DEVELOPMENTS IN THERMOCHEMICAL BIOMASS CONVERSION

DEVELOPMENTS IN THERMOCHEMICAL BIOMASS CONVERSION

Volume 1

Edited by

A.V. Bridgwater

Director of the Energy Research Group,
Department of Chemical Engineering and Applied Chemistry,
Aston University, Birmingham, UK

and

D.G.B. Boocock

Chair of the Department of
Chemical Engineering and Applied Chemistry,
University of Toronto, Toronto, Canada

SPRINGER-SCIENCE+BUSINESS MEDIA, B.V.

First edition 1997

© 1997 Springer Science+Business Media Dordrecht
Originally published by Chapman & Hall in 1997
Softcover reprint of the hardcover 1st edition 1997

ISBN 978-94-010-7196-3   ISBN 978-94-009-1559-6 (eBook)
DOI 10.1007/978-94-009-1559-6

A Catalogue record for this book is available from the British Library

Library of Congress Catalog Card Number: 96–86524

∞ Printed on permanent acid-free text paper, manufactured in accordance with ANSI/NISO Z39.48-1992 (Permanence of Paper).

CONTENTS

VOLUME 1

Laboratory experimentation

Pilot and demonstration

Analysis and characterisation of pyrolysis liquids

Combustion of pyrolysis liquid

Chemicals from pyrolysis liquid

Upgrading pyrolysis products

VOLUME 2

GASIFICATION

Fundamentals

Laboratory experimentation

Pilot and demonstration

COMBUSTION

Overview

Fundamentals

Pilot and demonstration

Environment

SYSTEM STUDIES

WORKSHOPS

PREFACE

There have been many developments in the science and technology of thermo-chemical biomass conversion since the previous conference on *Advances in Thermochemical Biomass Conversion* in Interlaken, Switzerland, in 1992. This fourth conference again covers all aspects of thermal biomass conversion systems from fundamental research through applied research and development to demonstration and commercial applications to reflect the progress made in the last four years. All aspects of bioenergy systems are covered from pretreatment through to end-user applications with increased consideration paid to the environmental benefits and problems of implementing bio-energy systems.

There was an excellent response with over 200 papers offered and over 180 delegates from 29 countries attending the conference. The programme was divided into five main areas covering pyrolysis, pretreatment, gasification, combustion and system studies and this division is reflected in the structure of these conference proceedings. Each main section was preceded by a state-of-the-art review to provide a focus for the ensuing presentations and an authoritative reference. All the papers included have been subject to a full peer review process.

As with any international conference, an important aim was to exchange ideas and discuss problems with fellow researchers, as well as to hear about the latest research and development and applications. A workshop programme was included to encourage this interaction in areas of interest selected by participants. The resultant workshop reports provide a summary of topical problems and opportunities.

The major developments in the bioenergy field include the growing appreciation of the contribution that can be made to mitigating greenhouse effects through the increased awareness of the environmental responsibilities for implementing bio-energy systems and also the growing emphasis on commercialisation. The academic community will, of necessity, continue to provide the lead in open dissemination of research results to stimulate the development of new ideas. These will translate into new products and processes and provide the justification for continued support of centrally supported R&D. The value of these conferences lies in providing an opportunity for industry to hear about the latest developments and provide an opportunity for their representatives to meet the active researchers and develop better links. A further development is the increased commercial activity which results from overt governmental and institutional support for bio-energy. This industrial pull for newer and better products and processes will stimulate researchers in the private and public sector and should deliver at least some of the promised benefits of the bio-energy sector.

Tony Bridgwater and Dave Boocock
October 1996

ACKNOWLEDGEMENTS

We would like to express our sincere appreciation to those people and organisations who have provided support:

- IEA Bioenergy, through the operating agent represented by C Wallace; and the thermochemical related conversion activities themselves represented by S Babu, J Hustad, K Mackie, Y Solantausta and T Bridgwater;
- Natural Resources Canada, represented by E Hogan;
- National Renewable Energy Laboratory, USA, represented by R Overend;
- The Department of Trade and Industry, UK, represented by N Barker;
- VTT, the Technical Research Centre of Finland, represented by K Sipila.

The scientific committee were very supportive throughout the preparation and running of the conference. Their encouragement and efforts in publicising the meeting, refereeing papers, providing constructive feedback on the programme, organising workshops and chairing sessions was much appreciated.

M. Antal, USA	S. Babu, USA
N. Bakhshi, Canada	T. Beenackers, Netherlands
M. Connor, Australia	J. Corella, Spain
B. Delmon, Belgium	D. Elliott, USA
B. Graham, Canada	E. Hogan, Canada
W. Kaminsky, Germany	B. Krieger-Brockett, USA
J. Kuester, USA	K. Mackie, New Zealand
R. Maggi, Belgium	K. Maniatis, Belgium
D. Meier, Germany	T. Milne, USA
T. Nussbaumer, Switzerland	R. Overend, USA
E. Rensfelt, Sweden	C. Roy, Canada
D. Scott, Canada	K. Sipila, Finland
K. Sjostrom, Sweden	Y. Solantausta, Finland
H. Stassen, Netherlands	S. Yokoyama, Japan

At a personal level, particular thanks are due to Karen Dowden and Claire Humphreys who provided the Conference Administration. We also wish to thank the research staff from Aston University who provided considerable support before, during and after the meeting.

PYROLYSIS

PYROLYSIS

Overview

OVERVIEW OF FAST PYROLYSIS OF BIOMASS FOR THE PRODUCTION OF LIQUID FUELS

J.P. DIEBOLD
Biomass Thermochemical Conversion Consultant, 57 N. Yank Way, Lakewood, CO80228, USA
A.V. BRIDGWATER
Energy Research Group, Aston University, Birmingham B4 7ET, UK

Abstract
The objective of this paper is to provide a brief overview for the fast pyrolysis of biomass for the production of liquid fuels. The basic concepts are described that pertain to the realization of fast pyrolysis in research, bench, and pilot scales. A description of the products and potential uses is included. This paper serves as an introduction for many of the following papers related to pyrolysis and provides representative references for further study.
Keywords: Biomass, pyrolysis, liquid fuels, chemicals, applications

1   Introduction

1.1   Biomass constituents

Biomass has three major constituents: about half cellulose: a quarter hemicelluloses and extractives: and a quarter lignin for a typical woody biomass. The cellulosic fibers are held together in a matrix of lignin and hemicelluloses, similar in concept to fiber glass fibers in polyester resins.

Cellulose is a linear polymer with a degree of polymerization (DP) of up to 10,000 six-carbon anhydroglucose sugar units, containing 49 wt% oxygen. Hemicelluloses are chemically similar to cellulose, but have a lower DP of 100 to 200 heterogeneously linked six-carbon and five-carbon anhydro sugars [1]. The five-carbon anhydro sugars have an oxygen content of 54 wt%.

Lignins appear to be primarily random, three-dimensional polymers of 4-propenyl phenol (p-coumaryl alcohol), 4-propenyl-2-methoxy phenol (guaicyl alcohol), and 4-propenyl-2,5-dimethoxy phenol (syringyl alcohol). The ratios of these phenols in lignin is a function of the plant species and environmental conditions [2]. The oxygen content of these lignin model compounds is between 12 and 29 wt%, which is much lower than for cellulose and hemicelluloses.

The overall range for the oxygen content of woody and herbaceous biomass is 40 to 45 wt% on a moisture and ash free basis [3]. One result of the high oxygen content is the relatively low lower heating value (LHV) of 19 to 20 MJ/kg of dry biomass [3], compared to hydrocarbon fuels having a LHV of between 40 and 44 MJ/kg. The energy content of biomass on a volume basis compares even less favorably to hydrocarbon fuels. If a bulk density of 240 kg/m^3 (15 lb/ft^3) is assumed for dry sawdust, the volumetric LHV is 4.8 GJ/m^3. This compares to the volumetric LHV of 34 GJ/m^3 for cetane, or over 7 times the energy per unit volume of cetane compared to sawdust.

Clearly one of the problems with utilization of sawdust for fuel is its low volumetric energy content. The practical problems involved with solids handling, as compared to liquids make the use of biomass even more challenging. In addition, biomass normally has a significant amount of potassium in it, which can lead to ash deposition during combustion, especially with grasses. This alkali rich ash can lead to corrosion or metal wastage of boiler tubes, heat exchangers, and turbine blades [4, 5, 6].

1.2   Thermal biomass conversion

As biomass is heated, a sequence of physical and chemical changes take place.

Firstly, pyrolysis occurs with heating in the absence of oxygen or air to produce a mixture of solid char, condensable liquids, and gases. For the purposes of this paper, the term fast pyrolysis is used to describe a process designed to optimize the formation of condensable organic vapors, with a minimum of gas and char. Many larger scale fast pyrolysis reactor designs add a small amount of air to the recycled char or to recycled pyrolysis gases to provide process heat by partial combustion but do so in such a manner that the condensable organic yields are not compromised.

Gasification is designed to produce non-condensable gases, usually with the addition of a small amount of oxygen or air (sub-stoichiometric) directly to the reactor to provide the process heat for the gasification reactions. Pyrolysis is a precursor to the gasification process.

Combustion implies the addition of air or oxygen directly to the reactor in sufficient quantity to completely oxidize the biomass (stoichiometric), usually with an excess of oxygen to ensure burn-out. During combustion of biomass, the biomass is first pyrolyzed to gases and organic vapors, which are then burned in flaming combustion. The char burns in glowing combustion after the pyrolysis step.

1.3   Drying

Drying occurs as the temperature of the biomass is slowly increased. Free moisture is reversibly evaporated from the biomass particles as the water vapor pressure on the particle surface exceeds the partial pressure in the surrounding atmosphere. At temperatures a little above 100°C, water that is loosely bound to the biomass is removed irreversibly to make what is known as kiln-dried lumber, if the humidity is low. If the humidity is high, the bound moisture is not lost at these low temperatures, but serves to soften and aid in the melting of the lignitic fraction of the biomass. This allows the cellulose fibers to slip within the matrix, as is practiced in the steam bending of wood for furniture. Upon cooling, the lignin solidifies and the bent wood appears to retain most of its original strength, indicating that the cellulose fibers are not significantly damaged.

As the temperature of the biomass increases, low molecular weight extractives are volatilized giving rise to "blue haze" from biomass dryers. However, at temperatures even below 100°C, pyrolysis begins very slowly with the exothermic dehydration reactions that form primarily char, water, and carbon dioxide over extended periods time, e.g. hours to months [7]. That these gaseous products are not combustible explains the self-extinguishing nature of thick pieces of burning wood, which heat very slowly deep inside.

In adiabatic situations, these slow chemical changes appear to be responsible for the transition from the self-heating of thermophilic bacterial decay of wet biomass to the chemical self-heating that leads to more rapid pyrolysis and spontaneous combustion.

1.4   Faster pyrolysis

Faster pyrolysis occurs during rapid heating conditions in which higher temperatures are reached prior to decomposition. These higher temperatures favor a different mechanism, involving depolymerization to what is believed to be a melt [8]. At lower temperatures, this melt has a high viscosity and may not flow except for a plastic shrinkage in the particle's outer dimensions by up to one-third as it slowly dehydrates to form char [9]. Thermal gravimetric analysis (TGA) of biomass with heating rates of several degrees per minute show that weight loss is very slow until temperatures reach about 150°C. Most of the weight loss of the biomass during this relatively slow heating occurs between 200 and 380°C [10].

Under these faster heating conditions and higher temperatures, the dehydration reactions to make char lose their predominance as organic volatiles are formed by cracking reactions. In the absence of catalysts, pure cellulose pyrolyses predominately to the monomer levoglucosan. However, in the presence of alkali catalysts that are naturally found in biomass materials, the carbohydrate rings are split open with the production of hydroxyacetaldehyde, organic acids, furfurals, and other oxygenated compounds. The pyrolysis of hemicelluloses gives rise to analogous families of chemical compounds [11, 12, 13].

Lignins in the presence of alkali catalysts pyrolyze to monocyclic aromatics and non-condensed bicyclic aromatic materials with a high phenolic content. These phenolics have zero, one, or two methoxy groups on them, depending upon the plant species. However, lignins also pyrolyze to form 40 to 50 wt% char, as they cross-link and condense to form three-dimensional, thermally stable aromatic polymers, known as char [14, 15].

1.5   Fast pyrolysis

Fast pyrolysis with high heating rates of thousands of degrees per minute, local temperatures as high as 465°C are thought to exist prior to the pyrolysis of dry biomass. At these very high heating rates, larger biomass particles will have large temperature gradients within them, such that the inside of a particle may still be at its initial temperature prior to being fed into the reactor. However, at temperatures above 450°C the melt has a very low viscosity and can be readily wiped from the biomass surface. If the melt is mechanically removed from the pyrolyzing surface, the pyrolysis front can be forced through the biomass at rates up to 3 cm/s [16,17]. This phenomena is the basis for ablative pyrolysis reactors, with the vapor forming reactions occurring primarily in the melt after it is removed from the biomass surface [8]. If the melt is not removed as it is formed, it will rapidly polymerize and retain the pore structure of the biomass and pyrolyze to form char, water, gases, and organic volatiles. Under the conditions of fast pyrolysis, the production of condensable organics has been optimized in a variety of different reactor designs with different heating methods and feedstock particle sizes [18].

1.6   Particle size

The particle size of the biomass has an important bearing on the ability to be heated quickly in a given heat flux environment. Biomass particles are rarely spherical or even cubicle, but tend to have a length that is several times larger than the thickness or the width, i.e., a pin chip. The commonly used hammer mill produces a comminuted, fibrous biomass material that contains a lot of very fine dust, regardless of the screen hole size used. The holes in the screen affect the maximum length of the larger particles primarily. There is a misleading tendency to report this maximum particle dimension as the nominal particle size.

However, the parameter that is normally of interest with respect to the heating rate of a particle and to fast pyrolysis is the surface area per unit weight. Assuming a small isothermal particle:

$$Q = h_r A_p (T_r - T_p) = m_p C_p (dT_p/dt)$$

$$dT_p/dt = (A_p/m_p)(h_r/C_p)(T_r - T_p)$$

where h_r is the overall heat transfer coefficient in the reactor to the particle, T_r is the effective temperature of the reactor, T_p is the temperature of the particle, m_p is the mass of the particle, and C_p is the heat capacity of the particle.

For an idealized pin chip, with the length equal to 5 times the width and with the width equal to the thickness it is easy to show that:

$$A_p/m_p = 22/(5t_p r)$$

where t_p is the particle thickness and r is the biomass density. This relationship is shown in Figure 1 using a typical biomass (wood) density of 640 kg/m^3. The surface area per unit weight and the heating rate both increase by an order of magnitude as the average particle thickness is reduced from 1 mm to 0.1 mm.

If the particle is too thick to be isothermal during heating and if that particle is in an ablative reactor, the situation is very different as is discussed later.

2   Heat Transfer

Heat transfer mechanisms used to move heat into fast pyrolysis reactors have included: hot preheated gases; hot preheated gases and heated reactor walls; heated reactor walls; hot sand; and preheated gases and hot sand [18].

2.1   Gas Heat Transfer

Hot carrier gas is one of the simplest ways to introduce heat into a pyrolysis reactor. The most common way to do this has been to stoichiometrically combust a fuel gas and to temper that with excess inert gas to control the temperature. Depending upon the hydrogen content of the fuel gas, this can introduce additional water vapor into the pyrolysis stream. This extra water will condense in the pyrolysis oil unless the condensers are operated at fairly high temperatures, so that the water and volatile organic solvents are not collected. The concurrent loss of the volatile organic solvents with the water vapor will decrease the organic oil yield and increase the viscosity of the oil. It has been found that if the carrier gas is too hot, it tends to crack the organic vapors to permanent gases [19, 20]. This led to the use of cooler gases and a relatively large amount of carrier gas of 6 to 8 kg carrier gas per kg dry feed to supply the necessary heat [20, 21, 22].

A variation on the addition of heat to the reactor by hot carrier gases is to heat them indirectly in a heat exchanger rather than directly by partial combustion. This has the advantage of not adding combustion moisture to the pyrolysis stream and avoids the possible hazards of introducing air or oxygen into the reducing atmosphere of the pyrolysis reactor. This has been demonstrated at the bench scale with a fluidized bed of sand that was heated by recycled pyrolysis gases heated indirectly by electric heaters [23] and the combustion of excess pyrolysis gases and auxiliary propane at the pilot scale [24].

2.2   Heated reactor walls

The use of hot carrier gas plus heated reactor walls has also been used to transport heat into pyrolysis reactors. The advantage of this is to reduce the amount of carrier gas required, as the temperature of the gas to reduce vapor cracking, or both. This has been practiced with fluidized sand beds at the 3-kg/h bench scale [23] and at the 125-kg/h pilot scale [25], with entrained flow in coiled tubes at the 4-kg/h bench scale [26] and also at a 1900 kg/h demonstration scale [27, 28], and with entrained flow in a special vortex tube at a 20 kg/h bench scale [16, 29].

Heat transfer from an externally heated reactor wall to the entrained biomass particles has been demonstrated at a research scale to be extraordinarily high, as the heat is transferred by conduction across a thin liquid film of pyrolysis intermediates to the biomass. The heat transfer rate to the particle is increased as the biomass particle is pushed harder onto the hot reactor wall, causing the liquid film to become even thinner. If the pyrolysis liquids are wiped from the surface of the biomass, the biomass can be ablatively pyrolysed at measured rates up to 3 cm/s [16,17].

In the vortex reactor, the particles are entrained at very high velocities to create high centrifugal forces. The thicker the particle, the harder the particle is pushed onto the reactor wall and the faster the heat transfer and the resultant rate of movement of the ablative pyrolysis front through the non-isothermal particle. However, the thinner the particle, the more of them that are present for a given reactor throughput, so the total heat transferred is independent of particle thickness in the vortex reactor [30].

Non-entrained ablative pyrolysis has been demonstrated between a rotating plate and a heated stationary plate at a 0.2 kg/h lab scale [31] and at a 3 kg/h bench scale [32], but with the addition of a significant amount of inert gas to aid in feeding the biomass and to reduce the gaseous residence time.

Frictional heat created by pushing biomass against a rapidly spinning metal disk or cylinder (with raised ribs for the escape of the vapors) was utilized to produce smoke aerosols for preserving meat products [33].

A much slower heat transfer rate to biomass occurs in a multi-hearth reactor, where the biomass is slowly pushed from one level to the next. However, in the presence of sub-atmospheric pressures to rapidly remove products, the pyrolysis products appears to be fairly similar to those attained with faster heating rates [34].

2.3   Rotating cone
Heat transfer from hot sand that is mechanically circulated by a inverted cone spinning on its axis has been demonstrated for fast pyrolysis of finely ground biomass. The heated sand is introduced near the bottom (apex) of the cone along with the biomass particles. The centrifugal forces generated move the hot sand, pyrolyzing particles, char, vapors, and gases upwards and out of the reactor [35].

2.4   Transported bed
Heat transfer by hot sand entrained with hot transport gas is quite similar to the riser reactor or the fluidized bed of the Fluidized Catalytic Cracker used in petroleum refining. In this concept, sand is heated by combustion of the pyrolysis gases or the char, or both. The hot sand is then transported by gravity or carrier gas to the pyrolysis reactor, where it is mixed with the biomass feed. After transferring heat to the biomass, the cooled sand and char are returned to the combustor where the char is combusted and the sand is reheated. Historically, this was demonstrated at the 1.4 kg/h and 135 kg/h scales and was attempted with refuse derived fuel in the riser/regenerator mode at 3500 kg/h using spent FCC catalyst or sand as the thermofor [36, 37]. More recently it has been demonstrated with biomass feeds using pneumatically transported sand as the heat transfer medium at scales ranging from 0.3 to 1000 kg/h in scale [38, 39, 40, 41].

3   Pyrolysis Products

The pyrolysis products from biomass consist of condensable, oxygenated organic liquids, permanent gases, solid char, and water. The ratios of these products varies depending upon the feedstock and the time and temperatures in the reactor system. The char increases with alkali and ash content of the feed. The yield of char and organic liquids both decrease with increased severity in the reactor. At more severe conditions, the organic vapors are cracked primarily to permanent gases and some water vapor.

3.1 Liquid Characteristics

Pyrolysis oil consists of water and organic compounds that are condensed and collected after the pyrolysis step, usually after the char is removed from the hot gas and vapor stream. This liquid is a very complex mixture that contains molecular fragments of the cellulose, hemicellulose, and lignin polymers that were able to escape the pyrolysis environment by volatilizing or by being part of a liquid droplet small enough to be entrained from the reactor [8]. Extensive reviews on the chemical [42, 43] and physical properties of pyrolysis oils have recently been published [43].

3.1.1 Molecular weight

The molecular weight of pyrolysis oil is not well defined. While still in the initial vapor state, molecular beam mass spectrometry shows that there are very few molecules having a molecular weight over 200 Daltons [15]. However, after the oil vapors have condensed, the liquid oil appears to have a much higher weight average molecular weight of over 500 Daltons, as determined by size exclusion chromatography (SEC). Upon re-evaporation of the oils, the oil vapors are still seen as having molecular weights below primarily below 200 Daltons. The molecular weight of the oils by SEC does increase during aging of the oil, particularly at elevated temperatures up to weight average molecular weights of 900 Daltons [44]. One explanation for the difference in apparent molecular weights by the two techniques is that upon condensation that weak, reversible bonds are formed which are broken during re-evaporation.

3.1.2 Composition

The elemental composition of the pyrolysis oil is a function of its pyrolysis history and water content. To better understand the nature of the oxygenated organics present, it is useful to consider the pyrolysis oil on a "dry" basis without the water present. In the case of low pyrolysis severity, the oxygen content of the dry organic liquids approaches that of the biomass feedstock. As the severity of pyrolysis increases, the yield of organic liquids decreases through reactions that selectively remove carbon oxides and water; this decreases the oxygen and hydrogen content of the remaining condensable organics and increases its heating value as it becomes more hydrocarbon in nature [44].

In pyrolysis processes designed to optimize the yield of pyrolysis oil, the pyrolysis severity is sufficiently low that the dry oil contains over 30 wt% oxygen [44]. This oxygen content causes the oil to be very polar in nature, with the ability to hold a considerable amount of water in solution. Water contents in whole oil are typically in the range of 15 to 35 wt%, although values outside of this range have been reported. Factors that affect the water content of the oil are the initial moisture content of the feed, presence of water formed by combustion in the carrier gases, char yields and the quantity and temperature of carrier gases that leave the condensation train saturated with water.

3.1.3 Heating value

The heating value of the pyrolysis oil increases as its oxygen content decreases due to higher pyrolysis severity. However, the pyrolysis oil yield decreases faster than the heating value increases from the loss of oxygen, to result in a net energy yield loss in the oil.

The heating value of the pyrolysis oil decreases as the water content increases. Although the higher heating value (HHV) decreases simply as the water dilutes the pyrolysis oil, the more significant lower heating value (LHV) is also reduced by the energy required to evaporate the water in the oil. With a water content of about 20 wt%, pyrolysis oil has about the same volumetric heating value as 95% ethanol. It is expected that the fuel value of pyrolysis oil will be based on the lower heating value, as this properly penalizes the oil for its water content.

3.1.4   Viscosity

A wide range of viscosities have been reported for pyrolysis oil that reflect differences in the average molecular weight or extent of cracking of the oligomers and monomers to monomer fragments, the relative success in recovery of the volatile organics from the carrier and pyrolysis gases, the water content, and possibly the content of char fines. Aging studies have shown that viscosity and molecular weight increase together with time. [44]

The addition of alcohols to pyrolysis oils greatly decreases the viscosity [46]. Adding water to pyrolysis oils decreases the viscosity about as effectively as methanol on a volume basis. Whole pyrolysis oils have been reported having a viscosity well within the range specified for ASTM #4 fuel oils [47].

3.1.5   Ash

The ash content of pyrolysis oils has been shown to be directly related to the char content of the oils. Through the use of hot-gas filtration to remove char fines, the ash content has been reported to be lowered to below 0.01 wt% [47], which meets the ash requirement for even the best quality diesel fuel [48, 49]. After the char fines and ash are removed by hot-gas filtration, the resultant oil has been found to have its alkali content lowered down to as low as 2.2 ppm [49], which is very close to that recommended for gas-turbine fuels [6, 48]. Liquid filtration of char-laden pyrolysis oils through a 0.7-mm filter only reduced the alkali metal content from an original 325 ppm down to 190 ppm [50].

3.1.6   Storage

The storage stability of fast pyrolysis oils is widely perceived to be poor with respect to viscosity increases and to separation into a thin water-rich phase and a thick organic-rich phase. However, it has been recently shown that pyrolysis oils which had been hot-gas filtered to remove char fines, had a much better storage life in accelerated aging tests [47. Supporting this finding, it has also been shown that the addition of char fines to pyrolysis oil greatly accelerated the polymerization reactions [51].

Using gel permeation chromatography (GPC), aging of fast pyrolysis oil was seen to decrease the amount of lower molecular weight compounds present and increase the higher molecular weight compounds, i.e., polymerization after aging 15 h at 90°C Analysis by Fourier Transform Infrared (FTIR) has shown an increase in the absorbance at frequencies associated with water (in agreement with Karl Fischer moisture results), ethers, and esters. The esters would be formed by the reactions of organic acids and alcohols, although no trend in pH was observed with aging [52]. Other reactions are speculated to involve phenols reacting with aldehydes in polymerization reactions, which also produce water as a by-product.

The elucidation of the reaction mechanisms is made difficult due to the large number of chemical species present and the large number of reactions possible. In addition, because the reactions could involve large molecules, all that is required is one extra bond to form a dimer and double the molecular weight.

3.2   Liquid Applications
The commercial uses of fast pyrolysis oils are thought to be as a source of high valued,

11

specialty chemicals in the short term, and as petroleum fuel substitutes in the long term.

3.2.1 Chemicals

The production of highly valued specialty chemicals, such as hydroxyacetaldehyde [53] and food flavors from fast pyrolysis liquids for meat preservation has made a 1000 kg/h pyrolysis system into a commercial operation, that would have otherwise been only a large pilot scale for fuel production purposes [54].

The recovery of levoglucosan from pyrolysis oils has been the target for several research groups [12, 55, 56, 57, 58] with the fermentation of pyrolysis derived sugars to ethanol said to be competitive with acid hydrolysis [54,59].

The recovery of other specialty chemicals from pyrolysis oils has been of interest over the years [56, 60, 61, 62,]. The simultaneous production of specialty chemicals and fuel oils from pyrolysis oils has also received attention [63].

The recovery of commodity chemicals, such as organic acids in the form of calcium salts from pyrolysis liquids has also received attention for road deicing[64] and also as a reactive carrier for calcium for scrubbing flue gases containing nitrogen oxides and sulfur oxides [65]. Recovery of other commodity type chemicals has centered around phenolics for use in phenol-formaldehyde resins [66, 67].

3.2.2 Fuels

The fuel uses of fast pyrolysis oils have been demonstrated in furnace applications. In general, the pyrolysis oils burn quite readily in a previously heated furnace, but with slightly higher carbon monoxide, nitrogen oxides, and particulate emissions than ASTM #2 fuel oil [68, 69] and medium fuel oil [70]. Compared to an ASTM #6 fuel oil, pyrolysis oil burned with lower NO_x emissions, but with particulate emissions that correlated with ash content (about the same or higher than #6 oil) [71]. The NO_x emissions were found to correlate with the nitrogen present in the pyrolysis oils [72].

Due to the wide spread use of high-efficiency very large gas turbines for power production, there is interest in using pyrolysis oils in them. However, the alkali content of pyrolysis oils is higher than is normally recommended for aero-derivative turbines [6], although the alkali content with hot-gas filtering is now approaching the suggested upper limit [49]. The use of pyrolysis oils having higher alkali contents is now being investigated in a 2.8 MW turbine designed for the high alkaline environment of salt water spray in marine applications, that is expected to demonstrate a thermal efficiency of 28% [73].

Diesel engines achieve higher thermal efficiencies in small to medium applications than do gas turbines, so there is also interest in the use of pyrolysis oil in diesel engines. Preliminary tests revealed a difficulty in achieving autoignition of pyrolysis oil in the diesel engine tested, so the research adopted the use of pilot ignition with a small amount of high cetane diesel fuel in a medium speed 84-kW tractor engine. The fuel injectors were found to very rapidly corrode or erode in the presence of the acidic pyrolysis oil containing char fines. However, the pyrolysis oil produced less NO_x, less smoke, and less hydrocarbon emission than the diesel fuel tested for comparison. After the catalytic converter, the carbon monoxide emissions for diesel fuel and for pyrolysis oil were identical [74]. A special self-cleaning filter was developed to remove the char fines, which were nearly all less than 25μm in size [74, 75, 76]. In later tests, a ceramic coated, special stainless steel fuel injector in one of the 18 cylinders of a low speed, 1.4 MW diesel engine was successfully tested with liquid-filtered pyrolysis oil for four hours with no corrosion problems and demonstrated a thermal efficiency of 45% [77].

A 250kWe modified dual fuel engine has been successfully tested on raw pyrolysis liquid with and without dilution with methanol and ethanol. Run durations above 10 hours continuous operation have been achieved on raw oil and at full power output with generation of electricity [78]. It is possible that the char is sufficiently soft to be ground in the pumping process but longer term trials are needed to establish the erosion effects.

Emulsions made of 65% pyrolysis oil, 20% ethanol, and 15 % diesel fuel have been evaluated in a small 2-cylinder diesel engine. The pyrolysis oil was filtered to remove the char fines. The emulsion was sufficiently stable for the tests, but would separate after a few hours. Autoignition was attained with this mixture, but with long ignition delays (the water and volatiles probably evaporated before the easier to ignite diesel fuel) [79].

Systematic combustion bomb tests have more recently shown that hot-gas filtered pyrolysis oil (NREL Run 175) can be quite readily ignited in diesel engines without pilot ignition or ignition aids, by operating at higher compression ratios and inlet air temperatures. This pyrolysis oil was much easier to ignite than either methanol or ethanol; it was nearly as easy to ignite as the diesel fuel tested for comparison [80]. It is not currently known if the pyrolysis oil used in the combustion bomb tests had different autoignition properties than the pyrolysis oil used in the above engine studies.

3.3   Gas products
Gases are also produced during fast pyrolysis of biomass. During the formation of char, a very small amount (less than 5 wt% of the dry feed) of primary gases are formed, which contain about 53 wt% carbon dioxide, 39 % carbon monoxide content, 6.7 % hydrocarbons (including methane), and than 0.8 % hydrogen. During the operation of practical fast pyrolysis reactors, a part of the organic vapors is cracked to secondary gases, which contain a different ratio of gaseous products: 9 wt% carbon dioxide, 63 % carbon monoxide, 27 % hydrocarbons (including methane), and 1.4 % hydrogen. Thus, the gaseous products may be considered to be a binary mixture of the primary and the secondary gases [81].

The LHV of these primary gases is 11 MJ/nm^3 (280 BTU/SCF) and that of pyrolysis gases formed after severe secondary cracking of the organic vapors is 20 MJ/nm^3 (510 BTU/SCF). When operating fast pyrolysis reactors for the production of liquids with a minimum of vapor cracking, a LHV for the gases of around 15 MJ/nm^3 (390 BTU/SCF) would be expected [81].

The design of most pyrolysis systems feature the use of pyrolysis gases for process energy. Unless the gas yields are higher than is preferred for oil yields, some form of auxiliary fuel such as char needs to be used in addition to the pyrolysis gases. Of course, the addition of inert purge gases or flue gases to the pyrolysis gas stream would lower the specific heating value, the resultant flame temperature, and the available heat of the pyrolysis gases.

The use of pyrolysis gases as a synthesis gas would require extensive reforming and shifting to result in the desired gas composition. Due to economy of scale, it probably will not be feasible to produce synthesis gas from the usually low yields of pyrolysis gases formed in the fast pyrolysis for liquids. If the pyrolysis vapors are deeply cracked and steam reformed to form gases, then the economics of synthesis gas production from are improved [82].

3.4   Char Product
Char is the other major pyrolysis product. The particle size of the char formed is highly dependent upon the particle size of feed used, the relative attrition of the char by the pyrolysis reactor system, and the mechanism of char formation. The char particles do not appear to agglomerate in the pyrolysis reactor, although they may do so in product transfer lines that are cooled to below the tar dew point. Many of the pyrolysis reactors act as particle classifiers, retaining the char in the recycle loop until it is ground down fine enough to not be collected with the recycled sand, larger char, and/or partially pyrolyzed materials.

The small particle size and high volatility of char made in fast pyrolysis cause it to be very flammable, similar to powdered coal. Qualitative tests of dropping sawdust or fine pyrolysis char through the flame of a propane torch showed the sawdust to be difficult to ignite, compared to the very rapid ignition of the char and its subsequent glowing combustion outside of the propane flame. The autoignition temperature of biomass charcoal is typically between 200°C and 250°C, so hot char directly from the pyrolysis process must be properly handled to avoid unexpected ignition. In fact, commercial charcoal must be stored for several days exposed to air before it can be shipped, to avoid accidental ignition from exothermic oxygen adsorption during transit [83].

The LHV of chars have been reported to be about 32 MJ/kg (14,000 BTU/lb) and volatilities have been between 15 and 45 wt%, but depending upon the completeness of pyrolysis the heating value could be lower and the volatility higher. This material has many similarities to powdered, high-volatile bituminous coal, which has an extensive history of commercial use in furnaces and boilers. Most fast pyrolysis flow sheets show the combustion of char for process energy, but in some site specific locations the char may have a premium value for other applications, e.g. charcoal, etc. From a simple mass balance, the ash content of the char is about 6 to 8 times greater than in the original feed. Because the ash in biomass is concentrated in the char and has a considerable alkali metal content, it has the potential of having the same problems in combustion as do high ash, biomass materials, due to the low temperature of ash slagging and alkali vapor deposition and corrosion [4, 5].

Pyrolysis char is normally not graphitized and appears not to be electrically conductive.

3.5   Product Yields
The yields from fast pyrolysis are a function of the biomass feed and the reactor conditions, but generally are between 40 and 65 wt% organic condensate, 10 to 20 % char, 10 to 30 % gases, and 5 to 15% water based on dry feed. Most pyrolysis reactors are fed with biomass containing between 5 and 15 wt% water.

In comparing the products from different reactors, it is important to compare the total or gross unit yield of recovered water (including moisture in the feed) and dry oil per unit of dry feed to compare the yield and quality of the oil product. Of course, the other parameter of economic interest is the amount of usable energy in the products, based on the LHV of the products per unit of dry biomass. In the misguided interest of selling technology, it has been observed that gross wet oil yields have been numerically reported without proper clarification as to the moisture present in them. This has led to unreasonable expectations for fast pyrolysis systems.

4   Insights and discussion

4.1   Moisture
The adverse role of moisture introduced in the feed is to increase the amount of high-temperature energy or "premium" energy required to be transferred into the pyrolysis reactor. All of the moisture present in the feed is evaporated in the reactor and then heated to the exit temperature of the reactor. The energy required to evaporate and heat steam at one atmosphere pressure to 482°C (900°F) is 3.43 MJ/kg (1480 BTU/lb) relative to its condensed state at 25°C. For comparison, the energy required to heat dry biomass to its pyrolysis temperature and evaporate the products has been estimated to be about 2 MJ/kg [84]. So, if a pyrolysis reactor throughput is limited by heat transfer to the biomass, every extra weight percent of moisture in the feed requires the premium energy input that could otherwise pyrolyze an additional 1.7 wt% of dry feed. A similar extra heat duty is placed on the condensers by the moisture present in the feed.

The net energy available in the pyrolysis oil is lowered as the water content increases. Nearly all of the moisture present in the feed and the water of combustion that may be introduced into the pyrolysis stream normally ends up in the pyrolysis oil product. The water in the oil must be evaporated a second time when the oil is finally burned, usually with premium heat again. The net energy available in the pyrolysis oil decreases by 0.023 MJ/kg of dry wood for each percent of water left in the feed (on a dry basis). It would seem much better to evaporate the moisture only once during drying, where low temperature energy could be utilized.

The problem with attaining low moisture contents in biomass appears to be the use of co-current or stirred-tank dryers, in which the wood attains equilibrium with saturated air. Contrary to experience with such dryers, wood will dry very quickly to very low levels, providing the hot drying air has a low humidity as is the case with counter-current drying. The caveat here is that dry wood is flammable, so the temperature of the drying air must be carefully controlled.

There are beneficial effects of added water vapor in the condensation train. Although the cooling and condensation of water vapor requires a considerable removal of heat in the condensation train, water would be expected to act as an absorption oil to aid the capture of water soluble, oxygenated organic volatiles, such as alcohols, ketones, esters, aldehydes, etc. Another benefit to moisture in the pyrolysis oil appears to be in promoting microexplosions after atomization to achieve an extremely fine atomization that aids rapid and complete combustion [85].

4.2   Collection of liquid product

Aerosols consist of submicron liquid droplets and they present a severe problem in the successful recovery of the pyrolysis oils. These aerosols appear visually as smoke. Some of these aerosols could be formed in the pyrolysis reactor, especially from a submicron biomass particle that would be rapidly depolymerized, but which might be rapidly entrained out of the reactor before it vaporized. Another mechanism proposed for the formation of aerosols in the pyrolysis reactor involves the ejection of liquid droplets from internally pressurized cell capillaries of a pyrolyzing particle[8].

If a minimum of carrier gas is used, it could be saturated with respect to oligomers, which could condense to form aerosols upon the slightest cooling in the transfer lines. However, operational experience has been relatively successful in avoiding transfer line plugging and hot-gas filtering temperatures around 400°C have been reported [49]. So, it must be concluded that most of the aerosols are formed as they enter the condensation train. It is probable that they continue to be formed as the vapors are sequentially cooled.

These aerosols are only partially recovered by venturi scrubbing, wet spray scrubbing, heat exchangers, long settling times, cyclones, high-speed blowers, packed beds, and other commonly used impingement devices. Coalescing filters which utilize Brownian movement of the aerosols show considerable promise for the capture of aerosols, but only where the captured aerosols have a sufficiently low viscosity to drain from the filter material. Common practice has been to combine several of these techniques sequentially to capture the aerosols. Electrostatic precipitation was used historically in slow pyrolysis applications to recover aerosols [86] and are beneficial in fast pyrolysis applications.

4.3   Carrier gas

The role of carrier, diluent, and purge gas to reduce the partial pressure of vapors in the pyrolysis stream is not widely appreciated. In the most pyrolysis reactors, the extra gases aid the evaporation of higher molecular weight tar species, which would otherwise stay longer in the high temperature environment and crack to more volatile species. In the condensation train, the extra gases tend to strip off the volatiles, which have low viscosities, e.g., alcohols, ketones, aldehydes, water, etc. Thus, with the use of more gases in the pyrolysis reactor, the viscosity of the recovered liquids would be expected to higher due to the increased tar content and the decreased volatile content.

4.4 Gas temperature

The exit gas temperature from the condensation train has a definite effect on the properties of the recovered oil product. As the exit gas temperature is lowered, progressively more of the volatile organic vapors and water will be condensed to become the pyrolysis oil product. These volatiles should be thought of as solvents as they have viscosities of 1 cP or less and have a very large beneficial impact on lowering the viscosity of the oil, even at low incremental increases in the product yield. The recovery of these volatile, oxygenated organic solvents is thought to decrease the tendency of the oil to phase separate.

The use of high exit gas temperatures to effect a partial condensation of the water vapor to compensate for using wet feed, could result in poor recovery of the organic volatiles and a consequent loss in yields and the production of higher viscosity pyrolysis oils. There will be economic tradeoffs, because the lower exit temperatures will require expensive refrigeration to achieve increased organic oil yields of a higher quality oil.

4.5 Particle size

The effect of the required particle size on economics can be substantial. As was noted above, the thickness of the particle has an important role in the surface area per unit weight and subsequently the heating rate of the biomass in a particular heat flux environment. Biomass size reduction can be a significant cost of operating a pyrolysis unit, particularly if a small particle is needed to have high liquid yields and to avoid combustion of partially pyrolyzed particles as char. This cost as a function of final weight average particle sieve size has been calculated based on measured power requirements, estimated cost of grinding machines, and estimated operating costs. It was calculated that grinding from a minus 20 mm to a minus 1 mm nominal particle size could add between US$3 and 9 per tonne of feedstock, depending upon the cost of electricity and the type of grinding machine [29]. This estimated grinding cost is significant compared to projected biomass costs of US$40 to $60/tonne. Clearly, in comparisons of different pyrolysis reactors, the cost of feedstock preparation should be included.

4.6 Energy integration

Energy integration and balancing is one of the key requirements for an efficient pyrolysis system. The major heat requirements are for the pyrolysis reactor, carrier gas preheating, and biomass drying. The pyrolysis reactor and the carrier gases will require high temperature or premium heat. However, the drying of biomass could be accomplished quite rapidly in well designed systems at low temperatures around 100°C. Heat management in naphtha cracking processes to produce ethylene and propylene have been developed that are quite energy efficient. Much could be gained by applying their practices to energy management in biomass pyrolysis systems. An area that is often overlooked when comparing fast pyrolysis reactor concepts is the process energy required to compress and preheat the gases used for carrier, fluidizing, and purge gases. These gases are then cooled in the condensation train, where they lower the dew point and require lower temperatures for the same level of oil recovery. It has been observed that pyrolysis process developers often use very high relative rates of these gases, which impacts adversely on their economic viability. With careful use of process energy, it has been shown that there is enough excess high quality energy available for cogeneration of steam to produce electricity. This cogeneration has a significant beneficial impact on the economics of producing pyrolysis oil [87].

4.7 Health and safety

The toxicity, hazards, health, safety of a new product like pyrolysis oil need to be established prior to commercialization.

Petroleum derived fuel oils are known to contain carcinogens and are otherwise health hazards [88]. However, at low pyrolysis severity, pyrolysis oil does not contain polycyclic aromatics and appears not to be carcinogenic. The practice of wood distillation to recover pyrolysis liquids existed for over 100 years, but toxicological damage or danger to health was not observed [83].

As the pyrolysis severity increases and yields decrease, the surviving condensable organics become more phenolic in nature. Under severe pyrolysis conditions, the condensable organic yield is relatively low and contains benzene and polycyclic aromatic compounds, which are more carcinogenic than benzo(a)pyrene [45]. It is unlikely that the use of fast pyrolysis to produce liquids will try to operate in the severe pyrolysis mode where the yields of carcinogenic, and viscous oils are quite low.

More recent studies have shown that whole fast pyrolysis oils have mutagenic effects in the Ames tests with bacteria. The vapors and mists were considered to be mildly toxic with acute inhalation by test animals. Dermal application of pyrolysis oils did not cause any signs of systemic toxicity. The oils were noted to be very irritating to the eye and direct application to the eye caused corneal damage. It was concluded that pyrolysis oils can be routinely handled with commonly used safe work practices, e.g., safety glasses or goggles, gloves, and shielding clothes [89].

From a practical perspective, organic vapor face masks and goggles have been popular with operators to avoid exposure to the irritating pyrolysis oil volatiles during clean up procedures. Rubber gloves have also been popular with the operators to avoid brown stains and the lingering odor of pyrolysis oil on their hands. It must be noted in passing that wood smoke has been used for centuries for preservation and flavoring of meat, with only a very recent suspicion that it might be linked to ill health effects.

4.10   Economics
The economics of fast pyrolysis to produce pyrolysis oil liquids have been extensively studied and reported [87, 90, 91, 92]. A review of these independent studies with different assumptions reveals similar conclusions that pyrolysis oil can be produced for between US$0.13 and $0.16 per liter of wet oil ($6.50 to $7.00/GJ--LHV), with feedstock costing between $44 and $60 per dry tonne [90, 91, 92]. With cogeneration, the cost dropped to $0.11 per liter ($6.00/GJ) with feedstock costing $44 per dry tonne [87]. Assigning a zero cost for feedstock, as would be the case for waste biomass, the predicted production costs dropped to between $2.00 and $3.00/GJ. For comparison, the recent tax-free cost to the consumer for petroleum-derived fuel oil was said to be $0.20/L for # 2 diesel fuel and $0.15/L for #4 fuel oil, and $0.10/L for #6 fuel oil [95], or an estimated $5.75/GJ, $4.00/GJ, and $2.50/GJ respectively. Fuel taxes vary greatly from country to country, but generally are $3/GJ to $11/GJ for #2 fuel oil, zero to $10/GJ for #4 fuel oil, and zero to $0.48/GJ for #6 fuel oil [93].

The production of pyrolysis oil for fuel applications will be economically attractive with low cost, waste feedstocks. If fuel taxes can be avoided, the economics for pyrolysis oil appear to be excellent in many site specific cases, especially if the oil can be a replacement for a better fuel oil than #6 fuel oil.

These relative energy values make a very strong case for research into making pyrolysis oils having lower viscosities with low ash contents, as substitutes for higher quality fuel oil applications.

5   Conclusions

The fast pyrolysis of biomass can produce large yields of pyrolysis oils from wood and other biomass. The adverse effects of char fines to accelerate the aging of pyrolysis oils have recently been demonstrated, as well as the use of hot-gas filtration to remove them. Recent insights into the nature of pyrolysis oil have resulted in the manipulation of the viscosity through the addition of alcohols or water. The ongoing demonstration of these oils as diesel engine and gas turbine fuel opens a new market for them. The economics for pyrolysis production are encouraging for the higher valued applications that require a fuel better than #6 fuel oil, especially with waste feedstocks for applications that would avoid paying high fuel oil taxes by using biomass derived oils.

References

1.   Theander, O. (1985) "Cellulose, Hemicellulose, and Extractives," *Fundamentals of Thermochemical Biomass Conversion*, R.P. Overend, T.A. Milne, and L.K. Mudge, eds., Elsevier Applied Science Publishers, pp. 35-60.

2.   Glasser, W. (1985) "Lignin," *Fundamentals of Thermochemical Biomass Conversion*, R.P. Overend, T.A. Milne, and L.K. Mudge, eds., Elsevier Applied Science Publishers, pp. 61-76.

3.   Grabosky, M. and Bain, R. (1979) "Properties of Biomass Relevant to Gasification," *A Survey of Biomass Gasification. Volume II.-Principles of Gasification*, T.B. Reed, ed., Solar Energy Research Institute, Golden, CO, SERI/TR-33-329, p. II-21 to II-66.

4.   Dayton, D.C. (1996) "The fate of Alkali Metal During Biomass Thermoconversion," in these proceedings.

5.   Baxter, L.L.; Jenkins, B.M.; Milne, T.A.; and Dayton, D. (1996) "Ash Deposit Formation in Biomass Boilers," in these proceedings.

6.   Moses, C. (1994) "Fuel-Specification Considerations for Biomass Liquids," *Proceedings Biomass Pyrolysis Oil Properties and Combustion Meeting*, T.A. Milne, ed., National Renewable Energy Laboratory, Golden, CO, NREL-CP-430-7215, pp.362-382.

7.   Stamm, A.J. (1956) "Thermal Degradation of Wood and Cellulose," *Ind. Eng. Chem.*, Vol. 48, No. 3, pp. 413-417.

8.   Lédé, J.; Diebold, J.P.; Peacocke, G.V.C.; Piskorz, J. (1996) "The Nature and Properties of Intermediate and Unvaporized Biomass Pyrolysis Materials," in these proceedings.

9.   McGinnis, E.A.,Jr.; Harlow, C.A.; and Beall, F.C. (1976) "Use of Scanning Electron Microscopy and Image Processing in Wood Charcoal Studies," *Scanning Electron Microscopy/1976 (part 1) Proceedings of the Workshop on Plant Science Applications of the SEM*, IIT Research Institute, Chicago, IL.

10. Gaur, S. and Reed, T.B. (1995) *An Atlas of Thermal Data for Biomass and Other Fuels*, National Renewable Energy Laboratory, Golden, CO, NREL/TP-433-7965, 153 pp.

11. Antal, M.J., Jr. (1983) "Biomass Pyrolysis: a Review of the Literature. Part 1. Carbohydrate Pyrolysis," *Advances in Solar Energy*, K.W. Boer and J.A. Duffield, eds., Solar Energy Society, NY, pp. 61-111.

12. Radlein, D.; Piskorz, J.; and Scott, D.S. (1991) "Fast Pyrolysis of Natural Polysaccharides as a Potential Industrial Process," *JAAP*, vol. 19, pp. 41-63.

13. Richards, G.N. (1994) "Chemistry of Pyrolysis of Polysaccharides and Lignocellulosics," *Advances in Thermochemical Biomass Conversion*, A.V. Bridgwater, ed., Blackie Academic and Professional, London, pp. 727-745.

14. Petrocelli, F.P. and Klein, M.T. (1985) "Simulation of Kraft Lignin Pyrolysis," *Fundamentals of Thermochemical Biomass Conversion*, R.P. Overend, T.A. Milne, and L.K. Mudge, eds., Elsevier Applied Science Publishers, pp. 257-274.

15. Evans, R.J. and Milne, T.A. (1987) "Molecular Characterization of the Pyrolysis of Biomass. 1. Fundamentals," *Energy and Fuels*, Vol. 1, No. 2, pp. 123-137.

16. Diebold, J.P. (1980) "Ablative Pyrolysis of Macroparticles of Biomass,

Proceedings Specialists' Workshop on Fast Pyrolysis of Biomass, J.P. Diebold, ed., Copper Mountain, CO, Solar Energy Research Institute, Golden, CO, SERI/CP-622-1096, pp. 237-252.

17. Lédé, J.; Panagopoulos, J.; Li, H.Z.; and Villermaux, J. (1985) "Fast Pyrolysis of Wood: Direct Measurement and Study of Ablation Rate," *Fuel*, pp. 1514-1520.

18. Bridgwater, A.V. (1995) "Engineering developments in flash pyrolysis technology", Proceedings of conference on bio-oil production and utilisation, Estes Park, CO., USA, 24-26 September 1994 (NREL)

19. Maniatis, K.; Baeyens, J.; Peeters, H.; and Roggeman, G. (1994) "The Egemin Flash Pyrolysis Process: Commissioning and Initial Results," *Advances in Thermochemical Biomass Conversion*, A.V. Bridgwater, ed., Blackie Academic and Professional, London, pp. 1257-1264.

20. Kovac, R.J.; Gorton, C.W.; and O'Neil, D.J. (1988) "Production and Upgrading of Biomass Pyrolysis Oils," *Thermochemical Conversion Program Annual Meeting*, Solar Energy Research Institute, Golden, CO, SERI/CP-231-3355, DE88001187, pp. 5-20.

21. Kovac, R.J.; Gorton, C.W.; Knight, J.A.; Newman, C.J.; and O'Neil, D.J. (1991) *Research on the Pyrolysis of Hardwood in an Entrained Bed Process Development Unit*, Pacific Northwest Laboratory, Richland, WA, PNL-7788, UC-245, 162+pp.

22. Kovac, R.J. and O'Neil, D.J. (1989) "The Georgia Tech Entrained Flow Pyrolysis Process," *Pyrolysis and Gasification*, G.L. Ferrero, K. Maniatis, A. Buekens, and A.V. Bridgwater, eds., Elsevier Applied Science, pp. 169-179.

23. Scott, D.S. and Piskorz, J. (1984) "The Continuous Flash Pyrolysis of Biomass," *Can. J. Chem. Eng.*, vol. 62, pp. 404-412.

24. Cuevas, A. and Medina, E. (1993) "Leben, Galicia: Collection, Handling, and Thermochemical Energy Use of Resulting Products. A Final Report," E.C. Joule Contractors Meeting, June, Athens, Greece.

25. Flanigan, V.J.; Huang, W.E.; Clancy, E.; and Sitton, O.C. (1986) "Commercial Design of an Indirect Fired Fluid Bed Gasifier System," *Proceedings of the 1985 Biomass Thermochemical Conversion Contractors' Meeting*, Minneapolis, MN, October 15-16, Pacific Northwest Laboratory, Richland, WA, PNL-SA-13571, CONF-8510167, pp. 319-338.

26. Diebold, J.P. (1980) "Gasoline from Solid Wastes by a Noncatalytic, Thermal Process," *Thermal Conversion of Solid Wastes and Biomass*, ACS Symposium Series 130, pp. 209-226.

27. Brown, D.B. (1996) "Continuous Ablative Regenerator System," *Bio-Oil Production and Utilization. Proceedings of the 2nd EU-Canada Workshop of Thermal Biomass Processing*, A.V. Bridgwater and E.N. Hogan, eds., CPL Press, Berkshire, U.K., pp. 96-101.

28. Brown, D.B. and Black, J. (1992) "Method and Apparatus for Ablative Heat Transfer," International Patent Application WO 92/09671.

29. Diebold, J.P. and Scahill, J.W. (1995) "Improvements in the Vortex Reactor Design," in these proceedings. See also U.S. Patent 5,413,227.

30. Diebold, J.P. and Power, A.J. (1988) "Engineering Aspects of the Vortex Reactor to Produce Primary Pyrolysis Oil Vapors for Use in Resins and Adhesives," *Research in Thermochemical Biomass Conversion*, A.V. Bridgwater and J.L. Kuester, eds., Elsevier Applied Science, pp. 609-628.

31. Reed, T.B. (1988) "Contact Pyrolysis in a 'Pyrolysis Mill'," *Research in Thermochemical Biomass Conversion*, A.V. Bridgwater and J.L. Kuester, eds., Elsevier Applied Science, pp. 192-202.

32. Peacocke, G.V.C. and Bridgwater, A.V. (1995) "Ablative Fast Pyrolysis of

Biomass for Liquids: Results and Analyses," *Bio-Oil Production and Utilization. Proceedings of the 2nd EU/Canada Workshop on Thermal Biomass Processing*, CPL Press, Newbury, U.K., pp.35-48.

33. Rasmussen, H.R. and Rasmussen, H.J. (1961) "Method and Apparatus for the Production of Smoke for Food-Treating Purposes," U.S. Patent 3,009,457.

34. Roy, C.; Pakdel, H.; and Amen-Chen, C. (1996) "Enhanced Production of Fine Chemicals from Biomass by Vacuum Pyrolysis," in these proceedings

35. Wagenaar, R.M. and Prins, W. (1996) "Construction and Testing of a 50 kg/h Pyrolysis Test-Unit for Shenyang Agricultural University," in these proceedings.

36. Chang, P.W. and Preston, G.T. (1981) "The Occidental Flash Pyrolysis Process," in *Biomass Conversion Processes for Energy and Fuels*, S.S. Sofer and O.R. Zaborsky, eds., Plenum Press, NY, pp. 173-185.

37. Harrison, B. and Vesiland, P.A. (1980) "The San Diego Flash Pyrolysis Project," in *Design & Management for Resource Recovery. Volume 2. High Technology-A Failure Analysis*, P.A. Vesilind, ed., Ann Arbor Science Publishers, pp.61-84.

38. Graham, R.G.; Freel, B.A.; Huffman, D.H.; and Bergounou, M.A. (1994) "Applications of Rapid Thermal Processing of Biomass," *Advances in Thermochemical Biomass Conversion*, A.V. Bridgwater, ed., Blackie Academic and Professional Publishers, London, pp. 1275-1288.

39. Graham, R.G.; Freel, B.A.; Huffman, D.R.; and Bergounou, M.A. (1994) "Commercial-Scale Rapid Thermal Processing of Biomass," *Biomass and Bioenergy*, Vol. 7, Nos. 1-6, pp. 251-258.

40. Freel, B.A.; Graham, R.G.; and Huffman, D.R. (1996) "Commercial Aspects of Rapid Thermal Processing (RTP)," *Bio-Oil Production and Utilization. Proceedings of the 2nd EU/Canada Workshop on Thermal Biomass Processing*, CPL Press, Newbury, U.K., pp. 86-95.

41. Boukis, I. (1996) "Practical Implications During Operation of a CFB Air-Blown Pyrolyser," *Bio-Oil Production and Utilization. Proceedings of the 2nd EU/Canada Workshop on Thermal Biomass Processing*, CPL Press, Newbury, U.K., pp. 49-65.

42. Milne, T.A.; Agblevor, F.A.; Davis, M.; Deutch, S.; and Johnson, D. (1996) "A Review of the Chemical Composition of Fast Pyrolysis Oils from Biomass," in these proceedings.

43. Fagernäs, L. (1995) *Chemical and Physical Characterization of Biomass-Based Pyrolysis Oils. Literature Review*, Technical Research Centre of Finland, Espoo, VTT Research Notes 1706, 113+ pp.

44. Czernik, S. (1994) "Storage of Biomass Pyrolysis Oils," *Proceedings of Biomass Pyrolysis Oil Properties and Combustion Meeting*, Estes Park, CO, September 26-28, National Renewable Energy Laboratory, Golden, CO, NREL CP-430-7215, pp. 67-76.

45. Elliott, D.C. (1988) "Relation of Reaction Time and Temperature to Chemical Composition of Pyrolysis Oils," *Pyrolysis Oils from Biomass. Producing, Analyzing, and Upgrading*, ACS Symposium Series 376, pp.55-65.

46. Casanova-Kindelan, J. (1994) "Comparative Study of Various Physical and Chemical Aspects of Pyrolysis Bio-Oils Versus Conventional Fuels Regarding Their Use in Engines," *Proceedings of Biomass Pyrolysis Oil Properties and Combustion Meeting*, Estes Park, CO, September 26-28, National Renewable Energy Laboratory, Golden, CO, NREL CP-430-7215 pp. 343-354.

47. Diebold, J.P.; Scahill, J.W.; Czernik, S.; Phillips, S.D.; and Feik, C.J. (1996) "Progress in the Production of Hot Gas Filtered Bio-Crude Oil at NREL," *Bio-Oil Production and Utilization. Proceedings of the 2nd EU/Canada Workshop on*

Thermal Biomass Processing, CPL Press, Newbury, U.K., pp. 66-81.

48. Diebold, J.P.; Oasmaa, A.; Bridgwater, A.V.; Piskorz, J.; Huffman, D.; Cuevas, A.; Gust, S.; Czernik, S.; and Milne, T.A. (1996) "Proposed Specifications for Various Grades of Pyrolysis Oils," in these proceedings.

49. Scahill, J.W.; Diebold, J.P.; Czernik, S.; and Feik, C.J. (1996) "Removal of Residual Char Fines from Pyrolysis Vapors by Hot-Gas Filtration," in these proceedings.

50. Agblevor, F.A.; Besler, S; and Evans, R.J. (1994) "Inorganic Compounds in Biomass Feedstocks: Their Role in Char Formation and Effect on the Quality of Fast Pyrolysis Oils," *Proceedings of Biomass Pyrolysis Oil Properties and Combustion Meeting*, Estes Park, CO, September 26-28, National Renewable Energy Laboratory, Golden, CO, NREL CP-430-7215 pp. 77-89.

51. Agblevor, F.A.; Besler, S.; Montané, D. and Evans, R.J. (1995) "Influence of Inorganic Compounds on Char Formation and Quality of Fast Pyrolysis Oils," ACS 209th National Meeting, Anaheim, CA, April 2-5.

52. Czernik, S,; Johnson, D.K.; and Black, S. (1994) "Stability of Wood Fast Pyrolysis Oil," *Biomass and Bioenergy*, Vol. 7, No. 1-6, pp. 187-192.

53. Radlein, D. (1996) "Fast Pyrolysis for the Production of Chemicals," *Biomass Thermal Processing. Proceedings of the First Canada/European Community R&D Contractors Meeting*, October 1990, Ottawa, CPL Press, Newbury UK, 1991, pp. 113-123.

54. Underwood, G. (1992) "Commercialization of Fast Pyrolysis Products," *Biomass Thermal Processing. Proceedings of the First Canada/European Community R&D Contractors Meeting*, Eds. Hogan, E., Grassi, G. and Bridgwater, A.V October 1990, Ottawa, CPL Press, London, pp. 226-228.

55. Scott, D.S.; Radlein, D.; Piskorz, J.; and Majerski, P (1992) "Potential of Fast Pyrolysis for the Production of Chemicals," *Biomass Thermal Processing. Proceedings of the First Canada/European Community R&D Contractors Meeting*, October 1990, Ottawa, Eds. Hogan, E., Grassi, G. and Bridgwater, A.V., CPL Press, Newbury UK, pp. 78

56. Longley, C.J.; Howard, J,; and Morrison, A.E. (1992) "Levoglucosan from Pyrolysis Oils: Isolation and Applications," *Biomass Thermal Processing. Proceedings of the First Canada/European Community R&D Contractors Meeting*, October 1990, Ottawa, CPL Press, London, pp. 179-180.

57. Moeus, L. (1995) "Isolation of Levoglucosan from Lignocellulosic Pyrolysis. Oil Derived from Wood or Waste," U.S. Patent 5,432,276

58. Pernikis, P.; Zandersons, J.; and Lazdina, B. (1996) "Obtaining Levoglucosan by Fast Thermolysis of Cellolignin-Pathways of Levoglucosan Use," in these proceedings.

59. Scott, D.S.; Czernik, S.; Piskorz, J.; and Radlein, D. (1989) "Sugars from Biomass Cellulose by a Thermal Conversion Process," *Proceedings of IGT's Energy from Biomass and Wastes XIII*, February 13-17, New Orleans. See also the same authors in *Pyrolysis and Gasification*, Eds. Ferrero, G-L, Maniatis, K.; Buekens, A.; and Bridgwater, A.V., Elsevier Applied Science, pp. 201-208.

60. Solantausta, Y.; Diebold, J.P.; Elliott, D.C.; Bridgwater, A.V.; and Beckman, D. (1994) *Assessment of Liquefaction and Pyrolysis Systems*, VTT, Espoo, Finland, VTT Research Notes 1573, pp. 75-84.

61. Scott, D.S.; Piskorz, J.; Majerski, P.; and Radlein, D. (1996) "Fast Pyrolysis of Biomass for Recovery of Exotic Chemicals," in these proceedings.

62. Kaminsky, W. and Brolund, N. (1996) "Petrochemicals from Bark by Fluidized-Bed Pyrolysis," in these proceedings.

63. Freel, B. and Huffman, D. (1994) "Applied Oil Combustion," *Proceedings of Biomass Pyrolysis Oil Properties and Combustion Meeting*, Estes Park, CO, September 26-28, National Renewable Energy Laboratory, Golden, CO, NREL CP-430-7215, pp. 309-315.

64. Oehr, K.H. and Barrass, G. (1992) "Biomass Derived Alkaline Carboxylate Road Deicers," *Biomass Thermal Processing. Proceedings of the First Canada/European Community R&D Contractors Meeting*, October 1990, Ottawa, CPL Press, London, pp. 181-183.

65. Oehr, K.H. (1996) "Simultaneous SO_x/NO_x Emission Control with Bio-Lime Derived from Biomass Waste," in these proceedings.

66. Kelley, S.S; Meyers, M.; Johnson, D.K.; Scahill, J.W.; and Diebold, J.P. (1996) "Phenolic Resin from Biomass Pyrolysis Oils," in these proceedings.

67. Himmelblau, D.A. (1995) "Phenol-Formaldehyde Resin Substitutes from Biomass Tars," *Proceedings Second Biomass Conference of the Americas: Energy, Environment, Agriculture, and Industry*, August 21-24, Portland, OR, National Renewable Energy Laboratory, Golden, CO, NREL/CP-200-0898, DE95009230, pp. 1141-1150.

68. Shihadeh, A.; Lewis, P.; Manurung, R.; and Beér, J. (1994) "Combustion Characterization of Wood-Derived Flash Pyrolysis Oils in Industrial-Scale Turbulent Diffusion Flames," *Proceedings of Biomass Pyrolysis Oil Properties and Combustion Meeting*, Estes Park, CO, September 26-28, National Renewable Energy Laboratory, Golden, CO, NREL CP-430-7215, pp. 281-295.

69. Huffman, D.R.; Vogiatzis, A.J.; and Clarke, D.A. (1995) "Combustion of Bio-Oil," *Bio-Oil Production and Utilization. Proceedings of the 2nd EU/Canada Workshop on Thermal Biomass Processing*, CPL Press, Newbury, U.K., pp. 227-235.

70. Gust, S. (1995) "Flash Pyrolysis Fuel-Oil," *Bio-Oil Production and Utilization. Proceedings of the 2nd EU/Canada Workshop on Thermal Biomass Processing*, CPL Press, Newbury, U.K., pp. 223-226.

71.. Ligasacchi, S.; Mosti, A.; and Rossi, C. (1995) "Bio-fuel Oil Combustion in a 0.5 MW Furnace," *Proceedings Second Biomass Conference of the Americas: Energy, Environment, Agriculture, and Industry*, August 21-24, Portland, OR, National Renewable Energy Laboratory, Golden, CO, NREL/CP-200-0898, DE95009230, pp. 1110-1120.

72. Baxter, L. and Jenkins, B. (1994) "Baseline NO_x Emissions During Combustion of Wood-Derived Pyrolysis Oils," *Proceedings of Biomass Pyrolysis Oil Properties and Combustion Meeting*, Estes Park, CO, September 26-28, National Renewable Energy Laboratory, Golden, CO, NREL CP-430-7215, pp. 270-280.

73. Andrews, R.G. and Patniak, P.C. (1996) "Feasibility of Utilizing a Biomass Derived Fuel for Industrial Gas Turbine Applications," *Bio-Oil Production and Utilization. Proceedings of the 2nd EU/Canada Workshop on Thermal Biomass Processing*, CPL Press, Newbury, U.K., pp. 236-245.

74. Solantausta, Y,; Nylund, N.; Oasmaa, A.; Westerholm, M.; and Sipilä, K. (1994) "Preliminary Tests with Wood Derived Pyrolysis Oil as Fuel in a Stationary Diesel Engine," *Proceedings of Biomass Pyrolysis Oil Properties and Combustion Meeting*, Estes Park, CO, September 26-28, National Renewable Energy Laboratory, Golden, CO, NREL CP-430-7215, pp. 355-361.

75. Oasmaa, A. and Sipilä, K. (1995) "Pyrolysis Oil Properties: Use of Pyrolysis Oil as Fuel in Medium Speed Diesel Engines," *Bio-Oil Production and Utilization. Proceedings of the 2nd EU/Canada Workshop on Thermal Biomass Processing*, CPL Press, Newbury, U.K., pp. 175-185.

76. Gros, S. (1995) "Pyrolysis Oil as Diesel Fuel," *Seminar on Power Production from*

Biomass II, March 27-28, Espoo, Finland.

77. Jay, D.C.; Sipilä, K.H.; Rantanen, O.A.; and Nylund, N-0. (1995) "Wood Pyrolysis Oil for Diesel Engines," *ASME Fall Technical Conference*, Milwaukee, WI, September 24-27.

78. Leech, J., (1996) "Third Interim Report on Development of an Internal Combustion Engine for Use with Crude Bio-oil and Evaluation of Associated Processes", Report to EC, Contract RENA-CT94-0070, June.

79. Casanova-Kindelan, J. (1996) "Crude Bio-Oils in Small Diesel Engines," in these proceedings.

80. Suppes, G.J.; Natarajan, V.P.; and Chen, Z. (1996) "Autoignition of Select Oxygenate Fuels in a Simulated Diesel Engine Environment," *AIChE National Meeting*, February 26, New Orleans, paper 74e.

81. Diebold, J.P. (1985) *The Cracking of Depolymerized Biomass Vapors in a Continuous, Tubular Reactor*, Colorado School of Mines, Golden, CO, Thesis T-3007, 236 pp.

82. Chornet, E.; Czernik, S.; Wang, D.; and Montané, D. (1996) "Catalytic Steam Reforming of Biomass-Derived Fractions from Pyrolysis Processes," in these proceedings.

83. Brocksiepe, H-G. (1986) "Charcoal," *Ullmann's Encyclopedia of Industrial Chemistry, 5th edition, Vol. A6*, VCH, pp.157-162.

84. Reed, T.B.; Diebold, J.P.; and Desrosiers, R. (1980) "Perspectives in Heat Transfer Requirements and Mechanisms for Fast Pyrolysis," *Proceedings Specialists' Workshop on Fast Pyrolysis of Biomass*, Copper Mountain, CO, October 19-22, Solar Energy Research Institute, Golden, CO, SERI/CP-622-1096.

85. Shaddix, C.R. and Huey, S. (1996) "Combustion Characteristics of Pyrolysis Oils Derived from Hybrid Poplar," in these proceedings.

86. Stamm, A. and Harris, E.E. (1953) *Chemical Processing of Wood*, Chem. Publishing Co., NY, p. 454.

87. Gregoire, C.E. and Bain, R.L. (1994) "Technoeconomic Analysis of the Production of Biocrude from Wood," *Biomass and Bioenergy*, Vol. 7, No. 1-6, pp. 275-283.

88. Sax, N.I. (1979) *Dangerous Properties of Industrial Materials*, Van Nostrand Reinhold, NY, p. 895.

89. Gratson, D.A. (1994) "Results of Toxicological Testing of Whole Wood Oils Derived from the Fast Pyrolysis of Biomass," *Proceedings of Biomass Pyrolysis Oil Properties and Combustion Meeting*, Estes Park, CO, September 26-28, National Renewable Energy Laboratory, Golden, CO, NREL CP-430-7215, pp. 203-211.

90. Beckman, D.; Elliott, D.C.; Gevert, B.; Hörnell, C.; Kjellstrom, B.; Östman, A.; Solantausta, Y.; and Tulenheimo, V. (1990) *Techno-Economic Assessment of Selected Biomass Liquefaction Processes*, Technical Research Centre, Espoo, Finland, VTT Research Report 697, pp. 38-63.

91. Beckman, D. and Graham, R. (1994) "Economic Assessment of a Wood Fast Pyrolysis Plant," *Advances in Thermochemical Biomass Conversion*, A.V. Bridgwater, ed., Blackie Academic and Professional Publishers, London, pp. 1314-1324.

92. Cottam, M-L. and Bridgwater, A.V. (1994) "Techno-Economics of Pyrolysis Oil Production and Upgrading," *Advances in Thermochemical Biomass Conversion*, A.V. Bridgwater, ed., Blackie Academic and Professional Publishers, London, pp. 1343-1358.

93. Tippee, R. and Beck, R.J. (1995) "Politics has a Strong Say in Who Benefits from Oil," *Oil and Gas J.*, April 3, pp. 41-53.

PYROLYSIS

Fundamentals

THE NATURE AND PROPERTIES OF INTERMEDIATE AND UNVAPORIZED BIOMASS PYROLYSIS MATERIALS

Intermediate liquid products of pyrolysis

J. LEDE*
Laboratoire des Sciences du Génie Chimique, CNRS-ENSIC, Nancy, France
J.P. DIEBOLD
National Renewable Energy Laboratory, Golden, Colorado, USA
G.V.C. PEACOCKE
Aston University, Birmingham, UK
J. PISKORZ
Resource Transforms International Ltd, Waterloo, Ontario, Canada

*   Author for correspondence

Abstract

Many types of chemical pathways aimed at representing the primary process of biomass degradation have been published. These models include or ignore the possibility of the existence of an intermediate "Active" species or state. The purpose of this paper is to gather theoretical results and experimental observations intended to open the discussion on the possible existence and nature of such an intermediate.

The results of the modelling of the chemical and thermal behaviour of biomass undergoing a pyrolysis decomposition are first given. Simple experimental and visual observations associated with theoretical considerations based on heat transfer measurements lead then to the conclusion that the overall reaction is similar to a fusion with production of an intermediate liquid species. From kinetic rate constants derived from literature, it appears that it is not necessary to take into account such a liquid in thermogravimetric analysis (TGA) experiments, but that it cannot be ignored in high temperature ablative pyrolysis conditions.

At the end of the paper, a discussion is conducted on the possible chemical natures of these liquids, the evolved vapours and the condensed pyrolysis oils formed in a pyrolysis process. It is concluded that the primary liquids could be composed of dimers and higher oligomers derived from cellulose and lignin, while the vapours would be mainly monomers and monomer fragments from the cracking of oligomers in the liquid phase.

Keywords: ablation, biomass, depolymerization, fusion, heat transfer, kinetics, oligomer, pyrolysis

1   Introduction

Several hundred papers have been published during the last decades in the field of lignocellulosic material thermal conversion. The yields, selectivities and chemical compositions of the recovered phases (char, gas, vapour, oil) can be very different following the type of the overall process. They are the result of different interdependent parameters: hydrodynamics of the phases, heat and mass transfer efficiencies, physical and chemical nature of the feedstock and characteristics of the elementary chemical processes. Obviously, the nature of the first primary steps of decomposition is of major importance.

Unfortunately, the steps occur very rapidly and give rise to intermediate and unstable products. Their recovery is hence difficult and needs very efficient mass transfer conditions. This is a first requirement for ablative pyrolysis, the second being the necessity to heat the feedstock under high heat flux densities. These conditions are reached when biomass is heated by more or less contact with a hot moving surface under high applied pressure. Ablative pyrolysis is achievable by very different devices such as hot wire [1], spinning disk [2] and process reactors: vortex [1], cyclone [3], rotating blades [4] and the pyrolysis mill [5].

In addition to these reactors, a great quantity of measurements have been made in thermogravimetric analysis (TGA) devices [7]. TGA methods for kinetic measurements are used worldwide. Unfortunately, they operate under such conditions that the ablative pyrolysis requirements previously mentioned are not fulfilled. Moreover, they are unable to detect chemical transformations occurring through a fusion (non weight loss) step followed by a limiting step of volatilization and/or gas production. These problems connected with the study of fast chemical processes and of pyrolysis performed under intensive heating are discussed by Shlensky et al. [6].

A considerable number of chemical pathways aimed at representing the primary processes of cellulose degradation have been published. Most of the measurements are usually made in specific experimental conditions and it is difficult to derive general conclusions from them. The most reliable and general work in fundamental pyrolysis research typically used pure cellulose. The results from this work are then very often used in a first approximation for biomass in spite of the complexity of its composition (cellulose, hemicellulose, lignin, ash, ...). It is for example the case of Antal and co-workers who reported recently [13] a mathematical model (in agreement with previous similar approaches of Lédé [12, 25]) based on cellulose kinetic constants to explain the behaviour of biomass itself. The first conclusion of this recent paper [13] states for example: "Over the extraordinary range of conditions examined in this work, cellulose pyrolysis occurs in a relatively narrow temperature interval. Consequently, biomass pyrolysis strongly reassembles a phase change (fusion) phenomenon".

Most cellulose pyrolysis models consider competitive steps giving rise respectively to char and gases and to vapours with subsequent secondary consecutive and interphase reactions. However, these competitive reactions concern an intermediate "Active" liquid product or directly the solid macrobiopolymer (Note: for convenience, the term "Active", initially given by Shafizadeh, will be used). In a recent review, Antal et al. [7] suggest that a simple step giving rise to vapours could explain the experiments made in TGA. For the same authors, a primary step giving "Active" species is superfluous between 523 and 643 K [10].

Let us consider the two main relevant simplified models as shown in Figure 1.

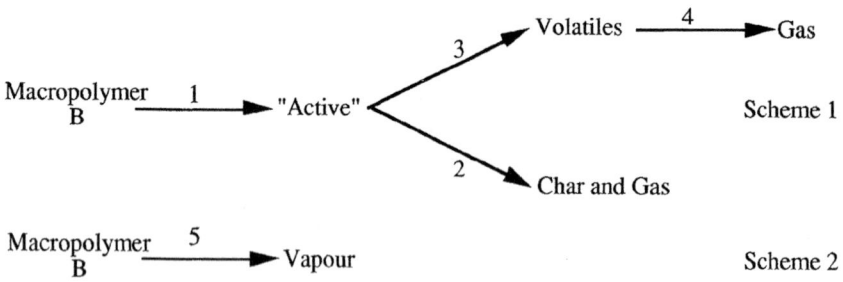

Fig. 1. Simplified biomass decomposition pathways

The first one is the so called Broido-Shafizadeh model [8, 9]. In both schemes, the vapour, when condensed, give rise to pyrolysis oils. These models represent primary steps and do not consider heterogeneous secondary vapour-solid reactions which could be the only source of char as suggested by Antal et al. [7]. From the fundamental chemical point of view, the difference in the assumptions of the two schemes is whether there exists random chain scission of the macrobiopolymer B to form "Active" intermediate or end chain scission giving rise to monomer or dimer vapours as discussed later. It must be noticed also that Narayan et al. [13] supposed that this liquid active intermediate observed for biomass would be attributed to lignin (not cellulose) pyrolysis, even if their modelling efforts, based upon cellulose, are used to explain the phenomenon of biomass fusion (and hence phase change giving a liquid).

The possible existence of "Active" is of capital importance from a fundamental point of view and one can question if it is of the same chemical nature as the volatiles and the condensed pyrolysis oils. It is also of major importance in process reactors. The occurrence of a high temperature liquid film can produce: a favourable lubrication between the feedstock and the hot walls; possible sticking with highly concentrated particles in entrained beds; possible source of aerosols; ... The aim of the present paper is to gather theoretical results and experimental observations intended to open the discussion on the controversial existence and nature of such an "Active" intermediate.

2 Reaction temperature of reacting biomass

2.1 Theoretical considerations
The mathematical modelling of the chemical and thermal behaviour of a solid (biomass) particle undergoing pyrolysis decomposition has been made in the case where the reaction occurs in competition with heat transfer resistances. The calculations have been made with an outside fixed heat source temperature [11, 12, 25] and also in conditions where this temperature increases with time in order to mimic dynamic TGA measurements [13, 25]. The reaction is supposed to be first order and of the type: solid (biomass) → fluids (liquids or gases).

The fluid products are supposed to be immediately eliminated with no possible subsequent reaction in the vicinity of the layer. In the case of a constant external temperature, two extreme cases have been considered: the ablation regime (reaction occurs in a very thin layer close to the surface of a large particle) and the chemical regime (reaction occurs in the whole volume of a small particle where the temperature is uniform). The evolution of the true biomass temperature as a function of the biomass conversion has been calculated from heat and mass balances and has taken into account many possible values of physical and chemical parameters (as for example heat of reaction). The results reveal, after an initial heating period, a very sudden stabilisation of solid temperature as soon as the reaction starts. In the ablation regime, this temperature is that of a thin peripheral layer of the particle: in the chemical regime it is the temperature of the whole volume of the particle. This reaction temperature is only weakly dependent on most of the parameters (heat source temperature, heat transfer coefficient, particle size, choice of enthalpy, ...). For example, an increase of wall temperature from 900 K to 1500 K will produce an increase of only 40 K of true sample temperature [12]. The only sensitive parameter is the activation energy constant in the Arrhenius kinetic rate equation.

Another important conclusion of the calculations [12] is that the true reaction temperature can be much lower than the heat source temperature (sometimes several hundred K difference). These results are confirmed in a recent paper [13]. It is concluded that biomass behaves as during a phase change phenomenon. However, it is not a true physical phase change as it is accompanied with some chemical decomposition.

Such a phenomenon is the result of two phenomena as demonstrated independently by Lédé [12, 14] and Antal [13]. The first one is a competition between heat flux required for heating the solid and heat flux required by the reaction occurring in a short time. If these theoretical results bring evidence for a phase-change-like phenomenon, they are however unable to bring conclusions as to a fusion (the case of step 1 in scheme 1, Figure 1) or to a sublimation (the case of step 5 in scheme 2, Figure 1). This approach does not consider the subsequent steps (2 and 3, (Figure 1)) as the model refers only to the transformation phenomena of the starting material B. Moreover, assuming kinetic constants (see Section 4), it is possible to calculate similar theoretical reaction temperatures [12, 14] for both possible steps 1 and 5 (Figure 1). It can be shown that the model also applies to the vaporisation step 3. Using the kinetic rate constants reported in Section 4.1 and a reaction temperature around 740 K (as discussed later) the ratio of rate constants k_1/k_3 is around 70. This result would mean that for scheme 1, step 3 would be rate limiting, showing that "Active" might be produced from cellulose in high yields before it is converted to volatiles. In these conditions, TGA could not be able to detect the step 1 and it would be concluded to the only occurrence of scheme 2. This point will be quantitatively discussed in Section 4.3.

2.2 Experimental considerations

The direct measurement of reaction temperature is very difficult for different reasons. As previously shown, very important errors can be made if one supposes that the particle temperature is similar to heat source temperature: it can be much lower. This is valid whatever the size of the sample. In ablative processes the reacting biomass surface is in contact with a hot surface and hence, direct pyrometric measurements of the reacting biomass temperature are impossible. Moreover, in ablation the reaction takes place in a very thin layer (a few tens of μm [2]) moving rapidly (several 10^{-2} m/s) towards the centre of the solid which is still at the original feedstock temperature. The direct measurement with a fixed and long-response-time thermocouple is also impossible. Consequently, only an indirect approach can be considered.

The experiments of ablation made with rods of wood pressed against a hot spinning disk have been made in quite similar conditions with true melting materials (rods of ice, paraffin and polyamide 11) [15]. Their behaviour appears to be quite the same as with wood. A very simple model has allowed the calculation of well known melting temperatures of these compounds. The same model applied to biomass experiments gave a theoretical phase-change-like temperature of 740 K [15]. Similar results are reported by Peacocke [4] who finds a reaction temperature of 710 K. These values are in very good agreement with the results derived from the modelling described in the previous section and related to both steps 1 and 5. Note that this good agreement concerns the results of calculations made with kinetic constants derived for cellulose, compared to experimental measurements of pyrolysis temperature performed with wood. If these results bring clear evidence of a phase-change-like reaction they are, however, unable to distinguish fusion (followed by further chemical and/or physical transformations) from simple sublimation. It could also be suggested at this point of the discussion, that in the assumption of scheme 2, vapour could be condensed inside the higher pressure interface between biomass and hot surface (as in an ablative pyrolysis reactor) leading to an apparent fusion behaviour. However, it is difficult to imagine that volatiles escaping from the cooler biomass surface would condense on the hotter heated surface.

2.3 Provisional conclusion

It can be said that biomass decomposes primarily in a narrow range of temperature to give fluid products. The results described above show also that biomass cannot be heated above this domain of temperature without its decomposition. A consequence is that many doubts can be expressed about the validity of kinetic constants derived from TGA experiments made under higher temperature conditions than the narrow temperature window described previously [12, 13, 14].

3   Physical nature of the primary products

3.1   Simple experimental observations

The friction under high velocity of a piece of wood against a red-heated surface gives a feeling of smooth lubrication. This is obviously the result of the presence of a fluid film at the interface. Unfortunately, a similar feeling would be produced either with a liquid or with gaseous volatiles, or both. That case corresponds to the leidenfrost phenomenon which is observable when a drop of water is placed on a very hot surface: the drop forms an insulative layer of steam that slows the rate of heat transfer to the droplet. This results in the droplet levitating above the hot surface for a longer time. In the case of ablative pyrolysis, there is a large increase in the rate of heat transfer compared to the case of a more or less direct contact between a solid and a heated surface. Unfortunately, it has not been widely obvious if this layer is a liquid, a gas or a gas/liquid layer.   It can be observed that the sliding resistance increases in two situations:

- By lowering the applied pressure and relative velocity: mass transfer resistances to the removal of the fluids increase and the liquid can undergo local further cracking to char.
- With lower wall temperatures: there is an increase of liquid viscosity and at the same time the step 2 of scheme 1, giving rise to char, becomes competitive compared to step 3. It must be noticed that no lubrication effect is felt with surface temperatures corresponding to the heat source temperature used in usual TGA measurements for the study of cellulose pyrolysis.

This lubricating effect is a basic assumption for the efficient working of entrained particle ablative reactors (vortex, cyclone, ...)[1,3].   Two major visual observations converge in favour of a liquid nature of these fluid products:

- Pioneering hot wire experiments of Diebold [1]. When, at a certain moment of the experiment, the electric current is turned off part way through a "cut", the products formed freeze on the wire to effectively glue the biomass to the wire. The kerf is thinner than the wire, indicating that the primary products were a melted phase, extruding around the wire as it passes through. Finally, a brownish varnish is left on the side of the cut. It is always dry to the touch, indicating that this high temperature melt is solid at ambient temperature, unlike condensed vapours known as pyrolysis oils.
- Spinning disk experiments conducted by Lédé [2] and later by Peacocke [4]. It is clearly observed a liquid trail left on the hot surface. This very short life time liquid product disappears very rapidly into secondary volatiles and smoke. This is a proof of the existence of a liquid even under atmospheric pressure.   The higher the temperature of the surface, the shorter the length of this trail [4, 16].   A final decomposition into solid char on the surface can be observed, mainly at low temperatures, disappearing slowly by direct oxidation if experiments are made in the air.   This simple observation shows that char can be formed from secondary reactions of the liquid intermediate product ("Active"), in competition with the formation of volatiles. This is in agreement with the scheme 1 (steps 2 and 3).

31

This shows also that secondary reactions of vapour and solid mentioned by Antal et al. [7] are not the only sources of char.

In the case of scheme 1, it can be assumed that "Active" gives volatiles by a boiling-like phenomenon (see Section 5.2). This process is a function of pressure and it follows that the pyrolysis temperature would be higher at higher pressures, with a subsequent decrease of "Active" viscosity. This could cause local pressure increases within the particle with possible expansion of the escaping products. This has been observed in the works of Lally [33] operating in a pressurisable thermobalance: the char formed had a considerable larger volume than the original wood chip. Moreover, photographs showed that the char had a foamed appearance. This clearly indicates that a melted material is intermediately formed, which is blown into a foam by the volatiles formed, before it polymerises into a solid.

Similar phenomena can be also observed in spinning disk experiments : when the applied pressure is very high a fraction of the liquid migrates upwards through the rod and finally escapes several cm higher with the appearance of a foam (liquid with volatile bubbles) through the open pores of the virgin wood. The difference with Lally's experiments is that here, the products are rapidly quenched at room temperature and easily recovered, instead of being further decomposed into char.

In a relatively old paper, Goring [17] reported glass transition temperatures of spruce enzyme lignin (364 to 465 K depending upon the presence or lack of steam) and of cellulose (505 K for dry Avicel accompanied with decomposition). It must be also noted that under heating, lignin shows a softening phenomenon beginning at about 363 K without significant decomposition. More recently, Piskorz et al. [9] pointed out the paper of Nordin et al. [18] which calculated a melting temperature of cellulose of 723 K from the glass transition temperature. This temperature is quite close to the values of apparent phase change, measured and calculated by Lédé [15] and Peacocke [4]. The same authors [18] report visual evidence that melting of cellulose is possible: cellulose was subjected to rapid heating using a CO_2 IR laser and then immediately cooled in a stream of liquid N_2. Photographs clearly show globules of solidified molten material. From the measurement of the degree of crystallinity they concluded that no chemical decomposition had occurred.

Diebold's hot wire experiments can be performed with cellulosic filter paper. The sheet cuts very rapidly leaving on the edges, microscopically visible globules of frozen melt. In very simple experiments made in CNRS, France, simple experiments to briefly contact microgranular cellulose powder and a piece of red hot material have been carried out. The microscopic observation of the products shows clearly that cellulose is transformed into a liquid phase causing agglomeration of the starting grains and resulting in the formation of a solid crust upon cooling to room temperature. The corresponding photographs shown by the authors of this paper during the conference definitely bring evidence that rapidly heated cellulose passes through a liquid intermediate which is solid at room temperature.

In other very simple experiments made at the University of Waterloo, cellulose particles were allowed to pass (free falling) through a vertical heated tube. The microscopic observation of the particles recovered at the bottom of the tube after rapid cooling reveal the presence of small spheres (evidence of a melted phase). The HPLC analysis of these products show several peaks corresponding to levoglucosan, cellobiosan and high oligomers up to a degree of polymerisation of about 6.

Thus it appears that the ability of lignin and cellulose to pass through a glass transition and melt is well known and easily observable. It seems likely that the phase change for lignin occurs at lower temperatures than with cellulose. This apparent fusion can be a simple physical process, or it can be accompanied with chemical decomposition, e.g. depolymerisation. Consequently, the short life time liquids formed during the ablative pyrolysis of wood can result from the fast depolymerization of either lignin and cellulose, or both, but not only from lignin, as has been previously assumed.

3.2 Theoretical considerations

Let us consider a cylindrical rod of wood (radius R) pressed against a hot surface. It is possible to write a very simple mass balance: the mass flowrate of disappearing solid equals the mass flowrate of fluids escaping out of the biomass/hot interface:

$$v_s \, \pi \, R^2 \, \rho_s = v_R \, 2\pi \, R \, e \, \rho \qquad \qquad \{1\}$$

v_s and v_R are respectively the rod ablation and radially escaping fluid velocities. For the sake of simplicity, this balance is written under atmospheric pressure: the mass density ρ and velocity v_R of the fluid products are not calculated inside the high pressure layer between the rod and the surface but just outside, when the fluids leave this interface. Assume that the heat transferred through the film is equal to $h(T_w - T_d)$, where T_w and T_d are the heat source and biomass decomposition temperatures. The thickness e of the fluid layer is equal to the ratio of the thermal conductivity λ of the fluid to the heat transfer coefficient h. The value of v_s with a 2 mm (2 x 10^{-3} m) diameter rod of wood has been measured to be close to 3 cm (3 x 10^{-2} m/s) with a corresponding h value of 40000 W m^{-2} K^{-1} [2, 19]. From very simple assumptions on the values of λ, ρ_s and ρ, it is possible to estimate the rate of ejection v_R in the two cases of a liquid and a gas. The starting feedstock density ρ_s is assumed to be 700 kg m^{-3}. With the assumptions of a liquid, ρ is the same as ρ_s and λ is chosen as 0.05 W m^{-1} K^{-1} (a mean value estimated from usual organic liquid after extrapolation to 740 K). With the assumption of a gas, ρ and λ are assumed to be respectively close to 3 kg m^{-3} and 0.01 W m^{-1} K^{-1}. For a gas v_R (calculated at atmospheric pressure) is found to be about 1000 times greater than for a liquid, with corresponding values close to 10000 m/s, e.g., much higher than the velocity of sound. This would be a very noisy rush of gas exiting from the reaction interface: however, ablative pyrolysis has been observed to be a quiet phenomena! Moreover, the calculated gas film would have a maximum thickness of only 0.2 µm. Such a film could lubricate only a very smooth surface. The possibility of a gas layer therefore seems to be quite unrealistic.

It is interesting to note that these values of e and v_R lead to mean values of the residence time of the liquid at the interface of the order of 0.1 ms, that is several orders of magnitude smaller than the characteristic chemical time of step 3 at 740 K (as discussed later). This means that there is little chance for the presence of a vapour layer between the liquid film and the surface of the hot wall.

The complete mathematical modelling of the ablation rate of a solid cylinder perpendicularly pressed against a stationary heated wall has been made [19]. Very good agreement with experiments has been obtained with true melting materials. The same comparison has been made with rods of beechwood. A good agreement is obtained after fitting the value of fluid viscosities. The calculated value is 72.5 cP (72.5 x 10^{-3} Pa s). This is a quite reasonable value for a liquid having a melting point of 740 K. This viscosity is in the range of a light oil and therefore this liquid would be expected to be a good lubricant. This viscosity is roughly 1000 times higher than the viscosity of a gas at that temperature. The viscosity of many liquids at their normal boiling point is around

0.3 cP (3.10^{-4} Pa s) [20]. This suggests that the liquid is not at its boiling point and would therefore undergo further decomposition to form volatiles.

3.3 Provisional conclusion

These experimental and theoretical considerations show that in ablative pyrolysis experiments, the layer of fluid separating the reacting biomass and the hot surface is a liquid. It is unrealistic to suppose that this liquid results only from lignin thermal degradation: both lignin and cellulose exhibit independently phase change phenomena when subjected to rapid heating. This liquid is probably present between the pyrolysis biomass particles and the walls of ablative reactors (vortex, cyclone, ...) where it plays an important role of lubricant in the trajectories of the particles. If subjected to an efficient thermal quench, this short life time liquid is a very stable solid at room temperature.

In the case of a reactor operating with highly concentrated particles these liquids can cause their sticking with possible further clogging of the reactor mainly during their trajectories in a colder gas environment when temperatures drop to lower values. Such a phenomenon occurs if the mean time between a two particles collisions (calculated from their mean free path) is smaller than the mean life time of the liquid product coating the particles. In addition, the endothermic aspect of the reaction of formation of anhydrosugars from biomass, will cause local cooling and then an increase in "Active" viscosity with an increased tendency for particles which contact each other to stick. It must be pointed out that this sticking phenomenon is more likely to be observed with pretreated (acid washed) feedstock.

Note that the measured heat transfer coefficients between the wood and hot surface have been measured to vary between 1000 and 40 000 W m^{-2} K^{-1} depending upon the applied pressure. Taking a thermal conductivity of the liquid of 0.05 W/m/K, it is possible to calculate a liquid thickness between roughly 1 and 50 μm. At this point, it is worthwhile recalling that the thickness of the ablation layer of the reacting solid has been estimated to be only slightly higher [2, 12]. In the assumption of a gaseous film, its calculated thickness is very small (down to less than 0.2 μm) and then only a very smooth surface could be lubricated, due to microsurface roughness.

4 Chemical kinetic considerations

4.1 Rate constants and experimental domains

Many published kinetic constants correspond to a global first order reaction [21, 7]. If such an approach can be sufficient for the global calculation of a reactor, it does not allow an easy determination of elementary steps. Moreover, if the measurements have been made under very specific experimental conditions (for example TGA) they are not necessarily valid for other more realistic and practical situations, i.e. process reactors. Fundamental research giving low activation energies that can be explained by a poor or limited knowledge of the true sample temperature are ignored at this point in the discussion. As a matter of fact, it has been shown recently [13] that an important thermal lag occurring during pyrolysis can explain the compensation effect observed. This is a probable explanation for the low activation energies derived from prior research [22].

As noted earlier in the paper (See Section 1), the kinetic constants used in the literature to represent biomass pyrolysis are mostly derived from published cellulose pyrolysis measurements. This is, of course, a great approximation as biomass is composed of cellulose, lignin, hemicellulose, ash, etc.. Moreover, the most usual constants are derived from measurements made with pure reactants while it is well known that ash, water, feedstock pretreatments, etc. have very important effects in the biomass pyrolysis [7]. Note, that in the case of the assumption of scheme 1, it is not established if these pretreatment act directly on the macrobiopolymer B (with modification of step 1) or on the liquid secondary decomposition (with modification of steps 2 and/or 3). If one consider that these competitive reactions are not limiting it is likely that these pretreatments would have only limited effect on step 1. The most generally recommended values for the Arrhenius parameters used in equation {2} of the elementary steps of schemes (1) and (2) are given in Table 1.

$$k = k_0 \exp \left\{ \frac{-E}{RT} \right\} \qquad \{2\}$$

The values for scheme 1 are derived from the original experiments of Bradbury et al. [8]. They are generally accepted today and worldwide recommended by numerous authors [e.g. 23, 24]. The values for scheme 2 correspond to the one step scheme proposed recently by Antal and co-workers [7, 10]. In the case of the scheme 1, the formation of volatiles and char are competitive.

Table 1. Kinetics constants for biomass pyrolysis steps corresponding to the two simplified pathways of Fig. 1.

		k_0 (s^{-1})	E (kJ/mol)
Scheme 1:			
Step	1:	$2.8 \ 10^{19}$	242
	2:	$1.3 \ 10^{10}$	151
	3:	$3.2 \ 10^{14}$	198
	4:	$4.3 \ 10^{6}$	108
Scheme 2:			
Step	5:	$1.3 \ 10^{18}$	238

The fraction of volatiles is favoured by higher temperatures. For example, the ratio k_3/k_2 exceeds 10 for temperatures higher than 723 K.

It is noted that scheme 2 has been fitted in TGA experiments for temperatures between 523 and 643 K. Recent calculations [25] using the scheme 1 have confirmed that in most TGA experiments performed under usual external heating rates (1.67 to 1.67 x 10^{-3} Ks^{-1}) the reaction is effectively achieved for sample temperatures between 573 and 633 K. Higher reaction temperatures could be theoretically reached in dynamic TGA measurements operating with higher heating rates. However, the results from reference [13] suggest that values much higher than 100 Ks^{-1} ought to be used to reach reaction temperatures much above 700 K (for example the fusion temperature of 740 K), with resulting important thermal lags between the sample and the heat source. These conditions would then be closer to what is encountered in ablative reactors operating with a fixed heat source temperature, where heating rates of several 1000 Ks^{-1} can be reached [12]. But, even in these extreme cases, TGA conditions could not be compared to ablative reactors because of the very poor mass transfer efficiencies.

The ratio of the rate constants k_1/k_3 for the steps 1 and 3 are always much greater than 1 inside the scope of sample temperatures considered in this paper. This means that in the assumption of the scheme 1, the step 3 would be rate limiting and hence, the step 1 occurring without mass loss could not be evidenced from TGA measurements.

Consequently, an "Active" liquid intermediate could not be observed in TGA. A consequence is that the simple scheme 2 is sufficient to interpret these experiments.

Therefore, it appears that the experimental conditions (true sample temperature and heating rate, mass transfer efficiencies, ...) are quite different in TGA and in ablative reactors. These differences are also evidenced if one considers that the mass flowrates of biomass that can be pyrolysed in usual TGA (around 2×10^{-8} kgs^{-1} [10] is more than 3000 times lower than in ablative pyrolysis where a 2×10^{-2} m diameter rod ablates at rates of around 3×10^{-2} ms^{-1} at 1073 K and an applied pressure of 4 MPa [2].

4.2 Calculation of cellulose true reaction temperature

As previously mentioned, it is possible to calculate the temperature at which biomass theoretically reacts, from the modelling of the competition between chemical and heat transfer processes. From the kinetic constant of step 1 the calculations give reaction temperatures between 693 and 753 K for constant heat source temperatures ranging between 873 and 1173 K and heat transfer coefficients between 2000 and 20000 W m^{-2} K^{-1}. The agreement is very good with measured values obtained for wood of 710 K and 740 K [4, 15].

It is worth noticing that a very good agreement is obtained between measured recent values of pyrolysis temperatures and calculations derived from weight loss measurements made 17 years ago in a vacuum at 200 Pa and with less sophisticated equipment compared to that presently available. It could be argued that Bradbury et al. [8] could not detect and then measure the constant of the step 1 if one considers that it is not rate limiting. But it is clear from Figure 5, page 3276 of reference [8], that this rate constant has been derived under lower temperatures (532 - 568 K). In these conditions, the ratio k_1/k_3 is close to only 4 at 532 K showing that the step 1 is more rate limiting than in the high temperature domain ($k_1/k_3 = 23$ at 643 K, and 70 at 740 K). This shows that "Active" can be observed in TGA weight loss experiments but only under very low temperatures where its formation is rate limiting compared to its volatilization. It is anticipated that it could be also observed in TGA under slow heating rates [32], where the sample is maintained a longer time at these low temperatures. Hence, these conditions: low temperatures and low heating rates, seem to be necessary conditions for observing the step 1 in TGA. However, they correspond to low concentrations of "Active" while in high temperature and high heating rate ablative reactors, the step 3 is rate limiting and concentration of "Active" very high. Finally we can assume that under very low temperatures, the "Active" may be a very high viscosity liquid or more likely a plastic deformable material that under the right conditions could react to form a few stable cross linking chemical bonds and hence become char.

It must also be stressed that the published kinetic constants, derived for cellulose are quite predictive for wood itself. Moreover it can be shown that the results of the model aimed at calculating the sample temperature [12, 14] is mainly dependent on the value of activation energy which is the most sensitive parameter in the calculations. The observed agreement is another proof of the validity of the high activation energy (242 kJ/mol) chosen to represent the step 1. It can be shown that the calculations lead to mathematical impossibilities if they are based on other small published activation energies [22].

4.3 Ability of schemes 1 and 2 to predict the rate of volatiles formation

Using the simple situation where a single particle of cellulose would react without heat and mass transfer limitation, and that the step 2 does not occur to a significant extent, the variations of the mole concentration of vapours V can be calculated as in a batch reactor as a function of time t [26]. For schemes 1 and 2 the corresponding expressions are :

Scheme 1:   (two consecutive reactions)

$$\text{Conc(Volatiles)} = \text{Conc(Macropolymer)}_0 \left[1 - \frac{k_3 \exp(-k_1 t) - k_1 \exp(-k_3 t)}{k_3 - k_1} \right] \qquad \{3\}$$

Scheme 2:   (one single reaction)

$$\text{Conc(Vapour)} = \text{Conc(Macropolymer)}_0 \left[1 - \exp(-k_5 t) \right] \qquad \{4\}$$

where k denotes kinetic constants for steps 1, 3 and 5.

The calculations show that both expressions lead to similar results (same volatile flowrates) in the domain of usual TGA reaction temperatures. It corresponds to ratios k_3/k_5 close to 1. However, for highest values and mainly for temperatures close to 740 K the differences are very important. For example, for a reaction time of 80 ms, and a reaction temperature of 740 K, the ratio of the two theoretical values of Conc(Volatiles) given by these two expressions (4) and (5) is only 0.28. In all the cases, the scheme 2 (represented by equation (4)) leads to a higher theoretical conversion than scheme 1 (represented by equation (3)) for a given reaction time. In the case of scheme 1, it is also possible to calculate the maximum achievable "Active" concentration and the corresponding time :

$$\text{Conc("Active")}_{max} = \text{Conc(Macropolymer)}_0 \left(\frac{k_3}{k_1} \right)^{\frac{k_3}{k_1 - k_3}} \qquad \{5\}$$

$$t_{max} = \frac{1}{k_3 - k_1} \, \text{Ln} \, \frac{k_3}{k_1} \qquad \{6\}$$

For a reaction temperature of 740 K, the calculations lead to a concentration of "Active" close to 94 %, this value being reached after about 18 ms. From equation $\{3\}$ it can be shown that these values correspond to a cellulose conversion close to 1. These results are in very close agreement with those of Diebold [27, 32].

4.4   Provisional conclusion

It is possible to predict very satisfactorily, the ablative pyrolysis temperature of biomass from the kinetic constants obtained by Bradbury et al. [8] from work on pure cellulose.

In the domain of temperatures usually explored in usual TGA experiments, both models give the same results for the rate of volatile formation. In these conditions, the simplest model (scheme 2) is quite sufficient. In ablation conditions where the temperatures, heating rates and mass transfer efficiencies are much higher, scheme 2 seems insufficient and an intermediate species "Active" must be taken into account to predict ablative pyrolysis.

The different behaviour for the rate of volatile formation appear to be theoretically observable for only small differences of sample temperature (a few 10 K) as it can be the case between TGA and typical ablative reactors. This shows also that the temperature measurements must be made with great accuracy in TGA experiments.

5 Chemical nature of the primary products

In the scheme 1, the "Active" gives rise to volatiles (step 3). If these volatiles are maintained at high temperature they will further crack to permanent gases. There seems to be good agreement about the corresponding homogeneous rate constants (step 4) for which Liden [24] and Diebold [27] obtained comparable values. It can be shown that the rate constant for step 4 represents quite accurately the variations of the fraction of condensables and gases obtained as a function of temperature in a cyclone reactor [3].

However, if these volatiles are rapidly cooled, they condense to give pyrolysis oils. A first unanswered question concerns the possible differences in the chemical composition of the three fluid species described by scheme 1: "Active" intermediate (liquid at high temperature and solid at ambient temperature), volatiles, and pyrolysis oils. A second question is how are these products formed from each other.

5.1 Practical difficulties in the measurement of the composition of "Active"

The measurement of the true reaction temperature of the starting sample is very difficult because of the rapidity of the phenomena. The direct determination of the chemical composition of "Active" and volatile is also a very difficult task. At 740 K, the characteristic time of "Active" formation (step (1)) is less than 10 ms. Expecting that the model developed for a solid decomposition is also valid for a liquid, it can be calculated that the "Active" would react (step 3) between 793 and 843 K with characteristic times of a few tens of ms. Even if this step 3 occurred at the fusion temperature of 740 K, the time would be of the order of 1 s. Thus, only indirect speculations can help to understand the chemical composition of "Active".

A first approach is to look at the possible molecular weight and/or degree of polymerisation of the species existing or likely to exist in these products. At the first moments of the process, the biopolymer could first melt and then depolymerise in the melt phase, or depolymerise in the solid phase to give liquid species. A first response comes from a previous section where it has been reported the ability of lignin and polymers of carbohydrates to pass through a glass transition before melting [17]. However it seems that cellulose oligomers are not chemically stable in the melted form. So, it is not clearly understood if "Active" is a short life-time melted form of the starting feedstock (as a true fusion) or if it represents a certain degree of depolymerization. It is likely that both situations can occur following the respective characteristic times of heating, residence and cracking.

5.2 How does step 3 (scheme 1) occur ? Possible nature of "Active" and Volatiles

Is it through a simple physical boiling phenomenon of the liquid "Active"? Are there fast depolymerization processes inside the liquid phase "Active" giving rise to smaller molecules? Are these small molecules gaseous at the reaction temperature, or are they producing volatiles after a final boiling phenomenon? Brule et al. [28] proposed an empirical relationship between molecular weight, boiling temperature and density of petroleum fractions. Assuming a simple physical vaporisation, it is possible to calculate estimates of the boiling points of lignin and cellulose monomers and oligomers. The calculation lead to 731 K and 866 K for hypothetical lignin dimer and trimer. As for cellulose, the predicted boiling points would be 612 K for the monomer levoglucosan, 854 K for the dimer cellobiosan and 1065 K for the trimer. These values would suggest that, as in TGA analysis of Avicel cellulose, the sample loses the majority of weight between 573 and 623 K, levoglucosan is vaporising as soon as it is formed and that the trimers and higher oligomers would not be very volatile in these conditions. It would also suggest that as the fusion temperature has been estimated around 740 K for biomass [15], the volatiles would be made mainly with monomer fragments, cellulose monomers and lignin dimers. These speculations suggest that "Active" would be primarily formed by higher molecular weight molecules undergoing cracking in the liquid phase to give monomer fragments, monomers and/or dimers that would evaporate as soon as they are formed. Such an assumption is supported by the results of Pouwels et al. [29] describing experimental evidence for the presence of oligomers in the pyrolysis of cellulose under vacuum.

5.3 How can the presence of higher molecular weight chemicals (tetramers and higher oligomers) found in the condensed pyrolysis oils be explained? Where are they coming from?

A first explanation would be that the low molecular weight volatiles can chemically recondense to give higher molecular weight compounds (reported by gel permeation chromatography). They would be formed by weakly bound monomers as supported by the similarity of the molecular beam mass spectrometry spectra for volatiles found at atmospheric pressure by pyrolysis and for re-evaporated pyrolysis oils, spectra of which show no significant peaks for trimers or larger species.

A second explanation comes from the true nature of these volatiles which contain gaseous vapour species, but also a large fraction of solid particles and liquid droplets, i.e. high molecular weight aerosols (their relative fractions depending on the reactor design). These aerosol droplets would be expected to be fairly soluble in the pyrolysis liquids and then form the high molecular weight fraction. These heavy species would then be issued directly under a liquid form from the first steps of biomass depolymerization without the occurrence of an intermediate vaporisation. This hypothesis about the primary nature of aerosols has been clearly demonstrated by Piskorz et al. [9] who showed that a considerable portion of the pyrolysis, especially the portion derived from lignin, is generated directly from an aerosol intermediate. However these aerosols are probably not made exclusively of "Active" which is solid at ambient temperature (as seen from hot wire experiments) while the trapping of aerosols in a cold filter gives a liquid phase. This could be due to the presence of water inside the droplets. In any case, the presence of these aerosols represents serious problems in pyrolysis processes and mainly in hot gas filtration where they can accumulate with subsequent plugging of the filter. Moreover, if the pyrolysis reactor produces a high proportion of aerosols that pass to the collection system, the resultant oil will have an undesirable higher viscosity. It is hence of major importance to understand how they are created: partial condensation in the gas phase from volatiles; direct production from the pyrolysing substrate by mechano-chemical processes [9] (for example from microexplosions of superheated steam and "Active" ejected out of the pores of the pyrolysing biomass) ; entrainment of small feed particles out of the reactor while they are still in the "Active" state; ...

Finally, one must recall the important observations made by Peacocke [4] who reported data on the mean molecular weight of several types of pyrolysis oils. The values were always lower when derived from ablative reactors (Aston pyrolyser and NREL vortex reactors) than from fast pyrolysis processes carried out in fluidised bed reactors (Waterloo). This is consistent with the formation of a liquid intermediate (issued from biomass depolymerization) undergoing further cracking or vaporisation on the hot surface of ablative reactors, but being directly ejected into the gas/vapour phase of fluidised beds.

6 Conclusions

The starting point of this paper has presented a certain number of theoretical and experimental data showing that biomass pyrolyses inside a narrow range of temperatures. Such a phenomenon is the consequence of the coupling between heat transfer and chemical processes. It leads to an apparent behaviour similar to a phase change phenomenon. The produced phase is shown to be a short life time intermediate that is liquid at the reaction temperature ("melting" behaviour of biomass pyrolysis). Very simple experiments have been described showing that pure cellulose passes through a liquid phase and hence, the liquid observed with biomass cannot be attributed only to lignin. The existence of such an intermediate liquid phase has been visually observed during spinning-disk experiments and is strongly inferred from hot-wire experiments. The rate of heat transfer calculated to exist in ablative pyrolysis, strongly suggests the existence of a thin film of liquid rather than a necessarily even thinner film of volatiles. This intermediate liquid has low viscosity at ablative pyrolysis temperatures, is a solid at room temperatures, and appears to have a higher molecular weight than the condensates made from pyrolysis volatiles. Boiling point predictions suggest that cellulosic and lignin-derived monomers boil well below the "fusion" temperature of biomass, which strongly suggest that the intermediate liquid is composed of dimers and higher oligomers.

In the domain of temperatures usually explored in TGA experiments, it is not necessary to take into account the existence of such an intermediate for representing the rate of volatile formation and hence mass loss. However in the higher temperature domain of ablative pyrolysis an intermediate unvaporised species cannot be ignored to predict biomass conversion specially for the modelling of process reactors.

It is possible to predict very satisfactorily the ablative temperature of biomass from the kinetic constants proposed by Bradbury et al. [8] 17 years ago with pure cellulose. Whatever the fact that these values have been obtained in rather crude experiments they are quite predictive for biomass and satisfy several modelling aspects of the problem. They are the only values available in the literature that take into account the existence of a liquid intermediate product.

Such an intermediate oligomeric liquid cannot be observed in TGA since its rate of volatilization (to give volatiles, step 3) is rate limiting compared to its rate of formation (step 1) occurring without weight loss. In any case it has been shown that it can be misleading and erroneous to compare results derived in usual TGA and in ablative pyrolysis because of the very different conditions in TGA of lower temperature domains, much smaller heating rates and very poor mass transfer efficiencies. The liquid intermediate can play a major role in process reactors: modelling of the process: lubrication (particle/particle, particle/wall, ...), sticking (particle/particle) causing possible blockages, migration through the pores followed by condensation and further cracking, source of aerosols, produce an increase of the molecular weight of the pyrolysis oil [4]. As pointed out by one of the reviewers of the paper, further experimental and theoretical work is needed. Any new progress in that field will improve the fundamental knowledge of these matters which are of major importance for designing any process of biomass thermal conversion. This is not a new problem, as it was already discussed by Diebold more than 15 years ago [30]. However, the specialists meeting held recently in Breckenridge [31] allowed new stimulating interchanges of ideas and thoughts. This resulting paper must be considered as the starting point of many new constructive discussions and fundamental works on this always controversial subject.

7 References

1. Diebold, J.P. (1980) Ablative Pyrolysis of Macroparticles of Biomass, *Proceedings of Specialists Workshop on Fast Pyrolysis of Biomass*, Copper Mountain Colorado, October 19-22, 1980, SERI/CP-622-1096, pp. 237-251.
2. Lédé, J., Panagopoulos, J., Li, H.Z. and Villermaux, J. (1985) Fast pyrolysis of wood: direct measurement and study of ablation rate. *Fuel*, Vol. 64, pp. 1514-1520.
3. Lédé, J., Verzaro, F., Antoine, B. and Villermaux, J. (1986) Flash pyrolysis of wood in a cyclone reactor. *Chem. Eng. Proc.*, Vol. 20, pp. 309-317.
4. Peacocke, G.V.C. (1994) *Ablative Pyrolysis of Biomass*. Ph.D. Thesis, The University of Aston in Birmingham (UK).
5. Reed, T.B. (1988) Contact Pyrolysis in a "Pyrolysis Mill". *Research in Thermochemical Biomass Conversion.*, (eds. A.V. Bridgwater and J.L. Kuester), Elsevier Applied Science Publishers, New York, pp. 192-202.
6. Shlensky, O.F., Aksenov, L.N. and Shashkov (1991) Thermal Decomposition of Materials. Effect of Highly Intensive Heating. *Studies in Modern Thermodynamics 12*. Elsevier, Amsterdam, Oxford, New-York, Tokyo.
7. Antal, M.J. and Varhegyi, G. (1995) Cellulose Pyrolysis Kinetics: the currrent State of Knowledge. *Ind. Eng. Chem. Res.*, Vol. 34, pp. 703-717.
8. Bradbury, A.G.W., Sakai, Y. and Shafizadeh, F. (1979) A Kinetic Model for Pyrolysis of Cellulose. *J. Appl. Polym. Sci.*, Vol. 23, pp. 3271.
9. Conference on *Frontiers of Pyrolysis : Biomass Conversion and Polymer Recycling* (1995) sponsored by NREL (Golden, Co, USA), Breckenridge, USA, June 25-30.
10. Varhegyi, G., Jakob, E. and Antal, M.J. (1994) Is the Broido-Shafizadeh Model for Cellulose Pyrolysis True? *Energy and Fuels*, Vol. 8, No. 6, pp. 1345-1352.
11. Villermaux, J., Antoine, B., Lédé, J. and Soulignac, F. (1986) A new Model for thermal volatilization of solid particles undergoing fast pyrolysis. *Chem. Eng. Sci.*, Vol. 41, No. 1, pp. 151-157.
12. Lédé, J. (1994) Reaction temperature of solid particles undergoing an endothermal volatilization. Application to the fast pyrolysis of Biomass. *Biomass and Bioenergy*, Vol. 7, No. 1-6, pp. 49-60.
13. Narayan, R. and Antal, M.J. (1996) Thermal Lag, Fusion and the Compensation Effect during Biomass Pyrolysis. *Ind. Eng. Chem. Res.*, Vol. 35, pp. 1711-1721.
14. Lédé, J. and Villermaux, J. (1993) Comportement thermique et chimique de particules solides subissant une réaction de décomposition endothermique sous l'action d'un flux de chaleur externe. *Can. J. Chem. Eng.* Vol. 71, pp. 209-217.
15. Lédé, J., Li, H.Z., Villermaux, J.and Martin, H. (1987) Fusion-like Behaviour of Wood Pyrolysis. *J. Anal. Applied Pyrolysis.* Vol. 10, pp. 291-308.
16. Lédé, J., Li, H.Z. and Villermaux, J. (1988) Pyrolysis of Biomass : Evidence for a Fusionlike Phenomenon. *Pyrolysis Oils from Biomass, Producing, Analysing and Upgrading*, ACS Symposium Series, 376, Washington D.C. (eds. J. Soltes and T.A. Milne), Chap. 7, pp. 66-78.
17. Goring, D.A.I. (1963) Thermal Softening of Lignin, Hemicellulose and Cellulose. *Pulp Paper Mag. Can.*, Vol. 64, No. 12, T517-T527.
18. Nordin, S.B., Nyren, J.O. and Back, E.L. (1974) An Indication of Molten Cellulose produced in a Laser Beam. *Textile Research J.*, pp. 152-154.
19. Martin, H., Lédé, J., Li, H.Z., Villermaux, J., Moyne, C. and Degiovanni, A. (1986) Ablative Melting of a Solid Cylinder Perpendicularly Pressed Against a Heated Wall. *Int. J. Heat Mass Transfer*, Vol. 29, No. 9, pp. 1407-1415.
20. Diebold, J.P. and Power, A.J. (1988) Engineering Aspects of the Vortex Pyrolysis Reactor to Produce Primary Pyrolysis Oil Vapors for Use in Resins and Adhesives.

Research in Thermochemical Biomass Conversion, (eds. A.V. Bridgwater and J.L. Kuester), Elsevier Applied Science Publishers, London, pp. 609-628.

21. Wagenaar, B.M., Prins, W. and Van Swaaij, W.P.M., (1993) Flash Pyrolysis Kinetics of Pine Wood. *Fuel Processing Technology,* Vol. 36, pp. 291-298.

22. Milosavljevic, I. and Suuberg, E.M. (1995) Cellulose Thermal Decomposition Kinetics: Global Mass Loss Kinetics. *Ind. Eng. Chem. Res.,* Vol. 34, pp. 1081-1091.

23. Di Blasi, C. (1994) Numerical Simulation of Cellulose Pyrolysis, *Biomass and Bioenergy,* Vol. 7, pp. 87-98.

24. Liden, A.G., Berruti, F. and Scott, D.S. (1988), A Kinetic Model for the Production of Liquids from the Flash Pyrolysis of Biomass, *Chem. Eng. Comm.,* Vol. 65, pp. 207-221.

25. Lédé, J. (1996) Influence of the Heating Conditions on the Thermal and Chemical Behaviour of Solid Particles Undergoing a Fast Endothermic Decomposition. *J. Thermal Analysis,* Vol. 46, pp. 67-84.

26. Villermaux, J. (1985) *Génie de la réaction chimique. Conception et fonctionnement des réacteurs,* Lavoisier, Technique et Documentation, Paris.

27. Diebold, J.P. (1985) *The Cracking Kinetics of Depolymerized Biomass Vapours in a continuous Tubular Reactor,* Masters Thesis T-3007, Colorado School of Mines Golden, Colorado.

28. Brule, M.R., Lin, C.T., Lee, L.L. and Starling, K.E. (1982) Multiparameter corresponding States Correlation of Coal-Fluid Thermodynamic Properties, *AIChE J.,* Vol. 28, No. 4, pp. 616-625.

29. Pouwels, A.D., Eijkel, G.B., Arisz, P.W. and Boon, J.J. (1989) Evidence for Oligomers in Pyrolysates of Microcrystalline Cellulose. *J. Anal. Applied Pyrolysis,* Vol. 15, pp. 71-84.

30. Diebold, J.P. (1980) Chairman's Report of the Specialists Workshop on Fast Pyrolysis of Biomass, Copper Mountain, Colorado, October 19-22. SERI/CP-622-1096.

31. Conference on *Frontiers of Pyrolysis : Biomass Conversion and Polymer Recycling* (1995) sponsored by NREL (Golden, Co, USA), Breckenridge, USA, June 25-30.

32. Diebold, J.P. (1995) A Unified Global Model for the Pyrolysis of Cellulose. *Biomass and Bioenergy,* Vol. 7, Nos. 1-6, pp. 75-86.

33. Lally, K.J. (1983) *The Design, Fabrication and Operation of a High Pressure, High Temperature Thermobalance for Wood Gasification Studies,* Thesis T-2780, Colorado School of Mines, Golden, Co, 80401.

COMPARISONS IN BIOMASS AND REFUSE DERIVED FUEL PYROLYSIS. MULTIVARATE ANALYSIS APPLIED TO AN EXPERIMENTAL DESIGN.

B. KRIEGER-BROCKETT*, W.C. LAI and W.C. CHAN
Research conducted at:
Dept. of Chemical Engineering, University of Washington
Box 351750, Seattle, WA USA 98195-1750

Abstract

At the University of Washington, biomass devolatilization or pyrolysis has been studied from a fundamental perspective by analysis of single reacting *macro*particles. Despite the steep temperature gradients inside macroparticles, large chips will likely be used in industrial processes for economic reasons owing to the high cost of biomass size reduction. To investigate particle size as well as other effects, numerous experiments have been made on actual and fabricated wood as well as densified refuse-derived fuel macroparticles under well-controlled, realistic pyrolysis conditions of industrial importance. A large number of quantitative measurements have been made on each devolatilizing macroparticle. These include: devolatilization product compositions (overall product fractions and detailed chemical species) as functions of time, internal particle temperature histories, as well as infra-red spectroscopy and surface area of the char residues. However, many of the reaction products and observables appear to vary together, in other words, they are highly correlated over a wide range of experimental conditions. In a previous paper, we examined the product slate correlation for wood devolatilization. In this paper, we explore the similarities and differences between biomass and municipal solid waste devolatilization of thermally thick particles. The large data base, obtained in the *same* experimental apparatus using the principles of statistical experimental design, now permits us to see if some devolatilization products are "surrogates" for others. By this we mean, can only a few gas or tar species be measured and be predictive of nearly the entire composition matrix? This paper presents preliminary efforts in this direction.

Keywords: devolatilization, biomass particle, refused-derived-fuel, pyrolysis product slate, fundamental pyrolysis

1 Introduction and Previous Work

Biomass and municipal solid waste are both heterogeneous, low quality, high oxygen content fuels and feedstocks for chemical production. Their low economic value and the fibrous and heterogeneous nature of both are likely to result in only a limited amount of size reduction or sorting as an affordable feed pretreatment. Thus, large macroparticles that experience steep temperature gradients during pyrolysis (thermally "thick" particles) are likely to be the feedstock used in both direct and indirect industrial scale pyrolysis reactors or combustors. The study of biomass macroparticle pyrolysis behavior can provide insight into pretreatments that enhance the product slate and reaction rate. Such studies can also suggest improved reactor designs with favorable temperatures and flow fields that tailor the macroparticle devolatilization process. Single particle measurements and insight will advance biomass thermal conversion in the same fashion that the study of single catalyst pellets advanced the field of catalytic reactor engineering in the 1950-1960's [1].

It has been shown that it is during pyrolytic conditions, even in high intensity combustors, that some of the final products or pollutants are formed and remain as refractory chemical species [2]. In a large pellet of low thermal conductivity such as biomass or municipal solid waste (MSW), the interior of the particle experiences pyrolytic, not combustion, conditions during much of its heating period. The particle interior also experiences a changing temperature dictated by particle size, particle composition, the presence of moisture, and local reactor heat flux. During heating, the complex porous structure inhibits the flow of volatiles from the particle core. There the volatiles can interact-react, form secondary reaction products, and pursue a different reaction trajectory, heat release, and devolatilization weight loss than would be observed from powdered small particles. For example, it has been speculated that heated large pellets may more readily form char by condensation reactions between reactive intermediates than small particles do, since the reactive intermediates easily escape from small particles [3]. Because more char may be formed, the weight loss as a function of temperature may be less steep for large than small particles. To quantify these differences, we study the thermally thick particle directly, and measure the internal temperatures as well as the instantaneous pyrolysis products.

The effects of reaction conditions such as particle size, heating rate, moisture, particle density, and to a degree, composition on biomass devolatilization have been published in papers by the authors. To quantitatively determine the effects of these reaction conditions or *independent* variables, x_j, on the product distribution, we have performed regression analysis to obtain a set of engineering predictive equations [4,5]. The nature of the equations published therein was

$$y_1 = f_1(x_1, x_2, ..., x_j, ... x_m)$$
$$y_2 = f_2(x_1, x_2, ..., x_j, ... x_m)$$
$$y_i = f_i(x_1, x_2, ..., x_j, ... x_m)$$
$$...$$
$$y_N = f_N(x_1, x_2, ..., x_j, ... x_m)$$

Eq. 1

where the y_i's, or the *dependent* variables, are the various products i such as CO, total gas, tar, and other product yields. In ordinary regression analysis, there must be an equation for each product i, and little summary information is apparent about the relationships *among* the y_i's. Previous papers and relationships focused on the *overall* product yields, and not the individual chemical constituents within the overall product fraction. In this paper, we examine the individual gas and low molecular weight or "light" tar product similarities between wood and MSW using correlation analysis. Correlation basically is the starting point [6] in determining a relationship *among* the dependent variables, the y_i's above; that is,

$$y_1 = g_1(y_2, y_3, ..., y_i, ... y_N).$$

Eq. 2

Other more detailed analyses such as factor analysis or principal component analysis involve subsequent steps after a correlation analysis. These steps are often very informative, but not as perceivable as the correlation analysis itself which can be visualized simply. Thus we limit our treatment to this step, and further analyses are in progress. A brief review of some related efforts at measuring product distributions from thermal decomposition of biomass appears below.

2.1 Small Particle Decomposition Studies - Modeling and Experiments

Early work on biomass is summarized in several papers [7-11] among others. A comprehensive bibliography and comments on published biomass thermal degradation kinetic parameters have been written by Krieger-Brockett [12] and only selected features are discussed here. Some studies have used pyrolysis mass spectrometry to characterize a number of whole biomass samples [13-17]. These three authors in particular have used statistical methods to reduce the large number of redundant peaks from mass spectrometry into a smaller number of highly "characteristic" peaks. In mass spectrometry, the parent compound produces fragmentation daughter peaks, and since all peaks are correlated to a high degree, frequently the most "characteristic" is the parent peak. It is usually determined using multivariate statistics [6,16]. Earlier work on MSW has been summarized by Lai and Krieger-Brockett [5,18] for both small and large particles.

2.2 Large Particle Studies - Modeling and Experiments

Some investigators have investigated large particle devolatilization using mathematical modeling studies [19-23]. Few modelers with the exception of Murty & Blackshear [24-28], Ohlemiller et al [29,30], and ourselves [31-35] have conducted *both* experiments and modeling studies of thermally-thick biomass or MSW particles during devolatilization.

The previous literature is dominated by studies at extremes. We found studies that measured too few decomposition products to regress against reaction conditions, and thus were unable to provide engineering correlations suitable for process modeling. We also found mass spectroscopic studies that measured so many decomposition products of such small concentration that they were difficult to correlate with manipulatable reaction conditions. Our work focuses between these extremes.

3 Experimental Details: Materials and Methods

In the work reported here the devolatilization behavior of wood or biomass was represented by lodgepole pine, a species of regional interest. Devolatilization of municipal solid waste was determined using pellets of densified refuse-derived fuel (d-RDF) made from four principal constituents in proportions bounded by typical ranges found in MSW, and dictated by a statistical experimental design to study composition effects [34]. The whole wood, shredded MSW, or ground and sieved biomass samples were compressed into nominal 1 cm diameter pellets with a constant measured density. All the materials were dried in an oven at 90°C for about 48 hours and well-mixed before pelletization. Samples with morphology removed were -80 mesh granules compressed in a cylindrical mold. Samples with morphology intact were lathed to fit the reactor. The whole wood macroparticles were prepared by choosing a uniform grain log and lathing 1 cm diameter dowels of varying lengths (thermal thickness in the direction of heating). Both four-component *fabricated* pellets of d-RDF, as well as reworked *actual* d-RDF pellets, were made from RDF obtained from Thief River Falls, MN under a collaboration with the National Renewable Energy Lab (NREL, 1987-1993). Finished pellets were kept in the air to reach equilibrium moisture content (about 5%). Additional moisture was added quantitatively using the procedure described in Lai [34]. Biomass and RDF pellets were prepared with the same sizes, lengths, moisture contents, and had thermocouples inserted at the same actual and dimensionless distances in the interior for meaningful comparisons. Surface temperature was measured somewhat less accurately using an infrared optical pyrometer mounted off-axis from the arc lamp beam.

A brief description of the single particle devolatilization reactor, analysis system and methodology is given below, and a more complete description is given in the theses of Chan [35] and Lai [34] as well as the papers by the authors. A funnel-shaped glass reactor held a single, fabricated pellet inside a second removable tube at the intersection of the "funnel" and the "neck". The double glass wall and intervening air space provided radial insulation and ensured 1-dimensional axial heating of the cylindrical sample. During the fixed duration pyrolysis run (12 min), the pellet front face was heated radiatively with an Oriel xenon arc lamp whose beam was barely attenuated by passing through the large diameter, fused silica window on the reactor "funnel" fitted with a baffle. The incident radiation was absolutely calibrated and, by spatial mapping, determined to be uniform across the pellet face. The fixed heating time was chosen short enough to examine only vigorous devolatilization (not "cooking" or gasification of the char) for all conditions. In a reactor design context, the fixed time was analogous to the transit time of a pellet in a reactor. Oxygen was excluded since we wished to study the feed polymer fragmentation patterns as well as pyrolytic conditions preceding gas phase combustion or other reactions.

An experiment consisted of heating the biomass particle at constant applied heat flux for a fixed time. The samples were prepared and pyrolyzed under specific conditions of a statistical experimental design partially shown in Table 1. The thickness ranged from 0.5 to 2.5 cm, the initial moisture from 5 to 110% dry basis, and macroparticles were heated in the apparatus at a constant applied heat flux ranging from 2 to 6 cal/cm^2-sec as shown in Table 1. The outflowing volatile products were quenched by helium carrier gas and swept from the pellet surface to the analysis system. The baffle directed the carrier gas flow; the negligible backmixing of these gases and the short transit time to analysis were quantified using tracer experiments. The volatile mixture passed through a dry-ice/acetone cold trap (at -40°C) to condense tars and water, collectively called "condensibles". Condensible samples were prepared for analysis by Liquid Chromatography (LC) or Gas Chromatography (GC). Downstream of the trap, a small portion of the non-condensable gas flowed through automated sampling valves into calibrated loops. The samples provided a semi-continuous history of the gas evolution rate and its composition as presented elsewhere [18,34-36]. The gas samples' composition was quantified after the experiment using a gas chromatograph employing authentic samples for compound identification and external standards for quantitation. GC analysis of the extracted tar sample was done using internal standards to determine water and the hydrocarbon composition. The fraction of the sample that reacted and its char yield were determined gravimetrically after the experiment. The char was easily discernable from the unreacted portion of the particle and was separated using a scalpel. Consistent with 1-dimensional heating, the interface between the char and unreacted biomass was flat across the pellet diameter. Char surface area was measured by dynamic gas adsorption and char composition by Fourier Transform Infra-red Spectroscopy. Time-integrated overall product yields were given in Chan et al [4] for lodgepole pine, Lai and Krieger-Brockett [5] for municipal solid waste/refuse-derived fuel. The analysis methods were described in detail in these papers. Other papers presented time-dependent results [18, 32] and comparisons to a mathematical model [31, 36].

4 Results and Discussion: Wood and RDF Pyrolysis Comparisons

Our experimental approach avoids performing simple comparative experiments because of the enormous particle variability known to exist in biomass and RDF. Changes in the pyrolysis product slate may result from sample-to-sample variability or "noise" rather than from the intended influence of the experimental conditions. Thus we use large numbers of experiments, hidden replication in experimental designs, regression analysis, and thereby develop *coefficients* that quantify the effects of the reaction conditions in spite of the sampling noise [37]. Hidden replication was a phrase used by Box et al [37] to describe repeated trials in *some* of the experimental conditions, while other experimental conditions are examined in a systematic (orthogonal) way, as will be illustrated in the example below. The particular statistical experimental design used for MSW devolatilization studies is given in Lai and Krieger-Brockett [5]; it is a combination of a mixture design and process variable fractional factorial design. The specific experimental design for the devolatilization of wood is given in Chan, et al [4] and Krieger-Brockett [38]. The rationale and practical importance of these reaction conditions is also given in the references by the authors. The ranges of reaction conditions are shown in Table 1. Sometimes the experimental design has few direct comparisons at identical conditions. However, a few direct comparisons will be made now.

Several moderate heating intensity (3 cal/cm^2-s) wood[2] pyrolysis runs were accomplished using the single particle pyrolysis reactor in order to establish similarities and differences between RDF and wood, and thus utilize the extensive data on wood pyrolysis for initial modeling work on RDF pellet pyrolysis. The temperature histories of lodgepole pine during the moderate heating are shown in Fig. 1, with the temperature histories of RDF (which has the same heating rate, moisture content, and particle size) indicated for comparison. The temperature profiles at long times in RDF appear to be lower (about 80°C) than those in pine at the same process conditions. Nor does the temperature rise as rapidly in RDF (initially about 250°C/min at the 2 mm, 50°C/min at 6 mm depth). However, the temperature increment between the 2 and 6 mm traces for both samples is similar; it might indicate that at higher temperature the thermal conductivity of pine is larger than that of RDF although they have similar thermal conductivities at lower temperature. This might be accounted for by the more porous product char from RDF compared to pine at higher temperature due to structure differences. Consistent with the higher temperatures shown during pyrolysis, the pine samples appear to react to a larger extent (54%) in the 12 minutes pyrolysis time than identically heated RDF samples (45%). The char surface area, as determined by CO_2 adsorption at 20°C, is lower for RDF (182 m^2/g) than for lodgepole pine (310 m^2/g). These results are expected because the RDF experiences a lower temperature and slower heating rate than pine [3].

Figure 2 presents the mass flux of non-condensible gases from the pellet surface as a function of time. It shows that the volatiles from RDF leave the particle less rapidly than for pine, consistent with its lower temperature, and potentially lower reactivity. All these observations imply that the RDF might have a smaller thermal conductivity than pine after all, although they have similar magnitudes of thermal conductivity at room temperature. This finding is important to optimumizing the temperature distribution in a reactor designed to handle either RDF or biomass or both. However, as stated earlier, we wish to avoid comparative pairs of experiments because of the great particle to particle variability in both wood and d-RDF. Therefore, we will compare larger sets of experimental design results from now on.

The devolatilization product slates from both biomass substrates have been published elsewhere [4,5,34,35]. To summarize them, there are over 15 time-integrated product yields, that is, the y_i's in Eq. 1. They consist of gross product fractions (char, tar, gas, and total %reacted in a fixed pyrolysis time) as well as individual species' yields. The species' yields were obtained from gas chromatography and the use of authentic samples for compound and peak identification. From experiments designed to elucidate the effects of cellulose, newsprint, and plastics [34,35] we can obtain Figure 3 which shows a spectrum of percent yields of 9 selected products from cellulose pyrolysis

---

[2] The lodgepole pine used in these pyrolysis experiments has the properties as follows: parallel grain direction, density (0.374 g/cm^3), thermal conductivity (3.82x10^{-4} cal/cm-s-°C), and heat capacity (0.6 cal/g-°C)

at two different heat fluxes, and RDF at a comparable heat flux. The experimental conditions are similar but not identical. We can see that the carbon oxide yields are quite different for the RDF pyrolysis than from either of the cellulose pyrolyses. If we expand and form the ratio of the product slate, Fig. 4, we see that the product yield ratio for both cellulose pyrolyses (despite the different conditions) fall near one. When the RDF and cellulose pyrolysis product spectra are ratioed, we see that that celllulose produces a greater fraction of carbon oxides, low molecular weight hydrocarbons, and char, but RDF produces considerably (over 2.5 times) more tar. Although RDF has a large fraction of newsprint (see Table 1) which contains considerable cellulose as does pine (about 60-75%), RDF contains plastic which evolves hydrocarbons during pyrolysis [18]. Apparently the greater reaction rate, reaction extent, and higher temperatures found in pine (see Figs. 1 and 2) are responsible for the lower product yield ratios in Fig. 4. The greater tar product from RDF may be due to the lower temperature and insufficient intraparticle residence time for the tar molecules to crack to lighter gases [39].

In addition to the identified light tar products such as acetic acid, we obtained typical chromatograms such as those in Ref. [38] for the gas and the high molecular weight tar product. In the high molecular weight tar chromatogram, over 30 peaks appear, only about 10 of which are positively identified. Reasonably, when faced with the approximately 20 additional tar components not yet analyzed nor positively identified, we pose the question - Are all the peaks independent of each other? or are some "correlated" with other peaks, and therefore need we only measure a few to compactly describe the tar composition?

It is useful to explain what we intend by "correlation" techniques used to analyze the devolatilization data. When product slates from biomass or MSW pyrolysis contain many measurements, and the measurements are obtained over a wide variety of known conditions (but in the same apparatus), it is revealing to correlate or determine a measure of the linear relationships among the components of the product slate, e.g., Eq. 2. A concrete example of correlation might be the way *weight and height* of primary school children increase together, or are strongly *correlated* with each other, despite individual variability. Weight and height are so-called "dependent" variables, that is, similar to the y_i's in Eqs. 1 and 2. In fact the weight and height of primary school children are also correlated with their ages, genetics, nutrition, and many other "independent" variables, that is, x_j's in Eq. 1. However, the weight and height "dependent" variable *correlation*, analogous to g_1 in Eq. 2, is a consistent, obvious, visible measure of student growth and we have an expectation of its range. For example, we easily recognize outliers, i.e., very thin, tall students, or very short, stocky students. Similarly in the realm of biomass or MSW devolatilization, we expect the heating value of a particular fuel to be strongly correlated with fuel carbon content *no matter what the conditions and devices (independent variables) are that measure the heating value.*

Since correlation among many dependent variables such as pyrolysis product slates depends on many independent variables, it is useful to *visualize* correlation behavior with a simple, but multidimensional example. One visualization is described by Malinowski [40]. He lucidly discusses the the correlation coefficient as being related to the cosine of the angle between two vectors of either dependent or independent variables. We will proceed in a related way. Though factor analysis or other techniques can reveal relationships among the dependent variables such as the product yields, they are less visual and correlation analysis is implicit in all of them.

A hypothetical data set was generated from simple values, but orthogonal or uncorrelated values of independent variables x_1 and x_2. They are shown as the 2 leftmost columns in Table 2. Correlation analysis and plotting of x_1 against x_2 will show a zero correlation and a plot such as Fig. 5a. If we generate synthetic data with a linear function, z_{12}, of these two independent variables, we will also choose the function such that the angle between z_{12} and *both* x_1 and x_2 is 45° (recall the cosine (45°) = .7071). Thus generate z_{12} as follows:

$$z_{12} = .7071\, x_1 + .7071\, x_2$$

as shown in column 3, Table 2. As is the case in real experiments, we add noise to these hypothetical data,

$$z_{12,n} = .7071\, x_1 + .7071\, x_2 + noise$$

which in this case is chosen arbitrarily to be 20% random noise. The resulting synthetic data are shown in Table 2 as column 4. A statistical analysis will show that the noise has a mean of zero and standard deviation of about 20% of z_{12}. Incidentally, Table 2 illustrates hidden replication. There are two *obvious* replicates of the conditions ($x_1 = -1$ and $x_2 = -1$). However, there is hidden replication in that the condition $x_1 = -1$ is investigated *four* times, albeit at different conditions of x_2. If x_2 turns out to have no impact on the outcome, we will have four estimates to the influence of x_1 [37]. The hidden replication feature has greater importance the more variables we investigate at once in an experimental design.

If we plot z_{12} against x_1 and x_2 in 2-D plots, such as Fig. 5b and 5c, the simple (known, synthetic) relationship with each of the independent variables x_1 and x_2 is not revealed. The added noise shows up as a slightly different pattern for each plot. The angle between x_1 and x_2 and z_{12} in each of Figs. 5b and 5c is about 45°, but is not easily discerned, both because only one of the variables x can be plotted in a 2-D plot, and because the noise is significant enough to obscure the relationship.

A more revealing 3-D plot is shown in Fig. 5d. Here the cross-hatched surface is the plane fitted to the data $z_{12,n}$. The angle of the fitted plane appears to be 45° relative to both x axes, and the noise is somewhat discernable as locations of the points above and below the fitted plane, although the viewing angle is not perfect. Performing correlation analysis on these four variables gives Table 3, where the correlation coefficient between pairs of variables is the intersection of a row and column. Thus, we see the correlation between x_1 and x_2 is indeed zero (orthogonal), and the correlation coefficient between z_{12} and each of x_1 and x_2 is .7071 as designed. The correlation between x_1, x_2 and the dependent variable $z_{12,n}$ (with noise) is reduced by only 2% despite the 20% noise added into the data! The correlation between z_{12} and $z_{12,n}$ (generated with noise equal to 20%) is nearly perfect at .9825!

This simple example illustrates the ability of correlation analysis to overcome effects of noisy data and reveal the intrinsic dependence of a complex entity like product slate on many independent variables. This simple example also illustrates how we might *visualize* the dependence of a pyrolysis product, say CO, on *many* independent variables such as the particle size, composition, moisture content, and heating rate experienced by a biomass or RDF pellet in a reactor. Were we to plot the CO produced against any one variable, the relationship would be obscured as in Figs. 5b and 5c. However, if we had a higher dimensional way of visualizing, such as Fig. 5d, we could see the covariation and correlation of CO yield on (BOTH or) all the independent variables. Correlation coefficients summarize this pairwise interdependence between two variables in higher dimensional spaces. However, we wish to apply it to the *dependent* variables, namely the pyrolysis chemical yields.

Table 4 presents the correlation coefficients among the gas and light tar products concerning all the conditions and compositions studied for lodgepole pine, and Table 5 presents the same for RDF. In what appears here, the correlation analysis is applied to 60 systematic experiments' product slates. Examination of the the two tables reveals a few selected relationships though there are more that could be mentioned. For pine, the carbon oxides are highly correlated with each other, the hydrocarbons are correlated with each other except C_2H_2, and surprisingly the hydrocarbons correlate highly with CO. For RDF, the methane yield is highly correlated with CO. These relationships are perhaps more easily seen if they are plotted.

Figure 6 presents the plotted correlation coefficients between reaction products for lodgepole pine pyrolysis. Displayed is the magnitude of the elements in a column of Tables 4 and 5 against the species row names on the abscissa. For example, CO is plotted with a square, and we see that CO has a high correlation coefficient with all but the acetic acid, acetone, and methanol products from pine pyrolysis. In contrast, CH_4 (methane - plotted with an upwards triangle) is negatively correlated with acetic acid and acetone production. A negative correlation means that when CH_4 increases, acetic acid decreases. If the two negatively correlated species were plotted against each other, a plane would result with about an angle measured from the positive abscissa of arccosine (-0.5), or about 120° counterclockwise. Since each column In Table 4 decreases in size from left to right, so do the lengths of the plotted lines in Fig. 6. Figure 7 presents a similar view for RDF. We see by comparing the two graphs that RDF product slates show no negative correlations, that is, apparently the product yields all go up or down in concert. CO and CO_2 appear to have a relatively low but positive correlation with acetylene production from RDF pyrolysis. Ethylene appears to have a high correlation with carbon oxides and methane production, but low with acetylene. Acetylene appears to be nearly uncorrelated with the tar compound production, e.g., acetic acid and other light tars.

A last comparison can be formed by taking a *ratio* of the respective correlation coefficients - one by one - in the columns and rows of Tables 4 and 5. Table 6 and Fig. 8 result. This is an "abstract" result to think about. It is most useful in revealing "differences". Since both RDF and pine correlation analyses were done with the same "order" and organization of compounds in the tables and matrices, there is much to be learned from Fig. 8. Most of the correlation coefficient *ratios* are between 1 and -1 as expected; in other words, each product appears to bear a similar relationship to its sister products in the product slate for both substrates, RDF and lodgepole pine. However, acetylene appears to have a completely different chemical origin from pine than from RDF, and that origin from pine bears a relationship to the production mechanism for ethylene and propane as well. Our next step is to examine the chemical mechanisms, such as those described by Richards [41], in order to fundamentally understand the similarities and differences in wood and RDF pyrolysis product slates.

5 Summary and Conclusions

We are continuing to examine quantitatively the similarities and differences between wood and RDF pyrolysis. Much more is known about wood. The wood substrate itself, though highly variable, is *less variable* than MSW or RDF. The single particle devolatilization results, recall, are obtained over a wide range of conditions (Table 1). Thus correlation analysis allows us to make some general observations about the interrelationships of the product slates over these widely varying conditions which will be quite useful in reactor design and optimization for heterogeneous feeds.

6 Acknowledgments

The author would like to acknowledge the financial support of the National Science Foundation for the work on wood, and the National Renewable Energy Lab (formerly Solar Energy Research Institute) for the work on wood (through the University

Programs) and RDF (through a research contract). The author would also like to acknowledge the helpful comments and support of MKB and the reviewers.

7 References

1. Peterson, E.E. (1965) *Chemical Reaction Analysis,* Prentice Hall, New Jersey, USA.
2. Tsang, W. (1990) Mechanisms for the Formation and Destruction of Chlorinated Organic Products of Incomplete Combustion. *Combust. Sci. and Tech., 74,* p. 99.
3. Brunner, P. H. and P. V. Roberts (1980) The Significance of Heating Rate on Char Yield and Char Properties in the Pyrolysis of Cellulose, *Carbon, 18,* pp. 217-224.
4. Chan, W. C. R., M. Kelbon and B. Krieger-Brockett (1988) Single-Particle Biomass Pyrolysis: Correlations of Reaction Products with Process Conditions, *Industrial and Engineering Chem. Res., 27,* pp. 2261-2275.
5. Lai, W-C. and B. Krieger-Brockett (1993) Single Particle Refuse-Derived Fuel Devolatilization: Experimental Measurements of Reaction Products, *Industrial and Engineering Chem. Res.* 32, pp. 2915-2929.
6. Davis, J.C. (1986) *Statistics and Data Analysis in Geology,* 2nd Ed., Wiley, N.Y, USA.
7. Roberts, A. F. (1971) Problems Associated with the Theoretical Analysis of the Burning of Wood, *Thirteenth Symposium (International) on Combustion,* The Combustion Institute, Pittsburgh, PA, pp.893-903.
8. Milne, T.A. (1979) Pyrolysis - The Thermal Behavior of Biomass Below 600°C, in *A Survey of Biomass Gasification Volume II - Principles of Gasification.* SERI/TR-33-239,*Solar Energy Research Institute,* (July 1979).
9. Antal, M. J., Jr. (1985) Biomass Pyrolysis: A Review of the Literature: Part 2 - Lignocellulose Pyrolysis, in *Advances in Solar Energy,* edited by Karl W. Boer and John A. Duffie, Vol 2, pp 175-255. Boulder, CO: American Solar Energy Society, Inc., and New York: Plenum Press.
10. Antal, M.J. (1989) Pyrolysis, in Kitani & Hall, *Biomass Handbook,* Gordon & Breach, Sci.
11. Kitani, O. and C.W. Hall, Eds. (1989) *Biomass Handbook,* Gordon and Breach Sci., 960 pages.
12. Krieger-Brockett, B. (1992) Analysis of Kinetics used for Biomass Combustion Modeling, *Nordic Seminar (Proceedings) on Biomass Gasification and Combustion,* Trondheim, Norway (Dec., 1992) 23 p.
13. Evans, R. J. and T. A. Milne (1986) Fundamental Pyrolysis Studies: Molecular Characterization of the Pyrolysis of Biomass. I. Fundamentals and II. Applications, *SERI/PR-234-3026, Solar Energy Research Institute,* (Sept. 1986).
14. Evans, R. J. and T. A. Milne (1987) Molecular Characterization of the Pyrolysis of Biomass. 1. Fundamentals, *Energy and Fuels, 1*(2), pp. 123-137.
15. Evans, R.J. and T.A. Milne (1988) Mass Spectrometric Studies of the Relationship of Pyrolysis Oil Composition to Formation Mechanisms and Feedstock Composition, in *Research in Thermochemical Biomass Conversion,* edited by A. V. Bridgwater and J. L.Kuester, New York: Elsevier Applied Science, 1193 p.
16. Meuzelaar, H.L.C., J. Haverkamp, and F.D. Hileman (1982) *Pyrolysis Mass Spectrometry of Recent and Fossil Material: Compendium Atlas,* Elsevier Sci. Pub., New York, 126 p.

17. Pouwels, A. D., A. Tom, G.B. Eijkel, and J. J. Boon (1987) Characterisation of Beech Wood and its Holocellulose and Xylan Fractions by Pyrolysis-Gas Chromatography-Mass Spectrometry", *J. of Analytical and Applied Pyrolysis, 11*, pp. 417-436.

18. Lai, W.-C., and Krieger-Brockett, B. (1992) Volatiles Release Rates and Temperatures during Large Particle Refuse Derived Fuel - Municipal Solid Waste Devolatilization. *Combust. Sci. and Tech.*, Vol. 85, pp. 133-149.

19. Roberts, A. F. and G. Clough (1963) Thermal Decomposition of Wood in an Inert Atmosphere, *Ninth Symposium (International) on Combustion*, The Combustion Institute, Pittsburgh, PA., pp. 158-166.

20. Roberts, A. F. (1967) An Analogue Method of Estimating Wood Pyrolysis Rates, *Eleventh Symposium (International) on Combustion*, The Combustion Institute, Pittsburgh, PA, pp. 561-565.

21. Roberts, A. F. (1970) A Review of Kinetics Data for the Pyrolysis of Wood and Related Substances, *Combustion and Flame, 14*, pp. 261-272.

22. Roberts, A. F. (1971) The Heat of Reaction During the Pyrolysis of Wood, *Combustion and Flame, 17*, pp. 79-86.

23. Grønli, M, and M. Melaaen (1993) Modeling the Convective Flow Inside Logs During Drying, Pyrolysis, and Combustion of Wood, *Nordic Seminar (Proceedings) on Biomass Gasification and Combustion*, Trondheim, Norway, Aug 30, 1993, 40 p.

24. Murty Kanury, A. and P. L. Blackshear, Jr. (1967) Pyrolysis Effects in the Transfer of Heat and Mass in Thermally Decomposing Organic Solids, *Eleventh Symposium (International) on Combustion,* The Combustion Institute, Pittsburgh, PA, pp. 517-523.

25. Murty Kanury, A. and P. L. Blackshear, Jr. (1970) Some Considerations Pertaining to the Problem of Wood-Burning, *Combustion Sci. and Technology, 1*, pp. 339-356.

26. Murty Kanury, A. (1971) Burning of Wood - A Pure Transient Conduction Model, *J. Fire and Flammability, 2,* p. 191.

27. Murty Kanury, A. (1972) Thermal Decomposition Kinetics of Wood Pyrolysis, *Combustion and Flame, 18,* pp. 75-83.

28. Murty Kanury, A. (1973) Rate of Charring Combustion in a Fire, *Fourteen Symposium (International) on Combustion,,* The Combustion Inst., Pittsburgh, PA, p. 1131.

29. Ohlemiller, T. J., T. Kashiwagi, and K. Werner (1985) Products of Wood Gasification, *National Bureau of Standards, NBSIR 85-3127*, U. S. Department of Commerce, April 1985.

30. Ohlemiller, T. J., T. Kashiwagi, and K. Werner (1987) Wood Gasification at Fire Level Heat Fluxes, *Combustion and Flame, 69*(2), pp. 155-170.

31. Chan, W. C. R., M. Kelbon and B. B. Krieger (1985) Modeling and Experimental Verification of Physical and Chemical Process during Pyrolysis of a Large Biomass Particle, *Fuel, 64*, pp. 1505-1513.

32. Krieger-Brockett, B. and D. S. Glaister (1988) Wood Devolatilization - Sensitivity to Feed Properties and Process Variables, in *Research in Thermochemical Biomass Conversion,* edited by A. V. Bridgwater and J. L. Kuester, pp 127-142, New York: Elsevier Applied Science.

33. Glaister, D. S. (1987) M.S. Thesis, The Prediction of Chemical Kinetic, Heat, and Mass Transfer Processes During the One- and Two-Dimensional Pyrolysis of a Large Wood Pellet, University of Washington, Dept. of Chemical Engineering.

34. Lai, W.-C. (1991) Reaction Engineering of Heterogeneous Feeds: Municipal Solid Waste as a Model. Ph.D. Dissertation, University of Washington, Dept. of Chemical Engineering.

35. Chan, W. C. Ricky, (1983) Ph. D. Thesis, Physical and Chemical Processes During Pyrolysis of a Single Wood Particle, University of Washington, Dept. of Chemical Engineering.

36. Lai and Krieger-Brockett (1996) An Experimental and Mathematical Simulation of a Large, Wet Pyrolyzing Refuse Derived Pellet, in preparation.

37. Box, G.E.P., W.G. Hunter, J.S. Hunter (1978) *Statistics for Experimenters*, Wiley, N.Y. 653 p.

38. Krieger-Brockett, B. (1994) Single Particle Pyrolysis and Combustion Research on Biomass, *IEA Biomass Combustion Conf. Proceedings*, Cambridge, UK, 29 Nov. 1994., 12 p.

39. Boroson, M. L., J. B. Howard, J. P. Longwell, and W. A. Peters (1989) Product Yields and Kinetics from the Vapor Phase Cracking of Wood Pyrolysis Tars, *AIChE Journal, 35*, pp. 120-128.

40. Malinowski, E. (1991) *Factor Analysis in Chemistry*, Wiley-Interscience, NY.

41. Richards, G.N. (1993) Chemistry of Pyrolysis of Polysaccharides and Lignocellulosics, Bridgwater, A.V. Ed., *Advances in Thermochemical Biomass Conversion*, Blackie Academic and Professional, London pp. 727-745.

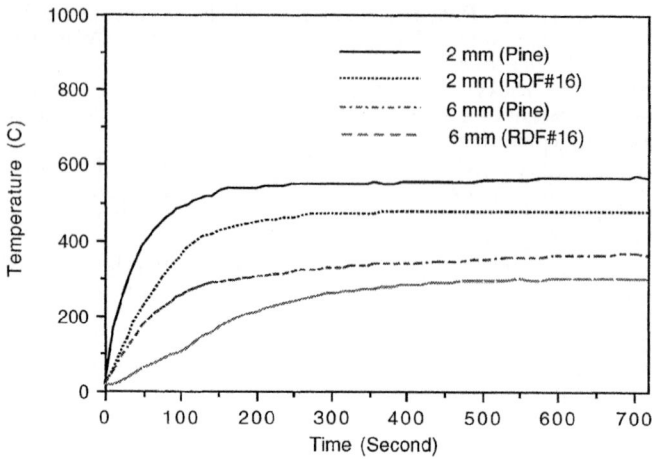

Fig. 1. Experimental temperature histories for 1.0 cm pine and RDF pellets under a 3 cal/cm^2-s heat flux at the axial surface

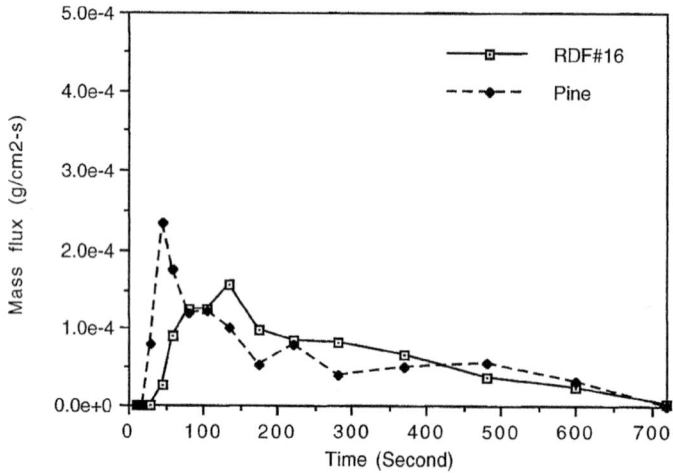

Fig. 2. Experimental total gas mass flux histories for pine and RDF pellets under a 3 cal/cm^2-s heat flux at the axial surface

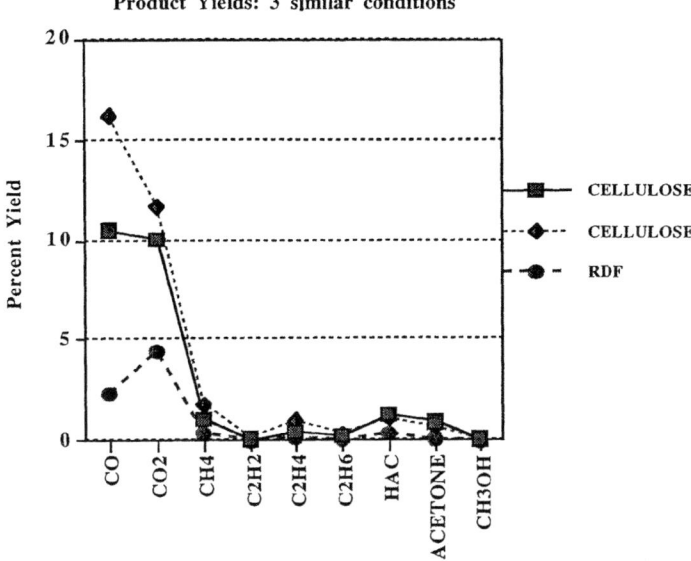

Fig. 3. Partial Product Spectrum for RDF and Cellulose pyrolysis under similar conditions.

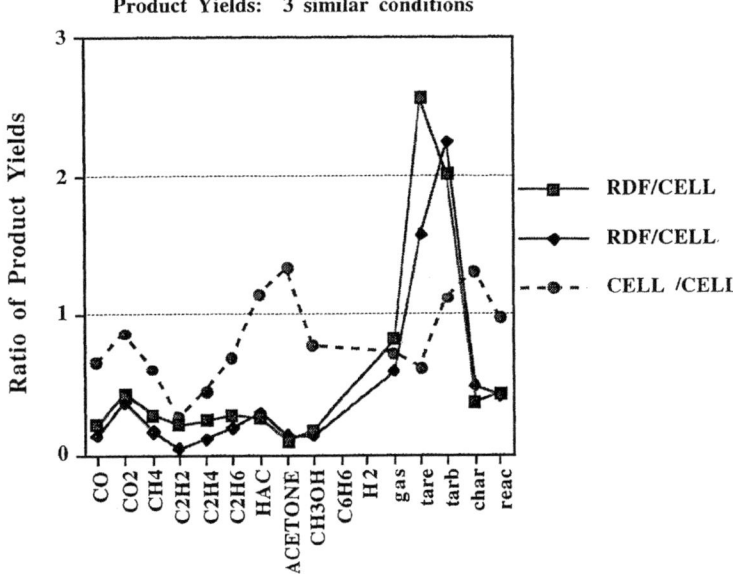

Fig. 4. *Ratios* of Product Slates: RDF yields/ Cellulose yields under similar conditions.

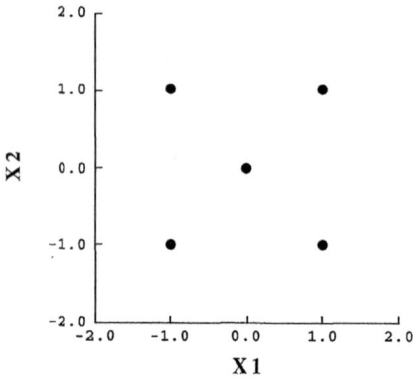

Fig. 5a - Visualization of Orthogonal Independent Variables

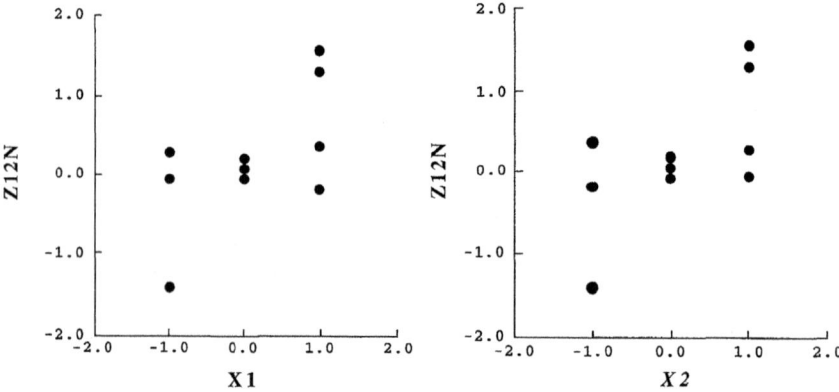

Fig. 5b. Data versus x_1 (independent)   Fig. 5c. Data versus x_2 (independent)

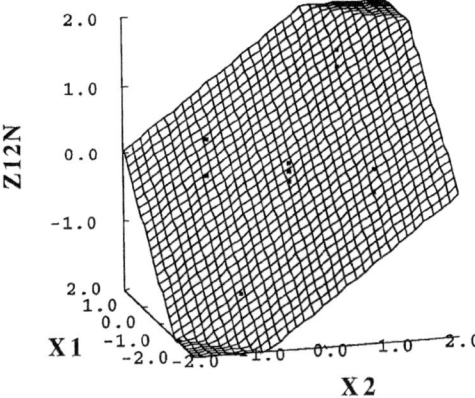

Fig. 5d. Visualization of Both Independent Variables' Influence on the Dependent Variable Z. Correlation Coefficient is 0.7 for each

Table 1. Experimental Ranges Studied

	Symbol	Substrate	Scaled conditions −1	0	+1	Comments
			Unscaled conditions			
Process variables						
Heat flux	q	d-RDF	3	4	5	cal/cm^2-s
		wood	2	4	6	cal/cm^2-s
Moisture content	MC	d-RDF	0.05	0.175	0.30	weight fraction, dry basis
		wood	0.1	0.55	1.10	weight fraction, dry basis
Particle thickness	L	d-RDF	1.0	1.5	2.0	cm
		wood	0.5	1.0	1.5	
Composition variables		d-RDF				
Paper	X$_1$		0.57	0.76	0.95	weight fraction
Plastics	X$_2$		0.00	0.10	0.20	weight fraction
Metal/Glass	X$_3$		0.05	0.10	0.15	weight fraction
Binder	X$_4$		0.00	0.04	0.08	weight fraction

Table 2. Simple Example of Correlation Analysis and Visualization

x_1	x_2	z_{12}	$z_{12,n}$
-1.00	-1.00	-1.414	-1.4102
1.00	-1.00	0.000	-0.1833
-1.00	1.00	0.000	-0.0585
1.00	1.00	1.414	1.5612
-1.00	-1.00	-1.414	-1.4060
1.00	-1.00	0.000	0.3664
-1.00	1.00	0.000	0.2710
1.00	1.00	1.414	1.3035
0	0	0.000	0.1795
0	0	0.000	-0.0882
0	0	0.000	0.0468

Table 3. PEARSON CORRELATION MATRIX for the synthetic data example

	$z_{12,n}$	z_{12}	x_1	x_2
$z_{12,n}$	1			
z_{12}	0.9825	1		
x_1	0.6912	0.7071	1	
x_2	0.6983	0.7071	-.3568E-16	1

Table 4. PEARSON CORRELATION MATRIX - Pine; Chan

	CO	CO2	CH4	C2H2	C2H4	C2H6	HAC	CH3CHO	CH3OH
CO	1.000								
CO2	0.922	1.000							
CH4	0.533	0.248	1.000						
C2H2	0.787	0.60	0.517	1.000					
C2H4	0.939	0.768	0.659	0.905	1.000				
C2H6	0.918	0.764	0.7972	0.689	0.894	1.000			
HAC	0.365	0.632	-0.462	-0.029	0.100	0.099	1.000		
CH3CHO	0.077	0.325	-0.636	-0.189	-0.122	-0.213	0.867	1.000	
CH3OH	-0.099	-0.087	0.183	-0.269	-0.113	0.042	-0.035	0.076	1.000

Table 5. PEARSON CORRELATION MATRIX - RDF; Lai

	CO	CO2	CH4	C2H2	C2H4	C2H6	HAC	CH3CHO	CH3OH	C6H6
CO	1.000									
CO2	0.915	1.000								
CH4	0.965	0.905	1.000							
C2H2	0.447	0.378	0.387	1.000						
C2H4	0.817	0.757	0.915	0.246	1.000					
C2H6	0.755	0.757	0.802	0.185	0.838	1.000				
HAC	0.509	0.562	0.507	0.118	0.488	0.304	1.000			
CH3CHO	0.747	0.701	0.699	0.164	0.586	0.583	0.550	1.000		
CH3OH	0.423	0.463	0.378	0.187	0.246	0.186	0.361	0.513	1.000	
C6H6	0.578	0.646	0.530	0.048	0.380	0.305	0.555	0.750	0.620	1.00

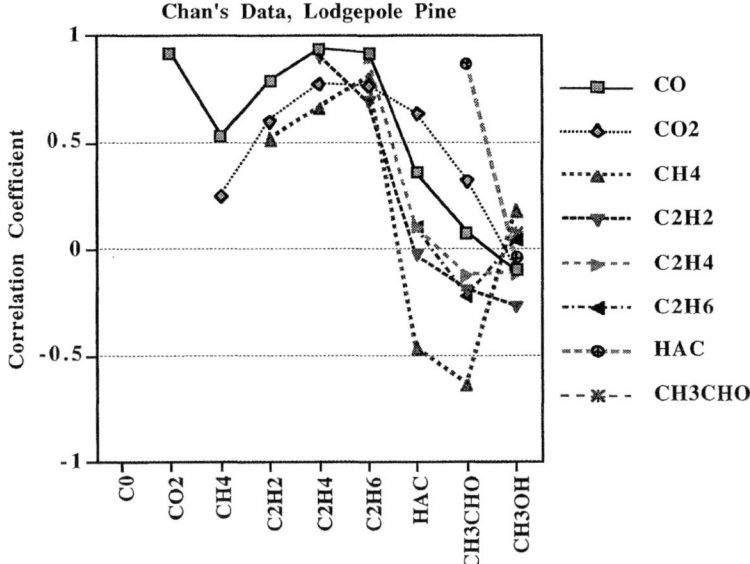

Fig. 6. Correlation among Pyrolysis Products for Pine

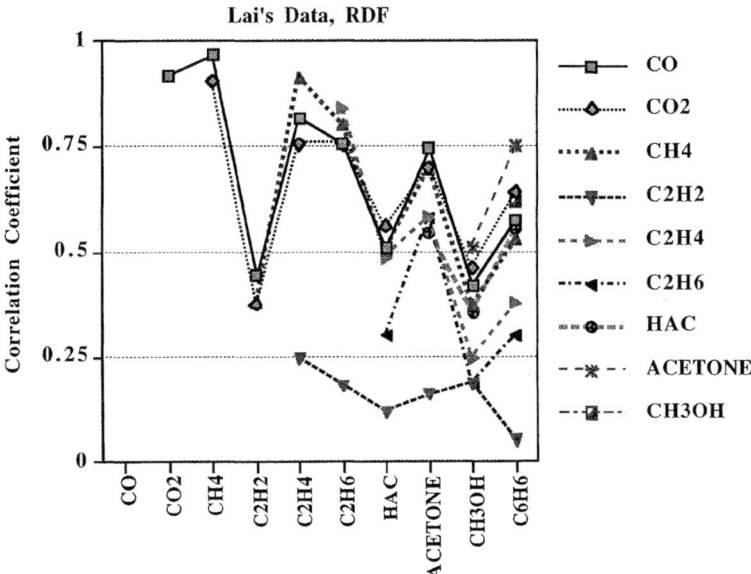

Fig. 7. Correlation among Pyrolysis Products for RDF

Table 6.   Ratio of PEARSON CORRELATION ELEMENTS - $C_{i,RDF}/C_{i,PINE}$

C(Pine) / C(RDF)	CO	CO2	CH4	C2H2	C2H4	C2H6	HAC	CH3CHO	CH3OH
CO	1								
CO2	1.007	1							
CH4	0.552	0.274	1						
C2H2	1.758	1.585	1.337	1					
C2H4	1.149	1.015	0.721	3.684	1				
C2H6	1.217	1.009	0.994	3.73	1.066	1			
HAC	0.717	1.125	-0.91	-0.242	0.206	0.327	1		
CH3CHO	0.103	0.464	-0.91	-1.151	-0.208	-0.37	1.58	1	
CH3OH	-0.23	-0.19	0.485	-1.439	-0.458	0.225	-0.1	0.149	1

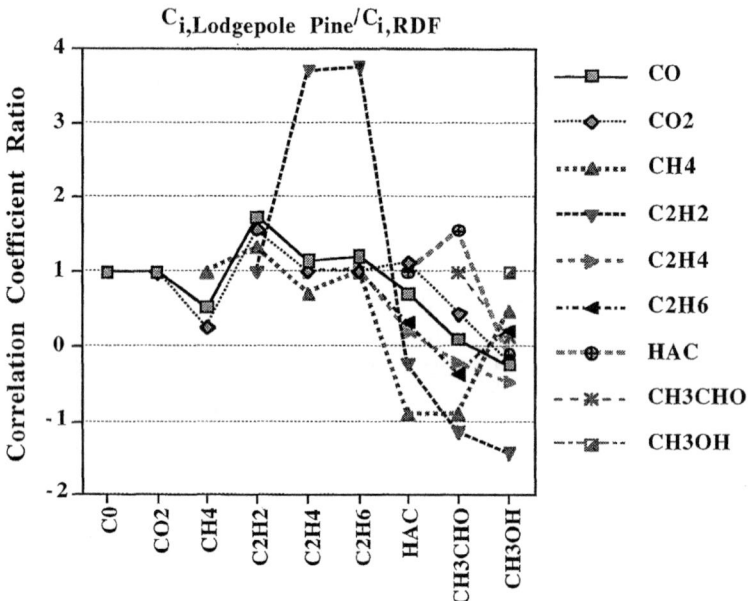

Fig. 8 Ratio of Respective Correlation Coefficients for RDF/Pine.

RAPID PYROLYSIS OF STRAW AT HIGH TEMPERATURE
Rapid High-Temperature Straw Pyrolysis

R. ZANZI, K. SJÖSTRÖM and E. BJÖRNBOM
Department of Chemical Engineering and Technology / Chemical Technology,
Kungl Tekniska Högskolan (KTH), Stockholm, Sweden

Abstract
Rapid pyrolysis of biomass is studied using a free-fall reactor. The biomass used in the study is untreated straw and straw pellets. The results are compared with those previously obtained in pyrolysis of wood. Before pyrolysis the straw is milled, sieved and wind classified to obtain fractions of uniform particle size.

The free-fall reactor has a heated length of 2.9 m and an inner diameter of 0.04 m. The char from the rapid pyrolysis is further pyrolyzed in nitrogen atmosphere in a thermobalance with a slow heating rate (20°C/min) up to 850°C in order to obtain complete pyrolysis.

The interest in the present work is focused on the effect of the treatment conditions such as heating rate and temperature on the distribution of products, gas composition and properties of the char obtained in pyrolysis.

The product distribution in rapid pyrolysis of wood, untreated straw and pelletised straw is similar. Wood gives a little less char than straw. The higher treatment temperature has led to lower yields of tar and higher yields of gaseous products. At higher temperature, the heat flux and the heating rate are higher. The higher heating rate favours also the decrease of char yield.

Keywords: Rapid pyrolysis, pyrolysis, biomass, wood, straw, agricultural residues, char.

1 Introduction

Utilisation of agricultural wastes is a suitable approach for substitution of solid fossil fuels for energy production. Straw is one of the most abundant agricultural residues, which can be used as a renewable energy source [1] [2] [3]. The present paper deals with a study on rapid pyrolysis of straw in a free-fall reactor. Rapid pyrolysis at high temperature is the first step in both combustion and gasification in fluidized bed reactors. The rate limiting step in steam gasification is the reaction between the char and steam. It is well known also that the conditions under the pyrolysis determine the char yield and its reactivity [4] [5]. Knowledge on rapid pyrolysis may contribute to the development of thermochemical processes such as gasification and combustion in fluidized bed reactors.

The interest in the present work is focused on the effect of the treatment conditions such as heating rate and temperature on the product distribution and gas composition. The results are compared to previous results obtained in pyrolysis of wood.

2 Experimental

The rapid pyrolysis of the fuel is conducted in the free-fall reactor shown in Figure 1.

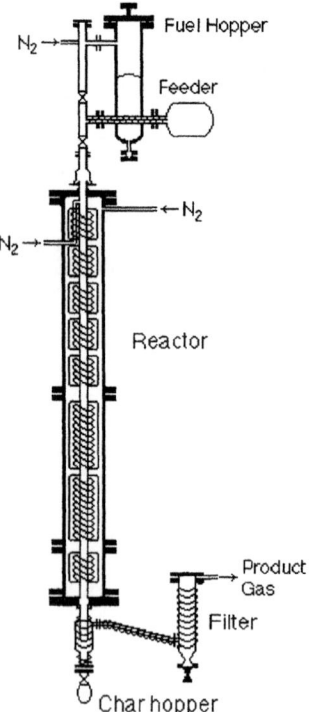

Figure 1. Free fall reactor for pyrolysis of solid fuels.

The free-fall tubular reactor has a heated length of 2.9 m and an inner diameter of 0.04 m. The carrier gas is preheated before it enters the reactor. The reactor tube is heated by eight independent electric heaters. The heated length of the reactor may be varied by changing the number of the heaters used for the experiments. The char is collected in a char receiver situated below the tube. The gas passes through a metallic filter. It is further cleaned in a tar condenser, dry ice traps and a cotton filter. The cleaned gas is analysed in a chromatograph for CO_2, H_2, CO, N_2, CH_4, saturated hydrocarbons (C_1-C_5) and aromatic hydrocarbons (benzene and toluene).

The char from the rapid pyrolysis is further pyrolyzed in nitrogen atmosphere in a thermobalance up to 850°C with a slow heating rate to obtain complete pyrolysis.

3 Raw material.

The biomass selected for this study is Danish wheat straw in untreated and in pelletised form. Before pyrolysis the straw is milled, sieved and wind classified to obtain a fraction of uniform particle size, for the experiments reported here 0.5-0.9 mm, with similar falling properties. The chemical composition of the raw materials, as well as their ash and moisture contents are shown in Table 1.

The low bulk density of the straw is a problem when feeding into the reactor (Table 2).

Table 1. Chemical composition of the raw materials used for pyrolysis

Raw Material	C (wt % maf)	H (wt % maf)	N (wt % maf)	O by diff. (wt % maf)	Ash (wt % mf)	Moisture (wt %)	H/C
Straw	46.5	6.3	0.9	46.3	3.2	7.1	0.135
Straw pellets	47.0	6.1	0.5	46.4	3.9	6.8	0.129
Birch wood	48.4	5.6	0.2	45.8	0.3	5.0	0.116

maf: moisture and ash free   mf: moisture free

Table 2. Bulk density of the biomass types used for the experiments

Raw Material	Bulk density (g/cm^3)
Straw	0.1
Straw pellets	0.3
Wood	0.3

4 Results and discussion

4.1 Effect of the heating rate

Table 3 compares the yields of chars produced by slow pyrolysis using a thermobalance with the yields of chars produced by rapid pyrolysis in free fall reactor. Rapid heating rate has resulted in less char formation than obtained at slow heating rates. The quick devolatilization of biomass in rapid pyrolysis results in formation of char with low density (Table 4).

The data in Table 3 are affected by the differences in the experimental conditions for slow pyrolysis in the thermobalance and for rapid pyrolysis in the free fall reactor. The slow removal of the volatiles from the bed of the fuel particles in the thermobalance facilitates the secondary pyrolysis reactions between the volatiles and the char, which favour the formation of secondary char. In the free-fall reactor, on the contrary, the removal of volatiles from the solids is facilitated because the fuel particles move as a stream of individual particles. Hence the secondary pyrolysis reactions are limited.

Though the conditions in the free-fall reactor and in the thermobalance are not completely comparable the differences in the results may be attributed mostly to the effect of the different heating rates.

Table 3. Char yield produced by slow and rapid pyrolysis of straw

Char yield from	Slow pyrolysis in thermobalance at 850°C (wt % maf)	Rapid pyrolysis in a free fall reactor at 800°C (wt % maf)
Straw	17.5	10.3
Straw pellets	18.0	11.3
Wood	16.5	5.5

nd: not determined   maf: moisture and ash free

Table 4. Bulk density of the char produced by slow and rapid pyrolysis of straw and wood

Bulk density of char from	Slow pyrolysis (g/cm^3)	Rapid pyrolysis (g/cm^3)
Straw	0.08	0.05
Straw pellets	0.25	0.10
Wood	0.18	0.09

4.2 Effect of the pyrolysis temperature

Table 5 shows the effect of temperature on the yield of products obtained in rapid pyrolysis of untreated straw, pelletised straw and wood. The higher treatment temperature has led to lower yields of tar and more tar cracking leading to higher yields of gaseous products. The distribution of products obtained from pelletised and untreated straw is quite similar. Under the selected experimental conditions the yields of char obtained in rapid pyrolysis of straw are higher than those produced in rapid pyrolysis of wood. The higher ash content of straw may strongly affect its reactivity under the pyrolysis.

The temperature markedly influences the heating rate. The heat flux is proportional to the driving force, the temperature difference between the particle and the environment. At higher temperature, the heat flux and the heating rate are higher. The higher heating rate results in decreased char yield.

The residence time of the particles in the free fall reactor is not enough for the final pyrolysis [6]. The char obtained by rapid pyrolysis contains a fraction that can be further volatilized by slow pyrolysis.

Higher treatment temperatures favour cracking of the hydrocarbons in the gaseous products and thus increases the yield of hydrogen. It has also decreased the content of CO_2 and increased the content of CO in product gas

Table 5. Effect of temperature on the yield of products obtained in rapid pyrolysis of straw in a free-fall reactor

Biomass	Straw		Straw pellets		birch, wood	
Temperature, °C	800	1000	800	1000	800	1000
Gas yield, wt % maf	75.8	86.0	75.4	85.5	77.7	87.0
Tar yield, wt % maf	0.9	0.1	0.8	0.1	1.1	0.2
Char yield, wt % maf	13.2	10.8	13.6	10.5	7.2	5.6
Water and losses	10.1	3.1	10.2	3.9	14.0	7.2
Final slow pyrolysis						
Char yield after total pyrolysis, wt % maf	10.3	8.9	11.3	9.3	5.5	5.0
Volatiles removed by secondary slow pyrolysis, wt % of original char.	22	17.6	16.9	11.4	23.6	12
Composition of the gaseous products, vol. %, mf, N_2 free						
H_2	35.0	43.9	24.2	38.8	16.8	34.0
CH_4	9.5	4.8	16.2	8.0	16.2	11.7
C_2H_2, C_2H_4	3.1	bdl	4.7	0.1	6.2	0.5
C_2H_6	0.1	bdl	0.5	bdl	0.3	bdl
Benzene	0.6	0.1	0.7	bdl	1.2	0.6
CO_2	23.7	5.0	19.3	4.7	8.3	7.5
CO	28.0	46.2	34.4	48.4	50.7	45.7

bdl: below detection limit,   maf: moisture and ash free   mf: moisture free

4.3 Effect of the fuel

The effect of the composition of the biomass is also shown in Table 5. The product distribution in rapid pyrolysis of wood, untreated straw and pelletised straw is similar. Wood gives a little less char than straw. The char yield after total pyrolysis is somewhat higher in pyrolysis of pelletised straw than untreated straw.

The differences in gas composition in pyrolysis at 800°C probably is due to the high ash content in straw. Alkali catalyses the water shift reaction and the yield of CO_2 and H_2 increases. In wood the content of alkali is low and the water shift reaction is not catalysed. The density of straw pellets is higher than that of untreated straw. Thus the residence time for straw pellets is lower than for straw.

Though the ratio H/C in the samples of straw is higher than in the wood (Table 1), which should favour the formation of volatiles, tar and gaseous products from straw, which is rich in H [7], the yield of char is higher in the experiments with straw than with wood. The higher content of inorganic compounds in the straw also may affect the formation of char.

4.4 Final pyrolysis

Previous experiments in varying the heated tube length in the free-fall reactor, i.e. the residence time of the particles, show that the fraction of char removed by slow pyrolysis in the thermobalance does not change significantly [6]. Pyrolysis proceeds in two steps: an initial fast step followed by a slower step including some chemical rearrangement of the char. The initial step is completed in a period of seconds. The secondary reactions need minutes to be completed. The residence time of the particles in the free fall reactor is not enough for the slower reactions. They are completed in the further slow pyrolysis in the thermobalance.

5 Acknowledgement

The present work is part of the project "Fixed bed gasification of agricultural residues" co-ordinated by Prof. Colomba Di Blasi, University of Napoli and is funded by the JOULE programme of the EU. Part of this research study was financially supported by the Swedish National Energy Administration and SAREC (Swedish Agency for Research Cooperation with Developing Countries).

6 References

1. Henriksen, U., Kofoed, E. and Christensen, O. (1994) Mass and energy distribution of the pyrolysis products from straw, in *Advances in Thermochemical Biomass Conversion*, papers from the International Conference in Interlaken, Switzerland, May 11-15, 1992, Edited by A.V. Bridgwater, Academic&Professional, London, Vol. 2, pp. 1110-1121.
2. Pedersen, K. (1994) Catalytic hydrocracking of tars from gasification of straw, in *Advances in Thermochemical Biomass Conversion*, papers from the International Conference in Interlaken, Switzerland, May 11-15, 1992, Edited by A.V. Bridgwater, Academic&Professional, London, Vol. 1, pp. 246-264.
3. Ergudenler, A. and Ghaly, A.E. (1994) A comparative study on the thermal decomposition of four cereal straws in an oxidising atmosphere, *Bioresource Technology*, Vol. 50, No. 3, pp. 201-208, Elsevier Science Limited.
4. Sjöström, K. and Chen G. (1990) Properties of char produced by rapid pressurised pyrolysis of peat, *Ind. Eng. & Chem. Research*, Vol. 29, No. 5, pp. 892-895.
5. Zanzi, R., Sjöström, K. and Björnbom, E. (1995) Rapid pyrolysis as the initial stage in combustion, *Biomass for Energy, Environment, Agriculture and Industry*, papers from the 8th European Conference held in Vienna, Austria, 3-5 October 1994, Edited by Ph. Chartier, A.A.C.M. Beenackers and G. Grassi, Pergamon, Elsevier Science Ltd, U.K., Vol. 3, pp. 1922-1929.
6. Zanzi, R., Sjöström, K. and Björnbom, E. (1994) Rapid pyrolysis of wood with application to gasification, in *Advances in Thermochemical Biomass Conversion*, papers from the International Conference in Interlaken, Switzerland, May 11-15, 1992, Edited by A.V. Bridgwater, Academic&Professional, London, Vol. 2, pp. 977-985.
7. Björnbom, E., Björnbom, P. and Sjöström, K. (1991) Energy-rich components and low-energy components in peat, *Fuel*, Vol. 70, No. 2, pp. 177-180.

THE PYROLYTIC DECOMPOSITION AND SUBSEQUENT COMBUSTION OF RICE HUSKS
Pyrolytic decomposition of rice husks

M.A. CONNOR, J.P. KISLER, N.I. ALESICH,
M.G. KANE, R.N. WATKINS and D.C. SHALLCROSS
Department of Chemical Engineering, University of Melbourne,
Parkville, Victoria, 3052, Australia.

Abstract
Waste rice husks are a potential but often little-used fuel source in many Third World countries. Rice husks contain up to 24% silica, in the form of a three-dimensional skeleton. Complete combustion is therefore harder to achieve than in the case of other, preferred fuels. To design more efficient rice husk combustion equipment, a better understanding of the pyrolytic and combustion steps in the overall combustion process is needed. To date, fundamental studies of these steps have usually employed thermogravimetric techniques. However work undertaken on the oxidation kinetics of crude oil suggested that an effluent gas analysis (EGA) technique could provide information both supplementary and complementary to that obtained by thermogravimetry. This EGA technique involves heating the material under study at a preset rate. At the same time gas with a known oxygen content is passed through the material, and the composition of the exit gas measured. A series of EGA experiments was carried out using rice husk. These not only demonstrated the potential of the EGA approach as a means for learning more about biomass pyrolysis but also provided insights into the effects of variations in pressure, oxygen partial pressure and particle size on rice husk combustion.
Keywords: Combustion, effluent gas analysis, fuel, pyrolysis, rice husk.

1 Introduction

Worldwide, the annual production of rice husks is around 100 million tonnes, 90% of which is generated in developing countries [1]. Sixty million tonnes are produced in the Asia/Pacific region, where much of the research into finding beneficial ways of using rice husks has taken place [2]. A diversity of uses has been found for rice husks. However, the extent to which they are made use of varies widely from country to country, being particularly dependent on the local availability of cheap and convenient fuels.

In the comparatively fuel-rich country of Indonesia, for example, a 1987 survey found that only 10% of the available husks were being used for economically productive purposes; uses included: as a fuel for brick kilns, as a clay extender in brickmaking, as animal litter and as a growing medium for mushrooms. Very little use was made of the fuel value of the husks [3]. This contrasts markedly with the situation in India, where, in the 1980's, 40% of husks were being used as boiler fuel in rice parboiling mills, while the remainder was almost all used as industrial fuel. In fact, by the end of the 1980's, rice husk in parts of India was commanding a price that effectively placed it out of the reach of fuel-starved potential domestic users [3].

The situation in India appears to have arisen because of that country's severe energy and fuel shortages and in spite of rice husk's many disadvantages as a fuel [3]. As emphasized by Beagle [4], there are many drawbacks to the use of rice husk as an energy source. An obvious problem is rice husk's low bulk density of around 100 kg/m^3, which makes it costly to transport any distance and also to store [5]; since rice is a seasonal crop, anyone wishing to use rice husk as a fuel at times outside the rice harvest period has to provide for husk storage. A further problem is the difficulty in achieving complete combustion of the oxidisable components of the rice husk. Rice husks contain 16 to 24% silica [4], present in the form of an internal silica skeleton that maintains its structural integrity during the pyrolysis and oxidation stages of the combustion process [5]. As a consequence, char formed within husks during the pyrolysis stage is not easily accessible to the oxygen in the air passing over the husks. The oxidation of this char is a process that becomes progressively more difficult as the surface of unreacted char moves inwards into the interior of the husk's silica skeleton. This prolongs the combustion process and, except where combustion temperatures are high, often leads to husks being discharged from furnaces or combustors in an incompletely burnt out state. This means that the potential heating value of the husks has been only partly realised. The problem of ash disposal is also aggravated since productive uses for rice husk ash usually require it to have a low carbon content.

Such problems, and others associated with the volume of ash produced and its highly abrasive nature, can be overcome economically where combustion of rice husks is undertaken on a large scale. An example of such a large-scale facility is the power plant operated by Agrilectric Power Partners Ltd, at Lake Charles, Louisiana. This is a highly automated plant burning rice husks ground to between 20 and 100 mesh and generating 10.6 MW of electricity [6]. However, for small-scale users, putting in place the equipment needed to overcome rice husks' peculiar problems is expensive; rice husk is therefore regarded as an inferior fuel and other fuels tend to be used where possible. Nevertheless, as populations grow, energy sources become harder to find, and problems associated with waste disposal intensify, greater use of rice husks in small-scale combustion units can be expected.

A better understanding of the pyrolytic and oxidation steps in rice husk combustion could help in developing more efficient small-scale processes for burning or gasifying rice husks. Rice husks, on an ash-free basis, have a composition not dissimilar to that of other materials of vegetable origin. Okeke and Obi [7] showed that, whilst there is significant variation between husks from different rice strains, cellulose, hemicellulose and lignin are present in proportions roughly similar to those found in wood, namely 2:1:1. Partial confirmation of these figures comes from Beagle [4], who reported a cellulose content for rice husk of around 50% on an ash-free basis, and from Williams and Besler [1], who determined the lignin content of their rice husks to be 18.6% (on a non ash-free basis).

On the basis of this similarity in composition, it might be expected that the pattern of weight loss shown by rice husk when heated would be similar to the now well-established pattern for wood and other materials [1][8]. This pattern involves the more or less sequential evolution of decomposition products of the hemicelluloses over the temperature range 220 to 320°C, followed by evolution of cellulose decomposition products over the range 250 to 360°C; lignin undergoes gradual decomposition over the range 180 to 500°C, with the bulk of the associated weight loss occurring at higher temperatures. The small thickness of individual rice husks, and the consequently large surface area to volume ratio, would suggest that conformity with the above pattern should be likely. However, the silica skeleton of the rice husk, and the increased resistance to heat transfer and mass transfer it introduces, would tend to offset this.

To date only limited studies have been undertaken of the fundamental aspects of rice husk decomposition on heating; most of the research carried out has been concerned with practical aspects of rice husk gasification, pyrolysis or combustion. Such fundamental work as has been carried out [1][9] has used thermogravimetric analysis (TGA) and differential thermogravimetric analysis (DTG) to determine patterns of weight loss from very small samples of rice husk.

Recent fundamental studies of the oxidation kinetics of crude oil, carried out at the University of Melbourne [10], suggested that effluent gas analysis (EGA) of gases leaving a rice husk combustion unit could provide information both supplementary and complementary to that obtained in TGA and DTG studies. To verify this, the project described in this paper was initiated.

2 Materials and Methods

2.1 Effluent gas analysis (EGA)

In the effluent gas analysis technique, a cell is loaded with a mixture of combustible material and sand. The cell is then heated so as to give a constant rate of temperature increase. At the same time, air or another oxidising gas of known composition is passed through the cell at a known flowrate. The pressure in the cell is maintained at a preset value. The gases leaving the cell have their composition analysed on a continuous basis. From the results of these analyses, insights can be gained into the rate and nature of the decomposition reactions undergone by the material being studied.

2.2 Rice husks

The rice husks used in this study had been stored for a number of years after being originally obtained from the rice growing area of southern New South Wales. Such husks typically have an ash content of around 22-24% on a dry basis; similar husks used by Minn [9] had an ash content of 24%. The cellulose, hemicellulose and lignin content of the husks was not determined.

2.3 Experimental equipment

As indicated above, the experimental EGA rig used in this investigation was originally developed to study the kinetics of crude oil oxidation. A full description of its development can be found in [10]. At the heart of the rig is the pressure cell, shown in Figure 1, into which is loaded a predetermined mix of sand and the material to be burned, in this case rice husk. The sand functions both as a flow distributor and as a thermal capacitance, damping out temperature fluctuations associated with endothermic or exothermic decomposition or oxidation reactions and ensuring that cell temperatures follow a linear rate of increase. Quartz sand with a size range of 425 to 500 μm was used; the bed porosity was 40 ± 0.5%.

The pressure cell is a cylindrical pressure vessel formed from a section of thick-walled stainless steel pipe. It is sealed at both ends with plugs, O-rings and screwed on end caps. The O-rings are of brass, to provide an effective high temperature pressure seal. Within the cell are two thin-walled stainless steel cups with perforated bases to allow gas to pass through. The lower cup is filled with sand and acts as both a flow distributor and a final preheater for the gas passing through the cell. Much of the preheating of the gas is accomplished in a gas preheating coil placed within the furnace immediately beneath the cell. The position of this coil can be seen more clearly on Figure 2, which is a schematic of the entire experimental set-up.

The upper cup is 27.0 mm in diameter and 70.7 mm high; this is the cup which held the mixture of 1 gram of rice husk and 40 grams of sand used in this study. The temperature of the contents of the upper cup is measured by means of a type-K thermocouple positioned vertically in the cell such that its tip is just above the cup's base. A second thermocouple, placed within the furnace but outside the cell, is used to control the furnace heating rate. A back-pressure regulator is employed to maintain a constant pressure in the cell. The regulator is operated manually allowing the pressure to be maintained to within ± 5 kPa of the desired value. A gas flow rate controller is used to ensure that the gas is supplied to the cell at a constant rate of 300 ml/min measured at standard conditions. The inlet gas flowrate was extremely stable, not varying by more than ± 0.2% over a 10 hour period.

Gases leaving the cell pass through a cleaning system comprising a condenser and a dryer. They then pass to three gas analysers (O_2, CO and CO_2) connected in series. The ranges of the CO, CO_2 and O_2 gas analyzers are 0-5 mol %, 0-20 mol % and 0-25 mol % respectively. The analysers are accurate to within ±0.5% of the full scale for changes up to 90% of the full scale. These gas analysers are calibrated before and after each run using a set of calibration gases of known compositions covering the range of measured values expected. A computer-controlled data acquisition system records the inlet gas flowrate, cell pressure, sample temperature and gas composition readings at 30 second intervals. The data is obtained by logging each channel ten times in rapid succession and then

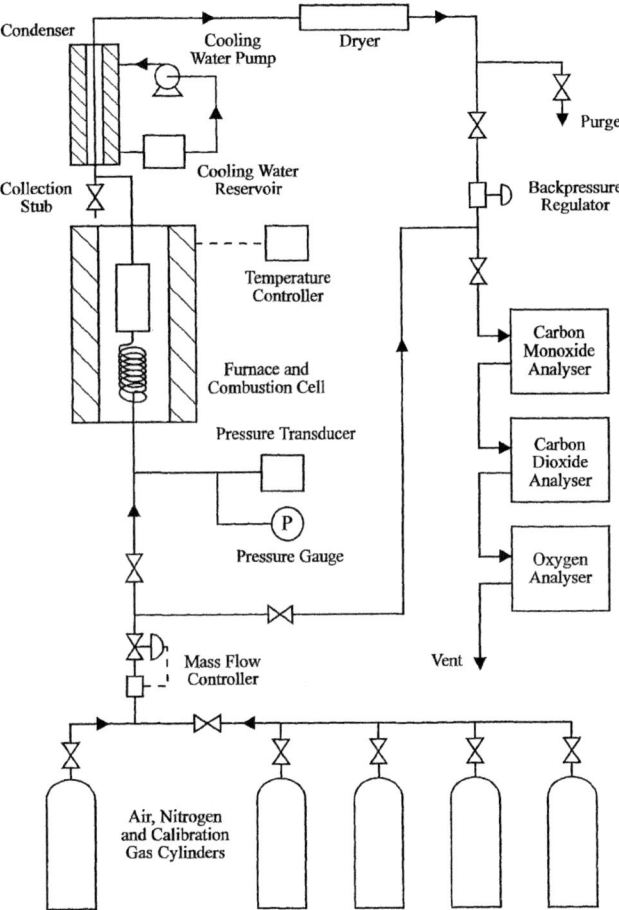

Fig. 1 Schematic of the EGA equipment

averaging the results.   Overall about six seconds are required to scan all the
channels ten times, average the values and apply the necessary calibration
equations. Provision is made in the data processing program to take account of the
time taken for gas from the cell to reach each of the gas analysers.

2.4 Experimental procedures
Once the cell has been loaded and placed in the furnace, the rig is pressure tested
to ensure that all connections are leak-tight. The furnace heater is switched to full
power to raise the furnace temperature to 200°C. While the furnace is heating, the
gas analysers are calibrated.   Once the cell temperature reaches 170°C (which
corresponds to a furnace temperature of 200°C) the heating program is adjusted to
give the desired heating rate (50°C/h or 75°C/h in the experiments discussed here)

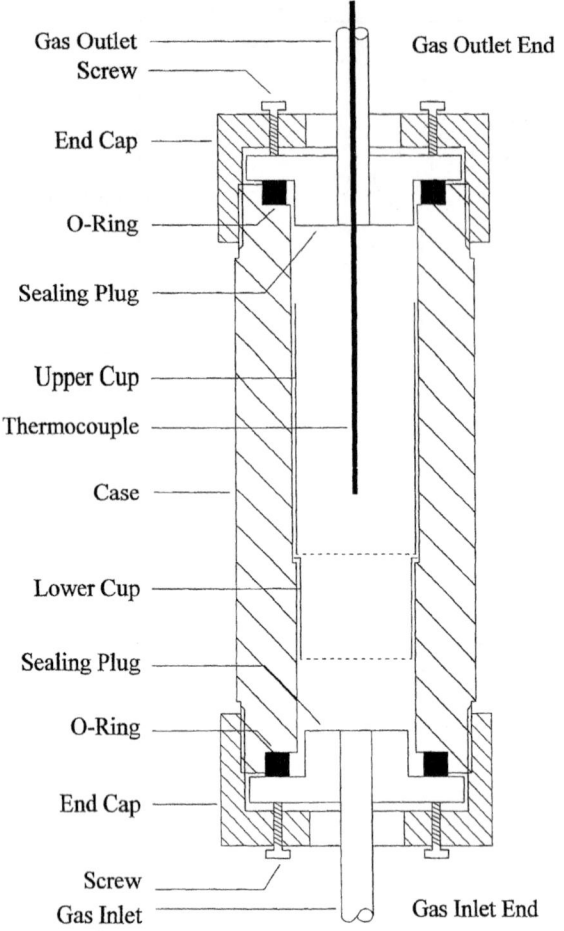

Fig. 2 The pressure cell

and the pressure regulator set to give the desired cell pressure. Flow of gas through the cell commences at this point, i.e. the experimental run begins. The run is allowed to continue until gas composition readings show that production of carbon oxides has ceased, i.e. combustion is complete. Thereupon the gas heater is turned off, the cell allowed to cool, and the gas analysers recalibrated.

3 Experimental Program

Using the equipment described above, a total of 21 rice husk combustion experiments was carried out. The first three were preliminary experiments used to establish an appropriate husk to sand ratio. These experiments showed that if too much husk were present carbon oxide levels in the effluent gas could exceed the upper limit of the range set on the measuring instruments. In addition, during periods when exothermic reactions were occurring, cell temperatures could rise above the desired level; this had the effect of increasing reaction rates, further accelerating rates of heat generation and leading to unwanted temperature and carbon oxide concentration "spikes". On the other hand, too low a rice husk content was found to give very low carbon oxide readings on the measuring instrument, thereby introducing larger than desirable inaccuracies into measured gas compositions. As a result of these experiments a mixture of 1 gram of rice husk in 40 grams of sand was settled on; for this mixture the rate of temperature increase in the cell remained linear while at the same time effective use was made of the measuring instrument ranges.

The remaining eighteen experimental runs were designed firstly to establish the reproducibility of the results obtained using the EGA technique and secondly to investigate the effect on rice husk decomposition of:

- total system pressure
- oxygen partial pressure
- heating rate
- the size of the rice husk particles
- the effect of soaking the rice husks in water prior to combustion
- the effect of rice straw particles present as a "contaminant" in the rice husk.

4 Experimental Results

The main product of each of the experimental runs was a plot like that shown in Figure 3. This plot shows changes in gas composition with temperature (and, effectively, with time, since a linear rate of increase of temperature occurred in all experiments from run 4 onwards). Actual carbon dioxide and carbon monoxide levels in the effluent gas are shown, together with the difference between the oxygen levels in the influent and effluent gases. These plots were all of similar form, having two main peaks, one in the 260-280°C region and the other in the 350-400°C region. Because of the slow rate of heating and the high frequency with which gas composition measurements were recorded (twice per minute), peak temperatures could be very accurately located; for run 4, the results for which are shown in Figure 3, the peak temperatures were 277.6°C and 372.4°C respectively.

The above plot shows marked similarities in form to those determined for wood using TGA techniques [8][11]. On the basis of these similarities the above two peaks can be taken to define periods during which pyrolytic decomposition products of first hemicellulose and then cellulose escape from the rice husks and undergo combustion. They are therefore often referred to from hereon as the hemicellulose and cellulose peaks, respectively. Also of interest are the minor peaks, less reproducible in form, that occur at around 380-430°C. Such peaks, though rarely commented on, also appear on weight-loss curves obtained by thermogravimetric analysis when wood is subjected to slow pyrolysis [8][11]. Little attention has yet been paid to these lesser peaks by researchers but the

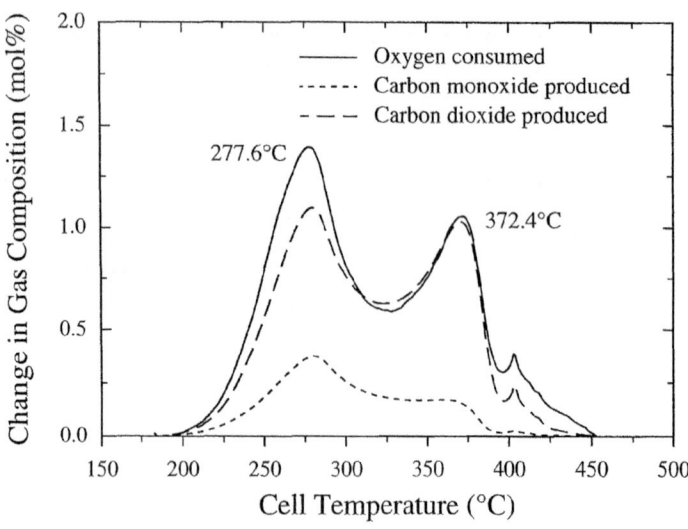

Fig. 3 A typical set of experimental results

temperature range over which they occur suggests they reflect the evolution of products of lignin decomposition. Their small size is assumed to be a result of lignin decomposing over a broad rather than a narrow temperature range.

4.1 Reproducibility of results

The reproducibility of results obtained by the EGA technique was checked by comparing the values of the peak temperatures obtained for runs carried out under similar sets of operating conditions. Peak temperatures were preferred to peak heights as a measure of reproducibility because they could be determined with considerably greater precision and accuracy. Peak temperatures determined for four runs carried out at a total system pressure, P_T, of 200 kPa and using air as the influent gas are shown in Table 1.

Table 1. Results for four runs carried out under identical conditions

Run number	Total pressure P_T (kPa)	Concentration of oxygen in gas (mol %)	Oxygen partial pressure (kPa)	Peak temperatures (°C)	
				Peak 1	Peak 2
4	200	21	42	277.6	372.4
5	200	21	42	277.6	369.3
9	200	21	42	279.2	371.1
16	200	21	42	276.2	368.7

For peak 1, the hemicellulose peak, temperatures lay within 1.5°C of the mean value of 277.7°C. For the cellulose peak, peak 2, temperatures lay within 2.0°C of the mean value of 370.4°C.

One additional run, run 10, was carried out under similar conditions to the four runs discussed above. However the rice husk sample used was observed to contain a number of sizeable pieces of rice straw, small amounts of which were present in the commercially obtained rice husk sample used in these experiments. For this particular run, the gas compositions varied in a slightly different way from those in runs lacking obvious contamination with rice straw, while the peak temperatures were slightly lower, namely 274.5°C and 368.8°C. To check whether this decrease could be attributed to the presence of rice straw, the experiment discussed in section 4.7 was carried out.

From the results in Table 1 it was concluded that good reproducibility of the hemicellulose and cellulose peak temperatures could be obtained provided that a rice husk sample of consistent composition was used.

4.2 Effect of system pressure

A major advantage of the effluent gas analysis technique over conventional thermogravimetric techniques is the ease with which the influence of pressure on decomposition patterns can be investigated. Pressurised thermobalances have been developed [12] but they are of necessity complicated pieces of equipment. In this study the effect of system pressures of from 2 to 7 atmospheres were investigated. The oxygen concentrations in the inlet gas were varied appropriately so that the effect of total pressure could be looked at independently of any effect of oxygen partial pressure. The results obtained are shown in Table 2. The lack of influence of system total pressure on peak temperatures is illustrated in Figure 4, which

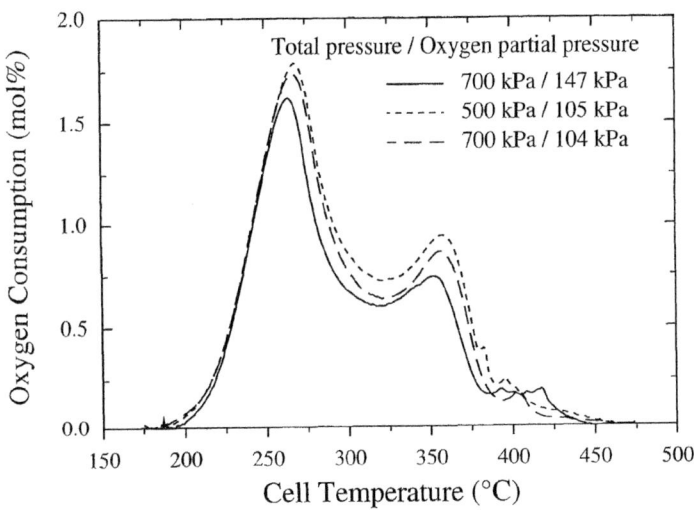

Fig. 4 The effect of changes in pressure and oxygen partial pressure on decomposition patterns

shows the peaks in oxygen consumption for runs 11, 12 and 13. This Figure also shows the small but significant effect of varying oxygen partial pressure (see section 4.3, below).

Table 2. The effect of system pressure on peak temperatures

Run numbers	Total pressure P_T (kPa)	Concentration of oxygen in gas (mol %)	Oxygen partial pressure (kPa)	Peak temperatures (^{o}C)	
				Peak 1	Peak 2
4,5,9,16	200	21	42	277.7 (±1.5)	372.4 (±2.0)
6,15	280	14.8	42	278.4 (±1.4)	370.1 (±0.1)
14	700	6.2	42	279.0	369.9
12	500	21	105	270	357.2
13	700	14.8	104	268.4	355.3

4.3 Effect of oxygen partial pressure

By employing suitable combinations of the system total pressure and the oxygen content of the inlet gas, experiments could be carried out at a variety of oxygen partial pressures. The results obtained are shown in Table 3.

Table 3. The effect of oxygen partial pressure on peak temperatures

Run numbers	Total pressure P_T (kPa)	Concentration of oxygen in gas (mol %)	Oxygen partial pressure (kPa)	Peak temperatures (^{o}C)	
				Peak 1	Peak 2
4,5,6,9,14,15,16	200-700	6.2 -21	42	278.1(±1.9)	370.2(±2.2)
7	280	21	59	275.8	367.3
17	350	21	73.5	271.0	361.4
12,13	500-700	14.8-21	105	269.2(±1.8)	356.3(±1.0)
8,11	700	21	147	264.5(±1.5)	352.9(±0.3)

It can be seen that, provided the oxygen partial pressure remains constant, the system pressure has no significant effect on measured peak temperatures. Given that the combustion of rice husk is currently believed to be affected by diffusion limitations, this finding was unexpected. Its implications are discussed in more detail in section 5.

Table 3 shows unequivocally how increases in the oxygen partial pressure lead to a decrease in the peak temperatures. The extent of the changes observed is such as to rule out any idea that they might be attributable to experimental inaccuracies or variations in other experimental conditions. Associated with these decreases in peak temperature at higher oxygen partial pressures was a substantial increase in the heights of the first (hemicellulose) peaks for CO_2 and CO, and a corresponding decrease in the heights of the second peaks. However very little change in gas composition occurred at temperatures of around 325°C, intermediate between the two peaks. The implications of these observations are also discussed in section 5.

4.4 Effect of heating rate

One experiment, run 19, was carried out under operating conditions identical to those used in runs 4,5,9 and 16, except that a heating rate of 75°C/hour was used instead of 50°C/hour (the rate used in all runs other than run 19). The peak temperatures measured for run 19 were 279.3°C and 374.1°C respectively. Comparison of these temperatures with corresponding values for the 50°C/hour experiments, as reported in Table 1, shows that the results for run 19 are slightly higher. The smallness of these increases suggests that at the slow rates of heating employed in the present study changes in heating rate have only an insignificant effect on decomposition processes; the small changes observed could well be linked to the existence of slightly larger temperature gradients within the rice husks at the higher heating rate and/or to response times of the temperature sensors.

4.5 Effect of particle size

If the process of rice husk combustion were to be diffusion limited, with the husk's silica skeleton providing the major mass transfer resistance, then grinding the rice husk could be expected to overcome much of this resistance. Certainly, such an approach has been taken at the Agrilectric plant in Louisiana [6]. To test the validity of this approach, a sample of rice husk was ground before being heated under conditions similar to those used for the runs of Table 1. The peak temperatures in this case were 270.5°C and 363.5°C, significantly below the corresponding mean values of 277.7°C and 370.4°C for the four runs in Table 1. At first sight, this would appear to confirm the hypothesis that the combustion process is diffusion controlled. However, grinding the rice husks also facilitates the transfer of heat into the husk interior and could be expected to increase the rate at which volatile pyrolytic decomposition products from within the husks would reach the husk surface, escape, and react with oxygen in the surrounding gases. Both of these effects would tend to reduce the peak temperatures and could account, at least in part, for the observed peak temperature decreases. These aspects are discussed further in section 5.

4.6 Effect of pre-soaking in water

The rice husk used in the above experiments was old and dry and probably not totally representative of conditions in husks from recently milled rice. To gain at least a partial insight into the effect of using a rice husk sample with a higher moisture content, rice husks were soaked in water for 24 hours before being heated. Again the same set of conditions was employed as those used for the runs of Table 1. In this case, the peak temperatures obtained were 283.7°C and 395.4°C, well above the corresponding mean values for dry rice husk of 277.7°C and 370.4°C. These changes show a similar trend to those observed by Jenkins et al [13] in studies of rice straw. In TGA experiments, their unleached rice straw showed a maximum rate of weight loss at a temperature of around 290°C whereas for rice straw leached in water the corresponding maximum was at about 320°C. Jenkins et al [13] showed that this change was linked to the removal of potassium and sodium from the rice straw during the leaching process; it was inferred that these elements are present in the rice husk in a form that has a pronounced catalytic effect on straw decomposition pathways. Whilst it was not confirmed that leaching of the rice husk resulted in removal of sodium and potassium, it seems highly probable that dissolution of these alkali metals during the 24 hour rice husk soaking process was responsible for the observed changes in measured peak temperatures.

4.7 Effect of rice straw contamination

As indicated in section 4.1, experience obtained with a rice husk sample containing a number of sizeable pieces of rice straw led to a run to test the effect of rice straw contamination of the rice husk. From the original purchased sample, pieces of rice straw were picked out and mixed with rice husk in a 44.6% straw: 55.4% husk mixture. When this mixture was heated under conditions similar to those used for the runs of Table 1, peak temperatures of 273.5°C and 361.4°C were obtained. These are significantly below the corresponding mean values of 277.7°C and 370.4°C obtained for the normal rice husk fraction but quite close to the temperatures of 270.5°C and 363.5°C obtained for crushed rice husk. Rice straw has a significant silica content of around 19% [14]. Although this is comparable to the ash content of rice husk, rice straw does not appear to have anything like as complex an internal silica skeleton as rice husk. For this reason, the above results imply that husk structure does exert a significant influence on the rice husk combustion process, but whether through hindering heat transfer, mass transfer or both, is uncertain.

5 Discussion

One as yet unresolved problem is how to find a reasonable explanation for the observations presented in sections 4.2, 4.3 and 4.5. The fact that total system pressure appears to have no influence on the rice husk decomposition patterns has several implications:

- that heat transfer into the rice husk is not materially affected by a change of gas pressure; this would be consistent with heat transfer occurring predominantly by conduction through the solid portion of the rice husk rather than by conduction and convection through the gases in the internal pores (a not unexpected result).
- that evolution of volatiles from the surface of the rice husk into the surrounding gas phase is not affected by the external pressure; this in turn implies that internal mass transfer resistances are small when the quantities of gas seeking to migrate to the surface are taken into account. This would be consistent with rates of decomposition within the husk being heat transfer limited.

The inferences drawn above are consistent with observations made when the rice husk was ground before heating. Such a pretreatment would have markedly increased the surface area to volume ratio of the rice husks and decreased the distance heat would have to travel to the most inward parts of the husk. The interior of the husk would reach the cell temperature more rapidly and evolution of volatiles (and their subsequent combustion) would occur at lower values of the cell temperature as measured by the inserted thermocouple.

There remains a need to explain why oxygen partial pressure should exercise an influence over peak temperatures. It could be expected that oxidation reactions would be predominantly gas phase reactions until the efflux of volatile, combustible decomposition products from the husks fell to a fairly low level, i.e. after the bulk of the cellulose decomposition products had been evolved and burned. Consequently the observed effect of varying the oxygen partial pressure can be inferred to reflect a change in the rate of the gas phase oxidation reactions. For such a change to manifest itself as a change in peak temperatures (and peak heights, as also occurred) some of the combustible products of husk decomposition must be escaping unburned. The experimental rig did not include a means of checking this and so the above explanation, though plausible, must remain unconfirmed. Extension of the rig to enable more complete carbon balances over the cell to be constructed is obviously desirable.

Also desirable is inclusion in the facility of a means of measuring the water content of the effluent gases. As Figure 5 shows, plotting the sum of the oxygen content of the carbon oxides together with the quantity of oxygen consumed has the potential to provide interesting insights into the nature of decomposition mechanisms. Where the oxygen consumed is less than that present in the carbon oxides, it is evident that carbon oxides are being evolved from the decomposing husk. However, lack of information on the amounts of oxygen leaving the cell as water vapour means that at present only qualitative conclusions about decomposition processes can be drawn from diagrams of this nature.

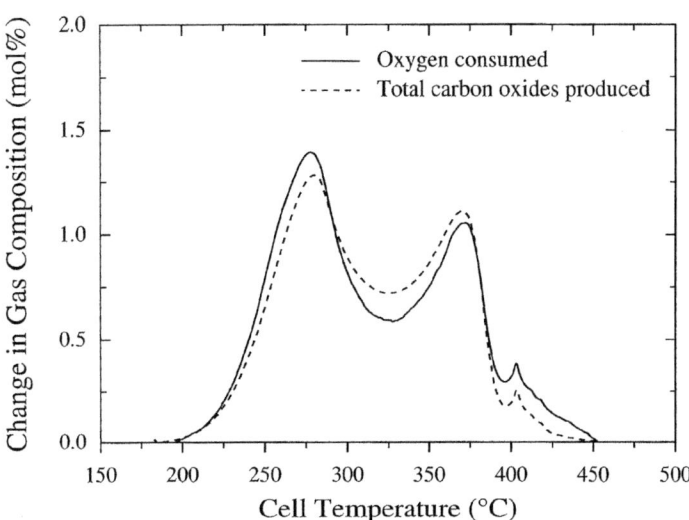

Fig. 5 Comparison between the amount of oxygen consumed and that leaving in the form of carbon oxides

6 Conclusions

The project described in this paper was initiated to see whether effluent gas analysis could provide information on the pyrolysis and subsequent combustion of biomass that was complementary to that obtained by thermogravimetry. It is concluded from the preliminary experiments described here that EGA has the potential to provide information that not only complements TGA-derived results but in some aspects is both qualitatively and quantitatively superior.

Though preliminary, the experiments described here were sufficient to show that total system pressure has no effect on patterns of rice husk decomposition whereas varying the oxygen partial pressure does. Also of importance is the size of the rice husk particles. Analysis of the results of experiments looking at the influence of the above parameters leads to the conclusion that over the range of heating rates studied, rice husk decomposition is heat transfer limited; the influence of oxygen partial pressure is concluded to be predominantly on gas phase oxidation reactions. Whether the situation is similar at the higher heating rates to be expected in actual rice husk combustors still remains to be established. It was also concluded that exposure of rice husks to water leads to dissolution of alkali metals in the husk and to slower rates of pyrolytic decomposition. This would appear to have significant implications as far as the storage and pretreatment of rice husks destined for combustors are concerned.

Whilst the EGA rig used in this project already has considerable potential for providing insights into biomass decomposition patterns, addition of instrumentation enabling determination of effluent gas moisture content and organic carbon content would further increase its usefulness.

References

1. Williams, P.T. and Besler, S. (1993) The pyrolysis of rice husks in a thermogravimetric analyser and static batch reactor, *Fuel*, Vol. 72, No. 2. pp. 151-59.
2. Coovattanachai, N. (1991) Gasification of husk for small scale power generation. *RERIC International Energy Journal*, Vol. 13, No. 1. pp. 1-17.
3. Mahin, D.B. (1990) *Energy from rice residues*, Winrock International Institute for Agricultural Development, Arlington, Va., USA.
4. Beagle, E.C. (1978) *Rice-husk conversion to energy*. FAO Agricultural Services Bulletin No. 31, Food and Agriculture Organization of the United Nations, Rome.
5. Kaupp, A. and Goss, J.R. (1983) Technical and economical problems in the gasification of rice hulls, physical and chemical properties, *Energy in Agriculture*, Vol. 1, No. 3. pp. 201-34.
6. Anon. (1985) Cogen plant burns rice hulls, avoids waste-disposal problem, *Power*, Vol. 129, No. 10. pp. 19-20.
7. Okeke, B.C. and Obi, S.K.C. (1994) Lignocellulose and sugar compositions of some agro-waste materials, *Bioresource Technology*, Vol. 47, No. 3. pp. 293-4.

8. Connor, M.A. and Salazar, C.M. (1988) Factors influencing the decomposition processes in wood particles during low temperature pyrolysis, in *Research in Thermochemical Biomass Conversion*, (eds. A.V. Bridgwater and J.L. Kuester), Elsevier Applied Science, London, pp. 164-78.

9. Minn, K. (1983) *Thermal effects influencing the pyrolytic decomposition of rice husks*, M.Eng.Sc. thesis, University of Melbourne, Victoria, Australia.

10. Kisler, J.P. (1995) *The application of in situ combustion to light Australian oils*, Ph.D. thesis, University of Melbourne, Victoria, Australia.

11. Salazar, C.M. (1987) *The influence of particle size and shape on the mechanisms of decomposition of wood during pyrolysis*, Ph.D. thesis, University of Melbourne, Victoria, Australia.

12. Richard, J.R. and Rouan, J.P. (1988) New apparatus for the study of the gasification of char in high temperature and pressure environments, in *Research in Thermochemical Biomass Conversion*, (eds. A.V. Bridgwater and J.L. Kuester), Elsevier Applied Science, London, pp. 370-83.

13. Jenkins, B.M., Bakker, R.R., Baxter, L.L., Gilmer, J.H. and Wei, J.B. (1996) Combustion characteristics of leached biomass. Paper presented at a Conference on Developments in Thermochemical Biomass Conversion, Banff, 20-24 May, 18 pp.

14. Barnard, G. and Kristoferson, L. (1985) *Agricultural residues as fuel in the Third World*, International Institute for Environment and Development, London.

MAC\PYRODECR.96

RELATIONSHIPS BETWEEN WOOD DENSITY, WOOD PERMEABILITY AND CHARCOAL YIELD
Wood properties and charcoal yields

M.A. CONNOR and M.H. VILJOEN
Department of Chemical Engineering, University of Melbourne,
Parkville, Victoria, 3052, Australia.
J. ILIC
CSIRO, Division of Forest Products,
Private Bag 10, Rosebank MDC, Clayton, Vic 3169, Australia

Abstract
Charcoal yields for woods from different tree species can vary widely. Charcoal is believed to consist of the combined residues of primary wood decomposition reactions and of secondary reactions undergone by volatile products en route to the wood surface. It might be expected that the harder it is for these volatile products to migrate to the wood surface, the more carbon will be laid down as a result of secondary reactions. Two wood properties that could be expected to influence volatiles' migration rates are density and permeability. Initially, experiments were conducted to establish whether density could be used as a predictor of charcoal yield. Early results seemed promising but extension of the investigation to other woods showed that links between density and charcoal yield were tenuous. A program to study the influence of wood permeability on charcoal yield was then initiated. Early results for samples from several Brazilian grown eucalypts confirmed the lack of a relationship between density and charcoal yield but suggested a stronger link between permeability and charcoal yield. However, extension of the investigation to woods of widely varying densities and permeabilities showed that no straightforward relationships exist between density, permeability and charcoal yield.
Keywords: Carbonisation, charcoal, density, permeability, pyrolysis, wood.

1   Introduction

It is well known to charcoal makers that the species of tree they obtain their wood from affects the yield and quality of the charcoal they produce. However, just why some woods yield more charcoal than others is still poorly understood. Factors having the potential to influence charcoal yield include chemical properties such as lignin content [1] or extractives content [2], as well as physical properties such as particle size and shape [3], and density [1] [4] [5].

Considerable evidence exists to suggest that the carbonaceous material we know as charcoal combines residual solid products from two different sets of reactions [3]. The first set comprises the primary decomposition reactions undergone by different wood fractions as the wood is heated. The second includes a variety of secondary carbon-forming reactions involving volatile products of the primary decomposition reactions. These secondary reactions are believed to occur whilst the volatile products migrate from their point of origin to the wood surface, where they disperse into the surrounding gas.

Given that both these mechanisms of charcoal formation are important, one would expect that the harder it is for the volatile primary decomposition products to move to the wood surface, the greater will be the charcoal yield. This is because the longer these products remain within the wood/charcoal matrix, the more likely they are to take part in carbon-forming secondary reactions. Support for this idea comes from the findings of Salazar [3][6]. His work showed that as the size of a piece of wood subjected to a standardised slow pyrolysis test routine increased, the charcoal yield also increased. However, this was true only as long as no cracks developed within the wood/charcoal matrix; if cracks formed, the charcoal yield was less than expected.

These observations are in complete accord with what would be predicted when secondary carbon-forming reactions contribute substantially to the overall charcoal yield. For example, in larger wood pieces the migration path is on average longer, increasing the time volatiles remain within the wood and hence their opportunities to undergo carbon-depositing reactions. In this case, a higher charcoal yield would be predicted. However, if cracks form, many volatiles migrate rapidly to the surface through them; this reduces their residence time in the wood/charcoal matrix and hence the likelihood of their taking part in the secondary reactions that contribute to char formation. In this case a reduction in charcoal yield would be expected.

One wood property that could be expected to affect the rate of migration of volatiles through wood is density. The chemical composition of woods from a majority of tree species is remarkably consistent. Given this, then the greater the wood density, the more closely packed the cellulose, hemicellulose and lignin molecules in the wood must be. This implies a corresponding reduction in the free volume within the wood. Such a reduction in free volume might be expected to increase the diffusion resistance for volatiles migrating to the wood surface, thereby prolonging their period of residence in the wood and leading to an increased secondary char yield.

The problem with this analysis, of course, is that wood is not a uniform material but has an anisotropic vascular structure. In hardwoods, for example, the vessels in the wood provide a means for transporting water, nutrients and other chemical substances from one end of the tree to the other. These vessels would be expected to serve as the primary conduits for migrating volatiles when wood is pyrolysed, and observation of small to medium sized branches in an open fire tends to confirm this; when such branches are heated in the middle, liquid droplets soon appear at the broken ends, followed by discharges of smoky vapours, which often catch fire and support a flame.

The resistance such vessels pose to the migration of volatiles appears dependent on three factors. The first of these is the percentage of the cross-sectional area (measured perpendicular to the direction in which the vessels run) which is occupied by vessels. This could be expected to be at least partly related to density. The second is the number of vessels per unit area and the size distribution of their diameters; the fluid flow resistance of a vessel depends on its diameter, and a mass of small diameter vessels will have a total flow resistance far greater than that of a single vessel of equivalent cross-sectional area. The third is the extent and nature of obstructions to fluid flow within the vessels. Such obstructions are necessary if fluid is to be transported upwards through the tree and their presence in sapwood is therefore essential. In heartwood these obstructions may be augmented by deposits of resins and other extractives. In addition, tree species groups such as the eucalypts develop tyloses, which are ray cells that grow into the vessels during heartwood formation. In the eucalypts and other species that form them, these tyloses constitute the major obstruction mechanism.

The ease with which fluids flow through wood has been of interest to the forest products industry for a long time. It is of particular concern to those involved in drying timber and in impregnating wood products with preservatives or fire retardants [7]. The fluid penetration properties of a given piece of wood are customarily expressed in terms of its permeability, which can be defined as the rate at which a fluid flows through a porous material under the influence of a pressure gradient [7].

From the above discussion it is apparent that both wood density and wood permeability are likely to affect volatiles' migration rates and hence charcoal yields. This has prompted a series of investigations into the effect of these properties on charcoal yield. Initially efforts were concentrated on characterising the influence of density, as this is such a simple property to measure. More recently, however, an extensive investigation into the effect of permeability on charcoal yield has been undertaken. The findings of these investigations are described below.

2   Materials and Methods

2.1   Pyrolysis equipment and measurement procedures

A detailed description of the pyrolysis equipment used to determine charcoal yields can be found in [6]. Its central feature is a heavily insulated stainless steel tube. The wood sample to be pyrolysed is suspended within this tube by a thin (0.4 mm diameter) nichrome wire. This wire passes through a 1.5 mm diameter hole in the roof of the tube and is attached to the base of a Sartorius Model 1219 MP electrobalance. Pyrolysis of the sample is achieved by passing heated nitrogen over it. The heating of the nitrogen is controlled so as to heat the sample slowly, following a preset pattern, to a maximum temperature of 420°C. As the sample is heated, measurements of the temperature of the gas close to the sample surface and the mass of the sample are taken at 1 minute intervals and stored in a computerised data acquisition system. Once the final temperature of 420°C has been reached, heating is continued until the rate at which the mass of the sample is decreasing has become very slow. The heating is then turned off and the sample allowed to cool. The charcoal yield is found by comparing the mass of dry charcoal obtained with the dry mass of the original sample. This mass is obtained by drying the sample at 105°C for 24 hours and allowing it to cool in a desiccator before weighing.

It should be noted that the carbonised material (charcoal) produced in this way only has a fixed carbon percentage of around 60%. Some researchers prefer to restrict the use of the term "charcoal" to products with a minimum fixed carbon percentage of 70%, and this is regarded as a positive step forward in resolving the confusion that presently exists about what the term "charcoal" signifies. Nevertheless, primarily for convenience, the carbonised products of the pyrolysis experiments described above have continued to be referred to as charcoal in this paper.

For historical reasons, much of the early work on the influence of density on charcoal yield was done using carefully machined cylindrical wood samples 10 mm in diameter and 20 mm long. Once the program of permeability measurements was initiated, wood samples 20 mm in diameter and 25 mm long were used; this size change was necessary to meet the requirements of the permeability apparatus.

2.2 Permeability apparatus and measurement procedures

The apparatus used to measure wood permeability is shown schematically in Figure 1. Details of the sample holder, with a sample in place, are shown in Figure 2. The outer shell of the sample holder consists of an open-ended stainless steel tube (B) with an inlet port (D) for the admission of sealing air. The sample (C) is held in the centre of the sample holder by two brass end-pieces (E), the inner sections of which have the same diameter as the sample (20 mm). Covering the sample, as well as the inner sections of the two end-pieces, is a piece of rubber tubing (H) of

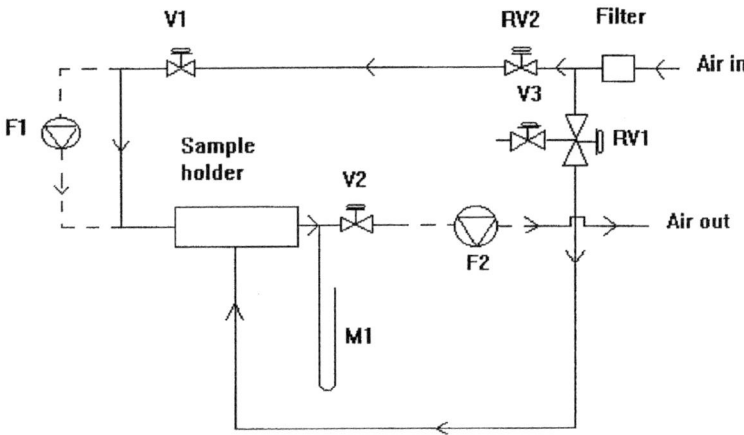

Fig. 1   Schematic diagram of the wood permeability measurement system

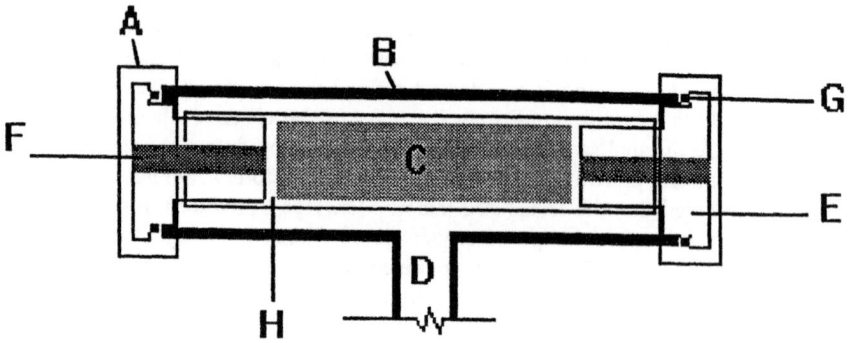

Fig. 2   Schematic diagram of the sample holder.   A: Screw-on end cap;
B: Stainless steel tube; C: Wood sample; D: Sealing air inlet; E: Brass end piece;
F: Measurement air inlet; G: O-ring seal; H: Rubber tubing

the same diameter (internally) as the sample. When the annular space between the
outside of this tubing and the inside of the stainless steel tube is pressurised with
sealing air, the tubing presses down on the outer surface of the sample and prevents
measurement air passing around instead of through the sample. The effectiveness
of this sealing mechanism was confirmed by substituting a 20 mm diameter solid
metal (and therefore impervious) cylinder for the wood sample; no flow around the
cylinder was detected when the sample holder was pressurised. The outer ends of
the sample holder are closed by the two brass end-pieces; O-ring seals (G) are used
to prevent leaks. The brass end-pieces are held in position by screw-on end caps
(A). Measurement air enters through a hole (F) drilled centrally in one of the end
pieces and leaves through a corresponding hole in the other end-piece. The gap
between the ends of the sample and the adjacent end-pieces is small, but large
enough for air introduced through the measurement air inlet to distribute itself over
the entire upstream face of the sample.

The air used in the permeability tests is first passed through a filter. This is
done to reduce the likelihood of dust particles accumulating on the sample surface,
or inside it, and obstructing the passage of air. Some of the air is directed through
pressure regulator RV1 and into the sample holder; as described above, this is done
to create a pressure seal around the outside of the sample and ensure that none of
the measurement air passes around instead of through the sample. The remainder
of the air, the measurement air stream, passes through pressure regulator RV2. If a
sample with a high permeability is being tested, the air is passed through flowmeter
F1 (Dwyer Instruments, range: 0.28 to 2.8 m^3/h (10-100 scfh)). The air then
continues through the sample and its outlet pressure is measured with a water-filled
manometer. For low permeability samples, for which gas flowrates are low, a
model M-5 mini-Buck Calibrator flowmeter, F2, (range: 1 to 6000 cm^3/ min.), is
used to measure flowrates. This is placed downstream of the sample holder
because of a requirement that the pressure in its flow cell remain below 6.7 kPa (1
psi). In this case a manometer containing a petroleum hydrocarbon of specific
gravity 0.785 is used, in order to measure the outlet pressure more precisely.

The procedure employed in measuring permeabilities can be summarised as follows. With valves V_1 and V_2 open, and the sample in place in the sample holder, regulator RV2 was opened just enough to give a pressure on the upstream face of the sample (P_{in}) of about 20 kPa. Regulator RV1 was then adjusted to give a seal pressure of around twice the maximum upstream pressure expected to be used in the experiment concerned. The outlet pressure, P_{out}, and the gas flowrate were noted once steady-state conditions had been achieved. RV2 was then adjusted to give a higher value of P_{in} and the above process repeated. The entire procedure was repeated for a total of 5 different inlet pressures.

2.3   Permeability calculation procedures

The approach used in calculating wood permeability from the above measurements is based on that described by Choong et al.[7]. This assumes that flows within the wood are laminar. Conventionally, when flows are laminar, the permeability of a porous medium is defined by Darcy's Law:

$$Q = \frac{KA(P_{in} - P_{out})}{\mu L} \tag{1}$$

where K is the permeability; A and L are the cross-sectional area and length of the sample, respectively; (P_{in}-P_{out}) is the pressure drop across the sample; μ is the fluid viscosity; and Q is the volumetric flowrate of fluid through the sample. Hence:

$$K = \frac{Q\mu L}{A(P_{in} - P_{out})} \tag{2}$$

As is pointed out in [7], equation (2) is valid only for incompressible fluids. For a compressible fluid such as air, the volumetric flowrate changes along the length of the sample. In such cases an apparent permeability, K_G, is what is actually measured, where

$$K_G = \frac{\bar{Q}\mu L}{A(P_{in} - P_{out})} = \frac{Q_0 \mu L P_0}{A(P_{in} - P_{out})\bar{P}} \tag{3}$$

Here $\bar{Q}$ and $\bar{P}$ represent the average volumetric flowrate and pressure in the test sample, while P_0 is the pressure at which the flowrate through the sample (Q_0) was actually measured.

A further complication discussed in [7] is that associated with the flow of gas in very small diameter channels, where so-called "slip" or molecular flow may occur. To take account of this problem when calculating an absolute permeability from the K_G values obtained from experimental data, use of the following equation, often known as the Klinkenberg equation, is recommended:

$$K_G = K(1 + b/\bar{P}) \tag{4}$$

where b is a constant.

To determine values of K from the measurements obtained as described in section 2.2, the procedure given below was followed. First, measured values of Q were plotted against corresponding values of $(P_{in} - P_{out})$. If this relationship was more or less linear it seemed safe to assume that no leaks or other equipment malfunctions had occurred. Then, from the readings taken at each set of pressures, the corresponding value of K_G was determined, using equation (3). Values of K_G calculated in this way were then plotted against $1/\bar{P}$; from equation (4), the intercept of this plot on the ordinate (K_G) axis gives the absolute permeability K.

3   The influence of density

3.1   Theoretical considerations

Earlier, the idea was proposed that a wood's density influences the charcoal yield when a sample of that wood is pyrolysed. It was suggested that the higher the density, the less free space there is likely to be within the wood material. This could well make it harder for the volatile products of primary decomposition reactions to migrate to the sample surface; opportunities for secondary carbon-forming reactions to occur would be increased and char formation promoted.

However, wood density changes also have the potential to affect charcoal yields in another way. As wood density increases it could also be expected that the wood's thermal conductivity would increase, since its components are being brought closer together. An increased thermal conductivity would facilitate heat transfer into a wood sample's interior during pyrolysis. The resulting increase in heat transfer rate would be reduced somewhat by the higher thermal capacity (per unit volume) associated with the wood's increased density. Nevertheless a more rapid rate of increase in internal temperatures (and a slower rise in surface temperature) would be likely to result [8]. If wood primary decomposition pathways are assumed to have a dependence on temperature similar to that put forward in the model of Kilzer and Broido [9], then the above change should lead to a decrease in char yield in the sample interior and an increase nearer the wood surface.

Secondary reactions would also be affected by changes in patterns of temperature variation such at those described above. However, as yet too little is known to predict how such changes might affect carbon decomposition. If these reactions show a typical dependence of reaction rate on temperature, there might even be an enhanced rate of carbon decomposition in the interior and a reduced deposition rate nearer the surface. This would have the effect of offsetting the changes in char yields from the primary decomposition reactions.

It is clear, therefore, that the influence of density on charcoal formation mechanisms is not a simple matter to establish and much work remains to be done to elucidate its impact.

3.2   Experimental observations

Initial studies of wood carbonisation at the University of Melbourne involved only the single species *Eucalyptus delegatensis*. However, comparison of charcoal yields for this species with the limited information then available for other species suggested that a correlation between charcoal yield and wood density was possible. This idea was supported by the observations of Cutter and McGinnes [5] that high density woods such as oak and hickory yield more char than low density woods such as basswood and cottonwood.

To test this hypothesis, charcoal yields for similar sized cylindrical samples of four very different woods were determined by Lin [4] [10]. These woods were chosen to span a wide range of wood types and characteristics. The woods selected for study were: alpine ash *(Eucalyptus delegatensis)*, the medium density hardwood used in previous pyrolysis experiments; jarrah *(Eucalyptus marginata)*, a high density hardwood already in use in Western Australia for producing industrial charcoal; blackwood *(Acacia melanoxylon)*, a low density hardwood used for furniture; and *Pinus radiata*, a common softwood. The piece of pine wood from which samples were cut varied considerably in structure; two sets of samples were therefore prepared, one from a region with closely spaced growth rings and one from a region where the rings were widely spaced.

When samples of these woods that were 15 mm in diameter and 20 mm long were carbonised under similar conditions, the charcoal yields obtained were as shown in Table 1. Also shown in this Table are the corresponding dry wood densities together with information, where known, on the presence of tyloses and deposits.

Table 1. Charcoal yields and densities (initial investigation)

Species	Average charcoal yield (%)	Wood density (kg/m^3)	Presence of tyloses (T) and deposits (D)
Eucalyptus marginata	43	751	T (abundant); D
Eucalyptus delegatensis	33	595	T (few)
Pinus radiata (closely spaced rings)	29	555	-
Acacia melanoxylon	26	423	D
Pinus radiata (widely spaced rings)	25	416	-

As the results from the project undertaken by Lin [4] [10] appeared to support the idea that wood density and charcoal yield might be linked, a second investigation was undertaken. This was oriented around woods available in South-East Asia; it was envisaged that if the link between density and charcoal yield were confirmed for a series of such woods, it would help charcoal makers of the region to select the most suitable woods for their kilns. The results obtained in this second investigation are shown in Table 2.

It is apparent that, with only one exception, the charcoal yields fall within the narrow range 41% to 46% despite the densities of the woods involved ranging from 339 to 912 kg/m^3. Particularly when the results in Table 2 are compared with those from Table 1, it is clear that density on its own is a quite inadequate predictor of charcoal yield. For this reason investigations into the influence of wood permeability on charcoal yield were initiated.

Table 2. Charcoal yields and densities (South East Asian woods)

Species	Average charcoal yield (%)	Wood density (kg/m³)	Presence of tyloses (T) and deposits (D)
Eucalyptus deglupta (1st sample)	41	546	T (abundant)
Eucalyptus deglupta (2nd sample)	46	579	T (abundant)
Gmelina arborea	42	503	T (abundant)
Rhizophora sp.	45	912	T (abundant); D
Paraserianthes falcatoria	42	339	Absent
Hevea brasiliensis	36	593	T (occasional)
Casuarina junghuhniana	41	770	T absent; D
Casuarina equisetifolia	43	831	T absent; D
Acacia mangium	42	437	Absent
Leucaena leucocephala	42	678	T absent; D

4 The influence of permeability

4.1 Previous investigations

Because of the relevance of wood permeabilities to the impregnation of wood with preservatives and flame retardants, there is a substantial literature on this subject. Permeabilities for woods from around one hundred tree species have been published by Smith and Lee [11]; these permeabilities range over six orders of magnitude, illustrating the marked variations in permeability that can occur between species. Significant variation also occurs between samples from the same species or even from the same tree; Smith and Lee [11], as well as Comstock [12], have pointed out that local variations in wood structure within the same tree can give rise to permeabilities that differ by an order of magnitude or more. Permeability is also highly anisotropic. For beechwood, for example, tangential, radial and longitudinal permeabilities are in the ratio of 0.02: 0.03; 13000 [11].

4.2 Experimental determinations

The investigation into permeability initiated at the University of Melbourne has involved extensive collaboration with the Division of Forest Products of the CSIRO (Commonwealth Scientific and Industrial Research Organisation). Measurements of wood permeability have all been undertaken at the CSIRO laboratories at Clayton (a suburb in Melbourne's south-east) while density and charcoal yield determinations were undertaken at Melbourne University.

The first phase of the permeability study was done on a series of *Eucalyptus* samples. These samples were cut from sections of eucalypts, of known age and provenance, kindly supplied by the Brazilian company ACESITA from their plantations in the vicinity of Belo Horizonte, Minas Gerais. Samples of both heartwood and sapwood were studied. Densities, longitudinal permeabilities and charcoal yields for three species, *E.saligna*, *E.cloeziana* and *E.urophylla* are shown in Table 3. For the sake of completeness, this Table includes a series of results obtained prior to the start of the permeability investigation, at a stage when only

Table 3. Densities, permeabilities and charcoal yields for three Eucalyptus species

Tree species	Density (kg/m^3)		Permeability		Charcoal yield (% of dry wood)	Presence of tyloses (T) and deposits (D)
	Air dried	Oven dried	Air dried	Oven dried		
E. saligna	600	550			46.5	T (moderately
(heartwood)	710	650			44.9	abundant)
	600	590	0.006	0.006	45.0	
	590	570	0.008	0.01	44.7	
E.saligna	660	610			38.5	Absent
(sapwood)	690	650			42.7	
	700	690	0.75	1.0	41.0	
	670	650	1.9	1.2	42.2	
E.cloeziana	730	660			51.0	T (abundant);
(heartwood)	710	650			49.5	D (abundant)
	780	740	0.004	0.004	49.2	
	760	720	0.004	0.004	50.1	
E.cloeziana	820	780			39.8	Absent
(sapwood)	820	770			40.8	
	810	780	0.64	0.63	44.8	
	820	790	0.77	0.55	44.5	
E.urophylla	650	620			37.0	T (occasional)
(heartwood)	640	610			37.1	
	620	590	7.15	9.06	41.6	
	610	580	3.43	3.30	39.2	
	620	600	0.71	1.37	42.6	
E.urophylla	710	680			38.1	Absent
(sapwood)	720	680			38.3	
	740	720	0.89	0.63	39.5	
	760	730	0.72	0.52	39.8	

the respective charcoal yields from the different *Eucalyptus* species were being studied.

The results in Table 3 confirm the inadequacy of density as a predictor of charcoal yield. For example, for all three species the heartwood density is lower than the sapwood density. It might therefore be expected that yields of charcoal would be greater in each case for the sapwood rather than the heartwood. However, whilst this is true, on balance, for *E.urophylla*, it is far from the case for *E.saligna* and *E.cloeziana*. There is also no consistency in the relationships between density and charcoal yield when inter-specific differences are considered.

When measured permeabilities (longitudinal) are compared with their corresponding charcoal yields, a much more consistent pattern is evident. Very low permeabilities were obtained for two sets of samples, the heartwoods from *E.saligna* and *E.cloeziana*, and these two woods gave charcoal yields well above the rest. The highest permeabilities found were for the heartwood from *E.urophylla*, and this was the wood for which the lowest charcoal yields (around 37%) were obtained. At intermediate permeabilities, the correlation between charcoal yield and permeability is not as good, however. Nevertheless, on the basis of these results, permeability did appear to be a much better predictor of charcoal yield than density.

Since the above results were obtained for woods from tree species of the same genus it was felt advisable to extend the test program to a very much wider range of woods. So a second phase of testing was initiated, using a selection of high and low density woods from the Division of Forest Products' H.E. Dadswell Memorial Wood Collection. Densities and permeabilities were obtained for these woods and the diversity of the resulting combinations of permeability and density is shown in Figure 3. (The results for the three *Eucalyptus* species discussed above are included on this Figure.) Charcoal yield determinations for a selection of the species being investigated in the second phase of testing are shown in Table 4.

The results shown in Table 4 make it clear that even with a knowledge of both the density and permeability of a wood, predicting the charcoal yield is still not possible. It would appear from the results for the *Nothofagus menziesii* and the

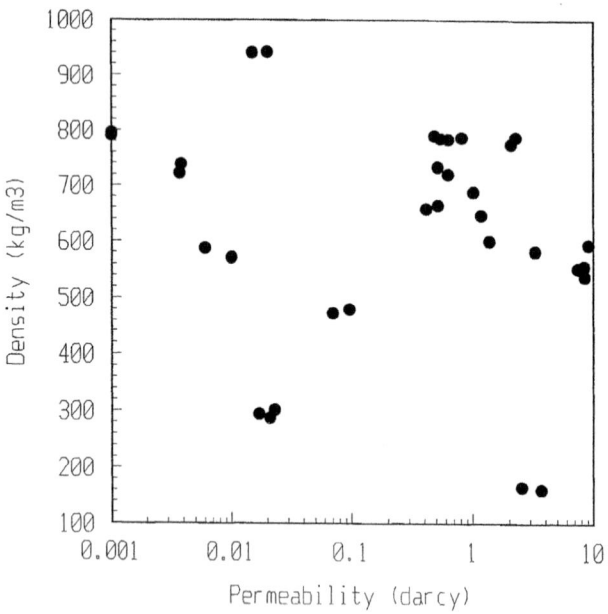

Fig. 3 Wood density/permeability pairings

4. Densities, permeabilities and charcoal yields (second phase results)

Tree species	Density (kg/m³)		Permeability		Charcoal yield (% of dry wood)	Presence of tyloses (T) and deposits (D)
	Air dried	Oven dried	Air dried	Oven dried		
Fraxinus	580	560	9.92	8.37	42.2	T (occasional
mandschurica	560	550	12.2	14.1	40.2	in heartwood)
Ochroma lagopus	170	160	2.4	2.53	40.6	T (occasional
	170	160	2.33	3.67	40.3	in heartwood)
Fagus sylvatica	820	790	0.55	0.49	38.5	T (frequent
	810	790	1.19	0.81	40.6	in heartwood)
Pterocarpus		660	0.16	0.52	47.1	T (absent);
indicus		660	0.14	0.42	46.7	D (abundant)
Pinus radiata		470	0.05	0.07	39.3	Absent
	510	490	0.05	0.05	39.0	
Shorea leprosula	300	290	0.02	0.02	37.4	T (occasional
(light red meranti)	320	300	0.02	0.02	39.5	in heartwood)
Shorea sp.	960	940	0.01	0.02	47.7	T (abundant
(dark red meranti)	960	940	0.01	0.05	47.1	in heartwood); D (abundant)
Nothofagus	820	800	0.001	0.001	48.8	T (abundant
menziesii	820	790	0.001	0.001	47.9	in heartwood); D (abundant)

Shorea sp. (a dark red meranti) samples that where permeabilities are very low and densities high, a high charcoal yield is likely. In addition, where permeabilities are at the higher end of the range, charcoal yields are at the lower end of their range. However, the lowest yields of charcoal were obtained for woods that had low permeabilities and densities at the low to medium end of the density range.

A concern that arose in connection with the above results was that the observed differences in charcoal yield could be merely a result of differences in the volatile matter content of the carbonised wood samples. To test this, proximate analyses of the charcoal produced in experiments on woods from five different tree species were carried out. (To obtain sufficient material for the analyses, for each species it was necessary to combine the carbonised product from two duplicate runs; the results of the analyses therefore represent an average over these two duplicates.) The species chosen were those whose charcoal yields were towards the extremes of the range obtained. The analyses were undertaken by an outside laboratory according to procedures in Australian Standard AS 1038, Part 3 (1989) (Proximate Analysis of Higher Ranked Coals).

Table 5. Proximate analyses (dry basis) of carbonised samples

Tree species	Charcoal yield (% of dry wood)	Ash (%)	Volatile matter (%)	Fixed carbon (%)
Pinus radiata	39.2	0.7	38.0	61.3
Eucalyptus saligna (sapwood)	41.6	0.7	41.5	57.8
Eucalyptus urophylla (sapwood)	39.7	0.6	38.7	60.7
Pterocarpus indicus	46.9	1.6	36.9	61.5
Eucalyptus cloeziana (heartwood)	49.7	0.1	36.5	63.4

The results obtained are presented (on a dry basis) in Table 5. As can be seen from Table 5, these results not only allay concerns that the differences in charcoal yield are associated with differences in volatile matter content but in fact accentuate the differences reported earlier. The highest fixed carbon percentage found was for samples of *Eucalyptus cloeziana* heartwood, which gave one of the highest charcoal yields, while the highest volatile matter content was that for *Eucalyptus saligna* sapwood, which has a comparatively low charcoal yield. Also reassuring are the low ash content values, showing that the results obtained were not affected significantly by differences in ash content between carbonised samples.

5 Discussion and conclusions

It is apparent from the results presented above that predicting charcoal yield on the basis of known wood densities and wood permeabilities is still not possible. Some trends are apparent, for example the tendency for charcoal yield to be high for woods of high density and very low permeability. A high wood permeability also seems to preclude a charcoal yield at the upper end of the range. However, away from these extremes, charcoal yields can vary widely.

An intrinsic problem with using wood permeability as a predictor of charcoal yield is that permeabilities can be expected to change as the carbonisation process proceeds. Marked changes in both the chemical [3] [8] [13] and physical [14] nature of the wood occur. This can involve an increase in the fraction of the wood cross-sectional area occupied by voids [14]. Changes would also be predicted in cases where low wood permeabilities are attributable in part to deposits of extractives in the internal flow passages. As the substances in these deposits are heated and decompose, they could well provide a reduced resistance to flow, increasing effective permeabilities. A pointer to the significance of such internal changes is provided by the permeability measurements given in Tables 3 and 4. These show that even the minor internal changes accompanying the oven drying of previously air dried wood can be quite significant.

Further progress in understanding charcoal formation would seem to require better characterisation of the internal structure of woods and of the manner in which this structure changes during the carbonisation process. Characterisation of how structure dependent properties like permeability alter during the carbonisation process would also be desirable.

6 Acknowledgements

The help provided by Dr Hoi Why Kong and Ms Vanya Pasa in obtaining wood samples is gratefully acknowledged. The contributions of Mr. T. Lin, Mrs. V. Daria and Dr. C.M. Salazar to investigations referred to in this paper are also much appreciated.

7 References

1. Brito, J.O. and Barrichelo, L.E.G. (1977) Correlações entre características físicas e químicas da madeira e a produção de carvão vegetal: I. Densidade e teor de lignina da madeira de eucalypto. *IPEF, Piracicaba*, Vol. 14, pp. 9-20.
2. Roy, C., Pakdel, H. and Brouillard, D. (1990) The role of extractives during vacuum pyrolysis of wood. *Journal of Applied Polymer Science*, Vol. 41, pp. 337-348.
3. Connor, M.A. and Salazar, C.M. (1988) Factors influencing the decomposition processes in wood particles during low temperature pyrolysis, in *Research in Thermochemical Biomass Conversion*, (eds A.V. Bridgwater and J.L. Kuester), Elsevier, London, pp. 164-178.
4. Connor, M.A., Lin, T. and Daria, V. (1994) Characterising the charcoal-producing potential of different woods. *Proceedings 2nd International Energy Conference: Energy from Biomass Residues*, Forest Research Institute of Malaysia, Kuala Lumpur, pp. 105-110.
5. Cutter, B.E. and McGinnes, E.A. (1981) A note on density change patterns in charred wood. *Wood and Fiber*, Vol. 13, No. 1. pp. 39-44.
6. Salazar, C.M. (1987) The pyrolysis of wood at low heating rates. *PhD thesis*, University of Melbourne.
7. Choong, E.T., Achmadi, S. and Tesoro, F.O. (1989) Variables affecting the longitudinal flow of gas in hardwoods, in *Cellulose and Wood - Chemistry and Technology*, (ed. C. Schuerch), John Wiley Interscience, New York, pp. 1175-1196.
8. Hirata, T., Kawamoto, S. and Nishimoto, T. (1991) Thermogravimetry of wood treated with water-insoluble retardants and a proposal for development of fire-retardant wood materials. *Fire and Materials*, Vol. 15, pp. 27-36.
9. Kilzer, R.J. and Broido, A. (1965) Speculations on the nature of cellulose pyrolysis. *Pyrodynamics*, Vol. 2, pp. 151-163.
10. Lin, T. (1988) Effect of wood variety on charcoal yield. *M.Eng.Sc.project report*, University of Melbourne.

11. Smith, D.N. and Lee, E. (1958) The longitudinal permeability of some hardwoods and softwoods. *Department of Scientific and Industrial Research, Forest Products Research: Special Report No. 13*, H.M.S.O., London, 13 pp.

12. Comstock, G.L. (1965) Longitudinal permeability of green Eastern Hemlock. *Forest Products Journal*, Vol. 15, pp. 441-449.

13. Browne, F.L. and Tang, W.K. (1962) Thermogravimetric and differential thermal analysis of wood and of wood treated with inorganic salts during pyrolysis. *Fire Research Abstracts and Review*, Vol. 4, pp. 76=91.

14. Connor, M.A., Daria, V. and Ward, J. (1994) Changes in wood structure during the course of carbonisation, in *Advances in Thermochemical Biomass Conversion* (ed. A.V. Bridgwater), Blackie, London, Vol. 2, pp. 846-858.

THE HIGH HEAT OF FAST PYROLYSIS FOR LARGE PARTICLES

Thomas B. Reed & Siddhartha Gaur
The Colorado School of Mines, Golden, CO 80401

ABSTRACT

The **heat FOR pyrolysis, h_p,** (including sensible heats required to raise the reactants to pyrolysis temperature; the heat **OF** pyrolysis, Δh_p, and the heat to raise the products to the surface temperature) is an important value required for the science and engineering of biomass thermal conversion.

A novel "water tracer" method of measurement of the **heat FOR pyrolysis** has been tested and the results are presented here. The method is based on measuring the time required for pyrolysis of a dry particle and an identical particle with high moisture content. The difference in time is attributed to the heat required to vaporize the water at the temperature of pyrolysis from which one can calculate the heat required for pyrolysis.

A dry biomass particle (birch dowel) was heated in the reducing flame of a Meeker burner while being slowly rotated around its axis and the time for pyrolysis was measured. The time for pyrolysis of an identical particle with high moisture content was also measured. The flame intensity is kept constant by monitoring its heat transfer to a thermocouple. The **heat FOR pyrolysis** can then be calculated from the difference in time according to

$$h_b = h_w \, m_w \, t_b \, / [m_b \, (t_{w,b} - t_b)]$$

Birch dowels with nominal diameters of 0.95, 1.27 and 1.90 cm X 2 cm long were pyrolysed in the flame of a Meeker burner maintained at a constant heat transfer rate of 28.1 watts/cm^2 with a thermocouple (reading 1000°C). The time for pyrolysis varied from 53 to 355 seconds and was reproducible within 1-3 sec. The heat for pyrolysis of the dowel was then calculated to be **3472, 3280, 2909 J/g** respectively. (These values are about 15% of the heat of combustion.) The charcoal yields were 3.1%, 7.8% and 10.7% reflecting higher yields in larger particles.

1. Introduction

1.1. Importance of the heat OF and FOR pyrolysis

There is a great deal of confusion about the value of the heat required **for** the pyrolysis of biomass under various conditions of heating. We have been surprised that workers in the field typically have no idea of even the magnitude of heat required for the pyrolysis step in pyrolysis, combustion and gasificaiton. In this paper we report the heat **for** pyrolysis of large birch particles measured using a water tracer technique and we relate this to other related values.

1.2. The heat OF pyrolysis

The heat **OF** pyrolysis, Δhp , for wood or other biomass is the heat absorbed (negative) or given off during the (pseudo second order) "phase change" of pyrolysis from solid wood to charcoal and pyrolysis gas and vapor. (We use a small "**h**" to designate this value, since there is no true molar value for biomass.) This is not a true heat of reaction, since pyrolysis involves many reactions occurring over a hundred or two hundred degrees, as shown in thermogravimetric (TG) tests on biomass..

Figure 1 shows the weight loss vs temperature of the average of 6 hardwoods heated at a rate of 10°C/min. The weight loss vs temperature curve is remarkably similar for different species of wood, but increases about 70°C with each factor of ten increase in heating rate.[1]

The importance of the heat of pyrolysis can be judged from the numbers of measurements that appear in the literature. Values of Δhp have been reported for wood and cellulose ranging from 240 kJ/mole (exothermic) to -2100 kJ/g (endothermic), so it is not surprising that workers in the field are confused.[2]

It has long been observed in slow charcoal making that when the wood reaches the temperature of about 275°C the wood pile temperature will exceed the surrounding temperature, indicating an exothermic reaction.

It can be shown with differential thermal analysis or differential thermogravimetry that the pyrolysis reactions can be endothermic or exothermic, depending on the yield of charcoal. Antal has measured Δhp for varying yields of charcoal (by varying pressure). He found that the heat of pyrolysis varied with pressure from 138 J/g with a charcoal yield of 22% to -270J/g with a charcoal yield of 9%.[3] The heat of pyrolysis for 36 data points was fitted satisfactorily by the equation

$$\Delta hp = -553 + 3142\ F_c\ \text{J/g} \tag{1}$$

(where F_c is the weight fraction of charcoal formed).

1.3. The heat FOR pyrolysis

A more useful quantity for engineers is the heat FOR pyrolysis, $\mathbf{h_p}$, the sum of the sensible heat required to bring the reactants to the temperature of pyrolysis; the heat **OF** pyrolysis, $\Delta \mathbf{h_p}$, and the sensible heats required to produce gas/vapor and charcoal in a specific process, ie

$$\mathbf{h_p} = \Sigma h(\text{reactants}) + \Delta\,h_p + \Sigma h(\text{products}) \tag{2}$$

the sums being taken over the range before and after pyrolysis. This quantity is required to answer the question "how much heat must I supply for pyrolysis in order to generate charcoal and gas in any given process of pyrolysis, gasification or combustion.

1.4.   Slow vs Fast Pyrolysis

The variation in products and heat required to produce them is partly due to the variation in heating rate. There is a broad boundary between "slow pyrolysis" and "fast pyrolysis" that can be defined by the Biot number of the heating conditions. The Biot number is the ratio of external to internal heat transfer,

$$N_B = (\text{External Heat Transfer})/(\text{Internal heat transfer}) = H(T)/\kappa\, r \qquad (3)$$

where $H(T)$ is the heat transfer coefficient at temperature T for the heat source (flame, radiation etc.), κ is the thermal conductivity of the body being heated and r is a characteristic radius of the particle.

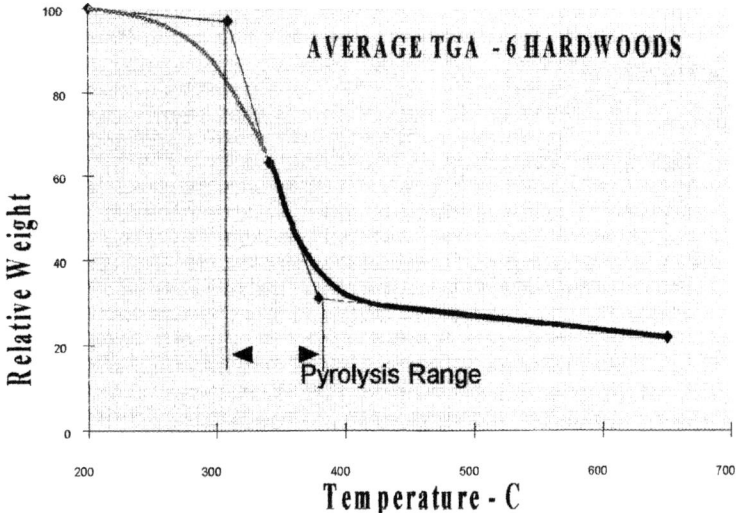

Figure 1. The Relative Weight vs temperature for 6 dry hardwoods

In slow heating biomass, the heat is transferred by conduction to the interior faster than it can be supplied and the whole particle and the particle is essentially isothermal for $N_b < 0.1$. For high heat transfer rates, $(N_b > 1.0)$, the surface is heated rapidly and water and gases are evolved at the surface while the inner portions remain cool.

In conventional "slow pyrolysis" manufacture of charcoal requiring days and weeks, the whole pile steams first as water is removed and then the pyrolysis gases come off. In the fast pyrolysis processes used in making pyrolysis oil very small particles are heated very fast and water and gas come off together. The relationship of energy required for slow and fast pyrolysis are shown diagramatically in Fig. 2.

The heat capacity of wood is on the order of 2.1 J/g and increases slowly with temperature up to the temperature where pyrolysis begins (as shown in Figs. 1 and 2). At pyrolysis temperature some of the reactions are endothermic (particularly vaporization) and some are exothermic (bonding to charcoal). Beyond that temperature the pyrolysis products pass out through the charcoal surrounding the particle and may react to form more secondary charcoal (slow pyrolysis) or may react with the charcoal (gasification) if the temperature is high enough.

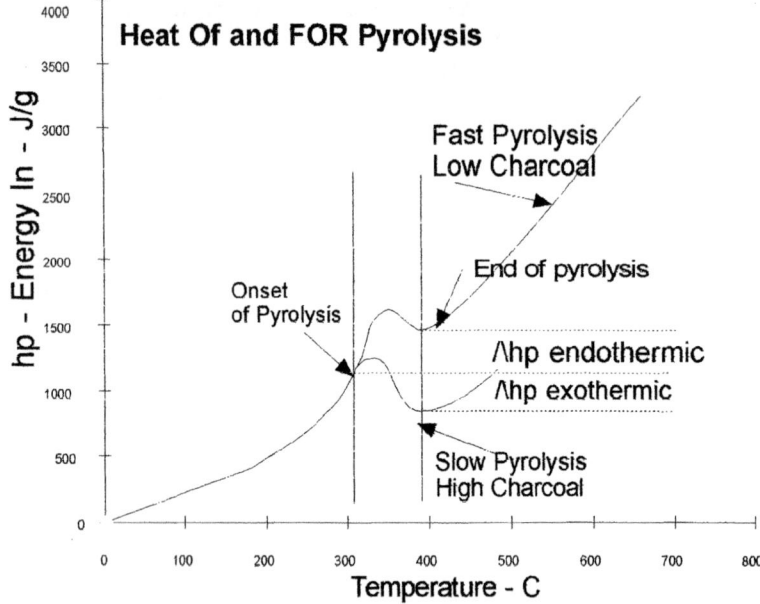

Figure 2. Energy required for slow and fast pyrolysis of larger particles of dry wood (biomass) showing relative contributions of sensible heat and heat for pyrolysis

1.5. Fast pyrolysis of large particles

There has been a great deal of interest in the last decade in "fast pyrolysis" where rapid heating of small particles of wood or biomass can produce a "pyrolysis oil" in yields of 60-70%. There is also an intermediate case that is of great interest in modern industrial processes where heat transfer can be quite high, but particle size (chips or logs) is large also significant. This paper deals with that case.

2. Experimental

We have found that we can measure the time required for pyrolysis, t_p quite accurately using a Meeker (oversize Bunsen) burner as a high heat transfer source, comparable in intensity to the flames or hot gases used in pyrolysis, gasification and combustion. During pyrolysis a bright luminous plume appears above the particle as shown in Fig. 3.

When pyrolysis is complete the plume suddenly disappears. The precision of measurement is 1-3 sec for pyrolysis requiring 60-500 sec.

Furthermore, we have found that the time required for pyrolysis is greatly affected by the moisture content of the particle. Since we know the thermal properties of water very well and the amount of heat required to remove the water, we can use the difference in time as a measure of the **heat OF pyrolysis** under these conditions. The only assumption is that the heat transfer from the flame remains constant for dry and wet biomass.

The experimental apparatus is shown in Fig. 3 and consists of a Meeker burner attached to a propane source, capable of generating a stable reducing flame 4 cm in diameter. A thermocouple is mounted 5 cm above the top of the burner and serves to record the relative heat transfer of the flame. Thermocouples do not give an accurate measurement of flame temperature because they radiate the received energy at a much lower temperature. However, they are an excellent indication of heat transfer, since an energy balance shows that

$$T_{tc} = (H_f/\pi \varepsilon \sigma)^{1/4} \tag{4}$$

(where T_{tc} is the temperature registered by the thermocouple immersed in the flame, H_f is the heat transfer by the flame to the projected area of the thermocouple, ε is the emissivity of the thermocouple (about 0.8) and σ is the Steffan Boltzman constant). At 1000°C the Meeker burner flame transfers about 0.18 W/cm²-K.

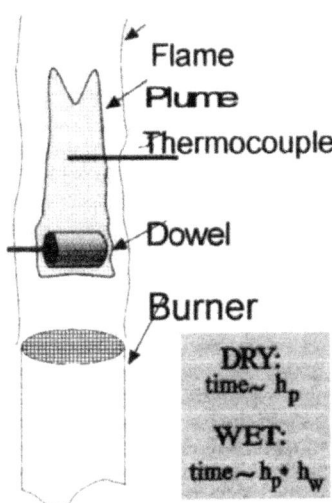

Figure 3. Apparatus for measuring the heat for pyrolysis using wet and dry dowels

The results are shown in Table 1. The particles used were birch dowels 0.95, 1.27, and 1.91 cm (3/8, ½ and ¾ in in diameter). Birch dowels were used because of their uniformity and all samples were cut from the same dowel. One set of dowels was oven dried at 100°C; the other set had high moisture content (~55%), made by boiling in water (to remove air). The sample was mounted on a steel threaded pin held in the jaws of an electric drill and rotated slowly in the flame to pyrolyse the sample evenly on all sides. The time required for the luminous pyrolysis tail to disappear was measured with a stopwatch. It was observed during heating that the outside of the dowel was non luminous during most of the heating period (T<650°C) but reached red heat (T>700°C) near the end of the period. It would be useful to record this surface temperature with an optical pyrometer to define the heat transfer better. However, since the flame temperature is near 2000°C, the heat transfer rate is only minimally affected by the surface temperature.

3.   Discussion

The time required for the pyrolysis of dry biomass is given by

$$t_b = h_b\, m_b\, /\, A\, H(T_{tc}) \qquad (5)$$

and for wet biomass it is

$$t_{b,w} = [h_b\, m_b\, +\, h_w\, m_w]/\, A\, H(T_{tc}) \qquad (6)$$

(where A is the area of the particle exposed to the flame; $H(T_{tc})$ is the heat transfer from the flame per unit area; t_b and $t_{b,w}$ are the times required to pyrolyse the dry and wet biomass respectively; h_b is the heat FOR pyrolysis of the biomass particle; m_b is the mass of the dry biomass particle; h_w is the enthalpy of water at the temperature of pyrolysis; and m_w is the mass of the absorbed water in the particle.)

These two equations can be solved for the **heat FOR pyrolysis** to give

$$h_b = h_w\, m_w\, t_b\, /(m_b\, (t_{w,b} - t_b\,) \qquad (7)$$

(Note that the area being heated and the heat transfer rate are constant for both times and so disappear from the final equation.)

The results of these measurements are shown in Table 1.

Table 1 - The heat for pyrolysis of $^3/_8$, $\frac{1}{2}$ and $\frac{3}{4}$ in dowels in a meeker burner flame

Run	Diam	m_b	$m_{b,w}$	m_w	MC(WB)	m_{char}	char	t_p	h
	cm	g	g	g	%	g	%	sec	J/g
DRY	0.95	0.83	-	0	0.0%	0.026	3.1%	53	3472
WET	0.95	0.83	1.83	0.99	54.4	0.026	3.2%	123	8060
DRY	1.27	1.38	-	0	0.0%	0.109	8.0%	75	3280
WET	1.27	1.38	3.21	1.83	57.1%	0.103	7.6%	192	8400
DRY	1.90	3.22	-	0	0.0%	0.316	9.8%	129	2909
WET	1.90	3.22	7.50	4.28	57.1%	0.376	11.7%	355	8610

The nominal heat transfer of the flame, $H(T_{tc})$, can also be calculated from the above equations as

$$H(T_{tc}) = h_b\, m_b\, /t_b A_{hb} \qquad (8)$$

For the three cases studied it was found to be 9.12, 7.54 and 6.07 W/cm^2. for the three cases.

The charcoal production, 3%, 8% and 10% for the three cases studied is surprisingly low relative to charcoal production in slow pyrolysis, typically 20%-25%. This is believed to be due to two possible factors. During fast pyrolysis there is less time for the pyrolysis products to undergo secondary pyrolysis as they pass out through the particle. Also during fast pyrolysis the temperatures in the outer regions are much higher and could even reach gasification temperatures so that the CO_2 and H_2O in the pyrolysis gases are partly reduced to steam by the outer charcoal layers.

4. Summary

The heat for pyrolysis of birch dowels (nominally $^3/_8$, $\frac{1}{2}$ and $\frac{3}{4}$ in in diameter) was measured using a water tracer technique and found to be 3472, 3280 and 2909 J/g respectively, about 15% of the heat of combustion.

5. References

[1] Gaur, S. and Reed, T. B., "An Atlas of Thermal Data for Biomass and Other Fuels", NREL/TP-433-7965, the National Renewable Energy Laboratory, Golden, CO 80401.
[2] Cowdery, C., "Measurement of the Mass and Energy Balance in Contact Fast Pyrolysis of Wood", MS Thesis T-3459, University of Colorado, 1987.
[3] Antal, M. J., "Biomass Pyrolysis: A Review of the Literature - Part 1 - Carbohydrate Pyrolysis", Advances in Solar Energy, ed K. W. Boer and J. A. Duffie (American Solar Energy Soc., Boulder, CO) Vol 1, pp 61-111, 1982.

CHEMICAL ENGINEERING ASPECTS OF SOLID (BIOMASS) PARTICLE PYROLYSIS : A REVIEW OF THE POSSIBLE RATE LIMITING FACTORS.

J. LEDE
Laboratoire des Sciences du Génie Chimique, CNRS-ENSIC, NANCY, France

Abstract

The yield and selectivity of chemical processes are function of several competitive chemical and physical parameters. For thermal multiphase processes, the problem is still much more complicated, mainly if the operation ocurs under short residence times and with phase changes. The possible rate limiting physical factors must be considered at the particle level (heat and mass transfer resistances) and at the reactor level (hydrodynamics of each phase). The present paper reviews all these points in the specific case of biomass pyrolysis.

Several results derived from the modeling of the competition between chemistry and heat transfer resistances are more specifically reported. They correspond to calculations made in several cases of fixed and increasing heat source temperature and for particles undergoing a thermal flash. Fundamental information is derived on the true temperature and heating rates of the solid in the cases of chemical and ablation regimes. The results suggest to extend the notion of fusion like phenomenon of solids to the idea of sublimation or boiling like phenomena for other types of elementary chemical processes.

The paper often emphasizes the important possible mistakes made in the measurement of chemical rate constants. It aso shows that using kinetic data derived from laboratory scale devices (TGA) for the calculation of process reactors can be very dangerous.

Keywords : chemical engineering, heat transfer processes, kinetic parameters measurements, mass transfer processes, particle, reactor, residence times distributions.

1 Introduction

Biomass pyrolysis is a chemical reation. A chemical reaction necessarily occurs in a chemical reactor. In every chemical reator, the yield and selectivity of the process is function of several operating parameters related to phenomena occuring at the particle and reactor levels. Evaluating each of these parameters is a very difficult task mainly for fast and complex thermal processes occuring with phase changes. This is the case for biomass pyrolysis. All these parameters are coupled with two main consequences:

- the practical determination of intrinsic chemical kinetics is very difficult. Their estimation must take into account transfer processes and reactor properties. They can be sometimes rate limiting. Unfortunately, they are often badly known and/or ignored. Consequently many published results are probably influenced by physical factors and do not represent pure chemistry.

- one must be very careful when using chemical data obtained in a given small scale laboratory device for scaling up calculations.

The present paper is divided into two parts, each one describing the main possible controlling factors located at two different levels:

- at the particle level: intrinsic chemical charateristics of the reaction; heat and mass transfer resistances inside the particle; heat and mass transfer resistances at the interface of the particle and external medium

- at the reactor level : hydrodynamic behaviour of the different flowing phases (reactor design)

Only the main ideas and results will be reported. More details can be found in related more detailed publications cited all along this paper.

2 Phenomena occuring at the particle level

2.1 Intrinsic chemistry

A great number of chemical pathways have been proposed in the literature for representing the pyrolysis of ligno cellulosic materials. Most of them consider competitive steps giving rise respectively to char and gases, and to vapours with subsequent secondary consecutive and interphase reactions. However, following the authors, these competitive reactions concern an intermediate "active" liquid product [1] or directly the starting solid material [2,3]. The first case is encountered for example in ablative pyrolysis when biomass is heated at relatively high temperatures. The second one is used to represent the results derived from TGA measurements that cannot be sensitive to the formation of an intermediate liquid produced without mass loss. The most commonly used model taking into account the formation of a short life time liquid during cellulose pyrolysis is the Broido-Shafizadeh type model. This model can fairly well represent most of the experiments made with cellulose, while the models ignoring the formation of a liquid are insufficient to explain the observations made in high temperature ablative pyrolysis [4]. Moreover, it has been shown recently that liquid species observed during biomass pyrolysis are issued from both cellulose and lignin pyrolysis [4]. Accordingly, it will be supposed in the present paper, that the Broido-Shafizadeh model can be extended to the most general case of biomass pyrolysis:

$$W \xrightarrow{\ (1)\ } A \quad \begin{array}{c} \xrightarrow{\ (4)\ } C+G \\[4pt] \xrightarrow[\ (2)\ ]{} V \xrightarrow[\ (3)\ ]{} G \end{array}$$

W: solid feedstock (biomass) ; A: "active" liquid intermediate; C: char; V: vapours; G: gases

All the elementary chemical processes are supposed to be first order decomposition reactions. Most of them occur with phase changes ((1) : solid $\rightarrow$ liquid; (2): liquid $\rightarrow$ vapour; ...). Each one has its own enthalpy and activation energy. They are organized in a network of competitive and consecutive reactions. Depending on the different temperatures, the complete process occurs in less than 1 second and some elementary reactions in a few milliseconds. These times are of the same order of magnitude as characteristic times of heat exchange (te = $\rho_s C_p L/h$). Thus it appears that the accurate determination of intrinsic pure chemical properties is a very difficult task. The overall apparent kinetics is controlled by several couplings between pure chemistry and heat transfer processes. In the same way, mass transfer resistances can also be rate limiting.

These transfer resistances can occur either inside the reacting particle and at its interface with the surrounding.

2.2 Mass transfer processes
These mechanisms are always intimately connected with heat transfer resistances.

2.2.1 Internal transfer

Resistances are located inside the feedstock reacting particles (W) and concern the ability for the products to escape outwards, to the outside surface. The solid can be considered as a porous medium, the porosity of which changes as the reaction proceeds. The products migrate by diffusion and/or convection through the pores. According to the relative values of transfer and elementary reactions times, the products arriving at the surface are liquids (A) or further cracked products (V). Possible chemical interactions (gasification; ...) with the solid may occur inside the pores. These products flow in counter current with internal heat transfer and are then interdependent. In the case of ablation regime (see section 2.3.2.2) the pyrolysis reaction occurs inside a very thin external layer while the heart of the solid is still cold. If the size of the pores is larger than this ablation layer, back transport of the fluids is then possible, with upstream flow of liquids (A) and/or condensation of vapours (V) in colder parts of the solid (W). These problems are likely to occur with big porous samples. They become negligeable in the case of small isolated particles reacting in chemical regime. In any case, if the reaction is very fast and/or produces high yields of products, local pressure increase may occur, causing fragmentation of the solid.

2.2.2 External transfer

Once arrived at the surface of the particle, the products must escape into the external flowing medium. If the removal is not efficient enough, further decomposition is then possible. In any case, these interface phenomena will interfere with local heat transfer resistances. Several cases must be considered following the nature of the products arriving at the surface (A, V or G), their possible local degradation and the nature of the outside medium.

2.2.2.1. The products are gases (V,G): the reaction (1) is very fast inside the solid or at its surface and is not rate limiting.

- The outside medium is a gas. The efficiency of elimination of the gaseous products is well represented by correlations proposed in the chemical engineering litterature [5,6]. They associate dimensionless numbers as the Sherwood number ($Sh = k_D d_p/D$) describing mass diffusion processes through the boundary layer and particle Reynolds number ($Re = u\rho_g d_p/\mu$). The products elimination is enhanced when the Reynolds increases. It follows that changing the gas residence times in the reactor for kinetics studies purposes may also influence mass transfer efficiencies. These laws are in very close analogy with the local heat transfer correlations (see section 2.3.1).

- The outside medium is a hot surface. The contacts between the two surfaces may be direct (true point contacts if the surfaces are not perfectly smooth) or through the intermediate of a more or less thin gas layer. The important parameters are the applied pressure and the relative velocity between the two surfaces, associated with the rate of production of the gaseous products. Modelling of the gas elimination efficiency is complex and no simple correlation exists in the litterature.

2.2.2.2 The products are liquids (A)

- The outside medium is a gas

The liquid (A) cannot be directly eliminated and will locally undergo further decomposition. At high temperature, the reaction (2) prevails, producing vapours (V), eliminated by convection (see mecanisms described previously). Steadystate conditions may be reached with the formation of a constant thickness of liquid layer. At lower temperatures, the reaction (4) takes place producing char (C), the elimination of which is very difficult. Its layer thickness has chances to increase continuously until slow pyrolysis conditions are reached.

- The outside medium is a hot surface

If its temperature is moderate, the two surfaces are always separated by a liquid (A) layer (through which heat transfer occurs at the same time). The thickness of this layer and rate of its elimination are functions of applied pressure, relative velocities, liquid viscosity and rate of production of the liquid [7]. This layer reduces friction (a kind of lubrication) but may cause possible sticking (in the case of moving and concentrated particles). After leaving the gap between the two surfaces, these liquids can stay on the particle where they undergo further decomposition.

For high temperatures, the liquid (A) decomposes partially at the interface to vapours (V). The solid particle is then isolated from the outside solid medium by a liquid and a gaseous layer (leidenfrost phenomenon). The elimination rate depends as previously on pressure, velocities, flux of production and viscosities of both fluids. Friction is less than with pure liquid and sticking is less probable. Conversely, heat transfer resistances are increased.

It is clear that the recovery of primary liquid products is always enhanced if the transfer medium is a solid surface. This is one of the conditions for ablation regime. The surface can be the hot wall medium as in a cyclone [8,9], a vortex reactor [10] and in spinning disk type experiments [11,12,13], or hot neutral particles impinging the biomass particles [14,15].

2.3 Heat transfer processes

As previously reported, they are always closely connected with mass transfer resistances and with the extend of secondary reactions at the vicinity of the solid surface.

2.3.1 Outside the particle

The solid particle may be heated by several possible mechanisms of exchange : convection with a hot flowing gas; interaction with a hot surface (radiation and/or more or less direct contact). Other sources of heat as plasmas, arcs and lasers will not be considered here. The heat flux ϕ transfered to the particle is usually written as the product of its exchange surface, temperature difference and heat transfer coefficient ($\phi = hS(T_w-T_s)$).

2.3.1.1 Gas phase (natural and forced) convection

The heat source is the gas flowing through the reactor. The correlations used to estimate the heat exchange are quite similar to those derived for mass transfer. They now associate a Nusselt number ($Nu = hd_p/k$) related to the heat diffusion through the boundary layer and the particle Reynolds number ($Re = u\rho_g d_p/\mu$). As for mass transfer, the heat transfer efficiency depends on gas phase residence time. Practically the values of h in forced convection are of a few hundreds W m^{-2} k^{-1} [16].

2.3.1.2   Particle/hot surface transfer

- Transfer by radiation

The heat transfer coefficient depends on surface temperature. For a reaction temperature of 740 K and an emissivity of 0.5, h varies between about 50 and 100 W m^{-2} k^{-1} for source temperatures T_w between 873 and 1173 K. This process is often neglected for low values of T_w. The liquid intermediate (A) produced by the primary process (1) has a very short life time. So, if it is not rapidly eliminated from the feedstock surface, it can further crack because it is continuously exposed to the incident radiation. Consequently, even if the radiative heat flux density is very high (as provided for example in solar or image furnaces) the ablation regime requirements (see section 2.3.2.2) is not ensured as long as no possible elimination of (A) is possible. If secondary cracking reactions are represented by process (4), char produced on the surface will finally insulate the feedstock from the incident radiation. This char is very slowly eliminated by secondary gasification processes only if the reaction is performed under reactive atmosphere (as steam or CO_2) [17]. As a conclusion, radiation is not an efficient process if primary products elimination is needed and if the surface is physically separated from the solid. As a conclusion, the fast pyrolysis requirements is fulfilled only during the very first times of exposure.

- Transfer by more or less close contact

* Simple short contacts during bouncings on a fixed surface (example : impinging of a two phase jet upon a hot surface [18]). The values of h are high but the contact times are short and hence, the heat exchanged may be small. The exchange occurs by direct contact points or by transfer through a thin fluid interface products layer.

* Impact between the reacting particles and neutral hot particles. This is encountered in fluidized bed reactors [19] or in impinging jet reactors [15]. The contact times are short but very efficient. Their frequency depend on mixing conditions ; level of turbulence ; inertia of particles ; porosity of the bed ; ... : this problem is directly related to reactor design. As previously seen in mass transfer problems, sticking phenomena can occur with possible advantageous increase of contact times. In the meanwhile, if they are too long, clogging of the reactor may occur.

* Continuous friction against a heated wall (solid convection). This is a very efficient exchange, closely coupled with mass transfer resistances and depending as seen previously on applied pressure, relative velocities, nature and viscosity of the interface layer. This case has been fundamentally demonstrated and studied in hot wire and spinning disk experiments [11,12,13,20]. The heat transfer coefficient is very high and may reach values up to 50 000 W m^{-2} K^{-1} under high pressure. Associated with excellent mass transfer efficiencies, these fundamental principles are at the basis of many actual ablative pyrolysis process reactors (vortex [10], cyclone [9], rotating blades [13], pyrolysis mill [21]). One of the advantages of ablative pyrolysis is that high rates of pyrolysis are achieved independently of the initial particle size of the feed material (this, significantly reduces the cost of feed preparation as during grinding for example). Most of these devices operate also in plug flow for the solid phase with the resulting advantage of a better selectivity than in more traditional fluidized bed reactors (see section 3.2). However, high applied pressures and important heat transfer surfaces are needed.

2.3.2   Inside the particle with outside fixed temperature

It will be supposed the simplified situation where a cold particle is suddenly immersed into a high and constant temperature medium. Assuming no intra particle mass transfer limitation, the conductive heat transfer resistances inside the particle are in direct competition with the primary chemical processes of solid decomposition.

2.3.2.1 Pure heating of a particle (no reaction)

When a non reacting solid particle is suddenly exposed to an external heat flux it heats, but not instantaneously. The rate of heating is governed by two factors: boundary layer resistance (heat transfer coefficient h) and rate of heat flow inside the solid [6]. The relative importance of these two factors is measured by the thermal Biot dimensionless number ($Bi = hL/k_s$). Two limit cases are usually considered:

- For low (<0.1) values of Bi, the main resistance is located in the outside boundary layer (high temperature gradient) while the temperature is uniform inside the solid
- For high (>40) values of Bi, the resistance to heat transfer is located inside the particle: a steep temperature gradient is observed in a peripherical thin solid layer

In both cases, the evolution of these profiles with time t is represented by the Fourier number ($Fo = \alpha t/L^2$).

2.3.2.2 Heating associated with an endothermal reaction

There are now strong couplings between heat transfer and chemical processes.

These couplings between both phenomena are mainly represented by two dimensionless numbers: the previously described thermal Biot number and the thermal Thiele number ($M = rL^2C_p/k_s$, (where r is the intrinsic chemical reaction rate) [22,23].

Note that M is the ratio of two characteristic times : L^2/α (heat penetration time, characterizing the conduction of heat inside the solid) and $1/k_r$ (first order reaction time characterizing the rate of the chemical reaction). A less influencing parameter is thermicity ($H = \Delta H/C_pT_W$) representing the enthalpy of the reaction.

- For small Bi and M numbers: the temperature and chemical rate are uniform in the whole volume of the particle. The overall rate of reaction is controlled by chemistry. This is the chemical regime.
- For large Bi and M numbers: the reaction occurs only inside a thin layer moving at a constant velocity towards the still cold heart of the solid. The overall rate is limited by a close coupling between heat transfer and chemistry. This is the ablation regime. The model has been derived for the general case of reactions of simple decomposition (no reaction with an external fluid) of solids giving rise to a fluid (no primary solid products). This is the case for the process (1) considered in this paper where only a liquid (A) is produced. It follows that this model cannot be directly compared to the shrinking core model. The ablation layer has a few microns thickness while the ablation velocity can exceed 10^{-2} m/s [11]. In this situation, a liquid (A) layer is formed outside the ablation layer. Steady state conditions are reached only if the products (A) are efficiently removed (see previous problems of mass transfer processes). If it is not the case, (A) can locally crack. If process (4) prevails, char produced will be much more difficult to eliminate. Consequently, the value of h decreases continuously and finally the chemical regime conditions will be reached. The efficient elimination of the primary products (A) is then a basic condition to observe a steady state ablation regime.

2.3.2.3 Consequences

- Solving heat and mass balances shows that the time required for pyrolysis is composed of a heating time and a reacting time [24]. Moreover, the temperature at which the solid particle decomposes during this reacting phase reaches rapidly a quasi-constant value T_R [23,25]. For a given kinetics, T_R is only weakly dependant on other operating parameters (T_W, d_p, h) and may be very different from that of the heat source temperature T_W : the difference $T_W - T_R$ may be sometimes much larger than 500 k !

This phenomenon, valid for every kind of reaction of the solid $\rightarrow$ fluids type leads to a phase change like phenomenon [25]. In the case of biomass first step of decomposition (process (1)) where very high yields of liquid (A) are expected [26], it looks like a fusion. Such a behaviour has been experimentally demonstrated [12,27]. The measured fusion temperature of 740 K (more or less a few 10 K following the conditions) is in close agreement with values issued from the model [23]. Besides these calculated temperatures, empirical relationships have been proposed for the estimation of reaction temperatures T_R [25]. The effect of stabilization observed for T_R associated with large differences with the heat source temperature lead to the conclusion that important errors can be made in the determination of the true solid temperature [23,28]. Consequently, many Arrhenius parameters obtained in such conditions must be very carefully considered and used.

- For given conditions, the reaction temperature T_R is lower (several 10 K differences) in ablation than in chemical regime. Consequently, in the case of solids decomposing following competitive reactions, the selectivity could be very different following the size of the sample [23].

- Another often considered parameter in pyrolysis in the heating rate. In the conditions of a constant outside temperature medium, its definition is difficult [29]. It depends on the type of regime (chemical or ablation) and must be defined before and during the reaction. In all the cases it is a function of time and it is important to state if one considers instantaneous (and hence variable) heating rates or mean heating rates. It can be shown that the heating rates are lower in ablation (a few 10^3 K s^{-1}) than in chemical regime and are always more dependent on operating parameters (ΔH, h, L, T_W) than the reaction temperature T_R [23,29]. Of course, the highest heating rates are observed before the beginning of the reaction (several 10^4 K s^{-1}).

- The fusion model has been derived for the reaction of a solid particle giving a liquid (process (1)). The heat and mass balances of the model can also be solved for another possible situation where the solid (W) would give rise directly to vapours. The calculations would similarly reveal a strong stabilization of solid temperature. In that case it would be concluded that the solid decomposes following a sublimation like phenomenon.

- Coming back to the Broido-Shafizadeh type model (assumed to be valid for biomass pyrolysis), the solid (W) gives rise to an intermediate liquid species (A). Following the mode of decomposition of (W), (A) can be obtained under the form of a droplet or of a liquid layer around (W). In both cases, (A) will undergo further decompositions. One of them is represented by the process (2). Similar heat and mass balances as for (W) primary decomposition can be written. The same conclusions can be derived: stabilization of the reaction temperature of (A). In that case, it can be concluded that the intermediate liquid (A) decomposes now following a boiling like phenomenon. So, these two consecutive chemical processes can be simulated by fusion and boiling like phenomena in series. The corresponding boiling like temperature may be close or higher than (W) fusion temperature according to the respective kinetic and thermodynamic parameters of reactions (1) and (2). Of course such a phenomenon does not occur if there is a true vaporization of (A) before its decomposition.

2.3.3 Inside the particle with outside variable temperature T_W

2.3.3.1 The heat source temperature increases with time

This is the case in dynamic thermogravimetric analysis (TGA). The heat and mass balances at the particle level can be solved after introducing a heating law for T_W (for example $T_W = T_0 + \beta t$) [28,29,30]. Under the conditions of usual values of β encountered in conventional TGA studies $(0.1 \leq \beta(\text{K min}^{-1}) \leq 100)$ it can be shown [29,30] that both the true temperature T_S and heating rate of the solid are close to T_W and β at every moment of the reaction. For example, the maximum difference $T_W - T_S$ has been calculated to be less than 9 K for $\beta = 6$ K min^{-1} [28]. In these practical situations, the thermal behaviour of the heat source is approximatively representative of that of the reacting sample although a few K differences can have important impacts in the kinetics of high activation energies processes. These results reveal a fundamental difference with the case of constant external heat source temperature T_W where the differences $T_W - T_S$ can reach values of several hundred K [30]. Such large differences can be also theoretically observed for very high heating rates of the heat source ($>> 100$ K min^{-1}) [28]. It can also be shown [30] that the true heating rates of the solid, observed with constant heat source temperatures ($> 10^4$ K min^{-1}) are always much higher than the usual values of β encountered in conventional TGA.

The calculations show also that for given heat transfer conditions, the temperature at which the solid reacts is always much lower in the case of increasing values of T_W than when it is fixed. Consequently for complex kinetic schemes, the selectivities can be very different in both cases. Let us consider for example, the processes (2) and (4) and let define an instantaneous yield of vapours φ ($\varphi = k_2/(k_2 + k_4)$). It has been shown [30] that in the case of variable T_W, φ increases as the reaction proceeds and strongly depends on β, while for fixed values of T_W, φ is constant during the reaction. Moreover, φ is always higher in conditions of fixed T_W.

These results show that the true reaction temperature of the reacting solid, the selectivity of the reaction and the relevant kinetic model can be very different if determined in TGA or in fixed external temperature conditions. Then, it appears that the results derived from TGA measurements must be very carefully used for the calculations of the behaviour and yield of a given process reactor operating with a constant temperature T_W.

2.3.3.2 Particles undergoing a thermal flash

It is supposed that the particle reacting in a heated medium is suddenly removed. The practical applications are: crossing of the particle through a highly concentrated radiation zone (focus of a solar furnace); crossing of the particle through a hot boundary layer after bouncing on a surface [18]; every process of quenching where a reacting solid is extracted from a hot medium and rapidly immersed into an other cold medium. The problem is to know whether and to which extent the reaction continues once the reacting particle arrives into the cold medium. The results [30] depend strongly on the enthalpy of the reaction.

For highly endothermal processes, the reaction is stopped quasi instantaneously.

For processes with small ΔH (the case for biomass pyrolysis), the reaction continues and the conversion extent may reach asymptotic values much higher (sometimes close to 1) than those reached at the moment of cooling. In these conditions, quenching is inefficient and final conversion observed is not representative. In an other hand, this result shows that a simple short flash heating could be sufficient for reaching a given high conversion.

3 Phenomena occuring at the reactor level

A particle is never isolated but is necessarily included inside a reactor. In most of the usual fast pyrolysis reactors, the recovered products are gases and vapours but also aerosols, char and ash particles (these condensed fractions are usually small enough to be entrained by the gas). It means that as these products leave the particle boundary layer, they can undergo further decomposition inside the reactor. These secondary processes are hence dependent on the reactor design and fluid dynamics. For a given process, the hydrodynamics of feedstock particles and of flowing gas are usually very different. It is then necessary to define two interdependent reactors: for the gas and for the solid phases.

For the sake of simplicity the reacting particles will be supposed diluted (no interactions) and possible heat exchanges between the species and the walls of the reactor will be ignored.

3.1 Gas phase

3.1.1 Hydrodynamics

In chemical reaction engineering, two basic continuous reactors are usually considered: the plug flow reactor (PFR) and the continuous stirred tank (or back mix) reactor (CSTR) [5].

In a PFR, all the molecules have the same residence times at a given cross section: a pulse of concentration delivered at the entrance would leave the reactor without any deformation. The parameters are only a function of space (or time): the chemical conversion is the same at any point of a cross section. Practically, PFR are encountered in tubular turbulent, packed beds and fluidized bed reactors. It is also idealized with a series of many CSTR.

In a CSTR, the composition, temperature and mixing level are the sames at any point of the reactor: the response to an entrance pulse of concentration would be an exponential, showing that all the residence times are possible (some molecules are by-passed, while other ones stay infinite time). Consequently, all the degrees of conversion yields are possible: from zero to one. CSTR are practically encountered in mechanically or jet stirred reactors; recycling processes; if turbulence or diffusion are high.

In practice, a real reactor can be globally represented by a network of one or several CSTR and/or PFR elements associated with by passes, dead zones, recyclings and exchanges. Each of these parts is characterized by fitted parameters [5]. For example, the flow of the gas in a cyclone reactor is similar to a PFR followed by a more or less by passed CSTR [31].

For known reactors the literature gives correlations to predict the hydrodynamics of the gas. For new types of reactors it is necessary to perform residence time distributions (RTD) measurements, to find correlations based on networks of CSTR and PFR or to use computation codes.

3.1.2 Conversion (in the assumption of isothermal operation and no volume change due to the reaction)

Because of their different residence time distributions, the CSTR and the PFR show different performances mainly for the secondary chemical processes (i.e. further cracking of gases and/or vapours). For a simple first order reaction and a given residence time, the conversion yield of a reactant is always higher in a PFR than in a CSTR. For two consecutive first order reactions, the yield of intermediate product is always higher in a PFR than in a CSTR for a given conversion of the starting reactant.

Its maximum possible yield is higher for a PFR but occurs for higher conversions of the reactant. The results depend on the relative activation energies of the two reactions. These are important points if the desired product is an intermediate species: for a given mean residence time and temperature, its recoverable yield will be different following the nature of the reactor.

3.2 Solid phase (particles)

3.2.1. Hydrodynamic

The aerosols and powder products are fine enough to be considered as perfect tracers of the gas. This is not the case for the reactant feedstock, made with big size particles. Their residence times may differ considerably from those of the carrier gas. For example, in a cyclone reactor, where the gas behaviour has been depicted previously [31], the particles behave as in a PFR [32]. Furthermore, it has been shown in that case that the solid residence time increases as the gas residence time decreases. For example, for a 12.5×10^{-2} m diameter cyclone, the residence time of biomass particles of a few millimeters can be five times larger than that of the carrier gas. In a fluidized bed reactor, the particles and the gas behave respectively as in a CSTR and a PFR [19]. These two examples show that for every process reactor, it is necessary to perform residence time distributions measurements for both phases: gas and solid.

If the experimental methods are relatively well known for the gas phase, though very hard to perform for residence times lower than 1 s, they are always much more difficult for the particles where size distributions must be taken into account. Preliminary measurements with model particles (known sizes, shapes, ...) undergoing known chemical reactions are sometimes recommended.

3.2.2 Conversion

As for the gas, the conversion of a reacting solid will depend on particles residence time distributions. If, for example, the particles behave as in a PFR, they will have the same conversion at the exit of the reactor. If they behave as in a CSTR, some will stay a long time and will be totally converted while others will immediately leave the reactor before any reaction.

For complex chemical pathways, the PFR gives better selectivities: this is the case for the solid phase in many ablative pyrolysis reactors.

3.3 Coupling between gas and particles behaviours

For a given chemical scheme, the overall yield for the primary solid decomposition and for secondary gas phase crackings, observed at the exit of the reactor, is then the result of couplings between the residence time distributions and chemical reactions of each phase. All the situations are of course theoretically possible following the type of the overall process. For example, the performance will be different in a process with short solid and large gas residence times (low solid conversion and important secondary crackings) compared to an other process with large solid and short gas residence times (complete solid conversion with no further cracking).

Modeling of the whole process becomes thus a very difficult task for complex chemical pathways (example: the Broido Shafizadeh model with simple, competitive and consecutive reactions concerning solid, liquid and gas phases) occuring in a reactor where each of the phases has its own residence time distributions and where the solid particles and/or liquid droplets have their own size distributions.

4 List of symbols

A   Intermediate liquid
C_p Heat capacity of the solid (J kg^{-1} K^{-1})
d_p Diameter of a particle (m)
D   Mass diffusivity of the gas (m^2 s^{-1})
Fo   Fourier number ($= \alpha t/L^2$)
G   Gas
h   External heat transfer coefficient (W m^{-2} K^{-1})
H   Thermicity
k_D External mass transfer coefficient (m s^{-1})
k_r Chemical kinetic constant (r = 1,2,... represents the chemical process number) (s^{-1})
k_s Solid thermal conductivity (W m^{-1} K^{-1})
k   Gas thermal conductivity (W m^{-1} K^{-1})
L   Characteristic length of the solid ($= d_p/6$ for a sphere) (m)
M   Thermal Thiele number ($= r\, L^2\, C_p/k_s$)
r   Chemical reaction rate (kg m^{-3} s^{-1})
Re   Reynolds number ($= u\, \rho_g d_p/\mu$)
S   External surface of the particle (m^2)
Sh   Sherwood number ($= k_D\, d_p/D$)
t   Time (s)
t_e Time of heat exchange ($= \rho_s\, C_p\, L/h$) (s)
T_R Reaction temperature of the solid (K)
T_S Temperature of the solid (K)
T_W External heat source temperature (K)
u   Slip velocity (difference between gas and solid velocities) (m s^{-1})
V   Vapour
W   Solid feedstock

Greek letters :

α Thermal diffusivity of the solid (m^2 s^{-1})
ß   External heating rate (K s^{-1})
ΔH Enthalpy of the reaction (J kg^{-1})
ϕ Heat flux transfered (W)
φ Yield of vapours
μ Viscosity of external flowing gas (kg m^{-1} s^{-1})
ρ_g Gas mass density (kg m^{-3})
ρ_s Solid mass density (kg m^{-3})

5 References

1. Piskorz, J., Scott, D.S. and Radlein, D. (1995) Mechanisms of the fast pyrolysis of biomass: comments on some sources of confusion, in *"Frontiers of*

Pyrolysis: Biomass Conversion and Polymer Recycling", Breckenridge, Colorado, June 1995

2. Varhegyi, G., Jakob, E. and Antal, M.J.A. (1994) Is the Broido-Shafizadeh Model for Cellulose Pyrolysis True ? *Energy and Fuels*, vol. 8, n° 6, pp. 1345-1352
3. Antal, M.J.A. and Varhegyi, G. (1995) Cellulose Pyrolysis Kinetics: the current State of Knowledge, *Ind. Eng. Chem. Res.*, vol. 34, pp. 703-717
4. Lédé, J., Diebold, J.P., Peacocke, G.V.C. and Piskorz, J. (1996) The nature and properties of intermediate and unvaporized biomass pyrolysis materials. Paper presented at the *Developments in Thermochemical Biomass Conversion*, Banff, Canada, May 20-24
5. Villermaux, J. (1985) *Génie de la Réaction Chimique. Conception et fonctionnement des réacteurs.* Lavoisier, Technique et Documentation, Paris
6. Levenspiel, O. (1984) *Engineering Flow and Heat Exchange.* Plenum Press, New York
7. Martin, H., Lédé, J., Li, H.Z., Villermaux, J., Moyne, C. and Degiovanni, A. (1986) Ablative melting of a solid cylinder perpendicularly pressed against a heated wall. *Int. J. Heat Mass Tranfer"*, vol. 29, n° 9, pp. 1407-1415
8. Lédé, J., Li, H.Z., Soulignac, F. and Villermaux, J. (1992) Le cyclone réacteur. Partie IV: Mesure de l'efficacité des transferts de chaleur entre les parois et les phases gazeuse et solide. *Chem. Eng. J.*, vol. 48, pp. 83-99
9. Lédé, J., Verzaro, F., Antoine, B. and Villermaux, J. (1986) Flash pyrolysis of wood in a cyclone reactor. *Chem. Eng. Proc.*, vol. 20, pp. 309-317
10. Diebold, J. and Scahill, J. (1988) Production of primary pyrolysis oils in a vortex reactor, in *Pyrolysis oils from Biomass, Producing, Analyzing and Upgrading*, (eds. J. Soltes and T.A. Milne), ACS Symp. Ser. 376, chap. 4, pp. 31-40
11. Lédé, J., Panagopoulos, J., Li, H.Z. and Villermaux, J. (1985) Fast pyrolysis of wood: direct measurement and study of ablation rate. *Fuel*, vol. 64, pp. 1514-1520
12. Lédé, J., Li, H.Z. and Villermaux, J. (1988) Pyrolysis of Biomass: evidence for a fusionlike phenomena. Same Ref. as 10, pp. 66-78
13. Peacocke, G.V.C. (1994) *Ablative Pyrolysis of Biomass*, Thesis, The University of Aston in Birmingham
14. Graham, R.G., Bergougnou, M.A., Mok, L.K. and De Lasa, H.I. (1985) *Fundamentals of Thermochemical Biomass Conversion* (ed. R.P. Overend), Elsevier Applied Science Publishers, New York, pp. 397-410
15. Tamir, A. (1994) *Impinging Stream Reactors. Fundamentals and Applications. Transport Processes in Engineering*, vol. 7, Elsevier, Amsterdam
16. Reed, T.B., Diebold, J.P. and Desrosiers, R. (1980) *Perspectives in Heat Transfer Requirements and Mechanisms for fast Pyrolysis* in Proceedings of Specialists Workshop on Fast Pyrolysis of Biomass, Copper Mountain Colorado, October 19-22 (1980), pp. 7-20
17. Lédé, J., Villermaux, J., Royère, C., Blouri, B. et Flamant, G. (1983) Utilisation de l'énergie solaire concentrée pour la pyrolyse du bois et des huiles lourdes du pétrole. *Entropie*, vol. 110, pp. 57-69
18. Lédé, J., Adam, P., Marcant, S. and Villermaux, J. (1993) A new method for studying reaction of solid particles sprayed by a gas jet onto a hot surface. *Trans. I. Chem. E.*, vol. 71, Part A, pp. 153-159
19. Kunii, D. and Levenspiel, O. (1991) *Fluidization Engineering.* Butterworth-Heinemann, Boston
20. Diebold, J. (1980) *Ablative Pyrolysis of macroparticles of Biomass.* Same Ref. as 16, pp. 237-251
21. Reed, T.B. (1988) *Contact Pyrolysis in a "Pyrolysis Mill".* Research in Thermochemical Biomass Conversion, Bridwater, A.V. and Kuester, J.L. (eds), Elsevier Applied Science Publishers, New York, pp. 192-202

22. Villermaux, J., Antoine, B., Lédé, J. and Soulignac, F. (1986) A new Model for thermal volatilization of solid particles undergoing fast pyrolysis. *Chem. Eng. Sci.,* vol. 41, n° 1, pp. 151-157

23. Lédé, J. (1994) Reaction temperature of solid particles undergoing an endothermal volatilization. Application to the fast pyrolysis of Biomass. *Biomass and Bioenergy.* Vol. 7, n° 1-6, pp. 49-60

24. Kothari, V. and Antal, M.J.A. (1985) Numerical studies of the flash pyrolysis of cellulose. *Fuel,* vol. 64, pp. 1487-1494

25. Lédé, J. et Villermaux, J. (1993) Comportement thermique et chimique de particules solides subissant une réaction de décomposition endothermique sous l'action d'un flux de chaleur externe. *Can. J. Chem. Eng.* vol. 71, pp. 209-217

26. Diebold, J.P. (1995) A unified global model for the pyrolysis of cellulose. *Biomass and Bioenergy,* vol. 7, n° 1-6, pp. 75-86

27. Lédé, J., Li, H.Z., Villermaux J. and Martin, H. (1987) Fusion-like Behaviour of Wood Pyrolysis. *J. Anal. Applied Pyrolysis.* vol. 10, pp. 291-308

28. Narayan, R. and Antal, MJ.A. (1996 Thermal Lag, Fusion and the Compensation Effect during Biomass Pyrolysis. *Ind. Eng. Chem. Res.,* In press

29. Lédé, J. (1995) Pyrolysis of solid materials: a chemical engineering approach. Same Ref. as 1

30. Lédé, J. (1995) Influence of the heating conditions on the thermal and chemical behaviour of solid particles undergoing a fast endothermic decomposition. *J. Thermal Analysis.* Vol. 45

31. Lédé, J., Li, H.Z. et Villermaux J. (1989) Le cyclone réacteur. Partie I: Mesure directe de la distribution des temps de séjour de la phase gazeuse. Lois d'extrapolation. *Chem. Eng. J..* vol. 42, pp. 37-55

32. Lédé, J., Li, H.Z., Soulignac, F. et Villermaux, J. (1989) Le cyclone réacteur. Partie II: Mesure directe de la distribution des temps de séjour de la phase solide. Lois d'extrapolation. *Chem. Eng. J..* vol. 42, pp. 103-117

SIMULTANEOUS HEAT, MASS AND MOMENTUM TRANSFER DURING BIOMASS DRYING

COLOMBA DI BLASI

Dipartimento di Ingegneria Chimica, Universitá di Napoli Federico II, P.le V. Tecchio, 80125 Napoli, Italy

Abstract
A mathematical model is presented of transport phenomena and moisture evaporation of biofuels exposed to radiative/convective heating. The medium is considered as a three-phase mixture: virgin solid with bound water to the FSP, capillary water that partially fills the pores, and bubbles containing inert gas and water vapor. Transport phenomena account for convection of capillary water, convection and diffusion of water vapor, surface diffusion of bound water, heat convection and conduction, liquid and gas phase pressure and velocity variations. The partial pressure of vapor is equal to its equilibrium value, which is a function of both temperature and moisture content. The high-temperature ($600K$) drying of $1 \times 10^{-2}m$ thick particles, with an initial moisture content of 50% on dry basis, has been simulated by varying the external heat transfer coefficient and the permeabilities to liquid and gas flow. The results of the simulations are applied to understand the dynamics of thermal drying and to assess the validity limits of simplified theories of moisture evaporation, usually introduced in the mathematical description of moist particle pyrolysis and gasification.
Keywords: biomass, bound water, capillary water, drying, modeling, transport phenomena.

1 Introduction

The moisture content of biomass feedstock, available in the form of wood chips, may reach values as high as 150% on dry basis. Moisture in solids can exist in three forms: water vapor and capillary (or free) water in the pores and bound (or hygroscopic) water in the solid matrix (when all available sites are occupied with water molecules, the medium is at the Fiber Saturation Point (FSP)). Sometimes, to reduce the amount of waste water, generated during thermochemical conversion, and the thermal load of pyrolysis/gasification reactors and incinerators, different devices, such as rotatory or silo driers [1], are used. Rotatory driers are usually provided with gases at about $600K$ (waste heat from air-cooled condensers and/or hot flue gas). After drying, the gases leave the driers at about $400K$. Air, slightly above the normal water boiling point, is blown up through a slowly moving bed of chips in silo driers. In other cases, when the efficiency of the reactor is not highly affected by the moisture content, or when this is not very high in the biomass feed-stock, drying and thermochemical conversion are made to occur simultaneously.

For slow heating rates and/or low temperatures (below the normal water boiling point), three main stages are seen during a drying process [2,3]: a constant rate period and two falling rate periods (funicular and pendular stages). When the moisture content is high, in the initial stage, moisture evaporates from the medium surface at a constant rate which is approximately the same as the evaporation rate from a continuous water surface under identical environmental conditions. This stage occurs at constant temperature (equal to the wet-bulb temperature of the environment). When the moisture level decreases below a certain critical value, the biomass temperature starts to increase. The rate of drying becomes successively slower because dry spots start to appear on the solid surface. However, inside the porous medium, there is still a continuous moisture distribution (this is the funicular stage). The decreasing drying rate is controlled by internal moisture transfer. When significant decrease in the drying rate is observed, the process becomes independent of environmental conditions. It may occur that the drying front moves from the surface towards the interior of the solid. This is the pendular stage, whose duration is strongly dependent upon the topology of the porous system.

Drying under fast external heat transfer rates and/or high temperatures (above the normal water boiling point) is characterized by the presence of large spatial gradients. The propagation of a rather steep drying front through the medium is observed [4-7] for the whole duration of moisture evaporation, so that process dynamics are profoundly different from those described above.

Although it is widely recognized that moisture largely affects reactor effciency and product quality, no extensive experimental investigation is available on the drying characteristics of biomass. Some analyses on the pyrolysis of moist wood [8-10] show that both the conversion time and the product yields are significantly dependent on the moisture content. Delay in the solid pyrolysis and the subsequent gas phase ignition are also observed in the combustion of wet wood [5,7,11,12].

As moisture evaporation from a porous solid exposed to high temperature is a non-isothermal process, accompanied by mass transfer from the liquid to the vapor phase and fluid motion, its mathematical description is constituted by a rather complicated set of partial differential equations. A local phase equilibrium is always assumed, that is the water vapor is in saturated equilibrium with the free water. The simplest treatment, usually adopted for high-temperature processes, is based on the application of the Clausius-Clapeyron equation, as for the drying of concrete [4,6], lignocellulosic materials [7,13,14] and coal [15], or on Arrhenius-type kinetics as in the case of pine wood [16]. Alves and Figueireido (1989) [17] describe moisture evaporation from degrading wood as a heat transfer controlled process. An empirical expression gives the equilibrium moisture content, U, as a function of temperature (above $373K$) as $T_b = T_b(U)$ (the variations in the vapor pressure are described as variations in the boiling point, T_b). This equation is used to compute T_b for $U \leq 0.144$, whereas for larger values of U, $T_b = 373K$. When $U > 0.144$, no vaporization occurs for $T < 373K$.

More general treatments have also been proposed for low-temperature drying. The equation of liquid-vapor equilibrium, known as sorption relation, is expressed as a family of curves giving the relative volumetric moisture content, U, as a function of the vapor pressure, p_v, for several temperatures, T: $U = U(p_v)$. The inclusion of the three variables U, p_v, and T into a single equation [3] consists of the sorption relation which gives the moisture content as a function of a characteristic pore length, r, $U = U(r)$. The Kelvin equation is then used to express r on dependence of T and p_v, and $U = U(r(p_v, T))$. Apart from the model of wood drying presented in [18], this approach has been applied mainly for construction products, such as clay bricks, cement paste. In the case of wood drying [19-21] the phase equilibrium relation is explicitly expressed as $p_v = p_v(T, U)$, which reduces to the Clausius-Clapeyron equation for $U > U_{FSP}$ or to specifically determined correlations of bound water sorption data for $U \leq U_{FSP}$. The same approach is also proposed in [22] for a parametric study of the drying of porous media.

Another important aspect in the modeling of solid drying is represented by the description of transport phenomena. In all cases, the assumption of local thermal equilibrium is considered valid, given that drying processes are usually slow. Thus enthalpy variations due to conduction, convection and moisture evaporation are in most cases taken into account. The main differences among the models are due to the different simplifications in the description of transport phenomena of the two-phase water. The most complete treatment is presented [19,21] for low-temperature drying of wood. Diffusion and convection of water vapor, convection of capillary water and surface diffusion of bound water are modeled by means of polyphase Darcy's law and empirical expressions for the capillary pressure and the diffusion coefficients. A rather detailed treatment of the problem is also proposed in [18,20,21], where the main simplification consists of neglecting the motion of the bound water. All movement in the porous system and pressure variations are limited only to the gas phase in the analyses of construction product drying presented in [3,4,6].

Highly simplified treatments of moisture transport phenomena are proposed when more or less sophisticated treatments of solid devolatilization are also introduced. A constant pressure formulation of the conservation equations of water vapor and enthalpy and no transport of liquid phase moisture have been proposed to simulate the drying of coal [15] and wood [13,16,17]. The assumption of constant gas phase pressure is removed in the wood pyrolysis model proposed in [7]. Furthermore, in all these cases, very few predictions of the drying characteristics have been presented.

The analysis of the literature, in relation to wood drying, shows that though rather detailed theories have been proposed, the applications deal with slow drying of very large samples. Therefore, the results are not applicable for the fast drying stage of biomass fuels fed to chemical reactors or driers. On the contrary, these conditions have been modeled only through highly simplified theories. Thus a detailed theory of fast drying is proposed in this study, to predict the dynamics of particle drying and to quantify the importance of the different processes.

2 Mathematical model

In the formulation of the mathematical model a phenomenological approach has been chosen, which applies the equations of the continuum physics to the porous medium, assuming that the porosity is fine and uniformly distributed (this approach gives the same equations as classical averaging techniques). The existence of a propagating evaporating front, which separates the material into a moist zone and an evaporation region is not explicitly assumed. On the contrary, the position of such a front is found as a part of the solution. Transport phenomena account for convection of capillary water, convection and diffusion of water vapor, surface diffusion of bound water, heat convection and conduction, liquid and gas phase pressure and velocity variations. The main assumptions are:

1. the porous system is one-dimensional,
2. the binary mixture of inert gas and steam behaves like an ideal gas (with a constant mean molecular weight),
3. there is a local thermal equilibrium among the solid, the liquid and the gas phases,
4. the Darcy law holds for the liquid and the gas phase,
5. the effects of gravity force are small for both the liquid and the gas phase,
6. the rule of mixture applies for thermal conductivity, and the thermal conductivity for each phase is constant,
7. specific heat of steam and inert are equal,
8. permeability to liquid and gas flow can be expressed in terms of relative permeabilities (these vary with the local moisture level),
9. the enthalpy for the three phases is a linear function of temperature,
10. there is no variation in the total volume initially occupied by the moist solid,
11. the liquid water is incompressible, though the water percentage continuously decreases during the drying process,
12. the liquid and the vapor phases remain in thermodynamic equilibrium at the local temperature (the Clausius-Clapeyron equation is valid until capillary water is present, otherwise the bound water content and the gas phase composition are related by a sorption isotherm, experimentally determined, at the local temperature),
13. the flow of bound water through the solid occurs by molecular diffusion, according to a gradient proportional to bound water concentration,
14. the viscosity of the liquid and the gas phase are constant,
15. no degradation of the solid occurs, so that the solid density remains constant.

Conservation equations are thus formulated for capillary water (1), bound water (2), water vapor (3), total gas mass (4), total enthalpy (5); gas and liquid phase flow is described by the Darcy law (6,7); the constraint of constant volume gives the porosity, ψ_g, (8) (the void fraction is that corresponding to dry material minus the contribution due to the presence of capillary water); thermodynamic relations (9-12) between gas phase concentrations, temperature and moisture levels, and definitions (13) make the drying process a well posed problem:

$$\varrho_l \frac{\partial \psi_l}{\partial t} + \varrho_l \frac{\partial(\psi_l u_l)}{\partial x} = -m_l, \quad \frac{\partial \varrho_b}{\partial t} = \frac{\partial}{\partial x}\left(\varrho_s D_b \frac{\partial(\varrho_b/\varrho_s)}{\partial x}\right) - m_b \qquad (1-2)$$

$$\frac{\partial(\psi_g \rho_v)}{\partial t} + \frac{\partial(\rho_v u_g)}{\partial x} = m + \frac{\partial}{\partial x}\left(\varrho_g D^* \frac{\partial(\varrho_v/\varrho_g)}{\partial x}\right), \quad \frac{\partial(\psi_g \rho_g)}{\partial t} + \frac{\partial(\rho_g u_g)}{\partial x} = m$$
$$(3-4)$$

$$(\varrho_s c_s + \psi_g c_g \varrho_g + c_l(\varrho_b + \varrho_l))\frac{\partial T}{\partial t} = m_b \Delta h_b + m_l \Delta h_l -$$

$$[u_g c_g \varrho_g + u_l c_l \varrho_l \psi_l]\frac{\partial T}{\partial x} + \frac{\partial}{\partial x}\left(\lambda^* \frac{\partial T}{\partial x}\right) \qquad (5)$$

$$u_g = -\frac{K_g k_g}{\mu_g}\frac{\partial p_g}{\partial x}, \quad u_l = -\frac{K_l k_l}{\mu_l}\frac{\partial(p_g - p_c)}{\partial x}, \quad \psi_g = \frac{V_g}{V} = \psi_{g0} - \psi_l \qquad (6-8)$$

$$p_v = p_{vs}(T) \ (U > U_{FSP}), \ p_v = p_v(T,U) \ (U \leq U_{FSP}), \ \varrho_v = \frac{p_v W_v}{RT}, \ p_g = \frac{\rho_g RT}{W_g}$$
$$(9-12)$$

$$U_l = \frac{\varrho_l}{\varrho_s}, \ U_b = \frac{\varrho_b}{\varrho_s}, \ U = U_l + U_b, \ \psi_l = \frac{V_l}{V}, \ m = m_l + m_b \qquad (13)$$

The initial temperature and moisture profiles must satisfy local equilibrium constraints. One side of the particle ($x = \tau$) is exposed to a flow with fixed characteristics (velocity which affects h_c), temperature (T_r) and relative humidity (Y_{v0})) and boundary conditions are written as:

$$\lambda^* \frac{\partial T}{\partial x} = -h_c(T - T_r) - \sigma\varepsilon(T^4 - T_r^4), \ D^* \frac{\partial Y_v}{\partial x} = -k_m(Y_v - Y_{v0}) \qquad (14)$$

The other side($x = 0$) represents a plane of symmetry.

The symbols have the following meaning: ϱ density, ψ liquid or gas void fraction, u_g and u_l gas and liquid velocities, Y mass fraction, T temperature, c thermal capacity, λ thermal conductivity, D diffusion coefficient, M mass, τ particle half-thickness, Δh heat of evaporation, K intrinsic permeability, k relative permeability, m evaporation rate, T_r external temperature, μ viscosity, p pressure, W_g mean gas molecular weight, σ Stefan-Boltzman constant, ε surface emissivity, h_c convective heat transfer coefficient, k_m mass transfer coefficient, p_c, capillary pressure, U, volumetric moisture content. Subscripts indicate: l capillary water, b bound water, s solid, g gas, v vapor, $v0$ saturated vapor, FSP Fiber Saturation Point, 0 initial or reference conditions.

3 Physical properties

Physical properties and their dependence on the moisture content and temperature should also be known to simulate the evaporation process. These have been taken from the literature and roughly describe a softwood. Though, in general, the properties of biomass are largely dependent on the type of material, it is believed that the choice here operated suffices for the main purpose of the work, that is the understanding of the role played by some key parameters on the evaporation process.

The equilibrium vapor pressure is expressed as in [20,21], assuming that capillary water evaporates first, followed then by bound water [19,22]:

$$\frac{p_v}{p_{vs}} = exp[(17.884 - 0.1423T + 23.63 \times 10^{-5}T^2) \times (1.0327 - 67.4 \times 10^{-5}T)^{92U}],$$

$$U > 0, \ T \leq 423K; \qquad p_v = p_{vs} \ \ U > 0, \ T > 423K \tag{15}$$

where the equilibrium vapor pressure of free water is given by:

$$p_{vs} = 10^{[-A/T_r + B - exp(-20(T_r - b)^2)]},$$

$$T_r = \frac{T}{T_c}, \ T_c = 647.3K, \ A = 3.1423, \ B = 8.3610, \ b = 0.163 \tag{16}$$

Equation (15) is valid for $U > 0$, as it is the case in slow heating, low temperature drying, where the moisture content is never rigorously brought to zero (values typically of the order 5% are established). In the simulations, values of the total moisture below 0.001 have been assimilated to zero, with a zero vapor pressure, as shown in Fig.1. It can be seen that, for moisture contents above 30% (that is above the FSP typical of softwoods), $p_v = p_{vs}$, and the Clausius-Clapeyron equation is valid. For lower moisture contents, significant vapor pressure depressions are predicted, mainly at the lower temperatures. It should also be noted that for $T > 423K$, the Clausius-Clapeyron equation is again assumed to be valid (though, for the conditions here considered, this limit in never achieved).

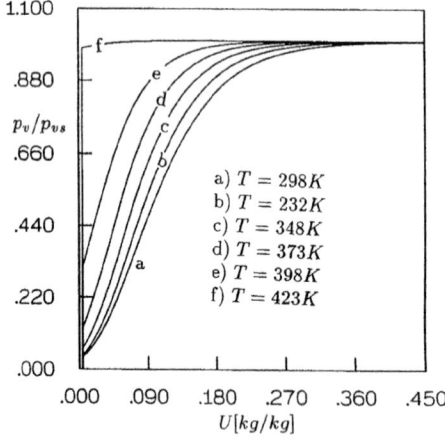

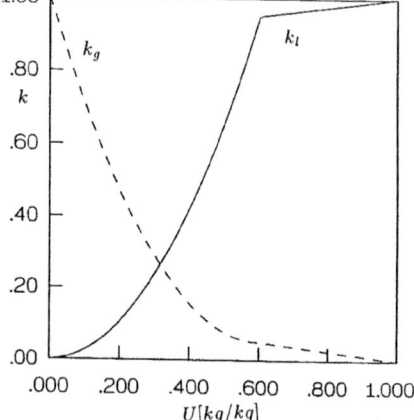

Fig. 1 - Depedence of p_v/p_{v0} on the moisture content and temperature.

Fig. 2 - Dependence of the relative permeabilities to liquid and gas flow on the moisture content.

The capillary pressure is again expressed as in [20,21] (where the surface tension, γ, at the liquid/gas interface is taken from [23]:

$$p_c = [1.364 \times 10^5 \gamma (U_l + 1.2 \times 10^{-4})^{-0.63}][Pa] \tag{17}$$

$$\gamma = 0.0769[1 - 0.00225(T - 273)] \times 10^{-2}[kg/m] \qquad (18)$$

The enthalpy of vaporization of capillary water is $\Delta h_l = 2260kJ/kg$, whereas, for bound water, it depends on the moisture content [17]:

$$\Delta h_b = 3348 - 13085U + 60262U^2 - 95778U^3[kJ/kg] \qquad (19)$$

It has been observed that, below a critical saturation point, the relative permeability goes to zero and liquid migration ceases due to a loss of continuity in the liquid phase [18]. The dependence of relative permeabilities on the moisture content is shown in Fig.2 and can be expressed [20,21] as ($U_{sat} = 1.33$ and $U_{cr} = 0.8$):

$$U_l = 0, \quad k_l = 0, \quad k_g = 1 \qquad (20a)$$

$$0 < U_l < U_{cr}, \quad k_l = 0.95(U_l/U_{cr})^2, \quad k_g = 0.95(1 - U_l/U_{cr})^2 + 0.05 \qquad (20b)$$

$$U_{cr} < U_l < U_{sat}, \quad k_l = 0.95 + 0.05(U_l - U_{cr}/(U_{sat} - U_{cr}),$$

$$k_g = 0.05(U_{sat} - U_l/(U_{sat} - U_{cr}) \qquad (20c)$$

Again, the diffusion coefficients of water vapor and bound water are assigned as in[20,21] (D_v is the diffusion coefficient of water vapor in air):

$$D^* = k_g D_v \times 10^{-4}[m^2/s], \quad D_b = exp(-9.9 - 4300/T + 9.8U_b)[m^2/s] \qquad (21-22)$$

Finally, introducing the volume fraction occupied by bound water as $\psi_b = \varrho_b/\varrho_s$ and according to the rule of mixture, the effective thermal conductivity is defined as:

$$\lambda^* = \lambda_s + \psi_g \lambda_g + (\psi_l + \psi_b)\lambda_l \qquad (23)$$

4 Results

The equations (1-23), after finite-difference approximations, are solved through two stages according to the method of operator splitting. The first stage accounts for the evaporation process and the second for transport phenomena.

The drying of a particle (half thickness $\tau = 0.5 \times 10^{-2}m$) with initial moisture contents equal to 50% on dry basis (boundary conditions (14) and $Y_v(x = \tau) = Y_{v0}$), exposed to ambient temperatures equal to $600K$ is considered. Properties are chosen to describe Southern pine (reference data): $\varrho_s = 600kg/m^3$, $\psi_{g0} = 0.6$, $\lambda_s = 0.14W/mK$ [24], $K_g = 5 \times 10^{-15}m^2$, $K_l = 5 \times 10^{-16}m^2$ [19]. Other properties are: $D_v = 0.22 \times 10^{-4}m^2/s$, $\mu_g = 2 \times 10^{-5}kg/ms$, $\mu_l = 1 \times 10^{-3}kg/ms$, $c_l = 4200J/kgK$, $\lambda_l = 0.610W/mK$. The reference value of the convective heat transfer coefficient is assigned as $20W/m^2K$ and then it has been varied in the range of those reported in the literature, $4 - 25W/m^2K$ [17]. The mass transfer coefficient is taken equal to $0.02m/s$, and simulations, in agreement with previous findings [20], show that, for high equilibrium temperatures, the drying rate is affected essentially by the heat transfer rate. A parametric study has also been conducted of the effects of the intrinsic liquid and gas permeability on the drying process.

The dynamics of high-temperature drying can be seen through Figs. 3-7, where the temperature, the moisture contents, the evaporation rate, the gas pressure, the gas velocity, the porosity and the liquid velocity profiles are shown for several times, as simulated for the reference data ($D_b = 0$, in this simulation). At a first glance, it can be seen that the evaporation rate at the surface is faster than the rate of internal liquid flow needed to maintain a continuous surface layer. Therefore, no constant period rate is observed, but soon the propagation of two drying fronts through the particle is seen, the first associated with the evaporation of the capillary water, the second with that of bound water (Fig.3).

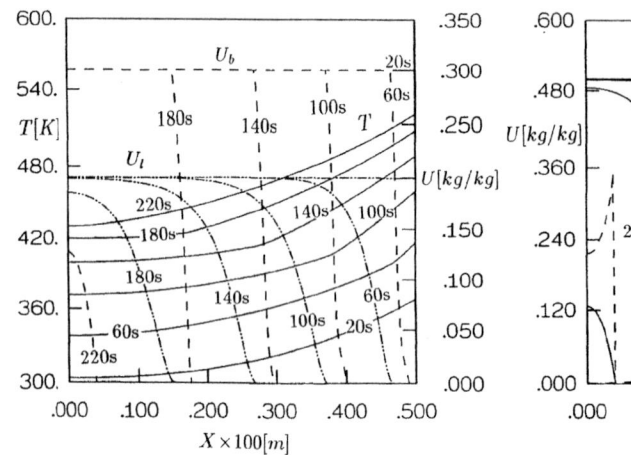

Fig. 3 - Spatial profiles of temperature, capillary and bound water contents (reference data).

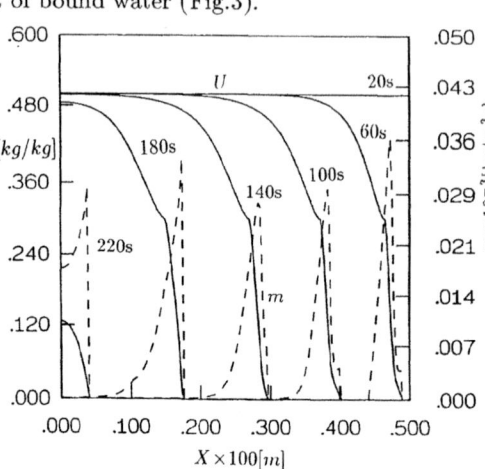

Fig. 4 - Spatial profiles of total moisture content and moisture evaporation rate (reference data).

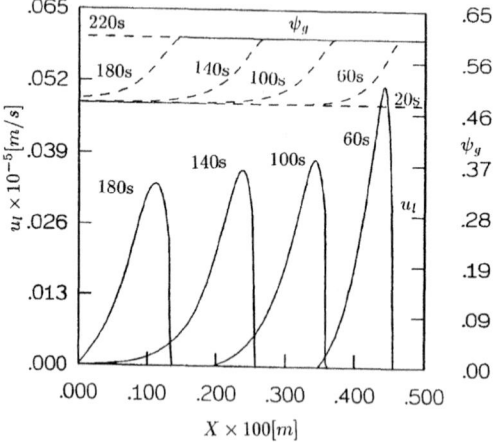

Fig. 5 - Spatial profiles of liquid velocity and porosity (reference data).

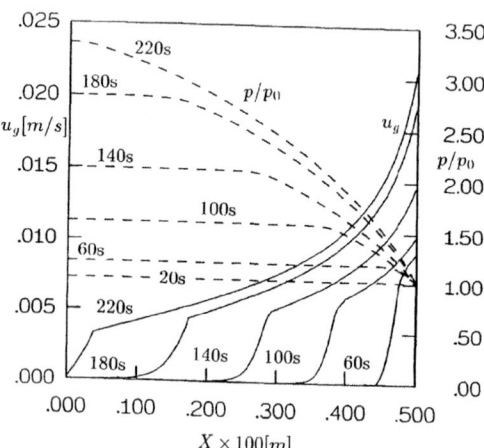

Fig. 6 - Spatial profiles of gas overpressure and velocity (reference data).

The position of the evaporating front is characterized by a change of slope in the temperature profile, due to the changes in the thermal properties and the endothermicity of water evaporation. Also, the temperature gradients are rather large, mainly between the drying front and the particle surface, indicating that the process is controlled by internal heat transfer. In the last stage of drying, the temperature gradients along the moist region are negligible and the temperature increases uniformly with time. As this temperature attains values suffciently high, evaporation starts and occurs along the last part of the particle, with very small temperature increase. The temperature, associated with complete evaporation of moisture slightly increases with the distance of the drying fronts from the heat exposed surface (from about $390K$ to about $420K$).

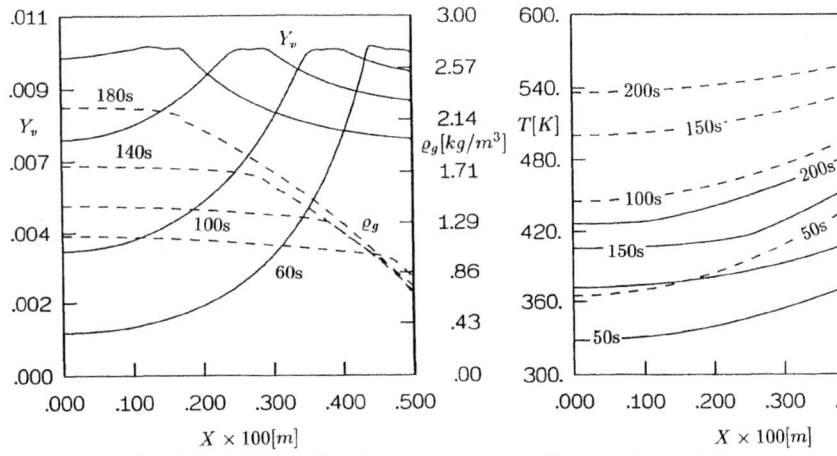

Fig. 7 - Spatial profiles of water vapor mass fraction and total gas density (reference data).

Fig. 8 - Spatial profiles of temperature for dry (dashed lines) and moist ($U_0 = 0.5$, solid lines) wood (reference data).

Figure 4 shows that the evaporation rate attains values different from zero only along a narrow region, except in the final stages of the process. Moreover, the right part of the profiles is rather steep, as it results from bound water, in contrast with the smoother left side, resulting from capillary water. Indeed, the profiles of bound water are also rather steep (surface diffusion, for the conditions here considered plays a negligible role). The movement of capillary water is significant in the neighborhood of its evaporating front, as shown by the liquid-phase velocity (Fig.5). This results from liquid phase pressure ($p_l = p_g - p_c$) gradients, which become significant at the evaporation front, and permeability to liquid flow, which goes to zero in the dry region. Associated with capillary water evaporation, an increase in the medium porosity is also observed.

The pressure profiles (Fig. 6) are the result of mass addition to the gas phase due to water evaporation, gas phase species transport and thermal gas expansion. At very short times, given the low temperatures and the localization of the evaporation process close to the heat exposed surface, the pressure remains near to atmospheric values. Then, at the drying front and along the wet region, it continuously increases above the ambient value, while the flow of vapor/gas out from the particle brings it to atmospheric values close to the surface. That is, similar to temperature, a discontinuity is seen near the evaporating front, as a consequence of property variation. For the reference values of the properties here used, the gas overpressures are rather high and reach, towards the end of the drying process, values about 3.3 times larger than the atmospheric value. It may be guessed that such high values can lead to structural failure of the particle, with consequent pressure reduction. Only at the beginning of the drying process, thus some convective transport of water vapor towards the moist region occurs (maximum negative flow velocities are of order $2 \times 10^{-6} m/s$). Therefore, pressure and water vapor concentration (Fig. 7) increase in the moist region mainly due to mass diffusion. This process is, on the other hand, enhanced by the successively large space gradients of vapor concentration. The spatial profiles of water vapor also show local maxima in correspondence to the evaporation front, followed by a decrease in the dry region, because of convective transport.

The flow of water vapor (Fig. 6) out from the particle is caused by the pressure gradient and the permeability to gas flow. It is worth noting that there are two distinct regions in the velocity profiles, the first (from the left) due to evaporation rates different from zero and the second to transport phenomena. For the reference values of the permeabilities here used, the rate of water vapor transfer to the sample surface is much faster than that of capillary water transport to the drying front (of several orders of magnitude).

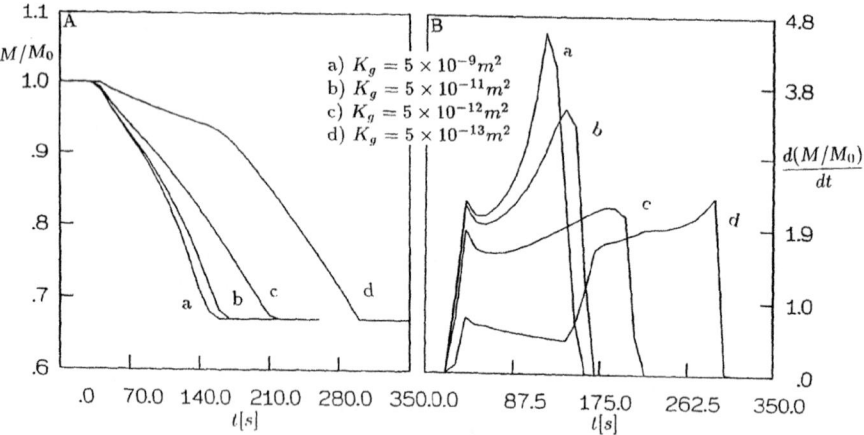

Figs. 9A-9B - Weight fraction (A) and time derivative of the weight fraction (B) as functions of time for several gas permeabilities.

The dynamics of gas overpressure explain the increase in the temperatures associated with complete evaporation of moisture. In fact, the increase of the gas pressure above the atmospheric value raises the boiling point of water, so that though the temperature becomes larger than $373K$, the evaporation rate is still low. Water vapor transport also affects the evaporation process. A decrease in the local mass concentration of water vapor (and vapor pressure) causes the water to evaporate at lower temperatures. The two effects tend to counteract each other, because the first would lead to an increase in the boiling point and the second would allow evaporation at lower temperatures. The dominance of the first or the second depends on the permeability to gas flow and the water vapor diffusion coefficient.

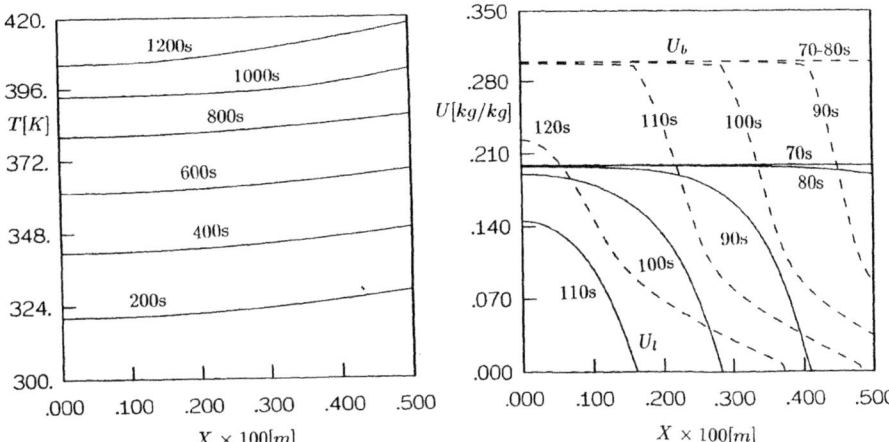

Fig. 10 - Spatial profiles of temperature for $h_c = 4.2W/m^2K$ (reference data).

Fig. 11 - Spatial profiles of bound and capillary water contents for $h_c = 4.2W/m^2K$ (reference data).

In conclusion, from the moisture and temperature profiles three main regions appear, corresponding to the moist zone (I), the evaporating zone (II) and the dry zone (III). In the first region the liquid phase is continue. Slow capillary convection causes motion of the liquid towards the drying front. The temperature in this region is almost constant because of the high thermal conductivity along the region II and the significant amount of heat required for the water evaporation (the rate is rather high). In the region II gas phase transport (diffusion) is from the high to the low temperature region. Thus in this region the moisture transport is directed away from the drying surface. The third region is the dry region. The vapor pressure is no longer dependent on temperature, as dictated by the sorption relation. The gradient in the partial pressure causes a gas phase moisture transport towards the particle surface.

As the drying process approaches completion, the pressure of the gas phase decreases, and all the variables of interest reach the ambient or the equilibrium values. The particle drying time is $220s$, which corresponds to a surface temperature equal to about $500K$, a value too low for the rates of solid thermal degradation to be significant. Consequently, particles smaller than $1 \times 10^{-2}m$ are completly dry before pyrolysis begins. A comparison of the temperature distributions between the cases of dry and wet particles (Fig.8), shows that moisture significantly delays the temperature rise. Indeed, though the effective thermal conductivity of the wet particle is larger than that of dry wood, the thermal capacity is also higher and plays a controlling role.

As the gas permeability is increased, the dynamics of moisture evaporation remain qualitatively the same, but the gas overpressures are successively reduced, the gas velocity increases and the temperature of complete moisture evaporation decreases. For $K_g = 5 \times 10^{-13} m^2$, the solution which, in practice, coincides with that obtained with a simplified version of the model, based on the assumption of constant gas phase pressure, predicts a significant reduction in the drying time ($160 s$ against $220 s$). A summary of the effects of the permeability to gas flow on the drying characteristics of wood particle is given by Figs. 9A-9B, where the particle mass and its time derivative are reported as functions of time (as there is no chemical reaction, these are representative of moisture dynamics). For all cases, after an initial period of negligible weight loss, a sharp rise in the global evaporation rate is simulated, due to the attainment of surface temperature near the normal water boiling point. This is followed by a decrease due to the process becoming heat transfer controlled, as drying occurs through the propagation of a rather steep front. A new local maximum in the rate of evaporation is simulated also for the conditions corresponding to the extention of the evaporation zone to the whole particle, that is when the temperature along the whole particle thickness reaches values large enough for water evaporation. As expected, due to the increase in the gas overpressure, the duration of the drying process increases as K_g is decreased.

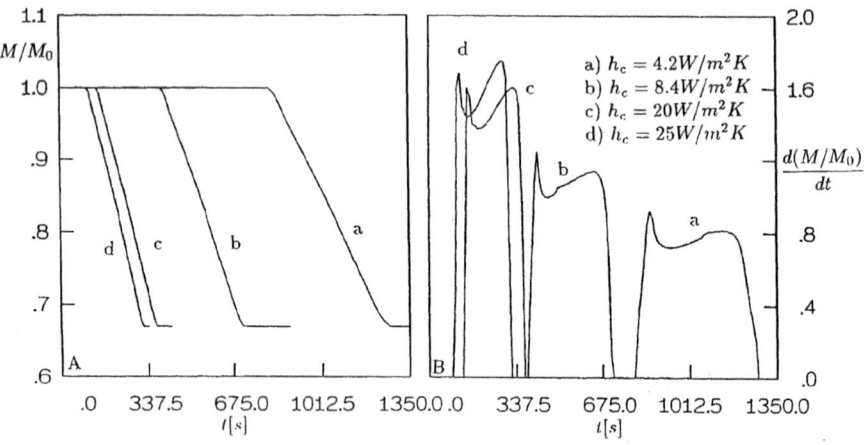

Figs. 12A-12B - Weight fraction (A) and time derivative of the weight fraction (B) as functions of time for several heat transfer coefficients.

As the external heat transfer rate is decreased through h_c ($\sigma = 0$), the spatial gradients of the variables of interest also decrease and the dynamics of moisture evaporation are modified, as shown in Figs. 10-11 ($h_c = 4.2W/m^2K$). An initial stage is seen where moisture evaporation occurs along the whole particle ($t = 800s$). For longer times, evaporation still occurs along the whole particle but its rate attains a maximum in a region near the high temperature surface. Consequently, a drying front appears. As fast evaporation rates are established along the whole moist region, a delay in the temperature increase is predicted. The gas overpressure is also significantly reduced (maximum $p/p_0 = 1.78$). Capillary transport is large along the whole moist region, with velocities continuously increasing as the drying front is approached and values comparable to those of Fig.5. On the contrary, the convective gas phase transport is noticeably reduced, with maximum velocities equal to $0.5 \times 10^{-2} m/s$. The drying time is about six times longer. On the whole, the drying characteristics are significantly dependent on h_c (Figs.12A-12B), but the dependence is successively less important as h_c is increased, because controlling mechanism shifts from external to internal heat transfer. It is worth observing that radiative heat transfer, for the external temperature here considered, significantly contributes to particle heating. In fact, in the absence of radiation, the drying time increases to $380s$ (against $220s$).

The role played by bound water diffusion appears to be negligible for high external heat transfer rates. At the low values of this, the movement of bound water towards the drying front makes the drying time shorted (for $h_c = 4.2W/m^2K$, this varies from $1310s$ ($D_b = 0$) to $1280s$).

Variations in the intrinsic permeability to liquid flow do not significantly affect drying, if the external heat transfer rate is sufficiently fast, because the process is controlled by heat and water vapor transport. However, liquid phase convection plays a role of increasing importance as h_c is decreased. For low h_c and high K_l, liquid phase convection becomes large compared to gas-phase convection, and moisture transport from the wet region to the evaporation region makes the drying times shorter and the duration of the initial stage, without a drying front, longer.

The influence of the sorption isotherm, compared to the Clausius-Clapeyron equation, is significant for conditions of relatively low temperatures of complete water evaporation, that is for high permeabilities to gas flow, low heat transfer coefficients, etc., and low moisture contents. Indeed, p_v/p_{v0} tends to 1, as the temperature and/or the moisture content are increased. Thus the final stage of water evaporation is effectively slower for low h_c and/or high K_g.

The dynamics of high temperature drying here described reproduce well the experimentally observed characteritics, with reference to the existence of a drying and an overpressure front [5,7,12], of a constant temperature evaporating region, close to the particle centerline [5,7,12,24], and of the shape of the drying curves [17]. Furthermore, though the large uncertainty on the transport properties and the lack of extensive experimental data for the drying rate and the temperature and moisture distribution do not allow for an extensive model validation, few simulations, made for moisture levels and particle sizes different from those here considered, predict drying times in the range of the few measured [17].

The characteristics of the drying process are also expected to depend on the initial moisture content, through variations in the medium properties and endothermic rate of moisture evaporation. Also, it is likely that variations in the particle size affect temperature profiles, so that different thermal regimes, already observed in pyrolysis, are established. Finally, it should be noted that high temperature drying of large particles can be characterized by simulatneous pyrolysis.

6 Conclusions

A generalized mathematical model of particle drying is applied to predict the drying behaviour of wood for external temperatures well above the normal water boiling point and intermediate moisture levels, for widely varying permeabilities to liquid and gas flow and heat transfer coefficients. Predictions presented exhibit realistic physical behaviour.

For high gas permeabilities, the pressure remains almost constant (equal to the atmospheric value) so that there is no increase in the moisture boiling point. If the vapor mass flux is high, convective transport due to mass addition to the gas phase may lead to a decrease in the local vapor pressure, so that evaporation takes place at low temperatures. For low gas phase permeabilities, the gas pressure may be rather high and the evaporation process is slowed. These two charactistics, that is the increase of the water boiling point with pressure and the decrease of the evaporation temperature, due to water vapor transport, may lead to large errors in simplified theories for conditions near to the atmospheric boiling point. In conclusion, gas pressure variations can be neglected only for permeabilities to gas flow above certain values. Also, they appear to be successively less important as the external heat transfer rates are reduced.

Bound water diffusion and capillary water convection do not affect the global characteristics of the drying process at high heat transfer rates. However, as these are decreased, liquid phase transport to the drying front significantly reduces the total drying time.

Finally, the effective liquid-vapor equilibrium sorption is an important factor which determines the shape of the drying curve at slow heating rates.

6 Acknowledgements

The work was supported by the European Economic Community under Contract AIR2-CT93-0889.

7 References

1. Solantausta Y., Diebold J., Elliot D. C., Bridgwater A. V., Beckman D. (1994) Assessment of liquefaction and pyrolysis sysytems, *VTT Research Notes* 1573.
2. McCabe W., Smith J. C., and Harriot P. (1985) *Unit Operations in Chemical Engineering*, McGraw-Hill, 4th ed. Singapore.
3. Harmathy T. Z. (1969) Simultaneous moisture and heat transfer in porous systems with particular reference to drying, *I & EC Fundamentals* Vol. 8, pp. 92-103.
4. Sahota M. S., Pagni P. J. (1979) Heat and mass transfer in porous media subject to fires, *Int. J. Heat and Mass Transfer* Vol. 22, pp. 1069-1081.

5. White R. H., Schaffer E. L. (1981) Transient moisture gradient in fire-exposed wood slab, *Wood and Fiber* Vol. 13 pp. 17-38.

6. Dayan A., Cluekler E. L. (1982) Heat and mass transfer within an intensely heated concrete slab, *Int. J. Heat and Mass Transfer* Vol. 25, pp. 1461-1467.

7. Fredlund B. (1993) Modelling of heat and mass transfer in wood structures during fire, *Fire Safety Journal* Vol. 20, pp.39-69.

8. Beaumont O. Schwob Y. (1984) Influence of physical and chemical parameters on wood pyrolysis, *Ind. Eng. Chem. Res.* Vol. 23, pp. 637-641.

9. Gray M. R., Corcoran W. H., Gavalas G. R. (1985) Pyrolysis of a wood-derived material. Effects of moisture and ash content, *Ind. Eng. Chem. Res.* Vol. 24, pp. 646-651.

10. Chan W. R., Kelbon M., Krieger-Brockett B.(1988) Single-particle biomass pyrolysis: correlation of reaction products with process conditions, *Ind. Eng. Chem. Res.* Vol. 27, pp. 2261-2275.

11. Simms D. L., Law M. (1967) The ignition of wet and dry wood by radiation, *Combustion and Flame* Vol. 11, pp. 377-388.

12. Lee C. K., Diehl J. R. (1981) Combustion of irradiated dry and wet oak, *Combustion and Flame* Vol. 42, pp.123-138.

13. Saastamoinen J. (1994) Model for drying and pyrolysis in an updraft gasifier, *Proc. of the Int. Conference on Advances in Thermochemical Biomass Conversion*, A. V. Bridgwater (Ed.), pp. 186-200, Blackie A & P.

14. Simmons W. W. (1983) Analysis of single particle wood combustion in convective flow, *Ph.D. Thesis*, University of Wisconsin - Madison.

15. Agarwal P. K., Genetti W. E., Lee Y. Y. and Prasad S. N. (1984) Model for drying fluidized-bed combustion of wet low-rank coals, *Fuel* Vol. 63, pp. 1020-1026.

16. Chan W. C. R., Kelbon M., Krieger B. B. (1985) Modelling and experimental verification of chemical processes during pyrolysis of a large biomass particle, *Fuel* Vol. 64, pp. 1505-1513.

17. Alves S. S., Figueiredo J. L. (1989) A model for pyrolysis of wet wood, *Chemical Engineering Science* Vol. 22, pp. 2861-2869.

18. Plumb O. A., Spolek A., Olmstead B. A. (1985) Heat and mass transfer in wood during drying, *Int. J. Heat and Mass Transfer* Vol. 28, pp.1669-1678.

19. Stanish M. A., Schajer G. S., Kayihan F. (1986) A mathematical model of drying for hygroscopic porous media, *AIChE J.* Vol. 32, pp. 1301-1311.

20. Nasrallah S. B., Perre P. (1988) Detailed study of a model of heat and mass transfer during convective drying of porous media, *Int. J. of Heat and Mass Transfer* Vol. 31: 957-967.

21. Perre P., De Giovanni A., Simulation par volumes finis des transferts couples en milieux poreux anisotropes: sechage du bois a basse et a haute temperature (1990) *Int. J. of Heat and Mass Transfer* Vol. 33, pp. 2463-2478.

22. Berger D., Pei C. T. (1973) Drying of hygroscopic capillary porous solis - A theoretical approach, *Int. J. of Heat and Mass Transfer* Vol.16, pp. 293-302.

23. Kollmann F. F. P., Cote W. A., Jr. (1968) *Principles of Wood Science and Technology*, Springer-Verlag.

24. Kanury A. M., Blackshear P. L., Some problems pertaining to the problem of wood-burning (1970) *Combustion Science and Technology* Vol. 1, pp. 339-355.

MODELLING AND SIMULATION OF MOIST WOOD DRYING AND PYROLYSIS

M. C. MELAAEN
Telemark Institute of Technology (HiT-TF), N-3914 Porsgrunn, Norway
M. G. GRØNLI
Norwegian University of Science and Technology, Division of Thermal Energy and Hydropower, N-7034 Trondheim, Norway

Abstract
A transient, one-dimensional model which can simulate drying and pyrolysis of moist wood is presented. The porous wood is divided into four phases: solid, bound water, liquid water and gas phases. Conservation equations for energy and mass together with Darcy's law for velocity and algebraic equations for the transport properties and physical properties are presented. The drying model is based on equilibrium between water vapor and bound or liquid water in the porous wood. For the thermal degradation process, two reaction schemes including a single one-step global and a multiple competitive reaction model which have been proposed in the literature, are included. A comparison between the two pyrolysis models has been made on dry wood and the results reveal that for the multiple reaction model, the heating rate and pyrolysis time have great influence on the ultimate char, tar and gas yields calculated. Simultaneously drying and pyrolysis of large wood particles are simulated. The effect of moisture content on the pyrolysis time is presented together with characteristic profiles for temperature, pressure and moisture distribution inside the particle. A temperature plateau can be observed at about 100°C where evaporation and condensation of liquid water takes place. The simulations show that the in-depth moisture content increases and exceeds the initial moisture content before evaporation. The reason for this increase is that when water evaporates in the front, some of it will be transported by convection and diffusion into the colder region and condensed.
Keywords: drying, pyrolysis, wood, modelling, simulation

1 Introduction

The environmental awareness related to CO_2 and other greenhouse gas emissions as well as concern over the ultimate availability of fossil fuels have increased the interest in using biomass as a renewable resource for chemical feedstocks and energy production. As produced, however, biomass has many disadvantages compared with fossil fuels. The physical form is rarely homogeneous and free flowing. Biomass has usually only a modest energy content and when harvested, the moisture content is very high. In its solid form, biomass is therefore difficult to use in many applications without substantial modification. To adapt the fuel to a particular end use, technologies for converting and upgrading the biomass into more convenient energy forms such as gaseous and liquid fuels may be introduced. One of the most promising ways to do so is pyrolysis. Pyrolysis is the thermal degradation of a feedstock in the absence of oxygen, leading to the formation of a mixture of liquid (tarry composition), gases and a highly reactive carbonaceous char of which the relative proportions depend very much on the method used. Temperature, pressure, heating rate and reaction time can be used to influence and determine the proportions and characteristics of the main products of the process. Pyrolysis can be used as an independent process for the production of useful energy holders or chemicals and is also the first step in the gasification and combustion process.

The development of thermochemical processes for biomass conversion and proper equipment design requires knowledge and good understanding of the several chemical and physical processes which are interacting in the thermal degradation process. Mathematical modelling and simulation of single particle pyrolysis represent a very useful tool for the understanding of these processes. Then one can focus on the solid phase internal processes such as heat and mass transfer that control the release of products, results which can be used in the design and control of large scale converters.

In this paper, we will present a transient, one dimensional (cartesian coordinates) model for the pyrolysis of large wood particles. Since wood is seldom completely dried before entering the thermal converter, we have focuseo on the complex mechanisms of drying which will occur simultaneously with the thermal degradation process. The effect of different moisture content on the pyrolysis time will be investigated. In addition, two models proposed in the literature for the thermal degradation process will be discussed and compared.

2 Formulation of the problem

Wood is a heterogeneous, porous substance consisting mainly of three major components, i.e; cellulose, the skeletal polysaccharide; hemicelluloses, which form the matrix; and lignin, the encrusting substance which binds the cells together. Additionally, wood contains many low-molecular-weight organic compounds known as extractives, small amounts of mineral matter and moisture. When a tree is harvested or felled it contains water in three phases; i.e; bound (hygroscopic or adsorbed) water which is believed to be hydrogen bonded to the hydroxyl groups of cellulose and hemicelluloses; free (liquid or capillary) water and water vapor which fills up the cell cavities or voids of the wood. The moisture content for which the cell walls are saturated with no free water in the cell cavities is called the fiber saturation point (M_{fsp}). This point is for most wood 30% of the ovendry weight, emphasising the large amount of water which can be adsorbed [1].

When moist wood is heated, most of the energy received by the wood is consumed by heating and evaporating water. If the initial moisture content is above the fiber saturation point ($M > M_{fsp}$), the pores contain liquid water and the internal moisture transfer is mainly attributable to capillary flow of free water through the voids. As drying proceeds, the surface moisture content reaches its maximum sorptive value ($M = M_{fsp}$), and the evaporation front begins to travel into the solid leaving behind a sorption zone ($M < M_{fsp}$). Inside the evaporation front, the material can still contain liquid water in the pores. Outside the evaporation front, no liquid water exists and the main transfer mechanisms are bound water diffusion and convective and diffusive transport of water vapor in the gas mixture phase. Evaporation takes place at the front as well as in the whole sorption zone. As time proceeds, the solid reaches the temperature at which the thermal decomposition takes place at the surface, forming a pyrolysing zone. After a certain period the pyrolysing solid loses all of its volatiles and the pyrolysing zone propagates into the interior of the solid, leaving behind a thermally insulating layer of char. As volatiles (tar, water vapor and gases) from the pyrolysis and water vapor formed during drying in the interior flow out through the high temperature char layer secondary reactions may occur both homogeneously and heterogeneously. Homogeneously in the gas phase where the heavy tar components crack into lighter hydrocarbons and heterogeneously which can be exothermic char gasification reactions of the oxygen-rich part of the volatiles. Wood loses up to 80% of its mass during the pyrolysis, obviously, this leads to changes in the physical structure such as internal shrinkage, increased porosity, surface recession and formation of surface fissures. Hence, drying and pyrolysis of wood involves the complex interaction in a porous media of heat and mass transfer with chemical reactions. Heat is transported by conduction, convection and radiation and mass transfer is driven by pressure and concentration gradients. The simulation of these processes involves the simultaneous solution of the conservation equations for mass and energy, Darcy's law for velocity together with kinetic expressions describing the rate of reaction.

The mathematical model which has been derived is based on earlier work on drying of non-hygroscopic and hygroscopic media done by Whitaker [2], Ouelhazi et al. [3], Perre et al. [4], and earlier work on pyrolysis of wood done by Chan et al. [5], Glaister [6], Krieger-Brockett et al. [7] and Di Blasi and Russo [8]. The most important assumptions of the model are: I) all the phases are at the same temperature and the partial pressure of vapor is equal to its equilibrium pressure (local thermodynamic equilibrium); II) the transport of bound water is modelled as a diffusion process given by a diffusion coefficient which is a function of the local moisture content; III) binary diffusion in the gas mixture phase. IV) Darcy's law for the bulk flow of gas mixture and liquid water; V) a linear variation between the virgin wood and char for the properties related to the solid structure; VI) wood shrinkage and crack formation are not considered. Compared to previous work, the present mathematical model has less assumptions, and the coupling between drying and pyrolysis is emphasised.

Early models of wood pyrolysis [9,10] used very simple one-step global reactions to describe the chemical kinetics. Among the shortcomings of employing such simple kinetics are that only total volatiles are predicted (by difference) and the inability to account for the variation in char yield on dependence of operating conditions, e.g. temperature, heat flux....etc. Another approach that has been more recently used is the so-called lumped parameter approach [5-8], where the reaction products have been grouped into classes of products such as gases, tars, and char. Two models for the thermal degradation, a one-step global model and a multiple, competitive reaction model proposed by Chan [5] and modified by Glaister [6] have been implemented in the mathematical model and will be discussed and compared here.

One-step global model

$$S \xrightarrow{k_1} a_1 C + (1 - a_1)[a_2 T + (1 - a_2)G]$$

Chan & Glaister's model

$$S \begin{array}{c} \xrightarrow{k_1} a_1 G_1 + (1 - a_1)V \\ \xrightarrow{k_2} T_1 \xrightarrow{k_4} a_3 G_2 + a_4 T_2 + (1 - a_3 - a_4)V \\ \xrightarrow{k_3} a_2 C + (1 - a_2)V \end{array}$$

In the one-step global model, the solid wood degrades to form a fixed fraction (a_1) of char (C) and volatiles. In this case, we have divided the volatiles into tar (T) and gases (G), and assumed a constant distribution between them. The model of Chan & Glaister includes three parallel and one consecutive reactions. Unreacted wood reacts through three competitive paths to form char (C) and vapor (V); primary gas (G_1) and vapor (V); and primary tar (T_1). The primary tar can further crack to form secondary gas (G_2), secondary tar (T_2) and vapor (V). The kinetic data, heat of pyrolysis and stoichiometric coefficients of the two models are listed in Table 1. The rate constants are all assumed to follow an Arrhenius type expression of the form:

$$k_i = A_i \exp(-E_i/R_0 T) \tag{1}$$

Table 1 Kinetic data, heat of pyrolysis and stoichiometric coefficients.

	One-step global model[*]					Chan & Glaister's model			
i	A_i	E_i	Δh_i	a_i	i	A_i	E_i	Δh_i	a_i
1	$7.4 \cdot 10^4$	88.0	200.0	0.2	1	$1.3 \cdot 10^8$	140.3	209.3	0.7
2				0.6	2	$2.0 \cdot 10^8$	133.1	209.3	0.9
					3	$1.1 \cdot 10^7$	121.4	209.3	0.65
					4	$1.5 \cdot 10^6$	114.3	-2009.7	0.25

[*] Data for spruce taken from Grønli [11]. The units of: A is [1/s], E is [kJ/mol], Δh is [kJ/kg].

3 Formulation of the model

According to Whitaker's [2] formulation, the equations of the model are obtained by averaging the classical mechanic and thermodynamic equations over a representative elementary volume, V, that contains all phases, as illustrated in Figure 1.

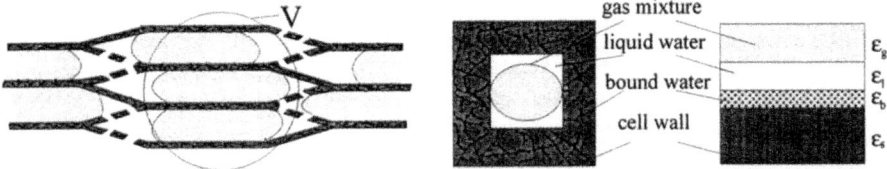

Figure 1 Averaging volume, V, and volume fractions of the different phases found in wood.

The phase average of a quantity φ associated with the phase γ, is defined as:

$$<\varphi> = \frac{1}{V} \int_{V_\gamma} \varphi \, dV \tag{2}$$

where V_γ is the volume occupied by phase γ in V. An intrinsic phase average of a quantity φ, which represents the averaged value inside phase γ is defined as follows:

$$<\varphi>^\gamma = \frac{1}{V_\gamma} \int_{V_\gamma} \varphi \, dV \tag{3}$$

The phase average and intrinsic phase average quantity are related by $<\varphi>=\varepsilon_\gamma<\varphi>^\gamma$, where $\varepsilon_\gamma=V_\gamma/V$ is the volume fraction occupied by phase γ. The volume fractions of the phases represented in moist wood, see Figure 1, and the relations between them are given by the following expressions:

$$\varepsilon_s = \frac{<\rho_s>}{<\rho_s>^s}, \quad \varepsilon_l = \frac{<\rho_l>}{<\rho_l>^l}, \quad \varepsilon_b = \frac{<\rho_b>}{<\rho_b>^b}, \qquad \varepsilon_s + \varepsilon_l + \varepsilon_b + \varepsilon_g = 1 \qquad (4)$$

where the intrinsic phase average densities are assumed to be $<\rho_s>^s=1500$ kg/m³ for the cell wall substance [1] and $<\rho_l>^l=<\rho_b>^b=1000$ kg/m³ for the liquid and bound water. By neglecting the contribution of the vapor fraction ($<\rho_v>$), the moisture content in wood is given by:

$$M = M_b + M_l = \frac{<\rho_b> + <\rho_l>}{<\rho_{SD}>} \qquad (5)$$

where M_b and M_l are the bound and liquid moisture contents, respectively.

3.1 Mass conservation equations

Mass conservation of liquid water:

$$\frac{\partial}{\partial t}<\rho_l> + \frac{\partial}{\partial x}(<\rho_l v_l>) = <\dot{w}_l> \qquad <\rho_l v_l> = <\rho_l>^l<v_l> \qquad (6)$$

Mass conservation of bound water:

$$\frac{\partial}{\partial t}<\rho_b> + \frac{\partial}{\partial x}(<\rho_b v_b>) = <\dot{w}_b> \qquad <\rho_b v_b> = -<\rho_{SD}>D_b\frac{\partial}{\partial x}\left(\frac{<\rho_b>}{<\rho_{SD}>}\right) \qquad (7)$$

Mass conservation of component i in the gas mixture phase:

$$\frac{\partial}{\partial t}(\varepsilon_g<\rho_i>^g) + \frac{\partial}{\partial x}(<\rho_i v_i>) = <\dot{w}_i>, \qquad <\rho_i v_i> = <\rho_i>^g<v_g> + <\rho_i U_i>$$

$$<\rho_i U_i> = -<\rho_g>^g D_{eff}\frac{\partial}{\partial x}\left(\frac{<\rho_i>^g}{<\rho_g>^g}\right) \qquad (8)$$

Mass conservation of solid:

$$\frac{\partial}{\partial t}<\rho_s> = <\dot{w}_s> \qquad (9)$$

3.2 Energy equation

$$\left(<\rho_s>C_{P,s} + <\rho_l>C_{P,l} + <\rho_b>C_{P,l} + \varepsilon_g<\rho_g>^g C_{P,g}\right)\frac{\partial<T>}{\partial t}$$

$$+\left(<\rho_l v_l>C_{P,l} + <\rho_b v_b>C_{P,l} + <\rho_g v_g>C_{P,g} + \sum_{i=1}^{N_g} C_{P,i}<\rho_i U_i>\right)\frac{\partial<T>}{\partial x} \qquad (10)$$

$$= \frac{\partial}{\partial x}\left(k_{eff}\frac{\partial<T>}{\partial x}\right) - \sum_i<\dot{w}_i>\Delta h_i - <\dot{w}_v>\Delta h_v + <\rho_b v_b>\frac{\partial}{\partial x}(\Delta h_{sorp})$$

3.3 Momentum equation

Since both gases and liquid water are flowing simultaneously through the porous wood, Darcy's law is extended by using relative permeabilities. These permeabilities are dependent on the relative saturation of each fluid. By neglecting gravitational effects, the superficial velocities for the gas phase and liquid phase can be written as:

$$<v_g> = - \frac{K_g K_r^g}{\mu_g} \frac{\partial}{\partial x} <P_g>^g, \qquad <v_l> = - \frac{K_l K_r^l}{\mu_l} \frac{\partial}{\partial x} <P_l>^l \qquad (11)$$

where the liquid pressure $<P_l>^l$ is connected to the gaseous pressure by capillarity:

$$<P_l>^l = <P_g>^g - P_c , \qquad P_c \sim P_c(M_l, T) \qquad (12)$$

3.4 Algebraic equations

In addition to the differential equations for mass, energy and momentum and an appropriate reaction scheme for the thermal decomposition, the mathematical model requires data for the thermo-physical and transport properties. Since some of these are dependent on the fiber direction, properties along the fibers (longitudinal direction) will consistently be used in the following. The gas mixture is assumed to be an ideal mixture of perfect gases, hence, the following relations exist:

$$<\rho_g>^g = \sum_i <\rho_i>^g, \quad <P_g>^g = \frac{<\rho_g>^g R_0 <T>}{M_g}, \quad M_g = \left(\sum_i \frac{<\rho_i>^g}{<\rho_g>^g M_i} \right)^{-1} \qquad (13)$$

Interpolation factor:
For the properties related to the solid structure, a linear variation between the virgin wood and char is assumed:

$$\varphi = \eta \varphi_{SD} + (1 - \eta)\varphi_c \qquad \eta = <\rho_s>/<\rho_{SD}> \qquad (14)$$

Effective thermal conductivity:
The thermal conductivity of wood varies with the direction of heat flow with respect to the grain, with temperature, with density and moisture content. The heat transfer takes place through a complex interaction between several transfer mechanisms (conduction, convection, radiation). The effective conductivity is introduced:

$$k_{eff} = k_{cond} + k_{rad} \qquad (15)$$

which is assumed to be made of the usual conductivity term given by Fourier's law:

$$k_{cond} = \varepsilon_g k_g + (\varepsilon_l + \varepsilon_b)k_l + \eta k_{SD} + (1 - \eta)k_C \qquad (16)$$

and a radiative conductivity term [12]:

$$k_{rad} = \kappa_{rad} \frac{\varepsilon_g}{(1 - \varepsilon_g)} \sigma \omega d_{por} T^3, \qquad d_{por} = \eta d_{por,SD} + (1 - \eta)d_{por,C} \qquad (17)$$

where d_{por} is the pore diameter which is assumed to be the characteristic length which the radiation can pass through.

Specific heat capacities:
The specific heat capacity of the gas mixture and the dry solid substance is given by:

$$C_{P,g} = \sum_{i=1}^{N_g} C_{P,i} \frac{<\rho_i>^g}{<\rho_g>^g}, \qquad C_{P,s} = \eta C_{P,SD} + (1-\eta)C_{P,C} \tag{18}$$

where the different heat capacities are assumed constant.

Fiber saturation point [1]:

$$M_{fsp} = \max[(M_{fsp}^0 + 0.298 - 0.001T),\ 0.2], \qquad M_{fsp}^0 = 0.3 \tag{19}$$

Vapor pressure:
The vapor pressure inside the porous wood structure is assumed to be in equilibrium (saturated) with the liquid or bound water:

$$
\begin{array}{lll}
M > M_{fsp} & <P_v>^g = P_v^{sat}(T) & \\
M < M_{fsp} & <P_v>^g = P_v^{sat}(T)\,h(M_b,T) &
\end{array} \tag{20}
$$

where $h(M_b,T)$ is the relative humidity given by sorption theory and $P_v^{sat}(T)$ is the equilibrium vapor pressure above the liquid free water surface. The following expression is used for the relative humidity and saturated vapor pressure [13]:

$$h(M_b,T) = 1 - \exp[-7.05(T-273)^{0.4}M_b^{1.5}]$$

$$P_v^{sat}(T) = 101325\exp(17.58 - 5769/T - 5.686\cdot10^{-3}T) \qquad N/m^2 \tag{21}$$

Heat of evaporation:

$$
\begin{array}{lll}
M \geq M_{fsp} & \Delta h_v = \Delta h_l & \\
M < M_{fsp} & \Delta h_v = \Delta h_l + \Delta h_{sorp} &
\end{array} \tag{22}
$$

The heat of evaporation of liquid water has been obtained by fitting the saturated steam data of [14]:

$$\Delta h_l = 3174.9 - 2.46T \qquad kJ/kg \tag{23}$$

Stanish et al.[15] assumed Δh_{sorp} to vary quadratically with the bound water content and at zero bound water content to be equal 40% of the heat of evaporation of liquid water:

$$\Delta h_{sorp} = 0.4\Delta h_l\left(1 - \frac{M_b}{M_{fsp}}\right)^2 \qquad kJ/kg \tag{24}$$

Gaseous diffusion [3,4]:

$$D_{eff} = \frac{D_{v-a}}{50}, \qquad D_{v-a} = 8.803\cdot10^{-5}\left(\frac{T^{1.81}}{P}\right) \qquad m^2/s \tag{25}$$

Bound water diffusion [3,4]:

$$D_b = \exp\left[-9.9 + 9.8M_b - \frac{4300}{T}\right] \qquad m^2/s \tag{26}$$

Capillary pressure:
Spolek and Plumb [16] derived the following expression for the capillary pressure in softwoods:

$$P_c = 1.364 \cdot 10^5 \sigma \, (M_1 + 1.2 \cdot 10^{-4})^{-0.63} \qquad N/m^2$$
$$\sigma = (128.0 - 0.185 \, T) \, 10^{-3} \qquad N/m \tag{27}$$

Permeabilities:
The intrinsic permeability of the gas mixture and the liquid water is given by:

$$K_g = \eta \, K_{g,SD} + (1 - \eta) K_{g,C}, \qquad K_1 = 5 K_{g,SD} \tag{28}$$

Perre et al. [4] derived the following expressions for the relative permeabilities in softwoods:

$$K_r^l = \overline{M}^8, \qquad K_r^g = 1 + (4\overline{M} - 5)\overline{M}^4, \qquad \overline{M} = M_1/M_{1,sat} \tag{29}$$

where $M_{1,sat}$ is the saturated liquid water content:

$$M_{1,sat} = \langle\rho_1\rangle^l \left(\frac{1}{\langle\rho_{SD}\rangle} - \frac{1}{\langle\rho_s\rangle^s} \right) - M_{fsp} \tag{30}$$

3.5 Initial and boundary conditions

Energy:
The left surface is subjected to a radiative heat flux (F) which balances the convective and radiative heat losses, the inward conduction and the evaporation and desorption of liquid and bound water:

$$F - h_T(\langle T_s\rangle - \langle T_\infty\rangle) - \omega_s\sigma(\langle T_s^4\rangle - \langle T_\infty^4\rangle) = -k_{eff}\frac{\partial \langle T\rangle}{\partial x} - \langle\rho_1 v_1\rangle \Delta h_1 - \langle\rho_b v_b\rangle \Delta h_b \tag{31}$$

where: $\qquad \Delta h_b = \Delta h_1 + \Delta h_{sorp}$

Moisture:

$$-(\langle\rho_1 v_1\rangle + \langle\rho_b v_b\rangle + \langle\rho_v v_v\rangle) = h_m(\langle\rho_v\rangle_s^g - \langle\rho_v\rangle_\infty^g) \tag{32}$$

Pressure:

$$\langle P_g\rangle_s^g = \langle P_g\rangle_\infty^g \tag{33}$$

Symmetry is assumed at the right surface. The initial conditions for temperature, density, pressure, moisture content and velocities have to be specified before the simulation can start.

The mathematical model with initial and boundary conditions is solved numerically. The convective terms are discretized by first-order upwinding, while central differencing is used for the diffusive terms. The time integrator must manage a system of differential and algebraic equations, and the numerical code DASSL has been chosen. A detailed description of the numerical solution procedure is given in [17].

4   Results and discussion

Pyrolysis of wood with a sample half-thickness of 2.5 cm has been simulated (due to symmetry). The moisture content has been varied between zero and 30%. The heat fluxes (F) applied on the left side surface (parallel with grain direction) are 40.0 kW/m^2, 80.0 kW/m^2 and 120.0 kW/m^2, and the total pyrolysis time is 600 s (10 min) for all the simulations. The initial and surrounding temperature and pressure are 298 K and 1 atm, respectively, and the constant thermo-physical data assumed are listed in Table 2. First, a comparison between the one-step global model and the model of Chan & Glaister (C&G) will be presented and discussed. These simulations have been done on dry wood where diffusion in the gas phase has been neglected. Next, C&G's model has been used in the simulation of moist wood pyrolysis with initial moisture contents of 10%, 20% and 30%. All the simulations have been carried out on a *90 MHz Pentium PC* with 100 grid points in space. The overall time of computation varied between 10 and 60 minutes, depending on the number of equations to be solved.

Table 2 Constant thermo-physical data for the model.

Symbol	Value	Reference	Symbol	Value	Reference
$C_{P,SD}$	2.3 kJ/(kg K)	[7]	$C_{P,C}$	1.1 kJ/(kg K)	[8]
$C_{P,1}$	4.2 kJ/(kg K) (at 60°C)	[13]	$C_{P,v}$	2.0 kJ/(kg K)	[13]
$C_{P,air}$	1.0 kJ/(kg K)	[13]	$C_{P,T}$	1.1 kJ/(kg K)	assumed
$C_{P,G}$	1.1 kJ/(kg K)	assumed			
μ_g	$3.0 \cdot 10^{-5}$ kg/(m s)	[13]	μ_l	$5.0 \cdot 10^{-4}$ kg/(m s)	[13]
k_{SD}	$3.5 \cdot 10^{-1}$ W/(m K)	[9]	k_C	$1.0 \cdot 10^{-1}$ W/(m K)	[9]
k_g	$25.77 \cdot 10^{-3}$ W/(m K)	[13]	k_l	$6.58 \cdot 10^{-1}$ W/(m K)	[13]
$d_{por,SD}$	$5.0 \cdot 10^{-5}$ m	[11]	$d_{por,C}$	$1.0 \cdot 10^{-4}$ m	assumed
$K_{g,SD}$	$1.0 \cdot 10^{-13}$ m^2	[11]	$K_{g,C}$	$1.0 \cdot 10^{-10}$ m^2	assumed
M_{air}	$2.9 \cdot 10^{-2}$ kg/mole	[13]	M_v	$1.8 \cdot 10^{-2}$ kg/mole	[13]
M_{T1}	$1.1 \cdot 10^{-1}$ kg/mole	assumed	M_{T2}	$1.1 \cdot 10^{-1}$ kg/mole	assumed
M_{G1}	$3.8 \cdot 10^{-2}$ kg/mole	assumed	M_{G2}	$2.8 \cdot 10^{-2}$ kg/mole	assumed
h_T	30.0 W/(m^2K)		h_m	$3.0 \cdot 10^{-2}$ m/s	
ω	0.9	assumed	ω_s	0.9	assumed
$<\rho_v>^g_\infty$	0.0		κ_{rad}	4.0	[10]

4.1   Pyrolysis of dry wood

Some of the results from the simulations of dry wood pyrolysis are presented in Figures 2 to 7, where the particle has been exposed to a heat flux of 120.0 kW/m^2. In Figures 2 to 6, the one-step model is compared with the model of C&G. The temperature evolution inside the particle as function of time and particle depth is shown in Figures 2 and 3, respectively. For both models, there is a steep increase in the surface temperature in the initial part of the simulations. At a certain time, however, the temperature produced by the model of C&G exceeds the temperature produced by the one-step model. This is due to the higher temperature caused by the secondary exothermic reaction in C&G's model. Since the radiative conductivity, eq. (17), is a function of temperature raised to power 3, the temperature effect on the effective conductivity, shown in Figure 4, is stronger than the effect of higher volume fraction

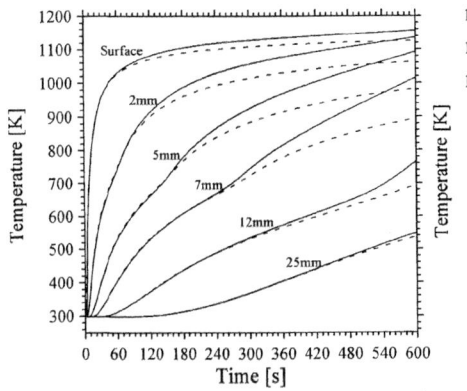

Figure 2 Temperature vs time. (——) C&G's model, (- - -) one-step model.

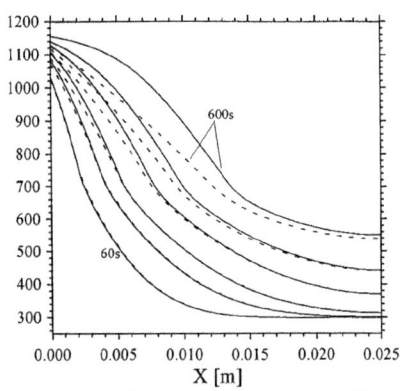

Figure 3 Temperature vs thickness. (At time: 60, 120, 180, 300, 420 and 600s)

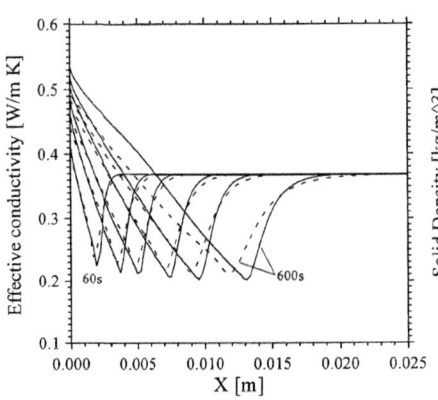

Figure 4 Effective conductivity vs thickness. (Notation: see Figures 2 and 3)

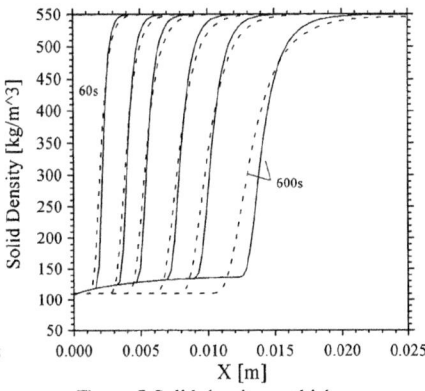

Figure 5 Solid density vs thickness. (Notation: see Figures 2 and 3)

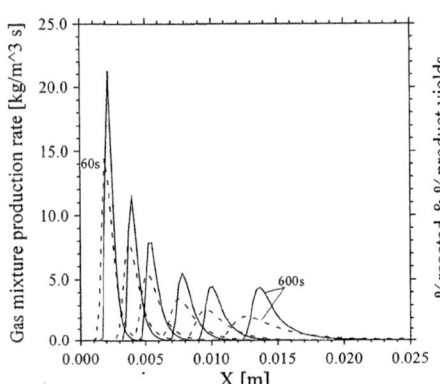

Figure 6 Gas mixture production rate vs thickness. (Notation: see Figures 2 and 3)

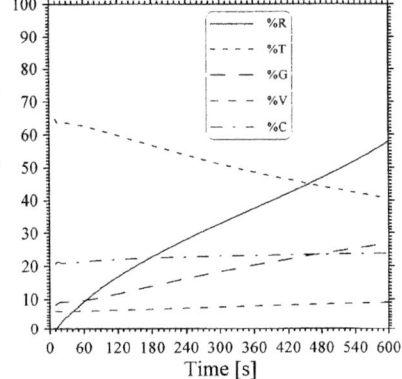

Figure 7 Percentage reacted and product yields vs time. (Notation: see Table 3)

of the gas mixture (porosity) produced by the one-step global model. One of the main advantages of C&G's model is that the char-yield is dependent on the temperature history and not assumed constant as in the one-step model. This is seen in Figure 5, where the solid density vs particle thickness is presented. While the one-step model gives a constant char density of 110 kg/m^3, the char density increases slightly from 110 kg/m^3 at the surface to approximately 135 kg/m^3 at a depth of 13 mm in the model of C&G. In Figure 6, the gas mixture (tar, gas, vapor) production rates of the two models are shown. Due to the higher temperature in the simulation of C&G's model and due to different kinetics, the gas mixture production rate is higher for C&G's model. This is also reflected in the much steeper pyrolysis front in Figure 5. The percentage reacted and time integrated product yields as a function of time are shown in Figure 7 for C&G's model. The char-yield increases slightly as the pyrolysis front propagates into the particle. The tar-yield decreases due to secondary cracking reactions causing the gas and vapor-yield to slightly increase. The results from all the dry sample pyrolysis simulations presented in percentage reacted (converted) and ultimate product yields are shown in Table 3. Obviously, as the heat flux is decreased from 120 to 40 kW/m^2, the percentage reacted decreases for both the models. While the ultimate product yields of char, gas and tar are independent of the incident heat-flux for the one-step model, more char and tar, and less gas and vapor is produced when the heat flux is decreased in the pyrolysis model proposed by Chan & Glaister.

4.2 Pyrolysis of moist wood

The results from the simulation of a particle with an initial moisture content of 30% exposed to a heat flux of 120.0 kW/m^2 are shown in Figures 8 to 13. The temperature variation in time and space are presented in Figures 8 and 9, where we also have included the dry sample simulation for comparison. A temperature plateau can be observed at about 100°C where evaporation and condensation of liquid water takes place. As expected, the moisture delays the process. The profiles for the moisture content in Figure 10 show that the moisture content increases and exceeds the initial moisture content before it evaporates. The reason for this increase is that when water evaporates in the front, some of it will be transported by convection and diffusion into the colder region and condensed. If the moisture content in Figure 10 and water evaporation rate in Figure 11 are compared, the region of evaporation and condensation can be seen. Figure 12 shows the gas overpressure. The highest pressure is found near the highest production rate of vapor. The increasing pressure at 420 and 600 seconds is due to the interaction between the production of water vapor and the symmetry boundary condition on right side. The symmetry condition produces a quicker increase of temperature than if the sample had been infinite, and the velocity at the symmetry boundary is zero. Both effects will increase the pressure. Depending on the pressure gradient, the superficial gas velocity in Figure 13 will have direction into or out of the sample, and hence water vapor can also be transported into the sample by convection. From Table 3, it is seen that the percentage reacted is decreasing when the moisture content in the wood is increased. The tar fraction increases while the gas and water vapor fractions from pyrolysis decrease. For the present moisture contents and heat flux, the char fraction only increases marginally.

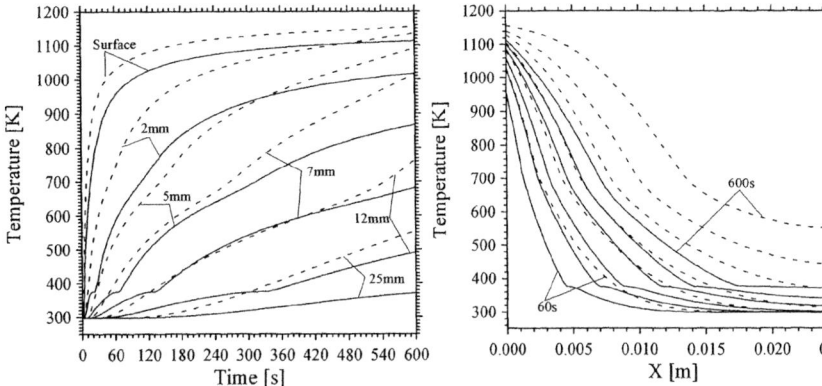

Figure 8 Temperature vs time. C&G's model
(——) M=0.3, (- - -) M=0.

Figure 9 Temperature vs thickness (At time: 60, 120, 180, 300, 420 and 600s)

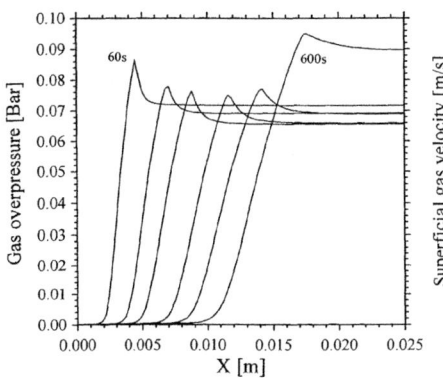

Figure 10 Moisture content vs thickness. (Notation: see Figures 8 and 9)

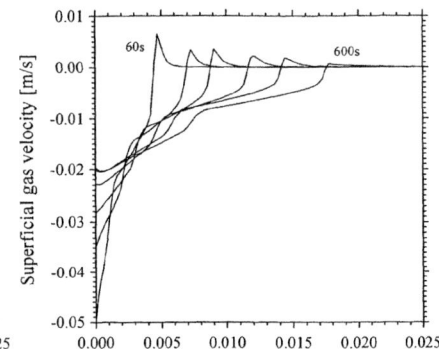

Figure 11 Water evaporation rate vs thickness. (Notation: see Figures 8 and 9)

Figure 12 Gas overpressure vs thickness. (Notation: see Figures 8 and 9)

Figure 13 Superficial gas velocity vs thickness. (Notation: see Figures 8 and 9)

143

Table 3 Percentage reacted and ultimate product yields after 10 minutes heating.

Model	%M	Heat flux [kW/m²]	%R*	%T*	%G*	%V*	%C*	%Evap
One-step	0	40	20.1	48.0	32.0	---	20.0	---
	0	80	40.9	48.0	32.0	---	20.0	---
	0	120	55.0	48.0	32.0	---	20.0	---
C&G	0	40	17.1	59.8	7.3	6.0	26.9	---
	0	80	38.3	55.0	13.4	6.9	24.7	---
	0	120	57.8	40.4	26.8	9.0	23.8	---
	10	120	41.1	49.5	18.9	7.8	23.8	88.4
	20	120	34.7	53.1	15.7	7.3	23.9	74.4
	30	120	31.0	55.2	13.9	7.0	23.9	64.8

%M=initial moisture content, %R=percentage reacted, %T=tar yield, %G=gas yield, %V=vapor yield, %C=char yield, %Evap= percentage moisture evaporated, *based on dry basis

5   Conclusions and further work

A mathematical model and simulation results for the pyrolysis of both dry and moist wood particles are presented. In the transient, one-dimensional model, the analysis is based on four phases: solid, bound water, liquid water and gas phases. Necessary algebraic and differential equations are presented. The first part of the simulations is done on dry particles, and a single one-step global and a multiple competitive pyrolysis model are compared for different heating rates. The results reveal that for the multiple reaction model, the heating rate and pyrolysis time have great influence on the ultimate char, tar and gas yields calculated. The other part of the simulations is done on moist particles. The multiple competitive pyrolysis model together with a drying model based on equilibrium between water vapor and bound or liquid water in the porous wood is used. The simulations show that an increase in moisture content delays the pyrolysis process and changes the ultimate product yields. The moisture content inside the wood increases and exceeds the initial moisture content before final evaporation takes place and the well known temperature plateau at about 100°C can be seen. Modelling predictions will be compared with experiments in [11].

Nomenclature

a_i coefficients in the pyrolysis model

A_i pre-exponential factor in the pyrolysis model, s^{-1}

C_p specific heat capacity at constant pressure, J/(kg K)

d_{por} pore diameter, m

D_b diffusivity of bound water, m^2/s

D_{eff} effective diffusivity in gas mixture phase, m^2/s

E_i activation energy in the pyrolysis model, J/mol

F   incident heat flux to surface, W/m^2

144

h	relative humidity	ε	volume fraction
h_m	mass transfer number, m/s	μ	dynamic viscosity, kg/(m s)
h_T	heat transfer number, W/(m^{2}K)	η	interpolation factor
k	thermal conductivity, W/(m K)	$<\rho_\gamma>$	(phase averaged) density of phase γ, kg/m^3
k_i	reaction rate constant, s^{-1}		
K	intrinsic permeability, m^2	$<\rho_\gamma>^\gamma$	intrinsic (phase averaged) density of component or phase i in phase γ, kg/m^3
K_r	relative permeability		
M	moisture content		
M	molecular weight, kg/mole	σ	surface tension, N/m; Stefan-Boltzmann constant, [= $5.67 \cdot 10^{-8}$ W/(m^{2}K^4)]
N_g	number of components in the gas mixture phase		
P	pressure, N/m^2	φ	general variable
P_c	capillary pressure, N/m^2	ω	emissivity in porous structure
$<P_g>^g$	pressure in the gas mixture phase, N/m^2	ω_s	surface emissivity
		κ_{rad}	radiation constant
$<P_l>^l$	pressure in the liquid phase, N/m^2		
R_0	universal gas constant, [= 8.3144 J/(mol K)]	*Subscripts and Superscripts*	
t	time, s	b	bound water phase
T,$<$T$>$	temperature, K	C	char
V	volume, m^3	fsp	fiber saturation point
$<v_i>$	superficial velocity of component i, m/s	g	gas mixture phase
		G	gas in the gas mixture phase
$<\dot{w}_i>$	Rate of production of component or phase i in phase γ, kg/(m^3 s)	i	component or phase i
		l	liquid water phase
		s	solid phase, boundary surface
x	cartesian coordinates, m	S	virgin or un-reacted solid during pyrolysis
Δh_i	heat of pyrolysis, J/kg		
Δh_l	latent heat of evaporation of liquid water, J/kg	SD	dry solid wood
		T	tar in gas mixture phase
Δh_{sorp}	heat of sorption of bound water, J/kg	v	water vapor
		∞	ambient

References

1. Siau, J. F. (1984) Transport processes in wood. *Springer-Verlag Series in Wood Science.*

2. Whitaker, S. (1977) Simultaneous heat, mass and momentum transfer in porous media: A theory of drying. *Advances in Heat Transfer.* vol. 13, pp. 119-203.

3. Ouelhazi, N., Arnaud, G and Fohr, J.P. (1992) A two-dimensional study of wood plank drying. The effect of gaseous pressure below boiling point. *Transport in Porous Media.* vol. 7, pp. 39-61.

4. Perre, P., Moser, M. and Martin, M. (1993) Advances in transport phenomena

during convective drying with superheated steam and moist air. *International Journal of Heat and Mass Transfer.* vol. 36, pp. 2725-2746.

5.   Chan, W.-C.R., Kelbon, M. and Krieger-Brockett, B. (1985) Modelling and experimental verification of physical and chemical processes during pyrolysis of a large biomass particle. *Fuel.* vol. 64, pp. 1505-1513.

6.   Glaister, D.S. (1987) The prediction of chemical kinetic, heat and mass transfer processes during the one- and two-dimensional pyrolysis of a large wood pellet". *M.Sc.-Thesis, University of Washington.*

7.   Krieger-Brockett, B. and Glaister, D.S. (1988) Wood devolatilization - sensitivity to feed properties and process variables. *In A.V. Bridgwater (ed): Research in Thermochemical Biomass Conversion.* pp.127-142.

8.   Di Blasi, C. and Russo, G. (1994) Modelling of transport phenomena and kinetics of biomass pyrolysis". *In A.V. Bridgwater (ed): Advances in Thermochemical Biomass Conversion.* pp. 906-921.

9.   Kung, H.-C. (1972) A mathematical model of wood pyrolysis. *Combustion and Flame.* vol. 18, pp.185-195.

10. Kansa, E.J., Perlee, H.E. and Chaiken, R.F (1977) Mathematical model of wood pyrolysis including internal forced convection. *Combustion and Flame.* vol. 29, pp. 311-324.

11. Grønli, M. (1996) A theoretical and experimental study of the thermal degradation of biomass. *Ph.D-Thesis, Norwegian University of Science and Technology.* (in press).

12. Panton, R.L. and Rittmann, J.G. (1971) Pyrolysis of a slab of porous material. *Thirteenth International Symposium on Combustion,* pp. 881-891.

13. Lai, W-C. (1991) Reaction engineering of heterogeneous feeds: municipal solid waste as a model". *Ph.D.-Thesis, University of Washington.*

14. Raznjevik. K. (1976) Handbook of thermodynamic tables and charts. *Hemisphere Publishing Corporation, McGraw-Hill Book Company.*

15. Stanish, M.A., Schajer, G.S. and Kayihan, F. (1986) A mathematical model of drying for hygroscopic porous media. *AIChE Journal.* vol. 32, pp. 1301-1311.

16. Spolek, G.A. and Plumb, O.A (1981) Capillary pressure in softwoods. *Wood Science and Technology.* vol. 15, pp. 189-199, 1981

17. Melaaen, M.C. (1996) Numerical analysis of heat and mass transfer in drying and pyrolysis of porous media. *Numerical Heat Transfer.* (in press).

A TRANSIENT, TWO-DIMENSIONAL MODEL OF BIOMASS PYROLYSIS

COLOMBA DI BLASI

Dipartimento di Ingegneria Chimica, Universitá di Napoli Federico II, P.le V. Tecchio, 80125 Napoli, Italy

Abstract

A two-dimensional, unsteady mathematical model of a thermally degrading cellulosic particle is presented. Heat mass and momentum transfer are modeled by the usual conservation equations for porous media. Chemical processes are described through a multi-step, lumped-parameter model. The dynamics of particle conversion, simulated for sizes above $0.5 \times 10^{-2}m$, show a process largely affected by the grain structure of the solid, with convective heat transfer predominating over conduction along the solid grain. A sensitivity analysis has shown that, among physical properties, the most important role is played by the char/solid thermal conductivities. The degradation process becomes slower as the these are made lower. However, in the first case, the reaction front enlarges and the lower reaction temperature causes an increase in the char yield. On the contrary, as the solid thermal conductivity is decreased, the thickness of the pyrolysis region shrinks and, as a consequence of the larger reaction temperatures, lower final char yields are obtained.

Keywords: biomass, heat conduction and convection, modeling, multi-step kinetics, pyrolysis, wood anisotropy.

1 Introduction

In recent years there has been a renewal of interest in thermochemical conversion [1-3] of cellulosic solid fuels, such as biomass and municipal solid waste, as potential sources of energy recovery. In most cases, the design of industrial chemical reactors is based on empirical approaches, because of the complexity of pyrolysis mechanisms and the poor understanding of the interaction between transport phenomena and chemical processes. As for pyrolysis, product yields, consisting of medium heating value gases, condensable organic components (oil) and charcoal, depend on the conversion conditions. Conventional pyrolysis of biomass fuels produces gas, oil and charcoal in approximately equal proportions. Flash pyrolysis, which is usually based on the conversion of small particles, subjected to high heating rates, moderate temperatures (400^0C-600^0C) and short volatile residence times, allows high yields of oil to be obtained. A variation of flash pyrolysis is represented by ablative pyrolysis, achieved by direct contact of the solid fuel with a high pressure, moving, hot surface. In spite of the optimism and interest created by recent estimates of biomass resources, their utilization by most industrialized nations remains small. Theoretical and experimental studies on biomass thermal degradation may improve the knowledge of this type of fuels and promote their utilization on a large scale.

An extensive critical review of the several attempts made to model thermal degradation of cellulosic materials has been already presented [4]. More recently, some transport models have also been proposed [5-10], where multi-step kinetics of solid pyrolysis are coupled with a rather detailed description of physical processes. Though such theories have contibuted to the definition of the different regimes and characteristics of cellulosic material pyrolysis on depedence of feedstock properties and reaction conditions, in all cases their validity is limited to one-dimensional heat and flow conditions. Thus, there is no adequate description of the anisotropic structure of lignocellulosic materials and no quantification of its implications in the pyrolytic degradation.

In this study a two-dimensional, unsteady mathematical model of a thermally degrading cellulosic particle is presented. Simulations have been made for different particle sizes to investigate the influences of medium anisotropy on product yields, conversion time and mass loss rate.

2 The mathematical model

Pyrolysis kinetics of cellulosic fuels are described, as in previous analyses [6-10], according to the Broido-Shafizadeh scheme, made by two competing reaction pathways, a) intermolecular dehydration, predominating at low temperatures, leading to char and gas and, in air, to smoldering combustion; and b) depolymerization reaction, predominating at high temperatures, leading to combustible volatiles (tars) and, in air, to flaming combustion:

$$SOLID \xrightarrow{K_1} ACTIVE\ SOLID \begin{array}{c} \overset{K_3}{\nearrow} TAR \\ \underset{K_2}{\searrow} \nu_C CHAR + \nu_G GAS \end{array}$$

Rates of endothermic reactions $(K_1 - K_3)$ can be adequately represented as first order in the mass of pyrolyzable material and having an Arrhenius type of temperature dependence. Secondary vapor phase tar decomposition and possible gasification of char are not taken into account because this study is mainly focused on physical processes.

The Broido-Shafizadeh model of cellulose pyrolysis has been widely applied in previous studies to predict the interaction between chemical and physical processes [6,7,8,9] and the choice has also been motivated. Though numerous experimental analyses on the thermal degradation of cellulose are available, when the chemistry of the process has to be coupled with transport phenomena, to predict the behavior of biomass in chemical reactors, the choice is somewath limited. In most cases, kinetic models consist of a one-step global reaction whose data are estimated to best fit experimental weight loss curves. The main drawback of these kinetic models is that they are generally based on the assumption of a constant ratio of the final volatile to char yield. Consequently, they cannot predict, even from the qualitative point of view, the dependence of product yields on reactor temperature and heating rate, which is of fundamental importance in reactor design and operation. Furthermore, in several cases, no kinetic data are reported. Only very few studies have been conducted to formulate semi-global kinetic schemes [4], which remove the previous assumption. Apart from the Broido-Shafizadeh mechanism, these schemes have not been supported by experimental studies different from those where they were proposed. Furthermore, the Broido-Shafizadeh has been shown [9,10] to predict well, at least from the qualitative point of view, the conversion of cellulosic samples under widely variable conditions, namely in the pure kinetic, thermally thin, thermally thick and ablative regimes.

The cellulosic particle is described as an anisotropic, porous medium with different properties along and across the grain (permeability to gas flow and thermal conductivity are larger along the grain). No particle shrinkage and/swelling are taken into account, so that the total volume V, remains constant. Physical processes described include:
1. convective and radiative heat transfer from the surrounding fluid/environemnt to the surface of the solid,
2. radiative, convective and conductive heat transfer interior to the solid,
3. momentum transfer to account for non zero pressure gradients and non uniform velocity,
4. variable properties (thermal conductivity, porosity and permeability),
5. accumulation of volatile enthalpy and mass in the pores of the solid,
6. volatile convection and diffusion through the pores of both the virgin solid and the charred region.

Thus the mathematical model is made by the conservation equations for solid phase species (eqns. (1-3), solid, active solid and char), gas phase species (eqn. (4) for total continuity), enthalpy (eqn.(5)), momentum (eqns.(6-7)), state equation (eqn.(8)) and solid volume variation (eqn. (9)):

$$\frac{\partial \varrho_W}{\partial t} = -r_1, \quad \frac{\partial \varrho_A}{\partial t} = r_1 - r_2 - r_3, \quad \frac{\partial \varrho_C}{\partial t} = \nu_C r_2 \qquad (1-3)$$

$$\frac{\partial(\varepsilon \varrho_g)}{\partial t} + \frac{\partial(\varrho_g u)}{\partial x} + \frac{\partial(\varrho_g v)}{\partial y} = \nu_G r_2 + r_3 \qquad (4)$$

$$(\varrho_C c_C + \varrho_W c_W + \varrho_A c_A + \varepsilon \varrho_g c_g)\frac{\partial T}{\partial t} + \varrho_g c_g u \frac{\partial T}{\partial x} +$$

$$\varrho_g c_g v \frac{\partial T}{\partial y} = \frac{\partial}{\partial x}\left(k_x^* \frac{\partial T}{\partial x}\right) + \frac{\partial}{\partial y}\left(k_y^* \frac{\partial T}{\partial y}\right) + Q_r \qquad (5)$$

$$u = -\frac{B_x}{\mu}\frac{\partial p}{\partial x}, \quad v = -\frac{B_y}{\mu}\left(\frac{\partial p}{\partial y} + \varrho_g g\right) \qquad (6-7)$$

$$p = \frac{\varrho_g RT}{W_g}, \quad \frac{V_s}{V_{s0}} = \frac{(\varrho_W + \varrho_C + \varrho_A)}{\varrho_{W0}} \qquad (8-9)$$

where

$$K_i = A_i exp(-E_i/RT) \; i = 1,3$$

$$r_1 = A_1 exp(-E_1/RT)\varrho_W \; r_i = A_i exp(-E_i/RT)\varrho_A, \; i = 2,3$$

$$\varrho_W = M_W/V, \; \varrho_C = M_C/V, \; \varrho_g = M_g/V_g = M_g/(\varepsilon V), \; \varepsilon = V_g/V, \; V_g = V - V_s$$

$$Q_r = K_1 \varrho_W [\Delta h_1 + (T - T_0)(c_W - c_A)] + K_2 \varrho_A [\Delta h_2 +$$

$$(T - T_0)(c_A - \nu_C c_C - \nu_G c_G)] + K_3 \varrho_A [\Delta h_3 + (T - T_0)(c_A - c_T)]$$

$$k_x^* = \eta k_{Wx} + (1-\eta)k_{Cx} + \varepsilon k_g + \sigma T^3 d_x/\omega, \; k_y^* = \eta k_{Wy} + (1-\eta)k_{Cy} + \varepsilon k_g + \sigma T^3 d_y/\omega$$

$$B_x = \eta B_{Wx} + (1-\eta)B_{Cx}, \; B_y = \eta B_{Wy} + (1-\eta)B_{Cy}$$

$$\eta = (\varrho_A + \varrho_W)/\varrho_{W0}$$

The symbols have the following meaning: A pre-exponential factor, E activation enrgy, ϱ density, ν stoichiometric coefficient, ε porosity, u and v velocity components, T temperature, c thermal capacity, V volume, Δh heat of reaction, B permeability, μ viscosity, p pressure, W mean molecular weight, σ Stefan-Boltzman constant, ω emissivity, d pore diameter. Subscripts indicate: W solid, C char, A active solid, G gas, T tar, s solid, g total volatiles, 0 initial or reference conditions.

From the physical point of view, the model here presented describes the transport phenomena occurring through the cross (square) section of a wooden, anisotropic sample exposed in a pre-heated, inert atmosphere furnace. To study the effects of anisotropy on the thermal degradation, different properties along the x and y directions are assigned. Because of the symmetry of the problem, only one fourth of the particle, initially at ambient temperature, is considered (the origin of the x and y axes is at the center of the cross section and the half-size, along the two directions, is τ). Boundary conditions are written as:

$$x = \tau: \quad k_x^* \frac{\partial T}{\partial x} = -\sigma(T^4 - T_r^4) - h_c(T - T_r), \quad p = p_0, \qquad (10)$$

$$y = \tau: \quad k_y^* \frac{\partial T}{\partial y} = -\sigma(T^4 - T_r^4) - h_c(T - T_r), \quad p = p_0 \qquad (11)$$

where T_r is the furnace temperature, h_c the convective heat transfer coefficient (at the other two sides, $x = 0$, and $y = 0$, symmetry conditions are imposed).

The numerical solution of eqns.(1-9) with initial and boundary conditions, is computed through a finite difference formulation of the model equations, based on the hybrid scheme, already applied for the simulation of the two-dimensional structure of smoldering combustion [11] and flame spread over solids [12].

3 Results

The properties used in the simulation roughly describe a cellulosic fuel. The kinetic data, the values of the heat of pyrolysis and the medium properties are the same as in [6-10]: $A_1 = 2.8 \times 10^{19} s^{-1}$, $A_2 = 1.3 \times 10^{10} s^{-1}$ $A_3 = 3.28 \times 10^{14} s^{-1}$, $E_1 = 242.4 kJ/mol$, $E_2 = 150.5\ kJ/mole$, $E_3 = 196.5\ kJ/mole$, $\nu_G = 0.65$, $\nu_C = 0.35$, $\Delta h_1 = 0$, $\Delta h_2 = \Delta h_3 = -418\ kJ/kg$, $\rho_{W0} = 400\ kg/m^3$, $c_W = c_A = 2.3\ kJ/kgK$, $c_g = 1.8\ kJ/kgK$, $c_C = 1.1\ kJ/kgK$, $d_x = d_y = 4 \times 10^{-5}\ m$, $k_g = 25.77 \times 10^{-3} W/mK$, $\mu = 3 \times 10^{-5} kg/ms$, $\omega = 1$, $h_c = 20\ W/Ks$. Thermal conductivities are taken from [13] and permeabilities have been estimated to give gas overpressures of the same order of those experimentally observed [13]. Values along the x and y directions are chosen to describe cross (low values) and parallel (high values) grain properties: $B_{Wx} = B_{Ax} = 1 \times 10^{-16}$, $B_{Cx} = 5 \times 10^{-13}$ $B_{Wy} = B_{Ay} = 1 \times 10^{-12}$, $B_c = 5 \times 10^{-12} m^2$; $k_{Wx} = k_{Ax} = 10.5 \times 10^{-2}$, $k_{Cx} = 7.1 \times 10^{-2}$, $k_{Wy} = k_{Ay} = 25.5 \times 10^{-2}$, $k_{Cy} = 10.46 \times 10^{-2} W/mK$. This set of data will be referred to as reference data. The furnace temperature is 900K (from $t = 0$) or it is increased from 500K, with a prescribed rate (10K/s), to 900K. The initial particle temperature is 300K. Simulations have been made for anisotropic (square) particles by varing their sizes, from $0.5 \times 10^{-2} m$ to $4 \times 10^{-2} m$. A parametric study of the effects of medium thermal conductivities and permeabilities on the thermal degradation has also been conducted and two-dimensional compared with one-dimensional results.

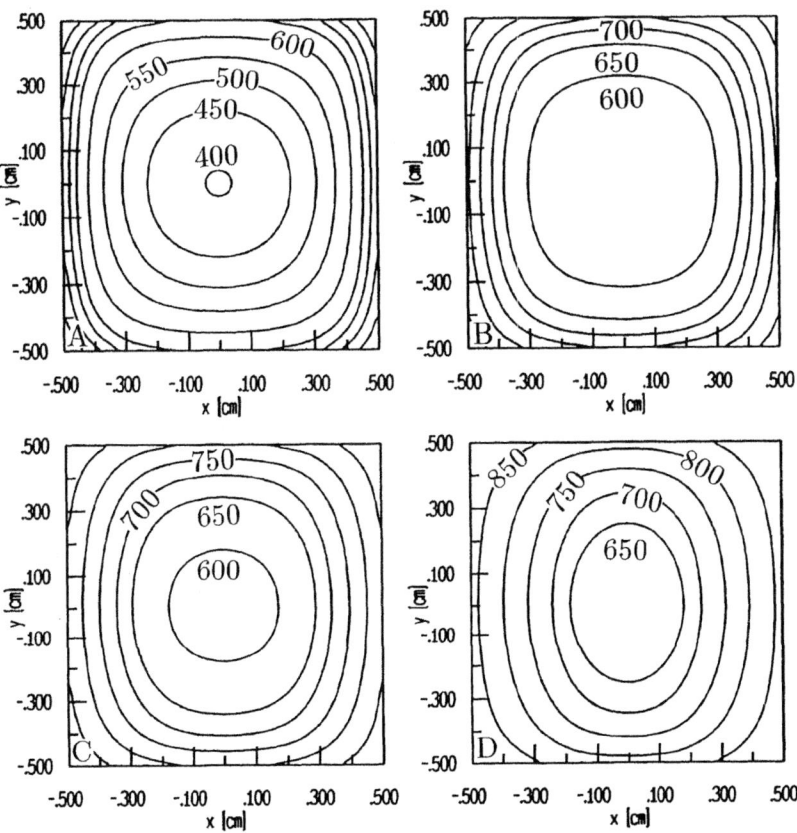

Fig. 1 - Isotherms [K] for t=15s (A), 30s (B), 45s (C) and 60s (D) and $T_r = 900K$ from $t = 0$ (plots refer to the whole particle).

For all particle sizes here considered, thermal conversion is a wave-like process, controlled by heat and mass transfer, and largely affected by the grain structure of the solid. Examples of process dynamics are given through the constant contour levels of temperature, gas overpressure, reaction rate (K_3) and conversion level η, the gas vector velocity and the conductive heat flux fields (Figs. 1A-1D, 2A-2D, 3A-3D, 4A-4D) for $\tau = 0.5 \times 10^{-2}m$ and $T_r = 900K$ (from $t = 0$). From the qualitative point of view, the process dynamics show the same characteristics as already simulated by one-dimensional models [5-11]. The usual regions of virgin solid, pyrolysis and charred residual are seen. The virgin solid is followed by an elevated pressure zone due to slow transport (convection and diffusion) of volatiles in a region of low porosity and permeability to gas flow. Heat conduction is the dominant mechanism of solid preheating, as convection and diffusion of volatiles do not significantly contribute in terms of heat exchange. The gas overpressure front sligthly precedes the pyrolysis zone, which becomes thicker and propagates with successively reduced rates, as its distance from the heated surface increases, because of the increasing resistance to heat transfer. The velocity field, which also affects heat transfer through the charred region and determines the intra-particle residence time of volatiles, is the result of both pressure gradients and

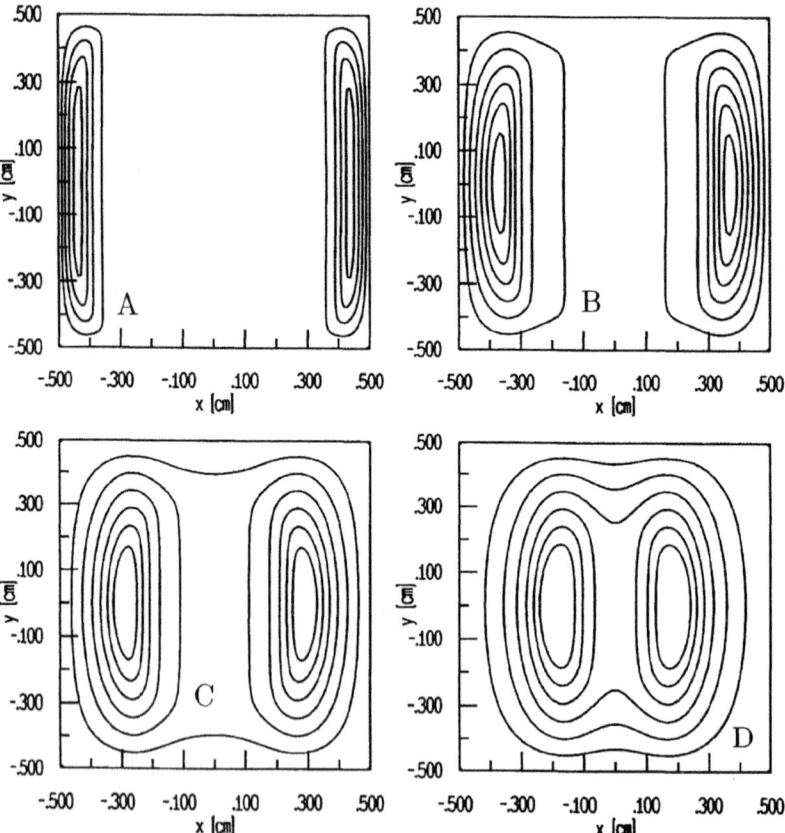

Fig. 2 - Constant contour levels of gas overpressure p/p_0 from 0.05 and then with step 0.05 for t=15s (A), 30s (B), 45s (C) and 60s (D) and $T_r = 900K$ from $t = 0$ (plots refer to the whole particle).

solid devolatilization rate. Similar to one-dimensional configurations, gas velocity reaches high values at the beginning and towards the end of particle conversion, because of the high initial devolatilization rate and the attainment of particle heat-up conditions, respectively. However, the two-dimensional simulations show that the shape of the temperature, pressure and reaction fronts, for fixed external conditions, is dictated by the anisotropy of the medium.

In the initial stage of particle conversion, because of the significantly lower thermal conductivity across the solid grain, the surface heat flux is larger (absolute value) along this direction (Fig.5A). Therefore, given the faster temperature rise (Fig. 5B), the reaction process begins across the grain. Since permeability is also lower, the volatile pyrolysis products mainly escape along the grain direction, through a thin section, corresponding to the thickness of the reaction zone across the grain direction. In fact, the flow velocity attains very large values along this narrow zone. As expected, the pressure front develops across the grain. The increase in the temperature across the grain causes a progressive decrease in the

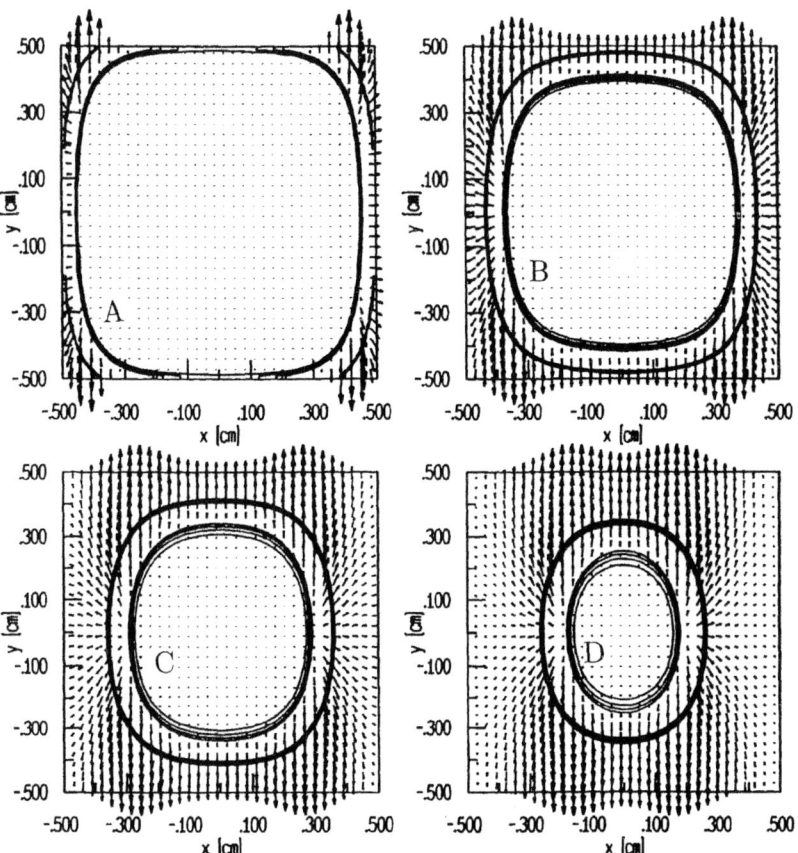

Fig. 3 - Constant contour levels of reaction rate $[kg/m^3 s]$ from 4 and then with step 3 and vector velocity field for t=15s (A), 30s (B), 45s (C) and 60s (D) and $T_r = 900K$ from $t = 0$ (plots refer to the whole particle). Maximum vector velocity [m/s]: 0.220 (A), 0.185 (B), 0.174 (C) and 0.180 (D).

surface heat flux along this direction, while the onset of the pyrolysis process along the grain and the consequent decrease in the medium thermal conductivity make comparable the two surface heat fluxes (that along the grain, however, remains always lower than that across the grain). On the contrary, interior to the particle, the vector heat flux field shows that larger values are attained along the grain. The lower reaction temperatures along the grain direction also give rise to higher char densities. Furthermore, the anisotropy of the solid, through the temperature distribution, also affects the shape of the reaction front. The thickness is larger along the grain, but a faster advancement is seen across the grain direction. Towards the end of the process, it can be seen that the reaction front assumes an ellipsoidal shape. Convective transport, with consequent larger solid cooling along the grain direction, and medium thermal conductivities are responsible for such a behavior.

A comparison of the numerical simulation results with experimental data is not possible because detailed measurements of the two-dimensional structure of pyrolysis fronts are not available. Very interesting experimental studies on the effects of the orientation of the anisotropic wood grain relative to one-dimensional heating have been conducted [13,14]. These results have already been compared with the one-dimensional version of the model here presented in [8].

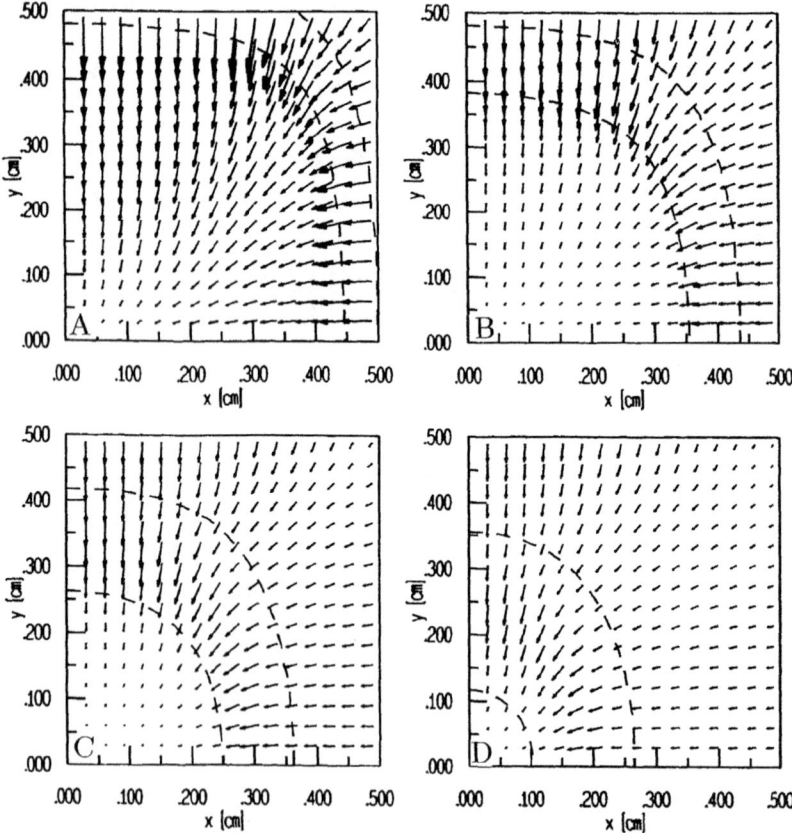

Fig. 4 - Vector conductive heat flux (boundary values excluded) and contours of the conversion level η (dashed lines) equal to 0.01 and 0.99 for for t=15s (A), 30s (B), 45s (C) and 60s (D) and $T_r = 900K$ from $t = 0$.

The influence of the anisotropy in terms of permeability to gas flow on the temperature distribution can be seen from Figs. 6A-6B, where the isotherms and the vector velocity velocity field ($\tau = 1 \times 10^{-2}m$, $T_r = 900K$ from $t = 0$) are compared, as simulated for the reference values of B_W, B_A and B_C and in the absence of pressure gradients. It can be seen that, in the second case, the velocity field becomes symmetric, so that convective transport of heat and mass is quantitatively the same along the two directions. The reaction zone is again thicker along the grain (larger thermal conductivity) but, contrary to the reference case and as expected from heat conduction effects, it is slightly more advanced along this same direction. This result indicates that the different permeability along and across the grain not only affects the magnitude of the convective flow and thus the activity of secondary reactions but also the temperature field and thus the rate of primary solid degradation.

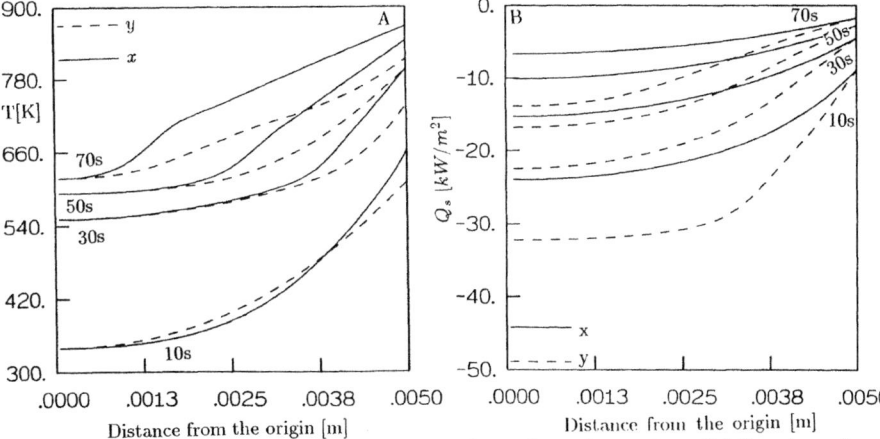

Fig. 5 - A) Temperature profiles along the axis x ($y = 0$, solid lines) and y ($x = 0$, dashed lines); B) surface heat flux profiles along the external boundary x ($y = \tau$, solid lines) and y ($x = \tau$, dashed lines); (times reported in the figure and $T_r = 900K$).

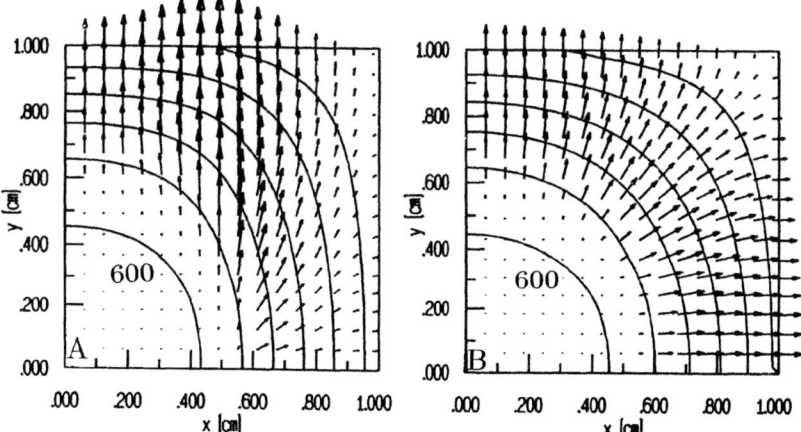

Fig. 6 - Constant contour levels of temperature [K] from 600 and then with step 50 and vector velocity field for t=150s with pressure gradients (reference data) (A) and in the absence of pressure gradients (high permeabilities along both x and y directions) (B) for $\tau = 1 \times 10^{-2}m$ and $T_r = 900K$. Maximum vector velocity [m/s]: 0.088 (A), 0.048 (B).

The temperature and char density profiles along the two axes (x (cross grain direction) and y (parallel grain direction)) simulated by the two-dimensional model in the absence of pressure gradients can be compared with the corresponding one-dimensional simulations through Figs. 7A-7D. For the two-dimensional case, temperatures for the char layer and the pyrolysis region are lower or hardly higher along the grain direction, whereas, apart from the beginning of the process, they are about equal for the virgin solid region. As expected, the lower reaction temperatures also give rise to larger char densities along the grain.

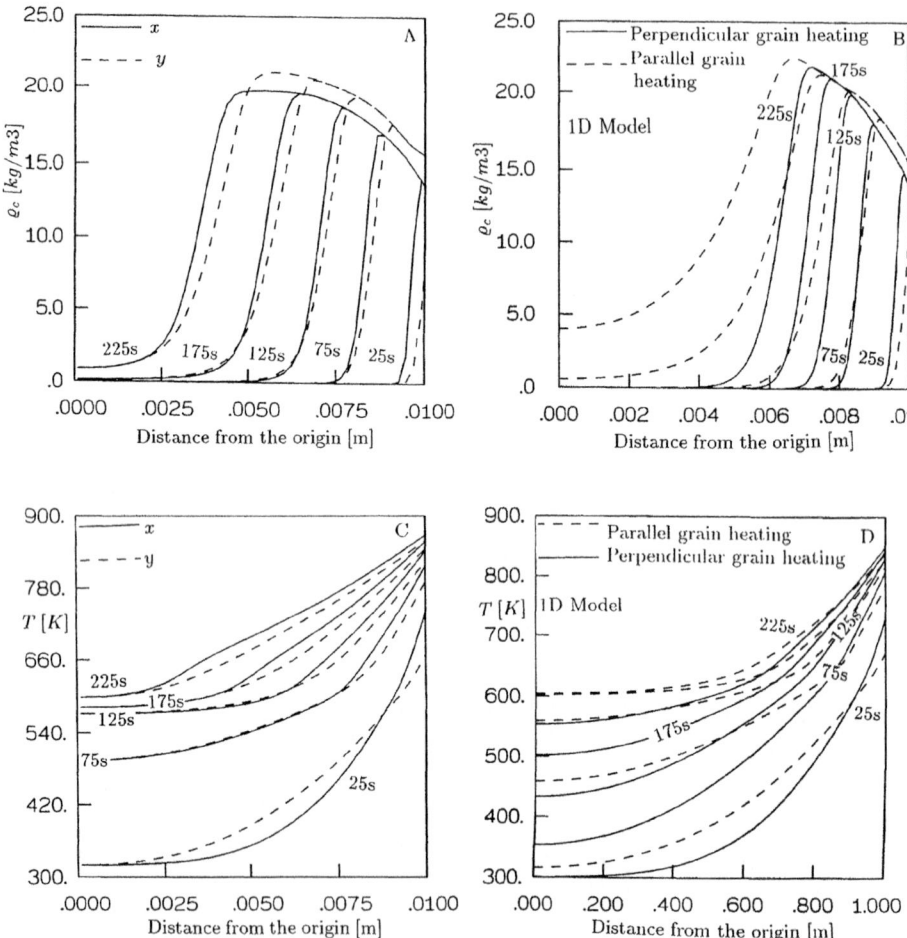

Fig. 7 - Char density profiles along the axis x ($y = 0$, solid lines) and y ($x = 0$, dashed lines) (A) two-dimensional and B) one-dimensional simulations); temperature profiles profiles along the axis x ($y = 0$, solid lines) and y ($x = 0$, dashed lines) (C) two-dimensional and D) one-dimensional simulations); (times reported in the figure and $T_r = 900K$).

For short times ($t < 75s$), the two-dimensional and the one-dimensional simulations produce results quantitatively close. Hower, for longer times, significant differences appear from both the quantitative and the qualitative point of view. The one-dimensional model predicts that temperatures, at the reaction front, for both parallel and perpendicular grain heating, are on average lower. Associated with this, there are other consequent effects, such as larger final char densities, a thicker reaction zone (mainly for the direction along the solid grain) and lower conversion levels (for times shorter than the one-dimensional particle heat-up time). It should also be noted that, along the grain direction, the one-dimensional model predicts a significant reduction in the particle heat up time, so that about half particle is pyrolyzed at temperatures well below the surface value. Another difference is that, for long times, the temperature for the parallel grain case becomes significantly larger for the pyrolysis and the virgin solid regions. These results indicate that multi-dimensional effects affect not only the detailed distribution of depedent variables but also global parameters, such as conversion time and product yields.

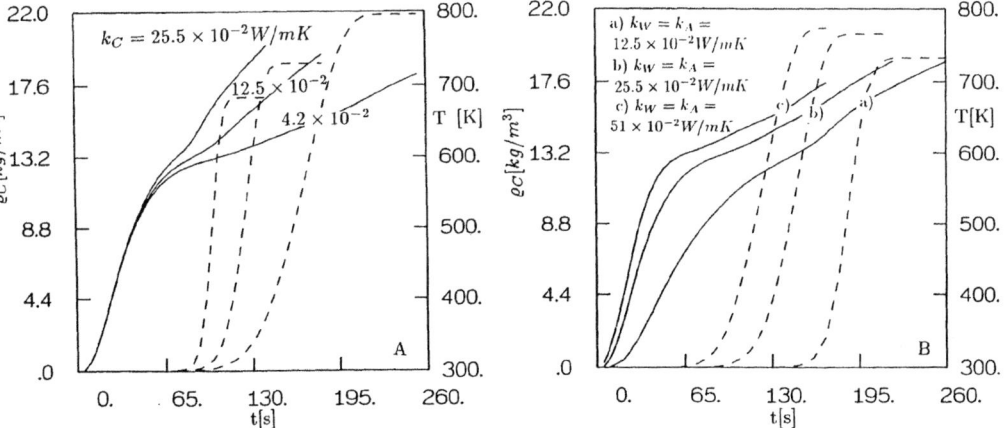

Fig. 8 - Time history of temperature and char density at a selected position (x=0.005 m, y=0.005 m) for an isotropic solid (parallel grain permeabilities) and for different values of the char (A) and solid (B) thermal conductivity. ($T_r = 900K$ and reactor heating rate equal to $10K/s$, A) $k_W = 25.5 \times 10^{-2}W/mK$, B)$k_C = 7.1 \times 10^{-2}W/mK$).

Given the importance of medium thermal conductivities, simulations have also been conducted, by varying these variables, for an isotropic two-dimensional solid with parallel grain permeabilities along both the x and y directions ($T_r = 900K$ and a reactor heating rate equal to $10K/s$). A first set of simulations has been made for $k_W = k_A = 25.5 \times 10^{-2}W/mK$ and varying k_C from $4.2 \times 10^{-2}W/mK$ to $25.5 \times 10^{-2}W/mK$. A second set of simulations has been made for $k_C = 7.1 \times 10^{-2}W/mK$ and varying k_W and k_A from $7.1 \times 10^{-2}W/mK$ to $125 \times 10^{-2}W/mK$. Results of the simulations are summarized in Figs. 8A-8B and 9A-9B. The curves of the mass loss and its time derivative as functions of time show a process successively slower, in terms of both conversion time and global devolatilization rate, as k_C is decreased. Indeed, heat transfer from the hot charred region to the virgin solid becomes more difficult. Similar to one-dimensional simulations, the reaction front enlarges and the lower reaction temperatures cause an increase in the char yield. Also, the convective transport of heat is lowered, however, the difference between the conductive and the convective contribution remains almost constant for the range of k_C tested. Even though sensitivities on volatile products are much lower, it can also be seen that tar yields increase at the expense of gas because tar formation is more favored and volatile residence times are reduced (higher solid devolatilization rates).

As the solid thermal conductivity is lowered, the onset of the pyrolysis process is anticipated but, on the whole, the degradation process becomes slower. The thickness of the pyrolysis region shrinks and, being on average the reaction temperatures higher, lower final char densities are predicted. The faster devolatization rates also lead to higher flow velocities and reduced extent of secondary reactions.

These results suggest that the lower char densities predicted for the cross grain direction, simulated in the reference case, are due not only to the reduced convective cooling but also to the lower solid thermal conductivity ($k_{Wx}/k_{Wy} = 0.41$) whose effects predominate over those due to the lower char thermal conductivity ($k_{Cx}/k_{Cy} = 0.68$), with the radiative contribution in the effective thermal condutivity playing a role of secondary importance.

To investigate the effects of char permeability on the thermal degradation of the solid, a set of simulations has been conducted for an anisotropic solid (reference data) and by varying B_{Cx} from $5 \times 10^{-13} m^2$ to $5 \times 10^{-15} m^2$. The maximum pressure increases as B_{Cx} is lowered, but , while the cross grain velocities decrease, those along the grain increase (Fig.10). Even though variations in the flow field cause variations in the convective transport along the x and y directions, there is, in some way, a compensation effect, and the devolatilization process (mass loss rate, conversion time and product distribution) does not undergo any significant change.

Convective transport becomes successively less important as the particle size is increased, because of the concomitant increase in the resistance to internal heat transfer and the reduction in the global devolatilization rate (Fig.11). The process is on the whole slowed and for very large particles, temperature distribution is essentially determined by heat conduction.

4 Conclusions

A two-dimensional model of a single particle pyrolysis has been presented, including the main physical process and the kinetics of primary degradation. Several mechanisms are responsible for the temperature (and reaction) field:

1. the convective transport has a cooling effect, thus, apart from the initial transients, higher temperatures are expected across the grain;

2. low char thermal conductivities do not favor heat transfer interior to the solid, through conduction. Therefore, associated with this mechanism, for a fixed time, lower temperatures and a slower advancement of the degradation front are ex-

pected across the grain (the reaction front enlarges and the lower reaction temperature causes an increase in the char yield);

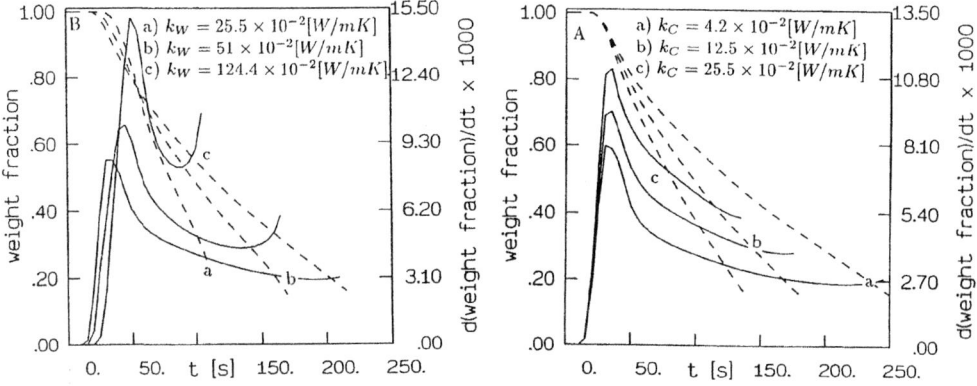

Fig. 9 - Weight fraction (dashed lines) and time derivative of the weight fraction (solid lines) as functions of time for several values of the char (A) and solid (B) thermal conductivity (data as Fig.8).

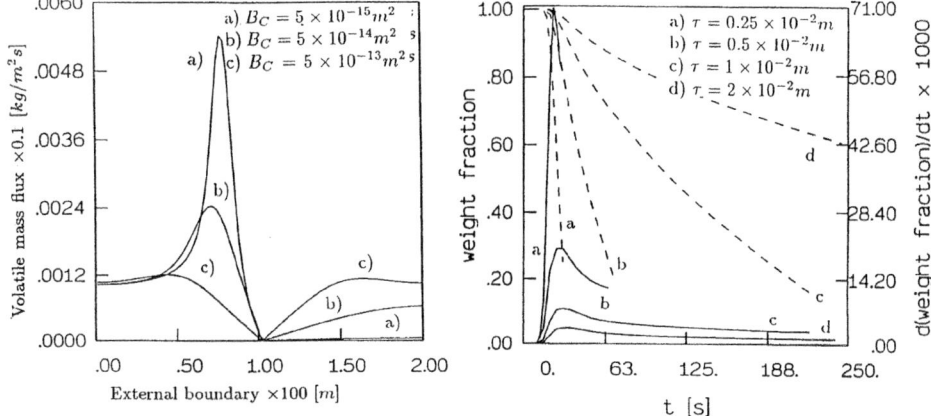

Fig. 10 - Volatile mass flux along the external boundary ($\tau = 1 \times 10^{-2}m$, $T_r = 900K$, $hr = 10K/s$).

Fig. 11 - Weight fraction and time derivative as functions of time ($T_r = 900K$, $hr = 10K/s$).

3. low solid thermal conductivities lead to a thinner reaction zone and higher reaction temperatures and thus a slower advancent of the degradation front is expected across the grain (the thickness of the pyrolysis region shrinks and, being the average reaction temperature higher, lower final char yields are obtained).

Numerical simulations of process dynamics for an anisotropic particle have shown that the larger heat convection (larger permeabilities) dominates over the larger heat conduction (larger medium permeabilities) and the reaction front progresses more slowly along the solid grain. For comparable convective transport, due to two-dimensional heat transfer effects, the propagation rates of the reaction front along the two directions become comparable even for different thermal conductivities along the two directions. On the whole, one-dimensional and isotropic, two-dimensional simulations show particle conversions significantly faster as the char and/or the solid thermal conductivities are increased.

5 Acknowledgements

The work was supported by the European Economic Community under Contract AIR2-CT93-0889.

6 References

1. Bridgwater A. V., Cottam M. L. (1992) Opportunities for biomass pyrolysis liquids production and upgrading, *Energy & Fuels* Vol. 6, pp. 113-120.
2. Bridgwater A. V. (1994) Catalysis in thermal biomass conversion, *Applied Catalysis A: General* Vol. 116, pp. 5-47.
3. Bridgwater A. V. (1995) The technical and economic feasibility of biomass gasification for power generation, *Fuel* Vol. 74, pp. 631-653.
4. Di Blasi C. (1993) Modeling and simulation of combustion processes of charring and non-charring solid fuels, *Progress in Energy and Combustion Science* Vol. 19, pp. 71-104.
5. Di Blasi C. (1993) Analysis of convection and secondary reaction effects within porous solid fuels undergoing pyrolysis, *Combustion Science and Technology* Vol. 90, pp. 315-339.
6. Di Blasi C. (1994) Numerical simulation of cellulose pyrolysis, *Biomass and Bioenergy*, Vol. 7, pp. 87-98.
7. Di Blasi C. (1996) Influences of model assumptions on the predictions of cellulose pyrolysis in the heat transfer controlled regime, *Fuel*, Vol. 75, pp. 58-66.
8. Di Blasi C. (1996) Heat, momentum and mass transfer through a shrinking biomass particle exposed to thermal radiation, *Chemical Engineering Science*, Vol. 51, pp. 1121-1132.
9. Di Blasi C. (1996) Kinetic and heat transfer control in the slow and flash pyrolysis of solids, *Ind. Eng. Chem. Res.*, Vol. 35, pp. 37-47.
10. Di Blasi C. (1996) Heat transfer mechanisms and multi-step kinetics in the ablative pyrolysis of cellulose, *Chemical Engineering Science*, Vol. 51, pp. 2211-2220.
11. Di Blasi C. (1995) Mechanisms of two-dimensional smoldering propagation through packed fuel beds, *Combustion Science and Technology* Vol. 106, pp. 103-124.
12. Di Blasi C. (1995) Predictions of wind-opposed flame spread rates and energy feed back analysis for charring solids in a microgravity environment, *Combustion and Flame* Vol. 100, pp. 332-340.
13. Lee, C., K., Chaiken, R., F., and Singer, J., M. (1976) Charring Pyrolysis of wood in fires by laser simulation, *Sixteenth Symposium (Int.) on Combustion*, The Combustion Institute, Pittsburgh, pp. 1459-1470.
14. Chan, W., R., Kelbon. M., Krieger-Brockett, B. (1988) Single-particle biomass pyrolysis: correlation of reaction products with process conditions, *Ind. Eng. Chem. Res.* Vol. 27, pp.2261-2275.

PYROLYSIS

Laboratory experimentation

LABORATORY EXPERIMENTS TO CHARACTERISE THE PYROLYSIS BEHAVIOUR OF SELECTED BIOMASS FUELS

A. MOILANEN, P. OESCH[*] and E. LEPPÄMÄKI
VTT Energy, P.O.Box 1601, FIN-02044 VTT, Espoo, Finland

Abstract
The aim of this work was to create fundamental characterisation data on the pyrolysis behaviour of various biomass fuels needed, e.g., for the development of gasification processes. Pyrolysis (Pyroprobe 1000) - Gas Chromatograph - Atomic Emission Detector (Py-GC-AED) technique was used to study the rapid pyrolysis of seven different biomass feedstocks: Kenaf, Miscanthus, pine bark, reed canary grass, sweet sorghum, wheat straw, and willow. The following fractions formed in pyrolysis were composed: light fraction, i.e. volatile gases (benzene and lighter gases), heavy fraction, i.e., volatile tars (heavier than benzene), and char residue. Carbon, hydrogen and oxygen content of the volatile fractions was determined with the AED. Pyrolysis temperatures (i.e., set in the pyrolyser) were 600, 800 and 1000 °C. Pine bark gave the highest amount of char residue and the lowest amount of volatiles at each temperature. However, each feedstock formed a considerable amount of condensable tars that was not analysable. The carbon content of the heavy fraction of the volatilised products decreased while the temperature increased. The oxygen of the volatilised products was mainly in the light fraction, and it increased only slightly with temperature. The char amounts measured in the Pyroprobe were compared with those obtained from the gasification reactivity measurements carried out under pressure in the pressurised thermobalance (PTG). In some biomasses, the char amount measured in the Pyroprobe followed the amount measured in the PTG in spite of the differences in experimental conditions. Pressure seemed to have no significance on the char amount measured in PTG. The oxygen content of the biomasses correlated with that of the volatiles.

The experimental conditions of the pyrolyser were found to affect the pyrolysis results due to the condensation of the heaviest volatilising products. The actual temperatures measured with a thin thermocouple inside the pyrolysable sample were 7-12% lower than the set temperatures. The actual heating rate of a sample was also much slower than the set value of 1000°C/s.
Keywords: biomass, pyrolysis, characterisation

---

[*] Present address: Imatran Voima Oy, Rajatorpantie 8, Vantaa, FIN-01019 IVO. Finland

1   Introduction

In the field of coal gasification, a lot of fundamental research has been carried out to create the scientific background required for the commercialisation of the new and fairly complex power production systems, as IGCC based on pressurised fluid-bed gasification. Gasification of biomass, especially under pressurised conditions is much less studied. Therefore, additional knowledge is needed to understand and control the phenomena taking place in biomass gasification processes. In literature, there are not many articles on the characterisation of biomasses used especially for thermal processes as gasification. Antal [1] has published a review about biomass pyrolysis. The pyrolysis literature is concentrated mainly on the characterisation of the volatilising product. The characterisation technique used for volatiles in this study, i.e., a gas chromatograph equipped with an atomic emission detector (AED), has been applied earlier for biomass pyrolysis [2].

In this work, the pyrolysis behaviour of various biomass feedstocks was characterised with laboratory instruments, considering, in particular, pressurised gasification. Biomass, when fed to a fluidised-bed gasifier, is first pyrolysed, and the amount of residual char limits the achievement of the total carbon conversion. The amounts of volatiles and their fractions are dependent on process conditions such as heating rate, final temperature, pressure, particle size, biomass type, gas atmosphere, etc. In this work, the rapid pyrolysis was studied, because it resembles the situation in which small particles and the surface of larger particles are pyrolysed when biomass is fed into a gasification reactor having a temperature range of up to 1000 °C. In larger particles, the pyrolysis is different due to the reasons that the particle centre heats up slower and the char layer is formed around it.

2   Experimental

The samples used in the study are given in Table 1, which shows the ultimate analysis of the samples used in this work. These were a selection of sub-samples from a larger set of biomass feedstock samples, which were used to characterise the gasification behaviour [3]. Pine saw dust has been included in earlier pyrolysis studies [2].

Table 1. The ultimate analysis (wt%, d.b.) of the biomass samples used in this study

Sample	Ash %	C %	H %	N %	$O_{diff.}$ %
Pine bark	2.0	55.0	5.9	0.3	36.7
Salix	1.2	49.7	6.1	0.4	42.6
Wheat straw	4.5	47.5	6.1	0.5	41.4
Reed canary grass	8.9	45.3	5.9	1.7	38.2
Miscanthus	3.3	47.8	6.2	0.4	42.2
Sweet sorghum	5.6	47.4	6.0	0.4	40.6
Kenaf	5.3	46.0	6.0	1.1	41.6

The pyrolysis experiments were carried out with a CDS Instruments Pyroprobe 1000 connected directly to a HP 5890 Series II gas chromatograph equipped with a HP 5921A atomic emission detector (AED). The scheme of the experimental set-up is shown in Fig. 1. The sample size was 2 mg and the particle size ranged 105-125 μm. The sample holder was a quartz tube. The microbalance Mettler AT 20 was used for weighing the sample before and after the pyrolysis. The pyrolysis conditions are seen in Table 2. The pyrolysis temperature and the heating rate indicated in the table are those set in the controller of the pyrolyser.

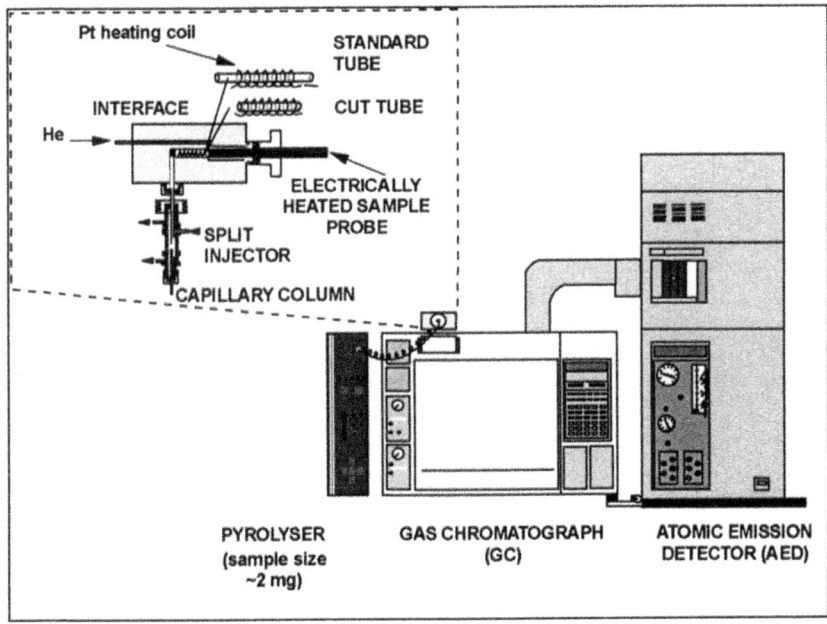

Fig. 1. The experimental set-up used in the study of the pyrolysis behaviour.

Table 2. The pyrolysis conditions used in the pyrolyser

Pyrolysis temperature (set points)	600, 800 or 1000 °C
Heating rate (set value)	1000 °C/s
Pyrolysis time	20 s
Pressure:	80 kPa (overpressure)

Pyrolysis took place in helium atmosphere while the pyrolysis chamber temperature was 250 °C. Helium was also used as carrier gas in the capillary column (HP 1.25 m, 0.32 mm, 0.17 μm). The oven program of the gas chromatograph was as follows:

Initial temperature   30 °C for 2 min
Heating rate   10 °C/min
Final temperature   300 °C for 5 min

Different model compounds such as carbon dioxide, butanal, phenol and chlorobenzene were used to determine the response factors for quantifying different elements with the GC-AED. Carbon, hydrogen and oxygen were measured and the mass balances of each feedstock were composed. For Miscanthus, the behaviour of chlorine compounds during pyrolysis was also detected by AED, the pyrolysis time being 20 seconds and the wavelength 479 nm [4, 5].

The standard length of the sample tube was 30 mm and the diameter 2 mm, and the length of the heating coil was 15 mm. The sample tube was placed inside the resistant heated platinum coil of the pyrolyser. Prior to the pyrolysis the sample in the coil was kept in the chamber at 250°C for about 3 minutes to equalise the conditions.

To find out possible effects of the experimental set-up on the pyrolysis products, a shorter sample tube and temperature measurements with a thin thermocouple were applied in the tests. The cut tube was used to find out the condensation and secondary reactions of the volatilising product by varying the residence time of the volatilising material in the sample holder. This tube was cut to the same length as the coil (see Fig. 1) and the distribution of pyrolysis products formed in the cut tube was compared to the products formed in the standard tube. The actual temperature values of the set points were measured using a thin thermoelement (diameter 0.25 mm) put into the quartz tube. In addition, the behaviour of the platinum coils with different calibration numbers was tested. Pine saw dust was used as the sample in these tests.

The char amounts from the pyrolysis tests were compared with those obtained from gasification tests carried for characterising the gasification behaviour in a pressurised thermobalance (PTG) [3]. In these tests, the char amounts were evaluated at the point just before the gasification of char started. The gasification behaviour was determined by measuring the gasification rates isothermally at temperatures of 700, 750 and 850 °C at atmospheric pressure and at 30 bar. The gasification gases were H_2O, CO_2 and gas mixtures H_2O-H_2 and CO_2-CO. For the evaluation of the char amounts, only the tests showing low reactivity values of char were used. In a test, the sample was lowered down to the reactor of the PTG where the reaction conditions were adjusted (i.e. temperature, pressure, gaseous environment). During the lowering the sample pyrolysed while heating up to the reaction temperature at an estimated rate of some 10 °C/s. The char amount was obtained from the weight signal at the point of 60 seconds from the beginning. If the reactivity in measuring conditions was high, it reacted too much during the first 60 seconds giving a small char amount.

3   Results and discussion

The main set of results was obtained from the tests carried out using the standard tube in the pyrolyser. Figs 2 and 3 show the mass balances of the feedstock, in Fig. 2 both for the total mass and the organic mass. In the results a so-called light fraction and a heavy fraction were separated. Benzene and lighter compounds (such as permanent gases) were included in the light fraction, and other pyrolysis products in the heavy fraction. The residue in the sample tube consisted of char. The results show that the increase in temperature decreased clearly the amounts of char and the heavy fraction of the

MASS BALANCE

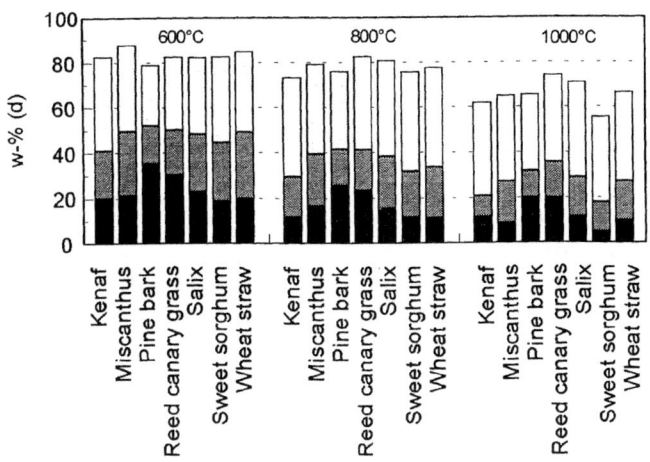

MASS BALANCE OF ORGANIC MATERIAL

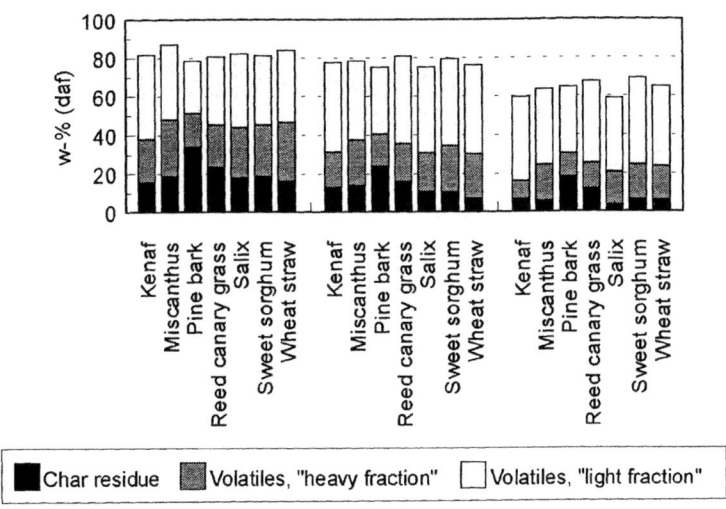

Fig. 2. Mass balances of the selected biomass samples (upper: wt% of dry sample; lower wt% of dry ash-free sample).

the volatiles, and increased only slightly the amount of the light fraction. At the set temperature of 600 °C, the average char amount was about 20% except for pine bark which had a significantly higher char amount of 35% than the other biomasses. At the set temperature of 1000°C the char amounts (daf) were less than 20% for all the samples. The light fraction seems to be almost stable at each temperature, being only slightly higher at 800 °C and above.

To a certain extent, the ash content of reed canary grass explains the char amount, but not for pine bark, when the mass balances of the organic material are examined. The amount of the heavy fraction of the volatilised products varied from one biomass to another being smallest for pine bark. The increase in pyrolysis temperature decreased the amount of char and the heavy fraction of the pyrolysis product, while the amount of the light fraction increased slightly.

Fig. 3 shows the release of carbon, hydrogen and oxygen from the volatilised products detected with AED for the samples at each pyrolysis temperature. To find out the distribution of the heating value of the fractions, the distribution of carbon and oxygen (wt% of dry ash-free sample) between the light and heavy fractions was plotted separately as seen in Fig. 4 (the hydrogen content was small in the volatilised products). Accordingly, the carbon content of the fractions varies with temperature so that the increase in temperature decreases the carbon content of the heavy fraction. The carbon content of the light fraction increased only from 600 °C to 800 °C, and at 1000 °C the changes were minor. The oxygen in the volatilised products was mainly in the light fraction and increased slightly with temperature.

The amounts of the volatilised fractions remained low for two reasons: Prior to the pyrolysis measurement there was a period of 3 minutes at 250 °C that caused an extra loss in the mass balances. This loss can be estimated to be about 10% including moisture. Another reason is that the GC-AED technique used did not enable to measure the heaviest volatilising products. The compound with a boiling point above approximately 500 °C condensed and could not be analysed.

The amounts of chlorine released from Miscanthus (Cl content 3030 ppm) were also determined. The highest yield obtained was less than 0.35% of the total chlorine content measured at 800 °C. The amount of the volatilised chlorine was minor detected with the AED, only less than 0.35% of the total chlorine content. This was probably due to the formation of hydrogen chloride (HCl) which could not be determined by the apparatus used. HCl is very reactive and does not elute in the capillary column. [2, 4].

Fig. 5 shows the comparison of the char amounts measured in the Py-GC-AED system and in the PTG. In general, the char amounts measured in the Pyroprobe were smaller than those measured in PTG. For pine bark and reed canary grass, however, the char amounts were almost equal in all tests. In the PTG tests, no systematic differences were obtained for the char amounts measured in atmospheric and in pressurised conditions. The differences observed can be explained partly by the fact that part of char reacted during the sample lowering, and partly by differences in the heating rates.

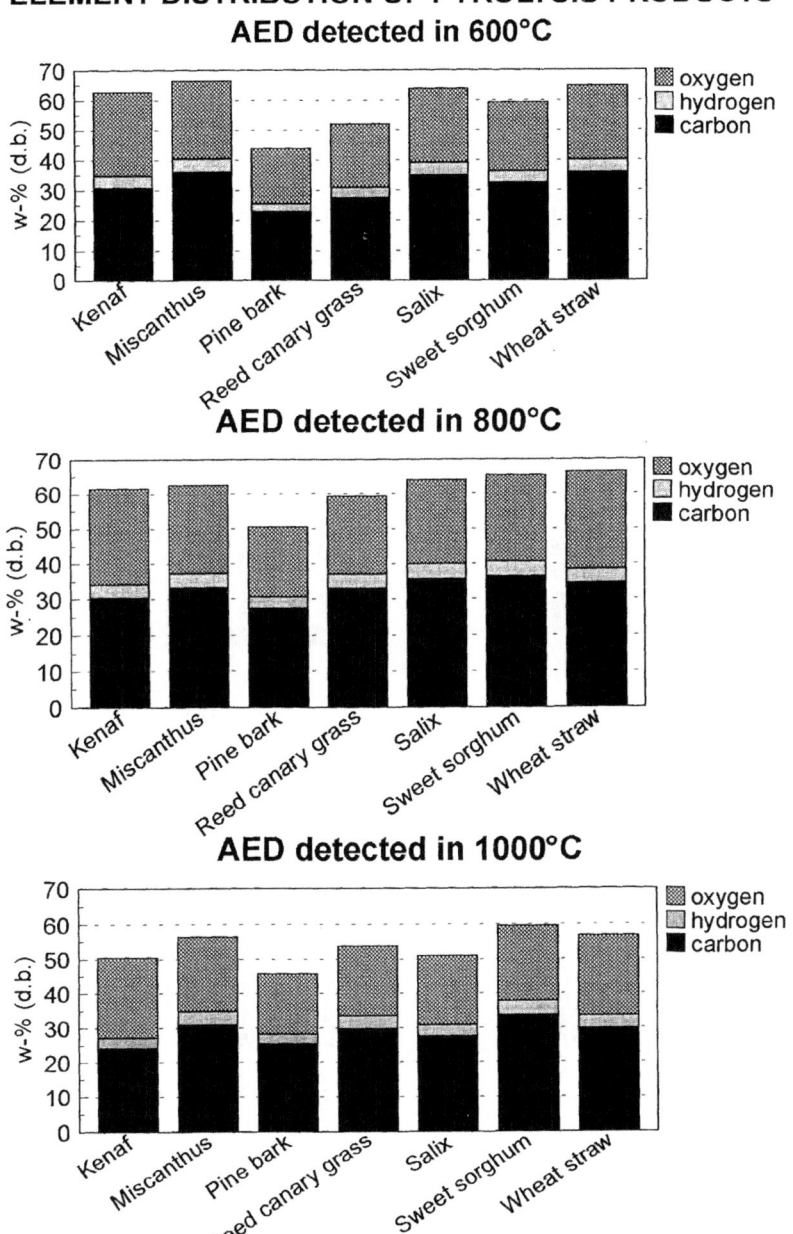

Figure 3. The release of carbon, hydrogen and oxygen in biomass pyrolysis detected by AED (wt% of dry sample).

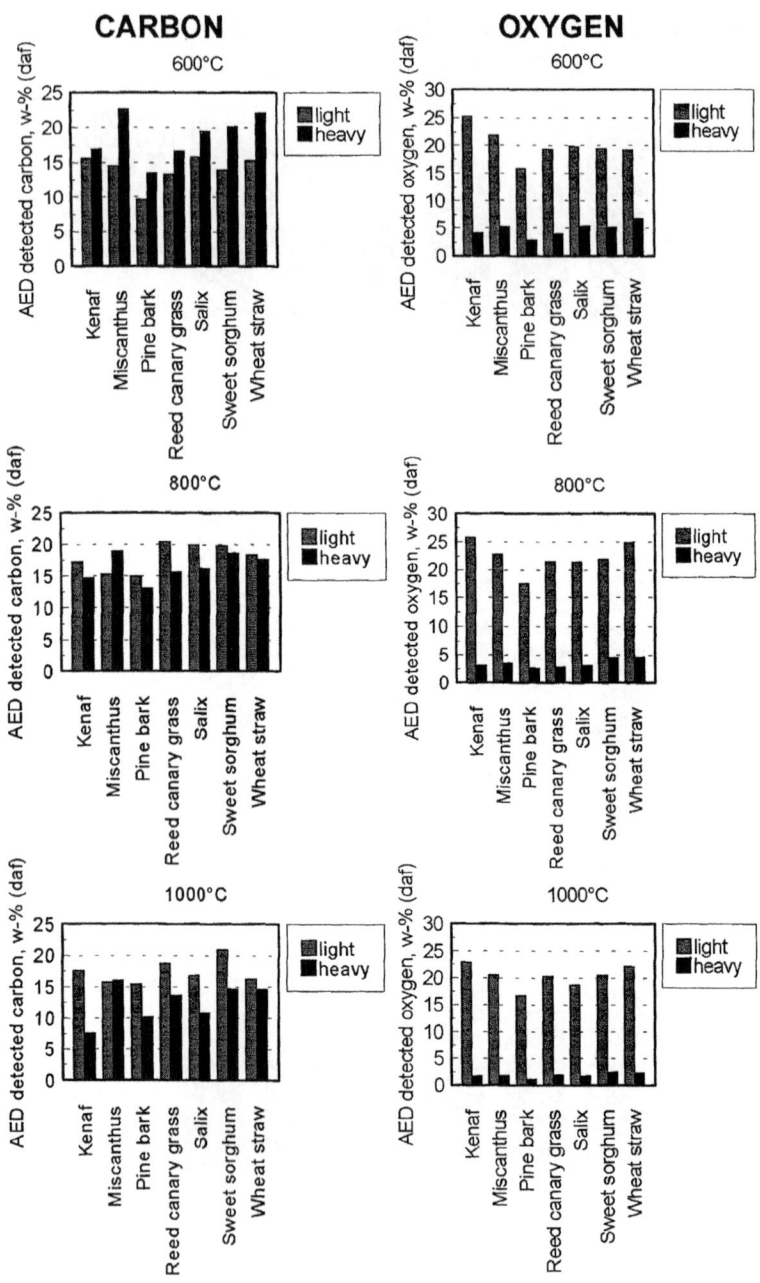

Figure 4. The carbon and oxygen content of the light and heavy fraction (wt% of dry ash free sample) at the set temperatures of 600, 800 and 1 000°C.

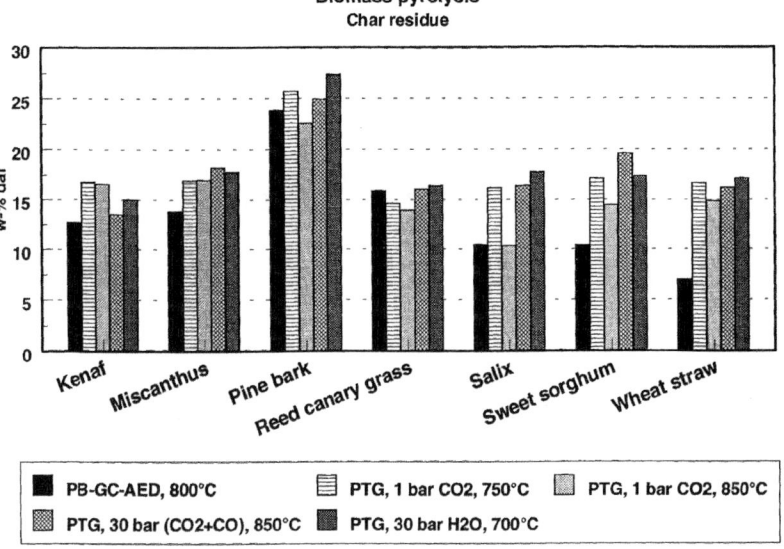

Fig. 5. The comparison of char amounts (organic material) in Py-GC and in PTG.

The effect of the experimental set-up on pyrolysis products is seen in Figs 6-9. Fig. 6 shows the effect of the tube length on the mass distribution in pine sawdust pyrolysis. Accordingly, the amounts of char were much lower in the cut tube than in the longer standard tube indicating that the char in the longer tube included condensed pyrolysis products. Fig. 7 shows, respectively, the released carbon, hydrogen and oxygen from pine sawdust determined by AED. The yields of the elements in the volatilising products were significantly higher in the cut tube at the temperatures below 1000 °C, but at 1000 °C temperatures the pyrolysis compounds detected decreased. Obviously, this was due to the formation of larger amounts of the compounds heavier than those detectable by the apparatus. When pyrolysing in the cut tube lighter compounds were obtained more than in the standard tube. There seemed to be more condensing reactions in the longer tube than was expected.

In the tests of the measuring conditions, the individual platinum coils of the Pyro-probe were also tested. Three coils were chosen for this study, each having its own cali-bration number related to the resistance of the coil. The temperature profiles for each coil were measured with the thin thermocouple put into the standard quartz tube. The results are shown in Fig. 8. Fig. 9 shows the sample temperature profiles measured dur-ing pyrolysis.

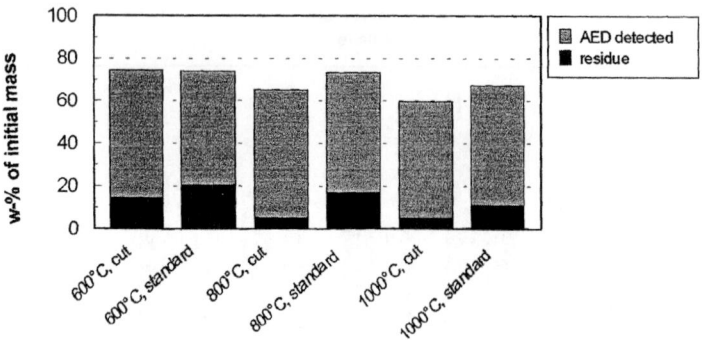

Fig. 6. Summary of the pyrolysis behaviour as a function of the pyrolysis tube length.

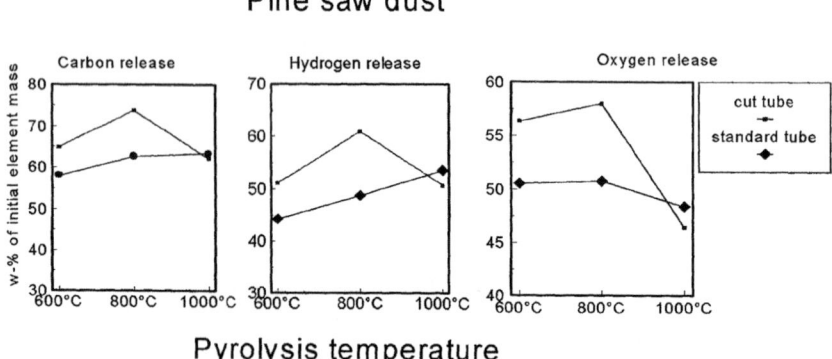

Fig. 7. Comparison of carbon, hydrogen and oxygen release in the cut and the standard tube for pine sawdust.

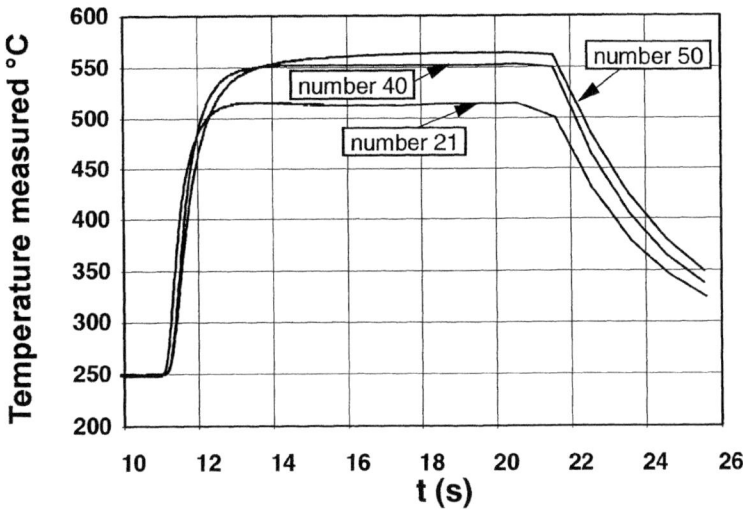

Fig, 8. Temperature profiles of three probes with different calibration numbers measured with the standard tube.

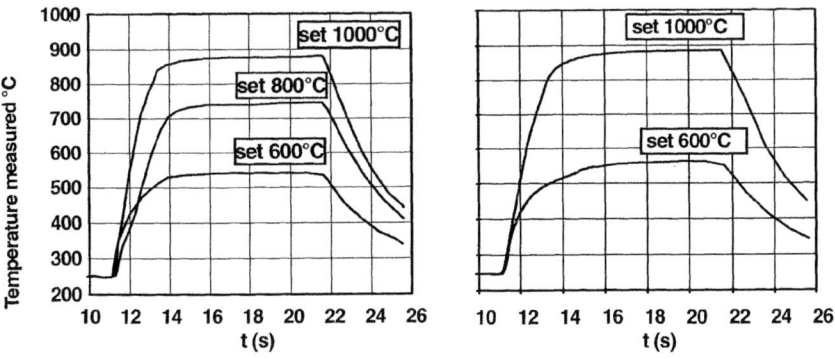

Fig. 9. Summary of the effects of the length of the tube on the temperature measured profile in pine saw dust pyrolysis. Left: the standard quartz tube (the length of the heating coil is half of that of the tube), right: the cut quartz tube (the length of the coil is the same as that of the tube).

The results showed clearly that the actual temperatures measured with a thin thermocouple remained 7-12% lower than the set temperatures. No significant differences were observed between the cut tube and the standard tube. At the set value of 600 °C the final temperature was only slightly higher with the cut tube than with the standard one, and the heating rate also reached a little higher value. The measured temperature also remained similarly lower when no sample was in the tube. This indicated that the helium flow and other heat losses decreased the actual temperature. The actual heating rates were considerably lower than the set value of the pyrolyser. For instance, in the pyrolysis at 800 °C, the heating rate in the beginning was only about 300 °C/s, and at the final stage it was about 100 °C/s, although the set value was 1000 °C/s. Kambara et al. [6] have also studied the temperature profiles for the probes and noticed some delay compared to the set values.

To see the correspondence between the oxygen content obtained from the ultimate analysis (determined as difference) and the oxygen content in the volatilising fractions detected with AED, they were plotted against each other, as seen in Fig. 10. The correlation was fairly good, especially for the values measured at 600 °C.

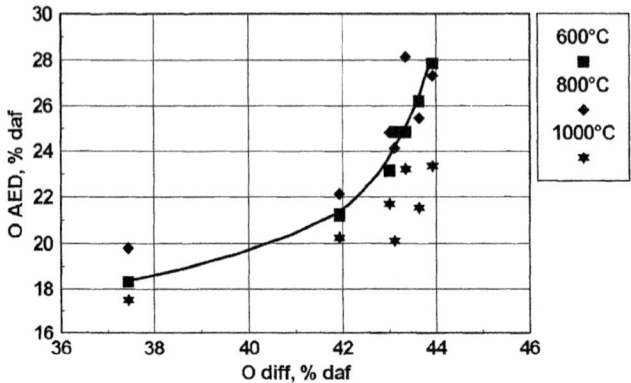

Fig. 10. The oxygen content of the volatilised products (detected by AED) vs. the oxygen content of the fuel (determined as difference). For clarity, the line is drawn for the points obtained at 600 °C.

4   Conclusion

In this study, laboratory experiments were carried out to characterise the rapid pyrolysis behaviour of selected biomass fuels. The following conclusions were drawn from the study:
- Increase in temperature decreased clearly the amounts of char and the heavy fraction of the volatiles, and increased only slightly the amount of the light fraction.
- The carbon content of the heavy fraction of the volatilised products decreased while the temperature increased. The oxygen of the volatilised products was mainly in the light fraction, and it increased only slightly with temperature.
- Experimental conditions affected the pyrolysis results (i.e. product distribution) significantly.

- For some biomasses, the char amount measured in the Pyroprobe followed the amount measured in the PTG in spite of the differences in experimental conditions. Pressure seemed to have no significance on the char amount measured in PTG.
- The oxygen content of the fuel correlated fairly well with that of the volatiles measured by AED .

5   Acknowledgements

The study was a part of the EU project AIR 2-CT93-1760 "Fundamentals of pressurised gasification of biomass feedstocks" carried out by VTT's Gasification Research Group. The project was financed by Academy of Finland, Ministry of Agriculture and Forestry and VTT.

6   References

1.   Antal, M.J. (1985) Biomass pyrolysis, A review of the literature, Part 2 - Ligno-cellulose pyrolysis. *Advances in Solar Energy*, Vol. 2, pp. 175-255.

2.   Alén, R., Oesch, P., Kuoppala, E. (1995) Py-GC/AED studies on the thermochemical behavior of softwood. *J. of Analytical and Applied Pyrolysis,* Vol. 35. pp. 259-265.

3.   Moilanen, A., Saviharju, K. (1996) Gasification reactivities of biomass fuels in pressurised conditions and product gas mixtures. Submitted to *Developments in Thermochemical Biomass Conversion*, Banff, Canada, May 20-24, 1996.

4.   Pitkänen, P. (1993) Thermochemical behaviour of chlorine compounds in biosludge, *Master thesis* (Finnish), Helsinki University of Technology, Espoo.

5.   Oesch, P., Alén, R., Leppämäki, E. (1996) Thermochemical behaviour of chlorine-containing material in biosludge during analytical pyrolysis, *J. of Pulp and Paper Science,* Vol. 22, No. 4. pp. J131-J134.

6.   Kambara, S., Takarada, T., Yamamoto, Y. & Kato, K.(1993) Relation between functional forms of coal nitrogen and formation of NO_x precursors during rapid pyrolysis, *Energy & Fuels*, Vol. 7, No. 6. pp. 1013-1020 .

OLIVE STONES PYROLYSIS: CHEMICAL, TEXTURAL AND KINETICS CHARACTERIZATION

Olive stones pyrolysis characterization

P.A. DELLA ROCCA, G.I. HOROWITZ, P. BONELLI, M.C. CASSANELLO
and A.L. CUKIERMAN

PINMATE - Departamento de Industrias - Facultad de Ciencias Exactas y Naturales, Universidad de Buenos Aires, Buenos Aires, Argentina.

Abstract

A study of olive stones pyrolysis over a wide range of temperatures, from ambient temperature up to 1123 K, has been carried out. Variations of chemical compositions and textural features taking place during olive stones pyrolysis have been determined. Porosity development with increasing pyrolysis temperature has been established from adsorption measurements employing nitrogen at 77 K and carbon dioxide at 298 K. Virgin olive stones and pyrolyzed samples structure modifications have been visualized by optical and scanning electron microscopy, respectively. Isothermal and non-isothermal thermogravimetric analysis has been carried out to examine pyrolysis kinetics. A recent model presented in the literature has been successfully applied to describe experimental data. Kinetic parameters have been estimated and compared with others reported for lignin. Both activation energy and preexponential factor have been found to increase with temperature.

Keywords: pyrolysis, biomass, char features, kinetics

1 Introduction

Olive stones, a copious agricultural by-product in several countries, have been recognized as an interesting source for activated carbons production. Two step thermal activation processes are commonly applied for activated carbons preparation: pyrolysis of raw materials under an atmosphere completely free of oxidant agents or with a limited supply of them, followed by in-situ gasification of the remaining solid with steam [1].

As pointed out in the literature [2-3], pyrolysis of lignocellulosic residues is complex since their major constituents, namely cellulose, hemicellulose and lignin, show different reactivities. During pyrolysis, numerous parallel and consecutive reactions take place. This complex network of reactions leads to significant variations in chemical composition and pore structure of the solid.

Although several studies of biomass pyrolysis have been reported in the literature [4-6], most of them have been concerned with different wood species, sugar cane bagasse, straw or biomass individual constituents [7-9]. Only a few have dealt with residues such as shells or fruit stones [2,10].

Studies to predict biomass pyrolysis rate in terms of its composition should be emphasized [5]. However, at the current state of knowledge, it is still difficult to predict the behavior of a particular type of biomass from data reported for others or based on its composition. In this context, the study of olive stones pyrolysis is important not only to improve activated carbons preparation but also because pyrolysis constitutes an attractive option for production of economic alternative fuels.

The present paper is concerned with several aspects involved in olive stones pyrolysis over a wide range of temperatures. It focuses on the reaction modeling together with the analysis of chemical compositions and textural features of the pyrolyzed samples.

2 Experimental section

2.1 Materials and samples preparation
Virgin olive stones free of pulp and mechanically cleaned with water were employed as raw materials. To analyze chemical and textural variations of olive stones during pyrolysis, chars from whole olive stones were prepared in a fixed bed reactor heated by an electric furnace under a nitrogen stream. Samples were heated at a rate of 10 K/min to different final reaction temperatures, 623 K, 873 K, 1123 K, and then kept constant for one hour. Afterwards, chars were cooled under flowing nitrogen to ambient temperature.

2.2 Chemical and textural characterization
Proximate and elemental analyses of milled virgin olive stones and chars prepared at the above mentioned temperatures were performed. Proximate analyses were carried out according to conventional ASTM techniques. A Carlo Erba EA 1108 elemental analyzer was employed to determine samples elemental compositions.

Specific surface areas of milled virgin and devolatilized samples were evaluated from adsorption isotherms based on the volumetric technique. These were obtained by employing a Micromeritics Accusorb 2100E sorption instrument with carbon dioxide at 298 K as adsorbate. Some experiments were also performed employing N_2 at 77 K.

Virgin olive stones and variations in sample structure at the three pyrolysis temperatures were visualized by optical and scanning electronic microscopy (SEM), respectively. Photographs were taken on cuts of whole virgin and pyrolyzed olive stones [11].

2.3 Kinetic measurements

Two sets of experiments under flowing nitrogen, 50 cm³/min, were performed by thermogravimetric analysis. A Netzsch STA 409 thermogravimetric balance coupled with a nitrogen flow device and a data acquisition system was used to carry out kinetic measurements. Virgin olive stones fractions of 37-45 μm particle diameter, to minimize diffusional effects, and 10 mg sample weights were employed.

Non-isothermal experiments were carried out at heating rates of 10 and 50 K/min from ambient temperature up to 1123 K. Experiments in isothermal conditions were performed at 100 K/min and twelve different final temperatures within the range of olive stones decomposition.

3 Results and discussion

3.1 Chemical and textural samples features

Tables 1 and 2 show results obtained from proximate and elemental analyses of virgin olive stones and chars prepared at the three pyrolysis temperatures.

Table 1. Proximate analysis (dry basis) of virgin olive stones and chars

Sample	Volatile matter (%)	Fixed carbon (%)	Ash (%)
Olive stones	74.84	24.74	0.42
Char (T = 623 K)	44.32	54.53	1.15
Char (T = 873 K)	22.80	75.62	1.58
Char (T = 1123 K)	22.14	76.24	1.62

Table 2. Elemental analysis (dry and ash free basis) of virgin olive stones and chars

Sample	% C	% H	% N	% O (∗)
Olive stones	46.59	6.04	—	47.37
Char (T = 623 K)	76.59	4.25	0.39	18.77
Char (T = 873 K)	93.32	1.95	0.41	4.32
Char (T = 1123 K)	94.39	0.53	0.77	4.31

(∗) by difference

Virgin olive stones have a considerable content of volatile matter (Table 1). Similar contents have been found for other lignocellulosic materials [7]. Ash percentage (Table 1) is low and agrees with the one reported for Spanish olive stones [1].

As expected, volatile matter, CO, CO_2, H_2O, hydrocarbons and other gases are released during pyrolysis accompanied by progressive increases in fixed carbon and ash contents. Increasing the final pyrolysis temperature leads to the decrease of volatile matter content. Successive decreases of this content between one temperature and the next become progressively smaller as temperature increases [12]. Above 873 K, changes are found to be minimal.

Results of the elemental analysis (Table 2) indicate that contents of carbon and nitrogen increase with pyrolysis temperature while those corresponding to hydrogen and oxygen decrease due to the decrease of volatile matter content. Similar variations of carbon percentages (Table 2) and fixed carbon (Table 1) of chars prepared at increasing pyrolysis temperature are found. As pointed out in the literature [12], the former is higher since part of the carbon is removed as volatile matter in the proximate analysis.

Losses in hydrogen and oxygen contents correspond to the scission of weaker bonds within chars structure and parallel results found for pyrolysis of some woods and coals. Relative increase of nitrogen with temperature may be explained considering that nitrogen is most probably present within heterocyclic ring structures in biomass and C-N aromatic bonds (sp^2 hybridized) are difficult to break [13].

Present trends with increasing pyrolysis temperature agree with others reported in the literature for different lignocellulosic materials [2,12,14].

In agreement with proximate analysis, no significant variations in char compositions are found between those prepared at 873 K and 1123 K. This indicates that for temperatures higher than 873 K almost complete pyrolysis is achieved.

In addition to composition changes, structural transformations of olive stones into chars at different pyrolysis temperatures were examined from adsorption isotherms and microscopy analyses.

The complementary use of N_2 at 77 K and CO_2 at 298 K has been strongly advised to characterize carbonaceous materials. The former provides information about meso and macropores whereas the latter is particularly suitable to investigate solids microporous textures or those in which micropores contribution is significant [1, 14-17].

Adsorption isotherms of N_2 were analyzed by applying the usual BET procedure. From carbon dioxide isotherms, specific surface area values were determined by applying the Dubinin-Radushkevich equation, which is given by [18]:

$$log\ (\phi) = log\ (\phi_0) - D\ log^2\left(\frac{p_s}{p}\right) \tag{1}$$

where ϕ is the amount adsorbed expressed as a liquid volume, ϕ_0, the ordinate at the origin that represents the total micropore volume of the sample, D, the Dubinin coefficient and p, p_s the equilibrium and saturation vapor pressures of the adsorbate, respectively.

In Figure 1, present experimental data plotted according to equation (1) are shown.

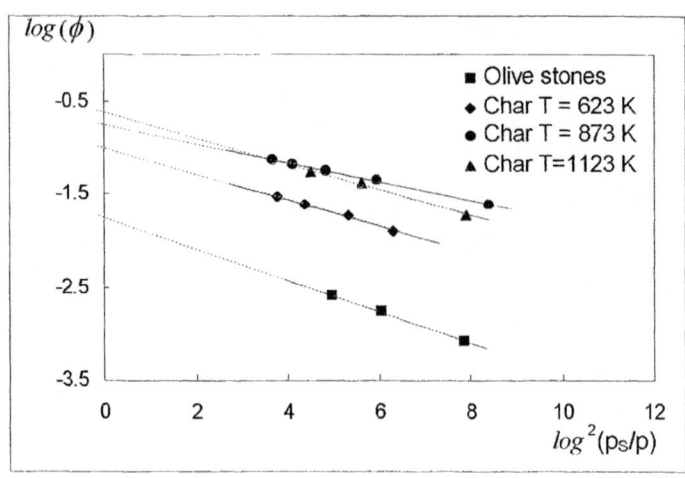

Fig. 1. Dubinin-Radushkevich plots for carbon dioxide adsorption
at 298 K

As may be observed, plots are linear over the relative pressure range with correlation coefficients larger than 0.996. Hence, the Dubinin-Radushkevich equation fits satisfactorily data obtained for virgin olive stones and chars. Then, for each sample, the value of ϕ_0 was determined and it was used to calculate specific surface areas following the evaluation procedure described in [15].

Table 3 details values of surface areas obtained by adsorption of nitrogen and carbon dioxide on virgin sample and examined chars.

Table 3. Specific surface area values

Sample	S_{N_2} (m^2/g)	S_{CO_2} (m^2/g)
Olive stones	0.6	53
Char (T = 623 K)	—	305
Char (T = 873 K)	85	548
Char (T = 1123 K)	—	737

As expected, nitrogen surface area values are lower than those evaluated from carbon dioxide adsorption, in agreement with numerous results reported for coals and biomass chars [14,19-21]. This behavior is attributed to the ability of carbon dioxide at 298 K to penetrate micropores whereas nitrogen adsorption at 77 K is mainly related to the contribution of larger pores, namely macropores and mesopores, to the total surface area [20]. Hence, present results points to microporous features of olive stones and their chars.

During olive stones pyrolysis, increases in both N_2 and CO_2 specific surface area values are found (Table 3), indicating porosity development. Higher pyrolysis temperatures induce larger areas. They suggest that increasing pyrolysis temperature leads to a larger amount of pores of any size, although micropores contribution is predominant. This increase may be attributed to development of new pores as a result of volatile matter release and to the widening of existing ones.

Virgin structure and typical microphotographs of cuts of olive stones pyrolyzed at the three temperatures, as obtained by microscopy and SEM analyses, are illustrated in Figure 2 (a-f).

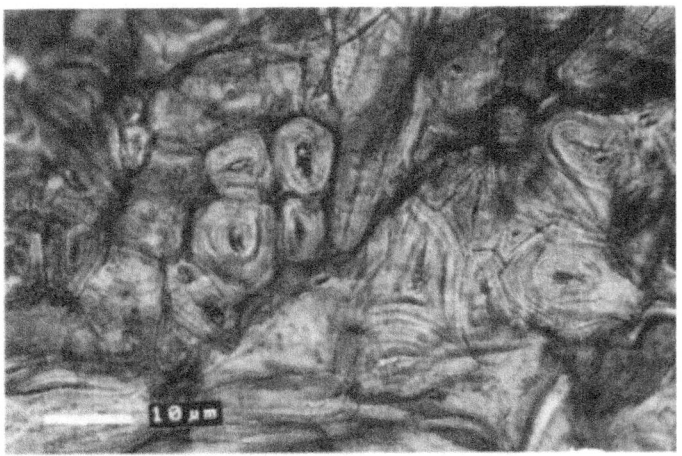

(a)

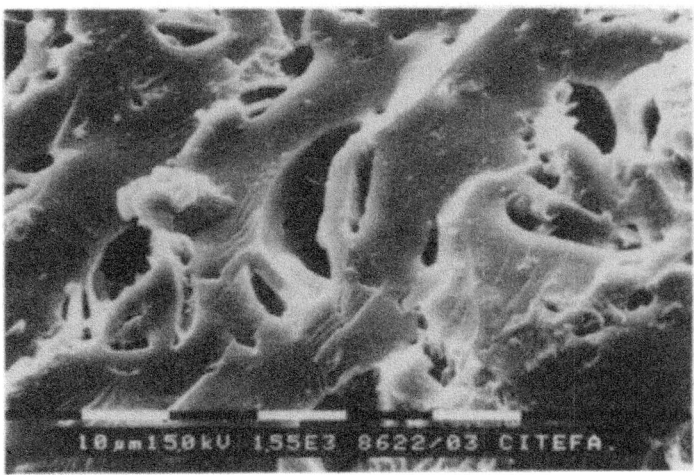

(b)

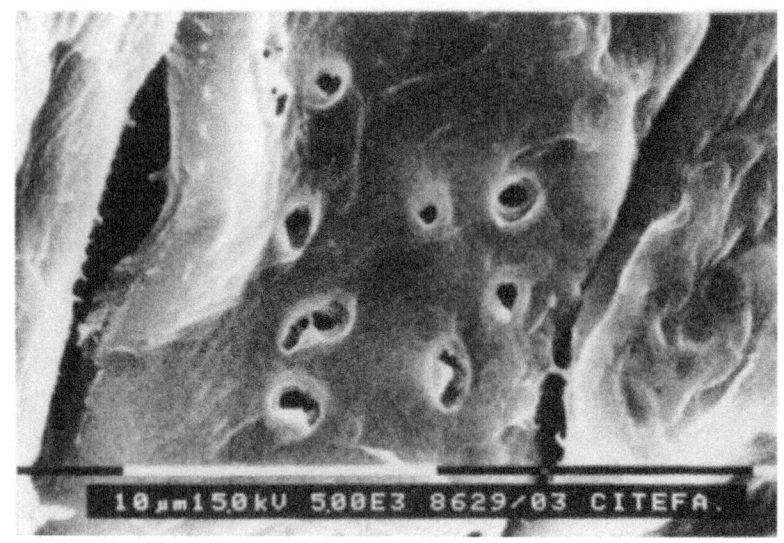

(c)

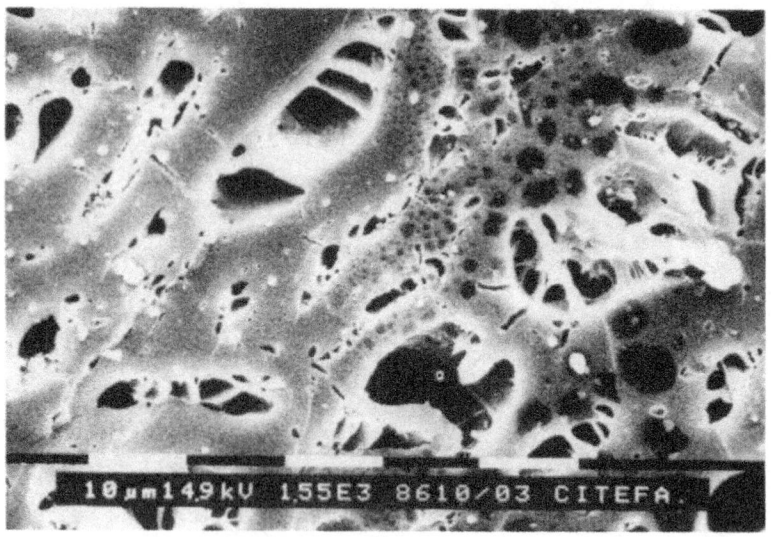

(d)

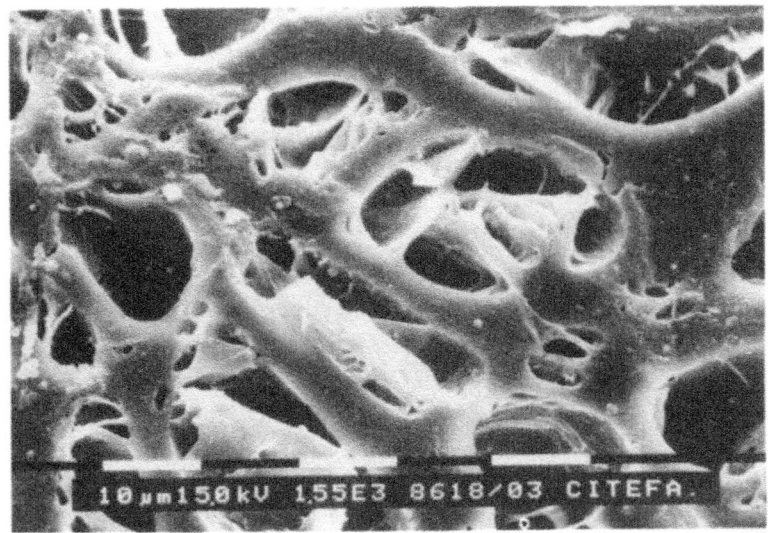

(e)

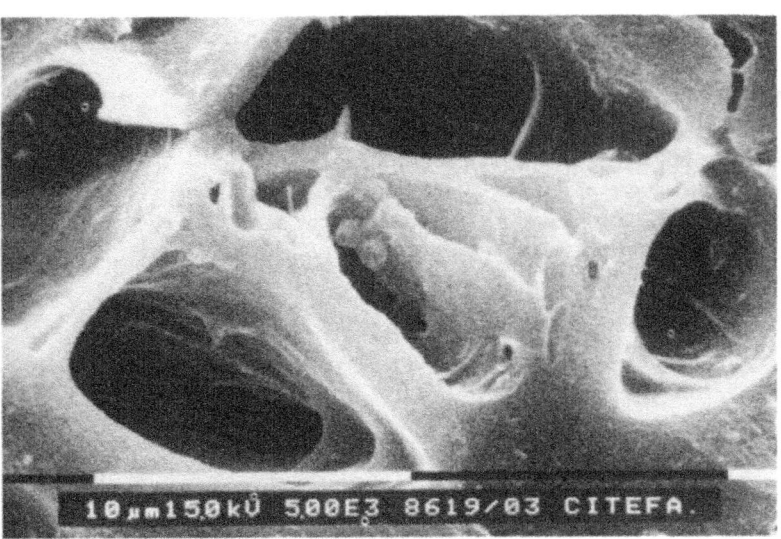

(f)

Fig. 2. (a) Optical micrograph of whole virgin olive stones cuts.
Scanning electron micrographs of chars prepared at: (b-c)
T = 623 K, (d) T = 873 K (e-f) T = 1123 K

Optical microscopy of virgin olive stones structure (Fig. 2a) suggests a close network of cells with almost polygonal cross-section, arranged in transverse and longitudinal directions.

As evidenced by SEM photographs (Fig. 2 b-f), removal of volatile matter during pyrolysis brings about an overall opening up of olive stone structure, which depends markedly upon final pyrolysis temperature. Development and widening of pores may be visualized. This result is consistent with that inferred from adsorption measurements.

At T = 623 K, the significant release of volatile matter is reflected in removal of cells contents, voids formation and development of new pores (Fig. 2b). Figure 2c details, on magnification, a section in which generated small pores may be visualized. Figure 2d portrays structural features of char prepared at T = 873 K. As may be seen, increasing temperature promotes creation of additional new pores. Numerous tiny pores may be observed. The microphotograph corresponding to olive stones pyrolyzed at the highest temperature (Fig. 2e) shows considerable voids, which may be attributed to the widening of tiny pores and coalescence between neighboring ones. Small pores may also be viewed. Figure 2f illustrates a larger magnification of the same char. The appearance of narrow pores developed on the walls of wider ones may be noticed.

Present results indicate that pyrolysis of olive stones leads to development of a wide variety of pores, which is largely affected by operating conditions. However, it should be kept in mind that microphotographs correspond to cuts from pyrolyzed whole olive stones. When these are crushed and milled to perform sorption measurements, only the smaller pores are mostly retained within the samples, explaining the predominant contribution of micropores found from adsorption measurements.

3.2 Kinetics and modeling

Typical thermogravimetric (TG) curves, namely values of sample weight loss in terms of temperature, obtained from non-isothermal experiments at heating rates of 10 K/min and 50 K/min are shown in Figure 3. It may be observed that TG curves are similar for both heating rates. The same behavior was reported by Balci et al. [2] for other lignocellulosic residues as woods and fruits shells.

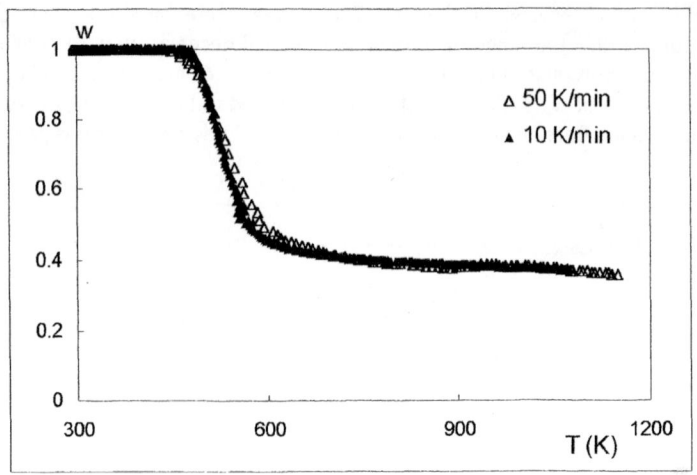

Fig. 3. TG curves of olive stones pyrolysis at different heating rates.

Pyrolysis modeling and evaluation of kinetic parameters are complex due to biomass heterogeneous composition and the numerous reactions that take place. One of the models usually employed to describe variations of pyrolysis rate is given by [6, 22]:

$$\frac{dw}{dt} = k\left(w - w_\infty\right)^n \qquad (2)$$

where k follows Arrhenius law and w_∞ is the residue fraction as $t \to \infty$. This fraction is often considered constant and n is commonly assumed to be 1.

As a first approximation eq. (2) was applied to fit experimental TG curves considering:
- a constant w_∞ and n as a fitting parameter;
- n = 1 and w_∞ as a fitting parameter;
- both n and w_∞ as fitting parameters.

A non-linear regression analysis was applied minimizing the following objective function:

$$\sum\left(w_{exp} - w_{calc}\right)^2 \qquad (3)$$

In the three cases, poor fits were obtained with standard deviations larger than 10%. This may be due to the fact that in heterogeneous solids such as lignocellulosic materials, the different components react at different temperatures. Then the assumption of a constant w_∞ may not held and a maximum pyrolyzable fraction for each temperature has to be determined as suggested in a recent model presented by Caballero et al. [22].

In this work, several isothermal TG curves were performed at different temperatures within the range of olive stones thermal decomposition to obtain values of maximum pyrolyzable fractions. Some of these curves are shown in Fig. 4 as an example. Values of w_∞ evaluated in this way were used to build the curve between the maximum residue fraction and temperature. In Figure 5 this relationship is presented.

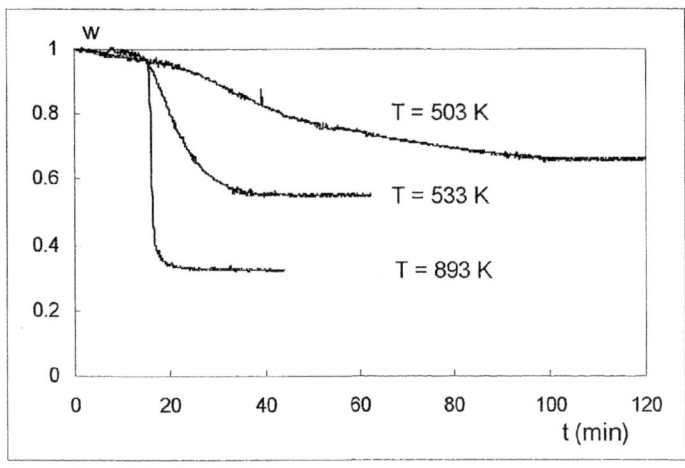

Fig. 4. Typical isothermal TG curves of olive stones pyrolysis at different temperatures performed to evaluate $w_\infty(T)$.

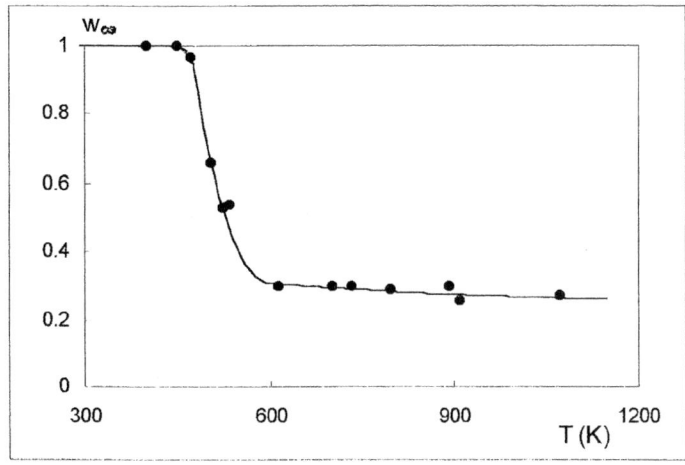

Fig. 5. Temperature dependence of the maximum residue fraction of pyrolyzed olive stones.

The continuous temperature dependence of w_∞ was obtained approximating different intervals by polynomials of degree two to four (solid line in Fig. 5). Considering the temperature dependence of w_∞, equation (2) becomes:

$$\frac{dw}{dt} = k_0 \ exp\left\{-\frac{E}{RT}\right\}\left(w - w_\infty(T)\right)^n \tag{4}$$

where k_0 and E may also be functions of the absolute temperature, T, and n may be different from unity.

Non-isothermal TG curves were fitted with eq. (4) using the temperature dependence of w_∞ obtained by the polynomial approximations. Different functions, including the constants, were tested for k_0 and E. The linear functions led to the best fitting. A comparison between experimental and calculated values are shown in Figure 6. It may be observed that present experimental data are satisfactorily predicted by eq. (4), with a standard deviation of 2%, and the following parameters:

$$k_0 \ \left(min^{-1}\right) = 6901 + 25.3 \ \left(T - 273\right)$$
$$\frac{E}{R} \ (K) = 891 + 13.4 \ \left(T - 273\right) \tag{5}$$
$$n = 1.8$$

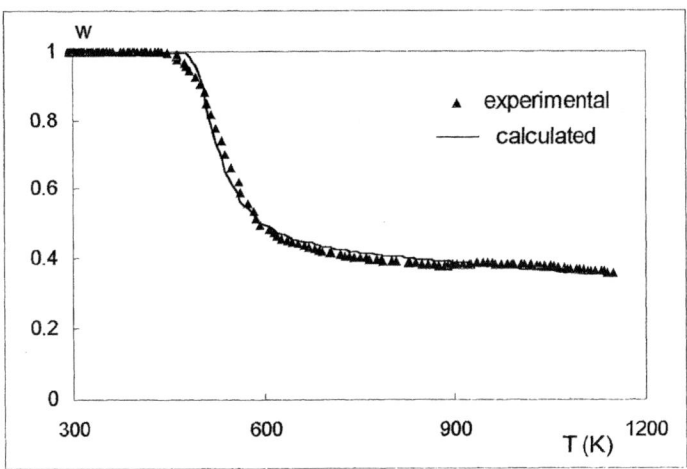

Fig. 6. TG curve of olive stones pyrolysis. Comparison between experimental and predicted values using eq. (4).

As pointed out in the literature [22], the preexponential factor, k_0, is related to entropy changes and solid surface modification. It has been shown in section 3.1 that solid surface area increases during pyrolysis, which may bring about the increase of k_0 with temperature. Increase of the activation energy with temperature may be attributed to breakage of weakest bonds that takes place at low temperatures while stronger ones break at higher temperatures. This behavior agrees with the effect of pyrolysis extent on activation energy found by Balci et al. [2] for other lignocellulosic residues.

Present activation energy values are lower than those found for pure Kraft lignin [22]. This is likely due to either the combined effect of biomass major components or to mineral matter content, which might catalyze the reaction.

4 Conclusions

Kinetics of olive stones pyrolysis over a wide range of temperatures together with variations of chemical compositions and textural features occurring in the solid residue have been analyzed.

Pyrolysis of olives stones in a fixed bed reactor are carried out at three final temperatures to investigate their variations. Proximate and elemental analyses are employed to characterize chemical compositions of virgin stones and pyrolyzed samples obtained in this way. Specific surface areas are evaluated from adsorption isotherms employing nitrogen at 77 K and carbon dioxide at 298 K. BET and Dubinin-Radushkevich equations have been applied, respectively. Virgin stones and samples structure modifications have also been visualized by optical and scanning electronic microscopy (SEM), respectively.

These techniques indicate that pyrolysis tends to open the structure due to volatile materials release depending strongly on pyrolysis temperature. Increasing temperature leads to generation of new pores and to pores widening.

For reaction study, two sets of experiments have been performed using thermogravimetric analysis of milled olive stones. Non-isothermal experiments have been carried out from ambient temperature up to 1123 K. Isothermal ones have been performed at twelve final temperatures within the range of olive stones decomposition to obtain the temperature dependence of the maximum pyrolyzable fraction.

A recent model reported in the literature has been applied to fit thermogravimetric data. Model has been found to describe satisfactorily olive stones pyrolysis with activation energy and preexponential factor that depend linearly with the absolute temperature.

5 References

1. González, M.T., Molina-Sabio, M. and Rodríguez-Reinoso, F. (1994) Steam activation of olive stone chars, development of porosity. *Carbon*, Vol. 32, No. 8, pp. 1407-1413.

2. Balci, S, Dogu, T. and Yücel, H. (1993) Pyrolysis kinetics of lignocellulosic materials. *Industrial & Engineering Chemistry Research*, Vol. 32, pp. 2573-2579.

3. Koufopanos, C.A., Papayannakos, N., Maschio, G. and Luchesi, A. (1991) Modeling of the pyrolysis of biomass particles. Studies on kinetics, thermal and heat transfer effects. *The Canadian Journal of Chemical Engineering*, Vol. 69, pp. 907-915.

4. Bridgwater, A. V. and Grassi, G. (1991) *Biomass Pyrolysis Liquids Upgrading and Utilisation*. Elsevier Applied Science, London and New York.

5. Antal, M. J. Jr. and Varhegyi, G. (1995) Cellulose pyrolysis kinetics: the current state of knowledge. *Industrial and Engineering Chemistry Research*, Vol. 34, No. 3, pp. 703-717.

6. Font, R. and Williams, P. T. (1995) Pyrolysis of biomass with constant heating rate: influence of the operating conditions. *Thermochimica Acta*, Vol. 250, pp. 109-123.

7. Magnaterra, M., Fusco, J. R., Ochoa, J. and Cukierman, A. L. (1994) Kinetic study of the reaction of different hardwood sawdust chars with oxygen. Chemical and structural characterization of the samples. *Advances in Thermochemical Biomass Conversion*, Edited by A.V. Bridgwater, Blackie A&P, London, Vol. 1, pp. 116-121.

8. Zanzi, R., Sjöström, K. and Björnbom, E. (1995) Rapid Pyrolysis of Bagasse at High Temperature. *Proceedings of the Third Asian-Pacific International Symposium on Combustion and Energy Utilization*, Hong Kong, Vol. 3, pp. 211-215.

9. Milosavljevic, I. and Suuberg, E. M. (1995) Cellulose thermal decomposition kinetics: global mass loss kinetics. *Industrial and Engineering Chemistry Research*, Vol. 34, No. 4, pp. 1081-1091.

10. Koufopanos, C. A., Maschio, G. and Lucchesi, A. (1989) Kinetic modeling of the pyrolysis of biomass and biomass components. *The Canadian Journal of Chemical Engineering*, Vol. 67, No. 1, pp. 75-83.

11. Della Rocca, P. *Pyrolysis and gasification of lignocellulosic residues*. PhD Thesis in preparation, Universidad de Buenos Aires, Argentina.

12. Figueiredo, J. L., Valenzuela, C., Bernalte, A. and Encinar, J. M. (1989) Pyrolysis of holm-oak wood: influence of temperature and particle size. *Fuel*, Vol. 68, pp. 1012-1016.

13. Wornat, M. J., Hurt, R. H., Yang, N. Y. C. and Headley, T. J. (1995) Structural and compositional transformations of biomass chars during combustion. *Combustion and Flame*, Vol. 100, pp. 131-143.

14. López-González, J. de D., Martínez-Vilchez, F. and Rodríguez-Reinoso (1980) Preparation and characterization of active carbons from olive stones. *Carbon*, Vol. 18, No. 6, pp. 413-418.

15. Gutierrez, M. C., Cukierman, A. L. and Lemcoff, N. O. (1988) Study of subbituminous coal chars: effect of heat treatment temperature on their structural characteristics. *Journal of Chemical Technology and Biotechnology*, Vol. 41, pp. 85-93.

16. Sousa, J. C., Torriani, I. L., Luengo, C. A., Fusco, J. R. and Cukierman, A. L. (1991) Microporosity and surfaces areas study of pine charcoal by SAXS and CO_2 adsorption techniques. *Journal of Applied Christalography*, Vol. 24, pp. 803-808.

17.   Amarasekera, G., Scarlett, M. J. and Mainwaring, D. E. (1995) Micropore size distributions and specific interactions in coals. *Fuel*, Vol. 74, No. 1, pp. 115-118.
18.   Gregg, S. J. and Sing, K. S. W., (1982) *Adsorption, surface area and porosity.* Academic Press Inc., London.
19.   Magnaterra, M. R. (1989) *Study of the combustion of different hardwood species devolatilized samples.* PhD Thesis, Universidad de Buenos Aires, Argentina.
20.   López-Peinado, A., Rivera-Utrilla, J., López-González, J. D. and Mata-Arjona, A. (1985) Porous texture characterization of coals and chars. *Adsorption Science and Technology*, Vol. 2, pp. 31-38.
21.   Gale, T. K., Fletcher, T. H. and Bartholomew, C. H. (1995) Effects of pyrolysis conditions on internal surface areas and densities of coal chars prepared at high heating rates in reactive and non-reactive atmospheres. *Energy & Fuels*, Vol. 9, No. 3, pp. 513-524.
22.   Caballero, J. A., Font, R., Marcilla, A. and Conesa, J. A. (1995) New kinetic model for thermal decomposition of heterogeneous materials. *Industrial & Engineering Chemistry Research*, Vol. 34, No. 3, pp. 806-812.

COMPARISON OF ABLATIVE AND FLUID BED FAST PYROLYSIS PRODUCTS: YIELDS AND ANALYSES
Biomass fast pyrolysis for liquids

G.V.C. PEACOCKE, C.M. DICK, R.A. HAGUE, L.A. COOKE and A.V. BRIDGWATER
Energy Research Group, Department of Chemical Engineering and Applied Chemistry, Aston University, Birmingham, England.

Abstract
A 3 kg/h ablative pyrolysis reactor and a 1.5 kg/h fluid bed reactor were operated using pine wood as the feedstock over a temperature range of 450-600°C. Mass balances and product analyses were performed and product comparisons made. A similar gas/vapour product residence time of around 1 s was used to reduce differences. Product yields followed the same general trends with a maximum organics liquid yield around 500-515°C for both reactors (59.4 wt% organics at 515°C and 1.19 s residence time in the fluid bed reactor and 62.1 wt% organics at 502°C and 1.1 s residence time for the ablative reactor]. Char yields increased for the fluid bed reactor above 515°C, and also in the ablative pyrolysis reactor. The volatile content of the char products were higher for the ablative, compared to the results for the fluid bed chars which showed a continual rapid decrease. Water yields were similar in the range of 11-16 wt%. Gas yields were markedly lower in the ablative pyrolysis reactor suggesting a less severe environment for the vapour products. The gaseous product composition variation with temperature in the fluid bed suggests that secondary vapour cracking is more prevalent in the fluid bed than the ablative pyrolyser.
Keywords: ablative pyrolysis, fluid bed fast pyrolysis, product yields and analyses

1   Introduction

Fast pyrolysis for the production of liquids in high yields for use as a fuel or as a source of speciality chemicals is becoming a more established and accepted technology. By controlling the key process parameters of reactor temperature, gas/vapour product residence time and temperature, the feedstock size and moisture content, high liquid yields can be obtained. The yields of these liquids and especially their chemical composition can be significantly changed by feedstock pretreatment, e.g. mild acid washing or additives.

The focus of this paper is a comparison of the effects of the reactor temperature on product yields and composition using a common feedstock in two different reactors. The ablative pyrolysis reactor described in this work has been in use since August 1992 and the fluid bed reactor since May 1995. A smaller 150 g/h fluid bed has also been commissioned and is used for small scale testing and feedstock evaluation. Feedstocks tested have included pine wood and pine bark, miscanthus and IEA Poplar to date, both pretreated and untreated in some cases.

2   Related Work

Prior work on fluid bed pyrolysis at similar reactor parameters to those investigated in this work has been extensively researched by the University of Waterloo [1, 2, 3] and on ablative pyrolysis by the National Renewable Energy Laboratory [4, 5]. It is not intended to review prior work, however, where relevant, comparisons and evaluations will be made.

3   Feedstock

The same pine wood feedstock was used in both reactors. The wood prepared for the ablative pyrolysis reactor had a particle size range of 1-3.35 mm and for the fluid bed reactor 0.6-1.7 mm. The feeds were sieved to remove fines. The particles for the ablative reactor were pin shaped, but the wood for the fluid bed was milled twice to remove pins which can cause feeding problems. These particles are roughly spherical in shape. The wood analysis on a wt% moisture free basis is: carbon 49.84, hydrogen 6.09, oxygen (by difference) 43.54, nitrogen 0.05, trace sulphur and ash 0.48. The ash content was determined using ASTM D1762-84. Elemental analyses were carried out to British Standard BS 5750 with oxygen calculated by difference. Feedstock for the ablative pyrolysis reactor was dried to near complete dryness, typically having a moisture content of less than 0.4 wt% (dry basis) by drying for 24 hours at 105°C in a fan assisted oven. Wood used in the fluid bed was air dried and had a water content of 9.25 wt% (dry basis). The wood had been stored since 1991 at Aston at ambient conditions.

4   Experimental equipment, methods and procedures

Results are presented for a 1.5 kg/h fluid bed and a 3-5 kg/h ablative pyrolysis reactor. The highest throughput to date with the ablative pyrolysis reactor has been 3 kg/h due to incomplete use of the contact area below the rotating blades and low biomass density as evaluated and described previously [6].

The conditions in the reactors were designed to imitate a demonstration or commercial fast pyrolysis process where it is likely that mean gas/vapour product residence times of 1-2 s will be used. For residence times of approximately 1 s, in conjunction with moderate reactor temperatures and low gas/vapour product temperatures (less than 500°C) to reduce secondary cracking reactions, the yields of organic liquids can be optimised. The aim of the work described here is to compare the products from different reactors to see if intrinsic features of the two reactors influence the product yield and composition at similar reactor parameters. For both reactors, the apparent gas/vapour phase residence time is the volume of gas and pyrolysis products at the average gas temperature from the reactor (taken as the temperature of the gases/vapours in the cyclone) divided by the empty reactor volume and the volume of the equipment to the liquids collection unit, discounting the volume of the reactor internals and sand for the fluid bed. The final gas exit temperature is measured as the dry gas at the gas meter and is measured to assist in the evaluation of volatiles losses from the reactors. For each reactor described below, the instrumentation, operation and product recovery procedures are described.

4.1 1.5 kg/h fluidised bed reactor

4.1.1 Instrumentation

A schematic of the fluidised bed reactor and the collection system is shown in Figure 1. Temperature and pressure measurements are made at the points indicated. Differential pressure measurements (not shown) are taken over the feed tube, the fluid bed reactor, the pyrolysis vapour products quench column and the cotton wool filter. The fluidising gas is electrically preheated by three heaters with additional energy supplied by two in-bed heaters. The temperature of the in-bed heaters is controlled independently of the fluidising gas heaters by measurement of the temperature in the sand bed in close proximity to the in-bed heaters. Two search thermocouples contained in a vertical sealed tube can be moved within the height of the reactor to check for any variation in reactor temperature in the sand bed. The liquids recycle rate is monitored with a volumetric flowmeter. The dry exit gas volume from the system is measured with a volumetric gas meter to 0.1 l in conjunction with gas temperature and pressure.

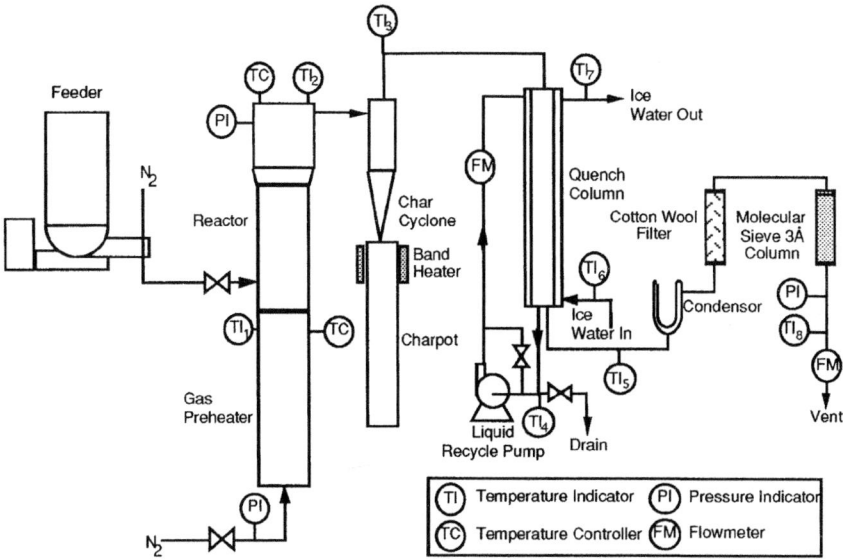

Fig. 1 Schematic of the fluidised bed reactor system

4.1.2 Operation

Nitrogen from a single supply is split into two streams; the fluidising gas stream to the reactor to fluidise the bed and the entrainment stream which purges the biomass feeder and entrains the biomass into the fluidised bed. The sand bed contains ~1.5 kg of 20-30 mesh sand (previously acid washed then washed with distilled water to neutrality).

The reactor operates in a blow through mode in that the product char, which has a low density relative to the sand and the biomass, is entrained out of the reactor with the hot gases and vapours after reaction. The char is then removed in the cyclone and collected in the char pot. The top of the char pot is heated by a band heater to 400°C to prevent the solids exit from the cyclone becoming blocked by organic vapours which may condense. Sufficient insulation reduces the extent of product condensation and collection on the transfer pipework within the run times used, rendering the use of trace heating unnecessary. The hot gases and vapours are quenched with either a solvent, propan-2-ol is used, or with a mixture of solvent and pyrolysis liquid from prior runs. For these experiments, FB22 used a solvent/pyrolysis liquid mixture from a previous run with the same feedstock, FB23 used a fresh batch of solvent, subsequent runs used a solvent/pyrolysis liquid mixture from the previous run. The quench column is externally jacket cooled with recycling ice cold water. The vapours are cooled and condensed by a recycle of solvent for the initial run: subsequent runs use the liquid collected from previous runs. The quench column liquid tends towards a pure pyrolysis liquid after four to five runs as the initial solvent is entrained from the column and collected downstream or removed as vapour in the gas stream. Exit gases from the quench column contain a significant quantity of aerosols and volatiles, including water vapour. Initial collection and further cooling is carried out in a dry ice/acetone condenser. Residual aerosols are removed in the cotton wool filter with gas drying in a molecular sieve 3Å column. Non condensable gas volume is measured by gas meter. Batch gas samples are taken at the gas meter for quantitative analysis by gas chromatography (GC).

4.1.3 Product recovery

All parts of the collection system are weighed before and after each run with the exception of the quench column. A weighed quantity of quench liquid is added to the quench column. The quench column is drained after each run and the quantity of liquid collected is weighed. Runs subsequent to the initial run use the liquid collected in the quench column as quench liquid. The water content of the quench liquid and all other liquids are determined by Karl-Fischer titration to an accuracy of ± 0.1 wt% absolute.

4.1.4 Results

Detailed results are presented for reactor bed temperatures from 449°C to 562°C as presented in Table 1 with experimental parameters and mass balance summaries. Product yields are quoted on a wt% of dry biomass fed basis. It is assumed that the moisture present in the feedstock is collected as water in the product liquid and takes no part in the pyrolysis reactions. The water yield quoted discounts this moisture and is thus the water of pyrolysis only and includes water recovered in the molecular sieve. Experiment FB22 used a demister pot to remove aerosols and volatiles from the quench column exit gases. Subsequent runs used a dry ice/acetone cooled condenser as described earlier and as shown in Figure 1. The use of the condenser reduced final gas exit temperatures by around 10°C leading to a reduction in the amount of volatile products lost in the exit gases. The optimum temperature for liquids production is between 500 and 515°C up to 59 wt% organics, consistent with the findings of Scott and Piskorz [7], but lower due to the longer gas/vapour product residence time. The water of pyrolysis yield is fairly consistent, between 11.5 and 15 wt%, dry basis, over the range of reaction temperatures.

4.1.5 Char yields and analysis

Char yields and analyses are summarised in Table 2. Elemental analysis is carried out to British Standard BS 5750. Ash and volatiles analysis are carried out using ASTM 1762-84. The char yield for the run FB25 is higher than anticipated-a result of 10-12 wt% was expected. At high reactor temperatures, greater than 550°C, excessive fluidising gas temperatures are require to supply sufficient heat for the pyrolysis process. In this case, the in-bed heaters were used to supplement the energy required to maintain the bed at the required temperature. Therefore, it was necessary to maintain the in bed heater setpoint at a higher temperature (30 to 50°C higher) than the desired reactor temperature in order to maintain the desired temperature in the bed. The char yield for FB23 and FB25 is believed to increase due to the in bed heaters cracking and charring the organic vapours and higher hydrocarbons (C_2+) within the reactor. Determination of the CHN contents of the chars showed variations of up to 30 wt% in some cases (60-90 wt% carbon for chars from FB25-results not shown). It is believed that this is due to the in-bed heaters causing secondary cracking, hence the variable carbon content as discussed in Section 5.2.1.

Table 1: Mass balances for the fluid bed reactor (product yields wt%, dry wood basis)

Run Number	FB22	FB24	FB26	FB23	FB25
Reactor Temperature (T_2, °C)	449	476	514	524	562
Apparent vapour residence time (s)	1.06	1.29	1.19	1.05	1.13
Gas/vapour product temp.(T_3,°C)	369	393	384	393	393
Nitrogen fed/Biomass fed (kg/kg)	6.41	7.43	8.78	8.23	9.08
Final gas exit temperature (T_8, °C)	23.9	16.6	13.8	15.6	12.6
Char	24.8	15.8	10.1	14.4	19.9
Organics	46.2	55.9	59.4	54.1	39.6
Gas	12.7	12.7	15.4	16.4	23.8
Water	11.9	11.6	12.4	14.8	13.7
Closure	95.6	96.1	97.3	99.7	97.0
Gas yields (wt% dry wood basis)					
CO	4.8	6.4	7.2	8.1	11.9
CO_2	7.5	5.6	7.1	7.1	9.6
CH_4	0.2	0.4	0.6	0.8	1.5
C_2H_4	0.1	0.1	0.3	0.3	0.6
C_3H_8	0.1	0.1	0.1	0.2	0.3
H_2	nd	nd	nd	nd	nd
Gas composition (vol%, nitrogen free basis)					
CO	47.9	58.7	54.7	56.3	55.9
CO_2	46.9	32.9	34.3	31.5	28.6
CH_4	4.1	6.4	8.6	9.5	12.0
C_2H_4	0.7	1.3	1.9	2.0	2.8
C_3H_8	0.3	0.7	1.5	0.7	0.8
H_2	nd	nd	nd	nd	nd

Table 2: Fluid bed char analyses (wt%, dry basis). Comparison with Scott et al. [7]

Run Number	FB22	FB24	FB26	FB23	FB25	Scott [7]
Reactor Temperature (°C)	449	476	514	524	562	541
Carbon	77.7	80.7	82.4	83.8	88.6	70.0
Hydrogen	3.8	3.5	3.3	3.5	3.1	2.8
Nitrogen	0.1	0.1	0.1	<0.1	<0.1	0.3
Oxygen (by difference)	13.6	12.0	9.5	8.4	4.3	20.4
Ash	4.8	3.7	4.7	4.2	3.9	6.5
Volatiles Content	24.9	22.0	14.3	14.2	13.2	nk

The carbon content of the char increases with reaction temperature with a corresponding decrease in volatiles content. This is to be expected with the increase in reaction severity, however the values quoted in Table 2 are considerably higher than those of Scott et al. [7] as shown in Table 2. This may also be due to the intrinsic nature of the feedstock.

4.1.6   Gas composition

The yield of non condensable gases increases at the expense of organic liquid production as the reaction temperature increases, as shown in Table 1. The composition of the gases also changes with increases in the relative amounts of higher hydrocarbons such as methane and ethylene. It is expected that other hydrocarbons are present in very low proportions in the fluidising gas but are not being reported by the GC system due to their very low concentration. Such variations in gas composition correspond with work of Scott and Piskorz [8] who noted that carbon monoxide yields increase more rapidly with reaction temperature than those of carbon dioxide while there is also an increase in the relative yield of hydrogen and higher hydrocarbons. The natural log of the major gases yields were plotted against the natural log of the reactor temperature (K) as shown in Figure 2 to identify trends and the degree of vapour product cracking.

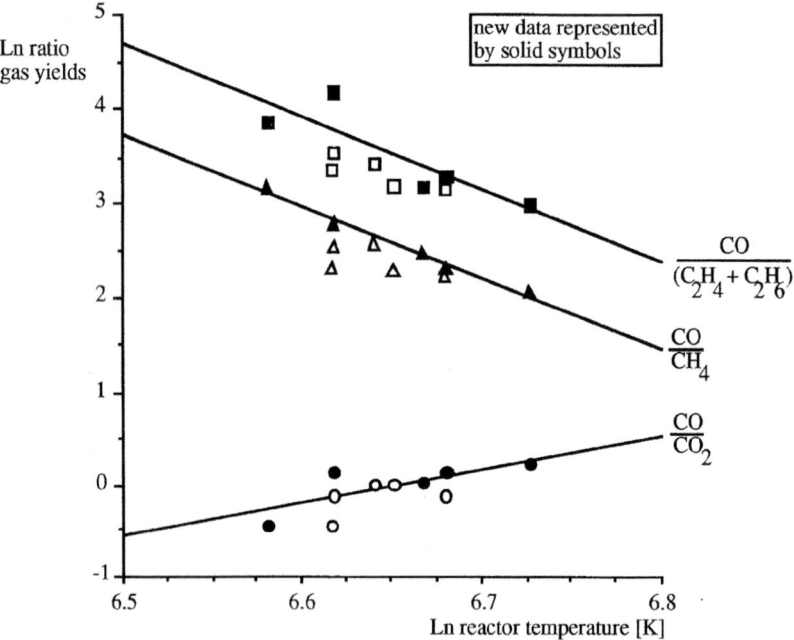

Fig. 2 Variation in gaseous product composition with reaction temperature

These show a good correlation with temperature as indicated, despite the evidence of some product cracking as discussed above. The correlation between CO/CH_4 is very consistent, however, more scatter is noticed for other results from unpublished data which have been included for comparison, also from pine wood. The ratio of CO/C_2+ hydrocarbons is more scattered due to the difficulty in measuring the low gas concentrations at lower reactor temperatures, as they are relatively minor products. It is intended to improve these correlations with further work. Diebold also reported that the ratio of CO/CO_2 in the pyrolysis gases would increase due to vapour cracking [9]. Differences in the gas composition are discussed in Section 5.2.3.

4.2 3-5 kg/h ablative pyrolysis reactor

4.2.1 Instrumentation
A schematic of the reactor and product collection system is shown in Figure 3. The reactor base is heated with 14 compact high wattage cartridge heaters, each fitted with an integral type K thermocouple. The reactor wall is heated with a band heater to 400°C. Two of the thermocouples in the heaters are used by the temperature controller which uses one as the control thermocouple and the other as an overload to prevent the maximum operational temperature of the heaters being exceeded. The controller maintains the reactor temperature to $\pm 1\%$ of the temperature setpoint. The gas pressure in the reactor is also measured. The gas exit temperature from the reactor and the cyclone is measured and also its temperature upon exit from the ice-cooled condensers and the cotton wool filter. The gas volume is measured to 0.1 l and is corrected for temperature and pressure.

4.2.2 Operation
The reactor and product collection system is shown in Figure 3 below. The biomass inlet slot from the feeder has a pressure sensor connected and is fitted with a nitrogen line for purging as shown. Ablation is achieved by means of four asymmetric blades rotating at 160 rpm which press and move the biomass particles relative to the heated surface.

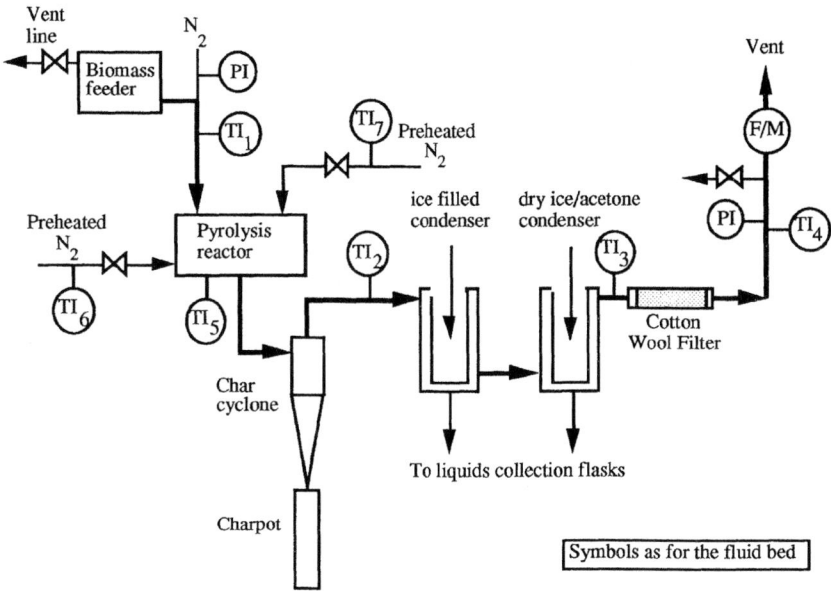

Fig. 3 Schematic of the 3-5 kg/h ablative pyrolysis reactor system

The blade clearance from the heated reactor surface is set at 0.5 mm. Preheated nitrogen is used to control the gas/vapour product temperature and the gas/vapour product residence time and assist in product removal from the reactor. During operation, biomass is rapidly ablated by the rotational action of the blades on the heated base. Fine char particles and vapour products exit through the reactor base where the char is removed in the trace heated cyclone and are collected in the char pot. The hot gas and vapour products enter the ice-cooled condensers-the first one containing ice and the second one a dry ice/acetone mixture. The remaining cooled and stabilised aerosols and non-condensable gases pass into the pre-dried cotton wool filter which removes any residual aerosols and water vapour. The molecular sieve column was not used as prior experience had indicated that water recovery in the column was very low, typically 1-2 g in 45 minutes, due to the dry-ice acetone mixture lowering the gas exit temperature from the condenser to less than 10°C. The non-condensable gases pass through a gas meter and are then vented to atmosphere. Batch gas samples are taken at the gas meter for analysis by GC. Similar procedures for the recording of data, determination of mass balances and the taking of gas samples are made on the ablative pyrolysis reactor as for the fluid bed reactor.

4.2.3 Product recovery
The product collection system is weighed before and after the run and then washed with ethanol to remove liquids from the condensers and the cotton wool. Propanol is not used as subsequent rotary evaporation of the pyrolysis liquids/solvent mixture is easier with ethanol and for subsequent HPLC analysis. The total liquid yield is then the mass difference of the product collection system, including the mass differences in the cotton wool, less the char recovered in a preweighed dried filter paper. Char is removed from the char pot and weighed. The filtered wet char is oven dried to remove the ethanol that was used to recover the liquid products from the collection system. The wood fed to the reactor is determined after the run by re-weighing the wood in the hopper. Typically, the ablatively produced char is defined as char less than the blade clearance from the heated reactor surface, i.e. material less than 0.5 mm in its minimum dimension. The gas yield is calculated from the dry non condensable gas fraction. Gases up to C_4 are analysed for.

4.2.4 Results
Results for four runs from 450 to 600°C with mass balances and product analyses are summarised in Table 3 overleaf. Similar results to that achieved in the fluid bed have been achieved with a maximum organic liquid yield of 62.1 wt% at 502°C. The residence time increases at lower temperatures due to the lower gas/vapour exit temperature, although the effects of gas/vapour phase temperature on product yields at 400-450°C is small. The char and gas fractions were similarly analysed to those for the fluid bed.

4.2.5 Char yields and analyses
Char yields and analyses are summarised in Table 4 overleaf. The CHN composition is relatively consistent, however, the carbon content is lower than previous work, suggesting a lesser severity of reaction and attrition of partially reacted material from the wood chips which is removed before complete reaction. The reaction front during ablation of the particles has been previously estimated to vary from 50-170 μm for reactor temperatures from 600-450°C [6]. The char particles range in size from 3.3 mm to micron size char, including unreacted material. Unreacted wood is removed by first sieving out material greater than 3.3 mm.

To determine the volatiles content of the char and to comply with the ASTM 1762-84, char particles of 250-500 μm were used, which represents about 10-20% of the total yield of ablatively produced char with the remainder less than 250 μm. The volatiles content of these particles is higher than expected for several reasons. Some wood/char will be attrited from the particles during ablation and will be partially reacted. The residence time of the fine particles can vary in the reactor and hence their degree of reaction. It is also probable that the "active" layer formed initially after rapid heating is not further heated after production of the fine particles, which then cools and does not complete depolymerisation and vaporisation from the particles.

Table 3. Ablative pyrolysis mass balances and results (product yields, wt% dry wood basis)

Run Number	CR24	CR25	CR26	CR27
Reactor Base Temperature (TI$_5$, °C)	450	502	555	604
Gas/Vapour Temperature (TI$_2$, °C)	330	343	374	383
Gas/vapour Residence Time (s)	1.53	1.11	0.86	0.80
Final dry gas exit temperature (TI$_4$, °C)	25.1	15.2	27.9	20.9
N$_2$/biomass flowrate (kg/kg)	4.06	3.08	0.91	0.52
Mass Balance (wt%, dry basis)				
Char	16.9	13.0	19.3	16.9
Organic Liquid	50.8	62.1	51.6	56.8
Water	15.8	12.3	12.5	11.4
Gases (N$_2$ free)	12.5	11.2	15.1	11.4
Closure	96.0	98.6	98.4	96.5
Gas yields (wt% dry basis)				
CO	4.11	3.30	5.68	4.05
CO$_2$	8.06	6.60	8.80	6.81
CH$_4$	0.27	0.80	0.48	0.35
C$_2$H$_4$	0.03	0.40		0.11
C$_3$H$_8$	0.03	0.12	0.09	0.06
H$_2$	nd	nd	nd	0.01
Gas composition, (vol. % nitrogen free basis)				
CO	42.1	35.2	46.0	43.3
CO$_2$	52.5	44.8	45.4	46.4
CH$_4$	4.9	15.0	6.9	6.5
C$_2$H$_4$	0.3	4.1		1.1
C$_3$H$_8$	0.0	0.9	0.5	0.4
H$_2$	nd	nd	nd	2.1

Table 4. Ablative Pyrolysis reactor char analyses (wt%, dry basis). Comparison with NREL data [10]

Run Number	CR24	CR25	CR26	CR27	CR10	NREL
Reactor Temperature (°C)	450	502	555	604	602	625
Carbon	74.4	74.6	75.9	76.2	82.5	81.8
Hydrogen	4.0	3.9	3.8	3.7	3.1	3.7
Nitrogen	0.1	--	0.1	0.1	0.1	0.0
Oxygen (by difference)	20.5	20.0	18.2	18.8	nk	11.5
Ash	1.1	1.4	1.4	1.2	nk	2.8
Volatiles Content	33.5	34.6	32.7	28.1	nk	nk

4.2.6 Gas analysis

The product gases were analysed by GC and the results given in Table 3 above. These results were then similarly plotted to the fluid bed results and are shown in Figure 4.

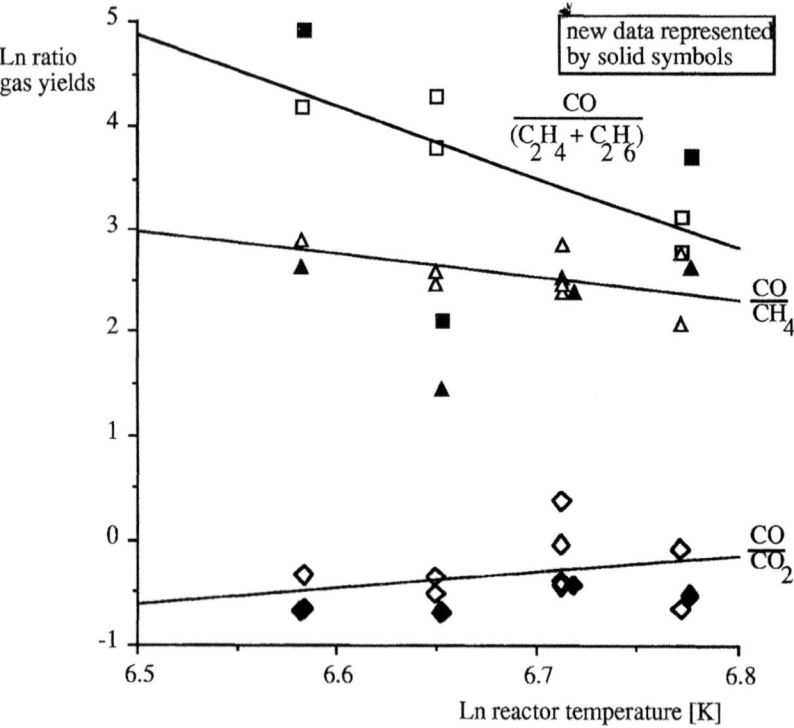

Fig. 4 Ratio of the product gas yields with reactor temperature: comparison with prior data [6]

The more recent data, shown using solid symbols in Figure 4, shows some significant differences. The CH_4 yield for CR25 is inconsistent with other data and appears to be due to a non-representative gas sample. Due to the shorter gas/vapour product residence times, the yields of higher hydrocarbons is lower, however the yield of CO has not increased, when compared with prior data [6], suggesting a low level of vapour cracking. The degree of secondary vapour phase reactions which gives rise to hydrocarbons has been reduced with the decrease in residence time to around 1 s. These results are comparable to those obtained by NREL [9]. It would appear that higher hydrocarbons (C_2+) and to a lesser extent methane are primary products, formed in the initial depolymerisation of the wood and are not formed extensively by secondary reaction of the liquid layer on the hot reactor surface.

5   Results comparison

5.1 Reactor parameters

In order to compare the results between the two reactors, their differences need to be noted in order to appreciate that a direct comparison to account for fundamental differences in the reactor systems is difficult.

The key differences in the two reactors are: degree of gas/vapour product dilution, biomass moisture content, particle size and shape, method of heat transfer, pyrolysis temperature, gas/vapour product residence time and temperature, and the biomass heating rate. It is not intended to discuss these in detail in this paper, only to highlight that there are intrinsic features in each system. Related process parameters are summarised in Table 5.

Table 5. Reactor comparison: process parameters

	Ablative	Fluid Bed
gas/vapour product dilution [#]	2.2-13.1	13.5-15.8
mass ratio N_2/biomass fed	0.5-4.1	6.2-8.6
biomass moisture content (wt%, dry basis)	0.36	9.25
particle size (mm) and shape	1 mm thick, 3.35 long	0.6-1.7, "cubes"
method of heat transfer	conduction	~90% conduction
pyrolysis temperature (°C)	450-600	450-560°C
gas/vapour product residence time (s)	0.80-1.53	1.06-1.29
gas/vapour product temperature (°C)	330-383	369-393
biomass heating rate (K/s)	not determined	not determined

[#] mass ratio of inert gas to pyrolysis products at reactor exit temperature

5.1.1   Product dilution
In the ablative pyrolysis reactor, the vapour products are diluted with some nitrogen to control the residence time and moderate the gas/vapour phase temperature. Higher gas/vapour dilution is required at low reactor temperatures as the reactor throughput is very low (~300 g/h) as this is at the lower temperature limit for effective ablative fast pyrolysis. In the fluid bed, the products are considerably more diluted. These values are similarly reflected in the mass ratio of nitrogen used in the two reactors.

5.1.2 Biomass moisture and particle size and shape

The particle sizes used and their shapes are suitable for the optimal production of liquids and are not expected to significantly influence the product yields for the conditions studied. The water content of the feed will affect the degree of gas/vapour product dilution and it is assumed that at moisture contents of less than 15 wt%, although the of biomass heating rate and product yields are affected as the heat requirement is increased.

5.1.3 Mode of heat transfer

In both reactors, the principal mode of heat transfer is by conduction, however, in the ablative reactor, this is done under high applied pressure between the biomass and the heated reactor base. Primary products are mainly deposited on the surface as a "melt" which then subsequently decomposes and gives rise to volatile products. The original top of the particle is eventually thinner than the clearance between the blade and the base which leads to the formation of char/partially reacted material which can be further heated before being carried out of the reactor.

In the fluid bed reactor, the products are directly ejected and released into the vapour phase. There are therefore different modes of vapour formation in the reactors which can lead to differences in the product composition and to a lesser extent yields.

5.1.4 Pyrolysis temperature

The pyrolysis temperature has been the source of continued debate over the past two years leading to discussion on the formation of products and the mode of volatiles formation . It has become apparent that biomass typically degrades below the reactor temperature, probably in the temperature range of 400-470°C, depending on the reactor temperature, rate of heating at the biomass reaction interface and mode of heat transfer. In the ablative pyrolysis reactor, after primary formation of the "active" layer, it continues to decompose and pyrolyse on the reactor surface, although there is some direct ejection of aerosols and vapours into the gas phase. A further paper on the nature of pyrolysis vapours generation has been submitted to this conference [11].

5.1.5 Gas/vapour product residence time and temperature

In both reactors, these parameters are fairly similar. The ablative pyrolysis reactor residence time is a function of the feedrate, reactor temperature and the degree of diluting nitrogen used, which is generally kept constant. The vapour phase temperature results for the fluid bed were lower than anticipated, possibly due to endothermic vapour forming freeboard reactions. This may be the source of differences in the gas product yields between the reactors in conjunction with the initial formation of volatiles. The gas/vapour product temperature is low enough to reduce secondary vapour phase decomposition reactions to a low level.

5.2 Mass balances

The product yields for FB26 and CR25 are the most similar, although the gas compositions are quite different due to the erroneous result for CR25. Despite the fact that a match in terms of even the four key reactor parameters of reactor temperature, gas/vapour phase residence time and temperature and heating rate was difficult to achieve, good comparable quantitative results have been obtained.

5.2.1 Char yields and analysis

The elemental analyses of the fluid bed char showed a much higher proportion of carbon, compared to other work and the ablative pyrolysis results. In the fluid bed, as previously noted, was due to the in-bed heaters causing some degree of coking at higher temperatures, which was evidenced in the wide range of CHN analyses obtained for the char samples. Ablatively produced char is high in volatiles, due to abrasion of some material from the biomass particles before complete reaction. The mechanism for the formation of these chars is very different and further analyses of the ablative pyrolysis char less than 250 μm is required.

5.2.2 Liquids analyses
Apart from Karl Fischer titration to determine the water contents of the various recovered liquid fractions from both reactor systems results of liquid analyses are not available for this paper.

5.2.3 Gas analysis
In looking at the gas yields for the ablative pyrolysis reactor, there is no clear trend of increasing gas yields with reactor temperature. It appears that the vapours are rapidly quenched after a secondary gas yield of 5-11 wt%. In the fluid bed reactor, however, the gas yields are more sensitive to the reactor temperature, an indication of the extent of vapour cracking. The lower gas yields in the ablative pyrolysis reactor suggests that the vapours "see" the reactor base temperature for a relatively short time interval-considerably shorter than the gas/vapour product residence time.

Gas yields for the reactors are compared in Figure 5. The use of ratios of gas yields allows a comparison to be made between different reactors to assess the degree of product cracking and other secondary effects. It is immediately apparent that gas yields in the fluid bed are much more sensitive to temperature than those of the ablative reactor. This suggests that secondary vapour cracking occurs to the fluid bed vapours to a greater extent as a consequence of the product vapours being exposed to higher temperatures for longer. Modifications are currently underway to measure the freeboard temperature directly, currently the gas/vapour product temperature is assumed to be that measured at the cyclone (T_3).

The vapours from the ablative reactor may be more rapidly quenched than the calculated vapour residence time due to the path taken by the vapours through the reactor. The preheated nitrogen purge to the reactor will tend to entrain product vapours from the reactor resulting in a reduced residence time and reducing the tendency to secondary vapour cracking. This hypothesis is supported by referring to Figures 2 and 4. The fluid bed shows a consistent trend for the hydrocarbons with CO at increasing reactor temperature, but in the ablative pyrolysis reactor, the ratio of CO/CH_4 is more constant, also the yield of higher hydrocarbons is lower in the ablative pyrolysis. CO formation increases more rapidly in the fluid bed reactor, with a corresponding decrease in CO_2, which occurs at a lower rate in the ablative pyrolysis reactor.

The lack of significant hydrocarbon formation in the ablative reactor combined with the relatively constant quantities of CO suggests a gas forming reaction with little sensitivity to temperature which implies that secondary vapour cracking occurs at an insignificant level. The opposite occurs in the fluid bed with the production of CO and hydrocarbons at the expense of carbon dioxide implying a significant secondary vapour cracking effect. Despite slightly longer calculated residence times in the ablative pyrolysis reactor, the formation of hydrocarbons is not significant suggesting that the primary products, formed as a "melt" are not subsequently cracked to permanent gases but depolymerise and vaporise to give condensable organic liquids.

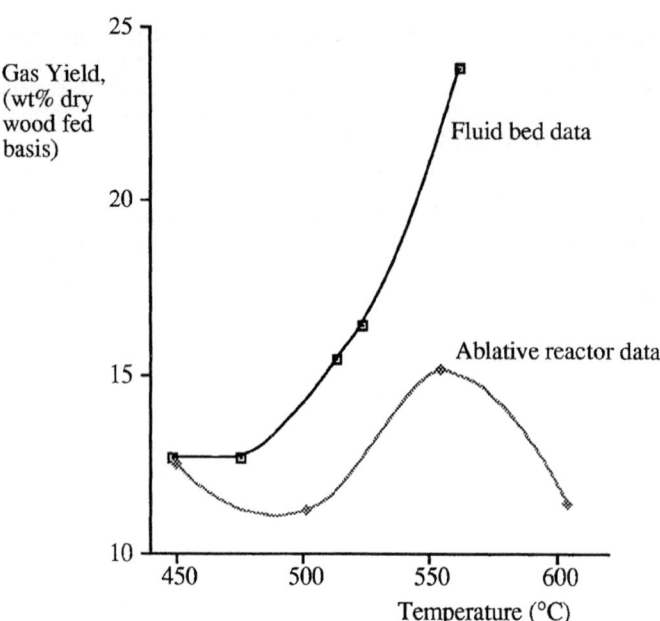

Fig. 5 Comparison of gas ratios for the ablative and fluid bed reactors: recent results

6   Conclusions

The preliminary data from the fluid bed which has been compared with the ablative pyrolysis reactor under similar conditions has yielded several results.   Similar mass balances can be obtained under very similar reactor parameters of reactor temperature, gas/vapour product residence time and temperature and biomass heating rate.

Differences in char products are due to the method of supplying heat to the biomass. In the ablative reactor, ablation leads to the formation of finely divided product char and attrition of the particles which also generates char fines.   The fluidised bed char also supplies heat by contact, but there is no applied pressure and the generation of fines is significantly less. Gas differences are due to the secondary formation of primarily CO and higher hydrocarbons in the fluid bed by secondary reaction of the volatiles in the biomass matrix and additional heat supply from the in-bed heaters causing additional vapour cracking to char and non condensable gases.   Also, vapours are more rapidly quenched from the ablative reactor due to the vapour path through the reactor leading to shorter than calculated residence times. Differences in the recent results presented here and prior data are due to residence time differences and lower mass balance closures.

7   Recommendations

Further data is required to confirm some of the possible reasons suggested for the results. Further experiments are planned and the liquid products are being extensively analysed by GC-MS and HPLC.

8   Acknowledgements

The authors wish to acknowledge the European Community AIR programme under contract no AIR2-CT93-0889 for providing the funding which has enabled this work to be carried out.

9   References

1. Scott, D.S. and Piskorz, J., Production of Liquids from Biomass by Continuous Flash Pyrolysis, *BioEnergy 84*, (eds. Egneus, H. and Ellegard, A.), Vol. 3, Elsevier Applied Science Publishers, London, 1985, p 15-23.
2. Scott, D.S., Piskorz, J. and Radlein, D., Liquid Products from the Continuous Flash Pyrolysis of Biomass, Ind. Eng. Chem. Process Des. Dev., Vol. 24, No. 3, 1985, p 581-588.
3. Scott, D.S., Piskorz, J., Radlein, D. and Czernik, S., Conversion of Lignocellulosics to Sugars by Thermal Fast Pyrolysis, *Proceedings of the 7th Canadian Bioenergy Research and Development Seminar*, 1989, Ottawa, Canada, pp. 713-718.
4. Diebold, J.P. and Power, A.J., Engineering Aspects of the Vortex Pyrolysis Reactor to Produce Primary Pyrolysis Oil Vapours for Use in Resins and Adhesives, *Research in Thermochemical Biomass Conversion*, (eds. Bridgwater, A.V. and Kuester, J.L.), Elsevier Applied Science Publishers, London and New York, 1988, pp. 609-628.
5. Diebold, J.P. and Scahill J., Production of Primary Pyrolysis Oils in a Vortex Reactor, *Pyrolysis Oils from Biomass, Producing Analyzing and Upgrading*, (eds. Soltes, Ed. J. and Milne, T.A.), ACS Symposium series 376, 1988, pp. 31-40.
6. Peacocke, G.V.C., Ablative Pyrolysis of Biomass, Ph.D. thesis, Department of Chemical Engineering, Aston University, October 1994, available from the British Library.
7 . Scott, D.S. and Piskorz, J., The Continuous Flash Pyrolysis of Biomass, *Can. J. Chem. Eng.* 1984, 62, pp. 404-412.
8   Scott, D.S. and Piskorz, J.,Flash Pyrolysis of Biomass, *Fuels from Biomass and Wastes*, (Eds. Klass, D.L. and Emert, G.H.) Ann Arbor Science Publishers; pp. 421-434 1981.
9. Diebold, J.P., The Cracking Kinetics of Depolymerized Biomass Vapors in a Continuous Tubular Reactor, MSc Thesis, Colorado School of Mines, Golden, CO, USA, 1985.
10. Gregoire, C.E. and Bain, R.L., Technoeconomic Analysis of the production of biocrude from Wood, *Biomass and Bioenergy*, Vol 7, No. 1-6, 1995, pp. 275-283.
11. Lede, J. Diebold, J.P., Peacocke, G.V.C. and Piskorz, J., The nature and properties of intermediate and unvaporized biomass pyrolysis materials, *Developments in Thermochemical Biomass Conversion*, Banff, May 1996, in press.

THE IMPACT OF WOOD PRESERVATIVES ON THE FLASH PYROLYSIS OF BIOMASS

Pyrolysis of wood and preservatives

S. WEHLTE, D. MEIER, J. MOLTRAN and O. FAIX
Institute for Wood Chemistry and Chemical Technology of Wood,
Federal Research Centre for Forestry and Forest Products, Hamburg, Germany

Abstract

The effect of wood preservatives on the amount and composition of pyrolysis products were investigated over a wide range of temperatures (300 to 700 °C). The experiments were carried out in a modified bench scale unit according to the Waterloo-Process. Of particular interest was the distribution of heavy metals in the different pyrolysis products.

Under strict German environmental laws, the combustion of wood waste in incinerators is restricted to those with gas cleaning systems which comply to the 17th Ordinance on the Implementation of the Federal Imission Control Act (17. BImSchV), because of the various wood preservatives, insecticides or coatings present in the wood. In this situation it is essential to find new and cheaper ways for wood waste management. After evaluating a series of experiments flash pyrolysis seems to be a promising approach of wood waste managing and the production of liquid fuel. Treated wood does not have a negative effect on the pyrolysis products. The amount of liquid remains high and the heavy metals from the preservatives accumulate in the char, not in the liquid. Furthermore, boron has a significant effect on the amount of levoglucosan, hydroxyacetone and hydroxyacetaldehyde. The levoglucosan yield in the pyrolysis liquid from treated wood increased whilst the hydroxyacetone and the hydroxyacetaldehyde yields decreased. The presence of creosote during the pyrolysis did not affect the products from the wood and the creosote did not change either.

Keywords: flash pyrolysis, heavy metals, wood waste management, wood preservatives

1 Introduction

The disposal of wood waste is a growing problem in Germany. The total amount of wood waste is around 3-3.5 million t/y. They contain various wood preservatives, insecticides or coatings, which may comprise heavy metals like Cr (C), Cu (C), As (A), Hg or and toxic substances like boron (B), fluorine (F), pentachlorophenol and lindane (Table 1).

Table 1. Hazardous potential and characteristics of several wood waste assortments in the former Federal Republic of Germany (West Germany) [1]

assortment	potential active ingredients	retention requirement	quantity of wood waste (t/y)
cross-ties	creosote, inorg. Salts	45-175 kg/m^3 ?	ca. 60,000-85,000
poles	CCB, CCF, CCA, creosote, HgCl$_2$	6-10 kg/m^3 90 kg/m^3 0.6-1.0 kg/m^3	ca. 15,000-25,000
timber from landscape	CCB, CCF, CCA Cu-HDO-salts creosote carbolineum chloronaphthalene LOS-preservatives	6-8 kg/m^3 3-4 kg/m^3 80 kg/m^3 250-400 g/m^2 250-350 g/m^2 200-280 g/m^2	ca. 220,000
hop-poles	CCB, CCF, CCA, creosote, HgCl$_2$	ca. 6-8 kg/m^3 ca. 90 kg/m^3 ca. 0.4-0.8 kg/m^3	current stock: 15,000-27,000
vineyard-posts	CCF, CCA, CCB creosote, HgCl$_2$ CFA	5-6 kg/m^3 ca. 10 kg/m^3 60-110 kg/m^3 0.6-1.0 kg/m^3 5-6 kg/m^3	ca. 9,000-14,000
cooling towers	CCB, CCF CC, CCA creosote	ca. 15 kg/m^3 ca. 12-13 kg/m^3 60-65 kg/m^3	ca. 5,500-8,300
wooden silos	CCB LOS-preservatives creosote	200-280 g/m^2	only small amounts expected
wood from demolition of buildings; building sites	all expect: creosote, chloronaphthalene and HgCl$_2$	no specification possible	ca. 500,000-2,000,000
bulky refuse	no specification possible	-:-	ca. 430,000
wood for packaging	seldom	-:-	ca. 470,000-970,000
cable drums	CCB, CCF, CC, (CCA)	6-8 kg/m^3	31,000-45,000

Under strict German environmental laws, the combustion of wood waste in incinerators is restricted to those with gas cleaning systems which comply to the 17. BImSchV.

The cost for landfill or burning wood waste in certified incinerators (according to the Federal Imissions Control Act, BImSchG) is generally between 200 to 700 US$ per tonne and in some cases up to 3000 US$. In this situation it is economical to find new and cheaper ways for the management of wood waste.

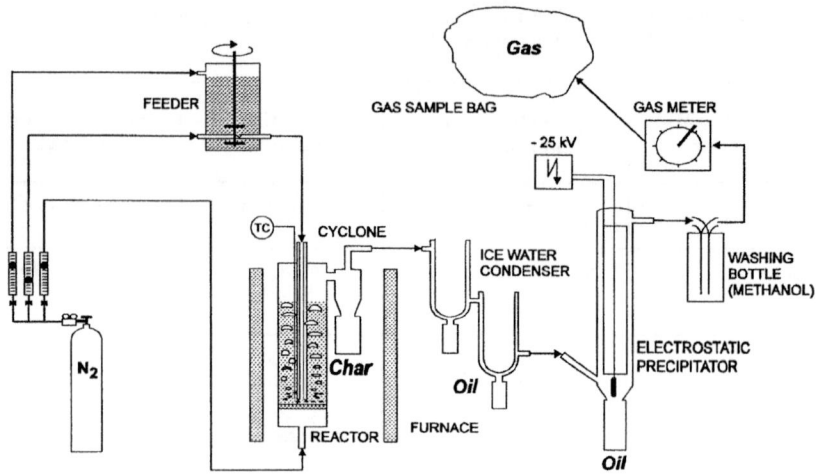

Fig. 1. Modified bench scale unit according to the Waterloo Flash Pyrolysis Process

Flash pyrolysis is a promising way. The heavy metals, which cause the main problems during incineration, remain to a large extent in the char. Compared to pyrolysis, incineration produces 5-20 times the gas which requires cleaning. Furthermore, the pyrolysis liquid may be used as a fuel or a source of chemicals.

In a research project funded by the German Foundation for Environment the utilization of waste wood for flash pyrolysis in a fluidized bed reactor is being studied.

Initial results from the flash pyrolysis of wood treated with CCB (chromium, copper, boron), which is the main wood preservative in Germany, and wood treated with creosote are presented. The runs were carried out in a modified bench scale unit (50-100 g/h) according to the Waterloo Flash Pyrolysis Process (Fig. 1) [2].

2 Experimental

2.1 Material

Wood: commercial wood meal (mixed softwood, beech)
Particle size: 0.3 - 0.5 mm
Moisture: ca. 12 % (dry basis)
Wood preservatives: CCB salt, creosote
Metal content (CCB salt):   Cu 2.18 mg/g, Cr 3.91 mg/g
Boron content (CCB salt):   high: 0.05 wt.%, low: < 0.01 wt.%

2.2 Reaction conditions

Temperature: 350 - 600 °C   Feed rate   40 - 50 g/h
Reaction pressure: atmospheric   Fluidization gas: N_2
Vapour residence time: ca. 1 s   Heat carrier: sea sand, 0.25-0.30 mm

2.3 Analytical methods

2.3.1 Pyrolysis liquids
- GC/FID: CHROMPACK (NL) CP 9000, detector: 280 °C, injector: 250 °C, carrier gas: He, pressure: 2 bar, oven program: 45 °C 4 min isothermal, 3 °C/min to 280 °C, 15 min at 280 °C, column: DB 1701 (J&W) 60 m x 0.25 mm, film thickness 0.25 μm, split ratio: 1:35, sample preparation: 30 mg liquid/1 ml MeOH + IS
- GC/MS: Hewlett Packard MSD 6890, carrier gas: He, oven program: 45 °C 4 min isothermal, 3 °C/min to 260 °C, 15 min at 260 °C, column: CP Sil 19 CB (Chrompack) 60 m x 0.25 mm, film thickness 0.25 μm, split ratio: 1:30, const. flow 1 ml/min

2.3.2 Pyrolysis gases
- CO, CO_2
  GC: CHROMPACK (NL) Micro-GC CP 2002, detector: TCD
  CO: oven: 50 °C isothermal, injection time: 250 ms, column: Molsieve 5Å, 4 m
  CO_2: oven: 70 °C isothermal, injection time: 250 ms, column: Hayesep A, 0.25 m
- Hydrocarbons
  GC: CHROMPACK (NL) Model 439 A, detector: FID at 250 °C, injector: 70 °C, oven program: 60 °C 10 min isothermal, 13 °C/min to 200 °C, column: PLOT (Al_2O_3+ KCl) 50 m x 0.32 mm, split ratio: 1:40

2.3.3 Quantification
- liquids: Internal standard method with a single calibration of a mixture of 24 thermal degradation products and fluoranthene as internal standard (Fig. 7 and Table 3). All compounds are proved with GC/MS and quantified with GC/FID [3, 4, 5, 6, 7].
- gases: External standard method with single point calibration of a mixture of 8 gaseous pyrolysis products.

2.3.4 Metals
Atomic absorption spectrometer: Perkin Elmer AAS 1100 B,
wavelength: Cr 357.9 nm, Cu 324.8 nm, spectral bandwidth: Cr 0.7 nm Cu 0.2 nm, flame: air/acetylene
Sample preparation: incineration at 440 °C for 12 h, digestion in a 4:1-mixture of H_2SO_4 (95-97 %) and $HClO_4$ (70 %) at 110 °C for 2 - 4 h under reflux.

2.4 Procedure
All parts of the pyrolysis unit were weighed at the beginning and the end of each experiment. During the run the gas was collected in a gas sample bag and analyzed later. After the run the whole unit was washed with water-free methanol. The solution of methanol and pyrolysis liquid was filtered through a paper filter that was subsequently Soxhlet-extracted with methanol. The extract was added to the solution and the methanol was evaporated at 45 °C and 150 mbar. The washing bottle (filled with water-free methanol) was analyzed for water (Karl Fischer titration) to complete the mass balance.

3 Results and Discussion

All yields presented are measured values and not calculated by difference. The results from the liquid analyses are presented on water-free and the gas compositions are on dry wood basis, whilst the product yields are given on a wet wood basis. The mass balance closure for all runs was between 90-95 %.

3.1 CCB-treated wood

The pyrolysis of softwood treated with CCB could be performed without any technical problems. The yields of the pyrolysis products are similar to those from untreated wood. Even in the presence of a high boron content there is only a small decrease in the amount of gas and a small increase of the char yield (Fig. 2). This was expected as boron salts are also used as flame retardants. The reduction of the liquid yield in combination with an increase of the char yield is a typical effect of flame retardants on cellulose [8]. Furthermore, Richards and Zheng [9] reported on similar effects due to the presence of Cu in wood.

The amount of liquid is high over a wide range of temperatures, with a maximum at 475 °C. Gas composition from untreated and CCB treated wood remained approximately the same (Fig. 3). The CO yield for untreated and treated wood showed similar trends for increase in temperature, however, CO yield with untreated wood increased from 425 °C whereas CO yield with treated wood remained relatively constant until 450 °C when it increased greatly with temperature.

The qualitative composition of the GC-detectable compounds is identical, but the quantification shows a significant difference between untreated and treated wood. While the lignin derived thermal degradation products remain practically unchanged (Fig. 5), the carbohydrate derived thermal products changed considerably.

The levoglucosan yield in the pyrolysis liquid from the CCB treated softwood with high boron content increased by a factor of around six compared to the untreated softwood (Fig. 6).

The hydroxyacetaldehyde yield in untreated and treated wood showed different trends, for example: untreated wood hydroxyacetaldehyde yield drops from 21 % at 400 °C to 4.5 % at 500 °C whilst treated wood remains at around 1.5 % between 400 °C and 550 °C. No data are available for untreated wood above 500 °C due to experimental problems.

The amount of hydroxyacetone in untreated and treated wood showed the same trend but in different quantities. The amount in untreated wood remains at 6 % whereas the amount in treated wood remains at 1 % in the temperature range between 375 °C and 550°C. CCB has no influence on the yield of acetic acid.

The amount of hydroxyacetone and hydroxyacetaldehyde represent between 30 and 70 % of the quantified liquid from untreated wood, but only around 10 % of the liquid from treated wood.

The elemental analysis shows a small decrease in the amount of carbon in the liquids from CCB treated wood (Table 2).

Of particular interest was the distribution of heavy metals in the different pyrolysis products. It was possible to recover the metals to a very high extent (Fig. 4). They are

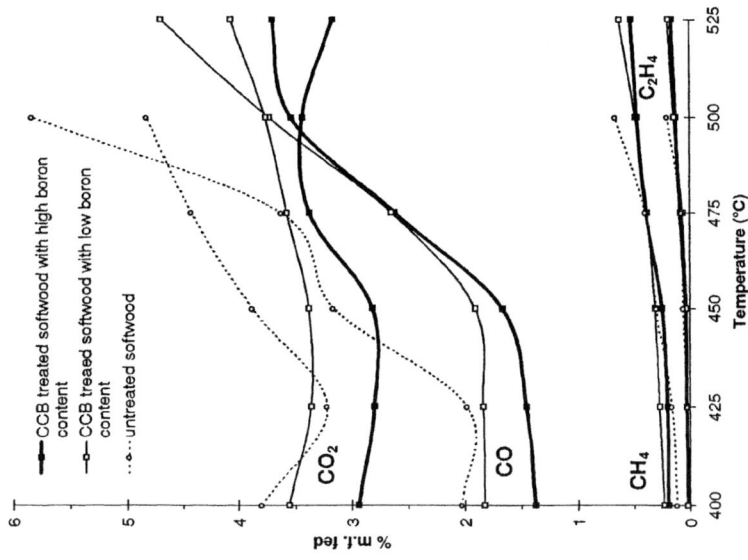

Fig. 3. Gas composition of untreated and CCB treated softwood

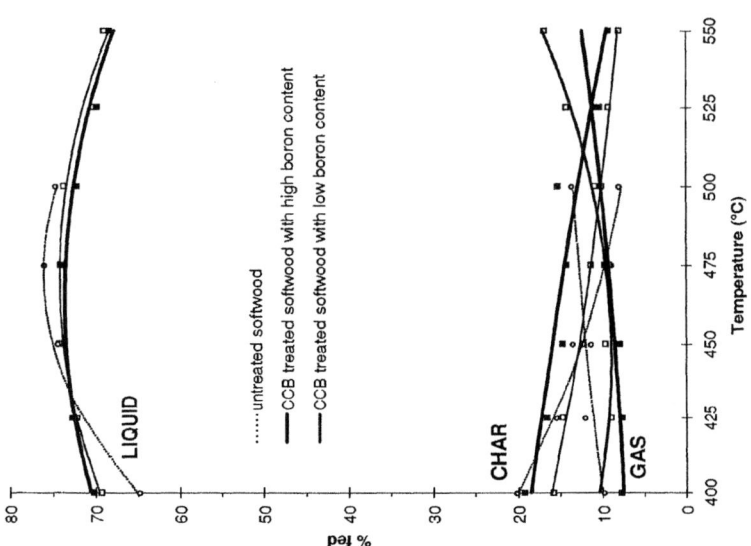

Fig. 2. Product yields of untreated and CCB treated softwood

211

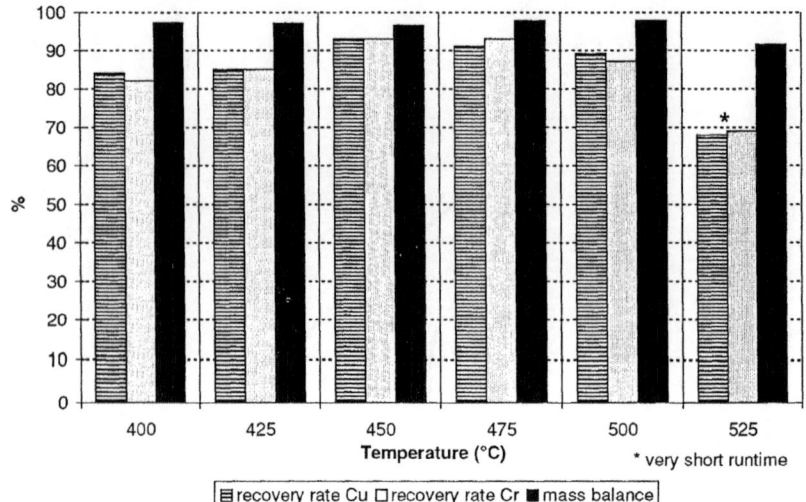

Fig. 4. Recovery rates and mass balances of CCB (high boron content) treated softwood

accumulated in the char. A small amount of metals remain in the thin char layer that coated the sand in the fluidized bed.

The amount of char on the sand shows a minimum at 450 °C. Visual inspection revealed that below 450 °C no char layer was formed (Fig. 8). The char detected below this temperature is due to char particles between the sand which are usually carried out of the reactor. Neither Cu or Cr were detected in the pyrolysis liquid.

Table 2. Elemental analyses of pyrolysis liquids from untreated and CCB treated softwood

	wt. % C		wt. % H		wt. % O*		wt. % N	
	treated	untreated	treated	untreated	treated	untreated	treated	untreated
425 °C	49.3	n.d.	7.2	n.d.	43.4	n.d.	0.2	n.d.
450 °C	49.6	51.1	7.1	7.2	43.1	41.5	0.2	0.2
475 °C	50.2	50.6	7.0	7.1	42.8	42.1	0.2	0.2
500 °C	49.6	51.5	7.2	7.2	43.1	41.1	0.2	0.2
525 °C	50.9	n.r.	7.2	n.r.	41.8	n.r.	0.2	n.r.

* by difference, n.d.: not determined, n.r.: no run available

212

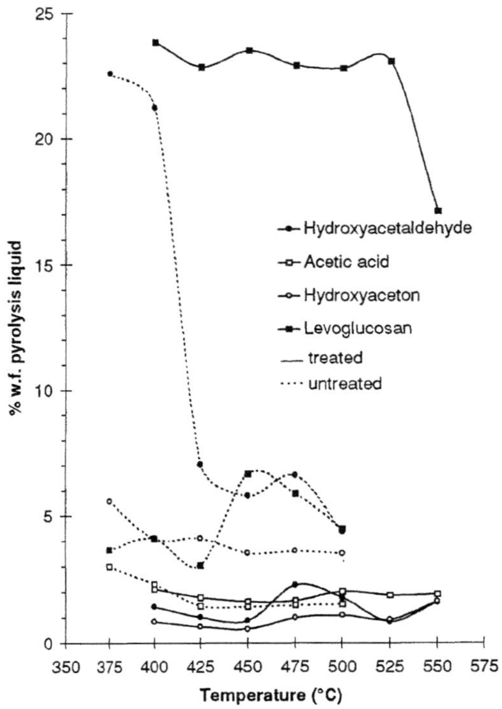

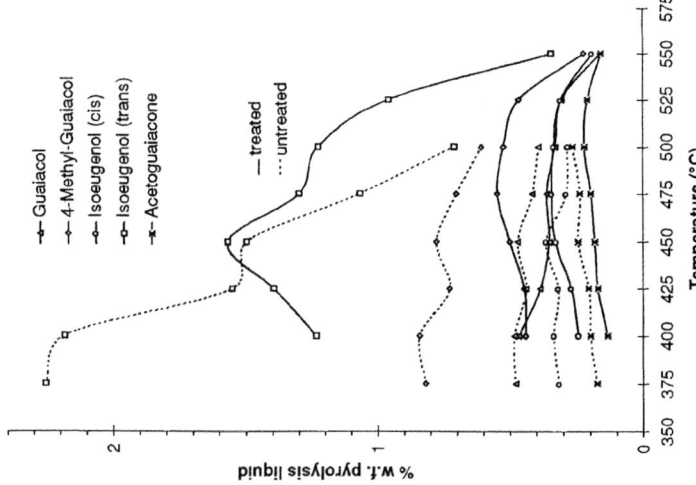

Fig. 6. Amount of some lignin derived products in pyrolysis liquids from untreated and CCB treated softwood

Fig. 5. Amount of some carbohydrate derived products in pyrolysis liquids from untreated and CCB treated softwood

Table 3. Yields of liquid compounds in pyrolysis liquids from untreated and CCB treated (high boron content) softwood (wt. % based on water-free liquid)

No.	Compound	400 °C		425 °C		450 °C		475 °C		500 °C	
		CCB	untreated	CCB	untreated	CCB	untreated	CCB	untreated	CCB	untreated
1	Hydroxyacetaldehyde	1.43	21.21	1.02	7.04	0.89	5.80	2.28	6.62	1.78	4.37
2	Acetic acid	2.12	2.32	1.80	1.47	1.63	1.44	1.67	1.50	2.03	1.54
3	Hydroxyacetone	0.84	4.11	0.65	4.12	0.56	3.54	1.01	3.61	1.10	3.52
4	2,5-Dimethoxy-tetrahydrofuran (cis)	0.04	0.27	0.05	0.38	0.06	0.41	0.10	0.31	0.14	0.37
5	Furfural-(2)	0.43	0.52	0.37	0.50	0.35	0.54	0.39	0.44	0.44	0.61
6	Furfurylalcohol-(2)	0.01	0.24	0.02	0.08	0.02	0.05	0.03	0.04	0.03	0.03
7	Furan-(5H)-2-one	0.14	0.69	0.13	0.59	0.15	0.62	0.23	0.55	0.26	0.55
8	2-Hydroxy-1-methyl-cyclopenten-(1)-3-one	0.19	0.29	0.18	0.19	0.16	0.21	0.15	0.17	0.14	0.18
9	Phenol	0.08	0.02	0.09	0.03	0.10	0.05	0.08	0.06	0.09	0.07
10	Guaiacol	0.47	0.48	0.38	0.44	0.35	0.47	0.36	0.41	0.33	0.39
11	o-Cresol	0.05	0.02	0.05	0.02	0.05	0.04	0.04	0.03	0.05	0.05
12	m-Cresol	0.02	0.01	0.06	0.02	0.06	0.02	0.01	0.02	0.05	0.05
13	4-Methyl-guaiacol	0.44	0.84	0.45	0.73	0.50	0.78	0.55	0.70	0.52	0.61
14	2,4-Dimethyl-phenol	0.05	0.03	0.04	0.02	0.04	0.04	0.04	0.04	0.04	0.05
15	4-Ethyl-guaiacol	0.12	0.12	0.14	0.15	0.08	0.11	0.13	0.10	0.08	0.11
16	Vinyl-guaiacol	0.19	0.41	0.24	0.09	0.26	0.11	0.22	0.09	0.29	0.08
17	Eugenol	0.20	0.29	0.22	0.27	0.26	0.28	0.27	0.25	0.26	0.23
18	5-(Hydroxy-methyl)-furaldehyde-(2)	0.40	0.73	0.37	0.38	0.31	0.46	0.29	0.42	0.32	0.41
19	Isoeugenol (cis)	0.24	0.33	0.27	0.32	0.33	0.36	0.34	0.29	0.33	0.28
20	Isoeugenol (trans)	1.23	2.18	1.39	1.55	1.57	1.50	1.30	1.07	1.23	0.71
21	Vanillin	0.23	0.20	0.26	0.24	0.26	0.26	0.23	0.24	0.29	0.28
22	Homovanillin	0.10	0.16	0.11	0.10	0.13	0.09	0.09	0.10	0.16	0.10
23	Acetoguaiacol	0.13	0.20	0.17	0.20	0.18	0.24	0.20	0.24	0.22	0.26
24	Levoglucosan	23.82	4.10	22.84	3.06	23.48	6.68	22.88	5.89	22.77	4.50
	total	32.97	39.77	31.30	21.98	31.78	24.09	32.87	23.17	32.96	19.35

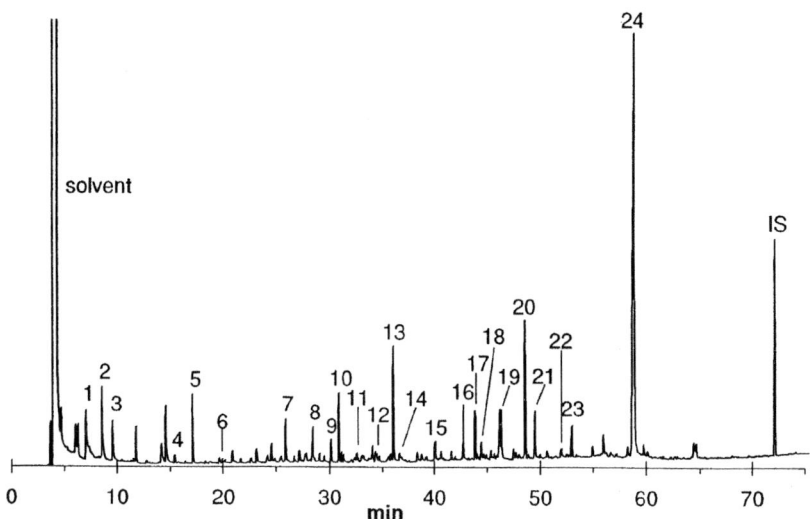

Fig. 7. Gas chromatogram of pyrolysis liquid from CCB treated (high boron content) softwood (numbers see Table 3)

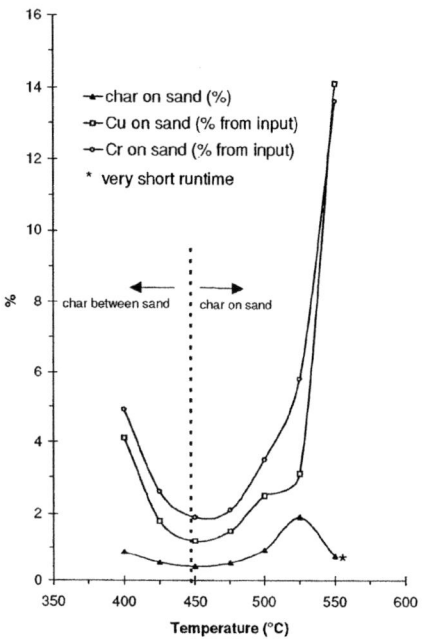

Fig. 8. Amount of char and heavy metals on/between the sand from the pyrolysis of CCB treated softwood

3.2 Creosote treated wood

The creosote treated hardwood pyrolysed well without any difficulties over a wide range of temperatures (375 to 700 °C). The yield of pyrolysis products from untreated and treated beech are practically the same. There is slightly more liquid in the treated wood pyrolysis products compared to untreated wood. This is due to the high amount of creosote (10-15 wt.% of feedstock) in the treated beech (Fig. 10).

From the GC chromatograms of creosote, untreated and treated wood (Fig. 9) it can be seen that the creosote has no real effect on the pyrolysis of wood. The treated wood chromatogram is the sum of the creosote and untreated wood chromatogram. The creosote is the cause of the higher carbon content in the treated wood pyrolysis liquid compared to the untreated (Table 4). The gas fraction showed no difference in compositions (Fig. 11).

Table 4. Elemental analysis of creosote and pyrolysis liquids obtained from untreated and creosote treated beech (pyrolysis temperature: 475 °C)

	pyrolysis liquid untreated beech	pyrolysis liquid creosote treated beech	creosote
wt. % C	50.6	62.0	89.1
wt. % H	7.3	6.7	6.3
wt. % O	42.1	30.6	3.6
wt. % N	0.2	0.7	1.1

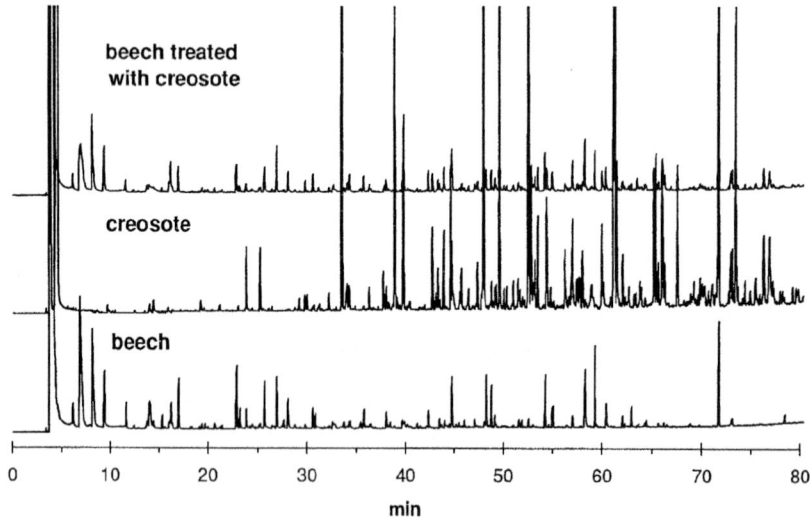

Fig. 9. GC chromatograms of creosote and pyrolysis liquids obtained from untreated and creosote treated beech (pyrolysis temperature: 475 °C)

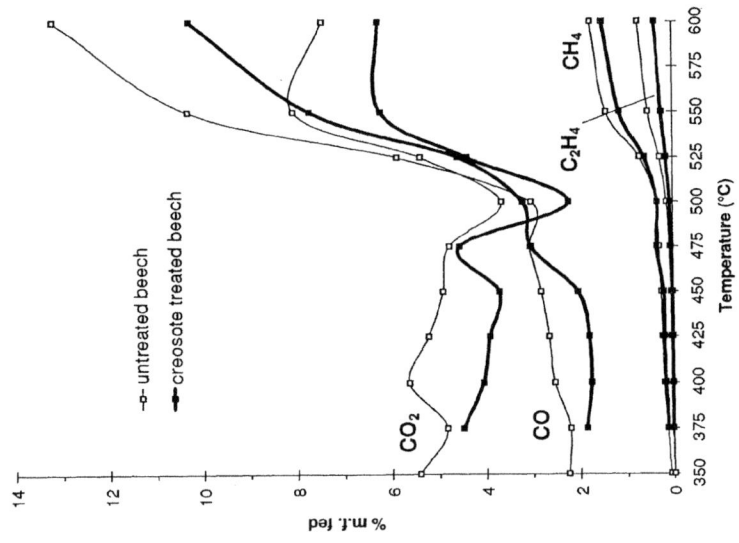

Fig. 11. Gas composition of untreated and creosote treated beech

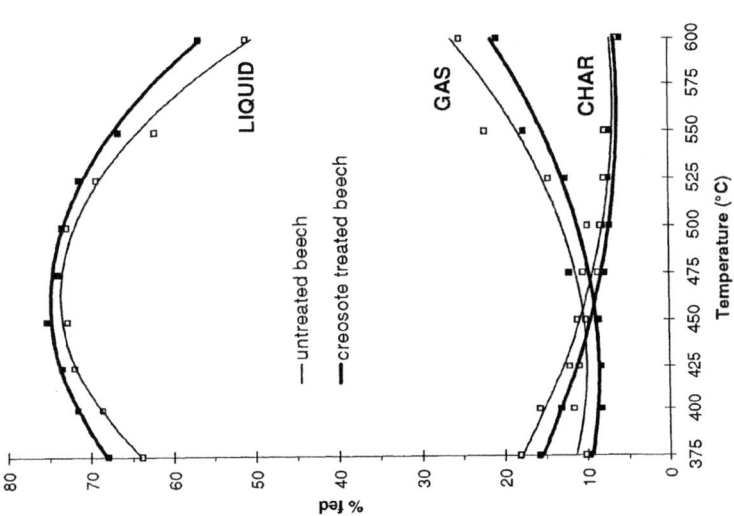

Fig. 10. Product yields from untreated and creosote treated beech

4 Summary and Conclusion

The results presented show that the presence of wood preservatives (CCB, creosote) does not have a negative effect on the fast pyrolysis. The amount of liquid remains high and the heavy metals from the preservatives are in the char. No heavy metals were detected in the pyrolysis liquid. Furthermore boron has a significant effect on the amount of levoglucosan, hydroxyacetone and hydroxyacetaldehyde which are the main carbohydrate derived degradation products of pyrolysis liquids. Boron is a flame retardant and favours the so-called transglycosylation reaction of cellulose which leads to higher yields of levoglucosan. In the case of untreated wood, the cycloreversion reaction is the main pathway yielding preferably hydroxyacetaldehyde [10]. There is no effect on the lignin derived products, they change only slightly.

The pyrolysis of creosote treated wood did not cause any technical difficulties. The presence of creosote during the pyrolysis does not affect the products from the wood and the creosote did not change either.

Flash pyrolysis of wood waste is a promising way of managing wood waste and producing cheap fuel, due to financial incentives in disposing of wood waste.

Following these results a pilot plant (3-5 kg/h) based on the HAMBURGER VER-FAHREN [11] will be constructed and operating this year.

Future research will be focused on the influence of other wood preservatives, contaminants and waste wood products, for example: particle boards, old furniture etc..

5 References

1. Voss, A and H. Willeitner (1995) Characteristics and quantity of impregnated wood waste in Germany. *International Research Group on Wood Preservation*, Document No.: IRG/WP 95-50041, 1995.
2. Scott, D. S. and J. Piskorz (1984) The continuous flash pyrolysis of biomass. *Can. J. Chem. Eng.*, Vol. 62, pp. 404-12.
3. Meier, D. and O. Faix (1992) Pyrolysis-gas chromatography-mass spectrometry of lignin, in *Methods in Lignin Chemistry* (eds. S. Lin and C. Dence), Springer Verlag, Berlin, Heidelberg, New York, London, Paris, Tokyo, pp. 177-99.
4. Faix, O; D. Meier and I. Fortmann (1990) Thermal degradation products of wood. gas chromatographic seperation and mass spectrometric characterization of monomeric lignin-derived products. *Holz Roh. Werkst.*, Vol. 48, pp. 281-89.
5. Faix, O; D. Meier and I. Fortmann (1990) Thermal degradation products of wood. A collection of electron-impact (EI) mass spectra of monomeric lignin-derived products. *Holz Roh. Werkst.*, Vol. 48, pp. 351-54.
6. Faix, O; D. Meier and I. Fortmann (1991) Thermal degradation products of wood-gas chromatographic separation and mass spectrometric characterization of polysaccharide derived products. *Holz Roh. Werkst.*, Vol. 49, pp. 213-19.
7. Faix, O; D. Meier and I. Fortmann (1991) Thermal degradation products of wood - A collection of electron-impact (EI) mass spectra of polysaccharide derived products. *Holz Roh. Werkst.*, Vol. 49, pp. 299-04.

8. Barker, R. H., Drews, M. J. (1985) Flame retardants for cellulosic materials, in *Cellulose Chemistry and its Applications* (eds. P. T. Nevell and S. H. Zeronian), Ellis Horwood Limited, Chichester, pp. 423-55.

9. Richards, G. N. and G. Zheng (1991) Influence of metal ions and of salts on products from pyrolysis of wood: applications to thermochemical processing of newsprint and biomass. *J. Anal. Appl. Pyrolysis*, Vol. 21, pp. 133-46.

10. Boon, J. J., I. Pastorova, R. E. Botto and P. W. Arisz (1994) Structural studies on the cellulose pyrolysis and chars by PYMS, FTIR, NMR and by wet chemical techniques, *Biomass and Bioenergy*, Vol. 7, pp. 25-32.

11. Kaminsky, W. (Ed.) (1990) *Entsorgung von ölhaltigen Sonderabfällen nach dem Hamburger Pyrolyseverfahren*, Erich Schmidt Verlag, Berlin.

RELEASE OF METALS DURING THE PYROLYSIS OF PRESERVATIVE IMPREGNATED WOOD

Pyrolysis of CCA treated wood

L. HELSEN and E. VAN DEN BULCK
Department of Mechanical Engineering, Katholieke Universiteit Leuven, Leuven, Belgium

Abstract

This paper describes an experimental facility for the pyrolysis of non-treated and CCA-treated wood particles. In a typical experiment, 10 grams of wood are loaded into the reactor. This reactor is a stainless steel circular tube with a diameter of 5 cm and a length of 20 cm. A flow of heated nitrogen is forced through the reactor. The reactor tube is thermally guarded with a heating tape with a controlled heating power such that the reactor outlet temperature equals the inlet temperature. At the outlet of the reactor, the nitrogen flow contains non-condensible pyrolysis gases as well as microscopic tar droplets. The gas flow is cooled in a highly compact coiled water cooled heat exchanger before passing through a filter.

The experimental results show that the loss of arsenic from the wood relative to the overall loss of mass is dependent upon the reactor temperature and the duration of the pyrolysis process, increasing with increasing temperature or duration of the heating period. It is shown that there exists an optimal combination of temperature and duration which yields a minimum loss of arsenic. At these conditions, the release of copper and chromium is negligible.

A qualitative study of the gaseous and liquid products, resulting from the pyrolysis process at these optimal conditions, is carried out by forcing the pyrolysis gas through a water scrubber and a glass fibre filter.

Keywords: Copper Chromium Arsenicum, filtration, gasification, pyrolysis, wood

1 Introduction

Waterborne chromated copper arsenate (CCA) has been a preferred treatment for the preservation of wood against fungal and insect attack for several applications. The treated wood is used in applications such as garden projects, utility poles and marine pilings.

At present there are no environmental regulations for the disposal of CCA treated wood for the end user. These wastes are considered to be non-hazardous and are disposed of in non-hazardous waste sites or burned in conventional incinerators with other combustibles [1]. However, there is a growing concern by utility companies as to who will be ultimately responsible for the disposal of CCA treated poles. The number of waste disposal sites is decreasing and redundant poles, piling and lumber, being a large volume material, may not be accepted at the limited number of sites in the future. Burning the CCA treated wood may cause a non-negligible pollution due to the release of the metals (copper, chromium and arsenic) in the flue gas. However, pyrolyzing the CCA treated wood can be done at lower temperature and there is no oxidising agent, resulting in a lower loss of metals.

To determine the amount of arsenic actually released to the atmosphere and the form of the arsenic when CCA wood is burned, McMahon et al [2] examined losses versus a range of temperatures and times at temperature. There results showed arsenic losses of about 22% at low temperature (400°C) but these losses could rise to 77% at high temperature (800-1000°C) and long times (6 hours). The major form of the volatilized arsenic was as particulates in the form of arsenites and arsenates. In addition some 11% copper and 15% chromium were found to vaporize.

In laboratory investigations by Dobbs and coworkers [3], the percent arsenic that was volatilized increased with increasing temperature and oxygen partial pressure. Only minor amounts of copper and chromium were volatilized. However, Pasek et al [4] found that the conditions at which arsenic volatilization approaches zero, are a limited air flow and high combustion temperatures (in excess of 1100°C). No volatilization of chromium or copper was observed.

The current literature clearly shows that the mechanism of arsenic volatilization during the burning of CCA treated wood is not yet completely understood.

The purpose of this study is to build an experimental, laboratory-sized test rig for the pyrolysis of CCA treated wood with the aim of maximizing the fraction of arsenic which is contained in the ash upon a maximal mass reduction of the wood. The resulting ash would subsequently be recycled to fresh CCA. The influence of the reactor temperature and the duration of the pyrolysis process on the metal content of the resulting ash and the mass reduction has been determined. The optimal combination of temperature and duration which yields a minimum loss of arsenic is found. At these conditions the process can be optimized to obtain a gas and liquid stream that can be used as biomass derived fuels in other applications.

2 Methods and materials

2.1 Feedstock
The feedstock used in the experiments is pine wood ('Pinus Sylvestris') impregnated with type C CCA salt: 32.5% $CuSO_4.5H_2O$, 41.1% $Na_2Cr_2O_7.2H_2O$ and 26.4% $As_2O_5.2H_2O$. The wood chips measure 5 cm in length and 2 mm in thickness. The impregnation has been carried out specially for these experiments to imitate the worst

case: the wood is treated twice for class 4 impregnation (wood may be in contact with groundwater and sweet water) and sap-wood is used instead of massive wood.

The metal concentration in the wood is determined after drying the wood, which is accomplished by forcing a heated (120°C) nitrogen stream through the wood. The end of the drying process is signaled by the end of the moisture release which is detected by a humidity sensor located in the nitrogen stream at the exit of the reactor. The percentage of moisture, calculated from the difference in mass before and after drying, varied between 9 and 14%. The average metal concentrations (average of 14 values) in the dried wood are given in table 1, together with the calculated spread. The high values of the spread indicate that the feedstock is not homogeneous.

Table 1. Average metal concentrations in the dried CCA wood.

	chromium (μg/g)	copper (μg/g)	arsenic (μg/g)
average value	8778	3499	8947
spread	8%	12%	12%

2.2 Pyrolysis process

An experimental facility for the pyrolysis of non-treated and CCA-treated wood particles has been built. In a typical experiment, 10 gram of wood is loaded into the reactor. This reactor is a stainless steel circular tube with a diameter of 5 cm and a length of 20 cm. A flow of heated nitrogen is forced through the reactor. The reactor tube is thermally guarded with a heating tape with a controlled heating power such that the reactor outlet temperature equals the inlet temperature.

To determine the influence of the reactor temperature and the duration of the pyrolysis process on the concentration of metals in the pyrolysis residue and the mass reduction, the reactor temperature is varied between 250 and 400°C and the duration of the pyrolysis process between 20 and 60 minutes. These experiments are carried out in an updraft system with a nitrogen flow rate of 5 Nm^3/h. The gas stream leaving the reactor is cooled to about 50°C in a water cooled spiral heat exchanger. Upon leaving the gas cooler, the gas stream is passed through 4 Whatman filters in series (Whatman nr. 42, 40, 43, 41), treated with tetra-n-buthyl ammonumhydroxide (TBAH) that is used to collect arsenic(III)oxides [2].

At the outlet of the reactor the nitrogen flow contains non-condensible pyrolysis gases as well as microscopic tar droplets. In the updraft system, this gas stream could not be fully determined because the liquid tar droplets and heavy gaseous components condense on the spiral heat exchanger. This condensate forms a dense sticky layer on the tubes that cannot be removed (water and tetrachloroethylene have been tried as solvents) nor analysed. The calculation of a complete metal mass balance was thus impossible. To overcome these problems, a water scrubber has been installed. Consequently a downdraft system was selected for further experiments. The gas flow, leaving the reactor, is cooled and washed in the water scrubber. The cold, wet gas flow at the outlet of the scrubber forms an aerosol of microscopic tar and water droplets. This aerosol is finally forced through a glass fibre filter with a 99.98% retention for droplets of 0.3 μm. The nitrogen flow rate is adjusted such that no water droplets penetrate through the filter, which occurs at a flow rate of about 800 l/h of dry nitrogen.

2.3 Metal analysis

The dried wood, the pyrolysis residue, the filters and the scrubber emulsion are analysed with an inductively coupled plasma - mass spectrometer (ICP-MS) for copper, chromium and arsenic concentrations. Since the elements to be analysed are injected in the ICP-MS as a liquid, an appropriate leaching method has to be applied to get the metals into solution starting from the dried wood, the pyrolysis residue and the filters. The scrubber emulsion can directly be injected after dilution and adding an internal standard (Indium) and nitric acid to obtain a uniform matrix.

A comparative study of leaching methods has been carried out to look for a method that is reproducible, easy to handle and that guarantees that all metals go into solution [5]. The conclusion of this study is that the BSI-method (British Standard Methods of analysis of preservatives and treated timber, part 3) [6] fulfills these requirements when starting from dried wood or filters for all of the three metals (copper, chromium and arsenic). The comparison with a method that brings all the solid matter into solution (reflux method), but which is very cumbersome, shows that the BSI-method fulfills these requirements when starting from the pyrolysis residue only for copper and arsenic, but not for chromium. Not all of the chromium, present in the residue, goes into solution, which means that the chromium is strongly bonded in the residue.

Prior to analysis with the ICP-MS, the leached solutions are filtered, diluted with a dilution factor dependent on the expected metal concentrations, acidified with nitric acid and Indium is added as an internal standard. Besides the test solutions, standard and blank solutions are analysed as well.

The actual ICP-MS measurements were carried out with a PlasmaQuad PQ2+ (Fisons Instruments) equipped with a conventional Meinhardt nebuliser for sample introduction with the following adjustments: 3 seconds measuring time, 120 seconds uptake time, 60 seconds wash time and autosampler introduction. The signal measured for the blank solution is subtracted from the signal measured for the standard and test solutions. The resulting concentration is the average value of three measurements. To the standard solutions of each metal to be measured (Cu, Cr, As) a least square approximation is applied by the software connected to the ICP-MS. The correlation coefficient of this straight line is at least 0.995, mostly 0.999 or 1. To eliminate interferences with other compounds (such as SO_2, SO_2H, C, ...) the concentrations of the isotopes ^{53}Cr, ^{63}Cu and ^{75}As are measured.

3 Results and discussion

3.1 Updraft system: influence of reactor temperature and duration of the pyrolysis process on the release of metals and mass reduction

The influence of reactor temperature and duration of pyrolysis process on the weight percentages of the metals in the residue after pyrolysis and on the mass reduction of the wood sample is illustrated in figures 1, 2, 3 and 4. The duration of the pyrolysis process is here defined as the time between the moment that the reactor temperature is 20°C lower than the pyrolysis temperature and the end of the pyrolysis experiment.

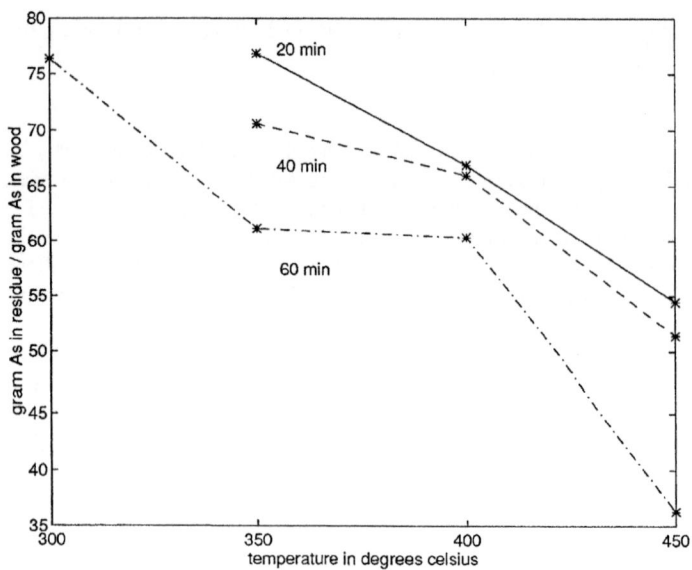

Fig. 1. Retainment of arsenic as a function of reactor temperature and duration of the pyrolysis proces.

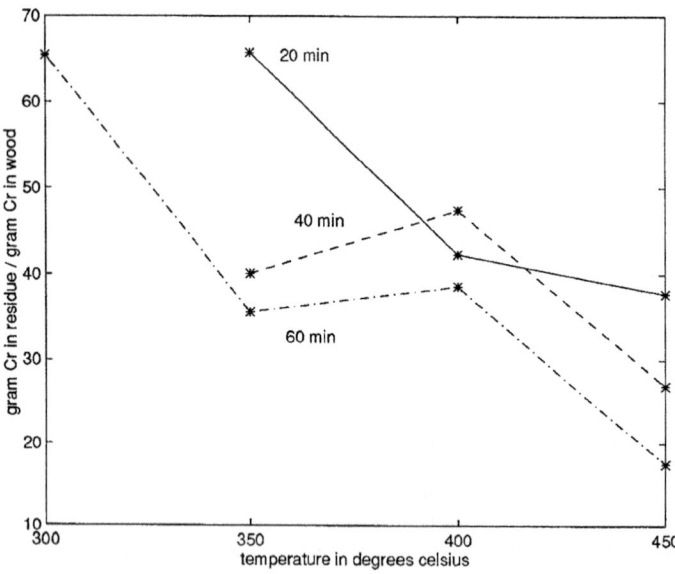

Fig. 2. Retainment of chromium as a function of reactor temperature and duration of the pyrolysis proces.

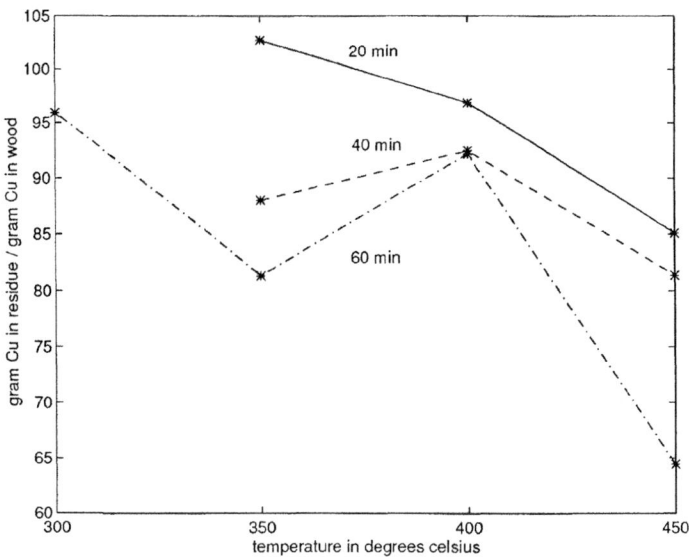

Fig. 3. Retainment of copper as a function of reactor temperature and duration of the pyrolysis proces.

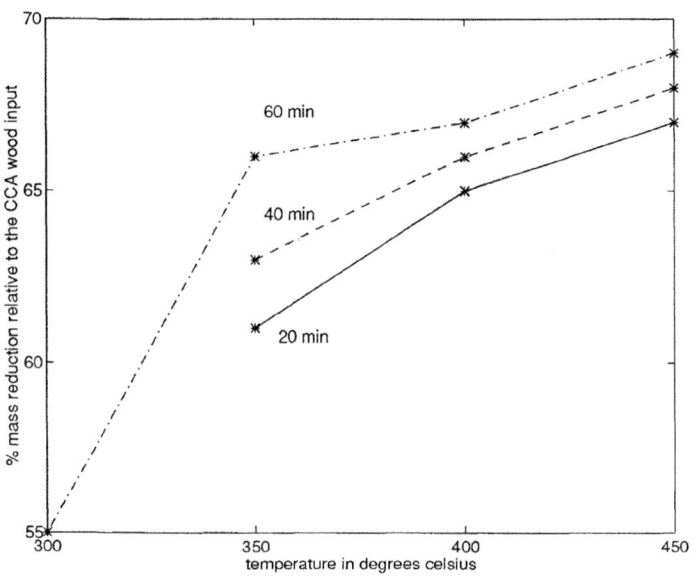

Fig. 4. Percentage mass reduction relative to the CCA wood input as a function of reactor temperature and duration of the pyrolysis process.

It should be taken into consideration that the chromium concentrations in the pyrolysis residue are determined using the BSI-method, which means that these values are too low. The values in figure 2 cannot be interpreted as absolute values, only a tendency can be seen.

These results show that the metal content in the pyrolysis residue decreases as the temperature or the duration of the process increases. The fact that for durations of 40 and 60 minutes a temperature of 400°C gives better results than a temperature of 350°C in the case of copper and chromium cannot be explained yet. This phenomenon does not appear for arsenic.

Besides the metal concentrations of the pyrolysis residue the metal concentrations in the filters are determined. These values, expressed as percentage metal in the leached filter solution relative to the metal concentration in the initial CCA wood sample, vary between 0.03 and 0.07% for chromium, 0.02 and 0.04% for copper and 5.0 and 5.7% for arsenic (2.5% in the 300°C - 60 minutes experiment). There is no explicit correlation between the metal content in the filter and the reactor temperature or the duration of the pyrolysis process. The filters may be saturated at the end of the experiment.

The optimal combination of reactor temperature and duration of the pyrolysis process, resulting from these experiments, is 350°C and 20 minutes. They retain most of the arsenic, the same amount as the combination 300°C - 60 minutes. The shortest duration is chosen with a view to industrial applications because the duration of the pyrolysis process corresponds to the residence time of the wood particles in a continuous flow system. A small residence time yields the smallest size of the high temperature pyrolysis chamber. The mass reduction is only 60%, resulting in 40% by weight wood char product. However this char is recycled to fresh CCA, which means that the equipment costs for char treatment and CCA recovery may equally be reduced, compared to an installation using directly the wood waste instead of the pyrolysis residue.

As already mentioned before, this updraft system is not optimal to calculate a complete mass balance or to recuperate the liquid product from the pyrolysis process. The gas flow, leaving the reactor, contains non-condensable pyrolysis gases as well as microscopic tar droplets. The condensable compounds precipitate on the water cooled spiral heat exchanger, polymerise and form a very viscous tar phase, which is very difficult to remove. If these tar compounds are mixed with water during the condensation process, polymerisation is much slower and the resulting liquid is much less viscous and can be recuperated.

The metal concentrations in the pyrolysis residue, resulting from a pyrolysis experiment in the downdraft system at a reactor temperature of 350°C during 20 minutes, have been determined.

To know exactly the metal content of the residue, the reflux method has been carried out on the solid matter that remained in the leached solution when the BSI-method has already been carried out. No additional arsenic or copper is found, but the chromium concentration found in the leached solution by the reflux method is not negligible. The results [5] indicate that the release of chromium and copper during the pyrolysis process at these optimal combination of temperature and duration is negligible. The release of arsenic can be accurately determined with the BSI-method.

3.2 Downdraft system: study of liquid and gaseous products

The pyrolysis gas, exiting the reactor , appears to exist more as an aerosol than a true vapour and thus collection is not a simple condensation problem. The vapours have similar properties to cigarette smoke; they can be characterised as a combination of true vapours, micron sized droplets and polar molecules bonded with water vapour molecules [7]. In the downdraft system the gas is cooled and washed in a water scrubber, resulting in an aqueous emulsion and an exiting cold, wet gas flow. The scrubber emulsion has been examined under a microscope, where micron sized (1 - 2 μm) droplets or particles were visible.

The cold, wet gas flow, which forms an aerosol of microscopic tar and water droplets, is finally forced through a filter. When Whatman filter paper is used at a nitrogen flow rate of 5 Nm^3/h, the filter paper becomes wet and bubbles appear at the backsurface of the filter section. Also with a Teflon membrane filter, which is hydrophobic, bubbles still appear at the external surface. None of these filters performed satisfactorily for retaining water and pyrolysis droplets. The best results are obtained with a glass fibre filter type A/E. There are still some bubbles visible, but to a lesser extent and the glass fibre filter retains more droplets or particles because the filter is much more coloured (yellow - brown) from the pollutants.

A small experiment is done to examine whether these droplets flow through the filter by mechanical entrainment. A heated nitrogen flow is forced through an empty reactor (no CCA wood) and cooled in the water scrubber when exiting the reactor. The flow rate is adjusted such that no water droplets penetrate through the filter, which occurs at a flow rate of about 800 l/h of dry nitrogen.

In future experiments, the pyrolysis experiment will be carried out with a nitrogen flow rate of 800 l/h and 10 gram of CCA wood in the reactor. This should result in an exhaust, which is a clear dry gas except for a short period in which a visible white smoke is emitted, as has already been seen in previous experiments. With this setup a complete arsenic balance can be determined.

4 Conclusions

CCA treated wood can be pyrolyzed with a negligible release of copper and chromium and a minimal release of arsenic (76.84% As in the pyrolysis residue) at a reactor temperature of 350°C during 20 minutes, using nitrogen at a flow rate of 5 Nm^3/h. The corresponding reduction in mass is 61%. However, the gas flow, leaving the system, after being cooled and washed in a water scrubber, through a filter, is not a dry clear gas, but rather contains liquid droplets (brown in colour).

Experiments with gas filtration have been carried out. A glass fibre filter at low flow rate proved to be the best setup for filtration of the pyrolysis gas.

Acknowledgements

The authors would like to thank the Department of Chemical Engineering of the K.U.Leuven for their valuable help in conducting the quantitative analysis of the samples. We are especially grateful to Prof. C. Vandecasteele from the chemical department for his valuable comments through the course of the study. We also thank Beaumartin S.A. and Mr. Hery in particular for the financial support, the helpful discussions and the wood samples.

5 References

1. Pasek, E.A. (1993) *Treatment of CCA waste streams for recycling uses*, Hickson Corporation, Conley, GA, USA.
2. MCMahon, C.K., Bush, P.B., Woolson, E.A. (1986) How much arsenic is released when CCA wood is burned, *Forest Products Journal*, 36, 45-50.
3. Dobbs, A.J., Phil, D., Grant, C. (1978) The volatilization of arsenic on burning copper-chromium-arsenic (CCA) treated wood, *Holzforschung* 32, 32-35.
4. Pasek, E.A., McIntyre, C.R. (1992) *Burning of CCA treated wood with minimal metals loss*, Treated wood life cycle management workshop, AWPI.
5. Helsen, L. (1996) *Comparative study of leaching methods for CCA treated wood*, Internal Report, K.U.Leuven, Belgium.
6. British Standard Institution (1979) BS 5666: *Part 3, British Standard Methods of analysis of wood preservatives and treated timber part 3: Quantitative analysis of preservatives and treated timber containing copper / chromium / arsenic formulations.*
7. Bridgwater, A.V., Peacocke, G.V.C. (1995) *Biomass fast pyrolysis*, Proc. 2nd Biomass Conference of the Americas, Portland, USA.

CO-PYROLYSIS OF BIOMASS WITH PLASTICS

FILOMENA PINTO, I. GULYURTLU, I. CABRITA and J. PINTO
ITE / DTC, Instituto Nacional de Engenharia e Tecnologia Industrial, Lisboa,
PORTUGAL
M. GONÇALVES
COALTEC e Ambiente, Lisboa, PORTUGAL

Abstract

The recycling of plastics for further uses as raw material is proving to be difficult due to reluctance of relatively conservative industry to receive them, particularly because of high costs involved in recycling and the lack of incentives. The incineration of these plastics is a controversial issue, mainly resulting from the concern in protecting the environment and in reducing the CO_2 emissions. Although PVC-origin plastics are getting less and less in use, their incineration gives rise to the formation of chlorinated hydrocarbons like furans and dioxins, some of which are highly toxic and very hazardous to human health.

An alternative approach to upgrade plastics wastes and their further use could be achieved through the application of liquefaction technology forming liquid and gaseous products.

Studies have been undertaken to pyrolyse plastics alone or together with biomass to produce liquid fuels in the above-named Department. The objective of the introduction of biomass was to make up the variations in the quality of plastics and to guarantee the security of supply in a future commercial unit. The biomass somewhat changed the nature of the liquids formed and there was some char left at the end.

Keywords: Pyrolysis, Liquefaction, Plastics, Biomass.

1 Introduction

The work undertaken at INETI investigates the production of hydrocarbons from biomass and plastic residues, which can be used both as liquid or gaseous fuels or as a raw material for petrochemical industry. Preliminary experiments have shown the feasibility of the process, although influences of several experimental conditions should be further studied in greater detail, in order to maximise total conversion and products quality.

Products obtained by this process involve a complex mixture of several hydrocarbons, which needs to be separated into several fractions before they can be analysed and identified for further application.

Presently, the most common solutions to deal with biomass and plastic residues, namely landfills and incineration, do not appear to be most suitable solution since they present various problems also related with the environment.

Landfilling plastic residues is not a solution, essentially because it has been increasingly difficult to find suitable places for building technically adequate landfills, due to the resistance imposed by the nearby populations and since there is the danger of leaching and soil impregnation, with the subsequent contamination of underground waters.

After their disposal in the landfill, residues suffer numerous biological and physical-chemical modifications simultaneously such as, minor biodegradation with the production of gases and organic liquids, oxidation with gas emission and diffusion through the landfill and finally, dissolution of organic and inorganic compounds on the nearby waters (superficial and underground waters). On the other hand, this process does not allow the recover of the organic content of plastic residues, that should be recycled in an organic system [3].

Incineration of biomass and plastic residues to produce heat may be a possibility, but its organic content would totally be destroyed and converted only into CO_2 and H_2O [14]. In addition, depending on its nature, combustion produces pollutants like light hydrocarbons, nitrous and sulfur oxides, dusts, dioxins and other toxins, that have highly negative impact on the environment.

Likewise incineration, biodegradation is also not the obvious solution for this type of wastes, since the synthetic products obtained in the petrochemical industries are generally resistant to this type of action. In fact, if plastics were all photo and/or biodegradable, their accumulation problem would automatically be solved. Although, some plastics have these characteristics (capsule covers, suture string degradable by organic action, etc.), they represent a very small percentage in the plastics universe, either due to the type of application for which they were projected, or due to economic reasons.

The degradation of plastics is a waste of valuable resources. Furthermore, it is still not well established what long term effects of their degradation could lead to in landfills. On the other hand, there is a strong possibility that the consumption of plastics and consequent accumulation of their wastes would increase rapidly, if the believe of their fast disposal is created.

An alternative approach to upgrade either biomass and plastic residues and their further use could be achieved through the application of liquefaction technology. The thermal decomposition of biomass and plastic residues at moderate temperatures without air, breaks their structure into smaller intermediate species, without damaging its intrinsic structure [12]. This process differs from the conventional destructive processes, because it allows the recovery of reaction products with added value, while at the same time decreases the carbon percentage as a secondary product.

The pyrolysis/liquefaction of a plastic/biomass mixture can be performed under moderate conditions of pressure and temperature (about 400 °C and 70 atm) and in the presence of an inert atmosphere. The presence of biomass is important, as due to its high contents of volatiles, high operation pressure are achieved even with lower initial pressures, thus contributing to lower operating costs [11]. Gases produced are proven to have high calorific value and thus could be burned to provide the heat necessary for liquefaction reactions. The pyrolysis process is auto-sufficient for its energetic needs, since gases produced have calorific capacity enough to supply energy for all reactions involved. Liquids produced could be used as fuel or as a raw material for petrochemical industry.

The co-pyrolysis reaction products exhibit higher quality than those obtained from the pyrolysis of biomass only, since they present higher calorific value and better stability properties. In fact, these liquids show a well-defined distillation point, stable physical properties, low acidity, good miscibility with conventional fuels, low contents of toxins. These liquids are hydrophobic, unlike biofuels that are oleophobic and hydrophilic [2].

Flash-pyrolysis of biomass can give high yields of liquids or *bio-oil* of up to 70 % (w) on a dry feed basis and laboratory scale. Three products are usually obtained: gas, liquid and char, theirs relative proportions, depend on the pyrolysis method and reaction parameters. This process involves very high heating rates combined with moderate temperatures of less than 600 °C, short residence times and rapid quenching of products. High heating rates of up to 1 000 °C/s or even 10 0000 °C/s, at temperatures below 650 °C, combined with rapid quenching, cause the intermediate liquid pyrolysis products to condense, before further reactions occur and break down these high molecular wheight species into gaseous products [22]. The liquid product known as *bio-oil*, may be readily burned and has been employed for this purpose, although some upgrading steps to remove water and oxigen are needed for stabilization and to give them full compatibility with conventional fuels [4].

This process has been tested at INETI with encouraging results, that requires further studies to optimize the global process and to characterize the reaction products. It is also fundamental to continue tests that allow clear understanding of the process mechanism and reaction kinetics.

2 Experimental

Experiments involving biomass only, were carried out in a tubular reactor with 1 cm diameter and 50 cm long. The transport gas used was N_2 with a 8,4 dm3/min flow. The biomass used was dried and then fed to the reactor with a mass flow of 30 g/h. Particle size was ranged between 105 and 250 µm. Tests were performed between 350 and 500 °C at the reactors end. Gases produced were cooled rapidly to achieve a fast condensation.

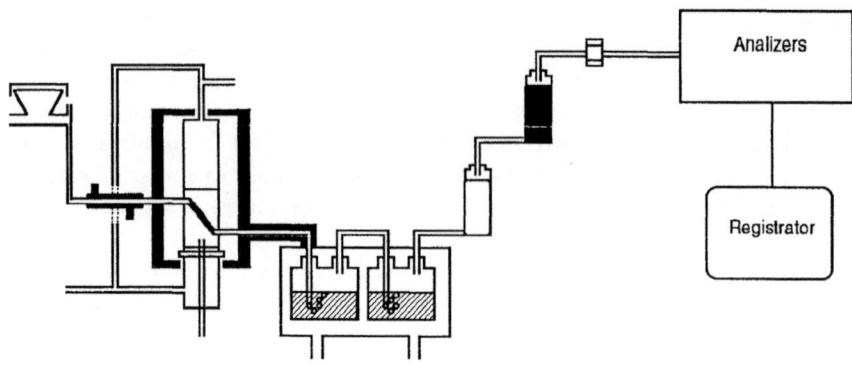

Fig. 1. Scheme of the biomass flash-pyrolysis installation.

Experiments done with plastic residues and mixtures of biomass and plastic residues, were carried out in an 1 liter autoclave, built in Hastelloy C276, by Parr Instruments. For monitoring each test run, the autoclave was connected to a PID programmable controller, which has a data acquisition system and was linked to a PC.

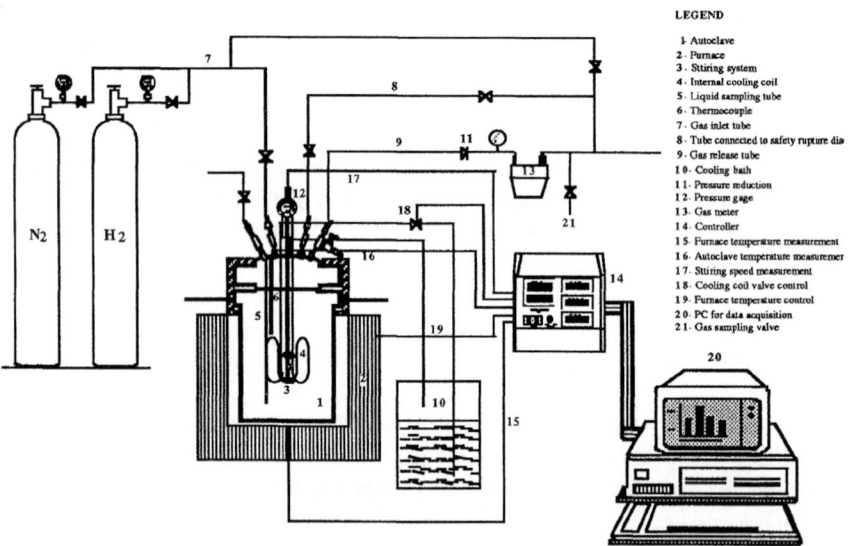

LEGEND

1 - Autoclave
2 - Furnace
3 - Stirring system
4 - Internal cooling coil
5 - Liquid sampling tube
6 - Thermocouple
7 - Gas inlet tube
8 - Tube connected to safety rupture disc
9 - Gas release tube
10 - Cooling bath
11 - Pressure reduction
12 - Pressure gage
13 - Gas meter
14 - Controller
15 - Furnace temperature measurement
16 - Autoclave temperature measurement
17 - Stirring speed measurement
18 - Cooling coil valve control
19 - Furnace temperature control
20 - PC for data acquisition
21 - Gas sampling valve

Fig. 2. Scheme of the plastics and biomass and plastic residues mixtures liquefaction installation.

After the autoclave was loaded, it was closed and pressurised with N_2 to purge the interior air. Then the autoclave was pressurised with N_2 to a pre-set value and was heated till the reaction temperature was reached, at which it was maintained for 30 minute. Then the autoclave was cooled down to room temperature. Gases were then collected, measured and analysed. The autoclave was opened and its content was weighted and analysed.

In each test, the amount of biomass was 20 g, for a biomass/plastic ratio of 1/2. Biomass particle size varied between 125 and 200 μm. Plastic particle size was 3 mm. Tests were performed between 380 and 450 °C. Initial pressure was ranged between 2 and 10 atm, whilst the average test pressure varied between 12 and 71 atm.

The biomass used in all runs had a forestry origin being pine its main component, since it is a common and widely distributed specie in Portugal.

All gaseous hydrocarbons were collected for direct analysis (by GC), whilst liquid hydrocarbons were extracted (separation of liquid components from residual solid constituted essentially by cellulose and inerts) and distillated. After these steps, liquid hydrocarbons obtained were also analysed by GC-MS with a capillary column associated to a mass spectrometry and FTIR, to identify their main compounds. Residual mass was analysed by thermogravimetry - TGA.

3 Results and discussion

3.1 Biomass and plastic liquefaction

To study the effect of biomass concentration on product yields, several experiments were carried out at 430 °C and at an initial pressure of 0.96 MPa. As shown in Figure 3, the liquid yield was observed to decrease with higher amounts of biomass added, whilst the gas quantity increased. When more biomass was used, more volatiles were released, which was responsable for greater amount of gas formed. The experiments involving plastic waste only, showed that for the same conditions, almost all the initial load was converted into liquids. This suggests that the nature of volatiles released from plastics are quite different from those of biomass origin and they do not appear to undergo cracking at temperatures in the range of 450 °C, thus giving rise to much lower gas yield.

Under the experimental conditions used the conversion of solid input to products was found to be always greater than 88 %. The highest value was obtained when biomass waste was added at an amount of 33 % (w). This was because the amount of plastic was still large enough to guarantee a high liquid yield whilst the amount of biomass ensured a release of gas of about 20 % (w).

For biomass concentrations higher than 33 %, the reduction in the quantity of liquids produced is negligible, whilst the amount of gas formed remained almost constant, hence no significant change in total conversion was observed. The biomass concentrations ranging between 0 and 33 % should be studied further in order to optimise the amount of biomass additions. This is important to determine the amount of gas that may be required to provide the heat necessary for the liquefaction reactions, so that the system could become auto-sufficient regarding its energy requirements.

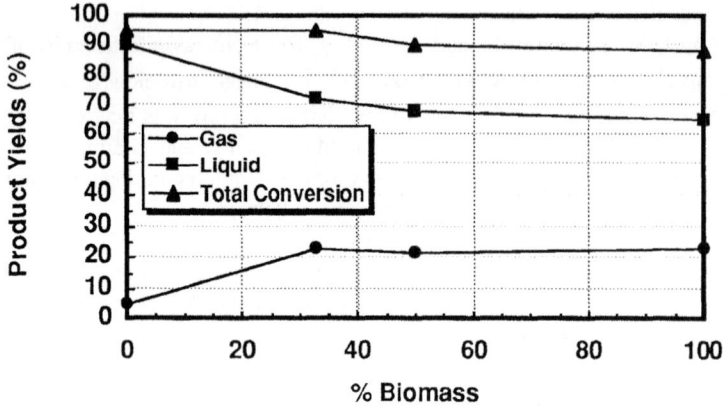

Fig. 3. Product Yields variation with percentage of biomass used for liquefaction runs.

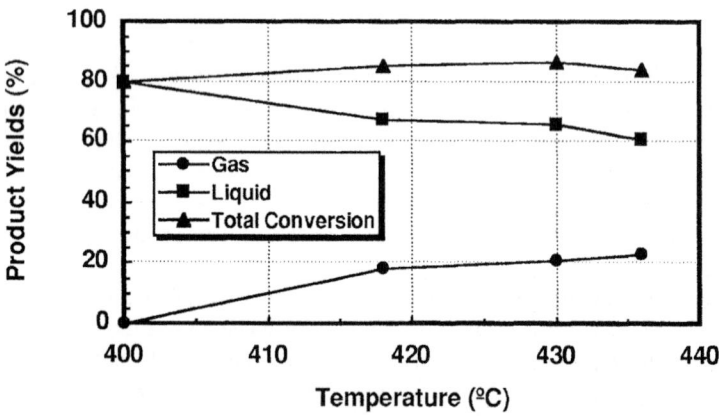

Fig. 4. Product Yields variation with Temperature for liquefaction runs when a plastic/biomass mixture was used.

In Figure 4, the effect of temperature on products yields and total conversion is shown, for a mixture of equal amounts of biomass and plastics' wastes. The initial pressure used in all these experiments was 0.96 MPa. The amount of gas formed and the overall conversion was observed to increase with temperature, whilst liquid yield decreased. This reduction in the liquid products could partly be due to the formation of higher amounts of gas, probably, resulting from cracking reactions of liquids at higher temperatures. However, because the total conversion did not remain constant, there could be other reactions taking place simultaneously. In fact, the overall conversion was found to increase with temperature, which suggests that there were several liquefaction reactions occurring in parallel that were favoured at higher temperatures. If the aim is to get higher yields of gas, the temperature should be fixed at about 430 °C, whilst for liquids, the ideal value is 400 °C.

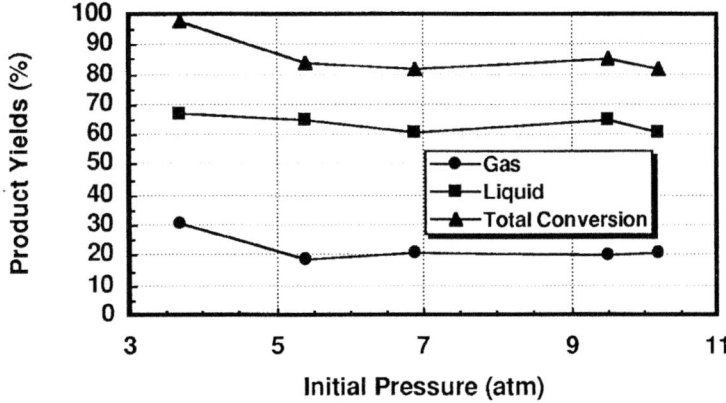

Fig. 5. Product Yields variation with Initial Pressure for liquefaction runs when a plastics/biomass mixture was used.

The influence of initial pressure on both product yields and total conversion was also investigated. The temperature in all these experiments was fixed at 430 °C. The results obtained are shown in Figure 5. The initial pressure does not seem to significantly affect the liquid yields. This is, however, not the case with the amount of gas formed which was found to decrease with the increase in pressure. This could be due to the effect of higher pressures making the release of volatiles more difficult as the pressure difference between the interior of the particle and the bulk phase could be too small so that the diffusional equilibrium was rapidly attained. The total conversion was higher at lower initial pressures because of higher gas yields. The results obtained so far suggest that the pressure should be kept at about 0.37 MPa (3,7 atm).

3.2 Biomass flash-pyrolysis

In Figure 6, the effect of temperature on product yields obtained from biomass only, is shown for flash-pyrolysis runs. The amount of gas formed was observed to increase with temperature, whilst solids yield decreased. Liquids yield was observed to increase from 350 to 450 °C and decreased from 450 to 500 °C. The liquid yield decrease observed between 450 and 500 °C, is probably the result of further cracking reactions of liquids, at these higher temperatures.

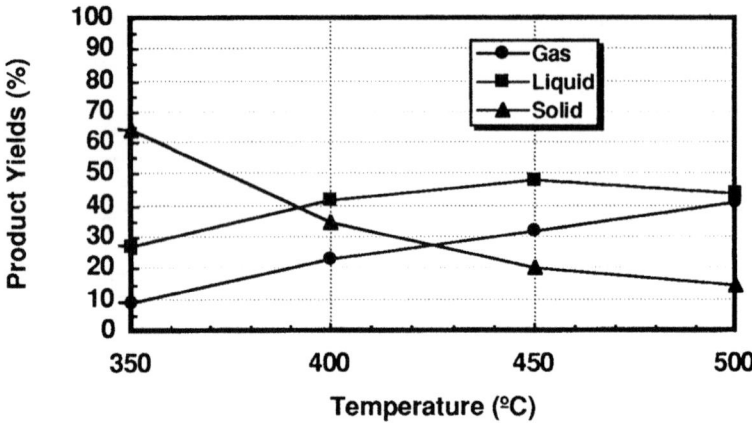

Fig. 6. Product Yields variation with Temperature for flash-pyrolysis runs.

The reduction of solids yields is partly due to the formation of higher amounts of gas and liquids, resulting from cracking reactions of solids at higher temperatures.

3.3 Plastics liquefaction

The effect of pressure was studied and the results obtained, demonstrated that pressure appeared to have little influence on products yields.

The results seem to show that for higher pressures, the cracking process was enhanced and the initial intermediate species were further broken down leading to the formation of smaller liquid hydrocarbon molecules. However, and probably due to the fact that higher initial pressures produced even greater run pressures, gas yields decreased. As demonstrated in Figure 7, higher pressures led to relatively lower gas yields and higher liquid yields. The total conversion was found to be about the same regardless the operating pressure.

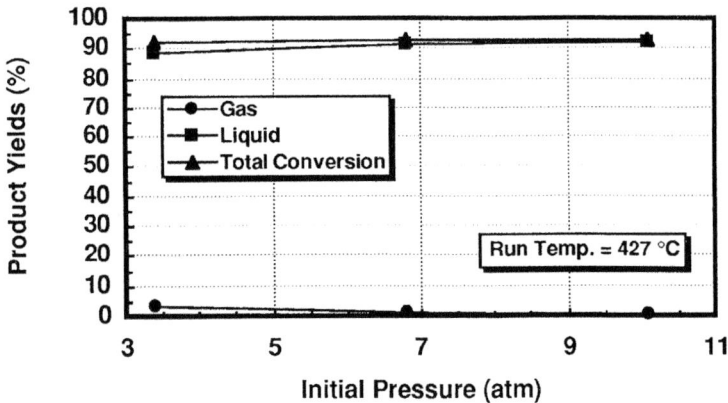

Fig. 7. Product Yields variation with Initial Pressure for liquefaction runs when only plastic was used.

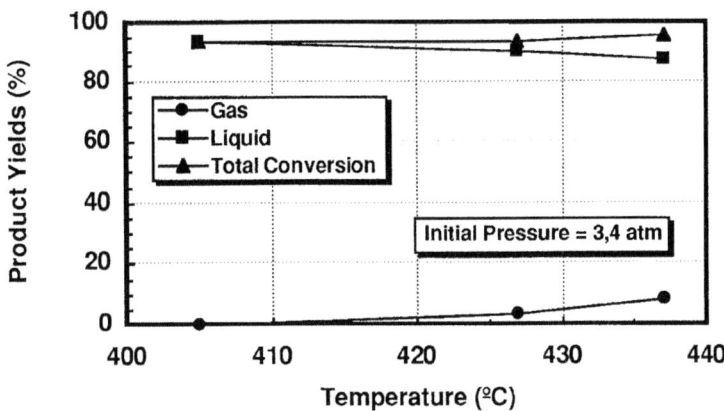

Fig. 8. Product Yields variation with Temperature for liquefaction runs when only plastic was used.

The test temperature appeared to influence both products yields and liquid quality, although total conversion remained almost the same. As shown in Figure 8, when the temperature increased higher gas yields and lower liquid yields were obtained.

Probably, at higher temperatures the intermediate species were further cracked, forming higher amounts of gaseous hydrocarbons and lighter liquid hydrocarbon molecules. Hence, the liquids obtained at 405°C have higher percentages of heavier hydrocarbon molecules, while those produced at 443°C have higher percentages of lighter hydrocarbon molecules. The experiment done at 423°C led to a mixture of both.

Liquids composition depended on the tupe of plastics presented in the waste. When only polypropylene (PP) and polyethylene (PE) were used, the aromatic yield obtained in the end products was very small, whilst high amounts of aliphatics were produced. The presence of polystyrene (PS) in plastic waste mixture increased the aromatic content of the end product and improved quality by rising calorific value and octane number.

For a mixture with composition similar to the one of plastics in municipal solid waste (68% (w) PE, 16% (w) PP, 16% (w) PS) the main liquid fraction produced was alcanes. Alcenes were produced in the least quantity and aromatics presented intermediate results. The individual compound analysis showed that alcanes yields increased from pentane to octane and then decreased proportionally when the number of atomic carbons rose. The heptane, octane and nonane yields presented similar values for all catalysts tested. The alcenes yield increased substantially from pentene to hexene, which presented the highest yields and then decreased proportionally to the rise of the number of atomic carbons. Ethylbenzene and toluene were the aromatic species produced in higher quantities. It was also detected the presence of compounds up to 20 carbon atoms and more, as well as other polyaromatic compounds such as: naphthalene and anthracene.

3.4 H/C Ratios

The elemental analysis of the plastic and biomass wastes were carried out and hence the H/C ratio was calculated, as shown in Figure 9. The H/C ratio of biomass was determined to be 1.5, whilst that of plastics alone was 2.17. The H/C ratios of the residue after the liquefaction reactions were found to be lower and varied from 0.6 to 1.0 as the variation in the amount of plastic used changed from 0 to 66.7 %. This could mean that when only biomass was used, the amount of hydrogen present in the reaction medium was lower and almost all appeared to be used up in the stabilisation of intermediate species formed, due to thermal cracking. As the amount of plastic increased, the amount of hydrogen present in the reaction medium was greater, leading to a final residue with a higher H/C ratio.

The H/C ratio of the biomass used in the flash-pyrolysis runs was 1.32, a little lower than the one used in the liquefaction runs. The H/C ratios of the products obtained are shown in Table 1.

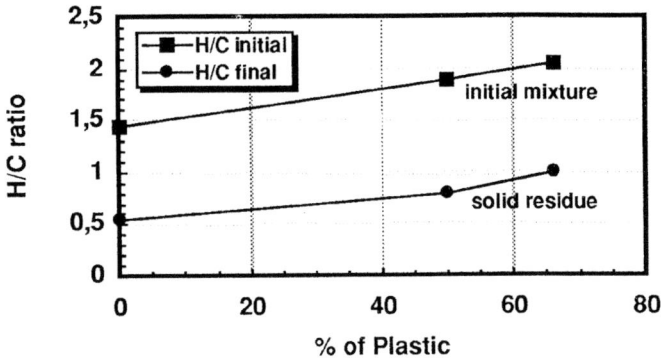

Fig. 9. H/C Ratio before and after liquefaction runs of biomass and plastic mixtures.

Table 1. H/C Ratio before and after liquefaction of plastics and flash-pyrolysis biomass runs.

Product / Initial H/C	Gases	Liquids	Solid residue
Biomass / 1.32	1.52	2.08	0.26
Plastic / 2.17	--	2.08	--

4 Conclusions

For the flash-pyrolysis of biomass, it could be concluded that:

i) The selection of the temperature was found to depend also, on the nature of the end product required. If the gas yield is to be maximised, temperature should be set at 500 °C or higher, whilst for liquids it should be set at about 450 °C.

ii) Flash-pyrolysis of biomass, was found always to produce higher amounts of gas than the liquefaction process using mixtures of biomass and plastic residues, or plastic wastes alone. The gas yields were observed to vary between 20 and 40% (w), depending on the temperature used. On the other hand, liquid yields obtained, were rarely superior to 40% (w), whilst total conversion (gas and liquids) was about 80% (w).

As regard to the liquefaction of plastic/biomass mixtures, it could be concluded that:

iii) It was observed that for higher pressures, the total conversion and the gas yield decreased, whilst liquid yield remained constant.

iv) The selection of the temperature was found to depend on the nature of the end product required. If the gas yield is to be maximised it should be set at 430 °C, whilst for liquids being at about 400 °C.

v) For the conditions used, it was found that the amount of biomass to be added should not exceed 33 % by weight of the original mixture, however further experiments should be carried out to investigate this variable.

vi) Co-pyrolysis of biomass and plastic residues mixtures gave rise to lower gas yields, around 20% (w), but much higher liquid yields, 70% (w), independent of the temperature used. Total conversion was found to vary around 90% (w).

For the liquefaction of plastic, it could be concluded that:

vii) Pressure appeared to have little influence on products yields. For higher pressures the cracking process was enhanced, leading to the formation of lighter hydrocarbon liquids.

viii) The selection of the temperature was found to depend also, on the nature of the end product required. If the gas yield is to be maximised, temperature should be set at 443 °C or higher, whilst for liquids it should be set at about 400 °C. Temperature had an enhanced action on products quality, producing lighter liquid hydrocarbons.

ix) Liquefaction of plastic residues only, present the lowest gas yields about 5% (w), however, liquid yields were very high, about 85% (w), as well as the total conversion being as large as 98% (w). The solid residue was found to be negligible and consisted mainly of plastic inert additives.

x) The co-pyrolysis and liquefaction products were observed to exhibit higher quality than those obtained from the pyrolysis of biomass only, since they have higher calorific value and better stability properties. In fact, these liquids show a well-defined distillation point, stable physical properties, low acidity, good miscibility with conventional fuels, low contents on toxins and hydrophobia, unlike *bio-oils* that are oleophobic and hydrophilic.

5 References

1. Ainsworth, S. J. (1992) C&E News, 3435.
2. Biofuels, European Commission DG XXII, 1994.
3. Brandrup, J., Immergut, E. H. (1966) eds. Polymer Handbook, New York.
4. Bridgwater, A.V., GRASSI, G. (1991) Biomass Pyrolysis Liquids Upgrading and Utilisation, Elsevier Applied Science.
5. Churin, E. (1991) Upgrading of Pyrolysis oils by hydrotreatment, Biomass Pyrolysis Liquids Upgrading and Utilisation, Ed. Bridgwater and Grassi.

6. Evans, R.J., Milne, T.A. (1985) Applied Mechanistic Studies of Biomass Pyrolysis, 17th Biomass Thermochemical Conversion Contractors Meeting, Minneapolis.

7. Fairbridge, C., Ross,R. and Spooner, P. (1975) A Thermogravimetric Study of the Pyrolysis of Jack Pine, Wood Science and Technology, No. 9. pp. 257.

8. Fouhy, K., Kim, I., Moore, S., Culp, E. (1993) Chemical Engineering, No. 30.

9. Furrer, R., Bakshi, N. (1988) Catalytic Conversion of Tall Oils to Chemicals and Gasoline Range Hydrocarbons, Research in Thermochemical Biomass Conversion, Ed. Bridgwater, Elsevier Applied Science.

10. Goldstein, I.S. (1983) Organic Chemicals from Biomass, CRC Press Inc.

11. Gulyurtlu, I., Pinto, F., Gonçalves, M., Cabrita, I., (1994) 8th European Conference on Biomass for Energy, Environment, Agriculture and Industry, Vol. 3, pp. 1908.

12. Gulyurtlu, I., Pinto, F., Lobo, L. S., Cabrita, I. (1995) Recycle'95, Vol. V, pp. 208.

13. Kanury, A.M. (1972) Thermal Decomposition Kinetics of Wood Pyrolysis, Combustion and Flame, No. 18. pp. 75 - 83.

14. Matthews, V. (1991) Recycle'91, V.7.1, Switzerland.

15. Phillips, M. (1991) Recycle'91, I.5.1, Switzerland.

16. Piskorz, J., Radlein, D., Scott, D. (1986) On the Mechanism of the Rapid Pyrolysis of cellulose, J. Anal. Appl. Pyrolysis, No. 9, pp. 121 - 137.

17. Ponder, G. R., Richards, G. N., Stevenson, T. T. (1992) Influence of Linkage Position and Orientation in Pyrolysis of Polysaccharides: a study of several Glucans, J. Anal. Appl. Pyrolysis, No. 22, pp. 217 - 229.

18. Potts, J. E. (1970) Continous Pyrolysis of Plastic Wastes, Industrial Water Eng. No. 7. pp. 32.

19. Richards, G. N. (1987) Glycolaldehyde from Pyrolysis of Cellulose, J. Anal. Appl. Pyrolysis, No. 10, pp. 251 - 255.

20. Richards, G. N. (1991) Influence of Metal Ions and Salts on Products from Pyrolysis of Wood: applications to thermochemical processing of Newsprint and Biomass, J. Anal. Appl. Pyrolysis, No. 21, pp. 133 - 146.

21. Shafizadeh, F., Mcginnis, G.D. (1971) Morphology and Biogenesis of Cellulose and plant cell walls, Advan. Carbohydr. Chem., No. 26. pp. 297.

22. Steinberg, M., Fallon, P., Sundaram, M.S. (1984) Flash Pyrolysis of Biomass With Reactive and Non-Reactive Gases, BNL-34510, New York.

23. Shafizadeh, F. (1982) Introduction to Pyrolysis of Biomass, Journal of Analytical and Applied Pyrolysis, No. 3. pp. 283 - 305.

24. Wampler, T.P., Levy, E.J. (1987) Reproducibility in Pyrolysis, Recent Developments, Journal of Analytical and Applied Pyrolysis, No. 12. pp. 75-82.

IMPROVEMENTS IN THE VORTEX REACTOR DESIGN

J.P. DIEBOLD and J.W. SCAHILL
National Renewable Energy Laboratory
Golden, CO USA

Abstract
Over the years, a variety of improvements have been incorporated into the vortex reactor that considerably increase its efficiency, durability, and operability. These improvements have included an internal rib to guide feed particles into a tight helix, an easily replaceable wear plate located at the vortex entry, a recycle loop that ensures complete pyrolysis of large feed particles, reorientation of the reactor from horizontal to vertical, better conservation of kinetic energy in the recycle loop with a new eductor design, and the ability to periodically remove inert particles from the reactor. A mathematical model that describes the vortex reactor was used to guide the design of a reactor with a nominal capacity of 32 dry tonnes of biomass/day. This reactor was fabricated with the internal spiral rib using large, but conventional, machining equipment. Other techniques of fabrication exist, but have not been proven. The National Renewable Energy Laboratory is interested in scaling up this technology in cooperation with industry.
Keywords: ablative pyrolysis, grinding cost, inert solids removal, recycle loop, vortex reactor, wear plate

1. Introduction

Fast pyrolysis with low char yields is an endothermic process. Reactor systems for the fast pyrolysis of biomass and organic wastes must be capable of rapid heat transfer to the feedstock, so that the char yields can be minimized and the organic vapors maximized. The yield of vapors will be reduced, if they are allowed to crack to gases. The method of supplying this heat transfer categorizes the various fast-pyrolysis reactors into relatively few groups: hot gas to feed particle by gaseous convection; hot particle to feed particle by conduction and thermal radiation; hot surface to feed particle by radiation; and hot surface to feed particle by conduction with or without sliding contact. Each method of heat transfer has a different range of heat transfer coefficients, that dictate the rate of heat transfer possible from a heat source to the biomass particles, the throughput possible for a given reactor size, the range of possible gaseous and solid particle residence times, and a range of acceptable feed particle sizes.

There are many different possible arrangements to accomplish fast pyrolysis, but most of them can be categorized as either entrained flow or fluidized bed. Except for vacuum pyrolysis on a rotating hearth [1], fast-pyrolysis reactors typically entrain the char out of the reactor to a gas-solid separation device (e.g., a cyclone separator or hot-gas filter).

The reactor design must provide sufficient time for the complete reaction of the feed particle before it leaves the reactor system. This can be accomplished by feeding small feed particles through the reactor in such a way that they are completely reacted during a single pass, or by recirculating them back through the reactor until they are fully reacted. If the heat-transfer rate to the larger particles is low, they will tend to form char rather than volatiles. With the high heat-transfer rates possible with ablative pyrolysis, the pyrolysis front can move through the particle at rates of over 3 cm/s, so even chips can be ablatively pyrolyzed, as has been widely seen with the hot-wire pyrolysis demonstration [2].

This potential for feeding large chips into an ablative-pyrolysis reactor can lead to significant cost savings, compared to the added cost of reducing the chipped feed to small slivers, granules, or powders. Note that for heat-transfer considerations, it is the smallest dimension of the particle that is of interest rather than the maximum dimension, as in particle sieving.

The vortex reactor was conceived and designed specifically for the ablative pyrolysis of biomass and organic wastes. It features the rapid entrainment of the feed into the vortex reactor with a supersonic eductor, and the transfer of heat from the externally heated walls through a thin liquid layer of oligomeric, intermediate pyrolysis liquids[3] to the rapidly moving biomass particles. The biomass particles take a helical path through the reactor. Partially pyrolyzed feed and large char particles are removed tangentially and recycled to the eductor. The vortex tube acts as a particle classifier, with char fines exiting the axial outlet of the reactor re-entrained with the product vapors and gases. Our experience has been that the vortex reactor works better with larger granules of feed than with finer powders. This paper discusses the advantages of the vortex reactor for the fast pyrolysis of biomass and the design improvements we have made over the years to this innovative reactor design, as revealed in our recent patent [4].

2. Feedstock Grinding Costs

In reviewing the detailed technoeconomics of fast pyrolysis [5], it is apparent that the cost of the pyrolysis reactor is a surprisingly small portion of the overall pyrolysis system. Peripheral equipment, such as feedstock receivers, driers, grinders, recycle gas compressors, condensation trains, and etc., add up to the majority of the estimated equipment and operating costs. One of the variables the process designer must choose is the feedstock particle size for the pyrolysis system. The smaller the feedstock, the more it costs to prepare it from the biomass. The following discussion will provide the reader with an appreciation for the cost of this grinding process.

The scaled-up version of the vortex reactor is expected to be able to ablatively pyrolyze as-received wood chips **without** requiring additional or secondary grinding of the feedstock. However, most other fast-pyrolysis concepts must use smaller feedstock particles to avoid slow heating and the resultant lower oil yields. Commercial wood-chipping machines used in the field (e.g., the "Brush Bandit" flywheel chipper by Foremost,

Inc.) produce chips with a nominal 20-mm (3/4-inch) dimension. The Brush Bandit, with a 48-kW (64-hp) gasoline engine, can chip 7 to 9 tonnes per hour of aspen poles having a 10- to 15-cm diameter. A study conducted some time ago at the National Renewable Energy Laboratory (NREL) was used as the basis for determining the cost of the secondary grinding of such aspen chips to smaller particle sizes[6].

That NREL study measured the shaft power required to operate a hammer mill or a knife mill having similar throughputs to grind aspen, straw, corn stover, or corn cobs. With aspen feed, various particle-size distributions were obtained by using screens with different hole sizes. The hammer mill was found to require a relatively large amount of energy for all screen sizes, apparently due to the large percentage of very fine particles produced (e.g., 50 wt% smaller than 0.72 mm with the 6-mm (1/4-inch) screen size). The knife mill required significantly lower power for the same screen sizes and produced a much smaller amount of fines (e.g., only 5 wt% smaller than 0.72 mm with the 6-mm screen). Only with the smallest screen, did the power requirement of the knife mill exceed that of the hammer mill.

The data for the power required for grinding was converted to estimated cost per tonne of ground feed by making the following assumptions:

a) Costs are in US$ (April 1995, Chemical Engineering Index = 381);

b) Electricity costs were $0.05/kWh;

c) Capital costs are based on a mill with a 10 tonne/h throughput (8,000 hours operation/yr);

d) the cost of a knife mill was assumed to be equivalent to that of a hammer mill for the same throughput;

e) Estimated costs of mill and electric motors were from charts in the literature [7];

f) Total installed costs are 4.15 times equipment costs;

g) Capital costs are for 15% interest and 20-year life (Annual Capital Recovery Factor = 0.16);

h) Annual maintenance costs are 5% of installed costs;

i) Labor costs are based on $25,000/yr/person with 0.25 persons per shift and four shifts, including overhead;

The weight-average particle size of the ground feed was determined from data presented in reference [6], based on the 50 wt% points from probability plots of screened particle sizes versus cumulative weight per cent.

Table 1 shows the cost breakdown by major category for use of the different screens with the knife mill and the hammer mill. The electrical cost is the largest component of the total cost. To illustrate the sensitivity of grinding costs to electrical cost, Figure 1 shows the estimated total amortized cost to grind to various particle sizes for electrical costs of $0.03 to 0.07/kWh. The total additional cost of grinding is seen to vary from $0.80/tonne for the 3.4-mm (0.13 inch) average particle size at $0.03/kWh with a knife mill to $11/tonne for an average particle size of 0.25 mm and $0.07/kWh with a hammer mill.

Table 1 also shows that with the hammer mill used in this study, the weight-average particle size was very small compared to the screen size (e.g. 0.72-mm average particle size with the 6-mm [1/4-inch] screen). It appears that the wood chips are quickly shattered by the hammers in the hammer mill, rather than a gradual attrition taking place that would allow the screens to better control the particle size. This is in agreement with our experience at NREL that feed produced with hammer mills has a large amount of powdery fines. In contrast, the weight-average particle size of the product was controlled much

Table 1. Feedstock grinding costs for 3/4" aspen chips for the knife mill (KM) and the hammer mill (HM).

Screen size, inches	Wt. Ave. Particle size, mm Himmel)	Power, kwh tonne (Himmel)	Elect. cost, $/tonne @ $0.05 kWh	Capital and Maint. Cost, $/tonne	Labor Cost, $/tonne	Total Cost, $/tonne
1/16 KM	0.80	132	6.60	1.43	0.31	8.34
3/32 KM	0.92	88	4.40	1.04	0.31	5.75
1/8 KM	1.05	53	2.64	0.67	0.31	3.62
1/4 KM	1.68	28	1.38	0.42	0.31	2.11
3/8 KM	2.42	16	0.82	0.30	0.31	1.43
1/2 KM	3.35	9	0.44	0.23	0.31	0.98
1/16 HM	0.24	137	6.85	1.25	0.31	8.41
3/32 HM	0.32	127	6.35	1.40	0.31	8.06
1/8 HM	0.32	127	6.35	1.40	0.31	8.06
1/4 HM	0.72	105	5.25	1.20	0.31	6.76

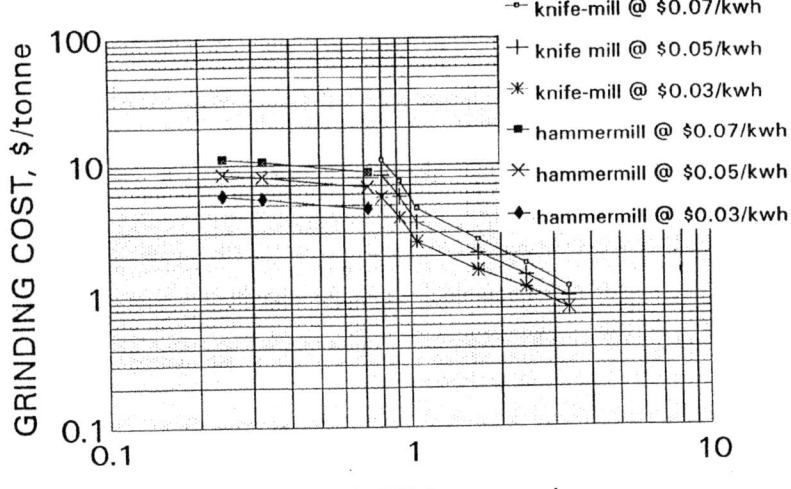

Figure 1. Grinding cost versus particle size

better by the screens in the knife mill used in the NREL study [6], which is also in agreement with our visual observations that knife milled material is low in fines.

As the average particle size of the knife-milled product is reduced past the 1.0-mm (0.039-inch) range, the power required is significantly more sensitive to the screen size and the resultant particle size, as seen in Figure 1. If the desired average particle size range is below 0.90 mm (0.035 inch), it would appear that the hammer mill would be more economical.

However, because the knives of a knife mill are subject to damage from stones and tramp metal that may be mixed in with the wood chips, the knife mill will be practical only with very clean feedstocks. Unless stones and tramp metal can be precluded or removed from the wood chip feed, it would appear that the hammer mill would be the preferred grinding machine to produce feed small enough for most fast-pyrolysis reactors. Thus, the predicted cost to convert 20-mm (3/4-inch) wood chips to a suitable feed for generic fast-pyrolysis reactors is in the range of $5 to $10/tonne (except for scaled-up ablative reactors, which are expected to require no grinding operation). Compared to the U.S. Department of Energy's cost goal of $42/ton of dry feedstock ($46/tonne), this grinding cost is significant.

3. Vortex Reactor

3.1 Design Considerations

The initial vortex reactor design was conceived to improve on the coiled-tube, entrained-flow fast-pyrolysis reactor developed at China Lake [8,9,10]. Cold-flow studies of entrained particle flow in transparent tubing showed that the particles were sliding only on that portion of the inside of the coiled tube, that was at the maximum distance from the axis of the coil. This meant that most of the tube's surface was not transferring heat to the pyrolyzing particle by the sliding contact. (The heat transfer by radiation was calculated to be relatively small in comparison.) It was thought that a better design would use all of the hot reactor wall surface for the sliding contact. In addition, in the entrained-flow tubular reactor, the gas and vapors have only a slightly lower residence time than the solids, whereas it was desirable for the particles to have a much longer residence time to ensure complete pyrolysis.

The use of a reactor modeled after a cyclone separator was considered, but not chosen because the heat-transfer surface area of a cylinder is greater than a cone for the same diameter and length. It was later reported that the solids' residence time in a cyclone separator is inversely proportional to the particle size [11], which is the opposite of the desired result for complete pyrolysis of the larger particles.

The use of a hot, rotating mechanical device to move the feed on the hot surface was thought to have developmental problems and was avoided. Likewise, the use of heat transfer from hot particles to the biomass particles was not chosen because of the increased ash expected in the char from the attrition of the thermofor material and the need to periodically add makeup thermofor.

3.2 Vortex Reactor Evolution

As shown in Figure 2, the original vortex-reactor design consisted of a tangential entrance into a horizontal 13-cm (5-inch) pipe. The outlet of the reactor was an axial tube located at the opposite end from the entrance that protruded about 5 cm (2 inches) into the reactor. The reactor was approximately 70 cm (27.5 inches) long. Initial operation of this reactor quickly revealed that the particles were taking a very coarse helical path through the reactor, rather than the desired tight spiral. This coarse path was found to have a pitch of about 1.3 times the reactor tube diameter, which was independent of entering velocity, temperature, gas molecular weight, and particle size. A review of the literature revealed that this coarse pitch is also seen in cyclone separators, Ranque-Hilsch tubes, and vortex tubes [12]. The solution to the coarse, natural spiral path of the entrained particles was the machining of a wide spiral groove into the wall of the reactor, leaving a narrow, raised spiral rib. The spiral rib was shown by high-speed movies at 4,000 fps to distribute the fast-moving solid particles uniformly on the inside surface of the reactor in cold flow. (Similar movies made during pyrolysis were not fast enough to photograph the better lubricated, faster moving particles.) [13,14]

Even with the use of the spiral rib to increase the residence time of the particles in the vortex reactor, adding a tangential solids recycle loop to the end of the reactor (as shown in Figure 3) was found to be necessary. For a time, these recycled solids swept the feed from the screw feeder for a common entry of fresh and recycled feed into the eductor [15]. However, it was found that over time the pyrolysis oil vapors would condense on the colder surfaces at the feed entry to create deposits that would solidify and then flake off. The recycled solids were then sent directly to the eductor, which resulted in the recycled solids and the fresh feed entering separately into the eductor [15].

During modeling efforts, it was recognized that the amount of energy required by the eductor to reaccelerate the recycled solids is significant. The eductor was then modified to bring the rapidly moving, recycled solids into the eductor at an acute angle and in the same general direction as the eductor jet to help conserve the kinetic energy of the recycled solids [4].

Although the vortex reactor used very high entering velocities, we have observed very little erosion with clean biomass feedstocks. However, feedstocks with tramp metal, dirt, and abrasives result in rapid, very localized erosion at the entrance of the vortex reactor. This erosion appears to be caused by high-velocity abrasive particles that enter on a chord rather than tangentially. In order to allow the vortex reactor to successfully operate with dirty feedstocks, we have modified the vortex reactor to have an easily replaceable, dove

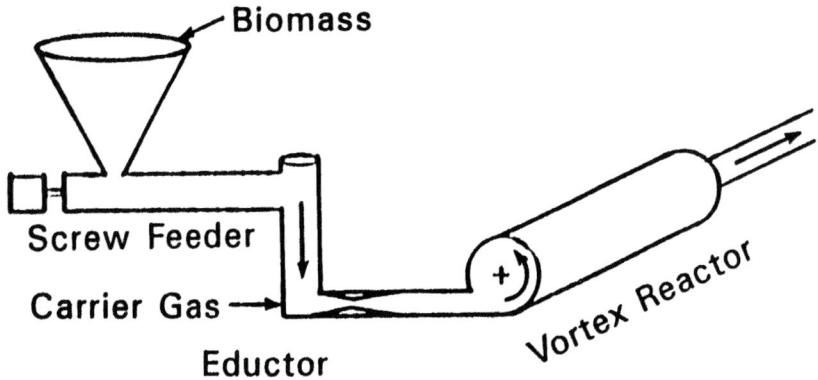

Figure 2. Original vortex reactor design

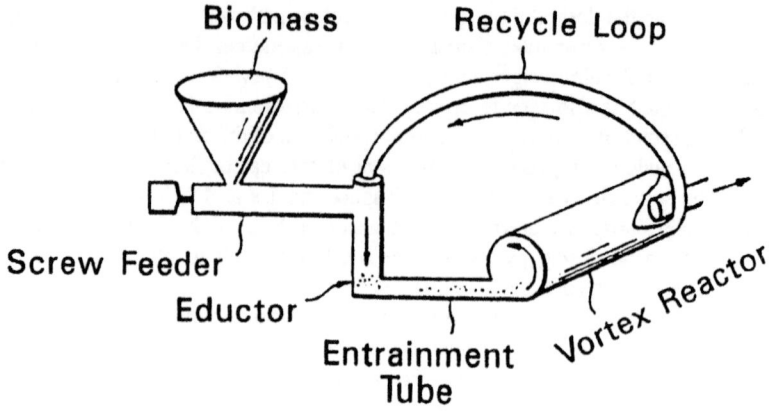

Figure 3. Vortex reactor with recycle loop

tailed wear plate that is accessible by removing the flanged head of the reactor, as shown in Figure 4. We have used wear plates made of stainless steel, as well as ceramic materials such as reaction-bonded silicon carbide [4].

While using organic feedstocks having a lot of tramp abrasives (e.g., refuse-derived fuel), we observed that large, slow-moving inert, solid particles could slow down and stop in the boundary layer of the bottom of the horizontal reactor. This led to an undesirable accumulation of solids in the bottom of the horizontal vortex reactor. The solution to this problem was to rotate the vortex reactor 90 degrees to a vertical orientation as shown in Figure 5, to allow the slow-moving, inert solids to fall to the bottom of the reactor, where they were re-entrained into the solids-recycle loop [4].

The change to a vertical vortex reactor apparently solved the solids-accumulation problem observed in the horizontal vortex reactor, but resulted in an increase in the erosion of the reactor by the recycled abrasives. The solution to this accumulation of abrasives in the vortex reactor was to add a special three-way valve to the recycle loop, as shown in Figure 6. This three-way valve normally allows the solids-recycle flow to pass unimpeded. Operationally, the feed is periodically shut off for a few seconds to allow the completion of the pyrolysis to take place. The three-way valve is then rotated to divert the entrained, inert solids into a small cyclone separator to remove the inert solids from the entraining recycled gases. The three-way valve is then returned to the straight-through position and the feeding is re-started. The vortex reactor is an inherently stable system that allows this type of stop-start operation with a minimum of problems [4].

3.3 Vortex Reactor Scale-up

At the request of a small private firm, we designed a 35 tons per day (TPD) vortex reactor, based on our previously presented computer model[13]. This NREL-designed reactor had the latest features described above, with the exception of the tramp-solids-removal valving. The reactor had a nominal inside diameter of 30.5 cm (12 inches) and was 4.6 m (15 feet) long. A raised, spiral rib was machined into the inside diameter of the reactor with a 10.2-mm (4-inch) pitch. This reactor was successfully machined on a large conventional lathe in

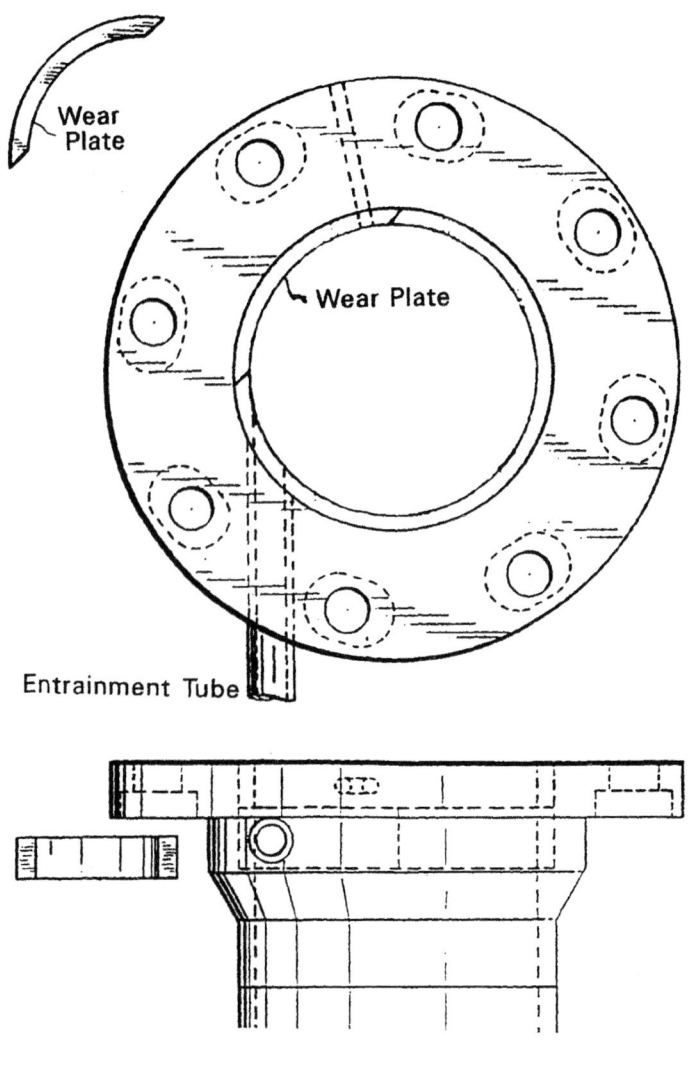

Figure 4. Replaceable wear plate in dove tailed slot

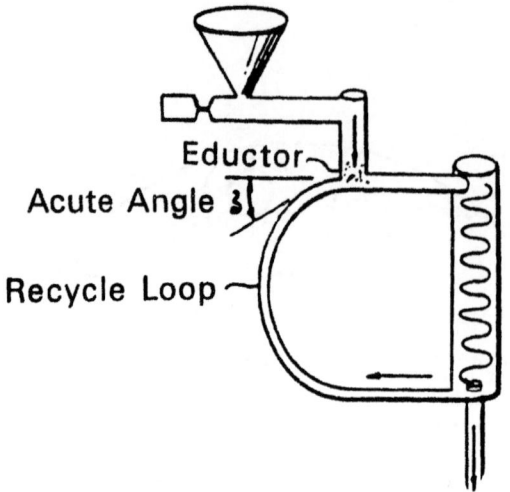

Figure 5. Vertical orientation of vortex reactor and recycle loop entering the eductor

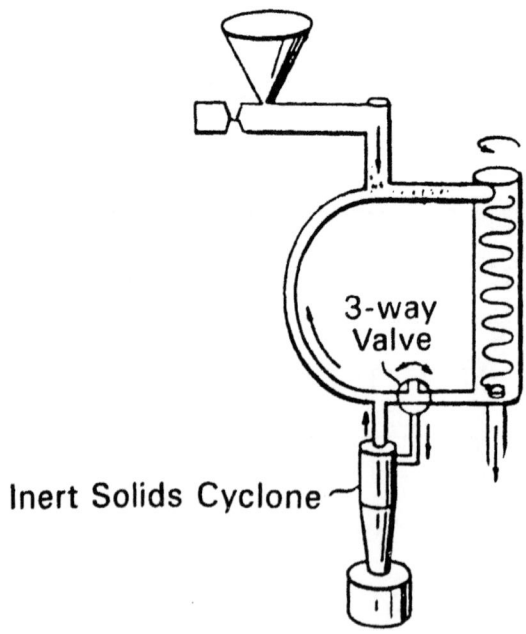

Figure 6. Special three-way dump valve to remove inert solids from vortex reactor

several sections, which were then welded together. The vortex reactor was to have been installed in a high-flux vortex furnace. The recycle loop was a 2-inch pipe [16]. Unfortunately, the small firm did not finish the installation of the completed vortex reactor, due to financial problems. NREL is no longer contractually involved with this firm and would like to transfer this technology to a new client.

4. Summary

The ability of the vortex reactor to ablatively pyrolyze large particles of biomass is seen as a distinct advantage over pyrolysis systems that do not rely on ablative pyrolysis and require ground feedstock. The grinding of the feedstock with a hammer mill was shown to add between $5 and $10 per ton, which is a significant part of the feedstock cost.

NREL's work on the vortex reactor has resulted in improved operation for a wide variety of feedstocks. The raised, spiral rib has been very effective in distributing the rapidly moving biomass particles in the reactor. Although we have successfully fed a wide variety of feedstock particles to this reactor, we have not demonstrated the use of large chips as feed because of the small size of the reactor. The use of replaceable wear plates has demonstrated that suitable abrasion-resistant materials can withstand this high-velocity environment. The removal of tramp, inert solid particles from the reactor system with very little downtime has shown that realistic waste biomass feedstocks containing dirt and abrasives can be processed.

The scale-up of the vortex reactor to at least 35 TPD appears to be quite feasible from the predictive model developed. Larger reactors also appear to be feasible, but we recommend that the risk of scale-up be minimized by gaining more pilot-plant experience.

5. References

1. Roy, C.; Blanchette, D.; de Caumia, B.; and Labrecque, B. (1994) "Conceptual Design and Evaluation of a Biomass Vacuum Pyrolysis Plant," *Advances in Thermochemical Biomass Conversion*, A.V. Bridgwater, ed., Blackie Academic and Professional, London pp. 1165-1186.
2. Diebold, J.P. (1980) "Ablative Pyrolysis of Macroparticles of Biomass," *Specialists' Workshop on Fast Pyrolysis of Biomass*, Copper Mountain, CO, Solar Energy Research Institute, Golden, CO, SERI-CP-622-1096, pp. 237-252.
3. Lédé, J.; Diebold, J.P.; Peacocke, G.V.C.; and Piskorz, J. (1996) "The Nature of Intermediate Unvaporized Pyrolyzed Biomass Materials," Developments in Thermochemical Biomass Conversion, Banff, May 20-24, (in press).
4. Diebold, J.P. and Scahill, J.W. (1995) "Improved Vortex Reactor System," U.S. Patent 5,413,227.
5. Solantausta, Y,; Diebold, J.P.; Elliott, D.C.; Bridgwater, A.V.; and Beckman, D. (1994) *Assessment of Liquefaction and Pyrolysis Systems*, Technical Research Centre of Finland, VTT Research Notes 1573, 123 pp. plus appendices.
6. Himmel, M.; Tucker, M.; Baker, J.; Rivard, C.; Oh, K.; and Grohmann, K.(1985) "Comminution of Biomass Hammer and Knife Mills," *Biotechnology and*

Bioengineering Symp. No. 15, pp. 39-58.

7. Garrett, D.E. (1989) *Chemical Engineering Economics*, Van Nostrand Reinhold, NY, pp. 291 and 292.

8. Diebold, J.P.; Benham, C.B.; and Smith, G.D. (1976) "R&D in the Conversion of Solid Organic Wastes to High Octane Gasoline," *AIAA Monograph Volume 20 on Alternate Fuel Resources*, Western Periodicals Co., pp. 322-325.

9. Diebold, J.P. (1980) "Gasoline from Solid Waste by a Noncatalytic, Thermal Process," *Thermal Conversion of Solid Wastes and Biomass*, J.L. Jones and S.B. Radding, eds., ACS Symposium Series 130, pp. 209-226.

10. Benham, C.B. and Diebold, J.P. (1980) "Pyrolysis in a Tubular Entrained Flow Reactor at China Lake," *Specialists' Workshop on Fast Pyrolysis of Biomass*, Copper Mountain, CO, SERI -CP-622-1096, pp. 271-286.

11. Lédé, J.; Verzaro, F.; Antoine, B.; and Villermaux, J. (1980) "Cyclone Reactor for Flash Pyrolysis of Solid Particles," *Specialists' Workshop on Fast Pyrolysis of Biomass*, J. Diebold, ed., Copper Mountain, CO, Solar Energy Research Institute, Golden, CO, SERI-CP-622-1096, pp. 327-346.

12. Diebold, J.P. and Scahill, J.W. (1985) "Ablative Pyrolysis of Biomass in SolidConvective Heat Transfer Environments," *Fundamentals of Thermochemical Biomass Conversion*, R.P. Overend, T.A. Milne, and L.K. Mudge, eds., Elsevier Applied Science Publishers, London, pp. 539-556.

13. Diebold, J.P. and Power, A.J. (1988) "Engineering Aspects of the Vortex Pyrolysis Reactor to Produce Primary Pyrolysis Oil Vapors for Use in Resins and Adhesives," *Research in Thermochemical Biomass Conversion*, A.V. Bridgwater and J.L. Kuester, eds., Elsevier Applied Science Publishers, London, pp. 609-628.

14. Diebold, J.P. and Scahill, J.W. (1988) "Production of Primary Oils in a Vortex Reactor," *Pyrolysis Oils from Biomass - Producing, Analyzing, and Upgrading*, E. Soltes and T.A. Milne, eds., ACS Symposium Series 376, pp. 31-40.

15. Diebold, J.P. (1980) *The Cracking Kinetics of Depolymerized Biomass Vapors in a Continuous, Tubular Reactor*, Master's Thesis T-3007, Colorado School of Mines, Golden, CO.

16. Johnson, D.A.; Maclean, D.; Feller, J.; Diebold, J.; and Chum, H. (1994) "Developments in the Scale-Up of the Vortex-Pyrolysis System," *Biomass and Bioenergy*, Vol. 7, Nos. 1-6, pp. 259-266.

REMOVAL OF RESIDUAL CHAR FINES FROM PYROLYSIS VAPORS BY HOT GAS FILTRATION

J. SCAHILL, J. P. DIEBOLD, and C. FEIK
National Renewable Energy Laboratory
Center for Renewable Chemical Technologies and Materials
Golden, Colorado, USA

Abstract

An NREL-designed vortex reactor fast pyrolysis process development unit (PDU) has been used to investigate hot gas filtration of biomass pyrolysis vapors. Most of the experimental work employed a conventional baghouse type of filter that used NEXTEL™ ceramic cloth filter bags as the filter medium.

A series of experimental runs demonstrated that hot gas filtered biocrude oils having less than 10 ppm of total alkali could be reproducibly made. Removal of the char cake from the filter elements proved to be a difficult problem. The char appears to become progressively more sintered to itself and the filter as a function of the cumulative biomass processed. Controlled oxidation does remove this dense char from the filters, but leaves residual ash on the filter cloth fibers. This ash may in turn cause subsequent biomass pyrolysis vapors that pass through the filter to produce additional char (coke) in the interstices of the filter cloth. Data are presented that suggest this char formation may contribute to a more rapid rise in the rate of filter blinding as measured by the increase in recovered filter pressure drop.

1 Introduction

The use of biomass pyrolysis oils in relatively sophisticated applications, such as power generation via gas turbines or diesel engine generators, has attracted increasing interest in recent years. These applications hold the promise of providing cost-effective markets for this form of renewable energy from biomass. Fuels used in these systems are required to meet specifications related to their physical, chemical, and combustion properties that affect the reliable long term operation of the application equipment. The more important properties are viscosity, ash, alkali metal content, heating value and Cetane number. Pyrolysis oils must be produced in such a manner that they will consistently meet the standard specifications required by the end-use application.

It has been long recognized that the physical properties of biomass pyrolysis oils change over extended periods of time. Recent studies have reported the rate of this change [1], which is characterized primarily by increases in average molecular weight and water content. When molecular weight increases, viscosity also increases. Other recent work has linked the presence of char fines in the oil to this viscosity increase [2]. Char has also been identified as the source of alkali metal, a contaminate that limits the use of pyrolysis oil in turbine applications [3]. Thus, the presence of char fines degrades the quality of pyrolysis oils and weakens their ability to penetrate the higher quality fuel markets. To meet the specification requirements of the high value end-user applications, high quality pyrolysis oils having uniform, reproducible properties must be free of char fines.

Char fines originate in the fast pyrolysis processes that produce high yields of these oils from solid biomass feedstocks. During processing, char and pyrolysis gases are produced simultaneously with oil vapors and are usually co-mingled with the vapors in the resulting process stream. Most reactor designs employed to pyrolyze biomass subject the char to some degree of attrition, either from the heat transfer medium (fluidized beds and some entrained flow designs) or during particle acceleration (vortex type). The exception to this is the staged-hearth, vacuum pyrolysis reactor developed at the University of Laval in Quebec, Canada [4].

The low density char from these processes, typically a powder as a result of this attrition, usually is entrained in the gas and vapor stream exiting the reactor. Cyclone separators have traditionally been used to remove solid char particles from pyrolysis vapor/ gas streams because of their low cost and reliable design. However, a small but significant portion of the char from pyrolysis processes falls below 5μm particle size and consequently is not captured by cyclones. Filtration and electrostatic precipitators (ESP) are both effective at removing this size particle from the gas stream. ESPs are more expensive from both a capital and operating standpoint.

NREL has been developing the use of high temperature filtration for separating char fines from a pyrolysis gas/vapor stream for the past three years [5,5a]. Advantages to filtering the gas/vapor stream prior to condensation are viewed as follows:

1) The char and oil are discretely separated to enable each to be marketed individually.

2) Separation prior to condensation of the vapors eliminates the detrimental influence of char on long term oil stability.

3) This process eliminates the monetary and environmental costs of disposing of the liquid /solid sludge produced by filtering char from the oil.

The major technical disadvantage of hot gas filtration is a small yield penalty resulting from vapor-cracking reactions in the baghouse filter at the higher temperatures necessary to prevent condensation of the vapors. It appears that a compromise will need to be made between a superior quality oil and slightly lower yields.

Hot gas filtration is a relatively new technology, and most of the advances in systems and materials have taken place since the early to mid 1980s [6]. Much of this work, which was focused on combustion and gasification processes, attempted to meet low particulate emission standards or specifications for the resulting gases. This early work, developed materials and designs that can withstand processing temperatures ranging up to 1000°C. Reliable long-lived filters and efficient particulate capture were the principal objectives of this development effort. Systems employing hot gas filtration and the operating principles involved have been presented in other publications and will not be redescribed here [7].

High temperature filtration in pyrolysis applications has different design constraints than those needed for combustion and gasification. The time/ temperature relationship of the vapors as they pass through the filter device is very important for pyrolysis. The optimum operating temperature is the lowest possible that will avoid condensing vapors on the filter cake or filter media. The dew point of this condensation is also a function of the amount of inert carrier gases present with the vapors. Residence time of the gas/vapor stream at elevated temperatures is generally not a concern for combustion or gasification processes, but it is a significant issue when the filtration is done on pyrolysis vapor streams. In conventional baghouse filter designs the volume adjacent to the filter elements is fixed at a minimum value based on the net gas throughput of the unit. If volume is too low, high gas velocities quickly re-entrain particles that have been dislodged from the filters following backpulsing. So, for any given system, the residence time is fixed, based on the design gas throughput. This gas throughput also establishes the face velocity or air to cloth ratio and the size (volume) of the vessel holding the filters [7]. For the gas/vapor velocities used in our process at NREL [5,5a], the residence time in the baghouse was approximately 5-6 seconds. In two separate runs using the same poplar feedstock, organic liquid yields increased from 36 wt% to 49 wt% (dry biomass basis) as the baghouse temperature was reduced from 410-450°C to 370-390°C. This may be the low end of the temperature range that can be safely used without condensing vapors on the filter, except for the case of relatively high carrier gas flow rates. However, it may be possible to reduce the residence time of the vapors by considering filter designs other than conventional baghouses.

2 Conventional Baghouse Filter

The filter used was a conventional baghouse filter designed with the minimum recommended volume and which employed flexible high temperature ceramic fiber cloth bags made by 3M as the filter medium. This cloth material (NEXTEL™), has a maximum operating temperature of 760°C (in the presence of alkali oxides). The oxides will melt above this temperature and form a glass that fuses to the fibers and cause them to become brittle [8]. The filter cloth has a nominal pore size of about 2μm, but in practice, it relies on the establishment of a cake of particles on the filter surface to actually do the separation of subsequent particles, especially the very small ones. The baghouse is fitted with four filter elements, each having a nozzle and venturi throat to provide a short pulse of gas inside the filter that imparts a rapid expansion front down the cloth bag to dislodge the cake when the pressure drop becomes excessive. This is the same baghouse filter that was used at NREL [5,5a], but the gaseous residence time is about half that previously reported.

3 Experimental Process Conditions

The following is a description of the relevant operating conditions used for the experimental activities conducted with the two types of filters described above.

3.1 Vortex Reactor

The pyrolysis operating conditions used in all experimental runs were essentially the same:

- Nitrogen carrier gas @ 650°C and 41 kg/hr flow rate (fixed for all runs)
- Vortex reactor wall temperature @ 625°C
- Vortex reactor exit gas/vapor temperature was initially at 520°C prior to feeding but equilibrated to 455°C at steady state feed conditions
- Biomass solids feed rate varied from 17 kg/hr to 24 kg/hr depending on the specific run
- Nitrogen carrier gas to oil vapor ratio: 3.6 to 2.6

3.2 Baghouse

The baghouse operating temperatures ranged from 385°C to 405°C at the entrance, which was controlled by the addition of liquid and/or gaseous nitrogen. This method of temperature control was quite good with deviations of only 3°- 4°C during a run. However, zone temperatures (bottom, middle, top) inside the baghouse typically varied by about 10°- 15°C; the top zone was consistently hotter.

3.3 Condensation Train

The condensation train was operated to reduce the gas/vapor stream to the lowest temperature possible. The initial stage of this temperature reduction occurs in the venturi scrubber where the gas/vapors enter at 360°-375°C and are quenched to 85°-90°C when passed through the throat area concurrently with recycled biocrude condensates. When starting up without the benefit of the recirculated condensates this quenching occurs by a combination of recirculating cooled pyrolysis gases and heat transfer to water-cooled walls. This typically resulted in an exit temperature of about 200°C. It usually required between 10-12 kg of feed to be processed before enough condensates were collected to begin the recirculation. Following the venturi scrubber the process stream passes through two conventional shell and tube heat exchangers where the temperature is reduced to between -5° and +5°C before entering the coalescing filters. This stream can rise to as much as 8°C at the exit of the filters. The gas/vapor stream usually starts out at the low end of this range and rises as the run progresses.

3.4 Feedstock

A hybrid poplar (*P. trichocarpa x P. deltoides*) was used for all these experiments [5,5a]. This biomass was obtained from the James River Paper Company and grown on the tree plantation that supplies feed to their pulp mill. The trees were harvested seven years into their growth cycle and debarked to a level of 1.0% bark remaining in the final dry chips. This feed is probably typical of that which would be obtained from a dedicated woody biomass energy farm. The chips were knife milled through a 3.2 mm screen and bone dried at 105°C before feeding to the process. Feedstock Analysis is shown in Table 1.

4 Process Operation Experience

Research collaboration and subcontract activities made it necessary to produce significant quantities of hot gas filtered biocrude oil. On a previous run with the baghouse filter (Run 175) we had demonstrated the ability to produce oils in reasonable yields with very low char levels. During this run two cyclones were used up-stream of the baghouse to remove the bulk of the char prior to the filter. This configuration pre-separated the coarser particles and sent only the very fine char to the filter which formed a denser cake that appeared to have sintered and was not removable by backpulsing. This phenomenon has also been observed in other hot gas filtering applications [9] where the cake is formed from a narrow particle size distribution of solids.

With the baghouse in place, five runs (numbered 6 through 10) were conducted with the objective of producing quantities of biocrude having low char and alkali metal content. The current system was configured without the cyclones upstream of the baghouse so that a broad distribution of char particles would be sent to the baghouse. This size distribution was expected to develop a char cake that would be less dense and less likely to sinter on the filter. It was also desirable to operate the process at steady state conditions with a narrow temperature range in the baghouse.

Good process temperature control and steady state operation was achieved on all runs except the first. Figure 1 shows trends of key process temperatures for run M2-8, which was typical for the other runs as well. However, it was not possible to reach an equilibrium recovered pressure drop across the filter on any of these runs except for a portion of run M2-10; instead, this pressure drop continued to increase at a steady rate throughout the run as can be seen in figure 2 for run M2-8. The recovered pressure drop is defined as the pressure drop measured immediately after going through a backpulsing cycle to dislodge the accumulated filter cake. In typical baghouse filter operation, the recovered pressure drop initially rises rapidly and then asymptotically approaches a relatively constant value. In all of these runs the pressure drop eventually rose to a level that caused excessive pressure in the eductor area of the process which then disrupted the ability to feed smoothly. This increase in pressure drop across the filter is caused by the char that could not be removed from the filter by backpulsing. From a practical standpoint, the long term continuous operation of the process was limited by this pressure drop across the filter and occurred after processing 80 to 100 kg of feed per m^2 of filter surface. During these runs, we looked at changing baghouse temperatures, backpulse pressure and duration to more effectively remove the filter cake. However, as can be seen from figure 3, which shows the recovered pressure drop as a function of the cumulative feed processed, these variables appeared to have no influence on the ability to remove the adhered char from the filters. The rate of pressure drop increase appears to be similar for all the runs regardless of the operating temperature or backpulse pressure and duration used.

Following the runs a dense permanent cake of char approximately 1-2 cm thick covered the entire surface area. On some runs, char bridged between the filter elements. The char could be scraped off in chunks or small sheets that easily crushed to a fine powder.

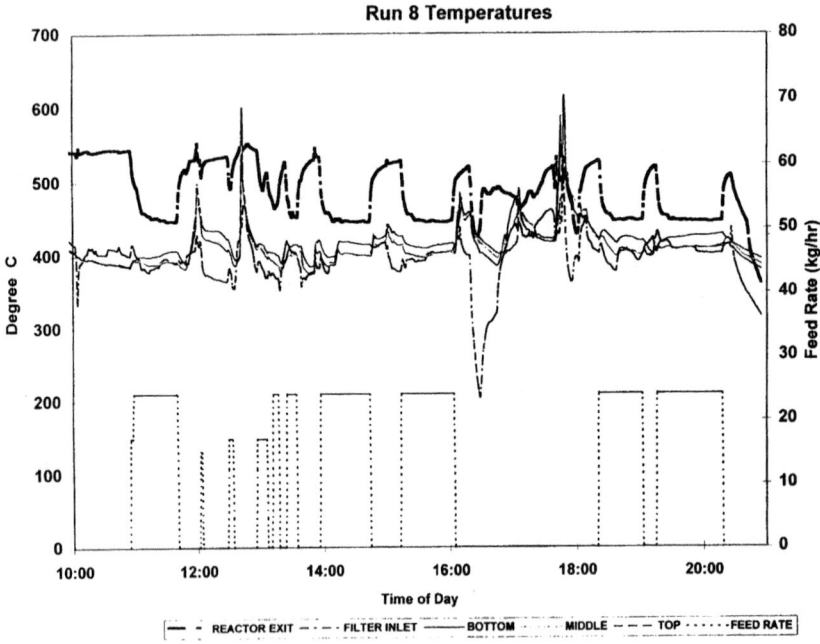

Fig. 1. Example of reactor exit and baghouse temperature profile for run M2-8

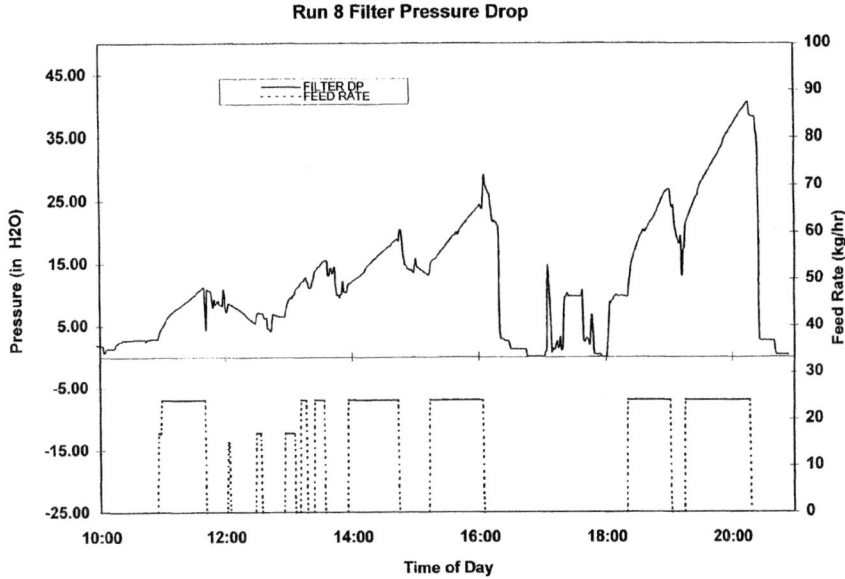

Fig. 2. Example of baghouse pressure drop profile for run M2-8

Once this cake was established, filters had to be removed from the baghouse and vacuum cleaned _ a very time consuming process. The possibility of using controlled oxidation as a means of regenerating the baghouse was explored before runs M2-7, M2-8, and M2-9. New filter bags were installed prior to run M2-10 instead of oxidatively regenerating after run M2-9. The NEXTEL™ filter bag temperature could not exceed 760°C in the presence of alkali metal oxides, because at these temperatures the metal oxides melt and fuse the fibers of the NEXTEL™ cloth together, making it brittle [8]. The baghouse was configured with thermocouples located 1 cm from the bags to monitor the gas temperature during the oxidation. Since it was not possible to measure the actual bag temperature, we carried out the regeneration while maintaining a 600°C temperature at the nearest thermocouple. The temperature was easily controlled by introducing measured air flows into the nitrogen carrier gas stream while the baghouse was at temperature. During the oxidation process the baghouse was periodically backpulsed to speed up the char removal rate. Although this procedure permitted complete recovery of the baghouse initial pressure drop, it took between 6-9 hours to complete the regeneration. Initially this appeared to be a viable method of regenerating the filters in the baghouse, however, further investigation indicated that this procedure may in fact lead to a faster rate of filter blinding (see discussion section).

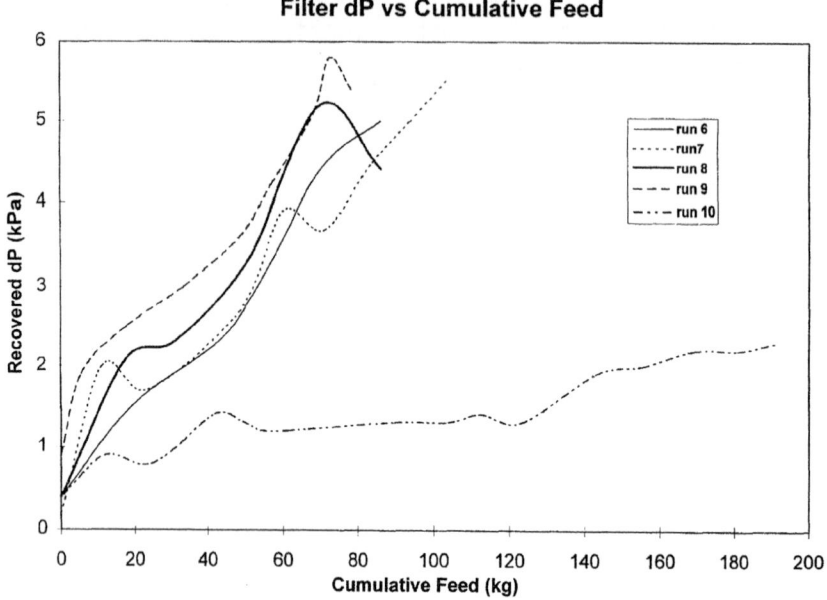

Fig. 3. Baghouse Recovered Pressure Drop vs. Cumulative Feed Processed. The recovered pressure drop is measured immediately after the filters have gone through a back pulse cycle.

5 Oil Properties and Yields

Ultimate analyses of the oils from each run are shown in Table 1. Elemental compositions of the oils are very similar except for a slightly lower oxygen content in oils from runs M2-7 and M2-9. Yields are presented in Table 2 and show slightly higher water yields from these two runs. Heating values for these runs are also slightly higher, which would be expected with less oxygen.

Trace element concentration was determined by instrumental neutron activation analysis and values are given in table 3. Alkali metal content of oils from runs M2-6, M2-7, M2-8, and M2-9 are very consistent but it is almost an order of magnitude lower in oil from run M2-10. M2-10 oil is the cleanest oil we have produced to date; the alkali metal levels almost achieving the 1 ppm maximum specified for a fossil-fuel derived turbine fuel. This low sulfur oil could probably be used in a turbine without harm to the blades. The ash levels in the ultimate analysis appear to correlate well with the alkali metal concentration as determined by instrumental neutron activation analysis.

The yields are not as high as those obtained on the original process operated at NREL [5,5a]. The coarser coalescing filters, which are not designed to recover all of the aerosols from the process stream, contribute to the yield losses but are necessary to minimize the overall system pressure drop. The higher relative carrier gas throughputs used in this process will also strip away more of the volatile compounds, in addition to some of the water, and send them to the process flare. However, these same gas throughputs have reduced the residence time in the baghouse by 41% (6 seconds to 3.5 seconds). Combined with the vortex reactor (0.3 sec) and transfer line (0.3 sec), the total gas/vapor residence time in the hot zone is about 4.1 seconds. At this lower residence time, losses due to cracking of the vapors should be reduced. The char yields appear to be linked to the biomass feed rates. On run M2-9 the actual feed rates were approximately 17 kg/hr whereas in the other runs the actual feed rate was closer to 24 kg/hr. At the higher feed rates, incomplete pyrolysis of the char was observed, which contributed to higher char yields.

Table 1. Ultimate Analysis of Biocrude Oil Made From Hybrid Poplar

Component wt%	Poplar Feedstock	M2-6	M2-7
Moisture		18.9	19.6
Carbon	49.0	45.8	47.4
Hydrogen	6.0	5.3	5.3
Oxygen	44.1	28.9	27.5
Nitrogen	0.1	0.25	0.13
Sulfur	0.03	0.03	0.02
Ash		0.016	0.016
HHV MJ/kg	19.5	18.9	19.2
LHV MJ/kg	18.2	17.3	17.5
Viscosity @40°C cSt		46.5	50.4

Table 1. Ultimate Analysis of Biocrude Oil Made From Hybrid Poplar (Continued)

Component wt%	M2-8	M2-9	M2-10
Moisture	20.0	19.6	16.8
Carbon	45.6	47.0	48.1
Hydrogen	5.4	5.1	5.3
Oxygen	29.0	28.1	29.6
Nitrogen	0.05	0.09	0.14
Sulfur	0.02	0.04	0.04
Ash	0.018	0.018	0.007
HHV MJ/kg	18.5	19.1	18.8
LHV MJ/kg	16.6	17.5	17.3
Viscosity @40°C cSt	27.3	34.0	42.7

Table 2. Pyrolysis Product Yields From Hybrid Poplar

Yields wt% of feed (moisture free)	M2-6	M2-7	M2-8	M2-9	M2-10
Organic Liq.	39.0	41.8	36.8	44.1	46.2
Water	9.1	10.2	9.2	10.7	9.3
Char	23.3	24.1	27.6	14.7	19.9
Gas	N/A	13.1	15.8	N/A	14.1
Volatiles, H_2O (by difference)	N/A	10.8	10.6	N/A	10.5

Table 3. Trace Element Analysis of Biocrude Oil Made From Hybrid Poplar Feedstock

Element, ppm	M2-6	M2-7	M2-8	M2-9	M2-10
Calcium	6	4	3	4	1
Potassium	5	4	4	4	1
Sodium	4	2	2	1	0.9
Magnesium	<3	3	1	4	0.7
Chlorine	8	3	3	3	11
Aluminum	2	1	1	1	0.3
Titanium	<.2	0.7	0.2	0.2	0.2
Vanadium	0.01	0.01	0.01	0.01	<0.01
Manganese	0.2	0.2	0.1	0.06	0.04

6 Discussion

The rate at which the filters blinded progressively worsened after each oxidative regeneration of the baghouse. The final baghouse pressure drop, measured after a full cycle of backpulses (4), was plotted against the net cumulative feed. It can be seen from these data, in figure 3, that the initial rate of increase is much steeper for the three runs that were preceded by an oxidative regeneration of the baghouse. To investigate the possibility that the temperature of the filter bag fibers exceeded 760°C we removed a small test sample from one of the filter bags after run 9. This sample was then oxidized under controlled conditions to simulate regeneration in the baghouse, and then cut in half. One piece was subjected to air pulses through the fabric to remove residual ash fines (as would be done in the baghouse); the other piece was thoroughly washed in deionized water to remove any residual ash. The samples were then examined by SEM and XPS microprobe techniques to evaluate their surface characteristics. The SEM micrographs in Figure 4 clearly show very small particles on the surface of the air-pulsed sample and none on the new bag sample. Results from the XPS analysis are shown in Figure 5 and indicate that both the air pulse and washed samples have more potassium and calcium on the surface relative to a new, never-used bag. The XPS results suggest that the particles are ash remaining from the char oxidation. Ash present during biomass pyrolysis has been reported to polymerize pyrolysis intermediates to produce more char [10]. A similar effect has also been reported for the char itself [11,12] when biomass pyrolysis vapors are passed over it. Thus, pyrolysis vapors may be catalytically reacting with the ash present on the surface of the filter and also with the char in the filter cake. These reactions may contribute to physical changes in the char cake that cause it to become more difficult to remove as the run progresses. The initial high rate of pressure drop increase may be due primarily to the ash on the fibers catalyzing char (coke) formation reactions that plug the interstices of the filter cloth. A similar phenomenon may also occur in the char cake, but at a slower rate because the alkali in the char is only partially exposed at the surface [12]. The relatively low rate of recovered pressure drop increase seen for run 10, particularly in the early part of the run, is thought to be due to the use of virgin filter bags. However, even at the later stages of this run the recovered pressure drop is beginning to climb. The humps seen in the curves followed periods of multiple backpulse cycles. During these cycles, small sheets of char could have been dislodged from the filters to give greater pressure drop recoveries but would more quickly reload with char because the gas/vapor stream would preferentially pass through this area of lower resistance. Once reloaded, the recovered pressure drop should follow its previous path. This explanation is consistent with the data shown in Figure 3. The recovered pressure drop also begins to rise more steeply at the end of runs 6,7,8,and 9 as would be expected when the reactive char cake thickness grows and cannot be removed from the filter.

The ash remaining on the filter cloth fibers may also be contributing to the differences in ash, and consequently alkali levels, in the oils produced from this series of experimental runs. Some of this ash could have passed through the bags during the heat-up phase of the run, before feeding began, and end up in the condensation train. This could explain the much lower levels of alkali metals in the run M2-10 oil, which was filtered with new filter bags.

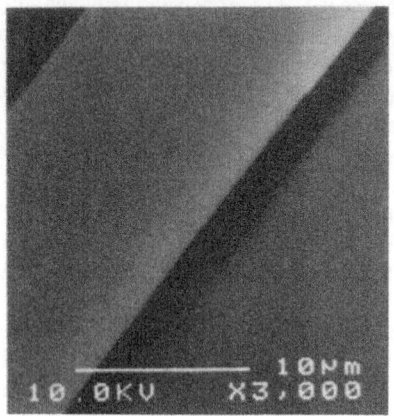 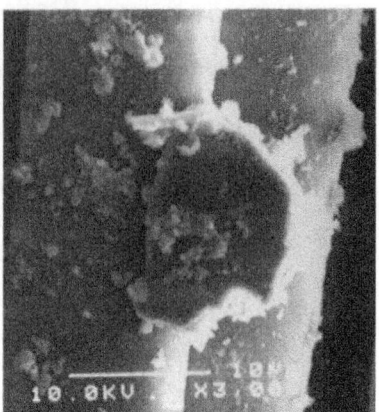

A   **B**

Fig. 4. SEM of NEXTEL™ filter cloth fibers. (A) a sample from a filter bag that has never been used. (B) the same material after three oxidative regenerations and subsequent air pulsing in an attempt to remove residual ash.

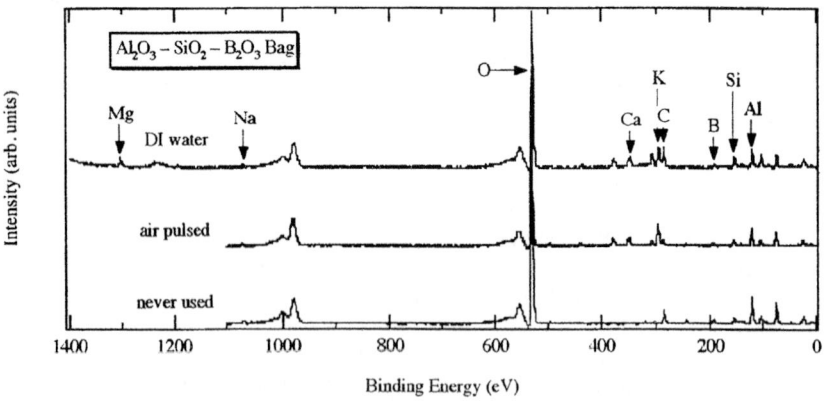

Fig.5. XPS analysis of the filter cloth samples shown in Fig.5. One sample was also washed in deionized water to evaluate the effects of rinsing on removing the residual ash.

7 Conclusions

Biomass pyrolysis oils having alkali metal concentrations below 10 ppm can be reproducibly made when employing hot gas filtration. However, it becomes progressively more difficult to dislodge the accumulated permanent char cake on the filter as more feed is processed. Standard baghouse filter operating techniques have proved ineffective in removing this char, especially when the cake thickness becomes greater than 1 cm.

Although it is possible to oxidatively regenerate the baghouse filter to recover the initial pressure drop, this technique also leaves residual ash on the surface of the filter cloth fibers that appears to exacerbate the rate of filter blinding. Given these observations and the known reactivity of biomass ash, oxidatively removing char from any type of hot gas filter is not recommended. The char cake itself also appears to promote a progressively higher pressure drop across the hot gas filter, although at a lesser rate than ash.

A more fundamental understanding of events within the char cake as biomass pyrolysis vapors pass through will be important in developing methods for long-term operation of hot gas filters.

8 Acknowledgments

We would like to acknowledge the U.S. DOE Office of Solar Thermal and Biomass Power Technology, Biomass Power Program, managed by Dr. Richard Bain, for supporting this work. Thanks also to Alice Mason and David Niles of NREL's Measurement and Characterization Center for providing the surface analysis of the NEXTEL™ filter cloth. Hazen Research Inc., and in particular Mr. Tim Quinn, must also be recognized for their technical support in making the production of these clean biocrude oils possible.

References

1. Czernik,S.(1995). "Storage of Biomass Pyrolysis Oils." *Proceedings, Biomass Pyrolysis Oil Properties and Combustion Meeting, September 26-28, Estes Park, Colorado*, Milne T.A. Ed. NREL-CP-430-7215, National Renewable Energy Laboratory, Golden, Colorado.
2. Agblevor, F.A., Besler, S., and Evans R.J. (1995). "Influence of Inorganic Compounds on Char Formation and Quality of Fast Pyrolysis Oils." ACS 209th National Meeting, Anaheim, California, April 2-5.
3. Agblevor, F.A., Besler, S., and Evans R.J. (1995). "Inorganic Compounds in Biomass Feedstocks: Their Role in Char Formation and Effect on the Quality of Fast Pyrolysis Oils." In *Proceedings, Biomass Pyrolysis Oil Properties and*

Combustion Meeting, September 26-28, Estes Park, Colorado, Milne T.A. Ed. NREL-CP-430-7215, National Renewable Energy Laboratory, Golden, Colorado, pp 77-89.

4.   Roy,C.,de Caumia, B., Pakdel, H. (1988). "Preliminary Feasibility Study of the Biomass Vacuum Pyrolysis Process" *Research in Thermochemical Biomass Conversion*, Bridgewater, A.V.,Kuester, J.L.,Eds., Elsevier Applied Science: New York, pp 585-596.

5.   Diebold, J.P., Czernik, S., Scahill, J.W., Phillips, S.D., and Feik, C.J. (1995). "Hot-Gas Filtration to Remove Char from Pyrolysis Vapors Produced in the Vortex Reactor at NREL." In *Proceedings, Biomass Pyrolysis Oil Properties and Combustion Meeting, September 26-28, Estes Park, Colorado*, Milne T.A. Ed. NREL-CP-430-7215, National Renewable Energy Laboratory, Golden, Colorado.

5a.   Diebold, J.P., Scahill, J., Czernik, S., Phillips, S.D., and Feik, C.J., (1996). " Progress in the Production of Hot-Gas Filtered Biocrude Oil at NREL", A.V. Bridgewater and E.N. Hogan, eds., CPL Scientific Information Services, ltd., Newbury , U.K., pp 66-81

6.   Bergman, L. (1993). The World Market for Hot Gas Media Filtration: Current Status and State-of-the-Art. *Gas Cleaning at High Temperatures*, Ed: Clift, R. and Seville, J.P.K., Published by Blackie Academic & Professional, Glasgow, pp 294-306.

7.   Croom, M., (July 1993). Effective Selection of Filter Dust Collectors, *Chemical Engineering*, pp 86-91.

8.   Gennrich, T.J.(1993). " High Temperature Ceramic Fiber Filter Bags",*Gas Cleaning at High Temperatures*, Ed: Clift, R. and Seville, J.P.K., Published by Blackie Academic & Professional, Glasgow, pp 307-320.

9.   Alvin, M.A., (1996). "Impact of Char and Ash Fines on Porous Ceramic Filter Life." *In Preprimts 211th ACS National Meeting*, New Orleans, LA, March 24-28, pp 672-675.

10.   DeGroot, W.F., Shafizadeh, F. (1984). "Influence of Exchangeable Cations on the Carbonization of Biomass." *Journal of Analytical Applied Pyrolysis*, 6(3), pp 217-232.

11.   Boroson, M.L., Howard, J.B., Longwell, J.P., Peters, W.A. (1989). "Heterogeneous Cracking of Wood Pyrolysis Tars over Fresh Wood Char Surfaces." *Energy & Fuels* Vol.3, pp 735-740.

12.   Agblevor, F.A. (April 1996) Personal communication, National Renewable Energy Laboratory

CO-COMBUSTION OF BIOMASS PRETREATED IN A FLUID BED PYROLYSIS UNIT

Hansen, M. W. and Henriksen, U., Department of Energy Engineering, Technical University of Denmark, DK-2800 Lyngby, Denmark.
Houmøller, S., dk-TEKNIK Energy and Environment, 15 Gladsaxe Möllevej, DK-2860 Söborg, Denmark.

Abstract

Keywords: Alkali, Biomass, Chlorine, Co-combustion, Fluid bed, Pyrolysis, Straw, Wood.

An atmospheric bubbling fluid bed pyrolysis unit has been developed to convert straw and wood into volatile matter to be combusted in a conventional boiler. This pre-treatment is a means of supplying a major proportion of the biomass energy to the boiler without the corrosive components.

The pyrolysis unit has been tested extensively as a means of pretreating the biomass before co-combusting straw and wood with fossil fuels in power plant boilers. The fluid bed pyrolysis unit converts the biomass into two fractions: a volatile fraction to be combusted in the power plant boiler and a solids fraction to be converted otherwise. The volatile fraction contains up to 90 per cent of the energy from the input biomass. The idea is to avoid feeding the corrosive alkali metals and chlorine present in the biomass into the boiler, as these cause corrosion damage to the superheater, yet still supply the boiler with most of the biomass energy. This principle keeps ash from biomass and ash from the fossil fuels separated, and thereby avoids contamination of residues.

1 The Idea

The original idea emerged from the notion of combining the two-stage fixed bed pyrolysis and gasification plant developed at the Department of Energy Engineering at the Technical University of Denmark with the more compact fluid bed technique. The two-stage process is well known; the Department has worked with it since 1988. The advantage of the process is a gas with a very low tar content and no slag formation as the tar is decomposed above the char layer. At present a 400 kW_{input} two-stage pilot plant has been built in co-operation between the Technical University of Denmark and the Danish boiler manufacturer REKA [1].

During the construction of the two-stage fluid bed unit, tests were made on the first stage, the pyrolysis unit [6]. These tests showed that about 90 per cent of the energy in the input biomass was relocated in the volatile matter. This could be of interest to the Danish utilities as a means of co-combusting straw with coal: The volatile fraction, containing about 90 per cent of the energy in the biomass, can be fed directly into a coal fired boiler, provided the corrosive alkali metals and chlorine can be retained in the solid fraction. The idea behind the project was to investigate the influence of important parameters on the energy content of the volatile matter produced in the pyrolysis unit, as well as the alkali and chlorine content of the char. The primary parameter is the temperature at which the reactor is run during an experiment.

2 Background

Political pressure has created a need for technologies to convert biomass into electricity. The Danish energy policy aims at reducing CO_2 emission in A.D. 2005 by 20 per cent under the 1988 level [2], [3]. This reduction is to be combined with a decrease in dependence on foreign fuels.

In 1993 the Danish Parliament directed the Danish utilities to increase the use of biomass for electricity production to 1.2 million tons of straw and 0.2 million tons of wood per year by 2000 A.D.. The utilities argue that the only way for them to meet this demand in time is by co-combusting the biomass with coal in existing power plants. Unfortunately the content of alkali metals and chlorine causes corrosion damage to the boilers. The fluid bed pyrolysis unit described in this paper could be one way of supplying the energy of the straw to an existing power plant boiler and of keeping the alkali metals and chlorine out of the boiler.

3 The Project

The purpose is to produce relatively pure volatile matter with as small a content of alkali metals and chlorine compounds as possible. Instead of feeding the straw directly into a coal fired power plant boiler, it is pre-treated in an atmospheric fluid bed pyrolysis unit in order to reduce the amounts of alkali and chlorine compounds. This is done to prevent corrosion of the boiler and to prevent contamination of the ash.

3.1 Principle of Operation

The working principle is shown in **Figure 1**. Straw is fed into the pyrolysis reactor filled with sand. The biomass is pyrolysed and divided in a cyclone into a volatile fraction and a char fraction. When pyrolysis takes place below a certain temperature, the bulk of alkali and chlorine in the input straw remains in the char. The volatile matter, containing limited amounts of alkali and chlorine, is partly led back into the bottom of the pyrolysis reactor for fluidization and partly led into a conventional boiler. A turbocharger working as blower is used to circulate the volatiles. The char from the cyclone is stored for further treatment which has not been addressed in the present investigation.

3.2 Alkali in Volatiles and Energy in Solids

Figure 2 illustrates the target of the investigation. At what temperatures do the alkali metals and chlorine compounds leave the solids fraction and transform into volatiles and how much of the original energy is found in the solids fraction at this point? **Figure 2** shows the result of some preliminary investigations combined with studies of literature. The alkali and chlorine compound content in the volatile phase increases with increasing temperature, while the energy content in the char decreases. This is undesirable, as the idea is to feed as much of the biomass energy as possible into the power plant boiler while keeping the alkalis and chlorines in the solids phase.

To investigate the alkali and chlorine content and the corresponding energy content in the char, the experimental set-up shown in **Figure 3** was used.

3.3 Experiments

During the above preliminary investigations, a large temperature gradient from the coldest to the hottest part of the unit was observed. As the effect of such large temperature differences is unknown but might be important, the first step was to reduce this gradient. This was done by improving the insulation round the unit and adding electrical tracing to the coldest parts. This reduced the gradient from about 500 K to under 50 K, depending on working conditions. It is important to note where the biomass is infed. Examples of temperature curves for experiments before and after the improvements are shown in **Figure 4**.

3.3.1 Preliminary Investigations with Wood

The first experiments were made with wood, as straw is known to be a difficult fuel. These clearly showed the effect of temperature, as can be seen in **Figure 5**. Four experiments conducted at the freeboard temperatures of 486°C, 582°C, 744°C and 801°C showed that the content of particles in the volatile sample, the total tar amount, and the water content all decreased. There were no problems with the turbocharger; it has been operating for about 24 hours with volatiles from pyrolysis and there has been no need for maintenance. The turbocharger is run reversed, which makes it possible to keep the high temperature in the turbocharger and thereby avoid deposition. The decrease in particle content is caused by a lower char production at elevated temperatures and an increased cyclone efficiency as the flow increases with temperature. The amount of tar is reduced as a greater part of the tar is thermally cracked and transformed into lighter hydrocarbons. The water content is reduced as chemical reactions consuming water become dominant.

The char content also decreases with increasing temperature, as can be seen from **Figure 6**. This is very important for the co-combustion concept as it means that with increasing temperature an increased part of the energy is found as volatiles and less as carbon in the char. The upward break of the curve can be explained by different variables attached to the experiment. All these observations are in agreement with the literature [4], [7].

Figure 7 shows the reactivity of char samples taken from the experiments. The reactivity, R_m, is mass-weighted. The intention is to give the reactivity, where the actual mass of the char sample is high, a larger weight, than where the actual mass is low.

$$R_m = \frac{\sum\limits_{t=t_0}^{t_n} [R(t) \times W(t)]}{\sum\limits_{t=t_0}^{t_n} W(t)}$$

Where

R(t)   : Actual reactivity versus time,
R(bo) : Reactivity versus burn off,
bo   : Burn off, defined as:

$$bo = 100 \times \frac{1 - W(t)}{W(0)}$$

W(t)   : Actual mass of convertible (ash free) char versus time,
W(0)   : Initial mass of convertible (ash free) char when the gasification starts,
t   : Time at the start of gasification,
t_0 : Time at 12% burn off,
t_n : Time at 92% burn off.

The reactivities of char from the experiments are compared with samples made during slow oven pyrolysis of both the wood used and a straw sample. It can be seen that the reactivity of char prepared by slow pyrolysis is much less than the reactivity of the fluid bed prepared char. This means that char produced under fast pyrolysis conditions (fluid bed) reacts faster with the gasification agent than char produced under slow pyrolysis (fixed bed). This is also in accordance with the literature [5].

The mass distribution for one of the experiments gives an example of the energy distribution as shown in **Table 1** below. An assumed lower heating value of 17 MJ/kg on an ash free basis with a water content of 10 per cent has been used for the input wood and for the char an assumed heating value of 27 MJ/kg was used. In general, only about 15 per cent of the input biomass energy reappears in the char fraction.

Table 1: Char fraction from the wood experiments.

Test No.	1	2	3	4
Temperature in freeboard	486°C	582°C	744°C	801°C
Wood in	2129g	2673g	2689g	4815g
Mass per cent	100%	100%	100%	100%
Char out	277g	281g	161g	347g
Mass per cent	13%	11%	6%	7%
Energy content in char	*21%*	*17%*	*10%*	*11%*

Table 2 shows the composition of the gas from the experiments. The temperature in the freeboard is important because it is the maximum temperature to which the volatiles are exposed. The composition is dominated by a large CO content. The compositions are corrected because of false air in the gas sample. The measuring procedures and subsequent analyses were made with the equipment at hand and the results should only be regarded as indicators, due to various uncertainties. The heating values are as expected when the biomass is wood and pyrolysed in a fluid bed.

Table 2: The gas composition.

Test No.	2	3	4
Temperature in freeboard	582°C	744°C	801°C
Hydrogen, H_2	7%	11%	14%
Methane, CH_4	13%	16%	16%
Carbon monoxide, CO	55%	60%	59%
Carbon dioxide, CO_2	27%	13%	12%
Heating value, MJ/Nm^3	8	12	14

3.3.2 Co-combustion Experiments with Straw

After the wood experiments the fuel was changed to straw as this is the fuel to be used in co-combustion. A series of experiments was performed with straw, of which six included analysis of the char. The six experiments were performed at freeboard temperatures ranging from 450°C to 740°C. In order to compare the results from fluid bed flash pyrolysis, with pyrolysis in a fixed bed, a reference experiment was carried out. Two analyses of the straw were made to determine the amount of harmful components retained in the straw and therefore not fed into the power plant boiler. Alkali and chlorine measurements in the gas stream have not been made, even though total balances would give more accurate figures. It was more important to get indications of the amount retained than to get accurate figures.

The mass distribution is determined and the energy distribution is shown in **table 3**. In a fixed bed about 40 to 50 per cent of the input energy is retained in the char. In the fluid bed this fraction is 20-40 per cent, making it possible to convert 60 to 80 per cent of the input biomass energy into energy bound in the volatile fraction.

Table 3: The char fraction from straw devolatilization experiments in a fluid bed.

Test no.	Fixed	6	5	4	3	2	1
Temperature in freeboard	600°C	450°C	500°C	550°C	580°C	660°C	740°C
Straw in	194g	917g	815g	865g	994g	1006g	811g
Mass per cent	100%	100%	100%	100%	100%	100%	100%
Char out	57g	252g	197g	180g	174g	166g	134g
Mass per cent	29%	27%	24%	21%	18%	17%	17%
Energy content in char	*46%*	*44%*	*38%*	*33%*	*28%*	*26%*	*26%*

Table 4 shows the amount of chlorine, potassium and sodium refound in the char after pyrolysis. The percentages are related to the amount in the straw infed. Values above 100 per cent are explained by uncertainties associated with the experiments as the laboratory tests are considered accurate. An accuracy of ± 5 per cent is attached to these tests.

Table 4: Chlorine, potassium and sodium contained in the char.

Test No.	Fixed	6	5	4	3	2	1
Temperature in freeboard	600°C	450°C	500°C	550°C	580°C	660°C	740°C
Chlorine	58%	82%	82%	58%	40%	14%	20%
Potassium	83%	111%	101%	73%	65%	58%	58%
Sodium	83%	104%	97%	73%	73%	63%	61%

The results of the experiments are shown in **figs. 8** and **9**. It can be seen that the alkali and chlorine retained in the char decreases with increasing temperature. As expected, more alkali and chlorine compounds are released when the temperature is increased, so to keep them in the solids fraction the temperature must be low.

4   Discussion

The idea of pretreating straw in a fluid bed pyrolysis unit before the volatile fraction is sent into a conventional boiler has proved to be viable. This has been the aim of the tests performed up till now.

A major advantage of this concept is that the residues are kept apart, in contrast to the usual co-combustion principle where straw and fossil fuel are combusted in the same boiler. When the residues are kept apart, the residue from the fossil fuels can be used for cement production as usual, and the straw residue can be led back to the field. When co-combusting traditionally, the residues contaminate each other and cannot be used for either purpose.

The results need to be improved to optimize the energy content of the volatiles and maximize the retention of harmful components in the solids phase. Char from the pyrolysis is the main component in this phase, and the conversion of the energy must be addressed to fully convert the biomass to energy. At present two possibilities exist: to gasify the char at a low temperature and supply the gas to the boiler or to combust the char at low temperatures.

Also, the uncertainties connected with the energy balances and balances of harmful components presented in this paper must be minimized. Especially the harmful components balances must be improved as the actual balances are only based on char analysis.

5   Further Work

The notion of pretreating biomass in a fluid bed pyrolysis unit emerged during initial tests on a two stage fluid bed pyrolysis and gasification unit. It was decided to postpone construction of the two stage fluid bed until more elaborate results on the pyrolysis unit were obtained. These results are reported in the present paper.

The work on the complete plant based on the two stage concept has now been resumed and is concentrated on testing this gasifier.

6   Conclusion

- It is possible to pyrolyse wood and straw in an atmospheric bubbling fluid bed, based on a principle of external heating and recycling of volatiles to be used for fluidization.
- The amount of alkali and chlorine retained in the char fraction decreases with rising pyrolysis temperature. If the pyrolysis is fast, this tendency is increased.
- The amount of energy contained in the char decreases with rising pyrolysis temperature.

Tests with the pyrolysis unit have revealed the relation between pyrolysis temperature, pyrolysis conditions and the amount of alkali metal and chlorine in the produced gas. An analysis of char produced under slow fixed bed conditions has been performed. The results of this analysis have been compared to the results of the analysis of char from fast, fluid bed pyrolysis operation. This showed that the content of alkali and chlorine in the char depended on the temperature and the speed of pyrolysis.

There is a contradiction between the alkali/chlorine and energy content in the gas: to get most of the energy in the gas phase a high temperature is needed, but the gas then contains a large portion of alkali/chlorine compounds. This correlation has been determined.

7   Acknowledgement

The Department of Energy at the Technical University of Denmark has been working for 10 years on pyrolysis and gasification. The Danish utility company ELKRAFT and the Danish energy research programme each provide half the funds.

The work described in this paper is mainly funded as a part of these activities while the expenses of Mr. Houmøller were paid by dk-TEKNIK. Since January 1996 the participation of dk-TEKNIK has been financed by the Danish UVE programme (Development Programme for Renewable Energy). This funding ends in October 1996 when the two-stage fluid bed concept will, one hopes, prove viable.

8   References

1. "Gasification of Straw in a two-stage 50 kW Gasifier", Henriksen, U., Christensen , O., Presented at the "8th European Conference on Biomass for Energy, Environment, Agriculture and Industry", Vienna, October 1994.
2. "Energy 2000", Programme from the Danish Energy Minestery 1990, ISBN 87-503-8379-5
3. "Energi 21", Programme from the Danish Energy Minestery April 1996, ISBN 87-7844-057-2
4. "Pyrolysis and Industrial Charcoal", Deglise, X., Magne, P, Chapter 10 in "Biomass - Regenerable Energy, John Wiley & Sons Ltd., Chichester, England, 1987, ISBN 0 471 90919 X.
5. "Nordic Seminar on Solid Fuel Reactivity", CTH and Nordic Energy Research Programme, Sweden, 24 Novembre 1993.
6. "Advanced Fluid Bed Gasification" - in Danish - Houmøller, S., Hansen, Martin W., M. Sc. Thesis, Institute of Energy Engineering, Technical University of Denmark, Denmark.
7. "Gasification of Straw - EFP88 - projekt 1 - Theory and Technical Basics", Jørgensen, L. B., Risø National Laboratory, Denmark.

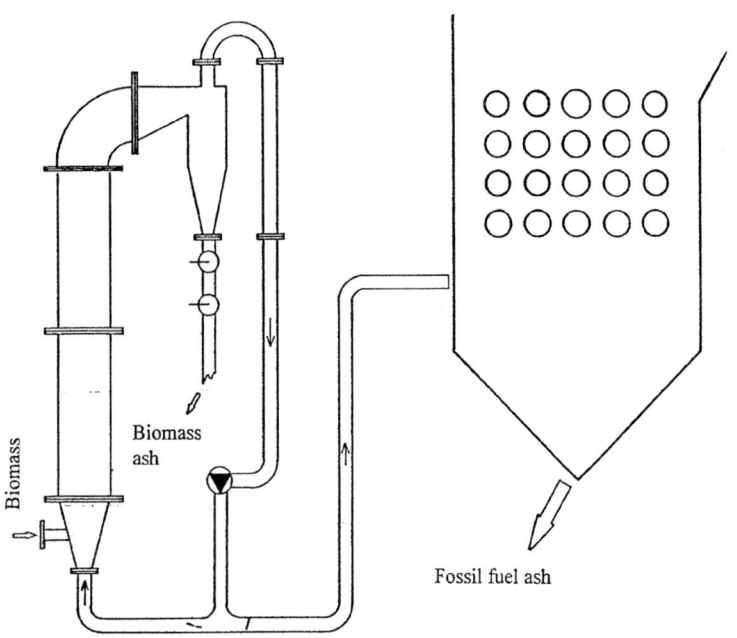

Figure 1. Working principle for the fluid bed pyrolysis unit.

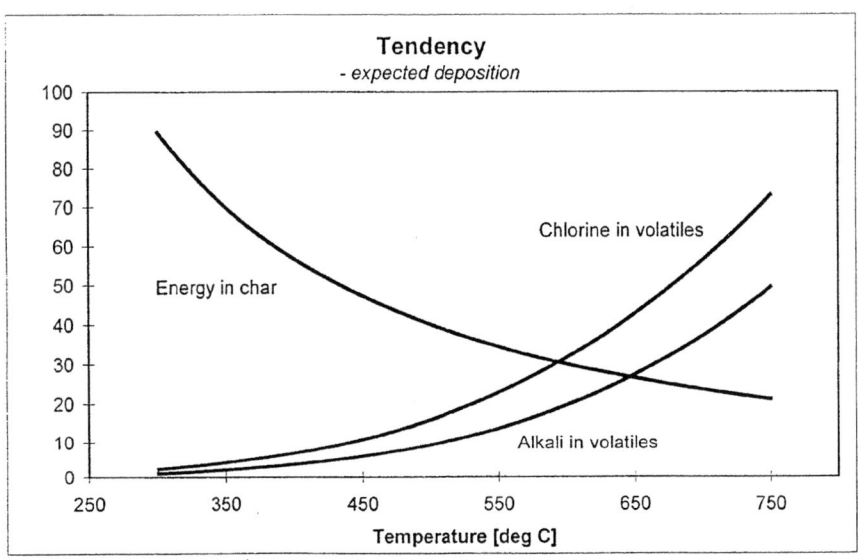

Figure 2. Alkali in volatiles and energy content in char - is there an optimum ?

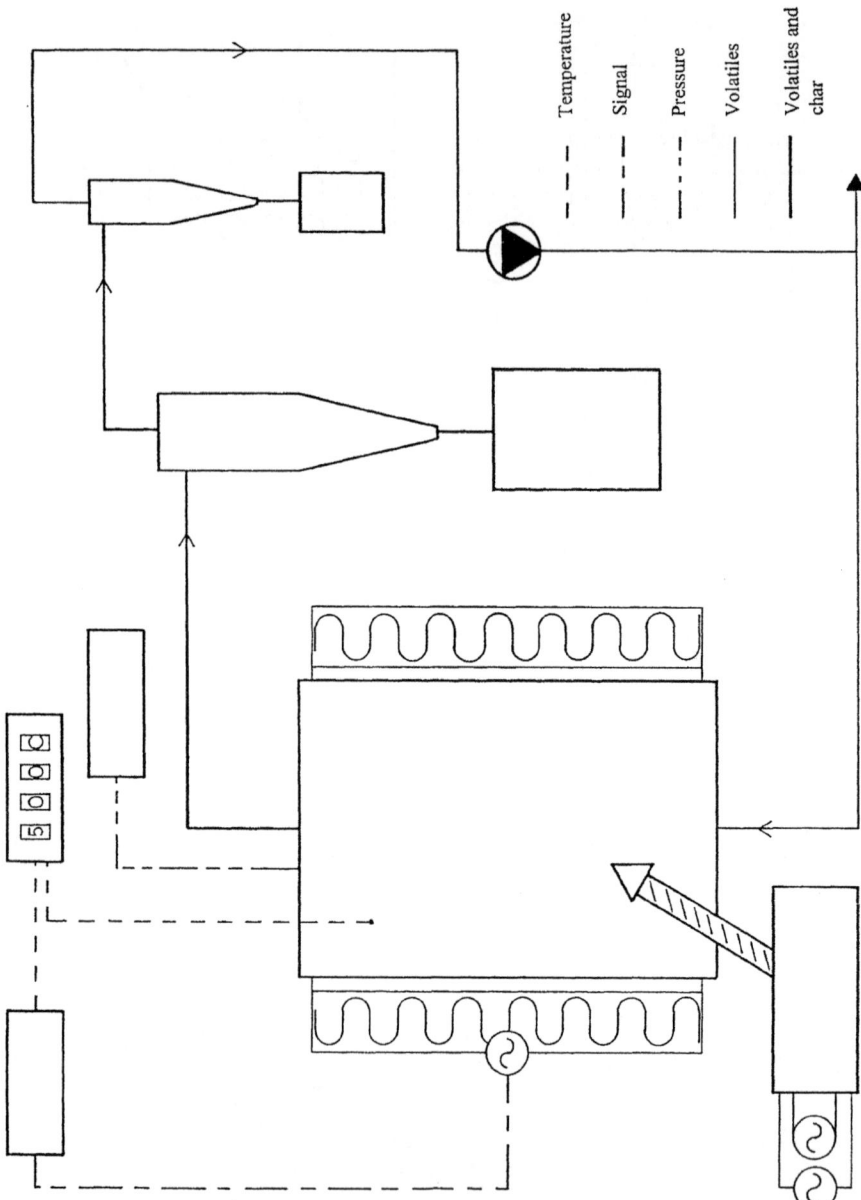

Figure 3. Experimental set-up.

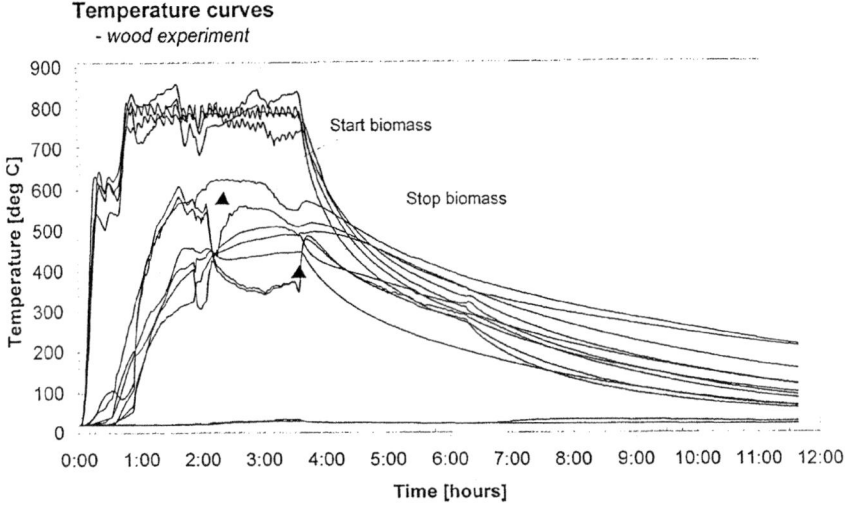

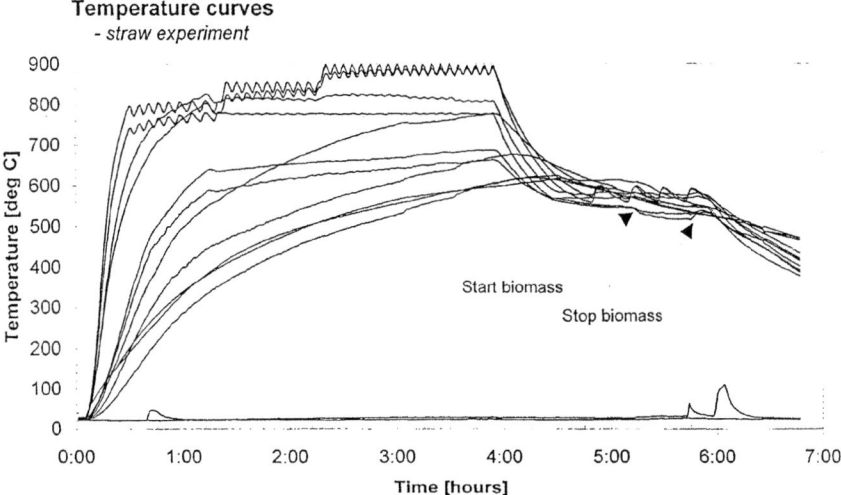

Figure 4. Graphs of temperatures before and after the improvements. The temperature gradient from the coldest to the hottest part of the unit is reduced from about 500 K to about 50 K. The first experiments were performed with wood; after the improvements the fuel was straw.

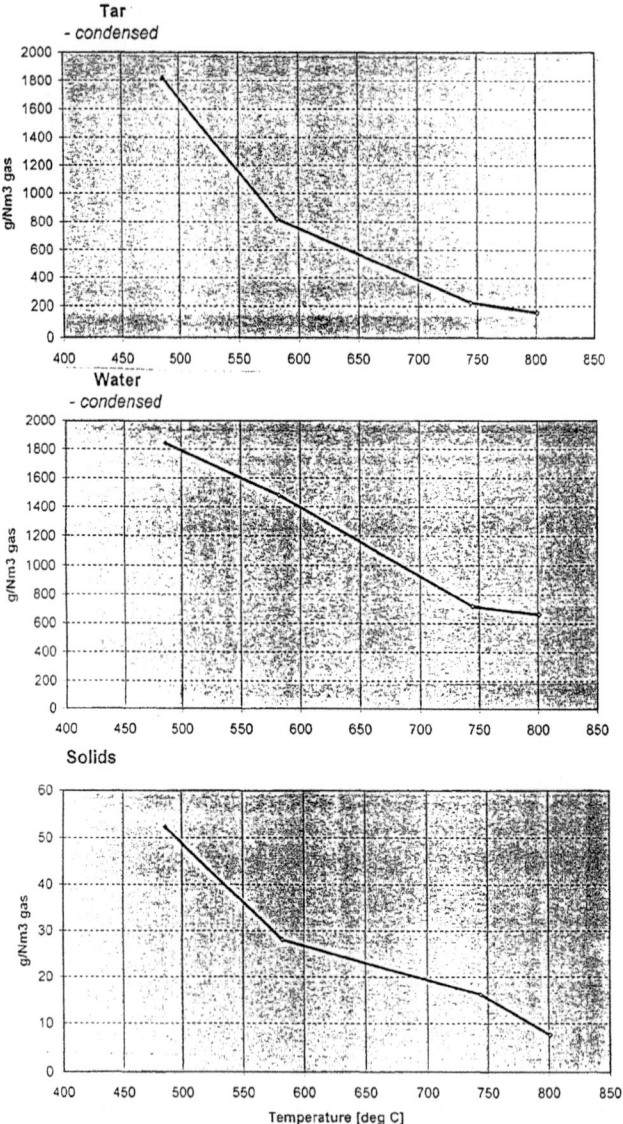

Figure 5. Influence of temperature on the tar, water and solids content of the gas. The tar content decreases as more tar components are cracked to lighter hydrocarbons with increased temperature; the water content decreases as water consuming processes become dominant. The decreased content of solids can be due to a higher rate of carbon conversion.

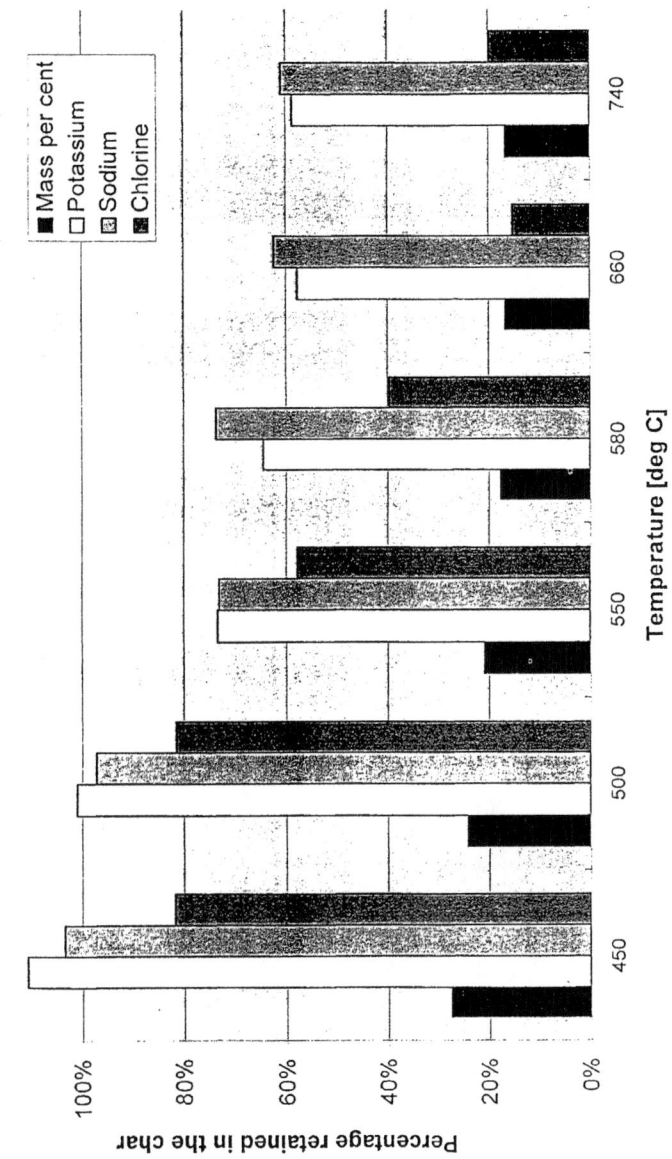

Figure 8. Chlorine and alkali content of the gas from the experiments with straw. As the temperature increases, retention of the corrosive compounds in the char is reduced. This is in conflict with the wish for a high energy content and a low alkali and chlorine content in the gas produced.

279

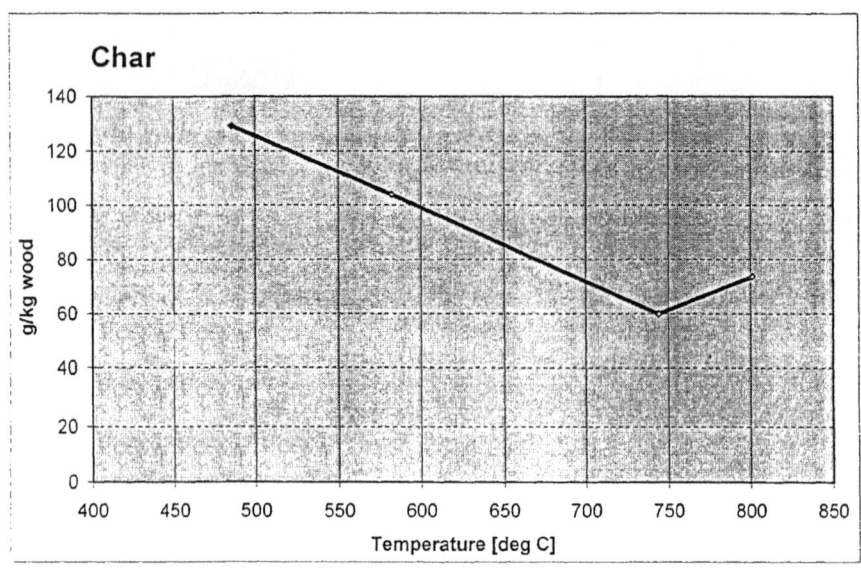

Figure 6. The char output decreases with increasing temperature. A smaller fraction of the infeed energy is therefore found in the char fraction.

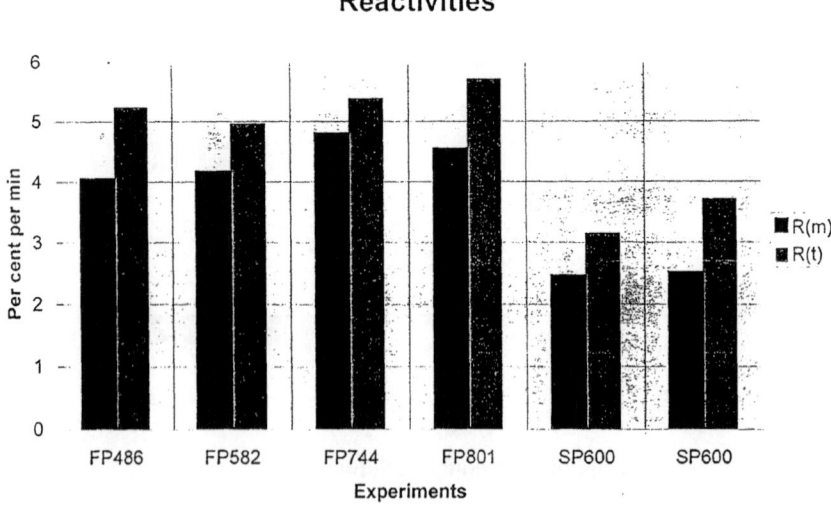

Figure 7. Reactivity of the char from the different experiments with wood. The char produced in the fluid bed is more reactive than the fixed bed TGA produced char.

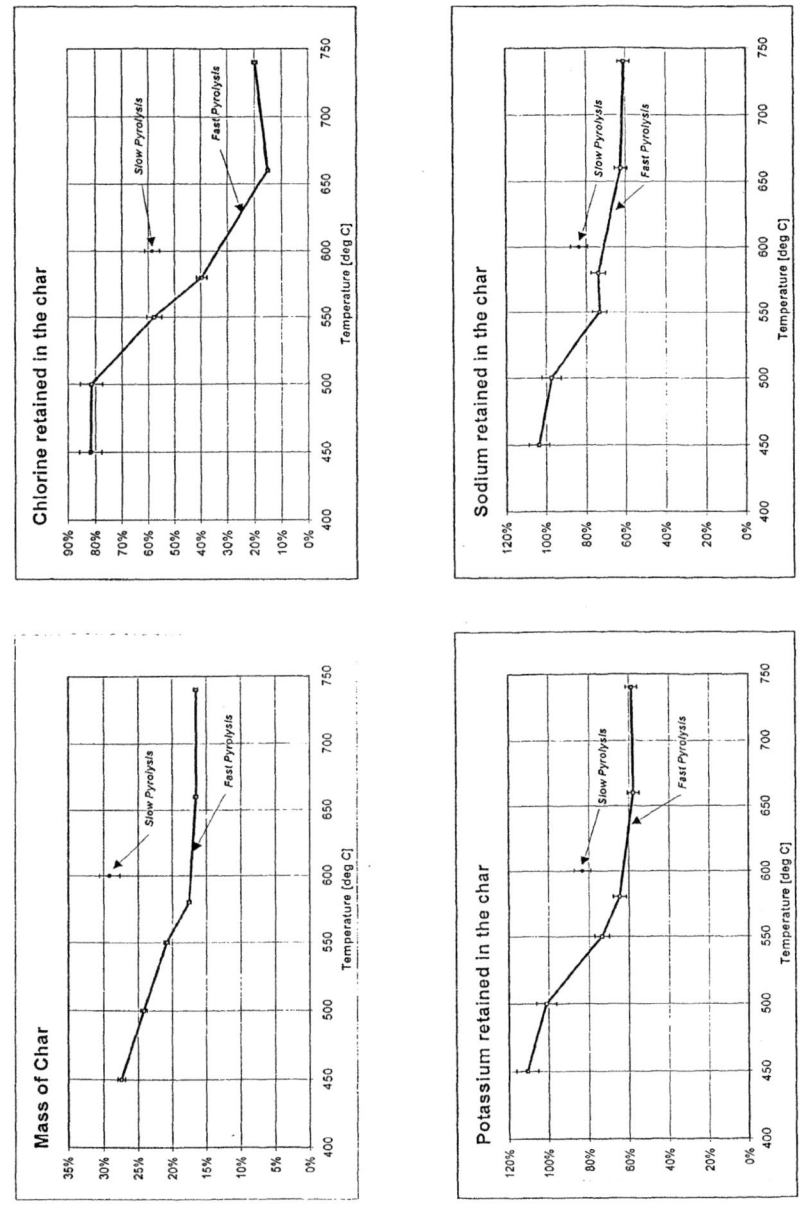

Figure 9. The essential results from the experiments with straw. The figures are the same as in Figure 8. The figures are given with an uncertainty of ±5 per cent. For the sake of comparison, the figures also include a single test based on slow pyrolysis at 600 °C.

THERMOCATALYTIC CONVERSIONS OF WOOD BIOMASS IN FLUIDIZED BED OF CATALYSTS

B.N. KUZNETSOV and M.L. SHCHIPKO
Institute of Chemistry of Natural Organic Materials Siberian Branch
of Russian Academy of Sciences, Krasnoyarsk, Russia

Abstract
The process of catalytic gasification of hydrolytic lignin and wood bark in fluidized-bed reactors was studied under different operating conditions.

The complete gasification of wood lignin was accomplished in the fluidizing circulated bed of catalyst. The composition of gaseous products was studied at the variation of gasification process operating parameters and types of wood biomass. The relationship has been established between the calorific value of produced gases and the gasifying mixture velocity.

The integrated process of gaseous fuel and carbon porous material producing from wood biomass in a fluidized bed of different catalytically active materials has been investigated. Some disposable catalysts like metallurgical slag particles with good catalytic and mechanical properties have been selected for fluidized-bed reactors.

Keywords: catalyst, char, gaseous products, gasification, fluidized bed, oxide catalyst, wood biomass.

1 Introduction

The thermal processing of plant biomass produces a mixture of gaseous, liquid and char products. The use of catalysts makes it possible to regulate an yield and composition of wood biomass pyrolysis and gasification products [1, 2]. But the catalyst application results in some additional technological problems, connected with a catalyst preparation, catalyst stability and re-using. The noted problems can be simplified when the thermocatalytic process is carried out at conditions of indirect catalysis - without direct interaction of catalyst and biomass material. Catalyst promotes the rate of gas-phase reactions of volatile substances emitted from biomass particles at the initial steps of their thermal conversion and some disposable catalytic materials can be successfully used instead of the synthetic catalysts.

The pyrolysis and gasification of wood-biomass combined with the cracking of volatile matter and tar substances in the presence of metal catalysts (Ni, Cu, Mo) can surve as examples of indirect catalysis [3, 4]. The principle of indirect catalysis was successfully used by authors for thermocatalytic processing of solid organic raw materials of different nature in a fluidized bed of oxide catalyst [5,6,7]. It was shown the catalyst can accelerate the reactions of volatile compounds emitted from heated fuel particles. The catalyst use makes it possible to support the demanded temperature of the process, to regulate the yield and composition of gaseous, liquid and solid carbon products and to reduce the harmful compounds formation.

This work was focused on the development of hydrolytic lignin and wood bark processing to gaseous and char products in the reactors with fluidized and circulated beds of catalysts. Hydrolytic lignin is produced on a large scale in Russia as a by-product of the ethanol, furfural and albumen yest production. Wood bark wastes are formed as a side product of the wood mechanical and chemical processing.

2 Experimental

The laboratory- and pilot-scale installations with a fluidized bed reactors were used for the wood biomass gasification. The reactors were made from stainless steel with the thermally resistant heat-isolation covering.

Two types of reactors described in [5] were applied in pilot experiments with wood-biomass partial gasification. For both reactors the height of the first section was 0.6 m and its inner diameter 0.3 m. This section is filled by a fluidized catalyst particles during the wood-biomass processing. The air-distributed grate was located in the lower part of the section with the charge of catalyst 12-16 kg. The second section has an inner diameter 0.3 m, a length of 1 m for short-type reactor and 2.8 m for long-type reactor. The latter also has a third section with height 0.8 m and an inner diameter of broad part of section 0.6 m. The variation of height and diameter of the reactors makes it possible to regulate the time of wood-biomass thermal treatment without changing the consumption of reactants.

The diesel fuel furnace and an electric heater were used for starting up the installation. The reactor was heated at first by a hot air from the electric heater. Then a diesel fuel and air were injected into the furnace. The air/diesel fuel ratio selected was capable of the combustion temperature 700-800°C. Hot gaseous products of diesel fuel combustion heated the reactor with a catalyst bed to a required temperature. The air and powdery lignin mixture was used as a fluidizing agent for catalyst bed. The main portion of lignin particles was moved through the fluidized bed of catalyst in entrained-flow conditions. Char particles with sizes less than 0.2 cm, a gaseous mixture of CO_2, H_2O, H_2, CO, CH_4, N_2 and tar were the main products of wood-biomass thermal conversion. Char products were trapped mainly in the cyclone and accumulated in the bunker. Gases and very small solid particles were burnt in the economizer, thus producing water-steam. The smoke-gas was purified in the multi-cyclone and in the hose filter.

A special device was applied for periodical sampling of reaction mixture from different points of the reactor. It consisted of cyclone, trap, cooler, filter and rotameter. During the sample selection the flow of gases in the sampling device was maintained at the same level as in the reactor. For the better reliability of the data three experiments were conducted at the same parameters and the average values were given in tables and figures.

For the complete gasification process the reactor was applied with vertical rectangular tube made from stainless steel with the inner size 0.22 x 0.22 m and the height 4.2 m [5]. The air-distributed grate with the free cross section 3% was placed in the lower part of the reactor. The water steam and air (or oxygen) were fed under the distributed grate. The powdery row material was injected above the air distributed grate.

The two different technology of lignin gasification process were studied. According to the one of them the char dust from the gas generator was separated from gaseous products in the cyclone, then collected and periodically removed from the installation. According to the other technology the char particles were recirculated through the combustion zone of the reactor.

Hydrolytic lignin from Krasnoyarsk biochemical plant and the mixture of coniferous bark from Krasnoyarsk pulping plant were used as initial raw materials. The particle size of lignin and wood bark was 0.03-0.12 cm. The lignin composition: moisture content from 4 up 12 wt.%, ash 3.3 wt.%, carbon (C^{daf}) 60.0%, hydrogen (H^{daf}) 6.2%, sulfur (S^{daf}) 0.4%, oxygen (O^{daf}) 33.4%. The initial lignin porosity was 1.1 cm^3/g, iodine sorption ability - 14%. The wood bark consists of 80-85% of silver-fir bark and 15-20% of spruce bark with moisture content up to 18 wt%, ash 2.4 wt.%, carbon (C^{daf}) 44.0%, hydrogen (H^{daf}) 6.4%, nitrogen (N^{daf}) 0.2%, oxygen (O^{daf}) 49.4% , volatile compounds 59.3%, porosity 0.7 cm^3/g, I_2 sorption ability -16%.

Al-Cu-Cr oxide catalyst was prepared by impregnation of γ-Al_2O_3 (surface area - 160 m^2/g) with $CuCr_2O_4$ in water solution, drying at 110°C and then calcining on air at 700°C for 6 hours. The catalyst contained 5.9 wt.% of $CuCr_2O_4$. Catalyst fraction with sizes 0.20 - 0.25 cm was used in a fluidized bed experiments.

Different slag materials were also tested as catalysts. Their catalytic activity was measured in the reaction of CO oxidation at 400°C. The reaction mixture CO-air of a stoichiometric composition was used. Experimental data were obtained in quartz reactor with the inner diameter 0.015 m and the length of 0.15 m, volume velocity 2000 h^{-1}. The catalyst charge was 10 cm^3 and catalyst particles sizes

0.5-3.0 mm. The concentration of carbon monoxide was measured before and after catalytic reactor using a gas chromatograph.

Gaseous products were analyzed by gas-liquid chromatograph with the thermal conductivity detector at 200°C. Argon was used as a gas-carrier. The analyses of gases was performed on columns with molecular sieves CaX and polysorb-1. In the first case H_2, O_2, N_2, CH_4 and CO were detected, in the latter one - CO_2 and low hydrocarbons.

Low temperature luminescence technique was used for 3,4-benzpyrene detection and for its concentration determination. This method is based on the measurement of intensity of 3,4-benzpyrene line 402.9 nm in fluorescence spectrum. The lower detectable limit was 0.0001 μg/ml, the analysis accuracy about 10 rel.%.

The ultimate analysis and γ-ray diffraction measurements were used for the char products study. Char porosity was determed by kerosene sorption. The meso- and micropores ratio in char samples was measured by automatic adsorbtion apparatus ASAP-2400, Micromeritics. Specific surface area was calculated by BET method from N_2 adsorbtion isotherms at -196°C. The capacity of solid carbon products for I_2 adsorbtion was tested using the photocolorimetry technique.

3 Results and discussion

3.1 The catalyst selection for wood-biomass processing in a fluidized bed

The mechanical resistance of different catalytically active materials in a fluidized bed was studied in the blank experiments. The obtained data are shown on Fig. 1. The synthetic $CuCr_2O_4/Al_2O_3$ catalyst and open hearth furnace slag particles had the most high mechanical resistance.

Some data on the composition of fluidized bed materials and their catalytic activity in CO oxidation are presented in table 1.

3.2 The wood biomass complete gasification

The different technological schemes of wood biomass gasification were experimentally tested for the selection of the optimal one. It was found that the process of one-stage air-steam gasification of hydrolytic lignin and other wood wastes produces a fuel-gas with the high amount of tar substances. Therefore a two-stage lignin gasification process with the lignin conversion to a char at the first stage should be preferably used to produce more pure gas.

The conventional methods of hydrolytic lignin pyrolysis yielded 15-30%wt of char product, 15-20%wt of tar insoluble in water, 5-7%wt of soluble tar, 25-57% of water, 20-25% of non-condensed gases [8]. The steam gasification of char in the fixed bed at 750-800°C and pressure 2.0-2.5 MPa produced syn-gas with the yield 1.7-1.9 m^3 per

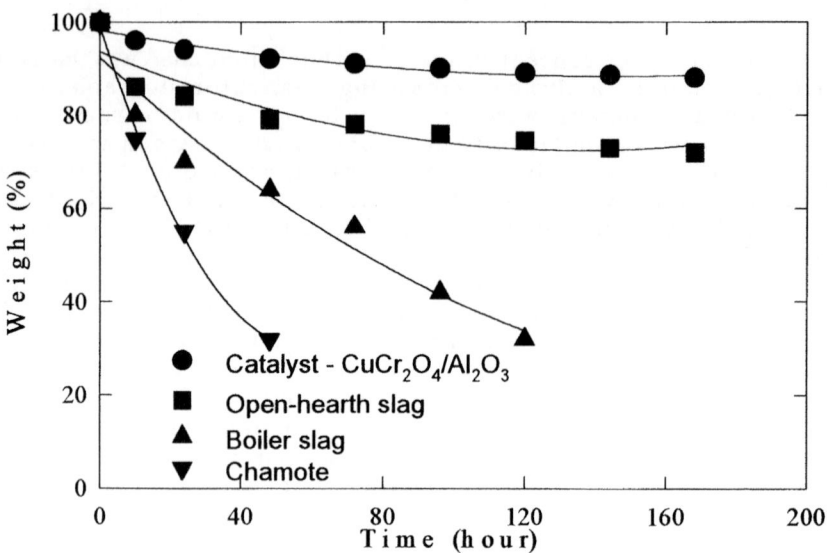

Fig. 1. The weight loss of different materials during their ittrition in a fluidized bed.

Table 1. Some characteristics of materials used in a fluidized bed

Indices	Shamote	Boiler slag	Open-hearth furnace slag
Content (%) :			
SiO_2	49.8	54.6	19.4
Al_2O_3	16.9	5.7	2.3
CaO	8.2	25.6	38.6
MgO	24.3	4.1	14.3
MnO	0.2	trace	5.7
FeO	0.4	8.2	6.7
Catalytic activity* (%)	15.5	25.5	54.4

* Degree of CO conversion at 400°C.

1 kg dry lignin and with the composition: H_2 63-67%vol., CO 9-13%vol., CO_2 21-25%vol., CH_4 - the remainder [8]. The same results were obtained in the present work for silver-fir bark char gasification in the fixed bed.

The use of a fluidized bed and powdery wood biomas results in the increase of the process intensity due to the improve mass- and heat transfer parameters. Fig. 2 demonstrates the influence of fluidized bed and circulated bed on temperature distribution in gas-generator for different methods of lignin and silver-fir bark chars gasification.

The initial composition of reaction mixture was (%wt): char - 27.0, O_2 -10.8, H_2O - 8.7. At first the char gasification process was carried out in the entrained flow conditions. In this set of experiments the narrow front of the flame was observed (Fig. 2A). The temperature about 1000°C was recorded inside the gas generator on the height of approximately 0.5 m above the air-distributed grate. After the addition of 12 liters of shamote particles (the fraction 1.0-1.4 mm) into the gasification reactor the length of high-temperature zone was increased (Fig. 2B). The extent of high-temperature zone was the highest for circulated bed of boiler slag particles (Fig. 2C). In the latter case the degree of char gasification was increased from 52 to 74% (relative to carbon content in the char) as compared to experiment with fluidized bed of shamote.

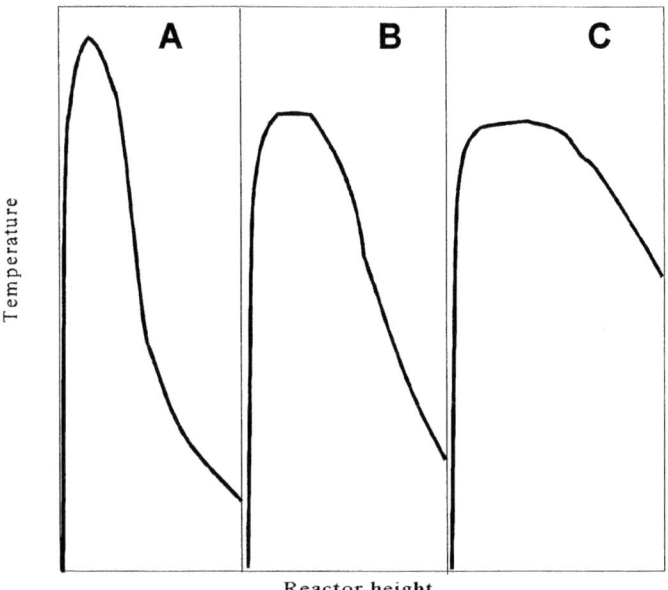

Fig. 2. The distribution of a temperature along the gas generator height during the process of char gasification by air-steam mixture: in entrained flow (A), entrained flow and fluidized bed of shamote particles (B), entrained flow and circulated bed of boiler slag particles (C).

The use of a fluidized bed should intensify the heat transfer from the combustion to the reduction zone of gas generator by the promotion of the reagents mixing. The observed variation in the temperature distribution along the height of the gas generator is confirmed above mentioned preposition.

The increse of the char conversion degree in the reactor with a fluidized bed of slag particles generally can be connected with the growth of fuel particles residence time in high temperature zone, with the intensification of reagents mixing and with the promotion of heat exchange between combustion and gasification zones of gas generator. But it is known that the complete reagents mixing is achieved at the fluidized bed height 50-100 mm above the powdery fuel injection point [9]. Therefore the intensification of the heat exchange between combustion and gasification zones of gas generator should be taken into account as main reason for the increase the char conversion degree in the presence of circulated bed of boiler slag particles.

In the investigated range of the gasification process parameters the degree of char conversion correlates with the height of a fluidized bed of boiler slag particles (see Fig. 3). Since the air-steam gasifica-

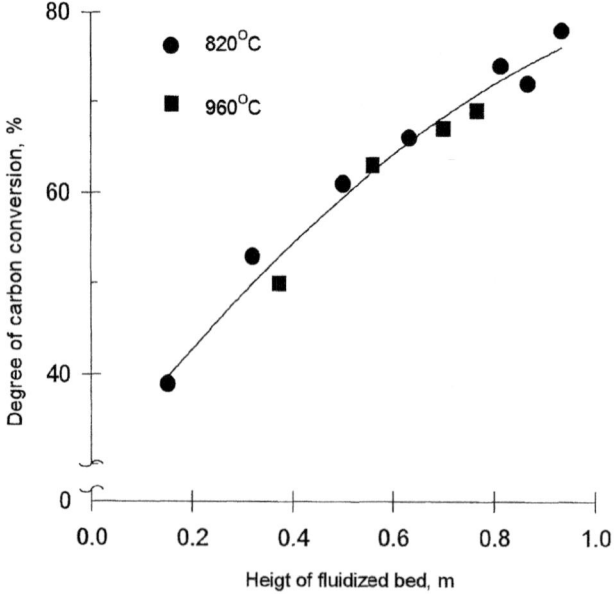

Fig. 3. The influence of the height of a fluidized bed of boiler slag particles on the degree of lignin char gasification by air-steam at temperatures 820°C and 960°C.

tion of lignin char proceeds with detectable rate already at 600°C [10], one can expect that the gasification process is limited by diffusion at the higher temperatures. This conclusion was supported by the fact that the variation of the temperature from 820 to 960°C had no strong influence on the degree of char gasification.

3.3 The wood biomass processing to a fuel-gas and char

The incomplete gasification of powdery hydrolytic lignin was studied by us in a fluidized bed of Al-Cu-Cr oxide catalyst under different operating parameters [10]. The process was carried out in the presence of limited amount of air in the autothermal conditions without the external heat supply. The stable regions of the autothermal processing of powdery lignin particles in a fluidized bed of oxidized catalyst were established by mathematical simulation and proved experimentally (Fig. 4).

The set of iron-containing catalysts was tested for powdery lignin oxidative processing. Obtained data are presented in table 2. At the same process parameters the temperature of a catalytic fluidized bed was about 40°C higher than the temperature of a fluidized bed of inert particles (compare experiments with boiler slag and oxide catalyst in table 2). This fact shows that the oxide catalyst accelerates the combustion reactions during the lignin thermal processing.

In a fluidized bed of catalyst the rate of volatile compounds oxidation is much higher than the rate of char oxidation [10]. This results in the decrease of O_2 concentration in the reaction mixture and prevents the char burning.

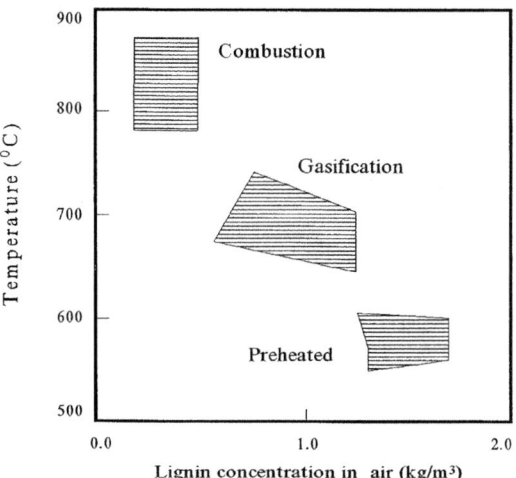

Fig. 4. Stable regions of lignin oxidative processing in a fluidized bed of Al-Cu-Cr oxide catalyst.

Table 2. Data on powdery lignin oxidative processing in a fluidized bed of different materials

Indices	Material of fluidized bed				
	Boiler slag	Oxide catalyst			
Lignin consumtion (kg/h)	24.7	26.6	69	102.0	162.3
Air consumtion (m³/h)	43.1	43.3	74.5	70.0	92.7
Ratio lignin/air (kg/m³)	0.6	0.6	0.9	1.5	1.8
Temperature of fluidized bed (°C)	740	780	760	670	620
Heat evolution (kW)	32.4	58.9	56.7	43.4	46.0
Yield of char (%)*	52.0	56.7	64.4	85.2	94.8
Concentratins of fuel components in dry gaseous products:					
H_2 (% vol.)	7.9	4.0	9.6	4.1	0.1
CO (% vol.)	6.7	4.7	7.8	2.2	0.2
CH_4 (% vol.)	2.4	0.9	2.1	0.7	-
tar (g/m³)	29	0.1	0.4	-	-

* relative to carbon in lignin

The advantage of a thermocatalytic lignin processing, besides increased yield of char, is the low level of harmful compounds in the gaseous products. The significant decrease in the formation of CO, CH_4 and tars makes this catalytic process environmentally friendly. Presented in Table 2 data show that CO and CH_4 concentrations in gaseous products of lignin oxidative processing are considerably reduced when Al-Cu-Cr oxide catalyst was used instead of the slag particles in a fluidized bed reactor (see in table 2). The gasification tar content depends on the fluidized bed temperature, the lignin and oxygen concentrations. It is known that the formation of tars is promoted by the increase of a volatile substances concentration in the pyrolysis gases. The oxide catalyst can reduce the content of these substances owing to their oxidation on the catalyst surface. The ratio of volatiles emission rate and the rate of their oxidation depends on the gasification process temperature. The application of the proper temperature, for instance, 670-620°C makes it possible to prevent almost completely the formation of gasification tars and polynuclear aromatic compounds like 3,4-benzpyrene during lignin oxidative processing in a catalytic fluidized bed.

The combined lignin processing in a fluidized bed of oxidation catalyst with the synchronous producing of fuel-gas and porous char materials was studied.

The process produces the combustible gases through the emission of volatile substances from lignin particles and by lignin-char gasification with steam and carbon dioxide. The concentration of volatile substances was rather high in the gas mixture because the content of volatile matter in lignin and wood-bark makes up 50%. The initially formed combustible gases can undergo to secondary reactions like oxidation, metanation, $CO-H_2O$ shift etc. In all studied runs the oxygen was almost completely consumed during gasification process and O_2 concentration in gaseous products did not exceed 0.6%. This way of gas formation is preferable for the producing of fuel gas, since the calorific value of pyrolysis gases is usually higher than that one of gases produced by the char air-steam gasification. One can expect that the amount of volatile substances, reacted with oxygen, depends on the velocity of reaction mixture flow. At the higher rates of gas flow the faster removal of volatile substances from lignin particles takes place. This promotes the oxygen diffusion to the char surface and accelerates the process of char oxidation. As a consequence the amount of oxygen, reacted with pyrolysis gases, is decreased and this results in the formation of fuel gas with higher calorific value (see Table 3).

Table 3. Data on lignin partial gasification in a fluidized bed of catalyst in the presence of air-steam mixture

Indices	Experimental values					
Temperature (^oC)	860	930	855	1060	980	950
Air consumption (m^3/h)	52.5	55.0	52.0	93.5	91.5	92.5
Lignin consumption (kg/h)	21.5	27.0	35.5	42.0	44.0	56.0
Water steam consumption (kg/h)	18.0	11.0	11.0	5.0	11.0	5.0
Yield of produced gas (m^3/kg)*	2.7	2.3	1.9	2.4	2.2	1.9
Calorific value of produced gas ($10^6 J/m^3$)	2.30	2.46	2.49	3.06	3.28	3.35
Yield of char (%)	17	15	24	19	18	25
Properties of char:						
-porosity, cm^3/g	1.6	1.8	1.6	1.7	1.8	2.1
-I_2 sorption ability, %	16	25	30	40	10	35

* Yield of produced gas m^3 at normal conditions per 1 kg of dry lignin, char yield - relative to carbon in lignin

4   Conclusion

Since the gasification processes are carried out at high temperatures the thermally stable catalysts based on slag materials, like open-hearth furnace slag, should be successfuly used in a fluidized bed. They have an acceptable level of catalytic activity and promote the oxidation of the wood-biomass volatile substances.   Finally this results in the additional heat supply for the gasification process. The composition of produce gases can be regulated within a broad limits by the variation of the expedentures of recirculated char particles, fresh wood biomass material and oxygen (or air).   The scheme of installation for wood-biomass gasification to fuel-gas and to syn-gas is shown on Fig. 5.   The advantages of the suggested technological scheme are connected with the following factors.   About 70-90% of heat, demanded for biomass gasification process, is supplied by recirculated char particles.   This allows to reduce the consumption of expensive oxygen.   Besides, the lower temperatures of the process and the catalyst action make possible to produce high calorific value fuel-gas or syn-gas with the higher ratio of H_2/CO, which favorite its following using in hydrocarbon syntheses. It is important that the produced syn-gas or fuel-gas contain low concentrations of tar substances and methane.   This simplifies the purification of a produced gas.

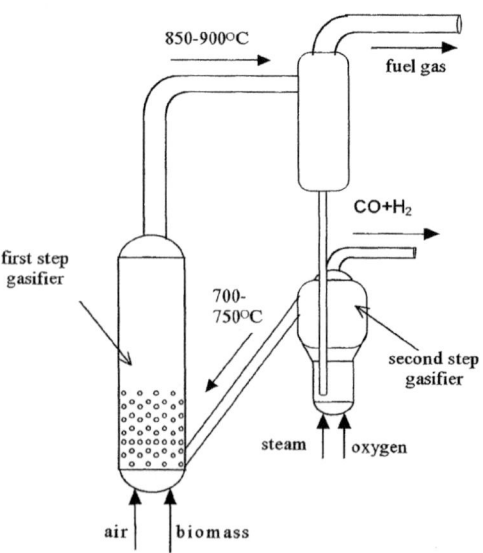

Fig. 5. The scheme of biomass gasification installation.

5 References

1. Robertus, B.J., Mudge, L.K., Weber, S.L., Sealock, L.J. and Mitchell, D.N. (1983) Use of catalysts in biomass gasification. *Alternative Energy Sources III.* Proc. 3 Int. Conf. Washington e.a., Vol. 3. pp. 409-24
2. Kuznetsov B.N. (1990) *Catalysis of coal and biomass chemical conversions*, Nauka, Novosibirsk (in Russian)
3. Yeno Koci. (1983) Wood gasification with the use of catalyst. *Kogaku to Koguo (Chemistry and Chemical Industrial).* Vol. 36, No. 9. pp. 670-1
4. Baker, E.G., Mudge, L.K. and Brown, M.D. (1987) Steam gasification of biomass with nickel catalysts. *Industrial and Engineering Chemistry Research*, Vol. 26., No. 7. pp. 1335-9.
5. Kuznetsov, B.N. and Shchipko, M.L. (1995) Environmentally friendly fuels from Kansk-Achinsk brown coal in *Proc 2nd Int. Conf. on Combustion Technologies for a Clean Environment*, Lisbon, Vol.2. pp. 94-102.
6. Shchipko M.L.,Kuznetsov, B.N. (1995) Catalytic pyrolysis of Kansk-Achinsk lignite for production of porous carbon materials. *Fuel.* Vol.74, No 5. pp. 751-5.
7. Kuznetsov, B.N. and Shchipko, M.L. (1995) The conversion of lignocellulosic materials to char products in fluidized bed of catalyst in Proc. 8th Int. Symp. on Wood and Pulping Chemistry, Helsinki, Vol. 3. pp. 349-54.
8. Yangolov, O.V., Kuznetsov, B.N., Volova, T.G., Gitelson, I.I., Kakhanov, Yu.G. and Konovalov, N.M. (1992) The gasification of waste wood raw materials with the purpose of receipt albumin cells. *Izvestiya Sibirskogo Otdeleniya Rossiiskoy Akademii Nauk*, Ser. Khim. (Communications of Siberian Branch of Russian Academy of Sciences, Chemistry), No. 6. pp. 40-4 (in Russian).
9. Makhorin, K.E. and Hinkis, P.A. (1989) *Fuel combustion in fluidized bed* (in Russian), Naukova Dumka, Kiev.
10. Kuznetsov, B.N. and Shchipko, M.L. (1995) The conversion of wood lignin to char materials in fluidized bed of Al-Cu-Cr oxide catalysts. *Bioresource Technology*, Vol. 52. pp. 13-9.
11. Yangolov, O.V., Ruznikov, S.G. and Schipko, M.L. (1992) The analyse start up process of reactors with fluidized bed for autothermal processing of raw organic solid. *Izvestiya Sibirskogo Otdeleniya Rossiiskoy Akademii Nauk*, Ser. Khim. (Communications of Siberian Branch of Russian Academy of Sciences, Chemistry), No. 6. pp. 59-63 (in Russian).

BLACK LIQUOR PYROLYSIS AND CHAR REACTIVITY
Black liquor pyrolysis and gasification

J.L. SÁNCHEZ, L. GARCÍA, M.L. SALVADOR, R. BILBAO and J. ARAUZO
Chemical Engineering & Environmental Department, University of Zaragoza, Zaragoza, Spain
Z. FANG
China National Centre for Rural Technology Development, Beijing, China

Abstract
Black liquor is the wastewater from the cooking of wood or straw in the production of pulp and paper. Nowadays new processes are being investigated as alternatives to the traditional recovery boiler used for black liquor treatment. One of the processes which appears to be more promising is gasification, for which further research is needed for its full industrial implementation.

Black liquor gasification occurs in several stages: drying of the black liquor droplets, devolatilization or pyrolysis, gasification of the resulting char and, depending on the temperature, coalescence and melting of the inorganic salts. The knowledge of each of these individual stages is important for a better understanding of the global process.

Black liquor pyrolysis and gasification have been studied in two different experimental systems:

a) In a bench scale reactor, where the composition of the gases has been analyzed by GC.

b) By means of a thermogravimetric system, experiments with dry black liquor and char have been conducted to analyze their reactivity. Different atmospheres N_2-O_2, (5-15%) and N_2-CO_2 (10-20%), have been used.

Keywords: black liquor, gasification, pyrolysis, soda, straw

1 Introduction

Black liquor is the residue of cooking wood or straw in the production of pulp and paper. Lignin, hemicellulose, other carbohydrates and, in the case of straw, inorganic compounds dissolve and form part of the black liquor. Recovery of the chemicals used, as well as of the energy contained in the organics, is important for the economics of the paper mill.

Alternative processes which in terms of energy can be more efficient, cheaper, safer and environmentally friendlier than the conventional recovery boiler are being investigated, and gasification is proving to be one of the more promising alternatives.

Important efforts have been made and are currently under development to increase the knowledge of the different stages in which the combustion and gasification of black liquor proceeds. These stages are drying, pyrolysis, char burning/gasification and inorganic reactions [1], pyrolysis and char gasification being the limiting stages.

Very interesting works can be found on the behaviour of Kraft black liquors from wood pulping in relation to pyrolysis [2] [3] [4] and char gasification [5] [6] [7] [8]. Similarly, studies on the combustibility and swelling of these black liquors under burning and gasification conditions have been reported [9] [10].

There is not much data about the behaviour of soda black liquors from straw cooking in the literature. Therefore the thermal decomposition and the reactivity of the product char with oxygen and CO_2 of one of these liquors have been studied.

2 Experimental

The black liquor used in this study was supplied by a Spanish paper mill where cereal straw is cooked with soda as the only chemical.

The original black liquor has a solid content ranging between 5 and 10% in weight. It is dried in an oven at 105 °C until completely dry. The black liquor solids (BLS) obtained are smashed and sieved to the size desired. The elemental analysis of the BLS, obtained in a CHNS Carlo Erba elemental analyzer (mod. EA 1108), is shown in Table 1.

Table 1. Elemental composition of BLS

Element	N	C	H	S
% in weight	1.29	45.85	4.49	0.55

The ash weight content of the BLS is about 22%, with the main components being 33.6% carbonate, 8% chloride, 28% sodium, 7.2% potasium, and 8.8% silica. The chloride comes from the water used for cooking while K, N, SiO_2 and sulphur come from the straw, and were disolved while cooking.

Different experimental systems were used. For the thermogravimetric experiments, two different thermobalances have been used, a Setaram TGA 92 and a Cahn Instruments. The gas composition and flowrate were set with mass flow controllers.

In order to get a sufficient amount of char for the reactivity studies another plant was used. It consists of a 8 cm i.d. tubular reactor, discontinuous for the solid which is introduced, held in a ceramic cup, vertically into an electrically heated furnace connected to a temperature and heating rate control system allowing us to program different heating rates up to 100 °C/min. The pyrolysis has been carried out under a N_2 flow controlled by a mass flow controller. The gas at the outlet is cooled and the tar condensed. Gas samples can be taken to be analyzed by Gas Chromatography. A scheme of the experimental system can be found elsewhere [11].

3 Results of the thermal decomposition of black liquor solid

Isothermal and dynamic experiments of the thermal decomposition of black liquor solid in a nitrogen atmosphere have been carried out. In these experiments the solid conversion values (X_s), defined as the ratio between the solid weight loss at a given time and the initial weight, were determined.

Examples of the results obtained in dynamic experiments using heating rates of 8 and 50 °C/min are shown in Figure 1. These experiments were carried out in a thermobalance with a particle size less than 150 μm and a nitrogen flowrate of 800 Ncm³/min. As with other materials, the solid conversion obtained at a given temperature is higher as the heating rate decreases. Figure 2 shows the variation of solid conversion with the time and temperature up to a temperature of 600 °C.

Two temperature ranges for which significant solid conversion variation occurs can be appreciated,both depending on the heating rate. The first range starts at about 200 °C and extends to about 325 °C. The second from about 400 °C to 500 °C. Significant higher values of dXs/dt are obtained when the heating rate is increased.

The kinetics of weight loss of black liquor solid during its thermal decomposition has been determined. As black liquor is a heterogeneous material, the methodology used in this work for the determination of the kinetic equations is the same as that used by this research group with other heterogeneous materials, such as lignocellulosic biomass [12] [13] [14].

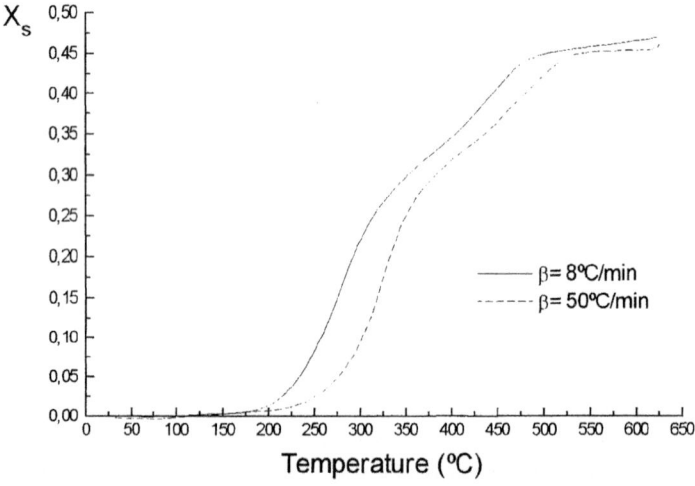

Fig. 1. Solid conversion for two heating rates in nitrogen atmosphere.

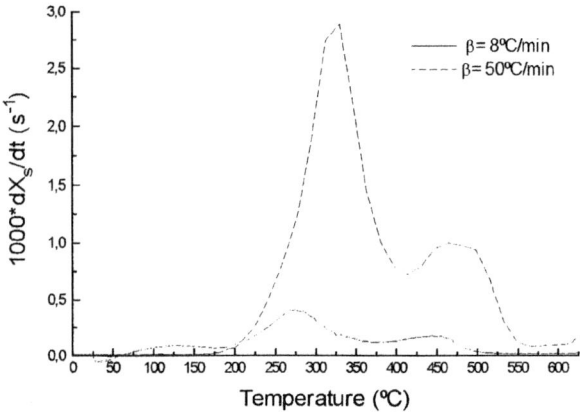

Fig. 2. Variation of solid conversion with time for two heating rates.

With this methodology the values of the pyrolyzable weight fraction (A_s) for different temperatures must be taken into account. This parameter is defined as:

$$A_s = \frac{W_0 - W_t}{W_0} \tag{1}$$

where W_0 is the initial weight and W_t is the final weight at a given temperature.

The A_s values have been determined from isothermal experiments of long duration in a nitrogen atmosphere and in the temperature ranges for which significant solid conversion is obtained (200-500 °C). These experiments started with a low heating rate

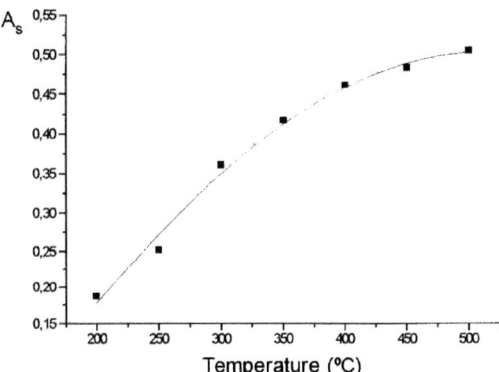

Fig. 3. Pyrolyzable weight fraction for different temperatures.

(10 °C/min) until the desired temperature is reached in order to avoid a temperature gap between the sample and its surroundings. The variation of A_s with temperature is shown in Figure 3.

It is considered that the solid conversion variation during the BLS thermal decomposition can be represented by a power-law equation:

$$\frac{dX_s}{dt} = k\left(A_s - X_s\right)^n \tag{2}$$

It has been assumed that, as for other black liquors [3] and other materials [15], $n \approx 1$, so the k_β values for each range can be calculated as:

$$k_\beta = \frac{dX_s / dt}{A_s - X_s} \tag{3}$$

Figure 4 shows the k_β values obtained for the two heating rates and the temperature ranges for which significant variation in the solid conversion is observed.

The kinetic constant values have been fitted to the Arrhenius equation.

For $\beta = 8$ °C/min, the constant values obtained are:

$200 \leq T \leq 285$ °C

$$k_{\beta=8}(s^{-1}) = 37.9 \exp\left(-\frac{5275}{T}\right) \tag{4}$$

$385 \leq T \leq 462$ °C

$$k_{\beta=8}(s^{-1}) = 5.3 \exp\left(-\frac{5631}{T}\right) \tag{5}$$

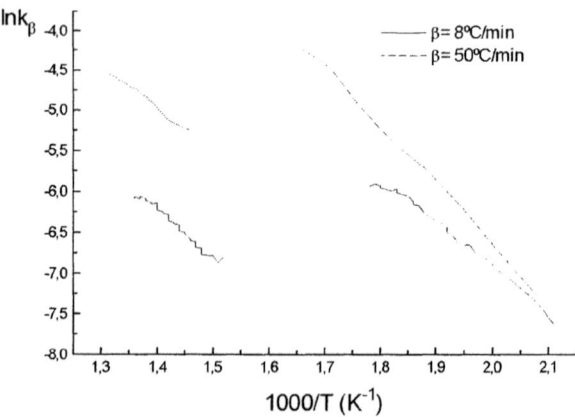

Fig. 4. k_β values obtained from dynamic experiments in nitrogen atmosphere.

For β= 50 °C/min, the constant values obtained are:

$208 \leq T \leq 329$ °C

$$k_{\beta=50} (s^{-1}) = 2578.3 \exp(-\frac{7245}{T}) \tag{6}$$

$412 \leq T \leq 496$ °C

$$k_{\beta=50} (s^{-1}) = 9.1 \exp(-\frac{5120}{T}) \tag{7}$$

In the bench scale plants, the product gas obtained in the thermal decomposition of BLS in a nitrogen atmosphere has been analyzed by Gas Chromatography, after cooling and condensing the tar. An example of the gas distribution obtained is shown in Figure 5. The conditions of this experiment were: final temperature 500 °C, β=50 °C/min, weight of BLS used 14.65 g, N_2 flowrate: 508,5 cm^3/min. The product distribution does not change significantly when varying the final temperature or heating rate. As can be seen, CO_2 is the main product.

4 Reactivity of the black liquor char

The char used was obtained by pyrolysis of BLS with a heating rate of 8°C/min and at a final temperature of 500 °C which was maintained during two hours, under a N_2 flow. The solid obtained was smashed and sieved to less than 75 μm, to minimize possible intra-particle diffusion during the reaction. The reactivity of this char with oxygen and with CO_2 has been studied. In each experiment a flowrate of 800 Ncm^3/min was used.

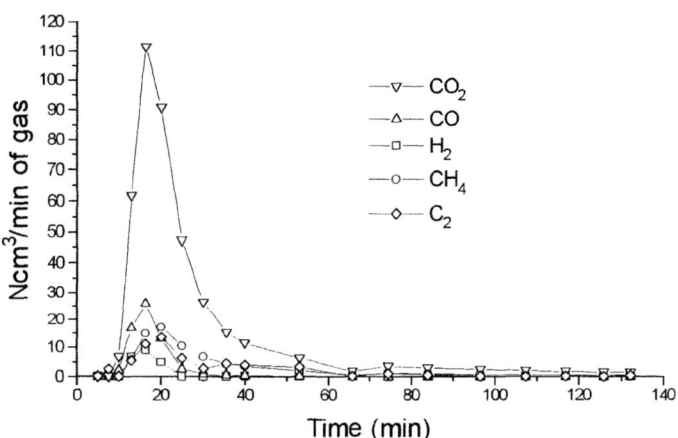

Fig. 5. Gas composition evolution during thermal decompositon of BLS.

4.1 BL char reactivity with different oxygen concentrations

Experiments in a thermobalance were carried out. The inlet gas composition ranged from pure N_2 to 14.7% O_2. The heating rate of these experiments was 8 °C/min. The solid conversion curves are shown in Figure 6, and the derivative of conversion with respect to temperature in Figure 7. It can be seen than most of the conversion occurs between 350 and 430 °C, with the maximum conversion rate at 380-390 °C. At 100°C a peak appears due to the water evaporation. The higher the oxygen content, the greater is the conversion rate and also the conversion reached for a given temperature.

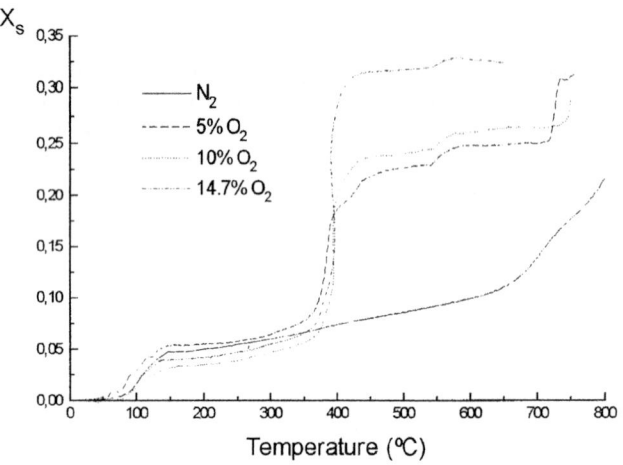

Fig. 6. Conversion of BL char for different oxygen concentrations.

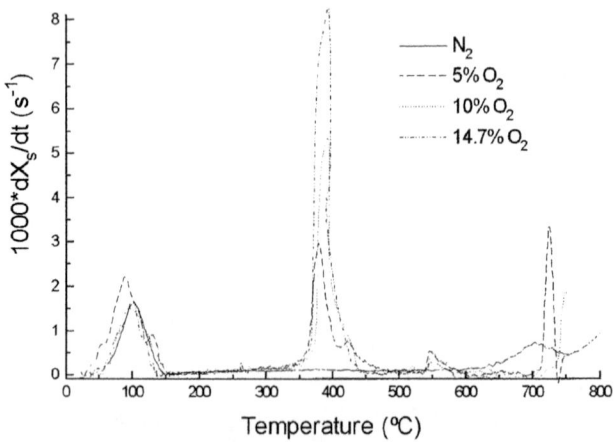

Fig. 7. Derivative of char conversion vs temperature with different oxygen concentrations.

300

It can also be observed that when the experiment is performed in a nitrogen atmosphere, an increase of the solid conversion from T=550 °C is obtained. When there is oxygen in the gas, this effect is not observed. This could be due to the fact that, in an inert atmosphere, alkaline carbonates in BL char decompose, which could be avoided with a small amount of CO or CO_2 in the gas [14] [16]. The fact that carbonate decomposition does not occur may be due to the fact that, as oxygen is present, CO and/or CO_2 are being generated on the surface of the char, in which case the effect would be similar to the introduction of CO and CO_2 in the gas.

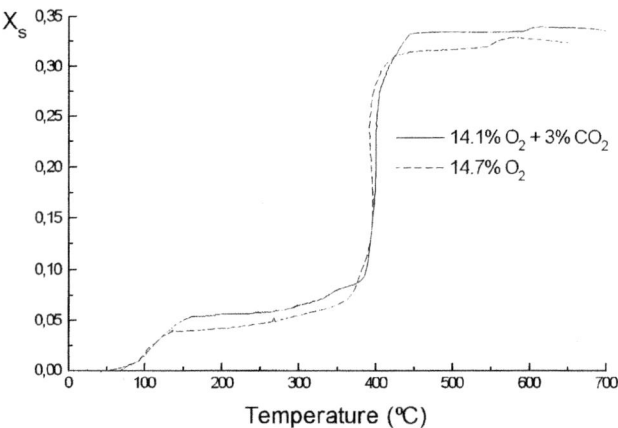

Fig. 8. BL char conversion with oxygen and CO_2.

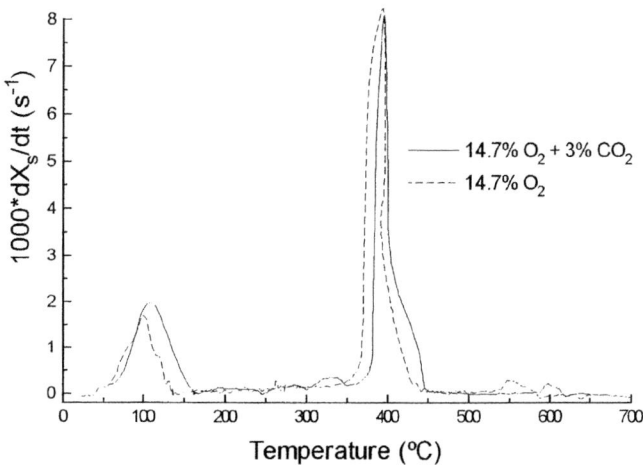

Fig. 9. Derivative of char conversion vs temperature with oxygen and CO_2.

In order to know if the presence of CO_2 affects the weight loss due to carbonate decomposition, an experiment was performed using 14.7% O_2 and with 3% CO_2 in the gas. As can be observed in Figures 8 and 9, the conversion curves are quite similar.

4.2 BL char reactivity with CO_2

Experiments in a thermobalance with a heating rate of 8 °C/min were performed. A gas flowrate of 800 Ncm³/min with CO_2 concentrations of 10 and 20% were used. Figure 10 shows solid conversion versus temperature and Figure 11 the derivative of the char conversion versus temperature.

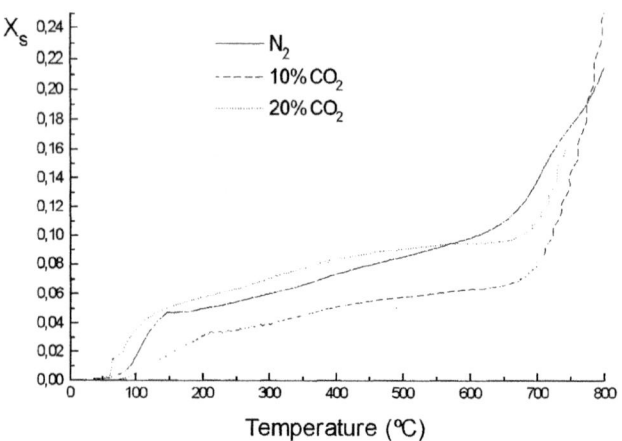

Fig. 10. BL char conversion with different amounts of CO_2.

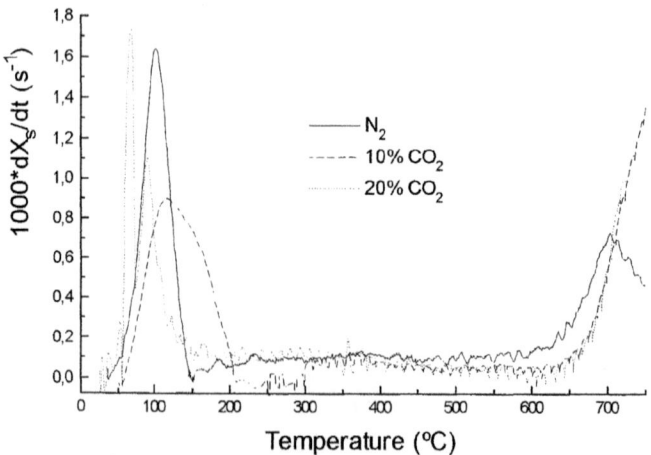

Fig. 11. Derivative of conversion vs temperature.

It can be seen that the conversion is not very different when 10% or 20% CO_2 or only pure nitrogen are used. Differences start at 550 °C, when the solid conversion with pure nitrogen is faster than with CO_2. The derivative of the weight loss is greater for the nitrogen curve until 700 °C is reached. At this temperature carbonate decomposition and CO_2 gasification can be significant [17], so both it is quite possible that both effects are present. It is interesting to observe that a sudden increase of the conversion also occurred at 700 °C when oxygen was used, as can be observed in Figures 6 and 7.

5 Conclusions

- Weight loss of black liquor solids (BLS) during its thermal decomposition can be represented by means of first order kinetic equations, with the constant being dependent of the range of temperatures.
- The concentration of oxygen significantly affects the reaction rate of the black liquor char and the final conversion achieved.
- In a nitrogen atmosphere a char weight loss is observed over 550 °C. This is attributed to the decomposition of alkaline carbonates. The presence of CO_2, even proceeding from the char oxidation in the presence of oxygen, inhibits the decomposition.
- BL char reacts significantly with CO_2 over 700 °C.

6 References

1. Hupa, M., Solin, P. and Hyöty, P. (1987) Combustion behavior of black liquor droplets. *Journal of Pulp & Paper Science*, Vol. 13, No. 2. pp. 67-72.
2. Alén R., Rytkönen, S. And McKeough, P. (1995) Thermogravimetric behavior of black liquors and their organic constituents. *Journal of Analytical and Applied Pyrolysis*, Vol. 31. pp. 1-13.
3. Bhattacharya, P.K., Shrinath, A.S. and Kunzru, D. (1985) Pyrolysis of black liquor. *Journal of Chemical Technology and Biotechnology*, Vol. 35A. pp. 223-33.
4. Nassar, M.M. (1984) Thermal studies on kraft black liquor. *Journal of Chemical Technology and Biotechnology*, Vol. 34A. pp. 21-4.
5. Frederick, W., Wåg, K. and Hupa, M. (1993) Rate and mechanism of black liquor char gasification with CO_2 at elevated pressures. *Industrial Engineering Chemical Research*, Vol. 32. pp. 1747-53.
6. Li, J. and van Heiningen, A.R.P. (1991) Kinetics of gasification of black liquor char by steam. *Industrial Engineering Chemical Research*, Vol. 30. pp. 1594-1601.
7. Li, J. and van Heiningen, A.R.P. (1989) Reaction kinetics of gasification of black liquor char. *Canadian Journal of Chemical Engineering*, Vol. 67. pp. 693-7.

8. Whitty, K., Frederick, W. and Hupa, M. (1992) Gasification of black liquor char with H_2O at elevated pressures. *International Chemical Recovery Conference*, Seattle, USA, pp. 627-39.

9. Noopila, T (1991) Laboratory studies on combustibility of black liquors, *Thesis for the Degree of Licenciate of Technology*, Åbo Akademi University, Finland.

10. W. Frederick, W., Noopila, T. and Hupa, M. (1991) Combustion behavior of black liquor at high solids firing. *Journal Pulp & Paper Science*, Vol. 17, No. 5. pp. 164-70.

11. Bilbao, R., Salvador, M.L., García, P. and Arauzo, J. (1993) Solid weight loss in the thermal decomposition of cellulose and pine sawdust. *Journal of Analytical and Applied Pyrolysis*, Vol. 24. pp. 257-71.

12. Bilbao, R., Millera., A. and Arauzo, J. (1987) Kinetics of thermal decomposition of cellulose. Part I. Influence of experimental conditions, *Thermochimica Acta*, Vol. 120. pp.121-31.

13. Bilbao, R., Millera, A. and Arauzo, J. (1989) Kinetics of weight loss by thermal decomposition of xylan and lignin. Influence of experimental conditions. *Thermochimica Acta*, Vol. 143. pp. 137-48.

14. Bilbao, R., Millera, A. and Arauzo, J. (1989) Thermal decomposition of lignocellulosic materials: Influence of the chemical composition. *Thermochimica Acta*, Vol. 143. pp. 149-59.

15. Williams, A. (1982) Urban and wildland fire phenomenology. *Progress on Energy Combustion Science*, Vol. 8. pp. 317-54.

16. Li, J and van Heiningen, A.R.P. (1990) Sodium emission during pyrolysis and gasification of black liquor char. *TAPPI Journal*, Vol. 73, No. 12. Pp. 213-9.

17. Sams, D.A. and Shadman, F. (1986) Mechanism of potassium-catalyzed carbon/CO_2 reaction. *AIChE Journal*, Vol. 32, No. 7. pp. 1132-7.

7 Acknowledgements

The authors express their gratitude to CICYT (Project AMB95-0575) for providing frame support for this work, and to MEC and EC for research grants awarded to J.L. Sánchez and Fang Z.

MASS BALANCE OF BIOCARBON ELECTRODES OBTAINED BY EXPERIMENTAL BENCH PRODUCTION

A.R.COUTINH
Methodist University of Piracicaba, Technological Center, 13400-901, Piracicaba-SP, Brazil
C.A.LUENGO
Applied Physics Department, Unicamp, 13083-970, Campinas-Sp, Brasil

Abstract

Biocarbon electrodes (BCE) for specialty applications were produced, bench scale, from biomass. The pyrolysis of wood logs at 1000°C yielded charcoal and volatile by-products which were condensed and later distilled to recover the biopitch, utilized as the binding agent. The biocoke was ground and mixed with the binder to obtain the bio-electrode paste. The green electrode was molded at 60 MPa and 150°C. Heat treatments include calcination at 1000°C followed by graphitization at 2700°C. The physical properties of BCE showed microcrystallite dimensions of $L_c = 124$ Å and $L_a = 565$ Å, electrical resistivities of $10^{-4} \Omega.m$ and the mechanical measurements yielded a Young's modulus near 3.0 GPa, rupture strength of 50 MPa and a thermal expansion coefficient of 6.10^{-6} °C^{-1}. The mass balance of the manufacture process indicates 31% of biocoke, 45% of condensed and 24% of volatile materials. The vacuum distillation of condensed fractions resulted in 11.2% of biopitch. Mixing and molding of biocoke with biopitch paste produced the green electrode which was calcined and finally, graphitized. Ultimately, the heat treatment process yields 28.9% of graphitized BCE.
Key Words: Biocarbon, Biopitch, Electrode, Biomass

1   Introduction

Electrodes are extensively used in aluminum production and for obtaining high grade metals [1-6]. A more restricted market consists of high purity graphites for transforming silica in quartz or semiconductors, for obtaining anodes of new batteries or manufacturing light structures of airplanes [7-9]. These specialty applications require pure carbons with appropiate physical and structural properties. Furthermore, they are sensitive to impurities such as sulfur, metals and others frequently found in fossil feedstocks. BCE may represent a viable option because of their inherent high purity, similar properties and, as shown below, production mass balances comparable to those currently obtained in conventional materials.

Previous reports described BCE preparation procedure and main physical and mechanical properties [10-11]. This paper outlines the corresponding mass balance.

2.1- Biocoke and biopitch production

Eucalyptus saligna (E.saligna.) wood logs were chopped to obtain the feedstock. The samples, cyllinders of 0.025m in diameter and 0.12m in height, were pyrolyzed up to 1000°C using heating rates of 3°C/min and an oxygen deficient atmosphere. Table 1 summarizes the characteristics of the feedstock and the product of this stage.

Table 1 *E.saligna* wood and charcoal characteristics

Properties	Wood	Charcoal
Lignin (%)	31.0	-
Density (g / cm^3)	0.55	0.18
Fixed Carbon (%)	34.0	95.5
Volatiles (%)	65.5	4.0
Ash (%)	1.0	0.5

The pyrolysis products were solid charcoal (biocoke), condensed liquids (tar) and gas. The mass balance of this stage is showed in Table 2.

Table 2 Mass yields of the pyrolysis of *E.saligna* wood

Product	Yield (%)
Biocoke	31.0
Tar	45.0
Gas	24.0

The biopitch binder was obtained by distilling the tar at 260°C, 25 kPa with a heating rate of 2°C/min. In Table 3 are compared the properties of the biopitch with those of conventional binders [12]. Note the much lower content of contaminants of the biopitch used for BCE .

Table 3 Analysis of biopitch (BP) and coal tar pitch (CTP)

Properties	BP	CTP
Softening Point (°C)	119.0	110.0
Toluene Insolubles (%)	50.0	35.0
Quinoline Insolubles (%)	0.02	11.0
Coking Value (%)	40.0	58.0
Density (g / cm^3)	1.3	1.3
Sulfur (%)	0.0	0.4
Ash (%)	0.01	0.2

2.2. The mass balance of BCE preparation procedure

The lay out of BCE pilot production normalized to 100kg *E.saligna* wood, is shown in Figure 1, note production stages and respective losses.

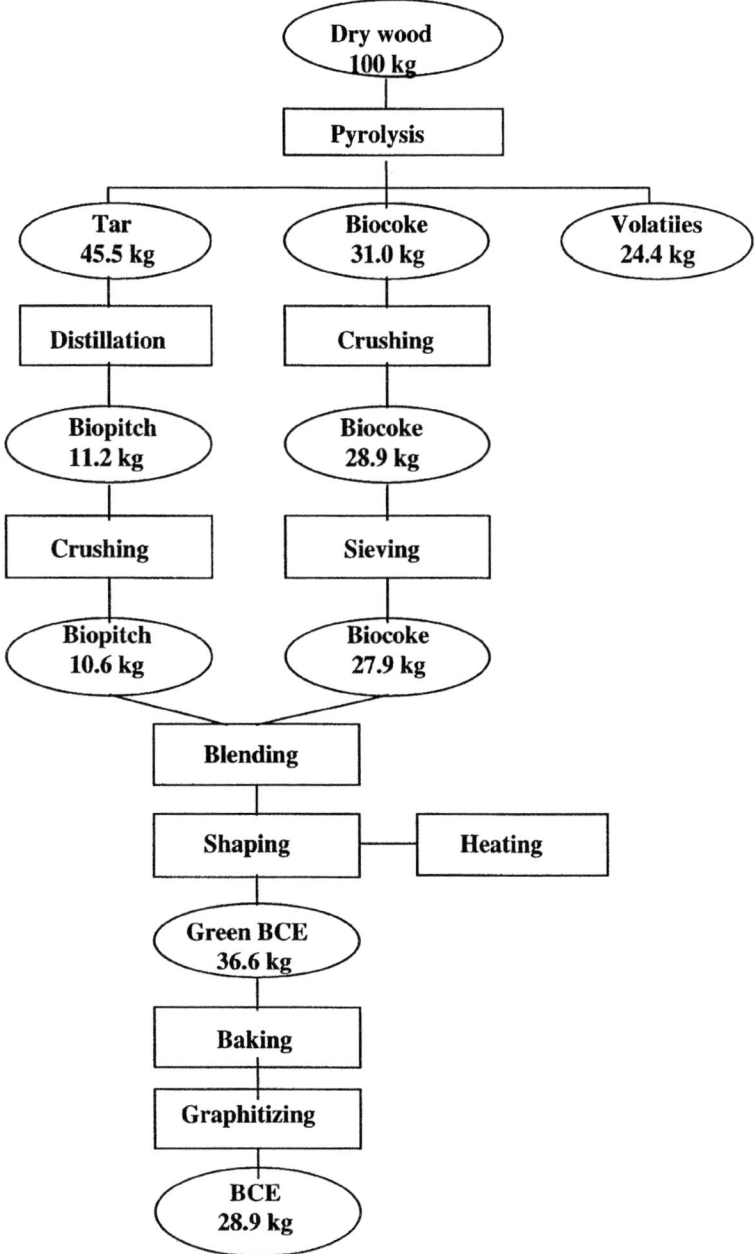

Figure 1 The mass balance of BCE production

After preparation and pyrolysis stages, appropriately sized biocoke and biopitch samples were blended and molded in a warm die (150°C) to a cylinder 0.025m in diameter and 0.12m length, i.e. the "green electrode" which has the properties listed in Table 4.

Table 4. Some properties of the green electrode

Electrical resistivity (Ω.m)	10^{10}
Apparent density (g/cm^3)	1.07
Fixed carbon (%)	73.8
Volatiles (%)	22.2
Ash (%)	4.0

Two options are possible for the heat treatment of BCE since, for applications where their electrical and mechanical properties are the most important, baking at 1000°C is enough. However, for applications in which the microscopic properties are important, an additional graphitization stage may be necessary.

It must be noted the comparison of mass balances for the production of BCE and its equivalent, reported in the literature for conventional electrodes, currently obtainedfrom coal and petroleum. From the layout shown in Figure 1 it is clear that the process efficiency of BCE production is close to 30% related to the *E.saligna* feedstock and 40% when related to the intermediaries biocoke and biopitch, comparing rather well with the 39% currently accepted for the case of fossil feedstocks [2-4].

2.3. Compositional analysis of the BCE

The thermogravimetric analysis (TGA) of the BCE as a function of the heat treatment temperature (HTT) indicates the evolution of the BCE material. Figure 2 shows the TGA of a sample from room temperature to 1100°C. The rate of mass removal goes through a steep maximum at 300°C and flattens out above, at a HTT of 1100°C indicating that 17% of the electrode mass has been removed, i.e. 70% of the biopitch added leaves the sample. It is thought that the remaining 30% is polymerized, acting as the binder.

The composition changes due to heat treatment may also be documented through proximate and porosity analysis of BCE samples. Representative data is shown in Table 5 where are displayed characteristics of calcined and graphitized (T$\cong$ 2700°C) samples.

Table 5   Analysis of the calcined and graphitized BCE

Properties	Calcined	Graphitized
Density (g/cm^3)	1.1	1.3
Porosity (%)	17.0	35.0
Fixed Carbon (%)	95.2	99.0
Volatiles (%)	3.8	0.9
Ash (%)	1.0	0.1

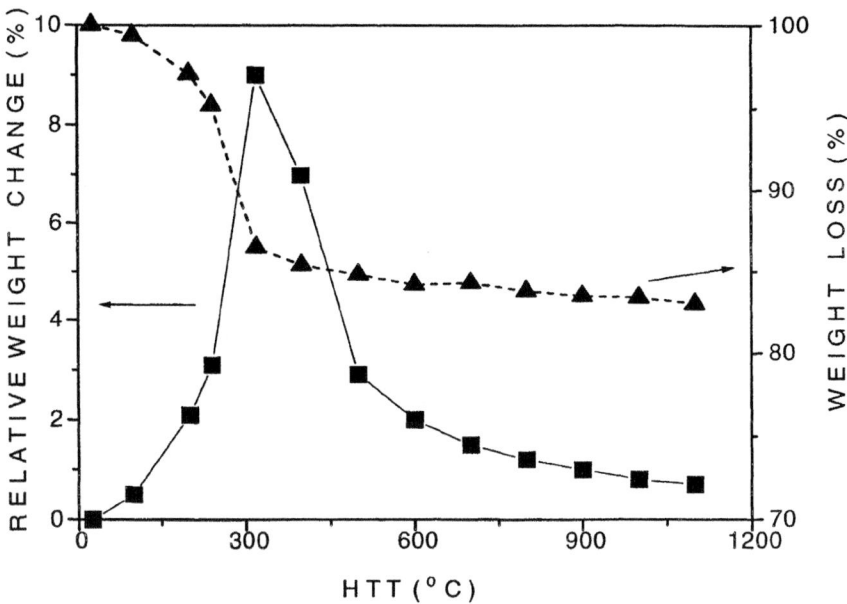

Figure 2   TGA analysis of the BCE

3   Microscopic properties

A Phillips diffractometer was used to obtain x-ray diffraction (XRD) spectra of several BCE as a function of their HTT. Narrowing of the lines evidenced the presence of a turbostratic structure. The parameters of microcrystallites, i.e. Lc, dimension of the parallel layer groups, and La, the layer dimension, may be calculated from the linewidth at half maximum intensity of (002) and (110) lines, respectively, using expressions from WARREN [13] and BISCOE & WARREN [14]. For BCE heat treated at 2700°C, the parameters Lc and La indicate values of 124Å and 565Å, respectively, similar to other non-graphitizing carbons [15].

3.1   Electrical resistivities

Measurements of electrical resistivity for BCE were with a two points technique for samples with a resistivity above 1.0 Ω.m., and a four points technique [16] in those with resistivities lower than that value. The resistivity decreases with HTT about nine orders of magnitude. Samples with HTT<400°C have values typical of non-conductor materials, then at HTT$\geq$800°C present a transition region to a value which remains nearly constant at $10^{-4}\Omega$.m, which is similar to values currently observed in commercial graphites. Moreover, the data correlates rather well with results recorded in the literature for other heat treated carbons [16-17].

3.2 Mechanical properties

The mechanical properties of BCE samples were measured at room temperature using a MTS instrument (model 810). The samples were positioned with their axis parallel to the molding force. The compression strength of BCE at room temperature, shown values of 50 MPa. The Young's modulus present a maximum of about 3.0 GPa for BCE heat treated near 1100°C, followed by an abrupt decrease with HTT, similar to observations in graphite artefacts [18-19]. BCE showed values lower than other biomass materials heat treated at high temperature, for example the babassu nut charcoal moduli are three times larger [20] like synthetic graphites, however the BCE behavior is similar to commercial electrodes derived from petroleum coke.

3.3 Thermal expansion

The room temperature coefficient of thermal expansion (CTE), was obtained using a Adamel Lomargy dilatometer (model LK02). Results of L-CTE for BCE, indicate values of $6.10^{-6}{}^{o}C^{-1}$ for samples heat treated at 2700°C, also observed in other heat treated carbonaceous materials [21-23].

4   Conclusions

Biocarbon Electrode (BCE), another end product from biomass, was obtained using a procedure initiating with the batch pyrolysis of E.*saligna* wood, a bio-source. The raw material showed a satisfactory behaviour in the manufacture process, resulting in a yield near 30%, from dry wood to final product. The solid fraction, charcoal, is used instead of petroleum coke, and biopitch substitutes the coal tar pitch both used in conventional synthetic electrodes. When compared, the alternative biocomponents of the BCE paste showed much lower inert and sulfur contents. The biopitch binder softening point, insolubles and fixed carbon are similar to those of conventional binders.

Blending carbon powder and biopitch with appropriate molding conditions and heat treatments (calcination followed by graphitization) resulted in prototypes mechanically resistant and very hard.

The BCE after graphitization monitored by XRD, showed the evolution towards turbostratic structure. The values of Lc and La, indicate growth in parallel direction, approximately four times larger than in perpendicular direction, resulting in the formation of long planes, due to lateral coalescence of the microcrystallites.

The BCE electrical resistivities are similar to those of graphites and other bio materials, like lignin coke and babassu nut charcoal, when heat treated at high temperatures. The mechanical properties, i.e. Young's modulus and rupture strength, compare favorably with commercial electrodes. Also, CTE is in the range of conventional electrodes. Thus, the BCE properties suggest a variety of possible applications in new devices such as batteries or electrodes for specialty processes, like fullerenes production and others. Some are being developed and will be reported elsewhere.

References

1   MANTELL, C.L. "The technology of the carbon electrode industry: History and development". *Chem. and Metallurgical. Eng.* (1922) 27, 109-112.

2   HARDER, R.N., GAMSON, B.L., BAILEY, B.L. "Graphite electrodes" *Industrial and Eng. Chemistry* (1954) 46, 2-11.

3   CUMMINAL, B., DE BOUAR, M. "Les graphites artificiels". *Chimie et Industrie Genie Chimique* (1972), 105, 1653-1657.

4   CHRIS, S.H.; "Graphite electrodes". *Metal Bulletin Monthly* (1981), Oct, 89-95.

5   PARISOT, C. "Graphite electrodes for arc furnace". *Metal Bulletin Monthly (*1981), n° 130, 99-109.

6   CLAUSE,J., CORNAULT,P. "Les graphites artificiels", *Chimie at Industie - Genie Chemique,* (1972), 105, 1569-1575.

7   MANTELL,C.L. **Carbon and Graphite Handbook**. (1968) Interscience Publisher, NY.

8   NIGHTINGALE,R.E. **Nuclear Graphite** (1962), Academic Press, NY.

9   DRESSELHAUS, M.S., DRESSELHAUS, G., "Intercalations compounds of graphite". *Advances in Physics,* (1981), 30, 139-326.

10   COUTINHO, A.R. "Síntese e caracterização de eletrodos grafíticos a partir da pirólise de *Eucalyptus saligna*". PhD Thesis (1992), IFGW, Unicamp, Campinas SP, Brasil.

11   COUTINHO, A.R., LUENGO, C.A. "Preparing and characterizing electrode grade carbons form eucalyptus pyrolysis products", *Advances in Thermochemical Biomass Conversion (*1994), vol 2, 1230-1240, Chapman Hall, London.

12   MARTIN, Y., GARCIA, R., ANDERSEN, J., LUENGO, C.A., SNAPE, C.; MOINELO, S.R. "Influence of the preparation condictions on the nature and thermal behaviour of coal tar pitches". *Proc.of 22 Int. Conf. on Carbon,* Granada, Spain (1994).

13   WARREN, B.E. "X-ray diffraction in random layer lattices". *Physical Review (*1951), 59, 693-698.

14   BISCOE, J, WARREN, B.E. "An x-ray study on carbon black". *J. Applied Physics,* (1942), 13, 364-371.

15   FRANKLIN, R.E. "Crystallite growth in graphitizing and non-graphitizing carbons". *Proc. of the Royal Soc.(*1951), 209 A, 196-219.

16   EMMERICH, F.G, SOUZA, J.C., TORRIANI, I., LUENGO, C.A.; "Applications of a granular model and percolation theory to the electrical resistivity of heat treated endocarp of babassu nut". *Carbon* (1987), 25, 417-424.

17   ERNANDES, G., CALDERON, .H., LUENGO, .A., TSU, . "Microscopic structure and electrical properties of heat treated coals and eucalyptus charcoal". *Carbon,* (1982*)*, 20, 201-205.

18   ANDREW, F, SATO, JF. "Studies of Young's modulus of carbons at high temperature". *Carbon (*1964), 1, 225-234.

19   ONDA, H, SANADA, Y., FURUDA, T. "Mechanical and thermal properties of heat treated carbons". *Carbon (*1966), 3, 421-428.

20   EMMERICH, F.G., LUENGO, C.A. "A theory for non-graphitizing carbons". *Carbon,* (1993), 31, 333-339.

21   HEINTZ, E.A. "The measurement of the coefficient of thermal expansion of graphite artefacts". *Carbon (*1990), 28, 233-234.

22   PRICE, R.J., BOKROS, J.C., KOYAMA, K. "thermal expansivities and preferred orientation of pyrolitic carbons". *Carbon (*1967), 5, 423-430.

23   MASON, I.B., KNIBBS, R.H. "the thermal expansion of polycrystalline carbon and graphite from -196°C to 2000°C". *J. of Nuclear Energy, parts A/B,*(1964*)*, vol 18, 311-329.

CARBONISATION OF FIVE SHORT ROTATION TROPICAL TREE SPECIES

J. A. FUWAPE
Federal University of Technology
Akure, Nigeria

Abstract

Five fast growing tree species : Acacia mangium, Cassia nodosa, Gliricidia sepium, Gmelina arborea and Leucaena leucocephala were converted to charcoal by pyrolytic process. Charcoal briquettes were prepared from charcoal fines and carbonized wood twigs. The charcoal yield and the combustion related properties of the fuelwood and charcoal, viz: moisture content, density, percentage fixed carbon, volatile matter and gross heat of combustion were determined.

The charcoal yeild of the tree species ranged from 29 to 42%. Although the nominal density of charcoal was less than that of wood, high density charcoal briquettes were produced by binding charcoal fines with starch. Average gross heat of combustion of charcoal briquettes, 32.5 MJkg-^{1}, was not significantly different from that of charcoal but was higher than that of wood, 22.14 MJkg-^{1}. The final carbonisation temperature and tree species had effect on charcoal yield and combustion characteristics. High charcoal yield was recorded at low carbonisation temperature. Charcoal produced from acacia mangium, Gmelina arborea and Leucaena leucocephala have desirable properties for domistic energy application.

Keywords: briquettes, carbonisation temperature, charcoal, firewood, thermal conversion, tropical trees, short rotation.

1 Introduction

Charcoal is mainly used as source of domestic fuel in developing countries. The increase in the demand for charcoal is due to its superior combustion characteristics, lower pollution problems and lower transport and handling cost when compared with firewood [1,2]. In addition to the domestic use of charcoal for cooking, it is used as source of heat for smelting by blacksmiths and as a raw material for chemical derivatives such as carbon disulphide, carbon tetrachloride, sodium cyanide, activated carbon, fertiliser and pharmaceuticals. In Brazil, charcoal supplies 40% of the energy needs of the steel industry while it is used as the major fuel for heat generation for cemect production in Uganda [3].

The quality and properties of charcoal is affected by the process of wood carbonisation. Pyrolytic conversion of wood to charcoal involves thermal separation of volatile matter from the solid charcoal residue. The proportion of fixed carbon and volatile matter in charcoal varies with the type of plant material used as feedstock and the pyrolytic technique. The pyrolytic process is affected by heating rate, residence time, particle size, temperature, chemical composition and moisture content of species and the final pyrolysis temperature [4]. The quality of charcoal in terms of percentage fixed carbon increases with increase in carbonisation temperature [5,6]. Charcoal containing high proportion of fixed carbon and low content of volatile matter are prefered for industrial applications while those with lower fixed carbon are acceptable for domestic use [7]. However, charcoal yield decreases with increase in carbonisation temperature: only charring takes place below 260°C while depolymerization of chemical components generally predominates at 275 to 400°C. Hemicellulose is readily converted to methanol and acetic acid at 200 to 280°C while lignin decomposes at temperatures above 280°C forming mainly tar and charcoal [8]. The main objective of this study was to investigate the effect of carbonisation temperature on the yield and combustion properties of charcoal from the selected five fast growing tropical wood species. The physical properties and heat of combustion of charcoal briquettes produced were also determined.

2 Materials and Methods

Twentyfive samples of five-year old trees of *Acacia mangium, Cassia nodosa, Gliricidia sepium, Leucaena leucocephala* and *Gmelina arborea which* were cut to 2m long billets were collected from the forestry plantation at The Federal University of Technology, Akure, Nigeria. Subsamples were randomly selected from the species for the determination of specific gravity and heating value of wood. Another set of subsamples were cut into regular dimensions (6 cm long by 2 cm width and 2 cm thick) and used for carbonisation experiment. Wood samples for carbonisation experiment were initially oven dried at 103 $\pm$ 2 °C for 24 h or until they attained constant weight. Carbonisation of sample from each treatment (and tree species) were replicated five times. Each specimen to be carbonised was weighed and placed in a crucible , the crucibles were then arranged in the furnace. A constant heating rate of 3°C per minute was used to heat the specimens to the final temperature. The wood samples were subjected to pyrolysis tests (slow pyrolysis) to the desired final carbonisation temperature of either 250,300, 350, 400, 500 or 600°C . The furnace was turned off as soon as it attained the desired temperature. The carbonised specimens were allowed to cool down in the furnace before they were removed and placed in the desicator. After the cooling process the carbonised specimens were weighed at room temperature to determine percentage charcoal yield. Charcoal briquettes were prepared from pulverized charcoal mixed with starch (cassava starch equivalent to 2.5% of oven-dry weight of charcoal was used as adhesive). The mixture was compressed in a briquetting machine according to the technique described in International Labour organisation handbook [9].

The charcoal samples were kept in the oven at 103 ± 2°C prior to proximate analysis and calometric tests to determine the heating value. The moisture content and specific gravity of both wood and charcoal produced from each tree species were determined according to ASTM D143-52(1979) procedures [10]. The heat of combustion was determined with the ballistic bomb calorimeter according to the ASTM D2015-79 procedures. Proximate analysis was conducted following the ASTM D1102-84 procedures. The percentage volatile matter, fixed carbon and ash content of each carbonised sample was determined.

3 Results and Discussion

The result of charcoal yield of selected species at different final carbonisation temperatures are presented in Table 1. Although there were significant differences in the charcoal yield between species at different final carbonisation temperature slight differences existed in the weight of carbonised material at a particular carbonisation temperature. The charcoal yield decreased with increase in pyrolysing temperature for all the species. The higher yield of carbonised material at 250-300°C may be attributed to limited thermal decomposition of extractives and hemicellulose in wood at 190° to 270°C [11]. The drastic reduction in charcoal yield per unit mass of wood is possibly due to the variation in the rate of thermal decomposition of the chemical components of wood. Cellulose which constitutes about 45% of tropical hard-wood, is thermally degraded at 270° to 325°C but lignin which makes up 25-30% of the chemical composition of wood is thermally stable below 270°C while 90% of it, is thermally degraded at 400°C [7,12].

Table 1 Charcoal yield and density of five selected wood species

Species	Wood	Final carbonisation temperature (°C)					
		250	300	350	400	500	600
Acacia mangium							
Charcoal yield (%)	-	94±2	53±1	48±2	36±1	24±2	23±3
Density	0.64	0.56	0.42	0.34	0.28	0.20	0.20
Cassia nodosa							
charcoal yield (%)	-	92±3	54±2	33±1	31±2	20±2	14±4
Density	0.35	0.30	0.26	0.2	0.2	0.18	0.18
Gliricidia sepium							
Volatile matter (%)	-	94±4	52±2	35±2	33±3	32±4	29±3
Ash content (%)	0.49	0.46	0.38	0.36	0.31	0.26	0.26
Gmelina arborea							
Charcoal yield (%)	-	96±3	44±3	39±1	32±2	27±2	19±3
Density	0.5	0.48	0.40	0.34	0.30	0.26	0.24
Leucaena leucocephala							
Charcoal yield (%)	-	94±2	47±1	42±2	34±1	31±1	26±4
Density	0.62	0.56	0.52	0.39	0.32	0.30	0.30

Each value is the average of five replicates

315

3.1 Proximate composition

The result of proximate composition of charcoal produced at different temperatures is presented in Table 2. The percentage volatile matter in the carbonised samples decreased with increase in final carbonisation temperature while there was increase in the percentage fixed carbon. The trend of the variations in the value of volatile matter and fixed carbon in the charcoal produced at different final carbonisation temperature in this study is similar to values reported in previous studies [13,14,15].

Table 2 Proximate composition of charcoal at different carbonisation temperature

Species	Final carbonisation Temperature °C						
	Wood	250	300	350	400	500	600
Acacia mangium							
Volatile matter (%)	77	71	64	50	38.4	23.3	21
Ash content (%)	1.2	1.4	2	3	3.6	4.7	5
Fixed carbon (%)	21.8	27	34	47	68	72	74
Cassia nodosa							
Volatile matter (%)	85	74	58	53	36	30.7	25
Ash content (%)	1.2	1.8	2.4	4	5	5.3	6
Fixed carbon (%)	14.8	24	39.6	43	59	64	69
Gliricidia sepium							
Volatile matter (%)	78	73	70	68	43	31	21
Ash content (%)	1.1	1.7	1.6	2	3	3.4	3.8
Fixed carbon (%)	23	26	28	30	54	62	75
Gmelina arborea							
Volatile matter (%)	83	79	74	60	38	29	20
Ash content (%)	.58	1.2	2.7	2.9	3.2	4.4	5
Fixed carbon (%)	15.8	19	23.3	36	58	66	75
Leucaena leucocephala							
Volatile matter (%)	74	69	59	55	41	30	26
Ash content (%)	.93	1.4	1.9	2.1	3	3.9	4.3
Fixed carbon (%)	25	29	39	43	56	67	70

Each value is the average of five replicates

The comparatively higher volatile matter content of charcoal produced at 300°C final carbonisation temperature was possibly due to evolution of water of constitution and the decomposition of volatile extractive components and hemicellulose in wood. The percentage fixed carbon and ash forming minerals in the charcoal increased with increase in carbonisation temperature for all the selected species. This may be due to the removal of volatile matter in the plant material during the pyrolytic process leaving the more stable carbon and ash-forming minerals.

There were significant differences in the apparent density of wood and charcoal from the selected tree species. For all the species, the density of charcoal decreased as final carbonisation temperature increased. The variations in the density of wood and charcoal derived from the three species may be due to the anatomical difference in the species. The decrease in the density of charcoal with temperature is possibly due to increase in pore volume with increasing carbonisation temperature. The pore volume increases with decomposition of cellulose and lignin component of wood at high temperature. The density of the charcoal briquettes were higher than those of the charcoal produced at high carbonisation temperature (Table 3).

Table 3 Density and heat of combustion of charcoal briquettes from five selected tree species

Wood species	Density (g/cm^3)	Heat of combustion (MJ/kg)
Acacia mangium	0.43 ± 0.01	32.67 ± 0.3
Cassia nodosa	0.38 ± 0.03	32.20 ± 0.5
Gliricidia sepium	0.40 ± 0.04	33.00 ± 0.3
Gmelina arborea	0.41 ± 0.02	33.86 ± 0.2
Leucaena leucocephala	0.43 ± 0.02	33.75 ± 0.2

3.2 Heat of Combustion

The values of the gross heat of combustion for the wood, charcoal and charcoal briquettes are presented in Table 4. The gross heat of combustion of charcoal were generally higher than those of wood. There was no significant difference in the heating value of charcoal and charcoal briquettes at different carbonisation temperatures but there was significant difference between heating value of wood and that of the charcoal (at 5% level of test). The greater gross heat of combustion recorded for some carbonised wood at final temperature when compared with wood or char produced at low pyrolytic temperature may possibly be due to higher percentage of fixed carbon in such charcoal produced at high carbonisation temperature. Since the heat of combustion of carbon is higher than that of volatile matter, high proportion of fixed carbon in sample is expected to increase the heating value. However, the increase in proportion of ash forming minerals in charcoal at temperature above 300°C tends to limit the extent of increase in gross heat of combustion of charcoal as the carbonisation temperature increased.

Table 4: Heat of Combustion of wood and carbonised material determined with bomb calorimeter on dry ash-free basis.

| CARBONISATION TEMPERATURE °C | HEAT OF COMBUSTION | | | | MJ/KG |
	ACACIA MANGIUM	CASSIA NODOSA	GLIRICIDIA SEPIUM	GMELINA ARBOREA	LEUCAENA LEUCOCEPHAI
0	22.3±.4	21.7±.5	23.2±.5	21.2±.3	23.3±.3
250	23.3±.2	22.9±.3	23.2±.4	22.5±.3	24.0±.2
300	32.3±.2	30.2±.1	33.1±.2	33.8±.2	33.8±.2
350	32.9±.2	32.1±.2	33.7±.2	34.4±.1	33.9±.2
400	32.6±.2	32.5±.2	32.5±.2	33.5±.2	33.7±.2
500	32.7±.1	33.0±.1	32.3±.2	33.6±.1	34.3±.2
600	32.7±.2	33.0±.1	32.9±.2	34.1±.2	34.1±.2

Each value is the average of five replicates.

4 Conclusion

The short rotation tree species studied had high percentage charcoal yield at low carbonisation temperature. Charcoal yield decreased with increase in carbonisation temperature. The volatiles in the charcoal decreased as temperature increased while there was increase in the percentage fixed carbon. Although the heat of combustion of charcoal produced at 600°C carbonisation temperature, was higher than that produced at 250°C there was no definite trend of increase in heating value of charcoal with final carbonisation temperature. Charcoal produced at low carbonisation temperature has high yield and greater heat of combustion than fire wood. The charcoal briquettes produced had higher density than that of charcoal and better combustion properties when compared with firewood.

References

1. Arnold, J.E.M. 1979. Wood energy and rural communities.
    Natural Resources Forum. Vol. 3 Nos. 3 UN. New York.
2. Fuwape, J.A. 1993. Charcoal and fuel value of agroforestry tree crops.
    Agroforestry Systems, Vol. 22: pp 55-59.
3. Smith, N. (1981) Wood: an ancient fuel with a new future.
    Worldwatch Paper 42. Worldwatch New York.
4. Baileys, R.T. and Blankenhorn, P.R. 1982. Calorific and porosity
    development in carbonised wood. *Wood Science*, Vol.15 No.1, pp19-28.
5. Lim, K.O. and Vizhi, S.M and Fong, S.K.(1994) Some physical properties
    and burning characteristics of Cocoa wood charcoal.
    Reric Energy Journal. Vol.16 No.1 pp 51-66.
6. Fuwape, J.A. (1996)Effects of carbonisation temperature on charcoal from
    some tropical trees. *Bioresource Technology* Vol. 56, pp1-7.
7. Connor, M.A. and Viljoen, M.H. (1995) Understanding the fundamental
    processes of wood carbonisation. IUFRO conference. August 1995.
    Tampere, Finland. 12pp.
8. Blankenhorn, P.R. and Barnes, D.P., Kine, D.E. and Murpher, W.K. 1978.
    Porosity and pore distributiuon of black cherry carbonised in an inert
    atmosphere. *Wood Science* Vol.11: No.1, pp23-29.
9. International Labour organisation (1985) Fuelwood and charcoal
    preparation. Geneva,48pp.
10. American Society for Testing and Materials 1979. ASTM D143-52
    standard method for testing wood. Part 22. Standard method for testing
    coal and coke Part 26. ASTM D 2015-72 Philadephia.
11. White, R.H. (1986). Effect of lignin content and Extractives on the higher
    heating value of wood. *Wood and Fiber Science*. Vol.19: No.1 pp 440-446.
12. Antal , M.J., Friedman,H.L and Rogers, F.E. 1980. Kinetics of cellulose
    pyrolysis in Nitrogen steam.*Combustion Science and Technology*. Vol. 21,
    pp 141-152.
13. Hoffman, O. and Fitz, A. 1968. Batch retort Pyrolysis of solid municipal
    wastes. *Environmental Science and Technology*, 2: No.11 pp. 1023-1026.
14. Brocksiepe, H. 1971. Products of wood pyrolysis, In Winacker K. and
    Kiiehler, L. edited "Chemische Technologie" Part III. Carl Hanser Verlag,
    Munchen.
15. Buekens, A.G. and Schoeters, J.G. 1987. Valueable products from the
    pyrolysis of biomass. In biomass for energy and industry. Proceedings of
    4th European Communities Conference. London pp 224-232.

DISTRIBUTION OF NITROGEN IN THE LIQUEFACTION OF ALBUMIN

Y. DOTE*, S. INOUE, T. OGI, and S. YOKOYAMA

National Institute for Resources and Environment, Tsukuba, Japan
*Present address; Miyazaki University, Miyazaki, Japan

Abstract
The oil obtained from protein-containing biomass such as sewage sludge by direct liquefaction contained nitrogen that is difficult to reduce nitrogen content in the liquefied oil by hydrotreatment. Albumin was liquefied to study the behavior of nitrogen in liquefaction. The maximum oil yield was 10%, much less than that for practical feedstocks(30-40 %). The maximum nitrogen content in the oil was 9%. Five percent of nitrogen in albumin was distributed to oil which is much less than that for practical feedstocks(30-50 %). No distribution of nitrogen to the oil under 150 °C occurred because of no oil yield. Most of the nitrogen in albumin, 80 %, was distributed to the aqueous phase above 200 °C. The distribution of nitrogen to oil was completed by 250 °C. Sodium carbonate used as catalyst prevented the distribution of nitrogen to oil. Albumin was decomposed to ammonia, not to amino acids.
Keywords: liquefaction, albumin, nitrogen distribution

1 Introduction

Direct liquefaction, one of the thermochemical conversion processes for biomass, can treat high-moisture content biomass without dewatering of the biomass and without the use of a reduction gas such as hydrogen or carbon monoxide. Dewatering processes consume high energy and reduction gas is costly. Thus, the advantages of this liquefaction are energy and cost savings.

Many researchers have studied the liquefaction of cellulosic biomass such as wood and peat, as summarized in the Reference 1. The authors have researched liquefaction of protein-containing biomass such as sewage sludge[2,3] in addition to cellulosic biomass[4]. For sewage sludge, a pilot plant for liquefaction (capacity of 5t/d) has been run[5]. For liquefaction of protein-containing biomass, however, the product oil contains nitrogen which is derived from protein in the biomass as will be shown later; this lowers the quality of the liquefied oil. In case of stillage, a beverage ethanol distillery residue, the upgrading of the liquefied oil has been attempted using hydrogen and Ni/Mo catalyst[6]. However, no low nitrogen-containing oil could be obtained; this suggested that it was difficult to reduce nitrogen in liquefied oil by hydrotreatment.

There is no information on the aqueous liquefaction of protein, alone. In the present paper, we clarify the effect of reaction temperature, holding time, and catalyst on the distribution of nitrogen in the reaction products for the liquefaction of albumin, a representative of protein. Moreover, we also summarize the distribution of nitrogen to oil using practical feedstocks in previous works.

2 Experimental

The protein employed was albumin from eggs(Wako Co., Japan), which is water soluble and exists in cell, body fluid, and plant seeds. Liquefaction was performed in a 300 mL autoclave made of stainless SUS 316) as shown in Figure 1. Albumin(4 g) and distilled water(100 mL) were charged to the autoclave with or without catalyst(0 or 5%). The residual air was purged with nitrogen gas, and then the autoclave was charged with nitrogen gas at 2 MPa, sealed and heated by an electric furnace. The required temperature was maintained for 0.5 or 2 h, followed by cooling. Sodium carbonate was used as catalyst because it showed a marked catalytic effect on liquefaction of wood[7]. Catalyst loading was defined as weight percentage of catalyst to organics in albumin.

Figure 2 shows the procedure for separating the liquefaction products. After liquefaction, gas, reaction solution, and 'retained' material were obtained. The gas, mainly composed of carbon dioxide and same nitrogen, was collected in a sample bag. The reaction solution which was a mixture of water, water soluble, oil-like material, and suspended solid, was collected in a beaker. The 'retained' material, a solid material stuck on the inner wall and blades of the autoclave, was washed with water. The washings and the reaction solution were filtered using paper filter. The combined filtrate was referred to as *aqueous phase*. Oil-like material in the reaction solution could not pass through the filter. The water insoluble material remained on the filter, which was dried at 60 °C for 1day. The retained material, after washing with water, was washed with dichloromethane. The washings and the dried water insoluble material were filtered using the filter, and the residue on the filter and the filter were washed with dichloromethane. The dichloromethane was evaporated under reduced pressure to obtain *oil*. The residue was dried at 105 °C for 1 day to obtain *solid residue*. The retained material after washing with water and dichloromethane was referred to as *retained residue*. We could not analyze it, because we could not scrape it from the wall and blades.

The analysis of elemental composition(carbon, hydrogen, and nitrogen) in solid and oil was performed by a Perkin Elmer Elemental Analyzer(Model 2400). Ammonia nitrogen(NH_3-N) in the aqueous phase was analyzed by the phenate method[8]. Free amino acid nitrogen(Am-N) was measured by a high performance liquid

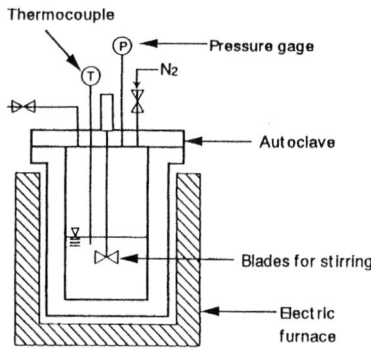

Fig. 1. Experimental apparatus

chromatography(Waters, PICO-TAG amino acid analysis system). The HPLC system was composed of a Model 600E solvent delivery system, a M486 UV/VIS detector, and column(Waters, Pico-Tag Column, 3.9x30 mm). Pre-column derivatization of amino acids with phenylisothiocyanate was performed[9], and the derivatized amino acids were detected at 254 nm. Kjeldahl nitrogen(Kj-N) in the aqueous phase was measured in the same way as ammonia nitrogen after decomposition by the macro-kjeldahl method[8] with selenium oxychloride catalyst instead of mercuric sulfate catalyst. Although organic nitrogen normally contains amino acid nitrogen, *organic nitrogen*(Org-N) in the aqueous phase was defined in the present paper as follows:

Org-N =Kj-N - (NH$_3$-N + Am-N)

Nitrogen in aqueous phase was composed of Org-N, NH$_3$-N, and Am-N.

The moisture content of the albumin was determined from the weight loss on heating at 105 °C for 1day. Organics in albumin were determined by ignition loss at 600 °C for 0.5 h.

Oil yield was defined as weight percentage of oil to organics in albumin. *Nitrogen distribution*(ND) was defined as follows:

ND(%)=(weight of nitrogen in product)/(weight of nitrogen in albumin) x 100

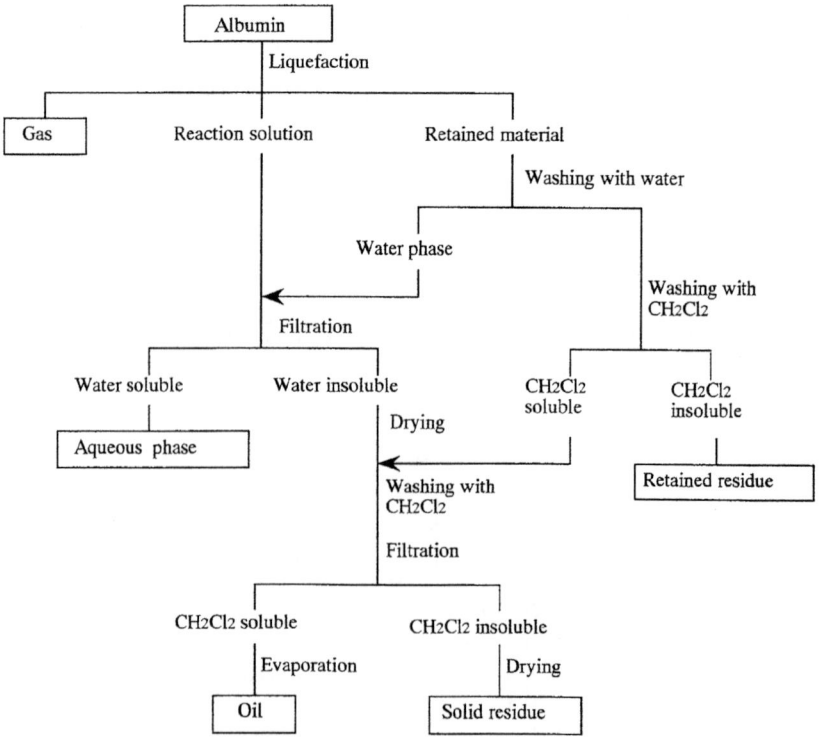

Fig. 2. Procedure for separating liquefaction products

3 Results and discussion

The characteristics of the albumin are summarized in Table 1. The nitrogen content was 15%. This was much higher than that in the practical feedstocks, where the protein content is less than 100 %.

3.1 ND to the oil from practical feedstocks

ND to the oils formed in previous studies is summarized in Table 2. Stillage is a beverage ethanol distillery residue as mentioned above. Liquefaction was performed with the same autoclave and in the same manner as in the present study. Oil yields were 30-40%. Nitrogen contents of the oils were 4-6%, much higher than those in petroleum oil. ND to the oil was 30-45%. The results will be used to compare with those from the present work.

3.2 Oil Yield

Oil yields at 150 °C were negligible and increased with temperature as shown in Figure 3. Oil yield using no catalyst for 2 h holding time remained constant above 300 °C. At other conditions, there was a little increase in oil yield above 300°C. This means that the most of oil production was completed by 300 °C. The maximum oil yields were 9% for 0.5h holding time and 10% for 2 h holding time. The maximum oil yield for albumin was much less than oil yields for the practical feedstocks(30-40%).

In general, oil yield increased with holding time. However, at higher temperatures, the oil yield was less affected by holding times.

The addition of catalyst decreased oil yield. Above 300°C, there was little effect of catalyst on oil yield for 0.5 h holding time, although the effect of catalyst on oil yield for 2 h holding time was apparent.

Table 1. Characteristics of albumin

Moisture content (%)	Organics content (%)	Elemental composition(%)			
		C	H	N	O*
2.9	92.7	45.6	6.4	13.9	34.1

*; calculated by difference

Table 2. DN to oil for practical feedstock

Feed stock	Stillage*				Sewage sludge **	
	Sweet potato	Barley	Rice	Wheat	A	B
Temperature(°C)	340	300	300	325	300	300
Holding time(h)	0	0.5	0	0.5	0	1
Catalyst(Na_2CO_3)(%)	5	2.5	None	None	***	***
N in feedstock(DS%)	4.0	3.9	7.4	7.0	5.4	5.4
Organics in feedstock(DS%)	95.6	89.7	98.7	96.6	83.6	83.6
Oil yield(%)	32.8	36.2	33.4	36.8	43.5	42.5
Nitrogen in oil(%)	5.9	4.2	6.4	5.8	6.1	5.8
ND to oil(%)	46.3	35.0	28.5	29.5	40.8	37.9

*; Ref. 11
**; Ref. 12
***; the results for sewage sludge were calculated as the average value between
catalyst loading of 0% and 5 %.

3.3 Nitrogen content in oil

Figure 4 shows nitrogen contents of oil. The maximum nitrogen content of 9% appeared at 250 °C for 0.5 h holding and at 200 °C for 2 h holding time. Above these temperatures, nitrogen content decreased with temperature. Nitrogen contents at 340 °C were 6-7 %. The amount of oil formed at 150 °C was too little to determine the elemental composition, except for the oil formed with catalyst at a 2 h holding time. Nitrogen contents of the oil formed from albumin at 300 and 340 °C were a little higher than those for the practical feedstocks.

No effect of catalyst on nitrogen content was apparent.

3.4 ND to the aqueous phase

ND to the aqueous phase, oil, and solid residue is shown in Figure 5. Most of the nitrogen in albumin, 80 %, was distributed to the aqueous phase at any condition except for the condition at 150 °C with no catalyst. Holding time had no effect on ND to the aqueous phase. The addition of catalyst above 200 °C reduced ND to aqueous phase, although the effect of catalyst on ND to the aqueous phase above 300 °C was only slight.

3.5 ND to the oil

ND to the oil is shown in Figure 6. ND to the oil increased with temperature and remained constant above 250 °C; the value with no catalyst for 2 h holding time was 5 %, and at other conditions 4 %. This means that the distribution of nitrogen to the oil was completed by 250 °C, although oil production was completed by 300 °C as shown above. In spite of a 2 h holding time, ND to the oil at 150 °C was negligible because of the negligible oil yield at 150 °C. This means that no distribution of nitrogen to the oil under 150 °C occurred.

ND to the oil increased with holding time. ND to the oil for 2 h holding at 200 °C was 2-3%, although at the 0.5 h holding time at 200 °C it was less than 0.5 %.

The maximum difference in ND to the oil between with catalyst and without catalyst above 250 °C was 0.8 %. The addition of catalyst decreased ND to the oil; this indicates that sodium carbonate prevented the distribution of nitrogen to the oil.

By comparison with the practical feedstocks, ND to oil for albumin was much lower. This was due to lower oil yields for albumin than those for the practical feedstocks. Because the practical feedstocks contain other components such as cellulose and lipid, the reason might be that these components either increase the amount of oil produced

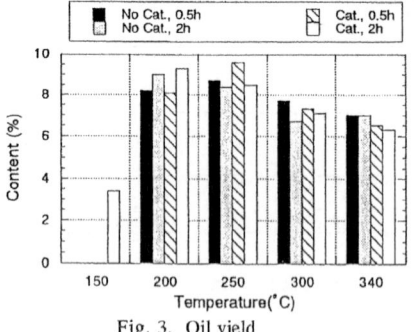

Fig. 3. Oil yield

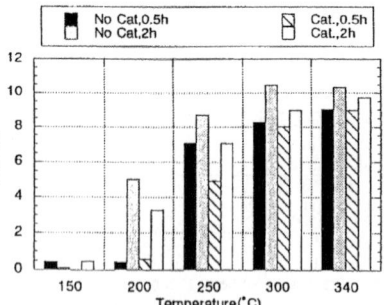

Fig. 4. Nitrogen content

from protein, or they react with nitrogen-containing compounds produced from protein during their conversion into oil.

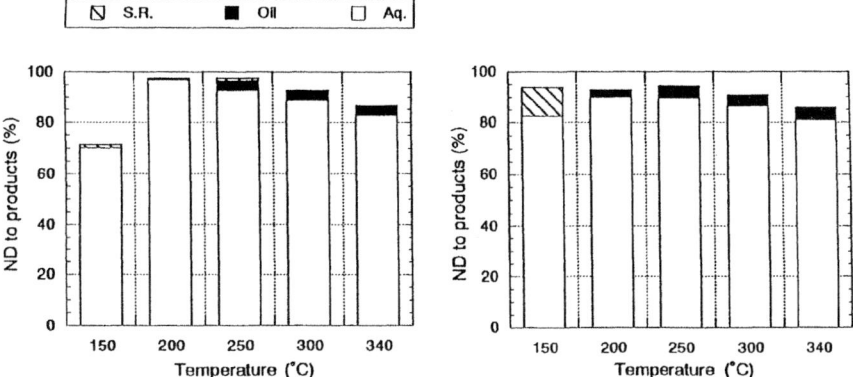

Fig. 5.(a) With no catalyst for 0.5 h holding time

Fig. 5.(b) With no catalyst for 2 h holding time

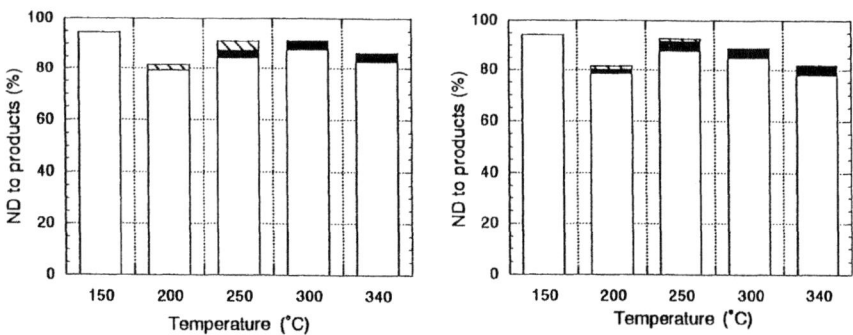

Fig. 5.(c) With catalyst for 0.5 h holding time

Fig. 5.(d) With catalyst for 2 h holding time

Fig. 5. ND to oil, aqueous phase(Aq.), and solid residue(S.R.)

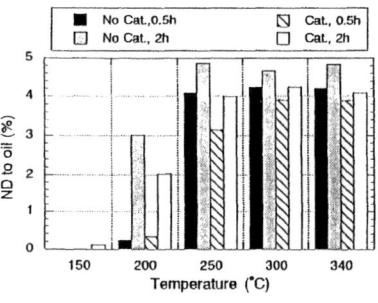

Fig. 6 ND to oil

325

3.6 ND to loss and solid residue

The loss of nitrogen occurred owing to the non-recovery of retained residue, in addition to the loss caused during product separation.

The retained residue produced at many conditions was a black material forming thin film on the inner wall and blades of the autoclave; the retained residue at 150 °C with no catalyst was a milk-white string-like material. Because denaturation of albumin begins at 70-80 °C, this result suggests that denaturation of albumin occurred at these conditions in liquefaction.

On the other hand, most of the nitrogen formed at 150 °C using catalyst was distributed to the aqueous phase as shown in Figure 5; thus, sodium carbonate used as catalyst made the denaturalized albumin soluble. Because sodium carbonate is base, the result suggests that alkali hydrolysis of albumin occurred at 150 °C with catalyst. This was supported by the fact that the pH of the reaction solution at 150 °C with catalyst was 8, and with no catalyst 7.

3.7 ND to NH₃-N and Am-N

ND to NH_3-N increased with temperature as shown in Figure 7. ND to NH_3-N increased with holding time. The effect of catalyst on ND to NH_3-N was negligible.

ND to Am-N had the maximum values at 200 °C (1% and 2 % for 0.5h and 2 h holding times, respectively) as shown in Figure 8. ND to Am-N above 300 °C was less than 0.3 %. ND to Am-N at 150 and 200 °C increased with holding time, although that at 250 °C decreased with holding time. The addition of catalyst increased ND to Am-N. At any condition ND to Am-N was less than that of NH_3-N. This suggests that organic nitrogen-containing compounds were decomposed to ammonia, not stopped at amino acids.

4 Conclusions

For the liquefaction of albumin, the conclusions were as follows:
1. The maximum oil yield was 10% , much less than that for practical feedstocks.
2. The maximum nitrogen content in oil was 9%.
3. ND to oil was 5% at most, much less than that for practical feedstocks.
4. No distribution of nitrogen to oil under 150 °C occurred.
5. The distribution of nitrogen to oil was completed by 250 °C.
6. Sodium carbonate used as catalyst prevented the distribution of nitrogen to oil.

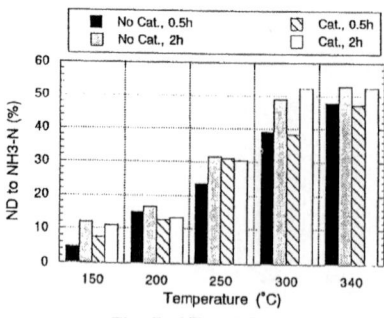

Fig. 7. ND to NH3-N

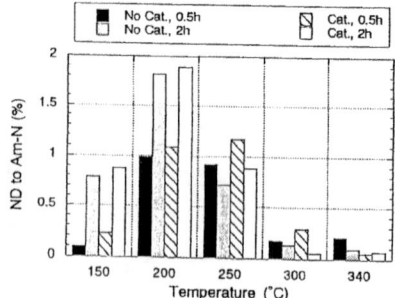

Fig. 8. ND to Am-N

7. Most of the nitrogen in albumin, 80 %, was distributed to the aqueous phase above 200 °C.
8. Albumin was decomposed to ammonia, not to amino acids.
9. At 150 °C with no catalyst, denaturation of albumin occurred.
10. At 150 °C with catalyst, alkali hydrolysis of albumin occurs.

References

1. Elliott, D.C. (1991) Developments in direct thermochemical liquefaction of biomass; 1983-1990, *Energy & Fuels*, Vol. 5, pp. 399-410.
2. Yokoyama, S., Suzuki, A., Murakami, M., Ogi, T., Koguchi, K., and Nakamura, E. (1987) Liquid fuel production from sewage sludge by catalytic conversion using sodium carbonate, *Fuel*, Vol. **66**, pp. 1150-1155.
3. Itoh, S., Suzuki, A., and Yokoyama, S. (1992) Direct thermochemical liquefaction of sewage sludge by a continuous plant, *Water Science and Technology*, Vol. **26**, pp. 1175-1184.
4. Ogi, T., and Yokoyama, S. (1993) Liquid fuel production from woody biomass by direct liquefaction, *Sekiyu Gakkaishi*, Vol. **36**, pp. 73-84.
5. Itoh, S., Suzuki, A., Nakamura, T., and Yokoyama, S. (1993) Production of heavy oil from sewage sludge by direct thermochemical liquefaction, *Proceedings of the IDA and WRPC world conference on desalination and water treatment*, Volume II, pp. 315-322.
6. Dote, Y., Yokoyama, S., Minowa, T., and Murakami, M. (1001) Liquefaction of barley stillage and upgrading of primary oil, *Biomass and Bioenergy*, Vol. **1**, pp. 55-60.
7. Ogi, T., Yokoyama, S., and Koguchi, K. (1985) Direct liquefaction of wood by alkali and alkaline earth salt in an aqueous phase, *Chem. Lett.*, pp. 1199-1202.
8. APHA (1992) *Standard methods for the examination of water and wastewater*, 18th ed., American Public Health Association, New York(1992)
9. Bidlingmeyer, B.A., Cohen, S.A., and Tarvin, T.J. (1984) Rapid analysis of amino acids using pre-column derivation, *J. Chromatography*, Vol. **336**, pp.93-104.
10. *Waters 600E multisolvent delivery system operator's manual*
11. Minowa, T., Murakami, M., Dote, Y., Ogi, T., and Yokoyama, S. (1994) Effect of operating conditions on thermochemical liquefaction of ethanol fermentation stillage, *Fuel*, Vol. 73, pp. 579-582.
12. Suzuki, A., Nakamura, T., and Yokoyama, S. (1990) Effect of operating parameters on thermochemical liquefaction of sewage sludge, *J. Chemical Engineering of Japan*, Vol. **23**, pp. 6-11.

APPLICATION OF THERMOCHEMICAL LIQUIDIZATION OF SEWAGE SLUDGE

Shin-ya Yokoyama, Tomoaki Minowa and Shigeki Sawayama
National Institute for Resources and Environment
16-3, Onogawa, Tsukuba, Ibaraki 305, Japan
Yutaka Dote
Department of Civil and Environmental Engineering, Miyazaki University
1-1, Gakuen, Kibanadai Nishi, Miyazaki 889-21, Japan
Akira Suzuki
Japan Organo Co. Ltd., 1-4-9, Kawagishi, Toda, Saitama 335, Japan

Abstract

Dewatered sewage sludge can be liquidized and converted to a free flowing material by treating between 150 and 200 C under pressure. This thermochemical liquidization phenomenon is due to the rupture of cells of the microorganisms that make up the major part of sewage sludge , and the water phase is then available to suspend the celluar debris thus reducing the viscosity. By this technology, organic waste such as dewatered sewage sludge which is solid in form can be easily transformed to liquid materials leading to a variety of technical application such as transport by pipelines, incineration drying, and methanation etc.

Keywords: Biomass, sewage sludge, solid waste, liquidization, transportation incineration, drying, methanation, anaerobic digestion.

1 Introduction

In Japan, more than 50 million cubic meters of sewage sludge (on a basis of 98% moisture content) is being generated annually by sewage treatment plants. Municipalitie not only have a financial burden concerning sludge treatment and disposal, but also difficulties in finding dumping sites for sewage sludge. Under such circumstances, a novel technology is expected from the viewpoint of economical and environmenta reasons.

For collecting sewage sludge, a pipeline is used for concentrated sludge which has abou 98% moisture content, or dump trucks are used for dewatered sludge which has abou 80% moisture content and is more solid in form. These methods, however, have problems. The transportation of the concentrated sludge is not very efficient, because the concentrated sludge has a high moisture content and only several percent of solid can be generally transported. The transportation of the dewatered sludge, however, is more efficient than that of concentrated sludge, because the dewatered sludge has a high solid concentration of 20%; therefore, the volume of the dewatered sludge to be transported becomes one-tenth that of the concentrated sludge. If it were possible to give fluidity to the dewatered sludge, the dewatered sludge could then be transported through a pipeline that is, efficient transportation of sewage sludge could be established.

The conversion of sewage sludge into liquid fuel has been investigated by some researchers [1,2,3]. We have also studied the conversion of dewatered sewage sludge into liquid fuel at about 300°C and under 10 MPa [4,5,6]. In the experiments, it was suggested that dewatered sludge could become fluid under milder conditions. This is referred to as thermochemical liquidization, a process used in the present paper to convert dewatered sewage sludge, which is hard to transport by pumping, into slurry and high fluidity. Furthermore, liquidized material can be dried or incinerated efficiently due to the homogenized character compared with that of bulk or solid materials. It has some possibility that liquidized material can be directly anaerobically digested to methane.

2. Experimental

2.1 Sewage sludge

Three kinds of dewatered sewage sludge, collected from different sewage treatment plants, were used for liquidization experiments. Concentrated sludge was used for comparison with the viscosity of liquidized sludge. Table 1 shows the properties of three dewatered sludges and a concentrated sludge. Moisture content was measured by drying the sludge at 105°C for 24h. Organic content was measured as ignition loss at 600°C for 1h.

Table 1 Properties of sewage sludge

No.	Type	Moisture content (%)	Organic content (%)*
A	Dewatered	87.3	80.0
B	Dewatered	74.2	56.9
C	Dewatered	77.0	79.7
D	Concentrated	98.8	—

* on a dry basis

2.2 Liquidization of dewatered sludge for measuring viscosity

Liquidization for measuring viscosity was performed in a 300 ml autoclave made of stainless steel (SUS 316) equipped with a pressure controlling circuit. The dewatered sludge (about 80g) was charged in the autoclave. After purging the residual air with nitrogen gas, nitrogen gas was added to 2 MPa plus the saturated vapor pressure of water at the operating temperature. The addition of 2 MPa of nitrogen gas was done to avoid vaporizing the water. The autoclave was heated with an electric furnace. After heating the autoclave up to the required temperature within 30 min, the temperature was maintained for the desired holding time and then the autoclave was cooled quickly to room temperature. The product removed from the autoclave was used for measurement.

2.3 Liquidization of dewatered sludge for measuring pressure loss

Liquidization to produce a large sample for measuring pressure loss was carried out by a continuously operating plant with a capacity of 15kg/h. The continuously operating plant consisted of an injection system, reactor, cooler, let-down system and boiler. Sludge was put into the bottom of the reactor, and about 1h was required to increase the temperature of sludge to the operating temperature. When the emperature was 250C, the pressure was 10 MPa. After the time at the operating temperature, the sludge was taken out from the top of the reactor and cooled quickly. Details of the operation can be found in Ref. 7. For the measurement of pressure drop in a pipe line, sludge A was liquidized at 200 or 250°C for a holding time of 0 min at a heating rate of 7-10°C/min.

2.4 Measurement of viscosity

Viscosity was measured on a rotational viscometer (TOKYO KEIKI, B-8M) with a sample adapter for small samples such as 5-10 ml. The rotation rate was varied from 0-60 rpm. Assuming a non-Newtonian fluid, the power law flow model was adopted in the analysis of data, the relevant equation being.

$$\tau = KG^n \tag{1}$$

where   is shear stress, G the shear rate, n   the flow behavior index, and K the consistency index. Apparent viscosity, μ_a was given by

$$\mu_a = KG \tag{2}$$

Parameters K and n were determined by use of least-squares regressions after transforming eqn.(1) to common logarithms;

$$\log \tau = \log K + n \log G \tag{3}$$

2.5 Measurement of pressure loss

The measurement of pressure loss was performed by use of a stainless steel pipe (i.d. 0.0107m) and a pump as described in Ref. 8.   Pressure loss was determined by measuring the difference in pressure between pressure gauges 1 and 2 set in the pipe. The flow rate of the sample was obtained by measuring the weight of the sample flowing out of the pipe against time. To compare with liquidized sludge A, the pressure loss of sludge D and sludge A was measured. For sludge A, a pipe 0.73m long was used, because the pressure loss of the sludge was too high to pump it for 10m.   The temperatures of the pipe and sample were about 6°C.

For a Newtonian fluid, the pressure loss $\Delta P/L$ is expressed by Fanning's equation:

$$\Delta P/L = 2fV^2\rho/D \tag{4}$$

where ΔP is pressure drop, L the length of the pipe, f the friction factor, V the average velocity, ρ the density, and D the inner diameter of the pipe. With regard to turbulent flow conditions, f is dependent on the Reynolds number Re, and the roughness of the pipe, which was 4.5×10^{-5} m in these experiments. The Reynolds number can then be given by

$$Re = DV\rho/\mu \tag{5}$$

where μ is viscosity.

Assuming that the velocity distribution in the pipe regarding laminar flow conditions was maintained in each experiment,

$$v = 2V\left|1 - (r/R)^2\right| \tag{6}$$

where r is radial distance and R the radius of the pipe, then the shear rate at the wall of the pipe was given by

$$G = - dv/dr \big|_{r=R} = 4V/R \qquad (7)$$

Although eqns (4) and (5) are valid exactly for Newtonian fluid and eqn(7) for laminar flow conditions, we used the equations for liquidized sludge, which was expected to be a non-Newtonian fluid, by substituting μ_a for μ on laminar and turbulent flow conditions. As will be seen later, the equations could primarily be used for calculating the pressure loss of liquidized sludge. The relationship between f and Re was obtained from Ref.9

3 Results and discussion

3.1 Liquidization of dewatered sludge

The change in the properties of the different sludges(A-C) on thermal treatment is shown in Table 2. The liquidization phenomenon was observed at 150-175°C in most cases. At much higher temperatures the sludge underwent a further change and would show phase separation by leaving a sediment on standing for 16h. In practice such a severe treatment is not required since sedimentation of transport lines and storage tanks has to be avoided. We consider that the liquidization phenomenon is due to the rupture of the cells of the microorganisms that make up the major part of the sewage sludge, and the water phase is then available to suspend the cellular debris thus reducing the viscosity.

Table 2 State of liquidized sludge

Liquidization condition (Temp. and holding time)	Sludge A	Sludge B	Sludge C
150°C and 0min	◯	✕	✕
150°C and 60min	◯	◯	✕
175°C and 0min	◯	◯	✕
175°C and 60min	△	◯	◯
200°C and 0min	◯	◯	✕
200°C and 60min	△	◯	◯
225°C and 0min	△	◯	◯
225°C and 60min	△	△	△
250°C and 0min	△	△	◯
250°C and 60min	△	△	△
300°C and 60min	△	△	△

✕ means that sludge was not liquidized.

◯ means that the sludge was liquidized.

△ means that the sludge was liquidized and that a solid sediment was deposited if left overnight.

Figure 1 demonstrates an example of the flow curve of liquidized sludge. A linear relationship was observed between log τ and G for all experiments: the correlation coefficient in each experiment was greater than 0.8 as shown in Table 3.   In all conditions, except when sludge C liquidized at 250°C for 60 min, the flow behavior index was less than unity. Therefore, liquidized sludge was regarded as a pseudoplastic fluid.

Figure 2 shows the effect of operating temperature and holding time on the consistency index at the viscosity measurement temperature of 20°C. The dotted line expresses the consistency index of sludge D for reference. The smaller the consistency index is, the lower the apparent viscosity is if the shear rate and flow behavior index are constant, as shown in eqn.(2). For each liquidized sludge, the consistency index increased as the operating temperatures increased. The long holding time tended to decrease the consistency index. At high temperatures especially, the consistency index was lower than that of sludge D which was concentrated sludge.

Table 4 expresses the effect of viscosity measurement temperature on the apparent viscosity of liquidized sludge A and sludge D. At a shear rate of 10 and 100 s^{-1}, the apparent viscosity tended to decrease as the viscosity measurement temperature increased.

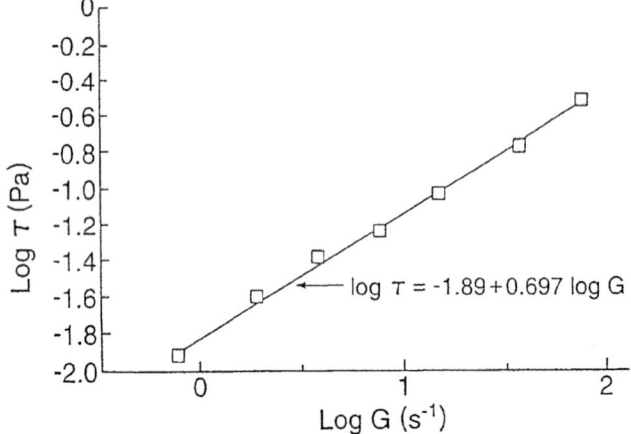

Fig. 1 Flow curve of sludge B liquidized at 225C for 60 min

For liquidized sludge at the shear rate of 100 s , the apparent viscosity at measurement temperatures of 20 and 30°C decreased by 20-30% relative to that at 6°C. The change in the apparent viscosity with temperature is less than one order of magnitude and is insignificant, although such a temperature dependence of the apparent viscosity is important when considering the effect of the seasonal variations of ambient temperature on pressure loss in buried pipelines.

Table 3 Correlation coefficient between log and G at a viscosity measurement temperature 20°C

Liquidization condition (Temp. and holding time)	Sludge A	Sludge B	Sludge C
150°C and 0min	0.984	—	—
150°C and 60min	0.982	—	—
175°C and 0min	0.970	0.9996	—
175°C and 60min	0.986	0.927	—
200°C and 0min	0.967	0.818	—
200°C and 60min	0.842	0.900	0.889
225°C and 0min	0.876	0.986	—
225°C and 60min	0.995	0.997	0.942
250°C and 0min	0.931	0.943	0.981
250°C and 60min	0.920	0.923	0.973
300°C and 60min	0.936	0.974	0.9993

— Means that viscosity could not be measured.

3.2 Pumping of liquidized sludge

Figure 3 shows the results of the measurement of pressure loss. The parameters n, K and used for calculations are listed in Table 5. The solid (sludge A liquidized at 250°C),

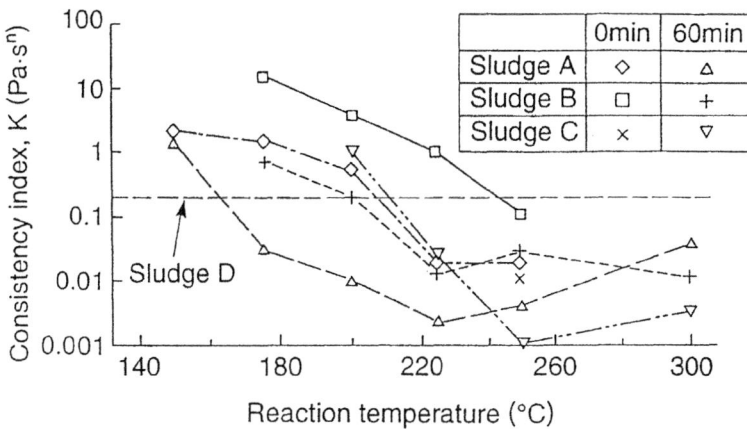

Fig. 2 Effect of reaction temperature on consistency index

dashed (sludge A liquidized at 200°C) and dotted (concentrated sludge D) lines were calculated using eqn(4). The reason why two slopes exist for sludge A liquidized at 250°C and sludge D is that flow conditions in the high viscosity region were turbulent and those in the low viscosity region were laminar by calculation. Liquidization could considerably reduce pressure loss. For example, at 1 m s^{-1}, the pressure loss of sludge A was 6 x 10^5 Pa m^{-1}, but that of sludge A liquidized at 250°C was 2 x 10^3 Pa m^{-1} and was equivalent to the pressure loss of sludge D (1 x 10^3 Pa m^{-1}). As compared with sludge D, the pressure loss of sludge A liquidized at 250°C was lower than that of sludge D at low viscosity. On the other hand, at high viscosity, both pressure losses were nearly equal.

Table 4 Dependence of apparent viscosity on viscosity measurement temperature

Temp. (°C)	Sludge A liquidized at 200°C for 0min		Sludge A liquidzed at 250°C for 0 min		Sludge D	
	μ_a^*	μ_a^{**}	μ_a^*	μ_a^{**}	μ_a^*	μ_a^{**}
6	656	213	14	3.5	41	10
10	615	178	16	3.0	47	9.7
20	675	177	6.6	2.8	37	7.0
30	560	160	9.0	2.4	36	19

μ_a^* : At a shear rate of 10s^{-1}.
μ_a^{**} : At a shear rate of 100s^{-1}.

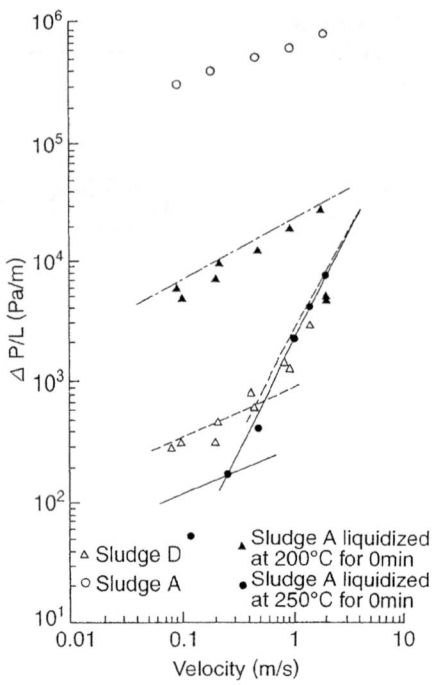

Fig. 3 Pressure loss of liquidized sludge

3.3 Drying and incineration of liquidized sludge

If the sludge is liquidized, it is expected that the sludge can be a homogeneous in terms of viscosity, particle size of suspended materials, and thermal property. According to the

Table 5 Parameters used for calculation of pressure loss

	$n \, (-)$	$K \, (\text{Pa·s}^n)$	$\rho \, (\text{kgm}^{-3})$
Sludge A liquidized at 200°C	0.386	2.018	1060
Sludge A liquidized at 200°C	0.512	0.059	1060
Sludge D	0.394	0.165	1000

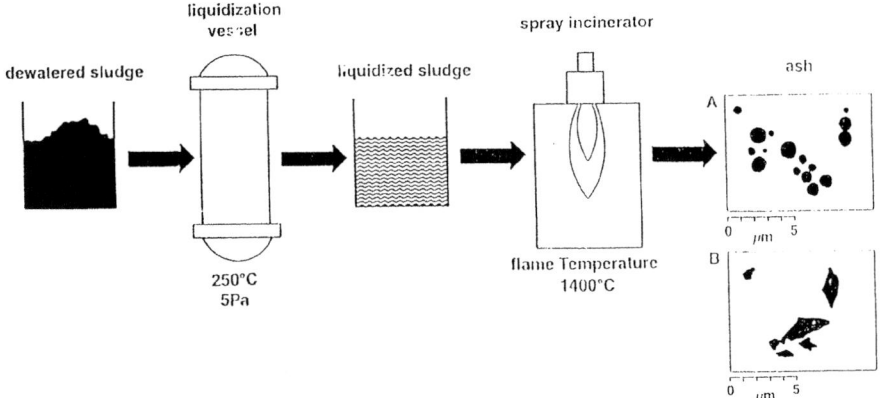

Fig. 4 Flow of spray incineration system of liquidized sewage sludge

report provided by Japan Organo Co. Ltd., the dewatered sludge could be successfully incinerated after liquidization. Figure 4 shows the flow of spray incineration of dewatered sludge. When the liquidized sludge is incinerated together with air, the sludge can self- combust without addition of auxiliary fuel. In the flame, liquidized sludge is dried and at the same time organic component is combusted and inorganic component is melted. As a result, the particle size of ash becomes smaller than that obtained in a traditional fluidized bed incinerator, which leads an efficient use of ash. The particle size distribution is shown in Figure 5.

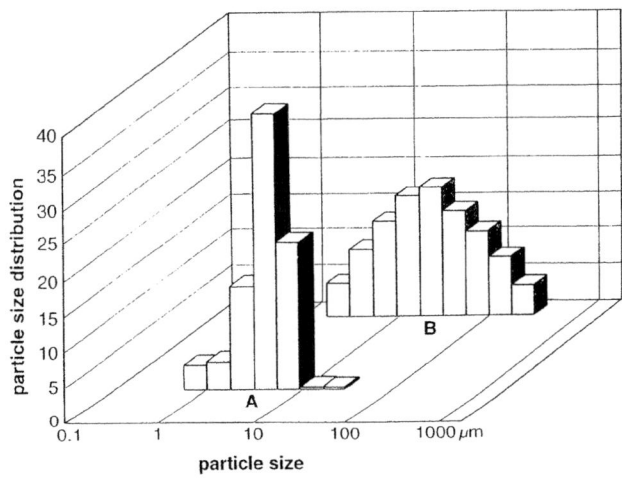

Fig. 5 Particle size distribution of ash from spray incineration
(A: Spray incineration, B: Fluidized bed incineration)

335

3.4 Anaerobic digestion for the production of methane

Dewatered sludge was thermochemically liquidized at 175 C and liquidized sludge was separated by centrifugation to supernatant and it was successfully digested. Biogas yield from the supernatant at organic loading concentrations of 1.9 to 2.2 g VS/l during 9 days incubation was 440 ml /g-added VS and digestion rate was 66% as shown in Figure 6.

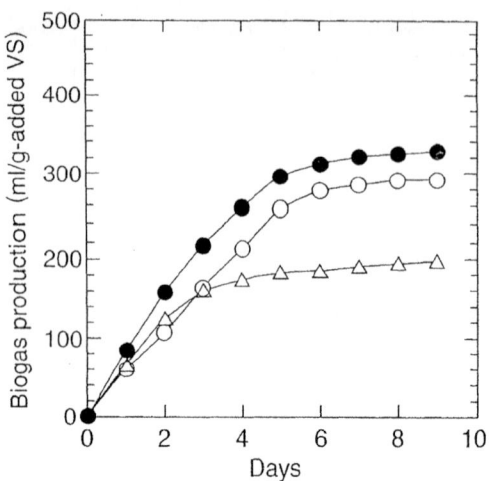

Fig. 6 Biogas yield from anaerobic treatments of dewatered sewage sludge, liquidized
sludge and supernatant
   (Open triangles, dewatered sludge; open circles, liquidized sludge; closed circles,
   supernatant)

Biogas yield in case of the dewatered sewage sludge was 257 ml/g-added VS and digestion rate was 45%. Digestion of the supernatant resulted in high biogas productivity and a high digestion ratio compared with that of the dewatered sewage sludge. Moreover, it was suggested that the precipitate can be incinerated without the need for any supplemental fuel.

4.Conclusions

The following conclusions were obtained:
(1) Dewatered sewage sludge could be liquidized at temperatures above 150-175°C.
 The viscosity of the liquidized sludge decreased by increasing the operating temperatures and holding time. The viscosity of the sludge liquidized under certain conditions could be less than that of concentrated sludge.
(2)Pseudoplasticity was found in liquidized sludges.
(3)The apparent viscosity of liquidized sludge depended on viscosity measurement temperature. At a shear rate of 100 s , the apparent viscosity measured at 20 and 30°C decreased by 20-30% of that at 6°C.
(4)The pressure loss of the sludge liquidized at 250°C for a holding time of 0 min could be nearly equal to that of concentrated sludge over 1 m s and lower than that under 1 m s.
(5)Fanning's equation can be used for calculating the pressure loss of liquidized sludge; therefore, design of a pipeline system is possible.

If the technology for liquidizing denatured sludge is put to practical use, its advantages are:

(1)For the transportation of liquidized dewatered sludge using a pipeline, the volume of sludge to be transported can be reduced to one-tenth of that of concentrated sludge; as a result, the construction and operating costs of the transportation facilities for liquidized sludge become cheaper than those for concentrated sludge.

(2)Liquidized sludge can be transported by tanker lorries or a vacuum car both of which are able to seal off the odor of the liquidized sludge, so that local inhabitants in the affected area will not be subjected to any unpleasant odors.

(3)The volume of storage of liquidized sludge is smaller than that of dewatered sludge since liquidized sludge is almost completely liquid, whereas dewatered sludge is more solid and its apparent density is low.

(4) Liquidized sludge can be self combust in a spray incinerator efficiently without an auxiliary fuel and the particle size of ash becomes smaller and uniform compared with those obtained from a current fluidized bed incinerator.

(5) Liquidized sewage sludge can be anaerobically digested to produce methane. Since the volume of dewatered sludge is generally one tenth of the sludge, the biogas digester can be greatly downsized by liquidization of the dewatered sludge.

Notation

D = inner diameter of pipe, m.
f = friction factor, dimensionless.
G = shear rate, s .
K = consistency index, Pa s .
L = length of pipe, m.
n = flow behavior index , dimensionless.
ΔP = pressure drop, Pa.
r = radial distance, m.
R = radius of pipe, m.
v = velocity in pipe, m s .
V = average velocity, m s .

Greek letters
μ = viscosity, Pa s.
μ_a = apparent viscosity, Pa s.
ρ = density, kg/m .
τ = shear stress, Pa.

References
1. E. Bayer and M. Kutubuddin, Proc.Int.Recycl.Congr., Volume 1(1982).
2. W. L. Kranich and A. E. Eralp, U.S. Environmental Protection Agency Project Summary, EPA-600/S2-84-010(1984).
3. P. M. Molton, A. G. Fassbender and M. D. Brown, U.S. Environmental Protection Agency, Project Summary, EPA-600/S2-86-036(1986).
4. S. Yokoyama, A. Suzuki, M. Murakami, T. Ogi, K. Koguchi, and E. Nakamura Liquid fuel production from sewage sludge by catalytic conversion using sodium carbonate. Fuel 66, 1150(1987).
5. A. Suzuki, T. Nakamura, S. Yokoyama, T. Ogi and K. Koguchi, Conversion of sewage sludge to heavy oil by direct thermochemical liquefaction J. Chem. Eng. pp 21, 288(1988).
6. A. Suzuki, T. Nakamura and S. Yokoyama, Effect of operating parameters on

thermochemical liquefaction of sewage sludge , J. Chem. Eng. Jpn 23,6(1990).

7. S. Itoh, A. Suzuki, T. Nakamura and S. Yokoyama, Direct thermochemica liquefaction of sewage sludge by a continuous plant.   Water Sci. Thechnol. 2€ 1175(1992).

8. Y. Dote, S. Yokoyama, T.Minowa, T. Masuta, K.Sato, S.Itoh, and A. Suzuk Thermochemical liquidization of dewatered sewage sludge, Biomass and Bioenergy, 4 243 (1993)

9. L. F. Moody, Trans. ASME 66,671(1944).

FUEL OIL PRODUCTION FROM HYDROCARBON-RICH MICROALGAE *BOTRYOCOCCUS BRAUNII*

T.OGI, S.INOUE and S.SAWAYAMA
National Institute for Resources and Environment, 16-3, Onogawa, Tsukuba,
Ibaraki 305 Japan
Y.DOTE
Miyazaki University, Nishi 1-1, Kibanadai, Miyazaki 889-21 Japan

Abstract

Botryococcus Braunii is a colonial green microalga which produces and accumulates oily hydrocarbons consisted mainly of botryococcenes($C_{34}H_{58}$), and it grows well even in secondarily treated sewage. Liquefaction of *B. braunii* with high moisture content was performed in the presence of Na_2CO_3 as a catalyst in order to recover hydrocarbons. A greater amount of oil was obtained in a maximum yield of 64 wt% at 300°C, 10MPa, for 1hr, with 5% Na_2CO_3 loading, which exceeded the content of original hydrocarbons in *B. braunii*(50wt% dry base). Analysis of the oil obtained showed the oil consisted mainly of three fractions; lower molecular weight hydrocarbons, botryococcenes, and polar substances. The recovery of botryococcences was about 75%.
Keywords:microalgae, *Botryococcus braunii*, hydrocarbon production, liquefaction

1 Introduction

Recently, serious environmental problems such as global warming have occurred. Carbon dioxide released in the atomsphere attributes to global warming and various technologies for fixation of CO_2 have been developed. Among these, promising is biological fixation by suitable photosynthetic microalgae cultured on a large scale.[1,2] *Botryococcus braunii* is a unique colonial microalga which produces and accumulates hydrocarbons, named botryococcenes, in itself with a dry weight range of 17~86%.[3-7] These hydrocarbons can be upgraded to transport fuels by hydrocracking[8] or catalytic cracking.[9]

Conventionally, hydrocarbons are recovered from microalgae by extraction with organic solvent after freeze-drying or filtering on the laboratory scale. These procedures, however, would not be suitable for a large-scale cultivation system, because they are complicated and costly. More effective method, therefore, is required.

The authors have investigated the conversion of wet biomass such as wood and sewage sludge into liquid fuel in high yield at ~300°C and 10 MPa using a catalyst such as sodium carbonate.[10,11] Advantages of this process are that no reducing gas such as carbon monoxide or hydrogen is necessary and that no drying is necessary, so wet biomass can be directly used as feedstock and energy is saved. If this process is applicable to microalgae, it can combine biological carbon dioxide fixation with the production of liquid fuel.

In this paper, we tried to liquefy *B.braunii* to recover hydrocarbon oil and analyze the oil obtained. The results are shown.

2 Experimental

2.1 Algal strain and culture conditions

The Berkeley strain of *Botryococcus braunii* was used for liquefaction. The cells were cultured in a 10 l bottle with a Chu 13 medium, aerating with 1% CO_2, at 3000 lx and 25°C. The algal cells of *B.braunii* cultured were filtered through a 20 m nylon and then used as a feedstock in experiments hereafter. Table 1 shows the properties of the algal cells.

2.1 Hydrocarbon extraction

The algal cells of *B.baunii* were filtered and then freeze-dried. The resulting algal biomass (943mg) was ultrasonically extracted with 50 ml of hexane for 30 min. The hexane was removed under reduced pressure to yield a yellow oil as hexane extracts(471 mg). In Table 1, also is shown the properties of the hexane extracts. The oil was loaded on a column of silica gel (20g, Wakogel C-200, Wako, Japan) using hexane as the eluent. The eluent was again removed under reduced pressure to yield 343 mg of botryococcenes (36% of dry cell weight) as a colorless oil. The recovery of botryococcenes was defined on an organic weight basis.

	Cells	Hexane-solubles[a]
Moisture (wt%)	90	–
Organics (wt% db)	98	–
Ash (wt% db)	2	–
Elemental analysis (wt% db)		
C	63.1	84.7
H	11.7	12.7
N	2.8	0.8
O[b]	22.4	1.8
Heating value (MJ kg^{-1})	–	47
Viscosity at 50°C (mPa s)	–	170

[a] 50 wt% of organics
[b] By difference

Table 1. Properties of algal cells of *B.braunii*

2.3 Liquefaction

Liquefaction was performed in a 300 ml autoclave as shown in Fig.1. Algal cells (~ 30g) were charged to the autoclave with or without a catalyst (sodium carbonate). Distilled water (~20 ml) was added, because the quantity of algal cells was too small for adequate stirring. After purging with nitrogen, the autoclave was charged with nitrogen at 2 MPa and then sealed and heated by an electric furnace. The required temperature was maintained for 1h, followed by cooling. The catalyst loading was defined as a percentage by weight of the organics in the algal cells.

Fig.2 shows the procedure for separation of the liquefaction products. The gas, mainly composed of CO_2, was collected in a sample bag. The reaction mixture in the autoclave was washed with dichloromethane. The solid residue was separated by filtration, washed with dichloromethane and dried at 105°C for 1 day. The mixture of filtrate and washings was extracted completely with the same solvent. The dichloromethane was evaporated at 35°C under reduced pressure to obtain the oil.

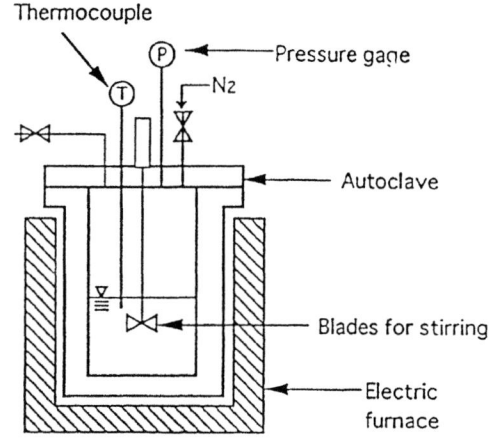

Fig. 1 Experimental apparatus for liquefaction

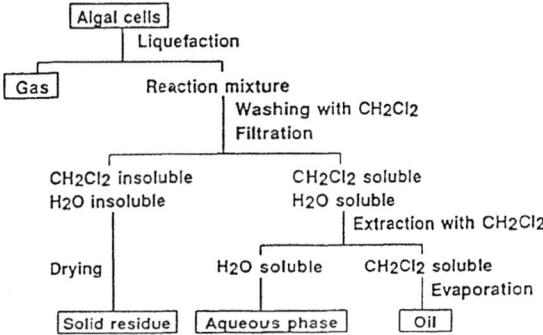

Fig. 2 Procedure for separation of liquefaction products

2.4 Fractionation and analysis

Thin-layer chromatography with hexane (TLC: Kieselgel 60 F254, layer thickness 0.2 mm, Merck, U.S.A) was used for a preliminary separation of the liquefied oil. Treatment with a color-producing reagent, ethanol-phosphomolybdic acid, revealed three spots. Based on the result, the liquefied oil was separated into three fractions by silica gel column chromatography. The method for fractionation is provided in Fig.3. Liquefied oil (~200 mg) was transferred to a silica gel column (30g, Wakogel C-300). The column was eluted with 200 ml (100 ml x 2) of hexane and then 200 ml of diethyl ether. The first 100 ml of hexane eluate removed under reduced pressure was collected as F1. In the same way, the next 100 ml of hexane eluate as F2, and the 200 ml of diethyl ether eluate as F3. The yield of fractions was defined on an organic weight basis. The fractions were analyzed using GC-MS technique. The mass spectrometer interfaced with gas chromatogragh was Hewlett Packard HP-5890-II-5971A. GC-MS analytical conditions are shown in Table 2. The mean molecular weight of the respective fractions was measured using the 117 Molecular Weight Apparatus(Corona Japan).

The yield of oil is defined as follows:

$$\text{Oil yields}(\%) = \frac{\text{Weight of oil}}{\text{Weight of organics in algal cells}}$$

The heating value of F1, F2 and F3 was calculated according to Dulong's formula:

$$Qh = 338.3C + 1442(H-O/8) \quad (J/g)$$

where C, H and O are the weight percentage of carbon, hydrogen and oxygen, respectively.

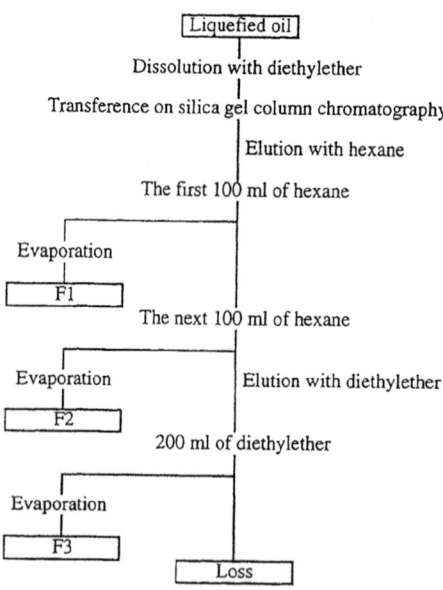

Fig. 3 Fractionation procedure of liquefied oil

HEWLETT PACKARD HP-5890II-5971A

Column:	DB-1 30m x 0.25mm
Column Temp:	100(10min) - 3°C/min - 280(25min)
Inj. Temp:	280°C
Det. Temp:	300°C

Table 2. Conditions of GC-MS analysis

3 Results and Discussion

3.1 Recovery of hydrocarbon rich oil

Fig.4 shows the effect of reaction temperature on oil yield. The yield was a maximum at 300°C, regardless of catalyst loading. The oil obtained at 200°C without a catalyst was not suitable for liquid fuel, because it was rubber-like. The maximum yield was 64 wt% at 300°C with the catalyst, exceeding the content of hexane solubles in the algal cells (50 wt% based on organics) Hence liquid fuel in greater amount than the hexane solubles in the cells can be obtained by liquefaction. The yield of oil at 300°C without a catalyst also exceeded the content of hexane-solubles in the cells. The yield at 340°C, however, was much less than the content of hexane-solubles in the cells. One reason for this may be that thermochemical degradation proceeded further to produce low-molecular-weight materials, which were lost while evaporating the dichloromethane to recover the oil

There was no apparent effect of catalyst loading on the yield of oil, although the catalyst was effective in the liquefaction of wood.[12] This difference may be due to the difference in feedstock: wood is mainly composed of holocellulose and *B. braunii* of hydrocarbons.

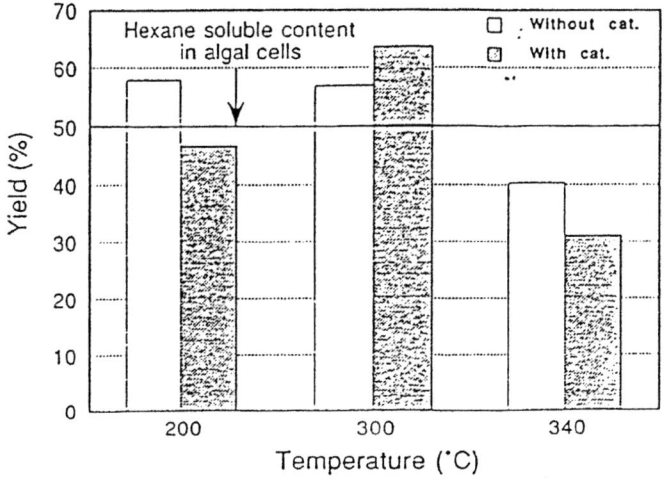

Fig. 4 Effect of reaction temperature on oil yield (catalyst loading 0 and 5 wt%)

Catalyst loading (wt%)	0			5		
Temperature (°C)	200	300	340	200	300	340
Heating value (MJ kg^{-1})	42	50	49	50	49	47
V΄ cosity at 50°C (mPa s)	–	79	77	160	64	71
Elemental composition (wt%)						
C	78.6	84.2	83.2	84.8	84.1	81.7
H	11.9	14.9	13.8	14.9	14.3	13.8
N	0.0	0.9	1.4	0.3	0.7	0.9
O^a	9.5	0.0	1.6	0.0	0.9	3.6

^{a} By difference

Table 3. Properties of oil obtained by liquefaction

Table 3 shows the properties of the oil obtained. The heating value was essentially the same under all conditions tested, ~50MJ kg^{-1} and superior to that of petroleum oil. The viscosity was low, 64-160 m Pas. All the oils contained ~1 wt% of nitrogen. The present results show that oil obtained a suitable for liquid fuel, e.g. for boilers, although flue gas treatment may be necessary to remove NOx.
To get more information about the oil, oil was fractionated and analyzed.

3.2 Analysis of oil

It was confirmed by TLC that the fractions F1, F2 and F3 represent the top, middle and bottom spots respectively on the silica gel TLC of the liquefied oil. The properties of F1, F2 and F3 and of the botryococcenes extracted from the algal cells are shown in Table 4. F1 and F2 were colorless oils, but F3 was brown.
Botryococcenes are terpenoid-type hydrocarbons having elemental composition of C30-36H48-64 and molecular weight of 408-496 (most typical : C34H58 shown below*).

	Mean molecular weight	Elemental composition			GC retention time (sec)
		C (%)	H (%)	O (%)	
Fraction 1	197–281	84.5–85.5	14.6–15.5	0.0	–3540
Fraction 2	438–572	85.0–86.5	13.2–14.0	0.0–1.0	4140–4380
Fraction 3	867–2209	73.3–77.3	12.2–13.3	10.1–14.3	4680–
Botryococcenes	472	85.7	14.3	0.0	4140–4380

Table 4. Properties of the fractionated oil and botryococcenes in the raw algal cells

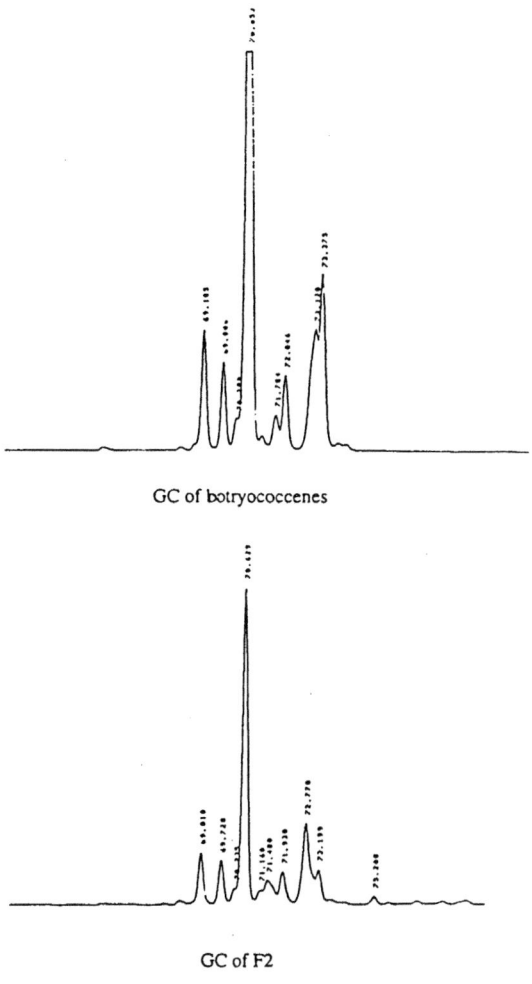

GC of botryococcenes

GC of F2

Fig. 5 Gas chromatograms of extracted botryococcenes and F2

The mean molecular weight of F2 was in the range of 438-572, and that of botryococcenes in the algal cells 472. The elemental composition of F2 was in good agreement with that of botryococcenes. In addition, the gas-chromatography retention time of F2 coincided well with the retention time of the botryococcenes(Fig.5). From these results, F2 was identified as the botryococcenes.

F1 mainly consisted of relatively light compounds having the mean molecular weight in the range of 197-281. F1 was determined as hydrocarbons from the elemental composition. On the other hand, F3 had a very high mean molecular weight (867-2209) and contained various oxygenated compounds. Gas chromatogram of F3 was very complicated. The retention times ranged from 2min. to 95 min. and more than 90 peaks were detected.

In Table 5, are shown the compounds identified as likely compounds in F1 and F3 of the oil obtained at 300 °C in the presence of Na_2CO_3. The compounds identified in F1 were C16-C22 hydrocarbons, which suggests that F1 is some kind of lightly degraded products of the botryococcenes. Compounds in F3 were considered to be produced from organic materials other than hydrocarbons through liquefaction. Phenolic compounds were identifeid. Microalgae containes little lignin which is usually considered to be source of phenolic compounds from lignocellulosic biomass. However, it is reported that aromatic compounds are produced also from cellulose and polysaccharides via degradation and condenzation process.[13] Phenolic compounds in F3 were produced from polymers which consist alga.

From these results, we could confirm that F1, F2 and F3 were low molecular weight hydrocarbons, botryococcenes and polar substances, respectively.

Fig. 6 shows the yield of each fraction of the liquefied oil and hexane extracts from the algal cells. The maximum recovery of the botryococcenes in the liquefied oil was obtained when the liquefaction was carried out at 200°C (78%) and 300°C (75%) both with the use of 5% Na_2CO_3.

Fraction 1
(Low molecular weight hydrocarbon)
5-Octadecene
Heptadecane
$C_{19}H_{38}$
$C_{20}H_{40}$
Octadecane
Tetramethyl hexadecene
Docosane

Fraction 3
(Polar substances)
Mesityl oxide
4-Hydroxy-4-methyl-2-pentanone
Phenol
Methyl phenol
Ethyl phenol
2,6-Bis(1,1-dimethylethyl)-4-ethyl phenol
Tetradecanoic acid
Trimethyl pentadecanone
Methyl tetradecanoic acid
Hexadecanoic acid
9-Octadecenoic acid
Octadecanoic acid
Eicosanoic acid
1-(7-Methyl-1-oxopentadecyl) pyrrolidine

Table 5. Likely components of liquefied oil in F1 and F3

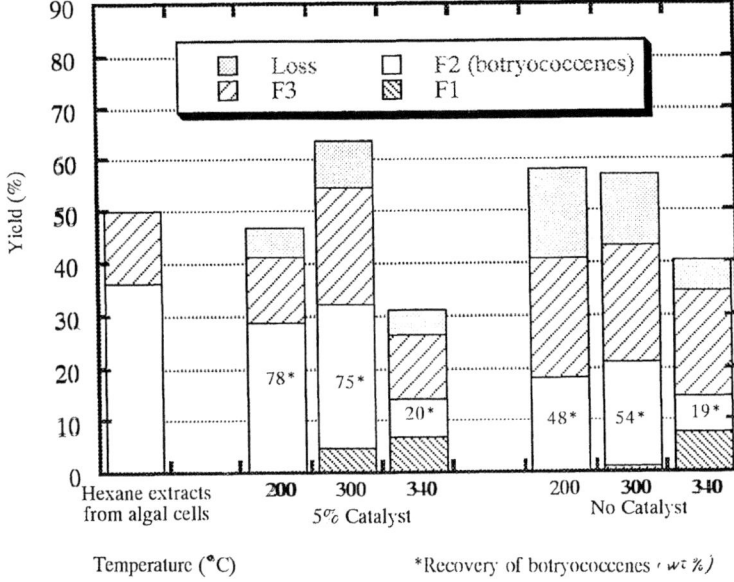

Fig.6 Yield of each fraction of liquefied oil and hexane extracts from algal cells.

F1 and F2, having high heating values of about 50 MJkg^{-1}, are suitable as an energy feedstock. The heating value of the oxygenate F3 (42 MJ kg^{-1}) was lower than that of F1 and F2. However, if the liquefied oil is to use as boiler fuel, it could be suitable.

4 Conclusion

Liquefaction is an effective method for converting algal cells of B. braunii into liquid fuel and recovering hydrocarbons. Maximum oil yield (64%) was attained at 300 C in the presence of Na2CO3.

The oil obtained from *B.braunii* consisted of three fractions:low molecular weight hydrocarbons, botryococcenes and polar substances.

Maximum recovery of the botryococcenes (75-78%) in the oil was achieved at 200°C and 300°C with the use of 5% Na2CO3 catalyst. Recovery of botryococcenes was improved by the addition of the catalyst.

5 References

1. Kodama, M. Ikemoto, H. and Miyachi, S., (1993) A new species of highly CO2-tolerant fast-growing marine microalga suitable for high-density culture. *J. Mar. Biotechnol.* 1 pp21-25

2. Sawayama, S. Minowa, T. Dote, Y. and Yokoyama, S., (1992) Growth of the hydrocarbon-rich microalga *Botryococcus braunii* in secondarily treated sewage. *Appl. Microbiol. Biotecnol.* pp135-138

3. Tornabene, T.G., (1982) Microorganisms as hydrocarbon producers. *Experimentia* 38, pp43-46

4. Wolf, F.R., Nonomura, A.M. and Bassham, J.A., (1985) Growth and branched hydrocarbon production in a strain of *Botryococcus brainii*. *J. Phycol.* 21 pp388-396

5. Derenne, S., Largeau, C., Casadevall, E. and Berkaloff, C., (1989) Occurence of a resistant biopolymer in the L race of *Botryococcus braunii* . *Phytochemistry* 28 (4) pp1137-1142

6. Metzger, P. Berkaloff, C., Casadevall, E. and Coute, A., (1985) lkadiene-and botryococcene-producing races of wild strains of *Botryococcus braunii* . *Phytochemistry* 24(10) pp2305-2312

7. Wolf, F.R., Nemethy, E.K., Blanding, J.H., and Bassham, J.A., (1985) Bioshnthesis of unusual cyclic isoprenoids in the alga *Botryococcus braunii* . *Phytochemistry* 24(4) pp733-737

8. Hillen, L.W., Pollard, G., Wake, L.V. and White, N., (1982) Hydrocracking of oils of *Botryococcus braunii* to transport fuels. *Biotechnol. Bioengng* XXIV pp193-205

9. Kitazato, H., Asaoka, S. and Iwamoto, H., (1989) Catalytic cracking of hydrocarbons from microalgae. *Sekiyu Gakkaishi,* 32(1) pp28-34

10. Yokoyama,S., Suzuki, A., Murakami, M., Ogi, T., Koguchi,K. and Nakamura, E., (1987) Liquid fuel production from sewage sludge by catalytic conversion using sodium carbonate. *Fuel* 66 pp1150-1155

11. Ogi, T. and Yokoyama, S., (1993) liquid fuel production from woody biomass by direct liquefaction. *Sekiyu Gakkaishi* 36(2) pp73-84

12. Ogi, T. Yokoyama, S. and Koguchi, K., (1985) Direct liquefaction of wood by alkali and alkaline earth salt, *Chem. Lett.,* pp1199-1202

13. Nelson, D.A., Russell, J.A., Molton, P.M., (1985) Formation of aromatic compounds from Condensation reactions of cellulose degradation products-II, "Fundamentals of Thermochemical Conversion", ed. Overend, R.P., Milne, T.A., Mudge, M.K., Elsevier Appl. Sci.Pub., pp1039

PYROLYSIS

Pilot and demonstration

DEVELOPMENT OF A NOVEL VACUUM PYROLYSIS REACTOR WITH IMPROVED HEAT TRANSFER POTENTIAL
Vacuum pyrolysis - Improved heat transfer

C. ROY, J. YANG, D. BLANCHETTE, L. KORVING and B. DE CAUMIA
Institut Pyrovac Inc.
1560, avenue du Parc Beauvoir, Sillery (Québec) Canada G1K 7P4

Abstract
    A novel reactor has been developed in our laboratory which addresses the heat transfer limitations usually encountered in vacuum pyrolysis technology. Conventional pyrolysis reactors such as multiple hearth furnaces, rotary kilns and screw type reactors exhibit overall heat transfer coefficients ranging from 10 to 60 $W \cdot m^{-2} K^{-1}$, depending primarily on the type of feedstock treated. The new reactor design includes a novel feedstock transport and agitation system which produces a forced exchange between the feedstock particles heated at the surface of the heating plate and the colder particles located at the core of the packed particle bed. Thus, the heat transfer between the reactor and the pyrolyzed material is dramatically increased. The other novelty of this reactor is the use of an indirect heating system involving commercial molten salts (Hitec®). Experimental and theoretical studies have been undertaken in order to tentatively correlate the reactor design parameters and the heat transfer coefficient. A new heat transfer model, based on Schlünder's heat transfer model, has been developed to model the heat transfer in the bed of particles as a function of the mechanical movement of particles created by the agitation. This model is validated by comparing theoretically calculated overall heat transfer coefficients in a batch reactor with experimentally measured values for gravel feedstock in the same reactor. The model is then used to predict overall heat transfer coefficients for various feedstocks in the novel continuous reactor. Coefficients ranging from 70 to 250 $W \cdot m^{-2} K^{-1}$ are obtained with this new system.
Keywords: Agitation, heat, transfer, model, Hitec®, packed, particles, surface, vacuum, pyrolysis.

1 Introduction

Over the last ten years, a vacuum pyrolysis process has been developed at Université Laval in Québec, Canada. This thermal decomposition process enables a large variety of solid and semi-liquid wastes to be transformed into useful products. Vacuum pyrolysis is typically carried out at a temperature of 400-500°C and a total pressure of 2-20 kPa [1]. These vacuum conditions allow the pyrolysis products to be rapidly withdrawn from the hot reaction chamber, thus preserving the primary fragments originating from the thermal decomposition.

A major limitation of vacuum pyrolysis technology is heat transfer. Previous studies have shown that the rate of heat transfer is essentially the rate limiting step for pyrolysis reactions [2]. Conventional pyrolysis reactors such as multiple hearth furnaces, rotary kilns and screw type reactors exhibit overall heat transfer coefficients ranging from 10 to 60 $W \cdot m^{-2} K^{-1}$ [3], depending on the type of feedstock treated. The low thermal conductivity of the feedstock materials partially explains why the heat transfer fluxes are low.

The multiple hearth furnace which has been used in the previous studies of this laboratory [4] can be used to provide an example of the factors which limit the heat transfer in this type of reactor. Since the heat source in the multiple hearth furnace is the external wall, the heat transfer is primarily effected by radiation. The plates on which the feedstock is placed inside the reactor are heated by radiation from the external wall and thus heat the feedstock from below. The feedstock is also directly exposed to the radiant heat source (the external wall) from above, as there is some distance between the different plates inside the reactor. One limitation of this radiative heat transfer mechanism is the view factor between the external wall and the plates inside the reactor. This factor is only slightly greater than 0.5 at the point where the plate meets the wall (assuming that the plate is a black body), and decreases considerably as one moves away from the external wall and approaches the middle of the plate in the center of the reactor, especially when the distance between neighbouring plates is small. Another limitation arises from the type of agitation system used in this reactor. Each plate inside the multiple hearth furnace is equipped with a series of radial metal bars which transport the feedstock as they move over the plates. The feedstock particles tend to accumulate along these bars, thus leaving almost 50% of the heated plates uncovered by feedstock and rendering the heat transfer inefficient, especially since there is very little mixing of the hot and cold particles. Finally, since the plates are only heated by radiation, the heat transfer coefficient is limited both by the unfavourable view factor and the minimal conduction between the heated plates and the feedstock. Both of these factors could be greatly improved, especially for small feedstock particles, by surface-to-particle contact conduction and radiation.

The new vacuum pyrolysis reactor, described herein, uses an indirect heating system involving commercial Hitec® molten salts which flow inside the heating plates, thus heating the feedstock on these plates both by conduction and radiation. In addition, the novel transport system in the reactor efficiently agitates the feedstock, thus greatly improving the heat transfer potential.

The importance of the influence of agitation on heat transfer has been investigated widely in previous studies [5-8]. These studies mainly dealt with the following two aspects: the contact heat transfer resistance between the heating surface and the feedstock particles, and the heat transfer in the feedstock as a function of the flowing dynamics of the particles. An important contribution was made by Schlünder [7] who developed a theory to quantitatively describe the surface contact heat transfer between a spherical particle and a heating surface. This theory was later further developed by Malhotra and Mujumdar [9] to describe the surface contact heat transfer of other particle shapes. The particle flowing dynamics have been studied by several authors [10-11]. However, there are still no general equations which correlate the particle movement with the actual design of the agitation system. The new heat transfer model proposed herein assumes that the heat transfer in a moving bed is only a function of the heat exchange at the heating surface which depends on the surface-to-particle contact resistance and the renewal of the heated bottom layer, as described in the theory section below. The influence of particle mixing in the bulk of the particle bed is ignored as agitation renders the interparticle heat transfer very efficient and therefore this heat transfer resistance component is negligible.

2 The new vacuum pyrolysis reactor

In our laboratory a semi-continuous horizontal pilot plant reactor with a length of 3 m, a diameter of 0.6 m and a capacity of about 100 - 400 kg·h^{-1}, depending on the feedstock treated, has been built (see Figure 1). The moving bed vacuum pyrolysis reactor uses a novel transport and agitation device which produces a forced exchange between the feedstock particles heated at the surface of the heating plate and the colder particles located at the core of the packed bed of particles. As a result, the heat transfer between the reactor and the pyrolyzed material is dramatically increased. The patents for the transport system have been claimed and hence further details cannot be given for the moment (12). In the reactor, the feedstock is conveyed over two horizontal heating plates, one on top of the other. The plates are 35 cm wide and together they have an effective length of 4.1 m. They are composed of rectangular tubes which are welded together in the longitudinal direction. Commercial molten salts (Hitec®) flow countercurrently with the feedstock through the interior of these tubes, thus indirectly heating the feedstock which is conveyed over the heating plates. The salts leaving the reactor are collected in a tank which is equipped with a vertical pump to circulate the molten salts through the system, to a heater to be reheated and then back to the reactor. The Hitec® molten salt is an eutectic mixture of potassium nitrate, sodium nitrite and sodium nitrate which can be used as a heat transfer medium between 150°C and 540°C. Below 150°C lies the melting point of the mixture (at 142°C) and above 540°C the salt will degrade too rapidly for commercial applications. The vapour pressure of the salt is negligible in the operating range. At 500°C, the Hitec® salts have a density of 1700 kg·m^{-3}, a specific heat capacity of 1549 J·kg^{-1}K^{-1} and a thermal conductivity of 0.61 W·m^{-1}K^{-1}.

During pyrolysis, the feedstock is heated under vacuum to a temperature of about 450-500°C. The pyrolysis vapours are rapidly evacuated from the hot reactor chamber by means of a vacuum pump which maintains a total pressure of 2-20 kPa in the reactor. The vapours are condensed in two spray towers. After pyrolysis, the solid residues leave the reactor under vacuum and are cooled with water in a quenching vessel. The sludge formed is directed towards a screen where the char is sieved and separated from the water. The cooling water is then recycled to the wetting vessel. The non-condensable gases can be burned to provide a significant portion of the energy necessary to heat the molten salts. The amount of heat which can be provided by burning the non condensable gases will depend on the nature of the feedstock. In applications involving plastics for example, three times more heat is available in the pyrolysis gas for each kg of feedstock treated than the actual amount required to achieve the pyrolysis reactions. For automobile shredder residues or auto fluff, this ratio is approximately two. Another example is contaminated soil containing 15% moisture where the "pyrolysis gas heat content per kg of feedstock / heat of pyrolysis per kg of feedstock" ratio is approximately 0.6. An induction heater can also be used as a make-up heat source to enable the molten salt to reach the exact set point temperature in the reactor.

In order to determine the best design and operation parameters for the new reactor, theoretical and experimental work has been carried out to investigate the correlation between the heat transfer coefficient and the design of the agitation system. The proposed new heat transfer model was validated using a smaller batch scale agitated bed reactor.

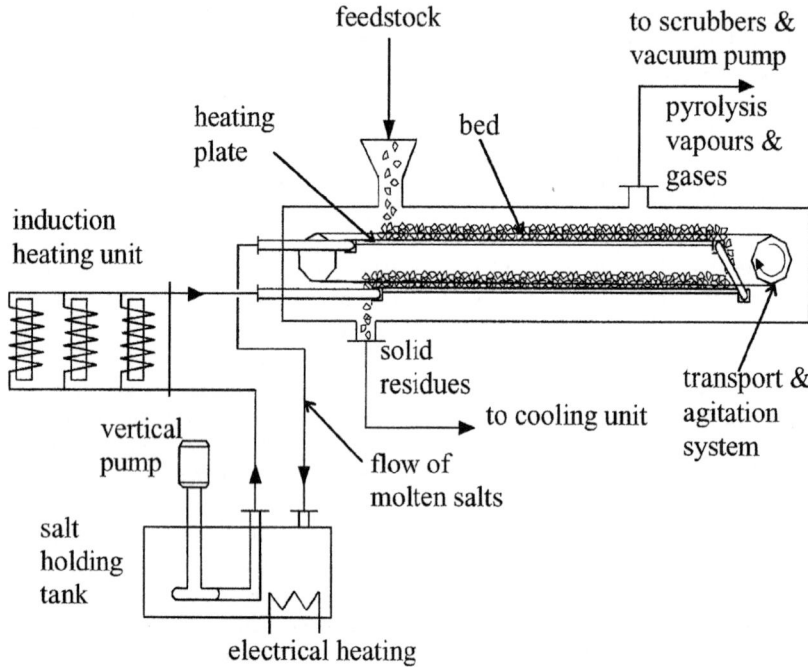

Figure 1. Pilot plant vacuum pyrolysis reactor

3 Theory

3.1 The surface renewal model

The proposed surface renewal model is based on the heat transfer model which Schlünder [13] developed for the indirect contact vacuum drying of solid particles.

In Schlünder's model, particle movement due to the agitation was described as a series of time periods during which the particle bed is assumed to be static, followed by a rapid mixing of the particles. Thus the average heat transfer coefficient during the static period can be calculated using the Penetration Theory, which results in the following equation:

$$\alpha_{bed} = \frac{\alpha_{ws}}{1 + \frac{\sqrt{\pi}}{2}\sqrt{\frac{\alpha_{ws}^2}{\rho C_p \lambda}\tau_R}} \tag{1}$$

In the equation above, α_{ws} is the surface-to-particle contact heat transfer coefficient, α_{bed} is the overall heat transfer coefficient in the agitated bed, ρ, C_p, and λ are the average density, specific heat capacity and the thermal conductivity of the particle bed and τ_R is the time period of the static period. In order to relate the time scale of the mixing device to the time scale of the time periods of the static bed, Schlünder introduced the following relationship into the formula above:

$$\tau_R = N_{mix} t_{mix} \tag{2}$$

Together these formulas yield the following formula:

$$\alpha_{bed} = \frac{\alpha_{ws}}{1 + \frac{\sqrt{\pi}}{2}\sqrt{N_{therm} N_{mix}}} \tag{3}$$

where N_{therm} is given by:

$$N_{therm} = \frac{\alpha_{ws}^2 t_{mix}}{\rho C_p \lambda} \tag{4}$$

In these formulas t_{mix} is the characteristic mixing time of the agitation device (normally the time necessary for one turn of the device). N_{mix} is defined as a "mixing number" whose value is a function of the geometrical and mechanical properties of the agitation system, *i.e.* the agitation design parameters. This parameter must be obtained from curve fitting of the measured α_{bed} as a function of the characteristic time of agitation, t_{mix}. Schlünder could not give a physical explanation for the value of this N_{mix}.

In order to improve Schlünder's model and to find the direct correlation between heat transfer and the mechanical properties of an agitation system, the heat transfer mechanism in an agitated bed has been analyzed. As shown in Figure 2, two heat exchange phenomena occur in an agitated bed: the wall-to-particle heat exchange at the heating surface and the interparticle heat exchange in the bulk. Particle agitation improves the heat transfer in the following two ways: (i) the hot particles are periodically removed from the heating surface, thus maintaining a large temperature difference between the particle bed and the heating surface and (ii) the hot particles are directed towards the bulk of bed, resulting in a solid convection which greatly improves the interparticle heat exchange. If the particle bed is not very thick, perfect mixing of the hot particles in the bulk should be obtained and the heat transfer resistance is considerably reduced. In this case, the heat transfer in an agitated bed can be simplified as a periodical wall-to-bed contact heat transfer problem, which is a function of the wall-to-bed contact heat transfer coefficient, the period of agitation and the quantity of particles removed after each agitation. The resulting new heat transfer model is based on the following assumptions:

1. The characteristic time of the interparticle heat transfer is shorter than the time required for the agitated hot particle to return from the bulk to the heating surface.
2. The overall heat transfer in the bed of particles is determined by the heat transfer between the bottom layer of particles and the heated plate.
3. Each agitation removes β percent of the bottom layer particles from the heating surface; β is defined as the bottom layer renewal efficiency.
4. The agitation results in a distribution of the residence time of the particles at the heating surface. The fraction of particles with a residence time kt_{mix} is $\beta(1-\beta)^{k-1}$; t_{mix} is the time that is necessary to complete one agitation.
5. During the residence time of a specific fraction, it is assumed that this fraction remains static (see Figure 2.a). Thus the heat transfer coefficient of this fraction can be calculated using the Penetration Theory.
6. The overall heat transfer coefficient is calculated by summation of the heat transfer coefficients of each fraction.

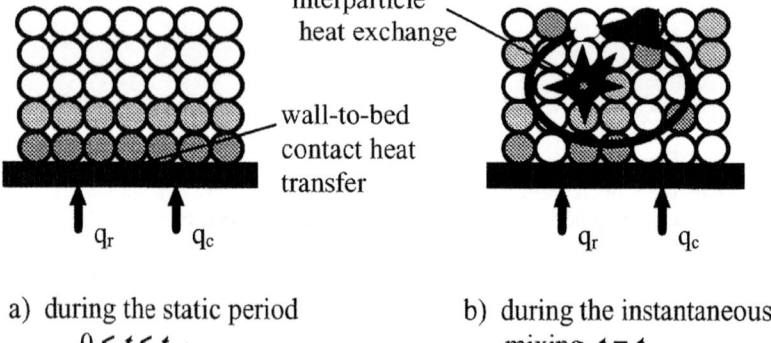

a) during the static period
$0 < t < t_{mix}$

b) during the instantaneous mixing, $t = t_{mix}$

Figure 2. Heat transfer mechanism in an agitated bed of particles

These assumptions result in the following equation:

$$\alpha_{bed} = \sum_{k=1}^{\infty} B_k \alpha_k \tag{5}$$

where α_{bed} is the overall heat transfer coefficient in the bed, and B_k and α_k are the quantity and the heat transfer coefficient of fraction k. The variables, B_k and α_k, are calculated by the following equations:

$$B_k = \beta \cdot (1 - \beta)^{k-1} \tag{6}$$

$$\alpha_k = \frac{\alpha_{ws}}{1 + \dfrac{\sqrt{\pi}}{2}\sqrt{\dfrac{\alpha_{ws}^2}{(\rho C_p \lambda)_{bed}} kt_{mix}}} \tag{7}$$

In this equation, ρ, C_p and λ are physical properties determined by the nature of the feedstock. Thus, the only parameters which depend on the reactor design are the surface-to-particle contact heat transfer coefficient, α_{ws}, and the bottom layer renewal efficiency, β. As β is directly influenced by the reactor design, this formulation allows a relationship to be established between the reactor design and the overall heat transfer in the particle bed.

In comparison with Schlünder's model, the major advantage of this formulation is that the parameter N_{mix}, which can only be obtained by curve-fitting, has been replaced by the experimentally measurable parameter β. The bottom layer renewal efficiency, β, has been defined as the particle fraction removed after each agitation. Thus, β has a direct physical meaning and can be easily obtained by a simple cold mechanical run using tracers.

3.2 Surface-to-particle contact resistance (α_{ws})

The so-called surface-to-particle contact heat transfer is defined as the heat transfer from the heating surface to the bottom layer of the particles. It consists of three components: the thermal conduction through the gas wedge between the particles and the heating surface (α_{wp}), the thermal conduction through the void space between the particles (α_{wg}) and the heat transfer by radiation (α_r). Therefore the contact heat transfer coefficient (α_{ws}) can be expressed by:

$$\alpha_{ws} = \alpha_{wp} + \alpha_{wg} + \alpha_r \tag{8}$$

According to Schlünder's theory, a minimal thermal contact resistance exists between a particle and a contacting surface, which is symbolized by α_{wp}. A formula for the calculation of α_{wp} has been developed for spherical particles [13]. It is expressed as:

$$\alpha_{wp} = \phi_A \cdot \frac{4\lambda_g}{d_p}\left[\left\{\frac{2(\sigma + S_r)}{d_p} + 1\right\} ln\left\{\frac{d_p}{2(\sigma + S_r)}\right\} - 1\right] \tag{9}$$

where ϕ_A is the area fraction of contact points. The heat transfer through the voids of gas (α_{wg}) can be formulated as:

$$\alpha_{wg} = \left(1 - \phi_A\right) \frac{2\lambda_g / d_p}{\sqrt{2} + \left(2\sigma + 2S_r\right)/ d_p} \tag{10}$$

The radiation between a particle and the heating surface (α_r) is expressed as [14]:

$$\alpha_r = \frac{0.2268}{\dfrac{2}{\varepsilon} - 0.264} \left(\frac{T}{100}\right)^3 \tag{11}$$

Since the equations above were developed for spherical particles, they are still not available to determine the surface-to-particle contact heat transfer coefficient in this work where the materials treated often have an irregular shape. The importance of these equations is that they indicate which parameters have a significant influence on the surface-to-particle heat transfer resistance. In this work, the surface-to-particle contact resistance is determined experimentally. The surface-to-particle heat transfer coefficient for irregular shaped particles is currently being further studied in our laboratory.

4 Experimental

4.1 Equipment
A batch scale reactor was developed and used as a tool to validate the new heat transfer model at a total pressure slightly lower than atmospheric pressure. The transport and agitation devices of both the laboratory and full scale reactors are equivalent. The heat transfer tests were carried out in order to determine the effect of the agitation on the rate of heat transfer. Once this influence is known, the results obtained at near-atmospheric-pressure can be corrected for vacuum conditions. The vacuum will only influence the thermal parameters, *i.e.* the surface-to-particle contact resistance, α_{ws}, and the conductivity in the bed λ_{bed}, and will not affect the agitation mechanism. Thus, the heat transfer relationship will remain the same even if the reactor pressure changes. So, the values of α_{ws} and λ_{bed} can be measured under vacuum and entered into the developed formulas so as to determine the overall heat transfer coefficient under the vacuum conditions found in a continuous feed reactor.

The batch reactor, shown in Figure 3, consists of a well insulated tank which contains molten salt and is equipped with heating elements in order to be able to heat the salt to the required temperature. A second insulated tank can be placed in the bath of molten salts. The feedstock enters the reactor through the feedpipe. The agitation mechanism transports and agitates the feedstock in a circular direction. The center of the reservoir is kept free of feedstock by a scraping mechanism. The diameter of the feedstock tank is 107 cm and the effective heat transfer area is 0.82

m². The total mass of salt in the salt-tank is 285 kg. During a pyrolysis run, the vapours formed are evacuated by a blower. The air tightness of the system and the capacity of the pump are able to maintain a total pressure slightly lower than ambient atmosphere which is sufficient to study all relevant heat transfer phenomena. For the tests described in this paper the blower was not used.

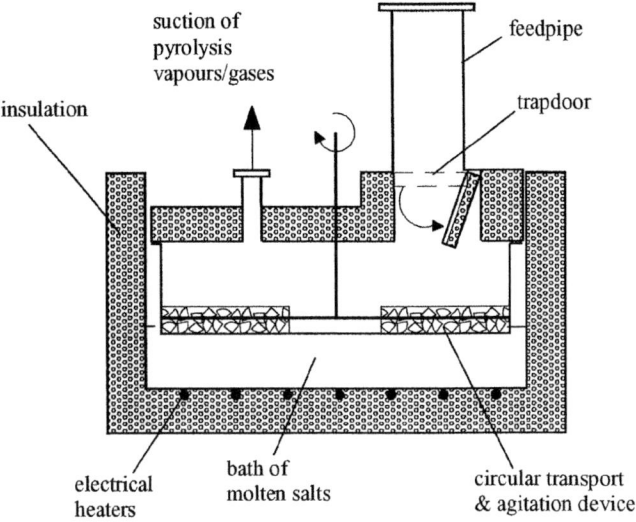

Figure 3. Batch vacuum pyrolysis reactor

Once the surface-to-particle contact resistance, α_{ws}, had been measured, the batch reactor was used to measure the bottom layer renewal efficiency, β for the circular agitation device and finally the overall heat transfer coefficient in the particle bed, α_{bed} (see section 4.2). The semi-continuous reactor was used to measure the bottom layer renewal efficiency, β, for the novel transport and agitation device (see section 4.2.2). The measured value of β was then used to predict the overall heat transfer coefficient in the new reactor, α_{bed} (see section 5.4).

4.2 Methodology

4.2.1 Determination of the surface-to-particle contact resistance (α_{ws})

The surface-to-particle contact heat transfer coefficient, α_{ws}, is measured with the apparatus shown in Figure 4. This apparatus consists of a stainless steel heating plate with a diameter of 400 mm, the particles to be heated, and a glass wool insulation plate with a thickness of 30 mm. The temperature of the heating plate, T_w, is increased from 30 to 200°C. The particles have an irregular shape. Their approximate dimension is 15 mm x 10 mm x 4 mm with a mean volume diameter d_p of 5 mm. The average of the maximum particle diameter and the volume average diameter is 10 mm. A thermocouple is placed in the center of the particle to measure the particle temperature, T_p.

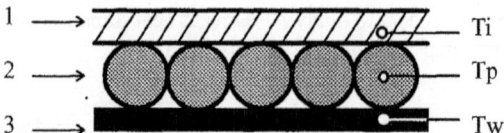

Figure 4. Measurement of the surface-to-particle contact heat transfer coefficient.
1. heating plate, 2. particles, 3. insulation material.

A computer data acquisition system is connected to record the temperatures, T_w, T_i and T_p. As the radial dimension of the particle bed is much larger than the axial dimension, the heat transfer in the radial direction can be ignored. Thus, an energy balance equation is developed for one single particle, which gives:

$$ mC_p \rho \frac{dT_p}{dt} = \alpha_w A(T_w - T_p) - \alpha_w A(T_p - T_i) \tag{12} $$

where m, C_p, ρ and A are the mass, specific heat capacity, density, thermal conductivity and surface area of the particle. The temperature of the heating plate, the particles and the insulation plate are represented by T_w, T_i and T_p, respectively.

4.2.2 Determination of the bottom layer renewal efficiency (β)

Tests with tracers have been carried out in order to determine the bottom layer renewal efficiency for the transport and agitation system in the batch reactor and in the semi-continuous pilot plant reactor. For both reactors the test-method was the following. For a small part of the bed the bottom layer of the bed was replaced by a monolayer of coloured particles. To determine β, the length of a transport unit was passed over the bed. For the batch reactor this was equivalent to one turn of the agitation and transport device; for the continuous system this would be a fraction of a turn depending on the number of transport units on the chain. Then several samples of the bottom layer were taken at various places in the bed. From these samples the total mass of coloured particles remaining in the bottom layer was calculated and divided by the original mass of coloured particles to give the fraction β.

For the batch reactor, β was determined for 10 mm gravel particles at seven different turning speeds (1, 3, 5, 10, 19, 26, 38 rpm). The tests with the transport and agitation device of the semi-continuous reactor were performed using gravel with different particle sizes (5, 10 and 14 mm diameter) at five different speeds of the transport and agitation device (0.10 m·s⁻¹, 0.11 m·s⁻¹, 0.16 m·s⁻¹, 0.21 m·s⁻¹ and 0.28 m·s⁻¹). For each test the bed height was 2.5 cm. More research is being carried out to determine the influence of the design of the agitation device and the mechanical feedstock properties on the β value [15].

4.2.3 Determination of the overall heat transfer coefficient in the agitated bed, α_{bed}

Using the batch reactor, four heat transfer tests were performed with gravel particles with a diameter of 10 mm. The conditions for each of these tests are listed in Table 1 and the thermal and physical properties of the feedstock are given in Table 2.

Table 1. Operating conditions for runs performed with the batch reactor

Test	Feedstock	Turning speed (rpm)	Load (kg)
G1	gravel	10	55.0
G2	gravel	19	55.3
G3	gravel	26	55.3
G3	gravel	38	53.5

About 15 minutes before the test, the heating of the salt (in the outer tank) was stopped. Immediately before the beginning of the test, the agitation mechanism was started and the feedstock was placed in the feedpipe. The test was then started by opening the trapdoor to allow the feedstock to fall into the inner feedstock tank. At that moment the temperature of the salt was approximately 520°C for each test. Several thermocouples were placed in the salt directly on the heat transfer surface and at different distances from this surface. During the tests both the temperature of the feedstock and the salt were recorded as a function of time.

Table 2. Physical and thermal properties of the feedstock

Physical / Thermal Property	Gravel 5 mm	Gravel 10 mm	Gravel 14 mm
Average particle size* (mm)	4.9	9.2	14.7
Apparent density (kg·m^{-3})	1260	1190	1100
True density (kg·m^{-3})	2590	2640	2664
Specific heat capacity (J·kg^{-1}K^{-1})	900	900	900
Thermal conductivity (W·m^{-1}K^{-1})	0.2	0.2	0.2

* Average of the maximum particle diameter and the volume average diameter

5 Results and Discussion

This section will describe how the new heat transfer model, based on the Schlünder heat transfer model, was validated. First, the results of the measured surface-to-particle contact resistance, α_{ws}, and bottom layer renewal efficiency, β, will be described. This will be followed by a description of how these two values were used to predict and then experimentally validate the value of the overall heat transfer coefficient, α_{bed} in the batch reactor using gravel feedstock. Finally, the measured values of β will be used to predict the overall α_{bed} in the semi-continuous reactor for various feedstocks.

5.1 Surface-to-particle contact resistance (α_{ws})

The surface-to-particle contact heat transfer coefficients of gravel have been measured. The measurements were carried out under atmospheric pressure. When the measured temperatures T_w, T_i and T_p are introduced into Equation 12, the contact heat transfer coefficient α_{ws} is obtained. The α_{ws} of the gravel is 250 W·m^{-2}K. The contact heat transfer does not depend on the material, but only depends on the particle dimension and shape, as well as on the fluid property between the heating surface and the heated particles (see Equation 9).

5.2 Bottom layer renewal efficiency(β)

Intuitively, it could be foreseen that the bottom layer renewal efficiency, β, would be a function of the moving speed of the transport unit, the design of the transport unit, the bed height and the physical properties of the feedstock. The values of β that were measured in the batch and semi-continuous reactor are summarized in Tables 3 and 4.

Table 3. Bottom layer renewal efficiencies for the circular agitation device in the batch reactor

Turning speed (rpm)	Characteristic time, t_{mix} (s)	Bottom layer renewal efficiency, β
1	46.2	0.67
3	19.4	0.72
5	11.8	0.73
10	6.1	0.56
19	3.1	0.51
26	2.0	0.46
38	1.6	0.47

Table 4. Bottom layer renewal efficiencies for the transport and agitation system of the semi-continuous reactor

Moving speed ($m·s^{-1}$)	Bottom layer renewal efficiency, β		
	Gravel, 5 mm	Gravel, 10 mm	Gravel, 14 mm
0.10	0.72	0.62	0.47
0.11	0.66	0.63	0.44
0.16	0.67	0.66	0.59
0.21	0.74	0.61	0.55
0.28	0.69	0.51	0.50

The value of the bottom layer renewal efficiency should ideally be as close to 1.0 as possible, for example a value of 0.9. However, a value of β greater than 0.5 is very satisfactory, since not much will be gained if the value of β is increased from 0.5 to 0.9. It can be shown that this increase will improve at best the rate of heat transfer by a factor of 1.3. However, the value of β should not be too low, since this will greatly affect the rate of heat transfer. It can be shown that a decrease of β from 0.5

to 0.1 will decrease the rate of heat transfer by a factor of 2.2. The results in Table 3 indicate that values of β greater than 0.7 could be obtained in the batch reactor at low turning speeds. The results also suggest that an increase in turning speed leads to a decrease in β. The bottom layer renewal efficiency is significantly influenced by the particle size, as indicated by the study carried out with the semi-continuous reactor (Table 4).

5.3 Overall heat transfer coefficient in the batch reactor (α_{bed})

In order to validate the heat transfer model which correlates the heat transfer coefficient in an agitated bed, α_{bed}, with the surface-to-particle contact heat transfer coefficient, α_{ws}, and the bottom layer renewal efficiency, β, heat transfer coefficients in the batch type agitation reactor have been measured with gravel feedstock and compared with the predicted values. Table 5 presents the experimental and theoretical results.

The theoretical results were calculated using Equation 5 and the thermal properties ρ, C_p and λ listed in Table 2. The measured surface-to-particle heat transfer coefficient, α_{ws} (250 W m^{-1} K^{-1}), and the bottom layer renewal efficiency, β (see Table 3), were used for the prediction. The value of β only changes slightly with the agitation speed.

Table 5. Measured and calculated overall heat transfer coefficients in the batch reactor for various feedstocks

Test	Feedstock	Turning speed (rpm)	Heat transfer coefficient, α_{bed} (W·m^{-2}K^{-1})	
			Calculated	Measured
G1	gravel	10	95	81
G2	gravel	19	112	117
G3	gravel	26	119	124
G4	gravel	38	132	156

The results in Table 5 indicate that the experimental measurements and the model predictions are in good agreement. Even though the model seems to overpredict the heat transfer coefficient at low turning speeds and underpredict this coefficient at high turning speeds, the difference between the experimental and theoretical values is less than 21%. This difference may be due to the fact that the model ignores the interparticle heat exchange (it was assumed that the agitation device renders this heat transfer resistance component negligible).

5.4 Predicted overall heat transfer coefficient in the semi-continuous reactor

The most important characteristic of the novel vacuum pyrolysis reactor is the significant increase in heat transfer between the reactor and the feedstock. The new heat transfer model proposed in this work enables a quantitative prediction of the overall heat transfer coefficient as a function of the design and operation parameters. As shown in Equations 5-7, the most important design parameters are α_{ws}, β and the agitation speed. When the design and operation parameters are fixed, the heat transfer coefficient will be a function of the thermal properties of the feedstock. Table 6 provides the thermal properties of various pyrolysis feedstocks and Table 7 presents the predicted heat transfer coefficient in the new pyrolysis reactor for different feedstocks. These values do not take into account the radiation effect above the bed.

Table 6. Feedstock physical and thermal properties used to predict overall heat transfer coefficients

Feedstock	Density $(kg \cdot m^{-3})$	Heat capacity $(J \cdot kg^{-1}K^{-1})$	Thermal conductivity $(W \cdot m^{-1}K^{-1})$	$\rho C_p \lambda$ $(\cdot 10^4)$
Gravel	1360	700	0.2	31.9
Charcoal	230	1000	0.07	1.6
Wood	240	1450	0.07	2.4
Rubber	600	1600	0.1	14.4

Table 7. Predicted overall heat transfer coefficients for various feedstocks

Feedstock	Predicted heat transfer coefficients $(W \cdot m^{-2}K^{-1})$ for varying surface-to-particle contact resistance values			
	$\alpha_{ws} = 250$ $W \cdot m^{-2}K^{-1}$	$\alpha_{ws} = 500$ $W \cdot m^{-2}K^{-1}$	$\alpha_{ws} = 750$ $W \cdot m^{-2}K^{-1}$	$\alpha_{ws} = 1000$ $W \cdot m^{-2}K^{-1}$
Gravel	143	200	232	252
Charcoal	70	82	87	90
Wood	81	97	104	107
Rubber	122	161	181	193

The values of β and t_{mix} used in the prediction were 0.7 and 1.7 second, respectively. As the tracer-tests show (see Table 3), $\beta = 0.7$ is a reasonable value for the agitation system. The moving speeds of the transport and agitation system of the semi-continuous reactor can reach 0.3 m·s^{-1}, a reasonable speed for full scale applications. With the current configuration of the mechanism this will give a characteristic mixing time of approximately 5 seconds. For larger scale reactors, the design of the transport and agitation system can easily be optimized to provide three times as much agitation, resulting in a mixing time of 1.7 s.

It is difficult to select one value of the surface-to-particle contact heat transfer coefficient, α_{ws}, since α_{ws} depends on both the reactor heating temperature and the feedstock particle diameter and shape (see Equations 8-11). Thus in Table 7, values of α_{ws} ranging from 250 to 1000 W·m^{-2}K^{-1} have been used, which include the most probable contact heat transfer coefficients in agitated reactors. The predicted results in Table 7 show that the overall heat transfer coefficient in the new vacuum pyrolysis reactor varies from 70 to 250 W·m^{-2}K^{-1}, and mainly depends on the value of the thermal properties of the feedstock, i.e. the type of materials treated.

These predictions were made with a proposed heat transfer model. The predictions do not take into account phase changes and pyrolytic reactions which will take place when feedstock such as rubber and biomass is pyrolyzed. Work is ongoing to study potential modifications of this model in cases where substrates such as biomass or rubber are used as feedstock.

6 Conclusions

A novel vacuum pyrolysis reactor has been described whose overall heat transfer coefficient α_{bed} ranges from 70 to 250 $W \cdot m^{-2} K^{-1}$, depending on the feedstock treated. This new reactor uses commercial molten salts (Hitec®) which flow through the heating plates in the reactor to indirectly heat the feedstock which is conveyed over the heating plates. The novel transport system agitates the feedstock, thus greatly improving the heat transfer potential in the reactor.

A new heat transfer model based on Schlünder's heat transfer model has been proposed. This model suggests that the two most important heat transfer parameters are the surface-to-particle contact resistance, α_{ws}, and the bottom layer renewal efficiency, β. Using a simple apparatus, α_{ws} was measured. Then β was measured in the batch and semi-continuous reactor and found to depend mainly on the feedstock particle size, although its value did slightly decrease at high moving speeds. The values of α_{ws} and β were then used to calculate the overall heat transfer coefficient, α_{bed}, in the batch reactor using gravel feedstock. These theoretical values were in very good agreement with measured values of α_{bed} in the batch reactor (at most a 20% difference). Having validated the proposed heat transfer model, the measured values of β and the previously obtained values of α_{ws} were used to predict the overall heat transfer coefficient in the semi-continuous pilot plant reactor for various feedstocks. The validated new heat transfer model will help optimize reactor design parameters and diminish the number of experiments required to determine the optimal parameters, thus facilitating the process of achieving optimal heat transfer in an industrial scale reactor.

Acknowledgements
This project has been supported by Pyrovac International Inc., Hydro-Québec and the Ministère des ressources naturelles du Québec. The collaboration of the Université Laval and Ecotechniek b.v. has been appreciated. The authors wish to thank Mrs. Carelle Malendoma and Dr. Annette Schwerdtfeger for their precious help in the preparation of this manuscript.

Notation
A	surface area, m^2
B	quantity of the agitated fraction k of the first layer of particles
C_p	specific heat capacity, $J \cdot kg^{-1} K^{-1}$
d	diameter, m
m	mass, kg
N_{mix}	experimental mixing number according to Schlünder
N_{therm}	dimensionless group variable according to Schlünder

S_r sum of particle and surface roughness

T temperature, K

t_{mix} characteristic time of the agitation device, s

Greek letters

α heat transfer coefficient, $W \cdot m^{-2} K^{-1}$

β bottom layer renewal efficiency

ϵ average emissivity of wall and particle

λ heat conductivity, $W \cdot m^{-1} K^{-1}$

λ_g continuum heat conductivity of the gas, $W \cdot m^{-1} K^{-1}$

ρ density, $kg \cdot m^{-3}$

σ modified mean free path of gas molecules, m

ϕ_A surface coverage factor

subscripts

i	insulation	ws	wall to surface
r	radiation	wp	wall to particle
p	particle	wg	wall to gas

7 References

1. Roy, C., Labrecque, B. and de Caumia, B. (1990) Recycling of scrap tires to oil and carbon black by vacuum pyrolysis. *Resour. Conserv. Recycl.*, Vol. 4, pp. 203-213.
2. Yang, J., Tanguy, P. A. and Roy, C. (1995) Heat transfer, mass transfer and kinetics study of the vacuum pyrolysis of a large used tire particle. *Chem. Eng. Sci.* Vol. 50, No. 12, pp. 1909-1922.
3. Perry, R.H. and Green, D. (1984) Perry's chemical engineers' handbook, sixth edition, McGraw-Hill Book Co., New York, pp. 11.48 - 11.49.
4. Labrecque, B. (1987) Etude du transfert de chaleur par radiation thermique dans un réacteur de pyrolyse sous vide des vieux pneumatiques. Mémoire de Maîtrise, Université Laval, Québec.
5. Lehmberg, J., Hehl, M. and Schugerl K. (1977) Transverse mixing and heat transfer in horizontal rotary drum reactors. *Powder Tech.*, Vol. 18, pp. 149-163.
6. Wes, G.W.J., Drinkenburg, A.A.H. and Stemerding, S. (1976) Heat transfer in a horizontal rotary drum reactor. *Powder Tech.*, Vol. 13, pp. 185-192.
7. Schlünder, E.U. (1985) Vacuum contact drying of free flowing mechanically agitated particle material. *DRYING'85*, (Eds) R. Toei and A.S. Mujumdar, Hemisphere/Springer-Verlag, New York, pp. 75-83.
8. Malhotra, K. and Mujumdar, A.S. (1989) Indirect heat transfer and drying in mechanically agitated granular beds - an annotated bibliography. *Drying Tech.*, Vol. 7, pp. 153-171.

9. Malhotra, K. and Mujumdar, A.S. (1990) Effect of particle shape on particle-surface thermal contact resistance. *J. of Chem. Eng. of Japan*, Vol. 23, pp. 510-512.

10. Weidenbaum, S.S. (1956) Mixing of solids, advances in chemical engineering, Vol. I, Academic Press, New York, pp. 209-324.

11. Cooke, H., Bridgwater, J., and Scott, A. M. (1976) Powder mixing - a literature survey. *Powder Tech.*, Vol. 15, pp. 1-20.

12. Roy, C., Blanchette, D. and de Caumia, B. (1996). Horizontal moving bed reactor. International patent pending.

13. Schlünder, E. U. (1984) Heat transfer to packed and stirred beds from immersed bodies. *Chem. Eng. Process.*, Vol. 18, pp. 31-53.

14. Waoka, N. and Kaguei, S. (1982) Heat and mass transfer in packed beds. Publisher: Gordon and Breach Science Publishers, New York.

15. Malendoma, C. (1996) Physical and mechanical parameters which influence the transport and agitation of a bed of particles. Institut Pyrovac Inc., Internal report.

DEVELOPMENT OF A SMALL INTEGRATED PILOT PLANT FOR FLASH PYROLYSIS OF BIOMASS

A.M.C. Janse, W. Prins, W.P.M. van Swaaij
Twente University, Department of Chemical Engineering, Enschede, The Netherlands

Abstract

A continuous test-unit for the flash pyrolysis of biomass to bio-oil, with a throughput of 10 kg pulverised biomass per hour, has recently been developed at the University of Twente. The apparatus is now under construction; it is built by Royal Schelde in Vlissingen, The Netherlands.

Central part is a sophisticated and compact reactor configuration including a rotating cone, two fluid beds and a riser. Sand is used as a heat carrier and continuously circulated inside the system. The main features of this novel flash pyrolysis reactor are a high degree of heat integration by internal combustion of produced char, a high biomass throughput, a high heating rate for the biomass particles, and little need for transport gas.

The design of the reactor will be discussed in this paper, and presented together with some results of design-supporting experiments.

Introduction

In the past, several reactor types have been developed for flash pyrolysis of biomass to bio-oil. Examples are the bubbling fluid bed (Scott et al., 1988), the circulating fluid bed (Graham, 1988), and the entrained flow reactor (Gorton et al., 1990). An overview of applied reactor technologies is given by Bridgwater and Bridge (1991). The products of flash pyrolysis are gas, liquids and char which are formed in ratios depending on reactor type and operating conditions. The yield of bio-oil is maximal when the heating rate of the biomass particles is high, and the residence time of the produced vapours is short. The reactors mentioned above can be operated at the appropriate conditions and have the advantage of a simple construction. A main disadvantage of these reactors is however the required amount of carrier gas, which is quite large and dilutes the produced vapour stream considerably. It makes an extra heat input necessary , and causes the need for relatively large downstream equipment. Another drawback may be the limited mixing capacity of these reactors: the possible occurrence of cold zones near the feed entry points could cause non-optimal flash pyrolysis conditions.

From 1989 to 1993 a new reactor type, called the rotating cone reactor, has been developed at the University of Twente (Wagenaar et al., 1994). This rotating cone reactor enables a high solids throughput, without requiring any transport gas (except for secondary equipment). Inside the rotating cone, biomass particles are mixed intensively with an excess of hot sand particles. The sand particles are added as an inert heat carrier, and also to prevent fouling of the cone wall.

In a prototype set-up built at the University of Twente, the sand particles were fed from a vessel positioned above the cone reactor. After a single pass through the reactor they were collected at the bottom of the oven placed around the rotating cone. In a future continuous installation, the sand particles must be trapped and re-entered to the cone reactor, for instance by applying an external riser system.

During the last few years, research at the University of Twente has been focussed on developing a considerably improved version of the rotating cone reactor. The new concept includes an internal recycling of the sand, allowing for a higher sand flow rate. In this way, the sand flow around the cone reactor is decoupled from the overall heat balance over the pilot plant. Besides, the improved reactor system is now fully heat integrated and can be operated autothermally. The energy content of the produced char is used directly by burning it inside the reactor system in a special combustion zone.

In the ensuing section, the design of the improved rotating cone reactor will be discussed extensively. The other sections report on the results of experiments concerning the hydrodynamic and heat-transfer characteristics of the novel concept.

The cold-flow performance of the reactor must be checked preceedingly to the construction of a hot test-unit to be built for continuous flash pyrolysis of pulverized wood. Meanwhile, the design of this small pilot plant installation for continuous pyrolysis of 10 kg per hour has been completed. The equipment and boiler manufacturer AKF in Goes, a daughter company of the Royal Schelde Group in Vlissingen, has offered to build the reactor system. It will be delivered next May and experimentation is expected to start in July 1996.

Design and Operation

The new cone reactor has no bottom plate. By inserting such a bottomless cone in a fluidized sand bed (the so-called cone bed), the sand particles will move upwards by centrifugal forces induced by the rotating cone along the cone wall, and fall back over the wall into the fluid bed. In this way an *internal* circulation for sand plus biomass is realized, indicated as "solids loop 1" in figure 1a, which shows a simplified picture of the pilot plant configuration. The sand particles of loop 1 are cooled while providing the energy for heating of the biomass particles and the (probably endothermic) pyrolysis reactions. To compensate for this heat loss a second fluid bed is positioned around the cone reactor and the inner fluid bed and used for the combustion of produced char. Sand and char particles can be transported from the inner cone bed to the combustor bed through an orifice, which connects the two fluid beds. The transport of reheated sand from the combustor to the top of the cone reactor is realized by means of a riser, partly submerged into the combustor bed. This circulation through the orifice and riser is called the *external* sand circulation and indicated as "solids loop 2" in figure 1a.

The energy content of the produced char is seldom used in pyrolysis processes. The char particles are usually separated from the pyrolysis vapours in a cyclone. In our new integrated concept, the energy content of the char is used to operate the installation autothermally. This implies the coupling of a reducing atmosphere (where the pyrolysis process will take place) with a fluid bed operating under oxidizing conditions. In the latter bed, the combustion of char takes place and the sand particles are reheated. A continuous exchange of sand between the pyrolysis zone and the combustor bed must be maintained. The coupling of two fluid beds operating in an oxidizing and reducing atmosphere can be realized by using the interconnected fluid bed principle (Masson, 1989, Korbee et al., 1994). This interconnected fluid bed concept has been simplified for application in our pilot plant. Flow of solids between the two fluidized beds through an orifice is induced by difference in bed density and height.

The connection of two fluid beds, one at reducing and the other at oxidizing conditions, is only applicable under the following stringent condition. To assure a safe operation, the pressure in the cone fluid bed (at orifice height) must always be higher than the pressure in the combustor bed. In this way, the percolation of air from the combustor to the flammable reducing atmosphere of the cone bed will be avoided. In a commercial installation, the cone fluid bed will preferably be fluidized by the non-condensible gases produced in the pyrolysis process (CO_2, CO, H_2, CH_4), and the combustor bed by air. To avoid the risk of explosion and prevent the formation of a local hot spot near the orifice, the exchange of gas between the cone bed to the combustor must be minimized (the leakage must be smaller than a few percent of the incoming cone gas flow). In our laboratory set-up the cone fluid bed will be fluidized with nitrogen although in a commercial installation it is certainly possible to recycle the non-condensible gases to the cone bed or the combustor.

The mini plant will be operated in a temperature range from 500 to 600 °C, where the bio-oil yield is expected to be maximal (Bridgwater and Bridge, 1991). The operating pressure is slightly above atmospheric. A heat balance over the two beds, based on a throughput of 10 kg wood per hour, revealed that a minimum external sand circulation rate of 108 kg/hr is required to realize an acceptable temperature difference of 180 °C. A riser with a diameter of 2 cm is large enough to satisfy this demand. A low temperature difference corresponds with a high external sand circulation rate. This situation requires large amounts of carrier gas to operate the riser, which is energetically unattractive. On the other hand, however, a large temperature difference can lead to excessive cracking of the bio-oil. Figure 1b shows a 3D-view of the pilot plant which main dimensions are presented in table 1.

Table 1: Characteristic dimensions of the continuous pyrolysis pilot plant

cone[1] top diameter (m)	0.32
cone[1] top angle (°)	60
cone[1] height (m)	0.22
outer diameter combustor bed (m)	0.39
outer diameter cone fluid bed (m)	0.16
inner diameter cone fluid bed (m)	0.09
total height combustor bed (m)	0.40

[1] : rotating cone

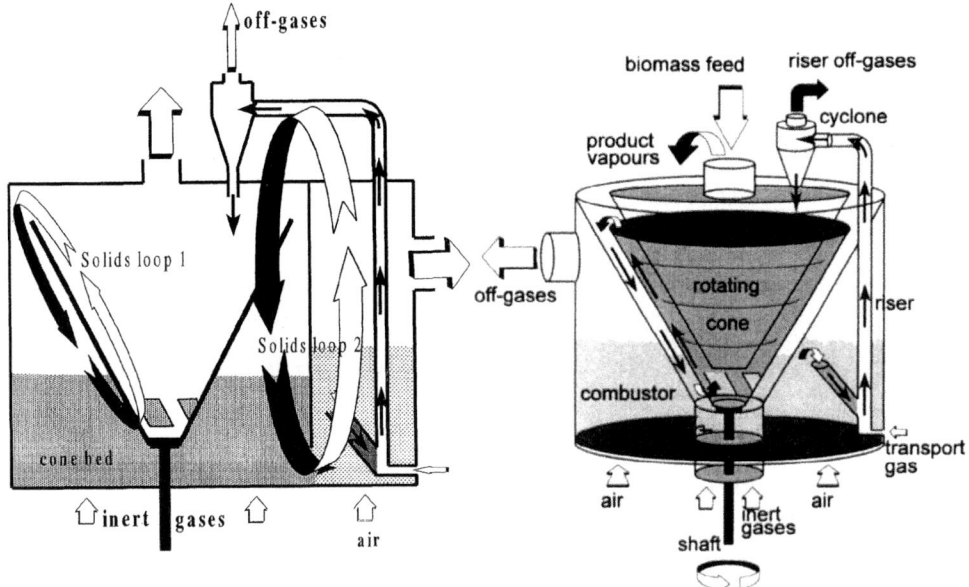

Figure 1a and b: Scheme (left) and 3D-view (right) of the continuous pyrolysis pilot plant

To assure stable and safe operation with a high throughput of wood, the pilot plant should be operated at the following conditions:
- large sand flow through the cone reactor (typically 20 kg/hr sand is required for 1 kg/hr biomass throughput).
- short residence time of the biomass particles and the produced bio-oil vapours (less than 1 s)
- complete coverage of the cone wall with sand particles
- high rate of heat transfer to the biomass particles
- small gas leakage between the two fluid beds (less than a few percent)
- stable external sand circulation

The experimental work, which is presented below, focusses on the determination of the above mentioned characteristic parameters.

Hydrodynamics

Cold flow experiments have been carried out in a special set-up (without a surrounding combustor bed) to determine the sand transport capacity and the residence time of particles inside the cone. A schematic presentation of the set-up is visualized in the left picture, while the right picture shows a top view of the cone. The main dimensions are different from the cone reactor of the future pilot plant. The cone top diameter is larger (0.68 m instead of 0.3 m) and the cone angle is 90° instead of 60°.

For this type of experimental work it is desirable to use a cone which is easily accessible from the top. Therefore, the cone is provided with a closed bottom to which the rotating shaft is connected. In this special case, the sand is forced into the rotating cone reactor from the surrounding fluidized sand bed through four large holes in the bottom part of the cone wall. A problem with this design is the sealing of the rotating shaft from the solids-gas mixture: a quite complex labyrinth, flushed with an inert gas, is applied for this sealing.

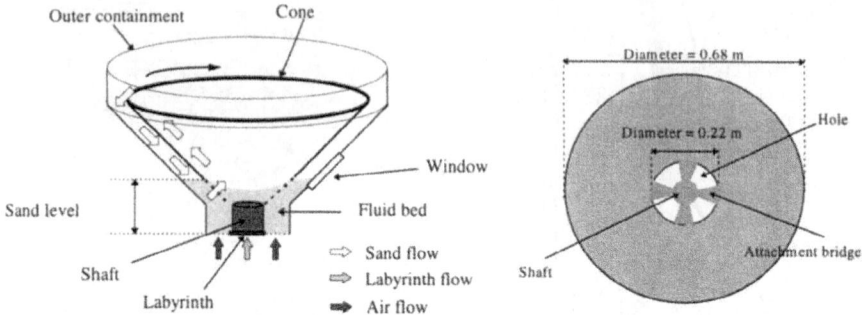

Figure 2: Schematic diagram (left) of the experimental set-up and the top of the cone (right)

In the cold flow set-up described here, the sand transport rate through the cone could be determined by collecting a well defined part of the sand flow over the top during a certain fixed time period (see figure 3).

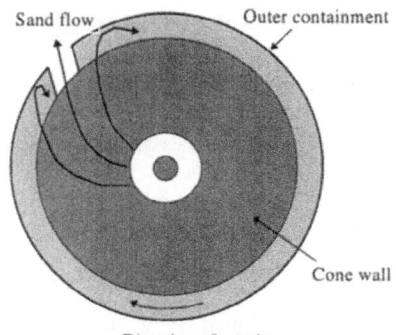

Figure 3: Measurement of the sand transport capacity

Preliminary investigations had already shown that the parameters determining the sand transport capacity are, besides the cone dimensions, the cone rotational frequency, the fluidizing gas velocity and the sand hold-up in the system.

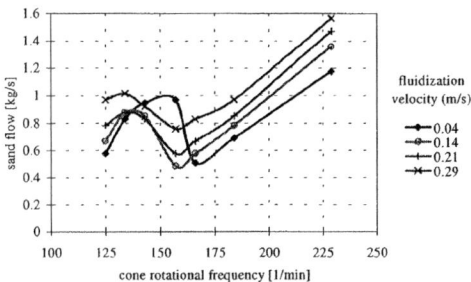

Figure 4: Sand transport capacity as function of cone rotational frequency (sand particle diameter: 200-500μm, minimum fluidization velocity: 0.08 m/s)

A superficial look at figure 4 shows that, regarding the size of the equipment, very high sand transport capacities (up to 1.5 kg/s) are possible. A second observation is that an increasing fluidization velocity leads to higher sand flows. Somewhat surprising is the course of the curves in figure 4. All of these curves pass through a maximum and a subsequent minimum, and finally end as inclining lines with approximately the same constant slope. As a possible explanation, it is likely that the attachment bridges (see figure 2) will induce complex flow behaviour. It is clear that sufficiently high transport capacities can be expected in the cone reactor of the pyrolysis plant which is now being built. Values of around 1 kg/s (corresponding to the maxima in figure 4) could be reached if the influence of a smaller top angle remains limited.

The residence time of particles in the cone has been determined by following fluorescent particles with a videocamera. Typical residence times, measured at a low gas fluidizing velocity are shown in figure 5 as a function of the cone rotational frequency. Experiments revealed a minor influence of the gas velocity or the hold-up on the particle residence time. An increase of the fluidizing gas velocity or hold-up, enhances the sand transport capacity, but only through an increased thickness of the sand layer, and not by a higher particle velocity.

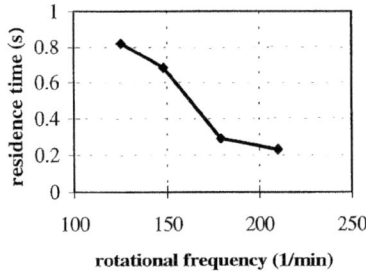

Figure 5: Typical residence times as function of cone rotation frequency (fluidizing gas velocity 0.04m/s)

For single pass conversion, the required residence time of a pyrolyzing wood particle is determined by its smallest dimension. Modelling of the flash pyrolysis of a single spherical wood particle indicates that a residence time of 1 s is sufficient for 95% conversion of a particle with a diameter of 300 μm at 600 °C. According to this conservative estimate it seems likely that such biomass particles can be largely pyrolysed in a single pass through the reactor (at 600 °C).

Heat transfer

In case of a coarse biomass feedstock (particle diameter > 500 μm) the flash pyrolysis process may be limited to a large extent by the external heat transfer process. Prediction of the reactor performance (in terms of biomass conversion) for the external heat transfer controlled conditions is only possible if the related heat transfer coefficient α is known. Therefore, experiments have been carried out to determine this heat transfer parameter for particles moving upwards within a sand flow along the rotating cone wall. They have been measured by the contactless fluoroptic technique, recently developed by Wagenaar (Wagenaar et al., 1995). Small samples of fluorescent particles (d_p is 106-212 μm, ρ_p 6000 kg/m³, c_p 318 J/kg/K) with a temperature of 300 °C were fed near the bottom of the (cold) cone wall (with the dimensions as depicted in figure 2) along which cold sand is circulated. The temperature of the cooling particles was recorded as a function of the radial coordinate. It is possible to calculate the external heat transfer coefficient from the particle temperature drop between two radial positions and the belonging time increment. Table 2 shows the determined external heat transfer coefficients, as a function of the rotation frequency.

Table 2: External heat transfer coefficients[1]

Rotational frequency (1/min)	Heat transfer coefficient (W/m²/K)
180	1154
226	982
273	723
317	619

[1] sand particle diameter: 200-500 μm; hold-up: sufficient for complete coverage of cone wall

The particle Reynolds number (and therefore Nusselt number) is not included in this table because the slip velocity between the particle and the gas phase cannot be determined properly. Table 2 shows a decreasing external heat transfer coefficient with an increasing rotational speed. Probably this can be ascribed to an increase in the contact efficiency between the hot fluoroptic particles and the cold sand particles at a lower rotation speed. The sand particles will then flow quietly upwards through a thick stable sand layer on which the fluorescent particles are observed to float. With an increase in the speed, the cold sand layer becomes more diffuse (c.q. less dense). From fluid bed heat transfer literature (Botterill, 1975) it is known that the amount of particles per unit volume is a crucial factor. A decrease of the contact frequency and/or the number of contact points per unit volume will immediately result in lower values of the heat transfer coefficient. This effect of a lower contact efficiency, however, may be partly counterbalanced by an increase in the slip velocity between solids and gas at higher porosities of the solids/gas mixture.

In comparison with fluid bed heat transfer literature data (for example Kunii and Levenspiel, 1991) the rotating cone reactor showed good heat transfer characteristics. Typically values for the heat transfer coefficient in a bubbling fluid bed for (sand) particles with a diameter of 300 μm are 300-600 $W/m^2/K$, while the average sand particle diameter used here was 390 μm.

Finally, preliminary calculations indicated that the observed heat transfer values are sufficiently high for single pass conversion of biomass particles with a characteristic dimension of less than 300 μm. But this is subject for future theoretical work.

External sand recirculation

To determine the sand transport capacity of the external sand circulation (solids loop 1 in figure 1), and the gas leakage from the cone bed to the combustor bed, cold flow experiments have been carried out in a precise copy of the miniplant to be operated later. The cone used in the miniplant and for the experiments described in this part, is smaller than the one applied for the studies of the preceding sections; its dimensions are given by a top diameter of 0.3 m and a cone angle of 60 °. Preliminary tests showed again that high sand transport rates through the cone can be achieved. Also, the cone wall appeared to be always covered completely with a dense sand layer.

Three different riser diameters have been considered in the sand transport capacity measurements, viz. 1, 2 and 3 cm. A riser diameter of 2 cm was sufficient for the demand of 30 g/s (108 kg/hr). The maximum throughput appeared to be more than 40 g/s (144 kg/hr), and could be influenced by adjusting both the gas flow and the pressure drop over the standpipe. The limiting step in the external sand circulation is the sand throughput of the riser and not the sand flow through the orifice.

After having established that the sand throughput is always sufficient to enable steady operation, the attention was focussed on the leakage of gas through the orifice and the corresponding pressure difference. This leakage consists of two contributions:

1: an interstitial part, consisting of gas, dragged with the sand particles through the orifice without slip. This contribution is a function of the mass-flow rate and porosity of the solids-gas mixture flowing through the orifice

2: a percolation part caused by the pressure difference over the orifice (Kunii and Levenspiel, 1991)

In the experiments, the cone-bed gas flow has been varied to investigate its influence on the gas leakage, while the gas flow through the combustor bed and the riser were kept constant. Four different orifice diameters have been considered: 1, 2, 3 and 4 cm. The leakage (expressed in m^3/hr) has been determined by "tracer" gas (CO_2) analysis. This tracer gas was added to the cone bed feed gas and its concentration was measured at the outlet of the cone-fluid bed, riser and combustor. Figure 6 shows the influence of the gas flow through the cone bed on the sand circulation rate and gas leakage for a constant combustor gas flow (1.77 #U_{mf}, while U_{mf} was determined on 0.08 m/s).

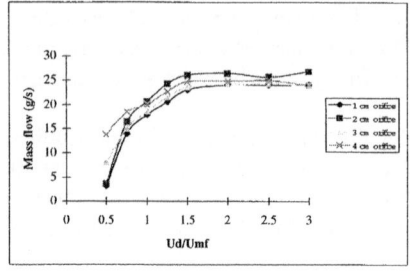 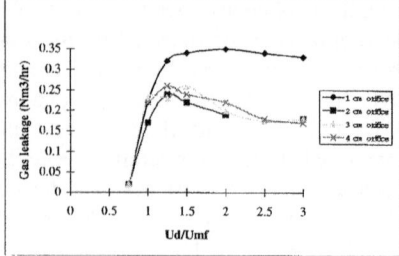

Figure 6: Absolute gas leakage and orifice pressure drop as function of the cone gas flow (gas flow rate through combustor is 1.77 U_{mf})

The sand recirculation rate is enhanced with increasing cone-bed gasflow. The reason for this, at first instance, quite unexpected behaviour, is the change in hold-up distribution between the two fluid beds. As the cone-bed is better fluidized, the hold-up of the cone bed will decrease. As a consequence, the combustor bed hold-up will increase, which will cause a higher pressure drop over the standpipe of the recirculation riser, inducing a higher mass flow through the riser (at constant riser gas flow). Varying the orifice diameter, not much influence is observed on the mass flow. Because of the large size of the combustor bed with respect to the cone bed, a severe change in the cone-bed hold-up will induce only slight change in the combustor bed hold-up.

At an orifice diameter of 1 cm the gas leakage reaches the highest values. While the pressure in the combustor bed at the level of the orifice is kept constant, the pressure in the cone bed increases until the minimum fluidization velocity is reached. Beyond that point, the cone bed pressure at the height of the orifice remains also nearly constant.

The interstitial part of the gas leak will follow the trend of the mass flow (to approximately a few percent of the cone-bed gas flow) and the percolation part will increase until a constant value of the pressure drop is achieved. The experiments show however that an increase in the cone-bed gas flow beyond U_{mf} induces a slight decrease in the gas leakage (and pressure drop). Probably, this is due to decreasing friction forces between particles near the orifice.

The orifice with a diameter of 1 cm shows the highest value for the gas leakage. While the mass flow at constant gas flows at varying orifice diameter is nearly constant, the orifice with the smallest diameter will exhibit the highest mass flux. A high mass flux corresponds with a high required pressure drop and consequently a high (percolation) gas leakage.

The pressure in the cone fluid bed must always be higher than the pressure in the combustor bed to avoid the inflow of oxidizing gases into the reducing atmosphere of the cone fluid bed. The experiments have shown that as long as a sand circulation is maintained, this condition is always fulfilled.

The pilot plant to be operated at high temperatures is currently under construction and will be started up in spring 1996.

Conclusions

A continuous pilot plant for the flash pyrolysis of biomass is being developed. It is intended to integrate the pyrolysis process of biomass to bio-oil with the combustion process of the by-product (char), to enable autothermal operation.

With the novel rotating cone reactor, high sand throughputs can be realized with minor amounts of carrier gas. The residence time of particles is short, and ranges from 0.2 to 0.8 second. The external heat transfer coefficients are high, in the order of 500-1000 $W/m^2/K$.

A high sand circulation rate between the rotating cone fluid bed and the combustor can be maintained without much leakage of gas between the two reactor beds. During operation the pressure in the cone fluid bed will always be higher than the pressure in the combustor bed, which is a safe situation with respect to the direction of the gas leakage.

Acknowledgments

This investigation was supported by the CEC-AIR programme (contract: AIR II-CT93-0889). We also acknowledge H.Weerdenburg, X. de Jong, E.J. Ransdorp, M. Biesheuvel and J. Nijmeijer for their assistence in the experimental work

Literature

1. Bridgwater, A.V., Bridge, S.A. (1991), A Review of Biomass Pyrolysis and Pyrolysis Technologies, In: *Biomass Pyrolysis Liquids Upgrading and Utilisation*, Elsevier Science Publishing Co., New York

2. Botterill, J.S.M. (1975), *Fluid bed Heat Transfer*, Academic Press, London

3. Gorton, C.W., Kovac, R.J., Knight, J.A., Nygaard, T.I. (1990), Modelling Pyrolysis Oil Production in an Entrained Flow Reactor, *Biomass*, Vol. 21, 1

4. Graham, R.G., Freel, B.A., Bergougnou, M.A. (1988), The Production of Pyrolytic Liquids, Gas and Char from Wood and Cellulose by Fast Pyrolysis, Proceedings in: *Research in Thermochemical Biomass Conversion*, Eds. Bridgwater, A.V., Kuester, J.L., Elsevier Applied Science, NewYork, 629

5. Korbee, R., Snip, O.C., Schouten, J.C., van den Bleek, C.M. (1994), Rate of Solids and Gas Transfer Via an Orifice Between Partially and Completely Fluidized Beds, *Chem. Eng. Sci.*, Vol. 49, pp. 5819

6. Kunii, D., Levenspiel, O. (1991), *Fluidization Engineering*, 2nd ed., (Edited by H.Brenner), Butterworth-Heinemann, Boston

7. Masson, H.A. (1989), A Twin Fluid Bed Pyrolyser Combustor System, *Fluidization VI*, Banff, Canada, pp. 383

8. Scott, D.S., Piskorz, J., Bergougnou, M.A., Graham, R., Overend, R.P. (1988), The Role of Temperature in the Fast Pyrolysis of Cellulose and Wood, *Ind. Eng. Chem. Res.*, Vol. 27, 8

9. Wagenaar, B.M., Prins, W., van Swaaij, W.P.M. (1994), Pyrolysis of Biomass in the Rotating Cone Reactor: Modelling and Experimental Justification, *Chem. Eng. Sci.*, Vol. 49, pp. 5109

10. Wagenaar, B.M., Meijer, R., Kuipers, J.A.M., van Swaaij, W.P.M. (1995), A Novel Method for Noncontact Measurement of Particle Temperatures, *A.I.Ch.E.J.*, Vol. 41, pp. 773

PLANS FOR THE PRODUCTION AND UTILIZATION OF BIO-OIL FROM BIOMASS FAST PYROLYSIS

G.TREBBI, C.ROSSI and G.PEDRELLI
ENEL-Thermal Research Centre, Pisa, I

Abstract

Within the framework of the applied research carried out by ENEL, aimed at utilising non-traditional fuels derived from the renewable resources of the territory, the Thermal Research Centre of Pisa has realized a R&D plant for bio-oil production, through a flash pyrolysis process of vegetable biomass. The Project, that is being carried out in close collaboration with the Umbria Region, is partially financed by EU. The plant has been erected at the ENEL thermoelectric power plant of Bastardo (Perugia). The process unit (RTP-Rapid Thermal Processing) has been developed by the Canadian Ensyn Company, whereas ENEL has designed and constructed the auxiliary facilities. The plant, with a capacity of processing 15 tonnes/day of dry feedstock and capable of producing around 10 tonnes/day of liquid fuel, is the largest plant of this type in Europe. The paper describes the technical characteristics of the plant and the aims of the tests that will be performed with hardwood sawdust and other sorts of feedstocks. The tests are finalized to assess the technical-economic fesibility of the flash-pyrolysis process, a step necessary for the commercialisation of the technology.

Keywords: Fast pyrolysis, bio-oil

1 Introduction

Biomass is a renewable resource spread in the territory, generally with a high water content and a very low mass and energy density. These characteristics affect its use for energy recovery, in particular as a consequence of the transportation costs. The combustion of biomass as received has been and is yet the simplest energy conversion process, but unfortunately it exhibits technological limits and low efficiencies, that have induced researchers to find out more efficient systems. In general these new processes require a proper pretreatment of biomass, in order to convert it into a more valuable fuel, that can be handled and utilized more easily. As a matter of fact, the biomass conversion, accomplished through proper thermochemical processes, into liquid, solid and gaseous fuels, characterised by higher specific energy, easier storage and handling, with the possible utilisation in energy conversion plants at higher efficiency, permits to overcome the limits posed by direct combustion. In particular, the most advanced biomass pyrolysis processes, oriented to the production of an organic liquid fuel (referred to as bio-oil, bio-fuel-oil or bio-crude-oil), appear to be very interesting for the several possible energy applications that can be envisaged for this special fuel.

The progress in this field allowed to pass from the traditional process of slow pyrolysis for the charcoal production to the most advanced fast pyrolysis processes, characterized by high yields production of liquid bio-fuel. Also if the operating conditions of these processes are not extremely severe as regards pressure and temperature, the high heat transfer rate to the biomass and the as much quick cooling (quenching) of the pyrolytic vapours permit to reach yield conversion of the order of 70 wt% on dry feed. The European technology is not so advanced in this field as in Canada and North America, but in any case to reach the commercialization of this technology still requires a lot of R&D activities, both as regards the bio-oil production process and the utilization of bio-fuel in plants for electricity production, such as boilers, steam and gas turbines, diesel engines etc.

2 The biomass pyrolysis in the Umbria Region

"Agricultural overproduction of food biomass within the EU framework" and "Renewable energy resources" are two subjects of great importance, to which EU is devoting a lot of human and financial resources. In Italy, Umbria Region is in particular actively interested in this sector, through its regional Agency ARUSIA, in charge of the agricultural development and innovation. Also the Research Department of ENEL, that operates in the field of thermoelectric production through the Thermal Research Centre of Pisa, is actively involved in the biomass energy utilization.

In agreement with the EU's trend to foster the use of renewable resources, not only in view of the benefits related to the saving of fossil fuels and to the reduction of petroleum-derived products import, ENEL is participating, in close collaboration with ARUSIA, in a Project for the construction and operation of a flash pyrolysis R&D plant. The Project, which is partially financed by EU, consists of two separated phases:

1) biomass production, chipping, drying and transportation;
   cost evaluation and energy consumption assessment of the treatment steps necessary to produce feedstock with the suitable characteristics required for the conversion into bio-oil;
2) thermochemical biomass conversion into bio-oil, by means of a large scale demonstration plant, based on a fast pyrolysis process, with the aim of producing significant amounts of bio-oil for technical and economic assessment of the technology.

The feedstock considered is hardwood sawdust, but there is the possibility of testing also different sorts of energy crops, such as robinia pseudo acacia, according to the new technique of "rapid rotation of crops", sweet and fiber sorghum, etc. The feedstock drying necessary to reach the low moisture content required by the pyrolyser, will be carried out in solar driers developed by ARUSIA.

The crude bio-oil will be directly utilized for experimental research activities in plants for production of heat and/or electricity or will be sent to experimental upgrading plants, in order to improve its chemical-physical characteristics for particular applications.

The costruction of the plant, that is the largest in Europe, is performed in the framework of bioenergy biomass utilization, i.e. an alternative to the biochemical sector, aimed at the production of valuable chemicals, such as resins, dyeing products, vitamins, fertilizers etc, as well as as fuels. Whereas in Canada and in USA the fast pyrolysis processes have been first developed just for the production of chemicals, the different energy situation of Europe favours the use of bio-oil as a fuel.

3 The Bastardo (Perugia) pyrolysis plant

3.1 Introductory remarks

The pyrolysis plant consists of the pyrolyser unit, supplied by Ensyn Technology Inc. All the auxiliary facilities necessary to complete the plant, have been designed and constructed by ENEL. The choice of the supplier has been suggested and recommended by the EU's experts, because at the present time Ensyn is the owner of the most advanced technology in this field. [1] The pyrolyser is based on the so called Rapid Thermal Process (RTP), partially patented.

The plant is capable of processing 15 tonnes per day of dry feedstock, corresponding to a production of 10 tonnes per day of raw bio-oil. The plant is located in Umbria Region, in a dedicated area inside the ENEL thermoelectric power plant of Bastardo, Province of Perugia, with the possibility of utilizing the power station auxiliary utilities. The biomass necessary for tests is supplied by ARUSIA, according to a formal agreement signed by the partners, that regulates the plant operation not only during the first experimental phase in order to meet the EU contract duties, but also for the future.

3.2 Description of the process

The RTP process is based on the principle of very fast pyrolysis, capable of producing the thermal degradation of biomass in an environment with no oxidizing agents, at atmospheric pressure and with a moderately high temperature.

With a very quick heat transfer to the biomass, carried out in less than 1 second, it is possible to reach yields of bio-oil around 70 wt%. It is important the choice of a right residence time, because a very short residence time results in incomplete depolymerisation of the lignin and in less homogeneous liquid product, while longer residence times can cause secondary cracking of the primary products, reducing yield and adversely affecting bio-oil properties. The quenching of the pyrolysis vapours must be performed very quickly, so that the decomposition products of the feedstock can maintain the typical molecular structure of a liquid, by preventing the possibility of reacting to form gaseous products in equilibrium.

The schematic of the process is shown in Fig.1. The feedstock, with a maximum size of 6 mm and moisture content of 8%, is transported to the plant via trucks, from which it is pneumatically transferred to a storage silo; the biomass is then sent to a surge bin, with

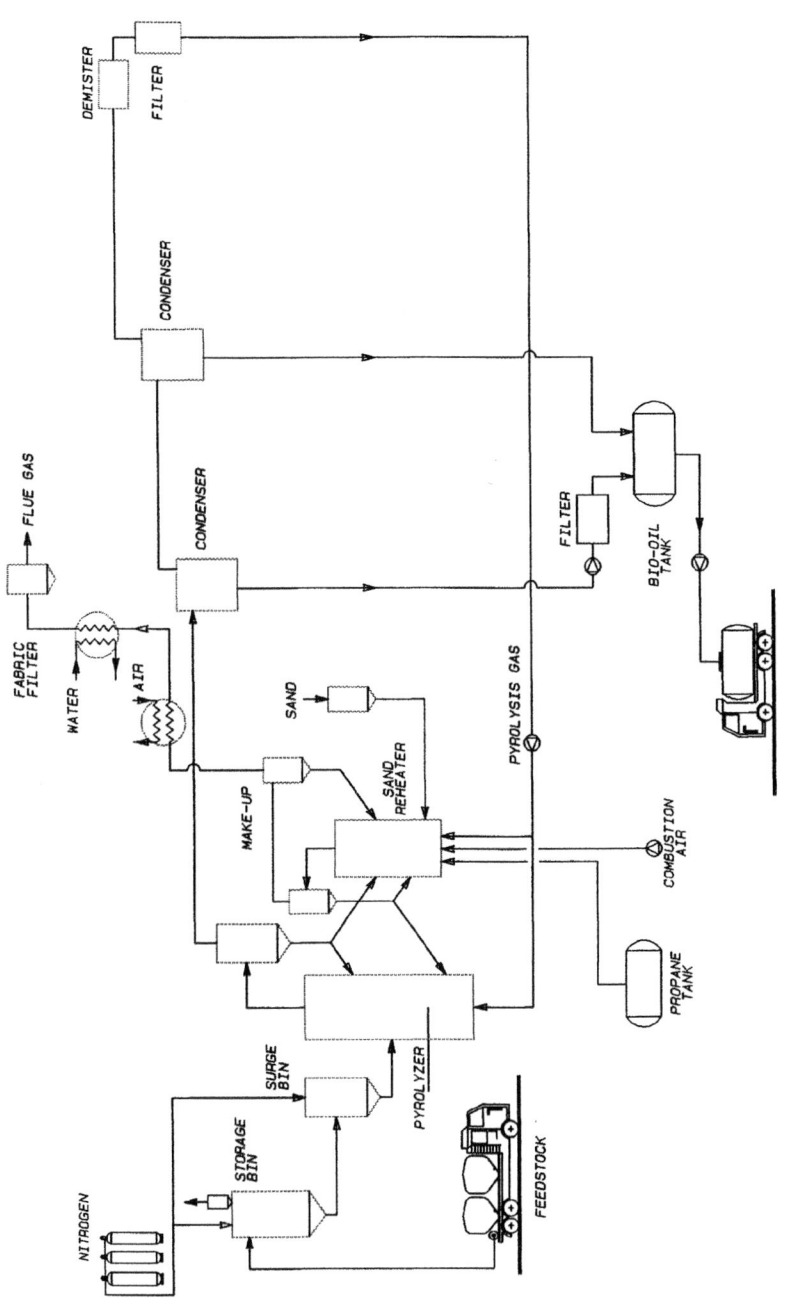

FIG. 1   FLOW SHEET OF THE ENSYN RTP FAST PYROLYSIS PLANT

a capacity sufficient for 5 hours of operation, and afterwards to a feed bin equipped with a variable speed screw feeder, that permits continuous biomass feeding to the pyrolysis reactor. The biomass feeding system is blanketted with nitrogen, in order to prevent possible explosions.

The pyrolysis process occurs inside the pyrolyser, which is a circulating fluidized bed, where the biomass enters radially and is put in contact with the heat carriers, namely hot sand and recirculated pyrolysis vapours. The pyrolysis reactor is divided into three main zones:

> a rapid mixing zone
> a transportation zone
> a separation zone

The conversion of the solid feedstock to a raw product vapour begins in the mixing zone and continues through the other two zones. The separation of the solids, namely sand and char, entrained by the gaseous stream coming out from the pyrolyser, is accomplished by means of a short residence time high-efficiency cyclone. The operating temperature inside the pyrolyser is 525°C. The sand is reheated in a direct contact heat exchanger, where the heat produced by the combustion of char and a fraction of pyrolysis gas is utilised. The temperature inside the sand reheater is around 760 °C. The char in the flue gas of the sand reheater is removed with high efficiency cyclone and fines are returned to the reheater. A suitable valves system permits to recirculate the solid products from the mentioned cyclones both to the pyrolysis reactor and to the sand reheater. It is also possible to recirculate sand and char in the pyrolysis reactor. The pyrolyser is a closed system, where a fraction of the pyrolysis gas is used as transportation fluid.

The gaseous stream from the pyrolyser cyclone, made up of condensible and non-condensible products and with a temperature of around 500 °C, is sent to the condensation sections, which consists of two separate columns.

The condensed liquid collected at the bottom of the two columns is the useful product that is sent to the storage tanks. The fraction of bio-oil coming from the first column, around 70 wt% of the total production, needs to be filtered, before being mixed with the liquid fraction drawn from the bottom of the second condenser.

The gas from the first column is sent to the second condenser and after a slight heating up, the gaseous stream is sent to a demister and afterwards to a filter located upstream of the blower that sends it in part to the pyrolyser and partly to the sand reheater. Before the immission into the atmosphere, flue gas from sand reheater are cleaned by a filter-bag.

The effluents from the plant consist of a small flow of ash mixed with sand. One of the most important feature of the plant is the possibility of utilizing the gaseous and solid pyrolytic products inside the process, with the advantage of reducing the disposal problem of waste. For start up it is necessary to provide some bio-oil for the condensing columns and propane for heating up the system; afterwards the plant is self sustaining and fully automated.

3.3 Design data

3.3.1 Functional warranties

The plant has been designed for a reference biomass, namely hardwood sawdust, for which Ensyn guarantees the performances indicated in the contract. In a ny case it will be possible to extend the experimental activity to other sorts of feedstocks with suitable modifications to the plant, in particular to the the biomass feeding system.

The performances guaranteed by the supplier with the reference biomass characterized by a maximum size of 6 mm and maximum moisture of 8%, are:

dry biomass processed	625 kg/h
bio-oil yield production	70 wt%
min HHV	15 MJ/kg

The plant is designed for continuous operation, on the basis of 330 days per year (7920) hours. It will be also possible to run the plant at intervals, both for the set up and the experimental research needs.

3.3.2 Energy balance

An overall accurate energy balance of the plant will be performed during the experimental campaigns; in any case the typical mass and energy balance of RTP plant anticipated by the supplier indicates an energy efficiency of 64%, defined as the ratio between the energy content of bio-oil produced and the energy input from biomass.
The major heat losses from the RTP process are from bio-oil condensation and from the flue gas coming out from the reheater.

3.4 Plant construction characteristics

The RTP plant supplied by Ensyn is mounted in modular skids, preassembled at the factory, that can be easily transported and reassembled on site.
The facilities prepared by ENEL are te follwing:

- compressed air system
- nitrogen supply for biomass silo blanketing
- biomass feeding system from feedstock arrival to the surge bin supplied by Ensyn
- cooling system of water for bio-oil refrigeration
- bio-oil storage tanks
- civil works and RTP building

The RTP plant is in fact lodged inside an indusrial building, which is suitably ventilated in order to avoid collection of dangerous vapours. The control room has been placed inside an existing building close to the plant area, where the offices for personnel have also been located. The major part of the components has been realized with stainless steel, in order to take into account the particular corrosivity of bio-oil, from its acidity. Particular attention has been devoted to the safety aspects, according to the severe standards in force in Italy.

3.5 Bio-oil characteristics

Bio-oil produced through the hardwood flash pyrolysis is an "off-spec fuel" completely different from petroleum-derived oils.

Its chemical, physical and thermogravimetric analyses show the very complex nature of this combustible product, which is characterized by high heterogeneity, due to the presence of light and heavy fractions spread over a wide range.

Bio-oil complexity arises from degradation of lignin, cellulose and hemicellulose, resulting in a broad spectrum of oxygenated compounds. The major components of bio-oil are depolymerised lignin, carbonyls, such as aldheydes and ketones and water. Other oxygenated constituens include carboxylic acids, carbohydrates and alcohols. Bio oil has an average water content of around 23% and exhibits a specific gravity higher than that of water. A typical composition of bio oil produced through the RTP process is shown in Table 1.

TABLE 1 CHARACTERISTICS OF TYPICAL RTP BIO-OIL

Bio-fuel properties	Range of values	Typical values
Moisture (%)	15-31	23
Specific gravity	1.15-1.25	1.20
HHV (MJ/kg)	15-18	17.5
LHV (MJ/kg)		16.2
Viscosity at 40 °C (cSt)	35-53	40
Acidity (pH)	2.8-3.8	3.2
Elemental (%)		
Carbon	51.5-58.3	54.5
Hydrogen	5.5-6.8	6.4
Nitrogen	0.07-.0.40	0.2
Sulfur	0.00-0.07	0.0005
Oxygen (by difference)	34.4-42.9	38.9
Ash (%)	0.13-0.21	0.16

Bio-oil composition is determined extrinsically by the feedstock composition and intrinsically by the temperature, rate of reaction, vapour residence time and temperature, rate of reaction, vapour residence time and temperature-time cooling and quenching process which control the extent of secundary reactions. Bio-oil is polar, therefore it is not miscible with solvents such as toluene and benzene, whereas the solvents that can be used are acetone, methanol and ethanol. Viscosity shows steep variations vs. temperature and is significantly affected by by the moisture content. The several experimental activities reported in the technical literature and the experiments carried out by ENEL itself on bio-oils similar to bio-fuel that will be produced at Bastardo plant, have demonstrated the burnability of this fuel, but at the same time have pointed out the problems posed by its use, as a consequence of acidity, corrosion potential, instability to polymerisation that can occur also at ambient temperature, viscosity and other rheological properties variation with the time etc.

Bio-oil has a number of special features and characteristics which require consideration in any application, including production, storage, transport, upgrading and utilisation. All these aspects need further accurate investigations, in view of wider use of bio-oil for energy applications. Just the analytical characterization of bio-oil poses some difficulties, due not only to its complex nature, but also to a lack of standard analytical methods; just the in depth investigation of these subjects could justify a wide research activity. Even the simple measurement of moisture content is not so easy as for conventional mineral fuel oil. RTP bio-oil contains a significant quantity of light volatile components with boiling point below that of water. Therefore simple distillation produced by oven-drying removes these organic volatiles as well as the water, giving an erroneous overestimated value for apparent water content. Drying operation at 105 °C can remove as much as 20 wt% of non-water organic volatiles.

The presence of a solid phase of char fines suspended in the bio-oil and of alkalis, which in particular are concentrated in the solid phase, from which they can leach, into the liquid, can pose severe limits to some bio-oil applications, such as gas'turbine and diesel engines.

Nevertheless the characteristics of raw bio-oil can be improved through a further upgrading process, based on hydrogenation at high pressure and temperature aimed at producing hydrocarbons with a lower oxygen content, or in zeolites cracking to high aromatic product with suitable catalysts. Unfortunately these upgrading processes are generally complex and expensive, so that production cost is nearly doubled.

For that reason the trend is to directly utilize crude bio-oil, or to improve its characteristics through modifications at the pyrolysis process level. A possible solution could be for example the hot filtration of pyrolytic vapours at a temperature higher than the dew point, upstream of the condensation stage.[2]

4 Experimental activity

4.1 Scope of tests

The general aim of the tests is to assess the RTP pyrolysis process in terms of reliability, flexibility and quality of bio-oil with respect to different sort of feedstocks.

Another important goal is the economic evaluation of this technology, to evaluate the potential penetration into the market.

It is also of paramount importance to collect experimental data concerning the plant operation conditions, in order to make further process improvements, taking into account that this advanced technology offers margins for a proper optimization in terms of bio-oil quality, energy efficiency of the process and environmental impact.

Another goal is to produce significant quantities of bio-oil at relatively low cost, in order to carry out exhaustive experimental activities, necessary for investigating several possible applications of bio-fuels.

These goals will be reached through accurate mass and energy balances, thermodynamic analysis, chemical-physical characterization of the feedstock and by measuring the pollutant emissions into the atmosphere.

The experimental campaigns at Bastardo plant will be carried out on the basis of the direct experience that ENEL has gained in the field of bio-oil laboratory characterization and through combustion tests of bio-oil in a 0.5 MWt furnace and bio-oil/ethanol mixtures in a small 40 kWe gas turbine, whose results have been already reported in previous papers.[3,4,5]

4.2 Factory acceptance test

According to the supply conditions, the RTP plant had to undergo an acceptance test at the supplier's factory, that will be followed by a final acceptance test, to be performed on-site at Bastardo power plant. The first acceptance test was carried out in the period August to September 1995. Personnel of ENEL and ARUSIA has attended the test, whose aim was to verify the proper operation of all plant components, with particular attention to the reliability of the process. The tests have been executed on the basis of testing procedures carefully prepared by ENEL and discussed with Ensyn.
The pyrolysis unit performed sufficiently well, so that the final approval was given.

4.3 Test time schedule

After the factory acceptance test, the RTP has been disassembled and shipped to Italy, where it has arrived at the end of November 1995. The works of reassembling started in December 1995, so that the plant will be ready for the final on-site acceptance test in the first half of 1996. Taking into account the general plant complexity as regards in particular handling and transportation of solids, a proper set up of about 3 months period will be necessy before being able to continuously run the plant for 24 hours per day for carrying out the planned experimental campaigns.

5 References

1.   Graham R.G, Freel B. and Bergougnou M.A. *Rapid Thermal Processing (RTP): Biomass fast pyrolysis overview R&D.* Contractors Meeting on Biomass Liquefaction, Ottawa 23-24 October 1990.

2.   Diebold J.P., Czernik S., Scahill J.W., Phillips S.D. and Feik C.J. *Hot-gas filtration to remove char from pyrolysis vapors produced in the Vortex reactor at NREL.* Proceedings of Biomass Pyrolysis Oil Properties and Combustion Meeting, Estes Park, Colorado, September 26-28 1994, pp.90-109.

3.   Rossi C. *Bio-oil combustion tests at ENEL* Proceedings of Biomass Pyrolysis Oil Properties and Combustion Meeting, Estes Park, Colorado, September 26-28 1994, pp.321-328.

4.   Ardy P.L., Barbucci P., Benelli G., Rossi C. and Zanforlin S. *Development of gas turbine combustor fed with bio-fuel oil.* Second Biomass Conference of the Americas. Portland, Oregon, USA 21-24 August 1995.pp. 429-438.

5.   Barbucci P., CostanziF., Ligasacchi S., Mosti A. and Rossi, C. *Bio-fuel-oil combustion in a 0.5 MW furnace.* Poster Session of the Second Biomass Conference of the Americas. Portland, Oregon, USA 21-24 August 1995. pp.1110-1120.

PYROLYSIS

Analysis and characterisation of pyrolysis liquids

PROPERTIES OF FAST PYROLYSIS LIQUIDS: STATUS OF TEST METHODS

Characterisation of fast pyrolysis liquids

D. MEIER
Federal Research Centre for Forestry and Forest Products, Hamburg, Institute of Wood Chemistry and Technology of Wood, Germany
A. OASMAA
VTT Energy, Technical Research Centre, Espoo, Finland
G.V.C. PEACOCKE
Energy Research Group, Aston University, Birmingham, England

Abstract
Analysis and characterisation of liquids from fast pyrolysis processes is an important area of research. Data on the physical and chemical properties of the liquids may give important indications about process parameters, quality, toxicity and stability of the pyrolysis liquids. Based on the growing interest in the utilisation of pyrolysis liquids as fuel for heat and power production as well as a source for chemicals, several analytical methods have been applied and modified to meet the special requirements of pyrolysis liquids. This paper will describe new achievements and developments in the physical and chemical characterisation of pyrolysis liquids. Two international working groups, Pyrolysis Network for Europe (PyNE) and the IEA activity on pyrolysis (PyRA) are currently active in reviewing, assessing, harmonising and developing standard procedures for pyrolysis liquids analysis. These activities will be presented and discussed.

The unusual characteristics of pyrolysis liquids will be noted with respect to their physical properties and, where relevant, data presented. Most standard test methods for conventional fuel oils may be applied to fast pyrolysis liquids, although there are modifications to some tests and new tests require development in some areas.

Key words: biomass pyrolysis liquid analysis, physical and chemical characterisation, standard test methods

1   Introduction

There is limited information available on the physical and stability characteristics of fast pyrolysis liquids which is essential if they are to be utilised as a transportable fuel. There are a few comprehensive reviews of liquids analyses [1, 2, 3] and the physical properties of pyrolysis liquids for use as a fuel [4, 5, 6]. Some success has been made in the chemical characterisation of fast pyrolysis liquids, with major peaks having been identified, by GC-MS (Gas Chromatography-Mass Spectroscopy) and HPLC (High Performance Liquid Chromatography).

Pyrolysis liquids are relatively unstable compared to orthodox fossil fuels and various reactions between the components of the pyrolysis liquids and with the storage environment can occur: a feature which needs to be taken into consideration in their subsequent storage, transport, testing and analysis. With the development of fast pyrolysis for the production of a liquid fuel, testing and characterisation of these liquids has become a key area of research. Biomass pyrolysis liquids have become an accepted source of energy, mainly by direct combustion in boilers and more recently in dual fuel diesel engines.

The characteristics of the liquids must therefore be carefully defined, in terms of chemical composition and physical properties for the relevant application. The IEA Pyrolysis Activity (PyRA) and Pyrolysis Network for Europe (PyNE) are addressing various aspects of the production and utilisation of fast pyrolysis liquids, chemical characterisation, physical properties and related health and safety issues over the next two years.

It is intended to review the development of methods for each physical property where appropriate. This is then followed by a review of the chemical analysis. The standards required of fast pyrolysis liquids for fuel applications has been reviewed by Diebold et al. from standards for conventional fuel oils [7]. The range of tests required for liquid fuels has been previously detailed by Rick and Vix [8]. They highlighted that a range of properties of fast pyrolysis liquid had not been done: it is intended that a thorough update is provided here. The first detailed work in determination of physical properties of pyrolysis liquids by standard methods were detailed by Garrett and Mallan who investigated the density, viscosity and water content of pyrolysis liquids and their ability to be distilled for use as a fuel [9]. Limited work on the physical characterisation was done until 1990 when interest in fast pyrolysis liquids for fuel applications was renewed.

2   Storage handling and sampling

2.1 Characteristics of biomass pyrolysis liquids
Pyrolysis liquids are both hygroscopic to a limited extent and highly oleophobic. Pyrolysis liquids, depending upon the initial biomass, can absorb up to 50 wt% before the formation of two phases occurs-a highly viscous phase derived from the lignin components of the original biomass and a fluid phase containing the bulk of the oxygenated chemicals and water. The liquids are highly acidic, having a pH of usually less than 3.

2.2 Sampling of pyrolysis liquids
The first stage in any method of characterisation or analysis is to ensure that an adequately representative sample of the pyrolysis liquid has been taken. As the liquids may have several phases, it would therefore be useful to assess the degree of phase separation (if any) initially by sampling from the top and bottom of the container, prior to homogenisation. This will allow an initial assessment of the degree of homogenisation required. Layering and phase separation may take place for some oils soon after preparation or for other liquids during storage. Even when phase separation is believed to occur, this is very difficult to observe visually and water content at different points in the sample should be measured for assuring the homogeneity of the liquid. In some cases, solids in the pyrolysis liquid may sediment gradually on the bottom of the barrel forming a thick sludge. The degree of sedimentation depends on the density difference of the liquid and particles.

Mixing and sampling methods depend on the type and size of the pyrolysis oil container. The temperature of the oil should be kept as low as possible (below 30 °C) and the mixing time as short as possible for avoiding changes in the material through physical or chemical interaction or evaporation of light components. Handling, storage and pumping may all affect the liquid properties in different ways.

In some cases pyrolysis liquids exist in two or more phases. One reason for this is high moisture content (above 30 wt%) of some pyrolysis liquids can cause the formation of a lignin-derived rich fraction, as noted above, and an aqueous phase. In such cases, homogenisation of the liquid is impossible. Reasons for this include: the viscosity of the lower part of the oil is too high for recirculation pumping, the phase separation is very fast even though the oil has been mixed and the sampling is impossible. This type of liquid cannot be homogenised and their use as a fuel is questionable and may prove difficult.

Mixing of pyrolysis oil in small containers (below 10 litres) can be provided by typical laboratory methods (mixers). Mixing of pyrolysis liquid in large storage drums, or tanks (1 m^3) is easiest to carry out by pumping at room temperature for one hour or so if the liquid has originally been a single-phase liquid. Samples should be taken from different depths and analysed for moisture and solids for assuring the homogeneity of the oil. The suitable pumping time should be optimised for each batch of oil. . More attention should be paid on proper mixing of solids.

3   Standards and Test Methods

The purpose of the tests noted below are to characterise fast pyrolysis liquids for their use as a fuel, either in an engine or other suitable equipment and to develop tests for their characterisation. Tests have been developed for conventional fuel products and these methods were logically applied to fast pyrolysis liquids, however, not all of these test methods are suitable for the reasons discussed below.

Where appropriate, typical property values are quoted from the available literature. The test methods are reviewed qualitatively in five key areas: physical/chemical properties; combustion technology; safety technology; composition and new tests. Areas where new test methods are required are noted.

3.1 Physical/Chemical Properties

3.1.1   Density
The standard test method for liquid density are: ASTM D4052 at 15°C using a digital density meter; ASTM D941 using a graduated bicapillary pycnometer; or ASTM D1298-85 using a hydrometer. Pyrolysis liquid density varies with time, temperature and water content, each of which can be easily measured. Char particles can be problematical with unfiltered viscous liquids in ASTM D4052 and ASTM D941 [10]. Hydrometer methods such as ASTM D1298-85 are also useable, provided the pyrolysis liquid has not been extensively exposed to the air or has been heated for a prolonged period above 50°C. Heating above 50°C can lead to changes in its composition by the loss of water vapour and low molecular weight chemicals such as acetic acid, formaldehyde and acetaldehyde. The use of electronic density meters are now becoming more widely used. Wood derived pyrolysis liquid has density of typically 1.2 g/cm^3 for a water content of 25 wt% at 20°C.

3.1.2   Lower heating value
The standard test method for lower heating value (LHV) is ASTM D2382/DIN 51900 using a Parr adiabatic bomb. The high water content pyrolysis liquids can lead to poor ignition (DIN 51900) and a fine cotton thread is often used as a wick. The heat content of

the thread is subtracted from the result. The lower heating value is calculated from the higher heating value (HHV) and hydrogen content by equation{1}:

$$LHV (J/g) = HHV (J/g) - 218.13 * H \% (wt\%) \qquad \{1\}$$

Typical lower heating values for fast pyrolysis liquids are in the range of 12.8-17.8 MJ/kg [11]. No other problems are known with these methods.

3.1.3   Viscosity- kinematic

The standard test method for viscosity is ASTM D445-88. The liquid viscosity is highly variable from one liquid to another depending on the water content, production conditions and whether it is the whole pyrolysis liquid or a recovered fraction. More recently electronic viscometers are becoming common, using a rotating cup viscometer which has been calibrated using reference fluids and measuring the dynamic viscosity. These systems typically are closed cup and allow the sample temperature to be varied and allows a more rapid estimation compared to the methods detailed in ASTM D445-88. The latter method may be subject to volatiles losses at temperatures above 50°C.

3.1.4   Viscosity- dynamic

This is calculated from the kinematic viscosity divided by the density. Calculated dynamic viscosity values have been found to vary from 30-2000 cP at 25°C due to changes in storage time, temperature and water content [11, 12].

3.1.5   Thermal conductivity

There is no standard ASTM or DIN method for the determination of thermal conductivity. Typical methods are comparative to a known reference material. Limited work has been done to determine the thermal conductivity of pyrolysis liquids with the only known work by Peacocke et al. who found that values varied significantly within samples which had been allowed to stand for several months [13]. Typical values ranged from 0.35-0.43 W/mK. The thermal instability of the liquid lead can to difficulties in determining accurate values above 50°C, in conjunction with variability of water content, char content and sample temperature. Due to the high thermal conductivity of water compared to organics, the thermal conductivity of the pyrolysis liquid will be expected to vary considerably with the water content. No satisfactory method has yet been devised for fast pyrolysis liquids.

3.1.6   Specific heat capacity

There is no standard ASTM or DIN method for the determination of specific heat capacity. Preliminary work on specific heat capacity has been attempted by Peacocke et al. who found determination of this property problematical [13]. Due to the high specific heat capacity, the specific heat capacity of the pyrolysis liquid is also expected to be a function of the water content. Results obtained for Ensyn liquids varied ranging from 2.55-3.75 J/gK over the temperature range of 25-60°C.

Researchers at Aston University are presently using Differential Scanning Calorimetry to determine the characteristics of the liquids and the specific heat capacity. Over the temperature range of 20-50°C, the average specific heat capacity has been found to be 2.75 J/gK for Aston fluid bed pyrolysis liquid, water content 16.5 wt% (run FB14). No other satisfactory alternative method has been devised for fast pyrolysis liquids.

3.1.7   Setting point, Pour point and Cloud point

The standard test method for the determination of the pour point is ASTM D97-87 and DIN 51583 for the setting point. Cloud point is determined by IP 219/82. The setting point indicates at which temperature the liquid ceases to flow as a result of an increase in viscosity. The pour point is the lowest temperature, expressed in multiples of 3°C, at which the liquid id observed to flow under prescribed conditions. The cloud point is the temperature at which a cloud or haze of wax crystals appears at the bottom of the test jar when the liquid is cooled under prescribed conditions. Limitations of the test method are that it is only applicable to liquids which are transparent in layers 38 mm thick and with a cloud point below 49°C. The determination of the cloud point, therefore, is not valid for wood derived pyrolysis liquids as they do not contain paraffins and the liquid is opaque.

Setting point is determined to the DIN method although for the ASTM method, the pour point is determined which may be 2-4 K above the setting point, depending on the liquid. Typical values for the pour point are -20 to -28°C.

3.2 Combustion Technology

3.2.1 Boiling curve
The standard test method for determining the boiling curve of liquids is ASTM D86-82. Due to the instability (due to the varied chemical composition) of pyrolysis liquids, they do not exhibit a true boiling curve. Upon heating above 50-60°C, the liquids begins to react and produce a viscous phase which readily turns to char as the temperature is raised to 100°C. Reactive chemical components within the pyrolysis liquid may react with each other to form polymeric products which eventually form char and water. Approximately 50 wt% can be volatilised but this recovered fraction is not representative of the original pyrolysis liquids due to thermal degradation and chemical reaction. ASTM D86-82 is therefore not suitable for fast pyrolysis liquids and no other test method has been developed.

3.2.2 Coke residue
The standard test methods for determining the coke residue are ASTM D189-88-Conradson carbon and ASTM D524-88. The preferred method for determining the coking behaviour of pyrolysis liquids is ASTM D189-88. ASTM D189-88 is for heavy fuel oils which are not mobile below 90°C. Comparison of results from the two methods should not be performed, due to the different conditions employed. Pyrolysis liquids typically have a coke residue of around 20 wt% with recent results reporting a value of 17.8 wt% [14]. Due to the high water content of pyrolysis oils, which causes foaming, the water needs to be evaporated carefully away before ASTM D189-88 or for subsequent ash (EN 7) determination. This test should not be confused with the determination of the char content of the pyrolysis liquids by filtration (see Section 3.4.2).

3.2.3 Miscibility
There is no standard test method for liquid fuels. Liquids derived from eucalyptus can adsorb only 35 wt% water before two phases form, although higher absorbencies for other wood derived liquids have been reported. Pyrolysis liquids are immiscible with conventional light fuel oils and only 1% soluble in hydrocarbon solvents such as hexane, increasing to 14 wt% soluble in toluene [15]. It has been observed that there may be significant miscibility with aromatic heavy oils [9]. Bahkshi et al. have also found similar behaviour in a more extensive series of tests [16].

3.2.4 Lubricity
The standard test methods for lubricity are IP 300-82 and IP 239/69T-four ball test method. The lubricity test for liquids is to assess the tendency of liquids to produce fatigue failures in rolling contacts. The test is to determine the suitability of a liquid as a lubricant between contacting metal surfaces, e.g. bearings. Limited work has been carried out on the lubricity of fast pyrolysis liquids. Initial results would suggest that IP

300-82 is not suitable for pyrolysis liquids, due to the loss of volatiles during the test. The problem in using common lubricity tests is the instability and volatility of pyrolysis liquid when exposed to air and especially when heated. VTT is currently testing the use of IP 239/69T which is a closed system but not totally air-tight, however, it may be used for room temperature measurements. The results are compared to a reference liquid possessing similar viscosity and known lubricant properties [10].

3.2.5 Corrosion

The standard test methods for determining the corrosion rates of metals in liquids is ASTM D130-88-copper strip method and ASTM D665A. Pyrolysis liquids typically have a pH of 2-3, making them highly acidic, mainly due to acetic acid. The liquids may be adjusted to neutrality by the addition of a hydroxide solution, however the liquid may be subject to polymerisation reactions as a result of the change in pH. Corrosion tests have been carried out to ASTM D130-88, showing no sign of weight loss [17]. The standard corrosion method, ASTM D665A also has shown weight loss in mild steel, but with no rust formation. 304 stainless steel has been found not to be resistant to fast pyrolysis liquids.

Materials such as teflon, polypropylene, copper and 316 stainless steels are acceptable. Rubber seals can swell if exposed to pyrolysis liquids and careful selection is required. Viton rubber, neoprene, aluminium, mild steels and buna rubber do not appear to be suitable materials [11, 17, 18].

3.3 Safety Technology

3.3.1 Flash point

The standard test method for determining the flash point of liquids is ASTM D93-90-Pensky Martens. The flash point of fast pyrolysis liquids appears to be problematical using ASTM D93-90, due to the release of water vapour causing the flashed vapours to extinguish before the instrument can take a reading leading to erroneous results at temperatures above 70°C [19]. Selection of the "correct" temperature can be difficult. Values typically to range around 55-70°C, however water evaporation begins to disturb the analysis at 70-75°C [10, 14, 18]. No other suitable alternative method has been devised.

3.3.2 Ignition limit

The standard test method for determining the ignition limits of fuels is DIN 51603. The ignition limit describes the limiting concentration of fuel vapour in air within which combustion can occur without further addition of air when ignition is caused by a suitable source [8]. Rick and Vix quote a value of 110-120°C for pyrolysis liquids [8]. No other results are known.

3.3.3 Ignition temperature

The standard test method for determining the ignition temperature is DIN 57194. The ignition temperature, or spontaneous ignition temperature is the lowest temperature at which a fuel-air mixture ignites itself without an external ignition source. The test method gives values which tend to be lower than those achieved in practice. Limited work has been carried out on the determination of the ignition temperature of pyrolysis liquids and no comprehensive results are known.

3.4 Composition

3.4.1 Water content

The standard test methods for determining the water content of conventional fuel oils are ASTM D1744, E203 and D95. ASTM D95 is not suitable for pyrolysis liquids. The recognised method of water determination is ASTM D1744 by Karl Fischer titration. ASTM D1744 which uses a 3:1 methanol:chloroform mixture, gives highly reproducible results (± 0.01 wt%). A suitable buffer is required in the solution to prevent interference from aldehydes and ketones present in the pyrolysis liquid. Confirmation of the results may be performed by successive water additions to the pyrolysis liquids and re-measurement of the water content.

3.4.2 Char content

Char is usually present in the pyrolysis liquids, due to inefficiencies in char removal equipment such as cyclones leading to carryover of char into the liquid collection system. The char content appears to be a problematical test and as yet, no standard method has been proposed. Pyrolysis liquids when dissolved in different organic solvents give different char results upon filtration [2, 10]. Some pyrolysis liquids, such as those produced from bark, are high in phenols which are only partially soluble in ethanol, which upon filtration gives erroneous results for the char content.

Extensive work has been performed by VTT [10]. For some pyrolysis oils differences on the amount of acetone and ethanol insolubles were observed. In all cases the smallest amount of solids was obtained by using ethanol as a solvent. For one wood-derived pyrolysis liquid, a solids content of 0.3 wt% was obtained by using ethanol and 0.9 wt% by using acetone. With a straw-derived pyrolysis liquid solids content of 0.8 wt% (in ethanol) and 10 wt% (in acetone) were obtained. Acetone does not dissolve straw pyrolysis liquid properly or it causes a precipitate to form. No suitable method has yet been devised.

3.4.3 Vapour pressure

The standard test method for determining the vapour pressure of conventional fuel liquids is IP 69/89. Reported work on pyrolysis liquids has been performed by Bakhshi et al. on fast pyrolysis liquids [16]. Typical values ranged from 5.2 kPa at 33.5°C to 62.5 kPa at 75.4°C, depending also on the pyrolysis liquid. The vapour pressure is closely related to the liquid composition and the temperature due to volatiles release and the vapour pressure of the water present in the liquid.

3.4.4 Surface tension

Surface tension is typically measured using a tensiometer, calibrated with liquids of known surface tension. Reported work on pyrolysis liquids has been performed by Bakhshi et al. on fast pyrolysis liquids [16]. Typical values averaged at 29.2 mN/m for Ensyn fast pyrolysis liquids [16].

3.4.5 Elemental Composition

Elemental analysis is quite straight forwarded. There are standard methods (ASTM D5291-92) for CHN analysis. The sample size for the method is typically small and attention should be paid on proper sampling and number of duplicates. ASTM D4239 may be used for sulphur determination for sulphur contents above 0.1 wt% (detection limit). However, there are typically only small amounts of sulphur in wood pyrolysis oils so more sophisticated methods like ICP may be used for obtaining exact sulphur levels. Oxygen (dry basis) is typically estimated by difference. Chlorine may be analysed by combusting the sample in a bomb according to IP 244/71 (88).

Alkaline metals may be analysed by AAS or ICP, although the level of accuracy at low concentrations (less than 100 ppm) is significantly reduced. For low concentrations (less than 100 ppm), other techniques such as neutron activation may be preferred. Wet combustion is an easy, fast and accurate pre-treatment method for avoiding losses caused by evaporation of alkalis. Dry combustion may be also used if the ashing temperature is low enough (520 °C). For other metals ICP is ideal because of obtaining several metals in one run. The composition of pyrolysis liquids has not been discussed here in detail as there are facilities capable of analysing and determining the elemental composition.

3.5 New tests
There are several identified areas where new tests are required for fast pyrolysis liquids to account for their unusual properties, as identified by the IEA PyRA Group in addition to the four key areas above.

3.5.1 Odour
There is no standard test method for the odour of a liquid fuel. The aldehydes, ketones, acids and phenolics present in pyrolysis liquids have low vapour pressure which gives the liquid its characteristic strong and acrid smell. For the purpose of safety and health, an indication of the odour of pyrolysis liquids is required. At present, there is no accepted test method for pyrolysis liquids.

3.5.2 Stability
The standard test method for determining the stability of distillate fuels is IP 378/87/ASTM D4625-86. Pyrolysis liquids are unstable, primarily due to slow chemical reactions that produce more polymeric compounds which gradually increases the molecular weight average of the pyrolysis liquids and consequently the liquid viscosity. A stability test for pyrolysis liquids has also been referred to as an "ageing" test. The rate at which the liquid appears to age is dependent upon the water content of the liquids, storage temperature, exposure to oxygen and light.

Stability tests have been conducted by Czernik to effectively "age" the liquid under controlled conditions and monitor the changes in liquid viscosity, molecular weight, density and water content [20]. The tests noted that 8 hours at 90°C is roughly equivalent to 6 months at 25°C for hot vapour filtered oil. Testing however, does need to be done at constant water content. Bahkshi et al. have monitored changes in similar properties with storage time [21]. The test method noted above may be used for pyrolysis liquids, although it is for long term tests (4-24 weeks) for research purposes and should not be used as a basis for quality control.

4   Summary of Pyrolysis Liquid Physical Properties

The status of test methods is summarised in Table 1 overleaf. It can be clearly see that some of the tests are not suitable and that the development of new tests is required to comprehensively characterise the liquids.

Table 1. Summary of status of physical properties of pyrolysis liquids

Property	ASTM/DIN/IP standard suitable	Alternative test method	No test developed	Under development
Physical/Chemical				
Density	√			
LHV	√			
Viscosity-kinematic	√			
Thermal conductivity	X	X		U
Specific heat capacity	X	X		U
Pour point	√			
Combustion Technology				
Boiling curve	X		X	X
Coke residue	√			X
Miscibility	X		X	
Lubricity	U	X		
Corrosion	√			
Safety Technology				
Flash point	X			
Ignition limit	?			
Ignition temperature	√			
Composition				
Water	√			
C, H, O, N, S, Cl	√			
Ash	√			
pH	√			
metals	√			
Others				
Odour	X			√
Stability	√			√
Vapour pressure	√			
Surface tension	X	√		

Key
√   yes   X   no   U   under development   ?   not known

5   Chemical Analysis of Fast Pyrolysis Liquids

5.1 Background

Pyrolysis liquids from woody biomass contain all the condensable thermal degradation products of the three main wood constituents: cellulose, hemicelluloses and lignin. Additionally, minor amounts may derive from extractives such as polyphenols, resins and fatty acids. Because of the non-specific thermal degradation mechanism of these macromolecules, hundreds of low molecular weight compounds and molecular aggregates of higher molecular weights are formed which constitute the pyrolysis liquid.

Therefore, analysis of pyrolysis liquids is a challenge and the direct application of standard methods known for example from petroleum liquid analysis is not possible.

Basically, there are several physical and chemical methods available for the characterisation of the oil. In the course of the development of thermochemical biomass conversion processes a whole bundle of chemical analysis procedures have been developed and tested comprising fractionation, gas chromatography, liquid chromatography and spectroscopic methods. A comprehensive overview on chemical and physical characterisation methods for biomass-based pyrolysis oils is given by Fagernäs [2, 22]. For most of the applied techniques fractions of the oils were used and only few data are available on direct GC analysis. Therefore, this section is aimed at presenting latest developments in direct flash pyrolysis liquid analysis by GC.

5.2 Separation of components

For chemical analysis high resolution capillary gas chromatography (HRGC) is required. In contrast to physical and spectroscopic methods which give mostly overall properties, HRGC in combination with flame ionisation detectors (FID) or mass spectrometric (MS) detection allows a very close view to the chemical composition of the volatile compounds in pyrolysis liquids. One essential aspect in pyrolysis liquid analysis by GC is the selection of the appropriate conditions. Special attention should be paid to the capillary column.

The compounds in the liquids have a wide range of boiling points and a wide range of different chemical characteristics ranging from strong and week acids (acetic acid, and phenols, respectively) to neutral compounds like alcohols, aldehydes, lactones and anhydrosugars. In the literature often low polar or unpolar columns, like DB-1 or DB-5 are used [1, 23]. They are commonly selected because a lot of separations can be performed with them. However, one major disadvantage in pyrolysis compound separation with this type of phases is the poor peak shape for very polar compounds, especially acids and levoglucosan which belongs to one of the most abundant degradation products in the liquids. Moreover, they show several peak overlapping of lignin and carbohydrate derived compounds.

A more suitable column is DB-1701 which was used for degradation products of carbohydrates [24], and lignin [25, 26]. This phase shows also a very good separation behaviour for flash pyrolysis liquids. Figure 1 demonstrates the differences between DB-5 and DB-1701 in the peak shape of levoglucosan. The following conditions were used for DB-1701:

column dimensions: 60 m x 0.25 mm, film thickness: 0.25 μm
GC conditions: inlet system: split-injector, 200 kPa He, 250 °C,
                      oven temp.: 45 °C isotherm for 4 min, 3 °C/min to 280 °C
                      detection: FID, 280 °C
                      equipment: CHROMPACK 9000 with autosampler injection
                      volume 1 μl of diluted pyrolysis liquids (3 wt% in methanol)

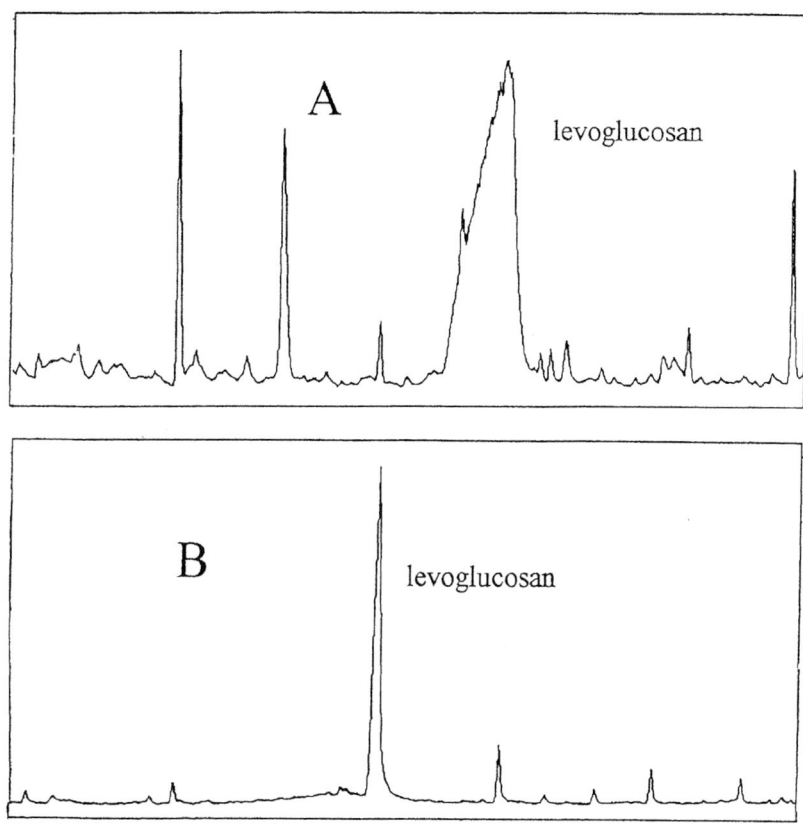

Fig.1 Peak shape for levoglucosan for DB-5 (A) and DB-1701

5.3 Identification of components

For peak identification mass spectrometry is the method of choice. Commercial mass spectral libraries contain in general those common chemical compounds which are generally found in drugs, pesticides, fungicides and other environmentally relevant areas. Only a small number of wood derived thermal degradation products are included in the standard libraries. Therefore, it is necessary to create a special library and include those specific compounds which can be detected in pyrolysis liquids. A good approach to do this is the use of hyphenated techniques which allow both pyrolysis and detection of products at the same time, for example analytical pyrolysis-GC/MS or analytical pyrolysis/MS. In the literature there are several papers dealing with this type of equipment for pyrolysis products from biomass, especially wood [27, 28, 29, 30]. Extremely helpful for compound identification is the separate analysis of the single wood constituents or oligomeric model compounds which helps to improve spectra assignment [31, 32, 33].

Table 2 shows a list of identified products found both in analytical pyrolysis of soft- and hardwood and in flash pyrolysis liquids produced from three different laboratories. Aston oil was produced from pine in a fluidised bed reactor (pilot plant scale), BTG oil from softwood comes from a rotating cone reactor (pilot plant scale) and IWC oil was from beechwood produced in a fluidised bed reactor (lab scale). The peak assignments are mostly based on published results and verified by the pyrolysis groups at the Institute of Wood Chemistry (IWC) [34, 35].

Table 2. Quantitative results from direct GC analysis of flash pyrolysis liquids from different laboratories (wt% based on water-free liquid)

No.	Compound	Aston	BTG	IWC
1	Hydroxyacetaldehyde	12.62	12.92	7.78
2	Formic acid	*	*	*
3	Acetic acid	3.22	3.04	4.65
4	Acetol	7.02	4.24	3.90
5	1.2-Ethanediol	*	*	*
6	1-Hydroxy-2-butanone	*	*	*
7	3-Hydroxypropanal	*	*	*
8	2.5-Dimethoxytetrahydrofuran (cis)	0.14		0.52
9	3-Furaldehyde		*	
10	Butanedial		*	
11	2.5-Dimethoxytetrahydrofuran (trans)	*		*
12	2-Furaldehyde		0.41	
13	2-Furfuryl alcohol		0.01	
14	1-Acetyloxypropane-2-one		*	
15	2-Acetylfuran		*	
16	Isomer of alpha-Angelicalactone			*
17	Isomer of alpha-Angelicalactone			*
18	(5H)-Furan-2-one	0.77	0.40	0.67
19	Methyl-(5H)-furan-2-one			*
20	4-Hydroxy-5,6-dihydro-(2H)-pyran-2-one	*	*	*
21	2-Hydroxy-1-methyl-1-cyclopentene-3-one	0.27	0.15	0.16
22	Phenol	0.06	0.15	0.03
23	Guaiacol	0.53	0.24	0.15
24	o-Cresol	0.05	0.09	*
25	Methyl-butyraldehyde derivative	*	*	
26	Furan derivative	*	*	
27	m-Cresol	0.31	0.10	*
28	p-Cresol	0.02	0.09	*
29	gamma-Lactone derivative	*	*	
30	4-Methyl guaiacol	0.88	0.36	0.14
31	2,4- and 2,5-Dimethyl phenol	0.05	0.08	0.41
32	4-Ethyl guaiacol	0.14	0.09	0.08
33	4-Vinyl guaiacol	0.06	0.03	0.05
34	Eugenol	0.22	0.16	0.06
35	4-Propyl guaiacol		*	
36	5-Hydroxymethyl-2-furaldehyde	0.45	0.33	
37	gamma-Lactone derivative	*	*	
38	Syringol			0.32
39	Isoeugenol (cis)	0.25	0.15	0.07
40	2-Hydroxymethyl-5-hydroxy-2,3-dihydro-(4H)-pyran-4-one	*	*	
41	Isoeugenol (trans)	0.67	0.24	0.30
42	4-Methyl syringol			0.30
43	Vanillin	0.26	0.34	0.10

44	Homovanillin	0.20	0.15	0.09
45	Acetoguaiacone	0.19	0.20	0.08
46	4-Vinyl syringol			*
47	Guaiacylacetone	*	*	*
48	4-Ally- and 4-Propyl syringol			0.21
49	1,6-Anhydro-beta-D-mannopyranose	*	*	
50	Propioguaiacone	*	*	
51	4-Hydroxy acetophenone	*	*	
52	Isomer of coniferyl alcohol	*	*	
53	4-Propenyl syringol (cis)			*
54	Levoglucosan	5.42	5.29	3.52
55	4-Propenyl syringol (trans)			*
56	Dihydroconiferyl alcohol	*	*	*
57	Syringaldehyde			0.25
58	Coniferyl alcohol (cis)	*	*	
59	Homosyringaldehyde			*
60	Anhydrosugar unknown	*	*	
61	Acetosyringone			*
62	Coniferylaldehyde	*	*	*
63	Propiosyringone			*
64	Dihydrosinapyl alcohol			*
65	Sinapaldehyde			*
	Total	33.79	29.29	23.86

*identified

5.4 Quantitation

Beside identification of single compounds within the complex mixture, quantification is another challenge for the analyst. Only with quantitative data, the observation of chemical changes or the determination of valuable chemical feedstocks is possible.

The most important aspect in quantification with GC-FID is the determination of individual response factors of each compound to be quantified, e.g. how much signal is produced from the detector for each compound eluting from the column. The area of a peak in a chromatogram is ideally a linear function of its amount. The response depends mainly on the chemical structure and the volatility of the compound. For example, oxygenated compounds generally have a lower response than hydrocarbons. Furthermore, the system performance can also influence the response.

The first step in calibration is the preparation of standard solutions which should include as much identified components as possible. Some can be found in catalogues for speciality chemicals other can be synthesised in the lab. Because of the large heterogeneity of functional groups in the pyrolysis liquids, the components for the standard solutions must be carefully combined to avoid mutual interactions like condensation reactions during storing. For example, if one would mix together acetic acid and furfural, the sensitive furfural would condense and a correct calibration would be impossible. Therefore, several solutions of standards should be prepared.

In general, every standard software for chromatography contains the calculation procedure of response factors using the internal standard method. The internal standard should be selected in such a way that it does not interfere with components from the pyrolysis mixture. For example fluoranthene has been proved to be a suitable standard compound because it does not occur in flash pyrolysis liquids.

Table 3. List of relative response factors used for quantification

Compound	relative response factor
Hydroxyacetaldehyde	7.90
Acetic acid	4.87
Acetol	6.6
2.5-Dimethoxytetrahydrofuran (cis)	2.50
3-Furaldehyde	2.06
2.5-Dimethoxytetrahydrofuran (trans)	2.50
2-Furaldehyde	2.06
2-Furfuryl alcohol	2.10
(5H)-Furan-2-one	2.19
2-Hydroxy-1-methyl-1-cyclopentene-3-one	1.2
Phenol	1.08
Guaiacol	1.41
o-Cresol	1.05
m-Cresol	1.06
p-Cresol	1.06
4-Methyl guaiacol	1.49
2,4- and 2,5-Dimethyl phenol	1.08
4-Ethyl guaiacol	1.26
4-Vinyl guaiacol	1.26
Eugenol	1.25
4-Propyl guaiacol	1.40
5-Hydroxymethyl-2-furaldehyde	3.54
Syringol	1.44
Isoeugenol (cis)	2.55
Isoeugenol (trans)	2.55
4-Methyl syringol	1.84
Vanillin	1.92
Homovanillin	1.90
Acetoguaiacone	1.90
4-Vinyl syringol	1.45
4-Ally- and 4-Propyl syringol	2.00
Levoglucosan	6.85
Syringaldehyde	2.14
Fluoranthene (internal standard)	1.00

Table 3 demonstrates the results of relative response factor (RRF) determination of various chemicals. RRF was calculated as shown in equation {2}:

$$RRF = \frac{amount_{Sample}}{area_{Sample}} \times \frac{area_{IS}}{amount_{IS}} \qquad \{2\}$$

Internal standard basis was the weight and peak area of fluoranthene. For quantitation the area of the corresponding compound must be multiplied by its RRF. GC conditions were as mentioned above. It is important to note that the factors cover a wide range. The larger the factor the worse is the response. Taking into account these results one can begin to quantify those peaks which could be detected by GC. A typical result is presented in Table 1. It reflects the current state of knowledge with respect to identification and quantification. It is intended to extend this list.

As Table 2 shows about one third of the organic matter of the pyrolysis liquids can be quantified. There are several peaks which are not yet quantified, because the right response factor is missing. However, it should be possible to take a factor from a known substance which has a similar chemical structure. Furthermore, one can assume that there might be some other compounds which had been vaporised but did neither pass the injector nor the columns. These high polarity compounds could possibly be determined by liquid chromatography. The remaining portion of the pyrolysis liquids consists of substances with a higher molecular weight. They are kept in solution by the lower monomeric compounds. The presence of larger molecules has been already described in the literature [36] and can easily be demonstrated by the addition of water which causes a lignin-like substance to precipitate.

The direct analysis of the crude pyrolysis liquid by HRGC has the advantage that no preparation of the sample is needed which might lead to changes of the sample's composition or even losses during fractionation. However, for a comprehensive qualitative analysis of all GC detectable compounds the oil should be fractionated for components enrichment. Fractionation is best achieved by liquid chromatography in an open column on silica gel as stationary phase, also called flash chromatography if a small stream of nitrogen is used to increase the flow rate of the solvents. This technique was first applied for the characterisation of coal derived liquids [37] and was used and modified by several researchers also for the liquid products from thermochemical biomass conversion processes [38, 39, 40]. Problems with losses arise when the solvents are being vaporised together with sample components and high molecular weight compounds are adsorbed permanently on the stationary phase.

6   Discussion

Pyrolysis liquid is relatively unstable with a unique range of properties which are not found in conventional petroleum-derived fuels. Their unusual characteristics are gradually becoming more understood and accepted. The detailed quantification and identification of the chemicals in the liquid is gradually being expanded as advances are made in analytical software and hardware. New tests for the physical characterisation need to be developed and one important area which was not addressed concerns transport and shipment of these liquids, for which there is no international designation.

7   Conclusions

Pyrolysis liquid physical properties have been shown to change over time. Both the viscosity and the water content increase with the time and temperature of storage. The stability characteristics of pyrolysis liquids produced from different feedstocks and processes are significantly different, based upon published work.

In addition, techniques for modification of certain physical properties of the pyrolysis liquid need to be developed such as the addition of water on viscosity to achieve the requirements for utilisation in an engine.

A three-year project on pyrolysis within the IEA Bioenergy Agreement, Task XIII, has been initiated 1995. One major task in the IEA work is product characterisation. The objective of the task is to review, assess and develop methods for characterisation and analysis of pyrolysis liquid and derived products. Basic fuel oil analyses will be tested. Recommendations for sampling and comments on methods will be summarised. The most important analytical methods in diesel engines, gas turbines etc. will be tested. Tests related to oil use tests for ageing, and stability will also be developed.

8. Recommendations

NREL (National Renewable Energy Laboratory) has suggested the second IEA Round Robin to be carried out and a similar study has been proposed within the PyNE group. Several pyrolysis oils from different processes would be included. Suggestions for handling and sampling procedure would be provided as well as some guidelines on minimum specifications or requirements for standard systems to help those entering the analysis and characterisation area to understand what is required and how it should be operated.

9. Acknoledgements

The authors gratefully acknowledge the contribution of the IEA Pyrolysis Group (PyRA) and especially: J. Diebold, S. Czernik, T. Milne (NREL, USA), J. Piskorz (RTI, Canada), D. Huffman (Ensyn, Canada) and S. Gust (Neste Oy, Finland).

10. References

1. Elliott, D.C., Sealock, L.J. and Butner, R.S. (1988) Product analysis from direct liquefaction of several high-moisture biomass feedstocks, *Pyrolysis Oils from Biomass - Producing, Analyzing, and Upgrading,* (eds. Soltes, E.J., Milne, T.A.), ACS, Washington, pp. 179-188.
2. Fagernäs, L. (1995) Chemical and physical characterization of biomass-based pyrolysis oils - Literature review. *VTT Research Notes,* Vol. 1706.
3. McKinley J. (1989), *Biomass Liquefaction Centralized Analysis,* prepared for Science Branch, Science Procurement Department of Supply and Services, Hull, Quebec, DSS file No. 23216-4-6192, Final report, Project No 4-03-837.
4. *Pyrolysis Liquids Upgrading and Utilisation,* (eds. A.V. Bridgwater, and G. Grassi), Elsevier Applied Science, p 177-218, 1991.
5. *Bio-oil Production and Utilisation- Proceedings of the 2nd EU-Canada workshop on Thermal Biomass Processing,* (eds. A.V. Bridgwater, and E.N. Hogan), CPL Press, Newbury, 1996.
6. *Proceedings Biomass Pyrolysis Oil Properties and Combustion Meeting,* (ed. Milne, T.A.),US DoE, 1994, available from NTIS.
7. Diebold, J.P. et al. (1996) Specifications of fast pyrolysis liquids for use as a fuel, paper to be presented to this conference.
8. Rick, F. and Vix, U. (1991) Product Standards for Pyrolysis Liquids for use as fuel in industrial firing plants, *Pyrolysis Liquids Upgrading and Utilisation* (eds. Bridgwater, A.V. and Grassi, G.), Elsevier Applied Science, pp. 177-218.
9. Garrett, D.E. and Mallan, G.M., *Pyrolysis Process for Solid Wastes*, US Patent no 4153514, approved 8 May 1979.
10. Oasmaa, A., Leppämki, E., Koponen, P., Levander, J., Tapola, E. (1996) Physical characterization of biomass-based pyrolysis oils. Application of standardised fuel oil analyses for pyrolysid oils, to be published in VTT Research Notes.
11. Diebold, J.P., Scahill, J.W., Czernik, S., Philips, S.D. and Feik, C.J., (1996) Progress in the production of hot-gas filtered biocrude oil at NREL, *Bio-oil*

Production and Utilisation- Proceedings of the 2nd EU-Canada workshop on Thermal Biomass Processing, (eds. A.V. Bridgwater, and E.N. Hogan), CPL Press, Newbury, pp. 66-81.

12. Tiplady, I.R., Peacocke, G.V.C. and Bridgwater, A.V. (1996) Physical properties of fast pyrolysis liquids from the Union Fenosa pilot plant, *Bio-oil Production and Utilisation- Proceedings of the 2nd EU-Canada workshop on Thermal Biomass Processing*, (eds. A.V. Bridgwater, and E.N. Hogan), CPL Press, Newbury, p 164-174.

13. Peacocke, G.V.C., Russell, P.A., Jenkins, J.D. and Bridgwater, A.V. Physical Properties of Flash Pyrolysis Liquids, *Biomass and Bionergy*, Vol. 7, No. 1-6, 1994, p 169-177.

14. Gros, S. (1995) Pyrolysis oil as diesel fuel, Wartsila Diesel International Ltd., Diesel Technology, Finland.

15. Graham, R.G., Ensyn Technologies Inc., Greely, Ontario, Canada, personal communication, 1991.

16. Bakhshi, N. and Adjaye, J.D., Properties and Characteristics of Ensyn bio-oil, in *Proceedings Biomass Pyrolysis Oil Properties and Combustion Meeting*, (ed. T.A. Milne),US DoE, 1994, pp. 54-66.

17. Oasmaa, A. and Sipilä, K., Pyrolysis oil properties: use of pyrolysis oil as fuel in medium speed diesel engines, *Bio-oil Production and Utilisation- Proceedings of the 2nd EU-Canada workshop on Thermal Biomass Processing*, (eds. A.V. Bridgwater, and E.N. Hogan), CPL Press, Newbury, 1996, p 175-185.

18. Andrews, R., Patnaik, P.C., Liu, Q. and Thamburaj (1994) Firing Fast Pyrolysis Oils in Turbines, *Proceedings of the Biomass Pyrolysis Oil Properties and Combustion Meeting*, (ed. T.A. Milne), NREL/CP-430-7215, pp. 383-391.

19. S. Gust, Neste Oy, Finland, personal communication, March 1996.

20. Czernik, S. (1994) Storage of Biomass Pyrolysis Liquids, *Proceedings of the Biomass Pyrolysis Oil Properties and Combustion Meeting*, (ed. T.A. Milne), NREL/CP-430-7215, pp. 67-73.

21. Adjaye, J.D., Sharma, R.K. and Bakhshi, N.N. (1992) Characterisation and stability analysis of wood-derived bio-oil, *Fuel Processing Technology*, vol. 31, no. 3, pp. 241-256.

22. Milne, T.A., Agblevor, F., Davis, M., Deutch, S. and Johnson, D. A review of the chemical composition of fast pyrolysis oils from biomass, *Developments in Thermochemical Biomass Conversion (ed. A.V. Bridgwater)*, 1996, in press.

23. Maggi, R. and Delmon, B. (1994) Comparison between 'slow' and 'flash' pyrolysis oils from biomass. *Fuel*, Vol. 73, No. 5. pp. 671-677.

24. Kelly, J., Mackey, M. and Helleur, R.J. (1991) Quantitative analysis of saccharides in wood pulps by quartz-tube pulse pyrolysis-polar phase gas chromatography. *J. Anal. Appl. Pyrolysis*, Vol. 19, pp. 105-117.

25. Meier, D. and Faix, O. (1992) Pyrolysis-gas chromatography-mass spectrometry [of lignin in solid state], in *Methods in Lignin Chemistry*, (eds. Lin, Y., S. Dence, W., C.), Springer, Berlin, pp. 177-199.

26. Faix, O., Bremer, J., Meier, D., Fortmann, I., Scheijen, M.A. and Boon, J.J. (1992) Characterization of tobacco lignin by analytical pyrolysis and Fourier transform-infrared spectroscopy. *J. Anal. Appl. Pyrolysis*, Vol. 22, pp. 239-259.

27. Bates, A.L., Hatcher, P.G., Lerch, H.E.I., Cecil, C.B. and Neuzil, S.G. (1991) Studies of a peatified angiosperm log cross section from Indonesia by nuclear magnetic resonance spectroscopy and analytical pyrolysis, *Org. Geochem.*, Vol. 17, pp. 37-45.

28. van der Hage, E.R.E., Mulder, M.M. and Boon, J.J. (1993) Structural characterization of lignin polymers by temperature-resolved in-source pyrolysis-mass spectrometry and Curie-point pyrolysis-gas chromatography/mass spectrometry. *J. Anal. Appl. Pyrolysis,* Vol. 25, pp. 149-183.

29. Pouwels, A.D. and Boon, J.J. (1990) Analysis of beech wood samples, its milled wood lignin and polysaccharide fractions by Curie-point and platinum filament pyrolysis-mass spectrometry. *J. Anal. Appl. Pyrolysis,* vol. 17, pp. 97-126.

30. Boon, J.J. (1992) Analytical pyrolysis mass spectrometry: new vistas opened by temperature-resolved in-source PYMS. *Int. J. Mass Spectrom. Ion Processes,* vol. 118-119, pp. 755-787.

31. Faix, O., Meier, D. and Fortmann, I. (1988) Pyrolysis-gas chromatography-mass spectrometry of two trimeric lignin model compounds with alkyl-aryl ether structure. *J. Anal. Appl. Pyrolysis,* Vol. 14, pp. 135-148.

32. Faix, O., Meier, D. and Grobe, I. (1987) Studies on isolated lignins and lignins in woody materials by pyrolysis-gas chromatography-mass spectrometry and off-line pyrolysis-gas chromatography with flame ionization detection. *J. Anal. Appl. Pyrolysis,* Vol. 11, pp. 403-416.

33. Meier, D., Berns, J., Grünwald, C. and Faix, O. (1993) Analytical pyrolysis and semicontinuous catalytic hydropyrolysis of Organocell lignin. *J. Anal. Appl. Pyrolysis,* Vol. 25, pp. 335-347.

34. Faix, O., Meier, D. and Fortmann, I. (1990) Thermal degradation products of wood. Gas chromatographic separation and mass spectrometric characterization of monomeric lignin-derived products. *Holz Roh-Werkst.,* vol. 48, pp. 281-289.

35. Faix, O., Fortmann, L., Bremer, J. and Meier, D. (1991) Thermal degradation products of wood. A collection of electron-impact (EI) mass spectra of polysaccharide-derived products. *Holz Roh- Werkst.,* Vol. 49, pp. 299-304.

36. Piskorz, J., Scott, D.S. and Radlein, D. (1988) Composition of oils obtained by fast pyrolysis of different woods. *ACS Symp. Ser.,* Vol. 376, pp. 167-178.

37. Farcasiu, M. (1977) Fractionating and structural characterization of coal liquids. *Fuel,* Vol. 56, pp. 9-14.

38. Achladas, G.E. (1991) Analysis of biomass pyrolysis liquids: separation and characterization of phenols. *J. Chromatogr.,* Vol. 542, pp. 263-275.

39. Pakdel, H. and Roy, C. (1991) Hydrocarbon content of liquid products and tar from pyrolysis and gasification of wood. *Energy Fuels,* Vol. 5, pp. 427-436.

40. Meier, D., Wulzinger, P. and Faix, O. (1995) Chemicals recovery from flash pyrolysis oils, *Biomass for Energy, Environment, Agriculture and Industry,* (eds. Chartier, P., Beenackers, AACM., Grassi, G.), Pergamon, pp. 1875-1880.

A REVIEW OF THE CHEMICAL COMPOSITION OF FAST-PYROLYSIS OILS FROM BIOMASS

T. MILNE, F. AGBLEVOR, M. DAVIS, S. DEUTCH, and D. JOHNSON
National Renewable Energy Laboratory
Center for Renewable Chemical Technologies and Materials
Golden, Colorado, USA

Abstract

As part of our thermochemical program for biofuels, power and chemicals from biomass we have undertaken a comprehensive review of the past literature on the detailed chemical composition of "fast" or "flash" pyrolysis oils produced under conditions that minimize secondary cracking. Such a review is complicated by the great variety of pyrolysis techniques and conditions used to produce the oils. In addition, almost every researcher in the past twenty years has used a different feedstock, the Canadian standard aspen being a notable exception.

This paper will summarize our early findings, including a list of compounds reported, and their quantitation. Where quantitative yields are available, an attempt has been made to relate yields to feedstocks and conditions of pyrolysis.

Keywords: Biocrude, Biomass, Bio-Oil, Chemical Composition, Fast Pyrolysis, Quantitation.

1 Introduction

The purpose of this paper is to annotate a thorough review of the literature on the chemical composition of so-called "fast-pyrolysis" oils, often referred to as "bio-oils" or "bio-crudes". The choice of oils is limited to those produced under: mild pyrolysis severity (generally a few seconds at 500° C); high oil yield (generally greater than 50%); and resulting in high oxygen content (>30% on a moisture-free organic oil basis). We had intended to include a discussion of the derivatization or fractionation of oils for analysis and of the various analytical techniques that have been used for these complex liquids. Fortunately, while this work was in progress, an excellent report by Leena Fagernäs of the Technical Research Centre of Finland was issued, which discusses these aspects in detail [1].

What follows are some highlights of a more comprehensive technical report to be issued in 1996 as an NREL document [2]. This forthcoming report will contain: the complete, briefly annotated bibliography of source references; a listing of all chemical species that have been reported qualitatively and quantitatively; a table of those that have been quantitated; and the temperatures, residence time, yield, feedstock and oxygen content of dry-oils when such information is given. Regarding the listing of chemical species: each species will be identified by its Chemical Abstract name; common names; Chemical Abstracts Index number and chemical structure. Hopefully this will save future readers the time-consuming task of cross-comparing the many names used by researchers in reporting their results.

2 Reported Compositions

In this paper, mainly studies reporting quantitative results of "fast pyrolysis" of untreated biomass feedstocks will be discussed. The Table shows the suite of compounds that have been quantified at levels greater than 0.1% of the wet oil. (The short-hand citation in the table gives the first author, the year of publication and the first page number). Any missing data will be included in the comprehensive NREL report. There has sometimes been ambiguity as to the basis of reported yields. Many data are reported on a weight percent of feed but for our purposes an attempt was made to reduce the data to weight percent of wet oil, assuming a single-phase oil. (this is usually true for "primary" bio-oils). Also included in the table are ranges of compositions reported in the thorough review of fast pyrolysis by Solantausta, et al. [3].

It will be noted that only ten groups are represented in the list of quantified results. Some of these groups have been analyzing bio-oils for over a decade, and with a variety of separation techniques and improved methods. Partial analyses are common, depending on the focus of the investigators at the time. The results are shown as reported, but with our conversion to a weight percent wet-oil basis.

3 Discussion

A goal of the review is to find correlations between the yield and distribution of chemical components reported and such variables as pyrolysis severity [f(T and residence time)] nature of the feedstock; oxygen content of the organic portion of the oil; total oil yield; and method of oil production. It also was hoped that a "typical oil composition" could be offered, such as was done by Solantausta, et al. [3] through the use of ranges of component compositions reported. Such an overall composition is difficult to define, in view of gaps in closing mass balances in most work and the large variability in reported yields from any one feedstock. As seen in the table one must "patch" together results from several runs, using different analytical techniques and pyrolysis system, to obtain an expression of the whole range of compounds that have been quantitated. We suspect that the problem of comparing oils is further compounded by the different collection schemes that have been used for condensed vapors and by the age and storage conditions of the different oils (e.g. those submitted for centralized analysis: [MCKN-88; ELLI 88 and 86].

4 General Notes from the Table

A visual inspection of the Table reveals the variable scope of analyses and quantified compounds. (Only compounds reported at $\geq$ .1 are listed. A number of other species were seen and reported at the < .1 or trace level and these are starred.) The number of species identified qualitatively, especially in the older "pyroligneous liquor" literature, and in recent mass spectral studies, probably increases the number of chemical species from 100 to the order of 300. Whether minor (or some major) species will be of technical or toxicological concern must await further studies of varied fuel applications.

No one compound has been quantified by all ten laboratories. Equally surprising is the observation that only 19 or so components, of the 108 listed in the Table, have been reported by four or more laboratories. This fact, plus the use of at least nine different pyrolysis systems, makes overall comparisons uncertain. The reader is referred to the papers cited to view variables such as particle size, temperature and feedstock effects in more detail.

Some qualitative observations from a visual inspection of the tabulated results are as follows.

A. Studies on aspen/poplar:
   - BC Research (McKinley) consistently reports lower acid concentrations than Laval/Waterloo (Roy, Scott). Western Ontario results are lower still but process temperatures are higher.
   - Sugars show no clear trend, but are quite variable.
   - Waterloo results show a strong increase in aldehydes with temperature.
   - Western Ontario oils show high phenols (94 and 108). These oils also show lower furfural than others.
   - For most other species data are too limited, or scattered, to document clear trends.

B. Variation with type of wood.
   Most of the results reported involve hardwoods. Spruce, perhaps hog fuel, Western Hemlock, Pine bark and sawdust, and Fir are the exceptions. There are few trustworthy differences except that Syringols are only reported for hardwoods.

C. Variation with type of pyrolysis.
   Differences between vacuum pyrolysis, fast-fluid bed pyrolysis and ablative or vortex pyrolysis are not great, either in yields or chemical content.

D. Variation with storage history of oils.
   Not available, but the BCR and PNL results probably involved older oils.

5 Conclusions and Observations

A. Only a small number of species are nearly, universally reported. These include formic, acetic and propanoic acids; levoglucosan, glucose, xylose; methyl-cyclopentenone, methanol, phenol, and furfural. Part of the reason for this is the emphasis that different investigators have placed on products of particular value to them for chemicals from biomass.

For fuel use, minor species may assume enhanced importance. As an example, specifications for fatty acid methyl esters from vegetable oils include concern for levels of glycerol at the 1% level for transportation "bio-diesel" use [4].

B.   Slight increases in cracking severity lead to the appearance of benzene, toluene, and polynuclear aromatics that may arouse toxicity concerns. (ELLI 88 and 86); Williams and Horne [5].

C.   Variation among woody feedstocks appears to be modest, though much of the work has been done with Aspen/Poplar.

D.   Combined effects of feed, pyrolysis method and pyrolysis conditions (assuming collection, storage and methods of analysis are comparable) do lead to reported differences of factors of 5-10 (See the Table). Chemometric techniques may help de-convolute the several variables in the extant data.

6 References

1.   Fagernäs, L. (1995) "Chemical and Physical Characterization of Biomass-Based Pyrolysis Oils - Literature Review". VTT Research Notes 1706. VTT Technical Research Center of Finland, Espoo, Finland. pp 115.

2.   Milne, T.A.; Agblevor, F.; Davis, M.; Deutch, S.; Johnson, D. (1996). "A Review of the Chemical Composition of Fast Pyrolysis Oils from Biomass." NREL report number to be assigned. National Renewable Energy Laboratory, Golden, CO, U.S.A.

3.   Solantausta, Y.; Diebold, J.; Elliott, D.C.; Bridgwater T.; Beckman, D. (1994). "Assessment of Liquefaction and Pyrolysis Systems," Research Notes, PB–95-129797/XAB, VTT/RN-1573, pp. 206.

4.   Peterson, C.L. (1994) "Conference Proceedings: Commercialization of Biodiesel: Establishment of Engine Warranties." Ed. C. L. Peterson. University of Idaho, Moscow, Idaho.

5.   Williams, P.T.; Horne, P.A. (1995) "Analysis of Aromatic Hydrocarbons in Pyrolytic Oil Derived from Biomass." *J. Analytical & App Pyrolysis.* Vol. 31, pp. 15-37.

TABLE OF BIO-OIL COMPOSITION

COMPOSITION	POPL ASPN	POPL ASPN	POPL ASPN	POPL ASPN	POPL ASPN	POPL ASPN	HYBD POPL	ASPN BROK	ASPN BROK	HYBD POPL	POPL BROK	WHIT SPRC	RED MAPL	IEA POPL	ULEX	BIO SHEL	HOG FUEL	WEST HEML	PINE BR/S
TEMPERATURE (Celcius)	450	500	550	600	600	650	450	500	550	504	504	500	508	504	506	504	500	490	
CITATION (First Author)	SCOT						SCOT			PISK	PISK							PISK	ELDR
YEAR	82-						84-			86-	88-							93-	80-
PAGE NO.	666						404			121	167							1432	217
RUN NO.	44	37	39	42	43	45	33	31	32	59	58	43	63	A-2	142	90	126	R-78	-17
WET OIL (WT. % Moisture-free wood)	60	69	65	54	53	42	61	70	67	78	73	78	78	78	71	66	59	66	
CHAR (WT. % Moisture-free wood)	15	13	10	8	8	6	25	13	11	12	17	12	14	7.7	13	20	32	18	15
GASES (WT. % Moisture-free wood)	7	10	13	18	18	25	8.6	12	19	11	12	9.8	9.8	11	14	13	13	12	
WATER (WT. % Moisture-free wood)	9	10	8	8	8	6	5.3	4.3	5.3	10	10	12	10	12	7.7	13	12	9.4	
ORGANIC LIQUID (WT. % Moisture-free wood)	51	59	57	46	45	36	56	66	62	66	63	67	68	66	63	53	47	57	
APPARATUS	BFB	BFB	BFB	BFB	BFB	BFB	PPFB	PPFB	PPFB	PPFB	PPFB	PPFB	PPFB	PPFB	PPFB	PPFB	PPFB	PPFB	GASF
RESIDENCE TIME (Seconds)	~0.5																		
PRESSURE (torr)	~760																		
MOISTURE IN WOOD (WT. %)	5.0	5.0	5.0	5.0	5.0	5.0	6.3	6.3	6.3	4.8	5.2	7.0	5.9	4.6	8.0	8.6	12.0	5.0	
ASH IN WOOD (WT. %)	0.4	0.4	0.4	0.4	0.4	0.4	1.3	1.3	1.3	0.5	1.2	0.5	0.4	0.5	0.4	2.2	2.2	0.6	
COMPOSITION WT. % WET OIL																			
ACIDS																			
46 FORMIC (METHANOIC)	3.8	1.5	3.1	2.0	1.8	4.7	8.8	7.5	7.6	5.6	7.4	9.1	8.1	4.0	5.8	5.0	1.8		
60 ACETIC (ETHANOIC)							1.7	1.5	1.8		8.6	4.9	7.4	6.9			4.5	1.5	
74 PROPANOIC (PROPIONIC)																			
76 GLYCOLIC (HYDROXYACETIC)																			
88 BUTANOIC (BUTYRIC)																			
102 PENTANOIC (VALERIC)																			
116 4-OXOPENTANOIC																			
116 HEXANOIC (CAPROIC)																			
122 BENZOIC																			
130 HEPTANOIC																			
SUGARS																			
150 D-XYLOSE																			
162 1,6-ANHYDROGLUCOFURANOSE														3.1					
162 LEVOGLUCOSAN (1,6-ANHYDRO-beta-D-GLUCOPYRANOSE)											3.3	5.1	3.6	3.8	7.5	4.2	3.7	2.7	
180 alpha-D-GLUCOSE (alpha-D-GLUCOPYRANOSE)											0.8	1.3	0.8	0.5		0.6	0.6	0.4	
180 FRUCTOSE											1.8	2.9	1.9	1.7	1.7	1.2	0.7		
CELLOBIOSAN											1.5	3.2	2.1	1.7		1.2	0.6	0.6	
ALCOHOLS																			
32 METHANOL	0.8	0.9	1.2	✻	✻	0.6	0.4	0.4	0.7	1.4	0.9	1.4	1.0	0.5	1.7				
46 ETHANOL	✻	✻	✻	✻	0.6	1.4													

413

TABLE OF BIO-OIL COMPOSITION

FEED	POPL ASPN	POPL ASPN	POPL ASPN	POPL ASPN	POPL ASPN	POPL ASPN	HYBD POPL	ASPN BROK	ASPN BROK	HYBD POPL	WHIT SPRC	RED MAPL POPL	IEA POPL	ULEX	BIO SHEL	HOG FUEL	WEST HEML	PINE BR/S
TEMPERATURE (Celcius)	450	500	550	600	600	650	450	500	550	504	500	508	504	506	504	500	490	
CITATION (First Author)	SCOT						SCOT			PISK	PISK						PISK	ELDR
YEAR	82-						84-			86-	88-						93-	80-
PAGE NO.	666						404			121	167						1432	217
RUN NO.	44	37	39	42	43	45	33	31	32	59	58	63	A-2	142	90		126	R-78
KETONES																		
70 2-BUTENONE																		
72 2-BUTANONE																		
84 CYCLO PENTANONE																		
96 3-METHYL-2-CYCLOPENTEN-1-ONE																		
112 2-ETHYLCYCLOPENTANONE																		
112 DIMETHYLCYCLOPENTANONE																		
124 TRIMETHYLCYCLOPENTENONE																		
ALDEHYDES																		
30 METHANAL (FORMALDEHYDE)	1.0	1.3	3.0	5.6	4.7	8.5							1.5		1.6	1.4		
44 ETHANAL (ACETALDEHYDE)													✱					
84 2-METHYL-2-BUTENAL																		
PHENOLS																		
94 PHENOL																		0.2
108 2-METHYL PHENOL o (o-CRESOL)																		0.1
108 3-METHYL PHENOL m (m-CRESOL)																		0.2
108 4-METHYL PHENOL p (p-CRESOL)																		
122 2,3-DIMETHYLPHENOL (2,3-XYLENOL)																		
122 2,4-DIMETHYLPHENOL (2,4-XYLENOL)																		
122 2,5-DIMETHYLPHENOL (2,5-XYLENOL)																		
122 2,6-DIMETHYLPHENOL (2,6-XYLENOL)																		0.1
122 2-ETHYLPHENOL																		
136 2,3,5 TRIMETHYLPHENOL																		
GUAIACOLS																		
124 GUAIACOL (2-METHOXYPHENOL)																		0.3
138 4-METHYL GUAIACOL																		0.8
152 4-ETHYL GUAIACOL																		0.2
164 4-PROPENYL GUAIACOL																		
164 EUGENOL (PHENOL, 2-METHOXY-4-(2-PROPENYL))																		0.4
164 ISOEUGENOL (PHENOL, 2-METHOXY-4-(1-PROPENYL))																		0.7
166 4-PROPYLGUAIACOL																		0.1
PYROLYTIC LIGNIN, WATER INSOLUABLE										34.0	27.0	27.0	21.0	31.0	34.0	47.0	21.0	
SYRINGOLS																		
154 SYRINGOL (2,6-DIMETHOXY PHENOL)																		
168 METHYLSYRINGOL																		
182 4-ETHYL SYRINGOL																		
182 SYRINGALDEHYDE																		
194 4-PROPENYLSYRINGOL (4-ALLYLSYRINGOL)																		
196 4-HYDROXY-3,5-DIMETHOXYPHENYL ETHANONE																		

414

TABLE OF BIO-OIL COMPOSITION

MW	Compound	POPL ASPN 450	POPL ASPN 500	POPL ASPN 550	POPL ASPN 600	POPL ASPN 600	POPL ASPN 650	HYBD POPL 450	ASPN BROK 500	ASPN BROK 550	HYBD POPL 504	ASPN BROK 504	WHIT SPRC 500	RED MAPL 508	IEA POPL 504	ULEX 506	BIO SHEL 504	HOG FUEL 500	WEST HEML	PINE BR/S 490
	TEMPERATURE (Celsius)	450	500	550	600	600	650	450	500	550	504	504	500	508	504	506	504	500		490
	CITATION (First Author)	SCOT						SCOT			PISK	PISK							PISK	ELDR
	YEAR	82-						84-			86-	88-							93-	80-
	PAGE NO.	666						404			121	167							1432	217
	RUN NO.	44	37	39	42	43	45	33	31	32	59	58	43	63	A-2	142	90	126	R-78	
FURANS																				
68	FURAN																			
82	2-METHYLFURAN																			
96	2-FURANONE						·													
96	FURFURAL (2-FURALDEHYDE)							0.4	0.3	*	0.7	0.6	0.4							
98	3-METHYL-2(3H) FURANONE																			
98	FURFURAL ALCOHOL																			
110	5-METHYLFURFURAL										0.3	0.3	0.1							
126	5-HYDROXYMETHYL-2-FURALDEHYDE													0.5						
MIXED OXYGENATES																				
58	GLYOXAL (ETHANEDIAL)											1.9	3.2	2.2	2.8	1.5	0.9	0.9	1.1	
60	HYDROXYETHANAL (HYDROXYACETALDEHYDE; GLYCOALDEHYDE)										11.5	8.9	9.6	9.7	13.0	8.1	8.3	6.4	11.4	
62	1,2-DIHYROXYETHANE (ETHYLENE GLYCOL)																			
72	PROPANAL-2-ONE (METHYL GLYOXAL, 2-OXOPROPANAL)											1.2	1.1	0.8	1.4 / 0.8	2.0	1.2	0.7		
74	1-HYDROXY-2-PROPANONE (ACETOL)										3.8	2.3	1.5	1.5	1.8		2.7	2.6	3.6	
74	2-HYDROXYPROPANAL (METHANOLACETALDEHYDE)																			
86	BUTYROLACTONE (gamma or beta)																			
100	2,3-PENTENEDIONE																			
110	1,2-DIHYDROXYBENZENE (CATECHOL)																			
110	1,3-DIHYDROXYBENZENE (RESORCINOL)																			
110	1,4-DIHYDROXYBENZENE (HYDROQUINONE)																			
112	2-HYDROXY-3-METHYL-2-CYCLOPENTENE-I-ONE																			
112	METHYLCYCLOPENTENOLONE																			
126	2-METHYL-3-HYDROXY-2-PYRONE																			
152	4-HYDROXY-3-METHOXYBENZALDEHYDE (VANILLIN)																			
	Notes	a,b									*									

TABLE OF BIO-OIL COMPOSITION

COMPOSITION	OAK	WOOD	WOOD	POP TREM	POP TREM	POP TREM	POP TREM	POP DELT	POP DELT	POP DELT	POP DELT	SOFT WOOD	SOFT WOOD	ASPN	POP DELT	POP DELT	IEA POPL	IEA POPL	MAPL	MAPL
TEMPERATURE (Celcius)	~500	383	476	540	450	450	525	450	450	450	450	525	525	500			500	504	508	508
CITATION (First Author)	CZER	PEAC	DICK	MENA	ROY	MENA		PAKD				PAKD		MCKN			WLFP	WLFP	WLFP	WLFP
YEAR	96-	96-	96-	82-	83-	84-		87-				96		89						
PAGE NO.	TBP	PC	PC	331	1147	418		155				124		SERI	LVVP	LVVP				
RUN NO.	154	CR27	PC	-			F.B.	24	25	23	66	643	644	01	06	03	01	02	08	05
WET OIL (WT. % Moisture-free wood)	66	68	68	71	~62			65	65	66		74	70	64	67	65	79	77	77	24
CHAR (WT. % Moisture-free wood)	12	17	16	18	~25							16	15	12	21	25	12	12	14	14
GASES (WT. % Moisture-free wood)	12	11	13	11	~13							11	12	16	11	9.7	20	11	10	9.8
WATER (WT. % Moisture-free wood)	10	11	12	23	~16	17		17	15	16				17	1.9		11	19	19	12
ORGANIC LIQUID (WT. % Moisture-free wood)	56	57	56	48	46	48		48	50	50										
APPARATUS	VORT	ABLT	FB	VAC	VAC	VAC	FB	VAC	VAC	VAC	VAC	VAC	VAC							
RESIDENCE TIME (Seconds)		0.8		<10	-2	10	10													
PRESSURE (torr)		760					760	40	30	12	10	4	4							
MOISTURE IN WOOD (WT. %)			9.3					5.9	5.9	5.9	5.9	4.8	5.4							
ASH IN WOOD (WT. %)	0.1	0.4						0.6	0.6	0.6	0.6	0.0	0.0							
COMPOSITION WT. % WET OIL																				
ACIDS																				
46 FORMIC (METHANOIC)	5.0			3.2	3.1	3.9	1.4	3.7	4.9	5.2				1.0	0.5	2.5	1.3	1.5	1.5	1.0
60 ACETIC (ETHANOIC)	7.6	9.9	4.9	9.3	8.8	9.7	4.3	8.2	8.5	7.9				4.1	0.4	3.7	2.3	3.1	2.9	1.0
74 PROPANOIC (PROPIONIC)				0.4	0.7	1.0	1.0	0.2	0.2	0.2				0.3	0.1	0.3	0.1	0.6	0.1	0.2
76 GLYCOLIC (HYDROXYACETIC)				0.9	0.8			*						0.5	0.7	0.7	0.6	0.7	0.7	0.2
88 BUTANOIC (BUTYRIC)					0.2	0.1	0.1	*	0.5					0.4	0.3	0.3	*	0.2	0.2	0.3
102 PENTANOIC (VALERIC)					0.7	0.8	0.2	0.1	*	0.1				0.3	0.3	0.2	0.1	0.1	0.1	0.2
116 4-OXOPENTANOIC																				
116 HEXANOIC (CAPROIC)				*	*	0.2	0.1	*						0.3	0.3	0.3	0.1	0.1	0.1	0.3
122 BENZOIC														*		0.3				
130 HEPTANOIC				*	*									*		0.3				
SUGARS																				
150 D-XYLOSE	1.4																			
162 1,6-ANHYDROGLUCOFURANOSE																	0.1	0.1		0.2
162 LEVOGLUCOSAN (1,6-ANHYDRO-beta-D-GLUCOPYRANOSE)	5.8	7.9	15.0	0.8		0.8	1.1							2.0	5.4	1.2	0.9	1.0	1.5	
180 alpha-D-GLUCOSE (alpha-D-GLUCOPYRANOSE)						*	0.5										*			
180 FRUCTOSE																				
CELLOBIOSAN			0.7																	
ALCOHOLS																				
32 METHANOL																				
46 ETHANOL																				

416

TABLE OF BIO-OIL COMPOSITION

COMPOSITION	OAK	WOOD	WOOD	POP TREM	POP TREM	POP TREM	POP TREM	POP DELT	POP DELT	POP DELT	SOFT WOOD	SOFT WOOD	POP ASPN	POP DELT	POP DELT	IEA POPL	IEA POPL	MAPL	MAPL
TEMPERATURE (Celsius)	-500	383	476	540	450	450	525	450	450	450	525	525	500			500	504	508	508
CITATION (First Author)	CZER	PEAC	DICK	MENA	ROY	MENA		PAKD			PAKD		MCKN						
YEAR	96-	96-	96-	82-	83-	84-		87-			96		89						
PAGE NO.	TBP	PC	PC	331	1147	418		155			124		SERI	LVVP	LVVP	WLFP	WLFP	WLFP	WLFP
RUN NO.	154	CR27		-		VAC	F.B.	24	25	23	643	644	01	06	03	01	02	08	05
KETONES																			
70 2-BUTENONE																			
72 2-BUTANONE																		0.1	
84 CYCLO PENTANONE																			
96 3-METHYL-2-CYCLOPENTEN-1-ONE						X	0.6												
112 2-ETHYLCYCLOPENTANONE																			
112 DIMETHYLCYCLOPENTANONE																			
124 TRIMETHYLCYCLOPENTENONE																			
ALDEHYDES																			
30 METHANAL (FORMALDEHYDE)	3.3																		
44 ETHANAL (ACETALDEHYDE)																			
84 2-METHYL-2-BUTENAL																			
PHENOLS																			
94 PHENOL			0.2	0.5		0.4	0.4				X		0.7	0.3	0.5	0.3			
108 2-METHYL PHENOL o (o-CRESOL)			0.1								X	X	X	0.1	0.1	0.1			
108 3-METHYL PHENOL m (m-CRESOL)			0.1				0.2				X	0.1	0.1	0.3	0.1		X		X
108 4-METHYL PHENOL p (p-CRESOL)						0.3	1.0												
122 2,3-DIMETHYLPHENOL (2,3-XYLENOL)			0.1																
122 2,4-DIMETHYLPHENOL (2,4-XYLENOL)							0.1												
122 2,5-DIMETHYLPHENOL (2,5-XYLENOL)							0.4							0.1					
122 2,6-DIMETHYLPHENOL (2,6-XYLENOL)							1.3												
122 2-ETHYLPHENOL							X												
136 2,3,5 TRIMETHYL PHENOL																			
GUAIACOLS																			
124 GUAIACOL (2-METHOXYPHENOL)			1.1	0.2		0.2	0.3				0.2	0.4			0.4				
138 4-METHYL GUAIACOL			1.9								0.4	0.6							
152 4-ETHYL GUAIACOL			0.3								0.1								
164 4-PROPENYL GUAIACOL																			
164 EUGENOL (PHENOL, 2-METHOXY-4-(2-PROPENYL))			0.6			0.2	0.2				0.3								
164 ISOEUGENOL (PHENOL, 2-METHOXY-4-(1-PROPENYL))			4.6			0.1	0.1				0.6	1.4							
166 4-PROPYLGUAIACOL																			
PYROLYTIC LIGNIN, WATER INSOLUABLE	38.0																		
SYRINGOLS																			
154 SYRINGOL (2,6-DIMETHOXY PHENOL)			0.2				0.4						0.7		0.7	0.3		0.2	
168 METHYLSYRINGOL																			
182 4-ETHYLSYRINGOL																			
182 SYRINGALDEHYDE						0.2	0.7						0.4	0.1	0.3	0.4	0.4	0.1	1.5
194 4-PROPENYLSYRINGOL (4-ALLYLSYRINGOL)																		X	
196 4-HYDROXY-3,5-DIMETHOXYPHENYL ETHANONE																			

417

TABLE OF BIO-OIL COMPOSITION

	OAK	WOOD	WOOD	POP TREM	POP TREM	POP TREM	POP TREM	POP DELT	POP DELT	POP DELT	SOFT WOOD	SOFT WOOD	ASPN	POP DELT	POP DELT	IEA POPL	IEA POPL	MAPL	MAPL
TEMPERATURE (Celcius)	~500	383	476	540	450	450	525	450	450	450	525	525	500			500	504	508	508
CITATION (First Author)	CZER	PEAC	DICK	MENA	ROY	MENA		PAKD			PAKD		MCKN						
YEAR	96-	96-	96-	82-	83-	84-		87-			96		89						
PAGE NO.	TBP	PC	PC	331	1147	418	F.B.	155			124		SERI	LVVP	LVVP	WLFP	WLFP	WLFP	WLFP
RUN NO.	154	CR27		-		VAC		24	25	23	643	644	01	06	03	01	02	08	05
FURANS																			
68 FURAN																			
82 2-METHYLFURAN																			
96 2-FURANONE			1.1																
96 FURFURAL (2-FURALDEHYDE)			1.1	0.7		0.6							0.7		0.5	0.4	0.5	0.4	0.5
98 3-METHYL-2(3H) FURANONE							0.5								0.1				
98 FURFURAL ALCOHOL			0.1	0.1		0.2	0.1						0.1						
110 5-METHYLFURFURAL													0.1						
126 5-HYDROXYMETHYL-2-FURALDEHYDE			2.2																×
MIXED OXYGENATES																			
58 GLYOXAL (ETHANEDIAL)	4.6	1.7																	
60 HYDROXYETHANAL (HYDROXYACETALDEHYDE; GLYCOALDEHYDE)	6.5	12.0	3.0																
62 1,2-DIHYROXYETHANE (ETHYLENE GLYCOL)																			
72 PROPANAL-2-ONE (METHYL GLYOXAL, 2-OXOPROPANAL)																			
74 1-HYDROXY-2-PROPANONE (ACETOL)	2.7	7.1	6.7			1.2	3.2						4.1	0.8	3.6	1.7	2.9	1.8	
74 2-HYDROXYPROPANAL (METHANOLACETALDEHYDE)						2.2	1.1												
86 BUTYROLACTONE (gamma or beta)						0.7	0.2						0.3		0.7	0.3	0.2	0.2	0.9
100 2,3-PENTENEDIONE																			
110 1,2-DIHYDROXYBENZENE (CATECHOL)						0.5	0.3				0.7	0.3							
110 1,3-DIHYDROXYBENZENE (RESORCINOL)						0.1	0.3					0.3							
110 1,4-DIHYDROXYBENZENE (HYDROQUINONE)						0.1	0.1					0.1							
112 2-HYDROXY-3-METHYL-2-CYCLOPENTENE-I-ONE			0.5				0.2								0.5				
112 METHYLCYCLOPENTENOLONE													0.4	0.3					
126 2-METHYL-3-HYDROXY-2-PYRONE						0.2	0.3						0.3	0.2					
152 4-HYDROXY-3-METHOXYBENZALDEHYDE (VANILLIN)			1.1																
Notes	k		l	w			c				d	d	e	e,n	e,n	e,o	e	e	e,p

TABLE OF BIO-OIL COMPOSITION

COMPOSITION	SPRU	ASPN POPL	IEA POPL	IEA POPL	IEA POPL	IEA POPL	IEA POPL	POPL	ASPN	OAK	OAK	FIR	FIR	ASPN	EUCL	WOOD
TEMPERATURE (Celcius)	500	500	800	700	750	750	750			450	485	400	550			
CITATION (First Author)	WLEP	WLFP	WO	WO	WO	WO	WO	ELLI	WAPO	ELLI	SERI	HILE		OASM		SOLA
YEAR	06	07	09	13	15	16	17	88		86		76-49	76-49	96		94-
PAGE NO.	78	73						1062		5943				175		VTT
RUN NO.								VAAS		GATE	SERI					RANG
WET OIL (WT. % Moisture-free wood)	78															
CHAR (WT. % Moisture-free wood)	12	16														
GASES (WT. % Moisture-free wood)	7.8	12										~10	~30			
WATER (WT. % Moisture-free wood)	26	19	19	7.7	21	5.2	64	17			16	18	15			
ORGANIC LIQUID (WT. % Moisture-free wood)																
APPARATUS														CETB	FB	LIT
RESIDENCE TIME (Seconds)			0.055	0.47	0.297	0.128	0.088									
PRESSURE (torr)																
MOISTURE IN WOOD (WT. %)																
ASH IN WOOD (WT. %)																
COMPOSITION WT. % WET OIL																
ACIDS																
46 FORMIC (METHANOIC)	1.5	1.3	0.4	0.4	0.8			1.2						1.1	3.5	.3-20
60 ACETIC (ETHANOIC)	1.2	2.9	0.1	1.6	1.8			12.0	7.0			0.6	0.7	6.6	3.4	5-32
74 PROPANOIC (PROPIONIC)				0.2	0.2									0.3	0.1	.1-3
76 GLYCOLIC (HYDROXYACETIC)	0.8	0.6	0.2	0.2	0.2									0.1	1.4	.1-7
88 BUTANOIC (BUTYRIC)	0.2	*	0.1	0.3	0.3				*					*		.1-4
102 PENTANOIC (VALERIC)	0.1	0.1		0.2	0.2											.1-3
116 4-OXOPENTANOIC									*			0.1	0.4			0.4
116 HEXANOIC (CAPROIC)	0.1	*														.1-3
122 BENZOIC							0.3									.2-3
130 HEPTANOIC																
SUGARS																
150 D-XYLOSE	0.1	0.2														0.1-2
162 1,6-ANHYDROGLUCOFURANOSE																3.1
162 LEVOGLUCOSAN (1,6-ANHYDRO-beta-D-GLUCOPYRANOSE)	2.2	1.5	0.8	1.4	0.9	2.3	0.9	>0.6		>0.6	>0.7					4-3.9
180 alpha-D-GLUCOSE (alpha-D-GLUCOPYRANOSE)										0.5						
180 FRUCTOSE																1.7
CELLOBIOSAN																1.7
ALCOHOLS																
32 METHANOL								2.4	1.2			0.3	0.4			1.2-2.4
46 ETHANOL																

TABLE OF BIO-OIL COMPOSITION

COMPOSITION	SPRJ 500 WLFP 06	ASPN POPL 500 WLFP 07	IEA POPL 800 WO 09	IEA POPL 700 WO 13	IEA POPL 750 WO 15	IEA POPL 750 WO 16	IEA POPL 750 WO 17	POPL ELLI 88 1062 VAAS	ASPN WAPO	OAK 450 ELLI 86 5943 GATE	OAK 485 SERI	FIR 400 HILE 76-49	FIR 550 76-49	ASPN OASM 96 175	EUCL	WOOD SOLA 94- VTT RANG
KETONES																
70 2-BUTENONE		*														
72 2-BUTANONE								0.3	*							.1-.2
84 CYCLO PENTANONE									0.9							.1-.9
96 3-METHYL-2-CYCLOPENTEN-1-ONE				0.1	0.2				*							.1-.5
112 2-ETHYLCYCLOPENTANONE								0.3								0.2
112 DIMETHYLCYCLOPENTANONE								0.3								.1-.5
124 TRIMETHYLCYCLOPENTANONE								0.2								
ALDEHYDES																
30 METHANAL (FORMALDEHYDE)												0.1	0.9			1.5
44 ETHANAL (ACETALDEHYDE)									0.2			0.2	1.4			0.1-.4
84 2-METHYL-2-BUTENAL										0.4	0.3		0.1			0.4-.5
PHENOLS																
94 PHENOL		0.3		0.4	0.9	1.0		0.4	0.2	0.2	*					3-3.8
108 2-METHYL PHENOL o (o-CRESOL)		0.1		0.3	0.5	0.6		0.1	0.1	0.1	*					.1-.3
108 3-METHYL PHENOL m (m-CRESOL)				0.2	0.3	0.4		0.1	*	0.1	0.1					.1-.2
108 4-METHYL PHENOL p (p-CRESOL)		*		0.2	0.5	0.5										.1-.3
122 2,3-DIMETHYLPHENOL (2,3-XYLENOL)																.1-.3
122 2,4-DIMETHYLPHENOL (2,4-XYLENOL)				0.1	0.2	0.2										.1-.3
122 2,5-DIMETHYLPHENOL (2,5-XYLENOL)						0.1										.2-.4
122 2,6-DIMETHYLPHENOL (2,6-XYLENOL)																
122 2-ETHYLPHENOL				0.1	0.1	0.7										.1-.2
136 2,3,5 TRIMETHYLPHENOL						*										
GUAIACOLS																
124 GUAIACOL (2-METHOXYPHENOL)								0.2	0.1	0.1	0.2	0.2	0.3			.1-.2
138 4-METHYL GUAIACOL								0.3	0.1	0.1	0.1	1.0	0.6			.1-.6
152 4-ETHYLGUAIACOL								0.1	*	*	*	0.4	0.1			
164 4-PROPENYL GUAIACOL								0.1	0.1	0.1	0.2					
164 EUGENOL (PHENOL, 2-METHOXY-4- (2-PROPENYL))	0.1															.1-2.3
164 ISOEUGENOL (PHENOL, 2-METHOXY-4-(1-PROPENYL))		*														2.4-7.2
166 4-PROPYLGUAIACOL												0.3				.1-.3
PYROLYTIC LIGNIN, WATER INSOLUABLE																
SYRINGOLS																
154 SYRINGOL (2,6-DIMETHOXY PHENOL)		0.3						0.4	0.2	0.2						.7-4.8
168 METHYLSYRINGOL								0.3	0.1	0.1						
182 4-ETHYLSYRINGOL								0.2	*	*						
192 SYRINGALDEHYDE		0.3						0.1	0.2	0.2						4-1.5
194 4-PROPENYLSYRINGOL (4-ALLYLSYRINGOL)								0.3	0.2	0.1						
196 4-HYDROXY-3,5-DIMETHOXYPHENYL ETHANONE								0.1	0.1	0.1						.1-.3

420

TABLE OF BIO-OIL COMPOSITION

COMPOSITION	SPRU	ASPN POPL	IEA POPL	IEA POPL	IEA POPL	IEA POPL	IEA POPL	POPL	ASPN	OAK	OAK	FIR	FIR	ASPN	EUCL	WOOD
TEMPERATURE (Celcius)	500	500	800	700	750	750	750			450	485		550	400		
CITATION (First Author)	WLFP	WLFP	WO	WO	WO	WO	WO	ELLI	ELLI		SERI	HILE		OASM		SOLA
YEAR								88	86			76-	76-	96		94-
PAGE NO.								1062	5943			49	49	175		VTT
RUN NO.	06	07	09	13	15	16	17	VAAS WAPO	GATE							RANG
FURANS																
68 FURAN												0.1				0.2
82 2-METHYLFURAN								0.2	0.1							0.1-0.2
96 2-FURANONE								0.9	0.6							.1-.9
96 FURFURAL (2-FURALDEHYDE)	0.2	0.4		0.1								0.6	0.5			.3-.7
98 3-METHYL-2(3H) FURANONE					0.1				✳			0.1				
98 FURFURAL ALCOHOL						✳						0.2	0.3			5.2
110 5-METHYLFURFURAL						0.1						0.1				
126 5-HYDROXYMETHYL-2-FURALDEHYDE												0.4	0.3			2.-2.6
MIXED OXYGENATES																
58 GLYOXAL (ETHANEDIAL)																2.8
60 HYDROXYETHANAL (HYDROXYACETALDEHYDE; GLYCOALDEHYDE)								5.7	11.4			0.9	0.9			12.9
62 1,2-DIHYROXYETHANE (ETHYLENE GLYCOL)																1.3
72 PROPANAL-2-ONE (METHYL GLYOXAL, 2-OXOPROPANAL)												0.9	0.6			1.2
74 1-HYDROXY-2-PROPANONE (ACETOL)	1.5	2.2		1.1	1.3	2.2		3.2	3.3			1.0	0.7			1.5-7.4
74 2-HYDROXYPROPANAL (METHANOLACETALDEHYDE)								0.9	1.4							
86 BUTYROLACTONE (gamma or beta)	0.1			0.2	0.5											.3-.7
100 2,3-PENTENEDIONE												0.1	0.1			
110 1,2-DIHYDROXYBENZENE (CATECHOL)																.1-.5
110 1,3-DIHYDROXYBENZENE (RESORCINOL)																0.1
110 1,4-DIHYDROXYBENZENE (HYDROQUINONE)				0.1												.1-1.9
112 2-HYDROXY-3-METHYL-2-CYCLOPENTENE-1-ONE									0.1	0.1		0.1	✳			
112 METHYLCYCLOPENTENOLONE												0.1				.1-1.9
126 2-METHYL-3-HYDROXY-2-PYRONE										0.3	0.3		✳			.2-.4
152 4-HYDROXY-3-METHOXYBENZALDEHYDE (VANILLIN)									0.8							
Notes	e	e	e	e	e	e	e	e,m	e,m	e,f,q	e,g,r	l,s	i,t	j,u	j,u	v

KEY TO REFERENCES IN TABLE

CZER-96-PC

    Czernik, S.; Wang, D.; Montane D.; Chornet, E. (1996). "Catalytic Steam Reforming of Biomass-Derived Fractions from Pyrolysis Processes." Paper presented at Developments in Thermochemical Biomass Conversion, Banff, AL, Canada, May 19-25, 1996.

DICK-96-PC

    Dick, C. (1996) Private communication. Aston University, U.K.

ELDR-80-217

    Elder, T.J.; Soltes, E.J. (1980). "Pyrolysis of Lignocellulosic Materials; Phenolic Constituents of a Wood Pyrolytic Oil," *Wood and Fiber*, pp. 217–226.

ELLI-88-1062

    Elliott, D.C. (1988). "Volume 4: Analysis and Upgrading of Biomass Liquefaction Products," Pacific Northwest Laboratory, DOE/NBM—1062, Richland, WA, pp. 5–89.

ELLI-86-5943

    Elliott, D.C. (1986). "Analysis and Comparison of Biomass Pyrolysis/Gasification Condensates: Final Report," Pacific Northwest Laboratory, PNL-5943, Richland, WA (USA), June, 100 pp. (This report superseded PNL-5555)

FONT-86-491

    Font, R.; Marcilla, A.E.; Verdú; Devesa, J. (1989). "Chemicals from Almond Shells by Pyrolysis in Fluidized Bed," *Pyrolysis and Gasification*, G.L. Ferrero, K. Maniatis, A. Buekens, A.V. Bridgwater, eds., Elsevier Applied Science, London and New York, pp. 230–237.

HILE-76-49

    Hileman, F.D.; Wojcik, L.H.; Futrell, J.H.; Einhorn, I.N. (1976). "Degradation Products of ∝-Cellulose and Douglas Fir Under Inert and Oxidative Environments". *Thermal Uses and Properties of Carbohydrates and Lignins* ed. F. Shafizadeh, K.V. Sarkanen, D.A. Tillman, Academic Press.

MCKN-89-SERI et al.

    McKinley, J. (1989). "Biomass Liquefaction: Centralized Analysis," Final Report and Appendices, July, Contract OSV84-00223, Science Branch, Science Procurement, Dept. of Supply and Services, Hull, Quebec. Project No. 4-03-837. B.C. Research.

MENA-82-331

    Ménard, H.; Roy, C.; Gaboury, A.; Belanger, D. (1982). "Analyse Totale de Pyroligneux Provenant de *Populus tremuloides* Par HPLC," *Proceedings Fourth Bioenergy R&D Seminar*, 29-31 March, Winnipeg, Manitoba, Canada, pp. 331-335.

MENA-84-418

    Ménard, H.; Belanger, D.; Chauvette, G.; Gaboury, A.; Khorami, J.; Grise, M.; Martel, A.; Potvin, E.; Roy, C.; Langlois, R. (1984). "Characterization of Pyrolytic Liquids from Different Wood Conversion Processes," Fifth Canadian Bioenergy R&D Development Seminar, S. Hasnain, ed., National Research Council of Canada, Ottawa, Ontario, Canada, Cuadra/Elsevier, p. 418–434.

OASM-96-175

    Oasmaa, A.; Sipila, K. (1996). "Pyrolysis Oil Properties: Use of Pyrolysis Oil as Fuel in Medium-Speed Diesel Engines." *Bio-Oil, Production and Utilization*. Ed. A. V. Bridgwater; E. N. Hogan, CPL Scientific Information Ltd., Newburg, U.K.

PAKD-96-124

Pakdel, H.; Amen-Chen, C.; Chang, J.; Roy, C. (1996) "Phenolic Compounds from Vacuum Pyrolysis of Biomass." pp 124-136 in *Bio-Oil, Production and Utilization.* Ed. A. V. Bridgwater; E. N. Hogan, CPL Scientific Information Ltd., Newburg, U.K.

PALM-93-947

Palm, M.; Piskorz, J.; Peacocke, C.; Scott, D.S. (1993). "Fast Pyrolysis of Sweet Sorghum Bagasse in a Fluidized Bed," *Proceedings Volume II: First Biomass Conference of the Americas: Energy, Environment, Agriculture, and Industry,* August 30-September 2, 1993, Burlington, Vermont, pp. 947–963.

PEAC-96-PC

Peacocke, C. (1996). Private Communication. Aston University, U.K.

PISK-88-167

Piskorz, J.; Scott, D.S.; Radlein, D. (1988). "Composition of Oils Obtained by Fast Pyrolysis of Different Woods", *Pyrolysis Oils from Biomass—Producing, Analyzing, and Upgrading, Chapter 16,* ACS Symposium 376, American Chemical Society, Washington, D.C., pp. 167–178.

PISK-92-64

Piskorz, J.; Scott, D.S.; Radlein, D.; Czernik, S. (1992). "New Applications of the Waterloo Fast Pyrolysis Process, *Proceedings of the First Canada/European Community R&D Contractors Meeting October, 1990, Ottawa, Canada,* E. Hogan, J. Robert, G. Grassi, A.V. Bridgwater, eds., pp. 64–73.

PISK-93-1432

Piskorz, J.; Radlein, D.; Scott, D.S. (1993). "Thermal Conversion of Cellulose and Hemicellulose to Sugars," *Proceedings: Inter. Sympos. on Advances in Thermochemical Biomass Conversion,* A.V. Bridgwater, ed., Blackie Academic & Professional, New York, pp. 1432–1440.

PISK-86-121

Piskorz, J.; Radlein, D.; Scott, D.S. (1986). "On the Mechanism of the Rapid Pyrolysis of Cellulose," *Journal of Analytical and Applied Pyrolysis,* Vol 9, pp. 121–137.

ROY-83-1147

Roy, C.; deCaumia, B.; Ménard, H. (1983). "Production of Liquids from Biomass by Vacuum Pyrolysis—Development of Data Base for Continuous Process," *Energy from Biomass and Wastes VII,* Institute of Gas Technology, Chicago, IL, pp. 1147–1170.

SCOT-85-581

Scott, D.S.; Piskorz, J.; Radlein, D. (1985). " Continuous Flash Pyrolysis of Biomass," *Industrial & Engineering Chemistry Process Des. Development,* Vol. 24, No. 3, pp. 581–587.

SCOT-82-666

Scott, D.S.; Piskorz, J. (1982). "The Flash Pyrolysis of Aspen-Poplar Wood," Canadian *J. Chem. Eng.,* Vol. 60, No. 5, pp. 666–674.

SCOT-84-404

Scott, D.S.; Piskorz, J. (1984). "The Continuous Flash Pyrolysis of Biomass," *The Canadian Journal of Chemical Engineering,* Vol. 62, Institute of Gas Technology, Chicago, Illinois, pp. 404–412.

SOLA-94-VTT

Solantausta, Y.; Diebold, J.; Elliott, D.C.; Bridgwater T.; Beckman, D. (1994). "Assessment of Liquefaction and Pyrolysis Systems," Research Notes, PB–95-129797/XAB, VTT/RN-1573, pp. 206.

NOTES TO TABLE

*	Less than 0.1 wt.%
a.	BFB = Bench-scale fluidized bed; PPFB = Pilot-plant fluidized bed; GASF = Updraft gasifier; VORT = Vortex reactor; VAC = vacuum pyrolysis reactor; ABLT = Ablative reactor; FB = Fluid bed reactor; LIT = Literature review; CETB = ENSYN system;
b.	Wood basis varies
c.	Sherbrooke analysis of Waterloo oil #32. 0.2 percent of gamma-valerolactone and trace of 2, 4, 6-trimethyl phenol also reported
d.	Phenolic faction only
e.	Centralized analysis of submitted oils
f.	Oil from Georgia-Tech, Tech-Air process
g.	SERI (NREL) ablative, vortex oil
i.	Analytical pyrolysis directly into a mass spectrometer. Results are on a wt.% feed basis.
j.	Analysis of submitted oils
k.	Also reported: xylitol, 1.7; 2-furoic acid, 0.4; and cyclotene, 0.6 wt. % of wet oil
l.	Also reported: 3-Methoxycatechol, 0.6; Homovanillin, 0.6; and Acetoguaiacone, 0.8; wt.% of wet oil
m.	Also reports: 0.8 wt.% $\propto$-oxy-propioguaiacone; 0.2 wt.% trimethylcyclopentenone; 1.9% methyl formate; 0.1% 2-furanyl; 0.1% 1-acetyloxy-2-propanone; and 0.7% dimethylcyclopentene in VAAS oil and 0.1% methylformate; traces of 2-butenal; 2-furanylethanone; 1-acetyloxy-2-propanone; and 3-pentanone in the WAPO oil.
n.	Also reports 0.1 wt.% of D-arabinose
o.	Also reports 0.3 wt.% of 2, 4, 6-trimethyl phenol
p.	Also reports 0.6 wt.% of 1,2,3-trihydroxy benzene
q.	Also reports 0.2 wt.% naphthalene; 0.1% acenaphthalene, and traces of phenanthrene, fluoranthene, and acephenanthrene.
r.	Also reports 0.2 wt.% of naphthalene and 0.3 wt.% of phenanthrene.
s.	Also reports traces of cyclohexanone and 2-methylpropene plus 0.1 wt.% of 2,3-pentenedione; 0.4 wt.% 2-methoxy-4-methylanisole; 0.2 wt.% 4-hydroxypentanoic acid; and 0.6 wt.% acrolein
t.	Also reports 0.1 wt.% of 2-methyl-2-butenal, 0.4 wt.% of 2-methylpropene, 0.1 wt.% 2,3-pentenedione, 0.2 wt.% 4-hydroxypentanoic acid; 0.9 wt.% acrolein; 0.1 wt.% 2-methoxy-4 methylanisole
u.	Also reports 8.3 and 8.4 wt.% of "total acids" in the aspen and eucalyptus respectively
v.	Also reports 0.1 wt.% D-arabinose; 0.1-1.9% methylformate; 0.4% cyclohexanone; 0.1-1.2% alpha-angelicalactone; 0.1% 1-acetyloxy-2-propanone; .1-.2% acetal; .1-2% benzene, 0.1% toluene, 2.8% acetone, .2-.4% trimethylcyclopentanone
w.	Also reports 0.5 wt.% pentanal

CHARACTERISTICS OF PYROLYSIS OIL AND CHAR FROM OIL PALM SHELLS

F. N. ANI
Fakulti Kejuruteraan Mekanikal
Universiti Teknologi Malaysia
Sekudai, Johor Darul Takzim, Malaysia
R. ZAILANI
Engineering Department - American Degree Program
MARA Institute of Technology
Shah Alam, Selangor, Malaysia

Abstract

Oil Palm cultivation in Malaysia has provided the world's largest producer of crude palm oil with more than 7 million tonnes in 1994. Besides producing the crude oil, it also generate the solid waste ie. palm shells, empty fruit bunches and fruit fibres. Part of these waste are used to generate the energy to run the palm oil mill. Surplus amount of hard palm shells are available and could be converted into activated carbon after the conversion to palm char and pyrolysis oil.

Preliminary studies were made on the characteristics of palm oil shells in terms of size distribution, physical and chemical properties and the thermal behaviours using TGA. Later an investigation was conducted on the char and liquid derived fuel from oil palm solid waste via fast pyrolysis. For this purpose, fluidised bed pyrolysis were conducted in an inert bed at varying temperature from 400 to 600°C. The liquid product was analysed for its properties and compared with other biomass pyrolysis oils and petroluem fuels. The influence of some process conditions on the relative proportions of the liquid and solid products together with its properties and characteristics are presented. Some data on palm shell char characteristics were also given.

Keywords: Fluidised bed pyrolysis, oil palm shell, palm char, pyro-crude oil, thermal analysis(TGA).

1 Introduction

Palm oil is one of the main commodity crop cultivated in Malaysia which produced more than 7 million tonnes of crude palm oil in 1994. Besides producing oil, the palm oil mills also produce solid waste as well as effluent which needed to be taken care off. The effluent at present are treated by means of ponding system which later with permissible BOD, it was channel to the downstream of the water source. The solid waste which consist mainly of palm fruit shells, fruit fibers and empty fruit bunches were incinerated for power and fertiliser purposes respectively. Burning of palm shells are less favourable due to combustion and emissions problems. The fibers alone are sufficient for the power generation of the mills. Other means of utilising the waste shells should be exploited and developed, thus increasing the value-added products for the maximum utilisation.

The recovery of pyro-crude oil from the solid waste have centred on the thermochemical process and amongst which fast pyrolysis have received increasing attention since the process conditions may be optimised to produce high energy density oil and char for activated carbon. The pyrolysis oil on the other hand could be converted to high grade fuel and chemical feedstocks which is easily transported or stored and has a high flexibility in its uses, Besler et. al. 1992. For this study, fast pyrolysis has been projected as the most competitive method of producing liquid fuel as well as activated carbon from the palm oil shells.

2 Characteristics of palm fruit shells

Physical and chemical characteristics of palm shells are essential for the designing of the conversion process. The typical elemental composition, proximate analysis and gross calorific value of palm shell are given in Table 1.

Table 1. Composition of palm fruits shells

Elemental composition % wt (dry ash free)		Proximate analysis % wt (air dry)		Gross CV (MJ/kg)
C	55.35	Volatile	68.8	
H	6.27	Fixed Carbon	20.3	19.56
N	0.37	Moisture	8.4	
O	38.1	Ash	2.5	
S	-			

Palm fruit shells contain a high percentage of volatile matter and since palm shells are the waste by-product of palm oil milling, the size of the waste does not varies much due to the nature of the fruit's shape. The average bulk density is about 440 kg/m^3 for < 18mm size. Palm shells are also granular in size and the distribution after the milling process is given in Figure 1. This reduces the initial preparation cost prior to the thermal processing since the size are smaller and there are excess energy from the mill for the purpose of further reducing the size. In this research work, the palm shells were grounded and sieved to study the effect of size and also to suit the size of the thermal conversion system that have been constructed.

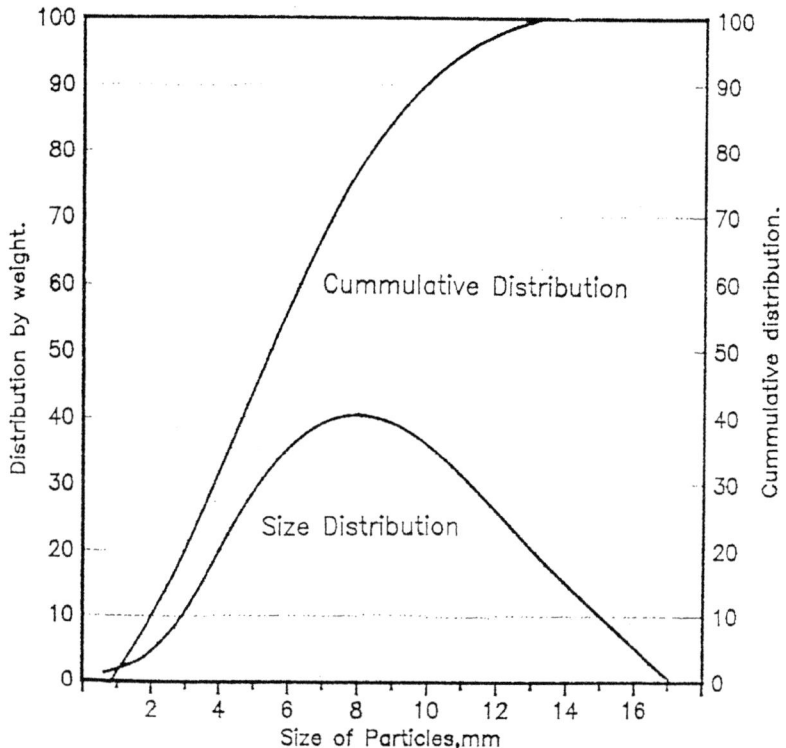

Fig. 1. The distribution of oil palm shell residues.

The thermal characteristics of palm shell was analysed by thermogravimetric analyser(TGA). Palm shell samples were pyrolysed in an inert atmosphere of nitrogen gas over a temperature range from ambient to 900°C at various heating rates. The loss in weight of the sample was continuously recorded as a function of temperature and time. The thermograms give information about the temperature at which pyrolysis is initiated, when the rate of devolatilisation is at its maximum and finally the temperature at which the process is completed. They also give an indication of the fractional weight of the volatile in the sample.

Figure 2 shows the thermogram from the TGA produced when the palm shells sample was subjected to a heating rate of 25°C/min. The inherent moisture is initially driven out at temperature less then 110°C which represent about 8% of the total weight. The devolatilisation initiates at about 230°C and approaches completion by 600°C. From the DTG curve, two distinct peaks at temperature around 315°C at a rate of develotilisation of 0.18 weight loss/°C and at 385°C at a rate of 0.15 weight loss/°C, which are common to all agricultural wastes ie. the decomposition of hemicellulose and cellulose respectively.

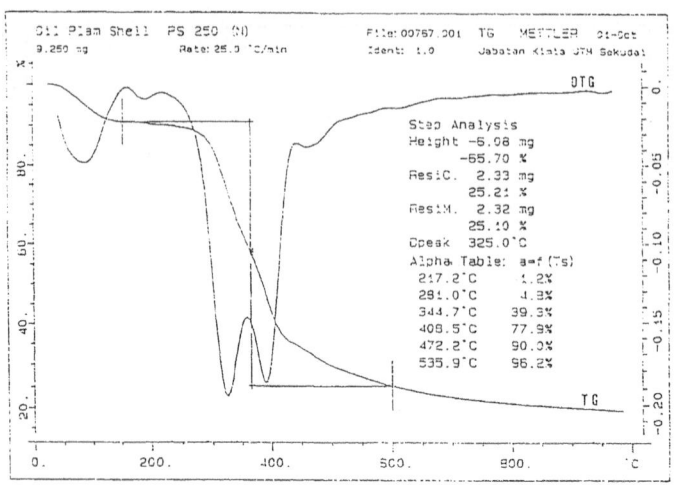

Fig. 2. TGA/DTG of palm shell subjected to heating rate of 25°C/min in nitrogen gas.

3 Experimental set-up

The pyrolysis process was performed using a fluidised bed reactor employing nitrogen as the fluiding gas and sand as the bed particles. The system consists of a stainless steel reactor of 50 mm diameter and 300 mm length, a fluid bed gas pre-heat chamber, a feed container with screw feeder, a solid -gas cyclone separator, a liquid condenser and liquid collectors. The heat container is supplied by two ring heaters of 1kW each around the reactor and the gas pre-heater respectively. The reaction tempertaure is controlled by a thermocouple within the fluid chamber bed which is connected to a PID temperature controller.

Based on studies by Scott and Piskorz(1985), the fluid bed with capacity of 0.5 kg/hr (dry raw feed) was designed to entrain the char product by blowing it from the bed while retaining the sand. This was done by careful selection of sand size, feed particle size and gas fluidising velocity. In one of the investigations, the sand particle size used was +212 μm to -350 μm with feed size of +150 μm to -420 μm and the fluidising gas flow rate set to 18 lit/min.

Initially the sand bed was heated to a desired temperature ranging from 400°C to 600°C. Palm shell of selected particles size were fed into the bubbling hot sand in the reactor at a controlled feed rate. The reaction products pass through a cyclone where the char was removed. The vapours and the gaseous product passed through the condenser and the condensed liquid was collected by the liquid collectors. The schematic diagram of the fluidised bed rig is shown in Figure 3.

The liquid product was analysed for their properties as potential fuel in comparison to petroluem fuel. The fuel characteristics compared were the physical properties, heating value, elemental analysis and the chemical composition. The pyrolysis oil was also analysed using Fourier Transform infra-red, (FT-ir) spectroscopy to determine the basic functional group of the compositions.

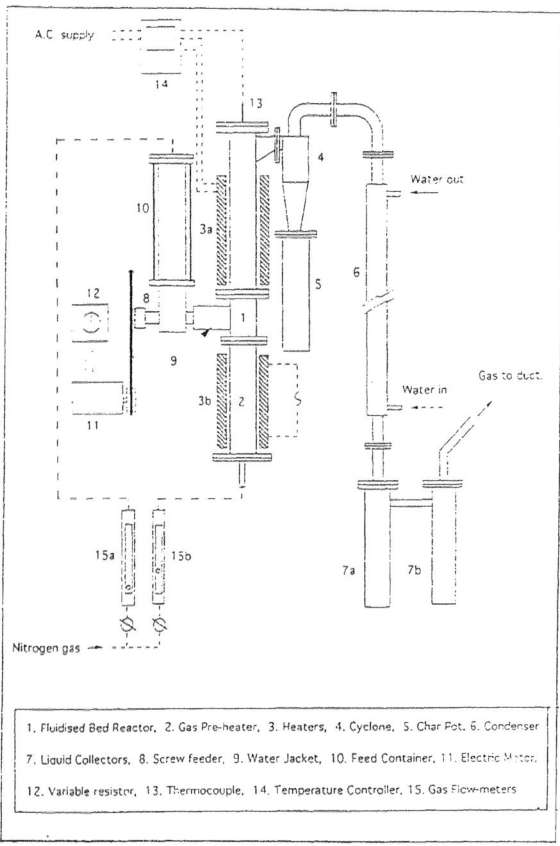

1. Fluidised Bed Reactor, 2. Gas Pre-heater, 3. Heaters, 4. Cyclone, 5. Char Pot, 6. Condenser
7. Liquid Collectors, 8. Screw feeder, 9. Water Jacket, 10. Feed Container, 11. Electric Motor,
12. Variable resistor, 13. Thermocouple, 14. Temperature Controller, 15. Gas Flow-meters

Fig. 3. The schematic diagram of the fluidised bed pyrolysis rig.

4 Results and discussions

From the fluidised bed pyrolysis studies, the percentage weight yield of liquid and char products are influenced by the process temperature. Figure 4 shows the percentage yield of liquid products and the char at various process temperature from 400 to 600°C. The liquid product shows only one phase of low viscosity liquid. The maximum liquid yield of about 58% of dry feed weight with feed particles size of 212 - 420 μm obtained at a temperature around 510°C. At bed temperature less than 450°C the char yield is high which is more than 25% of feed weight. This is due to the incomplete devolatilisation during the pyrolysis process. The liquid yield decreases while the char yield remains constant at the bed temperature above 525° C.

From the FT-ir spectroscopy analysis, the absorption frequency spectra representing the functional group of the pyrolysis is shown in Figure 5. The presence of alkanes is indicated by the strong absorbance peak of C-H vibrations between 3000 and 2800 cm^{-1} and the C-H and deformation vibration between 1465 and 1350 cm^{-1}.

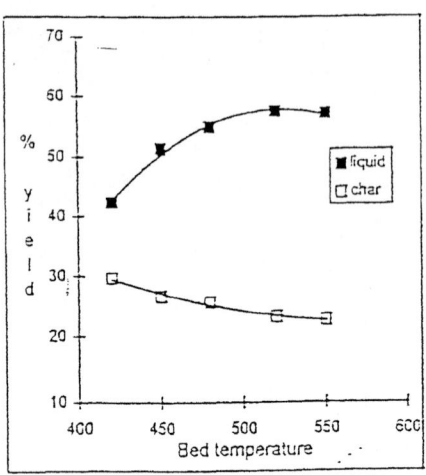

Fig. 4. Pyrolysis products yield againt bed temperature.

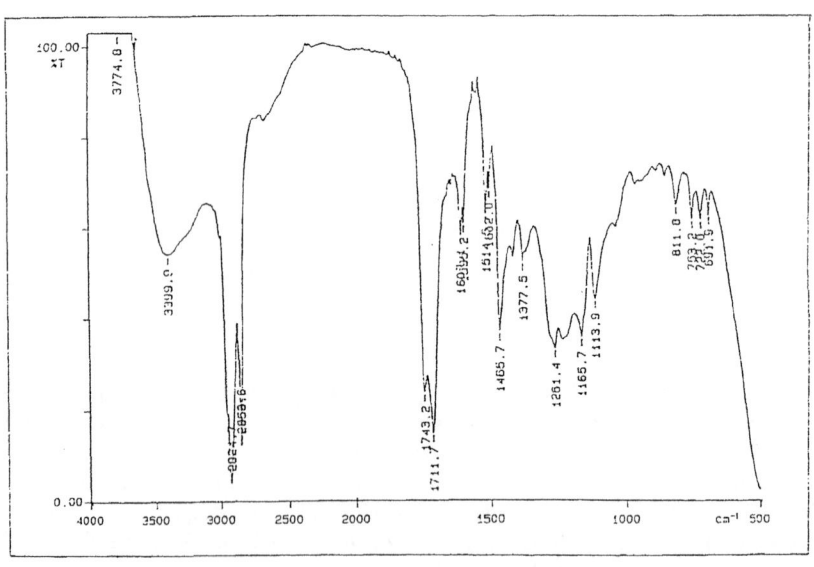

Fig. 5. FT-ir spectra of palm shell pyrolysis oil.

The presence of water impurities and other polymeric O-H are indicated in the pyrolysis liquid product by the broad absorbance peak of O-H stretching vibration between 3600 and 3200 cm^{-1}. The absorption peaks between 1780 and 1640 cm^{-1} represents C=O stretching vibration, indicates the presence of ketones and aldehydes. The presence of both O-H vibration and C=O stretching vibration are also indicate the presence of carboxyclic acids.

The possible presence of alkenes is indicated by the absorbance peaks between 1680 and 1580 cm^{-1} represent C=C stretching vibrations. The overlapping peaks between 1000 and 1300 cm^{-1} are suggested due to the presence of alcohols and phenols. Single and polycyclic and substituted aromatic groups may also be indicated by the absorption peaks between 900 and 650 cm^{-1}. These functional groups and the indicated composition have been identified in the pyrolytic derived oil from palm shells from fixed bed reactor by Williams et al.(1992) and Ramlan Z. (1995).

Table 2 Characteristics of the pyrolysis liquid compared to petroluem oil

Pyro-crude oil	Palm shells	Hard-wood (Solantausta et.al. 1993)	Brockville Poplar (Bridgwater et. al. 1991)
Elemental analysis			
C	53.48	55.5	54.7
H	6.69	6.7	6.90
N	0.44	0.1	-
O	39.37[a]	37.7[a]	38.4[a]
S	0.02	0.0	-
Calorific value (MJ/kg)	22.1	17.7	23.2
Viscosity @ 50°C cst	14.6	13	-
Density @ 15°C	1.20	1.22	1.20
Moisture content % wt	10.0	20.5	18.7

a - by different

Table 2 shows the characteristics of the derived oil in comparison with others found in literatures. The elemental analysis of the derived oil shows that it is highly oxgenated and acidic with PH value of 2.7 which influences the calorific value of the oil. The calorific value of the derived oil is lower than those of the petroluem fuels which is only 22.1 MJ/kg as compare to 42.4 MJ/kg on medium fuel oil and 45.6 MJ/kg on diesel fuel.

The char on the other hand was investigated in another study by Ani and Gibbs (1992). Pore size and the surface structures of palm shell char were examined using scanning electron microscope(SEM) and gas adsorptometer (BET method). The average pore sizes and the specific surface area are 0.8 μm and 253.6 m^2/gm

respectively. For activated carbon from palm shells, the average pore sizes and the specific surface area were found to be 19.8 μm and 1300 m^2/gm respectively, Ku Halim, Ramlan and Normah (1992).

5 Conclusions

Fluidised bed pyrolysis of palm fruit shell shows a significant recovery of pyrolysis oil. The derived oil have low calorific value due to its high oxygen content. Catalytic upgrading of the oil should be considered for producing a deoxygenated and a high carbon/hydrogen ratio fuel. FT-ir analysis of the pyrolysis oil indicates the presence of some compounds which can be found in petroleum fuels. Thus, these show high potential of fast pyrolysis of palm oil shells, which are available abundantly and as a source of liquid hydrocarbon fuel which is an attractive proposition in waste utilisation.

6 Acknowledgement

The authors wish to thank the Ministry of Science, Technology and Environment under the IRPA research programme for the research grant awarded (Vot No 62503) and Universiti Teknologi Malaysia for the support given to the research project.

7 References

1. Besler, S., Horne, P.A. and Williams, P.T. (1992), The Fuel Properties of Biomass Derived Pyrolytic Oils and Catalytically Upgraded Products, in Renewable Energy Technology and Environment (Sayigh, A.A.W (ed)) :Pergammon Press, pp 1341 - 1345.
2. F.N. Ani and Gibbs B.M.(1992), The Thermal Characteristics of Oil Palm Solid Wastes, Second International Energy Conference, Energy from Biomass Residues, 17 - 19 August 1992, Kuala Lumpur.
3. Ku Halim Ku Hamid, Ramlan Abdul Aziz and Normah Mulop(1992), Activated Carbon Plant for Cottage Industry, Second International Energy Conference, Energy from Biomass Residues, 17 - 19 August, Kuala Lumpur.
4. Ramlan Zailani (1995), Fluidised Bed Pyrolysis of Organic Solid Waste, MEng Thesis, Universiti Teknologi Malaysia.
5. Bridgewater, A.V. and Bridge, S.A.(1991), A Review of Biomass Pyrolysis and Pyrolysis Technologies in Biomass Pyrolysis Liquids, Upgrading and Utilisation, Bridgwater A.V. and Grassi G.(eds) : pp 11 - 92, Elsevier Applied Science, London.
6. Scott, D. S. and Piskorz J. (1985), Production of Liquids from Biomass by Continuous Flash Pyrolysis, Vol III, Bioenergy 84 (H. Egneous and A. Ellegard, eds) Elsevier Applied Science, London, pp 15 - 18.
7. Solantausta, Y., Nylund, N.O., Westerholm, M., Koljonen, T. & Oasmaa, A. (1993), Wood Pyrolysis Oil as Fuel in a Diesel-Power Plant'. Bioresource Technology, 46, pp 177-188.

PROPOSED SPECIFICATIONS FOR VARIOUS GRADES OF PYROLYSIS OILS

J.P. DIEBOLD, T.A. MILNE, and S. CZERNIK
National Renewable Energy Laboratory, Golden, CO, U.S.A.;
A. OASMAA,
Technical Research Centre, Espoo, Finland
A.V. BRIDGWATER
University of Aston, Birmingham, U.K.
A. CUEVAS
Union Fenosa, Madrid, Spain
S.GUST
Neste Oy, Porvoo, Finland
D. HUFFMAN
Ensyn, Greely, Ontario, Canada
J. PISKORZ
Resource Transforms International, Waterloo, Ontario, Canada

Abstract
A handicap facing both the producer and the user of fast-pyrolysis oils is the lack of a description of these oils that is adequate for commercial applications. These oils are highly oxygenated and are relatively immiscible with petroleum oils. Under the current IEA Biomass Energy Agreement, the new Pyrolysis Activity (PYRA) has taken on the task of establishing a useful description of a series of pyrolysis oils. This series roughly parallels that of petroleum fuel oils already described, so that with as few changes as possible to the users' equipment, a bio-oil could be used in place of the equivalent petroleum-derived oil. The specifications for biomass pyrolysis oils differ in the density, heating value, water content, and corrosiveness. These proposed specifications are presented for discussion by the biomass conversion community and feedback to the Pyrolysis Activity.
Keywords: pyrolysis oil, bio-oil, physical properties, fuel, specifications

1 Introduction

The new Pyrolysis Activity of the International Energy Agreement (IEA) under Task XIII has taken on the task of proposing a series of specifications for fuel oils made from biomass materials by the rapid pyrolysis technique. These pyrolysis oils are also referred to as bio-oils or biocrude oils. Over the years, the petroleum industry has been marketing fuel oils derived from petroleum for a variety of specialized applications. The fuel oil user has equipment that has been designed to properly feed, atomize, and burn a specific type of fuel oil having well established physical characteristics.

In general, the less viscous the fuel oil, the easier it is to move from the fuel tank to the fuel pump, to atomize it, and to combust it. Low sulfur levels are desirable to minimize SO_x emissions, or to eliminate the requirement to scrub the SO_x from the exhaust gases. The ash level should be low enough to minimize the particulate content in the exhaust gas to preclude the need for baghouse filtration, as well as the resultant wear and tear on pumps, injectors, etc. Water can exist as an emulsion in petroleum fuels and is undesirable as water contains soluble salts and promotes bacterial growth that creates scum. In summary, the premium valued petroleum-fuel oils have a lower viscosity and contain lower sulfur, lower ash, lower particulate, and lower water levels.

The ASTM specifications for petroleum-derived fuel oils and diesel fuel oils are very similar, with the exception of a Cetane specification and lower ash contents for diesel fuels. For gas turbines (non-aero), an additional ASTM requirement for low alkali metal content of less than 0.5 ppm is suggested. Although the gas turbine fuel grades 1 and 2 have many similarities to the #1 and #2 fuel oil grades, the gas turbine fuels #3 and #4 are more like #5 and #6 fuel oils respectively. [1][2]

In order to take advantage of the less expensive fuel oils, equipment manufacturers can design their products to function with higher viscosity and dirtier fuels. However, this omnivorous equipment is more complex and more expensive to build and to maintain. So, the fuel user traditionally has had the option of more expensive equipment costs in order to use less expensive fuels.

We now want to expand the options available to the fuel users by another dimension, that of using liquid fuels made by the pyrolysis of biomass materials (bio-oil). In addition, it is desired to make this bio-oil option available to an international market. It has been concluded by many that there is a need to develop specifications for bio-oil [3][4][5].

We will first review the specifications that have evolved for petroleum derived fuel oils in North America and Europe. We will then examine the properties of bio-oils and then attempt to categorize them into a several different grades, based on the properties that are important to the user.

2 Specifications for Petroleum-Derived Fuel Oils

In Tables 1 through 4 are listed the specifications for light, medium-light, medium, and heavy fuel oils, based on specifications obtained from the United States [1][2][6], Canada [7][8], Finland [9], Denmark [10], United Kingdom [11], and Spain [12]. It is entirely possible that specifications other than those referred to in these tables would result in a closer match of specified properties for the different grades of fuel oil, but this appears to be the best match up of the specifications from the six countries now being compared. In today's world of international trade, there are smaller differences in the specifications from the various countries than might have been expected.

In the texts of the ASTM specifications for fuel oils [2] and for diesel fuels [6], the general characteristics and uses for the different grades of fuels are summarized as follows:

#1 **Fuel Oil or #1 Diesel Fuel**---volatile fuel oils from kerosene to middle distillates having a low pour point and particularly adapted to vaporizing burners or for high speed diesel engines with varying loads and speeds. High volatility is required to produce vapors by contact with a hot surface or by radiative heating with a minimum of residue. Currently whole bio-oils are more viscous than this material

Table 1. Specifications for Light Fuel Oils

Property	U.S. ASTM#2	Canada CGSB#2	Finland POK5	Denmark GO 5	U.K. GO	Spain GO A
Flash point, °C minimum	38-FO 38-GT 52-D	40-FO	56	61	56	55
Water and sediment, max	0.05	0.05	0.025	0.015	0.05v 0.01w	0.1v
90% distillation temp, °C	282	----	360	----	350 (85%)	350 (85%)
Max. Distillation temp., °C	338	360	370 (95%)	385	----	380
Viscosity, cSt @ 40°C	1.9- 3.4 FO 4.1 GT 4.1 D	1.6- 3.5 FO 4.1D-B	2.0- 3.3	1.9- 3.7	1.5- 5.5	4.3- 5.2
RB Carbon residue on 10%, max. wt%	0.35	0.35	0.2	0.2 Conrads. carbon	0.2	0.2
Ash, wt% maximum	0.01GT 0.01 D	----	0.01	0.01	0.01	____
Sulfur, wt% max.	0.5 FO .05LSD 0.5 D	0.7	0.05	0.2	0.2	0.3
Density @ 15°C, kg/m³ maximum	876 FO 876 GT	900	820- 860	845	----	825- 860
Pour point, °C minimum	-6 FO -6 GT	varies	-15	-5	----	----
Cloud point, °C	varies D	varies	≤-5	2	-4 to -12	-1 to 4
Cetane number minimum	40 D	----	49	47	45	45
Heating value, MJ/kg, min.	----	----	40.1 LHV	42.9 LHV	----	44.0 HHV

Note: D = Diesel fuel; GT= gas turbine; FO = fuel oil

Table 2. Specifications for Light-Medium Fuel Oil

Property	U.S. #4 FO	Canada #4 FO	Finland POK 15	Spain GO C
Flash point, °C minimum	55	54	60	60
Water and sed., maximum vol%	0.5	0.5	0.025 wt%	0.1
90%. dist. Temp, °C	----	----	398	>65% <250; >80% @ <390
Max. dist. temp., °C	----	----	----	----
Viscosity, cSt @ 40°C	5.5-24	5.5-24	3.0-7.5	< 7.0
Carbon residue on 10% dist. residue, wt%	----	----	0.2	0.35
Ash, max. wt%	0.1	0.1	0.01	----
Sulfur, wt% max.	varies	varies	0.20	0.3
Density @ 15°C, kg/m^3 maximum	----	----	890	900
Pour point, °C minimum	-6	varies	-15	-6
Cloud point, °C minimum	----	varies	≤ 8	< 4
Cetane min.	30 D	----	45 D	----
Heating value, MJ/kg min.	----	----	40.3 LHV	43.12 HHV

and evaporate leaving a substantial residue, so we currently do not plan to have an equivalent #1 bio-oil at this time.

#2 Fuel Oil or #2 Diesel Fuel---Volatile fuel oils from kerosene to middle distillate, but less volatile than #1 fuel oils or diesel fuels. Suitable for high speed diesel engines with high loads and uniform speeds or for engines not requiring the higher volatility of #1 fuels. The #2 fuel is atomized into the combustion zone and the droplets typically burn while in suspension.

Table 3. Specifications for Medium Fuel Oils

Property	U.S. #5 FO #3 GT	Canada #5	Finland PORL 100	Denmark HFO 45	U.K. Shell F	Spain FO 1
Flash point, °C minimum	55	54	65	68	66	65
Water and sed., wt% maximum	1.00v	1.00	0.7	0.1 and 0.15	0.75v and 0.1w	1v
90%. dist. temp., °C	----	----	----	----	----	----
Max. dist. temp., °C	----	----	----	----	----	----
Viscosity, cSt	5.0-14.9FO 50 GT @ 100°C	17.1-80 @ 50°C	80 - 100 @ 50°C	38 - 45 @ 80°C	<20 @ 100°C	<25 @ 100°C
Carbon residue, wt%	----	----	≤ 14	----	----	----
Ash, max. wt%	0.15	----	0.1	0.1	0.05	----
Sulfur, wt% max.	varies	varies	0.85	1.0	3.5	2.7
Density @ 15°C, kg/m^3 maximum	----	----	900-1000	960	990	----
Min. pour point, °C	----	----	-15	----	----	----
Cloud point, °C	----	varies	----	----	----	----
Heating value, minimum MJ/kg	----	----	40.4 LHV	40.6 LHV	43.2 HHV	40.2 LHV

Table 4. Specifications for Heavy Fuel Oils

	U.S. #6 FO #4-GT	Canada #6 FO	Finland PORL 180; 380	Denmark HFO typical Elkraft	U.K. HFO SG	Spain FO-2
Flash point, °C minimum	60 FO 66 GT	60	65	80	66	65
Water and sediment, wt% maximum	2.0 FO 1.0 GT	2.0	0.7	----	1v and 0.25w	1v
90%. dist. temp, °C	----	----	----	----	----	----
Max. dist. temp., °C	----	------	----	----	----	----
Viscosity, cSt	<50 @ 100°C	93- 638 @ 50°C	140 - 380 @ 50°C	77 @80°C; 14.5 @ 130°C	40 @ 100°C	37 @ 100°C
RB Carbon residue on 10% residue, wt%	----	----	----	----	----	----
Ash, max. wt%	----	----	0.1	0.1	0.2	_____
Sulfur, wt% max.	varies	varies	1.0	1.0	3.5	1(BIA) 3.5
Density @ 15°C, kg/m^3 maximum	----	----	900- 1020	990	----	----
Pour point, °C minimum	15 FO	----	15	27	----	----
Hot filtration, max. wt%	----	----	0.15	----	----	----
Cetane minimum	----	----	----	----	----	----
Heating value min., MJ/kg	----	----	----	40.7	----	39.4

#4 Fuel Oil or #4 Diesel Fuel---This is usually a blend of middle distillate fuels with residual fuels to meet the desired viscosity value. These fuels are suitable to be burned in industrial applications designed to atomize the more viscous fuels. These fuels are also used in low and medium-speed diesel engines under sustained loads and relatively constant speed, e.g., train locomotives. Fuel oils in the lower range of the acceptable viscosity values (#4 light) are normally pumped and atomized without preheating. Slightly more viscous #4 fuel oil is also not normally preheated for handling, except in extremely cold weather.

#5 Fuel Oil is a residual fuel having a higher viscosity than #4 Fuel Oils and preheating may be necessary for some types of combustion equipment and for handling in colder climates.

#6 Fuel Oil or Bunker C is used mostly in commercial and industrial heating. It requires heating of the storage tank to permit pumping and additional preheating at the burner to permit atomization. The extra equipment required to properly burn this fuel normally precludes its use in small applications.

Historically, diesel engines have been used to burn heavy fuel oils, e.g., ASTM #5 fuel oil. In this application, the diesel engine is outfitted with two fuel tanks, a small one for light diesel fuel and a large one for heavy diesel main fuel. The associated plumbing is installed to allow the engine to be switched to the light diesel fuel for startup and shutdown. The engine is switched to the heavy diesel fuel for normal operation after warmup. In addition, the heavy diesel oil is normally heated to as high as 120°C to reduce its viscosity to between 7 and 13 cSt to achieve good atomization during injection [13]. In some diesel engines, a high Cetane diesel fuel is used to pilot ignite the low Cetane fuels [14].

Even engines using #2 diesel fuel use heat exchangers to preheat the fuel in the winter between the fuel pump and the injectors, typically using the engine coolant at 70-80°C as the heat source. A portion of the preheated fuel is returned to the fuel tank to keep the fuel in the tank from freezing or forming waxy precipitates.

There are also specifications for marine fuels, which are characterized by relatively high viscosities and high sulfur contents. However, because biomass-derived pyrolysis oils have low sulfur contents, it is thought that the bio-oils will compete economically with the more expensive, low-sulfur petroleum fuels. Therefore, we will not add to the complexity of this comparison by considering the marine fuels in detail.

Currently, whole pyrolysis oils have been produced which meet certain #4 fuel oils specifications, but not the #2 nor the #1 fuel oil viscosities. Consequently, the #1 fuel oils will not be compared in detail. However, the #2 fuel oil specifications remain as a goal, which perhaps may be reached by solvent dilution, or by fractionating the bio-oil by distillation, or other means to reduce the residual portion of the oil.

3 Significance of the Specified Physical Parameters

Flash Point This test measures the liquid temperature necessary for the vapors above a pool of the fuel to ignite by passing a flame through the vapors. This is a measure of the volatility of the oil, as well as its ease of ignition. The higher this number, the safer the oils are to handle because the risk of accidental vapor ignition is reduced. Note that all of the petroleum oils mentioned above are considered to be flammable.

Water and Sediment In petroleum fuels, the water and sediments are phases that differ from the non-polar hydrocarbon fuel. In petroleum fuels, water can cause corrosion of the fuel system and allow biological activity leading to deposits. Sediments can accumulate on filter screens or burner parts and obstruct proper flow of the fuel. In addition, the sediment can cause erosion of pumps and injectors.

Distillation Temperatures The distillation tests give an indication of the volatility of the fuel. A fuel with a low volatile content may be safe to handle but difficult to ignite. The complete distillation of the sample is important for those applications which require the volatilization of the fuel prior to burning. Equipment designed for the burning of droplets can utilize oil that does not completely evaporate, e.g., pyrolysis oil. The distillation temperatures are normally not specified for fuel oils heavier than ASTM #2.

Viscosity The lower the viscosity of the oil, the easier it is to pump and to atomize and achieve finer droplets. This is the major criterion upon which the oils are graded.

Ramsbottom or Conradson Carbon This test gives an indication of the residue left after the bulk of the oil has evaporated and the residue has been pyrolyzed to a char.

Ash Content Excessive amounts of ash can cause high wear in pumps and injectors and lead to deposits in combustion equipment. Gas turbines are very susceptible to corrosion by alkali metals, so gas turbine fuels have a very low suggested limit of 0.5 ppm (sodium + potassium)[1].

Sulfur Content The presence of sulfur in the fuel leads to SO_x emission in the exhaust, which can poison the catalyst in catalytic mufflers, as well as, directly pollute the atmosphere. Petroleum oils containing both sulfur and water have an increased corrosivity.

Density The density of the oil is a measure of its aromaticity in hydrocarbon oils, but not in biomass-derived oils. It is a necessary parameter used to calculate the volumetric output of pumps and injectors needed to supply a given rate of delivered energy, because the heat of combustion is determined on a weight basis.

Pour Point This parameter is an indication of the minimum temperature at which the oil can be pumped without heating of the storage tank.

Cloud Point As fuel oil is cooled, wax crystals precipitate forming an opaque or cloudy oil. This is very difficult to observe in heavier oils, which may not be very transparent even when heated. This parameter is used to estimate the lowest temperature at which the fuel can be filtered effectively or passed through small orifices.

Cold Filter Plug Point (CFPP) This test designed for diesel fuel and home heating oil determines the lowest temperature at which a quantity of the oil can be filtered through a reference filter element having a 45μm pore size with a 20 millibar (200 mm W.C.) pressure difference in one minute elapsed time[15]. However, because the oil sample is prepared by first filtering through a paper filter, most of the char fines remaining in the bio-oil would be removed prior to the CFPP test. So, this test as written will not predict the ability to filter the bio-oil (with char fines) prior to combustion. Another problem with the test as written is that the prescribed pressure difference must provide the driving force to push the oil through the filter **and** to push the filtered oil up a 200-mm column to the mark on the volumetric flask. Bio-oil that has a density higher than water has a lower driving pressure to push the oil through the filter and is unfairly handicapped. This test would need to be modified before it can be useful for pyrolysis oils. It is suggested that this test not be used for pyrolysis oils.

Cetane Number This is a measure of the ease of autoignition of the fuel in a diesel engine. In a manner analogous to the more familiar Octane number, the reference fuels are n-hexadecane (100 Cetane) and methyl naphthalene (0 Cetane).

Heating Value This is the heat that is released during combustion in oxygen under pressure, as measured by the temperature rise of a small bomb. The gross or higher heating value (HHV) is measured directly and includes the amount of heat released when the water present in the bomb condenses. The net or lower heating value (LHV) is calculated by subtracting the heat of vaporization of the water found in the bomb after the test.

4 Reported Physical Properties for Biocrude Oils

Chemical Composition Biocrude oils have some properties that are quite different from those of petroleum-derived oils. As can be seen from an inspection of Table 5, biocrude oils contain a large amount of oxygen. This oxygen is present as polar and non-polar carbonaceous compounds, as well as water in the wet oils. The polar nature of many of the oxygenated compounds allows a very significant amount of water to dissolve in the oils. The viscosity is rapidly reduced as the water content is increased. However, too much water in the oil causes the biocrude to separate into a thinner aqueous-rich phase and a thicker organic-rich phase. This phase separation is temperature dependent and can be difficult to detect visually, because the phases are both quite dark.

The oxygen content of the bio-oils (excluding water) is a function of the oxygenated feedstock from which they are made, as well as the residence time and temperatures of the pyrolysis step. An increase in pyrolysis severity reduces the organic liquid yield, with the production of gases rich in carbon oxides, but which leaves the remaining organic oil with less oxygen and a higher heat of combustion.

The sulfur content of bio-oils is naturally low, due to the low sulfur content of biomass. This low sulfur content is one of the positive aspects of bio-oils.

One of the variables with bio-oils will be the nitrogen content of the oil, which will reflect the variable protein content possible with biomass. Biomass which has a significant content of green, living plant cells (e.g., green grasses) will have a higher nitrogen content than biomass having a large content of dead plant cells (e.g., straw and wood). The nitrogen contained in the bio-oils will contribute to NO_x emissions.

Corrosion The other unusual aspect of these oils is their carboxylic acid content, e.g., formic and acetic acids, which causes the oils to have a pH of between 2 and 3 (similar to vinegar). This acidity causes the pyrolysis oils to be corrosive to mild steel,

Table 5. Example Properties of Biomass Pyrolysis Oils (Wet oil basis)

Feedstock {char removal method}	Poplar {hot-gas filtered}	Maple and oak {char cyclone}	Eucalyptus {char cyclone}
Water, wt%	18.9 (0)*	23.3 (0)	27.9 (0)
Elemental			
Carbon	46.5 (57.3)	44.8 (58.5)	39.4 (54.6)
Hydrogen	7.2 (6.3)	7.2 (6.01)	7.3 (5.8)
Oxygen	46.1 (36.2)	47.8 (35.4)	53.4 (39.6)
Sulfur	0.02 (0.02)	<0.01(<0.01)	<0.01 (<0.01)
Nitrogen	0.15 (0.18)	0.1 (0.1)	<0.1 (<0.1)
K + Na, ppm	9.9	NA	NA
Cl, ppm	7.9	NA	NA
Ash, wt%	0.01	0.09	0.09
HHV, MJ/kg	18.7 (22.3)	18.1 (23.5)	15.6 (21.6)
LHV, MJ/kg	17.4 (21.2)	16.6 (22.2)	14.0 (20.3)
Density, kg/m³	1200	1230	1260
Flash point, °C	64	> 106	> 70
Pour Point, °C	NA	-9	-12
Viscosity @ 50°C, cSt	13.5	70	50
Ethanol insoluble filtered solids, wt%	0.045	0.3	0.9
pH	2.8	2.8	2.3
CFPP, °C	50	NA	NA

*Numbers in parentheses are on a moisture-free oil basis.

aluminum, etc. Aldehydes also contribute to the low pH.

Aging The pyrolysis oils are chemically reactive with themselves and will polymerize with time, usually with the formation of additional water as a byproduct of the reactions. After prolonged storage, the oils tend to increase their molecular weight owing to chemical reactions, viscosity, and also the tendency to separate into a thin oil phase and a thick tar phase. It is very important that the bio-oils be used prior to this phase separation. An accelerated aging test at elevated temperatures would be useful to predict the storage behavior of an oil sample. Aging tests with hot-gas filtered bio-oil (Run 175 oil at NREL) in closed containers have shown that accelerated aging at 90°C for 8 hours increases viscosity about the same amount as would occur during storage at 20°C for a 6-month period [16]. Other bio-oils may have a different correlation of accelerated aging rate to room temperature aging rate, due to their specific chemical character and to the amount of char fines present.

Flash Point Pyrolysis oil often has a reported flash point of between 50°C and over 100°C, reflecting a wide variation in the content of volatiles. These flash points bracket what is commonly required for petroleum derived fuel oils. However, above temperatures of 70 to 75°C, water vapors from the pyrolysis oils start to disturb the analysis and a reproducible value is difficult to obtain.

Distillation Temperatures Pyrolysis oil distillation involving slow heating typically leaves about 40% polymerized residue in the bottom of the distillation flask. It is not reasonable to expect that crude, whole pyrolysis oil will pass the requirements imposed on petroleum distillates. Since we are not currently proposing the use of pyrolysis oil for those applications that require complete evaporation prior to combustion and, therefore, the use of ASTM #1 equivalent oils, we will not specify the distillation temperatures.

Density The density of these oxygenated oils is much higher, 1150 to 1300 kg/m³, than petroleum-derived hydrocarbon oils at 0.85 to 1.0 kg/m³. Unlike hydrocarbon oils, the higher density of the bio-oils reflects the high oxygen content, rather than a high polycyclic aromatic content. The bio-oils having relatively higher densities typically have lower water contents. The density will be needed if the pyrolysis oil is to sold be on a volume basis, as well as to ensure the proper sizing of fuel pumps and injectors.

Water Content In biomass-derived oils, water is soluble in the polar organic fuel to a considerable extent. This water content decreases the available energy from the oil. However, water in the biocrude reduces its viscosity, produces a good working gas after evaporation, lowers flame temperatures and NO_x, and appears to be a beneficial factor in promoting the microexplosions of atomized fuels that promote better burning [17]. The organic acid and phenolic content of whole bio-oil appear to act as bactericides to retard bacterial or fungal growth.

Sediment The sediment in the bio-oil is fine char, dirt, and attrited heat transfer medium that was not removed during its manufacture, as well as organic precipitates that can form during storage (however, some of these precipitates have a low density and float on top of the oil). The alkali content of the bio-oil resides in the fine char and other solids. It appears that the fine char may act as a catalyst for the formation of gummy tars that can readily plug filters and for much of the increase in viscosity with time or aging [18][19]. Bio-oil that was hot-gas filtered to remove char fines undergoes microexplosions during combustion that completely vaporize the droplets and reduce their burn time to half that of similar droplets of #2 diesel fuel [17]. Conversely, bio-oil which was not hot-gas filtered and had residual char fines, formed a solid, hollow spherical shape which shattered into burning solid particles that had burning times similar to #2 diesel fuel [20].

Heating Value Due to the variable water and oxygen content of these oils, it is expected that they will be sold on a net energy basis. This energy basis is expected to be the LHV, in which the HHV is measured and then corrected for the latent heat of evaporation of the water collected after the test. Included in this water is the water that was in the original sample as moisture, as well as the water formed during combustion. This measure of net energy properly penalizes the producer who has high wet oil yields because of the moisture in the biomass feed or other water sources.

5 Proposed Specifications for Biomass Derived Oils

In reviewing the specifications for petroleum derived fuel oils, it is apparent that the different grades are structured around viscosity and ash content of the oil. Consequently in Table 6, the proposed specifications for biomass-derived oils are also grouped in this manner with light oil (with an ash content, filtered particulates, and viscosity similar to ASTM #2), medium-light (similar to ASTM #4), medium oil (similar to PORL 100) and heavy oil (similar to ASTM #6). If the oil does not contain fine dirt from the feed or attrited thermofor, the ash content is a convenient method to estimate the amount of char fines in the oil, because the ash is apparently contained in the residual char fines, rather than in the liquid oil [19].

The ability of the bio-oil to flow at low temperatures does not need to be specified for universal applications, as the requirement for this property is site specific. For example, a user in Finland may have to heat the storage tank in order to pump the oil on a cold winter's day, whereas a user in Spain may have no problem with the same oil without heating. Therefore, it is concluded that this type of physical property needs to be measured and reported for the convenience of the user.

The accelerated aging rate is very important to the user, as it helps to establish guidelines as to how long the bio-oil can be stored before its viscosity increases to that of a less valuable oil grade or the oil-rich and aqueous-rich phases separate. In addition, an accelerated aging test gives the user an idea of how long the pyrolysis oil can be preheated before it becomes too viscous to atomize well. It is desirable to have at least six months of storage life, although some applications will not require more than a few days of storage prior to use. However, it is too early in the development of bio-oil to specify this rate, so it too will need to be measured and reported.

The allowable water content of the bio-oil will seem surprisingly high to many potential users. However, a yield of 70 wt% wet bio-oil from a biomass having a moisture content of 8 wt% could result in as much as 32 wt% water in the bio-oil product, if the assumed yield of water by pyrolysis is 14 wt% of the dry feed. This water content can be reduced by predrying the feed, only partially condensing the water vapor from the pyrolysis gas stream (but volatiles would be concurrently lost), or increasing the organic condensate yield. Consequently, an upper limit of 32 wt% water in the bio-oil seems reasonable, although lower water contents are desirable to avoid phase separation during storage and to increase the available energy content of the bio-oil. The maximum allowable water content without phase separation will be a function of the feedstock, pyrolysis severity, and product collection techniques (the volatile polar organic compounds apparently act as co-solvents for water). This upper limit for water needs to be carefully examined and adjusted as future experience dictates. It would be a shame to reject a low viscosity pyrolysis oil because its water content was above an artificially low value.

Table 6. Proposed Specifications for Biomass Derived Oils

	Light bio-oil (~ASTM #2)	Light-Medium bio-oil (~ASTM #4)	Medium bio-oil (~PORL100)	Heavy bio-oil (~Can. #6)
Viscosity, cSt	1.9-3.4 FO 1.9-4.1 D 1.9-4.1 GT @ 40°C	5.5-24 @ 40°C	17-100 @ 50°C	100-638 @ 50°C
Ash, wt%	0.05 FO 0.01 D 0.01 GT	0.05 FO 0.01 D	0.10 FO	0.10 FO
Pour point, °C min.	report	report	report	report
Conradson carbon, wt%	report	report	report	report
Max. 0.1μm filtered ethanol insol. Solids wt%	0.01 FO	0.05	0.10	0.25
Accelerated aging rate @ 90°C, cSt/h	report	report	report	report
Water, wt% of wet oil, max.	32	32	32	32
LHV, MJ/L min., wet oil	18	18	18	report
C, wt% dry H, wt% dry O, wt% dry S, wt% dry N, wt% dry K + Na, ppm	report " " 0.1 max. 0.2 max. report 0.5 GT	report " " 0.1 max. 0.2 max. report	report " " 0.2 max. 0.3 max. report	report " " 0.4 max. 0.4 max. report
Phase Stability @ 20°C after 8 h @ 90°C	single Phase	single phase	single phase	single phase
Flash point, °C minimum	52	55	60	60
Density, kg/m³	report	report	report	report

The elemental analysis, density, and higher heating value all need to be reported in order to calculate the volumetric lower heating value used to determine the value of the bio-oil as a fuel. A minimum volumetric lower heating value of 18 MJ/L was selected, which was considered to be extremely low by developers of the use of bio-oil as a diesel engine fuel [21].

The flash point of the oil is important to maintain at about the same value as the petroleum fuels, to avoid the need for expensive additional safety precautions. However, it should be noted that although flammable vapors can accumulate in the space above the bio-oil, liquid bio-oil is very difficult to ignite, presumably because of its high water content.

6 Summary and Conclusions

Petroleum-derived fuel oil specifications from six countries in North America and Europe were compared and used as the bases for analogous fuel oil specifications for bio-oils made by the fast pyrolysis of biomass materials. Notable exceptions to the petroleum fuel oil tests and specifications must be made with respect to moisture content, heating value, density, filtration, and distillation properties. Because of the variable water content reported for pyrolysis oils, it is expected that they will be sold on the basis of their lower heating values, rather than on a weight or volume basis like petroleum derived fuels. However, the concept of using viscosity and ash content as the primary determinants of the class of oil product appears to be applicable to bio-oils.

These proposed oil specifications are presented with the hope of receiving feedback from the potential pyrolysis oil generators and users. It is expected that these tests and acceptable values will evolve with time. Through the widespread utilization of these or similar specifications, the biomass thermal conversion industry will be able to market its pyrolysis oil products with more assurance that the user will successfully utilize them.

7 References

1. ASTM D 2880-94. (1994). "Standard Specification for Gas Turbine Fuel Oils." American Society for the Testing of Materials. Philadelphia, PA.
2. ASTM D 396-92. (1992). "Standard Specification for Fuel Oils." Oct. 15, American Society for the Testing of Materials, Philadelphia, PA.
3. Moses, C. (1994). "Fuel Specification Considerations for Biomass Liquids." *Proceedings Biomass Pyrolysis Oil. Properties and Combustion Meeting, T.A. Milne, ed., Estes Park, CO, September 26-28*, National Renewable Energy Laboratory, Golden, CO, NREL-CP-430-7215, pp. 362-382.
4. Milne, T.A. (1994). "Preface," *Proceedings Biomass Pyrolysis Oil. Properties and Combustion Meeting, T.A. Milne, ed., Estes Park, CO*, NREL-CP-430-7215, p. iii.
5. Solantausta, Y.; Nylund, N-O; Oasmaa, A.; Westerholm, M.; and Sipilä, K. (1994). "Preliminary Tests with Wood-Derived Pyrolysis Oil as Fuel in a

Stationary Diesel Engine," *Proceedings Biomass Pyrolysis Oil. Properties and Combustion Meeting, T.A. Milne, ed., Estes Park, CO, September 26-28,* National Renewable Energy Laboratory, Golden, CO, NREL-CP-430-7215, pp. 355-361.

6. ASTM D 975-94. (1994). "Standard Specification for Diesel Fuel Oils." American Society for the Testing of Materials. Philadelphia, PA.

7. Can/CGSB-3.2-M89. (1989). "Fuel Oil, Heating." Canadian General Standards Board, Ontario, Canada.

8. Can/CGSB-3.6-M90 (1990) "Automotive Diesel Fuel." Canadian General Standards Board, Ontario, Canada.

9. Neste Oy Product Data Sheets. (1994). Espoo, Finland

10. Elkraft Product Data Sheets. (1995). Copenhagen, Denmark.

11. British Standard 2869 part 2. (1988). "Fuel Oils", British Standards Institution London, United Kingdom.

12. Real Decreto 1485. (1987). "Productos Petrolíferos-Especificaciones," B.O.E. no. 291 de fecha 5 Deciembre, Spain. Ministry of Industry and Energy, Madrid.

13. Perry, R.H.; and Green, D. (1984) *Perry's Chemical Engineer's Handbook, 6th. ed.,* McGraw-Hill, p. 24-15.

14. Gros, S. (1995). "Pyrolysis Oil as Diesel Fuel," *Seminar on Power Production from Biomass II, March 27-28, Espoo, Finland, VTT Energy.*

15. DIN EN 116. (1983). "Diesel and Domestic Heating Fuels-Determination of Cold Filter Plug Point." European Committee for Standardization, Brussells, Belgium

16. Diebold, J.P. and Czernik, S. (1996) "The Use of Additives to Improve the Aging Properties of Stored Pyrolysis Oils." to be submitted to Biomass and Bioenergy

17. Davis, K.A.; Huey, S.P.; Davis, J.E. (1995). "Microexplosive Behavior of Biomass-Derived Pyrolysis Oils during the Combustion of Isolated Droplets." *Proceedings of Central and Western States (USA) sections and Mexican National Section of the International Combustion Institute and American Flame Research Committee, San Antonio, TX, April 23-26.*

18. Agblevor, F.A.; Besler, S.; Montagne, D.; and Evans, R.J. (1995). "Influence of Inorganic Compounds on Char Formation and Quality of Fast Pyrolysis Oils." *ACS 209th National Meeting, Anaheim, CA,* April 2-5.

19. Diebold, J.P.; Scahill, J.W.; Czernik, S.; Phillips, S.D.; and Feik, C.J. (1996) "Progress in the Production of Hot-Gas Filtered Biocrude Oil at NREL," *Bio-Oil Production and Utilization,* A.V. Bridgwater and E.N. Hogan, eds., CPL Scientific Information Services, LTD, Newbury, U.K., pp. 66-81.

20. Wornat, M.J.; Porter, B.G.; Yang, N.Y.C. (1994). "Single Droplet Combustion of Biomass Pyrolysis Oils." *Energy and Fuels,* 8, pp. 1131-1142.

21. Jay, D.C.; Sipilä, K.H.; Rantanen, O.A.,; and Nylund, N-O. (1995) "Wood Pyrolysis Oil for Diesel Engines," ASME Internal Combustion Engine Division's Fall Technical Conference, Milwaukee, WI, Sept. 24-27.

WOOD TAR PITCH: ANALYSIS AND CONCEPTUAL MODEL OF ITS STRUCTURE
Characterization of *Eucalyptus* Tar Pitch

V. M. D. PASA and F. CARAZZA
Departamento de Química - Universidade Federal de Minas Gerais - Belo Horizonte, Minas Gerais, Brazil
C. OTANI
Departamento de Física, Instituto de Tecnologia Aeroespacial - ITA, São José dos Campos, Brazil

Abstract

Wood tar is obtained as a by-product in *Eucalyptus* carbonization activities in Brazil. Efforts have been made to upgrade this raw material as source of fine chemicals and a precursor for polymeric materials. The distillation process of wood tar has been done on laboratory, pilot-plant and industrial scales. The first step in *Eucalyptus* tar distillation yields 20% of a watery phase, 30% of crude oil and 50% of pitch (bio-pitch).

The physico-chemical characterization of bio-pitch, obtained as a fusible distillation residue, is being carried out. The results show that bio-pitches are different from coal tar pitches, with lower aromaticity (C=64-75%, H=5.9-6.4%, N=0.4-1.3%, O= 20.7-29.3%) and have infrared absorptions, similar to lignin. ^{1}H NMR and ^{13}C NMR in solution and ^{13}C in the solid state give peaks representative of syringyl and guaiacyl units.

The softening point can be correlated with molecular weight and assumes values of 85-120 °C. The acetone insoluble contents (A.I.= 2%, for crude pitch) and HRSEC results indicated that bio-pitches contain macromolecular structures. HRSEC using polystyrene standards and tetrahydrofuran (THF) as solvent indicate an average molecular weight of 2000-5000 daltons, depending on distillation conditions. Bio-pitches show thermoplastic behaviour. A conceptual model of the macromolecular network is proposed.

Key-words: Bio-pitch, distillation, eucalyptus tar, HRSEC, infrared spectroscopy, macromolecular structure, model, NMR.

1 Introduction

Charcoal manufacture has been prominent as a Brazilian energy source, mainly for industrial use. The predominant application is for iron-making in charcoal fedblast furnaces, which consume over 32 million tons of wood per year, converted into 8 million tons of charcoal. This process produces 35% of the pig iron and 98% of all the ferroalloys in this country [1].

During wood slow carbonization the maximum temperature is 400-500 °C. The vapors generated were washed to recover about 600 kg of various chemical products per 1 ton of charcoal. The watery by-product is pyroligneous acid and the oily phase is the wood tar, rich in phenolic molecules. Efforts have been made to upgrade these by-products [2] as fine chemicals [3-5] or polymeric materials [6,7]. The biggest challenge for consolidation of biomass carbochemistry activities in Brazil is market development, in view of the high potential volume of by-products that can be obtained.

Fractionation of *Eucalyptus* wood tar produces 50% of solid residue as pitches. The characterization of these bio-pitches has been done and its properties compared with those of coal tar pitches. Attempts to develop commercial applications have been made [8].

The properties of bio-pitches depend on the processes used for the wood tar recovery and mainly on the distillation conditions. Steam distillation under severe conditions can generate infusible residue, which can be used as fuel or as carbon precursor [8]. Vacuum distillation is the most indicated process for wood tar and produces pitches with different softening points, depending on the final temperature and pressure [8,9]. Flash distillation permits a separation of volatile material from pitch precursor molecules, in a few seconds of processing, but the equipments manuntance costs are high.

This paper presents the results of *Eucalyptus* tar pitch analysis. This raw material is obtained in pilot plant units [2,8], using different techniques.

The bio-pitch characterization was carried out by elemental analysis (E.A.), infra-red spectroscopy (I.R.), proton and carbon-13 nuclear magnetic resonance in solution and ^{13}CNMR in solid state, high resolution size exclusion chromatography (HRSEC), softening point (S.P.), acetone insoluble content (A.I.) and residual carbon content (R.C.). The results permit a discussion of influence of processing on the properties of the bio-pitche. A simplified model of their macromolecular structure is thereby proposed.

2 Experimental

Samples: Four samples of pitches obtained in pilot plants using different distillation processes were studied: two were obtained as residues from steam distillation (SD1 and SD2), one was obtained by a vacuum process (V2) and the fourth was obtained by flash distillation (F1). Sample SD2 was acetylated to enhance its solubility in organic solvents.

*Acetylation:*1.0 g of SD2 was added to a mixture of acetic anhydride (5.0 mL) and pyridine (5.0 mL). The reaction mixture was stirred and refluxed for 48 h. The mixture was poured in a beaker containing ice. The acetylated pitch product was precipitated and filtered. The product was dried and stored in a vacuum desiccator.

Softening Point (S.P.):. The samples were analysed using the "Ring and Ball Method", ASTM D2398-73- Standard Test [10]. This test gives the temperature at which the sample becomes soft with coalescent aspect in a glycerol bath, which is an important property of coal tar pitches, that can be correlated to its average molecular weight and molecular mobility.

Residual Carbon Content (R.C.): samples were heated to 550 °C for 2.5 hours, in a reducing atmosphere, using ISO 6998 Standard Test [11]. The residual carbon obtained were weighted and correlated to the initial mass. This property is important for applications in which the pitch is used as a carbon source.

Acetone Insoluble Content (A.I.): the powdered samples were wrapped in filter paper and extracted with acetone in a soxhlet for several hours, to constant refractive index. Subsequently, the acetone was evaporated, the residue was weighted and compared to the initial mass, according to DIN 53.700 Standard Test [12]. The acetone insoluble content is an indication of the degree of crosslinking; the higher the acetone insoluble content, the more cured is the resin.

Elemental Analysis (E.A.): the samples were analysed on a Perkin Elmer 240 B Elemental Analyser to determine % C, % H and % N. The H/C ratio is correlated to the aromaticity of the sample, and the O/C ratio is related to the degree of oxygenation.

Infrared Absorption Spectroscopy (I.R.): samples were prepared as KBr pellets and analysed on a Mattson Instruments spectrometer. KBr was previously dried at 120 °C and the pitch concentration used was 1%. Conditions were: resolution = 4 cm^{-1} and scans = 8. The assignments were made by comparison to a lignin spectrum [13].

Nuclear Magnetic Resonance Spectroscopy - ^{13}C-NMR in Solution: the acetylated sample of SD2 was dissolved in $CDCl_3$ and analysed in the presence of 0.05 M chromium triacetilacetonate. Spectra were taken on a VARIAN VRX-300 MHz spectrometer operating at 75.426 MHz under the following conditions:pulse angle, 80°; pulse width, 10 μs; delay time, 12 s; gated broad band proton decoupling; number of pulses, 17000; sample temperature, 22 °C.

Nuclear Magnetic Resonance Spectroscopy - ^{1}H-NMR in Solution : The spectrum was taken on a VARIAN VRX - 300 MHz; number of pulses: 250. The acetylated sample of SD2 was dissolved in $CDCl_3$.

Nuclear Magnetic Resonance Spectroscopy - Solid State ^{13}C NMR: spectra were taken on a VARIAN VRX-300-FT-NMR spectrometer, operating at a carbon frequency of 75.426 MHz. Experimental conditions were: gated broad band proton decoupling, single pulse, "magic-angle"spinning - MAS (54°44") 6500 Hz; delay time,100 s; number of repetition, 500-2050 (depending on the sample), sample temperature, 22 °C.

High Resolution Size Exclusion Chromatography - HRSEC: the samples were analysed on a Waters Associates Chromatographer, coupled to an ultraviolet detector (440/010) at 254 nm. The elution was carried out in tetrahydrofuran (THF), 1mg of sample / mL of THF, flow speed 1.0 mL/min and 1000 psi. Injected volume: 12 mL. Samples were previously filtered and eluted into a PL-gel (pre-column , 5μm), coupled to three PL-gel columns (1000 Å, 5 μm; 500 Å, 10 μm and 100 Å, 10 μm). The calibration curve was built using polystyrene standards (molecular weight - M_W : 168-68000 daltons).

3 Results and Discussion

The physico-chemical characterization analysis is presented in table 1. The results permit discussion of the influence of the distillation process on the pitch properties. For the same fractionation technique (steam distillation) a large variation of softening points for the pitches was observed (85.5 °C and 155 °C); the higher the softening point, the more efficient the volatile separation. The higher softening point presented by SD1 is due to the severe distillation conditions employed. The average molecular weights for both samples as well as their elemental analysis are comparable.

Table 1. Physico-chemical analysis of bio-pitches obtained on pilot plant units

	Steam Distillation		Vacuum Distillation	Flash Distillation
	SD1	SD2	V2	F1
Softening Point -S.P. (°C)	155.0	85.5	91.0	84.0
Residual Carbon - R.C. (%)	45.0	35.0	33.0	38.0
Acetone Insoluble- A.I. (%)	30.7	-	20.3	-
Elemental Analysis -E.A.(%)				
C	72.0	70.0	68.5	64.0
H	6.0	6.3	6.4	5.9
N	1.3	1.2	0.4	0.8
O	20.7	22.5	24.7	29.3
H/C	1.00	1.08	1.11	1.11
O/C	0.22	0,24	0.27	0.34
Molecular Weight - M_W	2369	2426	3663	2287 *
Molecular Weight - M_Z	471	484	507	427 *
Polydispersion	5.0	5.0	7.2	5.4 *

* Corresponding 85% of sample (THF soluble portion).

V2 presented a higher Mw than SD1 or SD2, indicating that vacuum distillation favored polymerization more than steam distillation did. It is interesting to note that this is so in spite of V2 being less aromatic than SD2 (higher H/C and O/C). The residence time is 3 h for steam distillation and 15 h for vacuum distillation.

F1 had a higher Mw than the remaining samples, and was not completely soluble in THF. The residence time for flash distillation is 3-6 s. Its higher degree of oxygenation is probably due to the presence of lignin moieties which were not thermally degraded during the distillation process because of the short time spent by the material in the thin film evaporator.

It was observed that the higher the O/C ratio, the higher is the average molecular weight of the bio-pitches. This indicates that the macromolecules have oxygen fixed in their structures, which causes a reduction in aromaticity.

Figure 1 shows the infra-red absorption spectra of lignin and bio-pitch V2. The assignments are listed in table 2. The similarity between the samples becomes apparent by a comparison of their spectra. Bio-pitches, however, do not show absorptions at 1300 cm^{-1}, characteristic of O-H in primary and secondary alcohols.

Figures 2-4 show the ^{1}H-NMR and ^{13}C-NMR spectra for the acetylated sample of SD2; the corresponding assignments are listed in tables 3 and 4. Results from table 3 clearly indicate that the sample was acetylated and presents a significant content of methoxylic protons, confirming the results obtained by infra-red. It also confirms the lower aromaticity in relation to mineral pitches, as indicated by the elemental analysis.

The results presented in table 4 were obtained by comparison with studies carried out for lignin [13] and show that pitch from *Eucalyptus* tar retains in its structure the basic units present in the tar, which in turn originate from the pyrolitic conversion of lignin. They also show that these units are linked by methylene bridges (30-28 ppm), not found in lignin.

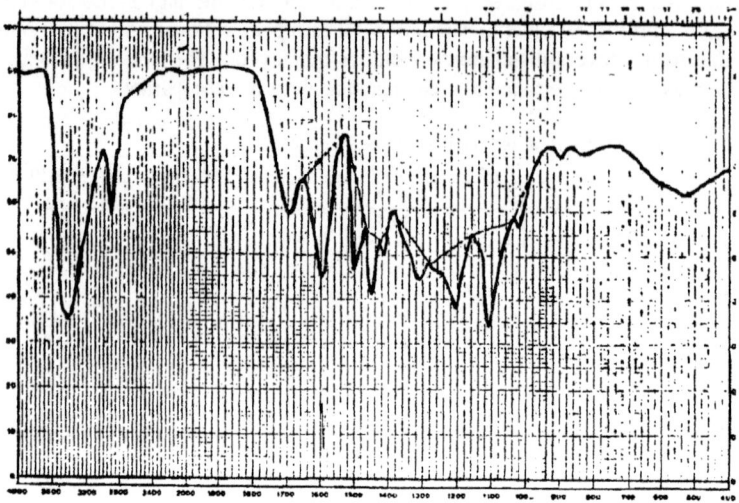

(a) Lignin

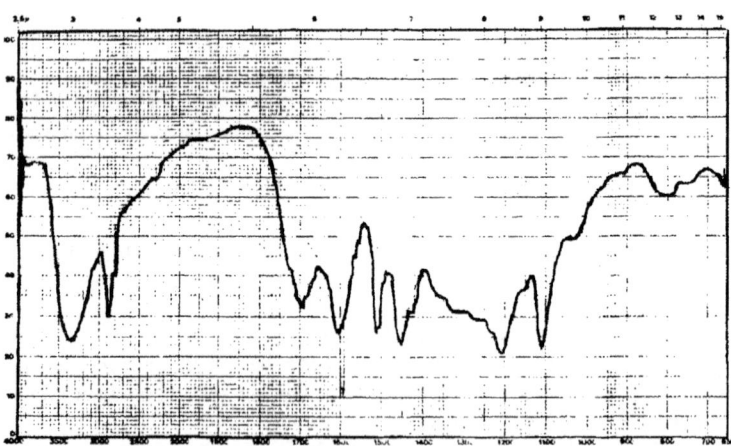

(b) Bio-pitch

Fig. 1. Infrared absorption spectra for lignin (a) and bio-pitch (b).

Table 2. Assignments for the main infrared absorptions of lignin and bio-pitches

Lignin (cm^{-1})	Pitch (cm^{-1})	Assignments
3400	3450-3400	Hydroxyl Groups (O-H)
3000-2850	2940-2830	C-H aromatic and aliphatic
1720-1690	1720-1700	C=O not conjugated
1660-1650	1660-1650	C=O conjugated
1600	1600	C-C aromatic rings
1505-1510	1500	C-C aromatic rings
1470-1460	1470-1440	C-H methylic and methylenic groups
1430-1415	1430-1410	C-C aromatic rings and C-H methylic groups
1330-1325	-	C-O syringyl units and O-H primary and secondary alcohol
1275	-	C-O guaiacyl units
1240-1230	1240-1220	C-O syringyl and guaiacyl units
1140-1120	1120	C-H syringyl and guaiacyl units
1085	-	C-O of secondary alcohol and of alkyl-aryl eters
1035-1030	1035-1030	C-O of alkyl eters, alkyl-aryl eters, primary alcohols and guaiacyl rings
915-815	815-915	C-H aromatic

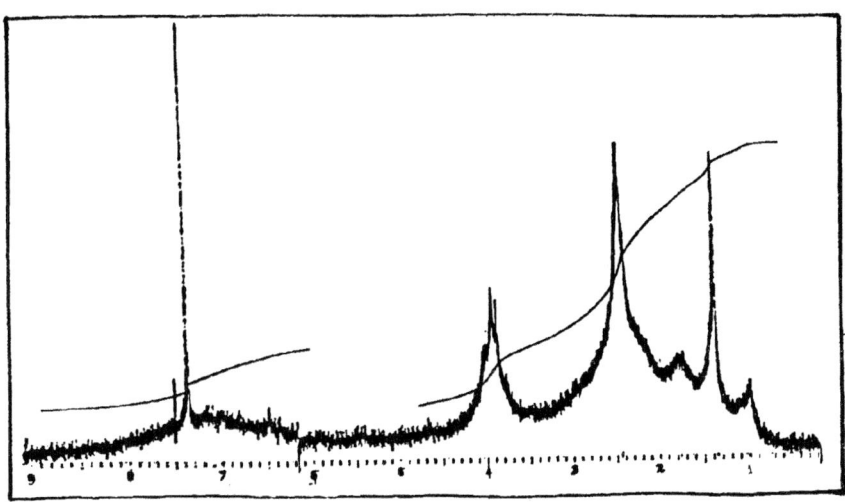

Fig. 2. ^{1}H NMR spectrum of SD2 acetylated pitch in CDCl$_3$ solution - 300 MHz

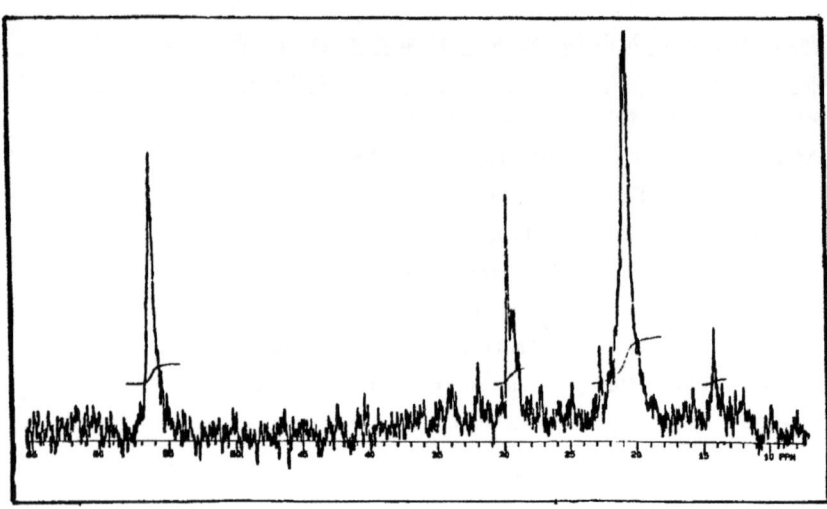

Fig. 3. ^{13}C NMR spectrum of SD2 acetylated, solution of CDCl$_3$ - 75 MHz, region of aliphatic carbon.

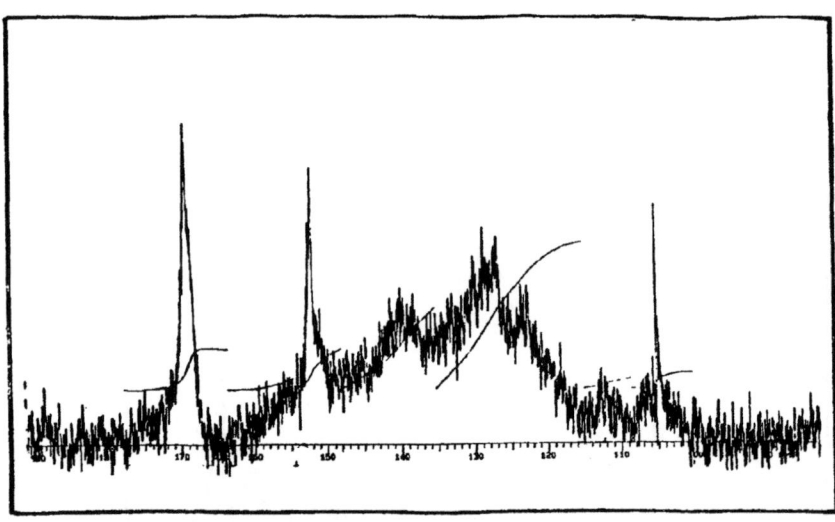

Fig. 4. ^{13}C NMR spectrum of SD2 acetylated, solution of CDCl3 - 75 MHz, region of aromatic carbon

Table 3. Assignments for chemical shift of protons - SD2 acetylated sample

Shift (ppm)	Assignments	%
9.0 - 6.0	Aromatic Protons	18.2
4.2 - 3.7	Methoxylic and Aryl Methylenic Protons	
2.6 - 2.3	Aromatic Acetoxylic Protons	
2.1 - 1.6	Acetoxylic Aliphatic and Acetoxylic Aromatic (ortho biphenyl linkage) Protons	81.8
1.6 - 0.5	Aliphatic highly shielding Protons	

Table 4. Assignments for chemical shift of carbons - SD2 acetylated sample

Chemical Shift (ppm)	Assignments	%
175.5-164.0	C=O acetyl groups of phenols and of primary and secondary alcohols	9.5
163.5-154.5	Aromatic carbon bridgehead	2.3
154.0-148.0	C-3 of acetylated guaiacyl (α-OR), C-3 and C-5 of acetylated syringyl (α-OAc), C-3 and C-5 of acetylated syringyl (α-OR).	9.4
148.0-135.5	C-4 of acetylated guaiacyl (α-OAc), C-1and C-4 of acetylated guaiacyl (α-OR) and C-1 of acetylated syringyl(α-OR)	18.9
135.0-115.0	C-4 of acetylated syringyl (α-OAc), C-4 of acetylated syringyl(α-OR), C-5 of acetylated guaiacyl, C-6 of acetylated guaiacyl (α-OR)	34.0
115.0-108.5	C-2 of acetylated guaiacyl (α-OAc) and C-2 of acetylated guaiacyl (α-OR)	2.4
108.5-100.0	C-2 and C-6 of acetylated syringyl (α-OAc) and (α-OR)	3.5
58.0-54.0	methoxyl (O-CH$_3$)	4.2
30.0-28.0	CH$_2$ between phenolic nuclei	3.4
21.0-18.0	CH$_3$ of acetyl groups (CH$_3$COO-)	11.1
15.0-13.5	CH$_3$ aliphatic	1.3
TOTAL		100

The ^{13}C-NMR/MAS spectra are presented in figure 5. A gated mode was used to avoid nuclear Overhauser effect, as well a single pulse, sample spinning at the magic angle and intervals between pulses (D1) of 100s, sufficient for total relaxation of the existing carbons, following a suggestion by Collin and co-workers [14]. Depending on the sample, 500 to 2050 transients were conected. Table 5 presents the types of carbons present in the pitches as also indicates that bio-pitches have lower aromaticity than coal tar pitches [15], confirmed by the H/C ratio. The content of aromatic carbon presented agrees with the H/C ratios presented in table 1.

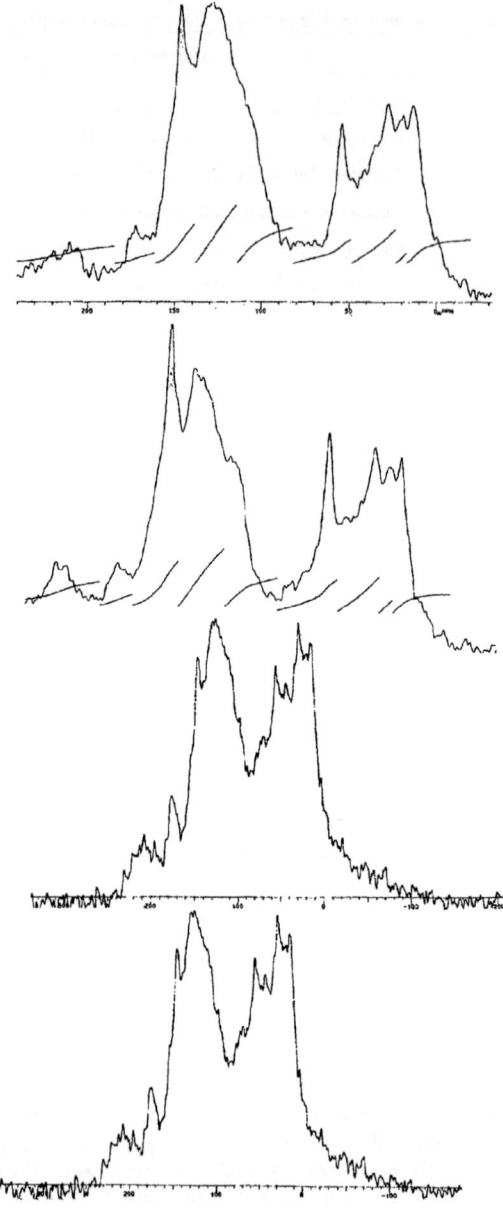

Fig. 5. ^{13}C-NMR spectra -solid state, 75 MHz, single pulse, D1=100 s, a) SD1 and b) SD2, transients =550 c) V2, transients= 2050 and d) F1, transients =515.

Table 5. Main types of carbon and their percentual content present in the samples: SD1, SD2, V2, F1 . Results of ^{13}C NMR - solid state, single pulse, D1 = 100s

Carbon	SD1 (%)	SD2 (%)	V2 (%)	F1 (%)
Saturated (0-80 ppm)	36,1	37,3	47,9	46,0
Insaturated: aromatic + olefinic (100-165 ppm)	52,9	50,1	44,0	45,1
Carbonylic (165-250 ppm)	11,0	12,6	8,10	8,90

The results in table 6, in which the functional groups were identified by using Trewhella's methodology [16], allow a good estimation of the chemical structure of pitches from *Eucalyptus* tar, which are less condensed than coal tar pitches, and posses different oxygenated chemical groups [15].

Table 6: Assignment of carbons presented in bio-pitch samples: SD1, SD2, V2 and F1, solid state ^{13}C-NMR, "single pulse", D_1 = 100s, using Trewhella method [24].

Chemical Functional Group	Region	Shift (ppm)	SD1 %	SD2 %	V2 %	F1 %
Methyl aliphatic -CH₂-CH₂-C*H₃	9	14	9.4	7.8	-	-
Methyl aromatic φ-C*H₃	8	20	4.0	4.9	13.2	14.1
Methylene -CH₂-C*H₂-CH₂ Methine and -CH₂-C*H-CH₂ Quaternary carbon - C* -	7	29 - 50	13.2	13.0	16.3	11.6
Oxymethylene C-C*H₂-O-C Oxymethine C-C*H-O-C	6	50 -70	9.5	11.6	18.4	20.3
Olefinic H₂C*=CH- -HC*=CH- and -C-C*=C-	5	110 -115	14.1	11.6	14.4	17.1
Non-substituted aromatic φ*-H	4	128	23.1	21.8	19.5	15.9
Condensed rings Alkyl substituted aromatic φ*-R Oxyaromatic φ*-OR	3	130 -160	15.7	16.7	10.1	12.1
Carbonyl (ester / acid) -COOR -COOH	2	178	4.2	4.1	4.0	3.2
Carbonyl (others) C=O	1	185 - 250	6.8	8.5	4.1	5.7

Pitch V2, obtained by slow distillation, a condition that favors polymerization reactions, presented the most intense signal in the region of methylenic bridges (29-50 ppm).

Despite the higher molecular weight of pitch F1, it suffered negligible polymerization, due to the short time spent in the evaporator (thin film). The analysis for this pitch showed that it presents:

- a higher content of aromatic and aliphatic methyl groups (region 9 and 8),
- a higher content of methoxyl (region 6) and olefinic (region 5) groups, implying it has a more oxygenated and less condensed structure, similar to lignin and
- a lower content of methylene bridges, indicating that these are formed during slow distillations, similar to V2.

4 Conceptual model of bio-pitch structure

Sample V2 is the most representative, considering that steam and flash distillation present high operacional costs. Pitch obtained by vacuum fractionation is thermoplastic and can be used in the development of polymeric materials, as phenolic resins [8].

Sample V2 gives Mw = 3663 daltons and about 20% of its mass is insoluble in acetone. This portion corresponds to macromolecular tridimensional units with similar behaviour to cured phenolic resins [6]. The soluble portion part (80%) was submitted to HRSEC, which gave MW = 2076 daltons.

Following these data, we suggest a model for the macromolecular network, which is an adaptation of that proposed for mineral coal by Haenel [17]. It is basically a two-component system consisting of a mobile phase and a stationary phase. The stationary phase is practically insoluble in organic solvents, consisting of a highly crosslinked macromolecular structure and with an average molecular weight about 10,000 daltons [8]. This network has some interstices, containing a mobile phase composed of relatively small molecules (500-2,000 daltons). These molecules can be extracted by organic solvents and also can be incorporated into the stationary phase by polymerization. The phase is mainly formed by alkyl substituted guaiacyl and syringyl groups, interconnected by aliphatic (methylene) bridges or bridges of the ether type (C-O-C). An schematic representation of this model is shown in Figure 6.

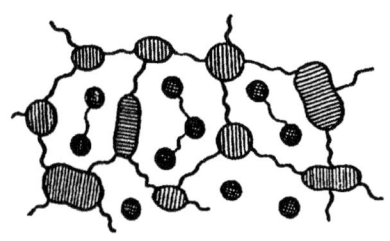

Macromolecular three-dimensional network (Immobile phase) Mw=10000 daltons	Methylene or oxymethylene bridges	Soluble molecules (mobile phase) Mw=2000daltons

Fig. 6. Conceptual bio-pitch model: two - component system

5 Conclusion

The bio-pitches have elemental analyses, infra-red absorption spectra and ^{13}C and ^{1}H NMR spectra similar to lignin. They preserve guaiacyl and syringyl units in their macromolecular structure. Their residual carbon (about 35%) is lower than coal tar pitches (50%) due to their low aromaticity and high oxygen content which limits their use as a carbon source.

^{13}C NMR spectra shows that bio-pitches have about 50% of aromatic and olefinic carbons. The spectroscopic results (IR and NMR) show that more than 30% of the carbon atoms are linked to oxygen. These oxygenated groups are mainly carbonyl, hydroxyl, carboxyl and methoxyl.

Chemical analyses of bio-pitches indicate that this macromolecular material is a kind of "fusible lignin" with thermoplastic behaviour.

Bio-pitch obtained by flash distillation is more similar to lignin than its counterpart obtained by vacuum fractionation. Slow distillation (vacuum) permits macromolecular fragmentation and rearrangements that can connect guaiacyl and syringyl units by methylene or oxymethylene bridges. The resulted pitches have an average molecular weight of about 2000-5000 daltons. The two-component model proposed for pitch macromolecular structure rationalises the phenolic resin behaviour of these materials.

6 Acknowledgements

The authors are grateful to Ms. S. Menezes (CENPES-Petrobrás) for her assistance with NMR spectra. This work was supported by ACESITA and CNPq.

7 References

1. ABRACAVE (1994) Anuário Estatístico, Associação Brasileira de Carvão Vegetal, Belo Horizonte, Brazil.
2. Carazza,F., Rezende,M.E.A., Pasa, V.M.D. and Lessa, A.(1994) Fractionation of Wood Tar, in *Advances in Thermochemical Biomass Conversion*, (ed. A.V. Brigwater), Blackie Academic & Professional, London, v. 2, p.1465-74.
3. Caraza,F., Pereira, M.O.S.and Oliveira, R. (1993) Derivatives From 2,6 Dimethoxyphenol, a Basic Constituent Of Eucalyptus Tar, (ed. D.P. Veloso and R. Ruggieiro), *Proceedings of The Third Brazilian Symposium on The Chemistry of Lignins and Other Wood Components*, Belo Horizonte, v. IV, p. 175-178.
4. Caraza, F., Machado, A.M.R., and Pereira, M.O.S. (1993) Effects of Methoxyl and Acetoxyl Groups Upon The Reaction Of NBS With Mono and Trioxygenated 4-Alkyl Phenols, (ed. D.P. Veloso and R. Ruggieiro), *Proceedings of The Third Brazilian Symposium on The Chemistry of Lignins and Other Wood Components*, Belo Horizonte, v.IV, p. 179-181.
5. Resende, A. M. M.(1992) *Análise por CGAR -EM/C de Misturas de Derivados Bromados de Constituintes Fenólicos do Alcatrão Vegetal*, Universidade Federal de Minas Gerais, dissertação de mestrado, Belo Horizonte.
6. Pasa, V.M.D., Otani, C. and Carazza, (1993) F. Wood Tar as Phenolic Resin Precursor, (ed. D.P. Veloso and R. Ruggieiro), *Proceedings of The Third Brazilian Symposium on The Chemistry of Lignins and Other Wood Components*, Belo Horizonte, v.IV, p. 291- 294.
7. Otani, C., Pasa, V.M.D. and Carazza F. (1990) The Structure and Chemical Characteristics Variation of Wood Tar Pitch During Its Carbonization, *Proceedings of The International Symposium on Carbon*, Tsukuba, v. 1. p. 546-549.
8. Pasa, V.M.D. *Piche de Alcatrão de Eucalyptus: Obtenção Caracterização e Desenvolvimento de Aplicações* (1994) Universidade Federal de Minas Gerais, Tese de Doutoramento, Belo Horizonte.
9. Otani, C., Polidoro, H. H., Rezende, L.C., Giana, H.E. e Pasa, V.M.D. (1994) Obtenção de Fibras Curtas de Carbono Ativadas a Partir do Piche Vegetal, *Metalurgia e Materiais*, Julho, p. 660-666, (ABM), São Paulo.
10 American Society for Testing and Materials. (1983) *Standard Test for Softening Point of Bitumen in Ethylene Glicol (ring and ball)*. ASTM, Phyladelphia, D3436-86.
11. International Standard Organization. (1984) *Carbonaceous Materials for The Production of Aluminium - Pitch for Electrodes - Determination of Coking Value*. ISO 6998.
12 Deutsche Industrie Normen. *Determination of Acetone Soluble Material in Polymeric Products*. DIN 53700.
13. Morais, S. A. L (1992) *Contribuição ao Estudo Químico e Espectroscópio da Lignina de Madeira Moída do Eucalyptus grandis: Isolamento, Quantificaçio e Análise Estrutural*. Tese de Doutoramento, Universidade Federal de Minas Gerais, Belo Horizonte.
14. Collin, E.S., Axelson, D.E., Botto, R.E. Delpuech, J.J., Tekely, P., Gerstein, B., Pruski, Marek, Maciel, G.E. and Wilson, M.A.(1989) 13 C NMR Solid State, a Discussion. *Fuel 68*, p. 548-60.

15. Rand, B., Hosty, A. J. and West, S.(1989) Physical Properties of Pitch to The Fabrication of Carbon Materials, *Introduction to Carbon Science,* Butterworths, chap.3.

16. Trewhella, M.J., Poplett, I.J.F. and Grint, A.(1986) *Fuel 65,* p. 541-46.

17. Haenel, M.W. (1992) Recent Progress in Coal Structure Research, *Fuel 71,* p.1211-1223.

PYROLYSIS

Combustion of pyrolysis liquid

COMBUSTION CHARACTERISTICS OF FAST PYROLYSIS OILS DERIVED FROM HYBRID POPLAR[1]

Combustion of poplar pyrolysis oils

C.R. SHADDIX and S.P. HUEY
Combustion Research Facility
Sandia National Laboratories
Livermore, California, USA

Abstract

A laminar-flow, single-droplet combustion facility has been used to evaluate the fundamental combustion properties of biomass pyrolysis oils produced by a fast, ablative vortex reactor at the National Renewable Energy Laboratory (NREL). Earlier single-droplet combustion experiments with pyrolysis oils derived from several biomass feedstocks (pine, oak, & switchgrass) revealed that droplets of these oils undergo violent microexplosions after relatively little mass loss, followed by disruptive, sooty burning and, for some oils, coke particulate formation. The present research focuses on the combustion behavior of three oils produced using the same hybrid poplar feedstock, but slightly different reactor and hot-gas filtration process variables. Surprisingly, the burning rates, degree of sooting, and coke particulate generation of these three poplar oils are found to differ considerably, presumably as a result of varying extents of fuel vapor cracking during the production of the oils. These results suggest that pyrolysis process conditions play a dominant role in the combustion behavior of the collected oils. A further implication is that, with adequate pyrolysis process control and possibly water or alcohol addition to the collected oil, a biomass pyrolysis oil may be produced which is optimized for fuel handling and combustion characteristics.

Keywords: biomass, burning rates, coke formation, droplet combustion, microexplosions, poplar, pyrolysis oils.

---

[1] This work was sponsored by the U.S. DOE's Office of Utility Technologies, Office of Solar Thermal and Biomass Power Technology, Biomass Power Program, and through direct support of the National Renewable Energy Laboratory, Golden, CO.

465

1 Introduction

Biomass pyrolysis oils (i.e. those produced from condensation of the pyrolysis vapors of biomass feedstocks) differ significantly in their chemical composition from traditional, petroleum-based fuel oils widely used in the energy marketplace. In particular, the biomass oils contain significant quantities of water and oxygenated organic compounds over a wide range of molecular weights (esp. organic acids, aldehydes, ketones, furans, alcohols, and phenols) [1–4]. These chemical differences are manifested as high densities, high viscosities, high surface tension, low pH, low heating values (energy content), and a much wider range of boiling points for the biomass oils [5–8]. In addition, significant quantities of char, with absorbed alkali metals, are often present in the biomass oils and pose potentially severe fuel handling and ash deposition/corrosion concerns in combustion environments [7–10].

As a consequence of the unique chemical and physical properties of pyrolysis oils, their combustion behavior may be expected to differ from that of conventional oils. Over the past few years a number of experiments [7,8,11,12] have evaluated the combustion properties of selected biomass oils. In all of these cases, the combustion characteristics of one or two selected biomass oils were compared in a given investigation, and for the most part the experimental burners utilized complicated, poorly characterized spray combustion systems. While these experiments have given insight into some of the potential problems of burning biomass oils, particularly with respect to boiler and furnace applications, they have provided little information on the differences in combustion behavior of biomass oils produced using different feedstocks or with the same feedstock under different pyrolysis conditions.

In contrast, an ongoing research program at Sandia's Combustion Research Facility has been investigating the fundamental droplet combustion behavior of biomass oils using a laminar-flow, single-droplet combustion facility. Oils produced by the fast, ablative pyrolysis reactor at the National Renewable Energy Laboratory (NREL) [13] are being investigated and compared to standard petroleum fuel oils. An initial investigation of oils produced from the pyrolysis of pine and oak feedstocks [14] revealed that biomass oil droplets generally undergo four successive stages of combustion: (a) quiescent burning, accompanied by a blue flame, which begins shortly after droplet injection, (b) microexplosion, accompanied by extensive flame luminosity, (c) subsequent disruptive burning of the remaining droplets, and (d) cenosphere formation and oxidation. These different combustion stages differ from those of distillate fuel oil droplets, which show quiescent, luminous burning throughout their lifetime.

Subsequent combustion experiments with a switchgrass oil and a poplar oil revealed a long initial quiescent burning period followed by violent microexplosion that completely fragmented the droplets and resulted in short burnout times, in spite of slow droplet burning rates during the quiescent burning stage [15]. These latter two NREL oils were produced after the pyrolysis unit cyclone system was replaced with a high-temperature baghouse for improved char removal from the pyrolysis vapors [16]. An inadvertent consequence of adding the baghouse to the pyrolysis unit was to increase the overall extent of thermal cracking of the fuel vapors before condensation into an oil. The lower char concentrations and the larger extent of thermal cracking in these latter oils were postulated to be responsible for the differences observed in the timing and intensity of the initial microexplosion in comparison with the earlier results for the oak and pine oils

[15]. However, the relative influence of different pyrolysis oil feedstocks could not be ascertained from these results.

In the present paper, the combustion behavior of three oils derived from the same hybrid poplar feedstock is evaluated. Production of these three poplar oils occurred with similar vortex reactor conditions and the same hot-gas filtration system, with some variation in the average hot-gas filter temperatures and the manner in which these temperatures were achieved [16,17]. In spite of the only minor differences present during production of these oils, they demonstrate a wide range of combustion properties. These results clearly illustrate the dominant effect of pyrolysis reactor cracking conditions on the microexplosion behavior of the resultant oils and suggest that addition of water or simple alcohols to some biomass oils may improve their combustion properties.

2 Experimental

2.1 Oil production and characterization

The three pyrolysis oils considered here were all produced from the same hybrid poplar (*P. trichocarpa x P. deltoides*) feedstock, a cross between Black and Eastern Cottonwood. These three oils will be referred to as poplar oil #1, #2, and #3, corresponding to NREL runs 160, 174, and 175, respectively. For all three runs, the vortex reactor wall temperature was maintained at 625 °C and the gaseous residence time in the reactor was ≈ 300 msec [16,17]. The vapor exit temperature was ≈ 575 °C during production of poplar oil #1 [16], and ≈ 525 °C during production of poplar oils #2 and #3 [17]. Flexible woven Nextel ceramic filter bags were used for hot-gas removal of entrained char particulate in the pyrolysis vapors. During production of oil #1, the baghouse, with a gas residence time of several seconds, was maintained at a mean temperature of ≈ 430 °C [16]. The baghouse temperature was reduced somewhat for production of oil #2 by spraying some of the condensed oil back into the transfer line upstream of the baghouse [17]. For production of oil #3 the baghouse temperature was further reduced by placing two cyclones directly upstream of the baghouse, eliminating the need for condensate spraying [17]. As indicated by the relative yields of pyrolysis liquids and uncondensable gases in Table 1, the extent of thermal cracking of the pyrolysis vapors was clearly greatest during production of oil #1 and was smallest for oil #3.

The combustion characteristics of a sample of diesel No. 2, a distillate fuel oil, were evaluated for comparison with the biomass oils. Proximate and ultimate composition as well as heating values were evaluated for all of the fuels and are shown in Table 1. The chemical analysis shown for oil #1 was performed on the raw oil by NREL shortly after its production. For the fuel oil and poplar oils #2 and #3, the chemical analysis results shown in Table 1 are duplicate sample averages of analyses performed by Hazen Research, Inc., on the oils at the time of their use in the combustion facility. All of the fuels were forced through a 0.45-μm filter at Sandia before use in the single-droplet combustion facility (to eliminate the potential for plugging of the droplet generator from residual char in the oils), but showed no noticeable buildup on the filter surface.

Three different heating values are given in Table 1. For fuels such as biomass oils that contain a substantial fraction of water, heating values on a dry basis can differ significantly from those evaluated on an "as received", or wet, basis. All of the values shown here are for the more meaningful wet basis. The higher heating value is most often measured and usually is the benchmark value for economic evaluations of a fuel. The lower heating value is a more practical estimate of the potential usable energy content of a fuel insofar as it accounts for the fact that in most combustors no useful heat transfer occurs at the dew point of the exhaust. The higher heating values of the biomass poplar oils are just over a factor of two lower than that of the No. 2 fuel oil, whereas the lower heating values show a difference of a factor of ≈ 2.5. Since these traditional heating values are defined in terms of the mass of the fuel, the ratio of lower heating values indicates the relative mass flowrates of fuel required to maintain a given thermal input to a combustor. However, the high density of biomass oils (≈ 1.2 g/ml, compared to 0.85–0.90 g/ml for distillate fuel oil) reduces the relative *volumetric* flowrate requirements for use of these oils in comparison with the indicated mass flowrate ratios. For fuels with significant oxygen content (e.g. explosives, coal, and biomass-derived fuels), neither the traditional higher heating value nor the lower heating value gives an accurate representation of the relative flame temperature or thermal

Table 1. Chemical compositions and pyrolysis run yields of investigated oils

	Poplar oil #1 (NREL 160)	Poplar oil #2 (NREL 174)	Poplar oil #3 (NREL 175)	No. 2 fuel oil
Elemental analysis (wt-% dry)				
C	64.4	59.3	57.1	87.3
H	7.0	6.6	6.6	12.9
O (by diff.)	28.3	33.8	36.3	-0.3
N	0.3	0.19	0.04	<0.01
S		0.00	0.04	0.1
Proximate analysis (wt-%)				
ash	0.005	<0.001	<0.001	<0.001
water	29.7	27.2	20.6	0.0
volatiles	64.5	67.4	71.8	99.8
fixed carbon (by diff.)	5.8	5.4	7.6	0.1
HHV (MJ/kg) [a]	19.0	17.5	17.6	43.8
LHV (MJ/kg) [b]	17.2	15.8	16.0	41.0
NLHV (MJ/kg) [c]	2.46	2.42	2.38	2.66
Viscosity (cP, at 40 °C) [d]	27	13	22	2
Pyrolysis run yields (wt-%)				
dry oil	35.8	41.0	49.3	
water	16.1	14.6	12.1	
char	11.7	16.4	12.2	
gases	23.6	21.7	18.2	

[a] HHV = higher heating value, wet basis (note: 23.3 MJ/kg = 10^4 Btu/lb)
[b] LHV = lower heating value, wet basis
[c] NLHV = normalized lower heating value (see text), wet basis
[d] viscosity was determined by NREL, with the exception of the No. 2 fuel oil, whose listed value is typical for this fuel

efficiency that can be expected with use of that fuel. A more useful heating value for these purposes is given by the so-called "normalized" heating value, also shown in Table 1. Whereas the traditional heating values are based on the ratio of combustion heat release to the mass of fuel, the normalized heating value reflects the ratio of combustion heat release to the total mass of fuel and stoichiometric air [18–20] and is derived from the measured, traditional heating value and an elemental analysis of the oil. These normalized values show that the thermal efficiency of a combustor should only drop slightly when switching from use of No. 2 fuel oil to the poplar biomass oils. Note that the relative heating value between two closely matched fuels can change when comparing traditional heating values with the normalized values, as illustrated here with poplar oils #2 and #3, due to the relative mass contributions of the water content of the oil and the amount of combustion air required for burning.

2.2 Droplet combustion facility

Sandia's Biomass Fuels Combustion System (BFCS) was used to evaluate the combustion properties of isolated biomass oil droplets. This facility has been described in detail elsewhere [14,15], so only a brief description will be given here. A stream of uniformly sized droplets is formed using a concentric nitrogen gas flow to aerodynamically shear the liquid fuel emanating from a capillary tube. As shown in Figure 1, the droplet stream then flows down the center of a laminar, quartz-walled flow

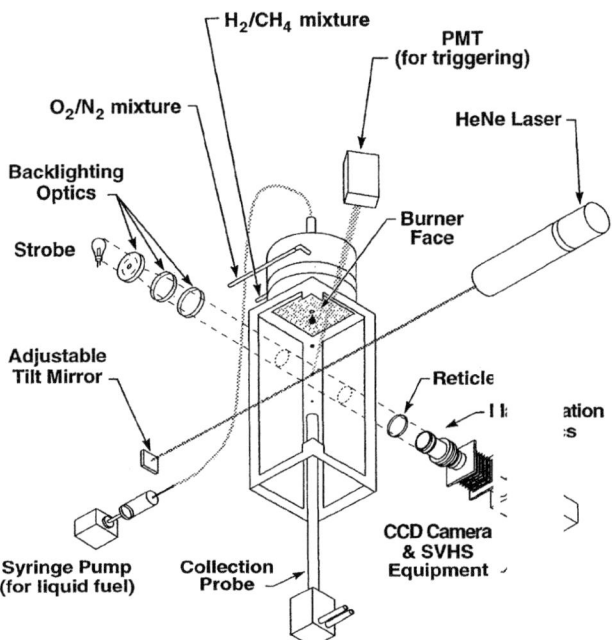

Fig. 1. Schematic of the Biomass Fuels Combu tion System (BFCS), an isolated-droplet combustion facility lesigned for analysis of biomass pyrolysis oils.

reactor. A flat, multiflamelet diffusion flame burner combusts a mixture of hydrogen, methane, oxygen, and nitrogen at the top of the reactor to produce a product stream of the desired temperature, velocity, and chemical composition. The droplet injection velocity is set to approximately match the burner effluent velocity, such that the droplet Reynolds number always remains less than 2 and convective effects on the droplet combustion are minimal. Imaging of the burning droplet shape and size is performed by laser triggering of a back-lighting strobe (1-μs pulse duration) and a high-resolution video imaging system. The flow reactor is mounted on a vertical translation stage surrounded by fixed optics, such that droplet imaging can be performed up to 30 cm from the burner face, corresponding to a droplet residence time of $\approx$ 220 ms. Droplet diameters are determined at a given measurement location by measuring the projected droplet area and assuming sphericity. Prior to the onset of any disruptive burning, measured droplet diameters have a standard deviation of $\approx$ 2% about the mean, with an average ratio of major to minor axis of 1.02. For the diameter values shown here, typically 6–10 individual measurements have been averaged. Local velocities are measured by setting the strobe to double-pulse at a separation of 500 μs, capturing two sequential images of a single drop on a single frame. From these velocity measurements the integrated droplet residence time can be determined as a function of axial position in the reactor.

For the experiments reported here, the liquid fuel and shear gas flows were adjusted to provide initial drop sizes of 300–450 μm. Drop-to-drop spacing was maintained at > 50 diameters in order to eliminate any potential interaction during combustion of successively injected drops. The gas mixture supplied to the flat-flame burner was adjusted to produce post-flame gases with 24.0 mole-% O_2, 13.2% H_2O, 2.4% CO_2, and 60.4% N_2 at a flowrate of 52 slpm (standard liters per minute), to which was added the $\approx$ 1 slpm N_2 flow through the droplet generator. Under these conditions, radiation-corrected thermocouple measurements show that the reactor centerline temperature rapidly rises to a peak of $\approx$ 1600 K several centimeters below the burner face and then slowly falls to 1500 K by the bottom of the reactor [15].

3 Results

3.1 Visual observations

Time-exposure photographs of the burning droplets of poplar oils and No. 2 fuel oil are shown in Fig. 2. Adjacent to the photograph of oil #3 combustion is a descriptive schematic to aid in the interpretation of these grayscale images. The photograph labeled as poplar oil #2 is actually that of an earlier switchgrass oil produced by NREL that was observed to have very similar visual combustion properties as the poplar oil #2. An attempt at photographing the combustion of poplar oil #2 several months after its detailed combustion characterization revealed a change in its combustion behavior to more closely resemble that of poplar oil #3. Consequently, the existing photograph of oil #2 droplet burning is not representative of the combustion behavior at the time its chemical and combustion characterizations were performed. Presumably this change in qualitative combustion behavior is a consequence of chemical degradation of the oil during storage.

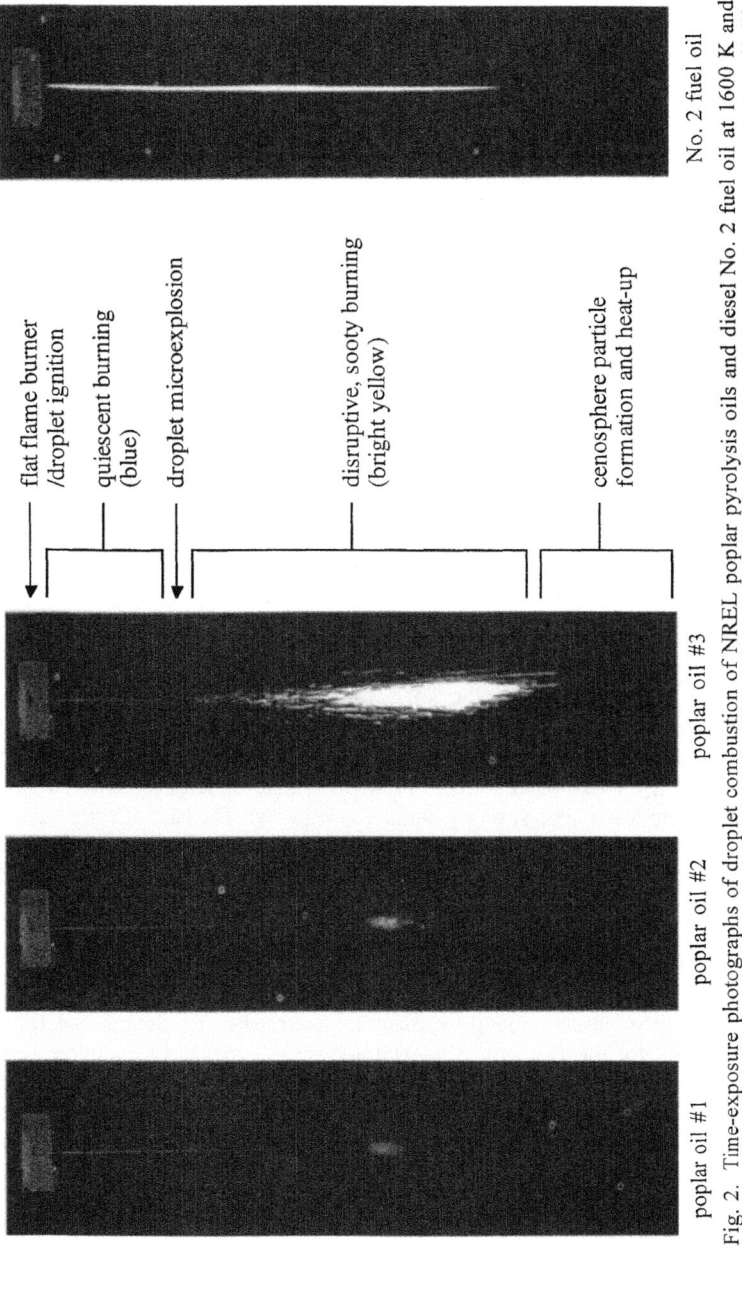

No. 2 fuel oil

poplar oil #1   poplar oil #2   poplar oil #3

flat flame burner
/droplet ignition

quiescent burning
(blue)

droplet microexplosion

disruptive, sooty burning
(bright yellow)

cenosphere particle
formation and heat-up

Fig. 2. Time-exposure photographs of droplet combustion of NREL poplar pyrolysis oils and diesel No. 2 fuel oil at 1600 K and 24 mole-% O_2. Initial droplet sizes are ≈ 350 μm. See note in text concerning photograph representing poplar oil #2. Exposure time of biomass oil photographs is ≈ 10 sec (or 300 drops), whereas the duration of the No. 2 fuel oil photograph is about 3 sec (100 drops). Chemiluminescent emission during the biomass oil quiescent burning stage has been artificially enhanced for illustrative purposes.

471

We have observed similar behavior with poplar oil #1 after storage for over 1 year and this aging phenomenon has been well documented for other biomass oils [17,21].

Coincident with these changes in combustion behavior of the aged oils, a marked increase in the pressure required to force the oils through a 0.45-μm filter (an indication of liquid viscosity) has been noted. However, proximate and ultimate analyses of poplar oil #2 have been performed after observation of the change in combustion behavior and reveal no significant differences from the original chemical analysis (reported in Table 1). Consequently, if chemical aging is indeed responsible for our observed change in combustion properties, the extent of gross chemical evolution (e.g. as reflected in the water content) is small and essentially undetectable with the current level of uncertainties in the proximate and ultimate analyses [22]. In support of these observations, Czernik et al. [21] found that aging of an oak oil resulted in over a factor of two increase in viscosity, but less than a 10% increase in water content.

As has been reported elsewhere [15], the combustion of poplar oil #1 is characterized by a relatively long quiescent period followed by a rapid, violent microexplosion that results in an audible popping noise and near-instantaneous evaporation of the drop. Combustion of poplar oil #2 shows the same general characteristics as that of oil #1, except for a slightly less efficient microexplosion event, resulting in a longer soot cloud tail associated with the microexplosions. The combustion of poplar oil #3 exhibits markedly different behavior than that for oils #1 and #2. In particular, the primary microexplosion occurs much earlier in the droplet lifetime and is much less effective at shattering the initial droplet. As a result, large fragmentary drops often remain after the first microexplosion and undergo a period of disruptive, sooty burning before ultimately forming solid coke particles.

In contrast to the biomass oils, the No. 2 fuel oil drops burn quiescently throughout their lifetime and display only a brief period of blue flame after ignition, before broadband (yellow) luminosity from soot dominates the flame appearance. After about half of the fuel oil burnout time, the luminous flame undergoes minor, transient shrinkage before resuming a strongly luminous character. The lack of coke formation for diesel No. 2 fuel oil and poplar oils #1 and #2, as well as the existence of coke particulate in the combustion of poplar oil #3, has been verified by probe sampling in the reactor.

3.2 Diameter measurements

Fig. 3 shows the square of the measured droplet diameters (d^2) as a function of the droplet residence time in the flow reactor. Asterisks (∗) denote the time of microexplosion for the two oils that exhibit complete or nearly complete droplet shattering, whereas a simple 'x' denotes the location of the original, primary microexplosion for oil #3, after which disruptive burning of the fragments occurs and no meaningful average droplet diameter can be determined. In order to place the curves for the different fuels on the same basis, the d^2 values are normalized by the square of the initial diameter (d_0^2). The initial diameter is assumed to be the first diameter measured in the reactor.

These initial diameters, droplet evaporation rates (see below), microexplosion times, and droplet burnout times are shown in Table 2. The residence times for the primary microexplosions vary over $\approx$ 10 ms for different drops of a given poplar oil, so the values shown in Fig. 3 and Table 2 represent averages. The burnout times in Table 2

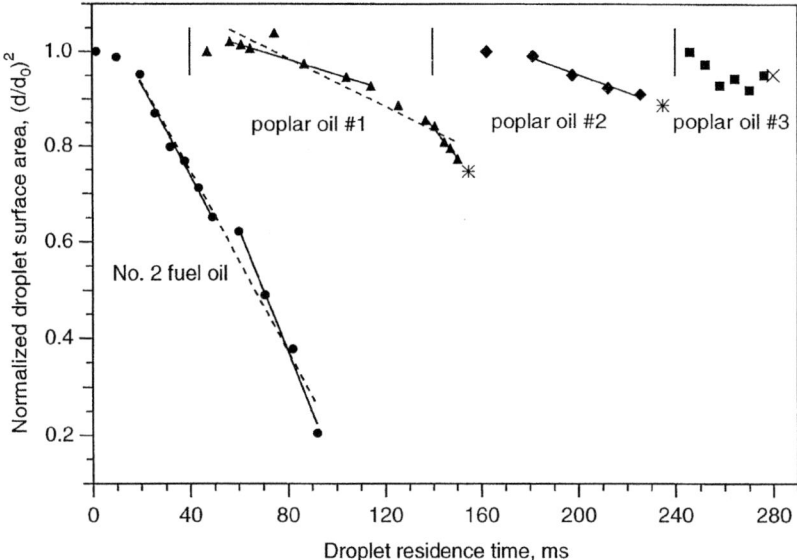

Fig. 3. Measured droplet diameter variation with residence time in the BFCS at 1600 K and 24 mole-% O_2. The initial diameters of the droplets are indicated in Table 2. For clarity the residence time zero has been set to 40 ms for poplar oil #1, 140 ms for oil #2, and 240 ms for oil #3. The lines shown are linear fits of the data between the data points corresponding to the ends of the fits. Solid lines denote "local" fits to the data, whereas dashed lines are used for post-heat-up average fits. For the early portion of the poplar oil #1 profile, the apparent outlier point was not included in the line fitting.

were derived for poplar oils #1 and #2 by measuring the height at which final luminous emission occurred and extrapolating the residence time from the location of the last velocity measurement. For poplar oil #3 the stated burnout time corresponds to a height at which the vast majority of the droplets and cenospheres appear to have been consumed. For diesel No. 2 the burnout time is extrapolated from the final diameter measurement and the evaporation rate. Also shown in Table 2 are the effective droplet burnout rates (K_{eff}) and the relative heat release rates ($\dot{H}_{eff}$) of the different oils, as will be discussed in section 4.1.

The general shape of the d^2 vs. time plots in Fig. 3 is consistent with the findings from droplet combustion of a number of pure compounds as well as mixtures [23]. The initial period in which the droplet diameter remains approximately constant (and may even increase slightly) is known as the droplet heat-up period, during which the droplet temperature rises rapidly and the resultant thermal expansion more or less compensates for surface regression due to vaporization. After this initial transient period, the droplets typically enter a quasi-linear "d^2-law" regime, wherein the rate of droplet evaporation is predominantly controlled by the rate of heat transfer back to the droplet from the flamefront. The slope of this portion of the d^2 vs. time plot is known as the evaporation constant (K_{evap}) or burning rate, for those instances in which a flame surrounds the droplet. In fact, due to the transient nature of fuel vapor accumulation and depletion within the surrounding flame, the rate of mass loss from the droplet surface cannot be assumed to accurately represent the instantaneous rate of fuel consumption at the droplet flame [24], so use of the term "burning rate" in this context can be misleading. Furthermore, for multicomponent droplets such as those investigated here, the d^2 vs. time trace can display a continuously sloping curve or can show one or more "plateaus" between quasi-linear sections, depending on the ambient environment, mass diffusion resistance of the droplet components, and the distribution of volatilities of the droplet components [23].

After an initial heat-up period of ≈ 10 ms, the drops of No. 2 fuel oil appear to follow two distinct d^2-law regions separated by a brief period with little change in diameter. In contrast, the d^2 plot for poplar oil #1 reveals a strongly sloping character before microexplosion, but two linear fits appear to match the data well for intermediate and late times. Poplar oil #2 undergoes a relatively early microexplosion, with a single linear d^2 section subsequent to the initial droplet heat-up. The d^2 plot for poplar oil #3 does not show the characteristic heat-up section, but appears to decay at approximately the same rate as oil #2 before reaching an early microexplosion. The absence of a heat-up region may be indicative of a skewed droplet trajectory entering the reactor, particularly since the trajectory of this oil in the reactor is difficult to determine due to its early microexplosion. This oil also shows evidence of an increase in diameter just before microexploding, which could result from preliminary vapor formation in interior regions of the droplet.

Table 2. Burning rate characteristics (1600 K, 24 mole-% O_2)

	Poplar oil #1 (NREL 160)	Poplar oil #2 (NREL 174)	Poplar oil #3 (NREL 175)	No. 2 fuel oil
d_0 (μm)	358	463	436	307
K_{evap} (mm^2/s) [a]	0.20, 0.80	0.39	----	0.92, 1.19
$\overline{K}_{evap}$ (mm^2/s)	0.33	0.39	----	0.88
microexplosion time (ms)	115	95	40	----
burnout time (ms)	120	105	≈ 180	110
K_{eff} (mm^2/s)	1.1	2.0	≈ 1.1	0.86
$\dot{H}_{eff}$	0.73	1.3	≈ 0.67	1.0

[a] values of local fits to the d^2 data

3.3 Microexplosion imaging

Examples of the characteristic types of microexplosions imaged for the different poplar oils are shown in Fig. 4. Poplar oil #1 exhibits two characteristic types of microexplosions that can be broadly categorized as "bubble blowout" or "complete atomization" [15]. A third, "wispy", microexplosion is also occasionally evident.

Poplar oil #2 also undergoes some microexplosions via the "bubble blowout" mechanism, but no evidence of the "complete atomization" mechanism is observed. Instead, this oil often undergoes fragmentary "splatter" types of microexplosion, with the production of many ultrafine droplets as well as some larger fragments. For both poplar oil #1 and #2, double-strobe imaging reveals that the microexplosion event occurs in less than the 500-μs probe flash separation, resulting in effective and rapid dispersal of the droplet mass.

In contrast to the other poplar oils, oil #3 exhibits a bubbly, "eruption" type of microexplosion which is similar to that imaged earlier in the investigation of the NREL oak and pine oils [14]. This type of microexplosion also appears to have a very rapid initial onset, but only results in partial fragmentation of the parent drop.

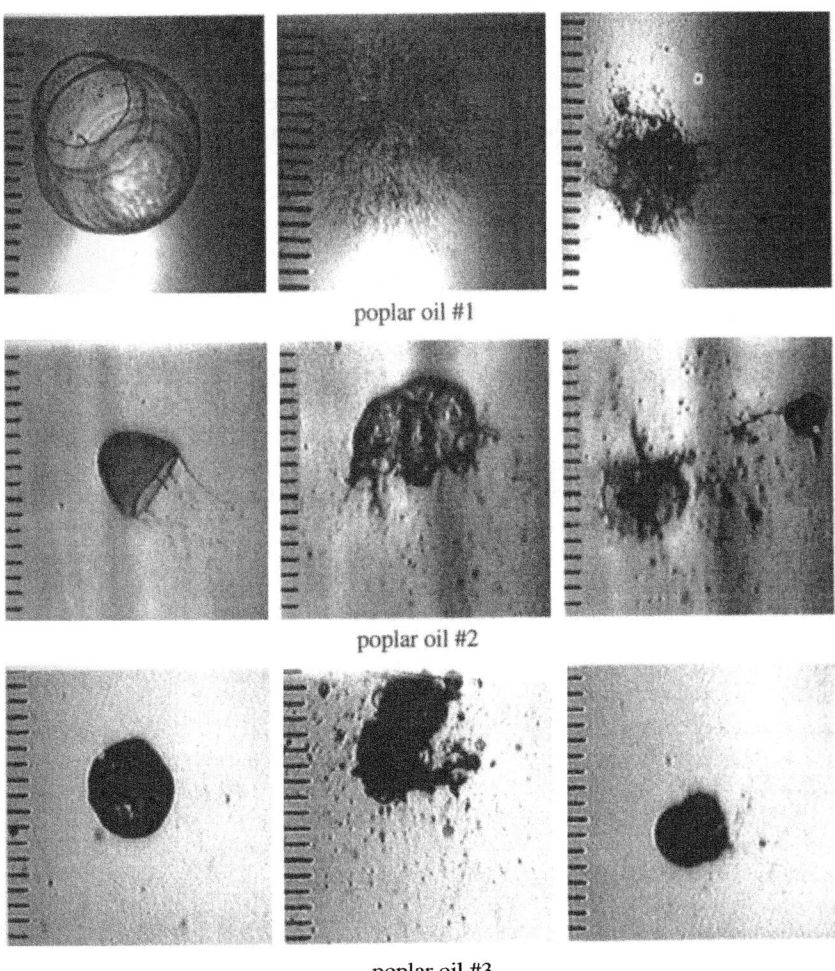

poplar oil #1

poplar oil #2

poplar oil #3

Fig. 4. High-magnification backlit video images of primary droplet microexplosion during the combustion of poplar oils in the BFCS entrained flow reactor at 1600 K and 24 mole-% O_2. Tick spacing on the left of each image is 100 μm. Initial droplet diameters are ≈ 350 μm.

4 Discussion

4.1 Burning rate comparisons

The relative rates of liquid fuel droplet consumption are usually evaluated in terms of the d^2-law linear fits of the data. Table 2 shows that for the flow reactor conditions investigated here, the No. 2 fuel oil vaporizes at a rate of approximately 1 mm^2/s, similar to values which have previously been published [25,26]. The plateau in the d^2 plot (Fig. 3) separating the two linear sections occurs at the same location that minor, transient luminous flame shrinkage is observed (slightly evident in Fig. 2), and therefore is in all likelihood not an artifact of the limited temporal density of data. Similar profiles have been measured for some binary mixtures [23], wherein the plateau represents an intermediate heat-up region between vaporization of mixtures with two characteristic volatilities.

The poplar oil #1 evaporation rate rapidly increases as the microexplosion time is approached, but even then is somewhat below the evaporation rate of the diesel. The *average* evaporation rate after the first 15 ms is 0.33 mm^2/s, about 3 times lower than that for diesel. Similarly, the evaporation constant for poplar oil #2 is 0.39 mm^2/s. However, poplar oils #1 and #2 undergo such complete microexplosions that their droplet lifetimes are just a few milliseconds longer than their microexplosion times and are at least as short as for comparably sized droplets of No. 2 fuel oil. Consequently, a more meaningful evaluation of droplet combustion for microexploding fuels is the effective droplet burnout rate (K_{eff}), defined as the square of the initial droplet diameter divided by the droplet lifetime. Values of this parameter are shown in Table 2.

For the experimental conditions investigated here, it is apparent that the effective burning rate of poplar oil #2 is approximately twice that of No. 2 fuel oil, and the effective rates of oils #1 and #3 appear to be slightly greater than that of the fuel oil. For poplar oil #3, however, it is impossible to determine a definitive or even statistical burnout time due to the wide shower of droplet fragments and variation in coke particle sizes formed, so any calculations based on oil #3 burnout time are suspect. Also, the calculation of K_{eff} implicitly assumes that the burnout time is proportional to the square of the initial droplet diameter for the range of initial droplet sizes considered here. This assumption is probably a fair one for the No. 2 fuel oil, but may be in error with respect to the biomass fuels, whose burnout time is dictated by the microexplosion event. We are presently making measurements to ascertain the effect of initial droplet diameter on the burning behavior and microexplosion times of biomass oils.

Regardless of the relative K_{eff} values, the actual droplet heat release rate is probably of most practical interest. The relative heat release rates can be estimated as shown below:

$$\dot{H}_{eff,2/1} = \frac{K_{eff,2}}{K_{eff,1}} \cdot \frac{\rho_{l,2}}{\rho_{l,1}} \cdot \frac{LHV_2}{LHV_1} \tag{1}$$

where ρ_l is the liquid fuel density (determined at room temperature) and K_{eff} and LHV have been previously defined. These heat release rates, evaluated with respect to the No. 2 fuel oil, are shown in Table 2. The ratio of mass consumption and therefore heat release rates is only meaningfully evaluated for a given initial droplet diameter, so, as before, K_{eff} is assumed to be constant for these fuels over the investigated range of initial droplet diameters. The estimated values of heat release rates show the poplar oil #2 droplets release energy 30% faster than the No. 2 fuel oil, whereas the two other poplar fuels release energy 30% slower than the diesel. If these results could be extrapolated to practical spray combustor conditions, they would suggest that the poplar oil #2 would form a tighter, more intense spray flame than the No. 2 fuel oil, whereas the other poplar oils would form a somewhat longer flame than the diesel. However, this speculation neglects any flowfield effects created by the significantly higher fuel flowrate required by the biomass oil compared to the diesel to maintain a given thermal output.

4.2 Effect of pyrolysis conditions

For a given biomass feedstock, the effect of an increase in thermal vapor cracking during pyrolysis is to convert relatively large, oxygenated primary pyrolysis products such as levoglucosan and coniferyl and sinapyl alcohols to carbon monoxide and light hydrocarbons (uncondensable gases), water, furfurals and acetaldehydes, lighter aromatic alcohols (phenols and guaiacols), and larger aromatic oxygenates [2]. In short, with increasing residence time at high temperatures during biomass pyrolysis, the oxygen content in the organic matter decreases, the water production increases, and an initial suite of molecules of intermediate molecular weight becomes more broadly distributed to both lower and higher molecular weights. These trends for oxygen concentration and water content in the poplar oils are evident in Table 1.

The droplet combustion images, d^2 plots, and microexplosion imaging all clearly demonstrate that even relatively minor variations in pyrolysis conditions (as experienced by the poplar oils examined here) can have dramatic effects on the combustion characteristics of biomass oil droplets. The sensitivity of the gross combustion behavior of biomass oils to the pyrolysis conditions probably arises from the dominance of the primary microexplosion to the later history of the droplet, and, in turn, the sensitivity of the microexplosion characteristics to the distribution of chemical species in the oil. In particular, the variations in oxygen concentrations and water contents of the oils can account for the microexplosion trends which are observed.

Biomass oils, with their high oxygen content, are known to polymerize over time, resulting in an increase in viscosity at a rate which is a strong function of temperature [17, 21]. Similarly, the high percentage of non-volatile residue formed from the distillation of biomass oils [27] is presumably due to these same polymerization reactions. One might expect, then, that at the elevated droplet temperatures experienced during combustion these reactions will be fast enough to result in a noticeable increase in droplet viscosity as the droplet evaporates. This increased viscosity would result in an enhanced resistance to the shattering effects of an internal vapor bubble and thus a relatively ineffective microexplosion. On the other hand, the production of polymerized, high-molecular-weight components during droplet combustion would cause an increase in the droplet surface temperature (= wet-bulb temperature of the surface mixture) and thereby hasten the occurrence of microexplosion. The tendency to polymerize presumably increases with an increase in oxygen content of the dry oil and thus with a decrease in vapor cracking during production of the oil. In fact, the microexplosion images for the previous NREL oak and pine oils [14] as well as the current poplar oil #3, all of which experienced a small amount of vapor cracking during production, suggest a very viscous solution exists at the time of microexplosion. These oils also undergo an early, weak microexplosion, resulting in the formation of large secondary

477

droplets and the production of coke particulate. This same combustion behavior of *aged* poplar oils #1 and #2 (i.e. "pre-polymerized oils") is consistent with this reasoning.

As the water content of the oil increases, the residence time to microexplosion should increase, because the higher water content will result in a longer period of surface-layer devolatilization before the droplet temperature rises considerably. Furthermore, assuming the microexplosion results from vapor nucleation in a miscible mixture, as is thought to be the case [15], the increased concentration of water will have a strong effect on the initial bubble vapor pressure, and thus on the speed with which the bubble expands through the droplet, due to the very high saturation pressure of water at superheat temperatures.

For the poplar oils considered here, it is difficult to deconvolve the relative influence of increased water content from that of decreased dry-oil oxygen content, since these factors are intimately linked through the pyrolysis process. For example, while our existing data would seem to provide some evidence for faster bubble expansion during microexplosion for the higher-water-content oils #1 and #2, the explanation for this phenomenon could also reside with the oxygen-content-induced increased viscosity evident during combustion of poplar oil #3. Similarly, decreased polymerization during combustion of oils #1 and #2 could result from the effects of either a high water content or a low oxygen content. In all likelihood, these microexplosion characteristics result from a combination of the effects of water and oxygen content in the oil.

4.3 Practical implications

Of the NREL poplar oils investigated, the one which experienced the most vapor cracking during production (oil #1) exhibits the most complete microexplosive atomization. The intermediately cracked oil (oil #2) demonstrates an effective microexplosion at an earlier residence time, resulting in very rapid net consumption of fuel – greater than that of No. 2 fuel oil, even on the basis of heat release. Finally, the least cracked poplar oil (oil #3), with the lowest water content and highest liquid yields during production, experiences an early, but ineffective, microexplosion, followed by sooty burning and the production of coke. At first glance these results may seem to suggest that the combustion behavior of biomass oils are ill-matched to the economics of production, which clearly favor a minimum of pyrolysis vapor cracking. However, our interpretation of the present results implies that the combustion behavior of the uncracked oils may be significantly improved simply by the addition of water to the oil (until the water content is 25–30 wt-%). Of course, the addition of water results in a decrease in the already-low heating value of the fuel, so adding methanol or ethanol (which can be derived from biomass) may be a preferable alternative, assuming a similar improvement in the combustion behavior results. In fact, for practical combustion applications addition of water or alcohol to the pyrolysis oils is desirable in order to reduce the inherently high viscosities of the oils, and researchers at NREL have investigated the effect of methanol and/or water addition on the viscosity of biomass oils [28]. We are presently investigating the effects of water or alcohol addition on the combustion properties of the biomass oil droplets, as well as the spray flame behavior of the biomass oils.

5 Conclusions

The single-droplet combustion characteristics of three hot-gas-filtered poplar pyrolysis oils have been evaluated and compared to those of No. 2 fuel oil. As was previously found for other biomass oils, the combustion behavior and effective droplet burnout times of the poplar oils are dominated by the characteristics of a primary microexplosion event. The timing and atomization quality of the microexplosion vary considerably with relatively minor variations in the pyrolysis vapor cracking experienced during production of these oils. The optimally combusting poplar oil demonstrates an effective droplet mass consumption rate (and even a heat release rate) that is significantly larger than that for the No. 2 fuel oil. The oil with the least thermal cracking during production exhibits the worst combustion characteristics – sooty burning and coke formation. These trends may be explained in terms of the effects of water content and dry-oil oxygen content on the microexplosion behavior, and the relation of these parameters to the extent of thermal cracking of biomass vapors. This reasoning suggests that the combustion behavior of low-cracking-severity biomass oils might be significantly improved by the addition of water, or perhaps methanol or ethanol, to the oil before combustion.

6 Acknowledgment

The authors thank Tom Milne, Jim Diebold, and John Scahill of NREL for supplying the NREL biomass oils and their characterization data, as well as information about the pyrolysis reactor conditions during their production. Jeff Davis assisted with data collection for poplar oils #1 and #2 and Kevin Davis (REI) directed the project during investigation of oil #1. Donald Hardesty has provided effective management of the project.

7 References

1. Radlein, D., Piskorz, J, and Scott, D.S. (1987) Lignin derived oils from the fast pyrolysis of poplar wood. *J. Anal. Appl. Pyrolysis*, 12: 51–59.
2. Evans, R.J., and Milne, T.A. (1987) Molecular characterization of the pyrolysis of biomass. 1. Fundamentals. *Energy & Fuels*, 1: 123–137.
3. Evans, R.J., and Milne, T.A. (1987) Molecular characterization of the pyrolysis of biomass. 2. Applications. *Energy & Fuels*, 1: 311–319.
4. Maggi, R., and Delmon, B. (1994) Comparison between 'slow' and 'flash' pyrolysis oils from biomass. *Fuel*, 73: 671–677.
5. Peacocke, G.V.C., Russell, P.A., Jenkins, J.D., and Bridgwater, A.V. (1994) Physical properties of flash pyrolysis liquids. *Biomass and Bioenergy*, 7: 169–177.
6. Bakhshi, N.N., and Adjaye, J.D. (1995) Characteristics of a fast pyrolysis bio-fuel and its miscibility with oxygenated and conventional fuels. *Proceedings: Second Biomass Conference of the Americas: Energy, Environment, Agriculture, and Industry*, Portland OR, Aug. 21–24, NREL, Golden CO 80401, CP-200-8098, pp. 1079–1088.
7. Barbucci, P., Costanzi, F., Ligasacchi, S., Mosti, A., Rossi, C. (1995) Bio-fuel oil combustion in a 0.5 MW furnace. *Proceedings: Second Biomass Conference of the Americas: Energy, Environment, Agriculture, and Industry*, Portland OR, Aug. 21–24, *op.cit.*, pp. 1110–1120.

8. Solantausta, Y., Nylund, N.-O., and Gust, S. (1994) Use of pyrolysis oil in a test diesel engine to study the feasibility of a diesel power plant concept. *Biomass and Bioenergy*, 7: 297–306.

9. Agblevor, F.A., Besler, S., and Evans, R.J. (1994) Inorganic compounds in biomass feedstocks: their role in char formation and effect on the quality of fast pyrolysis oils. *Specialists Workshop on Biomass Pyrolysis Oil Properties and Combustion*, Estes Park CO, Sept. 26–28, NREL, Golden CO 80401, CP-430-7215, pp. 77–89.

10. Elliott, D.C. (1994) Water, alkali and char in flash pyrolysis oils. *Biomass and Bioenergy*, 7: 179–185.

11. Ardy, P.L., Barbucci, P., Benelli, G., Rossi, C., and Zanforlin, S. (1995) Development of gas turbine combustor fed with bio-fuel oil. *Proceedings: Second Biomass Conference of the Americas: Energy, Environment, Agriculture, and Industry*, Portland OR, Aug. 21–24, *op.cit.*, pp. 429–438.

12. Gust, S. (1994) Flash pyrolysis fuel oil. *Specialists Workshop on Biomass Pyrolysis Oil Properties and Combustion*, Estes Park CO, Sept. 26–28, *op.cit.*, pp. 316–320.

13. Czernik, S., Scahill, J., and Diebold, J. (1995) The production of liquid fuel by fast pyrolysis of biomass. *ASME Journal of Solar Energy Engineering*, 117: 2–6.

14. Wornat, M.J., Porter, B.G., and Yang, N.Y.C. (1994) Single droplet combustion of biomass pyrolysis oils. *Energy & Fuels*, 8: 1131–1142.

15. Shaddix, C.R., Huey, S.P., and Davis, K.A. (1996) Microexplosive behavior of biomass-derived pyrolysis oils during the combustion of isolated droplets. To be submitted.

16. Diebold, J.P., Czernik, S., Scahill, J.W., Phillips, S.D., and Feik, C.J. (1994) Hot-gas filtration to remove char from pyrolysis vapors produced in the vortex reactor at NREL. *Specialists Workshop on Biomass Pyrolysis Oil Properties and Combustion*, Estes Park CO, Sept. 26–28, *op.cit.*, pp. 90–109.

17. Diebold, J.P., Scahill, J.W., Czernik, S., Phillips, S.D., and Feik, C.J. (1995) Progress in the production of hot-gas filtered biocrude oil at NREL. Presented at the Second E.U./Canada Workshop on Bio-Oil, Toronto, Ontario, May 8–9, NREL, Golden CO 80401, TP-431-7971.

18. L.L. Baxter, personal communication.

19. Hottel, H.C. (1983) The relative thermal value of tomorrow's fuels. *Ind. Eng. Chem. Fundam.*, 22: 271–276.

20. Goodger, E.M. (1995) Light distillate fuels for transport. *J. Instit. Energy*, 68: 199–212.

21. Czernik, S., Johnson, D.K., and Black, S. (1994) Stability of wood fast pyrolysis oil. *Biomass and Bioenergy*, 7: 187–192.

22. McKinley, J.W., Overend, R.P., Elliott, D.C. (1994) The ultimate analysis of biomass liquefaction products: the results of the IEA round robin #1. *Specialists Workshop on Biomass Pyrolysis Oil Properties and Combustion*, Estes Park CO, Sept. 26–28, *op.cit.*, pp. 34–53.

23. Wang, C.H., Liu, X.Q., and Law, C.K. (1984) Combustion and microexplosion of freely falling multicomponent droplets. *Comb. and Flame*, 56: 175–197.

24. Law, C.K. (1982) Recent advances in droplet vaporization and combustion. *Prog. Energy Combust. Sci.*, 8: 171–201.

25. Wood, B.J., Wise, H., and Inami, S.H. (1963) Combustion of fuel droplets. *AIAA J.*, 1: 1076–1081.

26. Godsave, G.A.E. (1953) Studies of the combustion of drops in a fuel spray – the burning of single drops of fuel. *4th Symposium (International) on Combustion*, The Combustion Institute, Pittsburgh PA, pp. 818–830.

27. Moses, C., (1994) Fuel-specification considerations for biomass liquids. *Specialists Workshop on Biomass Pyrolysis Oil Properties and Combustion*, Estes Park CO, Sept. 26–28, *op.cit.*, pp. 362–382.

28. Agblevor, F., Guran, S., Diebold, J., Scahill, J., Feik, C., and Graham, J. (1995) Pyrolysis for turbines, in *NREL Biomass Power Program Quarterly Progress Report: Third Quarter, FY 1995*, NREL, Golden CO, pp. 15–29.

COMBUSTION EXPERIENCES OF FLASH PYROLYSIS FUEL IN INTERMEDIATE SIZE BOILERS

Flash pyrolysis liquids combustion in boilers

STEVEN GUST
Neste Oy,
Porvoo, FINLAND

Abstract

Flash pyrolysis liquid was combusted with simple pressure atomisation equipment commonly used with light fuel oils in intermediate size boilers. With a number of modifications to the combustion system, carbon monoxide (CO) and nitrous oxide (NOx) could be reduced to acceptable levels: CO < 30 ppm and NOx < 140 ppm. On the other hand, particulate emissions were >2 times higher than from light fuel oil. Contributions to high particulate emissions were from high solids content (0.5-0.7 wt%) and poor atomisation caused by high viscosity of the samples (> 20 cSt at 80 °C).

The changes to the combustion system/conditions compared to light fuel oil were: acid resistant progressive cavity pump, higher oil preheat temperature and higher oil pressure, refractory section between burner and boiler warmed up to at least 800 °C. In addition, it was necessary to store pyrolysis liquid samples under inert conditions and to rinse nozzles with solvent after shutdown to prevent coking. The complexity and cost of these modifications are considered to be too great for current grades of flash pyrolysis liquid to be sold as a light fuel oil replacement.

Keywords: boilers, combustion, emissions, flash pyrolysis liquid.

1. Introduction

The aim of this work is to determine the technical feasibility of combusting flash pyrolysis liquids in intermediate size boilers where light fuel oil is currently used. Intermediate size boilers are roughly classified as larger than residential boilers (15-20 kW) but smaller than the lower range of heavy fuel oil boilers (1 MW). Equipment that is commonly used here is simple pressure atomisation where the liquid is pumped under pressure through a nozzle to form a fine spray which mixes with air and combusts. Typical customers could be small industry, schools, hospitals, block of flats, greenhouses etc. The total market potential in Finland is estimated to be about 1-1.2 Mtoe or about one-half of current light fuel oil use. The practical potential will depend on biomass availability and cost, government programmes for supporting renewable energy and ability of technology to meet the customer requirements.

Flash pyrolysis has been demonstrated as an efficient way (65-70 weight percent yields) to convert solid biomass to a liquid. The advantages over solid biomass systems are: lower shipping and storage costs, cheaper and simpler combustion systems and lower emissions. The main disadvantages come when flash pyrolysis liquid is compared to the light and heavy fuel oils it is meant to replace. (see below and Table 4).

Pyrolysis liquids are currently in the early stages of application testing. It is being tested as a fuel for gas turbines, stationary diesels and boilers. Present quality levels are intermediate between that of a low sulfur heavy fuel oil and a light fuel oil. For economic reasons, replacement of a light fuel oil is preferred since the market price of light fuel oils is normally 50-100% higher than heavy fuel oils. Based on the present properties of flash pyrolysis liquid, replacement of a residential heating fuel is not considered to be possible; but large, light fuel oil customers could utilise flash pyrolysis liquid providing a number of fuel characteristics are improved.

The main fuel properties to be improved are: high viscosity and storage stability. Other properties such as poor ignition and acidity can be handled with modifications to the combustion system.

2. Petroleum Fuel Oil Grades

The most important property which distinguishes between different fuel oil grades is viscosity. The lower the viscosity, the easier it is to store, pump and atomise a fuel. Light fuel oils or distillate fuels have room temperature viscosity of 3-6 cSt while heavy fuel oils are 200-500 cSt at 50 °C. Due to their high viscosity, heavy fuel oils are normally shipped and stored above 50 °C. Medium fuel oils can be blended to any desired viscosity between light and heavy. Other important fuel properties are: flash point, heating value, cloud point, cold filter plug point. These properties are used for the design of combustion systems and the choice of pumps, storage tanks etc. Contaminants in the form of heavy metals, sulfur, nitrogen and solids content are important for emission control and system reliability.

When considering fuel specifications for flash pyrolysis liquid, a good starting point is the specs for different petroleum products. Some of can be directly applied while others will have to be modified. Additional tests will be needed to measure for example stability.

3. Liquid Fuel Combustion

Atomisation or the forming of a fine spray is considered to be the most critical stage for fuel combustion since ignition delay, combustion efficiency and production of pollutants are all dependent on the size of the drops. In general, the smaller the drops, the better the fuel/air mix and the shorter the burn out time. A range of technologies exist for atomisation of fuels from simple pressure atomisation used for light fuel oils to steam, air or rotating cup atomisation for heavy fuel oils.

For low viscosity light fuel oils, a pressure swirl or simplex nozzle operating at 7-10 bars is normally used. Under these conditions, small drops on the order of 20-60 micrometer diameter are produced. The drop is heated, light compounds evaporate, mix with oxygen, are ignited and combust. Heavier compounds form cenospheres or soot and combust giving a yellow flame. The combustion system is composed of a storage tank, burner and boiler or furnace. The burner contains a fan, oil pump, a nozzle, ignition coil and burner retention head. The nozzle imparts circular or swirl motion to the oil and when combined with swirling air produced from the burner retention head, a good fuel/air mixture results. This turbulent mixing allows combustion to take place in a compact volume. Heat is transferred through the boiler walls to either air or water which is used for heating and hot water systems. The system is designed so that the flue gas temperature is kept well above the dew point at 150-170 °C.

Using this simple technology, relatively low emissions of carbon monoxide (CO), nitrogen oxides (NOx) and soot or smoke number are produced. Typical values are: CO 10-20 ppm, NOx 80-120 ppm, and soot as measured on the Bacharach smoke scale <1. Burners are also available in which fuel oil is vaporised prior to combustion. This lowers NOx emissions to under 50 ppm. Oil preheaters may be used to ensure good atomisation with heating of the light fuel to 40-50 °C. In order to achieve higher thermal efficiencies, condensing boilers are also available in which water vapour in flue gases are cooled allowing water to condense. Depending on fuel sulfur content and NOx emissions, the condensate must be neutralised prior to disposal.

Heavy fuel oils, due to their higher viscosity, require more extreme measures to ensure good atomisation and combustion. Atomisers range from the simple, pressure type, to air, steam or rotating cup. Here oil is preheated to 100-140 °C depending on the atomiser type and initial viscosity. Oil pressure ranges from 10 to 30 bars. These measures reduce drop size to those approaching light fuel oils. Emissions normally found are CO < 20 ppm, NOx 200-300 ppm due to fuel nitrogen and particulates 50-150 mg/m^3 due to ash content of the fuels (Table 4). For very viscous fuels, combustion air may also be heated to assist in fuel evaporation. These options increase investment costs and are not used on small scales.

In the case of pyrolysis liquids combustion, it will be critical to ensure good atomisation and to construct the system to minimise ignition delay. All pumps, valves, storage tanks, nozzles must be able to withstand the corrosive nature of the fuel.

4. Flash Pyrolysis Combustion

There is limited published work pertaining to the combustion of flash pyrolysis liquids in practical systems. Early work on slow pyrolysis liquid is of limited value due to the large differences in fuel properties. Recent work in flame tunnel tests at MIT, US [1], ENEL, Italy [2], CANMET, Canada [5] and together with single droplet studies at Sandia [3,4] have provided general insights into the combustion process. These works have shown that although the actual process of flash pyrolysis combustion may differ from petroleum oils, the overall combustion time required for complete burnout is similar provided good atomization and timely ignition has taken place. This means that flash pyrolysis liquids can be combusted in existing boilers and engines provided suitable atomizers and fuel delivery systems can be constructed.

In the combustion tests at MIT and CANMET, technology commonly used for heavy fuel oils was employed such as air or steam atomization, combustion air preheating. In addition, flame tunnels were used for the tests which had an initial refractory section that was warmed up with a normal fuel prior to firing pyrolysis liquid. All emissions were reportedly lower than for heavy fuel oil.

5. Samples Tested

The samples tested are shown in Table 1. In addition some preliminary tests were done on a sample from Ensyn with 25 wt% water but high solids content of 2 wt%. Some filtering trials were done on this sample but problems occurred due to combination of char and lignin fraction forming a paste and blocking of filters.

Table 1. Flash pyrolysis samples at Neste

	Ensyn		Union Fenosa
feedstock	hardwood		eucalyptus
at Neste	11/94	11/95	8/94
water wt%	18	23	19
identifier	En4/18	En5/23	UF4/19
solids wt%	0.5	0.7	0.7

6. Equipment & Methods

Oxygen was measured using a Hartmann & Braun (H&B) Magnos 3 paramagnetic analyzer, carbon monoxide on a H&B Uras 3G IR-absorption, nitrogen oxides on a AAL 443 chemiluminescence instrument. Particulate levels were determined by drawing a known volume of flue gas through a filter and optically measuring the grayness level and reporting as a Bacharach smoke number. Equipment used for pyrolysis liquid combustion is given below.

7. Results

7.1 Preliminary Combustion Tests
Initial combustion testing was performed on Ens93 (25 wt% water, 2% solids) in a district heating type 2.5 MW Danstoker boiler fitted with a dual fuel burner using normal pressure atomisation. Initial tests with unfiltered flash pyrolysis liquid caused pump pressure to fluctuate by ±3 bar and stable combustion could not be obtained. The liquid was then filtered to <35 microns. In the first series of combustion tests, the burner was used in the

dual fuel mode and different light fuel oil to flash pyrolysis liquid ratios were tested. A variety of nozzle types, excess air amounts, liquid preheat temperature and pressures were tested. It was found that flash pyrolysis liquid was not able to combust on its own under these conditions. The pyrolysis flame extinguished immediately after switching off the mineral oil.

In the next set of tests, a sheet metal tube was inserted into the boiler and warmed up with light fuel oil. This was done in order to assist in evaporation of water and to provide heat for ignition. This allowed combustion of pyrolysis liquid without the mineral oil but was found to slowly cool which lead to the flame extinguishing. During these tests it was noted that the pyrolysis liquid flame was composed of two sections. The section close to the nozzle was presumably from the combustion of light, volatile components of the pyrolysis liquid while the second section was of the char combustion.

Finally, the insert was wrapped with insulating material. The flash pyrolysis liquid was now found to be able to combust on its own without an auxiliary fuel. Emissions (CO, NOx and particulates) were high with the best results: NOx 140 ppm, CO 90 ppm and particulates with a Bacharach number of 6-7 . The particulates were both higher and due to their brown colour of different composition than for fuel oils.

7.2 System Modifications and Effect on Results

Based on the results of the initial tests, an existing combustion system was modified. A refractory section of dimensions approximately 1 meter long by 0.4 meter inner diameter was placed between an Arimax Eetta 200 kW boiler and a Oilon KP24H dual fuel burner. Burner control was modified so that two fuels could be combusted simultaneously or independently. A 1 kW fuel preheater was used to adjust temperatures in the range 60-90 °C. A variety of commercial nozzles were tested with results for a Danfoss 1.75 gal 60°S nozzle shown in Table 2. Nozzle capacities for mineral oil and pyrolysis liquid were chosen to enable rapid warming up of the refractory and to help in air adjustments when switching between fuels. The original oil pump was replaced with a progressive cavity pump. Different pump materials were tested. Typical results are shown in Table 2 together with comparison to Tempera 15 which is a light fuel oil..

Table 2. Flash pyrolysis liquid emissions (1 meter refractory)

	O_2 vol%	CO ppm	NOx ppm	Bach. value	oil press. bars	oil T C
Tempera 15	3	14	128	1	10	24
	6	10	111	1,2	10	24
En4/18	4	32	142	5	15	90
	6	28	137	5	15	90
UF4/19 +3%EtOH	5	11	100	7,4	16	85
+3% H2O	8	11	90	7,2	16	85

It was found necessary to warm the refractory up to at least 800 °C in this configuration before the pyrolysis liquid samples would ignite and combust on their own. After switching off the system, it was found necessary to rinse the nozzles out with alcohol to prevent blocking. This blocking was caused by the hot refractory radiating heat back to the nozzle raising its temperature to above 150 °C causing light compounds to evaporate leaving behind a sticky, highly viscous material. Rinsing out with mineral oils was found to be not adequate. Combustion of pure Union Fenosa sample was not stable. Ethyl alcohol and water were used to reduce viscosity and improve atomisation but pump pressure was not stable which lead to poor flame quality and high emissions.

During these tests it was noted that the flue gas temperature for this configuration could not be raised above 120 °C. This value is close to the dew point temperature for flue gases and lead to higher particulate emissions (Bach. value). The refractory was then shortened which raised the flue gas temperature to above 180 °C which reduced particulates as shown in Table 3.

Table 3. Flash pyrolysis liquid emissions (modified refractory)

	O_2 vol%	CO ppm	NOx ppm	Bach. value	oil press. bars	oil T C
Tempera 15	3	12	60	<1	10	24
En4/18+ 4% EtOH	5	40	170	2,5	18	80
	6	20	150	2,8	16	85

Both Danfoss and Monarch nozzles were tested. Monarch nozzles were found to be more acid resistant than the Danfoss nozzles and less susceptible to blocking but showed surface corrosion. Results for a Monarch 2.25 US gal./hour nozzle are shown in Table 3. Note that the alcohol added to En4/18 was used to reduce the viscosity to it's original value since the viscosity of the original sample roughly doubled in the one year storage time. Results for En5/23 were similar to that of En4/18 except for higher CO emissions.

8. Discussion of Results

1. Flash pyrolysis liquids can be efficiently atomised using commercial pressure swirl nozzles. Quality of atomisation (drop size) will depend on initial fuel viscosity and pump pressure. Raising pump pressure and fuel temperature were found to be effective ways to reduce drop size but insufficient to reduce average drop size to the level of light fuel oils.
2. Current materials must be replaced with more acid resistant types. Also fuel lines, storage tanks, pressure valves will be made from acid resistant material. To avoid blocking of nozzles after system shut off, rinsing with alcohol was necessary. This is not considered to be practical in small systems.
3. Provided flash pyrolysis liquid is atomised effectively and proper care is taken to ensure timely ignition, flame length does not increase compared to light fuel oils. This means that existing boiler size is adequate. Present dual fuel burners can be used with suitable modification of the control programme. Existing progressive cavity pumps with suitably chosen materials can be used.
4. In the combustion system configurations tested to date, it was found that flash pyrolysis liquid would not ignite and achieve self-sustaining combustion until a refractory section, placed between the burner and boiler, was warmed up to at least 600-800 °C depending

on the particular system. The failure to ignite at lower temperatures is most likely due to the high water concentrations and the need to evaporate this water before ignition and combustion. Modifications to the refractory design indicate that further improvements are possible. Burner retention heads normally used with natural gas or light fuel oil were tested with the natural gas model showing improvements in flame stability.

5. A summary of flash pyrolysis results is shown in Table 4 with comparison to wood chips, a light to medium fuel oil Tempera 15 and a heavy fuel oil Master 180. Other grades of heavy fuel oil are also available. Note in the table that the heating value is for dried wood chips. Fresh cut wood has moisture content of 40-50% and woodchips have lower heating value of about 3 MJ/l. Emissions vary considerably depending on what equipment is employed and a range of values is given in the table. Flash point for pyrolysis liquid varies depending on water content. The higher solids content for pyrolysis liquid is without filtering while the lower value is what can be obtained with hot gas filtering. The Sauter diameter values give an indication of drop size in the spray. For flash pyrolysis liquids, the lower the water concentration, the higher the viscosity and the larger the drop size.

Table 4. Flash pyrolysis liquid properties and emissions compared to alternative fuels.

	Tempera 15	Flash pyrolysis liquid	Wood chips	Mastera 180
Water content wt%	0.02	18...25	15...20	0.5
Heating value MJ/kg	42.4	17.....15	16...12	41
MJ/l	36.9	21...18	8...6	39.4
Viscosity cSt 30 °C	9	900....150		600
50 °C	4	150...20		180
80 °C	2	24...6		50
130 °C				12
Flash point °C	90	66...100		80
Solids	-	0.7 ... 0.01		<0.05
Ash wt%	<0.010	0.1...0.2	2...4	0.03
Sulfur wt%	0.15	0	0	1
Nitrogen wt%	0	0.03....0.1	0.1...0.2	0.3
Sauter diameter um	30-40	60-40		50-70
	10 bar,	20 bar,		30 bar, 130 C
	20 C	85 C		
Typical emissions				
CO (ppm)	15-30	30-50	500-6000	5-30
NOx (ppm)	80-120	120-150	80-160	200-400
particulates(Bach.)	0.2-1	2.5-5	+tars, PAH	1-4
mg/Nm3				50-150
Pour point C	15	-27		0

6. In the boiler configurations tested to date, particulates, NOx and carbon monoxide are all higher from flash pyrolysis liquid than from light fuel oil under similar combustion

conditions but are within acceptable limits except for particulates. Our results also compare favourably with those of CANMET /5/ and MIT/1/ where either air or steam atomisers were used. NOx emissions will have contributions from fuel nitrogen and will depend on feedstock. Particulate emissions will be strongly influenced by fuel solid content. The brown color of the particulates indicates presence of tars which can be reduced with further improvements of atomisation and/or of liquid quality.

7. Viscosity of flash pyrolysis liquid was found to slowly increase with time even when stored under inert conditions (under nitrogen). Viscosity doubled during the year of storage. This increase in viscosity is believed to be due to condensation polymerisation. Since viscosity of current samples is at present higher than the fuel oil, this increase in viscosity could lead to higher emissions, blocking of filters and problems in pumping. Mixing of old and new product will give an produce a sample of unknown viscosity.

9. Conclusions

1. Flash pyrolysis liquid can be combusted using simple pressure atomisation equipment and emissions reduced to acceptable levels with proper optimisation of combustion conditions.

2. Due to the number and extent of problems encountered with storing, pumping, igniting, and keeping nozzles operable after shutdown, it is not likely that present quality levels of flash pyrolysis liquid would meet the requirements of a typical light fuel oil customer.

10. References

1. Shihdah, A., et al. Combustion characterisation of wood-derived flash pyrolysis oils in industrial scale turbulent diffusion flames. Paper presented at the International Workshop "Properties and Applications for Pyrolysis Oils" Estes Park, CO, USA, September 1994.

2. Rossi, C. Bio-oil combustion tests at ENEL" Paper presented at the International Workshop "Properties and Applications for Pyrolysis Oils" Estes Park, CO, USA, September 1994.

3. Baxter, L. Jenkins, B. Baseline NOx emissions during combustion of wood-derived pyrolysis oils. Paper presented at the International Workshop "Properties and Applications for Pyrolysis Oils" Estes Park, CO, USA, September 1994.

4. Wornat, M., et al., Single droplet combustion of biomass pyrolysis oils. Energy & Fuels 8 (1994) 1131-1142.

5. Huffman, D.R., Vogiatzis, A.J. and Clarke, D.A. Combustion of Bio-Oil in Bio-Oil Production and Utilisation, Proceedings of the 2nd EU-Canada Workshop on Thermal Biomass Processing eds. Bridgewater & Hogan.

6. Solantausta, Y., Nylund, N-O., Gust, S. Use of Pyrolysis Oil in a Test Diesel Engine to Study the feasibility of a Diesel Power Plant Concept. Biomass and Bioenergy 7 (1994) 1-6, 397-306.

RTP™ BIOCRUDE: A COMBUSTION / EMISSIONS REVIEW

D.R. HUFFMAN and B.A. FREEL
Ensyn Technologies Inc.
Greely, CANADA

Abstract

The thermochemical conversion of solid biomass wastes to liquid fuels via fast pyrolysis is now firmly established with a number of research, pilot plant and commercial systems producing sufficient quantities for applied combustion tests. Biocrudes produced with Ensyn's RTP™ technology have undergone characterization and combustion tests in flame tunnels, furnaces, combustors and boilers, including work at MIT (USA), ERL (Canada), SANDIA (USA), ENEL (Italy), NESTE (Finland) and ARSTA (Sweden).

Data on flame temperatures, heat release profiles and emissions have been collected for systems using mechanical, air and steam atomization nozzles. These tests have conclusively established that biocrudes can be burned with steady-state self-sustaining flames that closely resemble those from petroleum-based fuel-oils.

Results have confirmed that conversion to a liquid permits better combustion control resulting in significantly lower emission levels than those typically obtained by direct combustion of the solid biomass wastes. Emission levels for carbon monoxide, nitric oxide, and total suspended particulate are close to those obtained from No. 2 fuel-oil and lower than from No. 6 fuel-oil. Sulphur emissions are dramatically lower than for either fuel-oil. It is therefore reasonable to expect that existing combustion systems for fuel-oils can be modified to burn biocrudes with emission levels that will allow permitting in normal situations.

Keywords: Biocrude, combustion, emissions.

1 Introduction

Fast pyrolysis, a thermochemical conversion technology that converts solid biomass to liquid fuels, is now considered a viable component in the field of bioenergy. Research, pilot plant and commercial systems are now operating on a variety of biomass feedstocks providing quantities of biocrude liquids for laboratory and industrial combustion tests. Biocrudes produced with Ensyn's RTP™ technology have undergone characterization and combustion tests in the USA, Canada, Italy, Finland and Sweden. Table 1 provides a listing of the tests carried out utilizing biocrude derived from hardwood feedstocks. They cover a wide range of combustor sizes, from fundamental single droplet tests to large-scale firing in utility boilers. All tests have shown that biocrudes can be burned with steady-state, self-sustaining flames that closely resemble those from petroleum-based fuel-oils.

Table 1. Combustion tests on RTP™ biocrudes

Location	Facility	Size	Reference
Red Arrow Manitowoc, Wisconsin	Fire-Tube Boiler (Natural Gas)	$0.6\,MW_t$	1
CCRL, CANMET Ottawa, Ontario	Flame Tunnel	$0.7\,MW_t$	2
ENEL Livorno, Italy	Test Furnace	$50\,KW_t$	3
CCRL, CANMET Ottawa, Ontario	Domestic Furnace	$33\,KW_t$	4
MIT Cambridge, Massachusetts	Flame Tunnel	$1.0\,MW_t$	5
SANDIA Livermore, California	Multifuel Test Combustor	$30\,KW_t$	6
	Single Droplet Test Combustor		7
Red Arrow Manitowoc, Wisconsin	Silo Combustor	$5.9\,MW_t$	8
Neste Oy Porvoo, Finland	Danstoker Boiler (Light fuel oil)	$2.5\,MW_t$	9
	Arimax Etta Boiler (Modified, light fuel oil)	$200\,KW_t$	9
M.P.U. Manitowoc, Wisconsin	Stoker Utility Boiler (Coal)	$50\,MW_t$	8
ARSTA Stockholm, Sweden	Water-Wall Utility Boiler (Heavy fuel oil)	$10\,MW_t$	10

490

2 Emissions

A major concern has been the determination of air-borne emissions from biocrude combustion. Data have been collected on CO, NOx, SOx, THC (as methane equivalent) and particulates during several of the tests. As shown in Table 2, there is a considerable range in the test values, generally resulting from significant differences in the quality of the biocrudes, particularly with respect to the content of char, ash and sulphur. While it is not possible to obtain meaningful averages as a result of this extreme variability, it is possible to select a probable emissions level based on recurring values from different tests carried out using relatively clean biocrudes. These values, 30 ppm for CO, 160 ppm for NO, 5 ppm for SO_2, 2 ppm for THC and 70 mg/Nm^3 for TSP generally reflect much lower emission levels than those typically obtained from the direct combustion of solid biomass wastes. Emission levels are close to those obtained for light fuel-oil and lower than for heavy fuel-oil. Sulphur emissions are dramatically lower than for either fuel-oil.

Table 2. Biocrude combustion emission values

Biocrude Test	CO ppm	NOx ppm (NO)	SOx ppm (SO_2)	THC ppm (CH_4)	Particulate mg/Nm^3
CCRL, 1992	<50,<160, 220,442	160,168, 176,212	2,3,17,38	NDR*	255,264, 399
ENEL, 1992	NDR	220-280	NDR	NDR	30,32
CCRL, 1993	40,60,90	60,88,103	NDR	2,6,7	NDR
MIT, 1994	18	165	NDR	NDR	NDR
SANDIA, 1994	<40	230-270	NDR	NDR	NDR
NESTE, 1995	28,72	137,215	NDR	NDR	NDR
CCRL, 1995	24,26	164,169	3.8, 46.1	NDR	31.7, 271
ARSTA, 1996	32,32,67	195,198, 208	NDR	0.8, 1.0, 1.4	105,144, 161
RANGE	18-442	60-280	2-46	0.8-7	30-399
Probable Typical Value	30	160	5	2	70

*NDR: No Data Reported

3 Specific observations

While biocrude combustion experience has varied as a result of differences in test equipment and biocrude quality, there are a number of common observations expressed by the authors of the test reports. These include:

* The fuel delivery system must be acid resistant.
  Most biocrudes have a significant content of organic acids (acetic, formic) resulting in a pH value below 3. As a result, storage tanks, fuel lines, pumps, heaters and nozzles must be constructed from acid resistant materials such as stainless steel, copper or various plastics.
* Pre-heating is beneficial.
  While it is possible to atomize biocrudes at ambient temperatures, their high viscosity can be reduced substantially by preheating to temperatures of 70-85ºC. This eases pumping and flow control, improves ignition and gives much better atomization thereby reducing emissions (CO and THC).
* Extended heating is detrimental.
  Components within the biocrude interact slowly over extended storage time leading to noticeable increases in viscosity. Heating greatly increases the rate of these reactions. Therefore, a high proportion of pre-heated biocrude return to storage should be avoided. Extended heating in open tanks can cause loss of water and organic volatiles leading to eventual solidification.
* Biocrude films must not be left exposed.
  When thin layers are exposed to the atmosphere, evaporation of volatiles occurs causing the biocrude to thicken, become sticky and eventually solidify. Lines, pumps and nozzles can be cleaned with mild caustic or alcohols but flushing with fuel-oil does not appear to be satisfactory. Pipes and pumps can be left undrained for extended periods at ambient temperatures, provided they remain completely filled with biocrude.
* Ignition is difficult.
  Spark ignition (as used for light fuel-oil) is not normally successful. Typically, the combustion chamber must be pre-heated or a pilot flame used to achieve reliable ignition.
* A thermal sink is required for stable combustion.
  Biocrudes have proven difficult to burn (without supplemental fuel) in a "cold" furnace, for example a full water-wall boiler. Stable combustion has been achieved by using a refractory or alloy flame stabilization chamber.
* Static biocrude will coke at elevated temperatures.
  On shutdown a nozzle must be flushed, cooled or removed from the hot zone. Operation in an off/on mode is therefore more difficult than continuous firing for base load.
* High char/ash levels must be avoided for small systems.
  Particulate emissions are directly related to the ash content of the biocrude. Low ash

contents are necessary to meet emission limits for small-scale combustors that do not employ particulate collection systems. Char particles cause sparklers which produce CO and THC emissions and cause plugging in filters, lines and nozzles.

4 Conclusions

In general, there are no fundamental impediments that would act as barriers to utilization of biocrudes in industrial-scale combustion systems. Tests have shown that, after ignition, a quality biocrude will burn with greater rapidity, higher luminosity and better heat release than common petroleum fuel-oils. As a result, there is no need for re-configuration of the combustion chamber dimensions or re-building of the combustor structure. There is no inherent de-rating as full capacity can be achieved. Fast pyrolysis biocrudes can be burned using standard atomization techniques with emissions controlled to acceptable levels.

However, it must be recognized that biocrude is not petroleum and therefore the burner must be set-up to accommodate the unique characteristics of the liquid biomass. The control strategy and combustion parameters must be appropriate to and optimized for the biocrude.

Combustion performance is controlled by the quality of the bio-crude which is determined by the pyrolysis conditions and the feedstocks. The content of N, S and ash affect emissions while the cracking severity influences viscosity, moisture content and specific combustion characteristics.

While current commercial biocrudes can easily substitute for heavy fuel-oils, improvement in quality is required for consideration as a replacement for light fuel-oils. This would include reduced viscosity and lower levels of char and ash.

5 References

1. Freel, B.A., Graham, R.G. and Huffman, D.R. (1990) The Scale-Up and Development of Rapid Thermal Processing (RTP) to Produce Liquid Fuels from Wood, Ontario Ministry of Energy Report (CF), Toronto, Canada.
2. Banks, G.N., Wong, J.K.L. and Whaley H. (1992) Combustion Evaluation and Heat Transfer Characterization of a Fast Pyrolysis Fuel Product, Division Report ERL 92-35 (CF), CANMET, Energy Mines and Resources Canada, Ottawa, Canada.
3. Barbucci, P., Costanzi, F., Ligasacchi, S., Mosti, A. and Rossi, C. (1995) Bio-Fuel Oil Combustion in a 0.5 MW Furnace Proceedings; Second Biomass Conference of the Americas NREL/CP 200-8098, Golden, USA.
4. Lee, S. Win (1993) Preliminary Combustion Evaluation of Wood-Derived Fast Pyrolysis Liquids Using a Residential Burner, Division Report ERL 93-29 (CF) CANMET, Energy Mines and Resources Canada, Ottawa, Canada.

5. Shihadeh, A., Manurung, R., Lewis, P. and Beér, J. (1994) Proceedings; Workshop on Biomass Pyrolysis Oil Properties and Combustion, Estes Park, USA.

6. Shaddix, C.R. and Huey, S.P. (1996) Combustion Characteristics of Pyrolysis Oils Derived from Hybrid Poplar, Proceedings; Developments in Thermochemical Biomass Conversion, Banff, Canada.

7. Wornat, M.J., Porter, B.G. and Yang, N.Y.C. (1994) Single Droplet Combustion of Biomass Pyrolysis Oils, Proceedings; Workshop on Biomass Pyrolysis Oil Properties and Combustion, Estes Park, USA.

8. Freel, B.A. and Huffman, D.R. (1994) Applied Bio-Oil Combustion, Proceedings; Workshop on Biomass Pyrolysis Oil Properties and Combustion, Estes Park, USA.

9. Gust, Steven (1994, 1995, 1996) Combustion Experiences of Flash Pyrolysis Fuel in Intermediate Size Boilers, Project Progress Reports (CF), Porvoo, Finland.

10. Hallgren, B. (1996) Personal Communication, Stockholm Energi AB, Stockholm Sweden.

FEASIBILITY OF FIRING AN INDUSTRIAL GAS TURBINE USING A BIO-MASS DERIVED FUEL

R.G. ANDREWS, S. ZUKOWSKI AND P.C. PATNAIK
Orenda Aerospace Corporation
Gloucester, Ontario, Canada

Abstract

This paper describes a development program aimed at determining the technical feasibility of utilizing a biomass derived fuel in an industrial gas turbine engine. The fuel addressed is a flammable liquid derived from wood waste through flash pyrolysis or "bio-fuel". The fuel has a heating value of approximately 18 MJ/kg, a density of 1.2 kg/l and a moisture content of 25%.

The turbine engine selected as the test vehicle, is a 2.5 MW class - GT2500 engine designed and built by Mashproekt in the Ukraine. The standard operating conditions and layout of this engine provide flexibility in optimization of the combustion system to accept lower than conventional grade fuels.

The characteristics of the fuel, the fuel handling system and the results of preliminary component and engine tests are presented. The GT2500 industrial gas turbine engine was successfully operated using bio-fuel from idle to full power.
Keywords: Bio-Fuel, biomass, flash pyrolysis, gas turbine, combustion and atomization.

1. Introduction

The rapid rate at which our usage of fossil fuels is releasing carbon dioxide emissions into the atmosphere has raised international concern and has resulted in efforts to develop alternative, renewable sources of energy with minimal detrimental impact to our environment.

Biomass derived fuels when substituted for fossil fuels could reduce carbon dioxide emissions that are thought to be contributing to the greenhouse warming, since the biomass absorbed the same amount of carbon dioxide in growing as it releases when consumed as fuel. In addition, bio-fuels contain only low levels of sulphur and hence, they do not contribute to sulphur related acid-rain problems.

1.1 Bio-Fuel

The bio-fuel used for this program was produced by Ensyn Technologies through their Rapid Thermal Process (RTP™) using a wood waste feedstock to ensure high fuel quality [1,2]. Ensyn has proven through combustion testing and commercial experience that their fuels can burn readily in a self sustaining mode [3,4,5]. In addition, a practical fuel production process and combustion design can lead to costs that are competitive with conventional methods of power production [6,7].

1.2 Potential Markets

Markets exist worldwide for the application of renewable and environmental fuels, and energy systems. In Europe the high cost of electricity combined with environmental concerns, which include incentives, provide an ideal market. The new costs for waste disposal and growing environmental concerns in North America combined with the need to identify renewable energy sources should provide a significant future market. In addition, Asia and the Pacific rim have abundant agricultural waste and limited power facilities offering another substantial market.

1.3 Objective

The objective of our program is to demonstrate the feasibility of gas turbine engine operation using bio-fuel. The tasks for engine development have been identified as: fuel property characterization, test plan development, fuel handling and engine fuel systems development, engine instrumentation, endurance testing and engine performance.

2. Fuel Property Characterization

A thorough characterization of the bio-fuel has been carried out and compared to diesel fuel in Table 1.

2.1 Bio-Fuel Property Technical Limitations

Some technical limitations resulting from the bio-fuel properties have been identified, including: high fuel viscosity, fuel acidity, ash content, lower heating value and alkali metal content [7]. Techniques exist in both the tailoring of fuel process parameters and the modification of gas turbine engine systems to deal with these limitations. The bio-fuel is not a petroleum fuel and cannot be practically modified through process conditions to provide petroleum fuel properties. Therefore, gas turbine engines may need modification to burn bio-fuels. A detailed discussion of the potential technical limitations are covered in detail by Moses et al., 1994 [7].

To overcome high viscosity, the bio-fuel must be preheated just prior to combustion, however, it cannot be held at temperature for extended periods and the preheat temperature should not exceed 90°C [3,4]. This limit is a result of bio-fuel decomposition which can occur at or above this temperature if held for more than a short period [8]. The effects of bio-fuel acidity can be minimized by the selection of applicable materials for the fuel systems and coating selection for the turbine hot section components. The lower heating value of the bio-fuel will require starting on conventional fuels and an increased fuel flow resulting in possible nozzle and combustor modifications. Ash levels have been significantly reduced using filtering techniques during processing [1] which in turn reduces the alkali metal content. The remaining alkali metal content needs to be addressed through improvements in fuel processing, reduction by fuel additives and minimization of component degradation by the use of advanced protective coatings [7].

Table 1: Bio-fuel oil Properties and Comparison to Diesel

Analyses		Standard	Units	Bio-Fuel Sample #1	Bio-Fuel Sample #2	Bio-Fuel Sample #3	Diesel
Calorific Value		LECO AC-300	MJ/kg	19.3	19.8	18.0	45.18
Flash Point		ASTM D93-90	°C	58.0	59.0	58.0	53
Kinematic Viscosity							
	at 20°C	ASTM D445-88	cSt	362.0	480.2	425.5	2.61
	at 40°C			119.2	66.99	58.3	-
	at 60°C			20.46	20.99	21.07	-
Moisture		ASTM D-1744	wt%	22.0	20.4	18.1	79 ppm
Ultimate							
	Carbon	LECO CHN-600	wt%	46.28	46.24	46.27	86.58
	Hydrogen	LECO CHN-600	wt%	7.29	7.55	7.70	13.29
	Sulphur	ASTM D-3120	wt%	0.029	0.030	0.029	0.11
	Nitrogen	ASTM D-4629	wt%	0.138	0.141	0.147	65 ppm
	Ash	ASTM D-482	wt%	0.04	0.04	0.07	0.0
	Oxygen	PE-240	wt%	46.2	46.0	45.8	0.01
Ash Analysis							
	Na	AASA	ppm	34	23	31	-
	K	AASA	ppm	135	128	106	-
	Ca	AASA	ppm	44	82	59	-
Density							
	at 20°C	ASTM D4052-91	g/ml	1.1980	1.1990	1.1976	0.8271
	at 30°C			-	-	1.1795	-
	at 40°C			1.1802	1.1802	-	-
	at 50°C			-	-	1.1630	-
	at 60°C			1.1636	1.1636		-
Surface Tension		Fisher Surface Tensiomat at 25°C	mN/m	36.4	35.3	36.3	29.2
pH		Sargent - Welch pH meter	pH	2.7	2.8	2.9	-
Particulate Content		PTFE Millipore					
	5 μm	Membrane	mg/l	17600	15400	15200	2.2
	1 μm	Filtration	mg/l	190	30	320	0.4

The moisture content is not contamination but part of the fuel make up and therefore cannot be compared directly. In addition, a lower moisture content results in a significant increase in viscosity. The high viscosity and moisture content will be affected by changes in ambient temperature. Temperatures below -20°C could result in a significant increase in viscosity, while temperatures exceeding 30°C for long periods could result in fuel decomposition. The pH was found to be below specified diesel levels, however the results of corrosion tests have shown that even at 70°C the bio-fuel does not corrode stainless steel, which is the standard fuel system material on the GT2500 engine.

3. Test Vehicle

Initially, the main concern of the program was to review and select a small gas turbine, which was best suited to achieve the main objective. The Mashproekt GT2500 engine was been identified as the ideal candidate for the bio-fuel program test vehicle.

Technically the GT2500 offers distinct advantages over other engines. Unlike the aircraft engine candidates, which use jet fuel, this small industrial engine uses diesel, as a standard operational fuel, which has physical properties closer to bio-fuel. The diesel fuel nozzles use a dual passage system which allows more flexibility for liquid fuel operation. A positive flame ignition system offers potential for bio-fuel starting. Advanced coating systems are used as standard for hot section durability and may provide sufficient protection against contaminants. The critical characteristic is that the engine offers the ability to easily modify the entire combustion system, since the design incorporates "Silo" type combustion chambers which are located above the engine, as shown in Figure 1.

In addition to the suitability of the engine for bio-fuel operation, the participation of the original engine manufacturer is necessary to provide design information and operational characteristics needed for testing and design of engine modifications.

Figure 1: Orenda/SPE Mashproekt GT2500 Industrial Gas Turbine Engine

4. Summary Of Test Requirements

The test requirements identified accuracy targets compatible with industry standards and defined the procedures to prepare accuracy estimates; the goal was to ensure confidence in the parameters used in the analyses and evaluations for both back to back comparisons and absolute assessments. The nominal instrumentation must be capable of meeting the minimum accuracy requirements outlined by the ASME Performance Test Code PTC-22.

4.1 Combustion Requirements

The combustion requirements were identified as a demonstration of engine ignition and stable operation on both diesel and bio-fuel over the normal operating range corresponding to the respective fuel heating value and the rated fuel flow. In addition, an efficiency comparison would be made which required the combustion chamber inlet and exit temperatures and pressures will be monitored during both diesel and bio-fuel operation. The commercial viability of the GT2500 operating on bio-fuel will be partially related to the combustion emissions and the related local requirements, therefore, the monitoring of combustion emissions was also defined as a requirement.

4.2 Engine Durability Requirements

The engine durability directly affects the cost effectiveness of a power utility due to the related down time, repair costs and warranty requirements. Component degradation will be continuously monitored by additional instrumentation which will include: combustion liner surface temperature and gas temperature in the gas collector prior to turbine inlet. In addition, detailed inspections will be carried out to qualitatively evaluate the durability of individual components.

4.3 Engine Performance

The engine performance evaluation will discern steady state changes in the following parameters for diesel and various bio-fuel test runs, using industry standard instrumentation: speed, power, airflow (calibrated), fuel flow, inlet pressure, compressor discharge pressure, combustor outlet pressure, lube oil pressure (various locations), fuel pressure, inlet temperature, compressor discharge temperature, combustor exit temperature, exhaust gas temperature, lube oil temperature, three locations of vibration. All parameters are to be measured at idle, part and full power. These parameters will be used to generate the performance information and curves necessary for diesel fuel and bio-fuel operational comparisons.

5. Engine Fuel Delivery Systems

5.1 Atomization Testing

The diesel fuel nozzle was tested first using water, followed by diesel and finally using bio-fuel. The main atomization test facility and fuel delivery systems are shown in Figure 2. A Malvern laser was used to measure the Sauter Mean Diameter for each of the test conditions and for the diesel and water tests. In addition, a sheet laser was used for imaging of atomization patterns.

The water tests provided a baseline for comparison without the complexity of working with fuels. Also in the case of the bio-fuel with its high molecular water content, the results for water were likely to be closer to the bio-fuel than the diesel. The water tests showed some marginal inconsistency in the concentricity of the nozzle spray, which was found in more than one nozzle but tended to be positioned in the same location. Therefore, an average location was selected which could be used for the diesel fuel and bio-fuel tests.

The atomization tests showed a wider cone angle for the bio-fuel and water than for diesel fuel due to the reduced viscosity and surface tension of diesel as well as a definite interaction between the primary and secondary flows. The combined diesel and bio-fuel results are close to diesel while the bio-fuel droplets were found to be twice the size of the diesel.

Figure 2: Atomization Test Facility

5.1.1 Prototype Bio-Fuel Nozzle

It was determined that, based on the atomization tests discussed earlier and the flame tunnel combustion tests presented in a later section, for the demonstration to be representative of commercial operation a prototype bio-fuel nozzle should be developed which would allow sufficient bio-fuel flow to achieve full power operation. The prototype development was used to reduce the complexity of the diesel to bio-fuel operational transition.

The nozzle developed is similar to the standard diesel fuel nozzle but with two distinct changes. The nozzle uses three concentric passages rather than two, where the third passage was designed for air to assist in atomization of the diesel fuel during ignition and the bio-fuel during normal operation. The second difference is that the secondary passage was enlarged for full power bio-fuel flow, which allows the diesel and bio-fuel flows to be controlled separately without mixing prior to atomization.

The prototype GT2500 full power nozzle was atomization tested by Mashproekt prior to combustion testing. The results suggested that this nozzle could adequately atomize the bio-fuel at the much higher flow required for equivalent heating value. The nozzle tests were verified using the atomization facility discussed earlier and the results indicated that the nozzles were operating within required specifications.

5.2 Diesel Fuel Supply System

The engine mounted high pressure diesel fuel system was used for the engine tests but required a low pressure test cell diesel fuel system which was installed tested and qualified for the full range of normal operation.

The prototype bio-fuel nozzle requires a diesel fuel preheat to avoid both a bio-fuel temperature drop in the nozzle and high thermal gradients in the thin walls of the nozzle passages. A steam heat exchanger and manual fuel tank return line were installed into the diesel fuel system to allow both preheating of the fuel in the tank and on line heating of the diesel fuel. The tank preheat is limited to below the fuel flash point of 65°C, with any additional required preheat to occur on line during engine operation.

5.3 Bio-Fuel Supply System

The low pressure bio-fuel system, shown in Figure 3, was installed, tested and was fully operational within the specified limits. The system includes a fully instrumented, closed loop, steam heat exchange system for accurate control of the bio-fuel temperature prior to entering the high pressure system. In addition, two pressure relief loops have been used to protect the low pressure pump and to avoid stagnation of the bio-fuel in the heat exchanger. The low pressure fuel system also includes an integral high pressure system bypass and complete fuel system rinse. The rinse system interrupts the bypass and introduces a Caustic Soda rinse solution to remove the residual bio-fuel from the system following a shutdown. The rinse tank is then replaced with a water feed which was used to wash out the Caustic.

As a result of the prototype bio-fuel nozzle development, the transition from diesel to bio-fuel can be completed using the two separate systems to remove the diesel while increasing the bio-fuel flow to compensate. The key to the control of the high pressure bio-fuel system is the computer controlled metering valve. The metering valves, which were designed using a computer model and verified by water flow tests, are capable of metering the much higher flows associated with prototype nozzle at a level of precision far exceeding the engine requirements.

Figure 3(a): Low Pressure Bio-Fuel System

Figure 3(b): Bio-Fuel Heating and Rinse System

Figure 3(c): High Pressure Bio-Fuel System

The atomization air for the prototype nozzle required the installation of on-line, closed loop air heating and pressure regulation systems. The heating was accomplished through the use of heat tapes and a PID controller on a high pressure feed line, while a differential pressure regulator, controlled pressure by tracking a compressor delivery bleed line.

6. Flame Tunnel and Furnace Exposure Testing

Flame tunnel and furnace exposure tests were performed to evaluate the hot corrosion characteristics of bio-fuel on the turbine components of the GT2500 engine. The tests used engine components to ensure that all material parameters were consistent with those used in service. The components being considered include: the liner, collector, first to third stage turbine nozzles, and first to third stage turbine blades.

The Flame Tunnel Test was carried out at the CANMET Energy Research Laboratory of Natural Resources Canada. The flame tunnel was used to simulate engine service conditions with the exception of velocity. This included a cooling air manifold used to achieve Mashproekt specified gas and metal surface temperatures. The atmospheric combustion of bio-fuel, although not as efficient will produce combustion products similar to that seen in the engine, which in turn will result in similar degradation mechanisms.

In addition to the component exposure provided by the flame tunnel, combustion and emissions data are given in Table 3 and 4 respectively. The combustion data was significant in that it provided the basic combustion information which was required for the development of the prototype bio-fuel nozzle discussed earlier. The emissions data provided an opportunity to evaluate potential problems associated with the combustion of bio-fuel, for example: the levels of NO_x were relatively high and therefore some combustion liner modifications will be required for a commercial product.

Following the exposure in the flame tunnel the components were placed in a furnace for 1000 hours. The liner, collector, first stage nozzle, first stage blade and second stage nozzle were held at 850°C and the remaining parts at 650°C. These temperatures represent the highest activity points for the alkali elements likely to be a source of corrosion and the components were assigned according to which temperature was closest to their normal operating surface temperature as supplied by Mashproekt.

The components from the flame tunnel showed no signs of damage, but were found to be coated by a bio-fuel combustion deposit, which flaked off easily even during exposure. The combustion liner, first stage blade and first stage nozzle, however, have shown some indication of attack following the furnace exposure. This attack was expected based on the aggressive nature of the tests, where the conditions were designed for maximum damage. During engine operating conditions it is expected that the deposits will build up and flake off similar to the flame tunnel results.

7. Bio-Fuel Performance Tests

The GT2500 engine installed in the National Research Council of Canada's, Institute for Aerospace Research Propulsion Laboratory test cell, is shown in Figure 4. The preliminary bio-fuel tests identified a series of problems including: operation of emergency systems, metering valve performance and diesel system hot gas back flow. Each problem required only minor modifications to the sub-systems or operational procedures.

The engine was then tested throughout it operational range, form idle to full power, using a atomization mixture of diesel and bio-fuel up to a proportion of 80% bio-fuel to 20% diesel. After completing these tests and proving the operation of all engine and accessory systems an experimental 100% bio-fuel engine test was carried out which was successful over the complete operational range of the engine.

7.1   Engine Stripping and Inspection

The engine was partially stripped and inspected before and after the diesel baseline runs to provide a baseline evaluation point for after the bio-fuel tests. The stripping included the removal of the fuel nozzles, combustion chambers and liners followed by a complete inspection of all removed components. In addition, all remaining hot section components were inspected using a borescope to determine if any deposits had formed or if any other damage may have resulted from the commissioning and baseline diesel tests.

Table 3: Flame Tunnel Bio-Fuel Combustion Parameters

Parameters	Bio-Fuel	Diesel Fuel
Fuel		
Flow Rate (kg/h)	85.1	32.3
Moisture (wt%)	23.4	0.0
Temperature (°C)	75.1	27.6
Thermal Input (kW)	435	408
Atomization Air		
Flow Rate (kg/h)	51.4	33.9
Temperature at Burner (°C)	111.5	134.8
Pressure at Burner (psig)	55	32
Combustion Air		
Flow Rate (kg/h)	604	636
Temperature at Burner (°C)	38.4	40.5
Heat Transfer Rate (kW/MJ)	0.175	0.172

Table 4: Flame Tunnel Bio-Fuel Combustion Emissions

Parameter	Bio-Fuel Oil	Diesel Fuel
Furnace Exit (°C)	480	511
Flowrate (Nm^3/MJ)	0.362	0.285
Particulate (g/Nm^3)	0.096	0.006
NO (g/MJ)	0.086	0.037
SO_2 (g/MJ)	<0.01	0.104
O_2 (% volume)	4.6	4.0
CO_2 (% volume)	15.0	12.3
CO (% volume)	20	18

Figure 4: GT2500 Engine Installed in NRC Test Cell

After the initial bio-fuel experimental tests inspections were carried at each stoppage. It was found that when the engine was shutdown following bio-fuel operation that a black bio-fuel tar deposit could be found on the nozzle, liner and gas collector, however, no deposits were found on the hot gas path airfoils. During these tests the shut down was modified to include a high temperature diesel operational period just prior to complete shutdown. This short period of diesel operation removed the black bio-fuel tar deposits. Once the black deposits were removed ash deposits were found on some gas path components, however no attack or damage was found on any engine components.

8. Conclusions

1. The Mashproekt GT2500 engine has been identified as an ideal candidate for the bio-fuel operation based on its standard operating conditions and flexibility in combustion system modification.
2. A comprehensive characterization of the bio-fuel was carried out with respect to its properties and constituents.
3. The development and testing of methods for effective handling, heating and storage of the bio-fuel were successfully completed. A prototype GT2500 full power bio-fuel nozzle has been developed which allowed the performance tests to be carried out up to full power.
4. The flame tunnel and furnace exposure was completed, which is being used to evaluate the gas turbine component durability. The results of these tests have shown damage in three engine components: the first stage blade, the first stage nozzle and the combustion liner.
5. The Mashproekt GT2500 engine was successfully performance tested using bio-fuel under varying operational conditions from idle to full power.
6. Potential exists for the direct use of bio-mass derived fuels in gas turbine engine power generation packages. Although bio-mass derived fuels are currently used in boilers, making them adaptable to gas turbines through the approach outlined here is expected to lead improved efficiencies particularly in applications that could benefit from cogeneration.

9. Acknowledgments

This work is being carried out as an international cost shared program supported by the following participants:

> SPE Mashproekt; Ensyn Technologies,
> Government of Canada, Department of Natural Resources: CANMET, Bioenergy R&D, Forestry Canada, Green Plan; National Research Council Canada,
> Government of Ontario, Ministry of Environment and Energy.

10. References

1. Bakhshi, N. "Properties and Characteristics of Ensyn Bio-Oil", Biomass Pyrolysis Oil Properties and Combustion Workshop Proceedings, Estes Park, Colorado, September 1994.

2. Elliot, C.D., "Chemical Analysis of Biomass Fast Pyrolysis Oils", Biomass Pyrolysis Oil Properties and Combustion Workshop Proceedings, Estes Park, Colorado, September 1994.

3. Shihadeh, A., "Combustion Characterization of Wood Derived Fast Pyrolysis Oils in Industrial Scale Turbulent Diffusion Flame", Biomass Pyrolysis Oil Properties and Combustion Workshop Proceedings, Estes Park, Colorado, September 1994.

4. Rossi, C., "Bio-Oil Combustion Tests at ENEL", Biomass Pyrolysis Oil Properties and Combustion Workshop Proceedings, Estes Park, Colorado, September 1994.

5. Jasas, G. Kaskper, J. and Trauth, R., "Gas Turbine Demonstration of Pyrolysis-Derived Fuels", Technical Report, DDE/ET/3333--T2, Report No. 1901, June 1983.

6. Gregoire, C., "Pyrolysis Oil Economics - Issues and Impacts", Biomass Pyrolysis Oil Properties and Combustion Workshop Proceedings, Estes Park, Colorado, September 1994.

7. Moses, A.C. and Bernstein, H., "Impact Study on the Use of Biomass-Derived Fuels in Gas Turbines for Power Generation", NREL/TP-430-6085, January 1994.

8. Czernik, S., "Storage of Biomass Pyrolysis Oils", Biomass Pyrolysis Oil Properties and Combustion Workshop Proceedings, Estes Park, Colorado, September 1994.

PYROLYSIS

Chemicals from pyrolysis liquid

EFFECT OF VARIOUS PYROLYSIS PARAMETERS ON THE PRODUCTION OF PHENOLS FROM BIOMASS

H. PAKDEL[a], C. ROY[a,b], and X. LU[a]
[a] Institut Pyrovac Inc.
1560, avenue du Parc-Beauvoir
Sillery (Quebec) Canada G1T 2M4

[b] Université Laval, Department of Chemical Engineering
Sainte-Foy (Québec) Canada G1K 7P4

Abstract

North American wood species have been pyrolysed in a vacuum-pyrolysis, laboratory reactor under various conditions. The pyrolysis oils were analyzed in detail for their content in phenolic compounds after derivatization to their acetyl derivatives. Pyrolysis process conditions such as temperature, heating rate, sample amount, particle size, vapour residence time, biomass moisture content and additives and/or catalysts leading to high yield of valuable phenolic compounds have been investigated. The influence of such parameters on the formation of syringylic and guaiacylic compounds is discussed.

Syringylic and guaiacylic compounds are among the most valuable compounds in the oils produced. Demethoxylation of syringylic and guaiacylic compounds occurs under prolonged heating and high temperature. Furthermore, dihydroxy aromatics contribute to char formation under severe pyrolysis conditions. The role of various additives including ash to catalyze lignin depolymerization is also investigated. Pyrolysis performance was evaluated in terms of total phenolic yield and pyrolysis oil composition.

Keywords: Vacuum, pyrolysis, wood, bark, fine, chemicals, phenols, extraction, analysis, separation.

1 Introduction

The forest industry produces large volumes of biomass wastes such as bark, waste, wood, lignin and lignosulfonates. Such wastes have a low or even negative value and are usually landfilled, incinerated or burned as a cheap fuel. A small amount is converted to compost. Lignin is one of the principal constituents of wood, bark and lignosulfonate. Lignin is an organic phenolic polymer which produces phenols upon depolymerization. In general phenols have good industrial value. Phenolic compounds can also be deoxygenated or converted to methyl aryl ethers. The "lignol" process has been developed earlier for the conversion of lignosulfonate or kraft lignin to gasoline with compatible octane rating to petroleum gasoline [1].

Various thermal conversion technologies are being developed which produce complex pyrolysis oils and wood charcoal. Depending on the pyrolysis process and biomass constituents, the oil composition considerably changes and usually reflects the

509

reactor environment [2]. Wood carbonization technology to produce charcoal and pyrolysis oil is a very old process. Carbonization techniques have low energy efficiency, but the pyroligneous liquor by-product has a commercial value.

Several studies have been conducted over the last decade in the area of thermochemical conversion of biomass to liquid fuels. There are various articles dealing with the upgrading of pyrolysis oils to produce fuels, the feasibility of which is rather questionable [3]. Fuel production in absence of a tipping fee or governmental support does not seem to be economical under current crude oil prices. The production of valuable oxygenated chemicals, such as phenols, could improve the process feasibility.

Vacuum pyrolysis process involves heating under reduced pressure of the complex polymeric structures of lignocellulosic materials. The process produces high yields of a complex primary oil rich in oxygen-containing compounds such as phenols and phenolic type components together with a good quality wood charcoal as by-product. However the complexity of the pyrolysis oils is an obstacle to their utilization as chemicals. There are two general routes to make use of the fine chemicals that are present in these complex liquid mixtures: i) the pyrolysis liquid products can be fractionated into simpler mixtures of components with similar chemical and/or physical properties, and ii) the pyrolysis products can be refined into more useful chemicals and/or chemical intermediates. In both cases, knowledge of the composition of the pyrolysis products is a prerequisite to their utilization as a chemical feedstock.

The recovery of pure chemicals from pyrolysis oils and the production of activated carbon by modern thermal processes has received little attention over the last decades while the demand for these products has increased on a worldwide basis. In the past, isolation of chemicals at the industrial scale has been performed to recover commodity compounds such as methanol, acetone, acetic acid and mixtures of phenols [4]. Phenols are valuable compounds and have been used as food aromas, pharmaceuticals or as intermediates for chemical synthesis. Limited work has been performed so far on the various potential applications of phenolic fractions separated from wood pyrolysis oils. Further exploitation of the chemical value of biomass-derived compounds would improve the overall pyrolysis process profitability.

1.1 Biomass-derived phenols

In their attempt to formulate new adhesives, Chum and Black [5] isolated a phenolic/neutral fraction from the fast pyrolysis of pine sawdust-derived oil. The authors used a series of liquid-liquid extraction steps. The phenols were partitioned into the ethylacetate phase from the pyrolysis oil phase. The ethylacetate phase was then treated with an aqueous sodium bicarbonate solution. Their results showed that about 50% of the phenol used in the formulation of phenol-formaldehyde resin adhesives could be replaced with a phenolic fraction of pyrolysis oils originating from wood. The performance of this material was comparable to Cascophen 313 adhesive brand. A preliminary feasibility study revealed a production cost of half that of phenol. Suzuki *et al.* [6] prepared a wood tar-based phenol adhesive by isolating the phenolic content of the tar at 1.3 kPa reduced-pressure distillation. The phenolic distillates could be blended with commercial phenol with a wood phenol/petroleum phenol ratio of 3:1. This blend matched the Japanese standards for minimum tensile shear-strength for adhesive evaluations. To take advantage of the well known germicidal activity of phenols, Guha *et al* [7] fractionated rice husk pyrolysis oil by applying a low pressure distillation. It was shown that the germicidal activity increased as composition of the volatile phenols increased in the fraction. Carazza *et al.* [8] distilled a wood tar under reduced pressure in two steps to isolate guaiacylic and syringylic compounds which had potential market value in the pharmaceutical industry and for the synthesis of new compounds.

Isolation of phenols from an Eucalyptus wood tar obtained by carbonization has been carried out with the objective to recover pure phenolic compounds such as syringol and guaiacol [9]. The approach involved the primary conversion of the raw wood tar into a lighter oil using a bench-scale vacuum pyrolysis unit followed by vacuum distillation and liquid-liquid extraction using alkaline and organic solvents. The extraction essentially yielded mixtures of pure phenols. Extraction efficiency increased at pH higher than 12. Process optimization is currently under investigation in this laboratory.

The objective of the work reported in this paper is to assess the potential of the vacuum pyrolysis technology as a new recycling process to transform wood residues to useful products such as pyrolysis oil and wood charcoal. Parameters such as the source material, pyrolysis temperature, pyrolysis pressure, vapour residence time and pyrolysis heating rate, feedstock bed thickness in the reactor and moisture content of the biomass influence the yield and composition of pyrolysis oils, gas and charcoal [10-13]. They do influence also the yield and type of chemicals obtained during pyrolysis. This paper will focus on factors leading to the production of high yields of phenolic compounds and phenolic type fractions, which could be used as a source of pure phenols or for commercial phenol formaldehyde resin replacement. Thermosetting polymers such as phenol-formaldehyde resins, are widely used as adhesives in the wood particle board and plywood industry [14].

2 Experimental

2.1 Feedstock
Table 1 shows various biomass feedstocks that were used during this investigation. Runs A074 and A075 were performed with dry and ground birch wood (*Betula alba*) sample under two different heating rates. All other runs A076 to A084 were performed with wild-cherry (*Cerasus serotina*) under different sets of conditions. A076 and A077 were performed to illustrate the effect of grinding. Run A078 shows the effect of wood natural moisture content on the pyrolysis products and phenol yields. As-received green, wild-cherry wood sample was ground and soaked in water for 48 h, then dried and pyrolysed (run A079). A similar sample as A079 was soaked in alkaline solution with pH 10 for 48 h then rinsed with distilled water and pyrolysed (run A080). Runs A079 and A080 exhibit the effect of ash, extractives and alkaline cations on pyrolysis product and phenol yields. Wood particles were dispersed in a support of glass wool, then pyrolysed (run A081). The effect of sample quantity, bed thickness and configuration and reactor pressure was studied during runs A081 to A084.

Table 1. Pyrolysis Conditions and Product Characterization (wt. %, anhydrous wood basis)

Run #	Feedstock	Sample weight (g)	Moisture wt. %	Ash wt. %	Heating rate °C/min.	Total pressure kPa	Yields wt. %			Yield of phenols wt. %
							Liquid	Solid	Gas	
AO74	Dry and ground birch	100.9	3.74	n.m	10	0.72	57.0	31.9	11.1	1.00
AO75	Dry and ground birch	100.9	3.74	n.m	20	0.72	55.2	32.0	12.8	1.39
AO76	Green wild-cherry	113.7	42.9	0.27	20	1.33	68.1	20.5	11.4	1.86
AO77	Green and ground wild-herry	138.7	38.1	0.30	20	1.47	66.9	18.7	14.4	1.77
AO78	Dry and ground wild-cherry	101.2	3.40	0.28	20	1.47	68.4	20.4	11.2	1.24
AO79	Water extracted, dry and ground green wild-cherry	87.2	4.40	0.19	20	1.47	69.9	21.0	9.1	1.51
AO80	Alkaline water extracted, dry and ground green wild-cherry	92.4	3.80	0.26	20	1.60	63.9	26.4	9.7	1.47
AO81	Green wild-cherry dispersed in glass wool	56.0	41.8	0.27	30	0.19	65.3	21.8	12.9	1.88
AO83	Green wild-cherry	56.1	41.8	0.27	20	0.20	66.7	21.7	11.6	2.72
AO84	Green wild-cherry	56.1	41.8	0.27	20	29.33	53.2	23.9	22.9	1.90

n.m.: not measured

The wood sample feedstocks were kept frozen for all the experiments reported in this work except sample A083 which was left at room temperature before pyrolysis. All wood samples originated from northern province of Québec. The wood sample used for runs A076-A084 contained a lower proportion of bark compared with those from runs A074 and A075.

2.2 Vacuum pyrolysis tests
Pyrolysis was performed with approximately 50 to 120 g of biomass feedstock in a 1 L batch reactor. Schematic view of the reactor and vacuum system used will be found elsewhere [12]. The pyrolysis oil was condensed in two condensers installed in series. Both condensers were maintained at -60°C in a mixture of dry ice and limonene (acetone was used instead of limonene in Ref. 12). The noncondensable gases were recovered in a 10 L metallic empty cylinder. Process conditions and pyrolysis product yields are shown in Table 1. Pyrolysis final temperature was 450°C for all runs. The mass balance closed to 99% for runs A074 and A075. In Table 1, the loss was assumed to be gases. For the other runs the gas phase was not analyzed and the mass balance calculations could not be performed. In those cases, the amount of gas produced was calculated by difference (see Table 1).

2.3 Phenol extraction
The content of the two condensers was recovered and mixed and the oil phase was analyzed for phenolic content after derivatization to the acetyl derivatives applying the following procedure:

Between 50 to 300 mg of oil was mixed with approximately 300 mg of acetic anhydride as derivatization reagent and two drops of pyridine as the reaction catalyst. The mixture was well mixed and stirred at 60°C for one hour. The reaction product was then eluted on about 2 g of silica gel using ca 50 ml of 20% dichloromethane in n-pentane. The eluate was concentrated and analyzed in a gas chromatograph coupled to a mass spectrometer (GC/MS).

A mixture of standard phenolic compounds such as phenol, cresol, 2,5-xylenol, guaiacol, catechol, resorcinol, hydroquinone, syringol, eugenol, vanillin, isoeugenol, trihydroxybenzene, allylsyringol, and syringaldehyde was prepared and derivatized. Anthracene was added as the internal standard.

2.4 GC/MS analysis
The purified fractions were analyzed in a Hewlett-Packard model 5890 gas chromatograph. The separation was performed on a 30 m x 0.25 mm i.d. HP-5MS fused silica capillary column with 0.25 mm film of crossed-linked 5% phenol methyl silicone. The GC oven temperature was held at 50°C for 2 min., then programmed to 200°C at 5°C min^{-1} and then to 290°C at 30°C min^{-1}. The oven temperature was maintained at the final temperature during 10 min. The injector temperature was 270°C with split mode (1/30 split ratio). Helium was the carrier gas with a flow of about 1 ml min^{-1}. The end of the column was directly introduced into the ion source of a Hewlett-Packard model 5970 mass selective detector operated with electron impact ionization mode. The mass spectrometer typical operating conditions were as follows: transfer line 270°C, ion source 250°C, electron energy 70 eV. Data acquisition was done with

HP-UX chemstation software using a HP-UNX computer and NBS mass spectra library data base. The mass range m/z: 30-350 Dalton was scanned every 0.8s. The computerized match was manually evaluated to ensure quality of identification. Identification of selected target compounds was also confirmed by matching mass spectra and retention times with standard compounds.

3 Results and discussion

3.1 Phenol precursors
Lignin is an abundant organic phenolic polymer produced by vascular terrestrial plants. The ultimate lignin precursors are CO_2 and H_2O which produce coumaryl, coniferyl, and sinapyl alcohols by biosynthesis. These three alcohols are the sole monomeric precursors of lignin. Coniferyl alcohol predominates. Lignins can be divided into several classes according to their structural elements. The so-called "guaiacyl lignin" which occurs in almost all softwoods is largely a polymerization product of coniferyl alcohol. The "guaiacyl-syringyl lignin", typical of hardwoods, is a copolymer of coniferyl and sinapyl alcohols, the ratio of which lies between 4:1 and 1:2 for the two monomeric units [15]. Depolymerization of lignin by thermal degradation produces various phenolic compounds.

As large volumes of lignocellulosic wastes exist, it would appear that a method to convert them to high value products would have a significant economical impact. The yield and composition of such products depend however on the pyrolysis conditions and the source material. Characteristic key compounds obtained by thermal decomposition of lignin, either isolated or in wood itself, are the phenolic ethers typified by guaiacol, in the case of softwood and mixture of guaiacol and syringol in the case of hardwood. Although quite an array of phenols and phenolic ethers have been identified and reported by many researchers in the pyrolytic products of lignin and wood, accurate quantitative analysis are rare, due to the great difficulties to separate and analyze the mixtures obtained. It may safely be said, however, that guaiacol, syringol and their homologs and derivatives predominate. The substituent groups are largely methyl, ethyl, propyl, vinyl, allyl, and propenyl and it is notable that the side chains do not exceed three carbon atoms and are almost always in the para position to the phenolic hydroxyl, hence reflecting the structure of the precursor monomers.

3.2 Phenol distribution
Phenolic content of the pyrolysis oils was determined and is listed in Table 2. Among those, guaiacol, catechol, resorcinol, eugenol, vanillin, isoeugenol, syringaldehyde and syringol and its derivatives have a relatively high commercial value. Their main application is in the food and fragrance industries. Phenols are mainly derived from the depolymerization of lignin at about 400-450°C [16]. Unlike hardwood, softwood-derived phenolic compounds have lower economic values as their syringylic type phenolic content is very low.

514

Table 2. Phenolic compounds obtained under various experimental conditions

#	Compound	Run #											
		A074	A075	A076	A077	A078	A079	A080	A081	A083	A084
		% wt., anhydrous wood basis									
1	Phenol	0.05	0.05	0.01	0.02	0.01	0.01	0.02	0.01	0.02	0.02
2	o-Cresol	0.02	0.02	0.01	0.02	0.01	0.01	0.01	0.01	0.02	0.01
3	m-Cresol	0.02	0.02	n.d	0.02	0.01	0.01	0.01	0.01	0.02	0.02
4	p-Cresol	0.06	0.05	0.01	0.01	0.01	0.01	0.01	0.01	0.02	0.01
5	Guaiacol	0.07	0.08	0.06	0.06	0.05	0.04	0.05	0.25	0.08	0.06
6	Propenylphenol	0.01	0.02	n.d	n.d	n.d	n.d	n.d	n.d	n.d	n.d
7	2-Methoxy-4-methylphenol	0.07	0.01	0.09	0.08	0.08	0.11	0.10	0.09	0.13	0.09
8	Catechol	0.08	0.18	0.13	0.13	0.09	0.08	0.09	0.13	0.15	0.12
9	2,3-Dimethoxyphenol	n.d	n.d	n.d	0.01	0.02	0.02	0.01	n.d	0.02	n.d
10	Hydroquinone	0.01	0.03	0.02	0.02	0.02	0.01	0.01	0.02	0.02	0.02
11	Methoxybenzenediol	0.04	0.24	0.16	0.19	0.07	0.10	0.09	0.14	0.16	0.13
12	Syringol	0.11	0.20	0.24	0.23	0.16	0.19	0.18	0.24	0.34	0.28
13	Dimethoxyphenol	0.05	0.02	0.03	0.03	0.02	0.04	0.03	0.03	0.04	0.03
14	Methylbenzenediol	0.03	0.07	0.04	0.03	0.09	0.08	0.10	0.09	0.13	0.04
15	Eugenol	0.01	n.d	0.01	0.01	0.01	0.02	0.02	0.01	0.01	0.01

515

16 Vanillin	n.d	n.d	n.d	0.01	0.01	0.01	0.01	0.01	0.01	0.01
17 Methoxyhydroxybenzaldehyde	0.09	0.08	n.d	0.02	0.01	n.d	0.01	n.d	0.01	n.d
18 Methoxymethylphenol	0.02	0.03	0.07	0.03	0.01	0.02	0.01	0.02	0.02	0.03
19 Dimethoxymethylphenol	0.00	0.01	n.d	n.d	n.d	n.d	n.d	n.d	0.01	n.d
20 Dimethylbenzenediol	0.00	0.02	0.02	0.01	0.01	0.02	0.02	n.d	0.02	0.02
21 Methylsyringol	0.04	n.d	0.21	0.20	0.15	0.19	0.18	0.19	0.27	0.24
22 Isoeugenol	0.05	n.d	0.06	0.07	0.04	0.06	0.05	0.03	0.13	0.08
23 Ethylhydroxymethoxybenzaldehyde	0.04	0.09	0.12	0.10	0.08	0.13	0.12	0.12	0.23	0.12
24 1,2,3-Trihydroxybenzene	0.01	0.02	0.11	0.04	0.02	0.03	0.03	0.07	0.08	0.04
25 Allylsyringol	0.01	0.02	0.01	n.d	n.d	n.d	n.d	n.d	n.d	n.d
26 Propylsyringol	0.01	n.d	n.d	0.02	0.01	0.01	0.01	0.01	0.03	0.02
27 Syringaldehyde	0.01	n.d	0.04	0.02	0.02	n.d	0.03	0.03	0.04	0.02
28 Dimethoxyhydroxybenzaldehyde	n.d	n.d	0.11	0.10	0.06	0.05	0.05	0.08	0.16	0.14
29 1,2,4-Trihydroxybenzene	0.01	0.01	n.d	0.03	0.01	0.03	0.02	0.03	0.06	0.03
30 Propenylsyringol	0.08	0.12	0.30	0.26	0.16	0.23	0.22	0.25	0.49	0.31
Total	1.00	1.39	1.86	1.77	1.24	1.51	1.47	1.88	2.72	1.90

n.d.: not detected

3.3 Optimun phenolic yield - process conditions

Analysis of the phenolic content of two hardwood feedstocks under various vacuum pyrolysis conditions was performed and the results are discussed below. The ultimate goal is to operate the pyrolysis under conditions leading to high yields of compounds of good commercial interest.

3.3.1 Heating rate

Table 1 shows the effect of pyrolysis operating conditions on the production of total phenolic compounds. The first two experiments, A074 and A075, involved birch wood waste and provided fairly similar pyrolysis solid and liquid yields (including both the pyrolysis oil and the pyrolysis water) under heating rates ranging between 10 and 20°C min^{-1}. However, their total phenolic yield is quite different. Run A075 was performed at a faster heating rate and produced higher phenolic yield. Fast heating facilitates the heat transfer between the hot heating plate of the reactor and the particles. The secondary degradation reactions of phenolic compounds are prevented when the primary vapours are quickly removed from the hot chamber. Usually low heating rates tend to produce higher yields of solid residues due to increased amounts of recondensed organic matter. Hot wood charcoal enhances the degradation of primary phenols. Any constriction in the expansion of vapours and their fast condensation will increase the residence time and will lead to secondary decomposition reactions in the hot reactor possibly following free radical mechanism. Detailed analysis of the phenolic compounds listed in Table 2 shows significant amounts of benzene diols and triols for runs A074 and A075. These are believed to be totally or partially the demethylation products of monomethoxy and dimethoxyphenolic derivatives. In Table 2, allylsyringol and propenylsyringol exhibit an increase at higher heating rate but concentration of the other dimethoxyphenols is reduced. 1,2,3-benzenetriol which is found in both pyrolysis oils is a decarboxylation product of gallic acid by pyrolysis [17]. Gallic acid originates from the hydrolyzable tannins occurring in bark [15]. Trihydroxyphenols have not been reported earlier in cellulose-derived vacuum pyrolysis oil, while Richards [18] identified these compounds in sodium chloride treated cellulose after pyrolysis.

Similar results have been reported in a recent article published by this laboratory using different wood residues rich in bark [19].

3.3.2 Pyrolysis final temperature

The effect of pyrolysis temperature on phenol yield has been studied earlier during a series of experiments using maple wood bark [19]. The difference in phenol yield is not large and to some extent there is an increase in phenolics at lower final temperature. Their structures will be preserved at lower temperatures under lower vapour pressure inside the reactor, since some phenols are primary pyrolysis products and are produced mainly by the cleavage of β-O-4 aryl ether bonds in lignin.

3.3.3 Particle size

Run A076 was performed with an as-received, green, wild-cherry sample which contained a small amount of bark with fine to large particle size. Run A077 was conducted with the same feedstock as A076 but was ground to 2 mm x 10 mm. Total phenolic content of the ground feedstock dropped by about 12 % which is considered to be a significant reduction. Fine particle size accelerates heat exchange reactions, but seems to be an obstacle to the quick removal of phenolic vapours even under vacuum. This might be again attributed to the compact wood charcoal bed which may catalyses the degradation of the evolving primary phenolic compounds. Previous results obtained in this laboratory had indicated higher pyrolysis oil yield using ground samples than large particle size [20].

3.3.4 Moisture content

The A078 pyrolysis experiment was performed with air-dried, ground wild-cherry with 3.4 % residual moisture. Both the pyrolysis solid and liquid yields slightly increased compared with the green sample (A077). However, the overall phenolic content was reduced by 30% with the dry wood sample. The concentration of the valuable syringol and its derivatives also drastically decreased upon drying. Heating rate and reactor pressure had been kept constant during both tests

The role of moisture in pyrolysis mechanism is not yet well understood. Gray *et al.* [21] suggested that water decreases wood crystallinity and viscosity at fusion temperature which in turn requires a lower volatilization temperature. The pretreatment of wood in solvent also allows the wood to swell into its full range, which subsequently favours the rapid diffusion of degraded products formed inside the wood structure during the pyrolysis process [22].

3.3.5 Wood pretreatment

Potassium, calcium, sodium, silicon, phosphorous and chlorine are the main constituents of wood ash. Wood extractive compounds are soluble in organic solvents and water. During biomass pyrolysis these compounds, especially potassium and calcium, have been reported to catalyze biomass decomposition and char-forming reactions [23]. In this study wood samples were separately treated with demineralized water and alkaline water. The results are discussed below.

Water extraction

Run A079 was performed with the same feedstock as the one used for runs A076 - A078 except the biomass feedstock was soaked in water during 48 h (Table 1). Water pretreatment reduces ash content as well as extractives and produces a higher pyrolysis liquid yield compared with the other experiments. Similar results have been published by the authors earlier [12]. Water extraction reduced the sample mass by 3.4 %. The loss included part of the wood extractives (total extractives have not been determined in this experiment) and approximately 32 % of the total ash content. Water treatment resulted in an increase of the phenolic yield with respect to run A078. Syringylic phenols such as propenylsyringol increased more than guaiacylic phenols such as guaiacol and methylguaiacol and diols such as catechol which is an interesting approach to inhibit diol production. It is not known yet whether ash and/or extractive components catalyze wood lignin decomposition reactions. Not understood also is the role of extractives on phenol production. The hydrophilic fraction of wood bark, *i.e.* the organic extractives, contain large amounts of phenolic constituents, mainly flavonoids which are polyphenols and will be partially extracted by water.

Alkaline water extraction

Neutral and alkaline salts have been reported to promote char formation and dehydration reactions [24]. Experiment A080 was performed using ground wild cherry sample which had been soaked in alkaline water. The pyrolysis liquid yield decreased and the yield of wood charcoal increased with respect to run A079. The phenolic content and its distribution was not significantly affected compared with A079.

3.3.6 Sample bed configuration

Run A081 was performed using a small quantity of wood particles (56 g) which were dispersed in about 2 g of glass wool and transferred to the reactor. The reactor pressure however was slightly lower than the other runs. There was an increase of the pyrolysis liquid yield compared with A080 experiment but the yield was lower than A076. The total phenol production was also lower than A076. A 1% increase in the yield of phenolic compounds based on an anhydrous feed using a thin bed of about 2 cm compared with a thicker bed of 20 cm was reported earlier [19]. Here the presence of glass wool seemed to increase the wood charcoal yield and reduce the yield of phenols. In principle heat transfer to the particules and vapour product evacuation is more efficient using a thinner feed bed or lower sample quantity due to reduced occurrence of secondary reactions. The distribution of all phenols changed during experiment A081. Surprisingly guaiacol production was favoured under run A081 experimental conditions. However additional work is needed before reaching a definite conclusion concerning the bed configuration parameter.

3.3.7 Sample initial temperature

Run A083 was performed using a sample half the size the one used during run A076. The wood sample was left at ambient temperature during a few hours before pyrolysis. Heat transfer to inner bed particles was faster under such pyrolysis conditions compared with the other runs (runs A076 to A081 and A084) that were performed with frozen particles. In such case, approximately 1/2 hour was needed to reach the temperature equilibrium in the reactor. The reactor pressure was also slightly lower for run A083 than A076. Phenolic yield was found to be higher for run A083 compared with the other experiments, including run A081 which was conducted at a very low vacuum. Run A083 also produced exceptionally high proportion of propenylsyringol compared with the other experiments.

Three experimental variables, namely the sample initial temperature, the pressure and the sample volume seem to contribute to phenol production. Further study is needed to distinguish the role of each variable. These three sets of pyrolysis parameters seem to be important for the enhancement of specific phenols. This observation however should be further investigated.

3.3.8 Pyrolysis pressure

Run A084 was performed using a similar mass of feedstock as for the previous run but under a higher pressure (see Table 1). As expected the pyrolysis yielded a lower quantity of liquid. Similarly the production of phenolics was reduced. However, the phenols distribution was not drastically modified when comparing runs A076 and A083. As two variables (the pyrolysis pressure and the sample initial temperature) were simultaneously studied during run A084, a well designed experimental plan needs to be implemented before any definite conclusion is drawn.

4 Concluding remarks

The objective of this work was to enhance the phenolic compound production in wood pyrolysis oils. The phenol distribution in the pyrolysis oils was investigated as well as their identification and their quantification.

Vacuum pyrolysis of birch and wild-cherry woods under various operating conditions was performed. Thirty phenolic compounds present in the pyrolysis oils were identified and quantified after derivatization to their acetyl derivatives. Pyrolysis yielded high amounts of syringol derivatives and benzenediols and triols. Pyrolysis operating conditions and feedstock pretreatment significantly influence the phenolic composition and yields. Green wood produced higher yields of phenolic compounds. Production of diols and triols was lower under the former pyrolysis conditions. Water and alkaline treatment of the feedstock did not increase the yield of phenols. Fast heat transfer between the hot reactor wall and the feed particles increases the phenolic content if the vapours are removed fast enough from the hot reactor chamber. Ground wood produced lower yield of phenols than the initial non ground wood.

Reproducibility tests have not been conducted systematically during this work and the data presented are often based on one pyrolysis test for each experimental variable investigated. This study is preliminary and exploratory. However, a similar trend of results has been reported earlier by this laboratory using different feedstocks and bed configurations. Further pyrolysis tests and analyses are in progress and the results will be reported in the near future.

As phenolic compounds have a high economical value, they merit more extensive efforts in the future to increase their yield and selectivity. It will be necessary also to improve the efficiency of the extraction methods as well as the purification of the single phenols obtained.

5 Reference

1. Oshima, M., Kashima, K., Kibo,T., Tabata, H. and Wafanabe. H. (1966) Studies of the hydrocracking of lignin. Bull. Chem. Soc. Jpn. Vol. 39, No.12. pp. 2750-2767.
2. Shafizadeh, F. (1983) *Saccharification of Lignocellulosic Materials*, Pure Applied Chem. Vol. 55, No. 4. pp. 705-720.
3. Cottam, M-L. and Bridgwater, A. V. (1994) Techno-Economic of Pyrolysis Oil Production and Upgrading, in *Advances in Thermochemical Biomass Conversion*, (ed. A. V. Bridgwater), Blackie Academic and Professional, New York, pp. 1343-1359.
4. Soltes, Ed. J. and Elder T. J. (1983) Pyrolysis, in *Organic Chemicals from Biomass*, (ed. I. S. Goldstein), CRC Press, Boca Raton, Fl, pp. 63-99.
5. Chum, H. L. and Black, S. K. (1990) *Process for Fractionating Fast-Pyrolysis Oils and Products Derived from*, USA patent 4, 942, 269. July 17.
6. Suzuki, T., Hiroshi, N., Yamada, T. and Homma, T. (1992) *Preparation of Wood Tar-Based Phenol-Resin Adhesives*, Mokuzai Gak. Vol. 38, pp. 321-324.
7. Guha, R., Das, D., Grover, P. D. and Guha, B. K. (1987) *Germicidal Activity of Valuable Compounds from Wood Tar-Derived Materials*, Biol. Wastes. Vol. 21, pp. 93-100.
8. Carazza, F., Rezzende, M. E. A., Pasa, V. M. D. and Lessa, A. (1994) Fractionation of Wood Tar, in *Advances in Thermochemical Biomass Conversion*, (ed. A. V. Bridgwater), Vol 2, Blackie Academic and Professional, New York, pp. 1465-1474.
9. C. Amen-Chen. (1995) *Preparative Separation of Valuable Compounds from Wood Tar-Derived Oils*, M.Sc Thesis, Université Laval, Québec, Canada.
10. Roy, C., Lemieux, R., de Caumia, B. and Blanchette, D. (1988) Processing of Wood Chips in a Semicontinuous Multiple-Hearth Vacuum-Pyrolysis Reactor, in *Pyrolysis Oils Biomass-Processing, Analyzing, and Upgrading* (eds. Ed. J. Soltes and T. A. Milne), ACS Symp. Series 376, ACS, Washington DC, pp. 16-30.
11. Shafizadeh, F., Furneaux, R. H., Cochran, T. G. Scholl, J. P. and Sakai,Y. (1979) *Production of Levoglucosan and Glucose from Pyrolysis of Cellulosic Materials*. J. Appl. Polym. Sci. Vol. 23, pp. 3525-3539.
12. Roy, C., Pakdel, H. and Brouillard, D. (1990) *The Role of Extractives During Vacuum Pyrolysis of Wood*. J. Appl. Polym. Sci., Vol. 41, pp. 337-348.
13. Domburg, G., Sergeeva,V., Kirshbaum, I., Sharapova, T. and Skripchenko, T. (1977) *Interaction of Wood Components During Thermal Treatment*. J. Therm. Anal. (Proc. 5th. Int. Conf.), pp. 304-307.
14. Chiu, S.-T. U.S. patent, 4, 433, 120, 21Feb. 1984 (filed 30 Sep. 1981)
15. Sjöstrom, E. (1981) *Wood Chemistry-Fundamentals and Applications*, Academic Press, Toronto.
16. Goheen, D.W. (1983) Chemicals from Lignin, in *Organic Chemicals from Biomass* (ed. I.S. Goldstein), CRC Press, Boca Raton, Fl, pp. 143-161.
17. Pakdel, H. and Roy, C. Unpublished results.

18. Richards, G.N. (1993) Chemistry of Pyrolysis of Polysaccharides and Lignocellulosics, in *Advances in Thermochemical Biomass Conversion* (ed. A.V. Bridgwater), Blackie Academic and Professional, New York, pp. 727-745.

19. Pakdel, H., Amen-Chen, C., Zhang, J. and Roy, C. (1996) Phenolic Compounds from Vacuum Pyrolysis of Biomass, in *Bio-Oil Production and Utilization* (eds. A. V. Bridgwater and E. N. Hogan), CPL Press, Newburg, UK, pp. 124-136.

20. Roy, C., de Caumia, B., Plante, P. and Menard, H. (1983) Production of Liquids from Biomass by Vacuum Pyrolysis-Development of Data Base for Continuous Process, in *Energy from Biomass and Wastes VII*, January 24-28, Lake Buena Vista, Fl, pp. 1147 - 1170.

21. Gray, M. R., Corcoran, H. and Gavalas, G. R. (1985) *Pyrolysis of a Wood-Derived Material - Effect of Moisture and Ash Content*. Ind. Eng. Chem. Process, Des. Dev. Vol. 24, pp. 646-651.

22. Lawson, J. R. and Klein, M. T. (1985) *Influence of Water on Guaiacol Pyrolysis*. Ind. Eng. Chem. Fundam. Vol. 24, No. 2, pp.203-208.

23. Agblever, F. A. and Besler, S. (1996) *Inorganic Compounds in Biomass Feedstocks. 1. Effect on the Quality of Fast Pyrolysis Oils.* Energy and Fuels, Vol. 10, pp. 293-298.

24. Zaror, C. A., Hutchings, I., Leo, I. P., Stiles, H. N. and Kandiyoti, R. (1985) *Secondary Char Formation in the Catalytic Pyrolysis of Biomass*, Fuel, Vol. 64, pp. 990-994.

FAST PYROLYSIS OF BIOMASS FOR RECOVERY OF SPECIALTY CHEMICALS
Specialized chemicals by fast pyrolysis

D.S.SCOTT and R.L.LEGGE
Dept. of Chemical Engineering, University of Waterloo, Waterloo, Ontario, Canada
J.PISKORZ, P.MAJERSKI and D.RADLEIN
Resource Transforms International Ltd, Waterloo, Ontario, Canada

Abstract
Many plant materials contain minor amounts, often only in the 100 ppm range, of complex compounds which may have a considerable present or potential application in biological or pharmaceutical areas. Extraction and concentration of these specialty chemicals by conventional technology can be a laborious and costly process.

Examples are given in which fast fluid bed pyrolysis has been used in our research to obtain pyrolysis liquids considerably enriched in such compounds, even when they have very low volatilities. One example involves the recovery of an alkaloid from plant leaves. Another feedstock gives complex terpene-based compounds (taxanes) which can be precursors for synthesis of new antitumor agents, or that can be used as a potential plant fungicide. Additional examples of complex compounds obtained in significant yields in fast pyrolysis oils can be found in the high molecular weight lignin fragments, largely aromatic in character, which are a part of the "pyrolytic lignin" fraction of a pyrolysis oil; and in the anhydrosugar monomers, dimers and oligomers obtained from the carbohydrate fraction of biomass on fast pyrolysis.

Some unique features, and some speculations on mechanisms, of such specialized pyrolytic processes are discussed.
Keywords: exotic chemicals, mechanism, pyrolysis speciality or fine chemicals.

1. Introduction

In the last fifteen years, the development of the
technology of fast pyrolysis processes to obtain high
yields of liquids by the rapid thermal decomposition of
biomass has reached a considerable degree of maturity. The
rather unusual composition of these fast pyrolysis liquids
has been investigated in considerable detail, and the
presence in unexpectedly high concentrations of some unique
components documented [1] [2] [3]. Some of these compounds
had not previously been identified in the thermal
decomposition products obtained from biomass prior to work
done in our laboratory over ten years ago [4].
The optimum temperature for high yields of fast
pyrolysis liquids is around 500°C for most biomass species.
At this temperature, thermal cleavage of many chemical
bonds occurs readily, and volatile compounds of low
molecular weight may be expected to form, even at the short
reaction times necessary for fast pyrolysis processes.
These volatile products are usually collected by
conventional cooling and scrubbing/condensation methods.
It is very surprising, therefore, to find in the pyrolytic
condensate also considerable amounts of compounds having
high molecular weights and of limited thermal stability.
The low volatility and thermally unstable nature of many of
these chemicals makes it difficult to accept that simple
distillation, or that some kind of secondary vapour phase
reactions, accounts for their presence in such significant
amounts in the pyrolytic condensate. Nonetheless, it is an
experimental fact that such complex high molecular weight
materials can be recovered in the condensate of volatiles
obtained in the fast pyrolysis process. This present work
gives some examples of how we have tried to exploit this
fact in our experimental program.

2. Previous work

It is now well established that lignin decomposition
products, largely aromatic in character, in amounts up to
15% to 30% of the biomass reacted can be found in fast
pyrolysis liquids - of the order of 80% of the original
lignin content. This product fraction has been defined for
convenience in our earlier studies as "pyrolytic lignin",
and was measured as the amount of organic material of low
solubility in water which precipitated on addition of
excess water to the pyrolysis liquid. Because water in
pyrolysis liquids in amounts less than about 40% by weight
acts as a bridge solvent, precipitation of pyrolytic lignin

only occurs when the water content exceeds this value. We have defined this lignin fraction as the amount of organic precipitate filtered off, dried and weighed when sufficient additional water was added to the pyrolysis liquid to bring the water content to 70% to 75% by weight.

This pyrolytic lignin fraction has been analyzed by Radlein et al [5] by ^{13}C NMR, and the spectra compared to that of other lignins. An example of these spectra is shown in Figure 1. The similarity of the spectra for the three poplar lignins (milled wood lignin, steam-exploded lignin, and pyrolytic lignin) is apparent, and shows that the various lignins are structurally similar. The pyrolytic lignin, therefore, does not appear to arise by secondary vapour-phase reactions.

The molecular weight distribution of the Waterloo poplar lignin has been measured by Johnson and Chum [6] using size exclusion chromatography, and the result is shown in Figure 2. It is apparent that most of this material has molecular weights >200, and the greater part has molecular weights from 500 to 1000 Da. As temperature or reaction time increases, the lignin fraction decreases, and is broken down to a heavy tar/coke and gas by secondary decomposition reactions. This decomposition behaviour is in contrast to that of cellulose, also a major natural biomass polymer. As Figure 2 shows clearly, cellulose fragments into lower molecular weight components, some of which are monomeric anhydrosugars, but the majority are low molecular weight carbonyl compounds.

It has also been found in our laboratory that if a biomass feedstock, free of alkaline cations, is subjected to fast pyrolysis, the cellulose or hemicellulose content is much more selectively depolymerized to form various monomeric anhydrosugars. These compounds typically have melting points from 150° to 200°C, and are thermally somewhat more stable than their hydrolyzed parent sugar, but are of very low volatility. Normal monomeric sugars e.g. glucose, cannot be successfully volatilized in fast pyrolysis, whereas anhydrosugars e.g. (levoglucosan) can be recovered in pyrolytic liquids in yields from 50% to 80% of the biomass cellulose content [7]. What is more interesting is that the anhydrosugar formed from the glucose dimer, cellobiose, that is, anhydrocellobiose or cellobiosan, can also be found in fast pyrolysis liquids from pretreated biomass in significant yields - 2% to 10% of the cellulose content. Yields of cellobiosan from a pure cellulose are shown as a function of temperature in Figure 3, as reported in one of our earlier publications [7]. Cellobiosan has a melting point of 285°C, is not thermally stable or volatile, and has a molecular weight of 324. It is surprising, therefore, to find such large amounts in the volatile products of cellulose fast

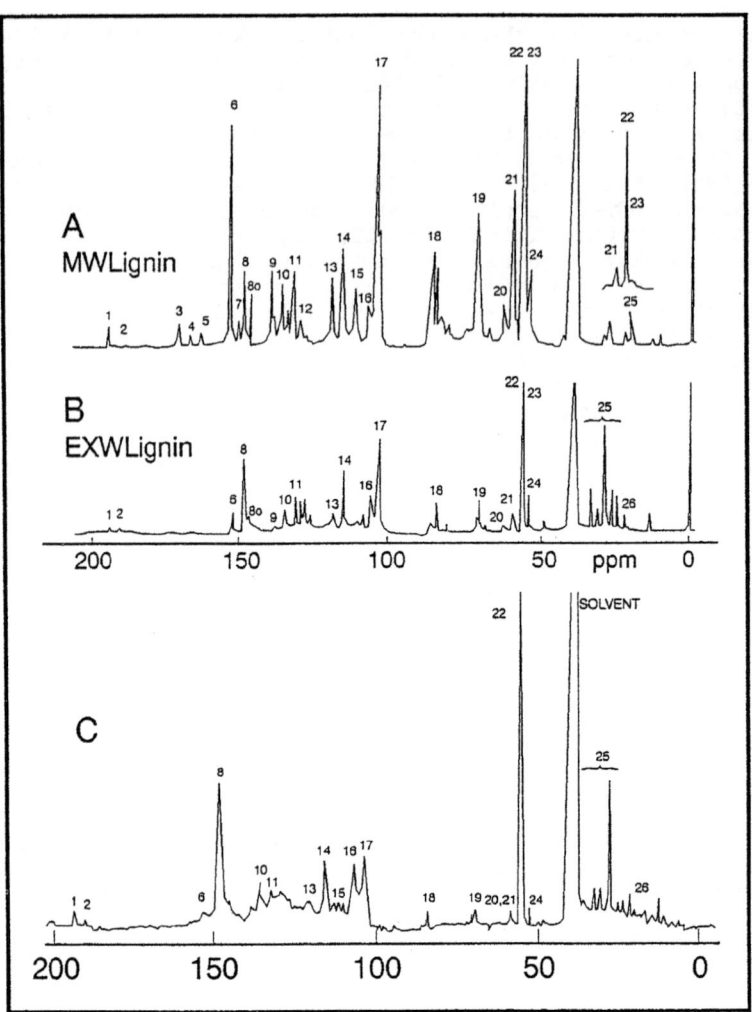

Fig. 1 NMR spectra of various lignins from poplar wood

A. Milled wood lignin
B. Steam exploded lignin
C. Waterloo pyrolytic lignin

pyrolysis.

Biomass materials often contain lesser amounts of high molecular weight compounds of limited volatility and thermal stability, but having a high commercial value. Because many of these compounds are of considerable

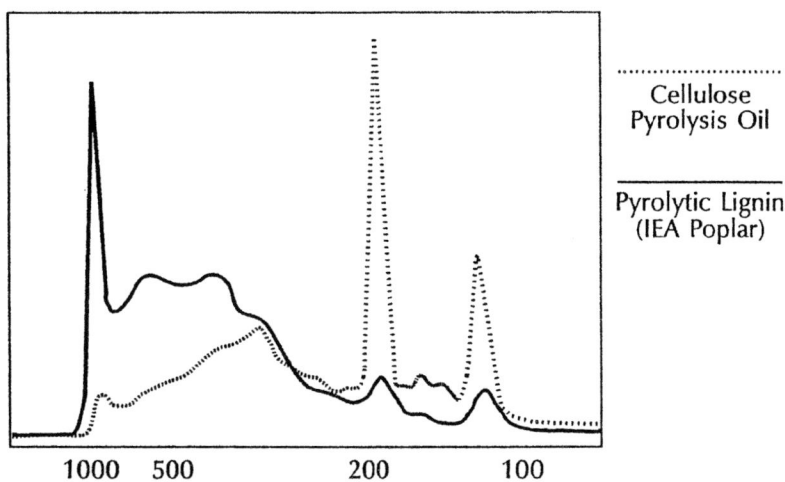

Fig. 2. Molecular weight distributions for cellulose pyrolysis oil and for poplar pyrolytic lignin (6)

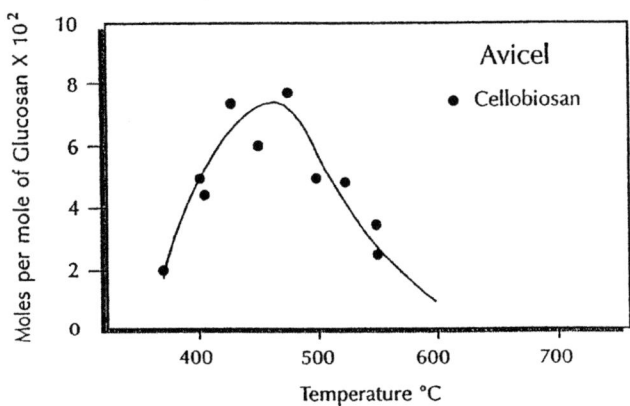

Fig. 3. Cellobiosan pyrolysis yield from cellulose [7]

interest for biological, pharmaceutical or flavouring uses, it was decided to test fast pyrolysis as a potential means of recovering such chemicals, or their complex derivatives.

3. Experimental

Two exploratory tests were done to evaluate the concept that desirable complex molecules of potentially high value could be recovered from plant materials by fast pyrolysis techniques. The first of these used Ontario flue-cured tobacco leaves (variety "Candell") as a feedstock, with the objective of recovering an alkaloid, nicotine. A second set of tests explored the possibility of recovering paclitaxel, (taxol), or related taxanes from prunings (needles and terminal twigs) from an Eastern Canadian native yew, taxus canadensis. Taxol has recently been shown to be an active anti-tumor agent for both ovarian and breast cancers, and has been approved for use in Canada.

The molecular structure of these materials is shown in Figures 4 and 5, together with some of their properties. It is apparent from these configurations that it would be surprising to find them in fast pyrolysis liquid products. Nevertheless, the presence of nicotine in tobacco smokes is well known, although the yields are low. Taxol is presently recovered by a multistep solvent extraction/ chromatography process, which uses large quantities of solvents, and is expensive and lengthy.

The tobacco leaves and yew prunings were both dried at ≈50°C and then ground to -0.5 mm particle size. Pyrolysis tests were carried out in the Waterloo Fast Pyrolysis Process bench scale continuous fluidized bed unit, which has been described in detail elsewhere [1]. Briefly, this apparatus is a small fluid bed unit using sand as the heat transfer medium, and operating at essentially atmospheric pressure. Capacity varies from 10 to 100 g/h. These controlled low solids feed rates are made possible by use of the specially designed RTI Ltd. pneumatic feeder. Volatile products were condensed and collected using surface condensers. Runs were normally from 20 to 60 minutes long, and complete material balances were made, usually with closures of 95% or better. A schematic of the complete pyrolysis apparatus is shown in Figure 6.

Analysis of the acidic pyrolysis condensate was carried out by GC/MS and HPLC (with an Aminex HPX-878 column at 65°C). Analyses for taxol and related taxanes was carried out by the Natural Products Branch, National Cancer Institute (USA). Gas analyses were done by conventional GC.

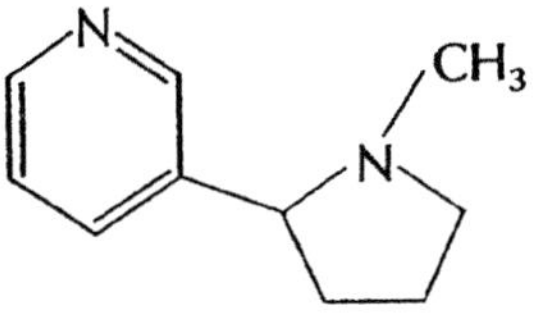

Fig.4   Nicotine

3-(1-methyl-2-pyrrolidinyl) pyridine

MW= 162.23   bp (745) = 247°C   (partial dec.)

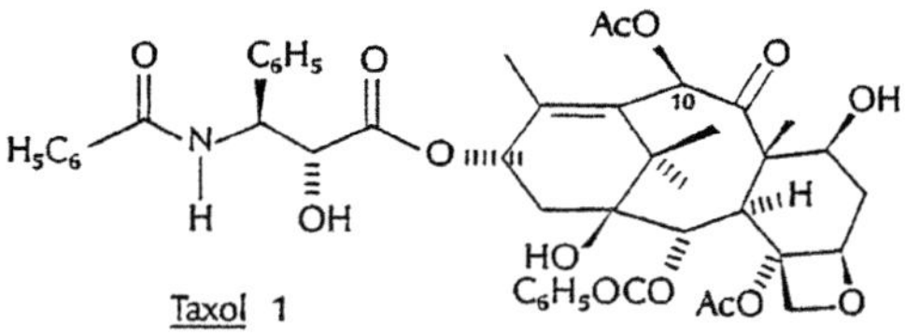

Taxol 1

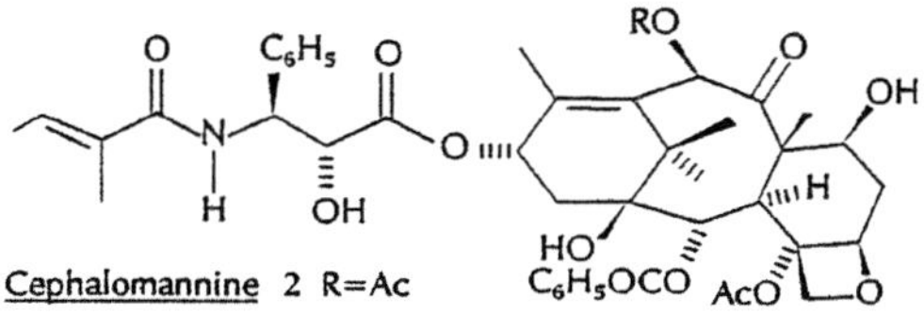

Cephalomannine 2 R=Ac
   10-deacetyl   3 R=H
Cephalomannine

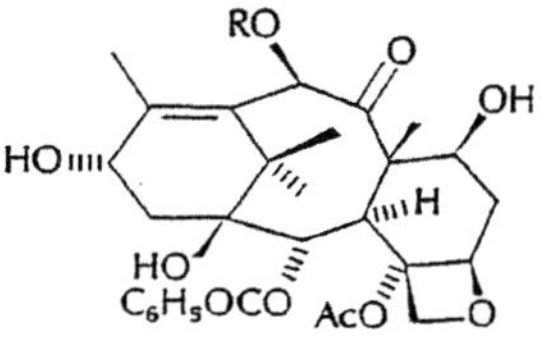

Baccatin III   4 R=Ac
10-deacetyl   5 R=H
Baccatin III

Fig. 5   Structure of some taxanes

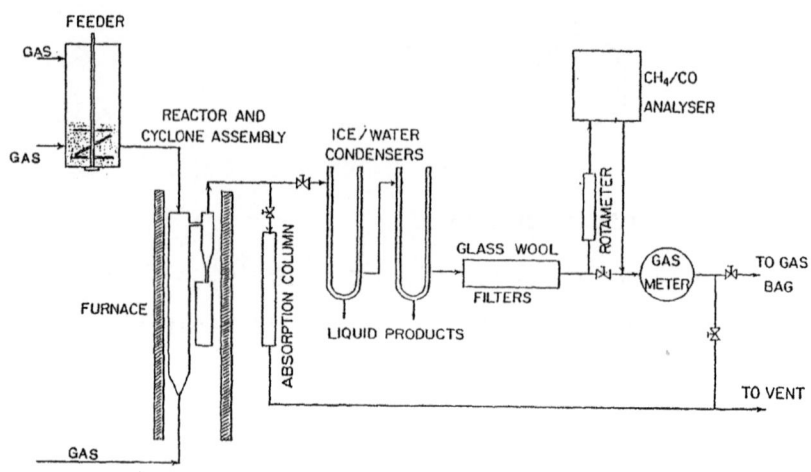

Fig. 6   Schematic diagram of the continuous bench
          scale fluid bed pyrolysis apparatus

4. Results

Details of the composition of the two feedstocks used and
of typical pyrolysis conditions are given below in Table 1.

Table 1.   Feedstock properties and pyrolysis conditions

	Tobacco leaves	Yew Prunings
Particle size	-0.5 mm	-0.5 mm
Moisture, wt %	8.0	19.0
Ash, wt % mf.	16.3	4.6
Temperature, °C	487	475.
App.vapour res.time	≈0.64 s	0.43 s

The yields from the two samples listed in Table 1 are
summarized in Table 2.

Table 2. Yields of pyrolysis fractions

	Tobacco leaves Yields, maf feed		Yew prunings Yields, mf feed	
Product	Exp	Normalized	Exp	Norm.*
Gas	19.8	21.0	8.3	10.8
Water	11.9	12.6	7.4	18.3
Char	25.2	25.2	21.4	21.4
Organic liquid	38.8	41.2	49.5	49.5
Recovery	95.7	100.0	86.6	100.0

*Normalization assumed all losses were gases or water due to small leak in gas collection system.

The gas product from the pyrolysis of the tobacco leaves was unusual in that it contained about 84% by weight of carbon dioxide. The liquid product from the tobacco (tar + aqueous liquor) was analyzed both by HPLC and by GC/MS. The nicotine is a weak base, and readily forms salts with acids (in the leaves it is usually present as a salt with citric or malic acid), and therefore does not readily elute in HPLC analyses. However, the nicotine nucleus is readily detected by GC/MS. The GC/MS trace for the pyrolytic liquid is shown in Figure 7. By far the largest concentration of material which could be chromatographed by the GC was nicotine, in an amount estimated to be about 6% by weight of the total liquid, or about 2.5% yield based on the dry moisture-free leaves. Since dry tobacco leaves contain from about 2% to 8% of nicotine, this recovery represents a minimum yield of about 35% to a maximum of 90%

The pyrolysis liquid recovered from the Taxus canadensis sample was analyzed for taxol and other taxanes by the US National Cancer Institute, using an HPLC system with a PDA (200-400 nm) detector to give three dimensional coloured chromatograms. Pyrolysis liquids are acidic, with pH values of 2.5 to 3.0 normally, and taxol is known to be unstable at these low pH levels. Our samples were stored for some time before analysis, and therefore only taxol precursors could be expected to exist in the samples at the time of analysis. A conventional solvent extraction of the T. canadensis prunings sample followed by chromatographic partitioning was also done as a part of this study, using standard methods described in the literature [8]. Briefly, this procedure consisted of a hexane wash of the dried sample, followed by extraction with 50/50 dichloromethane/methanol solution. The extract was evaporated to dryness, the residue dissolved in methanol, and this solution was then partitioned with a 50/50 dichloromethane/water mixture. The aqueous phase was re-extracted with dichloromethane, and the combined extracts were then

GC/MS CHROMATOGRAM FOR PYROLYSIS LIQUID FROM TOBACCO LEAVES

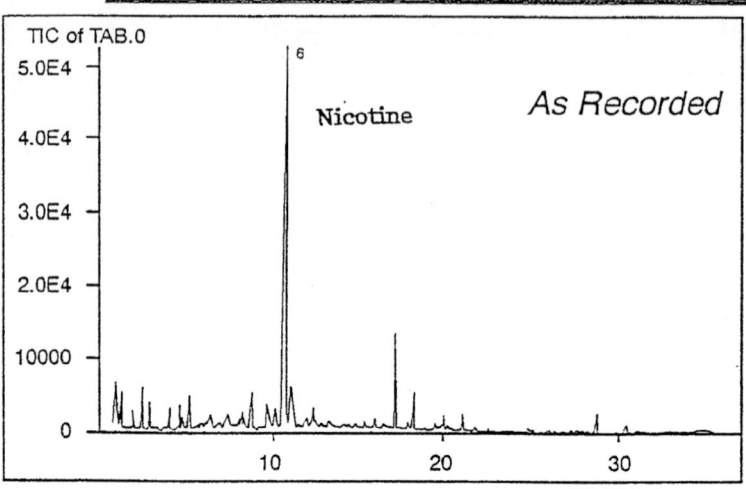

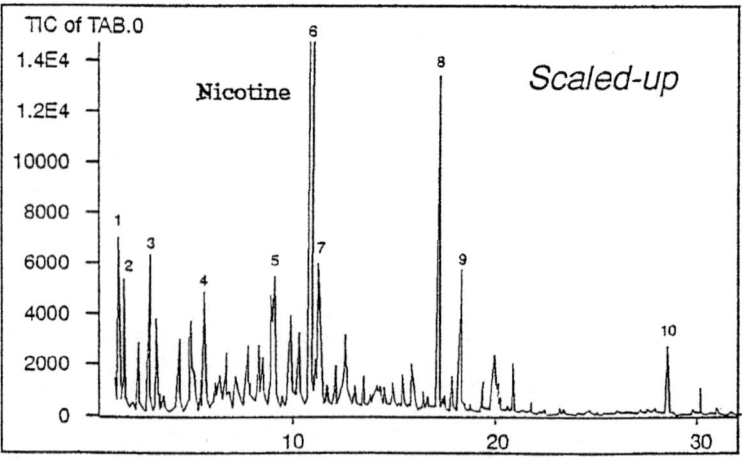

1. ethylene glycol
2. methyl acetate
3. ethylene glycol, diacetate
4. 1,2-cyclopentanedione, 3-methyl
5. 1,2-benzenediol

6. nicotine ⟵——
7. 1,3-bezenediol, 4-ethyl
8. 6-octen-1-ol, ?

9. tetradecanoic acid ?
10. ?

Fig. 7

evaporated to dryness, the residue taken up in methanol, and after filtering analyzed by HPLC using the NCI method. The pyrolysis liquid was subjected to the same extraction/partioning procedure as was the sample of yew.

This procedure gave values for the content of taxol and related compounds (see Figure 5), as determined by presently recommended analytical methods. The results of these analyses are shown in Table 3.

Table 3. Analyses for taxanes in T. canadensis prunings

Yield, wt. % dry yew	Conventional extract	Pyrolysis liquid
Taxol	0.0105	nil
Cephalomannine	0.0013	0.0003
Baccatin 3	0.0050	0.0149

As expected, no taxol was found in the pyrolysis liquid. A different product collection procedure would be needed to preserve this compound in the pyrolytic condensate. However, the taxol precursor, baccatin 3, was present in an amount approximately equal to the sum of the more complex taxanes identified by the US NCI in the yew prunings extract. As the analysis of the solvent extracted components showed only 50 ppm of baccatin 3 in the yew sample, one can speculate that the 149 ppm from the analysis of the pyrolysis liquor (196 ppm in the liquid) may be largely derived from the taxol content or its analogues.

5. Discussion

The normal extraction/partitioning procedure for the recovery of taxol requires several extraction/evaporation steps followed by chromatographic separation. Large volumes of volatile solvents are used. The fast pyrolysis method requires only one step, and yields a liquid in which it may be possible to readily concentrate the content of taxanes. However, methods for the economical recovery of taxanes from the pyrolysis liquids are still a subject of on-going research. Syntheses are already published which allow the production of taxol from precursors such as baccatin 3. Current research is now in progress in our laboratory in an attempt to exploit this method for the production of taxol, or of its precursors which could be the starting point for the synthesis of taxol or other biologically active analogues.

In a more general sense, the question arises of what is the operating mechanism whereby these low volatility, thermally unstable molecules, admittedly sometimes transitional species, can be recovered intact by high temperature fast pyrolysis? In an earlier presentation [9], the authors have proposed that a large fraction of the aerosols formed in fast pyrolysis are in fact generated directly as primary products, that is, the volatile product of fast pyrolysis is a bimodal mixture of true vapour and an aerosol containing non-volatile fragments or components from the biomass. Arguments are given based on the reported results of several researchers to support this conclusion. One might add to the examples already given here that of the presence of cannabinoids in the smoke from marijuana, principally tetrahydrocannabinol (see Figure 8), with a molecular weight of 314.45, $bp_{0.02} = 200°C$ (we did not experiment with this feedstock in this work!).

As one possibly plausible mechanism, it is suggested that that proposed in our earlier publication can serve at least as a useful working hypothesis which explains the results reported here, as well as the well-documented presence of similar molecular entities in most fast pyrolysis liquids. It also gives a justification for a planned exploitation of fast pyrolysis processes as a method of recovering from plant materials many complex chemicals of high value for the pharmaceutical or synthetic chemical industries.

Fig. 8

Tetrahydrocannabinol (THC)

Tetrahydro-6, 6-9-trimethyl-3-pentyl-6H-dibenzo (b,d)pyran-1-ol
or (-)Delta-9-trans-tetrahydrocannabinol
MW = 314.45

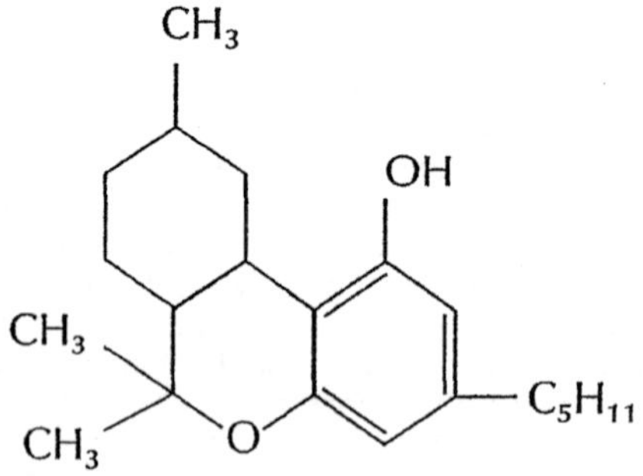

6. Acknowledgment

The authors would like to thank the many students and colleagues who have contributed to our research studies on biomass conversion over the years. We also extend our appreciation to the Natural Sciences and Engineering Research Council of Canada, and to Natural Resources Canada, who supplied some of the funding for the work.

7. References

1. Scott,D.S., and Piskorz,J. (1984) Continuous flash pyrolysis of wood for production of liquid fuels. Canadian Journal of Chemical Engineering, Vol.26,pp.404-412.
2. Scott,D.S., Piskorz,J. and Radlein,D. (1985) Liquid products from the continuous flash pyrolysis of biomass,Industrial & Engineering Chemistry Process Design & Development, Vol.24, pp. 581-588.
3. Piskorz,J.,Radlein,D., Scott,D.S. and Czernik,S. (1989) Pretreatment of wood and cellulose for production of sugars by fast pyrolysis. Journal Analytical & Applied Pyrolysis, Vol. 16, pp.127-142.
4. Radlein,D., Grinshpun,A., Piskorz,J.and Scott,D.S. (1987) On the presence of anhydro-oligosaccharides in sirups from the fast pyrolysis of biomass. Journal Analytical and Applied Pyrolysis, Vol.12, pp.39-49
5. Radlein, D., Piskorz,J. and Scott,D.S. (1987) Lignin derived oils from the fast pyrolysis of poplar wood, Journal Analytical Applied Pyrolysis, Vol.12, pp.51-59
6. Johnson,J. and Chum,H.L. (1987) Some aspects of pyrolysis oils characterization by HPSEC. Preprints ACS Symposium on Production,Analysis and Upgrading of Pyrolysis Oils from Biomass, Denver, CO, April
7. Piskorz,J., Radlein,D., Scott,D.S. and Czernik,S. (1988) Liquid products from the fast pyrolysis of wood and cellulose. Research in Thermochemical Biomass Conversion, (eds. A.V.Bridgwater and J.L.Kuester), Elsevier Applied Science, (London), pp. 557-571
8. Wheeler,N.C., Jeck,K., Masters, S., Brobst,S.W., Alvarado, A.B., Hoover, A.J. and Snader, K.M. (1992) Effects of generic, epigenetic and environmental factors on taxol content in T.brevifolia and related species. J. Natural Products, Vol. 55, pp.432-440
9. Piskorz,J., Scott,D.S. and Radlein,D. (1995) Mechanisms of the fast pyrolysis of biomass: Comments on some sources of confusion. Presentation at Symposium on Frontiers of pyrolysis biomass conversion and polymer recycling. Breckenridge, CO, June (National Renewable Energy Laboratories, Golden, CO)

OBTAINING OF LEVOGLUCOSAN BY FAST PYROLYSIS OF LIGNOCELLULOSE. PATHWAYS OF LEVOGLUCOSAN USE

R. PERNIĶIS, J. ZANDERSONS and B. LAZDIŅA
Latvian State Institute of Wood Chemistry, Riga, Latvia

Abstract
In the present work, the results of laboratory studies and pilot setup experiments are reported. The yield of levoglucosan depends upon the temperature, the particle size of lignocellulose and modulus of heat carrier. A temperature of 370 ... 410°C and a modulus of heat carrier of 4...5 kg/kg may be regarded as optimum parameters. The purification of levoglucosan by selective dissolution and cristallization of levoglucosan using 90...96% ethanol ensure its content in the purified product 95...96%. Levoglucosan can be used as a raw material in polymer chemistry, biotechnology and medicine.
Keywords: Lignocellulose, fast pyrolysis, levoglucosan, use of levoglucosan.

1 Introduction

1,6-anhydro-β,D-glycopyranose - levoglucosan (LG) is one of the products of the thermodestruction of cellulose. During vacuum-pyrolysis, the LG yield exceeds 70% of the cellulose mass. Since this carbohydrate anhydride is of practical interest, the mechanism of its formation as well as the technology of the obtaining and separation of LG during the vacuum-pyrolysis of cellulose and cellulose-containing materials was under study.

However, in late 60-s, it was found that LG with a good yield is formed during the process of the fast pyrolysis of sawdust, linter and other cellulose-containing materials at atmospheric pressure [1]. It has been shown by us that the fast pyrolysis of lignocellulose (LC) for obtaining LG can be carried out under pilot setup conditions. A good LG yield has been achieved (20...26% from the mass of oven dry LC or 47.5...63.0% from the mass of cellulose) [2].

The yield of the end products as well as the choice of the equipment during the fast pyrolysis of finely-disperse cellulose-containing raw materials in the heat carrier flow are determined by the particle sizes of the raw materials, the temperature in the reaction zone, the particles velocity, the thermodestruction reaction time as well as the mass ratio of the heat carrier and the destructed material (modulus of heat carrier). The studies of the process of fast pyrolysis of wood [3] have shown that the character of the mixture flow of the heat carrier and the material particles exerts a negligible effect upon the thermodestruction process.

At temperatures exceeding 300°C, the reaction of the formation of volatile products of cellulose thermodestruction prevails over carbonization reactions. In accordance with the data by G.E.Domburg and co-authors [4], LG formation begins at 320°C. Therefore, the fast heating of cellulose at low-temperature stages of thermal treatment prevents the development of dehydration reactions. The problem of fast heating was solved by the use of the superheated steam with a sufficiently high thermal capacity, promoting the removal of volatile products from the high temperature region and facilitating their recovery in the condensation system. During the fast pyrolysis at atmospheric pressure, the superheated steam ensures also the removal of oxygen from the system and lowers the partial pressure of the volatile products vapour.

The developed capillary structure of the material particles under other equal conditions of thermodestruction improves the heat mass exchange and increases the LG yield. A higher yield of LG can be expected when porous LC of hardwood, milled up to particle sizes of 2...3 mm, are used. Such particles are heated in 1.3...2.8 sec in the heat carrier flow up to 400°C.

The ever increasing commercial interest shown in the thermodestruction of the plant biomass made many scientists to investigate the cellulose pyrolysis chemism. A good review of the results of the studies of the last decade has been published recently by M.J. Antal and G Varhegyi [5].

It is known that the formation rate and levoglucosan yield are affected not only by the presence of ions of metals and hydrogen in the cellulose-containing raw material [6], but also by the composition of the gaseous medium and the partial pressure of volatile products vapors in the reactor. The heat carrier during the fast pyrolysis process ensures not only the heating, but also the transportation of the partially carbonized material as well as the evacuation of the volatile products from the reaction zone. The choice of the mass ratio of the heat carrier and the material (modulus of heat carrier) as well as the temperature of the heat carrier are connected with taking into account both the heat and hydrodynamic requirements of the process. In the present work, the results of the laboratory studies of the lignocellulose particle sizes, the reaction temperature as well as the mass ratio between the heat carrier and the cellulose-containing material upon the yield of the products of cellulose destruction in a cyclone type thermal reactor (see the flow sheet) are set forth.

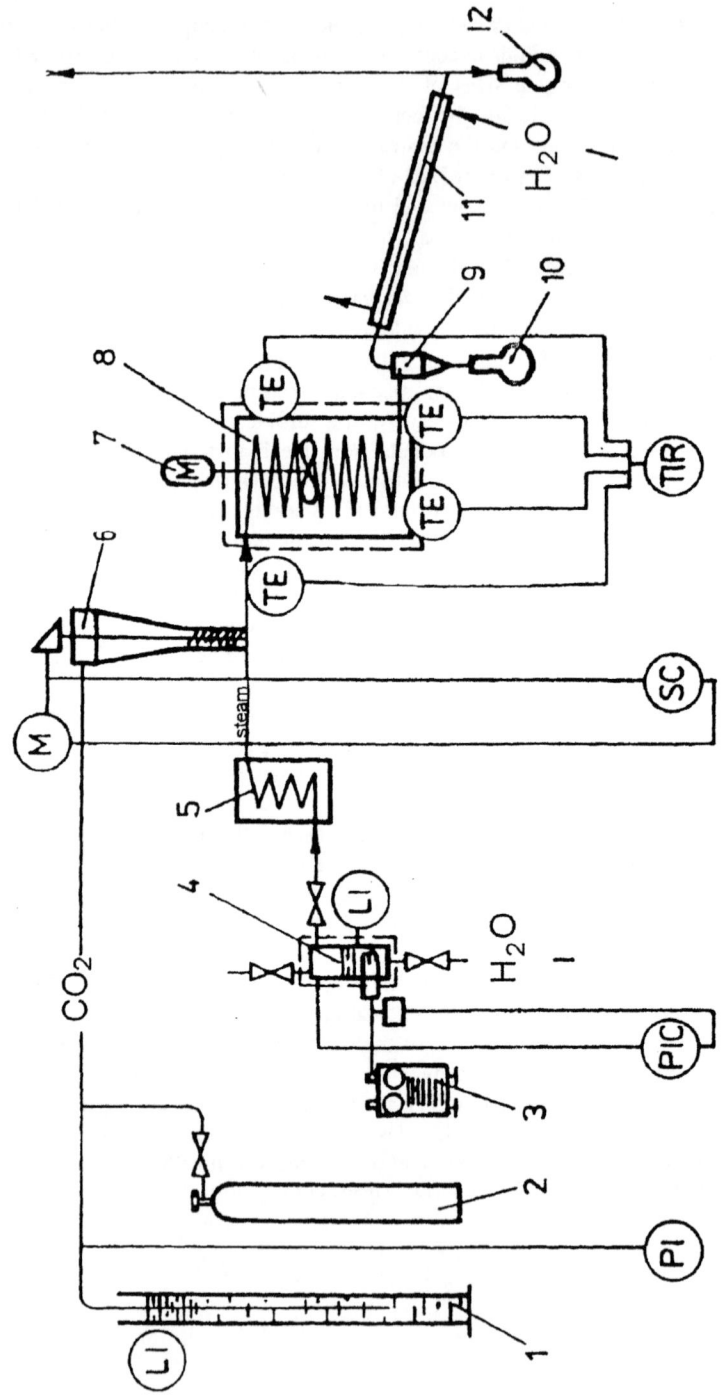

Flow-sheet of laboratory unit for fast pyrolysis: 1 - monostat; 2 - vessel with inert gas; 3 - autotransformer; 4 - steam generator; 5 - steam superheater; 6 - feeder; 7 - fan; 8 - thermoreactor; 9 - cyclone; 10 - char collector; 11 - condensator; 12 - condensate collector

2 Laboratory experiments

In our experiments, the duration of the presence of LC particles in the thermal reactor was 0.7...2.3 sec. (Table 1). We used birch lignocellulose obtained upon furfural production by the method developed at the Latvian State Institute of Wood Chemistry by Prof. N.A. Vedernikov [7].

Table 1. Main parameters of experiments

Modulus of heat carrier	Temperature (° C)	Coefficient of sliding	Movement rate in reactor (m/sec)		Time of passing reaction zone (sec)	
			steam-gaseous mixture	solids	steam-gaseous mixture	solids
	320	1.118	9.4	8.4	2.0	2.3
2.5	360	1.129	10.0	8.9	1.9	2.1
	400	1.138	10.7	9.4	1.8	2.0
	320	1.159	18.4	15.9	1.0	1.2
5.5	360	1.166	19.7	16.9	1.0	1.1
	400	1.179	20.9	17.8	0.9	1.1
	320	1.178	27.5	23.3	0.7	0.8
8.5	360	1.193	29.4	24.6	0.6	0.8
	400	1.206	31.2	25.8	0.6	0.7

It has been shown by special experiments on the carbonization of pulverized fuel in the gaseous flow that particles with a size of 1.0 and 0.5 mm are heated to 400°C in 0.36 and 0.16 sec, respectively, while 1.3...2.8 sec. are required for the heating of particles with a size of 2...3 mm [8]. In our work, the heating of the whole volume of lignocellulose particles up to the reaction temperature was ensured for fractions with a particle size up to 1 mm, and only at a modulus of heat carrier of 2.5 kg/kg, the time of passing the thermal reactor channel chosen by us ensured the heating of fraction particles of 1...2 mm. This can be seen if comparing the yields of the condensated organic products of lignocellulose thermolysis (thermolysis tar) and LG from different fractions. As can be seen from the data of Table 2, as the particle sizes increase, the products yields decrease. At a temperature of 400°C, this dependence is less expressed. The LG concentration in lignocellulose thermolysis condensates was determined by the known methodology [9], by the difference of the content of reducing substances before and after the hydrolysis of the sulphur acid sample analysed. A parallel determination of the quantative content of levoglucosan by the GLC method has shown that the results of the estimations coincide. It means that the condensates , apart from LG, do not contain any other sugar or oligosaccharide anhydrides, as it can be expected taking into account the data obtained by other

Table 2. Yield of products during fast pyrolysis of lignocellulose

Lignocellulose fraction (mm)	Temperature (°C)	Yield, % from o.d. LC		Yield, % from cellulose mass in LC					RS/LG	LG/TT
		Carbonaceous residue	TT	Organic acids	TT	LG	RS	Other dry substances		
				Modulus of heat carrier 2.5 kg/kg						
<1	320	58.4	15.9	1.7	38.5	18.5	9.7	10.4	0.53	0.48
	360	38.8	18.9	2.4	44.5	21.0	12.6	10.9	0.60	0.46
	400	26.0	24.4	2.5	59.2	26.8	16.7	15.7	0.62	0.45
1...2	320	65.6	11.1	1.5	24.5	13.2	6.4	4.9	0.49	0.54
	360	36.5	15.0	2.1	33.1	15.0	11.0	7.1	0.73	0.45
	400	27.0	24.3	2.7	53.8	24.1	16.5	13.2	0.68	0.45
2...3	320	64.4	9.7	1.4	20.5	10.6	5.5	4.4	0.52	0.52
	360	43.4	15.1	2.1	32.1	13.3	10.6	8.2	0.80	0.42
	400	30.2	23.7	2.8	50.1	23.9	16.4	9.8	0.69	0.48
				Modulus of heat carrier 5.5 kg/kg						
<1	320	65.2	18.9	1.7	45.6	22.9	9.9	12.8	0.43	0.50
	360	34.8	20.7	1.7	50.0	26.1	13.0	10.9	0.50	0.52
	400	28.9	24.8	2.7	60.1	30.5	16.2	13.4	0.53	0.51
1...2	320	66.1	15.3	1.8	33.7	17.2	7.5	9.0	0.44	0.51
	360	44.7	20.1	2.2	44.3	22.7	11.7	9.9	0.51	0.51
	400	27.0	26.7	3.4	58.9	31.5	20.4	7.0	0.65	0.53
2...3	320	69.7	13.9	1.7	29.4	12.3	5.9	11.2	0.48	0.42
	360	44.7	18.2	1.6	38.4	19.1	9.6	9.7	0.51	0.50
	400	30.3	26.1	3.1	55.2	25.7	18.2	11.3	0.71	0.46
				Modulus of heat carrier 8.5 kg/kg						
<1	320	74.5	14.0	1.8	33.7	21.7	7.7	4.3	0.35	0.64
	360	52.7	24.4	2.3	58.9	28.5	12.3	18.1	0.43	0.48
	400	31.9	27.1	3.9	65.4	29.9	17.6	17.9	0.59	0.46
1...2	320	87.3	7.8	1.3	17.2	11.0	4.4	1.8	0.40	0.64
	360	49.9	21.5	2.2	47.4	23.6	9.9	13.9	0.42	0.50
	400	30.3	26.9	3.2	59.4	28.5	18.1	12.8	0.63	0.48
2...3	320	78.3	8.2	0.9	17.3	7.8	4.2	5.3	0.53	0.45
	360	61.4	15.4	1.8	32.6	18.4	8.9	5.3	0.48	0.56
	400	24.2	21.7	2.8	45.9	26.2	14.6	5.1	0.56	0.60

*TT - thermolysis tar, LG - levoglucosan, RS - reducing substances

authors [10, 11], although obtained in differing experimental conditions, in particular, at a higher temperature and in the gaseous medium.

In the interval under study (320...400°C), the increase in temperature in all the variants causes the increase in the yield of the volatile products of cellulose thermodestruction. Owing to rather a short reaction time only at a temperature of 400°C and a modulus of heat carrier of 5.5 and 8.5 kg/kg, the levoglucosan yield has reached approximately a half of the yield obtained if the thermodestruction reaction time is 8...10 sec [1, 2]. The reducing substances (RS) yield is rather high. For clearness, it is presented in Table 2 as the RS/LG ratio. It makes up 0.5...0.6, and, in some cases, even 0.7...0.8. Normally, if the LC thermodestruction time is 8...10 sec, the RS/LG ratio is 0.2...0.3. We had no oportunity to study the component composition of thermolysis condensates in greater detail. The presence in them of sugar derivatives containing the aldehyde group, as it has been found by F. Shafizadeh and co-authors [12], or glyoxal, hydroxyacetaldehyde formaldehyde found in condensates of fast thermolysis of cellulose and wood by D.S. Scott and co-authors [13] can be only supposed. In our case, the presence of similar admixtures could be the reason for the elevated content of reducing substances in thermolysis tar. Carrying out the process on a laboratory as well a pilot-scale unit for fast pyrolysis, no water-insoluble residual tar was formed. Special studies of the mechanism of wood thermodestruction have shown that, irrespective of the surrounding temperature, a constant temperature is preserved in the wood bulk [14]. Before the accomplishment of the endothermal reactions of LG formation, the temperature inside the LC particles is maintained at a level of 320...350°C, although the surrounding temperature is higher and corresponds to the temperature of the beginning of lignin thermodestruction. However, as it has been shown by the experimental data obtained by D.S. Scott and co-authors [13], at a higher temperature (approx. 500° C), in another medium and another fast thermolysis apparatus a considerable splitting of lignin still takes place, and a water-insoluble material (according to the author, "pyrolytic lignin") is formed. The latter tend to complicate considerably the crystalline LG separation process.

The increase of the modulus of heat carrier increases the yield of all the products of thermolysis. This increase compensates the decrease of the reaction time when increasing the modulus of heat carrier in our experiment, which is clearly seen when comparing LG yields at a modulus of 2.5 and 8.5 kg/kg (Table 2). The favourable effect of the decrease of the partial pressure of LG vapors upon its yield manifests itself.

3 Experiments under pilot setup conditions

From the practical point of view, the prospects of using LC, formed during the acid-free production of furfural from oak wood extraction chips for LG production is of interest in some regions. As can be seen from Table 3, in this case, the preliminary treatment of the LC with sulphuric acid solution results in a 3-5-fold increase of the LG yield on condition that the surplus of acid is washed off. The surplus of acid (more than 0.1% from the mass of oven dry wood) decreases the yield of LG sharply [6]. It has been established that the yield of LG increases, the ash content of the

Table 3. Yield of products of fast pyrolysis of lignocellulose (LC) formed during acid-free production of furfural

No. of sample	Conditions of LC pretreatment at 20 °C			TT	LG	RS	Other substances	RS/LG	LG/TT
	Concentra-tion of H_2SO_4 (%)	Time (h)	Washing						
1	-	-	-	13.8; 31.6	4.8; 11.0	6.0; 13.7	3.0; 6.9	1.25	0.35
2	1	2	up to pH 2.0	33.1; 75.8	16.0; 37.7	5.1; 12.3	12.0; 27.5	0.28	0.41
3	1	2	-	12.4; 28.4	3.9; 9.1	5.0; 1.8	3.5; 8.0	1.74	0.31
4	1	15min	-	8.9; 20.4	1.8; 4.2	2.9; 6.8	4.2; 9.6	1.61	0.20
5	5	2	up to pH 2.0	33.7; 77.2	20.4; 48.3	7.4; 17.5	5.9; 13.5	0.36	0.61

* TT - thermolysis tar, LG - levoglucosan, RS - reducing substances
** Here and in Table 4 the first cipher denotes the yield from the mass of oven dry lignocellulose (percentage); the second cipher denotes the yield from cellulose mass (percentage) containing in LC.

542

Table 4. Yield of products of fast pyrolysis of lignocellulose (LC) in a pilot scale unit

Modulus of heat carrier (kg/kg)	Temperature of thermoreactor (°C)	Moisture content of LC (%)	Supply of oven dry LC (kg/h)	TT	LG	RS	Other substances	RS/LG	LG/TT
				LC of corn-cob stems					
2	340	27.6	33.5	16.1; 29.5	9.1; 16.6	5.9; 10.8	1.1; 2.0	0.70	0.54
3	370	27.2	33.3	23.3; 42.6	15.6; 28.5	5.7; 10.4	2.0; 3.7	0.38	0.66
4	400	27.5	26.0	28.5; 52.1	18.7; 34.2	6.2; 11.3	3.6; 6.6	0.33	0.65
5	410	27.5	27.0	28.0; 51.2	19.2; 35.1	6.0; 11.0	2.8; 5.1	0.32	0.70
7	360	15.5	15.5	35.1; 64.2	21.1; 38.6	8.4; 15.4	5.6; 10.2	0.40	0.60
6.3	460	23.7	23.7	31.0; 56.7	23.3; 42.6	7.3; 13.4	0.4; 0.7	0.31	0.75
				Hardwood LC					
2.9	375	7.3	41.7	28.3; 58.2	20.8; 42.8	4.2; 8.6	3.3; 6.8	0.20	0.73
3.5	380	10.0	63.0	26.9; 55.3	19.2; 39.5	5.5; 11.3	2.2; 4.5	0.29	0.71
3.7	350	27.4	40.4	18.6; 38.3	13.7; 28.2	3.3; 6.8	1.6; 3.3	0.24	0.74
4.7	420	7.7	...	33.8; 69.5	24.3; 50.0	8.2; 16.9	1.3; 2.7	0.34	0.72

* TT - thermolysis tar, LG - levoglucosan, RS - reducing substances

cellulose-containing material being decreased. We observed also some increase in the levoglucosan yield owing to the decrease in the ash content in LC (Table 3, samples 2 and 5) - from 1.11 to 0.90% from the mass of oven dry LC. The decrease in the ash content is achieved by increasing the concentration of the sulphuric acid solution used to wash the LC from 1 up to 5% during the preliminary treatment. The concentration of sulphuric acid in LC after the washing with tape water is equal (about 0.1% from the mass of oven dry LC), since both the samples are washed off up to pH 2.

Corn-cob-stump LC was subjected to thermolysis in a pilot-scale unit with a nominal capacity of 30 kg LC (in terms of oven dry lignocellulose), developed by the Latvian State Institute of Wood Chemistry. As can bee seen from Table 4, with the increase in temperature and the modulus of heat-carrier, the LG yield and its amount in volatile products of cellulose thermodestruction increase (the ratio of reducing substance/LG decrease, while that of levoglucosan/tar increase). A temperature of 400...410°C in the thermoreactor and a modulus of heat carrier of 4...5 kg/kg may be regarded as optimum parameters. The same modulus of heat carrier can be recommended also for the fast pyrolysis of hardwood LC. However, in this case, the optimum temperature of thermodestruction should be somewhat lower - 370...380°C. We observed no decrease in the LG yield during the increase of the moisture level of the initial LC, observed by A.V.Prokhorov and co-authors [6] during the vacuum-thermolysis of chips LC. The use of LC with a moisture content of 25...30% instead of 5...10% required only the increase in the degree of overheating of the heat carrier owing to the additional increase in the consumption of energy for water evaporation.

Table 5. Yield of technical levoglucosan (TLG) from concentrated condensate (moisture content 8 %, LG content 63% from mass of oven dry tar)

Crystallization time (min)	Production of TLG (% from oven dry thermolysis tar)	Amount of LG in TLG (%)		Amount of LG in residual solution (%)	
		from TLG mass	from LG content in concentrated condensate	from mass of oven dry residual tar	from LG content in concentrated condensate
15	29.3	90.7	42.2	51.6	57.8
30	33.6	91.6	48.8	48.6	51.2
45	36.4	92.1	53.3	46.4	46.7
60	37.1	92.0	54.1	46.0	45.9
75	36.2	92.0	52.9	46.6	47.1
90	35.4	92.0	51.5	47.3	48.5
105	38.1	91.9	55.6	45.2	44.4
120	37.5	91.8	54.6	45.8	45.4

Table 6. Results of purification of technical levoglucosan (TLG) by recrystallization from ethanol

Volume fraction of ethanol (%)	Temperature of solution (°C)	Concentration of solution (g/ml)	Amount of solution (ml)	Amount of TLG, passed into solution		Yield*		Crystals of cleaned LG		
				g	% from sample	g	% from TLG mass	Melting temperature (°C)	LG content (%)	Color
90	20	0.0951	300	29.48	98.27	11.77	39.9	174...177	97.3	White
90	30	0.0934	295	28.95	96.50	10.06	34.7	167	94.7	Graywhite
96	30	0.0877	295	27.19	90.63	16.00	58.8	167...169	95.8	White
90	40	0.0928	300	28.77	95.90	12.07	41.9	160	96.0	Browngray
96	40	0.0956	290	29.64	98.80	19.42	65.5	166...169	96.2	Gray

* LG yield is given after single crystallization without the use of residual solution.

4 Purification of levoglucosan

We propose to separate technical LG by the acetone method [15], consuming 2.5 ml acetone per 5 g concentrated thermolysis condensate with a moisture content not exceeding 10%, containing 45...70% LG (from the mass of oven dry tar). The study of the process of LG crystallization from a mixture of concentrated thermolysis condensate and acetone (Table 5) has shown that crystallization takes not more than 2 hours, since 54% LG (from its content in the tar) is separated from the mixture in the form of technical LG already in 60 min. and at a maximum yield of 55.6% in 105 min. The residual tar obtained after the separation of technical LG crystals is a remarkable raw material for the production of rigid polyurethane foams [16].

The acetone method of LG separation is easy and provides the obtaining of technical LG, containing 91...92% of the title substance from the mass of technical LG. However, fire hazard and a modest coefficient of separating the main compound (54...55%) require further work on this problem. To separate dying substances and oligomers, crystals of technical LG, dried at room temperature, are dissolved in ethanol to a 3...8% LG concentration at room temperature. It has been established that the dying components of thermolysis tar, polluting LG, are insoluble in 90...96% ethanol at 15...30°C. As a result, selective dissolution of LG is possible. 13...17 ml ethanol per 1 g technical levoglucosan is required so that LG could be dissolved completely [17].

Table 6 shows the results of the purification of technical LG with a moisture content of 3%, containing 90.9% of the main substance (from oven dry mass of the technical product) by re-crystallizing from ethanol with different concentrations after the extraction at a temperature of 20...40°C. LG solutions, purified by selective extraction, tend to crystallize excellently. Even at rather a high content of admixtures in technical LG, purified LG with a yield of 70...75% from its content in the technical product is obtained, containing 95-96% LG and having a melting temperature of 173...175°C. Later, we have developed a method for the separation and purification of LG by the silylation of thermolysis tar with subsequent distillation under vacuum and hydrolysis [18]. The yield of LG with a melting temperature of 173-175°C comprises 72% from the mass of LG in thermolysis tar.

5 Pathways of levoglucosan use

Due to the presence of 1,6-anhydrorings capable of breaking and three secondary hydroxyl groups, LG can be used for the synthesis of different low- and high-molecular compounds. The following reactions of LG should be mentioned: reactions accompanied by anhydroring breakage with the formation of high-molecular compounds; reactions by hydroxyl groups; reactions of the selective esterification of hydroxyl groups; complexing reactions. LG derivatives, synthesized by hydroxyl groups are used mostly for practical applications.

Esterification reactions made it possible to obtain simple and complex polyesters, polyurethanes and films, glues, plastic foams. It should be mentioned that the introduction of LG fragments (links) into polyesters enables to improve a number of operation characteristics, such as heat stability and rigidity; to extend the source of raw materials for polyatomic alcohols; to reduce the expenditure of food-stuffs to produce, i.e. glycerol and xylitol, applied for the preparation of polyesters and polyurethane [19]. Oligocarbonate methacrylates are characterized by high rates of polymerization under the action of UV-radiation, which allows their use in different photopolymerizing compositions. Polyfunctional polycarbonate methacrylates are of particular interest [20]. Their properties can be changed within a wide range of the oligomer block length. The viscosity of the product can be changed within a wide range depending on the oligomer block length, which is of primary importance for their further processing. The remarkable feature of these oligomers is their capacity of rotating the plane of polarization. The degree of conversion at curing reaches 70-85%. The increase in the oligomer block length results in the fall of glassing temperature from 90 to 20°C owing to the increase in the bridge links O-O-C with a low potential of internal rotation and a decrease in the density of polymer cross-linking.

Distinctive features of the given polymers should be mentioned. They are characterized by an insignificant volume shrinkage and a low index of birefringence. The combination of these properties makes it possible to use these polymers as a component of carriers for multiplying optical discs. Methacrylate and oligocarbonate methacrylates are offered and tested for the synthesis of high-modulus resins for the manufacture of compressing parts. Oligourethanepoxide and oligourethanacrylates are offered as modifiers and cross-linking agents for the production of polyvinylchloride and other synthetic polymers. Therefore, the thermal processing of cellulose-containing materials allows the production of a new monomer, levoglucosan, which can be used successfully as a raw material in polymer chemistry, biology and medicine.

6 References

1. USA Pat. 3,298,928. Esterer, A.K. Pyrolysis of cellulosic materials in concurrent gaseous flow.
2. Plūmiņš, M.A. and Zandersons, J.G. (1984) Production of levoglucosan from fast pyrolysis of lignocellulose. *Khim. Drev.,* No 5. pp. 89-92.
3. Bohn, M.S. and Benham, C.B. (1984) Biomas pyrolysis with an entrained flow reactor. *Ind. Eng. Chem. Process Des. Dev.,* Vol. 23, No 2. pp. 355-363.
4. Domburg, G.E., Skripchenko, T.N., Bērziņa, I.J. and Treimanis, A.P. (1978) The influence of cellulose supermolecular structure on thermal destruction. *Khim. Drev.,* No 5. pp. 6-14.
5. Antal, M.J., Varhegyi, G. (1995) Cellulose pyrolysis kinetics: The current state of knowledge. *Ind. Eng. Chem. Res.,* Vol. 34, No 3. Pp. 703-717.
6. Prokhorov, A.V., Alsups, I.A. and Sergeyeva, V.N. (1976) Preparation of cellolignin for vacuum thermolysis to obtain levoglucosan. *Khim. Drev.,* No 6. pp. 91-95.
7. Vedernikov, N.A., Karliwan, W.P., Trautmane, I.A. (1980) Furfurolherstellung bei der Komplexen Verwertung des Laubholzes. III Internationales Symposium

"Grundlagenforschung zur Komplexen Holznutzung". Vortrage, Bd. I, S. 274-279.

8. Shubeko, P.Z. and Enik, G.I. (1974) Continuous Process of Carbonization, Moscow.

9. Alsups, I.A., Zaķis, G.F., Kļava, Z.Ž. (1974) Determination of cellolignin fitness for obtaining levoglucosan by thermolysis. *Khim. Drev.*, No 1, pp. 110-114.

10. Radlein, D., Grinshpun, A., Piskorz, J., Scott, D.S. (1987) On the presence of anhydro-oligosaccharides in the sirups from the fast pyrolysis of cellulose. *J. Amal. Appl. Pyrolysis*, Vol. 12, pp. 39-49.

11. Arisz, P.W., Lomax, J.A., Boon, J.J. (1990) High-performance liquid chromatography / chemical ionization mass spectrometric analysis of pyrolysates of amylose and cellulose. *Anal. Chem.*, Vol. 62, No 14, pp. 1519-1522.

12. Shafizadeh, F., Furneaux, R.H., Cochran, T.G., Scholl, J.P. and Sakai, Y. (1979) Production of levoglucosan and glucose from pyrolysis of cellulosic materials. *J. Appl. Polymer Sci.*, Vol. 23, No 12. pp. 3525-3539.

13. Scott, D.S., Piskorz, J., Radlein, D. (1993) Yields of chemicals from biomass-based fast pyrolysis oils. *Energy, Biomass, Wastes,* Vol. 16, pp. 797-809.

14. Lede, J., Panagopoulos, J., Huai Zhi Li and Villermaux, J. (1985) Fast pyrolysis of wood: direct measurement and study of ablation rate. *Fuel*, Vol. 64, No 11. pp. 1514-1520.

15. Alsups, I.A., Medne, B.A. and Bērziņa, I.J. (1980) On isolation of levoglucosan from lignocellulose thermolysis products. *Khim. Drev.*, No 4. pp. 88-92.

16. Stirna, U.K., Pernikis, R.J., Alsups, I.A., Lazdiņa, B.O. and Stirna, J.A. (1974) Method for production of rigid polyurethanes. Authorship Certificate 431193 USSR, in *Bulleten' Izobretennij*, No 12, p. 79.

17. Zandersons, J.G., Plūmiņš, M.A., Auniņš, E.A. and Āze, D.F. (1985) Method for purification of levoglucosan. Authorship Certificate 1155604 USSR, in *Bulleten' Izobretennij*, No 18. p. 91.

18. Pernikis, R.J., Zaķis, I.N., Davidovich, J.A., Zandersons, J.G. (1990) Method for isolation of levoglucosan. Autorship Certificate 1574608 USSR, in *Bulleten' Izobretennij*, No 24. p. 102.

19. Pernikis, R.J. (1976) Oligomers and polymers on the basis of sugar anhydrides. Zinatne, Riga.

20. Sivergin, J.M., Pernikis, R.J. and Kirejeva, S.M. (1988) Polycarbonate(meth)-acrylates. Zinatne, Riga.

PETROCHEMICALS FROM BARK BY FLUIDIZED BED PYROLYSIS
Petrochemicals through Pyrolysis

W. KAMINSKY and N. BROLUND
Institute for Technical and Macromolecular Chemistry
University of Hamburg, Germany

Abstract

Coniferous bark, composed of 46,3 wt% carbon, 5,7 wt% hydrogen, 47,2 wt% oxygen and 0,8 wt% nitrogen, was pyrolyzed within a temperature range of 700 to 920 °C. The influence of the process parameters on the product spectrum was investigated. The goal was to receive high amounts of a synthetic gas, aromatics and charcoal. A laboratory size fluidized bed with a capacity of 1 to 5 kg/h was built and used for the continuous process. The main fractions are gases, making up 56 to 62 wt%, and charcoal, affording 17 to 22 wt%.

The gas fraction contains about 49 % CO, 27 % H_2, 15 % CH_4, and 9 % CO_2 (in vol.%) and is by this a valuable synthetic gas. The defunctionalisation of liquid wood components (5 wt%) is high; main products are benzene and naphthalene.

The dependence of the product composition on the temperature is not very strong. The amount of CO_2 decreases with increasing temperature. The charcoal has a good adsorption quality. The investigations have shown that the pyrolysis of bark, under exclusion of air, in a fluidized bed process is able to produce simple and useful products.

1 Introduction

As described earlier [1-4], the original goal of the Hamburg Pyrolysis Process was the production of gas with a high heating value and oil containing mainly benzene, toluene and xylene (BTX aromatics) from the pyrolysis of plastic wastes. In recent years other process variants have been investigated such as the thermal conversion of sewage sludge into petrochemicals [5-7]. Different biomasses have been investigated as feed for the fluidized bed pro-cess because of problems in waste treatment and to find alternative processes.

High amounts of coniferous bark are available from wood

sawing industries and are mainly disposed by composting or
incineration. A fluidized bed process is an alternative
possibility, using bark as a renewable source for petrochemicals. Germany imports high amounts of charcoal, aromatics and tar.

Tar is used to impregnate wood in houses and boats.
Charcoal is used in the metallurgical industry and as domestic coke. Aromatics play an important role in the petrochemical industry.

The Hamburg Pyrolysis Process uses an indirectly heated
fluidized bed process. The reactor is heated by incineration of a part of the produced gas in a cyclon burner or
fire tubes.

Different sizes of pyrolysis labora-tory plants have
been installed in the Institute. These have throughputs of
60 to 3000 g/h; a pilot plant for plastic and biomass with
a throughput of 10 to 40 kg/h for plastics and biomass has
also been built.

Currently, three strategies are being pursued:
1. Low temperature cracking of polymers at 500 to
   650 °C to produce mainly oil, charcoal and less gases
2. Pyrolysis in an inert gas stream at temperatures of
   650 to 800 °C in order to achieve a high content of
   simple products such as monomers
3. Pyrolysis at 600 to 800 °C with a fluidizing gas consisting of liquid-free pyrolysis gases. This yields
   gases of a high heating value, BTX-rich oils and charcoal from biomasses.

The feature of the third strategy will be discussed in
detail for the pyrolysis of bark.

2 Experimental

A 1 to 3 kg/h laboratory scale system at the University of
Hamburg was used in the experiments (Fig. 1).

The fluidized bed reactor with a diameter of 154 mm is
heated indirectly by burning propane in a concentric
combustion chamber. In the starting phase, the reactor is
filled with fine quartz sand of a size of 0,2 to 0,5 mm.
By pyrolysis the sand is exchanged by small particles of
charcoal.

The fluidizing pyrolysis gas is preheated in a copper
coil at temperatures close to 400 °C. Before entering the
fluidized bed, the gas can be further heated to temperatures of 500 °C. The coniferous bark, with a composition
of 46,3 wt% carbon, 5,7 wt% hydrogen, 47,2 wt% oxygene and
0,8 wt% nitrogen, is dried until a water content of 6,1
wt% is reached and then enters the reactor through a screw
system. After reaction and vaporization, the gas passes a
cooling and separation system. After precipitation of fine

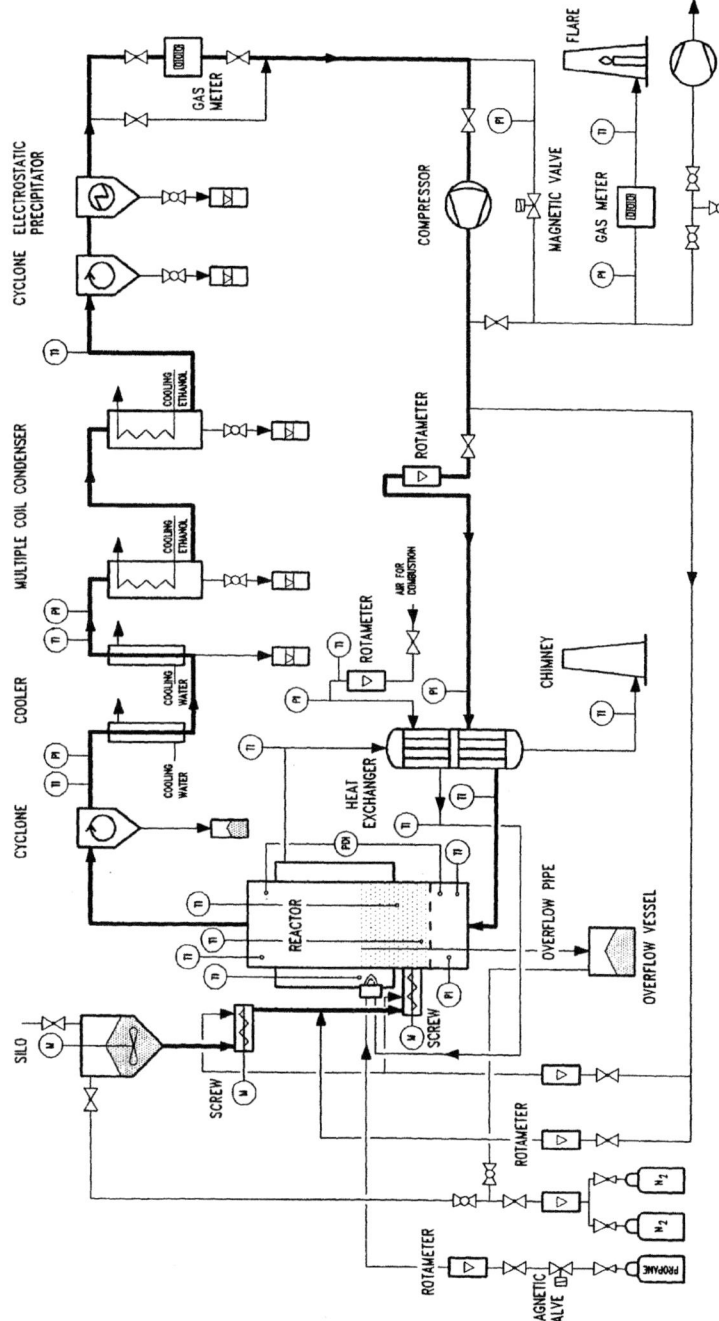

Fig. 1. Flow diagram of the laboratory process for the pyrolysis of bark

551

charcoal in a cyclone, water is condensed in a water cooling system. In the subsequent coolers, the gas is cooled
to temperatures of below -10 to -20 °C in order to condense the liquid products. A second cyclone and a second
electrostatic precipitator removes hydrocarbon oil droplets from the pyrolysis gas. In the laboratory scale system, the pressure is controlled by a compressor and the
gas flow measured by a gas meter and the gas itself burned
in a flame. Some parameters of the experiments are listed
in Table 1.

Table 1. Parameters of the pyrolysis of bark in a
         fluidized bed

Temperature (°C)	700	700	840	875	920
Input (kg)	4,86	7,76	8,97	4,99	6,87
Duration (h)	7,0	1,7	4,2	1,1	2,8
Feed rate (kg/h)	0,69	4,61	2,16	4,68	2,44
Fluidizing gas (m^3/h)	35,2	34,7	42,4	42,7	45,9
Residence time (s)	0,22	0,19	0,20	0,19	0,20
Heating energy (KW)	25	28	43	54	41

The first two experiments show the influence of the
throughput, the others the influence of the temperature.
The bark was collected by a sawing factory, was dried and
cutted into pieces of 0,5 to 5 mm length. The bulk density
of the feed was 0,3 to 0,4 g/cm^3.

3 Results

In all of the experiments, gases, water/oil mixtures, tars
and solid residues were obtained as product fractions.
Samples of the liquid pyrolysis products were fractionated
first by separation in a separating funnel and further by
distillation giving a water fraction and an oil fraction
and a distillation residue. The fractions were analyzed by
GC, GC/MS and elementary analysis.
    The product fractions are summarized in Table 2, the
main fraction being 49 to 64 wt% of gas, followed by 17 to
25 wt% of charcoal and 5 to 10 wt% of tar. The throughput
has a significant influence on the product composition.
Higher throughput gives more tar and charcoal and less
gas. The influence of the temperature on the fraction a-
mounts is not as large. On average, the gas fraction increases and the water amount is lowered. More information
of the composition of the gas gives Table 3.

Table 2. Product fractions of the pyrolysis of bark at different temperatures; [1] low feed rate, [2] high feed rate

Pyrolysis	700[1]	700[2]	840	875	920
Gas	49,4	57,2	64,1	58,9	62,5
Water	16,1	12,3	11,3	16,1	9,3
Tar	9,83	8,76	5,02	7,84	7,31
Charcoal	24,7	21,7	19,5	17,3	20,6

Table 3. Composition of the pyrolysis gas from bark in wt%; [1] low feed rate, [2] high feed rate

Temperature (°C)	700[1]	700[2]	840	875	920
Hydrogen	1,2	1,6	2,4	2,1	2,7
CO	57,1	52,5	63,3	64,1	70,4
CO_2	21,1	25,2	16,0	14,3	10,6
Methane	11,9	10,9	10,9	10,9	10,3
Ethane	1,4	1,2	0,21	0,19	0,09
Ethene	4,3	4,7	3,4	4,2	2,2
Ethine	0,09	0,09	0,52	0,77	0,76
Propene	0,43	0,70	0,05	0,07	0,03
Propadiene	0,01	0,02	0,02	0,02	0,03
Propine	0,02	0,05	0,02	0,03	0,02
Butadiene	0,21	0,34	0,06	0,11	0,04
Acetaldehyde	0,04	0,09	-	-	-
Others	2,2	2,6	3,1	3,2	2,8
Density (g/cm^3)	1,00	0,99	0,88	0,91	0,85
Heating value (Hu, MJ/m^3)	16,7	16,0	15,5	16,0	15,0

With rising temperature, the amount of CO and hydrogen grows while the amount of CO_2 and ethene decreases. Defunctionalization is very high. At a temperature of 800 °C no aldehyds or ketones are found. The methane concentration is constant over the whole temperature range. The heating value of the gas, which can be used to heat up the reactor, is about a third of natural gas.

Table 4 shows the full balance of the experiments. The water fraction contains 0,8 - 1,6 wt% of water soluble compounds (methanol, acetic acid). Main products in the tar are benzene, styrene, indene, and naphthalene. The part of phenols is less than 3 wt%. The analysis of the charcoal shows a fix-C-content of 88 wt% indicating a good quality for metallurgical or domestic use.

For selected compounds a regression calculation was made to find optimal process parameters of the fluidized bed reactor: pyrolysis temperature and residence time in the fluidized bed reactor. The standardized temperature and residence time are:

Table 4. Product composition of the pyrolysis of coniferous bark in a fluidized bed process

Temperature (°C)	700	700	840	875	920
Feed Rate (kg/h)	0,69	4,61	2,16	4,68	2,44
Hydrogen	0,58	0,89	1,5	1,2	1,7
CO	28,2	30,0	40,6	37,7	43,9
CO_2	10,4	14,4	10,3	8,4	6,6
Methane	5,9	6,2	7,0	6,4	6,5
Ethene	2,1	2,7	2,2	2,5	1,4
Other gases	1,3	1,9	0,66	0,79	0,66
Benzene	1,1	1,4	2,1	2,3	1,9
Styrene	0,25	0,33	0,09	0,27	0,08
Indene	0,41	0,51	0,33	0,45	0,21
Naphthalene	0,39	0,54	1,3	1,5	1,4
Acenaphthylen	0,14	0,19	0,35	0,4	0,42
Phenanthrene	0,06	0,11	0,30	0,4	0,39
Other tar components	8,3	6,7	2,5	4,3	4,6
Water	16,1	12,3	11,3	16,1	9,3
Charcoal	24,7	21,7	19,5	17,3	20,6

1) $t = \dfrac{T - 810\ °C}{110\ °C}$

2) $\tau = \dfrac{\tau m - 0,204\ sec}{0,013\ sec}$

The results for the methane concentration in the gas phase are received from:

3) CH_4-concentration =
   $6,371 + 0,314\ t - 0,074\ \tau + 0,201\ t \cdot \tau$

Similarly, the results for the benzene-concentration were calculated as follows:

4) benzene-concentration =
   $1,600 + 0,322\ t - 0,389\ \tau - 0,217\ t \cdot \tau$

The Figures 2 and 3 show the plots of the calculation. The result is that with increasing temperature and residence time the methane-concentration reaches an optimum value. A high concentration of benzene is obtained at a high temperature and a low residence time.

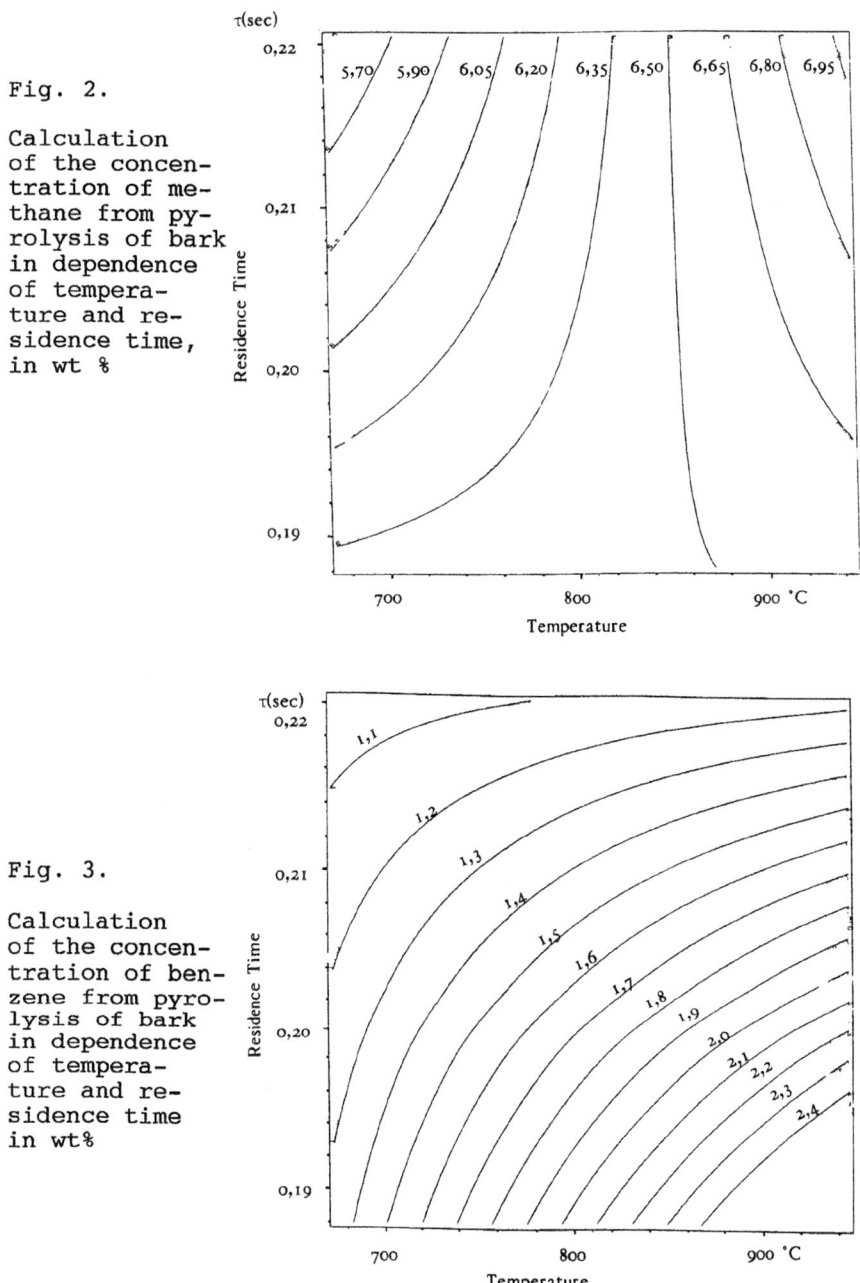

Fig. 2.

Calculation of the concentration of methane from pyrolysis of bark in dependence of temperature and residence time, in wt %

Fig. 3.

Calculation of the concentration of benzene from pyrolysis of bark in dependence of temperature and residence time in wt%

4 Conclusions

Laboratory scale experiments on biomass pyrolysis for the production of gas, tar, and charcoal were performed. Coniferous bark was used as biomass feed.

Pyrolysis was carried out in an indirectly heated fluidized bed process. Under exclusion of air, simple and useful products were obtained t a short residence time of 0,2 seconds.

The gas could be used as synthesis gas or for heating. The tar contains high amounts of small aromatics and naphthalene. The charcoal is similar to that obtained by other processes. In order to cover the needed energy for the fluidized bed reactor, about one half of the gas must be burned.

5 References

1. Kaminsky, W. (1985) Recycling of Polymers, *J. Anal. Appl. Pyrolysis*, No. 8, p. 439.
2. Kaminsky, W., Schlesselmann, B., and Simon, C. (1995) Olefins from Polyolefins and Mixed Plastics by Pyrolysis, *J. Anal. Appl. Pyrolysis*, No. 32, p. 19.
3. Kastner, H., Kaminsky, W. (1995), Recycle Plastics into Feedstocks, *Hydrocarbon Processing*, No. 74, p. 109.
4. Kaminsky, W. (1995) Pyrolysis with Respect to Recycling of Polymers, *Angew. Makromol. Chem.*, No 232, p. 151.
5. Kaminsky, W., and Kummer, A.B. (1989) Fluidized Bed Pyrolysis of Digested Sewage Sludge, *J. Anal. Appl. Pyrolysis*, No. 16, p. 27.
6. Kaminsky, W. (1989) Hamburg Leads Sludge to Oil Conversion, *World Water*, Vol. January, p. 24.
7. Heinrich, R., Kaminsky, W., and Ying, Y. (1993) Chemicals by Biomass Pyrolysis in a Fluidized Bed, in: *Advances in Thermochemical Biomass Conversion*, Bridgwater, A.V. (ed.), Blackie Acad. Professional, Cambridge, p. 1222.

USE OF BIOMASS PYROLYSIS OILS FOR PREPARATION OF MODIFIED PHENOL FORMALDEHYDE RESINS.

STEPHEN S. KELLEY, XIANG-MING WANG, MICHELE D. MYERS, DAVID K. JOHNSON, AND JOHN W. SCAHILL

National Renewable Energy Laboratory
1617 Cole Blvd.
Golden, CO 80401
USA

ABSTRACT

Fast pyrolysis can be used to convert a wide variety of biomass feedstocks into a liquid oil. A phenolic-rich (PN) component can be extracted from this oil and used as a low-cost replacement for petroleum-derived phenol in phenol formaldehyde (PF) resins. The National Renewable Energy Laboratory has developed an extraction process to recover this PN fraction, and it has demonstrated the performance of these PN oils in modified PF wood adhesives. Experience has shown that the performance of PN-PF wood adhesives depends on the chemical features of the PN.

This work investigates how PN's chemical reactivity and molecular architecture affected the properties of the PN-PF resin. Differences in the chemical features of the PN fraction have a distinct impact on the performance of PF resins. Differences in the reactivity and the functionality of the PN, relative to trifunctional, monomeric phenol must be considered during the preparation of PN-PF resins. When 25 weight percent of the monomeric phenol was replaced with PN the performance properties of properly formulated PN/PF resins compared favorably with those of commercial PF resins.

Key Words: Pyrolysis oils, Phenolics and Neutrals, Gel Theory, Wood Bonding

INTRODUCTION

Chemicals from Biomass

Production of value-added chemicals from biomass has long been a goal of scientists and entrepreneurs. Methanol was commercially produced from the slow pyrolysis of oak in the 1920s and 1930s, and more recently rosins and naval stores chemicals have been extracted from pine.[1,2] Since the 1950s there ongoing interest in utilizing waste lignin, produced by pulping processes, as one component in the production of phenol formaldehyde (PF) resins.[3,4] There has also been interest in using tannins and other polyphenolic compounds isolated from the extraction of wood bark in PF resins.[5] Owing to the relatively low cost of petroleum derived phenol in North America attempts at producing biomass-derived phenolics have met with limited commercial success. Recently, however, concerns with environmental sustainability and the long-term supply of petroleum-derived chemicals have renewed interest in the production of chemicals from biomass.[6]

Low molecular weight chemicals can be produced from biomass by fast pyrolysis. This advanced technology converts a solid biomass feedstock to a liquid oil that is a complex mixture of hundreds of individual chemicals. Common classes of chemicals present in pyrolysis oils [7-9] include phenolics; e.g., phenol, guaiacol, syringol and para substituted derivatives, carbohydrate fragments; e.g., levoglucosan, and polyols, organic acids; e.g., acetic acid, aldehydes; e.g., formaldehyde, acetaldehyde, and furfuraldehyde, and a host of oligomeric products. Recovery of pure compounds from this complex mixture is technically feasible but economically unattractive because of the high cost of recovering the chemical and the low yield (commonly below 10 wt%.) of any one specific compound. An alternative to production of pure compounds from this complex mixture is the utilization of classes of compounds in applications that do not require high purity. Classes of compounds can be isolated with common, low-cost fractionation or distillation techniques. One example of this approach is the use of liquid/liquid extraction to isolate a phenolic-rich stream from biomass pyrolysis oils.[10] This phenolic rich stream, or phenolics and neutrals (PN) fraction, which amounts to 18-20 wt% of the dry biomass is useful for the production of PF resins.[11-13]

The PN fraction appears to result primarily from the thermal decomposition of lignin in biomass. The exact chemical structure of the PN depends on the source of the biomass feedstock [11], the conditions used for the fast pyrolysis reactions, and the conditions used for the liquid/liquid fraction process. For example, use of a hardwood feedstock produces a PN with a higher methoxyl content than a PN produced from a softwood. A schematic representation of a native softwood lignin and the associated PN is shown in Figure 1.

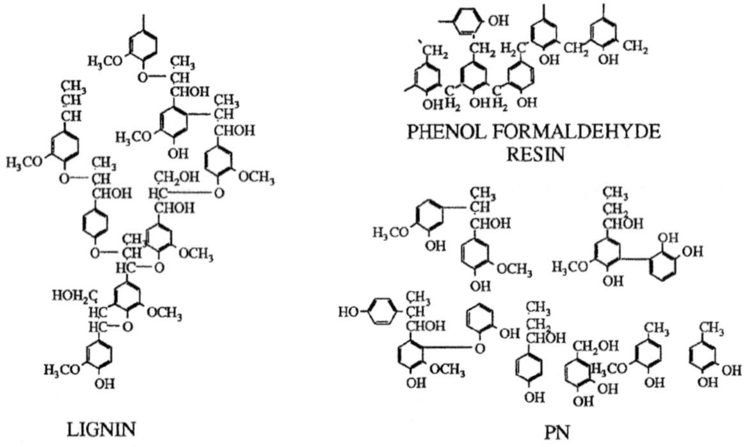

Figure 1. Schematic representations of native softwood lignin, a PN, and a phenol formaldehyde resin.

Addition of Biomass Derived Phenolics to PF Resins.

Because lignin and PF resins are structurally similar (Figure 1) there have been numerous attempts to incorporate lignin into PF resins. Although, lignin is frequently used as a filler or extender for PF resins, it has been difficult to develop methods for incorporating lignins into PF resins as a reactive component. In this context a 'reactive component' is both a reagent that is added to the resin during the initial resin synthesis and a component that is chemically incorporated into the resin backbone. Some efforts to incorporate lignin into PF resins have been technically successful but economically unattractive. Recently, modified PF resin containing about 25 wt% of a fractionated lignosulfonate have been commercialized.[14,15] However, most other attempts to incorporate lignin into PF resins have not been successful.

This lack of success in developing lignin-PF resins results from the performance requirements of the PF resins and the chemical features of lignin. Performance requirements for PF resins include, high mechanical strength, controlled viscosity and penetration into the wood substrate, rapid curing at elevated temperature, and reasonable shelf-life. Addition of polymeric lignin can increase the viscosity of the PF resin and lead to premature gelation and incomplete network formation, which in turn lowers the inherent mechanical strength of the lignin-PF resin and limits penetration of the lignin-PF resin into the wood substrate. A low strength wood composite results. Higher viscosity lignin containing resins can also limit diffusion during the curing process and require longer curing times. Finally addition of high molecular weight lignin to PF resins can reduce the ambient temperature shelflife by causing premature gelation of the high molecular weight fraction of the PF resin.

The phenolic-rich (PN) fraction of fast pyrolysis oils can overcome several of the inherent limitations common to lignin-based PF resins. Oligomeric phenolics present in PN [11,16] does not cause the same increase in solution viscosity commonly seen in with the higher molecular weight lignin. The PN is typically more reactive than lignin owing to a higher phenolic hydroxyl content and a lower aromatic methoxyl content (see below). However structural differences between PN and monomeric phenol require that the conditions under which the resins are synthesized be appropriately modified in order to successfully prepare PN-PF resins.

Goal of this Work

Because of the chemical features and functionality of phenol and PN differ significantly, the initial goal of this work was to determine how these differences affect the formation of PN-PF resins. The ultimate goal of this work is to use this information to prepare high-quality PN-PF resins.

EXPERIMENTAL

Fast Pyrolysis Reaction Conditions

NREL has conducted hundreds of fast pyrolysis runs during the past 10 years. These runs have allowed us to establish pyrolysis reaction conditions that are optimal for the generation of pyrolysis oils suitable for the production of chemicals.[17] The pyrolysis oils are then upgraded to yield the PN fraction used for the preparation of PN-PF resins.[10] These conditions are described in detail elsewhere [11,12,17] and are only generally described here.

Vortex Reactor:
- The pyrolysis operating conditions used in this run are listed below:
- Oven-dry, ground softwood,

- Nitrogen carrier gas @ 650°C,
- Vortex reactor wall temperature @ 625°C, and
- Vortex reactor exit gas/vapor temperature equilibrated to 455°C at steady state feed conditions

Condensation Train:
The condensation train reduced the gas/vapor stream to the lowest temperature possible. The initial stage of this temperature reduction occurred in the Venturi scrubber where the gas/vapors enter at 360-375°C and are typically quenched to 85-90°C when passing through the throat area concurrently with recycled pyrolysis oil condensates. Following the Venturi scrubber this stream passed through two conventional shell and tube heat exchangers where the temperature was reduced to -5 to +5°C before it entered the coalescing filters.

Upgrading of the Raw Oil:
For this extraction, one weight equivalent of water was added to the pyrolysis oil, which was then neutralized to a pH of 7. Following neutralization the pyrolysis oil was processed through a three-stage mixer settler extraction system using ethyl acetate as the organic solvent. Yields of PN from the raw oil are typically between about 35 wt% .

Synthesis Conditions

Similar reaction conditions were used for both the monomer reactivity studies and the gelation studies. These reactions are described in more detail elsewhere.[18] The phenolic model compounds studied in the reactivity experiments were phenol (**1**), 2-methoxy-4-methylphenol (**2**), 2-methylphenol (**3**), and 4-methylcatechol (**4**). For the monomer reactivity studies three principle variables were studied: reaction temperature (40°C, 60°C), NaOH to phenol (NaOH/P) mole ratio (0.05, 0.25), and formaldehyde to phenol (F/P) mole ratio (1.20, 1.80). The solvent was a 3:1 (wt./wt.) mixture of methanol and water.

Monomeric model compounds used in the gelation studies were o-cresol (2-methylphenol) (**2**), phenol (**1**), bisphenol A (4,4'-isopropylidene diphenol) (**5**), trisphenol (1,1,1-tris(4-hydroxyphenyl)ethane) (**6**), which have 2, 3, 4, and 6, respectively, reactive sites on each molecule. (Potential reactive sites for all of the phenolic model compounds are denoted with asterisks.)

Other reaction variables included several ratios of reactive site ratios of formaldehyde to monomeric model compounds (fF/fP=0.60, 0.80, 1.00, 1.20, 1.40). Reaction temperature of 80°C and the ratio of sodium hydroxide to phenolic hydroxyl groups (fNaOH/fP=0.083), were held constant.[1]

The monomeric model compounds were reacted in a 500-ml three-neck round-bottomed flask equipped with a reflux condenser, a thermometer, and a mechanical stirrer that was used in all the experiments. Temperature was maintained at ±1°C. In a typical synthesis the flask was charged with phenol or the phenolic model compound, paraformaldehyde (95% solid content). (For the reactivity studies, 20 wt% methanol was added as a co-solvent and 1,4-dioxane was added as an internal standard for the carbon nuclear magnetic spectroscopy (^{13}C NMR) experiments.) A 50 wt% aqueous NaOH solution and the required amount of water were then added with stirring. The solids content for the reactivity experiments and gelation studies were 20 wt% and 40 wt%, respectively.

For the reactivity experiments the first sample was taken when the solution was homogeneous, but prior to heating. The reaction mixture was then heated to the desired temperature, and additional samples were taken every half hour for the first two hours and then every hour for the next six hours. To prevent any further chemical reaction prior to analysis, all the samples were frozen immediately after removal from the reactor. Samples for the gelation experiments were taken every 30 minutes until a viscosity rise was detected and then samples were taken every 10-15 minutes.

Preparation of PN-PF resins was conducted using a modified PF resin preparation [19] replacing 25 wt% of the monomeric phenol with PN. Reagents were charged to a 500-ml three-neck round-bottomed flask equipped with a reflux condenser, a thermometer, and a mechanical stirrer. The mixture was heated to 60°C for varying lengths of time, from 60 to 120 minutes, and then heated to reflux until the solution viscosity was equivalent to stages to W-X as measured with Gardner-Holt viscosity tubes. (This equates to a solution viscosity of 1,000 to 1,200 cP.) The resin was then quenched and stored in a refrigerator until needed for the bonding experiments.

NMR measurements;

The reaction of the phenolic model compounds was followed with ^{13}C NMR using a Varian Unity 300 NMR spectrometer. Spectral conditions included a 90° tip angle pulse and 1-second recycle delay. The middle peak of the CD3OD was used as a reference (49.0 ppm). The relative concentration of the reactive phenolic models was measured by comparison to an internal standard (1,4-dioxane) that had been added to the resin mixture at the beginning of the reaction. The ratio of peak heights was used to monitor the loss of starting materials and the formation of products.

Gelation measurements:

The condensation reactions between the monomeric model compounds and formaldehyde were monitored by measuring viscosity as a function of reaction time. The gel point was taken as the point at which the reactive solution would not pour. The viscosity of the solution was monitored using Gardner-Holt viscosity tubes. The

---

[1] Sodium hydroxide to model compound ratios were determined by keeping the moles of NaOH and the number of phenolic hydroxyls constant. Formaldehyde to monomeric model compounds reactive site ratios were calculated assuming two reactive sites on formaldehyde and the appropriate number of sites on monomeric model compound, e.g., f=2 for o-cresol, f=3 for phenol, f=4 for bisphenol A, and f=6 for trisphenol.

samples were completely equilibrated at 25°C and the true solution viscosity was measured with a Brookfield viscometer using the appropriate spindle.

Wood Bonding Tests:
The properties of the PF-PN resins were tested according to the British Standard [20] using hard maple wood as the substrate. Between 100-200 mg of PN-PF resin or commercial PF resin were applied to a 25.4 x 25.4 mm (1" x 1") area at one end of each stick. After an open assembly time of 10 minutes at ambient conditions, the sticks were lapped over the length of their coated ends. The over-lapped stick assembly was then hot pressed at 200 psi and 160°C for 3.5 or 6 minutes, using a Carver Laboratory Press, Model-C 12 ton capacity. After bonding, all specimens were reconditioned at 51% relative humidity for at least 24 hours prior to mechanical testing. Mechanical properties were measured with an Instron Testing machine. Maximum tensile load was recorded for each specimen and averaged for the five specimens prepared for each resin-press time combination. A commercial plywood PF resin obtained from Georgia Pacific Resins was used as a standard.

EXPERIMENTAL RESULTS

Chemicals Features of Fast Pyrolysis Oils

Several chemical features of pyrolysis oils and PN permit the successful incorporation of biomass-derived phenolics into PF resins. These features include the number of free phenolic hydroxyl groups, the number of open reactive sites ortho and para to a free phenolic hydroxyl, and the molecular weight distribution of the reactive fragments. In combination these chemical features control the number of reactive sites on an individual molecule and the progression of the condensation reactions that occur between biomass-derived phenolic and formaldehyde during the resin synthesis. Wide variations in these chemical features with varying feedstocks and pyrolysis conditions demonstrate why a single set of PF resin synthesis conditions cannot simply be applied to all PNs.

Preparation of biomass modified PF resins depends on the reactivity and number of sites that can potentially condense with formaldehyde. These two chemical features are dominated by the number of free phenolic hydroxyl groups and the aromatic methoxyl content of the biomass-derived phenolic. These two important chemical features are shown in Figure 2 for phenol, a typical PN, and two types of lignin. This figure clearly demonstrates that PNs more closely resemble monomeric phenol than common lignins. A typical PN has more phenolic hydroxyls and fewer methoxyl groups than common lignins. These features allow for more chemical bonds to form between formaldehyde and the phenolic compounds present in PN, which results in a high quality PN-PF resin.

The number average molecular weight (M_n) of PN is between 250 and 350 [16], much lower than the M_n reported for most organosolv lignins and several orders of magnitude lower than the M_n reported of many kraft lignins [21]. The lower M_n of the PN allows for a gradual increase in molecular weight as the PN is condensed with formaldehyde under reaction conditions commonly used for preparation of PF resins. This gradual increase in molecular weight makes PN-PF resin synthesis easier to control and more reproducible than a comparable lignin-PF resin synthesis.

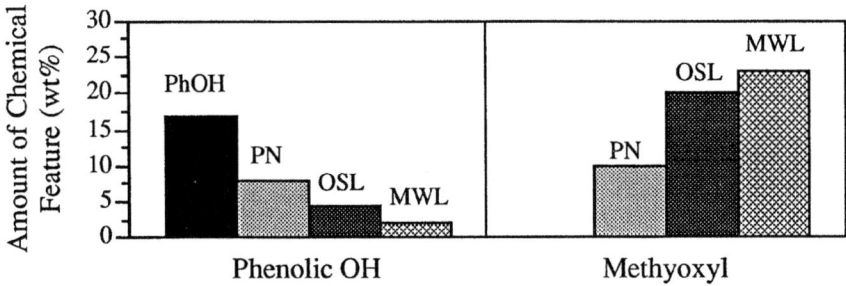

Figure 2. Key chemical features of monomeric phenol (PhOH), a typical PN, organosolv lignin (OSL), and milled wood lignin (MWL).

Reactivity of PN Model Compounds

A series of well defined model compounds were studied to determine if the structures commonly found in PN would react with formaldehyde under the conditions commonly used for preparation of PF resin, and be chemically incorporated into the PF resin network.

For each phenolic model compound, a plot of the logarithm of the product of compound concentration and formaldehyde concentration versus the reaction time shows a linear correlation at each reaction condition. This indicates that the reaction of phenol, or the model compounds with formaldehyde followed second-order reaction kinetics. The overall rate expression is shown in equation 1:

$$kt = (1 / a - b) \ln [b(a - x) / a(b - x)] \qquad (1)$$

where

k	$=$	second-order rate constant (L / mole-sec)
a	$=$	initial concentration of phenol or phenolic model compound (mole / L)
b	$=$	initial concentration of formaldehyde (mole / L)
x	$=$	amount of phenol or phenolic model compound reacted (mole / L) at different time intervals (s)

The effect of reaction temperature and the amount of formaldehyde on the reactivity of 2-methoxy-4-methylphenol is shown in Figure 3. Similar linear plots were derived for 4-methylcatechol, 2-methylphenol, and phenol.

These results, normalized for the reactivity of phenol, are summarized in Table 1. It shows that the reactivity of all model compounds was highest at high temperature ($60^{\circ}C$) and high base ratios (NaOH/P=0.25). Increasing both reaction temperature and amount of base significantly increased the reactivity of phenol and phenolic compounds with formaldehyde; temperature showed the greatest effect. These results are consistent with the kinetic studies by Malhotra et al. [22-24], who studied the reaction of a series of cresols with formaldehyde. He found that cresols followed second-order kinetics under basic conditions, and that the reaction rate increased with increasing temperature and pH.

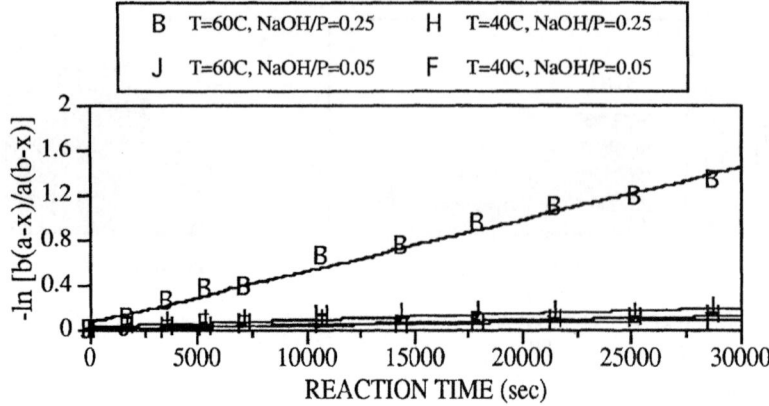

Figure 3. Reaction rates for 2-methoxy-4-methylphenol with formaldehyde at varying temperatures and pH.

These results suggest that the types of functional groups commonly found in PN are more reactive than monomeric phenol. This high reactivity will allow the PN to rapidly react with formaldehyde, under typical PF resin synthesis conditions, and become chemically incorporated into the PF resin network.

Table 1. Relative rate constants for each compound under the different reaction conditions (The relative rate constant for each model compound were obtained by dividing the actual rate constant by the rate constant of phenol for that set of reaction conditions.)

Temp (°C)	NaOH/P	F/P	2-Methoxy-4-methyl phenol	2-Methyl-phenol	4-Methyl-catechol	Phenol
40	0.05	1.80	1.53	1.31	2.02	1.00
40	0.25	1.80	1.34	1.63	4.49	1.00
60	0.05	1.80	2.05	3.43	2.02	1.00
60	0.25	1.80	3.52	2.75	6.04	1.00

Effects of Monomer Functionality

Flory's gel theory [25,26] has been applied to many crosslinked polymer systems, including PF resins.[27] However, to the best of our knowledge this theory has not been applied to phenol formaldehyde systems with more than three reactive sites on the phenolic component. Flory's gelation theory is shown in Equation 2.

$$X_n = \frac{f(1-\rho+1/r)+2\rho}{f(1-\rho+1/r-2p_a)+2\rho} \quad (2)$$

where

X_n	=	number average molecular weight
f	=	average functionality of the system
ρ	=	the ratio of **A** groups attached to branch points to the total number of **A** groups
r	=	the ratio of **A** to **B** groups, i.e., (moles of phenolic i)*(number of reactive sties on i)/(moles of formaldehyde)*2
Pa	=	the extent of reaction of **A** groups (proportional to reaction time)

Application of Flory's theory to PN-PF resins should be useful in predicting the effects of varying the pyrolysis and fractionation conditions to influence the average functionality of the system and the amount of monofunctional and difunctional phenolic monomers.

Using Flory's theory, the predicted effect of increasing the formaldehyde/phenol molar ratio from 0.8 to 1.6 is shown in Figure 4a. In this case increasing the ratio decreases the extent of reaction required for gelation, as shown by the rapid increase in the number average degree of polymerization (X_n). The effect of increasing the average functionality on the X_n is shown in Figure 4b. As the average functionality of the phenolic components increase from 3, to 4, to 6 (e.g., phenol, bisphenol A, and trisphenol), the extent of reaction at which gelation occurs decreases. Finally, the effects of adding difunctional or monofunctional phenolic model compounds to mixture a of tetrafunctional phenolic model compounds, while keeping the overall ratio of phenol and formaldehyde reactive sites constant, are shown in Figure 4c. Addition of monofunctional and difunctional monomers increases the extent of reaction required for gelation.

These predictions were tested for a series of phenolic model compounds. PF type resins were prepared using the different phenolic model compounds with varying numbers of reactive sites, varying amounts of formaldehyde, and varying amounts of difunctional phenolic monomers. The measured gel times are shown in Table 2. (For convenience the reaction time was used instead of the extent of reaction.)

It is obvious that increasing the functionality of the phenolic model compound decreases the reaction time required for gelation. In the case where equimolar amounts of the phenolic model compound and formaldehyde were used, increasing the functionality of the phenolic component from 3.0 to 6.0 decreases the gel time from 207 to 74. Increasing the amount of formaldehyde also decreases the gelation time. For example, in the case of bisphenol A, increasing the F/P ration from 0.60 to 1.40 decreases the gel time from 256 minutes to 82 minutes. As predicted by Flory's equation, the effects of varying the amount of formaldehyde are more important than changes in the average functionality of the phenolic component. The effects of adding difunctional phenolic model compounds is shown in Table 2. The addition of o-cresol to bisphenol A decreases the average functionality of the system (f=2.7), but it also adds difunctional reactive groups that are not capable of contributing to the formation of the network, decreasing ρ to 0.5 and increasing the time required for gelation.

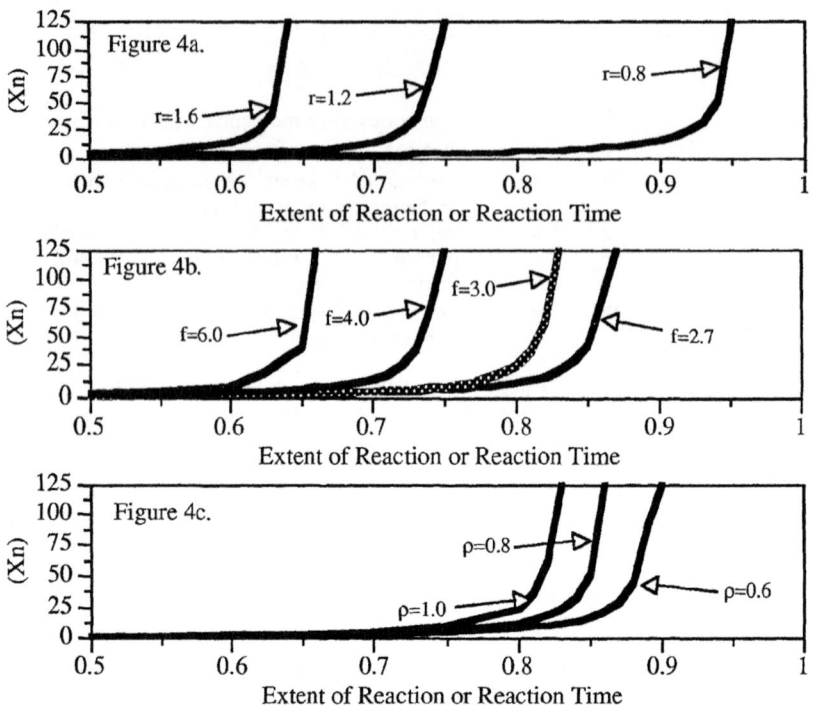

Figure 4. Effect of reaction time on the number average degree of polymerization as a function of f (Figure 4), r (Figure 4b), and ρ (Figure 4c) as predicted by equation 2.

These results suggest that Flory's theory can be used to predict the effects of changes in the overall functionality of the phenolic components of the system, in the ratio of phenolic reactive sites to formaldehyde, and in the presence of difunctional and monofunctional phenolics on the properties of PN-PF resins.

Table 2. Gelation time of phenolic resins with varying monomer functionalities and molar ratios of formaldehyde to phenolic model compound.

Molar Ratio (F/P)	Gelation Time (min.)			
	f = 6.0, ρ=1	f = 4.0, ρ=1	f = 3.0, ρ=1	f = 2.7, ρ=0.5
0.60	129	256	635	1253
0.80	--	185	294	362
1.00	74	130	207	250
1.20	--	101	148	--
1.40	53	82	117	151

Preparation of PN-PF Resins

A series of PN-PF resins were prepared under several reaction conditions which varied the levels of NaOH catalyst, the resin cook times, and the levels formaldehyde added. These PN-PF resins were then used to bond small wood composite samples. The performance of the PN-PF resins is shown in Figure 5.

In general, the mechanical properties of wood composites bonded with PN-PF resins were very similar to those bonded with a commercial plywood resin. At longer press times wood composites bonded with PN-PF resin were not significantly different from those bonded with the commercial PF resin. At the shorter press time wood composites bonded with the PN-PF resin did not perform as well as the samples bonded with the commercial PF resin. These results indicate that the PN-PF resins can be effectively used to bond wood composites, although for this PN feedstock they may be slightly less reactive than commercial PF resins.

Several different reaction conditions were used to prepare the PN-PF resins. Two replicate resins performed in a very similar manner. A PN-PF resin prepared under the same conditions, but cooked for longer times to give a higher solution viscosity, did not perform as well as the standard PN-PF resins. A PN-PF resin prepared with low levels of NaOH also performed very well.

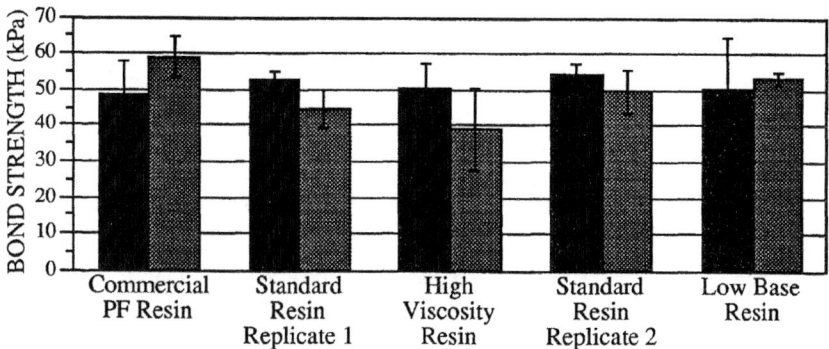

Figure 5. Mechanical properties of wood composites prepared with PN-PF and commercial PF resins. Press times of 6 minutes (solid black) and 3.5 minutes (gray).

DISCUSSIONS

Preparation of high quality, PF wood adhesives requires that the reactive components be properly balanced to insure the formation of a uniform, highly crosslinked network. Improper balance of the reactive components can lead to heterogeneity of the molecular weight distribution and the presence of a high level of unreacted monomer. Both of these features will lower the ultimate mechanical strength of the PF network.

Effects of Reactivity

Reactivity is the first consideration when one tries to use phenolics from a new source to replace phenol used in PF resins. The new phenolics must be as reactive or more reactive than monomeric phenol to insure that they are chemically incorporated into the PF resin network. In comparison, many naturally derived phenolics, such as tannins from wood bark or nut oils, the added phenolics are highly reactive and are chemically incorporated into the PF resin network. In other cases, such as most lignins, the new phenolic is less reactive than phenol and is not chemically incorporated into the PF resin network. If the reactive components are not as reactive as phenol and are not incorporated into the PF network they act more as a mechanical filler rather than as a integral part of the network. While fillers are commonly added to PF resins, they are generally used at low levels and typically lower the mechanical strength of the PF network. Thus, low reactivity replacements for monomeric phenol will lower the overall mechanical properties of the PF network.

Effects of Functionality

Flory's gel theory [25,26] predicts that the increase in molecular weight and the gel point of common condensation reactions will be influenced by both the overall functionality of the system and the ratio of the two reactive species, i.e., phenol and formaldehyde. In the case of a standard PF resin the average functionality of the system is fixed by the ratio of the two monomers. Thus, the molecular weight and gel point can be controlled by fixing the formaldehyde to phenol ratio, as is commonly done in industry. Addition of molecules with functionality greater than three, i.e., bisphenol A, PN, or lignin, to a PF resin will cause the resin to gel prematurely. Premature gelation can cause poor mechanical properties, since mechanical properties are closely tied to the number of cross-link points and the presence of unreacted or "dangling" chain ends. Unreacted chain ends can be caused by an imbalance in the ratio of the reactive components or the presence of monofunctional compounds that cannot propagate the PF condensation reactions.

The mechanical strength of a crosslinked network is controlled by the number of chemical bonds that can support mechanical load and the uniform distribution of these bonds within the network. Unreacted chain ends can produce stress concentrations that can limit the strength of the network. Flory and others [25-27] have shown that careful control over the polymerization conditions is required to produced a gelled system with low levels of unreacted, extractable oligomers. These unreactive oligomers are not chemically bound into the network and thus do not contribute to the ultimate mechanical strength on the network

The average functionality of the polymer system is controlled by the functionality of the individual monomers and the relative amount of the monomers. In the case of PF resins, phenol has three reactive sites while formaldehyde has only two reactive sites. If one maintains a formaldehyde to phenol mole ratio of one (i.e., r=1.5), then increasing the number of reactive sites on the phenolic component decreases the extent of reaction (i.e., the number of reactions) that must take place prior to gelation. For example, if the mole ratio of the two reactive components is held constant then increasing the number of reactive sites on the phenolic component from two (o-cresol or p-cresol), to three (phenol), to four (catechol or bisphenol), to six (1,1,1-tris(4-hydroxyphenyl)ethane, PN or lignin fragments) decreases the extent of reaction required for gelation. This relationship was predicted in equation 2 and Figure 4 and validated for a series phenolic model compounds shown in Table 2. An additional important implication is that the amount of unreactive monomeric compounds present in the resin at the gel point increases as the functionality increases.

PN-PF Resin Performance

In commercial wood adhesives the ratio of the two monomers, generally expressed as the F/P ratio, varies between 1.0 and 1.6. Although the F/P ratio may be as low as 0.75 and gelation will still occur. Low F/P ratios will produce weak resins with few crosslinking points and poor mechanical properties. Simply applying these commercial formulations to PN-PF resins will result in several undesirable changes in properties of the resin.

The chemical features of PN will affect all three of the variables in Flory's gelation equation. The average functionality of the phenolic system, the ratio of reactive phenolic sites to formaldehyde sites and the presence of difunctional and monofunctional phenolics can all be changed by the addition of PN. Perhaps the easiest example is the preparation of a PN-PF resin that maintains the same weight ratio of "phenol" and formaldehyde, and then replaces some of the phenol with PN. This change will effect the network formation process in three ways.

The first effect is on the ratio of phenolic reactive sites to formaldehyde reactive sites; r in equation 2. As shown in Figure 2, on a weight basis PNs have more reactive sites than lignins, but fewer reactive sites than monomeric phenol. So simply replacing phenol with PN will give a F/P ratio that is higher than anticipated. This higher F/P ratio will result in premature gelation of the PN-PF resin and will not allow for the formation of a high-quality network.

The second effect of using PN as a replacement for monomeric phenol is changes in the average functionality of the phenolic components; f in equation 2. On a weight basis the PN has fewer reactive sites than phenol, i.e., one gram of PN has fewer total reactive sites than one gram of phenol. However, on a molecular basis PN has more reactive sites than phenol. This is due to the higher average molecular weight of the PN, which suggests that an "average" PN molecule may have 3-4 phenolic rings that are all linked through covalent chemical bonds. This means that the "average" PN may be similar to trisphenol with a few added methoxyl groups. For this reason the simple replacement of phenol, functionality of three, with PN, functionality of three or more, will also reduce the extent of reaction needed for gelation.

The third effect of replacing phenol with PN, while apparently maintaining the same F/P ratio, is the effects of difunctional and monofunctional species; ρ in equation 2. Addition of difunctional and monofunctional species uses formaldehyde without contributing to the formation of a network. Monofunctional species are particularly troublesome since they lead to dangling chain ends that do not increase the mechanical integrity of the network. A modest number of difunctional species, (10 wt% to 20 wt%) simply increase the molecular weight between branch points and do not significantly decrease the mechanical properties of the network. Unlike the effects of increasing f and r which leads to premature gelation, the addition of difunctional and monofunctional species increases the extent of reaction needed for gelation.

It is clear that the recipe used for preparation of a PN-PF resin must be modified for the PN of interest. Simply replacing monomeric phenol with PN and cooking the mixture in the standard fashion is not likely to result in a high quality PN-PF resin.

CONCLUSIONS

The goals of this work were to develop an understanding of how the chemical features and functionality of PN and monomeric phenol differ, and to use this understanding to prepare high quality PN-PF resins. These goals have been met.

Chemical analysis of PNs produced from a large number of pyrolysis oils has shown that PNs have fewer free phenolic hydroxyl groups and more methoxyl groups than monomeric phenol, but more free phenolic hydroxyl groups and fewer methoxyl groups than lignins. The molecular weight of PNs is significantly higher than that of monomeric phenol but lower than that of lignin. At the current time it is not possible to completely analyze the properties of PN in such a way as to precisely predict the distribution of functionalities.

Studies on the reactivity of PN model compounds with formaldehyde indicate that the chemical structures commonly found in PNs are more reactive than monomeric phenol under the conditions used for production of commercial PF resins. These studies show that the typical PN structures will be rapidly incorporated in the PF resin network.

Flory's gelation theory was used to predict the effects of changing the average functionality of the phenolic components, the ratio of phenolics to formaldehyde, and the effects of difunctional and monofunctional species on the gelation properties of PF resins. The theoretical predictions agreed well with the results of phenolic model compound studies.

Finally, using analytical information on the chemical features and functionality of PNs and the theoretical insights gained from Flory's theory, a series of high quality PN-PF resins were prepared. These PN-PF resins then were used to prepare wood composites whose mechanical properties were comparable to those prepared using a commercial PF resin. Currently, additional work is focused on optimizing the performance of PN-PF resins prepared using PNs from a variety of biomass sources.

ACKNOWLEDGMENTS

The support of the United States Department of Energy, Office of Industrial Technologies, Mr. Charlie Russomanno, and Mr. Alan Schroeder are gratefully acknowledged. The support and technical advice of Drs. Helena Chum, Mark Davis, and Mr. James Diebold, and other researchers at the National Renewable Energy Laboratory, is also gratefully acknowledged. Additional support and advice from the member companies of the Pyrolysis Materials Research Consortium is also greatly appreciated.

REFERENCES

1. Buchanan, M.A. (1975) Extraneous Components of Wood, Chapter 7 in *The Chemistry of Wood*, (B.L. Browning ed.) Robert E. Krieger Publishing Company, New York, 313-67.
2. Hillis W.E. (1962) *Wood Extractives and Their Significance to the Pulp and Paper Industry*, Academic Press, New York.
3. Hemingway, R.W., Conner, A.H. and Branham, S.J. (1989) *ACS Symposium Series 385, Adhesives from Renewable Resources*, American Chemical Society, Washington, DC, pp. 13-151.
4. van der Klashorst, G.H. (1983) Lignin Formaldehyde Wood Adhesives, Chapter 6 in *Wood Adhesives, Chemistry and Technology*.
5. Hemingway, R.W., Conner, A.H. and Branham, S.J. (1989) *ACS Symposium Series 385, Adhesives from Renewable Resources*, American Chemical Society, Washington, DC, pp. 155-270.

6. Bozell, J.J., and Landucci, R. (1993) Alternative Feedstocks Program Technical and Economic Assessment, Thermal/Chemical and Bioprocessing Components, U.S. Department of Energy Report.

7. Soltes, E.J. and Milne, T.A. (1988) *ACS Symposium Series 376, Pyrolysis Oils from Biomass, Producing, Analyzing and Upgrading*, American Chemical Society, Washington, DC

8. Bridgwater, A.V. and Grassi, G. (1991) *Biomass Pyrolysis Liquids, Upgrading and Utilisation*, Elsevier Applied Science, London.

9. Bridgwater, A.V. (1994) Advances in Thermochemical Biomass Conversion, Vol. 2, Blackie Academic & Professional, London.

10. Chum, H.L. and Black, S.K. (1990) Process for Fractionating Fast-Pyrolysis Oils, and Products Derived Therefrom, U.S. Patent 4,942,269.

11. Chum, H.L., Diebold, J.P., Scahill, J.W., Johnson, D.J., Black, S.K., Schroeder, H. and Kreibich, R.E. (1989) Biomass Pyrolysis Oil Feedstocks for Phenolic Adhesives, Chapter 11 *ACS Symposium Series 385, Adhesives from Renewable Resources*, (Hemingway, R.W., Conner, A.H. and Branham, S.J. eds.) American Chemical Society, Washington, DC, pp. 135-53.

12. Chum, H.L. and Kreibich, R.E. (1992) Process for Preparing Phenolic Formaldehyde Resole Resin Products Derived From Fractionated Fast-Pyrolysis Oils, U.S. Patent 5,091,499.

13. Chum, H.L., Black, S.K., Diebold, J.P. and Kreibich, R.E. (1993) Resole Resin Products Derived from Fractionated Organic and Aqueous Condensates made by Fast-Pyrolysis of Biomass Materials, U.S. Patent 5,235,021.

14. Doering, G.A., (1992) Lignin Modified Phenol Formaldehyde Resin Giving Adhesive Compositions Useful in Bonding Wood Chips, Veneers and Sheets of Plywood, U.S. Patent 5,202,403.

15. McVay, T., Baxter, G.F. and Dupre, F.C., Jr. (1992) Reactive Phenolic Resin Modifier, Canadian Patent 2,070,500.

16. Johnson, D.K. and Chum, H.L. (1988) Some Aspects of Pyrolysis Oils Characterized by High-Performance Size Exclusion Chromatography, Chapter 15 in *ACS Symposium Series 376, Pyrolysis Oils from Biomass, Producing, Analyzing and Upgrading*, (Soltes, E.J. and Milne, T.A. eds.), American Chemical Society, Washington, DC, pp. 156-66.

17. Diebold, J.P. and Scahill, J.W. (1988) Production of Primary Pyrolysis Oils in a Vortex Reactor, Chapter 4 in *ACS Symposium Series 376, Pyrolysis Oils from Biomass, Producing, Analyzing and Upgrading*, (Soltes, E.J. and Milne, T.A. eds.), American Chemical Society, Washington, DC, pp. 31-40.

18. Wang, X-M, Davis, M.F. and Kelley, S.S. (1996) Kinetic Studies of Phenol and Related Model Compounds by ^{13}C NMR Spectroscopy, submitted to Holzforschung.

19. Kim, M.G., Amos, L.W. and Barnes, E.E. (1990) Study of Reaction Rates and Structures of a Phenol Formaldehyde Resole Resin by Carbon-13 NMR and Gel Permeation Chromatography, Ind. Eng. Chem. Res., Vol. 29, pp. 2032-37.

20. Anonymous, (1992) Adhesives, Phenolic and Aminoplastic, for Load Bearing Timber Structures: Classification and Performance Requirements, European Standard EN 301

21. Goring, D.A.I. (1971) Polymer Properties of Lignin and Lignin Derivatives, Chapter 17 in *Lignins, Occurrence, Formation, Structure and Reactions*

(Sarkanen, K.V. and Ludwig, C.H. eds.), Wiley-Interscience, New York, pp. 695-761.

22. Malhotra, H.C. and Gupta, V.K. (1978) Kinetics of alkali-catalyzed m-cresol-formaldehyde reaction. J. Appl. Polym. Sci. 22 , pp. 343-351.

23. Malhotra, H.C. and Kumar, V. (1979) Kinetics of the alkali-catalyzed o-cresol-formaldehyde reaction. J. Macromol. Sci. A13 (1), pp. 143-152.

24. Malhotra, H.C. and Tyagi, V.P. (1980) Kinetics of alkali-catalyzed 2,5-dimethylphenol-formaldehyde reaction. J. Macromol. Sci. A14 (5), pp. 675-686

25. Flory, P.J. (1953) Principles in Polymer Chemistry, Cornell University Press, pp. 347-56.

26 Flory, P.J. (1941) Molecular Size Distribution in Three Dimensional Polymers. II and III, J. Am. Chem. Soc., Vol. 63, pp. 3083-4002.

27. Uragami, T. and Oiwa, M. (1972) Studies on Formaldehyde Condensation Resins, XV. Gelation Theory for Phenolic Resins, Die Markomol. Chemie., Vol 153, pp. 255-67.

PYROLYSIS

Upgrading pyrolysis products

UPGRADING OVERVIEW

R. E. MAGGI
Université Catholique de Louvain, Louvain-la-Neuve, Belgium
D. C. ELLIOTT
Pacific Northwest National Laboratory, Richland, USA

Abstract
In the context of alternative sources of energy, many routes have been explored for using biomass. Direct combustion remains the most energy efficient use of biomass but liquids, rather than solids or gases, are preferred. Liquids have many advantages: high energy density, easy storage, handling and transportation and flexibility of use. Nevertheless, these liquids present some unwanted characteristics such as high viscosity, acidity, particulates content and chemical instability. Some upgrading is necessary before utilisation, specially for feeding turbines or internal combustion engines. Three different types of upgrading can be envisioned: physical, chemical/catalytic and the recovery of chemicals. Physical methods such as those to improve viscosity, offer potentially low cost steps which can be applied when the oil is used as soon as produced. The more expensive chemical/catalytic processes offer long term stabilisation and a range of improvement extending to high quality products. This paper reviews the main characteristics of the flash pyrolysis oils, their influence during the utilisation step and the possible solution to overcome this situation. The most important upgrading methods, both physical and chemical/catalytic, are summarised.
Keywords: upgrading, fast pyrolysis oils, deoxygenation, filtration, corrosion, bio-crude, viscosity, ash, char content.

1 Introduction

The petroleum crisis and the political instability that petroleum producing countries experienced, as well as the realisation of the limited amount of fossil energy reserve and of the environmental problems linked to their exploitation incited the authorities of developed countries to start scientific programs for the development of renewable energies.

In this frame, biomass holds a place of growing importance because of the socio-economical and environmental benefits its intensive use could bring. The main transformation routes envisaged are thermochemical or biochemical. Gas, liquid or solid can be produced by various more or less complicated multi-steps processes. Solid ligno-cellulosic biomass, especially plants which give a high yield of dry matter, are considered the best candidates for an energetic exploitation. Intensive researches are under way for the development of new prolific crops. On the other hand, a low energy consuming, simple and preferably one step transformation process is desired. Pyrolysis appears to fulfil these requirements.

Pyrolysis is the thermal decomposition of the ligno-cellulosic matter either in the complete absence of oxidising agent, or with a limited supply in order to avoid gasification. The products obtained include gas, liquids and solid char, their relative proportion varying with the pyrolysis method and the reaction conditions. Liquids are particularly interesting as energetic vector because they have, in comparison to gas and solid, a high energy density and offer advantages in transport, storage, flexibility of use and retrofitting. Another important point is the possibility to use these liquids in existing facilities such as boilers, diesel engines or turbines.

Since the early eighties, several groups have developed processes of fast pyrolysis for the production of high yield of liquids from solid biomass or coal. The underlying principle is the necessity that the decomposition reactions be very rapid and, hence, that the heat exchange processes be fast. An extremely quick cooling step preserves the valuable intermediary products before further repolymerisation. High liquid yields from biomass have been obtained using different processes related to fast pyrolysis: fluid bed (University of Waterloo (1) and, at pilot scale at Union Fenosa (2)), entrained bed (Georgia Institute of Technology (3)), ablative pyrolysis (National Renewable Energy Laboratory (NREL) (4) and Aston University (5)), transported bed (Ensyn (6)) and vacuum pyrolysis (Université Laval (7)). They maximise the transformation of wood into a liquid obtaining relatively constant yields of liquid around 65 %wt on a dry basis. These liquids, called fast pyrolysis oils, bio-oils or bio-crudes have the aspect of a tar, are viscous and not completely volatile, present a high oxygen content and are not miscible with fossil fuels.

The end use of fast pyrolysis oil must still be assessed in accordance with their physical and chemical properties. Actually, the high chemical and thermal instability of pyrolysis oils, as well as the presence of char particulates and dissolved water undermine the prospect of their direct energetic utilisation. One has to consider an intermediate stage of upgrading or refining. This upgrading could correspond to a physical treatment in order to improve the physico-chemical properties such as viscosity or water or solid content. The second possibility is a chemical/catalytic treatment in order to totally or partially eliminate the oxygenated functions which are largely responsible for the unwanted properties of bio-oils. Within the chemical treatments, two alternatives may be envisaged: partial upgrading aiming at the chemical stabilisation of the oils improving the storage and handling for the use in internal engines or turbines, or full refining to high grade products such as transportation fuels.

Figure 1 shows the different utilisations of oils which need some upgrading, this upgrading going from simply heating or dilution in a solvent to full deoxygenation involving expensive and complex chemical processes For the production of electricity and heat, which represents a short-term possibility of utilisation, simply physical upgrading or mild catalytic process are enough according to quality requirements. For the production of transportation fuels, a long-term utilisation, high severity chemical/catalytic processes are needed since the oils must be completely deoxygenated. The other possible utilisation is the recovery of special or high value chemicals contained in the oils, an interesting perspective is the recovery of these chemicals prior to further catalytic upgrading. In fact, high value molecules such as hydroxyacetaldehyde and levoglucosan are extremely reactive and cause coke deposition leading to the deactivation of the catalytic system. On the other hand, catalytic processes could be necessary prior to recovery in order to increase the percentage of the molecule in the oils.

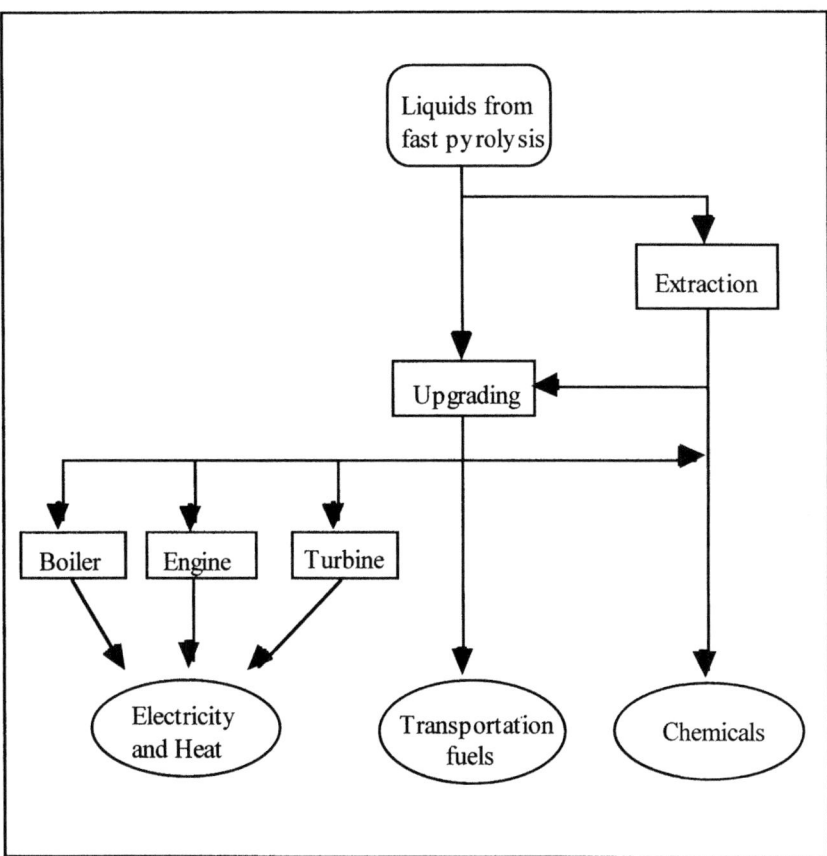

Fig. 1. Different possibilities of utilisation of upgraded oils

This paper reviews the characteristics of fast pyrolysis oils, their influence during storage and utilisation and the main physical and chemical/catalytic upgrading methods.

2 Why upgrade pyrolysis oils ?

All kinds of bio-crude quality improvement appear to be costly and add one or several steps to the primary process. Then, the question mentioned in this section arises naturally. Table 1 (8), comparing the characteristics of typical flash pyrolysis oils produced by Union Fenosa and those of a heavy petroleum cut, answers this question.

Table 1. Typical characteristics of flash pyrolysis oils (Union Fenosa, produced at the Meirama Pilot Plant) compared to a heavy petroleum fuel

Characteristic	Fast pyrolysis oil	Heavy petroleum fuel
Water content (wt%)	15-25 %	0,1 %
Microcarbon (1,6 mm)	0,5-0,8 %	0,01 %
C%	44,14-46,37	85,2
H%	6,60-7,10	11,1
O%	47,03-48,93	1,0
N%	0	0,3
S%	n. d. (*)	2,3
HHV (MJ/kg)	16,5-17,5	40
Density (kg/dm^3)	1,23	0,94
Viscosity (25°C)	400-1200 cp	-
Viscosity (50°C)	55-150 cp	180 cp
pH	2,4	-
Ash	0,01-0,14 %	-

(*) n. d.: non detected, limit of detection 0,05%

Typical characteristics of fast pyrolysis oils are compared to those of a heavy petroleum fuel in Table 1. Fast pyrolysis oils, are viscous, dark and smell tars. They contain a high quantity of oxygen (up to 49%) present as water and organic groups. It ensues that the calorific values of pyrolysis oils are less than a half of those of a petroleum cut. The density of pyrolysis oils is always higher than one which means that the heating values expressed on a volume basis are closer. The viscosities are quite comparable but it should be noted that, for pyrolysis oils, the viscosity depends largely on the water content. On the other hand, an important property of fast pyrolysis oil that was never quantified is their chemical instability. They show a high tendency to polymerise with light, heat or time. The polycondensation reactions become rapid at temperatures higher than 100 °C. This last property is actually the main reason why an intermediate upgrading stage is required before their energetic exploitation.

Table 2 summarises special features and characteristics of bio-crudes requiring some improvement or consideration before any storage or utilisation. Some of these problems originated by the oil characteristics are easily soluble while others are more difficult to solve. The key point is to recognise the problem and to establish the exact specifications and quality requirements.

Table 2. Unwanted characteristics of bio-crudes and methods for modification

Characteristic	Problem encountered	Possible solutions
Low pH	• corrosion	• adequate materials • neutralisation • catalytic upgrading
High viscosity	• difficulty in handling • difficulty in pumping	• addition of water • addition of solvent
Char and solid content	• combustion problems • equipment blockage • erosion	• liquid filtration • hot gas filtration
Alkali metals	• deposition of solids in boilers, engines and turbines	• feedstocks pre-treatment • hot gas filtration • catalytic upgrading
Water content	• complex effect on heating value, viscosity, pH, homogeneity and other characteristics	• problem recognition • optimisation and control of water content according to application
Instability and temperature sensitivity	• storage problems • increase of viscosity • phase separation • decomposition and gum formation on hot surfaces	• avoid contact with hot surfaces addition of water or solvents • stabilisation/refining by catalytic treatment
Incompatibility with fossil fuels		• emulsions • catalytic refining

We can conclude from this table that even if the upgrading processes are costly, complex and add one or several steps to the primary process, they improve the quality of the oils, they add value to the final product and, specially, they are necessary. Sometimes, as in the case of corrosion by acidity, the adaptation of the utilisation, such as the choice of adequate materials, can easily solve the problem. Nevertheless, this solution limits the utilisation of existing facilities or can be too expensive.

3 What kind and degree of upgrading?

It is clear that some quality improvement is necessary and this for any application. On the other hand upgrading processes are expensive and complex. Then, it is extremely important to determine the degree of upgrading in order to choose the most adapted and simple process.

The degree of upgrading is directly linked to the end use of the oil, **the first key point is to properly define the final objective**. According to Figure 1 these final objectives are:

- recovery of chemicals
- fuel for traditional boilers
- fuel for turbines
- fuel for internal engines
- transportation fuels

Concerning the recovery of chemicals the upgrading consists in increasing the concentration of one or more interesting molecules prior to the fractionation step. This kind of upgrading is generally carried out in the pyrolyser before the condensation of the vapour phase which is chemically treated with an adapted catalyst.

For energetic purposes the upgrading processes vary from simple dilution in a solvent and filtration to obtain a fuel for boilers to full deoxygenation to obtain tranportation fuels. Another important point is to determine if the oil will be used as soon as produced or if it will be stored. If it is to be stored, some chemical stabilisation is necessary in order to avoid polymerisation and linked problems such as increase of viscosity, phase separation or formation of gums. After the final objective for the utilisation is defined, **the second point is to properly define the specifications and quality requirements**.

The knowledge of the physico-chemical characteristics and the chemical composition of the oils is also very important prior to any kind of upgrading, specially in the case of chemical/catalytic processes.

Two major types of upgrading can be distinguished: **physical and chemical/catalytic**. Among the **physical methods** we can mention: control of the viscosity by heating, control of the viscosity by addition of water or organic solvents, liquid filtration, hot gas filtration, centrifugation, control of the acidity by steam stripping, by extraction or by neutralisation. Various of these methods are not interesting because they are not applicable in large scale (centrifugation, extractions), produce a negative secondary effect such as loss of volatiles with increasing of viscosity (steam stripping), or cause emmissions during burning (neutralisation). We will further describe two physical methods which appear useful and interesting: addition of water or organic solvents and hot gas filtration.

Concerning **chemical/catalytic methods** leading to the partial or total deoxygenation, we can distinguish two main categories. The first category includes processes at atmospheric pressure, without reducing gases over zeolite catalysts. The second category involves processes at high pressure of hydrogen using typical hydrotreating catalysts. In both categories we can distinguish different processes which will be further described in this paper.

4 Physical upgrading

4.1 Water or solvent addition

The control of the viscosity by addition of water or organic solvents appears to be a very interesting and simple upgrading method since the dilution of the crude oils reduces their viscosity, stops the increase of viscosity in time and facilitates handling in general and filtration in particular.

The addition of water and different organic solvents was considered by Bridgwater et al. (9). They added water to an oil produced by Union Fenosa with an initial water content of 17% in order to obtain total water contents of 20, 25 and 30%. Two main observations were reported: 1) a small amount of water produces a large decrease in oil viscosity (from 1127 cp for the initial oil to 590 cp for the oil containing 20%W), and 2) after 4 months storage there was no increase of viscosity for the oil containing 30%W and very low increase for the oil containing 25%W. They obtained similar results with organic solvents (methanol, ethanol and methanol/water mixture). UCL carried out similar tests (10) using water and solvents such as methanol, ethanol, butanol and octanol, the results being in complete agreement with those obtained in Aston University. At UCL the most relevant result was obtained with an oil produced by Fenosa with addition of 10% of butanol: the viscosity (205 cp) was unchanged after 8 months storage. This phenomenon has not yet been elucidated, a possible explanation could be the colloidal interactions in the mixture. At UCL we also observed that while the addition of organic solvent is very easy, the addition of water presents some inconvenience since bio-oils have limitd solubility in water, the oil will separate into two phases when enough water is added.

This method, easy and cheap (possibility to use bio-ethanol), presents nevertheless some inconvenience: decrease of heating value with the addition of water and increase of volume and weight for transportation and storage. In addition, it does not eliminate the chemical functions causing polymerisation.

4.2 Hot gas filtration

In the pyrolysis process a residual char product is formed. A positive aspect of the char in the pyrolysis oil product is that it increases the overall energy yield and the heat of combustion of the oil. However, there are several negative aspects of having the char in the oil, including sedimentation and sludge formation during storage, increased viscosity, plugging of small orifices, higher alkali content in the char and different burnout rates for the char compared to the oil droplets. Much of the char can be separated in cyclone separators in which centrifugal forces cause the separation.

Important considerations in the cyclone design are that the efficiency of solids removal decreases with decreased solids diameter and increases with the square of the velocity. Overall, there is poor efficiency of separation of particles less than 10 μm so that a small but significant portion of the pyrolysis char is not removed from the pyrolysis oil in cyclone separators.

One option for removing the char is to filter the oil product after condensation. This step forms a cleaner oil but also an oil/char sludge with high viscosity and ash. In addition, this cleanup allows the acids in the condensate to leach alkali from the char into the oil product. Filtration will remove only the solids, and dissolved alkali would remain in the oil product.

Hot vapour filtration is a better option which is under development at NREL (11). In this process the char is filtered prior to condensation so, no leaching of alkali into the oil is possible. All the char is collected as a separate product stream. Submicron filtration is possible with this technology which can be scaled up with a constant volume flow rate per filter area.

In hot vapour filtration a delicate temperature balance must be maintained. The temperature of the vapours must be high enough to prevent condensation which would plug the filter medium. But high temperature also leads to vapour cracking which reduces the oil yield. A fairly small temperature window exists to achieve both goals.

Commercial hot vapour filtration mediums have been developed. Typically they have multiple elements and the filter cake acts as a part of the filter. The thicker the filter cake, the better the filter; however, higher pressure drops also result. Pressure drop varies over the processing cycle between filter cake removals, from 1000 to 5000 Pa. High temperature filter mediums include rigid sintered metal powders, rigid porous ceramic foams and flexible ceramic cloth. The flexible bag elements have been used effectively at NREL. Other work with rigid metal (inconel) elements has not proved as successful. As a result of this work we know that pyrolysis oil is not black or cloudy brown, but is actually a clear reddish brown liquid when free of char particles. Comparative tests with switchgrass feedstock show that the hot vapour filtered (rigid metal) product contained only 100 ppm ash compared to 9500 ppm ash with cyclones only to recover char. With poplar wood feedstock, the hot vapour filtered product contained only 50 ppm ash using flexible ceramic cloth elements.

In conclusion, hot vapour filtration can be used to remove char particles and to lower the ash content of the oil. As a result, the apparent rate of ageing, sludge formation and burn time should all decrease. Concurrently, the value of the oil is increased as well as char recovery. Filtering removes fine particulates 10 to 100 times better than cyclones.

5 Chemical/catalytic upgrading

5.1 Catalytic cracking

The upgrading of pyrolysis oils by catalytic cracking is currently studied by different

European and North-American laboratories: CPERI (Greece), VTT (Finland), IWC (Germany), UCL (Belgium), University of Leeds, Aston University (UK), NREL (USA), University of Saskatchewan (Canada) and University of Waterloo (Canada) among others. This popularity is due to the fact that this process is carried out without reducing gases considered expensive and prior to the condensation of the pyrolysis vapours, this means that no supplementary step is added to the overall process. This process of catalytic cracking became so popular that the IEA (International Energy Agency) within its Bioenergy Agreement, Assessment of Liquefaction and Pyrolysis Systems Activity carried out a techno-economic assessment (12).

This method proposes deoxygenation at atmospheric pressure of pyrolysis liquids without reducing gases through simultaneous dehydration-decarboxylation reactions over acidic zeolite catalysts. This deoxygenation occurs with simultaneous synthesis of gasoline-type compounds. This zeolite catalyst is well-known for the production of gasoline from methanol. Its activity for the deoxygenation of other small oxygenated compounds has also been proved several times. At a typical temperature of 450 °C, oxygen is rejected as H_2O, CO_2 and CO (13). In the case of pyrolysis oil vapours, it is desired that the oxygen is eliminated as CO and CO_2 in order to preserve the hydrogen and maximise the yield of gasoline. Papers published in the scientific literature (14) indicate that the yield in hydrocarbons is low (a maximal theoretical yield of 42%) because of the loss of carbon on the catalyst (up to 15% of coke deposition on the catalyst and supplementary 15% as suspended carbon) and as CO and CO_2 and because of the low conversion of the phenolic compounds (15). Moreover, the large molecules present in pyrolysis vapours have no access to the pores of the catalyst and, consequently, are not selectively converted.

The catalytic reactor (fixed or fluidized bed) has to be placed just at the exit of the pyrolysis reactor to avoid the condensation of the vapours. On the other hand, as coke forms rapidly on the catalyst (up to 15%), a fluid bed reactor with constant catalyst regeneration is very likely to be required. This kind of reactor must be of large scale because of its investment cost and consequently is likely incompatible with the pyrolysis reactor scale. An additional problem arising with pyrolysis liquids is the alkali metals content, these alkali metals being a strong poison for the acidic zeolites.

Recently this catalytic cracking process has been applied to pyrolysis liquids (16) in order to explore the possibility to feed pyrolysis oils in typical FCC reactors. Results were disappointing since the yield in upgraded liquids, containing high quantity of water, were poor. In addition, authors reported a high coke deposition (15%) and a high char formation (17%).

Summarising, we can draw the following conclusions about catalytic cracking of pyrolysis liquids:
- process is well developed (Mobil)
- low yield (max. 42%)
- high coke deposition (up to 15% on the catalyst and 15% as suspended char)
- continuous catalyst regeneration
- zeolite poisoning by alkali metals

- requires development of new catalysts or modification of standard zeolites
- present economics: not favourable

5.2 Slurry phase reactor

This process is under development in Germany by Veba Oel (17). This upgrading process would be similar to a coke liquefaction process patented by Veba. There is very little information about this process which is carried out without catalyst under high hydrogen pressure (up to 200 bar) and using an additive. No results have been published yet, but Veba representatives have expressed satisfaction with process potential.

5.3 Catalytic hydrotreating

Catalytic hydrotreating of pyrolysis liquids is currently studied by Battelle-Pacific Northwest National Laboratory (PNNL) (USA), Università di Sassari (Italy), Université Catholique de Louuvain (UCL) (Belgium) and Institute of Wood Chemistry (IWC) (Germany). Veba Oel (Germany) stopped all activities in this field during the past year.

This process has been developed taking into account more than 40 years experience of the petroleum refining industry and proposes the deoxygenation of fast pyrolysis oils under high hydrogen pressure using traditional hydrodesulphurisation catalysts (NiMo or CoMo supported on alumina). Under these reaction conditions, oxygen is eliminated as water with simultaneous hydrocracking of large molecules and hydrogenation of double bonds.

In spite of the petroleum industry experience, the first experiences carried out with fast pyrolysis liquids were unsuccessful because the extensive polymerisation of the oil occurring at usual hydrotreating conditions (350-400°C). In order to avoid thermal degradation, Elliott et al. (18) developed a two stage treatment: pre-treatment, currently called stabilisation, at lower temperatures aiming to the elimination of more reactive molecules such as aldehydes, carbonyls, ketones and olefins followed by full refining with elimination of stable oxygenated molecules such as phenols and furans. With such two stage treatment, this process can be easily modulated from simple stabilisation (stopping the reaction at the first stage) to full refining. Later, Churin improved the yields obtained by Elliott using tetralin as hydrogen donor (19). These first experiments performed with real oils were followed by a systematic study using model compounds leading to the elucidation of reaction mechanisms and to the evaluation of the influence of reaction parameters such as total pressure, temperature, competition between molecules and partial pressure of H_2O, H_2S and NH_4 (20-23).

All studies above mentioned were performed using industrial catalysts and, even if the traditional sulphided CoMo or NiMo supported on alumina presented good activity, they were quickly deactivated by coke deposition and this using both the real oils or the model mixture. Therefore, other important modification required compared with the traditional hydrodesulphurisation process is the catalytic system. In fact, petroleum contains less than 5% of heteroatoms (sulphur, nitrogen, metals, some oxygen) to be eliminated while pyrolysis liquids contain more than 50% of heteroatom (oxygen) and, even if it is possible to use the same active phases (Mo, Co, Ni) the catalytic system must be adapted to this high heteroatom content. Centeno et al. demonstrated that the cause of the catalyst deactivation was the acidity of the alumina (24). Results obtained with catalysts supported on neutral supports such as activated carbon are promising concerning the selectivity and the resistance to deactivation but the activity of these catalysts is still poor and must be improved (25).

5.3.1 Full catalytic hydrotreating

This process aims to completely deoxygenate the oils in order to obtain hydrocarbons to be used as transportation fuels. The technical feasibility of full hydrotreating has been demonstrated at laboratory scale with real oils as well as with model mixtures (26,27). Upgraded oils containing less than 0,5% of oxygen were produced by Veba Oel (28) and UCL (19). The maximal theoretical yield of hydrocarbons is 50% since oxygen (50%) is eliminated as water. However two major inconvenieces must be underlined: the high hydrogen consumption and the production of water which must be treated in order to eliminate pollutants. The hydrogen consumption is estimated between 700 and 800 Nm^3/t, this value being close to the hydrogen consumed for the refining of heaviest petroleum cuts (bottom of the barrel).

The full catalytic hydrotreating is a feasible process to obtain transportation fuel from biomass but it remains still expensive and not competitive with fossil fuels. However it must be kept as a long term opportunity and further laboratory scale research as well as a pilot plant are necessary to improve the overall process. Summarising:

- technically feasible
- max. yield
- high hydrogen consumption (up to 800 Nl/kg)
- not necessary for electricity production
- expensive and not yet competitive with fossil fuel
- must be studied for the production of bio-transportation fuels

5.3.2 Mild catalytic hydrotreating

As it was mentioned above catalytic hydrotreating can be easily modulated from partial to complete deoxygenation. Mild hydrotreating, also refereed as stabilisation, appears as a very interesting process producing partially deoxygenated fuels which can be used for the generation of electricity and heat. If a stabilised oil with a partial quality improvement is enough for the feeding of boilers, engines or turbines, the severity of the treatment as well as the hydrogen consumption can be drastically decreased.

This process is under development in several laboratories using real oils and model compounds. Preliminary results are very promising: Conti successfully treated an oil produced by Union Fenosa obtaining 60% of deoxygenation (29), Elliott (30) reported in the present conference promising results obtained with an oil produced by NREL which has been filtered prior to condensation (process described in section 4.2), Laurent (31) obtained from 60 to 70% of deoxygenation and reported hydrogen consumptions from 38 to 163 Nl/kg, Meier recently reported results obtained with extremely mild hydrotreating using Pd catalysts and hydrogen partial pressures as low as 2-3 bar (32).

Using this process the hydrogen consumption is limited because only hydrogenation of double bonds and elimination of more unstable "oxygen" (carbonyl and carboxyls) are performed. The catalytic system is a key point in this process since the reactions are extremely selective, for ex. the carboxylic group which is extremely reactive can be stabilised without elimination of oxygen by the formation of the intermediary alcohol by simple hydrogenation avoiding further dehydration. This limits the hydrogen consumption and in addition, alcohols are good fuels and decrease the viscosity of the oils. New catalysts supported on carbon seem to play an important role in the development of this process since they avoid or limit the coke deposition and are very selective, their activity must still be improved.

This stabilisation process appears as a very interesting short term opportunity for the production of fuels with a degree of quality equivalent to heavy petroleum fuels. Summarising the mild hydrotreating process, we can draw the next conclusions:

- technically feasible
- only hydrogenation or elimination of "most reactive oxygen"
- very selective reaction
- elimination of molecules causing instability, then long term solution for storage
- important decrease of hydrogen consumption and severity compared to full hydrotreating
- the catalytic system plays a key role

6 Conclusions

Liquids produced by flash pyrolysis present a high potential as fuels, specially and in the short and medium term for the production of electricity and heat. These liquids are intended to be used as such but some quality improvements are necessary in order to overcome unwanted characteristics. In this context the upgrading methods appear as a key step for the utilisation of these bio-fuels.

On the other hand all upgrading methods are expensive and add one or several steps to the primary process. It is then essential to have exact specifications and quality requirements in order to choose the most adapted and, of course, least expensive process. At the present time, a large range of upgrading processes are available: from simple dilution to full deoxygenation. However, the different state of development between pyrolysis processes (demonstration/commercial scale) and all upgrading processes (laboratory/small bench scale) must be underlined. It is extremely important to continue research and further development of these upgrading methods, both physical and chemical.

7 References

1. Scott D. S. and Piskorz J. (1982), *Can. J. Chem. Eng.*, 60, p. 666.
2. Medina E. and Cuevas A. (1991), in "6th EC Conference on Biomass for Energy, Industry and Environment", (ed. G. Grassi et al.), pp. 1200-1205.
3. Knight, J. A., Gorton, C. W., Kovac R. J. and Newman, C. J. (1985), in "1985 Biomass Thermochemical Conversion Contractors Meeting.
4. Diebold, J. and Scahill, J. (1987), in "ACS Symposium on Production, Analysis and Upgrading of Pyrolysis Oils from Biomass", p. 21.
5. Bridgwater, A. V. and Peacocke, G. C. (1994), in proceedings "Biomass Pyrolysis Oil Properties and Combustion Meeting, available from NTIS, US Department of Commerce, p. 109.
6. Graham, R., Freel, B., Huffman, D. and Bergougnou, M. (1992), in AITBC, (ed. A. V. Bridgwater, p. 1275.
7. Roy, C., de Caumia, B., Brouillard D. and Ménard, H. (1985), in "Fundamentals of Thermochemical Biomass Conversion", (ed. Overend, R. P., Milne, T. A. and Mudge, L.K.), Elsevier, New York, p. 237.
8. Maggi, R. (1994), in 1st Progress Report Contract AIR-CT93-1086
9. Bridgwater, A. V. et al. (1995), in 4th Progress Report Contract AIR-CT92-0216
10. Maggi, R. et al. (1996), in 1st Progress Report Contract JOR3-CT95-0025 (in preparation).
11. Diebold, J. (1996), in "PyNE, Minutes of the second meeting".
12. Solantausta, Y., Diebold, J., Elliott, D., Bridgwater, T. and Beckman, D. (1994), in Assessment of liquefaction and pyrolysis systems, (ed. Technical Research Center of Finland (VTT)), p. 44.
13. Chang, C. and Silvestri, A. (1977), J. of Catalysis, 47, p. 249.
14. Diebold, J. (1996), in "PyNE, Minutes of the second meeting".
15. Renaud, M., Grandmaison, J., Roy, C. and Kaliaguine, S. (1988), in "Pyrolysis Oils from Biomass: Producing, Analysing and Upgrading", (ed. Soltes, E. and Milne, T.), p. 290.
16. Samolada, M. and Vasalos, J. (1996), in the present DITBC conference.
17. Baldauf, W. (1996), Private report.
18. Elliott, D. and Baker, E. (1983), in "Energy from Biomass and Wastes X", (ed. Klass, D. L.), p. 765.
19. Churin, E. et al. (1988), in "Research in Thermochemical Biomass Conversion Conference", (ed. Bridgwater, A. and Kuester, J.), p. 896.
20. Laurent, E., and Delmon, B. (1993), Ind. Eng. Chem. Res., 32, p.2516.
21. Laurent, E., and Delmon, B. (1994), Appl. Catal., 109, p. 77.
22. Laurent, E., and Delmon, B. (1994), Appl. Catal., 109, p. 97.
23. Laurent, E., and Delmon, B. (1994), J. Catal., 146, p. 281.
24. Centeno, A., Laurent, E., and Delmon, B. (1995), J. Catal., 154, p. 288.

25. Laurent, E., Centeno, A., and Delmon, B. (1994), in "Proceedings of the 6th International Symposium of Catalysts Deactivation " (eds. B. Delmon and G. F. Froment), Elsevier, Amsterdam, p. 573.

26. Baker, E. and Elliott, D. (1988), in "Research in Thermochemical Biomass Conversion Conference", (ed. Bridgwater, A. and Kuester, J.), p. 883.

27. Gagnon, J. and Kaliaguine, S. (1988), Ind. Eng. Chem. Res., 27, p. 1783.

28. Baldauf, W. (1994), in Final Report Contract JOUB-CT90-0055.

29. Conti, L. and Scano, G. (1996), in the present DITBC conference.

30. Elliott, D. (1996), in the present DITBC conference.

31. Laurent, E., Maggi, R. and Delmon, B. . (1994), in Final Report Contract JOUB-CT90-0055.

32. Meier, D. (1996), in "9th European Bionergy Conference" held in Copenhagen last June 1996.

BEHAVIOUR OF CATALYSTS SUPPORTED ON CARBON IN HYDRODEOXYGENATION REACTIONS

Hydrodeoxygenation carbon supported catalysts

A. CENTENO, O. DAVID, CH. VANBELLINGHEN, R. MAGGI and B. DELMON
Unité de Catalyse et Chimie des Matériaux Divisés
Université Catholique de Louvain
Louvain-la-Neuve, Belgium

Abstract

Liquids produced by flash pyrolysis of biomass contain a high quantity of oxygenated molecules determining unwanted properties. Oxygen can be eliminated by catalytic hydrotreating. Typical hydrodesulfuration catalysts supported on alumina have shown good activity for the hydrodeoxygenation of model compounds representing the real oils but they are immediately deactivated by coke deposition. This paper deals with the utilisation of a new catalytic system supported on activated carbon. Results are presented comparing two carbon supported catalysts to two other supported on alumina. Activity constants and selectivities have been calculated for each model compound for all catalysts. Reaction pathways and final products are presented for each reaction. The influence of the acidity and the catalytic mechanism are discussed.

Keywords: model compounds, pyrolysis oils, CoMo/C, NiMo/C, CoMo/Al$_2$O$_3$, acidity, reaction pathway, selectivity, activity constant.

1 Introduction

Liquids produced by flash pyrolysis have a high oxygen content (30-40% dry basis), as is reflected by the presence of most of the oxygenated functions causing unwanted characteristics such as high viscosity, thermal instability, corrosivity and low heating value (1). The quality of these flash pyrolysis liquids can be improved by partial or total elimination of the oxygenated functions.

One method proposes the deoxygenation without reducing gases through simultaneous dehydration-decarboxylation of the pyrolysis vapours over zeolites, oxygen being eliminated as CO and CO$_2$. This method is adapted for the conversion of small oxygenated molecules (alcohols, esters, ketones,...) into hydrocarbons (2) but is not very efficient in converting phenols (3).

The other method is hydrotreating under hydrogen pressure in the presence of a catalyst. The reactions occurring are hydrogenation of double bonds, elimination of oxygen as water and hydrogenation-hydrocracking of large molecules (4-6). This process was developed taking into account the experience in petroleum refining and using the same catalytic systems, since the elimination of oxygen occurs through the

589

same mechanism as for the heteroatoms in the petroleum cuts (sulphur, nitrogen). But these studies showed that, even if it is possible to use the same active phases (Mo, Ni and Co), the catalytic system must be adapted to be used with pyrolysis oils, considering that the total amount of heteroatom is much higher in pyrolysis liquids (30-40%) than in petroleum cuts (2-3%).

Another important point is the deactivation of catalysts by coke deposition (7-10). We demonstrated in preliminary studies (11,12) that the alumina support plays a key role in the formation of coke, principally due to molecules containing two oxygenated functions such as guaiacol and catechol, these observations are in agreement with the literature (13-16)

In this case, the use of a neutral support such as activated carbon is necessary. In recent years, the interest in carbon as a catalytic support has been increased. This interest is due to the flexibility of this neutral material as well as the facility to recuperate metals after complete deactivation. Activated carbon is well known as a support of noble metals catalysts for the hydrogenation of double bonds C=C (17, 18) but its utilisation in the petroleum refining is still limited (19). It has also been used as support of Co, Mo, CoMo, NiMo, Fe and other transition metals (20-25). These catalysts show higher resistance to coke deposition than those supported on alumina (20-22, 24-29). In addition several authors have obtained higher hydrodesulphuration activity with carbon than with the traditional alumina (20-22, 24, 25, 27, 28, 30, 31).

In this paper, we present and compare results obtained with CoMo and NiMo supported on carbon and typical CoMo alumina supported catalysts. An important difference in the behaviour of the two catalytic systems is the high selectivity in the production of phenol from guaiacol with carbon supported catalysts, which could be explained by the direct production of phenol avoiding the intermediary product (catechol). In order to study the reaction pathway, we carried out two series of experiments using catechol or guaiacol as reactants. We also performed experiments at high temperatures using p-cresol in order to study the hydrogenolysis of the $C_{aromatic}$-O bond.

2 Experimental

2.1 Catalysts

Four catalysts were used: CoMo/C, NiMo/C and CoMo/γAl$_2$O$_3$-1 prepared by ourselves and an industrial CoMo/γAl$_2$O$_3$-2 (HR-306 from Procatalyse). The precursors were Co(NO$_3$)$_2$.6H$_2$O or Ni(NO$_3$)$_2$.6H$_2$O and (NH$_4$)$_6$Mo$_7$O$_{24}$.4H$_2$O, both from Merck, always added by dry impregnation in order to obtain 15% MoO$_3$ and 3% CoO or NiO in the final catalysts. The support was an activated carbon from Norit RO 0.8 with the following characteristics: specific area 1300 m^2/g, total pore volume 0.78 cm^3/g, micropore volume 0.50 cm^3/g and final particle size 0.3-0.5 mm. For the alumina supported catalyst prepared by ourselves, the support was a γAl$_2$O$_3$ from Procatalyse with the following characteristics: specific area 240 m^2/g, total pore volume 0.68 cm^3/g and final particle size 0.3-0.5 mm. All laboratory made catalysts were prepared using the standard procedure (11). All catalysts were activated by reduction-sulphidation procedure (7,11).

The total acidity of the final sulphided catalysts was measured by TPD of ammonia. The procedure can be found elsewhere (11).

2.2 Catalytic tests

The catalytic tests were carried out using model compounds in a 570 ml stirred batch reactor at 280°C and 340°C and 70 bar. The volume of the liquid feedstock was 170 ml. Details of the reaction can be found elsewhere (11).

The model molecules were chosen on the basis of an in-depth chemical characterisation of the oils (1). Table 1 shows the different model compounds, their concentration, the type and weight of catalyst and the reaction temperature. The solvent was always p-xylene except in the tests where the model molecule was methylphenol (n-dodecane), the internal standard was always n-pentadecane. CS_2 (0.018 mol/liter), which decomposes in H_2S under the reaction conditions, was added in order to keep the sulphided state of the catalysts. Table 2 shows the composition of the model compounds mixture.

Table 1 Reactants and reaction conditions for different catalytic tests

Catalyst	Weight (g)	T (°C)	Reactant	Concentration (mol/liter)
CoMo/C	3.0	280	Methoxyphenol	0.237
CoMo/C	1.5	280	Hydroxyphenol	0.076
CoMo/C	1.5	280	Phenol	0.089
CoMo/Al-1	1.5	280	Methoxyphenol	0.237
CoMo/Al-1	1.5	280	Hydroxyphenol	0.076
CoMo/C	1.2	280	Mixture	table 2
CoMo/Al-2	1.2	280	Mixture	table 2
NiMo/C	1.2	280	Mixture	table 2
CoMo/C	1.5	340	Methylphenol	0.272
CoMo/Al-1	1.5	340	Methylphenol	0.272

Table 2 Composition of the model compounds mixture

Compound	Molecule	% Weight	Mol/liter
4-MA	4-methyl acetophenone	3	0.218
EDEC	ethyl decanoate	3	0.147
GUA	2-methoxy phenol	3	0.237
4-MP	4-methyl phenol	3	0.110
Internal standard	pentadecane	2	0.070
Solvent	xylene	-	-

2.3 Analytical follow-up

Liquid samples were taken every hour and analysed in a Chrompack CP-9001 gas chromatograph equipped with a FID detector at 250°C and a 25 m DB-5 capillaryy column. The peak integration was done using the software Maestro 1 from Chrompack. Each peak was then identified by a Finnigan-Mat GC-MS spectrometer.

3 Results

3.1 Expression of results

3.1.1 Catalytic activity

The catalytic activity is defined as the kinetic constant (k) in the pseudo first order reaction. This kinetic equation, which considers the reactant consumption and takes into account the decrease in volume caused by the sampling, has been reported by Gevert et al. (16).

$$- \ln \frac{C_i}{C_0} = k \, W \, f(t/V) \qquad [1]$$

$$f(t/V) = \sum_{i=1}^{n} \frac{t_i - t_{i-1}}{V_{i-1}} \qquad [2]$$

3.1.2 Selectivity

The selectivities are defined on the basis of the reaction pathways elucidated in former works (8,11).

Figure 1 shows the ethyldecanoate (EDEC) hydrodeoxygenation reaction pathway. Two different reactions occur in competition: i) hydrogenation leading to two alcohols which are then dehydrated to give olefins and, later, alkanes by hydrogenation, or ii) direct decarboxylation leading to an alkane. Figure 1 also shows a third reaction, the formation of the corresponding carboxylic acid which follows the same pathway as the ester.

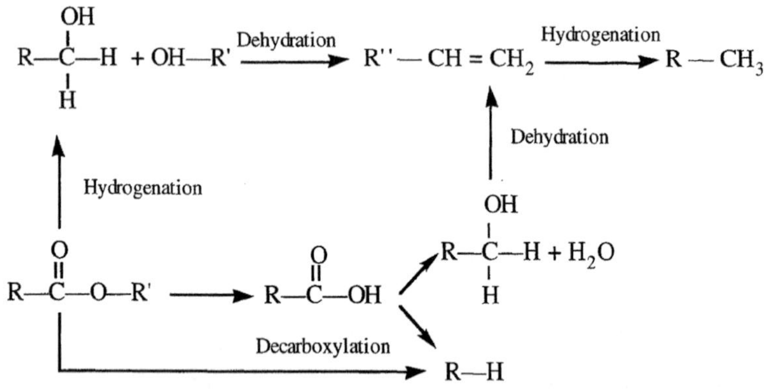

Figure 1. Hydrodeoxygenation reaction pathway for ethyldecanoate (EDEC)

Two selectivities can then be defined from the EDEC mechanism: the decarboxylation selectivity and the dehydration capacity.

The decarboxylation selectivity is defined as the direct carboxyl elimination from the ester as well as from the acid formed during the reaction:

$$S_{dec} = \frac{C_{non}}{C_{dec} + C_{decanol} + C_{decene} + C_{non}} \times 100 \qquad [3]$$

The dehydration capacity is defined as the ratio between the olefins and alkanes produced and all hydrogenation products.

$$C_{dehy} = \frac{C_{decane} + C_{decene}}{C_{decane} + C_{decanol} + C_{decene}} \times 100 \qquad [4]$$

Figure 2 shows that the hydrodeoxygenation of guaiacol occurs through the demethylation of the methoxy group, giving catechol which is then converted into phenol. If the reaction temperature is high enough (340°C), phenol is converted into hydrocarbons. A second possibility could be the direct conversion into phenol. This mechanism has not been elucidated yet and cannot be explained at the present time.

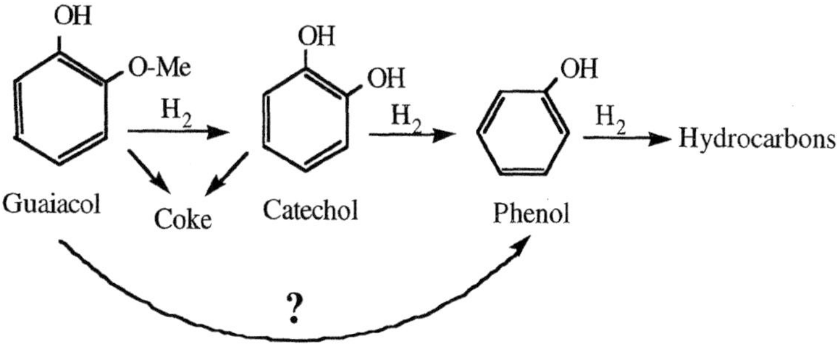

Figure 2. HDO reaction pathway for guaiacol

The phenol/catechol ratio defines the guaiacol hydrodeoxygenation, since the reaction products are catechol and phenol which can be later transformed in cyclohexane and benzene.

$$\frac{phe}{cat} = \frac{C_{phenol} + C_{benzene} + C_{cyclohexane}}{C_{catechol}} \times 100 \qquad [5]$$

Figure 3 presents the two competitive HDO reaction mechanisms for the methylphenol: i) Hydrogenolysis of the C-O bond leading to toluene and, ii) hydrogenation of the aromatic ring giving methyl-cyclohexane as main product and small quantities of methyl-cyclohexene and dimethylpentane.

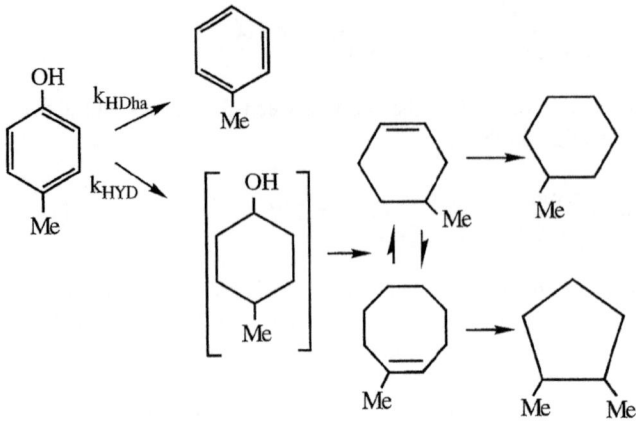

Figure 3. Hydrogenation and hydrogenolysis of p-cresol reaction pathways

The activity constant k is then defined as:

$$k_{p\text{-cresol}} = k_{HDY} + k_{HDha} \qquad [6]$$

The hydrogenolysis selectivity is defined as the ratio between toluene molar concentration (C_{Tol}) and the sum of the concentration of all hydrogenation products ($C_{pro\ hyd}$).

$$S_{HDha} = \frac{k_{HDha}}{k_{HDY}} = \frac{C_{Tol}}{C_{pro\ hyd}} \times 100 \qquad [7]$$

3.2 Reaction products

3.2.1 HDO reaction of 4-methylacetophenone (4-MA)
Methylethylbenzene was the only reaction product, no intermediary products were detected. The conversion of 4-MA reaches 100% after 50 to 90 minutes of reaction, depending on the catalysts.

3.2.2 HDO reaction of ethyldecanoate (EDEC)
The EDEC carboxyl group is more difficult to hydrodeoxygenate than the carbonyl contained in 4-MA. The conversion after 180 minutes varies from 27% using CoMo/C and 67% using CoMo/Al-2. All products presented in figure 1 were detected. Figures 4, 5 and 6 show the distribution of reaction products as a function of the conversion for three different catalysts.

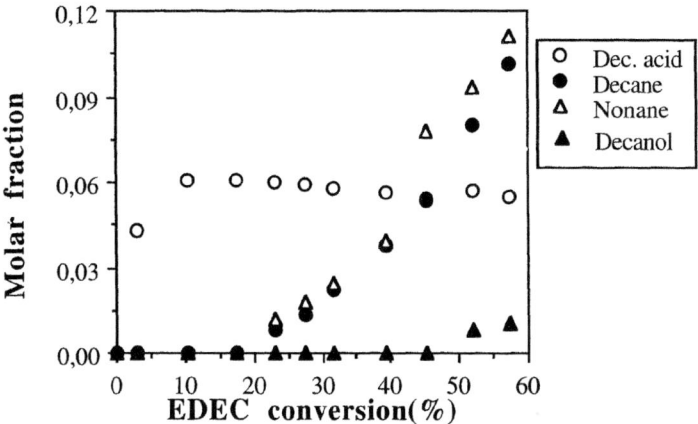

Figure 4. Main products obtained from EDEC using CoMo/Al-2 in the HDO of the mixture at 280°C

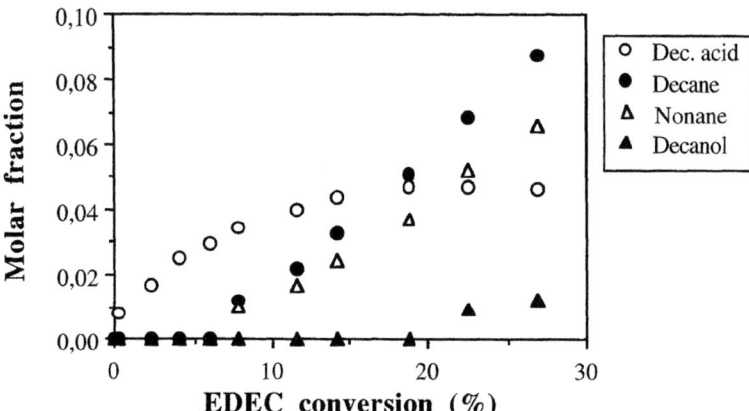

Figure 5. Main products obtained from EDEC using CoMo/C in the HDO of the mixture at 280°C

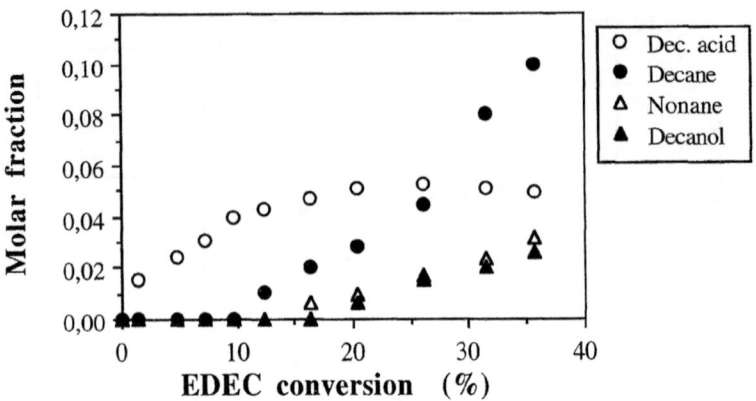

Figure 6. Main products obtained from EDEC using NiMo/C in the HDO of the mixture at 280°C

These figures show that the decanoic acid is the first compound produced, and this for all catalysts. Later, its concentration stabilises and, at the end of the reaction, a decrease is observed.

The production of decane as well as decanol is two times higher for carbon supported catalysts while alumina supported catalysts show a high decarboxylation selectivity.

3.2.3 HDO reaction of guaiacol (GUA)

Carbon supported catalysts present a low activity in the conversion of guaiacol compared to alumina supported catalysts. Nevertheless, they show a very high phenol selectivity with small production of catechol and hydrocarbons.

3.2.4 HDO reaction of p-cresol

P-cresol does not react at 280°C. Toluene, methylcyclohexane and small quantities of secondary products were produced at 340°C.

3.3 The mass balance

Mass balances are other important results. In fact, these balances reach 100% for 4-MA and EDEC and this for all catalysts. Concerning guaiacol, the mass balance is very different for catalysts supported on alumina and for those supported on carbon. Figure 7 shows the guaiacol mass balance as a function of the reaction time for three different catalysts (CoMo/C, NiMo/C and CoMo/Al).

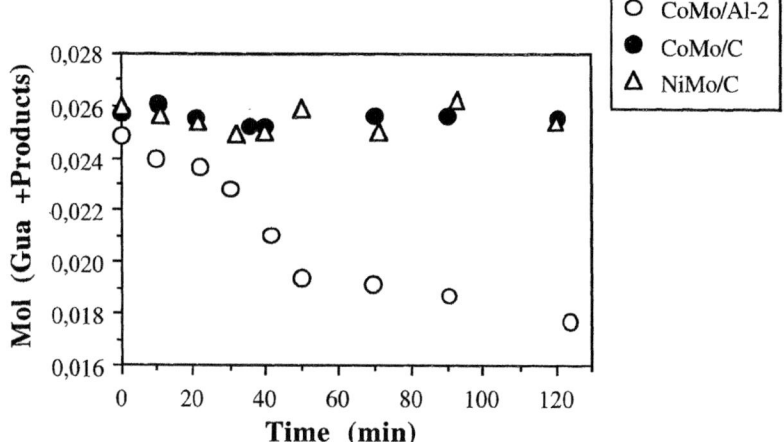

Figure 7. Mass balance as a function of time for guaiacol using three different catalysts

3.4 Acidity of catalysts

Table 3 presents acidity values for different catalysts and the two supports, all in the sulphide state.

Table 3. Acidity of different catalysts and supports

Catalyst	Carbon	g-Al$_2$O$_3$	CoMo/C	NiMo/C	CoMo/Al-2
Acidity (meq/g. cat)	22	359	134	211	522

The acidity of the carbon support is very low, as is shown in the table. The acidity of the catalysts supported on carbon is due to the sulphided phases.

3.5 Guaiacol reaction mechanism

In order to elucidate the guaiacol reaction pathway, we carried out different tests with guaiacol, catechol and phenol as individual reactants at the same reaction pressure and temperature using CoMo/C and CoMo/Al. Figure 8 shows the evolution of the catechol molar fraction as a function of the guaiacol conversion. Different behaviours for the two catalysts can be noted. With CoMo/C, there is no catechol formation at the beginning of the reaction and phenol is detected by GC-MS analysis, while with CoMo/Al catechol is detected in the first sample.

Table 4 presents values of activity constants (k), phenol selectivities (phe/cat) and activity constants for catechol as intermediary product ($k_{gua/cat}$) which have been calculated from figure 8. We can observe that $k_{gua/cat}$ is higher when catechol is the reactant than when it is an intermediary product using alumina supported CoMo. These values are very similar when the reaction is performed with CoMo/C.

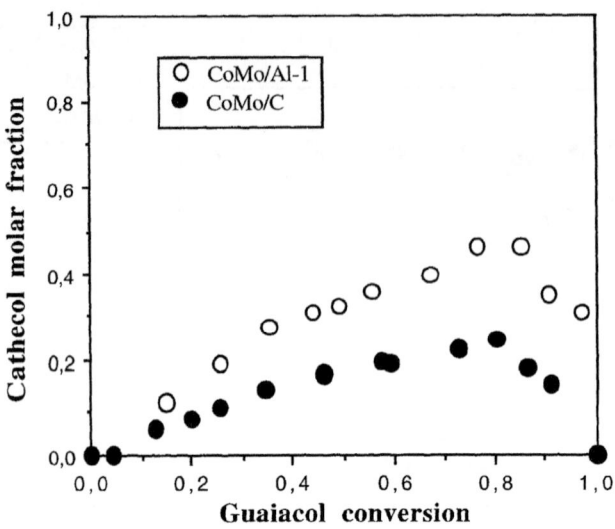

Figure 8. Catechol molar fraction as a function of the guaicol conversion at 280°C

When guaiacol is the reactant, the activity constant k is much higher than $k_{gua/cat}$ using CoMo/Al, while the values are almost identical for CoMo/C.

There is no activity for the conversion of phenol into hydrocarbons at this reaction temperature (280°C).

Table 4.

| Catalyst (weight) | Reactant | Activity | | |
		k_i (*)	$k_{gua/cat}$ (*)	phe/cat
CoMo/Al-1 (1.5 g)	Catechol	3.20		
CoMo/Al-1 (1.5 g)	Guaiacol	1.53	0.76	0.38
CoMo/C (1.5 g)	Catechol	0.53		
CoMo/C (3.0 g)	Guaiacol	0.45	0.4	0.95
CoMo/C (1.5 g)	Phenol	0.05		

(*)cm^3/min. g. cat

3.6 C_{arom}-O bond hydrogenolysis selectivity

Table 5. Activities and selectivities for the conversion of p-cresol

Catalyst	$k_{p-cresol}$ (*)	S_{HDha}	k_{HDha}	k_{HDY}
CoMo/Al-1	3.63	0.66	1.453	2.179
CoMo/C	0.65	0.98	0.32	0.33

(*)cm^3/min. g. cat

Table 5 presents the hydrogenation and hydrogenolysis activity and selectivity values for the conversion of p-cresol calculated from equations [1], [6] and [7]. In order to compare the two catalysts, values were calculated at the same conversion (65%).

Activity constant k is five times higher for CoMo/Al-1, nevertheless C_{arom}-O bond hydrogenolysis selectivity is higher using CoMo/C.

4 Discussion and conclusions

4.1 Catalytic activity and selectivity

In general, the catalytic activity for the hydrodeoxygenation of carbonyl, carboxyl and methoxy groups are higher for CoMo supported on alumina than for CoMo supported on activated carbon. NiMo/C follows the same trend as CoMo/C. This behaviour is due to a better dispersion on the support surface. The good interaction between molybdenum and alumina is well known while the interaction between carbon and molybdenum is very weak. In fact, the carbon surface groups are completely different in quality and quantity from those of alumina. HDO of the hydroxyl group in phenols cannot be performed at a temperature lower than 340°C, and these for all types of catalysts.

Figures 4, 5 and 6 show that the reaction mechanism for the HDO of carboxyl esters (figure 1) is the same for all catalysts. There is first production of the corresponding carboxylic acid and, later, two competitive reactions: decarboxylation or hydrogenation. Table 2 and figures 4, 5 and 6 show that the decarboxylation selectivity, leading to the formation of nonane, is higher for catalysts supported on alumina. In the other hand, table 3 shows that the CoMo/Al acidity is two and four times higher than for NiMo and CoMo supported on carbon. This could be explained by the fact that the decarboxylation reactions take place in acid sites formed by **metal sulphides** because both alumina and carbon (without active phases) do not have any decarboxylation capacity in spite of the difference of acidity (11). NiMo/C and CoMo/C present very different decarboxylation selectivities. This could be explained by the fact that interactions between Ni and Mo are different from those between Co and Mo. Concerning the alcohol dehydration capacity, the dramatic difference between CoMo/C and the other catalysts is also explained by the acidity which catalyses the dehydration reactions.

Concerning the HDO of guaiacol, the results presented in this work are in complete agreement with prior works (11,12): high phe/cat selectivity for catalysts supported on carbon in spite of the low activity. The selectivity values for both catalysts supported on carbon are very similar.

4.2 Guaiacol reaction mechanism

There are three different possible pathways for the hydrodeoxygenation reaction of guaiacol: i) prior demethylation to produce catechol and then phenol and hydrocarbons, ii) a possible direct production of phenol by breaking of the C_{arom}-O bond, and iii) parallel formation of coke. It is evident that the catalytic mechanisms are different for carbon and for alumina, as can be seen in table 4 and figure 8, but is is not yet possible to draw conclusions. One possibility could be that there are two different sites in both catalysts but that the distribution is different on the two surfaces.

Concerning the activity constant for the HDO of guaiacol, there is a great difference between catalysts supported on alumina and those supported on carbon. The k value obtained with CoMo/Al-1 is three times higher than k for CoMo/C. Nevertheless, much more phenol is produced when using the second catalyst as is proved by GC-MS analysis. Another important point is that the mass balance is close to 100% for both carbon supported catalysts while it reaches only 75% for CoMo/Al-1. The 25% difference corresponds to the coke formation.

4.3 C_{arom}-O bond hydrogenolysis selectivity

Results from table 5 show that catalysts supported on carbon have a higher capacity to break the C_{arom}-O bond by hydrogenolysis than those supported on alumina. This higher hydrogenolysis capacity can be compared to the direct breaking of the bond between the aromatic carbon and the methoxy group of guaiacyls, then the catalytic mechanism could be very similar.

5 Acknowledgments

The financial support of EU DG XII under contracts JOUB-CT90-0055 and RENA-CT94-0070 is gratefully acknowledged.

6 References

1. Maggi, R., and Delmon, B. (1994), Fuel, 73, p. 671.
2. Chang, C. D., Silvestri, A. J. (1977), Journal of Catalysis, 47, p. 249.
3. Diebold, J. P., Scahill, J. W.and Evans, R. I. (1985), in the proceedings of "Biomass Thermochemichal Conversion Contractors' Meeting", p. 31.
4. Churin, E., Grange, P., Delmon, B. (1988), in "Research in Thermochemical Biomass Conversion" (eds. Bridgwater, A. V., and Kuester, J. L.), pp. 896.
5. Soltes, E., Lin, S-C K. (1984), in "Progress in Biomass Conversion", Academic press, pp. 1.
6. Baker, E. G., Elliott, D. C. (1988), in "Research in Thermochemical Biomass Conversion" (eds. Bridgwater, A. V., and Kuester, J. L.), pp. 883.
7. Laurent, E., and Delmon, B. (1993), Ind. Eng. Chem. Res., 32, p.2516.
8. Laurent, E., and Delmon, B. (1994), Appl. Catal., 109, p. 77.
9. Laurent, E., and Delmon, B. (1994), Appl. Catal., 109, p. 97.
10. Laurent, E., and Delmon, B. (1994), J. Catal., 146, p. 281.
11. Centeno, A., Laurent, E., and Delmon, B. (1995), J.Catal., 154, p. 288.
12. Laurent, E., Centeno, A., and Delmon, B. (1994), in "Proceeding of the 6th International Symposium of Catalysts Deactivation " (eds. B. Delmon and G. F. Froment), Elsevier, Amsterdam, p. 573.
13. Train, P. M., and Klein, M. T. (1991), Fuel Sci. Tech. Int., 9, p. 193.
14. Petrocelli, P. F.,and Klein, M. T. (1987), Fuel Science & Techology Int., 5, p. 25.
15. Hurff, S., and Klein, M. (1983), Ind. Eng. Chem. Fundam., 22, p. 426.
16. Gevert, B. S., Otterstedt, J. E., and Massoth, F. E. (1987), Appl. Catal., 31, p. 119.
17. Schmitt, L., and Walker, P. L. Jr. (1971), Carbon, 9, p. 791.
18. Schmitt, L., and Walker, P. L. Jr. (1972), Carbon, 10, p. 87.
19. Corbett, R. A. (1989), Oil Gas J., p. 49.

20. Duchet, J. C., van Oers, E. M., de Beer, V. H. J., and Prins, R. (1983), J. Catal., 80, p. 386.
21. De Beer, V. H. L., Derbyshire, J. F., Groot, C. K., Prins, R., Scaroni, A. W., and Solar J. M. (1984),, Fuel, 63 , p. 1095.
22. Vissers,J. P. R., Groot, C. K., Van Oers, E. M., De Beer, V. H. J., and Prins, R. (1984), Bull. Soc. Chim. Belg., 93, p. 813.
23. Ledoux, M. J., Michaux, O., Agostini, G.and Panissod, P. (1986), J. Catal., 102, p. 275.
24. Vissers, J. P. R., Scheffer, B., De Beer, V. H. J., Moulijn, J. A., and Prins, R., (1987), J. Catal., 105, p. 277.
25. Schheffer B., Arnoldy P., and Moulijn J. A. (1988), J. Catal., 112, p. 516.
26. Scaroni, A. W., Jenkings, R. G., and Walker, P. L. (1985), Appl. Catl, 14, p. 173.
27. Vissers, J. P. R., de Beer, V. H. J., and Prins, R. (1987), J. Chem. Soc. Faraday Trans., 1 83, p. 2145.
28. Laine, J., Severino, F., and Labady, M. (1993), J. Catal., 147, p. 355.
29. Boorman, P. M., Kydd, A. R., sorensen, T. S., Chong, K., Lewis, M. J., and Bell, W. S. (1992), Fuel, 71, p. 87.
30. Groot, C. K., de Beer, V. H. J., Prins, R., Stolarski, M., and Niedzwledz, S. W. (1986), Ind. Eng. Chem. Prod. Res. Dev., 25, p. 522.
31. Vissers, J. P. R., Lensing, T. J., De Beer V. H. J., and Prins, R. (1987), Appl. Catal., 30, p. 21.

INFLUENCE OF THE PREPARATION PROCEDURE ON CATALYSTS SUPPORTED ON CARBON IN HYDRODEOXYGENATION REACTIONS

Preparation of hydrodeoxygenation catalysts

A. CENTENO, CH. VANBELLINGHEN, O. DAVID, R. MAGGI and B. DELMON
Unité de Catalyse et Chimie des Matériaux Divisés
Université Catholique de Louvain
Louvain-la-Neuve, Belgium

Abstract

The development of an adequate catalytic system is a key point in the optimisation of the valorisation of flash pyrolysis oils by catalytic hydrotreating. The catalysts preparation procedure and the modification of the support play a decisive role on the reaction selectivity and activity. In this paper we discuss the structural characteristics of the catalytic solids obtained by different procedures and the influence of the preparation method on the catalytic behaviour in hydrodeoxygenation reactions of model compounds. Thermal treatments between impregnations as well as intermediary sulphidation do not modify the textural and structural characteristics of the supports but lead to the formation of more stable active phases which increase the catalytic activity. Interesting results were obtained by changing the impregnation order. Catalysts prepared by cobalt impregnation in first place show a better dispersion and a higher Co/Co+Mo ratio than those obtained by molybdenum impregnation in first place. These characteristics lead to an increase of activity in the elimination of methoxy and carboxyl groups and favour the hydrogenation of carboxyls.

Keywords: hydrodeoxygenation, pyrolysis oils, carboxyl, carbonyl, methoxy, CoMo, sulphidation, thermal treatment, impregnation.

1 Introduction

Catalytic hydrotreating is a very interesting way to eliminate heteroatoms under high hydrogen pressure in the presence of a catalyst. This process, well known in the petroleum industry, was developed to refine petroleum cuts by elimination of sulphur and nitrogen atoms, cracking of large molecules and hydrogenation of aromatics. Elliott and Baker (1-3) demonstrated the technical feasibility of this process for the upgrading of flash pyrolysis oils by elimination of oxygenated functions, Churin (4) improved the yields obtained by Elliott by addition of a hydrogen donor solvent. Later Laurent (5) elucidated reaction pathways and studied the influence of reaction parameters using model compounds representing the most abundant functions contained in the oils.

All these studies were carried out using industrial CoMo/alumina and NiMo/alumina hydrodesulphurization catalysts and, although they presented good activity, they were quickly deactivated by coke deposition.

In our laboratory, Centeno (6-7) has recently demonstrated that the acidity of the alumina support was responsible of the polymerisation reactions leading to the formation of coke on the catalysts surface, the utilisation of a neutral support such as activated carbon should limit or avoid the coke deposition. Nevertheless, although these catalystssupported on carbon present the same activity for the elimination of the carbonyl group as those supported on alumina, the activity for the elimination of other oxygenated groups such as carboxyl and methoxy remains poor (6). An important point is the very high selectivity in the formation of phenol by deoxygenation of guaiacol which is one of the causes of the polymerisation reactions (7).

The objective of our work is the development of an active and adequate carbon supported catalytic system for the hydrodeoxygenation of flash pyrolysis oils. The activity of these catalysts should be increased by modification of the preparation procedure which can lead to a better dispersion of the active phases and a better accessibility of the catalytic sites. Concerning the support, several characteristics which differ completely from carbon to alumina certainly have an influence on the catalytic activity and can be modified in order to increase it. These characteristics are: i) textural characteristics such as specific area, total pore volume, micropore volume and average pore size, ii) type and quantity of surface groups which can interact or not with the active phases increasing or decreasing the dispersion and, iii) the thermal stability of the support.

In this paper, we present results of textural and structural characterisation and catalytic tests obtained with different catalysts prepared with different procedures. Three main modifications were tested: thermal treatment after each impregnation, modification of the impregnation order and reduction-sulphidation activation between the two impregnations.

2 Experimental

2.1 Catalysts

Six different catalysts were prepared, one by the standard method and the others by different modifications of the standard method. The precursors were $Co(NO_3)_2$ $.6H_2O$ and $(NH_4)_6Mo_7O_{24}.4H_2O$, both from Merck, always added by dry impregnation in order to obtain 15%MoO_3 and 3% CoO in the final catalysts. The support was an activated carbon for gas chromatography from Merck with the following characteristics: specific area 1100 m^2/g, total pore volume 0.58 cm^3/g, micropore volume 0.41 cm^3/g, average pore size 8.6 Å and final particle size 0.3-0.5 mm. The different steps of the standard preparation procedure are: dry impregnation of the cobalt precursor, drying at 130°C under argon, dry impregnation of the molybdenum precursor, drying at 130°C under argon, final activation by reduction-sulphidation (heating at 400°C during 3 h under 15% H_2S/H_2). Figure 1 shows the different modifications (thermal treatments, intermediary sulphidation and changing of the impregnation order) and the six catalysts obtained.

2.2 Textural characteristics

The textural characteristics were measured by adsorption of nitrogen at 77°K in a Micromeritics ASAP 2000 analyser. The samples (~0.1 g) were prior submitted to vacuum during 12 h at 180°C and 4 µHg pressure. All catalysts were activated by reduction-sulphidation before analysis.

As carbon is a microporous solid and nitrogen is adsorbed as a monolayer (8-9) the specific area is calculated by the Langmuir formula. The total pore volume was calculated with the Gurvitsch method and the micropore volume with the Dubinin-Raduskewitch method (10). Finally, the pore size distribution was calculated using the method of Horvath and Kawazoe (HK) which has been developed for microporous solids containing parallel plate shaped pores.

2.3 X ray photoelectron spectrometry (XPS)

The XPS spectra were obtained in a 4 eV resolution SSI X-100 spectrometer. The energy scale was calibrated using the $Au_{4f7/2}$ peak which has an energy bond of 83.98 eV. In addition to the general spectrum, we registered C_{1s}, O_{2p}, Mo_{3d} and Co_{2p} peaks. The C_{1s} (284.8 eV) peak with was taken as the reference in order to calculate the energy bonds. All catalysts were sulphided before XPS analysis.

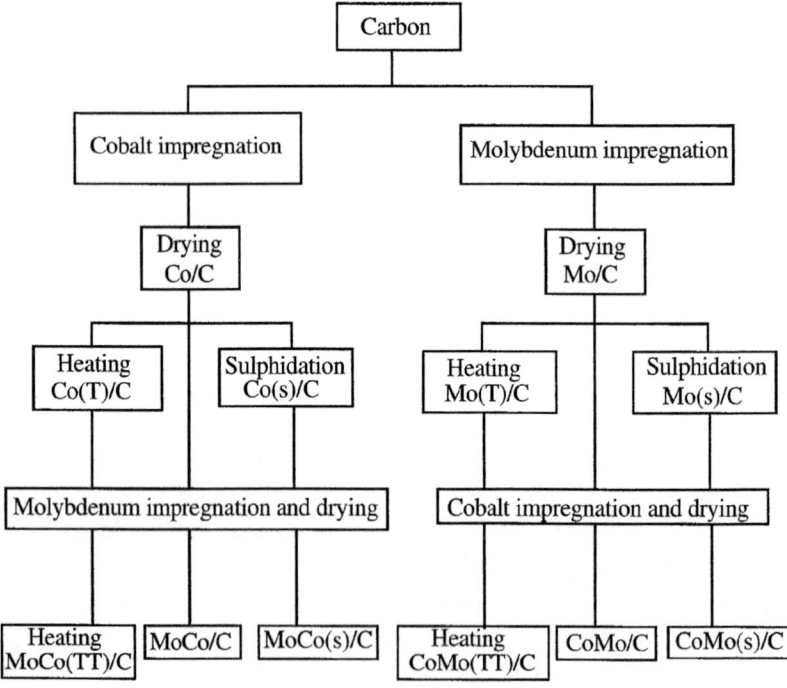

Figure 1. Different catalysts obtained by modifying the standard procedure

2.4 Catalytic tests

The catalytic tests were carried out using model compounds in a 570 ml stirred batch reactor at 280°C and 70 bar. The catalysts were activated by reduction-sulphidation procedure (6).

Table 1 Composition of the model compounds solution

Compound	Molecule	% Weight	Mol/liter
4-MA	4-methyl acetophenone	3	0.218
EDEC	ethyl decanoate	3	0.147
GUA	2-methoxy phenol	3	0.237
4-MP	4-methyl phenol	3	0.110
Internal standard	pentadecane	2	0.018
Solvent	xylene	-	-

The model molecules were chosen on the basis of an in-depth chemical characterisation of the oils. These groups must be eliminated during the stabilisation step (10). Table 1 shows the composition of the model compounds solution. CS_2, which decomposes in H_2S under the reaction conditions, was added in order to keep the sulphided state of the catalysts.

2.5 Analytical follow-up

Liquid samples were taken every hour and analysed in a Chrompack CP-9001 gas chromatograph equipped with a FID detector at 250°C and a 25 m DB-5 capillary column. The peak integration was done using the software Maestro 1 from Chrompack. Each peak was then identified by a Finnigan-Mat GC-MS spectrometer.

3 Results

3.1 Textural characteristics

Table 2 shows the textural characteristics of all catalysts after activation by the reduction-sulphidation procedure. We can note that the different modifications on the preparation procedure do not have a great influence on the micropore structure. The average micropore diameter remains very similar.

Table 2. Specific area, total pore volume, micropore volume and micropore diameter of different catalysts after activation procedure

Catalysts	S.A.(m^2/g)	Vp (cm^3/g)	Vµp (cm^3/g)	Dµp (Å)
Carbon	1094	0.58	0.41	8.5
CoMo/C	683	0.38	0.23	8.6
CoMo(TT)/C	634	0.36	0.21	8.7
CoMo(s)/C	560	0.32	0.19	8.8
MoCo/C	783	0.42	0.29	8.7
MoCo(TT)/C	643	0.36	0.27	8.7
MoCo(s)/C	678	0.38	0.23	8.6

For both CoMo and MoCo families, catalysts obtained by thermal treatments and intermediary sulphidation show lower textural characteristics than those obtained by standard preparation procedures.

We can observe that the MoCo catalyst obtained by changing the impregnation order shows higher textural characteristic values than the traditional CoMo, but both families present very similar values when thermal treatments are included.

3.2 XPS spectrometry
3.2.1 Qualitative results
The Mo_{3d} peak have been decomposed in two peaks $Mo^{4+}3d5/2$ (229.4 eV) and $Mo^{4+}3d3/2$ (232.5 eV). These values are in agreement with the literature ($Mo^{4+}3d5/2$ = 229.2 ± 0.3 eV and $Mo^{4+}3d3/2$ = 232.3 ± 0.3 eV) (11) and show the efficiency of the sulphidation procedure.

3.2.2 Quantitative results
Figure 2 shows the Co/C and Mo/C ratios on the surface of the catalysts. All MoCo catalysts present Co/C and Mo/C ratios much higher than the CoMo family and this for all treatments. For each series, the intermediary treatments (heating and sulphidation) do not influence the Co/C and Mo/C values.

Figure 3 shows the atomic Co/Co+Mo ratios for all catalysts. We can see that the MoCo series presents values two times higher than those for CoMo catalysts.

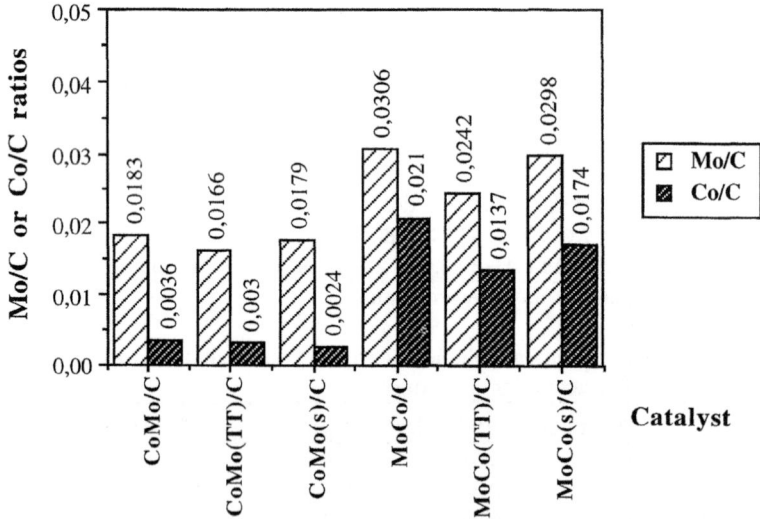

Figure 2. Co/C and Mo/C atomic ratios on the surface of activated catalysts (XPS)

606

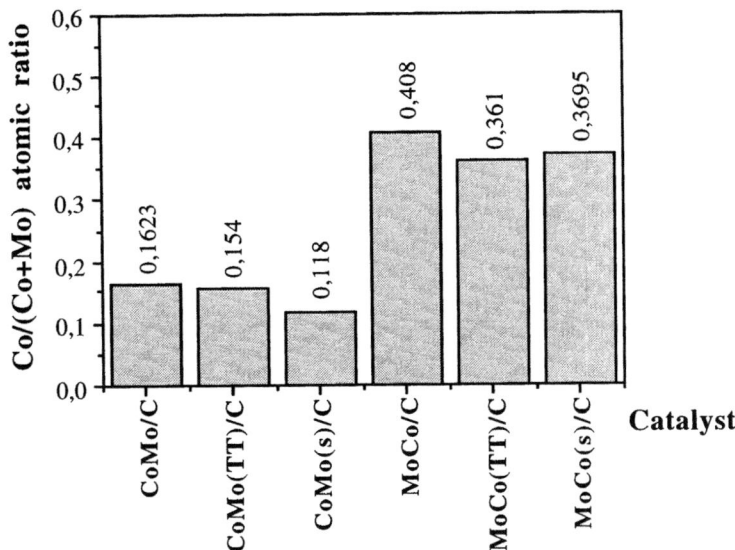

Figure 3. Co/Co+Mo ratios on the surface of all catalysts (XPS values)

3.3 Catalytic activity and selectivity

The reaction pathways for the different model compounds and the formulation of activity and selectivity have been published elsewhere (6,12). The catalytic activity is defined as the kinetic constant (k) in the pseudo first order reaction, considering the reactant consumption.

In the ethyl decanoate (EDEC) hydrodeoxygenation we considered the selectivity as the direct carboxyl elimination from the ester as well as from the acid formed during the reaction:

$$S_{dec} = \frac{C_{non}}{C_{non} + C_{dec} + C_{decene} + C_{decanol}} \times 100$$

The phenol/catechol ratio defines the guaiacol hydrodeoxygenation since the reaction products are catechol and phenol which can be later transformed in cyclohexane and benzene.

$$\frac{phe}{cat} = \frac{C_{phe} + C_{benzene} + C_{cyclohexane}}{C_{catechol}} \times 100$$

Table 3 presents activity and selectivity results for the hydrodeoxygenation of different model molecules using all catalysts prepared. We do not report results concerning 4-methyl phenol since it is not converted in such reaction conditions.

For the CoMo series, the intermediary sulphidation produces an important increase of activity and this for all molecules studied, while the thermal treatment produces a decrease of activity. Both treatments increase the decarboxylation selectivity.

For the MoCo system the intermediary sulphidation does not influence the catalytic activity in the hydrodeoxygenation of 4-MA and EDEC. We note a small activity increase but a selectivity decrease for GUA. Thermal treatments decrease the activity.

The impregnation order is the most important parameter since it has a strong influence on the catalytic activity. We note a small decrease in the conversion of 4-MA, but an important increase for the other molecules. The decarboxylation selectivity decreases while the phe/cat ratio increases.

Table 3. Activity and selectivity values for all catalysts prepared (*: cm^3/min. g-cat)

Compounds Catalysts	4-MA k (*)	EDEC k(*)	Sdec	GUA k(*)	phe/cat
CoMo/C	8.9	0.33	22.6	0.38	36.0
CoMo(s)/C	10.5	0.35	39.0	0.47	46.4
CoMo(TT)/C	7.9	0.30	36.2	0.29	39.1
MoCo/C	6.6	0.40	15.1	0.50	44.2
MoCo(s)/C	6.2	0.40	14.3	0.55	35.5
MoCo(TT)/C	5.5	0.34	11.0	0.29	49.6

4 Discussion

The modification of the preparation procedure can greatly influence the final physico-chemical characteristics of the catalysts as well as their catalytic behaviour. These two points will be discussed separately.

4.1 Influence on the physico-chemical characteristics

The thermal treatments have a moderate influence on CoMo catalysts (only 5% difference between CoMo/C and CoMo(TT)/C specific area values) while this difference is much more important in the MoCo family. This can be explained by the influence of the thermal treatment on the distribution of the active phases on the support surface and not on the support itself (13).

The CoMo(s)/C catalyst has lower textural characteristics than its reference (CoMo/C). This decrease, produced after the cobalt impregnation, can be explained by the fact that the sulphided molybdenun phase favours the cobalt adsorption which increase the micropore blocking. The MoS_2 could act as cobalt adsorption sites (11, 14-16). The MoCo family shows the same behaviour but the differences are less marked. We conclude that a sulphided phase on the surface increases the interaction between the support and the active phase.

MoCo catalysts have higher textural characteristics than CoMo catalysts. The average pore size is very similar for all preaparation methods, while the micropore volume is higher in MoCo than in CoMo. This means that some micropores are blocked while others are completely empty.

4.2 Catalytic activity and selectivity

We observed activity and selectivity changes in the ethyldecanoate conversion. The activity increases with the modification of the impregnation order. In fact, MoCo catalysts have better active phases dispersion and higher Co/Co+Mo ratio which favour the global catalyst activity.

We observed strong selectivity variations with a high increase of hydrogenating capacity and decrease of decarboxylation selectivity for MoCo catalysts. Concerning the intermediary thermal treatments and sulphidations, the two catalyst families show different selectivity behaviour. For MoCo catalysts, the values are very similar while we observed an increase of decarboxylation with treatments in CoMo family. This can be explained by the combined action of the active phases atomic ratio on the surface and the accessibility of the catalytic sites. In fact, carbon is a very complex microporous structure which can be modified by the different treatments with a redistribution of the active functional groups. When cobalt is impregnated in first place, it is adsorbed on the micropores, forming an independent stable sulphided phase. This could have a synergetic action increasing the hydrogenating capacity of the molybdenum sulphide on the surface.

Activity values for the conversion of 4-MA are similar for all catalysts, nevertheless we observed a small decrease for MoCo catalysts. In fact the limiting step in this reaction is the first hydrogenation leading to an alcohol which is influenced by the presence of the hydrogen "spill-over" generated by the cobalt sulphide. The high Co/Co+Mo ratios presented by MoCo catalysts should be located in the decreasing part of a typical synergy graph. This small decrease does not influence the general efficiency of the catalysts, since the activity of the 4-MA conversion is very high in all cases.

5 Conclusions

In the last years it has been demonstrated that an adequate catalytic system was the key step in the development of the pyrolysis oils hydrodeoxygenation process since typical hydrotreating catalysts are quickly deactivated. Catalysts supported on carbon are very promising but their activity must still be improved. The modifications in the preparation procedure of these catalysts, discussed in this paper, constitute a scientific basis for the development of an adequate catalytic system: active, selective and resistent to deactivation. The main conclusions are:

- The activated carbon textural characteristics are not modified by intermediary thermal or sulphidation treatments;
- Both intermediary treatments increase the stability of the molybdenum and cobalt sulphides on the catalyst surface, but they do not influence the dispersion of the active phases on the surface.
- The modification of the impregnation order produces a better dispersion of the active phases leading to an increase of hydrodeoxygenation activity for methoxy and carboxyl groups. The activity decrease for the carbonyl group, nevertheless, still remains very high. The impregnation order also has a strong influence on the decarboxylation selectivity.

- The modification in the hydrodeoxygenation reactions activity and selectivity can be explained by the interaction of separate stable phases and the accessibility of the catalytic sites. This last point is very important since flash pyrolysis oils contain large quantities of large and complex sharp molecules.

6 Acknowledgments

The financial support of EU DG XII under contract AIR-CT92-0216 is gratefully acknowledged.

7 References

1. Elliott, D. C., and Baker, E. G. (1985), in proceedings of 20th IECEC, SAE paper N° 859096.
2. Elliott, D. C., and Baker, E. G., United States Patent, 795, 841.
3. Elliott, D. C., and Baker, E. G. (1987), in "Energy from Biomass and Wastes X", (ed. D. L. Klass), Chicago, p. 765.
4. Churin, E., Maggi, R., Grange, P., and Delmon, B. (1988), in "Research in Thermochemical Biomass Conversion, Phoenix, 1988", (ed. A. V. Bridgwater and J. L. Kuester), Elsevier, London/New York, p. 896.
5. Maggi, R., Laurent, E. and Delmon, B. (1993), Final report contract JOUB-0055-CT90, EU DG XII.
6. Centeno, A., Laurent, E., and Delmon, B. (1995), J. Catal., 154, p. 288.
7. Laurent, E., Centeno, A., and Delmon, B. (1994), in "Proceedings of the 6th International Symposium of Catalysts Deactivation, (ed. B. Delmon and G. F. Froment), Elsevier, Amsterdam, p. 573.
8. Gregg, S. J., and Sing, K. S. W. (1982), The Physical Adsorption of Gases by Microporous Solids, in " Adsorption, Surface Area and Porosity", Academic Press.
9. Lecloux, A. (1971), in "Memoires Societé Royale des Sciences de Liège", 6e serie, Tome I, p. 169.
10. Laurent, E., Grange, P., and Delmon, B. (1991), in "Biomass for Energy, Industry and Environment", 6th EC Conference, (ed. G. Grassi, A. Collina and H. Zibetta), Elsevier, London and New York, p. 672.
11. Duchet, J. C., Van Oers, E. M., De Beer, V. H. J., and Prins R. (1983), J. Catal., 80, p. 386.
12. Laurent, E., and Delmon, B. (1994), Appl. Catal., 109, p. 77.
13. Gandia L. M., and Montes, M. (1994), J. Catal., 145, p. 276.
14. Vissers, J. P. R., Bouwens, S. M. A. M., De Beer, V. H. J., and Prins, R. (1987), Carbon, 25, p.485.
15. Vissers, J. P. R., Scheffer, V. H. J., De Beer, V. H. J., Moulijn, J. A., and Prins, R. (1987), J. Catal., 105, p. 277.
16. Crajé, M. W. J., De Beer, V. H. J., and Van der Kraan, A. M. (1991), Bull. Soc. Chim. Belg., 100, p. 953.

LIQUID FUELS BY LOW-SEVERITY HYDROTREATING OF BIOCRUDE

D. C. ELLIOTT and G. G. NEUENSCHWANDER
Pacific Northwest National Laboratory, Richland, Washington, USA

Abstract
Biocrude (fast pyrolysis oil from wood) was hydrotreated to minimize the negative aspects of this fuel. The instability of the oil was reduced by reaction of the most unstable functional groups. Concurrently, the oxygenated component of the oil was also reduced, resulting in an improved energy density. Changes in the physical handling properties were also modified. All of this change was accomplished at less severe processing conditions (lower temperature, shorter residence time) than that required for the earlier processing for gasoline production. Improved conversion was achieved by the use of a downflow reactor system. The experiments reported include those performed with clean (filtered hot vapor) biocrude just recently available from processing systems in the U.S.

Key words: biocrude, catalysis, fast pyrolysis, fuel oil, hydrotreating, upgrading

1 Introduction

Low-severity hydrotreating is one means to improve the fuel properties of biocrude (fast pyrolysis oils from wood). As produced, biocrude has some undesirable properties as fuel, such as: some thermally unstable components which can lead to gum formation, low energy density because of dissolved water and highly oxygenated compounds, a corrosive organic acid component, and phase instability with a tendency toward phase separation. High-severity hydrotreating, involving complete hydrodeoxygenation, some hydrocracking and minimal hydrogenation to produce an aromatic hydrocarbon fuel, is an expensive way to remedy these problems. As an alternative, we are developing low-severity hydrotreatment methods to improve the fuel oil properties. Low-severity hydrotreating involves partial hydrodeoxygenation, minimal hydrocracking, and effective hydrogenation to stabilize the biocrude and improve its energy density and handling properties.

Hydrotreating of biocrude has been under development for a decade. Initial efforts were stymied by the instability of the biocrude which required development of process modifications to specifically address the instability. A low-temperature catalytic hydroprocessing has been identified as an important first step in the overall process [1]. But low-temperature catalytic hydroprocessing alone does not yield a useful fuel oil product from the biocrude [2] [3] [4] [5]. The product is a viscous black tar which is more thermally stable and can even be distilled; but it still contains high levels of chemically combined oxygen and dissolved water.

Hydrodeoxygenation by catalytic processing has been developed for biocrude based on the initial stabilization by low-temperature hydroprocessing. Hydrocarbon products with gasoline and light fuel oil distillation ranges were produced from the biocrude. The hydrogen consumption for this processing was high with overall product yields in the range of 30% by weight of the biocrude [4] [6]. Initial estimates of the processing costs were high [7] [8]. Catalyst stability and gum formation in the biocrude feed lines were identified as process uncertainties.

As a result of the high costs calculated for gasoline production from biocrude, emphasis has shifted to lower cost processing to produce heavier products such as turbine fuel for electricity generation or as a first step toward cofeeding to a conventional petroleum refinery. Low-severity hydrotreating is an attempt to upgrade biocrude to a useful fuel oil product with a minimum of cost through catalytic hydroprocessing. Low-severity hydrotreating for upgrading biocrude has been under investigation for the past several years. Laurent et al. [9] pointed out the better economic possibilities in low severity processing to heavy fuel oils. A resulting design for a pilot plant considered turbine fuel production by low-severity hydrotreating [10]. Initial attempts show that the low-severity product can be produced at a lower cost [11], but the level of upgrading required versus the product properties needs to be further investigated.

Experimental attempts to produce low-severity hydrotreating products at Pacific Northwest National Laboratory (PNNL) have been reported with limited success [12]. Non-steady-state operations have plagued the experiments. Product quality has been lower than expected with higher viscosities (far outside the turbine fuel specification range) and oxygen contents of 20% and higher. The results led the researchers to redesign the continuous-feed reactor system to the down-flow configuration reported here. This research is also making use of the recent developments in biocrude cleanup via hot vapor filtration as well as catalyst improvements for hydrodeoxygenation of biocrude.

2 Bench-scale systems for hydroprocessing biocrude

The biocrude feedstocks used in these experiments was provided to us by Union Electrica Fenosa SA (UF) and the National Renewable Energy Laboratory (NREL). The UF oil was produced from eucalyptus in a fluidized bed flash pyrolysis system. The sample was over a year old at the time of our experiments. It was not filtered in the production process. The NREL oil was produced from hybrid poplar in an ablative fast pyrolysis system (M2-Run 6 at Hazen). The hot pyrolysis vapors were cleaned in a filter bag house prior to condensation and collection. The biocrude was only two months old at the time of the hydrotreating experiments. A sample from Ensyn Technologies, Inc. (RTP3) was also tested for comparison. The sample was at least 4 years old at the time of our testing. Analyses of these oils in given in Table 1.

Table 1. Biocrude Feedstocks for Low-Severity Hydrotreating

property	NREL	UF	RTP3
carbon, %	46.7	44.8	39.4
hydrogen, %	7.6	7.2	7.9
oxygen, %	45.7	48.1	52.7
nitrogen, %	0.2	0.2	0.2
sulfur, %	0.032*	NA	NA
ash, %	0.016*		0.1*
HHV, MJ/Kg	NA	NA	16.7*
density, g/mL	1.19		1.19
viscosity, cps	127	1510	43
suspended solids, %			
moisture, %	18.9*	NA	~24
K, ppm	5*	NA	30*
Ca, ppm	6*	NA	260*
Na, ppm	4*	NA	43*
Cl, ppm	8*	NA	NA

* analysis provided by NREL

2.1 Continuous-feed fixed-bed reactor system

The reactor system included two vessels containing fixed catalyst beds. The system was run in a down-flow (trickle bed) configuration in which the top bed (100 mL) served as the low-temperature stabilization reaction vessel and the bottom bed (425 mL) served as the main hydrotreater at higher temperature.

As shown in Figure 1, the oil was fed by a reciprocating high-pressure pump and the hydrogen entered the oil feed line from a gas manifold prior to the two reactor vessels. Pressure was controlled in the system by a dome-loaded back-pressure regulator (Teflon diaphragm) in the product line. Liquid and gaseous products were separated in a two- stage system of coolers and gas/liquid separators before they were recovered and analyzed.

The system was constructed of stainless steel (300 series) throughout all wetted parts. It is rated to 21 MPa. Feed lines were 6 mm (¼"), and the reactor effluent lines were 13 mm (½"). The feed lines were insulated and heat traced. Line heating up to 50°C was used with the Union Fenosa oil, while the other oils required no line heating for smooth pumping. Temperatures in the two reactor beds were controlled independently. The first bed temperature was controlled in two levels. Strong exothermic reactions occurred at times in the reactor beds. These were believed to result if pyrolysis condensation reactions overwhelmed the hydrogenating function of the catalyst. Coke formation in the catalyst bed resulted. The exotherms were controlled by sufficient low temperature hydrotreating to stabilize the biocrude before allowing the oil to be heated above 300°C.

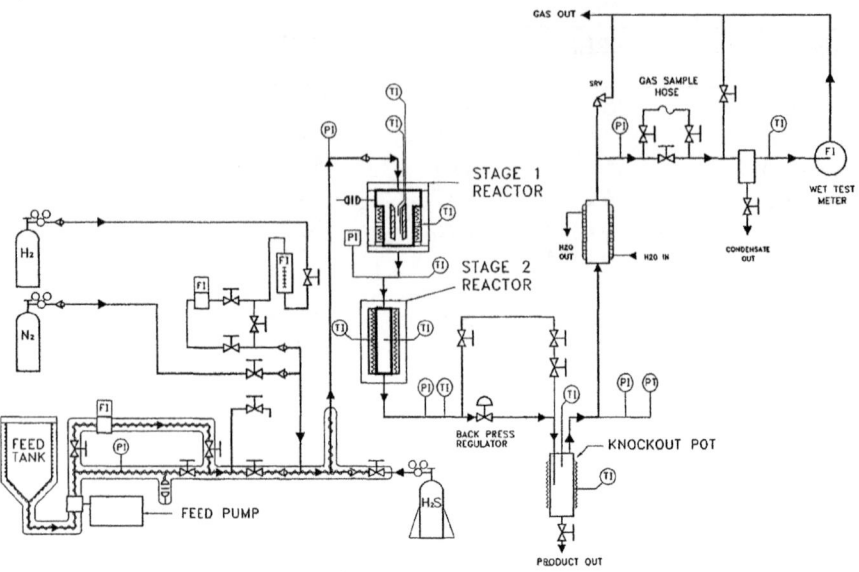

Fig. 1. A schematic of the bench-scale, continuous-feed, fixed-bed reactor system.

3 Experimental Results

The goal of this research was to determine the types and amounts of product oils which could be produced from biocrude at low-severity conditions. The work included both preliminary tests with highly active hydrogenation catalysts at low temperatures to evaluate the types of products which might be produced and developmental tests in the continuous-feed reactor to verify production rates and yields at a range of low severity conditions with conventional hydrotreating catalysts. The batch reactor results are presented in a separate report.

The results presented in this chapter were produced in recent experiments in the continuous feed catalytic hydrotreater at PNNL operated in a downflow configuration. The tests were made with three different biocrudes as described in Table 1. The bulk of the tests were made with a single batch of NiMo on alumina catalyst (Haldor Topsoe TK-751, 1 mm extrudates). The catalyst was placed in both reactors and presulfided with hydrogen sulfide 10% in hydrogen with a temperature ramp up to 400°C. A single charge of the NiMo catalyst was replaced in the first stage reactor after the initial test in which temperatures went outside the acceptable operating range and plugging occurred. The same catalyst charge was used in the balance of the NiMo tests. A second catalyst was also tested (BASF K8-11). This CoMo catalyst is unlike the NiMo in that the support is a spinel not the conventional high surface area γ-alumina. Better chemical stability of the support was envisioned, but the catalyst exhibited much lower activity overall.

The results of these tests are summarized in Table 2 below. Higher levels of conversion were seen in the downflow configuration compared to earlier upflow tests. High levels of hydrogen consumption and high product quality were seen even at much higher space velocity. Product oils were analyzed to quantify the effect of space velocity on the product properties. The two catalysts had significantly different activities. The different feed oils also had different levels of reactivity. The results are compared with literature values below.

Table 2. Biocrude hydrotreating results

	NREL -- downflow -- NiMo/alumina catalyst							CoMo/spinel	
1st temperature, °C	150	148	148	150	150	148	148	148	157
2nd temperature, °C	380	375	354	349	349	362	355	400±15	435±45
LHSV, L/L/hr	0.28	0.30	0.29	0.29	0.29	0.43	0.38	0.38	0.31
WHSV, g/g/hr	0.52	0.54	0.53	0.53	0.54	0.78	0.70	0.64	0.51
Yield, g/g biocrude	0.42	0.39	0.38	0.44	0.49	0.47	0.53	0.21	0.21
Deoxygenat'n, %	98.6	97.9	96.3	95.1	95.0	94.5	95.8	97.6	97.1
Density, g/mL	0.82	0.84	0.84	0.86	0.86	0.86	NA	NA	NA
Gasificat'n, %Carbon	27	20	22	25	33	29	29	23	22
H_2 Consumpt'n, L/L	881	746	727	808	813	791	779	494	313
Carbon balance, %	92	79	79	91	107	101	109	54	53

4 Discussion and Conclusions

4.1 Space velocity effects

The experimental reactor results are evaluated in terms of deoxygenation which incorporates the effects of temperature, pressure and catalyst. The deoxygenation is defined as the percent of chemically combined oxygen in the biocrude removed when compared to the oil product. Deoxygenation as a function of space velocity shows a dramatic effect. In Figure 2 space velocity is presented in terms of volume (liquid hourly space velocity, LHSV). By this measure the new downflow results in this paper stand out clearly compared to all of the earlier published results. Earlier, we agreed with VEBA that the difference between downflow and upflow operation was small considering the overall reaction and the little difference between our earlier upflow results and those of VEBA (on a LHSV basis). However, Figure 3 provides a data comparison on a weight basis (weight hourly space velocity, WHSV) wherein the effect of the diluted CoMo catalyst bed used by VEBA is dramatic. By using the diluted CoMo catalyst bed, VEBA achieved much higher processing rates based on the weight of catalyst. The one unexplained result is the poor showing of the NiMo catalyst in the downflow experiments at VEBA. Again, our downflow results are much improved compared to our earlier upflow results. In all of these tests the catalysts are supported on an alumina base. The differences in the results comparing the use of CoMo with NiMo in our earlier tests are relatively small in light of the differences in results achieved with downflow operation compared to upflow.

615

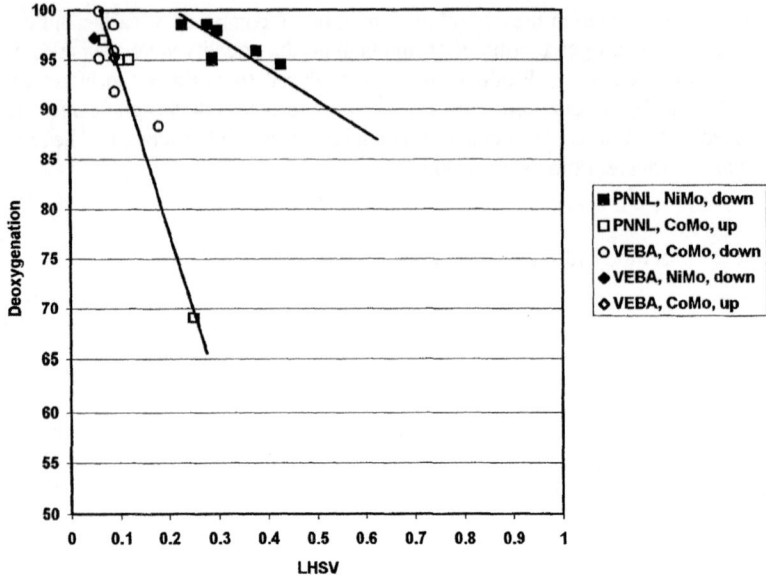

Fig. 2. Deoxygenation on a liquid hourly space velocity basis

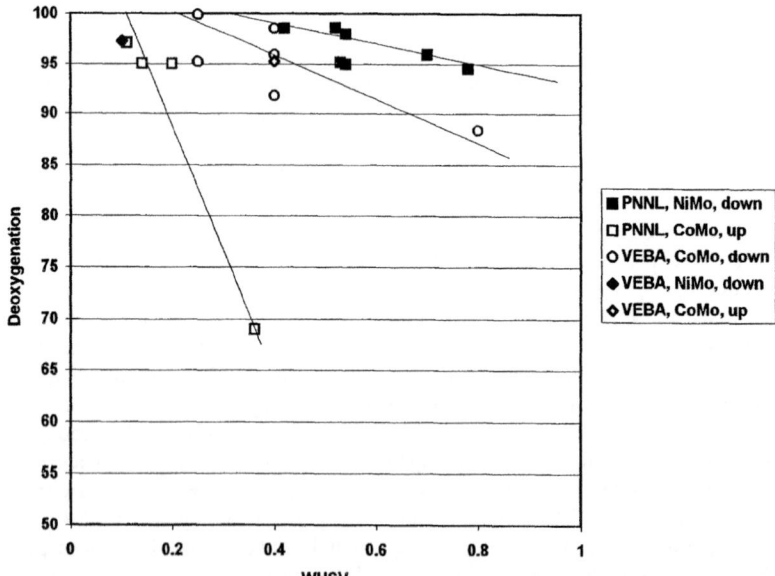

Fig. 3. Deoxygenation on a weight hourly space velocity basis

4.2 Product oil properties

The product oil density is clearly a function of product oil oxygen content as seen in Figure 4. The molecular weight may also be correlated with the oxygen content as they may both be changing coincidentally as a function of processing severity. Figure 4 shows that the function of oxygen content and density seems to apply to the full range of products reported in the literature as well as those from these tests. An important conclusion to draw from these data is that the product oil density approaches 1 at a relatively low oxygen content, about 10%. Products with oxygen contents around 10-15% and densities around 1 tend to form mixtures (emulsions?) with the water byproduct and can not be easily separated from the water. This lack of separation defeats one important purpose of the hydrotreating which is to remove the water and, thereby, dramatically improve the energy content of the oil.

Viscosity of the product oil is also a function of the oxygen content, as seen in Figure 5. The range shown goes all the way from the low viscosity required for turbine fuels of less than 5 cps to the heavy tar products with high oxygen contents and viscosities >100,000 cps whose pour points would be around room temperature. These results suggest that only the highly upgraded oils with oxygen contents of 5% or less have potential for direct use as turbine fuels because of viscosity limitations. Figure 4 also includes the data for the 3 biocrudes. The raw biocrudes show decreasing viscosity with increasing oxygen content because of increasing water content in the raw biocrude. These biocrude numbers suggest the use of oxygenated solvents to decrease the biocrude viscosity to the useful range for turbines thus allowing its direct use without chemical processing. However, use of water as the solvent for viscosity reduction of the biocrude does not appear feasible because phase separation would occur before the viscosity would be reduced to a level useful in turbines.

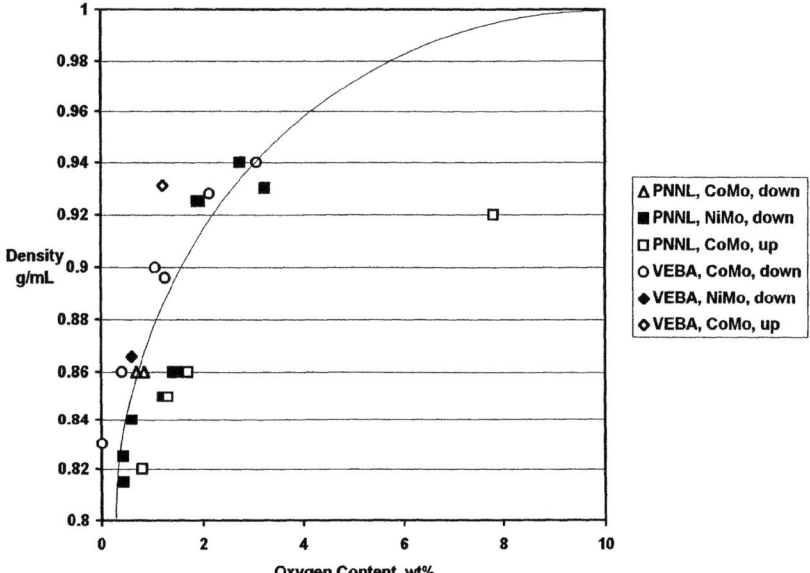

Fig. 4. Relation of product oil density to oxygen content

617

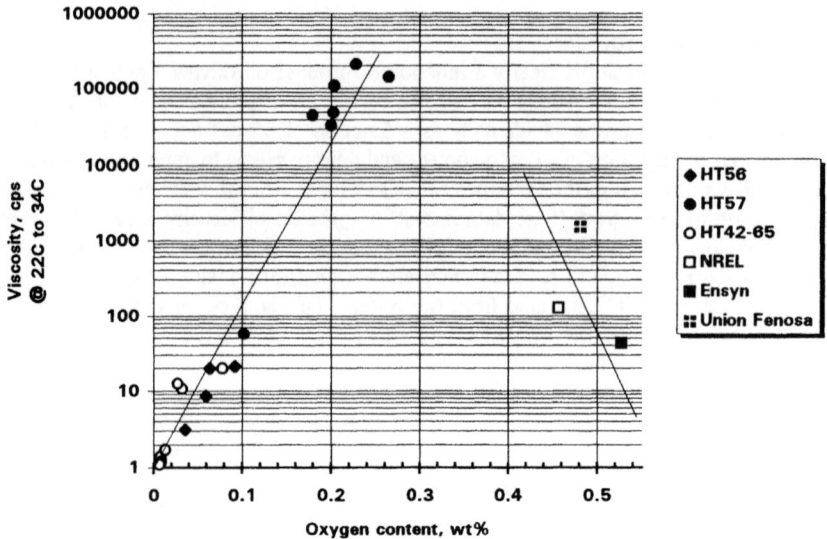

Fig. 5. Relation of oil viscosity to oxygen content

Figure 6 shows the effect of temperature on viscosity of several of the heavy oil products. These results suggest that the highly oxygenated products could not be used as turbine fuels because of high viscosity even with preheating. The 10% oxygen oil might be used as turbine fuel with preheating to 50°C or higher.

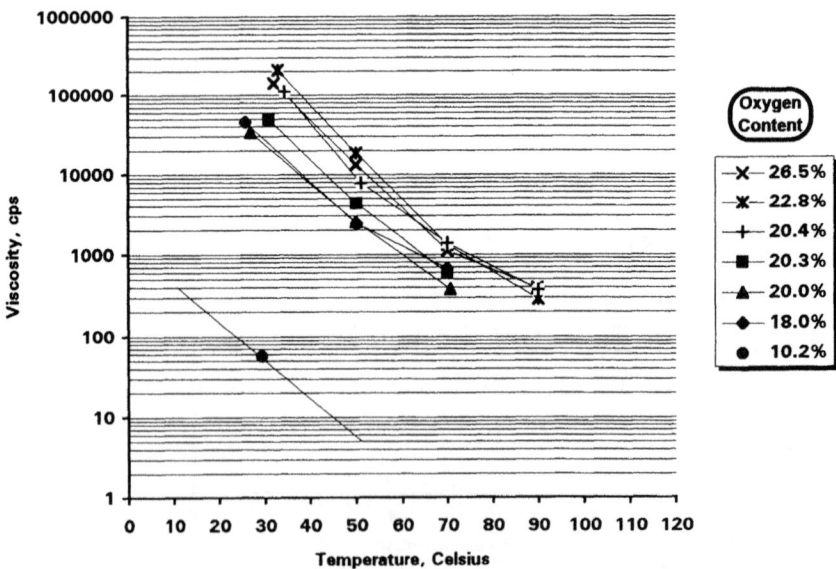

Fig. 6. Effect of temperature on oil viscosity

4.3 Byproduct water contamination

Organic contamination of the water byproduct from hydrotreating is a concern relative to the overall wastewater treatment requirements for the plant. Figure 7 clearly shows that the contamination of the byproduct water as represented by the carbon content increases with the oxygen content of the product oil up to a range of 5 to 10 wt% when the product oil oxygen content is >10%. At that level the oil water separation is difficult and oil phase contamination of the water samples leads to great variation in the results. Samples representing part and all of the oil product for 2-stage upflow tests are given.

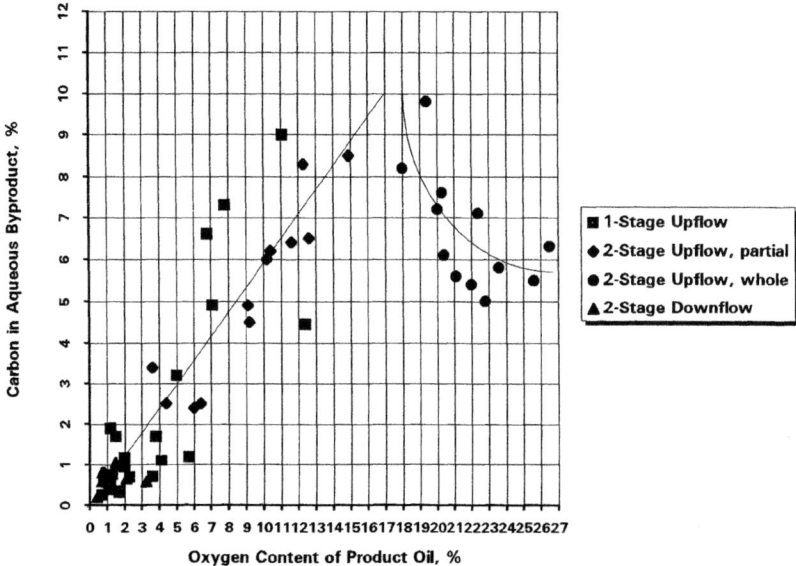

Fig. 7. Relation of carbon in the aqueous phase to the oxygen content of the oil

4.4 Biocrude feed effects

The effect of biocrude feed properties on the reactivity can be significant. Whereas the work in our lab in earlier years used a number of biocrude oils including several hardwood oils from different fluid-bed pyrolysis reactors, pine oil from ablative pyrolysis, hardwood oil from vacuum pyrolysis, and poorly-humified-peat oil from fluid-bed pyrolysis without identifying any major differences in reactivity; the current work shows large differences between poplar oil from ablative pyrolysis and eucalyptus oil from fluid-bed pyrolysis. Table 3 provides some of the results. Although the eucalyptus oil test was performed at higher temperature and lower space velocity, the product oil quality was significantly lower, i.e., higher oxygen content and density. Further substantiating the conclusion of lower reactivity is the lower gas yield and also the lower hydrogen consumption. The combination of high viscosity and poor pumping performance with the lower reactivity make the eucalyptus oil more difficult to hydrotreat.

Table 3. Hydrotreating results with different biocrudes

	NREL	Union Fenosa
Temperature, °C	355	365
WHSV, g oil/g catalyst/hr	0.70	0.54
Yield, g/g biocrude	0.53	0.41
Deoxygenation, %	96	92
Product density, g/mL	0.86	0.94
Gasification, % carbon	29	17
Hydrogen consumption, L/L oil	779	554
Carbon balance, %	109	85

4.5 Effect of catalyst support
The tests with the spinel supported CoMo catalyst (see Table 2) show the dramatic effect of lower catalyst activity. Without a strong hydrotreating catalyst effect in the reactor, exothermic pyrolysis condensation reactions caused major temperature excursions and lead to coke buildup on the catalyst. Buildup of heavy tar products in the reactor and effluent lines caused plugging and premature experiment terminations. Product fractionation occurred, similarly to the upflow experiments, in which the excess hydrogen gas exiting the reactor carried a substantial portion of the light products out while leaving most of the biocrude still as a heavy tar product. Total carbon balances were difficult to measure, but very low space velocities would be required to effectively hydrotreat the biocrude over this catalyst..

4.6 Hydrotreating economics
The four major cost factors in hydrotreating biocrude are 1) Biocrude cost, 2) Capital cost, 3) Hydrogen cost, and 4) Relative product value. The cost of the biocrude is the largest component in the hydrotreated product costs, therefore the product yield is a primary consideration for process optimization. The cost of capital is significant for this high-pressure process, therefore volume space velocity is a critical factor to maximize in order to reduce the capital cost per unit processed. Hydrogen consumed in the process is a significant cost. It typically can be generated by steam reforming of the byproduct gases. To minimize the cost of the reforming systems or the cost of procured hydrogen, the hydrogen consumption should be efficiently focused on the reactions which provide the product properties of importance, i.e., removing the unstable functional types, like olefinic and carbonyl groups; and removing oxygen. In the end, the process needs to be optimized based on the product oil value relative to the expenditures for the other three factors.

5 References

1. Elliott, D.C. and Baker, E.G. (1989) Process for upgrading biomass pyrolyzates. U.S. Patent #4,795,841, issued January 3, 1989.

2. Elliott, D.C. and Baker, E.G. (1987) Hydrotreating biomass liquids to produce hydrocarbon fuels, in *Energy from Biomass and Waste X*, Institute of Gas Technology, Chicago, pp. 765-784.

3. Gagnon, J. and Kaliaguine, S. (1988) Catalytic hydrotreatment of vacuum pyrolysis oils from wood. *Industrial and Engineering Chemistry Research*, Vol. 27, pp. 1783-8.

4. Baldauf, W. And Balfanz, U. (1992) *Upgrading of Pyrolysis Oils from Biomass in Existing Refinery Structures*, VEBA OEL AG, Gelsenkirchen. Final Report JOUB-0015.

5. Conti, L., Scano, G., Boufala, J., and Mascia, S. (1995) Experiments of bio-oil hydrotreating in a continuous bench-scale plant, in

6. Baker, E.G. and Elliott, D.C. (1988) Catalytic upgrading of biomass pyrolysis oils, in *Research in Thermochemical Biomass Conversion*, (ed. A.V. Bridgwater and J.L. Kuester), Elsevier Applied Science, London, pp. 883-895.

7. Baldauf, W. And Balfanz, U. (1991) *Upgrading of Pyrolysis Oils from Biomass in Existing Refinery Structures*, VEBA OEL AG, Gelsenkirchen. Progress Report (June 91- November 91) Joub - 0015/0055-C (MB).

8. Elliott, D.C., Baker, E.G., Beckman, D., Solantausta, Y., Tolenhiemo, V., Gevert, S.B., Hörnell, C., Östman, A., and Kjellström, B. (1990) Technoeconomic assessment of direct biomass liquefaction to transportation fuels. *Biomass*, Vol. 22, pp. 251-269.

9. Laurent, E., Grange, P., and Delmon, B. (1991) Evaluation of the upgrading of pyrolytic oils by hydrotreatment, in *Biomass for Energy, Industry and Environment, 6th EC Conference*, (ed. G. Grassi, A. Collina, and H. Zibetta) Elsevier, p.672.

10. Elliott, D.C. and Silva, L.J. (1992) *Conceptual Process Design and Cost Estimate for a Bio-Oil Hydrotreating Pilot Plant*. report to Aston University from Battelle Pacific Northwest Laboratories, contract #18621.

11. Baldauf, W. And Balfanz, U. (1991) *Upgrading of Pyrolysis Oils in Existing Refinery Structures*, VEBA OEL AG, Gelsenkirchen, Germany. Contractors Coordination Meeting, Gent, October 29-31, 1991. JOUB - 0015/0055-C (MB).

12. Elliott, D.C., Hart, T.R., Neuenschwander, G.G., McKinney, M.D., Norton, M.V., and Abrams, C.W. (1995) *Environmental Impacts of Thermochemical Biomass Conversion*. National Renewable Energy Laboratory, Golden, Colorado. NREL/TP-433-7867.

BIO-CRUDE OIL HYDROTREATING IN A CONTINUOUS BENCH-SCALE PLANT

L. CONTI, G. SCANO, J. BOUFALA, S. MASCIA
University of Sassari, Dipartimento di Chimica, Sassari (Italy).

Abstract

At University of Sassari, in the EC ambit AIR contract, the hydrotreatment of a bio-oil produced by a RTP pyrolysis process was investigated in a continuous reactor containing a packed bed of sulphided NiMo catalyst.

The process was found to decrease particularly some undesired physical and chemical properties of the bio-oil as neutralization number, density, acetone insolubles and oxygen levels and to increase the hydrogen content of the oil. These improvements have been achieved by a mild hydrotreatment step using a particular reactor temperature profile (140-280 °C) at 15 MPa hydrogen partial pressure. A stabilized oil, well separated from the water, with a yield of 72 wt % on dry bio-crude oil was obtained.

The operating life of the reactor was experimented up to ca. 120 hours and no apparent deactivation of the catalyst was observed. This experiment demonstrates the feasibility of the process.

The analyses of the mass balance of the whole test showed in particular a reduction of the oxygen content of 60 wt% and a consuption of hydrogen of 264 l/kg bio-crude oil.

The product is suitable for further hydrorefining processes.

Keywords: bio-oil, fixed bed, hydrotreating, NiMo catalyst, upgrading.

1 Introduction

Biomass contributes around 14 % to global primary energy input but it is an inefficiently utilized resource, often exploited unsustainably. A wider commercial exploitation on a sustainable basis awaits the development of modern technology to enable biomass to compete with conventional energy carriers. Pyrolysis seems one of the most suitable way to reach this aim.

High yields of crude bio-oils can be obtained from biomass by pyrolysis, but the quality of the product is relatively low. This is mainly due to the high content of oxygenate compounds and water. Bio-oils have relatively low heating value, have high viscosity, are unstable, are corrosive and incompatible with petroleum feedstocks. The upgrading of the crude bio-oils seems to be a good way to overcome these problems.

Hydrotreating technology can be regarded as promising for the conversion of bio-oils in valuable high quality fuel, but it requires long residence time and high temperatures and pressures. In such conditions thermal degradation of the bio-oil occurs because the polymerization rate is higher than the rate of the hydrogenation reaction.

Elliott suggested a hydrotreatment of the bio-oil at middle conditions before the upgrading in more severe conditions, in order to stabilize the bio-oil [1].

In this paper the plant description and some results obtained in the first stage of the process are reported.

2 Experimental

2.1 Analyses of the bio-oil
The upgrading tests were performed on a bio-oil supplied by ENSYN (wood derived). Some chemical and physical analyses of the oil are summarized in table 1.

Table 1. Chemical and physical Analyses of Ensyn Bio-oil

Acetone insolubles (%)	2.98
Ash (%)	0.20
Neutr. Number (mg KOH/g)	100
pH	3.17
Moisture (%)	24.8
Density (g/cm^3 20°C)	1.21
Viscosity (mPa·s 20°C)	179
HHV (MJ/kg m.f.)	19.4
Elemental Analysis (m.f.)	
C (wt%)	52.9
H (wt%)	6.2
N (wt%)	0.2
O (wt% by difference)	40.7

2.2 The plant
The schetch of the plant is shown in figure 1. It is mainly constituted by a high pressure piston pump for bio-oil feeding, a feed system for the hydrogen, a fixed bed reactor and two section for the collection of liquids and gaseous products. The reactor was

operated in upflow mode. A more detailed description is reported elsewhere [2].

2.3 Catalyst.
A NiMo catalyst (extrudate Ketjenfine 840-1.3Q) supplied by AKZO, was used. The catalyst was presulfided *in situ* with a process performed by us using dimethyldisulfide (DMDS) as sulfiding agent and a partial pressure of hydrogen of 3 MPa. A particular programmed temperature profile was carried out and the 10 wt% of sulfur was adsorbed on the catalyst.

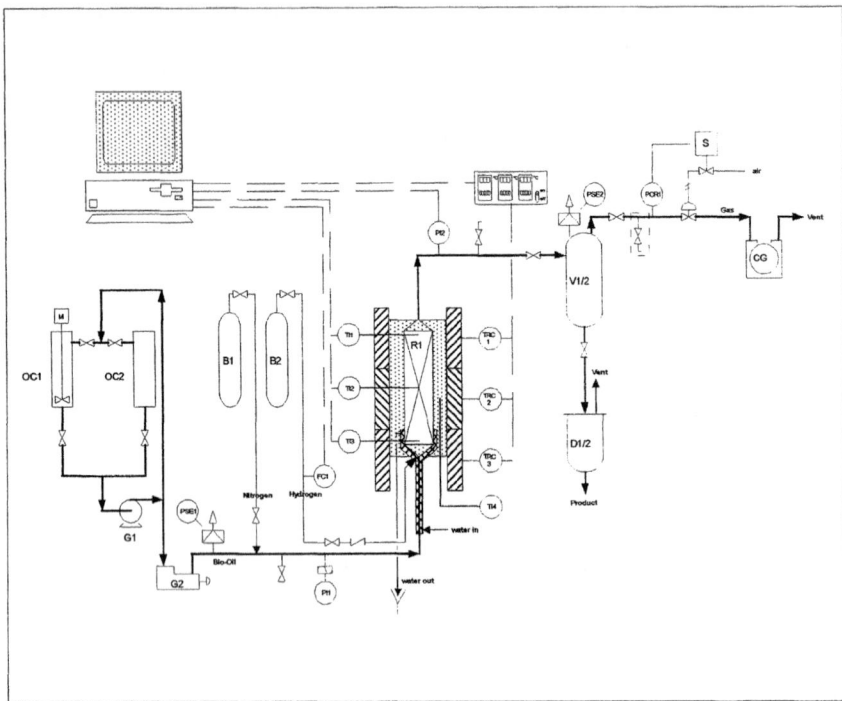

Fig. 1. Continuous hydrotreating plant scheme.

3 Results and discussion

3.1 Hydrotreating of the bio-oil
From the beginning, a lot of difficulties were encountered during the hydrotreating processing runs due to [3]:
- the separation of water from oil. Gum like deposits were found in the feeding pipes already at the ambient temperature [4]. This was ascribed to an annular flow regime caused from the different velocities of the gas and liquid. Probably the stream of hydrogen dries the bio-oil [5]. This problem was overcome by feeding the hydrogen in the catalytic bed directly.
- the decomposition of the bio-oil in the preheater and/or on the catalytic bed.

These problems were overcome by using a particular temperature profile of the catalytic bed; cooling of the feeding pipe in order to maintain the temperature less then 100 °C up to the reactor and the catalytic bed entrance at a temperature of about 140 °C . In these conditions the run lasted up to 120 hr which was voluntarily shut down.

The conversion of the bio-oil to upgraded product was determined for several balance periods and the gas sampled for the chromatographic analyses. The water content of each liquid sample was also determined.

The experimental conditions are listed in table 2 and the data summary sheet of the whole test is shown in table 3.

Table 2. Experimental conditions (nominal)

Inlet reactor temperature (°C)	115
Middle reactor temperature (°C)	250
Exit reactor temperature (°C)	275
Space velocity (h^{-1})	0.5
Total pressure (MPa)	15
H_2/feed (v/v)	2000
Throughput (ml/h)	50

Table 3. Data Summary Sheet

Hours on-stream *	18	33	49	63	87	95	107
Balance period (h)	8	8	8	6	8	8	4
Pressure (MPa)	15	15	15	15	15	15	15
Hydrogen rate (l/h)	105	105	105	105	105	105	105
LHSV (h^{-1})	0.58	0.57	0.48	0.57	0.58	0.49	0.51
Liquid feed (g/h)	63.1	62.0	51.9	61.7	63.7	53.0	55.3
Temperatures (°C)							
Bed entrance	137	138	136	125	142	148	148
Bed middle	243	246	256	254	268	265	260
Bed exit	262	265	277	275	291	288	281
Liquid products (g/h)	60.8	58.2	47.1	47.9	60.1	50.5	52.5
Product gas rate (l/h)	91.0	92.1	88	93.3	89.5	89.0	95.8
Liquid recovery (wt%)	96.4	94.0	90.7	77.6	94.4	95.3	94.8

* End of balance periods

The viscosities of the upgraded oil sample obtained at the end of each balance period were measured at 60° C; the values are plotted versus the time (fig. 2). The viscosity is in some extent, indicative of the catalyst deactivation. The catalyst was very active at the beginning; then the activity decreased rapidly and kept fairly constant after about 60 hours from the start.

The initial decline in activity has often been attributed to the coke formation [6].

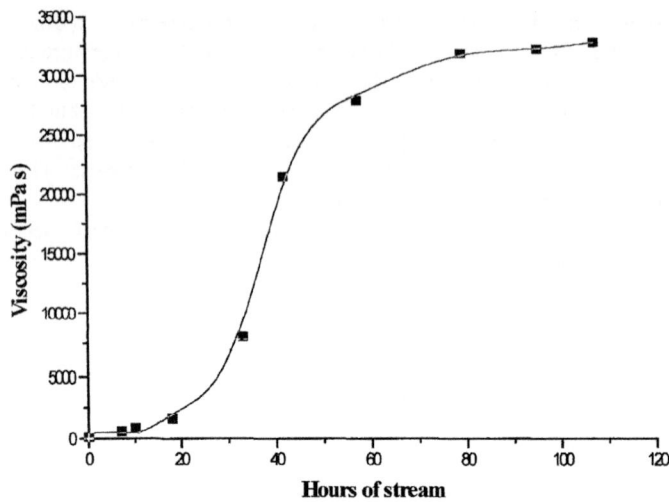

Fig. 2. Viscosity at 60 °C of the hydrogenated bio-oil sampled during the run (shear rate 4.87 s⁻¹).

Table 4 reports the mass balance of the whole test, data calculations are based entirely on recovered liquid products. The hydrogen balance value was obtained by comparing hydrogen concentrations at the inlet and outlet of the reactor. A totally indipendent elemental hydrogen consumption was calculated from the increased combined hydrogen in the product vs. the feed.

Table 4. Mass balance of the whole test.

		Feed				Products				
	Bio-oil m.f. (*) 4107 gr		Hydrogen 781 gr		Total 4888 gr	Upgr. oil (m.f.) 2961 gr		Water (**) 997 gr		Gas product 930 gr	
	%	gr	%	gr	gr	%	gr	%	gr	%	gr
C	52.3	2148			2148	68.8	2037			11.9	111
H	6.2	255	100	781	1036	7.7	228	11	110	75.1	698
O	41.3	1696			1696	23.2	687	89	887	13.0	121
N	0.2	8			8	0.3	8				

(*) 2000 ppm of DMDS added
(**) mainly separated by settling

The water produced in the hydrogenation of the compounds contained in the bio-oil is the addition of the separated water and that one left in the stabilized oil. The analysis of the mass balance shows:

- 72 wt% and 6.5 wt% yields of hydrotreated oil and gas respectively on the basis of the fed bio-oil m.f. were obtained.

- The hydrogen consumption was calculated as 264 litres per kg of fed bio-crude oil. These results do not agree with the ones obtained by Elliot (30 l/l bio-oil) [7]. This fact could be due to the use of a different catalyst (CoMo) which shows a smaller activity on the removal of the oxygen or to the different conditions of the reaction.
- 32.5 wt% of the utilized hydrogen is contained in the gaseous hydrocarbon, 67.5 % removes the oxygen from the oil producing water (75.7 wt% of the total). Additional water (24.3 wt%) derives from the cracking of the bio-oil. Oxygen is also eliminated as CO_2.
- On the whole about 60 wt% of the oxygen is removed from the bio-crude oil.
- In these experimental conditions no reduction of the nitrogen was observed.

The weighted average composition of the gaseous products, determined by gaschromatography, is shown in table 5. As it was expected a great amount of carbon dioxide, probably due to decarboxylation reactions, was produced. Referring to starting bio-oil, a limited production of gaseous hydrocarbons was obtained, olefinic materials were not found in these products.

Table. 5. Gas products composition (vol %)

Carbon dioxide	45.8
Methane	37.5
Ethane	6.6
Propane	3.3
C3+	6.8

3.3 Characteristics of upgraded bio-oil
The samples of the upgraded oils collected during the run were put together to obtain a averaged product. In table 6 the chemical and physical analyses of resulting hydrogenated oil are shown.

Table 6. Chemical and physical analyses of upgraded oil

Acetone insolubles (wt %)	0.21
Ash (wt %)	0.06
Neutr. Number (mg KOH/g)	5.0
pH	6.5
Moisture (%)	3.6
Density (g/cm^3 20°C)	1.07
HHV (MJ/kg m.f.)	30.2
Elemental Analysis (m.f.)	
C (wt%)	68.8
H (wt%)	7.7
N (wt%)	0.3
O (wt% by difference)	23.2

The extent of the upgrading was also evaluated in terms of the forming of benzene and n-pentane solubles, which increased with respect the starting bio-crude by 0.25 to 12.83 and 4.27 to 34.85 respectively (table 7). The elemental analyses of the different fractions show that the oxygen content is equally distributed.

Pentane solubles were separated in the saturated and the polar plus aromatic fractions using a properly modified ASTM D2700-75 method [8]. The saturated hydrocarbons fraction constitutes about 1.2 % of pentane solubles and their components are mainly linear paraffines. The GC analysis showed mainly the presence of C10-C26 chains, the structure was confirmed by I.R. analysis.

Table 7. Upgraded oil fractions

	yield (wt%)	C %	H %	N %	O%
n-pentane solubles	12.83	65.2	9.1	0.7	25.0
saturated hydrocarbons	1.20	84.9	15.1	---	---
polar and aromatics	11.63	63.1	8.5	0.8	28.0
benzene solubles	34.85	67.6	9.1	0.2	23.1
tetrahydrofurane solubles	52.32	70.5	6.4	0.3	22.8

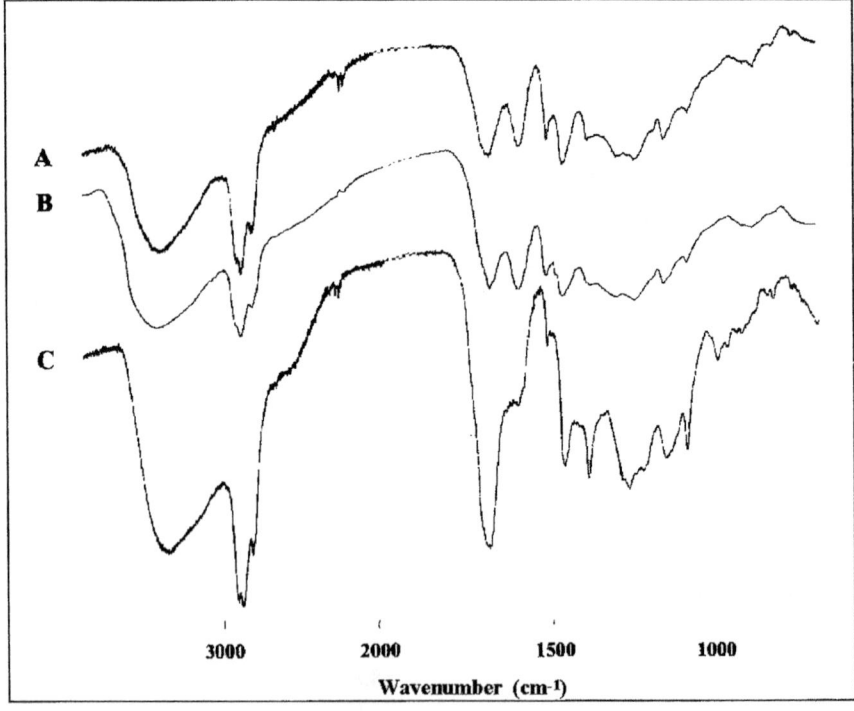

Fig. 3. Infrared spectra of A, benzene solubles, B, benzene insolubles and C, polar aromatic fractions of upgraded bio-oil.

As expected, the polar aromatic fraction has a high oxygen content; the calculated empirical formula is $C_{5.25}H_{8.50}N_{0.05}O_{1.75}$.

In the infrared analyses (fig. 3) typical signals of functions can be observed, a strong band around 1700 cm^{-1} due to the presence of carbonyl compounds; the bonded OH stretching at 3300 cm^{-1}; the aromatic CO stretching at 1150-1330 cm^{-1}, the pure aromatic skeletal stretching at 1600 and 1500 cm^{-1}, and a remarkable substitution of the aromatic ring in the 900-700 cm^{-1} region. The presence of a such a large amount of carbonyl compounds was unexpected in consideration of the reactivity scale proposed by other authors [9].

The I.R. spectra of the benzene solubles fraction and the insoluble one shows very large bands and a rather bad resolutions. This is due to the overlapping of absorptions of many high molecular weight compounds. In these fractions also carbonyls and strong bonded OH groups can be noted.

The stability and the reological properties of the upgraded bio-oil were investigated as well.

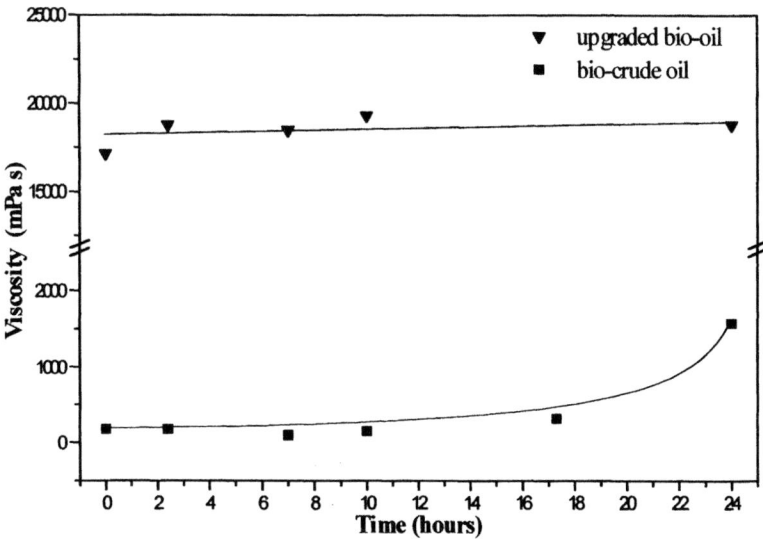

Fig.4. Stability of starting bio-crude oil and upgraded oil as a function of time at 100 °C

In figure 4 the trends of viscosity after heating at 100 °C for several hours of the starting bio-crude oil and the upgraded oil measured at 20 and 60 °C respectively are shown. The hydrogenated product has a higher viscosity then the starting bio-crude, but it shows an improved stability; infact, under the used conditions, no appreciable increasing of viscosity was registered.

The rheological behaviour of the upgraded oil can be described by the so-called power law [10], written as

$$\log \eta_{app} = A \log D + K$$

where η_{app} is the apparent viscosity, A measures the dependence of the viscosity on the shear rate D and K is the consistency index and gives the viscosity at D =1.

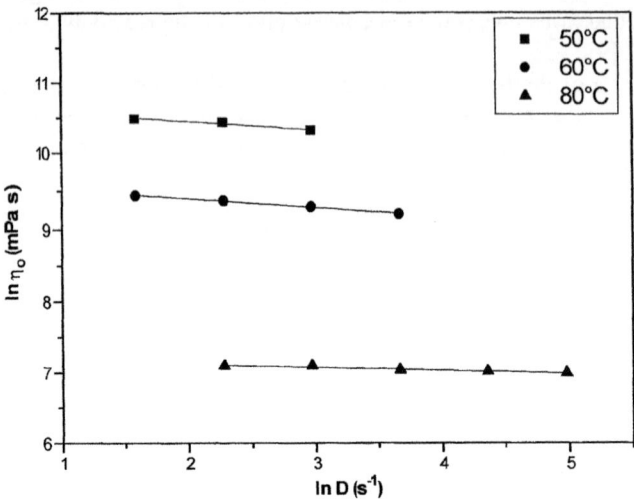

Fig.5. Rheological behaviour of hydrotreated bio-oil at different shear rates.

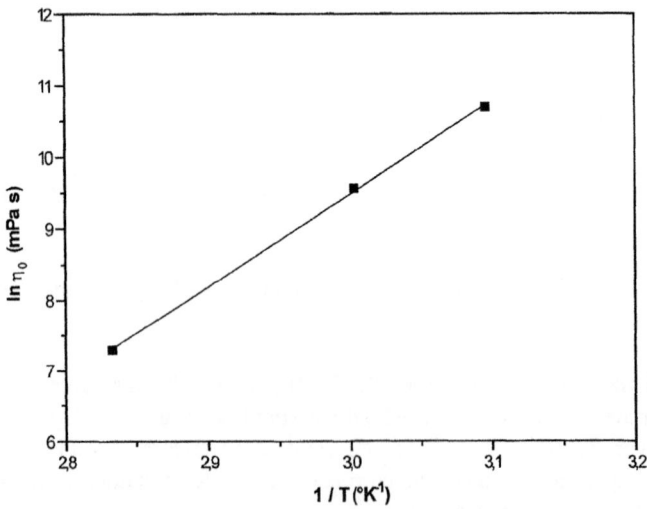

Fig. 6. Trend of the viscosity versus temperature of hydrotreated bio-oil (D= 1 s⁻¹).

The upgraded bio-oil appears slightly pseudoplastic at low temperatures and Newtonian at higher ones (figure 5).

Figure 6 shows the viscosity-temperature relations for the oils, as expressed by the Arrhenius-Guzman equation[10], at $D = 1$ ($\eta_{app} = \eta_0$):

$$\eta_0 = C\exp(B/RT)$$

where C is related to the molecular weight and molecular volume of the fluid [12] and B is the energy of activation of viscous flow.

B is related to the heat of vaporisation of the fluid, which is in turn highly dependent on the association between the constituent molecules. The upgraded bio-oil shows a linear relation between log η_0 and temperature, hence B is constant. This one means that the high viscosity of the oil may be due to the high molecular weight rather than molecular associations involving mainly polar groups[13].

4 Remarks and Conclusions

The process performed has been successful in significantly improving the physical and chemical properties of flash-pyrolysis bio crudes to produce stabilized oils for further upgrading by conventional hydrotreating processes. These improvements have been achieved by a mild hydrotreatment step as well as a particular temperature profile.

This experiment demonstrate the feasibility of the process. Even though no apparent deactivation of the catalyst had been observed during runs, additional work needs to be done in a pilot plant to study the long-term stability of the catalyst and new catalysts have to be developed to take into account the peculiarities of bio-oils, as well as the kinetic of the process.

5 References

1. Baker, E. G. & Elliot, D. C. (1988) Catalytic Upgrading of Biomass Pyrolysis Oils, In *Research in Thermochemical Biomass Conversion*, (eds A. V. Bridgwater and J. L. Kuester), Elsevier Applied Science, London and New York. pp.883-95.
2. Conti, L., Scano, G., Boufala, J., Trebbi, G. and Rossi, C.(1995) Upgrading of Pyrolysis Oils from Biomass in a Bench-Scale Continuous Plant, *8th European Conference on Biomass for Energy, Environment, Agriculture and Industry*, (eds P. Chartier, A.A.C.M. Beenackers and G.Grassi), Vol.3, Pergamon Press, Oxford. pp. 1901-7.
3. Conti, L., Scano, G. and Boufala, J. (1994) Continuous Hydrotreating Research at University of Sassari, *Proc. Contr. Meeting AIR CT92-0216*, Gelsenkirchen (D).
4. Baldauf, W. and Balfanz, U., Upgrading of Pyrolysis Oils from Biomass in Existing Refinery structures, *Progress Report: June 91-November 91, Contract JOUB-0015/0055-C (MB)*.
5. Kirk-Othmer, Vol. 10, *Fluid Mechanics*, p. 593
6. Thakur D.S. and Thomas M.C.(1985) *Appl. Catal.*, 15. pp. 191-225
7. Elliot, D. C., Beckman, D., Bridgwater, A. V., Diebold, G. B., Gevert, S. B. and

Solantausta, Y. (1991) Developments in Direct Thermochemical Liquefaction of Biomass: 1983-1990. *Energy&Fuel*, pp. 390-410.

8. Conti, L. and Rausa, R. (1987) Flash Pyrolysis of an Italian Low Rank Coal. *Fuel Processing Technology*, 17, pp. 107-15.

9. Laurent, E., Pierret, C., Grange, P. and Delmon, B.(1991) Control of Deoxygenation of Pyrolytic Oils by Hydrotreatment, *Proc. of 6th E.C. Conference Biomass for Energy, Industry and Environment*, (eds Grassi G., Collina A. and Zibetta H.), Elsevier Applied Science.

10. Wilkinson, W. L. (1960) *Non-newtonian fluids*, Pergamon Press, Oxford.

11. Weltman, R. N. (1960) *Rheology, Theory and Applications*, (ed. F.R. Eirich), Academic Press, New York, Vol. III.

12. Eyring, H. (1936) *J. Chem. Phys.*, 4, p. 283.

13. Conti, L., Boatto, G. and Passarini N. (1985) Coal-oil mixture from a synthetic oil obtained by hydroliquefaction of a South African coal. *Fuel*, 64, pp. 1317-19.

CATALYTIC CONVERSION OF CANOLA OIL IN A FLUIDIZED BED REACTOR

S. P. R. KATIKANENI, J. D. ADJAYE, R. O. IDEM AND N. N. BAKHSHI

Catalysis and Chemical Reaction Engineering Laboratory
Department of Chemical Engineering
University of Saskatchewan, Saskatoon, Saskatchewan, Canada

Abstract

Studies were conducted in a fluidized-bed reactor at atmospheric pressure, reaction temperatures in the range 400-500°C and fluidizing gas flow rates ranging from 175-275 mL/min to study the product distribution obtained from the conversion of canola oil over HZSM-5, silica-alumina and HS-Mix (a physical mixture containing 20 wt% HZSM-5 and 80 wt% silica-alumina).

Canola oil conversions increased with reaction temperature and were in the range 73-98 wt%. The product distribution mostly consisted of hydrocarbon gases in C_1-C_5 range, a mixture of aromatic and aliphatic hydrocarbons in the organic liquid product (OLP) and coke. Also, the product distribution varied widely and depended strongly on the catalyst type and operating conditions.

It was interesting to observe that C_2-C_4 olefins fraction in the gas product depended on reaction temperature and type of catalyst but was not affected significantly by changes in fluidization gas flow rate. On the other hand, C_5^+ hydrocarbon gases increased markedly with both an increase in the fluidization gas flow rate and a decrease in reaction temperature.

The OLP obtained over HZSM-5 contained mostly aromatic hydrocarbons whereas those for silica-alumina and HS-Mix contained considerable amounts of aliphatic hydrocarbons in addition to small amounts of aromatic hydrocarbons. Furthermore, both the reaction temperature and catalyst type determined whether the fractions of aromatic hydrocarbons in the OLP were higher in the fluidized-bed or fixed-bed reactor.

Keywords: Fluidized-bed, canola oil conversion, product distribution, HZSM-5, silica-alumina.

1 Introduction

As conventional energy sources deplete, the need for developing alternative energy resources becomes more imperative. Biomass is attracting increasing interest [1-14] as a renewable source for the production of fuels and useful chemicals.

In an earlier work [12-17], we studied the catalytic upgrading of canola oil over various acid catalysts such as HZSM-5, HY, H-mordenite, aluminum-pillared clays, silica-alumina and silico-alumino phosphates using a fixed-bed reactor in order to determine the effects on oil conversion and product distribution of catalyst characteristics such as acidity and acid site distribution, crystallinity, pore size and pore size distribution as well as to study the thermal stability of the catalyst under various operating environments. These studies showed that there were dramatic variations in both the overall oil conversion and yields of individual product for different catalysts used. For example, high yields of aromatic hydrocarbons together with high oil conversions were obtained with HZSM-5 catalyst whereas the reverse trend was observed for silica-alumina.

Also, in a recent study [15], the use of HS-Mix (catalyst consisting of a physical mixture of HZSM-5 and silica-alumina) for the upgrading of a wood derived bio-oil resulted in a very high selectivity for isooctane. In contrast, the use of HS-Mix catalyst for canola oil conversion resulted in a high selectivity for C_2-C_4 olefins [14]. Also, it has been shown [12-17] that apart from the catalyst, operating conditions such as space velocity and reaction temperature also affect conversion and the product distribution.

As the contacting pattern also has a tremendous affect on the product distribution, it was therefore decided to use a fluidized-bed reactor for converting canola oil over various catalyst systems of importance such as HZSM-5 and silica-alumina. Such studies using a fluidized-bed reactor for canola oil conversion have not been reported in the literature. These results are presented and discussed in this paper.

2 Experimental

2.1 Chemical composition of canola oil
The canola oil feed material used in this study was obtained from CSP Foods, Saskatoon, Canada. It consisted of about 60 wt % oleic acid, 20 wt % linoleic acid, 10 wt % linolenic acid, 2 wt % stearic acid, 4 wt % palmitic acid and small amounts of eicosenic and erucic acids. These fatty acids were present in the oil as their triglyceride molecules.

2.2 Catalyst preparation and characterization
The catalysts used were HZSM-5, silica-alumina and HS-Mix (a physical mixture containing 20 wt% HZSM-5 and 80 wt% silica-alumina). They were selected based on previous studies [12-17] in a fixed-bed reactor which showed that the product distributions obtained with these three catalysts were widely different.

HZSM-5 was prepared according to the procedure reported in the literature [13, 14] whereas silica-alumina catalyst was obtained from Union Carbide (Danbury, CT, USA). Each of the above catalysts was ground and sieved into particle sizes ranging from 104 to 150 µm. HS-Mix catalyst was obtained by physically mixing 20 wt % HZSM-5 powder with 80 wt % silica-alumina powder (each in the particle size range of 104-150 µm). All the three catalysts were thoroughly characterized using XRD, BET surface area, TPD of ammonia, FT-IR of pyridine adsorption and solid state NMR studies. The Dubinin-Astakhov program was used for measuring the micropore area of HZSM-5 whereas the BET program was used to determine the surface area of silica-alumina. HZSM-5 contains mostly micropores. Thus, the surface area of HZSM-5 is only a measure of the surface area based on these micropores. The average pore size of silica-alumina was determined using the BJH equation (for catalysts having mesopores) while the median pore size of HZSM-5 was determined using Horvath-Kawazoe equation (used for micropore catalysts).

Both are considered to be equivalent for the purpose of comparison. Details concerning the characterization of these catalysts are given elsewhere [13, 14, 17].

2.3 Experimental setup and procedure

The performance evaluation of the catalysts was carried out in a fluidized-bed reactor at atmospheric pressure at temperatures in the range 400-500 °C. This temperature range was selected based on previous studies [12-17] in a fixed-bed reactor which showed that large amounts of the products of interest were obtained within this temperature range.

Figure 1 shows the schematic diagram of the experimental set-up. The fluidized-bed reactor consisted of a top cylindrical portion (25.4 mm i.d. and 170 overall length) which tapered into a bottom cylindrical portion (6.3 mm i.d. and 250 mm length). The vertical height of the tapered section was 10 mm. All the reactor components were made of stainless steel (SS 316). The reactor was equipped with a distributor of circular cross section (25.8 mm dia and with size of openings ≈ 100 μm) which was located at a distance 150 mm from the top flange. The reactor was also equipped with a feed inlet tube as well as a products outlet tube as shown in Figure 1. On the other hand, the reactor did not contain a cyclone separator. Thus, to minimize catalyst carry over, the reactor was provided with a large free board section. Details concerning the fluidized-bed reactor are given elsewhere [18].

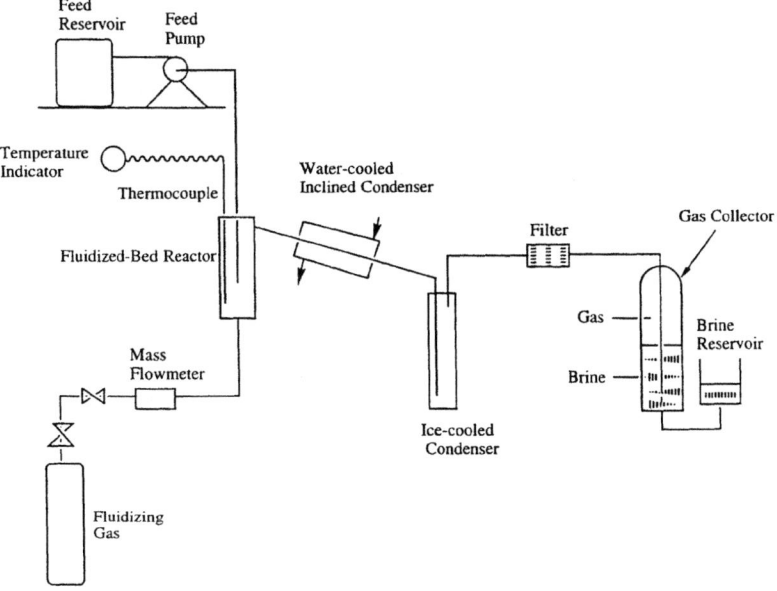

Figure 1: Schematic Diagram of Experimental Set-up

The reactor was placed in an electrically heated furnace whose temperature was controlled by a Series SR22 microprocessor-based autotuning PID temperature controller (supplied by Shimaden Co. Ltd., Tokyo, Japan) using a K-type thermocouple inserted into the furnace. A separate thermocouple was used to monitor the temperature of the fluidized catalyst bed as shown in Figure 1. The thermocouple was placed at 24 mm above the static bed height. When the catalyst was fluidized, the thermocouple was at the center of the fluidized bed. This arrangement was capable of ensuring an accuracy of $\pm$ 2 $^{\circ}$C for the catalyst bed temperature. The fluidization gas entered the reactor from the bottom.

The minimum fluidization velocity determined experimentally was 0.37 mm/min. The actual velocities used for the experimental runs were 0.37, 0.48 and 0.58 m/min. The corresponding fluidization gas flow rates were 175, 225 and 275 mL/min (at atmospheric pressure and 25 $^{\circ}$C (i.e., STP)).

A typical fluidized bed run was performed as follows: The reactor was loaded with 10 g of the desired catalyst placed on top of the distributor. The reactor was then heated in the flowing argon gas (175 mL/min) to the desired reaction temperature and then the argon flow was adjusted to the desired fluidization flow rate. Canola oil was then fed to the reactor from the top by an Eldex micrometering pump (Model A-60-S) at the rate of 0.6 g/min. The product mixture from the reactor was cooled in a chilled water-cooled condenser. The liquid product was separated in a gas/liquid separator and then collected in an ice trap. On the other hand, the gas product was collected over brine in a gas collector. Each experimental run lasted for 20 min.

2.4 Collection and chemical analysis of products

The liquid product was distilled at 200 $^{\circ}$C at a vacuum of 172 Pa using a Buchi GKR-50 distillation unit. The distillate obtained was referred to as "the organic liquid product (OLP)" and the residue was termed "residual oil". Details concerning the collection of all products are given elsewhere [13, 14]. The chemical composition of gas and organic liquid product (OLP) fractions were analyzed using a Carle GC (Model 500) equipped with FID and TCD detectors in conjunction with GC-MS analysis. Details concerning these analyses are also given elsewhere [12, 13]. The entire procedure was repeated a number of times for some runs in order to check for reproducibility. The error was less than $\pm$ 2 wt % in terms of both canola oil conversion and yields of products.

3 Results and discussion

3.1 Catalyst selection and characteristics

The catalytic properties of HZSM-5 are attributed to both its strong acid sites and 3D system of intersecting straight channels and near circular zig-zag channel. The acid sites in HZSM-5 lie on the intercrystalline system of the zeolite and are all accessible. Molecules such as cyclohexane, xylenes, trimethyl benzene have been found to diffuse through the pores of HZSM-5. In addition, HZSM-5 exhibits a high shape selective property. On the other hand, silica-alumina has an amorphous pore structure and does not exhibit any shape selective property. Also, the acid sites are buried in inaccessible locations, thus leading to a low acidity. Hence, these two catalysts were selected to determine the effect of crystallinity, shape selectivity and type of acidity on product distribution. The characteristics of the three catalysts are given in Table 1. surface area and XRD results confirmed that HZSM-5 was crystalline and had a uniform pore structure with an average pore size of 0.54 nm whereas silica-alumina was amorphous with an average pore size of 3.15 nm. Also, TPD of NH_3, solid state NMR and FT-IR results showed that HZSM-5 contained mostly strong Bronsted acid sites whereas silica-alumina contained Lewis acid sites.

Table 1: Characteristics of the catalysts

Catalyst	Surface area, m²/g	Pore size, nm	silica/alumina ratio	Acid density*, mm²m⁻²
HZSM-5	329	0.54	56.0	95
Silica-alumina	320	3.15	0.79	45
20% HZSM-5 + 80% silica-alumina	322	2.63	11.8	55

* measured by TPD of ammonia

In the literature, the acid site density for HZSM-5 and silica-alumina has been expressed either in terms of mm^2/m^2 or $mmol/m^2$ [19, 20]. Although the later expression gives the acid site density in absolute terms, we decided to use mm^2/m^2 as a convenient and accurate method to evaluate the relative acid strengths of the three catalysts. Table 1 shows the acid site density for HZSM-5 is larger than that of silica-alumina. The acid site density for the HS-mix catalyst was found to be in between that of HZSM-5 and silica-alumina. These results are consistent with those reported in the literature [12-17] for the three catalysts. The effects of these characteristics on canola oil conversion and product distribution in a fluidized-bed reactor are discussed below.

3.2 Catalytic conversion of canola

Table 2 shows the canola oil conversion and the yields of gas, OLP, coke and residual oil obtained for the catalytic conversion of canola oil in a fluidized-bed reactor as a function of type of catalyst, reaction temperature and fluidization gas velocity. It was interesting to observe that there were drastic changes in conversions and product distributions due to the changes in the operating variables. Catalysis literature indicates that such changes are due to the effects of the operating variables on the reaction mechanism. In an earlier study [13, 14, 16], a possible reaction scheme was developed to show the reaction sequence for the formation of various products from canola oil conversion. This sequence is given in steps 1-7 (i.e., Equations 1-7).

step 1 canola oil $\rightarrow$ heavy oxygenated C_xH_y (thermal) (1)

step 2 heavy oxygenated C_xH_y $\rightarrow$ heavy C_xH_y + H_2O + CO_2 + CO (thermal + catalytic) (2)

step 3 heavy C_xH_y $\rightarrow$ paraffins + olefins (short and long chain) (catalytic) (3)

step 4 light olefins $\rightarrow$ C_2-C_{10} olefins (catalytic) (4)

step 5 C_2-C_{10} olefins $\Leftrightarrow$ aliphatic C_xH_y + aromatic C_xH_y (catalytic) (5)

step 6 canola oil $\rightarrow$ coke (thermal) (6)

step 7 n (aromatic C_xH_y) $\rightarrow$ coke (catalytic) (7)

Here, steps 1 and 2 involve the primary cracking of canola oil molecules to give a mixture of heavy hydrocarbons and heavy oxygenated hydrocarbons. This mixture is regarded as "residual oil" which may undergo further deoxygenation and/or secondary cracking (step 3) to give olefins and paraffins. The light olefins oligomerize (step 4) to yield olefins in the C_2-C_{10} range. Aromatic hydrocarbons are eventually produced (step 5) as a result of the aromatization reactions of the C_2-C_{10} olefins. Coke is formed either due to polycondensation of canola oil molecules (step 6) or polymerization of large aromatic hydrocarbon molecules (step 7). The above sequence of steps are used in discussing the effects of operating variables on canola oil conversion and product distribution as follows.

3.2.1 Type of catalyst

Table 2 shows that canola oil conversions under similar operating conditions for the three catalysts decreased in the order: HZSM-5 > HS-Mix > silica-alumina. This order follows the same decreasing order as acid site density, crystallinity and shape selectivity as shown in Table 1. It was observed from earlier studies that the initial reaction in canola oil conversion was the thermal cracking of triglyceride molecules (step 1). Subsequently, the cracked molecules undergo various reactions on the surface/pores of the catalysts. The secondary reactions and the final product distribution is strongly dependent on the above catalyst characteristics. The literature [16] shows that some of the forward reaction steps (Equations 2-5) given in the reaction sequence are catalyzed by Bronsted acid centers. Thus, the greater the acid site density, the greater the shift in equilibrium in favor of the forward reactions, and consequently, the greater the conversion.

It was interesting to observe that both the gas and OLP fractions decreased in the same order as canola oil conversion (i.e., an increase in both gas and OLP fractions with acid site density (see Table 2)). In fixed-bed reactor operation [12-16], a higher Bronsted acid density in the catalyst resulted in a greater yield of the OLP fraction and a smaller gas yield. Thus, an increase in both gas and OLP fractions with acid site density obtained with fluidized-bed reactor may be attributed mainly to the contacting pattern that exists in the fluidized-bed reactor. The contacting pattern in the fluidized-bed reactor may have been responsible for large amounts of gas being formed, in the first place, and subsequently for a substantial fraction of the gas being converted to OLP irrespective of the acid density of the catalyst. However, at this time, this behavior is not fully understood. Further studies are being carried out to elucidate this behavior. On the other hand, it was observed that coke formation increased in the order: HZSM-5 < HS-Mix < silica-alumina. This is consistent with results reported in the literature [13, 14] which shows that coke formation decreases with an increase in Bronsted acid site density.

Usually, the components of interest in the gas phase products are C_2-C_4 olefins, as well as the C_4 and C_5 hydrocarbons. Table 2 shows that there were drastic variations in the concentrations of some components with the type of catalyst used. For example, the concentrations of ethylene and propylene (olefins) decreased in the order HZSM-5 > HS-Mix > silica-alumina while a reverse trend was observed for ethane and propane (paraffins). Also, the concentration of low molecular weight hydrocarbons such as methane decreased in the order HZSM-5 > HS-Mix > silica-alumina while a reverse trend was observed for high molecular weight hydrocarbon gases such as n-butane, isobutylene and C_5^+ hydrocarbons. These variations can be explained as follows. Our previous work [13-16] indicates that Bronsted acid centers promote hydrogenation reactions. This implies that the extent of hydrogenation which results in the formation of paraffins increases with the Bronsted acid site density in the catalyst whereas the extent of dehydrogenation to form olefins increases with a decrease in the catalyst acid site density.

Also, it is known that some of the reaction steps involved in canola oil conversion (steps 2-5) are catalyzed by acid centers. This implies that the extent of

Table 2: Mass balances and compositions of gas and OLP fractions for canola oil conversion in a fluidized-bed reactor

Catalyst	HZSM-5						HS-mix			Silica-alumina					
Temp., °C	400	400	400	450	500	500	400	450	500	400	400	400	450	500	500
Flow rate, mL/min	175	225	275	225	225	275	225	225	225	175	225	275	225	225	275
Gas	28	30	32	42	60	62	22	39	52	18	20	22	35	55	57
OLP	44	35	29	32	20	13	28	32	20	30	22	29	29	17	14
Residual oil	7	12	16	7	2	5	15	12	10	15	19	21	14	9	10
Coke	5	5	4	4	2	4	15	6	5	16	18	20	5	5	5
Unaccounted	16	18	19	15	16	16	20	11	13	21	21	8	17	14	14
Conversion	93	88	84	93	98	95	85	88	90	85	81	79	86	91	90
						Gas composition, wt % of Gas									
Methane	0.3	0.2	0.1	0.2	0.6	1	0.2	0.8	2	1	0.1	0.2	1	3	3
Ethylene	7	5	5	0.5	10	11	4	5	7	2	1	2	2	3	3
Ethane	1	1	0.4	8	1	1	0.4	1	2	1	0.2	1	2	2	2
Propylene	16	14	16	19	21	30	16	24	29	14	8	12	12	12	13
Propane	16	17	12	7	3	4	6	5	5	4	3	6	4	6	4
n-butane	9	8	9	10	10	12	6	8	13	4	2	3	5	5	6
Isobutane	9	7	5	7	5	1	18	10	2	12	11	11	7	9	9
Isobutylene	10	11	10	10	8	7	9	10	11	7	6	6	6	9	7
C_2-C_4 olefins	35	34	33	39	43	52	29	39	47	27	15	22	22	25	26
						OLP composition, wt% of OLP									
Aromatics	89	87	88	88	90	81	34	47	66	33	29	37	44	63	46
Aliphatics	1	1	-	1	2	2	28	20	13	29	37	25	26	17	16

cracking of each molecule increases with the acid site density. Thus, the higher the acid site density, the higher the extent of cracking, and consequently, the lower the molecular weights of the hydrocarbon products formed. Hence, cracking of molecules over HZSM-5 occurs to a greater extent than that over silica-alumina resulting in the formation of larger amounts of high molecular weight hydrocarbons (such as n-butane, isobutylene and C_5^+ hydrocarbons) with the latter catalyst than with the former catalyst.

Table 2 shows that differences in catalyst characteristics affect the liquid phase products (OLP). It shows that the OLP obtained over HZSM-5 (catalyst with strong Bronsted acid sites and high shape selectivity) contained mostly aromatic hydrocarbons whereas those obtained over silica-alumina (an amorphous catalyst containing a low acid site density) contained a considerable amount of aliphatic hydrocarbons. The amounts of aromatic and aliphatic hydrocarbons obtained for HS-Mix were somewhere between those obtained for HZSM-5 and silica-alumina. These results are consistent with those reported for fixed-bed reactor [13, 14].

3.2.2 Reaction temperature

Table 2 shows that there were drastic variations in the concentration of some components in the gas product with reaction temperature. For example, the concentrations of ethylene and propylene (olefins) increased with reaction temperature while a reverse trend was observed for ethane and propane (paraffins). Also, the concentrations of low molecular weight hydrocarbons such as methane increased with reaction temperature while a reverse trend was observed for high molecular weight hydrocarbon gases such as n-butane, isobutylene and C_5^+ hydrocarbons. These variations are explained below. Literature [22] indicates that cracking is an endothermic process and thus, the severity of cracking increases with temperature. Consequently, the higher the cracking temperature, the lower the molecular weights of the products formed and hence, the increase in methane formation and the decrease in the formation of n-butane, isobutylene and C_5^+ hydrocarbons with increasing temperature.

On the other hand, the opposing trends exhibited by ethane and ethylene and by propane and propylene show that the extents of hydrogen transfer reactions such as hydrogenation (as in the formation of ethane from ethylene or $C_2H_5^+$ ions) increase with a decrease in temperature whereas those for dehydrogenation reactions for ethylene formation from ethane or $C_2H_5^+$ ions increase with temperature. Similarly, the extents of hydrogenation reactions for propane formation from propylene or $C_3H_7^+$ ions increase with a decrease in temperature while those for dehydrogenation reactions for propylene formation from propane or $C_3H_7^+$ ions increase with temperature.

The yield of coke decreased as the cracking temperature increased. This can be explained on the basis that an increase in temperature facilitates coke gasification which results in a net decrease in the coke product. This result shows that high temperature operation can be advantageous since it results in lower coke formation and hence in lower catalyst deactivation. Previous studies in a fixed-bed reactor using HZSM-5 did not indicate any deactivation upto 90 min of operation. In addition, the used catalyst was regenerated upto 4 times without affecting the catalyst activity appreciably. One advantage of the fluidized-bed is that it can be generated in a reaction/regenerator mode similar to a typical fluid catalytic cracking (FCC) unit in a petroleum refinery.

Table 2 shows that for the three catalysts and fluidization gas velocities used, there is a maximum in the relationship between the yield of OLP and reaction temperature. This can be explained as follows. It was mentioned earlier that OLP is the distillate fraction obtained from the vacuum distillation of the total liquid product. The residue is called residual oil. Also, as was shown previously, a low reaction temperature results in the formation of large amounts of relatively long chain compounds (total liquid) at the expense of the gas product. This implies that both OLP and residual oil will increase as the temperature decreases. However, due to very limited cracking below a certain temperature, most of the long chain compounds end up as residual oil instead of OLP. Thus, below this temperature the yield of OLP starts to decrease and hence the maximum. Based on the variables studied in the present investigation, it appears that the maximum OLP is obtained at $450\,^{\circ}C$.

The variation of the concentrations of various aromatic hydrocarbons in OLP with reaction temperature also is shown in Table 2. It is seen that the concentration of aromatic hydrocarbons increased with cracking temperature. This shows that the extent of aromatization reactions increased as the cracking temperature increased. A similar trend was observed in an earlier study using a fixed-bed reactor [13, 14, 16].

It is seen in Table 2 that the concentration of C_6^+ aliphatic hydrocarbons decreased with temperature. This decrease is explained as follows. Our product classification indicates that OLP is composed mainly of C_6^+ aromatic hydrocarbons and C_6^+ aliphatic hydrocarbons. Also, it was shown earlier (Table 2) that aromatic hydrocarbon formation (i.e., cyclization and aromatization reactions) increased with cracking temperature. Thus, the implication is that as the cracking temperature increases, a larger fraction of C_6^+ hydrocarbons end up as aromatics instead of aliphatic hydrocarbons. The converse is also true and hence, the decrease in C_6^+ aliphatic hydrocarbons with an increase in cracking temperature.

3.2.3 Fluidization gas velocity

Argon gas was used as the fluidization gas. The range of fluidization gas velocities used was from 0.37 m/min to 0.58 m/min. Canola oil flow rate was constant for all the runs and the feed vapor was assumed to be entrained in the fluidization gas since the mass balances obtained with the fluidized-bed reactor were close to those obtained with the fixed-bed reactor.

Table 2 shows typical effects of fluidization gas velocity on canola oil conversion and product distribution at 400, 450 and $500\,^{\circ}C$ for HZSM-5, HS-Mix and silica-alumina. It is seen that canola oil conversion decreased with an increase in fluidization gas velocity. This is typical of cracking reactions. As was mentioned earlier, an increase in fluidization gas velocity results in a decrease in the contact time between the feed and the catalyst, and consequently, in a smaller extent of the cracking reaction.

It was rather interesting to observe that the gas yield increased with an increase in fluidization gas velocity (see Table 2) for both HZSM-5 and silica-alumina catalysts. This can be explained on the basis of the possible reaction sequence for canola oil conversion to various products (Equations 1-7). This sequence indicates that the reaction steps that result in the formation of gas products (Equations 2 and 3) occur much earlier than those that lead to subsequent consumption of the gas produced to form OLP (Equations 4 and 5). Thus, within the range of fluidization gas velocities used, an increase in fluidization gas velocity does not affect gas production whereas it adversely affects subsequent conversion of gas to OLP. Usually, this should result in a corresponding decrease in the yield of OLP with an increase in fluidization gas velocity. Our results show that there was a decrease in the yield of OLP with fluidization gas velocity in the case of HZSM-5. However, OLP yield appeared to be independent of fluidization gas velocity in the case of silica-alumina. This result for silica-alumina is not fully understood at this time. A more detailed investigation is underway.

Table 2 shows that in the gas phase products, the amounts of n-butane and C_5^+ hydrocarbon gases (high molecular weight gases) increased with fluidization gas velocity whereas the amount of methane (low molecular weight gas) decreased with fluidization gas velocity. As shown in the reaction sequence (Equations 1-7), gas phase hydrocarbons are formed from secondary cracking and deoxygenation of heavy oxygenated hydrocarbons (Equations 2 and 3). As was discussed previously, an increase in fluidization gas velocity (which implies a decrease in the contact time between the catalyst and the feed vapor) results in a less extensive cracking of each individual heavy molecule and consequently, in the presence of hydrocarbon components with relatively high molecular weights in the gas product.

The concentration of the total C_2-C_4 olefins appeared to be independent of fluidization gas velocity. As was discussed earlier, within the range of fluidization gas velocities used, an increase in fluidization gas velocity is not detrimental to gas formation. However, it does appear from the results that the net rate of formation of olefins in the gas phase and the subsequent conversion of these olefins to form other products such as aromatics in OLP are independent of fluidization gas velocity. On the other hand, there are no definite trends in the variations of both C_6^+ aromatic and aliphatic hydrocarbons with fluidization gas velocity as seen in Table 2. This result can be attributed principally to the wide variety of reactions involved in the formation and depletion of C_6^+ aromatic and aliphatic hydrocarbons. These reactions appear to be affected by fluidization gas velocity in a rather erratic manner.

3.3 Comparison with fixed-bed reactor studies

As was discussed earlier, there are various factors that can influence the cracking performance of a catalyst in both fluidized- and fixed-bed reactors. These are: (i) the average contact time between the catalyst and the feed, (ii) the contacting pattern, (iii) the axial and radial temperature distributions, and (iv) the possibility of loss of catalyst by entrainment in a fluidized-bed reactor [21-24]. The individual and/or gross effects of these factors on canola oil conversion and product distributions are discussed below. A comparison between the performance of the catalysts in a fluidized-bed reactor and that in a fixed-bed reactor is given in Tables 3.

The mass balances shown in Table 3 indicate that, generally, there were relatively more losses in fluidized-bed operation than the fixed-bed. These losses are referred to as "unaccounted fraction". The slightly larger losses in fluidized-bed reactor operation show that there was some entrainment of catalyst particles in some of the runs. In such cases, the amounts of coke and residual oil carried along with the catalyst particles were not included in the mass balance. However, the results show that the fraction of catalyst particles carried over was small and was less than 1 wt % for most runs and less than 5 wt % for runs performed at 500 °C (for a fluidization gas flow rate of 275 mL/min).

Table 3 shows that the canola oil conversions obtained in the fluidized-bed reactor were somewhat lower than those obtained in a fixed-bed reactor. This was attributed principally to the shorter average contact times between the feed and the catalyst in the fluidized-bed reactor.

The table also shows that, in general, gas yields were higher for fluidized-bed reactor operation than for fixed-bed reactor operation. On the other hand, a reverse trend was observed for OLP yields. This can be explained using the contact times employed for the two types of reactors in conjunction with the reaction sequence given in Equations 1-7. These results show that in the fluidized-bed reactor, the shorter contact times were sufficient for reactions that lead to the formation of gas products (Equations 2 and 3) to take place but not sufficient for the subsequent reactions of the gas to form OLP (Equations 4 and 5). Consequently, only small amounts of OLP were formed. In contrast, the longer contact time used in the fixed-bed reactor was sufficient for both types of reactions to take place. Consequently, larger amounts of OLP were formed at the expense of the gas products.

Table 3 shows that coke yield depended not only on the type of reactor but also on the reaction temperature. For example, larger amounts of coke were formed with a fluidized-bed reactor than with the fixed-bed reactor at 400 °C. At this time, this behavior is not fully understood.

Table 3: Comparison between the performance of catalysts in fluidized- and fixed-bed reactors

Temp., °C	400				450				500			
Catalyst	HZSM-5		Silica-alumina		HZSM-5		Silica-alumina		HZSM-5		Silica-alumina	
*Reactor configuration	1	2	1	2	1	2	1	2	1	2	1	2
Gas	25	30	11	20	34	42	29	35	50	60	45	55
OLP	63	35	40	22	56	32	51	29	41	20	42	17
Residual oil	3	12	7	19	1	7	4	14	0	2	1	9
Coke	3	5	36	18	4	4	13	5	5	2	10	5
Conversion	97	88	93	81	99	93	96	86	100	98	99	91
C_2-C_4 olefins	16	33	25	15	29	39	28	22	34	43	29	25
Total aromatics	82	87	28	29	88	87	34	44	96	90	53	63
Aliphatic hydrocarbon	1.6	1.0	30	37	0.8	0.6	22	26	0.3	2	14	17
[+]aromatic hydrocarbon conc'n ratio	0.94		0.97		1.01		0.77		1.07		0.84	

* 1 = fixed-bed reactor; 2 = fluidized-bed reactor

[+] Fixed-bed/fluidized-bed aromatic hydrocarbon concentration ratio

An entirely different scenario was observed in the case of the gas phase selectivity for total C_2-C_4 olefin production. Table 3 shows that the selectivity for production of these olefins was in the order: HZSM-5 > silica-alumina > HS-Mix which is different than that observed for a fixed bed reactor. The trend for total olefin selectivity for fixed-bed reactor was: silica-alumina > HS-Mix > HZSM-5.

On the other hand, Figure 2 shows that the concentrations of isobutane and isobutylene (branched C_4 hydrocarbons) in the gas product were higher for the fluidized-bed reactor than with the fixed-bed reactor. This implies that there is a greater degree of isomerization but without subsequent conversion of the iso-products to OLP or other products in fluidized-bed than the fixed-bed reactor.

The yields of aromatics in OLP for operations using the two types of reactors for HZSM-5 and silica-alumina is shown in Figure 3. As expected, the yields of aromatic hydrocarbons were higher with HZSM-5 catalyst than with silica-alumina. Also, aromatic hydrocarbon yields were higher for fixed-bed reactor operation than for fluidized-bed reactor operation. This is consistent with the longer contact times (between the catalyst and feed) that exists in the fluidized-bed reactor compared to that in the fixed-bed reactor.

On the other hand, both the reaction temperature and catalyst type determined whether the concentration of aromatic hydrocarbons in the OLP were higher in the

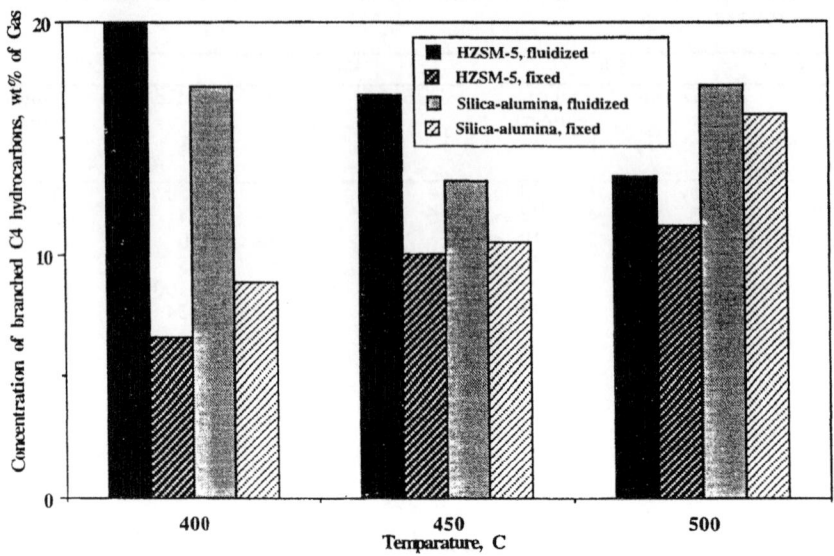

Figure 2: Comparison of the yields of branched C4 hydrocarbons in fixed- and fluidized-bed reactors

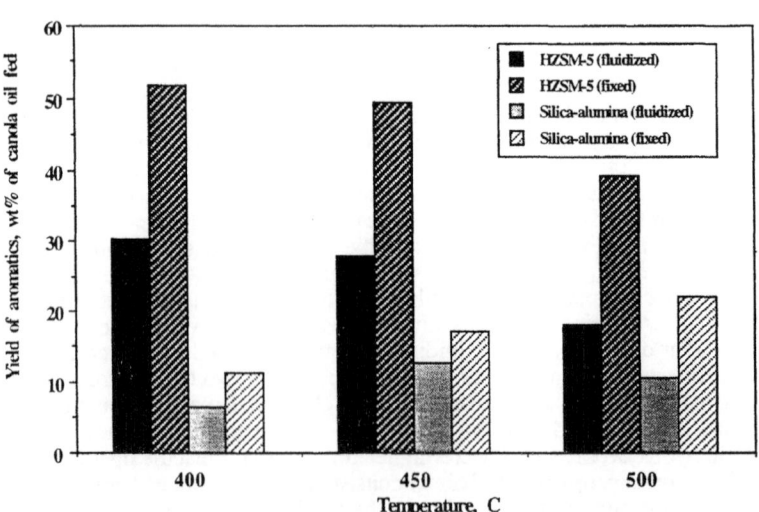

Figure 3: Comparison in the yield of aromatic in fixed- and fluidized-bed reactors

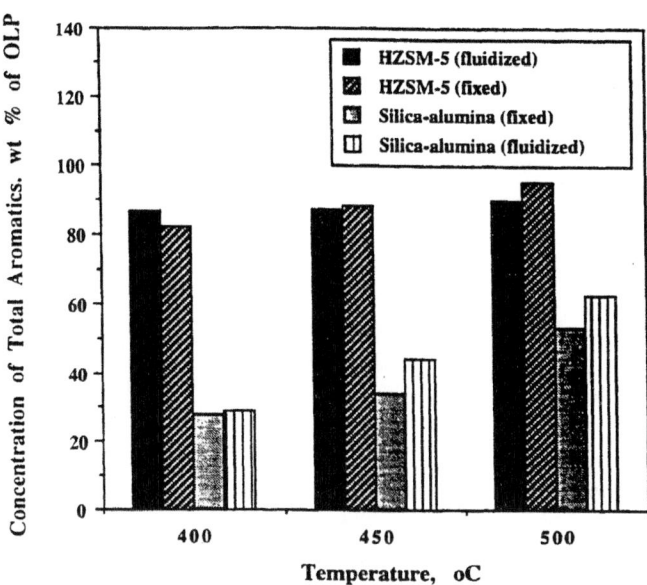

Figure 4: Variation of aromatic hydrocarbon concentration in OLP fraction as a function of temperature and type of catalyst

fluidized-bed or fixed-bed reactor. This is shown in Figure 4. In the case of HZSM-5 catalyst, aromatic hydrocarbon concentration was higher with the fixed-bed reactor than with the fluidized-bed reactor at 500 °C, as expected. However, the fixed-bed/fluidized-bed aromatic hydrocarbon concentration ratio decreased as the temperature decreased, and at 400 °C, this ratio was less than unity indicating a higher aromatic hydrocarbon concentration with the fluidized-bed reactor. On the other hand, in the case of silica-alumina, aromatic hydrocarbon concentration was lower with the fixed-bed reactor than with the fluidized-bed reactor at all temperatures. This behavior is not fully understood at the present time. Studies aimed at obtaining better understanding of the behavior are underway.

4   Conclusions

1. The yield C_4 and C_5 hydrocarbon gases was greater for the fluidized bed reactor operation than for the fixed-bed operation.

2. C_5 hydrocarbon gases increased with both an increased in fluidization gas velocity and a decrease in reaction temperature.

3   The highest yields of C_5^+ hydrocarbon gases were obtained with silica-alumina. This decreased in the order: silica-alumina > HS-Mix > HZSM-5 which is the same as the order of increasing catalyst acidity.

4   The selectivity for the production of C_2-C_4 in the gas phase products for the
     fluidized-bed reactor decreased in the order: HZSM-5 > silica-alumina > HS-
     Mix. This is different than the trend exhibited for the fixed-bed reactor which is
     silica-alumina > HS-Mix > HZSM-5.

5   For silica-alumina, both gas and OLP fractions increased with fluidization gas
     velocity. On the other hand, for HZSM-5, an increase in fluidization gas
     velocity resulted in an increase in OLP and a decrease in gas.

6   A greater selectivity for branched C_4 hydrocarbon was obtained with fluidized-
     bed reactor than with fixed-bed reactor.

5   References

1. Campbell, I.M. (1983) *Biomass, Catalysts and Liquid Fuels*, Holt, Rinehart and
   Winston, London.
2. Parkinson, G. (1992) April, *Chemical Engineering,* pp.35
3. Baker, E.G. and Elliot, C.D. (1987) Catalytic upgrading of biomass pyrolytic
   oils, in *Research in Thermochemical Biomass Conversion*, (ed. A.V.
   Bridgwater), ACS Division of Fuel Chemistry, Elsevier Appl. Sci., London,
   pp. 883-895.
4. Boocock, D.G.B., Konar, S.K., Mackay, A., Cheung, P.T.C., Liu, J. (1992)
   Fuels and chemicals from sewage sludge, *Fuel*, vol 71, pp 1291-1297.
5. Sharma, R.K. and Bakhshi, N.N. (1989) Catalytic conversion of crude tall oil
   to fuels and chemicals, Report of Contract File No. 0582-23283-8-6116,
   Renewable Energy Branch, Energy, Mines and Resources Canada, Ottawa,
   Canada.
6. Adjaye, J.D. and Bakhshi, N.N. (1985) Production of hydrocarbons by
   catalytic upgrading of fast pyrolysis bio-oil: part 1. conversion over various
   catalysts, *Fuel Proc. Tech.,* vol 45, pp. 161-183.
7. Perterson, C.L., Haines, H and Chase, C. (1993) Aug. 30-Sept. 2, Rapeseed
   and safflower oils as diesel fuel, in *Proc. 1st Biomass Conf. of the Americas:
   Energy, Environment, Agriculture and Industry,* Burlington, VT, (NREL,
   Golden, CO) pp. 922-927.
8. Craig, W. and Coxworth, E. (1987) Feb. 16-18, Conversion of vegetable oils to
   conventional liquid extenders, in *Proc. 6th Canadian Bioenergy R&D Seminar,*
   (ed. C. Granger), Richmond, BC, Canada, Elsevier Appl. Sci., London, pp.
   407-411.
9. Sharma, R.K. and Bakhshi, N.N. (1991) Catalytic conversion of crude tall oil
   to fuels and chemicals over HZSM-5: Effect of co-feeding steam, *Fuel Proc.
   Tech.,* vol 27, pp. 113-130.
10. Chantal, P., Kaliaguine, S., Grandmaison, J.L. and Mahay, A. (1984)
    Production of hydrocarbons from aspen poplar pyrolysis oil over HZSM-5,
    Appl. Catal., vol 10, pp. 317-332.
11. Sharma, R.K. and Bakhshi, N.N. (1993), Catalytic upgrading of pyrolysis oil,
    Energy and Fuels, vol 7, pp. 306-319.
12. Prasad, Y.S., Bakhshi, N.N., Mathews, J.F. and Eager, R.L. (1986) Catalytic
    conversion of canola oil to fuels and chemical feedstocks, part I. effect of
    process conditions on the performance of HZSM-5 catalyst, *Can. J. Chem.
    Eng.,* vol 64, pp. 278-284.
13. Katikaneni, S.P.R., Adjaye, J.D. and Bakhshi, N.N. (1995) Catalytic
    conversion of canola oil to fuels and chemicals over various catalysts, *Can. J.
    Chem. Eng.,* vol 33, pp. 484-497.
14. Katikaneni, S.P.R., Adjaye, J.D. and Bakhshi, N.N. (1995) Studies on the

catalytic conversion of canola oil to hydrocarbons: influence of hybrid catalysts and steam, *Energy and Fuels,* vol 9, pp. 599-609.

15. Adjaye, J.D., Katikaneni, S.P.R. and Bakhshi, N.N. (1995) Catalytic conversion of a bio-fuel to hydrocarbons: effects of mixtures of HZSM-5 and silica-alumina on product distribution, *Fuel. Proc. Tech.,* submitted.

16. Katikaneni, S.P.R., Adjaye, J.D., Idem, R.O. and Bakhshi, N.N. (1995) Catalytic conversion of canola oil over potassium-modified HZSM-5 catalysts: C_2-C_4 olefins production and model reaction studies, *Ind. Eng. Chem. Res.,* submitted.

17. Katikaneni, S.P.R., Adjaye, J.D. and Bakhshi, N.N. (1996) Hydrocarbon characteristics for conversion of a plant oil over Pt/HZSM-5 bifunctional catalyst, *Can. J. Chem. Eng.,* submitted.

18. Katikaneni, S.P.R., Adjaye, J.D., Idem, R.O. and Bakhshi, N.N. (1996) Performance of shape selective and non-shape selective cracking catalysts in the conversion of canola oil to fuels and chemicals in a fluidized-bed reactor, *Can. J. Chem. Eng., submitted.*

19. Prasad, Y.S. and Bakhshi, N.N. (1985) Effect of pretreatment of HZSM-5 catalyst on its performance in canola oil upgrading, *Applied Catalysis.,* vol 18, pp 71-85.

20. Nayak, V.S. and Choudhary, V.R. Isomerization of m-xylene on HZSM-5, *Applied Catalysis,* vol 4, pp 333-352.

21. Gary, J.H. and Handwerk, G.E. (1984) *Petroleum refining, technology and economics,* Marcel Dekker, Inc., New York.

22. Kunii, D. and Levenspiel, O. (1991) *Fluidization Engineering,* Butterworth-Heinemann, Massachusetts. ·

23. Rase, F.R. (1977) *Chemical Reactor Design for Process Plants: vol 1. Principles and Techniques,* John Wiley, Toronto.

24. Froment, F.F. and Bischoff, K.B. (1979) *Chemical Reactor Analysis and Design,* John Wiley, Toronto.

6 Appendix A

The definitions for the terms conversion, yield and selectivity as used in this work are given by the following equations:

$$\text{conversion (wt \%)} = (1-(R \text{ (wt \%)})/(100-UA(\text{wt \%})) \text{ X } 100 \tag{A1}$$

$$\text{yield (wt \%)} = (P \text{ (g)/canola oil fed (g))} \text{ X } 100 \tag{A2}$$

$$\text{selectivity, } S = (P(\text{wt \%}))/(100 - P(\text{wt\%}) - UA \text{ (wt \%)}) \tag{A3}$$

where   UA = unaccounted fraction
P = products (i.e., coke, gas, OLP, hydrocarbons, etc.)
R = residue

The Influence of Steam on the Zeolite Catalytic Upgrading of Biomass Pyrolysis Oils

Patrick A. Horne, Nittaya Nugranad and Paul T. Williams
Department of Fuel and Energy
The University of Leeds, Leeds, U.K.

Abstract.
Zeolite catalytic upgrading of highly oxygenated biomass pyrolysis liquids has been proposed as a means of producing hydrocarbon oils and gases. During catalysis the oxygen from the pyrolysis liquids is removed as water, carbon monoxide and carbon dioxide. Biomass derived pyrolysis oils have a relatively low hydrogen content which limits the potential yield of hydrocarbon products particularly when hydrogen is used in the removal of oxygen through the formation of water. The addition of steam to the pyrolytic vapours prior to upgrading may provide a potential source of hydrogen during catalysis to increase hydrocarbon formation. In this work steam was added to the vapours produced from the pyrolysis of biomass. The steam/pyrolytic vapour mixture was passed over a static bed of the zeolite catalyst, HZMS-5. The presence of steam led to the formation of large amounts of gaseous products with a yield in excess of 70% at 550°C catalyst temperature. The catalytic oil yield was reduced as the catalysis temperature was increased from 400 to 550°C.
Keywords. pyrolysis, zeolite, catalysis, steam

1. Introduction

The flash pyrolysis of biomass feedstocks has been shown to yield a liquid product which has a much greater energy density than the original biomass [1]. However, these liquids have limited potential as fuels due to their chemical and physical characteristics. They are highly oxygenated, corrosive, relatively unstable, especially at elevated temperatures, and chemically complex [1,2]. Therefore, the use of catalysts for the upgrading of pyrolysis liquids derived from biomass has received increased attention. One upgrading process which has been investigated by a number of workers [3,4,5,6] is catalytic cracking using zeolite catalysts, in particular the zeolite ZSM-5.

The zeolite catalyst ZSM-5 has a strong acidity, high activity and shape selectivity which converts the original feedstock into hydrocarbons predominately in the C_1 to C_{10} range. In the upgrading of biomass derived pyrolysis liquids the oxygenated compounds are catalytically cracked to give water and carbon oxides as by-products, a highly aromatic liquid product and hydrocarbon gases, in particular the alkenes, ethene and propene.

Evans and Milne [3] used molecular beam mass spectrometry to investigate the upgrading of biomass derived flash pyrolysis vapours. They found that the maximum total hydrocarbon yield was approximately 18% of the original biomass feed and that considerable amounts of coke were formed on the catalyst.

Williams and Horne [4,7,8] have investigated a number of catalytic parameters in regard to the use of zeolite catalysts for upgrading biomass derived pyrolytic liquids. They have shown that the zeolite catalyst is deactivated relatively quickly when upgrading biomass derived pyrolytic liquids and that temperatures in excess of 500°C are necessary to optimise the formation of hydrocarbon products. Due to the high process temperatures the upgraded oils produced were shown to contain large quantities of polycyclic aromatic hydrocarbons (PAH) which are undesirable due to their toxic nature.

Diebold and Scahill [9] upgraded wood pyrolysis vapours over the zeolite catalyst ZSM-5, using steam as the carrier gas. They found that with a catalysis temperature of 400°C a yield of 10% oil was obtained and the major gaseous products were carbon oxides and C_2 and C_3 alkenes.

Sharma and Bakhshi [10] have used the zeolite catalyst ZSM-5 to upgrade a bio-oil produced from the high pressure liquefaction of aspen wood. They co-fed the bio-oil to the catalyst reactor with steam, methanol and tetralin. They found that co-feeding with tetralin gave the highest concentration of phenols and aromatic hydrocarbons in the upgraded liquid condensate.

The pyrolysis liquids derived from biomass have a relatively low concentration of hydrogen. Churin [11] calculated that the maximum theoretical yield of hydrocarbons from the upgrading of biomass derived pyrolytic liquids is best achieved by the removal of oxygen as carbon oxides preferably carbon dioxide. Churin [9] also calculated that the greater the amount of hydrogen used in the expulsion of oxygen during catalysis the lower the potential hydrocarbon yields.

In this work steam was added to the pyrolytic vapours produced from the flash pyrolysis of biomass prior to their upgrading over the zeolite catalyst ZSM-5 which was in its hydrogen form (HZSM-5). Steam was added to the process to act as a hydrogen donor. This could take place by either char gasification and/or water gas shift reactions, with the hydrogen produced reacting further to increase the formation of hydrocarbon products.

2. Experimental

2.1 Biomass.

The biomass used was a mixture of wood types obtained from a wood working company. Table 1 shows the proximate and ultimate analyses of the biomass feed.

2.2 Catalyst.

The catalyst was the hydrogen form of the zeolite ZSM-5 and was obtained commercially from Merck LTD, Poole, UK. The ZSM-5 was in the form of spheres (2 mm) supported on a clay binder. The silica:alumina ratio of the catalyst was 50 and the elliptical pore aperture was 5.4-5.6 Å diameter.

2.3 Pyrolysis/Catalytic Reactor System.

The reactor system comprised of a fluidised bed pyrolysis reactor coupled directly to a fixed bed catalytic unit. Figure 1 shows the reactor system set-up. The pyrolysis reactor had a length of 50 cm and an internal diameter of 7.5 cm. The fluidised bed consisted of 250 μm quartz sand with a depth of 10 cm at rest. The fluidised bed was operated at three times minimum fluidising velocity. The char from the pyrolysis process remains in the reactor during the run. A gauze filter was placed in the freeboard of the pyrolysis reactor to stop any carry over of char particulate to the catalytic reactor. The operational temperature of the fluidised bed was 550°C for all experiments. The steam was added to the fluidised bed prior to the catalytic reactor using a HPLC pump. The amount of steam added to the reactor was selected to give approximately half the mass of the pyrolytic vapours/gases produced. The experimental conditions used in this work are shown in table 2.

The catalytic reactor was directly coupled to the fluidised bed reactor eliminating the need for condensing and revapourising of the pyrolytic liquid. The catalytic unit had the same dimensions as the pyrolysis reactor with fully independent temperature control. A fixed bed of catalyst was held in the base of the reactor.

Table 1. Proximate and Ultimate Analysis of the Biomass (wt. % as received)

Wood	Proximate Analysis
Volatiles	91.0
Moisture	7.50
Ash	1.50
Element	Ultimate Analysis (Wet basis)
Carbon	46.9
Hydrogen	5.71
Oxygen	46.6
Nitrogen	0.18

Table 2. Experimental Parameters.

Experimental Conditions	
Carrier Gas Flowrate (N_2) (kg h^{-1})	1.27
Wood feed (kg h^{-1})	0.2
Weight of Catalyst (kg)	0.2
Pyrolysis Temperature (°C)	550
Catalysis Temperature (°C)	400, 450, 500, 550
Weight of Steam (kg h^{-1})	0.08

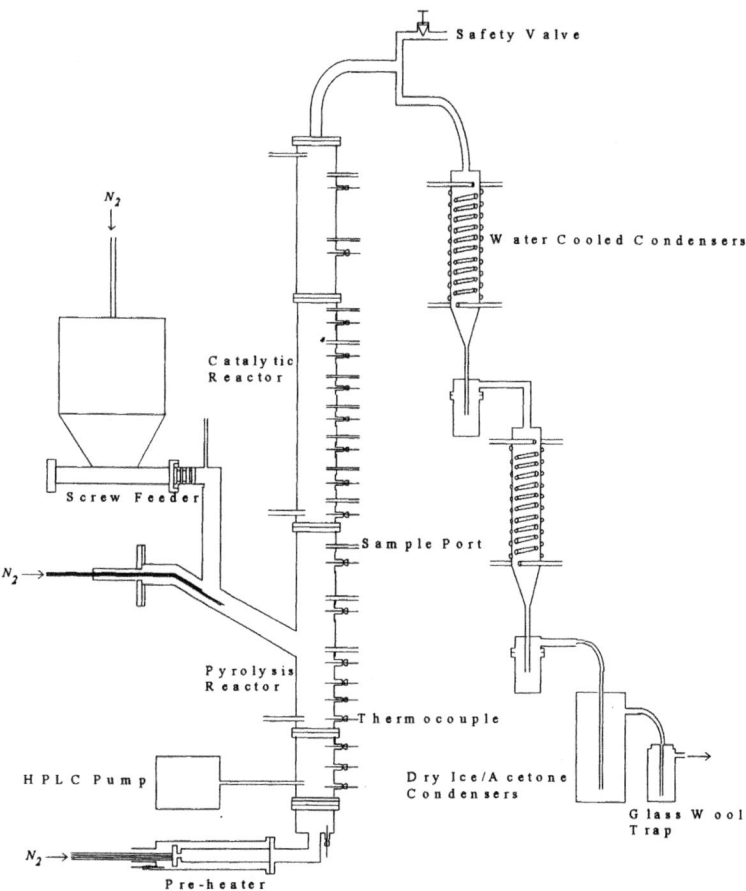

Figure 1. Schematic diagram of the pyrolysis and catalytic reactor system.

2.4 Gas Analysis.

A slip stream of the gaseous products from the upgrading process was sampled into a 100 litre Teflon gas bag. The individual gaseous components were then identified and quantified off-line by packed column gas chromatography. The gases analysed for were CO, CO_2, H_2, CH_4, C_2H_6, C_2H_4, C_3H_8 and C_3H_6. Details of the full chromatographic procedures have been given previously [4,7,8].

2.5 Oil Analysis.

The major single ring aromatic hydrocarbons present in the oils were identified and quantified using gas chromatography/mass spectrometry (GC/MS) and gas chromatography/flame ionisation detection (GC/FID). The GC/MS system was a Carlo-Erba, Vega HRGC with cold on-column injection, coupled to a Finnigan Mat ion trap detector. (ITD) via a heated transfer line. A 30 m × 0.3 mm fused silica DB-5 column was used with helium as the carrier gas. Further details on oil analysis techniques have been given previously [4,7,8].

3.0 Results & Discussion.

3.1 Product yields

The authors have already investigated the effect of catalysis temperature on the upgrading of biomass derived flash pyrolytic vapours with no steam addition [7]. These results are given in table 3 for comparison with steam/catalyst results of this work which are shown in table 3A. The mass balance closures in this work were typically 90-95% the results in table 3A were normalised.

For comparative purposes the amount of water added to the system was subtracted from the aqueous catalytic product yield. Less water was collected for the catalysis runs of 500 and 550°C than was initially added to the system for steam generation, hence the negative values recorded for these catalysis temperatures.

The addition of steam to the catalysis process had a profound effect on the product yields. The formation of gaseous products was vastly increased by the presence of steam during upgrading and there was a large decrease in the amount of aqueous phase produced compared to the catalyst only runs. The initial wood feed contained some moisture and the pyrolysis process also gives water as a product [7]. The authors [7] have shown that the amount of water present in the products from the pyrolysis of wood was approximately 14%. This pyrolytic water plus the water added will all pass over the catalytic bed. The catalytic process in the absence of steam has been shown to give water as one of its major products particularly at low temperature (table 3). However, the addition of steam significantly reduces the amount of water produced by the catalytic process with the higher catalysis temperatures giving the greatest reduction. Therefore, the addition of steam into the gas/vapour stream during catalysis causes either or both catalytic and/or thermal reactions to take place with the steam reacting with other species. The oil yield produced from the steam/catalytic process was less than that observed at the corresponding temperatures for the catalysis only runs. The steam/catalytic oils were a deep translucent amber colour compared to an opaque dark brown colour for the catalyst only runs. The formation of char was constant regardless of the presence of steam, but the formation of coke was slightly increased with steam present in the gas/vapour stream.

Table 3. Yield of catalytic products with no steam addition (%).

Temperature	Oil	Aqueous	Gas	Char	Coke
400	8.8	34.5	20.2	16.1	12.3
450	6.2	35.7	24.2	16.0	11.8
500	5.7	30.1	31.8	16.3	11.4
550	5.2	24.0	38.7	15.1	13.1

Table 3A. Yield of catalytic products with steam addition (%).

Temperature (°C)	Oil	Aqueous	Gas	Char	Coke
400	7.74	24.3	35.7	17.3	14.8
450	2.45	8.76	56.2	16.9	15.1
500	1.93	-1.80	68.1	17.8	13.9
550	0.37	-11.9	80.1	17.1	14.0

3.2 Gas Products.

Tables 4 and 4A show that gaseous products from the upgrading process with and without steam addition respectively. The presence of steam regardless of catalytic temperature gave an increase in the yield of each gaseous product compared to catalysis carried out with no steam addition. The higher the catalytic temperature so the more marked the increase in the formation of gaseous products, particularly the formation of carbon oxides and hydrogen. It is well documented that under certain process conditions water can react with carbon or carbon oxides as shown in equations 1 and 2.

$$C + H_2O \rightarrow H_2 + CO \qquad \text{carbon steam reaction} \qquad (1)$$
$$CO + H_2O \leftrightarrow H_2 + CO_2 \qquad \text{water shift reaction} \qquad (2)$$

During the upgrading of the pyrolytic vapours the steam present could react with the coke deposits on the catalyst as shown in equation 1 giving carbon monoxide and hydrogen as products. However, the coke yields observed in this work were greater than those in the absence of steam and indicate that equation 1 is not a major reaction pathway. The carbon monoxide present in the vapours produced during catalysis could react with steam to give carbon dioxide and hydrogen as shown in equation 2. Both equations 1 and 2 are well known thermally driven reactions which normally occur at high temperatures. However, the presence of the zeolite could shift these reactions to the right at lower temperatures contributing to the large observed increase in the formation of carbon oxides and hydrogen. As well as reactions 1 and 2 there could be numerous other reaction mechanisms taking place during the upgrading process between the steam, pyrolytic vapours and catalyst. Oxygen expulsion from the pyrolytic vapours leads to the formation of carbon oxides and water. The authors [7] have shown previously that the upgrading of pyrolytic vapours in the absence of steam leads to the formation of both water and carbon oxides with water being the favoured route of oxygen expulsion at low temperatures (400-450°C) and carbon oxides at higher temperatures. The steam/catalytic runs carried out in this work show a marked decrease in the yield of water and a comparative increase in the formation of carbon oxides suggesting a shift in the reaction mechanism for oxygen expulsion from the pyrolytic vapours.

The presence of steam gave a large increase in the formation of hydrocarbon gases especially at 500 and 550°C, however the formation of oil was reduced with temperature. For the 550°C steam/catalyst experimental run the total yield of hydrocarbon gases was 17.8% compared to 7.7% without steam. The increase in hydrocarbon gas formation could be due to the steam reducing the catalysis residence time and competing with the pyrolytic vapours for the active sites. Thus limiting the opportunity for the hydrocarbon gases to react further to produce liquid hydrocarbons.

Table 4. Yield of gaseous catalytic products with no steam addition (%).

Temperature (°C)	CO	CO_2	H_2	CH_4	C_2H_6	C_2H_4	C_3H_8	C_3H_6
400	9.50	8.80	0.02	0.58	0.12	0.37	0.10	0.75
450	12.2	8.70	0.04	0.91	0.14	0.69	0.24	1.10
500	15.0	10.6	0.05	1.54	0.22	1.18	0.12	1.98
550	19.0	12.0	0.16	2.53	0.31	1.96	0.11	2.64

Table 4A. Yield of gaseous catalytic products with steam addition (%).

Temperature (°C)	CO	CO_2	H_2	CH_4	C_2H_6	C_2H_4	C_3H_8	C_3H_6
400	14.8	15.0	0.12	1.06	0.13	1.53	0.27	2.63
450	24.6	22.4	0.27	2.79	0.62	2.37	0.35	2.74
500	29.1	24.1	0.40	4.30	0.60	3.76	0.75	5.55
550	35.6	25.2	1.05	5.96	0.83	4.39	0.85	5.82

3.3 Oil Analysis.

Previously the authors [7] have shown that for the upgrading of biomass pyrolytic vapours in the absence of steam that the higher the catalysis temperature (400 to 550°C) the lower the oil yield but the greater the formation or aromatic hydrocarbons. The oil yield for the experimental runs carried out with no steam addition has been shown to vary from 8.81 to 5.21% as the catalyst temperature was increased from 400 to 550°C.

In this work the addition of steam to the catalysis process gave oil yields of 7.74 to 0.37% as the temperature was raised from 400 to 550°C. These yields are considerably lower than those recorded in the absence of steam especially at higher temperatures. Detailed analysis of the catalytic steam oils has shown them to have a highly aromatic product slate with extremely high concentrations of single ring aromatics. The benzene, toluene and dimethyl benzene concentrations for steam/catalyst oils of 400, 450 and 500°C are shown in table 5. Even the 400°C steam catalyst oil contains high concentrations of single ring aromatic hydrocarbons, whereas low temperature oil produced at 400°C with no steam addition has been shown to contain only relatively low amounts of these compounds [7]. Indeed the 400°C steam/catalysis oil contained benzene, toluene and dimethyl benzene in concentrations similar to those found in upgraded oils produced with no steam addition at temperatures in excess of 500°C. Therefore, the addition of steam appears to promote the formation of aromatic hydrocarbons at 400°C, but at higher catalysis temperatures the formation of hydrocarbon gases becomes predominate.

Table 5. Concentration of single ring aromatic hydrocarbons in the steam/catalytic oils (%).

Temperature (°C)	Benzene	Toluene	Dimethyl Benzenes
400	2.40	12.7	21.7
450	0.50	13.2	26.2
500	22.2	25.2	19.7

4.0 Conclusions.

The oil yield was decreased with increased catalyst temperature. However, there was an large increase in the formation of hydrocarbon gases as the catalysis temperature was raised. The addition of steam to the upgrading process gave a marked increase in the formation of all the gaseous products compared to upgrading in the absence of steam, in particular the formation of carbon oxides. The increased formation of the carbon oxides was in contrast to the formation of water which was reduced compared to experimental runs carried out in the absence of steam. There was a significant increase of hydrogen with catalysis temperature with a maximum concentration of 1% at 550°C which corresponds to 0.5 mols for every 100 grams of wood fed. This indicates that the steam is reacting during the pyrolysis/catalysis process.

The concentration of single ring aromatics in the oils was increased with increasing catalysis temperature. However, in terms of wood fed to the system the greatest concentration of aromatic hydrocarbons was produced at the lowest temperature of 400°C. The upgraded oil product produced at 400°C was almost exclusively single ring aromatic hydrocarbons, which was in contrast to oil produced by the authors [7] previously under similar conditions but in the absence of steam which contained partially converted and unconverted pyrolysis products. The 400°C steam run gave an oil and gas yield similar to that achieved at 500 to 550°C in the absence of steam. Therefore, the addition of steam appears to be beneficial in lowering the operational temperature of the catalysis process. This is important as a lower catalysis temperature could also aid in the production of a quality oil product which contains low quantities of PAH. Previous work by the authors [7] in the absence of steam have shown that as catalysis temperature was increased so the formation of PAH increased which is undesirable as they are toxic and carcinogenic.

Acknowledgements

This work was supported by the U.K. Science and Engineering Research Council under grant numbers GR/F/06074 and GR/F/87837. The authors gratefully acknowledge this support.

References.

1. Bridgwater A. V. and Bridge S. A. (1991) A Review of Biomass Pyrolysis and Pyrolysis Technologies. In (A. V. Bridgwater and G. Grassi eds.). *Biomass Pyrolysis Liquids, Upgrading and Utilisation.* Elsevier Applied Science, London. pp. 11-92.
2. Stoikos T. (1991) Upgrading of Biomass Pyrolysis Liquids to High Value Chemicals and Fuel Additives. In (A. V. Bridgwater and G. Grassi eds.). *Biomass Pyrolysis Liquids, Upgrading and Utilisation.* Elsevier Applied Science, London. pp. 227-242.
3. Evans R. J. and Milne T. (1988) Molecular Beam Mass Spectrometric Studies of Wood Vapour and Model Compounds over HZSM-5 Catalyst. In (J. Soltes and T. A. Milne eds) *Pyrolysis Oils from Biomass Producing, Analysing and Upgrading.* ACS Symposium Series 376, American Chemical Society, Washington DC. pp. 311-327.
4. Horne P. A. and Williams P. T. (1995) The Effect of Zeolite ZSM-5 Catalyst Deactivation During the Upgrading of Biomass-Derived Pyrolysis Vapours. *Journal of Analytical and Applied Pyrolysis,* Vol. 34, pp. 65-85.
5. Diebold J. P., Chum H. L., Evans R. J., Milne T. A., Reed T. B., and Scahill J. W. (1987) Low Pressure Upgrading of Primary Pyrolysis Oils from Biomass and Organic Wastes. In (D. Klass ed.), *Energy, Biomass and Wastes,* Vol. 10, pp. 801-830.
6. Sharma R. K. and Bakhshi N. K. (1991) Upgrading of Wood Derived Bio-oil over HZSM-5. *Bioresource Technology.* Vol. 35, pp. 57-66.
7. Williams P. T. and Horne P. A. (1994) Characterisation of Oils from the Fluidised Bed Pyrolysis of Biomass with Zeolite Catalyst Upgrading. *Biomass and Bioenergy,* Vol. 7, No 1-6, pp. 223-236.
8. Williams P. T. and Horne P. A. (1995) Analysis of Aromatic Hydrocarbon in Pyrolytic Oils Derived from Biomass. *Journal of Analytical and Applied Pyrolysis,* Vol. 31, pp. 15-37.
9. Diebold J. and Scahill J. (1988) Biomass to Gasoline: Upgrading Pyrolysis Vapours to Aromatic Gasoline with Zeolite Catalysis at Atmospheric Pressure. In (J. Soltes and T. A. Milne eds) *Pyrolysis Oils from Biomass Producing, Analysing and Upgrading.* ACS Symposium Series 376, American Chemical Society, Washington DC, pp. 264-276.
10. Sharma R. K. and Bakhshi N. N. (1991) Upgrading of Wood-Derived Bio-Oil over HZSM-5. Bioresource Technology, Vol. 35, pp. 57-66.
11. Churin E. (1990) Catalytic Treatment of Pyrolysis Oils, Cat. No. CD-NA-12480-EN-C. Commission of the European Communities.

CATALYTIC CRACKING OF BIOMASS FLASH PYROLYSIS LIQUIDS

M.C. SAMOLADA and I. A. VASALOS
Chemical Process Engineering Research Institute
P.O. Box 361, 570 01 Thermi, Thessaloniki, GREECE

Abstract

This paper reports the results of investigations for producing conventional liquid fuels via catalytic cracking of biomass flash pyrolysis liquids (BFPL). A microactivity fixed bed testing unit (MAT) was applied for the selection of the most appropriate catalyst. A special feed system was designed for the bio-oil feed in a 1/1 wt solution with methanol. Optimum conditions for testing were selected using HZSM-5, a common de-oxygenation catalyst. A high residue conversion was measured (~95 wt%) accompanied by a high coke production (~15 wt %), corresponding to a minimum coke on catalyst of 7.5 wt %. The performance of typical FCC catalysts in the catalytic cracking of biomass flash pyrolysis liquids was also studied in order to explore the possibility of feeding this feedstock in an ordinary FCC unit. A number of commercially FCC catalysts were tested. These catalysts were found to be inactive unless ZSM-5 was included. Moreover, the achieved degree of de-oxygenation was very poor due to the low activity of the catalyst. Thus, the obtained liquid product contained a high water content (~30 wt %) and low molecular oxygenated compounds. Only the formation of some benzenes indicates the catalytic de-oxygenation activity of ZSM-5. The direct introduction of such a feedstock in ordinary FCC units seems to be at this time impractical due to extensive coking.

Keywords: α-Al_2O_3, biomass flash pyrolysis liquids, catalytic cracking, coke production, FCC catalysts, HZSM-5, liquid fuels, MAT

1 Introduction

Biomass flash pyrolysis liquids cannot be used as fuels directly without a prior upgrading step due to their high oxygen content (40-50 wt %) and the especially low H/C ratio (1). Likely routes for the upgrading of biomass pyrolysis liquids include: catalytic hydrotreatment and catalytic cracking. Catalytic hydrotreatment of bio-oils is not an economical approach till now, because of the high hydrogen consumption and the high pressures required to achieve an adequate degree of deoxygenation and the high water content of the obtained liquid product [1]. These drawbacks can be overcome if catalytic cracking is selected. HZSM-5 has previously been tested for the catalytic cracking of bio-oils. This catalyst has been initially developed by Mobil researchers for the effective conversion of oxygenates, e.g. CH_3OH, to a hydrocarbonaceous fuel [2]. It is a medium pore (5.4 Å) shape selective zeolite, selectively producing a C_5^+ gasoline. The obtained low coke yields are directly related to the achieved low conversion of the feedstock heavy compounds, controlled by the unique structure of the catalyst [3].

HZSM-5 was tested over a wide range of experimental conditions and feedstocks as well. Low temperatures (300-500°C) and low WHSVs (0.6-6 h^{-1}) under reactive and non-reactive atmospheres were generally applied. These conditions were found to be appropriate for the production of an aromatic gasoline with a high xylene/benzene ratio [4]. A wide variety of feedstocks comprising of fresh pyrolysis vapours [4-6], atmospheric flash pyrolysis liquids [7, 8], high pressure liquefaction liquids [9, 10], and oil-seeds [11, 12] have so far been tested.

High pressure liquefaction bio-oils reduce their non-volatile content (40 wt %) by 70-80 wt % when they are upgraded over HZSM-5 [9]. The high water content (20-30 wt %) of biomass flash pyrolysis liquids (BFPLs) is responsible for the enhancement of decarbonylation and decarboxylation reactions over the dehydration reactions [8].

The phenolic content of bio-oils (10-20 wt %) is especially reduced in the cracked product particularly in the presence of a "hydrogen donor" compound, such as tetralin, methanol or steam [7-9]. The phenolic fraction of bio-oil primarily contributes to coke formation [5, 7-9]. A promising upgrading rout could be the catalytic cracking of the non-phenolic fraction of a BFPL for the production of a fuel gasoline, using phenols as a starting material for the production of valuable chemicals [7]. Co-feeding "hydrogen donor" compounds in the cracking reactor, particularly tetralin, brings a positive effect in minimizing coke formation [5, 9]. Coke yields of 10 wt % were observed with the use of "hydrogen donor" compounds [7-9]. Otherwise, coke yields as high as ˜30 wt % resulted [9]. The role of the used hydrogen donor compounds in the catalytic cracking of bio-oils is primarily related with their stabilisation rather being a hydrogen source [9, 13], as it occurs during the classical hydrogenation of ordinary hydrocarbonaceous feedstocks. Moreover, the atmospheric pressure used for the catalytic cracking of BFPLs assures for the minor dehydrogenation extent of tetralin [9]. Co-processing steam reduces coke, while facilitates the separation efficiency of gasoline. Maximum conversion of pyrolysis liquids to gasoline of up to ˜35 wt % was reported [4, 6-9].

The behaviour of catalysts other than HZSM-5 was also studied using a high pressure liquefaction bio-oil [10]. It was found that catalysts of increased acidity and small pores are especially suitable for this process. Coke production is reduced with a concomitant increase of the conversion to hydrocarbons. Zeolites with especially high SiO_2/Al_2O_3 ratios (›35) and of medium pore size (5.7 Å) are selective in gasoline production from cellulose pyrolysis vapours [14]. Ordinary FCC catalysts were tested for the upgrading of a hydrogenated high pressure liquefaction bio-oil [15]. Gasoline yields of ˜30 wt % accompanied by high coke yields (˜20 wt %) were observed under modified micro-activity test (MAT) conditions.

Coke formation during catalytic cracking over HZSM-5 irreversibly depends on the Effective Hydrogen Index (EHI) of the feedstock [3, 16], its definition given by equation A_1 in the Appendix. The especially high coking observed using low EHI («1) feedstocks, e.g. BFPLs, can be reduced using mixtures with higher EHI (»1) compounds, like methanol.

In this study, it was tried to explore the possibility of using ordinary BFPLs as a feedstock in a common FCC unit. First, the performance of HZSM-5 in a broad range of experimental conditions was studied in order to establish a modified MAT procedure for BFPLs and find out the main operational variables affecting coke production. Extensive coking affects the proper operation of the regenerator part of the FCC unit and should be minimised as much as possible. Finally, various commercial FCC type catalysts were evaluated for the catalytic cracking of a BFPL in terms of coke production.

2 Experimental

2.1 The Bench Scale Reactor Unit

The modified MAT reactor system shown in Fig. 1 was constructed for the performance of the catalytic cracking experiments with bio-oil. The reactant liquid mixture was fed at the top of the fixed bed reactor using an ISCO 500D pump, capable of achieving an accurate small feed rate (0.1-10 cm^3/min). It was introduced on the top of the catalytic bed passing through a narrow stainless steel injector. A stream of N_2 (35 cm^3/min), introduced at the top of the reactor for the continuous withdrawal of the reaction products from the reaction zone. An additional N_2 flow (15 cm^3/min) was introduced after the end of the run to purge the transfer lines. A pressure indicator connected to the feed line was used as the basic tool for the effective detection of any plugs.

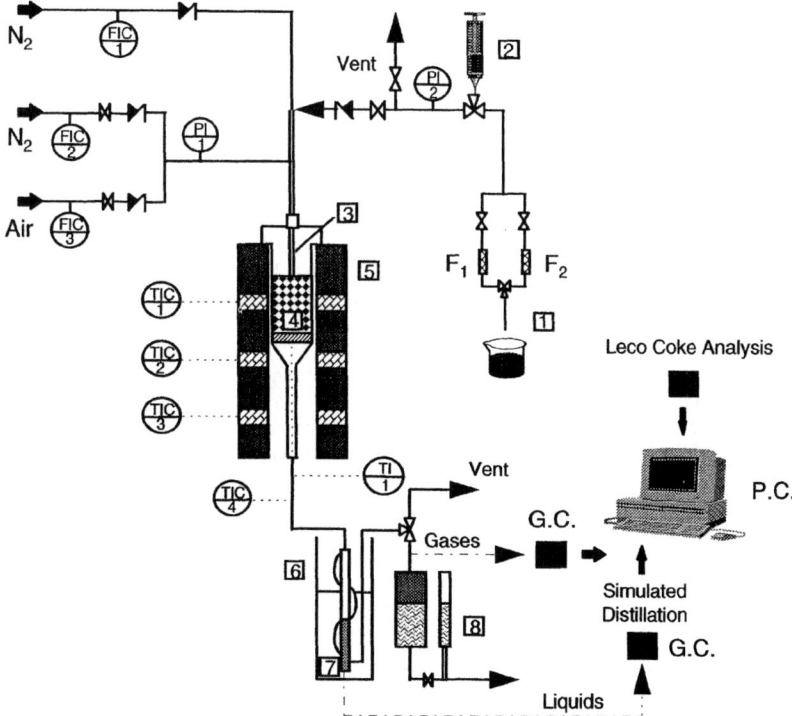

Fig. 1. The catalytic fixed bed reactor system (1: liquid feedstock; 2: syringe pump; 3: injector; 4: catalyst bed; 5: reactor furnace; 6: cold liquid bath; 7: liquid product receiver; 8: gas collection system)

The fixed bed reactor (12.1 cm height x 1.25 cm D_i), made of stainless steel 316, was heated to the desired temperature using a 3-zone radiant furnace (Fig. 1). The temperature of each zone was controlled using a standard temperature controller. The temperature of the reactor exit was controlled at 290°C in order to prevent the condensation of the liquid reaction products inside the transfer lines and to prevent additional thermal cracking. After the end of the experiment N_2 purges the system for 25 min in order to achieve the maximum liquid product recovery.

The liquid products were collected in a cold liquid bath (-15°C) and their amount was weighted in a glass receiver. The gaseous products were collected and measured by paraffin oil displacement, subtracting the volume of the carrier gas N_2. After the end of the experiment, the reactor was washed with methanol. The amount of liquid remaining after methanol evaporation was considered as unconverted liquid. The amount of coke deposited on the reacted catalyst was measured by burning .

A fixed amount of catalyst (HZSM-5: 2g; FCC type: 3g) was introduced into the reactor. A statistical experimental design table (Table 1) was selected in order to establish the effect of the primary cracking parameters (T, WHSV, Catalyst/Oil (C/O) ratio) on the coke on catalyst content. This response is of primary importance especially for a pilot scale operation. It should be minimised as much as possible ($\leq$ 1 wt %) for the effective operation of the FCC unit. HZSM-5 was used as a catalyst for this study. The optimum experimental conditions for minimum coke production were applied for the evaluation of various FCC catalysts in the modified MAT unit.

Table 1. Experimental design for the BFPL solution (BFPL/MeOH : 1/1 wt) catalytic cracking over HZSM-5

Run	T (°C)	WHSV (h^{-1})	C/O ratio	L (wt %)	Gas (wt %)	Coke (wt%)	Conv$_r$[a] (wt %)	Deox.[b] (%)	Y[c] (wt %)
1	315	9	1	62.5	10.1	25.9	91.7	99.8	23.6
2	375	9	1	64.3	13.5	19.8	90.4	95.4	18.8
3	315	30	1	61.6	7.7	28.9	88.2	92.5	28.4
4	375	30	1	64.8	8.4	24.3	82.9	82.9	23.1
5	315	9	1.3	64.5	9.7	24.3	98.4	98.4	17.6
6	375	9	1.3	65.3	12.9	17.0	90.3	97.6	14.2
7	315	30	1.3	60.3	6.6	30.7	97.7	100.0	23.1
8	375	30	1.3	65.2	8.4	22.8	46.2	97.1	17.2
Effect	-4.71±1.99	4.39±1.99	-5.60±1.99						

where: [a]: residue conversion (wt %); [b]: deoxygenation (%); [c]: coke on catalyst (wt %)

An eucalyptus flash pyrolysis liquid (B-EU-01-A) produced in the Union Fenosa Flash Pyrolysis Pilot Plant Unit (Spain) was used as a feedstock. The elemental analysis and the group chemical composition of the used BFPL is listed in Table 2. A BFPL mixture with CH_3OH (1/1 wt) was used in order to overcome the operating difficulties related with pure bio-oil feeding and assure for a high EHI feedstock [3].

Table 2. Elemental analysis and group composition of the Union Fenosa BFPL (wt %)

Sample	C	H	O	EHI
BFPL	46.5	6.8	46.7	0.27
Pl* (49.0 wt %)	59.4	6.2	34.4	0.38
Aqueous (25.8 wt %)**	51.6	5.9	42.5	5.1
Phenols (11.2 wt %)**	61.5	6.8	31.7	0.55
Acids (3.8 wt %)**	59.4	6.3	34.3	0.40
Neutrals (3.1 wt %)**	67.0	6.9	26.1	0.65

where: *: Pyrolytic Lignin (H_2O/BFPL: 3/1 wt); **: Liquid-liquid extraction method [17]

2.2 Analysis of products

The density of the feed used in each cracking experiment was measured just prior entering the fixed bed reactor using a DMA-48 PAAR densitometer in order to accurately estimate the amount of the feed entered the reactor during the experiment. The water content of the liquid product was measured by applying the Karl Fisher titration method using a Crison Micro TT 2050 instrument. The content of compounds with boiling points within the range of gasoline, diesel and heavier was measured by performing a simulated distillation in an HP 5890 GC (ASTM D2887-78). The qualitative composition of the liquid products was determined by applying GC/MS analysis. CHN analysis was also performed in order to find the elemental composition of the upgraded liquid (Leco CHN 800). The Micro Carbon Residue (MCR) of the liquid feedstock was measured by applying the standard ASTM D 4530-85 procedure, being equivalent to the so called Conrandson Carbon Residue test (ASTM D 189). The cracked gases were analysed in an HP 5890 refinery gas analyser equipped with both a TCD and a FID. Their composition used for the estimation of the total gas yield.

2.3 Catalysts

A high Si/Al (60/1) HZSM-5 catalyst was received from Amoco Oil USA in a powder form. This catalyst was not a commercial product and was prepared for laboratory use only according to well known procedures [18]. The catalyst was pelletized and crushed. The fraction of 425-600 μm particles was used for the cracking experiments.

Table 3. Physical properties of the catalysts tested

Catalyst	Company	Zeolite	PV_{Hg} (ml/g)	BET (m^2/g)	Area[a] (m^2/g)
α-Al$_2$O$_3$	Amoco	–	0.300	3.0	26.1
HZSM-5	Amoco	ZSM-5	0.525	460.0	–
Z-100	Engelhard	ZSM-5	0.359	43.7	7.0
ELDA	–	Y	0.303	262.8	36.7
FCC-A	Engelhard	Y	0.249	349.7	82.2
FCC-B	Engelhard	Y	0.153	341.7	29.4
FCC-C	Engelhard	Y	0.575	98.8	96.0
SP5061	Grace/Davison	ZSM-5	0.348	44.8	16.7
SP5062	Grace/Davison	REUSY	0.300	46.1	14.0
SP5063	Grace/Davison	REHY	0.344	188.6	17.4

where: [a]: estimation of mesopores area based on PV_{Hg} data

A number of commercial FCC catalyst samples were received from various companies. Their coded names and physical characteristics are listed in Table 3 (BET Surface Area: Quantachrome Autosorb 1; PV_{Hg}: Micromeritics Autopore II 9220). These catalysts are likely to be the most appropriate for BFPLs cracking. A fresh catalyst sample, ELDA (Table 3), used in a commercial FCC unit (Hellenic Aspropyrgos Refinery) was also tested. All catalysts were tested without deactivation, since some preliminary tests with the equilibrium form of ELDA catalyst exhibited especially high coking and minor activity. Their >75μm fraction was used, in order to avoid high pressure drops along the catalyst bed.

3 Results

3.1 HZSM-5 Performance

The design table was repeated and the mean values of all product yields and the considered experimental response (Y: coke on catalyst) are listed in Table 1. The factor effects (Table 1) were calculated based on standard statistical design formulas [19]. In order to establish the most significant effects, the confidence interval of each effect was estimated (Table 1) based on replicate runs. It was found that all factors are significant according to the basic statistical design criteria [19].

The sign of the effects suggests that for minimum coke on catalyst high reaction temperatures, low WHSVs and high C/O ratios should be applied. The fuel product is the organic phase of the derived liquid. A constant degree of deoxygenation (~95 wt %), given by equation A_2 in the Appendix, was always achieved (Table 2) due to the especially high activity of HZSM-5. This response is mainly related with the catalyst characteristics and thus cannot be dramatically affected by the system variables. The achieved residue conversion, defined by equation A_3 (see Appendix), was very high (88–98 wt %) in almost all the experiments. Low residue conversions were obtained at high temperatures and WHSVs due to the catalyst extensive deactivation.

The organic liquid product from selected experiments was analyzed by GC/MS (Fig. 2). It was found that the main compounds fall in the gasoline boiling range and are substituted benzenes. An entirely aromatic gasoline was produced at the selected optimum operating conditions using HZSM-5 as a catalyst. The following operating parameters: T=370°C, WHSV=8 h^{-1}, C/O ratio =2 (Q_1= 0.4 cc/min, C=3g) should be selected in order to evaluate catalysts for BFPL catalytic cracking following a modified MAT procedure.

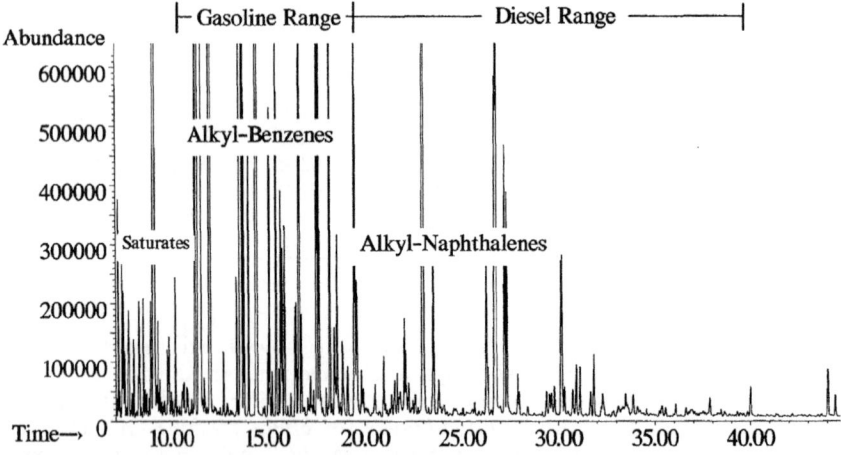

Fig. 2. GC/MS analysis of the organic liquid product obtained from BFPL catalytic cracking over HZSM-5

3.2 Performance of FCC catalysts

These catalysts could not be effectively tested by applying the selected operating conditions. Their high bulk density combined with the great sensitivity of BFPL under heating, were responsible for the complete plugging of the catalyst. The on-line catalyst regeneration was then required. It was found that for an optimum system performance FCC catalysts should be placed between two layers of α-alumina (0.8g at the bottom, 1.9 g at the top of the catalyst), in order to avoid plugging and to use a constant bed height in all the experiments, equivalent to HZSM-5 (5.7 cm). Moreover, lower WHSVs (4 h^{-1}) should be applied in order to get reliable mass balances. This result is consistent to the results from HZSM-5 study, where these low WHSVs could not be practically operated.

The experimental results derived from the evaluation of various commercial catalysts are listed in Table 4. All catalysts give a small amount (~50 wt%) of a colorless liquid product containing primarily water (80-100 wt % H_2O). Only catalysts containing HZSM-5 give a coloured liquid product at an acceptable yield (~70 wt %, Table 4), which was characterized by a high water content (30-40 wt %) as well. This is an indication that deoxygenation occurs primarily via dehydration reactions, thus minimizing hydrocarbons formation. Dimerisation (dehydration) of methanol is the primary reaction at these conditions proved by the particular high yields of dimethyl ether (4-10 wt%) and water. Dimethyl ether is the MTG process intermediate product converting methanol to hydrocarbons over HZSM-5 [20]. GC/MS analysis of this liquid product verifies the existence of low molecular oxygenated compounds (e.g. phenols) compared to those contained in the original BFPL. The existence of some alkyl-benzenes proves that catalytic cracking takes place in a minor extent. The overall process looks more like a precoking rather than a catalytic cracking process. The observed low reactivity of these catalysts, commonly used as additives in the FCC process, is due to their low surface area (Table 3).

Table 4. Performance (wt %) of commercial FCC Catalysts in the catalytic cracking of BFPL (T: 370°C; WHSV: 4h^{-1}; C/O ratio: 2)

Catalyst	Run	Liquid	Gas	Coke	Mass Balance	Coke/Cat	CH$_3$OCH$_3$ (wt %)
Z-100	BIC9F-1	69.9	13.2	12.4	78.1	6.8	5.2
" "	BIC9F-2	70.9	13.3	11.6	80.0	6.3	5.3
ELDA	BIC8F-3	48.3	16.6	30.9	86.7	15.4	-[a]
" "	BIC8F-4	48.5	13.5	35.1	87.5	15.6	-[a]
FCC-A	BIC12F-1	54.9	16.8	25.6	71.4	13.4	4.6
" "	BIC12F-2	56.1	16.4	24.2	73.6	12.5	3.8
FCC-B	BIC13F-1	57.5	13.6	22.0	71.1	11.5	4.6
" "	BIC13F-2	55.0	16.6	25.6	74.5	13.4	4.5
FCC-C	BIC14F-2	54.9	18.7	23.1	73.1	12.2	4.1
" "	BIC14F-3	58.0	15.6	23.4	73.6	12.2	4.3
SP5061	BIC15F-1	77.6	5.7	10.9	82.6	6.5	3.7
" "	BIC15F-3	75.9	7.2	12.6	72.9	6.2	5.4
SP5062	BIC16F-1	53.0	18.4	24.8	64.5	12.9	8.4
" "	BIC16F-2	53.2	18.7	24.9	67.8	12.9	5.6
SP5063	BIC17F-1	51.7	20.7	24.4	63.2	12.7	10.4
" "	BIC17F-2	52.1	19.2	26.0	62.6	13.7	9.6

where: [a]: not analysed

HZSM-5 containing catalysts (Table 3) give the least coke yield (¯12 wt %) and the optimum mass balance closure (75-85 wt %). A constant coke yield of ¯10 wt % (Fig. 3) originating from thermal cracking reactions was always deposited over the upper layer of α-Al₂O₃. If this layer did not exist, coke would deposit on the top of the catalyst bed thus plugging the bed eventually. Common FCC catalysts give an even higher coke yield of ¯24 wt % (Fig. 3).

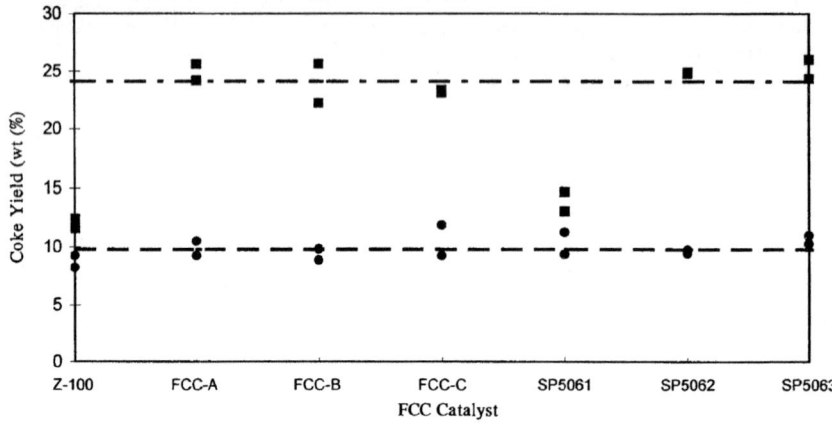

Fig. 3. Coke formation during BFPL catalytic cracking over commercial FCC catalysts (■: FCC catalyst; ●: α-Al₂O₃ upper layer)

A heavy aromatic refinery product (HAO) was also tested to compare its behaviour with that of BFPL (Table 5). Experimental conditions used in the catalysts evaluation study of BFPL and the standard MAT procedure were both applied using ELDA catalyst. It was found that HAO operates better at high WHSVs, where low yields of gases and coke are obtained. Operation at low WHSVs (4 h⁻¹) supports with high coke yields and poor mass balances (¯70 wt %) because of the higher feed conversion [21]. These classical trends observed during the catalytic cracking of various hydrocarbonaceous feedstocks cannot be observed using BFPL. The especially high yields of coke and gases are the main reasons for the poor mass balances (Table 5). Low conversion conditions (WHSV: 20 h⁻¹) are not operational. The unreacted BFPL is converted to coke under heating, thus plugging the reactor bed almost completely (Table 5).

Testing both feedstocks over α-alumina (Table 6) their thermal cracking performance was obtained. Coke yield is slightly affected by WHSV variation in case of BFPL. An almost constant coke yield (¯15 wt %) was always produced. Coke yield originating from HAO thermal cracking was very low (0.24 wt %). It is thus verified that coke yield formed during BFPL catalytic cracking results mainly from thermal cracking reactions and cannot be reduced beyond 15 wt % by further optimization of the process conditions.

Table 5. Distribution of product yields (wt %) from the BFPL solution and HAO catalytic cracking over ELDA catalyst

| | BFPL/CH$_3$OH (1/1 wt); C/O ratio: 2 | | | | |
| | WHSV: 4h^{-1}, T: 710°F | | WHSV: 4h^{-1}, T: 932°F | | WHSV: 20h^{-1}, T: 932°F |
Run	BIC8F-3	BIC8F-4	BIC8F-7	BIC8F-9	BIC8F-12
Liquid	48.3	48.5	40.3	45.3	22.7
Gas	16.6	13.5	31.1	32.2	8.9
Coke	30.9	35.1	25.6	19.2	8.3
Unconverted	4.2	2.3	3.0	3.3	2.4
Mass Balance	86.7	87.5	79.7	81.4	42.3 (reactor blocked)
Coke/Cat.	15.5	17.4	12.8	14.6	

| | HAO (C/O ratio: 2) | | | | | | | |
| | WHSV: 4h^{-1}, T: 710°F | | WHSV: 20h^{-1}, T: 710°F | | WHSV: 4h^{-1}, T=932°F | | WHSV: 20h^{-1}, T=932°F | |
Run	HA8F-6	HA8F-9	HA8F-12	HA8F-13	HA8F-2	HA8F-3	HA8F-10	HA8F-11
Liquid	61.6	60.7	66.7	65.1	51.9	44.2	65.6	64.0
Gas	11.4	12.1	4.5	4.7	25.8	25.4	10.3	12.0
Coke	14.9	15.6	11.4	11.8	12.6	16.9	11.2	12.1
Unconverted	12.1	11.5	17.4	18.3	10.1	13.6	12.2	12.0
Mass Balance	70.3	68.4	92.4	92.1	70.5	67.8	96.6	96.6
Coke/Cat.	7.5	7.8	5.3	5.9	6.3	8.5	5.6	5.8

Table 6. Performance of α-alumina (425-600μm) in the catalytic cracking of BFPL/CH₃OH and HAO (T: 370°C; C/O ratio: 2)

Feed	BFPL/ CH₃OH				HAO
WHSV (h⁻¹)	4	6		8	4
Run	BIC2F-5	BIC2F-1	BIC2F-2	BIC2F-3	HAC2F-1
Liquid	79.4	77.9	74.7	76.1	85.0
Gas	3.5	3.5	3.6	2.3	2.5
Coke	13.3	12.5	13.7	15.8	0.24
Unconverted	3.8	6.1	8.0	5.8	12.2
Mass Balance	109.8	94.7	101.6	106.2	110.9
Coke on Catalyst	7.3	6.3	6.8	7.9	0.1

4 Discussion

The obtained HZSM-5 performance is quite comparable with the results reported in the recent literature with similar feedstocks [7, 8]. Since methanol is also reactive at the used experimental conditions its effect could not be subtracted and thus only the overall yields can be compared.

The nature of the feedstock and the experimental conditions applied are essential for the final coke yield. The Ensyn BFPL (E-BFPL) tested in the literature [7, 8] has a considerably different composition than the Union Fenosa BFPL (UF-BFPL) used in the present study (Table 7). It was characterised by a higher oxygen and water content, a lower viscosity and a lower content of polar organic compounds. Even though tetralin was co-fed during the E-BFPL catalytic cracking a high coke yield of 15 wt % was obtained, accompanied by a 16 wt % char formation at 370°C [8]. The addition of these two figures gives a "total coke" of 30 wt %, quite higher to the minimum coke yield obtained in the present study (17 wt %). The higher coke yields reported for E-BFPL are related with the lower C/O ratios (0.74) applied, verified to enhance coke formation. Lower WHSVs (1-3 h⁻¹), applied during E-BFPL cracking, were also tried in this study but they were abandoned due to extensive coking problems. UF-BFPL was expected to form more coke than E-BFPL under the same cracking conditions. This fact is strongly supported by the high tar content of UF-BFPL, considering its specific low solubility in CH₂Cl₂ (Table 7).

The EHI of the feedstock was so far claimed to be the primary property affecting coking during catalytic cracking over HZSM-5 [3, 16]. This conclusion was obtained based on results from volatiles catalytic cracking (e.g. acetic acid) and can be hardly extended to BFPLs due to the production of a lot of "thermal coke" during their cracking (Table 6). Thermal coke is equivalent to the MCR content of the BFPL solution used (Table 6, 8). Tetralin is characterised by a zero MCR content and an EHI of 1.2, equivalent to that of the UF-BFPL methanol solution used in the present study (Table 7). Moreover, tetralin is almost unreactive under atmospheric, low temperature (≤400°C) catalytic cracking conditions, forming no coke [9].

Tar, containing in pyrolytic lignin and carbohydrates, containing in the aqueous fraction of BFPL, are the main groups contributing a lot to the MCR content of a BFPL (Table 7, 8). These two fractions are characterised by an increased polarity, a high oxygen content (Table 2) and a great instability under heating conditions. Sharma and Bakhshi [7] studied also the performance of the aqueous fraction of E-BFPL (50 wt % H₂O) over HZSM-5. The minimum coke+char yield reported was 17 wt %, even though the EHI of H₂O is infinite. Moreover, this coke reduction was accompanied by a ⁻60 %

reduction of the obtained fuel product yield. Thus the accomplished benefit is uncertain considering the fact that ordinary FCC units would have significant operating problems with such an aqueous feedstock, containing water more than ≥ 5 wt % H_2O.

Table 7. Basic physical characteristics of various BFPLs

Property	UF-BFPL*	E-BFPL**
Elemental Composition (wt %)		
C	46.5	38.1
H	6.8	8.4
O	46.7	52.8
H_2O (wt %)	14.8	27.0[b]
viscosity, 25°C (cp)	395[a]	107[b]
PL[c] (wt %)	67.4	34.4
MCR (wt %)	26.9	15.2
EHI[d] (BFPL/CH_3OH: 1/1 wt/wt)	1.12	1.28
CH_2Cl_2 insolubles (wt %)	47.0	20.0

where: *: Union Fenosa BFPL; **: Ensyn RTP III process BFPL; [a]: [17]; [b]: [22]; [c]: pyrolytic lignin (H_2O/BFPL: 1/1 wt); [d]: defined according Chen et al. 1988 [16]

It is believed that the MCR content (ASTM D 4530-85) of a feedstock is the most representative measurement for the prediction of the coking tendency of a BFPL under catalytic cracking rather than EHI. BFPLs MCR content is in the range of 15-30 wt %. HAO is characterised by an almost zero MCR content (Table 7, 8). The MCR content is improved considerably by mixing original BFPLs with methanol (Table 8). The particularly high MCR value of BFPLs is the main drawback for their inability to be used in a common FCC unit. The polar high molecular compounds contained in these liquids cannot be effectively evaporated under the existing reaction conditions and are thermally converted to coke undergoing polymerisation reactions. This coke gradually plugs the catalyst. This phenomenon was even more essential when the FCC catalysts were tested. A lot of coke was then formed at the upper end of the catalyst bed. Because of this particular problematic behaviour of BFPLs, the ordinary MAT operating conditions (T:520°C, WHSV: 19 h^{-1}, C/O reatio: 3-7) cannot be applied. For these reasons, hydrogenated bio-oils with low amounts of polar oxygenated compounds (O: 0.8 wt %) were so far tested using FCC type catalysts [15]. Catalytic cracking of biomass flash pyrolysis vapours at high temperatures (500°C) over HZSM-5 give comparable coke yields (15 wt %) to those obtained in the present study, due to the zero thermal coke formed [23]. The use of methanol decreases the amount of coke formed, while increases the amount of the aqueous compounds considerably [23], as it was also observed in the present study.

The low coke yield obtained over HZSM containing FCC additives is related to the low conversion achieved due to the low surface area of these materials (Table 3). Ordinary FCC catalysts support a higher coke yield related to their higher BET surface area (Table 3) and their broader pore structure (8.4 Å) than HZSM-5 (5.4 Å). Medium pore zeolites (HZSM-5) produce lower coke yields compared to the large pore zeolites (HY), by preventing cracking of the high molecular compounds [10]. This result is valid only when a completely volatile feedstock is used. This is not the case with BFPLs. A different coke formation mechanism was proposed for the medium and large pore zeolites [10]. Moreover the bulk of coke yield proved to originate from thermal cracking reactions of BFPLs and cannot be effectively reduced.

Table 8. MCR content (wt %) of various feedstocks

Feedstock	MCR (wt %)
BFPL	26.9
BFPL + CH_3OH (1/1 wt)	13.6
BFPL + CH_3OH (1/1 wt)	16.4 (at 400°C)
Aqueous Phase (75 wt % H_2O)	1.9
Pyrolytic Lignin	49.6
Tar	51.8
HAO	0.01

Three are the most plausible solutions to the problem of the effective catalytic cracking of biomass flash pyrolysis liquids in ordinary FCC units: (a) the development of new active FCC catalysts for this application, (b): the isolation of the polar oxygenated compounds of BFPL to obtain a low MCR feedstock and (c) the use of upgraded bio-oils. The first approach is not a short term solution to the problem, while the second was found to be unrealistic. Tar compounds are separated along with water, due to their strong polarity. The resulted material is still characterized by a high MCR content (~30 wt %). The final result is that original flash pyrolysis oils cannot be upgraded in ordinary FCC units. The main reason for their problematic behaviour is their very high tendency to coke formation under heating (MCR: 26.9 wt %). Thus the only possible approach could be the use of upgraded BFPLs as a feedstock.

5 Conclusions

The main conclusions drawn from the present experimental study could be summarised to the following:

- The performance of HZSM-5 was studied for the catalytic cracking of UF-BFPL, a typical biomass flash pyrolysis liquid. The optimum conditions supporting minimum coke on catalyst were selected to evaluate various commercial FCC catalysts for the same reaction, thus formulating a modified MAT procedure.
- The use of commercial FCC catalysts results in the production of 7-15 wt % coke on catalyst and cannot be used for the catalytic cracking of BFPLs in the ordinary FCC process. Their activity in de-oxygenation is of minor importance, even in the best case when HZSM-5 was contained in their structure. The liquid product contains a large quantity of water (30-40 wt %) and has a comparable composition to a thermal cracking product.
- The high MCR content of BFPLs (26.5 wt %) is responsible for their high conversion to coke under heating.
- There is not an easy and effective separation procedure for the effective isolation of the tar components from original BFPLs.
- The catalytic cracking of ordinary BFPLs is at this time not practical. The use of upgraded BFPLs with improved MCR values is the most likely solution to the problem

Acknowledgements

The authors would like to acknowledge the financial support of EU under Contract: AIR2-CT93-1086 for the realisation of this work. The excellent co-operation with Union Fenosa should be mentioned here, particularly for supplying us with the necessary amounts of BFPL for our experiments. Engelhard and Grace Davison, the well known catalysts manufacturers are gratefully thanked for their kindness to provide us with adequate quantities of commercial FCC catalysts for our experiments. The contribution of Mrs E. Stergioula (technician) for the performance of the experimental runs included in this work is also acknowledged. Finally, The scientific contribution of Professor R. Bertolacini (University of Delaware, USA) and Dr. M. Goula in the preparation of HZSM-5 is also greatly appreciated.

Appendix

1. Effective Hydrogen Index (EHI) definition [16]

$$EHI \ or \ (H \ / \ C)_{effective} = \frac{H - 2O - 3N - 2S}{C} \tag{A_1}$$

where: H, C, O, N and S are the atoms per unit weight of the sample of hydrogen, carbon, oxygen, nitrogen and sulphur respectively.

2. De-oxygenation (%) was defined as:

$$Deoxygenation \ (\%) = \frac{(\% \, O_f - \% \, O_{org,p})}{\% \, O_f} \ x \ 100 \tag{A_2}$$

where: O: oxygen content; $_f$: feedstock; $_{org,p}$: organic product

2. Residue Conversion (wt %) was defined as:

$$Conversion_r \, (\%) = 100 - \frac{R_p}{MCR_f} \, x \ 100 - \frac{R_p}{100} \, (Y_{org,p}) \tag{A_3}$$

where : R: resid content ; $Y_{org,p}$: yield of the organic layer ; MCR_f : the Microcarbon Residue Content of the feedstock .

Literature

1. Baldauf W. and U. Balfanz, "Upgrading of Pyrolysis Oils from Biomass in Existing Refinery Structures", in (A.V. Bridgwater and G. Grass (eds)): Energy from Biomass Thermochemical Conversion, Proc. of the EC Contracors' Meeting, 29-31 October 1991, Gent- Belgium, p 147.
2. Weisz P. B., Haag W. O. and P. G. Rodewald, "Catalytic Production of High Grade Fuel from Biomass Compounds", *Science*, 1979, **206**, p57.
3. Chen N. Y., Degnan Jr. T. F. and L. R. Koenig, "Liquid Fuel from Carbohydrates", *Chemtech*, 1986, **August**, p 506.
4. Milne T. A., Evans R. J. and J. Filley, "Molecular-Beam Mass-Spectrometric Studies of HZSM-5 Activity During Wood Pyrolysis Product Conversion", in A. V. Bridgwater

and J. Kuester (eds), "Thermochemical Biomass Conversion", Proc. 2nd Int. Conf., Phoenix, AZ, USA, 2-6 May, 1988, p 910.

5. Evans R. J. and T. Milne, "Molecular-Beam Mass-Spectrometric Studies of Wood Vapor and Model Compounds over an HZSM-5 Catalyst", in J. Soltes and T. Milne (eds), "Pyrolysis Oils from Biomass: Producing Analysing and Upgrading", ACS Symposium Series 376, Washington DC 1988, ch 26, p 311.

6. Scahill J. and J. Diebold, "Engineering Aspects of Upgrading Pyrolysis Oil Using Zeolites", in A. V. Bridgwater and J. Kuester (eds), "Thermochemical Biomass Conversion", Proc. 2nd Int. Conf., Phoenix, AZ, USA, 2-6 May, 1988, p 927.

7. Sharma R. K. and N. N. Bakhshi, "Catalytic Upgrading of Pyrolysis Oil", *Energ. & Fuels*, 1993, **7(2)**, p306-314.

8. Sharma R. K. and N. N. Bakhshi, "Catalytic Upgrading of Fast Pyrolysis Oil Over HZSM-5", *Can. J. of Chem. Eng.*, 1993, **71**, p383.

9. Sharma R. K. and N. N. Bakhshi, "Upgrading of Wood-Derived Bio-Oil Over HZSM-5", *Bioresource Technology*, 1991, **35**, p 57.

10. Adjaye J. D. and N. N. Bakhshi, "Catalytic Conversion of a Wood-Derived Oil to Fuels and Chemicals: Effect of Process Conditions and Catalysts", in Proc. of the 1st Biomass Conf. of the Americas, Vol II, Aug. 30-Sep. 2, 1993, Burlinghton, Vermont, USA, p 1215.

11. Prasad Y. S. and N. N. Bakhshi, J. F. Mathews and R. L. Eager, "Catalytic Conversion of Canola Oil to Fuels and Chemical Feedstocks Part I. Effect of Process Conditions on the Performance of HZSM-5 Catalyst", *Can. J. of Chem. Eng*, 1984, **64**, p 278.

12. Prasad Y. S. and N. N. Bakhshi, J. F. Mathews and R. L. Eager, "Catalytic Conversion of Canola Oil to Fuels and Chemical Feedstocks Part II. Effect of Co-feeding Steam on the Performance of HZSM-5 Catalyst", *Can. J. of Chem. Eng*, 1986, **64**, p 285.

13. Laurent E., Maggi R. and Delmon B., "Partial and Full Hydrorefining of Pyrolysis Oils: Overview and Feasibility Assessment", Proc. 8th Eur. Conf. on Biomass for Energy and Environment, Agriculture and Industry (Ph. Chartier, A.A.C.M. Beenackers and G. Grassi, eds), Vienna, Oct. 1994, Pergamon (Elsevier Sci. Ltd), 1995, vol 2, p 1485.

14. Frankiewicz T. C., "The Conversion of Biomass Derived Pyrolytic Vapors to Hydrocarbons", Proc. of the "Specialists Workshop on Biomass Fast Pyrolysis", Golden, Colorado, 1980, p 123.

15. Gevert B. S. and J. E. Otterstedt, "Upgrading of directly Liquefied Biomass to Transportation Fuels: Catalytic Cracking", *Biomass*, 1987, **14**, p 173.

16. Chen N. Y., Walsh D. E. and L. R. Koening, "Fluidised Bed Upgrading of Wood Pyrolysis Liquids and Related Compounds", in Soltes E. D. J. and T. A. Milne (eds), "Pyrolysis Oils from Biomass", ACS Symposium Series 376, 1988, p 277.

17. Vasalos, I.A., Samolada M.C., Iatridis D., Ikonomou D., Patiaka D. and M. Goula, "Production Treatment and Utilization of Bio-oils from Pyrolysis, for Energy and Alternative Fuels and Chemicals", 1st AIR2-CT93-1086 Progress Report, CPERI contribution, November 1994.

18. Chen N. Y., N. Miale and W. J. Reagan, US Patent 4,112,056, Sep. 5, 1978.

19. Box G. E. P., Hunter W. G. and J. S. Hunter, "Statistics for Experimenters. An introduction to design, data analysis and model building", John Wiley & Sons Inc., USA, 1978, ch 20, p306.

20. Bhatia S., in "Zeolite Catalysis: Principles and Applications", CRC Press, 1989, p266.

21. Gary J. H. and Glenn E. H., in "Petroleum Refining", Marcel Dekker, NY, 1975, ch 7, p87.

22. Maggi R. and B. Delmon, "Production Treatment and Utilization of Bio-oils from Pyrolysis, for Energy and Alternative Fuels and Chemicals", 1st AIR2-CT93-1086 Progress Report, UCL contribution, November 1994.
23. Horne P. A., Nugrand N. and P. T. Williams, "Catalytic coprocessing of biomass-derived pyrolysis vapours and methanol", *J. Anal. Appl. Pyrol.*, 1995, **34**, p87.

CATALYTIC STEAM REFORMING OF BIOMASS-DERIVED FRACTIONS FROM PYROLYSIS PROCESSES
Catalytic steam reforming of bio-oils

S. CZERNIK, D. WANG, D. MONTANÉ[1], and E. CHORNET[2]
National Renewable Energy Laboratory, Golden, Colorado, USA

Abstract
Biomass conversion via hydrolytic, solvolytic, and pyrolytic processes generates liquid streams that, after separation of marketable products, can be used to produce either syngas or hydrogen, a strategy being considered in this work. Catalytic steam reforming of model oxygen-containing compounds, their mixtures, bio-oil, and its fractions has been studied using Ni-based catalysts. Tests performed on a microreactor interfaced with a molecular beam mass spectrometer showed that, by proper selection of the process variables: temperature, steam-to-carbon ratio, gas hourly space velocity, and contact time, almost total conversion of carbon in the feed to CO and CO_2 could be obtained. These tests also provided possible reaction mechanisms where thermal cracking competes with catalytic processes. Bench-scale, fixed bed reactor tests demonstrated high hydrogen yields from model compounds and carbohydrate-derived pyrolysis oil fractions. Reforming bio-oil or its fractions required proper dispersion of the liquid to avoid vapor-phase carbonization of the feed in the inlet to the reactor. A special spraying nozzle injector was designed and successfully tested with an aqueous fraction of bio-oil. The techno-economic assessment showed that the process could be economically viable if the lignin-derived oil fraction was sold for adhesives and only carbohydrate-derived fraction was converted to hydrogen.
Keywords: Hydrogen, biomass, fast pyrolysis, pyrolysis oils, catalytic steam reforming, model compounds, oxygenates.

1 Introduction

Steam reforming of hydrocarbons, partial oxidation of heavy oil residues, and gasification of solid fuels to yield syngas (a H_2/CO mixture), followed by shift conversion to produce H_2 and CO_2, are well established processes. Renewable lignocellulosic biomass has been considered as a potential feedstock for gasification to produce syngas. However, economics of current processes favor the use of hydrocarbons (natural gas, C_2-C_5, and naphtha) and inexpensive coal. An alternative approach to the produc-

---

[1] On leave from Universitat Rovira i Virgili, 43006 Tarragona, Catalunya, Spain.
[2] Also affiliated with Université de Sherbrooke, Québec, Canada J1K 2R1.

tion of H_2 from biomass is the catalytic steam reforming of liquid streams generated from various biomass conversion processes such as pyrolysis, hydrolysis, and solvolysis that also yield valuable oxygenate products. This latter approach has the potential to be cost competitive with the current commercial processes for hydrogen production.

The proposed technology for producing hydrogen would be comprised of two stages: fast pyrolysis of biomass to generate bio-oil and catalytic steam reforming of the oil or its fractions. Recent advances in pyrolysis processes allow for conversion of 75% of biomass into a liquid product (bio-oil). This bio-oil is made of oxygenated organic compounds [1,2]: acids, alcohols, aldehydes, ketones, furans, substituted phenolics and complex oxygenates derived from carbohydrates and lignin, as well as water. Very little ash and char are present in the bio-oil when appropriate filtration technology is used in the pyrolysis process [3].

An efficient steam reforming of the oil using no external heat can result in a hydrogen yield of 12.6 wt.% of dry biomass. This yield is similar to that from the biomass gasification/water-gas shift process (11.5 wt.%). The yield of hydrogen for the stoichiometric reaction between wood (poplar) and steam using externally supplied heat is 17.2 wt.%. These values were calculated using current yields for fast pyrolysis [4-6] and gasification [7] processes. An advantage of the pyrolysis process is that it is less severe and does not require an oxygen supply as in some gasifiers.

This paper will describe the concept of our approach, thermodynamic calculations and experimental results to support it, and results and assumptions of techno-economic analysis to estimate the cost of hydrogen production. Our objectives are to gain insight into the chemistry involved in the catalytic steam reforming reactions of biomass-derived oxygenates and to achieve a high yield of hydrogen with a long catalyst life time. These shall lead to a successful process for making renewable hydrogen.

2 Fast pyrolysis: bio-oil yields and composition

Several fast pyrolysis technologies, at pilot and demonstration levels, have been reported to give high yields of bio-oil, including fluid beds [4-6], transport reactors [8,9], and cyclonic reactors [10,11]. The common feature of all these technologies is rapid heating of biomass particles at temperatures between 450 and 550 °C, and short residence time of volatile products in the reaction chamber (a fraction of a second to a few seconds). For example, a 76 wt.% yield of bio-oil can be obtained from fluid bed fast pyrolysis of dry poplar ($CH_{1.47}O_{0.67}$) [4]. The organic fraction of the bio-oil represents 85 wt.% and its elemental composition is $CH_{1.33}O_{0.53}$. About 15 wt.% of the bio-oil is water from dehydration reactions. Additional water will result from the moisture in the feedstock. The water is soluble within the organic fraction of the bio-oil due to the hydrophilicity of carbohydrate-derived compounds that constitute the major fraction of the bio-oil. Vacuum pyrolysis is another technology that offers high bio-oil yield [12].

Reliable values of oil, water, char and gas yields are limited in the literature; most published work contains unclosed and/or unreported material balances. The composition of bio-oil has been quantitatively reported from fluidized bed studies conducted at the University of Waterloo [4-6]. It is shown for several feedstocks in Table 1, together with the composition of the bio-oil derived from the cyclonic (vortex type) reactor located at NREL [11].

Table 1. Composition of pyrolytic oils derived from different feedstocks by two processes (Yields in wt.% of dry biomass)

Product	Fluidized bed (Univ. of Waterloo)[c]			Vortex (NREL)[d]
	poplar 504 °C	maple 508 °C	spruce 500 °C	oak ~500 °C
acetic acid	5.4	5.8	3.9	5.0
formic acid	3.1	6.4	7.2	3.3
hydroxyacetaldehyde	10.0	7.6	7.7	4.3
glyoxal	2.2	1.8	2.5	3.0
methylglyoxal		0.65		
formaldehyde		1.2		2.2
acetol	1.4	1.2	1.2	1.8
ethylene glycol	1.1	0.6	0.9	
levoglucosan	3.0	2.8	4.0	3.8
1,6-anhydroglucofuranose	2.4			
fructose	1.3	1.5	2.3	
xylose				0.9
glucose	0.4	0.6	1.0	
cellobiosan	1.3	1.6	2.5	
oligosaccharides	0.7			
pyrolytic lignin[b]	16.2	20.9	20.6	24.9
unidentified	11.9	17.1	12.9	5.8
oil[a]	65.8	67.9	66.5	55.3
water[a]	12.2	9.8	11.6	10.4
char	7.7	13.7	12.2	12.4
gas	10.8	9.8	7.8	12.2
mass closure	96.5	101.2	97.7	90.3

[a] Oil + water = bio-oil.   [c] From [6].
[b] Material precipitated by addition of water.   [d] Private communication.

Variations in the composition of pyrolysis oil as shown in Table 1, should be expected as a function of raw material, pyrolysis treatment severity (temperature, residence time, and heating rate profiles) and using or not using catalysts during the pyrolytic step. However, the information presented in Table 1 indicates that the bio-oil is essentially a mixture of two acids (acetic and formic), numerous aldehydes and alcohols plus a significant fraction of lignin (denoted as pyrolytic lignin, a low to medium molecular weight material). The unidentified compounds range from a large number of carbohydrate-derived components [13] to monomeric lignin-derived products having alcohol characteristics [1].

3 Thermodynamics

Thermodynamic simulations of reforming model oil components as well as bio-oil have been performed in order to determine equilibrium yields of hydrogen. The results are listed in Table 2, together with stoichiometric yields (moles and wt %). In general, the lignin-derived phenolics have a higher potential for producing hydrogen than the products from carbohydrates. This is because, for an oxygenate with a chemical formula of

$C_nH_mO_k$, the stoichiometric yield of hydrogen is $2+m/2n-k/n$ moles per mole of carbon in the feed (eq. 1), and k/n is usually in the fractions for the aromatic phenolics from lignin, while k/n is close to 1 for most carbohydrate-derived products such as sugars.

$$C_n H_mO_k + (2n-k) H_2O = n CO_2 + (2n + m/2-k) H_2 \qquad (1)$$

Table 2. Hydrogen yields from steam reforming, assuming necessary external heat

Sample	Formula	Stoichiometric H_2 Yield		Equilibrium H_2 Yield	
		moles[a]	% (by wt.)[b]	% (of st.)[c]	% (+ WGS)[c,d]
methane	CH_4	4.00	50.3	85.6	95.0
methanol	CH_4O	3.00	18.9	87.1	96.8
acetone	C_3H_6O	2.67	27.8	85.2	96.2
ADP[e]	$C_{11}H_{14}O_3$	2.36	27.0	84.7	96.4
guaiacol	$C_7H_8O_2$	2.29	26.0	84.6	96.5
syringol	$C_8H_{10}O_3$	2.25	23.5	84.8	96.6
lignin	$C_7H_9O_3$	2.21	22.1	84.9	96.7
furfuryl alcohol	$C_5H_6O_2$	2.20	22.6	84.9	96.7
poplar oil[e]	$CH_{1.33}O_{0.53}$	2.14	19.7	85.2	96.9
poplar	$CH_{1.47}O_{0.47}$	2.07	17.2	85.5	97.1
furfural	$C_5H_4O_2$	2.00	21.0%	84.6	96.9
5-HMF[e]	$C_6H_6O_3$	2.00	19.2	84.9	97.0
xylan	$C_5H_8O_4$	2.00	15.3	85.8	97.3
cellulose/levogluosan	$(C_6H_{10}O_5)_n$	2.00	14.9	85.8	97.3
cellobiose	$C_{11}H_{22}O_{11}$	2.00	13.4	86.3	97.5
HAc, HAA[e]	$C_2H_4O_2$	2.00	13.4	86.3	97.5
formic acid	CH_2O_2	1.00	4.4	87.7	98.9

[a] Moles of H_2 produced per mole of carbon in the reactant being reformed. [b] Amount of H_2 formed divided by the sample molecular weight. [c] Equilibrium moles of H_2 predicted at 750°C and S/C=5 divided by the stoichiometric yield. [d] Assuming all of CO present under 750°C and S/C=5 being completely shifted to H_2 in a downstream water-gas shift (WGS) reactor.
[e] ADP: 4-Allyl-2,6-dimethoxyphenol; 5-HMF: 5-Hydroxymethylfurfural; HAc: acetic acid; HAA: hydroxyacetaldehyde. Poplar oil is on water-free basis.

Table 3. H_2 yields from steam reforming of poplar oil ($CH_{1.33}O_{0.53}$)

Temperature (°C)	Steam-to-carbon ratio (S/C)	% of st. H_2 yield at equilibrium (reforming only)
500	5	57.0%
500	10	76.3%
600	5	76.9%
600	10	90.4%
700	5	84.6%
700	10	92.8%
800	5	84.7%
800	10	91.7%
900	5	82.8%
900	10	89.9%

Thermodynamic calculations indicate that the equilibrium yield of hydrogen (as percentage of the stoichiometric yield) under typical steam reforming conditions (750°C and S/C=5) does not vary much for these oxygenates, averaging around 85%. The small variation in this yield reflects the true ratio $(5+k/n)$ of steam-to-carbon in the catalytic reformer: external steam supplied (S/C=5) plus that derived from the feed (S/C=k/n). A further water-gas shift (WGS) operation to convert all the CO formed in the reformer would improve the overall yield to about 97%. If coking could be avoided, the conversion of oxygenates to hydrogen would be almost quantitative through the combination of reforming and water-gas shift.

The equilibrium yields of hydrogen from a typical bio-oil (of poplar) [4] as a function of temperature and steam-to-carbon ratios (S/C) were also calculated (Table 3). Results suggest that a wide range of combinations of the two key operating variables can be chosen to approach stoichiometric yields. Temperatures higher than 700°C and S/C greater than 5 are recommended for high efficiency conversion.

4 Catalytic reforming of oxygenates

4.1 Literature review

Processes of steam reforming C1-C5 hydrocarbons, naphtha, gas oils, and simple aromatics are well known and commercially practiced. When the objective is to maximize the production of H_2, the stoichiometry describing the overall process is:

$$C_n H_m + 2n\, H_2O = n\, CO_2 + (2n + m/2)\, H_2 \qquad (2)$$

In a hydrocarbon reformer, the following reactions take place concurrently:

$$C_n H_m + n\, H_2O = n\, CO + (n + m/2)\, H_2 \qquad (3)$$
$$CH_4 + H_2O = CO + 3\, H_2 \qquad (4)$$
$$CO + H_2O = CO_2 + H_2 \qquad (5)$$

At normal operating conditions, reforming of higher hydrocarbons (eq. 3) is irreversible [14], whereas the methane reforming reaction (eq. 4) and the shift conversion reaction (eq. 5) approach equilibrium. A large molar ratio of steam to hydrocarbon will ensure that the equilibrium for reactions (4) and (5) is shifted towards H_2 production.

The main problem in the practical steam reforming of hydrocarbons is carbon (coke) formation. Three types of carbon may be formed: whiskers, encapsulating deposits ("polymers"), and pyrolytic carbon [15]. The carbon formation is related to the kinetics, and hence the selectivity, of the catalyst. Significant improvements have been made in controlling coke formation by promoting the classical Ni-alumina or Ni-silica/alumina formulations with Ca and/or K and by the use of magnesia as magnesium-aluminum spinel [16].

The reforming operation requires relatively high steam-to-carbon molar ratios (S/C = 3.5 - 5.0 for methane), adequate temperature profiles in the tubular reactor (low inlet temperature increased along the tube length to a maximum of 750-775°C due to the structural constraints of the Ni-Cr alloys used), and a Ni-based catalyst (20-30 wt.% as NiO) on a refractory support having appropriate steam adsorption characteristics. The reforming operation is not very sensitive to pressure which is essentially dictated by the applications of the product H_2 [17]. Space velocities (G_{C_1}HSV, as volume of C_1 equivalent/h/volume of catalyst) used in tubular reformers are typically in the order of 1500-2000 h^{-1}.

Aznar et al [18] have demonstrated that steam reforming raw, tar-laden syngas obtained from the gasification of lignocellulosics is possible using commercially available Ni-based catalysts (Topsoe R-67 and RKS-1). Rapid deactivation of the catalyst takes place after a few hours only. Tars produced during gasification contain a high percentage of polycyclic aromatics, while pyrolysis oil is essentially a mixture of carbohydrate-derived fragments and lignin-derived phenolics [1].

Literature on the steam reforming of oxygenates falls within two categories: (a) reforming simple alcohols, such as methanol and ethanol; and (b) reforming oxygenated aromatics, i.e. cresols. A detailed review on this subject has been given elsewhere [19]. The steam reforming of methanol is a special case and a large body of information has been accumulated [20]. A variety of catalysts are capable of carrying out the conversion at low temperatures and the product slate is markedly affected by the type of catalyst and the experimental conditions used. A different mechanism, involving methyl formate as an intermediate, was initially proposed by Takahashi et al [21] and based on the formation of methylformate ester on the catalyst surface between two adsorbed species, H_3CO and HCO. The ester is hydrolyzed to methanol and formic acid. The latter decomposes directly to CO_2 and H_2 with 100% selectivity, i.e. no CO is formed. Ethanol reforming has also been studied to produce acetic acid [22] and to hydrogen by Garcia and Laborde [23]. Steam reforming of oxygenated aromatics has been directed towards the dealkylation of cresol.

When considering the bio-oil composition shown in Table 1, it is clear that challenges in catalytic steam reforming will be linked to the ability to handle both the "pyrolytic lignin" fraction and the carbohydrate-derived fraction: acetic acid, aldehydes, sugars and anhydrosugars, and oligomers. It is likely that the oligomeric fragments, either lignin- or carbohydrate-derived, once condensed as bio-oil, cannot be revolatilized without significant char formation. Thus, our strategy will be either (a) to separate the "pyrolytic lignin" (which has a potential market of its own as a phenolics substitute for the production of adhesives) from the condensed bio-oil and steam reform the remaining *aqueous fraction* which, while still containing some substituted, but monomeric phenolics, will predominantly be constituted of the carbohydrate-derived components; (b) to carry out a fractional condensation of the pyrolytic vapors to separate the oligomers and thus directly recover a "light" fraction comprised essentially of mixed monomers; or (c) try to reform the entire uncondensed vapors by coupling the reformer to the pyrolysis unit.

Given the chemical diversity of the bio-oil *aqueous fraction*, and according to the literature reviewed, Ni-based catalysts operating in the 700-800°C range ought to be able to carry out the steam reforming conversion. An appropriate temperature profile through the reactor will have to be developed experimentally to carefully control coke formation.

4.2 Rapid screening and mechanistic studies
A series of model oxygen-containing compounds, biomass and its main components (cellulose, xylan, and lignin), and bio-oil and its various fractions were screened under identical conditions using a commercial catalyst, G-90C, from United Catalyst Inc. (UCI). We also tested a number of research and commercial steam reforming catalysts and a WGS catalyst and determined H_2 yields using four model compounds (methanol, acetic acid (HAc), an aqueous solution of hydroxyacetaldehyde (HAA), and a methanol solution of 4-allyl-2,6-dimethoxyphenol (ADP)) under the same operating conditions. Methanol steam reforming was used as a standard test for checking the performance of the system as well as catalyst deactivation.

These experiments were performed in a vertical dual-bed microreactor coupled to a molecular-beam mass spectrometer (MBMS) [24,25]. This system has the advantage of simultaneously detecting, unconverted reactants, intermediates and products at the exit of the microreactor by rapidly sampling through a supersonic, free-jet expansion nozzle. This expansion cools the reaction products and forms a molecular beam that is ionized and analyzed by a quadrupole mass spectrometer. The microreactor was housed in a tubular furnace with four independently controlled temperature zones. The dual bed configuration of this reactor enabled us to study either the differences between thermolysis and catalysis or to compare the performances of two catalyst under the same temperature conditions. Mass spectra in the m/z range of 1-350 were obtained for pyrolysis and catalysis products with the MBMS operated under 25 eV EI ionization conditions. The detection limit of the MBMS corresponded to conversions of approximately 99.95% for most of the samples, i.e., from the sample signal without catalyst at more than 1.0×10^6 counts/s (cps) to less than the instrument lower limit of 500 cps (noise = ± 250 cps) with catalyst. Among the most important parameters for steam reforming are catalyst bed temperature(s), molar steam-to-carbon ratio (S/C), gas hourly space velocity ($G_{C_1}HSV$), and residence time (t, calculated from the void volume of the catalyst bed divided by the total flow rate of gases at the inlet of the reactor; void fraction = 0.4).

Under conditions of 600 °C, S/C=10-13, $G_{C_1}HSV = 180$-1680 h^{-1}, and t=0.1s, essentially complete conversion of all oxygenates was achieved using the UCI G-90C catalyst, and in most cases, the yield of hydrogen was high, and close to stoichiometric values. However, there were several cases where the hydrogen yield was far below that predicted by stoichiometry and thermodynamic calculations, because of the formation of carbonaceous materials during vaporization or pyrolysis of the feed prior to entering the catalyst bed.

Other commercial and research catalysts tested were capable of reforming the model compounds and high conversions (>99%) were observed [25]. The H$_2$ yields for all catalysts and model compounds were high, averaging 90% (± 5%) of the stoichiometric. Within our experimental error limit, there is no clear indication of one catalyst being better than the others. The yields were reproduced very well (typically within ± 2-3%; the worst within ± 5-6%). Among the different model compounds, we observed that methanol and HAA yield slightly more H$_2$ than HAc and ADP.

Among the various operating parameters studied, temperature has the most profound effect on steam reforming reactions. Methanol can be steam reformed at 300 °C; HAc and HAA are reformed in high conversions (>99.95%) at above 350 °C. The lignin model compound, ADP, was more difficult to reform with steam, requiring temperatures above 600°C for complete conversion. Within experimental error limits, varying residence time from 0.04 to 0.15 s and increasing S/C from 4.5 to 7.5 showed no significant effects on the yield of hydrogen under the conditions of 600°C and $G_{C_1}HSV = 1680$ h^{-1}; however, it did affect the yield of CH$_4$.

The microreactor-MBMS system also allows the study of reforming mechanisms through the continuous monitoring in real time of the intermediate products present in the gas phase at low conversions. Acids, ketones, alcohols and aldehydes were represented by HAc, acetone, ethylenediol, glycerol, and HAA. The phenolic series included phenol, anisole, cresols, resorcinol, 2,6-dimethylphenol, guaiacol, syringol, and ADP. The furan family of model compounds consisted of furan, 2-methylfuran, 2,5-dimethylfuran, 2-furfuraldehyde, furfuryl alcohol, 5-methylfurfural, and 5-hydroxymethylfurfural. A complete report on HAc and HAA has been published [26], and details of these studies are to be published in separate papers being prepared. HAc produced significant amounts of carbon on the catalyst during steam reforming. No intermediate was detected during steam reforming of HAA and phenol above 400 °C. Phenol was the only intermediate detected at both 400 and 600 °C from *o*-cresol steam reforming, while significant amounts of cresols, phenol, benzene, and toluene were observed from thermolysis of *o*-cresol at both 400 and 600 °C. The results obtained for guaiacol and syringol are consistent with the weakening effect of MeO substituents on ArO–Me and ArO–H bonds (Ar = aromatic group).

The results obtained for these model compounds suggest that there are significant differences in steam reforming mechanisms of hydrocarbons and oxygenates. Most of the oxygenates found in bio-oils are thermally unstable. At the operating conditions of a typical steam reformer, these oxygenates undergo homogeneous (gas-phase) thermal decomposition, as well as cracking reactions on the acidic sites on the catalyst support. These reactions compete with the catalytic steam reforming reaction to hydrogen. However, a complete conversion of both the oxygenate feed and its decomposition products to hydrogen can be achieved with commercial Ni-based catalysts under reasonable operating conditions, if char formation prior to reaching the catalyst bed and coking on the catalyst can be eliminated, or at least controlled. This will be further discussed below in our bench-scale studies.

4.3 Bench-scale reformer studies

A bench-scale, fixed-bed reactor system was used to establish the global and elemental mass balances and the carbon-to-gas conversion, to quantify the distribution of gas products under conditions of complete conversion of the pyrolysis oil feedstock, and to study catalyst lifetime and regeneration. This apparatus has been described in detail in our previous publication [25]. The reactor main body is a stainless steel tube (1.65 cm id x 42.6 cm length) housed in a tubular furnace equipped with three independently controlled heating zones. Steam was generated in a boiler and superheated up to 850 °C. The organic feed from a diaphragm metering pump was sprayed with ambient N_2 gas followed by mixing with superheated steam in a triple nozzle inlet. Products exiting the reactor were passed through a condenser. Every 4-5 min, the condensate (just water in most cases) weight, volume and compositions of the permanent gas output were recorded. An on-line IR gas analyzer was used to monitor CO/CO_2 concentrations and a MTI-QUAD GC was used to measure concentrations of H_2, N_2, O_2, CO, CO_2, CH_4, and other light hydrocarbons. The reformer system was interfaced with a computer to monitor temperatures and other important parameters (once every 30 s).

We initially used model compounds (methanol, acetic acid, syringol and *m*-cresol, both separately and in mixtures), then real bio-oil and its aqueous fraction. Most studies were carried out using the UCI G-90C catalyst; in several latest testings we used a dual-catalyst bed of 46-1 and 46-4 from ICI Katalco. Representative results are listed in Table 4 for the steam reforming of acetic acid, syringol/methanol mixture, a three-component feed made of acetic acid, *m*-cresol, and syringol, and an aqueous fraction obtained from a poplar oil. Profiles of the output gas composition are shown for the 3-component mixture in Fig. 1 and for the poplar oil aqueous fraction in Fig. 2.

Acetic acid. Results show that the temperature of the reforming section of the bed determines the extent of the carbon conversion to gas products. At temperatures below 500 °C, only 73% of the carbon contained in the feed was converted to gas products. It was observed that a significant amount of carbon was formed during these experiments.

Table 4. Summary of results for catalytic steam reforming experiments on the bench-scale reformer

Feed	Catalyst	S/C[a]	G_{C_1}HSV[b]	Temperatures (°C)			Yield (mol/100 mol of carbon fed)				Carbon-gas conversion	% st. yield of H_2 (+WGS)[c]	Time on stream (h)
				Top	Middle	Bottom	H_2	CO_2	CO	CH_4			
acetic acid	UCI G-90C	4.7	1973	685	716	833	145.8	50.3	49.7	0.055	101	73 (98)	6
	"	12.8	777	710	789	830	171.8	74.9	29.0	0.005	104	86 (100)	8
syringol (MeOH solution)	"	6.3	2454	702	745	830	195.0	45.4	53.9	0.2	100	75 (96)	4
	"	7.4	1985	750	803	863	197.3	46.0	54.9	0.1	101	76 (97)	4
3-component mixture	ICI 46-1/46-4	6.5	1053	738	na	833	167.6	67.4	28.6	0.00	96	78 (91)	11
	"	4.9	1053	782	753	834	187.8	79.8	27.4	0.03	105	86 (98)	17
poplar oil, aq. fraction	"	19.3	1110	480	730	818	206.7	85.2	9.6	1.7	97	103 (108)	2
	"	30.0	1000	530	744	821	205.8	86.6	8.4	6.9	102	103 (107)	4

[a] Molar ratio of steam to carbon. [b] Gas hourly space velocity on C_1 basis (h[-1]). [c] Assuming all CO being converted to H_2 in a down stream WGS unit.

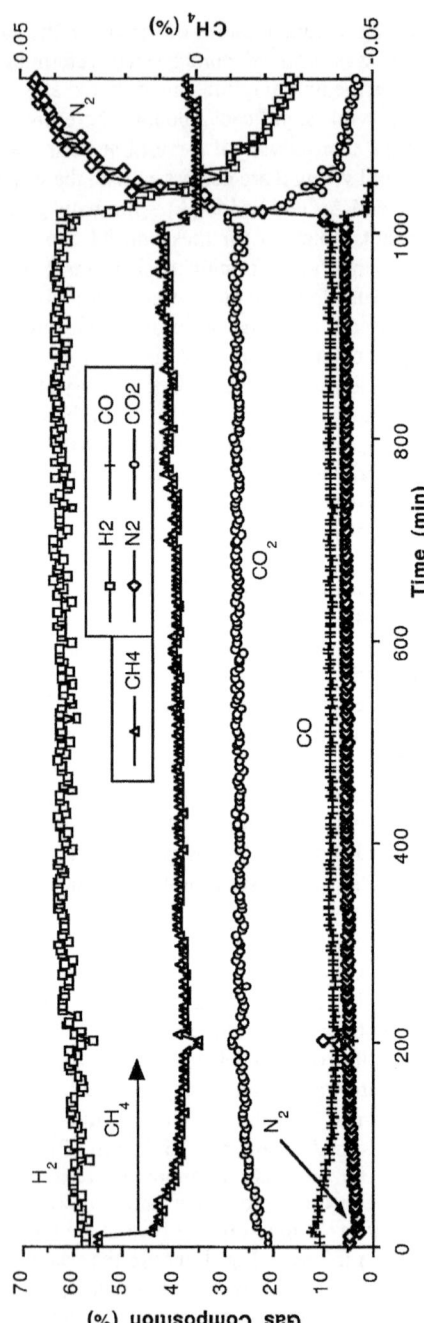

Fig.1. Composition of gaseous products during steam reforming of a 3-component mixture using ICI 46-series catalysts.

Part of this carbon was entrained from the bed by the gas stream and was collected in the condenser. The accumulation of carbon in the bed caused the pressure at the reactor entrance to increase continuously during the experiment. An increase in temperature to 535 °C in the reforming zone of the catalyst bed improved the recovery of carbon in the gas products up to 80% of that in the feed, and although carbon was still formed, it was not entrained by the gas but retained in the upper section of the catalyst bed. We observed that the carbon deposited in the catalyst bed during the steam reforming of acetic acid was gasified, at least partially, by steam and that, after the steam treatment, the pressure at the reactor entrance recovered to the initial value. Further increase in the reforming temperature resulted in a lower extent of carbon deposition on the catalyst. Formation of carbon deposits was minimal when the temperature was raised to 665 °C. At this temperature, the average carbon recovery was 98% and the pressure drop (~50 kPa) across the catalyst bed was stable throughout the experiment (90 min).

Below 600°C, hydrogen yield improves with rising reforming temperature due to the increase in the amount of carbon converted to gas products. Above 600°C, where carbon conversions approaching 100% are obtained, hydrogen yield tends to decrease as the temperature increases. This behavior is due to the thermodynamics of the reforming process, since WGS reaction (3) is displaced towards the formation of CO and water at high temperature. This results in a lower hydrogen concentration at the reformer exit. Thus, a unit for the steam reforming of acetic acid using commercial nickel-based catalysts does not differ conceptually from the process for the steam reforming of naphtha or methane, and must include high- and low-temperature shift reactors to obtain an optimal yield of hydrogen. In this context, higher temperatures in the reformer are favorable because they reduce not only the extent of carbon deposition on the catalyst bed but also the yield of methane, thus resulting in a higher yield of hydrogen after the conversion of carbon monoxide in the shift reactors.

Syringol. The steam reforming of syringol was studied at conditions similar to those employed for the reforming of acetic acid. Syringol was dissolved in methanol at a weight concentration ratio of 40:60, which corresponds to a mixture in which 53% of the carbon is provided by syringol. Methanol was selected since it was verified that this solvent was readily reformed in the range of experimental conditions to be tested without leaving any appreciable amount of carbon on the catalyst surface. As a general trend, syringol does not react as well as acetic acid during steam reforming using the UCI G-90C catalyst. At 670 °C, the conversion of the carbon in syringol to gas products was estimated to be only 90%. Raising the catalyst temperature to 700 °C and above resulted in almost complete carbon conversions (Table 4). Similar to the reforming of acetic acid, the increase in temperature during the steam reforming of syringol also results in the displacement of the gas composition: a decrease in the contents of CH_4, H_2,, and CO_2, together with an increase in the amount of CO.

Three-component mixture. Two long-duration runs using a 3-component mixture were performed in order to compare performances of catalysts from two manufacturers. This mixture contained approximately 67% acetic acid, 16% *m*-cresol and 17% syringol, to represent the components of the carbohydrate fraction and the lignin fraction in bio-oil. We observed *some coke deposits* on the top portion of the UCI G-90C catalyst bed. The overall mass balance (carbon, hydrogen, and oxygen) was 98.8% and the carbon conversion to gas was 96.1%. The other catalyst tested for steam reforming of the 3-component mixture was the 46-series from ICI Katalco (46-1/46-4). This dual catalyst bed showed an excellent and steady performance *without any coke deposition* on the catalyst. Figure 1 shows the reactor temperature profiles, the output gas flow rate (very steady), and the gas composition, which remained constant throughout the whole run. The overall mass balance (including carbon, hydrogen, and oxygen) was 104.3%, and for carbon 105.1%, indicating that there may be a systematic error in our measurement. An excellent hydrogen yield of 85.6% was obtained, and the total hydrogen potential may be as high as 97.6% with a second water-gas shift reactor. These results confirm that both the UCI G-90C and especially the ICI 46-series catalysts can efficiently convert oxygenates to hydrogen.

Aqueous fraction of bio-oil. Steam reforming of bio-oil or its fractions was found to be more difficult than that of model compounds. The main problem that needed to be solved was feeding the oil to the reactor. Bio-oil cannot be totally vaporized; significant amounts of residual solids are often formed that block the feeding line and the reactor. Thus, the simple injection system used for model compounds had to be modified to allow spraying bio-oil and its fractions in to the catalytic reactor without prior char formation. Because the process economics favors the separation of lignin-derived phenolics as co-products (discussed below), we only used the aqueous, mostly carbohydrate-derived fraction to generate hydrogen.

A poplar oil generated in the NREL vortex reactor system was extracted using ethyl acetate (EA) and water (weight ratios of 1:1:1 for oil:EA:water). The resulting aqueous fraction (55% of the whole oil) contained 25% organics and 75% water. It was successfully fed to the reactor using a triple-nozzle spraying system with minimal accumulation of char in the reactor inlet. A large excess of steam (S/C = 20-30) was used, together with a high flow rate of nitrogen, to allow for proper oil dispersion and heat transfer required to maintain a sufficiently high temperature (>500 °C) at the reactor entrance. A portion of water and other volatiles in the sprayed droplets evaporate during mixing with the superheated steam and the remaining will contact the catalyst surface directly. The ICI 46-series catalysts performed satisfactorily with no coke formation. We observed a stable gas production rate and composition throughout the whole 4-hour-long experiment (Fig. 2).

The carbon conversion of the aqueous fraction to gas products was almost quantitative in both runs that used the same catalyst bed (Table 4). We observed similar levels of mass balances as in the experiments using model compounds: global 99%, carbon 105%, and hydrogen 97% (assuming a formula of CH_2O for the organics in the aqueous fraction). The methane concentration (with N_2 excluded) increased from 0.56% in the first run (2 hrs, t=0.03 s) to 2.2% in the second run (4 hrs, t=0.02 s), and both values were much higher than that (0.01%) obtained from the 3-component model compound mixture (17 hrs, t=0.09 s). This was likely caused by the shorter residence time forced by the large flow rate of steam and nitrogen used in the experiment, and also possible incomplete reduction or reoxidation of the catalyst.

Summary. It can be concluded that oxygenates present in bio-oil can be effectively steam-reformed using commercial catalysts under conditions similar to those used for the steam reforming of naphtha: S/C=5, >700 °C, and G_{C_1}HSV around 2000 h^{-1}. Excellent mass balances and high yields of hydrogen have been obtained in most cases. A catalyst formulation, such as the UCI G-90C and the ICI 46-series, used for the reforming of naphthas, can be effectively employed in converting bio-oil to hydrogen.

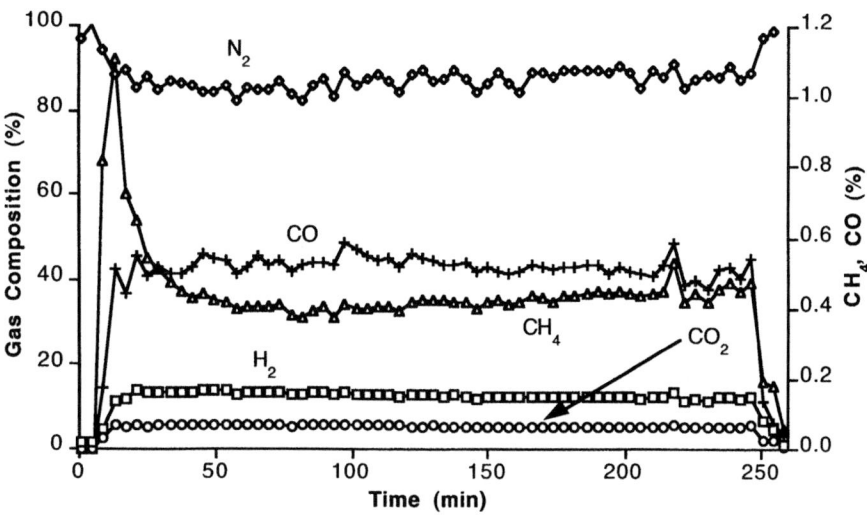

Fig. 2. Composition of gaseous products during the steam reforming of bio-oil aqueous fraction using ICI 46-series catalysts.

5 Process design and economics

Two scenarios for the production of hydrogen from pyrolysis oil can be envisioned: i) A *regionalized* system of hydrogen production with small and medium-sized pyrolysis units providing condensed bio-oil to a centralized reforming unit. At the reforming plant, the separation of the lignin-derived oligomers will generate a "clean" aqueous stream with simple carbohydrate-derived soluble organics which will be fed to the reformer. ii) Biomass is transported to a larger pyrolysis unit and the uncondensed vapors are fed directly to an *integrated* adjacent reforming unit located in the same plant premises. In this case, we can also condense the heavier fraction of the pyrolysis vapors via fractional condensation and feed the remaining uncondensed vapors to the reformer. The first case has the advantage of the availability of cheaper feedstocks, which include biomass residues. In the integrated system, however, the costs of entirely condensing the vapors to produce the bio-oil and transporting it to the hydrogen facility are avoided.

We have estimated the selling price of hydrogen from the process i) above through a detailed technoeconomic analysis [27]. The process analyzed consists of fluid bed pyrolysis of biomass, water and EA extraction to recover a phenolic substitute coproduct, and steam reforming to produce H_2 in a process based on that used for natural gas reforming. The cost of the pyrolysis plant was taken from Beckman and Graham [28]. A pressure swing adsorption system is used to separate H_2. Steam is produced through detailed heat integration and is intended to be sold as a by-product.

The current selling price of H_2 in industry is between \$5/GJ and \$14/GJ, depending on the size of the production facility. This range is for hydrogen as produced by the plant: purified, but not compressed or stored. Results indicate that the necessary hydrogen selling price for the base case analysis is \$7.70/GJ. The process to produce a phenolics substitute and hydrogen from pyrolysis via extraction and steam reforming has significant potential to be economically feasible. A high rate of return can be obtained and the process can sustain a high drop in coproduct selling price before the hydrogen becomes more expensive than current markets will allow. Relative to the market value of hydrogen, this analysis found that the process is able to sustain large changes in capital cost, hydrogen production capacity, and feedstock cost.

6 Conclusions

Fast pyrolysis of biomass is an advanced technology to produce a bio-oil in high yields (70-75 wt.% of dry biomass). This bio-oil is a complex mixture of simple aldehydes, alcohols and acids together with more complex carbohydrate- and lignin-derived oligomeric materials and water. Fractional condensation of the pyrolysis vapors could separate the simple monomeric materials, which comprise about one third to one half of the bio-oil from the complex oligomeric fraction. Steam reforming of the simple monomeric materials, the entire bio-oil or its fractions is thermodynamically and chemically feasible. Our screening tests have shown that such catalysis is possible using commercially available Ni-based catalysts. Improvements in their formulation may be needed to optimize the activity/selectivity/lifetime stream relationships when reforming biomass-derived oxygenates. Results obtained at the bench-scale level on the steam reforming of model compounds and their mixtures, as well as bio-oil fractions have substantiated our findings at the microscale level. The challenge seems to be the introduction of the bio-oil or its fractions to contact the catalyst prior to gas-phase thermal decomposition to form char. In addition, progressively increasing the catalyst activity by either temperature ramping or catalyst formulation is an appropriate strategy to minimize coke formation on the catalyst. We have achieved a reasonable comprehension of the spray nozzle design and operating conditions to fulfill these goals.

The preferred implementation strategy consists of small- to medium-size regional fast pyrolysis units which will produce the bio-oil from either dedicated crops or plantations, or will use waste lignocellulosics as feedstocks. The bio-oil produced will be transported to a central separation/reforming unit serving a given region where H_2 and co-products will be produced. The process concept is simple: a desulfurization unit is not needed, and the bio-oil can be atomized in a flow of steam which is then processed through a reforming unit. By a proper choice of conditions the near equilibrium design can be driven to maximize H_2 production. A pressure swing adsorption unit will purify the gas stream. An economic analysis of the process indicates that the necessary selling price of hydrogen is well within current market values. This analysis studied the co-production of hydrogen and a phenolics substitute for resin manufacture. The process studied was based on pyrolysis oil production at distributed locations followed by extraction and steam reforming at a centralized facility. Relative to the current market price of hydrogen, the economics of this process have been shown to be insensitive to changes in the assumptions made.

An alternate strategy, but less favorable economically, consists of transporting the biomass to a central conversion site where the fast pyrolysis and reforming operations can be conducted jointly. Fractional condensation of the vapors can be carried out in order to recover, for instance, the depolymerized lignin fraction which has value as a mixture of phenolics to be used in resin formulations.

7 Acknowledgments

We thank the US DOE Hydrogen Program for financial support of this research project under contract DE AC 36-83CH10093, and Margaret Mann of NREL for providing results of the techno-economic analysis.

8 References

1. Elliott, D.C. (1988) Relation of reaction time and temperature to chemical composition of pyrolysis oils, in *Pyrolysis Oils from Biomass: Producing, Analyzing and Upgrading* (eds. E.J. Soltes and T.A. Milne), ACS Symposium Series 376, ACS, Washington, D.C. pp. 55-65.
2. Fagernas, L. (1995) *Chemical and Physical Characterization of Biomass-Based Pyrolysis Oils: Literature Review*, VTT Research Notes, No. 1706, Tech. Res. Center of Finland, Espoo, Finland.
3. Diebold, J.; Czernik, S.; Scahill, J.; Philip, S.D. and Feik, C.J. (1995) Hot gas filtration to remove char from pyrolysis vapors produced in the vortex reactor at NREL, in *Proceedings of Biomass Pyrolysis Oil Properties and Combustion Meeting*, Estes Park, CO. pp. 90-108.
4. Radlein, D.; Piskorz, J. and Scott, D.S. (1991) Fast pyrolysis of natural polysaccharides as a potential industrial process. *Journal of Analytical and Applied Pyrolysis*, Vol. 19, pp. 41-63.
5. Scott, D.S.; Piskorz, J. and Radlein, D. (1985) Liquid product from continuous flash pyrolysis of biomass. *Industrial & Engineering Chemical Process Design & Development*, Vol. 24, pp. 581-86.
6. Piskorz, J.; Scott, D.S. and Radlein, D. (1988) Composition of oils obtained by fast pyrolysis of different woods, in *Pyrolysis Oils from Biomass: Producing, Analyzing and Upgrading* (eds. E.J. Soltes and T.A. Milne), ACS Symposium Series 376, ACS, Washington, D.C. pp. 167-78.
7. Probstein, R. and Hicks, R. (1982) *Synthetic Fuels*, McGraw Hill, New York, pp. 235.
8. Graham, R.G., Freel, B.A. and Bergougnou, M.A. (1988) The Production of pyrolysis liquids, gas, and char from wood and cellulose by fast pyrolysis, in *Research in Thermochemical Biomass Conversion*, (eds. A.V. Bridgwater and J.L. Kuester), Elsevier Applied Science, London, pp. 629-41.
9. Kovac, R.J., Gorton, C.W. and O'Neil, D.J. (1988) Entrained flow pyrolysis of biomass, in *Proceedings of Thermochemical Conversion Program Annual Meeting*, Solar Energy Research Institute, Golden, CO, SERI/CP-231-2355, pp. 5-20.
10. Diebold, J. and Scahill, J. (1988) Production of primary pyrolysis oils in a vortex reactor, in *Pyrolysis Oils from Biomass: Producing, Analyzing and Upgrading*, (eds. E.J. Soltes and T.A. Milne), ACS Symposium Series 376, ACS, Washington, D.C. pp. 31-40.
11. Czernik, S., Scahill, J. and Diebold, J. (1995) *Journal of Solar Energy Engineering*, Vol. 117, 2-6.
12. Roy, C.; de Caumia, B.; Plante, P. Performance study of a 30 kg/hr vacuum pyrolysis process development unit, In *5th European Conference on Biomass for Energy and Industry*; Grassi, G.; Goose, G.; dos Santos, G., Eds.; Elsevier Applied Science: London, 1990, pp 2595.
13. Antal, M.J. (1982) Biomass pyrolysis: a review of the literature. part 1–carbohydrate pyrolysis, in *Advances in Solar Energy*, (eds. K.W. Boer and J.A. Duffield), Solar Energy Society, New York, pp. 61.
14. Tottrup, P.B. and Nielsen, B. (1982) Higher hydrocarbon reforming. *Hydrocarbon Processing*, March, pp. 89-91.
15. Rostrup-Nielsen, J.R. and Tottrup, P.B. (1979) in *Proc. Symp. on Science and Catalysis and its Applications in Industry*, FPDIL, Sindri, pp. 379.
16. Rostrup-Nielsen, J.R. (1984) Catalytic steam reforming, in *Catalysis Science and Technology*, (eds. J.R. Enderson and M. Boudart), Springer-Verlag: Berlin, Heidelberg, Vol. 5, Ch. 1.

17. Patil, K.Z. (1987) Hydrogen, *Chemical Age of India* Vol. 38, No. 10, pp. 519-27.

18. Aznar, M.P., Corella, J., Delgado, J. and Lahoz, J. (1993) Improved steam gasification of lignocellulosic residues in a fluidized bed with commercial steam reforming catalysts. *Industrial Engineering Chemical Research*, Vol. 32, pp. 1-10.

19. Wang, D., Czernik, S., Montané, D., Mann, M. and Chornet, E. (1996) Biomass to hydrogen via fast pyrolysis and catalytic steam reforming of the pyrolysis oil or its fractions. (submitted to *Industrial & Engineering Chemical Research*).

20. Jiang, C.J., Trimm, D.L., Wainright, M.S. and Cant, N.W. (1993) Kinetic mechanism for the reaction between methanol and water over a Cu-ZnO-Al$_2$O$_3$ catalyst. *Applied Catalysis A: General*, Vol. 97, pp. 145-58.

21. Takahashi, K., Takezawa, N. and Kobayashi, H. (1982) The mechanism of steam reforming of methanol over a copper-silica catalyst. *Applied Catalysis* Vol. 2, pp. 363-66.

22. Iwasa, N. and Takezawa, N. (1991) Reforming of ethanol - dehydrogenation to ethyl acetate and steam reforming to acetic acid over copper-based catalysts. *Bulletin of Chemical Society of Japan*, Vol. 64, pp. 2619-23.

23. Garcia, E.Y. and Laborde, M.A. (1991) Hydrogen production by steam reforming of ethanol: thermodynamic analysis. *International Journal of Hydrogen Energy*, Vol. 16, No. 5, pp. 307-12.

24. Evans, R. and Milne, T.A. (1987) Mass spectrometric studies of the pyrolysis of biomass. 1. Fundamentals. *Energy and Fuels*, Vol. 1, pp. 123-37.

25. Chornet, E., Wang, D., Montané D. and Czernik, S. (1995) Hydrogen production by fast pyrolysis of biomass and catalytic steam reforming of pyrolysis vapors, in *the Proceedings of the 2nd EU-Canada Workshop on Thermal Biomass Processing*, Toronto, Canada, May 8-9, pp. 246-62.

26. Wang, D., Montané, D. and Chornet, E. (1996) Catalytic steam reforming of biomass-derived oxygenates: acetic acid and hydroxyacetaldehyde. *Applied Catalysis A: General* (in press).

27. Mann, M.K.; Spath, P.L.; Kadam, K. (1996) Technical and economic analysis of renewables-based hydrogen production, in *Proceedings of 11th World Hydrogen Energy Conference*, Stuttgart, Germany, June 23-28.

28. Beckman, D. and Graham, R. (1993) Economic assessment of a wood fast pyrolysis plant, in *Advances in Thermochemical Biomass Conversion*, (ed. A.V. Bridgwater) Blackie Academic & Professional, London, pp. 1314-24.

LINEAR MONO-OLEFINES BY THERMAL DEOXYGENATION OF LIPIDS OVER ALUMINA

D. G. B. BOOCOCK, S. K. KONAR, A. LEUNG and E. VHONGIA
Department of Chemical Engineering and Applied Chemistry, University of Toronto, Toronto, Canada.

Abstract
When lipids, in the form of triglycerides and fatty acids, are passed over a suitably demoisturized activated alumina catalyst at 450°C and a weight-hourly space velocity of approximately 0.5, high yields of deoxygenated liquids, containing mostly linear mono-olefines, are obtained. Fatty acids are intermediates in one of the two major pathways by which triglycerides are deoxygenated, the other pathway being the direct production of olefines having three less carbon atoms than the triglyceride side chains. The fatty acids are converted to symmetrical ketones, which then form methyl ketones and mono-olefines, presumably by a γ-hydrogen transfer mechanism. The methyl ketones can 1) reduce to alcohols, which then dehydrate to olefines, and 2) oxidize to carboxylic acids, probably via isomerization to aldehydes. The reduction and oxidation are likely coupled because the alumina does not have reductive or oxidizing properties. The new fatty acids ultimately yield mono-olefines which possess one more carbon atom than the original methyl ketones. Catalyst deactivation appears to be caused mainly by the deoxygenation of the methyl ketones. This can be overcome by a simple strategy designed to maintain the balance of acidic sites and/or provide some required hydrogen atoms. The results are relevant to the utilization of waste fats and oils, possibly as detergent precursors, as well as to the biogenesis of petroleum.
Keywords: Catalytic Deoxygenation of Lipids, Petroleum Biogenesis, Linear Mono-Olefines.

1 Introduction

Lipids occur in all organisms as structural components of cell membranes. Triglycerides are not important membrane lipids, but are stored in most plants and animals as a metabolic energy reserve. In higher plants, triglycerides are found in the seeds, and are the source of vegetable oils. Triglycerides are triesters of 1,2,3-propanetriol (glycerol) and fatty acids (see Fig 1). In vegetable oils, many of the long chains of the fatty acids contain one, two, and even three double bonds, but in some oils, such as coconut and palm, the chains are mostly saturated . Chain lengths of 16 and 18 carbon atoms are most common, but in coconut oil are somewhat shorter, lauric acid (C_{12}) being the main acid moiety. In 1993-1994, the split-year world production of vegetable oils was almost 62 million tonnes [1]. Soybean oil accounted for 17.6 million tonnes,and palm for 13.8 million tonnes   Other major oils (in millions of tonnes) are sunflower (7.8), rape/canola (8.8), cotton (3.6), peanut (3.5) and copra (3.0). Palm oil shows the greatest increase in world production.[1] The world average disappearance of fats and oils per capita is 15.5 kg per annum. In Canada, the U.S.A. and the E.E.C. the values are 35.4, 44.2 and 40.9 kg per annum respectively. Considerable amounts of waste fats and oils are generated in food operations and are available as a waste stream. Other waste streams are generated in the actual processing of oils.   These waste streams, often contain fatty acids as well as triglycerides, and, therefore, any process for the conversion of the waste to useful products must be capable of handling both substrates. In some jurisdictions,   facilities exist for refurbishing waste fats and oils.

In 1991 we discovered that raw sewage sludge lipid could be converted in high yields to a hydrocarbon mixture by passing it over activated alumina at 450°C at a weight-hourly space velocity of 0.46 [2].   The key to this deoxygenation was the removal of 15 mass % water from the catalyst before use. Other workers had reported the partial deoxygenation of triglycerides over activated alumina, but had failed to identify the requirement for water removal [3]. A patent reporting the deoxygenation of lipids over aluminosilicates is not supported by other scientific literature[4].

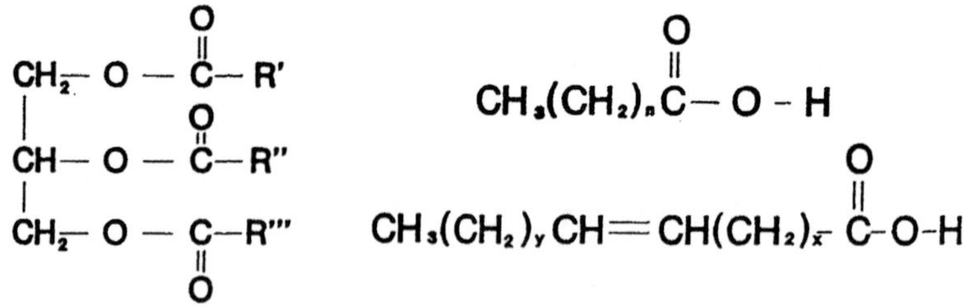

Fig. 1 General structures of triglycerides (left) and fatty acids (right).

The products from the deoxygenation were mostly linear mono-olefines. The sewage sludge contained 65 mass % fatty acids and only 7 mass % triglycerides, the balance being unsaponifiables. Because the process, converted both fatty acids and triglycerides, it was therefore suitable for waste fats and oils. Triglycerides in the form of soybean oil, canola oil and coconut oil, as well as pure triglycerides in the form of glycerol trioleate and glycerol trilaurate have all been converted. More recently, waste oils in the form of fast-fry oils and general food wastes have also been converted.

2 Experimental

2.1 Pyrolysis unit

The pyrolysis unit has been described elsewhere [5], but a brief description follows for the sake of clarity: The unit consists of an angled insulated 316 stainless steel preheater tube (1.3 cm i.d. x 50 cm length) which extends 2.54 cm into a 316 stainless steel fixed bed tubular reactor (2.5 cm i.d. x 46 cm length) oriented vertically. The reactor is heated by a cylindrical block heater. Two type J (iron-constantan) thermocouple probes are used to monitor the internal catalyst bed temperature as well maintain a constant reactor wall temperature in combination with a temperature controller. A syringe pump is connected to the front end of the preheater. An angled water-cooled condenser is connected to the base of the reactor which in turn is connected to the product receiving flask (replaceable). A line from the vacuum head leads to a brine displacement vessel where gas can be measured and collected. A nitrogen inlet is provided between the feed line from the syringe pump and the preheater. A removable stainless steel screen at the mid-section of the reactor supports the catalyst (usually 40 g of Alcan AA 200 activated alumina, BET surface area 270-290 m^2/g, pore volume 40 cm^3/g, bimodal pore distribution with 66% in pores less than 30Å radius and the remainder in larger pores). The reactor and preheater are operated at 180-190 and 450°C, respectively.

2.2 Operation of the pyrolysis unit

Prior to a run, fresh catalyst (usually 40 g) is demoisturized at 450°C for a two- hour period by intermittently passing nitrogen over the catalyst bed. This step is critical to the successful performance of the unit, even though water is a product of the process and the feed may contain some moisture. Fifteen to eighteen percent moisture is removed by this process so the catalyst , as used, only has a mass of approximately 34 grams. A syringe pump delivers the substrate to the preheater, usually at a weight-hourly space velocity (WHSV) of 0.46 based on the original catalyst or approximately 0.54 based on the demoisturized form. The substrate is delivered to the catalyst bed by gravity down the angled preheater. The liquid product is cooled in the condenser and collected in receiving flasks.

Various substrates such as canola oil, soybean oil, coconut oil and waste fastfry oil and food fats have been used [6]. The waste triglycerides must be filtered, prior to use. Other substrates, including lauric acid (dodecanoic acid), glycerol trilaurate, various ketones, methyl ketones, dodecanal and 1-dodecanol, were used to probe the functional group pathways by which the deoxygenation occurs [5,7].

3 Results and discussion

3.1 Liquid products and yields

For fats and oil, which contain mostly triglycerides, the actual organic liquid yield is close to 75 mass % [6]. When allowance is made for the loss of the glycerol and ester unit, presumably as gas and water, the yield is approximately 94 mass %. Mechanistic considerations based on the products formed, show that the CH_2 next to the carbonyl group is lost, in which case the yield is close to 100 mass %. The organic liquid clearly separates from a small lower aqueous layer. Carbon dioxide is a major gas phase product showing that oxygen is removed in this form as well as water. Elemental analysis confirms that the oxygen content of the liquid is close to zero. Infrared spectroscopy of the liquid confirms the absence of the carbonyl group, as well as other oxygenated functionalities such as C-O and O-H.

The liquids have the characteristic aroma of olefines, which to some people also manifests itself as an aftertaste. Gas chromatography (Fig 2) shows a broad distribution of liquid components across the C_6-C_{17} range, with individual components clearly clustered around the retention time of each n-alkane standard.

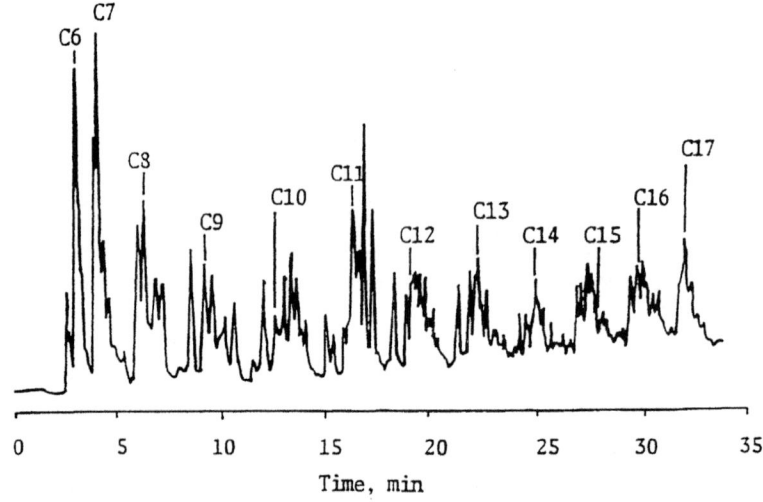

Fig. 2 Gas chromatograph of canola oil pyrolysis liquid.

Carbon-13 nmr clearly shows that the products are mainly linear mon-olefines although the presence of n-alkanes is not ruled out.. Significant sharp signals in the range 114-140 ppm show the presence of carbon atoms forming double bonds. Two clear signals at 114.5 and 139.5 ppm confirm carbon atoms in terminal double bonds. This is significant because terminal double bonds are thermodynamically least stable, which thus gives information relevant to the mechanism of olefine formation. Mass spectrometry has also been used to confirm the mono-olefinic nature of many of the components. Unfortunately, mass spectrometry can not clearly identify specific double bond isomers, because their fragmentation patterns are very similar. Hydrogenation of the liquids shows that in the case of products deriving from canola and soybean oils, the number of carbon atoms per double bond is 14-17. If all the products are mono olefines then gas chromatographic data suggests this number should be close to 11. In the case of trilaurin, in which the side chains are all saturated C_{12} units, the number of carbon atoms per double bonds (10.1) corresponds closely to that predicted by consideration of the most likely mechanism, as well as the gas chromatographic data.

The boiling range of a typical liquid is 60°C-325°C, the upper value being the 95% distillation point. With the removal of some light and heavy ends the product meets all the requirements for diesel fuel, although the products have more value as chemical feedstock [8].

4.2 Conversion routes

The functional group pathways for the conversion of triglycerides and fatty acids to mono-olefines have been investigated by the use of substrates in which the carbon chains are saturated. Although double bonds obviously play a role, such as in chain cleavage, the deoxygenation can be explained solely in terms of the other functionalities.

4.2.1 Olefine formation by γ-hydrogen transfer

The catalytic conversion of glycerol trilaurate, the triglyceride in which the side chains are C_{12}, including the carbonyl group, yields product which contains mostly linear decenes (C_{10}) [6] (see Fig. 3). Initially, this was surprising given that cleavage at the carbonyl group might be considered more likely. However, there is precedent for such a fragmentation in mass spectrometry (McClafferty Rearrangement) and photochemistry (Norrish Type 2 Rearrangement) of carbonyl-containing compounds, where in both cases the fragmentation is preceded by activation of the carbonyl group. In mass spectrometry it is proposed that the carbonyl group oxygen carries the positive charge following electron impact. In the pyrolysis it is proposed that the carbonyl oxygen has a partial positive charge after binding to a Lewis acid site on the alumina. A γ-hydrogen transfer mechanism on the ester linkage of the trilaurate initially forms 1-decene and the enol form of a methyl ester of the glycerol moiety (Figure 4). This proposed mechanism could be confirmed by preparation of the suitably deuterium labelled substrate, but at this time this has not been attempted. This mechanism is also proposed for the major fragmentations of the ketones which form from the fatty acids, which in turn derive from triglycerides (see below). For example, diundecyl ketone

691

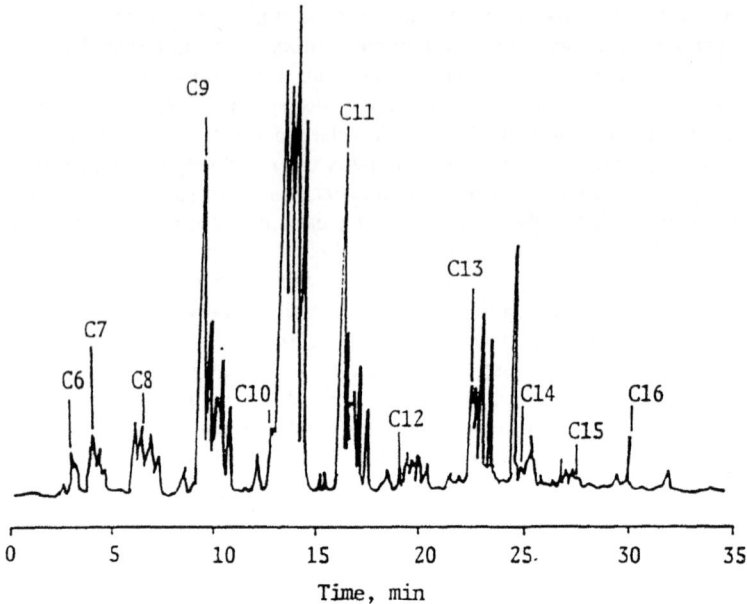

Fig. 3 Gas chromatograph of glycerol trilaurate pyrolysis liquid.

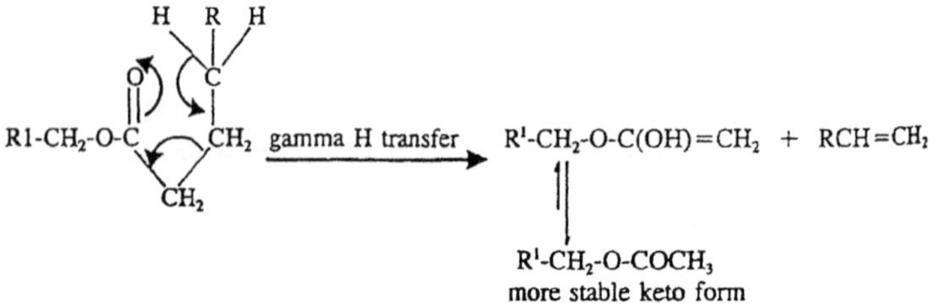

Fig. 4 γ-hydrogen transfer mechanism, (R¹= two remaining carbon atoms of glycerol moiety with relevant attached atoms)

forms decenes and, if not allowed to react completely, forms some 2-tridecanone, a methyl ketone [7]. The latter also forms decenes, although these are not the major products. The γ-hydrogen transfer mechanism does not appear to be a major factor in the fragmentation of fatty acids. The product slates and compositions resulting from the pyrolysis of decanoic acid and diundecyl ketone are virtually identical, in which case the conversion of fatty acids to the ketones must be fast.

4.2.2. Fatty acid formation by β-elimination

β-Elimination to form fatty acids involves removal of a complete glycerol side chain, including the glycerol oxygen, and the hydrogen atom from an adjacent carbon atom. The clue that the this occurs in triglycerides, is the presence of significant amounts of C_{13} alkenes in the product from trilaurin [6]. These alkenes, along with the decenes, are the major components of the liquids formed from dodecanoic acid and diundecyl ketone [7]. When glycerol triacetate, which can not undergo the γ-hydrogen transfer mechanism, is pyrolysed, only β-elimination occurs with the formation of the two possible propenediol diacetates, acetic acid and acetone. The latter is formed from the acetic acid [5].

4.3.3. Ketone formation

The formation of symmetric ketones from carboxylic acids by the elimination of carbon dioxide and water over heated metal oxides is well known [9]. The pyrolysis of a number of carboxylic acids over the alumina catalyst verified that this was the case in this system [7]. The similarity of product slates and compositions in the case of trilaurin and lauric acid confirm that the ketone formation is fast compared to the rest of the deoxygenation process.

4.3.4. Methyl ketone reduction/dehydration

The product from the pyrolysis of 2-undecanone contains 41 mass % undecenes which are the major components [5]. This substrate also causes faster deactivation of the catalyst. The question arises as to how the required reduction (to alcohol, followed by dehydration) occurs. One clue is the presence of C_{12} olefines in the product. These products contain one more carbon atom than the original substrate. It appears that some kind of isomerisation/disproportionation can occur. In the extreme this would involve isomerization of the methyl ketone to aldehyde, followed by disproportionation to alcohol and carboxylic acid. Recycling of the carboxylic acid through the ketone route explains the presence of the C_{12} olefines. Only two dodecene isomers are formed, and because neither of them is the 1-isomer, then they are almost certainly the cis and trans 2-isomers. On the other hand, there are at least four undecenes, which mass spectrometry suggested were the 1-, cis and trans 2- and 4-isomers. The pyrolysis of 1-decanol yielded at least five dodecene isomers, which is again inconsistent. We have not performed the key experiment of using a pure olefine isomer, such as 1-dodecene as substrate to see if it isomerizes beyond the 2 position. Pyrolysis of 2- undecanone with an equimolar amount of acetone causes an increase in the amount of dodecenes at the expense of the undecenes. This implies that the smaller methyl ketone is preferentially reduced.

4.3.5 Other conversion routes

From the gas chromatograph of the liquid product from glycerol trilaurate (Figure 3), it is clear that significant amounts of C_9 and C_{11} compounds are formed. Of these the n-alkanes form a significant fraction. We propose that the n-undecane is formed by free radical cleavage adjacent to the carbonyl group, followed by the attachment of a hydrogen atom to the resulting undecyl radical. The n-nonane is probably formed by cleavage of ethylene from undecyl radicals followed by hydrogen abstraction by the resulting nonyl radicals.

5 Catalyst deactivation

As in every catalytic process, deactivation of the catalyst is always a concern. In the case of the alumina catalyst and vegetable oils, the catalyst could handle only 2-3 times its own weight of substrate. Even though the catalyst is cheap, the deactivation was too rapid to be useful. the deactivation manifested itself by the appearance of solid at the entrance to the condenser. The deactivation can be monitored by observing the reappearance of the carbonyl group in the infrared spectrum. With the infrared spectrum set in the usual absorbance mode, the solid appears just before the carbonyl absorption reaches the same magnitude of the largest C-H bending absorption at 1465 cm^{-1}. This feature could be used as an on-line monitor of the process. The catalyst appeared unevenly coloured, some granules being grey while some were still white. Attempts to reactivate the catalyst by decoking met with only partial success and it was not clear that coking was the major problem. Attempts to reactivate the catalyst by treatment with acid met with only limited success.

Recently we have found a very simple method which prolongs the life of the catalyst. As a result, the catalyst has been run for up to 10 hours whilst maintaining activity. Beyond this point the catalyst, when deactivated, could be discarded or recycled. The method used for prolonging the catalyst life, as well as results from the use of an improved replacement catalyst (Alcan AA300), will be reported in the near future.

6 Conclusions

The use of demoisturized Alcan AA 200 activated alumina catalyst, under conditions which prolong catalyst life, allows the thermal conversion of waste or purified fats and oils to linear mono-olefines in high yields. This atmospheric pressure process operates at 450°C and weight hourly space velocities of approximately 0.5. The process converts fatty acids, as well as triglycerides. In addition linear ketones, aldehydes and alcohols are also deoxygenated. The conversion of methyl ketones, intermediates in the process, has been identified as the major step responsible for catalyst deactivation. The process has now been run for ten hours, beyond which time the inexpensive catalyst could be discarded or, if necessary, recycled. Results from the use of the superior replacement catalyst, Alcan AA 300, will be reported in the near future.

7 Acknowledgements

Operating grants, a Strategic Grant and a scholarship to E. Vhongia from the Natural Sciences and Engineering Research Council (Canada) supported this research.

8 References

1.   Fats and Oils in Canada, Annual Review (1993) Agriculture Canada, Ottawa, Canada, K1A 0C5.
2.   Boocock, D.G.B., and Konar S.K. (1994) Fuels and chemicals from sewage sludge III: Liquid hydrocarbons by the pyrolysis of raw sewage sludge lipid over activated alumina, **Fuel**, Vol. 73, pp. 642-646.
3.   Anjos, J.R.S.D., Lan, Y. and Frety, R. (1981) **R.. Bol. Tec. Petrobras,** Vol. 24, p 139.
4.   Chow, P.W. (1993) U.S. Patent 5,233,109.
5.   Vonghia, E., Boocock D.G.B., Konar, S.K. and Leung A. (1995) Pathways for the deoxygenation of triglycerides to aliphatic hydrocarbons over activated alumina, **Energy and Fuel,** Vol. 9, pp. 1090-1096.
6.   Boocock, D.G.B., Konar, S.K., Mackay, A., Cheung, P.T.C. and Liu, J. (1992) Fuels and chemicals from sewage sludge II, Alkanes and alkenes by the pyrolysis of triglycerides over activated alumina, **Fuel,** Vol. 71, pp. 1291-1297.
7.   Leung, A., Boocock, D.G.B. and Konar, S.K. (1995) Pathways for the catalytic conversion of carboxylic acids to hydrocarbons over activated alumina, **Energy and Fuels,** Vol. 9, pp. 913 -920.
8.   Bahadur, N.P., Boocock, D.G.B. and Konar, S.K. (1995) Liquid hydrocarbons from the catalytic pyrolysis of sewage sludge lipids and canola oil: Evaluation of fuel properties, **Energy and Fuels,** Vol. 9, pp. 248-256.
9.   March, J. (1977) Advanced Organic Chemistry: Reaction Mechanisms and Structure, 2nd Edition, Mcgraw Hill Book Co., New York, p 448.

LIQUID-PHASE THERMAL TREATMENT OF TALL OIL SOAP INTO HYDROCARBON FUELS

A. OASMAA, P. McKEOUGH, E. KUOPPALA and H. KYLLÖNEN
VTT Energy, P.O. Box 1601, 02044 VTT, Espoo, Finland

Abstract

Liquid-phase thermochemical processing of tall oil soap has been experimentally investigated. The organic matter of tall oil soap, which is a by-product of kraft pulping, originates mainly from wood extractives. Conventional processing of tall oil soap involves acidulation to yield crude tall oil and subsequent distillation of the oil at centralised refineries. Because tall oil originating from birch wood is far less valuable than that from pine, there is an economic incentive in the Nordic countries to develop alternative conversion processes for the tall oil soap produced at pulp mills where birch is widely used as feedstock.

Both catalytic hydrotreatment and liquid-phase thermal treatment of a mixed pine/birch soap have been investigated in the laboratory. Liquid and gaseous hydrocarbons were the principal products formed in both types of treatment. Although the liquid product formed under hydrotreatment conditions contained a higher proportion of aliphatic compounds, it was clear that the thermal treatment process would be, economically, the more favourable option. In the thermal process, a predominantly hydrocarbon oil product was obtained at 450 °C using a residence time of at least 60 min. About 50% of the energy content of the tall oil soap was recovered as oil product, about 30% as gases, and about 20% as organic material contained in an aqueous slurry phase. When the oil product was carefully separated from the aqueous slurry phase, only trace amounts of impurities, such as water, sodium and organic acid salts were present in the product. A techno-economic evaluation of the process concept has been performed. The economics of the thermal treatment process appear to be favourable in comparison with those of the conventional acidulation process.

Keywords: Fuel, kraft pulping, liquid-phase, tall oil soap, thermal treatment, thermochemical conversion.

1 Introduction

The Technical Research Centre of Finland (VTT Energy) has studied alternative energy uses for the waste/side streams of the kraft pulp mill since the early eighties [1]. One such stream is tall oil soap. The organic matter of tall oil soap originates mainly from wood extractives. Conventional processing of tall oil soap involves acidulation to yield crude tall oil (Table 1) and subsequent distillation of the oil at centralised refineries. Because tall oil originating from birch wood is far less valuable than that from pine, there is an economic incentive in the Nordic countries to develop alternative conversion processes for the tall oil soap produced at pulp mills where birch is widely used as feedstock. Conventional acidulation of tall oil soap with sulphuric acid also results in a significant input of sulphur into the recovery cycle. Efforts to, on the one hand, close water circuits of pulp mills and, on the other hand, apply sulphur-free alkaline pulping to hardwoods have thus led to increased interest in alternative processes for treating tall oil soap.

Table 1. Typical composition of Finnish tall oils [2]

	Pine	Birch[a]	Pine/birch about 50:50
Acid number, mg KOH/g	160	110-125[b]	130
Unsaponifiable,%	7	30	15
Resin acids,%	40	0	25
Fatty acids,%	53	70	60

[a] Extracted from black liquor. [b] Acid number of a tall oil soap from a mill where birch was used as a feed. Traces of resin acids are left from use of pine.

Previous efforts to thermochemically upgrade tall oil soap have been very limited [3], the main reasons obviously being the high contents of water (about 35%) and inorganics (sodium content of about 4%). Upgrading of tall oil, i.e. acidulated tall oil soap, has, on the other hand, been very widely researched [4-46]. Catalytic hydrotreatment of tall oil has received most attention.

2 Material and methods

2.1 Material
The raw material for the experiments was a mixed tall oil soap (Table 2) from a kraft pulp mill using both pine and birch wood. A gas chromatographic analysis of the organic material in the tall oil soap is given in Fig. 1. The acids are present as their sodium salts in the tall oil soap.

2.2 Experimental methods
The experiments were performed in a 1-litre batch reactor equipped for stirring. The typical amount of feed was 150 g. The reactor free volume was purged with nitrogen (0.1 MPa). The rate of mixing was 1500 rpm.

The product was recovered manually from the autoclave and separated by centrifugation into different product phases: oil, aqueous and heavy phases. The material remaining in the autoclave and in the stirrer was washed with acetone, weighed after evaporation of the solvent, and added into mass balances. The experimental conditions for liquid-phase thermal treatment runs are shown in Table 3.

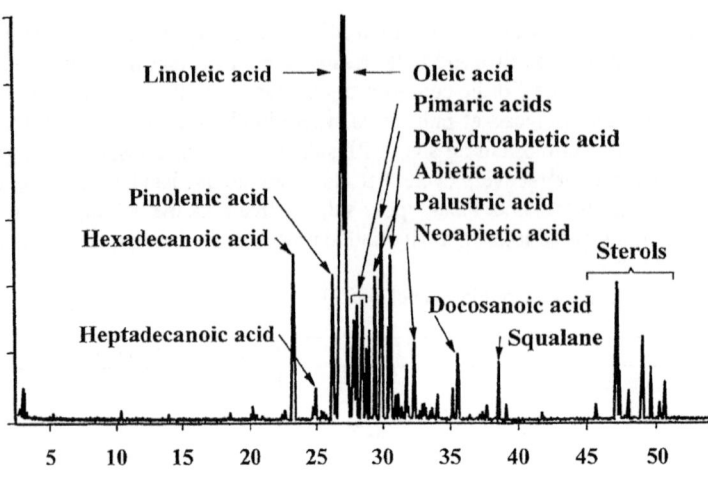

Fig. 1. Gas chromatographic analysis of the organic material in the tall oil soap.

Table 2. Feedstock analyses

	Tall oil soap
Water, wt%	35.1
Elemental analysis, wt% dry basis	
Carbon	69.5
Hydrogen	10.5
Nitrogen	0.1
Sodium	6.1
Sulphur	0.4
Oxygen (diff)	13.4
Heating value, MJ/kg	
HHV	22
LHV	20
Density (15 °C), kg/dm^3	0.845

Table 3. Thermal liquid-phase experiments

	1/95	5/94	8/95	4/95	16/94	24/94	29/94	7/95	18/94	21/94	9/94	25/93
T, °C	420	435	435	435	450	450	450	450	450	450	450	450
t (heating), min	75	85	75	75	75	85	90	85	90	85	75	70
t (reaction), min	190	5	70	130	5	25	65	70	120	130	130	140
P, MPa	16	16	16	17	12	17	19	17	20	11	13	11

2.3 Analytical methods

The water content was analysed by Karl-Fischer titration according to ASTM D 1744. Sodium and sulphur were analysed by atomic absorption spectroscopy (AAS). Elemental analysis was performed according to ASTM D 5291-92 using a microelemental analyser (Leco CHN-600) for carbon, hydrogen, and nitrogen.

For component identification, a mass-selective detector was used with a gas chromatograph containing HP-1 column (25 m, 0.32 mm, 0.17 µm). Helium carrier gas at 80 kPa was used in a split injection mode. The derivatized sample (0.5-1 mg) was prepared for gas chromatography (GC) by heating for 30 min at 75 °C with 0.5 ml pyridine (dried by KOH) and 0.1 ml BSTFA [bis(trimethylsilyl)trifluoroacetamide] including 1% TMCS (trimethylchlorosilane) in a sealed class tube to produce the trimethyl (TMS) derivatives. The oven temperature was started at 100 °C. It must be noted that for the GC analyses the samples were silylated using an acidic reagent and thus any acids were liberated from their sodium salts.

3 Preliminary upgrading experiments

In preliminary experiments at 450 °C, three types of liquid-phase treatment of tall oil soap were investigated: (1) catalytic hydrotreatment, (2) non-catalytic hydrotreatment and (3) thermal treatment. In each case, similar product groups were obtained but in different proportions (Fig. 2). The proportional amount of straight-chain hydrocarbons to other product groups (naphthalenes, anthracenes, phenanthrenes) was highest in the catalytic hydrotreatment experiment and lowest in the non-catalytic thermal treatment. The corresponding yields of oil were 44 wt% and 40 wt% of the organic feed material, respectively. The oils, being composed predominantly of hydrocarbons, were entirely soluble in hexane. Although the liquid product formed under hydrotreatment conditions contained a higher proportion of aliphatic compounds, it was clear that the thermal treatment process would be, economically, the more favourable option. Hence the main part of the study was directed towards the thermal treatment of tall oil soap.

4 Liquid-phase thermal treatment of tall oil soap

The main parameters affecting the cracking of tall oil soap into good-quality oils (free of acids and inorganics) are temperature and time. At 420 °C after a residence time of 180 minutes the oil product (Fig. 3a) still contains fatty acids (as salts), water (12-23 wt%) and sodium (above 3 wt%). Note that any acids appearing in the gas chromatograms presented in this paper were present as their sodium salts in the original samples (refer Analytical Methods). At 435 °C after a residence time of 120 minutes a low-viscosity oil product (oil yield 42 wt%) with traces of water and sodium is obtained (Fig. 3b). The main compounds in the oil product at 435 °C are listed in Table 4.

After a residence time of 60 to 120 minutes at 450 °C the product oil (oil yields of 39-40 wt%) contains only traces of moisture, sodium, and fatty acids (as salts). A smaller more viscous oil phase ("jelly") containing fatty acid salts is formed below the main oil layer. After 120 minutes residence time the straight-chain hydrocarbons start to crack

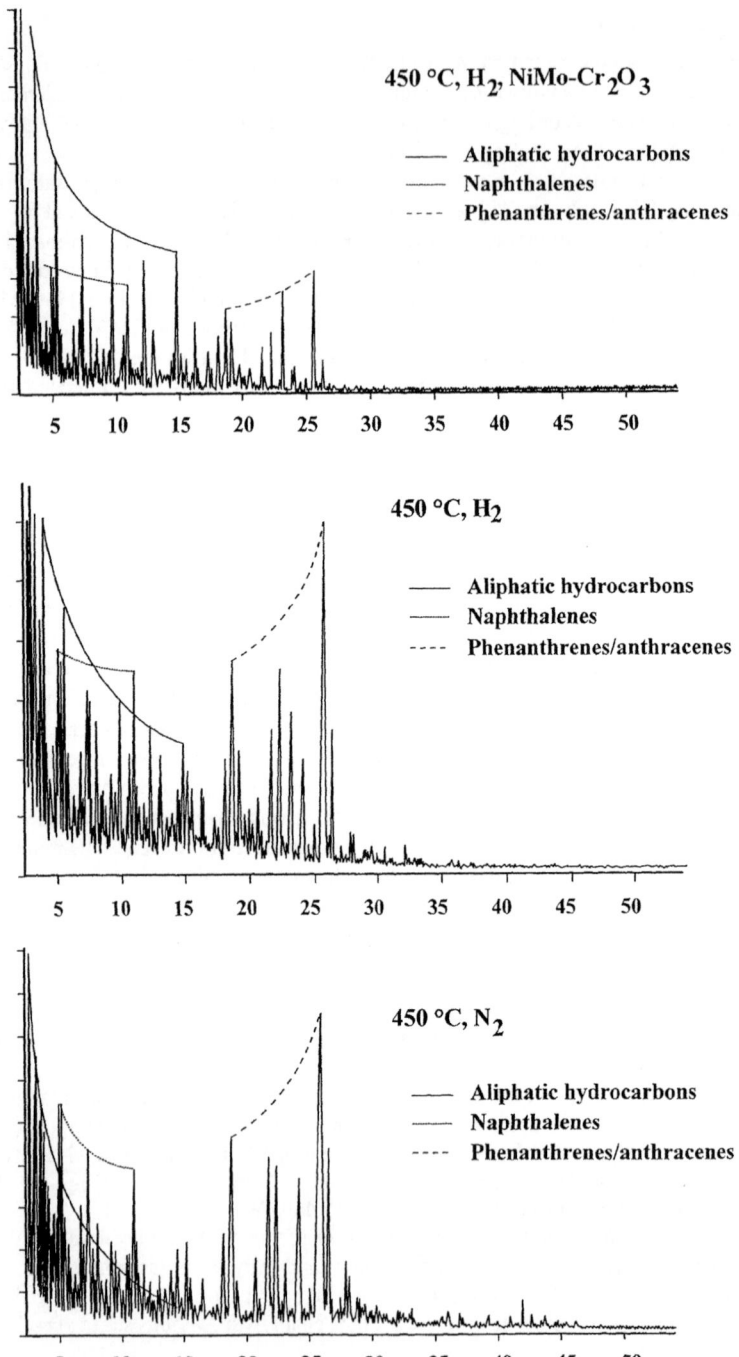

Fig. 2. Gas chromatographic analyses of oil products from tall oil soap.

further and the oil yield decreases. Fig. 4 shows the distribution and composition of the different product phases formed by thermally treating tall oil soap at 450 °C for 65 min. The major part of the sodium is found in the aqueous (40-50 wt% of the feed sodium) and solid phases (50-60 wt% of the feed sodium). These phases also contain significant amounts of fatty acids (as salts). It seems that the thermal treatment severs the chains of the acid salts present in the tall oil soap, resulting in the formation of hydrocarbons, on the one hand, and shorter-chain acid salts, on the other hand. At some point during the course of a sufficiently severe treatment, the aqueous phase and the acid salts become totally immiscible in an oil layer composed largely of hydrocarbons. The effect of the residence time at 450 °C on the product composition is shown in Fig. 5.

Fuel properties of one product oil are presented in Table 5. The amount of inorganics in the oil can be further reduced by additional centrifugation or by washing the oil with water. A sodium content as low as 3.5 ppm has been achieved. The lowest possible levels have not yet been established.

Table 4. Major components of product oil (435 °C, 120 min) identified by mass spectro-metry

Alkanes, saturated, unsaturated, alkyl substituted (about 30 wt%)
butane (C_4H_{10}), pentane, hexane, heptane, octane, nonane, decane, undecane, dodecane, tridecane, tetradecane, pentadecane, hexadecane, heptadecane, octadecane, nonadecane, eicosane, docosane ($C_{22}H_{46}$)
Cyclic hydrocarbons (about 5 wt%)
One-ring aromatic hydrocarbons (about 15 wt%)
toluene, alkyl benzenes
Two-ring aromatic hydrocarbons (about 20 wt%)
naphthalene, alkyl naphthalenes
Three-ring aromatic hydrocarbons (about 25 wt%)
phenanthrenes, anthracenes, alkyl phenanthrenes, alkyl anthracenes
Other (about 5 wt%)

Table 5. Fuel properties of the oil product obtained at 450 °C and 120 min

Expt.no.	18/94
Water, wt%	0.08
Elemental analysis, wt%	
Carbon	87.8
Hydrogen	10.5
Nitrogen	0.1
Sodium	0.2
Sulphur	0.14
Oxygen (diff)	1.3
Heating value, MJ/kg	
HHV	42.1
LHV	39.8
Density (15 °C), kg/dm^3	0.943
Viscosity, cSt	
20 °C	4
50 °C	2

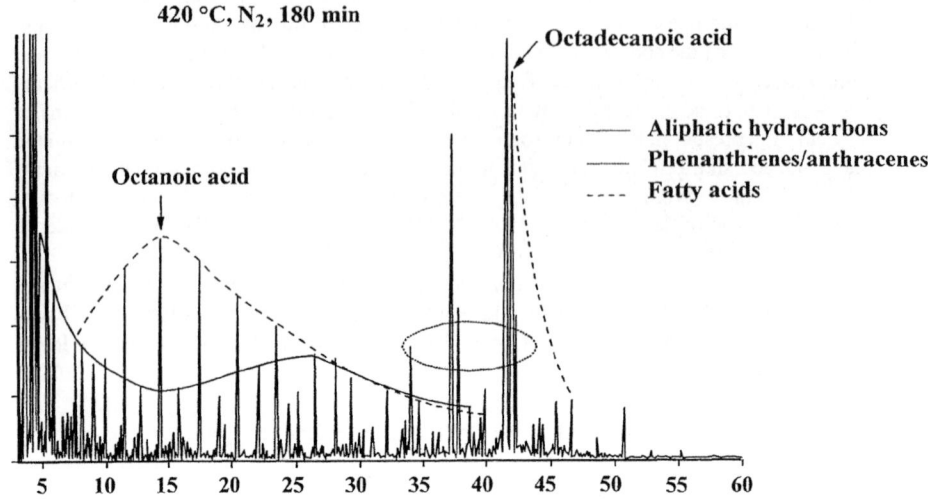

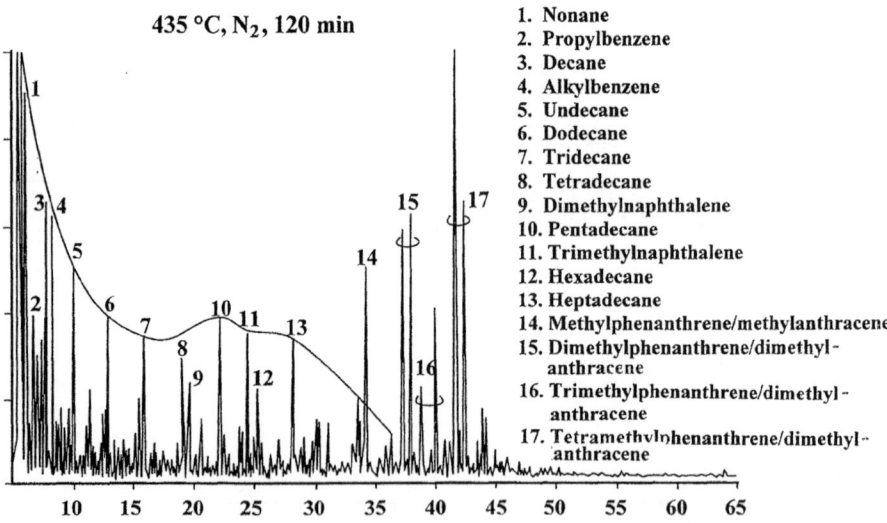

Fig. 3. Gas chromatographic analyses of oil products obtained at 420 °C and 180 minutes (3a) and at 435 °C and 120 minutes (3b).

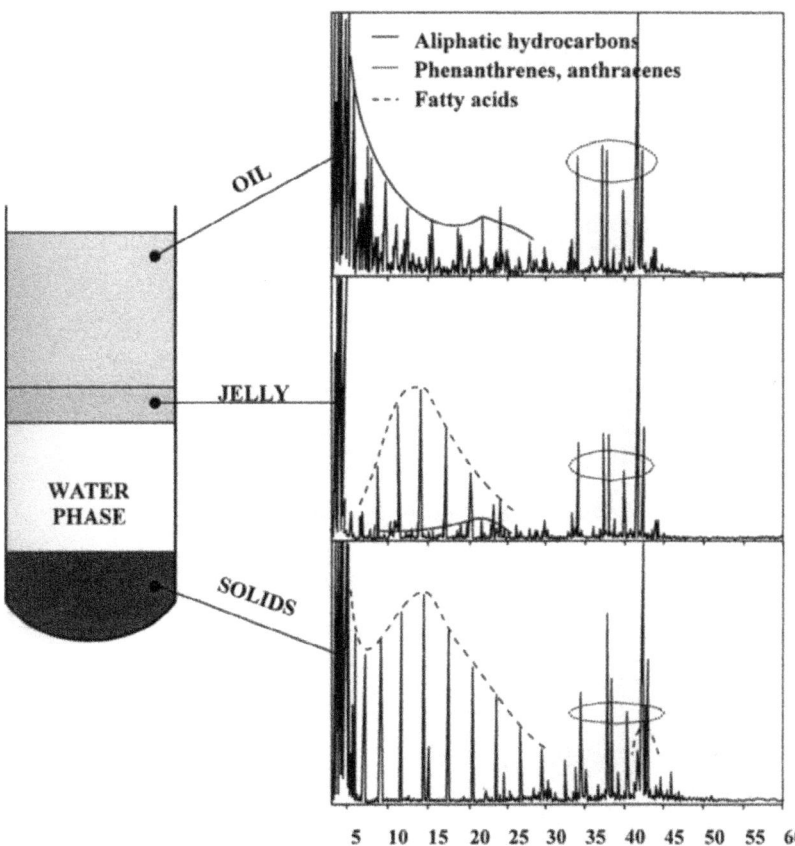

Fig. 4. The distribution and composition (gas chromatographic analyses) of different product fractions (obtained by centrifugation) after thermal treatment at 450 °C for 65 min. Product fractions: oil (above), jelly, aqueous, and solids (lowest) fraction. Water content of phases: oil (0.1 wt%), jelly (8.8 wt%), solids (36.0 wt%).

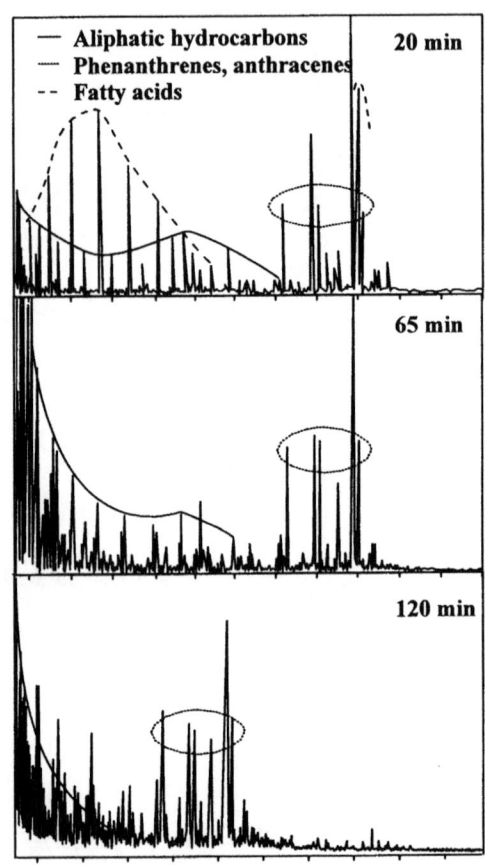

Fig. 5. The effect of residence time on the product oil composition (gas chromatographic analyses) at 450 °C. GC/MS column length: 50 m (Ultra 1) in short residence time runs, and 25 m (HP-1) in 120 min run.

5 Techno-economic evaluation

The process concept depicted in Fig. 6 was evaluated. The design capacity of the plant is 30 000 t/a of tall oil soap (35% moisture). This amount roughly corresponds to that produced by a Finnish kraft mill manufacturing 400 000 t/a of pulp.

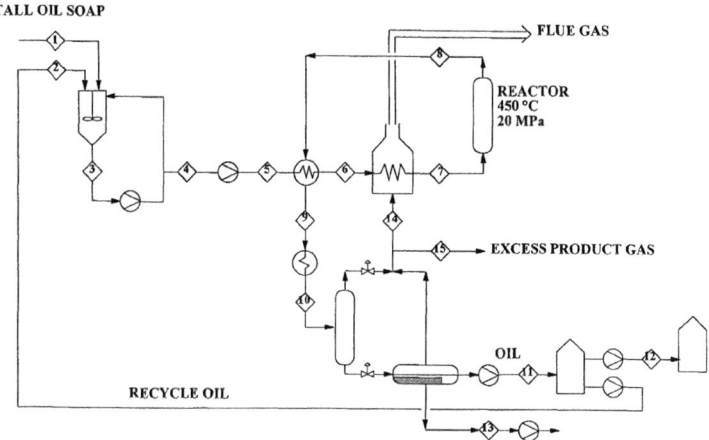

Fig. 6. Flow diagram of the process for thermally treating tall oil soap.

The thermal-treatment mass and energy balances, used as a basis of the evaluation, are given in Table 6.

Table 6. Mass and energy balances of thermal treatment

	Mass, % of tall-oil-soap solids	Energy (HHV basis) % of tall-oil-soap energy
Product oil	40	50
Product gas	25	30
Dry matter in aqueous slurry	30	20
Product water	5	-

In the evaluation, the investment costs and the operating costs of the thermal treatment process were estimated and then compared to the corresponding costs of the conventional acidulation process. Certain bases of the economic evaluation are given in Table 7.

The total capital requirement for the thermal treatment plant was estimated to be FIM 17.9 million, significantly greater than the FIM 5.2 million required for the acidulation process. On the other hand, the operating costs of the acidulation process were estimated to be higher than those of the thermal process, mainly because of the costs of sulphuric acid. The precise cost of tall oil soap is unknown and so a range of values was employed in the evaluation. Table 8 presents a summary of the estimated production costs for two

Table 7. Bases of the economic evaluation

Capital charges factor	0.118*
Value of excess product gas	60 FIM/MWh (LHV)
Cost of electricity	200 FIM/MWh
Cost of steam	72 FIM/MWh
Cost of sulphuric acid	400 FIM/t

*annuity method, 10%, 20 years

different processes of tall oil soap. In the thermal treatment process, about 80% of the product gas is available for utilisation elsewhere in the pulp mill, e.g. in the lime kiln.

The results in Table 8 indicate somewhat higher production costs for the thermal treatment alternative. On the other hand, the product oil of the thermal process should be a more valuable fuel oil than tall oil, which can, at best, be used as a substitute for heavy fuel oil. In Finland, the value of low-quality tall oil, originating in part from birch wood, is similar to that of heavy fuel oil (exclusive of taxes) - about 65 FIM/MWh. Thus, the present results suggest that the thermal treatment process would be competitive with the conventional acidulation process in cases where the tall oil produced by acidulation is of limited quality.

Table 8. Estimated oil production costs for thermal treatment process and for acidulation process. Capacity: 30 000 t/a of tall oil soap.

		Thermal treatment 7 800 t/a of oil		Acidulation 13 000 t/a of tall oil	
Cost of tall oil soap	FIM/t	150	200	150	200
	FIM/MWh (LHV)	27	36	27	36
Capital charges	1000 FIM/a	2 100	2 100	600	600
Feedstock costs	1000 FIM/a	4 500	6 000	4 500	6 000
Other operating costs	1000 FIM/a	1 300	1 300	2 400	2 400
Credit for product gas	1000 FIM/a	-2 500	-2 500	-	-
Production costs	1000 FIM/a	5 400	6 900	7 500	9 000
	FIM/t	690	880	580	690
	FIM/MWh LHV)	63	80	58	69

The economics of light fuel oil (LFO) production through thermal treatment of tall oil soap have also been compared to the economics of other processes for producing liquid biofuels in Finland [47]. The other biofuels were: ethanol from barley (and further conversion to gasoline additive ETBE), rapeseed methyl ester (RME) for diesel substitute, rapeseed fuel oil for LFO substitute, ethanol from wood (ETBE), methanol from peat and wood (and further conversion to gasoline additive MTBE), and pyrolysis fuel oil from wood (as a LFO substitute). The estimated production costs of these biofuels were compared to their market values in Finland (January 1996). The cost ratios are shown in Fig. 7. It appears that only one of the other alternatives - MTBE production from peat and wood - is as economically promising as the present tall-oil-soap conversion process.

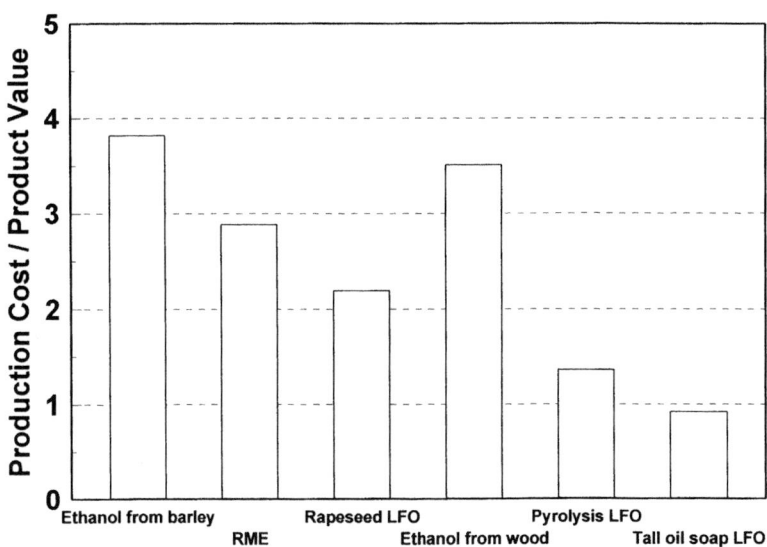

Fig. 7. Ratios of estimated production costs to product values for various liquid biofuels in Finland [47].

6 Conclusions

The main conclusions of this study are:
- Tall oil soap can be thermally processed at 435-450 °C to yield a low-viscosity oil (yield 32-42 wt% of feed organics) having a heating value (HHV) of 42-44 MJ/kg. The required residence times are 180 minutes at 435 °C and 60 minutes at 450 °C.
- During the treatment, the chain lengths of the tall-oil-soap acid salts are reduced, and the remnant acid salts become immiscible in the product oil layer.
- Nearly all the sodium is recovered in the aqueous slurry phase. The sodium content of the oil product can be further reduced down to a few ppm by water wash or by additional centrifugation. The lowest possible level of inorganics in the oil has not yet been established.
- Liquid-phase thermal treatment of tall oil soap has the potential to be competitive with the conventional acidulation process in cases where the tall oil produced by acidulation is of limited quality.

Acknowledgements

The authors wish to thank technicians Mrs. Eija Tapola and Ms. Johanna Levander for the product handling and Mr. Reijo Häkkinen for carrying out the experiments. Thanks are also due to the personnel of the analytical department for the product analyses. The initiative work of Prof. Raimo Alén (University of Jyväskylä) for the project is acknowledged.

References

1. Oasmaa, A., Alén, R., McKeough, P. & Johansson, A. (1992) Thermochemical conversion of black liquor organics into fuels. VTT Publications 102, Espoo.
2. Holmbom, B. (1972): Nyare synpunkter på förädling av tall- och björkolja. *Kem. teoll.* Vol. 29, No. 10. pp. 721-22, 724-26. (in Swedish).
3. Segessemann, E. and Molnar, N.M. (1949) Tall-oil treatment. U.S. 2,481,356, Sept. 6, 1949.
4. Walden, E. (1950) Manufacture of lubricating and emulsifying oils from tall oil. *Paper and Timber* (Finland), Vol. 32. pp. 140-1.
5. Manning, L.A. (1951) Cracking natural resins and their derivatives. Fr. 977,671, April 4.
6. Enkvist, T. and Kahila, S. (1950) Catalytic production of lubricating oil from crude tall oil and tall oil pitch. *Paper and Timber,* Vol. 32B. pp. 26-32
7. Teregerya, N.V., Gundobin, N.V. and Mednikov, F.A. (1970) Hydrogenation of fatty acids of tall oil. *Izv. Vyssh. Ucheb. Zaved., Les. Zh.,* No. 13(4). pp. 100-4 (in Russian).
8. Teregerya, N.V. and Mednikov, F.A. (1972) Hydrogenation of tall oil resin acids. *Sb. Nauch. Tt., Ivanov. Energ. Inst.,* No. 14. pp. 263-7 (in Russian).
9. Teregerya, N.V., Pasechnik, M.S. and Mednikov, F.A. (1969) Preparation of anhydrous sats of hydrogenated tall oil fatty acids and their testing as industrial lubricants in the drawing of steel wire. *Nauch. Tr., Leningrad. Lesotekh. Akad.,* No.114. pp. 22-5.
10. Pasechnik, M.S., Mednikov, F.A. and Teregerya, N.V. (1969) Preparation of the optimum concentration of stable emulsions of hydrogenated tall oil fatty acids for the cold rolling of steel. *Nauch. Tr., Leningrad. Lesotekh. Akad.,* No. 114. pp. 26-8.
11. Pasechnik, M.S., Mednikov, F.A. and Teregerya, N.V. (1969) Laboratory research on the use of highly saturated tall oil fatty acids in a solid state for cold rolling steel. *Nauch. Tr., Leningrad. Lesotekh. Akad.,* 1969, No. 114. pp. 37-41
12. Teregerya, N.V. (1969) Preparation of an active catalyst for the low-temperature hydrogenation of a fatty-acid fraction of tall oil. *Nauch. Tr., Leningrad. Lesotekh. Akad.,* No.114. pp. 29-32.
13. Teregerya, N.V., Pasechnik, M.S. and Mednikov, F.A. (1966) Tall oil hydrogenation for production of special technical lubricants. *Tr. Leningr. Lesotakh. Akad.,* No.105. pp. 82-8.
14. Secor, R. & Lewko, L. (1984) Tall oil as fuel for mobile equipment in the kraft pulp industry. In: *Can. Bioenergy R&D Semin., Proc.,* (ed. S. Hasnain), 5th, Elsevier Appl.Sci., London, UK, pp. 526-30.
15. Furrer, R.M. and Bakhshi, N.N. (1988) Catalytic conversion of tall oil to chemicals and gasoline range hydrocarbons. In: *Research in thermochemical biomass conversion,* (eds. A.V. Bridgwater and J.L. Kuester), pp. 956-973.
16. Furrer, R. M. & Bakhshi, N. N. (1990) Catalytic conversion of tall oil and plant oils to gasoline range hydrocarbons using shape selective catalyst HZSM-5. *Energy Biomass Wastes,* Vol.13. pp. 897-914.

17. Sharma, R.K. and Bakhshi, N.N. (1991) Catalytic upgrading of biomass-derived oils to transportation fuels and chemicals. *Can. J. of Chem. Eng.* (Canada), Vol. 698,Oct. pp. 1071-81.

18. Sharma, R.K. and Bakhshi, N.N. (1990) Efficient utilization of waste tall oil from the kraft pulp industry. Tappi Journal, Vol. 73, Sept. pp. 175-80.

19. Sharma, R.K. and Bakhshi, N.N. (1991) Upgrading of tall oil to fuels and chemicals over HZSM-5 catalyst using various diluents. *Can. J.of Chem. Eng.* (Canada), Vol. 69 Oct. pp. 1082-86.

20. Sharma, R.K. and Bakhshi, N.N. (1991) Catalytic conversion of crude tall oil to fuels and chemicals over HZSM-5: effect of co-feeding steam. *Fuel Processing Technology* (Netherlands), Vol. 27, No. 2, Apr. 1991. pp. 113-30.

21. Bakhshi, N.N., Furrer, R.M. and Sharma, R.K. (1989) Catalytic conversion of tall oil and plant oils to fuels and chemicals. In: *Proc. of 7th Canadian Bioenergy R and D Seminar*, (ed. E. N. Hogan), Canada Centre for Mineral and Energy Technology, Ottawa, ON (Canada) (1186420), pp. 687-92.

22. Sharma, R.K. & Bakhshi, N.N. (1991) Catalytic conversion of de-pitched tall oil to fuels and chemicals over H-ZSM-5: single and dual reactor system. *Appl. Catal.*, Bol. 76, No. 1. pp. 1-17.

23. Craig, W. K. & Soveran, D. W. (1991) Production of hydrocarbons with a relatively high cetane rating. US 4992605 A 12 Feb.

24. Wong, A. (1991) *Naval Stores Review*, July/August. pp. 14.

25. Wong, A. (1995) Tall-oil-based cetane enhancer for diesel fuel. Arbo-TaneTM meets demands for diesel. *Pulp & Paper Canada*, Vol. 96, No.11. pp. 37-40.

26. Ross, J. and Turck, Jr., J.A.V. (1944) *Hydrogenation of fatty acid soaps.* U.S. 2, 363, 694, Nov. 28.

27. Turck, Jr., J.A.V. (1944) *Hydrogenation of tall-oil material.* U.S. 2,309, 483, Jan. 26.

28. Colgate-Palmolive-Peet Co. (1943) *Hydrogenation of unsaturated fat acid salts.* Brit. 550,356, Jan. 5.

29. Turck, J.A.V. and Ross, J. (1945) *Selectively hydrogenated tall oil.* U.S. 2,389,284, Nov. 20.

30. Dressler, R.G. and Vivian, R.E. (1943) *Hydrogenation of tall oil.* U.S. 2,336, 472, Dec. 14.

31. Dressler, R.G. and Vivian, R.E. (1945) *Tall-oil hydrogenation.* U.S. 2,369, 446, Feb. 13.

32. Gordon, E. and Ragot, S. (1948) Hydrogenation of tall oil. *Bull. mens. ITERG*, No.4. pp. 32-4.

33. Chernova, I.K., Bychkov, B.N., Koshel, G., Budnii, I.V. (1986) Catalytic activity of some metals during hydrogenation of C18 unsaturated acid esters. *Izv. Vyssh. Uchebn. Zaved., Khim. Khim. Tekhnol.*, Vol. 29, No. 8. pp. 31-4.

34. Bergström, H.O.V., Trobeck, K.G. and Heijmer, G.B. (1946) *Treatment of tall oil.* Swed. 117,543. Nov. 5.

35. Braude, G.L. (1967) *Polymeric plasticizers from tall oil fatty acids.* U.S., 3pp. US 3337594, 22 Aug.

36. Jacini, G. Tall oil. (1950) *Chimica e industria (Milan)*, Vol. 32. pp. 328-34.

37. Hoffman, A.N. and Montgomery J.B. (1955) *Tall-oil alcohols*. U.S. 2,727,885, Dec.20.

38. Dressler, R.G., Vivian, R.E., and Hasselström, T.(1946) *Refining and fractionation of tall oil*. U.S. 2,396,646, March 19.

39. Harwood, J. and Binkerd, E.F. (1947) *Hydrogenation of tall oil*. U.S. 2,423,236, July 1.

40. Armour and Co. (1949) *Hydrogenation of tall oil*. Brit. 630,686, Oct. 19.

41. Numashima, M. and Takemoto, K. (1986) Clear epoxy coatings. *Jpn. Kokai Tokkyo Koho*, 3pp. JP 61103964 A2 22 May Showa.

42. Kulkarni, M.V. and Scheribel, R.L. (1970) Hydrogenation of fatty acid polymers. *Ger. Offen.* 28 p.

43. Ishigami,M. and Inoue, Y. (1976) Emulsifiers for emulsion polymerizations. *Japan. Kokai*, 5pp. JP 51025586 2 Mar Showa.

44. Rendko,T., Mistrik, E.J., Filadelfy, I. and Mikestik, A. (1978) Tall oil and its hydrogenation. *Petrochemia*, Vol. 18, No. 4. pp. 129-38.

45. Soveran, D.W., Sulatisky, M., Robinson, W.K.Ha, Stumborg, M. (1992) The effect on diesel engine emissions with high cetane additives from biomass oils. *Prepr. Pap. Am. Chem. Soc. Div. Fuel Chem*, Vol..37, No.1. pp. 74-85.

46. Secor, R.J. (1987) *Preparation of tall oil fuel blend*. US 4634452 A 6 Jan. 4 pp.

47. Solantausta, Y., McKeough, P. and Sipilä, K. (1996) Upgrading and liquefaction of domestic fuels. VTT Energy, Report to the Ministry of Trade and Industry, February.

PyNE: PYROLYSIS NETWORK FOR EUROPE

A. V. BRIDGWATER and K. DOWDEN
Energy Research Group
Chemical Engineering and Applied Chemistry Department
Aston University
Birmingham B4 7ET
UK

Abstract

A network of active researchers and enthusiasts in pyrolysis of biomass has been established in Europe. The overall objective of the network is to establish a forum for the discussion and exchange of information on new scientific and technological developments on biomass pyrolysis and related technologies across Europe. In particular, it will review progress in pyrolysis technologies, in analysis and characterisation of products, in upgrading of products by any method, and in applications of products. The purpose of these reviews are to identify and prioritise R&D needs on technical and economic grounds including identification of technically and economically sensitive components in relation to the agro-industrial sector and the renewable energy sector.

The work is being accomplished by establishment of regular meetings to review each area of interest, by organising seminars and technical visits and through the publication of a regular newsletter - PyNE News. This latter activity will be the main dissemination outlet for all the information collated and for the results of discussions and analyses.

Keywords: Biomass, Pyrolysis, Network, Upgrading, Pyrolysis products, Characterisation, Analysis, Newsletter.

Structure

The overall objective of the EC sponsored Pyrolysis Network for Europe is to provide a forum for a review of fast pyrolysis and related processes for production of fuels and chemicals from agricultural materials.

The specific goals are:

- to establish a European network of active technology researchers and developers of biomass pyrolysis processes and related technologies,
- to establish a forum for the discussion and exchange of information on new scientific and technological developments on biomass pyrolysis and related technologies between national programmes within the EU and beyond; with international programmes such as the IEA Bioenergy Agreement, FAO and UN etc., EC sponsored researchers and private companies and individuals;
- to review progress in important technical areas including technology, analysis and characterisation, upgrading and applications; identify significant obstacles to their implementation; and produce authoritative reports for distribution,
- identify R&D needs on technical and economic grounds including identification of technically and economically sensitive components in relation to the agro-industrial sector and the renewable energy sector,
- to encourage and provide for interaction by exchange of personnel, information and ideas between laboratories and with industry to improve the rate of development and knowledge base of relevant technologies,
- to encourage the active involvement of industry in development and exploitation of the R&D being undertaken,
- to establish a database of relevant activities for information dissemination and improved co-operation,
- to disseminate information on activities from the database and results of meetings and common research activities by newsletter, reports and other publications, to those in the network and other interested parties in regional, national and international programmes.

Management

The co-ordinator has the overall responsibility of organising, administering and maintaining the project assisted by a steering committee of the convenors of the four specialist working groups. They will agree structures and procedures and maintain a continuous monitoring role on each specialist sub-group and each participating member to ensure that declared objectives and targets are met.

Membership

Most EU countries have at least one representative as shown in Table 1 below, who provides a centralised facility for collation and dissemination of information. Enquiries for further information or interest in participation should be directed to the national representative to ensure that this interest is properly co-ordinated. In case of difficulty the Co-ordinator may be contacted directly.

Table 1. Members of PyNE

Coordinator			Greece	Y Boukis
	Tony Bridgwater			C.R.E.S. Biomass Department
	Energy Research Group			19th km Athinon
	Aston University			Marathona Ave
	Birmingham B4 7ET, UK			Pikermi - Attikis
Austria	J Spitzer			GR-190 09, GREECE
	Institut fur Energieforschung		Italy	L Conti
	Joanneum Research			University of Sassari
	Elisabethstrasse 11			Dipartimento di Chimica
	Graz A-8010 , AUSTRIA			Via Vienna 2
Belgium	R Maggi			Sassari I-07100, ITALY
	UCL		Holland	W Prins
	Place Croix du Sud 2/17			Twente University
	Louvain-la-Neuve 1348			Chemical Engineering Lab.
	BELGIUM			PO Box 217
Denmark	K Pedersen			7500 AE Enschede
	Danish Technological Institute			NETHERLANDS
	Technology Park		Norway	M Fossum
	Aarhus DK-8000, DENMARK			Norwegian Inst. of Technology
	E Winther			Trondheim 7034, NORWAY
	Elkraft Power Co Ltd		Portugal	F Pinto
	5, Lautruphoj			Faculty of Engineering
	Ballerup DK-2750, DENMARK			University of Porto
Finland	Y Solantausta			4099 Porto Codex, PORTUGAL
	VTT		Spain	J Arauzo
	PO Box 1601			University of Zaragoza
	FIN-02044 VTT, FINLAND			Chemical & Environmental Eng.
France	P Girard			Maria de Luna 3
	CIRAD Forêt			Zaragoza 50015 , SPAIN
	73 rue d.f. Breton			
	BP5035			A Cuevas
	Montpellier Cedex 34032			Union Electrica Fenosa SA
	FRANCE			Capitan Haya 53
				Madrid 28020 , SPAIN
Germany	D Meier		Sweden	E Rensfelt
	BFH/Institute Wood Chemistry			TPS (Studsvik Energiteknik)
	Leuschnerstrasse 91			Nykoping S-611 82,
	Hamburg -Bergedorf D-21031			SWEDEN
	GERMANY			
	W Baldauf			
	Veba Oel AG			
	Pawiker Str. 30			
	Gelsenkirchen D-4650			
	GERMANY			

Additional members are encouraged to participate to aid wider involvement and dissemination, especially where there are particular skills that will complement those available in the Network.

Methodology

The Network functions through the basic structure of the participants who provide good representation of geographical regions, subject areas, and research/industrial orientation. These individual participants contribute in a number of ways:

1 by acting as a focus in their geographical region and technical subject area to collate information on relevant activities on recent and ongoing projects in universities, research institutes or industry for assembling into a database. This will be related to current and future needs of industry and provide indicators for future developments. A comprehensive report of this database will be provided with the following contents:

- list of ongoing and completed projects with a short description containing:
  - the biomass feedstock used,
  - the technology used, benefits, advantages, disadvantages,
  - main results with analysis and evaluation,
- present state of research,
- goals, volume and time horizon of national R&D programmes including a short description of support and market introduction programmes,
- market evaluation of the opportunities to use biomass pyrolysis,
- present and future potential costs and performance of biomass pyrolysis processes,
- present and future economic evaluation of pyrolysis,
- list of active groups (universities, research centres, manufacturers, consultants),
- list of most important literature.

The data collection will be integrated and related with other ongoing activities both within Europe and elsewhere in the world, in particular by establishing a link with the corresponding IEA working group, and by reporting on pyrolysis activities in countries other than the IEA area respectively including a summary report of the activities in the rest of the world in co-operation with the IEA group.

2 by actively contributing to one of the Specialist Groups that will examine and review a small number of discrete topics that are critical for the successful development and implementation of biomass pyrolysis and related technologies. Again this will focus on the current status and future needs and opportunities for the successful exploitation of developing technologies. Each group will provide an authoritative review of the relevant sub-group topic with recommendations for further research, development and industrial opportunities, and identify significant obstacles to their implementation. Reports will be produced for dissemination.

3 by providing and encouraging communication channels for dissemination of information to relevant organisations in their geographical region and technical subject area and thereby encouraging interaction and improving opportunities for exploitation by industry,

4 by actively contributing to the publication of a newsletter (see later),

5	by task sharing when, as an example, laboratories will carry out comparative blind testing to establish preferred methodologies. The results will be reported and disseminated.
6	through personnel exchange when researchers will visit other laboratories in universities and industry for up to two weeks for example to learn new techniques or carry out parallel comparative experiments.

Output

The deliverables will include the following:
1	Publication of a newsletter (1),
2	Establishment of a database of recent and current relevant activities in Europe,
3	Country reports,
4	Authoritative reviews of each sub-group covering pyrolysis, analysis and characterisation, upgrading and applications/utilisation as well as other selected topics such as:
    •	R&D activities around the world,
    •	R&D needs and a proposed R&D strategy,
    •	commercial pyrolysis and related technologies and opportunities for integrated systems,
    •	strategies for an improved flow of information within the IEA countries,
    •	Organisation of meetings/forums/seminars for discussions and presentations,
5	Reports on information exchange, mobility experiences and establishment of closer links between research organisations and industry
6	Overview of the opportunities and possibilities for biomass pyrolysis and related activities,
7	Recommendation of a strategy for implementation of biomass pyrolysis systems in Europe.

PyNE Newsletter
The Newsletter is the most significant means of collecting, reviewing, analysing and disseminating information and is thus the main focus of the Network.

Newsletter Contents

The first edition of PyNE News (1) was published in March 1996 and includes the following contents:
•	Editorial
•	Diary of events
•	Publications and reports - annotated bibliography
•	Profiles of companies and research groups
•	New developments
•	Book reviews
•	Calls for proposals from national, EC and other sources

- Incentives available for biomass RD&D in EU countries
- Meeting and conference reports
- Current projects
- Survey of energy costs in EU countries
- Pyrolysis Technology Group report
- Characterisation and Analysis Group report
- Upgrading Group report
- Applications Group report
- News from Canada
- News from the USA

The next newsletter is planned for publication in October 1995, then at six monthly intervals. Material for publication should be sent at least two months previously.

Database
A database is being established of recent and current relevant activities in Europe to improve interaction and exploitation. This has been distributed to all Network members with regular updates and information made available to enquirers as hard copy and from an on-line system through the Internet. A database of expertise and technology will be established to provide a unique record of all relevant activities in the areas of biomass pyrolysis and related processes.

Country Reports
Each PyNE member will review the status of fast pyrolysis and related technologies in his/her own country and identify opportunities for further development and implementation.

Specialist Subject Groups
In order to complement the activity of the Network, a number of topics is being examined in depth by small working groups made up from members of the Network and led by a convenor with the responsibility of formulating a specific programme of work and producing an authoritative review and recommendations for RD&D. The subjects for the Specialist Sub-Groups are:
1. Pyrolysis technology
2. Upgrading
3. Characterisation and analysis of pyrolysis products
4. Applications of pyrolysis products

Pyrolysis technology

Fast pyrolysis is now on the brink of becoming widely commercialised. Technology has advanced considerably in the last 15 years, but there are still aspects of the process and product that require improvement such as char removal and stabilisation respectively. A further commercially inhibiting factor is the high cost of the product that, without some fiscal incentive, requires a low or zero cost feedstock to produce an economically viable liquid fuel. This specialist group is considering the following:
- reviewing options available for the production of a clean liquid for use as a fuel,

- examining current problems including removal of char and improvement of physical properties,
- reviewing the current status of fast pyrolysis in the world - reviewing actual processes and assessing results and information which may be derived for general use by the pyrolysis community,
- reviewing generic processes and unit operations in the pyrolysis system from feedstock reception to products exiting the process,
- sharing experiences, problems and possibilities, and devising solutions,
- identifying areas where further developmental work is required and making suitable recommendations.

Characterisation and analysis

Knowledge of the properties of the feedstock for pyrolysis is important as it has an impact on the process performance and the product quality. Important characteristics include:
- source and species of feedstock,
- chemical composition (content of lignin, cellulose, hemicellulose, and extractives),
- ash content and composition,
- moisture content,
- particle size,
- elemental composition,

The following characteristics of liquid products are important and a range of physical and chemical methods have been identified which are being evaluated, modified where necessary and new methods are being devised. The work includes cross-checking by inter-laboratory round robin tests and exchange of information:
- viscosity at different temperatures
- water content by Karl Fischer titration, solvent methanol
- pH at different temperatures
- density
- elemental composition
- Stiazny no.
- lower heating value
- ash content
- microcarbon content
- surface tension
- solubility in different solvents
- ageing
- pour point
- flash point

Pyrolysis liquids from woody biomass contain all the condensable thermal degradation products of the three main wood constituents: cellulose, hemicellulose and lignin as well as minor contributions from extractives such as polyphenols and resins. Because of the non specific thermal degradation of the macromolecules, hundreds of low molecular weight compounds and molecular aggregates of higher molecular weights are formed building up the pyrolysis liquid.

Chromatographic and spectroscopic methods are preferred for chemical analysis of the whole oil. Chemical analysis may be more effective if the pyrolysis liquid is first fractionated before. Three methods are currently being used and developed:

1. flash chromatography with sequential elution by different solvents with increasing polarity. This gives fractions containing components with similar chemical characteristics,
2. steam stripping. This also gives fractions with similar chemical characteristics,
3. phase separation by addition of water to give water soluble and water insoluble compounds

The same chemical analysis methods mentioned above can be applied to the individual fractions. Additionally, HPLC with an ion-exchange column can be applied to separate water soluble products, and GPC can be used for the determination of the average molecular weight of the high molecular weight compounds. Several short term tasks have been agreed:

- a resource package for analytical methods will be established.
- analysis and characterisation methods will be harmonised
- a material and safety data sheet will be established
- a round robin will be performed in collaboration with the IEA Bioenergy Pyrolysis Group.

Upgrading

Upgrading refers to any physical, chemical and/or catalytic process that improves any property of the pyrolysis liquid product.

It is necessary to carefully consider the degree of upgrading of pyrolysis oil for each application. It is fundamental to identify and define the potential customer as well as a realistic price for the upgraded liquids. In all cases specifications such as standard fuel properties. stability, acidity and solid content must be known.

It is necessary to look for new catalysts and new processes or combinations of them, for example removing acids which are responsible for catalyst poisoning by coke deposition before mild chemical upgrading. There are many ongoing methods being evaluated including fractionation and vacuum distillation which give low yields and are very expensive. A more radical approach may be needed once the requirements of the users of the bio-oils can better define the limiting characteristics of the pyrolysis liquids.

Applications

The liquid product currently only has one truly commercial application which is as a flavouring for the food industry. Other speciality chemicals can be derived in technically feasible quantities but for which there is currently no established market. The liquid has been used successfully as a boiler fuel, as a furnace fuel and as a diesel engine fuel, but the quality is not yet sufficiently controllable to consider widespread application, nor are the economics sufficiently attractive.

This group will work closely with the Characterisation Group in defining properties and tests and with potential users of the bio-oil to determine limiting characteristics and properties.

A key part of the implementation of liquid bio-fuels is the development of efficient and cost effective systems and the development of cost effective integrated systems is also being considered.

Conclusions

A European-wide network of active researchers, developers and enthusiasts has been successfully established and is contributing to the development of the technology through promotion and collaboration and dissemination. The PyNE newsletter is a regular publication containing a variety of topical news and information and is available free on request. Contribution to the Newsletter and participation in the meeting is warmly welcomed.

Acknowledgements

The PyNE network is funded by the EC AIR Programme. Special thanks are due to Claire Humphreys at Aston University.

Reference

1   PyNE Newsletter No. 1, March 1996, Energy Research Group, Aston University, Birmingham B4 7ET, UK

PRETREATMENT

AN OVERVIEW OF DEVELOPMENT OF STANDARD ANALYTICAL PROCEDURES FOR BIOMASS FEEDSTOCKS

F.A. AGBLEVOR, S.P. DEUTCH, C. EHRMAN, AND R. RUIZ
National Renewable Energy Laboratory, Golden, USA

Abstract

Research and development of bioenergy as a renewable energy source has made significant progress during the past few decades and as bioenergy matures into a commercial energy source the need for biomass standards cannot be overemphasized. The bioenergy community recognized this need more than a decade ago and organized the "Biomass Energy Meeting" to define what feedstocks to use as standards for biomass analysis and conversion processes. Four biomass feedstocks were identified by this workshop: wheat straw, sugar cane bagasse, pinus radiata, and populus deltoides. These feedstocks have been developed into standard research materials and can be obtained from NIST at a nominal cost.

In addition, standard methods of analysis of the reference feedstocks were developed to complement the feedstocks preparation. These analytical methods were tested in a worldwide round robin in 1991. After the round robin test, researchers at NREL have been collaborating with the American Society of Materials and Testing (ASTM) Subcommittee E-48.05 on Biomass Conversion to convert to ASTM standards some of the compositional analysis methods developed during the round robin. Six methods (E1755-95, E1756-95, E1757-95, E1758-95, E1721-95, and E1690-95) have been approved for publication as ASTM standard methods. Other methods are undergoing review for consideration as ASTM standards.

Additionally, NREL researchers are collaborating with the ASTM E48.05 Subcommittee on Biomass Conversion to develop standard analytical methods for various biomass conversion processes and products including simultaneous saccharification and fermentation (SSF), rapid analytical pyrolysis of biomass, and moisture analysis of biocrude oils.

Keywords: ASTM standards, analytical methods, biomass standards, renewable energy.

1 Introduction

Research and development of biomass technologies have made significant progress during the past two decades. The current state of the various biomass conversion technologies indicates that some are approaching commercialization and others have been commercialized for niche markets. These technologies include thermochemical conversion (gasification, pyrolysis, and liquefaction), biochemical conversion (gasification and saccharification and fermentation) and direct combustion. Because most of these processes are still in their developmental stage, few standards exist for products, product analyses, and process parameters. However, for biomass technologies to make significant impact on both the fuel and chemical industries, standards must be developed to guide both producers and consumers. The goal of this paper is to give an overview of the development of standard reference materials for biomass feedstocks, existing methods of biomass analysis, new standard analytical procedures being developed for biomass feedstocks analysis, as well as to discuss future directions for biomass standards development.

2 Development of whole-biomass reference materials

In 1984, the bioenergy community organized a workshop in Gaithersburg, Maryland [1] to define among a whole range of other issues, biomass feedstocks to use as standard reference materials for both conversion and analytical processes. Four feedstocks were identified: wheat straw, sugar cane bagasse, monterey pine, and eastern cottonwood. In a collaborative effort between the National Renewable Energy Laboratory (NREL), National Institute of Standards and Technology (NIST), the International Energy Agency (IEA) Bioenergy Agreement Biomass Conversion Annex VII-Voluntary Standards Activity Group, and Energy, Mines, and Resources Canada, standard reseaarch materials were prepared for the four biomass feedstocks (details of the preparation of the feedstocks are published elsewhere [2]). These feedstocks were used for a worldwide round-robin analytical test in 1991 [3]. The test validated the uniformity of the materials and also the analytical methodology. These packaged feedstocks are listed in the NIST Standard Reference Materials catalogue [4] and they are available for sale to the public at nominal cost by NIST.

Because standard ASTM procedures exist for higher heating values (HHV), carbon, hydrogen, nitrogen, oxygen for solid fuels (coal) as well as refuse derived fuels (RDF) these procedures were not tested during the round robin. However, there is a need to develop specific methods for biomass feedstocks instead of the wholesale adoption of the coal methods as is currently practised.

2.1 Development of feedstock analytical methods

Organizers of the 1991 round-robin test discovered that there were no uniform analytical methods for whole biomass feedstocks for fuels, fiber, chemicals, and other technical products. Different methods were used by the pulp and paper, wood, forage, and food industries. The pulp and paper and wood industries use the Technical Association of Pulp and Paper Industry (TAPPI) methods (T249-cm-85, T222 om-88, and T264 om-88), and ASTM methods (D 1106-84 (1990), D 1105-84 (1990)) for extractives, acid-insoluble lignin, and carbohydrate determinations. These methods produce very accurate results for woody feedstocks but are generally unsuitable for forage and other herbaceous materials. The major shortcomings of the TAPPI and ASTM standards for herbaceous feedstock analysis are interference from proteinaceous material, ash, cutins, suberin, starch, and other minor components during the lignin determination step. Lignin (Klason or sulfuric acid lignin) is determined as the acid-insoluble residue after hydrolysis with sulfuric acid. The proteinaceous components partially condense with the lignin and some of the ash is also retained with the lignin. This method results in very high lignin values for forage and herbaceous feedstocks because they contain significantly higher quantities of ash and crude protein than does wood. Because the compositions of various forms of herbaceous biomass vary significantly with respect to these minor nonstructural cell wall components, no single correction factor can be applied to the lignin values thus determined. Further, sample quantities for analysis are usually very large (5 g), which is unsuitable for situations where the test material is very small.

The forage and food industries have several competing methods for polysaccharides and lignin analysis. The oldest method for determining the amount of cell wall component in a given feedstock was the Weende method [5]. This method uses sulfuric acid and sodium hydroxide for treating the petroleum ether-extracted feedstock at elevated temperatures. Although this method is recognized by the Association of Official Analytical Chemists (AOAC) [5], it has a major deficiency in that some non-cellulosic components and some of the lignin are lost during the procedure.

The most widely used method for forage characterization is the detergent method introduced by Van Soest and Goring [6]. It is very useful for the analysis of individual species, but it is not very useful for comparing composition of different species because various plant materials behave differently in the detergent. A major shortcoming of this procedure is the overestimating of the cellulose and hemicellulose components of the feedstock and its dependence on feedstock composition.

The hemicellulose component is estimated by the difference between the neutral detergent fiber (NDF) and the acid detergent fiber (ADF) and cellulose is determined by the difference between the ADF and lignin. Lignin content is determined by the permanganate degradation of the ADF. It is assumed that NDF estimates the total structural cell wall components of the feedstocks, and that ADF estimates mostly lignin and cellulose. However, chemical analyses of NDF fractions from timothy grass, alfalfa, and wheat straw have shown that they contained 1–6% crude protein in addition to the cellulose, hemicellulose, and lignin [7].

Similarly, chemical analysis of ADF fractions of the above feedstocks has shown that these contain 7–14% hemicellulose and 1–4% crude protein in association with the cellulose and lignin [7]. Furthermore, the cellulose fraction also contained 8–13% hemicellulose and 2–7% lignin [7]. Clearly, for detailed analysis of biomass feedstocks for relating products yields to feedstock composition, this method is inadequate.

In order to address some of the shortcomings of the above methods, a new method was proposed by Theander [8]. This method, known as the Uppsala method for cell wall analysis, is based on the gas liquid chromatography (GLC) of the alditol acetates of individual cell wall polysaccharides released by sulfuric acid hydrolysis of the feedstock.

The results for individual sugars are very accurate and quantitative, but the lignin data suffer from the same shortcomings as the TAPPI and ASTM methods. However, it has been shown to be applicable to both woody and herbaceous feedstocks especially for the cell wall polysaccharides.

A summary of other methods available for the analysis of biomass materials have been published in a sourcebook by Milne et al. [9].

3 New ASTM standard methods for the analysis of biomass feedstocks

In converting biomass feedstocks to value-added products, it is desirable to compare the physical and chemical characteristics of individual feedstocks to one another and to correlate these with the yields of various products during the conversion process. The performance of each feedstock during the process can then be assessed and operating parameters changed to suit the feedstock. For meaningful comparison of analytical data, the methods of analysis of each feedstock must either be the same or similar. It is therefore of paramount importance to select an analytical procedure which is applicable to both woody and herbaceous biomass feedstocks.

The Uppsala method is closest to meeting the above criteria. Therefore, it was adopted, with some modifications, for the round-robin biomass analysis in 1991 because it was considered to be more versatile than the other methods in the determination of carbohydrates in biomass. The main changes introduced were the extractives determination, the preparation of the alditol acetates, and the use of considerably smaller sample sizes for the analyses. The extractives determination was modified from that in the Uppsala procedure to maximize accessibility to as many users as possible in the international community of biomass analyzers. Whereas the Uppsala method called for the ultrasonication or a combined homogenization and extraction using an Ultra-Turax with 80% (v/v) aqueous ethanol, the modified method used the following procedure. The sample was air-dried at room temperature and Soxhlet extracted with 95% (v/v) ethanol for 6 h. Extractives content was determined after evaporating the extract to dryness in a rotary evaporator under reduced pressure. The alditol acetate derivatization followed the procedure originally proposed by Blakeney et al. [10]. Whereas the original Uppsala method required sample sizes of 1.0–3.0 g, the modified method requires only 300 mg samples for the analysis. This makes it possible for small samples to be analyzed in situations were sample size is limited.

The modified Uppsala method is also similar to the TAPPI method T 249 cm-85 which also uses GLC analysis of alditol acetates to determine carbohydrates in wood and pulp samples. The main difference between the two procedures is the use of catalyst. Whereas the TAPPI method requires the reaction mixture to be heated for 1 h at 60 °C, the modified Uppsala method uses 1-methyl imidazole to catalyze the reaction and the process takes only 10 min. without heating.

In addition to the above changes, it was discovered that during the round-robin analysis of the whole biomass feedstocks, some laboratories used a high performance liquid chromatography (HPLC) method instead of the prescribed GLC method. However, when the results from the two methods were compared, the results were very similar. The largest differences- in glucan and xylan contents- were 1.0% and 1.2% respectively. Consequently, the HPLC method was also included in the new standards.

Ash analysis is a very important component for the herbaceous feedstocks and also contributes to the high values obtained for the determination of acid-insoluble lignin in these feedstocks. Because of the high potassium content of herbaceous feedstocks and the tendency of potassium salts to vaporize during the ashing process [11], a more suitable method was developed for both woody and herbaceous biomass feedstocks.

In 1993, the ASTM Subcommittee E-48.05 on Biomass Conversion evaluated its standards for biomass conversion processes and found that there were no suitable standards for the potentially emerging biomass energy industry. The Biomass Conversion subcommittee therefore initiated collaborative efforts with NREL researchers to convert already existing standard procedures for biomass feedstock analysis to ASTM procedures under the jurisdiction of the ASTM E-48 Committee on Biotechnology. The modified Uppsala method used for the round-robin was examined by the subcommittee and found suitable as a standard reference method in view of the fact that the method was tested in several laboratories worldwide and found to be reproducible. The subcommittee therefore decided to divide the procedures into several methods and reformat them to meet ASTM format. The wet chemical methods were divided into several procedures as listed in Table 1 with their ASTM designation numbers.

3.1 Standard practice for preparation of biomass for compositional analysis

This practice gives a reproducible way to convert hardwoods, softwoods, herbaceous materials (such as switchgrass and sericea lespedeza), agricultural residues (such as corn stover, wheat straw, and bagasse), wastepaper (such as office waste, boxboard, and newsprint), feedstocks pretreated to improve suitability for fermentation, and fermentation residues, into uniform material suitable for compositional analysis.

3.2 Standard method for the determination of total solids in biomass

The total solids content is used to adjust the mass of biomass so that all analytical results may be reported on a moisture-free basis. This method is intended to determine the amount of total solids remaining after drying a sample. Materials suitable for this procedure include samples prepared according the standard practice for preparation of biomass for compositional analysis.

3.3 Standard method for ash in biomass

This test method covers the determination of ash expressed as the mass percent of residue remaining after dry oxidation (oxidation at $575\pm25°$ C) of hard and softwoods, herbaceous materials (such as switchgrass and sericea lespedeza), agricultural residues (such as corn stover, wheat straw, and bagasse), wastepaper (such as office waste, boxboard, and newsprint), acid and alkaline pretreated biomass, and the solid fraction of fermentation residues. All results are reported relative to the 105°C oven-dried basis.

3.4 Standard method for the determination of ethanol extractives in biomass

This test method covers the determination of ethanol soluble extractives expressed as the percentage of the oven-dried biomass. It is applicable to biomass feedstocks

Table 1. ASTM standard methods for the analyses of biomass feedstocks

ASTM #	Title	Status
E1757-95	Standard practice for preparation of biomass for compositional analysis.	Approved for publication in 1995
E1756-95	Standard method for the determination of total solids in biomass.	-do-
E1755-95		-do-
E1690-95	Standard practice for preparation of biomass for compositional analysis.	-do-
E1721-95	Standard test method for determination of acid-insoluble residue in biomass	-do-
E1758-95	Standard method for determination of carbohydrates in biomass by high performance liquid chromatography.	-do-
No Number	Standard method for determination of carbohydrates in biomass by gas liquid chromatography.	At subcommittee approval stage
No Number	Standard method for determination of acid-soluble lignin in biomass by ultra violet spectroscopy	In preparation at NREL

prepared by the standard practice for the preparation of biomass materials for compositional analysis.

3.5 Standard test method for determination of acid-insoluble residue in biomass

This method covers the acid-insoluble residue of hardwood, softwood, herbaceous materials, agricultural residues, wastepaper, acid and alkaline pretreated biomass and solid fraction of fermentation residue. The residue collected contains acid-insoluble lignin, ash and any condensed proteins from the original sample. An independent nitrogen and ash analysis would be required to determine the acid-insoluble lignin content separate from the condensed protein and ash fractions. For the ash content, the method for ash determination described above is used.

3.6 Standard method for determination of carbohydrates in biomass by high performance liquid chromatography

This method describes the determination of the major cell wall polysaccharides of biomass. The polysaccharides are hydrolyzed to their sugar monomers (glucose, xylose, arabinose, galactose, and mannose) by sulfuric acid in a two-stage hydrolysis. The monosaccharides are then quantified by ion-moderated partition HPLC. The method is applicable to all biomass samples listed under the standard preparation method above.

3.7 Standard method for determination of carbohydrates in biomass by gas liquid chromatography

This method is similar to the HPLC method except that the monomeric sugars are derivatized as alditol acetates and analyzed by gas liquid chromatography. The results from the two methods have been shown to be similar with no statistically significant differences between them. This method is currently undergoing review by the various ASTM committees for eventual approval as a standard method.

3.8 Standard method for determination of acid-soluble lignin in biomass by ultra violet (UV) spectroscopy

This method describes the determination of acid soluble lignin by UV. It is particularly relevant to hardwood, herbaceous materials, acid and alkaline pretreated materials and agricultural residues. It is not useful for softwoods because the amounts of soluble lignin in these feedstocks are extremely small. This method is being prepared for ASTM approval.

3.9 Other Standards

Unlike the analysis of the whole biomass feedstocks in which significant progress has been made during the past few years, standards for conversion and product analyses are lacking. A concerted effort from the biomass research community is required to establish the necessary standards and practices. In addition to the wet chemical procedures listed in Table 1, ASTM subcommittee E-48.05 on Biomass Conversion is currently discussing the potential for developing standards for thermochemical and biochemical conversion processes such as pyrolysis, gasification, and simultaneous saccharification and fermentation (SSF). Product specification is another area the ASTM committee hopes to pursue. For instance, the subcommittee would like to develop specifications for various grades of biomass pyrolysis oil to ensure uniformity of products on the market. However, all these procedures are still in their developmental stages.

4 Conclusions

Significant progress has been made towards the development of standard methods for analysis of whole biomass feedstocks. However, no standard methodologies exist for thermochemical or biochemical conversion processes or for products characterization. They need to be developed as soon as possible. All the standards discussed above have been published in the Annual Book of ASTM Standards, Volume 11.05 under the jurisdiction the E-48 Committee on Biotechnology. These standards are subject to revision every five years.

5 Acknowledgements

The authors acknowledge R. Overend, H. Chum, T. Milne, E. Chornet, N. Hinman, A. Wiselogel, D. Johnson, NREL and L. Eitel, Jacobs Services Company, Commerce City, CO for their continuing interest and support for this project, and the US DOE Biofuels System Division for funding the project.

6 References

1. Milne, T.A. (1984). *Proceedings, Workshop on Standards in Biomass for Energy and Chemicals.* National Bureau of Standards, Gaithersburg, Maryland, August 1-3, 1984. SERI/CP-234-2506.

2. Agblevor, F.A., Chum, H.L. and Johnson, D.K. (1992). Compositional analysis of NIST biomass standards from the whole feedstock round-robin, in *Energy from Biomass and Wastes XIV,* (ed. Klass, D.L.), Institute of Gas Technology (IGT), Chicago, IL. pp. 395-421.

3. Chum, H.L., and Gellerstaedt, G. (1991). *Proceedings, IEA Presymposium on Methods of Analysis of Wood, Annual Plants and Lignins,* November 29 - December 1, 1991, Monteleone Hotel, New Orleans, LA. National Renewable Energy Laboratory, Golden, CO.

4. NIST Standard Reference Materials (1996). NIST special publication 260, U.S. Department of Commerce, Technology Administration, National Institute of Standards and Technology, Gaithersberg, MD.

5. Theander, O. and Westerlund, E. (1993). Quantitative analysis of cell wall components, in *Forage Cell Wall Structure and Digestibility,* (eds. H.G. Jung, D.R. Buxton, R.D. Hatffield, and J. Ralph), American Society of Agronomy, Crop Science Society of America, and Soil Science Society of America, Madison, WI, pp. 83-99.

6. Goering, H.K. and Van Soest, P.J. (1970). Forage Fiber Analysis. *USDA Agric Handbook 379.* U.S. Gov. Print Office, Washington, D.C.

7. Theander, O. and Aman, P. (1980). Chemical composition of some forages and various residues from feeding value determinations. *J. Sci. Food Agric.* Vol. 31, pp. 31-37.

8. Theander, O. (1991). Chemical analysis of lignocellulose Materials. *Animal Feed Sci. Tech.,* Vol. 32, pp. 35-44.

9. Milne, T.A., Brennan, A.H., and Glenn, B.H. (1990). *Sourcebook of Methods of Analysis of Biomass and Biomass Conversion Processes.* Elsevier, New York.

10. Blakeney, A.B., Harris, P.J., Henry, R.J. and Stone, B.A. (1983). A simple and rapid preparation of alditol acetates for monosaccharide analysis. *Carbohydrate Research,* Vol. 113, pp. 291-299.

11. Misra, M.K., Ragland, K.W. and Baker, A.J. (1993). Wood ash composition as a function of furnace temperature. *Biomass and Bioenergy,* Vol. 4. No. 2 pp. 103-116.

FUEL CHARACTERISTICS OF BRIQUETTES FROM WOOD WASTE AND RECYCLED PAPER

G.DANON, T. STEVANOVIC JANEZIC, B. BUJANOVIC AND G.STANOJEVIC
Faculty of Forestry, University of Belgrade, Belgrade, Yugoslavia

Abstract

The procedure is developed for light briquette production from wood and bark materials of different technical origin, regardless of their dimensions and moisture content. Cohesiveness of the light briquette is achieved by the addition of pulp and/or recycled paper waste. The mixture of materials is completely saturated with water and then shaped in molds at low pressure (3 to 5 bars).

The briquettes produced by the described procedure are characterized by high porosity and therefore low density (240 to 331 kg/m^3).

The light briquette structure does not decline upon moisture absorption or water suction, which is the common problem with the wood briquettes produced by conventional methods.

The samples of light briquettes were determined to have: low ash content (up to 2.5%), high volatile content (between 70% and 78%), charcoal content (between 15% and 22%) and higher heating value about 20,000 kJ/kg.

Combustion characteristics of briquette samples were examined in the furnace. The testing was carried out in the constant temperature conditions. The measurements were repeated in the range between 400°C and 800°C, advancing at constant intervals.

The chamber temperature was determined to have the influence upon the volatility of briquettes and the ignition delay time of the volatiles. The influence of the chamber temperature upon combustion time of the volatiles was determined to be insignificant.

Keywords: Combustion, forestry wood waste, fuel characteristics, higher heating value, ignition, light briquettes production, pulp, recycled paper.

1 Introduction

The idea of briquetting wood waste is not new. The briquetting was performed with more or less success during the 1920's. These briquettes were used as solid fuel in mills and households. Increased production and processing of oil, together with its low cost, made the briquetting technology uninteresting.

Because of oil prices and the need for environmental protection, utilization of wood wastes is coming of age again.

The briquettes are commonly produced according to the following principles [1]:

- Starting material is primarily wood waste. Utilization of bark is to be avoided, as it affects adversely the mechanical properties of briquettes.
- Starting material must be ground subsequent to briquetting (granular size below 10 mm [2]).
- The moisture content of the briquetting material should be around 12%. This infers drying of the material before briquetting.
- Compactness of the briquettes is achieved by compressing. Commonly applied pressures range between 1000 and 1500 bar. Pressure application contributes to the increase of briquette density to 1.1-1.4 t/m^3.

The resistance of briquettes to air humidity or to direct contact with water is improved by thermal treatment. Thermal treatments are usually performed at temperatures between 200 and 300^0C before briquetting, or at 60-80^{0}C during the briquetting procedure.

Common pelleting procedure yields products with excellent handling and storage characteristics, with four times the energy concentration of firewood, thus greatly reducing transportation costs and improving boiler efficiency.

2 Low density bio-briquette production

Light bio-briquette presented here is seen as an alternate and complementary solution for forest waste (residue) utilization.

Wood residues available for energy purposes are characterized by differences in regard to chemical composition, morphological origin, moisture content and size [6]. Practicable utilization of this waste as fuel, for energy purposes, depends primarily on its calorific value.

The subject of our previous research was to estimate the energy potential of forest and wood processing wastes and to correlate them with their chemical composition in order to enhance the utilization of waste [3, 4, 5].

Light bio-briquette production is based on various technical-technological principles that enable [6]:

- Utilization of materials with different moisture contents, without the need for prior moisture equalization.
- Briquetting of heterogeneous materials including particles coarser than 10 mm.

- Compressing by pressure ranging between 3 and 5 bars.
- Possibility of outdoor storage of briquettes, since the swelling or splitting of the layers is not observed with the increase of internal moisture content.

The briquettes produced within the concept presented here are sufficiently compact and water resistant. Cohesiveness of the bio-briquette is achieved by the addition of groundwood pulp or recycled paper [7]. The mixture of materials is completely saturated with water and then shaped in molds at low pressure (3 to 5 bars).

During compression, free evacuation of excess water from the mold is provided. Physico-chemical bonding between different materials is taking place gradually, during the entire drying process, even after the moldings has been finished.

Air seasoning of bio-briquettes is possible in summer. During the winter period, however, forced air drying must be provided. Because of its simplicity, the described procedure is applicable in households and small-scale enterprises.

3 Materials and methods

The light bio-briquettes were produced from various types of residues of different wood species. At the Faculty of Forestry of the University of Belgrade, the briquette samples were produced from different constituents (Table 1).

Table 1. Light bio-briquette sample contents

Sample A			Sample B			Sample C		
Components	size [mm]	[%][1]	Components	size [mm]	[%][1]	Components	size [mm]	[%][1]
Beech sawdust	4-10	85	Beech sawdust	4 -10	51	Conifer sawdust	4-10	42
Pulp and recycled paper		15	Pulp and recycled paper		15	Pulp and recycled paper		15
Water	-	-	Pine needles[2]	5 -15	34	Conifer bark	5-30	43
-		-	Water			Water		-

[1] percentage of dry mass
[2] residue after water vapor distillation

Preliminary examination of the manufactured bio-briquettes was performed at the Faculty of Forestry and the Faculty of Mechanical Engineering in Belgrade. The investigations included determination of density, moisture content, water capacity, compressibility, volatile content, charcoal and ash contents and higher heating values of briquette samples. In order to research the above briquette characteristics, 5 samples of each type of briquette were measured in each type of research.

Density was determined by standard method JUS D.A1.044 which includes the determination of sample volume and mass in the dry state.

Moisture content was determined by TAPPI T 12 wd-82 method by drying the briquette samples at the temperature of 105 ± 2 °C up to a constant mass.

Water capacity of briquette samples was determined after the briquette samples were soaked in water for 12 h and then their mass was measured and compared to the initial mass.

Compressibility of bio-briquettes was determined by the standard method JUS D.A1.045 which is usually applied in compression testing of wood samples. The compression strength measurements were performed on the 30x30x20 mm specimens.

Volatile content and charcoal were determined by JUS B.X8.317 method by annealing the briquette samples at the temperature of 900 °C, during 7 minutes, without the presence of air.

Ash content of investigated briquette samples was determined by the method TAPPI T15 wd-80 by annealing the briquette samples at the temperature of 600-850 °C during 1h.

Higher heating values of investigated briquette samples were determined by standard method JUS.B X8.318 in the calorimetric bomb.

Combustion characteristics of briquette samples were determined in the furnace of 3000 W with maximum temperature up to 1000 ^{0}C. The testing was carried out in the conditions of constant temperature in the furnace place. The measurements were repeated in the range between 425 ^{0}C and 800 ^{0}C, advancing at 25 ^{0}C intervals. The temperature in the fire place was maintained within the limits of 5 ^{0}C. The samples were prepared to be cubic in form, length 10-12 mm. The exact sizes of the samples were difficult to achieve, due to the non-homogeneous briquette composition and different granulation of the briquette components.

The time from loading the samples in the furnace till the end of the combustion of the volatiles was measured. The time of the first flame appearance was also recorded. In the period between sample loading and the appearance of the flame, the fuel is heated and thermally decomposed and the mixture of the volatiles and air is formed. This period is called ignition delay time of the volatiles and it is connoted by t_1. It is affected by fuel sample composition and size, as well as by temperature, pressure and oxygen concentration in the fireplace.

The period of combustion of the volatiles starts with their ignition and lasts as long as the visible flame. The length of this period depends on volatile quantity, temperature, pressure in the fireplace, as well as on sample particles. This period is called combustion time of the volatiles and it is denoted by t_2.

As it has already been mentioned, the analysis was carried out (1) in a 3000 W furnace. So as to enable a quick and uniform input of briquette samples (3) in each trial, a movable platform (2) was equipped along with fireproof glass (4) and sample holder (5), also with a video camera holder (6). During the testing of lightweight briquette combustion characteristics, the furnace door was removed and replaced with a fireproof door (4) for the visual observation of the course of analysis. Sample holder (5) was made of steel thermoresistant to furnace temperature (900 °C). The camera (7), video recorder (9) and monitor (10) enabled the direct visual monitoring of the combustion process with simultaneous video recording. The problem of visibility was solved by a reflector (11). Sample combustion time was measured by a digital stopwatch (12), precise to 0.1 sec., able to memorize the transit time.

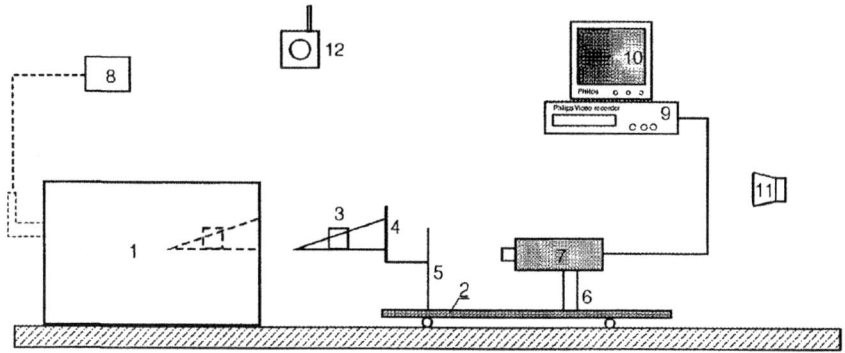

Fig. 1. Fuel characteristics experimental setup

The components of the experimental setup for fuel characteristics measurement (see Fig. 1.) are:

(1) furnace of 3000 W,   (7) video camera,
(2) movable platform   (8) gas analyzer,
(3) sample   (9) video recorder,
(4) fire proof glass,   (10) monitor,
(5) sample holder,   (11) reflector and
(6) video camera holder,   (12) digital stopwatch.

The above experimental installation was constructed at the Faculty of Mechanical Engineering in Belgrade [9]. Its improvement with a gas analyzed (8) will enable the determination of the total combustion time.

3 Results and discussion

The obtained results of measurements of light bio-briquette properties have been presented in Table 2.

All the examined samples of light briquettes have good energy characteristics which is in accordance with combustion characteristics of the type of waste used in their manufacture [2] and their percentage (Table 1). It is known that the upper heating value of lignocellulosic materials depends on the percentage of the main components of wood, as well as on the content and nature of the extractives. Cellulose and lignin contribute the values of about 16,000 kJ/kg and 25,000 kJ/kg, respectively [5]. The share of recycled paper and pulp in the analyzed samples, taking into account the high content of cellulose, reduces the briquette heating value compared to that of the corresponding species. Simultaneously, the share of bark and needles which are characterized by higher contents of extractive substances ($H_g = 40,000$ kJ/kg, [2]) in the briquette samples B and C, increases their heating value compared to that observed for sample A (see Table 2). The real energy potential of the briquettes is somewhat lower because of their moisture content (6.4 - 7.9 %).

The volatiles include volatile carbon, hydrogen, oxygen and possibly nitrogen, i.e. predominantly hydrocarbon compounds. Volatile content and charcoal were determined at 900 °C during 7 minutes, without the presence of air. The quantity of the volatiles in the analyzed briquette samples, which varies depending on furnace temperature [9], affects ignition delay tome t_1 and reduces the briquette combustion time (see Fig. 2. and 4.).

Coke residue is an unevaporable part of the fuel and in lignocellulosic substances it consists mostly of unevaporable carbon. The ratio of the volatile and the coke residue contents was described by the equation of reactivity index:

$$IR = V_g/C_{fix}.$$

In the construction of the furnace for light bio-briquette combustion, the differences in reactivity indices of lignocellulosic briquettes and different types of coal should be considered, as the volatile content in coals is very low.

Table 2. Light bio-briquettes properties

Sample	Higher heating values	Volatile	Char-coal	IR	Ash	Density	Moisture content	Water potential
	kJ/kg	%	%	-	%	kg/m³	%	%
A	18,600	77.70	14.70	5.24	0.62	331	7.60	241
B	20,675	75.34	18.25	4.13	2.26	325	6.41	229-279
C	19,400	71.67	21.41	3.35	2.51	240	6.92	337-394

The briquettes produced by the described procedure have high porosity and therefore low density. Oven-dry (bone dry) density of the briquettes was determined to range from 240 to 331 kg/m³. Great porosity of lightweight briquettes improves the speed of combustion.

The share of ash in briquette samples is considerably lower compared to the content of ash in solid fossil fuels. For example, the content of ash in the lignite of the coal mine Kolubara ranges between 5-24 %. However, as the wood waste from which the analyzed briquette samples were manufactured are dirty due to different phases of processing, the content of ash in the briquettes is higher than the ash content in the corresponding constituents of the briquettes [7].

All the examined types of briquettes were determined to absorb high quantities of water without a structure decline which facilitates handling and storage. The determined water potential was 241 % for sample A, from 229 to 279 % for sample B, and up to 394 % for sample C.

The compression strength measurements were performed on the 30x30x20 mm specimens, with a 50 % decrease of height relative to the initial height, after the loading. Disintegration of specimens was not observed in these experiments.

The obtained results of the measurements of ignition delay time for all samples are presented in Fig. 2.

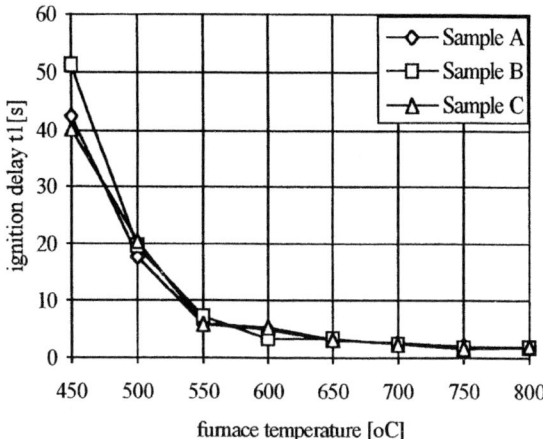

Fig. 2. Relation between ignition delay time of the volatiles and furnace temperature

The temperature in the furnace has a great influence on the ignition delay time. By increasing the furnace temperature, the quantity of the volatiles also increases, and the time till the ignition is reduced. The type of briquette and its properties, e.g. its density, do not affect substantially the time t_1, especially at higher temperatures.

Relations of combustion time of the volatiles and temperature in the stove for all samples are presented in Fig. 3..

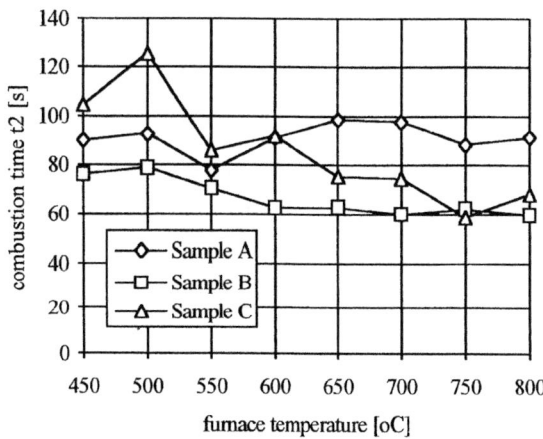

Fig. 3. Relation between combustion time of the volatiles and furnace temperature

The results of the volatile combustion time t_2 measurement do not reflect any laws. The observed differences in the combustion time of the analyzed briquette samples cannot be correlated to their characteristics. From Table 2 it can be seen that the share of the volatiles is approximately equal for all the three briquettes. The conditions of testing were always approximately identical.

The possible causes of the observed differences are the different sizes of the tested samples. It should be taken into account that during the analysis it was impossible to provide that the masses of the analyzed briquette samples be identical for each individual measurement. As it has already been stated, combustion time of the volatiles depends on the share of the volatiles in the total fuel volume, conditions of the fire place, and also on sample sizes, i.e. mass.

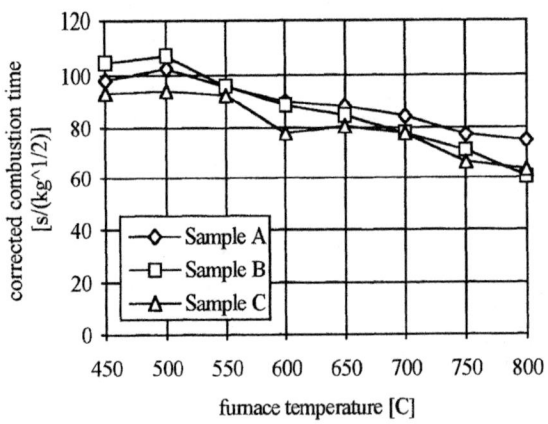

Fig. 4. Relation between corrected combustion time of the volatiles and furnace temperature

Fig. 4. presents the dependence of the corrected combustion time of the volatiles t_2 and furnace temperature. After several trials, the correction was made by dividing the measured values of t_2 with the square root of sample mass. The differences of sample mass are the result of the deviations in their sizes, as well as their different densities. This was influenced by non-uniform granulation of briquette constituents. The correction resulted in the more uniform change of corrected t_2 with the change of testing temperature in the chamber, and smaller differences between different briquette samples.

5 Conclusions

Wood residues are characterized by differences in regard to chemical composition, morphological origin, moisture content and size. Practicable utilization of this waste as fuel, for energy purposes, depends primarily on its calorific value.

The manufacture of lightweight bio-briquettes based on wood waste, pulp and recycled paper can provide the optimum utilization of these materials, in spite of their varying characteristics, sizes and moisture contents. Cohesiveness of light briquettes is achieved by the addition of pulp and /or recycled paper waste.

The described procedure enables the formation of briquettes under low pressure application (3 to 5 bar) and the possibility of air seasoning of briquettes to 10...15 % moisture content, which can be performed in yards. Because of its simplicity, the described procedure is applicable in households and small-scale enterprises.

All the examined samples of light briquettes have good energy characteristics (H_g is between 18,600 and 20,675 kJ/kg), which is in accordance with the combustible characteristics of waste types used in their manufacture.

The temperature in the furnace has great influence on the rate of volatile extraction and therefore the reduction of ignition delay time. The type of briquette and its properties, such as density, do not affect substantially the time t_1, especially at higher temperatures.

Combustion time of the volatiles t_2 depends on the share of the volatiles in the total fuel volume, conditions of the fire place, and also on sample sizes, i.e. mass. By introducing the corrected time of the parameter t_2, which is obtained by dividing the measured values of t_2 with square root of sample mass, the above effects are excluded successfully.

The light briquette structure is strong enough and does not decline upon moisture adsorption or water suction, which is a great advantage in briquette handling, storage and transport.

Low density of briquettes (between 240 and 331 kg/m^3) influences unfavorably the transportation and storage costs. The boiler fireplace capacity may also be a source of problems. The results of compressibility measurements are, however, encouraging, even though further investments are necessary.

6 References

1. Carre J.,L. Lacrosse and Schenkel, Y. (1989) PRODUCTION D'ENERGIE A PARTTIR DE LA BIOMASSE DES RESIDUS AGRICOLES ET AGRO-ALEMENTAIRS, Annales de Gembloux , 95, pp 199-223.
2. FAO PUBLICATION (1990): Energy Conservation in Mechanical Forest Industries, Forestry Paper No. 93, Rome, pp 71-89.
3. Stevanovic Janezic, T., Bujanovic, B. (1990) Comparative Analysis of Chemical Composition of Beech and Fir Barks, Sumarstvo (Forestry) XLIII (6), pp.31-34(In Serbian).
4. Stevanovic Janezic, T. (1991) Determination of Lignin in Barks, Mitteilungen der BFH, 168, Hamburg, pp.391-396.

5. Stevanovic Janezic, T., Danon, G., Bujanovic, B. Dedic, A.(1993) Correlation Between Chemical Composition and the Higher Heating Value of Some Domestic Wood Species, Drevarsky Vyscum, 36 (3),pp. 1-7.

6. Stevanovic Janezic, T. et al. (1995) Wood Residues In Energy Production For Domestic Use In Serbia , XX IUFRO Tampere Finland, 6 - 12. O8 .1995, Abstracts of Invited Papers, p 39

7. Stevanovic Janezic, T., Kolin, B., Jaic, M., Danon, G. (1995) Enhancement Of Wood Technologies In Correlation With Properties of Wood Chemical Constituent, Monograph, Faculty of Forestry, pp 115-126 (in Serbian).

8. Rice R.W., Willey, R.M. (1995) Higher Heating Values for Pellets Made From Woodwaste And Recycled Newsprint, Forest Product Journal, Vol 45, No.1, pp 84 - 85.

9. Radovanovic M., Stanojevic G., Stojiljkovic D., Jerinic N. (1995) Light Briquettes-New Technology, EKO-Ek 95 Biomass-Energy- Ecology- Economy, Belgrade, Yugoslavia, 03.06.1995, Proceedings of the Conference, pp 177-192. (in Serbian)

10. Leko A. (1995) Finalyear Project, Faculty of Mechanical Engineering Belgrade. (in Serbian)

BIOMASS FEEDSTOCK VARIABILITY AND ITS EFFECT ON BIOCRUDE OIL PROPERTIES

F.A. Agblevor, S. Besler-Guran, D. Montane, and A.E. Wiselogel
National Renewable Energy Laboratory, Golden, USA

Abstract

These studies employed similar varieties of switchgrass (*Panicum virgatum L.*) grown at three different locations and three hybrid poplar clones grown at one location. The feedstocks were pyrolyzed in a fluidized bed reactor at 500 °C. The gas products of pyrolysis were analyzed on-line, and the liquid products were analyzed for elemental composition and higher heating values.

Apart from small differences in the yield of char, the yields of pyrolysis liquids (biocrude oils) and gases were similar for switchgrass feedstocks grown at all three locations. The char yields ranged from 21.1% to 22.9% by weight; total liquids (organic liquids and water) yields ranged from 59% to 60.5%; and the gas yields ranged from 11% to 12%. The higher heating values (HHVs) of the oils were similar (24.3-24.6 MJ/kg).

For the hybrid poplar feedstocks, total liquids (65%-69%), char (10%-11%), and gas yields (15.6%-17%) were similar for all three poplar clones. However, the elemental composition and the HHVs of the biocrude oils had statistically significant differences. The NC5260 biocrude oils had lower HHV (22.0±0.5 MJ/kg) compared to the DN17 and DN182 poplar clones (23.2±0.3 MJ/kg).

The yields of total liquids and organics for the three poplar clones were higher than those for the switchgrass feedstocks. The gas yields for the hybrid poplar clones also were higher than those for the switchgrass, but had compositions similar to those of the switchgrass feedstocks. The char yields of the switchgrass were twice those of the hybrid poplar clones.

Keywords: biomass feedstocks, fluidized bed pyrolysis, feedstock variability.

1 Introduction

Converting biomass feedstocks to fuels via thermochemical routes is a major focus of renewable energy research at the National Renewable Energy Laboratory. Among several thermochemical routes is the fast pyrolysis process, which produces liquid fuels from woody and herbaceous biomass feedstocks. There is large variability in the composition of biomass feedstocks due to plant variety and environmental factors and it is important to assess how these differences in chemical composition affect the properties of thermochemical liquid fuels produced from these feedstocks.

Most published studies [1-7] have not addressed these variabilities. This work assesses the influence of plant environment and species variety on the thermochemical conversion process and on the properties of biomass pyrolysis oils (biocrude oils).

2 Methods

2.1 Biomass Feedstocks

For this study, hybrid poplar feedstocks were selected from the following clones: DN17 (*deltoides x nigra*), DN182 (*deltoides x nigra*), and NC5260 (*tristis x balsamefera*). The selection was aided by summative analysis of the poplar clones that showed significant differences in the composition of the feedstocks [8]. The poplar clones were harvested from sites in Ashland, Wisconsin, by the U.S. Department of Agriculture Forest Service (USDA-FS) North Central Experiment Station, Rhinelander, Wisconsin.

Switchgrass (*Panicum virgatum L.*) samples were harvested from plots in Mead, Nebraska, West Lafayette, Indiana, and Ames, Iowa. The samples from each plot were too small for triplicate runs to be conducted in the fluidized bed reactor; therefore, samples from several plots were combined to create enough sample for triplicate runs. This approach was justified because the summative analysis studies revealed no significant difference in the composition of switchgrass from different plots at the same location [9].

Switchgrass and hybrid poplar samples were prepared for pyrolysis by knife milling in a Wiley mill (model 4) until all the samples passed through a 2-mm screen. The milled materials were then sieved to -20/+80 mesh size and dried at room temperature to a moisture content of 4%-6% before pyrolyzing in the fluidized bed reactor. This particle size was selected to facilitate comparison of summative analysis and pyrolytic conversion results. The summative analysis procedure cannot handle particle sizes less than 80 mesh because the samples clog filters during the acid digestion process. The Sauter mean diameter of the feedstocks was 345-400 μm.

Both the switchgrass and hybrid poplar feedstocks were analyzed for their elemental composition and higher heating values at Huffman Laboratories, Inc. in Golden, Colorado.

2.2 Fluidized Bed Pyrolysis of Feedstocks

To assess the influence of plant location and species variety on thermochemical conversion products, the feedstocks (switchgrass and hybrid poplar) were pyrolyzed in a fluidized bed reactor.

The reactor consisted of a 50 mm (2 in) schedule 40 Inconel steel pipe that was 500 mm (20 in) long (including a 140 mm (5.5 in) preheater zone below the gas distribution plate) and equipped with a 100-μm porous metal gas distributor. The fluidizing medium was silica sand, and the bed was fluidized with nitrogen. The reactor was externally heated with a three-zone electric furnace. The reactor tube contained a bubbling fluid bed with back-mixing of the feed and sand. The static sand bed height was 90 mm (3.5 in), expanding to 120 mm (4.7 in) when fully fluidized. The silica sand particles had a Sauter mean diameter of 350 μm.

Reactor temperature was measured by three K-thermocouples inserted into thermal wells inside the reactor. The first thermocouple measured the temperature of the fluidized bed, the second measured the temperature above the fluidized bed, and the third thermocouple measured the temperature of the exiting pyrolysis gases and vapors. The temperatures in all three zones were controlled by the MKS™ data acquisition system (described in [1]).

The biomass was loaded into a feed hopper (batch-wise) and conveyed by a twin-screw feeder into an entrainment compartment. Here, high-velocity nitrogen gas entrained the feed and carried it through a jacketed air-cooled feeder tube into the fluidized bed. Although both the switchgrass and the hybrid poplar feedstocks were ground to the same average particle size and they had similar moisture contents, the switchgrass was more difficult to feed compared to the hybrid poplar. The hybrid poplar flowed very readily whereas the switchgrass flowed with difficulty. For hybrid poplar feedstock, we could feed the reactor at 230 g h⁻¹ without any blockage of the feeding tube, but at similar flow rates for the switchgrass, the feeding tube was blocked. However, at 105 g h⁻¹ the switchgrass feedstock did not block the feeding tube. To ensure uniform processing conditions for comparison of their pyrolysis performance, both the switchgrass and the hybrid poplar feedstocks were fed at 105 g h⁻¹.

Because a twin screw-feeder was used to feed the biomass to the entrainment compartment, at 100 g h⁻¹ and higher feeding rates, the feeding was relatively uniform, but below 85 g h⁻¹ the feeding was intermittent.

The pyrolysis temperature was maintained at 500 °C and the apparent pyrolysis vapor residence time was about 0.4 s. The apparent vapor residence time in the reactor was estimated using the Waterloo Fast Pyrolysis Process methodology [5]. In this approach, the apparent residence time of gases and vapors is defined as the free reactor volume (the empty reactor volume minus the true volume of the sand) divided by the entering gas flow rate expressed at reactor conditions. Runs lasted from 2.0-3.5 h, and the feed rate was 105 g h⁻¹ for both feedstocks and the total gas flow rate (fluidizing gas plus feed gas) was 19 L min⁻¹ at STP. The total gas to biomass ratio was 14:1 for both feedstocks. The feed rate, gas flow rate, and reactor temperature were kept constant during each run.

Pyrolysis gases and vapors exiting the reactor passed through a heated cyclone to separate char and any entrained sand. The cyclone and char pot temperatures were maintained at 400 °C to avoid condensation of the pyrolysis vapors. The pyrolysis gases and vapors passed through a condensation train consisting of a chilled water condenser, an ice/salt mixture condenser, an electrostatic precipitator, and a coalescence filter, all connected in series. The gas outlet temperature was 25 °C. The electrostatic precipitator was maintained at 18-20 kV throughout each run. The reactor temperatures, gas flow rates, pressure drop across the reactor, and electrostatic precipitator voltage were controlled and/or monitored by an MKS™ data acquisition unit.

The gaseous products were analyzed using an on-line HP 5890 Series II gas chromatograph (GC). The gaseous products were sampled and analyzed every 30 minutes during the run. In addition, gas samples were collected in syringes between GC runs, and these were analyzed after the run. Total gas volume was measured by a dry test meter. To ensure good mass closure, the entire setup (excluding the dry test meter) was weighed before and after each run. Biocrude oils were recovered (after weighing the pyrolysis unit) by rinsing the condensers with acetone. The oil-acetone mixture was filtered through a 40-60 μm fritted glass filter, and the moisture content of the filtrate was determined by Karl Fischer titration. The mass of the acetone-insoluble fraction of the filtrate was added to the mass of the char recovered from the char pot and the fluidized bed reactor. These were recorded as total char produced from each run. The acetone was evaporated under vacuum (40 °C and 61.3 kPa), and the recovered oils were analyzed for elemental composition and HHV.

Elemental composition and HHVs of the biocrude oils were determined in triplicate at Huffman Laboratories in Golden, Colorado. For the char samples, only single determinations of elements and HHVs were carried out.

2.3 Analysis of Gaseous Products
The pyrolysis gases were sampled and analyzed on an on-line HP 5890 Series II GC. Two packed columns (Porapak N and molecular sieve 13X) connected in series were used to analyze the gases. The oven was temperature programmed and a thermal conductivity detector was used.

3 Results

Elemental composition and HHVs of the switchgrass feedstocks are shown in Table 1. There is no statistically significant difference between the composition of the various feedstocks at the 95% confidence level, except the nitrogen content. The Ames switchgrass feedstock has a lower nitrogen content than did the other two feedstocks. However, the difference between the nitrogen contents of the switchgrass feedstocks from West Lafayette and Mead was insignificant.

3.1 Switchgrass Feedstocks Pyrolysis Products
Assessing the influence of plant location on the pyrolysis products of the switchgrass feedstocks was based on the following criteria: total liquids (organic liquids and pyrolysis water), gas yield, char yield, elemental composition, and HHV of the biocrude oils.

The yields of total pyrolysis liquids from feedstocks grown at all three locations (Mead, Ames, and West Lafayette) were very similar and ranged from 59% to 60% (see Table 2). However, there were significant differences between organic liquid yields

Table 1. Characteristics of switchgrass feedstocks[1]

Composition (wt%)	West Lafayette, IN	Mead, NE	Ames, IA
Carbon	44.02±0.14	44.08±0.11	43.47±0.16
Hydrogen	6.13±0.01	6.13±0..03	6.18±0.05
Oxygen[2]	44.64±0.31	44.37±0.08	44.82±0.08
Nitrogen	0.93±0.03	0.91±0.01	0.76±0
Sulfur	0.12±0	0.12±0	0.09±0.01
Ash	5.42±0.15	6.06±0.35	5.41±0.25
Glucan	33.5	33.5	32.0
Lignin	13.7	12.1	11.0
Protein	5.3	6.9	6.3
HHV(MJ/kg)	17.5±0.1	17.5±0.1	17.3±0.05

[1]Error bars are standard deviations on triplicate analysis (all data are on moisture-free biomass basis).
[2]Oxygen by difference.

from the three sites. The West Lafayette sample had a slightly lower organic liquids yield (49.9±2.0%) than the other two sites, which had average organic liquid yields of 53.2%. The biocrude oils were dark brown, viscous, single-phase, acidic liquids (pH 2.5-3.0). The char content of the oils, before filtration through a 40-60 μm filter, was 1%-2%.

The condensation train segregated the oils into a light fraction with high moisture content and a heavy fraction with low moisture content. The light fraction was condensed in the chilled water condenser. The heavy fraction was composed mostly of aerosols that formed during the pyrolysis process; these aerosols did not condense in the chilled water condenser but were captured in the electrostatic precipitator. The heavy oil fraction contained very little water and was extremely viscous and sticky. The sticky oils accumulated on the electrostatic precipitator electrodes and often led to voltage leakage and arcing. The heavy oil fraction conducted electrical current, probably because it contained char that had about 1% potassium. The oils stuck to the walls of the glass containers and, once they dried, did not dissolve readily in acetone or methanol. The glass equipment usually was cleaned with an alcoholic potassium hydroxide solution, the only solvent that could dissolve these dry oils.

The char was defined as the acetone-insoluble material recovered from the cyclone separator, fluidized bed reactor, and 40-60 μm filter. About 94-95% of the char was recovered from the cyclone separator, 1%-2% from the filtration of the oil, and the rest from the reactor. The char contained about 50% carbonaceous material, 20% oxygen, and small amounts of hydrogen (3%). Most of the ash in the original feedstock is

Table 2. Material balance and gas analysis of West Lafayette, Mead, and Ames switchgrass pyrolysis products[1]

	West Lafayette, IN	Mead, NE	Ames, IA
Mass Balance (wt.% of dry feed)			
Gas	11.5±0.8	11.8±0.3	11.0±2.0
Water[2]	9.3±2.2	6.1±0.9	7.3±0.3
Organics	49.9±2.0	53.2±0.6	53.2±0.3
Char[3]	21.8±0.4	22.9±0.6	21.2±0.2
Total	92.6±0.6	94.0±1.5	92.6±2.5
Gas Analysis (wt.% of dry feed)			
CO_2	8.5±0.4	8.6±0.2	8.0±1.9
CO	2.1±0.2	2.5±0.3	2.4±0.4
CH_4	0.2±0.1	0.2±0.1	0.3
C_2H_4	0.2±0.05	0.4	0.1
C_2H_6	0.1±0.05	0.1	0.2
C_3H_8	0.2±0.05	0.1	0.1
C_4H_8	0.3±0.15	0.3±0.1	0.3±0.1
Total	11.5±0.8	11.8±0.3	11.0±2.0

[1] The oil yields were determined gravimetrically and were not washed with acetone before the determination. Errors are standard deviations from three determinations from three pyrolysis runs (all data are on moisture-free biomass basis).
[2] Water determined by Karl Fischer titration of the biocrude oil.
[3] Char is the mixture of carbonaceous material and sequestered inorganic material recovered from the char pot, fluidized bed reactor, and acetone insoluble material recovered after filtration.

associated with this carbonaceous material; hence, the material had a high ash content (28%; see Table 3).

Char yields from the different sites showed statistically significant differences at the 95% confidence level. Feedstock from the Mead site had the highest char yield (22.9±0.6%), and the Ames site feedstock had the lowest (21.2±0.2%).

Gas yields from all three sites were similar, and differences were not statistically significant (see Table 2). Most of the gas was carbon dioxide and carbon monoxide, with small amounts of hydrocarbons. The hydrogen and water vapor contents of the gases were not analyzed. Multiple analyses of the gas fraction during each run, which lasted 2-3 h, revealed no significant differences in the composition of the gases over the

Table 3. Elemental composition and HHVs of switchgrass biocrude oils[1] and pyrolysis char

Composition (wt%)	West Lafayette, IN	Mead, NE	Ames, IA
Carbon	55.8±2.3	57.6±1.1	57.6±1.5
Hydrogen	7.4±0.6	6.8±0.2	6.9±0.1
Oxygen[2]	36.1±1.6	35.3±1.8	33.9±0.6
Nitrogen	1.24±0.04	1.40±0.14	1.25±0.10
Sulfur	0.07±0.01	0.16±0.08	0.13±0.06
Chlorine	<0.1	<0.1	0.10
Ash	<0.01	<0.01	0.16±0.07
HHV(MJ/kg)	24.3±0.8	24.6±0.8	24.6±0.8
Char Analysis (wt% of char)			
Carbon			50.45
Hydrogen			2.76
Oxygen[3]			19.72
Nitrogen			1.23
Sulfur			0.14
Ash			28.2

[1]Error bars are standard deviations of three determinations from three pyrolysis runs.
[2]Oxygen by difference. [3]Direct oxygen determination.

duration of the run, indicating that very little or no cracking of pyrolysis vapors took place on the fresh char retained in the reactor. If there were cracking on the fresh char, the carbon monoxide and carbon dioxide contents of the gases would have increased significantly [10].

The elemental compositions and HHVs of the biocrude oils appeared to be independent of the origin of the feedstock. There was a small, statistically insignificant, difference in the HHV of the oil from the West Lafayette feedstock compared to the other oils (see Table 3).

3.2 Hybrid Poplar Feedstock Pyrolysis Products

There were no statistically significant differences in the elemental compositions and HHVs of the three hybrid poplar clones at the 95% confidence level (see Table 4). However, very small differences exist between the carbon content of the NC5260 clone and the other clones.

Table 4. Elemental composition and HHVs of hybrid poplar feedstocks[1]

Composition (wt%)	DN17	DN182	NC5260
Carbon	46.43±0.03	46.74±0.05	46.90±0.05
Hydrogen	6.39±0.03	6.37±0.03	6.39±0
Oxygen[2]	48.75±0.03	48.61±0.11	49.37±0.18
Nitrogen	0.10±0.01	0.11±0.01	0.10±0.01
Sulfur	0.01±0	0.02±0	0.01±0.00
Ash	0.53±0.03	0.63±0.03	0.50±0.02
Glucan	43.7±0.5	43.9±2.3	46.2±1.7
Lignin	23.3±0.5	23.5±0.2	21.3±0.4
HHV(MJ/kg)	18.3±0.1	18.5±0	18.2±0.0

[1]Error bars are standard deviations on triplicate analysis (all data are on moisture-free biomass basis).
[2]Oxygen by difference.

Assessing the influence of clonal variation on the pyrolysis products of the hybrid poplar feedstocks was based on criteria similar to those for the switchgrass feedstocks. The yields of biocrude oils for all three poplar clones were very similar and ranged from 65% to 69% (see Table 5). The oils were dark brown, viscous, single-phase acidic liquids (pH 2.5-3.0). These oils were less sticky than the switchgrass biocrude oils. Their moisture contents were similar to those of the switchgrass biocrude oils, but the char content before filtration was about 0.5%. Unlike the switchgrass feedstocks, there was no significant difference between the organic liquid yields from the different clones.

The behavior of the poplar biocrude oils in the condensation train was similar to that of the switchgrass biocrude oils. However, the poplar biocrude oils were less viscous and less sticky than the corresponding switchgrass biocrude oils, and therefore did not accumulate on the electrostatic precipitator electrodes. Thus, no voltage leakage or arcing occurred during the pyrolysis runs. The oils were much easier to handle than the corresponding switchgrass oils; they dissolved readily in acetone and did not stick tenaciously to the glassware.

Char yields ranged from 10% to 11% (see Table 5). About 95% of the char was recovered from the cyclone separator, 0.5% from the filtration of the oil, and the rest from the reactor. The char from the reactor and the cyclone was shiny black lightweight material, whereas the sample recovered from the filtration was fine and dull black.

The char is mostly carbonaceous material, with small amounts of oxygen and hydrogen (see Table 6). Gas yields ranged from 15.6% to 17.2% (see Table 3) and were higher than those for the switchgrass feedstock. The composition of the gases was similar to that of the switchgrass feedstocks.

Elemental composition of the biocrude oils showed statistically significant differences between the clones (see Table 5). The pyrolysis oil from the NC5260 clone

Table 5. Material balance and gas analysis of hybrid poplar pyrolysis products[1]

	DN17	DN182	NC5260
Mass Balance (wt.% of dry feed)			
Gas	16.6	15.6	17.2
Water[2]	6.2	6.2	7.0
Organics	61.1	59.3	62.3
Char[3]	10.1	11.0	10.7
Total	94.0	92.1	97.2
Gas Analysis (wt.% of dry feed)			
CO_2	10.8	10.1	9.8
CO	4.5	4.8	6.5
CH_4	0.6	0.3	0.3
C_2H_4	0.2	0.1	0.1
C_3H_8	0.1	0.1	0.1
C_4H_8	0.4	0.3	0.4
Total	16.6	15.6	17.2

[1]The oil yields were determined gravimetrically and were not washed with acetone before the determination (all data are on moisture-free biomass basis).
[2]Water determined by Karl Fischer titration of the biocrude oil [3]Char is the mixture of carbonaceous material and sequestered inorganic material recovered from the char pot, fluidized bed reactor, and acetone insoluble material recovered after filtration.

feedstock had a slightly lower (but statistically significant at the 95% confidence level) HHV compared to the biocrude oils from the DN17 and DN182 clones. The DN17 and DN182 feedstock biocrude oils had similar HHVs. The differences in the HHVs are reflected in the carbon content of these biocrude oils. The NC5260 feedstock pyrolysis oil had a lower carbon content than the other hybrid poplar clone feedstock biocrude oils.

4 Discussion

4.1 Influence of Plant Location on the Pyrolysis Products of Switchgrass

The yields of pyrolysis products from switchgrass varieties grown at the Ames, Mead, and West Lafayette sites were similar except for the char yields, which showed significant differences. The Mead site switchgrass gave the highest yield of char compared to the switchgrass from the other sites, and the Ames switchgrass gave the

Table 6. Elemental composition and HHVs of hybrid poplar biocrude oils[1] and pyrolysis char

Composition (wt%)	DN17	DN182	NC5260
Carbon	55.8±0.8	55.4±0.4	53.6±1.1
Hydrogen	6.7±0.2	6.4±0.3	6.2±0.6
Oxygen[2]	37.9±0.1	38.9±0.7	41.1±1.3
Nitrogen	0.20±0.04	0.20±0.05	0.17±0.02
Sulfur	0.04±0.01	0.02±0.00	0.03±0.00
Chlorine	<0.1	<0.1	<0.1
Ash	<0.01	0.01	0.01
Higher heating value (MJ/kg)	23.2±0.3	23.0±0.2	22.0±0.5
Char Analysis (wt% of char)			
Carbon	73.80		73.90
Hydrogen	3.46		3.44
Oxygen[3]	19.16		19.31
Nitrogen	0.31		0.30
Sulfur	0.06		0.06
Ash	6.7		5.9

[1] The oils were recovered by rinsing the apparatus with acetone, filtering, and then vacuum rotary evaporation of the acetone. Error bars are standard deviations of three analyses.
[2] Oxygen by difference.
[3] Oxygen by direct method.

lowest char yield. The differences in the yields were significant at the 95% confidence level.

The differences in char yields were attributed to the protein and lignin contents of the feedstocks rather than the ash contents. Although the ash in biomass feedstocks promotes char formation [11], the ash contents of the feedstocks from the three sites were similar. It is therefore unlikely that the ash caused the differences in the char yields. The lignin and nitrogen compounds (proteins) also promote char formation [12], and are a probable cause of the observed differences in char yields. The data show that the crude protein content, which is obtained by multiplying the Kjeldahl nitrogen content by a factor of 6.25 [13], was higher in the Mead switchgrass (6.9%) than in the West Lafayette switchgrass (5.3%). Because proteins tend to denature and also react with carbohydrate decomposition products at high temperatures [14], this is a likely source of the higher char yield for this switchgrass.

The small but statistically insignificant differences in the char yields of the Ames and West Lafayette feedstocks were probably caused by the differences in their lignin and protein contents because the ash contents of these feedstocks were similar.

The difference in the yield of char from the different switchgrass feedstocks suggests that the time of harvest and degree of fertilization may play some role in the quality of herbaceous thermochemical biomass feedstocks. In temperate-climate regions, it has been observed that the yield of dry matter is directly proportional to nutrient intake [15-16], especially nitrogen. In the case of nitrogen, yields of forage increase in direct proportion to application rates of 336 to 404 kg/ha annually; however, above these values, no significant dry matter yield increases were observed. As the amount of nitrogen available to the forage grass is increased, the intake of other elements, particularly potassium, also increases.

Because nitrogen is essential for the formation of protein [15], a heavily fertilized field will produce feedstocks with higher crude protein and nitrates as well as higher ash contents. Further, the crude protein content is inversely proportional to the crude fiber content of the plant because nitrogen fertilization tends to decrease the plant's soluble carbohydrates (polysaccharides) levels and, in young growing plants, crude fiber content tends to be very low [16]. Because most of the biocrude oils originate from the structural component (crude fiber) of the feedstock, pyrolysis of feedstocks from heavily fertilized fields may result in high char yields and subsequently lower the liquid and gaseous product yields.

Additionally, the high crude protein content of the feedstock will lead to a significantly higher nitrogen content of the biocrude oils because, unlike potassium, calcium, and phosphorus, which are sequestered in the char during the pyrolysis process, only 60% of the nitrogen is sequestered [17]. This is undesirable because studies have shown that fuel-bound nitrogen is more efficiently converted to NOx than thermal NOx from nitrogen in the air [18].

The crude protein and crude fiber contents of herbaceous feedstocks such as switchgrass also are strongly influenced by season. Thus, the time of harvest of the feedstock may be a crucial factor in the production of high-quality biocrude oils. During spring, when the plant is actively growing, the crude protein content tends to be high and the fiber content low. As the herbaceous plant approaches maturity and starts to set seed, the lignin content rapidly increases and the crude fiber content increases while the crude protein content decreases. During the winter, when the grass is dormant and weathered, the crude protein content drops significantly [16]. In studies carried out at the Oklahoma Agricultural Experiment Station, the crude protein content of switchgrass dropped from 10.5% in May to 2.3% in December [16]. On the other hand, the dry matter content increased from 42.5% in May to 94.2% in December. Thus, a judicious agricultural practice can minimize the protein content and maximize fiber content for herbaceous biomass feedstocks for the production of high-quality biocrude oils. Current forage harvesting practices, which are geared toward maximum crude protein content will have to be modified to produce suitable feedstock for pyrolytic conversion processes.

The same studies at the Oklahoma Experimental Agricultural Station found that the ash content of switchgrass did not decrease as drastically as did the protein content. The ash decreased from 6.6% in May to 5.4% in December during the same year the studies for the crude protein were conducted. Thus, although the crude protein content of the feedstock can be reduced significantly by the time of harvest, the same cannot be said of ash. The high ash contents of the herbaceous feedstocks will lead to high concentrations of potassium and chlorine in the biocrude oils. These elements are detrimental to many combustion applications due to potential fouling, slagging, and hot corrosion of the alloy steel components.

The influence of the ash on the pyrolysis products is of equal importance to the overall process efficiency and economics of potential bioenergy facilities. The potassium and calcium in the ash catalyze char formation reactions, which, in combination with the nitrogen in the feedstock, will increase the char yield at the expense of the desirable liquid products. The ash-catalyzed char formation reactions level off above a total potassium and calcium content of 2.5% of the feedstock [19]. However, because the potassium and calcium concentrations of the switchgrass feedstocks (1.3%) are high, char yield on an ash-free basis is significantly higher for switchgrass than for woody biomass. Our studies have shown that potassium, calcium, and phosphorus are completely sequestered in the char, while chlorine, sulfur, and nitrogen are only partially sequestered [17]. Thus, complete removal of char from the biocrude oil will remove the harmful elements from the oil, even though the total liquid yield will decrease significantly as a result of the ash-catalyzed char formation reactions.

4.2 Influence of Clonal Variation on Hybrid Poplar Pyrolysis Products

The hybrid poplar clones showed no significant differences in the yields of various products from pyrolytic conversion; however, statistically significant differences were observed in the carbon contents (and consequently the HHVs) of the biocrude oils. The pyrolysis oil from the NC5260 poplar clone had a significantly lower HHV compared to the DN poplar clones. This can be attributed to the differences observed in the summative composition of the feedstocks (see Table 4 and ref. [8]). The NC5260 poplar clones were significantly lower in lignin (21%) and higher in cellulose content (46%) compared to the DN poplar clones (23% lignin and 43% cellulose). This difference in composition implies lower carbon content and higher oxygen content. The conversion of carbon in feedstocks to liquid products appeared to be more efficient for the DN clones than the NC clone. For the NC clone, 77% of the carbon was converted compared to 82% and 78%, respectively, for the DN17 and DN182 clones.

The extractives component of woody species tends to have higher energy content than the lignocellulosic component. Because no statistically significant differences were detected between the extractives components of the poplar clones [8], the differences in the HHVs of the biocrude oils were attributed solely to the differences in the lignin/cellulose content of the original feedstocks.

4.3 Merits of Switchgrass and Hybrid Poplar as Thermochemical Feedstocks

Although there were significant differences in the chemical composition and physical properties of the hybrid poplar and switchgrass biocrude oils, their HHVs were similar. The carbon conversion efficiency for switchgrass grown at the three different locations ranged from 75 to 80%, which is comparable to 77%-82% for the hybrid poplar clones. The hybrid poplar clones had higher liquid yields and lower char yields compared to the switchgrass from the three sites. The switchgrass feedstocks, because of their higher nitrogen and ash contents, produced higher char than the hybrid poplar clones on an ash-free basis. Further, because char particles are not very efficient at capturing nitrogen and sulfur compounds, the nitrogen and sulfur contents of the switchgrass biocrude oil was 6-7 and 2-5 times greater, respectively, than those found in hybrid poplar biocrude oils. Ash levels in the two oils are similar because they were each subjected to the same separation process, which reduced the ash concentration to the same level.

Although the energy content and the carbon conversion efficiency of the two feedstocks were comparable, the hybrid poplar biocrude oil appears to be a better quality oil. It contains lower concentrations of both sulfur and nitrogen, which are potential sources of NO_x and SO_x during combustion.

Because all the ash in the biomass are sequestered by the char during pyrolysis, the high ash content of the switchgrass feedstock implies that the average ash content of the pyrolysis chars will always be high (>20%); with their low carbon contents (50%), the HHVs of the chars will always be low. Alternatively, the average ash content of the hybrid poplar pyrolysis chars will always be low (<10%) and, due to their high carbon contents (>70%), the HHVs will be relatively high. The hybrid poplar chars can therefore find ready application in the production of charcoal briquettes, whereas the switchgrass chars exceed the upper limit of 12% ash in charcoal [20]. The high concentration of potassium, calcium, and phosphorus and the high porosity of the switchgrass pyrolysis chars, may make them marketable as soil conditioners or fertilizer.

Thus, from the pyrolytic conversion process point of view, hybrid poplars appear to be better feedstocks than switchgrass. However, other factors such as production cost, handling, and drying may make switchgrass a more attractive feedstock than hybrid poplar.

5 Conclusions

For all practical applications, it appears that the environment does not significantly affect the properties of biocrude oils generated from switchgrass. However, for the hybrid poplars studied here, clonal variation had a significant effect on the HHVs of the biocrude oils. Feedstocks with high cellulose/low lignin content (such as NC5260) had lower HHVs. The overall oil yield was similar for all poplar clones. Biocrude oil and gas yields were not significantly affected by the environment, but the small differences in the crude protein and lignin contents of the switchgrass due to the differing plant environment did affect char yields. The protein, ash, and lignin contents appear to promote char formation reactions and caused increased char yields. Because most protein in the switchgrass appeared to be converted into char, it had very little influence on the properties of the oils.

Agricultural practices may be an important factor in the production of herbaceous thermochemical feedstocks. Low-protein, high-fiber switchgrass varieties may be more suitable than high-protein switchgrass for pyrolytic conversion processes. Overall, hybrid poplar feedstocks appear to be more suitable, technically, than switchgrass feedstocks for pyrolytic conversion processes.

6 Acknowledgements

The authors acknowledge N. Hinman, NREL Biofuels Program Manager, and E. Chornet, H.L. Chum, and R.P. Overend for their continued interest in and support of the Biofuels program. We also thank the U.S. Department of Energy, Biofuels Systems Division, for funding the project.

7 References

1. Agblevor, F.A., Besler, S., and Wiselogel, A.E. (1995). Fast pyrolysis of stored biomass feedstocks. *Energy & Fuels*, Vol. 9, No. 4, pp. 635-640.

2. Agblevor, F.A., Evans, R.J., and Johnson, K.D. (1994). Molecular-beam mass-spectrometric analysis of lignocellulosic materials I. Herbaceous biomass. *J. Anal. Appl. Pyrol.*, Vol. 30, pp. 125-144.

3. Agblevor, F.A. and Boocock, D.G.B. (1989). The origin of phenol produced in the rapid hydrothermolysis and alkaline hydrolysis of hybrid poplar lignins. *J. Wood. Chem. Tech.*, Vol. 9, No.2, pp. 167-188.

4. Piskorz. J., Scott, D.S., and Radlein, D. (1988). Composition of oils obtained by fast pyrolysis of different woods, in *ACS Symposium Series 376: Pyrolysis Oils from Biomass, Producing, Analyzing, and Upgrading*, (eds. E.J. Soltes, and T.A. Milne), American Chemical Society, Washington D.C., pp. 167-178.

5. Palm, M., Piskorz, J., Peacock, C., Scott, D.S., and Bridgwater, A.V. (1993). Fast pyrolysis of sweet sorghum bagasse in a fluidized bed, in *Proceedings Vol. II, First Biomass Conference of the Americas: Energy, Environment, Agriculture, and Industry*. National Renewable Energy Laboratory, Golden, pp. 947-963.

6. Desbene, P.-L., Desmazieres, B., Lange, C., and Basselier, J.-J. (1992). Pyrolytic Behavior of lignin and cellulose in various wood species using direct pyrolysis MS-MS., in *Biomass for Energy, Industry and Environment, 6th E.C. Conference*, (eds. G. Grassi, A. Collina, H. Zibetta), Elsevier Applied Science, New York, pp. 776-781.

7. Hallgren, A. and Wanzl, W. (1993). Screening of pyrolysis behavior of different biomass. *ACS Division of Environmental Chemistry Preprints*, Vol. 33, No.2, pp. 48-51.

8. Davis, M.F., Johnson, D.K., Deutch, S., Agblevor, F., Fennell, J. and Ashley, P. (1995). Variability in the composition of short rotation woody feedstocks, in *Proceedings, Second Biomass Conference of the Americas: Energy, Environment, Agriculture, and Industry*. NREL/CP-200-8098, DE95009230, National Renewable Energy Laboratory, Golden, pp. 216-225.

9. Johnson, D.K., Ashley, P.A., Deutch, S.P., Davis, M.F., Fennell, J.S., Wiselogel, A. (1995). Compositional variability in herbaceous energy crops, in *Proceedings, Second Biomass Conference of the Americas: Energy, Environment, Agriculture, and Industry*. NREL/CP-200-8098, DE95009230, National Renewable Energy Laboratory, Golden, pp. 267-277.

10. Boroson, M.L., Howard, J.B., Longwell, J.P., Peters, W.A. (1989). Heterogeneous cracking of wood pyrolysis tars over fresh wood char surfaces. *Energy & Fuels,* Vol. 3, pp. 735-740.

11. DeGroot, W.F., and Shafizadeh, F. (1984). Influence of exchangeable cations on the Carbonization of Biomass. *J. Anal. Appl. Pyrol.,* Vol. 6, No. 3, pp. 217-232.

12. Agblevor, F.A., Rejai, B., Evans, R.J., and Johnson, K.D. (1993). Pyrolytic analysis and catalytic upgrading of lignocellulosic materials by molecular beam mass spectrometry, in *Energy from Biomass and Wastes XVI,* (ed. D.L. Klass). Institute of Gas Technology (IGT), Chicago, pp. 767-795.

13. Fisher, D.S., Burns, J.C., and Moore, J.E. (1995). The nutritive evaluation of forage, in *Forages Volume 1: An Introduction to Grassland Agriculture (Fifth edition),* (eds. R.F. Barnes, D.A. Miller, C.J. Nelson), Iowa State University, Ames, pp. 105-116.

14. Theander, O. and Nelson, D.A. (1989). Aqueous high temperature transformation of carbohydrates relative to utilization of biomass. *Adv. Carbohydrate Chem. Biochem.,* Vol. 46, pp. 273-326.

15. Miller, D.A. and Reetz, H.F. (1995). Forage fertilization, in *Forages Volume 1: An Introduction to Grassland Agriculture (Fifth edition),* (eds. R.F. Barnes, D.A. Miller, and C.J. Nelson), Iowa State University, Ames, pp. 71-88.

16. Waller, G.R., Morrison, R.D., and Nelson, A.B. (1972). Chemical composition of native grasses in Central Oklahoma from 1947-1962. *Oklahoma State University, Agricultural Experiment Station, Bulletin B-697.*

17. Agblevor, F.A. and Besler, S. (1995). Inorganic compounds in biomass feedstocks I: Effect on quality of fast pyrolysis oils. *Energy and Fuels,* Vol 10, No. 2, pp. 293-298.

18. Moses, C.A. and Bernstein, H. (1994). *Impact of the use of biomass-derived fuels in the gas turbines for power generation.* NREL/TP-430-6085; UC Category: 247, DE94000261. National Renewable Energy Laboratory, Golden.

19. Agblevor, F.A., Besler, S., and Evans, R.J. (1995). Inorganic compounds in biomass feedstocks:Their role in char formation and effect on the quality of fast pyrolysis oils. In *Proceedings, Biomass Pyrolysis Oil Properties and Combustion Meeting,* (ed. T.A. Milne). NREL-CP-430-7215, National Renewable Energy Laboratory, Golden, pp. 77-89.

20. Johnson, D.A. et al. (1995). Economic development through biomass systems integration in Northeastern Kansas. In *Proceedings, Biomass Pyrolysis Oil Properties and Combustion Meeting,* (ed. T.A. Milne). NREL-CP-430-7215, National Renewable Energy Laboratory, Golden, pp. 329-337.

EXTRACTION OF HEMICELLULOSES FROM POPLAR USING A TWIN-SCREW REACTOR: INFLUENCE OF THE MAIN FACTORS

Fractionation of biomass using a twin-screw reactor

S. NDIAYE and L. RIGAL
Laboratoire de Chimie Agro-Industrielle
Ecole Nationale Supérieure de Chimie, 31077 Toulouse, France
C. GOYETTE and P.F. VIDAL
Department of Chemical Engineering
University of Sherbrooke, Sherbrooke, J1K 2R1 Canada

Abstract

Fractionation of poplar wood chips, *Populus tremuloides*, was carried out using a modified Clextral BC 45 twin-screw reactor. The effect of the main factors: temperature (40-70°C), liquid flow rate (17.5-47.5 kg.h^{-1}), NaOH concentration (3-7% w/w), solid flow rate (1.5-4.5 kg.h^{-1}), and screw rotation speed (100-250 rpm), on the solubilization of the hemicelluloses was studied according to the Doehlert's method for the design of experiments. The main results indicate that the reactor behaves either as a batch reactor when the sodium hydroxide concentration in the extracting solution is above 5%: yield of the solubilized hemicelluloses increases as the NaOH concentration is increased, or behaves as a Thermo-Mechano-Chemical (TMC) reactor when this concentration is under 4%: the yield of solubilized hemicelluloses increases as the NaOH concentration is reduced.

Keywords: Design of experiments, Fractionation, Hemicelluloses, Poplar, Populus tremuloides, Twin-screw reactor.

1 Introduction

The growing shortage in softwood resources is causing the pulp and paper industry to explore the possibility of new sources of the raw material, such as annual plants and short rotation forestry, for the production of fiber according to new fractionation processes. Twin-screw extruders, used as pulping reactors, have been studied extensively and the developments have led to many industrial applications [1]. As with other TMC pulping processes, the use of NaOH induces the solubilization of hemicelluloses. However, these polysaccharides can be extracted continuously - in the form of a liquid solution - as long as the twin-screw reactor is equipped with a filtration unit located at the end of the barrel, facing the reverse screw element.

A Clextral BC45 modified in this manner was used to study the effect of: i) the barrel temperature, ii) the input liquid flow rate, iii) the NaOH concentration in the input liquid stream, iv) the input solid (wood chips) flow rate, and v) the screw speed, on i) the concentration of hemicelluloses in the filtrate, ii) the yield of hemicelluloses, iii) the production rate of hemicelluloses, and iv) the thrust pressure.

The effect of these five factors on the four responses was studied according to the Doehlert's method for the design of experiments. Results based on the analysis of the isoresponse curves will be discussed thereafter.

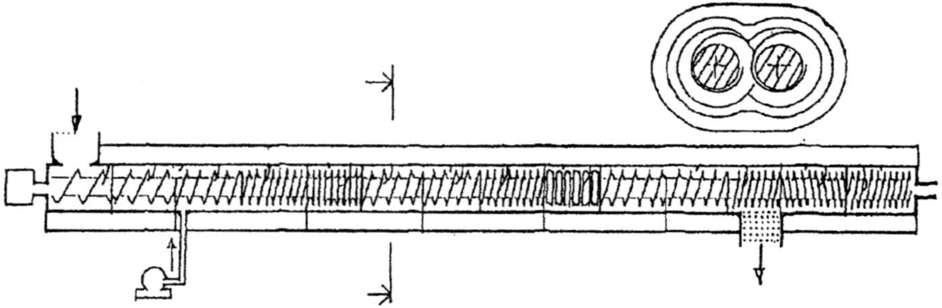

	A		B		C			D		E		F		G	
TYPE	T2F	C2P	C2F	C2F	B	C2F	C2F	C2F	C2F	DM	C2F	C2F	C2F	C2FC	C2F
PITCH (mm)	66	50	33	25		33	33	33	25		33	33	25	-25	25
LENGTH (mm)	100	100	100	100	100	100	100	100	100	100	100	100	100	100	100
MIXING	+	+	+	+	++++	+	+	+	+	++	+	+	+	+++	+
SHEARING	+	+	+	+	++	+	+	+	+	+++	+	+	+	+++	+
CONVEYING	+++	+++	+++	+++	0	+++	+++	+++	+++	+	+++	+++	+++	---	+++
COMPRESSION	AXIAL	AXIAL	AXIAL	AXIAL	RADIAL +++	AXIAL	AXIAL	AXIAL	RADIAL + AXIAL	RADIAL + AXIAL	AXIAL	AXIAL	AXIAL	AXIAL +++	AXIAL

T2F : Twin Flight Trapezoidal   **C2F** : Twin Flight Self Wiping   **B** : Bilobal Paddles   **DM** : Eccentric Monolobal Paddles
C2FC : Twin Flight Reverse Pitch Self Wiping
+ : very low positive action ; **++** : low positive action ; **+++** : average positive action ; **++++** : high positive action ;
---- : high negative action ; **0** : none

Fig. 1. Clextral BC 45 twin-screw reactor: configuration and screw profile.

757

2 Material and methods

2.1 Feed stock

Debarked chips of *Populus tremuloides* from Quebec's Eastern Townships region were ground to an average size of 15x2x2 mm using an Electra type VS1 industrial grinder.

The composition of the wood chips was as follows: alpha-cellulose: 44.92% (ASTM D-1103) ; Klason lignin: 17.67% (ASTM D-1106), pentosans: 17.82% (ASTM D-1787); and ash: 019% (ASTMD-1106). The moisture content of the chips during the experimental batch and extrusion programs was 41% wt [2].

2.2 Twin-screw extruder

Experiments were performed using a modified Clextral BC45 (Framatome, France) twin-screw extruder (Fig 1). Four sections (B, C, E, and F), heated by induction belts and cooled by water circulation, form the 1.4 m long barrel. All experiments were carried out with a reverse screw element which has peripheral slots grooved in the screw flight for leakage flow. Conical holes (1 mm entry, 2 mm exit) form the filter element added at the end of the barrel in order to extract the liquid phase from the slurry. The die consists of a 25 mm long cylindrical hole, 12 mm in diameter .

2.3 Recovery and analysis of the solubilized hemicelluloses

Liquid filtrates from the twin-screw reactor were treated as follows in order to recover the hemicelluloses: centrifugation, acidification to pH 5, precipitation with ethanol, filtration, and, finally, washing with acetone and air drying [2].

The concentration of hemicelluloses in the liquid filtrate at the twin-screw reactor exit, C_{HC}, is calculated by weighing the amount of organic soluble matter in an aliquot.

The yield of hemicelluloses, R%, and the production rate, P, are expressed as:

$$R\% = \frac{\text{weight of hemicelluloses precipitated}}{\text{weight of wood dry matter introduced x 0.1782}} \text{ x } 100$$

$$R\% = \frac{C_{HC} \cdot L_{output}}{S_{input} \text{ x } 0.1782} \text{ x } 100$$

$$P = \frac{C_{HC}}{L_{output}} \text{ (kg / h)}$$

2.4 Design of experiments.

Experimentation was carried out according to a Doehlert's experimental design. The experimental domain is represented in Table 1.

Table 1. Experimental domain for the study of the effect of the main factors.

Studied factors		Actual variable		Normalized variable [1]
Symbol	Definition	Center U_I^0	Variation ΔU_I	
T	Reactor temperature (°C)	55	15	X1
L_{input}	Input liquid flow rate (kg/h)	32.5	15	X2
[NaOH]	Sodium hydroxide concentration (w %)	5	2	X3
S_{input}	Input solid flow rate (kg/h)	3	1.5	X4
SS	Screw rotation speed (rpm)	175	75	X5

$$(1) \quad X_i = \frac{U_i^0 - U_i}{\Delta U_i}$$

The NEMROD software [3] was used for the determination of the coefficients of the second order polynomial equation of the model as well as for the printing of the isoresponse curves.

3 Results and discussion

The five factors, [T], [L_{input}], [NaOH], [S_{input}] and L, were studied according to the method of Doehlert for the design of experiments. The resolution of the system allows the determination of the coefficients b_i, b_{ij} of a second order polynomial model which links the studied response, R_n, to the normalized values, X_j, of the four factors:

$$R_n = b_0 + b_1 X_1 + b_2 X_2 + b_3 X_3 + b_4 X_4$$
$$+ b_{11} X_1^2 + b_{22} X_2^2 + b_{33} X_3^2 + b_{44} X_4^2 +$$
$$b_{12} X_1 X_2 + b_{13} X_1 X_3 + b_{14} X_1 X_4 + b_{23} X_2 X_3 + b_{24} X_2 X_4 + b_{34} X_3 X_4$$

with $X_i = \dfrac{U_i^0 - U_i}{\Delta U_i}$, X_i: normalized variable, U_i: actual variable, U_i^0: actual value at the center of the experimental domain, and ΔU_i: variation of the variable U_i.

The following conclusions can be drawn from the analysis of the isoresponse curves predicted by the model for the experimental domain:

- An increase in the solid flow rate, S_{input}, (wood chips feeding) and a decrease in the liquid flow rate, L_{input}, (NaOH extracting solution) result in an increase in the concentration of hemicelluloses in the filtrate (Figure 2).

- At high solid flow rates, an increase in L allows a slight increase in the production rate of hemicelluloses, P (Figure 3).

- However, the higher yields of hemicelluloses, R%, are obtained essentially at low solid flow rates (Figure 4), the liquid flow rate, L_{input}, being of little effect, if any. This last result shows that the liquid/solid separation (filtration) remains efficient, whatever the L_{input}/S_{input} ratio, since the quantity of hemicelluloses in the solid stream (processed fiber) does not vary.

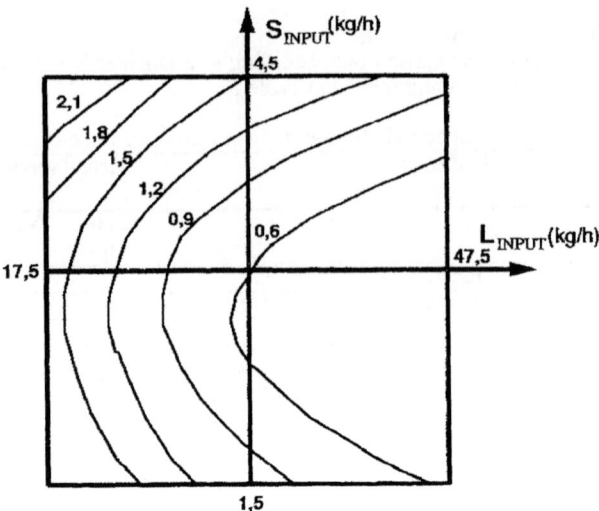

Fig. 2: Isoresponse curves for the concentration of hemicelluloses in the filtrate, C_{HC}, as a function of the input liquid flow rate, L_{input} and the input solid flow rate, S_{input}. Section of the experimental domain at $X_1=0$ (T=55 °C), $X_3=0$ (NaOH=5%) and $X_5=0$ (SS=175 rpm).

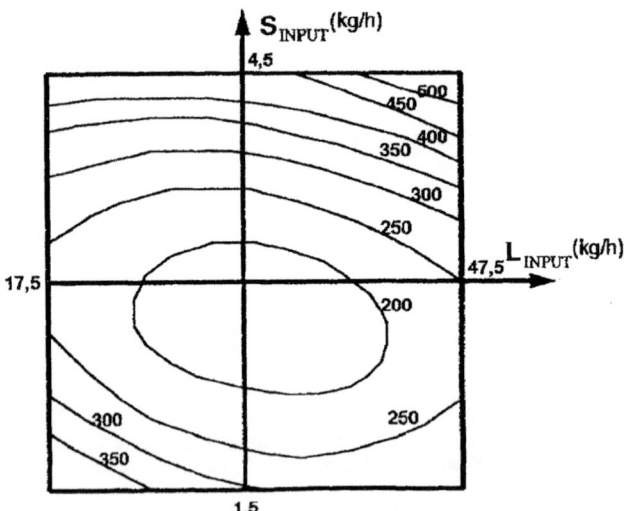

Fig. 3: Isoresponse curves for the production rate of hemicelluloses , P, as a function of the input liquid flow rate, L_{input} and the input solid flow rate, S_{input}. Section of the experimental domain at $X_1=0$ (T=55 °C), $X_3=0$ (NaOH=5%) and $X_5=0$ (SS=175 rpm).

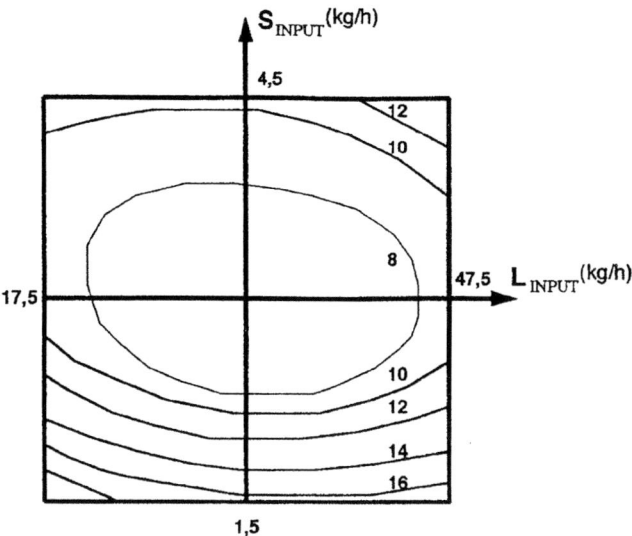

Fig. 4: Isoresponse curves for the yield of hemicelluloses , R%, as a function of the input liquid flow rate, L_{input} and the input solid flow rate, S_{input}. Section of the experimental domain at $X_1=0$ (T=55 °C), $X_3=0$ (NaOH=5%) and $X_5=0$ (SS=175 rpm).

- At high solid and liquid flow rates, a decrease in the screw rotation speed appears to be favorable to the concentration, C_{HC}, the yield, R%, and especially the production rate, P, of hemicelluloses (Figure 5). The decrease in the residence time of both the liquid and the solid due to the increase in their respective flow rates is therefore offset by the increase in their residence times due to a slower rotation speed.

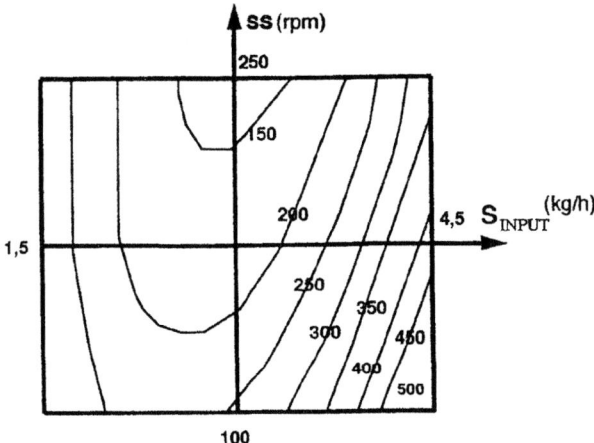

Fig. 5: Isoresponse curves for the production rate of hemicelluloses , P, as a function of the input solid flow rate, S_{input}, and the screw rotation speed, SS. Section of the experimental domain at $X_1=0$ (T=55 °C), $X_2=0$ ($L_{input}=32.5$ kg/h) and ($X_3=0$ (NaOH=5%).

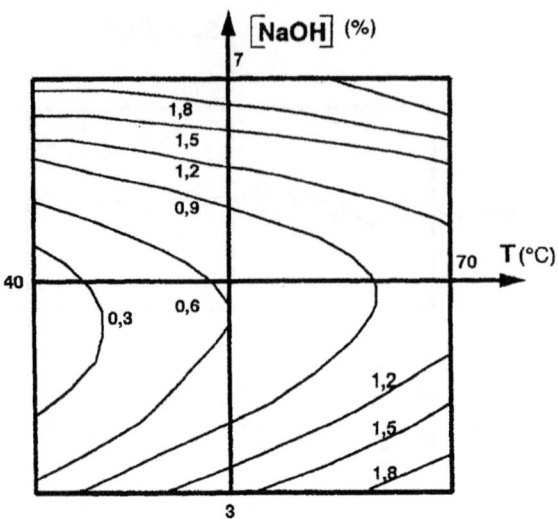

Fig. 6: Isoresponse curves for the concentration of hemicelluloses in the filtrate, C_{HC}, as a function of the temperature, T, and the sodium hydroxide concentration, [NaOH]. Section of the experimental domain at $X_2=0$ ($L_{input}=32.5$ kg/h), $X_4=0$ ($S_{input}=3$ kg/h) and $X_5=0$ (SS=175 rpm).

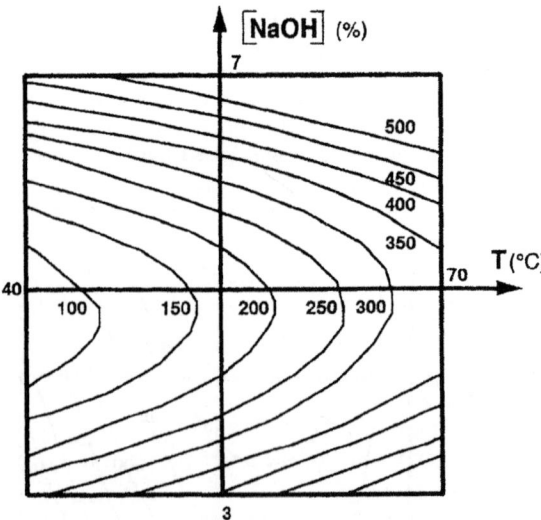

Fig. 7: Isoresponse curves for the production rate of hemicelluloses , P, as a function of the temperature, T, and the sodium hydroxide concentration, [NaOH]. Section of the experimental domain at $X_2=0$ ($L_{input}=32.5$ kg/h), $X_4=0$ ($S_{input}=3$ kg/h) and $X_5=0$ (SS=175 rpm).

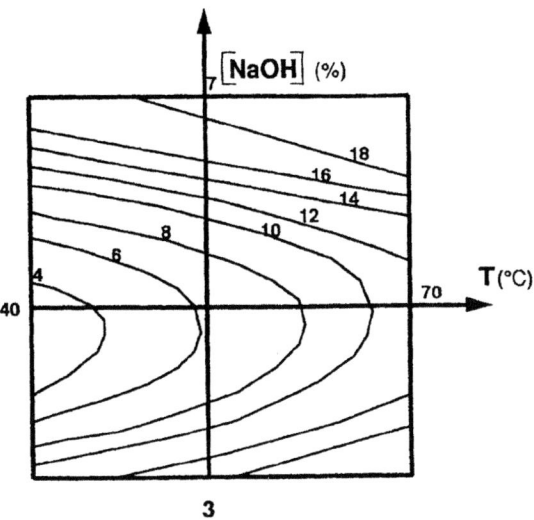

Fig. 8: Isoresponse curves for the yield of hemicelluloses, R%, as a function of the temperature, T, and the sodium hydroxide concentration, [NaOH]. Section of the experimental domain at $X_2=0$ ($L_{input}=32.5$ kg/h), $X_4=0$ ($S_{input}=3$ kg/h) and $X_5=0$ (SS=175 rpm).

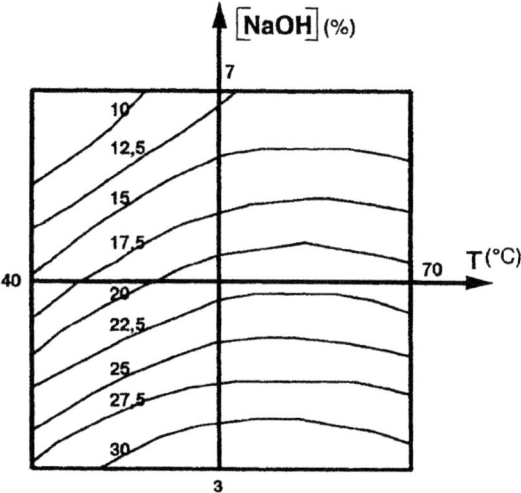

Fig. 9: Isoresponse curves for the thrust pressure, X, as a function of the temperature, T, and the sodium hydroxide concentration, [NaOH]. Section of the experimental domain at $X_2=0$ ($L_{input}=32.5$ kg/h), $X_4=0$ ($S_{input}=3$ kg/h) and $X_5=0$ (SS=175 rpm).

- As it has already been shown in the case of the extraction of hemicelluloses in a batch reactor ( ), an increase in the reaction temperature, T, favors the concentration, C_{HC}, the yield, R, and the production rate, P, of hemicelluloses (Figures 6,7 and 8). Moreover, since the three studied responses present a minimum located between 4 and 5% NaOH, whatever the temperature, T, the values of the NaOH concentration of the extracting solution clearly define two operating domains for the twin-screw reactor:

- *for NaOH concentrations lower than 4%*, the thrust pressure , which is a measure of the axial counter-pressure on the screws shafts due to the material processing, increases as the NaOH concentration is diminished (Figure 9). As well, the dry matter content of the solid extrudate which is a function of the efficiency of the squeezing effect at the counter-screw element and, as a consequence, of the L_{output}/S_{output} (hemicelluloses-rich liquor/processed fiber) separation, varies in the same way. The mechanical action due to the screws elements becomes more significant at lower NaOH concentrations and favors the accessibility of the hemicelluloses: the extraction yield increases despite the decrease in the sodium hydroxide concentration;

- *for NaOH concentrations above 5%,* the processed slurry is more fluid and the twin-screw reactor behaves as a stirring reactor for which an increase in the NaOH concentration favors the extraction of the hemicelluloses. According to the isoresponses curves, concentration, C_{HC}, yield, R%, and production rate, P, of 1.8 g/L, 16 %, and 500 g/h, respectively, could be obtained at a temperature of 55 °C and a NaOH concentration of 6.5%. Under these conditions, the residence time of the solid is 30 times shorter, the L_{input}/S_{input} ratio is 3 times lower (and consequently the NaOH/S_{input} ratio is lower) as compared to a batch reactor. These predictions from the model were confirmed experimentally.

4 Conclusion

The statistical analysis according to the method of Doehlert of the main operating parameters governing the extraction of hemicelluloses from poplar using a TMC twin-screw reactor allowed us to clearly demonstrate that the reactor behavior depends mainly on the NaOH concentration within the extracting solution.

5 References

1. De Choudens, C.; Angelier, R. and Combette, P.H. (1984) Pâtes mécaniques de résineux. Pâtes chimicomécaniques de feuillus. Nouveau procédé de fabrication. *Revue ATIP*, Vol. 38, No 8, pp 405-416.
2. N'Diaye, S.; Larocque, P; Vidal, P.F. and Rigal, L (1996) Extraction of hemicelluloses from poplar, Populus tremuloides, using an extruder-type twin-screw reactor: a feasibility study. *Biomass Bioengineering*, in press.
3. Mathieu, D. and Phan-Tan-Luu, R. (1992) *New Efficient Methodology for Research using Optimal Design (NEMROD) Software*, LPRAI, Centre St-Gérome, University of Aix-Marseille, France.

Acknowledgments
Financial support of Ministère des Affaires internationales-Direction générale France (Québec) and Ministère des Affaires étrangères (France) within the France-Québec collaborative project "Fractionnement des résidus lignocellulosiques et conversion des hémicelluloses" is gratefully acknowledged.

HOT LIQUID WATER PRETREATMENT OF LIGNOCELLULOSICS AT HIGH SOLIDS CONCENTRATIONS
Aqueous pretreatment of lignocellulosics

S.G. ALLEN, M.J. SPENCER and M.J. ANTAL, JR.
Hawaii Natural Energy Institute and the Department of Mechanical Engineering,
University of Hawaii at Manoa, Honolulu, USA
M.S. LASER and L.R. LYND
Thayer School of Engineering, Dartmouth College, Hanover, USA

Abstract
Rapid (120 s), batch, hot-liquid-water pretreatment of biomass (up to 30 g) at high solids concentrations has been studied using feed materials representative of woody and herbaceous species, namely aspen and sugar-cane bagasse. Although the current pretreatment apparatus was not designed to minimize water consumption, a mean reaction temperature of 220 °C was still achieved at solids concentrations of up to 18 %. The yield of pretreated biomass was insensitive to the mean reaction temperature over the range studied (205 to 220 °C), but increased slightly with increasing solids concentration (ie. fractionation decreased). A minimum of 10 % of the biomass was fractionated into the hydrolyzate at 18 % solids. After pretreatment at 9 % solids concentration, less than 10 % of the hemicellulose content of the feed material remained within the pretreated lignocellulosics. The production of possible inhibitors such as furfural and hydroxymethyl furfural accounted for less than 3 % of the original carbohydrates. Initial results indicate that the pretreated biomass can be readily converted to ethanol by SSF. A device capable of pretreating 1 kg of biomass at high solids concentrations is now being designed and constructed.
Keywords: Pretreatment, lignocellulosics, liquid water, solids concentration, SSF

1 Introduction

Pretreatment of some kind is needed for bioconversion of naturally-occurring lignocellulosics to ethanol. Otherwise, the hydrolyzing enzymes do not have sufficient access to the cellulosic fibers to produce fermentable sugars at high rates or yields. For a pretreatment process to be practical, however, the pretreated lignocellulosics and hydrolyzate must be available in a concentrated form. Pretreatment at high solids concentration is needed, in part, to prevent the ethanol produced by any subsequent bioconversion step from being too dilute to recover economically. The size of the pretreatment equipment and operating costs are also reduced by avoiding dilute process streams.

A variety of pretreatment processes are available [1]. A particularly simple pretreatment process consists of a brief exposure of lignocellulosics to hot, liquid water at elevated temperature and pressure. A rapid (45 s to 4 min.) immersed percolation using only hot (190 to 230 °C), compressed (P > P_{sat}), liquid water [2] was found to be a highly effective pretreatment at low solids concentrations. A range of lignocellulosics pretreated in this way were readily converted to ethanol by simultaneous saccharification and fermentation [3].

In the absence of experiment, it was not clear if the efficacy of the dilute hot-liquid-water pretreatment could be retained at high solids concentrations. Specifically, we wanted to know the effect of increasing the solids concentration on important process parameters such as the yield of pretreated lignocellulosics and the reactivity of the fiber. The yield of pretreated lignocellulosics indicates the level of fractionation between the more readily-soluble components (hemicellulose and lignin) and the relatively insoluble cellulose fibers. Fractionation of these components is desirable in that it increases the accessible surface area for bioconversion of cellulose to ethanol. Fiber reactivity is increased as a result. Indeed, a recent review of the literature suggests that increasing the accessible surface area is a key determinant of pretreatment processes [1].

2 Experimental

2.1 Feed materials
Whole sugar-cane bagasse was obtained from Hawaiian Commercial & Sugar Company, Puunene, HI. Except where noted, the bagasse was depithed prior to use. Aspen chips (smallest dimension 3 mm) were supplied by the United States Forest Products Laboratory, Madison, WI.

2.2 Pretreatment
The feed materials were pretreated "as is" by a batch, immersed percolation in hot, compressed, liquid water. Typically, 15 to 30 g (oven dry basis) of biomass was packed within a perforated metal cannister and pretreated using a custom built 250 mL reactor [2]. After loading and pressurizing the preheated reactor (walls at 280 to 300 °C), compressed liquid water preheated to between 235 and 265 °C was immediately displaced into this vessel. Enough hot, liquid water was displaced to fully immerse the cannister. The reaction pressure (6 to 8 MPa) was sufficient to prevent formation of an aqueous vapor phase. After 2 minutes at temperature, the hydrolyzate was drained from the reactor (still pressurized). Subsequent flushing of the reactor cold water cooled the lignocellulosic residue and removed any residual solubilized material. After depressurizing the apparatus, the cannister was removed and the lignocellulosic residue recovered. The solids concentration achieved was defined as the mass of the feed material pretreated (oven dry basis) expressed as a percentage of the mass of the

hydrolyzate. The reaction temperature was taken as the mean temperature of the water within the reactor inlet and outlet connections.

The yield of pretreated lignocellulosic was determined from the amount of the pretreated feed material recovered from the liquid products. To that end, aliquots (4 to 50 mL) of the hydrolyzate and flush were evaporated to dryness. The mineral content of the tap water used for pretreatment was corrected for by also periodically evaporating samples of the feedwater to dryness. This correction typically lowers the yield of pretreated lignocellulosics by no more than 3 % [4]. The yield of pretreated lignocellulosics was not determined by drying the solid product, as in earlier work [2], because drying adversely affects the subsequent bioconversion to ethanol.

The hemicellulose content of the feed material and pretreated lignocellulosics was determined by quantitative saccharification [5,6]. A representative sample was first dried to less than 20 % moisture and comminuted to -40 mesh. Duplicate subsamples of the comminuted lignocellulosic (0.5 g each) were dried @ 105 °C for 6 h to establish its moisture content. Approximately 200 g of this powder was soaked in 2 mL of 72 % sulfuric acid at 30 °C for 1 h with intermittent stirring. This slurry was then diluted with 54 mL of water and heated for 1 h at 120 °C. A filtered aliquot of the hydrolyzate was analyzed by high-performance liquid chromatography (HPLC). Trifluoracetic acid (0.01 M) was delivered at a flowrate of 0.5 mL/min to a Bio-Rad Aminex HPX-87H column heated to 60 °C and connected to a differential refractometer. The hemicellulose content of the substrate was calculated from a composite HPLC peak for xylose, mannose and galactose plus the arabinose peak, using the loss factors of Mok and Antal [7].

The furfural and hydroxymethyl furfural content of selected hydrolyzates was also determined using the HPLC methodology described above. An aliquot of fresh hydrolyzate was filtered and analyzed directly. The amount of hemicellulose and cellulose degraded to furfural and hydroxymethyl furfural during pretreatment was calculated from their HPLC peaks using multiplicative factors of 1.4 and 1.3 respectively [8].

2.3 Bioconversion

Batch simultaneous saccharification and fermentation of the pretreated lignocellulosics was carried out in a peptone (20 g/L)/yeast extract (10 g/L) medium at 37 °C in shaken 250 mL serum vials. The initial solids concentration corresponded to approximately 20 g/L cellulose. Cellulase (Genencor Cytolase) at 12 to 17 FPU/g was supplemented with β-glucosidase (Novozyme) at a 5:1 activity ratio relative to the cellulase. The fermenting organism was D_5A, supplied by the National Renewable Energy Laboratory, Golden CO. For selected samples, the peptone/yeast extract was replaced with a corn-steep-liquor medium supplemented with β-glucosidase (Novozyme) at a 1:1 activity ratio relative to the cellulase. Conversion based on ethanol production was expressed as a percentage of the theoretical ethanol yield, assuming a theoretical ethanol yield of 0.51 g ethanol/g cellulose. Conversion based on cellulose was determined by taking the percentage of the final cellulose content relative to the initial cellulose content, as determined by quantitative saccharification [3].

3 Results and discussion

Lignocellulosics (aspen and sugar-cane bagasse) were pretreated by an immersed percolation in hot, liquid water at high solids concentrations (dry biomass feed/hydrolyzate w/w, greater than 5 %). Although the pretreatment apparatus used was originally designed (and used) for "proof of concept" studies of hot, liquid water pretreatment under dilute conditions [2,3], a mean reaction temperature of 220 °C was achieved at solids concentrations of up to 18 %. This temperature was realized by: increasing the temperature of the liquid water input from 235 °C to as high as 265 °C, increasing the reactor wall temperatures from 260 °C to between 280 and 300 °C and minimizing the thermal mass of the cannister used to load the feed material. The solids concentration was increased by: increasing the amount of feed material pretreated/batch, only displacing enough hot, liquid water into the reactor to fully immerse the feed material and flashing the hydrolyzate while still at temperature (i.e. before the reactor contents were cooled by flushing it with cold water).

Increasing the solids concentration caused a small change in the yield of pretreated lignocellulosics, as evident from Fig. 1. Indeed, with the exception of whole bagasse (pith included), the standard deviation of the yield measurement for each lignocellulosic over the entire dataset is less than twice the expected deviation obtained from multiple measurements made under the same experimental conditions ($\sigma_x = 3$ % [4]). In the case of whole bagasse, the decrease in the fractionation achieved with increased solids concentration results from the pith. This fine powder simply raised the packing density of the feed material within the cannister enough to hinder heat and mass transfer, raising the yield of pretreated lignocellulosics. In all cases, however, significant fractionation of the lignocellulosics was still possible. The minimum amount of feed material solubilized (measured by the reduction in the yield of pretreated lignocellulosics) was 10 % at the highest solids concentration studied (18 %).

As implied by the small change in the yield of pretreated lignocellulosics with solids concentration, the ability of an immersed percolation in hot, liquid water to readily solubilize hemicellulose was retained at high solids concentrations. After pretreatment at 9 % solids concentration, less than 10 % of the hemicellulose content of the feed material remained within the pretreated lignocellulosics from both aspen and bagasse. Admittedly, the presence of residual hemicellulose does represent a decrease in performance. When operated as a dilute system, complete removal of the hemicellulose was achieved [2,7].

Excellent fiber reactivity was also retained at high solids concentrations. Over 90 % of the cellulose in pretreated bagasse and aspen was converted to ethanol by simultaneous saccharification and fermentation, as shown in the data summarized in Table 1. Of particular note is the finding that the leaner, less-expensive, corn-steep-liquor fermentation medium performed as well as the yeast extract/peptone medium with respect to fiber conversion. With the yeast extract/peptone fermentation medium, 90 % of the final conversion was achieved within 3 days. Understandably, fermentations conducted in the much leaner corn-steep-liquor medium did take longer to complete (300 h for 90 % of the final conversion).

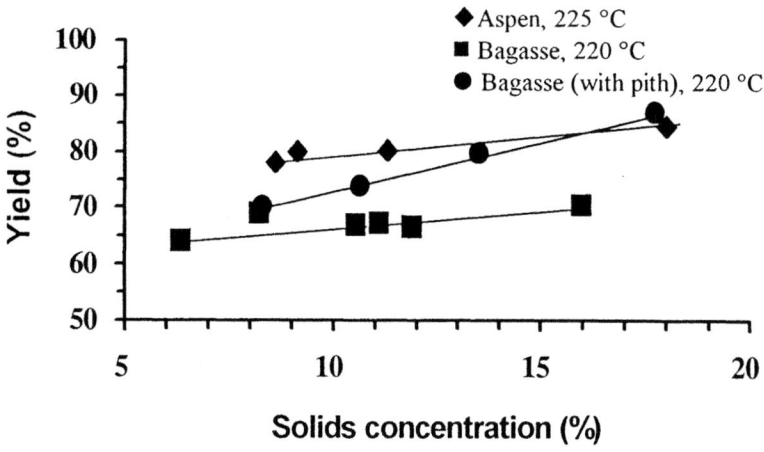

Fig. 1. Yield of lignocellulosics pretreated at different solids concentrations.

Table 1. Reactivity of lignocellulosics pretreated for 2 minutes at a mean reaction temperature of 220 °C

Feed material	Solids conc. (%)	Fermentation medium	Final conversion	
			By cellulose (%)	By ethanol (%)
Aspen	1	Yeast extract/peptone	94	103
Aspen	9	Yeast extract/peptone	91	110
Aspen	9	Corn steep liquor	89	101
Bagasse	1	Yeast extract/peptone	87 [3]	99 [3]
Bagasse	7	Corn steep liquor	88	102
Bagasse	10	Yeast extract/peptone	93 [3]	104 [3]

Of equal importance, initial results indicate that the hydrolyzate from hot, liquid water pretreatment at high solids concentrations is only slightly inhibitory. Fermentation of glucose to ethanol in the presence of the hydrolyzate exhibited some inhibition with respect to the fermentation rate but had no effect with respect to the final conversion achieved [3]. In keeping with this finding, only small amounts of sugar degradation products (known to be inhibitory) were detected in hydrolyzates from the hot, liquid water pretreatment of aspen and bagasse at 9 % solids concentration. The production of furfural and hydroxymethyl furfural accounted for less than 3 % of the original carbohydrates. Unlike pretreatment at low solids concentrations [2], hydroxymethyl furfural was present as well as furfural. The presence of hydroxymethyl furfural is a consequence of the higher water and reactor temperatures needed for pretreatment at 220 °C at high solids concentrations.

Modifications to the existing pretreatment equipment are now underway to reduce the formation of inhibitory, sugar-degradation products. Specifically, saturated steam at 220 °C will be used to preheat the feed material to 200 °C. Hot, liquid water will then be displaced into the reactor to bring the lignocellulosics to temperature and provide the liquid reaction medium desired. Exposure of the lignocellulosics to the high temperatures which degrade the sugars will thus be avoided, while retaining the benefits of a liquid water pretreatment.

The feasibility of lowering the pretreatment temperature is also being determined. The yield of pretreated lignocellulosics was insensitive to the mean reaction temperature over the range studied (205 to 225 °C), as evident from Table 2. Provided the efficacy (i.e. high fiber reactivity, low hydrolyzate inhibition) of the pretreatment is also unchanged, this finding is encouraging. Lower reaction temperatures (and pressures) could then be used in a commercial-scale operation.

To more closely approximate a commercial operation, the current system is being scaled-up approximately 50-fold. The preliminary design for the scaled-up pretreatment device is complete and complies with the ASME code. The apparatus consists of six vessels, a feedwater tank, a high-temperature hot-water boiler (with blowdown vessel), the reactor (with blowdown vessel) and the hydrolyzate reservoir. A schematic diagram is provided (Fig. 2.). The reactor will be loaded and unloaded through a hinged closure, with the lignocellulosics retained in a perforated metal cannister. Also, the liquid hydrolyzate will be removed from the reactor safely and quickly by means of the reactor blowdown vessel. Hot, liquid water can be generated at temperatures up to 280 °C.

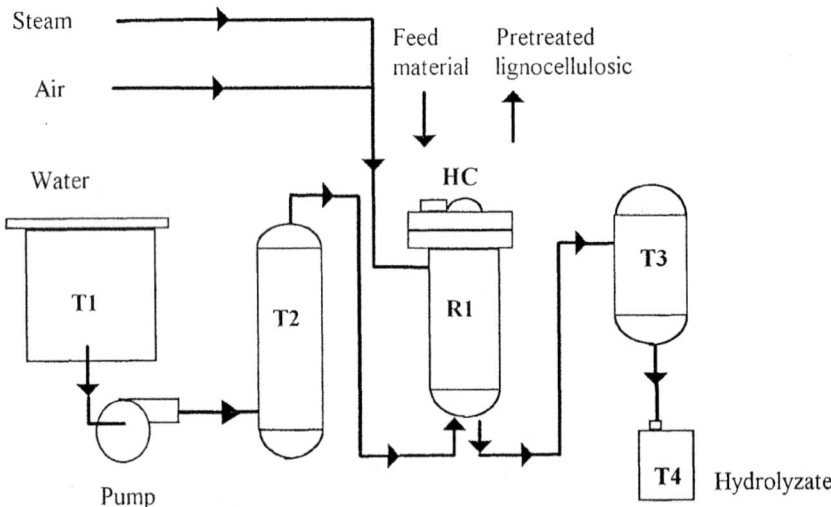

Fig. 2. A schematic diagram of the 1 kg/batch liquid-hot-water pretreatment apparatus under construction, showing the (T1) feedwater tank, (T2) high temperature-hot water generator (associated blowdown vessel not shown), (R1) reactor (with a (HC) hinged closure for rapid loading/unloading, air line for pressurization and a steam line for preheating of the feed material), (T3) reactor blowdown vessel and (T4) product reservoir.

Table 2. Yield of pretreated sugar-cane bagasse at different mean reaction temperatures

Mean reaction temperature (° C)	Solids conc. (%)	Reaction time (min)	Yield of pretreated lignocellulosic (%)
205	8	2	70
210	8	2	67 [3]
215	9	2	72
220	8	2	69

which is more than sufficient to meet our current and expected needs for pretreatment of lignocellulosics. Lastly, since only hot, liquid water is used, the boiler, the reactor and their associated blowdown vessels are being constructed from carbon steel pipe.

4 Conclusions

1. Rapid (120 s), batch, hot-liquid-water pretreatment of representative woody and herbaceous species (aspen and sugar-cane bagasse) is possible at a mean reaction temperature of 220 °C at solids concentrations of up to 18 %. The minimum amount of feed material solubilized (measured by the reduction in the yield of pretreated lignocellulosics) was 10 % at the highest solids concentration studied (18 %).
2. The yield of pretreated lignocellulosics was slightly sensitive to the solids concentration.
3. The yield of pretreated lignocellulosics was insensitive to the mean reaction temperature over the range studied (205 to 225 °C).
4. Hemicellulose was readily solubilized by hot-liquid-water pretreatment at high solids concentrations. After pretreatment at 9 % solids concentration, less than 10 % of the hemicellulose content of the feed material remained within the pretreated lignocellulosics from both aspen and bagasse.
5. Excellent fiber reactivity was achieved by hot-liquid-water pretreatment at high solids concentrations. Over 90 % of the cellulose in pretreated bagasse and aspen was converted to ethanol by simultaneous saccharification and fermentation.
6. The production of possible inhibitors to fermentation was low. Only small amounts of sugar degradation products (known to be inhibitory) were detected in hydrolyzates from hot, liquid water pretreatment of aspen and bagasse at 9 % solids concentration. The production of furfural and hydroxymethyl furfural accounted for less than 3 % of the original carbohydrates.

5 Acknowledgements

This research was supported, in part, through The Consortium for Plant Biotechnology Research, Inc. by the United States Department of Energy (DOE) cooperative agreement No. DE-FC05-92OR22072. This support does not constitute an endorsement by the DOE or by The Consortium for Plant Biotechnology Research, Inc.

of the views expressed in this article. Additional support was provided by The Linbergh Foundation, The Link Foundation, The University of Hawaii Foundation (Coral Industries Endowment) and the National Renewable Energy Laboratory as part of the Hawaii Integrated Biofuels Research Program. We wish to thank Francis Keany (Hawaiian Commercial & Sugar) and Andy Baker (U.S. Forest Products Laboratory) for supplying the feed materials, William Mok for assistance in the design of the pretreatment process and Eric Croiset and Ben Respicio for construction of the process equipment. We also thank Peter van Walsum for advice on SSF procedures.

6 References

1.   Weil, J., Westgate, P., Kohlmann, K. and Ladisch, M.R. (1994) Cellulose pre-treatments of lignocellulosics substrates. *Enzyme Microbiology and Technology*, Vol. 16, No. 11 pp 1002-1004.

2.   Allen, S.G., Kam, L.C., Zemann, A.J. and Antal, M.J. Jr. (in press) Fractionation of sugar cane with hot compressed liquid water. *Industrial and Engineering Chemistry Research.*.

3.   van Walsum, P.G., Allen, S.G., Spencer, M.J., Laser, M.S., Antal, M.J. Jr. and Lynd, L.R. (in press) Conversion of lignocellulosics pretreated with hot compressed liquid water to ethanol. *Applied Biochemistry and Biotechnology*.

4.   Spencer, M. J. (1995) An investigation into chemical-free pulping of biomass using only hot liquid water. M. Sc. Thesis, University of Hawaii.

5.   Saeman, J.F., Bubl, J.L. and Harris, E.E. (1945) Quantitative saccharification of wood and cellulose. *Industrial and Engineering Chemistry Research*, Vol. 17, No. 1 pp 35-37.

6.   Moore, J.F. and Johnson, D.B. (1967) *Procedures for the Chemical Analysis of Wood and Wood Products*, Forest Products Laboratory, Forest Service, U.S. Dept. of Agriculture.

7.   Mok, W.S.-L. and Antal, M.J. Jr. (1992) Uncatalyzed solvolysis of whole biomass hemicellulose by hot compressed liquid water. *Industrial and Engineering Chemistry Research*, Vol. 31, No. 4 pp 1157-1161.

8.   Kaar, W.E., Cool, L.G., Merriman, M.M. and Brink, D.L. (1991) The complete analysis of wood polysaccharides using HPLC. *Journal of Wood Chemistry and Technology*. Vol. 11, No. 4 pp447-463.

ALKALINE OXIDATION OF SUGAR : THERMOCHEMICAL CONVERSION OF XYLOSE FROM HEMICELLULOSE INTO LACTIC ACID

S. RAHARJA, L. RIGAL*
Laboratoire de Chimie Agro-Industrielle
Ecole Nationale Supérieure de Chimie de Toulouse, 118, Route de Narbonne, 31077 Toulouse, FRANCE.
P.F. VIDAL
Departement of Chemical Engineering
University of Sherbrooke, Sherbrooke, JIK 2R1, CANADA.

Abstract

Hemicelluloses are the first extracted fraction in numerous fractionation processes of lignocellulosic basic materials. Acid hydrolysis of hemicelluloses give a mixture of carbohydrates in which xylose is often in a majority. Transforming these carbohydrates into lactic acid is an interesting objective due to its numerous applications.

The behaviour of the principal monosaccharides of hemicelluloses (xylose, arabinose, mannose, galactose and glucose) was compared in alkaline solution (2 N NaOH, 40°C). Kinetics of the formation of four carboxylic acids (lactic, glyceric, acetic and formic) were studied. Increase temperature (100°C) and NaOH concentration (6N) favour the formation of lactic acid. The mass yield neighbouring to the maximum theoritical yield was achieved from xylose (54 %) and arabinose (52 %). These results were confirmed for high concentration in sugar (10-20 %) in the Hemiox Process Development Unit implemented with a gaz/liquid contactor and high temperature (235°C). The conversion of extracted hemicellulose was studied in Hemiox PDU.

Keywords : Alkaline oxidation, pentose, hexose, hemicelluloses, lactic acid, conversion.

1. Introduction

Hemicelluloses are the most accessible and the first extracted fraction in numerous fractionation processes of lignocellulosic materials. Our work has focussed on upgrading of the hemicellulose liquor extracted from poplar using a twin-screw reactor (1, 2). Acid hydrolysis of this hemicelluloses give a mixture of carbohydrates, in which xylose is in a majority (3). Transforming this carbohydrate into lactic acid is an interesting objective due to its numerous applications (4) : i) in food industry as acidulent or preservative, ii) in pharmaceutical industry as intermediate in drugs synthesis (5) and iii) in polymer industry for special applications as implants and melting sutures (6). Poly (lactic acid) has also been considered for biodegradable plastics and packaging material (7, 8, 9, 10).

Alkaline treatment of sugars has been extensively studied under different experimental conditions (11, 12, 13, 14, 15, 16, 17, 18). It produce various hydroxy carboxylic acids : formic, acetic, oxalic, lactic, glyceric, glycolic, saccharinic acids... (19, 20, 21). The objective of this work was to study the thermochemical conversion of hemicelluloses into lactic acid.

---

*To whom all correspondence and reprint requests should be addressed

2. Material and methods

2.1. Material
Chemicals : D-Xylose, L-Arabinose, D-Glucose, D-Mannose, D-Galactose, Cellobiose, Xylan, NaOH, used in this study were of laboratory grade, from Prolabo, Jansen Chimica and Aldrich. Hemicelluloses are obtained as decribe in (1). Chemicals used for the analytical identification were of HPLC grade and laboratory grade from Dionex, SDS and J.T. Baker.

Batch experiments : Alkaline sugar oxidation was conducted in a 2 L thermostated reactor equipped with mechanical stirer and reflux condensor. A volume of 2000 mL NaOH solution was brought to temperature in the agited reactor. Introduction of sugar is time zero. Aliquots of the reaction mixture were taken for HPLIC analysis of sugars and acids.

Continous process : The HEMIOX Process Development Unit is schematically represented in figure 1. Premixed sugar/sodium hydroxyde solution at room temperature is pumped at a flow rate of 40 mL/min and injected in the gaz/liquid mixer. Heat up of the solution was made by tangential injection of steam into the lower chamber the mixer after the liquid solution has passed throught the capillary. The reaction is cooled at the exit of the reactor by means of a condensor. All the reaction mixture is collected for HPLC analysis of acids and sugars.

2.2. Analysis
HPLIC analysis were conducted as described in (21), with Dionex BioLC system. Condition for sugars analysis were : precolumn PA1 Guard ; column Carbopac PA1 ; eluents NaOH 100 mM and H2O (> 18 MOhm) flow rate 1 mL/min ; temperature 20°C ; postcolumn addition NaOH 300 mM, flow rate 0,8 mL/min ; Pulsed amperometric detector. Condition for acids analysis were : Column HPICE AS1 ; eluants Octane Sulfonic Acid 1mM- Propanol-2 2 % - water, flow rate 0,3 mL/min ; Regeneration of Anion Micromembrane Suppresor with TetrabutylAmmoniumhydroxid, flow rate 0,5 mL/min and Conductrimetric detector.

HPLC analysis were conducted with the following system : Beckman, model 100A, pump (0,6 mL/min of 0,01 mM H_2SO_4), rheodyne manual injector (20 mL loop), Animex HPX 87H column (7,8 mm ID X 300 mm), Biorad model HPLC Column heater (65°C), Perkin Elmer model LC85 UV detector (210 nm) and Hewlett Packard model 3392 A integrator.

The yield of the acids of interest, R_n are expessed as :

$$R_n = \frac{\text{weight of the acid in the treated solution}}{\text{weight of initial sugar}} \times 100 \%$$

Continuous Alkaline Oxidation of Sugars
The UdeS Process

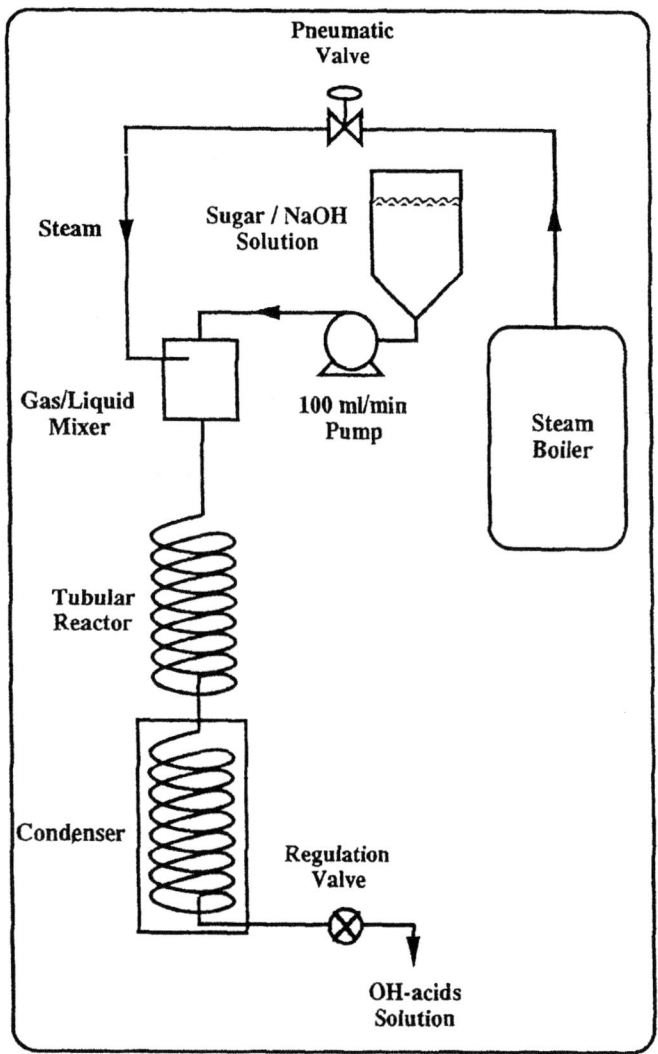

Fig. 1. The HEMIOX Process Development Unit reactor

3. Result and Discussion

Reaction pathways of alkaline degradation of hexoses and pentoses are complex (13). Its conduct to numerous hydroxycarboxylic acids with variable proportions depend on operating conditions. At 40°C, with 2 N NaOH, four acids have been analysed in the case of five monosaccharides (fig. 2a, 2b, 2c, 2d) :

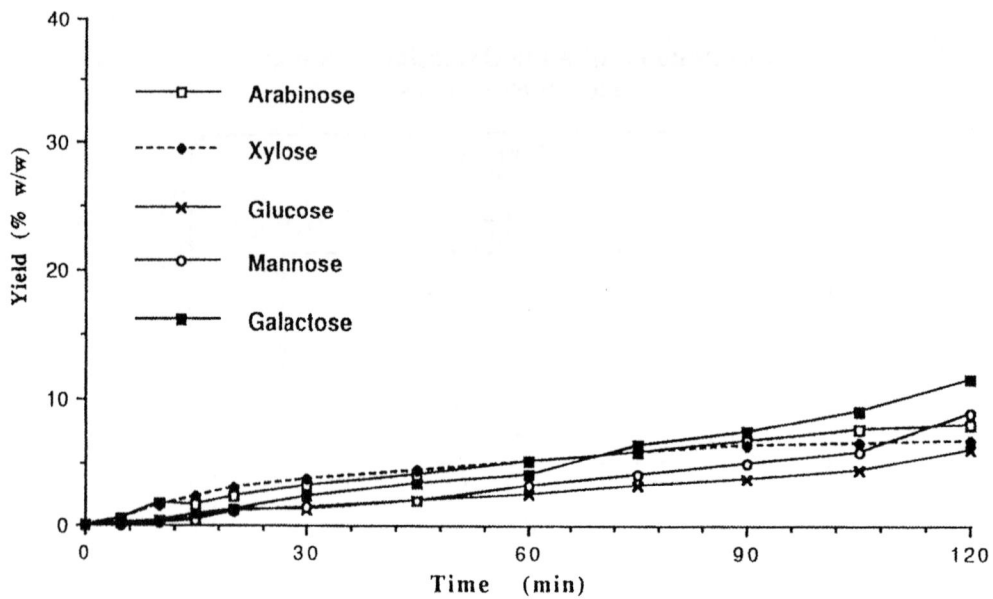

Fig. 2a. Formation of glyceric acid

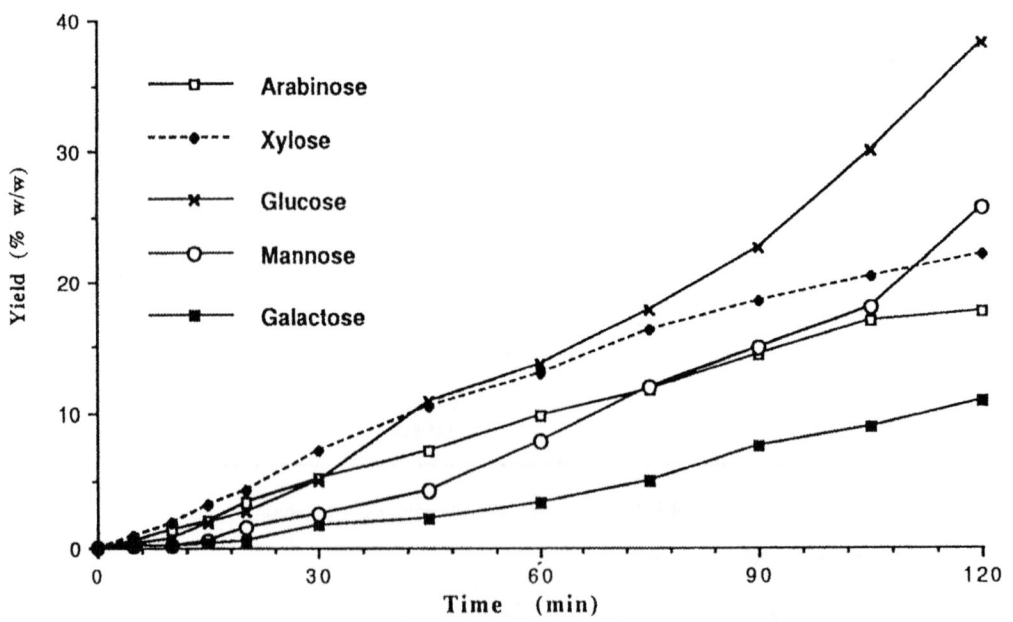

Fig. 2b. Formation of lactic acid

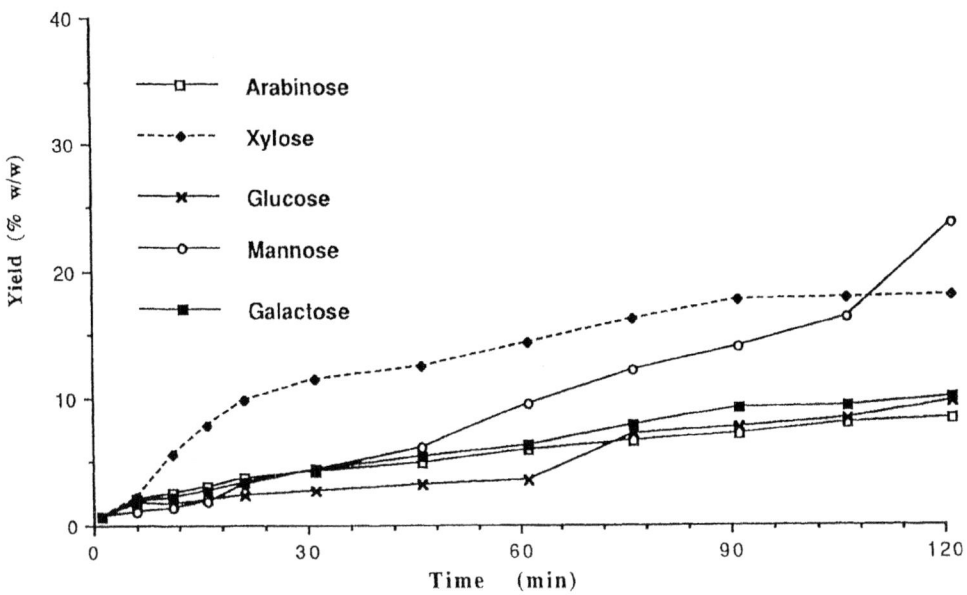

Fig. 2c. Formation of formic acid

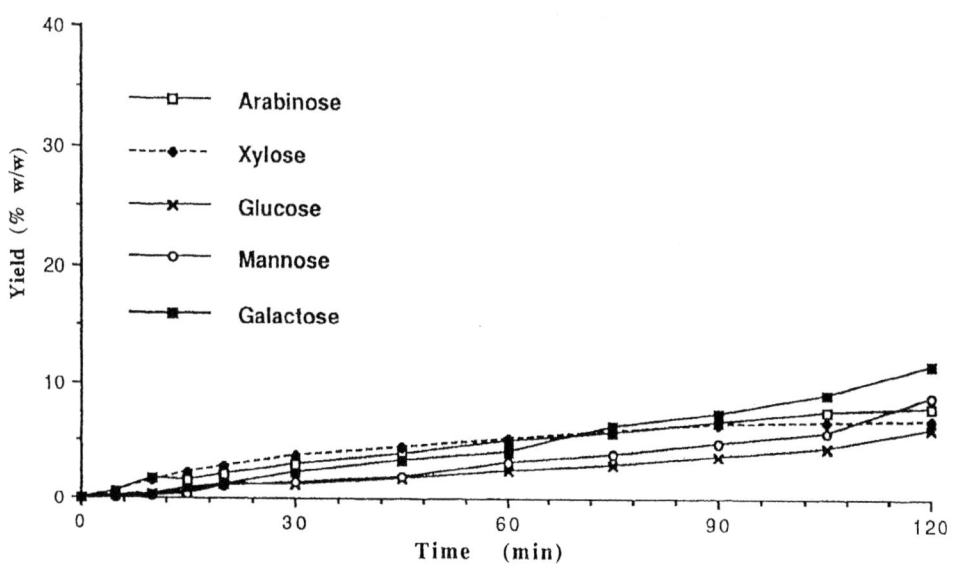

Fig. 2d. Formation of acetic acid

Operation conditions : batch reactor, 40°C, NaOH 2N, [sugar]o = 1 %

Fig. 2. Formation of carboxylic acids from arabinose, xylose, glucose, mannose
and galactose by alkaline oxidation

(i). Acetic acid is in a minority in all cases.
(ii). Lactic, formic and acetic acids were formed in equivalent proportion from galactose.
(iii). Formic acid is obtained in more important proportion from mannose and xylose.
(iv). The rate of lactic acid formation is higher from xylose and glucose.

Simultaneous augmentation of temperature and NaOH concentration minimized these differences of behavior, except for galactose (table 1). At 100°C and 24 % of NaOH, neighbouring or upper 50 % lactic acid yields were acheived from the four other sugars within five minutes xylose and arabinose have the same behaviour and gave lactic acid mass yield near to the maximum theoritical yield (100 x 90/150 = 60 %).

Table 1. Alkaline oxidation of pentoses and hexoses in highest temperature and NaOH concentration

	Acid Yield (% w/w)											
	Glyceric			Lactic			Formic			Acetic		
time (min)	5	15	60	5	15	60	5	15	60	5	15	60
Arabinose	7.1	7.2	7.4	50.4	50.9	51.9	1.3	1.3	1.4	0.2	0.2	0.3
Xylose	7.0	7.3	7.5	52.1	52.7	54.5	1.5	1.5	1.5	0.3	0.3	0.4
Glucose	6.8	6.6	6.3	55.1	55.9	57.6	1.0	1.1	1.2	0.5	0.5	0.5
Mannose	13.0	12.9	11.9	45.0	45.5	46.9	1.2	1.2	1.2	0.4	0.4	0.5
Galactose	8.3	8.7	9.1	17.1	17.6	17.8	1.1	1.1	1.1	0.2	0.2	0.3

Operating conditions : batch reactor, 100°C, NaOH 6 N, [sugar]$_o$ = 1 %

Acid hydrolysis of extracted hemicelluloses from poplar wood or fiber sorghum (2) lead to xylose in a majoritary and arabinose in small portion (table 2). Glucose, mannose and galactose were in a minority. Nevertheless, at 100°C and NaOH 6N, direct conversion of theses xylanic hemicelluloses into lactic acid is very limited (fig. 3). Beta-1-4 glycosidic linkage between xyloses in main chain of polysaccharides blocked the reactions of enolization and beta- elimination which are in general accepted as the first stage of the alkaline degradation mechanism (22, 23). However, the reaction may be take place on sugars in terminal position of the polysaccharides chain. Conversion of cellobiose, similar dimer of xylobiose give an example (fig. 4). Increase of temperature could favor the rupture of osidic linkage and the conversion into lactic acid of released monomers.

Table 2. Mass yield (% w/w) of sugars from hemicelluloses acid hydrolyzate

	Xylose	Arabinose	Mannose	Glucose	Galactose
Poplar(1)	60.4	1.06	0.2	0.5	0.6
Sorghum (2)	48.2	13.3	/	3.9	3.9

Operating conditions : batch reactor, 100°C, 60 min, H_2SO_4, 5 % (1) 4 % (2)

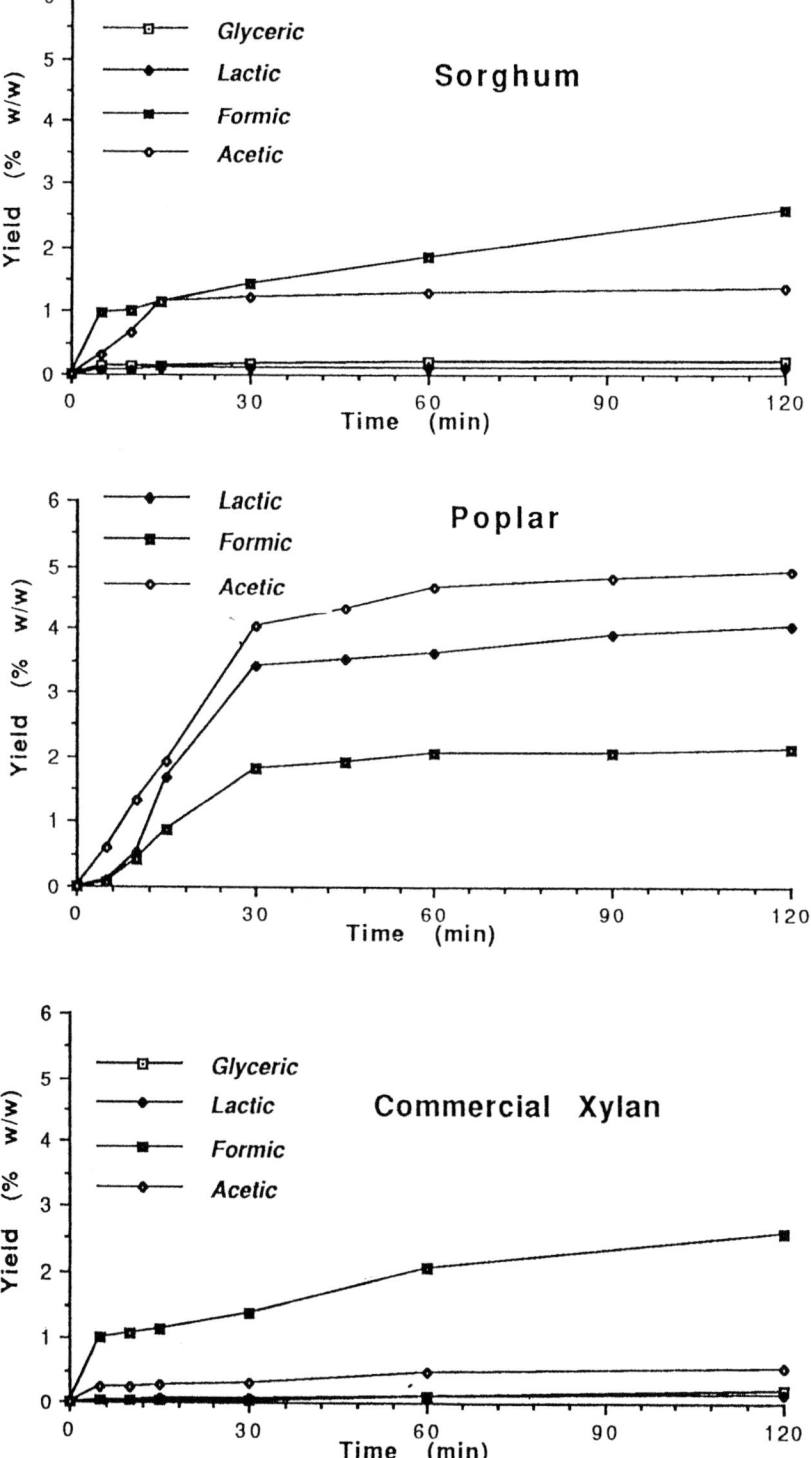

Operating conditions : batch reactor, 100°C, NaOH 6N, [hemicelluloses]$_o$ = 1 %

Fig. 3.   Alkaline oxidation of hemicellulose extracted from sorghum and poplar, and commercial xylan

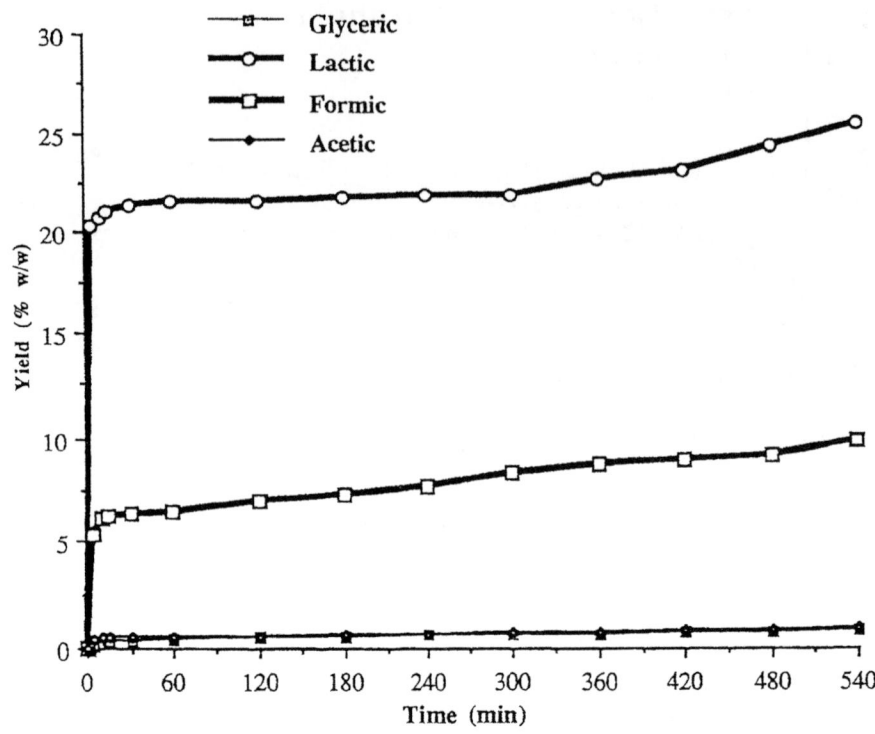

Operating conditions : batch reactor, 100°C, NaOH 6N, [cellobioses]o = 1 %

Fig. 4. Alkaline oxidation of cellobiose

At 235°C, in HEMIOX Process Development Unit (fig. 1) fited out with a gaz/liquid contactor, the conversion of xylose into lactic is very fast, with high yield (table 3). First trials for transformation of hydrolyzate hemicelluloses in such conditions were very encouraging : 40 % mass yield of lactic acid was achieved.

Table 3. Alkaline oxidation of xylose in HEMIOX PDU

NaOH (%)	Xylose (%)	T (°C)	Rlactic acid (%)
10.0	7.0	241	47.6
10.0	10.0	254	49.3
12.4	8.5	227	45.5
12.5	11.45	262	44.0
12.5	5.6	235	56.0
12.5	14.3	235	45.0
15.0	10.0	235	48.5

4. Conclusion

Thermochemical conversion of xylose using continuous HEMIOX Process Development Unit at high temperature (230-260°C) and concentration of soda (10-15 %) conduct to good yields of lactic acid, even with high initial concentration of sugar (5 to 15 %). Treatment of prehydrolyzate hemicelluloses in such conditions gave similar results. Our objective is now to study the operating conditions for simultaneous depolymerisation of hemicelluloses and conversion of monosaccharides into lactic acid.

5. References

1. N'Diaye, S. ; Rigal, L. ; Larocque, P. ; Vidal, P.F. (1996) Extraction of hemicelluloses from poplar *(Populus tremuloides)* using an extruder type twin-screw reactor : a feasibility study. Bioresource Technology. In press.
2. N'Diaye, S. (1996) Fractionnement de la matière végétale : mise au point d'un procédé thermo-mécano-chimique et modélisation du fractionnement d'un réacteur bi-vis. Thèse de doctorat INP, Toulouse.
3. Qiabi, A. ; Rigal, L. ; Gaset, A. (1994) Comparative studies of hemicelluloses hydrolysis processes : application to various lignocellulosic wastes Industrial Crops and Products (3) 95-102.
4. Lipinski, E.S. ; Sinclair, R.G. (1986) Is lactic acid a commodity chemical ? Chem. Eng. Prog. Augt. 26-32.
5. Peckman, G.T. Jr. (1944) The commercial manufacture of lactic acid. Chem. Eng. News. 22 (6) 440-443.
6. Moo-Young (1995) Lactic acid in comprehensive biotechnology : the principle, applications and regulations of biotechnology in industry, agriculture and industry, M. Moo-Young Eds.
7. Mc Carthy-Bates, L. (1993) Biodegradable blossom into field of dreams for packagers. Plastic World. March 1993, pp. 22-27.
8. De Vries, K.S. (1989) Preparation of polylactic acid and copolymers of lactic acids. US Patent 4.797.468.
9. Gruber, P.R. et al. (1992) Continuous process for manufacture of lactide polymers with controlled optical purity. US Patent 5.142.023.
10. Suzuki, K. (1995) Lactic acid polymers composition molded product and film. E.P. Patent 0.683.207 A2.
11. Ishuzu, A. ; Lindberg, B. ; Theander, O. (1967) Saccharinic acids from D-xylose and D-fructose-6-14C. Carbohydrate Res. 23. 207-215.
12. Harris, J.F. (1972) Alkaline decomposition of D-xylose-1-14C, D-glucose-1-14C and D-glucose-6-14C. Carbohydrate Res. 23. 207-215.
13. De Bruijn, J.M. (1986) Monosaccharides in alkaline medium : isomerisation, degradation and oligomerisation. PhD Thesis, Delf University of Technology, The Netherlands.
14. Shaffer, P.A. ; Friedemann, T.E. (1930) Sugar activation by alkali I. Formation of lactic acid saccharinic acids. J. Biol. Chem. 26. 345-374.
15. Braun, G. (1935) Process for the production of lactic acid. US Patent 2.024.565.
16. Krochta, J.M. ; Tillin, S.J. ; Hudson, S.S. (1988) Thermochemical conversion of polysaccharides in concentrated alkali to glycolic acid. Appl. Biochem. Biotechnol. 17. 23-32.
17. Niemela, K. (1988) The conversion of cellulose into carboxylic acids by a drastic oxygen-alkali treatment. Biomass. 15. 223-231.
18. Niemela, K. (1990a) Conversion of xylan, starch and chitin into carboxylic acids by treatment with alkali. Carbohydrate Res. 204. 37-49.

19. Niemela, K. (1990b) the formation of hydroxy monocarboxylic acids and dicarboxylic acids by alkaline thermochemical degradation of cellulose. J. Chem. Tech. Biotechnol. 48. 17-28.
20. Sjöström, E. (1991) Carbohydrate degradation products from alkaline treatment of biomass. Biomass Bioenergy. 1(1) 61-64.
21. Dubois, M.F. ; Rigal, L. ; Gaset, A. (1995) Dosage comparatif des produits issus de la dégradation alcaline des hexoses. Can. J. Chem. 73. 1582-1592.
22. Kieboom, A.P.G. ; Van Bekkum, H. (1984) Aspects of the chemical of glucose, Rec. Trav. Chim. Pays-Bas, 103. 1-12.

Acknoledgments

Financial support of Ministère des Affaires Internationales - Direction Générale France (Québec) and Ministère des Affaires Etrangères (France) within the France-Québec collaborative project "Fractionnement des résidus lignocellulosiques et conversion des hémicelluloses" is gratefully acknowledged.

EMISSIONS OF BIOMASS DRYING

L. FAGERNÄS and K. SIPILÄ
VTT Energy, P.O.Box 1601, 02044 VTT, Espoo, Finland

Abstract
Different dryers based on steam or flue gas drying have been or are being developed for biomass. In the thermal drying organic compounds are released from biomass and cause environmental effects as gaseous emissions to the ambient air or as liquid effluents into natural waters, and, in addition, operational problems by contributing to the formation of deposits. When preparing an Environmental Impact Assessment for new biomass-based plants the impacts of drying are to be considered.

The paper presents a review of the recent studies on emissions from biomass drying. The work covers the emissions from different feedstocks (wood, bark, peat, forest residues), and from different dryers and drying tests, both atmospheric and pressurised ones, and the factors affecting the emissions. The formation and effects of the emissions, and methods to reduce them are also discussed. Emissions of biomass drying consist of gaseous and condensable compounds. The emissions are affected by dryer type, drying conditions and feedstock material. The drying temperature has an important role in the formation of emissions. The emissions from low-temperature drying comprise mainly lipophilic compounds; monoterpenes and fatty and resin acids. At higher temperatures the emissions contain to a higher degree compounds from thermal degradation of biomass; acetic acid, aldehydes, furfurals, carbohydrates and carbon dioxide.

Keywords: Biomass, drying, effluents, emissions, flue gas drying, organics, steam drying

1 Introduction

Biomass materials contain a high amount of water. The moisture content of biofuels like wood, bark, forest residues and sludges, is typically 50-65%. Raw peat contains still more water, up to 90%, which is generally reduced to 40-60% by drying on the production field. The biomass should generally be dried before feeding to various thermochemical conversion processes like combustion, gasification or pyrolysis processes, to a 10-30% moisture content.

Dryers used for biofuels are based on direct or indirect heating. In direct dryers, convection is created by hot gas contacting the wet material [1]. In indirect dryers, conduction is carried out from a hot surface contacting the material. Flue gas drying or steam drying techniques are commonly used for biofuels. The temperatures and pressures applied on the dryers can vary considerably. The steam drying technique has drawn interest increasingly in the drying of moist fuels, mainly for environmental reasons. In the 1980s, an energy-efficient steam drying technique utilizing back-pressure steam or drying in superheated steam under pressure was developed in Sweden for drying porous materials such as paper pulp, different hog fuel and agricultural products, biomass and peat [2]. The use of steam drying technique for biomass is assumed to be increased in the future. Different new dryers based mainly on steam drying have been developed for drying biomass [3, 4, 5, 6]. When planning the integrated gasification combined-cycle (IGCC) technology of biomass, pressurised steam drying technique has been found to be suitable for wet biomass [7].

In flue gas and steam drying processes, biomass is commonly dried at low temperatures (100-450 °C), high heating rates and in short residence times (0.5 - 60 s). However, the thermal treatment brings changes in the organic material. Even under mild drying conditions, organic compounds are released from biomass. These compounds may cause environmental problems in the form of gaseous emissions to the ambient air when the flue gases are led uncondensed into the stack, or as liquid effluents into natural watercourses, if they are condensed or scrubbed [8, 9]. Malodorous emissions from dryers and power plants have also been detected [10]. The organic compounds released may also cause operational problems in drying processes by contributing to the formation of deposits in down-stream sections [9].

The liquid and gaseous emissions formed in flue gas and steam drying of biomass have been studied mainly in Finland and Sweden in recent years [8-19]. The biomass materials have comprised wood, bark, peat and forest residues. The dryers have been different flue gas and steam dryers, atmospheric or pressurised, industrial or experimental ones, using different drying temperatures and residence times. The emission studies on drying of wood and wood products have mostly focused on veneer drying in the panelboard industry and on air pollution from mechanical pulp mills [20].

The aim of the paper is to present a review of the recent studies on the emissions of biomass drying. The work covers the emissions, both aqueous and gaseous, and the factors affecting the quantity and the composition of the emissions. In addition to emissions, their formation and effects, and methods to reduce them are discussed.

2 Dryers for biomass

Flue gas processes used for biomass are based on flash, rotary drum, fluidized-bed and mill drying. There are several bark flue gas drying systems in operation, for example, the Hot Hog, the vibratory hot conveyor, the rotary dryer and the Bahco, a cascade fluidizing dryer [2]. Flash dryers are also used for bark, and for peat at peat power plants. The temperature of the flue gases fed to the dryer ranges generally from 250 to 450 °C, and the outlet temperature from 90 to 110 °C. The residence time of the fuel in the dryer ranges 1-20 min. A. Ahlstrom Corporation has recently developed a fixed-bed dryer able to utilise low-temperature flue gas (<150 °C) from the heat recovery heat generator in the IGCC plant, and has built a PDU dryer in Karhula, Finland [6]. Wood chips, peat and bark have been tested in the dryer. In the IGCC demonstration plant of Sydkraft AB in Värnamo, Sweden, the fuel received to the site is dried in a rotary drum dryer in an external drying and fuel preparation facility [21]. The flue gas used is, prior to being discharged in the stack, scrubbed with water.

Steam drying techniques comprise atmospheric and pressurised processes including pneumatic, fluidized-bed, flash and rotary dryers. The steam dryers applied for biofuels are Modo-Chemetics AB and Niro A/S pressurised dryers. The Modo dryer is a closed system where the material to be dried is exposed to indirectly heated superheated steam at a pressure of 2 to 6 bar [2]. The pneumatic dryer consists of transport pipes, heat exchangers, cyclone and fans. The heat for drying is transmitted from the outside of the tubes, the shell side, through condensing steam. Dried material and steam are separated and the steam is recirculated. The excess steam generated during drying is continuously bled off from the system. The dryer is in use in Sweden for bark at Husum pulp mill, for peat at Sveg briquette factory and for pulp at Rockhammars mill [3].

The Niro dryer is a superheated steam fluid bed dryer [3]. The material is fed into the first of 16 cells placed around the superheater in a closed vessel. In the cells, the material is kept suspended by superheated steam blown through the perforated base plate by the impeller. The material moves successively through each of the 16 cells and is finally discharged by the screw conveyor. The maximum steam pressure is 4 bar inside the dryer and 25 bar in heating side. The outlet temperature of the steam is about 150 °C. The first wood-fuel application was drying wood chips in Borås Energi, Sweden [22]. Other biomass references are bark drying in Mönsterås, Sweden, sludge drying in Mackolsheim, France, and sugar beat pulp drying in several countries.

Imatran Voima Oy (IVO), Finland, has developed a pressurised steam dryer for moist fuels [4]. The dryer operates at pressures up to 27 bar and temperatures up to 400 °C. The material is dried in direct contact with superheated steam. The dryer has been planned to be integrated to the process to produce injection steam for the gas turbine of the process called IVOSDIG, which is the combination of a pressurised steam dryer, a pressurised air-blown gasifier, and a gas turbine. To examine drying at high pressures, a pilot plant with capacity of 1 000 kg/h of wet feedstock was built in 1991 by IVO and VTT. 1 000 hours of test runs with peat and wood biomass have been carried out.

Later on, IVO developed a bed mixing dryer, which is suitable for power plants utilizing fluidized-bed combustion for combined heat and power generation (CHP) or mere power generation [5]. The idea of the dryer is to use the heat of the fluidized-bed directly for drying in a dryer operating at steam atmosphere at ambient pressure, and the latent heat of evaporation can be recovered. The first application is the peat and wood fired power plant of Kuusamo, Finland, commissioned in 1994.

Dryers used for wood or wood products comprise particle dryers, veneer dryers and timber kilns [20]. The particle dryers are flash, rotary and fluidized bed dryers for milled wood. The rotary drums are the most common drying systems in the particle board industry.

3 Emissions from biomass dryers

3.1 Research on emissions
The organic compounds released in the drying of peat and bark were studied at VTT in co-operation with Åbo Akademi University over the years 1987-1992 [9, 12, 13]. The

objectives were to determine the organic compounds released from peat and bark in different conditions of flue gas and steam drying, and to determine the effect of these compounds on the environment and on the formation of deposits in processes. The laboratory-scale experiments were carried out with fluidized-bed steam dryers at temperatures of 150-350 °C. The condensable compounds and non-condensing gases were analysed. Bark, condensate and deposit samples were taken from a pressurised (6 bar) steam dryer in a pulp mill. Peat and deposit samples and particle samples from the gas flows were taken from a peat power plant with flue gas drying system.

Organic compounds released from peat and biomass in pilot and PDU scale drying processes [8, 23, 14, 6] have also been analysed at VTT. The organic load of the condensates formed from peat and bark was studied in pneumatic and fluidized-bed PDU drying test rigs at 140-170 °C [8] and in a pressurised PDU steam drying test rig at 8-20 bar with the inlet steam temperature of 200-350 °C and outlet steam temperature of 140-220 °C [23]. To study the emissions from the pressurised steam drying process of IVO, the steam flow from the pilot plant was analysed for organic compounds [14]. The feedstocks were peat, wood chips and saw dust. The operating pressure was 23 bar with the saturation temperature 220 °C. The steam temperature varied from 300 to 350 °C. The emissions from the fixed-bed PDU test rig dryer of Ahlstrom were studied using bark as feedstock [6].

Studies on emissions from different flue gas and steam dryers were carried out by Borås Energi and Startekno AB, when a dryer was sought for drying of wood chips at the power plant of Borås Energi in Sweden [3]. The Niro dryer was chosen for the plant.

When planning IGCC plants based on biomass, an Environmental Impact Assessment (EIA) has to be done. A wide EIA was carried out in Sweden when Vattenfall AB was planning to build a biomass-based IGCC plant in Eskilstuna [15, 16]. In connection with this VEGA project Vattenfall tested drying of the steam dryers of biomass from the suppliers MoDo Chemetics, Stork Friesland and Niro [16] and analysed the condensates of the dryers. Potential environmental effects were also evaluated in 1995 by VTT [17] for an IGCC plant planned to be built at the paper mill of Enso Publication Papers Ltd Oy in Summa, Finland, and to use wood waste from the mill and surrounding forests [18].

The formation of organic compounds during steam drying of bark chips in the exhaust steam was investigated as a function of time, temperature and the age of the bark chips by Björk and Rasmuson [24]. Fresh and stored bark chips were dried at 140 and 160 °C for 3, 11, 30 and 60 min.

A literature review of the emissions of wood, bark and wood product dryers was carried out by Wastney [20] in the Environmental Systems Activity of the International Energy Agency (IEA) Bioenergy Programme [19]. Bongers [25] has, recently, studied organic emissions from pressurised steam drying of Australian brown coal.

3.2 Composition of emissions
Emissions of biomass drying consist of aqueous and gaseous emissions. Aqueous emissions are formed when steam liberated from the dryer is condensed and hence consist of condensable compounds. The emissions that remain volatile in ambient conditions, the non-condensing gases, form the gaseous emissions.

The compounds released from biomass comprise many different organic compound groups. The detailed analysis of the compounds is presented in the studies of VTT [9]. The compounds comprise both hydrophilic and lipophilic, gaseous and condensable ones. The hydrophilic compounds were analysed with different chromatographic methods, and the lipophilic ones by capillary gas chromatography and mass spectrometry (GC/MS). The condensable hydrophilic compound groups are volatile carboxylic acids such as formic and acetic acids, alcohols mostly methanol and ethanol, aldehydes, furfurals, and carbohydrates such as anhydrosugars. The dominant condensable lipophilic compound groups are fatty acids, hydroxy fatty acids, fatty alcohols, resin acids and triterpenoid alcohols. The most volatile lipophilic compounds consist of monoterpenes, the major components being α-pinene and β-pinene. Sesquiterpenes are also volatilised to some degree [3]. The non-condensing gases consist mainly of carbon dioxide and to a lesser amount of hydrogen, carbon monoxide, methane and C_2-C_4 hydrocarbons [9].

The condensates from the pressurised steam drying experiments with Australian coals have been analysed by Bongers et al. [21] in addition to gas chromatography (GC) and GC/MS with nuclear magnetic resonance spectroscopy (^{1}H NMR). The main compounds identified were substituted aromatics such as different benzenes and phenols.

4 Factors affecting the emissions

The emissions formed in drying are affected by many factors such as dryer type, drying conditions such as temperature and residence time, and feedstock material.

4.1 Dryer type

4.1.1 Flue gas dryers

The emissions from flue gas dryers have recently been studied in connection with the planning of new biomass plants, the Borås Energi plant [3] and the VEGA plant [15]. The emissions from a cascade dryer and a drum dryer were reported by Munter & Bäcker [3]. The substances were analysed with flame ionization detector (FID) and an absorption method followed by a spectrophotometric analysis. The analytical results obtained with these methods differed from each other. In the absorption method, organic acids and alcohols were not analysed. The emissions from the cascade dryer were considerably lower compared with those from the drum dryer. The reason for this might be the longer residence time of the material in the drum dryer. On the other hand, drying to a low moisture content increased the emissions considerably. The estimated values for organic emissions from these flue gas dryers are presented in Table 1. As to separate compounds some values were given for the flue gas after drying: <3 g monoterpenes or <6 g acetic acid or <10 g formic acid per kg dried fuel.

According to Setzman et al. [15] the emissions from direct and indirect dryers differ from each other. In the direct dryers, part of the gaseous compounds is led to combustion, the rest being led into the stack. The outlet flue gas contains terpenes and other non-condensing gases. The water-soluble compounds can be separated as condensates. In indirect dryers the gaseous compounds are burned and the emissions into the air are

Table 1. The estimated emissions of organic substances from drying of wood fuel in the flue gas dryer [3]

FID (mg CH_4-eq/m^3 (n) flue gas)	100[1]- 450
FID (mg CH_4-eq/kg dried fuel)	770-3 450[2]
FID changed to mg C/kg dried fuel	600-2 700

[1] drying to 30% moisture content
[2] obtained from the amount of flue gas and flow at Borås Energi

lower. The water-soluble compounds can be condensed. The emissions of terpenes, other volatile organic substances as TOC, and dust in outlet gases or suspended solids in condensates, when using direct or indirect dryers in the planned VEGA plant were estimated. The emissions of terpenes were 0.15-0.57%, those of TOC 0.24% and of solids 0.04% of dry wood. The terpene emissions were lower from the indirect dryer than from the direct dryer.

The emissions obtained for the fixed bed dryer of Ahlstrom were as follows: pH 3.6-3.7, monoterpenes 0.2-0.7%, TOC 0.10-0.23%, aldehydes 0.05-0.13% and volatile acids 0.05-0.16% of dry bark [6]. The dust content was under 10 mg/m^{3}n for bark and wood chips, but higher or under 40 mg/m^{3}n for peat. The amounts of monoterpenes and TOC are of the same magnitude as the values estimated above for the emissions of the VEGA plant.

4.1.2 Steam dryers

The emissions from steam drying are generally obtained as condensates, and the emission values are thus often given as wastewater parameters. The basic wastewater analysing results obtained for condensates from different steam dryers are compiled in Table 2. The emission values obtained for the condensates differed considerably from each other. In the table the dryers are classified in accordance with the type of the dryer: pressurised, atmospheric, industrial and test rig dryers. For condensates obtained from pressurised industrial dryers with different biomass materials, pH ranged from 3.3 to 7.5, BOD 200-1 700 mg/l, COD 300-2 900 mg/l, TOC 310-880 mg/l, suspended solids 5-1 500 mg/l, phosphorous 0.1-0.4 mg/l, total nitrogen 14-160 mg/l and ammonium nitrogen 11-14 mg/l. In one separate case the TOC and solids contents were as high as 9 700 mg/l and 1 500 mg/l, respectively. The concentrations per kg dry feedstock were presented in some cases, ranging then for COD 510-1 250 mg/kg. In the condensates from pressurised test rigs, the pH varied 3.3-6.8 and TOC 70-5 000 mg/l (90-9 530 mg/kg). Other analyses were determined only for some condensates. For the samples from atmospheric test rigs the organic load was higher than for those from pressurised drying. The pH varied 3.4-4.1, COD 880-2 630 mg/l (850-2 620 mg/kg), TOC 300-56 000 mg/kg, solids 80-8 000 mg/kg and total nitrogen 1-3 mg/kg.

The values for the condensate of the dryer of Borås Energi were estimated as follows: pH 3-5, BOD 200-800 mg/l (110-460 mg/kg), COD 400-1200 mg/l (230-690 mg/kg) and suspended solids 5-50 mg/l [3].

The organic composition has mainly been analysed for condensates from different test rigs. Hydrophilic and lipophilic compounds have been analysed for the condensates from atmospheric drying in the studies of VTT [8, 9, 12, 13]. In addition, lipophilic compounds have been analysed in the condensates from the pressurised PDU steam drying test rig [23], from the pressurised pilot process of IVO [14] and from the pressurised apparatus by Björk and Rasmuson [24].

In the condensates from atmospheric drying experiments, the amount of identified organic compounds ranged from 1 to 10% of dry peat, from 0.1 to 11% of pine bark and 0.3 to 17% of birch bark [9]. The compound groups for peat were carboxylic acids, alcohols, aldehydes, aromatic compounds, furanoic compounds, terpenoids and sterols. The carboxylic acids consisted of short-chain C_1-C_6 acids and long-chain C_{14}-C_{32} fatty acids. The main alcohol was methanol. Glycolaldehyde was the dominant aldehyde and levoglucosan the dominant carbohydrate. The furanoic compounds were dominated by furfurals. The dominant compound groups released from pine bark were carboxylic acids, aldehydes and carbohydrates, whereas those released from birch bark included carboxylic acids and terpenoids. The lipid compounds of peat were dominated by fatty acids, those of pine bark by fatty and resin acids and those of birch bark by betulinol.

The condensates from pressurised drying experiments contained lower amounts of lipophilic extractives than those from atmospheric drying experiments. The condensates from the IVO flash dryer contained extractives (ethyl ether-solubles) 600-3 600 mg/l for peat and 1 400-5 500 mg/l for wood biomass [4] (Fig. 1). The release of extractives from wood biomass in this dryer has been reported to be about 0.2-0.5% from dry ash-free feedstock [14]. The dominant lipid groups were fatty and resin acids, which covered 70-90% of the total lipids. The condensates from the pressurised PDU test rig [23] with peat contained 0.1-0.6% of dry peat lipophilic extractives (dichloromethane-solubles), of which identified lipids 0.04-0.26% of peat and fatty acids 0.02-0.16%. Björk and Rasmuson [24] have found different types of terpenes and sesquiterpenes among the lipophilic compounds. Munter and Becker [3] have reported the emissions for steam drying as monoterpenes to be <0.05%, as acetic acid 0.2% and as formic acid as 0.5% of dry feedstock.

4.2 Temperature
The emissions of biomass drying, both the amount and composition of the compounds, are highly affected by the drying temperature. The emissions from peat and wood biomass drying have been found to increase with the rise of temperature. In atmospheric fluidised-bed PDU peat drying experiments the average BOD, COD, and TOC values increased to nearly threefold, as the temperature rose from 110 °C to 130-140 °C [8]. At higher temperatures in laboratory-scale fluid-bed test dryers, the amount of identified organic compounds released from peat increased from 1 to 10% of dry ash-free peat, when the temperature rose from 190 to 350 °C, and that released from pine bark and birch bark at 150-350 °C were 0.1-11% and 0.3-17%, respectively (Fig. 2) [9]. The amounts of lipid compounds were 0.1-1% of peat, 0.1-0.9% of pine bark and 0.1-10.5% of birch bark. For peat, and for pine bark at higher temperatures (300-350 °C), hydrophilic compounds comprised the bulk of the organic substances released. At lower drying temperatures (250 °C or below), most of the organic materials released from pine bark were lipophilic compounds.

Table 2. Basic wastewater analyses of condensates from steam dryers, compiled from references [8, 9, 14, 16, 23, 24]

Dryer type	pH	BOD mg/l	BOD mg/kg dry feed	COD mg/l	COD mg/kg dry feed	TOC mg/l	TOC mg/kg dry feed	Solids content mg/l	Solids content mg/kg dry feed
Industrial dryers [8, 9, 16]									
Pneumatic, MoDo, Stork									
- wood	4.5-5	600	340	900	510	9 700	10 800	5-10	
- bark	3.3-3.6	800-1 400		1 900-2 200	1 100-1 250			10-15 (1 500)	1 700
- forest residues	4.5-7.5	1 100-1 700		1 900-2 400	440-560	580	130-150	9 (5 800)	
- peat	3.3-4.5			300-1 350		310-450		170-180	30-70
Fluidised-bed, Niro									
- forest residues	4.5	1 400		2 900		880		20	
Test rigs									
Pressurised rigs									
Pneumatic, pilot [8]									
- peat	6.7-6.8	130-180	140-190	470-570	500-630	90-140	90-160	220-350	230-390
Flash, PDU [23]									
- bark	3.3					390	9 500		
- peat	3.3-6.2					70-450	600-5 900	10-150	
Flash, pilot, IVO [14]									
- wood chips						1 100-1 600			
- saw dust						450- 00			
- peat						700-5 000			
Laboratory [24]									
- bark, 3 min/60 min			270-990/ 2 800-12 800		600-2 040/ 9 100-27 200		180-580/ 1 900-6 500		
Atmospheric rigs									
Fluidised-bed, VTT, PDU [8]									
- bark	3.4			2 630	2 620	440	460	390	390
- peat	4.1	530		880	850	310	300	80	80
Fluidised-bed, VTT, labor. [9]									
- peat	3-4						1 000-56 000		1 000-8 000
- bark	4						27 000		

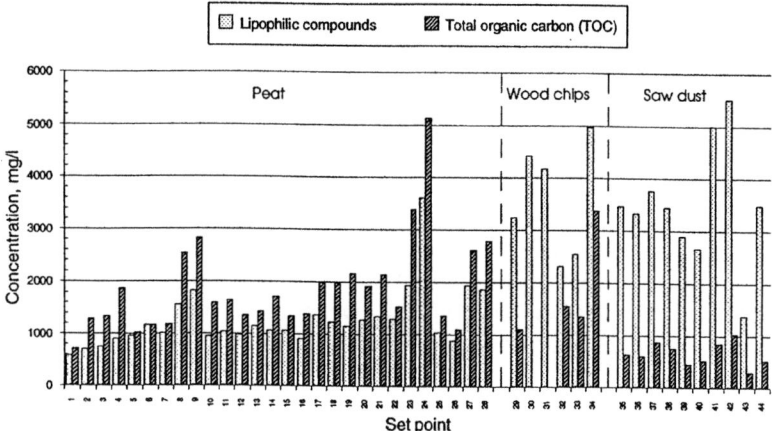

Fig. 1. Concentrations of TOC and extractives (lipophilic compounds) in the condensates from pressurised drying of peat, wood chips and saw dust at different drying conditions [4].

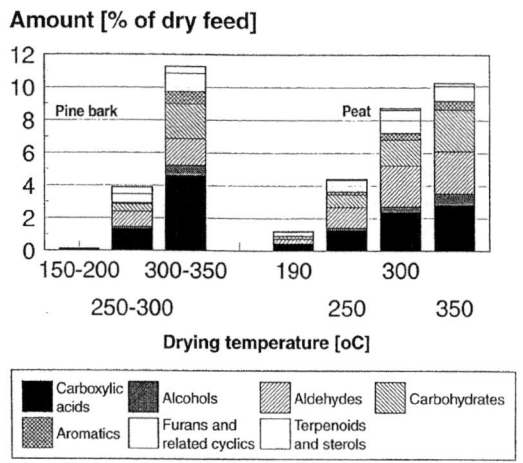

Fig. 2. Amounts of organic compound groups released from pine bark and peat in laboratory-scale fluidised-bed atmospheric steam dryers at different reactor temperatures [9].

When drying forest residues in industrial pressurised dryers the organic load in the condensate was lower at lower temperatures. The COD was 1 900 mg/l at 160-180 °C (Stork dryer) and 2 400-2 900 mg/l at 175-200 °C (Niro and MoDo dryers) [16]. For bark in an industrial dryer the TOC was 1.1% of bark and the release of lipids 0.5% of bark at 180 °C [9]. These lipids were dominated (80%) by fatty and resin acids. Björk and Rasmuson [24], when studying the formation of organic compounds in exhaust steam from bark drying, found that the formation rate strongly depends on the temperature of the superheated steam. The rates of BOD, COD and TOC formations were at least twice as high at 160 °C as at 140 °C.

In the experiments with the pressurized PDU test rig, the organic load in the outlet steam increased, as the pressure (2.5-20 bar) and the temperature of the outlet steam (140-220 °C) rose [23]. This steam temperature was assumed to be close to the temperature reached by the biomass particles in the drying tests. The TOC and the amounts of identified lipids in the condensates from peat with different outlet temperatures are shown in Fig. 3. The fatty acids were the dominant compound group among the lipids comprising on average 65% of the lipids. The TOC of the bark-based condensate obtained at 20 bar was 1% of dry bark and the amount of the dichloromethane-solubles 0.6% of dry bark. In the experiments with the pressurised IVO flash dyer, the differences in the concentrations of the lipophilic extractives in condensates between the set points have been reported to be mainly due to different drying temperatures [4].

Concerning the wood and veneer dryer exhaust streams, there are also indications that increased drying temperatures lead to increased emissions [20]. The boiling points of monoterpenes range from 150 to 180 °C. According to Marutzky [26] nearly all of the terpene content in wood should be removed during particle drying since the boiling point is lower than the typical drying temperatures. In Marutzky´s study on average 74% of the original terpene content of wood particles was removed. The other extractives like fatty and resin acids have significant vapour pressures at the veneer dryer temperatures (up to 260 °C) leading to their vapourisation.

4.3 Feedstock

Biomass feedstock has a great effect on the emission quantity and composition. However, the effect is not so straightforward, because it depends to a high degree on the drying temperature.

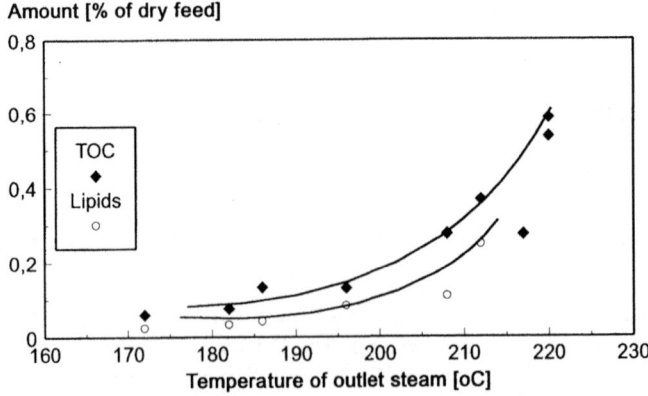

Fig. 3. The amounts of TOC and lipids in the condensates obtained from pressurised drying of peat at different outlet steam temperatures [23].

The organic content of the condensates obtained from steam drying of different raw materials like wood, bark, peat and forest residues, varies between the feedstocks (Table 2). In industrial pressurised dryers (3-6 bar) with the temperature range of 160-200 °C, the load of the condensates has been considerably higher for bark and forest residues than for wood and peat [8, 9, 16]. The forest residues also contain bark and needles with a higher extractive content than wood. In the pressurised pilot test rig of IVO at the pressure of about 23 bar, the TOC values of the condensates were higher for peat than for wood chips and saw dust [4]. In the condensates from atmospheric fluid-ized-bed PDU drying test runs, the organic load as TOC was higher for bark than for peat at low temperatures (<150 °C) [8]]. The considerably lower proportions of carbo-hydrates and aldehydes in the material condensing from bark at 250 °C than from peat may be due to the lower content of readily hydrolysable matter (hemicelluloses) in pine bark than in slightly decomposed peat [9].

Emissions from veneer drying have been presented to be derived almost entirely from the flash-off of extraneous compounds in the wood [27] and vary thus between and within the wood species due to the different extractive contents.

The formation of organic compounds in bark drying has been found to depend on the age of the chips, too [24]. The formation rate for fresh bark dried at 160°C was more than three times that for stored bark at 140 °C. In addition, drying time and the particle size and final moisture content of the biomass particles affect the composition and rate of emissions. In the drying process the smaller particles dry faster. The drying of small particles in a high temperature gas flow, the factors influencing the drying, including drying of a single particle at low temperatures (50 - 200°C) and the drying of a single particle in steam under various pressures have been studied by Saastamoinen [28].

5 Formation of emissions

Emissions of biomass drying consist of compounds that are either native constituents of raw materials and are released as such, or are formed from raw materials as a result of thermal degradation [9, 20, 3].

At lower temperatures (<200 °C) the emissions consist mainly of lipophilic extrac-tives, which are released from biomass (Fig. 4). These lipids comprise volatile mono-terpenes that are already naturally emitted from vegetation. Drying at temperatures above 100 °C liberates almost all of the monoterpene content from wood [20]. In addi-tion to monoterpenes, other lipid compounds such as fatty and resin acids are released in drying. At temperatures above 150 °C these compounds have significant vapour pressures leading to their vapourisation (Figs 3 and 4). In the atmospheric drying ex-periments of VTT [9] the major lipid groups released from pine bark consisted of fatty and resin acids, those from birch bark triterpenoid alcohols, and those from peat fatty acids. As the temperature rose from 150 to 350 °C, 10-80 wt% of the free fatty acids and 10-50 wt% of the resin acids in the pine bark were released. The main fatty acids were linoleic acid and oleic acid and the main resin acid dehydroabietic acid.

Thermal degradation starts already at the temperatures above 100 °C liberating volatile carboxylic acids from the biomass, but is more considerable at above 200 °C. Hemicelluloses, cellulose and lignin degrade in the temperature range of 200-500 °C.

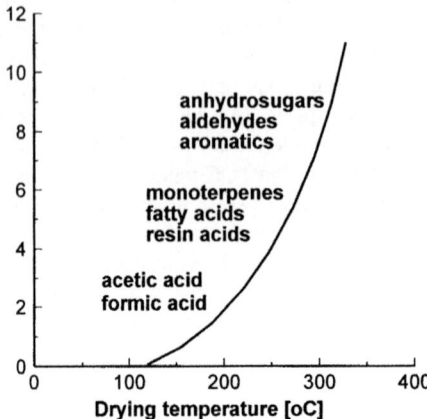

Amount [% of dry feed]

anhydrosugars
aldehydes
aromatics

monoterpenes
fatty acids
resin acids

acetic acid
formic acid

Drying temperature [oC]

Figure 4. The total amount and nature of organic compounds released in atmospheric drying of biomass at different temperatures.

As a result different compound groups like volatile carboxylic acids (formic and acetic acid), alcohols (methanol, ethanol), aldehydes (formaldehyde, glycolaldehyde), furfurals and carbon dioxide are formed and released from biomass. Their amounts increase strongly, as the temperature rises (Figs 2 and 4).

6 The effects of the emissions

The emissions may have an environmentally detrimental effect by causing organic load as condensates on natural waters or gaseous emissions to the air.

Due to their organic load the aqueous effluents require the adjustment of pH and wastewater treatment. According to Liinanki and Karlsson [16] condensates from the steam drying of forest residues need mechanical, biological and chemical purification prior to be led to the environment. The organic load in condensates from the dryer planned to the VEGA plant (25 m^3/h) corresponds a communal wastewater load of 15 000 persons. The condensates were found to be toxic, and were suggested to be diluted with twenty parts with other wastewaters corresponding to those from 30 000 persons before the wastewater plant [15]. A high amount of bark and needles in biomass was found to increase the toxicity of the condensate.

Blue haze is the visual discoloration of exhaust gases, which has been detected from wood dryers and considered by Wastney [20]. Blue-haze is formed when gases from the exhaust stack of dryers condense to form submicron aerosols in ambient conditions. Blue-haze has been found to be dependent on the amount of condensable rather than volatile emissions.

The lipids released may occur in the steam or in the gas flow as vapour phase or in a liquid state as aerosols or on the surface of particles [9]. The tacky fatty and resin acids released have been assumed to contribute to the formation of deposits on the surfaces of peat and bark dryers. These compounds have been found to be enriched in the deposit samples from the heat exchanger tubes, flue gas blowers and recycling fans.

Malodorous emissions from power plants and dryers have been detected [9, 10]. These emissions may be caused by monoterpenes, aldehydes and low-molecular-mass carboxylic acids. Monoterpenes and formaldehyde are strong-smelling irritants.

Monoterpenes have been found to react with nitrogen oxides and produce ozone in the presence of solar radiation. Wastney [20] has widely considered the role of monoterpenes in the formation of photo-oxidants like ozone. In the VEGA project [15] the ozone production was also considered, but was not regarded to become a problem, because the planned power plant was not to be used in summertime.

There is some concern that some drying fumes are carcinogenic [20]. Kurttio et al. [29] have investigated the mutagenicity of fumes from the heating of freshly cut spruce and birch chips. No mutagenicity was found in wood extractives, but during drying of spruce and birch at 170 °C mutagenic compounds were emitted. One of the mutagens was suspected to be Δ^3-carene. The mutagenic compounds appear to be volatile, low-molecular mass organic compounds.

7 Conclusions

The emissions from thermal drying of biomass comprise gaseous emissions to the air and liquid effluents to natural waters. These emissions have been studied in recent years. The studies have focused mainly on those from steam drying processes due to the increased interest in steam drying and in new biomass-based conversion and combustion plants.

The emissions formed in drying processes are affected by many factors, such as dryer type, drying conditions and feedstock material. The comparison of the emissions from different studies with each other is, however, not so straightforward, because in the studies carried out the drying conditions such as temperatures, superheat, and residence time, and biomass material; its composition, particle size and moisture content before and after drying, vary greatly.

The drying temperature seems to have a very important role in the formation of emissions. The emissions from low-temperature (<200 °C) drying comprise mainly lipophilic compounds; high volatile monoterpenes and fatty and resin acids. At higher temperatures the emissions also contain considerable amounts of compounds from thermal degradation of biomass. The amounts of the organic compounds released are then affected by the biomass raw material; its organic composition.

The emissions have environmental, health and operational effects. Environmentally they affect by producing organic load and toxicity in the wastewaters, malodorous odours and ozone formation. The significance of these effects is, however, dependent on local conditions. Operational effects consist of the contribution to the formation of deposits on the surfaces of the dryers.

To reduce emissions from biomass drying it is advisable to use lower drying temperatures and lower drying times, to minimise the gaseous emissions by employing steam drying processes or burning the terpenes containing outlet gases, and by using gas cleaning devices and wastewater treatment after pH adjustment and dilution for the condensates formed.

8 Acknowledgments

The review work was funded by Technology Development Centre Finland (TEKES) through the Bioenergy Research Programme (BIOENERGIA) and by VTT Energy.

9 References

1. McCormick, P.Y. (1988) The key to drying solids, *Chemical Engineering*, August 15. pp. 113-122.
2. Svensson, C. (1979) Back pressure drying, a new system for hogged fuels, *Svensk Papperstidning*, Vol. 82, No. 10, pp. 281-287.
3. Munter, M. & Bäcker, K. (1992) *Biobränsletork för Borås Energi*, Ryaverket, Startekno AB, NUTEK Projektrapporter, VK-93/6, 80 p. + app. 8.
4. Hulkkonen, S., Äijälä, M. & Raiko, M. (1993) Development of an advanced gasification process for moist fuels, in *Proc. 12th EPRI Conference on Gasification Power Plants*. Oct. 27-29, 1993. San Francisco, CA. 16 p.
5. Hulkkonen, S., Parvio, E. & Raiko, M. (1995) An advanced fuel drying technology for fluidized bed boilers, in *Proc. 13th International Conference on Fluidized Bed Combustion*, (ed. K.J. Heinschel), ASME, pp. 399-403.
6. Lundqvist, R. & Österman, J. (1994) A fixed-bed dryer research, in *Bioenergia Res. Progr. Yearbook 1994*, VTT Energy, Jyväskylä. pp. 49-55. (in Finnish)
7. Sipilä. K., Solantausta, Y.. & Kurkela, E. (1994) New options for biomass-based power production by IGCC: a Finnish national research programme- JALO, in *Proc. Advances in Thermochemical Biomass Conversion*, (ed. A.V. Bridgwater), Blackie Academic & Professional, London. pp. 77 - 91.
8. Fagernäs, L. & Wilén, C. (1988) Steam drying processes for peat and their organic condensates, in *Proc. VIII Intnl Peat Congr.*, Vol. II. International Peat Society, Moscow. pp. 261-271.
9. Fagernäs, L. (1992) *Peat and bark extractives and their behaviour in drying processes*. Technical Research Centre of Finland, Espoo., 51 p. + app. 95 p. VTT Publications 121.
10. Moilanen, E., Partonen, S., Moring, K. & Äijälä, M. (1990) Management of emissions in peat drying and burning, in *Proc. Low-grade Fuels*, (ed. M. Korhonen), Technical Research Centre of Finland, Espoo. pp. 243-260. VTT Symposium 108.
11. Salin, J.-G. (1988) Steam drying of wood for improved particle board and lower energy consumption. *Pap.Puu*, No. 9, pp. 806-809.
12. Fagernäs, L. (1993) Formation and behaviour of organic compounds in biomass dryers. *Bioresource Technology*, Vol. 46, pp. 71-76.
13. Fagernäs, L., Sipilä, K. & Ekman, R. (1994) Behaviour of organic compounds in biomass drying, in *Proc. Advances in Thermochemical Biomass Conversion*, (ed. A.V. Bridgwater), Blackie Academic & Professional, London, pp.1533-44.
14. Hulkkonen, S., Heinonen, O., Tiihonen, J. & Impola, R. (1993) Drying of wood biomass at high pressure steam atmosphere; experimental research and application, *Drying Technology*, Vol. 12, No. 4, pp. 869-887.

15. Setzman, E., Brännström-Norberg, B.-M., Rosén-Lidholm, S. & Sundell, P. (1993) *Miljökonsekvensbeskrivning "Från vaggan till graven - fallstudie VEGA"*, Vattenfall AB, Stockholm, 108 p. + 3 app.Forskn.rapport 108. (In Swedish)
16. Liinanki, L. & Karlsson, G. (1994) VEGA - Test och verifikationer. Trycksatt förgasning av biobränsle. *Vattenfall-rapport*, No.12, 79 p.
17. Mroueh, U.-M., Thun, R., Fagernäs, L. & Wihersaari, M. (1996*) Environmental impact assessment of bioenergy systems*, SIHTI 2 Research Programme, Final report, VTT, Espoo, 50 p. (in Finnish). Confidential report.
18. Salo, K. & Keränen, H. (1995) Biomass IGCC, in *Proc. Power production from biomass II*, (ed. K. Sipilä & M. Korhonen), VTT Energy, Espoo, pp. 23-40. VTT Symposium 164.
19. Bridgwater, A.V., Elliott, D.C., Fagernäs, L., Gifford, J. S., Mackie, K.L., Rivard, C. & Toft,A.J. (1995) *The nature and control of solid, liquid and gaseous emissions from the thermochemical processing of biomass*, Report No. ES 94/2, IEA, Task X, Activity 7- Environmental Systems, New Zealand Forest Research Institute, Rotorua, 162 p.
20. Wastney, S.C. (1994)Emissions from wood and biomass drying: A literature review, IEA, Biomass Agreement, Task X, Environmental Systems Activity, New Zealand Forest Research Institute, Rotorua, 68 p.
21. Palonen, J., Lundqvist, R.G. & Ståhl, K. (1995) IGCC technology and demonstration, in *Proc. Power production from biomass II*, (ed. K. Sipilä & M. Korhonen), VTT Energy, Espoo, pp. 41-54. VTT Symposium 164.
22. Liinanki, L., Horvath, A., Lehtovaara, A. & Lindgren, G. (1994) The development of a biomass based simplified IGCC process, in *Proc. 13th EPRI Conference on Gasification Power Plants*, San Fransisco, CA, Oct. 19-21, 1994. 18 p.
23. Fagernäs, L., Ståhlberg, P. & Sipilä, K. (1995) *Drying experiments with a pressurized steam dryer*. Confidential research report. 18 p. + app. 22 p. (In Finnish).
24. Björk, H. & Rasmuson, A. (1996) Formation of organic compounds in superheated steam drying of bark chips, *Fuel*, Vol. 75, No. 1, pp. 81-84.
25. Bongers, G.D., Woskoboenko, F., Jackson, W.R., Patti, A.F. & Redlich, P.J. (1995) The properties of the coal and steam condensate from pressurised steam drying, in *Coal science*, Vol. 1, (ed. J.A. Pajares & J.M.D. Tascón), Elsevier, London, pp. 901-904.
26. Marutzky, V.R. (1978) Emissions of terpenes from drying wood particles, *Holz als Roh-und Werkstoff*, Vol. 36, pp.407-411.
27. Cronn, D.R., Truitt,S.G & Campbell, M.J. (1983) Chemical characterisation of plywood veneer dryer emissions, *Atmospheric Environment*, Vol. 7, No. 2, pp. 201-211.
28. Saastamoinen, J. & Impola, R. (1994) Drying of solid fuel particles in hot gases. *Drying 94. Proc. 9th International Drying Symposium*, (ed. V.Rudolp, R.B.Keey & A.S.Mujumdar), pp. 623-630.
29. Kurttio, P., Kalliokoski, P., Lampelo, S., Jantunen, M.J. (1990) Mutagenic compounds in wood-chip drying fumes, *Mutation Research*, No. 242, pp. 9-15.

Printed by Printforce, the Netherlands

DEVELOPMENTS IN THERMOCHEMICAL
BIOMASS CONVERSION

DEVELOPMENTS IN THERMOCHEMICAL BIOMASS CONVERSION

Volume 2

Edited by

A.V. Bridgwater
Director of the Energy Research Group,
Department of Chemical Engineering and Applied Chemistry,
Aston University, Birmingham, UK

and

D.G.B. Boocock
Chair of the Department of
Chemical Engineering and Applied Chemistry,
University of Toronto, Toronto, Canada

SPRINGER-SCIENCE+BUSINESS MEDIA, B.V.

First edition 1997

© 1997 Springer Science+Business Media Dordrecht
Originally published by Chapman & Hall in 1997
Softcover reprint of the hardcover 1st edition 1997

ISBN 978-94-010-7196-3   ISBN 978-94-009-1559-6 (eBook)
DOI 10.1007/978-94-009-1559-6

A Catalogue record for this book is available from the British Library

Library of Congress Catalog Card Number: 96–86524

∞ Printed on permanent acid-free text paper, manufactured in accordance with
ANSI/NISO Z39.48-1992 (Permanence of Paper).

CONTENTS

VOLUME 1

vii

Chemicals from pyrolysis liquid

Upgrading pyrolysis products

Chemicals from pyrolysis liquid

Upgrading pyrolysis products

COMBUSTION

Overview

Fundamentals

SYSTEM STUDIES

WORKSHOPS

PREFACE

There have been many developments in the science and technology of thermo-chemical biomass conversion since the previous conference on *Advances in Thermochemical Biomass Conversion* in Interlaken, Switzerland, in 1992. This fourth conference again covers all aspects of thermal biomass conversion systems from fundamental research through applied research and development to demonstration and commercial applications to reflect the progress made in the last four years. All aspects of bioenergy systems are covered from pretreatment through to end-user applications with increased consideration paid to the environmental benefits and problems of implementing bio-energy systems.

There was an excellent response with over 200 papers offered and over 180 delegates from 29 countries attending the conference. The programme was divided into five main areas covering pyrolysis, pretreatment, gasification, combustion and system studies and this division is reflected in the structure of these conference proceedings. Each main section was preceded by a state-of-the-art review to provide a focus for the ensuing presentations and an authoritative reference. All the papers included have been subject to a full peer review process.

As with any international conference, an important aim was to exchange ideas and discuss problems with fellow researchers, as well as to hear about the latest research and development and applications. A workshop programme was included to encourage this interaction in areas of interest selected by participants. The resultant workshop reports provide a summary of topical problems and opportunities.

The major developments in the bioenergy field include the growing appreciation of the contribution that can be made to mitigating greenhouse effects through the increased awareness of the environmental responsibilities for implementing bio-energy systems and also the growing emphasis on commercialisation. The academic community will, of necessity, continue to provide the lead in open dissemination of research results to stimulate the development of new ideas. These will translate into new products and processes and provide the justification for continued support of centrally supported R&D. The value of these conferences lies in providing an opportunity for industry to hear about the latest developments and provide an opportunity for their representatives to meet the active researchers and develop better links. A further development is the increased commercial activity which results from overt governmental and institutional support for bio-energy. This industrial pull for newer and better products and processes will stimulate researchers in the private and public sector and should deliver at least some of the promised benefits of the bio-energy sector.

Tony Bridgwater and Dave Boocock
October 1996

ACKNOWLEDGEMENTS

We would like to express our sincere appreciation to those people and organisations who have provided support:

- IEA Bioenergy, through the operating agent represented by C Wallace; and the thermochemical related conversion activities themselves represented by S Babu, J Hustad, K Mackie, Y Solantausta and T Bridgwater;
- Natural Resources Canada, represented by E Hogan;
- National Renewable Energy Laboratory, USA, represented by R Overend;
- The Department of Trade and Industry, UK, represented by N Barker;
- VTT, the Technical Research Centre of Finland, represented by K Sipila.

The scientific committee were very supportive throughout the preparation and running of the conference. Their encouragement and efforts in publicising the meeting, refereeing papers, providing constructive feedback on the programme, organising workshops and chairing sessions was much appreciated.

M. Antal, USA	S. Babu, USA
N. Bakhshi, Canada	T. Beenackers, Netherlands
M. Connor, Australia	J. Corella, Spain
B. Delmon, Belgium	D. Elliott, USA
B. Graham, Canada	E. Hogan, Canada
W. Kaminsky, Germany	B. Krieger-Brockett, USA
J. Kuester, USA	K. Mackie, New Zealand
R. Maggi, Belgium	K. Maniatis, Belgium
D. Meier, Germany	T. Milne, USA
T. Nussbaumer, Switzerland	R. Overend, USA
E. Rensfelt, Sweden	C. Roy, Canada
D. Scott, Canada	K. Sipila, Finland
K. Sjostrom, Sweden	Y. Solantausta, Finland
H. Stassen, Netherlands	S. Yokoyama, Japan

At a personal level, particular thanks are due to Karen Dowden and Claire Humphreys who provided the Conference Administration. We also wish to thank the research staff from Aston University who provided considerable support before, during and after the meeting.

GASIFICATION

GASIFICATION

Fundamentals

CHEMISTRY OF TAR FORMATION AND MATURATION IN THE THERMOCHEMICAL CONVERSION OF BIOMASS

ROBERT J. EVANS and THOMAS A. MILNE
National Renewable Energy Laboratory
1617 Cole Blvd
Golden, Colorado 80401, USA

Abstract
An understanding of the molecular details of tar formation and maturation in thermochemical processes is fundamental to the development of gasification and gas cleaning systems for high efficiency applications, such as in internal combustion engines or in gas turbines. Tars are functionally defined as the condensible organic fraction from gasifier effluents, and hence, are generally considered to be aromatic in nature. This definition does not allow a distinction between classes of compounds which originate under different reaction regimes, such as the primary pyrolysis products, which may be in the gasifier effluent because of low temperature operation or due to process upsets, and high molecular weight polynuclear aromatic hydrocarbons, which are produced under higher reaction severity and are the precursors of particulate matter formation. This paper describes the effect of time, temperature, and oxygen on product composition and the maturation of tars through distinct classes of products. The effect of oxygen at temperatures of 600 °C to 700° C is shown to accelerate the destruction of primary pyrolysis products but has no significant effect on benzene.

Keywords: Biomass, Gasification, Mass Spectrometry, Oxidation, Pyrolysis, Tar Formation

1 Introduction

Past work using molecular beam mass spectrometry (MBMS) [1] has suggested that a systematic approach to the classification of pyrolysis products as primary, secondary, and tertiary products can be used to compare products from the various reactors that are used for pyrolysis and gasification [2]. These gas-phase reactions are not usually explicitly studied in the comparison of gasifier operating modes even though most of the mass of the wood is volatilized in the initial thermal reactions. In indirect gasification systems where thermal cracking, usually in the presence of steam, is used to produce synthesis gas, this reaction pathway is the principal controlling factor in tar composition.

The added effect of partial oxidation of these different product suites has not been addressed in terms of the effect on chemistry or on the relative destruction rates of primary, secondary and tertiary products. Gas-phase partial oxidation is relevant to direct gasification systems, such as the oxygen or air blown fluid bed gasifiers.

Several workers have looked at the kinetics of the vapor-phase pyrolysis of biomass-derived primary pyrolysis products. Antal [3] has studied the vapor phase cracking kinetics of cellulose conversion to permanent gases. Diebold [4] studied the kinetics of wood to gases in the vortex reactor with a tubular vapor cracker attached. These studies have focused on the kinetics of indirect gasification and not on the effect of reactor severity on the formation and composition of the remaining condensed products. The thermal cracking of tars with the addition of steam and oxygen in the cracking zone was reported by Jonsson [5]. Both additives increased the cracking rate over the temperature range studied from 950-1250 °C.

The composition of biomass pyrolysis products and gasifier tars from different processes has been studied by Elliott [6]. He has shown the transition as a function of process temperature from primary products to phenolic compounds to aromatic hydrocarbons. The MBMS was used to rapidly screen tars and oils [2] and this technique allows the products to be placed in the primary, secondary and tertiary product ranges that laboratory MBMS studies have shown [1].

The goal of this paper is to show the relevancy of vapor-phase reactions to biomass gasification and to show how process chemistry can be used to address the nature of the tar that is formed. This information is relevant to the development of catalytic tar cracking since the chemical nature of these tars, as well as the amount, affect the performance of a catalyst, such as dolomite. It is possible that a process utilizing downstream catalytic processing may be better off using a lower temperature to control tar composition even if more tar is made since catalyst performance may be better for the more primary or secondary tars than for the more refractory polynuclear aromatics.

2 Experimental Approach

The data reported in this paper was collected using molecular beam mass spectrometry (MBMS). This technique has been described in a previous paper [1] and in references within. Hence the experimental system is only briefly described below. Through MBMS one samples all light gases and heavy vapors, while preserving reactive and condensible species through rapid quenching and wall-less conversion to molecular flow before line-of-sight introduction into the mass spectrometer ion source. Sampling is in real time with millisecond time resolution for single ions or complete, repetitive scans of 300 amu/second. This allows the simultaneous observation of time-variant processes and the total product distribution by integrating the scans over the whole pyrolysis time interval.

Studies of the effect of thermal cracking and gas-phase oxidation on wood pyrolysis product composition were performed in a tubular reactor. Wood pyrolysis was carried out in flowing gas which had been heated to the desired temperature. The temperature of the up-stream, vapor-phase, cracking zone was controlled independently and the gas-phase residence time was controlled by the flow rate of carrier gas. In a typical experiment, a particle of wood is inserted through the sample introduction port after the mass spectrometer starts scanning. The primary pyrolysis products are mixed with the carrier gas and flow through the vapor cracking zone of the reactor. The effluent from the reactor is sampled through the orifice of the MBMS. To study oxidation, oxygen can be mixed with the carrier gas before the sample introduction port or it can be injected just downstream of the sample so that char oxidation products can be eliminated from the product slate and only gas phase oxidation observed.

3 Results and Discussion

This section will describe three experiments that address the effect of time and temperature on product composition (3.1), the effect of oxygen and temperature on product temperature (3.2), and the results of primary zone gasification where oxygen level and residence time is fixed and temperature varied in the primary product zone form 600 to 670 °C (3.3).

3.1 Major product classes as a result of gas-phase thermal cracking
The temperature was varied from 500 to 1000 °C and three gas phase residence times were used: 150, 300 and 750 ms. The carrier gas for this experiment was helium. This data set was previously used [1] to show the changes that occur in gas phase pyrolysis of wood pyrolysis products. Three reaction zones were identified and labeled primary, secondary and tertiary and the 33 variations in temperature and residence time showed smooth transition between these three main product classes. In that paper, principal component analysis was used to show these changes graphically and how the primary and tertiary products were mutually exclusive. That is, the destruction of the primary products occurs before the tertiary products appear. The tertiary aromatics were formed from cellulose as well as lignin although higher molecular weight aromatics were formed faster from the lignin derived products.

This data set was reanalyzed using multivariate analysis and using self-modeling software [7, 8]. To determine the underlying independent components that are of importance in this data set. Four mathematically-extracted sub-spectra were calculated and the results are shown in Figure 1. The product spectra of each of the original experiments were recalculated to show the relative amounts of these four components. The four components shown in Figure 1 are:

1. Primary products characterized by cellulose-derived products at m/z 126 98,85, 73 60,43, hemicellulose-derived products at m/z 114, etc, and lignin-derived products at 180, 164, 150, 137, 124. These products are described in detail in Reference [1].
2. Secondary products characterized by the phenolic peaks at m/z 136, 120, 110, 108 and olefins at m/z 56, 42.
3. Alkyl tertiary products that include methyl derivatives of aromatics at m/z 166

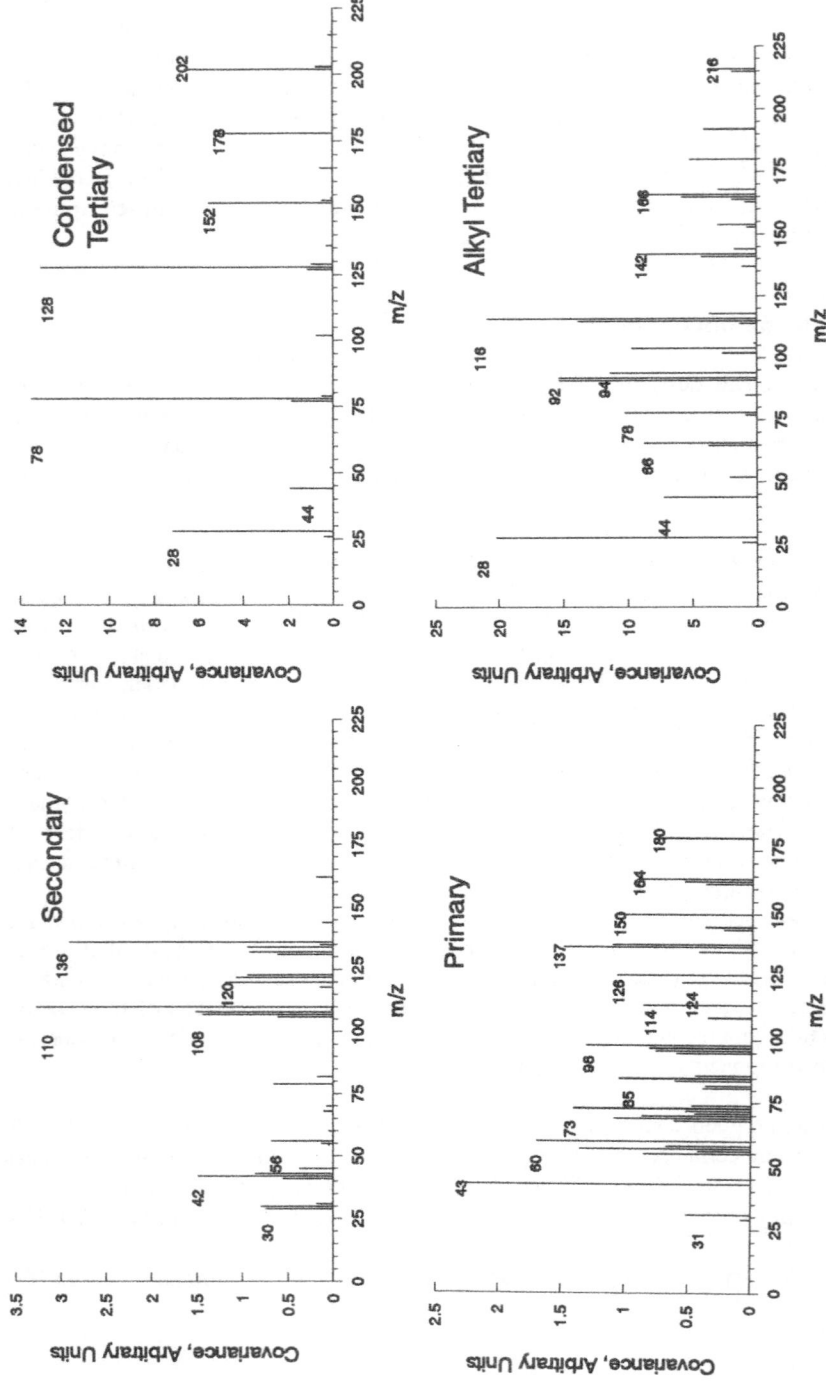

Fig. 1. Mathematically reconstructed component spectra of the compound classes in the thermal cracking data set derived by the methods of Windig et al. [7].

(methyl acenaphthylene), m/z 142 (methyl naphthalene), m/z 92 (toluene), m/z 116,(indene), as well as m/z 94 (phenol) and benzene, m/z 78.

4. Condensed Tertiary Products, that show the polynuclear series without substituents: benzene (m/z 78), naphthalene (m/z 128), acenaphthylene (m/z 152), anthracene/phenanthrene (m/z 178), and pyrene (m/z 202).

The distribution of the four components are shown in Figure 2 for the 300 ms residence time data set. This plot shows how the secondary and alkyl tertiary products grow as the primary products are removed and are then in turn removed as the condensed tertiary products grow. These data show composition of the tars and not tar yields, i.e., the numbers sum to 1 at any given temperature, hence nothing can be said directly about total tar cracking from these data.

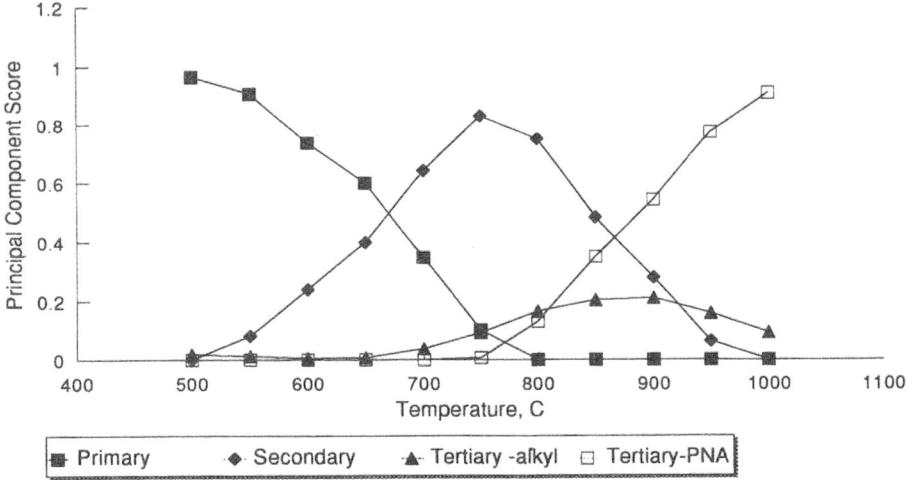

Fig. 2. The distribution of the four components shown in figure 1 as a function of temperature for the 300 ms residence time data set.

The assumption is often made that tars crack to CO, H_2 and other light gases with temperature. This is obviously true with the cracking of the primary products and yields of 50% by weight of CO is possible by thermal cracking [3,4]. However, this is not likely for the condensed tertiary products. These are more likely to be growing in molecular weight with reaction severity. Figure 3 shows the ratio of m/z 202/128 as a function of temperature and residence time. This is an illustration of the increase in molecular weight of the products with reaction severity. Again these data do not show what the absolute changes in the two products are, but rather that the ratio is changing. This raises the question of the ability to analyze for higher molecular weight components as they become more refractory with reaction severity and possibly evolve into particulate matter.

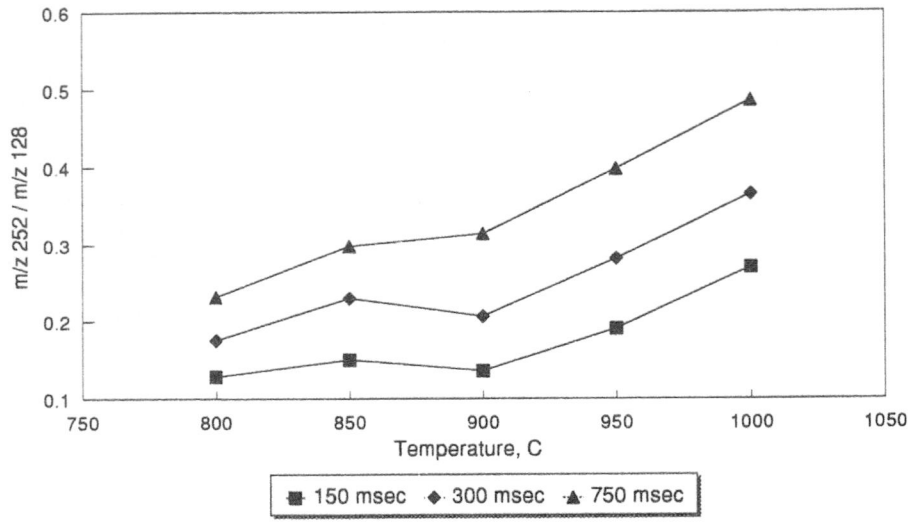

Fig. 3. The ratio of m/z 202/128 as a function of temperature and residence time.

3.2. The effect of temperature and oxygen partial pressure

In this experiment, the temperature was varied from 500 to 800 °C and the oxygen varied from 0 to 20 % of the carrier gas. The 20% oxygen level was approximately equal to 300% of the calculated stoichiometric air requirement. The plots of the % of total abundance for selected ions are shown in Figure 4:

1. m/z 28, primarily due to CO,
2. m/z 124, primarily due to guaiacol and methyl catechol,
3. m/z 60, primarily due to hydroxyacetaldehyde, acetic acid and fragment ions;
4. m/z 128, primarily due to naphthalene.

These plots clearly show that oxygen under these reaction conditions can influence the cracking of these products before complete oxidation occurs. Under these conditions, CO levels rise and reach a maximum at 700 °C and 5% oxygen. The decrease of primaries can be followed by m/z 60 which shows regular decrease in relative abundance as a function of both oxygen and temperature. The trend for m/z 124 shows an initial rise in abundance with temperature for straight pyrolysis and for the 1.2% oxygen level, but decreases regularly with temperature for higher oxygen levels. At 650 C, low levels of guaiacol are observed where m/z 28 is still rising. The conclusion is that lower temperature oxidation may be possible utilizing conditions that avoid the formation of the more refractory tertiary products. The trends for m/z 128 illustrates the behavior of the tertiary products. Naphthalene begins to form at around 650 °C (lower temperature data is due to primary products which occur at m/z 128 and should be ignored here).

At 700 °C the effect of oxygen at levels below 10 % is to increase the naphthalene abundance. This trend continues up to 800 C. Oxygen concentrations of 11 and 20% lead to destruction of naphthalene under these conditions.

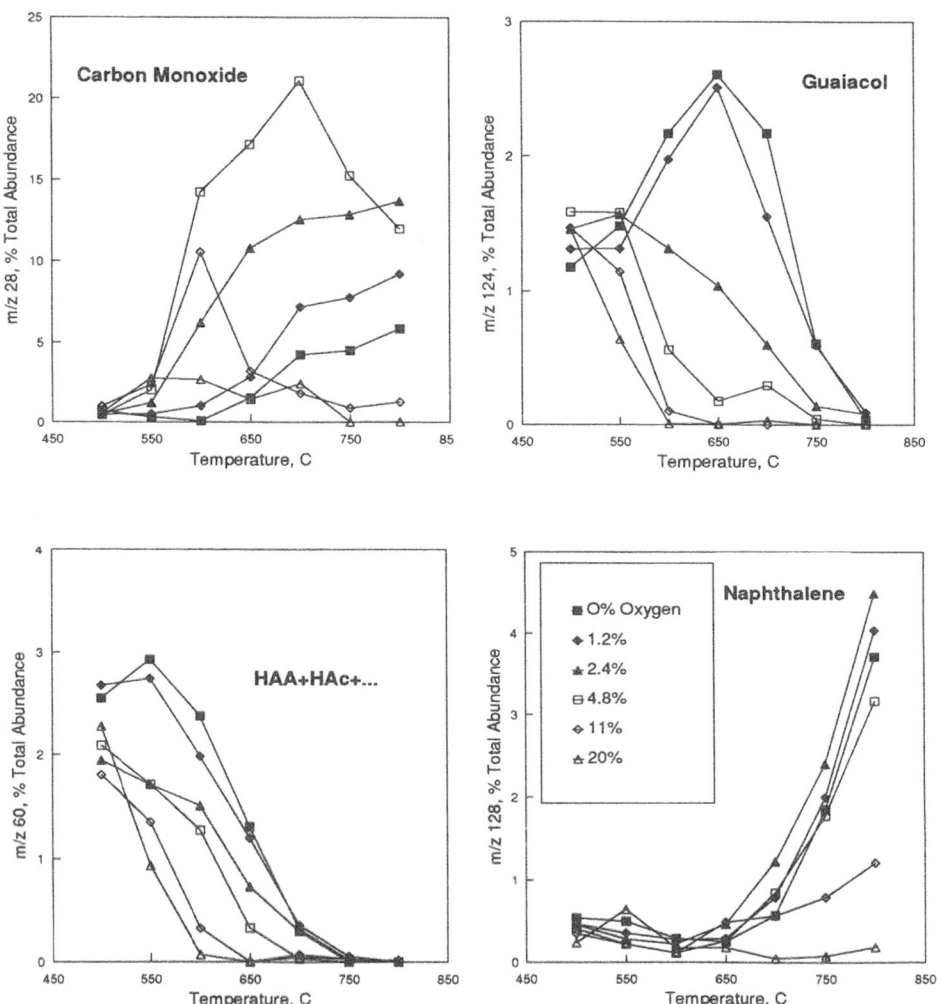

Fig. 4. The effect of temperature and oxygen partial pressure for selected products at a gas phase residence time of 1000 ms.

3.3 Primary Zone Gasification

These oxidation results illustrate the importance of understanding the oxidation of primary, secondary, and tertiary products so that the performance of gasifiers can be mapped according to the relevant zone of operation. That is, the reaction severity and the available oxygen can be used to predict tar composition. However, it is important that realistic conditions be simulated as closely as possible in these tubular reactor studies so as to increase the relevancy of this data. It is also desirable that the data be compared on the basis of yield so that product trends can be followed in a more meaningful way.

An experiment was performed to explore the primary reaction zone by varying vapor cracking temperature from 600 to 670 C. Steam was used as the carrier gas with steam /biomass ratios of 1 to 2 to simulate more realistically the partial pressure of vapors. The wood pyrolysis pulse lasted 30 seconds and the flow of steam was adjusted so an equal weight of steam flowed during that time resulting in a S/B = 1.

Oxygen concentrations were explored that gave incomplete oxidation and this turned out to be from 0 to 10% of the total flow (steam + wood vapor). The 5% oxygen level corresponded to enough oxygen during the pyrolysis pulse to be equal to 45% (including the oxygen in the wood) of the amount needed for complete combustion. The vapor phase residence time was 800 ms for all runs. Runs were performed in duplicate and 33 different conditions were compared in this data set. While quantitative analysis was not performed which would have allowed yields of products to be determined, argon was added to the carrier gas as an internal standard to allow comparisons between runs. Therefore, the data discussed below was corrected for variations in intensity due to changes in the sampling process from run to run and trends in the data reflect changes in yields.

Figure 5 shows three mass spectra from the data set for pyrolysis at 600 C, partial oxidation of the pyrolysis products at 600 °C and 5% oxygen, and partial oxidation at 672 °C and 10 % oxygen. The mass spectrometer was tuned to favor low masses so that hydrogen could be monitored at m/z 2, although it not shown in Figure 5. The pyrolysis spectrum at 600 °C shows a slight degree of cracking as judged by the increase in m/z 28 and 44 and the predominance of m/z 110 over higher molecular weight lignin products, although this is in part due to mass spectrometer conditions. The numbers on the different Abundance Axis are comparable for the three spectra so comparing 0 and 5% oxygen shows increase in CO at m/z 28 and CO_2 ant m/z 44 and decreases in the organic products at other masses. Note that CO and CO_2 are not the only contributions to m/z 28 and 44, since pyrolysis products and secondary hydrocarbon products also are present. This is particularly true in the 600 °C pyrolysis spectrum. The main conclusion to be drawn from this comparison is that the CO is increasing while the organics are being reduced through oxidation.

The spectrum of products at 672 °C and 10% oxygen shows that CO_2 is higher than CO, but traces of the organics are still present. Note that benzene at m/z 78 is now evident as a predominant product indicating a change to secondary products at this higher cracking severity.

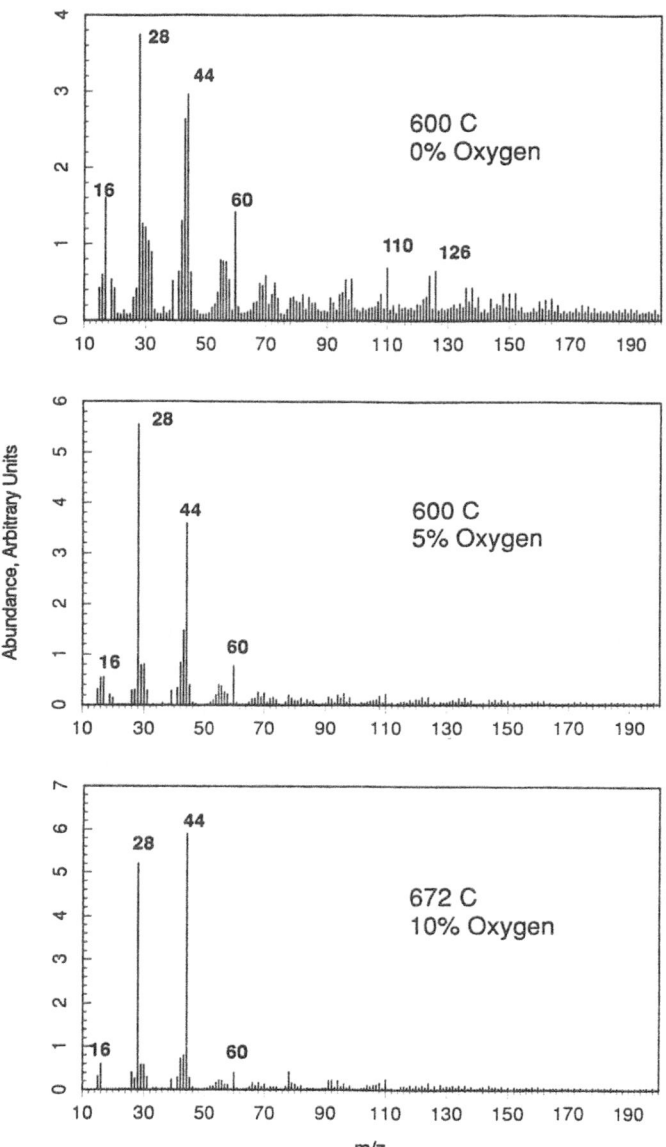

Fig. 5. Mass spectra of the gas phase products from wood pyrolysis and oxidation at different temperatures and oxygen concentrations. Results have been normalized to argon as an internal standard. Argon (m/z 40) and water (m/z 18) have been removed from the graphs.

811

The relative rates of oxidation among different products is further demonstrated in figures 6 and 7, which show results for duplicate runs at two temperatures (600 and 672 C) and three oxygen levels (0, 5, and 10%). Hydrogen shows a significant reduction between 600 °C at 0% oxygen and both 672 °C at 0% oxygen and 600 °C at 5% oxygen. The differences between all other conditions were insignificant.

The difference between the 0 and 5% oxygen level can be explained by H_2 oxidation, but the difference between 600 and 672 °C with no added oxygen is more difficult to rationalize unless it reflects hydrogen consumption by oxygen functionality in the vapor phased organics. Given the low sensitivity for hydrogen compared to other ions in this experiment, this observation should be confirmed by further study. The important observation is that hydrogen shows its own unique response to these conditions compared to the other products to be discussed next and selective oxidation is occurring under these conditions.

The response of m/z 28, CO, to these conditions is to increase with temperature and with oxygen level, however there is no significant difference between temperatures at 5 and 10 % oxygen levels. The important observation is that under realistic conditions of organic vapor partial pressure, the net effect is that the CO level is not reduced by oxidation for the levels of oxygen used here. This confirms the results for the previous experiments shown in Figure 4 although in that experiment the oxygen was introduced before the sample introduction port so char oxidation contributes to the CO signal. In this experiment, the oxygen is injected after the sample port so changes in CO are due only to vapor phase reactions.

The ions from m/z 46-200 were summed together to give the organics shown in Figure 6. The effect of changes in temperature and oxygen level is similar to the results for hydrogen with about the same level of reduction achieved by increasing the pyrolysis temperature to 672 °C as by raising the oxygen level to 5%. Nothing significant is observed between temperatures at 5% oxygen or between 5 and 10 % oxygen.

This effect of total organics is not always true for each individual product and three individual products are shown in Figure. 7 to illustrate. The signal at m/z 124 is largely due to guaiacol, but could also have some contribution from methyl catechol. Nevertheless, the reduction is more pronounced then that demonstrated by the total organics plot. The results for m/z 60, which could represent hydroxyacetaldehyde, acetic acid and a fragment ion of levoglucosan, does not show as pronounced an effect of oxygen addition as the total organics, but does show a decrease due to temperature. The lower response to oxygen than the other products is probably due to acetic acid formation as an oxidation product. Nevertheless, 5% oxygen at 600 °C decreases the level while at 672 °C there is no significant effect. Benzene is quite different, showing an increase at the higher temperature which reflects its classification as a secondary product (it is also present in the tertiary products as well, reflecting several pathways of formation and reaction). The effect of oxygen on benzene is not significant. This is a minor product under these primary product conditions and so conclusions about the trend with temperatures should be studied in a higher temperature zone.

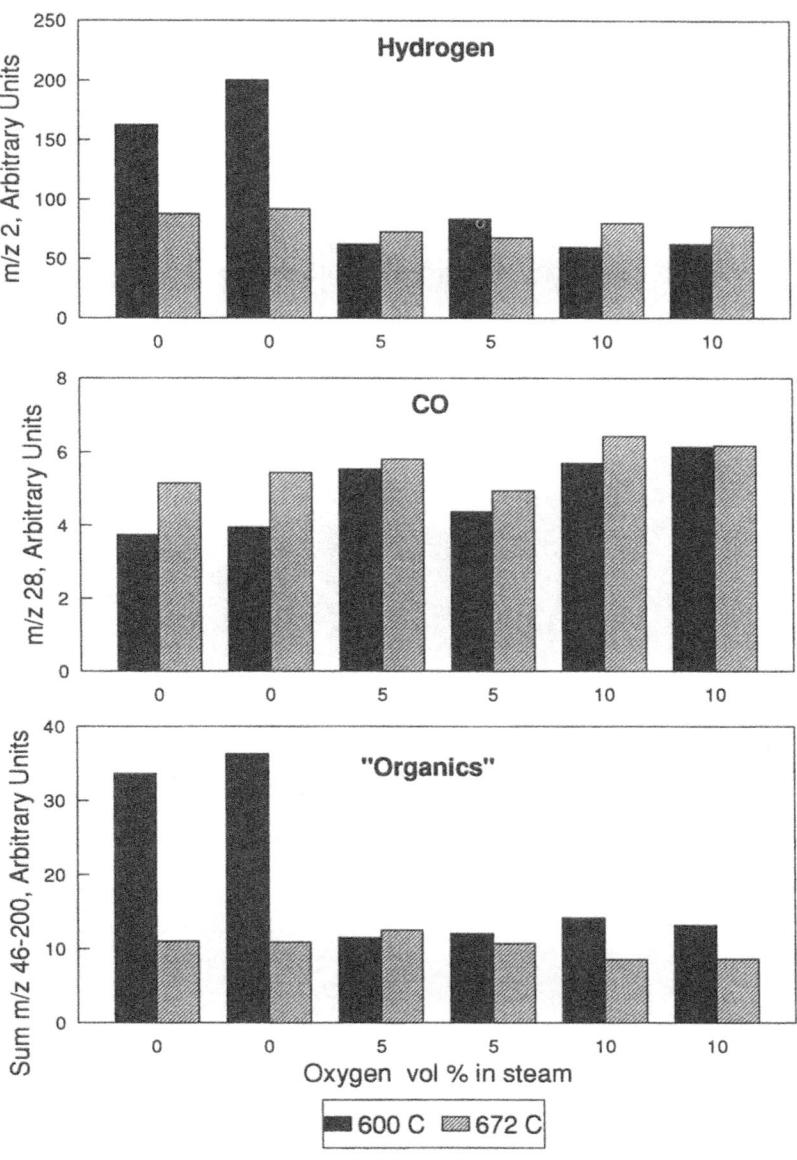

Fig. 6. Comparative yields of products from partial oxidation of wood derived vapors under different conditions of temperature and oxygen concentration

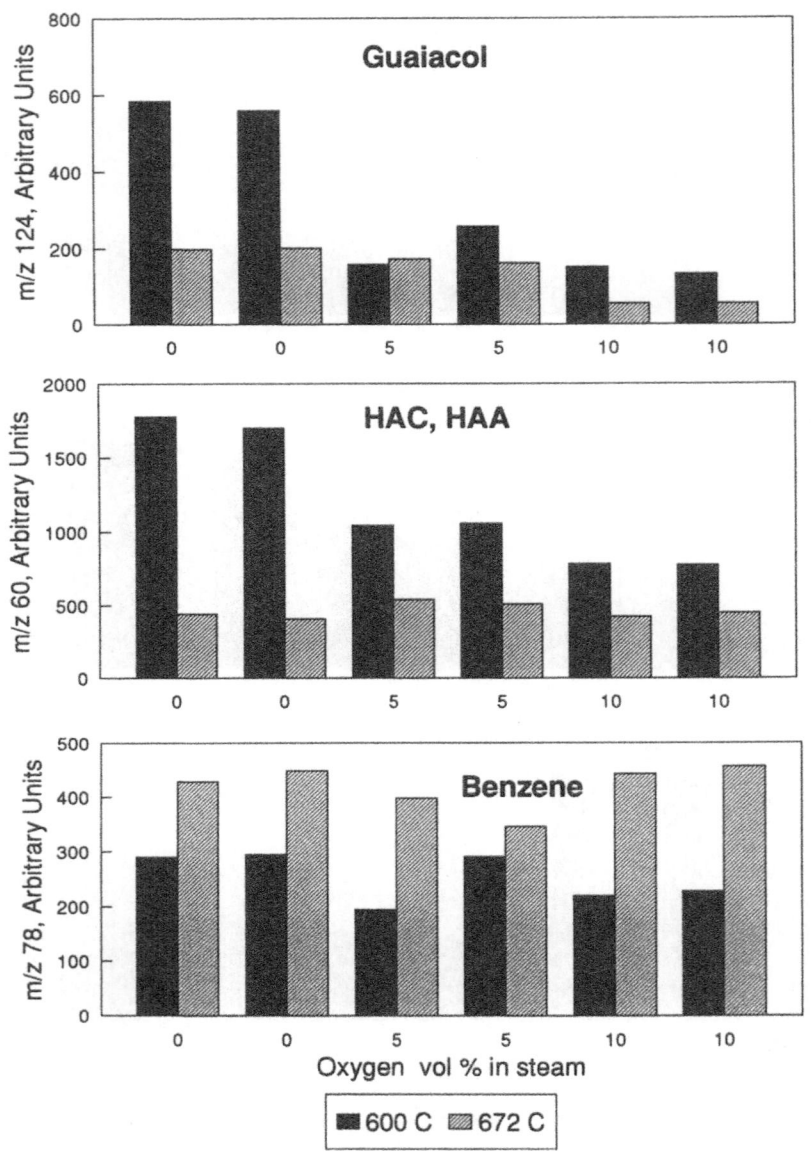

Fig. 7. Comparative yields of products from partial oxidation of wood derived vapors under different conditions of temperature and oxygen concentration.

4 Conclusions

Pyrolysis tar maturation pathways have been shown to be relevant to gasification and understanding the maturation pathway can help to characterize reactor performance. For example the presence of primary and tertiary tars in the same tar sample would indicate non uniform conditions such as channeling or the occurrence of process upsets. The decision to run a gasification system at high severity to crack tars should be balanced by a consideration of the composition the tars that remain. The condensed aromatics in these tertiary tars may prove harder to remove by downstream, catalytic cracking than the larger amount of primary or secondary tars produced under less severe gasification conditions. Results from the reaction severity study showed a temperature where secondary tars are maximum and primary and tertiary tars are minimum, which may be a balance between decreasing the amount of material and controlling the composition so that catalytic materials can function effectively. These results show that the molecular weight of polynuclear aromatic hydrocarbons increases through the tertiary cracking zone and hence maturation of tar to soot or particulates is a process that should be kept in mind both in running gasifiers and in performing chemical analysis to determine the effectiveness of tar cracking. This latter point is based on the possibility that these materials may become difficult to dissolve and mobilize in chromatographic systems as the molecular weight increases.

Oxidation of the vapor phase material is an area that deserves more consideration since most of the mass of the feed will react by this route and not by char oxidation. These results show that partial oxidation increases the tar cracking rate and causes a net increase in CO levels. Secondary and tertiary products are less susceptible to oxidation than primary products and each primary product appears to have its own reaction rate indicating a selective bimolecular process.

Primary zone gasification shows the possibility of selective oxidation of the tar components without sacrificing CO yield and minimizing the formation of secondary aromatic products. The effect of oxygen level on hydrogen is more complicated and hydrogen appears to be more reactive to oxidation than CO under these conditions, but it is not completely removed by the presence of oxygen. A possible process implication of these observations is that primary zone gasification may be a useful step prior to catalytic tar removal even though higher tar loads are present, since the catalyst should function more effectively with theses species rather than the polynuclear aromatics commonly produced by running the gasification process with high reaction severity.

The partial catalytic oxidation of tars was performed by Mudge et al .[9] where oxygen was used to remove the coke from the catalyst. This approach could be applied after a gas-phase partial oxidation to remove the remaining tar and use up the remaining oxygen in addition to oxidizing the carbon deposited on the catalyst. Common gasifier operating conditions of high severity may be selected to optimize char gasification reactions, such as steam gasification, which requires higher temperatures. The net effect on process economics of alternative reaction conditions leading to different chemical pathways is worthy of consideration.

5 References

1. Evans, R.J. and Milne, T.A. (1987) Molecular characterization of the pyrolysis of biomass: I. Fundamentals. *Energy and Fuels*, Vol. 1, pp. 123-137.
2. Evans, R.J. and Milne, T.A. (1987) Molecular characterization of the pyrolysis of biomass: II. Applications. *Energy and Fuels*, Vol. 1, pp. 311-1319.
3. Antal, M.J. (1982) A review of the vapor phase pyrolysis of biomass-derived volatile matter, in *Fundamentals of Thermochemical Biomass Conversion*, (eds. R.P. Overend, T. A. Milne, L.K. Mudge), Elsevier Applied Science, London, pp. 511-536.
4. Diebold, J.P. (1985) *The Cracking Kinetics of Depolymerized Biomass Vapors in a Continuous, Tubular Reactor*, Masters Thesis T-3007, Colorado School of Mines, Golden, CO., USA.
5. Jonsson, O. (1985) Thermal Cracking of tars and hydrocarbons by addition of steam and oxygen in the cracking zone, in *Fundamentals of Thermochemical Biomass Conversion*, (eds. R.P. Overend, T.A. Milne, L.K. Mudge), Elsevier Applied Science Publishers, London, pp. 733-746.
6. Elliott, D.C.(1987) Relationship of reaction time and temperature to chemical composition of pyrolysis oils, in *Pyrolysis oils from Biomass: Producing, analyzing and Upgrading*, (eds. E.J. Soltes and T.A. Milne), ACS Symposium Series 376, ACS, Washington, D.C. pp.55-65.
7. Windig, W., McClennen, W.H., Meuzelaar, H.L.C. (1987) Chemometrics Intelligent Lab. Systems, Vol. 1, pp. 151-165.
8. Agblevor, F.A., Evans, R.J., and Johnson, K.D. (1994) Molecular-beam mass - spectrometric analysis of lignocellulosic materials. I. Herbaceous biomass, *Journal of Analytical and Applied Pyrolysis*, Vol. 30, pp. 125-144.
9. Mudge, L.K., Baker, E.G., Brown, M.D. and Wilcox, W.A. (1988) Catalytic destruction of tars in biomass-derived gases, in *Research in Biomass Thermochemical Conversion*, (ed. A.V. Bridgwater and J.L. Kuester), Elsevier Applied Science, London pp. 1141-1155.

FUNDAMENTALS OF PRESSURIZED GASIFICATION OF BIOMASS
Pressurized biomass gasification

C. ROSÉN, E. BJÖRNBOM, Q. YU and K. SJÖSTRÖM
Department of Chemical Engineering and Technology / Chemical Technology,
Kungl Tekniska Högskolan (KTH), Stockholm, Sweden

Abstract
This is an investigation on tar yield and nitrogen components from gasified birch. The experiments have been carried out in a laboratory development unit (LDU) with a pressurized fluidized bed reactor. The objective is to study, under different process conditions, the variations in product gas, tar and nitrogen compounds from gasified wood fuels. The test reactor is a top-fed bubbling fluidized bed with a working range up to 3 MPa and 900°C. A well-defined Swedish birch has been used for the experiments and two different sand bed materials, silver sand and olivine sand. The two sand bed materials were tested to see if they have any process influence, which could cause misinterpretations of the test results. A mixture of nitrogen and oxygen was the main agent to fluidize the sand bed and gasify the fuel. The main variables for this study were temperature and pressure in the range 700-900°C and 0.4-1.0 MPa.
Keywords: Biomass, fluidized bed, gasification, pressurized gasification.

1 Introduction

The gasification research at the Department of Chemical Engineering and Technology, KTH, started in the middle of the seventies with basic gasification studies [1]. The purpose from the start was to build a gasifier, a fluidized bed, for production of synthesis gas from biofuels. The result came to be a pressurized equipment with a fluidized bed reactor. The LDU, i.e. Laboratory Development Unit, was constructed around 1977 and a secondary reactor was integrated with the system some years later. The research studies on basic pressurized gasification have advanced and now include also co-gasification [2] and environmental aspects.

2 Experimental equipment

Several reconstructions have been made to give today's design with maximum operating pressure of 3.0 MPa. The LDU is constructed for a feeding rate up to 15 kg/h of biomass fuel and with different kinds of fluidizing and gasifying agents such as air, steam and carbon dioxide. Maximum temperature is 900°C for the reactor and secondary reactor, and for the filter 500°C. The fuel hopper and feeder has a fuel capacity of 120 liters.

The feeding system is a stirred equipment which transports the fuel downwards and through the feeder's outlet. After the feeder, the fuel falls freely and passes a feeding measurement device before entering the reactor. The reactor is fed from the top directly into the fluidized bed through a cooled and insulated pipe. The 1.5 meter long reactor is split in two different zones, the reaction zone in the bottom and the freeboard in the upper part. The diameters of the reaction zone and of the freeboard are 0.144 m and 0.2 m, respectively. The inside reactor wall and bottom, designed like a perforated cone, are made of Inconel steel. The electric preheater, in the reactor bottom below the cone, increases the bed temperature up to 600°C. The remaining 100 to 300°C in increased temperature is obtained and controlled by burning added hydrogen with added oxygen. The hydrogen and the preheater are shut off before the experiments and fuel feeding begin. The reactor temperatures and pressure drops are continuously monitored. Thermocouples measure the temperature at nine levels and pressure pipes control the pressure drops in the bed and over the whole reactor. Through the perforated cone, the preheated fluidizing agents pass up and fluidize the bed material.

The product gas exits the reactor at the top and is transported in a heated pipe to the filter. In the high-temperature filter the product gas is cleaned from ash and char dust particles by three 1 m long felted Inconel steel tube socks, heated to prevent condensation of volatiles and tar. The gas finally passes the so-called secondary reactor, or reformer, in which it is possible to reform and crack gas and tar thermally or with a catalyst. The pressure valve is the last heated part, placed after the secondary reactor, before the product gas is cooled in two water coolers at atmospheric pressure. The water coolers condense most of the tar and the produced water, remaining aerosol is trapped in two 5 μm filters. The product gas has to be carefully cleaned and dried before analysis and flow measuring to avoid disturbances.

The whole unit is controlled manually and computerized only for recording temperatures, pressures and flows. The analysis of the product gas is made on a gas chromatograph, connected on-line with analyses every ten minutes. The pressurized fluidized bed reactor with the connecting high-temperature filter and secondary reactor are shown in figure 1.

Figure 1. Pressurized fluidized bed reactor (LDU).

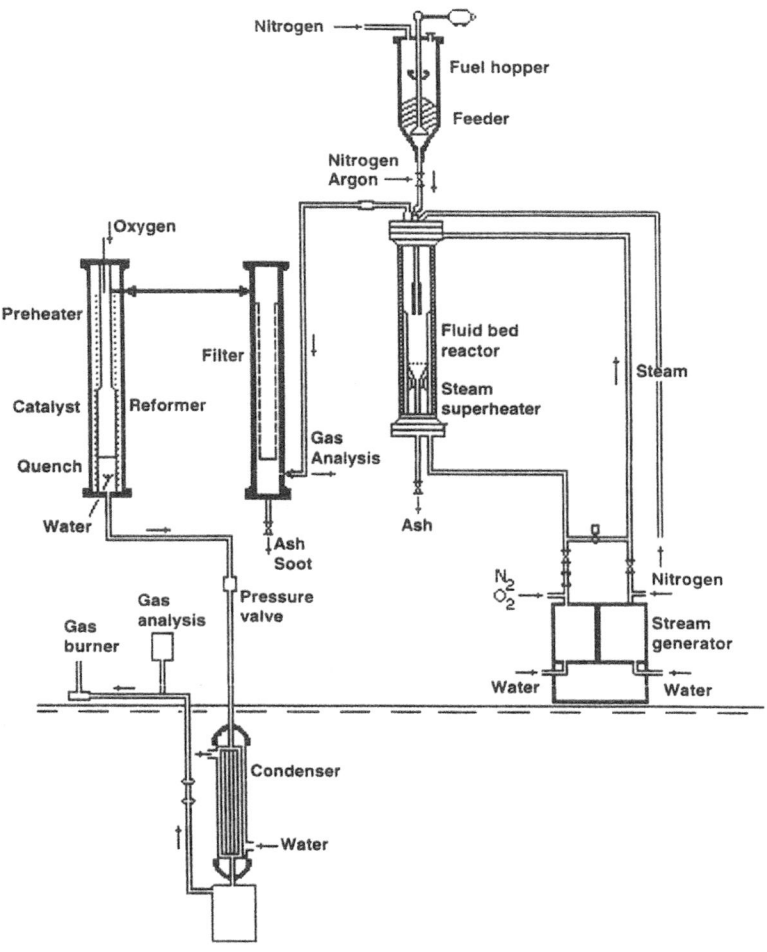

3 Experimental procedure

3.1 Materials and methods
To investigate how bed materials, fluidization gase as well as temperature and pressure levels influence composition of product gases, tar and nitrogen compounds, experiments have been carried out in the bubbling fluidized bed reactor (LDU-unit) (figure 1). The variations in experimental conditions are shown in table 1. The experimental parameters were planned in a factorial design and the order of the experiments was randomized.

Table 1. Treatment conditions for gasification of birch.

Temp. (°C)	Pressure (MPa)	Feeding rate [1] (kg/h)	Particle size (mm)	Sand bed (liters)
700, 800, 900	0.4, 1.0	~3-3.5	1-3	1, 2.5 [2], [3]

[1] calibrated, [2] silver sand, [3] olivine sand

3.1.1 Raw material
The fuel for the experimental work is a well-defined Swedish birch with less than one percent bark. The room-dried birch fuel, with a moisture content of 6-8 %, was milled and sieved to a particle size of 1-3 mm. If the moisture content becomes higher it will influence the feeding system. The problem with too high moisture content is the decreased fuel flow and flow behavior in a feeder with no screw-feeding device. The ash content, which was controlled after each run, had some fluctuations depending on how well the bark was mixed into the fuel for each analysis sample. The chemical composition with average ash and energy content of the wood raw material are shown in table 2.

Table 2. Data on birch for gasification in the LDU unit.

Material	(wt%)	C	H	N	O	S	Ash	MJ/kg (HHV)
Birch		49	7.0	0.1	n.d.	<0.1	0.2	19.3

3.1.2 Reactor bed material
Two different bed materials in different sizes were tested and compared, silver sand and olivine sand (table 3), to see if any effect on the process occurs from and between the sand types. Existing data, from our previously evaluated test runs at the same conditions using olivine sand, made silver sand the mainly used bed material during these tests. To fit our equipment and to keep the fluidization gas at manageable flow levels, the sand fraction size was calculated based on flow, pressure and temperature. To get a well fluidized bed, we need four times the calculated minimal fluidization rate [3]. The fluidizing and gasifying gas used was a mixture of nitrogen and oxygen.

Table 3.   Data on sand bed materials for gasification of birch in the LDU unit.

Material	MgO (wt%)	SiO$_2$ (wt%)	FeO+Fe$_2$O$_3$ (wt%)	Al$_2$O$_3$ (wt%)	CaO (wt%)	Compact density (kg/m^3)
olivine sand *	47-48	42-43	8.0	1.0	0.5	3300
silver sand *	-	98.4	-	-	-	2300

* standard analysis values from manufacturer

3.1.3 Experimental sampling and analysis

The sampling for tar and nitrogen compounds was taken after the reactor and before the filter during one hour, at stable test conditions.   A conduit of product gas, split into two lines, supports the tar and nitrogen compounds sampling. The device for the tar sampling is water-cooled condensers and cyclone-shaped dry ice traps.   For trapping the nitrogen compounds, ammonia and hydrocyanide, the product gas was led through bubble bottles with sulfuric acid and sodium hydroxide.   The total sampling flow is around 3 l product gas a minute.   When the gas has passed the sampling equipment, it is analyzed by an on-line NO indicator as a last step.   The amounts of tar and tar composition in the product gas are analyzed by chromatography, though we know that heavier components cannot be analyzed.   The ammonia is analyzed with an ISE (Ion Selective Electrode) and hydrocyanide by a wet chemical procedure.

The main product gas flow is analyzed every 10 minute in an on-line gas chromatograph with two detectors.   Analyses of CO$_2$, H$_2$, Ar, N$_2$, CH$_4$, CO and C2 hydrocarbons are made on the TC-detector and hydrocarbons up to C7 on the FI-detector.

3.1.4 Mass balance

For control of the experimental reliability an integral carbon balance is made over the system for every run (table 5).   Carbon out in gas, tar, reactor char and filter dust is compared with total carbon in fed wood.   Normal values for carbon balances in our unit are 90-100% for tests at 800 and 900°C and somewhat lower figures for runs at 700°C.   The largest error in the carbon balance comes from product gas losses.   This rather big LDU unit is difficult to keep totally leakage free.   To ascertain the size of the leakage during runs, some cold pressure tests have been performed.   This kind of tests give an idea of leakage levels.   To confirm and to get more precise leakage data, a small flow of argon was added to the equipment as an internal leak finder during the runs.   The argon content in the product gas was then analyzed on-line on the gas chromatograph, and any losses could easily be accounted for as product gas leakage.   Besides the argon balance a comparison with the nitrogen balance is made over the system.   The second source of error in the carbon balance is remaining soot and tar in the reactor, which adhere especially at lower reactor temperatures.   The top part of the free-board in the reactor is principally heated up by the out-going product gas.   The reactor gets a decreasing temperature profile from the reaction zone to the outlet in the top which may cause, at low temperature tests, that the top flange becomes too cold. With too low temperature in that area an accumulation of soot and some condensed volatiles get stuck on the flange surface. This accumulated carbon will be missing in the test evaluation. The soot can easily be burned out with increasing temperature, which can cause a carbon balance of over 100% for a subsequent high temperature test.

4 Results

Thirteen tests have been carried out with analyses of char, tar, product gas and nitrogen gas compounds. Tar analysis results are shown in table 4 and remaining data from the tests in tables 5 and 6.

4.1 Char

After the experimens, char samples were taken for analysis from the reactor bed material. The bed material of sand and char contained more or less char depending on the temperature and pressure. The char production in the high temperature tests was low but increased at increased pressure. The pressure influence on char production is low up to 1.0 MPa but increases between 1.0 and 1.5 MPa. Visual inspection of the char particles show that sand from the sand bed was stuck on the char surfaces and impossible to separate. The measured char ash content of around 70% has same error because of the sand particles. The main part of the fuel nitrogen is converted to nitrogen gas compounds. The remaining nitrogen content in char is low, somewhat lower than literature data [4]. Elementary analyses on the reactor chars indicate only small increase of nitrogen content, compared with the raw material.

4.2 Tar

Tar yields in the LDU-unit are generally low compared with amounts of tar in the literature. There may be something in our equipment or operating parameters which promotes decomposition. For the two tests B and C (table 5), made at identical

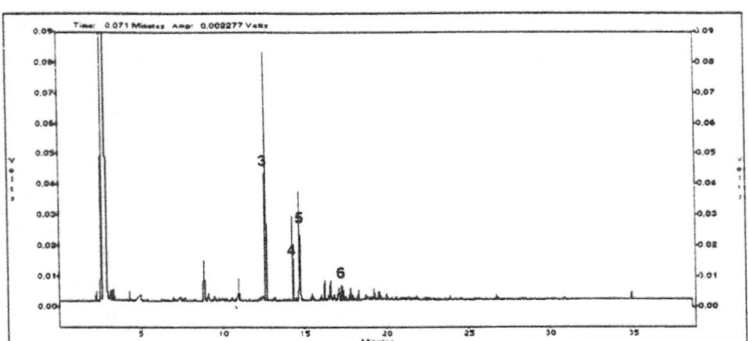

Figure 2. Capillary GC of crude tar produced at 700°C and 0.4 MPa.
Peaks: 1= Benzene; 2= Toluene; 3= Phenol; 4= Indene+o-Cresol; 5= p-Cresol;
6= Naphthalene; 7= 1-Methylnaphthalene+2-Methylnaphthalene;
8= Acenaphthylene; 9= Fluorene; 10= Phenanthrene+Anthracene;
11= Pyrene

822

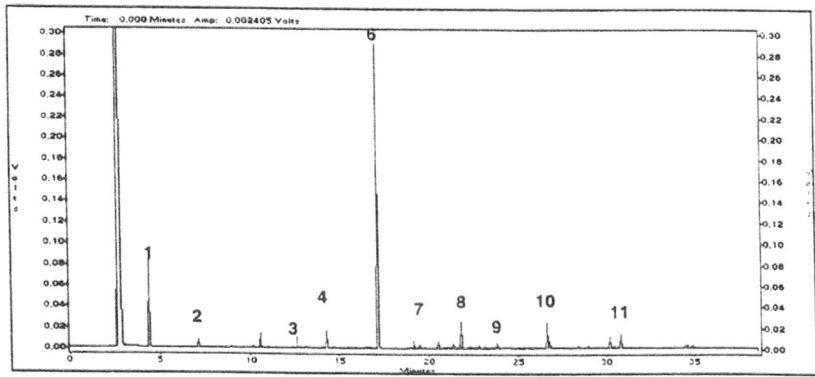

Figure 3. Capillary GC of crude tar produced at 900°C and 0.4 Mpa.
Peaks as in figure 2.

conditions exept for the amount of sand bed material, the tar yield is actually lower
without sand bed. The test B has a relatively high feeding rate and the higher pressure
level which produces more char. The bed temperature decreases the char production
but during the run the reactor contained a momentarily large char amount. A possible
explanation of the decreased tar amount in test B could be the gas residence time in

Table 4. Test data and composition of chromatography-identifiable tar from LDU
tests. Tar analysis results in wt.% of total tar.

Test nr.	1	2	3	4	A	B	C	E	F	G	5	6	7
Temp. [°C]	690	850	660	880	700	880	880	890	690	670	800	700	790
Pressure [Mpa]	0.4	0.4	0.4	0.4	0.4	1.0	1.0	0.4	0.4	1.0	0.4	1.0	1.0
Sand type	Silver	Silver	Silver	Olivine	Silver	-	Silver	Silver	Olivine	Silver	Olivine	Silver	Silver
Sand bed [liters]	2.5	1.0	1.0	1.0	2.5	-	2.5	2.5	2.5	2.5	2.5	2.5	2.5
Phenols	71.0	4.2	67.1	3.8	60.8	5.2	4.1	4.0	62.8	66.3	55.3	76.5	51.3
Neutrals	29.0	95.4	31.8	96.0	38.6	94.6	95.9	96.0	37.2	33.7	43.7	21.3	47.6
Bases	- [1]	0.4	1.1	0.2	0.6	0.2	- [1]	- [1]	- [1]	- [1]	1.0	2.2	1.1
Tar [2]	4.1	3.7	1.6	4.0	4.6	3.4	5.3	3.2	3.2	1.2	1.5	1.6	2.3

[1] value under calibration level, included in total tar

[2] g tar/kg moisture free fuel

823

contact with the reactive char. The amount of tar (g/kg raw material) which is analyzable by chromatography is nearly constant in the temperature-interval 700-900°C. The results give a low conversion to tar of fuel carbon [5]. Our method to define the total amount of tar and tar composition by chromatography analysis does not give a complete result, but the low total tar amounts could only be larger by the amount of the heavier molecular component losses. When increasing the tar amounts with around 10-15% [6] of higher molecular weight components, the obtained amounts are still at quite low levels. Benzene and toluene are here calculated as belonging to the gas phase [7] (table 6). The analysis method is under development to fit our small tar-water samples, with low tar concentration. The evident structure change in the tar between 700 and 900°C can be seen in table 4 and figures 2 and 3. The tar composition and amounts for a given temperature are not appreciably influenced by pressures up to 1.0 MPa.

4.3 Nitrogen compounds

The fuel nitrogen converts principally to nitrogen gas compounds, with a low remaining concentration in the char. With the low conversion to NH_3, HCN and NO from the fuel nitrogen, a large proportion must convert to nitrogen gas. The concentration of ammonia in the product gas seems follow the earlier results [2] by increasing with temperature, with no evident influence of varying sand bed materials and amounts. Increased pressure seems to decrease production of ammonia (figure 4).

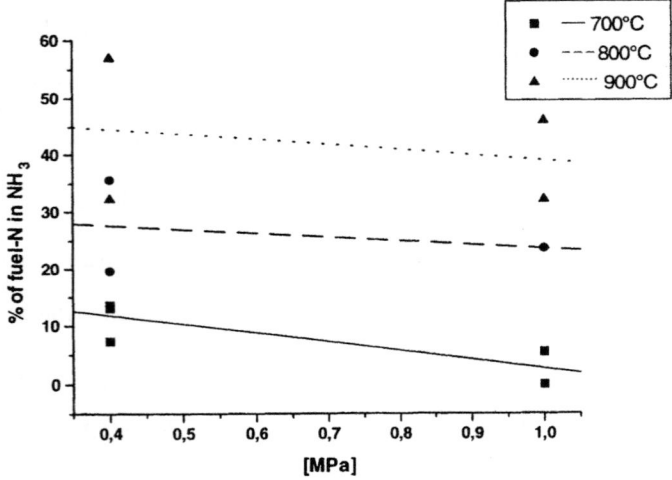

Figure 4. Diagram of fuel nitrogen conversion to NH_3 vs. pressure at various temperatures.

The fuel nitrogen conversion to ammonia in the LDU reactor is very low. The conversion, in percent, is around half compared with other similar reactors [5, 8]. Something in the reactor either impedes ammonia formation or cracks the produced ammonia molecules by some catalytic effect or radical reactions, which may occur.

Table 5. Data and analysis results from gasification of birch.

Test	1	2	3	4	A	B	C	E	F	G	5	6	7
Sand type:	Silver	Silver	Silver	Olivine	Silver	No sand	Silver	Silver	Olivine	Silver	Olivine	Silver	Silver
Sand bed [liters]:	2.5	1.0	1.0	1.0	2.5	-	2.5	2.5	2.5	2.5	2.5	2.5	2.5
Average sand size [mm]:	0.27	0.27	0.27	0.27	0.27	-	0.2	0.3	0.25	0.18	0.27	0.18	0.2
Feeding rate [kg mf*/h]:	2.3	3.1	1.7	2.6	2.7	5	2.5	3	2.7	2.4	2.6	1.1	2.0
Bed temperature [°C]:	690	850	660	880	700	880	880	890	690	670	800	700	790
Pressure [MPa]:	0.4	0.4	0.4	0.4	0.4	1.0	1.0	0.4	0.4	1.0	0.4	1.0	1.0
Gas yield [Nm^3/kg mf* fuel]:	0.91	0.86	0.92	0.86	0.78	0.62	0.95	0.94	0.85	0.72	0.85	1.16	0.93
Mass balance C (out/in) [%]:	77	115	84	101	87	83	85	92.5	75	76	93	87	98
Input-C in product gas [%]:	69	97	73	97	80	82	82	91	63	60	91	83	97
Input-C in tar [%]:	0.77	0.70	0.31	0.75	0.86	0.64	3.68	0.60	0.60	0.23	0.29	0.30	0.69
% of fuel-N in NH_3:	13.2	35.6	13.6	56.9	13.1	32	45.9	32.2	7.3	5.44	19.6	-[2]	23.6
% of fuel-N in NO:	6.6	3.7	10.2	2.5	4.72	0.95	3.0	-[1]	3.76	2.61	4.7	11.1	5.9
% of fuel N in HCN:	0.07	0.72	-[2]	2.48	-[2]	0.09	0.18	-[2]	-[2]	-[2]	-[1]	0.05	0.19
% of fuel N in char:	3.2	-[3]	0.54	-[3]	6.2	-[3]	-[3]	-[3]	5.2	6.3	-[3]	0.73	-[3]
Tar[4] [g/kg mf * fuel]:	4.10	3.70	1.65	4.00	4.55	3.40	5.30	3.17	3.17	1.24	1.54	1.59	2.32
Nitrogen in maf ** char [%]:	0.11	-[3]	0.34	-[3]	0.42	-[3]	-[3]	-[3]	0.5	0.3	-[3]	0.34	-[3]

[1] sampling incorrect, [2] value under calibration level, [3] no char, [4] tar analyzed by GC

* mf: moisture free, ** maf: moisture and ash free

Table 6. Composition of gaseous products obtained in gasification of birch (vol.%, nitrogen and steam free)

Test nr.	1	2	3	4	A	B	C	E	F	G	5	6	7
Temp. [°C]:	690	850	660	880	700	880	880	890	690	670	800	700	790
Pressure [Mpa]:	0.4	0.4	0.4	0.4	0.4	1.0	1.0	0.4	0.4	1.0	0.4	1.0	1.0
CO:	33.8	42.1	33.5	38.0	46.9	41.9	41.1	40.4	41.2	41.3	41.0	34.4	39.2
CO_2:	39.7	27.3	39.9	30.4	33.1	26.6	30.7	31.6	28.8	32.8	30.7	45.0	38.7
H_2:	11.9	11.6	11.4	13.0	6.9	11.1	9.5	10.0	13.2	9.2	9.5	7.2	7.9
CH_4:	10.8	13.9	10.8	13.5	9.4	15.4	13.3	12.8	11.6	12.2	13.4	9.4	10.4
C_2H_2,H_4:	2.5	3.6	2.9	3.6	2.9	3.1	3.9	3.7	3.2	2.6	3.9	2.6	2.7
C_2H_6:	1.00	0.29	0.94	0.21	0.55	0.76	0.69	0.30	1.34	1.33	0.69	0.91	0.52
C_6H_6:	0.30	1.10	0.30	1.19	0.23	1.02	0.66	1.02	0.34	0.42	0.64	0.23	0.52
C_7H_8:	-	0.11	0.26	0.11	-	0.15	0.13	0.14	0.26	0.18	0.13	0.13	0.10
Gas yield*:	0.91	0.86	0.92	0.86	0.78	0.62	0.95	0.94	0.85	0.72	0.85	1.16	0.93

* Nm^3 nitrogen and steam free gas/kg moisture free fuel at NTP

The abilities of silver sand and olivine sand to convert wood fuel nitrogen to different nitrogen gas compounds, or influence the study of the main test variables are equally low and not determining factors in our unit. Somewhat increased formation of nitrogen oxide when using silver sand is apparent. NO decreases at increasing temperature which was expected and fluctuates at low concentration levels. The very low concentrations of HCN, with values commonly under calibration level, follow the same temperature behavior as ammonia.

5 Discussion

The amounts of produced nitrogen compounds from fuel nitrogen and their distribution are influenced mostly by the temperature. The main part of the fuel nitrogen converts to nitrogen gas and only a small amount remains in nitrogen compounds and in the char. Why the fuel nitrogen conversion to ammonia only reaches relatively low levels in the LDU unit, overall lower than in other similar reactors [5] is difficult to explain. One possible explanation is that some radical reactions or catalytic wall effects occur in the reactor.

Our tar results from the LDU unit show no large difference between the tests but with lower carbon conversion to tar compared with literature data. A tendency is that tar yield seems follow trends in literature. The experimental conditions with large dilution by nitrogen of the produced gases gave little tar in low concentration in water. The small tar water samples are impossible to balance out, therefore chromatography analysis is used to obtain total tar amount. This is done although higher molecular weight compounds such as asphaltenes cannot be detected which may lead to somewhat low amounts of total tar, especially at lower temperatures. But we are not sure that tar components from the LDU reactor with higher molecular weight actually do get lost and should be added to the total tar amount. Several parameters are involved and could influence the tar yield. The tar analysis method is under development.

6 Acknowledgement

The project was initiated at the Department of Chemical Engineering and Technology and the research is conducted within the framework of the AIR2 Programmme of the European Commission, but it is presently financed by the Swedish Energy Agency (NUTEK). This financial support is gratefully acknowledged.

7 References

1. Rensfelt, E. et al. (1978) Basic Gasification Studies, IGT-conference on *Energy from Biomass and Wastes*, Washington, DC.
2. Chen, G. et al. (1994) Co-Gasification of Biomass and Coal in a Pressurized Fluidized Bed Reactor, *Biomass for Energy, Environment, Agriculture and Industry*, 8th E.C. Conference, Pergamon, Oxford, Vol. 3, pp. 1830-1835.
3. Wen and Yu (1966) *A.I.Ch.E. Journal*, Vol. 12, 610.
4. Keller, R., Nussbaumer, T. (1993) Fuel Nitrogen in Wood, in *Advances in Thermochemical Biomass Conversion* (ed. A. V. Bridgwater), Blackie Academic & Professional, London, pp. 549-562.
5. Kurkela, E., Ståhlberg, P. and Laatikainen, J. (1993) Pressurized fluidized-bed gasification experiments with wood, peat and coal at VTT in 1991-1992, Part 1: *Test facilities and gasification experiments with sawdust*, VTT Publications, Technical Research Center of Finland, Espoo.
6. Brage, C., Yu, Q. and Sjöström, K. (1995) Characteristics of evolution of tar from wood pyrolysis in a fixed-bed reactor, *Fuel*, Vol. 75, 213.
7. Aldén, H., Espenäs, B.-G. and Rensfelt, R. (1988) Conversion of tar in pyrolysis gas from wood using a fixed dolomite bed, in *Research in Thermochemical Biomass Conversion* (eds. A.V. Bridgwater and J.L. Kuester), Elsevier, London, pp. 987-1001.
8. Leppälahti, J., Kurkela, E., Simell, P. and Ståhlberg, P. (1993) Formation and Removal of Nitrogen Compounds in Gasification Processes, in *Advances in Thermochemical Biomass Conversion* (ed. A. V. Bridgwater), Blackie Academic & Professional, London, pp. 160-174.

GASIFICATION REACTIVITIES OF BIOMASS FUELS IN PRESSURISED CONDITIONS AND PRODUCT GAS MIXTURES

A. MOILANEN and K. SAVIHARJU[*]
VTT Energy, P.O.Box 1601, FIN-02044 VTT, Espoo, Finland

Abstract

Measuring data required for describing the reactivity of biomass-based fuels for pressurised gasification, like various straws (wheat, barley, reed canary grass) and wood-based feedstock (pine, willow, forest residues, miscanthus, sweet sorghum, kenaf), are discussed. The measurements were carried out isothermally in a pressurised thermobalance by determining the gasification rates of char. The main variables were temperature and partial pressures of H_2O and H_2, and CO_2 and CO. The reaction rates were determined in binary gas mixtures H_2O-H_2, and CO_2-CO as well as in product gas mixtures H_2O-H_2-CO_2-CO. The total pressure range was 1-30 bar and the temperature range 650-950 °C. The char samples were produced by pyrolysing the samples in a thermobalance in the gasification conditions by placing the sample in the reactor in the reaction conditions adjusted.

In the paper, the characteristic gasification behaviour of the samples in the presence of product gases is discussed. For wood kinetic parameters were determined using the Langmuir-Hinshelwood kinetics, which takes into account the effect of the product gases on the reaction rate. The results indicated that the H_2O-H_2 system was well in conformity with this kinetics, while the reaction rates measured for the CO_2-CO system required the description of the catalytic effects of the ash-forming material. In the gasification of wood as well as of other biomasses, the behaviour of catalytically active ash components is more complex during char gasification.

Keywords: biomass, gasification, reactivity, pressure, product gas

1 Introduction

The basis of this study was in the need of adding knowledge of the behaviour of various biomass-based fuels, especially, for the development work on pressurised gasification processes based on fluidised-bed technique. In coal gasification, a lot of fundamental research has been carried out to create the scientific background required for the commercialisation of new and quite complex power production systems such as pressurised IGCC technique. Characteristics of various biomasses have been studied much less.

---

[*] Present address: Ahlstrom Machinery Corporation, P.O. Box 5, FIN-00441 Helsinki, Finland

In the present study, the gasification behaviour of biomass feedstocks was characterised for pressurised fluidised-bed gasification. The testing conditions chosen were typical of a pressurised fluidised-bed gasification process, i.e. temperature range max. 1000°C and pressure max. 30 bar. The aim was to produce measuring data for the assessment of carbon conversion in pressurised gasification of solid fuels. The factor affecting the carbon conversion is the reactivity of the char residue after the feedstock material is pyrolysed in the reactor. There are fairly many articles in literature on the characteristics of biomass gasification reactivity and the role of the inorganic material in catalysing it, e.g., [1, 2, 3].

To evaluate the behaviour of biomass in the fluidised-bed gasification, the conditions involved in the reactivity of char material are those existing in the bed and in the freeboard. In this context, particularly the presence of the product gas components must be taken into account, especially for the char material circulating in the freeboard. The product gas components have been found to significantly reduce the gasification rate for coal and peat [4, 5, 6, 7]. The biomass-derived fuels differ from coal, i.e., with regard to the high content of volatile substances (i.e. small amount of residual char) and other type of ash-forming material, which can catalyze the char gasification reactions.

Gasification rates of residual char formed in pyrolysis in the presence of the product gas of gasification were determined on a pressurised thermobalance (PTG) for pine sawdust. The most significant variables were temperature, pressure and H_2O - H_2 and CO_2 - CO ratios. The measurements were carried out for these gas mixtures as a function of partial pressures of the components and of temperature. Kinetic parameters were determined on the basis of the reaction rate values obtained with the aid of Langmuir-Hinshelwood formulas. Gasification reactivities of other biomass types determined in the PTG were compared to the results obtained for pine sawdust.

2 Experimental

The measurements were carried out in the pressurised thermobalance system. The principle of the testing method has been presented previously [7]. The tests were carried out in the temperature range of 700-900 °C and in the pressure range of 1-30 bar (abs.). Pine sawdust was mainly used as sample, the average particle size being 0.1 mm. For characterising the gasification behaviour of different biomasses the samples presented in Table 1 were used [8].

The sample was pyrolysed by lowering it by the winch of the thermobalance direct into the reactor, into gasification conditions. The aim was to get rapid pyrolysis that to some extent corresponds to process conditions. The average heating rate of the sample obtained in this way was some 10 °C/s. The conditions for measuring the char gasification were stabilised after 60 seconds from the input of the sample, including the pyrolysis stage. This part of the weight-time signal recorded by the thermobalance system was removed in the data evaluation. The results obtained for mass change and for the rate of mass change with this treatment are shown in Fig. 1. The rate of mass change due to char gasifying and pyrolysis was given as a so-called instantaneous rate of reaction r" (%/min). It was calculated by dividing the rate of mass change of the sample by the residual ash-free mass. X (%) is the conversion of the total ash-free sample.

Table 1. The characteristics of the feedstock samples, wt%, dry basis

Sample	Volatile matter, %	Fixed carbon, %	Ash %	C %	H %	N %	$O_{diff,}$ %	S %
Pine saw dust	83.1	16.8	0.1	51.0	6.0	0.1	42.8	nil
Pine bark	73.1	25.3	1.7	52.5	5.7	0.4	39.7	0.03
Forest residue (pine)	79.3	19.4	1.3	51.3	5.8	0.4	40.9	0.02
Salix	79.9	18.9	1.2	49.7	6.1	0.4	42.6	0.03
Wheat straw	77.7	17.6	4.7	47.5	5.9	0.6	41.5	0.07
Barley straw	76.1	18.0	5.9	46.2	5.7	0.6	41.5	0.08
Reed canary grass	73.5	17.6	8.9	45.0	5.7	1.4	38.9	0.14
Miscanthus	78.5	18.2	3.3	47.9	6.0	0.6	41.6	0.6
Sweet sorghum	77.2	18.1	4.7	47.3	5.8	0.4	41.7	0.1
Kenaf	79.4	17.0	3.6	46.6	5.8	1.0	42.8	0.1

Ash chemical composition, wt%									
	SiO_2	Al_2O_3	Fe_2O_3	CaO	MgO	K_2O	Na_2O	TiO_2	SO_3	P_2O_5
Pine saw dust	8.3	2.0	1.8	41.8	11.8	12.3	0.3	0.12	1.9	5.2
Pine bark	1.3	5.3	0.3	40.6	4.5	7.6	0.5	0.12	2.0	4.8
Forest residue (pine)	38.5	4.7	3.7	15.4	4.0	8.3	0.4	0.5	1.6	3.2
Salix	0.4	0.3	0.2	30.8	5.1	26.5	0.3	0.02	3.0	11.5
Wheat straw	59.9	0.8	0.5	7.3	1.8	16.9	0.5	0.04	1.1	2.3
Barley straw	62.0	0.2	0.2	4.5	2.2	19.3	0.5	0.02	1.4	2.5
Reed canary grass	89.8	1.4	1.1	3.5	1.5	3.1	0.1	0.05	1.1	4.1
Miscanthus	42.8	0.5	0.4	7.6	4.8	25.3	0.7	0.03	2.1	5.3
Sweet sorghum	57.8	0.7	0.5	9.0	2.7	8.2	1.5	0.05	3.0	3.0
Kenaf	6.6	1.8	1.2	30.8	6.0	13.3	1.3	0.08	5.7	2.7

The kinetic parameters were calculated on the basis of the minimum reaction rates of the gasification rate vs. fuel conversion curve, see point A in Fig. 1.

3 Results

Fig. 2 shows the dependence of the reaction rate of wood residual char on temperature, measured in 1 bar pressure of CO_2 and H_2O. The activation energy was 217 kJ/mol for steam gasification and 229 kJ/mol for CO_2 gasification. These values are of the same magnitude range as those measured earlier [9].

The dependence of gasification of wood char on pressure can be seen from Fig. 3, measured in the pressure range of 1-15 bar CO_2 and H_2O at temperatures of 750 and 850 °C. When the pressure was raised in the range under study, the rate of CO_2 gasification was reduced slightly at both temperatures, while the rate of H_2O gasification increased clearly.

The effect of the product gas on the gasification reactivity of wood residual char was studied using binary gas mixtures CO_2- O and H_2O-H_2. The dependence of reaction rate on the partial pressure of product gas components were described with Langmuir-Hinshelwood formulas as used for coal [5].

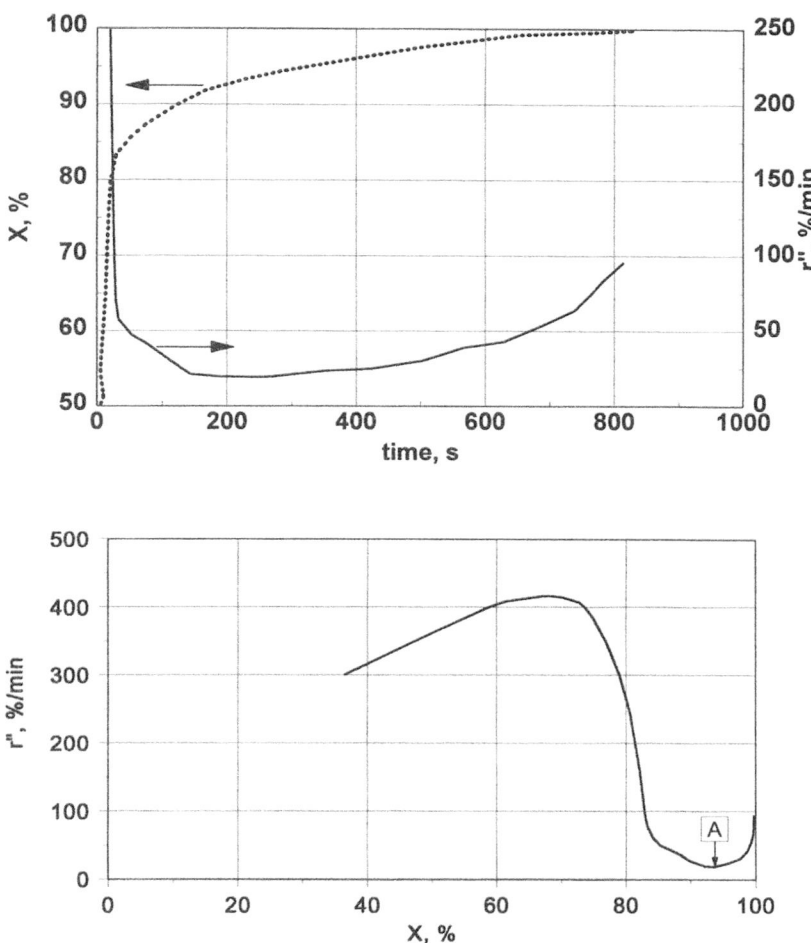

Fig. 1. Fuel conversion (X, %), and gasification rate (r", %/min), as a function of time (upper graph) and r" as a function of fuel conversion (X, %) (lower graph). Point A shows the minimum gasification rate.

According to the Langmuir-Hinshelwood kinetic expression, the reciprocal of the reaction rate should have a linear correlation with the ratio of the partial pressures, e.g., P_{CO}/P_{CO2}. The measurements indicated that the reciprocal of the reaction rate in the gasification of wood residual char in the CO_2-CO system was dependent linearly fairly well on the ratio of partial pressures, while the level of dependence was dependent on the total pressure (Fig. 4).

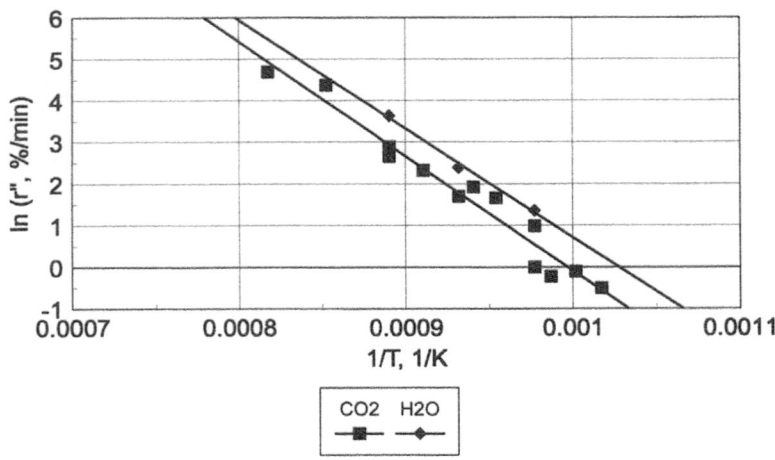

Fig. 2. Dependence of the gasification rate of wood char on temperature at 1 bar CO_2 and H_2O pressure.

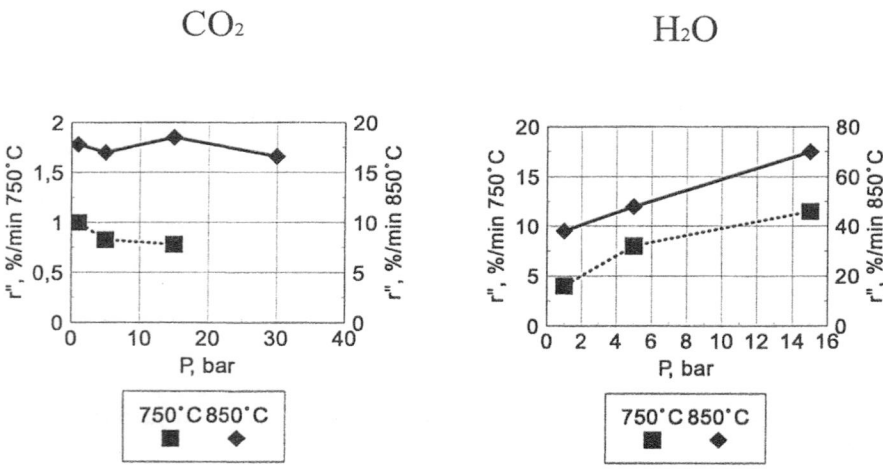

Fig. 3. Dependence of the gasification rate of wood char on CO_2 and H_2O pressure at 750 and 850 ^{O}C.

In the H_2O-H_2 mixtures the corresponding dependencies were significantly better and the total pressure was of no great significance, as can be seen from Fig. 5.

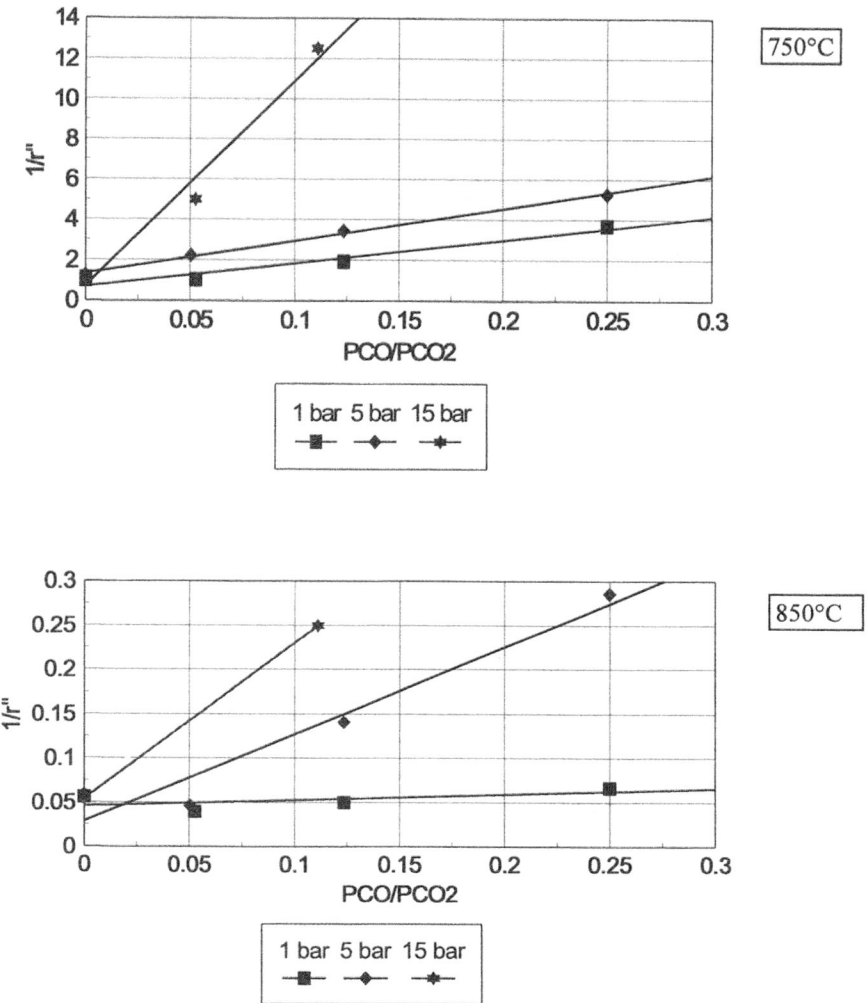

Figure 4. Dependence of the gasification rate of wood char and the P_{CO}/P_{CO2} ratio at 750 and 850 °C in the pressure range of 1-15 bar.

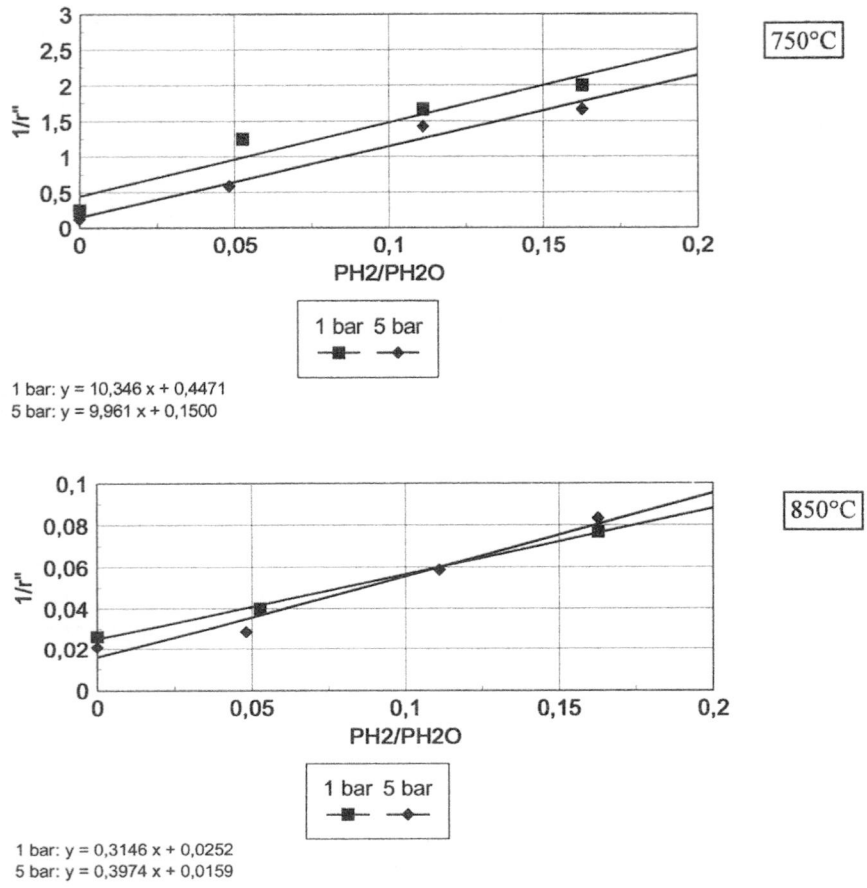

1 bar: y = 10,346 x + 0,4471
5 bar: y = 9,961 x + 0,1500

1 bar: y = 0,3146 x + 0,0252
5 bar: y = 0,3974 x + 0,0159

Figure 5. Dependence of the gasification rate of wood char and the H_2/H_2O ratio at 750 and 850 °C in the pressure range of 1-15 bar.

The reason for the differences observed seems to be the complexity of pressure dependence, which is possibly due to the behaviour of ash-forming substances during gasification. The catalytic effect of the ash-forming substances on the gasification reactivity of solid fuels is well-known, e.g., for lignites [10], and the gasification reactivity behaviour of low-rank coals is dependent on the behaviour of these substances [11, 12, 13] . Metals contained in biomasses have also been found to have a significant effect on their gasification reactivity [1-3, 7]. Thus, the behaviour of these substances during gasification can affect also the reactivity behaviour.

The behaviour of ash-forming substances during gasification can be very complicated. Possible effective behaviour models are an increase in diffusion resistance caused by ash or a decrease in catalytic activity of the ash-forming substances. In fuels with a low silicon content, the reaction rate correlates with the K + Ca content [2]. Biomasses, however, can contain very variable, even high, amounts of silicon [14], as shown in Table 1. Silicon occurs in plants dominantly as amorphous silicahydrate SiO_2nH_2O or as polymerised silicic acid [15]. Silicon distribution is highly dependent on the plant species. Silicon flocculates in cell structures protecting and supporting the plant. In fuel use, silicon can affect the reactivity by reducing it. The effect of silicon has been observed in the gasification of rice husk as a reduction in carbon conversion [3]. Research on silicate formation in gasification indicates an increase in diffusion resistance as the gasification reactions progress, when silicate ash sinters with alkali metals and carbon is encapsulated inside it. Kannan et al. [2] found out the reactivity-reducing effect of silicon by adding fine quartz powder to well-gasifiable biomass, when its gasification reactivity collapsed. Ash-forming substances may also react with each other or with the gas phase, when new compounds, different in catalytic effect can form [16].

The reactivity results obtained with other biomasses indicate great scattering compared to the results obtained with wood. As can be seen from Fig. 6, the pressure dependence between various biomasses in steam and CO_2 gasification can be rather different and unpredictable. As an example, the effect of the presence of 3 bar hydrogen in steam on the gasification rate was determined at 30 bar total pressure. The results show that the presence of hydrogen reduces the gasification rate, and differently for different biomass types.

4 Conclusions

The most significant conclusion from this study is that the determination of characteristics that describe gasification requires detailed knowledge of fuel structure and ash chemistry. A hypothesis is that the deviations found out in this study, such as the reduction of reaction rate at increasing pressure and differences in the gasification behaviour of CO_2-CO and H_2O-H_2 mixtures are due to the behaviour of ash-forming substances in gasification.

5 Acknowledgements

This research was part of the gasification research programme of VTT Energy. The study was funded by the Finnish National Research Programme LIEKKI 2, Academy of Finland, Ministry of Agriculture and Forestry and VTT.

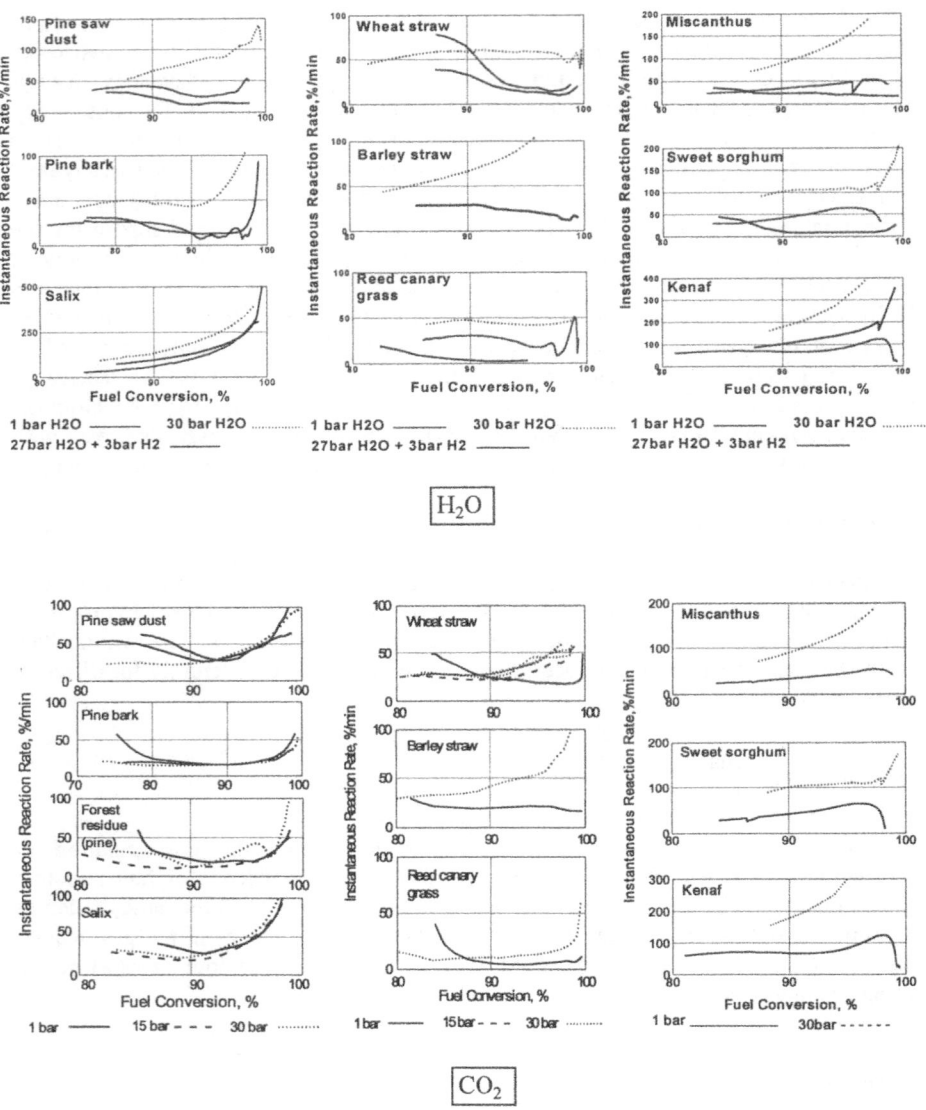

Fig. 6. Gasification behaviour of various biomasses measured in 1-30 bar pressure and at 850°C.

6 References

1. DeGroot, W.F. & Richards, G.N. (1988) Influence of pyrolysis conditions and ion-exchanged catalysts on the gasification of cottonwood chars by carbon dioxide. *Fuel*, Vol. 67. pp. 352-360.
2. Kannan, M.P., & Richards, G.N. (1990) Gasification of biomass chars in carbon dioxide: dependence of gasification rate on the indigenous metal content. *Fuel*, Vol. 69. pp. 747-753.
3. Ganesh, A. & Grover, P.D., Ramachandra Iyer, P.V. (1992) Combustion and gasification characteristics of rice husk. *Fuel*, Vol. 71. pp. 889-894.
4. Mühlen, H.-J., van Heek, K.H. & Jüntgen, H. (1985) Kinetic studies of steam gasification of char in the presence of H_2, CO_2 and CO. *Fuel*, Vol 64. pp. 944-949.
5. van Heek, K.H. & Mühlen, H.-J. (1991) Chemical kinetics of carbon and char gasification, in *Fundamental Issues in Control of Carbon Gasification Reactivity*, (ed. J. Lahaye & P. Ehrburger), Kluwer Academic Publishers, Netherlands, pp. 1-34.
6. Hüttinger, K.J. & Merdes, W.F. (1992) The carbon steam reaction at elevated pressure: formation of product gases and hydrogen inhibitions. *Carbon*, Vol. 30 No. 6. pp. 883-894.
7. Moilanen, A. & Mühlen, H.-J. (1996) Characterisation of gasification reactivity of peat char in pressurised conditions - effect of product gas inhibition and inorganic material. *Accepted for publication in Fuel*.
8. Moilanen, A. & Kurkela, E. (1995) Gasification reactivities of solid biomass fuels, in *Am. Chem. Soc. Div. of Fuel Chem., Vol. 40, No. 3, Preprints of Papers Presented at the 210th ACS National Meeting Chicago, Aug. 20-24, 1995*, pp.688-693.
9. Moilanen, A., Saviharju, K. & Harju, T. (1994) Steam gasification reactivities of various fuel chars, in *Proc. Advances In Thermochemical Biomass Conversion*, (ed. A.V. Bridgewater), Blackie Academic & Professional, London. pp. 131-141.
10. van Heek, K.H. & Mühlen, H.-J. (1987) Effect of coal and char properties on gasification. *Fuel Processing Technology*, Vol. 15. pp. 133-133
11. Takarada, T., Tamai, Y. & Tomita, A. (1985) Reactivities of 34 coals under steam gasification. *Fuel*, Vol 64. pp. 1438-1442.
12. Miura, K., Hashimoto, K. & Silveston, P.L. (1989) Factors affecting the reactivity of coal chars during gasification, and indices representing reactivity. *Fuel*, Vol. 68. pp. 1461-1475.
13. Miura, K., Makino, M. & Silveston, P.L. (1990) Correlation of gasification reactivities with char properties and pyrolysis conditions using low rank Canadian coals. *Fuel*, Vol 69. pp. 580-589.
14. Osman, E.A. (1982) A study of the effects of ash chemical composition and additives on fusion temperature in relation to slag formation during gasification of biomass. *Dissertation*, University of California, Davis, 182 p.
15. Mengel, K. & Kirkby, E.A. (1982) *Principles of plant nutrition*. 3rd ed., International Potash Institute, Bern, 654 p.
16. Meijer, R., van der Linden, B., Kapteijn, F. & Moulijn, J.A. (1991) The interaction of H_2O, CO_2, H_2 and CO with the alkali-carbonate/carbon system: a thermogravimetric study. *Fuel*, Vol. 70. pp. 205-214.

GAS QUALITY FROM BIOMASS GASIFICATION; AN EXTENSIVE PARAMETRIC EQUILIBRIUM STUDY

Parametric equilibrium study of gasification

A. NORDIN
Energy Technology Centre, Department of Inorganic Chemistry, Umeå University, Umeå, SWEDEN
P. KALLNER
Department of Heat & Power, Royal Institute of Technology, Stockholm, SWEDEN
E. JOHANSSON
Umetri AB, Umeå, SWEDEN

Abstract
Chemical equilibrium relations between the major gas components in gasification are valuable for reactor design and process optimisation, as well as for a generally increased understanding of the processes. However, most previous equilibrium calculations have been limited to more or less narrow ranges of operating conditions and specific fuel characteristics. In the present work, an extensive systematic parametric study was performed utilising a D-optimal experimental design to structure the equilibrium calculations. The sensitivity analysis of designed calculation data, covered equilibrium relations between the responses CO, H_2, CH_4, CO_2, H_2O, NH_3, HCN, higher heating value and all reasonable combinations of gasification conditions and fuel properties. The effects were evaluated by partial least squares (PLS) analysis and the results were illustrated with concentrations and higher heating values as functions of the most influential factors. The data set is available for interested readers.

Keywords: Biomass, gasification, chemical equilibrium, D-optimal design, mixture design, parametric study, sensitivity analysis.

1 Introduction

Interest in biomass gasification has increased during recent years because of the environmental benefits associated with these fuels. Presently, gasification in combined cycles with gas and steam turbines is the thermochemical method considered to be the most economical and with the highest potential for efficient fuel to electricity conversion. Also for other process cycles with special fuel needs, biomass gasification is considered. Much research and development work is therefore focused on designing biomass gasifiers of various types and sizes. Here, chemical equilibrium calculations may be used to illustrate predicted effects of changes in potentially influential parameters. Although not considering reaction rates and mechanisms of the chemical reactions, equilibrium model calculations always identify the possible directions and driving forces for a complex system. Furthermore, at gasification temperatures above 700 °C the reaction rates are relatively high, and most of the gaseous products have been reported to be rather close to their equilibrium levels for most practical gasifiers [s.f.1]. Equilibrium relations are therefore valuable for reactor design and process optimisation, as well as for generally increasing knowledge of gasification processes.

In the area of coal gasification, predictions based on equilibrium calculations have been reported in a significant number of publications. Equilibrium modelling of biomass gasification has been less actively exploated, but includes the work by Cousins [2], Sesh and Sunavala [3], Kinoshita et al. [4], Buekens and Schueters [5], Shand and Bridgwater [6] and Desrosiers [7]. However, most previous chemical equilibrium studies have been limited to more or less narrow ranges of operating conditions and specific fuel characteristics. To our knowledge, an extensive systematic parametric study, covering all relevant fuel and process variables still can not be found in the literature. Such a sensitivity analysis may preferably be performed by utilising a combined mixture (fuel variables) and process statistical experimental design. For the analysis of mixture data, the partial least squares (PLS) method [8] has recently been found to be superior to the traditional multiple regression method [9], mainly due to the mixture constraint. In particular when both mixture and process variables are involved, it offers a flexible and simple approach which works well in practice.

The objective of the present work was, therefore, to perform a sensitivity analysis by PLS analysis of designed calculation data, covering the equilibrium relations of all reasonable combinations of gasification conditions and fuel properties. The applicability of the PLS model is further illustrated with some concentration and higher heating value (HHV) data for the product gas as functions of the most influential factors.

2 Model calculation procedure

The equilibrium calculations were performed according to a D-optimal experimental design to structure the work [10]. D-optimal designs are mathematical schemes in which all potentially important factors are changed systematically to maximise the determinant $(X'X)$, thereby significantly facilitating the identification of process relations and the extraction of general sets of accurate regression models. In the present work, the D-optimal design combines both mixture and process factors and it was used to allow a model including linear, interaction and quadratic model terms to be established. The modelling and design program MODDE [11] was used to generate the design, calculate the model and illustrate the results. In Fig. 1, a schematic illustration of the process, as well as the factors and their design levels are given. The factors considered were temperature (Te) and pressure (P) of the process, oxygen to fuel ratio (OFR), nitrogen to oxygen ratio (NOR), concentrations of C, H, O, N in the fuel and the amount of water in the process, added as either fuel moisture or steam (St).

A total number of 100 different calculations were performed by using the computer code CHEMSAGE [12]. Thermodynamic data were taken from the data bank of the Scientific Group Thermodata Europe (SGTE), mostly originating from the tabulations of JANAF [13].

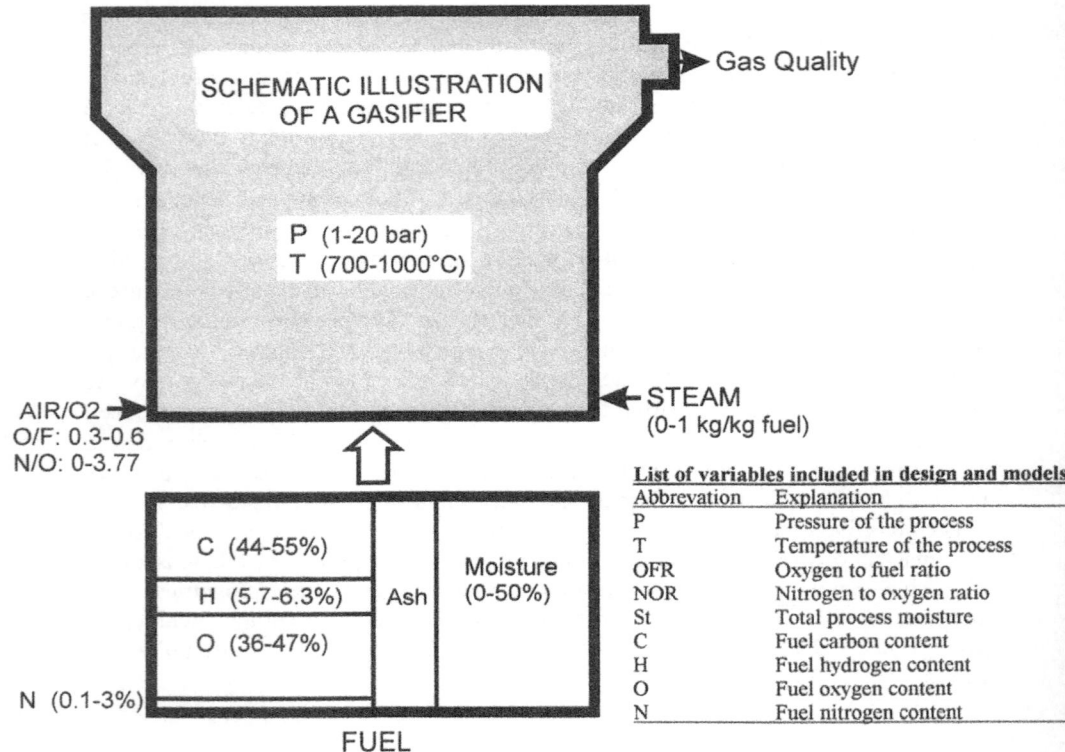

Fig. 1 Illustration of studied process and variable ranges

3 Results and discussion

The effects of the nine factors on all responses were evaluated by statistical methods, according to Box et al. [14], Eriksson et al. [8] and Kettaneh-Wold [9]. One PLS model was determined for all the dependent variables (responses), using the same underlying statistical design. The responses were the formed equilibrium concentrations of CO, H_2, CH_4, CO_2, H_2O, NH_3, HCN and the calculated higher heating value (HHV). All of the other species considered (~200) formed in only negligible concentrations. To facilitate the calculation of moles or weights, the total amount of gaseous species formed from 100 gram (dry) fuel, was also included as a response.

The effects of the different factors on the gas concentrations and HHV of the gas are illustrated in Figs. 2-9, where the importance of each of the different factors is given as scaled regression coefficients. The scaling makes the coefficients directly intercomparable, and a large absolute value indicates a strong influence, positive or negative. The general model equation can be written as:

$$y = \beta_0 + \beta_1 x_1 + \beta_2 x_2 + \beta_3 x_3 + \ldots +$$
$$\beta_{11}(x_1)^2 + \beta_{22}(x_2)^2 + \beta_{33}(x_3)^2 + \ldots +$$
$$\beta_{12}x_1 x_2 + \beta_{13}x_1 x_3 + \beta_{23}x_2 x_3 + \ldots + \beta_{123}x_1 x_2 x_3 + \ldots \qquad (1)$$

where y is the response, β_0 the constant term, $\beta's$ the regression coefficients and x_j the varied factors. To obtain an optimal model, insignificant model terms (>> 95% confidence interval) were eliminated.

As shown in Table 1, good agreements between calculated and model data were obtained. Large R^2 indicates that a major fraction of the response is explained by the model [14], whereas large Q^2 (> 0.7) indicates that the model has a good predictive ability [8,11]. The goodness of fit of the model is also illustrated in Figure 10, where the observed versus the predicted values are depicted for the responses with the best (closed symbols) and the least good fit (open symbols).

Table 1. Summary of model fit

Response									
Total model	CO	H2	CH4	HHV	CO2	H2O	NH3	HCN	Total amount
R^2 0.950	0.958	0.958	0.977	0.965	0.904	0.930	0.940	0.940	0.975
Q^2 0.855	0.860	0.889	0.909	0.890	0.776	0.818	0.828	0.741	0.922

From Figs 2-9, it can be concluded which factors are the most influential for the different responses and in the following some of these relations are illustrated.

None of the major gas components were significantly influenced by the different characteristics of different biomass fuels. However, the N-content may effect the concentrations of NH_3 and HCN, but only marginally. CO was not significantly influenced by any interactions and the major effect had the amount of water in the process, the process temperature, OFR and NOR. An increase in NOR will result in a linear decrease in CO (by dilution) and in Figs. 11 and 12, the effect of the other factors are illustrated.

The concentrations of H_2 is influenced significantly by several process factor interactions, although the major linear effects are by OFR and NOR. Increase in these ratios result in a proportional decrease in the H_2 concentration. The effect of temperature and steam, for given values of OFR and NOR is illustrated in Fig. 13.

In Fig. 14, the most important factors for the concentration of CH_4 is illustrated, although both OFR and steam significantly influences the concentration. CH_4 will only form at low temperatures, and is enhanced by high pressure, low OFR and water content. The calculated HHV of the product gas is most significantly influenced by the OFR and NOR (dillution) but an increased water load will also decrease the HHV. With the dillutive factors constant, the HHV may also be influenced somewhat by changing the temperature and pressure, as illustrated in Fig. 15.

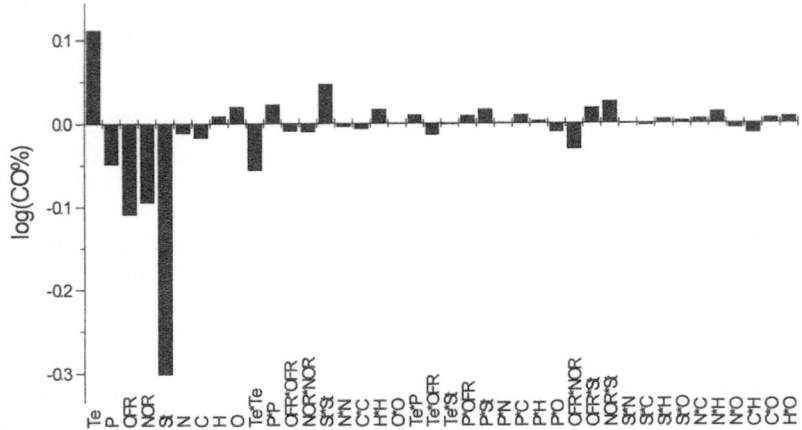

Fig. 2 Relative effects of the different factors on the equilibrium concentration of CO.

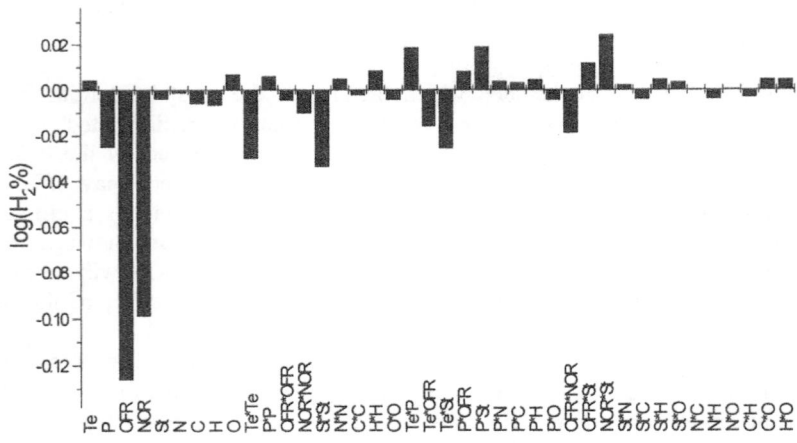

Fig. 3 Relative effects of the different factors on the equilibrium concentration of H₂

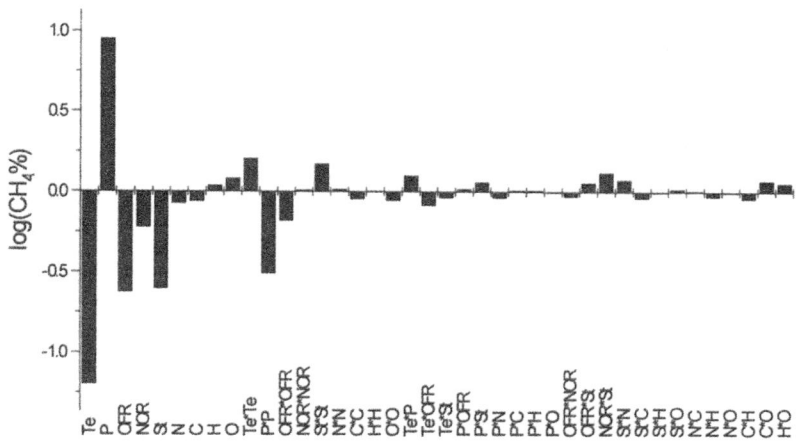

Fig. 4 Effects of the different factors on the equilibrium concentration of CH₄.

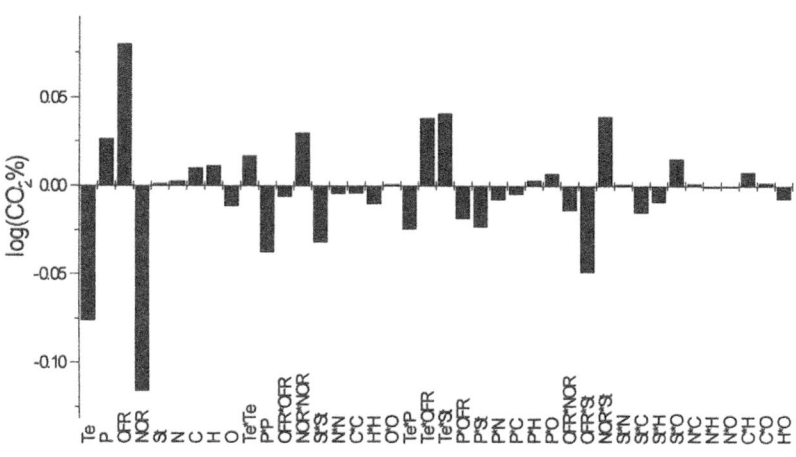

Fig. 5 Relative effects of the different factors on the equilibrium concentration of CO₂

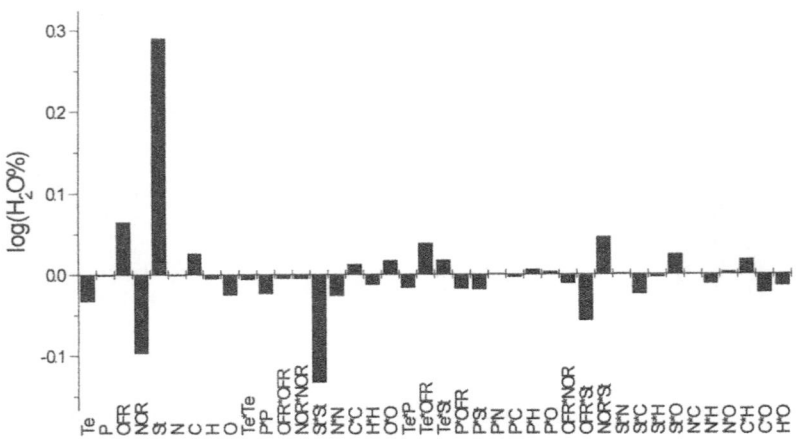

Fig. 6 Relative effects of the different factors on the equilibrium concentration of H₂O

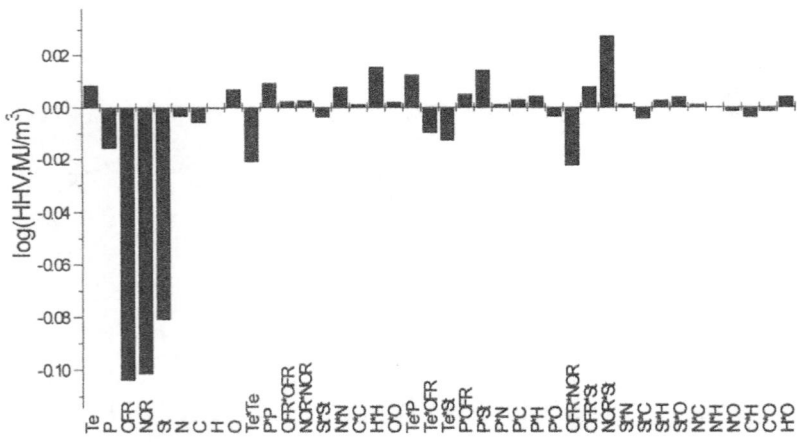

Fig. 7 Relative effects of the different factors on the higher heating value of the product gas

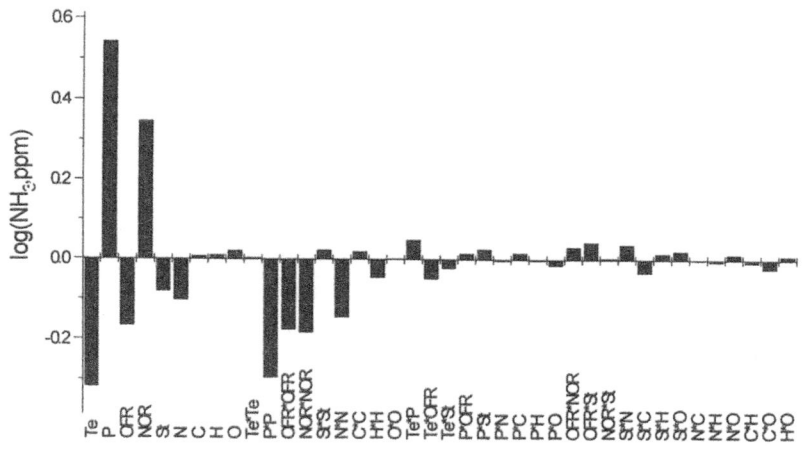

Fig. 8 Relative effects of the different factors on the equilibrium concentration of NH₃

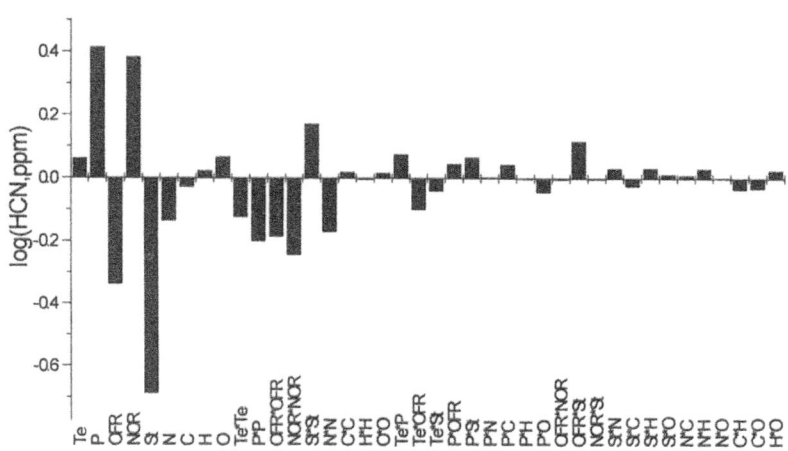

Fig. 9 Relative effects of the different factors on the equilibrium concentration of HCN

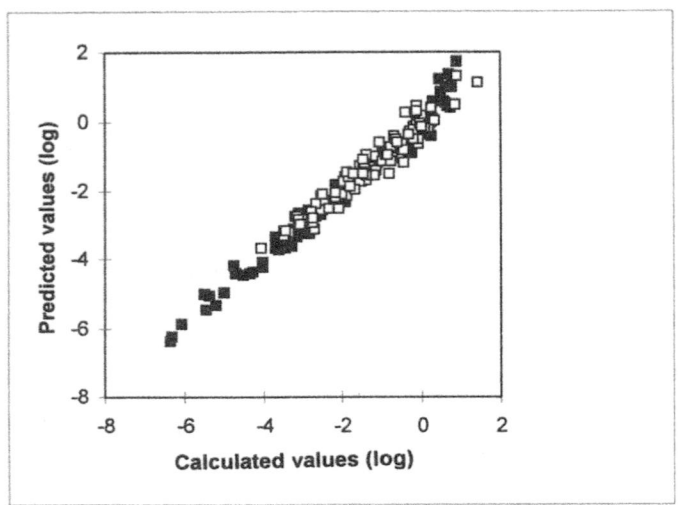

Fig. 10 Calculated versus predicted values for CH₄ (open symbols) and HCN (closed symols)

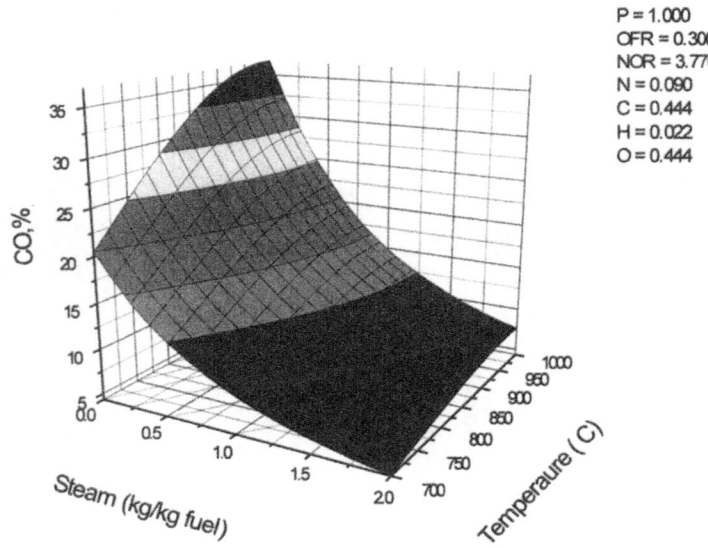

Fig. 11 Illustration of the effect of total water load and temperature on the CO concentration for a low OFR (0.3)

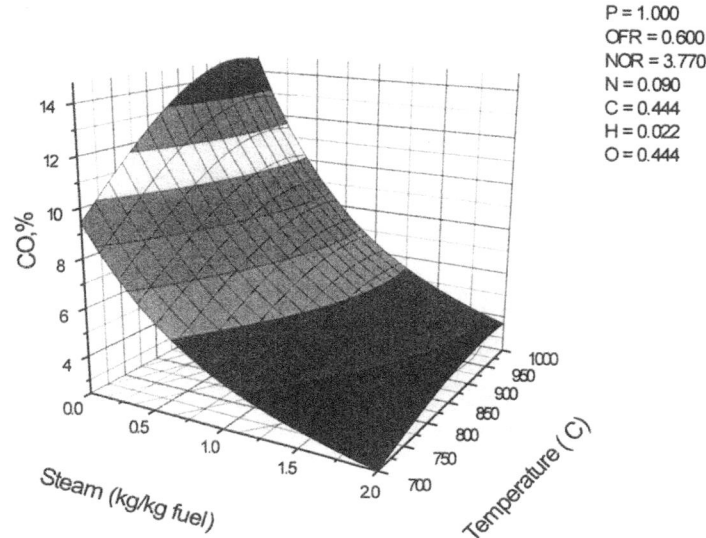

Fig. 12 Illustraton of the effect of total water load and temperature on the CO concentration for a high OFR (0.6)

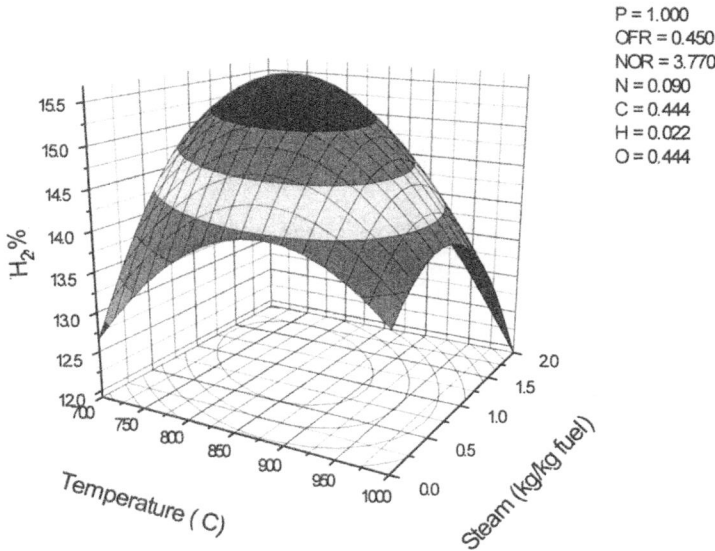

Fig. 13 Effect of temperature and water load on the concentration of H_2.

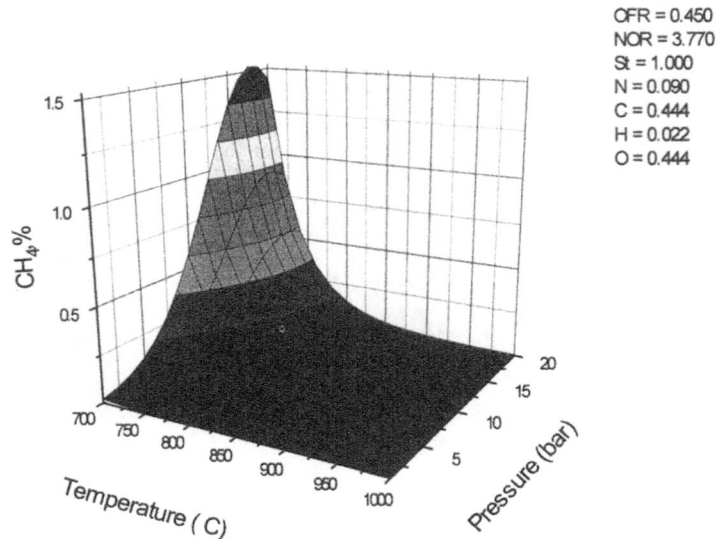

OFR = 0.450
NOR = 3.770
St = 1.000
N = 0.090
C = 0.444
H = 0.022
O = 0.444

Fig. 14 The effect of temperature and pressure on the concentration of CH₄.

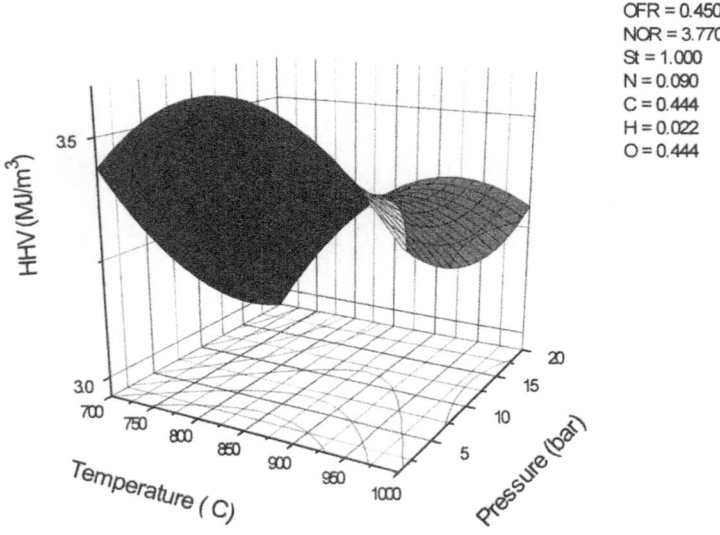

OFR = 0.450
NOR = 3.770
St = 1.000
N = 0.090
C = 0.444
H = 0.022
O = 0.444

Fig. 15 Effect of temperature and pressure on the HHV of the product gas

NH_3 is influenced by interactions between all process factors and also somewhat by the fuel nitrogen content. In Fig. 16, the effect of pressure and NOR is illustrated with all other process factors at their worse levels (most ammonia). The concentrations of HCN are very low for all conditions except for the most extreme dry, high pressure and air based processes (maximum 25 ppm).

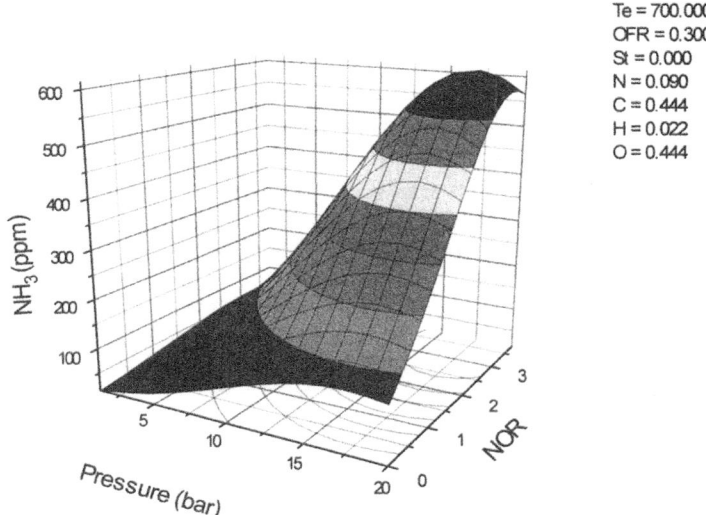

Te = 700.000
OFR = 0.300
St = 0.000
N = 0.090
C = 0.444
H = 0.022
O = 0.444

Fig. 16 Maximum NH_3 equilibrim concentration

One of the objectives of the present work was to compile the equilibrium data into a more compressed and accessible form. This was accomplished by utilising the PLS method, which also implies that the interested reader must have access to a multivariate program. A demonstration version of the MODDE program, including a PLS module, as well as the data set from the present work are, therefore, available on request from the authors (A.N.). In addition, an extended set of illustrations are available as a word document on internet (etc@etc.postnet.se).

4 Conclusions

The results of the parametric study showed that the maximal variations in elemental characteristics of different biomass fuels do not significantly influence the major gas compositions or heating values of the gas, although a small effect of the N-content may influence the NH_3 and HCN concentrations somewhat. The major conclusion of the present work was that the approach of utilising statistical experimental designs for parametric and sensitivity studies based on chemical equilibrium calculations is very useful. A complete compiled set of accurate equilibrium relations have hereby been made available for interested scientists, students etc.

In addition to effects based on chemical equilibrium shown in this study, the processes may be strongly influenced by gasification residence times and heating rates. Obviously, these aspects can not be evaluated with an equilibrium approach as they are kinetic in their nature. However, despite the fact that the results shown here are ideal and do not account for kinetic effects they do set possible directions and limits for gas qualities that can be attained. Above all, the results indicate those combinations of gasification settings that should be chosen as a starting point for experimental optimisation of a gasification process.

References

1. Watkinson, A.P., Lucas, J.P. and Lim, C.J. (1991) A prediction of performance of commercial coal gasifiers. *FUEL*, Vol. 70, pp 519-527.

2. Cousins, W. J. (1978) A theoretical study of wood gasification processes. *New Zeeland Journal of Science*, Vol 21, pp 175-183.

3. Shesh, K. K., Sunavala, P.D. (1990) Thermodynamics of pressurized air-steam gasification of biomass. *Indian Journal of Technology*, Vol. 28, pp 133-138

4. Kinoshita, C. M., Wang, Y. and Takahashi, P. (1991) Chemical equilibrium computations for gasification of biomass to produce methanol. *Energy Sources,* Vol. 13, pp 361-368.

5. Buekens, A. G. and Schoeters, J. G. (1984). Mathematical modelling in gasification (Keynote paper), in *Thermochemical processing of biomass,* (ed. Bridgwater, A. V.), Butterworths, London.

6. Shand, R. N. and Bridgwater, A. V. (1984), Fuel gas from biomass: Status and new modelling approaches. in *Thermochemical processing of biomass,* (ed. Bridgwater, A. V.), Butterworths, London.

7. Desrosiers, R. (1981) Thermodynamics of gas-char reactions, in *Biomass gasification, principles and technologies,* (Ed. T.B. Reed), Noyes Data Coorp., New Jersay, pp 119-153

8. Eriksson, L., Hermens, J. L. M., Johansson, E., Verhaar, H. J. M. and Wold, S. (1995) Multivariate analysis of aquatic toxicity data with PLS. *Aquatic Sciences*, Vol. 57, pp 1015-1621

9. Kettaneh-Wold, N. (1995) Analysis of mixture data with partial least squares. *Chemometrics and Intelligent Laboratory Systems*, **14**, 57-69

10. DuMouchel, W., and Jones, B. (1994) A simple bayesian modification of D-Optimal designs to reduce dependence on an assumed model. *Technometrics*, Vol. 36, pp 37-46.

11. MODDE, (1995) "*Modelling and design, users guide*", Vol 3.0, Umetri AB, Umeå.

12. Eriksson, G. and Hack, K. (1990) Chem-Sage - A computer program for the calculation of complex chemical equilibria. *Metal Transactions*, Vol. 21B, pp 1013-1023.

13. JANAF (1985) Thermochemical Tables. Journal of Physical and Chemical Reference Data, 3rd ed. Vol 14.

14. Box, G. E. P., Hunter, W. G., and Hunter, J. S. (1978) *Statistics for experimenters; An introduction to design, data analysis and model building.* Wiley, New York.

THEORETICAL AND EXPERIMENTAL INVESTIGATION ON HEAT TRANSFER IN FIXED CHAR BEDS

Heat transfer in fixed beds

S.B. PETERSEN, L. TH. PEDERSEN, U. HENRIKSEN
Department of Energy Engineering, Technical University of Denmark
Lyngby, Denmark

Abstract

The heat transfer by thermal radiation and conduction in char has been investigated as a part of an effort to model the pyrolysis and gasification of biomass.

A test stand for measurements was built based on a modification of the ASTM C201 standard [1] to permit investigation of flammable materials at temperatures up to 1000°C.

Experimental values of thermal conductivity for char are presented for three bulk densities of straw, 18.5 [kg/m^3], 20.2 [kg/m^3] and 36 [kg/m^3] respectively and one bulk density of wood chips, 146.4 [kg/m^3]. The experimental results are approximated in second-order polynomials over a range of temperatures from 300°C to 1000°C.

Experimental results show that the thermal conductivity depends strongly on the bulk density of the material. There is a strong decrease in the heat transfer with increasing bulk density. This is caused by a decrease in thermal radiation with increasing bulk density as described by the theoretical model.

Using a model for heat transfer in packed beds developed by Yagi and Kunii, the heat transfer through char was calculated for a range of values of bulk density, void fraction and particle size.

The measurements on char from pyrolysed straw have shown close agreement between experimental results and values from the theoretical model. It is concluded that the model is valid for computing the thermal conductivity of char from pyrolysed straw for different bulk densities and fluids in the packed bed.

Keywords: Biomass, char properties, combustion, gasification, pyrolysis, straw, thermal conductivity

Nomenclature

h_{rv}	effective heat transfer coefficient	$W/(m^2 \cdot K)$
h_{rs}	heat transfer coefficient for radiation at contact surface	$W/(m^2 \cdot K)$
k	thermal conductivity	$W/(m \cdot K)$
k_e	effective thermal conductivity	$W/(m \cdot K)$
k_f	thermal conductivity of the fluid	$W/(m \cdot K)$
k_s	thermal conductivity of the solid	$W/(m \cdot K)$
l_s	equivalent thickness a layer of solid should have to represent the same thermal resistance as the sphere	m
l_v	equivalent thickness a layer of fluid should have to represent the same thermal resistance as the fluid film	m
$\dot{q}_s$	conduction through solid	W
D_p	Particle diameter	m
T	temperature	K
ε	emissivity coefficient	--
η	void fraction	-
Δl	characteristic length between two particles	m
ΔT	temperature difference	K
ΔT_{ics}	temperature drop over the contact surface	K
ΔT_s	temperature drop through the solid	K

Indices

e	effective
f	fluid
cf	contact surface
p	particle
r	radiation
s	solid
v	void

1 Introduction

This paper contains the results of a study conducted at the Department of Energy Engineering at the Technical University of Denmark, concerning the thermal conductivity of char. This study addresses only heat transfer by conduction and radiation, i.e. the measured results do not apply to heat transfer by forced convection. In earlier studies on the properties of char no data on the temperature dependence of thermal conductivity on temperature was available [2].

For the present investigation a 400 kW, two-stage fixed bed gasifier has been developed and is currently being tested. Investigations of the heat balance of the gasifier have shown that not all the heat transfer can be explained by attributing it solely to convection. Therefore this study was initialised to determine the importance of neglecting heat transfer by conduction and radiation in a fixed bed reactor. Besides obtaining experimental results for the thermal conductivity of char, the aim of the study was to develop a correlation that could describe the dependency of thermal conductivity on the physical properties of the char, i.e. bulk density, structure.

A search for literature on the specific subject of thermal conductivity of char was negative, but a few references were found on theoretical models for heat transfer in chemical packed bed reactors. This paper contains the description of a test apparatus for measuring the thermal conductivity of char. The design of the apparatus is based on the ASTM C201 standard, which describes an apparatus for measuring the thermal conductivity of thermal refractories. The design was subsequently altered to accommodate measurements on flammable materials. A theoretical model was developed. The test results that have been obtained are presented and compared with the predictions of the theoretical model.

2 Experimental methods

For the purpose of measuring the thermal conductivity of char an apparatus has been designed and built. The design of the apparatus is based on the ASTM C 201 standard. The standard describes a testing method for measuring the thermal conductivity of thermal refractories. The idea of the apparatus is that good insulation of the walls and a large cross-sectional area of the test sample creates one-dimensional heat transfer in the centre, thus allowing easy calculation of thermal conductivity. The original design does not permit measurement on flammable materials, so in order to make measurements on char possible, the design was modified by introducing an inert gas in the heating chamber.

The design of the apparatus is shown in Figure 1. The apparatus consists of an electrically heated chamber capable of reaching temperatures up to 1100°C. A plate of iron above the test material provides uniform heat distribution over the test sample. Beneath the test sample two independent water-cooled calorimeters are placed concentrically around the central calorimeter. The purpose of the outer and innerguard calorimeters is to provide a uniform temperature beneath the test sample, making a one-dimensional heat transfer over the centre calorimeter possible.

The assumption made in the ASTM C201 standard concerning the one-dimensional heat transfer through the test sample has been investigated by computer modelling of the test set-up. The results showed that the deviation from the assumed one-dimensional heat transfer is less than 10%. Therefore no correction of the obtained results is needed.

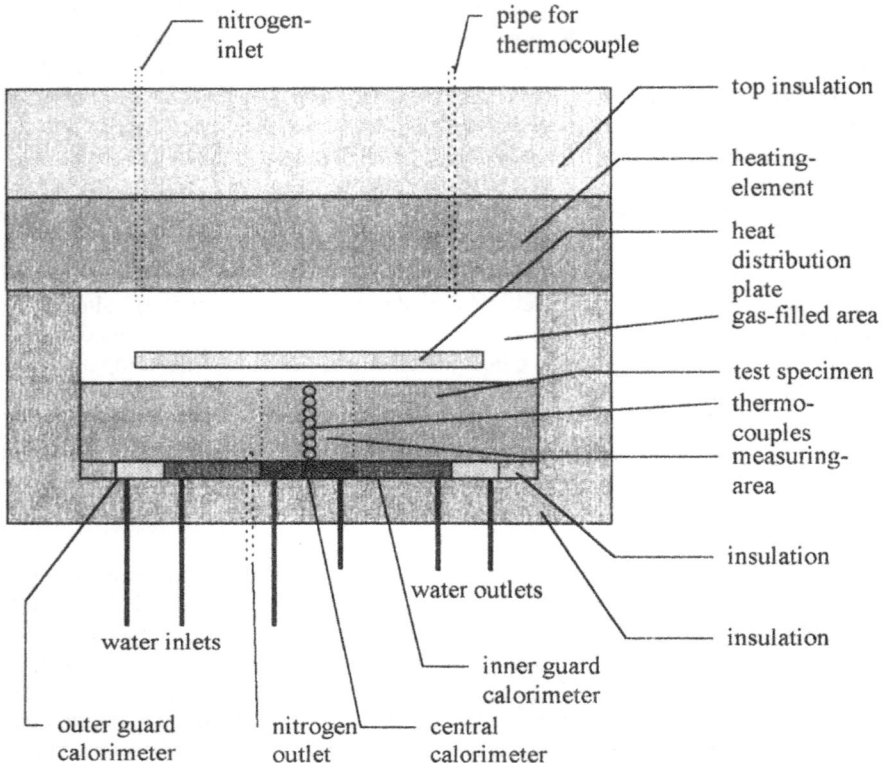

Figure 1. X and Y cross sections of the experimental set-up. The set-up is rectangular
with dimensions (height*width*length) [mm]= 660*690*805

To achieve steady-state conditions the temperature in the heating chamber is controlled
by a thyristor connected to the electrical heating element. A steady flow of cooling
water through the calorimeters is achieved by maintaining a stable water pressure
(waterfeed from a reservoir). The water flow through the calorimeters can be changed
by microvalves attached to each pipe. To ensure that no heat from the surroundings is
transferred to the cooling water through the bottom of the test set-up, the cooling water
must have a temperature within a few degrees centigrade of the room temperature. The
temperature rise over the calorimeters is to be less than 2-3°C to ensure a uniform
temperature in the bottom of the set-up.

2.1 Validation of test results

In the ASTM C201 standard, the inter-laboratory precision for a 95% significance level, is a 25% difference in the grand average of measured thermal conductivity of a given test sample. To ensure that the measurements made with the apparatus were valid a series of calibration tests with KaoWool as test material were made. The manufacturer of KaoWool, Thermal Ceramics Ltd. also uses an apparatus built according to the ASTM C201 standard for measuring the thermal conductivity of their products and therefore the results obtained could be considered as "inter-laboratory tests". These tests shows a difference in the values given by Thermal Ceramics Ltd., and the values measured with the presently constructed apparatus that are of the order of 10-12%, and therefore well below the 25% difference given by the ASTM C201 standard.

Therefore it can be concluded that the test results that are achieved with this apparatus are valid.

2.2 Test Preparation

Before a measurement is made a test sample of char is weighed and the volume is measured. Then the char is placed in the apparatus and "stacked", so that the char is uniformly distributed, also between the thermocouples. The volume of the char is then measured again. The apparatus is then sealed off and for the next 12 hours there is a flow of nitrogen through the apparatus at a rate of 3 l/min. After 12 hours the flow of nitrogen is reduced to 0.5 l/min, the electric power is switched on, and the water flow through the calorimeters is activated.

When the apparatus has reached steady state conditions, the temperature rise over the calorimeters are checked, and adjustments are made to the water flows if the increases in temperature over the calorimeters exceeds the values prescribed by the standard. The apparatus then remains switched on until all temperatures show values that vary less than 0.01°C for one hour. When steady state conditions have been reached, the power is switched off, and the apparatus is left to cool. As temperatures reach room temperature, the flows of nitrogen and water are turned off, the seal is broken, and the apparatus is opened. The volume and weight of the char is then measured.

3 Experimental results

3.1 Size and shape of the char

The char consists of straw or wood chips pyrolysed at 600°C. The char particles from straw used in the tests are near-cylindrical with diameters of 3-4 mm and lengths ranging from 20-100 mm, typically around 50 mm. Orientation of the char particles is random, but since the length of the particles is larger than the diameter most particles are orientated horizontally.

Bulk density of the packed bed was varied by gently pressing the packed bed, making sure the particle structure of the char was kept. This is equivalent to the increased pressure on a "layer" of char particles at different heights in a real-life reactor.

The char from wood chips has a wide variety of sizes between 15*10*2 [mm] and 25*15*5 [mm].

3.2 Results

Using the modified test apparatus the thermal conductivity is determined for a char packed bed at different bulk densities. The results obtained are shown in Figure 2 below. They are represented by the thermal conductivity that is defined as the heat transferred divided by the cross-sectional area and the temperature gradient.

The test data has been curvefitted with second-order polynomials.

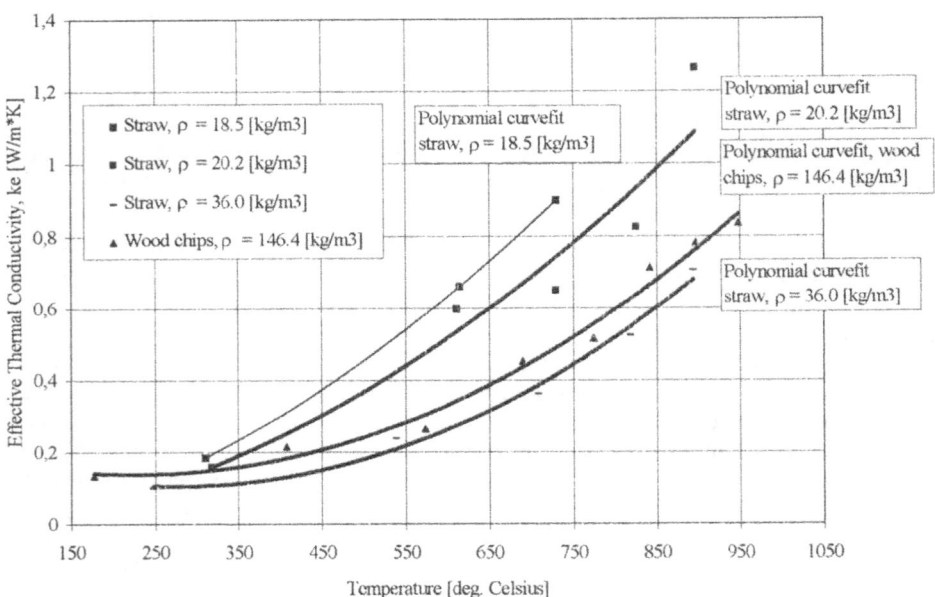

Figure 2. Diagram of Experimental Results and Second-order Polynomial Approximations. Nitrogen used as inert gas.

The experimental results are approximated by Forsythe-polynomials and listed in Table 1. The listed equations can be used to calculate the thermal conductivity of the char as a function of the temperature. The uncertainties show the largest deviation of a single point with respect to the second-order polynomial approximation.

Material	Curvefit (Temperatures must be inserted in degrees Celsius)
straw, ρ=18.2 [kg/m³]	k=1.25*10⁻⁶*T²+4.07E-04*T-0.061 [W/m*K] (±5%)
straw, ρ=20.5 [kg/m³]	k=2.27*10⁻⁶*T²-1.07*10⁻³*T+0.287 [W/m*K] (±15%)
straw, ρ= 36.0 [kg/m³]	k=1.51*10⁻⁶*T²-8.40*10⁻⁴*T+0.222 [W/m*K] (±10%)
wood chips, ρ=146.4 [kg/m³]	k=1.40*10⁻⁶*T²-6.40*10⁻⁴*T+0.211 [W/m*K] (±15%)

Table 1. Thermal conductivity of char as a function of temperature.
NOTE: Valid Temperaturerange is 300°C to 1000°C.

3.3 Discussion of experimental results

The thermal conductivity of the packed bed depends strongly on temperature, but the dependency is not linear, as with most solid materials. This is due to the fact that the heat transfer is composed of radiation and conduction. Since conduction is a linear function of the temperature whereas radiation is a power function of temperature thermal conductivity will increase as the temperature and therefore the percentage of the heat transfer by radiation increases.

3.3.1 Straw

It can also be observed that the effective thermal conductivity of the straw packed bed decreases with increasing bulk density. This effect is due to the fact, that if the bulk density of the straw packed bed increases, the voids between the char particles becomes smaller and the contact area increases. Since radiation mostly occurs in the voids, the heat transfer by radiation will decrease and the heat transfer by conduction increases thus making the effective thermal conduction smaller.

3.3.2 Wood chips

A single test on char from wood chips was carried out, and the result is shown in Figure 2. It is remarkable, that a wood chip packed bed with bulk density of 146.4 [kg/m³] has an effective thermal conductivity close to the conductivity of a straw packed bed with bulk density in range 25-35 [kg/m³]. This contradicts the earlier conclusion, that thermal conductivity decreases with increasing bulk density.

This effect is caused by the differences in size and structure of straw and wood chips. The wood chip particles have a far greater bulk density than the straw particles, and therefore the wood chip packed bed consists of more solid material per unit volume and less voids. This makes the heat transfer by conduction larger and heat transfer by radiation smaller.

4 Theoretical model

4.1 Description

Two theoretical models of the heat transfer have been investigated with respect to their ability to predict the heat transfer through char as a function of temperature and physical properties such as bulk density, fraction void, fluid thermal conductivity etc.

The two models investigated were very different, one being based on the principle of radiation shields, and the other being a literature model developed for chemical packed bed reactors. The literature model wasby far better able to describe the heat transfer in the packed char bed. The model was developed by Yagi and Kunii [3], and later improved by Kunii and Smith [4]. The model was originally developed for predicting heat transfer in chemical packed bed reactors, and has, in an investigation by Vortmeyer [5], been identified as the best model for that purpose, by comparison with other models.

Originally the model was derived for a bed consisting of spheres, but good results are also achieved if applied to packed beds filled with Raschig rings, Berl saddles etc.

The model identifies two basic heat transfer mechanisms between two spheres with diameter D_p, lying in two different layers with a temperature difference ΔT.

Transfer through voids in the packed bed (mechanism 1) and transfer through the particles (mechanism 2). The two mechanisms are further sub-divided into the mechanisms sketched in Figure 3.

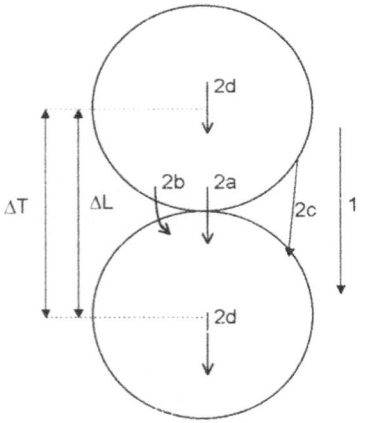

1. Conduction and radiation through voids filled with fluid.
2a. Conduction through contactpoints.
2b. Conduction through fluidfilm. The fluidfilm is a sort of thermal boundary layer between two particles [3].
2c. Radiation between particles.
2d. Conduction through solids.

Mechanism 2a is negligible in comparison with the other mechanisms, except for very low pressures.

Figure 3. The theoretical model.

The two basic heat transfer mechanisms, 1 and 2, are considered to be connected in parallel. In the subdivision of mechanism 2, mechanism 2b and 2c are connected in parallel, and as a group serial connected with mechanism 2d.

The effective thermal conductivity k_e is then given as Equation 1

$$-k_e \cdot \frac{\Delta T}{\Delta l} = [\text{transport through mechanism 1}] + [\text{transport through mechanism 2}] \quad (1)$$

4.2 Mechanism 1

Introducing the expression for the transport through mechanism 1, and denoting transport through solid, mechanism 2, as $\dot{q}_s$, equation 1 becomes:

$$-k_e \cdot \frac{\Delta T}{\Delta l} = \left[-k_f \cdot \eta \cdot \frac{\Delta T}{\Delta l} + h_{rv} \cdot \eta \cdot (-\Delta T) \right] + [\dot{q}_s] \quad (2)$$

- k_f is thermal conductivity of the fluid
- η is the void fraction
- $\dot{q}_s$ is conduction through solid.
- Δl is the characteristic distance between two particles. For spheres $\Delta l = Dp$, but for a packed bed of char it is an average value that depends on the packing, size and shape of the particles.
- h_{rv} is the effective radiation heat transfer coefficient of the voids, and is given as:

$$h_{rv} = 0.1952 \cdot \left(1 + \frac{\eta}{2(1-\eta)} \cdot \frac{1-\varepsilon}{\varepsilon} \right)^{-1} \cdot \left(\frac{T}{100} \right)^3$$

- ε is the emissivity coefficient.

The reason that the dependency of the temperature is only a function of T^3 and not T^4 is that the radiation heat transfer is approximated with a Taylor-series expansion, neglecting higher-order terms [3].

4.3 Mechanism 2

The heat transfer through mechanism 2, $\dot{q}_s$, is a function of the temperature drop over a particle, ΔT, and $\Delta T = \Delta T_s + \Delta T_{cs}$.

- ΔT_s is the temperature drop through solid.
- ΔT_{cs} is the temperature drop over the contact surface.

ΔT_s and ΔT_{cs} can be written as functions of the heat transfer via mechanism 2 ($\dot{q}_s$):

$$\Delta T_s = \frac{-\dot{q}_s}{\dfrac{k_s}{l_s}} \cdot \frac{1}{1-\eta} \quad (3) \text{ and } \quad \Delta T_{cs} = \frac{-\dot{q}_s}{\dfrac{k_1}{l_v} + h_{rs}} \cdot \frac{1}{1-\eta} \quad (4)$$

- k_s is the thermal conductivity of the solid.
- l_s is the equivalent thickness a layer of solid should have to represent the same thermal resistance as the sphere.
- h_{rs} is the heat transfer coefficient for radiation at the contact surface
- l_v is the equivalent thickness a layer of fluid should have to represent the same thermal resistance as the fluid film. This is the most difficult parameter to determine, but in the paper by Yagi and Kunii [3], diagrams show the relations. The relation can be correlated in the following equation:

- $$l_v = 0,15912 \cdot \Delta l \cdot \left(\frac{k_{air}}{k_{fluid}}\right)^{0,3716} \cdot \eta^{1,7304}$$

The radiation heat transfer coefficient at the contact surface, h_{rs}, is defined as:

$$h_{rs} = 0.1952 \cdot \left(\frac{\varepsilon}{2-\varepsilon}\right) \cdot \left(\frac{T}{100}\right)^3 \quad . \text{ For derivation see [4]}.$$

4.4 Conclusion, theoretical model

By combining equation 2 through equation 4 and solving for $\dot{q}_s$ the following equation is obtained

$$k_e = \eta\left(k_f + h_{rv} \cdot \Delta l\right) + \frac{(1-\eta) \cdot \Delta l}{\dfrac{1}{\dfrac{k_f}{l_v} + h_{rs}} + \dfrac{l_s}{k_s}} \quad (5)$$

It can be seen that the effective thermal conductivity calculated by equation 5 is not very sensitive to changes in the void fraction , whereas it is very sensitive to changes in the emissivity. In the model a large part of the total heat transfer is by radiation, therefore lower values of the emissivity, ε, results in a lower effective heat transfer in the packed bed. This is equivalent to a lower effective heat transfer coefficient, k_e, for the same temperature level and vice versa.

5 Validation of theoretical model

5.1 Comparison between experimental and theoretical results

In Figure 4 plots of the effective thermal conductivity calculated by equation 5 for different values of Δl and the experimental results are shown. The inert gas in the packed bed is nitrogen and the solid is modelled as consisting of carbon. The emissivity for the char, ε, is given as ε = 0.80 [2] and is practically independent of temperature and conversion level. The fraction void, η, has been measured to be η = 0.85.

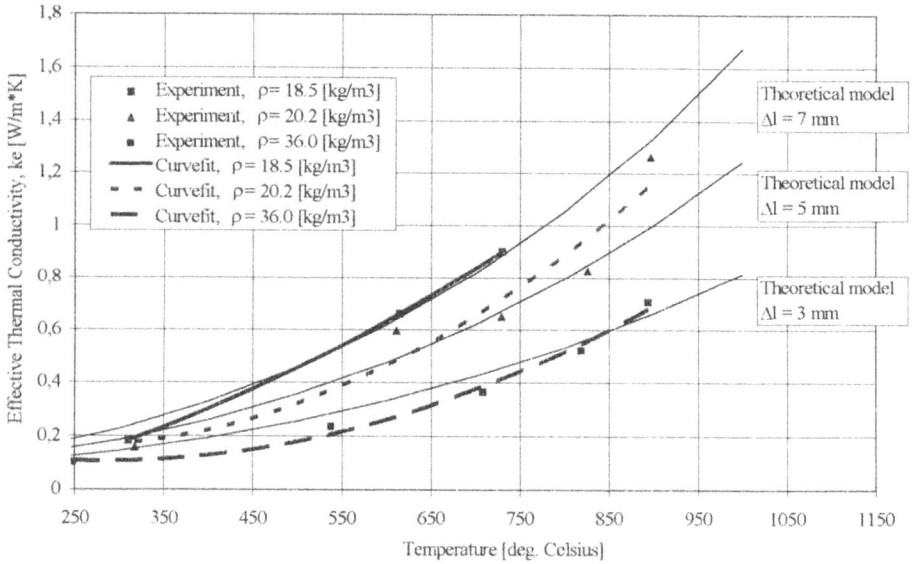

Figure 4. Plot of values of the effective thermal conductivity of the packed bed, calculated by equation 5 versus the experimental results.

It can be seen from Figure 4, that the curves from the theoretical model closely resemble the shape of the polynomial fit of the experimental results obtained, both with respect to shape and values.

There is a certain deviation in shape between the theoretical model and the polynomial curvefits of the experimental results at high temperatures. These deviations can be caused by the measurement of the temperature in the top of the packed bed. The thermocouple causing the "deviating" measurement may have been able to "see" the radiation from the heating-element. It is placed only a few centimetres deep in the packed bed, and the random packing of the bed may have been such, that the thermocouple was not shielded completely by the surrounding particles. Radiation from the heating element would cause the thermocouple to measure a higher temperature than that of the surrounding charparticles, causing the observed deviation. Secondly, a curvefit of relatively few measurements (in this case five) will always be approximate.

5.2 Discussion

The curvefit of the experimental results are scattered round a characteristic length of the straw particles in the interval between 3 and 7 mm. The characteristic length is defined as the average distance between the centres of two particles. The diameter of the charparticles is between 3 and 4 mm.

Figure 4 shows that an increase in bulk density of the packed bed corresponds to a decrease in the characteristic length of the theoretical model. This correlation between bulk density and characteristic length corresponds very well with the physical reality. A higher bulk density is achieved by "packing" the char particles gently, i.e. by applying a pressure on the bed. Since the structure of the individual particle is unchanged the pressure has the effect of reducing the voids between the charparticles. Therefore the average distance, equivalent to the characteristic length in the theoretical model must decrease.

The overall agreement between the experimental results obtained and the theoretical values proves that the theoretical model is valid for predicting the thermal conductivity of char particles from straw as a function of temperature under quiescent conditions.

6 Conclusion

The apparatus described by the ASTM C201 standard, was modified for measuring heat transfer in flammable materials. The result was successful.

Measurements of the heat transfer related to radiation and conduction in char from straw with different bulk densities has been carried out and reliable results have been obtained.

By inserting characteristic diameters of 3-7 mm which seems realistic with regard to the bulk density and particle size the theoretical model shows corresponds closely to the measured conductivities.

A single value of the thermal conductivity has been measured using char from wood chips.

7 Acknowledgements

This work was supported by The Danish Department of Energy and ELKRAFT.

8 References

1.   American Standards, *Standard Test Method for Thermal Conductivity of Refractories*, ASTM C201/177

2.   Ragland, K.W., Aerts, D.J. (1991) *Properties of Wood for Combustion Analysis*, Bioresource technology 37.

3.   Yagi, S. and Kunii, D. (1957) *Studies on Effective Thermal Conductivities in Packed Beds*, A.I.Ch.E.J. Vol. 3, pp.373.

4.   Kunii, D and Smith, J.M. (1960) *Heat Transfer Characteristics of Porous Rocks*, A.I.Ch.E.J., Vol.6, pp.71.

5.   Vortmeyer, D. (1978) *Radiation in packed solids*, 6th int. Heat Transfer Conference, Toronto, Canada 7-11/7 vol.6

HIGH-PRESSURE CARBON DIOXIDE REMOVAL IN SUPERCRITICAL WATER GASIFICATION OF BIOMASS

Y. MATSUMURA, T. MINOWA[†], X. XU, F. W. NUESSLE, T. ADSCHIRI[‡], and M. J. ANTAL, JR.
Hawaii Natural Energy Institute, University of Hawaii at Manoa, Honolulu, USA

Abstract
The utilization of high-pressure water to separate hydrogen from carbon dioxide is described for a biomass gasification process in supercritical water. Based on the higher solubility of carbon dioxide than hydrogen in water, the addition of excess water to the product gas mixture under high pressure results in hydrogen with a purity above 90 mol% in the gas phase. Power consumption for the compression of water needed to dissolve the carbon dioxide is negligible. Hydrogen dissolved in the water together with carbon dioxide is utilized for process heat. An evaluation of this process was conducted by both experimental and theoretical approaches. Energy flow analysis conducted on the gasification process with this separation shows the significance of the heat exchanger efficiency.
Keywords: Biomass, gasification, gas separation, hydrogen production, solubility, supercritical water.

1 Introduction

Wet biomass (water hyacinth, cattails, banana tree stem, kelp, green algae, etc.) grows rapidly and abundantly around the world. For example, water hyacinth (*Eichhornia crassipes*) produces over 150 t/ha/yr of dry organic matter, and cattails (*Typha* sp.) can deliver 97 t/ha over a 7 month growing season [1]. Nevertheless, wet biomass is not regarded as a promising feedstock for conventional thermochemical gasification

---

[†] *Visiting scholar, present affiliation: Biomass Laboratory, National Institute for Resources and Environment, Tsukuba, Japan.*

[‡] *Visiting scholar, present affiliation: Department of Chemical Engineering, Tohoku University, Sendai, Japan*

processes because the cost of drying the material is too high. This problem can be circumvented by employing water as the reaction medium. Namely, gasification of wet biomass in supercritical water is accomplished without having to dry the material, thus avoiding the high costs associated with drying process. The foci of our interest [2-4] are therefore the steam reforming reactions which convert biomass to hydrogen and carbon dioxide. Using glucose as a model compound for complex biomass feeds, the steam reforming stoichiometry is given by:

$$C_6H_{12}O_6 + 6H_2O \rightarrow 6CO_2 + 12H_2 \tag{1}$$

Prior work in this laboratory [3-6] showed that *low* concentrations of glucose and various wet biomass species (water hyacinth, algae) could be completely gasified in supercritical water at 600°C and 34.5 MPa after a 30 s residence time. But higher concentrations of glucose evidenced incomplete gasification. In these studies, both the extent of the conversion to gas and the gas composition were observed to depend upon both the chemical composition and the condition of the reactor's wall [4]. These results suggested that heterogeneous catalysis might be employed to increase the extent of gasification of concentrated feeds. Consequently, reactors were fabricated which could accommodate packed beds of catalysts and research was initiated on heterogeneous catalysis of the steam reforming reactions of biomass materials in supercritical water.

Succeeding studies [7] showed that a wide range of carbons effectively catalyze the gasification (steam reforming) reactions of concentrated feeds (22% by weight) of glucose in supercritical water at 600 °C and 34.5 MPa. Carbon gasification efficiencies near 100% were easily achieved, irrespective of the available surface area of the carbon. Coconut shell activated carbon was shown to effectively catalyze the gasification of cellobiose and various whole biomass feeds, including depithed bagasse liquid extract, and sewage sludge. No significant deactivation of the catalyst was detected for runs lasting six hours once a swirl flow turbulence generator was placed into the entrance region. These results give cause for optimism that a practical process can be developed for hydrogen production at high pressures by the catalytic steam reforming of abundant wet biomass feedstocks.

One of the problems to be considered after complete gasification is achieved is the separation of the product gases. Ideally, the product gases are entirely hydrogen and carbon dioxide according to Eqn 1. We are only interested in hydrogen, however, which is valuable for electricity production via fuel cells, heat generation by combustion, or in reducing ores in metallurgical processes. Separation of the gases in order to obtain pure hydrogen is thus essential for biomass gasification processes. From an environmental point of view, the carbon dioxide from this process originates from biomass, and thus there is no net production of carbon dioxide. Separation of carbon dioxide from hydrogen can decrease the amount of carbon dioxide released into the atmosphere.

Conventionally, hydrogen and carbon dioxide are separated via an absorption process such as the Girbotol amine process, the Sulfinol process, and the Rectisol process. The Girbotol amine process[8] employs the reversible reaction:

$$2HOC_2H_4NH_2 + H_2O + CO_2 \rightleftharpoons (HOC_2H_4NH_3)_2CO_3 \tag{2}$$

Carbon dioxide is uniquely absorbed by a monoethanolamine solution at a low temperature (27-65°C), and is then released in a regeneration tower by heating up the solution to 100-150°C. However, these separation processes use toxic or expensive solvents. We thereby propose to separate the gases by preferentially dissolving carbon dioxide into water at the high pressure of the supercritical gasification process (34.5 MPa).

Solubility of gases into liquid is determined by the Henry constant, which is expressed by the partial pressure of a gas over the solution divided by molar fraction of the gas in the solution. Lower values of the Henry constant result in higher solubilities of the gas. As is shown in Table 1, the Henry constants for carbon dioxide dissolution into water is much higher than hydrogen. Henry constants of these gases approach together with increase in pressure, but even under high pressures such as those for supercritical gasification (34.5 MPa), the Henry constant of carbon dioxide is still one eighth of that of hydrogen. Since under these high pressures the amount of carbon dioxide that will dissolve into water is significant, this difference in Henry constants can be exploited to selectively remove almost all of the carbon dioxide in the produced gas mixture.

This paper proposes and evaluates employment of water as solvent for carbon dioxide in the supercritical water gasification process. Energy flow analysis is also conducted for the proposed process.

Table 1. Henry constants of pure carbon dioxide and hydrogen for dissolution into water.

Pressure	Henry constant [MPa]	
[atm]	CO_2	H_2
1	40	8210
5	50	8216
10	61	8222
50	152	8278
100	263	8348
500	1067	8908

2 High-pressure carbon dioxide separation process

2.1 Proposal of the separation process

Actual product gas of glucose gasification contains carbon monoxide, methane and traces of higher hydrocarbons in addition to hydrogen and carbon dioxide. As a first step, however, we treat the idealized problem of hydrogen and carbon dioxide mixture. In this case, after the supercritical gasification process, the reactor effluent, composed of hydrogen, carbon dioxide and water, is readily cooled to room temperature at 34.5 MPa. It is expected that this effluent is composed of two phases; a gas phase composed mainly of hydrogen and carbon dioxide, and a liquid phase of principally water. Some fraction of the carbon dioxide is dissolved in the water phase, which is determined by the pressure of the system and the mole fractions of each component--water, carbon dioxide, and hydrogen. It is desirable that all of the carbon dioxide be dissolved in the liquid phase, resulting in pure hydrogen as the gas phase product.

The total pressure is predetermined by the process gasification pressure, 34.5 MPa for this work; the amount of water in the feed is also fixed by the concentration of the biomass feed, which should be higher than 18 wt% to make the system economically feasible. A back-of-envelope calculation shows that the water present in the reactant alone is not sufficient to dissolve all of the produced carbon dioxide; therefore, an additional water source needs to be provided at this high pressure. A supplementary amount of cold water is thus added to the effluent in order to dissolve most of the carbon dioxide. After the absorption of carbon dioxide into this additional water, the effluent is delivered to a high-pressure gas-liquid separator. The gas phase, mostly composed of hydrogen, is utilized for downstream processes. Depressurization of the liquid phase releases the dissolved gases, and the water can then be re-pressurized and recirculated to redissolve the carbon dioxide. The energy needed to compress liquid water is negligible, and hardly affects the energy efficiency of the whole system.

2.2 Estimation of phase equilibrium

The amount of each gas dissolved in the liquid phase can be calculated by a flash calculation, given the amounts of carbon dioxide, hydrogen, and water at 25°C, 34.5 MPa[9]. The SRK equation of state was utilized here. This equation is a cubic equation of state, which is a modified Redlich-Kwong equation [10] by Soave [11]. Although the SRK equation is not necessarily ideal for the precise prediction of a system that has polar molecules like water, no other well established estimation are available for this type of system. Thus, for rough calculations, the SRK equation should be sufficient. Thermodynamic properties required for the calculation were obtained from the literature, which is shown in Table 2. The composition of each phase is determined so that the fugacities of each compound in each phase become equal. The program shown in the literature [12] was modified so that binary interaction parameters would be considered.

To predict the behavior of such mixtures, binary interaction parameters are of great importance. Binary interaction parameters between water, hydrogen, and carbon dioxide were measured by Goto [13] at 10 MPa, 40 and 60°C based on measurements of both liquid and gas phase compositions. These values are listed in Table 3. The applicability of these values was cross-checked by comparing the solubility data of carbon dioxide in water in the presence of hydrogen at 30 MPa, 25°C[14] and the predicted solubility using these binary interaction parameters. Values obtained by extrapolating to 25°C were used for this calculation. As shown in Fig. 1, the predicted solubility does not agree well with the literature data.

Given these discrepancies, we formulated the interaction parameters by a least square fitting technique to the literature data[14]. The literature provided only liquid phase data, based on which these parameters were determined. Solubility data from the literature was well reproduced using the values we obtained. These interaction parameters are also shown in Table 3. These values are not significantly different from the values by Goto; they predict the solubility accurately according to Fig. 1. Thus, we utilized our least square fit values for further calculations.

Table 2. Thermodynamic properties of compounds used in the calculation

	Critical temperature [K]	Critical pressure [MPa]	Critical specific volume [cm^3/mol]	Acentric factor [-]
H_2O	647.3	22.12	57.1	0.344
H_2	33.2	1.30	65.1	-0.218
CO_2	304.1	7.38	93.9	0.239
CO	132.9	3.50	93.2	0.066

Table 3. Binary interaction parameters

	Binary system		
	H_2-CO_2	H_2-H_2O	CO_2-H_2O
By Goto 40°C	0.3536	-1.4314	-0.1171
By Goto 60°C	0.3536	-1.1324	-0.0918
This work	0.921	-1.36	-0.08

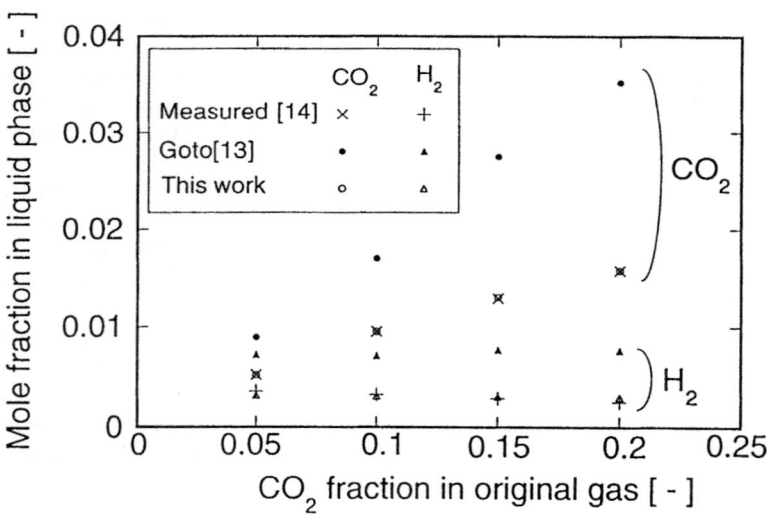

Fig. 1. A comparison between solubility data from the literature and estimated solubility of carbon dioxide in the presence of hydrogen.

2.3 Experimental

To check the validity of the estimation, gasification of 2 M and 6 M formic acid solutions at 34.5 MPa, 600°C was conducted. The gas effluent was sampled at the operating pressure after being cooled down to room temperature. Previous experiments in our laboratory have shown that formic acid decomposes completely into hydrogen and carbon dioxide with only a small amount of carbon monoxide under these conditions. No methane or other hydrocarbons were generated. Thus, formic acid is a suitable compound for measuring the behavior of a pure hydrogen-carbon dioxide system.

Figure 2 displays the experimental apparatus used in this work. The reactor was made of Inconel 625 tubing with a 9.53 mm OD and a 4.75 mm ID. Formic acid solutions were fed to the reactor by a Waters 510 HPLC pump at a flow rate of 1 mL/min. The temperature of the reactant flow was abruptly raised to the desired value by a cooling water jacket juxtaposed with an entrance heater. The reactor was maintained at isothermal conditions by a furnace and a down-stream heater. The product flow was then rapidly quenched by a cooling water jacket at the exit of the reactor. An annulus was employed inside the reactor to measure the temperature of the fluid inside the reactor; it was placed all the way through the entrance region with its end one half inch past the entrance heater. A coiled wire was affixed to the exterior of the annulus to create swirl flow in the entrance region of the reactor, increasing the turbulence of the flow, and thus improving heat transfer to the reactant.

The heated zone and the downstream cold section of the reactor was filled with sintered aluminum oxide (Aldrich Chemical Co.). The axial temperature profile along the reactor's functional length of approximately 0.48 m was measured by 15 fixed, type K thermocouples mounted on the reactor's outer wall, with two of them in ground out areas under the entrance and guard heaters. Furthermore, a movable type K thermocouple inside the annulus provided a temperature profile inside the reactor. Pressure in the reactor system was measured by an Omega PX302 pressure transducer. A Grove Mity-Mite model 91 back-pressure regulator reduced the pressure of the cold products exiting the reactor from 34.5 MPa to ambient. After leaving the back-pressure regulator, the reactor effluent passed through an in-house fabricated glass gas-liquid separator. The gas flow rate was measured using a wet test meter.

Gas samples are taken at both high and low pressures. A high-pressure gas sample vessel with a 1 cm^3 volume was evacuated, and then the high-pressure effluent gas was allowed to accumulate inside by the operation of valves. After the holder was filled with effluent gas, the valve at the bottom was closed, and the sample was released into a pre-evacuated gas sampling tube. Low pressure gas samples were collected with a gas-tight syringe from the gas sampling port of the gas-liquid separator. Since the amount of gas dissolved in the liquid phase under atmospheric pressure was negligible, the low-pressure sample was used to determine the actual composition of the generated gas.

The gaseous products were analyzed by a Hewlett-Packard model 5890 gas chromatograph (GC) equipped with flame ionization and thermal conductivity detectors. A 800/100 mesh carbosphere molecular sieve packed column was used, operating at 35°C for 4.2 min, followed by a 15°C/min ramp to 227°C, a 70°C/min ramp to 350°C, and a 5.3 min hold at 350°C. A mixture of 8% hydrogen in helium from AIRCO was used as the carrier gas. A standard gas mixture, obtained from AIRCO, was used to calibrate the gas chromatograph.

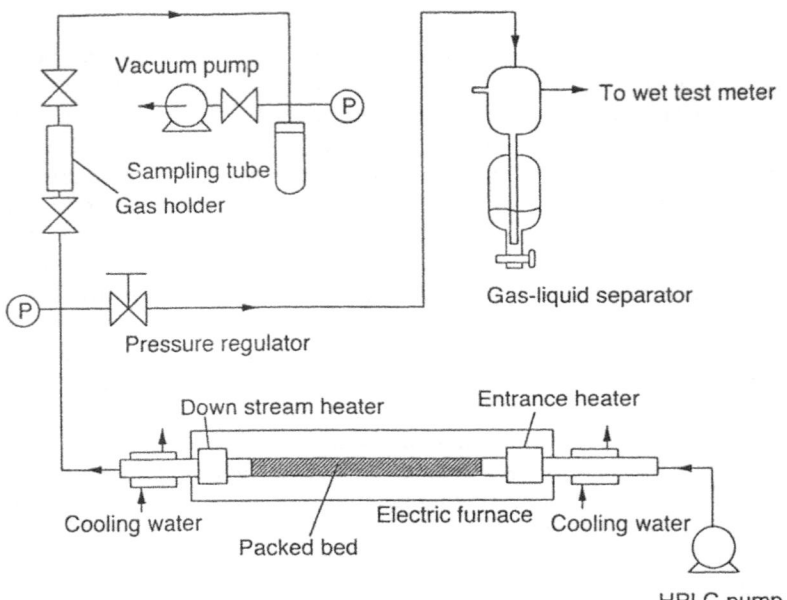

Fig. 2. Experimental apparatus.

2.4 Results and discussions

The low pressure gas sample was composed of 44 mol% hydrogen, 44 mol% carbon dioxide and 11 mol% carbon monoxide from the 6 M formic acid gasification, and 45 mol% hydrogen, 47 wt% carbon dioxide, and 8 mol% carbon monoxide from the 2 M formic acid gasification. These values are reasonable considering the gasification reaction of formic acid produces water and carbon monoxide via the stoichiometric equations:

$$HCOOH \rightarrow H_2O + CO \tag{3}$$

and/or:

$$HCOOH \rightarrow H_2 + CO_2 \tag{4}$$

takes place. Thus, generated carbon dioxide and hydrogen should be equimolar, and observed molar amounts of hydrogen and carbon dioxide are in good agreement. A carbon balance was calculated using these gas compositions and the total gas generation rate measured by a wet test meter. The disturbance of the system caused by the high-pressure sampling does temporarily perturb the gas generation rate during the experiment. However, the amount of carbon in the product gas agreed with the amount of carbon in the feed solution within 8% error before taking the high-pressure gas samples. Thus, the measured gas compositions of low pressure sample are utilized to determine the initial composition of the flash calculation.

Table 4 shows the comparison between the experimental results and the estimated values for the high-pressure gas composition. Considering the relative uncertainty of the

SRK equation for a system including water, good agreement is obtained. In this calculation, the binary interaction parameters regarding carbon monoxide were taken as zero to simplify the calculation. Since the mole fraction of carbon monoxide is small, this approximation does not affect the results significantly.

Lowering the concentration of the feed solution has the same effect as adding water to the effluent. The higher purity of hydrogen observed in the 2 M formic acid gasification indicates the increased effectiveness of high-pressure carbon dioxide separation when more water is present to dissolve the carbon dioxide.

This agreement enjoyed here between the theoretical estimates and the experimental results assures the applicability of this prediction method to real systems.

Table 4. Comparison of the experimental values and estimated values of the molar fraction of product gases at 34.5 MPa.

Gas		Formic acid concentration [mol/L]					
		6.0			2.0		
		H_2	CO_2	CO	H_2	CO_2	CO
Data	Sample 1	0.54	0.27	0.14	0.57	0.32	0.10
	Sample 2	0.49	0.38	0.11	0.59	0.30	0.11
	Sample 3	0.48	0.37	0.12	0.53	0.36	0.11
Prediction		0.50	0.39	0.11	0.59	0.26	0.15

2.5 Estimation of effect of additional water

The chemical composition of real biomass varies widely, depending on the species and other factors, like pretreatment. Because of this variation, theoretical calculation of the resulting gas composition is difficult. Here, glucose is used as a model compound for biomass because cellulose, the main constituent of most biomass, is a polymer of this hexose. Ideally, decomposition of 1 mol of glucose produces 12 mol of hydrogen and 6 mol of carbon dioxide according to Eqn 1. Although methane, carbon monoxide and other hydrocarbons are produced in the real system, this ideal situation is used for the calculation. A concentration of 1.2 M, 20.0 wt%, is model operating concentration. Under this condition, the separation of the produced hydrogen and carbon dioxide is estimated.

Figure 3 shows the effect of adding excess water on the resulting purity of hydrogen in the vapor phase and the ratio of hydrogen dissolved in water phase to the total hydrogen gas. When no supplementary water is added, the high-pressure gas phase composition is 69.6% hydrogen and 30.0% carbon dioxide. The rest is water vapor, which is negligible. The hydrogen concentration in the pressurized gas phase is higher than that at low pressure, which is theoretically 66.7%. The effect of the preferential dissolution of carbon dioxide into water is evident. However, the amount of water in the reactant feed is not sufficient to dissolve all of the carbon dioxide.

The addition of water increases the purity of the hydrogen gas. This estimation shows that to obtain 97.6% purity hydrogen, 19 L of cold water should be added to the effluent produced from 1 L of 1.2 M glucose. The water has to be supplied at high pressure, and thus compression of the water to 34.5 MPa is necessary. This work can be estimated by integrating the volume change of water along an isotherm, which is 0.26 kJ/kg. The compression of 19 L of water therefore requires 4.94 kJ. Considering the hydrogen produced from 1 L of solution is as much as 14.4 mol, with a heat of combustion as high as 242 kJ/mol, the work required to compress the additional water is negligible.

The additional water also results in the dissolution of a larger amount of hydrogen

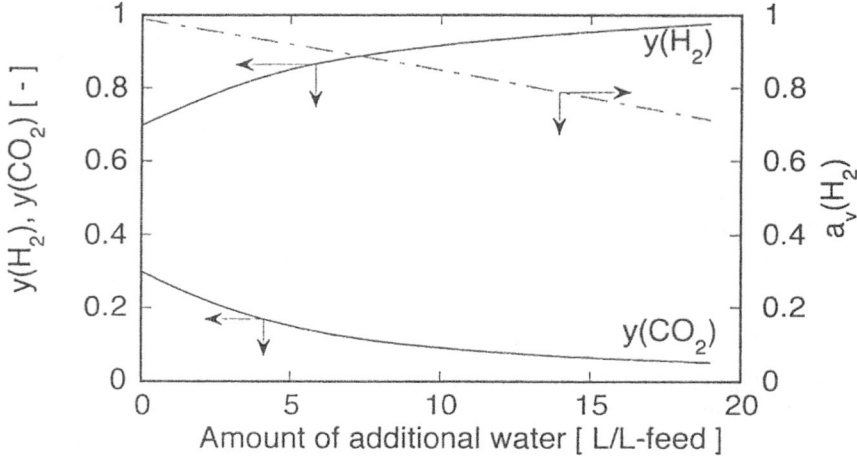

Fig. 3. Effect of adding water on the purity of hydrogen in the vapor phase.

into the liquid phase. The ratio of the dissolved hydrogen to total hydrogen is also shown in Fig. 3. In the case of no additional water, only 2% of the produced hydrogen is present in the liquid phase. When 19 L of water is added to obtain 97.6% pure hydrogen, 29% of the produced hydrogen is dissolved in the liquid phase. Since the yield of hydrogen directly affects the efficiency of the system, this loss of hydrogen is a drawback for this high pressure separation system. Burning this dissolved hydrogen to produce the heat necessary to increase the temperature of the incoming flow helps to increase the system efficiency. However, if the amount of hydrogen dissolved as a result of the additional water is greater than required for this purpose, it will adversely affect the efficiency of the system.

2.6 Gasification process flow diagram

Figure 4 shows a biomass gasification system for hydrogen production with the proposed gas separation system. Biomass is made into a slurry by a grinder. The biomass slurry is then pressurized to 34.5 MPa, heated up to 600°C via a heat exchanger and an additional heater, and is then fed into the supercritical water reactor. In the reactor, ideally, the biomass is catalytically converted entirely into product gas composed of hydrogen and carbon dioxide in the packed bed of carbonaceous materials. The effluent is cooled down by a heat exchanger, returning heat to the incoming flow.

After passing through the high-pressure absorption column, hydrogen is separated and stored for end uses. Since the hydrogen is already at a pressure as high as 34.5 MPa, the pressure can be utilized to conduct additional work, or to increase the efficiency of the downstream turbine. The liquid phase includes dissolved carbon dioxide and some part of the hydrogen. Depressurization of this liquid allows these gases to be released. Since the amount of the dissolved hydrogen is not negligible, recovery of this dissolved hydrogen should be considered. Here, the released hydrogen is burned to compensate the heat unrecovered by the heat exchanger. The efficiency of the heat exchanger is 50-75% for the supercritical oxidation process. When the additional heat supplied by the liquid phase hydrogen is not insufficient, part of the hydrogen in the gas phase is also used.

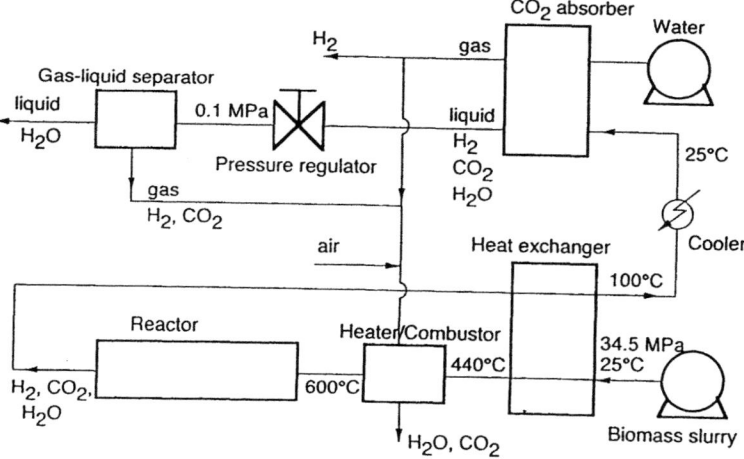

Fig. 4. Biomass gasification process with carbon dioxide separation by adding water.

3 Energy balance calculation

Figure 5 shows the energy flow analysis of the gasification system for 1 mol of glucose in a 1.2 M solution when the efficiency of the heat exchanger is 75%. The combustion heat of glucose, 2.54 MJ/mol, and the heat required to heat up the reactant slurry is supplied to the reactor. The heat capacity of the solution is assumed to be the same as for water. Most of the sensible heat requirement is provided by the heat exchanger. Supplemental heat is provided by combustion of the dissolved hydrogen and a part of the produced hydrogen. The gasification reaction itself is slightly endothermic; the heat of reaction is 0.37 MJ/mol-glucose. Thus, the heat of combustion of the produced hydrogen is higher than the heat of combustion of glucose by this amount. The final amount of hydrogen after subtracting the heat supply to the reactant solution (7.69 mol), has a heat of combustion of 1.86 MJ. The thermal efficiency of this process is 73%. Although 27%

of the heating value of the original biomass is lost, considering the fact that no heat is available from wet biomass (glucose solution) itself, a recovery of 73% is quite remarkable. Lower heating values are used in the calculation.

Figure 6 shows the same energy flow diagram in the case of a 50% efficient heat exchanger. It is clear that a much larger part of the produced hydrogen is consumed to heat up the reactant solution, reducing the final amount of hydrogen. The thermal efficiency for this case is as low as 47%. This result indicates the importance of the efficiency of the heat exchanger.

Although in this calculation hydrogen and carbon dioxide are considered to be the only products, carbon monoxide, methane and larger hydrocarbons are generated as well in the gasification of glucose[7]. The presence of these gases is important. If we are able to use these other combustible gases as a part of the downstream process, for example in the case of using the products as a heat source, the presence of these gases is not problematic. If, on the other hand, we need pure hydrogen, as for the case of supplying the product gas to a fuel cell, some downstream treatment is required. To reduce carbon monoxide, the water-gas shift reaction needs to be enhanced. We are currently trying to increase the degree of the shift reaction in the supercritical reactor, and are now working on this subject experimentally. For conversion of methane and other hydrocarbons, a downstream reformer would be necessary. Conversion of methane inside the gasification reactor is impossible because the chemical equilibrium at 34.5 MPa indicates the stability of methane.

Increasing the concentration of the reactant solution decreases the heat requirement for heating up the solution. This decreases the amount of the produced hydrogen needed to cover the heat lost by the heat exchanger. Thus, for a real biomass feed, a highly-concentrated biomass slurry is desirable. High-concentration slurries tend to clog the reactor, but we are currently developing a high-concentration paste-like slurry which can be fed smoothly into the reactor. We expect that we can achieve a slurry concentrations higher than 20 wt%.

4 Conclusions

- Based on the higher solubility of carbon dioxide than that of hydrogen in water at the high pressure of the reactor effluent, a carbon dioxide separation process using only water is proposed. By just adding water to the product gas mixture under high pressure, most of the carbon dioxide can be dissolved into the liquid phase, leaving much purer hydrogen in the gas phase. Power consumption in compressing water to dissolve the carbon dioxide is negligible.
- The estimated result by a flash calculation using the SRK equation agreed with the experimental results by 34.5 MPa. Binary interaction parameters for this equation were obtained by fitting calculated results to literature data.
- The calculated result shows that the addition of 19 L of water to the effluent per 1 L of 1.2 M glucose solution feed gives 97% hydrogen as the gas phase product.

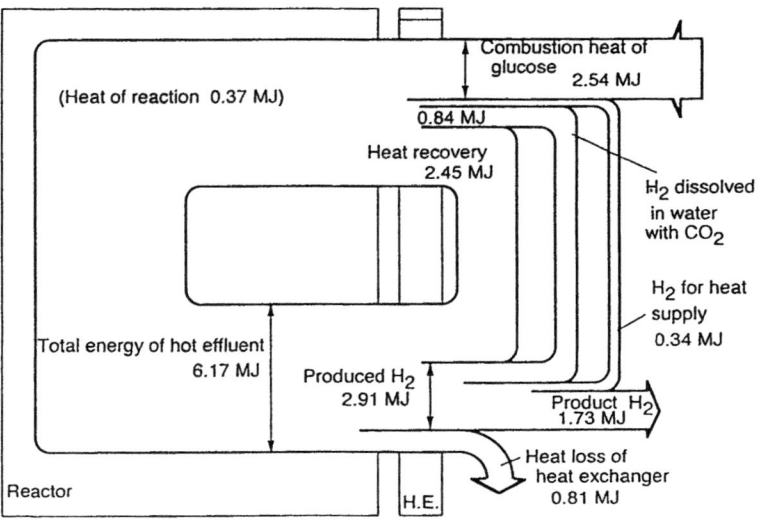

Fig. 5. Energy flow analysis of the gasification system for 1 mol of glucose in a 1.2 M solution when the efficiency of the heat exchanger is 75%.

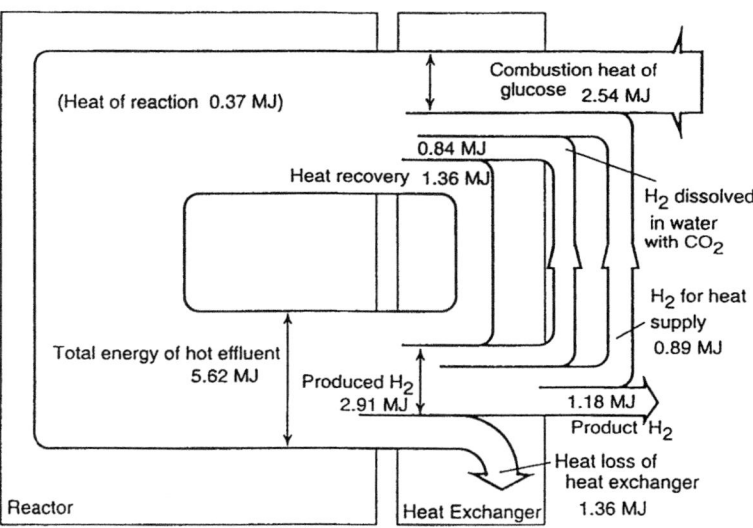

Fig. 6. Energy flow analysis of the gasification system for 1 mol of glucose in a 1.2 M solution when the efficiency of the heat exchanger is 50%.

- Part of the produced hydrogen is dissolved into the liquid phase, but is released by depressurizing the liquid effluent, and is then utilized to compensate the heat unrecovered by the heat exchanger. The water can be recycled after releasing the dissolved gases.
- Energy flow analysis was conducted for the gasification process of 1.2 M glucose, using this separation unit. It shows the significance of the heat exchanger efficiency.

5. Acknowledgment

This work was supported by NREL/DOE under contracts DE-FC36-94AL85804, XCF-5-14326-01, and the Coral Industries Endowment.

6. References

1. Wolverton, B.B. and R.C. McDonald (1980) Vascular plants for water pollution control and renewable sources of energy. in *Bio-Energy '80* Atlanta, Georgia, U.S.A.: The Bio-Energy Council.
2. Antal, M.J., Jr. (1983) Effects of reactor severity on the gas-phase pyrolysis of cellulose- and kraft lignin-derived volatile matter. *Ind. Eng. Chem. Prod. Res. Dev.*, Vol.22, pp.366-75.
3. Antal, M.J., Jr., T. Leesomboon, W.S. Mok, and G.N. Richards (1991) Mechanism of formation of 2-furaldehyde from D-xylose. *Carbohydr. Res.*, Vol.217, pp.71-85.
4. Yu, D., M. Aihara, and M.J. Antal, Jr. (1993) Hydrogen production by steam reforming glucose in supercritical water. *Energy & Fuels*, Vol.7, pp.574-7.
5. Manarungson, S. (1991) MSE Thesis, Department of Mechanical Engineering, University of Hawaii at Manoa.
6. Yu, D. (1993) MSE Thesis, Department of Mechanical Engineering, University of Hawaii at Manoa.
7. Xu, X., Y. Matsumura, J. Stenberg, and M.J. Antal, Jr. (submitted) Carbon catalyzed gasification of organic feedstocks in supercritical water. *submitted to Ind. & Eng. Chem. Res.*
8. Bottoms, R.R., (1933) US Patent No.18,958 (reissue of 1,783,901), Sept. 26 (transferred to The Girdler Corp.).
9. Reid, R.C., J.M. Prausnitz, and B.E. Poling (1987) *The properties of gases & liquids.* 4th ed. , New York: McGraw-Hill.
10. Redlich, O. and J.N.S. Kwong (1949) On the thermodynamics of colutions V. An equation of state. Fugacities of gaseous solutions. *Chem. Rev.*, Vol.44, pp.233-44.
11. Soave, G. (1972) Equilibrium constants from a modified Redlich-Kwong equation of state. *Chem. Eng. Sci.*, Vol.27, No.6. pp.1197-203.
12. Saito, S. (1983) *Toukei netsurikigakuni yoru heikou bussei suisanno kiso.* Chemical Engineering Series, No. 6: Baifukan.
13. Goto, S. (1989) B.S. Thesis, Dept. of Chem. Eng., Tohoku Univ.

14. Linke, W.F., ed. (1958) *Solubilities: inorganic and metal-organic compounds; a compilation of solubility data from the periodical literature.* 4th Ed. ed. , American Chemical Society: Washington.

GASIFICATION

Laboratory experimentation

RELATIONSHIP BETWEEN GASIFICATION REACTIVITY OF STRAW CHAR AND WATER SOLUBLE COMPOUNDS PRESENT IN THIS MATERIAL

Henriksen, U. Jacobsen, M.J. Lyngbech, T. Hansen, M.W., Department of Energy Engineering, Technical University of Denmark, DK-2800 Lyngby.

Abstract

Keywords: Alkali, Biomass, Char, Gasification, Catalysis, Straw, Washing.

By gasification of char from straw the reactivity depends of the content of water soluble compounds in the straw. The dependency was studied by extracting the water solubles and by reabsorption of these compounds into a previously washed sample.

Washing reduced the reactivity significantly, and it was seen that the original high reactivity could be re-established by absorption of the water soluble compounds back in the sample. It was also seen, that this effect on the reactivity was caused by the presence of water solubles during the gasification, while their presence during pyrolysis had no measurable effect on the gasification reactivity. A linear correlation between the reactivity and the concentration of the water solubles was found.

1   Background and Purpose

It has been observed that the gasification properties of straw change when the straw is exposed to rain. In order to meet the need for data on the reactivity to be used in mathematical models for the design of gasification reactors, the relationship between water soluble compounds and reactivity was investigated experimentally. The purpose was to give an idea of the extent of the effect of water solubles on char reactivity, but also to describe the phenomena qualitatively without giving detailed chemical explanations.

1.1   Method

1.1.1   TGA test

A large scale TGA test plant was used to measure the reactivity of char during gasification (figures 1 and 2). The char was produced in a separate pyrolysis oven at low heating rate (about 10 °C per minute), and a residence time of about 2 hours at the final pyrolysis temperature, which for all tests was 600°C. Pyrolysis was carried out in a nitrogen atmosphere.

The mass of the char samples in the TGA was 5 g. The samples were not pulverized, but were processed still having the shape and structure of straw. All samples were dried at 104°C before pyrolysis, before washing and before reabsorbtion of water solubles.

The gasification tests started by heating the reactor with a char sample to a temperature of about 120- 130°C for drying. After 1 hour the temperature was raised at a rate of 24°C per minute in an atmosphere of nitrogen (1 litre per min). During this heating process, the sample was finally pyrolysed, as the gasification temperature, which was 800°C for all tests, was higher than the pyrolysis temperature at which it was pyrolysed in the oven. When the desired gasification temperature of 800°C was attained, the temperature was kept constant for 1 hour. After this period the nitrogen atmosphere was changed to an atmosphere consisting of 100 per cent preheated steam. (The rate of steam addition was 1500 g per hour).

The gasification was carried out at 800°C for all tests. A typical gasification run is shown in figure 3.

1.1.2   Washing

Washing was carried out by mixing a sample in 3000 ml pure water at 32°C for two hours by continually stirring. All samples for the washing experiments were based on 50 g dry straw.

All experiments were carried out with samples from one well mixed bale of barley straw. A chemical analysis of the straw is shown in table 1.

1.1.3 Absorption

Experiments in which samples of straw or char were washed, were carried out. Other in which samples were submerged in an extract from the used washing water in order to return the water solubles to the sample, were carried out.

The washing water from a sample (50 g dry straw or char from 50 g dry straw) was boiled from 1500 ml to an concentrate of about 300 ml. The sample (50 g dry straw or char from 50 g dry straw) was then submerged until this was saturated. The rest of the extract was removed and the sample was dried. This was repeated 3 to 5 times until all the water solubles from the extract were absorbed in the sample.

1.1.4 Experiments

Six tests were conducted:
1. TGA gasification test of char from straw (see figure 4).
2. TGA gasification test of char from washed straw (see figure 5).
3. TGA gasification test of char from washed straw, but where the extract from the washing water was reabsorbed in the pyrolysed char (see figure 4).
4) TGA gasification test of washed char (see figure 5).
5) TGA gasification test of char from straw which has been washed, but where the extract of the washing water was reabsorbed in the straw before it was pyrolysed (see figure 4).
6) TGA gasification test of char from straw which had been washed, but where the extract from the washing water resulting from the washing of 100 g dry straw was absorbed in 50 g dry washed straw before it was pyrolysed. This char now had approximately about twice the original content of water solubles (see figure 6).

1.2 Results

1.2.1 Chemical analysis

A chemical analysis of the straw was carried out, and some of the relevant data are shown in table 1.

Table 1. chemical analysis of the investigated barley straw.

	Mass Per cent on dry basis	Uncertainty Per cent on dry basis
Ash content, total (550°C)	4.2	± 0.05
Chlorine, total	0.80	± 0.01
Chlorine, water soluble	0.81	± 0.02
Potassium, total	1.55	± 0.05
Potassium, water soluble	1.01	± 0.1
Sodium, total	0.08	± 0.01
Sodium, water soluble	0.056	± 0.01
Silicon (Si), total	0.99	± 0.05

1.2.2 Reactivity

Reactivity for char gasification is defined as

$$R(t) = \frac{1}{W(t)} \times \frac{dW}{dt}$$

It can be seen from figures 4 to 6 that for many runs, reactivity increases with increasing burn off. In order to obtain one characteristic number for the reactivity for one test run, a mass-weighted reactivity R_m was defined. The aim of this was to give the reactivity where the actual mass of the sample was high, a larger weight than the reactivity where the actual mass of char was low.

R_m was calculated between 12 per cent and 92 per cent sample weight conversion (burn off).

$$R_m = \frac{\sum_{t=t_0}^{t_n} R(t) \times W(t)}{\sum_{t=t_0}^{t_n} W(t)}$$

$R(t)$: The reactivity at time t.
$R(bo)$: The reactivity versus sample weight conversion (burn off).
bo : Burn off, defined as bo = 100 (1-(W(t)/ W(0))).
$W(t)$: The actual mass of convertible (ash free) char at time t.
$W(0)$: The initial mass of convertible char at the moment when the gasification starts.
t : The time elapsed from the start of the gasification.
t_0 : The time at 12 per cent burn off.
t_n : The time at 92 per cent burn off.

In table 2 the R_m value for the six experiments (**1** to **6**) is showed.

Table 2. Mass-weighted reactivity R_m.

Test no.	R_m per cent per minute
1	15
2	7
3	14
4	6
5	14
6	25

1.2.3   Original straw compared to washed straw

It was found that the reactivity R_m for char from original straw was about 15 per cent per minute.   It was also found, that when the straw was washed before pyrolysis, the reactivity R_m for the char gasification was reduced to about 7 per cent per minute.(see table 2, test **1** and **2**).

From test **1** it was found, that for char from the original straw, the reactivity R(bo) increased significantly with increasing burn off (bo).   It was also found, that if the char was produced from washed straw, the increase of the R(bo) with burn off (bo) was reduced significantly by a factor of 2 to 3 (see figures 4 and 5).

1.2.4   Absorption of water solubles

As described above, washing of straw resulted in decreased reactivity R_m of the char from about 15 to about 7 per cent per minute. However it was also demonstrated that in spite of the straw sample having been washed, it was possible to reobtain the high reactivity by absorbing an extract from the washing water back in to the straw sample (see table 2, test **1** and **5**). Also the reactivity R(bo) as a function of the sample weight conversion (burn off) was reobtained (see figure 4).

These tests indicate that the reactivity-reducing effect of the washing, was not a consequence, of changes in the structure of the sample caused by the washing, but as a result of active compounds being removed from the sample. The effect of these compounds could be reobtained when they were reabsorbed in the sample.

In the literature [1, 2] potassium and sodium have been described as having a favourable influence on char reactivity. As the results of chemical analysis (see table 1) shows presence of water soluble potassium and sodium, this explanation is reasonable.

1.2.5   Presence of water solubles in the sample during pyrolysis

It was observed, that when the char was washed after it was pyrolysed, the reactivity R_m was reduced to about 6 per cent per minute (see table, 2 test **4**). This is very similar to the result that we get from testing char derived from washed straw (see table 2, test **2**). Also, the progress of R(bo) versus burn off is very similar for these two tests.

This indicates that the presence of water solubles during pyrolysis does not influence reactivity of the char gasification.

This was also demonstrated by testing a sample from which the water solubles had been removed from the straw before pyrolysis, and reabsorbed in the char after pyrolysis before gasification (see table 2, test **3**). Test **3** showed that both the R_m and the progress of R(bo) were similar to results from test **5**, where the water solubles from the washed straw were re-absorbed in the washed straw before pyrolysis.

The fact that the presence of water solubles during pyrolysis did not have a measurable influence on the reactivity of the char indicated that water solubles does not have any significant effect on the structure in the char (active spots, pore structure, surface, etc.) during pyrolysis.

By comparing tests **1**, **3** and **5** (see figure 4), It was observed that there were no significant differences in the char reactivity, whether the water solubles were absorbed into the straw structure, absorbed into the structure of pyrolysed char, or just remained in the original straw.

1.2.6 Correlation between reactivity and content of water solubles

In test **6** water solubles washed out of 100 g dry straw were absorbed in 50 g washed dry straw. Thus the content of water solubles in this sample was about double that of the original straw. In figure 6 the reactivity R(bo) is shown, and the R_m increased significantly (see table 2).

It was assumed that the water solubles remained in the sample during gasification, that all water solubles was washed out of the sample during washing, and that all water solubles were reabsorbed during absorption. As no analysis of the actual concentration of water solubles in the char were carried out for these tests, all concentrations were related to the concentration in char from the original straw. Let $C(0)_x$ designate the concentration of water solubles in the char before gasification, expressed as mass of water solubles per unit of mass of convertible char for test x. Then we have:

$$C(0)_1 = C(0)_3 = C(0)_5 = 1,$$
$$C(0)_2 = C(0)_4 = 0 \text{ and}$$
$$C(0)_6 = 2.$$

In figure 7 the R_m values is presented as a function of the content of water solubles in the sample (test nos. **1** to **6**). A nearly linear correlation was seen. This was observed for biomass with a small content of silicon e.g. in reference no. 1. In the present work this observation is being extended to biomass with a high silicon content.

During a test run the reactivity was seen to increase with increasing burn off. This could be explained by the increase in the concentration of water solubles with increasing burn off. The following correlation was suggested : $R(bo)_x = R_0 + A_0 C(bo)_x$, where $C(bo)_x$ was the actual concentration of water solubles during gasification, expressed as mass of water solubles per unit of mass of convertible char for test x. The R_0 can be regarded as a basic reactivity, and was suggested to be a best line through the reactivity curve of the washed samples tests **2** and **4**, (see figure 8). It is now possible to calculate the concentration of water solubles, $C(bo)_x$ versus burn off:

$$C(bo)_x = \frac{C(0)_x}{1 - \frac{bo}{100}}$$

and the reactivity as:

$$R(bo)_x = R_0 + A_0 \cdot \frac{C(0)_x}{1 - \frac{bo}{100}}$$

The proposed correlation was compared to the experiments in figure 8 with $A_0 = 4.5$. For burn off below 80 per cent the correlation described the experimental data satisfactorily.

1.3 Conclusion.

It was possible to study the reactivity-promoting effect of the water soluble compounds on straw char by washing samples and by reabsorption of the water solubles back into the samples.

Washing reduces the reactivity radically (up to a factor 5), and it was seen that the original high reactivity could be re-established by absorption of an extract from the original washing water by the sample. It was also seen that the effect on the gasification reactivity was caused by the presence of water solubles during gasification, while their presence during pyrolysis had no measurable effect on the gasification reactivity of the char.

It was seen, that there were no significant differences in the char reactivity, whether the water solubles were absorbed into the straw structure, absorbed into the structure of pyrolysed char, or just stayed in the original straw. It was found that there were a linear correlation between the reactivity and the concentration of water solubles in the char.

1.4 References:

1. DeGroot, W.F., Kannan, M.P., Richards, G.N., Theander, O., (1990), Gasification of agricultural residues (biomass): Influence of inorganic constituents. *Journal of Agricultural and Food Chemistry USA*, Vol. 38(1), pp. 320-323.
2. Kannan, M.P., Richards, G.N., (1990), Gasification of biomass chars in carbon dioxide: dependence of gasification rate on the indigenous metal content. *Fuel - UK*. Vol 69(6), pp. 747-753.

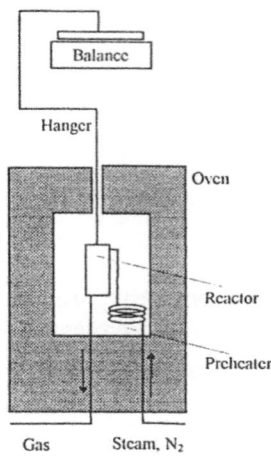

Figure 1. Test equipment for the TGA tests.

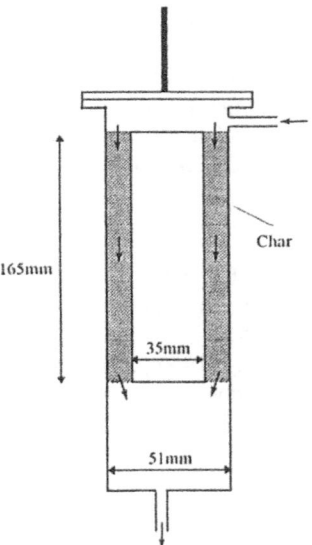

Figure 2. Reactor for gasification in the TGA tests.

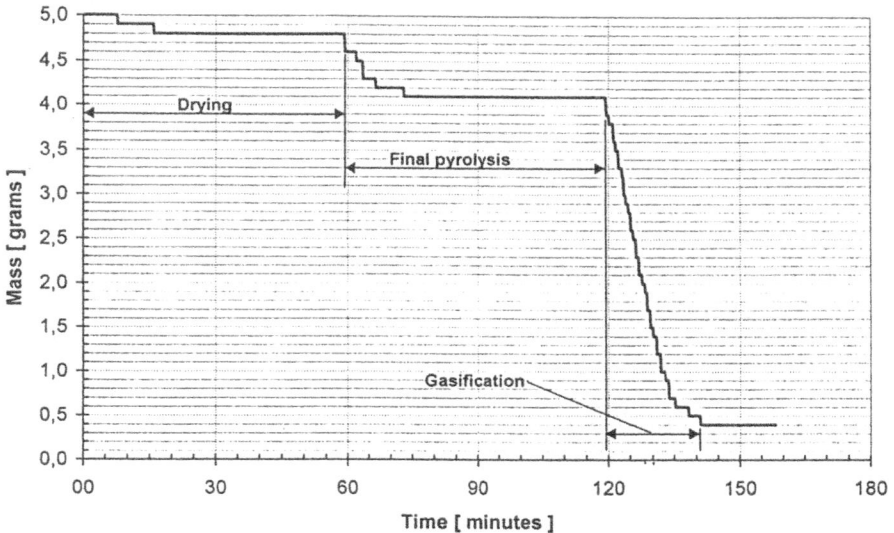

Figure 3. The mass of a char sample versus time during gasification in a TGA test.

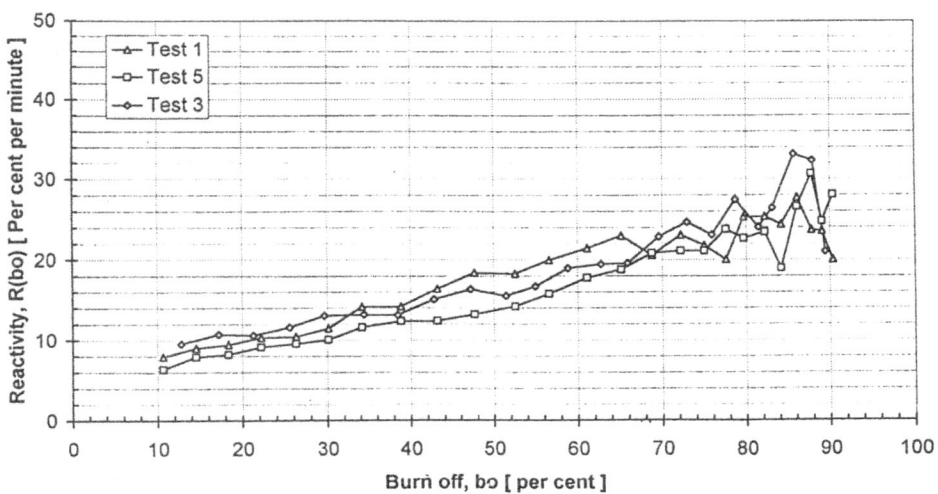

Figure 4. Reactivity versus burn off for tests 1, 3 and 5. All the samples used in these tests were containing water solubles.

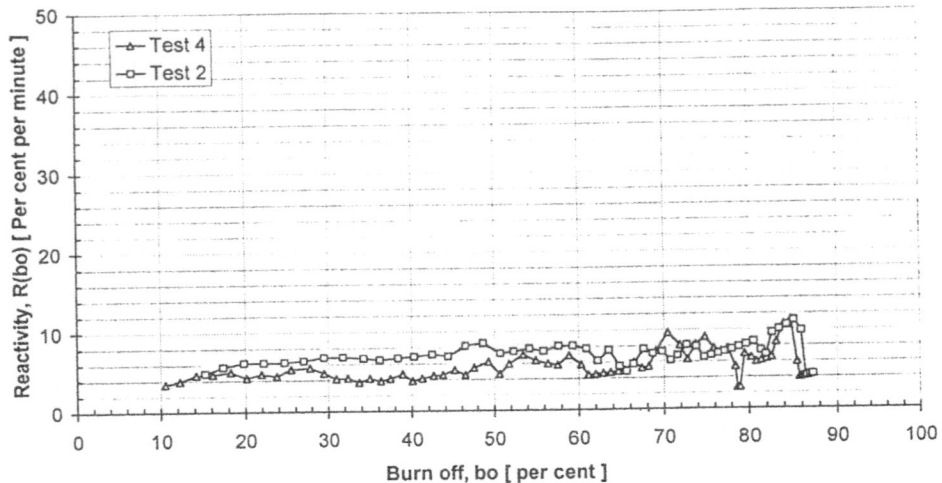

Figure 5. Reactivity versus burn off for tests 2 and 4. The samples used for these tests were not containing water solubles.

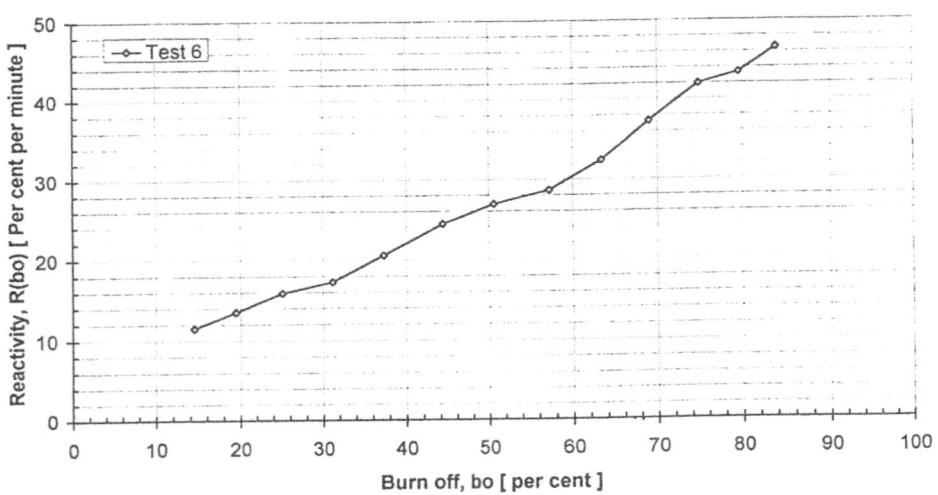

Figure 6. Reactivity versus burn off for test 6. The sample contained the double amount of water solubles as original.

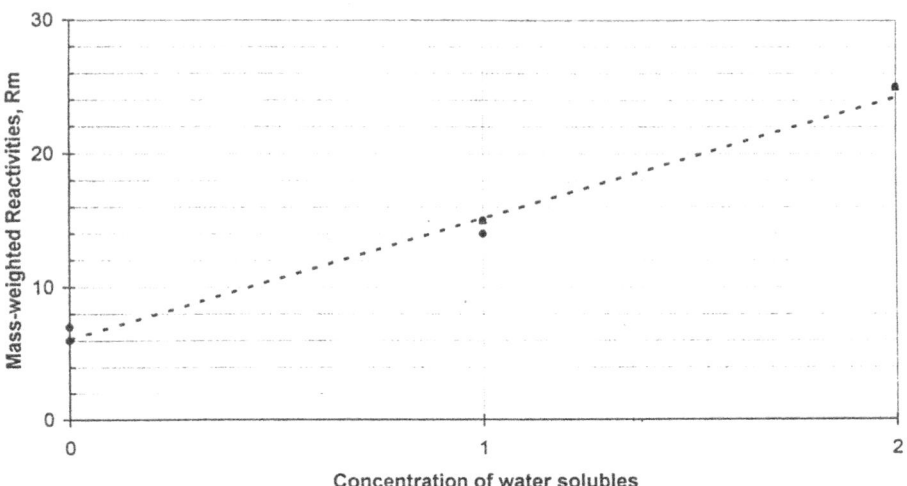

Figure 7. Mass weighted reactivity R_m versus content of water solubles. The content of water solubles are related to the content in the original straw.

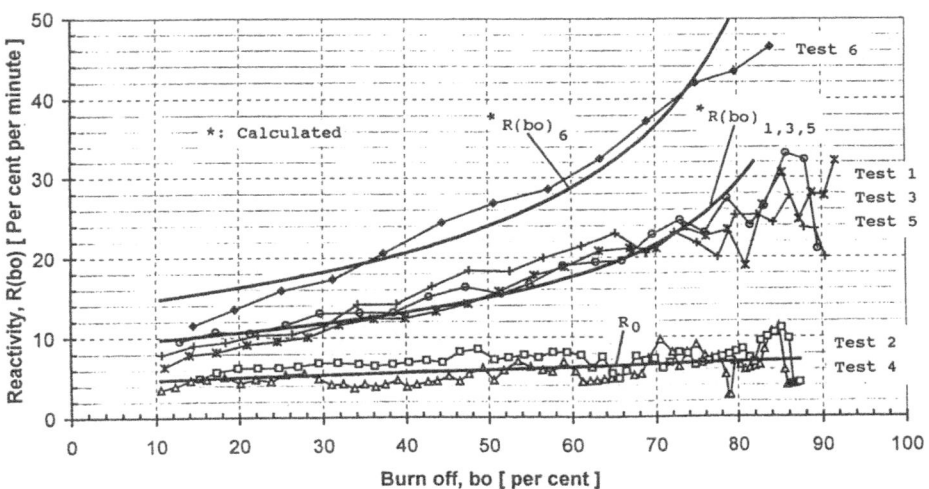

Figure 8. Reactivity versus burn off related to the actual concentration of water solubles in the sample.

RESEARCH ON STRAW WASTE GASIFICATION AND APPLICATION IN STRAW PULP MILL

M. XU, Z.Z. GU, L. SUN, D.Y. GUO and T. HAN
Energy Research Institute of Shandong Academy of Sciences
Jinan, P.R. China

Abstract

Wheat straw is main raw material for pulp mill in China. To produce 1 ton of paper, 2.5 tons wheat straw have to be used, of which 10 ~ 15% is unsuitable to paper-making and has to be screen out as waste, so resulting in environmental pollution.

The Energy Research Institute of Shandong Academy of Sciences developed a gasification system to convert such waste material which is characterized by low density, high ash and moisture to fuel gas which could be burnt in a boiler. This technology properly solves the waste treatment problem, reduces pollution to environment and greatly save fuel consumption in pulp mill.

This paper describes the design of gasification equipment, experimental data, and analyses its economic and environmental advantages.

1 Introduction

Wheat straw is one of the main raw material for pulp mill in China. About 6 million tons of wheat straw are consumed every year, of which 10 ~ 15% are grass joints, leaves and dust and have to be screened out before feeding into the digester. It is a difficult problem to treat these wheat straw waste for many pulp mills because its poor quality, loose, dirty and so on. Stacking it in the mill will occupy large space and influence environmental sanitation. Furthermore, it will ferment and release off harmful gases after a period stacking. Actually this is another largest pollution source besides the black liquor.

Energy Research Institute of Shandong Academy of Sciences conducted a testing program on the gasification of the wheat straw waste, in order to make use of this material efficiently and eliminate environmental pollution caused by the waste. A type of stratified downdraft gasifier was designed and tested, a experimental system was put into run in a pulp mill. The gas produced from the waste was sent into a boiler as fuel combined with coal. Test data and experiment indicate that the technology is a cost-effective and practical way to use industrial biomass wastes with quality as poor as the straw waste.

This paper presents the chemical and physical properties of the wheat straw waste and their influence on gasification process, the design of the stratified downdraft gasifier and the gas clean and burn system. The technology evaluation and feasibility analyses are also given here.

2 Description of the testing system and equipment

The principle testing system is shown in Fig. 1. The wheat straw waste screened from the grass cutter is fed into the gasifier by a screw feeder. In the gasifier it undergoes a succession of processes including drying, pyrolysis, oxidation and reduction, and finally convert into combustible fuel gas containing CO, H_2, and CH_4. The ash moved down through the grate and then discharged away by a ash screw. The fuel gas is cleaned by the removal of tar and dust by passing through a gas clean system consisted of a glover tower, a gas-liquid separator and a subsider while its temperature drops down to ambient from about 500℃. Leaving the gas clean system, the gas is transported to the boiler and jetted into the furnace to burn combined with coal. To run continuously, the gasifier and gas clean system was designed to operate under weak negative pressure, e.g. the fan placed at the rear of the system. A water seal trough was installed at the bottom of the gasifier and gas clean equipment to prevent air from leaking into the system.

3 Properties of wheat straw waste

Wheat straw waste is a kind of biomass material with very poor quality. Ultimate analysis data, moisture content and low heat value for typical samples taken from three pulp mill are shown in Table 1. It can be seen from it the total convertible part (C, H, O) of the waste is only about 60%, and the low heat value is only about 9 kJ / kg. But the ash and moisture content are as high as 20% individually. So it is difficult to believed that this feed stock can be suitable for gasification process. In the convertible part, however, relative content of C, H, O is closed to other biomass materials, nearly equal to $CH_{1.8}O_{0.8}$. Therefore from viewpoint of chemical reaction, it will be possible to gasify the wheat straw waste if the considered ash as inert matter and overcome the trouble caused by high moisture content.

The wheat straw waste is very loose compared with other biomass materials. According to our test data, the bulk density of the waste is only 40~50 kg / m^3. This brings a large amount of feed calculated by volume of material and a small gravity by which the fuel move down in the gasifier. And what is more important, this means a significantly reduced heat capacity of the oxidation zone. As we known, a steady operation of the gasifier rely upon the steady oxidation reaction because that is the only

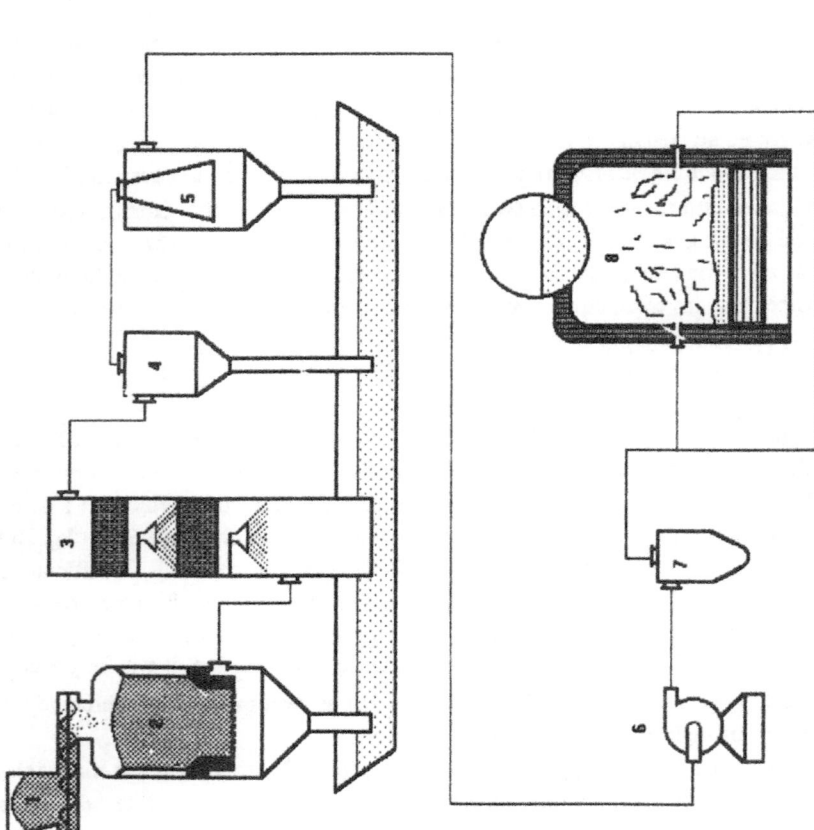

1. Feeder
2. Gasifier
3. Glover tower
4. Gas-liquid separator
5. Subsider
6. Fan
7. Water collection tank
8. Boiler

Fig 1. The principle system of wheat straw waste gasification and combustion

894

Table 1. Ultimate analysis data for three wheat straw waste samples

Sample No.	C	H	N	S	O	Ash	Moisture	Low heat value kJ / kg
1	26.63	3.43	0.4	0.11	27.26	29.2	16.7	9325.4
2	26.94	3.56	0.4	0.11	23.99	19.6	25.4	9124.7
3	25.62	3.29	0.38	0.11	22.40	16.7	31.5	8403.5

exothermic reaction and it is the heat produced by that which makes the other reactions to proceed. Therefore the little bulk density will influence the steadiness of operation.

The size and shape of the feed stock particles are important for determining the difficulty of moving and delivering the fuel, as well as the behavior of the fuel in the gasifier such as bridging and channeling. Also the size and shape determine the thickness of the oxidation zone, the pressure drop through the bed and the maximum hearth load for satisfactory operation. Unfortunately, the wheat straw waste particles possess fine and ununiform size and shape. The length distribution range is 0 ~ 6 mm and the shape likes a thin slice with width 0 ~ 3 mm. The experimental data have shown that the rate of oxidation reaction is controlled by the rate of the oxygen diffusing to the surface of solid particles. In this situation, the thickness of oxidation zone is only 3 ~ 4 times of the equivalent diameter of the particle. So the finer particles means a thinner and more unstable oxidation layer.

After pyrolysis of wood, its volume is reduced by 15 ~ 25%, forming an excellent charcoal layer with good ventilation ability, high internal porosity and high reaction activity. However, the volume of wheat straw is reduced by about 80% after pyrolyzing because the straw is soft and has high volatile content. The charcoal formed in this way has a poor ventilation ability and its pellet is very small and easy to be swept off by the gas stream.

The ash content of the waste is much high than that of pure wheat straw. That is because extra ash always added into the straw in harvesting and transporting and when the waste is screened out from the straw in pulp mill the extra ash also is screened out with the waste. High ash content will intensify the slagging behavior in the gasifier, much greater attention will be required to grate design.

Water must be spray onto the wheat straw when cutting it in pulp mill to prevent raising of dust. So the moisture content of the waste is high and variable significantly. Actually the moisture of biomass materials always reduce the heating value of the gas that will be produced in a gasifier. When the moisture is higher than a specified value, such as 20 ~ 25 %, it will affect the steadiness of gasification process because it absorb a mount of heat to vaporize.

To summarize the physical properties of the straw waste, indicate that the feedstock is difficult for the gasification process. The design of the gasifier must be made carefully to keep a steady condition of the gasification process.

4 Design of the stratified downdraft gasifier

Based on several experiments, a type of stratified downdraft gasifier has been designed and constructed, which can gasify 250 kg the straw waste per hour. About 500 m³ / h of gas with low heating value of 3800 kJ / m³ are produced.

As shown in Fig. 2, the gasifier has a simple cylindrical furnace construction without hearth constriction and air nozzles located at the size of the furnace commonly. The whole furnace was made of refractory material with a excellent heat insulation. Together with the fuel, the air enter the gasifier through the open top. During operation of the stratified downdraft gasifier, air and fuel pass uniformly downward through each reaction bed zone, and the wheat straw waste is converted into combustible gas after undergoing a series of processes consisted of reactions such as drying, pyrolysising, oxidation and reduction. A shaking grate was installed at the bottom of fuel bed. Shaking the grate periodically, the ash and slag will vibrate and fall down through the grate and the fuel bed will be loosen. A water cooling jacket located above the grate to

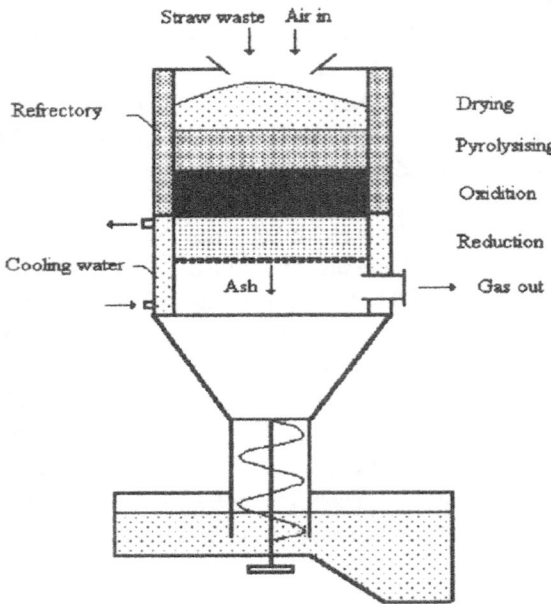

Fig 2. The stratified downdraft gasifier

prevent slagging of ash. Under the grate, there was a inverted cone ash trap with a thick tube extended into the water seal trough. Inside the tube an ash screw was installed to draw the ash down and remove it from the gasifier.

The gasifier is operated under a small negative pressure, that is a suction operation with the fan placed at the rear of the gasifier. The most significant advantage of the design is that open top permit feed stock to be fed continuously by simple feeder and operators can observe the condition in the gasifier directly and to eliminate the bridging and channeling by poking.

The stratified downdraft gasifier has some important advantages for treating the biomass materials with poor quality. The cylindrical furnace without constriction section decrease the possibility of bridging and channeling, so the limit for size and shape of feed stock particles is softened. The uniform passage of air on the whole cross section of reaction layer overcome the disadvantage of poor ventilation caused by the thinner charcoal particles. The heat insulated furnace wall maintain a quietly constant temperature field to improve the reaction conditions. And the refractory material of the furnace wall functions as heat accumulator to increase heat capacity of the reaction layer.

The gasifier has successfully gasified the straw waste with a basically steady operation condition. The composition and low heat value of the gas produced are shown in Table 2., the test number is corresponding to the fuel sample number shown in Table 1., and the basic properties of the gasifier are shown in Table 3. Although the low heat value of the gas is only 3800 kJ / m^3, the conversion efficiency is as high as 75.5%.

5 Gas combustion in boiler

A diffusion type of gas burner without any pre-mixed air was selected for the gas combustion combined with coal in a 10 t / h boiler. One advantage of this burner is its

Table 2. Main gas composition and low heat value

Test No.	Gas composition, %						Low heating value
	CO	H_2	CH_4	O_2	CO_2	N_2	kJ / m^3
1	12.7	15.3	1.6	1.8	13.4	55.2	3829
2	11.5	16.2	1.7	1.9	14.2	54.5	3810
3	12.1	14.4	1.7	2.1	14.0	55.7	3726

Table 3. Properties of the gasifier (Fuel moisture: 25.4%, ash: 19.6)

Fuel input kg / h	Gas output m^3 / h	Gas low heating value, kJ / m^3	Gasification efficiency, %	Gas-fuel ratio m^3 / kg
260	470	3810	75.5	1.8

897

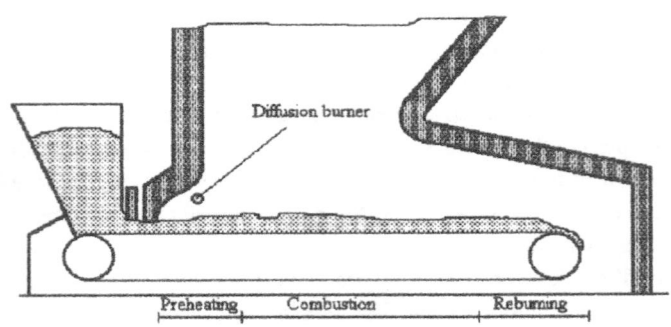

Fig 3. The placement of the burners in furnace

flame steadiness. This is most important for the system operation considering the lower gas heat value and the unsteady factors occurred in the gasification process caused by poor physical properties of the straw waste.

Along the length direction of the chain grate which is the main combustion device of the boiler, there are three sections. First one is the coal preheating section which is about 800 ~ 1200 mm of length changed with sorts of coal and load, the second is main combustion section and the third is residual carbon reburning section. Two gas burners were placed at the center of first section and oppositely on the sides of the furnace as shown in Fig. 3. Where there is always a amount of air passing upwards through the grate and coal bed to satisfy the need of gas combustion. And there is also a quite higher temperature caused by the heat reflection of front arch to ensure a good ignition condition of the gas. With the characteristics of large chamber space and the feature of the type of gas burner, a steady and complete gas combustion can be obtained.

Before the testing system was built up, the boiler consumed about 1.4 t / h of coal whose low heat value was 21.5 MJ / kg with a efficiency as low as 78 %. Total energy input was about 30,000 MJ / h. When the system was put into running, about 1800 MJ / h of energy (equal to 6 % of total energy input) was carried by the gas into the boiler. According to the statistics data of three months running, average coal consumption decreased to 1.25 t / h, e.g. a 10.7 % of energy saving was obtained and the efficiency of the boiler was increased to 81.6 %. This situation means that the addition of the gas produced from the straw waste not only supplied a equivalent energy into the boiler but also increased the efficiency of the boiler and a extra energy saving was obtained. That is because the combustion of gas exhausted the surplus air present on the coal preheating section of the chain grate, as a result of that decreased the heat lose of flue gas.

6. Conclusion

The wheat straw waste is a kind of biomass material with very poor quality such as high ash and moisture content, and especially its poor physical properties which bring many unsteady factors for gasification operation.

The data from testing and running indicated that a stratified downdraft gasifier can gasify this kind of biomass material with a efficiency as high as 75.5%. A extra energy saving was obtained by the system composed of waste gasification process and gas combustion in a boiler.

The testing system presented in this paper provide a cost-effective and practical way to treat industrial biomass waste with the quality as low as the wheat straw waste. There are many pulp mill and other factories that produce a lot of biomass waste such as sugarcane waste, wood dust and so on. So this technology has a extensive commercial potentiality.

7 References

1. Reed, T. B. (1980) A survey of biomass gasification, SERI/TR 33-239.
2. Reed, T. B. (1988) Handbook of biomass downdraft gasifier engine systems , SERI/SP-271-3022.
3. Sun, L., Xu, M., Gu, Z.Z. and Guo, D.Y. (1995) Study on cropstraw gasification for village cooking gas. *Preceedings of Asia-Pacific regional seminar on technology for utilization of rice husks and other agricultural waste.* pp. 30-41.

CO-GASIFICATION OF BIOMASS AND COAL IN A PRESSURISED FLUIDISED BED GASIFIER

Co-gasification of biomass and coal

J. ANDRIES and K.R.G. HEIN
Laboratory for Thermal Power Engineering
Department of Mechanical Engineering and Marine Technology
Delft University of Technology
Delft, The Netherlands

Abstract

During a 3 year (1996 - 1998) project, partly funded by the EU as part of their JOULE 3 programme, experimental and theoretical research will be done on co-gasification of biomass and coal in a pressurised fluidised bed gasifier.

Delft University of Technology will determine experimentally the influence of feedstock and operating conditions on the characteristics of the gasifier. Pelletized straw and Miscanthus will be used as biomass feedstock.

Experimental studies, using state of the art, laboratory scale, methods will be executed by other partners in the project (KTH, TPS and IC) to determine the extent and the origins of synergistic effects and to provide background data for the assessment of the experimental results obtained from the Delft test rig.

Besides the assessment of the time-averaged properties of the fuel gas, the time-dependent characteristics will be determined experimentally. The results will be compared with the acceptability range which will be provided by Nuovo Pignone, one of the other partners in the project. The results of this evaluation will be used to implement and test an optimised control strategy for operating the pressurised fluidised bed gasifier.

A detailed description is given of the objectives of the project, the test rig to be used and the time schedule of the project.

Keywords: biomass, coal, co-gasification, pressurised, fluidised bed, straw, miscanthus, gas turbine

1 Introduction

The use of coal-biomass mixtures to produce heat and electricity has the following advantages when compared with the use of pure biomass:

- seasonal and annual variations in the availability of biomass can be met by varying coal-biomass ratio's.
- the combined use of coal and biomass enables a wider range of system sizes compared with pure biomass-based power production systems, thus increasing the optimisation possibilities.
- coal and biomass can display synergistic effects with regard to char reactivity, tar formation and emission of harmful components.

The Laboratory for Thermal Power Engineering of the Delft University of Technology is participating in an EU funded, international, R&D project which is designed to aid European industry in addressing issues regarding co-utilisation of biomass and/or waste in advanced coal conversion processes. The project comprises three main programmes, each of which includes a number of smaller subprograms. The three main programmes are:

- coal-biomass systems component development and design.
- coal-biomass environmental studies.
- techno-economic assessment studies

The first of these programmes will focus on the most critical issues of the coal-biomass fuelled gasification combined cycle process: feeding of biomass, carbon conversion, gas quality, control of harmful emissions, reliability of gas treatment, impacts of co-gasification on the requirement for the gas turbine and utilisation or disposal of solid residues of the co-gasication process (partners VTT /FI, Enviropower /FI, Schumacher /DE, CTDD /GB, TUD /NL and Nuovo Pignone /IT).

The second programme, coal-biomass environmental studies, will concentrate on the use of laboratory scale experimental techniques. The overall intention is to provide a high level of fundamental understanding of those factors which influence the gasification behaviour of a range of fuels in direct support of the industrial scale studies (partners CRE /GB, TPS /SE, KTH /SE, IC /GB).

The third programme, techno-economic assessment studies, done by the University of Ulster /GB, will provide the means to assess the potential benefits and the technical issues that need to be addressed.

2 Objectives

The aims of the first programme are to establish coal-biomass systems components development criteria and the design of such components through R&D including:

- coal-biomass preparation and feeding systems.
- the evaluation of the impact of co-utilisation on fuel gas quality in fluidised bed gasification systems.
- evaluation of the impact on component performance through such co-utilisation.
- evaluation of the impact of changes in the derived fuel gas quality and composition on gas turbine performance.

The contribution of the Laboratory for Thermal Power Engineering to this project will mainly consist of the experimental evaluation of the impact of co-utilisation on fuel gas quality in fluidised bed gasification systems.

3 Installation

Since a number of years the Laboratory for Thermal Power Engineering has been doing research on pressurised fluidised bed combustion using the test rig shown in figure 1.
The existing test rig is being modified in phases to enable experiments with pressurised fluidised bed gasification of coal using flue gas recirculation and oxygen injection. The modifications consist of changes to the reactor vessel and the addition of an oxygen-

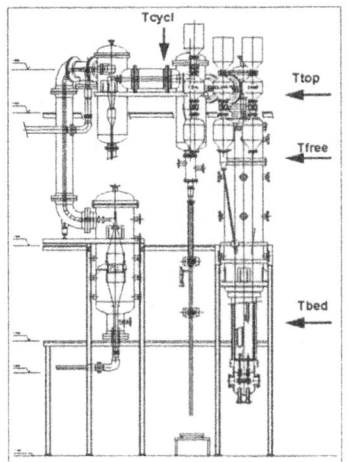

Fig. 1. The Delft PFBC/G reactor.

and steam supply system, a high temperature ceramic filter, a high temperature ammonia removal system, a pressurised topping combustor, a scrubber, a booster compressor and an atmospheric combustor. The main design data of the modified fluidised bed reactor are given in table 1.

Table 1. Main design data of the Delft PFBC/G test rig.

diameter bed	0.4 m
max pressure	10 bar
max bed height	2 m
freeboard height	4 m
fluidization velocity	0.8 m/s
max thermal capacity	1.6 MW

The former PFBC test rig contained a cylindrical fluidised bed reactor with an internal diameter of 0.49 m enabling a maximum coal feed rate of 150 kg/h. The design of the present PFBC/G reactor is based on the same maximum coal feed rate to enable the use of the existing coal feed- and heat removal infrastructure. The internal diameter of the lower part of the reactor vessel which contains the bed has been decreased to 0.38 m to maintain the same fluidisation velocity as in the former PFBC reactor and the internal diameter of the freeboard region has been kept at 0.49 m.

The parameter, which is used to control the temperature of the fluidised bed reactor, depends on the operating conditions: during combustion the amount of heat to be extracted from the bed is strongly dependent on the amount of coal supplied, while during gasification the amount of heat to be extracted is mainly determined by the amount of oxygen in the fluidization gas. When the process conditions change from combustion to gasification (for instance by increasing the coal feed rate while keeping the fluidization velocity constant) the amount of heat to be extracted from the reactor increases to a maximum at stoichiometric conditions and decreases with a decrease of the stoichiometric ratio. The maximum amount of heat, which has to be removed from the reactor at the design coal feed rate of 150 kg/h, is 800 kW.

Oxygen is supplied to the laboratory in liquid form by road tankers and kept in a cryogenic storage tank equipped with an evaporation system. When the test rig is operating at full load, 320 kg/h of oxygen is needed. During continuous operation at these condition the storage tank has to be filled every two days. The use of pure oxygen demands special care for the system lay-out and the choice of material. The detailed design and construction have been done in close cooperation with the oxygen supplier (Air Products /GB). The modified fluidised bed reactor has a gas distribution plate which is equipped with a central nozzle with small, radially outward directed, orifices. The oxygen, which is supplied to the bed through the central nozzle, is mixed with either air (when enriched air is used as fluidization gas) or with recirculated flue gas (when a CO_2/O_2 mixture is used as fluidization gas). The recirculated flue gas must be free of combustible gases and oil to ensure a safe mixing process. This implies the use of a booster compressor which delivers an oil free gas. The maximum allowable local oxygen concentration in the central nozzle is 40 %.

Recirculation of the flue gas implies recirculation of water vapour formed during the combustion of the coal and/or the fuel gas. The water concentration can become as high as 30 vol % at 8 bar. Because of the high dew point of the resulting flue gas this could result in severe corrosion in the system. Furthermore the water concentration in the flue gas strongly influences the composition of the fuel gas produced by the gasifier. To control the water vapour concentration a scrubber has been designed and installed. By injecting water into the scrubber the gas is cooled to about 50 °C and the saturated gas leaving the scrubber contains only a few percent water vapour. The scrubber will also remove SO_2 and remaining dust particles from the gas stream thus protecting the booster compressor from erosion, corrosion and deposition.

The booster compressor, which recirculates the flue gas to the fluidised bed gasifier, must supply a sufficient pressure increase to compensate for the pressure losses in the system. The most important pressure drops are those across the bed, the gas distribution plate and the valve controlling the gas flow to the bed. The compressor must be able to deliver a 2 bar pressure increase at system pressures ranging from 3 to 8 bar. The amount of recirculated flue gas, which is used for fluidising the gasifier and cooling the pressurised combustor depends on the operating conditions of the test rig. To simplify the control of the operation of the compressor, the excess capacity of the compressor is recirculated via the scrubber. The compressor chosen, is a centrifugal one with a vaneless diffusor which can be used to compress CO_2 - rich flue gas as well as air. At a system pressure of 8 bar it can supply 275 Nm^3/h of recirculated flue gas with a pressure of 12 bar. A schematic of the final configuration of the modified test facility is shown in figure 2.

The components to be used during phase 2 (pressurised topping combustor and heat

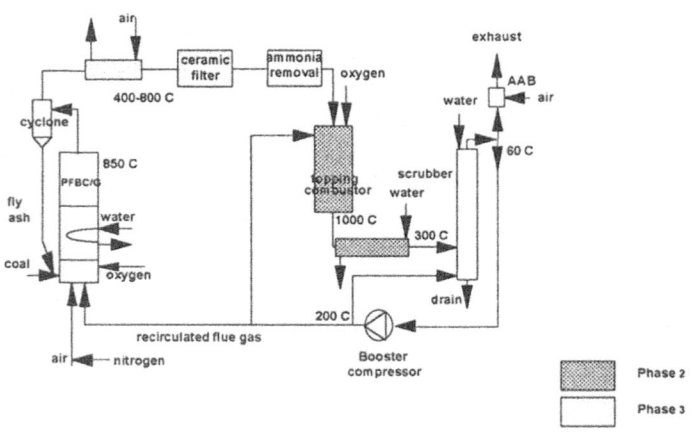

Fig. 2. The Delft PFBC/G test facility.

exchanger) will be operational at the end of March 1996, the components to be used during phase 3 (steam supply, heat exchanger, ceramic filter and ammonia removal system) will be operational in September 1996.

These modifications will enable gasification as well as combustion experiments using coal, biomass and coal-biomass mixtures as fuels. The gasification can be achieved using air, air-steam mixtures, oxygen enriched air and oxygen enriched recirculated flue gas as the fluidisation/gasification medium. The low calorific value fuel gas can be combusted in a pressurised, high temperature combustor using air or pure oxygen and recycled flue gas as oxidant and cooling gas. It is also possible to use steam to control the flame temperature and hence the NO emissions.

A more detailed description of a related EU-funded project on pressurised, high temperature combustion of biomass-derived, low calorific fuel gas is given in [1].

4   Work programme

The Delft contribution to the **first** programme of the JOULE3 project consists of 3 test series of 50 effective running hours each. During a test series, 4 sets of steady state operating conditions will be maintained for at least 10 hours each. The 3 test series will consist of experiments aimed at:

- Co-gasification of coal-straw mixtures. The coal type, mixing ratio and operating conditions will be determined after consultation with the partners in the project. The results of the experiments will be used to assess the influence of site-specific parameters on the fuel gas quality by comparing them with results obtained in a previous EU-funded project . The experiments will be done during the first half of 1997.

- Co-gasification of coal-miscanthus mixtures. The gasification temperature will be varied between 700 °C and 900 °C and the amount of biomass (based on thermal input) will be varied between 25 and 75 %. The detailed test programme will be based on available information from other experimental programmes and earlier experiments in this project. The results of the experiments will be used to assess the feasibility of using this energy-crop in a pressurised fluidised gasifier. Besides the assessment of the time-averaged properties of the fuel gas, dynamic characteristics will be determined experimentally. The experiments will be done during the second half of 1997.

- Gasifier control improvements. The dynamic characteristics of the fuel gas quality will be compared with the acceptability range with regard to time-dependent behaviour provided by Nuovo Pignone, one of the partners in the project. The result of this evaluation will be used to design and implement an optimised control strategy for operating the pressurised fluidised bed gasifier. The control strategy will be based on the possibilities which are available in the Delft facility, such as fuel feed, steam/fuel ratio, pressure and fluidisation velocity. The choice of biomass type and mixing ration will be made on the basis of the preceding experiments. The influence of the resulting gas quality on gas turbine performance and the consequences of the chosen control strategy on the system design will be assessed in close cooperation with Nuovo Pignone. The experiments will be done during the first half of 1998.

Work done in the **second** programme by KTH, TPS and IC will provide valuable background data for the larger scale complementary studies being carried out by TU Delft. Studies will be done by KTH to evaluate the factors which may influence the synergistic effects. Experiments using lab-scale facilities will be done to assess the influence of the fuel composition (coal and biomass), the effect of coal-biomass ratio and the effect of process pressure. TPS will evaluate individual steps in the overall gasification reaction like char gasification, devolatilisation rates and pyrolysis product distribution. Imperial College will assess the pyrolitic behaviour of coal-biomass mixtures under a wide range of process conditions in order to identify the origins of any synergistic behaviour.

Information, obtained from work done in the **third** programme by the University of Ulster, will be used during the course of the project to focus both the larger scale and laboratory scale programmes carried out as part of other two main activity areas.

5   Acknowledgement

This research is funded in part by the European Commission in the framework of the JOULE3 R&D programme (contract JOF3-CT95-0018).

6   References

1.   Andries,J., Hoppesteyn,P.D.J., and Hein, K.R.G., 'Pressurised combustion of biomass-derived, low calorific value, fuel gas', paper presented at *Developments in Thermochemical biomass conversion*, Banff, Canada, May 20 - 24, 1996

CLEANING OF HOT PRODUCER GAS IN A CATALYTIC ADIABATIC PACKED BED REACTOR WITH PERIODIC FLOW REVERSAL

L. VAN DE BELD, B.M. WAGENAAR and W.PRINS
BTG Biomass Technology Group BV
Enschede, the Netherlands.

Abstract
To use the product gas of a biomass gasification unit in, for instance, gas engines, its tar content must be reduced to less then 100 mg/Nm^3 [1]. A novel technology has been tested which is characterised by periodic reversal of the gas flow through a packed-bed tar removal reactor. Because the entrance and outlet sections of the reactor act as efficient heat exchangers, the producer gas leaves the reactor at relatively low temperatures. A small fraction of the producer gas is burned in situ to supply the energy required for the tar conversion reactions. This way of operation provides a high degree of heat integration.
The main parts of the pilot plant installation are a 100 kW$_{th}$ wood gasifier and a packed bed tar converter, with a diameter of 0.22 m and is filled with approximately 50 kg of dolomite catalyst. Depending on the operating conditions the direction of the flow is reversed every 10 to 60 min. Tar conversions above 98% were achieved, without any notable on the caloric value of the treated gas. Optimization of the system should focus on improvement of the switching valves and application of a more active catalyst.
Keywords: tar removal, gasification, reverse flow operation, catalytic gas cleaning.

1 Introduction

Biomass materials are attractive as feedstocks for thermal conversion processes as e.g. gasification. The gasification process is a two stage process: in the first stage the biomass is burnt partially to produce gases and charcoal; during the second stage the product gases (mainly CO_2 and H_2O) are reduced by the charcoal to form carbon monoxide and hydrogen. The process also yields some methane and other higher hydrocarbons depending on the design and operating conditions of the gasifier.

Unfortunately, a small fraction of the biomass is converted into what is called "tar".

Tar is a substance without a generally accepted exact definition: its is a complex mixture of more or less easily condensable substances. This tar will impose serious limitations in the use of gasifier product gases due to the fouling of downstream process equipment. Corrosion and build-up of tar deposition obstructing the flow in gas ducts, are typical examples of these fouling problems. When the fouling problem is used as a basis for defining the complex tar mixture, tar can be identified as a mixture of components, which condense on surfaces at 20 °C (room temperature).

For the use of gasifier product gas in e.g. gas engines the tar concentration must be reduced to an acceptable level; according to Stassen *et al.* [1] the maximum allowable tar content is 100 mg/Nm^3. However, an exact figure is difficult to give, because it is not only the absolute tar content, but also the *nature* of the remaining tars causing the fouling and corrosion problems.

This paper describes a novel tar conversion technology which is based on a catalytic fixed bed reactor with periodic flow reversal. The thermal behaviour of the reactor and results of the first tar removal experiments will be discussed.

2. Reverse flow operation of a fixed bed reactor

In Novosibirsk, Russia, some decades ago Boreskov and coworkers [2] developed a new technology which is characterised by the periodic flow reversal of the feed flow. The basic idea was that operating a packed bed reactor under transient conditions can improve the performance of the reactor: in this particular case unsteady-state conditions are created by periodically reversing the direction of the feed flow. The benefits of operating a reactor under unsteady-state conditions are summarised by Matros [3]. For the tar removal system the high degree of heat integration is the major advantage.

2.1 Basic characteristics
A schematic drawing of a reverse flow reactor is shown in Fig. 1. During start-up the reactor bed is preheated to the desired reaction temperature of, for instance, 800 - 1000 °C. Subsequently the contaminated gas is fed to the reactor at a much lower temperature. The gas flow is then heated by the hot solid phase, and, consequently the reactor bed will cool down. As a result, a heat front travels through the reactor towards the outlet. (see Fig. 2) The reaction starts in the region of the bed where the temperature is high enough; at this position the reaction heat is released. Because the ratio of the heat capacities of the solid phase and the gas phase is large, it will take a rather

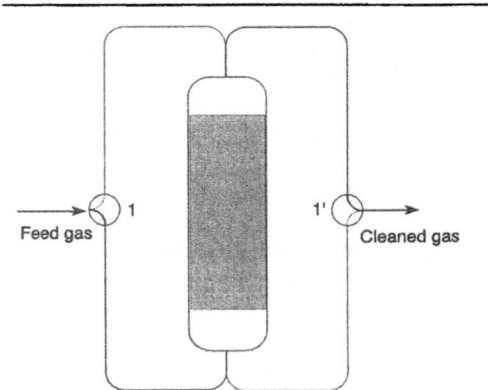

Figure 1: Schematic representation of a reverse flow reactor

908

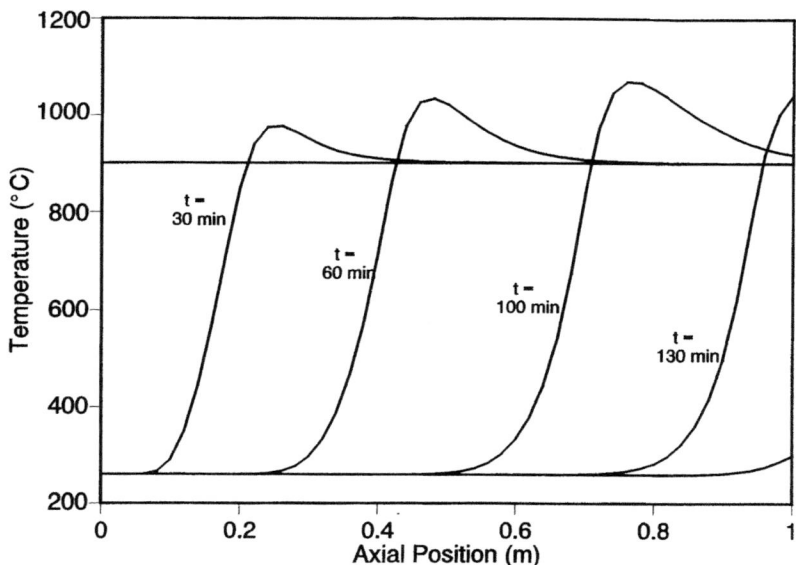

Figure 2: Typical axial temperature profiles for packed bed reactor without flow reversal

long time before the heat front reaches the reactor outlet. However, if no action is taken the final situation will be a completely cold catalyst bed in which reaction is impossible (see Fig. 2). To prevent the occurrence of this undesirable situation the direction of the feed flow is reversed, and as a result the heat front will travel in the opposite direction. This process of flow reversal is repeated continuously. In this way it is possible to keep the heat inside the reactor, provided that the heat generation by the reaction is high enough to compensate for the convective heat removal by the gas phase. After a large number of flow reversals a so-called Pseudo-Steady-State (PSS) is achieved. The maximum temperature in the PSS remains constant and only the heat of reaction is removed from the reactor by an increase in the temperature of the outlet stream. The moving temperature profiles are kept in between two extremes. Because the in- and outlet sections of the bed act as regenerative heat exchangers only, they can be filled by inert material.

In literature, many papers have been published concerning the behaviour of reverse flow reactors, and several processes were studied. Known examples are:
- SO_2 - oxidation, see Boreskov et al. [2].
- Selective Catalytic Reduction (SCR) of NO_x, see Bobrova et al. [4].
- Partial oxidation of methane for production of syn-gas, see Blanks et al. [5].
- Methanol synthesis, see Neophytides and Froment [6].
- Purification of air polluted with traces of organic compounds, see e.g. Eigenberger and Nieken [7], Van de Beld [8 - 9].

All these processes have in common that the reactions considered are exothermic. To guarantee stable operation in a reverse flow reactor, a net positive heat effect is required. The tar conversion process considered in this work is endothermic and thus heat consuming. To achieve an overall positive heat effect a second exothermic reaction is introduced to counterbalance the endothermic tar conversion reaction. This is realised by injecting small amounts of air into the converter, where the temperature is very high. The oxygen is then used for the combustion of a small part of the produced H_2 and CO. By adjusting the air flow into the tar converter, the temperature level in the reactor can be controlled. In literature, *a reverse flow process in which the combination of two independent reactions, one being exothermic, the other being endothermic is carried out, has not been demonstrated before [10].*

3. Experimental set-up

To enable a proper design of the reverse flow tar converter, several catalyst screening experiments in a small laboratory scale set-up were performed. These experiments have also provided the information required to determine the operating conditions.

A flow-sheet of the experimental pilot-plant installation is given in Fig. 3. Five major sections can be distinguished: the gas supply system, the biomass feeder, a co-current downdraft gasifier, the tar converter and the analysis section.

The air flows to the gasifier and the tar conversion reactor are passed through a control valve and a flow indicator. In case of emergency, and for start-up/shut-down

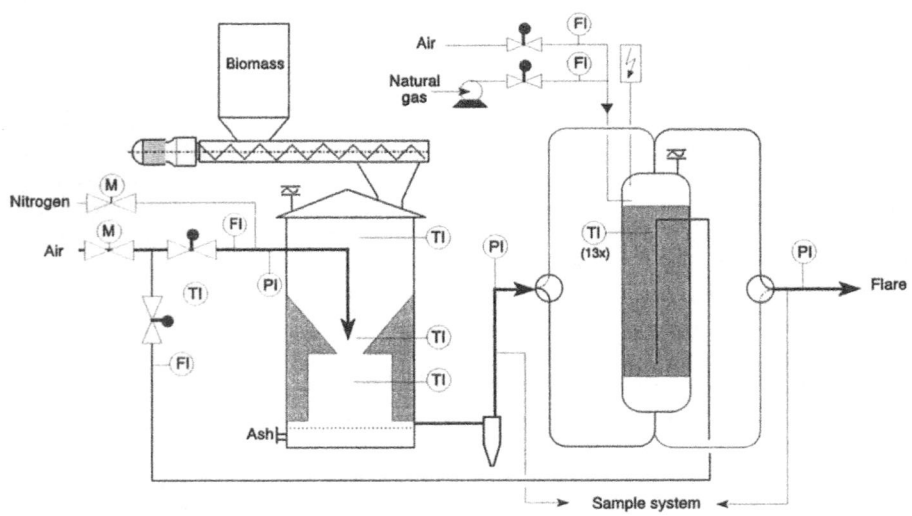

Figure 3: Flowsheet of the experimental installation

purposes the air flow is replaced by nitrogen. The system will switch automatically to nitrogen if i) the nitrogen pressure becomes too low (P < 0.2 bar) or ii) if the pressure in the gasifier becomes too high (P > 0.3 bar).

A screw feeder with a maximum capacity of 60 kg/hr has been installed to supply beech-wood to the gasifier.

Heat losses through the wall of the co-current downdraft gasifier are limited by efficient insulation. The gasifier throat has a diameter of 0.10 m, the diameter of the reduction zone is about 0.2 m. Thermocouples are inserted at three different axial positions in the gasifier. Ash is collected in the bottom section of the gasifier and needs to be removed after a few runs. For safety reasons a rupture disc, withstanding a pressure of at most 0.5 bar, has been placed on top of the gasifier In the downstream piping of the gasifier a cyclone is included to remove dust from the producer gas. An electrical heat tracing element has been wound around the connection line between the gasifier and tar conversion system to prevent condensation of tar in the piping.

Before start-up the space below the throat (i.e. the reduction zone) is filled with charcoal. The gasifier is then ignited by bringing a small amount of glowing charcoal into the gasifier just above the throat (i.e. the oxidation zone). Then a layer of beechwood is dumped on top of it, after which the air flow to the gasifier is switched on.

The tar converter is an adiabatic packed bed reactor, with a length of 1.2 m, and a diameter of 0.22 m; the total diameter including the insulation layer is 0.5 m. Temperatures inside the reactor are measured at 13 different axial positions. Pressure indicators have been placed in the inlet and outlet of the reactor. This reactor is also equipped with a rupture disc, since the producer gas is flammable and explosive. The bottom part (0.2 m) of the reactor is filled with bauxite with a negligible tar cracking activity. On top of this inert layer 0.8 m of dolomite is poured. The top section (0.2 m) is filled again with bauxite. The composition and some relevant properties of the dolomite and bauxite are listed in Table 1.

Initially, before the tar conversion process can be started the reactor must be heated to the required reaction temperature (900 - 1000 °C). This is realized by burning a mixture of natural gas and air in top of the reactor, where it is ignited by a spark. The hot flue gases are then passing through the bed while heating it. The heating is

Table I: Properties of dolomite and bauxite

Dolomite (supplier Duwa Kalk en Dolomiet BV, the Netherlands)

			Chemical composition	
particle diameter:	2 - 7	mm	$CaCO_3$	54 %
density (calcined)	1200	kg/m^3	$MgCO_3$	44 %
			SiO_2	< 0.5 %
			Fe_2O_3	< 0.4 %
			Al_2O_3	< 0.2 %

Bauxite

particle diameter	3 - 6	mm
density	2500	kg/m^3

controlled by adjusting the mixing ratio of air and natural gas. Disadvantageously, the combustion reaction produces significant amounts of CO_2, which reacts with the calcined dolomite. The dolomite must be recalcined, before it can be used as a cracking catalyst; this causes some complications which will be discussed later.

A special sampling device is developed to measure the tar content of the gas before and after the tar converter. This device consists of a cellulose filter for dust removal and a condenser to collect tar and water.

The composition of the collected gas samples is measured with a Shimadzu gas chromatograph (GC 14B). A mol-sieve column is applied to separate hydrogen, oxygen, nitrogen, methane and carbon monoxide, while carbon dioxide, water and C_3 are separated on a Porapak Q column. A Thermal Conductivity Detector (TCD) is used for detection of the components, and its signal is integrated by a CR-5A integrator.

The experimental work carried out with the system is divided in three parts: i) characterisation of the thermal wave, ii) preheating of the tar converter and iii) the tar removal experiments.

4. Results and Discussion

4.1 Characterisation of the thermal wave.

An important phenomenon occurring inside the tar converter is the movement of the heat front; it is therefore important to first study the heat characteristics of the reactor by a cooling experiment. Most likely, the dynamic behaviour of the reactor is fully

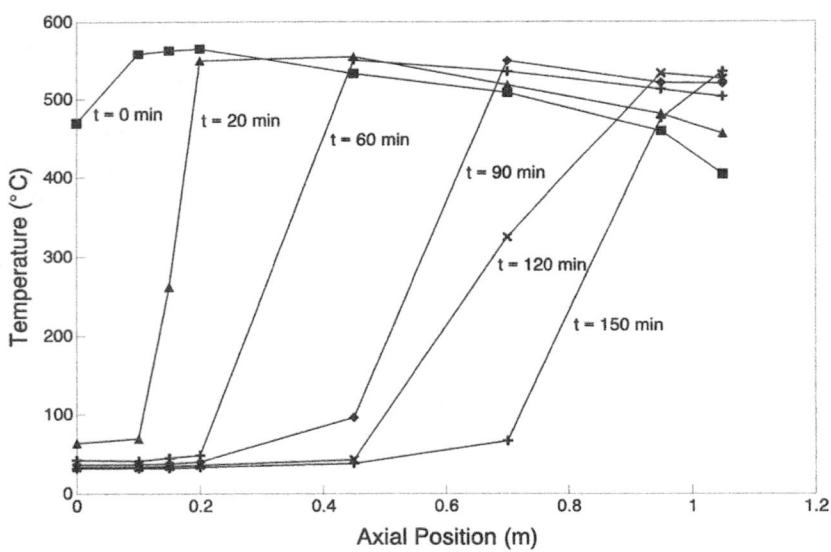

Figure 4: Development of the axial temperature profiles during a cooling experiment. Air flow = 25 Nm³/hr; T_{in} = 25 °C.

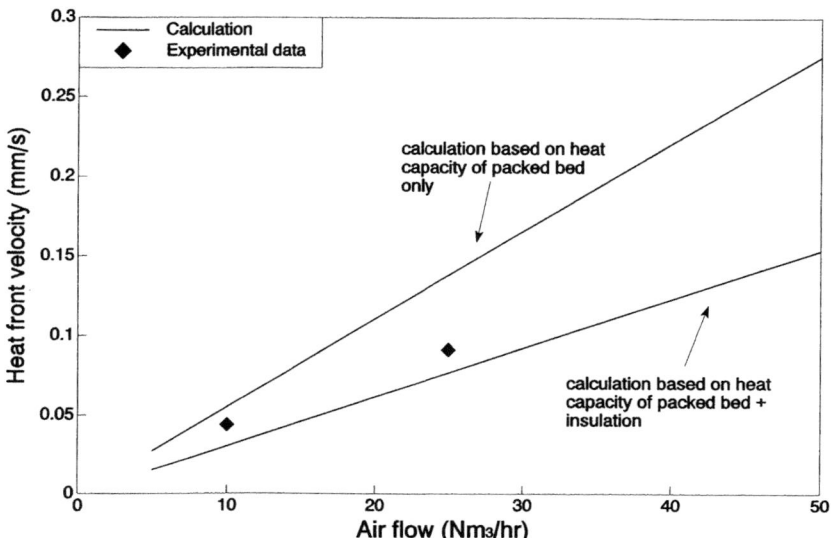

Figure 5: Comparison of calculated and measured heat front velocities as a function of the air flow. Calculations according to Eqn. 2 and 3.

determined by the heat capacity of the system (including the insulation layer).

A cooling experiment is carried out as follows: the reactor bed is first heated with hot flue gas until the steady state is obtained. In this steady state the whole reactor content has almost an uniform high temperature. Hereafter, the actual experiment is started by switching from hot flue gas to cold air, which almost results in a step change in the inlet temperature. The air will cool the hot solid phase, and a "cold" front" will pass through the reactor. As an example the result of such an experiment is shown in Fig. 4, where the temperature is plotted as a function of the axial position in the reactor. The residence time of the gas in the reactor bed (based on normal conditions = STP) is in the order of a few seconds, whereas "the residence time" of the heat front exceeds 3 hours.

From the cooling experiments the heat front velocity can be easily calculated:

$$v_{heat} = \frac{\Delta z}{\Delta t} \tag{1}$$

In this equation, Δz represents the displacement of the heat front during a time period Δt. From theoretical considerations the heat front velocity can be predicted. The simplest expression is given by the following relation:

$$v_{heat} = \frac{\varepsilon(\rho C_p)_g}{(1 - \varepsilon)(\rho C_p)_s} \cdot u_{g,i} \tag{2}$$

Herewith, it has been assumed that the only heat sink in the system is the packed bed itself; no additional heat buffer is taken into consideration. However, in reality the tar converter is insulated extensively to achieve adiabatic operation. Significant heat fluxes to and through the insulation will affect the dynamic behaviour of the reactor seriously [8]. A modified expression for the calculation of the heat front velocity, accounting for the heat capacity of the insulation, can be deduced:

$$v_{heat} = \frac{\varepsilon(\rho C_p)_g}{(1 - \varepsilon)(\rho C_p)_s + A_{ins}/A_R \cdot (\rho C_p)_{ins}} \cdot u_{g,i} \tag{3}$$

A comparison of the calculated and measured heat front velocities as a function of the air flow is given in Fig. 5. On basis of a comparison with two experimental data points, it seems that for the system studied the heat sink of the insulation will cause a reduction in heat front velocity of about 30%.

4.2 Preheating the tar converter

Before a tar conversion experiment is started, the catalyst bed must be at a sufficiently high temperature. For the dolomite catalyst used in this work, the minimum temperature required is approximately 900 °C. The preheating process is carried out as follows: air and natural gas are premixed and fed to the top section of the reactor, where it is ignited by spark ignition. Hot flue gas is then passing through the packed bed. Because the heat capacity of the system is rather large, it will take quite some time to heat the whole reactor bed. In the low temperature regime calcined dolomite reacts with CO_2 (which is

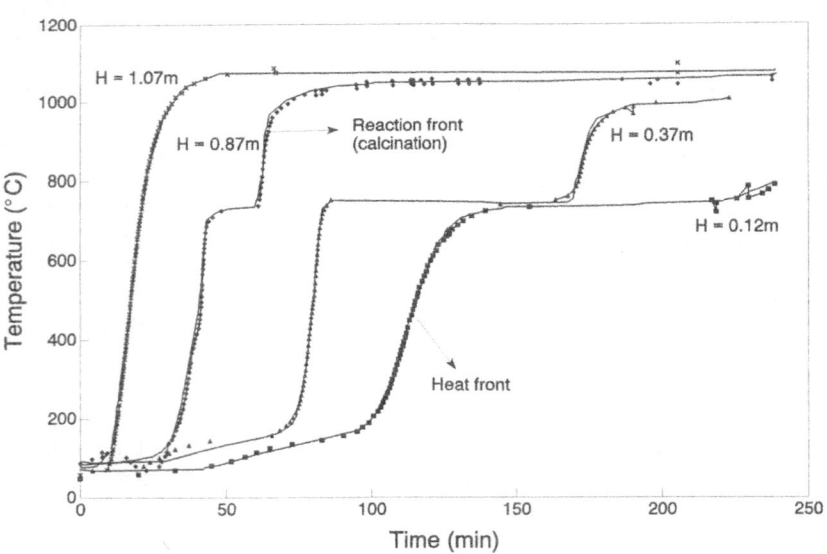

Figure 6: Temperature - time profiles measured in the reactor during preheating at different axial positions.

a substantial part of the hot flue gas) to form $CaCO_3$ and $MgCO_3$. At high temperatures (> 700 °C) recalcination of the dolomite occurs. As a consequence energy is not only required for heating the solids, but also for the (reversible) calcination reaction. This phenomenon is illustrated by Fig. 6. It can be observed that:

- the thermocouple at H = 1.07 m is positioned in the bauxite layer. Because recalcination will not occur in this part of the packed bed, a normal break-through curve is observed.
- the thermocouples at H = 0.87 m and H = 0.37 m are positioned in the dolomite bed. For these temperature-time lines two break-through curves are observed: the first one forms the real thermal front (probably enhanced by the exothermic adsorption of CO_2), the second is caused by recalcination (at 750 °C) of the dolomite, and is called the reaction front. Due to the difference in front velocities, the time lag between both break-through curves will increase further downstream the reactor.
- The thermocouple at H = 0.12 m is again located in the bauxite bed. Although recalcination does not take place here, the shape of the curve is still the result of both phenomena occurring in the dolomite bed.

Obviously, the recalcination process causes a considerable delay in the preheating of the packed bed. However, this problem is only related to start-up and not to continuous operation.

4.3 Tar removal experiments

Several experiments have been carried out with the pilot-plant installation, mainly to

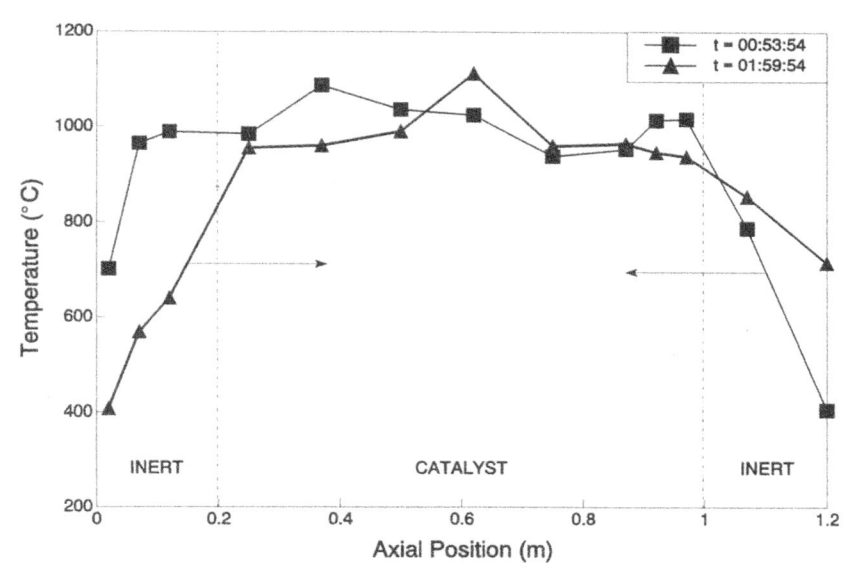

Figure 7: Axial temperature profiles in the tar conversion reactor. Conditions and results see Table 2, experiment no. 4.

Table II: Results of the experiments with the tar conversion pilot-plant.

Nr	1		2		3		4		5	
Temperature (°C)	1050		1050		1050		1050		1050	
$\phi_{gasifier}$ (Nm^3/hr)	8		8		8		8		8	
ϕ_{sec} (Nm^3/hr)	2		1.5		1.9		1.5		1.9	
ϕ_{out} (Nm^3/hr)	18.5		12.3		15.6		19.8		18.1	
Biomass cons. (kg/hr)	8		8		8		8		8	
Duration of run (hr)	3:00		3:44		3:36		3:00		3:00	
Cycle period (min)	40		40		40		40		40	
u_{mean} (m/s)	0.5		0.4		0.4		0.5		0.5	
τ_{mean} (s)[a]	1.6		2.1		1.8		1.5		1.7	
$C_{tar, in}$ (g/Nm^3)	23.5		21.2		36.2		20.9		16.7	
$C_{tar, out}$ (g/Nm^3)	0.23		0.50		0.18		0.29		0.20	
Conversion (-)	99.0		97.6		99.5		98.6		98.8	
	Gas Composition[b]									
	in	out	in	out	in	out	in	out	in	out
CH$_4$ (vol%)	2	1	1	2	2	1	3	2	2	3
CO (vol%)	17	15	17	13	18	15	7	8	22	24
CO$_2$ (vol%)	7	8	7	9	7	9	7	7	9	11
H$_2$ (vol%)	10	11	10	11	11	13	6	7	1	11
H$_2$O (vol%)	12	12	12	12	15	15	18	18	19	19
O$_2$ (vol%)	1	1	1	1	2	1	2	1	3	1
LHV (MJ/Nm^3) [c]	3.9	3.6	3.8	3.9	3.2	2.8	3	2.8	4.0	4.6

[a]   based on the total length of the reactor (including inert parts)
[b]   nitrogen balance
[c]   based on dry gas

study the behaviour of the tar conversion system. The gasifier is only used as a tar generator and, deliberately, operated at non-optimized conditions. This resulted in large amounts of tars in the producer gas, providing the opportunity to test the tar converter at very severe conditions.

Results of the experiments are listed in Table II. The cycle period (see Table II) is defined as half a cycle upflow and half a cycle downflow. By this definition the flow is reversed twice in a single cycle period.

In all experiments a high conversion of the tar components was found; the outlet concentration of tars could be reduced to about 200 - 300 mg/Nm^3. The addition of only small amounts of air (i.e. secondary air) were sufficient to maintain the temperature in the catalytic part of the reactor at 1000 - 1100 °C. Table II shows that a high degree of tar conversion has been reached for all five experiments. However the outlet concentration was never below the critical value of 100 mg/Nm^3 for engine application due to the high inlet values of more than 20000 mg/Nm^3. The heating value of the tar-free producer gas is not seriously affected.

In Fig. 7 two typical examples are given of the measured axial temperature profile. It can be observed that the in- and outlet temperature are relatively low compared to the temperature in the catalytic part of the reactor, where a high temperature zone exists due to secondary air injection at some discrete points. The temperature level in the reactor is so high that the oxygen in the air is consumed immediately by combustion reactions. Consequently, the reaction heat is released quite locally and as a result some scatter in the measured temperature profiles is found.

The temperature-time profiles at both ends of the reactor are plotted in Fig. 8. Due to the switching of the flow, the in- and outlet sections are periodically heated and cooled. When one end of the reactor is heated, the other end is cooled at the same time. The period of this heating and cooling cycle is set by the cycle period. If a cyclic steady state is obtained, the minimum and maximum temperatures at the end of the cycle will be the same for successive cycles.

In this section the testing of novel tar cracking technology has been discussed. The objective was to achieve a very high tar conversion at a minimum energy requirement. However, since the tar inlet concentration was quite high, the tar concentration in the outlet gas of the cracker was still a little too high with respect to the engine application. Two possible explanations can be put forward. The first one is related to the design of the tar converter, which is based on the results of small scale experiments providing the

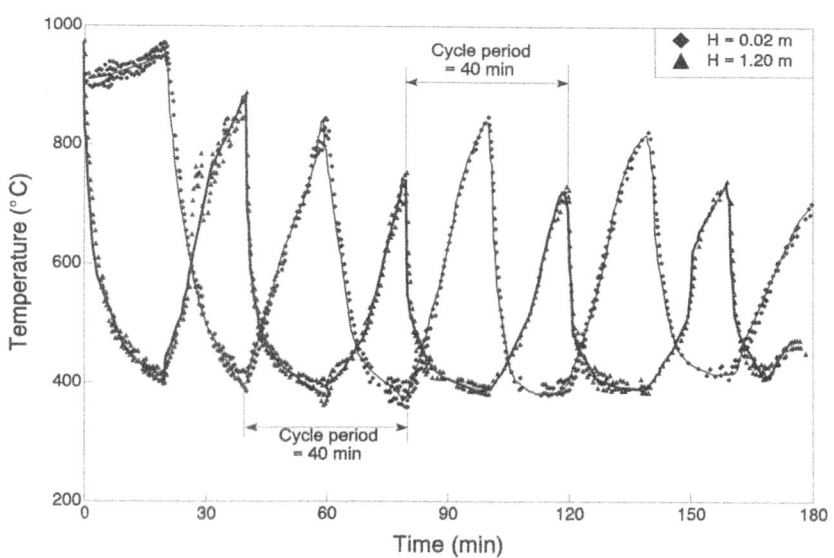

Figure 8: The temperature in both ends of the reactor as a function of time. Conditions and results, see Table II, experiment no. 5.

reaction kinetics. In these small-scale experiments, the dolomite was heated by an electric oven. In the pilot-plant experiments however, the dolomite was exposed to a CO_2/H_2O containing atmosphere during preheating. This may have changed the dolomite activity, by which the design specification can not be reached any more.

A second explanation is the possible by-passing of producer gas through the switching valves; even a small leakage has already a dramatic effect. Let us, for instance, consider a producer gas containing 30 g_{tar}/Nm^3 is fed to the tar converter. Assuming that in the converter a tar conversion of 100% is realised and only 1% of the total flow bypasses the reactor due to leakage of the valves, the overall result will be 99% conversion. This gives an outlet concentration of 300 mg/Nm^3.

The problem can be solved by application of dedicated valves. BTG has currently designed such valves; the construction is not extremely complicated or more expensive than commercially available valves.

5. Conclusions

A novel tar conversion reactor has been tested, which is characterised by periodic reversal of the gas flow direction. The reactor consists of a single packed bed built up of a heat exchanging section (inert bauxite), a catalytic section (active dolomite) and again a heat exchanging section (inert bauxite). In such system a heat exchanger and catalytic reactor are integrated in one apparatus. While the in- and outlet sections of the reactor are relatively cool, the temperature in the active central part is high enough for sufficient tar conversion.

Reverse flow operation enables cleaning of the producer gas at a low energy demand. Since the tar conversion reactions are slightly endothermic, a small part of the producer gas must be combusted in situ to achieve an overall positive heat effect. The tar converter was operated successfully, although the outlet concentration of tar was still somewhat too high. Probably this was caused by leakage through the closed valves by which a small part of the flow by-passes the cracking reactor. The heating value of the treated gas hardly changed. The efficiency of the system will be further increased by using other, more active catalysts, which may enable lower operating temperatures and/or a smaller reactor.

6. Future work

In the near future some long-run experiments will be carried out. Furthermore, the pilot-plant installation will be extended with a gas engine and electricity generator. Obviously it is the ultimate goal to prove the applicability of the proposed tar conversion technology for cleaning producer gas. The development of this technology is continued in a EU research project (JOR 3-CT95-0053) for the next 3 years.

7. Notation

A	area	m^2
a$_p$	specific area	m^2/m^3
C$_p$	Specific heat capacity	[J kg^{-1} K^{-1}]
C$_{tar}$	Tar concentration	[g/Nm^3 or mole/Nm^3]
d$_p$	Particle diameter	[m]
LHV	Lower Heating value	[MJ/Nm^3]
t	time	[s]
T	Temperature	[K]
u$_{g,i}$	interstitial gas velocity	[m s^{-1}]
u$_{mean}$	mean gas velocity in the reactor	[m s^{-1}]
v$_{heat}$	heat front velocity	[m s^{-1}]
z	Axial position	[m]

greek symbols

ε	bed porosity	[-]
ρ	density	[kg m^{-3}]
τ$_{mean}$	residence time based on the mean interstitial gas velocity in the cracker	[s]
φ$_{gasifier}$	gas flow to gasifier	[Nm^3/hr]
φ$_{out}$	gas flow after tar converter	[Nm^3/hr]
φ$_{sec}$	secondary air flow	[Nm^3/hr]

sub- and superscripts

g	gas
in	inlet tar converter
ins	insulation
out	outlet tar converter
s	solid or superficial
0	inlet conditions

8. References

1. Stassen, H.E.M., Venendaal, R., Knoef, H.A.M. (1993) UNDP/WB small-scale biomass gasifier monitoring report, Vol. 1 Findings.
2. Boreskov, G.K., Bunimovich, G.A., matros, Yu.Sh., Ivanov, A.A. (1982) Catalytic processes under non-steady-state conditions. I. Switching the direction for the feed of the reaction mixture to the catalyst bed. Experimental results., *Kinet.Catal.*, Vol. 23, pp. 335-338.
3. Matros, Yu.Sh. (1990), Mathematical modelling of chemical reactors - development and implementation of novel technologies, *Angew.Chem.Int.Ed.Engl.*, Vol. 29, pp. 1235.
4. Bobrova,L.N., Slavinskaya, E.M., Noskov, A.S., Matros, Yu.Sh. (1988), Unsteady-state performance of NO$_x$ catalytic reduction by NH$_3$, *React.Kinet.Catal.Lett.*, Vol.

37, pp. 267-272.

5. Blanks, R.F., Wittrig, T.S., Peterson, D.A. (1990), Bidirectional adiabatic synthesis gas generator, *Chem.Eng.Sci.*, Vol. 45, pp. 2407-2413.

6. Neophytides, S.G., Froment, G.F. (1992), A bench scale study of reversed flow methanol sysnthesis, *Ind.Eng.Chem.Res.*, Vol. 31, pp. 1583-1589.

7. Eigenberger, G., Nieken, U. (1988), Catalytic combustion with periodic flow reversal, *Chem.Eng.Sci.*, Vol. 43, pp. 2109-2115.

8. Beld, L. van de (1995), Air purification by catalytic oxidation in an adiabatic packed bed reactor with periodic flow reversal, PhD thesis, University of Twente, the Netherlands.

9. Beld, L. van de, Borman, R.A., Derkx, O.R., Woezik, B.A.A. van, Westerterp, K.R. (1994), Removal of volatile organic compounds from polluted air in a reverse flow reactor: An experimental study, 1994, *Industrial & Engineering Chemistry Research*, Vol 33, pp. 2946-2956

10. Biomass Technology Group BV (1995), Patent no. 1001555, the Netherlands

Acknowledgement

The financial support of the Netherlands Agency for Energy and the Environment (NOVEM: 355200/2040) and the European Commission program EC/AIR (AIR2.CT93.1436) is gratefully acknowledged.

ALKALI SEPARATION IN STEAM INJECTED CYCLONE WOOD POWDER GASIFIER FOR GAS TURBINE APPLICATION

FREDRIKSSON, C. AND KJELLSTRÖM, B.
Division of Energy Engineering
Luleå University of Technology
S-971 87 Luleå, Sweden

Abstract
Cyclone gasification of wood powder at atmospheric pressure has been studied. The cyclone gasifier works as a particle separator as well and the wood powder is injected into the cyclone with air or air/steam as transport medium. The effects of stochiometry and steam injection on the gasification temperature and separation of char particles are investigated. The experimental results are compared with theoretical predictions. The amount of Potassium (K) and Sodium (Na) that can be separated together with the char is also studied since this is of special interest if the gas is used to operate a gas turbine. The fuel flow has been 26 kg/h corresponding to a thermal input of 140 kW. The equivalence ratio was varied between 0.15 and 0.4. Wood powder has been injected with air only or air/steam with steam mass flows of 50-80% of the fuel flow. It has been possible to separate 20-40% of the potassium and 10-20% of the sodium supplied with the wood. The resulting alkali content in the gas entering a turbine at a temperature of 850°C is between 4 and 8 mg/kg gas.
Keywords: Alkali, cyclone, gasification, gas turbine, steam injection, wood powder.

Introduction

Environmental aspects has increased the interest in biomass as an alternative fuel for power generation. To maximise the electricity output from a power plant, a combined-cycle power generating system is necessary. This means that both a gas turbine and a steam turbine are used in series. Economical studies in Sweden have shown that one promising way to use biomass in a co-generating power plant is to burn wood powder directly in a modified gas turbine combustor. The use of such a process is however associated with problems that are caused by the inorganic materials in the fuel. In the case of a direct wood powder fired gas turbine the problems are possible erosion, caused by the particles in the gas and deposition of alkali compounds on the turbine blades which may cause flow blockage, vibrations and possibly corrosion [1,2]. The compounds of concern are sulphates, chlorides and hydroxides of potassium and sodium. The problems can be expected to be more severe if a high inlet temperature to the gas turbine is chosen in order to improve cycle efficiency. Experiences with direct fuelling of a gas turbine with wood powder [3] show no erosion or corrosion but severe deposits when the inlet temperature exceeded 800°C.

A possible approach to the problem is two-stage combustion of the wood powder where the first stage is a cyclone gasifier/separator. In the cyclone, the wood is gasified at temperatures below 900°C. Alkali metals are supposed to be in the fuel mainly as carbonates which has melting temperatures between 850-890°C [4]. These carbonates has been reported to dissociate and vaporise above 900°C [5]. If the gasification takes place at sufficiently low temperature, the corrosive ashes remain condensed in the char particles and can be separated from the gas in the cyclone. The wood powder can be injected into the cyclone with air or steam as transport medium. The gas from the cyclone gasifier can be burned in a modified gas turbine combustor.

In previously reported experiments with atmospheric cyclone gasification the fuel was injected with air [6,7]. This has several draw-backs from a safety point of view which should be eliminated with the use of steam for wood powder injection.

The aim of this work has been to investigate the effect of steam injection and stochiometry on the gasification temperature and separation of alkali in the cyclone gasifier which forms the first stage of a two-stage gas turbine combustor for wood powder.

Experiments with cyclone gasification

Experimental equipment

The experimental set-up is a two stage atmospheric gasification/ combustor rig (Fig.1). The fuel system consists of a fuel bin, screw feeders and two chutes that are connected to the cyclone inlets. The inlets are two opposite injectors driven by pressurised air or steam. The injectors are entering the cyclone in a tangential direction. From the screw feeders down through the chutes an air-fuel mixture is sucked into the injectors where it mixes with the transport medium (steam or air) and is blown into the cyclone. This cyclone gasifier which is designed as a standard cyclone separator [8] works as a particle separator as well and the separated char is collected at the bottom outlet. The top outlet of the cyclone is connected to a combustor where the gas is completely burned with air.

The cyclone is made of temperature resistant stainless steel and is mounted vertically standing on a rotary feeder where the separated char is collected. The cylindrical part of the cyclone is 250 mm in diameter and 560 mm in height. Overall height with the conical part included is 1060 mm. The cyclone and the pipe connecting with the secondary combustor are insulated with a thin layer of heat resistant insulation surrounded by a 200 mm layer of rockwool. For heating during start-up, a propane burner can be introduced in the cyclone.

During the experiments, cyclone gas- and wall temperature and the flows of fuel-, air- and steam are measured. Separated char is collected at the cyclone bottom each two minutes. It is collected at the rotary feeder and is stored in an air tight collecting vessel. Six wall temperatures of the cyclone are measured with Type K sheathed thermocouples attached vertically three and three to the outside of the wall. A type N sheathed thermocouple with radiation shield is used for gas temperature measurements at the cyclone outlet. Steam- and air flows are measured with orifice plates according to ISO 5167-1:1991(E). The fuel screw feeders were calibrated manually to determine the fuel flow as a function of screw speed.

All the measured variables are logged at 1 Hz. Experiments have been made at different equivalence ratios. This was achieved mainly by changing the air mass flow that is sucked into the injectors together with the fuel.

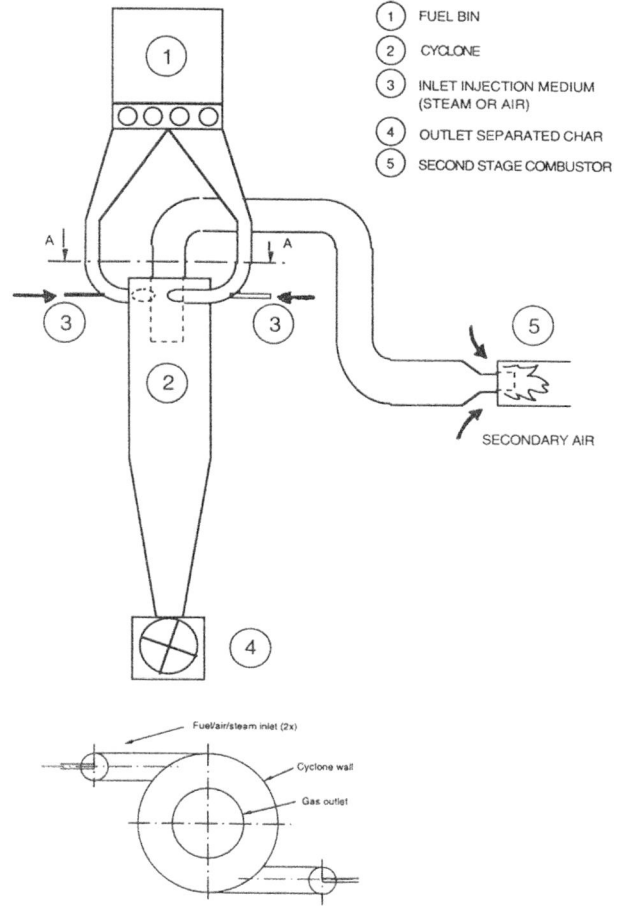

1	FUEL BIN
2	CYCLONE
3	INLET INJECTION MEDIUM (STEAM OR AIR)
4	OUTLET SEPARATED CHAR
5	SECOND STAGE COMBUSTOR

SECONDARY AIR

Fuel/air/steam inlet (2x)

Cyclone wall

Gas outlet

Fig. 1. Side view and top view A-A of the experimental set-up.

Ash analysis

The ash has been analysed at SGAB Analys in Luleå, Sweden. Ash samples of 0.5 g were dissolved in HF/HNO3/HClO4 and final analysis was made by Inductively Coupled Plasma Atomic Emission Spectroscopy, ICP-AES [9]. The ash samples have been produced from the fuel and char by decarbonisation at 550°C.

Experimental conditions

The range for different parameters covered in the experiments are listed in Table 1.

Ideally the experimental equivalence ratio should cover the same range for both air and air/steam injection but this was not possible with this experimental set-up. The lower level for this parameter corresponds to the level at which the gasifier function turns unstable, leading to decreasing temperatures. The higher level is determined by how much air it is possible to inject together with the fuel or steam/fuel.

Table 1. Experimental parameter ranges

Parameter	Specification/ Range
Fuel feed rate	26 kg/h
Thermal input	137 kW
Injection medium	Air or Air/Steam
Equivalence ratio	
-Air injection	0.16-0.32
-Air/Steam injection	0.30-0.40
Steam/Fuel ratio	0.5 and 0.8 kg steam/kg fuel
Cycl. inlet velocities	10-25 m/s
Cyclone pressure	Atmospheric
Gas temperature	
-Air injection	800-950°C
-Air/Steam injection	900-1000°C

Table 2. Fuel and ash analysis and particle size distribution

Fuel analysis	
Substance	weight-%
C	50.2
H	6.4
N	0.2
S	<0.04
Cl	<0.01
O (diff.)	38.8
Ash	0.3
Moisture	4.1

Ash analysis	
Substance	w-% of fuel
Al	0.003
Ca	0.070
Fe	0.002
K	0.020
Mg	0.008
Na	0.002
P	0.016
Mn	0.006

Table 3. Wood particle size distribution

Size distribution	
Particle size	%
>1 mm	0.3
0.5-1	18.4
0.25-0.5	38.9
0.125-0.25	17.7
0.100-0.125	6.3
0.074-0.100	1.4
0.063-0.074	9.3
0.040-0.063	6.5
<0.040 mm	1.6

Table 4. Inorganic elements in injected steam

Water analysis	
Substance	ppm-w
Al	1.05
Ca	3.98
Fe	3.3
K	2.48
Mg	0.65
Na	3.17
P	0.135
S	1.26
Mn	2.38

The wood powder that has been used is a commercial fuel. Ultimate analysis of the fuel, ash analysis and the size distribution of the powder can be seen in Table 2 and 3. Inorganic elements in the steam that is injected together with the fuel are listed in Table 4.

Operational experience

The behaviour of the char is very much dependent on the temperature in the cyclone. When temperature is decreased the char build-up on the cyclone walls seems to increase. Pieces of char are collected at the cyclone bottom when they fall off the cyclone wall. When the wall temperature at the conical part of the cyclone is below 600°C and the gas temperature is under 830°C, tar is coming out of the rotary feeder. A similar observation was made by [7] where a sticky char was generated at gas temperatures below 800°C. When the cyclone has been inspected visually after the test, there has been char build-up at the top of the cyclone even if the wall temperature has been above 800°C. At the top of the cyclone there is some turbulent regions close to the inlets where small particles can rest and start a build-up of char. This seems to happen even if the particles does not have any sticky characteristics. The amount of build-up is however very small and should not have any effect in these experiments.

The amount of char that is collected at the cyclone bottom each two minutes is slightly varying due to variation in the fuel flow. Irregular combustion in the second stage combustor leads to abrupt pressure changes in the cyclone which will also cause a varying amount of separated char. As the char is collected for 15-20 minutes, these variations are considered to be negligible.

The char is stored in a collecting vessel with a air tight cap to minimise weight loss due to combustion of the hot char.

Theoretical predictions

For the application of this cyclone gasification technique in a power plant it is of interest to be able to predict the resulting temperature and gas composition from the gasifier.

The measured temperatures are therefore compared with equilibrium calculation for the same condition performed by the ChemSage computer program [10].

The computer program predicts the equilibrium composition by minimisation of Gibbs free energy. The output are the adiabatic gas temperature and the molal amounts of all reaction products in gas-, condensed- and solid phase at equilibrium. In this work the predictions are focused on the adiabatic gas temperature and the amount of alkali in diffe-rent phases.

Results from experiments and equilibrium predictions

Temperature

The gas temperature is a function of equivalence ratio and is increasing as equivalence ratio is increased, see Fig. 2 and Fig. 3. In the same figures the results from theoretical predictions are plotted as continuous lines. As can be seen the measured temperatures differ from the predicted temperatures. When only air is injected the difference in temperature is significant. When the wood powder is injected with air/steam the difference is smaller. This has earlier been explained by the fact that there is an amount of carbon that does not participate in the reactions due to the separation of char and short residence time for the particles in the cyclone [11].

Char separation

The amount of char that is separated and collected at the cyclone bottom is plotted versus gas temperature in Fig. 4. The general trend is that the char separation decreases as temperature is increased. There does not seem to be any difference in collection efficiency when the amount of steam that is injected is changed. Note that the temperature is increased by increasing the equivalence ratio in the cyclone. A higher temperature is therefore corresponding to a higher volume flow and inlet velocity.

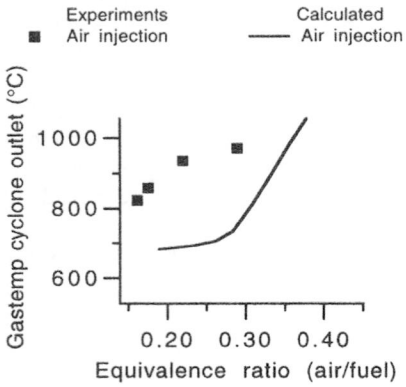

Fig. 2. Gas temperature when only air is used as injection medium.

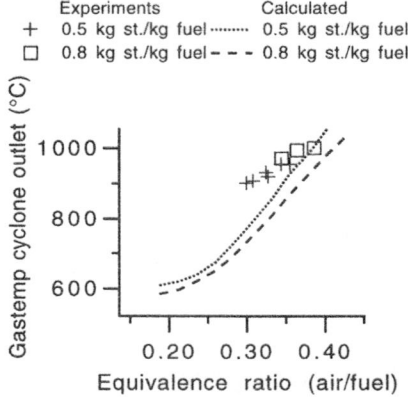

Fig. 3. Gas temperature when air/steam is used as injection media.

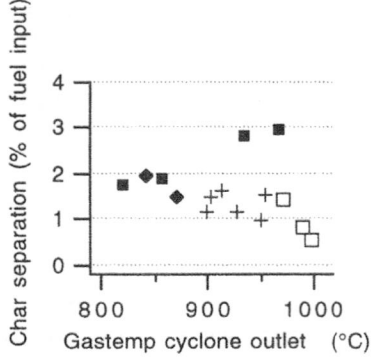

Fig. 4. Amount of char collected at the cyclone bottom.

926

Ash composition

The char has been analysed with respect to the total ash contents and the distribution of ash components in the ash. In Fig. 5 the quantity of ash in the char is plotted and in Fig. 6 the concentration of calcium is shown. Figure 7 to 9 shows the contents of potassium, sodium and magnesium relative calcium contents in the ash plotted versus gas temperature. Calcium is considered to stay condensed in the ash and is therefore used as reference. For the comparison between Fig. 7 to 9, the range for the left axis is chosen as ±40% of the average value in the plot . The amount of ash in the char is increasing as temperature is increased while the distribution of different ash element are almost constant independent of gas temperature and injection media. In Fig. 10 the predicted amount of condensed alkali are plotted versus temperature.

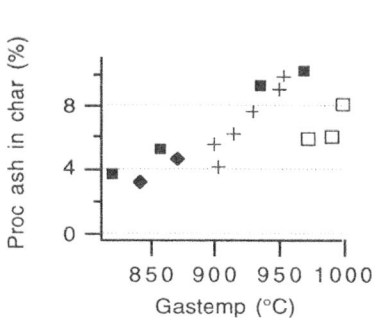

Fig. 5. Ash content in char.

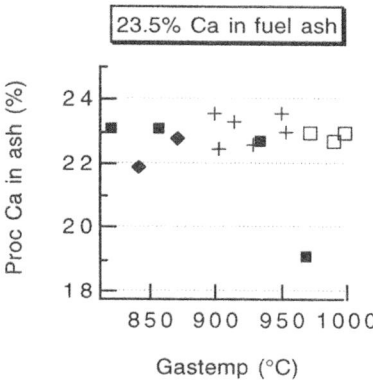

Fig. 6. Calcium (Ca) content in separated ash.

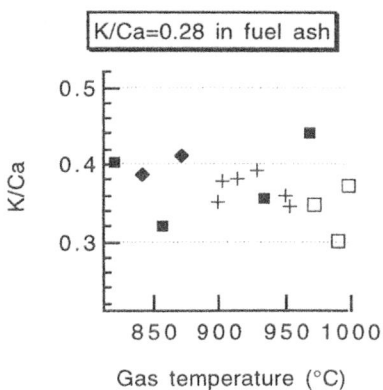

Fig. 7. Normalised concentration of potassium (K) in separated ash.

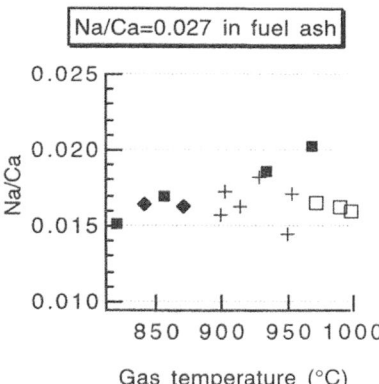

Fig. 8. Normalised concentration of sodium (Na) in separated ash.

927

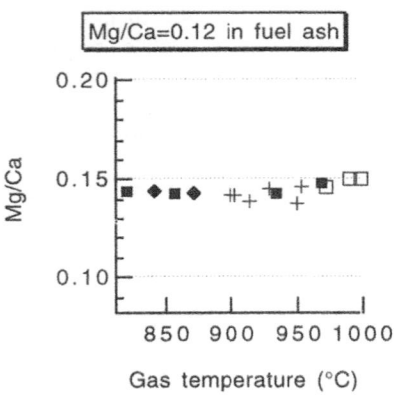

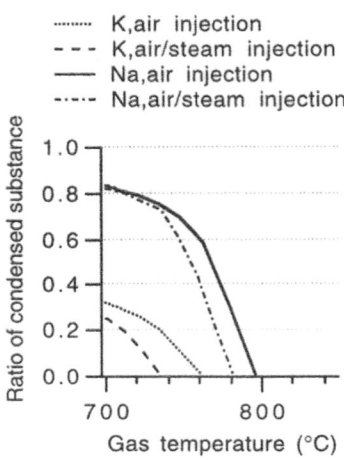

Fig. 9. Normalised concentration of magnesium (Mg) in separated ash.

Fig. 10. Predicted ratio of condensed K and Na at equilibrium as a function of temperature.

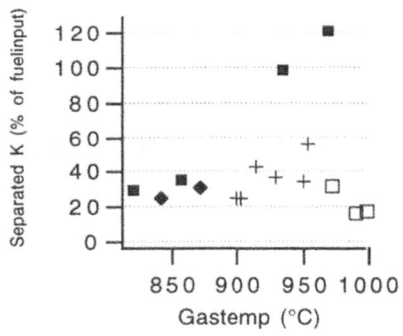

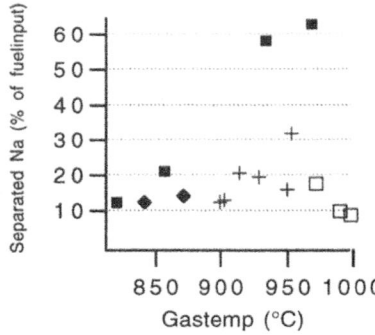

Fig. 11. Separated potassium (K).

Fig. 12. Separated sodium (Na).

Separation of alkali metals

The total amount of alkali metals that has been separated in the cyclone gasifier/separator can be seen in Fig. 11 and 12. The quantity of alkali that can be separated in the cyclone is a measure of the cyclone performance. In practice it is of secondary importance as the main concern is the amount of alkali and the form of alkali compounds entering the gas turbine. The amount of alkali relative to gas volume leaving the cyclone is shown in Fig. 13. The alkali content in the gas from the cyclone are calculated as the difference between input alkali with fuel and separated alkali with char. The temperature is raised by adding more air to the cyclone and the increase in volume flow is the main reason for the decrease in alkali content in the gas from the cyclone. The total alkali content in the gas to a gas

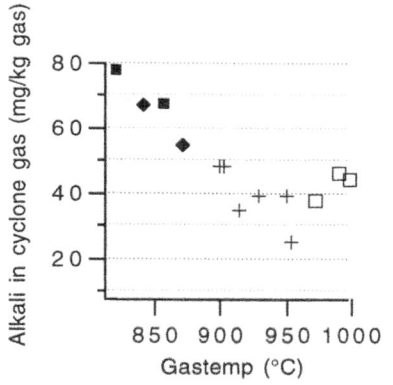

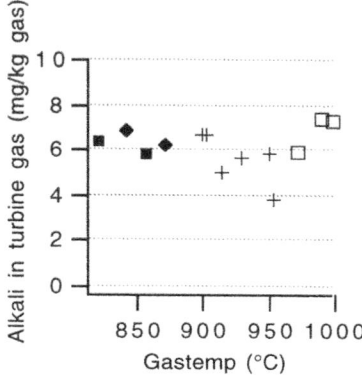

Fig. 13. Alkali content in gas from cyclone.   Fig. 14. Alkali content in gas to turbine inlet with inlet temperature 850°C.

turbine can be seen in Fig. 14 where the gas flow is calculated for a turbine inlet tempera-ture of approximately 850°C. The alkali in the steam is supposed to vaporise and leave the cyclone with the gas. 2-3% of total alkali in the gas to the turbine origins from the injected steam.

Discussion

In the case of cyclone gasification with separation of char with approximately 95% carbon content, the loss of carbon may explain why the predicted temperature is lower than the measured temperature shown in Fig. 2 and Fig. 3 [11]. This difference will increase because there is an error in the measured gas temperature due to radiation from the surrounding surfaces. As the surrounding surfaces are colder than the gas temperature, the actual tempera-ture will be higher than measured due to radiation losses. This will further enhance the difference between predicted and measured gas temperature. This error is decreased by surrounding the thermocouple with a radiation shield. Measurements shows that the shielded thermocouple underestimates the gas temperature between 10 and 20°C compared to a sucking pyrometer.

For the application of this cyclone gasifier connected to a gas turbine, it is of interest to know how much alkali metals that can be separated in the first gasification/separation stage. This is shown in Fig.11 and 12. Measurements show an increasing separation of alkali when air is used as injection media. If this effect really exists or if it is caused by experimental inconsistency is not clear. Until now it seems reasonable to calculate with a separation of potassium to between 20 and 40 % of input and a separation of 10 to 20% of input sodium. The level of alkali separation are almost constant as temperature is increased. This is a result of decreasing char separation, increasing ash content in the char and an almost constant distribution of ash elements in the ash. Ash analysis in Fig. 6 to 9 shows that the ash composition remains almost the same independent of temperature or injection media. This can be compared with equilibrium calculations shown in Fig.10 which predicts that all alkali metals are vaporised between 750° and 800°C. Experimentally it has been shown by Misra et al. [5] that the potassium carbonates dissociate and oxidise to CO_2 and K_2O as gas above 900°C. Both the equilibrium predictions and experimental results are only valid for long reaction time where all transformations can be completed. The wood ash samples in the experiments in [5] have been held at the specified temperature for over an hour. This indicates that even if the temperature are above the temperature of vaporisation for the alkali compounds, the residence time in the cyclone is to short for the alkali compounds to vaporise. If a plug-flow model of the cyclone is used [12], the particle residence time in the cyclone can be estimated to about 0.4s. From this it follows that the reactions in the cyclone appear not to reach equilibrium conditions due to the short residence time. As the temperature do not influence the ash composition, the separation of alkali metals is only a matter of separation efficiency and carbon burnout (i.e. ash content in char).

Figure 6 shows that the level of K/Ca is higher in the char than in the fuel. Figure 9 shows the opposite relation for Na/Ca i.e. the level of Na/Ca is lower in the char than in the fuel. As calcium is considered not to volatilise from the char, it is not clear why the amount of potassium is higher in the char than in the fuel. Besides possible errors in char analysis, it might be explained by different levels of calcium in wood powder particles of different size. If there is less Ca in larger particles that are separated while the content of K remains the same, the level of K/Ca in the collected char can be higher than in the original fuel. The loss of Na may be caused by release of easily volatilised compounds such as NaCl. The remaining sodium can be in the form of carbonates which remain condensed in the particles. These observations are in conflict with the equilibrium calculation which predict that potassium is more easily volatilised than sodium. In these calculations it is assumed that the alkali metals are bound in the fuel as carbonates which may not be the case.

Of even more interest than the separated alkali is the amount of alkali in the gas from this first stage cyclone. The content of alkali in the gas from the cyclone is decreasing when gas temperature is increased, see Fig. 13. This is mainly a result of an increase in mass flow as the temperature level is controlled by the amount of air that is injected together with the fuel. If steam is used as injection media, the relative amount of alkali will of course decrease further. The calculated amount of total alkali content in the gas to the turbine inlet is shown in Fig. 14. It is calculated for an inlet temperature of approximately 850°C and the amount is between 4 and 8 mg alkali/kg gas and more or less constant as the temperature in the cyclone is increased. Kallner et al. [7] has for the same turbine inlet temperature presented levels of alkali between 6 and 10 mg/kg gas. The lower level was measured at a cyclone gas temperature of 800°C and the level was increasing as temperature is increased.

The results obtained in these experiments show that the effect of steam injection on the relative alkali content in the gas is positive due to the increased mass flow. In a co.-generation plant there will be high pressure steam available which is reported to also have positive effects on cycle efficiency [13]. It has also been concluded that the ash composition is more or less constant as temperature is increased. The particle residence seems to be too short for any volatilisation of alkali compounds to take place in the cyclone. For the use of the gas in a gas turbine application it is necessary to investigate the chemical composition of the unseparated char particles and the faith of the alkali compounds as they pass through the cyclone, second stage combustor and turbine. Special attention should be taken to the form of the alkali compounds in the gas and char particles. The melting temperature and temperature of vaporisation of the compounds relative to the material temperature in the

930

turbine are important for the deposition mechanism. This work does not include such a study but is of importance in future research.

Acknowledgements

This work has been sponsored by the National Board for Industrial and Technical Development, NUTEK and an industrial group consisting of ABB Carbon, ABB Stal, Berglunds Rostfria AB, Elforsk, Luleå Energi AB, SCA, Vattenfall AB Värmekraft and Älvsby Fjärrvärme AB. The authors are also grateful to the staff at ETC, Energy Technology Center in Piteå, Sweden and to our fellow co-workers.

References

1. Fredriksson, J. and Kjellström, B. (1992) Fastbränsleeldade gasturbiner i kraftvärme-tillämpningar, *Stiftelsen för Värmeteknisk Forskning*, Stockholm, Sweden. ISSN 0282-3772, report 458.
2. Moses, C.A. and Bernstein, H. (1994) Impact Study on the Use of Biomass-Derived Fuels in Gas Turbines for Power Generation, *U.S. Department of Energy*, Contract DE-AC02-83CH10093.
3. Hamrick, J.T. (1991) Development of Wood as an Alternativ Fuel for Gas Turbines, Battelle Memorial Institute, USA, PNL-7673/UC-245.
4. Miles, T.R.Sr. and Miles, T.R.Jr. (1994) Alkali Deposits Found in Biomass Power Plants: A Preliminary Investigation of Their Extent and Nature, *Summary Report for the National Renewable Energy Laboratory*, USA, Contract TZ-2- 11226-1.
5. Misra, M.K., Ragland, K.W. and Baker, A.J. (1993) Wood ash composition as function of furnace temperature, *Biomass and Bioenergy*, Vol.4, No.2, pp.103-116.
6. Cousins, J.W., and Robinson, W.H. (1985) Gasification of Sawdust in an Air-Blown Cyclone Gasifier, *Ind. Eng. Chem. Process Design & Development,* nr 24, pp. 1281-1287.
7. Kallner, P., Fredriksson, J., and Kjellström B. (1995) Wood Combustion and Ash Separation in a Cyclone for Gas Turbine Applications, International Gas Turbine Institute, Vol.9, ASME COGEN TURBO Congress.
8. Perry and Green (1984) *Chemical Engineers' Handbook*, sixth edition, McGrawHill.
9. Skoog, D.A. and Leary, J.J. (1992) *Principles of Instrumental Analysis*, Fourth edition, Harcourt Brace College Publishers, Fort Worth, USA.
10. ChemSage version 1.0 (1991) *Practical Manual and Handbook*, GTT-Technologies, Kaiserstrasse 100, 52134 Herzogenrath 3, Germany.
11. Fredriksson, C., Degerman, B. and Kjellström, B. (1995) Effects of Steam Injection and Stochiometry on Cyclone Gasification of Wood Powder, Submitted to the ASME Turbo Expo Conference, Birmingham, United Kingdom, June 10
12. Lede, J., Li, H.Z., Soulignac, F. and Villermaux, J. (1989) Le Cyclone Réacteur Partie 2: Mesure Directe de la Distribution des Temps de Séjour de la Phase Solide-Lois d'Extrapolation, *The Chemical Engineering Journal*, 42, pp 103-117.
13. Fraize, W.E. and Kinney, C. (1979) Effects of Steam Injection on the Performance of Gas Turbine Power Cycles, *Journal of Engineering for Power*, Vol.101, April, pp 217-227.

HYDROGEN PRODUCTION FROM LIGNOCELLULOSIC MATERIALS BY STEAM GASIFICATION USING A REDUCED NICKEL CATALYST

Hydrogen production from biomass

T. MINOWA, T. OGI and S. YOKOYAMA
National Institute for Resources and Environment, Tsukuba, Japan

Abstract
Cellulose and wood (Japanese oak) were gasified in the presence of water at a temperature of 350 °C and pressure of 17 MPa using a reduced nickel catalyst and sodium carbonate as a promoter. Main gas components were hydrogen and carbon dioxide, and a small amount of methane and carbon monoxide. The gas yield reached 94 wt% for cellulose and 55 wt% for wood at a nickel catalyst loading of 20 wt%, and hydrogen recovery as hydrogen gas was 74% and 49%, respectively. The effects of catalyst loading and operating pressure on gas yield and composition were examined.
Keywords: Catalytic gasification, hydrogen production, lignocellulosic materials, reduced nickel catalyst, steam gasification

1 Introduction

Using fuel cells, hydrogen can be used for clean power production. Biomass is a renewable resource, and hydrogen production from biomass has become attractive. However, biomass usually has a high moisture content and a low moisture feedstock is required for conventional gasification. Since drying is an energy intensive operation, a new method to convert wet biomass directly is beneficial.

Antal et al. of the University of Hawaii reported that wet biomass is completely gasified to hydrogen, carbon dioxide and carbon monoxide in supercritical water at 600 °C and 35 MPa [1 - 3]. Elliott et al. of the Pacific Northwest Laboratory gasified wet biomass directly to methane rich gas using a reduced nickel catalyst at relatively low temperatures of 350 - 450 °C and low pressures of 15 - 35 MPa [4 - 6]. In the latter process, three major reactions, steam reforming, water-gas shift and methanation, can occur.

$$CH_xO_y + (1-y)H_2O \rightarrow CO + (x/2+1-y)H_2 \qquad (1)$$
$$CO + H_2O \rightarrow CO_2 + H_2 \qquad (2)$$
$$CO + 3H_2 \rightarrow CH_4 + H_2O \qquad (3)$$

In our experiments, methanation was found to be suppressed by using liquid water instead of steam in a high pressure process with nickel catalyst. This way a hydrogen rich gas was obtained [7, 8].

This paper summarizes our findings.

2 Experimental

Cellulose, a major component of woody biomass, and Japanese Oak were used as a starting material. The cellulose sample was microcrystalline (E. Merck). Both were dried at 105 °C for 24 h prior to this study. **Table 1** shows the elemental and chemical compositions of the samples. The elemental composition was determined using an elemental analyzer (Perkin-Elmer, 2400 CHN), and the chemical composition was analyzed at the Forest Development Technical Institute, Japan.

Table 1 Analyses of cellulose and Japanese Oak

	Cellulose	Japanese Oak
Elemental composition / wt%		
Carbon	42.4	46.7
Hydrogen	6.7	6.2
Nitrogen	0.0	0.1
Oxygen [a]	50.9	46.6
Chemical composition / wt%		
Cellulose		45.21
Hemicellulose		27.96
Lignin		20.20
Alcohol & Benzene soluble extracts		1.64
Ash		0.39

a; calculated by difference

The reduced nickel catalyst (about 50 wt% nickel on kieselguhr) was prepared by ordinary precipitation [9, 10]. Sodium carbonate solution was added to a slurry of kieselguhr and nickel nitrate solution at 70 °C to obtain the precipitate, which was then dried at 105 °C for 12 h, crushed to 60 - 150 mesh, calcined at 350 °C for 4 h and reduced with hydrogen at 350 °C for 4 h to finally obtain the reduced nickel catalyst. A commercial nickel catalyst (Engelhard, Ni-3288, about 50 wt% nickel on silica alumina) was used for comparison, and crushing and reduction were similarly done. Hereafter, they are referred to as the prepared catalyst and the commercial catalyst, respectively.

The reaction was performed in a conventional stainless steel (SUS-F316L) autoclave (146 cm^3 capacity including dead-space of 26 cm^3) with a magnetic stirrer. Water, cellulose or wood, the reduced nickel catalyst and sodium carbonate were charged into the autoclave. Nitrogen was used as purge gas. Additional nitrogen was added to achieve the desired initial pressure. The reaction was started by heating the autoclave to the required temperature using an electric furnace. The pressure in the autoclave was

933

monitored with a pressure transmitter. On completion of the reaction, the autoclave was cooled to room temperature with an electric fan.

The sweeping of gaseous phase during the reaction was performed using the same autoclave. **Fig. 1** shows a schematic diagram of the reactor. The sweeping was conducted by operating valves, while the operating pressure was kept constant by adding nitrogen gas. The rate of the sweeping was about 2 l/min. Other operation was done in the same manner as above.

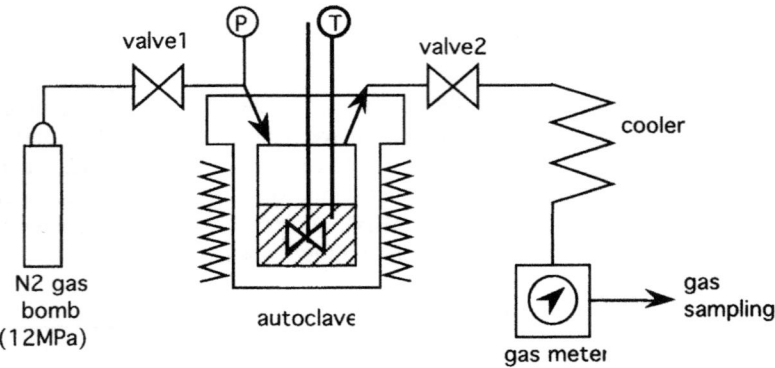

Fig.1 Schematic diagram of reactor apparatus

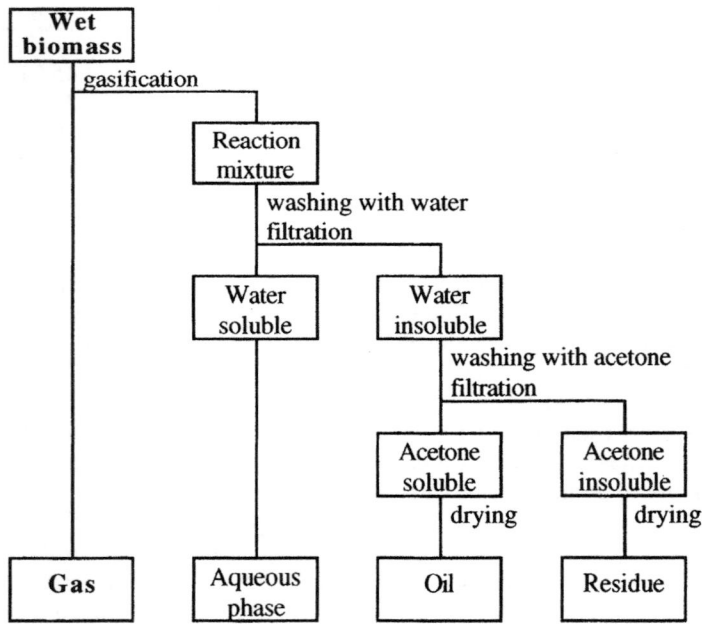

Fig. 2 Separation procedure of products

The gas was collected in a gas sampling bag and its volume was measured with a gas meter (Shinagawa Seiki, W-NK-0.5Bf). Gas composition with respect to methane, ethane, ethylene, propane and propylene were determined by a gas chromatograph (Shimadzu, GC-9A) with a hydrogen flame ionization detector and a packed column (Squalane on Alumina). Hydrogen, nitrogen, carbon dioxide and carbon monoxide were measured by a gas chromatograph (Shimadzu, GC-12A) with a thermal conductivity detector and a packed column (MS-5A or Porapak Q).

The autoclave was then opened to remove the reaction mixture. **Fig. 2** shows the procedure using for separation of the different reaction products. The aqueous phase was separated by washing with water and by filtration. Remaining material was washed with acetone and the acetone solution was separated by filtration. Acetone was evaporated from the filtrate at 70 °C to obtain an oil. The residue on the filter paper was dried at 70 °C for 24 h. The amount of carbon in the aqueous phase was determined using a total organic carbon meter (Yanaco, TOC-8L). The elemental composition of the oil and residue was determined as for the starting materials.

3 Results and discussion

3.1 Effects of initial pressure

The effects of initial pressure on gas production rate and gas composition were examined using cellulose as a starting material at 350 °C and 1 h reaction time, for results see **Fig. 3**. The initial pressure was varied from 0.1 to 5 MPa to get various operating pressures. The operating pressure, defined as the pressure at 350 °C, is also indicated in Fig. 3. The gas consisted mainly of hydrogen, carbon dioxide and methane, with minor amounts of carbon monoxide and hydrocarbons. The gas composition changed dramatically at the initial pressure of 1 MPa. Hydrogen yield was 40 - 50 mmol at initial pressures exceeding 1 MPa, decreasing rapidly at lower initial pressures. Similar effects were observed for the production of carbon dioxide. Methane yield reached a maximum value of 10 mmol at an initial pressure of 0.5 MPa. The prepared catalyst had enough activity for methanation reaction at the reaction condition of 400 °C and 13 MPa [11]. Therefore, much hydrogen production was considered due to the reaction condition.

The operating pressure reached values of 17 - 18 MPa at initial pressures above 1 MPa, decreasing rapidly at lower initial pressure by the effect of the dead-space of the autoclave. At 350 °C the saturated pressure of water is about 16.5 MPa, thus, water around the catalyst should be in a liquid state at initial pressures higher than 1 MPa and in a steam state at lower initial pressures. From the experimental results, we therefore conclude that methanation is suppressed when water is in the liquid state. When the catalyst is in liquid water, hydrogen gas produced through steam reforming and water-gas shift may move from the liquid phase to the gas phase, and the catalyst would not be in contact with much hydrogen, thus, preventing methanation.

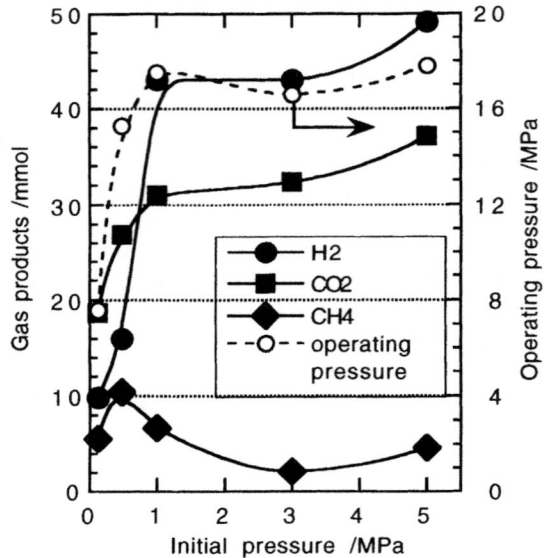

Fig.3 Effects of initial pressure on gas products
(350 °C, 1 h, cellulose/water/prepared Ni cat.=3/30/0.6 g)

3.2 Effects of catalyst loading

To study the role of the nickel catalyst, the effect of catalyst load on the distribution of gas and by-products such as oil, residue and aqueous phase were examined. The results are shown in **Fig. 4**. The yield was defined as the weight percentage of the organics in each phase to the weight of cellulose charged. The weight of the organics in the residue was calculated by subtracting the weight of the catalyst. The carbon amount measured by TOC meter was regarded as the weight of the organics in the aqueous phase, that is, the amounts of hydrogen and oxygen were neglected, because the weight of the organics in the aqueous phase cannot be measured due to the presence of water. The gas products are expressed as a mole ratio in the gas in Fig. 4. Without a catalyst, about 20 wt% cellulose was converted to gas (almost all carbon dioxide), about 40 wt% to the residue, and about 40 wt% was lost. With increase in catalyst loading, the gas yield increased linearly, the sum of the yields of by-products decreased linearly, and the loss also decreased linearly. The gas yield reached 91 wt% at the catalyst loading of 20 wt%, and this gas consisted of 50 vol% hydrogen, 40 vol% carbon dioxide and 10 vol% methane. Hydrogen recovery as hydrogen gas reached 70 wt%.

To determine the reason for the loss, the balances of carbon, hydrogen and oxygen were estimated. Carbon recovery was more than 90 wt% as shown in **Fig. 5**. Therefore, carbon loss is negligible. The mole ratio of hydrogen loss to oxygen loss was about 2:1 as shown in **Fig. 6**. The loss is thus probably due to water production. The addition of catalyst decreased the water production, because it is consumed by reactions such as steam reforming and water-gas shift. As evident from Fig. 6, the

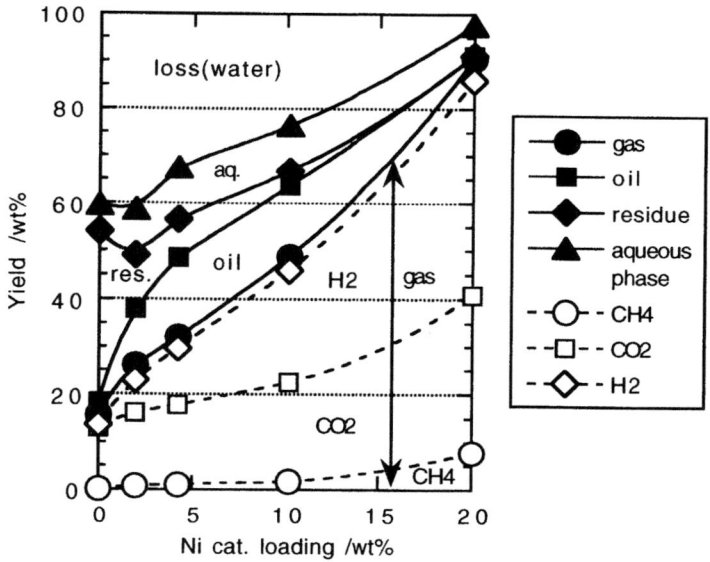

Fig. 4 Effects of prepared Ni cat. loading on product distribution
(350 °C, 18 MPa, 1 h, cellulose/water=5/30g)

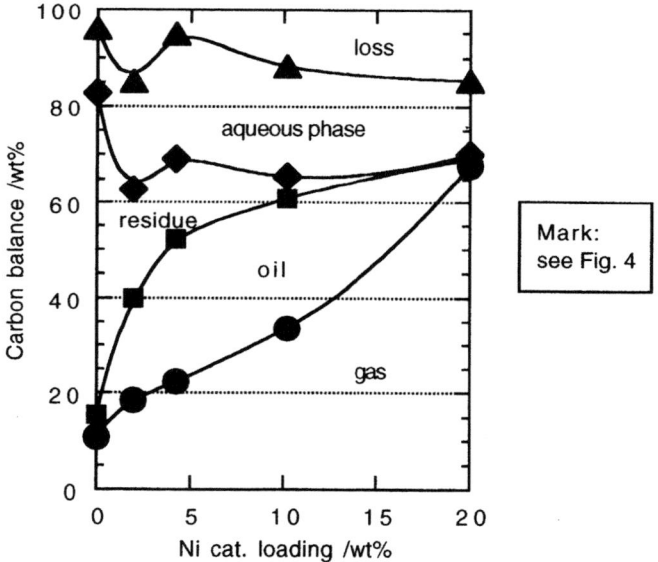

Fig.5 Effect of prepared Ni cat. loading on carbon balance
(reaction condition; see Fig. 4)

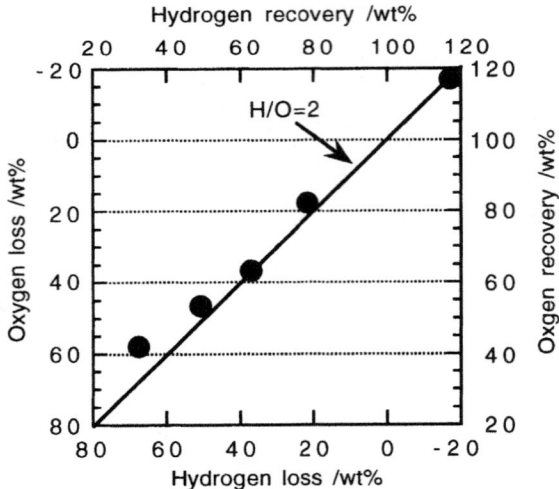

Fig.6 Relationship of hydrogen loss (recovery) to oxgen loss (recovery)
(reaction condition; see Fig. 4)

highest values of hydrogen and oxygen recovery exceeded 100%. Thus, water is probably reactant under the conditions of our experiments.

Fig. 4 shows the residue yield to decrease, and the oil yield to increase with catalyst load. Nickel catalyst is also used for the liquefaction of biomass to produce oil [12]. The oil formation would be one of the effects of the catalyst, and the oil may be an intermediate in the gasification.

3.3 Effects of kinds of catalyst

A preliminary gasification experiment was carried out using the commercial catalyst. **Table 2** shows the results of gasification using the prepared and commercial catalysts. With the prepared catalyst, hydrogen yield was in excess of 120 mmol and only 20 mmol methane was obtained. With the commercial catalyst, the amount of hydrogen decreased to 35 mmol, while the amount of methane doubled. The amount of carbon moved to the gas was similar in both cases. Therefore, about 20 mmol carbon dioxide and about 80 mmol hydrogen should react stoichiometrically to form about 20 mmol methane.

$$CO_2 + 4H_2 \rightarrow CH_4 + 2H_2O \qquad (4)$$

The carrier of the prepared catalyst was kieselguhr, and that of the commercial catalyst was silica alumina. The difference in the activity may be due to the carrier. Although no detail analysis of the catalyst was measured, this result suggests that the methanation may be suppressed or accelerated by modification of catalyst.

Table 2 Comparison of prepared catalyst and commercial catalyst

	Prepared catalyst	Commercial catalyst
Gas yield / wt%	93.7	79.4
H_2 / mmol	124.9	35.2
CO_2 / mmol	91.7	70.3
CH_4 / mmol	20.5	40.4
Carbon to gas / %	63.2	68.9
Hydrogen to gas / %	102.4	78.4
Oxygen to gas / %	115.5	88.3
By products		
Oil / wt%	1.9	1.1
Residue / wt%	3.8	0.0
Aqueous / wt%	6.2	4.0
Loss / wt%	-5.6	15.5

cellulose/water/prepared Ni cat./Na_2CO_3=5/30/1/0.5g
at 350 °C, 1 h, and initial pressure of 3 MPa
(operating pressure of about 18 MPa)

Table 3 Gasification of Japanese Oak

	Japanese Oak	Cellulose
Gas yield / wt%	55.4	93.7
H_2 / mmol	55.1	124.9
CO_2 / mmol	57.1	91.7
CH_4 / mmol	4.5	20.5
Carbon to gas / %	33.7	63.2
Hydrogen to gas / %	43.9	102.4
Oxygen to gas / %	78.9	115.5
By products		
Oil / wt%	14.3	1.9
Residue / wt%	3.3	3.8
Aqueous / wt%	9.8	6.2
Loss / wt%	17.2	-5.6

feedstock/water/prepared Ni cat./Na_2CO_3=5/30/1/0.5g
at 350 °C, 1 h, and initial pressure of 3 MPa
(operating pressure of about 18 MPa)

3.4 Gasification of Japanese Oak

Japanese Oak was used as lignocellulosic material in a gasification experiment. The results are summarized in Table 3. The gas yield was only 55 wt%, consisting mainly of hydrogen, carbon dioxide, and small amount of methane.

For comparison, the results of the gasification of cellulose are also shown in Table 3. Cellulose was gasified in the high yield of 94 wt%. Since wood consists not only of cellulose, but hemicellulose and lignin as well, the low gas yield for Japanese Oak was considered due to hemicellulose or lignin. Lignin is able to decompose into phenoxy radicals in hot water and the radicals can condense and/or repolymerize quickly to a solid material [13, 14].

3.5 Effects of gaseous phase sweeping during the reaction

Uemiya et al. of Waseda University carried out the steam reforming of methane in a hydrogen-permeable membrane reactor, and found that the methane conversion is increased by removing hydrogen from reaction system [15]. For our process, if hydrogen can be removed from the reaction system, methanation could be suppressed. The effects of gaseous phase sweeping during the reaction were examined. The results are shown in Fig. 7. Hydrogen production increased significantly by sweeping, and methane production decreased. Without sweeping, methanation proceeded during the entire holding time; hydrogen production decreased and methane was obtained. Methanation can be thus suppressed by sweeping; removing the produced hydrogen from reaction system.

In this experiment, all gas products were swept. If hydrogen gas could be removed selectively using a membrane reactor operable at pressures of 10 - 20 MPa, hydrogen production with higher efficiency may be expected.

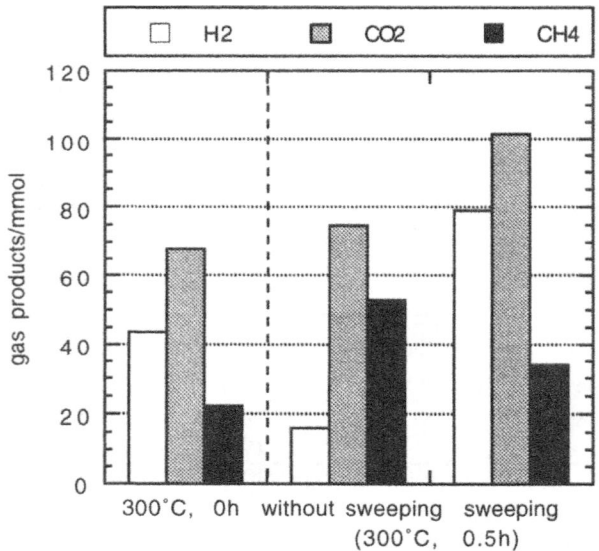

Fig.7 Effects of sweeping on gas products
(cellulose/water/commercial Ni cat./Na2CO3=5/30/2/0.5 g)

4 Conclusion

"It is clear from the present study that wet biomass, cellulose and Japanese Oak, can be steam gasified to yield hydrogen rich gas using a reduced nickel catalyst at 300 - 350 °C and 10 - 20 MPa. Low reaction temperature means less energy consumption. This gasification, therefore, is expected as one of the promising methods for hydrogen production from wet biomass. In addition, we showed that hydrogen production can be improved by modification of catalyst and sweeping of gas products. "

5 References

1. Yu, D., Aihara, M. and Antal, M. J., Jr. (1993) Hydrogen production by steam reforming glucose in supercritical water. *Energy & Fuels*, Vol. 7, No. 5, pp. 574 - 577.
2. Antal, M. J., Jr., Manarungson, S. and Mok, W. S. (1994) Hydrogen production by steam reforming glucose in supercritical water. *Advances in Thermochemical Biomass Conversion*, Vol. 2, pp. 1367 - 1377.
3. Antal, M. J., Jr., Matsumura, Y., Xu, X., Stenberg, J. and Lipnik, P. (1995) Catalytic gasification of wet biomass in supercritical water. *Prepr. Pap. - Am. Chem. Soc., Div. Fuel Chem.*, Vol. 40, No. 2, pp 304 - 307.
4. Elliott, D. C., Butner, R. S. and Sealock, L. J., Jr. (1988) Low temperature gasification of high-moisture biomass. *Res. Thermochem. Biomass Convers.*, pp. 696 - 710.
5. Sealock, L. J., Jr. and Elliott, D. C. (1991) Method for the catalytic conversion of lignocellulosic materials. *U. S. Patent 5019135; Chem. Abstr., 115, 94724g* .
6. Elliott, D. C., et al. (1993, 1994) Chemical processing in high-pressure aqueous environments 1 - 4. *Ind. Eng. Chem. Res.*, Vol. 32, pp. 1535 - 1541, pp. 1542 - 1548, Vol. 33, pp. 558 - 565, and pp. 566 - 574.
7. Minowa, T., Ogi, T. and Yokoyama, S. (1995) Effect of pressure on low temperature gasification of wet cellulose into methane using reduced nickel catalyst and sodium carbonate. *Chem. Lett.*, pp. 285 - 286.
8. Minowa, T., Ogi, T. and Yokoyama, S. (1995) Hydrogen production from wet cellulose by low temperature gasification using a reduced nickel catalyst. *Chem. Lett.*, pp. 937 - 938.
9. Catalyst Society of Japan (1985) *Syokubaikouza No. 5*, Koudansha, Tokyo.
10. Shirazaki, T. and Toudou, N. (1974) *Syokubaichosei*, Koudansha, Tokyo.
11. Minowa, T., Ogi, T. and Yokoyama, S. (1994) Methane production from cellulose by catalytic gasification. *Renewable Energy*, Vol. 5, pp. 813 - 815.
12. Boocock, D. G. B., Mackay, D. and Lee, P. (1982) Wood liquefaction: extended batch reactions using Raney nickel catalyst. *Can. J. Chem. Eng.*, Vol. 60, pp. 802 - 808.
13. Chornet, E. and Overend, R. P. (1985) Biomass liquefaction: an overview. *Fundamentals of Thermochemical Biomass Conversion*, Elsevier Applied Science Publishers, pp. 967 - 1002.
14. Minowa, T., Ogi, T., Dote, Y. and Yokoyama, S. (1994) Effect of lignin content on the direct liquefaction of bark, *International Chemical Engineering*, Vol. 34, No. 3, pp. 428 - 430.
15. Uemiya, S., Sato, N., Ando, H., Matsuda, T. and Kikuchi, E. (1991) Steam reforming of methane in a hydrogen-permeable menbrane reactor. *Applied Catalyst*, Vol. 67, pp. 223 - 230.

GASIFICATION

Pilot and demonstration

GASIFICATION OF WOOD IN A PILOT SCALE SPOUTED BED GASIFIER

A.K. JANARTHANAN
Department of Chemical Engineering, University of Nebraska-Lincoln, USA
L.D. CLEMENTS
Department of Biological Systems Engr., University of Nebraska-Lincoln, USA

Abstract

A pilot scale biomass gasification unit, known as the Spouted Bed Gasifier (SBG), has been set up to demonstrate and characterize a novel approach to reactor design. The SBG couples indirect radiant heating with a spouted bed reactor configuration. This paper describes the design of the 0.25 tonne per day pilot plant which is an intermediate step in the development of a 1+ meter diameter commercial module.

Performance results for operation at a feed rate of 1 kg/hr of pine chips (+1,-2 mm) are presented for operating temperatures of 500, 600 and 700 $^{\circ}$C, with and without steam addition. Gas yields (w/w) of 80-90% were attained under pyrolysis conditions and 40-58% under steam gasification conditions. The mass balance closures averaged 95% for the eight runs reported and heat balance closures averaged 94%. At 700 $^{\circ}$C reaction temperature, the pyrolysis gas product had a calorific value of 3.6 kcal/g and the steam gasification product had a calorific value of 3.4 kcal/g.

Keywords: Gasification, indirect radiant heating, pilot plant, pyrolysis, spouted bed.

1 Introduction

This work is the first description of the design and performance of a novel type of reactor. The Spouted Bed Gasifier (SBG) combines the fluid mechanics of a spouted bed with indirect heating technology [1] to achieve pyrolysis and gasification.

The SBG utilizes the action of a spouting bed [2] to achieve a high degree of gas-solid contact and to provide more control over solids residence time and heat flux effects in pyrolysis and gasification. A spouted bed may be considered a variation of a fluidized bed, with a central, ascending gas-solid jet and an annular, descending moving solids bed. In this reactor, the fluidizing media, along with the feed particles, is introduced through a small central opening in the conical bottom of the reactor. The reactor is typically cylindrical and characterized by large height and relatively small diameter. The fluid entering the bottom inlet carries the solid particles upwards. An individual particle travels upward until it elutriates from the bed or until its velocity drops to zero. The particles that do not elutriate then travel downward in the annulus between the wall of the reactor and the high-velocity, high-voidage central core. A particle can enter the spout again or get discharged at the bottom.

Heating zones can be placed in the fountain region or within the solid bed, or both to provide the heat required for pyrolytic decomposition and gasification. The heaters may be electrically powered or they may be gas burners, utilizing a portion of the syn-gas product.

The inherent advantages of this approach are as follows:

1) The high degree of control of reaction temperatures and residence times helps to better separate the effects of drying, heating and reaction within a single reactor.

2) The recycling action inherent in the spouted bed increases the total residence time of the solids and obviates the need for external recycling.

3) The short residence times and the ability to maintain high temperatures within the heating zone permit selective decomposition of the feed constituents, when desired.

4) Indirect heating is desirable for pyrolysis and gasification because there are no combustion products in the final synthesis gas.

5) Radiant heating as a component in the indirect heating regimen allows for much higher heating rates than those possible by convection or direct contact heat transfer alone.

6) The SBG also allows for the addition of catalytic components, if desired, to the reactor system. The reaction zone chemistry can be controlled by the addition of steam, CO_2, or other reactive gases, as desired.

2 The pilot plant

The major effort of the current work involved the design and construction of a process development unit capable of processing about 250 kg/day of organic feed stock [3]. All of the heaters have not been installed at this time and the operating capacity is consequently lower at 125 kg/day. Also, current limitations with the solids feeder caused the plant to be tested at a feed rate under 1.5 kg/hr. The system is designed to operate in the range of 300-800 $^{\circ}$C and at pressures not exceeding 3.04×10^{5} Pa (3 atm). The process flow diagram for the plant is shown in Fig. 1.

2.1 Overall process flow

At the core of the pilot plant is the spouted bed reactor. The fluidizing media is a mixture of steam and nitrogen. This mixture is preheated to about 250 $^{\circ}$C in a preheater and charged into the gasifier. Solid organic feed particles are introduced, along with the fluidizing media. The electrical heating elements within the reactor provide the thermal energy for the pyrolysis and gasification processes to occur. The chief product is a gas stream rich in carbon monoxide and hydrogen (other gaseous constituents are CH_4, CO_2, and low molecular weight hydrocarbons).

The downstream section includes a cyclone separator to remove fine solids carried over in the gas stream. The gas then passes through condenser-separator units, wherein cooling occurs and the condensible organics (tars) are removed from the gas. The product gas stream is metered before exiting the facility. Ash and char produced in the reactor are collected in the ash collector at the base of the unit.

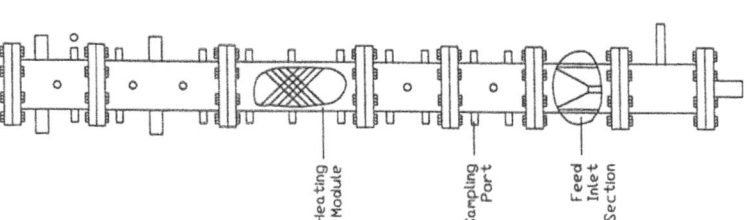

Fig. 2. The spouted bed reactor

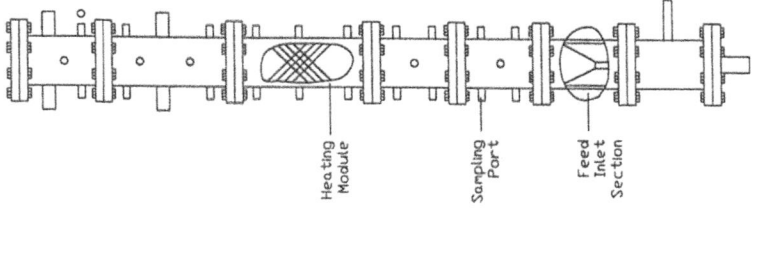

Fig. 1. Process flow diagram for gasification

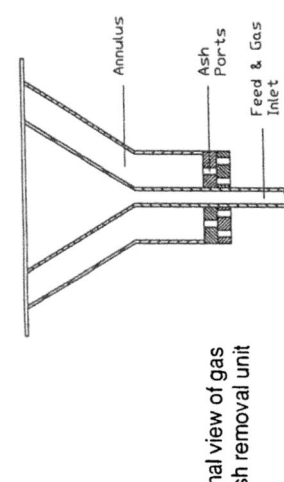

Fig. 3. Sectional view of gas
inlet/ash removal unit

2.2 Design and construction of the spouted bed reactor

The spouted bed reactor shown in Fig. 2 was constructed of a 0.203 m (8 inches) dia., schedule 80 steel pipe and was manufactured by Haynes Engineering Corporation, Norcross, Ga. The height of the reactor could be varied from 2.43 to 3.65 m (8 - 12 feet) by adding various modular sections as depicted in Fig. 2. The modular design allows for a high degree of flexibility in terms of location of the drying, pyrolysis, gasification and solids disengagement zones. The amount and composition of the feed stream and the desired product distribution have an important bearing on the placement of these zones. The modular units were coated with a special ceramic coating, which provides high reflectivity to the internal surfaces and increases internal reflection of the radiation from the heating elements. The reactor was insulated with a four-inch layer of Duraback (Carborundum Co., Niagara Falls, N.Y.) alumina-silica fiber insulation.

The reactor has two heating modules. Each module consists of two banks of Chromalox TRI-1612 WExx (Chromalox Industrial Div., Pittsburgh, Pa) electrical heating elements, with a total of 41 heating elements. The elements are set in an 'X' configuration at an angle of 60° to the horizontal and spaced 2.5 cm from each other. Each heating element is rated at 350 W and 240 V, for a maximum of 14.35 kW per module. The tubular heating elements (0.64 cm dia.) are Incoloy sheathed and can withstand sheath temperatures up to 870 °C. Each element has a length of 41 cm and a heated length of 23 cm.

For biomass pyrolysis, a heat requirement of 2000 J/g has been estimated [4]. Considering the 14.35 kW of rated heating capacity available in the first module, the projected throughput works out to 25.8 kg/hr. This is a solids loading of about 800 kg/hr.sq.m and is comparable to the 260-860 kg/hr.sq.m loadings [5] achieved with the GROW reactor.

A feature of the reactor design is the stainless steel, conical feed unit at the base of the reactor (Fig. 3). It consists of a double cone arrangement. The feed, along with the fluidizing media, enters through the apex of the inverted cone, while dense ash particles drop through the annulus between the cones. The conical entrance establishes the flow jet that creates the spouted bed action in the reactor. Ash and char fall through the ash ports and then finally into the ash collector at the bottom of the gasifier.

2.3 Fluidizing media and the solids feed system

Steam was generated in a Sussmann ES-18 electric boiler (Sussmann Electric Co., Long Island City, N.Y.) rated at 17 kW. The steam was available at a maximum pressure of 90 psig and 100 °C. A Brooks Hi-Pressure Armored Flow Indicator (Series 3600 rated at 1500 psig, 400 °F; Brooks Instruments, Hatfield, Pa), with an accuracy of ±10% full-scale, was used to measure the steam flow rate.

Nitrogen gas with a purity of 99.5% was sourced from liquefied nitrogen gas packs available from commercial suppliers (Linweld Co., Lincoln, Nebr.). The N_2 flow was measured using an RMC-102 variable area flowmeter (Dwyer Instruments Inc., Michigan City, IN). It has an accuracy of ±2% full scale and a maximum operating pressure of 35 psi. Dwyer bourdon tube type pressure gages (Spirahelic 7100-G60, 0-60 psi) with an accuracy of ±2% were used to monitor the pressures.

The steam and nitrogen, after metering, were mixed and preheated to a predetermined temperature. Sixteen Chromalox TRS-1612 WExx heating elements supplied up to 5.6 kW. The preheater was designed in 4 modules for simplicity and ease of removal/maintenance and to allow later modification.

Solid feed system: The solid feed in pelletized form (or directly, if of suitable physical form and size) was fed into the reactor by a Vibra Screw SCR-20 solids (Vibra Screw Inc., Totowa, N.J.) screw feeder. The feeder is equipped with an electronic variable speed drive and allows precision metering of solids. The feed hopper was purged with nitrogen to prevent backflow of fluidizing media.

2.4 Downstream product and byproduct collection systems

The downstream section of the plant has as its primary objective removing the solid and liquid co-products and cooling the product gases to room temperature.

A Kice CE-9 high-efficiency cyclone separator (Kice Industries, Wichita, Kan.) removes the ungasified char and carried-over fine ash. The cyclone was thoroughly insulated to ensure that the gas temperatures do not drop below 250 $^{\circ}$C and to prevent the organic condensibles from condensing on the walls.

In the next stage, a series of separator-condenser units removes the condensibles (water and tars) from the gas stream. A combination of condensation and impingement action help achieve this. The units are double jacketed. Either cooling water or steam may be passed thorough them to maintain each of these units in a desired temperature range. This allows for the collection of different fractions of the organic condensibles, based on their boiling points. The downstream product gases are metered prior to atmospheric venting. Dwyer RMC-108 variable area flowmeters measured the flowrate.

2.5 Temperature control and monitoring

The key parameter in the pyrolysis/gasification reactions and in the operation of the SBG is the temperature. The peak operating temperature is maintained by on/off control of the electric heaters in the heating module. The temperature controllers were Omega CN-76122 automatic process controllers that could operate in on-off or PID control modes. AC solid state relays were used as the final control elements.

The total energy consumed for preheating the fluidizing media and that consumed by the elements in the heating module were measured by two Schlumberger J5S cyclometer type watthour meters (Schlumberger Electricity, Norcross, Ga.).

The temperature was continuously monitored at various points in the system, including the feed streams, the downstream sections and within the reactor itself. K type, grounded thermocouples (Omega KQIN-14G-18) provide input to a Fluke Hydra electronic data bucket (John Fluke Manuf. Co. , Everett, Wash.). The data bucket can monitor up to 21 channels and was linked to a personal computer for continuous on-screen monitoring of the temperature profiles.

3 Preliminary performance studies

In this part of the work, studies were performed to characterize the hydrodynamics and the thermal stabilization/control aspects of the system. An understanding of these factors is essential for further operation of the plant.

3.1 Hydrodynamic studies

The studies were performed in the unheated unit with pine chips (+1,-2 mm size). The observations were made by viewing the bed condition from the top of the reactor. Bed heights in steps of 5 cm were used. The findings are presented in Fig. 4. The gas flow rates were monitored using flowmeters as described earlier and corrected for temperature and pressure variations. The pressure drop across the bed was monitored using 'U' tube water manometers.

The incipient spouting velocity (U_{is}) is the superficial gas velocity just prior to the break in bed resistance and the onset of fluidization/spouting. The minimum spouting velocity (U_{ms}) gives the minimum fluid requirement, below which spouting ceases. The velocities represent the limits of spouting as shown in Fig. 4.

While both the velocities are bed history dependent, they were found to be definitely reproducible (to the order of $\pm 10\%$). It was also found that these velocities did not vary appreciably (about 5-10% higher when closed) with the reactor top being open or closed to the atmosphere. This allowed for open-top viewing of the bed condition.

At the incipient spouting velocity, the fluidization was very heavy and spouting indistinguishable. The spouting was also accompanied by spurts of erratic spout formation reaching almost to the top of the reactor. The spout was better defined and highly stable at superficial gas velocities lower than U_{is}. The region between curves B and C in Fig. 4 represents the best operating range. In a 'hot' reactor, the incipient spouting velocity would be observable from the sharp drop in manometric pressure, and then it would be possible to adjust the gas velocity to lie in the optimum zone.

At bed heights greater than 35 cm, the bed reached the heaters and the bed resistance was too high to permit spouting without exceeding the design pressure at the gas inlet. For deeper beds it may be necessary to move the heating module higher up the reactor. At the lower end, there was no spout formation at a bed height less than 10 cm. The pressure drop across the bed, under stable spouting, was a constant for any given bed height. Information on pressure drops and on pressure drop v/s superficial velocity hysteresis loops can be found elsewhere [3].

3.2 Thermal stabilization and control

The characteristics of the system such as the time taken to attain thermal steady state, the temperature response to changes in gas flow and the temperature controller performance were studied.

Steady state is assumed to have been attained when the temperature changes by less than 10 °C per hour at all points in the system. For all the flow rates studied (200-800 SCFH), it took about 7 hours or more to attain steady state, though the peak temperature in the heating zone was attained in about an hour.

The system response to change in flow was observed by suddenly increasing the gas flow rate and recording the response. The controller held the temperature at its set point to $\pm 1\%$. However, the reactor exit gas temperature changed immediately and took about 6-7 hours to reattain steady state. The long time required to attain equilibrium is due to the large thermal mass of the system.

4 Pyrolysis and gasification of wood

The final part of the current work involved studies on the gasification of wood. The objectives of this part were 1) to test the working of the pilot plant and its subsystems, 2) to obtain good material and energy balances and 3) to study the effect of some of the controllable parameters (namely, temperature and steam addition) on the pyrolysis and gasification processes.

4.1 Feed material: Preparation and properties

Pine wood chips were obtained from a pallet manufacturer in Nebraska. According to the manufacturer, the logs were sourced from the U.S. Northwest and from Canada. It was estimated that the chips contained about 95% pine, with the balance being oak.

An Aggra-Test CL-400 (Soil Test Inc., Evanston, Ill.) sieve shaker was employed to obtain feed particles in the size range +1, -2 mm. Table 1 lists the particle size distribution. The material was processed four times (resulting in about 65% rejection of raw material) to remove as much of the oversize particles as possible. The oversize particles were usually needle shaped. They tended to clog the screw feeder and affect the solid feed rates. The prepared feed was then weighed and charged into the feed hopper. Random samples of the feed were taken before each run and analyzed for particle size distribution, moisture, carbon and ash contents. Table 2 presents the characteristics of the feed stock. The moisture content was measured by recording the weight loss of a sample maintained at 105 °C for 24 hours. The ash, carbon and nitrogen contents were analyzed at Midwest Laboratories Inc., Omaha, Nebr.

4.2 Reactor operation

A typical run consisted of three stages of operation, namely 1) the approach to steady state, 2) the steady state tests and 3) plant shutdown.

1) Approach to steady state: Initially, nitrogen alone was allowed to flow through the reactor with all of the heaters turned on. The gas was preheated to a desired set point (in this case, 250 °C) prior to entering the reactor. The temperature limit was set at the Peak Operating Temperature (POT) required for the run. The POT is defined as the maximum measured temperature just above the heater zone. Higher

Table 1. Feed particle size distribution

US. Std. Sieve	Size Range mm	Mass Percent Size Distribution						
		Run 5	6	7	8	9	10	Average
8	>2.28	0.6	0.4	0.4	0.5	0.1	0.1	0.4
10	+2,-2.28	1.3	2.4	1.2	3.0	0.3	0.4	1.4
14	+1.41,-2.0	37.1	37.8	43.0	36.9	28.3	31.5	35.8
16	+1.18,-1.41	27.0	26.0	26.3	24.0	30.0	29.4	27.1
18	+1.0,-1.18	20.9	19.6	17.2	20.4	20.3	21.0	19.9
35	+0.5,-1.0	12.8	13.6	11.6	13.7	18.8	17.0	14. 6
pan	<0.5	0.3	0.3	0.17	1.6	2.2	0.5	0. 8

Table 2. Feed characteristics

Bulk density of pine chips	:	0.16 g/cc
Size	:	1-2 mm
Moisture content	:	4 - 10 % (by weight of as-obtained feed)
Ash content	:	0.78% (by weight of as-obtained feed; AOAC 942.05)
Carbon content	:	52.5 % (dry weight basis; method ASA 29-2.2.4)
Nitrogen content	:	0.04 % (by weight of as-obtained feed)

temperatures exist only on the surface of the reactor and in between the heaters.

Steam flow calibration was also conducted during the first phase of operation. After the temperature of the gas entering the condensers exceeded 200 °C, steam was introduced and preheated along with nitrogen. Steam rates were measured by condensing and weighing steam in 10 minute intervals, over an hour.

Similarly, the gas flow rates at the entrance and exit of the plant were monitored to determine and calibrate the gas leakage. The design of the partition moving levers in the solid fines collection boxes caused some gas to leak (about 5%).

2) *Steady state tests:* When the reactor reached steady state (after about 6-7 hours), feed introduction was begun. Another hour was allowed for the reactor to attain a further steady state. No other procedures were necessary. At three 30- minute intervals, steady state data were taken. The data included the power consumed, pressure drop across the bed, inlet and exit gas flow rates and the liquids collected. The ash box and cyclone fines box partitions were moved to collect steady state material and then moved back after the end of all steady state periods. The solids could be collected only for the entire steady state period and were time averaged for half hour intervals. Also, at the three half hour intervals, product gas samples were obtained in gas collection bags for composition analysis. Longer steady state tests were not possible at this time due to limited feed hopper capacity.

The operational behavior of the reactor itself was highly satisfactory under the tested conditions. However, a few problems were encountered in the operation of some subsystems. The most critical problem encountered was the carryover of organic liquids and water vapor beyond the water cooled condensers, even though the gases were cooled down to room temperature. The carried-over vapors tended to condense on the walls of the rotameter, rendering it inoperable. This was successfully rectified by installing additional condensers kept immersed in an ice bath. This arrangement cooled the gases to around 10 °C and helped condense out almost all of the carried-over vapors.

The solids feeder had a maximum capacity of about 1.5 kg/hr. At this low feed rate, it was not possible to form and sustain a bed within the reactor because the feed immediately pyrolyzed and disintegrated. Consequently, fully developed spouted bed operation was not studied in this work. The low-capacity feeder was used because it was already available and the decision had been made to install a higher capacity unit after performing preliminary operational studies.

All other subsystems performed very well. The nitrogen feed system, the preheater and the solids feeder operated exceptionally well. The cyclone separator was efficient in removing about 99% of the fines. The digital temperature controls and data acquisition components operated reliably and flawlessly.

4.3 Pyrolysis and gasification

The main variables analyzed were the peak operating temperature and the effect of steam. The solids feed rate and the inlet gas temperatures (at 250 °C) were kept constant for all the runs. Initially, experiments were carried out at three temperatures (500, 600 and 700 °C), in the absence of steam. The experiments were repeated later with steam in the approximate ratio of 1.25 parts of steam to 1 part of feed (by weight). The temperatures were chosen to provide a spectrum of data in between the limits (400 - 800 °C) of pyrolysis and the initiation of char gasification. Runs were repeated at 700 and 500 °C to confirm the repeatability of performance.

Table 3 presents the mass balance data for the runs. Mass balance closures averaging around 95% were achieved. Minor gas leaks and the accuracy of the gas flow meters at $\pm 2\%$ contribute to the current closure level. The repeat runs 11 and 12 (Table 3) provide evidence for good reproducibility of product distribution and composition.

Figures 5 and 6 show the effect of temperature on the gas, char and organic liquid yields. The gas yields were calculated using the product gas as obtained by difference (Table 3). As expected, the highest gas yields occurred at the highest operating temperatures, with correspondingly lower liquid yields. However, it was observed that the addition of steam profoundly increases the liquid yield, at the cost of gas production. This indicates that the steam-char reactions do not occur at the temperatures studied. This is consistent with the findings of Garrett [6] who found that temperatures higher than 800 °C are necessary for steam-char reactions. Antal [7] found that with gas phase residence times of 0.5 to 1 s and at temperatures of 500-750 °C, the gasification process is dominated by cracking reactions and not by

Table 3. Material balance data (in grams/ hour)

Component	Run 5	6	7	8	9	10	11	12
Input								
Nitrogen	14040	17182	15580	14451	15390	14244	14669	15426
Feed	1266	1076	1078	1053	1158	1054	1009	1010
Steam	0	0	0	1336	1462	1470	0	0
Output								
Nitrogen	14040	17182	15580	14451	15390	14244	14669	15423
Oil[a]	79	152	69	1592	1906	2013	63	247
Char	23	71	63	173	159	42	67	70
Cyclone fines	8	16	3	11	5	3	2	2
Product gas	1249[b]	1054	1043	692	586	300	852	488
	(1156)[c]	(837)	(943)	(613)	(550)	(466)	(877)	(691)
Closure[d] %	107	120	109	103	101	93	98	80
Temperature[e]	600	500	700	700	600	500	700	500

a) Includes pyrolytic tars and water.
b) The values are calculated from observed exit gas flow rates and gas composition.
c) Values in parentheses refer to the mass obtained by difference of all other outputs from the inputs.
d) Output x 100 / input; Nitrogen free basis; Observed exit gas flow rates used to calculate output.
e) Refers to the peak operating temperature in °C.

steam reforming. This conclusion is further supported by the higher char residue in the runs with steam. The effect of decreased gas yields with steam addition was also observed by Beck et al. [8], who postulated that the steam served as a heat sink and reduced the reaction temperature of the particles.

The low gas yields and the higher liquid yields suggest that steam addition under the current operating conditions is beneficial for maximizing pyrolytic oil yields. Gas residence times were of the order of 15 s within the reactor. Shorter residence times may further increase tar yields at the cost of gas yields, but this remains to be experimentally observed in our reactor. The operating conditions of the present trials were such that the feed was subjected to 'flash pyrolysis' (450-650 $^{\circ}$C with heating rates of the order of 100 to 200 $^{\circ}$C/s) and to low temperature steam gasification (700 $^{\circ}$C). Flash pyrolysis conditions may be optimized for tar production with much shorter residence times and higher heating rates (both of which are achieved by high fluidizing media flow rates). Piskorz et al. [9] have reported that temperatures of 450-550 $^{\circ}$C and short residence times of 0.5 s are optimal for tar maximization.

The product yields in relation to flux are shown in Fig. 7. Tar yields of the order of 5-15% were obtained as shown. These values agree very well with those predicted by Clements [1] for tar yields as a function of heating rate. No analysis of the pyrolytic liquids was carried out, though preliminary indications are that the pyrolytic liquids contain significant amounts of water soluble components.

The product gas was analyzed for its components (H_2, CO, CO_2 and CH_4) by gas chromatography (with TCD). The analysis was conducted by Midwest Laboratories Inc. , Omaha, Nebr. Higher hydrocarbon gases were not analyzed for at this time. It was assumed in all subsequent calculations that these four gases made up all of the product gas. Compositions are listed in Table 4.

High concentrations of CO were obtained in the runs without steam. Addition of steam, however, increases H_2 and CO_2 concentrations. The results indicate that the reaction temperature is a significant parameter in determining the gas compositions. A consistent trend observed in both sets of runs (with and without steam) is the higher change in concentration of CO and H_2 between 500 - 600 $^{\circ}$C, as compared to the change between 600 and 700 $^{\circ}$C.

A better understanding is obtained by comparing yields (g per g feed; N_2 free basis) as in Fig. 8 and 9. Carbon monoxide and CO_2 yields show a high degree of dependence on the temperature. Hydrogen yields increase with temperature and with addition of steam. Carbon dioxide is an undesirable component in the product gas, and the total non-CO_2 gas yields are clearly higher in the fast pyrolysis runs than those obtained with steam addition at all temperatures.

In Table 4, a comparative listing of data from other studies is included. The product yields and compositions depend to a great extent on numerous factors such as feed size and nature, heating rate, solid and gas residence times, temperature, and fluid media. This makes it very difficult to compare the results. In Table 4, the work of Brink and Flanigan come closest to the operating conditions of the current work. The gas yields from this study were higher than those of Brink for pyrolysis, but were less than those of Flanigan for steam gasification.

Energy Balance: Table 5 lists the energy balance data for the system. The heat loss data was obtained by separate calibration runs [3]. The estimated heat loss (H)

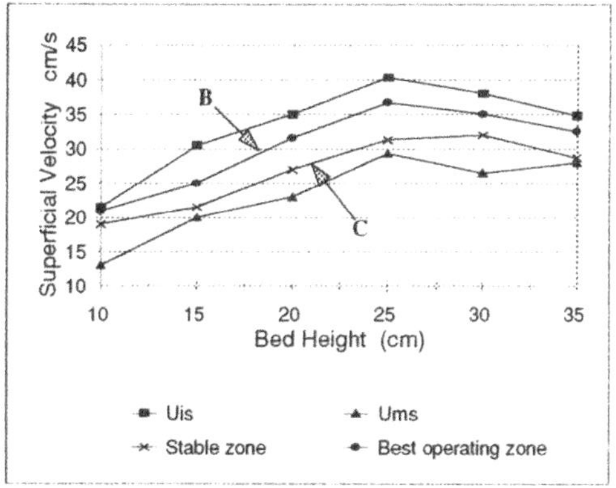

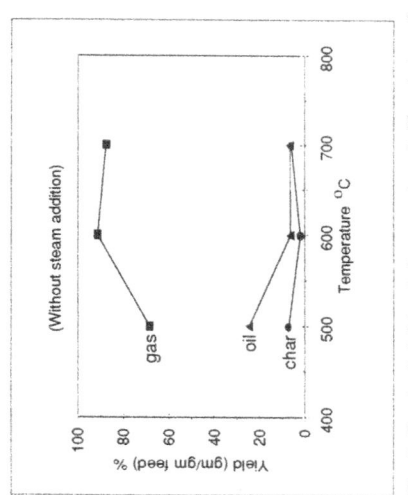

Fig. 5. Product yields v/s temperature (without steam)

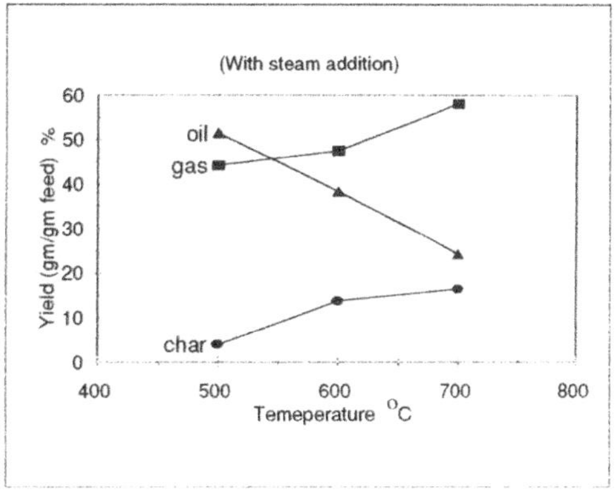

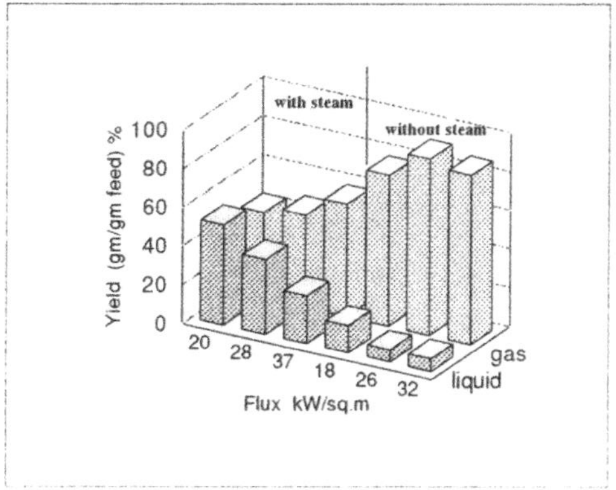

Fig. 7. Effect of flux on product yields

Fig. 4. Spouted bed performance v/s bed heights

Fig. 6. Product yields v/s temperature (with steam addition)

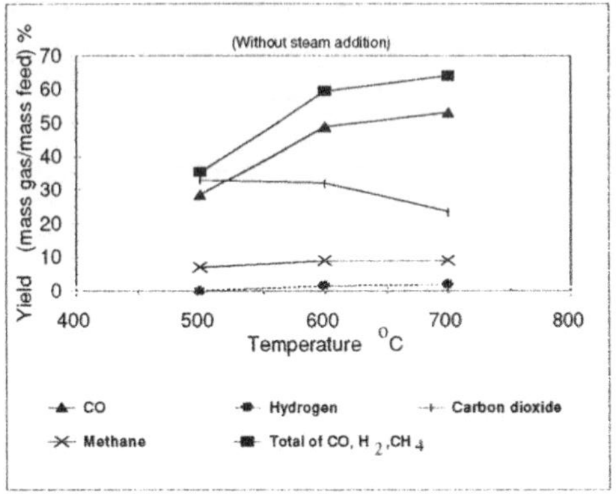

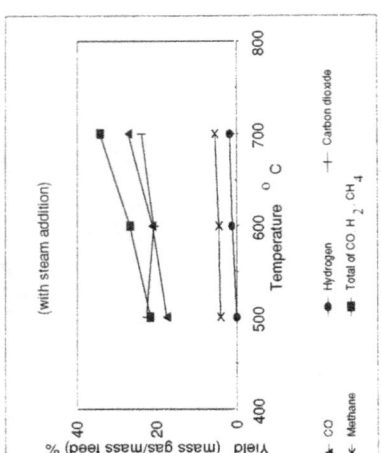

Fig. 9. Gas component yields v/s temperature

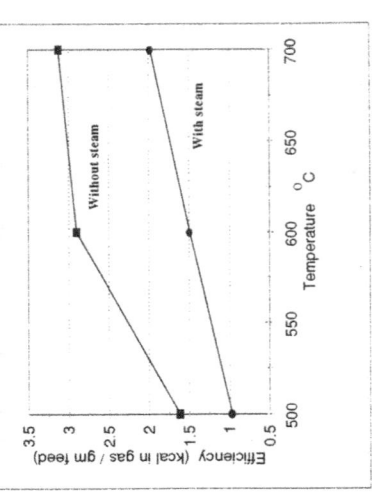

Fig. 11. Effect of temperature on gasification efficiency

Fig. 8. Gas component yields v/s temperature

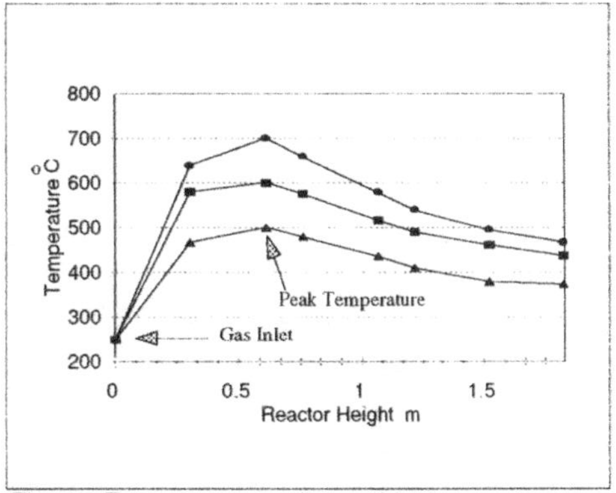

Fig. 10. Temperature profiles within the reactor

Table 4. Comparison of gasification studies*

Author	Heat Rate °C/s	Res. Time s	Feed Size mm	Temp °C	Yield %			Gas Composition %				kcal gas/gram feed
					Gas	Oil	Char	CO	H_2	CO_2	CH_4	
Pyrolysis This work	100-200	15	1-2	500	80	14	7	46	0	34	20	1. 6
				600	90	6	2	46	20	19	15	2. 9
				700	87	6	6	48	24	14	14	3. 1
Brink [10] flow reactor		3-5	. 4 - .8	488	60	20	8					
				604	70	20	6	65	0	22	13	2. 7
				704	80	2	12	50	34	3	11	3. 8
Hajaligol [11] static sample			0. 1	500	5	25	15					
				600	15	60	10	68	0	25	3	
				700	25	65	<5	65	5	18	6	
Steam Gasification This work	100-200	15	1-2	500	44	52	4	45	0	37	18	1. 0
				600	40	38	14	36	28	23	13	1. 5
				700	58	24	16	35	32	20	13	2. 0
Flanigan [5],indirect heat, fluidized bed				600	73	14	10	37	15	24	14	2. 0
				735	77	10	11	42	26	16	9	2. 4
Hightower [12] steam fluidized				690			1.5	45	14	30	7	2.8

Values are approximate as they have beenadapted and rounded off to facilitate ease of comparison

from the reactor and the cyclone ranged from 8,000 to 14,000 kJ/hr, depending on temperature. It is described by H = 27.75T - 5638, where H is in kJ/hr and T is in °C. The heat balance was calculated around the reactor, including the cyclone. Char was assumed to leave the reactor at the temperature of the inlet gas. The cyclone fines, gases and other liquids were assumed to be at the cyclone exit temperature. The methods used to estimate the heat capacities of the char and the pyrolytic oils are explained elsewhere [3]. Heat balance closures in the range of 87-100% were obtained. Though it was expected that the heat balance data would show evidence of exothermic reactions, that was not the case. It is believed that the low resolution of the energy meters, along with the estimation methods used for calculating the heat capacities of the solid and liquid products, led to this result.

In Fig. 10, the temperature profiles existing within the reactor are presented. The heating zone is present in between the 0.3 and 0.6 m level of the reactor. The rapid fall between the 0.6 and the 1.2 m level is normal and expected due to heat losses to the surroundings. The value at the 0.3 m level indicates the temperature within the bed, or within the spouting zone, depending on the bed height. In all of the runs, the temperature at this point dipped slightly upon feed introduction and then gradually increased to a higher value. This leads us to believe that, first, the temperature falls as the solids initially take up heat and later, the temperature rises due to exothermic heat of reaction. This agrees with the exothermic heat of reaction noted by Hightower [12].

Table 5 Energy balance data (in kJ/hr)

Component	Run 5	Run 6	Run 7	Run 8	Run 9	Run 10	Run 11	Run 12
Input[a]								
Nitrogen	3260	3995	3620	3360	3580	3310	3410	3585
Feed	0	0	0	0	0	0	0	0
Steam	0	0	0	3973	4348	4372	0	0
Heaters	14040	11160	16560	16560	14760	10800	15480	9000
Total	17300	15155	19820	23893	22688	18482	18890	585
Output[a]								
Nitrogen	3630	3815	4277	4120	3886	3087	3704	3003
Product gas	1632	1087	1406	736	820	539	1427	813
Liquids[b]	66	109	61	4343	4746	4710	51	156
Char +fines	17	48	36	101	90	25	38	40
Heat loss	11167	8160	13710	13710	11167	8160	13710	8160
Total	16512	13219	19490	23010	20709	16521	18930	12172
Closure[c] %	95	87	98	96	91	89	100	97
Temperature°C	600	500	700	700	600	500	700	500

a) Reference temperature : 25 °C   b) Includes water and organics   c) output*100/input

Gasification efficiency: A primary motive for this work is to maximize conversion with maximum gas quality (heating value). Gasification efficiency is defined by us as the energy content of the product gas per unit mass of 'as-obtained' feed. It is an important factor with critical influence on the process economics. The highest efficiency was observed for pyrolysis (Fig. 11) 700 °C. The gasification efficiency is comparable to those attained by other researchers (Table 4). Table 6 lists the calorific value of the product gas. It should be noted that for a given temperature the calorific values are relatively the same, irrespective of steam addition, though the gasification efficiencies are not (due to differing gas yields).

Table 6.   Calorific value (HHV) of product gas (kcal/g; N_2 free basis)

Temperature °C	Product Gas Calorific Value	
	Pyrolysis	Steam Gasification
500	2. 36	2.19
600	3. 18	3.14
700	3. 57	3.4

5   Acknowledgments

The financial support of the Nebraska Energy Office, the Western Area Power Association, the Southeast Regional Biomass Energy Program, HEC Environmental Inc. of Norcross, Ga and the University of Nebraska Agricultural Research Division is gratefully acknowledged.

6 References

1. Clements, L.D. (1985) *Indirect, Radiant Heating for Biomass Pyrolysis and Gasification*, Biotechnology and Bioengineering Symp. No.15, John Wiley & Sons, Inc., pp. 91-98.
2. Mathur, K.B. and Gishler, P.E. (1955) *A Technique for Contacting Gases With Coarser Particles*, AIChE Journal, 1(2), pp.157-164.
3. Janarthanan, A.K. (1996) *Design, Construction and Operation of a Pilot Scale Spouted Bed, Biomass Gasification Plant*, M.S Thesis, University of Nebraska-Lincoln, U.S.A.
4. Reed, T.B., Diebold, J.P., and Desrosiers, R.E. (1980) *Perspectives in Heat Transfer Requirements and Mechanisms for Fast Pyrolysis*, Proceedings of the Specialists' Workshop on Fast Pyrolysis of Biomass, SERI/CP-622-1096.
5. Flanigan, V.J., Punyakumleard, A., Sineath, H.H. and Sitton, O.C. (1984) *20 Fire Tube Gasification System*, Proc. of the 16th Biomass Thermochemical Conversion Contractor's Meeting, pp.83-102.
6. Garrett, D.E (1977) *Factors in the Energy Efficiency of Biomass Processing and Disposal*, 70th Annual AIChE Meeting, N.Y.
7. Antal, M.J. (1981) *The Effects of Residence Time, Temperature and Pressure on the Steam Gasification of Biomass*, Biomass as a Nonfossil Fuel Source, ACS Symposium Series 144, pp.313-334.
8. Beck, S.R., Wang, M.J. and Hightower, J.A. (1981) *Gasification of Oak Sawdust, Mesquite, Corn Stover, and Cotton Gin Trash in a Countercurrent Fluidized Bed Pilot Reactor*, Biomass as a Nonfossil Fuel Source, ACS Symposium Series 144, pp.334-349
9. Piskorz, J., Radlein, D.ST.A.G. Scott, D.S, and Czernik, S. (1988) *Liquid Products from the Fast Pyrolysis of Wood and Cellulose*, Research in Thermochemical Biomass Conversion, ed. Bridgewater, A.V. and Kuester, J.L., Elsevier Applied Science, pp. 557-571.
10. Brink, D.L and Massoudi, M.S. (1978) *A Flow Reactor Technique for the Study of Wood Pyrolysis*, Journal of Fire & Flammability, Vol.9, pp.176.
11. Hajaligol, M.R., Howard, J.B., Longwell, J.P. and Peters, W.A. (1982) *Product Compositions and Kinetics for Rapid Pyrolysis of Cellulose*, Industrial and Engineering Chemistry Process Design and Development, 21, pp.457-465.
12. Hightower, J.A. (1979) *Fluidized Bed Gasification of Agricultural Residues*, M.S Thesis, Texas Tech University, U.S.A.

APPLICATION OF GASIFICATION TO THE CONVERSION OF WOOD, URBAN AND INDUSTRIAL WASTES

N. ABATZOGLOU[1,2], J.-C. FERNANDEZ[2,3], L. LARAMÉE[2,4], P. JOLLEZ[1,2] and E. CHORNET[1,2,5]

[1]Kemestrie, Inc., Sherbrooke (Québec), Canada
[2]Université de Sherbrooke, Sherbrooke (Québec), Canada
[3]Ph.D. student, on leave from Instituto de Tecnologia y Modelización Ambiental, UPC, Terrassa (Barcelona), Spain
[4]Ph.D. student at the Université de Sherbrooke, Sherbrooke (Québec), Canada
[5]Also affiliated with the National Renewable Energy Laboratory, Golden (Colorado), U.S.A.

Abstract

Gasification is widely accepted as a technological option for the production of synthesis gas (SG) via partial oxidation of heterogeneous organic matter such as, residual biomass, classified urban wastes (RDF), autofluff, residual non-recyclable plastics, rubbers as well as other industrial organic wastes. The "producer gas", after an appropriate conditioning step, whose nature depends upon its final intended utilisation, can be used either directly or mixed with natural gas (NG) for the generation of electricity and/or heat.

The energy and environmental performances of the BIOSYN Atmospheric Fluidized Bed Gasifier (AFBG) will be presented and discussed in this paper.

These advances have led to the construction and operation of a demonstration unit for the simultaneous recovery of aluminum and energy from post-consumer residues (i.e. packaging material). These developments allows us to consider gasification as a valuable option for the similtaneous in situ or ex situ disposal of wastes and energy recovery.

Keywords: Gasification, Wood, Waste, Fluidized bed, Gas conditioning, Wastewater treatment.

1 Introduction/Objectives

The vision of the gasification of organic residues within the frame of the sustainable development and the "zero effluent" principle is well represented in Illustration 1. Our approach is based on the following choices:

1. Offer both wet scrubbing and hot filtration/reforming options for increased flexibility.

2. Minimization of wet scrubbing tarry liquid effluents by:
- efficient tar/water separation;
- water insoluble heavy tar recycle back to the gasifier;
- scrubbing water reuse until saturation with dissolving organics;
- use of wet oxidation (OXYJET) to eliminate 80-90% of the organic charge in the tarry waters (1);

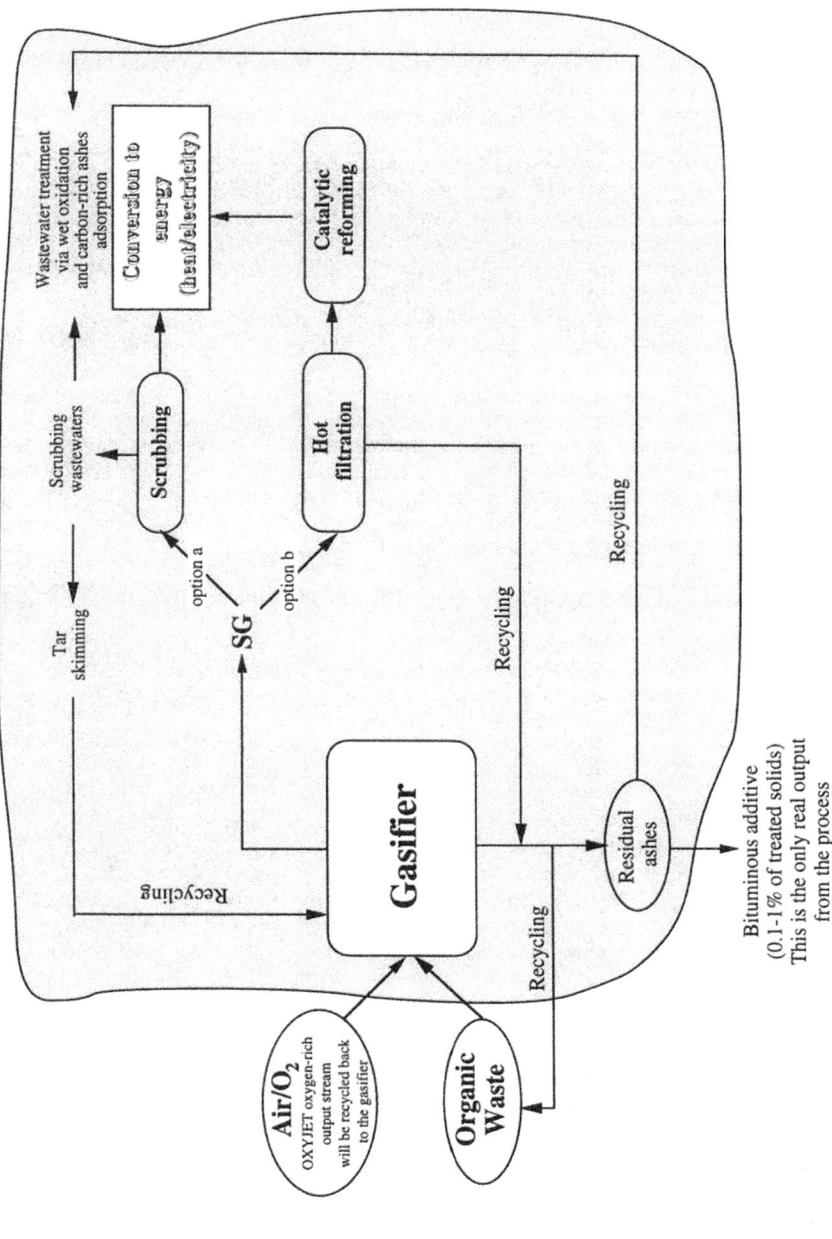

Illustration 1 : The gasification option

- use of the carbon-rich gasification ashes to complete the detoxification treatment of the tarry scrubbing effluents (2); 10 % of the ashes being purged from the system in order to avoid metals and inorganic anions concentration built-up in the gasifier.
- recycling of the carbon-rich ashes back to the gasifier after their use in the adsorptive treatment of the scrubbing/OXYJET effluents.

The Process Development Unit (PDU) used for this work is based on the BIOSYN technology and has been previously described (3, 4, 5).
Schematic representations of new gas conditioning and co-combustion modules are given in Illustration 2.
The work presented in this paper has focused in the following objectives:
- Performance of a 50kg/h (d.b.) atmospheric fluidized (bubbling) bed gasification unit, using both air or oxygen-enriched air as fluidizing and gasifying agents. Mass and energy balances are presented. Namely the High Heating Value (HHV) as well as the energetic efficiency of the AFBG are presented as function of the operational variables and parameters of the gasification runs.
- Design characteristics and performance of the wet scrubbing system.
- Tarry liquid treatment using wet oxidation (OXYJET) and adsorption on carbon-rich ashes.
- Energy and exergy performances of a commercial natural gas (NG) burner/boiler used for heat cogeneration with mixtures of synthetic gas and natural gas.
- Isokinetic sampling at: a) the gasifier and commercial burner/boiler stacks and b) before and after the producer gas conditioning module, in order to evaluate the environmental performances of the proposed gasification option.

The organic feedstocks tested so far have been:
- Woody biomass;
- RDF composed of paper, wood, plastic (PE and PVC) and putrescible organic matter;
- Rubber residues.

Moisture in the different feedstocks varied between 3 and 30% w/w.
Composition and analysis of all feedstocks are available but not presented here due to space limitations.

2 Methodology

2.1 The AFBG unit
The AFBG PDU, which has a nominal capacity of 50kg/h (d.b. solids), is based on the BIOSYN technology. This unit was designed, constructed and initially operated by researchers at Hydro-Quebec. Since 1990 our research group has been adapting the technology to waste conversion in the range of 0.5-5.0 tonnes/h. The gasifier operation has been successfully simulated using a mathematical model (6). The model predicts satisfactorily the behaviour of the unit for various feedstocks with different granulometries and calorific values; improvement of the model, mainly in the expressions for the intrinsic kinetics, is under way. A schematic of the gasifier has already been discussed in the past (2).
The new gas conditioning module and SG/NG combustion system, which are the focus of this paper are presented in Illustration 2.

2.2 The wet scrubbing system
The producer gas, after passing through two cyclones in series, where about 90% of its particulate matter is removed, is directed to a wet scrubbing system composed of a cooling tower in series with a venturi scrubber and a cyclonic demister (7, 8); in this conditioning step the gas is quenched to about ambient temperature while removing tar,

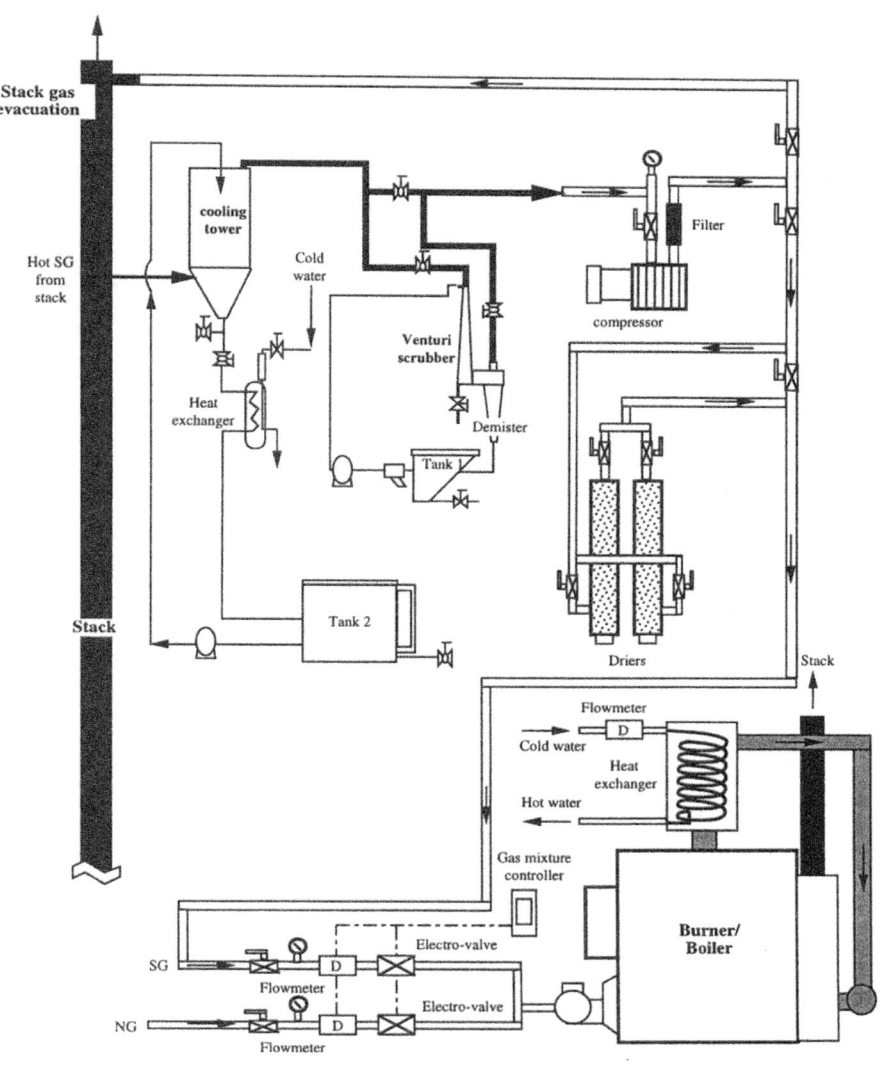

Illustration 2. Synthesis gas conditioning (scrubbing) and co-combustion

condensable light organics and inorganic particulate matter, including alkalis and volatile toxic metals (i.e. Hg, Pb). When gasifying organic matter with high acidic gas production potential (i.e chlore, sulphur), caustic alkali scrubbing can be applied. The wet scrubbing system is kept under vacuum induced by a "Liquid Ring" Compressor; it is designed to allow controlled recycling of the whole or part of the scrubbing water (see also Illustration 2).

2.3 The compressor and burner system
The scrubbed producer gas is just below atmospheric pressure when leaving the scrubber and, needs to be compressed prior to its mixing with natural gas, if co-combustion is the final purpose. In our unit a HIBON, A 01, Liquid Ring Compressor (2) is used to bring the SG at a pressure equal to that of the NG feeding line (about 32.4 kPa). This compressor, operating in series with two upstream parallel dessicator units (bed of hydroscopic alumina), is able to compress up to about 30 m3/h (at the suction side conditions; minimum absolute pressure of about 400 mmHg).

After the appropriate train of the gas mixing devices, including tubing, flow mass meters and controllers (2) the SG/NG mixture is fed to a commercial, Series "HXB-G-C3 & C4" FP Industries, burner, designed to operate with either natural gas or fuel no 2. The input capacity range of this burner is 450-1100 MJ/h. More details are given in (2). This burner is coupled with a Series "ES" FP Industries boiler, used to recover the heat generated by the combustion.

The system gas mixing/burner/boiler is equiped with the necessary thermocouples and pressure probes, connected with a HP data acquisition system; a data collection frequency of $0.1s^{-1}$ is chosen for the needs of our experimentation.

2.4 The experimental approach for emissions determination
Three isokinetic sampling trains, appropriately designed following EPA's standard methods (9, 10, 11, 12), and installed in the stack of the burner (3), are used.

The sampling combined with the appropriate analysis methods allows to:
- Calculate the gross composition of the stack gas (the H_2O, H_2, CO, CO_2, N_2, and O_2 content) using Gas Chromatography (combination of Porapak Q & Molecular sieve column packings) as well as the particulate matter load of the stack gas.
- Detect and quantify toxic and acid components via colorimetric methods (Matheson-Kitagawa's Kits) in the stack gas (i.e. SO_X, NO_X, HCl) (3).
- Identify and quantify the main organic components dissolved in the scrubbing water by means of GC/MS analysis.
- Measure the inorganic anions retained by scrubbing water using ion chromatography.
- Identify and quantify the distribution of metals in the output streams of the gasifier and its modules using X-rays fluorescence, AES/ICP, SEM and AA analyses.

Other analyses were used in order to gather additional information. These are as follows:
- Calorimetry for the different feedstocks tested.
- TGA for the study of the thermal stability/instability of the different feedstocks tested.
- TOC, COD and BOD_5 analyses for the scrubbing water before and after treatment
- BET analysis for the porosity of the carbon-rich gasification ashes used for the adsorption tests.

3 Results and discussion

3.1 Performance when using oxygen-enriched air as the gasifying agent
Tables 1 and 2 present typical results of the experimental runs conducted with both wood and RDF (RDF with either PE or PVC as plastic component). From these results it can be concluded that:
1. HHV is an increasing function of the oxygen (%) present in the gasifying air whithin the range of 20-50%.

Table 1 : Wood runs mass balances; the effect of O_2-enriched air

Runs Results	Wood (21% O_2) T=748°C	Wood (30% O_2) T=766°C	Wood (40% O_2) T=785°C	Wood (50% O_2) T=805°C
1A HHV (MJ/Nm^3)	5.6	8.6	9.3	11.2
2A Gas (Nm^3/kg)	2.2	1.8	1.3	1.1
3A Cold efficiency (%)	62.7	74.6	61.4	60.8
4A Tar+VOC (g/Nm^3)	5	10	10	90
5A Ashes (kg/h)	0.4	0.5	1	1
6A Water (g/Nm^3)	173	135	80	100
7A Carbon balance (%)	98	108	95	100
8A Solids feed rate (kg/h)	30	39	39	39
9A Humidity (%)	11	13	13	13
1B N_2 (% v/v, d.b)	58.7	43.1	35.7	28.3
2B H_2 (% v/v, d.b.)	2.1	2.7	3.1	3.3
3B CO (% v/v, d.b.)	17.8	26.2	28.1	33.9
4B CO_2(% v/v, d.b.)	14.1	17.5	21.5	20.4
5B HC (% v/v, d.b.)	6.2	9.7	10.7	12.9

Table 2 : RDF runs nass balances; the effect O_2-enriched air

Runs Results	RDF (21% O_2) T=690°C	RDF (30% O_2) T=770°C	RDF (40% O_2) T=765°C	RDF/PVC (30% O_2) T=770°C
1A HHV (MJ/Nm^3)	5.0	9.5	12.0	7.3
2A Gas (Nm^3/kg)	1.7	1.4	1.33	1.5
3A Cold efficiency (%)	41.0	83.5	78.7	54.0
4A Tar+VOC (g/Nm^3)	35	12	15	10
5A Ashes (kg/h)	0.4	0.2	0.5	0.2
6A Water (g/Nm^3)	310	180	230	285
7A Carbon balance (%)	99	105	115	104
8A Solids feed rate (kg/h)	31	49	45	37
9A Humidity (%)	10	10	12	13
1B N_2 (% v/v, d.b)	63.9	42.8	30.8	47.0
2B H_2 (% v/v, d.b.)	1.1	2.6	4.2	1.9
3B CO (% v/v, d.b.)	15.4	26.1	31.4	25.9
4B CO_2(% v/v, d.b.)	13.2	16.6	18.9	16.4
5B HC (% v/v, d.b.)	5.3	11.0	13.8	7.7

P.S. :   1. All runs at equivalence ratio λ=30% (+/- 2%)
2. A lines: Gasification results from mass and energy balances
3. B lines: SG composition

2. Wood and RDF runs gave similar results but RDF runs are conducted at lower temperatures and consequently give higher amounts of tars; this is mainly due to the lower calorific value of RDF compared to the woody biomass.
3. In spite of the higher HHV, SG obtained when the oxygen % in the gasifying air increases, it seems clear in both wood and RFD cases that the cold gas efficiency of the gasification, defined as the percentage of dry feedstock HHV recovered as HHV of the dry cold SG, attains a maximum at around 75%.

3.2 The effect of the equivalence ratio on the gasification performance

Wood and RDF feedstocks were also used for this study. Table 3 presents the effect of the equivalence ratio (λ) on the gasification performance using air as the oxidizing agent. It can be seen that the temperature of the run is a direct function of this λ. These results suggest the following:

1. HHV of SG increases as λ approaches around 30%. For higher λ values (42% was the maximium used) the HHV decreases. However the $\lambda=42\%$ point shows a very good efficiency (around 80%) despite the lower HHV (see also point 3.2.c).

2. Temperature also increases with λ leading to lower tar production.

3. The cold gas efficiency reaches 62.7% at $\lambda=0.31$ and it is enhanced to nearly 80% when $\lambda=0.42$. More work is under way to explore the $\lambda=0.30-0.45$ ratio where mixing effects can lead to an enhancement of the efficiency.

3.3 Co-combustion using mixtures of NG/SG

Wood and RDF feedstocks were used for this study. Table 4 presents typical energy and exergy balances for the system burner/boiler. It must be indicated that the producer gas used for these co-combustion experimental runs had an average HHV of about 5.6 MJ/Nm3 and was produced from gasification runs using air as the gasifying and fluidizing agent.

It can be concluded that:

1. One can easily consider SG/NG=1/1 co-combustion with energetic and exergetic efficiencies higher than 50%.

2. One can even consider SG/NG=3/1 co-combustion using commercialy available equipment (burners/boilers). Nevertheless, if one intends to use low HHV SG, as produced when using just air as gasifying agent, and SG/NG ratios higher than 1/1, some modifications are needed in order to improve the efficiencies and yields of the burner/boiler system, while decreasing the high CO production, typical of low HHV/low temperature combustion regimes.

Table 3: Wood runs mass balances; the effect of the equivalence ratio (λ)
Note: the temperature is fixed by the equivalence ratio

Runs	Results	Wood $\lambda=50\%$ T=822°C	Wood $\lambda=42\%$ T=780°C	Wood $\lambda=31\%$ T=748°C	Wood $\lambda=27\%$ T=680°C
1A	HHV (MJ/Nm^3)	2.7	5.4	5.6	5.3
2A	Gas (Nm^3/kg)	3.1	3.0	2.2	1.9
3A	Cold efficiency (%)	41.2	79.8	62.7	51.1
4A	Tar+VOC (g/Nm^3)	4	5	5	15
5A	Ashes (kg/h)	0.4	0.4	0.4	1.8
6A	Water (g/Nm^3)	172	159	173	200
7A	Carbon balance (%)	104	126	98	105
8A	Solids feed rate (kg/h)	35	35	30	20
9A	Humidity (%)	11	11	11	10
1B	N$_2$ (% v/v, db)	69.1	59.6	58.7	58.9
2B	H$_2$ (% v/v, db)	1.1	2.0	2.1	1.4
3B	CO (% v/v, db)	9.8	17.3	17.8	17.8
4B	CO$_2$(% v/v, db)	15.9	14.1	14.1	12.2
5B	HC (% v/v, db)	2.7	5.9	6.2	5.7
6B	O$_2$ (% v/v, db)	1.3	1.2	1.2	4.5

P.S.: All runs at equivalence ratio $\lambda=30\%$ (+/- 2%)

Table 4: Energy and exergy balances; the effect of the ratio natural gas over synthetic gas (NG/SG)

Runs	Results	NG/SG 100/0	NG/SG 75/25	NG/SG 50/50	NG/SG 25/75
1A	Boiler energy efficiency ε (%)	82.0	69.0	64.3	49.2
2A	Boiler exergetic yield η (%)	73.0	61.4	58.2	49.5
3A	Combustion exergetic yield η' (%)	53.1	51.9	46.8	27.3
1B	Flame temperature (°C)	853	805	640	256
2B	Energetic degree of combustion β (%)	99.9	99.3	98.9	98.8
3B	Exergetic degree of combustion δ (%)	99.3	99.4	99.0	98.8

3.4 Stack gases analyses as function of the ratio NG/SG

Table 5 presents the stack gas analyses for different co-combustion NG/SG ratios and for SG derived from either woody biomass or RDF gasification runs. Runs were chosen at similar SG HHV and λ in order to eliminate the influence of other variables in the final results.

The results indicate that:

1. The CO content of the burner stack gas is increasing as the ratio SG/NG increases. It is something expected since the flame temperature is decreasing as SG/NG is increasing.

2. The quantity of particulate matter contained in the boiler stack gas is, in each case, much lower than the upper environmentally acceptable limit (12, 13). Consequently the amounts of alkali metals as well as other harmful and toxic volatile metals (i.e. Hg, Pb) are negligible and certainly lower than the upper environmentally acceptable limits (12, 13). This is mainly achieved due to the high efficiency of the gas scrubbing system.

3. The amounts of SO_x, NO_x et HCl are not detectable and, in every case, lower than the upper environmentally acceptable limit (12, 13).

4. There are no results for dioxins, furans and HAP; plans for these tests are under way.

3.5 The case of gasification of residual rubber material

When residual rubber is used as feedstock the following results were observed (see Table 6):

1. The high calorific value of this feedstock (34 MJ/Nm^3) allows gasification regimes at higher temperatures (up to 900° C). The producer gas HHV varies between 4.9 and 7.3 MJ/Nm^3.

2. Feedstocks with moisture as high as 30% can be gasified with acceptable performances without pretreatment (drying) step.

3. Cold gas efficiencies are lower than in biomass and RDF gasification runs; this is mainly due to the high ash content (7% w/w) as well as the existence of a thermally stable carbon content of the feedstock (around 50% w/w). The elemental analysis as well as the TGA of this feedstock clearly supports this result. In almost all of our experimental runs with rubber residues as feedstock the gasification solid residue production (ash+carbon) is higher than 25% of the input solids.

4. The influence of the equivalence ratio (λ) in the range between 20 and 37% does not seem to be as important as in the wood and RDF gasification cases.

5. The use of oxygen-enriched air as gasifying agent cannot be clearly considered beneficial for gasification. Despite the gasification temperature increase, which led to

Table 5: Gasifier and boiler stack gas analysis; the effect of the O_2-enriched air and the ratio NG/SG, (Wood & RDF runs).

	Runs	WOOD NG/SG 75/25	WOOD NG/SG 50/50	WOOD NG/SG 25/75	RDF NG/SG 67/33	RDF NG/SG 75/25
	Results					
1A	N_2	53.1	57.3	52.4	65.0	63.9
2A	O_2	1.1	1.5	0.6	1.4	1.1
3A	H_2	6.3	2.9	4.0	2.6	1.1
4A	CO	23.0	17.7	20.9	12.7	15.4
5A	CO_2	10.8	14.1	14.7	13.8	13.2
6A	HC	5.8	5.9	7.4	4.5	5.3
1B	N_2	79.5	79.9	84.6	85.2	83.6
2B	O_2	7.0	6.9	7.3	7.5	10.5
3B	CO_2	13.5	13.2	8.1	7.4	6.0
4B	CO	0.2	0.3	0.7	0.3	0.5
5B	SO_2	-(*)	-	-	-	-
6B	NO_x	-	-	-	-	-
7B	HCl	-	-	-	-	-
8B	Particles/Metals (mg/Nm^3)	0.04	0.05	0.01	0.03	0.01
1C	Gasification Equivalence ratio λ (%)	31	30	30	29	30
2C	HHV (MJ/Nm^3)	5.8	5.9	7.4	4.5	5.0

A : Main results of gasifier stack gas analysis

B : Main results of boiler stack gas analysis: combustion was carried out at λ=130-150%

C : Additional information on gasification conditions and SG

(*) : Below detectable limit and higher acceptable limit

lower tar production, cold gas efficiency and producer gas HHV were not improved; in fact, the nitrogen content decrease has been balanced by a CO_2 increase. This result is in agreement with the TGA results.

3.6 The tarry waters treatment analysis

Two methods have been used to treat the tarry waters resulting from the scrubbing operation: wet oxidation using OXYJET technology, and adsorption tests using the carbon-rich gasification ashes. The results are presented briefly below:

1. 2600 ppm TOC skimmed tarry waters were treated using an OXYJET PDU having a nominal capacity of 30l/h. The efficiencies obtained vary between 70 and 80% TOC reduction. Typical conditions used are: T= 310 C, air at 13-14 MPa, oxidation time 3-6 min.

2. Adsorption treatment of the same tarry waters using carbon-rich residues from gasification of rubber proved that:

- the adsorption efficiency of these ashes is of the order of 150 mg-organic carbon/g-residue; according to the literature (14, 15) this value is similar to low grade commercial active carbons;

- this efficiency and the amount of residue produced are high enough to treat the tarry waters; the water effluent is then clean enough and non-toxic to be accepted by urban wastewater biological treatment facilities.

Table 6 : Rubber residues as feedstock: mass balances

Runs	Results	NA-18 λ=31.9	NA-18 λ=36.7	NA-19 λ=32.4	NA-19 λ=20.3	NA-19 λ=20.1	NA-20 λ=34.6	NA-20 λ=30.2
1A	Solids feed rate (kg/h)	20	20	18.4	26.4	30	17.4	17.4
2A	Humidity (%)	11.5	11.5	26.4	26.4	26.4	3	3
3A	Temperature (°C)	880	900	720	728	700	887	882
4A	(%) of oxygen in the air	21	30	21	30	21	21	21
5A	HHV (MJ/Nm^3)	5.50	5.45	6.62	5.60	7.27	4.88	5.67
6A	Gas (Nm^3/kg)	3.22	2.59	3.32	1.61	2.17	3.52	3.02
7A	Cold efficiency (%)	52.0	41.6	64.6	26.5	46.4	58.8	43.4
8A	Tar+VOC (g/Nm^3)	2+2	2+2	5+4	40+30	30+20	2+1	3+5
9A	Ashes (kg/h)	7	7	5	9	8	5	6
10A	Water (g/Nm^3)	135	135	164	346	300	90	98
11A	Carbon balance closure (%)	104	104	105	100	104	104	98
1B	N_2 (% v/v, d.b)	70.2	62.2	69.0	67.3	65.6	69.4	70.6
2B	H_2 (% v/v, db)	1.7	2.0	1.7	1.4	2.0	2.0	1.8
3B	CO(% v/v,db)	5.4	6.3	8.4	6.5	7.3	5.6	5.6
4B	CO_2(% v/v,db)	11.1	18.1	9.2	14.3	12.0	12.5	13.1
5B	HC (% v/v, db)	9.5	9.2	10.1	8.3	11.2	9.6	8.0
6B	O_2 (% v/v, db)	2.2	2.3	1.6	2.2	1.9	1.0	0.9
7B	SO_2 (%v/v,db) (3)	0.1-0.06	0.15-0.09	0.15-0.08	n.a.	n.a.	0.07-0.04	0.07-0.04
8B	NO_x (ppm)	15-5	20-5	0	n.a.	n.a.	n.a.	n.a.
9B	HCl (ppm)	0	0	0	0	0	0	0
10B	Particulates (3) (g/Nm^3)	20-0.004	20-0.004	4.2-0.003	4.2-0.003	4.2-0.003	n.a.	n.a

P.S. : 1. A lines: Gasification results from mass and energy balances
2. B lines: SG composition
3. Before and after gas conditioning

4 Conclusions/Work continuation

This R&D program aiming at the application of AFBG for the disposal of organic wastes generated by both urban and industrial activities, and generation of heat or electricity, is showing promising results. Our approach is being implemented at the demonstration level for aluminum recovery from post-consumer packaging wastes containing aluminum foil.

The results of this paper have demonstrated that:

• Gasification can be made compact, energy efficient and environmentally friendly organic waste treatment; the technology is well positionned within the frame of the sustainable development and the principle of "zero effluent" industrial operations; it is particularly suitable for small capacity solid waste management (0.5-5 tn/h = 0.5-10 MW$_{elec.}$) and is addressed to both urban and industrial wastes producers.

- The system is able to operate continuously and constantly either with air or oxygen-enriched air as gasifying and fluidizing agent.
- The system produces a synthetic gas clean enough to be co-burned with natural gas for heat cogeneration using commercially available burners/boilers; ratios of SG/NG of 3/1 are possible when producing SG with HHV as high as 11-12 MJ/Nm^3; and this is possible when operating the gasifier with oxygen-enriched air.
- The technology provides a synthetic gas whose co-combustion gives stack gases environmentally acceptacle according to the existing environmental regulations.

Our research group is pursuing its R&D efforts focusing on:
1. Hot gas filtration/Catalytic reforming of the tars to enhance the energetic content of the SG. A new reforming catalyst formulation (patent pending in Canada and U.S.A.) has been developed by our group and tested with success in both laboratory and PDU.
2. The simultaneous treatment of the tarry liquids produced during the gas scrubbing using the technology of the jet-reactor wet oxidation (OXYJET) also developed by our group. The results obtained so far are positively conclusive. The use of carbon-rich residues as a polishing strategy can also be coupled to the wet oxidation process.

5 Nomenclature

λ:	Equivalence Ratio
AFBG:	Atmospheric Fluidized Bed Gasifier
ash:	Ashes
BET:	Brunauer, Emmett and Teller surface area
BOD5:	Biological Oxygen Demand (5 days)
COD:	Chemical Oxygen Demand
GRTPC:	Groupe de recherches sur les technologies et procédés de conversion
Hg:	Mercury
HHV:	Higher Heating Value
NG:	Natural Gas
OC:	Organic Carbon
PAH:	Polycyclic Aromatic Hydrocarbon
Pb:	Lead
PE:	Polyethylene
PVC:	Poly-Vinyl-Chloride
RDF:	Refuse Derived Fuel
PDU:	Process Development Unit
SG:	Synthetic Gas
TGA:	Thermo-Gravimetric Analysis
TOC:	Total Organic Carbon
VOC:	Volatile Organic Carbon

6 Definitions

OXYJET: Oxidation technology patented in Canada and U.S.A. and belonging to the Université de Sherbrooke. Exclusive licence held by KEMESTRIE Inc.
BIOSYN: Gasification technology patented in Canada and U.S.A. and belonging to the Centre québécois de valorisation des biomasses et des biotechnologies (C.Q.V.B.). Licence are held by KEMESTRIE Inc. and BIOTHERMICA Ltd.

7 Acknowledgments

The authors are indebted to KEMESTRIE Inc., the Centre québécois de valorisation des biomasses et des biotechnologies (C.Q.V.B.), Ressources Naturelles Québec, Ressources Naturelles Canada, Gaz Métropolitain, the Industries FP and Hydro-Québec for their financial contributions.

Stimulating discussions with W. Yaïci and D. Elkaïm from the Centre de Technologies du Gaz Naturel (C.T.G.N. du Gaz Métropolitain), J. Baladi from EnerQuest Inc., C. Guy from École Polytechnique de Montréal and F. Tremblay from Alcan are also acknowledged.

The technical contribution of J. Bureau, G. Phaneuf, J. Gagné et M. Trottier have been essential in the implementation and execution of the R&D program.

8 References

1. S. Czernik, P.G. Koeberle, P. Jollez, J.F. Bilodeau and E. Chornet (1993) "Gasification of Residual Biomass via the Biosyn Fluidized Bed Technology", in "Advances in Thermochemical Biomass Conversion", Edited by A.V. Bridgewater, vol.2, Blackie Academic & Professional, ISBN 0 7514 0171 4.
2. W. Yaïci et D. Elkaïm (1995) Rapport d'étape "Gazéification: Co-combustion du gaz naturel-synthétique", N. 92-030.1, Centre des technologies du gaz naturel, Gaz Metropolitain, Montréal, Québec, Canada.
3. N. Abatzoglou et K. Belkacemi (1995) "Gazéification de Biomasse et de RDF: Co-combustion du gaz de synthèse avec du gaz naturel", Rapport final, GRTPC/Université de Sherbrooke.
4. S. Gassó, J.-M. Baldasano, M. Gonzalez, N. Abatzoglou, J.-P. Lemonnier and E. Chornet (1992) "Wet Oxidation via Two-Phase Flow Reactors and High Mass Transfer Regimes", Ind. Eng. Chem. Research, vol. 32.
5. M. Stenzel (1993) "Remove Organics by Activated Carbon Adsorption" Chemical Engineering Progress, vol. 89, N. 4, pp 36-41.
6. J.-F. Bilodeau (1993) "Gazéification de la biomasse en lit fluidisé: Modélisation mathématique et expérimentation", Mémoire de Maîtrise, UdeS, Génie chimique.
7. J. E. McCarthy (1980) "Scrubber Types and Selection Criteria", A.I.Ch.E., pp. 58-62, CEP.
8. H. Haller, E. Muschelknautz and T. Schultz (1989) "Venturi Scrubber Calculation and Optimization", Chemical Engineering Technology, vol. 12, pp. 188-195.
9. D. L. Brenchley, C. D. Turley, R. Yamac (1974) "Industrial Source Sampling", Ann Arbor Science Publishers, Ann-Arbor, Michigan, USA.
10. T. A. Milne, A.H. Brennan, B.H. Glenn (1990) "Sourcebook of Methods of Analysis for Biomass and Biomass Conversion Processes", Elsevier Applied Science.
11. Environnement Canada, Rapport EPS-1-AP-74-1 (1979) "Méthodes de référence normalisées en vue d'essais aux sources: mesure des émissions de particules provenant de sources fixes".
12. L. Laramée (1994) "Epuration des particules fines émises lors da la conversion thermique de résidus et de biomasse", Mémoire de Maîtrise, UdeS, Génie chimique.
13. Le Conseil Canadien des Ministres des Ressources et de l'Environnement (1988) "Directives Canadiennes sur le fonctionnement des incinérateurs de déchets urbains solides et sur les émissions".
14. R. D. Vidic, M. T. Suldan and R. C. Brenner (1993) "Oxidative Coupling of Phenols on Activated Carbon: Impact on Adsorption Equilibrium", Environmental Science Technology, vol. 27, N. 10, pp. 2079-2085.

15. E. Diamantopoulos, P. Samaras, G. P. Sakellaropoulos (1993) "The Effect of Activated Carbon Properties on the Adsorption of Toxic Substances", Water Science Technology, vol. 25, N. 1, pp. 153-160.

DEMONSTRATION SYSTEMS OF COOKING GAS PRODUCED BY CROP STRAW GASIFIER FOR VILLAGES

L. SUN, Z.Z. GU, D.Y. GUO and M. XU
Energy Research Institute of Shandong Academy of Sciences
Jinan, P. R. China

Abstract

Several demonstration systems were designed, built, tested and put into use in order to develop a new way of producing cooking gas from crop straw for villages by biomass gasification technology. A type of crop straw gasification unit composed of a down draft gasifier and gas clean equipment was also developed. The gas produced from crop straw gasifier was sent to each farmer's house of the villages by a distributing network on a scale of 100 ~ 300 households.

The paper indicated that the annual production of 600 million tons of cropstraws, a reasonable amount of which are burnt up directly in the fields, comprises the largest biomass resource in China and provides the practical significance of the project, and the demonstration systems introduced a biomass conversion method to use the cropstraws efficiently and to improve the farmer's cooking way and the rural environment.

The design of the crop straw gasification unit and the demonstration systems are presented in the paper, and the economic evaluations and the development prospect are also given here.

Keywords: Biomass, gasification, crop straw, rural energy.

1. Introduction

Energy Research Institute of Shandong Academy of Sciences has conducted a research and development program on the crop straw gasification and cooking gas supplying system for villages. The aim of this program was two-fold. The first objective was to make efficiently use of various kinds of agricultural residues especially cropstraws that were wasted in a huge amount and the second objective was to provide an advanced method of using energy for farmers to improve their cooking way therefore they will live a more comfortable and civilized life.

Experiments were made upon the down-draft gasifier to gasify many kinds of cropstraws. A low-Btu gas was produced from the cropstraws with the efficiency as high as over 70%. Two models of crop straw gasification unit were designed and tested. Five demonstration systems have put into run and other two systems are building. The results presented in this article show both of technological and economic feasibility of this system.

Each of these activities was designed to move the technology toward commercialization.

2. Process description

2.1 System

The principle of the village biomass gas supplying system is shown in Fig. 1. Biomass materials such as corn straw and wheat straw with moisture content below 20% are, after broken into the pellet of the length of 10 ~ 15 mm, fed into a downdraft gasifier by the screw feeder. The gas produced in the gasifier is removed the tar and ash dust by passing through the cyclone, cooler and filter, and its temperature drops down to ambient from about 450℃. Afertwords the gas is sent to the gas holder by a fan. The gas holder functions as a balance for the pulsation of consumption and a manostat with its pressure maintained on a specific value (2000 ~ 4000 Pa) to overcome the deliver resistance of the network and to provide a steady pressure for the combustion of stoves. On the inlet of the gas holder, a water seal tank is installed to prevent the gas flowing back to the shop when the gasifier stopped. The gas from the gas holder is sent into the network of pipelines to be distributed to every household for cooking.

2.2 Crop straw gasification unit

Main materials practically used in commercial down-draft gasifier are charcoal, wood and small amount of rice husk. All kinds of cropstraws are considered as not suitable for the feed of the gasifier.

Compared with those of woody materials, the chemical properties of cropstraws have no notable differences expect their lower calorific value resulted from the higher ash content. The relative content of their main elements, carbon, hydrogen and oxygen are basically the same, as shown in Fig. 2, almost equating to $CH_{1.4}O_{0.6}$.[1][2]

On the other hand, the straws are greatly different from wood in physical properties. (1) The bulk densities of typical biomass materials are shown in Table 1. It can be seen that the bulk density of corn straw is only equal to 1/5 ~ 1/4 of that of wood, and as for wheat straw it become 1/10. Such a little bulk density brings much

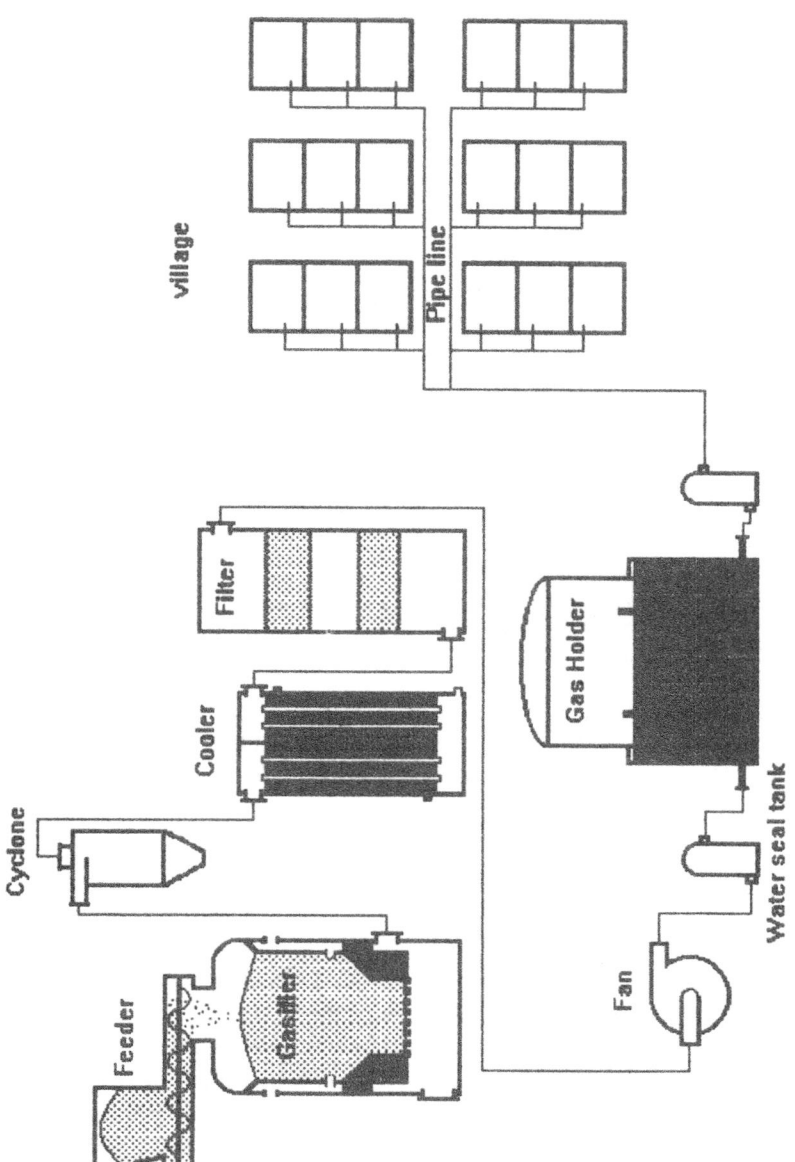

Fig. 1 Principle process of the cooking gas supplying system

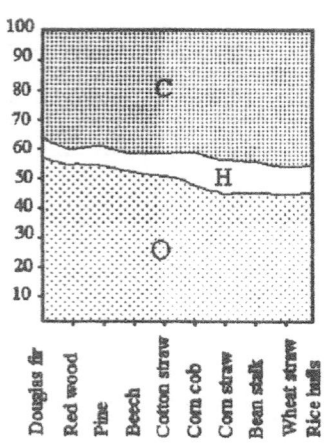

Fig. 2 The relative content of C, H, O
for selected biomass materials

Table 1. Bulk Density for Selected Biomass Materials

Material	Bulk Density kg / m^3
Hard wood	330
Soft wood	250
Charcoal	150 ~ 230
Corncob (11% moisture)	304
Cotton straw (23% moisture, powder)	340
Corn straw (10 ~ 15 mm length)	67
Wheat straw (10-15 mm length)	25

difficult to the collection and storage of the feed. And what is more important is this means that the heat capacity in the oxidation zone will be reduced greatly with the result of a unsteady gasification reaction. (2) The repose angle of corn straw with the length of 10 ~ 15 mm is close to 90° , and for wheat straw, this angle exceed 90° and become an abtuse angle. This will make the moving down of straw by gravity in the gasifier difficulty, with forming of bridging and rat-holdings. (3) After pyrolysising, the volume of corn straw is reduced by 50 ~ 55%, and wheat straw by about 80%. This is because the straws is soft and the mechanical strength of charcoal formed can not maintain its initial shape and volume. The charcoal formed in this way has a poor ventilation ability, and its pellet is small and easy to be swept off by the gas stream. (4) The slagging behavior of various cropstraws are much severer than that of woody materials because their higher ash content.[3]

Table 2. Main Gas Composition and Low Heat Value

Feed	Gas composition, %					Low heat value
	CO_2	O_2	CO	H_2	CH_4	kJ / m^3
Corncob	12.5	1.4	22.5	12.3	2.32	5302.8
Cotton straw	11.6	1.5	22.7	11.5	1.92	5585.2
Corn straw	13.0	1.65	21.4	12.2	1.87	5327.7
Wheat straw	14.0	1.7	17.6	8.5	1.36	3663.5

Table 3. Properties of the Model-XFF Crop straw Gasification Units

	XFF-1000	XFF-2500
Gas output, m^3 / h	216	524
Gas low heat Value, KJ / m^3	5327.7	5215.6
Energy output, MJ / h	1151	2733
Conversion efficiency, %	73.92	73.10
Gas-feed ratio, m^3 / kg	1.90	1.92

Researches, experiments and improvements have been conducted to overcome the disadvantages of the straws in physical properties. Two models of crop straw gasification units named XFF-1000 (energy output equal to 1000 MJ / h) and XFF-2500 (2500 MJ / h) have been developed. Each unit comprises a down-draft gasifier, a set of gas clean equipment (cyclone, cooler, filter, etc.), and a fan. The gasifier is operated under weak negative pressure, that is a suction operation with the fan placed at the rear of the gasifier. So the upper part of the gasifier will not need to be sealed and can be operated openly. This is more important for the material with extremely low bulk density because it make the continual feeding possible, and the operator can observe the conditions in the gasifier and eliminate the bridging and rat-holes by poking. An thermal accumulative section is set up in the furnace using the refractory, so the heat capacity of oxidation zone is increased by several dozens of times and a steady temperature field maintained.

The units produce a low-Btu product gas with the tar and dust content below 100 mg / m^3. The experimental data of the units using four kinds of feed are shown in Table 2 and 3. Main material practically used in the demonstration systems is the corn straw.

2.3 Demonstration systems

Five demonstration systems of cooking gas supplying have built and put into use, and other two systems are building. The first system supplying gas for 94 households was built up in Oct. 1994 in Huantai County, Shandong Province and has run for 16 months. Brief situation of these systems is shown in Table 4.

Table 4. The Situation of Seven Demonstration Systems

Sys. No.	1	2	3	4	5	6	7
Scale of system, households	94	90	156	130	132	186	268
Gas supplied per day, m³ / d	800	540	950	780	780	1100	1500
Unit installed, XFF-	1000	1000	1000	1000	1000	2500	2500
Volume of gas holder, m³	80	250	280	250	250	250	250
Pressure of gas holder, Pa	2250	2050	2800	2800	2800	3200	3000
Farthest distance of pipeline, m	230	180	360	380	370	680	520
Straw consumed per day, kg	410	280	440	400	400	580	800
Date put into run	94.10	95.12	96.01	96.05	96.05		

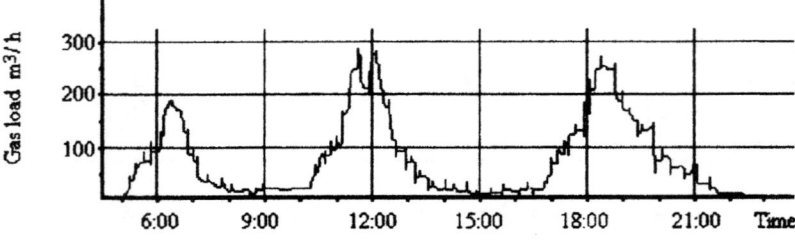

Fig. 3   Gas consuming curve of Sys. No. 2

In each system, there are a crop straw storage field, a gasification station in that all the equipment are installed, and a network of pipelines. Several liquid collector located in the pipelines to remove automatically the water, that formed in the pipe when temperature changed, from the gas. All of the pipelines are made up of polyvinyl chloride or polypropylene pipes and placed under ground.

General speaking, an average family in rural area of Shandong Province has 3.82 peoples. According to the data from the systems running and testing, every family consume about 6 m³ the gas with low heat value of 5.2 MJ / m³ per day. The gas consuming curve for a typical winter day tested in the system No.2 is shown in Fig. 3. It can be seen that there are three load peaks every day, one for breakfast at 6:00 to 7:00, one for lunch at 11:00 to 12:30 and another for dinner at 18:00 to 19:30. From 22:00 to next 5:00, no gas would be consumed. Therefore the reasonable design of the systems is to make the gasification unit run when cooking and stop on other time. The extra peak load when cooking lunch and dinner and the valley load are filled up by the gas holder. In this case, the gas holder can be designed much smaller than that the unit run all day continuously. So the total investment can be decreased much by this way.

3 Technology evaluation

3.1 Fuel flexibility

The researchers in the biomass gasification field have tested on almost all kinds of solid biomass materials and got a mount of experimental data. In China's rural area the largest biomass resource available for energy is composed of the all kinds of agricultural residues, especially the cropstraws so-called "soft firewood".

The ultimate analysis data of the main agricultural residues are listed in Table 5. The authors have tested these materials in a down-draft gasifier whose construction is not changed and obtained satisfactory results except the rice straw. Because the rice straw is too "soft" and higher ash content that make the operation unsteady.

Also some woody wastes such as wood chips, barks and wood shavings have tested by the unit and more satisfactory results have obtained when the moisture content is controlled under 20%.

To evaluate the biomass gasification, a important feature of the process is its excellent feed flexibility.

3.2 utilization of the low-Btu gas

A vague conception is that the higher the heat value of a gas, the more flourished is the fire. Many people say the gas produced from cropstraws by the gasifier is not suitable for using as cooking gas because its heat value is too low to burn in a gas stove. However, what is burnt actually in the stove is the gas-air mixture instead of the gas only. Commonly, the gas with higher heat value has to mix with proportionally more air for complete combustion and the heat value of the mixture is not so high as that many people think. Table. 6 shows the low heat value of typical fuel gas, the amount to air mixed for complete combustion and the heat value of their mixture. We can see that from it, although the low heat value of the natural gas is seven times higher

Table 5. Ultimate Analysis Data for Selected Agricultural Residues
(Dry Basis, Weight Percent)

Material	C	H	N	S	O	Ash	High heat value kJ / kg
Corncob	46.3	5.6	0.57		42.19	5.34	20,620
Rice hulls	38.5	5.7	0.5		39.8	15.5	15,720
Corn straw	45.43	6.15	0.78	0.13	41.11	6.4	17,794
Sorghum straw	45.18	5.59	0.62	0.11			18,020
Cotton straw	45.50	6.01	0.98	0.23	30.08	17.2	18,330
Bean stalk	46.65	6.12	0.89	0.12	36.02	10.2	18,120
Wheat straw	45.30	5.89	0.68	0.19	40.54	7.4	17,769
Rice straw	40.44	5.31	0.66	0.12	34.07	19.4	16,329

Table 6. Low heat value of typical fuel gas and the gas / air mixture

	Gas low heat value kJ / m^3	Air amount ($\alpha=1$) m^3 / m^3	Mixture low heat value kJ / m^3
Natural gas	36586	9.64	3438
Coke gas	17615	4.21	3381
Mixed gas	13856	3.18	3315
Generator gas	5735	1.19	2618
Biogas	21223	5.65	3191
Biomass gas	5316	0.9	2798

than that of the biomass gas produced by the down-draft gasifier, the low heat value of the gas-air mixture is increased no higher than 23%. And the low heat value of mixed gas which is civil used in city proper is twice times higher than that of the biomass gas, but the low heat value of the gas-air mixture is only 18% higher.

The heat value will greatly influence the power of a internal combustion engine if the gas were used as its fuel.[1] And the heat value will influence the investment of a pipeline network if the gas were transported for a long distance. But in a conversion of gas to heat, there is no notable influence caused by the heat value. Only thing we must do is to design a special gas stove , especially to calculate exactly the diameter of the nozzle through which the gas is jetted into the stove.

It is general considered that it is not economical to deliver the low-Btu gas by pipeline, because the volume is higher than medium-Btu gas for delivering a specific energy. None of low-Btu gas is delivered for civil use in city proper. But another factor will influence this prediction to a great extent, that is the concentration degree of gas consumption load or the residence density for the cooking gas supply network. Chinese farmers live concentratively in many natural villages, and the residence density in a village is over 10000 / km^2, well exceeding that for ordinary cities. Such as for Jinan City, the capital of Shandong Province, the residence density is 2987 / km^2. Moreover, the delivering distance of pipelines is short, the flow rate of the gas is small, and the delivering pressure head is only several thousands Pa, when treating a natural village as a unit. The using of plastic pipe in the pipelines makes its cost well below that of urban gas network. So it is the residence feature of Chinese farmers that compensates the disadvantage of low-Btu gas and makes it feasible economically.

3.3 Efficiency of the gasification unit and the system
The energy balance tests was conducted to evaluate the energy conversion efficiency and find the energy loses of the gasification unit. Table 7 shows the results of using corn straw. The efficiency of the crop straw gasification unit is usually over 70%.

The test on the gas stove indicates that a efficiency of 50 ~ 53% can be obtained

Table 7. Energy balance tests results of the crop straw gasification unit

Test No.	1	2	3
Feed rate, kg / h	113.7	105.4	118.1
Straw low heat value, kJ / kg	13692	13692	13692
Energy input, kJ / h	1556791	1443137	1622776
Product gas output, m^3 / h	216	199	226
Gas low heat value, kJ / m^3	5327.7	5280.1	5247.3
Energy output, kJ / h	1150783	1050604	1187873
Gasification efficiency, %	73.92	72.80	73.20
Amount of ash, kg / h	9.8	9.2	10.1
Carbon content in ash, %	42	43	39
Carbon heat value, kJ / kg	33500	33500	33500
Energy lose in ash, kJ / h	37886	132526	131957
Energy lose rate in ash, %	8.86	9.18	8.13
Amount of tar, kg / h	5.1	4.7	5.7
Tar heat value, kJ / kg	26790	26790	26790
Energy lose in tar, kJ / h	136629	125913	152703
Energy lose rate in tar, %	8.78	8.72	9.40
Temperature at gasifier outlet, ℃	382	392	405
Temperature at cooler outlet, ℃	28	31	32
Gas specific heat, kJ / m^3	1.5	1.5	1.5
Apparent heat lose, kJ / h	114696	107758	126447
Apparent heat lose rate, %	7.36	7.46	7.80
Surface area of gasifier, m^2	9.1	9.1	9.1
Average heat flow, W / m^2	210	204	220
Heat radiation lose, kJ / h	6878	6683	7207
Heat radiation lose rate, %	0.42	0.46	0.44

when using the biomass gas as cooking fuel. Therefore the total energy efficiency of the demonstration system is over 35%. Comparing with that the energy efficiency of a old style stove burning straw directly is as low as 10%, and that of new style stove developed recently is between 15 ~ 20%.

4 Economic evaluation

4.1 Investment of the systems

The careful designs have been doen to reduce the investment and unit cost of building of the demonstration systems. Commonly say, a farmer's family will pay average 2000 yuan for the building of the system. A investigation indicates that the families whose annual income reach or over the level of 1500 yuan each body are willing to pay for it to improve their cooking way.

4.2 Running Cost

As shown in Table 8, the cost of the demonstration system operation includes: (1) The fee for collection of straws; (2) The electricity fee for cutting the straws and running the fan; (3)The worker's salary; (4) The fee for equipment maintenance. Because the system is run as a public welfare service of village, the cost includes none profit and tax.

According to the statistic data, a average household consume 6 m^3 of gas per day. So the gas charge will be 19.8 yuan monthly. When they use honeycomb briquette, month fuel cost is 18-24 yuan, and using liquefied gas the cost is about 30 yuan. Compare with the latter fuels, the unit price of biomass gas is quite cheap.

Table 8. Running cost of demonstration system No. 1

Item	Data
1. Annual straw consumption, t	90
2. Unit cost of straws, yuan / t	60
3. The fee of straws, yuan	5400
4. Annual electricity consumption, kw.h	6570
5. Unit price of electricity, yuan / kw.h	0.5
6. Annual electricity fee, yuan	3285
7. Salary of workers, yuan	7200
8. Maintenance fee, yuan	3000
9. Total runing cost, yuan	18885
10. Annual gas supplying, m^3	172,000
11. Unit cost of gas, yuan / m^3	0.11

4.3 Economic scale of the system

As mentioned above, Chinese farmers live in many natural villages . For example, there are 67.02 million people live in 89393 villages in Shandong province. A average scale of village has 197 farmer's households and about 800 tons straws yield each year. In fact a natural village is a social unit in which agricultural and industrial production and other social activities are managed. If we treat a village as a cooking gas supplying system, some conveniences will be got. (1) The collection of straws is easy and cheap. (2) The gas pipeline is short, commonly less than 800m, and the delivering pressure is not higher than 5000 Pa. (3) The management of the system: storage of straws; gas measure and charge; maintenance of equipment; etc. It is difficult and expensive to collect straws from and transport gas to other villages.

By a careful calculation, we plan to develop three model of crop straw gasification unit whose energy output are 1000, 2500 and 4000 MJ / h and a series of auxiliary equipment such as gas holders. One unit can supply cooking gas for 100, 200 and 300 households individually and the combination of two units can meet the need of larger villages.

5 The development prospect

Crop straw is the largest biomass resource available in China. The approximate 600 million tons of all kinds of straws are produced in China every year. The yields of straws are shown in Fig. 4. Mainly the corn straw and wheat straw are produced in northern regions of China, and the rice straw in southern regions. These are the valuable renewable resources and the traditional cooking energy resources. But their utilization condition is extremely unreasonable for a large amount of straw is burnt directly in the stove for cooking with low efficiency.

Recently, Chinese government is puzzled by a serious problem that is the surplusage of straws in a huge amount. At each harvesting season, the farmers put the excessive straws in the field and by the road to be burnt up directly. This not only wasted the resource but also lead to air pollution. According to a rough estimate, the surplus amount is about 35 ~ 40% of total yield of straws, including 50 ~ 60% of corn straw, 40% of cotton straw, 25% of wheat straw and considerable part of rice straw. That means about 200 million tons of straws are wasted every year. In contrast to this , the supply of commercial energy in rural area such as honeycomb briquette and liquefied gas is in shortage and high-priced.

It is obvious that the situation of main cooking energy coming from agricultural and forest residues will go on. Especially considering their renewability and cleanliness, the

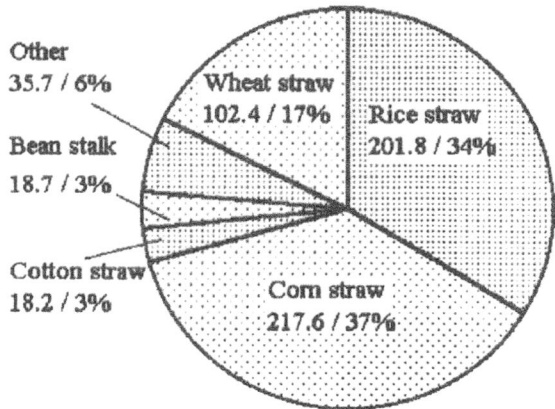

Fig. 4 Crop straw resources in China

biomass should be one kind of sustained energy resource. Very important thing is to improve the conversion technology of biomass energy. As mentioned above, the gasification process can use almost all kinds of the residues, so it should be a cost-effective and practical way.

In 1992, the total living energy consumed in China's rural area is 320.47 Mtce, 71.45% of which is biomass, e.g. 228.97 Mtce. The cropstraws used for cooking energy is 316.16 million tons, equal to 135.5 Mtce.

As the rapid development of China's agricultural economic, the component of living energy has changed. Before 20 years, over 90% of the cooking energy came from biomass, and the straws were almost burnt up with 2 ~ 3 months shortage. Now the ratio has decreased to about 70%. The main reason of this situation is that the farmers want to use high-grade energy of convenience and cleanness. With the increasing of farmer's income, the conditions of living and housing in rural area have improved greatly, farmers now put it high on their agenda to get rid of strenuous, time-consuming and smoky cooking method they traditionally used for many years.

It will be difficult and expensive to solve this problem by relying on commercial energy, considering the fact of 800 million farmers living in rural area in China. Therefore, the urgent need is, in the point of view of the authors, to change the conversion method of biomass materials. The data from the crop straw gasification and the demonstration system shows that this technology is one of feasible way for its higher conversion efficiency, better fuel flexibility, simpler equipment and lower investment.

There are over 100 million villages in China. The renewability of biomass resources and the strong desire of farmers for improvement their cooking way indicate the excellent prospect and potential market of this technology.

6 References

1. Reed, T.B. (1980) A survey of biomass gasification, SERI / TR 33-239.
2. Reed, T.B. (1988) Handbook of biomass downdraft gasifier engine systems, SERI / SP-271-3022.
3. Sun, L., Min, X., Gu, Z.Z. and Guo, D.Y. (1993) Crop straw gasifier and village biomass gas supply network. *Proceedings of the 5th international energy conference*, Vol. III, pp 290-9.
4. Sun, L., Min, X., Gu, Z.Z. and Guo, D.Y. (1995) Study on crop straw gasification for village cooking gas. *Proceedings of Asia-Pacific regional seminar on technology for utilization of rice husks and other agricultural waste.* pp.30-41.

A WOOD-GAS STOVE FOR DEVELOPING COUNTRIES

T. B. Reed and Ronal Larson
The Biomass Energy Foundation, Golden, CO., USA

Abstract

Through the millennia wood stoves for cooking have been notoriously inefficient and slow. Electricity, gas or liquid fuels are preferred for cooking - when they can be obtained. In the last few decades a number of improvements have been made in **wood-stoves**, but still the improved wood stoves are difficult to control and manufacture and are often not accepted by the cook.

Gasification of wood (or other biomass) offers the possibility of cleaner, better controlled gas cooking for developing countries. In this paper we describe a **wood-gas stove** based on a new, simplified wood gasifier. It offers the advantages of "cooking with gas" while using a wide variety of biomass fuels. Gas for the stove is generated using the "inverted downdraft gasifier" principle. In one mode of operation it also produces 20-25% charcoal (dry basis).

The stove operates using natural convection only. It achieves clean "blue flame" combustion using an "air wick" that optimizes draft and stabilizes the flame position.

The emissions from the close coupled gasifier-burner are quite low and the stove can be operated indoors. Keywords: Inverted downdraft gasifier, domestic cooking stove, natural draft

1. Introduction

1.1. The Problem

Since the beginning of civilization wood and biomass have been used for cooking. Over 2 billion people cook badly on inefficient wood stoves that waste wood, cause health problems and destroy the forest. Electricity, gas or liquid fuels are preferred for cooking - when they can be obtained, but they depend on having a suitable infrastructure and are often not available in developing countries.

In the last few decades, many improved wood stoves have been developed (the Chula, the Hiko, the Maendeleo, the Kuni Mbili, etc.), but the new wood stoves are often more difficult to manufacture, often more heat goes to the stove than to the food, and they do not offer good control of cooking rate. They are not often accepted by the cooks for whom they are developed.[1]

Because of the problems of wood cooking, people often cook over charcoal. However, charcoal manufacture is very wasteful of energy and very polluting, so the problems of the wood stove are externalized but not solved.

1.2. The Solution

Gas is preferred for cooking wherever it is available. Gas can be made from wood and biomass in gasifiers developed in this century, but these gasifiers are generally too big for home use. A downdraft stove for domestic cooking is now being manufactured in China.[2]

We have developed a new "inverted downdraft gasifier" stove shown in Fig. 1. It operates using only natural convection. The rate of gas production and heating is controlled by the primary air supply to the gasifier. As an option, the gasifier can make charcoal with a 20-25% yield.[3]

The wood-gas stove consists of an "inverted downdraft gasifier" (shown in Fig. 2) plus a burner to mix air and gas and burn cleanly (Fig. 3). These sections are discussed below.

The stove has been started and operated indoors with no exhaust fans and no odor of burning wood. However, we believe that there is still much work to be done in optimizing the stove for various fuels, adapting it to various cooking situations and developing other uses. For that reason we are publishing our preliminary results and hope that others will help adapt these principles to improve world cooking and wood conservation.

2. Construction of the wood-gas stove

2.1. The Inverted Downdraft Gasifier

Wood gasifiers can be classified as: Fixed bed (updraft and downdraft); and fluidized bed. Fluidized bed gasifiers Require high power input, and exact controls and are suitable only for large installations.

Updraft gasifiers produce large amounts of tar while consuming the charcoal residue and are not suitable for cooking.

Downdraft gasifiers in the 5-100 kW level were widely used in World War II for operating vehicles and trucks because of the relatively low tar levels. [4] In operation, air is drawn down through a bed of burning wood, consuming the volatiles. The resulting gas then passes over the resulting charcoal and is reduced to a low energy fuel gas. However, since hot gases naturally rise, it is necessary to supply power to draw the gases DOWN through the gasifier.

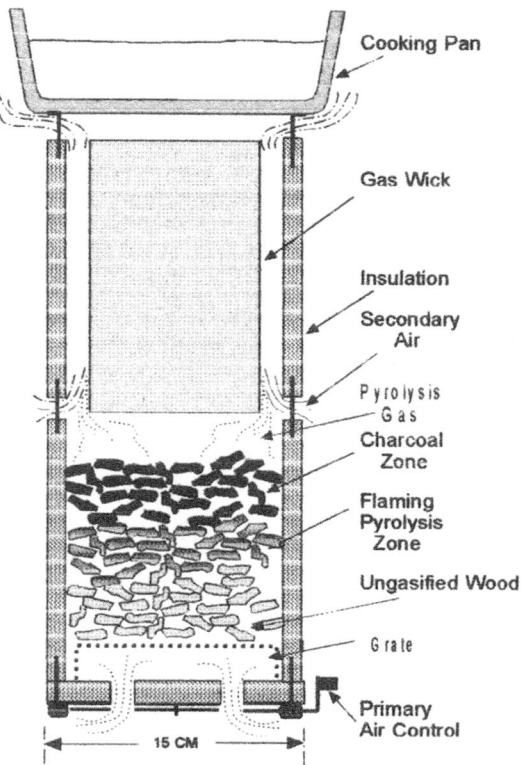

Cooking Pan

Gas Wick

Insulation

Secondary Air

Pyrolysis Gas

Charcoal Zone

Flaming Pyrolysis Zone

Ungasified Wood

Grate

Primary Air Control

15 CM

Fig. 1 - Wood-gas cooking stove showing lower gasifier section, upper burner section and pan heating

3. Construction of the wood-gas stove

3.1. The Inverted Downdraft Gasifier

Wood gasifiers can be classified as: Fixed bed (updraft and downdraft); and fluidized bed. Fluidized bed gasifiers Require high power input, and exact controls and are suitable only for large installations.

Updraft gasifiers produce large amounts of tar while consuming the charcoal residue and are not suitable for cooking.

Downdraft gasifiers in the 5-100 kW level were widely used in World War II for operating vehicles and trucks because of the relatively low tar levels. [4] In operation, air is drawn down through a bed of burning wood, consuming the volatiles. The resulting gas then passes over the resulting charcoal and is reduced to a low energy fuel gas. However, since hot gases naturally rise, it is necessary to supply power to draw the gases DOWN through the gasifier.

In 1985 we developed the "inverted downdraft gasifier" (also called "upside down-draft, or pyrolysing gasifier) operating on natural draft. The name comes from the fact that the fuel charge is lit ON THE TOP, and forms a layer of charcoal there; the flaming pyrolysis zone is below that; the unburned fuel is on the bottom of the pile, and primary air for pyrolytic gasification enters at the bottom and moves UP, forming gas in the flaming pyrolysis zone, as shown in Fig. 2.

At that time we built a clean, efficient stove using a jet of compressed air to mix secondary air with the gas and a venturi burner to hold the flame. However, developing country households typically do not have compressed air, so we began development of a natural draft, close coupled cooking gasifier. In 1991 we described a cooking stove based on the inverted downdraft gasifier with natural draft secondary air entering the gasifier above the charcoal zone. The combustion in this stove was relatively clean, but the poor air-gas mixing resulted in a unstable, partly yellow flame. The stove is marketed under the name "GAS-I-FIRE". [5]

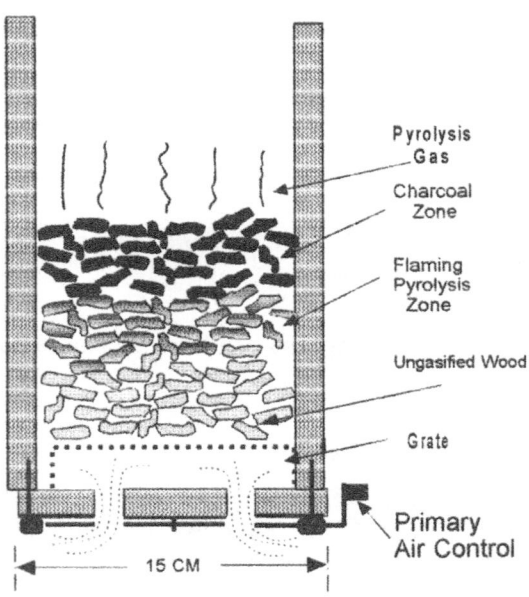

Fig. 2 - Inverted downdraft gasifier made from "riser sleeve", showing primary air inlet, fuel zone, flaming pyrolysis zone and charcoal zone.

The inverted downdraft gasifier is operated in batch mode, appropriate for cooking meals. (The gasifier can also be operated continuously by addition of an auger feed for the fuel at the bottom and an auger to remove charcoal at the top. However, this complicates construction.)

We have built several dozen gasifiers from tin cans, pails, stove pipes and "riser sleeves". We have operated gasifiers varying in size from 10 cm (4 in) to 25 cm (10 in) inside diameter. We have built several dozen burner combinations while looking for clean combustion from the burner.

The simplest gasifiers are made from 2 lb coffee cans using a "church key" can opener to punch holes in the bottom. Heat losses are high in can stoves, but they are very simple to build. We use metal shears for cutting the metal parts, and sheet metal screws for outside attachments and burner and pan supports.

"Riser sleeves" are particularly useful for construction of these stoves. They are made for use in casting molten iron and bronze, so are not affected by the temperatures involved in gasification and combustion. They are available in nominal 3 in to 6 in inside diameter in ½ in steps and in 6 to 10 in inside diameter in one inch steps. All stock sleeves are 30 cm (12 in) tall. As purchased, they are relatively soft and can be cut with saw or razor knife. However, they can also be "rigidized" by application of amorphous silica. They provide excellent insulation, and this is particularly important in small diameter gasifiers. 15 cm OD, 12.5 cm ID sleeves 30 cm tall capable of containing molten steel retail for about $3 in the U.S.[6] We believe that our stove can be built for under US$10 in the U.S. or developing countries.

A major advantage of the inverted downdraft gasifier is that the rate of gas production depends on the amount of primary air admitted to the bottom. For this reason it is very important to put a tight sealing valve on the bottom which permits a wide range of air adjustments. A simple valve made from 24 gauge sheet metal is shown in Fig. 3.

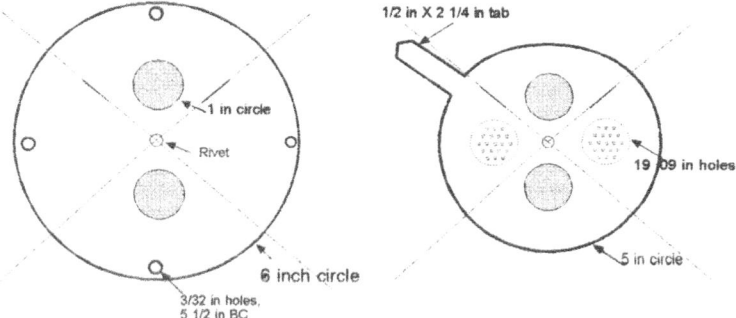

Fig. 3. Details of one type of air valve for coarse and fine control of primary air.

We have also constructed very satisfactory stoves from tin cans with riser sleeve liners. Use of the outer can permits addition of simple handles to the gasifier and burner sections.

3.2. The Wood-Gas Burner

Many burner designs have been tested for the wood-gasifier. The simplest is a series of holes in the gasifier above the level of the charcoal (as discussed for the Hottenroth stove above). However, this does not give good mixing and complete combustion. A better burner consists of a second can mounted a distance of about 1 cm above the first can. Air enters through this annular ring and mixes with the gas and burns.

We have developed a "blue flame" burner (Fig. 4), using a "gas wick" to burn the gas in a very clean manner. In order for the amount of combustion products to have maximum draft, the area must be adjusted. (Conventional gas stoves typically have a ring of fire about 70-120 mm in diameter and about 5 mm across the ring.)

3.3. Wood-gas Stove Fuels

We have successfully gasified 1-3 cm softwood chips, 1-2 X 10 cm hardwood sticks and 5 mm diameter canes from bushes. The hard woods produce a relatively dense charcoal; the softwoods produce a lighter more friable charcoal.

Much of the tests made on the stove used 2 X 1 X ½ cm air dry (about 7% moisture in Denver) Aspen chips in order to have reproducibility . However, we have find the stove operates at least as well on sticks standing vertically. Smaller fuel particles do not permit sufficient air to pass through the bed, and so we have not had success gasifying 0.5 mm wood pellets or corn.

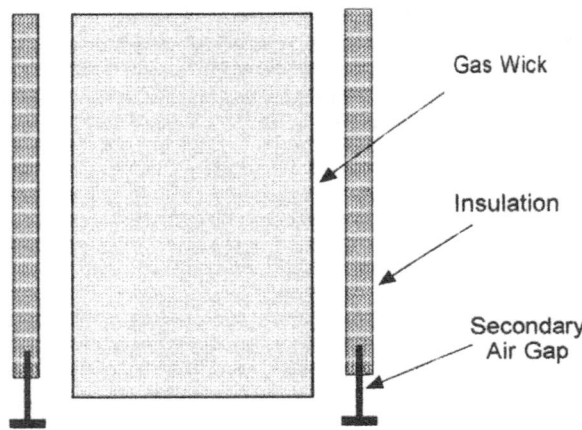

Fig. 4. Annular burner for "blue flame" combustion of wood gas using a "gas wick"

3.4. Wood-Gas Stove Operation

The gasifier is filled to within 2 cm of the top. It is desirable to have the fire spread rapidly laterally across the surface to provide gas over the whole area, so an easily combustible material, such as dry pine needles, fibers soaked in animal fat or vegetable grease or other "tinder" is placed on top. The tinder is ignited in several places with a match and allowed to burn for a minute or so. When a stable flame is established all around, the burner is placed slowly on top of the gasifier to permit the inside surfaces to become hot.

A continuous (smokeless) blue flame was established in 2 minutes and ran continuously for 37 minutes when it went out in Run # 5, p. 92, Book II. The weight of stove (plus the change of weight of 1 liter of water in a pan, added at minute 11) is shown in

Fig. 5. It is seen that the stove burned about 12 g/min with full primary air. At 20 min primary air was turned to 20 % and the stove then burned at a rate of 4.5 g/min. 155 g of charcoal remained at the end of the run.

3.5. Wood-gas stove emissions

Wood-gas cooking has very low emissions and can be operated indoors with proper precautions once established. However, if the fire is extinguished for any reason the production of smoke and CO continues, so that it is important to be able to extinguish it completely or take it outside. The stove is difficult to extinguish and the bottom air valve must be AIR TIGHT. A pan of water can then be placed on top of the gasifier and there will be no subsequent production of smoke.

In a preliminary measurement we have measured the CO level 80 cm above the stove as 22 PPM. The level in the room had no measurable rise. During gasification of wood, if the fire is extinguished the acrid tars in the wood make it unlikely that one would breathe too much CO. However, after the wood is gasified the stove contains hot charcoal and so can be a major source of CO. For this reason it is important to either continue to burn the charcoal or to cut off ALL air to the stove and preserve the charcoal as discussed below.

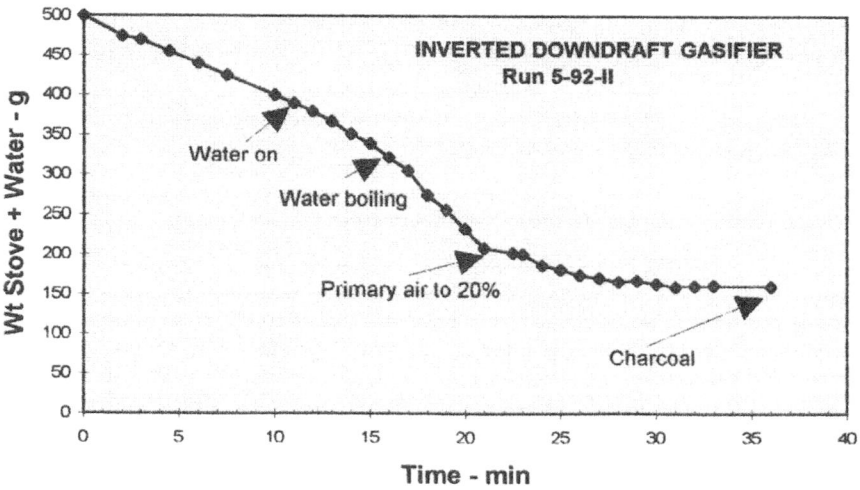

Fig. 5. Typical inverted downdraft gasifier run, No. 5, p. 92, Book II, showing initial stable combustion at 12 g/min, pan on at 11 min, water boiling at 19 min, primary air turned to 20% for simmering at 20 min, and a yield of 150 g charcoal.

3.6. Charcoal manufacture

We find that in a typical run we have a charcoal yield of 10-15% (dry basis). If it is desired to keep the charcoal, it is important that it be possible to close off ALL air to the charcoal after the volatiles are consumed. The valve shown in Fig. 3 should be made of relatively stiff metal to permit a tight closure. The top of the gasifier is closed with a pan over the top. In addition it is useful to spray water on the top pan and the lower parts of the stove to cool it down to prevent continuing oxidation of the charcoal.

4.   STOVE TESTING

Fuel consumption in most of the development runs on the stove were monitored by placing the stove, chimney-burner etc. on a digital balance and recording the weight every minute during the run. In this way it was possible to determine the rate of consumption as a function of primary air setting. A typical chart is shown in Fig. 5.

A mass-energy balance was made in a small gasifier connected to a flowmeter and an integrating total flow (Singer) meter. The flow was fixed at 10 scfh from 3 to 30 minutes. The gas was burned in a 10-15 cm high diffusion flame above the gasifier. 135 g of aspen chips 7.8% moisture produced 26.3 g of charcoal (yield =19.5%, 24.2% dry basis). The total air flow was 160 g of air or 30.8 g O_2. The mass and energy balances are then:

Chips	+	Water	+	Air	➔	Gas	+	Charcoal	
124.5 124.5		10.5		160		269		26.3	grams
2614						1957		657	kJ

The air/fuel ratio is 1.28. The air/fuel ratio for complete combustion is 6.36, so the equivalence ratio (air consumed/air required for complete combustion) for this test is 0.20 (compared to .25 for equilibrium gasification of dry wood. The gas probably has a higher energy content than conventional gas because no air is required for char conversion. The hat gas energy content will be particularly high because it contains the sensible heat and volatiles as well as the gas.

5.   Preliminary modeling of the wood-gas stove

A major advantage of the wood-gas stove is that the gasifier section is stratified and so can be modeled using a simple one dimensional flow scheme. We modeled the conventional stratified downdraft gasifier ("open-top" gasifier, "topless gasifier") and are using the same approach for the inverted downdraft gasifier. The following guidelines are useful for predicting fuel consumption, burn time, stove power and stove size.

Power: The heat content of the gas is ~ 18kJ/g, so consumption of 4-10 g/m produces 1.2-3.0 kWt . (U.S. Electric stoves usually have a 1.5 and a 2.5 kW burner.). The production rate is approximately proportional to cross sectional area. Larger cooking jobs will require larger stoves.

Primary Air: The heat for pyrolysis is about 2,000 j/g. Pyrolysis requires 2/3 g air/per g wood or 600 ml air/g wood. 3 kWt requires 6 l air/min

Primary draft: The pyrolysis gas temperature is about 500 °C and so generates a draft of 100 μm-water pressure

Secondary draft: The gas flow around the gas wick is quite complex and work is in progress on calculation of the gas and air flow

6. Conclusions

The wood-gas stove has been started and operated indoors with no exhaust fans and no odor of burning wood. However, we believe that there is still much work to be done in optimizing the stove for various fuels, determining the effect of moisture, adapting it to various cooking situations and developing other uses. For that reason we are publishing our preliminary results here and hope that others will help adapt these principles to improve world cooking and wood conservation.

7. References

1. See for instance: Barnes, F. Openshaw, K., Smith, K. and van der Plas, R., "What Makes People Cook with Improved Biomass Stoves", World Bank Technical Paper No. 242, Energy Series, 1994.
Kammen, D. M., "Cookstoves for the Developing World", Scientific American, July, 1995, p. 72.
Baldwin, Sam, "Biomass Stoves" Engineering Design, Development and Dissemination, Volunteers in Technological Development (VITA), Arlington, VA. 1987.
Many Papers in "Boiling Point", Intertechnology Development Group, (ITDG) Myson House, Railway Terrace, Rugby, CV21 3HT, UK. (Tel 0788 560631) or the German Assocation for Technical Cooperation (GATT).
Many papers in Publications of Stockholm Envirnment Institue, SEI, Box 2142, S-103 14 Stockholm, Sweden
WOODHEAT FOR COOKING (eds:K.Krishna Prasad and P.Verhaart), Indian Academy of Sciences, 1983.
2. Gao Xiansheng, "Biomass Domestic Cooking Gasifier Stove", in "Advances in Thermochemical Biomass Conversion" Conference, Interlaken, Switzerland, A. Bridgwater, Ed., 1992.
3. La Fontaine, H. and Reed, T. B., "An Inverted Downdraft Wood-Gas Stove and Charcoal Producer", in Energy from Biomass and Wastes XV, D. Klass, Ed., Washington, D. C.,
4. Reed, T. B. and Das, A., "Handbook Of Biomass Downdraft Gasifier Engine Systems", (SERI-1988) BEF Press, Golden, CO.
5. Hottenroth, F,"Gas-I-Fire Stove by ZMART STOVES", Z.Z.Corp., 10806 Kaylor Street, Los Alimitos, CA. 90720.
6. Kalmin 70 riser sleeves, obtained from United Western Supply Co., 4401 E 46th Ave., Denver CO 80216-3261.

INTEGRATED GASIFICATION COMBINED CYCLE BASED ON PRESSURIZED FLUIDIZED BED GASIFICATION

KARI SALO
J.G. PATEL

ENVIROPOWER INC./CARBONA INC.
P. O. Box 610
FIN-33701 Tampere, Finland

Abstract

Enviropower Inc. has developed a modern power plant concept based on an integrated pressurized fluidized bed gasification and gas turbine combined cycle (IGCC). The work is continued since March 1996 by Carbona Inc.

The process is capable of maximizing the electricity production with a variety of solid fuels - different biomass and coal types - mixed or separately.

The development work is conducted on many levels. A gasification pilot test program with a paper mill residue was finished in 1995 and is highlighted in this paper.

The first commercial size IGCC-plant based on Enviropower technology is being built in India. The main fuel of the plant is lignite but mixing biomass i.e. bagasse as feedstock has also been considered.

1 Introduction

The IGCC process is an efficient way to convert bioenergy into electricity and heat. The high efficiency of power and heat generation is due to the combined cycle configuration of the process, the optimized integration of the gasification plant into the gas/steam process and in the case of biomass, the optimized integration of the fuel dryer into the system.

Enviropower has also made a significant effort on the development of hot gas cleanup systems[1,2]. The development work concentrates on sulfur removal for coal gasification and on tar and ammonia destruction and removal for biomass gasification. In cooperation with acknowledged research institutions like the Institute of Gas Technology (IGT) and Technical Research Center of Finland (VTT) significant progress was achieved in this field[3]. The R&D work related to the cleanup of biomass derived gas led to the development of a new type of gas cleaning train (hot gas cleanup) which incorporates gas cooling and dust filtration.

The feasible size of a biomass IGCC plant seems to be between 30-100 MWe[4]. The plant has to be large enough for a reasonable investment/kW, but also biomass must be available in large quantities, which seems to set the upper limit in size in most of the countries.

The development work has led to the commercial phase of business in form of the first nominal 60 MWe IGCC-plant to be built in India.

2 Process description

Since simplified IGCC systems derive one of their main advantages from the close coupled integration of the various subsystems, an optimized integration is a key feature to reach high plant efficiency. The flow diagram of the basic process is shown in the Figure 1.

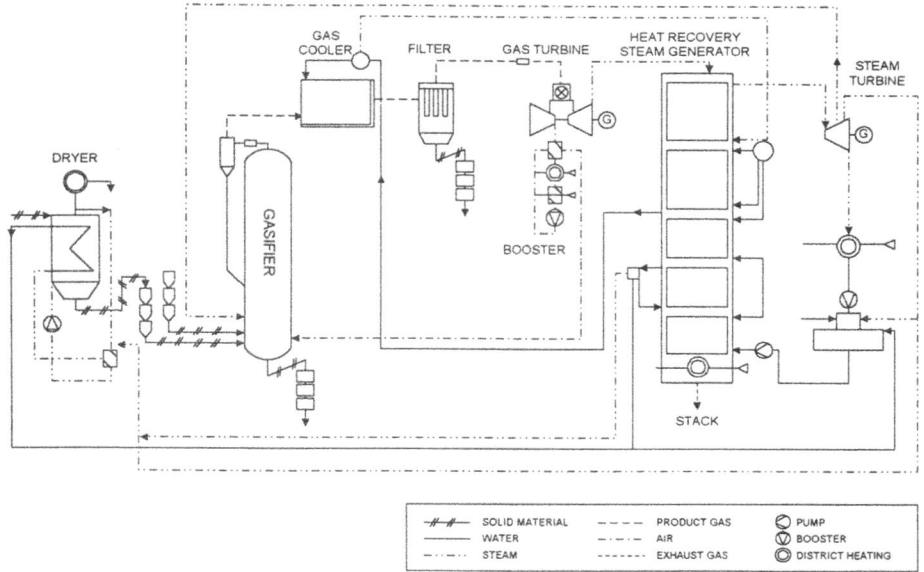

Figure 1.   Enviropower's IGCC Concept

Fuel Drying and Gasification are inseparable parts of the IGCC process for biomass. In Enviropower's gasification system biomass is gasified with air. Biomass has to be dried to 20-30% moisture content to generate a gas of reasonably high heating value. By applying steam drying an efficient integration of the drying and steam process can be realized. The dried fuel is fed through feeders into the fluidized bed of the gasifier. Since the ash content of biomass is low, inert bed material such as sand or dolomite is used.

The gasifier operates at the temperature level of 800-950°C and generates a low calorific value gas of 4-6 MJ/kg lower heating value. The gasifier pressure is in the range of 15 to 25 bar determined by the gas turbine requirements. The product gas leaving the gasifier is cooled in a simple fire tube type of gas cooler to 400-550°C, which is an acceptable temperature for the gas turbine and other downstream equipment.

Hot Gas Cleanup is one of the key parts of a simplified IGCC process. In biomass gasification tar formation, destruction and removal are the main issues. In Enviropower's IGCC concept tar cracking occurs in the gasifier at high temperature and pressure, (using dolomite as long residence time catalyst) efficient mixing in the bed and recirculation of entrained fines. All of these are essential features for efficient gasification and for cracking.

The final gas cleaning step before the gas turbine is the high temperature/high pressure filtration where the dust, heavy metal and alkali metal loading is reduced to a negligible amount. The filter can be of ceramic or metal candle type.

The low calorific value gas application of the commercially available Gas Turbines requires certain modifications to the gas turbine hardware. The gas turbine has to be equipped with modified combustion chamber for burning the biomass derived gas efficiently and with low emissions (low-NOx burner). Air has to be extracted from the gas turbine after the compressor for the gasification process. The pressure and temperature of the gasification air is adjusted in a booster compressor/heat exchanger system to the appropriate level.

The Steam Cycle section of the process (gas turbine/steam turbine) includes mainly commercial technologies, as a usual combined cycle plant. The product gas cooler, operating as an evaporator, is coupled to the high pressure section of the HRSG. The steam process coupled to the gasification plant/gas turbine complex is designed to maximize the electricity production of the process with high efficiency allowing also the required heat production.

3 Process performance

The size of the IGCC plant is determined by the selection of the gas turbine. The number of suitable gas turbines is very limited today. The main issue is the development of the GT-combustion systems to burn LHV-gas. The turbines of main manufacturers who quote their turbines for biomass IGCC today are listed below, with consumption and performance data.

Table 1. Gas Turbines for Biomass IGCC

Gas Turbine	IGCC Power, MW	Net Efficiency, LHV, %	Biomass Consumption	
			tn/a	m³/a
EGT, Typhoon	7.5	42	64000	200000
Mitsubishi, MW151	39	45	300000	940000
GE, F6B	62	48	465000	1.5 million
Westinghouse, CW251	72	48	540000	1.7 million
Siemens, V64.3	97	48	865000	2.7 million
GE, F6FA	110	49	921000	2.9 million

Assumptions: 8000 h/a, wood waste of 50% moisture AR, condensing plant

Aeroderivative gas turbines with small combustors and without possibility of low-NOx combustion (i.e. low ammonia to NOx conversion) are not considered as a potential choice for this kind of process application.

The power output and the calculated efficiency of IGCC in condensing mode are also presented in the Table 1. Also biomass amounts (wood based) needed for each size of IGCC are also given.

The IGCC process incorporates advanced control systems that will significantly reduce the level of gaseous pollutants being emitted. Furthermore, the better efficiency of the plant as compared to conventional and competing technologies will also reduce the amount of so called greenhouse gases. The anticipated emissions, when gasifying wood based residue are as follows:

SO_2	< 10 mg/MJ
NOx	<50 mg/MJ
Particulate	< 5 mg/MJ
CO_2	<780 gCO$_2$/kWh

4 Feasibility of Power Production from Biomass

The economical feasibility of the application of a power generating system has to be examined case by case because of the different conditions in each country and case. The trends of power generation price formation discussed below are for Finnish conditions. In Finland and in the Scandinavian countries the biomass and electricity prices are low since biomass is available in large quantities and power is produced from many sources which influences significantly the economical feasibility of a biomass fueled IGCC.

The *specific investment* of the biomass IGCC power plant (based on Enviropower's process concept) and a conventional fluidized bed (FBC) combustion technology based power generation process are compared in Figure 2, ref 4,5.
The specific investment cost of and IGCC plant declines much more rapidly than that of a conventional plant when the power plant size is increased. The investment cots of the IGCC plant is the estimated cost of a commercial plant after demonstration phase.

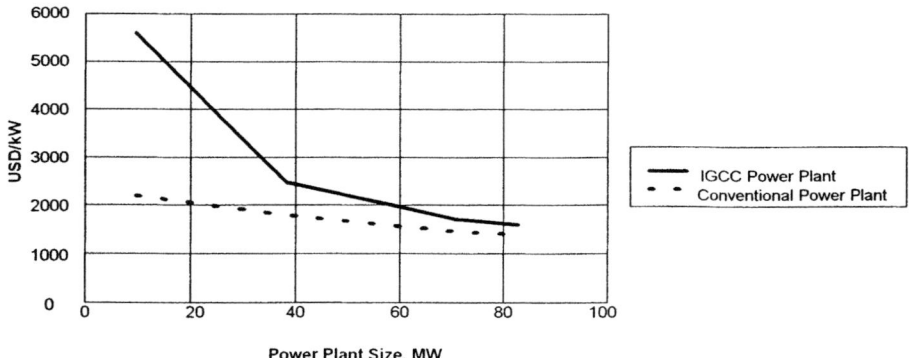

Figure 2 - Specific Investment Costs

The *cost of electricity* was calculated for condensing power generation and for typical process heat generation in the pulp and paper industry based on 8000 hour annual operating time, 20 years life time, 12% interest rate, 3% inflation and price escalation, assuming similar fuel prices and Internal Rate of Return (IRR) for each case. Typical efficiencies, availabilities and operation costs are considered for each case.

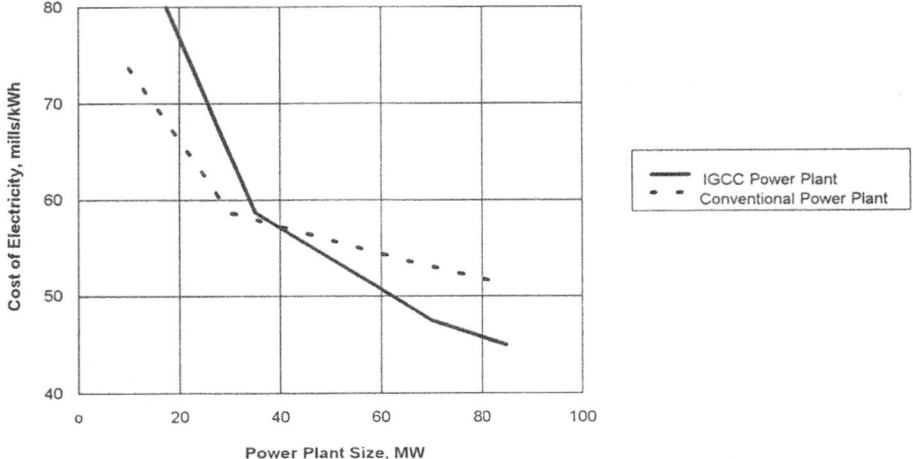

Figure 3 - The Cost of Electricity, Condensing Power Plants

Figure 3 shows the cost of electricity (COE) produced in a biomass IGCC power plant and in a conventional biomass fluidized bed combustion (FBC) power plant, considering both as a condensing power plant. The price of biomass fuel (forestry wood waste) was assumed to be \$2.6/million Btu. Power plant sizes up to 80 MW size were investigated. Power plants over 80 MW sizes are not realistic in most of the cases due to the limited availability of biomass and the limitations on fuel supply logistics (refer to Table 1) and related fuel price. COEs and the IGCC plants decrease more rapidly compared to the conventional power plant when the plant size increases. The COE of the first of the kind IGCC power plant will be equal to the COE of a fluidized bed conventional power plant at about 35 MW plant size. Thus this study can be concluded that at Scandinavian conditions the plant size of the biomass fueled IGCC plants will be between 35 and 80 MW and that the application of the biomass IGCC is economically more feasible than that of a conventional plant, even for only power generation.

Figure 4 shows the development of COE and electricity for both power plant types at *industrial applications*. In these applications the back pressure heat supply is main requirements, thus the electricity generation and purchase of the industrial power plant is also determined by the heat demand. Typical process conditions of a paper mill of thermomechanical pulp production were considered assuming load control with a condensing steam turbine. The same heat price was assumed for both generating technologies.

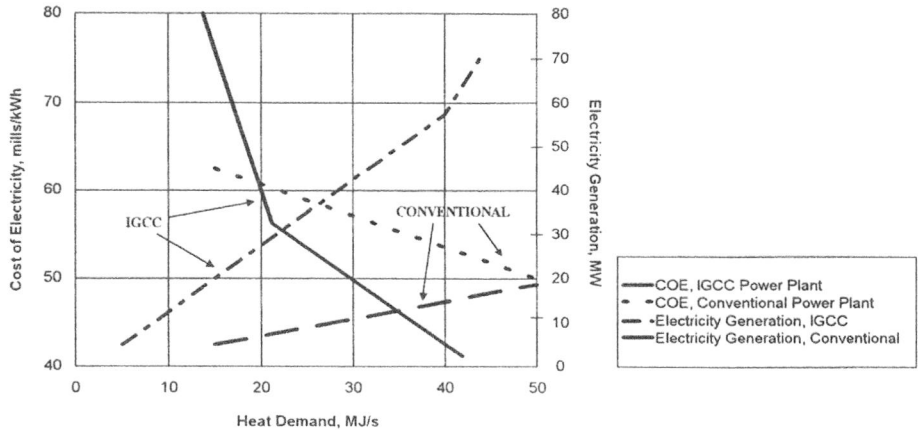

Figure **4** - Process Heat Generation in Pulp and Paper Industry

The diagram shows that the application of IGCC power plant is more feasible at a heat demand over 20 MJ/s (68 MMBtu/h) and it produces more than twice as much electricity as a conventional, FBC technology based power plant.

5 Potential of Biomass IGCC Power Production

Biomass availability and potential as well as IGCC market share was investigated by using information from the World Energy Council, the Commission of the European Communities and the U.S. Department of Energy, ref. 4.

In short term the largest potential for biomass IGCC applications seems to be in countries where biomass waste is generated by large industrial units using biomass as raw material. The two major industries are pulp and paper industry with the related forestry producing wood waste and the sugar industry producing bagasse.

The investigation concluded that about 10% of the national biomass potential of different countries could be used as fuel for biomass IGCC power plants in the size range of 30-80 MW. The estimated short term biomass IGCC potential by 2010 and the potential areas of applications are presented in Figure 5.

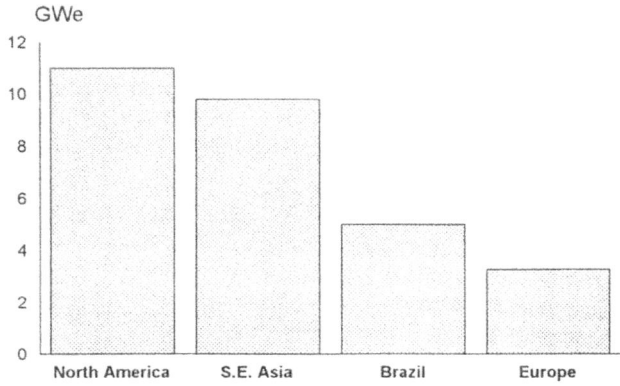

Figure **5** - Estimated Biomass IGCC Potentials by 2010

999

6 Biomass IGCC process development

Enviropower's biomass fueled IGCC process is based on the extensive gasification know-how and experience of IGT and further development by Enviropower[1,4]. In order to minimize the technical and economical risks related to design and operation of a demonstration plant, a development and test program was established to verify the critical process steps in the IGCC concept. A test program has been accomplished covering all aspects of a biomass based IGCC process. The program has included pilot or full scale testing of the following IGCC subprocesses.
- drying of biomass in three different types of steam dryers
- feeding of biomass to a pressurized system
- pressurized gasification and gas cleaning
- combustion of low calorific value gas in a gas turbine combustor.

Gasification test results with a paper mill residue, mainly bark, are highlighted in the following.

6.1 Pilot plant

The Pressurized fluidized bed Gasification Pilot Plant is located in Tampere, Finland. The pilot plant includes all essential modules for research, component testing and to complete the development of the gasification process for IGCC applications. The pilot plant has a maximum thermal input of 15 MW and it can be operated at up to 30 bar pressure and 1100°C temperature. The process flow diagram of the pilot plant is shown in Fig. 6.

The gasifier is a single stage pressurized fluidized bed gasifier. Metallurgical coke, which is used as fuel for cold start up is fed through the coal lock hopper system to the gasifier. Biomass feed to the gasifier is accomplished through a separate lock hopper system and a screw feeder which enters the bed above the fluidization grid. Nitrogen is used for pressurizing the lock hoppers and for screw purge. The dolomite feeding system, based on the lock hoppers, is also pressurized by nitrogen.

No external steam is used in the gasifier during biomass gasification. The excess steam produced in the system (gas coolers) is directed to the auxiliary condenser. The gasifier temperature is controlled by air and biomass flow rates, and the pressure is controlled according to the freeboard pressure measurement.

The gasifier ash (bed additive) is removed through the bottom of the gasifier while the filter fines through the filter ash removal system, both using lock hopper systems. In all cases nitrogen is used for lock hopper pressurization and ash conveying from the lock hopper to the weight hopper.

The bulk of the fines elutriated from the fluidized bed are separated from the product gas in an external cyclone. The fines from the cyclone are returned to the fluidized bed. During this test program two different cyclones were tested separately. The product gas leaving the gasifier cyclone is cooled in two steps: first to 400-600°C, then after hot gas cleanup 200-350°C. Both gas coolers are fire-tube type of boilers. After the first gas cooling stage, the remaining dust particles are removed from the gas stream by a ceramic candle filter cleaned by nitrogen pulsing. The filter unit temperature can be controlled in a certain range by the gas cooler by-pass.

The pilot plant is equipped with an extended process measuring system which is connected to the data information system. Over 1000 parameters (gas flows, temperatures, pressures, etc.) are measured continuously, product gas and flue gas quality is analyzed on-line and all solids flows are sampled.

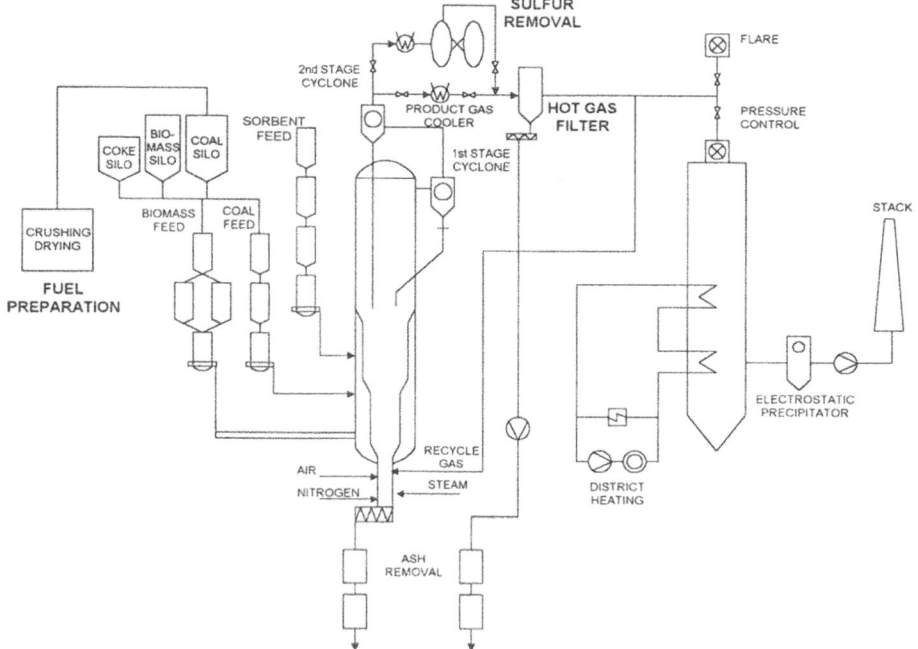

Figure 6 - The Process Flow Diagram of the Pilot Plant

6.2 Biomass gasification tests

The biomass gasification test program was started in 1993 as a part of the joint development project of Vattenfall and Enviropower to demonstrate the IGCC technology for forest based biomass fuels using air blown gasification and hot gas cleanup. A test program for paper mill residue was conducted in 1995[14]. The main objective of the test runs was to verify and demonstrate the ability of the gasifier process and hot gas cleanup system under commercial scale IGCC plant operating conditions.

A total of 3000 tons (metric) of wood based biomass (as received) was gasified during 900 test hours during 1993. Long, stable set points were achieved under commercial plant operating conditions verifying the suitability of the process for forest residues.

In 1995 a total of 1600 tons of paper mill residue and also willow was gasified with extremely good results as described in the next chapter[15]. A summary of biomass fuels tested is presented in the Table 2 below.

Table 2 - Biomass Pilot Test Runs

Operation pressure	14 - 18 - 22 bar
Operation Temperature	800 - 950 °C
Plant Capacity	15 - 17 MJ/s (MWth), 100 ton/day

FUELS GASIFIED (1993 - 1995)

	Amount/tn
Wood Chips	1630
Forest Residue	1750
Paper Mill Waste (bark, paper, sludge)	1180
Willow	400
Straw with Coal	20 (+120 tn coal)

6.3 Gasification of paper mill residue

Three test runs were carried out at the Enviropower Pressurized Gasification Pilot Plant during spring 1995 using paper mill specific fuels. The main objectives of the test runs were to demonstrate the ability of the gasifier and hot gas cleanup system for gasification mill specific fuels and to provide firm technical basis for the engineering of a commercial project.

The test runs were carried out using paper mill specific fuel, and mixtures with forest residues. Mill wood waste consisted of bark and sludge (bark, fiber and bio), and some amount of waste including paper, wood residue and plastics. Forest residue was trunk wood including also branches, bark and needles, representing a wide variety of wood types of Finnish origin including for example spruce, pine, birch, alder and aspen.

More than 5000 m³ (as received) biomass fuel was gasified during 360 hours of biomass gasification. The main test variables were fuel type, fuel moisture, freeboard temperature, pressure, filter temperature, fluidization velocity (part load performance), gas residence time, bed material feed rate, bed material type and bed size, temperature of filter pulsing gas, gasifier controls, and gasifier gas flow arrangements.

During these test runs the operation of the gasifier was stable and reliable. The gasifier was easily operated establishing different kind of temperature distributions in the freeboard area and varying the air distribution. Biomass was gasified successfully without external steam supply, without freeboard air and using low dolomite feed rate (only 0,5 % of biomass feed rate). Gasifier and biomass feeding system allowed the feeding of biomass of large particle size (max. particle length about 50 mm). Gasifier was also operated successfully below 50 % part load under constant pressure. The gasifier can tolerate high momentarily variations of bed temperature and fluidization velocity resulting in a wide operation window and it can be easily recover from process disturbances. Operation with the cyclone resulted in reliable and efficient cyclone behavior. The pilot plant was also started up successfully with pure wood fuel from warm conditions.

The carbon conversion was in the range of 97-99 %. MAF fuel conversion (MAF = moisture and ash free) was thus about 98-99+ %. The gasifier bottom ash discharge flow contained only small amounts of carbon originated mainly from dolomite. The carbon conversion increased with increasing air ratio and steam/carbon-ratio.

The LHV of wet product gas, was in the range of 3.4 - 5.0 MJ/Nm³ when the fuel moisture varied between 18-33 %-w. The product gas LHV of 4.2 MJ/Nm³ (wet) was obtained at 30 % fuel moisture content. This level of LHV provides proper limit for momentarily moisture or quality variations of the fuel because LHV of 3.5 MJ/m³$_n$ is still suitable for gas turbines. The product gas LHV will still increase in large scale applications due to the heated gasification air, reduced feeding of inert gas and lower relative heat losses.

The total amount of heavy tars, which may cause problems in gas cooling and filtration, were in the range of 25-160 mg/mm³, which is a low level. The total light tar concentrations including benzene were in the range of 5-10 g/m³$_n$. Benzene and naphthalene represents about 95 %-vol of the total light tar concentration. The sum of the light tar compounds (from pyridine to pyrene) was low in every set point, varying between 1.2-2.5 g/m³$_n$. The concentration of benzene was in the range of 4.1 - 7.9 g/m³$_n$. About 25 % decrease in light tar yield was obtained when the total gas residence time was raised and alternatively by increasing the freeboard temperature about 50°C. The most obvious mechanism of the tar destruction is the thermal cracking.

The concentrations of ammonia and HCN were in the range of 1450-2810 ppm-vol (dry) and 10-30 ppm-vol (dry), respectively. The conversion of fuel bound nitrogen to ammonia with mill specific fuel was in the range of 55 and with wood about 40 %. The measured TRS (Total Reduced Sulfur) concentrations in wood gasification were in the range of 80-100 ppmv (dry) and in mill specific fuel gasification 180-300 ppmv (dry). The ammonia and TRS concentrations in the product gas depend mainly on the nitrogen and sulfur content of the feedstock.

The fate of fuel alkali metals in gasification was examined by carrying out vapour-phase alkali measurements after the ceramis filter and by analyzing the alkali content of the solids streams. The measured vapour-phase alkali concentrations (Na+K) were mainly below 0.05 ppm-wt. Main part of the alkalines (typically 80-95%) was found

Table 3 - Pilot Test Program/ Test Results

		Forest residue	Bark
Pilot plant fuel input	MJ/s	15	15
Product gas LHV (dry)	MJ/m³	4.5 - 5.6	4.5 - 5.6
Fuel conversion	%	97 - 99	97 - 99
Product gas dust content after ceramic filter	mg/m³N	< 5	< 5
Alkalines (K + Na)	ppm(w)	0.01	0.01
Product gas H₂S content	mg/m³N	15 - 50	-200
Product gas NH₃ content	mg/m³N	500 - 1500	-2200

from the ceramic filter fines. The HC1-concentration in the product gas stayed at a low level (<30 ppm-wt) and most of the concentrations were under the analyzing limit.

The ceramic filter unit functioned well. Dust concentrations after the filter were mainly below 5 mg/Nm^3. This level of dust loading is accepted by gas turbine manufactures, and it is well below the value required by environmental regulations. In all test runs the filter dp-baseline stayed low and there were no difficulties to clean the candles by nitrogen pulsing. Filter operation with slightly warmed pulsing gas (85-110°C) resulted in normal filter behavior. The filter unit was operated in the temperature range of 450-570°C. The variation of filtration temperature caused no major changes in the filter behavior.

7 First IGCC cogeneration plant to India

IBIL Energy Systems (IES) of Madras, India is developing an innovative nominal 60 MW integrated combined cycle (IGCC) cogeneration plant based on Enviropower gasification technology. Work has already begun on the plant which will supply power and steam to a private 2.6 million ton per year cement plant being built by Sanghi Industries in the Kutch region of Gujarat State in India. The combined cycle part of the plant has already been purchased from General Electric Co., USA.

The IES/Sanghi project will be the first commercial scale IGCC plant in India and the first in the world to use lignite as the main fuel. The captive or "inside-the-fence" project will deliver power and process steam on a build, own, operate, transfer (BOOT) basis to a green-field cement plant. IES will own and operate the plant for the first ten years, after which time ownership will be transferred to Sanghi Cement. The IES/Sanghi plant will conclusively demonstrate the commercial viability of IGCC technologies in India and will encourage rapid diffusion of this environmentally-friendly technology throughout the country.

The IES/Sanghi cogeneration plant will provide peak power of 52.5 MW and an average of 30 MW to the cement plant. Surplus power will be sold to the grid. The project will also provide 40 tons per hour of steam for a captive desalinization plant which will provide process water to the cement plant and domestic water to the local community. The plant will initially use naphtha to generate power while the gasification island is being completed. Upon completion of the gasification island, the plant will gradually shift to lignite as the fuel.

The project features a simplified air blown, pressurized fluidized bed gasification technology developed by the Institute of Gas Technology (IGT) of Chicago, Illinois and perfected by Enviropower. The power island includes a 38 MW gas turbine, a 90 ton per hour heat recovery steam generator (HRSG), and a 26 MW steam turbine. General Electric will provide the gas and steam turbines for the power island. Ignifluid Boilers India, Ltd. (IBIL) will manufacture the gasifier and the HRSG.

Erection of power island equipment and civil works construction at the site has begun. The first stage of the power island (the gas turbine) will be operational by October 1996 and the second stage (combined cycle operation with the steam turbine) by March 1997. The gasification island contract with Enviropower was signed in March

1996. The gasifier island will be operational in 1998.

The IES/Sanghi project will demonstrate the technical and commercial feasibility of IGCC technologies in industrializing countries. The project is especially important because it will clearly demonstrate the technical and economic viability of mid-size IGCC cogeneration plants which are ideal for captive industrial projects and distributed generation applications.

8 Conclusions

Power production by an IGCC process is a niche market area with high market entry barriers but a lot of expectations. The development of the gasification technology has taken a long time and is highly capital intensive, which limits the number of potential developers.

The relatively high investment cost of an IGCC plant means that the efficiency and reliability of the plant must be maximized to reach a low specific cost (per kilowatt). This is possible only by using pressurized gasification technology, hot gas cleanup and advanced heavy duty gas turbines.

The minimum size of an IGCC plant is determined by the plant economy which however, differs from country to country. In small size up to 20-30 MWe the conventional boiler/steam turbine process seems to be the right choice in most cases. The maximum size of the plant seems to be determined by the availability of huge amounts of biomass at a reasonable price level. In countries with large forestry and pulp and paper industry plant sizes of 80-100 MWe seem to be feasible.

The IGCC technology is entering the phase of demonstration and commercialization. Pilot testing of different biomass has shown that wood based fuels are most suitable for gasification. The first large scale IGCC-plant based on Enviropower technology is being built in India by Ignifluid Boilers India and Carbona Inc.

9 References

1. 13th EPRI Conference, USA, (1994), "The Development of a Biomass Based Simplified IGCC Process", Liinanki, Horvath, Lehtovaara, Lindgren
2. Power production from biomass, 2nd. seminar, March 1995, Finland " Biomass IGCC", Salo, Keranen
3. Conference on New Power Generation Technology, USA (1995), "Demonstration of Biomass Gasification for IGCC Application", Horvath, Lehtovaara, Salo
4. CEPEX Conference, Poznan, Poland, April 1995, "Biomass Gasification", Keränen, Salo
5. 14th EPRI Conference, USA (1995), "Demonstration of Biomass Gasification for IGCC application", Horvath, Lehtovaara, Salo

IGCC POWER PLANT FOR BIOMASS UTILISATION VÄRNAMO, SWEDEN

K. STÅHL
Corporate R & D, Sydkraft AB, Malmö, Sweden

M. NEERGAARD
Sydkraft Konsult AB, Malmö, Sweden

J. NIEMINEN
Foster Wheeler Energia Oy, Varkaus, Finland

P. STRATTON
European Gas Turbines Ltd., Lincoln, United Kingdom

1 Introduction

It is expected that increasingly heavy demands will be made on future power plants in terms of efficiency, impact on the environment, fuel flexibility, power production costs etc.

Biomass fuels are in most countries a domestic source of fuel and are often found as waste products in different kinds of industries e.g. agriculture, forestry, pulp- and paper. Further, a lot of biomass garbage such as e.g. packaging is presently landfilled and in future there will probably be more stringent requirements to recycle garbage, thereby reducing landfilling.

Carbon dioxide, methane and freons are examples of greenhouse gases that absorb infrared radiation from the earth and contribute to the net increase of radiation energy in the atmosphere, which results in a temperature increase. In the debate on the greenhouse effect carbon dioxide is generally considered to be the main culprit, simply because its concentration in the atmosphere by far exceeds that of other gases, although it is a weaker greenhouse gas than the other gases. By utilising biomass fuels as feedstock there will be no net increase in carbon dioxide levels from power production over a relatively short time period, in contrast to a situation where fossil fuels are used.

Integrated gasification combined cycles (IGCC) have been developed and demonstrated for power generation using fossil fuels as feedstock. The main features are the possibility of cleaning the gas produced to remove from impurities, such as particulates, sulphur, etc. under pressure before the gas enters the combustor of the gas turbine, and also the relatively high electrical efficiency. Higher efficiencies also leaves relatively lower emissions.

IGCCs can generate electricity and heat out of solid fuels and also out of liquid and gaseous fuels by using the combined cycle alone.

On the basis of these considerations, Foster Wheeler Energy International, Inc. (formerly A. Ahlstrom Corp., Pyropower sector) and Sydkraft AB of Sweden have been developing the pressurised IGCC for biomass fuels since 1991.

A demonstration plant was built at Värnamo, Sweden and has been undergoing commissioning since mid in 1993. During the autumn in 1995, the plant was fully integrated and the gas turbine was successfully operated on gas produced in the gasifier. An aerial view is shown in Figure 1.

2 The Bioflow concept

Bioflow Ltd. a joint venture company was formed in 1992 for the purpose of marketing the technology. Bioflow Ltd. will exploit Foster Wheeler´s and Sydkraft´s common experience from the development work.

The Bioflow concept effectively utilises biomass fuels through pressurised gasification and a combined cycle, i.e. a combination of gas turbine and steam turbine. In the co-generation mode, it has a high ratio of electric power - to - heat generated, typically 0.8 - 1.2. Considering that the net electrical efficiency is 40 - 45 %, this gives an efficient way of generating electricity and heat out of solid biomass fuels. In the condensing mode, the electrical efficiency is 44 - 50 %. Of course, the efficiency is highly dependent on the gas turbine being used.

The fuel is dried in an integrated flue gas dryer to a moisture content of 10 - 20 %. The fuel particle size is typically approximately 30 mm. The dried fuel is pressurised to a value which is determined by the pressure ratio of the gas turbine, and is fed into the gasifier. This is an air blown circulating fluidized bed (CFB) gasifier with an operating temperature of 950 - 1000°C.

The fuel is pyrolized immediately on entering the gasifier. The gas transports the bed material and the remaining char towards the cyclone. In the cyclone, most of the solids are separated from the gas and are returned to the bottom of the gasifier through the return leg. The recirculated solids contain some char which is burned in the bottom zone where air is introduced into the gasifier. The combustion maintains the required temperature in the gasifier.

The formation of tars in the gas is minimised by thermal cracking due to the high gasification temperature and the catalytic effect of the bed material. The gas produced is cooled in a gas cooler where steam is generated.

The gas is cleaned in a hot gas filter where particulates are removed. Ash is removed from the process both via the bottom of the gasifier and via the hot gas filter.

The purified gas flows to the combustion chambers of the gas turbine. Since the lower calorific value of the gas is only about 5 MJ/m^3n, many of the commercially available gas turbine combustors require modifications in order to ensure efficient combustion and acceptable flame stability.

Air is extracted from the gas turbine compressor and is pressurised further in a booster compressor to supply the gasifier.

The flue gas from the gas turbine enters a heat recovery steam generator (HRSG) which contains a superheater, evaporator and economiser. Steam is generated and fed to the steam turbine along with the steam from the product gas cooler. The steam turbine cycle could be of condensing type or of back-pressure type for process steam or heat for district heating.

3 The demonstration plant at Värnamo

In June 1991, Sydkraft took the final decision to build a co-generation plant at Värnamo, Sweden, to demonstrate the technology. The plant generates 6 MW of electricity and 9 MW of heat for district heating.

The Värnamo demonstration plant is the first of its kind in the world. The plant is aimed at demonstrating the complete integration of a gasification plant and a combined cycle plant, fuelled by biomass. The basic idea is to demonstrate the performance of the technique rather than to run a fully optimised plant. Flexible and conservative solutions were chosen for the plant layout and design, to ensure the success of the project and to make the plant suitable for R & D activities.

4 Planning and procurement

Detailed plans were submitted, and a positive attitude of the local authorities paved the way to all the necessary permits and licences being granted after 6 months of negotiations.

The engineering work was done jointly by Foster Wheeler and Sydkraft Konsult (a wholly-owned subsidiary of Sydkraft). Procurements has started immediately after the decision to build, and construction work began on site in September 1991.

The beginning of commissioning was scheduled for the spring of 1993. Because of the rather short time schedule, it was necessary to split up the deliveries in parts which were optimal for different suppliers. As a result, Sydkraft have made more than one hundred procurements and undertook the co-ordination between all the suppliers. Sydkraft has also assumed the main burden of functional responsibility.

5 Fuel preparation plant

The Värnamo plant consists of two separate parts i.e. a fuel preparation plant and the co-generation plant. The fuel preparation plant upgrades the wood fuel to a moisture content of 10 - 20 % and a definite size distribution. Drying takes place in a rotary drum dryer and the flue gas, which is used as the drying medium, is produced in a small gratefired boiler. Upstream of the stack, the flue gas is cleaned in two cyclones and a scrubber.

Fuel is crushed into two different fractions in the plant. The larger fraction is used in the gasifier and the smaller fraction (< 2 mm) is delivered to another facility in a neighbouring city where it is used as a substitute for oil in an old boiler that was previously oil fired.

One reason for not integrating the fuel drying into the IGCC is the fact that two thirds of the fuel produced is sold for external use.

6 Power plant

A simplified process diagram is shown in Figure 2 and a view of the gasification plant is shown in Figure 3.

The dried and crushed wood fuel is pressurised in a lock-hopper system and is fed by screw feeders into the gasifier located a few meters above the bottom. The operating temperature of the gasifier is 950 - 1000°C and the pressure is approximately 18 bar (a).

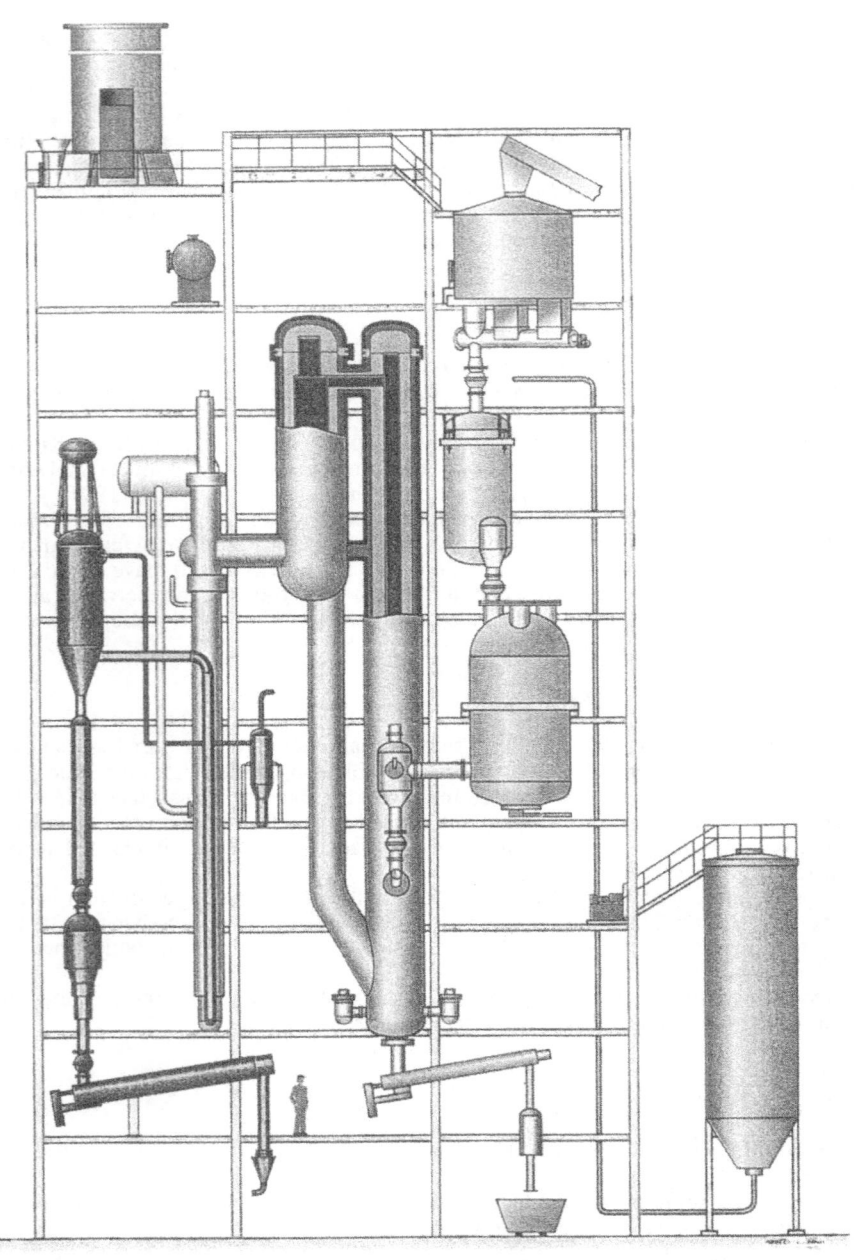

**BIOFUEL GASIFICATION PLANT
VÄRNAMO, SWEDEN**

PROCESS-SCHEMA

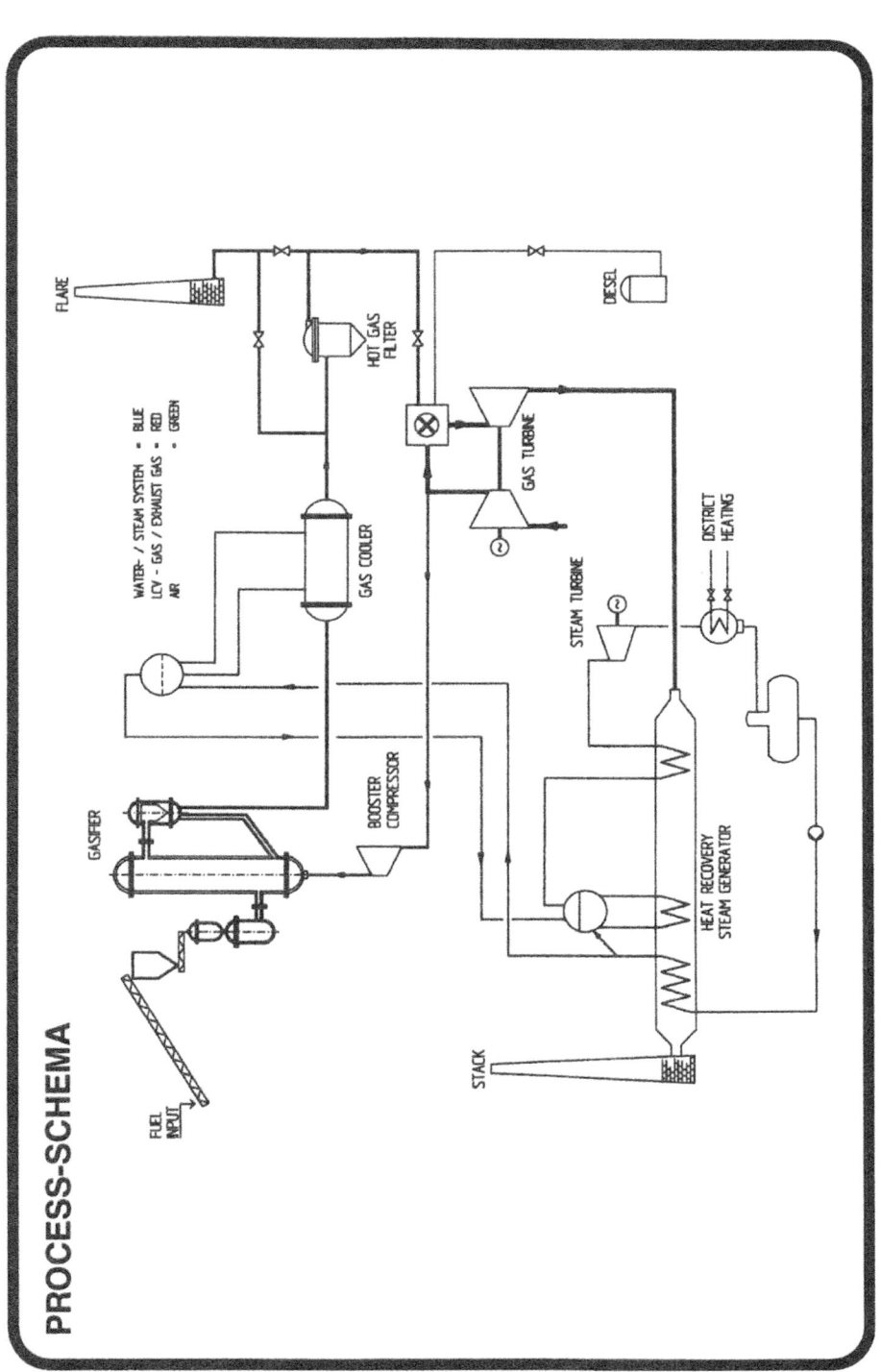

FLARE

HOT GAS FILTER

DIESEL

WATER- / STEAM SYSTEM = BLUE
LCV - GAS / EXHAUST GAS = RED
AIR = GREEN

GAS COOLER

GAS TURBINE

STEAM TURBINE

DISTRICT HEATING

GASIFIER

BOOSTER COMPRESSOR

HEAT RECOVERY STEAM GENERATOR

FUEL INPUT

STACK

The gasifier is of a circulating fluidized bed type and consists of the gasifier itself, cyclone and cyclone return leg. The three parts are totally refractory lined. After the cyclone , the gas produced flows to a gas cooler and a hot gas filter. The gas cooler is of a fire tube design and cools the gas to a temperature of 350 - 400°C before it entering the ceramic filter vessel where the particulate clean-up occurs. Ash is discharged from the ceramic filter and from the bottom of the gasifier, and is cooled before entering the depressurization system. The gasifier is of air-blown type and about 10 % of the air is extracted from the gas turbine compressor and is compressed further in a booster compressor, and then flows to the bottom of the gasifier.

The gas generated is burned in the combustion chambers and expands through the gas turbine, generating 4 MW of electricity. The gas turbine is a single-shaft industrial gas turbine. The fuel supply system, fuel injectors and the combustors have been re-designed to suit the low calorific value gas. Because of the large volume of gas, the control valve and stop valve are housed in a special fuel module outside the gas turbine enclosure. The new combustors are designed for the low calorific value gas but can also burn diesel oil which will be used during plant start-up. The hot flue gas from the gas turbine is ducted to the heat recovery steam generator (HRSG). The steam generated, along with steam from the gas cooler, is superheated in the HRSG and is supplied to a steam turbine (40 bar, 455 °C). The steam turbine is of a simple design and does not have a particularly high efficiency, which negative affects the electrical efficiency as well as the power-to-heat ratio. As mentioned earlier, the purpose is to demonstrate the complete pressurised IGCC in terms of reasonable size and cost rather than top performance.

The plant is equipped with a flare on the roof of the gasification building. The flare is used during the start-up procedure and when testing less well-known conditions, in order to protect the gas turbine.

Electric power is generated both by the gas turbine (4 MW) and by the steam turbine (2 MW). The combined cycle can be run as a co-generation plant and be cooled by a district heating system or by separate air coolers. The reason for installing air coolers is to achieve independence of the heat load when testing. The technical data is summarised in Table 1.

The Värnamo plant is dominated by the 40-metre high gasifier tower. The walls of the gasifier tower consist of overlapping horizontal metal sheets. The gaps between the sheets are big enough to allow gas to be ventilated away in the event of leakage. The gasification plant is classified as an outdoor installation.

The combined cycle plant is compact and placed in a conventional building with metal panels on a steelwork frame. Offices, workshops and storage space are provided in a separate building, where the control room is also located.

7 Plant commissioning

Commissioning of the plant started late in 1992 by start-up of the fuel preparation plant. Commissioning of the combined cycle was completed on liquid fuel during March 1993. The first gasification test on wood chips at an elevated pressure was performed in June 1993. Combustible gas with the predicted composition was produced and burned in the flare. Until February 1996, the plant has been in operation for more than 4000 hours.

Commissioning of the combined cycle part of the plant was carried out according to plan, which has to be expected since the design is almost conventional. However, commissioning of the gasification plant as well as the total integration of the two parts has been slightly more time consuming than planned.

TABLE 1 TECHNICAL DATA OF VÄRNAMO PLANT

Plant rating	18 MW_{fuel} (85%ds)
Fuel	Wood chips
Lower calorific value of Product Gas	5 MJ/m^3n
Gasification pressure	18 bar.
Gasification temperature	950 - 1000 °C
Total net efficiency (LCV)	83%
Power generation	6 MW_e
Heat for district heating	9 MW_{th}
Steam pressure	40 bar
Steam temperature	455°C
Plant owner	Sydkraft AB

Suppliers:

• Engineering	Sydkraft Konsult/ Foster Wheeler
• Gasifier	Foster Wheeler
• Gas cooler	Foster Wheeler
• Ceramic filter	Schumacher GmbH
• Gas turbine	European Gas Turbines Ltd.
• Booster compressor	Ingersoll- Rand
• Steam turbine	Turbinenfabrik Nadrowski GmbH
• Heat recovery steam generator	Foster Wheeler

For cost reasons, very few redundant components and systems are installed and , considering the high degree of integration and the complexity, the risk of delay in commissioning always exists.

This and the new technology of the gasification process were considered when planning the commissioning. We have to admit that what could be considered reasonably conventional technology has caused in some instances unexpected delays. This in particular applies to the handling of solid materials.

Furthermore, different bed materials heve been tested to find out how they affect the process. During tests with different materials, temperatures and pressure levels, deposits sometimes occurred. Deposits and fouling have verified the importance of controlling the process and ensuring suitable design of components.

It is of great importance to operate the whole process under suitable conditions to achieve the best results, i.e. minimum posible deposits and best possible gas quality. What we have achieved at present is to run the process quite satisfactorily, but we are confident that we can optimise it further. However, before performing this optimisation work on the gasification process, it was decided to demonstrate the possibility of running the gas turbine with product gas and full plant integration.

Accordingly, in the late spring of 1995, the product gas fuel supply system, fuel injectors and combustors were installed in the gas turbine. In the beginning of September, after the summer shut-down, the modified gas turbine commissioning on liquid fuel was completed, and the first test runs on product gas were performed in October. In order to minimise the risks, the first test runs were very short and product gas was introduced gradually with a corresponding reduction in liquid fuel, which eventually resulted in operation solely on product gas. Observations during the fuel changeover phase and during operation on 100 % product gas showed no adverse effect on either combustion performance or turbine behaviour. Unfortunately, leakage of liquid fuel caused a fire inside the gas turbine enclosure in early November, and the plant had to be shut down for a couple of months, as the gas turbine had to be shipped to UK for repairs. Today, the gas turbine is in operation and the commissioning has almost been concluded. Inspection of the LCV combustion system during this time showed that the combustors and fuel injectors are in excellent condition.

An extensive demonstration/development programme will be carried out during 1996-97. The objectives are to confirm the technical and economic viability of the technology and to provide engineering data for the design of commercially rated power plants. Plant performance and process characteristics will be evaluated in order to achieve high cycle efficiencies, high ratios of electric power to heat, and compliance with environmental requirements by low emission levels in future plants.

There is also a need for further basic research work on the process itself and in related areas. Foster Wheeler and Sydkraft participate in several programmes where institutions such as the Technical Research Centre (VTT) of Finland, Åbo Academy, Finland and the Lund Institute of Technology, Sweden, conduct theoretical and experimental work, financed by industry as well as by governments. For several years, the VTT Laboratory of Fuel and Process Technology has been conducting research and development work on biomass gasification in fluidized beds. A large number of tests on biomass, peat and coal have been run in the laboratory. The Lund Institute of Technology is operating a 60 kW pressurised gasification CFB pilot plant aimed at studying biomass gasification at pressure levels up to 25 bar and temperatures up to 1050 °C.

The first complete IGCC plant in the world to utilise biomasses as fuel and pressurised CFB gasification is in operation at Värnamo, Sweden. The demonstration plant is an important step forward and gives opportunities for testing a concept for high-efficiency power generation from renewable solid fuels, for further development of the technology in order to increase efficiency, optimise process performance and improve the hardware and process design.

The FICFB - Gasification Process

H. HOFBAUER; G. VERONIK T. FLECK; R. RAUCH
Vienna University of Technology, AUSTRIA
H. MACKINGER; E. FERCHER
Austrian Energy & Environment, Graz, AUSTRIA

Abstract

A novel fluidized bed gasification reactor has been developed to get a product gas with a high calorific value (up to 15 MJ/Nm³) and nearly free of nitrogen. The gasification process is based on an internally circulating fluidized system and consists of a gasification zone fluidized with steam and a combustion zone fluidized with air. The circulating bed material acts as heat carrier from the combustion to the gasification zone. Gas mixing between these two zones is avoided by construction measures. Furtheron, the apparatus is characterized by a very compact design.

The development of the gasification reactor has been carried out step by step. First, a cold flow model was operated to study the fluid mechanics of the fluidization system. The second step was a laboratory scale test rig to study the main features of the reactor by varying different operating and geometrical parameters. After this step a pilot plant was constructed and has been successfully operated. The results attained came fully up to the expectations.

Keywords: Biomass, steam gasification, fluidized bed, compact reactor design, medium energy gas, low nitrogen content.

1 Introduction

In Europe, Austria is one of the leading countries in using bioenergy. The most common utilization of biomass for energy is the combustion for heating applications. Gasification could become a second important route especially for power production (Sipilä, 1995; Solantausta et. al, 1996).

Usually, biomass gasification is carried out using fixed or fluidized beds. As the overall gasification reactions are endothermic, the gasification process must be supplied with heat. The easiest way is to use air as gasification agent and to burn the biomass partially within the gasification reactor. In this case the product gas has a low calorific value (around 4-6 MJ/Nm³) and a high nitrogen content of 45-55 %.

A gas with a low nitrogen content and a higher calorific value (about 12 MJ/Nm³) can be produced with pure oxygen as gasification agent but the costs for the oxygen production are high. Another possibility is to supply heat with heat exchangers but here material problems due to the high temperature level will arise. The dilution of the product gas by nitrogen can also be avoided by using a dual fluidized bed system. In this case no oxygen generator is necessary and also no serious material problems due to high temperatures will appear. A good overview is given by Bridgewater (1995).

2 The Gasification Concept

The basic idea of the gasifier concept is to divide the fluidized bed into two zones, a gasification zone and a combustion zone. Between these two zones a circulation loop of bed material is created but the gases should remain separated. The circulating bed material acts as heat carrier from the combustion to the gasification zone (Hofbauer, 1983). The priciple is shown graphically in Figure 1.

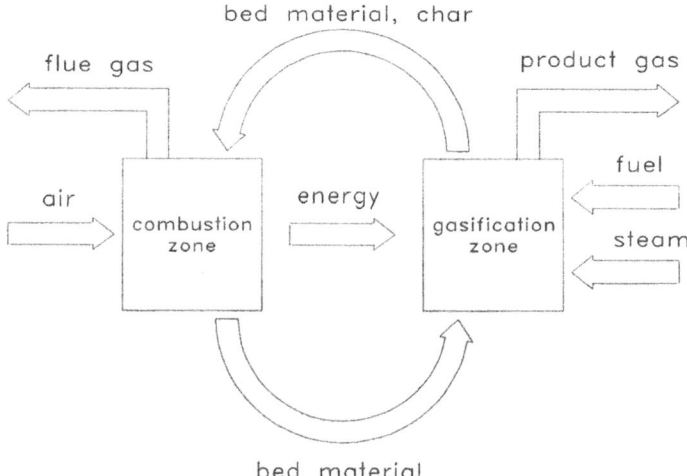

Fig. 1.   Basic idea of the gasification process.

The fuel is fed into the gasification zone and gasified with steam. The gas produced in this zone is therefore nearly free of nitrogen. The bed material, together with some charcoal, circulates to the combustion zone. This zone is fluidized with air and the charcoal is partly burned. The exothermic reaction in the combustion zone provides the energy for the endothermic gasification with steam. Therefore the bed material at the exit of the combustion zone has a higher temperature than at the entrance. The flue gas will be removed without coming in contact with the product gas. With this concept it is possible to get a high-grade product gas without use of pure oxygen. This process can be realized with two fluidized beds connected with transport lines

(Schiefelbein, 1989; Paisley, 1993) or with an internally circulating fluidized bed. The latter one has the following advantages:
- simple reactor design
- low investment cost because of compact construction
- reduced energy losses because of efficient thermal household

3 The Test Rig and Pilot Plant

The gasification process has been developed step by step. Three steps are carried out already and a demonstration step will still follow:
- cold flow model
- laboratory test rig
- pilot plant

3.1 Cold Flow Model

At the beginning of the development a cold flow model was built to study and optimize the fluid mechanics of an internally circulating fluidized bed with a draught tube and a surrounding annular bed. Circulation rates of the bed material and gas leackage between the two zones were measured. The circulation rates are important for the heat transport. Furtheron the gas leackage must be minimized. The results were very promising therefore a laboratory scale test rig for gasification tests was constructed.

3.2 Laboratory Test Rig

The laboratory test rig was designed for a thermal output of 10 kW (Zschetzsche et. al. 1995). Fuel is fed into the annular bubbling bed by a double screw feeding system. The annular bed is fluidized with steam which also acts as gasification agent. The bed material together with the charcoal moves down towards a riser which is situated in the centre of the reactor. The particles are transported up by air through the riser where charcoal is partly burned. At the top of the riser the particles are separated from the gas. The particles fall down and come back to the gasification zone via an annular gap. This gap and the location of the steam and air inlets ensure, that the gas leackage between the two zones is lower than 5 % of the total gas input. Both zones have separated gas exits.

3.3 Pilot Plant

The next step was the construction of a pilot plant with a thermal output of about 100 kW based on the experience of the laboratory test rig. The circular-symmetric geometry was changed to a rectangular cross-section because of scale up consideration. A scheme of the pilot plant can be seen in Fig. 2. Now the riser (2) and the bubbling gasification bed (1) are arranged side by side. To avoid large amounts of gas mixing a siphon (3) was introduced in the line from the combustion zone to the gasification zone. The bed particles are splitted from the riser gas stream using a U-beam separator (4).

The fuel feeding system (5) consists of a hopper and a multi-screw-conveyor. Air is supplied by blowers into the riser and during the start up period also into the gasification zone. Steam is produced by an electrical steam generator and overheated by an electrical heater. The product gas and the flue gas have separated exits from the reactor but they are mixed together after taking gas samples for analyzing. A gas burner ensures that the product gas is completely combusted before entering a cyclone (6) and a flue gas cooler (7). The flyash can be returned continuously into the gasifier with the aid of a pneumatic flyash recycle system. Table 1 contains characteristic data and dimensions of the pilot plant.

Table 1.   Characteristic data of the pilot plant

thermal output	100 kW
fuel	wood chips
reactor	80x12 cm
riser	7.5x12 cm
riser height	250 cm
bed material	quartz
bed mass	37.5 kg
mean diameter	0.6 mm

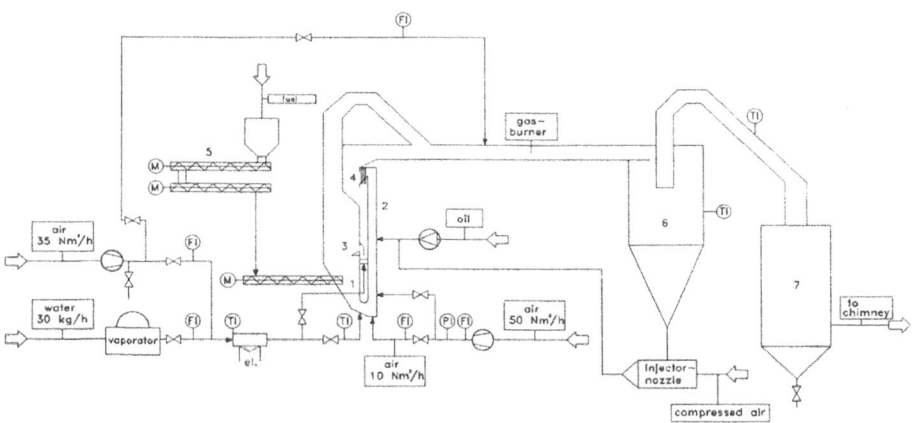

Fig. 2.   Fast internally circulating fluidized bed gasifier

The reactor is manufactured with stainless steel and is insulated. The warm up is carried out with electrical preheating of all air streams and lasts about 1.5 hours. During gasification tests all electrical air preheaters are switched off. A diesel oil feeding into the riser is installed which gives the possibility to change the temperature level of the system without varying other operation parameters. With this installation parameter studies can be carried out very easily.

4 Results of Gasification Tests

Gasification tests were carried out using wood chips (<20 mm) as fuel. Table 2 contains the analysis of the wood chips.

Table 2.   Fuel analysis (Analysis after DIN standards, dm ... dry matter)

		wood chips
water	w.-%	12.1
ash	w.-%	0.6
volatile matter	w.-%	73.3
fixed carbon	w.-%	14.0
C	w.-%, dm	51.5
H	w.-%, dm	6.3
O	w.-%, dm	41.3
N	w.-%, dm	0.22
S	w.-%, dm	<0.05
calorific value	MJ/kg	15.6

For start up purposes electrical air preheaters are installed which are able to heat up the fluidization gases up to 700 °C. During start up instead of steam air is blown into the gasification zone. As soon as the bed temperature exceeds 400 °C fuel feeding is started. After reaching the desired temperature the fluidization gas in the gasification zone is changed over to steam and the preheaters are switched off.

Figure 3 shows typical results attained with wood chips gasified in the 100 kW pilot plant. A lot of temperatures, pressures, volume streams and gas concentrations are measured continuously during gasification. This figure contains only a few of these data to improve overview namely the bed temperature in the combustion zone (T3), the bed temperature in the gasification zone (T6) and the concentrations of CO and H2 in the dried product gas.

In this example the warm up was not recorded. The gasification was started when the fluidization gas was switched over from air to steam. Immediatly the concentrations of CO and H_2 raise to a level of about 22 and 32 % respectively. The temperatures in the combustion as well in the gasification zones decrease slightly and remain nearly constant after 25 minutes. According to the concept explained above the temperature in the combustion zone is about 100 °C higher than in the gasification zone. This means that the circulating bed material transports heat from the combustion to the gasification zone. After 1400 s a decrease in CO and H_2 concentrations can be observed because of changing the analysis channel for tar and water analysis of the product gas. The test was finished after 7000 s of operation because the whole system reached complete equilibrium.

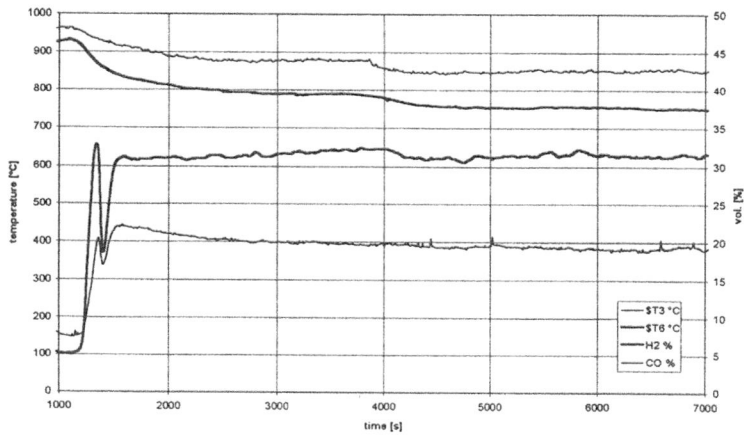

Fig. 3.
Pilot plant test run with wood chips (further data see Table 3)

In addition to the continuous analysis of the gas composition gas samples were taken and analyzed with gas chromatography. A small part of the product gas is drained for analysis. This sample gas is cooled down to about 10 °C where most of the condensate and tar is removed. Further gas cleaning stages are a glasswool filter and a high-grade paperfilter. Typical results of gas analysis from selected gasification tests for the wood chips can be seen in Table 3.

The nitrogen content is very low in the case of the test rig (< 2.5 %) and something higher in the product gas of the pilot plant (< 5 %). However, the content is small compared with air blown gasifiers (45 - 55 %). The calorific value is therefore much higher and lies above 13 MJ/Nm³ in any case.

The amount of tar produced during steam gasification has been quantified. In all experiments the tar amount did not exceed 1 g/Nm³ dry product gas although no catalytic bed material has been used up to now. Compared to air blown gasification the tar content is low. Experiments show that the tar content in this case is between 10 to 15 g/Nm³. This results are in agreement with literature (e.g. Rei et al., 1986; Baker et al. 1987) and can be explained by steam reforming of the tar.

Table 3.   Example of detailed analysis of the product gas (dry) using gas chromatography

operation parameter

installation	-	pilot plant
type of fuel	-	wood chips
feed	kg/h	30
steam	kg/h	14.89
feed/steam	-	2.01
gasif. temp.	°C	754
comb. temp.	°C	845

product gas analysis

gas	vol. [%]	calorific value [kJ/Nm³]
H_2 [1,31]	31.50	3402
O_2 [1,63]	0	0
N_2 [1,71]	2.79	0
CO [2,54]	22.66	2865
CH_4 [2,83]	11.21	4023
CO_2 [5,87]	27.46	0
ethene [6,21]	3.52	2.095
ethane [7,07]	0.55	353
propene [9,90]	0.27	243
propane [10,35]	0.02	14
i-butene [11,67]	0.02	20
i-butane [12,16]	0.00	0
n-butane [12,97]	0.00	0
sum :	100.00	13015

The following variations of parameters were carried out in several measurings:
(Figure 4,5,6)

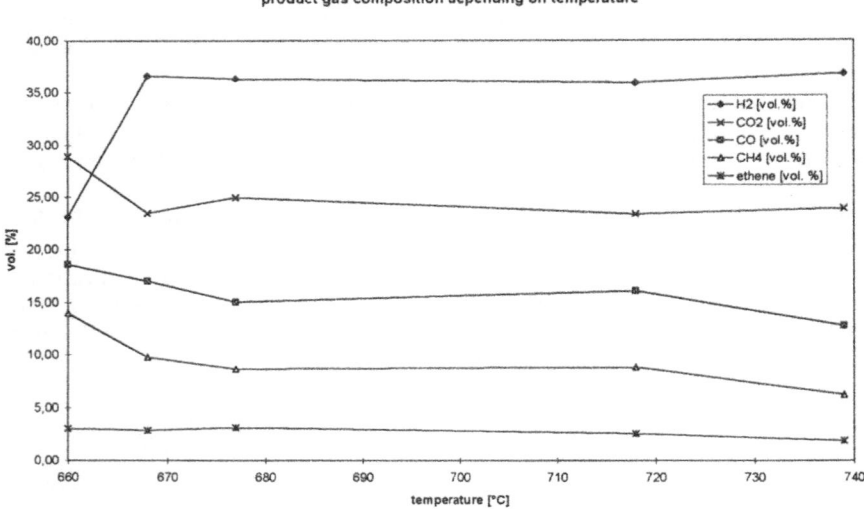

Fig. 4. Product gas composition depending on temperature

Figure 4 shows the product gas composition depending on temperature. In these
experiments the fuel to steam ratio was held constant at about 1,1 as far as possible. It
can be seen that the product gas composition does not change worth mentioning at
temperatures above 670 °C.

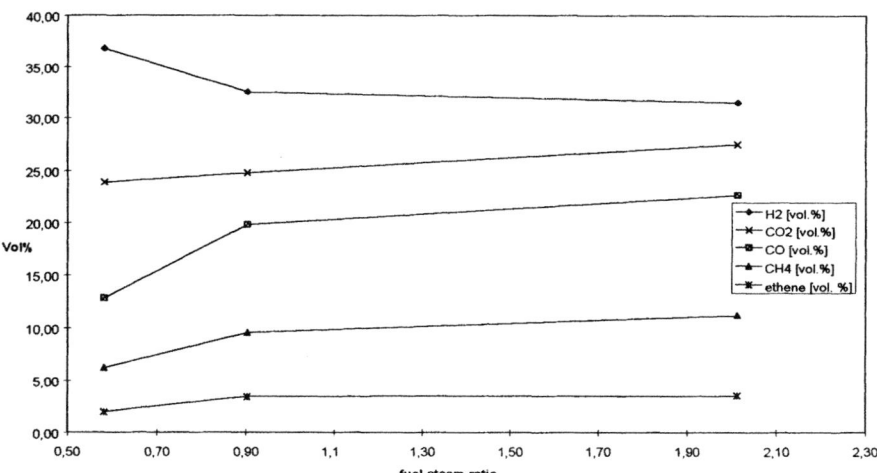

Fig. 5.   Product gas composition depending on fuel-steam-ratio

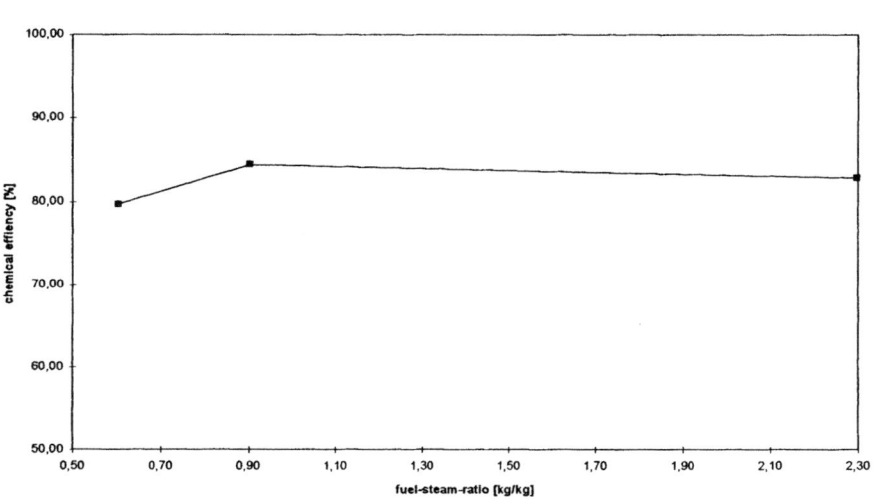

Fig. 6.   Chemical efficiency of the process depending on fuel-steam-ratio

Figure 5 shows the product gas composition depending on the fuel to steam ratio. In these experiments the temperature was held constant at about 745 °C as far as possible. It can be seen that a decrease of H_2 goes hand in hand with an inreasing CO value by increasing the fuel to steam ratio. All other components in this diagramm increase sligthly with the fuel to steam ratio too. We are aware that too much steam is necessary to run this reactor and therefore the variation of the fuel to steam ratio shown in figure 5 is of less importance. In this reactor that high amount of steam is necessary to keep up fluidization but in the next step of development the reactor will be reconstructed to avoid this fact and to reach higher fuel to steam ratios.

Figure 6 shows the chemical efficiency of this process depending on the fuel to steam ratio. In these experiments the temperature was held constant at about 740 °C. It can be seen that the chemical efficiency is almost constant at a high level though changing the fuel to steam ratio.

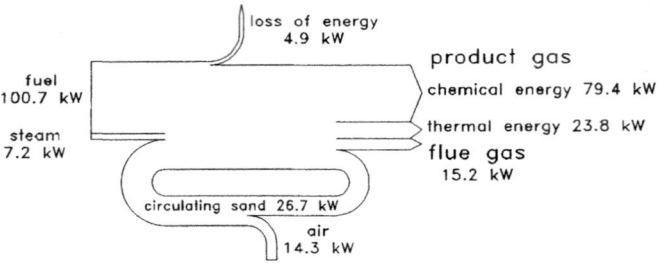

Fig. 7. Energy diagram of the FICFB pilot plant

Figure 7 shows an energy diagram of the pilot plant with energy input, energy output, energy losses and the amount of energy, which is circulating within the gasifier.

5. Conclusion

A novel gasifier was presented which has the following main advantages compared with an air blown gasifier:

- product gas nearly free of nitrogen
- calorific value higher than 13 MJ/Nm³
- very low tar content due to steam gasification
- gas quality is independent of water content in biomass feed
- the apparatus is very compact
- a wide range of feedstock can be gasified
- possibility to use a catalyst as bed material (regeneration of catalyst in combustion zone) to influence the gas composition and gasification kinetic in a more positive way

6. References

Baker, E. G.; Mudge, L. K.; Brown, M. D.; (1987). „Steam Gasification of Biomass with Nickel Secondary Catalyst". Ind. Eng. Chem. Res., Vol. 26, No. 7, pp. 1335 - 1339.

Bridgewater, A. V.; (1995). „The Technical and Economic Feasibility of Biomass Gasification for Power Generation". Fuel, Vol. 74, No 5, pp. 631-653.

Hofbauer, H.; (1986). „Wirbelschicht mit innerer Zirkulation". In „Möglichkeiten und Grenzen der Wirbelschichttechnik". Berichte der Akademie für Umwelt und Energie, Heft 5, pp. 116-126.

Hofbauer, H.; Stoiber, H.; Veronik, G.; (1995). „Gasification of Organic Material in a Novel Fluidization Bed System", Proc. of the 1st SCEJ Symposion on Fludization, Tokyo, pp. 291-299.

Rei, M. H.; Lin, F. S.; Su, T. B.; (1986). „Catalytic Gasification of Rice Hull. (II) The Steam Reforming Reaction". Applied Catalysis, Vol. 26, pp. 27 - 37.

Schiefelbein, G. F.; (1989). „Biomass Thermal Gasification Research: Recent Results from the United States DOE's Research Program". Biomass, Vol. 19, pp. 145-159.

Siplä, K.; (1995). „Research into Thermochemical Conversion of Biomass into Fuels, Chemicals and Fibres". Proc. of the 8th European Biomass Conference on Biomass for Agriculture and Industry, Vol. 1, pp. 156-167.

Solantausta, Y.; Bridgewater, T.; Beckman, D.; (1996). „Electricity Production by Advanced Biomass Power Systems". VTT Research Notes 1729, ISBN 951-38-4884-1, pp. 115.

Zschetzsche, A.; Hofbauer, H.; Schmidt, A.; (1994). „Biomass Gasification in an Internally Circulating Fluidized Bed". Proc. of the 8th European Conference on Biomass for Agriculture and Industry, Vol. 3, pp. 1771 - 1777.

PRESSURISED COMBUSTION OF BIOMASS-DERIVED, LOW CALORIFIC VALUE, FUEL GAS

Combustion of LCV fuel gas

J. ANDRIES, P.D.J. HOPPESTEYN and K.R.G. HEIN
Laboratory for Thermal Power Engineering
Department of Mechanical Engineering and Marine Technology
Delft University of Technology
Delft, The Netherlands

Abstract

During a 3 year (1996 - 1998) project, partly funded by the EU as part of their JOULE 3 programme, experimental and theoretical research will be done on the pressurised combustion of biomass-derived, LCV, fuel gas.

European Gas Turbines Ltd will design, manufacture and supply a pressurised, high temperature combustor for the biomass derived, LCV, fuel gas matched to the Delft gasifier. The combustor will be installed in the Delft test rig and experiments carried out to gather experimental data on the steady state and dynamic behaviour of the combustor. Mathematical models which simulate the steady state and dynamic behaviour of the combustor will be developed.

The refined and validated steady state and dynamic combustor models will be used by European Gas Turbines Ltd to develop a gas turbine model which will be incorporated in a plant layout for an advanced biomass-fuelled IGCC plant.

A detailed description is given of the objectives of the project, the test rig to be used and the time schedule of the project. Some preliminary eexperimental results are described.

Keywords: biomass, gasification, combustion, gasturbine, low calorific value gas, pressurised

1   Introduction

A potentially attractive option for using biomass in power production is the advanced biomass-fuelled IGCC system. Each of the relatively few IGCC systems built using biomass derived, low calorific value (LCV), fuel gases has shown that it is very difficult to reliably predict the behaviour of the system as a whole and of the gas turbine in particular. This has resulted in a number of operational problems.

The steady-state characteristics of the combustor and the influence of this on the design of the gas turbine as a whole require considerable design modifications. A sound understanding of the interaction between the dynamic characteristics of the gasifier and the combustor is critical for successful operation of an IGCC system. Reliable experimental and theoretical information on the characteristics of the gas turbine as a whole and the combustor as such leading to this understanding is needed prior to commercialization of these IGCC systems.

The combustion of 'conventional' gaseous fuels in gas turbine combustors has been studied extensively. The combustion of low calorific value, biomass-derived, fuel gas however has been studied in much less detail.

The Laboratory for Thermal Power Engineering of the Delft University of Technology is participating in an EU-funded, international, R&D project which is designed to aid European industry in addressing issues regarding pressurised combustion of, biomass-derived, low calorific value fuel gas. The groups participating in this 3-year project are Delft University of Technology /NL, European Gas Turbines Ltd /GB and Fluent Europe Ltd /GB.

2   Objectives

The objectives of the project are:

- to design, manufacture and test a pressurised, high temperature gas turbine combustor for biomass derived, LCV fuel gas, obtained from the Delft 1.5 MW_{th} pressurised fluidised bed gasifier.
- to develop a steady-state and dynamic model describing a combustor using biomass-derived, low calorific value, fuel gases.
- to gather reliable experimental data on the steady-state and dynamic characteristics of a pressurised, high temperature, combustor by analysing the fuel gas quality and the combustor behaviour and use these results to refine and validate the models.
- to study the steady-state and dynamic plant behaviour using a plant layout which incorporates a model of a gas turbine suitable for operation on low calorific value fuel gas.

This information obtained should be available in a form which can be used in industrial design practice.

3   Installation

Experimental research at a semi-technical scale using installations with capacities in the range from 0.5 to 1.5 MW_{th} offers major advantages when compared with laboratory-scale experiments as well as with full industrial scale plants, such as:

- ability to make measurements under well defined, controllable process conditions.
- ability to determine the fate of gaseous components by doing measurements at different locations along the gas path.
- experimental results are direct applicable to large scale installations.

Such a rig for pressurised fluidised bed gasification is available at the Laboratory for Thermal Power Engineering of the Delft University of Technology. The rig has been used for about 9 years to conduct combustion experiments and has been extensively nodified during 1995 and 1996. A schematic of the final configuration of the modified installation is shown in figure 1. A detailed description of the test rig is given in [1].

The components to be used during phase 2 (pressurised topping combustor and heat exchanger) will be operational at the end of March 1996, the components to be used during phase 3 (steam supply, heat exchanger, ceramic filter and ammonia removal system) will be operational in September 1996. The modifications will enable gasification as well as combustion experiments using coal, biomass and coal-biomass mixtures as fuels.The gasification can be achieved using air, air-steam mixtures, oxygen enriched air and oxygen enriched recirculated flue gas as the fluidisation/gasification medium. The low calorific value fuel gas can be combusted in a pressurised, high temperature combustor using air, pure oxygen and recycled flue gas as oxidant and/or cooling gas. It is also possible to use steam to control the flame temperature and hence the NO emissions. The modified installation has a maximum thermal capacity of 1.5 MW_{th} and will be operated at pressures up to 10 bar and temperatures of 900 °C. A 2 m high bed zone with a diameter of 0.4 m is followed by an adiabatic freeboard approximately 4 m high with a diameter of 0.5 m. The freeboard is equipped with a possibility to inject secondary air.

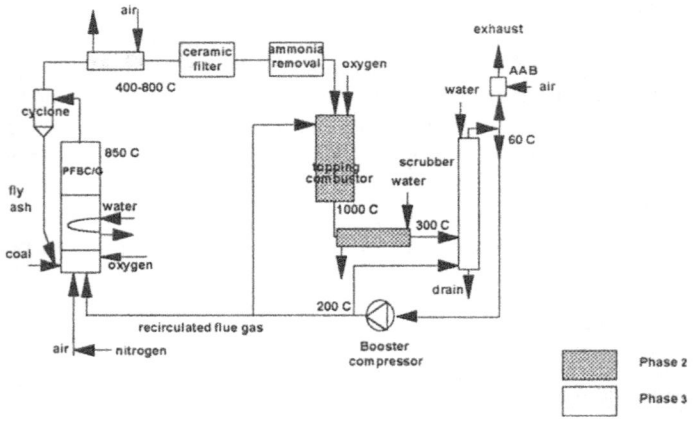

Fig. 1. The Delft PFBC/G test facility.

4   Work programme

European Gas Turbines Ltd will design, manufacture and supply a pressurised, high temperature combustor for the biomass derived, LCV, fuel gas matched to the Delft gasifier. The combustor will be installed in the Delft test rig and experiments will be carried out to gather experimental data on the steady-state and dynamic behaviour of the combustor. The experimental results will be compared with simulation results obtained from the mathematical models describing the steady-state and dynamic behaviour of the combustor and used to refine and validate these models.

The mathematical model which simulates the steady-state behaviour of the combustor will be developed by Delft University of Technology in collaboration with Fluent Europe Ltd. The mathematical model which simulates the dynamic behaviour of the combustor will be developed by European Gas Turbines Ltd.

The refined and validated steady-state and dynamic combustor models will be used by European Gas Turbines Ltd to develop a gas turbine model which will be incorporated in a plant layout for an advanced biomass-fuelled IGCC plant. This plant layout will be used for the simulation of the steady-state and dynamic behaviour of the system. The time schedule of the project is given in figure 2.

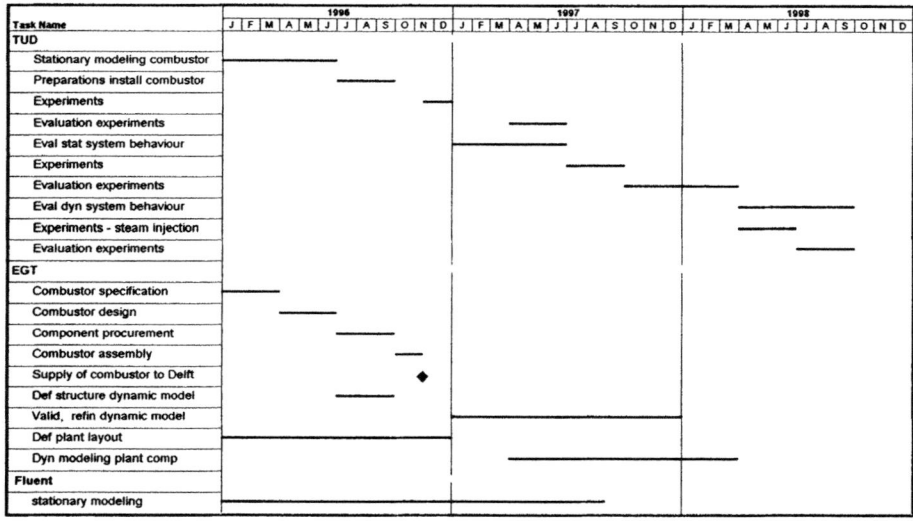

Fig. 2. Time schedule of the project.

5   Preliminary experiments

Preliminary combustion experiments have been done using an LCV fuel gas obtained from gasification of coal in a 60 kW_{th} atmospheric fluidised bed gasifier. The fuel gas was combusted in a cyclone combustor using air. The calorific value of the fuel gas was controlled by varying the operating conditions of the gasifier. It was found that ignition, combustion and flame stability posed no problems with calorific values down to 1.5 MJ/Nm^3.

As part of an EU-funded project aimed at gasification of coal, using recirculated flue gas and oxygen, a combustion chamber has been designed and built to combust the low calorific fuel gas obtained from the gasification of coal using recirculated flue gas and pure oxygen. The combustor uses pure oxygen as oxidant and recirculated flue gas as coolant gas. The burner, which uses a non-swirling flow, is designed in cooperation with Air Products UK. The combustor will be tested during the second quarter of 1996. The characteristics of the combustor have been simulated using the Fluent cfd package and the experimental results will be used to validate and refine the model.

During the third quarter of 1996 the burner will be replaced by a so-called winnox vortex burner. This very compact burner uses a very highly swirling flow to enhance the mixing properties close to the burner and to realize a very short low-NO_x flame. This burner will be tested using pure oxygen as oxydant and recirculated flue gas as coolant gas.

At the end of 1996 the combustion chamber will be replaced by a gas turbine combustion chamber designed and built by EGT. This combustion chamber will operate on LCV fuel gas obtained from the gasification of pure biomass using air and steam. Air will be used as oxidant and cooling gas.

6   Acknowledgement

This research is funded in part by the European Commission in the framework of the JOULE2 and JOULE3 R&D programmes (contracts JOU2-CT92-0154 and JOR3-CT95-0027).

7   References

1   Andries,J. and Hein, K.R.G., 'Co-gasification of biomass and coal in a pressurised fluidised bed gasifier', paper presented at paper presented at *Developments in Thermochemical biomass conversion*, Banff, Canada, May 20 - 24, 1996

ALKALI REMOVAL AND BED MATERIAL AGGLOMERATION STUDIES APPLIED TO BIOMASS GASIFICATION

N. PADBAN, Z. YE and I. BJERLE
Department of Chemical Engineering II
Lund Institute of Technology(LIT)
University of Lund, Sweden

Abstract

TGA and TDA methods were employed to investigate the alkali removal and its effect on the bed material agglomeration in the PFB biomass gasification process. Several silica-alumina based sorbents and bed materials as well as the bed ash from the biomass gasification process were screened for their alkali removal capacity at a temperature range of 400 - 1200 °C. Model alkali compounds were KCl, KNO_3, K_2SO_4 and K_3PO_4 in these measurements. There seems to be an optimum temperature for alkali absorption and the highest alkali removal was measured at around 890 °C. Fyle sand was found capable of capturing 0.25 mol K/mol SiO_2 and is the best of the tested sorbents. However the agglomeration of the sand took place at a level much lower than this ratio. The superficial gas flow only slightly affected the capturing capacity of the sorbents.

The results from agglomeration studies show that the lowest agglomeration temperature is decided by the alkali sort and experimental environment.

Keywords: Agglomeration, alkali, alkali removal, bed material, biomass, gasification, high temperature.

1 Introduction

The alkali induced problems during biomass gasification and combustion do not differ much from those observed in the conventional processes of coal utilisation. In the Integrated Gasification Combined Cycle technology (IGCC) process, the corrosion and erosion effects of alkali are of large concern. The volatilized alkali deposit in different parts of the plant. The life time of gas turbines which are relatively costly investments decreases drastically with increasing alkali concentration in the fuel gases.

Agglomeration and sintering of the bed material caused by alkali salts or their low melting eutecticume can disturb fluidization of fluidized bed utilisation. The lowering of the ash melting point is in most cases related to the amount of the alkali present.

The alakli problem in coal processes that has been extensively studied is not applicable for predicting the behaviour of alkali in biomass gasification or combustion. The reason is that the sodium and potassium content in coal are present in stable states such as chlorides for sodium and alumino-silicates for potassium.

The inorganic compounds in biomass can vary depending on not only the sort of biomass but also the harvesting place and season. Even local varieties in climate may have some effect.

Several sorbents used for the removal of alkali have been tested and up to now the main components in various sorbents are SiO_2 and Al_2O_3. Two sorbents frequently mentioned in the literature are activated bauxite (SiO_2 7.0-10.0%, Al_2O_3 81.5-88.0%) and diatomaceous earth (SiO_2 92.0%, Al_2O_3 5.0%)[1]. On a weight basis diatomaceous earth is more active than bauxite, but on a volume basis bauxite is more active. Removal efficiency can be affected by the impurities in the sorbents. It seems that bauxite has high removal capacity for NaCl and K_2SO_4 while diatomaceous earth is more active for KCl. A pigment that contains 44.8-45.3% SiO_2 and 37.5-39.7% Al_2O_3, a clay containing 68% SiO_2, 12% Al_2O_3 and 10.5% MgO, and a silica gel of pure SiO_2 are found capable of removing alkali from the gas phase [2] but not as active as bauxite and diatomaceous earth. Sulfur dioxide content, less oxygen and reduction conditions in the flue gas may be helpful for alkali removal. J. McLaughlin et al [3] tested calcium montmorillonite with a general formula of $Ca_{0.2}Al_{16}Mg_{0.4}Si_4O_{10}(OH)_2$ at temperatures 827-927 °C. The concentration of HCl was found to strongly affect the alkali capture. Compounds such as silica, alumina and graphite, commercially available such as calcined limestone and bauxite, and a bituminous ash collected from the bottom of a combustion furnace were tested in a temperature range of 800-850 °C by F.Shadman et al [4]. Bauxite, Kaolin and Al_2O_3 were found to be effective sorbents for both NaCl and KCl. The bituminous ash surprisingly captured KCl as much as bauxite, while the reactivity of SiO_2, limestone and graphite were low. Emathlite is another known alkali sorbent at temperatures lower than 900 °C and might be a better sorbents than activated bauxite [5].

Since 1994 a pressurized fluidized bed biomass gasifier has been run at LIT and some 30 experiments have been carried out. Alkali induced bed material agglomeration and consequent shut-down of the plant were frequently observed [6]. Therefore the investigation of hot fuel gas alkali removal with adequate sorbents that are compatable with bed material is rather necessary. In this study several alkali compounds that most likely are a part of the biomass fuel were selected as alkali resources. Three potential bed materials for fluidized bed biomass gasification were taken for the preliminary measurements.

2 The thermodynamic study - HSC analysis

The thermodynamic study was conducted by means of a HSC [7] Chemistry program developed by Outokumpu Research Oy in Finland. The program can supply various thermodynamic data such as reaction heat, equilibrium constants, heat and material balance, phase stability diagram and so on utilising an extensive thermochemical database which contains enthalpy (H), entropy (S) and heat capacity (C) data for more than 5600 chemical compounds. The reaction of SiO_2 with KCl, NaCl and KNO_3 are considered in this study since SiO_2 is the main content of the candidate bed materials.

NaCl and KCl can react with O_2 under oxidized conditions forming K_2O and Na_2O which can easily further react with SiO_2. Alkali silicates can exist in several forms: $M_2O*3SiO_2$, $M_2O*2SiO_2$ and M_2O*SiO_2 with M substituted for the alkali metal. To be comparable the following reactions are all taken at 800 °C that is considered as the practical temperature for alkali removal. From a viewpoint of reaction heat and the equilibrium constant the reaction were compared in order to estimate the possible reaction route even this may not be true.

$$2NaCl + 0.5\,O_2 \longrightarrow Na_2O + Cl_2 \qquad\qquad K= 5.9E\text{-}18 \qquad (1)$$

$$Na_2O + nSiO_2 \longrightarrow Na_2O*nSiO_2 \qquad n=1 \qquad K= 1.3E11 \qquad (2)$$
$$n=2 \qquad K=3.9E11 \qquad (3)$$
$$n=3 \qquad K=1.5E12 \qquad (4)$$

$Na_2O*3SiO_2$ is probably the most stable one and can be the dominated product. However, the formation of the Na_2O is obviously difficult. Instead, a mechanism with the formation of gaseous NaCl might be more practical.

$$NaCl \longrightarrow NaCl(g) \qquad\qquad K=3.52E\text{-}4 \qquad (5)$$

$$2NaCl(g) + 0.5O_2 + 3SiO_2 \longrightarrow Na_2O*3SiO_2 + Cl_2 \qquad K=72.3 \qquad (6)$$

For KCl the potassium silicates in forms of $K_2O*2SiO_2$ and $K_2O*3SiO_2$ probably do not exist since they can not be found in the data base while $K_2O*4SiO_2$ melts at 767 °C, therefore the stable form seems to be the K_2O*SiO_2.

$$KCl \longrightarrow KCl(g) \qquad\qquad K=6.6E\text{-}4 \qquad (7)$$
$$2KCl(g) + 0.5O_2 + SiO_2 \longrightarrow K_2O*SiO_2 + Cl_2 \qquad K=4.1E\text{-}3 \qquad (8)$$

In comparison with NaCl, KCl does not easily react with SiO_2. Under definite conditions the removal of KCl may take more time.KNO_3 is relatively easy to release K_2O:

$$2KNO_3 \longrightarrow K_2O + N_2 + 2.5\,O_2 \qquad\qquad K=1.4E\text{-}6 \qquad (9)$$

The presence of SiO_2 favors the decomposition of KNO_3:

$$2KNO_3 + SiO_2 \longrightarrow K_2O*SiO_2 + N_2 + O_2 \qquad K=4.4E8 \qquad (10)$$

Bed matrial contained Al_2O_3 can react with $M_2O*nSiO_2$ and form alkali aluminu-silicates. The possible stable form for Na can be $Na_2O*Al_2O_3*3SiO_2$ and for K $K_2O*Al_2O_3*6SiO_2$.

A fuel gas contains H_2O and H_2, and the reaction of alkali with SiO_2 can be different. Further more the phase balance may also control the formation and the stability of the final products [8]. A complete discussion should therefore take into account all these effects, but this means a more complicated work.

The stability of the alkaline compounds is affected by the environmental conditions. Reducing conditions speeds up the decomposition of the nitrates drastically. For pure potassium nitrate, K_2O, $K(g)$ and $KO(g)$ are the stable decomposition products at temperatures higher than $650°C$. In the presence of SiO_2 and at high temperatures K_2O*SiO_2 and $K_2O*4SiO_2$ are the most favourable compounds independent of environmental conditions.

For potassium chloride there is no environmental condition effect but the presence of SiO_2 prolongs the evaporation time of the liquid chloride.

In the neutral environment the K_3PO_4 is the most stable form of the K-P-O composition but both reducing and oxidising conditions give formation of $K_4P_2O_7$ and KOH (g) at high temperatures. In the presence of SiO_2, an equilibrium between K_3PO_4, $K_4P_2O_7$, K_2O*SiO_2 and $K_2O*4SiO_2$ exists independent of environment.

K_2SO_4 dissociates easily under reducing conditions and if the reduction medium is CO, then the carbonate will be the most stable form at temperatures up to $900°C$. Over this temperature other potassium compositions such as K_2SO_4 (l), $K_2S(g)$, $KS(g)$ and $K(g)$ will dominate. In the presence of SiO_2, the major alkali compounds in the system are K_2O*SiO_4 and $K_2O*4SiO_4$.

3 Experimental

The measurements of alkali capture were carried out in a Cahn 2000 TGA balance. A sample basket Pt-basket was hanging in a quartz reactor which is 19 cm long and 1.5 cm in diameter. The reactor is surrounded by a furnace which is able to raise the temperature up to 1250 °C. The alkali compounds with or without the sorbent particles were put in the basket. The sorbents formed a layer on the alkali compound in the basket. The sample was heated gradually from room temperature until the pre-set temperature. The alkali compound evaporates under air gas sweeping of 42 Nl/h. Three alkali compounds, KCl, NaCl and KNO_3 was measured. Three absorbents, Quartz sand, Baskarp sand and Fyle sand, that were employed as the candidates for the bed material in the HPHT biomass gasifier, were tested for their capacity of absorbing the vapour phase alkali compounds. The tested temperatures was 680, 850 and $1100°C$.

The weight loss during the evaporation was monitored with a strip recorder and after every run the final weight of the sample confirmed with a electric balance.

In the studies of the alkali caused agglomeration and sintering of bed material a Simultaneous Thermal Analysis (STA 409) instrument made by NETZSCH (Germany) was used. The experiments have been done in the temperature range of 20 to 1200°C.

The instrumentation is able of doing both Thermo-Gravimetric Analysis (TGA) and Differential Thermal Analysis (DTA and DSC). It is possible to distinguish the endothermic from exothermic reactions by studying the DTA curves. Both the DTA and DSC analyses are based on the constant sample volume. The chemical reactions between bed material and alkali compounds normally take place at high temperatures. However, the high temperature constant-volume sample holders are not available at LIT and therefore the quantitative thermal analysis has not been possible.

The experiments were run in three different environments: nitrogen , air and reducing. The reducing environment was created by using a mixture of nitrogen and carbon monoxide (80% N_2 + 20% CO).

The experiment includes tasks with both pure potassium salts and the mixtures of these salts with bed material. The studied bed material is Fyle Sand. The alkali salts were finely ground before mixing with Fyle Sand (dp_{mean} 180 μm).

In analysing the experimental results both visual observation and water solubility of the agglomerates have been considered. Most of potassium salts are water soluble, consequently water solubility test can give valuable information about the agglomeration mechanisms and even the kind of products in the case of chemical reaction between SiO_2 and alkali. The results from experiments have been compared with those from thermodynamical calculations (HSC).

4 Results and discussion

4.1 Alkali capture

Three potential bed materials were studied with their chemical compositions shown in Table 1. It can be seen that the quartz and Fyle sand are almost pure SiO_2, the only difference is perhaps that the alkali content in the Fyle sand is a bit higher than the quartz sand. The Baskarp sand, on the other hand, contains less SiO_2 but more Al_2O_3 and alkali compounds.The evaporation of three alkali compounds (KNO_3, KCl and NaCl) were tested at various temperatures.

Generally KNO_3 releases quickly while KCl and NaCl had a comparable evaporation rate. Figure 1 shows the NaCl evaporation profile affected by the various sorbents at 1090 °C. It seems that all sorbents can more or less delay the evaporation of NaCl. Fyle sand strongly hinders the evaporation by prolonging the time needed to complete the evaporation. Besides, both Fyle sand and Quartz sand can capture, about 2-3%, the alkali compound. While Baskarp sand as shown in the picture was less active with NaCl. It is obvious from the figure that all these three sorbents can hardly keep alkali from releasing in gas phase due to the high temperature they encountered.

Table 1. Chemical compositions of three tested sorbents

components, %	Quartz	Baskarp sand 150	Fyle sand 180
SiO_2	99.4	87.7	>99.5
Fe_2O_3	0.08	0.51	-
Al_2O_3	0.16	6.0	-
CaO	-	0.51	-
MgO	-	0.11	-
Na_2O	-	1.32	8.9 ppm
K_2O	-	2.48	40 ppm
total	99.95*	98.63	>99.5
particle size/μ	100, estimated	150	180

* Quartz sand also contains small amount TiO_2.

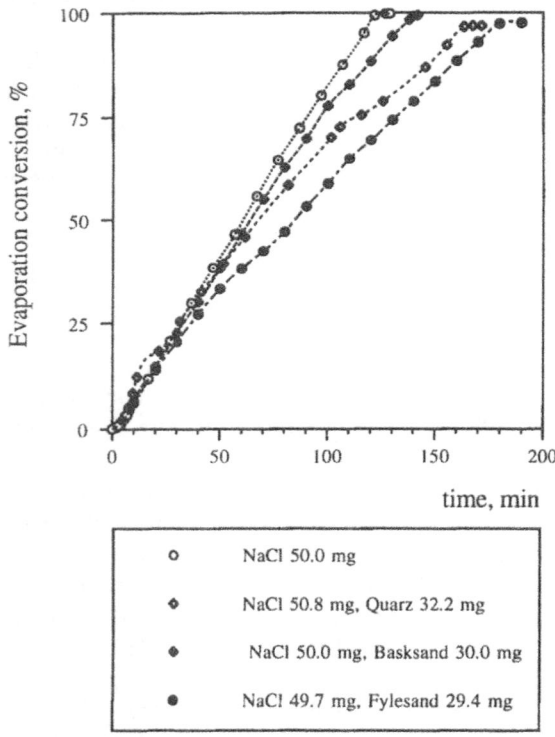

Fig. 1. The evapoeation of NaCl affected by three sorbents at 1090°

○	NaCl 50.0 mg
◇	NaCl 50.8 mg, Quarz 32.2 mg
◉	NaCl 50.0 mg, Basksand 30.0 mg
●	NaCl 49.7 mg, Fylesand 29.4 mg

At lower temperature of 890 °C, indeed, Baskarp sand and Fyle sand captured 17% and 23.5% of the evaporated KNO_3 respectively, shown in Figure 2. One interesting phenomenon is that the KNO_3 sample with Baskarp sand sorbent evaporates faster than the sample without sorbent. One possible explanation is that the alkali content in the sorbent evaporates at the same time when the alkali sample evaporates while the alkali absorbed onto the sorbent surface is different in structure from those original alkali contained in the sorbents. The later is easy to release at relevant temperature. This effect can be related to the alkali content in the sorbent. At even lower temperatures, e.g. 680 °C, the evaporation of the alkali itself could not be complete. KNO_3 evaporates until 88% of the content was converted and then levelled off. With Baskarp sand as sorbent the released KNO_3 was 80%, this means that 8% of the alkali was captured.

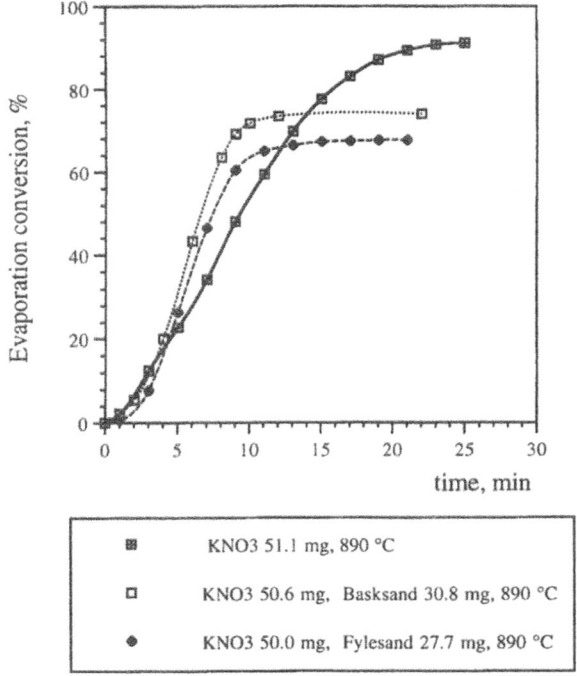

⊞	KNO3 51.1 mg, 890 °C
❑	KNO3 50.6 mg, Basksand 30.8 mg, 890 °C
◆	KNO3 50.0 mg, Fylesand 27.7 mg, 890 °C

Fig. 2. Sorbent effect on the evaporation of KNO$_3$ at 890°C

There may exist an optimum temperature that gives the highest capture of alkali. From the above experimental results a temperature around 890°C might be a good choice. It is believed that the alkali capture by sorbents undergoes the physisorption first and then chemcal reaction. The former is affected negatively by increased temperature while the latter is favoured. Thus an adequate temperature must be found for various sorbents in order to improve their alkali capture capacity.

It is also noticed that the alkali can strongly agglomerate the sorbent particles. A very sticky cake was always found in the sample basket if the sorbent was added to the alkali compounds. The absorbed alkali in a form of silicates on the sorbent surface may be responsible for the agglomeration. The KNO_3 captured by Fyle sand 180 at 890 °C showed a alkali/sorbent weight ratio of 0.16 mg/mg. Since the sorbent in this case was saturated with alkali compounds, a ratio much lower than 0.16 can be enough to agglomerate the sand particles. This is very important in the hot biomass gasification test rig where the alkali from biomass material accumulates on the bed material. The agglomeration finally can block the bed.

The Quartz sand were tested for its capacity of capturing KCl at 850, 900, 1000, 1100 °C respectively. Unlike the KNO_3, the evaporation of KCl seems not to be affected by quartz addition. Neither the evaporation rate nor the evaporation conversion was clearly changed. In the measurement of all three sorbents with KCl at 850 °C, only Baskarp sand can obviously delay the evaporation time while Fyle sand, that was most active to NaCl (see Figure 1), was inactive to the KCl evaporation. The chemical composition of the sorbents might not be able to explain the different behaviour with KCl and NaCl.

4.2 Bed material agglomeration

4.2.1 KNO3
A weak particle adhesion tendency was observed for the mixtures of KNO_3 and SiO_2 at the temperatures below 500°C. The reason can be the low melting point of the potassium nitrate. At higher temperatures (> 700 °C) a water insoluble cake was formed. The weight loss of the mixture is corresponding to the reaction:

$$2KNO_3 + nSiO_2 \longrightarrow K_2O*nSiO_2 + (N_2 + 2.5\, O_2)$$

The thermodynamical explanation for silicate forming is that:
 - The potassium nitrate dissociates at the temperatures higher than 400°C;
 - The potassium oxides are not stable and can not exist alone.
No environment effects were observed in this case. The temperature is believed to have the determining role on dissociation of the nitrates.

4.2.2 KCl
At temperatures higher than 800°C water soluble agglomerates were found. The agglomeration was a result of evaporated KCl condensation that attached to the SiO_2 particles together. The effect of the environment is not noticeable probably due to the fact that the KCl is the most stable form among potassium salts. On the other hand, the comparison between the TG analysis for pure KCl and a mixture between KCl and SiO_2 shows a delay in temperature during the whole evaporation. A pure KCl sample

evaporates almost completely from ~780 to 950°C (heating rate: 10°C/min). In the case of mixture and with the same heating rate more than 60% of the sample is left. This difference in evaporation rate combined with water solubility of the agglomerates can be an indication of the physisorption of KCl on to SiO_2 particles.

4.2.3 $K_3PO_4 * 7H_2O$

At temperatures higher than 1100°C, independent of reaction environment, water insoluble agglomerates were observed. Although the melting point of the K_3PO_4 is much higher than 1100 °C the agglomeration phenomena are difficult to explain. The weight loss for the mixture takes place at around 200°C and corresponds to dehydration of the phosphate. In the case with the sand placed upon the alkali phosphate inside the sample holder a water insoluble agglomerate was observed only in the boundary layer between the salt and SiO_2. It is possible that some eutecticum composition of K-P-O-Si is involved in the agglomeration phenomena.

4.2.4 K_2SO_4

In the air and nitrogen environments no agglomeration tendency was observed for the samples containing K_2SO_4 at the temperatures below 1000°C. For these samples the weight loss started at temperatures above ~1150°C. The weight loss at this temperature and neutral or oxidising conditions is very small and can be explained by smoothly evaporation of the K_2SO_4 (Figure 3).

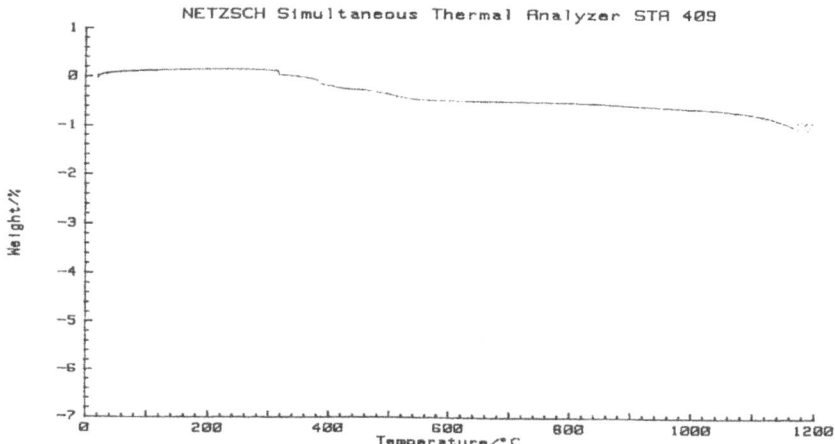

Fig. 3. TG analysis for a mixture with the composition: 89.0% SiO_2, 11.0% K_2SO_4 in air

In the reducing environment the weight loss started at much lower temperature (~800°C)(Figure4). Water insoluble agglomerate is formed at temperatures between 900°C and 1200°C. the weight loss corresponds to the reaction:

$$K_2SO_4 + nSiO_2 \longrightarrow K_2O*nSiO_2 + (S + 1.5 O_2)$$

The thermodynamical calculations show that at high temperatures and reducing environment K_2SO_4 is not stable and dissociates easily. Although the dissociated potassium compounds are not thermodynamically stable, they will react with SiO_2 and form different alkali silicates. The most part of the potassium will form K_2O*SiO_2, while a minor part is presented in $K_2O*4SiO_2$. The gaseous potassium sulphide will begin to form at temperatures higher than 1000°C.

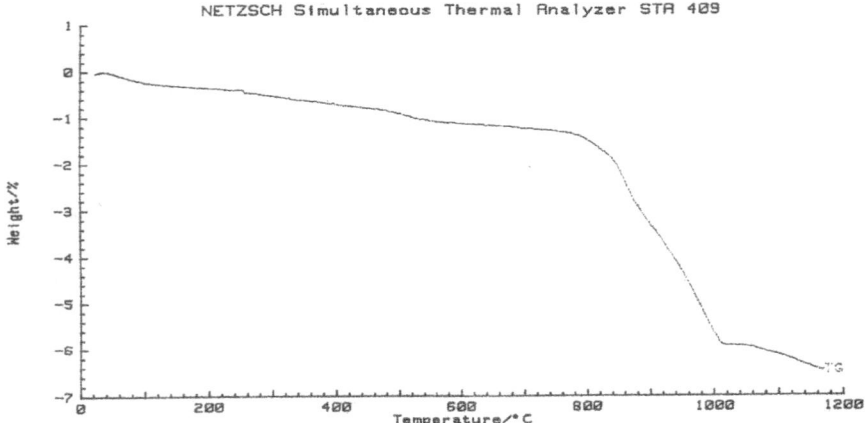

Fig. 4. TG analysis for a mixture with the composition: 88.7% SiO_2, 11.3% K_2SO_4 in reducing atmosephere (20% CO + 80% N_2)

5 Conclusions and perspective

Both physisorbtion and chemisorption are responsible for the alkali capture from gas phase. There may exist an optimum temperature that gives the highest capture of alkali. The experimental results show that a temperature around 890°C might be a good choice.

The stability of the alkali compounds has a decisive importance on their reaction with bed material. In the case of stable alkali compounds such as chlorides, the agglomeration is caused by the condensation of evaporated alkali on the surface of bed material particles. On the contrary the unstable alkali compounds react chemically with SiO_2 resulting in the formation of low-melting point eutectics. The effect of environment is crucial because it affects the stability of the alkaline salts.

The next step of our hot gas alkali removal involves the pressurized TGA measurement and a TDA analysis. Some more alkali compounds such as Na_2SO_4 and K_2SO_4 will be included. The candidate sorbents will expand to take in more actively researched compounds such as bauxite and emathlite. Pure Al_2O_3 will also be tested as a single sorbent. The bed materials used in the hot test rig and the ash of the biomass gasification process should be included, too. Since the normal temperature range for removing alkali compounds locates at around 800 °C, a temperature interval between 600-900 °C will be considered. The pressure is limited by the instrument and will not be higher than 5 bar. The catalytic effect of Fe_2O_3 and the influence of SO_2 on the alkali removal will be investigated.

6 References

1. J. Nykänen(1994), "Alkali metal measurement - a literature review", Thermal Power Engineering, Faculty of Mechanical Engineering and Marine Engineering, Delft University of Technology, Netherlands, EV-1742

2. S.H.D. Lee, I. Johnson(1980), "Removal of gaseous alkali metal compounds from hot flue gas by particulate sorbents", J. of Engineering for Power, 102, 397-402

3. J. McLaughlin, R. Schultz, R. Clift(1993), "Reaction of getter minerals with alkali salt vapours" Proceedings: 2nd Int. Symp. on Gas Cleaning at High Temperatures, Guildford, UK, Sep 27-29, pp556-573, Edited by R. Clift, J.P.K. Seville, ISBN 0-7514-0718-1

4. F. Shadman, W.A. Punjak(1988), "Thermochemistry of alkali interactions with refractory absorbents", *Thermochimica Acta*, 13, 141-152

5. P.R. Mulik, M.A. Alvin, D.M. Bachovchin(1982), "High-temperature removal of alkali vapours in hot-gas cleaning system", Westinghouse Electric Corporation, Pittsburgh, PA 15235, USA

6. Z. Ye, N. Padban, M. Mirazovic and I. Bjerle(1995), "Biomass gasification activities at LIT", presented at the Nordic Seminar on Thermochemical Conversion of Soilid Fuels, Dec. 13-14, 1995, Trondheim, Norway

7. Outokumpu Research Oy. Finland

8. N. Padban(1994), "Alkaliproblematiken vid termisk bioförgasning-Termodynamiska beräkningar - Rapport nr. 1 : Litteratur" (in Swedish), Dept. of Chemical Engineering II, Lund University, Sweden, Sept. 1994

GASIFICATION

Commercial

NEW DEVELOPMENTS IN BIOMASS UTILIZATION FOR ELECTRICITY AND LOW ENERGY GAS PRODUCTION, ON THE GASIFICATION PLANT OF GREVE IN CHIANTI - FLORENCE

Development in exploitation of L.E.G. from B.F. and R.D.F.

G.L. BARDUCCI
TAVOLINI Environmental Engineering, Florence, IT
P. ULIVIERI and G.C. POLZINETTI
S.A.F.I. S.p.A. Florentine Area Environmental Services, Florence, IT
with contribution of:
A. DONATI and F. REPETTO
ANSALDO INDUSTRIA÷ANSALDO RICERCHE, Genoa, IT

Abstract

The paper reports the technical results carried out at the plant of Greve in Chianti - Florence (an industrial plant with two C.F.B. Gasifiers of 20 MWt each and with a electric power production capacity of 6.7 MWe) and the new proposal for industrial utilization of purified L.E.G. (from R.D.F. and B.F.) with fossil fuel (N.G.) in the same improved power cycle. The aim is to demonstrate the economic convenience and the environmental acceptance of the electricity production using an improved steam cycle based on L.E.G. and N.G. At the same time the feasibility and the operability of a high temperature L.E.G. cleaning will be demonstrated on an industrial scale.

Keywords : B.F. (Biomass Fuel), L.E.G. (Low Energy Gas), N.G. (Natural Gas), R.D.F. (Refuse Derived Fuel)

1 Introduction

The proposed activities will be performed at the existing gasification plant at GREVE in CHIANTI - Italy that is the largest industrial gasification plant operating in Europe with RDF (Refuse Derived Fuel). The plant is located in an industrial area close to a cement factory with the aim of utilizing the positive energetic co-operation between the two plants [1]. (See the enclosed pictures 1 - 2- 3).

Picture 1 - General view of the plant location

Picture 2 - The Greve Gasification plant seen from the front

Picture 3 - The Gasifiers from above

2 Text

2.1 The status of art

The Greve in Chianti gasifier represents one of the three proposed plants of the Florentine area for treatment, recycling and disposal of Municipal Solid Waste; the other two plants are a selection plant and a landfill. (see scheme 1). These three plants together will be able to treat up to 1300 T/d of M.S.W. (Municipal Solid Waste), partially transforming it into energy and recyclable products and disposing of it.

In this integrated system the gasifier plays the role of electric power generator starting from RDF (the light combustible part of M.S.W.) through the thermal transformation of this fuel into low energy gas (L.E.G.).

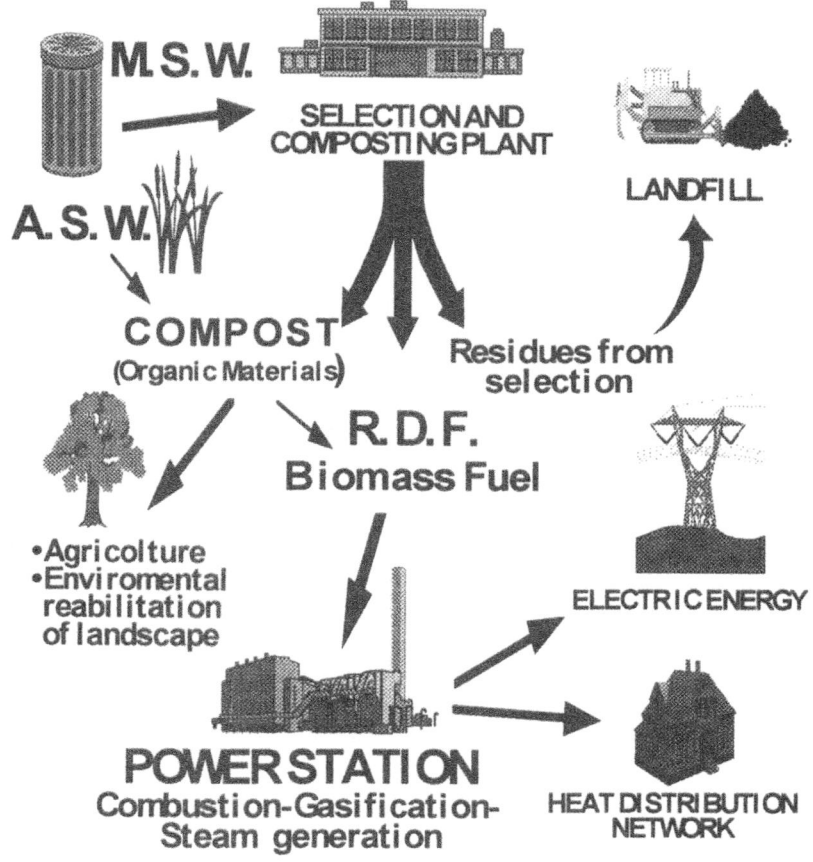

Scheme 1 - Integrated system for treatment, recycling and disposal of M.S.W. of Florentine Area

The R.D.F. production plant is located nearby Florence; the present capacity of the plant is about 600 T/d of MSW corresponding to about 180 T/d of RDF).

The R.D.F. produced will be sufficient for feeding the Greve gasifier, that will be able to transform it into L.E.G. Nevertheless the actual configuration of the Greve plant and the problems encountered during the running of the first boiler, make the management, first of all of the Greve gasification plant and generally of the whole integrated system of M.S.W. treatment of Florentine Area not economic as we expected.

The objective of the actual development is the production of electricity through a combined combustion of L.E.G. and Natural Gas in a new dual fuel boiler. The produced steam is then used in an existing power system. The aim is to demonstrate the economic convenience and the environmental acceptance of the electricity production using an improved steam cycle based on L.E.G. and N.G. At the same time it will be demonstrated on an industrial scale the feasibility and the operability of a high temperature L.E.G. cleaning.

The capacity of the plant is about 200 T/d of pelletised fuel (R.D.F. from M.S.W. and Biomass fuel from A.S.W. - Agricultural Solid Waste) using two reactors of 20 MWt each (C.F.B.G. - Circulating Fluidized Bed Gasifier) [2] Of course, regarding the environmental impact angle, the plant has a advanced flue gas treatment system and automatic monitoring and control system, for the evaluation of the environmental pressure on the surrounding area (an area of about 20 Km^2 is continuously monitored by special stations directly connected with the central electronic control-system).

Speaking about Biomass fuels (coming from A.S.W.), and Bioelectricity production, the Sorghum bagasse, is the residual and fibrous part of sorghum, derived from the industrial transformation of sweet sorghum for sugar production. In the test campaign of 1994 this kind of vegetal biomass was used as combustible with the following objectives:

- experiment of the gasification plant functionality using biomass as fuel instead of R.D.F. (the project fuel);
- measurement of the gasifier running parameters in various operational conditions;
- chemical - physical characterization of the lean gas; (L.E.G.).
- use of the gasifier in the case of sending gas to the cement woks, near the Testi plant (the cement factory has an advanced dry rotary kiln with pre-calcinator and tertiary air).

The chemical and physical characteristics of R.D.F. and sorghum are presented together with the lean gas chemical composition, lower heating value and flow (tables 1 and 2).

Table 1 - Main characteristics of the R.D.F. and sorghum bagasse [3]

Type		RDF	Sorghum Bagasse
		pellets-briquette	pellets
Form		small cylinders or others	small cylinders
Dimensions (average) mm.	diameter	10-15	5-10
	length	50-150	20-30
Density kg/m3		500-700	500-600
Composition % by weight	Moisture	6,45	10,66
	Volatile Matter	71,13	65,60
	Fixed Carbon	11,42	16,59
	Ash	10,99	7,25
Net calorifc value MJ/Kg		17,217	14,262
	Kcal//Kg	4100	3400

Table 2 - Flow rate, analysis and gas composition

Average experimental variants	Lean gas flow (Nm³/h)	Particulate (g/Nm³)	Condensate (g/Nm³)	Lean gas LHV (KJ/Nm³)
Sorghum	3,940	41,90	66,91	5,347
RDF	4,502	48,02	84,11	7,245

Average experimental variants	H₂O	CO	H₂	CO₂	N₂	CH₄	CₓHᵧ	H₂S (ppm)
Sorghum	7,54	14,44	11,94	15,90	44,28	4,31	1,41	75,85
RDF	9,49	8,79	8,61	15,65	45,83	6,51	4,88	48,61

During the test we encountered some technical problems for the adjustment of the "running parameters" of the gasifiers, but, fortunately, they were resolvable.
On the contrary, no particular problems were detected during the use of sorghum lean gas in the preclinkering furnace of the nearby cement factory, in spite of it not being cleaned before entering the coal burners (see scheme 2).

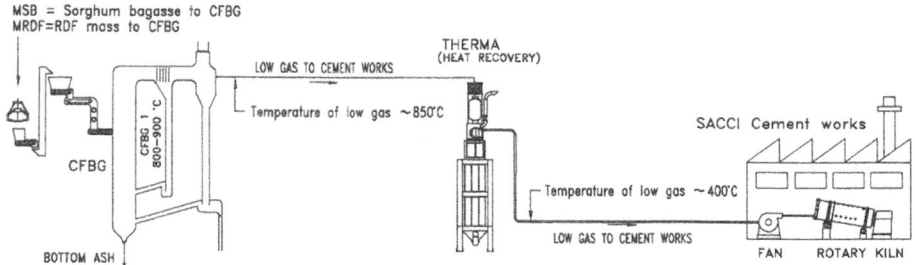

MSB = Sorghum bagasse to CFBG
MRDF=RDF mass to CFBG

A) Sorghum emitted in the CFBG = 2600 kg/h
   Sorghum P.C.I. = 14232 kJ/kg
   Hourly thermal power emitted in the CFBG = 37000 MJ/h
   Hourly thermal power arriving to the SACCI cement works oven = 17000 MJ/h

$$\eta = \frac{17000 \ MJ/h}{37000 \ MJ/h} = 46 \ \%$$

B) RDF emitted in the CFBG = 2400 kg/h
   RDF P.C.I. = 16744 kJ/kg
   Hourly thermal power emitted in the CFBG = 40185 MJ/h
   Hourly thermal power arriving to the SACCI cements works oven = 26260 MJ/h

$$\eta = \frac{26260 \ MJ/h}{40185 \ MJ/h} = 65 \ \%$$

Scheme 2 - Functioning test [3]

With this configuration we are able to feed a cement factory with low energy gas. This is new in the technical world because it is the first time that we have tested industrial production, such as cement production, with an alternative fuel, and especially with a fuel coming from agricultural waste like sorghum bagasse.

At the moment a new campaign is being carried out, using C.D.F. (Coppice Derived Fuel) coming from forestry residues (chestnut; oak). This new test uses about 2000 ton of C.D.F. with the shown characteristics. (see table 3)

Table 3 - Expected characteristics of C.D.F.

Type		Chips
Form		Small pieces and splinters
Dimensions (average)	mm.	25x25x10
Maximum size tolerable		
(22% of material)	mm.	120
Bulk density	Kg/mc	250-350
Moisture contents		
Before Drying	%	35
After Drying	%	8-10
Net calorific value	MJ/Kg	17-18
	Kcal/Kg	4000-4300
Ash content	% by weight	1-3

2.2 The future plant

The new developments are the completion of the existing gasification plant in Greve in Chianti. At present in fact the plant consists of two C.F.B.G. but in a single boiler (fuelled by the gas coming from one gasifier each time - See scheme 3) and in a steam turbine plus generator running at about 50% of nominal power (2.8-3.3 MWe respect to 6.7 MWe). The new plant is the coupling of the existing gasifier n. 2 with a new section consisting of an L.E.G. cleaning, a dual fuel boiler with external superheater, in a flue gas cleaning system and in a steam cycle. The plant is fuelled with N.G. and L.E.G. produced by R.D.F. gasification.

The R.D.F. gasification takes place in a existing CFBG. The produced gas is dedusted in a special section and then partially burned in a dual fuel boiler and partially (from 10 to 20 %) sent to the high temperature gas cleaning system for dechlorination and TAR cracking. (see scheme 4)

The gas cleaning will be carried out with a first step of dedusting (axial centrifugal system with compensation of flow variations) and a second step consisting in a thermal process. The thermal process will be arranged with a high temperature dechlorination unit (using calcium carbonate) and a TAR cracking (using dolomite) (scheme 4). In the dechlorination unit a basic reactant (Calcium Carbonate in a renewed bed) is dissociated at high temperature (about 900°C) in CaO that reacts with HCl forming salts and eutectics, that (a certain condition) firmly remove the HCl avoiding dissociation and further reactions [4]. This unit makes possible increasing the life time of the dolomite bed the following process of TAR cracking, that has downstream a final dedusting step before the co-combustion of clean L.E.G. with and N.G. in the Super Heater section.

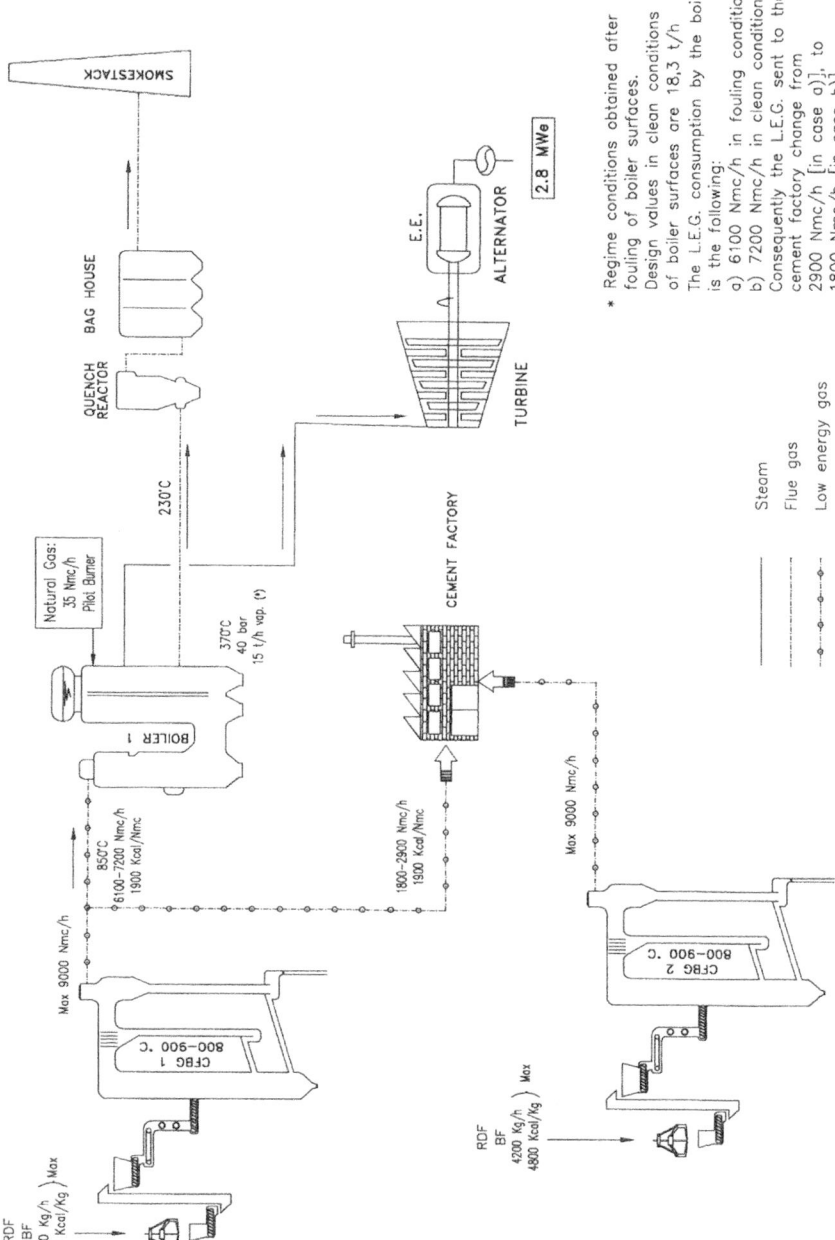

SMOKESTACK

BAG HOUSE

QUENCH REACTOR

230°C

Natural Gas: 35 Nmc/h Pilot Burner

370°C
40 bar
15 t/h vap. (*)

BOILER 1

Max 9000 Nmc/h

850°C
6100-7200 Nmc/h
1900 Kcal/Nmc

Max 9000 Nmc/h

1800-2900 Nmc/h
1900 Kcal/Nmc

CEMENT FACTORY

CFBG 1
800-900 °C

RDF
BF
4200 Kg/h } Max
4800 Kcal/Kg

CFBG 2
800-900 °C

RDF
BF
4200 Kg/h } Max
4800 Kcal/Kg

E.E.

ALTERNATOR   2.8 MWe

TURBINE

* Regime conditions obtained after
fouling of boiler surfaces.
Design values in clean conditions
of boiler surfaces are 18,3 t/h
The L.E.G. consumption by the boiler
is the following:
a) 6100 Nmc/h in fouling conditions
b) 7200 Nmc/h in clean conditions
Consequently the L.E.G. sent to the
cement factory change from
2900 Nmc/h [in case a)], to
1800 Nmc/h [in case b)].

———— Steam

–·–·– Flue gas

–◇–◇– Low energy gas

Scheme 3 – Process diagram actual configuration

1053

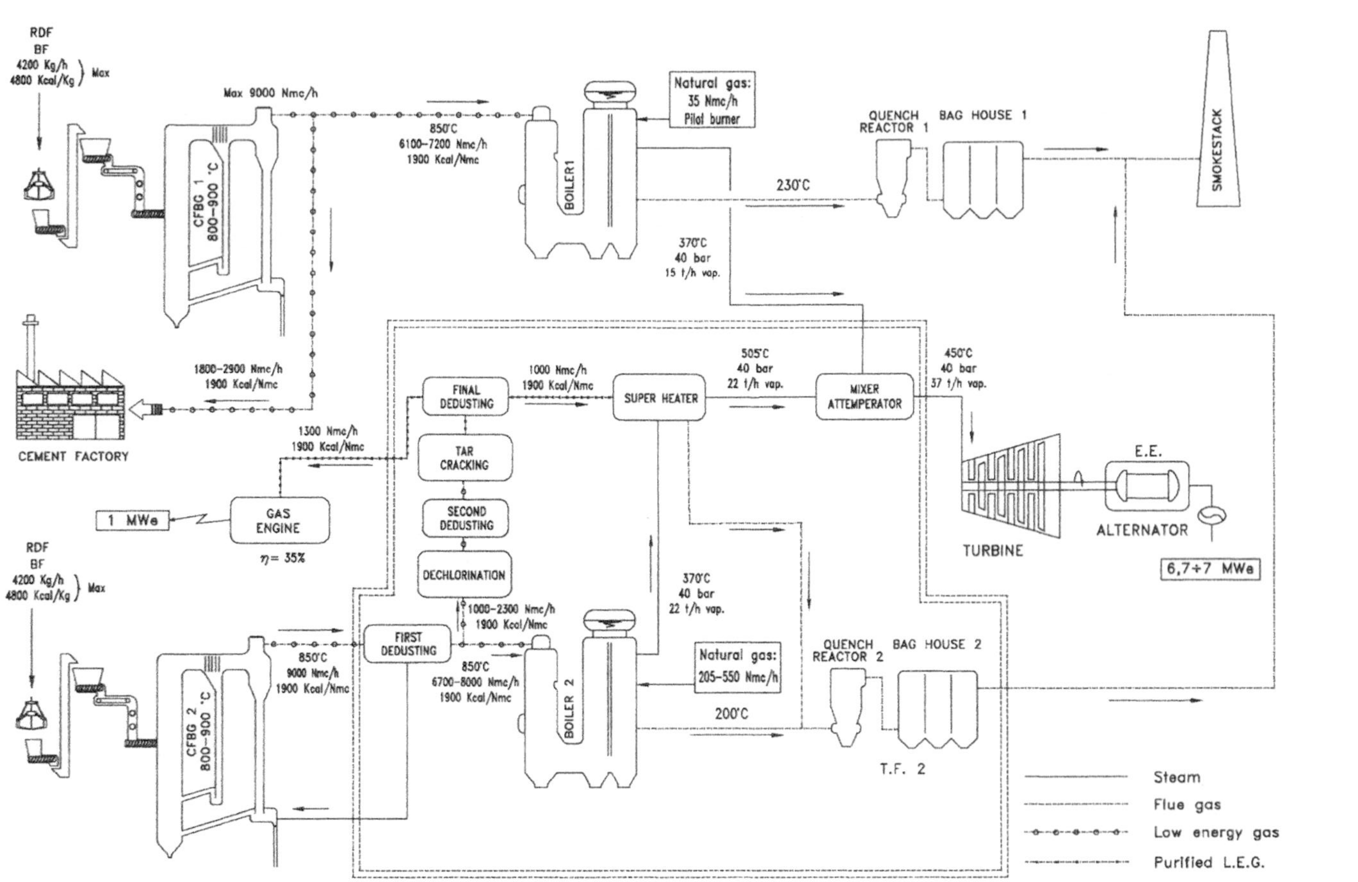

Scheme 4 – Process diagram project situation and future development

1054

The improved steam cycle is based on an external superheater using both N.G. and L.E.G. coming from the gas cleaning system. This configuration is already used in some W.T.E (Waste To Energy) plants using N.G. only.

The project will demonstrate that it is convenient to substitute the N.G. with L.E.G. The plant is designed and will operated as an industrial plant. This means that its continuous operation in compliance with the environmental limits will be the most important point of the project.

As we have explained the dual fuel boiler using L.E.G. and N.G. produces super heated steam at 370°C and 40 bar. The steam temperature is then increased in the aforementioned external SuperHeater up to 505°C. The external superheater is fuelled at the same time by the purified gas and N.G. The steam coming from the external SuperHeater is mixed with the steam coming from the existing boiler and, with an average temperature of 450°C used in a existing turbine to produce up to 7 MWe. The external SH is based on the existing Ansaldo application and will be designed for the specific L.E.G. that will be used in this development of Greve Gasifier.

The first deduster section separates the inlet gas into two streams having a different concentration of particles: the cleaner part (90 %) goes to the utilisation while the dirty part is recirculated to the gasifier.

Table 4 - Flue gas analysis [5] [6]

Parameters	Average
TEMPERATURE (°C)	120 - 145
DRY SMOKE FLOW (Nmc/h)	30000 - 50000
Macropollants:	
• TOTAL DUSTS (mg/Nmc)	3 - 7
• Pb (Lead, mg/Nmc)	max 0.005
• Cd (Cadmium, mg/Nmc)	Less than 0.0004
• Hg (Mercury, mg/Nmc)	0.008 - 0.05
• HCl (mg/Nmc)	3 - 20
• HF+HBr (mg/Nmc)	Less than 0.1
• SO2 (Sulphur dioxide, mg/Nmc)	5 - 15
• NOx (Nitrogen oxides, mg/Nmc)	200 - 300
• HCT (Loose carbon, mg/Nmc)	0.5 - 2
• CO (Carbon monoxide, mg/Nmc)	2.5 - 5
Micropollants:	
• PCDD and PCDF (Total, mg/Nmc)*[1]	13,1E- 6*[3]
• PCB (Total, mg/Nmc)*[2]	163,0E-6
• IPA (Total, mg/Nmc)*[2]	14,0E- 6

* Concentration of Oxygen that is in 11% of smokes
*[1] Minimum amounts took with instruments: 0.005 ng/Nm.
*[2] Minimum amounts took with instruments: 0.2 ng/Nmc
*[3] Isomeric classes observed: 4F, 4D, 5F, 5D, 6F, 6D, 7F, 7D, 8F, 8D
(whose classes: 4D, 5D, and 6D aren't detectable)

The low emissions of the gasifier are (see table 4) further reduced by two integrative sections: the first (deduster) is located at the exit of the gasifier before the boiler and the second (gas purification) on the by-pass of the lean gas feeding the external S.H. The advantages in terms of emission could be remarkable also in terms of suppression of De Novo Synthesis, due to hot gas clean up devices, able to remove HCl and particulates. (These particulates contain metals that catalyze the De Novo Synthesis reactions).

Another secondary objective is to obtain a feasibility analysis of the energy production by using a purified L.E.G. in a diesel engine in order to improve the efficiency of the electrical production. (see scheme 4)

Considering the use of a gas-fed Diesel engine, today's experiences are limited to the cold gas feeding system:

- downstream of the gasifier a tar cracking treatment is used for reducing tar concentration;
- then a centrifugal cyclone and a bag filter are employed to clean the gas stream;
- a scrubber cools the gas to 50°C and washes away the remaining tar which at this temperature is in a liquid phase.

In order to entirely keep the tar contribution in the gas stream, it is proposed to feed the Diesel engine with a gas temperature higher than the tar condensation value. In this way the diesel engine components are preserved by tar liquid particles and the gas maintains its original L.H.V. (Low Heat Value)

The new proposed gas line consists of: (see scheme 5)

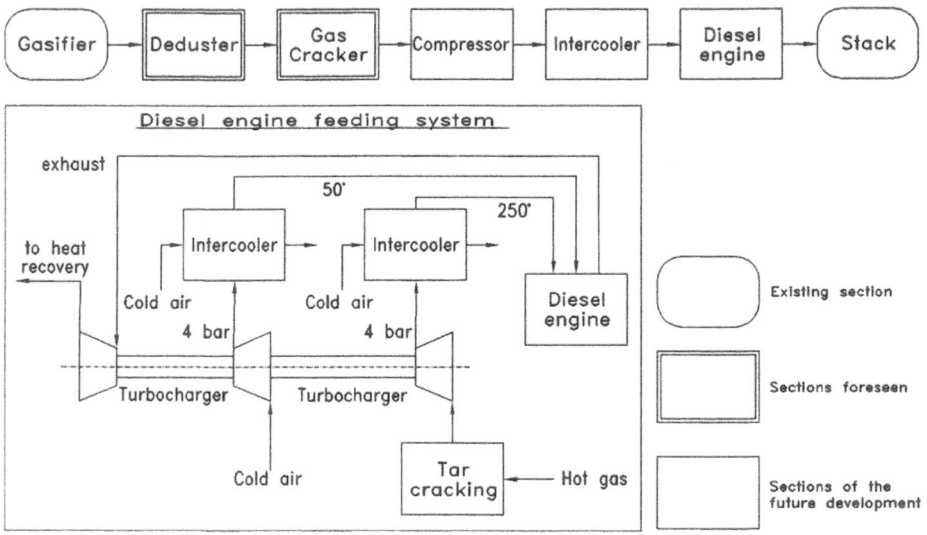

Scheme 5 - Future development based on diesel engine use

From the preliminary comparative study it emerges that the proposed gas line could be characterised by:

- important advantages in terms of total and electrical conversion efficiency
- strong simplicity in the gas treatment line requested by the hot fuel gas feeding in the engine
- probably lower costs due to a simpler gas treatment line

3 Conclusion

The thermal treatment of the Municipal Solid Waste and of the renewable fuels in general presents, in the Mediterranean area, a small utilization because of the low social acceptance of this kind of plant. Consequently the most widely adopted solution for the disposal of the municipal, industrial and agricultural waste is the landfill with a strong impact on the environment and on the quality of life.

It is now a common opinion that the solution for the reduction of this impact will be a recycling policy mainly connected with thermal treatment with energy recovery.

That means an increase in the R.D.F. production with high calorific value.

The best way for the energy recovery from R.D.F. materials and from Biomass fuels (like C.D.F. or Sorghum Bagasse) is the realization of a high efficiency power cycle and the gasification process is a good and reliable way for this purpose.

Consequently the perspectives for a wider application of this kind of technology will be very important if the technological, economical and environmental aspects have positive solutions.

4 References

1. Barducci G.L., Campagnola G.C. (1991) *R.D.F. pellet gasification gas clean up development programme*, In 6th European Conference on Biomass for Energy, Industry and Environment, Athens, Greece

2. Rensfelt, E. (1991) *Gasification for Power Production*, 1st European Forum on Electricity Production from Biomass and Solid Wastes by Advanced Technologies, Florence, Italy

3. Barducci G.L., Daddi P., Polzinetti G.C., Ulivieri P. (1995) *Thermic and electric power production and use from gasification of biomass and R.D.F.: experience at C.F.B.G. plant at Greve in Chianti*, Second Biomass Conference of the Americas: Energy Environment Agriculture and Industry - Portland - Oregon

4. Gori V., Colucci M., Capecchi G., Frontini L., Repetto F., (1995), *New technologies for exhaust treatment deriving from precious metals thermal recovery plant*, Second European Precious Metal Conference - Lisbon - Portugal

5. Barducci G.L. (1993) *The R.D.F. gasifier of Florentine Area*, First Biomass Conference of the Americas: Energy Environment Agriculture and Industry, Burlington, U.S.A.

6. Stassen H.E.M. (1994) *Gasification of waste Evaluation of the waste processing facilities of the Thermoselect and TPS/Greve*, In the NOVEM Programme E.W.A.B. (Energy from Waste and Biomass) 9420 - Utrecht, Holland

IISc-DASAG biomass gasifiers:
Development, technology, experience and economics

H. N. Sharan [1], H. S. Mukunda [2], U Shrinivasa [3], S. Dasappa [2],
P. J. Paul [2], N. K. S. Rajan [2]

Abstract

The open top gasifier technology developed by the R&D team at the Combustion, Gasification and Propulsion Laboratory, CGPL, of the Indian Institute of Science, Bangalore, successfully ran irrigation pumping sets, where it replaced diesel oil by producer gas. The technology was subsequently further developed for power generation, also with promising results. A laboratory - industry co-operation between IISc and DASAG, established to engineer the technology to rigorous power plant standards before commercialising it, has progressed into a successful South - North technology partnership being supported by the Swiss and the Indian governments.

The paper describes the gasifier and presents details of its performance both in the laboratory and in the field, including experience of a plant running in an Indian village, results of tests carried out jointly by Indo-Swiss teams in Bangalore and the results of the initial operation of a plant in Chatel-St-Denis, Switzerland. Details of a commercial 80 kWe cogeneration plant supplying electricity and process heat for drying to a hand-made paper factory have also been presented.

The paper gives details of steps taken to commercialise the technology and puts forward the concept of Independent Rural Power Producers IRPPs for setting up decentralised power stations in villages. The paper presents data on the cost of power generation in biomass based Independent Rural Power Producers (IRPPs) and compares the economics of electricity supplied from such decentralised plants with those of coal based electricity supplied by the Indian power grid.

Key words: Biomass gasification; decentralised power plants; Independent Rural Power Producers IRPP; commercialisation of renewable energy; South-North technology transfer.

---

[1] DASAG Energy Engineering Ltd., Seuzach. Switzerland.

[2] Combustion, Gasification and Propulsion Laboratory (CGPL), Indian Institute of Science, Bangalore.

[3] Department of Mechanical Engineering, IISc, Bangalore. India.

Introduction

Early in 1979, the attention of two authors of this paper (US and HSM) was brought to the translation of the Swedish experience on gasifiers by the Solar Energy Research Institute, USA. A study of this document showed that the development of low power gasifiers posed problems of gas quality, particularly of the tar content, specially at part loads. The problems at lower power levels was analysed to be one of the relative values of heat generation Vs heat loss. The heat loss through the hardware (however well designed) will always be unfavourable for small systems. In order to ensure reliable operation with good gas quality it is imperative that energy conservation is ensured by providing an adiabatic thermal environment . A large number of tests on the classical closed top designs revealed that while at a flow rate close to the rated value one could get nearly tar-free gas, at reduced flow rates the performance was poor.

At this time, the authors came across the laboratory studies of Thomas Reed[4] describing the open top system from a mechanistic point of view (discussing the propagation of a flame front in a porous bed of charcoal / wood pieces) and attempted a convenient variant of the reactor. The reactor consisted of a twin stainless steel shell, the inner portion containing the wood chips, the annular space transporting the hot combustible gases (at 500 to 600 °C) and transferring heat to the incoming wood chips. The outer wall was insulated with light high temperature ceramic. An addition to the geometry of Reed was an air nozzle at a certain height from the grate to enable quick light up. This reactor worked like magic, so different from its closed top counterpart. Such a positive result led to the decision to begin rigorous studies to understand why this system behaved so much better in terms of gas composition and low tar content than the closed top designs.

These studies made clear the roles of the air flow from the top as well as from the air nozzles above the grate in stabilising the operation of the reactor and providing residence time for tar reduction. The significant engineering inputs which had gone into the cooling and cleaning systems of the closed top design were integrated into the new reactor design and the IISc gasifier system with the twin stainless steel shell was developed into a Mark I product for operation with diesel engines. A remarkable feature found during the dual fuel runs was that there was no derating in terms of power, and the diesel replacement was as high as 92 %.

A point to be stressed is that the twin shell design of the reactor with an outer insulation is thermodynamically the most efficient solution, even though it may cause problems related to the service life of some components in the zone immediately above the air nozzle which is at a high temperature, operating sometimes under reducing and at other times under oxidising environment. A design solution consisting of a twin shell for the top region and a ceramic walled single shell for the lower region has proved successful in practice.

---

[4] Reed, T. B. "Types of gasifiers and gasifier design considerations". "Biomass gasification: Principles and Technology". Noyes Data Corporation, New Jersey. 1981.

Initial Field Experience

Most of the initial experience was gained in the development of the 3.7 kWe systems for electrical generation and mechanical drive. The development programme received a very welcome boost by the introduction of a national programme on biomass gasification which subsidised qualified gasifiers for saving diesel consumption for water pumping. The qualification tests were carried out under a rigorous programme at Indian Institute of Technology, Bombay, (IITB), by Parikh and co-workers. Successful manufacturers of both closed top and open-top designs (the latter being based on technology transfer from IISc) which achieved a minimum diesel replacement of 60%, and tar and particulate contents not exceeding about a hundred and fifty ppm each, were each allowed to produce about three hundred and fifty units for dissemination to individual farmers.

Results of tests at the Indian Institute of Science IISc, Bangalore and the Indian Institute of Technology, Bombay, IITB.

Both sets of results were obtained under nearly similar conditions with a 3.7 kWe gasifier system connected to a dual fuel diesel engine. The lower calorific value of the gas measured was in the range of 4.5 to 5 MJ/kg in IISc tests and 5 to 5.8 MJ/kg at IITB. The gas composition measured in both the laboratories were within the range of ±2 % for the major components, CO, H_2, CO_2, and N_2. Values of methane are also within 2±1% at both the laboratories. The diesel replacement in the engine was high, exceeding 85 % as measured in both the laboratories. No effluent analysis was conducted at IITB. The results of the particulate and tar measurements are presented in the Table 1 below.

Table 1: Comparison of Results on Particulate and Tar

Gas Flow Rate g/s	Tar ppm		Particulate matters ppm	
	IISc	IITB	IISc	IITB
2.0	50 - 120	40 - 120	80 - 100	-
3.0	50 - 100	40 - 120	100 - 150	50 - 300

The techniques used for the measurements of the particulate matters and tar concentration were not identical at the two institutes and may account for some of the differences in the results. The differences, however, are not significant and have been discussed in an earlier paper by Dasappa et. al.[5].

The utility and impact of the dissemination process of the gasifier based pumps in villages, which lasted over a period of three years, were assessed through field visits by the national monitoring agency, the state nodal agencies and by the designers. These visits to the far-off sites were very cordial at the personal level, and very revealing and instructive on the technical side. Based on

---

[5] Dasappa, et. al.: "Five-kilowatt wood gasifier technology: evolution and field experience". Sadhana, J.I.A.Sc., 14, pp. 187-212,1989

the analysis of the field experience several modifications were introduced: e.g., the bottom metal cap was replaced by a water seal to achieve long hours of uninterrupted operation, the use of water from the pump for cooling the gas was introduced and the design was modified to enable a single individual to dismantle and re-assemble the plant.

Other Indian Projects

The success of the field experience led to an interesting project to provide power to an unelectrified village. The objective was to provide a total package of services - lighting, drinking water, grinding food grains, supplying irrigation water and running other minor electrical machines - and gain experience regarding project implementation as well as field problems. This project has been in operation for the last six years, been through several stages of development and in the process has provided invaluable experience on the operation and management of such systems.

During this period the CGPL team was awarded a contract to build a 100 kVA dual-fuel plant at Port Blair, Andaman and Nicobar islands. Based on the experience at that time, it was possible to give an assurance about the operational and performance behaviour of the entire plant except for the expected life of the inner reactor shell. The system design was custom made to suit the local conditions in the Chattam Island Saw Mill near Port Blair. The plant was built within ten months and its performance was demonstrated on the test bed before its shipment to Port Blair. The whole plant including an automatic fuel feed system, and instrumentation and control system for optimising the diesel replacement were made operational by 1990.

Another interesting project started two years ago involved the supply of irrigation water to a select group of willing partners in a village from a 25 kVA gasifier based power generation system. The plant operation has provided very useful operational experience.

The Indo-Swiss qualification testing at CGPL, Bangalore

The experiments on Indo-Swiss project was started when one of the authors (HNS) prevailed upon the Swiss scientists and the energy ministry to examine the technology developed at Indian Institute of Science for possible adaptation to Swiss conditions. Under a qualification test programme Swiss and Indian experts jointly established the test procedures, instrumentation and analytical methods. The test procedure for tar and particulate matters, based on the Finnish experience, was designed to extract the small amounts of tar and particulate matters by conducting eight to ten hours' run on the gasifier. Data was collected at different loads and three types of wood mixtures on particulate and tar contents in the hot and the cold gas, gas composition and temperatures, metal temperatures, wood consumption rates, cooling water analysis, analysis of water used in the water seals of the reactor and sand bed filters, etc. Six major tests and several repeats were carried out using causarina pieces.

The results have been published by Mukunda et. al.[6]: some of the important values are summarised here in Table 2a. Data on various parameters of the wash water which are needed for specifying the water treatment system either for recycling the water through the cooler, or for meeting the legal limits for discharging it into the municipal drainage system are given in Table 2b. The minimum amount of dilution required to reach the permissible limits for irrigation water can also be determined on the basis of such data.

Table 2a
Typical Results of Tests with Causarina

Parameter	Value	Comments
Gasification efficiency (Hot)	80% to 85%	At full load. At 30% load about 5 % reduction.
Lower calorific value	4.5 to 4.8 MJ/m^3	The calorific value stabilises at 1- 3h after start.
Hydrogen content (Important parameter for engine performance)	16 - 19 %	The H$_2$ content also stabilises at 1 - 3h after the start. H$_2$ is higher with 10 -15% moisture.
Particulate (Hot) (Cold)	700 mg/m^3 50 mg/m^3	
Tar (Hot) (Cold)	120 mg/m^3 20 mg/m^3	
Particle size distribution	0.7 - 2 microns	The particle size distribution meets the requirements of a gas turbine.

---

[6] Mukunda et. al.: "Results of an Indo-Swiss programme for qualification and testing of a 300 kWth IISc-DASAG gasifier". Energy for Sustainable Development, vol. 1, No. 4, pp. 46-49, November 1994.

Table 2b
Effluent per kg moisture free wood

Item	P+T	BOD	COD	Phenol	DOC	NH$_3$ / NH$_4$
g per kg mf-wood	1.45	0,14	1.9	0.077	2.32	1.72

An IISc-DASAG gasifier testing programme in Switzerland

Subsequent to the successful testing in Bangalore, a gasifier plant was ordered from India by a Swiss consortium supported by the Swiss Ministry of Energy and was installed in a Swiss village called Chatel-St-Denis, about 250 km south of Zurich. The complete plant was shipped in a container and was installed in about 4 weeks. Joint tests were conducted in Jan - February 1996 by ETH, Zurich and IISc, and further tests are still continuing. While all the results from the completed tests are still not available, the preliminary indications are that the performance of the gasifier is very similar to the results obtained during the qualification testing in India. A particularly interesting result, as pointed out later in Table 3a, is that pine wood which is quite light, is capable of being gasified at a high moisture content of 30 %.

The first Commercial plant in Ochhra, Central India.

The interesting point about this project is the fuel. Eipomia is a weed which after being dried produces branches of fairly light specific weight. Results on the test bed showed that pieces of this fuel can be gasified with normal efficiency and the dual fuel engine operated with 80% diesel substitution. The plant, which will go in commercial operation next month, is briefly described below.

Description of a IISc-DASAG gasifier power plant

Figure 1 shows schematically the different elements of a power generation system. The main elements of the gasifier are the reactor, and the cooling, cleaning and the control systems. The auxiliary systems upstream of the gasifier are the fuel preparation, storage, feeding and, if required, fuel drying systems. The downstream elements are an engine, either a dual-fuel diesel engine or a gas engine and, if required, an exhaust gas conditioning and a waste heat recovery systems. In Orchha two engines have been installed for operational reasons and they will run in parallel at high unit loads.

The reactor is an open top down-draught system. In its internal structure it is a simple cylindrical shell with the air induction from the top as well as from the radial air nozzles arranged in the lower section, which is a metal shell lined inside with ceramic. The upper section is a double walled stainless steel shell. The hot combustible gas from the bottom of the reactor is passed into the top section via a side recirculating duct. The hot gas transfers part of the heat into the wood pieces in the reactor and leaves at the top section. It then enters the cooling section where two direct contact cooling water sprays bring down the temperature to the ambient. The gas, drawn into the suction zone of a blower, is sprayed with water in the impeller section. The mist from the spray combined with dust and possibly with some fugitive tar is separated from the gas by the centrifugal action of the impeller, removed in a downstream cyclone and passes on into a sand bed filter. The gas velocities are kept low - a few cm/s so that the gas passing through the tortuous path of the sand bed leaves behind a part of the remaining fine particulate and

Fig. 1: **Schematic arrangement of a biomass based power-cum-heat plant using an IISc-DASAG gasifier and a dual-fuel engine.**

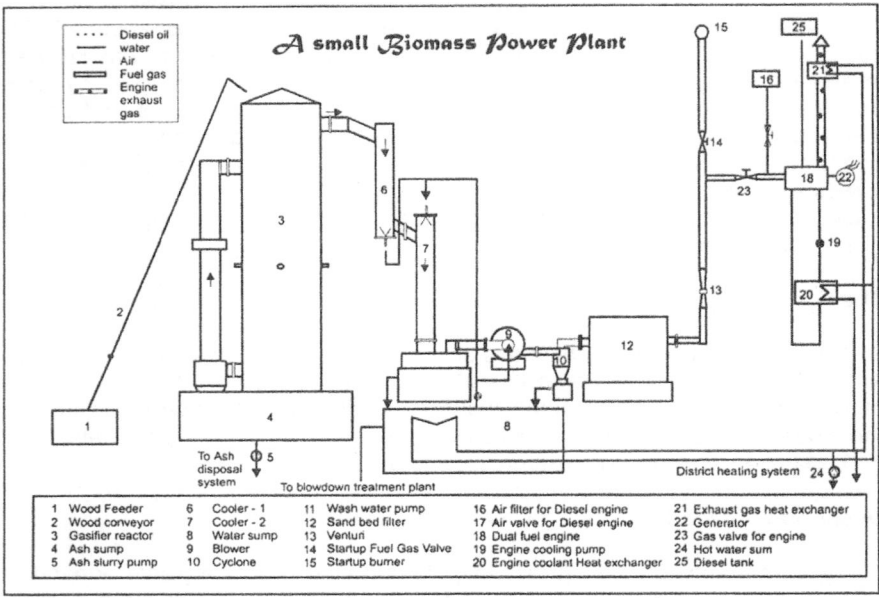

1	Wood Feeder	6	Cooler - 1	11	Wash water pump	16	Air filter for Diesel engine	21	Exhaust gas heat exchanger
2	Wood conveyor	7	Cooler - 2	12	Sand bed filter	17	Air valve for Diesel engine	22	Generator
3	Gasifier reactor	8	Water sump	13	Venturi	18	Dual fuel engine	23	Gas valve for engine
4	Ash sump	9	Blower	14	Startup Fuel Gas Valve	19	Engine cooling pump	24	Hot water sum
5	Ash slurry pump	10	Cyclone	15	Startup burner	20	Engine coolant Heat exchanger	25	Diesel tank

tarry matters. The various processes combine to produce clean gas of a good calorific value.

The fuel preparation system consists of manual cutting of wood or cutting it with suitable mechanical devices. The drying system can be air drying, supplemented , e.g., by a simple drum type drier if necessary. The dual fuel engine is a standard diesel engine modified at the air inlet section to take the mixture of fuel gas and air with appropriate valves which are motorised if automatic operation is needed. The C&I system consists of instrumentation to determine the diesel flow rate and other load parameters as well as normal measurements of temperatures and pressures. The electronic diesel flow rate indicator, e.g., can be used to optimise the diesel flow in the dual fuel operational mode. In another variant, the oxygen sensor in the exhaust gases can be the control variable to optimise the diesel replacement. This follows the discovery that at any given load, the reduction in diesel flow rate is limited by the amount of oxygen in the exhaust gas. A completely automatic system for the gasifier start-up and shut down sequence has been provided for the Châtel-St-Denis plant.

Types of wood gasified

The total gasifier running experience gained during the last ten years exceeds 40,000 hours. Hosahalli village has done about 5000 hours, 1500 hours have been clocked with the power plant for supplying irrigation water described above, about 200 to 400 recorded hours each have been done by more than a hundred water pumping systems, and the other plants have experience of some hundreds of hours. The kinds of wood used in the last ten years are listed in Table 3.

Table 3: Woody biomass tested in various gasifiers

Species	Density kg/m^3	Moisture content %	No. of hours	Size of wood chips
Causarina	550 - 650	15	200	- 75 mm - Mix: 75 mm + branches - Mix: 75 mm + 50 %: 10 to 15 mm + 10% sawdust
Eucalyptus	400 - 650	15	6000	50 - 75 mm
Phadauk	1050-1100	15	700	50 - 75 mm
Silver oak	250 - 300	20	150	50 - 75 mm
Pine (European)	200 - 250	20 - 30	100	40 - 75 mm
Mulberry stalk	300 - 350	15	1000	10 to 20 mm dia 30 - 50 mm long
Eipomia	200 - 250	15	50	10 - 20 mm dia 30 - 50 mm long
Jungle wood	300 - 600	20	100	50 - 75 mm
Arecanut husk	100	10 & 30		

Basic research in biomass combustion / gasification

About five years after the start of the programme, it was felt that basic work was essential in order to understand the processes that take place inside the reactor. Gasifier modelling was started so that all the other experimental and theoretical work could get integrated towards producing cohesive results. For instance, questions such as the relationship between geometric parameters of the reactor and power level, the role of moisture (in wood) on the producer gas composition, the prediction of the exit gas composition from known parameters of the reactor, or even finer questions such as the prediction of the tar level in the hot producer gas, needed to be answered. In addition to the prediction of gasifier performance, other general data such as the speed of propagation and the flammability limits of producer gas had also to be generated.

To answer these questions, it was necessary as a first step to have calibrated models of wood char reactivity with carbon dioxide, water vapour and oxygen in a nitrogen environment. Experimental and modelling work have been completed on most of these aspects and reported in a number of publications by Mukunda et. al.[7]. As far as the characterisation of producer gas is concerned, both experimental and computational aspects of flame propagation have been completed (Kanitkar et. al.[8], Mishra et al.[9], Chakrabarty et al.[10]). The results of flame propagational speed and the limits of propagation are presented for a composition CO : 20%, H_2 : 18 %, CH4 : 1%, CO_2 13 %, Rest nitrogen in figure 2. The figure contains experimental results and computational results from a code which accounts for full chemistry and realistic diffusion. The work on modelling the reactor operation is currently in progress.

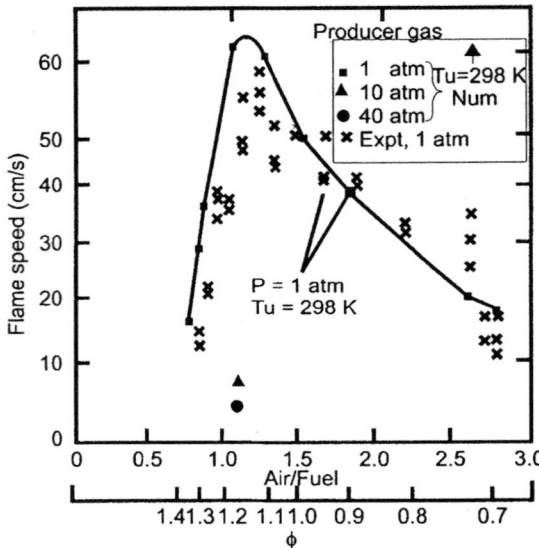

[7] Mukunda, Dasappa, et. al:
Proceedings of the 20th and 23rd Symposium (International) on Combustion, 1984 / 1990
Proceedings, the Third International Conference on small engines and their Fuels, Reading, UK, 1990
Third and Fourth National Conference on Biomass gasification, Baroda, 1991 / 1993

[8] Kanitkar et al: The Flame speeds, temperature, and limits of flame propagation for producer gas-air mixtures: Fourth National Conference on Biomass gasification, Baroda, 1993

[9] Mishra et al. : Flame speeds of wood gas at ambient and engine operating conditions: Third National Conference on Biomass gasification, Baroda, 1991

[10] Chakrabarty et al.: The theoretical calculations of the limits of flame propagation for producer gas mixtures. Fourth National Conference on Biomass gasification, Baroda, 1993.

One very important aspect of the IISc open top gasifier design is that its operation is close to being one-dimensional and hence the one-dimensional mathematical approximation is likely to be a good representation of reality.

IISc-DASAG Cyclone Gasifiers

Given the fact that the availability of agricultural residues is much more abundant in villages and the economics of biomass power plants are quite attractive, the technology for a reactor which is capable of gasifying pulverisable fuels like sawdust, sugarcane trash, rice husk, coconut coir pith, etc. is also being developed. Cyclone gasifiers of three ratings up to 2.75 MWth are in advanced stages of development and testing. The system elements downstream of the reactor have been kept similar to the current solutions.

Experiences of the village electrification system

Before discussing the commercialisation plans of the IISc-DASAG gasifier based power stations in the rural areas of India, it is considered important to review the experience of the village electrification project. The village consists of 42 households with a population of 209. The project was implemented jointly with the village community in the following five phases.

- Phase 1: Raising energy forest for supplying wood in a sustainable way / Installing a wood gasifier-diesel generator plant
- Phase 2: Providing electricity for lighting to all the households
- Phase 3: Pumping drinking water
- Phase 4: Installing a flour mill
- Phase 5: Pumping water for irrigation

It took about a year to stabilise the operations of Phases 1 and 2. In its sixth year, the project is currently running in Phase 4 while Phase 5 is under implementation.

The performance of the plant and the maintenance problems

The performance has been measured by the reliability of electricity supplied for lighting and the number of days on which water is supplied. The electricity supply has been maintained with a regularity of over 95 %, a dual-fuel operation being maintained during more than 90 % of this period. Typical diesel replacement values are about 78 %, the average for 32 months including all associated aspects being 67%. The amount of wood consumption, kg per kWh has been about 1.6 - 2.0, which is higher than the designed value, due to the use of wood with higher moisture content and the effective load being higher than designed because of power factor correction. Wood harvesting from the energy forest, with the dry wood productivity of 6.9 tonnes/ha/year, supplies the fuel on a sustainable basis.

As Table 4 shows, the major problems have not arisen from the gasifier but are the normal ones for conventional diesel engine operation.

Table 4
Engine operation and maintenance

Engine	Total hours	Hours on dual-fuel	Hours before maintenance	Items of maintenance
E1	1705	1534	1005	Overhaul + change of rings
E2	2004	1802	1117	Only overhaul; oil circulation pump problem; Pump rack worn out and replaced. Radiator leakage and scale formation.

Operators' training

Initially, two boys from a nearby village, who were trained at the IISc, operated the system for about 20 months. Witnessing the benefits of the technology and the operational experience, the villagers have now taken on the responsibility themselves and two local boys have been trained to run the system, in addition to gathering, chopping, drying and storing the wood. An indication of the social change is that a local woman has volunteered to operate the system.

Social benefits

Not all the social benefits can be financially quantified. Foremost is the impact of drinking water supply on the women who earlier had to haul water from open tanks. The water from the borewell, being potable, has improved the health and hygiene in the village. The cattle are also assured of drinking water now. Secondly, the villagers are able to extend their work at home till late at night, especially on silk culture, and the children can do their home work in the evenings. With a reliable supply of electricity, the scope for local entrepreneurial development and local employment has also improved substantially.

Commercialisation of the IISc-DASAG Biomass Gasifiers Technologies

Conferences, even UN ones, can sometimes be useful: for it was at one such workshop organised by UNCHS Habitat in December 1992 that the first contacts were established between the IISc team and DASAG, a Swiss engineering company engaged in the development and commercialisation of new energy technologies, including renewable. An assessment of the design concepts and solutions, the laboratory and field results, the facilities, and above all, the R&D team who are behind the open top gasifier indicated that at long last there was a highly realistic probability of building small biomass power plants on a commercial basis. A strategic business plan was evolved and included the following linkages:

- Combining the energy market in the rural sector in India with the increasing demand for reducing CO_2 emissions in Europe, which has sustainable supplies of wood.
- Combining the Indian and Swiss expertise with Indian manufacturing capabilities to produce affordable plants of high quality.

A license and technology co-operation agreement was signed in 1993, the 100 kWe unit was re-engineered, and the previously mentioned Indo-Swiss qualification test programme was carried out in Bangalore with the support of the Swiss government. The results showed that the performance of the gasifier in terms of its thermal efficiency, tar and particulate concentrations and ease of operation was one of the best tested by the Swiss experts [11].

These positive results led to the formation of a new Indo-Swiss joint venture company, NETPRO, with equity participation of IISc and the goal to commercialise the gasifier technology. Two orders were booked immediately: one to supply a gasifier plant for a Swiss testing and demonstration project in Châtel-St-Denis, and the other to install a decentralised power plant to supply electricity to a hand-made paper factory in Orchha, Central India. As mentioned earlier, the Swiss plant was commissioned early this year and is still being tested. The proposal to start the second phase in Châtel-St-Denis by adding an engine and feeding power into the local grid is presently under discussion.

The Orchha plant has been tested on the test bed and is expected to be commissioned in April. Since the cost of power generated by the Orchha plant is competitive with power supplied from the grid, DESI Power, the Indian Independent Rural Power Producer (IRPP) which has promoted the Orchha plant, has decided to install 10 such units in India during the next two years.

The Role of Independent Rural Power Producers IRPP based on Biomass

It is a sad fact that even after 50 years' of heavy financial investments, the centralised power system and its rural electrification programme have been unable to provide electricity, the most basic input required for economic and social development, to the rural areas and small towns in Developing Countries DCs. The creation of Independent Rural Power Producers [IRPP], a socially responsible market instrument to provide power and energy services to these areas, is therefore, long overdue. Similar in many ways to IPPs in the centralised power and industrial sectors, Commercial IRPPs, based on local sources of renewable energy, are competitive with the cost of power supplied from conventional fossil fuel power plants at the point of consumption, e.g., in India using biomass, Sharan and Khosla[12]. In addition, IRPPs will also save the

---

[11] Bühler: "State of technology in wood gasification" . 3. Holzenergie-Symposium, ETH, Zürich. Oct. 1994.
Kaufmann: "Pilot tests in India on a wood gasifier". Forum Bios-Energie, Châtel-St-Denis, June 1994.
Sharan: "Economic aspects of wood gasification" . Forum Bios-Energie, Châtel-St-Denis, June 1994.
[12] Sharan and Khosla: Power to the Rural Poor: A new Business Initiative for Sustainable Development. Development Alternatives Newsletter, January 1996

centralised power sector substantial amounts of money through reduced T&D losses and peak-load shaving.

The results of a comparison between a 600 kWe biomass based IRPP and two 210 MW coal fired power plants, one mine-mouth and the other load centre, are shown in Fig. 3. Values used in the calculations on which the bar charts are based are also given in Table 5 on the same page. The data for the coal fired centralised system are valid for current Indian conditions and the data for the biomass plant is based on the plant just built. The financial conditions used for the IRPP are the same as those currently used for the IPPs. The interest on capital during construction and the T&D losses of the centralised system have been taken into account in calculating the cost of electricity delivered in the villages.

Table 5: Base data used for comparing the cost of electricity.

Parameters	210 MW Mine Mouth	210 MW Load Centre	600 kW IRPP with Biomass
Capital Cost [13] Rs / kW	37000	34000	20000
Auxiliaries + T&D losses %	9.5 + 25	9.5 + 10	10.6 + 6
Annual operating hours	7008	7008	7008
Dividend on 30% equity %	16	16	16
Interest on Indian debt (40%) %	18	18	16
Interest on foreign debt (30%) %	12	12	12
Payback period years	12	12	12
Depreciation % / y	7.8	7.8	7.8
Coal price Rs / t	400	1200	400
Annual O&M cost, % of capital	2.5	2.5	6
Cost of electricity delivered in the village, Rs / kWh	2.68	2.74	1.87

---

[13] The total cost includes interest during construction and the cost of T&D lines. These two items of cost are much lower than for the IRPPs than for coal plants.

Fig. 3: Comparison of the cost of electricity supplied in a village

Break-up of Power Costs in a Village
Plant I. 210 MW Mine-mouth coal IPP   Plant II. 210 MW Load-centre coal IPP
Plant III. 600 kW Local biomass IRPP

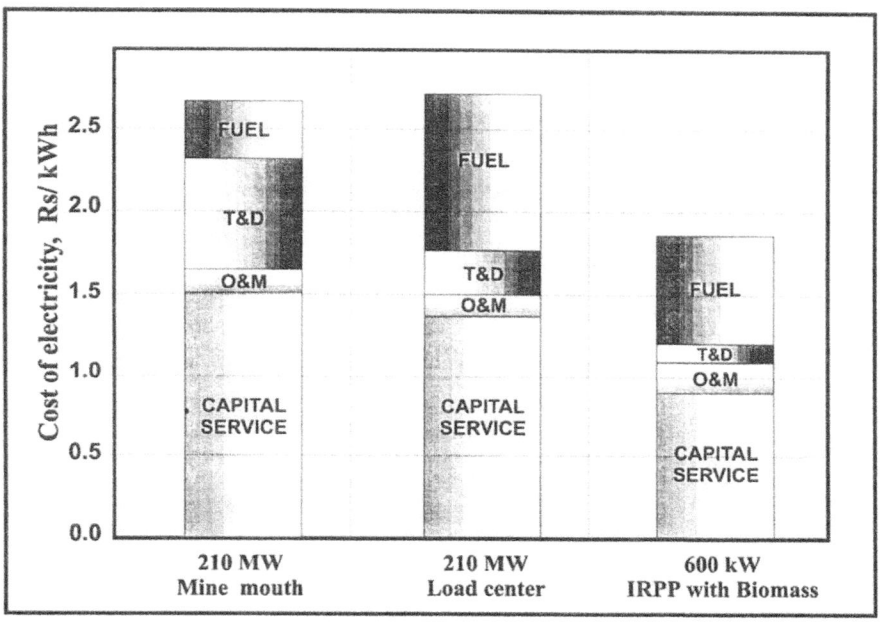

	Plant I	Plant II	Plant III
Capital Service	1.51	1.38	0.89
O & M (Operation and Maintenance)	0.15	0.13	0.20
T & D (Transmission and Distribution)	0.67	0.27	0.11
Fuel	0.36	0.95	0.67
Cost of electricity Rs / kWh	2.69	2.73	1.87

The economics of the commercial IRPP in Orchha described earlier is given below.

Table 6
The Economics of an Independent Rural Power Producer IRPP
A 80 kWe Power Plant in Orchha, Central India.
Decentralised Energy Systems India Pvt. Ltd (DESI Power), New Delhi.

Cost of the plant	1.6	million Rs.
(Civil works, gasifier system, 2 units of 40 kWe		
dual fuel engines, operator training, spares)		
Cost of the fuel (Eipomia, a local weed)	300	Rs/t
Diesel Replacement	80	%
Operating hours	6132	h/year
Financing:		
Equity	100	%
Loan	0	%
Dividend (as in Fossil IPPs)	16	%
Depreciation	7.8	%
Annual O&M charge (% of cap. cost)	9.1	% / y
Sales price of electricity	1.81	Rs/kWh
Comparable cost of grid supply in the village	2-2.5	Rs/kWh

Conclusions:

To sum up, technical experience and economic comparisons clearly show that IRPPs based on biomass fulfil all the criteria for successful commercialisation. Where biomass is available and decentralised plants can be operated with an adequate plant load factor, they can provide energy services reliably and cleanly, at rates competitive with centralised grid supply. In the process they also help the centralised sectors to save their commercial losses resulting from small non-remunerative loads and reduce the technical losses of their T&D systems. The only bottleneck, but undoubtedly the most difficult one, which remains to be overcome is the lack of a national and international policy framework which can make investments in the IRPPs attractive for the centralised sector operators and the private sector investors.

For IRPPs located in villages where there is inadequate electricity purchasing capacity due to lack of job opportunities, it may be necessary to start with semi-commercial IRPPs, in which a part of the electricity generated can be sold to the poorer sections of the villagers at prices lower than the commercial during the initial years. The accumulated losses can either be made up by marginally increasing the sales price of commercial electricity or by an innovative linkage to external financing sources such as an investment fund to be created by the centralised power sector who will save on T&D losses; ethical investments; CO_2 sales under Joint Implementation proposals of the Climate Convention (AIJ); yet-to-be-created funds based on CO_2 / energy levies / taxes, etc.

COGENERATION IN BRAZILIAN PULP AND PAPER INDUSTRY FROM BIOMASS-ORIGIN TO REDUCE CO_2 EMISSIONS

S.T.COELHO and S.G.VELÁZQUEZ
Institute of Electrotechnique and Energy - University of São Paulo - São Paulo - Brazil
D.ZYLBERSZTAJN,
Secretary of Energy of the State of São Paulo - São Paulo - Brazil.

Abstract

For developing countries, electricity cogeneration is quite important due to the economic difficulties of local utilities, avoiding investments for the construction of new plants.

Moreover, when electricity generation is from biomass origin, it allows significant reductions in CO_2 emissions due to carbon absorption during biomass re-growth. In Brazil, hydroelectric power plants correspond to 90% of electrical generation and, until now, they were not considered as a source of carbon emissions. However, recent studies evaluated that even hydroplants release greenhouse gases under the form of methane [1], although much less than fossil fuels.

In South-southeastern Brazil, biomass-origin cogeneration occurs mostly in sugar and alcohol industries [2]. Nevertheless, there is a high potential also in pulp and paper industries, using process by-products, through conventional or advanced technologies (biomass-fired gasifier/gas turbine systems), as evaluated in this paper.

Pulp and paper segment is in third place among the highest energy consumers in Brazilian industrial sector and most of its electricity is purchased from the grid. Fuel oil consumption is also quite high in this segment due to current low prices of fuel oil.

More efficient technologies allow an electricity surplus, besides the possibility of replacing fossil fuels. In this paper, existing data [3] for more advanced technologies are adapted to Brazilian plants and compared to conventional systems. Results show that electricity cogeneration could be more than twice the current amount generated by the chosen plants.

Although environmental benefits are evident, main difficulties remain in economic aspects because of low prices of electricity for industrial sector. Also, proposed price to purchase electricity from self-producers is not attractive and so more efficient technologies are not economically competitive. Because of that, special mechanisms are needed to improve cogeneration in Brazil.

Keywords: biomass, cogeneration, emissions, energy, environment, gasifier, gas turbine, pulp-paper industry.

1 Introduction

The consequences of carbon dioxide emissions on greenhouse effect are evident, despite the controversy among the specialists. Recent phenomena regarding the climate change over the world, as well as the melting of polar ice, reinforce the need of controlling carbon emissions, as already stated in ECO'92 Conference, in Rio de Janeiro. Although most emissions come from developed countries, also the developing ones can contribute to reduce them, improving their development through a sustainable way.

Brazil is quite efficient in carbon dioxide emissions: around 1% of total emissions, due to the large use of hydro and biomass-origin energy. Evaluations on CO_2 emissions in Brazil range from 242 Mt [4] to 269 [5] Mt[1] the amount of CO_2 emitted in 1990. Brazilian efficiency is high because most of energy consumption is from renewable sources. More than 90% of consumed electricity is produced in hydroelectric power plants; half of energy used in automotive vehicles is from sugar cane-ethanol. In sugar and alcohol segment, all energy needs come from sugar cane bagasse (a by-product from sugar cane), corresponding to more than 6% of total energy consumption in Brazil [5].

Those figures show that Brazil is more efficient than developed countries, regarding CO_2 emissions per kWh generated (15 times more efficient than USA) and also in terms of CO_2 emissions per capita (much less emissions than Germany, Great Britain, Japan, Italy, France and USA) [5].

On the other hand, recent studies [1] show that even hydroelectric power plants release carbon emissions (besides other well-known environmental impacts [4]), mainly as methane, from the flooded biomass, although much less than fossil fuels[2]. Tucuruí power plant, in Amazon, is estimated to release 103 Mt of CO_2 equivalent, under the form of methane, during 100 years [1].

Also, a significant amount of fossil fuels - mainly fuel oil - is still used in Brazilian industry due to low international prices of oil. Because of that, 1.3 Mt of CO_2[3]are estimated (from [5]) as having been produced by industrial sector in 1993, only from fuel oil. Brazilian pulp and paper industry is responsible for 10% of that amount; this is one of the highest energy consumers in Brazilian industry. By replacing fuel oil by process by-products (biomass residues and black liquor from Kraft process used for pulp production), and introducing more efficient technologies, those CO_2 emissions could be avoided (besides the other pollutant emissions), due to the carbon absorption during biomass re-growth.

Besides those environmental aspects, cogeneration improvements in industrial sector, through private investments, also present economic advantages for the

---

[1]*In this paper, metric tones were used.*
[2]*A thermoelectric power plant releases 88 times more carbon per megawatt than a hydroelectric one [1].*
[3]*Also 10,600 t of SO_2 and 708 t of NO_x were emitted, as evaluated from [6] and [7].*

Brazilian electric sector. Most of this sector is still state-owned - although there is nowadays a privatization program being started mainly in the state of São Paulo - and the utilities have significant financial difficulties. Therefore, private investments in more efficient cogeneration technologies would collaborate to the expansion of this sector, without the construction of new dams, and avoiding their subsequent economic and environmental problems [1], [8].

In pulp and paper plants (as well as in sugar/alcohol and iron/steel industries), thermal and power profiles are adequate for cogeneration systems, as evaluated by many studies [2], [9], [10], [11]. Therefore, this paper analyzes the cogeneration potential in a group of large Brazilian pulp and paper industries (corresponding to 40% of Brazilian production), aiming to generate an electricity surplus to be sold to the grid. To achieve this objective, more efficient technologies are proposed, since those ones commercially available in Brazil, until more advanced ones, such as gasifier/gas turbine systems [3]. On the other hand, fuels with lower pollutant emissions - mainly process by-products (biomass residues and black-liquor, from Kraft pulp production process) - are used to replace oil derived fuels.

The results are significant. Although conventional processes (condensing or back-pressure extraction-steam turbines) do not allow self sufficiency of all the chosen plants, natural gas-fired combined cycles can generate a surplus evaluated in more than 65,000 MWh/month to be sold (after supplying all energy needs of the plants). This surplus can become quite higher (81,000 MWh/month) when gasifier/gas turbine systems are used, fired only with the biomass residues and black liquor available from the plant. Almost 300,000 MWh/month can be produced, more than twice the current total amount generated by the existing systems (the current deficit of the chosen plants is more than 77,000 MWh/month), and corresponding to almost 40% of the electricity consumption in all Brazilian pulp and paper industries[4].

However, the introduction of these technologies depends on some mechanisms, like an adequate purchase price of electricity from self-producers, competitive with the generation costs. Electricity prices for industrial sector should be established on real levels (current low tariffs do not encourage investments on more efficient technologies); moreover, special policies are necessary to improve biomass-origin cogeneration (incorporation of externalities, such as carbon-avoided costs, carbon taxation, etc.[12], [13]).

2 The Brazilian pulp and paper industry

Brazilian pulp and paper industry has produced, in 1993, more than 10 million tones of pulp and paper, divided into: plants producing only cellulose-pulp (23.81%), paper plants (30.35%) and integrated plants (45.84%), producing both pulp and paper. In the last years this sector has shown a strong growth (1993's production was 10% more

---

[4]*792,000 MWh/month, 1993 data[14].*

than in 1991), with a subsequent rise on electricity demand (more than 9,500 GWh per year, around 8% of Brazilian industrial consumption [14]).

However, electricity self-production (hydroelectric self-generation and cogeneration systems) corresponds to only 40% of that global consumption. This percentage varies significantly according to the type of industry: pulp plants, because of by-products' large availability, cogenerate more than 80% of consumed electricity, while paper and integrated industries produce only 10% and 40% of their electric needs.

Because there is no availability of by-products in paper plants (currently paper industries' self-production is mainly from hydroelectric origin), improvements on cogeneration depend on the purchase of additional fuel (such as natural gas to be fired in gas turbine systems). In this case, generation costs are much higher than the established electricity tariffs [15].

On the other hand, in pulp and integrated plants, cogeneration is more feasible, under economic aspects, due to the availability of biomass residues and of black liquor, from pulp production process. Black liquor is a by-product from cellulose digestion in Kraft process, which is the most used for pulp production. It is consumed in recovery boilers, a technology commercialized since the beginning of this century, to recuperate chemical products and to produce high pressure steam. A chemical smelt is then produced, containing sodium carbonate and sodium sulfate, this one converted into sodium hydroxide; these chemical products are recycled for use in the pulp process ([16] apud [3]).

Besides 5 Mt of black liquor and 2.4 Mt of biomass[5] (1993 data), the consumption of other fuels is also quite high: fuel oil's consumption is 811,000 cubic meters per year, corresponding to 10% of the whole industrial sector, together with 257,000 tones of charcoal (15% of industrial sector) and 794,000 tones of oil equivalent (toe) in other oil-derived fuels (8% of industrial sector)[14].

Despite the fact that thermal and power profiles in pulp and paper plants are favorable to cogeneration systems, investments in more efficient technologies - even those already commercialized in Brazil (high pressure boilers and condensing extraction steam turbines) - are not accomplished by the industries due to the low prices of electricity, as established by Brazilian utilities.

However, with the lack of equilibrium between demand forecasts and electricity offers (forecasts show figures as 20% for deficit risks [9]), due to the financial difficulties of the utilities, cogeneration appears to be a promising option to Brazilian industrial sector, to guarantee the electricity needs without shortage risks.

---

[5] *Hog fuel and bark, including a small amount of purchased biomass, like firewood and sugar cane-bagasse*

3 Cogeneration potential in pulp and paper plants

The group of selected pulp and paper plants produces around 12,000 tones per day of pulp and paper, corresponding to 40% of Brazilian pulp and paper production. These industries are quite large, producing more than 1,500 tones per day of pulp and paper. Their current global electricity consumption is around 216,000 MWh/month, 65% of which is self produced (cogeneration systems), with a cogeneration efficiency higher than Brazilian average. Existing systems for that electricity generation, besides the production of process steam, include high pressure boilers and back-pressure steam turbines. All black liquor produced in pulp process is burned in high pressure boilers (recovery boilers) and produced steam is fed in the steam turbines. Complementary process steam is produced in other boilers, burning biomass residues and firewood (4,300 tones per day), fuel oil (1,800 tones per day) and some charcoal (less than 200 tones per day).

To evaluate the possibilities of increasing the electricity cogeneration for this group of pulp and paper plants, aiming their energy self-sufficiency and the generation of some electricity surplus, the following technologies already commercialized are considered as a first approach:

1. *System 1*: This configuration (figure 1) aims only *the thermal self-sufficiency of each plant; electricity production is defined by steam flow needed to process,* approximately as the current situation.

This system includes high pressure (and more efficient than the existing ones) boilers (60 bar, 470°C), burning the current consumed amount of black liquor and biomass, according to the availability of each plant.

When it is necessary, complementary firewood is utilized to replace the fuel oil previously consumed, to supply the thermal needs of each plant. For this group of plants, authors' evaluations (based on thermodynamics analysis) show that firewood demand in these additional boilers corresponds to 10% of current biomass consumption in the plants.

High pressure steam goes to steam turbines with steam-extraction to process (10 bar and 4 bar). A condenser is used when the amount of generated steam is higher than the plant's needs; otherwise, back-pressure-extraction steam turbines are employed[6] so all produced steam goes to process. With this system, energy balances (developed separately for each plant) show electricity deficits for some plants (more than 16,000 MWh/month for the group, around 8% of the group's global consumption).

---

[6]*Those differences among the plants are due to the (different) thermal needs and to the (different) availability of by-products of each plant.*

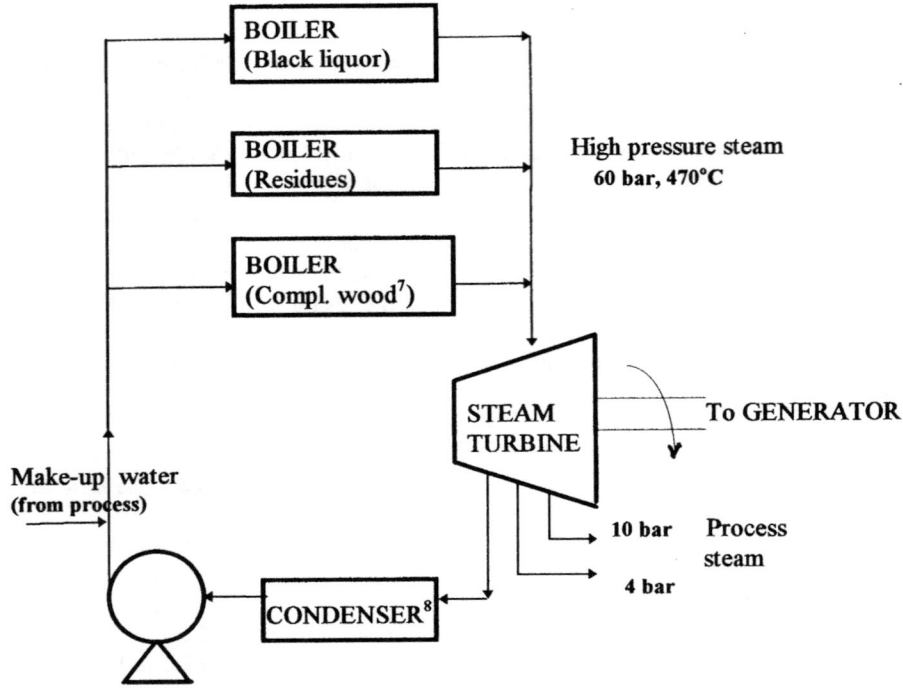

Figure 1: Rankine cycle (steam turbine cycle); boilers fueled with available black-liquor and residues from process and additional boiler firing complementary wood, if necessary (*system 1*).

2. *System 2*: This configuration aims *thermal and electrical self-sufficiency of each plant*. A natural gas-fired gas turbine (figure 2) is adapted to the steam turbine of system 1 and the additional firewood-boiler is replaced by a heat recovery steam generator (using exhaust gases from the gas turbine). With this combined cycle, authors' calculations evaluate that energy self-sufficiency of all selected plants can be obtained, with a small surplus (around 65,000 MWh/month for the group). However, plants' natural gas consumption is evaluated in more than 220 million cubic meters per year, more than twice the consumption of Brazilian pulp and paper sector, corresponding to almost 12% of our whole industrial consumption. Because the current availability of natural gas in Brazil is not so high (4,820 million of cubic meters in 1993 [14]), processes with high consumption of this fuel depend on the construction of the gas pipeline Brazil-Bolivia, still under negotiations.

---

[7] *If necessary.*
[8] *If necessary.*

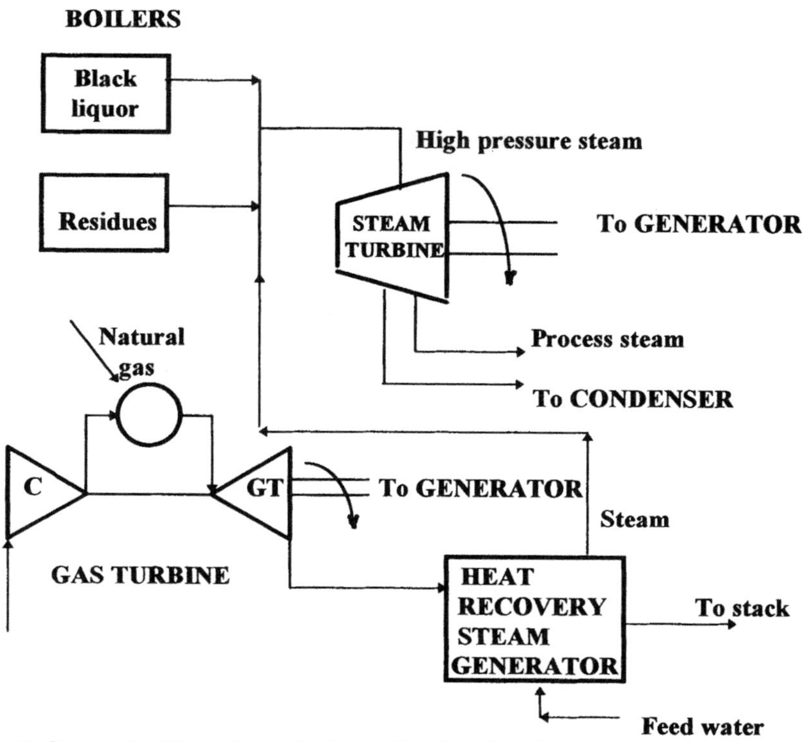

BOILERS

Figure 2: *System 2* - Natural gas fired-combined cycle with gas turbine adapted to the steam turbine of system 1

For systems 1 and 2, data related to the availability and the heating values of black liquor and biomass residues are assumed equal to the real ones, for each plant. Also process steam flows are assumed equal to the current ones, without any energy conservation measure. As it was considered in system 1, a back-pressure extraction steam turbine is utilized for the plants where there is no steam surplus (return water from process to the boiler is assumed at 120°C). In the same way, a condensing-extraction steam turbine is used when necessary.

For both proposed systems, thermodynamics efficiencies are assumed higher than the current ones, as follows: 80% for black liquor recovery boilers, 90% for biomass residues boilers[9], 90% for steam turbines' isentropic efficiency; 90% is the capacity factor. Condenser temperature is assumed equal to 40°C.

---

[9] It is assumed the most efficient boilers for biomass firing (high pressure boilers), under development now in Brazil [2]. There are recovery boilers and biomass boilers in some of chosen pulp plants working with efficiencies up to 86%.

4 Cogeneration potential from gasifier/gas turbine systems

To allow a higher efficiency on cogeneration process, achieving the generation of an electricity surplus, *system 3* is adopted (figure 3) here. This system includes a biomass-gasifier/steam injected gas turbine (BIG-STIG), as proposed in [3]. A steam injected gas turbine is fired with the gases produced in a fluidized bed gasifier, with two possibilities: one single gasifier is fed with black liquor or two gasifiers are fed, respectively, with both black liquor and biomass residues available in the plant. Results for black liquor gasification are estimated [3] from existing experimental data in wood gasifiers. In the same way, STIG efficiencies correspond to theoretical assumptions. Electricity generation with this technology is evaluated in 1,320 kWh per tone of pulp, if only black liquor is gasified, rising up to 1,703 kWh/t of pulp, if biomass residues are also gasified [3].

Most of development efforts occurs now on combined cycles [22], [23]; however, preliminary evaluations indicate that overall costs and performance characteristics of both technologies can be assumed equal[10]. Steam injected gas turbines are also interesting options due to economic[11] and maintenance aspects [10], [15], but further research is still needed to provide experimental results, mainly when fired with low heating value gases.

There are two main types of biomass gasifiers under development nowadays: fixed bed and fluidized-bed gasifiers, both using air-blown (instead of oxygen blown). Experimental results (with wood) are quite recent but they are promising and show that this technology shall be commercially available in a near future [3].

A pilot plant is being developed in Northeastern Brazil [17] for electricity generation from wood plantations using gasifier/gas turbine systems. In this project both types of gasifiers are being evaluated to choose the most convenient one[12]. Another pilot plant, for bagasse[13] gasification, has started, in 1995, in Hawaii (Manoa Island) [18], with a pressurized fluidized-bed gasifier (air blown) producing bagasse-gas to be fired into a gas turbine. In another system, biomass residues feed a pressurized gasifier for a combined cycle in pulp plants [19]. There are also significant researches under development now in black liquor gasification, whose results will allow a more accurate evaluation of the cogeneration potential for black liquor gasifier/gas turbine systems [24], [25].

The (theoretical) electricity generation data for system 3 correspond to a steam injected gas turbine (STIG), firing the gases produced in atmospheric fluidized bed

---

[10]*LARSON, E., 1996. **Personnal communication**.*
[11]*Gas turbine costs are not affected by scale factor, as it hapens to steam turbines.*
[12]*By the time this final version was written, the atmospheric gasifier system was chosen for the mentionned plant, in Brazil.*
[13]*Bagasse is a sugar cane by-product after juice extraction, that is fired in boilers for electricity cogeneration in sugar plants, in Hawaii, and in sugar/alcohol plants, in Brazil.*

gasifiers. These gasifiers are proposed to be fed with black liquor and biomass residues, available from the pulp plant (MTCI technology, in [3]).

Assuming BIG/STIG efficiencies for the selected plants, forecasts indicate an electricity generation up to 230,000 MWh/month (only with black liquor gasification) or 297,000 MWh/month (from black liquor and biomass residues), almost three times the plants' current self-production. There is then a surplus evaluated in up to 81,000 MWh/month, corresponding to more than 10% of the electrical demand of Brazilian pulp/paper sector, with the advantage of no fossil fuel's consumption.

However, to achieve those results, steam consumption in the plants must be reduced to 9.8 GJ/t of pulp [3], what requires significant improvements on energy conservation; nowadays, average steam consumption for the selected plants is 14.9 GJ/t[14] (16.3 GJ/t in a typical pulp plant in USA [3]). This reduction appears to be possible with the introduction of energy conservation measures [3].

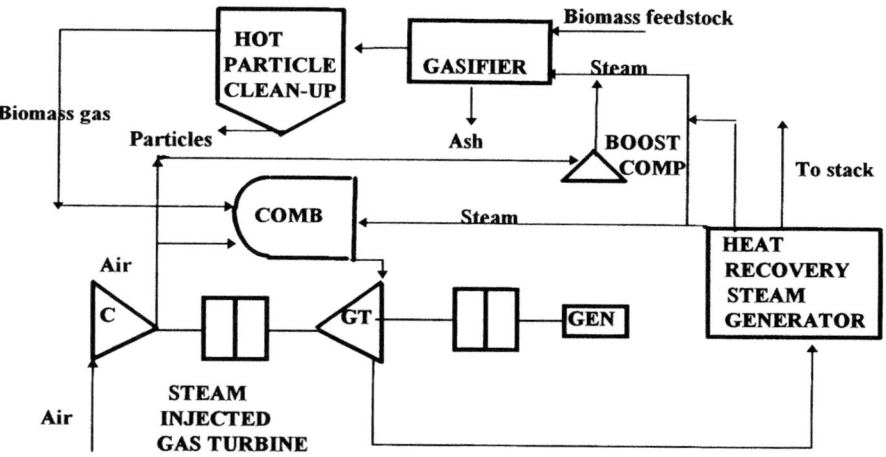

Figure 3: *System 3* - Biomass gasifier/steam injected gas turbine (BIG/STIG), fed with available black liquor and residues from the pulp plant [3].

5 Avoided carbon emissions with cogeneration improvements in pulp and paper plants

Because fuel oil is replaced by natural gas and biomass in the proposed systems, pollutant emissions can be reduced significantly. Carbon emissions are assumed here as follows [21]:

1. fuel oil: 73 g of CO_2 per MJ

---

[14]This consumption includes the steam for the paper production.

2. natural gas: 49 g of CO_2 per MJ

Carbon dioxide net emissions from biomass origin are almost null due to carbon re-absorption during biomass re-growth. Therefore, the replacement of current technologies by more efficient ones, as proposed here, is highly favorable also under environmental aspects.

Table 1 ahead illustrates the avoided carbon emissions with the introduction of the systems 1, 2 and 3.

From that table, we conclude that BIG/STIG systems, fired only with available by products from the plant, avoid around 104,000 tones of CO_2 emissions per year, besides the electricity surplus generated to be sold to the grid.

Table 1. Electricity cogeneration and carbon avoided emissions for the selected group of pulp/paper plants:

Proposed cogeneration system	Electricity production	Electricity deficit/surplus	CO_2 emissions (tones per month)	
	(MWh/month)	(MWh/month)	emissions	avoided emissions
1.Biomass fired steam turbine cycle	199,868	- 16,355	almost null	8,660
2.Natural gas fired combined cycle	281,368	+ 65,145	2,144	6,516
3.Black liquor and biomass residues-fired BIG/STIG	297,089	+ 80,866	almost null	8,660

It is important to notice that those figures correspond only to the chosen group of plants. Although those plants produce 40% of Brazilian pulp/paper production, their CO_2 emissions from fuel oil origin are almost 80% of the total CO_2 emitted by the sector from fuel oil. Avoided carbon emissions appear therefore to be quite significant with proposed technologies.

6 Economic aspects of electricity cogeneration in pulp and paper industry

Previous studies [8] have already analyzed economic feasibility of cogeneration in integrated pulp/paper plants. Obtained results for systems 1 and 2 show that these more efficient technologies are not economically competitive with the current electricity tariffs in Brazil (around US$ 20/MWh, in average, for this sector).

Such technologies would be economically feasible only for higher tariffs (US$ 50 or 60/MWh) and assuming the sale's price of electricity surplus (sold to the grid) equal to US$ 55/MWh. This should be the marginal cost for expansion/transmission in

South/Southeastern Brazil, as evaluated by many specialists, despite other higher evaluations[15] (up to US$ 70/MWh). However, the existing examples[16], in this region, for the purchase of surplus electricity by utilities are not promising. The (state-owned) utilities, in São Paulo, can offer only US$ 38/MWh for the purchase of electricity from self-producers, because this is the expansion/transmission marginal cost established by Federal Government for South-southeastern Brazil.

In the BIG/STIG system, it is assumed [3] an installation cost of US$ 1,360/kW, since the first commercial plant should cost from US$ 1,600/kW to US$ 1,700/kW [3]. This is an estimate for biomass-gasifier/gas turbine systems, because there is not yet commercial black-liquor fired systems. Similar pilot plants under development (bagasse gasification in Hawaii and wood gasification in Brazil, as already mentioned) estimate installation costs approximately the same[17]. From these assumptions, BIG/STIG generation costs are evaluated in US$ 40/MWh [3]. By consequence, neither this system would be economically competitive with current Brazilian electricity tariffs. For a private ownership, the internal rate of return is attractive (10%-13% per year), but only if the surplus electricity could be sold by US$ 50/MWh to local utilities[18], what is not feasible nowadays.

Those difficulties do not allow cogeneration improvements in Brazilian industrial sector. On the other hand, the financial situation of Brazilian utilities being quite hard, the privatization of the electric sector in Brazil (under development now) appears to be a first step to solve these problems. Also wheeling procedures (not allowed until today in Brazil) could improve electricity sales to other industries and would collaborate to make cogeneration economically feasible.

7 Conclusions

Despite the large benefits from environmental aspects (avoiding the carbon emissions responsible for the greenhouse effect) and despite the advantages of allowing the expansion of electricity offer, through private capital, biomass-origin cogeneration improvements are not economically feasible. Electricity tariffs are quite low for the industrial sector, specially for pulp/paper industries and therefore cogeneration costs are not competitive. Also the purchase's price established for the electricity surplus is not attractive to encourage this process.

---

[15]*HUKAI, R. (IEE/USP, 1992). **Personnal communication.***
[16]*Brazilian sugar and alcohol sector.*
[17]*A second plant, similar to that one in Hawaii, would have an estimated installation cost of US$ 2,000/kW (KINOSHITA, C., University of Hawaii, 1995. **Personnal communication**). A commercial plant for wood gasification/gas turbine system, in Brazil, shall present a cost equal to US$ 1,300-1,500/kW [18].*
[18]*Surplus electricity self-producers can be sold only to local utilities, according to current Brazilian legislation.*

Brazilian utilities present significant financial difficulties due to the low prices of electricity established during many years by the previous Federal Governments (used as a strategy against the high inflation rates of that period). Because of that, privatization process, as being now started in the Brazilian electric sector, appears to collaborate to reduce these problems. The participation of private capital in the sector's expansion and in the electricity distribution, as well as the wheeling, allowing an electricity surplus to be sold directly to other industries, are adequate mechanisms to achieve those proposed objectives.

On the other hand, the participation of international agencies could also collaborate to develop this process faster, reducing carbon emissions through the implementation of biomass-origin cogeneration, as discussed in [12].

8 References

1. Rosa, L.P. (1994) Methane and carbon dioxide emissions of hydroelectric power plants in the Amazon compared to thermoelectric equivalents, in *Workshop and Latin-American Seminar on Greenhouse Gas Emissions of Energy Sector and their Impacts*, Rio de Janeiro.
2. Coelho, S.T. (1992) *Evaluation of Electricity Cogeneration from Sugar-cane Bagasse-origin in Gasifier/gas turbine Systems*, Instituto de Eletrotécnica e Energia da Universidade de São Paulo. Master thesis (in Portuguese), São Paulo.
3. Larson, E. (1990) Biomass-gasifier/gas turbine applications in the pulp and paper industry: an initial strategy for reducing electric utility CO_2 emissions, in *Biomass for Utility Applications*, Tampa.
4. Moreira, J.R., Jannuzzi, G.M., Poole, A., Zylbersztajn, D. and Fries, J. (1992) Brazil, in *Global Warming Collaborative Study on Strategies to limit CO_2 Emissions in Asia and Brazil*, Tata McGraw-Hill Pub. Co. Ltd.
5. Rosa, L.P., Cecchi, J.C. (1994) The greenhouse effect and the burning of fossil fuels in Brazil (in Portuguese). *Ciência Hoje*, Vol. 17, No. 97, p.26.
6. Gustavsson, L. et al. (1992) An environmentally benign energy future for western Scania, Sweden. *Energy*, Vol. 17., No. 9, pp. 809-822.
7. Haas, R., Wirl, F. (1992) Market penetration of natural gas in Europe: prospects and impediments. *Energy Sources*, Vol. 14, No. 1, pp. 21-32.
8. Moreira, J.R., Zylbersztajn, D. (1990) The Brazilian Alcohol Program: performance and difficulties, in *The World Renewable Energy Congress*, Reading.
9. Nogueira, L.A.H. (1990) *Potential for Industrial Cogeneration in Brazil*. Escola Federal de Engenharia de Itajubá. Itajubá.
10. Coelho, S.T., Ieno, G.O. (1994) *Cogeneration Potential in Pulp and Paper Plants*. EEC/IEE/Unicamp Agreement (European Economic Community-Instituto de Eletrotécnica e Energia da USP - Universidade Estadual de Campinas). São Paulo.
11. Walter, A.C.S. (1994) *Feasibility and Perspectives for Cogeneration and Thermoelectric Generation in Sugar/alcohol Sector* (in Portuguese). Universidade Estadual de Campinas. Doctor thesis. Campinas.

12. Coelho, S.T., Zylbersztajn, D. (1996) A preliminary analysis of mechanisms to improve biomass-origin cogeneration in Brazil (accepted) in *9th European Bioenergy Conference.* Copenhagen.

13. Krause, G. (1994) *La Re-réglementation du Secteur Elétrique et laPlace de la Production Indépendante* (in French). E.H.E.S.S. Doctor thesis. Paris.

14. Ministério de Minas e Energia (1994) *National Energy Balance* (in Portuguese). Brasília.

15. Coelho, S.T., Ieno, G.O., Zylbersztajn, D. (1993) Technical and economic evaluation of cogeneration in gas turbine systems for paper plants, in *26°. Annual Congress of Pulp and Paper.* São Paulo (in Portuguese).

16. Grace, T.W., Malcolm E.W. (1989) Pulp and paper manufacture. Vol. 5. *Alkaline pulping,* TAPP, Atlanta.

17. Elliot, P., Booth, R. (1993) *Brazilian Biomass Power Demonstration Project - Special Project Brief.* Shell Group (based upon papers presented to the "Seminar on Power Production from Biomass", Finland, "Strategic Benefits of Biomass and Waste Fuels Conference", Washington).

18. Kinoshita, C. (1995) Gasification of sugar-cane bagasse for electricity cogeneration, in *Perspectives for Alcohol Fuel in Brazil.* São Paulo.

19. Nousiainen, I.K. (1995) Possibilities for increasing the power generation with biomass pressurized gasifier in pulp and paper mills, in *Second Biomass Conference of the Americas.* Portland.

20. Mansour, M.N., Durai-Swamy, K., Voelker, G. (1992) MTCI/Termochem steam reforming process for biomass, in *Second Biomass Conference of the Americas.* Portland.

21. Wilson, D. (1990) Quantifying and comparing fuel-cycle greenhouse-gas emissions. *Energy Policy,* Vol. 18, No. 6, pp. 550-562.

22. Berglin, N., Persson, L., Berntsson, T. (1995) Energy system options with black liquor gasification, in *1995 International Chemical Recovery Conference.*

23. Ihrén, C.N., Sverdberg, G. (1994) Simulation of combined cycles with black liquor gasification, in *International Gas Turbines and Aeroengine Congress and Exposition.* The Hague.

24. Swedish Natinal Board for Industrial and Technical Development (1992) *Black liquor gasification. Consequences for pulping process and energy balance. Final Report.* Stockholm.

25. Brown, C. (1993) Black liquor gasification update, in *AFPA Recovery Boiler Subcomittee.* Atlanta.

CYCLOHEXANE-STEAM CRACKING CATALYSED BY CALCINED DOLOMITE [CaMg(O)$_2$]

G. TARALAS[†]

Bioresource Technology Unit (BTU), Department of Chemical Engineering, Division IV, National Technical University of Athens, Zografou Campus, Athens, Greece.

Abstract

Experimental kinetic data on catalytic steam cracking of cyclohexane were obtained at cracking temperatures between 700 and 800 °C, atmospheric pressure 101.3 kPa and time factor values (0.11-0.16 kg·h/m^3). Various low-cost mineral rocks (dolomite, quicklime, dolomitic magnesium oxide) used as catalysts were examined.
Keywords: CaO, MgO, NiMo catalyst, calcined dolomite, steam-cracking, cyclohexane.

1 Introduction

Previous studies have shown that low-cost, naturally occurring, carbonate rocks such as dolomite [Ca·Mg(CO$_3$)$_2$] have been identified as a rather effective material to be used in hot gas upgrading at intermediate pressure gasification and pyrolysis of biofuels [1-6]. In low-pressure gasification, using cheap and disposable catalysts for gas upgrading is a key economical issue, since it is not cost effective to protect a catalyst bed from fly-ash by using high temperature rigid filters at low pressure drops (e. g., in diesel or spark ignited engine applications). Other results demonstrated tar yield reductions and quality improvement emanating from gasification of coal [7], oil shale, lignite [8] and municipal residues [9] by use of low-cost materials such as calcium oxide (CaO or calcined dolomite). Moreover, calcined dolomite and limestone have also been found to decompose tar nearly as effectively as commercial nickel-containing catalysts, which are more costly and intolerant to oxygen breakthrough [10 and references therein].

[†] *Current address: on leave from the Royal Institute of Technology (KTH), Department of Chemical Engineering and Technology/Chemical Technology, Stockholm, Sweden.*

To avoid problems posed by the structural complexity of tar (e. g., a collective term for the PAH form during the pyrolysis and gasification processes), and to elucidate the role of calcined (heated) dolomite as tar cracking catalyst, well-defined model compounds were chosen for cracking studies [2,11-14].

Early work [15] in this laboratory concedes that steam cracking of hydrocarbons on calcined dolomite proceeds via adsorption and/or dissociation to the catalytic surface which further reacts with steam. Later on, a mechanism for steam cracking of naphthalene over calcined CaO was proposed by Garcia and Huttinger [16] which had some features in common with the Morita's [15] mechanism, viz., the intermediacy of carbon residues to the surface. The main objection to these mechanisms concerns the initiation step, i. e., the hydrogen abstraction from the parent molecule (naphthalene), however, the hydrogen atom abstraction was considered not to be a rate determining step [12].

The present work deals with cracking of cyclohexane (C_6H_{12}, is a saturated hydrocarbon without π-electrons), chosen as a well defined reactant for catalytic steam cracking study. A commercial $NiMo/\gamma-Al_2O_3$ (for more information on their tar cracking conversion, the reader is referred to the work [13]), commonly available Swedish quarried calcined dolomites, limestone, quicklime and Norwegian dolomitic magnesium oxide (Norsk Hydro) have been used as cracking catalysts.

2 Experimental Section

2.1 Experimental setup

The experimental flow apparatus, down-flow continuous operation, and analysis arrangement used was similar to that described in previous studies [2,11,13]. Thermal and catalytic cracking experiments were run in a conventional fixed-bed reactor operated under plug flow conditions in the absence of diffusional intrusions in which the dispersion module (D_e/uL_R)<0.020 [11].

The work has been taken to determine the ability of samples of calcined mineral rocks (particle size of diameter between 1.4 and 1.5 mm) to achieve a cracking efficiency. However, after the course of a run, the catalyst was weighed, and in some cases was analysed for "coke" content [14].

2.2 Reagent and catalyst preparation

The feedstock (density 0.779 g/cm³) cyclohexane was of 99.63% w/w mass purity (KEBO AB, Stockholm) and used without further purification. The detailed pore size distributions in particles of calcined Limestone [$CaCO_3$], BET: 2.6 m²/g, low surface area quicklime [CaO], BET: 0.9 m²/g, dolomitic magnesium oxide [MgO], BET: 19.0 m²/g, Sala dolomite [$Ca\cdot Mg(O)_2$], BET: 8.8 m²/g, Glanshammar dolomite, [$Ca\cdot Mg(O)_2$], BET: 7.3 m²/g, Larsbo dolomite [$Ca\cdot Mg(O)_2$], BET: 7.5 m²/g and $NiMo/\gamma-Al_2O_3$, BET: 217.6 m²/g used as the catalysts are given in Figure 1. It is apparent that the mean pore radius ranges approximately between 225 and 300 Å for calcined dolomites, and 22.5 Å for $NiMo/\gamma-Al_2O_3$, respectively. Pore size distributions and (BET) surface areas were measured in a Micromeritics ASAP 2000, by nitrogen

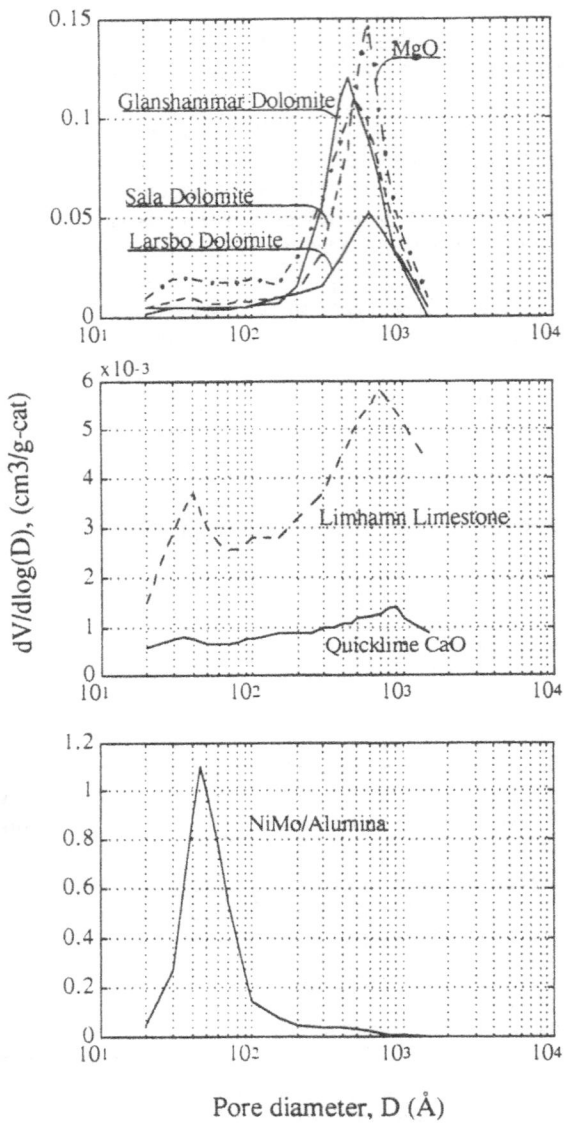

Fig. 1. Pore size distribution of fresh calcined mineral stones and NiMo/γ-Al₂O₃ catalyst.

adsorption, using a volumetric method. The commercially available NiMo catalyst (Katjenfine 153 S-1.5 E, nom. size 1.50 mm) was obtained from the Degussa AG, Germany. The catalyst was initially pre-treated in a hydrogen stream (500 °C) before used. The origin and chemical, as well as other physical properties of the mineral rocks used as a catalyst and NiMo/γ-Al$_2$O$_3$, have been reported previously [2,11,13].

Calcination of the mineral samples was carried out *in situ* at 10 to 20 °C/min to 850 °C in nitrogen (0.5 L/min, STP) for 14 h [2]. According to thermodynamic equilibrium, calcination of the samples should be complete in these experimental conditions [13]. At the calcination temperature of 850 °C, our previous results [13,14] supported the observation made by Borgward and Harvey [17] about the structure of fully calcined dolomite. It was shown that during the calcination of dolomites, Sala and Glanshammar dolomite, CaO cristallizes independently from MgO and forms very small grains attached to the surface of much larger MgO grains.

In this work, a scanning electron microscopy (SEM) study was also carried out to study the structural features of the calcined dolomite, Larsbo dolomite. It was found that during the calcination of virgin Larsbo dolomite samples, CaO cristallizes independently from MgO.

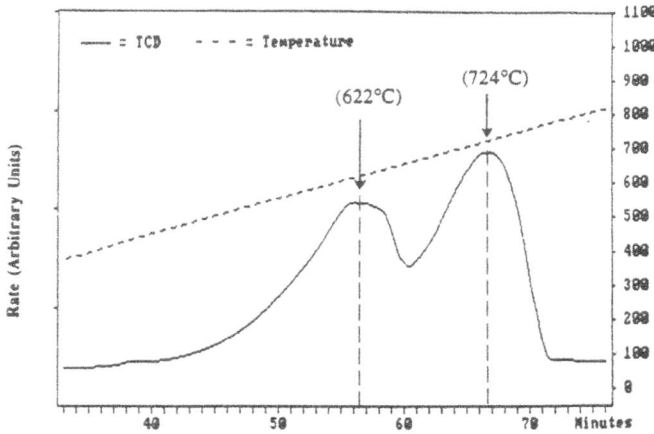

Fig. 2. Reduction profiles of calcined Larsbo dolomite from TPR analysis.

A Micromeritics TPD/TPR 2900 Analyzer was used to monitor the reduction of calcined Larsbo dolomite. In TPR (temperature programmed reduction) experiments, a mixture of H$_2$/Ar (10%/90% v/v), was introduced at 60 mL/min to reduce ~3 g of fully calcined dolomite sample in a fixed-bed reactor, while heating from 30°C to 825 °C, at a constant rate of 10 °C/min as shown in Figure 2. From this Figure, it is obvious that the reduction of dolomite is higher at a low (622 °C) temperature (big crystallites of MgO) and at a higher (724 °C) temperature (small crystallites of CaO).

3 Results and Discussion

3.1 Thermal and catalytic cracking of cyclohexane

Under the applied thermal and catalytic conditions, the following definitions were adopted in the present work: Yield (moles of product/100 moles of cyclohexane feed); Selectivity (moles of product/100 moles of cyclohexane decomposed). Some experiments were performed without catalyst to determine whether any conversion would take place. The space time=effective volume of reactor/molar flow rate of reactants (τ: sec.) at reaction conditions, was calculated by:

$$\tau = \frac{(S_R L_E)P}{[(F_{C_6H_{12}} + F_{N_2} + F_S)(RT_R)]} \tag{1}$$

where $F_{C_6H_{12}}$, F_{N_2}, F_S inlet molar flow rates of cyclohexane, carrier gas and water (mole/s), S_R the surface of the annular cross section of the reactor (m^2), P the atmpospheric pressure, R the universal gas constant. The space time for the thermal experiments, was calculated using a pseudoisothermal approach based on the equivalent length of the reactor [18]. The equivalent reactor length L_E is defined as the length which would give at the temperature of pyrolysis T_R, the same conversion as the actual reactor with its varying temperature T_i profile. Thus, we have:

$$L_E = \frac{\int_0^L [\exp(-E/RT_i] \, dl}{\exp(-E/RT_R)}. \tag{2}$$

In the trial and error calculation a start value of 192.5 kJ/mole was used. The inverse space velocity or time factor (θ: kg·h/m^3) for the catalytic runs was calculated by:

$$\theta = \frac{WP}{[(F_{C_6H_{12}} + F_{N_2} + F_S)(RT_R)]} \tag{3}$$

where W, the weight of calcined catalyst [2,11,13]. Two conversion values were proposed, the percentage conversion degree of cyclohexane defined as X_C obtained from the production of the gaseous components containing carbon (excluding C_6H_{12}) [13] calculated from the commencement of the run, i. e.,

$$X_C = [\frac{\sum_i \frac{C_{i(gas)out}}{6}}{C_{(C_6H_{12})in}}] \cdot 100 \quad (mol.-\%) \tag{4}$$

however, in order to accomodate this conversion the total percentage conversion degree, defined as X, was calculated from Eq. 5

$$X = [1 - (\frac{\text{moles of } C_6H_{12} \text{ leaving reactor}}{\text{moles of } C_6H_{12} \text{ entering reactor}})] \cdot 100 \quad (mol.-\%) \tag{5}$$

Table 1. Operating conditions and comparison of main product selectivities of the cracking of cyclohexane in presence of H_2O

Experiment	Empty Reactor (1)	Empty Reactor (2)	Empty Reactor (3)	Empty Reactor (4)	Calcined 700 °C Magnesium Oxide [MgO]	Calcined 700 °C Quicklime [CaO]	Calcined 700 °C Limestone Limhamn [CaO]	Calcined 700 °C Dolomita Lanbo Glanshammar [$Ca \cdot Mg(O)_2$]	Calcined 700 °C Dolomita Lanbo Glanshammar [$Ca \cdot Mg(O)_2$]	Calcined 700 °C $NiMo/\gamma\text{-}Al_2O_3$	Calcined 800 °C Dolomite Lanbo [$Ca \cdot Mg(O)_2$]	Calcined 800 °C $NiMo/\gamma\text{-}Al_2O_3$
Conversion, X_C (mol.-%)	3.1	4.5	5.2	14.5	13.0	6.3	18.8	22.0	20.0	28.0	66.6	67.7
Time factor, (kg·h/m³)	---	---	---	---	0.13	0.11	0.13	0.15	0.12	0.16	0.12	0.11
Space time, (sec.)	0.56	0.55	0.55	0.54	---	---	---	---	---	---	---	---
PC_6H_{12}, (kPa)	1.0	2.0	3.0	3.9	3.9	3.5	3.9	5.1	3.7	4.6	1.1	1.0
PH_2O, (kPa)	18.8	18.7	18.7	18.2	18.2	17.9	18.2	22.1	17.5	21.6	19.9	18.7
Elapsed cracking time, (min)	123	101	107	185	141	149	175	182	165	148	220	225
Coke on catalyst, (wt.-%)	---	---	---	---	ND	ND	ND	ND	0.03	8.00	0.48	12.0
Gaseous Products						Moles of Product/100 Moles of C_6H_{12} decomposed						
H_2	0.0	0.0	0.0	124.1	153.9	265.4	302.9	469.4	579.8	920.9	1281.9	1111.0
CH_4	0.0	0.0	0.0	0.0	0.0	0.0	34.9	33.3	27.1	26.2	37.9	21.4
C_2H_4	135.3	119.4	117.1	108.4	107.9	89.1	70.6	52.9	54.3	5.26	13.1	2.97
C_2H_6	0.0	0.0	5.66	5.77	5.80	8.17	15.8	18.3	16.0	1.94	4.39	0.0
$C_3H_6+C_3H_8$	0.0	24.5	25.2	23.2	24.9	26.9	32.8	20.4	20.4	0.62	0.18	0.0
$1\text{-}C_4H_8$	71.2	66.3	65.3	59.3	60.4	54.5	34.1	24.6	22.1	0.32	0.0	0.0
$n\text{-}C_4H_{10}$	0.0	0.0	0.0	0.0	0.0	0.0	1.14	1.49	1.22	0.0	0.0	0.0
$n\text{-}C_5H_{12}$	0.0	0.0	0.0	2.54	1.93	0.0	0.79	1.16	0.0	0.0	0.0	0.0
$n\text{-}C_6H_{14}$	0.0	0.0	0.0	0.0	0.0	0.0	1.55	3.99	1.79	0.0	0.0	0.0
$c\text{-}C_6H_{10}$	0.0	0.0	0.0	0.43	0.48	0.34	0.27	0.39	0.44	0.0	0.0	0.0
Benzene, (C_6H_6)	7.46	3.74	2.39	1.60	2.99	7.31	10.2	13.6	10.9	73.9	3.96	29.8
CO	0.0	0.0	0.0	0.0	0.0	0.0	0.0	37.3	67.2	45.6	223.9	202.8
CO_2	0.0	0.0	0.0	0.0	25.6	60.7	76.9	108.1	131.9	67.2	278.8	191.2
Total cracked gas	213.96	213.94	215.65	325.34	383.90	512.42	581.95	784.93	933.15	1141.94	1844.13	1559.17

ND: Not Determined.

Table 1 shows selectivities of gaseous products from thermal (τ=0.54-0.56 sec.) and catalytic (θ=0.11-0.16 kg·h/m^3) cracking of cyclohexane on calcined mineral rock, and NiMo catalyst samples, at 700 °C in presence of water vapor. The nitrogen-free product gas of the thermal cracking [empty reactor] of cyclohexane consisted of hydrogen, ethylen, ethane, propane+propylene, isobutene, n-pentane, n-hexane, cyclohexene and benzene as the major products; no carbon oxides were formed. The selectivity of benzene indicated that some dehydrogenation reactions may occur to a slight extent [19]. As shown in Table 1, the product selectivities of unsaturated hydrocarbon species obtained were lower when calcined dolomites were used as catalysts. Methane production was only observed when calcined limestone, dolomite and NiMo catalyst was used.

As can be seen from Table 1, the selectivities of cracked gas (carbon oxides-hydrocarbon gaseous product) increased with the order of calcined rock samples [MgO]<[CaO]<[CaCO$_3$]<[Ca·Mg(CO$_3$)$_2$]. The decrease in the "coke" deposition on the calcined dolomite, and increase in carbon oxides selectivity, are most likely due to the enchanced gasification of carbonaceous materials [13,14]. The amount of "coke" deposited was significantly higher on the NiMo catalyst, (8.0 wt.-%). Certainly, in the case of NiMo catalyst, the main gaseous products are hydrogen, carbon oxides, methane (other light hydrocarbons being negligible). Higher selectivities of benzene were also detected compared to mineral stones.

From Table 1, it can also be suggested that the overall selectivity of the main gaseous products in presence of dolomitic magnesium oxide [MgO] and quicklime [CaO] are almost identical with those obtained by thermal cracking [empty reactor]. Reference to the literature [11] indicates that [MgO] and [CaO] may affect the conversion of the feedstock without change in product selectivity.

A relatively large amount of oxygen (CO, CO$_2$) was measured in the product gas when calcined dolomites were used as a catalyst, especially at the initial stages of the cracking time. Precautions were made to prevent oxygen penetration into the experimental system, and the oxygen content obtained both in black and in thermal cracking runs was negligible. Oxygen may stem from the calcined dolomite samples released during the catalyzed runs. It should be noted, however, that the total amount of oxygen analysed in the product gas was less than 1.0% of the weight of dolomite sample used. This point should be further investigated.

3.2 Catalysability and effect of temperature

A series of runs at different temperatures was made with cyclohexane on calcined mineral rocks with the experimental conditions as shown in Figure 3. Increase in cracking temperature of cyclohexane from 700 to 800 °C, facilitates the decomposition of the C$_2$H$_4$, C$_3$H$_{6,8}$, and 1-C$_4$H$_8$ (Table 1 and Figure 3). Hence, for all subsequent experiments, the calcined dolomite samples affected the formation of unsaturated hydrocarbon species and diminished the "coke" deposition (0.48 wt.-%).

In previous studies [2,11,13], it was also reported that during steam cracking of n-heptane over calcined dolomites, the unsaturated hydrocarbon selectivities and yields ranges lower than saturated hydrocarbon selectivities and yields. In a current study

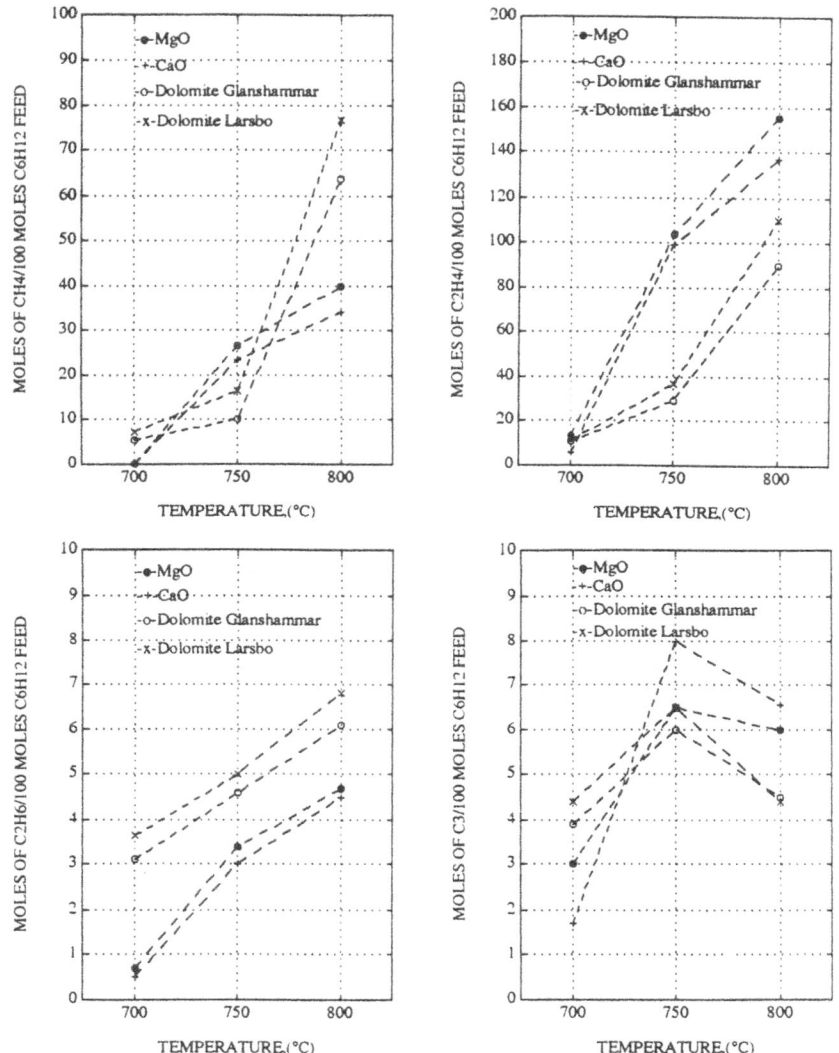

Fig. 3. Effect of temperature on hydrocarbon (CH$_4$, C$_2$H$_4$, C$_2$H$_6$, C$_3$H$_{6,8}$) yields from the catalytic cracking of cyclohexane. (Experimental Conditions: MgO: T=700-800 °C, X$_C$=13.0-63.2 (mol.-%), θ=0.12-0.15 (kg·h/m^3), CaO: T=700-800 °C, X$_C$=6.3-57.8 (mol.-%), θ=0.11-0.14 (kg·h/m^3), Calcined Dolomites:T=700-800 °C, X$_C$=21.7-77.7 (mol.-%), θ=0.11-0.14 (kg·h/m^3). Elapsed cracking time for the runs: 137-182 (min.).

[20], experiments have been conducted concerning the catalysability of dolomite and its influence on the steam cracking of the olefinic ethylene (C_2H_4, is an unsaturated hydrocarbon with π-electrons). Calcined dolomites were found to be the most effective of the catalysts tested. It may be suggested that this effect is related to the interaction of the Ca^{+2} electrons with the π-electrons of the hydrocarbon species.

The effect of temperature on steam cracking of cyclohexane can be described in terms of an apparent activation energy, in the Arrhenius equation

$$k=k_0 \exp(-E/RT_R) \tag{6}$$

derived by assuming overall first-order kinetics [21,22,23]. The straight lines were obtained by a least-squares fit of the natural logarithm of k, with the inverse absolute temperature as the independent variable. Table 2 summarizes the results for the mineral rocks. The coefficient of determination, r^2, indicates that better than 98.0% of the variability in the values of the conversion is accounted for by the linear regression model. Comparison of the absolute values of these apparent activation energies with literature data obtained from thermal cracking of cyclohexane may be instructive. The estimated value of 153.8 kJ/mole for calcined dolomite [Ca·Mg(O)$_2$], Glanshammar dolomite, is less than the literature values for thermal decomposition of cyclohexane of 270.0 kJ/mole [21], 163.5 kJ/mole [22], 192.5 kJ/mole [23].

Table 2. Activation energies for cyclohexane cracking on calcined [MgO], [CaO], and dolomites [Ca·Mg(O)$_2$]

Catalyst	Activation Energy (kJ/mole)	Correlation coefficient (r^2)
Dolomitic magnesium oxide, [MgO]	177.7	-1.00
Quicklime, [CaO]	202.8	-1.00
Glanshammar dolomite, [Ca·Mg(O)$_2$]	153.8	-0.99
Larsbo dolomite, [Ca·Mg(O)$_2$]	168.8	-0.98

3.3 Effect of partial pressure of cyclohexane

The effects of partial pressure of cyclohexane (H_2O/C=0.78-3.11 mole/mole) on calcined dolomite, Glanshammar dolomite and NiMo catalyst samples, are shown in Table 3 and Figure 4. On the basis of the experimental conditions (Table 3), the partial pressure of water vapor is considered to be constant, and, assuming plug flow reactor unde steady state conditions, the integration formula for a first-order reaction and zero-order reaction suggests that the conversion of cyclohexane on NiMo catalyst samples becomes constant to the change of partial pressure of cyclohexane ($X=1-\exp(-k\theta)$), k in m^3/kg·h, but varies inversely with the reactant partial pressure over calcined dolomite. This implies that the necessary time for a specific degree of conversion θ_X on NiMo catalyst is given as $\theta_X=\log(1-X)/k$. These results have also empirically been found to be of the same mechanisms during steam cracking of n-heptane over calcined dolomite and NiMo catalyst [13].

Table 3. Operating conditions and comparison of main product selectivities of the catalytic cracking of cyclohexane on $NiMo/\gamma$-Al_2O_3 and calcined dolomites at 700 °C in presence of H_2O

Catalyst	[$NiMo/\gamma$-Al_2O_3]				Calcined						
					Glanshammar Dolomite [$Ca\cdot Mg(O)_2$]				Sala Dolomite [$Ca\cdot Mg(O)_2$]		
Conversion, X (mol.-%)	91.6	100.0	94.0	100.0	56.2	47.0	39.3	36.9	43.6	26.0	27.0
Time factor, (kg·h/m³)	0.12	0.13	0.13	0.16	0.23	0.23	0.23	0.23	0.12	0.12	0.13
$P_{C_6H_{12}}$, (kPa)	1.0	2.0	2.8	4.6	1.0	2.0	3.0	4.0	3.9	2.9	1.9
P_{H_2O}, (kPa)	18.2	18.7	17.5	21.6	18.6	18.4	18.4	18.5	17.5	17.9	18.0
Elapsed cracking time, (min)	248	190	122	148	122	125	127	183	121	126	165
Gaseous Products	Moles of Product/100 Moles C_6H_{12} decomposed										
H_2	983.9	767.7	958.6	920.9	492.0	450.4	416.7	428.3	579.8	456.3	539.7
CH_4	0.0	15.9	25.5	26.2	16.8	32.2	29.8	30.1	27.1	23.8	23.9
C_2H_4	7.76	5.34	4.44	5.26	91.7	79.1	76.7	76.5	54.2	64.1	51.9
C_2H_6	0.0	0.71	1.23	1.94	15.7	17.1	17.0	17.3	16.0	13.9	13.7
$C_3H_6+C_3H_8$	0.0	0.74	0.23	0.62	18.9	21.2	20.9	20.8	20.4	19.0	18.6
1-C_4H_8	0.0	0.23	0.10	0.32	20.2	26.0	28.79	29.7	22.1	22.9	20.2
n-C_4H_{10}	0.0	0.0	0.0	0.0	0.0	0.0	0.99	1.23	1.22	1.25	0.0
n-C_5H_{12}	0.0	0.0	0.0	0.0	0.0	0.0	0.58	0.96	0.0	0.0	0.0
n-C_6H_{14}	0.0	0.0	0.0	0.0	1.09	1.41	1.50	1.53	1.79	2.34	2.44
c-C_6H_{10}	0.55	0.0	0.0	0.0	0.23	0.33	0.30	0.41	0.44	0.68	0.94
Benzene, (C_6H_6)	65.2	77.0	68.1	73.9	10.7	12.1	12.2	11.6	10.9	17.8	22.5
CO	88.6	39.9	54.7	45.6	0.0	0.0	8.90	15.4	67.2	27.7	22.0
CO_2	104.8	66.8	99.4	67.2	158.9	124.9	105.0	94.9	131.9	114.1	130.9
Total cracked gas	1250.81	974.32	1212.30	1141.94	826.22	764.74	719.36	728.73	933.05	763.87	846.78

Table 3 shows also the experimental results obtained with calcined dolomite samples, Sala dolomite ($H_2O/C=0.78-1.56$ mole/mole). In this case, the sequence of the catalyst treatment was reversed. The experiments were carried out starting with a high partial pressure of cyclohexane. The obtained conversion increased as the partial pressure of cyclohexane decreased.

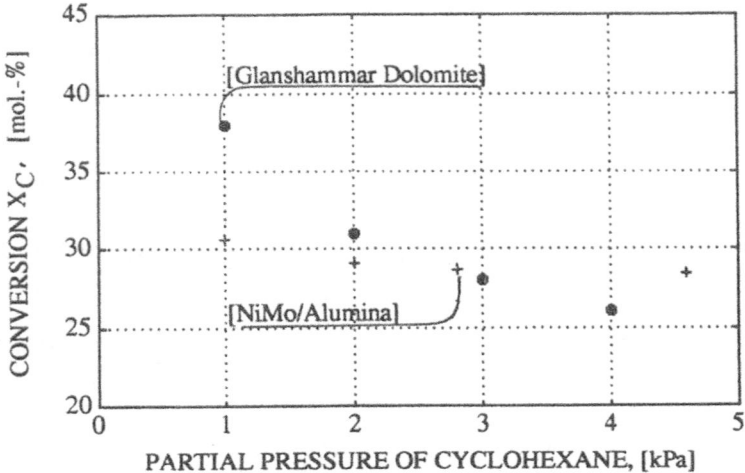

Fig. 4. Steam cracking of cyclohexane as a function of partial pressure and conversion on calcined Glanshammar dolomite [Ca·Mg(O)$_2$], and Ni catalyst samples (T=700 °C, operating conditions as in Table 3).

3.4 Effect of hydrogen addition

The influence of hydrogen on cyclohexane steam cracking was investigated by keeping the partial pressure of feedstock and water vapor constant, and varying the hydrogen pressure by interchange with nitrogen. In this way, the time factor could be kept constant, and any effect of varying total pressure was avoided. As can be seen from Table 4, the degree of total converted cyclohexane decreased by the addition of 17 kPa and 14 kPa hydrogen in the carrier gas in presence of calcined limestone [CaO], and calcined dolomite [Ca·Mg(O)$_2$], Larsbo dolomite. When H_2 is added, the total cracked gas (excluding hydrogen) and the product distribution varies; in particular, the yields of CH_4 and carbon oxides are decreased, but it in turn enhance the role of free-radical reactions with hydrogen leading to increase the yields of $C_2H_{4,6}$, $C_3H_{6,8}$, $C_4H_{8,10}$, C_5H_{12}, C_6H_{14}, c-C_6H_{10} and benzene (C_6H_6) hydrocarbon species. According to Billaud et al., [19 and references therein], ring-opening isomerization of cyclohexane, via biradical and/or monoradical mechanisms, leads to formation of open-chain alkenes. Likewise, the thermal cracking of cyclohexane studied by Aribike et al., [22], explained the formation of gaseous products, and proposed that during an induction period, the cyclohexane cracking becomes 1-hexene (·C_6H_{12}·) pyrolysis. Therefore, a reaction of radicals with the hydrogen *ad. hoc* cannot be excluded [24]. The results are in agreement with reported findings which observed that hydrogen inhibited the heterogeneous decomposition of naphthalene over CaO [16], as well as *n*-heptane and naphthalene on calcined dolomite [12,13].

Table 4. Influence of hydrogen on conversion and main product yields of the catalytic cracking of cyclohexane on calcined Limhamn $CaCO_3$, and Larsbo dolomite at 700 °C in presence of H_2O.

Experiment	Over calcined Limhamn [$CaCO_3$]	Over calcined Limhamn by the addition of (P_{H_2}=17 kPa)	Over calcined Larsbo dolomite [$Ca\cdot Mg(CO_3)_2$]		Over calcined Larsbo dolomite by the addition of (P_{H_2}=14kPa)
Operating Conditions		_____[CaO]_____		_____[$Ca\cdot Mg(O)_2$]_____	
Temperature			_____700 °C_____		
Conversion, X, (mol.-%)	57.4	77.1	35.0	71.9	87.1	50.0
Conversion, X_C, (mol.-%)	18.9	13.5	15.4	21.7	14.7	17.2
$P_{C_6H_{12}}$, (kPa)	3.9	3.9	3.7	5.1	4.5	3.9
P_{H_2O}, (kPa)	18.1	17.9	17.4	22.1	20.5	18.6
Time factor, (kg·h/m³)	0.13	0.13	0.13	0.15	0.13	0.12
Elapsed cracking time, (min)	175	67	65	182	82	83
Gaseous Products		----------Moles of Products/100 Moles Cyclohexane feed--------------				
CH_4	7.37	4.86	1.60	7.28	5.98	4.82
C_2H_4	14.9	10.1	8.77	15.0	7.47	8.23
C_2H_6	3.30	2.27	2.55	4.93	2.79	2.92
$C_3H_6+C_3H_8$	6.29	3.97	3.47	5.69	2.85	3.53
1-C_4H_8	7.33	4.73	5.72	7.31	3.22	5.42
n-C_4H_{10}	0.26	0.13	0.41	0.44	0.19	0.54
n-C_5H_{12}	0.23	0.09	0.35	0.23	0.03	0.35
n-C_6H_{14}	0.38	0.18	0.45	1.11	0.46	1.48
c-C_6H_{10}	0.07	0.02	0.20	0.13	0.06	0.24
Benzene,(C_6H_6)	2.77	0.98	2.15	4.30	1.30	3.53
CO	0.00	0.00	0.00	6.00	9.85	4.57
CO_2	15.0	12.4	2.30	27.4	18.6	3.67
Craked gas	57.90	39.73	27.97	79.82	52.80	39.30

4 Conclusions

1/. Using a first-order reaction model, the apparent activation energies for the cyclohexane-steam cracking in presence of calcined dolomitic magnesium oxide [MgO], quicklime [CaO] and dolomite [$Ca\cdot Mg(O)_2$] were found to be 177.7 kJ/mole, 202.8 kJ/mole, 153.8 kJ/mole for calcined Glanshammar dolomite, and 168.8 kJ/mole for

Larsbo dolomite. From the data obtained, it is seen that calcined dolomite samples affected the production of unsaturated hydrocarbon species.

2/. The alkaline oxides produced by calcination of limestone Limhamn [$CaCO_3$], quicklime [CaO] and dolomites [$Ca \cdot Mg(CO_3)_2$] reduced the "coke" deposition on the catalyst, whereas higher amounts of "coke" were found on NiMo catalyst samples under the applied conditions used. Hydrogen (P_{H_2}=14-17 kPa), when added to the systems, (calcined Limhamn limestone and Larsbo dolomite used as cracking catalysts), do suppress conversion of cyclohexane.

3/. Calcined quicklime [CaO] and dolomitic magnesium oxide [MgO] appeared to increase the conversion of cyclohexane without changing the distribution of gaseous products (700 °C), compared with cracking in an [empty reactor], thermal cracking.

5 Acknowledgements

The author is indebted to the J.-W. Gustafsson, N. Vorbrodt and B. Johansson for kindly supplying the samples of mineral stones (e.g., Sala and Glanshammar dolomites). Thanks go to Dr. K. Sjöström for his encouraging support. I also thank Prof. M. P. Aznar and Prof. J. Corella for contributing with helpful discussions.

6 References

1. Sjöström, K., Taralas, G., Liinanki, L. (1988) Sala Dolomite-Catalysed Conversion of Tar from Biomass Pyrolysis, in *Research in Thermochemical Biomass Conversion*, (ed. A. V. Bridgwater and J. L. Kuester), Elsevier Applied Science Publishers, London and N. Y, pp. 974-986.
2. Taralas, G. (1990) Effects of MgO, CaO and Calcined Dolomites on Model Substance Cracking and Conversion of Tar from Biomass Gasification/Pyrolysis Gas. *Lic. of Engineering Dissertation*, Royal Inst. of Technology (KTH), Stockholm, Sweden, Dept. of Chemical Engineering and Technology/Chemical Technology, ISBN 91-7170-043-9, pp. 70.
3. Vassilatos, V., Taralas, G., Sjöström, K., Björnbom, E. (1992) Catalytic Cracking of Tar in Biomass Pyrolysis gas in the Presence of Calcined Dolomite. *Can. J. Chem. Eng.*, 70, pp. 1008-1013.
4. Corella, J., Narvaez, I., Orio, A., Aznar, M.P., Saenz de Ynestrillas, A., Torres, A. (1995) Thermochemical Biomass Conversion: Upgrading of the Crude Gasification Product Gas. *Annual Progress Report, 1 Jan. 94-31 Dec. 1994*, University Comlutense of Madrid (UCM) Dept. of Chemical Engineering, pp. 44.
5. Taralas, G. (1994) Refuse Derived Fuel (RDF) Management and Atmospheric Fluidized Bed Combustion of RDF in Greece. Center for Renewable Energy Sources (CRES), Pikermi, Athens, Nov., pp. 22.
6. Arvelakis, S., Koullas, D.D., Taralas, G., Koukios, E.G. (1995) Biomass Gasification an Emerging Technology for Energy Production from Biological Raw Materials. Bioresource Technology Unit (BTU), Dept. of Chemical

Engineering, Division IV, National Technical University of Athens, Athens, Greece, May, pp. 46.

7. Yeboah, Y. D., Longwell, J.P., Howard, J.B., Peters, W.A. (1980) Effect of Calcined Dolomite on the Fluidized Bed Pyrolysis of Coal. *Ind. Eng. Process Des. Dev.*, 19, pp. 646-653.

8. Floess, J. K., Plawsky, J., Longwell, J. P., Peters, W.A. (1985) Effects of Calcined Dolomite on the Fluidized Bed Pyrolysis of a Colorado Oil Shale and a Texas Lignite. *Ind. Eng. Chem. Process Des. Dev.*, 24, pp. 730-737.

9. Rensfelt, E., Ekström, C. (1988) Basic Gasification Studies for Development of Biomass Medium-BTU Gasification Process, in *Energy from Biomass and Wastes XII*, Institute Gas of Technology, Chicago, IL, USA, paper No. 32.

10. Simell, P. A., Bredenberg, J.B.-son. (1990) Catalytic Purification of Tarry Fuel gas. *Fuel*, Vol. 69, pp. 1219-1225.

11. Taralas, G., Vassilatos, V., Sjöström, K., Delgado, J. (1991) Thermal and Catalytic Cracking of *n*-Heptane in Presence of CaO, MgO and Calcined Dolomites. *Can. J. Chem. Eng.*, 69, pp. 1413-1419.

12. Aldén, H., Björkman, E., Carlsson, M., Waldheim, L. (1994) Catalytic Cracking of Naphthalene on Dolomite, in *Advances in Thermochemical Biomass Conversion*, (ed. A.V. Bridgwater), Blackie A & P, Glasgow, Vol. I, pp. 216-232.

13. Taralas, G. (1996) Catalytic Steam Cracking of *n*-Heptane with Special Reference to the Effect of Calcined Dolomite, accepted for publication in Ind. Eng. Chem. Research.

14. Taralas, G., Sjöström, K., Björnbom, E. (1994) Dolomite Catalysed Cracking of *n*-Heptane in Presence of Steam, in *Advances in Thermochemical Biomass Conversion*, (ed. A.V. Bridgwater), Blackie A & P, Glasgow, vol. I, pp. 233-245.

15. Morita, Y.J. (1978) Calcined Dolomite. *J. Japan Petrol. Inst.,* Vol. 21, No. 1, pp. 2-9.

16. Garcia, X., Huttinger, K.J. (1990) Steam Gasification of Naphthalene. Catalyst: Calcium Oxide. *Erdöl und Kohle*, 43, 7/8, pp. 273-281.

17. Borgwardt, R. H., Harvey, R.C. (1972) Properties of Carbonate Rocks Related to SO_2 Reactivity. *Env. Sci. & Techn.*, Vol. 6, Nr 4, pp. 350-360.

18. Hougen, O.A., Watson, K.M. (1947) *Chemical Process Principles, III*, John Wiley & Sons, New York, 884.

19. Billaud, F., Duret, M., Elyahyaoui, K., Baronnet, F. (1991) Survey of Recent Cyclohexane Pyrolysis Literature and Stoichiometric Analysis of Cyclohexane Decomposition. *Ind. Eng. Chem. Res.*, 30, pp. 1469-1485.

20. Taralas, G., Sjöström, K., Järås, S., Björnbom, E. (1993) Thermal and Catalytical Cracking of Ethylene in Presence of CaO, MgO, Zeolite and Calcined Dolomite. *Interim report I*, Royal Inst. of Technology (KTH), Stockholm, Sweden. Dept. of Chemical Engineering and Technology, Chemical Technology, TRITA-KT-1991:1, ISSN 1101-9271, pp. 47.

21. Illés, V., Welther, K., Pleszkats, I. (1973) Pyrolysis of Liquid Hydrocarbons, II, Overall Decomposition Rates for Liquid Hydrocarbons. *J. Acta, Chem.*, (Boudapest), 78, 357.

22. Aribike, D.S., Susu, A.A., Ogunye, A.F. (1981) Mechanistic and Mathematical Modeling of the Thermal Decomposition of Cyclohexane. *Thermochim. Acta,* 47, 1.
23. Aribike, D.S., Susu, A.A. (1988) Kinetics of the Pyrolysis of Cyclohexane Using the Pulse Technique. *Ind. Eng. Chem. Res.,* 27, pp. 915-920 .
24. Korzum, N.M., Magaril, R.Z., Plyusnina, G.N., Semukhina, T.I. (1979) Influence of Hydrogen and hex-1-ene on the Thermal Decomposition of Cyclohexane. *Russ. J. Phys. Chem.,* 53, 631.

GASIFICATION

Upgrading

GASIFICATION GAS CLEANING WITH NICKEL MONOLITH CATALYST

P. SIMELL, P. STÅHLBERG, Y. SOLANTAUSTA, J. HEPOLA and E. KURKELA
Technical Research Centre of Finland (VTT), VTT Energy, Finland

Abstract
Particulate-containing gas derived from fluidized-bed biomass gasification can be efficiently purified from tars and ammonia by using a nickel monolith catalyst. Catalyst deactivation by H_2S and carbon deposition can be avoided at 20 bar pressure by using temperatures over 900 °C. Catalyst deactivation was not observed in a long-term test that lasted 500 h. According to a techno-economical evaluation catalytic hot gas cleaning is economically equal to ammonia removal by SCR in an 60 MWe IGCC plant.
Keywords: Catalytic gas cleaning, gasification gas, tar, ammonia, monolith

1 Introduction

Power production from biomass by gasification has been a topic of interest in the Nordic countries during the last years. The most attractive technology in the power scale of 50 - 150 MW_e has been the simplified integrated gasification combined-cycle (IGCC) process. Most of the process concepts are based on pressurised fluidised-bed air gasification and particulate and alkali removal by hot gas filtration followed by combined gas turbine and steam turbine cycle.

However, biomass-derived gasification gas contains always harmful impurities like tars and ammonia that may cause plugging problems in downstream process units and NO_x emissions. These problems can be avoided by using a catalytic gas purification unit for decomposing tars and ammonia. The catalyst should thus have high tar and ammonia decomposing activity and it should also tolerate gaseous catalyst poisons and the high particulate content of the gasifier product gas.

It is well known that nickel catalysts and carbonate rocks decompose tars and ammonia in gasification gas [1 -7]. The use of these materials has been studied at VTT using laboratory and PDU-scale facilities [8 - 12]. These options include a specially designed monolith catalyst reactor as well as the use of cheap bulk materials such as limestones and dolomites. Tests with these facilities showed that gasification gas can be efficiently purified from tars and ammonia also at 20 bar pressure with nickel catalysts.

The aim of this paper is to summarize the experimental work done with the monolith catalysts to develop a hot gas cleaning step for an IGCC process. In addition, the results of techno-economical evaluation of the process are briefly discussed.

2 Experimental

Monolith catalyst testing was carried out in a pressurized bench scale reactor. The feed was dust-containing slip-stream gas from a PDU fluid bed gasifier (Fig. 1). The reactor comprised a pressure vessel, a three-zone furnace and the reactor tube proper, within which the gas preheated and the monolith catalyst were sighted. Ceramic nickel monolith catalysts (Ni/Al_2O_3) were manufactured by BASF AG and had square channels. The total length of the tested monolith assembly was 600 mm, and a cross-sectional area of 31 x 31 mm was in effective use (part of the cross-section was lost for mounting and sealing). The inside temperatures of the monoliths were measured by three thermocouples placed in the center channels near the inlet, in the middle and at the bottom part of the monolith assembly. The temperature values presented in this study are a mean of these three thermocouple readings.

Various biomass fuels were used as gasifier feedstocks in the monolith catalyst tests: forest waste wood, bark, wood chips, eucalyptus chips and peat. These yielded feed gases of different compositions (typical gas with each is presented in table 3) that had tar 1 - 7 g/m^3_n, NH_3 600 - 6000 ppmv, H_2S 30 - 250 ppmv and dust 0,6 - 6 g/m_3n. The temperature range studied was 880 - 960 °C at 5 bar pressure. The total dry gas flow rate was varied in the range 0.3 - 1.9 m^3_n/h resulting in space velocities 1 000 - 5 000 1/h (residence times 0.5 - 2 s). Sampling and analytical methods have been described elsewhere [8, 10]. Catalyst testing was carried out in four separate runs that lasted 45, 95, 98 and 481 h.

The 500 h long-term test was carried out over four months in four parts, which lasted 100 - 170 h. Forest waste wood, bark and eucalyptus chips were used as gasifier fuels. Between the run periods the reactor was cooled down and the catalyst chamber was filled with nitrogen. The catalyst was not removed until the end of the test.

Catalysts were pre-reduced in situ for the first two test runs with H_2/N_2 mixture. Reduction procedure was as follows: heating rate 50 °C/h, total gas flow 10 l/min, heating to 200 °C to 700 °C in H_2/N_2 mixture (20 % H_2), from 700 °C to test temperature again in N_2 flow. Direct reduction by gasification gas was applied in the later test runs. In these cases the catalyst was heated up in N_2 flow (heating rate 50 °C/h) to the test temperature (900 °C) after which it was exposed to gasification gas.

Laboratory scale reactors were used in catalyst screening tests in the early phase of the studies and later to study the effect of H_2S on nickel catalysts. The apparatus and experimental methods used in the screening tests is described in detail in [8 - 10] and those used in the H_2S poisoning studies in [13, 14].

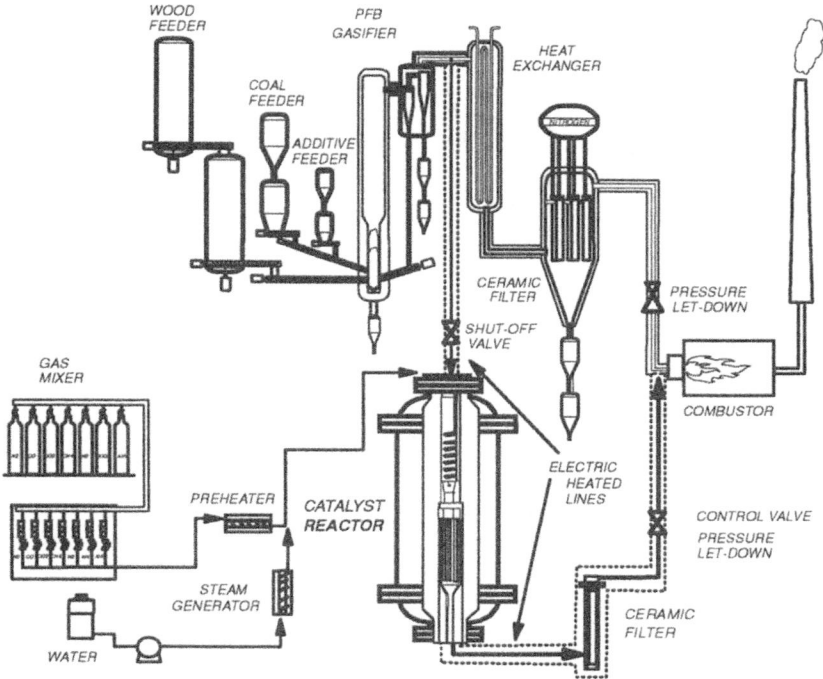

Fig. 1. The experimental set-up used for monolith catalyst studies.

3 Results and discussion

3.1 Effect of operation parameters

The effects of reactor temperature and space velocity on tar and ammonia conversion are presented in Figs 2 and 3. These graphs show results obtained during the four test runs and with the various feedstocks used.

Complete tar decomposition and about 80 % ammonia conversion were achieved at 900 °C temperature and at a space velocity of 2 500 1/h at 5 bar pressure. Changes in operation conditions affected more ammonia conversion than tar conversion in the condition range applied. At the highest temperatures ammonia conversion increased to 95 %.

With respect to the main gas components almost equilibrium gas composition was achieved in the applied conditions (Table 2). These results correspond to earlier results obtained for fixed-bed nickel catalysts. Thus, high yields of H_2 and CO were obtained whereas H_2O and CO_2 reacted resulting in a reduction of the contents of these compounds. However, with respect to NH_3 and CH_4 the equilibrium values were not reached as closely with the monolith as with the fixed-bed reactor.

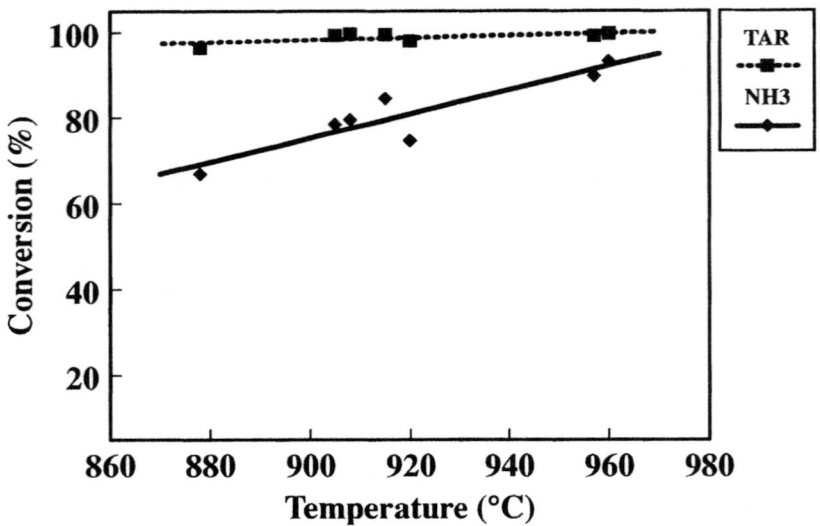

Fig. 2. Effect of temperature on tar and ammonia conversion. Several biomass feedstocks, reactor space velocity 2 000 1/h and pressure 5 bar.

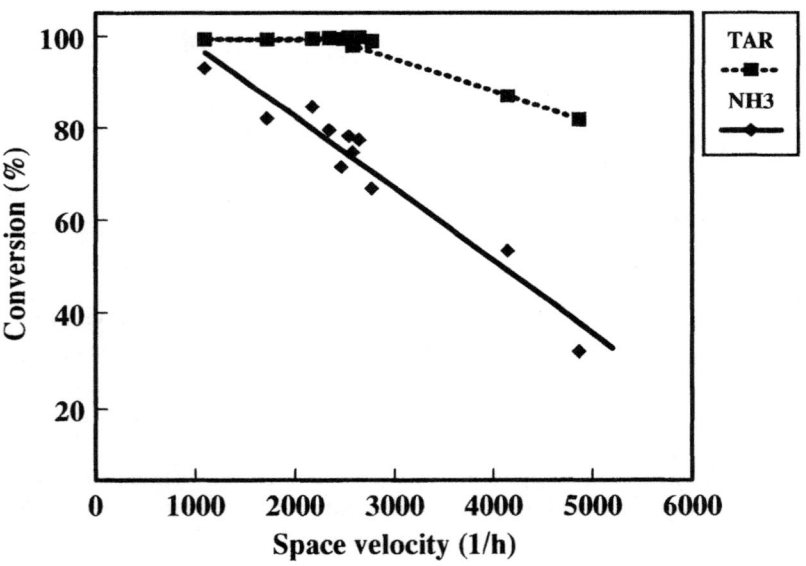

Fig. 3. The effect of space velocity on tar and ammonia conversion. Several biomass feedstocks, reactor temperature 900 - 910 °C and pressure 5 bar.

Table 2. Gas composition at reactor inlet and outlet and the calculated equilibrium composition. Fluid bed gasification gas as feed (wood chips as feedstock). Fixed bed conditions: temperature 900 °C, SV 1900 1/h, pressure 1 bar. Monolith reactor conditions: temperature 910 °C, SV 2 200 1/h, pressure 5 bar

Component	Fixed bed			Monolith		
	in	out	equil.	in	out	equil.
N_2. vol%					48.6	
CO.vol%	12.2	18.3	19.6	11.0	16.0	16.5
CO_2. vol%	13.7	10.1	8.8	12.3	9.5	8.9
CH_4. vol%	3.3	0.1	<<0.1	3.2	0.6	<<0.1
LHC. vol%	0.5	-	0	0	-	0
H_2. vol%	9.8	19.3	17.9	8.5	14.7	15.7
H_2O^a. vol%	12.0	8.3	9.8	12.7	10.6	10.6
NH_3. ppmv	3800	20	10	1880	290	45
Benzene + tar. ppmv	1300	<10	0	1160	4	0

a = calculated from material balance

Space velocities lower than 2 500 1/h were required for high tar and ammonia conversion. Increase in space velocity decreased tar and NH_3 conversion linearly. However, with the fixed-bed reactors operated at atmospheric pressure equally high tar and ammonia conversions were achieved at higher gas flow rates (residence time about 0.3 s) if the temperature was the same (900 °C).

The composition of gasification gas was dependent on feedstock. The greatest differences were found in the contents of hydrocarbons, ammonia, H_2S and water (Table 3). The gas derived from eucalyptus chips contained, in addition, a considerable amount of HCl (up to 90 ppmv). However, catalytically treated gas was almost equilibrated in the conditions applied and hence the residual amounts of ammonia and tar in the outlet gas were quite the same regardless of the feedstock.

Table 3. Gas composition at the reactor inlet and outlet with the tested fuels. Temperature 900 - 920 °C, SV 2000 - 2500 1/h, pressure 5 bar

Component	Gasifier fuel and content of gas component, vol%									
	wood chips		bark		forest waste wood		eucalyptus chips		fuel peat	
	in	out	in	out	in	out	in	out	in	out
CO	12.6	18.1	11.4	16.4	10.0	14.9	8.9	15.9	8.6	13.0
CO_2	10.8	7.9	12.4	9.6	12.6	8.5	14.1	9.4	12.8	10.3
CH_4	3.4	0.2	3	0.5	2.9	0.6	3.8	0.2	2.3	0.4
H2	8.9	17.1	8.8	14.3	10.3	18.0	8.4	18.2	9.1	15.2
H_2O^a	19.3	15.5	11.4	9.6	13.0	11.5	12.0	9.0	17.5	14.5
N_2	44.7	41.3	52.8	49.7	50.9	48.3	52.5	47.2	49.4	46.6
Benzene. ppm	910	1	780	6	960	24	2060	2	720	<1
Tar. ppm	590	<1	220	<1	450	<1	380	<1	160	<1
NH_3. ppm	560	140	1200	230	1160	150	660	150	2700	170
H_2S^a, ppm	30	30	90	80	90	80	35	30	170	160

a = calculated from material balance

The catalyst reduction process seemed not to affect the catalyst performance. Within the measuring accuracy the ammonia and tar conversions were of the same magnitude irrespective of whether the catalyst was reduced with H_2 or directly with gasification gas.

3.2 Long-term activity

The activity of the monolith catalyst was studied by conducting a 481 h long test run. The conversions of ammonia and tar during this test are presented in Fig. 4. Ammonia conversion ranged 70 - 80 % in the steady operating conditions and tar conversion was almost 100 %. During this test H_2S content of gas ranged 35 - 110 ppmv, the highest HCl content was 90 ppmv and the dust content varied in the range 0.2 - 5.8 g/m^3_n. The monoliths looked after the test quite intact and no blocking or carbon deposition in the channels was observed.

The reactor inlet pipe was blocked by particulates after 330 operating hours. This resulted in the decrease of pressure to 4 bar and hence, in an increase in the gas volumetric flow rate (increase of space velocity) in the reactor, which lowered the conversions at around 340 h exposure time. After the removal of the blockage the conversions were restored.

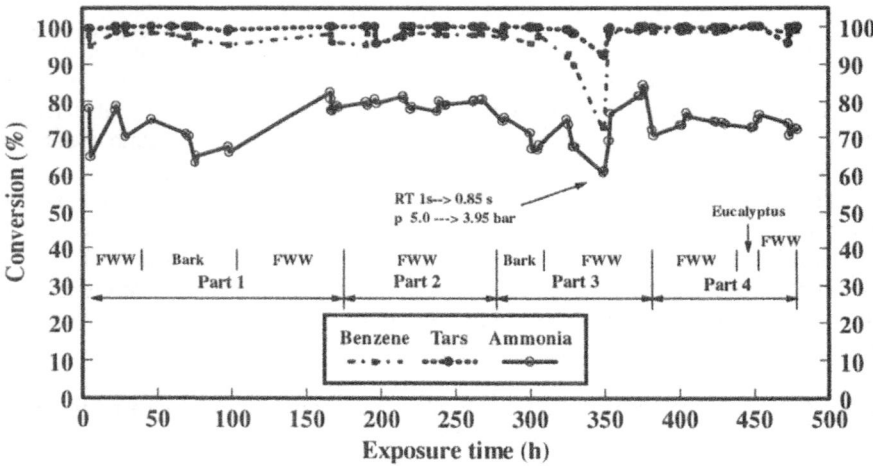

Fig. 4. Conversion of ammonia and tar during the long-term test. Average temperature 905 °C, SV 2 200 - 2 800 1/h, pressure 5 bar.

3.3 Reactions in catalytic hot gas cleaning

The reaction of the oxidizing components and the almost complete decomposition of the hydrocarbons suggest that the chemical reactions in the applied conditions very closely follow the schemes presented for steam reforming [15 - 17]. Hence, the hydrocarbons can be assumed to have reacted according to steam reforming reaction (1) forming H_2 and CO. Ammonia decomposition can in turn be described by the reverse of ammonia synthesis reaction (2). In addition to these reactions, water gas shift reaction (3) takes place in these conditions.

$$C_7H_8 + 7H_2O \rightarrow 7CO + 11H_2 \qquad \qquad -\Delta H^0 \, (900\ °C) = -876\ kJmol^{-1} \qquad (1)$$

$$2NH_3 \leftrightarrow N_2 + 3H_2 \qquad \qquad -\Delta H^0 \, (900\ °C) = -112\ kJmol^{-1} \qquad (2)$$

$$CO + H_2O \leftrightarrow CO_2 + H_2 \qquad \qquad -\Delta H^0 \, (900\ °C) = 33\ kJmol^{-1} \qquad (3)$$

However, it is also possible that the hydrocarbons reacted by CO_2 (dry) reforming reactions (4). Thermodynamically this reaction is slightly more favorable ($\Delta G = -789$ kJmol^{-1}) than steam reforming reaction ($\Delta G = -775$ kJmol^{-1}) in the operation conditions applied (900 °C). Studies carried out at VTT have shown that these two reactions occur at almost equal rate at 900 °C [18].

$$C_7H_8 + 7CO_2 \rightarrow 14CO + 4H_2 \qquad \qquad -\Delta H^0 (900\ °C) = -1105\ kJmol^{-1} \qquad (4)$$

The performance of a catalytic reactor with respect to a chemical reaction can be evaluated by the approach to equilibrium value. The approach to equilibrium at the exit of the catalyst bed is the difference between the gas temperature at the exit of the catalyst bed and the equilibrium temperature corresponding to the gas composition [15]. In Table 4 the equilibrium temperature corresponding to the gas composition calculated according to reactions (1) - (3) at the reactor outlet is presented. Of the hydrocarbons, only methane was considered here, as all the other hydrocarbons reacted almost completely. The $-\Delta T$ value is the difference between the reactor temperature and the equilibrium temperature.

Table 4. Approach to equilibrium with fixed bed and monolith nickel catalysts. Fluid bed gasification gas as feed (wood chips as feedstock). Fixed bed conditions: temperature 900 °C, SV 1900 1/h, pressure 1 bar. Monolith reactor conditions: temperature 910 °C, SV 2200 1/h, pressure 5 bar

Reaction		Fixed bed			Monolith		
		K	T_{eq} at outlet, (°C)	$-\Delta T$, (°C)	K	T_{eq} at outlet, (°C)	$-\Delta T$, (°C)
(1)	$CH_4*H_2O/CO*H_2^2$	$5.69*10^{-6}$	715	185	$9.99*10^{-5}$	730	180
(2)	$NH_3^2/N_2*H_2^3$	$1.27*10^{-11}$	800	100	$5.45*10^{-9}$	605	305
(3)	$CO_2*H_2/CO*H_2O$	0.814	760	160	0.822	895	15

The residual content of ammonia was larger than the equilibrium value even at the highest temperatures tested. With the fixed catalyst bed, however, a closer approach to equilibrium was obtained (Table 4). With respect to methane the approach was quite the same with the both reactor types indicating that the monolith reactor decomposed hydrocarbons as well as the fixed bed in the conditions applied.

Possible reasons for the differences in performance between the tested monolith and fixed-bed reactors can be partial deactivation of the catalyst surface by H_2S enhanced by pressure or differences in catalyst formulation. The lower ammonia conversion may also be due to a poorer heat and mass transfer to the catalyst surface in the monolith [19]. Furthermore, the highly endothermic hydrocarbon reforming reactions cool the catalyst surface considerably so that the heat transfer to the catalyst limits the reactor performance in practice [15, 20]. The monolith surface can thus have a lower temperature than the fixed bed in otherwise comparable conditions leading to lower conversions.

However, in these tests the monolith had a temperature gradient of about 20 °C which was lower than in the fixed beds (50 - 100 °C). This may be due to the placement of thermocouples. With the fixed bed, the thermocouples were inside the granular catalyst bed in good contact with the catalyst surface. This also allows them to cool to near the actual surface temperatures. On the other hand, with the monolith the thermocouples were placed in the channels, where the contact to the surface was not controllable.

3.4 Catalyst deactivation

On the basis of a literature study the deactivation of the high-temperature catalyst may be due to poisoning, fouling (i.a. carbon formation), sintering, vaporization (loss of active substance), or weakening of mechanical properties [21]. According to a thermo-dynamical study, sulfur adsorption or carbon formation in the catalyst may be significant reasons for the deactivation of nickel catalysts, depending on the process conditions applied.

In the gasification conditions, carbon deposition can take place by hydrocarbon decomposition reactions (5) and (6) or by Boudouard reaction (7). Carbon deposition can occur if gas does not contain enough oxidizing components (H_2O, CO_2) or the catalyst temperature is too low [15]. Thermodynamic calculations made for the gas compositions presented in Table 2 indicated that carbon will be present in equilibrium at temperatures lower than 600 - 640 °C at 1 bar pressure and 750 - 780 °C at 25 bar pressure (Fig. 5).

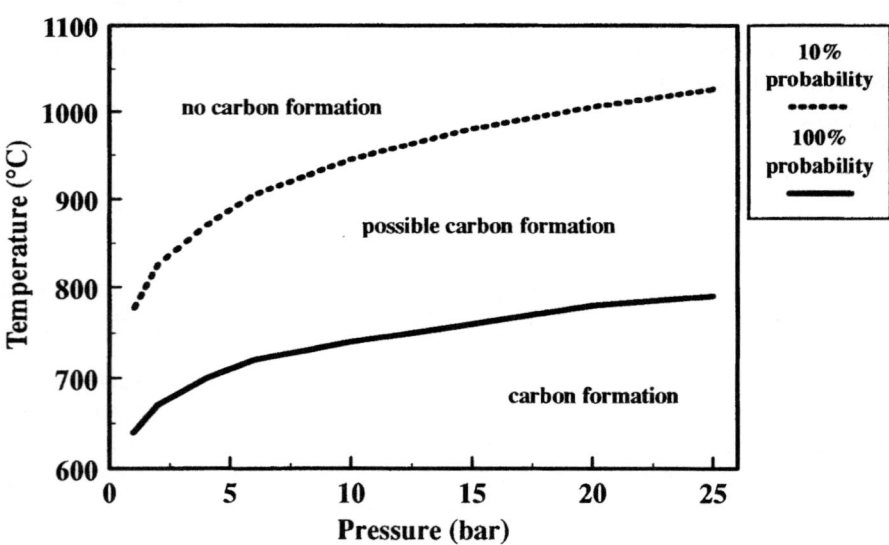

Figure 5. Limiting temperature for carbon deposition at 1 and 20 bar pressure at thermodynamic equilibrium. Average biomass derived gasification gas composition.

$$C_nH_m \rightarrow \text{"carbon"} + xH_2 \tag{5}$$

$$CH_4 \leftrightarrow C + 2H_2 \qquad\qquad -\Delta H^0\,(900\ ^\circ C) = -91\ \text{kJmol}^{-1} \tag{6}$$

$$2CO \leftrightarrow C + CO_2 \qquad\qquad -\Delta H^0\,(900\ ^\circ C) = 168\ \text{kJmol}^{-1} \tag{7}$$

In the catalytic gas cleaning conditions, the main sulfur compound in gasification gas is H_2S and only small quantities of COS exist [22]. The main reactions possible in these conditions can be described by equations (8) and (9). Thermodynamically, bulk nickel sulfide formation will occur in gasification conditions when H_2S concentration in the gas is above 900 ppmv. However, below this concentration the poisoning effect of sulfur must be interpreted in terms of sulfur adsorption on the surface of the metal (surface sulfide formation). It has been concluded by comparing the free energies of formation of surface sulfides and bulk sulfides that surface sulfides are more stable than bulk sulfides [23]. The adsorption of sulfur on most catalysts causes important modifications in the electronic and structural properties of the surface. Thus, the catalytic properties are also strongly affected. The presence of sulfur on the catalysts can either totally or partially inhibit the adsorption or the dissociation of molecular species and also inhibit the surface reaction between adsorbed species.

$$COS + H_2O = H_2S + CO_2 \qquad\qquad -\Delta H^0\,(900\ ^\circ C) = 34\ \text{kJmol}^{-1} \tag{8}$$

$$H_2S + (x/y)Ni = (1/y)Ni_xS_y + H_2 \tag{9}$$

Laboratory tests [14, 24, 25] indicated that the temperature and total pressure, and the hydrogen sulfide content of the gas have a strong effect on poisoning due to sulfur. At about 800 $^\circ$C and 20 bar pressure even a slight content of hydrogen sulfide in synthetic gasification gas spoiled the activity of the nickel catalyst with regard to tar and ammonia decomposition. The temperature at 20 bar pressure should be higher than 900 $^\circ$C to prevent deactivation of the catalyst with regard to ammonia and tar decomposition (Figs 6 and 7). In these conditions, sulfur prevented decomposition of ammonia more efficiently than that of tar. In particular, the deactivation sensitivity of nickel catalysts towards ammonia decomposition, was increased by pressure rise from 5 to 20 bar, while tar decomposition decreased only slightly. In certain conditions, it was possible to recover the activity of the catalyst by reducing the hydrogen sulfide content or by removing the hydrogen sulfide from the gas.

The laboratory-scale fixed-bed tests proved to be a fairly feasible way of determining the effects of hydrogen sulfide on the activity of the catalyst. The results are well comparable to those of the long-term tests with real gasification gas, and also give new interesting data on the behavior of sulfur in the process conditions concerned.

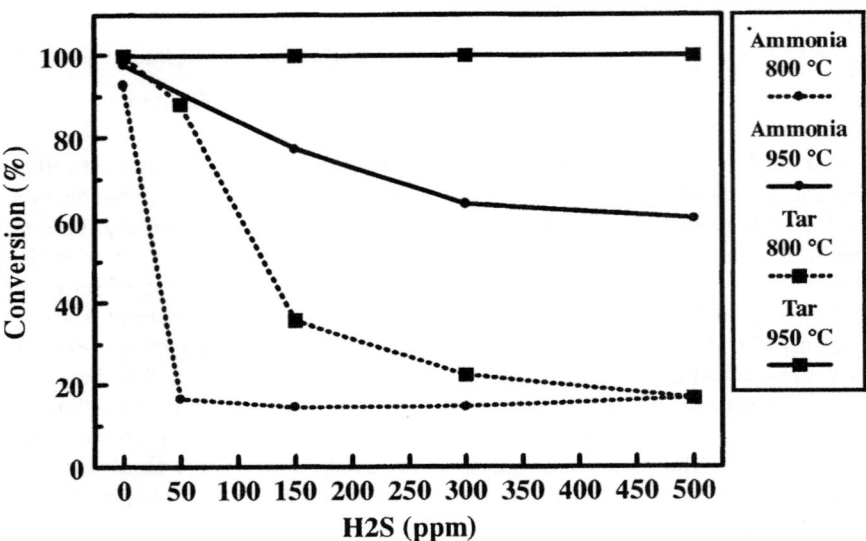

Fig. 6. Effect of H_2S on the conversion of ammonia and tar (toluene) with Ni/Al_2O_3. Synthetic gasification gas, temperature 800 - 950 °C, SV 15 000 1/h, pressure 20 bar.

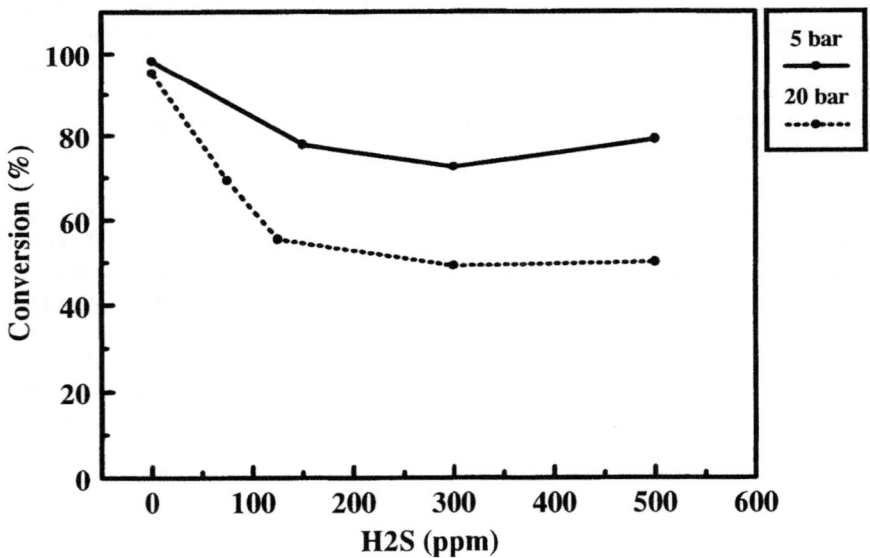

Fig. 7. Effect of H_2S on the conversion of ammonia with a Ni/Al_2O_3 catalyst. Synthetic gasification gas, temperature 900 °C, SV 7 500 1/h, pressure 5 and 20 bar.

4 Techno-economical evaluation

A assessment was carried out to study the feasibility of employing a monolith catalyst in the fuel gas stream of an IGCC using wood fuel. The application is shown in principle in Fig. 8. It was compared to an IGCC case, where a selective catalytic reduction (SCR) was included for the NO_X compounds in the gas turbine exhaust gas. Both concepts were based on an IGCC concept employing General Electric Frame 6B gas turbine having a nominal power output of 40.8 MW [26]. Co-generation of district heat and power was assumed. The plant characteristics are summarized in Table 5. A conventional fluidized-bed (FB) boiler power plant was employed as a reference. The capacity for the conventional plant was selected such that the district heat production was about the same in all cases. The capacity of a district heat power plant is typically determined by the heat demand. The power plant configurations with their performance (determined with Aspen-Plus™) and cost (determined in an IEA project) are reported in detail in [27]. It should be noted that the investment cost of IGCC has been estimated for a mature technology taking into account the cost reduction assumed when several plants have been built.

The economic feasibility was assessed for a utility with a district heat system in Finland with an additional nominal heat demand of 55 MJ/s. The production cost of power is calculated by assigning fixed and variable costs of production to both heat and power production according to their respective power-to-heat ratios. In addition, the price of heat used is the same as would be produced in a stand-alone district heat plant using woody biomass. The price of heat is also dependent on the number of operating hours per year. A software designed by the Finnish engineering-contractor Ekono Energy Ltd. was employed for calculations.

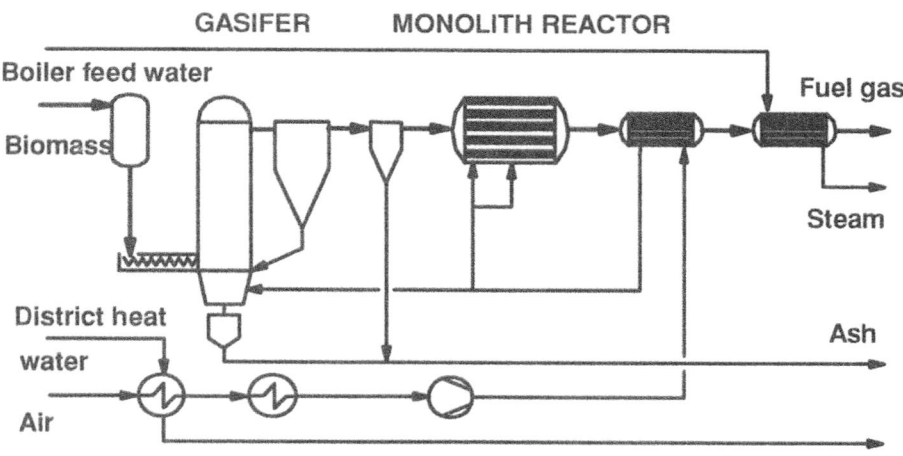

Fig. 8. A monolith catalyst installed at an IGCC power plant.

Table 5. Power plant cost and performance, two IGCC concepts and a conventional steam cycle plant. Efficiencies based on the lower heating value of wood (8.2 MJ/kg @ 50 % moisture)

	IGCC (SCR)	IGCC (monolith)	FB boiler
Net power output MW	56	55	28
District heat MJ/s	55	56	55
Power production efficiency %	43.3	42.9	25.5
Heat production efficiency %	42.8	43.3	59.9
Power-to-heat ratio	1.02	0.98	0.51
Total investment Million US$	90	90	50
Specific investment US$/kWe	1440	1470	1740

Annual savings for a utility, comparing IGCC cases and the conventional boiler plant against the Finnish electricity tariff, are shown in Fig. 9. Two lifetimes, 5 and 2 years, are assumed for the monolith catalyst. It can be seen that considerable savings, in the order of 2 to 3 million US$ per year, are possible with an IGCC compared to about 0.6 million US$/a for the conventional technology. The break even operating times are about 3 000 h/a for the IGCC cases and about 5 000 h/a for the conventional technology, above which co-generation would be more economic than buying electricity. Typically a co-generation plant of similar size in Finland may have a yearly operation time of about 5 000 h.

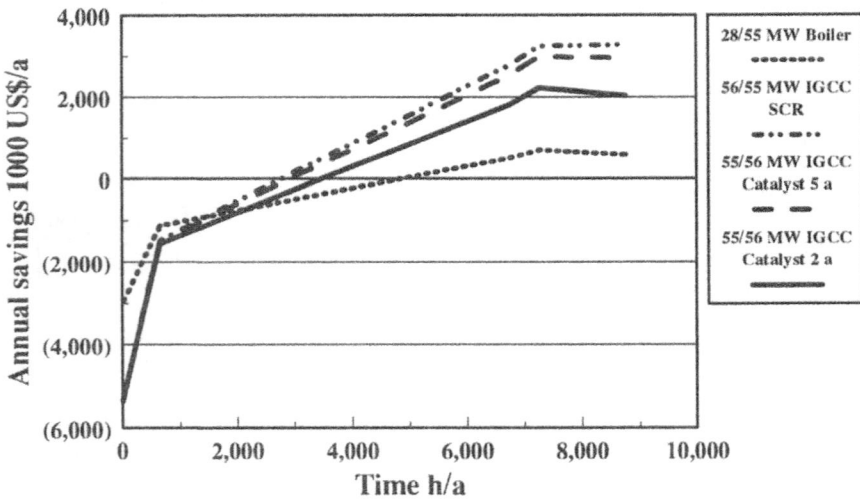

Wood 11 $/MWh, annuity for capital costs 0.08

Fig. 9. Annual savings in 1000 US$ for the utility employing different co-generation technologies. Savings calculated compared to a Finnish electricity tariff in place 1995.

It can also be seen that the catalyst lifetime of 5 years will yield about the same economics as the IGCC concept with SRC. It should be noted that the catalyst lifetime of 10 years will give precisely the same savings as the IGCC with SCR. Extending the catalyst lifetime from 2 to 5 years clearly improves the economy. However, extending the lifetime from 5 to 10 years has only a modest effect. It may be concluded that if the lifetime of the catalyst exceeds 3 - 5 years, the monolith has no negative effects on the economics of the IGCC.

5 Conclusions

With the nickel monolith catalyst the complete tar conversion and 70 - 80 % ammonia conversion were achieved at 900 °C with a space velocity of 2 500 1/h and at 5 bar pressure. Fluidised bed gasifier gas that contained particulates up to 6 g/m^3n was used as the feed.

Gas composition in the range studied did not have any marked effect on tar or ammonia conversion. The feed gas was varied according to the feedstock used in the gasifier. A range of biomass fuels including wood chips, forest waste wood and bark were tested.

The use of a monolith type of reactor seems to be the most promising way of avoiding plugging problems typical of fixed or moving bed reactors with high particulate loading. Catalyst deactivation or blocking due to dust was not observed during the 500 h long test run. However, tests in the range of thousands of operating hours are needed to demonstrate the applicability of this type of gas cleaning process for gasification applications.

Temperatures of over 900 °C are required at 20 bar pressure to avoid catalyst deactivation by H_2S or carbon deposition.

If the lifetime of the catalyst exceeds 3 - 5 years in an IGCC process, the monolith catalyst is economically competitive when compared to NO_x removal by SCR.

6 References

1. Alden, H., Espenäs, B.-G. and Rensfelt, E. (1988) In *Research in Thermochemical Biomass Conversion*, (eds A.V. Bridgwater and J.L. Kuester), Elsevier, New York, pp. 987 - 1001.
2. Corella, J., Herguido, J. Gonzales-Saiz, J. et al. (1988) In *Research in Thermochemical Biomass Conversion*, (eds A. V. Bridgwater and J. L. Kuester), Elsevier, New York, pp. 754 - 765.
3. Taralas, G., Vassilatos, V. and Sjöström, K. (1991) *Can. J. Chem. Eng.*, Vol. 69, pp. 1413.
4. Krishnan, G. N., Wood, B. J., Tong, G. T. and McCarty, J. G. (1988) *Study of Ammonia Removal in Coal Gasification Processes*, SRI International, California, USA, DOE/MC/23087-2667.
5. Baker, E. G. and Mudge, L. K. (1987) *Ind. Eng. Chem. Res.*, Vol. 26, pp. 1335.
6. Baker, E. G. and Mudge, L. K. J. (1984) *Anal. Appl. Pyrolysis*, Vol. 6, pp. 285.
7. Garcia, X and Hüttinger, K.J. (1990) *Erdöl und Kohle - Erdgas*, Vol. 43, pp. 273.

8. Simell, P. and Bredenberg, J.B-son (1990) *Fuel*, Vol. 69, pp. 1219.

9. Simell, P., Kurkela, E. and Ståhlberg, P. (1993) In *Advances in Thermochemical Biomass Conversion*, Vol. 1 (ed. A.V. Bridgwater), Blackie Academic & Professional, Glasgow, pp. 265.

10. Leppälahti, J., Simell, P. and Kurkela, E. (1991) *Fuel Processing Technology*, Vol. 29, pp. 43.

11. Leppälahti, Kurkela, E., Simell, P. and Ståhlberg, P. (1993) In *Advances in Thermochemical Biomass Conversion*, Vol. 1 (ed. A.V. Bridgwater), Blackie Academic & Professional, Glasgow, pp. 160

12. Simell, P., Leppälahti, J. and Bredenberg, J. B. (1992) *Fuel*, Vol. 71, pp. 211.

13. Simell, P., Leppälahti, J. and Kurkela, E. (1995) *Fuel*, Vol. 74, pp. 938.

14. Hepola, J., Simell, P., Kurkela, E. and Ståhlberg, P. (1994) In *Catalyst Deactivation 1994* (ed. B. Delmon and G.F. Froment), Elsevier, Amsterdam, pp. 499 - 506.

15. Rostrup-Nielsen, J. R. (1984) In *Catalytic Steam Reforming*, Springer-Verlag, Berlin.

16. Rostrup-Nielsen, J. R. (1993) *Journal of Catalysis*, Vol. 144, pp. 38.

17. Jennings, J. R. and Ward, S. A. (1989) In *Catalyst Handbook*, 2nd ed. (ed. Twigg, M.V.), Wolfe Publishing Ltd, Frome, pp. 384 - 440.

18. Simell, P. and Hepola, J. (1995) In *Book of Abstracts of Europacat II*, Maastricht, Sept 3 - 8, 1995, the Netherlands.

19. Cybulski, A. and Moulijn, J. (1994) *Catal. Rev. - Sci. Eng.*, Vol. 36, pp. 179.

20. Xu, J. and Froment, G. F. (1989) *AIChE Journal*, Vol. 35, pp. 97.

21. Hepola, J. (1993) *Usability of Catalytic Gas Cleaning in a Simplified IGCC Power System. Deactivation of Ni/Al₂O₃ Catalysts. Literature Review.* Espoo: Technical Research Centre of Finland, 79 p. (VTT Research Notes 1445).

22. Kurkela, E., Hepola, J., Ståhlberg, P and Lappi, M. (1991) *Sulphur Removal by Calcium-based Sorbents in Fluidized-bed Gasification.* 1991 International Symposium on Energy and Environment. American Society of Heating, Refrigerating and Air-Conditioning Engineeers, Inc., pp. 267 - 276.

23. Bartholomew, C. H., Agrawal, P. K. and Katzer, J. R. (1982) Sulfur poisoning of metals. *Advances in Catalysis*, Vol. 31, pp. 135.

24. Hepola, J., Simell, P., Kurkela, E. and Ståhlberg, P. (1994) In: *Abstract book of 6th Nordic Symposium on Catalysis.* Hornbaek: Danish Society of Chemical Engineers, 2 p.

25. Hepola, J. and Simell, P. (1995) In: *Abstract Book of Europacat II 1995.* Maastricht, pp. 661.

26. Palmer, C. and Erbes, M. (1993) *Performance Predictions of the Frame 6 Engine Fired on Low-Btu Fuels*, Enter Software, Inc., Menlo Park, CA, 32 p.

27. Solantausta, Y., Bridgwater, A. and Beckman, D. (1996). *Electricity Production by Advanced Biomass Power Systems.* Espoo: VTT Energy. 115 p.+ app. 79 p. (VTT Research Notes 1729).

STEAM REFORMING WITH NICKEL-BASED CATALYSTS ON GAS FROM BIOMASS GASIFICATION

Steam reforming on gasified biomass

M. BERG and L. WALDHEIM
TPS Termiska Processer AB, Studsvik, Nyköping, SWEDEN
J. KONINGEN and K. SJÖSTRÖM
Kungl Tekniska Högskolan, Department of Chemical Engineering and Technology, Chemical Technology, Stockholm, SWEDEN

Abstract

The work presented has focused on the conversion of tars and methane present in the raw gas from biomass gasification with special attention on the effect of sulphur poisoning. Experimental work on catalytic hydrocarbon steam reforming of biomass pyrolysis gas has been performed using a commercial nickel-based catalyst. The presented work is part of a larger project to experimentally and theoretically study hot gas cleaning for a product gas from biomass gasification for use in advanced applications, i.e. methanol synthesis and fuel cells.

Compared with the gas generated from coal, gasification gas produced from biomass contains more hydrocarbons (tars, methane etc.) and less sulphur. The advanced applications mentioned above require cleaning from sulphur in excess of that required by environmental restrictions. Since there seems to be no feasible method for the removal of sulphur at temperatures corresponding to the gasifier operating temperature, the steam reforming will be performed with a sulphur containing feed gas.

At KTH, a model gas mixture, free from tar, but containing all the other major components, was used to calculate a value for the activation energy describing both steam reforming kinetics and the effect of sulphur poisoning. At TPS, a pyrolysis gas, containing a relevant spectrum of hydrocarbons and other contaminants, was produced under reproducible conditions. Hydrogen sulphide and steam were added to this gas and apart from these two parameters, the temperature, space velocity and particle size were also varied. Finally, the results from both types of experiments were compared. In the light of the assumptions made, the agreement between the two series of experiments is remarkably good. This implies that with good approximation, the methane conversion in a biomass gasification gas can be calculated as a function of temperature and hydrogen sulphide concentration.

The work presented was performed within the framework of the European Unions Joule II Extension Programme and was funded by the Swedish National Board for Industrial and Technical Development (NUTEK) and TPS Termiska Processer AB.

Keywords: biomass gasification, nickel-based catalyst, steam reforming, sulphur poisoning.

1 Background

1.1 The TPS and KTH Joule II Extension project

During the world-wide oil crisis in the nineteen seventies, all over the world research activities in the field of energy resources other than oil were booming. In Sweden, among a number of other activities, an extensive programme concentrating on the gasification of biomass fuels was initiated. These materials have a high fraction of volatiles and reactive char, but have a lower heating value than solid fossil fuels. To compensate for this low energy content, adapted processes were developed for generating synthesis gas for upgrading to vehicular fuels. At the Royal Institute of Technology, KTH, these activities resulted in the construction of a Laboratory Development Unit, followed by a pilot plant at Studsvik, now TPS Termiska Processer, both based on fluidised bed, pressurised, oxygen-steam gasification. In the mid nineteen eighties, the confidence in oil as a relatively long-term energy source was reinstated, resulting in decreasing oil prices and consequently most of the alternative fuels projects based on biomass or coal were abandoned.

In recent years the awareness of the problems connected with carbon dioxide emissions has grown. Following the greenhouse debate and the Rio Declaration, it is clear that the carbon dioxide targets cannot be met without substitution of fossil fuels both in power production and for transportation fuels. This has resulted in renewed interest in alternative fuels, particularly from biomass. In 1994, a project was initiated by KTH and TPS, within the framework of the Joule II Extension Programme of the European Union. The purpose was to study the gas cleaning requirements for applications such as production of synthesis gas, hydrogen production and fuel cells. A number of problems, associated with the gas cleaning of biomass gasification gases is similar to those with relevance to coal based IGCC power generation, e.g., sulphur and chloride removal at high temperature. The progress made in the research efforts in the field of coal gas cleaning might be applicable, with minor or major adaptations, for processes with lower starting concentrations, such as biomass gasification gas. The project was intended to combine the knowledge and expertise, present at both the Royal Institute of Technology and TPS Termiska Processer in the field of gasification, catalysis and gas cleaning related to coal gasification.

The gasification of biomass operated at atmospheric or pressurised conditions, with or without the addition of steam, typically yields a broad spectrum of products. The main products are the permanent gases hydrogen, carbon monoxide, carbon dioxide, methane, ethene and ethane, but the gas also contains substantial amounts of heavier hydrocarbons and aromatics. Furthermore the gas contains small amounts of chloride, nitrogen, mainly as ammonia and sulphur, mainly as hydrogen sulphide, but to some extent also carbonyl sulphide. Although the concentrations of these species are low, their effect on the gas cleaning and the utilisation of the gas can be significant. The process scheme, which would be able to upgrade and clean this complicated mixture to produce gas in a quality, high enough for the before mentioned applications, can be thought to include the following units:

- a steam reforming unit, which is used to catalytically convert the hydrocarbons in the presence of steam into hydrogen and carbon monoxide.
- a shift conversion unit, in which carbon monoxide reacts over a catalyst with additional amount of steam to form hydrogen and carbon dioxide.
- a sulphur removal unit, in which the sulphur compounds released from the biomass during gasification are removed, preferably in a regenerative manner, from the original level of about 100 ppm to a level required for the catalytic units in the process or the downstream application.

To develop a process with a high (thermal) efficiency, the location of a hydrogen sulphide removal unit will be determined by the following considerations. Firstly, the catalysts that are used, both in steam reforming and in shift conversion, are poisoned by hydrogen sulphide, resulting in a significant decrease in their activity. Secondly, the temperatures in the different units constrain the applicability of low-temperature removal techniques. High temperature techniques offer advantages, but might be difficult to apply under the specific conditions in the process, for example because of the high amount of steam present.

The project therefore focuses on the catalytic systems and their behaviour in the presence of hydrogen sulphide. Furthermore, the different possibilities for the removal of hydrogen sulphide are investigated. The methods of cleaning will be comparatively evaluated by the end of 1996. The work presented here covers the steam reforming studies carried out within the project.

1.2 Steam reforming

Steam reforming is a common way to convert hydrocarbons at high temperatures in the presence of steam into hydrogen and carbon monoxide. In this way hydrogen is extracted from both water and hydrocarbon according to the equilibrium reaction:

$$C_nH_m + n\,H_2O \leftrightarrow (n+m/2)\,H_2 + n\,CO \tag{1}$$

Simultaneously carbon monoxide and water react, yielding carbon dioxide and extra hydrogen through the water-gas shift reaction:

$$CO + H_2O \leftrightarrow H_2 + CO_2 \tag{2}$$

As the reforming reaction is highly endothermic, it will be favoured by high temperature. To reach as high a conversion, which means approaching equilibrium as close as possible, at a specific temperature, requires the use of a catalyst. The metals of group VIII of the periodic system are active for the steam reforming reaction. In practice nickel-based catalysts are almost exclusively used in industry. The stoichiometric ratio between steam and carbon is one, but in practice higher values are used to favour the equilibrium and to prevent carbon formation [1,2].

The kinetics of the steam reforming reaction has been the subject of numerous studies [3,4,5,6]. Practically all studies report the rate of reaction of methane steam reforming to be first order in methane. On the effects of other reactants or products there is less agreement. Furthermore, a variety of experimental conditions, different catalysts and possible limitations concerning mass or heat transfer resulted in a substantial variation in the empirically determined values for the activation energy of the steam reforming reaction. All kinetic expressions can be written as a function of temperature, T, and partial pressures, p_i, of reactants and products:

$$r_{CH4} = k_0 \cdot \exp\left(\frac{-E_{A,CH4}}{RT}\right) \cdot f(p_i) \tag{3}$$

Values for the activation energy are reported in the range of 70 to 160 kJ/mole.

Sulphur as hydrogen sulphide is known to poison steam reforming catalysts considerably. Although the poisoning effect is reversible, the activity of the catalyst is decreased in the presence of only a few ppm hydrogen sulphide. However, if the surface of the catalyst is not completely covered by chemisorbed hydrogen sulphide, there are still active sites available for the steam reforming reaction. Rostrup-Nielsen [7] investigated the chemisorption of hydrogen sulphide on nickel at various temperatures and hydrogen sulphide concentrations and determined the following relation for relatively low degrees of sulphur coverage:

$$1 - \Theta_s = k_s \cdot \exp\left(\frac{-E_{A,S}}{RT}\right) \cdot \left(\frac{p_{H2S}}{p_{H2}}\right)^{-0.3} \tag{4}$$

From these chemisorption studies an activation energy for sulphur poisoning $E_{A,S}$ of 36 kJ/mole was calculated. This is in the same order of magnitude as the activation energy for H_2S chemisorption for nickel on α-alumina, reported by McCarty et al. [8] The rate of the partly poisoned catalyst, r_{sp}, can be related to the rate of a sulphur-free catalyst according to:

$$r_{sp} = r_{CH4} \cdot \left(1 - \Theta_s\right)^3 \tag{5}$$

The power coefficient implies that an ensemble of three nickel atoms is involved in the steam reforming reaction [7]. By combining equations (1) to (3) a general expression for sulphur-deactivated steam reforming can be found:

$$r_{sp} = k' \cdot \exp\left(\frac{-E_{A,tot}}{RT}\right) \cdot \left(\frac{p_{H2S}}{p_{H2}}\right)^{-0.9} \cdot f(p_i) \tag{6}$$

The value of $E_{A,tot}$ is then equal to $E_{A,CH4}$ plus three times $E_{A,S}$. The simplest kinetic expression for methane steam reforming is a first order reaction rate in methane partial pressure, $f(p_i) = p_{CH4}$, with $p_{CH4} = p_{0,CH4} \cdot (1-x)$. Using the general mass balance for a tubular reactor, an expression can now be obtained which relates methane conversion and temperature:

$$\frac{W_{cat}}{F_{CH4}} \cdot k'' \cdot \exp\left(\frac{-E_{A,tot}}{RT}\right) \cdot p_{H2S}^{-0.9} = \int_0^x \frac{1}{(1-x)} dx \tag{7}$$

The simplicity of equation (7) involves a number of assumptions:

- the temperature in the whole catalytic bed is constant and equal to the measured temperature
- the diluting effect of the non-equimolar steam reforming reaction is ignored
- the effect of the water-gas shift reaction on the gas composition is not taken into account
- any effects of other components on the reaction rate are neglected
- the hydrogen partial pressure is equal to the mean value of inlet and outlet value, and is assumed to be independent of the methane conversion within the range of the experiments

By integration of the right-hand term of equation (7), a value of $E_{A,tot}$, a combined activation energy of both the steam reforming reaction and the hydrogen sulphide absorption, can be calculated. This value is expected to lie in the order of 180 to 270 kJ/mole.

To validate the derived expression for sulphur-deactivated steam reforming, the results of the model gas experiments were used to plot the left-hand term of equation (7) against the integral on the right-hand side, which should give a linear correlation. Hereby experiments, performed at different temperatures, space velocities and hydrogen sulphide concentrations can be compared in a qualitative way. By plotting the results of the biomass pyrolysis gas experiments in the same figur, it is possible to compare them with the model gas experiments and determine if model gas experiments can be used to describe biomass pyrolysis gas experiments.

2 Experimental

2.1 Model gas experiments

A series of experiments was carried out in a laboratory set-up, which is schematically shown in figure 1. The centre of the set-up is a tubular fixed bed reactor with a bed of catalyst material. The catalyst used in these investigations was coded C-11-9-061, and is a proprietary catalyst produced by United Catalysts Inc. According to the specifications of the manufacturer it contains 11 to 20 wt% nickel oxide on refractory alpha-alumina. A gas mixture was used, which simulated gas, produced by a pressurised air-steam gasifier. The composition of the model gas was 15.0 vol.% H_2, 5.0 vol.% CH_4, 15.0 vol.% CO, 15.0 vol.% CO_2, 25.0 vol.% N_2 and 25.0 vol.% H_2O. The total inlet gas flow was 2 l/min wet gas for all experiments. The various gaseous components of the feed gas were supplied from high-pressure cylinders, and measured by mass flow controllers. Steam was produced in an evaporator, which consisted of a heated stainless steel tube filled with a highly porous inert material. Water was delivered to the evaporator by means of a high-precision rotary pump, providing a steady steam flow. All lines downstream from the steam generator were heated and insulated to prevent water condensation. Hydrogen sulphide was added to the gas stream from a high-pressure cylinder, containing 2077 ppm H_2S in nitrogen. The reactor, made of high temperature stable Inconell steel, had a length of 54 cm and an inner diameter of 15 mm. The catalyst particles, 11x8 mm three spoke wheels, were crushed to a size fraction of 0.7 to 1.4 mm. A catalyst bed was prepared by mixing 15.00 grams of catalyst material and 10.00 grams of inert material of the same size. In this way a bed height of about 9 cm was obtained, which was placed in the isothermal part of a tubular furnace. This corresponded to a space velocity of 5700 h^{-1}.

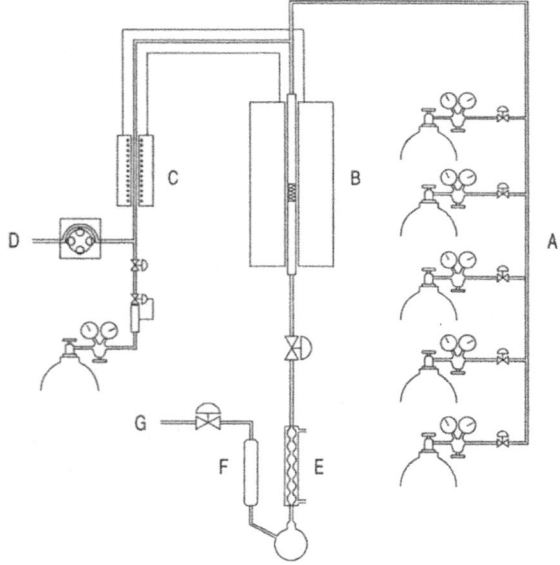

A Gas mixing section
B Furnace with tubular
   reactor and catalyst bed
C Steam generator
D Water pump
E Condenser
F Phosphor pentoxide bed
G Gas outlet to MS

Figure 1. Experimental set-up for the model gas experiments.

Experiments were carried out at atmospheric pressure and at temperatures between 750 and 1010 °C. The catalyst was pre-reduced before the experiments, using a total flow of 600 ml/min of 25 vol.% H_2 and 75 vol.% N_2. The pre-reduction was performed for two hours at a constant temperature of 900 °C.

The product gases were led through a cooling section, in which the water was removed in a countercurrent cooler, followed by a bed of phosphorpentoxide. Finally the product gas stream was analysed for H_2, CO_2, CO, CH_4, N_2 and H_2S, using a Balzers QMG421C Quadrupole mass spectrometer.

2.2 Biomass pyrolysis gas experiments

A second series of experiments was carried out in the apparatus shown in figure 2. The total height of the unit was 2.3 metres. It had two furnaces, one for the generation of pyrolysis gas and one for the pre-heating and catalytic treatment of the gas. The reactors were made of quartz glass except the fuel container that was made of stainless steel. The system provided a steady flow of product gases and made it possible to obtain reproducible experiments. The tar rich pyrolysis gas produced resembles a biomass gasification gas at all major points.

During all tests, aspen wood dried at room temperature was used as fuel. The fuel container was filled with 80 g fuel and 10 g char, produced in a previous identical experiment, was added. A weight was placed on top in order to compress the char formed during pyrolysis. The added char ensured that the gas composition was homogeneous even in the first stages of pyrolysis.

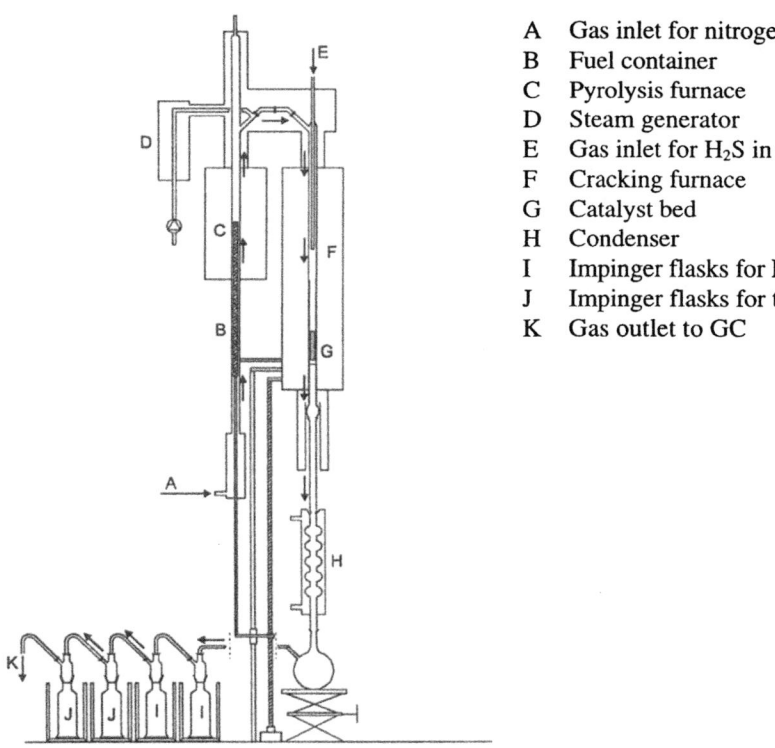

A   Gas inlet for nitrogen
B   Fuel container
C   Pyrolysis furnace
D   Steam generator
E   Gas inlet for H_2S in nitrogen
F   Cracking furnace
G   Catalyst bed
H   Condenser
I   Impinger flasks for NH_3
J   Impinger flasks for tar
K   Gas outlet to GC

Figure 2. Experimental set-up for the pyrolysis gas experiments.

The container was then introduced into the pyrolysis furnace (held at 700 °C) from below at a constant speed. The speed was adjusted to a given fuel feed of 40 g/hour. Nitrogen was used as a carrier gas at a rate of 676 ml/min. Pyrolysis gas passed through vents in the sides of the fuel container. Downstream of the pyrolysis reactor, steam and hydrogen sulphide could be added to the gas in the pipe that connected the pyrolysis reactor with the cracking reactor. This pipe was heated to 400 °C. The total dry gas flow was approximately 1.4 l/min. The cracking reactor had a central tube that forced the gas to pass out towards the reactor wall. This ensured the pre-heating of the gas to the selected cracking temperature. The gas then passed over the catalyst located in the quartz glass bed holder. For this experimental series two temperatures, 850 °C and 900 °C, were used. The same type of nickel-based catalyst as described for the model gas experiments was studied. The particle size was also the same except for one series of experiments where the effect of particle size was evaluated. The diameter of the fixed bed was 26 mm and the height 45 mm when 35 g of catalyst was used (23 mm for 17.5 g of catalyst). This corresponded to a space velocity of 3500 h^{-1} (1750 h^{-1} for 17.5 g of catalyst). Before each experiment the catalyst was pre-reduced in the cracking furnace, following the same procedure as previously described for the model gas experiments.

Ammonia and tar were collected downstream the cracking reactor in a water-cooled condenser in series with four impinger bottles held at 0 °C. The first two were filled with 50 ml 0.01 M H_2SO_4 for absorption of ammonia and the others with 50 ml acetone for tar collecting. The condensate was added to the H_2SO_4-solution and filtered. The condenser and impingers with H_2SO_4-solution were washed with distilled water to make sure that no losses of ammonia occurred. The aqueous solution was transferred to a bottle for later analysis of the ammonia content. The filter paper and all glassware were rinsed in acetone, and the acetone was collected for analysis of the amount of tar.

The gas exiting the impinger bottles was analysed on a Hewlett Packard 5890 Plus gas chromatograph for H_2, O_2, N_2, CO, CO_2, CH_4, C_2H_4 and C_2H_6. The gases were separated on Porapak Q column and Molecular Sieve 5A. The Molecular Sieve had a pre-column arrangement with a Porapak column. The pre-column arrangement was to limit the consumption of the Molecular Sieve by back blowing all the other gases except H_2, O_2, N_2 and CO. From the Porapak Q column CH_4, CO_2, C_2H_4 and C_2H_6 were separated and H_2, O_2, N_2 and CO resulted in one peak. The carrier gas was argon and a thermal conductivity detector was used.

The moisture content of the fuel varied between 5 and 7 wt% corresponding to about 5 vol.% in the gas. The addition of water was kept at 6 ml H_2O/h, corresponding to a total concentration in the gas of 14 vol.%. Hydrogen sulphide was added to the system 30 minutes after the start of pyrolysis. The hydrogen sulphide was taken from a bottle with a concentration of 0.3 vol.% H_2S in nitrogen.

3 Results and discussion

3.1 Model gas experiments

In table 1 and figure 3, the results of the model gas experiments are showed. Two levels of hydrogen sulphide (104 and 208 ppm) and a sulphur-free gas stream were used.

Table 1.   Results of the experiments at various hydrogen sulphide concentrations.

0 ppm H$_2$S		104 ppm H$_2$S		208 ppm H$_2$S	
Temp. (°C)	Conv. (%)	Temp. (°C)	Conv. (%)	Temp. (°C)	Conv. (%)
766	10.6	829	0.0	828	2.6
791	19.1	831	2.4	869	7.2
814	29.1	852	5.7	873	6.5
815	33.2	868	5.1	894	12.6
834	47.7	872	10.5	913	21.3
852	55.6	893	18.3	914	19.4
861	63.1	913	29.0	933	32.9
872	75.5	927	33.5	952	47.2
872	72.4	953	57.3	966	53.9
893	87.1	966	65.0	972	62.2
908	90.3	973	72.3	993	76.7
972	99.1	994	84.4	1010	82.3
1014	99.5	1010	90.7		
		1013	92.7		

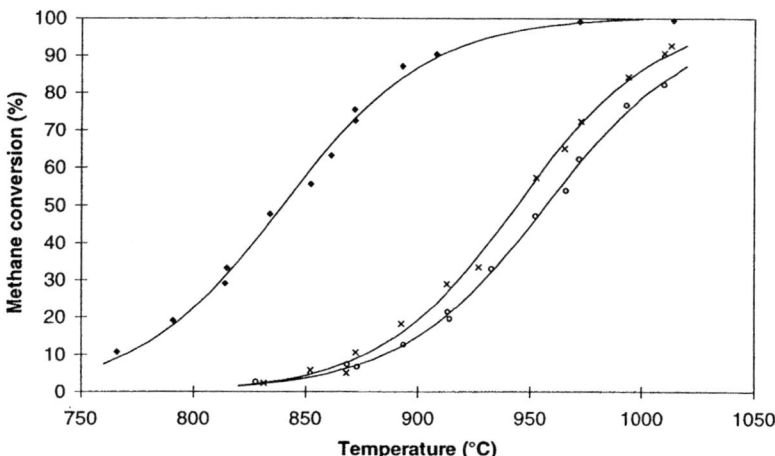

Figure 3. Methane conversion versus temperature for the various H$_2$S concentrations.
♦ 0 ppm H$_2$S, × 104 ppm H$_2$S, ○ 208 ppm H$_2$S

The curves in figure 3 show the effect of temperature and hydrogen sulphide concentration on the methane conversion. For conversion values over 90%, the reaction is controlled by diffusion instead of by kinetics, indicated by the decreasing slope of the conversion curves. At conversion values below 10%, the activity of the catalyst is low and the accuracy of the measurements is becoming uncertain.   Therefore, only conversion values between 10 and 90% were used for the calculations. From the figure, the poisoning effect of hydrogen sulphide is obvious; a temperature increase of 100 degrees is necessary to reach comparable levels of conversion for a sulphur containing feed. The difference between 104 and 208 ppm hydrogen sulphide is relatively small. This implies that an amount of 104 ppm hydrogen sulphide almost completely covers the nickel surface, and that doubling this amount hardly affects the degree of coverage and thus the poisoning of the catalytic surface by blocking of the catalytically active sites.

Equation (7) was used to calculate the activation energy of the sulphur-deactivated steam reforming reaction. This resulted in activation energies of 272 kJ/mole for 104 ppm and 282 kJ/mole for 208 ppm. Figure 4 shows the Arrhenius plots for the experiments with hydrogen sulphide in the gas stream.

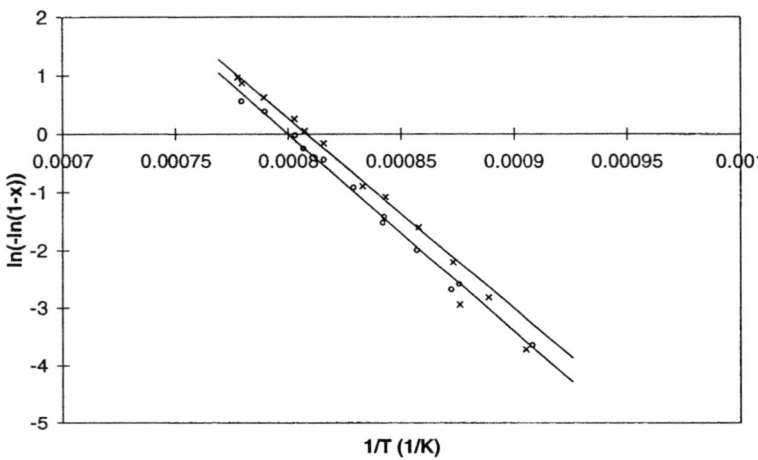

Figure 4. Arrhenius plot for the various H_2S concentrations.
× 104 ppm H_2S, ○ 208 ppm H_2S

The values for the activation energies are high, compared to those reported in literature [7]. As mentioned before, a number of assumptions were made to simplify the calculations and the comparison between model gas and biomass pyrolysis gas experiments. An estimate of the various errors as a result of these assumptions, showed that the effect of an inaccurate temperature measurement is large in comparison to the error made by approximating the hydrogen concentration. Furthermore, the effect of neglecting the volume increase caused by the steam reforming reaction, is opposite for hydrogen and hydrogen sulphide. If a more complicated rate equation is used, the effect of the partial pressures of the other reactants and products is difficult to predict.

3.2 Biomass pyrolysis gas experiments

Figure 5 shows the typical course of events for three pyrolysis gas experiments with varying sulphur addition.

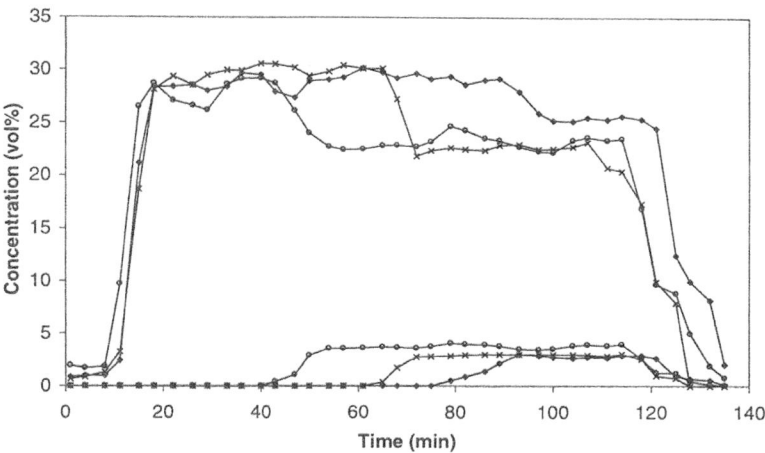

Figure 5. Results from biomass pyrolysis gas experiments at 900 °C. The high curves represent hydrogen concentrations, the low curves methane concentrations.
◆ 67 ppm total H_2S, × 115 ppm total H_2S, ○ 211 ppm total H_2S

When the pyrolysis starts, the hydrogen concentration increases to a stable concentration around 28 vol.% in the dry gas. The fresh catalyst is still active and no methane can be detected. With the addition of hydrogen sulphide the catalyst will with time be partly poisoned and the methane concentration increases to a stable level with a simultaneous decrease in hydrogen concentration. This methane level can according to the described theory be calculated as a function of the hydrogen sulphide concentration in the gas.

The second parameter affecting the methane conversion for the pyrolysis gas experiments was the inlet methane concentration. For all the experiments the outlet methane concentration from the cracking furnace was measured and experiments performed without any catalyst were used to determine the inlet concentration. However, due to the formation of methane from thermal cracking of higher hydrocarbons the inlet concentration to the catalyst increased with temperature. Therefore the methane concentration was measured for three different temperatures, 700, 850 and 900 °C. For 700 °C a methane concentration of 4.6 vol.% (dry gas) was determined. At this temperature, which was the same temperature as in the pyrolysis reactor, no formation of methane from heavier hydrocarbons was assumed to occur and 4.6 vol.% of methane was therefore used as the inlet concentration to the cracking furnace. Assuming that the formation of methane from heavier hydrocarbons follows first order kinetics, that the concentration of heavier hydrocarbons can be regarded as constant and that the homogeneous reforming of methane can be neglected result in a linear increase in methane concentration over the length of the cracking furnace. Based on these assumptions the inlet methane concentration to the catalyst could be calculated to 5.5 and 5.6 vol.% (dry gas) at 850 and 900 °C, respectively.

Table 2. Methane conversion for the pyrolysis gas experiments.

43 ppm H$_2$S			67 ppm H$_2$S			115 ppm H$_2$S			211 ppm H$_2$S		
Temp. (°C)	Weight (g)	Conv. (%)	Temp. (°C)	Weight (g)	Conv. (%)	Temp. (°C)	Weight (g)	Conv. (%)	Temp. (°C)	Weight (g)	Conv. (%)
			850	35	21.3	845	17.5	10.4	851	17.5	6.7
			853	17.5	19.5	849	35	19.5	851	35	15.9
894	35	64.3	893*	35	53.6	892	35	33.9	890	17.5	21.4
895	17.5	39.3	897	17.5	28.6	894	35	46.4	893*	35	26.8
			899	35	50.0	894	35	44.6	896	35	32.1
						894	17.5	19.6			
						894	35	30.4			
						896	35	48.2			
						897*	35	37.5			
						899	35	41.1			

*Particle size 1.4-2.0 mm

The results of the pyrolysis gas experiments are summarised in table 2. The evaluation of the results presented here focused on the conversion of methane. However, it should be noted that the corresponding tar levels after the nickel-based catalyst are very low, typically below 10 g/ton dry fuel for condensable tars. The addition of hydrogen sulphide was performed on four different levels, 24, 48, 96 and 192 ppm H$_2$S. These levels were the concentrations in the gas excluding the sulphur originating from the fuel. A fuel analysis showed that the sulphur content of the fuel was below the detection limit (about 50 mg/kg). However, measurements performed in the reactor in absence of any catalyst, using pre-calibrated absorption tubes, showed a sulphur level in the fuel of 61 mg/kg. The sulphur levels given in table 2 and figure 5 are the calculated total concentrations, based on this value and the measured gas flows.

3.3 Combining the two series of experiments
By using equation 7, all factors that influence the methane conversion can be plotted into one picture. In this way, the model gas experiments and the biomass pyrolysis gas experiments, with their different methane flows, catalyst amounts, temperatures and hydrogen sulphide concentrations can be compared in a qualitative way. The results for this comparison are plotted in figure 6. The hydrogen sulphide partial pressure was expressed in ppm, the units for catalyst weight and methane flow were grams and normal cubic meters per hour, respectively. A mean value for the activation energy of 277 kJ/mole was used.

The experiments using a model gas mixture show a remarkable agreement with the biomass pyrolysis gas experiments, with a much more complex composition. With a relative simple description of the steam reforming reaction and the deactivation of the catalyst by hydrogen sulphide, obtained by model gas experiments, the steam reforming of a biomass gasification gas can be described with satisfactory accuracy.

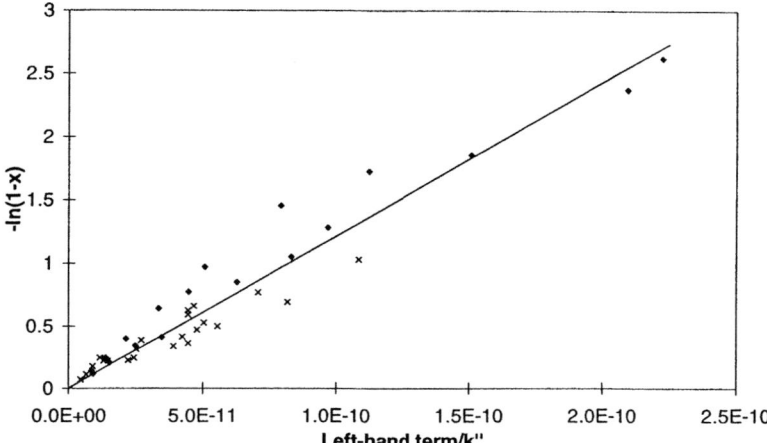

Figure 6. Graph combining the results of model gas and biomass pyrolysis gas experiments. On the x-axis, the left-hand term from equation (7) is plotted.
◆ model gas (KTH), × biomass pyrolysis gas (TPS)

The small deviation between the two experimental series is within the uncertainty of the measured parameters. Especially the temperature and hydrogen sulphide concentration have a strong influence on the results and for example an increase of the total hydrogen sulphide concentration of 15 ppm can explain the difference between the two experimental series. The value for the activation energy, which was used for the comparison is, as previously mentioned, high compared with other sources, but this does not influence the result of the comparison. From a scientific point of view, it is interesting to reveal the cause of this high activation energy. However, for future prediction of the conversion of methane, produced from biomass gasification, the accuracy is satisfactory, especially taking into account the approximate calculations on which the results are based.

Future activities at KTH will include analysis of the catalyst samples using scanning electron microscopy, to investigate the catalyst surface for any possible disturbing influences. Furthermore, more precise calculations will be made, to eliminate errors due to the assumptions made. Similar tests will also be performed on other industrial catalyst, which have been promoted to minimise the effect of poisons such as hydrogen sulphide. At TPS, pressurised experiments will be used to complement the previous results.

4 Conclusions

When gas from gasification of biomass will be used for advanced applications, i.e. methanol production or fuel cells, hydrocarbon steam reforming will be necessary. In the work presented here a kinetic expression for steam reforming of methane over partly sulphur-deactivated nickel-based catalysts has been derived and verified in model gas experiments. The expression takes the low concentration of sulphur into account and a combined activation energy for the partly sulphur-deactivated catalyst was calculated. The same kinetic expression has also shown to be a useful approximation for the evaluation of steam reforming of a biomass pyrolysis gas.

The low sulphur levels present in a gas from gasification of biomass strongly deactivate the nickel-based catalyst. However, it is clear from the results presented in this article that the partly sulphur-deactivated catalyst can be used for steam reforming of a biomass gasification gas. For the dimensioning of the steam reformer it is enough to base the calculation on the inlet concentration of methane and hydrogen sulphide since the heavier hydrocarbons will not have any strong influence on the reaction.

5 Acknowledgement

This project has been performed within the European Unions Joule II Extension Programme for Clean Coal technologies. Financial support to the work performed at KTH and part of the work at TPS Termiska Processer by NUTEK, the Swedish National Board for Industrial and Technical Development, was gratefully acknowledged. Additional financial support was received from the Board for Non-nuclear Research at Studsvik and this support was also gratefully appreciated.

6 References

1. Twigg, M.V. (1989) *Catalyst Handbook, 2nd edition*, Wolfe Publishing Ltd, London
2. Rostrup-Nielsen, J.R. (1984) *Catalytic Steam Reforming, Catalysis, Science and Technology*, Vol. 5, (ed. J.R. Anderson and M. Boudart), Springer-Verlag, Berlin
3. Bodrov, I.M., Apel'baum, L.O. and Temkin, M.I. (1964) *Kinet. Katal.*, Vol. 5, pp. 696
4. Bodrov, I.M., Apel'baum, L.O. and Temkin, M.I. (1968) *Kinet. Katal.*, Vol. 9, pp. 1065
5. Ross, J.R.H. and Steel, M.C.F. (1973) Mechanism of the Steam Reforming of Methane over a Coprecipitated Nickel-Alumina Catalyst. *J. Chem. Soc. Faraday Trans. 1*, Vol. 1, pp. 10
6. Nekrich, E.M. (1970) Kinetics of partial oxidation of natural gas by a small amount of steam. *J. Appl. Chem. USSR*, Vol. 32, No. 2, pp. 372
7. Rostrup-Nielsen, J.R. (1984) Sulphur-Passivated Nickel Catalysts for Carbon-Free Steam Reforming of Methane. *J. Catalysis*, Vol. 85, pp. 31
8. McCarty, J.G. and Wise, H. (1980) Thermodynamics of sulfur chemisorption on metals, I. Alumina-supported nickel. *J. Chem. Phys.*, Vol. 72, No. 12, pp. 6332

INVESTIGATIONS IN HIGH TEMPERATURE CATALYTIC GAS CLEANING FOR PRESSURIZED GASIFICATION PROCESSES

H. ALDÉN, P. HAGSTRÖM, A. HALLGREN, and L. WALDHEIM.
TPS Termiska Processer AB, SWEDEN

Abstract

One of the main paths for the development of thermochemical conversion processes involve systems with enhanced process pressures; to increase the overall process efficiency but also to improve the environmental performance of the system. Thus there is a need for fundamental experimental knowledge on chemical gas cleanup at elevated pressures.

A flexible and easy to operate pressurized apparatus has been installed for investigations in high temperature gas cleaning by means of thermal and catalytic or chemical procedures. In the current study tar decomposition of a pyrolysis gas from biomass fuels was investigated. A semi continuos fuel feeding concept allows a very constant formation of a gas product at 700 °C. The gas product is subsequently introduced into a fixed bed secondary reactor where the actual gas cleanup may take place. In this study two different catalysts were used as fixed bed material in this secondary reactor: a Swedish dolomite and a commercial nickel-based catalyst. Tests were performed at cracking temperatures between 800-900 °C and at pressures between 1-20 bars.

Novel technical solutions make the test rig an advanced and versatile platform for thermochemical conversion studies at high temperatures and at enhanced pressures.
Keywords: Tar decomposition, catalysis, dolomite, pressurized, pyrolysis.

1 Introduction

Gasification of solid fuels releases fuel impurities to the gas as well as smaller or larger fragments from the decomposed organic structure. These components, i.e. sulfur and nitrogen compounds; tars and inorganic salts, are harmful to the environment by formation of toxic, acidic or atmospheric active products during the combustion of the gas. Certain released compounds may also cause corrosion and fouling problems in reactors and tubing. To avoid such problems the gas product has to be cleaned thoroughly. Potential advantages are expected if the gas cleaning process additionally is accomplished at enhanced temperatures. There is a need for further investigations in this area and furthermore to expand activities to involve pressurized gas treatment processes for specific purposes. One important example of such purposes may be catalytic gas treatment (gas product reforming) integrated in fuel cell applications.

An endeavor in pressurized gas treatment is being made at TPS where a project is carried out within the framework of the European Commission AAIR Program. The objective of the initial stage of the project was to construct a flexible, easy to operate, pressurized apparatus for the study of high temperature gas cleaning by means of thermal and catalytic or chemical procedures. In this first phase the unit was planned to be used for the study of tar decomposition of a pyrolysis gas from biomass fuels.

2 Experimental

2.1 Equipment
The installed test rig consists of two externally heated pressure vessels (see figure 1). In the first pressure shell, the pyrolyser or more generally the gas producing unit is contained. The fuel container is introduced gradually at a controlled rate into the bottom of the pyrolyser by a piston that mechanically levers the container into the pyrolyser compartment by means of a DC motor. In cases where an increased fuel load is desired the solid fuel may be compressed to a pellet, before being put into the fuel container. In this way the density is enhanced and subsequently the required volume for the fuel material is reduced. In the pyrolyser itself the fuel devolatilization takes place and the gas produced is swiftly removed from the reactor by carrier gas. Additional technical information is presented in figure 1.

The produced pyrolysis gas is introduced into the secondary reactor. This reactor contains a fixed bed inside an internal gas tight pipe that is fitted to the bottom flange. This flange also includes connections for monitoring and measurements equipment. Gas sampling has been provided after the fixed bed. Sampling can be adapted to the desired components, e.g. tar, ammonia, cyanide, and hydrogen sulfide.

The gas product is analyzed with a GC where O_2, N_2, H_2, CO, CO_2, CH_4, C_2H_4, and C_2H_6 are analyzed on-line. Tar is absorbed, extracted, and analyzed with gas chromatography. This procedure has been described elsewhere [1]. Other components, like for instance ammonia, can also be absorbed and then measured, in this case with a NH_3 selective electrode.

The data acquisition including the control of the unit is performed with a separate PLC (Process Logical Controller) system connected to a PC (Personal Computer). The computer communicates with the PLC which enables the user to operate the unit easily.

The first experimental program featured slow pyrolysis and subsequent tar decomposition in a secondary gas treatment reactor. The intentions for the investigations were also to comparatively study the tar decomposition efficiency as a function of gas residence time, carrier gas composition, as well as the choice of catalyst. In a second stage of the investigation a Ni-based commercial catalyst will be studied.

2.2 Procedure
In the current study tar decomposition of a pyrolysis gas from biomass fuels was investigated. The semi continuos fuel feeding concept, at a maximum rate of 700 g/h, allowed a very constant formation of a gas product at 700 °C. The gas product was

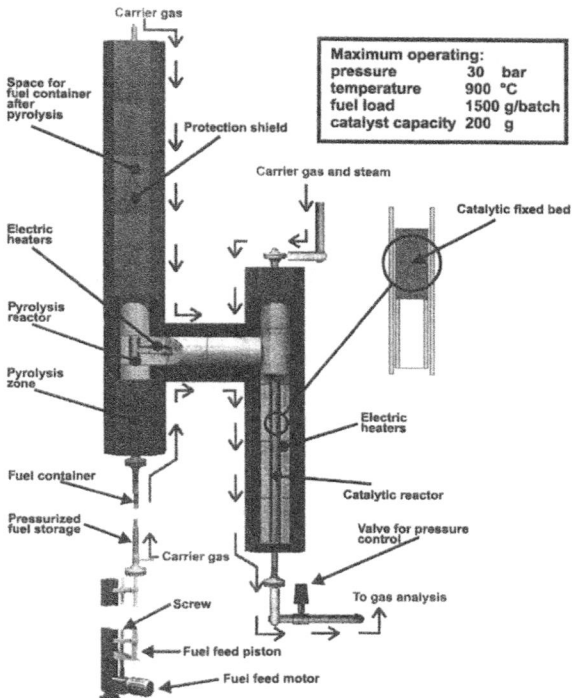

Maximum operating:
pressure   30   bar
temperature   900   °C
fuel load   1500 g/batch
catalyst capacity 200   g

Fig. 1. The experimental set up.

subsequently introduced into a fixed bed cracker where the actual tar decomposition took place.

The fuel used was Swedish Aspen chips (fuel analysis, please refer to table 1) which were totally devolatilized in the primary reactor forming gaseous, condensable, and solid pyrolysis products. The condensable pyrolysis products were of the main concern in this investigation where the high molecular weight hydrocarbons were aimed to be converted into lighter compounds by means of catalytic or/and thermal methods.

Two different catalysts, dolomite and a commercial Ni-based catalyst, were tested for tar cracking properties in atmospheric as well as pressurized conditions and at temperatures between 800 - 900 °C.

The dolomite catalyst (size fraction 0.7 - 1.4 mm) was, prior to the actual pyrolysis experiment, placed in the secondary reactor at 500 °C which also is the temperature at standby of the system. The temperature was thereafter ramped to the final calcination temperature. The dolomite was calcinated at 900 °C for 50 minutes in a nitrogen flow of 2.1 Nl/min. No CO_2 was detected in a subsequent gas analysis which otherwise would suggest insufficient calcination of the dolomite.

Regarding the nickel catalyst (size fraction 0.5 - 1.4 mm), similar procedure was undertaken apart from the calcination. The catalyst was introduced in the secondary reactor at 500 °C with a subsequent ramping to the set process temperature (800 - 900

°C). Prior to the experiment, the catalyst was pre-reduced *in situ* in a 25 vol% hydrogen atmosphere.

The experiments were carried out in mainly nitrogen atmosphere apart from a few test run with a mixture of steam and CO_2 additions to the carrier gas. Following each test run, the system was flushed with air at 500 °C to spare the reactor material and to clean the reactor parts from pyrolysis rest products. In a later stage of the experimental program, influences on the performance of the Ni-catalyst due to the oxidation procedure were anticipated. To avoid such influences, after the oxidation of the system, the internal parts of the test rig were flushed with nitrogen overnight. In this way the amount of oxygen in the system was reduced to a level where influences on the Ni-catalysts would be negligible.

2.3 Sampling

2.3.1 Gaseous products
The pressure of the main gas flow (maximum 20 Nl/min.) leaving the unit was reduced to atmospheric pressure. A slipstream of the main flow was taken for analysis of tar and the so called permanent gas components (H_2, O_2, N_2, CO, CO_2, CH_4, C_2H_4, C_2H_6). The gas which was sampled for tar analysis was kept at 350 °C to avoid condensation of liquid products, while the gas which was sampled for analysis for the permanent gas components was quenched to 10 °C to condense the water content in the gas.

After the quenching of the gas products the gas flow passes through a cotton filter to separate the rest of the tar from the gas phase.

The gas was dried and separated from any other liquid products and led to the gas analysis. This was made with gas chromatography (TC-GC) which admits a gas sample for analysis every three minutes. The results from the gas analysis were continuously transferred to the computer for evaluations.

2.3.2 Condensable products
The slipstream of the gas product was led from the test rig to the sampling equipment in preheated tubes. A manual system for tar sampling has been installed together with the sampling systems for other condensable products (HCN and NH_3).

The gas was separated from tar components through scrubbing with acetone in wash-bottles. After extraction to dichloromethane, the tar content was determined by FID-GC.

The gas volume passing through the gas wash-bottles was measured and the tar content was evaluated as gram tar/ton dry fuel. Three parallel lines for the analysis of the condensable product enable continuos sampling procedures. Alternatively, one tar sample may be taken accounting for an integral value while the other two lines were used for repeated sampling during the test run. The control system monitors the sampling.

2.3.3 Solid products
Following each test run the fuel container and the residues were weighted. The catalysts was removed from the cracker, weighted, and visually inspected with respect to the physical condition of the particles and the amount of soot captured in the catalyst bed.

3 Results

In the following, the results from the experiments are not presented individually. Instead they are discussed in an aggregated form to generalize trends. As the apparatus was commissioned in the autumn of 1995, the results of the tests (more than 40 tests in 3 months) are to a large extent checking on reproducibility etc. Additionally, the experimental basis does not merit sophisticated evaluation in terms of kinetics etc. However, the number of experiments shows the productivity of the unit.

3.1 Tar decomposition

With reference to atmospheric catalytic tar cracking studies earlier performed at TPS, it has been shown that the temperature has a distinct effect on the tar decomposition even in the narrow temperature interval 800 - 900 °C [1]. Similar tendencies were indicated in this study especially at elevated pressures. The first results clearly indicated influences from the catalysts on the tar decomposition but suggested as well a significant thermal contribution to the overall decomposition. However, the tar yields from these tests were higher than previously seen in similar tests in an atmospheric laboratory unit. The possible causes for this behavior, some of which can be linked to differing experimental conditions, are discussed in the next section.

3.1.1 Dolomite catalyst

The tar yield, with and without dolomite catalyst, versus the temperature and the pressure is shown in figure 2a-c. With the Glanshammar dolomite a reduction of approximately 70% in the amount of condensable tars, compared to the experiments with no catalyst, was achieved in the secondary reactor (800°C). At 900°C, a further 10 -15% reduction was accomplished (figure 2a).

The tar yield was even further decreased when the pressure was increased to 10 bar (figure 2b). A typical value of the amount of condensable tars in the product gas was found to be less than $1g/Nm^3$ at 900°C, 20 bar and with Glanshammar dolomite (figure 2c). The condensable tar components were defined as aromatic and polyaromatic hydrocarbons with higher molecular weight than o-xylene.

The increase in pressure did also affect the tar yield for the experiments without a catalyst in the same direction. An increase in the pressure, decreased the ratio between (condensable tars without a catalyst)/(condensable tars with a catalyst). Thus, under these conditions, the gain in using a dolomite catalyst is less on the refractory aromatics, if sufficient residence time can be allowed for in the design. Interesting to note was that a matching trend for total tar was not found, i.e. the BTX fraction did not, in general, decrease as much as the heavier compounds did with an increase in the pressure.

The primary release of volatile hydrocarbons is also influenced by the pressure as shown below. In these experiments a constant mass flow of carrier gas was used, such that the residence time in the catalyst bed was also increased at enhanced pressures. However, as the total system pressure was raised, also the partial pressure of CO_2 was raised, leading to partial recarbonisation of the catalyst in these experiments. In general the CO_2 concentration was 10-12 % (by vol. dry basis). During the non-catalytic and in the dolomite tests the partial pressure could be as high as 2 bar under pressurized conditions. As the most active catalyst is believed to be the fully calcined material, the approaching results between the non-catalytic tests and the dolomite tests, may also be caused by a gradual decrease in the activity of the dolomite catalyst at 10 and 20 bar.

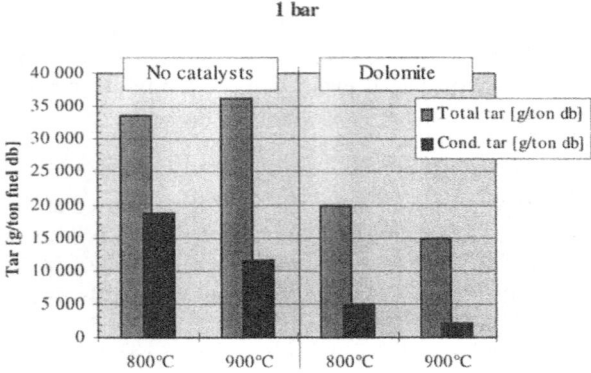

Fig. 2a. Conversion of pyrolysis tars without catalyst compared with Glanshammar dolomite at 1 bar.

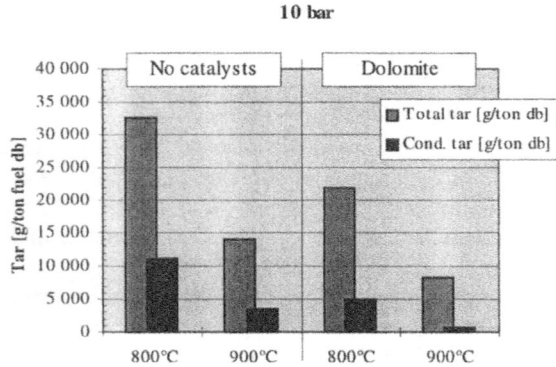

Fig. 2b. Conversion of pyrolysis tars without catalyst compared with Glanshammar dolomite at 10 bar.

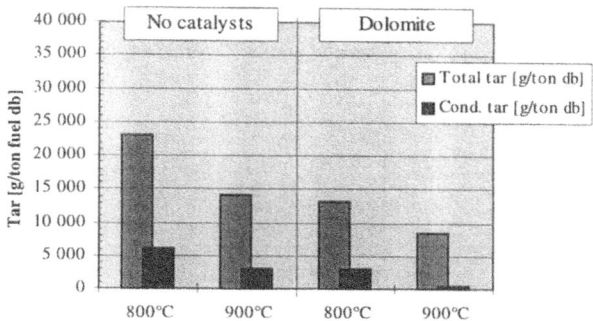

Fig. 2c. Conversion of pyrolysis tars without catalyst compared with Glanshammar dolomite at 20 bar.

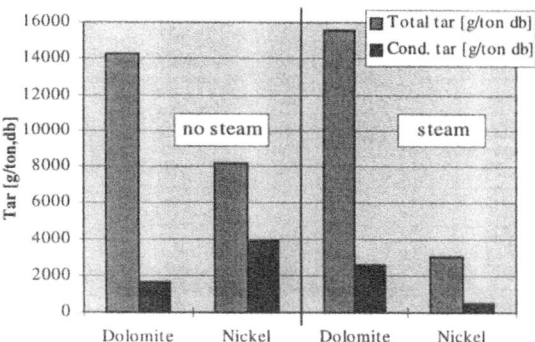

Fig. 3. Tar yields for two catalysts in experiments with steam and without addition of steam.

3.1.2 Comparison of selected catalysts

The dolomite catalyst was also compared to a commercial steam reforming nickel catalyst. The tar yields are shown in figure 3 for these two catalysts in experiments with steam and without addition of steam.

Without steam addition, the nickel catalyst was less efficient in decomposing the condensed tar than the dolomite. Some soot formation was suspected. However, when steam was added the condensed tar yield dropped to a level below the dolomite results. For dolomite no significant effect was present from steam.

To present the impact of the choice of the catalyst, in addition to effects on tar hydrocarbon, also influences on the bulk gas (H_2, CO, and CH_4) components were studied. Figure 4 shows the concentration of H_2, CO, and CH_4 in four different cases at 900°C and 1 bar. An increase in the hydrogen yield can be seen when dolomite is used, compared to the non-catalytic test. This can mainly be attributed to the effect of the dolomite on the water gas shift reaction as the methane content was not affected. With nickel catalyst, a decrease also in the CH_4 yield was seen ("Nickel catalyst" and "Nickel catalyst + steam"). The combined effect of the tar and methane steam reforming together with the water gas shift reaction resulted in higher concentrations of hydrogen in these cases.

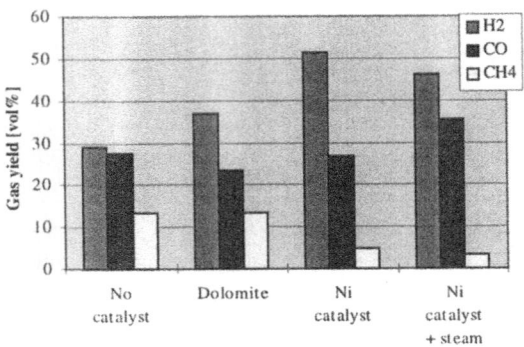

Fig. 4. H_2, CO, and CH_4 as a function of the choice of catalyst; dry basis, inert gas (N_2) excluded.

Table 1. Mass balances with respect to the choice of catalyst at 900°C/1 bar and 900°C/20 bar

	900°C/1 bar			900°C/20 bar	
Test run No.	31	15	44	37	38
wt%	No catalyst	Dolomite	Nickel catalyst	No catalyst	Dolomite
gas incl. water	68	68	66	54	50
char	22	23	22	29	29
tar	4	2	1	1	1
Total mass	94	93	89	84	80
Difference	- 6	- 7	-11	- 16	- 20

3.2 Mass balances

In table 1, the mass balances for a selection of tests runs are presented. Here the weight of gas is based on the gas analysis during the tests, and the weight of tar is based on the tar analysises. The weight of water and char are based on gravimetric analyses.

In the series at 900°C/1 bar, the best closure was shown in the case where no catalysts were involved. However, deviations in the atmospheric experiments were in general reasonably small.

The table shows that higher pressures gave a higher char yield, as was expected. This result has also been shown at other works [2]. However it also shows that the char yield is reproducible under similar conditions, which is of importance in studying the downstream reactions of the gas. It is also apparent that pressure influences the inlet gas and tar conditions for such downstream treatment. This has to be taken into account when evaluating the results of such testing.

The lack of closure in the balance can be attributed to several factors, as for example:

- soot formation
  The formation of soot from tar can occur under certain condition, and there is also a possibility for the Bouduard reaction to produce soot at elevated pressures during cooling of the gas, especially when little steam has been added. As the stop procedure included oxidation, soot has not been seen. Pressure drops etc. however, indicates that especially in nickel catalyst experiments this is a probable source of error in the mass balance. It can also be shown that the main component missing in the mass balances is carbon.

- recarbonisation of the dolomite
  This is predominately an effect at elevated pressures. The change in weight for the dolomite catalyst after the experiment at 20 bar showed an increase of an order that indicates a near half-calcined state. This explains about 7 % of the failing mass in this particular case. Thereby the results are brought more in line with the other experiment at pressure, but also with the atmospheric experiments.

4 Discussion

4.1 Thermal tar decomposition

Effects from the reactor material on the tar decomposition can not be excluded. The reactor material is made of steel which contains appreciable amounts of Ni that may catalyze the tar decomposition, even if the alloy matrix is protected by a chromium oxide surface film.

A combination of homogenous gas phase, thermal effects and influences from the reactor material constitutes the so called background conversion added to the actual conversion caused by the investigated catalysts. This behavior may be observed in figure 2a-c in the cases without catalyst involved. A significant decrease in the condensable tar yield is achieved in the temperature interval 800 - 900°C. At atmospheric pressure this does not translate in a similar decrease in total tar. On raising the pressure, the total amount of tar does not change appreciably at 800 °C, but at 900 °C a significant reduction can be seen. However, the residence time has also changed by a factor of 10 and 20 respectively. The project time has not allowed performing additional tests to isolate these two effects.

Presently it cannot be settled to what extent catalytic effects from the reactor material were influencing the non-catalytic tar decomposition. These effects, if any, may also be overshadowed by thermal effects. The extent of thermal effects was investigated in earlier experiments at TPS with a different reactor material at atmospheric pressures. They were found to be of a similar magnitude as in these experiments for the condensable fraction of the tar [3].

4.2 Dolomite tar decomposition

Dolomite has a proven ability to decompose tar produced in thermochemical conversion processes. In this series of experimental results, influences from the temperature as well as the pressure are demonstrated. As in the case without catalyst, the results from dolomite indicate a significant tar cracking effect from the temperature as well as from the pressure. At higher pressures (20bar) the function of the catalyst seems to be influenced more by the partial pressure of the gaseous components than by the temperature. An additional reduction of the tar content is obtained even at the lower temperature. Again a significant change in residence time is at hand when raising the pressure. Also, significant recarbonisation of the dolomite to the less active semi-calcined state occurred during the course of the experiments at elevated pressure. Thus, the approaching results to the results of the thermal cracking may well be caused by a less active catalyst being formed during the experimental conditions. The difference in the results could therefore be even larger if recarbonisation had not occurred. However, this is of academic interest, as carbon dioxide is always present in gasification gases to an extent that makes it impossible to maintain the calcined state of the dolomite at 900 °C and elevated pressures. A much higher temperature must be used to achieve the additional benefit of catalytic tar cracking. However, in that case also the thermal cracking will be more significant, thus the need for catalysts for this particular purpose can be argued.

The absolute yield of condensable tar in these experiments were found to be high when compared to the yields achieved in the atmospheric laboratory unit. To understand the reasons behind the high yield in the new unit, it was necessary to analyze the operating conditions more closely in these units and to find causes for this phenomena. One possible cause is influences of the reactor materials themselves. As concluded above, it is not however possible to distinguish the thermal effect from any such effect. Thermal cracking results seem consistent between the units, which makes this theory less plausible. The operating conditions were found to differ between the units. In the new unit, the residence time used at atmospheric conditions was only half of the residence time in the laboratory unit. Secondly, the scale-up to the new unit had made the catalyst bed longer, and this causes a less even temperature profile in the bed. An analysis of the magnitude of these effects shows that the results are consistent with the results from the laboratory unit when extrapolating to twice the residence time, i.e. a reduction of the tar yield to the order of 10-20 % compared to the reported yields in this report.

4.3 Comparisons of selected catalysts

When comparing nickel and dolomite catalysts, it is quite apparent that the operating mechanism differs from one catalyst to the other. First, in the case of dolomite the tar yield is not very influenced by the addition of steam, even suggesting a slight negative effect that has been pointed out in other studies previously at TPS and KTH [4]. In the case of nickel, however, the steam addition leads to tar yields that are extremely low. The mechanism of the nickel catalyst for steam reforming is to adsorb the hydrocarbon and the steam to nickel sites. The effect of steam seen in the experiments is consistent with such a mechanism. Dolomite, on the other hand has been shown to, as far as the mechanism has been elucidated, influence the exchange rate of hydrogen atoms between the gas phase and the hydrocarbon, presumably as radicals. The steam will as a result of reactions produce more hydrogen. Thereby is the probability shifted in a way that a hydrocarbon radical formed from splitting off a hydrogen radical will revert to the initial state rather than proceed into reactions finally breaking the aromatic carbon bonds. Contrary to this effect on the tar, the effect on methane concentration of dolomite is negligible whereas the effect when using the nickel catalyst is high, especially when steam is present. The commercial Ni-catalyst is a specific steam reforming catalyst which also can be observed in figures 3-4. The reaction:

$$CH_4 \quad + \quad H_2O \quad \Leftrightarrow \quad CO \quad + \quad 3H_2$$

is particularly responsible for the increased yields in hydrogen. The reaction is shifted to the right to convert CH_4 to H_2. An slight increase in the CO yield is also apparent. The shift reaction, see below, will change the observed concentrations from the values expected from the reaction given above. This indicates that from the point of view of the nickel catalyst, methane is just another hydrocarbon, whereas in the case of dolomite there is a qualitative difference between the aromatic tar hydrocarbons and the aliphatic type of hydrocarbons present in the gasification gas.

The difference seen in figure 4 between "No catalyst" and "Dolomite" reflects the equibrilium in the water shift reaction:

$$CO \quad + \quad H_2O \quad \Leftrightarrow \quad CO_2 \quad + \quad H_2$$

which is shifted to the right at these temperatures favoring the formation of hydrogen and carbondioxide. This is also seen in the results for the nickel catalyst. As discussed above, the ability of the dolomite to exchange hydrogen radicals between other components may well also be the cause for the activity of dolomite in the water gas shift reaction.

In these tests no significant poisoning effect of the nickel catalyst by hydrogen sulfide is seen, as is also the case in other experiments with the aspen wood used in these tests. At higher levels of sulfur, the nickel will decrease in activity through chemisorbtion of sulphur, up to a point where bulk sulfide formation will occur [5]. This deactivation will be studied more closely in a separate project. A part of this project will however be presented at this conference [6].

When comparing the high activity of the nickel catalyst to the dolomite activity, it must also be kept in mind that the nickel catalysts are specifically developed and optimized for this purpose and are very costly compared to a bulk mineral such as dolomite. A cost benefit analysis would probably result in that the added value of increased tar cracking from nickel, and not including synthesis gas manufacture, is too low.

5 Conclusions

- The new apparatus which has been installed for investigations in high temperature gas cleaning at enhanced pressures has shown to be flexible and easy to operate.

- The semi continuos fuel feeding concept allows a very constant formation of the gas product. This make the test rig an advanced and versatile platform for thermochemical conversion studies at high temperatures and at enhanced pressures.
- When no catalyst is used, the tar yield is decreased when both the temperature and the pressure are increased. This tendency is a result of both thermal decomposition of the tar, and catalytic effects from the reactor material.
- When the Swedish dolomite Glanshammar is used as catalyst the tar yield is decreased more, compared to when no catalyst is used, when the temperaure and the pressure are increased.
- When steam is added to the carrier gas, the nickel catalyst decompose tar to a larger extent than dolomite. This is a result of the character of the nickel catalyst as a steam-reforming catalyst. The gas yields at the experiments when the different catalysts are used show results in the same direction, that is the yield of hydrogen is increased and the methane content is decreased when the nickel catalyst is used, compared to when the dolomite is used. This comparison is based on experiments at atmospheric pressure.
- The massbalances show 6-11% difference at atmospheric pressure, while the differences are a little bit larger at enhanced pressures (16-20% at 20 bar). These larger differences can be explained by more soot formation and recarbonisation of the dolomite at the larger pressure.

6 Acknowledgement

This project was performed within the framework of the European Commission AAIR program, as a cooperation between Deutsche Montan Technologie (Germany), ENEL (Italy), VTT (Finland), KTH (Sweden), and TPS (Sweden). Financing was made available by NUTEK, The Board for Non-Nuclear Research at Studsvik and TPS. The apparatus will be used to study steam reforming of pyrolysis gas hydrocarbons (methane, tars) in the Joule IIx framework with new project partners.

7 References

1. Aldén, H., Espenäs, B.G., and Rensfelt, E., Conversion of tar in pyrolysis gas from wood using a fixed dolomite bed. In Research in Thermochemical Biomass Conversion, ed. Bridgwater, A.V., and Kuester, J.L., Elsevier Applied Science, New York, 1988, pp. 987-1001.
2. Bridgwater, A. V., Catalysis in thermal biomass conversion, Elsevier Applied Catalysis A: General 116 (1994) 5-47
3. Aldén, H., Carlsson, M., Nedbrytningsreaktioner på dolomitkatalysator - försök med modellsubstanser, Studsvik/EP-90/15, TPS Termiska Processer AB, Nyköping, 1990.
4. Aldén, H., Björkman, E., Carlsson, M., Waldheim, L., Catalytic cracking of naphtalene on dolomite, Advances in Thermochemical Biomass Conversion, Conference, Interlaken, Switzerland, May 11-15, 1992

5. Hepola, J., Usability of catalytic gas cleaning in a simplified IGCC power system, Deactivation of Ni/Al_2O_3 catalysts, Literature review, Laboratory of Fuel and Process Technology, Technical Research Centre of Finland, Espoo 1993

6. Berg, M., Waldheim, L., Koningen, J., Sjöström, K., Steam reforming with nickel-based catalysts on gas from biomass gasification, TPS Termiska Processer AB, Studsvik, Nyköping, SWEDEN, Kungl Tekniska Högskolan, Department of Chemical Engineering and Technology, Chemical Technology, Stockholm, SWEDEN

CHARACTERIZATION AND ACTIVITY OF DIFFERENT DOLOMITES FOR HOT GAS CLEANING IN BIOMASS GASIFICATION

A. ORÍO, J. CORELLA, I. NARVÁEZ

Department of Chemical Engineering, University 'Complutense' of Madrid, 28040 Madrid, Spain. Fax + 34-1-394 4164.

Abstract

The aim of this work is to identify if the type, origin or composition of the calcined dolomite has some influence on its activity for tar elimination in a hot flue gas coming from a biomass gasifier, bubbling fluidized bed type. For this purpose four different dolomites from four different quarries and Companies have been studied. Chemical analysis, adsorption isotherms, surface and pore size distributions both with nitrogen and by mercury porosimetry, etc,... have been made for three different samples of each dolomite. Activity tests for fresh tar destruction have been simultaneously carried out for each type of calcined dolomite in a fixed bed of 6 cm i.d. The tar elimination activity of the dolomite and the product distribution from it seem do not depend much on the composition or type of the dolomite used.

1. Introduction

People working on biomass gasification know quite well how calcined dolomites and related materials are active for the upgrading by gas cleaning of the raw gas from biomass gasifiers. In the last ten years research has been made on this subject at, at least, the University of Nancy 1 (France), KTH and TPS AB (Sweden), VTT (Finland), BTG A.V. (The Netherlands), and Universities of Zaragoza and 'Complutense' of Madrid (Spain). A detailed analysis on the work carried out till date on this area has been recently made (1) and will not be reported here. Some conclusions which emerge from the published data are: 1st) it is not known yet if the type or composition of the dolomite has some influence on its tar elimination capacity. For instance, it has been said that the Fe_2O_3 content in the dolomite could affect its activity. 2nd) Some physical properties of the dolomites are not well known and never have been related to their activity for tar elimination or gas upgrading in biomass gasification. The work here presented will intend to solve at least these two questions.

The overall mechanism for tar elimination over calcined dolomites is not well known yet but it includes reactions of steam reforming, steam cracking, thermal cracking, etc... (2). The activity of the calcined dolomite can depend not only on its physico-chemical properties but also on the reacting gas atmosphere. The content of steam, H_2,

... in the flue gas can have some effect on the overall kinetics for tar elimination. On the other hand, the tar composition can also have some effect on the kinetics of its disappearance. Both gas and tar compositions depend on how they were generated, that it is to say: on the gasifier design and operating conditions, like the equivalence ratio used (ER). Besides, gasifying with air does not produce the same tar and gas composition than when gasification is made with steam or (steam + O_2) mixtures. This communication will be only referred to a gasification process with air. Similar studies for gasification with steam or with (steam + O_2) mixtures will be presented elsewhere.

2. Types of dolomites studied

2.1 Origin (quarry) and chemical composition.
In Spain there are four different Companies producing or selling dolomites. Some of them have in turn several quarries under production. The quarries indicated are very distant, at least 300 km, between themselves. These Companies periodically (daily sometimes) made chemical analysis of their dolomites. Samples of about 100 kg each were obtained from such four companies. The averaged chemical analysis of such dolomites are shown in Table 1, together with a magnesite also here used. The geological characterization of these dolomites is also known but it is here omitted because does not give us some more useful information.

The main difference on their chemical composition can be their content in Fe_2O_3. Malaga and Sevilla dolomites have nothing Fe_2O_3, they are white. Dolomite from Norte has some amount of Fe_2O_3 (0.12 wt%) and are light brown colour. Chilches dolomite has a high content in iron (0.77 wt%) and, correspondly, it is dark brown-grey colour.

Table 1.- Chemical characterization of the dolomites used.

Origin (wt %)	Norte (Bueras, Cantabria)	Chilches (Peña Negra, Castellón)	Malaga (Coín)	Sevilla (Gilena)
CaO	32.2	29.7-31.3	30.6	30.5
MgO	18.7	17.5-19.0	21.2	21.5
Fe_2O_3	0.12	0.74-0.80	0.01	0.01
CO_2	45.5	47.4	47.3	47.2
Al_2O_3	0.06	1.19	0.40	0.60
MnO	0.09	0.04	n.d.	n.d.
Na_2O	0.01	0.05	n.d.	n.d.
K_2O	0.01	0.24	n.d.	n.d.
SiO_2	3.3	3.2	n.d.	n.d.

3. Physical characterization. Pore structure.

Since raw dolomites are heterogeneous minerals their correct sampling for further analysis is essential. Three samples of about 1 kg each one were taken for each dolomite, once crushed and sieved till -2.0 +1.0 mm. Such samples were calcined in an oven at near 0% CO_2, during 1h at 900°C. This temperature was measured in the surrounding air not on the dolomite bed. The (fixed) bed in the oven could have some temperature gradients which could produce further differences between the dolomite particles. Once calcined, they were kept in closed vessels to avoid their carbonation with the CO_2 in the air.

A few grams of each one of the three samples of each dolomite were taken for analysis by N_2 adsorption (in an ASAP 2000 apparatus) and by mercury porosimetry (in an 9320 Poresizer apparatus). The remaining solid was used for testing its chemical activity for fresh tar elimination. Due to the fact that this paper is only focused to this aspect only, these results will be shown in detail. A typical adsorption isotherm, for the calcined Norte Dolomite, is shown in figure 1. The low value for the saturation plateau already indicates a low surface area. From the right part of the figure it is deduced that the pores are cylindrical and in parallel.

The pore volume distribution (before its using) are shown in figures 2, 3, 4 and 5 for a sample of 'fresh' calcined dolomite from Sevilla, Malaga, Norte and Chilches, respectively. Sevilla and Malaga dolomite have here the same pore distribution (average pore diameter, dp, of about 900Å), but Chilches and Norte have still bigger macropores (dp ≈ 2000 - 4000Å).

Pore size distributions of calcined dolomites once used in biomass gasification experiments are shown in figure 5. They have a similar pore distribution but with pores somewhat smaller in diameter, indicating than some coke has been deposited on the pore walls decreasing something thus its diameter.

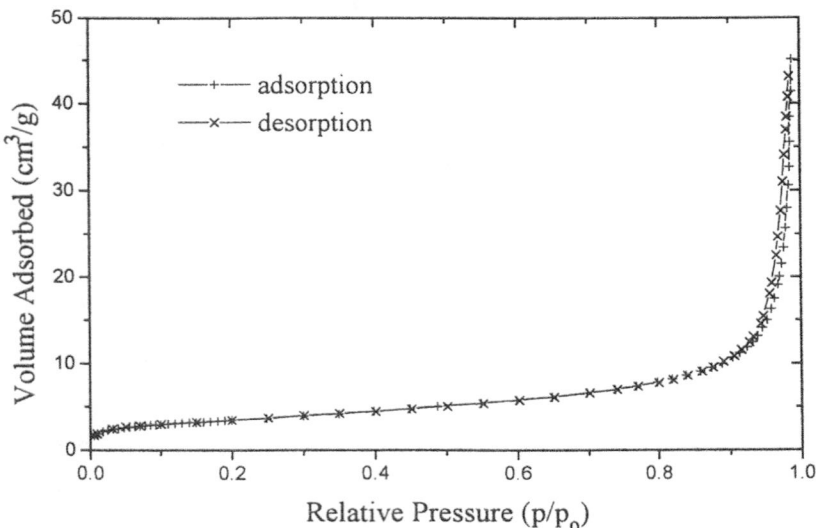

Figure 1.- Adsorption-desorption isotherm for the calcined Norte dolomite.

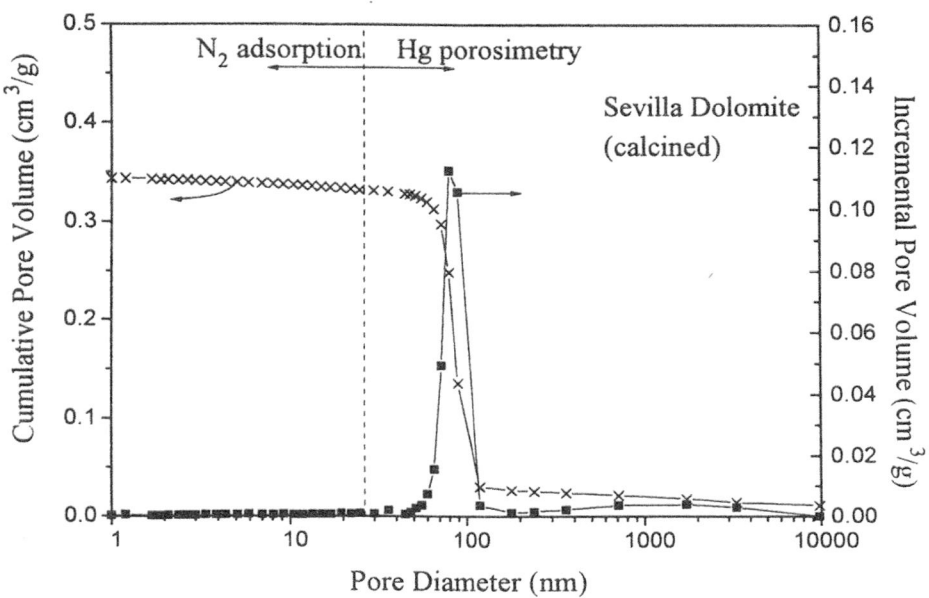

Figure 2.- Pore volume distribution for the calcined Sevilla dolomite. (Calcination conditions: T = 900°C, t = 1h, 0% CO_2, dp = -1.6 +1.0 mm)

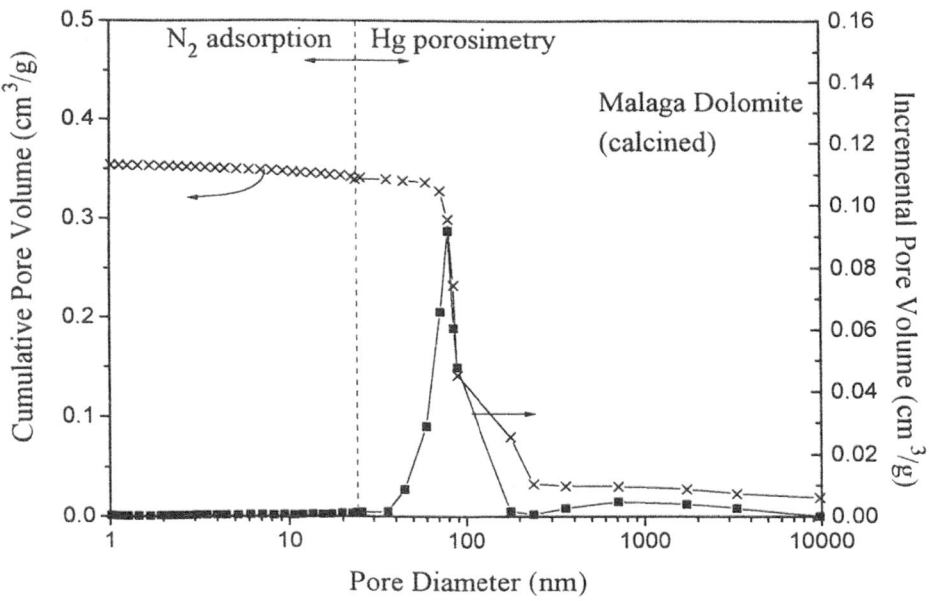

Figure 3.- Pore volume distribution for the calcined Malaga dolomite. (Calcination conditions: T = 900°C, t = 1h, 0% CO_2, dp = -1.6 +1.0 mm)

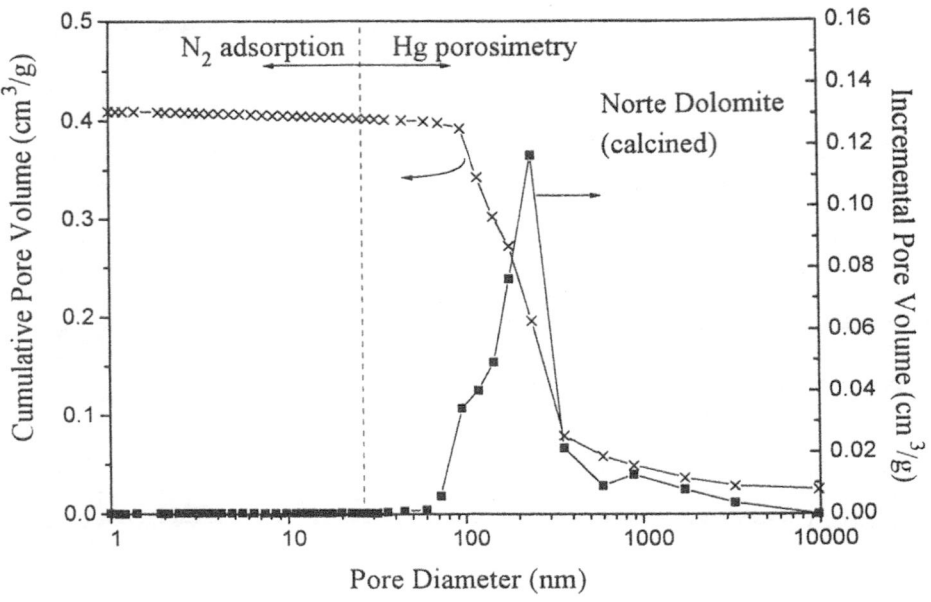

Figure 4.- Pore volume distribution for the calcined Norte dolomite. (Calcination conditions: T = 900°C, t = 1h, 0% CO_2, dp = -1.6 +1.0 mm)

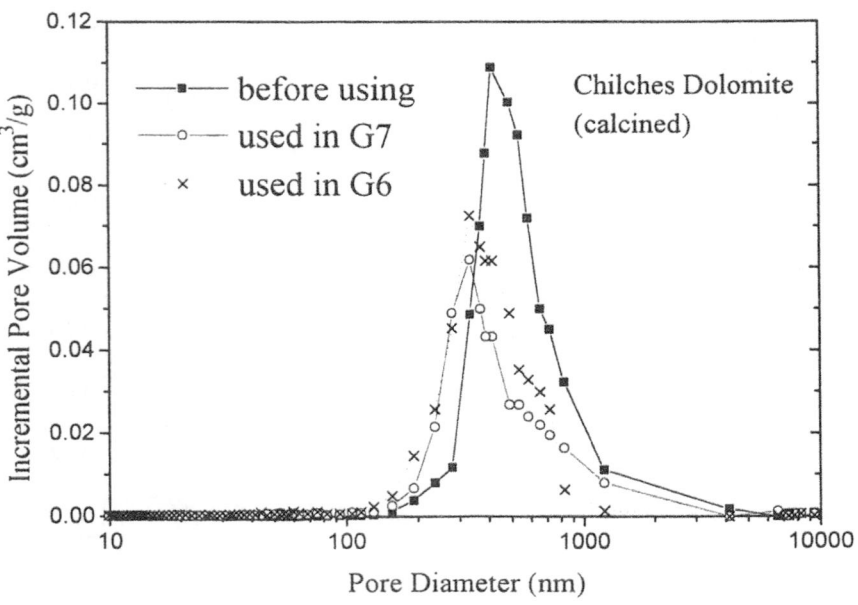

Figure 5.- Pore distribution (by Hg porosimetry) for the calcined Chilches dolomite. (Calcination conditions: T = 900°C, t = 1h, 0% CO_2, dp = -1.6 +1.0 mm)

The pore area distribution is also important for characterization of these solids. The incremental pore area distributions are shown in figures 6 and 7 for a sample of each one of the four dolomites. Again the Malaga and Sevilla dolomites are very similars. Macropores of around 900Å give the most important contribution to the pore area, but pores of about 20-40Å are also present in these solids and the micro and 'meso' porosities (till 500-600Å) have some importance: they mean the 50% of the overall pore area.

Chilches and Norte dolomites are somewhat different from the two other dolomites. They seem to have big macropores of two different sizes (diameters). Pores of 15-50Å also exist, and micro and 'meso' porosities produce the 50% of the total pore area.

Figures 1-7 were determined for each one of the 3 samples for each dolomite. Main results from them are summarized in Table 2.

Some important conclusions are: 1st) There can be a difference of about 50% in results from one sample to another one due to the facts that these solids are minerals, with some heterogeneity, and at the calcination temperature can be different for each part of the sample. 2nd) The macro porosity and the big pores of 800-3000Å are the most important ones. 3rd) The micro (below 17Å) porosity is near not existing. 4th). There are not major differences between the four dolomites excepting the Chilches one which has a clear lower BET surface area (3.9 m^2/g) instead of the 11-12 m^2/g of the Sevilla and Malaga ones. The Norte one has an intermediate value. 5th) A low surface area is associated with large pores: for the Chilches dolomite its mean pore diameter is the biggest one ($\approx$3000Å) instead of 800-1600Å of the other here tested dolomites.

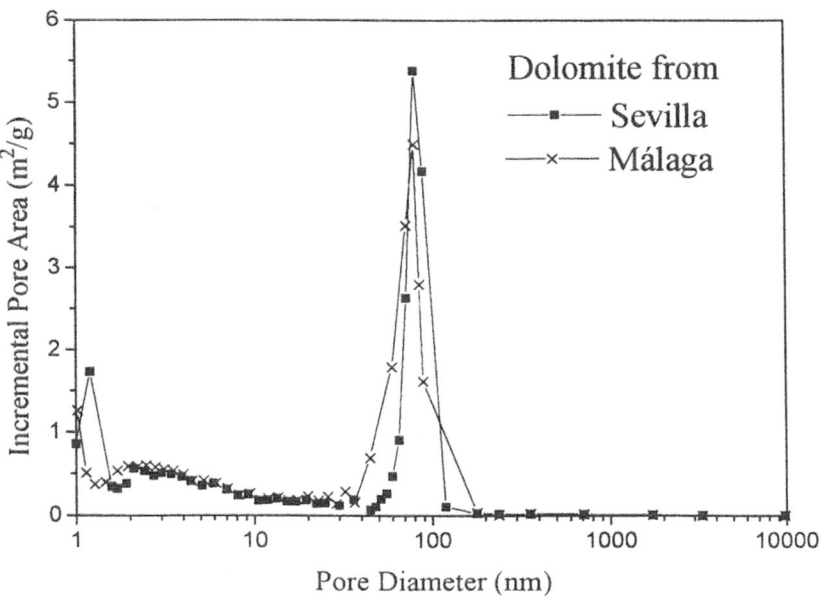

Figure 6.- Pore area distribution for calcined dolomites from Sevilla and Málaga. (Calcination conditions: T = 900°C, t = 1h, 0% CO$_2$, dp = -1.6 +1.0 mm).

Table 2.- Averaged parameters for the pore structure of the four types of the fresh dolomite once calcined 1h at 900°C.

		Norte	Chilches	Málaga	Sevilla
	BET Surface Area	m²/g	6.6	3.8	12	10.8
	Micropore Area	"	0.46	0.75	0.91	0.82
N_2	Area 17-3000 Å	"	7.5-8.7	3.4-3.9	18-20	18-20
ADSORPTION	Pore Volume 17-3000Å	cm³/g	0.024	0.009	0.070	0.069
	Micropore Volume	"	$3.1\ 10^{-4}$	$3.7\ 10^{-4}$	$4.9\ 10^{-4}$	$3.5\ 10^{-4}$
	Average Micropore Diameter	Å	95	74	144	154
	Pore Area	m²/g	9.9	5.4	16.1	17.7
Hg	Total Pore Volume	cm³/g	0.41	0.40	0.34	0.34
POROSIMETRY	Average Pore Diameter (4V/A)	Å	1630	2940	850	760
	Bulk density	g/cm³	1.35	1.38	1.52	1.51
	Apparent (skeletal) density	g/cm³	2.98	3.03	3.16	3.09

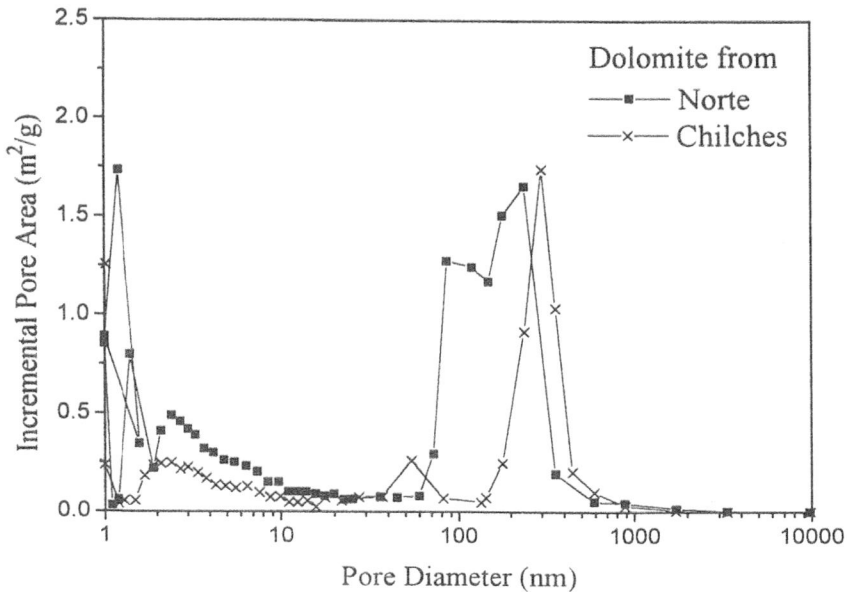

Figure 7.- Pore area distribution for calcined dolomites from Norte and Chilches. (Calcination conditions: T = 900°C, t = 1h, 0% CO_2, dp = -1.6 +1.0 mm).

4. Activities of the dolomites for tar elimination and reforming of the gas composition

4.1 Experimental conditions.

The dolomites were used in a fixed bed of 6.0 cm i.d. downstream the biomass gasifier. So, they received a fresh and hot tar there generated. The biomass gasifier was a bubbling fluidized bed of 6 cm i.d. continuously fed at 9-20 g biomass/min by three screws near the bed bottom. Some experimental conditions of the gasifier bed are indicated in table 3 for some typical experiments.

The raw gas generated in the biomass gasifier was first passed through a hot metallic filter at 500-600°C. All the gas flow rate entered to the bed of the calcined dolomite in which there was a previous bed of an 'inert' material (silica stones) for preheating purposes, as figure 8 shows. This secondary or downstream reactor, or bed of dolomite, was externally heated by an oven. The gas was sampled for tar and gas analysis before and after this bed of dolomite at different times-on-stream. Some experimental conditions in this bed of dolomite for some typical experiments are indicated in table 4.

To avoid its erosion if fluidized, the dolomite was used as fixed bed which is not isothermal. So, their axial and longitudinal temperature profiles were carefully measured in each experiment with two mobile thermocouples located in the bed axis and in the wall (inside). An example of the axial temperature profile is shown in figure 8, indicating important differences of temperature in the bed. The temperature here used as reference will be the averaged in the bed axis from its bottom to its top.

Table 3.- Main experimental conditions and results from the gasifier. (Some typical and representative experiments).

Run number		18	19	20	21	22
ER_1 (bed)		0.52	0.26	0.35	0.44	0.33
ER_1 (freeb)		0.000	0.000	0.000	0.000	0.000
H/C	mol/mol	1.61	1.65	1.60	2.13	2.11
O/C	mol/mol	1.78	1.30	1.44	1.88	1.67
gas composition at the bed inlet, vol %						
N_2		75.7	71.8	74.3	67.6	64.8
O_2		20.1	19.1	19.7	18.0	17.2
H_2O		4.2	9.1	6.0	14.5	17.9
$(H_2O/C)_{1,0}$	mol/mol	0.12	0.13	0.11	0.37	0.37
$T_{1,c}$	°C	730	800	800	810	810
$T_{1,freeb}$	°C	550	555	530	505	500
solid in bed		silica sand	silica sand	silica sand	silica sand	silica sand
dp_1	μm	-500+320	-500+320	-500+320	-500+320	-500+320
$H_{1,0}$	cm	20.0	20.0	20.0	20.0	20.0
u_{mf}	cm/s	10	10	10	10	10
$u_{1,0}$	cm/s	28	34	32	36	28
$u_{1,e}$	cm/s	33	44	46	46	38
$\tau_{1,0}$	s	0.65	0.52	0.51	0.49	0.60

RESULTS

gas composition at the exit, vol % (dry gas basis)						
H_2		6.5	7.0	9.5	8.0	13.6
CO		11.5	15.0	16.5	11.0	11.8
CO_2		13.0	13.0	14.0	16.0	12.8
CH_4		3.0	2.7	3.0	2.3	2.4
C_2H_4		0.0	1.3	1.2	1.1	0.9
N_2		66.0	61.0	55.8	61.6	58.5
H_2O (wet gas)*		9.5	6.6	8.2	14.9	9.5
$C_{tar,1}$	mg/Nm^3	16900	28030	3700	7200	3010
LHV_1	MJ/Nm^3	3.3	4.4	4.9	3.7	4.4
$Y_{tar,1}$	g/kg daf fuel	49.5	45.3	8.57	19.1	6.39
$Y_{gas,1}$	Nm^3/kg daf fuel	2.93	1.62	2.32	2.66	2.13

* from mass balances

Table 4.- Main experimental conditions and results from the downstream bed of dolomite.

Run number		18	19	20	21	22
solid in bed		Sevilla Dol.	Sevilla Dol.	Sevilla Dol.	Sevilla Dol.	Malaga Dol.
W_2 (calcined)	g	612	612	612	300	300
ER_2		0	0	0	0	0
$T_{2,c}$	°C	818	814	805	875	844
dp_2	mm	(-2 +1.6)	(-2 +1.6)	(-2 +1.6)	(-2 +1.6)	(-2 +1.6)
$H_{2,0}$	cm	30	30	30	15	15
u_{mf}	cm/s	55	55	55	55	55
$u_{2,0}$	cm/s	38	46	49	52	41
$u_{2,s}$	cm/s	52	48	63	54	48
$\tau_{2,0}$	s	0.66	0.64	0.54	0.28	0.34
$\tau'_{2,0}$	kg/(m³,wet/h)	0.132	0.128	0.108	0.056	0.066
$SV_{2,0}$	h⁻¹	5400	5600	6700	12700	10700
$(H_2O/C)_{2,0}$	mol/mol	0.38	0.21	0.25	0.55	0.36
RESULTS						
gas composition at the exit (dry gas basis)						
H_2	vol %	12.5	11.2	18.0	12.0	18.6
CO	vol %	15.5	13.5	15.0	10.5	13.5
CO_2	vol %	21.0	12.5	20.2	16.0	14.0
CH_4	vol %	2.8	2.4	2.4	1.6	2.4
C_2H_4	vol %	0.0	1.0	1.0	0.8	0.7
N_2	vol %	48.2	59.4	43.4	59.1	50.8
H_2O (wet gas)*		11.4	3.7	3.9	12.3	8.7
$C_{tar,2}$	mg/Nm³	2140	1040	580	1000	200
LHV_2	MJ/Nm³	4.4	4.4	5.3	3.7	5.0
$Y_{tar,2}$	g/kg daf fuel	8.6	1.7	1.7	2.8	0.49
$Y_{gas,2}$	Nm³/kg daf fuel	4.0	1.7	3.0	2.8	2.4
X_{tar}	%	87.3	96.3	84.3	86.1	93.3

* from mass balances

4.2. Some results on gas cleaning

The gas yield (Nm³/kg biomass daf) before and after the bed of dolomite is shown in figure 9 for several operating conditions (ER) of the gasifier. Note that the gas yield at the secondary bed inlet is roughly the one at the gasifier exit. A clear increase (of about a 10-20%) in this yield is observed due to the conversion of big molecules (like the ones of the tar) to several smaller others (CO, H_2, CH_4,...). From results shown in figure 9 there seems to have no effect of the type/origin of the dolomite on this variable, gas yield.

The low heating value (LHV) of the gas before and after the bed of dolomite is shown in figure 10 at different equivalence ratios of the gasifier. Though the experimental conditions in the bed of dolomite were not always the same, there is always an increase (of about a 15%) on the LHV of the gas and this increase seems do not depend on the origin of the dolomite.

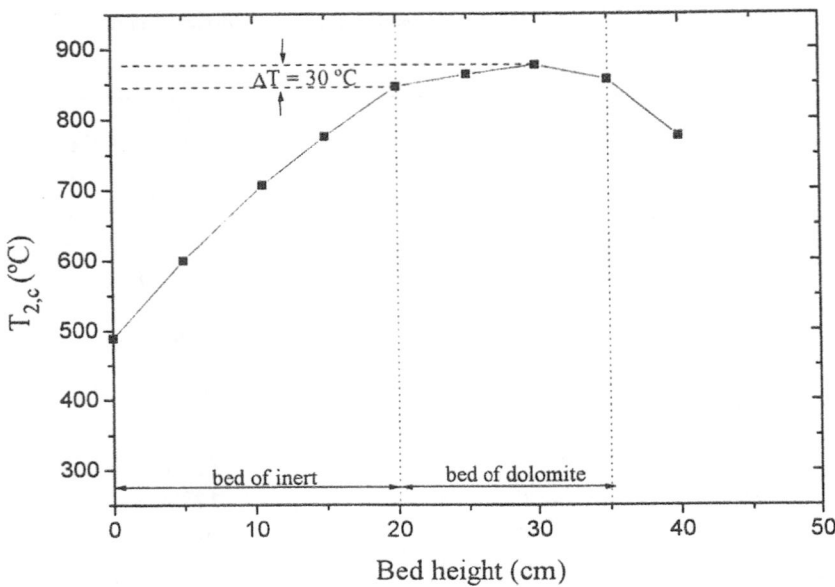

Figure 8.- Axial temperature profile in the bed of dolomite, from Málaga, during the
reaction. Run no G39.

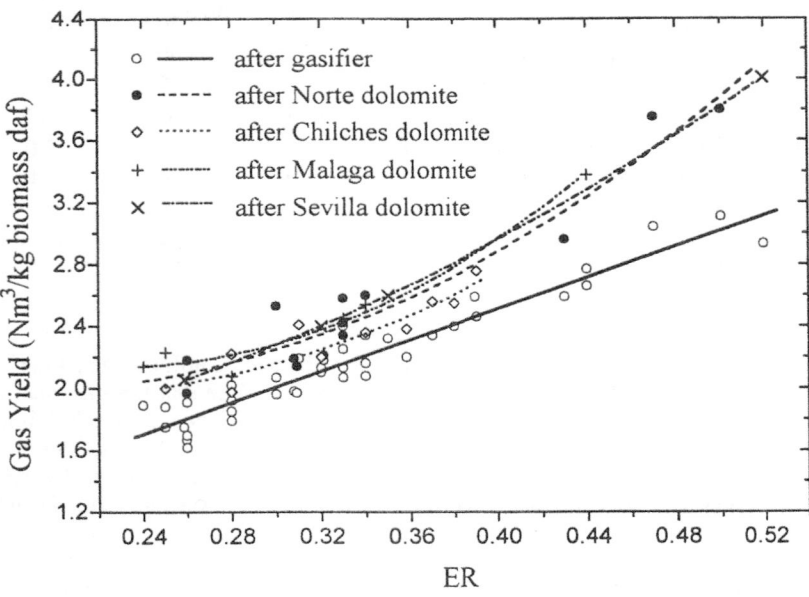

Figure 9.-Effect of the type of dolomite on the gas yield at different equivalence ratios.

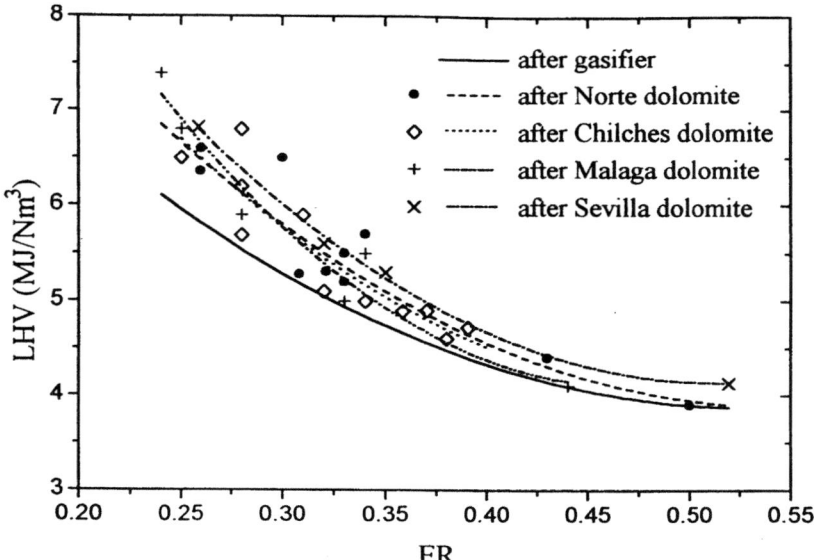

Figure 10.- Effect of the type of dolomite on the low heating value of the gas at different equivalence ratios.

Tar conversions (X_{tar}) for several space-times (τ) in the bed of dolomite are shown in figure 11. From these values an index for the activity of the dolomite has been calculated. This activity index is the apparent kinetic constant (k_{app}) for tar elimination supposing first order for the overall tar elimination reaction. Having worked at conditions close to piston flow, a mass balance gives:

$$-\ln (1-X_{tar}) = k_{app} \, \tau$$

This equation is used to calculate k_{app} for the secondary reactor or bed of dolomite ($k_{2,app}$). $k_{2,app}$ values so calculated are shown in figure 12 according to the Arrhenius law for the four dolomites here tested.

Results in figures 11 and 12 seem to indicate how the origin or type of dolomite affects something its activity for the tar elimination reaction. The Chilches and Norte dolomites seem to be something more actives. They are the dolomites with the biggest pore diameters and iron contents (tables 1 and 2). So, these variables seem to have some effect on the activity for the tar elimination. But notice also how: 1st) The tested dolomites had not a big difference between themselves to show an important difference on chemical activity. 2nd) Some possible minor or second-order (not more) effects of the type of dolomite can also be here masked by the not isothermicity of the fixed bed used. That means that the experimental error has an order of magnitude similar to the small differences in activity. To summarize, the most important conclusion of this work is perhaps that the type of dolomite, between the ones here used, of course, has only a minor effect on its activity for tar elimination in a gas from a biomass gasifier. And this small and positive effect can come from big pores or from high contents of iron in the dolomite.

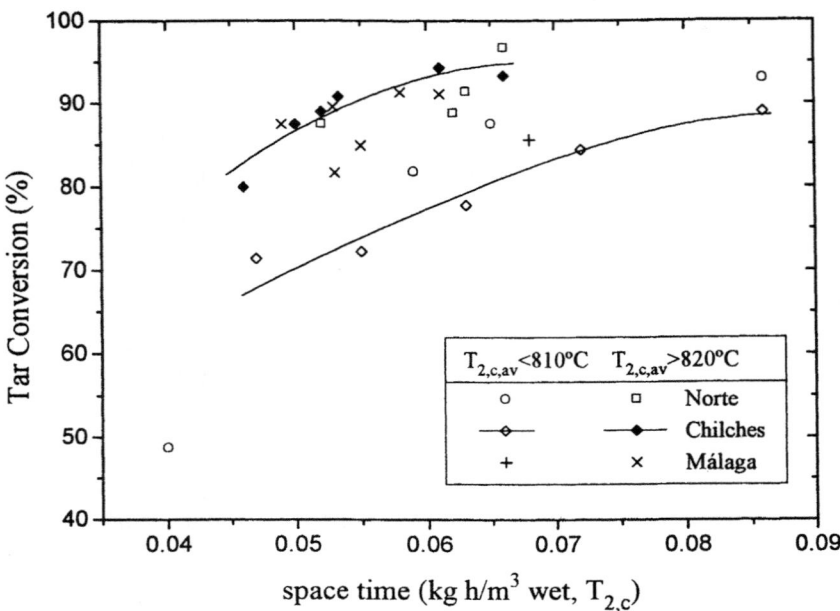

Figure 11.- Tar conversion vs. space time in the second reactor, for different types of dolomite at different temperatures.

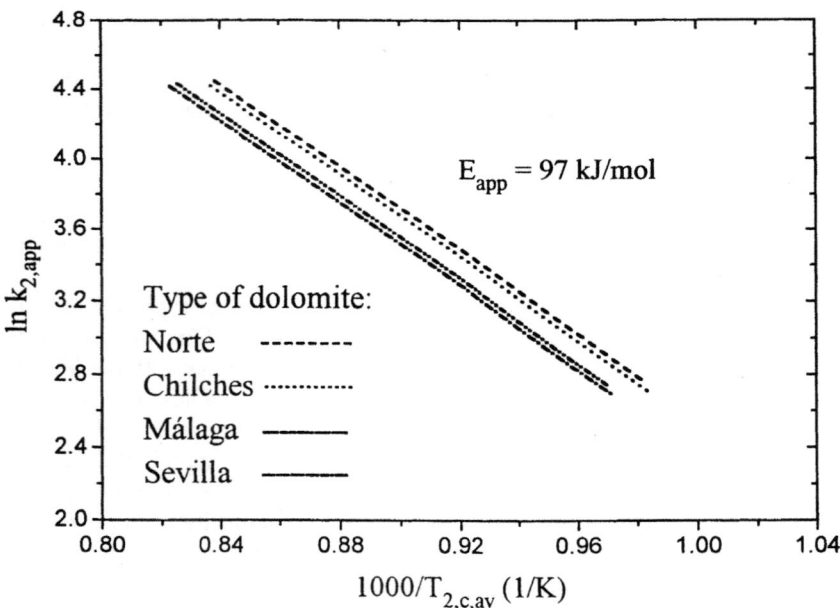

Figure 12.- Arrhenius representation for the different types of dolomite here tested.

5. Acknowledgments

This work has been made thanks to the financial support of the EU, DGXII, through the project of the Agroindustry Programme no. AIR2-CT93-1436. We also express our gratitude to J. Talamantes of 'Cales de la Plana, S.A.', to H. Varona of 'Dolomitas del Norte S.A.', to J. Macias of Prodomasa, and to 'Minera del Santo Angel S.L.' de Gilena (Sevilla) for the provision of samples of their dolomites.

6. References

1. Delgado, J.; Corella, J.; Aznar, M.P. Dolomites, Magnesites and Calcites for Raw Hot Gas Cleaning from Biomass Gasifiers. *Ind. Eng. Chem. Res.*, **1996**, in press.

2. Corella, J.; Narváez, I.; Orío, A. Criteria for Selection of Dolomites and Catalysts for Tar Elimination from Biomass Gasification Gas; Kinetic Constants. In *New Catalysts for Clean Environment*; Maijanen, A.; Hase, A., Eds.; Ed. by Technical Research Centre of Finland (VTT) in Espoo, Finland, **1996**, 177-183.

ASSESSMENT OF COPRECIPITATED NICKEL-ALUMINA CATALYSTS FOR PYROLYSIS OF BIOMASS

Biomass catalytic pyrolysis

L. GARCIA, J.L. SANCHEZ, M.L. SALVADOR, R. BILBAO and J. ARAUZO
Chemical Engineering & Environmental Department, University of Zaragoza, Zaragoza, Spain

Abstract
Various catalysts are being developed and tested for the pyrolysis of biomass using the Waterloo Fast Pyrolysis Process (WFPP) technology. The present paper describes pyrolysis tests with wood in a continuous bench scale fluidized bed reactor at 650 °C and at short gas contact times. A crystalline nickel aluminate catalyst of spinel structure prepared by coprecipitation is used.

Different techniques have been employed in order to characterize the catalysts. These are: X-Ray Diffraction (XRD), Temperature Programmed Reduction (TPR), Gas Adsorption, Mercury Porosimetry, Thermogravimetric analysis, elemental analysis, and X-Ray Photoelectron Spectroscopy (XPS).

The tar produced in the process can be significantly reduced using this kind of catalyst and the main objective is to develop a catalyst resistant to carbon deposition. This paper presents the results of the catalyst deactivation and the product distribution data using catalysts prepared with different calcination temperatures and activation conditions (time and hydrogen flow). The calcination temperature of the catalyst precursor influences the characteristic of the catalyst and the product distribution obtained. An increase in the calcination temperature implies more severe activation conditions.

Keywords: biomass, catalyst, coprecipitated, pyrolysis, nickel-alumina.

1 Introduction

Biomass gasification at low temperature is of great interest from an energy point of view, because the input of energy necessary for the process decreases and higher efficiencies can be achieved. The problem of working at low temperatures is that significant amounts of tars can be produced and the char reactivity decreases.

The use of catalysts is a posible solution to avoid these problems without an increase of temperature. These catalysts can be both primary and secondary.

The primary catalysts, mainly alkali carbonates, increase the rate of biomass char gasification. Such catalysts have been used in coal gasification, but due to the high reactivity of biomass char, their influence on biomass gasification is not so significant as it is for coal [1].

The function of the secondary catalysts is the transformation of tars which are obtained from primary pyrolysis. A suitable catalyst is a nickel catalyst supported on Al_2O_3 [2]. There are other metals (eg. Pt, Ru, Rh, ...) as active as Ni, but Ni is cheaper and sufficiently active. The Al_2O_3 acts as a support to encourage high metal dispersion and as an acid catalyst to crack tars.

The use of several Ni-Al_2O_3 catalysts presents good initial results for biomass gasification, but they reveal a problem of deactivation mainly caused by carbon deposition. The deposition of pyrolytic carbon is the most important cause of the deactivation, but, as a result of the nickel action, polymeric and whisker carbon can also be found on the catalyst.

In this context, the development of an adequate catalyst has been planned and a coprecipitated nickel-alumina catalyst selected. By calcinacing this precursor, spinels of nickel aluminate can be formed. Two types of nickel sites have been reported for these catalysts: monodispersed nickel atoms associated with aluminum atoms and metallic nickel resulting from the reduction of nickel oxide [3]. A suitable combination of these site types would allow us to obtain a catalyst of good activity and low deactivation.

A study of the behaviour of these catalysts in the biomass pyrolysis is very useful because pyrolysis is a previous process to gasification.

2 Experimental

2.1 Experimental System
The experimental system built makes use of the WFPP technology, which has been described previously [4] [5]. A scheme of the installation is shown in Figure 1.

This system is composed of a continuous biomass feeder capable of giving biomass rates of 10-100 g/h [6]. The sawdust is mechanically stirred and fed into the bottom of the reaction bed.

The gasifier, which has an internal diameter of 2.54 cm, operates under fluidized conditions using nitrogen as the fluidizing gas.

The product gas is cleaned by being passed through a cyclone, condenser traps and a cotton filter. The amount of tar condensed is determined by weight. After the cleaning system, the gas flowrate is determined using a dry test meter, and CO and CO_2 compositions are continuously analyzed. Moreover, gas samples are taken at regular time intervals and analyzed by chromatography. Reaction temperature, total flow gas and CO and CO_2 concentrations are registered by a data acquisition system.

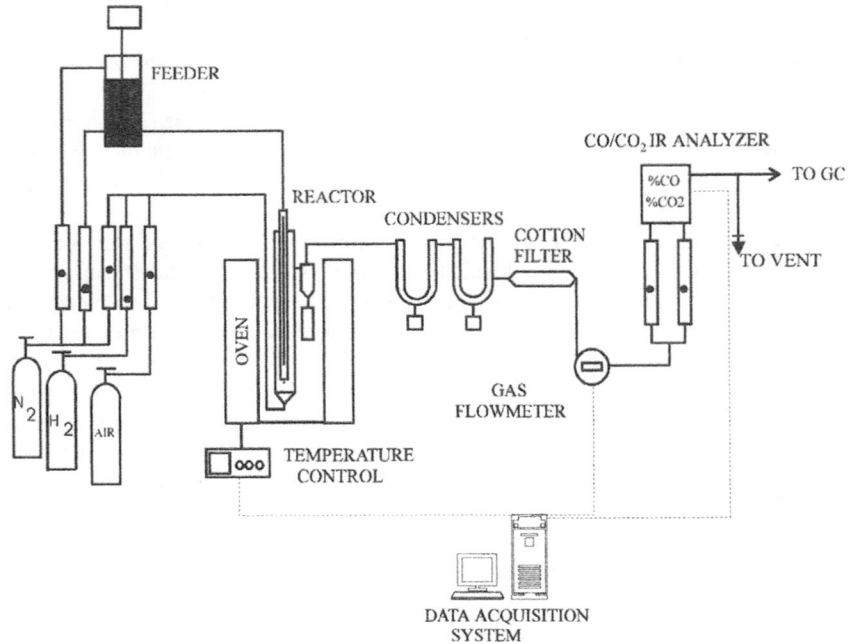

Fig. 1. Schematic drawing of the experimental system.

2.2 Catalyst preparation and characterization techniques

The catalysts used have a molar ratio 1:2, (Ni:Al). These precursors have been prepared by coprecipitation from a solution of 600 ml of distilled water with Ni $(NO_3)_2$ $6H_2O$ and Al$(NO_3)_3$ $9H_2O$ in appropriate proportions. This solution is kept at 40 °C and moderately stirred. Ammonium hydroxide is added until the pH reaches 7.9. The precipitate obtained is filtered and washed with 400 ml of distilled water at 40 °C, and dried for about 15 hours at 105 °C [3].

The calcination is carried out in an air atmosphere. The heating rate is low, particulary at the temperature range around 300 °C for which a maximum weight loss is observed. Satterfield [7] states that the catalyst should be heated under controlled conditions to a temperature at least as high as the reaction temperature. Other authors [8] suggest that the optimal calcination treatment involves slow heating with the removal of the gases. In our work the final temperatures of the calcination process were 650, 750 and 850 °C for 3 hours.

The calcined catalyst is reduced in the reactor using hydrogen just before the pyrolysis is performed. Two different hydrogen flow rates (1740 and 3080 Ncm^3/min) and reduction times (1 and 2 h) were used.

Characterization of the catalyst was carried out using different techniques: atomic emission spectrometry by inductively coupled plasma (ICP), gas adsorption, mercury porosimetry, thermogravimetric analysis, X-ray Diffraction (XRD), Temperature Programmed Reduction (TPR) and X-ray Photoelectron Spectroscopy (XPS).

2.3 Biomass

Pine sawdust with a moisture content of about 10 % and a particle size of -350 +150 μm has been used.

The results of the elemental analysis of pine sawdust are shown in Table 1.

Table 1. Elemental analysis of pine sawdust

element	%wt
carbon	48.27
hydrogen	6.45
oxygen*	45.19
nitrogen	0.09
sulphur	not detected

*Oxygen was calculated by difference.

3 Results of catalyst characterization

The analysis by atomic emission spectrometry (ICP) has shown an elemental composition of the catalyst precursor before calcination in accordance with the formulated stoichiometry.

The results obtained from the thermogravimetric analysis indicate that the weight loss produced during the calcination occurs following two steps. In the first step, at about 114 °C, molecular water is lost from the interlayer of the layer structure. In the second, at about 298 °C, the layer structure decomposes with the evolution of nitrogen oxides. In this latter step the maximum decompositon takes place. The results are similar to those obtained by Alzamora et al. [8].

The surface area of the catalyst after calcination has been determined. Nitrogen adsorption has been used to obtain these values. Isothermal data are fitted using the BET equation. The results of this analysis are shown in Table 2. As can be observed, increasing the calcination temperature decreases the surface area, in agreement with the results of other authors [8].

The results obtained by mercury porosimetry, Table 3, indicate that the total pore area decreases when the calcination temperature is increased. The pores existing in the catalyst are small in size 0.0044-0.0048 μm and similar values have been obtained for different calcination temperatures.

X-ray diffraction of the calcined catalysts have been also carried out. The spectra of XRD catalysts calcined at 650 °C and 850 °C for 3 hours are presented in Figures 2 and 3. The comparison between these spectra and the NiO and $NiAl_2O_4$, (spinel) standards shows a good accordance.

Table 2. Results of surface area using nitrogen adsorption.

Calcination temperature (°C)	Surface area (m^2/g)
650	185
750	154
850	124

Table 3. Results of catalyst characterization by mercury porosimetry

Calcination temperature (°C)	Total pore area (m²/g)	Median pore diameter (μm)
650	210	0.0044
750	188	0.0044
850	146	0.0048

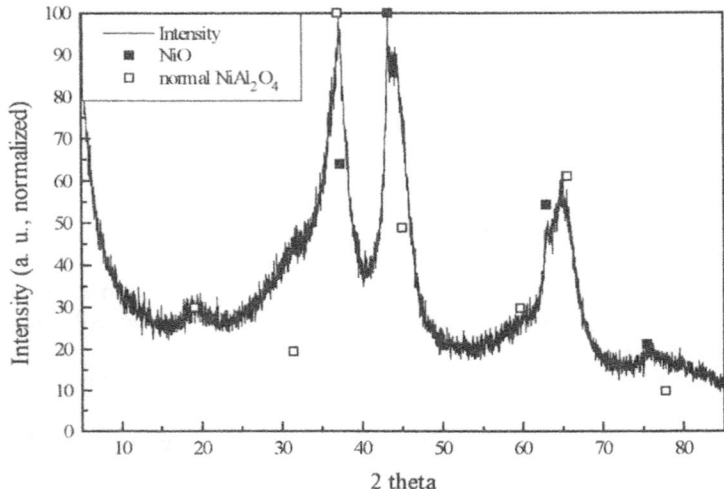

Fig. 2. XRD spectrum of a catalyst calcined at 650 °C.

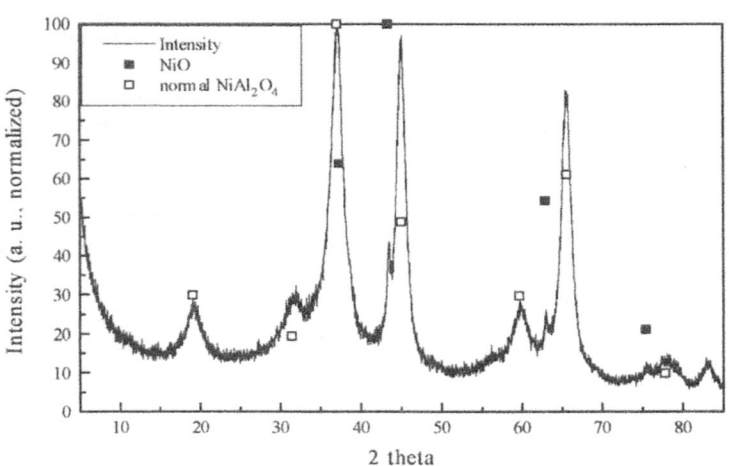

Fig. 3. XRD spectrum of a catalyst calcined at 850 °C.

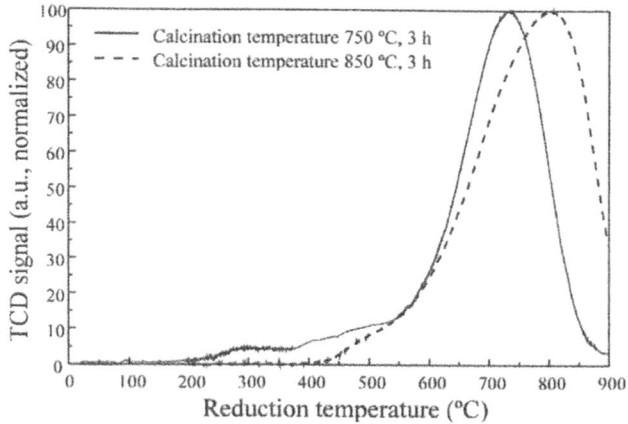

Fig. 4. Graphics of TPR for calcined catalysts at different temperatures.

When the calcination temperature is increased, the sample crystallinity increases, obtaining less nickel oxide phase and more spinel phase. These results agree with others presented in the bibliography [3]. The material calcined at relatively low temperatures contains nickel oxide rich crystallites with dissolved aluminum ions, but at higher calcination temperatures, a phase separation to pure nickel oxide and nickel aluminate is found [8]. The temperature value at which this phase separation occurs depends on the nickel content of the catalyst.

The reducibility of the calcined catalyst has been studied using the temperature programmed reduction technique. The results show that the temperature at which a maximum reduction is observed increases when the calcination temperature increases. The results obtained with catalysts calcined at 750 and 850 °C are presented in Figure 4. For a catalyst calcined at 750 °C, the temperature at which the maximum reduction velocity is observed is about 720 °C and the reduction is completed at 900°C. For a catalyst calcined at 850 °C, the maximum reduction appears at about 820 °C, the reduction not being completed at 900 °C.

The results of X-ray Photoelectron Spectroscopy (XPS) present a similar high atomic ratio for catalysts calcined at different temperatures, 750 and 850 °C. This suggests that nickel is well dispersed in the support surface. The atomic ratio Ni/Al shows a value close to the formulated ratio.

4 Results of catalytic pyrolysis

In order to evaluate the effect of the use of catalysts in the pyrolysis process, preliminary experiments without a catalyst have been carried out at a temperature of 650 °C. The reaction bed consisted of inert silica. The results obtained for experiments lasting between 0.25 and 0.50 hours are presented in Table 4. In these experiments tar formation is observed from the beginning of the experiments.

Table 4. Product distribution for a pyrolysis experiment without catalyst.

Liquid (tars)	46%
Gas	32%
Char	20%
H_2/CO	0.2- 0.4
CO/CO_2	4 - 5
$CH_4/(CO+H_2)$	0.15

Table 5. Product distribution data with catalyst at 650 °C.

Run	1	2	3	4	5
Catalyst calcination and reduction conditions					
Calcination temperature (°C)	750	750	850	750	850
Activation temperature (°C)	650	650	650	650	650
Hydrogen flow rate (Ncm³/min)	1740	1740	1740	3080	3080
Activation time (h)	1	2	1	1	1
Pyrolysis results					
Reaction time (h)	2.00	1.98	1.61	1.82	2.04
Sawdust feeding rate (g/h)	15.34	16.19	16.30	18.64	15.91
Results					
Liquid (tars)	8.3 %	3.4 %	23.4 %	26.8 %	11.0 %
Char	11.4 %	12.0 %	9.2 %	8.2 %	8.9 %
Gas	75.5 %	81.2 %	65.2 %	60.7 %	78.9 %
H_2/CO	1.28	1.15	0.84	1.06	1.00
CO/CO_2	9.75	17.40	7.52	7.60	10.33
$CH_4/(CO+H_2)$	0.01	0.001	0.04	0.03	0.03

Experiments of wood pyrolysis using catalysts have been performed at 650 °C. The catalyst weight was 20 g in each experiment.

The influence has been studied of the calcination temperatures and reduction conditions on the products obtained. Table 5 shows the products obtained in several experiments.

It is also interesting to analyse the variation of the products obtained along the reaction time. The sawdust feeding rate (g/h) has not been the same in all the experiments. In order to compare results corresponding to different experiments the ratio of sawdust processed/catalyst in the reactor (g/g) has been used.

4.1. Influence of reduction time

In general, (Runs 1 and 2, Table 5) lower tar yields and higher yields to gas are obtained with a catalyst reduction time of two hours rather than one. In the same way, higher ratios of CO/CO_2 are produced with a two hour reduction time.

Figure 5 presents the variation of the gas yield for different ratios of sawdust weight processed and catalyst weight in the reactor. Higher yields to gas are obtained when the reduction time increases. In each experiment, a decrease in the gas yields over the reaction time is observed. Figure 6 shows the H_2/CO ratio, being higher for the catalyst reduced for 2 hours.

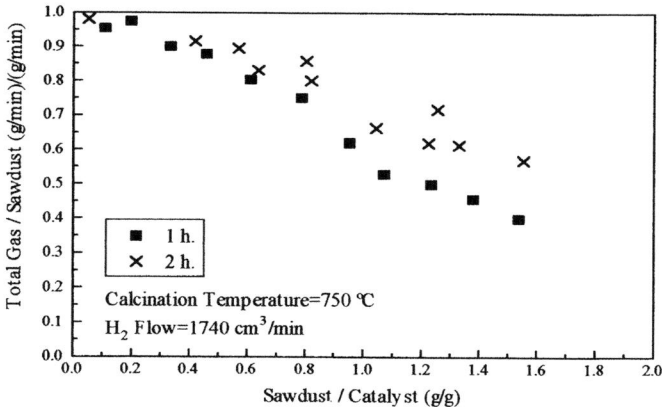

Fig. 5. Gas yield obtained for different reduction times.

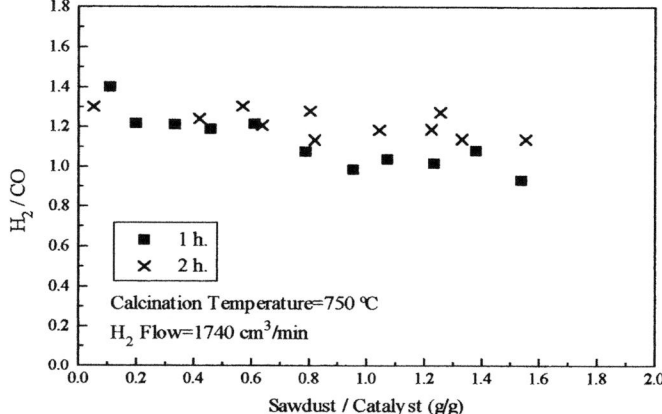

Fig 6. H₂/CO ratio obtained for different reduction times.

4.2. Influence of the catalyst calcination temperature

When the calcination temperature decreases (Runs 1 and 3, Table 5), higher values of gas yield and of CO/CO_2 ratio, and lower tar production, are observed.

Figures 7 and 8 show the gas yield and the H_2/CO ratios in function of the ratio between the sawdust processed and the catalyst introduced in the reactor. Higher gas yields are obtained at a calcination temperature of 750 °C. The H_2/CO values are similar for low sawdust/catalyst values at both calcination temperatures, Figure 8, but higher for a calcination temperature of 750 °C for a sawdust/catalyst value higher than 0.8.

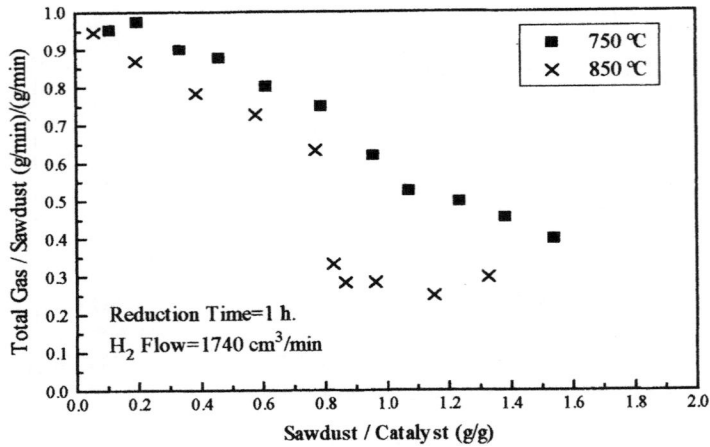

Fig. 7. Gas yield obtained for different catalyst calcination temperatures.

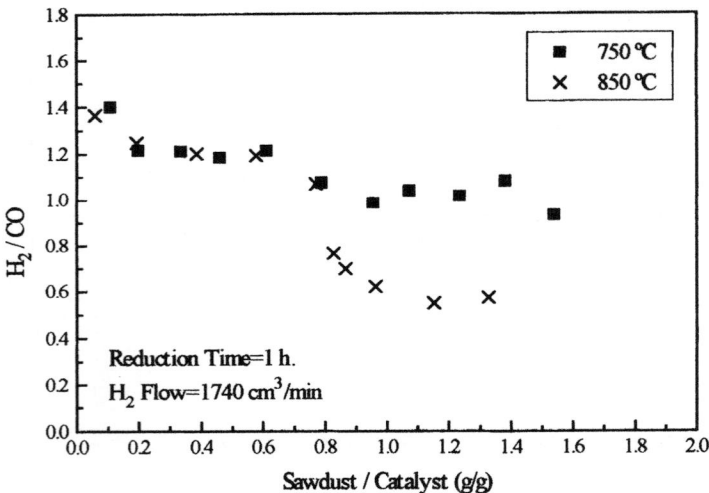

Fig 8. H_2/CO ratios for different calcination temperatures.

4.3. Influence of activation hydrogen flow rate

For a calcination temperature of 750 °C an increase in the hydrogen flow rate produces more liquid and less gas (Runs 1 and 4, Table 5). It can be concluded that for this calcination temperature there is an optimum in the activation conditions and that a more severe reduction produces an excess of metallic nickel. For a calcination temperature of

850 °C (Runs 3 and 5, Table 5), an increase of the hydrogen flow rate originates more gas and less liquid because an increase in the calcination temperature implies more severe activation conditions.

Figures 9 to 12 show the yield to gas and the H_2/CO ratios for different sawdust/catalyst weight ratios. Similar trends to those mentioned above can be observed.

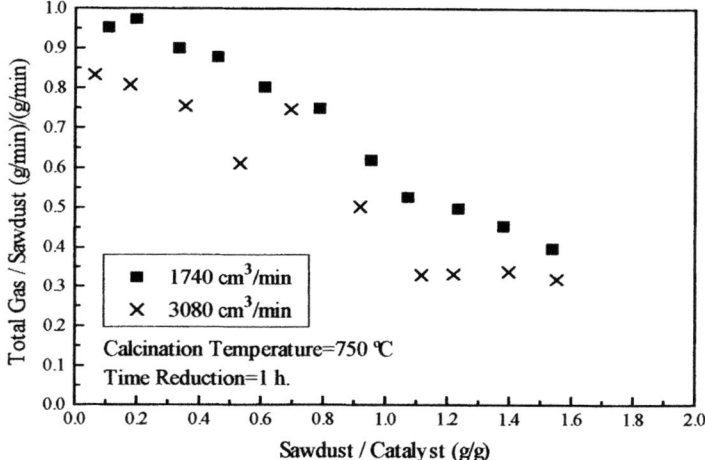

Fig. 9. Gas yield obtained for different hydrogen flow rates. $T_c = 750$ °C.

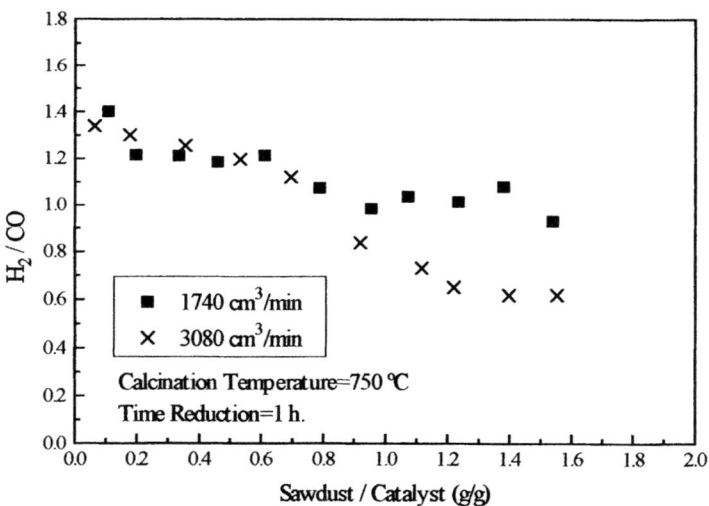

Fig. 10. H_2/CO ratios for different hydrogen flow rates. $T_c = 750$ °C.

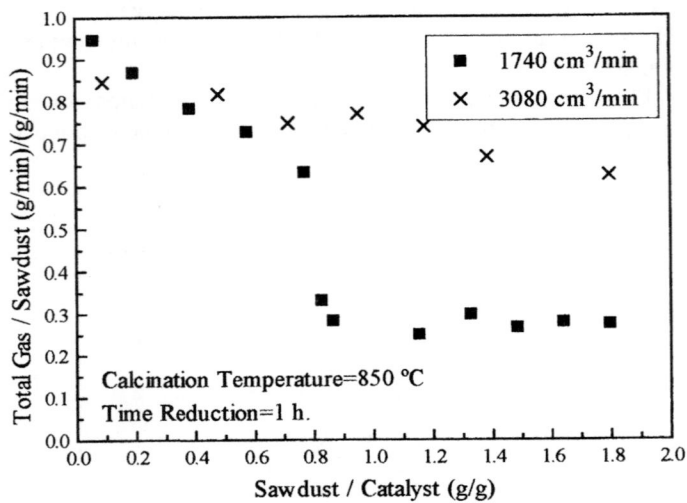

Fig. 11. Gas yield obtained for different hydrogen flow rates. $T_c = 850$ °C.

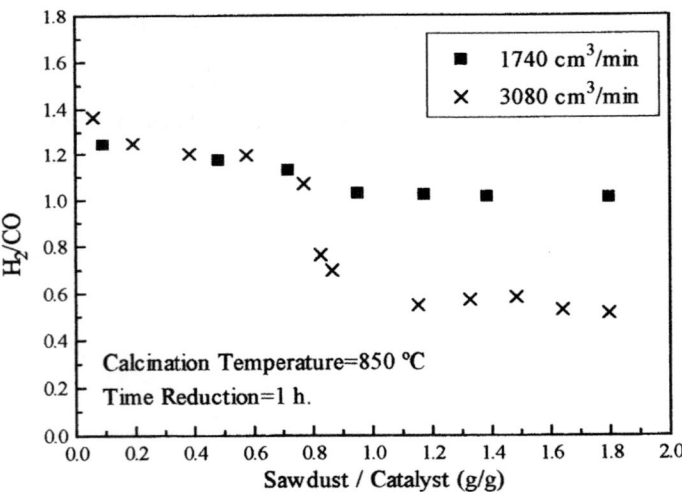

Fig. 12. H_2/CO ratios for different hydrogen flow rates. $T_c = 850$ °C.

5 Conclusions

The tar produced in biomass pyrolysis at low temperatures can be significantly reduced using nickel aluminate catalysts.

The calcination temperature of the catalyst precursor is an important variable because it influences the characteristics of the catalyst and the product distribution obtained in the biomass pyrolysis.

There is an optimum in the activation (reduction) conditions of the calcined catalyst. These conditions depend on the catalyst calcination temperature. An increase in the calcination temperature implies more severe activation conditions.

6 References

1. Baker, E.G. and Mudge, L.K. (1984) Mechanisms of catalytic biomass gasification. *Journal of Analytical and Applied Pyrolysis*, 6, pp. 285-297.
2. Tanaka, Y., Yamaguchi, T., Yamasaki, K., Ueno, A. and Kotera, Y. (1984) Catalyst for Steam Gasification of Wood to Methanol Synthesis Gas. *Ind. Eng. Prod. Res. Dev.*, Vol. 23, No. 2, pp. 225-229.
3. Al-Ubaid, A. and Wolf, E.E. (1984) Steam Reforming of Methane on Reduced Non-Stoichiometric Nickel Aluminate Catalysts. *Applied Catalysis*, 40, pp. 73-85.
4. Scott, D.S. and Piskorz, J. (1982) The Flash Pyrolysis of Aspen-Poplar Wood. *The Canadian Journal of Chemical Engineering*, 60, pp. 666-674.
5. Arauzo, J., Radlein, D., Piskorz, J. and Scott, D.S. (1994) A New Catalyst for the Catalytic Gasification of Biomass. *Energy & Fuels.*, Vol. 8, No. 6, pp. 1192-1196.
6. Scott, D.S. and Piskorz, J. (1982) Low Rate Entrainment Feeder for Fine Solids. *Ind. Eng. Chem. Fundam.*, Vol. 21, No. 3, pp. 319-322.
7. Satterfield, C.N. (1991) *Heterogeneous Catalysis in Industrial Practice*, 2nd ed., McGraw-Hill, Inc., New York.
8. Alzamora, L.E., Ross, J.R.H., Kruissink, E.C. and van Reijen, L.L. (1981) Coprecipitated Nickel-Alumina Catalysts for Methanation at High Temperature. *J. Chem. Soc., Faraday Trans. I.*, 77, pp. 665-681.

7 Acknowledgment

The authors express their gratitude to D.G.I.C.Y.T. for providing financial support for the work (Proyect PB93-0593) and also to the Ministerio de Educacion y Ciencia (Spain) for a research grant awarded to L. García. The authors also would especially like to acknowledge Dr. J.L. G. Fierro for his XPS analysis realized.

CATALYTIC UPGRADING OF THE CRUDE GASIFICATION PRODUCT GAS
Tar cracking

C. MYRÉN, C. HÖRNELL, K. SJÖSTRÖM, YU Q., C. BRAGE and E. BJÖRNBOM
Kungl Tekniska Högskolan (KTH), Department of Chemical Engineering and
Technology/Chemical Technology, Stockholm, SWEDEN

Abstract
Knowledge of necessary conditions for sufficient removal of tars from gasification gas
is the over-all goal. A tar-rich gas is produced through pyrolysis at 700°C. After steam
addition the produced gas is cracked thermally and catalytically, at temperatures
between 700 and 900°C using short contact times to allow study of tar decomposition.
The remaining tar is analyzed by chromatography.

Tars obtained from the agricultural residues Miscanthus and straw have been cracked
using dolomite and nickel as catalysts. Analyses of tars from Miscanthus and straw are
also compared with those of tars from hardwood. The amounts and compositions of the
tars reflect their origin.
Keywords: Biomass, catalytic tar cracking, dolomite, gasification.

1 Introduction

The tars produced in gasification must be removed at acceptable cost to allow efficient
use of the produced gas and to avoid condensation of tars further down the system
which will cause problems. One way of studying tar removal from gasification product
gas through thermal and catalytic cracking is to produce a tar-rich gas by pyrolysis and
then to expose the pyrolysis gas to various treatments.

The work presented here has been part of the EC AIR2 project "Thermo-chemical
biomass gasification: upgrading of the crude gasification product gas". It has been a
cooperation between the Universidad Complutense in Madrid, Rijksuniversiteit in

Groningen, Biomass Technology Group B.V. in Twente, Termiska Processer AB in Studsvik and KTH in Stockholm. Our part in this project was to find suitable catalysts for upgrading of tar from partial gasification of agricultural residues and to study the catalytic decomposition of biomass tar. A subtask was to perform reliable and fast analyses for determination of selected target compounds in the complex biomass tar.

2 Experiments

Tars from two types of pelletized agricultural residues, straw and Miscanthus, have been obtained. We have studied thermal and catalytic cracking of these tars with dolomite and nickel as catalysts. The results are also compared with previously obtained results using mixed hardwood [1]. Input rate of raw material together with carrier gas feed have been chosen so that the longest gas contact time in the cracker has been about 1 second.

3 Equipment

A scheme of the equipment used is shown in figure 1. A steady-flow of a tar-rich gas is obtained by pyrolysis of the chosen raw material. The input is fed by a rotating conveyor through a pyrolysis zone, heated up to 700°C, with N_2 as carrier gas. The char residue is separated from the tar-rich gas immediately after the pyrolysis zone and the gas is mixed with steam and led into an isothermal catalytic cracker. The temperature in the cracker is varied between 700 and 900°C. Contact time in the catalytic reactor may be varied by changing amount of catalyst in the reactor or by changing raw material input rate to N_2-flow. The cracking reactor has an inner diameter of 2 cm and its total length is about 40 cm. A net is placed about 10 cm above the outlet, on which the catalyst is placed. The temperature is measured at half total catalyst bed height. There are three separate heaters round the cracking zone to avoid any temperature gradients.

The residual tar in the leaving gas after the cracker bed is condensed in several steps passing first through water-cooled condensers and then through impingers cooled by dry ice-ethanol and its main components are analyzed. The leaving gas is analyzed for permanent gases and hydrocarbons up to toluene. Higher hydrocarbons are found in the condensates.

During startup of an experiment, as well as during the first half hour of feeding, an auxiliary cooling system is used for evolved products. After feeding has ended, evolved products are again switched to pass through the auxiliary system.

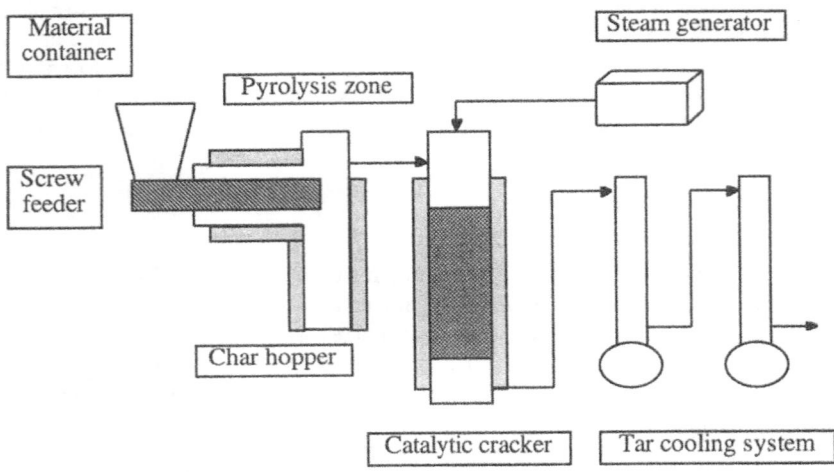

Figure 1 Experimental equipment

4 Tar analyses

After experiments the cooling system is weighed prior to washing out all the condensed tars with dichloromethane. Tar components are then separated by solid phase extraction: aliquots of tar sample are mixed with dichloromethane containing known amounts of an internal standard. An aliquot of the mixture is loaded onto an amino-bonded silica phase micro column. Neutrals are eluted from the column by dichloromethane. Phenols are then eluted by a mixture of dichloromethane-2-propanol followed by 2-propanol. The fractions are then subjected to capillary GC after trimethylsilylation of the phenols as silylation has been found to give the best GC resolution for isomeric compounds [2]. Besides identification of selected target neutral compounds and phenols, tar may also be analyzed for bases using a recently published method [3]. Basic compounds are separated from the organic extracts by extraction with sulphuric acid. The acidic water phase, after addition of internal standard and adjusting the pH, is then concentrated using reverse-phase (n-C_{18}) solid phase extraction. The retained solutes are eluted with a mixture of dichloromethane, acetonitrile and diethylamine, and then also analyzed by capillary gas chromatography.

Since not all tar components are analyzable by chromatography, total tar is also determined gravimetrically. The gravimetric tar determination is the basis for calculation of output water.

5 Materials and experimental conditions

Elementary analyses of the raw materials are given in table 1. Before use, the pellets of Miscanthus or straw have been ground and sieved. The fraction 0.5-1 mm has then been wind-sieved to obtain a raw material with uniform flow properties.

Table 1 Elemental analysis (moisture-free) in weight% of raw materials

Raw material	C	H	N	O	Ash
Straw	46.8	6.1	0.5	42.2	4.4
Miscanthus	46.4	5.9	0.5	41.5	5.7
Mixed hardwood	47.2	5.9	0.1	45.9	0.9

The dolomite used is from the Sala quarry, it has previously been shown to give good results for cracking [1]. The size fraction 1.2-1.7 mm was used. Prior to experiments, the dolomite was calcined in situ at 900°C for 8 hours. The nickel catalyst used was C11-9-061 from United Catalysts Inc. It was crushed, and the size fraction 0.5-1 mm was used.

Experimental conditions are given in table 2 . The runs with hardwood were made at approximately comparable conditions, with 202 g raw material/h, about 60 Nl N_2/h and a steam input corresponding to 90 g/h.

Table 2 Experimental conditions

Raw material	Miscanthus			Straw	
Temp in cracker (°C)	700	900	900	700	700
Catalyst weight[1] (g)	-	-	80	-	10
Raw material (g/h)	188	201	102	157	105
Moisture content (%)	4	5	4	5	5
Added N_2 (Nl/h)	56	56	28	56	56
Steam (g/h)	43	54	24	52	50
Experiment duration (min)	90	149	137	110	145
Duration of tar collection (min)	70	123	121	90	114
Space time (kg (calcined) catalyst*h/Nm^3) [2]			0.365		0.103

[1] Catalysts used here are dolomite for Miscanthus (uncalcined weight) and Ni for straw
[2] without N_2 and H_2O

6 Results

Comparatively short contact times have been used to follow the decomposition of the formed tars [4, 5].

For control of experimental reliability an integral carbon balance is calculated over the system for every run. Carbon out in gas and tar is compared with total carbon in feed. The carbon balances indicate well-functioning experimental equipment at temperatures over 700°C, with values for Miscanthus of 90-101%.

6.1 Tars

Some results can be seen in table 3, where the compositions of tars from Miscanthus are compared with those of tars from mixed hardwood, obtained previously at similar experimental conditions. Tar from hardwood at 700°C, without catalyst, contains more cresols than Miscanthus, but less phenols. Tar from straw (table 4) obtained at the same temperature appears to have an intermediate composition. Even after thermal cracking at 900°C, hardwood contains more cresols and less phenols than the Miscanthus, while aromatics are similar in composition, maybe except for indene. Miscanthus tar then persists in containing more phenols than hardwood after cracking at 900°C over 80 g of dolomite, which otherwise reduces tar to mainly naphthalene.

The proportion of identified compounds of total tar, as determined gravimetrically, increases with severity of processing conditions for both Miscanthus and hardwood. However, the unknown phenols and aromatic compounds were not included under their respective headings in the analysis of the hardwood experiments, so total amounts of each compound class are not directly comparable.

Composition of the tars from pyrolysis and cracking of straw at 700°C, without catalyst and using 10 g Ni are given in table 4. The presence of catalyst increases aromatics appreciably, while the decrease in total amount of tar per 100 g raw material is smaller.

6.2 Gases

Diagram 1 shows a typical variation of permanent gas composition over time. The diagram shows gas evolution during the period when tar is sampled for straw as raw material, with 10 g Ni, 700°C in the cracker. Zero time is thus when the gas has been switched from the auxiliary cooling system to the cooling system from which tar samples are taken. Calculations on gas composition for mass balance are based on values obtained once steady state for the permanent gases has been reached, which takes relatively long. An explanation for this may be that when evolved gases are switched to pass through the cooling system used for sampling of tars, the whole system is filled with inert gas.

7 Discussion

When long contact times in the cracker are employed, the resulting high conversion leaves almost only naphthalene in the condensed tar phase. We have used short contact time in the cracker which gives low conversion of tar and allows the study of decomposition patterns.

The gas produced by the experiments reported here contains more tar than a gas produced by gasification of the same raw materials would. The first step, however, also in gasification is pyrolysis. When heating is not extremely rapid, which never is the case in practical gasification, the pyrolysis step always produces the same types of

Table 3 Tars from Miscanthus and mixed hardwood (mg/100 g raw material)

Temperature (°C)	700	700	900	900	900 80	900 80
Dolomite[1] (g)	-	-	-	-	80	80
Raw material	Miscanthus	Hardwood	Miscanthus	Hardwood	Miscanthus	Hardwood
Phenols						
Phenol	241.4	106.9	144.2	75.3	5.6	0.4
o-Cresol	38.9	92.9	109.1	30.0	-	-
m-Cresol	47.2	114.0	25.0	50.3	-	-
p-Cresol	51.7	73.9	19.2	27.6	-	-
C2-Phenols	89.8	121.2	-[3]	62.8	0.6	0.3
Unknown	27.7	n.d.[2]	133.8	n.d.	-	n.d.
Total phenols (by GC)	496.7	509.0	431.3	250.0	5.7	0.7
Aromatics						
C1 - C3-Benzenes	245.5	170.5	187.9	119.1	22.7	13.8
Indene	103.6	64.9	96.5	142.5	-	n.d.
Naphthalene	136.5	34.5	179.3	172.2	65.8	92.9
2-Methylnaphthalene	36.9	24.2	30.6	45.6	-	0.3
1-Methylnaphthalene	25.3	16.4	20.2	28.9	-	-
Acenaphthylene	47.2	20.8	47.3	-	0.5	-
Fluorene	18.4	11.9	26.1	27.6	-	-
Phenanthrene	21.2	6.5	37.9	36.8	3.9	3.7
Anthracene	7.8	1.7	8.8	10.7	-	-
Pyrene	5.8	4.9	4.4	14.0	-	-
Unknown	339.0	n.d.	210.7	n.d.	7.9	n.d.
Total aromatics (by GC)	987.1	361.0	849.8	610.0	100.8	113.2
"Total tar" (gravimetry)	2679.3	4201.9	1487.5	1835.2	106.5	181.5

[1] uncalcined weight
[2] not determined
[3] below detection limit

Table 4 Tars from straw (mg/100 g raw material)

Temperature (°C)	700	700
Ni catalyst weight (g)	-	10
Phenols		
Phenol	161.9	175.3
o-Cresol	77.9	80.7
m-Cresol	73.2	78.9
p-Cresol	65.0	64.7
C2-Phenols	142.3	160.7
Unknown	34.2	93.9
Total phenols (by GC)	554.6	654.2
Aromatics		
C1 - C3-Benzenes	373.2	335.6
Indene	69.2	46.7
Naphthalene	66.1	40.4
2-Methylnaphthalene	37.5	28.2
1-Methylnaphthalene	28.6	20.8
Acenaphthylene	25.0	18.5
Fluorene	21.6	10.7
Phenanthrene	14.1	10.8
Anthracene	4.4	3.7
Pyrene	2.0	1.0
Unknown	85.0	588.6
Total aromatics (by GC)	726.8	1104.7
"Total tar" (gravimetry)	3319.5	3079.0

compounds - the differences arising from variations in raw material structure and composition.

In gasification, part of the tars produced by the pyrolysis step are consumed in the gasification step itself. Thus, the tars produced by pyrolysis alone may be considered as somewhat "heavier" than those remaining in a gas produced by gasification.

The higher concentration of tars in the experiments reported here, as compared to gasification tar concentrations, means that for tars from gasification, sufficiently low output concentrations are obtained at considerably lower conversions.

Our results indicate that almost any agricultural residue may be suitable as gasification raw material, once requirements on tar amount and composition have been established. A prerequisite, not studied within the framework of the present project, is that the ash composition of the raw material should allow gasification without clogging or melting problems.

8 Future work

This year our project continues with financing from the EC JOULE/THERMIE programme. We are continuing our research on tar characterization of new raw materials, new combinations of catalysts for tar cracking and overall kinetics of catalyzed tar elimination.

9 Acknowledgement

Mikael Lundgren is thanked for experimental assistance. Though this work has been part of an EC project, it has been financed by the Swedish Energy Agency (NUTEK). This financial support is gratefully acknowledged.

Diagram 1 Gas evolution vs time for straw during tar sampling, with 10 g Ni, 700°C in the cracker (without carrier-gas, N_2).

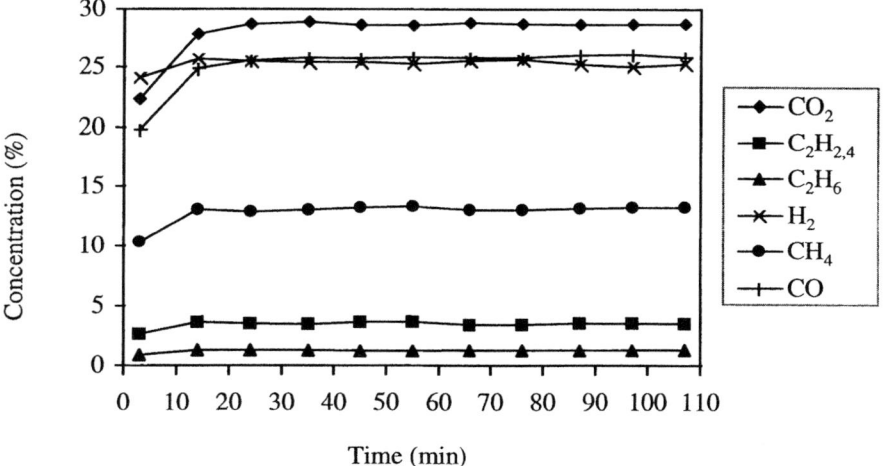

10 References

1. Vassilatos, V., Taralas, G., Sjöström, K. and Björnbom, E. (1992) Catalytic cracking of tar in biomass pyrolysis gas in the presence of calcined dolomite, *Can J. Chem. Eng.* (70) 1008.
2. Brage, C. and Sjöström, K. (1991) Separation of phenols and aromatic hydrocarbons from biomass tar using aminopropylsilane normal-phase liquid chromatography, *J. of Chromatography* (538) 303.
3. Brage, C., Yu, Q. and Sjöström, K. (1996) Characteristics of evolution of tar from wood pyrolysis in a fixed-bed reactor, *Fuel* (75) 213.

4. Henriksen, U., Kofoed., E. and Christensen, O. (1994) Mass and energy distribution of the pyrolysis products from straw, in *Advances in Thermochemical Biomass Conversion* (ed. A.V. Bridgwater), Blackie, London, pp. 1110-1121.
5. Aldén, H., Espenäs, B.-G. and Rensfelt, R. (1988) Conversion of tar in pyrolysis gas from wood using a fixed dolomite bed, in *Research in Thermochemical Biomass Conversion* (eds. A.V. Bridgwater and J.L. Kuester), Elsevier, London, pp. 987-1001.

CATALYTIC TAR REMOVAL FROM BIOMASS PRODUCER GAS WITH SECONDARY AIR

Tar removal with secondary air

G. LAMMERS and A.A.C.M. BEENACKERS
Department of Chemical Engineering, University of Groningen,
Groningen, The Netherlands
J. CORELLA
Department of Chemical Engineering, Universidad Complutense,
Madrid, Spain

Abstract

The effect of air addition on biomass tar conversion in catalytic packed bed crackers was studied using both an isothermal micro reactor and a fluidised bed bench scale biomass gasification set up with down stream tar crackers.

The micro reactor was applied for experiments with artificial biomass producer gas containing naphthalene as a model tar compound. Experiments were carried out with inert silica and catalytically active calcined dolomite bed material both with and without air addition. Kinetic results were modelled assuming three parallel first order naphthalene decomposition reactions:

$$R_{Naphth} = k_1^{eff}\, c_{Naphth,0}\, (1 - \zeta_{Naphth}) \text{ with: } k_1^{eff} = [k_1\, \varepsilon + k_1^*\, (1-\varepsilon) + k_1^{ox}]$$

k_1 : rate constant of the homogeneous thermal decomposition reaction [s^{-1}.]
k_1^* : rate constant of the catalytic heterogeneous decomposition reaction [s^{-1}.]
k_1^{ox} : rate constant of the oxygen induced decomposition reaction [s^{-1}.]

With secondary air the results show a significant increase in tar decomposition rate both with dolomite and silica. With dolomite, the naphthalene decomposition reactions on the catalyst surface dominate the overall conversion rate making the relative effect of air addition less significant. However, with dolomite, secondary air feeding has the positive effect of significantly decreasing the catalyst deactivation rate.

Experimental results with real tar from the fluidised bed bench scale gasification set up were in qualitative agreement with results from the micro reactor experiments.

Keywords: Gasification, catalytic gas cleaning, secondary air.

1 Introduction

The concept of catalytic tar removal from biomass producer gas is a promising step toward the use of biomass for power production especially in combined cycle systems. Various authors, Corella *et al.* [1] and Simell *et al.* [2], discussed the excellent tar removal in well designed tar cracker systems with suitable catalysts like dolomite on a laboratory scale. With the choice for the TPS atmospheric gasification process in the Brazilian Biomass power demonstration project, atmospheric catalytic tar removal from biomass producer gas is entering the demonstration scale.

In this report we discuss the effect of feeding secondary air to the catalytic cracker reactor on tar conversion and catalyst life time. For commercial cracking and steam reforming Nickel catalysts, the use of secondary air in the catalytic tar cracker was discussed by Mudge *et al.* [3]. Results showed good tar decomposition and an increase in energy of the product gas up to 20 %. Furthermore, at temperatures above 600 °C. catalyst deactivation was largely prevented which led to the conclusion that above this temperature air addition effectively removed carbon from the catalyst surface.

The combination of catalyst life time and catalyst costs is still unfavourable for use of Nickel catalysts in large scale tar cracking. The Finnish companies Tampella Power and Enviropower and the Swedish company TPS prefer calcined dolomite in their commercial scale Power from Biomass processes [4-5]. In this paper we therefore study the effect of secondary air with calcined dolomite as a catalyst using both real and artificial biomass producer gas. For comparison, the experiments with artificial biomass producer gas were repeated using inert silica and a BASF Nickel catalyst as bed material.

2 Experimental

Experiments were carried out at TPS research laboratory in Studsvik, Sweden and at the Chemical Engineering department of the Universidad Complutense in Madrid, Spain.

At TPS Studsvik the effect of secondary air addition on tar decomposition was studied using a micro cracker reactor, see Figure 1. The reactor consisted of a quartz tube placed in a thermostatted oven with a maximum operation temperature of 900 °C. Reactor temperature was measured using a K-type thermocouple placed outside the catalyst packed part the reactor.

Naphthalene was used as a model tar compound because it is one of the major compounds in high temperature fluidised bed biomass gasification [6] and is considered to be one of the most problematic compounds [7]. The naphthalene was fed to the reactor using a naphthalene evaporator in which a gas stream of artificial biomass producer gas was saturated with naphthalene. Naphthalene concentration in the gas could be controlled by adjusting the evaporator temperature. The composition of the artificial biomass producer gas is presented in Table 1.

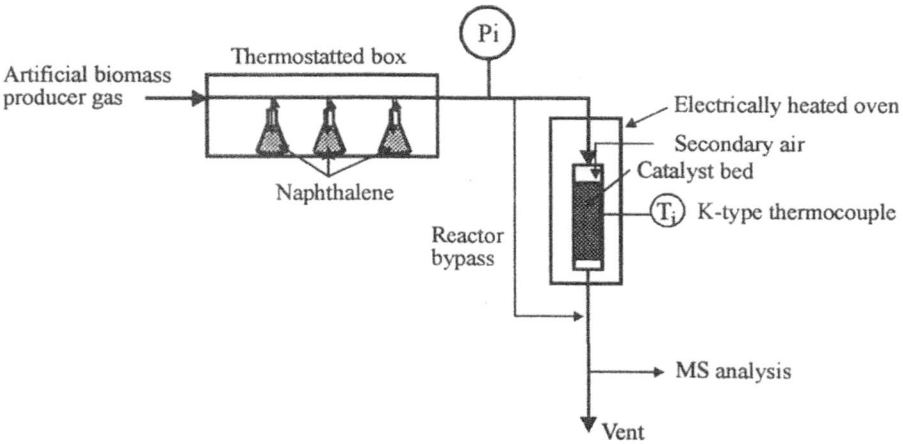

Fig. 1. Micro catalytic packed bed reactor for study of model tar compound decomposition kinetics.

Table 1. Artificial biomass producer gas composition.

Compound	Vol. %
H_2	11.8
CO	14.2
CO_2	15.0
H_2O	13.2
CH_4	5.0
C_2H_4	1.5
N_2	39.3

Secondary air was introduced using a quartz tube with d_i = 1 mm positioned at the front of the catalyst bed. In all experiments the air to artificial biomass producer gas volumetric flow ratio was 0.056. Flow conditions were laminar both in the secondary air tube and in the reactor with Re ≈ 10. However, fast mixing between secondary air and artificial biomass producer gas was assured by the mixing action of the packed bed.

The reactor outlet gas was analysed for naphthalene, benzene, toluene, H_2O, H_2, CO_2 and CH_4 using mass spectrometry (MS). The gas components CO, C_2H_4 and N_2 could not be measured separately because of their coinciding mass numbers. Data on the catalysts used in the three different experiments are presented in Table 2.

Table 2. Catalyst data for the three experimental runs.

Run	Catalyst type	Amount [g]	Sieve size [mm]	T [°C.]
1	Dolomite*	4.32**	0.50-0.71	850, 880
2	Silica	3.37	0.355-0.50	850, 880
3	BASF G 1-22***	3.43	0.50-0.71	450 - 800

* : Swedish 'Glanshammer' dolomite
** : prior to calcination, specific density 2850 kg/m³.
*** : 19-25 weight percent NiO on alumina carrier

The experiments at the University of Madrid (Universidad Complutense de Madrid, UCM) were carried out using a bench scale bottom fed fluidised bed gasifier with two downstream fixed bed tar cracker reactors in series, see Figure 2. In the first tar cracker Spanish dolomite (Malaga) was used as a catalyst, in the second bed a commercial Nickel catalyst (BASF G 1-25S) was used to obtain nearly complete tar decomposition.

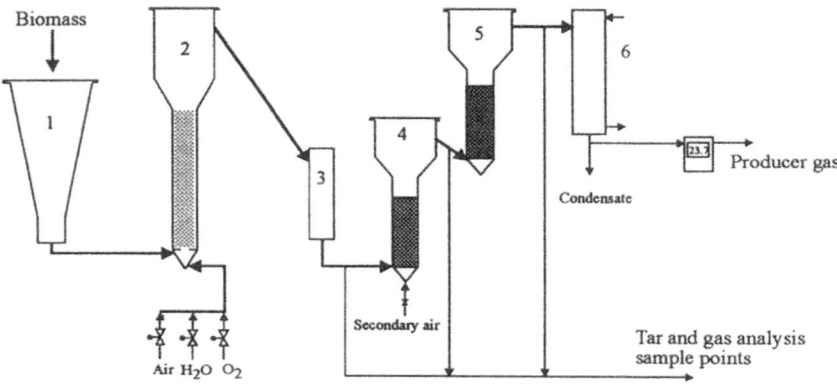

1: Biomass hopper; 2: Fluidised bed gasifier; 3: hot metallic filter; 4: dolomite bed;
5: Nickel catalyst bed; 6: Condenser

Fig. 2. Scheme of the bench scale biomass gasification unit at the University of Madrid.

To carry out an experiment, the electrical ovens of both gasifier and catalytic reactors were switched on and the hopper was filled with biomass (pine sawdust). After stable oven temperatures were reached feeding of biomass and air to the gasifier was started. After about one hour the gasifier and tar crackers reached stable operation conditions. Gas chromatographic analysis of the gas was carried out by sampling the outlets of both the gasifier and of the first and second catalytic cracker reactor. Tar sampling was carried out by condensing part of the gas at the three sample points. The condenser temperature was maintained at 0 °C. The tar content in the condensate was determined using a Total Organic Carbon (TOC) analyser. After determining the performance of the facility without secondary air, feeding of secondary air was started at the bottom of the first tar cracker reactor. The air was heated and radially dispersed in the inert part of the bed.

In order to get a realistic picture of the advantages of secondary air feeding on gas tar content the overall Equivalency Ratio (E.R.) was kept constant. This means that during secondary air feeding the air flow to the gasifier was lowered accordingly. After establishing stable reactor temperatures and gas compositions tar sampling was started at the three sample points.

The experiment was stopped by switching from feeding air to feeding nitrogen to the gasifier and simultaneously stopping the feed of biomass and secondary air.

3. Results and discussion

3.1 Micro packed bed reactor results

3.1.1 Cracking without air addition

The result of run #1 were obtained using calcined dolomite. Figure 3 shows the analysis result of the tar components naphthalene, benzene and toluene during the experimental runtime. Changes in operation conditions are indicated.

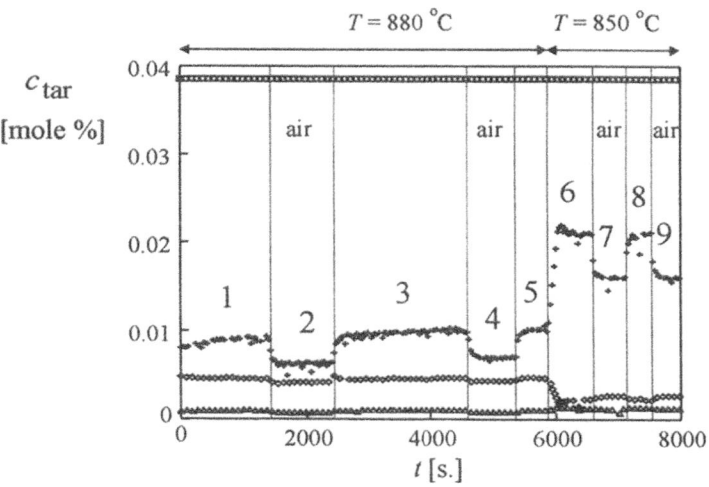

Fig. 3. Experimental naphthalene, benzene and toluene concentrations in the reactor outlet gas during run #1. Dolomite, reactor temperatures 850 and 880 °C.

Experiment #2 was carried out using a silica bed which generally is supposed to have no catalytic activity. Figure 4 shows analysis results for the concentrations of the tar components naphthalene, benzene and toluene in the reactor outlet gas. Changes in reactor operation conditions are indicated.

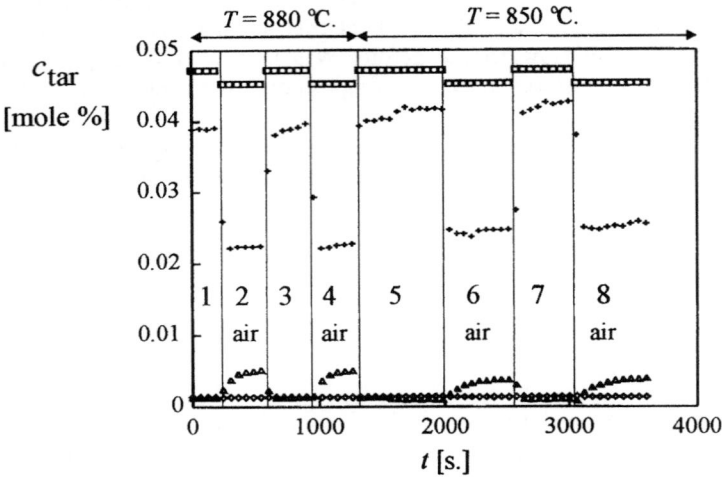

□ : Naphthalene inlet; + : Naphthalene outlet
◇ : Benzene; △ : Toluene

Fig. 4. Concentrations of naphthalene, benzene and toluene in the reactor exit gas during run #2. Silica bed, reactor temperatures 850 and 880 °C.

We will discuss the naphthalene decomposition process assuming a first order overall reaction mechanism. The overall naphthalene decomposition rate is the sum of the thermal homogeneous decomposition reactions and the catalytic decomposition reactions on the catalytic surface. The overall naphthalene conversion in the micro reactor follows from:

$$F_{\text{Naphth},0} \frac{d\zeta_{\text{Naphth}}}{dV} = R_{\text{Naphth}}$$

with: $R_{\text{Naphth}} = [k_1 \, \varepsilon + k_1^* (1-\varepsilon)] \, c_{\text{Naphth},0} \, (1-\zeta_{\text{Naphth}})$

The experimental rate constants can be calculated from:

$$[k_1 \, \varepsilon + k_1^* (1-\varepsilon)] = \frac{\phi_{\text{V}}}{V_{\text{bed}}} \, \ln \frac{1}{(1-\zeta_{\text{Naphth, end}})} \tag{1}$$

with: $[k_1 \, \varepsilon + k_1^* (1-\varepsilon)] = k_1^{\text{eff}}$ \hfill (2)

The rate constant of the thermal naphthalene decomposition reaction can e determined from experimental data of run #2. For the silica bed we can write: $k_1^{\text{eff}} = k_1 \, \varepsilon$ \hfill (3)

Table 3 presents calculated values of k_1.

Table 3. Experimental naphthalene conversion and calculated rate constant of the naphthalene decomposition reaction during run #2, silica bed material without oxygen addition, $\varepsilon_{bed} = 0.45$.

Run sequence*	T [°C]	Air	ζ_{Naphth} [-]	k_1 [s^{-1}]
1	880	off	0.173	2.45
3	880	off	0.172	2.44
5	850	off	0.120	1.61
7	850	off	0.106	1.41

*: indicated in Figure 4.

The temperature dependency of the rate constant of the thermal naphthalene decomposition reaction is described by the Arrhenius equation. Regression calculation of the experimental values of k_1 in Table 3 yields:

$$k_1 = A\, e^{-\frac{E_a}{RT}} \quad \text{with: } A = 1.89\ 10^8\ \text{s}^{-1} \text{ and } E_a = 174 \text{ kJ/mole } (r^2 = 0.965) \qquad (4)$$

The rate constant of the dolomite catalysed naphthalene decomposition reaction can now be calculated from experimental data of run #1 and the Arrhenius equation of the thermal decomposition reaction, see Table 4.

Table 4. Experimental naphthalene conversion, overall rate constant and calculated rate constant of the catalysed naphthalene decomposition reaction during run #1. Calcined dolomite bed, $\varepsilon_{bed} = 0.56$, no secondary air addition.

Run sequence	T [°C]	ζ_{Naphth} [-]	$k_1\,\varepsilon + k_1^*(1-\varepsilon)$	k_1^* [s^{-1}]
1	880	0.77	8.0	16.0
3	880	0.75	7.4	14.7
5	880	0.74	7.2	14.3
6	850	0.54	4.1	8.1
8	850	0.53	4.0	7.9

Expressing the temperature dependency of k_1^* by the Arrhenius equation gives:

$$k_1^* = A^*\, e^{-\frac{E_a^*}{RT}} \quad \text{with: } A^* = 2.58\ 10^{11}\ \text{s}^{-1} \text{ and } E_a^* = 227 \text{ kJ/mole } (r^2 = 0.98) \qquad (5)$$

3.1.2 Effect of air addition

In the discussion of the effect of air addition on naphthalene decomposition rate we propose that oxygen decomposes naphthalene via a parallel first order reaction route:

$$F_{Naphth,0}\, \frac{d\zeta_{Naphth}}{dV} = R_{Naphth}$$

with: $R_{Naphth} = [k_1\,\varepsilon + k_1^*(1-\varepsilon) + k_1^{ox}]\, c_{Naphth,0}\, (1 - \zeta_{Naphth})$

and: $[k_1 \, \varepsilon + k_1^* \, (1-\varepsilon) + k_1^{ox} \,] = k_1^{eff}$ \hfill (6)

Values of k_1^{eff} were calculated from the experimental naphthalene conversion data. k_1^{ox} was calculated from k_1^{eff} by subtraction of the contributions of the thermal naphthalene decomposition reaction $(k_1\varepsilon)$ in case of experiments with inert bed material and subtraction of both the contributions of the thermal naphthalene decomposition reaction $(k_1\varepsilon)$ and the catalytic naphthalene decomposition reactions $(k_1^* \, (1-\varepsilon))$ in case of the experiments with dolomite.

Results of micro reactor tests with air addition both with silica and dolomite are listed in Table 5.

Table 5. Experimental data on naphthalene conversion and calculated rate constants with air addition both with dolomite and silica bed material

bed solid	T [°C]	ζ_{Naphth} [-]	k_1^{eff} [s^{-1}]	k_1^{ox} [s^{-1}]
Dolomite	880	0.84	10.5	2.9
Dolomite	880	0.82	98	2.2
Dolomite	850	0.58	4.9	0.81
Dolomite	850	0.58	4.9	0.81
Silica	880	0.506	4.30	3.19
Silica	880	0.503	4.27	3.16
Silica	850	0.462	3.69	3.00
Silica	850	0.445	3.50	2.82

Results show that the relative contribution of k_1^{ox} to the value of k_1^{eff} is different for the experiments with silica and calcined dolomite. The experiments with inert bed material show hardly an effect of temperature on k_1^{ox}. This leads to the conclusion that the added oxygen reacts completely and that the selectivity of the reaction of naphthalene with oxygen is independent of temperature in the range of temperatures investigated. The relative error in the values of k_1^{ox} for the dolomite experiments is relatively large because experimental errors in k_1, k_1^* and k_1^{eff} accumulate in the calculated value of k_1^{ox} and the contribution of k_1^{ox} to the overall decomposition constant is relatively small.

In the discussion of the experimental results, isothermal operation of the micro reactor is assumed both for the situation with and without secondary air feeding. In agreement with this the thermocouple at the outside of the catalyst bed did not indicate a temperature rise during secondary air feeding. However, the relatively high contribution of radiation to heat transfer and the difference in radiation absorption coefficient for different materials like the metal surface of the thermocouple, the quartz reactor tube and catalyst bed material prevents accurate temperature measurements. It is thus conceivable that a relatively small increase in catalyst bed temperature would not have been indicated by the thermocouple at the outside of the reactor. Therefore, the expected temperature rise of the catalyst bed during secondary air feeding was calculated from the reaction enthalpy and the overall heat transfer from catalyst bed to surroundings. The reaction enthalpy was obtained from equilibrium calculations using AspenPlus™. The concentrations of CH_4 and C_2H_4 were prefixed because the concentrations of these compounds do not normally go to equilibrium at typical tar cracker conditions. Table 6 lists calculated and experimental results.

Table 6. Experimental and calculated component concentrations in the reactor outlet gas at 850 °C., for feed concentrations, see Table 1.

Component	calculated	experimental
H_2 [vol. %]	11.2	-
H_2O [vol. %]	12.9	13.2
CO [vol. %]	13.9	-
CO_2 [vol. %]	14.7	14.4
CH_4 [vol. %]	5.0	5.1
C_2H_4 [vol. %]	0.9	-
N_2 [vol. %]	41.4	-

-: not determined

$\Delta H_r = 4.7$ kJ/mole inlet gas

$\Delta T_{ad} = 103$ K

The temperature rise of the catalyst bed was calculated assuming heat transfer by radiation only:

$$P_{reac} - \varepsilon_{rad} A_{bed} \sigma [(T+\Delta T)^4 - T^4] = 0 \qquad (7)$$

With $\varepsilon_{rad} = 0.8$ (estimate based on [8]), $A_{bed} = 1.6*10^{-3}$ m^2. and P_{reac} (reaction heat generation rate) = 0.9 W. The result is:

$\Delta T = 2.2$ °C.

The reactor temperature required to account for the increase in naphthalene conversion in the inert bed experiments, assuming that no parallel reaction of naphthalene with oxygen occurs, can be calculated from the experimental data. With air at $T = 850$ °C. we get: $k_1 = 7.9$ s^{-1}. Substitution in eq.(4):

$T = 958.6$ °C.

This result, compared to the calculated adiabatic temperature rise of the gas mixture of 103 °C. and the calculated increase in bed temperature of 2.2 °C. leads to the conclusion that the increase in naphthalene conversion is caused by a parallel direct reaction of part of the oxygen with naphthalene. The change in product distribution of the naphthalene decomposition products, benzene and toluene, see Figure 4, gives further support to this conclusion.

The positive effect of secondary air addition on catalyst deactivation rate is evident from the experimental naphthalene exit concentration presented in Figure 3. The naphthalene exit concentration is significantly increasing during run sequences without secondary air feeding while the naphthalene exit concentration remains constant during secondary air feeding. This effect is caused by oxygen gasification of carbon on the catalyst surface.

To check the claims of Mudge *et al.* [3] the experimental procedure was repeated using BASF G1-22 catalyst. The catalyst was reduced in the reactor at 900 °C. using a 25 vol. % H_2 in N_2 mixture. Reactor operation at 800 °C, without secondary air feeding, resulted in conversion of the tar components naphthalene, benzene and toluene to a value below the detection limit. Tar component concentration in the reactor outlet remained below the detection limit while the reactor temperature was lowered in 50 °C. increments to a lowest operation temperature of 450 °C. The run was stopped when at 450 °C the pressure drop across the catalyst bed increased rapidly due to the formation of carbon in front of the bed. After this front the catalyst surface appeared visually unchanged. The result shows that a commercial Nickel catalyst like BASF G1-22 is a very efficient naphthalene decomposition catalyst in an artificial biomass producer gas mixture even without use of secondary air. In the paper of Mudge *et al.* [3], no direct comparison between tar conversion with or without secondary air was made.

3.2 Bench scale reactor results

We will now discuss results of the experiments with and without secondary air carried out at the University of Madrid using real biomass producer gas. The first experiment was carried out using dolomite in the first catalytic cracker while in the second experiment silica was used in this reactor. The aim was again to study the possible synergetic effect of catalytic surface and oxidative reactant on tar decomposition rate. Table 7 gives an overview of the experimental conditions for both experiments.

Table 7. Experimental conditions for secondary air tar cracking experiments at the University of Madrid.

Operational parameters	Dolomite		Silica	
	run period 1	run period 2	run period 1	run period 2
$\Phi_{M,bio}$ [kg/s]	$2.00*10^{-4}$	$2.00*10^{-4}$	$1.97*10^{-4}$	$1.94*10^{-4}$
m_{H2O} [-]	0.18	0.18	0.24	0.24
$\Phi_{Mair,1}$ [kg/s]	$3.03*10^{-4}$	$2.51*10^{-4}$	$3.03*10^{-4}$	$2.51*10^{-4}$
$\Phi_{Mair,2}$ [kg/s]	0	$0.52*10^{-4}$	0	$0.58*10^{-4}$
$E.R._{overall}$ [-]	0.28	0.28	0.30	0.30
$T_{g,c}$ [K]	1073	1073	1073	1073
$T_{g,f}$ [K]	823	823	823	823
W_{Solids} [kg]	0.30	0.30	0.63	0.63
$T_{r1,c}$ [K]	1123	1123	1123	1123
W_{Nickel} [kg]	0.08	0.08	0.264	0.264
$T_{r2,c}$ [K]	1073	1073	1073	1073

Analysis results are presented in Table 8.

Table 8. Analysis results with respect to gas composition and tar content after the gasifier and after the first and second catalyst bed.

	Dolomite					
Component	without secondary air			with secondary air		
	Gasifier	Crack_1	Crack_2	Gasifier	Crack_1	Crack_2
H_2 [vol. %]	6.0	13.2	15.5	7.3	14.0	16.1
CO [vol. %]	17.3	16.2	17.4	19.4	17.0	17.2
CO_2 [vol. %]	14.7	15.5	17.3	12.3	13.6	17.2
CH_4 [vol. %]	4.5	4.5	3.6	5.9	5.1	4.5
C_2H_4 [vol. %]	1.8	1.2	0.57	2.3	1.2	0.55
N_2 [vol. %]	55.7	49.4	45.6	52.9	49.1	43.8
Tar [mg/Nm^3]	16500	4700	100	27400	1000	55
ζ_{tar} [-]	-	0.72	0.98	-	0.96	0.94
LHV [MJ/Nm^3]	5.6	5.9	5.6	6.8	6.3	6.1
$\Phi_{v,gas}$ [Nm^3/s]	$3.53*10^{-4}$	$3.97*10^{-4}$	$4.35*10^{-4}$	$3.08*10^{-4}$	$3.31*10^{-4}$	$3.00*10^{-4}$
P [W]	1977	2342	2436	2094	2085	1830

	Silica					
	without secondary air			with secondary air		
Component	Gasifier	Crack_1	Crack_2	Gasifier	Crack_1	Crack_2
H_2 [vol. %]	5.2	7.4	22.7	6.2	8.1	19.1
CO [vol. %]	16.9	15.6	21.8	18.8	16.4	21.1
CO_2 [vol. %]	15.1	16.7	14.5	15.3	17.9	15.7
CH_4 [vol. %]	4.0	4.0	0.2	4.9	4.7	2.5
C_2H_4 [vol. %]	1.2	1.1	0.0	1.5	1.3	0.50
N_2 [vol. %]	57.6	54.7	40.8	53.3	50.9	40.5
Tar [mg/Nm^3]	9800	5020	10	13500	2100	40
ζ_{tar} [-]		0.488	0.998		0.844	0.981
LHV [MJ/Nm^3]	4.9	4.9	5.3	5.8	5.5	6.0
$\Phi_{v,gas}$ [Nm^3/s]	$3.20*10^{-4}$	$3.36*10^{-4}$	$5.10*10^{-4}$	$2.87*10^{-4}$	$3.00*10^{-4}$	$4.74*10^{-4}$
P [W]	1568	1646	2.703	1665	1650	2844

Table 8 shows the relatively large increase in gasifier exit gas tar concentration due to a decrease in primary air flow rate during the second part of the experiments.
Analogue to the discussion of the micro reactor data we calculate the kinetic constants k_1, k_1^* and k_1^{ox}. For results see Table 9.

Table 9. Values of k_1, k_1^* and k_1^{ox} calculated from experimental tar conversions in the first tar cracker of the bench scale equipment at UCM.

bed material	secondary air	k_1 [s^{-1}]	k_1^{eff}	k_1^* [s^{-1}]	k_1^{ox} [s^{-1}]
Silica	no	3.4*	1.9	-	-
Silica	yes	3.4	4.8*	-	2.8*
Dolomite	no	3.4	5.7*	8.3*	-
Dolomite	yes	3.4	12.5*	8.3	6.8*

*: determined in the experiment listed in this row

Again, the results show that the use of secondary air in a silica tar cracking bed results in a larger relative increase in overall tar decomposition rate constant than when using dolomite. Absolute values of the rate constant are, just as with the micro reactor experiments, smaller with silica than with dolomite both for the situation with and without secondary air.

Contrary to the results of the micro cracker experiments the absolute value of k_1^{ox} is larger with dolomite than with silica. This result can be explained by the temperature rise in the dolomite bed during secondary air feeding. Figure 5 presents experimental axial catalyst bed temperature profiles for the situation with and without secondary air feeding.

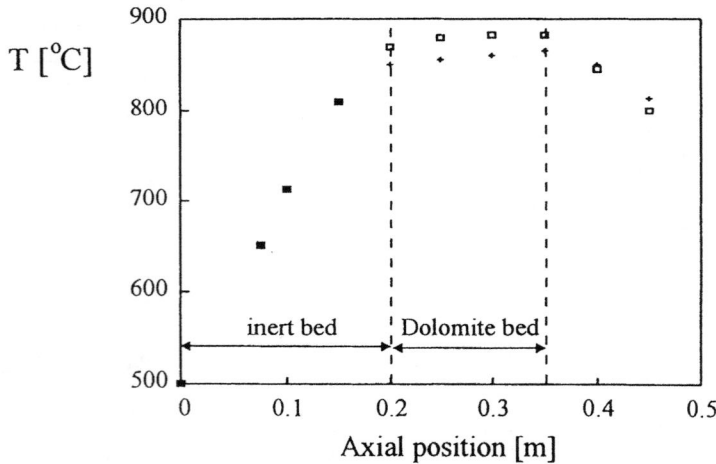

Fig. 5. Axial temperature profile in the centre of the dolomite bed for the situation without (+) and with ($\square$) secondary air feeding for experiment #1.

In the dolomite bed secondary air feeding resulted in a temperature increase from 851 °C. to 869 °C. The apparent high value of k_1^{ox} in the dolomite experiment is thus caused by the combined effect of temperature increase and direct reaction of oxygen with tar components. In the silica bed air addition did not affect bed temperature. This can be caused by the slower oxygen reaction rate with inert material. With dolomite the air introduced in the radial centre of the bed causes a local temperature increase. With inert material more radial dispersion of the air takes place before oxygen is depleted, leading to a more even radial heating of the bed. This way the temperature increase caused by air addition is compensated by a decrease in power output of the electric oven controlled by a thermocouple at the inside of the cracker wall. With dolomite build up of a radial temperature profile during air addition was observed experimentally.

We have calculated the equilibrium composition of the biomass producer gas at the conditions of tar cracker 1 using the AspenPlus™ flow sheet program ver. 9.1.3. In the calculations, the composition of the gas with respect to Methane and Ethylene were adjusted iteratively to meet the experimental concentrations. Concentrations of these two components do not normally go to equilibrium at typical catalytic cracker conditions. Calculations were carried out assuming complete carbon burn-out in the gasifier. Results are presented in Table 10.

Table 10. Experimental and equilibrium gas composition in mol %, for conditions see Table 7.

| Component | dolomite | | | silica | | |
	without air	with air	calculated	without air	with air	calculated
H_2	13.2	14.0	13.0	7.4	8.1	12.36
CO	16.2	17.0	16.1	15.6	16.4	12.97
CO_2	15.5	13.6	16.4	16.7	17.9	18.18
CH_4	4.5	5.1	4.8	4.0	4.7	4.4
C_2H_4	1.2	1.2	1.2	1.1	1.3	1.2
N_2	49.4	49.1	48.5	54.7	50.9	50.86

In agreement with [9] the results show that with calcined dolomite in the cracker reactor the water gas shift reaction more closely approaches equilibrium then with inert bed material.

4 Conclusions

The results from the experiments carried out using the micro cracker reactor at TPS lead to the conclusion that use of air in the catalytic cracker introduces an extra parallel naphthalene decomposition reaction. With dolomite the naphthalene decomposition reactions on the catalyst surface dominate the overall conversion rate making the relative effect of air addition less significant than with silica. However, with dolomite use of secondary air has the important effect of significantly reducing the deactivation rate of the catalyst.

Result obtained at the University of Madrid (UCM) have shown that, in agreement with results obtained at TPS, the biggest relative increase in overall tar decomposition rate constant with secondary air is observed with supposedly inert bed material. However, overall values of the naphthalene (TPS) and tar (UCM) decomposition constants are smaller with silica than with dolomite both for the situation with and without secondary air.

The significant improvement in observed tar conversion in the dolomite cracker with use of secondary air is the result of the combined effect of both the direct conversion of tar components with oxygen and the resulting temperature increase in the catalyst bed. At a constant overall equivalency ratio the addition of secondary air caused an increase of the gasifier outlet gas tar concentration. Overall however, the addition of secondary air resulted in a sharp decrease in tar concentration in the outlet gas of the first tar cracker. In commercial scale gasifiers the ratio of secondary air and primary air will be limited by the resulting decrease in carbon burn out in the gasifier.

Acknowledgements

This work was carried out under the EC AIR program project no. AIR2-CT93-1436. Bengt-Göran Espenäs of TPS, Sweden is thanked for his kind cooperation. Alberto Orio and Ian Narváez are thanked for carrying out the fluidised bed gasification experiments at UCM, Madrid.

Notation

A	=	Pre exponential Arrhenius factor [s^{-1}]
A_{bed}	=	bed surface area [m^2]
$c_{Naphth,0}$	=	naphthalene inlet concentration [mole/m^3]
d_i	=	inside tube diameter [m. or mm.]
E_a	=	energy of activation [J/mole]
E_{rad}	=	radiation heat emission [W/m^2]
$E.R._{overall}$	=	Overall equivalency ratio based on primary + secondary air [-]
$F_{Naphth,0}$	=	molar Naphthalene inlet flow rate [mole/s]
k_1	=	first order reaction rate constant of the homogeneous thermal naphthalene and tar decomposition reaction [s^{-1}]
k_1^*	=	first order reaction rate constant of the heterogeneous catalytic naphthalene and tar decomposition reaction [s^{-1}]
k_1^{ox}	=	first order reaction rate constant of the oxygen induced naphthalene and tar decomposition reaction [s^{-1}]
k_1^{eff}	=	first order reaction rate constant of the overall naphthalene or tar decomposition reaction [s^{-1}]
LHV	=	Lower Heating Value of the gas [MJ/Nm^3]
m_{H2O}	=	mass fraction H$_2$O in biomass [-]
P	=	$\Phi_{v,gas}$ LHV, cold gas power output [W]
P_{reac}	=	reaction heat production rate [W]
R_{Naphth}	=	Naphthalene decomposition rate [mole/m^3 s]

T	=	temperature [K]
$T_{g,c}$	=	gasifier centre temperature [K]
$T_{g,f}$	=	gasifier freeboard temperature [K]
$T_{r1,c}$	=	Dolomite bed cracker centre temperature [K]
$T_{r2,c}$	=	Nickel catalyst bed centre temperature [K]
ΔT_{ad}	=	adiabatic temperature rise [K]
$\Phi_{Mair,1}$	=	primary air mass flow rate [kg/s]
$\Phi_{Mair,2}$	=	secondary air mass flow rate [kg/s]
$\Phi_{M,bio}$	=	biomass mass flow rate [kg/s]
$\Phi_{v,gas}$	=	volumetric gas flow rate [m^3/s]
ε	=	bed void fraction [-]
ε_{rad}	=	radiation emission coefficient [-]
σ	=	Stefan-Boltzman radiation constant: $5.67 \ 10^{-8}$ [W/m^{2}K^4]
ζ_{Naphth}	=	Naphthalene conversion, calculated from reactor in- and outlet Naphthalene concentrations [-]
ζ_{tar}	=	tar conversion, calculated from reactor in- and outlet tar concentrations [-]

References

1. Corella, J., Aznar, M.P., Delgado, J., and Aldea, E. (1991) Steam gasification of cellulosic wastes in a fluidised bed with downstream vessels, *Ind. Eng. Chem. Res.*, **30**, pp. 2252-2262.
2. Simell, P.A. and Bredenberg, J.B. (1990) Catalytic purification of tarry fuel gas, *Fuel*, **69**, pp. 1219-1225.
3. Mudge, L.K., Baker, E.G., Brown, M.D. and Wilcox, W.A. (1988) Catalytic destruction of tars in biomass-derived gases, in: Research in Thermochemical Biomass Conversion; Eds. A.V. Bridgewater and J.L. Kuester, pp. 1141-1155.
4. Salo, K. (1995) Presentation of Enviropower Oy, Finland during the Seminar on Power Production from Biomass II, Espoo, Finland.
5. Waldheim, L. (1995) Presentation of Termiska Processer AB, Sweden during the Seminar on Power Production from Biomass II, Espoo, Finland.
6. Alden, H., Espenäs, B.G. and Rensfelt E. (1988) Conversion of tar in pyrolysis gas from wood using a fixed Dolomite bed, Research in Thermochemical Biomass Conversion, Int. Conference Phoenix, 2-6 May, p. 987.
7. Simell, P.A., Kurkela, E. and Ståhlberg, P. (1992) Formation and catalytic decomposition of tars from fluidized-bed gasification, in "Advances in Thermochemical Biomass Conversion" vol. 1. p. 265 Ed. A.V. Bridgewater.
8. Perry's Chemical Engineers Handbook, 'Heat transmission by radiation' ed. D. Green, p. 10-53, McGraw-Hill Inc., 1984.
9. Garcia, X. and Hüttinger, S.A., 1990, Erdöl und Kohle, **43**, p. 273.

BIOMASS GASIFICATION WITH STEAM AND OXYGEN MIXTURES AT PILOT SCALE AND WITH CATALYTIC GAS UPGRADING. PART I: PERFORMANCE OF THE GASIFIER

M.P. Aznar*, J. Corella+, J. Gil, J.A. Martín, M.A. Caballero, A. Olivares, P. Pérez, E. Francés.

Dept. of Chem. and Environm. Eng., University of Saragossa, 50009 Saragossa, and (+) Dept. of Chem. Eng. , University `Complutense´of Madrid, 28040 Madrid, Spain.

Abstract
Biomass gasification with steam + O_2 mixtures is studied at small pilot plant (10-20 kg/h) scale. The gasifier used is a turbulent fluidized bed of 15 cm. i.d. and 3.3 m high. The pilot plant has a downstream slip flow to study the catalytic upgrading of the raw gas. A guard bed with dolomite and a catalytic bed with a steam reforming catalyst are used and eight different commercial catalysts have been tested till date. Product distribution from the gasifier, including gas proposition and tar content in the gas, are here shown in detail at different (steam + oxygen)/ biomass and (H_2O/O_2) ratios and gasification temperatures (800-880°C).

1 Introduction

To say when a plant or facility can be called pilot is difficult in biomass gasification. In coal and oil processing it is quite well established the size or throughput which has to have a plant to be called pilot of small, medium of high scale. According to several rankings, the facility used in this work can be called (small) pilot plant. Apart from the well known pilots of TPS AB, VTT and Battelle Columbus [1] there are not many more pilot plants in the world for biomass gasification. Some facilities in some companies have been used sometimes for biomass gasification tests, but they are not devoted exclusively to gasification of biomass. The pilot here used was presented in the 8th EC Conference on biomass in Vienna, the first experiment in it was made in September 1994, and it can be called a 3rd generation or advanced pilot plant because two other ones were used before by the same people [2]. They were based on a bubbling fluidized bed and on a circulating and multisolid system, Battelle Columbus type [3].

Biomass gasification can be made with at least three main gasifying agents: steam, steam + O_2 mixtures, and air (including air with some steam). Oxygen enriched air has not been proved yet as gasifying agent for biomass. At University of Saragossa (UZ), Corella and co-workers studied the gasification with steam from 1984 till 1992 [ref. 4, for instance]. Biomass gasification with air is being studied in many Institutions, including a branch of this group at University of Madrid. Gasification with steam + O_2 mixtures is much less studied and known. So, the work here presented has been made gasifying with steam + O_2 mixtures only.

Steam gasification is highly endothermic and, to get an authothermal process, very complex circulating systems have to be used [3]. We decided thus to leave these circulating systems (used at UZ from 1989 till 1993) and to use some oxygen in the gasifying agent to provide the heat necessary for the steam gasification. Oxygen is expensive of course, but a more valuable gas is obtained when compared with the one from gasification with air.

2 Description of the Pilot Plant.

The pilot plant is shown in Fig. 1. It is based on a fast or turbulent fluidized bed gasifier of 15 cm i.d. and 3.3. m. high. It is continuously fed near the bed bottom (13 cm up from the distributor plate). It can gasify (as it as been tested in short intervals) till 40 kg/h but it usually works at 10 kg biomass/h, which is equivalent to a throughput of 564 kg biomass/h m^2. Higher throughputs have been achieved, but this presents a problem of the `disposal' of the produced gas (the pilot plant and stack is located in the centre of the campus, in the centre of the city).

The biomass feeding system consists of two hoppers, of 80 and 70 litres. The base of the lower one is connected to the dosifier screw. There are two lock valves (i.d. 15 cm), located at the top and at the bottom of the upper hopper. The lower hopper is never put in contact with the outside, by means of the alternating opening of the valves (one of them is always closed).

The screw feeder system consists of a dosifier screw (6 cm diameter, 90 cm length) connected to a 0.75 H.P. engine with a variant frequency equipment, which enables a wide range of solid flow rates (from 3 kg/h till 40 kg/h). The second and high speed screw (6 cm diameter, 60 cm length) leads the biomass into the gasifier and is connected to a 1 H.P. engine, which provides a high speed (690 r.p.m.) in order to lead biomass inside the fluidized bed in less than one second. This second screw has an external water-cooling system to avoid the pyrolysis of biomass inside this high speed screw.

Due to the flexibility required in this pilot plant for the gasifying agent, two separated lines were installed for steam and for air/O_2. So, this plant lets the study of different steam/O_2 (or steam/air) ratios. To obtain the required flow rate of steam, a pump leads

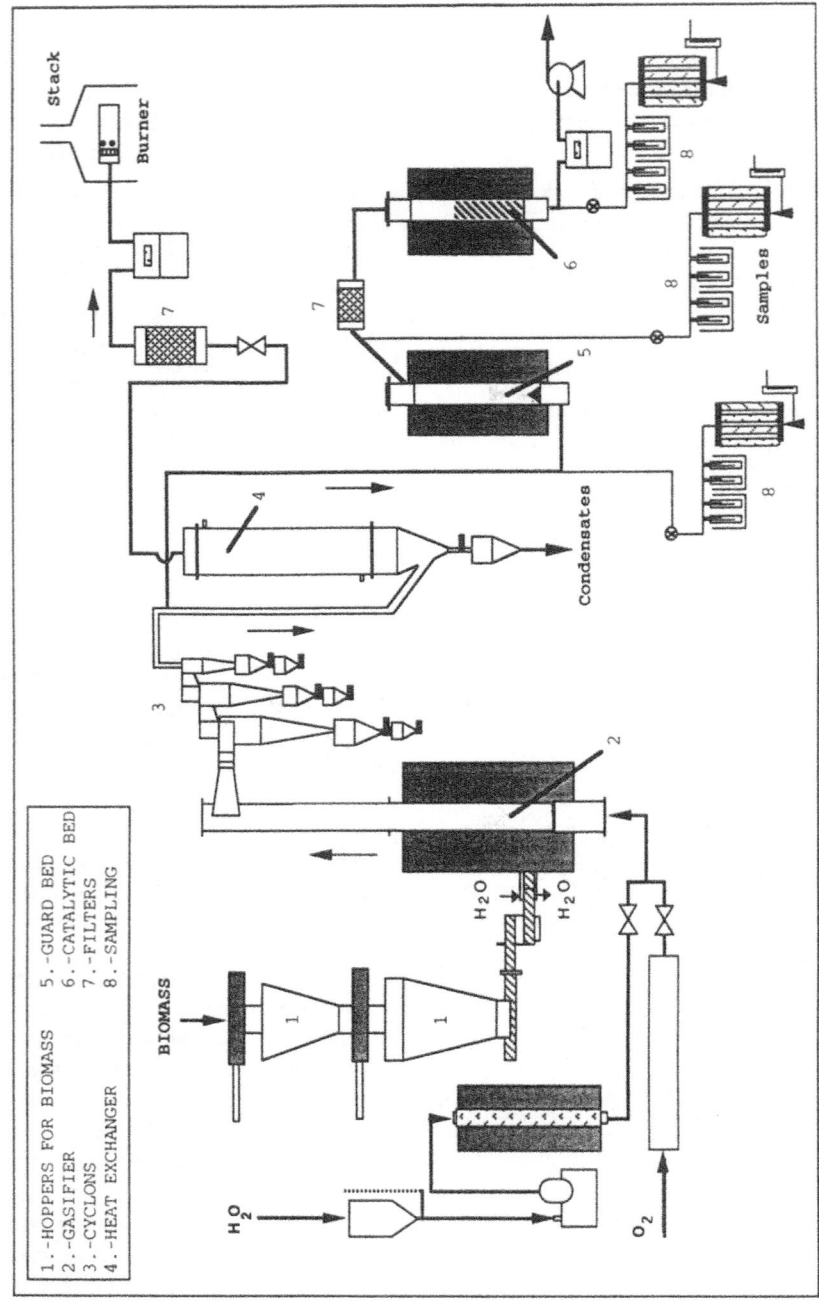

Fig. 1. AFB pilot plant for biomass gasification with downstream catalytic reactors in a slip flow

the water into a boiler of a 5 cm i. d. and 200 cm. length filled with 5 mm steel balls. The heat is provided by an external oven (4 kW) and two internal resistances (3 kW).

2.1 Gasifier: The gasifier is made of a refractory steel with 4 mm wall thickness. The total height is 3.3 m. with two parts of 150 cm each one, to provide an easy assembly/disassembly for cleaning. The gas distributor plate has 22 bubbling caps of different diameter (15, 20 and 50 mm) with 1 mm holes around their perimeter. The fluidized solid is silica sand (40-50 cm height as bulk fixed). The necessary heat to reach the starting temperature (500 °C) of the experiment, and to maintain it at stops, is delivered by an external oven (13 kW) at the lower part of the gasifier and by three resistances (3 kW) at its freeboard.

Biomass used as feedstock was pine (*Pinus pinaster*) wood chips of -6.0 + 1.0 mm with a moisture of 13 wt%. and whose characterisation was given elsewhere [4].

After the gasifier, there is a system of three cyclones in series of 16, 12 and 8 cm diameter, respectively. They separate the elutriated solid particles with an efficiency of 99% for particles of 122, 80 and 48 μm respectively.

A valve opens downstream a secondary line or slip flow in which a small and constant flow is processed in two catalytic beds. An heat exchanger (to cool the gases and to condensate water and tars) is placed in the main exit gas flow. Gas and condensates flow inside the tubes for a better further cleaning. The gas passes then through a glass wool filter and a flow meter. The gas is finally led to a burner and stack.

2.2 Catalytic or testing reactors: In the slip or by-pass flow, a sucking pump delivers a constant flow for the two catalytic reactors connected in series as Fig. 2 shows. Both reactors of 4 cm. i.d. are made of refractory steel. They have external and independent heating systems (ovens of 2.2 kW each one). They can work up to 950 °C. Their length is 108 cm. for the first or guard bed, and 69 cm. for the second or catalytic (Ni) one.

The first reactor is a (fixed or fluidized) guard bed, with calcined dolomite where a preliminary tar elimination (of about 80-95%) is made. Temperatures used were between 800 and 900°C. A glass wool filter separates the elutriated solid at the exit.

The second reactor is a fixed bed of 41 mm i.d. with a commercial reforming, nickel based, catalyst. A scheme of it is shown in Fig. 2.

After the catalytic bed the gas is cooled in a small heat exchanger in order to condensate the steam contained in the gas stream. Finally, the gas passes through a flow meter and the sucking pump and is led to a burner.

2.3 Tar sampling and analysis: There are three tar and gas sampling devices which work independently. One tar and gas sampling device is shown in Fig. 3. This sampling equipment and method is similar to those used by VTT in Finland [5] and by UCM in Madrid[6]. It consists of four impinger flasks in two cool (ice-water and dry ice) baths and of an equipment for gas sampling. There are also a flow meter and a vacuum pump in each sampling line. The first sampling point is located just before the first reactor (or guard bed), to sample the raw gasification gas entering the bed of

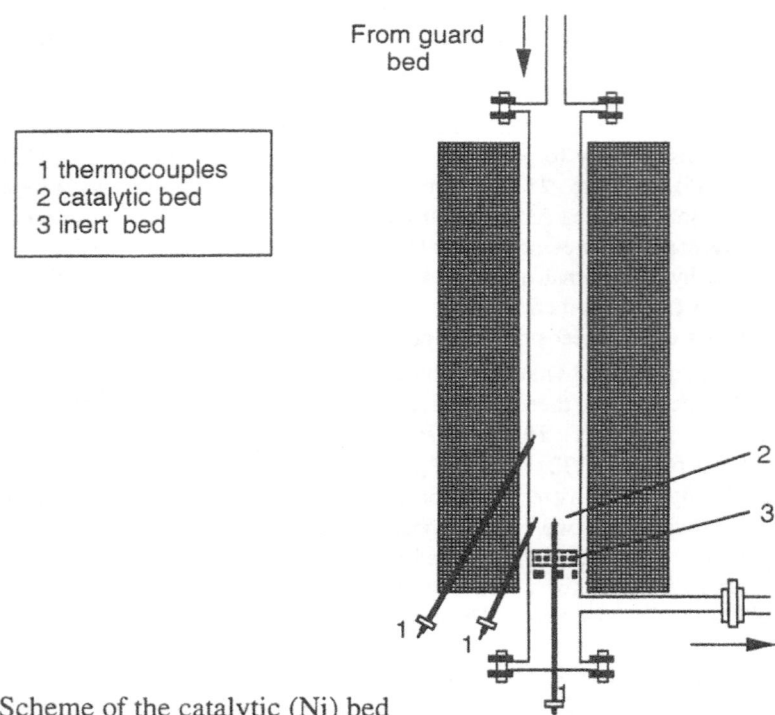

From guard
bed

1 thermocouples
2 catalytic bed
3 inert bed

2
3

1
1

Fig. 2. Scheme of the catalytic (Ni) bed

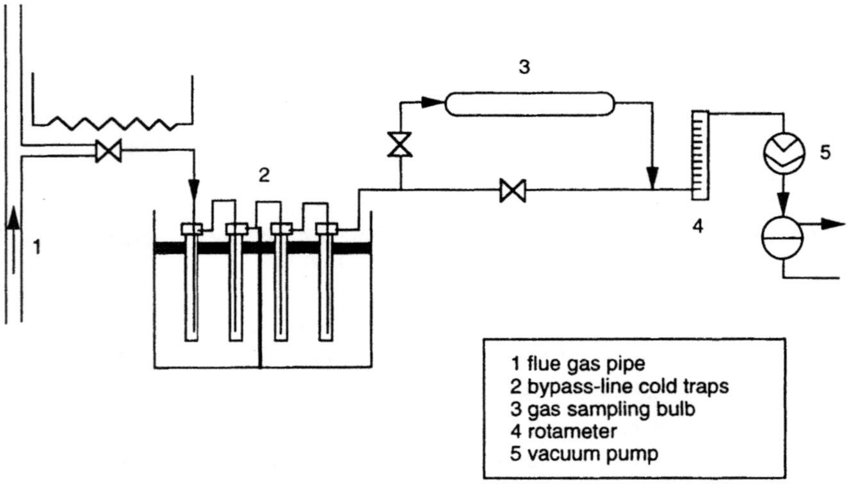

3

2

5

1

4

1 flue gas pipe
2 bypass-line cold traps
3 gas sampling bulb
4 rotameter
5 vacuum pump

Fig. 3.. Scheme of the tar and gas sampling system

dolomite. The second sampling point is located after the guard bed, which corresponds to the inlet of the catalytic bed. The third sampling point is located after the catalytic (Ni) reactor (see Fig. 3).

2.4 Control equipment: The most advanced technology available in this field has been used for the process control of the installation. Temperatures are controlled with several thermocouples located inside the bed, using Phase Angle Control (P.A.C.) technology for the electricity supply to the oven, proportionally to the 4/20 mLA control signal. Controllers have Proportional Integral Differential (P.I.D.) algorithms, with antiovershoot function. Thus, the temperature control is extremely stable, without using a more conventional control based on the oven wall temperature.

Oxygen flow rate is accurately measured with a mass flow controller with an answer speed of about 600 ms. Hopper seals have been solved with differential pressure measures and P.I.D. loops for the N_2 flow control, which is totalized to calculate the mass balance.

The control set has been built in 19" rack and it integers all the control actions and process automatisms. All the process signals are registered at real time by a data acquisition unit, which checks the 32 signals in 0.1s. Graphical information of the process evolution is delivered at real time.

2.5 Operating conditions: As an example, Table 1 shows some main experimental conditions in the three reactors involved in the pilot plant for some typical experiments.

3 Process development

Besides the chemical analyses for gas composition and tar content shown in following paragraphs, there are other important aspects or data related to correct operation of the pilot plant which are difficult to be given in figures. They are concerned with the continuous upgrading, revamping and/or improvement of the pilot plant. Due to its complexity (two parallel flows with different pressure drops in each one), to get a continuous and good long term operation was not an easy job at the beginning. A lot of not expected problems appeared which had to be solved before each experiment. Solving such problems made that each run or experiment needed by average one month. Some problems encountered were:
  i) Difficulties in keeping a steady feeding, due to the shape and low density of the biomass. The biomass even get static electricity in the hoppers.
 ii) During the start up and shut down of the plant, noticeable enlargements and contractions appeared in it, which caused gas leaks.
iii) In long-term experiments, obstructions or plugging in valves, bends and other cold points along the facility, because of the tar condensation, were sometimes observed.
 iv) ΔP in the slip flow can increase with time-on-stream by filter or beds plugging.

Table 1. Experimental conditions in some typical gasification experiments with steam and oxygen mixtures.

Run number		2	3	4	5	6
time on stream	h	1.2	1.8	7	9.2	12.4
FEEDSTOCK		pine sawdust	pine sawdust	pine sawdust	pine sawdust	pine sawdust
moisture	w%	13	13	9	13	13
dp	mm	(-4+0.5)	(-4+0.5)	(-4+0.5)	(-4+0.5)	(-4+0.5)
biomass feed rate	kg/h	10.0	5.0	7.4	7.6	6.4
throghput	kg/h m2	566	283	419	430	362
steam	kg/h	4.64	2.32	2.32	3.44	3.45
oxygen	kg/h	4.13	2.06	2.06	3.05	3.07
(steam+oxygen)/biomass daf	kg/kg	1.00	1.00	0.70	0.98	1.20
steam/oxygen	mol/mol	2	2	2	2	2
GASIFIER						
Tbed	°C	833	863	880	873	884
Tfreeb.-center	°C	801	813	871	800	839
Tfreeb.-top	°C	583	608	667	576	646
solid bed		sand	sand	sand	sand	sand
dp	mm	(-0.63+0.40)	(-0.63+0.40)	(-0.63+0.40)	(-0.63+0.40)	(-0.63+0.40)
L	cm	35	35	42	45	45
umf	cm/s	12	12	12	12	12
u	cm/s	33	59	35	39	42
u/umf		2.8	4.9	2.9	3.3	3.5
τ	s	1.06	0.59	1.20	1.15	1.08
GUARD BED						
catalyst		Malaga dolomite	Malaga dolomite	Malaga dolomite	Malaga dolomite	Malaga dolomite
W	g	236	236	247	293	232
Tbed	°C	835	884	839	880	839
Twall	°C	730	787	843	889	846
L	cm	25	25	17	28	23
dp	mm	(-2.0+1.0)	(-2.0+1.0)	(-2.0+0.8)	(-1.6+1.0)	(-1.6+1.0)
umf	cm/s	50	50	50	40	40
u	cm/s	17	29	130	50	49
u/umf		0.3	0.6	2.6	1.3	1.2
τ	s	1.47	0.86	0.19	0.56	0.47
CATALYTIC BED						
catalyst		BASF G25-1S	BASF G25-1S	UC C11-9-062	UC C11-9-062	UC C11-9-062
W	g	184	184	244	257	241
Tbed	°C	737	734	732	734	726
Twall	°C	735	750	772	724	713
L	cm	11	11	13	13	15
dp	mm	(-4+0.5)	(-4.0+0.8)	(-2.5+0.8)	(-0.8+0.2)	(-0.8+0.2)
u	cm/s	15	26	133	47	47
τ	s	0.66	0.42	0.10	0.28	0.31

4 Results in the whole process / plant.

Experience has been gained at least with three different. parts of the process: gasifier, bed of dolomite or guard bed, and catalytic (Ni) bed. This communication is mainly focused on results gained from the gasifier. Due to lack of space, results from the beds of dolomite and of catalyst will be shown in forthcoming papers.

Gas composition after each reactor: The gas composition obtained after the three reactors in run no. 5 is shown, as an example, in Fig. 4. It is firstly observed how both beds of dolomite and of nickel catalyst reform or modify quite a lot the gas composition. Major trends or observations from Figure 4 are:

i) The H_2-content increases from 18 vol% at the gasifier exit till 52% at the catalytic bed exit.

ii) The CO_2 decreases in both beds, indicating there is also dry-reforming (and not only steam-reforming).

iii) The CO decreases in the bed of dolomite mainly, perhaps by the shift reaction due to the high steam content and temperature in both reactors.

iv) The gas content in CH_4 and light hydrocarbons also decreases, in the catalytic (Ni) reactor mainly.

Tar content: The tar content in the flue gas after the three reactors involved in this process is shown in Fig. 5 for a typical experiment, run no. 6. The tar content after the gasifier, of about 7 g/Nm^3 (in this case), is reduced to 250 mg/Nm^3 after the bed of dolomite and to only about 10 mg/Nm^3 after the bed of the nickel catalyst. These figures, in this example, clearly indicate how this catalytic gas cleaning is effective and how very low tar contents can be (easily) achieved.

5 Composition of the raw gas after gasifier with steam + O_2 mixtures.

The amount of reactants fed to the gasifier will be described by two parameters or ratios:
- (H_2O + O_2) / Biomass (kg/h / kg daf/h)
- (H_2O / O_2) in the gas or gasifying agent.

5.1 Gas Composition: Gas composition at the gasifier exit is shown in Fig. 6 for several operating conditions. The CO amount decreases as the gasifying agent-to-biomass ratio increases and less oxygen is fed. On the other hand, CO_2 increases when such ratio or the oxygen fed increase. The hydrogen decreases slightly when this said ratio increases. Although it should decrease when there is more oxygen in the fed, it increases due to the increase of temperature caused by it. The hydrocarbons content decreases as the gasifying agent-to-biomass ratio and the oxygen fed increase, as it was expected.

The hydrogen-to-carbon monoxide ratio in the raw gas depends on the gasifying agent-to-biomass ratio, but it is very influenced by the steam/oxygen ratio in the fed, Fig.7.

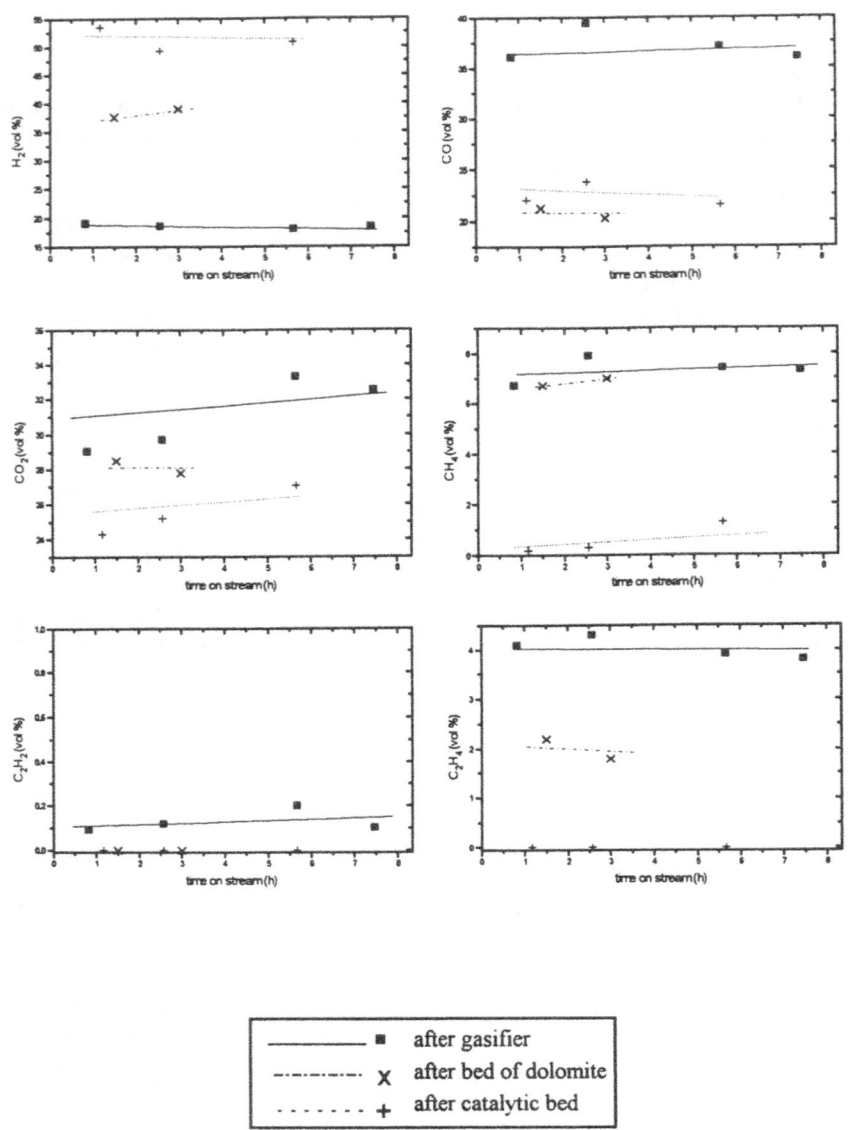

		after gasifier
	x	after bed of dolomite
	+	after catalytic bed

Fig. 4. Gas composition at the exit of each reactor vs. time on stream. Run no. 5.

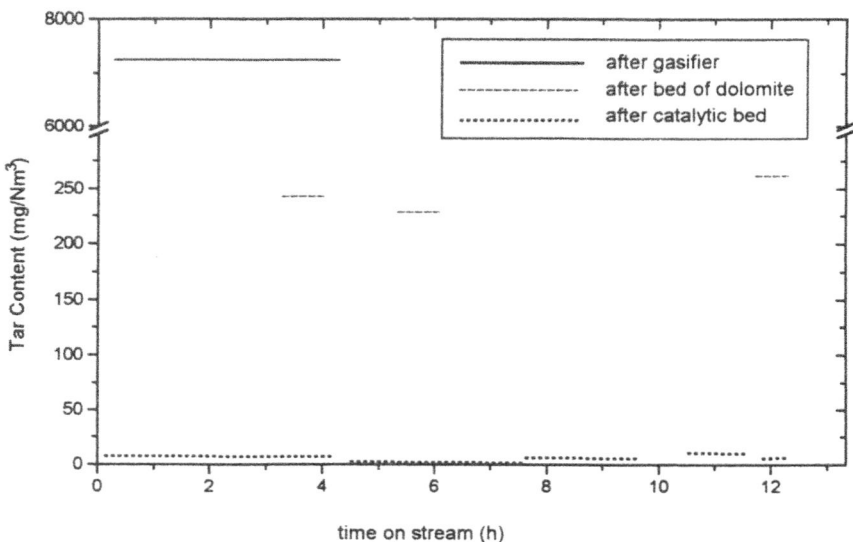

Fig. 5. Tar content in the gas at the exit of each reactor.
Run 6 (cat.: UC C11-9-062).

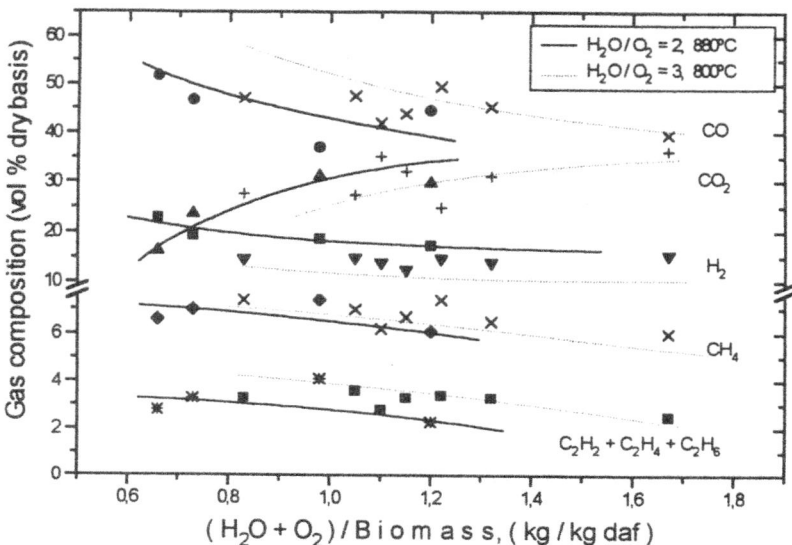

Fig. 6. Gas composition at the gasifier exit vs. gasifying agent to biomass
ratio. [(H_2O / O_2) = 2 and 3 (mol/mol)].

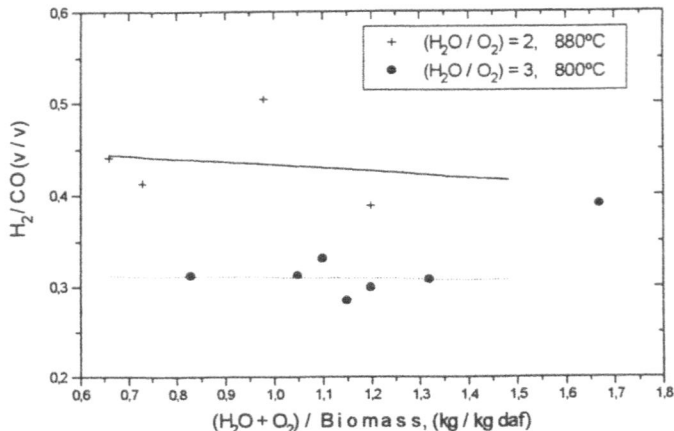

Fig. 7 . (H_2/CO) at the gasifier exit vs. gasifying agent to biomass

ratio. [(H_2O / O_2) = 2 and 3 (mol / mol)].

For further analysis on the activity of the dolomites and commercial catalysts used under these atmospheres is very important to know the steam content in the flue gas. Such amount is shown in Fig. 8. Note that it is higher than the typical 10-20 vol% steam content when the gasification is made with air (in which the steam comes from the biomass and air moisture content).

The low heating value (LHV) of the raw gas (dry basis) from the gasifier is shown in Fig. 9. Note how, i) it depends much on the $(H_2O + O_2)$ / Biomass and (H_2O / O_2) ratios used, and ii) it is clearly higher than the typical value of 4-6.5 MJ/Nm³ obtained for gasification with air.

5.2 Product distribution: Char yield. It decreases as the gasifying agent to biomass ratio increases, as it was expected and as Fig. 10 quantitatively shows for two (H_2O / O_2) ratios and two gasification temperatures.

The important **tar content** in the raw gas is shown in Fig. 11 at different (gasifying agent / biomass) and (H_2O / O_2) ratios (at two different gasification temperatures). These values can be compared with the ones obtained in gasification with air [6] or with steam [4]. Tar content or tar yield gasifying with (steam + O_2) mixtures is quite lower than gasifying with pure steam, and it is of the same order of magnitude than the one obtained gasifying with air.

Gas yield: The gas yield, as dry and wet gas, is shown in Fig. 12. An increase in the gasifying agent to biomass ratio produces a higher gas yield as it was expected.

5.3 Thermal efficiency: The apparent thermal efficiency (calorific value of the dry gas / calorific value of the biomass) is shown in Fig. 13. These values are quite high at lowest gasifying agent to biomass ratio, which would mean an useful gasification process (if it were not for the price of the O_2, which fed also has to be taken into account in an economical evaluation).

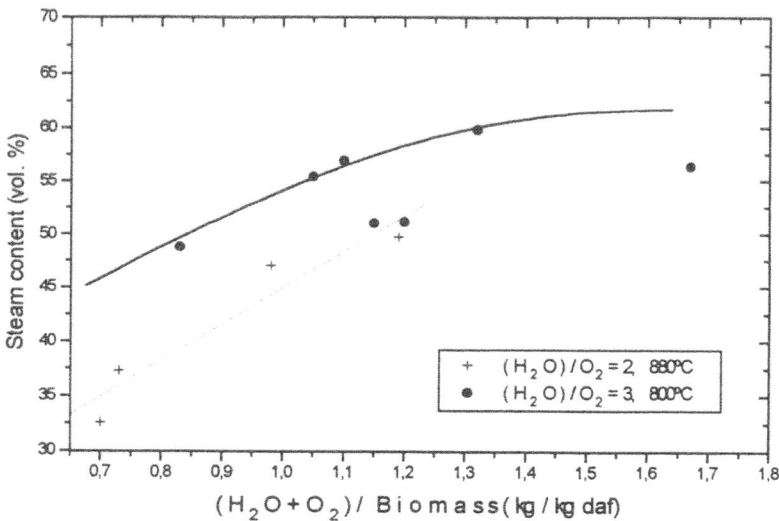

Fig. 8. Steam content in the gas at the gasifier exit vs. gasifying agent
to biomass ratio. [(H_2O / O_2) = 2 and 3 (mol / mol)].

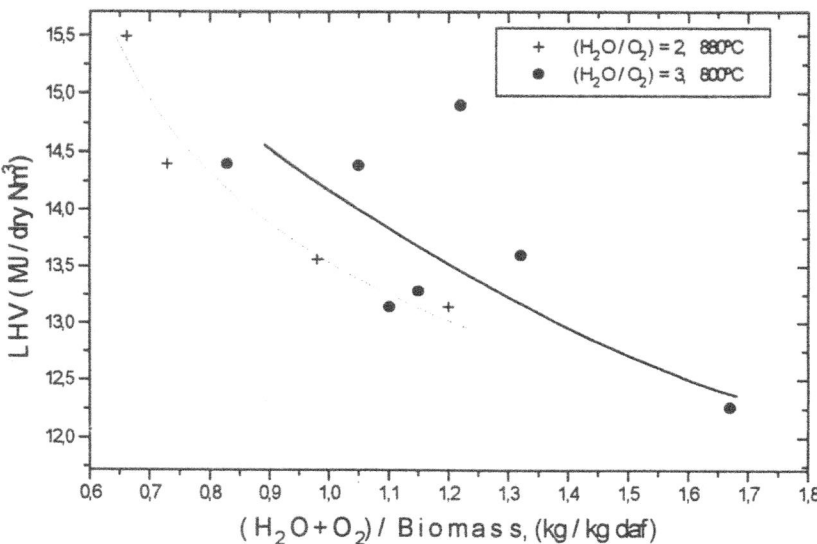

Fig. 9. Low heating value of the dry gas at the gasifier exit vs. gasifying
agent to biomass ratio. [(H_2O / O_2) = 2 and 3 (mol/mol)].

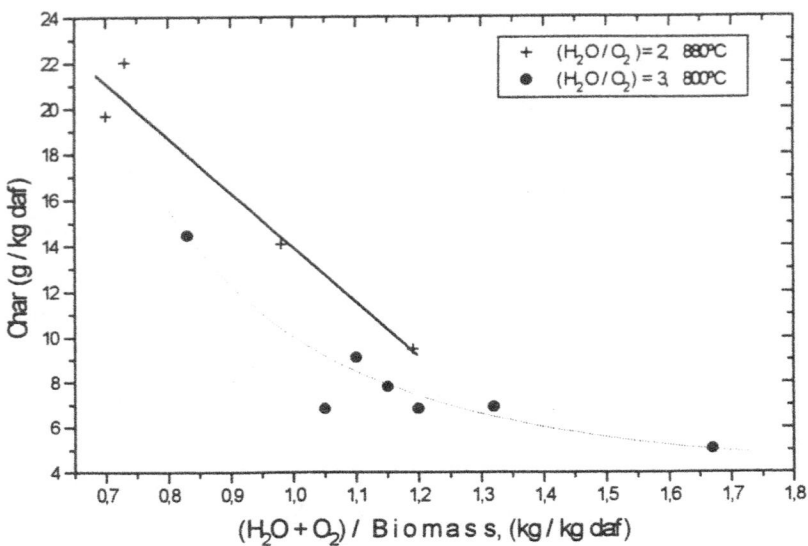

Fig. 10. Char produded (and collected in the cyclons) vs. gasifying agent to biomass ratio.

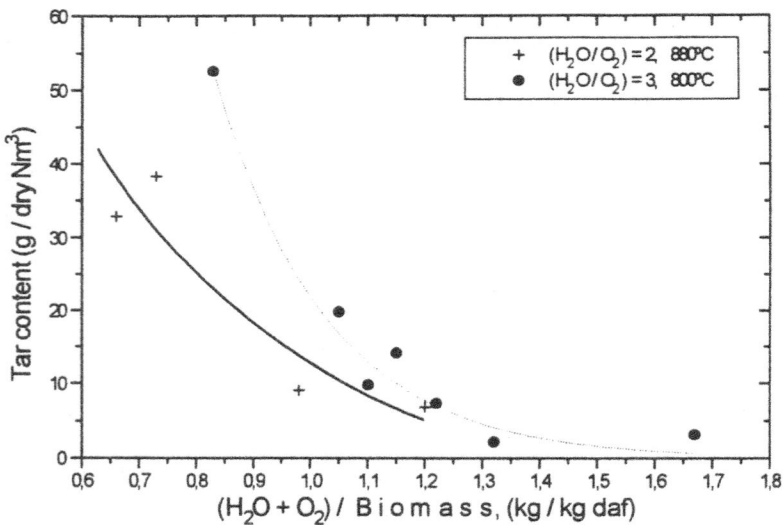

Fig. 11. Tar content in the gas at the gasifier exit vs. gasifying agent to biomass ratio. [(H_2O / O_2) = 2 and 3 (mol/mol)].

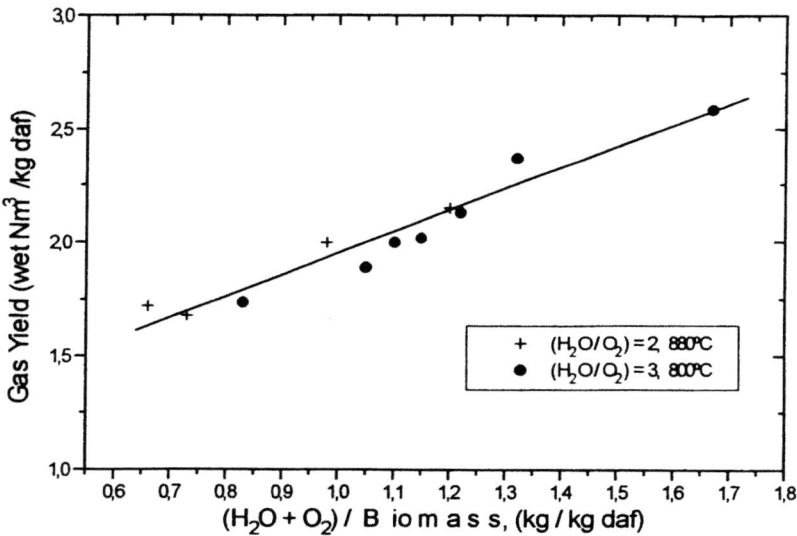

Fig. 12. Gas Yield at the gasifier exit vs. gasifying agent to biomass ratio. [$(H_2O / O_2) = 2$ and 3 (mol/mol)].

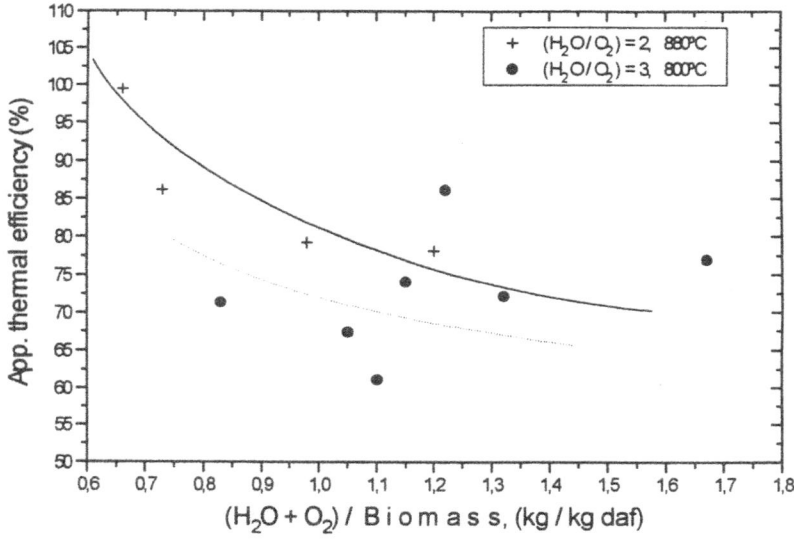

Fig. 13. Apparent thermal efficiency vs. gasifying agent to biomass ratio. [$(H_2O / O_2) = 2$ and 3 (mol/mol)].

6 Acknowledgement.

This work has been carried out thanks to the financial help of the European Commission, DGXII, under the projects of the AgroIndustry Programme, Contract no. AIR2-CT93-1436, and of the JOULE 2 Programme, Contract no. JOU2-CT93-0399.

7 References.

1. A. E. Brown, E.J.M.T. van den Heuvel. "Producer Gas Quality Requeriments for IGCC Gas Turbine Use", Biomass Technology Group BV, (1996).

2. M.P. Aznar, J.A. Borque, I.J. Campos, J.A. Martín, E. Francés and J. Corella. "New Pilot Plant for Biomass Gasification in Fluidized Bed and for Testing Downstream Catalyst", Biomass for Energy Enviroment Agriculture and Industry, (1995), 2, 1520-1527.

3. J. Herguido; J. Corella; J. Artal, J.E. Garcia-Bordejé, "Results with a multisolid circulating fluid bed pilot plant for the improved steam gasification of biomass". In Biomass of Energy Industry and Environment. Grassi, Collina, Zibetta (Eds.). Elsevier Appl. Sci.. (1992), pp 792-796.

4. J. Corella; M.P. Aznar; J. Delgado. E. Aldea, " Steam gasification of cellulosic wastes in fluidized beds with downstream vessels", Ind. Eng. Chem. Res., (1991), 30,2252-2262

5. P.A. Simell, J.B.-son Bredenberg. "Catalytic purification of tarry fuel gas." Fuel (1990), 69, 1219-1225.

6. I. Narvaez, J. Corella, M.P. Aznar, A. Orio. "Biomass gasification with air in an atmospheric bubbling fluidized bed ". Ind. Eng. Chem. Res. (1996) , in press.

Printed by Printforce, the Netherlands

DEVELOPMENTS IN THERMOCHEMICAL
BIOMASS CONVERSION

DEVELOPMENTS IN THERMOCHEMICAL BIOMASS CONVERSION

Volume 2

Edited by

A.V. Bridgwater
Director of the Energy Research Group,
Department of Chemical Engineering and Applied Chemistry,
Aston University, Birmingham, UK

and

D.G.B. Boocock
Chair of the Department of
Chemical Engineering and Applied Chemistry,
University of Toronto, Toronto, Canada

SPRINGER-SCIENCE+BUSINESS MEDIA, B.V.

First edition 1997

© 1997 Springer Science+Business Media Dordrecht
Originally published by Chapman & Hall in 1997
Softcover reprint of the hardcover 1st edition 1997

ISBN 978-94-010-7196-3   ISBN 978-94-009-1559-6 (eBook)
DOI 10.1007/978-94-009-1559-6

A Catalogue record for this book is available from the British Library

Library of Congress Catalog Card Number: 96–86524

∞ Printed on permanent acid-free text paper, manufactured in accordance with ANSI/NISO Z39.48-1992 (Permanence of Paper).

CONTENTS

VOLUME 1

Chemicals from pyrolysis liquid

Upgrading pyrolysis products

Laboratory experimentation

Pilot and demonstration

COMBUSTION

Overview

Fundamentals

SYSTEM STUDIES

WORKSHOPS

PREFACE

There have been many developments in the science and technology of thermo-chemical biomass conversion since the previous conference on *Advances in Thermochemical Biomass Conversion* in Interlaken, Switzerland, in 1992. This fourth conference again covers all aspects of thermal biomass conversion systems from fundamental research through applied research and development to demonstration and commercial applications to reflect the progress made in the last four years. All aspects of bioenergy systems are covered from pretreatment through to end-user applications with increased consideration paid to the environmental benefits and problems of implementing bio-energy systems.

There was an excellent response with over 200 papers offered and over 180 delegates from 29 countries attending the conference. The programme was divided into five main areas covering pyrolysis, pretreatment, gasification, combustion and system studies and this division is reflected in the structure of these conference proceedings. Each main section was preceded by a state-of-the-art review to provide a focus for the ensuing presentations and an authoritative reference. All the papers included have been subject to a full peer review process.

As with any international conference, an important aim was to exchange ideas and discuss problems with fellow researchers, as well as to hear about the latest research and development and applications. A workshop programme was included to encourage this interaction in areas of interest selected by participants. The resultant workshop reports provide a summary of topical problems and opportunities.

The major developments in the bioenergy field include the growing appreciation of the contribution that can be made to mitigating greenhouse effects through the increased awareness of the environmental responsibilities for implementing bio-energy systems and also the growing emphasis on commercialisation. The academic community will, of necessity, continue to provide the lead in open dissemination of research results to stimulate the development of new ideas. These will translate into new products and processes and provide the justification for continued support of centrally supported R&D. The value of these conferences lies in providing an opportunity for industry to hear about the latest developments and provide an opportunity for their representatives to meet the active researchers and develop better links. A further development is the increased commercial activity which results from overt governmental and institutional support for bio-energy. This industrial pull for newer and better products and processes will stimulate researchers in the private and public sector and should deliver at least some of the promised benefits of the bio-energy sector.

Tony Bridgwater and Dave Boocock
October 1996

ACKNOWLEDGEMENTS

We would like to express our sincere appreciation to those people and organisations who have provided support:

- IEA Bioenergy, through the operating agent represented by C Wallace; and the thermochemical related conversion activities themselves represented by S Babu, J Hustad, K Mackie, Y Solantausta and T Bridgwater;
- Natural Resources Canada, represented by E Hogan;
- National Renewable Energy Laboratory, USA, represented by R Overend;
- The Department of Trade and Industry, UK, represented by N Barker;
- VTT, the Technical Research Centre of Finland, represented by K Sipila.

The scientific committee were very supportive throughout the preparation and running of the conference. Their encouragement and efforts in publicising the meeting, refereeing papers, providing constructive feedback on the programme, organising workshops and chairing sessions was much appreciated.

M. Antal, USA	S. Babu, USA
N. Bakhshi, Canada	T. Beenackers, Netherlands
M. Connor, Australia	J. Corella, Spain
B. Delmon, Belgium	D. Elliott, USA
B. Graham, Canada	E. Hogan, Canada
W. Kaminsky, Germany	B. Krieger-Brockett, USA
J. Kuester, USA	K. Mackie, New Zealand
R. Maggi, Belgium	K. Maniatis, Belgium
D. Meier, Germany	T. Milne, USA
T. Nussbaumer, Switzerland	R. Overend, USA
E. Rensfelt, Sweden	C. Roy, Canada
D. Scott, Canada	K. Sipila, Finland
K. Sjostrom, Sweden	Y. Solantausta, Finland
H. Stassen, Netherlands	S. Yokoyama, Japan

At a personal level, particular thanks are due to Karen Dowden and Claire Humphreys who provided the Conference Administration. We also wish to thank the research staff from Aston University who provided considerable support before, during and after the meeting.

CATALYTIC HOT GAS CONDITIONING OF BIOMASS DERIVED PRODUCT GAS

MARK A. PAISLEY, PE
Battelle
Columbus, Ohio, U.S.A.

Abstract

Biomass gasification provides the potential to efficiently and economically produce a renewable source of a clean gaseous fuel suitable for power generation or synthesis gas (syngas) applications. An important side benefit of the use of biomass is the effective minimization of the primary greenhouse gas, carbon dioxide (CO_2), by providing a means to close-loop the CO_2 cycle. However, high molecular weight hydrocarbon constituents (tar) in the product gas from gasification can complicate the downstream uses of the gas. This paper discusses both the development of a low cost, disposable catalyst system that can eliminate these heavy hydrocarbons from the gas and the use of the catalyst in conjunction with the Battelle high-throughput gasification process for power generation and synthesis applications.

Keywords: biomass gasification, tar cracking, catalysis, power generation

1 Introduction

Biomass resources currently supply over 3 quads (3×10^{15} Btu) the nation's energy supply and are projected to provide between 17 and 55 quads of the nation's energy needs in the future [1]. Because they use the photosynthesis reaction for their growth, biomass resources provide a means to reduce the quantity of carbon dioxide (CO_2) emitted to the atmosphere since CO_2 is used in the growth cycle of the biomass feedstocks. These inherent characteristics along with their high chemical reactivity make biomass resources an attractive alternative to conventional fossil fuels.

Several high efficiency technologies are being developed by the U. S. Department of Energy and others to utilize these renewable, environmentally attractive biomass fuels. Chief among these developing technologies is gasification. Biomass gasification, while generally similar to coal gasification practiced commercially in the early part of the 20th century, presents a unique product slate that enables its ready acceptance for commercial applications. This unique product slate with biomass gasification also presents a unique set of challenges to commercial application of the technology.

In gasification, the biomass is converted into a mixture of gases that can later be used as a clean, gaseous fuel for heating, power generation, or as a feedstock for chemical synthesis. Chemical synthesis generally requires the use of a medium-Btu (non-nitrogen diluted) gas with minimal contaminants for optimum conversion to chemicals. Medium-Btu gas containing primarily CO and H_2 can be generated using oxygen as the gasifying medium in a single-vessel system, but the costs of pure oxygen are high. Alternatively, the gas can be generated by heating the biomass materials indirectly by using a circulating heat carrier. The resulting gas is nitrogen free, in contrast to air blown gasification, but still contains some level of hydrocarbons in addition to the CO and H_2.

Biomass derived fuel gases must behave as the fuels they are replacing not only in terms of heating value but in terms of other physical and chemical properties associated with their use. Both gas turbines and fuel cells require that the fuel gas used be essentially free of contaminants. The removal of these contaminants is a major barrier to the end use of biomass derived fuels, regardless of the conversion technology used to generate them.

The primary contaminants present in these biomass derived fuels are particulates from ash or reactor bed materials and condensible organic species (tars). This paper discusses removal of these tar materials through catalytic cracking .

1.1 Tar Characteristics

The characteristics of the specific conversion process will determine, to a large extent, the physical and chemical properties of the tars formed in a gasification process. Shorter reactor residence times and higher biomass heat-up rates will generally produce tars that are lower in molecular weight than those produced in gasifiers with longer residence times and slower heating rates.

The condensible organic materials produced during gasification have been shown to be of two basic types. These are commonly referred to as "secondary oils" and "tertiary oils" or tars. Biomass conversion systems can produce either or both of these types of materials depending on the specific process operating conditions. Chemically these materials can be aliphatic (straight chain) compounds or be highly aromatic (rings) compounds and include a host of polynuclear aromatic materials (PAH).

2 Hot-Gas Conditioning

For a biomass derived synthesis gas to be successfully utilized for power generation or chemical synthesis, the composition must be modified, the methane concentration reduced, and the high molecular weight hydrocarbons that make up the so-called "tar" must be eliminated. Purification and composition adjustment is collectively referred to as hot-gas conditioning. Hot-gas conditioning is done by passing the raw gasifier product gas over a solid catalyst in a fluidized or fixed bed reactor under temperature and pressure conditions that essentially match those of the gasifier. Similar conditions are used to minimize the heating requirements for the process. As the raw gas passes over the catalyst, the tar compounds react on the catalyst surface with excess steam from gasification to produce additional CO and H_2,. This is essentially *in-situ* catalytic steam reforming of tar, and generically can be represented by [2]:

$$C_nH_m + nH_2O \rightleftharpoons nCO + (n + \frac{m}{2})H_2$$

$$\Delta H_{298} > 0 \tag{1}$$

Simultaneously, the $H_2/(CO)$ ratio is adjusted via the water-gas shift reaction

$$CO + H_2O \rightleftharpoons CO_2 + H_2$$

$$\Delta H_{298} = -41.2 \ kJ/mol \tag{2}$$

The concentration of methane also must be reduced when the synthesis gas is to be used for methanol production -- the CH_4 concentration needs to be below 2-3 vol% for conventional gas phase methanol synthesis [2,3] because it accumulates in the recycle loop. Methane is acceptable for power generation systems because it is a fuel. Substantial methane is made during Fischer-Tropsch hydrocarbon synthesis so CH_4 removal is less important in this application.

The methane steam reforming reaction is equilibrium limited, and for catalysts that take the reaction to equilibrium, the temperature must be above about 800°C for complete methane elimination.

$$CH_4 + H_2O \rightleftharpoons 3H_2 + CO$$

$$\Delta H_{298} = +206 \ kJ/mol \tag{3}$$

Power generation requires high levels of gas cleanup especially in gas turbine systems. Gas turbines are particularly sensitive to particulate and condensible materials in the gas as well as alkali material. Alkali can be transported both on fine particulates and in the vapor phase [4]. Current improvements in gas turbine design for higher efficiency operation, will further increase the cleanup requirements for these power generation systems.

2.1 Additional Requirements of a Hot-gas Conditioning System

For advanced power generation cycles such as fuel cells, large H_2 concentrations are needed in the product gas necessitating modification of the gasifier output. As stated earlier, methane reforming is not as important for power generation, because it contributes significantly to the heating value of the product gas, however, tar destruction and the water-gas shift are still process requirements. Biomass derived product gas must be interchangeable with convention high-value fossil fuels such as natural gas or distillate oil, and therefore biomass gasification product gases must behave as the fuels they are replacing not only in terms of heating value but in terms of fuel cleanliness as well.

3 Development of a Biomass Gasification System

Battelle has developed an indirectly heated biomass gasification process [5]. Development efforts on the Battelle High Throughput Gasification Process were initiated in 1977. Detailed process development activities were initiated in 1980 with the construction and start-up of a process research unit (PRU) at Battelle's West Jefferson Laboratory. These PRU investigations, conducted during the mid-1980s demonstrated the technical feasibility of the gasification process and provided the basis for a detailed process conceptual design to be generated.

3.1 Process Description
The Battelle biomass gasification process produces a medium-Btu product gas without the need for an oxygen plant. The process schematic in Figure 1 shows the two reactors and their integration into the overall gasification process. This process uses two physically separate reactors: (1) a gasification reactor in which the biomass is converted into a medium Btu gas and residual char and (2) a combustion reactor that burns the residual char to provide heat

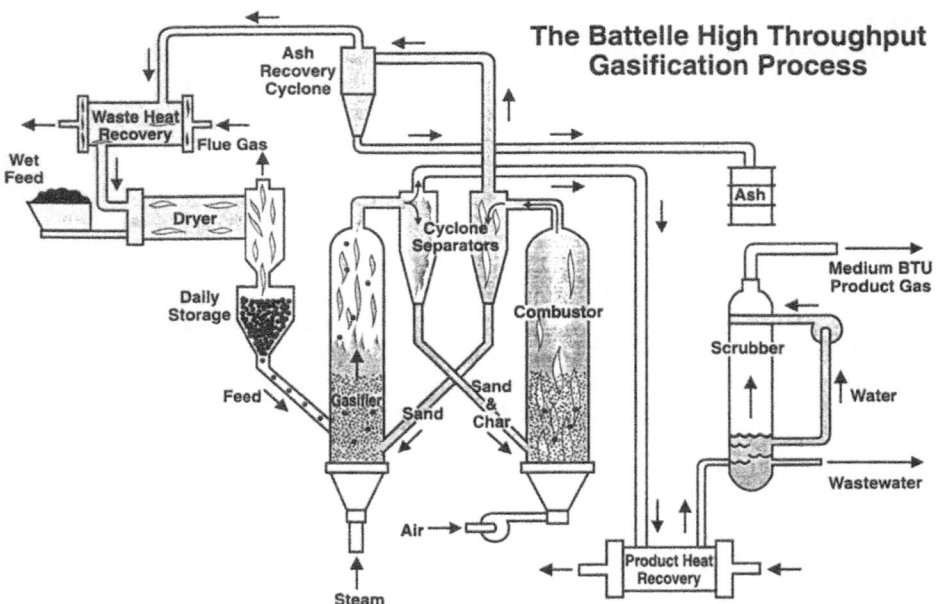

Figure 1. The Battelle Biomass Gasification Process

for gasification. Heat transfer between reactors is accomplished by circulating sand between the gasifier and the combustor.

The Battelle Process, unlike conventional gasification processes, utilizes circulating fluidized bed reactors to take advantage of the inherently high reactivity of biomass feedstocks. The reactivity of biomass is such that throughputs in excess of 3000 lb/hr-ft^2 can be achieved. In other gasification systems throughput is generally limited to less than 200 lb/hr-ft^2.

The high heatup rates inherent in the system by indirect heating of the incoming biomass with a circulating sand phase along with the short residence times in the gasification reactor reduce the tendency to form condensible tar-like materials.

Even though the production of tars is reduced in the process, the small quantities that are produced can present significant problems to downstream unit operations if not removed from the gas. The tar compounds found in raw product gas from the Battelle gasification process consist mainly of so called secondary and tertiary pyrolysis products [6].

3.2 Process Improvement Opportunities

Preparation of the medium-Btu gas prior to compression for power generation or chemical synthesis provides a significant opportunity for improvements in process operation and economics. Such preparation includes the hot-gas conditioning operations of hydrocarbon (tar) destruction, methane reforming, and water-gas shift reactions. By performing these operations as part of a hot-gas conditioning system, the overall process efficiency of the gasification system can be optimized. In this context, hot-gas conditioning refers to gas purification and composition adjustment that takes place at nearly the same process conditions as the gasification reactions, that is no significant heating or cooling of the gases prior to chemical change. The absence of heating or cooling operations reduces the overall cost of the hot gas conditioning options by eliminating process steps.

4 Product Gas Conditioning for Power Generation

The tars produced in biomass gasification are primarily polynuclear aromatic hydrocarbons that are generally considered difficult to destroy by any process other than combustion. Conventional scrubbing systems are generally the technology of choice for removal of these tars from the product gas. Scrubbing cools the product gas and results in a waste water stream that must ultimately be disposed of. A scrubber based system is shown schematically in Figure 2. In the Battelle process, between 0.5 and 1 percent of the dry wood weight exits the gasifier as condensible tar material. In conventional scrubbing systems, these tars are quenched producing a waste water stream laden with tars that must be cleaned before discharge from the plant.

In addition to the waste water cleanup issue, wet scrubbing systems can leave a small fraction (10 to 30% by weight of the original 0.5 to 1%) of the tars in the gas as a fine aerosol mist that is difficult to remove, but can create problems in downstream equipment such as compressors or turbines. Such tar deposition has been observed during integrated system testing of the biomass gasifier-compressor-turbine system at Battelle.

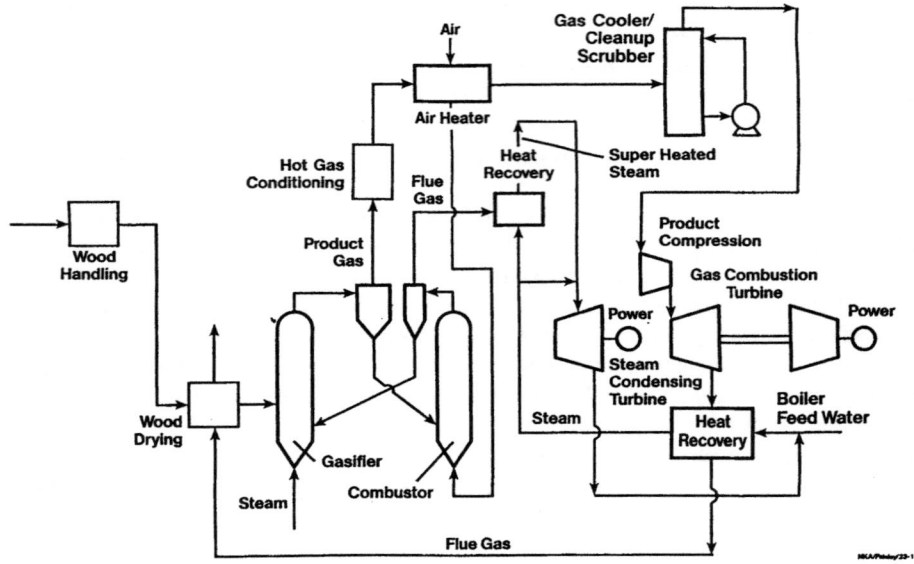

Figure 2. Schematic of a Battelle Gasification Combined Cycle Power Generation System

4.1 A New Catalyst for Hot-gas Conditioning

Battelle has identified and tested a low cost, disposable cracking catalyst (DN34) that can be used to condition the product gas prior to its use in a power generation system. This catalyst has been evaluated by Battelle at the PRU scale in slip-stream and full scale tests, and has been studied at the microscale at the National Renewable Energy Laboratory (NREL) to better understand catalyst activity and selectivity [7]. DN34 is a proprietary catalyst described in US Patent 5,494,653. Because of the proprietary nature of the catalyst, its precise formulation and properties are not discussed in this paper.

4.2 DN34 Testing with the Battelle PRU

DN34 has been evaluated under actual biomass gasification conditions in two separate reactor systems in the Battelle gasification PRU. These systems were, a 6-inch diameter slip stream reactor and a 36-inch diameter "full-flow" reactor designed to process the complete output of the PRU gasifier. Both reactors were operated as fluidized beds with superficial velocities of approximately 30.5 cm/sec. Gaseous hourly space velocities (GHSV), expressed as $m^3_{gas}/m^3_{catalyst}/hr$ (at operating conditions), were controlled between 1500 hr^{-1} and 2500 hr^{-1} for the experiments reported in this paper. The following paragraphs discuss results of the experimental operation and the results obtained during operation of these two reactor systems.

During the tests, tar concentration in the product gas was measured by sampling using a modified method 5 (MM5) train. The MM5 train consist of a series of 5 impingers place in an ice bath followed by a dry gas meter to measure the quantity of gas sampled. After sampling, the impingers were rinsed with toluene to remove the tars and water collected. Toluene and water were removed from the samples by heating in an oven at 65C overnight. Tar concentration at the inlet and outlet of the catalyst bed was then compared to determine tar destruction efficiencies.

4.2 Catalyst Operating Conditions

Catalyst DN34 was operated at conditions similar to the gasifier conditions (815C and 17 kPa) to simulate adiabatic operation of the catalyst system in both the slip stream and "full-flow" catalyst chambers. Data generated on tar destruction showed that high levels of tar destruction were possible with the catalyst and that following gas cooling in the PRU scrubbers, essentially all tars were removed from the product gas [5,8].

4.4 Results of Testing with DN34

Tar concentrations measured in the product gas from the Battelle gasifier are typically 1.6×10^{-2} kg/m^3. These tars are highly aromatic in character and are relatively insoluble in water. As discussed above, these tars can be removed by conventional water based scrubbing, however such scrubbing methods can leave significant amounts of tar behind in the product gas. Table 1 below shows tar removal rates as the tar laden product gas from the Battelle PRU passes through the venturi and spray tower scrubbers in the PRU.

Table 1. Tar Removal Efficiencies - PRU Scrubber

Gasifier Temperature, C	Tar Production, kg/m^3	Scrubber Outlet, kg/m^3
~700	3.2×10^{-2}	5.3×10^{-3}
~800	2.3×10^{-2}	1.5×10^{-3}

By providing hot gas conditioning of the product gas using DN34, significant improvement in the quality of the product gas can be realized. Table 2 shows the data generated using DN34 both in the slip stream reactor and in the "full flow" catalyst reactor in the PRU.

When using the "full flow" catalyst chamber, the gas exiting the reactor was passed through the existing PRU scrubbers prior to compression [9]. This process removed all remaining tars from the gas resulting in no measurable residual tars leaving the scrubber system. Figures 3 and 4 show the MM5 impingers at the exit of the gasifier and at the exit of the scrubber with the DN34 catalyst system in use. In the figures, the gas passes through each impinger in series, from left to right. As discussed above, the impingers are all placed in an ice bath to cool the gases. The final impinger contains a charge of silica gel to remove any residual water from the gas and provide an accurate dry gas volume measurement in the dry gas meter.

Figure 4 illustrates the cleanliness of the product gas possible using the DN34 catalyst system. The cleanliness of the product gas is further illustrated by measurements made of the water discharged from the PRU scrubber. Without the catalyst chamber in line the scrubber discharge water (single pass) showed an organic carbon content of 1200 mg/l but with the catalyst chamber in line this value was cut by over 90% to 110 mg/l.

Table 2. Hot Gas Conditioning Test Results

Gasifier Temperature, C	Tar Production, kg/m^3	Catalyst Unit	Catalyst Outlet, kg/m^3
~800	2.3 x 10^{-2}	Full Flow	1.4 x 10^{-3}
815	1.9 x 10^{-2}	Slip Stream	1.4 x 10^{-3}
815*	1.2 x 10^{-1}	Slip Stream	3.6 x 10^{-3}

* Green wood feedstock

The tar destruction activity of DN34 is, as might be expected, highly dependent on reactor temperature. Figure 5 shows the temperature dependence of the cracking reactions in the "full-flow" catalyst system. At normal gasifier operating temperatures such as those listed above, nearly complete destruction of the tars can be expected as shown in the figure. These temperatures correspond to an adiabatic reactor system placed immediately downstream of the gasifier.

4.5 DN34 as a Water Gas Shift Catalyst
A second benefit of hot-gas conditioning, when applied to advanced power generation cycles such as those including fuel cells, is that the hydrogen content of the gas can be increased. Fuel cell applications require high hydrogen content fuel gases. By using catalyst DN34, the hydrogen content of the product gas can be raised to a level so that no further water gas shift reaction is necessary in the fuel cell system. The product gas leaving most biomass gasifiers, including the Battelle gasifier, has a H$_2$ to CO ratio much less than 2:1 and in some cases less than 1:1. The ratio can be adjusted by the water gas shift reaction shown by equation (2) earlier. This reaction requires the presence of a catalyst to enhance reaction rates. Catalyst DN34 is effective in enhancing the water gas shift reaction to produce a gas with a high H$_2$ to CO ratio as shown in Table 3 below. With a hydrogen content in excess of 60% no further conversion of CO to H$_2$ would be necessary for fuel cell applications.

Catalyst temperature was elevated above the gasifier temperature by 50°C during these tests to enhance the water gas shift reactions. These results are of a preliminary nature but they show the significant potential for DN34 as a means of achieving gas compositions suitable for supplying fuel cell power systems.

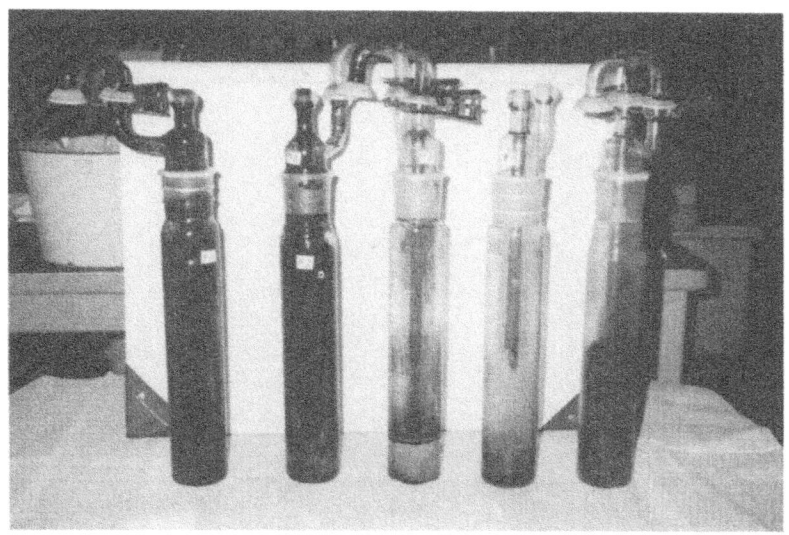

Figure 3. Modified Method 5 Impingers Showing Gasifier Outlet Tar Concentration

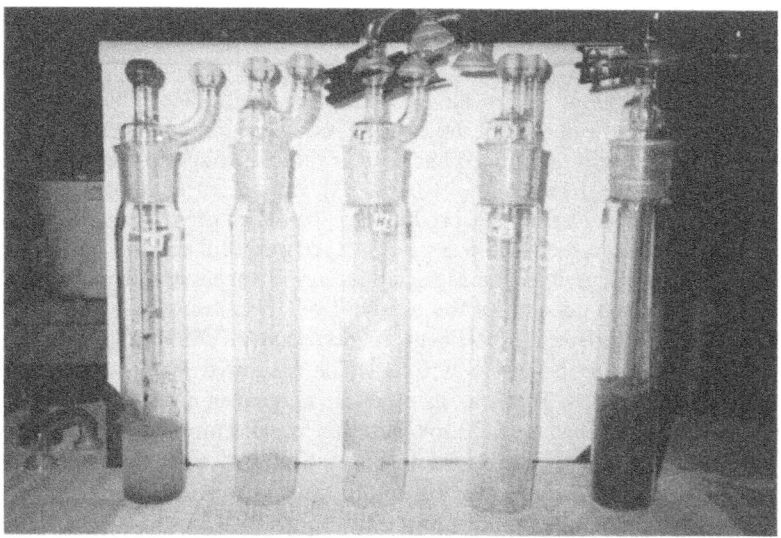

Figure 4. Modified Method 5 Impingers Showing Tar Concentration at Exit of DN34 Catalyst Reactor

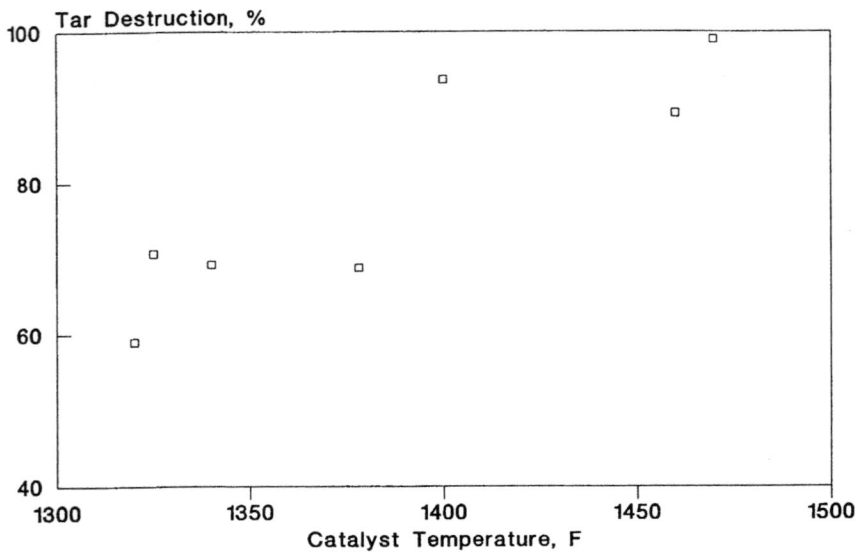

Figure 5. Temperature Dependence of Tar Destruction Activity of Catalyst DN34

5 Microscale DN34 Tests with Model Systems

NREL conducted two sets of microscale experiments using DN34. These were catalyst testing in a 25 gram fixed bed microreactor system using gas chromatography for product analysis and testing with a 4 gram fixed bed reactor directly coupled to a molecular beam mass spectrometer (MBMS). In both experiments, H_2, and a 3:1:1 mixture of CO, CO_2 and CH_4. were used to formulate a synthetic product gas. Two high pressure liquid pumps were used to feed water and benzene which was used as a surrogate for the tar component of the product gas. Benzene was used as a tar surrogate because it is present in the tar in significant quantities and is a difficult component to destroy. DN34 was tested using 50 vol% steam (similar to that measured in Battelle PRU experiments) using a GHSV of 1950 hr^{-1} at 765C. The inlet H_2/CO ratio for the test was 0.68. The measured H_2/CO ratio exiting the microreactor was 2.6 [10]. This was in excellent agreement with the thermodynamic equilibrium value of 2.5 obtained from the low methane steam reforming activity calculation.

The catalytic chemistry of tar destruction over DN34 was examined in microscale experiments where product analysis was done using the molecular beam mass spectrometer (MBMS). The temperature, inlet gas flow rate and composition were similar to those used described above, except that actual tar samples obtained from the Battelle PRU were used in place of benzene and the smaller reactor was close coupled to the sampling port of the MBMS to reduce the condensation of the high molecular weight polycyclic aromatic hydrocarbons. Lower conversions of benzene, naphthalene, phenanthrene and pyrene were observed compared to toluene, phenol, m-cresol, 1-methylnaphthalene, and 1,4-dimethylnaphthalene. This work is discussed in detail elsewhere [11].

Table 3. Water Gas Shift Results Catalyst DN34, Battelle PRU

Component	Raw Product, % dry basis	DN-34 OUTLET
H_2	24.64	60.45
CO_2	18.24	31.84
C_2H_4	4.61	ND
C_2H_6	0.46	ND
C_2H_2	0.56	ND
N_2	8.40	3.36
CH_4	10.46	2.46
CO	32.64	1.89

The microreactor experiments with toluene steam reforming over DN34 have shown that steam dealkylation is an important reaction path. In the case of toluene, benzene and methane are formed. Similarly, substituted naphthalenes will yield naphthalene and methane, water, etc., depending on the identity of the substituent group. The observation that substituted aromatic hydrocarbons are more easily destroyed over DN34 can be explained by higher rates of steam dealkylation than the rates of total steam reforming to produce additional syngas. This chemistry is consistent with observations from real time analysis of the Battelle PRU product gas before and after conditioning with DN34. In that work, the NREL transportable molecular beam mass spectrometer (TMBMS) showed that the raw product gas contained a large number of compounds, with significant concentrations of so-called secondary pyrolysis products (e.g. phenol, cresol, etc.) in addition to the more thermally stable tertiary pyrolysis products (e.g. benzene, naphthalene, etc.). Conditioning with DN34 essentially eliminated the secondary compounds, leaving the tertiaries behind. Details of the TMBMS work are reported elsewhere [12].

5.1 Methane Cracking Activity of DN34
DN34 exhibits relatively poor activity for methane steam reforming (< 20% conversion) as shown in Table 4. Low methane cracking activity is a drawback for synthesis gas applications but is actually a benefit in power applications. This is because the water-gas shift is moderately exothermic (equation 2) and the total concentration of tar is quite small compared to the amount of CO and H_2O undergoing the shift. While the steam reforming reactions for the tar compounds are endothermic (equation 1), the small amount of tar results in a relatively

small overall heat demand. In a system where CH_4 is totally steam reformed (e.g. using a supported Ni catalyst) there will be a significant heat requirement in the syngas conditioning reactor (see equation 3). The use of autothermal reforming by adding a small amount of O_2 has been suggested to avoid adding external heat to the catalytic hot-gas conditioning reactor.

NREL also preformed thermodynamic equilibrium calculations to determine the approach to equilibrium of tar destruction, the water-gas shift and methane steam reforming reactions. Equilibrium compositions were calculated for temperatures between 700C and 1000C in 5°C increments for various measured inlet gas compositions taken during PRU catalyst testing. The predicted H_2/CO ratios were between 2.3 and 5.7 at 815C with the exact value depending on the CH_4 steam reforming activity assumed for DN34 and the inlet gas composition (especially the steam content).

Table 4. Methane Cracking Activity of DN34

Component	Raw Product, % dry basis	DN-34 OUTLET
H_2	18.23	35.86
CO_2	11.74	24.84
C_2H_4	3.67	1.68
C_2H_6	0.21	0.20
C_2H_2	0.39	ND
N_2	28.22	22.76
CH_4	9.78	8.20
CO	27.75	6.47

Equilibrium calculations where all reactions are kinetically allowed gives the result that the concentration of CH_4 is essentially zero at T≥800C. This result is significantly different that the measured results using DN34. To simulate the low CH_4 steam reforming activity of DN34, the calculations argon was substituted for methane in the inlet gas. Additional study of the equilibrium in this system is necessary to completely explain the differences between the equilibrium predictions and the measured H_2/CO ratios shown in Table 3.

6 Power Production Impact Using DN34

A conceptual process design was developed and based on the following criteria: 1) electrical production of approximately 50 MW; 2) an industrial gas turbine system; 3) a dual pressure steam cycle; 4) gas turbine exhaust gas for chip drying. The gasifier to supply the gas turbine fires 818 dry tons per day of whole-tree chipped wood [13].

A gas conditioning system using DN34 provides a reduced burden on the waste water treatment system and provides a clean gas for compression and gas turbine firing. The system illustrated in Figure 2 provides the basis for evaluation of the economic potential of biomass gasification/ gas turbine power generation. The DN34 catalyst system in a plant of this scale, based on current PRU experimentation, will be a fluidized bed approximately 15 feet in diameter (slightly larger than the system combustor) and will operate at the same temperature as the gasifier.

Capital costs were estimated from preliminary designs for the gasifier and combustor vessels and are shown in Table 5. Costs for the gas and steam turbines were obtained from literature data [14]. The total capital investment for the 56MW plant is approximately $58 million or $1037 per installed kilowatt, a cost that is competitive with conventional power generation systems.

Operating costs for system are also shown in Table 5 along with the estimated cost of power from the system. A capital charge rate of 20 percent was used in the analysis which is equivalent to approximately a 10 percent ROI.

7 Conclusions

A new catalyst, DN34, has been identified that shows promise for tar destruction and water gas shift applications in biomass gasification hot-gas cleanup. DN34 shows significant activity for steam reforming substituted polycyclic aromatic tar compounds formed during biomass gasification, and to a smaller extent steam reforming benzene, naphthalene, etc. This catalyst is projected to be a low cost alternative to conventional scrubbing systems and can greatly reduce or eliminate the cost of waste water treatment systems in biomass gasification power generation systems.

Table 5. Cost of Power for a Gasification Cogeneration System at an Existing Site

Capital Cost Estimate	$ x 10^6	$ / kW
Gasifier Plant	15.0	267
Turbines	43.1	770
TOTAL	58.1	1037

Operating Cost Assumptions:
Load Factor = 0.9 (Base-Loaded Plant)
Fuel - Whole Tree Chips - Delivered Cost of $16/ton wet
Capital Charge Rate = 0.200
Operation and Maintenance - Personnel = 12 @ $15/hr
    Supplies Per Year for Operation and Maintenance
    GT/SP Plant - 1 % of Capital Cost; Gasification Plant - 5 % of Capital Cost
Power Delivered/Year = 441.5 x 10^6 kWh

Cost Component	Cost	
	$/Yr	Cents/kWh
Capital	11.6 x 10^6	2.63
Fuel	7.17 x 10^6	1.62
Operation & Maintenance		
Personnel	0.56 x 10^6	0.13
Purchased Supplies	1.18 x 10^6	0.27
Total	20.51 x 10^6	4.65

8 References

1. "Electricity from Biomass", Solar Thermal and Biomass Power Division, US Department of Energy, April, 1993.

2. Satterfield, C.N., Heterogeneous Catalysis in Industrial Practice, 2nd ed., McGraw Hill, New York (1991), chapter 10.

3. Twigg, M.V.(ed), Catalyst Handbook, 2nd ed., Wolfe Publishing Ltd., Frome, England, 1989, chapters 5 and 9.

4. Dayton, D.C. and Milne; Direct Sampling of Inorganic Vapors Released During Biomass Combustion, *Proceedings, Applications of Free-Jet Molecular Beam Mass Spectrometric Sampling*, Estes Park, CO., October 11-14, 1994. Evans, R.J. and Milne, T.A., "Molecular Characterization of the Pyrolysis of Biomass, 1 Fundamentals," *Energy and Fuels* 1 (1987a), 123.

5. Paisley, M, et.al, "Operation and Evaluation of an Indirectly Heated Biomass Gasifier, Phase Completion Report," USDOE, January, 1993

6. Evans, R.J. and Milne, T.A., "Molecular Characterization of the Pyrolysis of Biomass, 2 Applications," *Energy and Fuels* 1 (1987b), 311.

7. Rejai, B., et.al, "Catalyst and Feedstock Effects in the Thermochemical Conversion of Biomass to Liquid Transportation Fuels", Proc. Annual Conf. Af the American Solar Energy Society, Cocoa Beach FL, (1992).

8. Paisley, M, et.al, "Gas Turbine Power Generation from Biomass Gasification," ASME 1994

9. Paisley, M, and Overend, R, "Biomass Gasification for Power Generation," Thirteenth EPRI Conference on Gasification Power Plants, October, 1994

10. Jacoby, W.A., Gebhard, S.C. and Vojdani, R.L, (1995) unpublished results.

11. Gebhard, S.C.; Wang, D.; Overend, R.P.; and Paisley, M.A., "Catalytic Conditioning of Synthesis Gas Produced by Biomass Gasification," *Biomass and Bioenergy* 6 (1-6) (1994a)307-313.

12. Gebhard, S.C.; Gratson, D.A.; French, R.J.; Ratcliff, M.A.; Patrick, J.A.; Paisley, M.A.; Zhao, X.; and Cowley, S.W., *American Chemical Society, Division of Fuel Chemistry Preprints*, 39 (4) (1994b) 1048-1052.

13. Breault, R, "Design and Economics of Electricity Production from an Indirectly Heated Biomass Gasifier, " Report TR4533-049-92, 1992

14. Wiltsee, G., McGowin, C., and Hughes, E., "Biomass Combustion Technologies for Power Generation," First Biomass Conference of the Americas, August, 1993

COMBUSTION

COMBUSTION

Overview

OVERVIEW OF BIOMASS COMBUSTION

T. NUSSBAUMER
Verenum Research and Swiss Federal Institute of Technology, Zurich, Switzerland
J. E. HUSTAD
SINTEF Energy and Norwegian University of Sience and Technology, Trondheim, Norway

Abstract

The main combustion systems for biomass fuels are presented and the respective requirements are discussed. Wood stoves and stick wood boilers are used for individual house heating. Under stoker furnaces are used for wood chips from native wood with low ash content. Combustors for pulverized wood are used for dry wood residues (saw dust etc.) in industrial boilers or for co-combustion in power plants. For fuels with high ash content an efficient ash removal system is needed and mainly moving grate firings are used. Fluidized bed combustion is an option if a wide variety of fuels (including biomass fuels) is burnt. However, fluidised bed combustors are only used for large scale combustion. For biofuels with low ash fusion temperature like grass, straw, miscanthus etc., slagging on the grate and at the combustion chamber walls can cause severe operation problems usually above 850 °C – 1000 °C, depending on the fuel. For straw and similar biofuels, furnaces for whole bales like cigar burners can be used. The main functions of a biomass combustion system and the respective requirements are discussed in the paper.

Keywords: Biomass combustion, wood stove, wood boiler, under stoker, grate firing, cigar burner, fluidized bed combustion, co-combustion, emissions, slagging.

1   Introduction

There is a considerable unused potential for energy production by the use of biomass in the IEA-countries[1] through combustion, gasification and other conversion techniques, both for heat production, electricity generation and for liquid fuel production. Of the various techniques for biomass to energy conversion, biomass combustion is the oldest and most mature technique. However, there is still a great challange to develop new and more efficient and environmental acceptable biomass combustion systems both in small and large scale. The driving force for development are feedstock

1229

variations since new feedstocks like straw, grass and short rotation coppiece are to be more utilized in the future, environmental legislations, and increased energy recovery through development of new equipment and processes. The main functions of a biomass combustion system and the respective requirements can be summarized as follows and are discussed in detail in the paper:

- Complete combustion of the carbon -> Appropriate design of the combustion chamber, homogenous mixing of air and combustible gases, stable operation of the furnace at optimum excess air, accurate combustion control system.
- Separation of ash and gases -> I.e. ash removal on moving grate firings and low furnace exit gas temperatures for burning high alkali fuels.
- Energy recovery -> Flue gas condensation for biofuels with high water content and new electricity generation systems.
- Reduction of nitrogen oxide emissions -> Primary measures by staged combustion (air staging and fuel staging), deNO$_x$-techniques as secondary measures if primary measures are not sufficient.

When heated to temperatures above 300 °C biomass as wood, grass, straw or miscanthus decompose into volatile components (CO, H$_2$, CH$_4$ and others) and char. At 500 °C about 85% by weight of the wood substance is converted into gaseous compounds. There is only a small variation in the decomposition temperature between the various types of biomass [2]. However, the ash content and the ash behaviour (ash softening, melting point and sintering) of different biomass vary in a wide range. Therefore, the ash removal system, the grate and the boiler must be designed carefully for the type of biofuel used.

Wood combustion can be described as a two-stage process: Devolatilization/Gasification(dependent of the fuel/air ratio) of the solid substance and subsequent oxidation of the gases and charcoal. Thus, two main groups of pollutants from the combustion of native wood can be distinguished: unburnt pollutants: CO, HC, PAH, soot and oxidized pollutants: NO$_x$ and CO$_2$. Further, additional pollutants can be emitted if biomass containing Cl, metals etc. are burnt. Depending on the fuel composition, the design of the combustion equipment and the operation of the system, the combustion of biomass can lead to emissions of CO, HC, (VOC, UHC), PAH, tar, soot, particles, NO$_x$, N$_2$O, HCl, SO$_2$, salts, PCDD/F and heavy metals (Pb, Zn, Cd and others). The various groups of emissions can be identified as can be seen from table 1. The emissions from biomass combustion can be distinguished between:

- Emissions which are mainly influenced by the combustion equipment and process (design and operation of firing system; unburnt pollutants which can be avoided by complete combustion: CO, HC, PAH etc.).
- Emissions which are mainly influenced by the fuel properties(emissions which are formed from elements found in the biomass: NO$_x$ from N, HCl from Cl etc.).

Table 2 shows typical data for emissions from automatic wood furnaces (under stoker furnaces, grate firings, dust firings) which are mainly influenced by the combustion (CO, HC, PAH).

Table 1. Main groups of emissions from biomass combustion

	Origin, fuel	Emissions
1	Unburnt pollutants (all biomass)	CO, HC, Tar, PAH, unburnt particles
2	Oxidized pollutants (all biomass)	NO_X, N_2O, CO_2 in certain cases
3	From biomass containing Cl and S:	HCl, SO_2,
	Altholz = urban waste wood and demolition wood;	salts (KCl, K_2SO_4, NH_4Cl etc.)
	Short rotation biomass = straw, grass, miscanthus etc.	
4	Ash	Particles
5	From biomass containing heavy metals (Altholz)	Pb, Zn, Cd, Cu, Cr etc.
6	From native biomass (low content of Cl)	PCDD/F (low concentrations)
	From Altholz (high content of Cl, Cu)	PCDD/F (higher conc.)

Table 2. Emissions which are mainly influenced by the combustion. Comparison between poor and high standard design of the furnaces (typical values). Data from investigations in Switzerland [2,3,4,5], [6], [7], [8], [9], [10], [11].

Emissions at 11% O_2	Poor standard (mg/Nm³)	High standard (mg/Nm³)
Excess-air ratio, λ	2 - 4	1.5 - 2
CO	1000 - 5000	20 - 250
HC	100 - 500	< 10
PAH	0.1 - 10	< 0.01
Particles, after cyclone	150 - 500	50 - 150 *

* except dust firings usually > 150

These emissions have traditionally been in focus when optimizing combustion systems burning native wood. However, the emissions especially in traditional small combustion systems have been very high in the past. In the recent years new and better systems with high standard have been available on the market. Table 3 shows typical emissions which are mainly influenced by the fuel (NO_x, HCl, particles, Pb, Zn, Cd, PCDD/F). These emissions have in recent years been more focussed because of the new feedstocks utilized in biomass combustion.

Larger boilers (above 5 MW) developed for biomass combustion include conventional travelling grates with spreader stoker which is the most common boiler, dedicated folded boiler designs like straw-fired boilers, bubbling fluidized beds (BFB) and circulating fluidized beds (CFB). For the size range down to about 1 MW, the cyclone furnace is common used. Below 1 MW, the underfeed stoker and the fixed grate are the most common used firing systems in boilers. Below 20 kW, various types of wood log boilers, wood stoves and fireplaces exist on the market.

Table 3. Emissions which are mainly influenced by the fuel. Comparison between different fuel types (typical values). Altholz = urban waste wood and demolition wood.

Emissions at 11% O_2	Fuel type		Typical Data
NO_x (mg/Nm³)	Native wood (soft wood)		100 - 200
	Native wood (hard wood)		150 - 250
	Straw, grass, miscanthus, chip boards		300 - 800
	Altholz		400 - 600
HCl (mg/Nm³)	Native wood		< 5
	Altholz, straw, grass, miscanthus,		raw gas: 100 - 1000
	chip boards (NH_4Cl)		with HCl absorption: < 20
Particles (mg/Nm³)	Native wood	after cyclone:	50 -150
	Straw, grass, miscanthus,		
	chip boards	after cyclone:	150 - 1000
	Altholz	after cloth or electric filter:	< 10
Σ Pb, Zn, Cd, Cu (mg/Nm³)	Native wood		< 1
	Altholz	raw gas:	20 - 100
	Altholz	after cloth or electric filter:	< 5
PCDD/F (ng TE/Nm³)	Native wood	typical:	< 0.1
		range:	0.01 - 0.5
	Altholz	typical:	2
		range:	0.1 - 20

2 Complete oxidation of the carbon

The combustion of wood can be described as a two-stage process as can be seen in figure 1 employing air excess level (λ) less than 1 (gasification) in the first step and final oxidation of the gases in the second step:

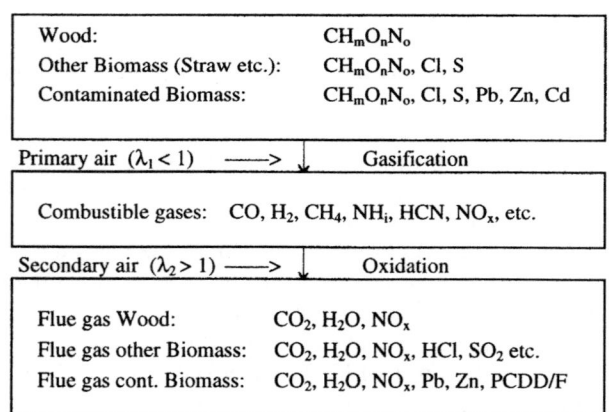

Wood:	$CH_mO_nN_o$
Other Biomass (Straw etc.):	$CH_mO_nN_o$, Cl, S
Contaminated Biomass:	$CH_mO_nN_o$, Cl, S, Pb, Zn, Cd

Primary air ($\lambda_1 < 1$) ——> $\downarrow$ Gasification

Combustible gases: CO, H_2, CH_4, NH_i, HCN, NO_x, etc.

Secondary air ($\lambda_2 > 1$) ——> $\downarrow$ Oxidation

Flue gas Wood:	CO_2, H_2O, NO_x
Flue gas other Biomass:	CO_2, H_2O, NO_x, HCl, SO_2 etc.
Flue gas cont. Biomass:	CO_2, H_2O, NO_x, Pb, Zn, PCDD/F

Fig. 1   Combustion of wood: 2-stage process with primary air for the gasification and secondary air for the burnout of the combustible gases [3].

By arranging the combustion process in this way, a low NO_x concept can be achieved for an underfeed stoker system. The unburnt pollutants can further be reduced effectively if the combustion air, especially the secondary air is mixed homogenously with the combustible gases and the burnout takes place in a hot combustion chamber. Furthermore, it is very important that the fuel/air ratio is optimized. The optimum excess air ratio is necessary to guarantee:

- a high combustion temperature (excess air as low as possible) and
- complete combustion (excess air > 1; usually 1.5 – 2 to avoid unburnt pollutants).

Each wood furnace shows a typical correlation between the CO-emissions and the excess air ratio. If the excess air ratio is too low, high CO-emissions can be found due to a (local) lack of oxygen as can be seen in figure 2. The shape of the curve and the numerical values depend on the furnace design. The graph shows typical data for a downdraft boiler (manual) and an underfeed stoker furnace (automatic). The downdraft boiler is shown schematically in figure 3. If the excess air ratio is too high, burnout of the gases is restricted because the combustion temperature is too low. Therefore an optimum of the excess air ratio can be found.

Since PAH's are formed during the devolatilization/combustion, a maximum at a certain combustion temperature can be observed. If the temperature is low, the formation of PAH is low, and if the combustion temperature is high enough, the formed PAH's are oxidized in the flame [4,12]. For batch fired wood stoves and fireplaces the CO-emissions decrease with increasing average wood consumption as can be seen in figure 4.

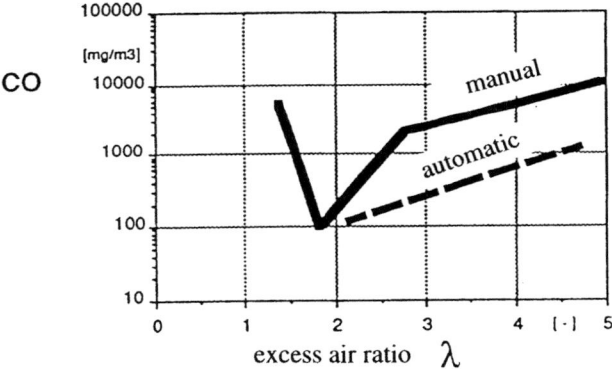

Fig. 2   Carbon monoxide emissions in function of the excess air ratio for automatic and manual wood furnaces [5].

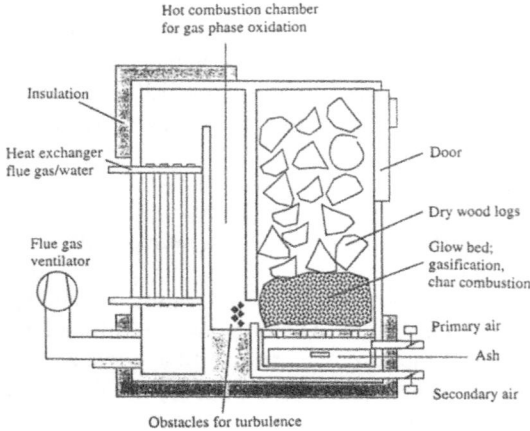

Fig. 3   Scematic view of a downdraft boiler.

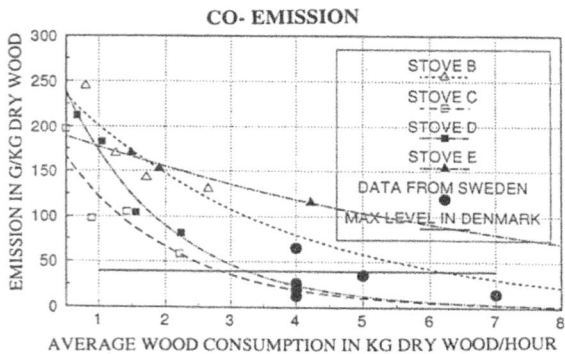

Fig. 4   Carbon monoxide emissions as a function of average wood consumption for wood stoves and fireplaces[12].

The optimum excess air ratio is influenced by the grate and the boiler construction, the actual load and the quality of the fuel (moisture content etc.). Combustion control systems are necessary to ensure a high combustion quality and high efficiency for biomass combustion. The most important control systems used today are control of the temperature in the combustion chamber *(combustion temperature control)*, control of the excess air ratio *(lambda control)* and control with a sensor which detects unburnt gases. A new control system combines lambda control with a setpoint optimization *(CO/lambda control)*. The application to a 1 MW underfeed stoker showed that at part load operation the efficiency could be improved by up to 5 % and CO emissions were reduced by more than 80 % compared to a traditional lambda control. Since emissions of CO, HC, tar and soot are caused by incomplete combustion of the gases, a certain furnace specific correlation between these parameters can be found in different wood firing systems and with different biomass fuels [5,12].

3 Separation of ash and gases

For biofuels with low ash content (native wood) underfeed stokers can be used for small scale systems as can be seen in figure 5. A certain separation between ash and gas takes place in the combustion chamber. However, a dust separator in the flue gas (multi-cyclone) is used to guarantee low dust concentrations in the flue gas (< 150 mg/Nm3). If biofuels with high ash content are used (bark, grass, straw, miscanthus, Altholz = urban waste wood + demolition wood) a more efficient ash removal system is needed.

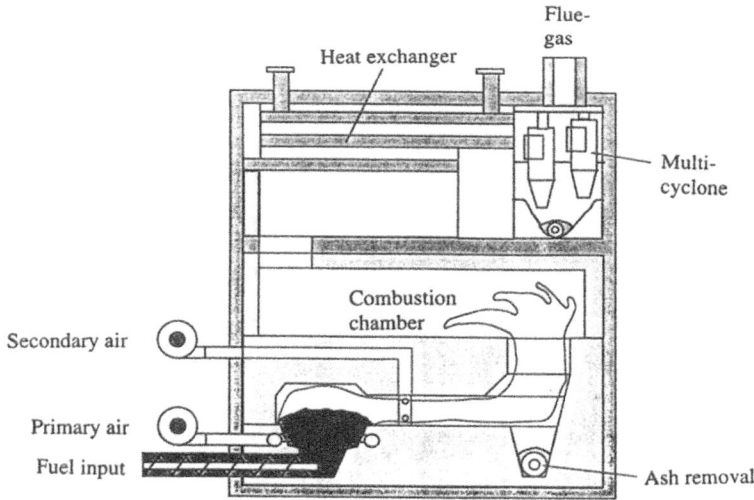

Fig. 5   Underfeed stoker for wood chips with automatic ash removal [4].

Different types of grate firings have been developed for solid fuels such as coal, municipal solid waste (MSW) and biofuels. To burn fuels with high water content(50 to 60%), counter-current flow on the grate is necessary, where the volatile flame is used to achieve a pre-drying of the wet fuel on the first part of the grate as can be seen schematically in figure 6.

For biofuels with high ash content mainly moving grate firings are used. If dry wood residues are burnt, co-current furnaces can be used(fig. 6). However, it is important to have a good mixing between combustible gases and secondary air to enable combustion with low excess air.

Co-current   Cross-current Counter-current

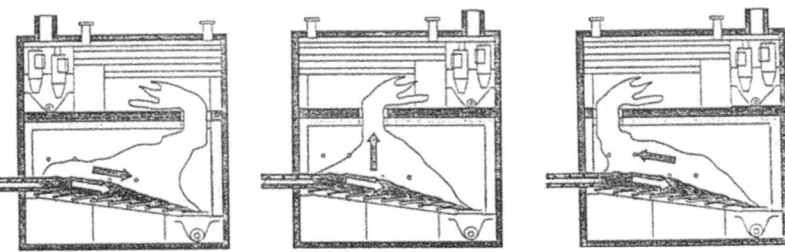

Fig. 6   Various furnace gas flow arrangements [4].

The aim of the grate is to guarantee a fuel bed with a constant thickness and pressure drop. At the same time the transport of the fuel bed and the removal of the ash at the end of the grate must be guaranteed. Poking, raking of the glow is another important function of the grate. Poking improves air flow, supports drying and increases load but also increase particle concentration in the flue gas and supports slagging on the grate. One of the main disadvantages of moving grate firing is the instability of the air distribution in the fuel bed due to changes of the fuel bed by the movement of the grate. Sometimes the main part of the primary air passes through a few holes in the fuel bed and this effect may cause slagging, and improved combustion with low excess air levels becomes impossible. Furthermore, part load operation is difficult to achieve in grate firings. A new type of grate design has been developed in Switzerland to avoid the above mentionned problems. The grate shows a slight and continous movement in the horizontal and the vertical direction.   With this 'forward/backward moving grate' a good fuel distribution on the grate is obtained, slagging can be reduced and the grate can be operated at a low primary excess air ratio (<1). A low primary excess air ratio is necessary to obtain high efficiency with low total excess air ratio, complete combustion and low NO_x-operation. The new design is shown in figure 7.

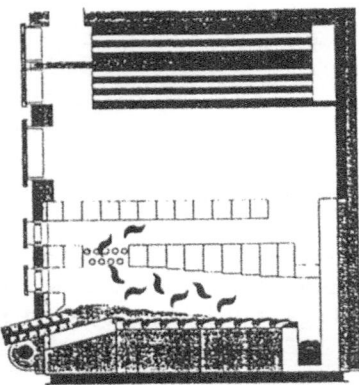

Fig. 7   Horizontal forward/backward moving grate firing (app. 0.3 MW – 3 MW) [4].

Cigar firing is a continous process developed for straw combustion in Denmark[13] where the straw bales are fed continously by a hydraulic piston through a feeding tunnel as shown in figure 8.

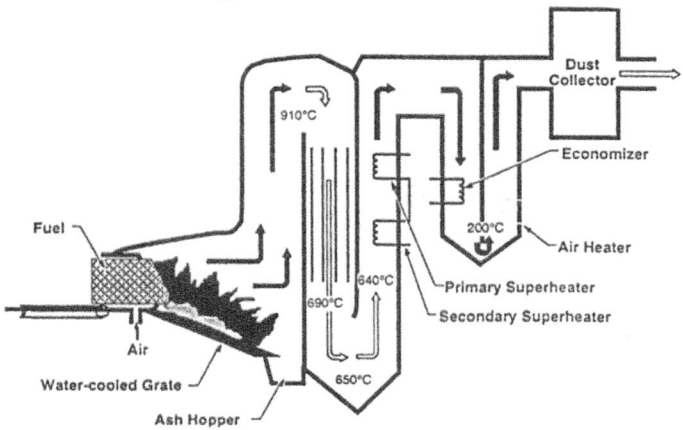

Fig. 8   Cigar firing [14].

Volatile gases are released and combusted by the secondary air. The unburnt charcoal and the ashes are pushed forward to a watercooled grate where the oxidation of the solid carbon is completed and the ash is transported to the end and removed. Another combustion system for straw is combustion of whole bales where the bales are conveyed through a feeding sluice to a gasification chamber and air is injected through a large number of nozzles to ensure a good air distribution. The volatile gases are subsequent burnt in the boiler. This combustion system is semi-continous with batch-wise operation giving peaks in both temperature and CO and the control system today is not satisfactory. Combustion systems for scarified and cut straw also exist and these systems are very much alike. The difference is whether the bale is scarified or the straw is cut and further fed by a hydraulic stoker which pushes the straw onto an air-cooled grate. In Denmark 22 scarified straw combustion systems were in operation in 1993. The number of whole bale systems were 16, and for cut straw and cigar firing about 12 systems existed each.

Both bubbling fluidized bed (BFBC) and circulating fluidized bed (CFBC) are suitable for combustion of biomass fuels. In BFBC a bed of usually silica sand is located in the lower part of a conventional furnace. The sand particle size is about 1mm and the height of the bed is approximately 1 m. The temperature in the bed can vary from 850 ^{0}C to 900 ^{0}C. Primary air is blown through small nozzles in an air distributor plate and the fluidizing velocity varies from 1 m/s to 2.5 m/s. The BFBC system as can be seen in figure 9 is a multifuel combustion system suitable for reactive fuels like wood, bark and other biofuels and the system is quite flexible regarding fuel size and moisture content.   Less reactive fuels can also be combusted to some extent and cofired with reactive fuels to avoid unburnt in carbon loss. Heat exchanger surfaces are not to be located in the bed section due to erosion of the tubes. In Finland approximately 30 BFBC systems are installed and the largest boiler is 300 MW.

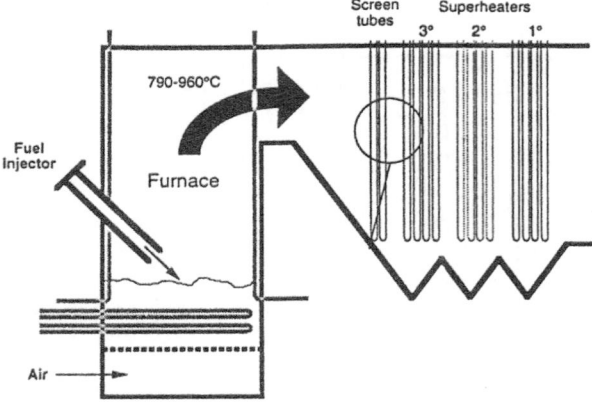

Fig. 9   Bubbling fluidized bed [14].

Increasing the fluidizing velocity up to approx. 6 m/s and using somewhat smaller sand particles, the sand will be transported with the flue gas, separated in a cyclone and fed back to the bed, a CFBC system is achieved. The flow in a CFBC is more turbulent with better mixing and heating surfaces can be placed in the upper part of the furnace or in a separate fluidized bed located below the cyclone. The CFBC system is shown in figure 10.

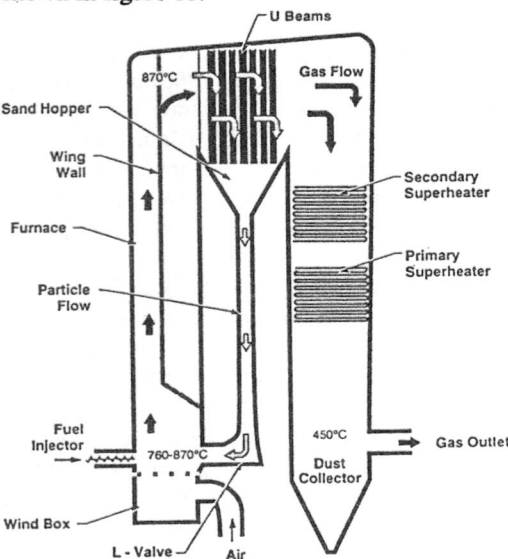

Fig. 10   Circulating fluidized bed [14].

1238

The advantage of the CFBC compared to BFBC is longer residence times so less reactive fuels can be burnt and the system is more efficient for desulphurization. The disadvantage is larger size and therefore higher investment costs. In addition, only small fuel particles (typically less than 10 mm) can be used often causing more investments in fuel pretreatment. Co-firing biofuels with fossil fuels, especially coal in large power plants, has been performed with good results in the Netherlands, Denmark and in the US. All annual growth like urban tree trimmings, annual crops or their residues like straw, energy crops, contain sufficient volatile alkali compounds to lower the ash fusion temperature so it melts in combustion or the elements vaporize and condense on boiler tubes or refractories. These mechnisms creates serious fouling and slagging in conventional boilers. In studies done in the USA [14], sintered or fused deposits were found on grates and in agglomerates in fluidized beds. Potassium sulfates and clorides were found condensed on the upper furnace walls where it mixed with fly ash. Convection tubes were coated with alkali clorides, carbonates and sulfates mixed with silica from fly ash or fluidized bed media. Deposits occur as a result of the boiler design, fuel properties and boiler operation. Conventional designed boilers are not suitable for burning high alkali fuels. Special boiler designs with low furnace exit gas temperature(less than 800 ^{0}C) are required for annual crops or residues, including grasses and straws. Design should include larger waterwall surface areas and parallell heat exchanger surfaces, regulation of the combustion air to control gas temperature and use of grates suitable for removing large quantities of ash and soot blowing systems to remove deposits. A maximum level of alkali compounds, defined as the sum of potassium(K) and sodium(Na), of 0.17 kg/GJ is recommended to avoid problems [14]. Fuel blending to obtain this value can also be adapted, but must be implemented with care.

For small scale (<20 kW) biomass combustion systems wood stoves and fireplaces are used. Traditional stoves and fireplaces have quite high emissions of CO and particles. To reduce these emissions various new techniques have been investigated like catalytic afterburners, 2 stage combustion(air staging), downdraft combustion systems and flue gas recirculation[12]. Best results have been obtained with downdraft combustion systems and catalytic afterburners[12].

4 Energy recovery

Energy can be recovered in the form of electricity and/or heat. The most common system for electricity generation from biomass combustion is the steam turbine, and in the United States about 7500 MW$_e$ is generated in this way. A simple steam cycle with a steam turbine is not very efficient for electricity generation, typical efficiencies are in the range of 20%. However, more than 70% of the biomass power in the Unites States is cogenerated with process heat which will increase the average annual efficiency of the total energy recovery up to 85 - 90%. Other possibilities for electricity generation include new versions of steam engines, stirling engine, indirectly fired gas turbine (hot air turbine) and pressurized combustion(fixed bed or fluidized bed) coupled to a directly fired gas turbine. A comparison of the costs of such systems with combined heat and power has been made as a part of our IEA-project and interesting results favorable to the stirling engine were obtained[15]. Similar comparisons have been done in Austria[16], and preliminary calculations and discussions to build a stirling engine prototype unit has been performed, the stirling engine is shown schematically in figure 11. The stirling engine system gave a calculated electricity price of 0.16 ECU/kW$_{el}$, lower than the steam turbine.

To achieve a high yield of heat recovery the flue gas temperature must be as low as possible. Typical data in practice which are achieved without having condensation problems in the chimney are about 120 °C – 150 °C. However, in practice particle depositions in the heat exchanger may lead to a higher flue gas temperature. Cleaning of the heat exchanger is then necessary. To reach a low flue gas temperature at full load and at part load, an optional heat exchanger can be used but only during full load operation. With this system an increase in efficiency of up to 5% can be reached.

If biofuels with high water content are burnt, flue gas condensation can be used to increase efficiency of app. 20%. Further, the flue gas condensation leads to a reduction of particle emissions and certain gas phase emissions. However, to use flue gas condensation, low temperature heating systems must be used (cold side app. 45 °C).

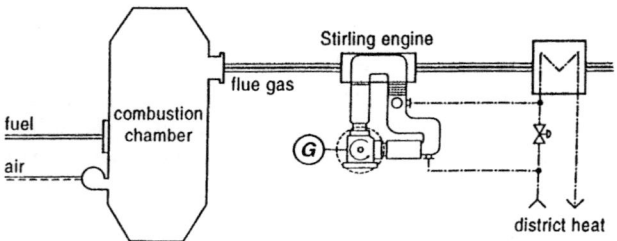

Fig.11   Biomass combustion with stirling engine [16].

5. NO$_x$ reduction

Nitric oxide emissions from biomass combustion originate mainly from the fuel bound nitrogen, thermal NO$_x$ emissions are only of minor importance. Since biomass combustion leads to higher NO$_x$ emissions than gas or fuel oil combustion, primary or secondary measures for NO$_x$ reduction are proposed. To minimize NO$_X$ emissions by primary measures, the fuel nitrogen must be reduced to molecular nitrogen in zones with an excess air ratio < 1. If primary measures are not sufficient, secondary measures can be employed.

Primary measures for NO$_x$ reduction
For the combustion of wood the following reactions are considered to be important for the NO$_x$ emissions:

- During the gasification HCN and NH$_i$ radicals (i = 0 ... 3) are formed. These components can react in different reactions to form molecular nitrogen:

$$NO + NH_2 \rightarrow N_2 + H_2O \tag{1}$$

- Where there is a shortage of oxygen, NO_x acts as an oxidizing agent for carbon monoxide, methane, hydrocarbons, hydrogen and carbon:

$$NO + CO \rightarrow CO_2 + 0.5\,N_2 \tag{2}$$

$$NO + CH_4 \rightarrow CO + 2\,H_2 + 0.5\,N_2 \tag{3}$$

$$NO + H_2 \rightarrow H_2O + 0.5\,N_2 \tag{4}$$

$$NO + C \rightarrow CO + 0.5\,N_2 \tag{5}$$

To show under which conditions the above mentioned reactions can reduce NO_x emissions a test reactor with fixed bed updraft gasification followed by a gas phase combustion was built [9]. The reactor can be operated as an underfeed stoker which was used as reference for conventional combustion. The following concepts of staged combustion were investigated:

- *staged combustion with seperate reduction chamber* between the gasification and the combustion chamber
- *air staging* in the gasification chamber
- flue gas recirculation
- combinations of the listed measures.

Due to the results of different experiments carried out in the reactor reaction (1)is supposed to be most important for fixed bed combustion of wood. Reaction (2) might be of importance together with catalytic effects on ash while for reaction (3) and (4) no effects were found during wood combustion in the test reactor. Reaction (5) is supposed to be of major influence during fluidized bed combustion of coal [17]. In the 25 kW test furnace a NO_x reduction of 50% compared to conventional fixed bed combustion (from 200 mg/m^3 to 100 mg/m^3 for native wood) could be reached [4] by a staged combustion with a (heated) reduction chamber between the gasification and the combustion if the following conditions were met: Primary excess air ratio 0.7(0.6 - 0.8), temperature in the reduction chamber 1160 °C - 1250 ^{0}C, residence time 0.5 s(>0.3 s). Fuel staging showed a comparable or even higher potential than air staging. However, fuel staging can be used only for larger plants while air staging can be used for small automatic wood furnaces as well as for large scale combustion. Further it was shown that flue gas recirculation was of minor influence on NO_x emissions in these experiments. The primary excess air ratio in the gasification chamber influenced the NO_x emissions to a great extent. The minimum NO_x emissions found at a primary excess air ratio of approximately 0.7 can be explained as follows:

- Primary air ratio << 0.7 (gasification): Only NH-components are formed during the gasification (no NO). The NH-components are oxidized in the combustion chamber to NO resulting in high NO-emissions typical for conventional wood furnaces.

- Primary air ratio > 1: The pyrolysis gases are burnt immediately in the gasification chamber (= combustion). The concentration of NH-components to reduce NO in the reduction chamber is close to zero.

Secondary measures for NO_x reduction

The injection of sal ammoniac, ammonia or urea can be used as secondary measure for NO_x reduction. For the *selective catalytic reduction (SCR)* with a catalyst at about 250 °C a NO_x reduction of 95% can be reached without significant slippage of ammonia (< 5 mg/m^3). For the *Selective non-catalytic reduction (SNCR)* a temperature between 850 °C and 950 °C and an amount of ammonia higher than the stochiometric ratio is necessary. Further, a separate reduction chamber is needed to achieve a relavant NO_x reduction without significant slippage of ammonia. For the SCR process the long term behaviour of the catalyst can be a major problem and for the SNCR process the accurate control of the temperature is a critical issue.

6. Conclusions

The driving force for development of new biomass combustion systems are utilization of new feedstocks like straw, grass and short rotation coppiece, environmental legislations and increased energy recovery. The main functions and requirements of a biomass combustion system are:

- Complete Combustion
- Ash separation
- Energy recovery
- NO_x reduction

Pulverized fuel- and cyclone furnaces are mainly used for dry wood residues (saw dust etc.). Fluidized bed combustion is mainly used for larger plants (> 10 MW). For straw-fired boilers new design with larger heating surfaces and thus lower furnace exit temperatures are developed to decrease problems with ash deposits and corrosion. A maximum level of alkali compounds of 0.17 kg/GJ is recommended to avoid these problems. The most common commercial biomass combustion systems in the range from 0.2 – 5 MW are:

- Low ash content: Underfeed stokers, grate firing, cyclone combustor
- High ash content: Moving grate firing
- High water content: Counter-current moving grate firing

New designs for wood log boilers, wood stoves and fireplaces (< 20 kW) include catalytic combustors, downdraft combustion systems and air staging which improve the combustion efficiency and reduce pollutant emissions.

References

1. Hustad, J.E. and Sønju, O.K. (1992). Biomass combustion in IEA countries, *Biomass and Bioenergy* vol 2 Nos 1-6, Pergamon Press Ltd. 192, 239-261.

2. Nussbaumer, Th. *1993: Sekundärmassnahmen zur Stickstoffoxidminderung bei Holzfeuerungen*, BWK Bd. 45 (1993) Nr. 11, 483-488

3. *Nussbaumer, Th.:* Emissionen von Holzfeuerungen, *Final Report NFP* 12 project 4.971.0.86. 12, Institute of Energy Technology, ETH Zürich, Februar 1988

4. Nussbaumer, Th: Overview of small scale combustion systems for biomass, *Joint IEA avtivity meeting on small scale cogeneration systems,* Vienna, Austria, May 1995.

5. Nussbaumer, Th.: *Schadstoffbildung bei der Verbrennung von Holz,* PhD Thesis ETH Nr. 8838, Zürich 1989

6. *Graf, S.:* Emissionsarme Holzschnitzelfeuerung; research report Nr. 9, *Institute of Energy Technology,* ETH Zürich, 1991

7. *Good, J. 1994:* Combustion control for automatic wood firings, in: Bridgwater (Ed.), *Advances in thermochemical biomass conversion,* Vol. 1, 1994

8. Kerschbaumer, D.:*Regelung einer stuckholzfeuerung,* PhD thesis, University of Neuenburg 1990.

9. Keller, R.:*Primarmassnahmen zur NOx minderung bei der holzverbrennung mit dem schwerpunkt der luftstufung,* PhD diss. ETH No 10514, Zurich 1994.

10. Hasler, Ph.; Nussbaumer, Th.; Bühler, R. *1993:* Dioxinemissionen von Holzfeuerungen. *Schriftenreihe Umwelt* Nr. 208, Bundesamt für Umwelt, Wald und Landschaft (BUWAL), 1993.

11. Hasler, Ph.; Nussbaumer, Th. 1994a: Dioxin- und Furanemissionen bei Altholzfeuerungen, *Bundesamt für Energiewirtschaft,* EDMZ-Nr. 805. 174 d, April 1994.

12. Karlsvik, E.K., Hustad, J.E. and Sønju, O.K.: Emissions from wood stoves and fireplaces, *Advances in thermochemical biomass conversion,* Vol. I, Blackie Academic 1993, ISBN 0 7514 0171 4.

13. Nikolaisen, L.: Utilization of straw in district heating and CHP plants, *Proceedings from Bioenergy 93 Conference,* Espoo, november 1993.

14. Miles, T.R., Miles, T.R. Jr, Baxter, L.L., Bryers, R.W., Jenkins, B.M. and Oden, L.L.: Alkali deposits, *Summary report for NREL Subcontract Tz-2-11226-1, April 1995.*

15. Jacobsen, H.H.: Technologies for small wood cogeneration systems, *Proceedings from IEA biomass combustion conference,* Cambridge November 1995.

16. Spitzer, J.: Small scale cogeneration systems in Austria, *Joint IEA activity meeting on small scale cogeneration systeme,* Vienna, Austria, May 1995.

17. Beér, J.: Advanced combustion methods for low grade coal utilization, in: Korhonen (Ed.), *Low-grade fuels,* Vol. 1, VTT Symposium 108, Espoo (SF) 1990, 83-112.

COMBUSTION

Fundamentals

INFLUENCE OF ASH DEPOSIT CHEMISTRY AND STRUCTURE ON PHYSICAL AND TRANSPORT PROPERTIES

Transport Properties' Relationship to Structure

L. L. BAXTER, T. GALE, S. SINQUEFIELD, AND G. SCLIPPA

Combustion Research Facility, Sandia National Laboratories, Livermore, CA USA

Abstract

Boiler ash deposits generated during combustion of coal, biomass, black liquor, and energetic materials affect both the net plant efficiency and operating strategy of essentially all boilers. Such deposits decrease convective and radiative heat exchange with boiler heat transfer surfaces. In many cases, even a small amount of ash on a surface decreases local heat transfer rates by factors of three or more. Apart from their impact on heat transfer, ash deposits in boilers represent potential operational problems and boiler maintenance issues, including plugging, tube wastage (erosion and corrosion), and structural damage.

This report relates the chemistry and microstructural properties of ash deposits to their physical and transport properties. Deposit emissivity, thermal conductivity, tenacity, and strength relate quantitatively to deposit microstructure and chemistry. This paper presents data and algorithms illustrating the accuracy and limitations of such relationships.

Keywords: ash, biomass, boilers, combustion, deposition, furnaces, inorganic material, transport properties

1. Introduction

Ash deposit properties in boilers depend on many factors, including deposit structure and composition. Thermal conductivity and emissivity, the two properties with the greatest impact on heat transfer, demonstrate strong and complex dependencies on both deposit structure and composition. The effects of deposit structure relate largely to the phases present in the deposit and the extent of sintering or contact between individual particles. This paper focuses on the effects of emissivity and porosity variations on heat transfer through boiler deposits.

Heat and mass transfer through porous media depend on macroscopic and microscopic structural properties of the media. Upper and sometimes lower bounds for transfer coefficients can be established based on easily measured structural properties, but precise expressions for transfer rates depend on a high level of structural detail, commonly beyond what could reasonably be expected to be available in practical applications. Our approach is to identify the limits and increase the level of sophistication of our models up to the point that we make the best use of available information.

2. Background

The thermal, radiative, and physical properties of ash deposits determine their effect on overall combustion performance. The properties of greatest interest include emissivity and absorbtivity, thermal conductivity, strength, tenacity, viscosity, composition, rate of accumulation, and porosity. The current status of understanding of each of these properties is described below.

The dependence of condensed-phase transport, physical, and chemical properties on porosity and composition of the deposit is an area of active research. Generally, upper bounds on properties such as thermal conductivity can be established from a knowledge of deposit composition and porosity. However, lower bounds and accurate predictions of actual properties are more difficult to establish.

Physical and transport properties of ash deposits depend on both the properties of the material from which they are formed and on the interactions of the material once it arrives at the surface. Recent research, sponsored by PETC and others, has provided new insight into the formation of fly ash and how fly ash properties depend on combustion conditions and fuel properties [1-10]. The formation of ash deposits from this fly ash is also under study [1, 3, 11-13]. However, no concerted effort in describing ash deposit properties has been initiated.

2.1.1 Emissivity and Absorbtivity of Deposits

Recent literature discussing deposit radiative properties indicates their dependence on chemical composition and structure [14, 15]. Most ash deposits show spectral variation in their emissivity as a function of wavelength. The spectral dependence is due in part to deposit composition and in part to deposit morphology. This variation gives rise to a temperature-dependent total or effective emissivity, as is illustrated for coal and char particles in our earlier work [16] and for ash deposits in more recent work [17, 18]. The FTIR emission spectroscopy diagnostic developed at Sandia for measuring surface species composition is also suitable for measuring ash deposit spectral emissivity over the range from about 3 to 20 μm. The strong dependence of emissivity on deposit morphology indicates that the most meaningful measurements will be obtained from an *in situ* device. That is, preparation of deposits for post mortem analyses often alters their structure. In addition, removing them from the combustion zone often alters the details of their chemistry and morphology.

Quantitative measurements of deposit emissivity are available in the literature. Deposit emissivity is shown to increase with increases in particle size in the deposit, up to at least 400 μm. Emissivity is also profoundly affected by the presence of atoms that form mixed silicates, sometimes referred to as coloring agents [19]. More formal relationships between composition and emissivity have been investigated recently [14], and experimental data illustrating the dependence of emissivity on structural properties [20] have been reconciled with published theoretical treatments indicating similar trends [21]. Systematic studies of specific constituents of ash deposits as a function of composition and temperature and of ash deposits directly have also been presented, although particle size and morphology effects are not typically addressed [22-27]. Industrial experience with deposits has documented the effect of deposit emissivity on pollutant production, boiler derating, convection pass fouling, and other operational issues and indicates that deposit-emittance based boiler diagnostics can be successful in anticipating the problems [28-32]. Deposit emissivity and absorbtivity depend strongly on both morphology and composition. Morphological considerations include porosity, shape, and thickness. Porosity is the dominant morphological factor if it is defined broadly, i.e., to include particle size and pore size information. The effect of porosity can be large. For example, weakly absorbing materials develop high hemispherical reflectivities when ground to a fine particle size and spread over a surface. Analytical approaches for describing such phenomena are available in the literature at several levels of approximation [14, 17, 21].

2.1.2 Thermal Conductivity of Deposits

Thermal conductivity of ash deposits represent the second major variable controlling heat transfer rates in boilers. The potentially complex chemical species formed in ash deposits do not all have conductivities with well known dependencies on temperature. However, the greatest source of uncertainty in predicting thermal conductivities is associated with the deposit porosity. Heat transfer through porous media can be over ten times less efficient than transfer through a nonporous material of the same composition.

Thermal conductivity has been observed to increase with increasing particle size and, in the case of fine, nonsintered dusts, to approach the value of air [19]. Initial studies of heat transfer through porous media have been completed at Sandia, in part in conjunction with researchers associated with Yale University [33-38]. Experimental data describing thermal conductivity dependencies on both morphology and composition are also available from practical systems [28, 29, 39]. Fundamental approaches to describing the thermal conductivity based on detailed knowledge of ash deposit structure are available [36, 37, 40, 41], although they appear to have been applied only to idealized systems to date.

2.1.3 Deposit Strength, Tenacity, and Resistance to Thermal Shock

Ash deposition on heat transfer surfaces is inevitable in essentially all coal combustors and is expected to be a major design and operation consideration in Combustion 2000 equipment. Successful management of these deposits in dry-walled units by soot blowers, wall blowers, or water lances is critically dependent on the deposit strength, tenacity, and thermal shock resistance. Definitions for these terms as used in this document are quite precise. As described earlier, strength relates to the bulk deposit and represents its ability to resist stress without plastic or catastrophic deformation. Tenacity is a similar property, but relates to the interface between the deposit and a surface. Thermal shock resistance is a combination of a thermal expansion coefficient, which indicates the magnitude of a stress generated in a deposit as a consequence of a temperature gradient, and the deposit strength or tenacity.

Deposit strength development is related to the physical microstructure of the deposit [35]. As individual particles in the deposit increase contacting efficiency with neighboring particles, strength increases significantly [40]. Contacting efficiency increases as particles sinter, as vapors condense or liquids accumulate around particles, and as deposits consolidate (smaller particles fill voids around larger particles). These trends have successfully been used to predict some aspects of deposit strength and tenacity development in commercial systems [12]. Deposit tenacity is similarly affected by sintering, condensation, and consolidation. First-order models of deposit tenacity have been developed based on these concepts in previous work [12].

2.1.4 Ash Viscosity

Slagging combustors offer the potential advantages of producing an environmentally more benign ash than dry-walled combustors and of increased ash capture and removal in the early stages of the combustion process. Some Combustion 2000 contractors recognize these advantages and are considering slagging combustors as part of their proposed systems. In a slagging combustor, ash viscosity plays the role of the dominant design consideration after the same manner as deposit strength, tenacity, and thermal shock resistance do in dry-walled systems.

Correlations of deposit viscosity have been proposed by several investigators [11, 42] based in large measure on the early work by Urbain [43]. These models are based on relationships and theory from the glass-making industry and represent correlations of viscosity with elemental composition.

2.1.5 Deposit Porosity

Deposit porosity plays a critical role in determining most of the physical and heat transfer properties of the deposit. The development of deposit porosity is influenced by ratio of particle to condensate in the deposit, the sintering of granules in the deposit, and the generation of gases in fluid material [40, 44]. The results below illustratre direct measurements of porosity and its effect on transport properties such as thermal conductivities. Prediction of such properties is not described in any detail.

3. Results

A useful idealization for illustrating the major effects of deposit structure on thermal conductivity is a solid of known porosity and thermal conductivity and with no conduction in the gas phase. Quanta of vibrational energy (heat) move randomly through this solid. A temperature gradient in the solid is represented by spatial differences in the population of phonons. We seek an expression relating the efficiency at which phonons can move through the porous material to its physical structure. In this simple model, heat transfer proceeds through the solid phase at its customary rate but stops when it encounters the void phase.

Spatial autocorrelation functions relate the probability of two locations being the same phase (solid or void) as a function of distance between them. Generally, autocorrelation functions are bounded by ±1 and are identically unity at displacements of 0. Characteristically for real materials, they also decay to a limiting values in a smooth but not necessarily monotonic fashion. For isotropic material, the limiting value is the volume fraction of the phase present at a displacement of 0. If the presence of void vs. solid phase is represented as a random event, there are fairly general conditions under which the autocorrelation becomes an exponential decay, with the spatial constant of the exponent a measure of average grain size.

In addition to the amount of solid vs. void volume in the material, the connectedness of the solid phase plays a large role in determining the heat transfer rate. There are higher order correlation functions and connected correlation functions that statistically give clues to the connectedness of a phase. The concept of tortuosity is the approach we have taken, where the tortuosity is defined as the shortest average path length through the solid phase between two points divided by the straight-line distance between the same points. As the solid phase becomes less connected, the tortuosity increases. Using these three most readily available characteristics of the solid phase, the solid volume fraction, the mean particle size, and the tortuosity, we have developed a model for the dependence of the average thermal conductivity

1250

on structural properties. We are currently pursuing means of extending the model to nonisotropic conditions more sophisticated descriptions of deposit structure. In its current state, the heat transfer model depends on material porosity and tortuosity of both the condensed and gaseous phases, in addition to the thermal conductivity of the two phases.

Aside from the anisotropies of the material, this approach largely ignores the efficiency of the connections between particles. Particles that connect at a single point or over a very small area typically conduct heat far worse than those that are connected over large fractions of their projected areas. In some analyses, the connection points dominate the heat transfer process. This connectedness is captured somewhat, but not entirely, in the concept of the tortuosity. We will examine this aspect of our model in the future. In its current state, it may somewhat over-predict the heat transfer rate in porous media.

The over-prediction is partially compensated by the effect of our initial assumptions. In the original model, the void space was assumed to be non-conducting and radiative heat transfer through the material is ignored. In reality, both intra-media radiative heat transfer and conduction through the gas phase occur. At present, we allow these two simplifications in the model and recognize that they are somewhat balanced by the incomplete descriptions of connectedness of particles in the condensed phase.

In its current state, the heat transfer model reveals some useful insights. These will be illustrated by models of heat transfer through artificially conceived by realistic deposits under boiler-like conditions. The deposits are assumed to exist on cylindrical surfaces and the analysis at this point is limited to one dimension, i.e., the radial dimension.

A one-dimensional, steady-state temeprature gradient through a cylindrical body with constant transport coefficients is described in this model by

$$T(r) = T_1 - \frac{q r_1}{k_{eff}} \ln\left(\frac{r}{r_1}\right)$$ (1)

where the effective thermal conductivity, k_{eff} is given as a function of the porosity and tortuosity of each of the n phases by

$$k_{eff} = \left(\sum_{s=1}^{n} \frac{\tau_s}{v_s(1-\xi_s)k_{ss}}\right)^{-1}$$ (2)

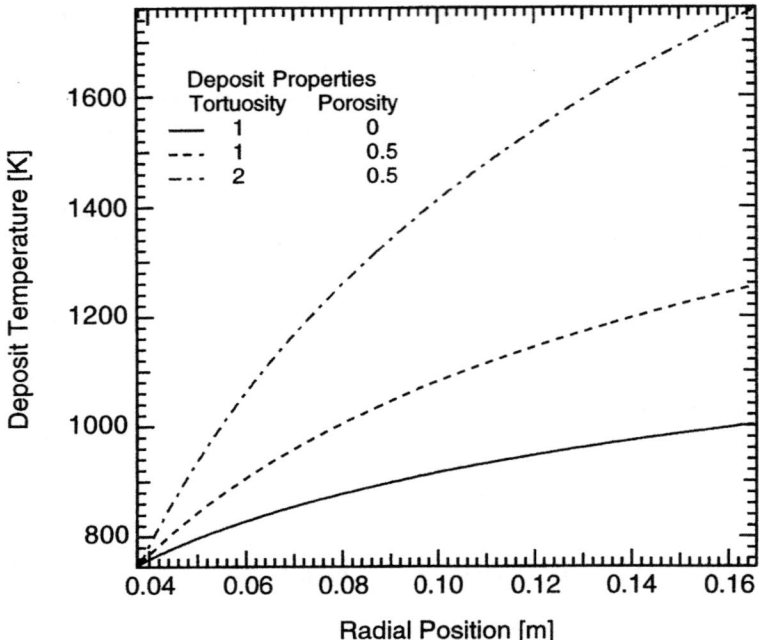

Figure 1   Parametric variation of deposit temperature as a function of position for various values of the solid volume fraction and tortuosity. See text for details of incident heat flux, etc.

This form reduces to a linear dependence of deposit temperature on distance in the limit of small deposit thickness relative to the radius of curvature. An example temperature profile is illustrated in Fig. 1 for the case of a five-inch deposit resting on a three-inch, outside-diameter steam tube with a 750 K surface temperature exposed to a heat flux of 10 kW/m² and with a thermal conductivity of 2.22 W/(m K). Both the porosity and tortuosity are considered to be unity in the base case, with both parameters being varied by a factor of two to illustrate the effects of deposit properties on the temperature profile. The temperature range depends linearly on the tortuosity and inversely on the porosity such that a change in either quantity changes the difference between deposit surface temperature and tube surface temperature by the same factor. The extent of curvature in the prediction is determined by geometry, not deposit physical properties. Deposits with solid volume fractions lower than (more porous than) 0.5 and tortuosities higher than 2 are common in many systems.

The previous predictions assumed that the incident heat flux, whether from radiation or convection, does not change as deposit surface temperature changes. In practice, incident heat flux is strongly coupled to deposit surface temperature. As an illustration, the heat transfer model predictions for the furnace section of a typical boiler are illustrated below. Only radiative heat transfer is considered, with an assumed black body radiative temperature of 2200 K, deposit thickness of 2 mm, deposit solid phase thermal conductivity of 2.22 W/(m K), and a waterwall composed of 750 K walls made of four inch OD tubes. Predictions of deposit surface temperature and heat flux are illustrated for a range of porosity and deposit emissivity values. Intra-deposit radiative heat transfer and intra-deposit conductive heat transfer through the gas phase are neglected and deposit tortuosity is assumed to be unity. None of these assumptions is generally accurate. They are made here to allow illustration of the impact of porosity and emissivity on heat transfer.

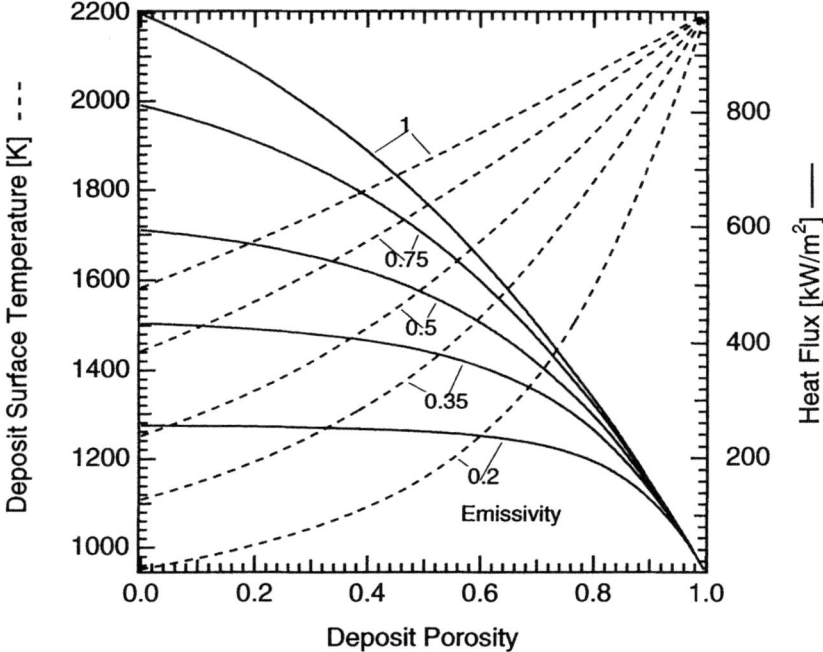

Figure 2   Deposit surface temperature and heat flux as a function of porosity and emissivity assuming no intra-deposit radiative heat transfer and a non-conducting gas phase. Tortuosity is assumed to be unity.

The parametric graph indicated in Fig. 2 belies the potential complexity of the relationships between deposit physical properties and heat transfer rates. While the trends in Fig. 2 indicate relatively smoothly varying and monotonic relationships between emissivity or porosity and heat flux, in practice the relationship may not be monotonic. In many cases of practical interest porosity and emissivity are correlated. Heat fluxes under such conditions may not vary monotonically with physical properties. Figure 3 illustrates the trends with an assumed linear relationship between porosity and emissivity, as read by the dual abscissae. As the relationships become more complex, and as factors such as intra-deposit radiative heat transfer and tortuosity are included, the relationships can become increasingly complex.

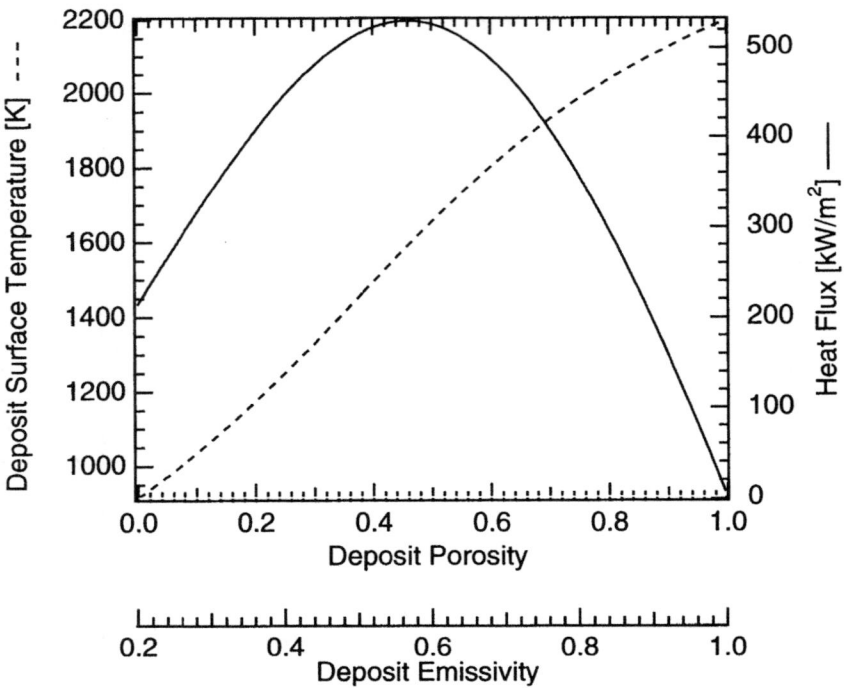

Figure 3   Deposit surface temperature and heat flux under the same assumptions as in Figure 2 but assuming a linear relationship between emissivity and porosity.

Structural properties of ash vary temporally, effecting changes in both porosity and tortuosity. A common example is sintering or melting of deposits, accompanied by increases in particle-to-particle contacting area and decreases in tortuosity and porosity. A simple example is illustrated in Fig. 4. In an idealized case of uniform spheres, a change in linear dimension of less than 15 % is accompanied by a change in contacting efficiency of theoretically zero in the initial case to 50 % in the slightly sintered case. This gives rise to proportional changes in tortuosity and the porosity changes from 0.48 to 0.17. Such changes lend themselves to mathematical treatments in predicting heat transfer through ash deposits. Similar treatments describe the effect of condensation or sulfation on deposit microstructure. These have been used in the past to explain the development of deposit properties ranging from tenacity to strength.

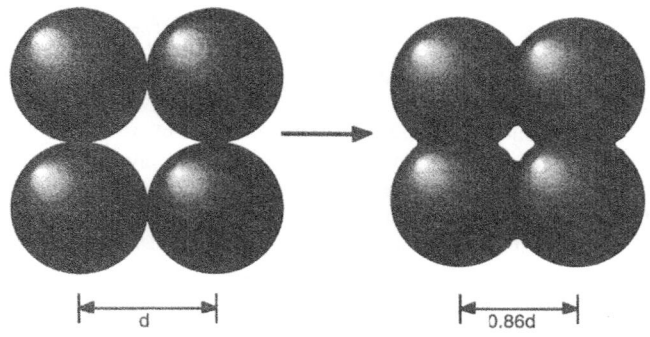

Figure 4   Conceptual illustration of the changes in contacting efficiency and tortuosity with sintering/melting.

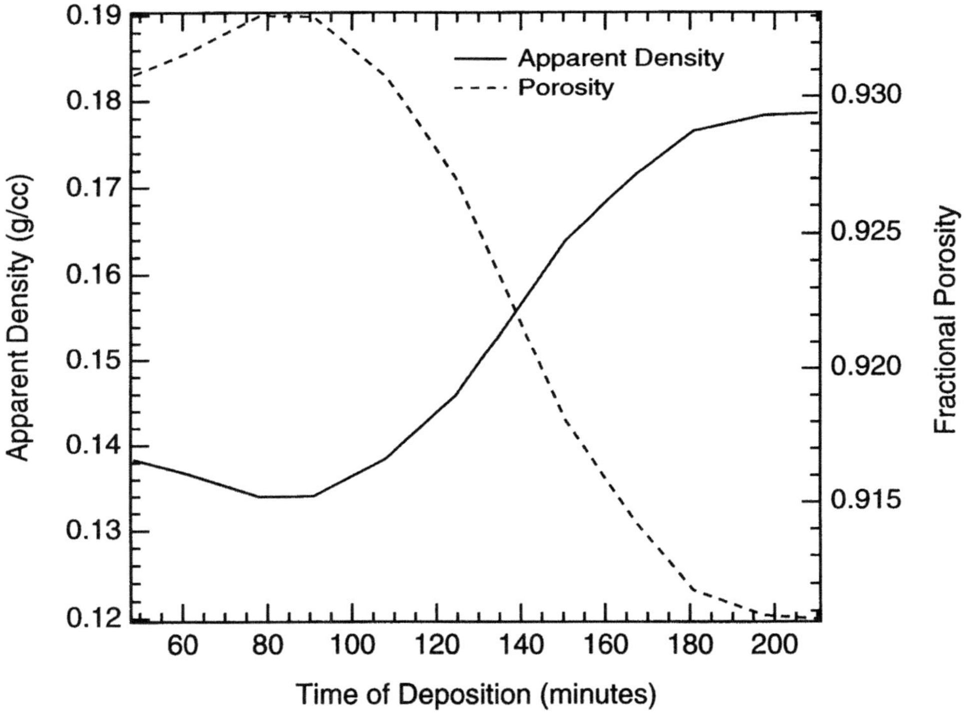

Figure 5.   Change in porosity and apparent density with time for an Illinois #6 coal ash accumulating on a tube in cross flow.

In situ, time-resolved, simultaneously measured trends in apparent density and porosity are indicated in Figure 5. The porosity decreases and apparent density increases with time, suggesting that the deposit is sintering. The extent of sintering is quite small, however, with less than a 3% decrease in porosity.

Thermal conductivity is determined from measured values of deposit surface temperature, probe temperature, overall heat flux through the probe as determined by change in gas temperature, and deposit thickness. Several measurements of deposit surface temperature and probe surface temperature are made, and since the probe dimensions and material are well characterized, the change in gas temperature can be used without probe surface temperatures to determine the thermal conductivity. In practice, this means there are several avenues available for the determination of thermal conductivity from our data. Two nearly independent analyses are illustrated Figure 6. They are nearly independent because they rely a few of the same measured values. As is indicated, the thermal conductivity is seen to increase by a factor of between 3 and 5, depending on the analysis. The data nicely illustrate the dependence of thermal conductivity on porosity, among other things.

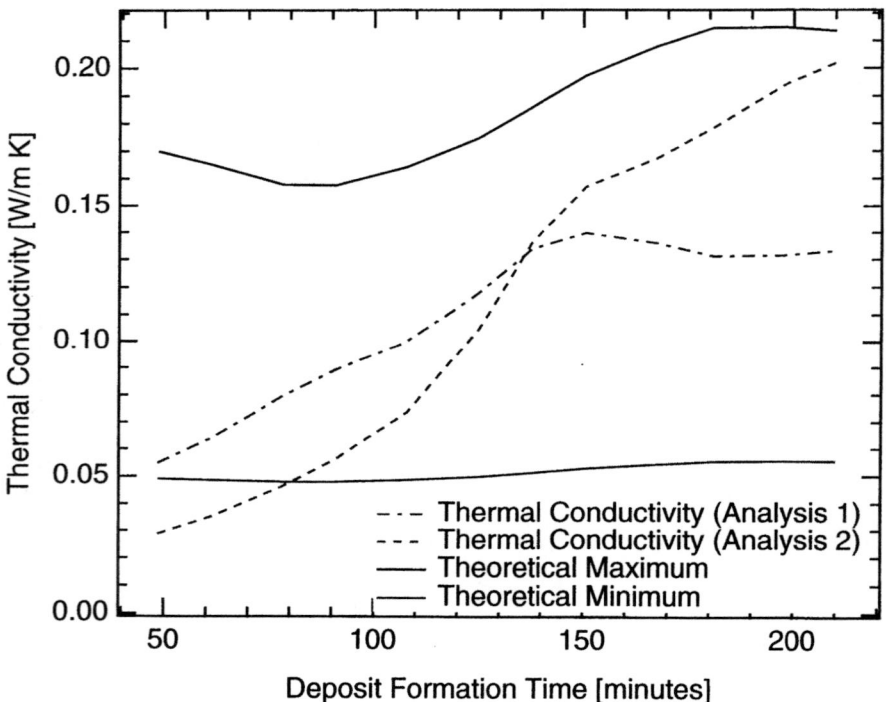

Figure 6   Measured and theoretical development of thermal conductivity in an ash deposit formed in the Multifuel Combustor.

Theoretical maxima and minima based on theoretical analyses of deposit properties are illustrated as a function of time for the same data. Over most of the range, they bracket the measured results. These maxima and minima also depend on deposit structure and therefore exhibit time dependencies.

Similar data for a Powder River Basin (Black Thunder) coal are illustrated in Figure 7, illustrating similar trends. Powder River Basin coals produce deposits rich in calcium sulfate, in some ways similar to deposits formed from woody biomass fuels. These materials are transparent in the infrared but form as small particulate on the surface, producing a highly reflective deposit with emissivities sometimes as low as 0.2. Biomass ash deposits formed from calcium sulfate, silica, potassium chloride, and potassium sulfate are qualitatively similar to those from low-rank coals with respect to their optical properties. As seen in the figure, these low rank coals also show indications of sintering.

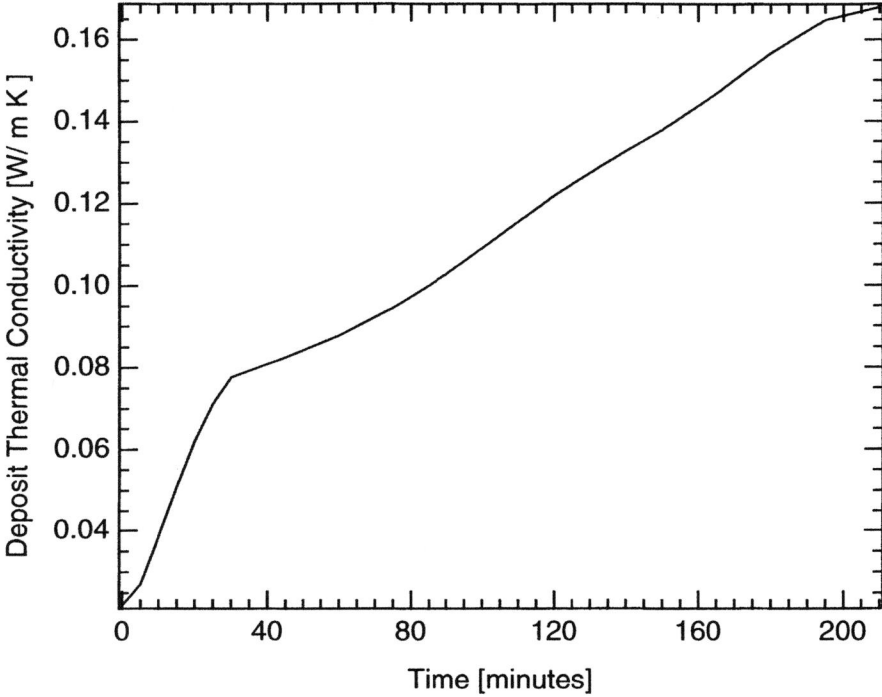

Figure 7   Measured development of thermal conductivity in an ash deposit from a Black Thunder coal.

In other experiments (not illustrated), the thermal conductivity in the earliest stages of deposit growth decreased from a value well above the theoretical maximum to a more reasonable limit, exhibiting the opposite trend in time as these data suggest. This behavior we attribute to the neglect of intra-deposit radiation on the heat transfer analysis. Nonabsorbing porous materials exposed to high incident radiative fluxes and at high temperatures can transfer as much heat by radiation as by conduction.

4. Conclusions

Ash deposit microstructure influences the mechanical and transport properties by impacting the degree of connectedness between particles and the tortuosity of heat transport through the deposit. Mathematical models are used to predict the impact of microstructural features on bulk deposit properties and on resulting boiler performance. Deposit surface temperatures can change many hundreds of degrees, depending on deposit thermal and structural properties.

Heat fluxes are also dominantly influenced by similar structural properties. Two properties that encapsulate much of the deposit microstructure effect are the porosity and tortuosity. Rational models of the dependence of thermal conductivity on these parameters are presented with predicted results. Experimental examples of how tortuosity and porosity develop in deposits, depending on deposit phase, are also presented.

5. Acknowledgments

Portions of this work were supported by U.S. Department of Energy through the Energy Efficiency and Renewable Energy Office's Biomass Power Program and through Pittsburgh Energy Technology Center's Direct Utilization Advanced Research and Technology Development Program. In addition, a consortium of industries with interest in biomass power financially contributed to this project. These include Mendota Biomass Power Ltd. and Woodland Biomass Power Ltd. (both associated with Thermo Electron Energy Systems), CMS Generation Operating Co. (formerly Hydra-Co Operations Inc.), Wheelabrator Shasta and Hudson Energy Cos., Sithe Energy Co., Delano Energy Co. Inc., the Electric Power Research Institute, Foster Wheeler Development Corp., and Elkraft Power Co. Ltd. of Denmark. Most of these companies also contributed fuels, use of facilities, and technical expertise in reviewing results of the project. The authors are grateful for the support and interaction with all of the individuals representing these organizations.

6. References

1. Beér, J.M., *et al. From Coal Mineral Matter Properties to Fly Ash Deposition Tendencies; a Modeling Approach.* in *The Seventh International Pittsburgh Coal Conference.* 1990. Pittsburgh, PA:

2. Boni, A.A., *et al. Mineral Matter Transformations During Coal Combustion: Fundamental Results and Relevance to Deposition.* in *ASME Seminar on Fireside Fouling Problems.* 1990. Provo, Utah:

3. Baxter, L.L. and R.W. DeSollar, *A Mechanistic Description of Ash Deposition During Pulverized Coal Combustion: Predictions Compared to Observations.* Fuel, 1993. **72**(10): p. 1411-1418.

4. Baxter, L.L. and D.R. Hardesty, *The Fate of Mineral Matter During Pulverized Coal Combustion: January-March.* 1992, Sandia National Laboratories:

5. Baxter, L.L. and D.R. Hardesty, *The Fate of Mineral Matter During Pulverized Coal Combustion: April - June.* 1992, Sandia National Laboratories:

6. Flagan, R.C. and D.D. Taylor. *Laboratory Studies of Submicron Particles from Coal Combustion.* in *Eighteenth Symposium (International) on Combustion.* 1980. Waterloo, Canada: The Combustion Institute.

7. Sarofim, A.F., J.B. Howard, and A.S. Padia, *The Physical Transformation of the Mineral Matter in Pulverized Coal Under Simulated Combustion Conditions.* Combustion Science and Technology, 1977. **16**: p. 187-204.

8. Srinivasachar, S., *et al., Mineral Behavior During Coal Combustion 2. Illite Transformations.* Progress in Energy and Combustion Science, 1990. **16**: p. 293-302.

9. Srinivasachar, S., *et al., A Kinetic Description of Vapor Phase Alkali Transformations in Combustion Systems.* Progress in Energy and Combustion Science, 1990. **16**: p. 303-309.

10. Zygarlicke, C.J., M. Ramanathan, and T.A. Erickson, *Fly Ash Particle-Size Distribution and Composition: Experimental and Phenomenological Approach,* in *Inorganic Transformations and Ash Deposition During Combustion,* S.A. Benson, Editor. 1992, American Society of Mechanical Engineers: New York. p. 525-544.

11. Srinivasachar, S., *et al. A Fundamental Approach to the Prediction of Coal Ash Deposit Formation in Combustion Systems.* in *The Twenty-Fourth Symposium (International) on Combustion.* 1992. The University of Sydney, Australia: The Combustion Institute.

12. Baxter, L.L. and L. Dora, *The Combustion Behavior of a Blend of Eastern and Western Coals: Comparisons Between a Blend and Its Individual Components.* ASME Paper No. 92-JPGC-FACT-14, 1992. .

13. Zygarlicke, C.J., *et al., Combustion Inorganic Transformations.* 1992, University of North Dakota Energy and Environmental Research Center:

14. Goodwin, D.G., *Infrared Optical Constants of Coal Slags*. 1986, Stanford University:

15. Wall, T.F., *et al. The Properties and Thermal Effects of Ash Deposits in Coal-Fired Furnaces*. in *Engineering Foundation Conference on Ash Deposition during Coal Combustion.* 1993. Solihull, UK:

16. Baxter, L.L., T.H. Fletcher, and D.K. Ottesen, *Spectral Emittance Measurements of Coal Particles*. Energy & Fuels, 1988. **2**: p. 423-430.

17. Wall, T.F., *et al.*, *Ash Deposits, Coal Blends, and the Thermal Performance of Furnaces*, in *Engineering Foundation Conference on Coal Blending and Switching of Western Low-Sulfur Coals*, R.W. Bryers and N.S. Harding, Editor. 1993, ASME: New York. p. 453-463.

18. Richards, G.H., *et al. Radiative Heat Transfer in PC-Fired Boilers — Development of the Absorptive/Reflective Character of Initial Ash Deposits on Walls*. in *Twenty-Fifth Symposium (International) on Combustion.* 1994. Irvine, CA: The Combustion Institute.

19. Boow, J. and P.R. Goard, *Fireside deposits and their effect on heat transfer in a pulverized-fuel-fired boiler: Part III. The influence of the physical characteristics of the deposit on its radiant emittance and effective thermal conductance*. Journal of the Institute of Fuel, 1969. **42**: p. 412-419.

20. Markham, J.R., *et al.*, *Measurement of Radiative Properties of Ash and Slag by FT-IR Emission and Reflection Spectroscopy*. Transactions of the ASME, 1992. **114**: p. 458-464.

21. Bohren, C.F. and D.R. Huffman, *Absorption and Scattering of Light by Small Particles*. 1983, New York: John Wiley & Sons. 530.

22. Mitor, V.V. and I.N. Konopel'Ko, *The Study of Emissivity of Solid Bodies*. Thermal Engineering, 1966. **13**(7): p. 92-97.

23. Mitor, V.V. and I.N. Konopel'Ko, *Experimental Investigation of Radiation and Reflection Characteristics of Some Solid Bodies*. Thermal Engineering, 1970. **16**(5): p. 47-52.

24. Mitor, V.V. and I.N. Konopel'Ko, *A Study of the Emissivity Factor of Ash Deposits and Certain Refractory Materials*. Thermal Engineering, 1970. **17**(19): p. 61-63.

25. Vassallo, A.M., *et al.*, *Infrared Emission Spectroscopy of Coal Minerals and their Thermal Transformations*. Applied Spectroscopy, 1992. **46**(1): p. 73-78.

26. Vassallo, A.M. and K.S. Finnie, *Infrared Emission Spectroscopy of Some Sulfate Minerals*. Applied Spectroscopy, 1992. **46**: p. 1477-1482.

27. Street, P.J. and C.S. Twamley, *Fuel Particle Emissivities*. Journal of the Institute of Fuel, 1971. **44**: p. 477-478.

28. Mulcahy, M.F.R., J. Boow, and P.R.C. Goard, *Fireside Deposits and their Effect on Heat Transfer in a Pulverized-fuel-fired Boiler. Part I: The Radiant Emittance and Effective Thermal Conductance of the Deposits*. Journal of the Institute of Fuel, 1966. **39**: p. 385-394.

29. Mulcahy, M.F.R., J. Boow, and P.R.C. Goard, *Fireside Deposits and their Effect on Heat Transfer in a Pulverized-fuel-fired Boiler. Part II: The Effect of the Deposit on heat Transfer from the Combustion Chamber Considered as a Continuous Well-stirred Reactor.* Journal of the Institute of Fuel, 1966. **39**: p. 394-398.

30. Goetz, F.J., N.Y. Nsakala, and R.W. Borio, *Development of Method for Determining Emissivities and Absorbtivities of Coal Ash Deposits.* Journal of Engineering for Power, 1979. **101**: p. 607-621.

31. Carter, H.R. and C.G. Koksal. *On-line Imaging and Emissivity Measurements to Determine Furnace Cleanliness.* in *International Joint Power Generation Conference.* 1991. San Diego, CA:

32. Carter, H.R., C.G. Koksal, and M.A. Garrabrant, *Furnace Cleaning in Utility Boilers Burning Powder River Basin Coals.* 1992, ASME:

33. Tassopoulos, M., J.A. O'Brien, and D.E. Rosner, *Simulation of Microstructure/Mechanism Relationships in Particle Deposition.* AIChE Journal, 1989. **35**(6): p. 967-980.

34. Tassopoulos, M. and D. Rosner, *Simulation of Vapor Diffusion in Anisotropic Particulate Deposits.* Chemical Engineering Science, 1992. **47**(2): p. 421-443.

35. Tassopoulos, M. and D.E. Rosner, *Microstructural Descriptors Characterizing Granular Deposits.* AIChE Journal, 1992. **38**(1): p. 15-25.

36. Tassopoulos, M. and D.E. Rosner, *The Effective Thermal Conductivity of Packings of Spheres I. Conduction through the Solid Phase.* International Journal of Heat and Mass Transfer, 1992. .

37. Tassopoulos, M. and D.E. Rosner, *The Effective Thermal Conductivity of Packings of Spheres II. Conduction through the Solid and Void Spaces.* International Journal of Heat and Mass Transfer, 1992. .

38. Tassopoulos, M., L.L. Baxter, and D.E. Rosner, *Comparison of Predicted and Measured Microstructure of Coal Ash Deposits.* AIChE Journal, 1994 (in press). .

39. Anderson, D.W., R. Viskanta, and F.P. Incropera, *Effective Thermal Conductivity of Coal Ash Deposits at Moderate to High Temperatures.* Journal of Engineering for Gas Turbines and Power, 1987. **109**: p. 215-221.

40. Jagota, A., P.R. Dawson, and J.T. Jenkins, *An Anisotropic Continuum Model for the Sintering and Compaction of Powder Packings.* Mechanics of Materials, 1988. **11**: p. 357-372.

41. Jagota, A. and C.Y. Hui, *The Effective Thermal Conductivity of a Packing of Spheres.* Transactions of the ASME, Journal of Applied Mechanics, 1990. **57**: p. 789-791.

42. Kalmanovitch, D.P. and M. Frank. *An Effective Model of Viscosity for Ash Deposition Phenomena.* in *Mineral Matter and Ash Deposition from Coal.* 1988. Santa Barbara, CA: The Engineering Foundation.

43. Urbain, G., *Viscosity of Silicate Melts.* Journal of the British Ceramic Society, 1981. **80**.

44. Raask, E., *Mineral Impurities in Coal Combustion.* 1985, Washington: Hemisphere Publishing Corporation. 484.

THE FATE OF ALKALI METAL DURING BIOMASS THERMOCHEMICAL CONVERSION

Alkali Metals and Biomass Combustion

DAVID C. DAYTON
National Renewable Energy Laboratory
Golden, CO USA

Abstract
The fate of alkali metal released during combustion of biomass and biomass-derived fuels has been studied by directly sampling the hot gases liberated from combustion of small samples in a variable-temperature quartz-tube reactor employing a molecular beam mass spectrometer (MBMS) system. This experimental technique has been successfully used to identify alkali species released during the combustion of 23 solid biomass feedstocks, 4 biomass chars, and 7 biomass-derived pyrolysis oils. This paper will focus on the results for oils and chars produced from switchgrass pyrolysis as representative examples. By studying the combustion of biomass pyrolysis oils and chars it is possible to determine how the alkali metal is partitioned during biomass pyrolysis. Multiple combustion conditions have been investigated to target those conditions that minimize alkali metal release. The nitrogen, sulfur, and chlorine content of a given pyrolysis oil is a function of the starting feedstock material.

The results of extensive laboratory studies indicate that during biomass combustion, initial feedstock composition has the most pronounced effect on alkali metal released. Depending on the feedstock composition, four mechanisms of alkali metal release involving potassium sulfate, potassium chloride, potassium hydroxide, and potassium cyanate have been identified. Alkali metals are sequestered in the char during biomass pyrolysis. The form of alkali metal released into the gas phase during combustion of the biomass char was similar to the alkali metals observed during biomass combustion. In general, the biomass pyrolysis oils studied had low alkali metal concentrations.

Keywords: Alkali metals, combustion, molecular beam mass spectrometry, pyrolysis

1 Introduction

Combustion, gasification, and combustion of oils produced by pyrolysis are all potential processes for producing electricity from biomass. The fate of the alkali metals initially present in biomass during each of these processes can affect the success and efficiency of electricity production. Alkali metals, in particular potassium, are thought to promote fouling and slagging of heat transfer surfaces in power generating facilities that convert biomass to electricity via direct biomass combustion [1-3]. When biomass is used as a fuel in boilers, the deposits formed reduce efficiency and, in the worst case, lead to unscheduled plant downtime [4]. Biomass pyrolysis oils have been targeted as future fuels in diesel engines, and ultimately, aeroderivative gas turbines. Alkali metals released during the combustion of pyrolysis oils can lead to fouling of fuel injectors in diesel engines and fouling and corrosion of turbine blades. Alkali metals released during biomass gasification also pose challenges for using the product gas in combustors, turbines, and fuel cells.

Biomass-derived pyrolysis oils must contain very low levels of alkali to become competitive fuels for use in turbine combustors that generate power. Based on standards set for traditional petroleum fuel oils, turbine manufacturers require total alkali (Na and K) levels in the sub-ppm range [5]. NO_x and SO_x released during pyrolysis oil combustion must be controlled to alleviate environmental concerns. Chlorine released during biomass and pyrolysis oil combustion will have important implications concerning corrosion of heat transfer surfaces in boilers and turbine blades in direct-fired facilities. The transformation of nitrogen, sulfur, and chlorine during the production of biomass-derived pyrolysis oils has not been studied in detail. When pyrolysis oil is produced, the majority of the alkali metal present in the starting biomass material remains sequestered in the char fines [6]. The alkali metal content of the oils, therefore, is thought to be predominantly a function of the amount of char fines suspended in the oil.

2 Experimental Approach

The release of alkali vapor species, NO_x, SO_x, and chlorine during the combustion of selected biomass feedstocks, biomass-derived pyrolysis oils, and chars was monitored and studied using a direct sampling, molecular beam mass spectrometer (MBMS) system [7,8] in conjunction with a quartz-tube reactor that has been described in detail in the literature [9,10]. The MBMS system is ideally suited for studying the high-temperature, ambient- pressure environments encountered during the present screening studies. The integrity of the sampled, high-temperature combustion gases is preserved by the free-jet expansion which effectively quenches chemical reactions and inhibits condensation. As a result, reactive and condensable species remain in the gas phase at temperatures far below their condensation point for long periods of time in comparison to reaction rates.

2.1 Oil and char samples

In the past several years, alkali metal released during the combustion of 23 different biomass feedstocks at multiple combustion conditions has been investigated [9-11]. A list of these feedstocks, their ultimate and ash analyses, and the results from the combustion studies are available in the literature [11]. In addition to the biomass feedstocks, seven different pyrolysis oils and three different biomass chars have been studied using similar experimental techniques.

Four of the oils were produced in the NREL vortex reactor [12]: switchgrass oil (high alkali, run #146A); oak oil (run #154); switchgrass oil (low alkali, run #157); and poplar oil (run #160). During the production of the first two oils (run #146A and run #154), the char fines were filtered out of the hot gas stream using a cyclone. A hot baghouse filter was used to remove particulates during the production of the second two oils (run #157 and run #160). The remaining three oils were produced in a laboratory scale (2" diameter) fluidized bed combustor (FBC) [6]. These oils were produced from switchgrass (time 0, run #101), poplar (time 0, run #4), and corn stover (time 0, run #2), respectively. In these cases, time 0 corresponds to freshly harvested biomass that has not been stored in the field.

The char samples investigated in this study [13] were collected from the chars remaining after biomass was pyrolyzed in the NREL vortex reactor. The following char samples were received: oak char (run #154); poplar char (run #156); and switchgrass char (run #157). The ash content of the pyrolysis oils studied is very low (less than 0.1% by weight). The biomass chars studied, however, can be as much as 10% ash. The switchgrass char is actually 22% ash by weight. Results from combustion studies using the switchgrass-derived oils and chars are presented below.

2.2 Molecular beam mass spectrometer/quartz-tube reactor apparatus

Ten to fifty milligrams of each sample were loaded into hemi-capsular quartz boats which were then placed in a platinum mesh basket attached to the end of a quarter-inch diameter quartz rod. This quartz rod was inserted and translated into the heated zone of a quartz tube reactor enclosed in a two-zone variable temperature furnace [9,10]. The experiments simulate the continuous combustion of the sample from initial heating to ignition to complete char burnout and ash "cooking." By the ash "cooking" stage most or all of the volatile material has been released, however, the ash was left in the high temperature furnace to insure that all of the volatile material was released. Furnace temperatures during these screening studies were maintained at 1100°C and 800°C, respectively. Gas temperatures near the quartz boat were measured with a type-K thermocouple inserted through the quartz rod. The actual boat, sample, and flame temperatures were not measured during the combustion event, however, the particle temperature during flaming combustion is expected to be higher than the furnace temperature. The atmosphere in the reactor consisted of a flowing mixture of 20% oxygen in helium at a total flow rate of 4.4 standard liters per minute. Helium instead of nitrogen was used as the inert gas because it is optimum for molecular beam formation. This also ensured that all of the nitrogen oxides formed during combustion originated from fuel-bound nitrogen. With the exception of thermal NO being formed from the nitrogen in the added gases, the use of nitrogen as the inert gas would not affect the combustion products.

The molecular beam sampling system consists of a three-stage, differentially pumped vacuum chamber. A conical, stainless steel molecular beam sampling orifice was positioned at the downstream end of the quartz tube reactor to sample the high temperature, ambient pressure combustion gases. The tip of the sampling orifice protruded into the furnace, keeping it at an elevated temperature (not measured) to prevent alkali metals and other species from condensing on the orifice. Sampled gases underwent a free-jet expansion into the first stage of the vacuum system. A molecular beam was formed by collimating the gas stream with a conical skimmer located at the entrance to the second stage of the vacuum system. The molecular beam was directed into the ionization region of the mass spectrometer located in the third stage of the vacuum system. Electron impact ionization (25 eV) of the species in the molecular beam yielded ions that were filtered by a triple quadrupole mass analyzer and detected with an off-axis electron multiplier.

3 An Overview of Biomass Combustion

Alkali metal release and speciation during biomass combustion was investigated by examining the mass spectral results recorded during biomass combustion in a 20% O_2 in helium atmosphere at 1100°C. As discussed in the literature [9-11], biomass combustion occurred in three distinct phases: the devolatilization phase, the char combustion phase, and the ash "cooking" phase.

Typical combustion products such as CO, CO_2, and H_2O, as well as SO_2 and NO, were produced during the devolatilization phase while O_2 was consumed. Most of the alkali metal was released into the gas phase during the char combustion phase and it was possible to identify the alkali species in the mass spectrum averaged over the char combustion phase. Although the volatile matter has been liberated by the beginning of the ash "cooking" phase, the remaining ash was left in the high temperature reactor to insure that all of the volatile matter had been released.

Of the multiple phases that occur during biomass combustion, the char combustion phase is the most important in terms of studying alkali metal release. The mass spectrum averaged during a given char phase of biomass combustion qualitatively reflects the feedstock composition as determined in the ultimate analysis. Although the ultimate analysis is important for determining the total amount of alkali metal in a given feedstock, it does not reflect how much alkali metal is released into the gas phase nor the form of the alkali metal-containing species released. For this reason, the MBMS technique continues to be valuable for directly studying alkali metal released during biomass combustion.

To date, 23 biomass feedstocks have been screened for alkali release and speciation under various combustion conditions [9,11]. The conditions were chosen to investigate the effect of temperature, oxygen concentration, and excess water vapor on the release of alkali vapor during biomass combustion. The focus of the alkali screening studies has been to identify alkali metal containing species released during biomass combustion, and conditions, if any, which reduce the amount of alkali released.

Changing the combustion conditions, over the range studied (800-1100°C, 5% - 20% initial O_2 in helium, and 0-20% added steam), has little effect on the release of alkali vapor species. The most significant parameter that affects alkali vapor release is the composition of the feedstock being combusted. The higher the chlorine content and fuel bound nitrogen of a given feedstock, the more gas-phase chloride and NO_x that is released during combustion. The combustion of a feedstock with a high potassium content does not necessarily translate into high levels of gas-phase potassium.

Although each individual feedstock appears to have its own unique combustion properties, these feedstocks can often be grouped together into classes of feedstocks. Woody feedstocks have comparatively low alkali metal content and low levels of chlorine. Consequently, combustion of woody feedstocks release very little alkali vapor.

Herbaceous feedstocks, grasses, and straws contain very high levels of alkali and chlorine compared to the woody feedstocks. Large amounts of alkali metal are released into the gas phase during combustion of herbaceous feedstocks. As a result, these feedstocks have a high fouling and slagging potential.

Chlorine content has been correlated with alkali vapor release [11]. It appears that high chlorine levels tend to facilitate alkali metal release; however, alkali metals can still be released in the absence of chlorine. The agricultural residues with little chlorine still release significant amounts of alkali vapor during combustion, however, the alkali metal is present as the hydroxide and potassium cyanate [11,14] instead of the chloride. The formation of potassium cyanate seems to dominate during the combustion of feedstocks that have high potassium and nitrogen contents. Adding excess steam to the combustion atmosphere substantially reduces the amount of potassium cyanate that is released during combustion of these feedstocks.

4 Combustion of Biomass Chars

The combustion behavior of the biomass chars was different than that of the biomass precursor. A devolatilization phase would not be expected during biomass char combustion; however, depending on how completely the biomass was pyrolyzed, some volatile matter can be left in the char. Multiple phases of combustion were observed during biomass char combustion but these phases do not necessarily correspond with the phases observed during combustion of the parent biomass material. In the following discussion, two phases observed during biomass char combustion will be discussed.

Figure 1 shows the temporal profile of the total ion current recorded during switchgrass char (run #157) combustion in 20% O_2 in helium at 1100°C. Each point corresponds to a complete mass spectrum from m/z = 10 to m/z = 130. The sample was introduced into the high temperature region of the furnace at approximately 0.5 minutes in the figure. The initial peak in the total ion intensity corresponds to phase 1 and the average mass spectrum over this phase is presented in Figure 2. The mass spectrum averaged over the background was subtracted from the mass spectrum averaged during phase 1. The signal intensities in Figure 2 were normalized to the $^{34}O_2^+$ signal measured before the sample was inserted in the hot zone of the reactor. The switchgrass starting material was shown to have a high alkali metal and chlorine content [10]. Therefore, the corresponding switchgrass char was also expected to have a high alkali metal and chlorine content.

The ultimate analysis of the switchgrass char, which is listed in the inset in Figure 2, confirms the high alkali metal and chlorine content of the switchgrass char. Compared to the parent switchgrass material, which is 0.83% K by weight (dry basis), the switchgrass char is 4.2% K by weight (dry basis). This clearly indicates that potassium was concentrated in the char when switchgrass was pyrolyzed. The mass spectrum averaged during phase 1 (Figure 2) indicates that this phase was dominated by the release of CO_2, CO, HCl, KCl, NaCl, and SO_2. Based on the measured intensities of the $NaCl^+$, KCl^+, and K_2Cl^+ (the KCl dimer fragment ion) signals indicated in Figure 2, a large fraction of alkali metal vapor was released during this phase of the switchgrass char combustion.

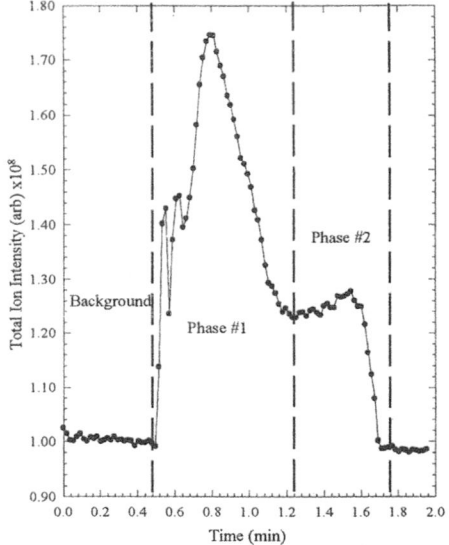

Figure 1: Temporal profile of the total ion current measured during switchgrass char combustion at 1100°C in 20% O_2 in helium.

The average mass spectrum during phase 2 of switchgrass char combustion is presented in Figure 3. In contrast to phase 1, the phase 2 mass spectrum indicates that little alkali metal vapor was released during this phase because of the absence of the KCl^+ and K_2Cl^+ peaks and only a small K^+ intensity at m/z = 39. NO^+ and SO_2^+ were still observed in the phase 2 mass spectrum. These results indicate that alkali metal was released during the combustion of switchgrass char primarily in the form of the chloride in the first phase and possibly as a sulfate in the second phase [9,10].

The effect of temperature on the combustion of the biomass chars was investigated by studying combustion of the switchgrass char at a furnace temperature of 800°C in a 20% O_2 in helium atmosphere. The duration of char burnout was about a factor of two longer at the lower furnace temperature. In the case of the switchgrass char, the majority of the alkali chloride released during combustion was shifted to the second phase. This suggests that the release of alkali metal release during the switchgrass char combustion was highly temperature dependent.

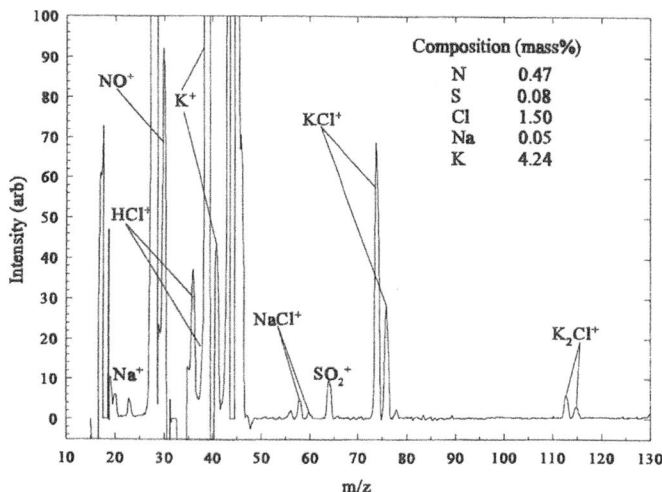

Figure 2: Average mass spectrum recorded during phase 1 of switchgrass char combustion at 1100°C in 20% O_2 in helium. Mass spectral intensity has been normalized to the background $^{34}O_2^+$ signal. Sample composition (dry basis) as determined in the ultimate analysis is listed in the inset.

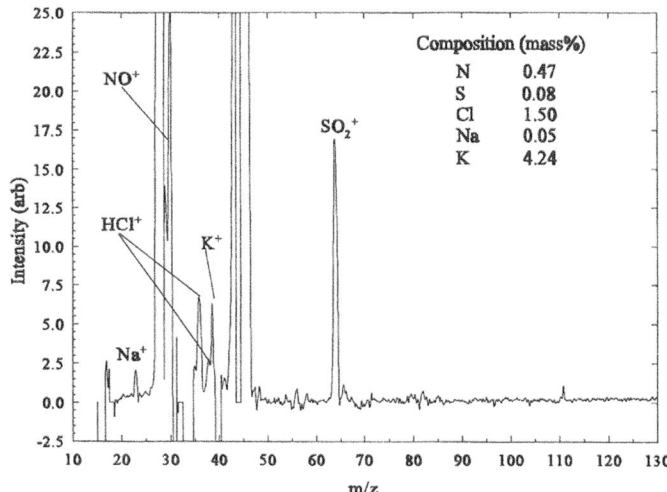

Figure 3: Average mass spectrum recorded during phase 2 of switchgrass char combustion at 1100°C in 20% O_2 in helium. Mass spectral intensity has been normalized to the background $^{34}O_2^+$ signal. Sample composition (dry basis) as determined in the ultimate analysis is listed in the inset.

5 Combustion of Biomass-Derived Pyrolysis Oils

In general, the combustion behavior of the oils was considerably different than that of solid biomass and chars. When the oil samples were inserted into the hot region of the reactor, the combustion produced a very luminous flame indicative of sooting that was visible from the rear of the reactor. The pyrolysis oil combustion event also appeared more violent than that of solid biomass because the pyrolysis oils sputtered during initial heating and combustion. As a result, a serious problem with particulate induced noise (PIN) was encountered. Particulates (char-ash particles and/or aerosols) released during combustion became entrained in the gas flow and collided with metal surfaces inside the mass spectrometer. These collisions randomly produced ions that were detected. This resulted in a time-dependent, mass-independent signal that saturated the detector and obscured any mass spectral data recorded during the PIN event.

The PIN problem made screening the pyrolysis oils more difficult than screening solid biomass material. The residence time of the combustion gases (and particles) in the reactor was increased to almost one second (compared to about 0.1 seconds for the solid biomass combustion) which reduced the PIN problem by allowing sufficient time for particle burnout. Although PIN continues to make it difficult to monitor pyrolysis oil combustion, the nature of pyrolysis oil combustion and the gases released can still be investigated.

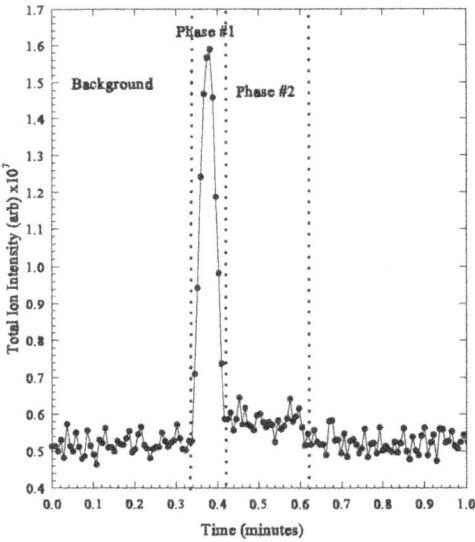

Figure 4: Temporal profile of the total ion current measured during FBC switchgrass oil (time 0) combustion at 1100°C in 20% O_2 in helium.

Several observations can be made about the combustion of biomass pyrolysis oils upon analysis of the mass spectral data. As seen in Figure 4, which shows the total ion current profile versus time measured during combustion of FBC switchgrass oil at 1100°C in 20% O_2 in helium, the duration of the pyrolysis oil combustion was considerably shorter than combustion of solid biomass, and there was little indication of a char combustion phase during pyrolysis oil combustion. The screening studies of biomass chars discussed above also indicated that most of the alkali metal in the parent material was sequestered in the char leaving relatively low levels of alkali metal in the oils. The preliminary screening of the three switchgrass pyrolysis oils indicated that, indeed, these oils contained little alkali metal compared to the biomass precursors. As an example, Figure 5 displays the mass spectrum averaged during phase 1 of FBC switchgrass oil combustion. Similar to combustion of the solid feedstocks, it was easy to identify the major combustion products, H_2O, CO, and CO_2, in the mass spectra recorded during the combustion of the pyrolysis oils. It was also possible to identify NO and SO_2 released during combustion of the pyrolysis oils. Individual alkali metal-containing species were not detected, but it was possible to identify the K^+ fragment ion from potassium-containing species.

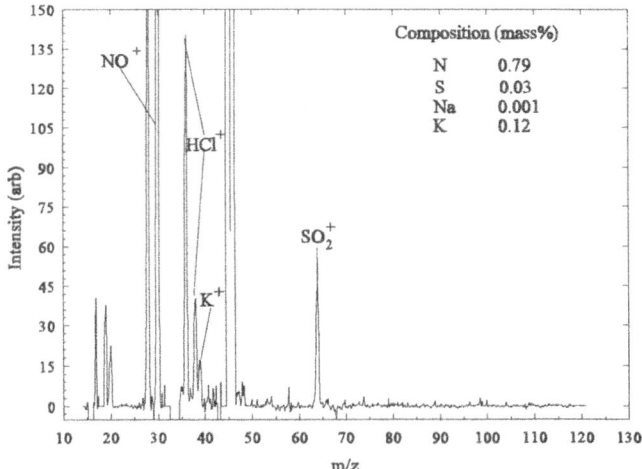

Figure 5: Average mass spectrum recorded during phase 1 of FBC switchgrass oil combustion at 1100°C in 20% O_2 in helium. Mass spectral intensity has been normalized to the background $^{34}O_2^+$ signal. Sample composition (dry basis) as determined in the ultimate analysis is listed in the inset.

The mass spectra averaged during the combustion of the pyrolysis oils produced from switchgrass indicated that a substantial amount of HCl was released during combustion. In general, the oils produced from switchgrass in the vortex reactor released less HCl during combustion than the switchgrass oil produced in the fluidized bed combustor. In fact, no HCl was detected during combustion of the vortex switchgrass oil (run #157), which was produced using the baghouse hot gas filter to remove the particulate matter.

The amount of NO observed during pyrolysis oil combustion was relatively constant and correlated with the fuel-bound nitrogen in the feedstocks used to produce the oils. The NO released during combustion of the switchgrass-derived oils was relatively consistent and does not appear to be a function of the way the oils were produced.

Reducing the furnace temperature to 800°C had little effect on the combustion behavior of the pyrolysis oils investigated during this study. Little change was observed in the mass spectra averaged during the combustion of the vortex switchgrass (run #157) oil compared to the higher temperature results (spectra not shown). The mass spectra averaged during combustion of the FBC switchgrass and the high alkali, vortex switchgrass oil (run #146A) at the lower furnace temperature suggest that less SO_2 and HCl was released compared to the higher temperature results. The ratio of the SO_2 to HCl signals, however, was not a function of temperature, indicating that the differences in the signal intensities may be just an artifact of scaling or differences in instrument response. Individual alkali metal-containing species were still not observed during combustion of these oils at a lower furnace temperature.

6 Quantification of the KCl Released During Combustion of Switchgrass, Switchgrass Oil, and Switchgrass Char

The amount of potassium chloride, the dominant alkali metal-containing species, released during the combustion of switchgrass char was quantified and compared to the amount of potassium chloride released during combustion of the parent feedstock. The amount of gas phase KCl released during combustion of switchgrass char at 1100°C and 800°C in 20% O_2 in helium was quantified by a previously described procedure [9,10]. KCl standards were prepared by drying 100 µl of a 4.654 mM aqueous KCl solution in a quartz boat in air. This resulted in 35µg of solid KCl in the sample boat. Standards were inserted into the furnace in the same fashion as the biomass samples with the underlying assumption that all of the KCl was volatilized, and that the KCl/reactor wall interactions were identical for the standards, the biomass, and char samples. The area under the m/z = 74 time profile was used to quantify the amount of gas phase KCl liberated during switchgrass and switchgrass char combustion. This insured that the KCl released during all phases of combustion was accounted for. The results for the amount of KCl released during switchgrass and switchgrass char combustion are summarized in Table 1.

The largest source of error from this calibration method arose from the variability of the mass spectral signals recorded from the KCl standards. The ratio of the areas under the m/z = 74 and m/z = 76 time profiles for the KCl standards, the switchgrass, and switchgrass char samples yields, within error, the characteristic 2.47:1 intensity ratio indicative of the natural isotopic abundance of KCl containing ^{35}Cl, ^{37}Cl, ^{39}K, and ^{41}K [15]. The ratio of the areas under the m/z = 39 and m/z = 74 time profiles for the KCl standards is a measure of the fragmentation of gas phase KCl in the ionizer at a given furnace temperature. This ratio is slightly temperature dependent [16]. The ratio of the areas under the m/z = 39 and m/z = 74 time profiles was significantly larger for the switchgrass and switchgrass char samples indicating that other potassium-containing species may have been liberated into the gas phase, fragmented, and contributed to the signal at m/z = 39. Although the majority of the alkali metal liberated during combustion of switchgrass and switchgrass char was in the form of KCl, other alkali metal-containing species (presumably sulfates or hydroxides) were also being released.

Table 1: Quantification of KCl Released from Switchgrass, Switchgrass Char, and Switchgrass Oils

Sample	Ratio of areas (39/74)	Ratio of areas (74/76)	% K released (as KCl)	Gas phase K from KCl (ppmw)
1100°C 20% O$_2$ in helium				
KCl Standard	3.6	2.5	100	----
Switchgrass	4.1±0.2	2.35±0.18	23.2±4.0	532±62
Switchgrass Char	8.0±0.3	2.6±0.1	36±2	1397±86
Vortex Switchgrass Oil	3.24±1.4	1.31±1.24	23±10	51±23
Vortex Switchgrass Oil	3.24±.23	1.46±1.65	63±26	18±7
FBC Switchgrass Oil (Time 0)[*]	5.16±2.9	1.07±0.11	179±11	36±2
800°C 20% O$_2$ in helium				
KCl Standard	3.9	2.6	100	----
Switchgrass	4.03±0.31	2.28±0.42	16.5±1.9	378±44
Switchgrass Char	5.8±0.7	2.4±0.1	45±7	1755±280
Vortex Switchgrass Oil	4.2±3.6	1.08±0.65	20±6	44±13
Vortex Switchgrass Oil	2.62±1.0	0.85±.27	65±12	23±4
FBC Switchgrass Oil (Time 0)[*]	3.19±0.93	1.54±0.31	145±62	29±13

The errors are one standard deviation.

[*]Alkali metal released during switchgrass pyrolysis oil combustion was quantified using the m/z = 39 peak (K[+]) assuming that this ion came solely from KCl fragmentation.

[**]Calculated as described in the text

Using the information from the ultimate analysis of the switchgrass and switchgrass char, it was possible to determine the average percent of K (as KCl) released into the gas phase. At 1100°C, 36±2% of the potassium in the switchgrass char was released into the gas phase as KCl and at 800°C, 45±7% of the potassium was released into the gas phase. Given the errors in the calibration method, these values are essentially the same. This is a larger fraction, however, compared to switchgrass combustion where only 23±4% of the available potassium was released into the gas phase as KCl at 1100°C and 16±2% of the potassium was released as KCl at 800°C. A more practical way of presenting the amount of alkali released is in parts per million inside a typical combustor. This quantity was calculated by assuming switchgrass and switchgrass char combustion in 20% excess air, conditions that are typical inside a boiler or combustor. The average calculated amount of potassium from KCl released during switchgrass char combustion at 1100°C is 1397±86 ppmw, and at 800°C is 1755±280 ppmw. These values are 2.6 to 4.6 times larger than the amount of gas phase potassium released from combustion of the parent switchgrass feedstock at 1100°C (532±62 ppmw) and 800°C (378±44 ppmw).

Similar attempts were made to quantify the alkali metal released during combustion of the switchgrass-derived pyrolysis oils (vortex switchgrass oil (run #146A), vortex switchgrass oil (run #157), and FBC switchgrass oil (time 0)). These results are suspect and less reliable than the results for KCl released during combustion of the solid samples discussed above. Based on the mass spectral results, KCl was the dominant alkali metal-containing species released during switchgrass combustion. Individual alkali metal-containing species were not identified in the mass spectral results recorded during switchgrass oil combustion (see Figure 5), however, it was possible to identify the K^+ fragment ion. In an attempt to quantify the amount of alkali vapor released during switchgrass oil combustion, several assumptions were made. The first assumption was that the dominant alkali metal-containing species released was indeed KCl even though it was not identified in the mass spectral results. For these three switchgrass oils, however, it was possible to identify K^+ in the mass spectrum, which was assumed to come solely from the fragmentation of KCl. Consequently, the $m/z = 39$ signal (instead of the $m/z = 74$ signal) was used to quantify the alkali metal released during switchgrass pyrolysis oil combustion.

The ratio of the $m/z = 74$ and $m/z = 76$ areas determined from the mass spectral results recorded during combustion of the switchgrass-derived pyrolysis oils are in error compared to the expected ratio of 2.47:1. The signal intensities at these two masses were very low or near zero (see Figure 5), and the areas under the time traces for these ions predominantly consisted of the noise. The ratio of $m/z = 39$ and $m/z = 74$ contain the same error. The shortcomings were most prevalent when values for the percentage of potassium released were compared for the different switchgrass oils. Twenty percent of the available potassium was released as KCl during combustion of the vortex switchgrass oil (run #146A). This value is similar to the results for combustion of the solid feedstocks. The results for the remaining two switchgrass oils, however, are less reasonable. In fact, for the FBC switchgrass oil, the calculations indicate that more than 100% K was released indicating that either KCl was not the dominant form of the alkali metal released during combustion of these oils, or that the amount of alkali metal released was below the detection limits of the mass spectrometer. Considering all of the potential errors in this quantification method, the amount of alkali metal released during combustion of these oils (assuming combustion in 20% excess air) is believed to be on the order of tens of parts per million. The values in Table 1 for the amount of gas phase K from KCl can be considered upper limits for these oils. Nevertheless, the switchgrass oils still contain sustantially more alkali metal than the recommended specifications (<20 ppb Na and K combined) for gas turbine fuels [5].

7 Conclusions

Investigating biomass combustion has revealed several dominant pathways for alkali-metal release depending on the composition of the feedstock being combusted. Low alkali-metal and chlorine feedstocks release alkali metals via vaporization or decomposition of potassium sulfate. High-alkali and low-chlorine feedstocks release alkali metal as potassium hydroxide; if the feedstock also has a high nitrogen content, release as potassium cyanate becomes important. Feedstocks with high alkali and high chlorine content release potassium chloride during combustion.

The combustion of biomass-derived pyrolysis oils and chars was also investigated to determine the fate of various components of the parent biomass feedstock during pyrolysis oil production. The nitrogen, sulfur, and chlorine content of a given pyrolysis oil is a function of the starting feedstock material. High nitrogen, sulfur, and chlorine-containing biomass feedstocks yield pyrolysis oils with corresponding high levels of nitrogen, sulfur, and chlorine. Based on the results from the present screening studies involving biomass-derived pyrolysis oils and chars, alkali metal is sequestered in the char fines during the production of biomass-derived pyrolysis oils which suggests that efficient filtering should yield a lower alkali metal-containing pyrolysis oil. This result is desirable if pyrolysis oils are to become a competitive fuel for use in turbine combustors for generating power.

Attempts were made to quantify the amount of alkali metal released during combustion of the switchgrass pyrolysis oils and chars. The results suggest that more alkali metal was released during combustion of switchgrass char than during combustion of similar amounts of switchgrass. The amount of alkali metal released during combustion of the switchgrass pyrolysis oil is at least an order of magnitude lower than the release during combustion of the parent switchgrass feedstock. Although the alkali metal levels in the switchgrass pyrolysis oils are lower than the parent switchgrass feedstock, the amount of alkali metal in the oils is still three orders of magnitude higher than recommended specifications for turbine fuels [5]. The levels of alkali metal in pyrolysis oils can be lowered by starting with biomass material having lower alkali metal levels, such as woody feedstocks. For example, the amount of potassium in the poplar oils (vortex poplar oil (run #160) and FBC poplar oil (time 0)) is 30 ppm and 27 ppm, respectively which is considerably lower than the potassium levels in the switchgrass oils. Even with these potential improvements, it will be difficult to produce biomass-derived pyrolysis oils with acceptable levels of alkali metal as required by manufactures of gas turbines for fossil-derived fuels. Whether such stringent criteria will apply to biofuels remains to be determined. For diesel applications, these strict requirements may also not apply.

8 Acknowledgements

The author is grateful for the support of the Solar Thermal and Biomass Power Division of the Department of Energy Office of Energy Efficiency and Renewable Energy. Special thanks go to Richard L. Bain, Thomas A. Milne, and Ralph P. Overend for both programmatic and technical support and guidance. Special thanks go to Stefan Czernik and John Scahill for providing the vortex oils and chars studied and to Foster Agblevor and Serpil Besler for providing us with pyrolysis oil produced in the fluidized bed combustor.

9 References

1. Bryers, R.W. (1978) *Ash Deposits and Corrosion Due to Impurities in Combustion Gases*, Hemisphere Publishing Corporation, New York, NY.
2. Bryers, R.W. and Vorres, K.S., Eds. (1990) *Mineral Matter and Ash Deposition from Coal*, Engineering Foundation Conference February 22-26, 1988 Santa Barbara, CA.
3. Baxter, L.L. (1993) Ash deposition during biomass and coal combustion: A mechanistic approach. *Biomass and Bioenergy*, Vol. 4, pp. 85-102.
4. Turnbull, J.H. (1993) Use of biomass in electric power generation: The California experience. *Biomass and Bioenergy*, Vol. 4, pp. 75-84.
5. Moses, C.A. and Bernstein, H. (1994) *Impact Study on the Use of Biomass-Derived Fuels in Gas Turbines for Power Generation*. NREL Technical Report (NREL/TP-430-6085). January 1994.
6. Agblevor, F.A.. Besler, S., and Wiselogel, A.E. (1995) Fast pyrolysis of stored biomass feedstocks, *Energy and Fuels*, Vol. 9, pp. 635-640.
7. Evans, R.J. and Milne, T.A. (1987) Molecular characterization of the pyrolysis of biomass: I. Fundamentals. *Energy and Fuels*, Vol. 1, pp. 123-127.
8. Evans, R.J. and Milne, T.A. (1987) Molecular characterization of the pyrolysis of biomass: II. Applications. *Energy and Fuels* Vol. 1, pp. 311-319.
9. French, R.J., Dayton, D.C., and Milne, T.A. (1994) *The Direct Observation of Alkali Vapor Species in Biomass Combustion and Gasification.* NREL Technical Report (NREL/TP-430-5597). January 1994.
10. Dayton, D.C., French, R.J., and Milne, T.A. (1995) The direct observation of alkali vapor release during biomass combustion and gasification. I. The application of molecular beam/mass spectrometry to switchgrass combustion. *Energy and Fuels*, Vol. 9, pp. 855-865.
11. Dayton, D.C. and Milne, T.A. (1995) Laboratory measurements of alkali metal containing vapors released during biomass combustion, to be published in the conference proceedings of the *Application of Advanced Technologies to Ash-Related Problems in Boilers* conference July 16-21, 1995 in Waterville Valley, NH.

12.   Diebold, J. and Scahill, J. (1985) Ablative pyrolysis of biomass in solid-convective heat transfer environments, in *Fundamentals of Thermochemical Biomass Conversion*, eds. R.P. Overend, T.A. Milne, and L.K. Mudge, Elsevier Applied Science, New York, pp. 539-555.

13.   Scahill, J. and Czernik, S. (1994) NREL personal communication

14.   Dayton, D.C. and Wang, D. (1995) CID studies of inorganic species released during biomass combustion, *43rd ASMS Conference of Mass Spectrometry and Allied Topics*, May 21-26, 1995, Atlanta, GA. Paper TPA 034.

15.   Weast, R.C. Ed. (1985) *66th Edition of the CRC Handbook of Chemistry and Physics*, CRC Press, Inc., Boca Raton, FL.

16.   Hastie, J.W., Zmbov, K.F. and Bonnell, D.W. (1984) Transpiration mass spectrometric analysis of liquid KCl and KOH vaporization, *High Temperature Science*, Vol. 17, pp. 333-364.

THERMOCHEMICAL EQUILIBRIUM AS AN INDICATOR OF ASH DEPOSITION PROBLEMS IN BIOMASS BOILERS
Equilibrium Relationships to Ash Deposition

L. L. BAXTER
Combustion Research Facility, Sandia National Laboratories, Livermore, CA, USA
M. BLANDER
Quest Inc., Argonne, IL, USA
D. DAYTON and T. A. MILNE
National Renewable Energy Laboratory, Golden, CO, USA

Abstract

Combustion of many biomass fuels, in particular herbaceous materials such as straws and grasses, often leads to unmanageable ash deposition problems. The inorganic material from such fuels deposits on heat transfer and structural surfaces and undergoes several chemical and physical changes. For many herbaceous fuels, the resulting deposits have proven to be unmanageable for long-term operation.

The formation of fly ash and its deposition on surfaces are now reasonably well understood. The complex thermochemical reactions that can occur after deposition and, to a lesser extent, during flight of fly ash particles, are less well understood. This paper presents rational means of using non-ideal thermochemical equilibrium as an indicator of the potential difficulty of managing ash deposits in furnaces. The application of this theory to practical systems and the role of changing the chemical composition of deposits to influence their manageability are discussed. The equally important roles of chemical kinetics and transport processes are also acknowledged.

Keywords: ash, biomass, boilers, combustion, deposition, equilibrium, furnaces, inorganic material

1. Introduction

Empirical indices of ash deposition behavior have been developed and used by both operators and designers of biomass and coal combustors for several decades [1]. While the success of the current power generation industry attests to the usefulness of these indices, most people who have dealt with biomass selection or boiler design recognize their limitations. In the past, many boilers were operated with a single fuel for most of their lifetimes, allowing experience to compensate for a lack of detailed knowledge about ash deposit formation mechanisms and their dependence on fuel type, operating conditions, and boiler design. Even so, most experts regarded ash deposition as the most significant factor controlling boiler operation and design [2]. The past luxuries of long-term commitments from single sources of fuel have largely disappeared. Responses to the Clean Air Act Amendments [3], development of new utilization technologies, and expanded biomass markets have all contributed to a more diverse set of ash-related problems in boilers. In almost all cases, these changes have worsened the ash deposition problems (lower-rank fuels, high temperatures, reducing conditions, etc.). The problem that many already viewed as the most daunting in the biomass and coal industry has, in many cases, become worse.

These changes challenge the capability of traditional ash indices to adequately anticipate problems in boilers. A significant amount of research has been performed in an attempt to reconcile observed behavior with fuel and boiler properties on a more scientific basis [4-9]. The discussion below presents a framework in which practical aspects of ash deposition can be discussed in terms that are scientifically sound and are common to engineers in both the power plant and research labs.

2. ASH CONSTRAINTS ON A COMBUSTOR'S OPERATING WINDOW

Formation of ash deposits in combustors depends on fuel properties, boiler design, and boiler operating conditions [10, 11]. Figure 1 illustrates some of the relationships among these factors. Deposit temperature and composition are both regarded as independent variables in this figure. At a given deposit composition, temperature increases generally lead to increasingly tenacious and strong deposits. The line labeled Tenacity Threshold indicates the dividing line above which deposits cannot be removed from combustor walls by means available to the operator (soot or wall blowing, etc.). For example, as deposit temperature increases, components of the deposit begin to sinter and, consequently, the deposit develops strength and density. At some point, the deposit can no longer be removed from the surface on which it is formed. This point is indicated by the tenacity threshold line. As deposits change composition, the tenacity threshold also changes, as indicated in Fig. 1. The specific shape of the line depends on deposit thermodynamics and chemistry, and will be discussed later. For ash deposits to remain manageable, the deposits must be maintained below the tenacity threshold line.

Combustors are also subject to other constraints during operation. The principal constraint is that of meeting the steam generation load. Ash deposition problems could be avoided if temperatures were dramatically reduced, for example, by reducing, firing rate. However, efficient power generation requires that the heat transfer surface, and hence the deposit temperature, be maintained above some minimum value that is indicated in the figure as a Load Constraint. Successful operation of the boiler requires operating in regions that are both below the tenacity threshold line and above the load constraint. This region is labeled Operating Window in Figure 1.

This simplified view of ash management during boiler operation captures most of the first-order considerations. Recent advances in technology enable much more careful characterization of the tenacity threshold, relationships between deposit temperature and gas and steam temperatures, and relationships between deposit composition and fuel composition, boiler design, and boiler operating conditions. Boiler design influences both the deposit composition and the deposit temperature for any given fuel at a specified location in the boiler. However, boiler design cannot be changed rapidly or easily. For this reason, it is not considered a tool at the operators disposal to manage the ash deposition in the boiler. (It is the principal tool at the boiler designers disposal to manage ash deposition). This discussion focuses on fuel composition as it relates to deposit composition and boiler operating conditions and the relationship to deposit temperature, structure, and composition.

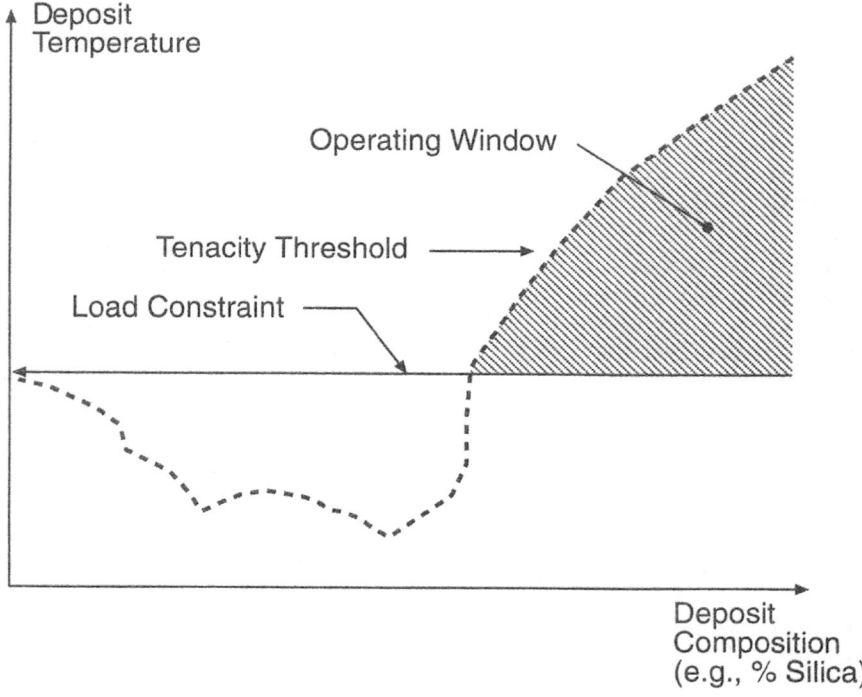

Figure 1   Schematic illustration of parameters that control successful operation of a boiler from the standpoint of ash management.

The representation illustrated in Fig. 1 is oversimplified in several respects. First, deposit composition cannot adequately be represented as a single parameter, as is suggested in Fig. 1. At least eight elements contribute significantly to overall deposit composition in biomass- and coal-fired boilers (O, Si, Al, Fe, Ca, Na, K, and S) with several others playing a minor role (Ti, Mg, Cl, P, and C). Furthermore, these elements may be present in various chemical forms, with silicates, oxides, and sulfates the most common. Also, the morphology of the deposit influences its removability. A complete description of ash management requires more than a single parameter to characterize deposit composition, and therefore more sophisticated descriptions are being developed. The single-parameter approach is used here primarily for illustration of what is inherently a multi-dimensional problem.

A second limitation of this approach to ash management is the specification of the tenacity threshold. Fundamentally, this threshold depends on the details of deposit structure — in particular the interface of the heat transfer surface and the deposit. The threshold varies with time, even in combustors operating under steady conditions, as chemical reactions drive deposits nearer to their equilibrium compositions and as the deposit structure matures. Reactions between gas-phase species, such as alkali-containing vapors, and deposits also alter compositions. In some circumstances, deposit tenacity is strongly affected by the attraction of small particles to the surface rather than physical bonding. Finally, a tenacity threshold clearly depends on the location and efficiency of soot blowers and other boiler-specific considerations. All of these factors are difficult to incorporate in a single line such as is represented in Fig. 1. The approach here is to incorporate the most significant and common features of deposits that render them tenacious. These factors include (1) the formation of molten phases, and (2) the formation of sulfates or other possibly non-molten materials in intimate contact with the surface.

A third consideration is that ash deposits are not necessarily at chemical equilibrium with their environment. It takes time for condensed-phase reactions to occur and for materials to reach equilibrium. While these time scales are very long compared to gas-phase reactions, they are typically comparable to controlling time scales for ash deposition. For example, between two soot blowing cycles (typically 8 hours minimum), high-temperature ash deposits can react significantly. Many reactions occur on time scales comparable to the residence time of fly ash [12, 13]. While nonequilibrium considerations may expand the operating window slightly larger than is implied in Fig. 1, equilibrium properties represent a good point of departure for many applications.

This approach can be used to describe most of the features of ash deposits observed in several field tests and in pilot and laboratory tests in practically useful ways. The remainder of this report focuses on how to derive quantitatively the parameters indicated in the figure. The discussion briefly describes two examples.

2.1 Tenacity Threshold

The underlying concept in establishing the tenacity threshold is illustrated in Fig. 2, with quantitative results in Fig. 3. The feature of a deposit that makes it difficult to separate from a surface is the contacting efficiency and strength at the surface-deposit interface. On the left, a typical particle in a granular deposit is illustrated. The particle is rigid and is in intimate contact with the surface in only a small area. By comparison, the figure on the right illustrates a situation where a deposit is composed of both some condensate and some particulate material. The area of contact for the deposit on the right is increased greatly compared to the particle on the left. A similar phenomenon occurs if the particles become partially molten. This increase in contacting area is a primary mechanism whereby deposits develop tenacity.

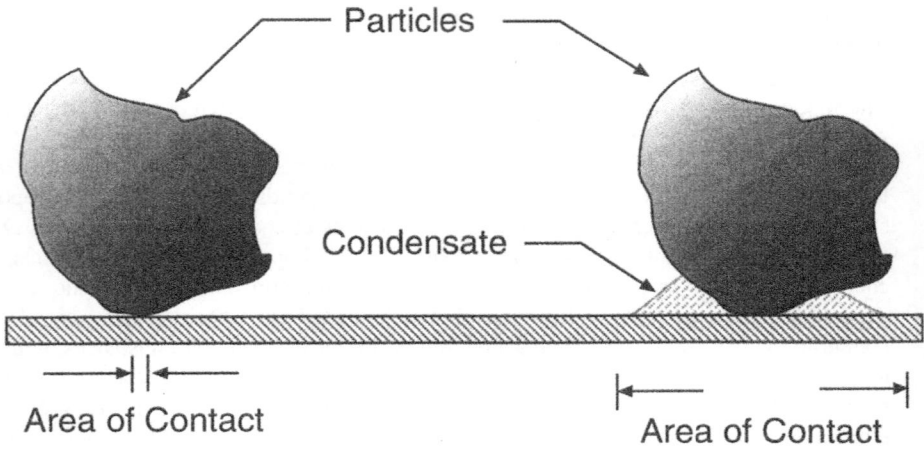

Figure 2   Illustration of the effect of condensate on deposit contacting efficiency. Formation of molten phases has a similar effect.

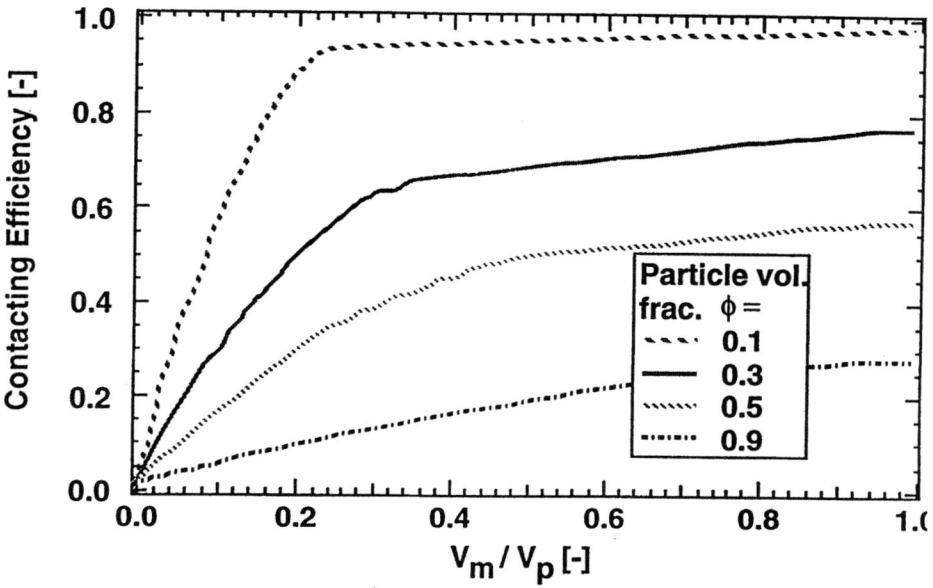

Figure 3   Quantitative dependence of contacting efficiency on relative proportions of molten on condensate volume to particulate volume.

The fraction of the surface that is intimate contact with the deposit is defined as the contacting efficiency. Figure 3 illustrates how the relative volumes of molten material to particulate material (V_m/V_p) influences this contacting efficiency [14]. The relative volume is the sum of the volumes deposited by condensation or chemical reaction plus the fraction of particulate that has become molten (V_m) divided by the volume of ash deposited by inertial impaction or thermophoresis that has not become molten (V_p). The details of these deposit formation mechanisms have been discussed elsewhere [15]. Deposits with contacting efficiencies above approximately 0.10 are difficult to remove by traditional means. Typical deposits have particle volume fractions (ϕ) of about 0.3, ranging from 0.1 to 0.5. Under these conditions, no more than 5 to 10 percent of the deposit volume can be condensate or molten if the deposit is to be manageable (maintain a contacting efficiency of less than ten percent).

2.2 Equilibrium Considerations

The development of contacting efficiency with a plate can occur through condensation (as previously discussed), formation of molten phases, or sintering. The formation of molten fractions of the deposit can be predicted by a combination of thermodynamic and chemical kinetic arguments. An illustration for interactions of sodium and silicon (components of most low-rank, US coals) is indicated in Figure 4. Similar interactions occur between potassium and silicon in biomass. The figure indicates the temperatures at which 10 and 100 percent of this mixture of oxides becomes liquid as a function of mixture (deposit) composition. As is indicated, if the deposit contains greater than 3 mole % sodium oxide, the maximum deposit temperature that can be allowed before forming more than 10 percent liquid is about 785 °C (1450 °F). The entire deposit becomes molten at temperatures of around 1600 °C (3000 °F). Deposits in most practical systems have more complex chemistry than can be easily illustrated in such a diagram, the addition of other components changes the results significantly. For example, small amounts of calcium can lower the melting temperatures illustrated in Fig. 4 even further, where as addition of aluminum typically increases the melting temperature. Both aluminum and calcium are important constituents of biomass and low rank coals. Despite the complexity of more realistic mulicomponent systems, the concepts illustrated in Fig. 4 are useful. Furnace exit gas temperatures in commercial biomass-fired boilers range from 700 - 900°C. Surface temperatures at the leading edge of deposits is one design point for determining the minimum deposit temperature that will allow economical operation of the unit. This is indicated as the lower limit for operating at the maximum continuous rating (MCR) in Fig. 4. The region above this minimum temperature and below the 10% liquid line represents the operating window for the boiler. This region is seen to be very small for the case of deposits composed entirely of silica and sodium oxides. It is larger for most systems that contain significant aluminum. The lowest melting point in the silica-alumina system, for example, is at 1584 °C (2883 °F).

Similar but more severe behavior is predicted for potassium silicates. Figure 5 illustrates the same data as are indicated by the dashed line in Fig. 4 for sodium together with potassium data for comparison. As is seen, the potassium silicates former slighter lower melting compounds and do so with fewer moles of potassium than is observed for sodium. Otherwise, the data are qualitatively similar. An analysis of the 10% melting point for potassium compounds would also produce results qualitatively similar to those shown in Fig. 4, with the potassium line tolerating less alkali at a given temperature and forming 10% solutions at lower temperature for a given molar concentration than sodium.

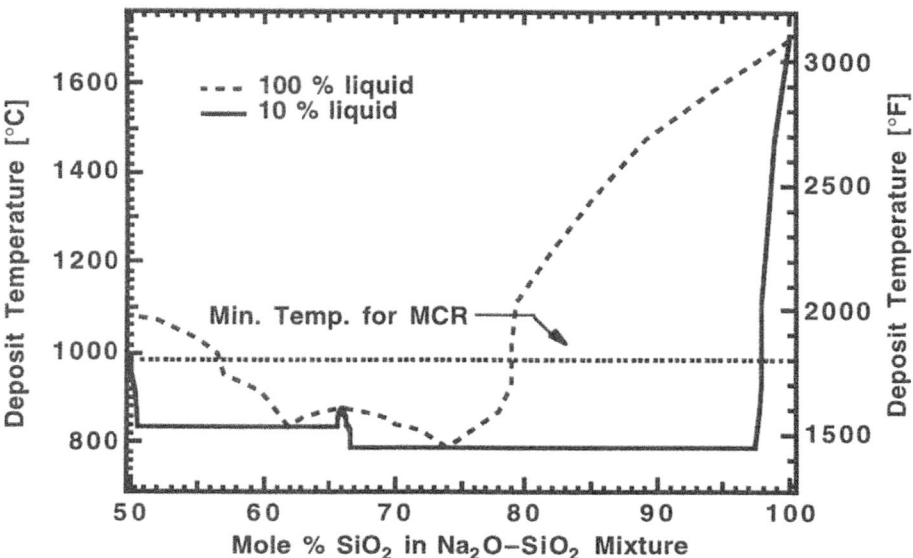

Figure 4   Melting behavior of a silica:sodium system.

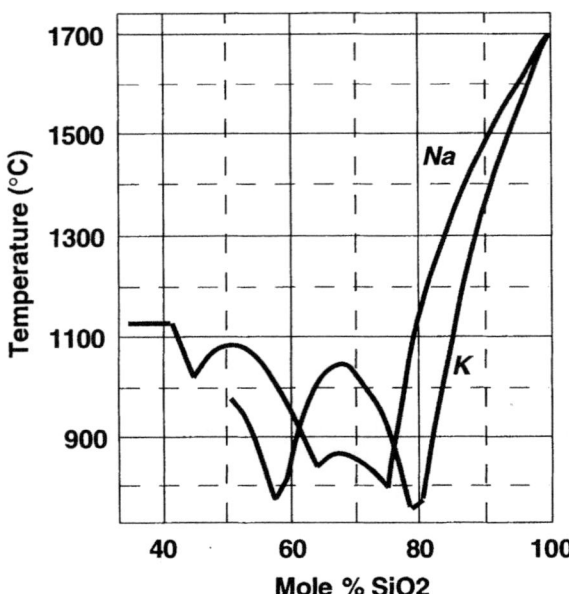

Figure 5   Melting behavior of a silica:sodium and silica:potassium systems commonly found in deposits generated during low-rank coal and biomass combustion, respectively.

Traditional indices of ash deposition behavior often fail to describe these issues. For example, laboratory tests on over thirty commercially significant coals and over twenty five biomass fuels have been performed in Sandia's Multifuel Combustor, with many complementary studies performed at commercial sites. We contrast two of these results in Table 1, where the ash fusion temperatures (reducing conditions) are compared for one coal and one biomass fuel (wheat straw). Both field and laboratory tests were performed for both of these fuels [15-18]. The coal fusion temperatures are uniformly more than 100 °F lower than those for wheat straw. The straw heating value is about 70% that of this particular coal (as-fired basis). In the laboratory tests, the fuels were fired under the same conditions of gas temperature, oxygen concentration, burnout, and surface temperature. In the field, the straw was burned in a traveling-grate stoker-fired boiler with 50-60% excess air whereas the coal was burned in a tangentially fired pc boiler with 20% excess air. In both laboratory and field tests, the straw formed molten deposits at much lower temperatures than the coal. The wheat straw ash is composed of about 60% silica and 20% potassium, with about 3.5% calcium and less than 2% aluminum. The silicon is almost entirely in the form of free silica (as opposed to silicates) in fuel. The Hanna Basin ash comprises about 40% silica, 16% aluminum, 25% calcium. Silicon in the coal is primarily in the form of silicates.

The fuels in both laboratory and field tests behave opposite from what the fusion temperatures imply. The fuel with the highest fusion temperatures is the most prone to form molten slags. There are at least three potential reasons for the discrepancy: (1) the fusion temperature test is performed on ash from the fuel that may differ significantly in composition from the ash deposit in the boiler because of selective deposition; (2) the fusion temperature tests are performed quickly compared to the time available for ash deposits to react and form molten phases; and (3) the fusion temperature procedure uses ash samples that may be depleted in alkali during preparation (due to vaporization) and does not recreate the alkali-laden gas stream to which the actual deposits are exposed. Our observation was that as the wheat straw deposits matured on the surface of the laboratory system, potassium became increasingly incorporated in the silica matrix of the deposit. Formation of molten phases in a silica-potassium mixture is very similar to that illustrated in Fig. 4 for sodium and silica.

Table 1. Fusion temperatures and observed behaviors for one coal and one biomass fuel.

Fusion Temperature	Hanna Basin Coal		Wheat Straw	
(Reducing Condit).	°F	°C	°F	°C
Initial Deformation	2186	1196	2464	1351
Spherical	2245	1229	2466	1352
Hemispherical	2276	1246	2467	1353
Fluid	2338	1281	2474	1357
Observed Behavior	Dry, Granular, Easily Managed Deposit		Molten, Running Deposits Leading to Unscheduled Outage	

As a practical example, Figure 6 is a scanning electron micrograph of a deposit produced during combustion of rice straw in Sandia's Multifuel Combustor. The deposit was generated on the wall of the MFC combustor and accumulated over a 3 hour test period. The wall temperature was 900 °C, the gas temperature was 1000 °C, and the gas composition is estimated to contain 6 % oxygen. Most of the deposit has a glassy appearance, with occasional nodules under the otherwise smooth surface.

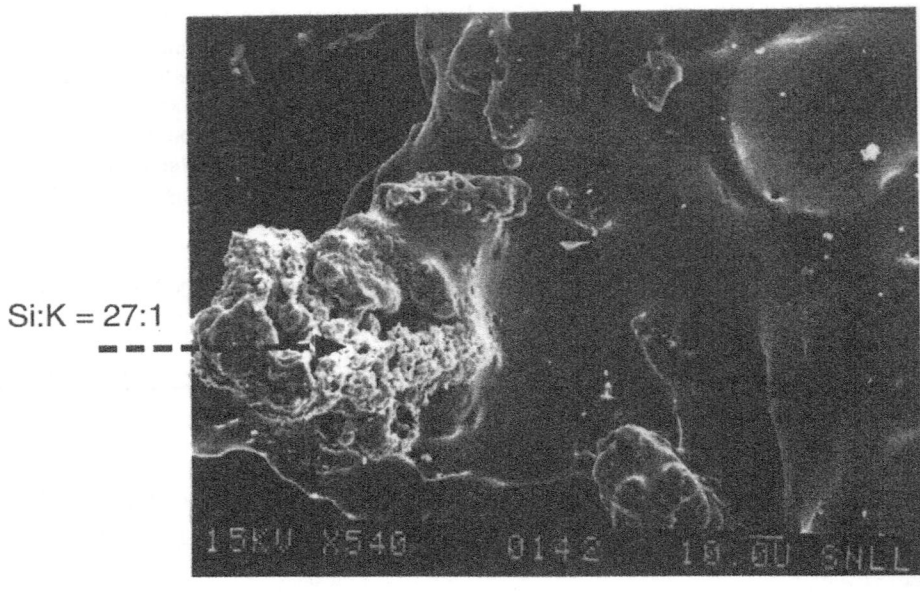

Si:K = 3.6:1

Si:K = 27:1

15KV X540   0143   10.0U SNLL

Figure 6   Scanning electron micrograph of a portion of a rice straw deposit
                collected from a ceramic surface in the MFC. The porous, silica
                based material was exposed by fracturing the deposit though one
                of the nodules evident in many locations on the smooth, glassy
                surface.

The deposit composition is determined as a function of location on the surface using electron dispersive spectroscopy in a scanning electron microscope. Both phases are composed principally of silicon. By comparison with the nodules, the glassy phase contains more nonsilaceous material. More than half of the nonsilaceous fraction is potassium.

For example, the melting point of silica decreases from about 1700 °C to less than 750 °C as potassium is introduced to form potassium silicates. Incorporation of additional materials, in particular other alkalis and alkaline earth materials, usually lowers the melting point further still. The silicon to potassium ratio observed in the glassy portions of the deposit illustrated in Figure 6 is about 3.4 on a mass basis, or about 81 % SiO_2 to 19 % K_2O. An equilibrium mixture of such material becomes completely molten at approximately 1300 °C. This is slightly above the temperature of deposit, but the addition of calcium and other heteroatoms to this mixture reduces the melting point significantly. The nodular material, on the other hand, has a much higher melting point. These changes in phase have obvious effects on the microstructure of the deposit and hence on its physical and transport properties.

A final consideration in using thermodynamic predictions of the sort illustrated in Figs. 4 and 5 is that the composition of the fuel ash is often not related in any simple way to that of the deposits that form on surfaces. An illustration of this point derived from a commercial biomass combustion test appears in Fig. 7. These data are derived from a test of a blend of wheat straw and wood (about 20 % wheat straw) in a bed-based boiler. As is seen, the composition of the deposit differs strongly from that of the ash in the fuel blend, and thermodynamic calculations based on the latter would be poor indicators of the properties of the former. The specific mechanisms giving rise to such chemically selective deposition processes are discussed elsewhere [17-19] and easily explain the reasons for the observed differences. The point to be made here is that thermodynamic calculations made on the basis of the fuel ash composition will commonly be of little more value than the indices they are intended to replace without some analysis of the relationships between fuel properties and deposit composition.

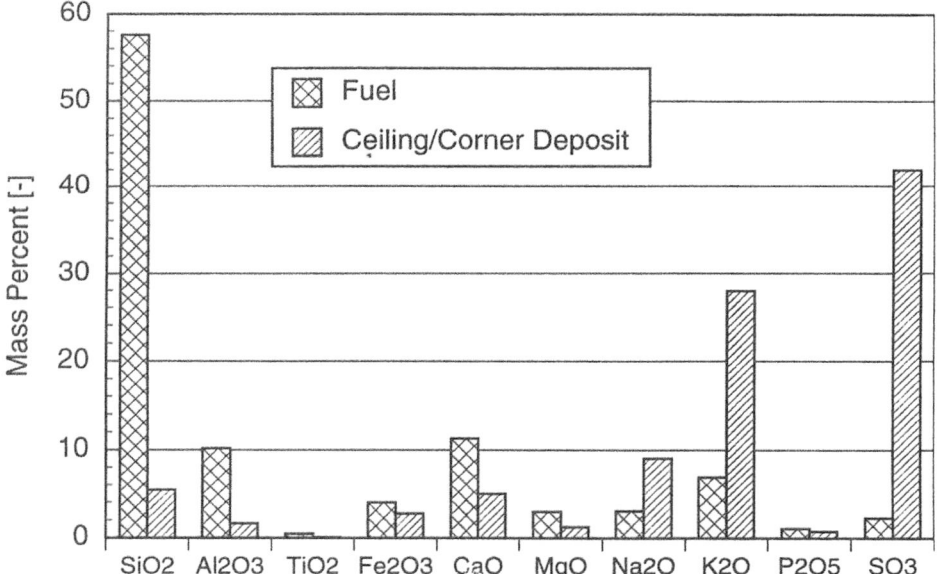

Figure 7   Elemental composition of fuel and ash deposits from a commercial biomass-fired boiler, illustrating that deposition processes can be highly selective in the species they involve.

can be used to predict deposit formation in conditions where traditional indices fail. Deposits are not in equilibrium initially, and the chemical reactions between ash constituents and gas-phase species that drive the composition toward equilibrium are shown to contribute significantly to critical regions of deposit strength, such as along heat transfer surfaces and between deposited particles. The combination of new diagnostics, sophisticated descriptions of materials, and careful experimentation is shown to improve understanding of deposit properties.

4. ACKNOWLEDGMENTS

Portions of this work were supported by U.S. Department of Energy through the Energy Efficiency and Renewable Energy Office's Biomass Power Program and through Pittsburgh Energy Technology Center's Direct Utilization Advanced Research and Technology Development Program. In addition, a consortium of industries with interest in biomass power financially contributed to this project. These include Mendota Biomass Power Ltd. and Woodland Biomass Power Ltd. (both associated with Thermo Electron Energy Systems), CMS Generation Operating Co. (formerly Hydra-Co Operations Inc.), Wheelabrator Shasta and Hudson Energy Cos., Sithe Energy Co., Delano Energy Co. Inc., the Electric Power Research Institute, Foster Wheeler Development Corp., and Elkraft Power Co. Ltd. of Denmark. Most of these companies also contributed fuels, use of facilities, and technical expertise in reviewing results of the project. The authors are grateful for the support and interaction with all of the individuals representing these organizations.

5. REFERENCES

1. Winegartner, E.C., *Coal Fouling and Slagging*. 1974, ASME:

2. Raask, E., *Mineral Impurities in Coal Combustion*. 1985, Washington: Hemisphere Publishing Corporation. 484.

3. *Federal Register: Part III; Air Contaminants*. 1989,

4. Abbott, M.F., *et al. A Modeling Strategy for Correlating Coal Quality to Power Plant Performance and Power Costs.* in *Engineering Foundation Conference on The Impact of Ash Deposition on Coal-Fired Plants*. 1993. Solihull, England:

5. Srinivasachar, S., *et al. A Fundamental Approach to the Prediction of Coal Ash Deposit Formation in Combustion Systems.* in *The Twenty-Fourth Symposium (International) on Combustion*. 1992. The University of Sydney, Australia: The Combustion Institute.

6. Beér, J.M., *et al. From Coal Mineral Matter Properties to Fly Ash Deposition Tendencies; a Modeling Approach.* in *The Seventh International Pittsburgh Coal Conference*. 1990. Pittsburgh, PA:

7.   Zygarlicke, C.J., M. Ramanathan, and T.A. Erickson, *Fly Ash Particle-Size Distribution and Composition: Experimental and Phenomenological Approach,* in *Inorganic Transformations and Ash Deposition During Combustion,* S.A. Benson, Editor. 1992, American Society of Mechanical Engineers: New York. p. 525-544.

8.   Baxter, L.L., M.F. Abbott, and R.E. Douglas. *Dependence of Elemental Ash Deposit Composition on Coal Ash Chemistry and Combustor Environment.* in *Engineering Foundation Conference on Inorganic Transformations and Ash Deposition During Combustion.* 1991. Palm Coast, Florida: The American Society of Mechanical Engineers.

9.   Baxter, L.L. and R.W. DeSollar. *Ash Deposition as a Function of Coal Type, Location in a Boiler, and Boiler Operating Conditions: Predictions Compared to Observations.* in *International Conference on Environmental Control of Combustion Processes.* 1991. Honolulu, HA:

10.   Baxter, L.L., *Ash Deposition During Biomass and Coal Combustion: A Mechanistic Approach.* Biomass and Bioenergy, 1993. **4**(2): p. 85-102.

11.   Benson, S.A., M.L. Jones, and J.N. Harb, *Ash Formation and Deposition,* in *Fundamentals of Coal Combustion for Clean and Efficient Use,* L.D. Smoot, Editor. 1993, Elsevier: New York.

12.   Srinivasachar, S., J.J. Helble, and A.A. Boni, *Mineral Behavior During Coal Combustion 1. Pyrite Transformations.* Progress in Energy and Combustion Science, 1990. **16**: p. 281-292.

13.   Srinivasachar, S., et al., *Mineral Behavior During Coal Combustion 2. Illite Transformations.* Progress in Energy and Combustion Science, 1990. **16**: p. 293-302.

14.   Baxter, L.L. and L. Dora, *The Combustion Behavior of a Blend of Eastern and Western Coals: Comparisons Between a Blend and Its Individual Components.* ASME Paper No. 92-JPGC-FACT-14, 1992. .

15.   Baxter, L.L. and R.W. DeSollar, *A Mechanistic Description of Ash Deposition During Pulverized Coal Combustion: Predictions Compared to Observations.* Fuel, 1993. **72**(10): p. 1411-1418.

16.   Baxter, L.L., et al. *Transformations and Deposition of Inorganic Material in Biomass Boilers.* in *Second International Conference on Combustion Technologies for a Clean Environment.* 1993. Lisbon, Portugal: Commission of European Communities.

17.   Miles, T.R., et al., *Alkali Deposits Found in Biomass Power Plants.* 1995, National Renewable Energy Laboratory:

18. Baxter, L.L., *et al.*, *The Behavior or Inorganic Material in Biomass-Fired Power Boilers -- Field and Laboratory Experiences: Volume II of Alkali Desposits Found in Biomass Power Plants.* 1996, National Renewable Energy Laboratory:

19. Baxter, L.L., *et al.*, *Ash Behavior in Biomass Boilers.* 1995,

THE INORGANIC CHEMISTRY OF BIOMASS COMBUSTION: PROBLEMS AND SOLUTIONS

Chemical problems of biomass combustion

M. BLANDER
QUEST Research, South Holland, Illinois, USA

Abstract

High temperature molten products of the combustion of biomass present special problems related to fouling and corrosion. We have performed calculations of the total inorganic chemistry of biomass combustion using a free energy minimization computer program with a large database for solids, liquids and gases (over 6000 species). The program also includes theories and models which predict the properties of multicomponent molten salts (including carbonates, sulfates and chlorides) and silicates from the properties of the subsidiary binaries (and also ternaries for the silicates). The calculations indicate two different types of low melting liquids: (1) a very low melting ($<500°C$-$700°C$) silicate containing aluminum, potassium, sodium and other oxides and (2) a low melting potassium rich reciprocal salt solution of K^+,Ca^{2+},Na^+/CO_3^{2-},SO_4^{2-}, Cl^-. Typical examples of these two types are: (1) wheat straw which produces a silicate with a liquidus which is probably less than $500°C$ and (2) aspen wood which produces a high temperature potassium sulfate rich liquid which crystallizes close to $1000°C$ as well as a lower temperature carbonate rich liquid between about $800°C$ and $900°C$ at fairly high pressures (4-10 atm). The computer program and our knowledge of the combustion chemistry enable us to determine additives which will minimize fouling and corrosion.

1 Introduction

Biomass provides renewable fuels that are alternatives to fossil fuels. They generate ash that may be returned to the soil as a fertilizer, and they do not lead to a net increase in the CO_2 contribution to the greenhouse effect. In this paper, I will discuss two different biomass materials, aspen wood and wheat straw. These two were chosen because they represent two distinct types of biomass. Aspen wood contains essentially no silica and is chemically basic after combustion. In an excess of oxygen, it tends to react with SO_2 and CO_2 to form molten salts. Wheat straw is relatively high in silicon and potassium which tend to form molten silicates during combustion. The molten salts or silicates thus formed could be a primary cause of problems with fouling and corrosion in combustion systems. The aim is to illustrate that calculations of the inorganic chemistry of combustion of biomass help to identify the problems and to find possible solutions which minimize the detrimental effects of fouling and corrosive liquids.

2 Calculations for Aspen Wood

Prior calculations [1] for the combustion of aspen wood have shown that, at the high pressures needed to power a turbine (4 to 10 atm), there are two successive liquid condensates, a molten sulfate rich phase which forms at high temperatures (and at 4 atm crystallizes at 996°C) and a molten carbonate rich phase (which forms at 853°C and crystallizes at 802°C at 4 atm). A similar liquid also forms at somewhat higher temperatures at 10 atm. This carbonate rich liquid is the apparent cause of fouling of a gravel bed which had been used at 800-900°C in a combustor at the University of Wisconsin. The operation of the combustor improved considerably by raising the gravel bed temperature above 900°C [1]. The prime cause for two separate liquids being formed, the sulfate at relatively high temperature and the more troublesome carbonate at lower temperature, is that the relative molar abundance of potassium is much higher than twice that of sulfur. Most of the potassium which has not condensed as sulfate is present as KOH molecules in the gas phase. After the gaseous sulfur compounds (SO_2 largely but some SO_3) are depleted in the gas by reactions such as,

$$2KOH + SO_2 + 0.5O_2 \Leftrightarrow K_2SO_4 + H_2O \tag{1}$$

essentially no sulfate can form while there is considerable KOH left over. Because CO_2 which is, of course, very abundant in a combustion process, is a weaker acid than SO_2 + $0.5O_2$, a carbonate rich phase (liquid in this case) forms at lower temperatures than the sulfates (which crystallize at about 1000°C). These facts suggest a possible method for ameliorating the problem of fouling in the combustion of aspen wood. What is needed is an acidic material which is acid enough to react with the KOH in the vapor to form solids at relatively high temperatures where the combustor and a turbine might be efficient and economically reasonable. Three such materials are SiO_2, Al_2O_3 and SO_3 (or S or SO_2 in an oxidizing gas). Silica and alumina together should have a synergistic effect because of the special stabilities of the alkali aluminosilicate compounds. However, the rate of reaction of gaseous KOH with these solids might be slow since many strong bonds (such as e.g., Si-O-Si bridge bonds) need to be broken to react. Additions of sulfur as a gaseous species such as SO_2 are more likely to fully react on the time scales of the residence of the gases in the combustor.

Calculations were performed at 4 atm at temperatures ranging from 1100°C to 600°C at 25° intervals for the combustion of 100 grams of aspen wood with an excess of air and with 0.115 grams of added sulfur. The composition of 100 gm of aspen wood, an excess of air and the sulfur addition, in grams, is 51.57 C, 6.27 H, 0.47 N, 0.02 S,

39.52 O, 0.0085 P, 0.0810 K, 0.1524 Ca, 0.0256 Mg, 0.00504 S, 0.00036 B, 0.001008 Mn, 0.001872 Fe, 0.001008 Al, 188 O, 619 N, and 0.115 S. We used the FACT computer system [2] which incorporates the CHEMSAGE [3] algorithm for free energy minimization. The FACT system has a database of over 6000 solid, liquid and gaseous species [4-7] and is capable of representing the thermodynamic properties of multicomponent solid salt, molten salt [8,9] and silicate solutions [10-13]. The results are exhibited in Tables 1-3.

Table 1. Molten and solid salt solutions produced in the combustion of 100 gm of aspen wood with 0.115 gm of S added at 4 atm. (mol%)

T($^{\circ}$C)	1100	1000	900	800	600
$CaSO_4$ (liq)	72.97	62.72	------	------	------
K_2SO_4 (liq)	26.05	36.99	------	------	------
$CaCO_3$ (liq)	0.72	0.19	------	------	------
K_2CO_3 (liq)	0.26	0.11	------	------	------
moles	3.22 E-3	2.69 E-3	0	0	0
K_2SO_4 (ss)	------	------	------[1]	99.81	99.95
$CaSO_4$ (ss)	------	------	------	0.19	0.04
K_2CO_3 (ss)	------	------	------	1.67 E-3	7.41 E-3
$CaCO_3$ (ss)	------	------	------	3.21 E-6	2.90 E-6
moles	0	0	0	1.04 E-3	1.04 E-3

liq refers to the liquid solution and ss refers to the solid solution
[1] Note that there is no solid solution at 900°C because the stable phases are $CaSO_4$ and $K_2Ca_2(SO_4)_3$

In Table 1 we see that a calcium sulfate rich liquid is present at 1100°C and persists to below 1000°C. The actual temperature of disappearance of the liquid is 908.3 where all the liquid has formed $CaSO_4$ and $K_2Ca_2(SO_4)_3$ (see Table 2) by a peritectic reaction [14]. The latter sulfate has a lower temperature limit of stability and decomposes to a K_2SO_4 rich solid solution with a composition given in Table 1 and solid $CaSO_4$ exhibited in Table 2. The solid solution is, in essence, slightly contaminated K_2SO_4. The total number of moles of potassium sulfate and calcium sulfate is essentially limited by the total amount of sulfur. Thus, we see that the addition of sulfur has raised the highest temperature below which no molten salt forms from 802°C to 908.3°C. This temperature of disappearance of the sulfate, 908.3°C is lower than that of the sulfate melt with no added sulfur (996°C) because there is more calcium sulfate in the solution with added sulfur than without and the liquidus temperature of the melt is then lower. This is also reflected by the total amount of sulfate which, at 1025°C, is 6 x 10^{-4} in the prior work with no added sulfur and 2.7 x 10^{-3} in this work. This increase comes about from the increase of both K_2SO_4 and $CaSO_4$ which are products of the reaction of most of the KOH vapor species, and the complete reaction of the CaO, which might otherwise have formed without the added sulfur. The added sulfur makes the reaction

$$CaO + SO_2 + 0.5O_2 \Leftrightarrow CaSO_4 \qquad (2)$$

go much further to the right and thus creates a higher activity and concentration of $CaSO_4$ in the molten salt. The mole fractions of $CaSO_4$ at $1025°C$ are 0.09 with no added sulfur and 0.65 with added sulfur. It should be noted in Table 3 that the total amount of SO_2 in the vapor at $600°C$ and $700°C$ is less than 10^{-10} even with the added sulfur. Because of the buffering effect of the assemblage, even an addition of

Table 2. Moles of solid compounds produced in the combustion of 100 gm of aspen wood with 0.115 gm of S added at 4 atmospheres.

T(°C)	1000	900	800	700	600
MgO	1.03 E-3	1.00 E-3	9.91 E-4	9.87 E-4	9.70 E-4
$CaSO_4$	1.62 E-3	1.27 E-3	3.33 E-3	3.33 E-3	3.33 E-3
$Ca_5HO_{13}P_3$	9.15 E-5	9.15 E-5	9.15 E-5	9.15 E-5	9.15 E-5
$Ca_3Al_2O_6$	---------	---------	---------	---------	---------
$Ca_2Fe_2O_5$	1.67 E-5	---------	---------	---------	---------
Mn_3O_4	1.56 E-6	---------	---------	---------	---------
$Mg_3B_2O_6$	7.20 E-6	1.47 E-5	1.58 E-5	1.62 E-5	1.65 E-5
$K_2Ca_2(SO_4)_3$	---------	1.03 E-3	---------	---------	---------
$CaFe_2O_4$	---------	1.68 E-5	1.23 E-5	1.27 E-5	---------
Mn_2O_3	---------	---------	4.83 E-6	6.70 E-6	8.17 E-6
$MgFe_2O_4$	---------	---------	4.50 E-6	4.08 E-6	1.67 E-5
$CaCO_3$	---------	---------	----------	---------	1.27 E-5

Table 3. Mole fractions of gaseous species formed during the combustion of 100 gm of aspen wood with 0.115 gm S added, at 4 atmospheres.

T(°C)	1000	900	800	700	600
NO	1.53 E-4	7.37 E-5	3.11 E-5	1.10 E-5	3.05 E-6
OH	3.71 E-6	1.00 E-5	2.13 E-7	3.29 E-8	3.32 E-9
KOH	1.61 E-6	2.01 E-7	3.96 E-8	4.44 E-9	1.32 E-10
SO_2	1.82 E-6	2.10 E-7	2.08 E-9	---------	---------
NO_2	1.58 E-6	1.22 E-6	8.96 E-7	6.20 E-7	3.95 E-7
KBO_2	3.75 E-7	1.68 E-8	9.64 E-10	---------	---------
K_2SO_4	2.58 E-7	4.54 E-8	4.07 E-9	1.18 E-10	---------
SO_3	1.09 E-7	2.73 E-8	6.88 E-10	---------	---------
HBO_2	7.07 E-8	9.31 E-9	8.26 E-10	---------	---------
O	1.08 E-8	1.40 E-9	1.25 E-10	---------	---------
H_3BO_3	1.68 E-7	1.00 E-7	5.37 E-8	2.50 E-8	9.58 E-9
N_2O	1.70 E-8	8.63 E-9	3.87 E-9	1.48 E-9	4.58 E-10
$Ca(OH)_2$	2.41 E-10	---------	---------	---------	---------
$Mg(OH)_2$	3.81 E-10	---------	---------	---------	---------
$Fe(OH)_2$	2.39 E-10	---------	---------	---------	---------

The fractional partial pressures and total number of moles of the major species are essentially constant at all temperatures at N_2 = 0.719, CO_2 = 0.140, H_2O = 0.101, O_2 = 0.041 and 30.769 moles. --------- denotes $<10^{-10}$ mole fraction in the gas phase.

0.2 grams of sulfur per 100 grams of aspen wood produces a gas effluent at 600°C with $<10^{-10}$ mole fraction of SO_2. Thus, the addition of sulfur would improve the performance and efficiency of a combustor-turbine system using aspen wood as a fuel by raising the operating temperature of a combustor. In order to use the effluent to drive a turbine one would have to clean up the solids in the combustor effluents with a cyclone and or filter. Further adjustments in the chemistry are possible.

3 Calculations for Wheat Straw

Table 4. Composition of wheat straw (elemental mol %)

H	49.0240	K	0.1923	Ca	0.0298
C	27.2655	Cl	0.0542	Al	0.0228
O	22.1420	S	0.0425	P	0.0115
Si	0.6528	Na	0.0406	Fe	0.0078
N	0.4747	Mg	0.0387	Ti	0.0008

The calculations of combustion of 100 moles of wheat straw in an excess of air included 296.6992 moles of N and 78.8694 moles of O. P, Fe and Ti, the least abundant elements, are not included in the calculation.

In combustion tests, problems arise from the inorganic products of combustion of wheat straw which lead to corrosion and fouling of combustors and power systems [15]. Experimental combustors had to be derated by operation at temperatures lower than design temperatures. We have performed calculations of the inorganic chemistry of the combustion of wheat straw at 25° intervals between 500°C and 1200°C using the FACT computer system. The composition of wheat straw is given in Table 4.

Table 5. Compositions (mol %) and moles of molten condens- ates in the combustion of 100 elemental moles of wheat straw in excess air at 1 atmosphere.

T°C	1200	1000	800	600	500
SiO_2	80.91	82.65	80.49	77.82	76.28
K_2O	9.66	11.23	11.15	11.79	11.71
CaO	2.92	0.49	0.64	0.82	0.85
Na_2O	2.57	2.81	4.24	4.89	5.66
MgO	2.46	1.11	0.37	0.06	0.02
Al_2O_3	1.47	1.61	2.67	3.78	4.51
Cl	0.61 E-3	4.46 E-3	65.59 E-3	0.50	0.50
SO_4	0.51 E-3	89.93 E-3	0.39	0.35	0.47
CO_3	0.38 E-5	0.33 E-5	1.75 E-5	23.08 E-5	13.20 E-4
Moles	0.7736	0.7083	0.4269	0.3014	0.2526

The most important result of the calculation is on the formation of molten silicates with compositions given in Table 5 at five temperatures. A molten silicate phase is present at all temperatures of the calculation down to at least 500°C. We performed calculations below this temperature and found that the liquid appeared to persist to lower temperatures. We can not be certain that these very low temperatures are realistic because the calculations depend on a long extrapolation of higher temperature input data. Because there is a measured eutectic temperature of 540°C in the SiO_2-K_2O-Na_2O ternary system [14] near the composition at 500°C in Table 5, it is likely that the calculated liquid (with many more solutes than in the ternary) will persist at least down to 500°C. The last three solutes in the molten silicate labeled Cl, SO_4, and CO_3 are the sums of a collection of chloride sulfate and carbonate salts of K, Na, Mg, and Ca. Since the amounts of these salts of the alkalis and alkaline earths are much smaller than the amounts of the corresponding oxides, we simplified Table 5 to make the magnitude of the salt contents clear with an insignificant loss of information on the overall composition. In any case, a persistent liquid silicate phase is formed consisting of 0.774 moles at 1200°C and decreasing to 0.253 moles at 500°C. These liquids are the likely cause of fouling and corrosion in the combustion of wheat straw.

In order to minimize the deleterious effects of the molten silicate phase, one must find a way to convert the liquid to a solid. Because of the stabilities and relatively high melting points of the alkali and alkaline earth aluminosilicates, alumina addition is one likely possibility. It would be particularly useful as an additive if there is a waste biomass which contains relatively large amounts of alumina. Mixed waste paper [4] is such a biomass with a composition for 80 moles of the elements given in Table 6 in which the most important inorganic material which can form is alumina.

Table 6. Moles of elements in 80 elemental moles of mixed waste paper.

C	24.59998	Si	0.23988	Ca	0.06853
H	40.49816	Al	0.52877	Mg	0.03005
N	0.06157	Ti	0.02752	Na	0.00874
O	14.17705	Fe	0.00517	K	0.00173
		S	0.02432		

Calculations were made for combustion of a mixture of 100 elemental moles of wheat straw, 80 elemental moles of mixed waste paper, an excess of air with 148.5 moles of O and 558.6 moles of N, and 0.5 moles of SiO_2. The silica was added because of a flaw in the calculation for molten silicates in a high alumina mixture. As a result, even though the calculations are correct, they may not be the optimum solution which can be achieved. In any case, they do indicate a considerable increase in the temperature range in which no liquid silicate is present. The equilibrium products of combustion were calculated at most of the temperatures at 25° intervals between 500°C and 1175°C. The temperatures, compositions and numbers of moles of the molten silicate phase is given in Table 7. The liquid completely crystallizes before 836.6°C which is considerably higher than the lowest temperature of stability of a molten silicate from wheat straw alone. It should be noted that we could not perform optimal calculations because of the limitations of the computer program which will be corrected in the near future.

Table 7. Molten silicate compositions (mol %) produced in the combustion of 100 elemental moles of wheat straw mixed with 80 elemental moles of mixed waste paper and 0.5 moles of SiO_2.

T($^{\circ}$C)	1175	1000	900	836.6
SiO_2	71.18	80.80	82.67	81.29
Al_2O_3	14.24	9.74	8.52	8.80
K_2O	5.16	5.33	4.83	4.85
Na_2O	1.32	2.23	2.63	3.68
CaO	1.61	1.48	1.21	1.22
MgO	3.50	0.42	0.14	0.12
Cl	7.77 E-7	8.91 E-5	1.59 E-4	5.13 E-4
SO_4	6.29 E-8	9.08 E-5	1.64 E-4	4.39 E-2
CO_3	1.62 E-7	--------	--------	--------
Moles	1.865	1.052	0.769	0.000

Table 8. Moles of stoichiometric solids formed in the combustion of wheat straw mixed with waste paper and 0.5 mole SiO_2.

T($^{\circ}$C)	1000	900	800	700	600	500
$KAlSi_2O_6$	0.0578	0.0774	0.1050	0.1014	0.0039	-------
$Mg_2Al_4Si_5O_{18}$	0.0321	0.0338	0.0343	0.0344	0.0344	0.0344
Fe_2O_3	0.0065	0.0065	0.0065	0.0065	0.0065	0.0065
$Al_6Si_2O_{13}$	0.0115	-------	-------	-------	-------	-------
Al_2SiO_5	-------	0.0395	0.0372	0.0912	0.1038	0.1040
SiO_2	-------	0.0624	0.3624	0.4127	0.3278	0.3241
$CaSO_4$	-------	-------	-------	0.0540	0.0666	0.0668

Table 9. Mole % and number of moles of a hypothetical ideal plagioclase solid solution formed in the combustion of 100 moles of wheat straw mixed with 80 moles of waste paper and 0.5 moles SiO_2.

T($^{\circ}$C)	1175	1000	900	800	700	600	500
$CaAlSi_3O_8$	92.32	75.77	63.52	41.55	23.79	11.69	11.46
$KAlSi_3O_8$	7.26	22.02	30.13	37.62	49.74	70.13	70.60
$NaAlSi_3O_8$	0.42	2.21	6.35	20.83	26.47	18.18	17.94
Moles	0.013	0.109	0.140	0.237	0.186	0.271	0.275

Tables 8 and 9 present the amounts and compositions of the solids produced at equilibrium. As can be seen in Table 8, most of the added silica reforms at low temperatures. Table 9 presents the mole fractions and number of moles of an ideal solid solution of the calcium, potassium, sodium plagioclases which was assumed to be ideal. The binary calcium-sodium plagioclase is known to be essentially ideal.

Table 10 presents the mole fractions and numbers of moles of the major constituents of the gas phase at four temperatures between 1000°C and 500°C. From the presence of SO_3, SO_2 and HCl, one can deduce that the assemblage is acidic and that the effluent gas might require some scrubbbing or other treatment with basic materials to avoid a potentially corrosive and polluting effluent.

Table 10. Mole fractions of important gaseous products of the combustion of 100 moles of wheat straw, 80 moles of mixed waste paper and 0.5 moles of SiO_2

T(°C)	1000	800	600	500
SO_2	1.65 E-4	1.45 E-4	1.61 E-7	1.33 E-10
NO	1.55 E-4	3.15 E-5	3.09 E-6	6.17 E-7
HCl	1.37 E-4	1.38 E-4	1.38 E-4	1.38 E-4
OH	5.62 E-6	3,23 E-7	5.03 E-9	2.81 E-10
SO_3	5.03 E-6	2.45 E-5	3.33 E-7	1.58 E-9
NO_2	8.14 E-7	4.63 E-7	2.03 E-7	1.15 E-7
Cl	4.82 E-7	9.08 E-8	7.99 E-9	1.49 E-9
KCl	4.75 E-7	4.42 E-8	1.89 E-9	1.34 E-10
CO	5.55 E-8	3.87 E-9	----------	----------
NaCl	4.54 E-8	2.75 E-8	7.38 E-10	----------
Moles	392.9	392.9	392.8	392.8

Mole fractions of the major ingredients of the gas in this range of temperature are N_2-0.712, CO_2-0.132, H_2O-0.114 and O_2-0.042

3 Conclusions

Calculations of the inorganic chemistry of the combustion of biomass identify liquids which are the probable causes of corrosion and fouling in combustor systems. For aspen wood at 4 atm, a sulfate forms at relatively high temperature and crystallizes at about 996°C. A second molten salt (carbonate) forms between 853°C and 802°C. This suggested that the observed fouling of a gravel bed in a combustor at the University of Wisconsin could be eliminated by raising its temperature >900°C. Performance of the gravel bed improved considerably when this was done [1]. The addition of S eliminates this molten carbonate phase and increases the total amount of molten sulfate which completely crystallizes before 908.3°C. Thus the sulfur addition permits one to raise the operating temperature of a combustor by about 100°C without a liquid precipitate below that temperature. This should improve the efficiency of the system by permitting one to possibly use the effluent gas to drive a turbine if hot gas cleanup of the solid condensates can be performed . The influence of non-equilibrium effects on the chemistry and further improvements for further raising the upper temperature of operation of the combustor 100°C are under examination for this case.

In the combustion of wheat straw, a low melting (<500°C) silicate forms and persists to low temperatures. In the combustion of a mixture of wheat straw and mixed waste paper containing significant amounts of alumina, the lowest temperature where a liquid silicate is present is 836.6°C which enlarges the possible temperature range of operation of the combustor by more than 336°. Thus, our calculations indicate the problem materials which are the root cause of the observed fouling of wheat straw combustors which have led to derating and are an aid in exploring possible solutions to fouling and corrosion in combustors which we have discussed. Our methods will probably be most effective when they are used in parallel with an experimental program.

4 References

1. M. Blander, K. W. Ragland, R. L. Cole, J. A. Libera and A. D. Pelton (1995) The inorganic chemistry of wood combustion for power production. *Biomass and Bioenergy*, Vol. 8, pp 29-38
2. C. W. Bale, A. D. Pelton and W. T. Thompson Facility for the Analysis of Chemical Thermodynamics (FACT). CRCT, Ecole Polytechnique de Montreal, P.O. Box 6079, Station "Downtown", Montreal, Quebec H3C 3A7, Canada
3. G. Eriksson and K. Hack (1990) Chemsage--a computer program for the calculation of complex chemical equilibria. *Metallurgical Transactions,* 21B, p 1013
4. I. Barin, O. Knacke and O. Kubaschevski, Thermochemical Properties of Inorganic Substances, Springer-Verlag, Berlin (1977)
5. I. Barin, Thermochemical Data of Pure Substances, VCH, Weinheim, Germany (1989)
6. JANAF Thermochemical Tables, 3rd Ed., J. Phys. Chem. Ref. Data (1985)
7. R. G. Berman, T. H. Brown and H. J. Greenwood, Atomic Energy of Canada Ltd., TR-377 (1985)
8. M. Blander and S. J. Yosim, J. Chem. Phys., **39** 2610 (1963)
9. M.-L. Saboungi and M. Blander, J. Am. Ceram. Soc., **58** 1-7 (1975)
10. A. D. Pelton and M. Blander, Metall. Trans., **17** 805 (1986)
11. M. Blander and A. D. Pelton, Geochim. Cosmochim. Acta, **51** 85 (1987)
12. A. D. Pelton and M. Blander, Calphad, **12** 97 (1988)
13. Ping Wu, G. Eriksson, A. D. Pelton and M. Blander, Iron and Steel Inst. of Japan, **33** 26 (1993)
14. Phase Diagrams for Ceramists, Volumes I-VII, Am. Ceram. Soc., Westerville, Ohio, U. S. A. (1969-1989)
15. T. R. Miles, T. R. Miles Jr., L. L. Baxter, R. W. Bryers, B. M. Jenkins and L. L. Oden (1996) Alkali Deposits Found in Biomass Power Plants: A Preliminary Investigation of Their Extent and Nature. Summary Report NREL/TP-433-8142, National Renewable Energy Laboratory, Golden, CO, U. S. A.

Acknowledgements

This work was supported by the National Renewable Energy Laboratory (NREL) under subcontract no. TCD-5-15623-01. The author is indebted to Tom Milne of NREL for many valuable suggestions and discussions.

A NO/N2O - CLASSIFICATION SYSTEM OF SINGLE FUEL PARTICLES

NO/N2O - classification system

F. WINTER, C. WARTHA, and H. HOFBAUER
Institute of Chemical Engineering, Fuel Technology and Environmental Technology
Vienna, AUSTRIA

Abstract

The formation rates of the nitrogen containing species of NO, N_2O, HCN and the emissions of hydrocarbons, CO, CO_2 of various fuels have been measured during devolatilization and char combustion. FT-IR, ND-IR, mass spectroscopy, GC-ECD, chemi-luminescence, a paramagnetic technique, and flame ionization have been used for gas analysis.

The fuels range from different wooden fuels (spruce-wood chips, beech-wood spheres, wooden pressboard and wooden hardboard chips), straw pellets to sewage sludge pellets, spherical sub-bituminous and bituminous coal particles.

Single fuel particles in a size range of 3 to 20 mm are fed into a laboratory-scale multimode fluidized bed combustor (MM-FBC) which is made of quartz-glass and exclusively electrically heated. The oxygen partial pressure is varied between 0 and 21 kPa, and the bed temperature between 600 and 900 °C. The bed material is silica sand with a mean diameter of 225 μm.

A classification system is developed based on the nitrogen content of the fuel and its conversion to NO, N_2O, and HCN. In a detailed analysis it is distinguished between the conversions of volatile-nitrogen and char-nitrogen. During devolatilization and char combustion the instantaneous conversion rates of the fuel nitrogen to NO, N_2O, and HCN are obtained as a function of carbon conversion.

Keywords: NO/N_2O, classification, single fuel particles, fluidized bed combustor.

1 Introduction

NO and N_2O are harmful emissions generated, beside other sources, during the combustion process. NO plays an important role in the formation of acid rain and ground level ozone. N_2O is a very effective greenhouse gas and causes depletion of the ozone layer in the stratosphere.

Low temperature combustion technologies, e.g. fluidized bed combustion, are generally low in their NO_x emissions because neither thermal nor prompt NO_x is formed in significant quantities. The nitrogen of the fuel is the source for NO_x and N_2O emissions. One important advantage of the fluidized bed combustion technology is its fuel flexibility. The type of the fuel can range from coke, bituminous to low-grade coals, and peat, to wood and wooden wastes, and different types of biomass, and sewage sludge. Due to these very different fuels, the amount and the type of fuel-bound nitrogen varies in a broad range, e.g. the amount of fuel-nitrogen can range from 0.05 up to 6 w-%.

Due to economical and ecological reasons, further progress in NO and N_2O reduction is necessary and primary reduction techniques are essential to achieve this goal. But the conversion of fuel-nitrogen to NO and N_2O is very complex. Homogeneous and heterogeneous formation and destruction reactions form reaction networks. And, depending on fuel, operation and local conditions, one or the other reaction path is preferred. To optimize primary NO and N_2O reduction, a deeper knowledge of the on-going chemistry and fuel-nitrogen conversion under combustion conditions is essential. The differences and the similarities of the various used fuels should be understood.

2 Experimental

Eight different fuels are studied under fluidized bed combustor conditions. The fuels range from different wooden fuels (spruce-wood chips, beech-wood spheres, wooden pressboard and wooden hardboard chips), straw pellets to sewage sludge pellets, spherical sub-bituminous and bituminous coal particles. The nitrogen content of these fuels varies from 0.11 w-% (beech-wood) up to 6.3 w-% (sewage sludge pellets), refer to Table 1 for further fuel characteristics.

Table 1. Proximate and ultimate analysis of the fuels

proximate analysis	bituminous coal as received (after devol.)	sub-bituminous coal as received (after devol.)	wooden pressboard as received (after devol.)	wooden hardboard as received (after devol.)	spruce-wood chips as received (after devol.)	beech-wood spheres as received (after devol.)	straw-pellets as received (after devol.)	sewage sludge pellets as received (after devol.)
calorific value [MJ/kg]	32	16	17	17.5	17	17.5	16	16
density [kg/m³]	1520 (832)	1250 (600)	730 (270)	880 (320)	500 (140)	670 (150)	n.a.	1140 (640)
volatile matter [w-%]	30	32.2	73.3	77.1	76.2	81.7	65.4	60.6
moisture [w-%]	6.0	25.3	8.0	6.2	9.5	5.6	8.1	7.0
ash content [w-%]	2.4	12.3	0.9	0.4	0.4	0.3	6.8	26.7
fixed carbon (C-fix) [w-%]	61.6	30.2	18.7	16.6	14.3	12.4	19.7	5.7
ultimate analysis	(after devol.)	water ash free						
carbon [w-%]	86.8 (85.65)	68.11 (73.1)	49.69	51.03	49.91	49.76 (87.31)	49.04 (76.3)	56.11 (38.45)
hydrogen [w-%]	4.27 (0.94)	2.84 (1.05)	5.68	5.88	5.42	5.94 (1.63)	5.65 (1.3)	7.5 (1.3)
nitrogen [w-%]	1.48 (1.76)	0.64 (0.9)	2.89	0.16	0.18	0.11 (0.24)	0.43 (0.52)	6.3 (3.65)
sulfur [w-%]	0.5 (0.32)	1.52 (0.73)	0.05	<0.2	<0.02	< 0.02 (<0.02)	0.06 (0.13)	0.8 (0.57)
oxygen (diff.) [w-%]	6.95 (11.3)	26.88 (24.2)	41.68	42.91	44.47	44.17 (10.83)	44.8 (21.8)	29.28 (56.04)

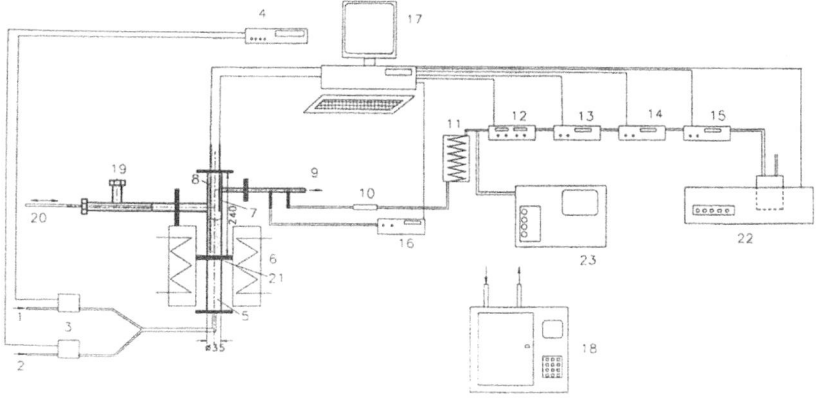

Fig. 1. The experimental set-up of the MM-FBC: 1, 2 inlets of the fluidizing gas (air, nitrogen), 3 mass-flow controllers, 4 mass-flow display, 5 preheating zone, 6 heating shells, 7, 8 thermocouples, 9 to the chimney, 10 filter, 11 cooler, 12-15 CO/CO_2, O_2, NO, CLD, 16 FID, 17 data acquisition, 18 GC-ECD, 19, 20 fuel inlet, 21 quartz frit, 22 FT-IR, 23 mass spectrometer.

The single fuel particles in a size range of 3 to 20 mm are fed into a laboratory-scale multi-mode fluidized bed combustor (MM-FBC). To minimize the catalytic reactivity and to allow insight into the combustor, the MM-FBC is made of quartz-glass. The inner diameter of the main quartz-tube is 35 mm and the height is 240 mm. To study exclusively the emissions of the single fuel particles, the MM-FBC is electrically heated by two heating shells. Ni/CrNi - thermocouples are used at different positions to control the combustion temperature between 600 and 900 °C. The MM-FBC has been optimized to study the formation rates of the species of interest, i.e. short residence times at high temperatures (typically below 140 ms). The bed material is silica sand with a mean diameter of 225 μm. The static bed height is about 40 mm. Air/nitrogen mixtures are used to vary the oxygen partial pressure between 0 and 21 kPa. The su-perficial gas velocity is 0.68 m/s and controlled by two mass-flow controllers. The detailed experimental set-up is given in Fig. 1.

To obtain reliable data of the instantaneous concentrations of the various species (CO_2, CO, CH_4, other hydrocarbons, total hydrocarbons (THC), NO, N_2O, HCN, and O_2), different gas analyzers with different analytical methods are tested. Non-disper-sive infrared analyzers (ND-IR) are used for CO_2, CO, and NO. The high performance fourier transform infrared spectrometer (FT-IR, BIO-RAD, FTS 60A) has been com-bined with a long-path, low volume gas-cell (Foxboro LV7, cell volume = 223 cm^3, optical path length = 7.25 m) to follow the concentration histories of CO_2, CO, CH_4, other hydrocarbons, NO, N_2O, and HCN. To optimize time resolution together with spectra resolution, a resolution of 0.5 cm^{-1} of the FT-IR (the maximum resolution is 0.1 cm^{-1}) has been used, i.e. four scans are added to one spectrum in 3.8 s. An analyzer (CLD) based on the chemi-luminescence technique is used to obtain the NO concentrations. A gas-chromatograph combined with an electron capture detector (GC-ECD) has been optimized for off-line N_2O measurements (For further details refer to [1]). Additionally, NO and HCN are measured with a chemical ionization mass spectrometer (CI-MS). The chemical ionization allows soft, i.e. low energy, ionization of the species. Oxygen is measured by using a standard paramagnetic technique. The total hydrocarbons are obtained with the flame ionization detector (FID). Comparing these different analyzers, it is concluded that the analytical set-up using the FT-IR and the oxygen detector gives very reliable data and an efficient data collection process. For further details compare [2]. The formation rates as well as the operating conditions are recorded during devolatilization and char combustion. Additionally, the ignition and the extinction of the flame of the volatile matter, the phenomenological behavior, fuel particle temperatures, and the fluid-dynamical behavior have been investigated. Mass-balances of the fuel carbon lead to kinetic data of devolatilization and char combustion. These results have been reported elsewhere [3-5] and are used whenever necessary to explain the NO and N_2O formation characteristics.

3 Results and Discussion

Figure 2 shows the conversion of fuel-nitrogen to NO, N_2O, and HCN versus the nitrogen contents of the fuels measured in the MM-FBC at 800°C bed temperature and 10 kPa oxygen partial pressure. Spruce-wood chips, beech-wood spheres, straw pellets, and wooden hardboard chips are low in their nitrogen contents but relatively high in their conversion of fuel-nitrogen to NO, N_2O, and HCN during devolatilization and char combustion. These fuels are grouped in class I. Class II consists of fuels with medium nitrogen contents (the sub-bituminous and the bituminous coal) corresponding to medium conversions. The wooden pressboard chips and the sewage sludge pellets are nitrogen rich fuels with low nitrogen conversions (class III).

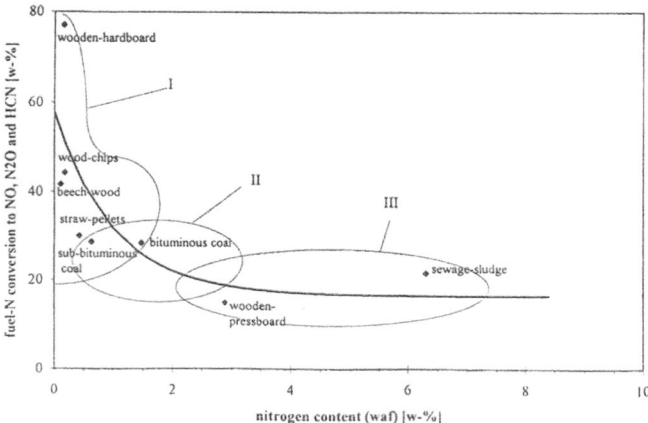

Fig. 2. Fuel-nitrogen conversion to NO, N_2O, HCN versus nitrogen content of the fuel.

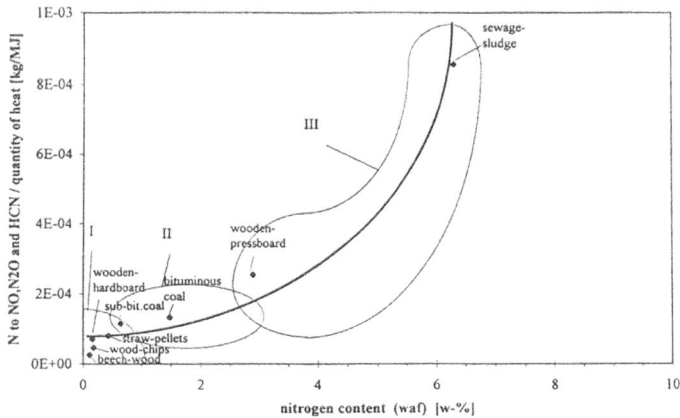

Fig. 3. The conversion of fuel-nitrogen based on the quantity of released heat of different fuels.

Figure 3 is corresponding to Fig. 2. But the fuel-nitrogen conversions are based on the quantities of released heat resulting in specific conversions. The high conversions of class I fuels are over-compensated by their low nitrogen contents and by their relatively high calorific values. Burning class I fuels under fluidized bed conditions should basically result in low NO_x and N_2O formation.

Although these specific fuel-nitrogen conversions of class I fuels are low, the conversions of the fuel-nitrogen are much higher than the conversions of class II and class III fuels. A deeper understanding of these formation characteristics is very important to develop effective primary NO and N_2O reduction techniques.

In Figs. 2 and 3 the net nitrogen conversions (conversions during devolatilization and

Table 2. Nitrogen conversion characteristics of selected fuels.

	bituminous coal (after devol.)	sub-bituminous coal (after devol.)	beech-wood spheres (after devol.)	straw-pellets (after devol.)	sewage sludge pellets (after devol.)
N/C [1]	0.017 (0.021)	0.0094 (0.012)	0.0021 (0.0027)	0.0088 (0.0068)	0.11 (0.095)
N/C - change [%]	124	128	129	77	86
volatile-N [w-%]:	28	26	68	74	76
char-N [w-%]:	72	74	32	26	24
devolatilization (N/vol.-N)					
N to NO [w-%]	38	40	28	27	10
N to N2O [w-%]	8	3	0.7	2.2	0.9
N to HCN [w-%]	4	7	10	3.6	n.a.
N to (NO, N2O, HCN) / fuel-N [w-%]	14	13	26	24	9
char combustion (N/char-N)					
N to NO [w-%]	16	18	31	14	21
N to N2O [w-%]	1.5	0.5	3.9	1.4	35
N to HCN [w-%]	1.4	3.5	13	6.7	n.a.
N to (NO, N2O, HCN) / fuel-N [w-%]	14	16	15	5.7	13

char combustion are combined) in terms of fuel-nitrogen are given. But devolatilization and char combustion are very different processes. In the following the conversion of the fuel-nitrogen is analyzed on a detailed basis separating the fuel-nitrogen conversion during devolatilization from the fuel-nitrogen conversion during char combustion.

In Table 2 the nitrogen conversion characteristics of five selected fuels (beech-wood and straw pellets representing class I, the sub-bituminous and the bituminous coal representing class II, and the sewage sludge pellets representing class III) are given. The values in parentheses indicate the nitrogen to carbon ratio of the fuel after devolatilization which increases in the case of the beech-wood (129 w-%) and the coals (124, 128 w-%) but decreases for the straw pellets (77 w-%) and the sewage sludge pellets (86 w-%). The splitting of the fuel-nitrogen into volatile-nitrogen and char-nitrogen is given in the next two lines. Two thirds or respectively three quarters of the nitrogen of beech-wood, straw pellets, and sewage sludge pellets is volatile-nitrogen. In the case of the coals volatile-nitrogen is below one third. The next part of Table 2 shows the different conversions of the volatile-nitrogen to NO, N_2O, and HCN. The last line of this part presents the total conversion to NO, N_2O, and HCN based on fuel-nitrogen. The third part of Table 2 presents the corresponding conversions for char combustion.

It can be seen from Table 2 that the conversions to NO are generally much higher than to N_2O. Thus, the NO conversions are mainly responsible for the emission characteristics given in Fig. 2. The high conversions of the fuel-nitrogen of beech-wood and the straw pellets are mainly based on the high conversions during devolatilization (26, 24 w-%, respectively). In the case of the coals the conversions of fuel-nitrogen during devolatilization (14, 13 w-%) and char combustion (14, 16 w-%) are of nearly equal importance. The sewage sludge pellets reveal a similar behavior (9 w-% during devolatilization and 13 w-% during char combustion). In terms of volatile-nitrogen the conversions of the beech-wood and the straw pellets to NO are lower (28, 27 w-%) than the conversions of the coal nitrogen (38, 40 w-%). During char combustion the high conversion to NO of the beech-wood nitrogen is paramount.

In Fig. 4 the instantaneous conversion rates of the volatile-nitrogen of the beech-wood spheres, the straw pellets, and the two coals to NO, N_2O, and HCN versus the volatile-carbon conversions are given. The oxygen partial pressure of the fluidizing gas is 10 kPa, the bed temperature is 800 °C.

The NO conversion rates are slowly increasing with carbon conversion in the case of the beech-wood, the straw pellets, and the bituminous coal. In the case of the sub-bituminous coal the rates steeply increase after the early stages of carbon conversion (8 %). The relatively high moisture content of the sub-bituminous coal (25.3 w-%) may account for that delay. N_2O formation also increases in the later stages of carbon conversion. A maximum can be found for the bituminous coal at about 80 % of volatile-carbon conversion. Similar effects, but not that obvious, are found for the sub-bituminous coal and the straw pellets. It may be concluded that N_2O formation increases with NO formation. NO is necessary to homogeneously form N_2O via the following reaction path (Refer to [6]).

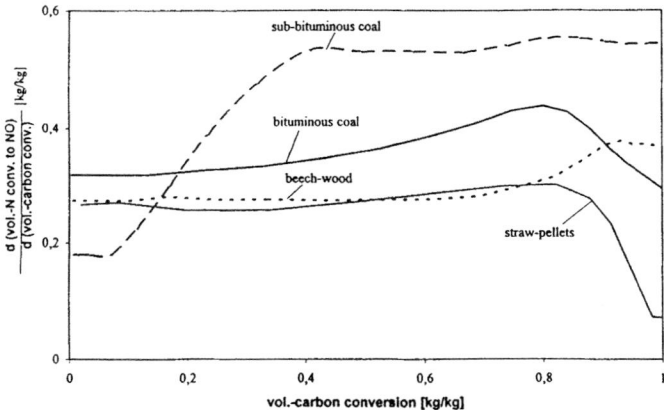

Fig. 4a. Instantaneous conversion rates of volatile-nitrogen to NO versus volatile-carbon conversion.

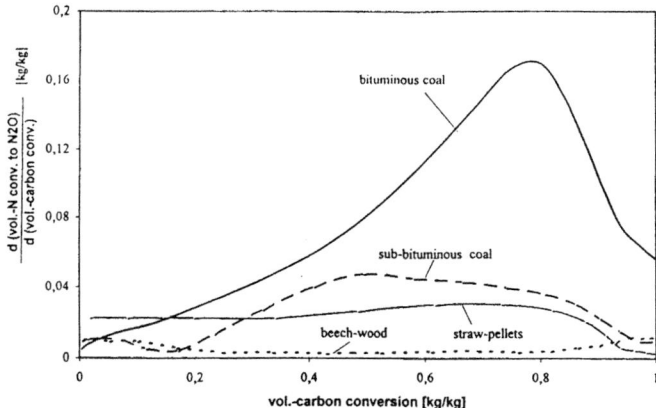

Fig. 4b. Instantaneous conversion rates of volatile-nitrogen to N_2O versus volatile-carbon conversion.

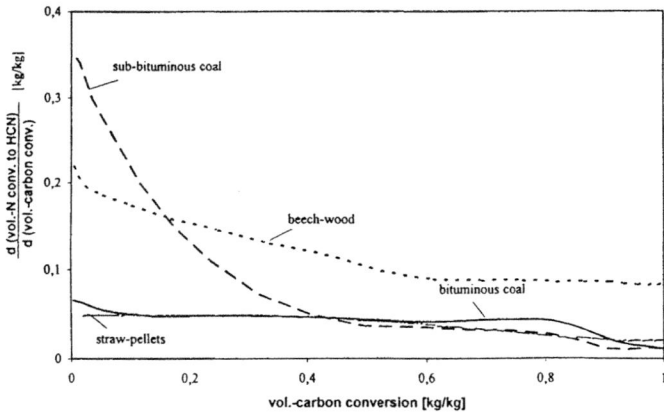

Fig. 4c. Instantaneous conversion rates of volatile-nitrogen to HCN versus volatile-carbon conversion.

$$NCO + NO <=> N_2O + CO \tag{1}$$

Conversion to HCN is continuously decreasing with carbon conversion. This may be due to an increasing oxidation of HCN via the following reaction path.

$$HCN + O <=> NCO + H \tag{2}$$

In Fig. 5 the instantaneous conversion rates of the char-nitrogen of the beech-wood spheres, the straw pellets, and the two coals to NO, N_2O, and HCN versus the char-carbon conversions are given under the same conditions (10 kPa oxygen, 800 °C).

The NO conversion rates of the beech-wood spheres stay constant throughout the char combustion at high levels whereas the NO conversion rates of the two coals are rather low. The low nitrogen content of the beech-wood may be the reason for the high conversion rates to NO. Once NO is generated the likelihood of the heterogeneous reduction to N_2 or N_2O is low (Compare also [7])

$$NO + (-CN) <=> N_2 + (-CO) \tag{3}$$
$$NO + (-N) \quad <=> (-N_2O) \tag{4}$$

where species in parentheses represent surface bound species. (-N) is any nitrogen containing site of the char surface, e.g. (-NCO) or (-CN). The increasing conversion to N_2O of the beech-wood can be explained by the increasing HCN concentrations around the char particle where N_2O is homogeneously formed mainly via reaction 1 and 2 (Compare [8]). The low NO conversion rates of the straw pellets (till 90 % carbon conversion) may be due to the catalytic reactivity of the straw-ash. The conversion to HCN of all four fuels increases with carbon conversion. This may be due to the decreasing diameter of the fuel particles and the decreasing residence time in the particle.

In Fig. 6 the conversions of the volatile-nitrogen to NO versus the bed temperature (Fig. 6a), to N_2O versus the bed temperature (Fig. 6b), and to N_2O versus the oxygen partial pressure in the fluidizing gas (Fig. 6c) are given relatively to the corresponding conversions at 800 °C bed temperature and 10 kPa oxygen. Four selected fuels are used, viz. the beech-wood spheres, the straw pellets, and the spherical bituminous and sub-bituminous coal particles.

Generally, NO conversion increases with bed temperature for all fuels except for the straw pellets where a maximum around 800 °C can be found. Around 800 °C maxima for conversions to N_2O exist for the straw pellets and the two coals. The slight maximum for the beech-wood spheres is shifted to lower temperatures (around 700 °C). The N_2O decrease at higher temperatures may be due to the increase of destruction reactions with the H and OH radicals and thermal decomposition (Refer to [6, 9]).

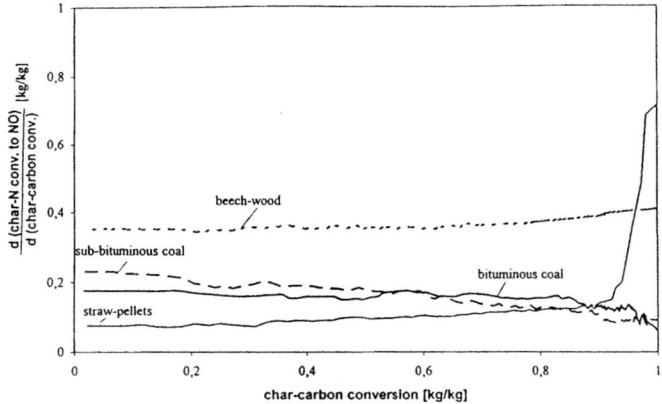

Fig. 5a. Instantaneous conversion rates of char-nitrogen to NO versus char-carbon conversion.

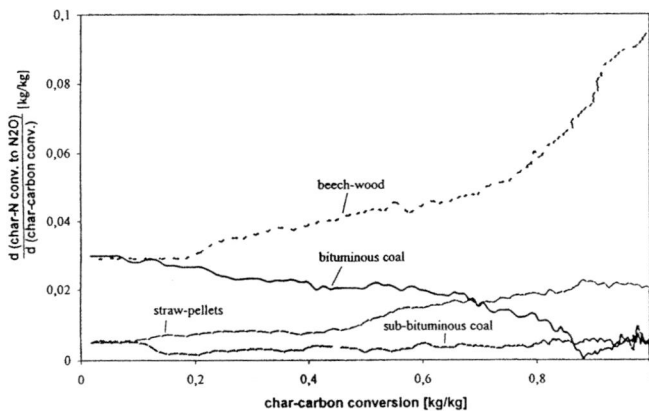

Fig. 5b. Instantaneous conversion rates of char-nitrogen to N_2O versus char-carbon conversion.

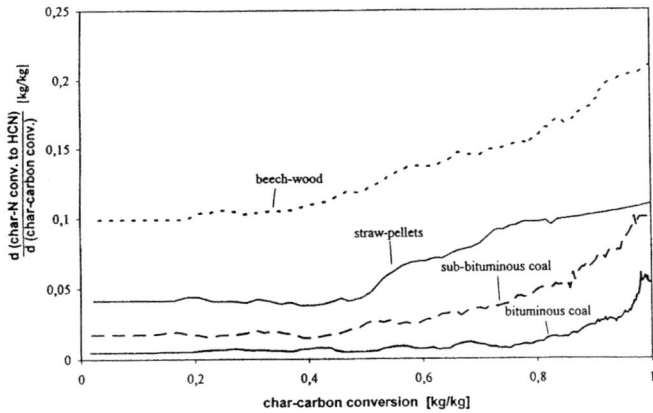

Fig. 5c. Instantaneous conversion rates of char-nitrogen to HCN versus char-carbon conversion.

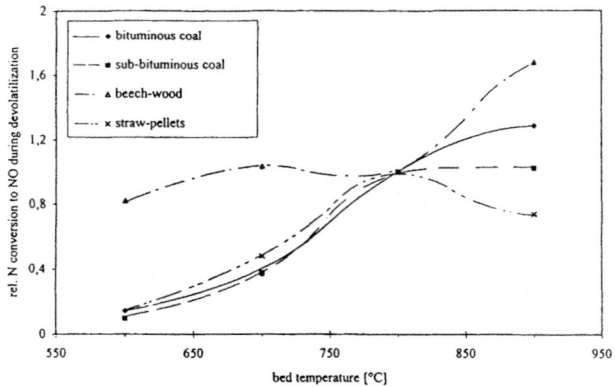

Fig. 6a. The relative conversion of volatile-nitrogen to NO versus the bed temperature.

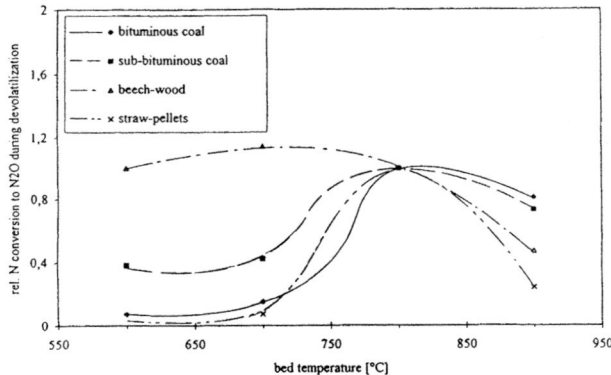

Fig. 6b. The relative conversion of volatile-nitrogen to N_2O versus the bed temperature.

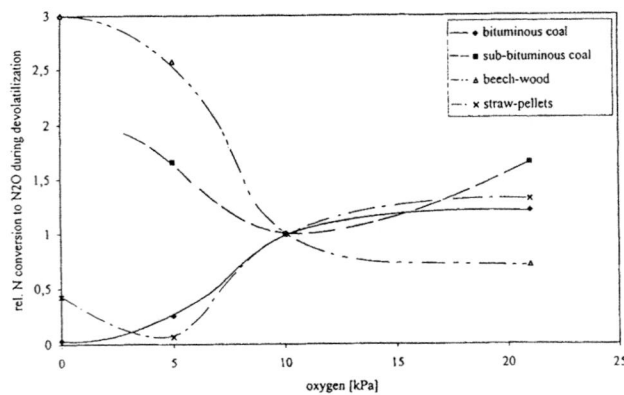

Fig. 6c. The relative conversion of volatile-nitrogen to N_2O versus the oxygen concentration.

$$N_2O + H \quad <\!=\!> N_2 + OH \qquad (5)$$
$$N_2O + OH <\!=\!> N_2 + HO_2 \qquad (6)$$
$$N_2O + M \quad <\!=\!> N_2 + O + M \qquad (7)$$

Increasing the oxygen partial pressure in the fluidizing gas, N_2O conversion increases in the case of the straw pellets and the bituminous coal whereas it decreases for the beech-wood spheres. The spherical sub-bituminous coal particles reveal a minimum around 10 kPa.

In Fig. 7 the conversions of the char-nitrogen to NO versus the bed temperature (Fig. 7a), to N_2O versus the bed temperature (Fig. 7b), and to N_2O versus the oxygen partial pressure in the fluidizing gas (Fig. 7c) are given relatively to the corresponding conversions at 800 °C bed temperature and 10 kPa oxygen. The same fuels are used as in Fig. 6. Regarding the NO conversion, maxima can be found at 800 °C for the straw pellets and the sub-bituminous coal particles whereas in the case of the bituminous coal NO conversion continuously increases. For the beech-wood spheres a minimum exists at 800 °C. Similar trends can be found for the N_2O conversions. Minima of N_2O formation are around 10 - 13 kPa oxygen in the fluidizing gas for all fuels except the beech-wood where N_2O conversion continuously decreases.

4 Conclusions

- To obtain the instantaneous formation rates of NO, N_2O, and HCN together with CO_2, CO, CH_4, and other hydrocarbons of single fuel particles, a high-performance FT-IR spectrometer in combination with a long-path, low-volume gas cell has been successfully proven.
- Three different classes of fuels, characterized by their nitrogen contents, can be found. Fuels of class I are low in their nitrogen contents but the conversions of the fuel nitrogen to nitrogen oxides (NO, N_2O) and precursors (HCN) are high. Class II fuels have medium nitrogen contents corresponding to medium conversion rates. Fuels in class III are nitrogen rich with low conversion rates (Refer to Fig. 2).
- If these conversion rates are based on the quantities of released heat, the high nitrogen conversion rates of class I fuels are over-compensated by their low nitrogen contents and by their relatively high calorific values. Burning class I fuels should basically result in low NO_x and N_2O formation rates (Refer to Fig. 3).
- The nitrogen to carbon ratio increases after devolatilization for the beech-wood, the bituminous and the sub-bituminous coal but decreases for the straw pellets and the sewage sludge pellets indicating the different chemical composition of these fuels.
- The majority of the nitrogen of high-volatile fuels such as beech-wood, straw and sewage sludge pellets is volatile-nitrogen (between 68 and 76 %).
- Conversion to NO is generally much higher than the conversion to N_2O.
- The high fuel-nitrogen conversions of beech-wood and straw pellets result from their high fuel-nitrogen conversion to NO during devolatilization. In the case of the two

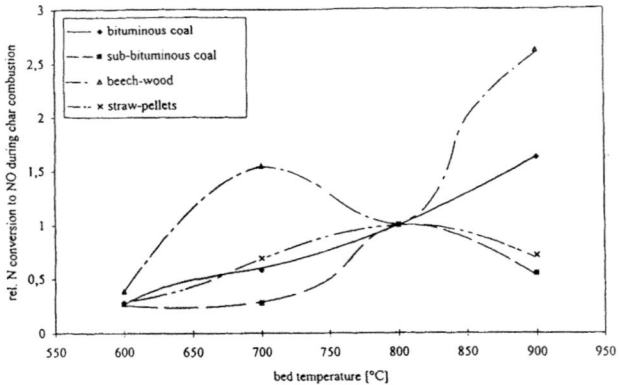

Fig. 7a. The relative conversion of char-nitrogen to NO versus the bed temperature.

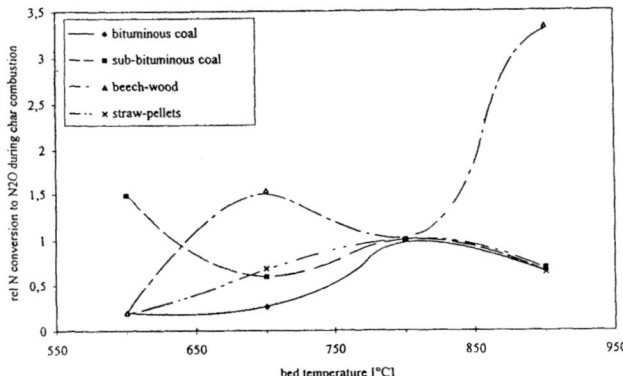

Fig. 7b. The relative conversion of char-nitrogen to N_2O versus the bed temperature.

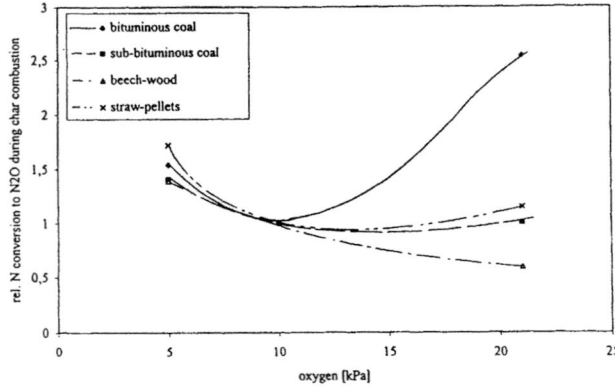

Fig. 7c. The relative conversion of char-nitrogen to N_2O versus the oxygen concentration.

coals the conversions during devolatilization and char combustion are of nearly equal importance.

- In terms of volatile-nitrogen the conversions of the beech-wood and the straw pellets to NO are lower (28, 27 %) than the conversions of the coal nitrogen (38, 40 %).
- NO and N_2O formation during devolatilization is mainly based on homogeneous reaction paths as well as N_2O formation during char combustion whereas NO formation during char combustion is mainly based on heterogeneous reaction paths. But although differences and similarities can be qualitatively explained more knowledge has to be gained to quantitatively describe formation characteristics.

5 References

1.   Buchtela, G., Hofbauer, H., Vitovec, W. and Hackl, A. (1990) Comparison of four Methods to Analyze N_2O in Flue Gases, *presented at the European Workshop on N_2O Emissions*, June 6-8, 1990, Lisbon, Portugal.

2.   Wartha, C. (1995) NO/N_2O - Formation Rates: An Analytical Approach, *Workshop on Fluidized Bed Combustion* (Eds. H. Hofbauer, B. Leckner), December 8-10, 1995, Vienna University of Technology, Vienna, Austria.

3.   Winter, F., Krobath, P. and Hofbauer, H. (1995) Comparison of a Circulating Fluidized Bed Combustor with a Stationary Fluidized Bed Combustor, *Thirteenth International Conference on Fluidized Bed Combustion,* ASME, Orlando, Florida, 1995, pp. 1477-1487.

4.   Winter, F., Wartha, C., and Hofbauer, H. (1995) Characterisation and Emissions of Single Fuel Particles under FBC Conditions, *Third International Conference on Combustion Technology for a Clean Environment*, Lisbon, Portugal, 1995, Sec. 15.2.

5.   Winter, F., Prah, M. E., and Hofbauer, H. (1995) Intra Particle Temperatures under various Fluidized Bed Combustor Conditions: The Effect of Drying, Devolatilization, and Char Combustion, submitted to *Combust. Flame* (1995).

6.   Kilpinen, P. and Hupa, M. (1991) Homogeneous N_2O Chemistry at Fluidized Bed Combustion Conditions - A Kinetic Modeling Study, *Combust. Flame,* Vol. 85, pp. 94-104.

7.   Goel, S.K., Morihara, A., Tullin, C.J. and Sarofim, A.F. (1994) Effect of NO and O_2 Concentration on N_2O Formation During Coal Combustion in a Fluidized Bed Combustor: Modelling Results, *Twenty-Fourth Symposium (International) on Combustion,* The Combustion Institute.

8.   Winter, F., Wartha, C., Löffler, G. and Hofbauer, H. (1996) The NO and N2O Formation Mechanism during Devolatilization and Char Combustion under Fluidized Bed Conditions, submitted to *Twenty-Sixth Symposium (International) on Combustion,* The Combustion Institute, Naples, Italy.

9.   Glarborg, P., Johnsson, J.E. and Dam-Johansen, K. (1994) Kinetics of Homogeneous Nitrous Oxide Decomposition, *Combust. Flame,* Vol. 99, pp. 523-532.

Combustion Characteristics of Leached Biomass

B.M. Jenkins and R.R. Bakker
Department of Biological and Agricultural Engineering
University of California, Davis, CA 95616 USA
L.L. Baxter
Combustion Research Facility
Sandia National Laboratories, Livermore, California USA
J.H. Gilmer
Department of Chemical and Materials Engineering
University of California, Davis, California
J.B. Wei
Department of Biological and Agricultural Engineering
University of California, Davis, California USA

Abstract

The simple leaching of straw, grass (e.g. switchgrass) and other types of biomass with water is shown to result in substantial improvements to the combustion behavior. Although the economic feasibility of leaching is case specific, leaching does extract large amounts of alkali metals and chlorine which improve the fouling, slagging, and burning characteristics of biomass fuels, especially grasses and straws. Increases in the fusion temperature of ash arise from extraction of alkali elements by leaching, and decreases are, in general, observed in the amount of volatile ash. Pilot combustion tests showed lower rates of fouling when burning leached rice straw compared to unleached fresh material. Pyrolysis rates obtained via dynamic thermogravimetric analysis are slower for leached materials, probably as a result of a reduction in catalytic effects due to alkali metal salts, although volatile matter emission is observed to terminate earlier under isothermal heating for leached compared to unleached fuel. For the highly refractory fuel--rice straw--improvements in the ignition and burning characteristics of single straws have been observed to accompany leaching, possibly as a result of lower chlorine activity in terminating free radical chain reactions.

Keywords: wet extraction, alkali metals, potassium, chlorine, straw, grass, wood, properties, fouling, slagging, kinetics, heating value.

1 Introduction

One of the key challenges in the thermochemical conversion of biomass is to overcome the problems of slagging, fouling, and agglomeration arising from transformations among fuel elements at elevated temperatures. These phenomena occur as a result of complex physical and chemical interactions, exacerbated by the relative concentrations of certain key elements, especially alkali and alkaline earth metals in combination with silica, chlorine, and sulfur [1 - 6]. Herbaceous and other annual growth biomass, including straws and grasses, in general exhibit severe slagging and fouling behavior compared to wood fuels in most thermal systems. Among the gramineae, high concentrations of silica in combination with the macronutrient potassium give the ash a character very similar to that of an ordinary glass recipe. Slagging is therefore a common occurrence when firing straws in most combustion and gasification systems. Much of the potassium is volatile at higher temperatures as inorganic or organometallic compounds, and, facilitated by the presence of chlorine, contributes to the fouling of heat exchange surfaces, filters, and other components of the energy conversion system. The rapidity with which straws lead to the fouling of ordinary combustion furnaces and boilers has greatly reduced the use of straw for power generation, even where such use could be beneficial for other purposes, e.g., in reducing air pollution from open burning of straw for disposal. These problems are by no means restricted to straw, as all biomass fuels contribute to fouling, just at varying rates, the slower rates being more manageable.

Recently, we have found that the fusion temperatures of straw ash can be remarkably increased through the simple process of wet extraction or leaching [7, 8]. By leaching with water, most of the potassium and chlorine are extracted from the fuel. A number of other elements are also removed in large amounts. This happens naturally when straw is left in the field and subject to precipitation, or may be accomplished artificially. Sugar cane bagasse is successfully used as boiler fuel largely because of the leaching of fouling elements occurring during the extraction of sugar from the parent material. The feasibility of leaching for a variety of biomass types as an integral process to power plant operation is currently under investigation. This paper addresses some of the combustion behavior of leached biomass, with specific attention to the compositional changes due to leaching, the effects on heating value, ash fusibility, and ash volatilization, pilot scale experiments on the combustion of leached straw, and changes observed in the kinetic behavior of leached materials relative to the untreated parent material.

2 Compositional changes due to leaching

Table 1 compares the properties of untreated and leached rice straw obtained via two different techniques: laboratory rinsing with distilled water, and field leaching under natural precipitation. The field leaching was accomplished by forming windrows of straw behind the grain harvester, and leaving them in the field for a period of 4 months over winter. In the first case, the changes in elemental composition and heating value are due to leaching the material with water alone. In the second case, the primary effects are caused by leaching with rain water, with additional influences due to contamination from dirt accumulating over the course of the winter, and loss of organic matter as a result of decomposition. The latter effects are relatively minor for this material. Listed in the table are the compositions and heating values for fresh, unleached straw, as well as for leached straw. The relative differences between the two are also listed. Aside from large relative differences in some of the minor species such as aluminum and titanium, leaching removes large amounts of alkali metals (K, Na) and chlorine. It also removes variable amounts of sulfur and phosphorous. In addition to inorganic elements, there is some extraction of organic materials, including sugars and phenolic compounds. The removal of macronutrients by field leaching is particularly valuable in nutrient recycling where straw is harvested as fuel, rather than incorporated into the soil. The heating value for both methods of leaching is increased by about 2% overall, which is due largely to the loss of inorganic diluents. Although Table 1 is specific to rice straw, similar effects have been observed for other biomass materials [7, 8].

The rate of leaching has been examined in several different ways. Figure 1 displays the incremental electrical conductivity (EC) of leachate solution obtained by leaching five different types of biomass: rice straw (California), wheat straw (California), switchgrass (Nebraska), and two different wood fuel blends obtained from commercial biomass fueled power plants (California). In each case 50 g of milled (2 mm) fuel was leached by successively pouring 500 mL of distilled water through the material. The incremental EC was determined for each batch of leachate. Although the peak magnitudes of EC vary depending on the type of fuel and the ion concentration, the location of the peak (1 L or 0.02 L g^{-1}) and the total volume of applied water to leach

Table 1. Compositions and heating values of fresh and leached rice straw.

	Laboratory rinsed			Windrowed, Left in field		
	Fresh	Leached	% Difference	Fresh	Leached	% Difference
Precipitation (mm)				0	254	
Elemental Analysis (% dry matter)						
Carbon	37.87	39.53	4	39.64	40.37	2
Hydrogen	4.61	4.76	3	4.64	4.62	-1
Nitrogen	0.63	0.53	-16	0.72	0.90	25
Sulfur	0.14	0.08	-43	0.11	0.09	-18
Ash	20.87	17.89	-14	18.27	17.51	-4
Chlorine	1.01	0.04	-96	0.26	0.02	-92
Ash analysis* (% dry matter)						
SiO_2	15.08	15.94	6	13.20	14.79	12
Al_2O_3	0.01	0.21	1,357	0.01	0.05	283
TiO_2	0.004	0.01	114	<0.002	0.004	>100
Fe_2O_3	0.05	0.08	42	0.04	0.06	39
CaO	0.43	0.44	2	0.30	0.39	28
MgO	0.43	0.24	-44	0.27	0.27	-2
Na_2O	0.57	0.06	-90	0.08	0.02	-73
K_2O	2.46	0.25	-90	3.03	0.66	-78
P_2O_5	0.37	0.09	-76	0.34	0.21	-38
SO_3	0.23	0.07	-68	0.16	0.11	-28
Reconstructed mass	19.65	17.39	-12	17.43	16.56	-5
Percent of Ash	94.17	97.22	3	95.40	94.57	-1
Higher heating value (dry basis)						
BTU/lb	6,342	6,514	3	6,489	6,625	2
MJ/kg	14.75	15.15	3	15.09	15.41	2
Water Soluble Potassium						
(% dry matter)	2.07	nd		2.29	nd	
(% of total K2O)	83.85			75.51		
Elemental Ash (% ash)						
SiO_2	72.28	89.12	23	72.23	84.47	17
Al_2O_3	0.07	1.19	1,600	0.07	0.28	300
TiO_2	0.02	0.05	150	<0.01	0.02	>100
Fe_2O_3	0.26	0.43	65	0.22	0.32	45
CaO	2.08	2.47	19	1.65	2.21	34
MgO	2.06	1.35	-34	1.49	1.52	2
Na_2O	2.71	0.32	-88	0.42	0.12	-71
K_2O	11.80	1.39	-88	16.60	3.77	-77
P_2O_5	1.78	0.49	-72	1.86	1.21	-35
SO_3	1.11	0.41	-63	0.86	0.65	-24
Cl	4.07			1.15		
CO2	0.10	0.12	20	0.14	0.17	21
Total	98.34	97.34	-1	96.69	94.74	-2
Undetermined	1.66	2.66		3.31	5.26	

*computed from elemental ash analysis (% ash). nd=not determined

the fuel (0.04 L g^{-1}) are nearly identical among the fuel types (the ash concentrations of the woods and the ion concentrations in the leachate are rather low; their EC values are read on the right scale of the figure). The leaching in these cases was done without recycling or water recovery (e.g. by membranes). The extent to which water use for leaching can be reduced by these techniques is currently under investigation. The rate of 0.04 L g^{-1} is equivalent to 25 mm of rainfall on spread rice straw in California, or 50 to 100 mm of rainfall on windrowed rice straw.

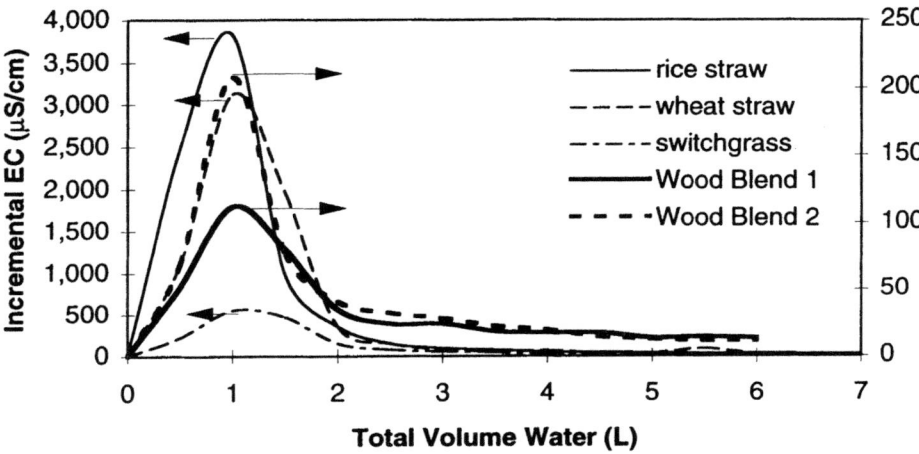

Figure 1. Incremental electrical conductivity (EC) for leachate from straw, grass, and wood (50 g samples milled to 2 mm particle size leached in distilled water). Wood blends refer to commercial wood fuels obtained from two separate biomass power plants.

The leaching rate has also been tested by immersing whole and milled fuel in water for different times. Soaking times of as little as two minutes with agitation have removed substantial amounts of potassium and chlorine [8]. The extraction rates by this method are currently being investigated.

Field leaching is of substantial interest because of its perceived greater practicality compared to plant site leaching and the beneficial impacts from nutrient recycling (nutrients in leachate from site leaching could also be recycled to the fields, but at added cost). Figures 2 and 3 illustrate the leaching rates obtained in the field for windrowed rice straw in California. The samples used for these analyses are field composites, so the ash and potassium concentrations vary somewhat from the compositions in Table 1.

Figure 2 shows the ash concentration in rice straw as a function of cumulative precipitation. Figure 3 gives the potassium concentration and cumulative precipitation over time. The ash concentration prior to the precipitation event of 13 Dec remains relatively constant, as does the potassium concentration (Figure 3). The curve connecting the concentrations for 26 Nov and 13 Dec in Figure 2 is arbitrary (assumed linear), as this single event deposited slightly more than 100 mm precipitation within 30 hours and the ash concentration during the event is unknown. The ash concentration reaches a minimum at 3 Jan as increasing precipitation continues to remove inorganic materials, but then begins to increase as microbial induced decomposition begins to remove organic matter at a rate faster than inorganic material extraction by leaching. The organic matter loss slows at later times as weather changes allow the straw to dry (not shown). Figure 3 shows the rapid loss of potassium with the first precipitation event. The total extraction of potassium is, however, lower than obtained for rice straw spread in a thin layer in the field (as is normally done behind the grain harvester to expedite straw drying and burning) [8]. The reduced extraction is attributed to the

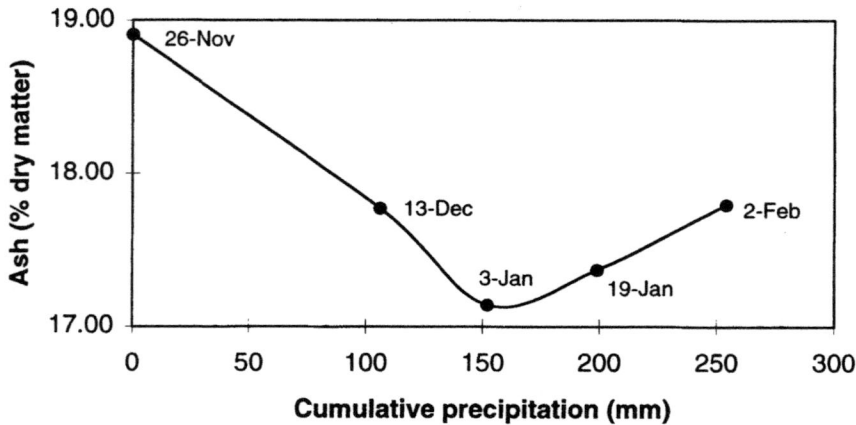

Figure 2. Total ash content of rice straw left windrowed in the field over the winter (variety M202, Sutter, California).

thatching effect of windrowed straw, whereby the upper part of the windrow is heavily depleted in potassium, but water penetration to the lower part of the windrow is inhibited by shedding of water from the upper windrow (in a manner similar to a thatched roof). Compositional assays of the upper and lower regions of the windrow confirm this result (after 330 mm rainfall, potassium concentration at the top surface of the windrow was 0.19% dry matter, and at the lower surface it was 1.12% dry matter).

These results show that potassium and chlorine are rapidly and quantitatively removed from biomass by water leaching. The results are for the most part consistent with chemical fractionation assays whereby biomass is sequentially leached with water, ammonium acetate, and hydrochloric acid to investigate the mode of occurrence of different elements [3] [9] [10], although in general the extraction of potassium by water is here shown to be larger than in the case of the standard method used for chemical fractionation, which may be due to incomplete extraction during chemical fractionation. The results for potassium are also consistent with the standard test for water soluble potassium, whereby the fuel is leached with water at 90°C. The extraction with cold water seems equally effective. The extraction of both alkali metals and chlorine, along with some of the other elements (particularly sulfur and phosphorous) are beneficial in terms of improving the fuel value of biomass materials, not only from the fouling standpoint but from those of corrosion and emission formation as well.

3 Increase in heating value due to leaching

As mentioned above, heating value tends to increase for leached material compared to

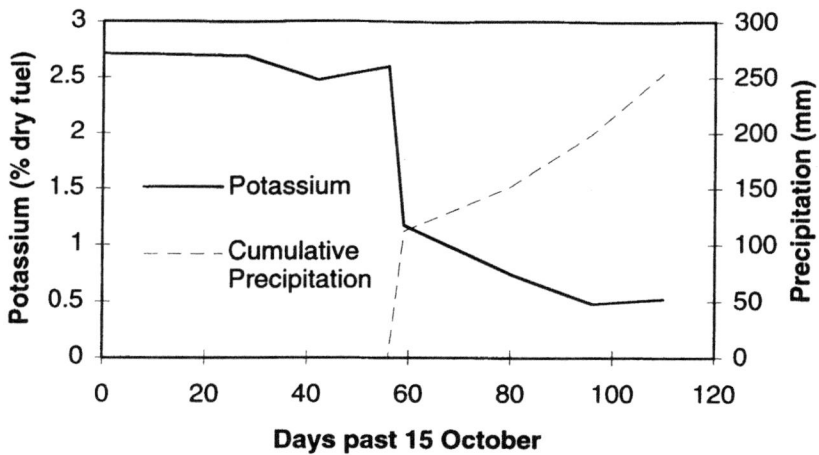

Figure 3. Potassium concentration in windrowed rice straw.

the original unleached material. Trends for rice straw leached naturally by rainwater in the field, and for untreated and laboratory washed switchgrass are shown in Figure 4 as a function of ash content. For each fuel, the point with highest ash content and lowest heating value is the untreated material. Leaching causes a decrease in ash content and an increase, in general, in heating value. Also shown in the figure is the correlation line developed by Jenkins [11] using 112 different types of biomass. This correlation represents the general decline in heating value with increasing ash content, largely as a result of ash displacing energetic organic matter. Although the rice straw and switchgrass data lie below the correlation line, the trends are similar, both across groups (indicating species specific variations), and within groups (indicating the effects of leaching).

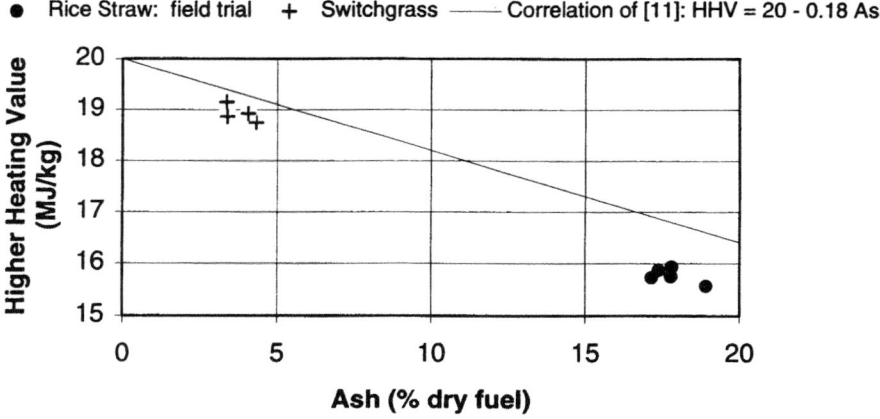

Figure 4. Heating values of rice straw and switchgrass. The line is a correlation for 112 different biomass types [11].

4 Effect of leaching on ash fusibility

Rather remarkable improvements in the fusibility of biomass ash can be demonstrated for leached materials. Figure 5 gives severity ratings of ash fusibility for three types of straw materials: rice straw (Sutter County, California), a low alkali wheat straw (Yolo County, California), and a high alkali wheat straw (Imperial County, California) [8]. The test used to evaluate ash fusibility is different than the standard ASTM pyrometric cone test (D1857/E953), and was developed [8] to overcome some of the perceived limitations of the standard test. This new test burns a pellet (a compressed cylinder roughly 12 mm diameter by 12 mm height and weighing 0.5 to 1 g) of milled (2 mm) fuel at various temperatures in a high temperature furnace. The appearance of the pellet is observed over time at each temperature. Typically the pellet will swell, expanding primarily in height, particles will eventually lightly sinter into a porous mass, then increasingly compact through strongly sintered, slagged and partially fused, and finally fully molten and fused glassy states. In some cases, the ash will eventually disappear through volatilization. Fuels which achieve higher states of fusion at lower temperatures are perceived to be more troublesome in terms of slagging and fouling in thermal converters. Figure 5 shows that for each of the three fuel types, leaching results in a increase in fusion temperature compared to the original unleached material. In all cases the increase is substantial.

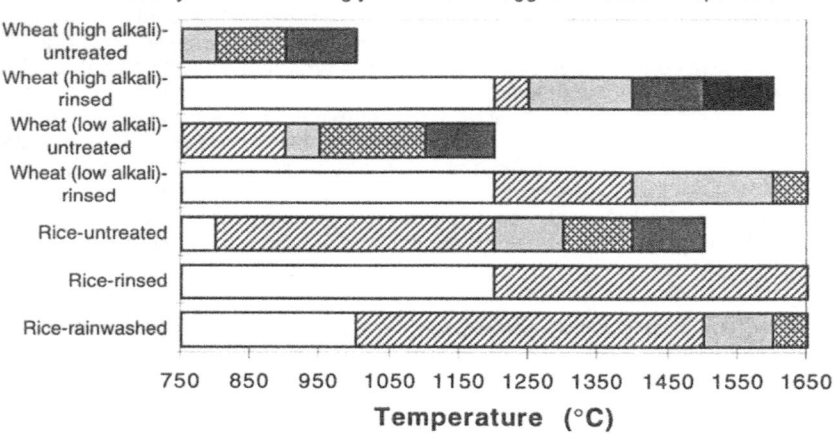

Figure 5. Fusibility ratings for rice and wheat straw [8].

This fusibility test was also applied to the field leached windrowed rice straw discussed above. The results are shown in Figure 6 for the windrow composite samples as a function of cumulative precipitation. The first rainfall of 106 mm substantially improves the fusion character of the straw due to the leaching of alkali metals and other elements. There is some light sintering at the lowest temperature tested (800°C) until 199 mm of precipitation. The improvement is similar to that obtained for spread straw after 65 mm precipitation (Figure 5).

Similar improvements can be observed for other materials. Figure 7 gives results for switchgrass in four treatments: 1) fresh unleached, 2) rain washed in the field, 3) rinsed with distilled water, and 4) soaked in distilled water. The latter two cases both use a water application rate of 0.04 L g^{-1}. Above 1000°C there is a reduction in severity rating at the same temperature for treated switchgrass compared to untreated fuel.

Soaking in this case appears to be less effective compared to rinsing the material by pouring water through a bed of fuel. The untreated switchgrass used in these experiments had an ash content of about 4% dry matter; leaching reduced this to about 3% dry matter.

Wood fuels from the two power plants mentioned above were blended with leached rice straw and tested in the same manner. Whereas the wood alone was observed to fuse at 1250°C, the blends with straw were observed to fuse at increasingly higher temperatures as the straw concentration increased, until at 50% straw, only sintering was observed up to 1500°C. As leached straw is heavily depleted of alkali and chlorine, blending with untreated wood serves to reintroduce these materials (more so potassium, not so much chlorine) into the fuel mix. The effect on fouling due to blending is currently under investigation.

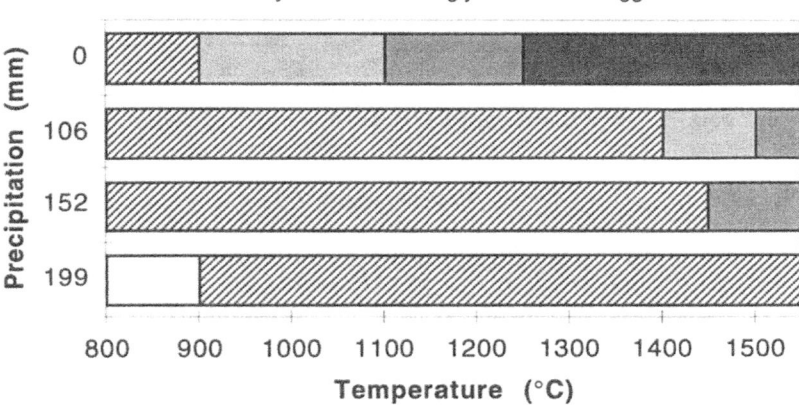

Figure 6. Fusibility ratings for rice straw left windrowed in the field over winter.

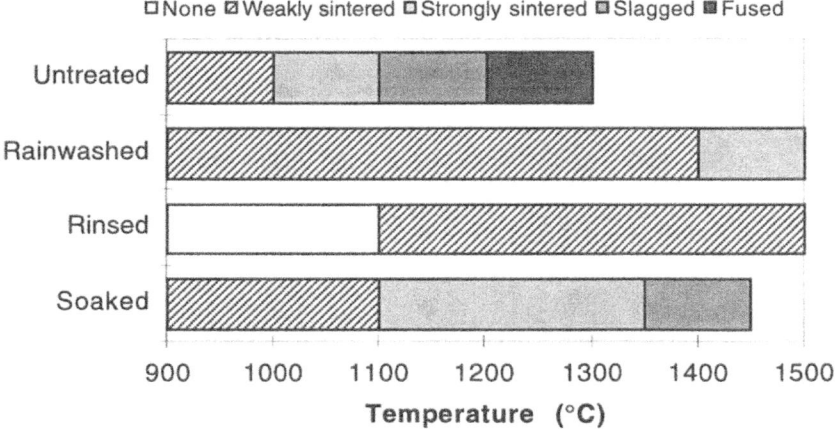

Figure 7. Fusibility ratings for untreated and leached switchgrass.

5 Effect of leaching on ash volatilization

Ash fusibility can be used to infer the slagging characteristics of the fuels, but is only partly indicative of the fouling behavior. Fouling involves volatilization of ash elements as well. To some extent, the fouling behavior can be evaluated on the basis of

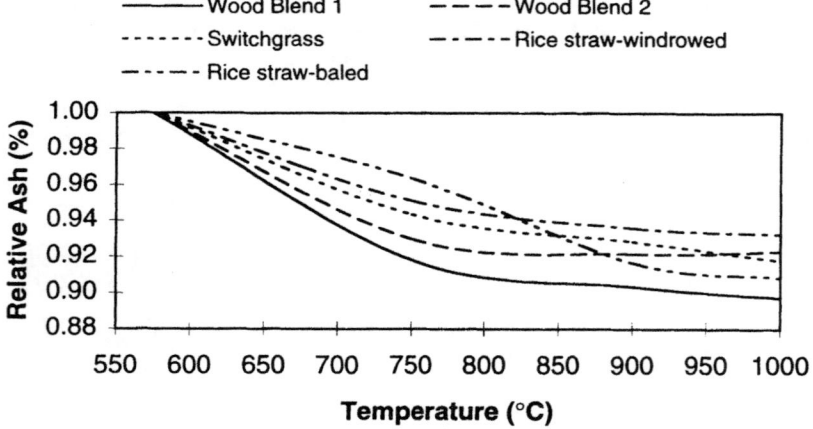

Figure 8. Relative ash contents of unleached fuels.

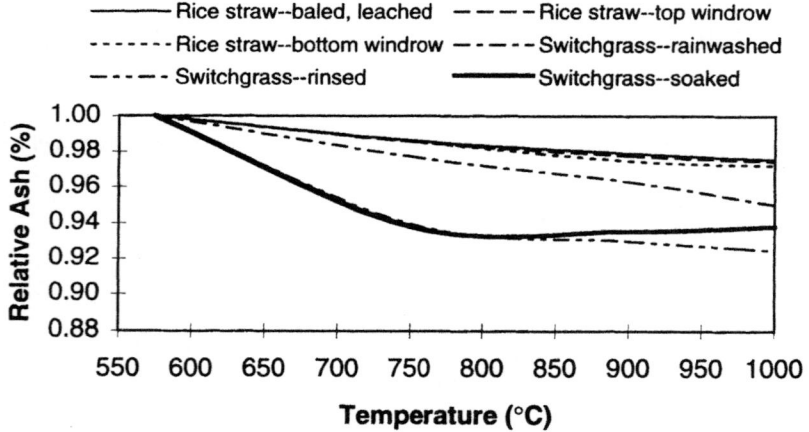

Figure 9. Relative ash contents of leached fuels.

the mass loss of ash with increasing temperature. Figures 8 and 9 give the relative ash contents for several fuel types as a function of ashing temperature in a muffle furnace under oxidizing conditions between 575 and 1000°C. The relative ash content is obtained by dividing the ash content at each temperature by the ash content at 575°C. The fuels shown in Figure 8 are unleached, and include autumn harvested rice straw obtained in baled form prior to any precipitation, windrowed rice straw from the field leaching trials as described above, switchgrass, and the two power plant wood fuel blends mentioned above. There is a substantial decline in measured ash content between 575°C and 750°C, due most likely to the volatile loss of alkali metals [12]. The composition of the ash at each temperature has not yet been determined. The same analysis appears in Figure 9 for various leached samples of rice straw and switchgrass. All of the rice straw samples show much lower volatile loss compared to the unleached samples. The rainwashed sample of switchgrass yields lower volatile loss compared to the laboratory leached samples. The reasons for the difference are not entirely clear and await compositional analyses of the ash. In part, the difference between switchgrass and rice straw may be due to the greater silica and lower alkaline earth metal concentrations of rice straw [8]. A reduction in volatile ash should accompany a reduction in fouling tendency.

6 Pilot combustor tests with unleached and leached straw

To test the effect of leaching on the fouling characteristics of the otherwise high fouling fuel, rice straw, pilot scale combustion experiments were carried out in the Multifuel Combustor (MFC) at Sandia National Laboratories, Livermore, California. This unit is an entrained flow combustor in which milled fuel (passing 16 mesh) is pneumatically injected into and burned in the air stream of an electrically heated down-flow tube furnace of 15 cm internal diameter and 4.3 m length. The wall temperature of the furnace can be controlled, as can the temperature of a vitiated air stream (obtained by burning natural gas in the air to increase the temperature) entering the top of the furnace. In the tests conducted here, the wall temperature was held at 900°C, and fresh inlet air (without natural gas preheat) was used. The furnace exhausts into an open test section, through which deposition probes of various types can be inserted. In these tests, air cooled steel tubes were used as deposit collection probes simulating superheater tubes in commercial power plants. The oxygen concentration in the exhaust was varied between 3 and 10%.

With unleached rice straw, deposits were observed to rapidly build on the furnace walls, eventually clogging the furnace and leading to early termination of the test (dislodgment and falling of furnace wall deposits also led to loss of deposit on the probes). Microscopic examination of these deposits shows a potash glass matrix surrounding nodules of silica, demonstrating the transformation of the fuel ash to a molten glass by reaction of potassium and silica at furnace temperatures. A similar deposit was observed to build on the deposition probes, but the rapid deposition on the furnace walls prevents any strong conclusions regarding the deposition rates on these probes.

When leached rice straw (the laboratory leached straw of Table 1) was fired, very light furnace wall deposits appeared, but these were continuously blown off by the furnace flow. Samples were collected at the furnace exhaust, and were observed to consist of loose, friable aggregates having almost no sinter strength. Microscopic examination of this material revealed no perceptible fusion of the ash particles, unlike the untreated straw. No measurable deposit mass was collected on the deposition probes for the two hour duration of the test. Qualitatively, these tests support the conclusions above regarding the likely improvements to fuel quality arising from leaching. Full scale experiments are pending.

7 Effect of leaching on rates of reaction

Pellet burning tests described above to evaluate ash fusibility characteristics suggested a difference in the volatile matter emission rates for leached compared to unleached materials (smoke emission was observed to stop earlier for leached materials) [8]. To explore this effect, dynamic thermogravimetric analysis (TGA) was carried out on 10 mg samples of rice and wheat straw (other materials are currently being investigated) in both air and argon at heating rates of 50 and 100 K min⁻¹. Typical results are shown in Figure 10. Two main stages of conversion (weight loss) can be observed: pyrolysis followed by char conversion. In every case, the rate of pyrolysis is delayed for the leached fuels relative to the unleached fuels. In the inert (argon) atmosphere, a distinct separation of evolving components can be observed for the leached materials. Although these results do not explain the difference in volatile emission rates observed during ash fusibility testing, they are consistent with the known effects of alkali metals on pyrolysis rates for carbohydrate materials [13 - 15].

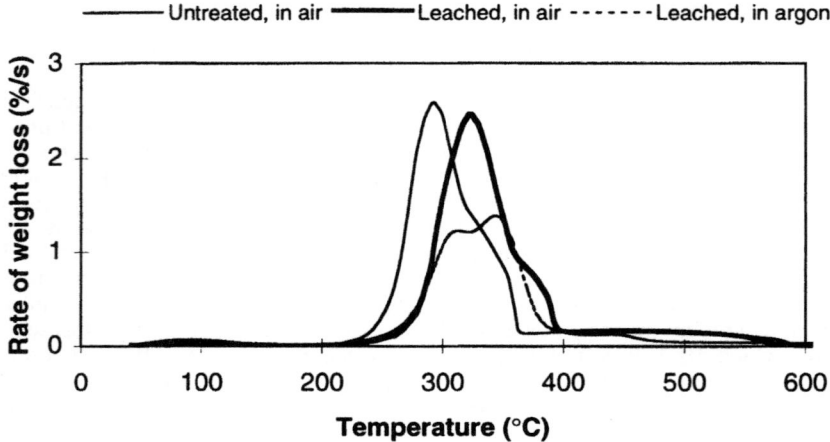

Figure 10. Rate of weight loss for unleached and leached rice straw in air and argon atmospheres by dynamic TGA (100 K min⁻¹).

To quantify the differences observed via TGA, coefficients of simple first order Arrhenius kinetic models for the two individual stages were derived. Defining the extent of conversion (or just conversion), x, of the fuel as,

$$x = \frac{m_o - m}{m_o - m_f}$$

(1)

where m_o (kg) is the original mass of fuel, m_f (kg) is the final mass of fuel, and m (kg) is the mass of fuel at any time, t (s), the rate of conversion can be modeled as

$$\frac{dx}{dT} = \frac{k_o}{q} \exp\left(-\frac{E}{RT}\right)(1 - x)$$

(2)

in which k_o (s⁻¹) is the frequency factor, E is the activation energy (kJ mol⁻¹), R is the universal gas constant (kJ mol⁻¹ K⁻¹), T is the absolute temperature (K), q is the constant temperature ramp, $q = dT/dt$ (K s⁻¹) and the expression 1-x represents the first order reaction. The coefficients k_o and E can be found from the TGA data through the direct numerical integration and linear regression (E from the slope and k_o from the intercept) on the ln-transform of equation (2).

The results of such analysis for rice straw, a high alkaline wheat straw, and a low alkaline wheat straw are shown in Table 2 for air. Several considerations apply to these data. The heating rates utilized are relatively high (for kinetic studies, not necessarily in comparison to actual heating rates in reactors and combustors), and the simple kinetic model utilized does not correct for heat transfer in the TGA nor account for competitive reaction pathways. Higher heating rates typically result in lower activation energies and primary char formation than lower heating rates [13]. Various techniques to account for the effects of heating rate have been suggested [13, 16]. The evaluation here is intended for comparative purposes, and although there exists some variability in the results, the trends are clear for this separate stagewise analysis of the pyrolysis and char conversion. In each case, leaching causes a reduction in the activation energy and frequency factor of pyrolysis, and an increase in the activation energy and frequency factor of char conversion for the imposed conditions. The reduction in activation energy would cause an increase in the reaction rate, but this is more than compensated by the reduction in frequency factor which results in an overall reduction in reaction rate.

Table 2. Activation energies, E (kJ mol^{-1}) and frequency factors, k_o (s^{-1}), within the temperature range ΔT ($^\circ$C) for first order Arrhenius kinetics of untreated and leached straw during two main stages of decomposition. Kinetic parameters determined by numerical integration of transformed TGA data in distinct linear regions. TGA in air on 40 mesh particle size at 100 K min^{-1}.

	Stage I: Pyrolysis			Stage II: Char conversion		
	ΔT	E	k_o	ΔT	E	k_o
Rice Straw						
Untreated	275 - 300	110.25	8.35×10^8	390 - 475	24.90	0.76
Sprayed	275 - 300	95.80	3.87×10^7	380 - 460	29.11	1.97
Soaked	305 - 330	94.98	1.13×10^7	420 - 480	40.14	12.31
Rainwashed	312 - 330	60.46	9.99×10^3	465 - 485	51.20	78.38
High Alkaline Wheat Straw						
Untreated	274 - 305	103.01	1.94×10^8	380 - 465	22.13	0.45
Leached	297 - 325	96.68	2.41×10^7	405 - 475	36.61	7.51
Low Alkaline Wheat Straw						
Untreated	295 - 315	117.09	2.04×10^9	385 - 490	23.45	0.51
Leached	315 - 340	93.60	5.37×10^6	450 - 505	43.86	16.82

Biomass consists of three main organic structural components: hemicellulose, cellulose, and lignin. An alternative approach to the evaluation of the kinetic coefficients is to model the overall conversion as the sum of the conversions of multiple components [13, 17, 18]. The component conversions are described by equations (1) and (2) based on individual component mass fractions, and then superimposed. The method has the conceptual advantage of modeling the conversion over the full temperature range without the need to segregate the reaction history as in the stagewise method described above. Simple superposition of individual component conversions fails to account for interactions among components leading to more complex behavior, however [13]. More sophisticated models are needed to account for the effects of heating rate. Recognizing these limitations, the simple superposition technique still offers the means to comparatively assess differences due to leaching. The results of such an analysis from a 3-component model are shown for unleached and leached rice straw in Table 3.

Table 3. Mass fractions (m/m_o) and kinetic coefficients derived from 3-component conversion model for untreated and leached rice straw via TGA at 100 K min^{-1}

Untreated Rice Straw

Component	1	2	3
m/m_o (--)	0.15	0.56	0.29
k_o (s^{-1})	1.4×10^{11}	2.2×10^{14}	4.2×10^{3}
E (kJ mol^{-1})	142	166	77

Leached Rice Straw

Component	1	2	3
m/m_o (--)	0.15	0.55	0.30
k_o (min^{-1})	6.7×10^{10}	1.8×10^{14}	4.2×10^{3}
E (kJ mol^{-1})	142	174	75

In generating these results, the model is calibrated against the experimental data for each test by fitting the mass fractions and kinetic coefficients (9 variables in all). The choice of three components is essentially arbitrary, and assumes that hemicellulose and cellulose form reactive materials (chars) during conversion and these materials react at the same rate as the third lignin component. The model fit is quite good in the pyrolysis region, less so in the char conversion region, as might be anticipated. The trends in kinetic coefficients differ somewhat from the results of the stagewise analysis of Table 2, although the overall result in terms of decreased reaction rate for leached material is maintained. In the case of component 1, the activation energy remains constant between the unleached and leached materials, but the frequency factor declines for the leached straw. For component 2, the activation energy increases for the leached materials, with a decrease in frequency factor. For component 3, the activation energy decreases slightly with leaching, while the frequency factor remains constant. Whether using the stagewise or multicomponent modeling technique, the overall relative kinetic behavior is consistent with a shift towards higher temperatures for the conversion of leached materials, most likely due to the decrease in catalytic effects attributable to alkali metals and alkali salts. Reasons for the earlier cessation of volatile matter emission during isothermal reaction of leached materials is still under investigation, and may have to do with transport limitations in pellets of unleached materials and differences in composition, in particular, chlorine.

8 Effects of leaching on ignition and burning

Another effect of leaching has been qualitatively observed in the case of rice straw, and is now also being investigated in more detail. Rice straw is a relatively refractory material due to its high ash content. We have observed that in the case of burning single straws, leaching facilitates the ignition and burning. This is readily observed by vertically suspending single straws of unleached and leached material, and igniting from below.

We have also observed differences in the spreading rates of fires in beds of straw during wind tunnel simulations of open burning [19]. We find consistent reductions in fire spreading rates with increasing chlorine concentration in the fuel. Differences in the burning characteristics of polymeric materials containing chlorine are known to be related to the role of chlorine in terminating free radical chain reactions (e.g., poly-vinylchloride). Halogens are commonly used as flame retardants. Where substantial quantities of chlorine are leached from biomass fuels, improvements in the ignition and burning characteristics of the fuels may be expected. Other effects, such as those on acid gas and toxic emission formation may also be anticipated. Further analysis in this regard is continuing.

References

1. Miles, T.R. Jr. and T.R. Miles. (1995) *Alkali deposits found in biomass power plants--a preliminary investigation of their extent and nature.* Summary report for the National Renewable Energy Laboratory, NREL Subcontract TZ-2-11226-1, Golden, CO.

2. Baxter, L.L. (1993) Ash deposition during biomass and coal combustion: a mechanistic approach. *Biomass and Bioenergy* 4(2):85-102.

3. Baxter, L.L., T.R. Miles, T.R. Miles, Jr., B.M. Jenkins, G.H. Richards, and L.L Oden. (1993) Transformations and deposition of inorganic material in biomass boilers. In M.G. Carvalho (ed), *Second International Conference on Combustion Technologies for a Clean Environment*, 1:Biomass II 9-15, Commission of European Communities, Lisbon, Portugal.

4. Jenkins, B.M., L.L. Baxter, T.R. Miles, T.R. Miles, Jr., L.L. Oden, R.W. Bryers and E. Winther. (1994) Composition of ash deposits in biomass fueled boilers: results of full-scale experiments and laboratory simulations. *ASAE Paper No. 946007*, ASAE, St. Joseph, Michigan.

5. Miles, T.R., T.R. Miles, Jr., L.L. Baxter, B.M. Jenkins, and L.L Oden. (1993) Alkali slagging problems with biomass fuels. *First Biomass Conference of the Americas*, Burlington, Vermont.

6. Miles, T.R. , T.R. Miles, Jr., R.W. Bryers, L.L. Baxter, B.M. Jenkins, and L.L. Oden. Alkalis in alternative biofuels. (1994) *FACT 18, Combustion modeling, scaling, and air toxins*, pp 211-220, ASME, New York.

7. Jenkins, B.M., R.R. Bakker and J.B. Wei. (1995) Removal of inorganic elements to improve biomass combustion properties. *Proceedings 2nd Biomass Conference of the Americas*, National Renewable Energy Laboratory, Golden, CO.

8. Jenkins, B.M., R.R. Bakker and J.B.Wei. (1995) On the properties of washed straw. *Biomass and Bioenergy* 10(4):177-200.

9. Miles, T.R., T.R. Miles, Jr., L.L. Baxter, R.W. Bryers, B.M. Jenkins, and L.L. Oden. (1995) *Alkali deposits found in biomass power plants: a preliminary investigation of their extent and nature.* National Renewable Energy Laboratory, Golden, CO.

10. Baxter, L.L., T.R. Miles, T.R. Miles, Jr., B.M. Jenkins, D. Dayton, T. Milne, R.W. Bryers and L.L. Oden. (1996) *The behavior of inorganic material in biomass-fired power boilers--field and laboratory experiences.* Vol. II, Alkali deposits found in biomass power plants, National Renewable Energy Laboratory, Golden, CO (draft report).

11. Jenkins, B.M. (1993) Properties of biomass. In: Wiltsee, G.A., Biomass Energy Fundamentals, Vol. 2: Appendices. EPRI TR-102107, Electric Power Research Institute, Palo Alto, California.

12. Misra, M.K, K.W. Ragland and A.J. Baker. (1993) Wood ash composition as a function of furnace temperature. *Biomass and Bioenergy* 4(2):103-116.

13. Antal, M.J., Jr. (1983) Biomass pyrolysis: a review of the literature. Part I--carbohydrate pyrolysis. In: *Advances in solar energy, Vol. 1*, Boer, K.W. and J.A. Duffie (eds.), American Solar Energy Society, Boulder, CO, pp 61-112. Also, Antal, M.J., Jr. (1985) Biomass pyrolysis: a review of the literature. Part 2--lignocellulose pyrolysis. In: *Advances in solar energy, Vol. 2*, Boer, K.W. and J.A. Duffie (eds.), American Solar Energy Society, Boulder, CO, pp 175-255.

14. Essig, M., T. Lowary, G.N. Richards and E. Schenk. (1988) Influences of "neutral" salts on thermochemical converison of cellulose and of sucrose. In

Research in Thermochemical Biomass Conversion, Bridgwater, A.V. and J.L. Kuester (eds.), Elsevier Applied Science, London, pp. 143-154.

15. Williams, P.T and P.A. Horne. (1994) The role of metal salts in the pyrolysis of biomass. *Renewable Energy* 4(1):1-13.

16. Gaur, S. and T.B. Reed (1994) Prediction of cellulose decomposition rates from thermogravimetric data. *Biomass and Bioenergy* 7(1-6):61-67.

17. Vovelle, C., H. Mellottee and R. Delbourgo (1992) Kinetics of the thermal degradation of cellulose and wood in inert and oxidative atmospheres. *19th Symposium (International) on Combustion*, pp. 797-805.

18. Bining, A.S. (1996) *A study of reaction kinetics for thermochemical conversion of rice straw.* Unpublished PhD dissertation, University of California, Davis.

19. Jenkins, B.M. (1996) *Atmospheric pollutant emission factors from open burning of agricultural and forest biomass by wind tunnel simulations.* Final Report, CARB A932-126, California Air Resources Board, Sacramento.

COMBUSTION OF A SINGLE WOOD LOG UNDER FURNACE CONDITIONS
Combustion of single wood logs

K. M. BRYDEN and K. W. RAGLAND
Department of Mechanical Engineering
University of Wisconsin - Madison
Madison   USA

Abstract
This paper presents the results of experimental combustion studies of large particle (>10 cm) thermally thick woody biomass under conditions representative of packed bed combustion. The combustion rate of thermally thick yellow poplar and paper birch tree segments (12 cm to 21 cm in diameter by 1.5 m long) in a typical combustion environment was measured in a special test furnace using load cells to measure mass versus time. Test conditions included a temperature range of 950°C to 1340°C and a variety of oxygen, carbon dioxide, and water vapor concentrations. Both green and dry woods were tested. The time to consume 90% of the mass of the wood ranged from 13 min to 55 min depending on wood and furnace conditions. The thickest char layer observed during these studies was 2 cm. A simple relationship for the buildup of the char layer within the range of conditions studied was developed. A numerical model to determine the combustion rate of single logs within the kiln, for the range of experimental values was developed using global reaction rates for combustion and gasification. Agreement between the numerical model and the experimental values is within 30% for nearly all of the 72 data points.
Keywords: Thermally thick combustion, Woody biomass, Wood logs

1   Introduction

Biomass is an attractive fuel for electric power generation because it is a locally available renewable energy source that reduces SO_2 and CO_2 emissions. Several proposed methods for efficient generation of electric power from biomass use large chunks of wood or whole tree segments rather than wood chips. For example, packed bed combustion of chunked wood 2.5 cm to 8 cm thick and 10 cm to 15 cm in diameter has been proposed [1]. In another proposal a deep bed combustor/gasifier is fed with large logs and whole trees rather than wood chips [2]. The advantages of this large fuel size are that the chipping costs are saved, the fuel may be stored on site without degradation, and the heat release per plan area of the combustor is greater than with wood chips. However, there is little information on how large-sized woody biomass fuel reacts under the intense conditions found in a deep fixed bed.

Although these proposed methods use stoker type technology, they are unique in that large tree segments rather than wood chips are used, and the fuel bed is very deep to allow sufficient residence time to dry and pyrolize the wood and to combust the char. Fresh wood is introduced at the top of the bed. Exposed to the hot combustion gases from lower portions of the bed, the wood rapidly forms a char layer. As the fuel travels downward in the bed, it undergoes drying, pyrolysis, and char consumption simultaneously. Before reaching the lowest section of the bed, all the moisture and volatiles and a significant portion of the char have been consumed, leaving char particles of reduced diameter that rapidly react with the incoming oxygen. This lower region quickly consumes the oxygen in the upwardly flowing gases that then gasify, pyrolize, and dry the fuel in the upper region. In effect the system is both a combustor and gasifier in which the energy to support the gasification reactions in the upper section is provided by the lower combustion region.

Wood combustion is a complex process that is not completely understood. In wood combustion the external gas phase reactions, the surface char reactions, and the internal pyrolysis reactions at the surface of the yet unaffected wood are coupled together. These three coupled reaction zones jointly determine the consumption rate, heat release rate, and the composition of the released gases. The pyrolysis products and moisture flow outward through the char layer driven by the transient heat transfer from the surface temperature. At the surface of the char, oxygen diffusing inward from the external flow and carbon dioxide and water vapor from both the internal pyrolysis reactions and diffusion from the free stream react with the char. The products of these char reactions combined with remaining pyrolysis products then enter the freestream and may or may not result in an attached or unattached flame zone that consumes the oxygen flux to the surface and feeds back radiant energy to the surface. In the case of the whole tree or chunkwood combustor the large particle size makes the fuel thermally thick in the sense that all three reaction zones are present simultaneously and at all locations in the combustor except at the bottom of the bed where only char remains. The overall bed heat release rate depends on the heat and mass transfer to the logs, the formation of the char layer, the reaction rate of the char with oxygen, carbon dioxide and water vapor, and the nature and rate of reaction of the pyrolysis products.

Modeling of wood combustion is an active area of research with many different areas being actively pursued. Flame spread and ignition of both thermally thin and thermally thick charring materials has been investigated by several researchers. Jannsens [3] provides a detailed review of piloted ignition of wood and notes briefly the differing models for thermally thin and thermally thick materials. Di Blasi [4,5] discusses the mechanisms of flame spread for thick and thin materials. Pyrolysis of thermally thin and thermally thick material has been examined by numerous researchers. Antal and Varhegyi [6] provide an in-depth review of cellulose pyrolysis kinetics. Limited data for large wood char particle combustion and gasification are available. Ragland et al. [7] report mass loss as a function of time for wood chunks up to 10 cm thick inserted in a spreader stoker fired with coal and present a transient conduction model to account for the mass loss and growth of the char layer. The chunkwood shrinking rate was 1.8 mm/min and the burnout time was 30 min. The maximum wood char thickness was 3 mm for green wood and 15 mm for air-dried wood. Drying, pyrolysis, and char burn occurred simultaneously. The temperature in the char layer was calculated and shown to drop rapidly from the surface temperature of 1250°C to the pyrolysis layer which was 1.5 mm thick and at 500°C. The wood temperature behind the pyrolysis layer was below 100°C. Albini and Reinhardt [8] have examined the time to ignition and mass loss as a function of time for cylindrical wood elements 2.5 - 10 cm in diameter with and without bark under conditions representative of those experienced in a wildland fire. Modeling of fixed bed combustion of large woody biomass is discussed by Bryden and Ragland [9].

The purpose of this paper is to present the experimental data and numerical modeling of the burning of a single log versus time for conditions representative of the interior of a deep bed, combustor-gasifier using large logs. This is a simpler case than the deep fixed bed reactor, and can provide insight into the processes involved.

2 Test Setup and Procedures

The furnace was designed by Energy Performance Systems Co. and installed at the U.S. Forest Products Laboratory in Madison, Wisconsin. The furnace had a working chamber 1.5 m long by 30.5 cm wide by 36.8 cm high and was heated by a 200 kW natural gas burner (Fig. 1). Single logs ranging in diameter from 12 cm to 21 cm by 1.4 to 1.5 m long were tested. Gases from the burner were mixed with additional oxygen depending on the conditions desired. The added oxygen was obtained from

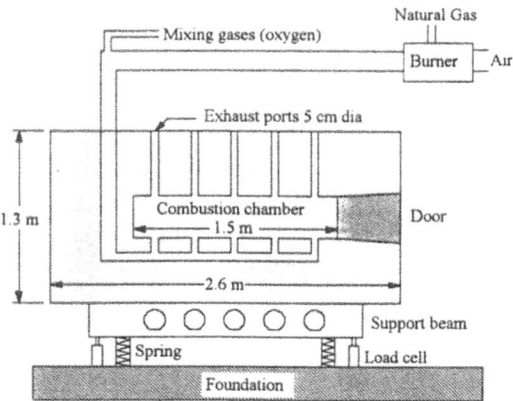

Fig. 1. Schematic of test furnace setup.

liquefied oxygen, which was vaporized in heat exchangers. The burner gases and mixed gases flowed down, beneath the working chamber and up into the chamber through five 5-cm diameter holes in the bottom of the chamber. Flue gases from the natural gas combustor were used to simulate the environment that a wood log would experience in a deep packed bed combustor.

The entire furnace and burner assembly was supported on five springs and four load cells. Load cell data versus time was stored on a computer at the site. In-place calibration of the load cells was performed at the beginning and end of each day of testing. Temperatures were measured in the working chamber with a thermocouple probe and an optical pyrometer before insertion of the log. The flow of the natural gas, and the oxygen and carbon dioxide concentrations in the mixed inlet gas were measured with continuous gas meters. From this information the gas flow rate and composition of the flow into the working chamber were determined. Temperatures up to 1340°C at a flow rate of 3.1 standard-m^3/min were achieved.

The logs tested were yellow poplar and paper birch from the same stand of trees. One-half of the logs were dried in a constant humidity room for several months. Before testing, the as-received moisture was determined from 15-cm lengths cut from each end, and each log was measured and weighed. Two different sets of tests were performed. In the first, a log was placed in the furnace at specified operating conditions, and the mass versus time was measured. Data from the four load cells were summed and normalized by the initial weight of the log, and an exponential curve fit was made to the data. Thirty-two separate logs were tested, and usable data were obtained from twenty-four of the tests.

In the second set of tests, a log was placed in the furnace at specified operating conditions and then removed after a specified period of time. Upon removal the log was extinguished using a CO_2 fire extinguisher. After the log cooled it was weighed and cross sectioned to measure the char layer thickness and moisture profile of the remaining wood. The purpose of these tests was to confirm the load cell data and to provide detailed char thickness and moisture profile data. Thirty separate logs were tested.

3 Test Results

A typical mass loss versus time curve is shown in Fig. 2. Typically the first 60% to 70% of the mass loss shows a nearly linear decrease with time. The wood was not completely pyrolyzed until after approximately 98% to 99% mass loss. Approximately the last 10% of the load cell data is unreliable due to system noise. For all of the tests the time to consume 90% of the mass of the log varied from 13 min to 55 min depending on the log diameter, moisture content, oxygen content, and gas temperature.

The first set of data in Table 1 is for relatively dry logs (10% to 23% moisture, as-received) and the second set of data is for relatively wet logs (34% to 45% moisture, as-received). The gas temperature was varied from 950°C to 1340°C (the maximum achievable with the test setup). The superficial gas velocity of 0.54 m/s to 0.76 m/s was also the maximum achievable. The inlet oxygen concentration was varied from 0

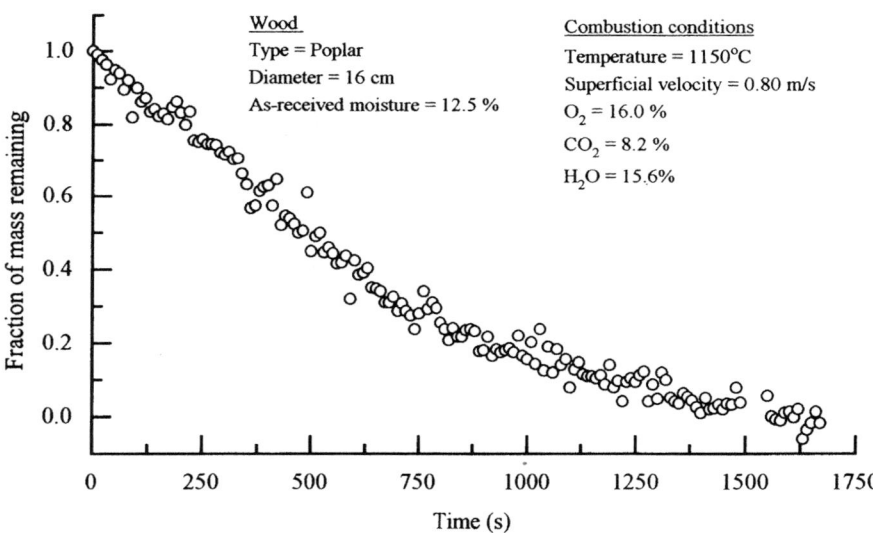

Fig. 2. Typical normalized mass loss vs. time curve.

to 21%. The inlet carbon dioxide concentration was 8% to 9% except when the burner was shut off or for the rich burner runs. The inlet water vapor concentration varied from 15% to 21% unless the burner was shut off. The mass loss rate during the first 30% of burnout varied from 0.5 kg/min to 1.9 kg/min depending on the wood and furnace conditions. From 30% to 60% and 60% to 90% burnout, the mass loss rate varied from 0.32 kg/min to 1.2 kg/min and 0.16 kg/min to 0.72 kg/min, respectively. The mass loss rate depends on the gas temperature; gas velocity, oxygen, carbon dioxide, and water vapor concentration; and size and moisture content of the log. In this test setup several of these variables were interrelated. To examine the relationship and impact of these variables a numerical model of the combustion of a single log was developed and compared with the data as discussed below.

The char layer grows with time in the furnace and did not exceed 2 cm for the logs tested. The growth of the char layer is a complex issue with a large number of variables. The goal for this portion of this project was to develop a simple expression for the growth of the char layer within the range of parameters studied. This expression was found by nondimensionalizing the parameters that may have an impact on the growth of the char layer and then using multiple regression. The proposed correlation is:

$$\frac{c}{d} = 0.0715 \cdot \left[\left(\frac{d_i}{d} - 1 \right) \cdot \frac{1}{y_{H_2O}^{0.25}} \right] \qquad (1)$$

where c is the char thickness, d is the average diameter of the unaffected wood, d_i is the initial diameter, and y_{H_2O} is the initial as-received moisture content. This

Table 1. Measured combustion rate for mass loss vs. time experiments.

#	type	d_{ave} (cm)	As-received moisture (%)	T_{ave} (°C)	V_s (m/s)	O_2 (%)	CO (%)	CO_2 (%)	H_2O (%)	0-30% burnout	30-60% burnout	60-90% burnout	Time to 90% burnout (min)
		Wood			**Furnace Conditions**					**Combustion Rate (g/s)**			
1	poplar	17	13.4	980	0.70	2.1	0.0	8.8	16.8	8	5	2.7	55
2	poplar	15	10.5	1090	0.70	7.7	0.0	9.0	17.2	10	7	4.5	21
3	poplar	16	12.5	1150	0.80	16.0	0.0	8.2	15.6	19	13	7.3	20
4	poplar	17	13.4	1080	0.77	17.5	0.3	7.8	15.5	18	12	5.9	27
5	birch	18	17.9	1250	0.91	17.8	0.0	8.0	15.3	24	18	12	21
6	birch	17	22.8	1240	0.79	8.2	0.0	8.9	17.7	24	16	8.6	25
7	birch	13	9.8	1290	0.78	1.3	0.0	9.1	17.4	21	14	6.4	17
8	poplar	21	12.1	1340	0.60	0.0	7.9	3.3	21.4	15	10	5.9	44
9	poplar	15	9.5	1340	0.60	0.0	7.9	3.3	21.4	12	8	4.1	29
10	poplar	20	19.3	1290	0.58	21.0	0.0	0.0	0.1	20	14	6.8	33
11	poplar	19	15.3	1300	0.58	21.0	0.0	0.0	0.1	18	12	6.8	31
12	poplar	13	10.1	1060	0.82	16.0	0.0	8.2	15.6	15	10	4.5	18
13	poplar	14	34.9	1040	0.70	2.5	0.1	8.5	16.4	11	7	3.2	41
14	poplar	13	43.6	1090	0.70	9.4	0.1	8.8	16.9	14	10	5.9	27
15	poplar	18	44.8	980	0.70	16.0	0.2	8.0	15.7	22	15	8.2	37
16	birch	13	34.8	990	0.68	12.3	0.2	8.3	16.4	19	13	6.8	22
17	birch	17	34.1	950	0.70	17.3	0.2	7.9	15.5	18	13	6.8	37
18	birch	15	39.0	1250	0.89	16.9	0.0	8.1	15.5	30	16	10	23
19	birch	18	39.3	1290	0.82	8.4	0.0	8.9	17.1	26	17	8.6	33
20	birch	12	34.6	1290	0.76	0.0	0.7	9.2	18.9	32	20	8.6	13
21	poplar	17	43.7	1280	0.58	20.9	0.0	0.0	0.1	14	11	9.1	35
22	poplar	14	42.6	1290	0.54	0.0	0.0	9.7	18.6	14	13	8.2	24
23	poplar	12	36.9	1120	0.68	1.2	8.0	4.2	20.3	15	10	4.1	25
24	poplar	14	38.2	1330	0.98	16.0	0.0	8.2	15.6	21	14	7.7	22

relationship is shown in Fig. 3; the correlation coefficient R^2 is 0.985. These results are similar to those reported by Ragland et al. [8], with the highest moisture content wood having the thinnest char thickness. Samples beyond the ratio of c/d = 1.6 could not be obtained because the remaining material would fall apart during removal from the furnace. Based on this, it appears that beyond the ratio of c/d = 1.6 the pyrolysis of the particle is complete. At this limit, burnout of the log is more than 99% complete.

As the char layer reacts and the log shrinks in size, the moisture content of the remaining wood gradually increased when the initial moisture was relatively low (10% to 15%), but did not show significant increase for higher initial moisture (35% to 45%). Drying, pyrolysis, and char oxidation occur simultaneously until approximately the last 1% of the mass loss when only the char remains. As shown in Fig. 4, for a typical 15-cm log the maximum char layer thickness will vary from 1.2 cm to 1.9 cm depending on the as-received moisture content. Additionally, it can be seen that the char layer grows slowly at the start of the process, with the rate of the char layer growth increasing slowly but steadily.

Interestingly, the char layer thickness was not dependent on temperature or oxygen concentration. Temperature does not appear as a variable due in part to the limited

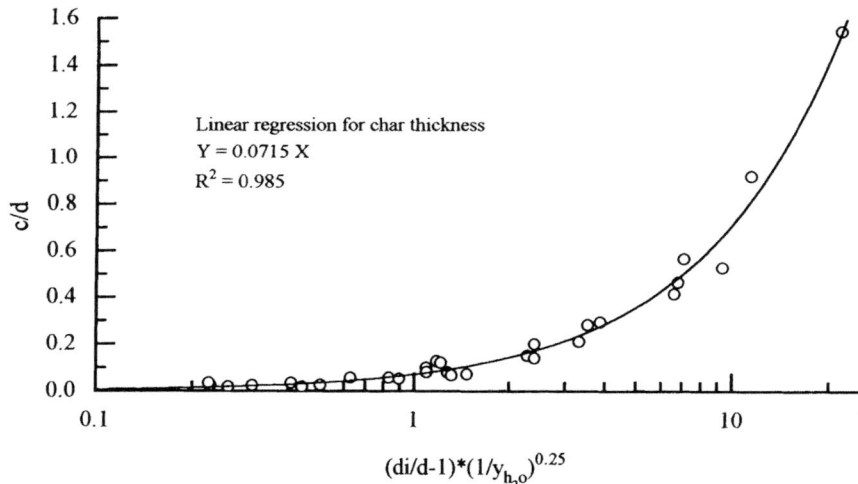

Fig. 3. Char layer growth in large wood segment combustion at furnace conditions. The X-axis has been expanded using log scale to provide the reader with a clearer understanding of the correlation through the entire range of the data.

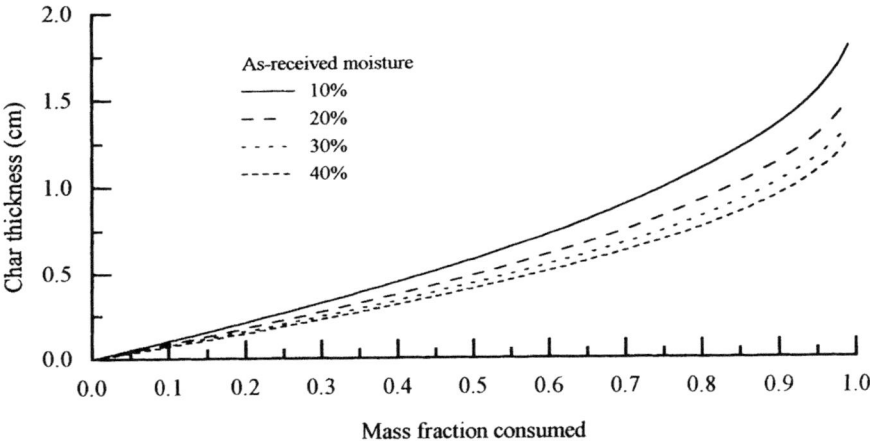

Fig. 4. Char layer growth as a function of mass consumption for a typical 15-cm poplar log.

range of temperatures over which the tests were performed (990°C to 1170°C). The upper temperature bound was set by concerns for personnel safety while manually removing the logs with a large pair of tongs from the furnace rather than the maximum working temperature of the furnace. Additionally, wood combustion at furnace-level temperatures is generally diffusion controlled, which minimizes the impact of temperature.

Although the inlet oxygen concentration was varied from less than 1% to 18% and had significant impact on the combustion rate of the log, oxygen concentration had little impact on the buildup of the char layer thickness as a function of the remaining undisturbed wood. The inlet carbon dioxide concentration, which was varied from 8% to 9%, and the inlet water vapor concentration, which was varied from 15% to 18%, also had no apparent effect on the char thickness.

4 Combustion Model of a Single Log in the Test Furnace

The model of combustion of a single log includes simultaneous heat and mass transfer, and chemical reaction of char with oxygen, carbon dioxide, and water vapor, and pyrolysis of the wood (Fig. 5). Conservation of energy for the solid is based on a surface energy balance at the char surface,

$$A_p h_{conv}(T_s - T_\infty) + A_p F \sigma(T_s^4 - T_\infty^4) = H_1 r_1 + H_2 r_2 + H_3 r_3 + \dot{m}_r h_r - \dot{m}_p h_p \qquad (2)$$

where h_{conv} is the convective heat transfer coefficient, A_p is the char surface area, T_s is the surface temperature of the char, T_∞ is the furnace temperature, H_i is the heat of reaction for reaction r_i, h_r is the enthalpy of the reactants, h_p is the enthalpy of the

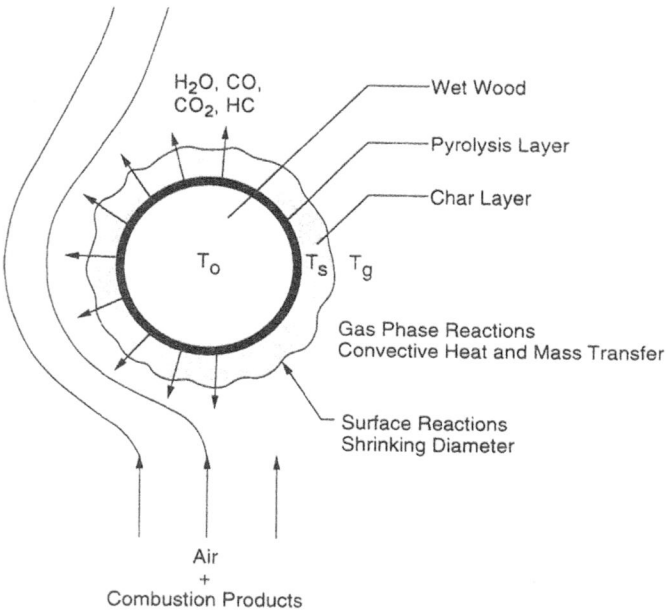

Fig. 5. Single reacting log.

Table 2. Oxidation and gasification kinetic rate constants used in the model.

$$k_{c(i)} = ATe^{-E/RT}$$

#	Reaction	A (cm/s K)	E/R (K)	Ref.
1	$C + 0.5\,O_2 \rightarrow CO$	86.8	9,000	12
2	$C + CO_2 \rightarrow 2\,CO$	342	15,600	11
3	$C + H_2O \rightarrow H_2 + CO$	1.67×342	15,600	13

products, $\dot{m}_p$ and $\dot{m}_r$ are the mass flow rate of the products and reactants, respectively, σ is the Stefan-Boltzman constant, and F is the view factor of the char to the combustion chamber. The reactions are given in Table 2. The convective heat transfer coefficient is obtained from a correlation for nonreacting cylinders in a high turbulence cross flow [10] with a screening factor to account for blowing and mass transfer at the surface.

The reaction rate for the pyrolysis reaction is

$$r_{pyr} = r_c \delta_{pyr} \tag{3}$$

where δ_{pyr} is the ratio of wood pyrolized to char consumed and is determined from Eq. 1. The reaction rate for char, r_c, is determined by

$$r_c = -r_1 - r_2 - r_3 \tag{4}$$

These reactions depend on diffusion from within the wood and diffusion from the gas stream to the char surface, as well as heterogeneous kinetics.

For a global reaction rate of order n with respect to gaseous component i, the burning rate is given by

$$r_i = \frac{dm_{c(i)}}{dt} = s_i \left(\frac{M_c}{M_i}\right) A_p k_{c(i)} \left(\rho_{i(s)}\right)^n \tag{5}$$

where s_i is the stoichiometric ratio of moles of product per mole of reactant i, i represents CO_2, O_2, and H_2O, M_i is the molecular weight of reactant i, $k_{c(i)}$ is the kinetic rate constant for the char reaction with gaseous species i, $\rho_{i(s)}$ is the reactant partial density of species i at the surface of the particle, and n is the order of the reaction. The reactant concentration at the log surface is not known; however, it can be eliminated by performing a mass balance on the reactant, equating the reactant consumed by reaction with the char with the reactant flow through the surface and the diffusion of the reactant through the boundary layer. Since a simplification occurs when the reaction order is 1, only this case is considered. Then,

$$\left(\rho_{i(\infty)} - \rho_{i(s)}\right)h_{d(i)} + \rho_{i(pyr)}h_{pyr} - \rho_{i(s)}h_{pyr} = k_{c(i)}\rho_{i(s)} \tag{6}$$

where $h_{d(i)}$ is the convective mass-transfer coefficient, $\rho_{i(pyr)}$, $\rho_{i(\infty)}$, and $\rho_{i(s)}$ are the density of reactant i in the pyrolysis products approaching the char surface from the pyrolysis region, in the freestream, and at the char surface, respectively, and h_{pyr} is the rate at which the pyrolysis products are presented to the surface of the char. The assumed pyrolysis product yield is given in Table 3. The increase in h_{pyr} leaving the char surface due to the gasification of the char is neglected. The convective mass-transfer coefficient is obtained from analogy to the heat transfer coefficient. Simplifying,

$$\rho_{i(s)} = \frac{\rho_{i(\infty)} h_{d(i)} + \rho_{i(pyr)} h_{pyr}}{h_{d(i)} + k_{c(i)} + h_{pyr}} \tag{7}$$

Hence Eq. 5 becomes

$$r_i = s_i \left(\frac{M_c}{M_i} \right) A_p k_{e(i)} \rho_{i(\infty)} \tag{8}$$

where

$$k_{e(i)} = \frac{k_{c(i)} \left(h_{d(i)} + h_{pyr} \dfrac{\rho_{i(pyr)}}{\rho_{i(\infty)}} \right)}{h_{d(i)} + k_{c(i)} + h_{pyr}} \tag{9}$$

and

$$h_{pyr} = \frac{r_{pyr}}{A_p \rho_{pyr}} \tag{10}$$

where ρ_{pyr} is the density of the pyrolysis products as they reach the char surface.

These equations are integrated with respect to time using a differential-algebraic implicit solver to provide mass loss versus time. The kinetic rate parameters used are given in Table 2. As noted by Brewster et al. [11], kinetic rate parameters for large particle coal oxidation and reduction are scarce to nonexistent. This is even more problematic for wood combustion and gasification. The oxidation kinetic rate parameters are based on wood charcoal at a stagnation point without the splitting and checking typical of large particle combustion [12]. The kinetic rate parameters of Baxter [11] for carbon dioxide reduction of small particle lignite are used. The molar reaction rate of water vapor with char is assumed to be 1.67 times the molar reaction rate of carbon dioxide [13].

Table 3. Distribution of pyrolysis products for dry wood under combustion conditions.

	Poplar	Birch
Mass fraction of char [measured]	0.166	0.150
Mass fraction of volatiles	0.834	0.850
	1.000	1.000
Composition of volatiles [14]		
Mass fraction of H_2O	0.312	
Mass fraction of CO_2	0.144	
Mass fraction of CO	0.229	
Mass fraction of H_2	0.006	
Mass fraction of hydrocarbons	0.309	
	1.000	

5 Discussion of Results

The single log model was run for each of the test cases in Table 1. As shown in Fig. 6 the majority of the calculated mass consumption rates fall within 30% of the measured mass consumption rates. The measured values for 60% to 90% mass are generally low compared to the model results. This is due to nonuniform flow towards the end of the run caused by the inlet jets, resulting in lower combustion rates. The accuracy of these results is believed to be primarily due to three issues. The first is the large variation in the rate parameters that are available and the lack of kinetic rate parameters directly applicable to the case of large biomass particle combustion. The other two issues are items not directly addressed in the model. The first is the impact of cracking and

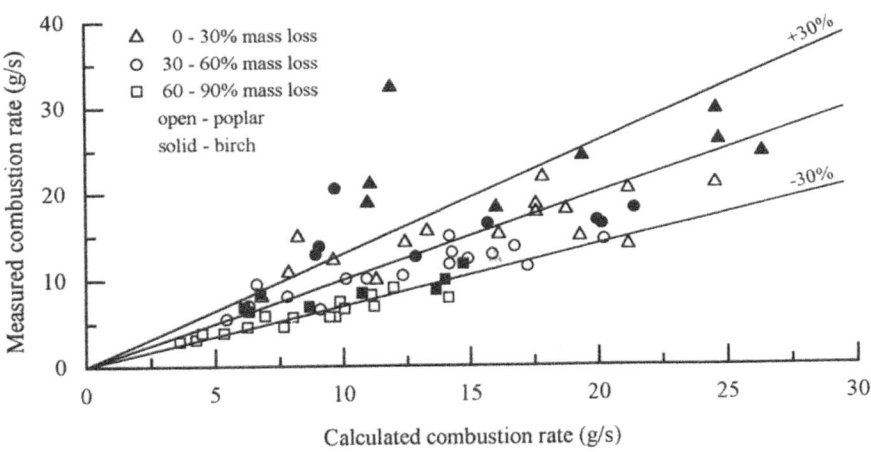

Fig. 6. Comparison of calculated and measured combustion rates.

checking of the char on heat and mass transfer to and from the surface of the wood and the use of apparent area to determine the kinetic reaction rate. Because the external surface area of the partially combusted wood showed considerable cracking and checking and because the carbon dioxide and water vapor are in intimate contact with this surface as they pass through the char layer, the area available for the reduction reactions (e.g., reactions 2 and 3) was increased. The optimum correlation was found when the area was increased by a factor of 8 for reduction reactions. The area for diffusion was not increased. The second additional factor is the impact of an attached or unattached flame zone on the surface reactions. The flame zone consumes the oxygen available for surface reactions and increases heat transfer to the char surface. The differences in fuel properties caused by the changes in moisture content were included directly as discussed in Section 4. The differences between poplar and birch are included via differing densities and char yields as shown in Table 3. The differences in char and density are small and have little impact on the results. Higher moisture content impacts the combustion of the log in two ways, increasing the length of time to consume the log. As the moisture content rises the thermal wave within the particle is slowed and the char layer is thinned. Additionally, the higher moisture content results in more shielding of the particle from convective heat and mass transfer.

Fig. 7 compares the measured normalized mass versus time with the results from the model for three typical logs with combustion times ranging from 13 min to 55 min. The model results show the same curve shape and trends as the measured data and thus give confidence to the results of Fig. 6. Fig. 8 shows the results of many model calculations for the variation of the average combustion rate as a function of inlet oxygen concentration and as-received moisture content for a 15 cm diameter by 1.5 m long poplar log. With rising oxygen levels the combustion rate rises, but not quite linearly, due to the impact of the reduction reactions. For higher moisture levels the mass consumption rate increases because of the greater mass of the logs with higher moisture content. However, it should be noted that the total time to consume the log increases. This occurs because the higher moisture levels provide more screening of the oxygen diffusion to the surface of the char. Also, the additional energy required to heat and vaporize the water lowers the surface temperature of the char, reducing the reaction rate and slowing the penetration of the thermal wave.

6 Summary

Combustion of a large single biomass particle (yellow poplar and paper birch logs) was investigated in a convective furnace. The time to consume 90% of the mass in a test furnace was found to vary from 13 min to 55 min depending on the as-received moisture content, the log diameter, the inlet gas mixture, and furnace temperature. The growth of the char layer was investigated and it was found that significant char consumption occurred prior to complete pyrolysis of the wood. A simple relationship for the growth of the char as a function of the undisturbed wood core was found for the range of the variables tested. A numerical model of a single log was constructed for the range of values investigated, and most calculated values were found to agree within 30% of the measured values.

	Wood			Combustion conditions				
Case	Type	Moisture (%)	Dia. (cm)	Temp. (°C)	Velocity (m/s)	O_2 (%)	CO_2 (%)	H_2O (%)
1	poplar	43.6	13	1090	0.70	9.4	8.8	16.9
2	birch	34.1	17	950	0.70	17.3	7.9	15.5
3	poplar	13.4	17	980	0.70	2.1	8.8	16.8

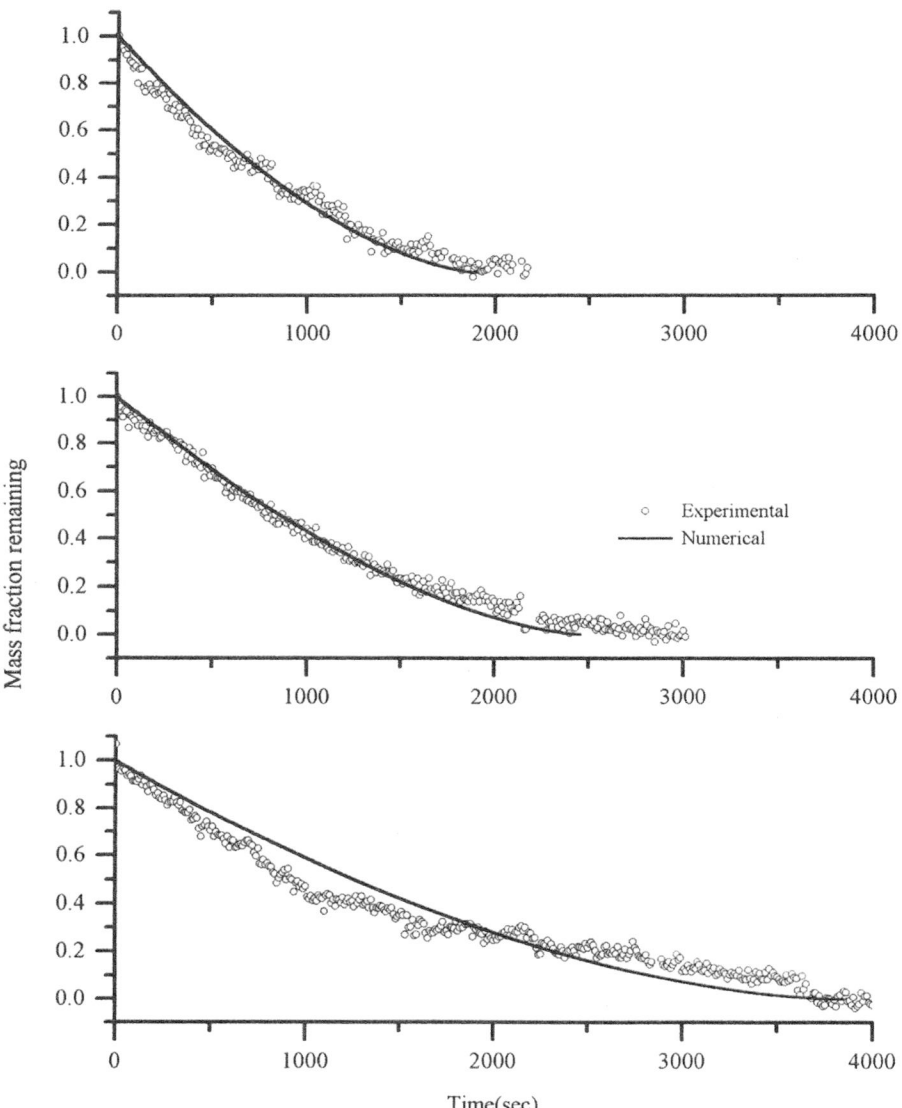

Fig. 7. Comparison of numerical and experimental results for a single log.

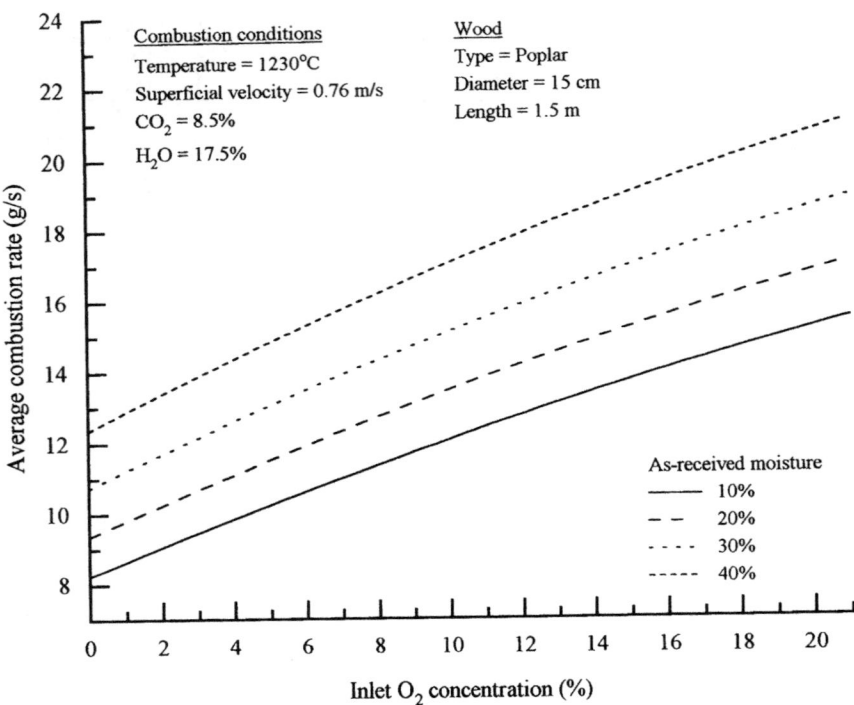

Fig. 8. Average combustion rate from 0% to 90% of a 15-cm poplar log as a function of inlet oxygen concentration and as-received moisture concentration.

The combustion of thermally thick, charring biomass material is significantly different from both thermally thin and non-charring materials. Coupling between the pyrolysis, char, and gas phase reactions add significant complication to the analysis. Important considerations in the wood combustion process include the gas phase reactions and flame zone, the screening of the heat and mass transfer to and from the surface, the reactions of the products of pyrolysis and drying of the wood core with the char surface, and the cracking, checking, and surface roughness. The analysis presented accounts for the carbon-dioxide and water vapor reaching the hot char surface internally from the pyrolysis zone and drying of the inner wood surfaces as well as externally from the combustion products.

Further research into the coupling of the gas phase, char, and pyrolysis reactions; the impact of cracking and checking of the wood; and the effect of outgassing on the surface heat and mass transfer is needed so that a complete picture of wood combustion can be drawn. Additionally, there is a need for basic kinetic rate information of large particle biomass combustion.

7 Acknowledgments

This work was funded by the Electric Power Research Institute under contract RP3407-07; Evan Hughes was the project manager. Dave Ostlie at Energy Performance Systems Co. designed and delivered the convective furnace and motivated the study. Andrew Baker at the U.S. Forest Products Laboratory provided the logs and helped with the test design. Doug Hall and Earl Geske from the U.S. Forest Products Laboratory assisted with the test setup and instrumentation. Ryan Roloff, a student at UW-Madison, helped with the testing and data reduction.

8 References

1. Arola, R. A. et al. (1983) Chunkwood production: a new concept. *Forest Product Journal*, Vol. 33, No. 7/8. pp. 43-51.
2. Lamarre, L. (1994) Electricity from biomass. *EPRI Journal*, Jan/Feb. pp. 17-24.
3. Jannsens, M. (1991) Piloted ignition of wood: a review. *Fire and Materials*, Vol. 15. pp. 151-167.
4. Di Blasi, C. (1994) Processes of flames spreading over the surface of charring fuels: effects of the solid thickness. *Combustion and Flame*, Vol. 97. 225-239.
5. Di Blasi, C. (1995) Predictions of wind-opposed flame spread rates and energy feedback analysis for charring solids in a microgravity environment. *Combustion and Flame*, Vol. 100. pp. 332-340.
6. Antal, M. J. and Varhegyi, G. (1995) Cellulose pyrolysis kinetics: the current state of knowledge. *Industrial and Engineering Chemical Research* Vol. 34. pp. 703-717.
7. Ragland, K. W., Boerger, J. C. and Baker, A. J. (1988) A model of chunkwood combustion. *Forest Products Journal*, Vol. 38, No. 2. pp. 27-32 .
8. Albini, F. A. and Reinhardt, E. D. (1995) Modeling ignition and burning rate of large woody natural fuels. *Int. J. Wildland Fires*, Vol. 5, No. 2, pp. 81-91.
9. Bryden, K. M. and Ragland, K. W. (1996) Numerical modeling of a deep, fixed bed combustor. Accepted for publication in *Energy and Fuels*.
10. Zukauskas, A. and Ziugzda, J. (1985) *Heat Transfer of a Cylinder in Crossflow*, Hemisphere Publishing Corporation, Washington.
11. Brewster, B. S. et al. (1993) Comprehensive modeling. Chapter 8 in *Coal Science and Technology, vol. 20, Fundamentals of Coal Combustion for Clean and Efficient Use*, edited by L. D. Smoot, Elsevier Publishing Corp., Amsterdam.
12. Evans, D. H. and Emmons, H. W. (1977) Combustion of wood charcoal. *Fire Research*, Vol. 1. pp. 57-66.
13. Yoon, H., J. Wei, and Denn, M. M. (1978) A model for moving-bed coal gasification reactors. *AIChe Journal*, Vol. 24. No. 5. pp. 885-903.
14. Adams, T. N. (1980) A simple model for predicting particulate emissions from a wood-waste boiler. *Combustion and Flame*, Vol. 39. pp. 225-239.

CO-COMBUSTION OF FORESTRY BIOMASS WITH TYRES IN A CIRCULATING FLUIDISED BED COMBUSTOR

I. GULYURTLU, E. PENHA and I. CABRITA
Departamento de Tecnologias de Combustão
ITE - INETI, Lisboa, PORTUGAL

Abstract:

Combustion studies of mixtures of forestry biomass and used tyres were undertaken in an atmospheric circulating fluidised bed. The combustor was operated over a temperature range of 700 to 1000 °C.

The main parameters that were investigated are i) where and how to feed both forestry biomass and used tyres to the combustor, ii) the ratios of amounts of forestry waste/used tyres, iii) air staging, and iv) excess air levels along the riser.

Whole waste tyres are usually incinerated in cement kilns. There are boilers designed to burn whole tyres but the experience has not been successful because the time for combustion is not long enough, the temperature is relatively lower for complete combustion and the distribution of air has been proven to be insufficient. The work of INETI investigates the combustion of waste tyres which are reduced in size prior to their burning in a fluidised bed. This type of combustion studies with waste tyres is new and as results obtained by far are encouraging this option could be an alternative to the eventual utilisation of waste tyres as a boiler fuel.

Introduction

The combustion of coals of different origin has been studied over many years in fluidised beds which has led to the compilation of valuable data about the behaviour of these fuels. In recent years, however, for several environmental reasons the combined use of coal with other solid fuels like biomass and other wastes is gaining particular interest. These new applications require further research studies to understand better the behaviour of these fuels when burned together to verify: i) the interaction between ashes of different fuels, ii) the control of emissions of NO_x, SO_2, and unburned hydrocarbons, iii) the mixture ratios of these fuels and iv) feeding techniques used for the fuels in question. In certain cases, when coal is not available but there is an abundance of biomass waste, these residues could become a prime fuel in co-combustion systems involving various types of solid wastes from indutrial processes.

Because of limitations mainly due to the high collection and transport costs, the amount of different types of biomass that could be used for any energy process will always be limited and from logistic point of view, their combined use with other residues could be beneficial, particularly for environmental reasons as biomass contains none or very small amounts of elements like nitrogen or sulphur thus leading to the dilution of the emission levels of pollutants such as NO_x and SO_2. Their use with low quality fuels like industrial wastes could especially be profitable because the high volatile nature of biomass could combine ideally with such wastes.

The main problem associated with using biomass as a fuel lies with difficulties involved in its preparation which includes size reduction, handling and feeding to the combustor. The highly irregular shapes of biomass particles turn difficult to select systems that could adequately handle them to be supplied to any type of combustor troublefree. Most of industrial wastes also present feeding problems and require some preparation prior to their combustion.

Elimination of waste tyres is a major problem as their quantities have grown almost exponentially in the last ten years. There are several options that are under study. Combustion of waste tyres has been tried, particularly in cement kilns in which the whole tyre is thrown into the kiln for burning (1 & 2). The results have been quite satisfactory, however, arrangements have to be made to feed whole tyres into the kiln. The use of whole waste tyres in other applications like boilers has, however, not been so successful because the time needed for complete combustion is extremely high and necessary mixing conditions between the fuel and air are not fulfilled. It is preferrable in such applications, to cut down the size of tyres to much smaller pieces which could easily be transported to the combustion chamber. The recent developments in equipment for shredding decrease the cost of size reduction and enables sizes with the dimension of about 50 mm x 50 mm to be prepared for the supply for burning. The controlled combustion temperature prevents the fusion of the metal lining which can be recovered for recycling. Combustion tests with waste tyres chips of 30 mm x 30 mm, already undertaken at INETI have confirmed these observations.

Studies at the above-mentioned Department aim at using co-combustion of waste tyres with biomass with the objective of eliminating them in a clean manner whilst taking advantage of their high calorific value which could positively supplement fairly low heating value of biomass fuels. The main parameters that have been studied by far are i) where and how to feed both forestry biomass and used tyres to the combustor, ii) the ratios of amounts of forestry waste/used tyres, iii) air staging, and iv) excess air levels along the riser.

Experimental

The fluidised bed combuster is square in cross sectionwith each side 300 mm long and its height is 5 000 mm. The combustor is built of refractory steel and is well insulated on the exterior with high temperature-resistant ceramic fibres. The same entry was used for feeding both biomass and waste tyres which were mixed prior to their combustion. They were supplied under gravity just to the top of the bed. The system was calibrated prior to combustion tests. The fuel feed rate is related to the air requirement for several excess air levels and the fuel/air ratio is controlled through a software developed by INETI. The total air needed is divided between that for fluidising and the rest being for the freeboard as secondary air and the INETI software regulates their amounts which are pre-fixed. The fuel feed rate is also correlated with the water flow rates passing through cooling coils placed both in the bed region and in the freeboard zone for controlling the temperature. The fluidising air enters the bed through a inverted cone type distributor plate. There are ports in the freeboard for the introduction of secondary air in stages. The temperatures in the bed and along the freeboard and that of the flue gases leaving the reactor were continuously monitored. There are gas sampling probes situated at various heights of the combustor to carry out instant measurement of CO, CO_2, N_2O, NO_x, and O_2. The combustion gases leaving the combustor are let go through a a high-efficiency cyclone to capture the particulates elutriated out of the combustor. The bed is heated up to the combustion temperature using propane as a start-up auxiliary fuel. The operating conditions are given in Table 1.

Table 1 - Operating conditions of the pilot-scale fluidised bed combustor

Bed temperature (oC)	750 - 1000
Fuel feed rate (kg/h)	5.0 - 8.0
Gas velocity (m/s)	1.0 - 4.5
Excess air levels (%)	80 -180
Particle size range (mm)	2.0 - 4.0
Bed height (m)	0.2 - 0.4
Average sand particle size used (mm)	0.5

The fuels used were the following: Eucalyptus and waste tyre. The fuel characteristics are given in Table 2:

Table 2 - The ultimate analysis of the fuels used

	EUCALYPUS	WASTE TYRE
Moisture (% by weight)	4.5	0.83
Ash (% by weight)	1.37	0.05
Carbon (% by weight)	45.2	82.1
Hydrogen (% by weight)	5.5	6.93
Nitrogen (% by weight)	0.06	0.66
Sulphur (% by weight)	0.00	1.54
Oxygen (% by weight)	43.37	7.6
Calorific value (kJ/kg)	17 600	34 400

Results and discussion

The results obtained are summarised in Tables 3 and 4. They are given for different excess air levels and for two different mixtures of biomass with waste tyre based on weight. The fuel feed rate used in all experimental tests was 7.5 kg/h. The feeding of waste tyre particles of about 4 mm offered the greatest challenge to attaining stable operating conditions. The elastic nature of these particles was found to be highly difficult material to handle and to transport. When mixed with biomass, the feeding was observed to be relatively easier and this technique was as a result employed for the tests carried out in this work. Some elutriation of very fine particles was observed during the experimental runs, however, this was not so serious to cause significant reduction in combustion efficiency.

The principal reasons for any decrease in the combustion efficiency were observed to be unburned hydrocarbons and CO. With the addition of biomass, the amount of volatiles liberated increased and most of which appeared to bypass the bed to burn in the riser. However, there first were some difficulties in achieving complete combustion of these volatiles in the freeboard when the mixing of the combustion air was inadequately carried out. In fact, a plug flow behaviour of gas in the freeboard was observed which at times led to unacceptably high emissions of CO resulting from the poor mixing. The amount of air that had to be supplied to the riser needed to be increased to reduce the levels of CO and unburned hydrocarbons. In addition, it was found necessary to stage this air to improve its mixing with volatiles. In this way it was possible to bring down the hydrocarbon emissions, however, in some cases CO concentrations still continued to be high. This could be due to sluggish nature of CO over the temperature range used in the riser as given in Table 4. The combustion efficiency is very much dependent on the temperature as demonstrated in Figure 1 for the reasons stated above.

Table 3 - The summary of the results of co-combustion of biomass with waste tyre

Bed temp (°C) (Bed/ Free.)	Fuel	Excess air (%)	Prim air/ Sec air	CO (ppm)	CO2 (% vol)	SO2 (ppm)	NOx (ppm)	Unburned HCs (ppm)	Unburned fuel particles (% by wt of input)
700/500	Biomass with 30% of waste tyre by wt	108	1.0/ 1.3	700	9.2	300	160	28	2.1
720/500	Biomass with 30% of waste tyre by wt	106	1.0/ 1.5	5000	11.0	400	180	40	2.5
790/570	Biomass with 30% of waste tyre by wt	115	1.0/ 1.6	2180	13.0	370	195	20	0.5
810/550	Biomass with 30% of waste tyre by wt	110	1.0/ 1.7	3000	9.0	350	160	26	1.8
860/630	Biomass with 30% of waste tyre by wt	70	1.0/ 1.2	1590	13.0	320	160	15	0.8

Increased amounts of biomass in the original mixture, as stated above, was found to influence the degree of burning in the riser due to the combustion of volatiles released primarily from biomass particles as shown in Figure 2. The temperature in the zone immeadiately above the bed increased when only biomass was used. Differences up to 50 °C were measured.

Table 4 - The summary of the results of co-combustion of biomass with waste tyre

Bed temp (°C) (Bed/ Free.)	Fuel	Excess air (%)	Prim air/ Sec air	CO (ppm)	CO2 (% vol)	SO2 (ppm)	NOx (ppm)	Unburned HCs (ppm)	Unburned fuel particles (% by wt of input)
700/500	Biomass with 30% of waste tyre by wt	140	1.0/ 1.6	7000	11.0	550	220	76	1.9
720/580	Biomass with 30% of waste tyre by wt	125	1.0/ 1.3	2500	10.0	550	200	70	1.9
730/580	Biomass with 30% of waste tyre by wt	110	1.0/ 1.2	3500	9.0	600	230	80	1.8
740/670	Biomass with 30% of waste tyre by wt	105	1.0/ 1.1	950	13.0	500	250	68	1.5
760/470	Biomass with 30% of waste tyre by wt	160	1.0/ 1.4	2590	11.0	470	185	76	1.1
800/550	Biomass with 30% of waste tyre by wt	115	1.0/ 1.3	4000	12.0	500	190	70	2.5

Figure 4 compares the emissions of NO_x for different cases which were: i) biomass mixed with 30% by weight of waste tyres ii) biomass mixed with 50% by weight of waste tyres, iii) biomass combustion only and iv) waste tyre combustion only. Due to the higher amount of fuel-N present in waste tyres, the amount of NO_x released increased. However, it is clear from the results in Fig. 3 that the effectiveness of the use of biomass mixed with waste tyres in achieving substantial reduction in the amounts of oxides of nitrogen. The level of NO_x released lowered with the increase in the amount of biomass in the fuel feed and this decrease was more than the reduction in the quantity of fuel-N entering the combustion system. This could be due to the fact that the biomass gave rise to conditions leading to a decrease in the amount of NO_x formed by influencing the reaction paths, particularly by modifying the nature of percursors responsible for the formation of both No_x. In fact, it was observed that most of the fuel-N in the biomass was released as NH_3 rather than HCN which is more dominantly responsible for the NO_x formation.

Experimental studies suggested that what was more important was the manner of introducing air to the combustion zone and not the quantity of excess air. It was preferrable to introduce the combustion air in stages so that the volatiles released from eucalyptus residues could mix well with the air to ensure complete combustion. The ports of secondary air were located at heights of 700, 1 200 and 1 800 mm above the distributor plate. Separate studies showed that if the total secondary air, regardless of its amount, was supplied only to the port 700 mm above the distributor plate, the effect of air staging was insignificant which suggested that at the top of the bed in the splash zone there was not an efficient mixing of air with volatiles. The advantages of both the use of biomass and the split of the air as primary and secondary between the bed and the riser could be combined to reduce the amounts of NO_x formed during combustion as shown in Fig. 4. If the stoichiometric conditions were maintained in the bed, and by increasing the amount of the air supplied to the riser the formation of NO_x was quite successfully suppressed in the bed section where most of the waste tyre was burned.

Conclusions

The combustion efficiency appeared to be unaffected by the addition of waste tyre to biomass.

With increased amount of waste tyre added to biomass, combustion in the freeboard became more significant, thus resulting in higher temperatures.

Temperature appears to be still very influential i) on the amount of NO_x formed, ii) in reducing the levels of hydrocarbons.

The air staging could produce even lower emissions of NO_x.

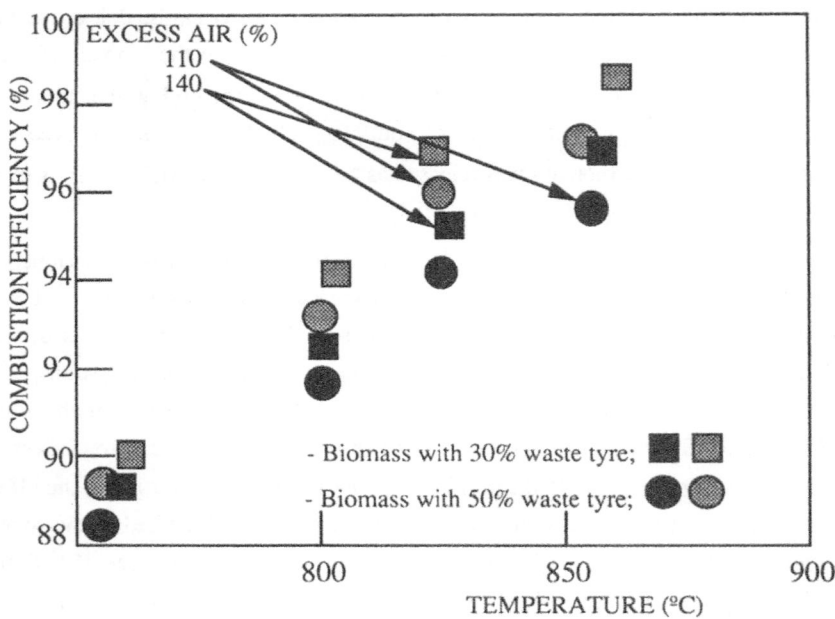

Figure 1 - The influence of the temperature on the combustion efficiency for different amounts of waste tyre mixed with biomass

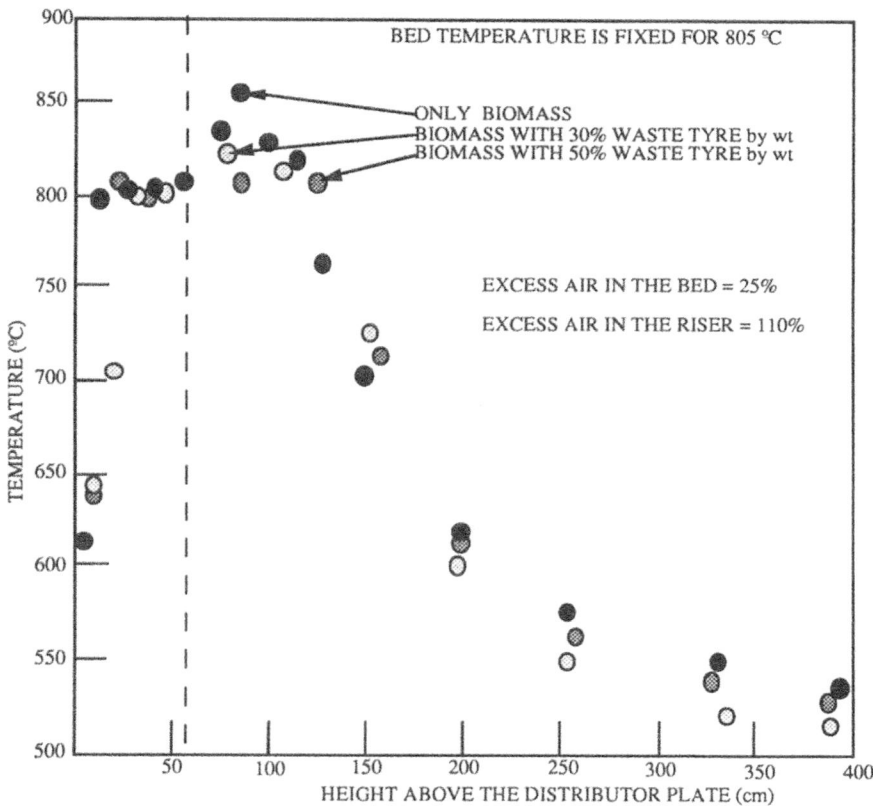

Figure 2 - The variation in the temperature along the combustor height

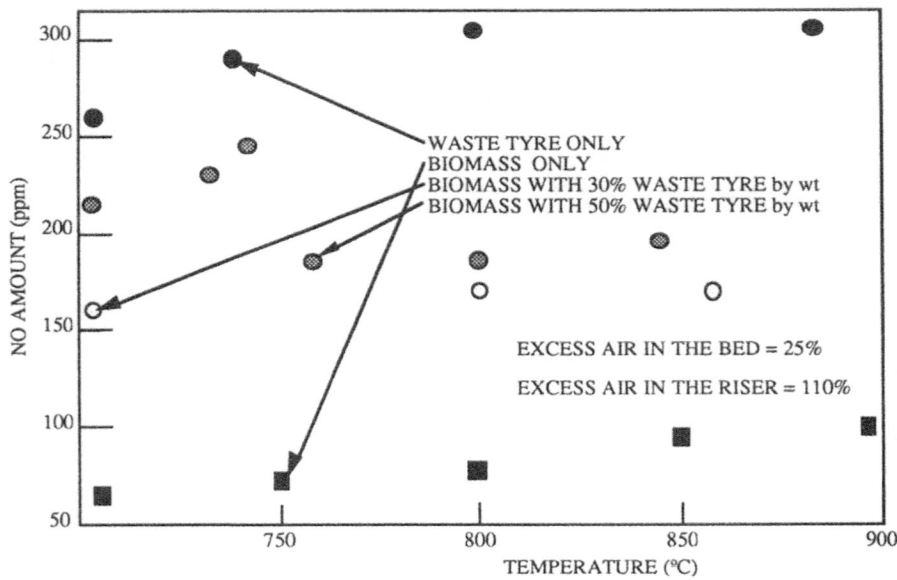

Figure 3 - The influence of biomass/waste tyre ratio on the amount of NO

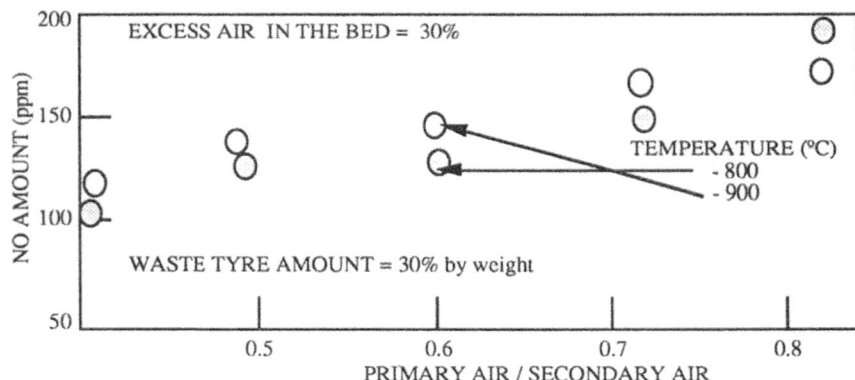

Figure 4 - The effect of air staging on the levels of NO formed

REFERENCES

1)   Dambrine, E., (1994) "Wastes burning and thermal destruction of organic compounds in cement kiln - Ciments Français trials", International Flame Days, Biarritz, March.

2)   Dupont-Wavrin, X., (1994) "Waste tyres used as supplement fuel in cement industry", International Flame Days, Biarritz, March.

3)   Kim, I., (1994) "Incinerators & Cement kilns", Chemeical engineering, p.41-45, April.

CIRCULATING FLUIDIZED BED COMBUSTION OF BROWN COAL DURING MIXING UP BIOMASS

W. NEIDEL; M. GOHLA; R. BORGHARDT; H. REIMER; M. KLEINDIENST
Otto-von-Guericke-University of Magdeburg
Institute of Apparatus Design and Environmental Technology
P.O.Box 4120, D-39016 Magdeburg, Germany

Abstract

Especially for large CFBC units it is possible to employ only the co-firing of biomass because of logistic problems. So it is interesting to get knowledge about burnout and emission behavior as well as best working parameters to use biomass as co-combustion fuel in already existing units. Therefore the co-combustion of coal/biomass were investigated in bubbling and circulating fluidized beds.

The experimental investigations has been realized by two test fluidized bed units,
- bubbling fluidized bed (diameter 100 mm, height 4000 mm) and
- circulating fluidized bed (diameter 100 mm, height 8000 mm)

at the combustion laboratory of the University of Magdeburg.

The kinds of biomass
- wood,
- straw and a
- fast growing species (china reed)

are mixed with brown coal.

The objects of the theoretical and experimental investigations are
- the determination of the burnout characteristics and of the kinetic datas of the fuel by special batch-tests in the bubbling fluidized bed and
- the influence of the biomass mixing rate and of the process parameters on the combustion and emission behavior in the CFBC.

The results of the burnout investigations shows that the burning time of a raw brown coal sample with a particle range of 0,35-0,5mm is comparable to a china reed or straw sample with a particle range of <3mm. So it is possible to modify the particle range of coal to get same burnout times. It is also possible to modify the burnout time of the biomass with special fuel preparation like compressing.

The NO$_x$ emissions don't increase with co-combustion of biomass during brown coal firing. The SO$_2$ emissions are lower with co-combustion of biomass during brown coal firing. The lower sulfur content of the biofuel corresponds to the SO$_2$ emissions in the case of china reed co-firing. The SO$_2$ emission level decreases more than the percentage of biomass feeding in the case of straw co-firing. The reason for this effect may be the high calcium content of the straw in connection with the low fuel sulfur content.

Fluidized beds are qualified for co-combustion of biomass during brown coal firing. The results of the investigations can be used for the biomass co-combustion in industrial units.

Keywords:
Biomass, Burning behavior, Combustion, Fluidized bed, Flue gas emission, Kinetic datas

1 Characterization of burning behavior by batchtest

The modelling of the combustion process needs fuel characteristics of burning behavior as an input parameter. The fuels can be characterized by special burn out tests [3,4]. These tests, so called batch-tests, make it possible to obtain a fuel and technology characterized reaction rate constant directly in a fluidized bed. Different investigators have used similar tests to obtain specific reaction rate constants [1,2,5,6]. The apparatus for the batch tests is the fluidized bubbling bed unit (Fig.1) of the Magdeburg University.

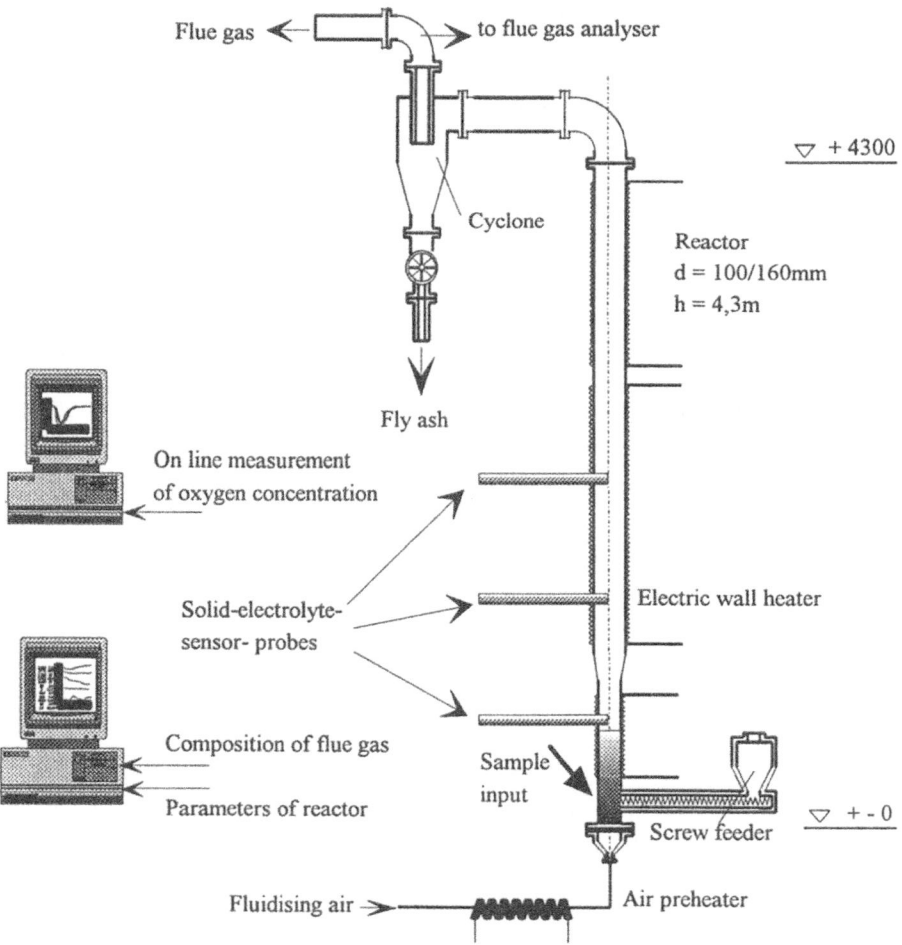

Fig.1:FBC experimental setup for determination of burn out behavior with solid electrolyte sensors probes (batch test)

Exactly prepared samples of the tested fuels (Table 1) are feeding rapidly into the bubbling bed by a two valve system. The measuring value is the oxygen concentration in the flue gas (Fig. 2). The oxygen concentration is measured with special sensors (solid electrolyte sensor probes) in the splash zone of the bubbling bed and in the freeboard. It is possible to follow the burn out process directly by a computer.

The combustion process can be described in tree parts: drying, combustion of volatile matter and combustion of fixed carbon. The combustion of volatile matter is characterized by very rapidly consumption of oxygen. On the other hand, the combustion of fixed carbon is relatively slow with a great dependence on fuel and particle diameter.

Table 1: Analysis of used fuels

	Brown Coal	China Reed	Wood	Straw
Volatiles (dry, ash free) [mass%]	57,7	81,3	83,5	80,1
Moisture (raw) [mass%]	52,7	9,0	43,3	11,2
Ash (raw) [mass%]	4,9	1,8	0,2	5,2
Lower Heating Value [kJ/Kg]	9140	15857	9298	14466
Carbon (dry, ash free) [mass%]	65,44	49,08	50,07	49,59
Hydrogen (dry, ash free) [mass%]	4,50	5,76	5,75	5,40
Oxygen (dry, ash free) [mass%]	29,35	44,76	44,18	44,40
Nitrogen (dry, ash free) [mass%]	0,48	0,31	0,00	0,46
Sulfur (dry, ash free) [mass%]	0,23	0,09	0,00	0,15
Ash Soften Temperature [°C]	1200	<1000	<1000	<815

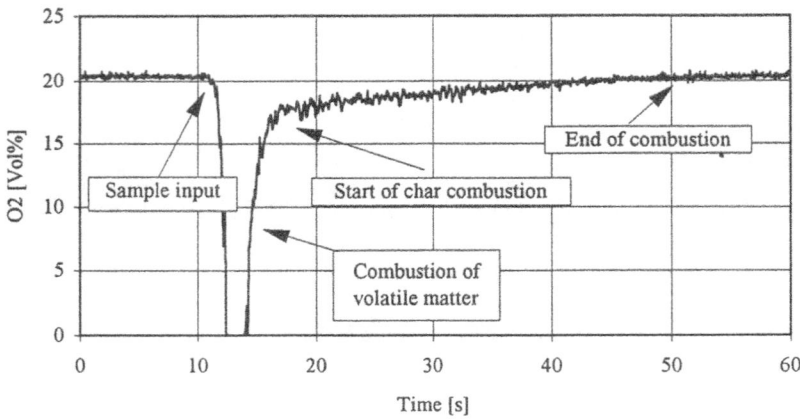

Fig. 2: Burn out behavior of a fuel sample in FCB (batch test)

Big differences in burn out behavior are shown in the following diagrams. Fig. 3 shows the influence of the fuel, especially the content of volatiles and char, on the burnout time and combustion behavior. All the biomasses have a volatile content of about 80% and the burnout times is nearly the same in case of biomass combustion. The burnout time of the brown coal is longer because of the higher char content.

Fig. 4 shows the burnout time of raw brown coal, china reed and ist mixtures. It is possible to modify the whole combustion time of a sample by various fuel mixtures.

The influence of particle size shows Fig. 5. The burning time of a raw brown coal sample with a particle range of 0,35-0,5mm is comparable to a china reed or straw sample with a particle range of <3mm. So it is possible to modify the particle range of coal to get same burnout times.

Longer burnout times are reachable with fuel compressing (Fig. 6). The combustion of volatiles is not obeserved directly at the beginning of the combustion. Small amounts of volatiles burnes over the whole time (see the number of the peaks) because of the compressing of the fuel.

Existing units, constructed for special residence times during coal combustion, can be used for biomasses, if the biomass is especially prepeared.

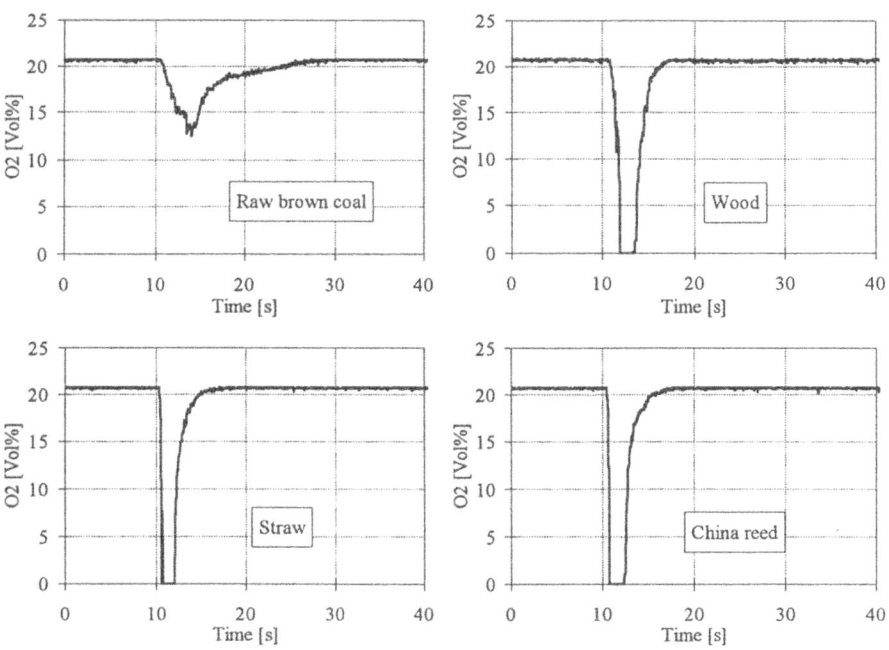

Fig. 3: Dependence of burn out behavior on fuel (raw brown coal, wood, straw, china reed; batch test in FBC; same particle size)

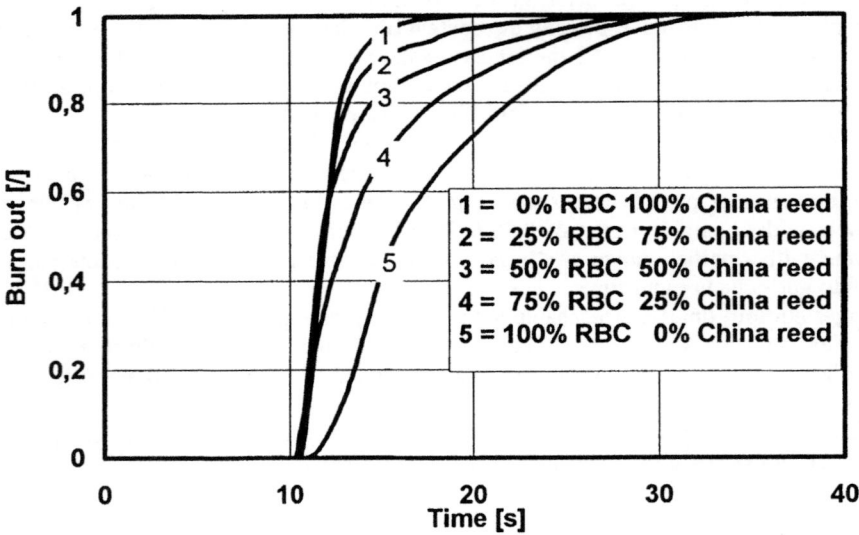

Fig. 4: Dependence of burnout behavior on mixed china reed / raw brown coal (RBC); bubbling bed; T_{bed}=800°C; u_0=1,9 m/s; carbon mass of the samples =1,0 g

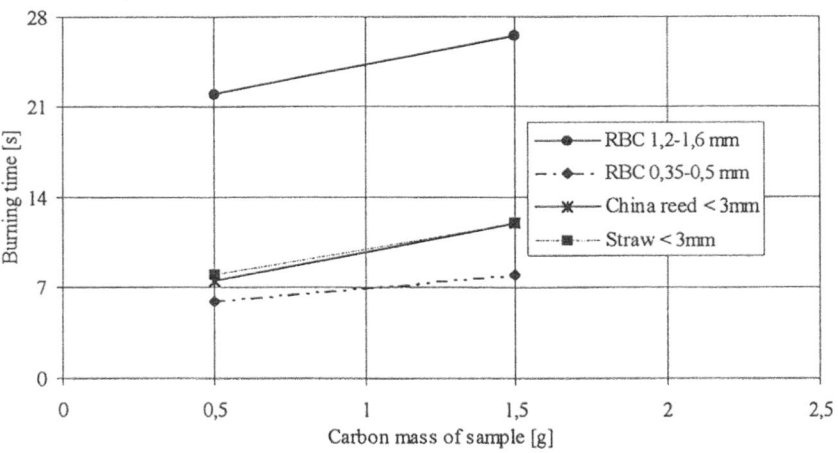

Fig. 5: Dependence of burnout behavior on different fuels (coal, biomass) with different particle sizes; batch test in bubbling bed

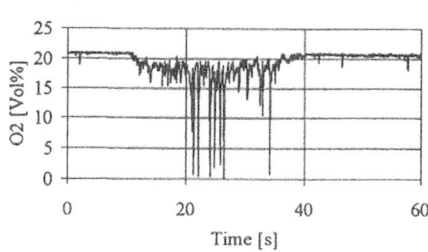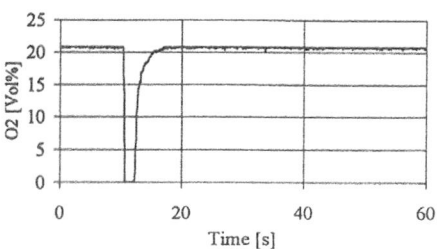

Fig. 6: Dependence of burn out behavior on biomass preparation (chaff/pellets); batch test in bubbling bed

2 Calculated kinetic datas by batch test results

The integral analysis of the calculated rate constants of devolatilization k_d and the rate constants for the char combustion $k_{c,Cfix}$ (Fig. 7) are shown in table 2.

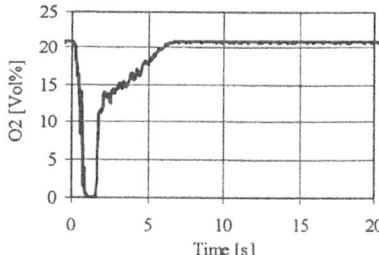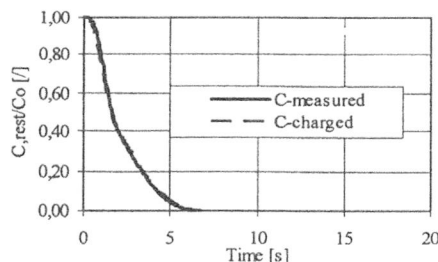

Fig. 7: Batch test function in comparison with measured and calculated burnout

Table 2: Rate constant of devolatilization and the rate constant for the char combustion

| Fuel | dp [mm] | k_d [g/m²s] | | | $k_{c,Cfix.1}$ [g^{0,5}/s] | | |
		k_d [/]	A [/]	E [J/mol]	$k_{int.}$ [/]	A [/]	E [J/mol]
Raw brown coal	0,42	5-7	5,81	(0)	0,3-0,4	0,856	8063
	0,71	8-10	31,10	12381	0,2-0,25	0,395	5066
	1,5	10	30,57	10318	0,09-0,12	0,134	2353
Wood	0,5	6-9	20,56	9686	0,29-0,35	0,15	0
	1,5	11-16	38,23	9559	0,2-0,25	0,16	0
	7	28-35	47,97	(3954)	0,08-0,14	0,01	0

The rate constants of devolitilization k_d is related to the surface of the particle. That´s why there is a low dependence on the mass of the probe and the particle diameter.

The velocity of the combustion of volatiles is prescribed by the velocity of devolitilization and is independend of the existing partial pressure of oxygen. The velocity of the combustion of volatiles is depending on the reactivity of the fuels and additional on the resistance of transport (boundary layer suspension-particle, surface boundary layer suspension-bubble), that means the velocity of oxygen supply is an important fact. The first fact mentioned could be important for bigger particles and the second for smaller particles with a fast devolitilization. In comparisan with literature datas there is a good agreement for the level of the activation energy (E).

The constant for the char combustion $k_{c,Cfix}$ is not suitable for afterwards modelling because there are strong dependences on the used fuel masses and on the diameter of particles. But it's really suitable to announce the char combustion velocity in the dependence of the temperature. If the reaction rate constant is known it is possible to compare directly probes of fuel with the same mass of carbon.

3 Influence of process parameters and mix ratios on flue gas emission

The influence of process parameters and the mix ratio of coal and biomass on the emission level were investigated in the circulating fluidized bed reactor (Fig. 8) with a thermal power of 60kW.

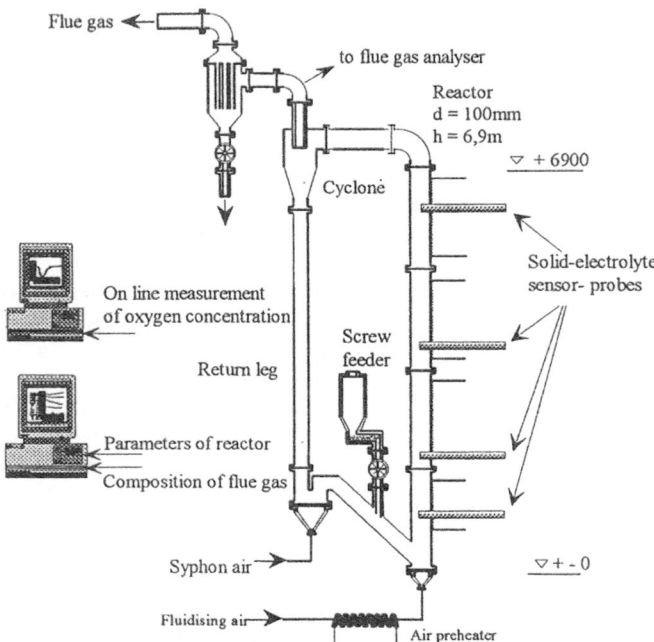

Fig. 8: CFBC experimental setup

Fig. 9 shows SO₂ emission level with increasing mass part of china reed. The mass part of calcium of the used china reed is 0,148% (water ash free). In contrast to china reed the mass part of calcium of the used straw is 0,364% (water ash free). So Fig. 10 shows the SO₂ emission level with increasing mass part of straw. It was determined a desulpherisation effect increasing straw part.

Fig. 11 and 12 shows the NO_x emission level in dependence of mix coal/biomass. There is no increase of NO_x emission during increasing of biomass fuel content.

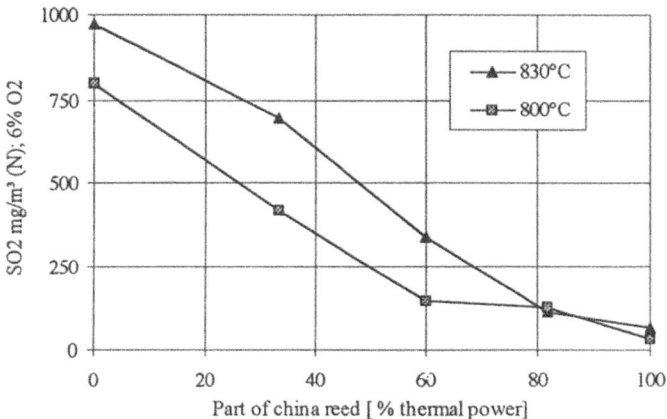

Fig. 9: Sulfurdioxide emission level of mixed china reed / raw brown coal; CFBC

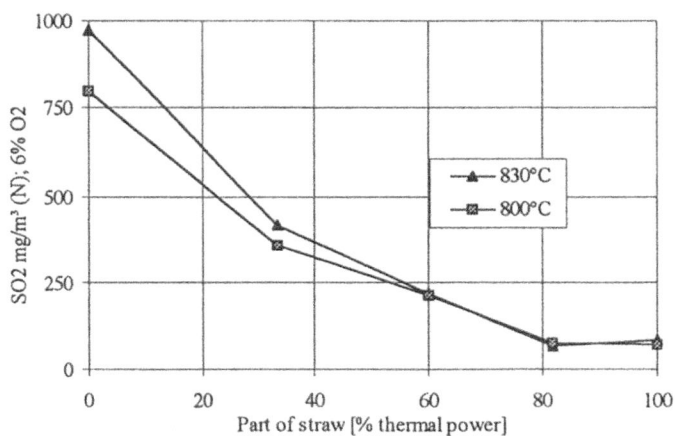

Fig. 10: Sulfurdioxide emission level of mixed straw / raw brown coal; CFBC

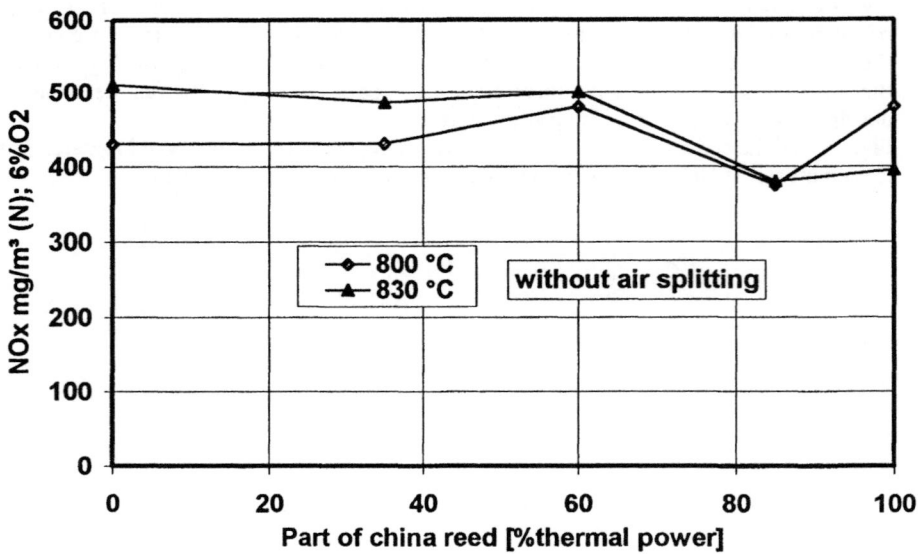

Fig. 11: Nitrogen oxide emission level of china reed / raw brown coal; CFBC; without air splitting

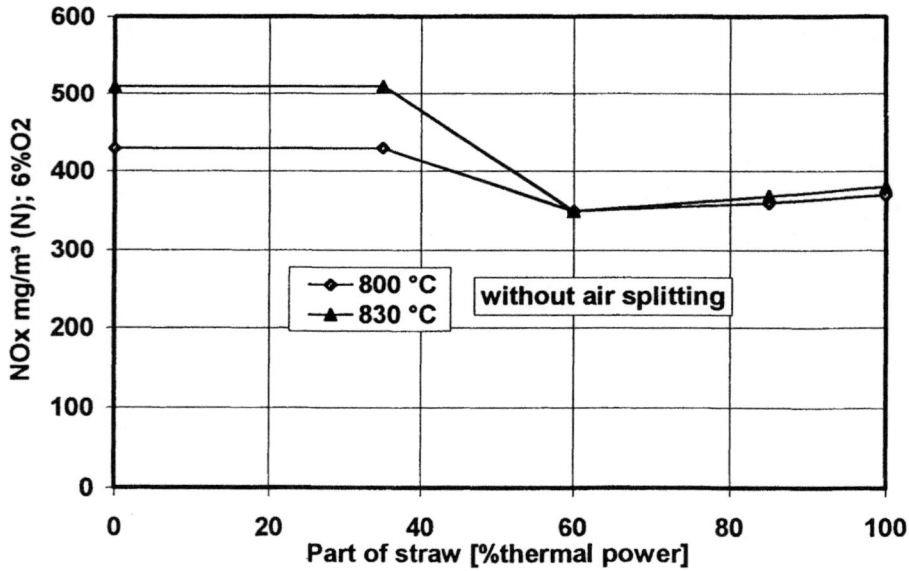

Fig. 12 Nitrogen oxide emission level of straw / raw brown coal; CFBC; without air splitting

4 Conclusions

The aim of the investigations was to describe the combustion process of mixed coal/ biomass in fluidized beds.

It was necessary to become more knowledge about the burning behavior of the different fuels. So the fuel characteristics of burning behavior had been determined by batch tests under fluidized bed conditions. The results of the batch tests are $O_2 = f(t)$ functions, describing the integrated burning behavior of the fuels. The kinetic datas of the burning process can be determined with special functions.

The burning time of the fuel depends on the particle size and on the content of volatiles and char [7].

It is also interesting to get more knowledge about the influence of process parameters and biomass mix ratio on the emissions. NO_x emissions don't increase with higher biomass fuel content.

There is a possibility of sulfur capture by some biomasses (in this case straw). This depends on the Ca/S molar ratio of the fuel [8]. General the low sulfer content of the biomass lowers the SO_2 emissions with higher biomass feeding rate.

The results of the investigations gives conclusions for the co-firing of biomass in CFBC industrial units.

References

1. Zelkowski, J.(1986): *Kohleverbrennung*, VGB-Kraftwerkstechnik, Essen.

2. Neidel, W.; Gohla, M.; Schanko, (1991*): Computer Simulation of Coal Combustion in CFB-Units*, D. Proc. of the 2nd Int. Symp. on Coal Combustion. Beijing; pp. 159 - 167.

3. Kraisha, Y.H.; Dugwell, D.R. (1992) 5: *Coal combustion and limestone calcination in a suspension reactor*, Chem.Eng.Sci. 47; pp. 993-1006.

4. Junk, K.W.; Brown, R.C. (1993): *A model of coal combustion dynamics in a fluidized bed combustor*, Combustion and Flame 95; pp. 219-228.

5. Jia, L.; Becker, H.A.; Code, R.K. (1993*): Devolatilisation and char burning of coal particles in a fluidized bed combustor,* The Can. Jour. of Chem. Eng. 71.

6. Adanez, J.; Abanades, J.C.; de Diego, L.F. (1993): *Determination of coal combustion reactivities by burnout time measurements in a batch fluidized bed*, Fuel 73.

7. Lorenz, H.; Rau, H.; Borghardt, R.; Neidel, W. (1993): *Untersuchungen zum Abbrandverhalten von Brenn- und Abfallstoffen in der Wirbelschicht mit Hilfe von Sauerstoff-Elektrolyt-Sensoren*, VDI-Berichte, No. 1090, pp. 431 - 438.

8. Gohla, M., Borghardt, R., Reimer, H., Neidel, W., Meschgbiz, A.; (1995): *Feuerungstechnische Untersuchungen zur Verbrennung von Braunkohlen und Biomassen in zirkulierenden Wirbelschichtfeuerungen unterschiedlicher Leistung*, Proceeding, 9. VGB-Konf. „Forschung in der Kraftwerkstechnik"Essen, pp. C3.

FRACTIONATED HEAVY METAL SEPARATION IN BIOMASS COMBUSTION PLANTS AS A PRIMARY MEASURE FOR A SUSTAINABLE ASH UTILIZATION

Heavy Metal Fractionation in Biomass Combustion Plants

I. OBERNBERGER and F. BIEDERMANN
Institute of Chemical Engineering, University of Technology Graz, Austria

Keywords
Biomass ash, combustion, heavy metals, fractionation, ash utilization.

Abstract
Previous research has shown that the elementary cycle of nature within the process of biomass combustion is disturbed by deposition of heavy metals on the forest ecosystem caused by environmental pollution. By separating a heavy metal rich side stream (filter fly-ash) it should be possible to recycle the major part of the ashes produced (usable ash). Consequently, the aims of technological development are to reduce the heavy metal concentrations in the usable ash and to upgrade them in the filter fly-ash. In order to find out the necessary requirements for a fractionated heavy metal separation, 11 large-scale tests in an Austrian biomass district heating plant were carried out. In addition, the influence of the specific surface of the ash particles on the heavy metal precipitation was examined by sieving experiments. The results showed that above 800 °C the temperature is the most important influencing variable for the condensation/desublimation of Zn, Cd and Pb. The reduction potential for these heavy metals and their compounds in fly ashes precipitated at high temperatures is considerable. Below 600 °C, the concentrations of these heavy metals increase with decreasing particle size. Consequently, at low temperatures the surface of the particles represents the dominating influencing variable. The low amounts of volatile heavy metals in the bottom ashes show that a reducing atmosphere around the ash particles also seems to play an important role.

1 Introduction

Coincident to forcing the thermal energy utilization of biomass (wood chips, sawdust, bark, straw, cereals), the amounts of combustion residues (ash) increase. Therefore it is necessary to find ways of utilizing the ashes produced in a sustainable manner.

Previous research has shown that the elementary cycle of nature within the process of thermal energy utilization from biomass is disturbed by dry and wet deposition of heavy metals on the forest ecosystem caused by environmental pollution. By separating a heavy metal rich side stream (the so-called filter fly-ash - precipitated in electrostatic filters, fibrous filters or flue gas condensation units) it should be possible to recycle the major part of the ash produced, the so-called "usable ash". The usable ash represents a mixture of bottom ash and fly-ashes collected before filter fly-ash precipitation takes place (see Fig. 1).

A necessary requirement for the usable ash is to meet the national limiting values for ash recycling in forests or on agricultural fields or, even better, to meet the respective guiding values for soils. Furthermore, the amount of usable ash should be as large as possible in order to be able to close the nutrient cycles of nature effectively and to keep the amount of filter fly-ash that has to be disposed of or industrially utilized at a low level.

Previous analyses carried out have shown that the concentrations of Cd, and in some cases of Zn in the usable ash mixture were close to the limiting values for an agricultural sewage sludge utilization [1]. For sawdust combustion plants the concentrations of Cd even exceed these limits.

Taking into account that in the near future the limiting values will gradually be reduced, the aims of technological development are to decrease the heavy metal concentrations in the usable ash and to upgrade them in the filter fly-ash fraction by designing combustion technology that makes an efficient heavy metal separation possible during the combustion process. Such a primary measure has the advantage that no further ash treatment is necessary which reduces the operating costs of a plant and meets the requirements for a decentralized closed-cycle economy within energy production from biomass.

2 Objectives

Based on the biomass research already carried out [2] and the necessary requirements for a sustainable ash utilization, comprehensive investigations on the heavy metal fluxes and the influencing variables in biomass combustion plants were carried out. The research work done covered the following points [3]:

1. Implementation of test runs in a state-of-the-art biomass combustion plant (Lofer, Austria) under consideration and variation of operating parameters:
   a. Tests with different kinds of bio-fuels (bark, wood chips).
   b. Variation of the temperatures in the combustion zones in order to check the influence on the heavy metal concentration in hot precipitated fly-ashes.
   c. Tests at different plant loads in order to check the influence of different fly-ash production rates on the heavy metal fluxes.
2. Determination of the concentrations of the heavy metals and nutrients in the different ash fractions produced and the bio-fuels used.
3. Calculation of material balances to test the plausibility of the analytical results and to determine the mass fluxes of the inorganic elements from the bio-fuel to the different ash fractions and the flue gas.

4. Research on the influence of the particle size of the fly-ashes on heavy metal precipitation.
5. Evaluation and interpretation of the results in order to examine the possibilities and the potential of a new combustion technology with integrated fractionated heavy metal separation.

3 Methodology

3.1 Sample taking and measurement of operating parameters
Altogether, 11 large-scale tests were carried out in a 4 MW_{th} biomass combustion plant with a moving grate furnace. During the test periods, samples of 7 different ash fractions precipitated at different temperatures between 1000 and 40 °C and of the bio-fuel used were taken at regular intervals and analyzed for their contents of nutrients and heavy metals. Moreover, the amounts of biomass fired and ashes produced, the ash precipitation temperatures, the amount of heat produced, the water content of the bio-fuel and the excess oxygen in the flue gas were measured.

Fig. 1 shows a diagram of the biomass combustion plant where the test runs were performed. Every test period lasted for 3 days. The first day was necessary to adjust the combustion unit to the specified side constraints. After 24 hours pre-run, the boiler, the combustion unit and the fly-ash precipitation units were carefully cleaned and a 48 hour test period followed.

In addition to the bottom ash, cyclone fly-ash and condensation sludge fraction, ash samples from the secondary- and tertiary- combustion zone, from the boiler and from the flue dust were taken (see Fig. 1). Furthermore, the condensate produced was examined.

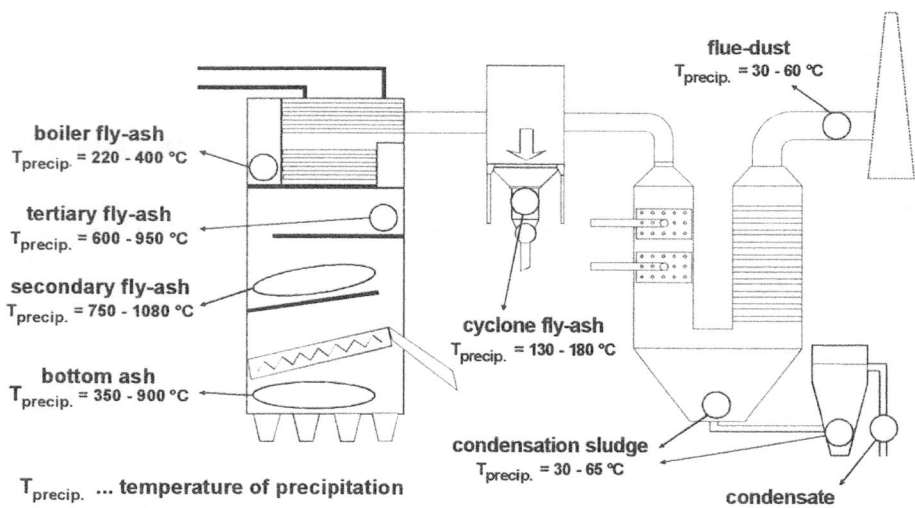

Fig. 1. Ash fractions sampled and their precipitation temperatures

3.2 Analysis of ashes and bio-fuels

Representative samples of the ashes and bio-fuels collected were analyzed for their contents of nutrients (Ca, Mg, Na, K and P), heavy metals (Fe, Mn, Cu, Zn, Co, Mo, Ni, Cr, Pb, Cd, V and Hg) and other inorganic elements (Si, Al, S, Cl and As). In addition the content of carbon and carbonate in the ash samples was determined. The analyses were done with special methods developed and tested at the Institute of Chemical Engineering, University of Technology Graz in cooperation with the Institute of Analytical Chemistry, Radio- and Microchemistry, University of Technology Graz [4, 5]. These methods guarantee a correct and complete detection of inorganic elements which is essential when material fluxes of trace elements have to be investigated.

3.3 Material balances

The plausibility of the analytical results was checked by the results of the material balances calculated in comparison to similar data already available [1].

Furthermore, the mass distribution of the inorganic elements among the different ash fractions was investigated.

The calculations were done with a spread sheet based program for mass and energy balances of biomass combustion plants [6].

3.4 Sieving experiments

To examine whether heavy metal precipitation depends on the specific surface of the fly-ash particles, sieving experiments with 3 different fly-ash fractions (boiler fly-ash, cyclone fly ash and flue-dust) were performed. The sieving experiments were carried out in acetone. 4 different sieves with a mesh width of 60, 40, 20 and 5 µm were used. In addition, the cyclone fly-ash fraction was sieved with 100 µm mesh width. This was necessary because of the larger particle sizes of this ash fraction. The sub-fractions obtained by sieving were dried at ambient temperature, weighted to determine the particle size distribution and analyzed for their contents of Zn, Cd, Pb, Cu, S and Cl.

4 Results

The discussion of the results achieved focuses on the heavy metals in bio-fuels and ashes that are environmentally most significant. These are primary Cd and secondary Zn [7, 8].

4.1 Concentrations of heavy metals in ash fractions produced

Fig. 2 and Fig. 3 show the average concentrations and the standard deviation of Zn and Cd in the different ash fractions sampled.

The concentrations of Cd in the fly-ash fractions precipitated at temperatures above 900 °C (secondary fly-ash) are similar to the concentrations detected in the bottom ash. Furthermore, there is a clear dependence of the Cd- concentrations in the secondary and tertiary fly-ash on the precipitation temperature (see section 5, Fig. 9).

At precipitation temperatures between 600 and 900 °C (tertiary combustion zone) the Cd- contents in the fly-ash are on average 3 times higher than in the bottom ash, but lower than in the cyclone fly-ash. Consequently, the fractionation potential of Cd at high temperatures is considerable.

The concentrations of Zn in the secondary- and tertiary- fly-ash are significantly higher than in the bottom ash (4 times higher in average), but they are lower than in the cyclone fly-ash. This means that the fractionation potential of Zn in dependence on the temperature seems to be low, due to the less volatile Zn-compounds formed in the flue gas [3].

The low concentrations of Cd and Zn in the bottom ash show that the reducing atmosphere on the grate seems to enhance the volatility of these elements by impeding the oxide formation (see section 5).

Moreover, the results presented in Fig. 2 and Fig. 3 show that the concentration of Cd and Zn in the fly-ash fractions significantly increase with decreasing precipitation temperatures and particle size. For this reason samples of fly-ash fractions classified by size were prepared and analyzed in order to obtain more information about the influence of particle size on heavy metal precipitation (see section 4.2).

The highest concentrations of Cd and Zn are found in the finest fly-ash fraction precipitated at normal conditions, in the condensation sludge (filter fly-ash). Consequently, the filter fly-ash has to be deposited or industrially utilized. The flue dust emitted with the flue gas exceeds the heavy metal concentrations in the condensation sludge but its amount compared with the amount of the other fly-ash fractions produced is low. This is due to the efficient dust cleaning facilities installed (see section 4.3, Table 1).

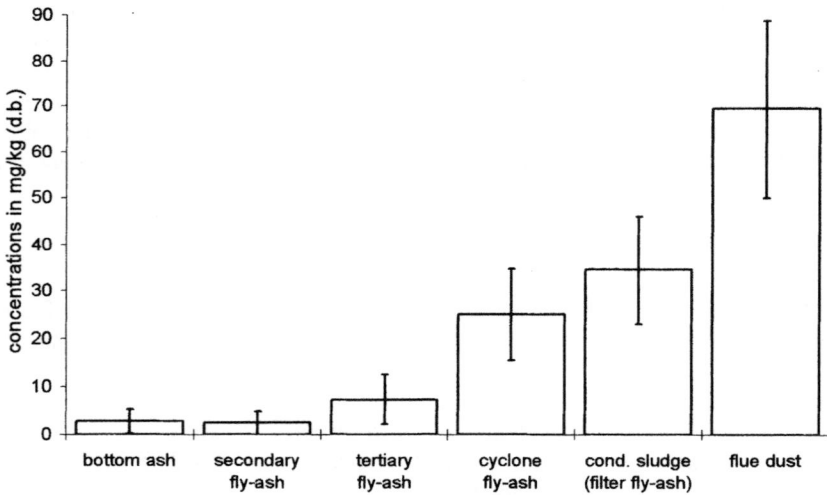

Fig. 2.   Average contents of Cd in the ash fractions examined

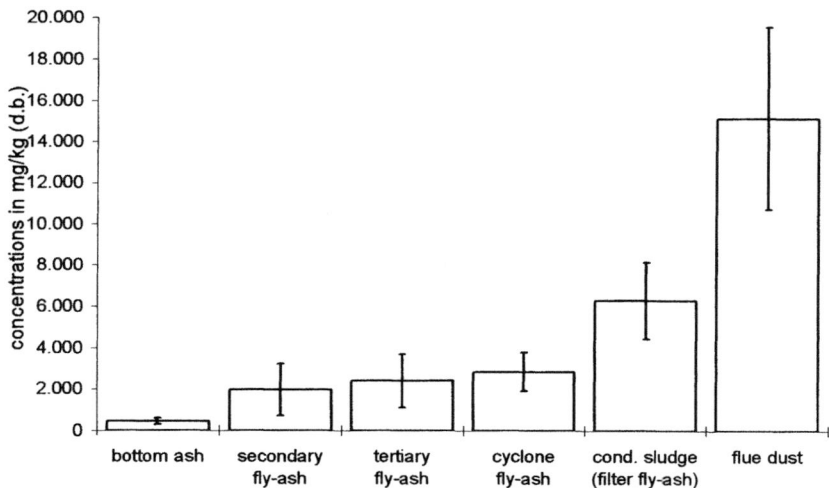

Fig. 3.　Average contents of Zn in the ash fractions examined

4.2　Influence of particle size on the heavy metal concentration in fly-ash fractions

Besides the temperature, the particle size (specific surface) seems to play an important role for heavy metal precipitation on fly-ash. For this reason, sieving experiments were carried out using 3 different fly-ash fractions precipitated at different temperatures.

4.2.1　Particle size distribution of fly-ashes

Fig. 4 shows the results of the sieving experiments undertaken. High amounts of particles of the cyclone fly-ash are larger than 100 μm (about 42 wt% of the total particle amount). This sub-fraction contains unburned carbon particles with larger diameters.

The average particle size of the boiler fly-ash is smaller than the particle size of the cyclone fly-ash. 60 wt% of the boiler fly-ash particles are between 5 and 40 μm. This is due to the fact that fly-ash sampling in the boiler was carried out with a filter bag which has a significantly higher precipitation efficiency for smaller particle sizes than the multi-cyclone, but does not work effectively for light particles that contain large amounts of unburned carbon.

The different particle size distributions measured are due to precipitation technologies and their serial connection. Consequently, the flue dust particles not precipitated in the condensation unit are very fine (81 wt% of the particles are smaller than 20 μm).

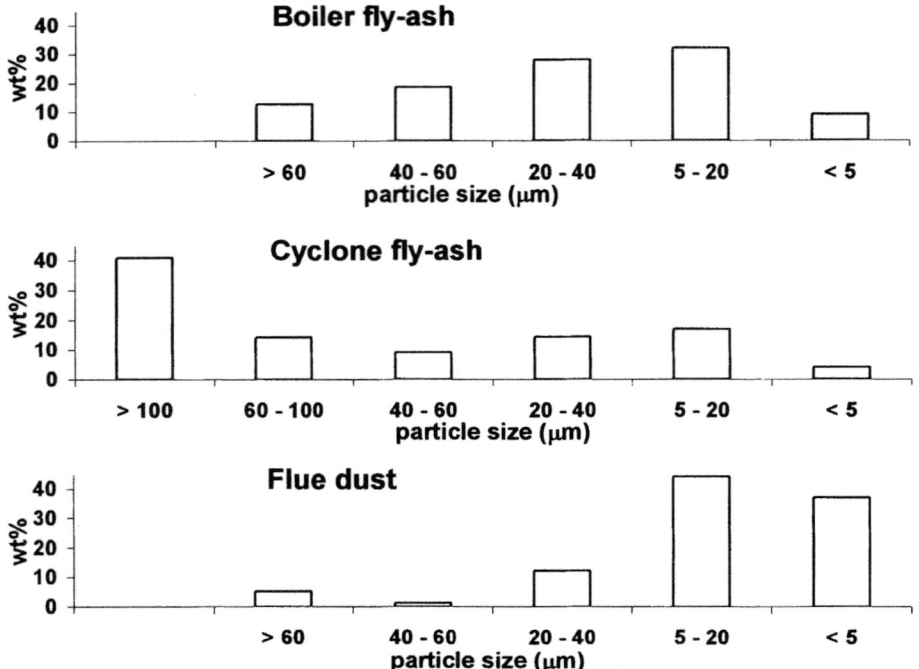

Fig. 4.   Particle size distribution of different fly-ash fractions

4.2.2   Concentrations of heavy metals in the sub-fractions classified by size

The concentrations of Zn and Cd in the sieve fractions of the boiler fly-ash and the cyclone fly-ash increase with smaller particle size (see Fig. 5 and Fig. 6). In the smallest sieve fraction (< 5 μm) the concentrations are three to five times higher than in the > 60 μm fraction (60 - 100 μm for the cyclone fly-ash). The smallest concentrations of volatile heavy metals were found in the > 100 μm sieve fraction of the cyclone fly-ash.

The concentrations of Cd and Zn in the sieve fractions of the flue dust do not depend on the particle size. This is probably due to a partial dissolution of Cd and Zn in the water drops formed in the condensation unit when coming into contact with the surface of the ash particles. The low pH- value of these drops (due to SO_2 and HCl uptake from the flue gas) force Cd and Zn to dissolve. The relatively low concentration of Cd and Zn in the fine dust fractions implies that the condensation unit does not only have a dust precipitation, but also a cleaning effect. The evidence that equal concentrations of Cd and Zn among the different sieve fractions are not due to agglomeration of the flue dust particles during collection and storage is the increasing concentration of Cu and Pb with decreasing particle size (the solubility of these elements is less pH-dependent) [3, 9, 10].

The concentrations of Cd and Zn in the sieve fractions of the boiler fly-ash examined are higher than in the corresponding sieve fractions of the cyclone fly-ash. The sieving experiments undertaken do not give information about the distribution of particles within one sieve fraction. It is likely that particle distribution within the sieve fractions of the boiler fly-ash - compared with the cyclone fly-ash - is shifted to smaller particle sizes, since the particle size distributions of these two fly-ash fractions show a similar tendency (see Fig. 4). Moreover, the sample taking method applied for the boiler fly-ash was more efficient with fine particles. This implies and explains the higher concentrations in the sieve fractions of the boiler fly-ash.

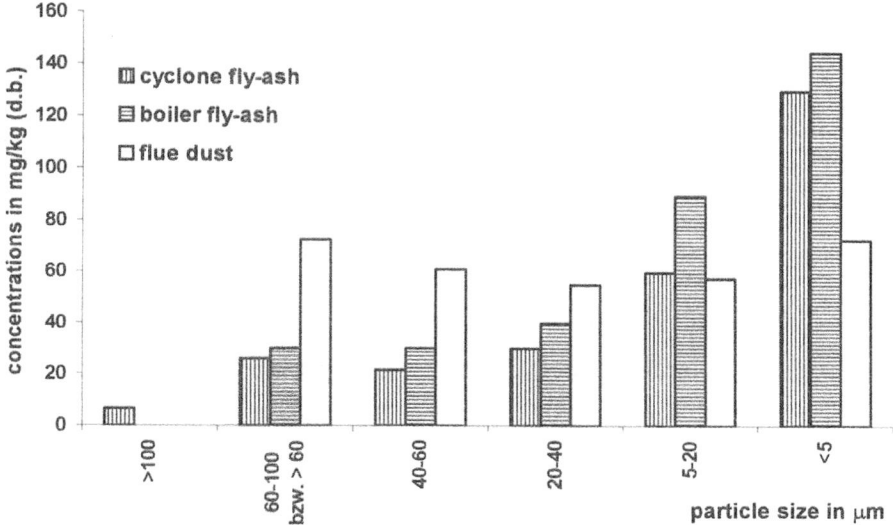

Fig. 5.   Concentrations of Cd in the size classified sub-fractions of different fly-ash fractions

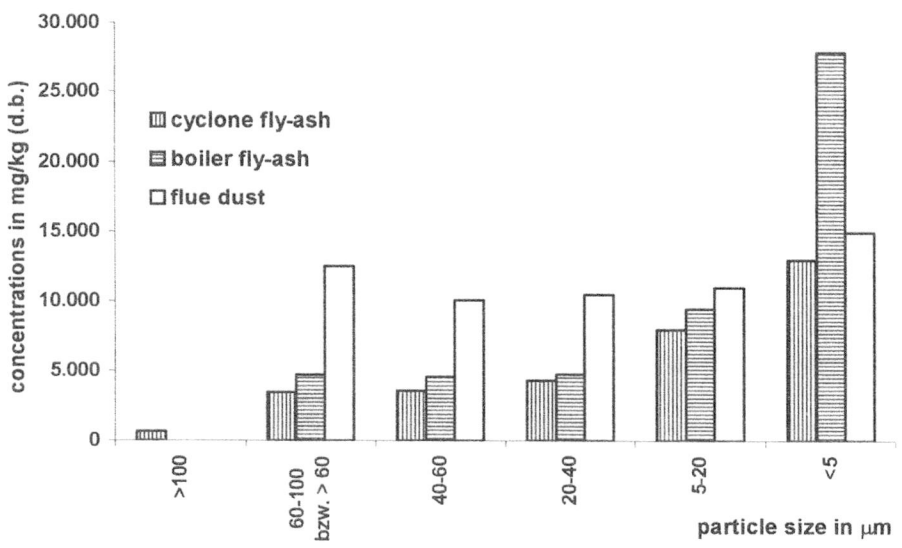

Fig. 6.   Concentrations of Zn in the size classified sub-fractions of different fly-ash fractions

Summing up, the results show that the particle size (particle surface) is the most important influencing variable for the Cd and Zn precipitation on fly-ash particles at temperatures below 600 to 700 °C.

4.3 Results of the mass balances for Cd and Zn

4.3.1 Amount of the ash fractions produced at different plant loads
The distribution of the total ash amounts among the ash fractions produced is influenced by the plant load (see Table 1). At full load the amount of cyclone fly-ash produced is significantly higher than at partial load. In contrast the quantities of bottom ash increase with decreasing plant load. This is due to the smaller amount of primary air at partial load and a longer retention time of the flue gas in the combustion zone. For this reason, less ash particles are whirled up from the grate.

Furthermore, the amount of condensation sludge increases with decreasing plant load, because the multi-cyclone works at a lower efficiency.

The ash content of wood chips is 50 to 70 % smaller than that of bark. This is due to lower concentrations of inorganic elements and lower amounts of mineral impurities (sand, earth) in the wood chips. Because of the smaller particle size of wood chips (in comparison to bark), the amounts of fly-ash produced are higher.

Table 1. Total ash amount and mass distribution among the different ash fractions

biofuel used: bark

	average output (kWth)	amount of ashes produced per MWh (kg/MWh)	mass distribution among the different ash fractions			
			bottom ash (wt%)	cyclone fly-ash (wt%)	cond. sludge (wt%)	flue dust (wt%)
full load	3277	9,2	57,1	37,3	2,9	2,7
medium load	-	-	-	-	-	-
partial load	1635	12,2	85,1	8,5	4,6	1,7

biofuel used: wood chips

	average output (kWth)	amount of ashes produced per MWh (kg/MWh)	mass distribution among the different ash fractions			
			bottom ash (wt%)	cyclone fly-ash (wt%)	cond. sludge (wt%)	flue dust (wt%)
full load	2831	4,1	48,9	42,6	3,7	4,8
medium load	1945	4,6	62,5	21,3	12,6	3,6
partial load	1420	4,4	77,1	12,7	7,0	3,2

4.3.2 Material fluxes of Cd and Zn during combustion
Material balances for Zn and Cd make a qualitative evaluation of fractionated heavy metal separation possible.

Fig. 7 and Fig. 8 show the average distribution of Cd and Zn among the different ash fractions produced at different plant loads and for different bio-fuels used.

The results show that there is a strong dependence of the mass distribution of Cd and Zn on the plant load. This is due to the varying distribution of the total ash amount among the different ash fractions at changing plant load. The concentrations of the heavy metals in the different ash fractions at different plant loads do not change significantly.

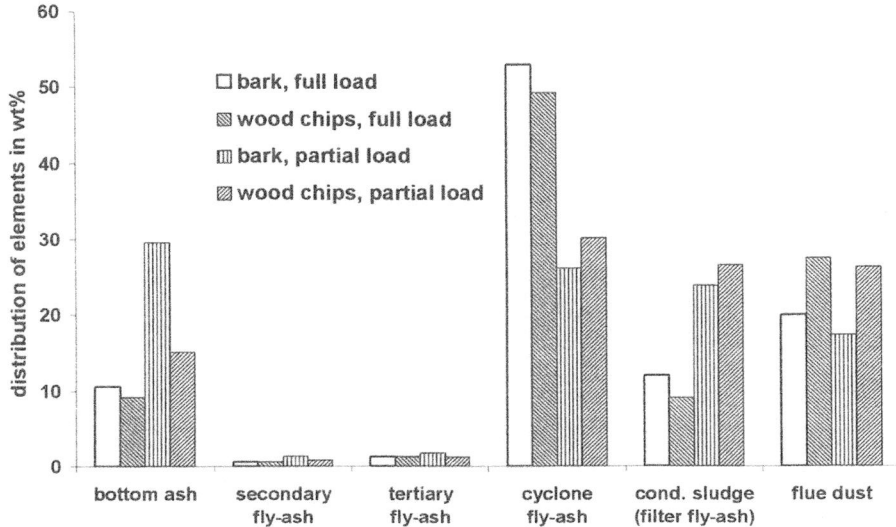

Fig. 7. Distribution of Cd among the different ashes in dependence on the plant load and the kind of bio-fuel used

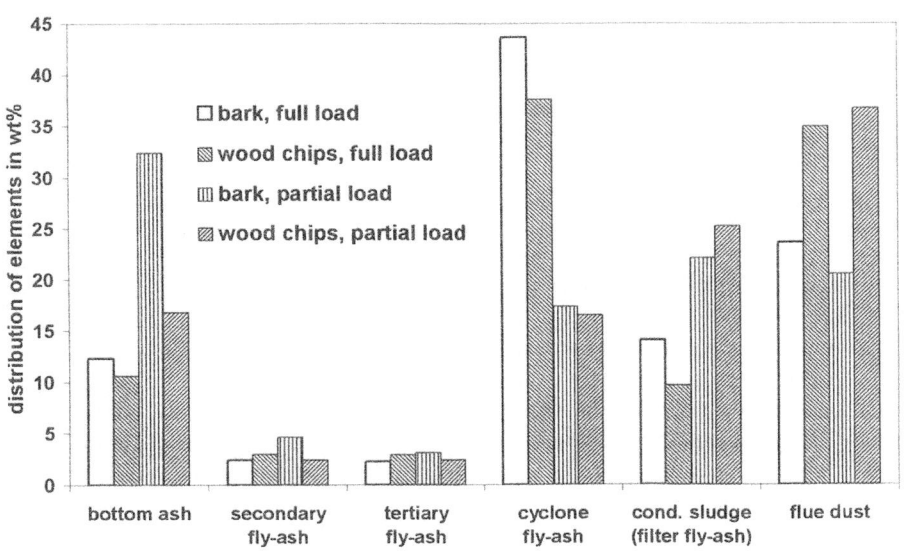

Fig. 8. Distribution of Zn among the different ashes in dependence on the plant load and the kind of bio-fuel used

The effect of the distribution of Cd and Zn among the ash fractions at different plant loads for a fractionated heavy metal separation is demonstrated by the calculation of the resulting concentrations of elements in the usable ash (see Table 2). The main goal of research on fractionated heavy metal separation is to reduce the amounts of volatile and environmentally important heavy metals in the usable ash. The significantly better results at partial load show that for an efficient fractionated heavy metal separation it is important to keep the fly-ash production as small as possible.

Table 2. Concentrations of elements in the usable ash mixture in dependence on the plant load and the kind of bio-fuel used

Explanation: usable ash - mixture of bottom ash and fly-ashes without the filter fly-ash

element	bark		wood chips	
mg/kg (d.b.)	full load	partial load	full load	partial load
Zn	1.118,0	768,0	1.358,0	896,0
Cd	8,5	4,1	9,1	5,8

5 Important influencing variables for an efficient fractionated heavy metal separation in biomass combustion plants

The test runs and examinations carried out show that the potential for fractionated heavy metal separation in biomass combustion plants is considerable.

According to the location where the ash precipitation takes place the following influencing variables are important (see Table 3):

1. The temperature of precipitation.
2. The reduction potential of the gaseous phase.
3. The particle size of the fly-ashes.

As shown in Fig. 9 and Table 3, the concentration of Cd in the ashes depends on one hand on the gaseous atmosphere around the ash particles (important for the bottom ash), and on the other hand on the temperature of ash precipitation. The temperature as an influencing variable is of importance in zones with oxidizing gaseous atmosphere and temperatures above 800 °C (secondary and tertiary combustion zone). According to high temperature equilibrium calculations, Cd and its compounds is mainly to be found in the gaseous phase at temperatures above 800 °C [11, 12, 13, 14], which ties in well with the results achieved. At lower temperatures, Cd starts to desublimate forming aerosols or precipitating on fly-ash particles. Consequently, at temperatures below 600 °C the only influencing variable on the Cd-concentration in the fly-ash particles is the particle size (specific surface).

Zn and its compounds are less volatile than Cd compounds. Therefore the reducing atmosphere on the grate in combination with small fly-ash production rates are of great importance for an efficient fractionated heavy metal separation. The reducing atmosphere on and above the grate, due to high amounts of unburned carbon and CO formed, in combination with high temperatures makes the volatilization of elementary

Zn and $ZnCl_2$ probable [11, 12]. As soon as an oxidizing atmosphere is present (in the secondary combustion zone) ZnO will be formed. According to high temperature equilibrium calculations, ZnO is solid in flue gas atmospheres of biomass combustion plants below 950 °C [11, 12]. Therefore the precipitation of Zn in the secondary and tertiary combustion zone does not depend on the temperature (see Fig. 10).

Summing up, ash precipitation at high temperatures (above 800 °C) in a reducing gaseous atmosphere enforces Cd and Zn volatilization and consequently low concentrations of Cd and Zn in the ashes precipitated there. The most important influencing variables are for Cd the temperature and for Zn the reducing atmosphere.

A fractionation of Cd and Zn at temperatures below 600 °C (in and behind the boiler) seems only to be possible through particle size- selective ash precipitation methods.

Table 3. Important influencing variables for an efficient fractionated heavy metal separation in dependence on the zone where ash precipitation takes place

influencing variables	precipitation zone		
	grate and primary-combustion zone	secondary-combustion zone	boiler and downstream units
reducing atmosphere	X		
temperature of precipitation	X	X	
size of the fly-ash particles			X
temperature range [°C]	400 - 900	800 - 1100	< 800

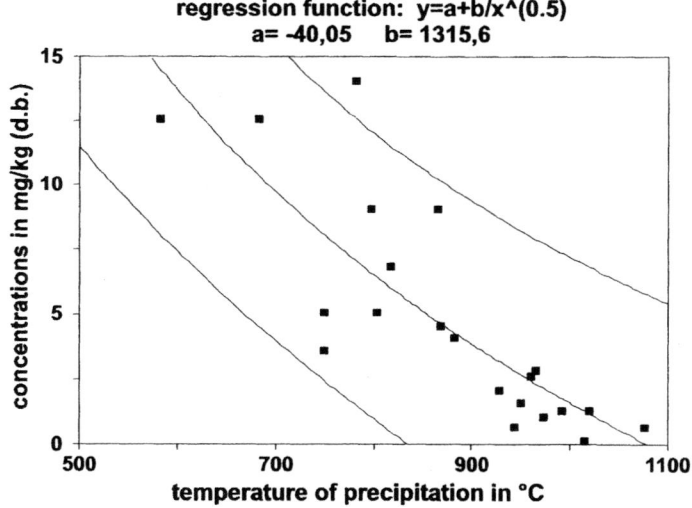

Fig. 9. Concentrations of Cd in the secondary- and tertiary- fly-ash versus temperature of precipitation

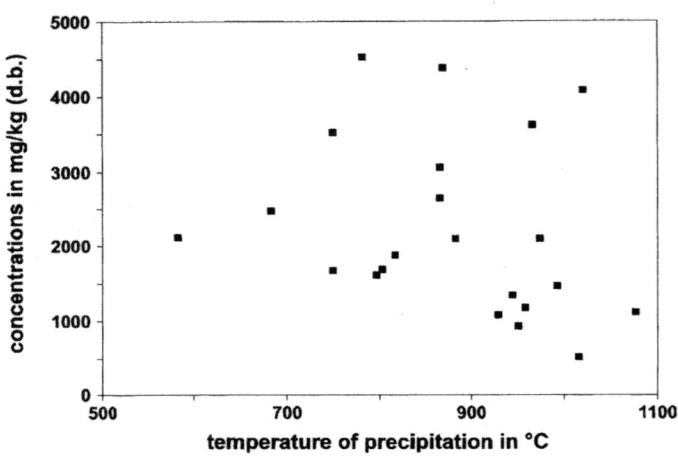

Fig. 10. Concentrations of Zn in the secondary- and tertiary- fly-ash versus temperature of precipitation

6 Conclusions for a new combustion technology with integrated fractionated heavy metal separation

Based on the research results achieved, the following points were considered for the design and development of a new combustion technology with integrated fractionated heavy metal separation:

1. The amount of bottom ash should be increased as much as possible.
2. The major part of the fly-ash (particle size >20 μm) should be precipitated at temperatures above 850 °C.
3. The precipitation of the fine fly-ash remaining in the flue-gas should be carried out in an efficient way to avoid an increase of heavy metal emissions with the flue gas.

To realize these objectives, the geometry of the furnace and the fly-ash precipitation technology was designed in a new way (see Fig. 11). The grate area was increased to reduce the specific primary air load (lower gas velocities). Consequently, less ash and fuel particles should be whirled up from the grate. Furthermore, a relaxation zone in the primary combustion zone will increase the retention time of the flue gas and enhance a fly-ash precipitation on the grate.

A second and bigger relaxation zone is situated in the secondary combustion chamber. Larger fly-ashes particles will be precipitated there at temperatures between 800 and 1100 °C.

The furnace is equipped with "air staging" technology, which means that the gasification of wood and the combustion of the gases produced take place in different combustion chambers. The reducing atmosphere in the primary combustion zone should increase the heavy metal evaporation as well as the reduction of nitrogen oxides formed.

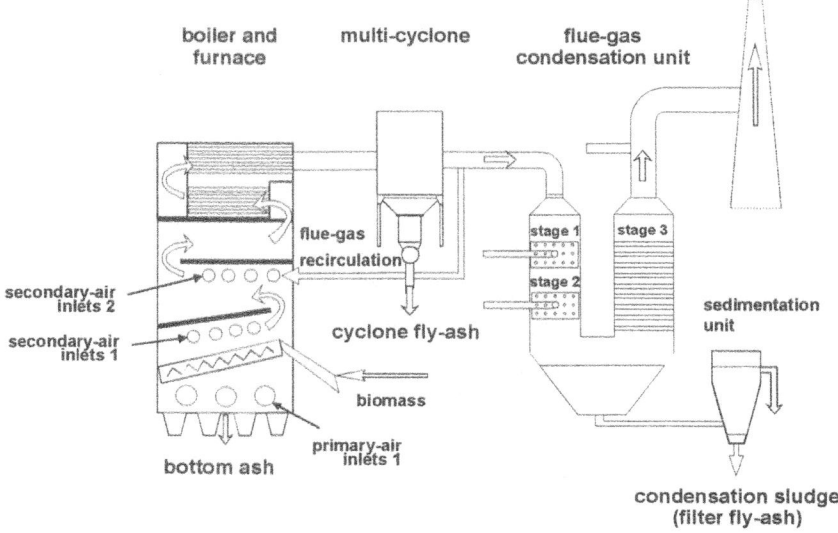

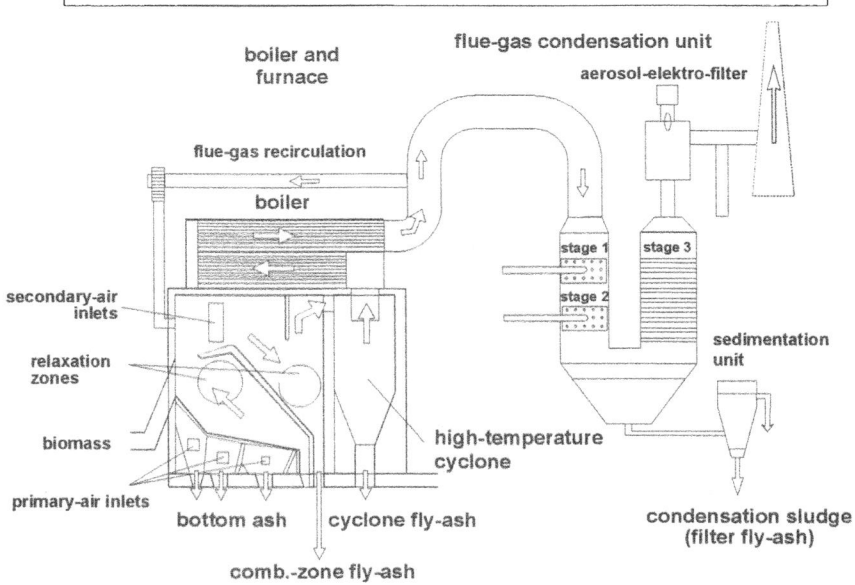

Fig. 11. Comparison of a state-of-the-art combustion plant with the new combustion technology with integrated fractionated heavy metal separation

A completely new unit is the high temperature cyclone placed behind the combustion zone. The precipitation temperatures in this cyclone will vary between 850 and 950 °C to guarantee low concentrations of volatile heavy metals (especially Cd) in the ash fraction produced.

A mixture of bottom ash and the hot precipitated fly-ashes will form the usable ash. The increasing amount of fine fly-ash particles leaving the boiler will be effectively precipitated in a flue-gas condensation unit with an aerosol electrostatic filter placed behind it at low temperatures (about 40 °C). This filter fly-ash fraction will be deposited or industrially utilized.

The fractionation potential for Cd and Zn expected by means of this technology is summarized in Table 4 for full and partial load. In conclusion, it should be possible to reduce the Cd concentration in the usable ash between 40 and 70 wt% of the present values which means that between 55 and 80 wt% of the total Cd input by the bio-fuel are expected to be concentrated in the filter fly-ash and therefore can be removed from natural cycles.

This new technology was realized in an Austrian biomass combustion plant (nominal boiler capacity 3 MW_{th}). Comprehensive test runs at this plant will take place in winter/spring 1996 to check the efficiency of this technology. Taking the results of these test runs into consideration within an approved EC JOULE THERMIE project, this new and further improved technology will be demonstrated in the Austrian biomass district heating plant Tamsweg in 1997.

Table 4. Concentrations of Cd and Zn expected in the usable ash in biomass combustion plants with and without fractionated heavy metal separation technology

element	without fractionation technology (mg/kg d.b.)	with fractionation technology (mg/kg d.b.)
scenario 1: full load		
Zn	1100-1300	700-900
Cd	9-11	3-5
scenario 2: partial load		
Zn	400-600	350-550
Cd	3-4	2-3

7 References

1. Obernberger I., Narodoslawsky M. (1995): *Material Fluxes of Inorganic Elements in Biomass Combustion Plants and Their Characteristics,* in Proceedings of the 3rd European Conference on Industrial Furnaces and Boilers, 18-21 April 1995, Lisbon, Portugal.

2. Obernberger I. (1995*): Characterization and Utilization of Ashes from Biomass Combustion Plants,* in Proceedings of the Earth Conference on Biomass for Energy, Development and the Environment, 10-13 Jan, 1995, Havana, Cuba.

3. Obernberger I., Biedermann F. (1995) *Forschungsbericht FRACTIO - Fraktionierte Schwermetallabscheidung in Biomasseheizwerken,* annual report 1995, Institute of Chemical Engineering, University of Technology Graz, Austria

4. Obernberger I., Pölt P., Panholzer F. (1994) *Charakterisierung von Holzasche aus Biomasseheizwerken, Teil I: Zur chemischen Analytik der Aschefraktionen,* in Umweltwissenschaften und Schadstoff-Forschung - Zeitschrift für Umweltchemie und Ökotoxikologie, Heft 6 (1994), Germany.

5. Panholzer F. (1996) *Homogenization, Digestion and Analysis of Biomass Fuels and Ashes,* to be published in Proceedings of the IEA activity meeting, Task XIII, Activity 6 "Integrated Bioenergy Systems"; 18-20 Sept, 1995, University of Technology, Graz, Austria.

6. Biedermann F. (1994): *EDV- gestützte Bilanzierung und Bewertung von Biomasseheizwerken,* diploma work, Institute of Chemical Engineering, University of Technology Graz, Austria.

7. Ruckenbauer P., Obernberger I., Holzner H. (1996) *Erforschung der Verwendungsmöglichkeiten von Aschen aus Hackgut- und Rindenfeuerungen,* Endbericht der Projektphase II, Institut für Pflanzenbau und Pflanzenzüchtung, BUKU Wien, Vienna, Austria.

8. Ericson S.O. (1994) *Salix cleans soil from cadmium,* in Proceedings of the 8th European Conference on Biomass for Energy, Agriculture and Industry, 3-5 Oct, 1994, Vienna, Austria.

9. Obernberger I., Panholzer F., Arich A. (1996) *System- und pH-Wert-abhängige Schwermetallöslichkeit im Kondensatwasser von Biomasseheizwerken,* Institute of Chemical Engineering, University of Technology Graz, Austria.
Hutchinson S.M. (1995) *Solubility - pH Relationship of Heavy Metal Compounds,* diploma work within the ERASMUS- program, Institute of Chemical Engineering, University of Technology Graz, Austria.

10. Scheffer, Schachtschabel (1992) *Lehrbuch der Bodenkunde,* 13th Edition, ENKE Verlag Stuttgart, Germany.

11. Nordin A. (1993) *On the Chemistry of Combustion and Gasification of Biomass Fuels, Peat and Waste; Environmental Aspects,* Department of Inorganic Chemistry, University of Umeå, Sweden.

12. Dahl J. (1995*): Thermodynamik Investigation of the Faith of Heavy Metals in Biomass Combustion,* diploma work within the ERASMUS- program, Institute of Chemical Engineering, University of Technology Graz, Austria.

13. Tillmann D.A. (1994*) Trace Metals in Combustion Systems,* Academic Press, Enserch Environmental, Sacramento, California.

14. Flemming F. (1995) *Trace Metals from Coal Combustion,* Tekst og Tryrk A/S, Vedbæk, Denmark.

8 Acknowledgment

The research presented has been financially supported by the Austrian Fund for Innovation and Technology and by the cooperation of the Federal and the State Governments of Austria for Environmental Research.

COMBUSTION

Pilot and demonstration

PYROLYSIS GAS FROM BIOMASS AND PULVERIZED BIOMASS AS REBURN FUELS IN STAGED COAL COMBUSTION

H. Rüdiger, A. Kicherer, U. Greul, H. Spliethoff, K.R.G. Hein
University of Stuttgart, Germany
Institut für Verfahrenstechnik und Dampfkesselwesen, IVD
(Institute for Process Engineering and Power Plant Technology)
Pfaffenwaldring 23, D-70569 Stuttgart,
Phone: ++49 711 685 3745, Fax: ++49 711 685 3491

Abstract

The possibility of a combined application of hard coal and biomass using two different co-combustion technologies (combined combustion of pulverized fuels / pre-pyrolysis of biomass and use as a reburn fuel) was investigated in a 0.5 MW_{th} test rig and in a 50 kW_{th} small scale test facility. Reburn investigations with three pulverized biomasses in the 0.5 MW_{th} facility resulted in NO_x-emissions of approx. 300 mg/m^3 (at 6% O_2 in the flue gas). Besides the high DeNOx-efficiency, the test runs showed that the feedstock burnout attained approx. 99% (depending on the biomass particle size) and that no problems arose caused by CO-emissions.

With pyrolysis gas as reburn fuel in the 50 kW_{th} facility, minimum NO_x-emissions of 200 mg/m$_n^3$ (100 ppm) at 6% O_2 in the flue gas are possible. The main parameters are pyrolysis gas composition, stoichiometry, and residence time in the reduction zone. When biomass was used as feedstock the pyrolysis temperature showed only a weak influence on the reburn behaviour. The nitrogen concentration, especially in the tar components of the pyrolysis gas, appears to have a positive effect on NO_x-reduction in the reburn zone of the combustion reactor.

The report will show the influence of different process parameters on the NO_x-reduction applying biomass as reburn fuel by using two different test facilities.

Key Words: Biomass, Combined Combustion, NO_x-Emissions, Reburn, Pyrolysis

1 Introduction

Biomass as a fuel in power generation shows high potential as a CO_2-neutral energy source, which receives increasing interest in combustion technology. Still, coal is one of the world's main sources to produce electricity and heat. Therefore, using available knowledge, combustion technologies and already existing power plants, a first step towards applying biomass in power generation could be a combined firing of coal and biomass using the biomass as reburn fuel. Beside the co-combustion with solid biomass, it also seems attractive to run a combined fired process, processing biomass in pre-gasification or pre-pyrolysis to produce a gas that can be applied as an additional fuel in a secondary combustion zone of a coal-fired boiler. Apart from advantages in the separation of the feedstocks' ashes, this technology promises benefits in NO_x-emission control.

This publication regards the combined combustion of hard coal and biomass. The co-combustion investigations were aimed at different objectives. Major points of interest were burnout and CO-emissions of flames using coal for primary fuel and pulverized biomass for reburn fuel. Furthermore, the feasibility of a co-combustion process using pre-pyrolysed biomass was examined. In this case the main reductive components were examined if pyrolysis gas as reburn fuel was used. In all cases low NO_x emissions were achieved without using SCR.

1.1 CO_2 reduction by combustion of biomass

Generally, biomass is regarded as a CO_2 neutral energy source in a relatively short carbon cycle, because the combustion of biomass releases the same amount of CO_2 as the plant has extracted during its growing phase. These reflections do not take into account e.g. the energy consumption by cultivation, fertilization, transportation or preparation of the plants for combustion.

Table 1 contains a CO_2-balance [1] when biomass is used as a fuel in power generation. The emissions are divided into the segments provision (which contains all the CO_2-emissions produced until the fuels are burned in the boiler) and combustion. The results are compared with CO_2-emissions caused by the use of hard coal in power generation.

Table 1. CO_2-emissions using hard coal and biomass (energy plant) in combustion

fuel	high volatile hard coal		biomass (miscanthus)	
	(t CO_2 /t fuel)	(kg CO_2 /GJ_{th})	(t CO_2 /t fuel)	(kg CO_2 /GJ_{th})
provision	0.10	3.4	0.11	6.03
combustion	2.73	93.2	--	--
total	2.83	96.6	0.11	6.03

The provision of hard coal for power generation consumes hardly less energy (approx. 3.4 kg CO_2 /GJ_{th}) compared with the use of biomass (6.03 kg CO_2 /GJ_{th}). But coal combustion emits approx. 93.2 kg CO_2 / GJ_{th} while biomass is CO_2 neutral. The total sum points out that the use of hard coal in combustion compared with biomass produces CO_2-emissions which are more than 15 times higher.

The CO_2-emissions shown in table 1 refer to the application of a cultivated energy plant in combustion. This reflection on a biomass use represents the most unfavourable case from the aspect of energy saving. If biomass residuals are used (e.g. forestry wood residuals or industrial wood residuals) the energy consumption for the provision might be lower.

1.2 NO$_x$ reduction by staged combustion

Formation of NO$_x$ in combustion is caused mainly by three different mechanisms (thermal NO$_x$ /Zeldovich, prompt NO$_x$ /Fenimore, fuel nitrogen), which are reported or described in various publications. Because of relatively low temperatures in the combustion zone in the BTS test facility the main NO$_x$-emissions are caused by fuel nitrogen. After the devolatilization of the raw fuel at the beginning of the combustion the combustion air converts the fuel nitrogen into NO$_x$. While the conversion of volatile nitrogen can be up to 60% in unstaged combustion [2] the conversion of char nitrogen by air is only between 20% and 30% [3]. The conversion of volatile nitrogen can be lowered to a range comparable with the conversion of char nitrogen by air-staged combustion. In fuel-staged combustion the NO$_x$ is reduced by the addition of a reburn fuel after the primary zone. Depending on the reburn fuel composition a conversion of NO$_x$ to N$_2$ takes place. Detailed descriptions on NO$_x$-formation and NO$_x$-decomposition mechanisms are given by different authors [4,5,6].

The NO$_x$-formation mechanisms show that a coal combustion process, which separates a part of the fuel nitrogen by a thermal pre-treatment, burns char with a low nitrogen content, or raw coal as primary fuels, at high air ratios and uses the volatile nitrogen as a reductive in a reburn zone has advantages regarding the NO$_x$-emission control. This staged combustion concept using a thermally generated reburn fuel can also be applied with biomass pyrolysis gas.

2 Experimental Section

2.1 Small scale test facility

The BTS (*B*rennstoff*t*renn*s*tufung / 'fuel splitting and staging') test rig is shown in figure 1. The facility was developed by the IVD, University of Stuttgart, Germany, and the Saarbergwerke AG, Saarbrücken, Germany, a mining company which also operates coal-fired power plants. The main objective of the test facility is to investigate the NO$_x$-formation and -decomposition during the combustion of hard coal. A coal-originated reburn fuel for NO$_x$-reduction should be used.

The process separates the fuel by pyrolysis into volatiles with a certain amount of fuel-nitrogen and the residual char with the remaining nitrogen. The gas is separated from the residual char in a hot gas filtration. Coal or char are burned in the combustion unit (PF) at high air ratios in the primary zone to achieve high burnout and efficiency. The pyrolysis gas is added in a second zone as a reburn fuel. The arrangement of the test facility allows a separate investigation of the three steps pyrolysis, hot gas filtration and combustion. Detailed results regarding NO$_x$-reduction by coal pyrolysis gas are published elsewhere [7,8].

The 50 kW$_{th}$ pyrolysis unit consists of a gravimetric feeding device, an inert gas pre-heater, an entrained flow pyrolysis reactor and a hot gas filtration. The pyrolysis reactor is an electrically heated furnace (30 kW$_{el}$) with an internal 2000 mm long reaction tube. The tube diameter is between 50 mm and 100 mm. The maximum furnace temperature is between 1200°C and 1300°C depending on the material of the reaction tube (CrNi-steel or ceramic). The main stream of the inert gas is pre-heated

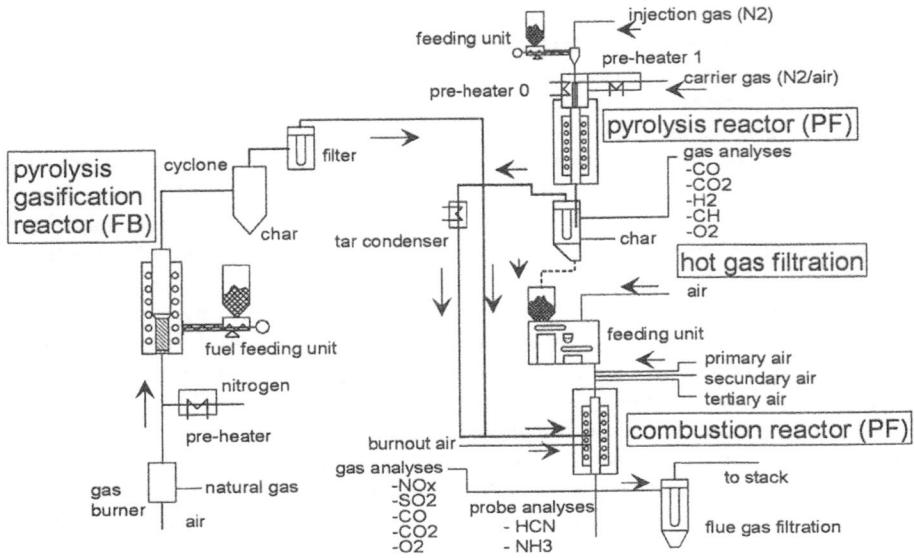

Fig. 1. Flow diagram of the BTS test facility

to ensure high heating rates of the fuel at the entrance to the reaction tube. The residence time in the reactor is between 2 s and 5 s. The hot gas filtration is also electrically heated to avoid tar condensation. A detailed description of the pyrolysis unit is given elsewhere [9]

The 50 kW_{th} combustion reactor is electrically heated, too. This concept avoids heat losses and allows, in contrast to conventional test rigs, the investigation on temperature influences on NO_x-formation and -decomposition at different locations in the reactor. The maximum reactor wall temperature is approx. 1350°C. The reaction tube (inner diameter: 200 mm) is made of ceramic. Fuel- and air-staging can be investigated with different residence times in each reaction zone. Gas measurements are carried out at the bottom of the reactor or along the centre line using a cooled suction probe. The combustion unit is connected with the hot gas filter by heated tubes to inject the pyrolysis gas into the reburn zone.

Additionally to the original BTS test rig a fluidized bed reactor (50 kW_{th}) has been installed and connected with the combustion unit. For a better handling and maintenance it is electrically heated, too and can be operated at temperatures up to 1100°C. The aim is to produce a reburn gas by pyrolysis or gasification and compare the reburn efficiency with gases produced in the entrained flow reactor.

2.2 Pilot scale test facility

A vertical 0.5 MW_{th} combustion unit (length: 7 m, inner diameter: 0.75 m - 0.85 m) has been installed at the IVD for experimental work on combustion processes in coal flames and to validate combustion modelling. Newer research is done using different biofuels or sewage sludge for feedstock in pulverized combustion.

The combustion unit is constructed for experiments under power plant conditions. The first reaction zone, refractory lined with a water-cooled shell, simulates the combustion zone of a boiler. The second part of the combustion chamber, only water-cooled, represents the convective part.

Three rows of ports over the whole length and staggered for 90° allows an application of different suction probes and optical measurement techniques.

3 Results

3.1 Feedstock
Different feedstocks were used in the experiments reported. The analyses of the fuels are shown in table 2 [9,10]. In pyrolysis experiments all fuels were dried at approx. 105°C to a water content between 2.5% and 10% to minimize reactions at higher temperatures and avoid influences on the mass balances of the pyrolysis products.

Table 2. Feed stock analyses for pyrolysis and combustion experiments

fuel	straw	miscanthus	wood	hard coal
proximate analyses (%)				
fixed C (daf)	18.9	18.8	16.9	63.4
volatile matter (daf)	81.1	81.2	83.1	36.5
ash (dry)	6.0	4.9	0.6	10.3
particle size				
d_{50} (µm)				58.8
sieve (mm)	1.5 [1]	1.5 [2]	4.0	
ultimate analyses (% daf)				
carbon	50.5	53.3	50.8	81.4
hydrogen	4.8	4.6	6.5	5.6
nitrogen	0.8	0.5	0.2	1.5
sulphur	0.1	0.3	0.1	1.0
oxygen	43.0	41.1	42.4	10.6
chlorine	0.8	0.2		0.24

[1] in co-combustion 0.75 mm - 4 mm; [2] in co-combustion 1.5 mm - 6 mm

3.2 Reburn experiments with biomass pyrolysis gas
The results present the combined combustion of hard coal as primary fuel and biomass pyrolysis gas as reburn fuel. The experiments were carried out at the BTS facility. Miscanthus, straw and wood (saw dust of conifers) were used to produce a pyrolysis gas. A high volatile hard coal was fired at the combustor. In all cases the residence time in the reduction zone was approx. 1.5 s. The temperature of the reduction zone was fixed at 1300°C. Some of the main results are given in advance:

- The residence time in the reduction zone of the BTS facility must be at least 1.0 s to realize a sufficient flue gas / pyrolysis gas blending in order to achieve a high NO_x-reduction [11].

1391

- High temperatures in the reduction zone improve NO_x-reduction [12].
- NO_x-reduction in staged combustion depends only slightly on the pyrolysis temperature when biomass pyrolysis gas is used. The reduction capability of coal pyrolysis gas strongly depends on the pyrolysis reactor temperature.
- A high N-content in the pyrolysis feedstock results in a better NO_x-reduction.

Detailed information on the influence of different process parameters on NO_x-reduction by fuel staging using pyrolysis gas for reburn fuel are presented in other publications [8,13].

3.2.1 Pyrolysis results

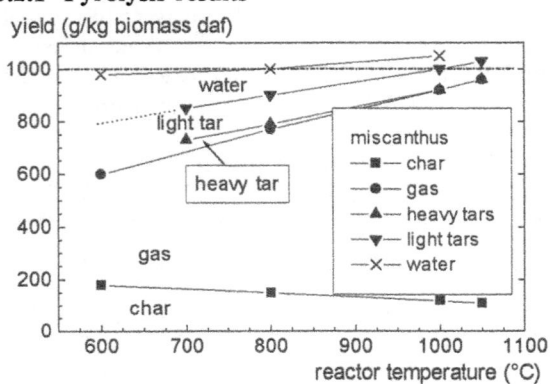

Fig. 2. Total balance in biomass pyrolysis

A view of the main biomass pyrolysis products is given in figure 2. Biomasses generally produce a higher yield of gaseous components in the whole investigated temperature range than hard coal does. The char production in biomass pyrolysis does not increase strongly between 600°C and 1050°C but the yield of all other the products changes. The reaction water content in the pyrolysis products is estimated by an oxygen balance. In contrast to coal pyrolysis, soot was not determined in relevant quantities in biomass pyrolysis. A detailed description of the measurement techniques used is given elsewhere [9,14].

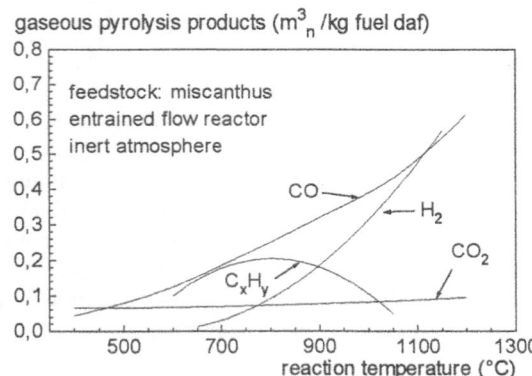

Fig. 3. Gaseous products in biomass pyrolysis

Figure 3 shows the gas production for the main components C_xH_y, H_2, CO, CO_2 in biomass pyrolysis. The investigated temperature range was between 400°C and 1150°C. The curves are designed by mean values, which are estimated by about 30 measurements in the whole temperature range for each gaseous component.

Carbon monoxide and hydrogen are generated in a higher share with increasing temperatures; hydrogen is produced in significant quantities at temperatures higher than 800°C. Gaseous hydrocarbons are produced at low temperatures, too, and reach a maximum somewhat over 800°C. The sum of gaseous hydrocarbons (C_xH_y) were measured using an on-line FID, which was calibrated with methane. The carbon dioxide production was not influenced by the pyrolysis temperature in the range examined.

3.2.2 Reduction results

First test runs were carried out using different synthesis gas mixtures for reburn fuel in the reduction zone of the combustor. The aim was to find the influences of nitrogen-free pyrolysis gas mixtures on NO_x-reduction.

Table 3. Synthetic reburn fuel composition

(vol.-%)	CO	CO_2	H_2	CH_4	C_2H_x
synthesis gas 1	26.5	5.35	7.75	53.1	7.3
synthesis gas 2	41.5	0	37.5	10.8	10.2

Table 3 shows the composition of two synthesis gas mixtures used [7]. The gas mixtures are comparable with coal pyrolysis gas produced at different reactor temperatures.

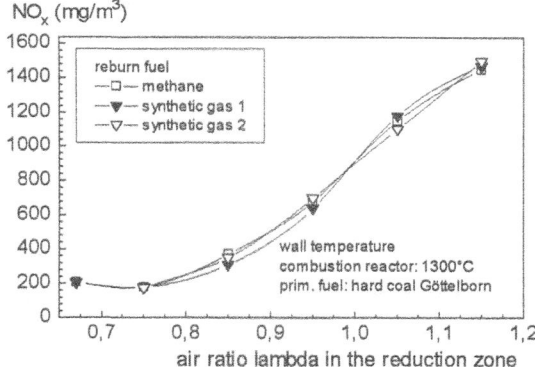

Fig. 4. Reburn with nitrogen-free gas mixtures

The reburn tests were carried out using the same hard coal feed rate for all set points. To achieve different air ratios in the reduction zone the synthesis gas stream was changed. Figure 4 points out that there is only a little difference in NO_x-reduction between the gas mixtures shown in table 3. The maximum deviation was less than 10% in all cases [7]. Like in the test runs using pyrolysis gas for reburn fuel, the residence time in the reduction zone was 1.5 s, the temperature was 1300°C. The unstaged combustion showed high NO_x-emissions of approx. 1450 mg/m_n^3 (6% O_2). At an air ratio – = 0.75 minimum NO_x-emissions of approx. 200 mg/m_n^3 could be achieved. With lower air ratios than 0.75 the NO_x-emissions increased. This behaviour points to NO_x-formation due to unreacted fuel radicals in the reburn zone

The pyrolysis temperature has only a small influence on NO_x-reduction if biomass (miscanthus) pyrolysis gas is used for reburn fuel (see figure 5). To set different air ratios in the reburn zone the feed rate of the pyrolysis reactor was varied. Compared with figure 4 (synthesis gas mixtures) the NO_x-reduction is shifted slightly to higher air ratios in the combustion unit reduction zone. The optimum NO_x-reduction is given at – approx. 0.85. The reburn capability of biomass pyrolysis gas might be improved by nitrogen components in the biomass pyrolysis gas. Especially nitrogen-containing

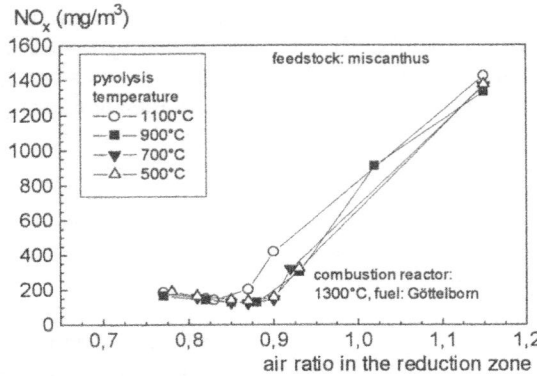

Fig. 5. Reburn with miscanthus pyrolysis gas

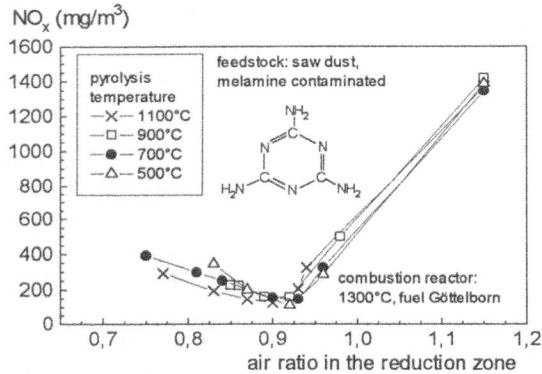

Fig. 6. Reburn with N-contaminated saw dust

tar species seem to have a positive influence on NO_x-decomposition. This was checked by using coal pyrolysis gas with and without tars. The gaseous nitrogen species NH_3 seems to have less influence on the NO_x-reduction in the described process, which was shown using CH_4/NH_3 mixtures for reburn fuel.

Figure 6 presents results using a nitrogen containing saw dust for feedstock in pyrolysis. The feedstock was a residual of industrial furniture manufacturing where a melamine-containing adhesive was used. The structure of melamine is shown in figure 6. The N-content of the saw dust was improved from 0.2% to 4.2% (daf).

Compared with miscanthus pyrolysis gas the NO_x-minimum is shifted to higher air ratios (− approx. 0.92) when the melamine contaminated saw dust is used in pyrolysis. The nitrogen components in the pyrolysis gas are also responsible for the strong increase of NO_x-emissions at lower air ratios in the reduction zone. Unreacted N-species of the pyrolysis gas might be oxidised by the injection of burnout air downstream of the reburn zone. At lower air ratios there seems to be a slight influence of pyrolysis temperature on the NO_x-emissions. Gas produced at low pyrolysis temperatures shows higher NO_x-emissions than pyrolysis gas generated at higher temperatures. This could also be explained by the nitrogen in the raw fuel. Higher temperatures might increase the decomposition of nitrogen containing pyrolysis products producing inert N_2 during the residence in the heated CrNi reaction tube and on the hot surface of the filter candles. At lower pyrolysis temperatures reactive nitrogen components pass the pyrolysis unit without decomposition and are injected into the reburn zone. The decomposition e.g. of NH_3 in the pyrolysis reactor depending on the reactor temperature was demonstrated using NH_3 test gas, and also noticed by other authors [15].

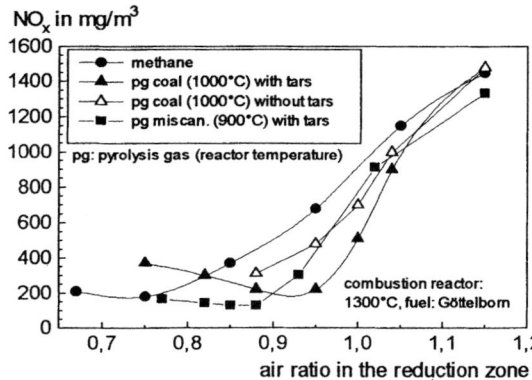

NO$_x$ in mg/m^3

Legend:
- methane
- pg coal (1000°C) with tars
- pg coal (1000°C) without tars
- pg miscan. (900°C) with tars

pg: pyrolysis gas (reactor temperature)

combustion reactor:
1300°C, fuel: Göttelborn

air ratio in the reduction zone

Fig. 7. Reburn with different pyrolysis gases

The influence of different gaseous reburn fuels is shown in figure 7 [14]. In contrast to miscanthus pyrolysis gas, the NO$_x$-reduction capability of hard coal pyrolysis gas strongly depends on the pyrolysis temperature [8] (not shown in figure 7), which might be caused by the different nitrogen release behaviour of coal compared with biomass and the decomposition of nitrogen species mentioned above. Using hard coal for feedstock in the pyrolysis unit the NO$_x$-minimum is shifted to 0.95. Best results were achieved with gas from coal which contains all pyrolysis tar species. Due to the nitrogen components in the coal pyrolysis gas the NO$_x$-emissions increase at lower air ratios.

The efficiency of coal pyrolysis gas without tars seems to be in the same range of biomass pyrolysis gas. A view on the total tar yield in the pyrolysis gas of both feedstocks shows that they are quantitatively in the same range. Therefore, the composition and the nitrogen content in the pyrolysis tars may influence the reburn behaviour.

3.2 Reburn experiments with pulverized biomass

All experiments using pulverized biomass in staged combustion were carried out in the 0.5 MW$_{th}$ test rig [16]. Like in the experiments described above, the primary fuel was hard coal (Göttelborn). The air ratio in the primary zone was 1.15 for all set points. In contrast to the experiments using pyrolysis gas as an additional fuel the thermal input in the whole 0.5 MW$_{th}$ chamber was constant in each case. The range of residence times in the reduction zone was between 1.0 s and 1.6 s. NO$_x$-emissions were measured at the outlet of the reactor, downstream from the burnout zone.

The main results using pulverized biomass for reburn are [10]:

- The NO$_x$-reduction mainly depends on the volatile content of the reburn fuel. The N-content of the reburn fuel has only a secondary influence on the NO$_x$-decomposition. Higher volatile fuels result in better NO$_x$-reduction (compare biomass with coal) at the available residence times. Longer residence times increase the DeNOx-capability of low volatile fuels [17].
- The extend of the NO$_x$-minimum depends on the N-content of the solid reburn fuel. A high N-content in the reburn fuel increases the NO$_x$-emissions at lower air ratios. This behaviour could be shown better in experiments with pyrolysis gas for reburn fuel.

- The N-content of the pulverized reburn fuel influences the position of the NO_x-minimum. Fuels with a comparable volatile matter content but higher N-content shift the NO_x-minimum towards higher air ratios in the reburn zone. This behaviour could be shown in experiments using pulverized saw dust for reburn fuel which was contaminated with melamine [1,18].

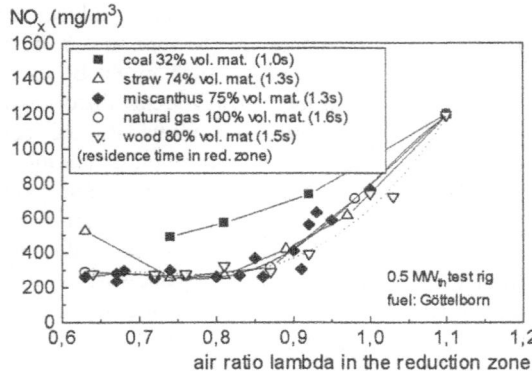

Fig. 8. Reburn with pulverized fuels

NO_x-emissions using different solid reburn fuels are presented in figure 8 [10]. Because of the short residence time in the reduction zone hard coal shows the lowest NO_x-reduction of all fuels tested. Compared with the high volatile fuels biomass (and natural gas) the NO_x-emissions only decreased to approx. 500 mg/m_n^3. Better results were achieved with longer residence times in the reduction zone, but due to the geometry of the combustion chamber the residence time in the burnout zone was to short in this experiment. All the other reburn fuels decreased the NO_x-emissions from 1200 mg/m_n^3 at unstaged combustion to values lower than 300 mg/m_n^3 for air ratios – between 0.75 - 0.85. Further results regarding the combined combustion of pulverized biofuels (biomasses, sewage sludge) and coal using pre-blended coal/biomass feedstocks or air-staging are reported elsewhere [19,20].

Burnout and CO-emissions can be a restriction in staged combustion. Experiments regarding the burnout showed that increasing particle sizes of miscanthus or straw result in a decreasing burnout at the end of the reactor [10]. Nevertheless the burnout was higher than 96% for air ratios > 0.7 in all cases and higher than 98 % for miscanthus sieved < 0.5 mm. The burnout lowering with increasing particle size may be caused by limitations of the residence time in the burnout zone downstream from the reduction zone. Therefore a fine grinding of pulverized reburn fuels is required.

4 Conclusions

The use of biomass in power generation can reduce CO_2-emissions. Applying existing know-how and technology, co-combustion of biomass and coal in large power plants has advantages regarding the high emission standards. The biofuels can be used pre-blended with coal or as separate fuels. The separate application as a reburn fuel (pulverized or gaseous by pre-pyrolysis/pre-gasification) shows a high DeNOx capability. A comparison of the NO_x-reduction with pulverized fuels and with pre-pyrolysed fuels is given in figure 9 [14].

Reburning with pulverized biomass reaches a higher NO_x-reduction than using pulverized coal for reburn fuel. This may be caused by the high volatile content of the biomass and the reduction of NO_x by the gaseous components produced by devolatilization in the reburn zone.

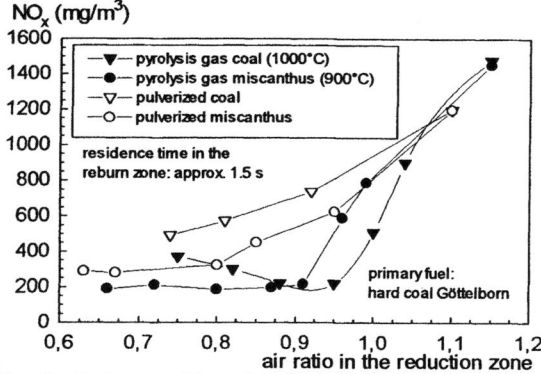

Fig. 9. Reburn with pulverized and pyrolysed fuels

Pyrolysis gas for reburn fuel seems to show an even higher DeNOx capability than pulverized fuels. But this behaviour might be caused by parameters specific to the test rig like mixing effects or temperatures in the reduction zone. This could be shown by the different NO_x-reduction of natural gas in the 0.5 MW_{th} pilot scale test rig (approx. 300 mg/m_n^3) and methane in the 50 kW_{th} facility (approx. 200 mg/m_n^3).

Coal and biomass are suitable to generate a reburn fuel by pyrolysis. Compared with pulverized coal, the use of coal pyrolysis gas results also in a high NO_x-reduction without problems regarding the burnout or CO-emissions at the end of the combustor.

Regarding only biomass, pulverized and pre-pyrolysed reburn fuels show nearly the same DeNOx behaviour. Therefore, high volatile fuels can be used as pulverized reburn fuels, whereas low volatile fuels should be pre-pyrolysed for a reburn process. But the application of pulverized biomass as reburn fuel may have disadvantages concerning the ash behaviour in the boiler.

References

1. Lewandowski, I. CO_2 and Energy Balance for the Production of Biomass and the Combustion in a Power Plant. 8th European Conference on Biomass for Energy, Environment, Agriculture and Industry. Paris: Ademe, 1994.
2. Song, Y.H., Pohl, J.H., Beer, J.M., Sarofim, A.F. Nitric Oxide Formation During Pulverized Coal Combustion. Combustion Science an Technology, Vol. 28, 1982, pp. 31-39.
3. Schulz , W. Formation of Nitrogen Oxides in Pulverized Coal Combustion and Its Avoidance (in German). VGB Kraftwerkstechnik, Vol. 66, 1986, pp 541-550.
4. Pohl, J.H., Sarofim, A.F. Devolatilization and Oxidation of Coal Nitrogen. 16th (Int.) Symposium on Combustion, The Combustion Institute, pp. 491-501.
5. Fenimore, C.P. Studies of Fuel-Nitrogen Species in Rich Flame Gases. 17th Symp. on Combustion, 1979, pp. 661-670.
6. Mechenbier, R., Kremer, H. Fuel Staging of Coal Dust/Methane to Reduce Fuel Originated NO_x-Emissions (in German). VDI-Report Nr. 645, Düsseldorf, 1987, pp. 87-98.

7. Spliethoff, H., Rüdiger, H., Greul, U. Combined Minimizing of NO$_x$-Production and Reduction of Formed NO$_x$ During Combustion of Coal Dust (in German). Final Report, Federal German Ministry of Education, Science, Research and Technology, BMBF, Dez. 1993.

8. Greul, U., Rüdiger, H., Spliethoff, H., Hein, K.R.G. Use of Pyrolysis Gas as Reburn Fuel. 3rd European Conference on Industrial Furnaces and Boilers, Lisbon, April 1995.

9. Rüdiger, H., Greul, U., Spliethoff, H., Hein K.R.G. Co-Pyrolysis of Coal/Biomass- and Coal/Sewage Sludge-Mixtures in an Entrained Flow Reactor. Final Report, APAS Clean Coal Technology Programme CT92-0001, Commission of the European Communities, Bruxelles, 1995.

10. Kicherer A. Biomass Combustion in Pulverized Fuel Boilers - Technical Possibilities and Pollutant Emissions - (in German). PhD Thesis, University of Stuttgart, Germany, 1995.

11. Greul, U., Rüdiger, H., Spliethoff, H., Hein, K.R.G. NO$_x$ Controlled Combustion in a Bench Scale Test Facility. 21st International Conference on Coal Utilization and Fuel Systems, Clearwater, USA, March 1996.

12. Spliethoff, H., Greul, U., Rüdiger, H., Hein, K.R.G. Basic Effects on NO$_x$ Emissions in Air Staging and Reburning at a Bench Scale Test Facility. submitted to: Fuel, Dez. 1995.

13. Spliethoff, H., Greul, U., Rüdiger, H., Magel, H.C., Schnell, U., Hein, K.R.G. (University of Stuttgart, IVD, Germany) and Nelson, P.F., Li, C.Z. (CSIRO, Australia) NO$_x$-Reduction Using Coal Pyrolysis Gas as Reburn Fuel: Effects of Pyrolysis Gas Composition. 8th International Conference on Coal Science, Oviedo, Spain, Sept. 1995.

14. Rüdiger, H., Kicherer, A., Greul, U., Spliethoff, H., Hein, K.R.G. Investigations in Combined Combustion of Biomass and Coal in Power Plant Technology. submitted to: Energy and Fuels, ACS-Journal, USA, Okt. 1995.

15. Li, C.-Z. Use of Pyrolysis Gas as Reburn Fuel. Joint Project of CSIRO and University of Stuttgart (IVD), First Year Report to the Rothmans Foundation, CSIRO, Australia, 1995.

16. Kicherer, A., Maier, H., Spliethoff, H., Hein, K.R.G. Possibilities of the Use of Regenerative Fuels in a Pulverized Fuel Fired Furnace (in German). VGB Conference Forschung in der Kraftwerkstechnik, Essen, 1993.

17. Kicherer, A., Spliethoff, H., Maier, H., Hein, K.R.G. The Effect of Different Reburning Fuels on NO$_x$-Reduction. Fuel 1994, Vol. 73, No. 9, pp. 1443-1447.

18. Kicherer A., Maier, H., Greul, U., Moersch, O., Spliethoff, H. Investigations on Staged Combustion (Fuel-Staging / Air-Staging) for NO$_x$-Reduction in Pulverized Coal Combustion (in German). DFG (German Research Association) Project DO 227/7-1, 1994.

19. Kicherer, A., Gerhardt, T., Spliethoff, H., Hein, K.R.G. Co-Combustion of Biomass/Sewage Sludge with Hard Coal in a Pulverized Fuel Semi-Industrial Test Rig. Final Report, APAS Clean Coal Technology Programme CT92-0002, Commission of the European Communities, Bruxelles, 1995.

20. Kicherer, A., Gerhardt, T., Görres, J., Spliethoff, H., Hein, K.R.G. Investigations and Calculations onto Biomass Co-Combustion in Pulverized Fuel Units. 3rd European Conference on Industrial Furnaces and Boilers, Lisbon, 1995.

FLUIDIZED BED COMBUSTION OF LOW GRADE COALS AND BIOMASS

L. ARMESTO, A. CABANILLAS and A. BAHILLO
CIEMAT, Madrid, Spain

Abstract

Fluidized bed is one of the most promising methods for combustion today. Its application to boilers is recognized primarily for its low sensitivity to fuel quality and its capacity to limits air pollution.

This technology is being used all over the world for biomass as well as for coal combustion. Nevertheless, there are no results available on the joint utilization of these fuels.

The utilization of biomass in blends with low grade coal is expected to improve the conversion of the fossil fuels due to the high volatile content of the biomass, thus reducing the emissions of SO_2 and NOx as well as these of CO_2 due to the renovable origin of the biomass.

The main objective of the proposed paper is to demonstrate the technical feasibility of the circulating fluidized bed as a clean technology for the combustion of low grade coal/biomass blends.

The paper collets and analyses combustion test data. The objective of the tests is to achieve high efficiency and low emission. The influence of biomass content in blends, bed tmperature and fluidization velocity are studied.

Keywords: Cocombustion, Circulating fluidized bed combustion, Biomass, Low grade coals.

1 Introduction

The circulating fluidized bed combustion has become recognized as a viable alternative to conventional combustion boilers. The features of this process that make it an attractive alternative are: combustion temperature, heat transfer, its capacity to limits air pollution and low sensitivity to fuel quality.

These characteristics permit the posibility to burnt fuels with high ash and sulphur content with a minimun environmental impact.

Ciemat was carring out a project in order to demostrate the technical feasibility of the circulating fluidized bed as clean technology for the combustion of low grade coal /biomass blends.

The use of low calorific value and/or high sulphur content coal mixed with biomass is interesting because of the complementary properties of the two fuels since, as opposed to the typical characteristics of poor coals, biomass has a low ash content (generally less than 3%) and high volatil content that is favourable, both from the technical and environmental point of view, to a clean combustion of the coals that are being taken into consideration.

The raw materials to be used as fuels in the tests are blends of different proportions of the following materials:

- Low grade coal: lignite with a heating value of 4970 Kcal/Kg (LHV, db), a sulphur content of 9.61 % (db) and 20.06 % ash content.
- Biomass: Forestry wastes from wood cleaning whose heating value is 4645 Kcal/Kg (LHV, db)

An absorbent, limestone, has been used for reducing the sulphur emission.

The combustion experiments have been carried out in a 0.5 Mw_{th} atmospheric circulating fluidized pilot plant.

The aim of this work was to optimice the effect of the different operating parameters on the SO_2 and NOx emissions and combustion efficiency. These parameters are: bed temperature, fluidization velocity and ratio of coal and biomass mixtures.

2 Tests facilities

The combustion tests have been carried out in a 0.5 Mwth atmospheric circulating fluidized bed pilot plant.

This pilot plant (Figure 1) has as main components of the system, the combustor, the cyclone and the standpipe, for solid recirculation, and the baghouse.

The combustor is a cylinder with 200 mm inside diameter and a total height of 6.5 m. It is covered inside with refractory ceramic.

The cooling systems are three heat exchangers. These systems are in different combustor heights. The cooling fluid is water.

Solid recirculation is carried out using a cyclone, a standpipe and N-valve. The solid circulating rate is controlled by the air in N-valve.

The flue gases leaving the plant are cooled in a water jacketed pipe and finally discharged into a self cleaning baghouse.

The solid feeding system is a screw mounted on a weighing scale to provide an accurate account of feed rate. The system is also equipped with on line analyzers for CO, CO_2, O_2, NO_x and SO_2.

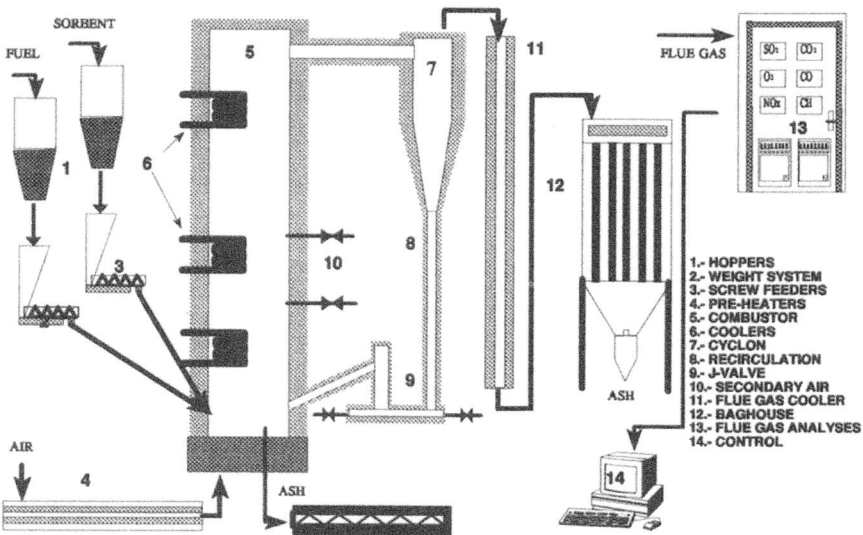

Figure 1.- Scheme of the circulating fluidized bed combustion pilot plant.

3 Fuel analysis

The raw materials to be used as fuels in the tests are blends of a lignite from the Teruel area (Central-east Spain) and pine chips as biomass fuel. This is the most common forest residue available in Spain.

Table 1 shows the analysis of the material used in the combustion tests.

The coal has medium volatile and ash content and very high sulphur content. The biomass has very high volatile and low sulphur content; the ash content is low. The K_2O and Na_2O contents in the pine chips biomass and lignite ashes are high. A high content of alkali metals in the ash can often indicate possible sintering problems.

Limestone, from the lignite area, has been used in order to reduce the SO_2 emission. This limestone has 38.3 % calcium content.

Table 1. Fuels analysis (as burnt)

	LIG	BIO		LIG	BIO
PROXIMATE ANALYSIS (ab)			ULTIMATE ANALYSIS (ab)		
H_2O (%)	6.90	12.50	Carbon (%)	50.33	45.13
V M (%)	38.92	66.69	Sulphur (%)	9.61	0.04
Ash (%)	22.06	1.57	Nitrogen (%)	0.51	0.34
Fixed Carbon (%)	32.12	19.24	Hydrogen (%)	4.32	5.23
			Oxygen (%)	6.26	35.18
HEATING VALUE (KJ/Kg)					
HHV	20349	18201	LHV	19199	16704
ASH ANALYSIS (%)					
SiO_2	43.9	19.51	MgO	2.59	7.35
Al_2O_3	12.16	3.62	MnO	<0.06	1.36
CaO	23.26	32.11	P_2O_5	<0.04	4.92
Na_2O	1.00	0.62	TiO_2	0.74	0.16
Fe_2O_3	6.87	1.78	K_2O	2.06	11.88

4 Tests Result

In order to study the influences of different parameters in combustion efficiency and low emission, the tests have been made with different operating conditions: Ca/S molar ratio (1.3-2.5), biomass weight percent (0-50).

For all tests the experimental measurements have been the following:

A.- Temperature measurement along the length of the combustor and standpipe as well as at several other important locations such as the cyclone exit, baghouse inlet, etc.

B.- On-line analysis for CO, CO_2, O_2, NO_x and SO_2.

C.- All input air flow rates, flue gas, rate of fuel feed and solid discharge rates from the bed and the baghouse.

Table 2 shows the main operation conditions and test results.

Table 2. Operation conditions and tests results

TE	B %	Ca/S	T °C	Vf m/s	Qm Kg/h	O$_2$ %	CO$_2$ %	CO	NOx	SO$_2$
								mg/Nm^3, 6% O$_2$ exc		
1	0	2.0	850	6.0	42.0	3.4	12.4	112	275	4724
2	10	2.0	850	6.0	38.2	4.1	9.0	111	253	4577
3	20	2.0	850	6.0	22.0	7.0	12.0	50	187	3064
4	20	1.3	850	6.0	25.2	6.5	11.6	100	205	3232
5	40	1.3	849	6.0	26.8	6.8	14.8	101	150	3158
6	40	1.7	837	5.0	29.0	3.4	13.6	303	243	2506
7	40	1.7	849	5.1	28.9	4.0	14.0	556	187	2489
8	40	1.7	853	6.2	35.6	3.6	13.6	134	249	3437
9	40	2.5	856	6.1	39.5	3.7	13.8	336	220	970
10	40	2.0	850	6.0	32.4	5.0	8.4	59	169	2011
11	50	2.0	850	6.0	33.6	6.5	10.0		160	2558
12	50	1.5	850	6.1	27.6	6.7	10.0		108	4050

4 Discussion

Most of the parameters shown in Table 2 and Figures 2, 3, 4 and 5 are self-explanatory, however, there are some with diverse comments.

4.1 NO$_x$ emission

The NO$_x$ emission remained at very low levels in all cases. The influence of primary air/secondary air ratio or the temperature on NO$_x$ emissions has not been found within the range of parameters used in the current investigation.

Figure 2 shown the influence of amounts of biomass in NO$_x$ emission.

When the percentage weight of biomass increases the NO$_x$ formed decrease.

This reduction was found to be more than just due to the dilution effect of the addition of the biomass. In fact, the bed temperature profiles (Figure 3) suggest the possibility of mixture segregation within the bed. The biomass is separated of lignite because of its different density. So the coal is burnt in the bottom of the bed and the biomass in the top. This phenomenon suggesting that the biomass fuel can act as reburning fuel, thus causing reduction in the NO$_x$ emission.

On the other hand, when the Ca/S ratio increases, the NOx emissions increase, because the CaO are active catalysts for the oxidation of nitrogen to NOx

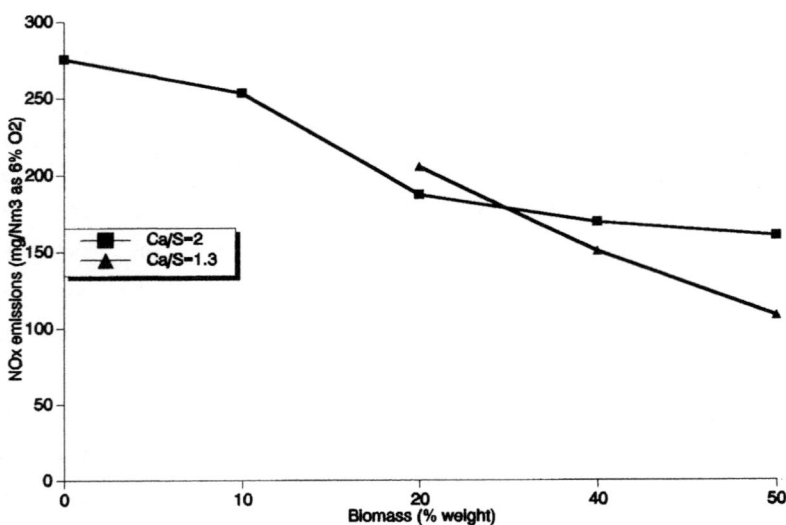

Figure 2.- Influence of coal/biomass fuels ratio on NOx emissions

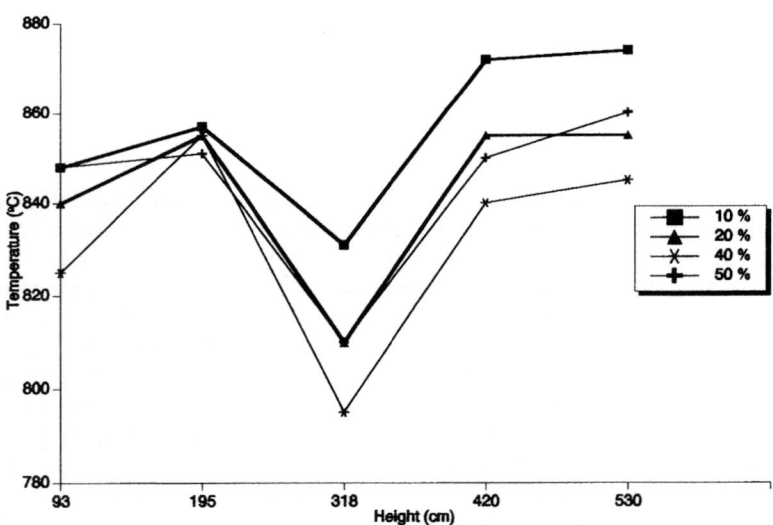

Figure 3.- Temperature profile

4.2 SO₂ emission

As might has been expected as the calcium/sulphur ratio increases the SO_2 emission decrease, because the sulphur retention increase (tests 5,8,9 and 10).

The amount of biomass in blends burned has very influence of SO_2 emission. When the biomass weight percentage increase between 10 to 40 %, the SO_2 emissions decrease, as shown in Figure 4. The reduction was found to be due to the dilution effect of the addition of the biomass.

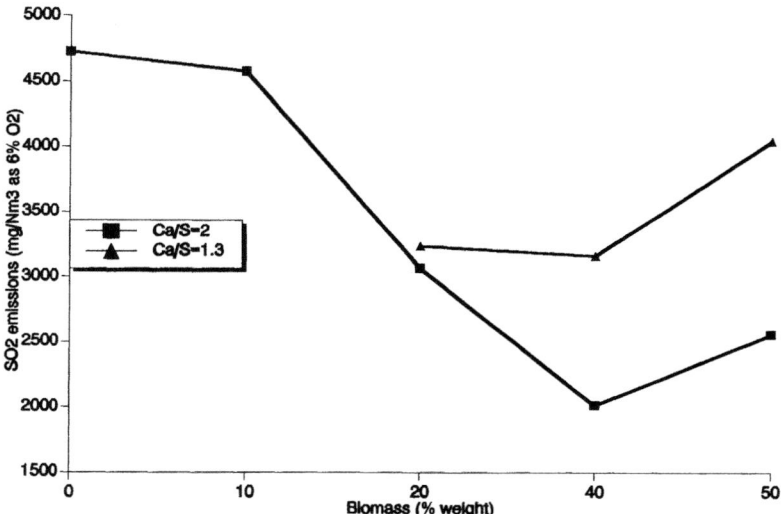

Figure 4.- Influence of coal/biomass fuels ratio on SO₂ emissions.

However, when the amount of biomass is major that 40 % weight, the SO_2 emissions increase. This phenomenon suggest that when the amount of biomass is greater the effect of mixture segregation within the bed has more influence of SO_2 emissions that the dilution effect.

4.3 Combustion Eficciency

When the amount of biomass added increases the unburned carbon content in the ash recovery in the baghouse increases, (as shown in Figure 5) and the combustion efficiency decreases particulary when the fluidization velocity increases.

This is due to the high volatile nature of biomass, the release of volatiles and their subsequent combustion was observed to take place at the top of tne bed, when the amount of biomass is greater, it is possible the the combustion of same volatiles did not occur, then the ash recovery in the baghouse presented high carbon content.

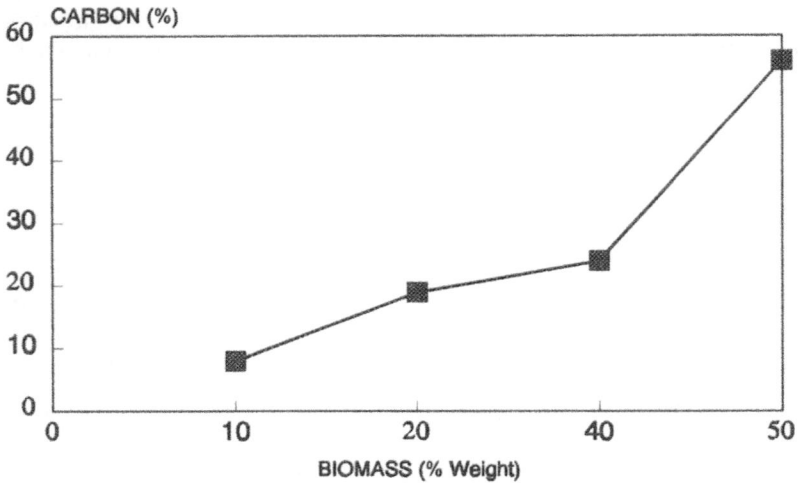

Figure 5.- Influence of coal/biomass fuels ratio on the amount of unburned carbon in the ash.

4.4.- CO_2 emission

Table 2 shown that CO_2 emission light increases when the Ca/S ratio increases, because the amount of limestone usage is major.

The influence of biomass content in blends is very small. However, if the net emissions to the environment are considered; the biomass is considered to be neutral with respect to CO_2, then the process shows a much lower level of CO_2 emissions.

5 Conclusions

In this experimental program different blends of biomass and lignite have been used as fuels. The main characteristics of these raw materials are: lignite has a high sulphur content (9.61 %) and ash content (22.06%), the biomass has high volatile matter content and low sulphur content.

The combustion tests have been carried out in an atmospheric circulating fluidized bed pilot plant.

The most significant findings resulting from the program are:

- The pilot plant tests shown that the co-combustion of lignite with biomass fuels was successfully carried out in a circulating fluidized bed.
- The levels of SO_2 emissions decrease with the increase of the amount of biomass fuel added.
- The addiction of biomass fuel causing reduction in NO_x emissions, because the biomass fuels acted as a reburning fuel.

6 References

1.- Carpenter, A.M. (1988), Coal Classification, *IEACR/12*, October (1988).
2.- Couch, G.R. (1988), Lignite resources and characteristics, *IEACR/12*, December (1988).
3.- D´Acierno, J.P. (1989), CFBC Pilot Plant testing of coals, coal wastes and others fuels. A comparison of performance, *10Th International Conference on Fluidized Bed Combustion*.
4.- Levi Hanson, J. (1991), Agricultural waste fired fluid bed combustor Delano, California. *11Th International Conference on Fluidized Bed Combustion*.
5.- McGowin, C.R. (1989), Fluidized Bed Combustion testing of coal/refuse-derived fuel mixtures, *10Th International Conference on Fluidized Bed Combustion*.
6.- Louhimo, J.T. (1993), Combustion of pulp and papermill sludges and biomass in BFB, *12Th International Conference on Fluidized Bed Combustion*.

Acknowledgements

The authors want to acknowledge to the European Commission for the financiation of this project.

ASH BEHAVIOR IN A CFB BOILER DURING COMBUSTION OF SALIX.

B-J. SKRIFVARS*, G. SFIRIS**, R. BACKMAN*, K. WIDEGREN-DAFGÅRD**,
M. HUPA*
*Åbo Akademi University, Åbo/Turku, Finland
**Vattenfall Utveckling AB, Stockholm, Sweden

Abstract

A study on the combustion characteristics of Salix Viminalis, a fast growing willow, was conducted at a 12 MW circulating fluidized bed boiler. The purpose of the study was to increase the understanding of the mineral matter behavior in the boiler and to foresee possible bed agglomeration or slagging and fouling problems that may occur during the combustion of this type of a fuel. Special focus was given to the impact of ash chemistry on the slagging, fouling and bed agglomeration. Samples from all in-going (bed material, fuel) and out-going solid material streams (secondary cyclone and bag filter) as well as from the bed and the return leg were collected and analyzed chemically. Selected bed samples and ash samples were also analyzed with a scanning electron microscope (SEM/EDAX). Deposit samples were collected at the cyclone inlet and from two different locations in the convective path using specially designed surface temperature controlled deposit probes. All collected probe deposits were photographed and characterized visually. Selected samples from both windward (front) side and leeward (back) side of the sampling probes were analyzed chemically as well as with SEM/EDAX. On top of these samples, the boiler operation was monitored carefully. This included collection of operational data (fuel feed, air distribution and total air), collection and monitoring of pressure drops in the furnace, fluegas temperature profiles and emissions. Multi-component multi-phase thermodynamic equilibrium calculations were then performed for predictions of the fly ash the thermal characteristics, using the fly ash chemical composition as input data. The thermal characteristics i.e. the melting behavior, was predicted for the different ash samples and compared with the results from the full scale fouling measurements. The paper discusses the impact of the ash chemistry on the bed agglomeration and fouling tendency, found during the combustion tests and draws conclusions about their relevance to the operation of the boiler.
Keywords: Ash behavior, biomass, fluidized bed combustion, Salix, slagging and fouling.

1 Introduction

Salix Viminalis, a fast growing willow, is for some countries an indigenous fuel which may become a valuable energy resource. Cultivation of Salix for energy production purposes serves also a possibility to increase the share of non food agricultural production. Combustion of Salix is further attractive from an environmental point of view since it doesn't increase the net CO_2 emissions into the atmosphere.

New combustion techniques such as fluidized bed combustion, have broadened the possibility to fire Salix in a more feasible way than before. Little is, however, known yet about the long term use of Salix in a fluidized bed boiler. Especially the mineral matter behavior is still to some extent unknown. There are indications from combustion of other types of biomasses that the mineral matter in those may sometimes cause ash related problems during combustion. [1 - 4].

We wanted to collect some initial data on the ash behavior of fluidized bed combustion of Salix. A study was conducted in a 12 MW circulating fluidized bed boiler. The purpose of the study was to increase the understanding of the mineral matter behavior in the boiler and to foresee possible bed agglomeration or slagging and fouling problems that may occur during the combustion of this type of a fuel. Special focus was given to the impact of ash chemistry on the slagging, fouling and bed agglomeration.

2 The field measurements

During a test period of two weeks Salix was combusted in a 12 MW circulating fluidized bed boiler. The boiler is a semi-full scale CFB, that produces hot water to the local district heating net. The boiler is also a very suitable research facility and has former been succesfully used when addressing for example emission questions from CFBC:s [5, 6]. The CFB boiler is equipped with a number of operational controlling systems, such as advanced air and flue gas recirculation, bed particle cooler, primary air preheater, and fly ash recirculation possibilities. This enables a wide range of operating options to be tested under controlled conditions in a wide load range (3-12 MW). The unit is also equipped with a large number of analyzing instruments for continuous monitoring and recording of the operation. Numerous instrument tappings enable measurements in many locations in the boiler. A schematic view of the boiler is presented in Figure 1.

Three different running modes were used during the tests. The three modes were related to the fuel mixes used. In the first mode, Salix was fired alone, in the two other cases, wood pellets made from forrest residue were fired along with Salix, to reach higher loads. The air and fuel feed was adjusted so that they corresponded to a gas velocity of approximately 5.5 - 6 m/s in the furnace. Other boiler operational data such as fuel feed, air distribution and total air feed, pressure drops in the furnace, fluegas temperature profiles and gaseous emissions were further recorded continuously. Selected data are summarized in Table 1.

Samples from all in-going (bed material, fuel) and out-going solid material streams (secondary cyclone and bag filter) as well as from the bed and the return leg were collected twice a day and analyzed quantitatively on the elements Si, Al, Fe, Ca, Mg, P, Na, K, Cl, and S by wet chemical analyses. Selected bed samples and ash samples were also analyzed semi-quantitatively with a scanning electron microscope (SEM/EDAX).

Deposit samples were collected at the cyclone inlet (T_{FG} = 850 °C) and from two different locations in the convective path (T_{FG} = 680°C and 250°C resp.) with a sampling time varying from 15 minutes to 21 hours. The sampling locations are indicated in Figure 1 as

Loc. 1, 2 and 3. The samplings were done with specially designed surface temperature controlled deposit probes. The probes were also equipped with a removable ring for later SEM/EDAX analyses.

The surface temperature in the tests were set at 450°C in the sampling locations 1 and 2 to simulate a superheater tube, while the probe in coldest sampling location 3 was held uncooled. All the probes were photographed and characterized with respect to their collected deposits. Selected samples from both front (windward) side and back (leeward) side of the sampling probes were analyzed quantitatively on the elements Si, Al, Fe, Ca, Mg, P, Na, K, Cl, and S as well as semi-quantitatively with SEM/EDAX.

3 Results and discussion

3.1 Fuel analyses

The fuel analyses are summarized in Table 2. The Salix moisture content was very high, >50 % by weight as received. The fuel was fired within a couple of days from harvesting. No extra drying of the fuel was done between harvesting and firing. The ash content in the Salix was approximately 2 % by weight. The corresponding value for the wood pellets, made of forrest residue, was 0.5 %. Sulfur and chlorine content in Salix was low (0.04 % and 0.02 % by weight resp.). In the wood pellets the chlorine content was low, 0.01 % by weight, and the sulfur content moderate, 0.12 % by weight.

The main ash element in both fuels was calcium. Salix contained roughly 25 wt-% calcium of the ash while the corresponding number for wood pellets was 29 %. In Salix also fairly high amounts of potassium and phosphor were found, 13 % and 8 % by weight respectively. Potassium was also in the wood pellets the second biggest element, 6.8 % by weight, although in clearly lower amounts than in Salix. The amount of phosphor in the wood pellets was low only 0.7 % by weight.

The bed material used in the combustion tests was quartz sand with a particle size of approximately 500 μm.

3.2 Bed, primary, secondary cyclone and bag house ashes

Outgoing ash materials were collected twice a day from four different locations in the boiler; i) from the bed at the bottom of the furnace, ii) from the return leg of the primary cyclone, iii) from the secondary cyclone after the heat exchanger surfaces in the convective part of the fluegas channel and iv) from the bag house. These materials were then analyzed quantitatively by wet chemical analyses on the elements Si, Al, Fe, Ca, Mg, P, K, Na, S, Cl and SO_4. Figure 2 summarizes the analyses results. In this figure the analyzed elements are recalculated to their corresponding oxides. A number of trends can be extracted from the figure.

One is the decreasing share of silicon and the increasing share of major ash elements in the samples as we move from the bed material and primary cyclone ash to the secondary cyclone and bag house ashes. The reason to this trend is obvious; the larger sized quartz bed material is separated at the primary cyclone and stays within the internal loop, furnace-primary cyclone, while smaller sized particles, such as condensed ash particles, small fragments of the bed and all gases work their way out to the convective part of the fluegas channel.

A second trend is the decrease in the total amount of analyzed elements in the different samples as we move from the bed and primary cyclone samples to the secondary cyclone and bag house ashes. In the bed and primary cyclone samples, the analyzed elements calculated to their corresponding oxides add up to approximately 100 %, which shows that most of the elements in the samples are countered for. In the secondary cyclone and bag house ashes, however, the sum of oxides remain clearly below 100 %, indicating some missing element. Assuming the analyzed calcium to calcium carbonate and estimating the amount of carbon as CO_2, one can correct the missing share.

A third noteworthy trend is the shares of sulfur and chlorine in the different samples. In the bed and primary cyclone samples the sulfur content is low but increases as we move out to the secondary cyclone and bag house samples. Chlorine is missing completely in the bed, primary cyclone and secondary cyclone samples but turns up in the bag house samples. It seems obvious that chlorine escapes out through the primary cyclone and is enriched in the fly ash fraction collected by the bag house.

One can also see shifts in the running conditions in these analyses. For example the share of silicon in the bed samples taken during the condition A is somewhat higher than those of the samples taken during conditions B or C. The trend is even more clear in the primary cyclone samples. Obviously a higher share of fuel ash elements were present in the bed during conditions B and C, maybe due to the higher temperature in the bed and furnace which would enhance the reaction between the bed material and ash elements.

SEM/EDAX analyses were also performed on selected samples. Figure 3 presents an image of a bed sample taken during the running condition C. The image is taken with a x100 magnification. The numbers indicated in the image are EDAX point analyses performed on that specific location in the sample. These analyses are shown in Table 3. The SEM/EDAX analyses clearly indicate the heterogeneousity of the bed. It consisted of original quartz bed particles that had reacted with Al, Ca and K (points 1, 4 and 5 in the SEM image) and further seemed to be coated with elements originating from the fuel ash (points 2 and 6 in the SEM image). Individual fuel ash particles seemed to be missing.

3.3 Deposit samples

The quantitative wet chemical analyses of the deposit samples collected during the Salix combustion tests are summarized in Figure 4. The deposits collected in the three different locations in the boiler (Loc. 1, furnace at cyclone inlet, Loc. 2, convective part before the heat exchangers and Loc. 3, convective part after the heat exchangers) were generally not very thick. One can, however, distinguish between four different types.

Very thin, shell-type deposits, thickness below 0.1 mm. This type of a deposit was mainly found on the front and back side of the probe in the sampling location 1, in the upper part of the furnace at the primary cyclone inlet at a fluegas temperature of approximately 810 - 870°C. The deposits were generally very thin and did not grow even if the sampling time was increased from 15 minutes to 21 hours. The deposits were also very hard to remove from the sampling probes. Hence, no deposit samples could be taken for quantitative wet chemical analyses. The semi-quantitative SEM/EDAX analyses on the probe rings detected the following elements in the front and back side deposits in amounts as shown below:

Front side SEM/EDAX, ave 3 overall analyses		Back side SEM/EDAX ave 3 overall analyses	
Ca	32.8 wt-%	Ca	21.5 wt-%
K	11.3 wt-%	K	10.9 wt-%
P	7.3 wt-%	P	4.8 wt-%
Mg	4.7 wt-%	Mg	3.6 wt-%
S	3.9 wt-%	S	3.1 wt-%
Cl	0.1 wt-%	Cl	1.7 wt-%
Si	1.3 wt-%	Si	0.6 wt-%
Al	0.2 wt-%	Al	0.1 wt-%
Na	0.1 wt-%	Na	0.8 wt-%

2. A second type of deposits was a moderately thin, brittle, shell typed deposit, thickness between 0.5 and 5 mm. This type was found on the front side of the probe in the sampling location 2, in the convective part of the fluegas channel at a fluegas temperature of approximately 600 - 640°C. This deposit grew slowly but continuously as the sampling time was increased from 15 minutes to 21 hours. The deposit was fairly easy to remove from the probe. The semi-quantitative SEM/EDAX analyses of the probe rings (average of 3 overall analyses) detected Ca (17.5 wt-%), K (26.7 wt-%), P (5.2 wt-%), Mg (4.1 wt-%), S (4.4 wt-%) and Cl (8.8 wt-%) as the six main elements. The corresponding quantitative wet chemical analyses show Ca (18.9 wt-%), K (15.1 wt-%), P (3.6 wt-%), Mg (1.7 wt-%), S (4.6 wt-%) and Cl (3.5 wt-%). On top of those elements detected by the SEM/EDAX analysis, the deposit contained also small amounts of Si (2.6 wt-%), Al (0.3 wt-%), Fe (0.5 wt-%) and Na (0.4 wt-%).

3. The third type of deposit was a loose, dust typed one, with a thickness of 0.1 - 1 mm. This type was found on the back side of the probe in the sampling location 2. It was very easy to remove from the probe. The semi-quantitative SEM/EDAX analyses detected here Ca (25.0 wt-%), K (9.2 wt-%), P (6.2 wt-%), Mg (5.1 wt-%), S (2.2 wt-%) and Cl (2.3 wt-%) as main elements. These analyses seem again to be fairly similar to those values determined by the wet chemical analyses (Ca 24.6 wt-%, K 10.1 wt-%, P 4.5 wt-%, Mg 2.8 wt-%, S 2.3 wt-%, Cl 1.3 wt-%). The deposit contained also Si (5.0 wt-%), Al (0.6 wt-%), Fe (0.7 wt-%) and Na (0.3 wt-%).

4. The fourth type of deposit found during the tests was a thick, grayish, deposit, thickness 1 - 5 mm. This type of a deposit was found on the front side of the probe in the sampling location 3, in the convective part of the fluegas channel after the heat exchanger surfaces at a fluegas temperature of approximately 280 - 360°C. This deposit was the thickest one found in all the sampling locations. The SEM/EDAX analyses indicate Ca (26.9 wt-%), K (10.2 wt-%) and Si (8.5 wt-%) as the main

elements and small amounts of P (3.0 wt-%), Mg (1.5 wt-%), and S (1.8 wt-%). The quantitative wet chemical analyses show the amounts of these elements being roughly the same, i.e., Ca (23.4 wt-%), K (9.2 wt-%), Si (8.5 wt-%), P (4.6 wt-%), Mg (3.0 wt-%), and S (2.9 wt-%). Cl was determined to be 0.1 wt-% in this deposit. The back side of the sampling probe in this location appeared visually to be clean. SEM/EDAX analyses showed, however a thin deposit layer attached to the ring surface, consisting almost completely of K (31.8 wt-%) and Cl (20.6 wt-%).

3.4 Evaluations of chemistry in the bed and deposits

Based on the quantitative wet chemical analyses, we evaluated the distribution of different ash elements in the boiler by using an enrichment factor, FE. We defined the factor as the share of a certain element, related to calcium in one sample divided by the share of that same element, related to calcium in the in-going fuel. The factor defined in this way gives a value of 1 for the fuel. A value above one indicates enrichment, below one depletion. The reason to relate the amounts to calcium is that we in this way don't get diluting effects caused by the bed. The analytical expression of the FE for potassium in the deposit is shown below

$$EF_{K, DEP} = \frac{K/Ca_{deposit} \ (mol/mol)}{K/Ca_{fuel} \ (mol/mol)} \tag{1}$$

In Figure 5 the enrichment factors for K, Cl and S are shown as a function of the different sampling locations in the boiler. For potassium some enrichment can been seen in the bed while the rest of the samples seem to have some depletion, compared to the fuel. Chlorine seems to be enriched in deposits and bag house samples. Very little chlorine is found in bed, primary cyclone or secondary cyclone samples. Sulfur is enriched in all samples.

Two clear implications from these distribution evaluations can be drawn. One is that alkali salts may be present in the deposits on heat exchangers right after the cyclone even if the fuel contains low amounts of chlorine and sulfur. This may cause fouling and corrosion. The second is that since potassium is absorbed in the bed, preferably as silicate, it may cause bed agglomeration through a formation of a molten potassium silicate phase.

It was therefore of interest to evaluate also the melting behavior or stickiness of the bed and the deposits. This was done partly by using thermodynamic multi component, multi phase equilibrium calculations [7] partly by extracting melting behavior data from existing phase diagrams. Inorganic mixtures like those found in fuel ashes do not melt at one certain temperature, but have a wide temperature range where both a solid and a liquid phase is present. If the sample, in our case a bed sample, a fly ash or a deposit, contains a high enough portion of molten phase, it will cause problems. In the bed this would be seen as bed agglomeration, in a deposit or a fly ash as a growing deposit. In certain types of boilers it has been found that a value of approximately 10 - 20 % melt in a fly ash will make it sticky [8, 9]. For silicate melts one has also to take into consideration the viscosity of the silicate melt [10].

The melting behavior evaluations of a mixture with a composition similar to that which was analyzed in the bed particles during the tests proved to be very sensitive to the content of aluminum. A pure potassium silicate has a first melting temperature, T_0, of 742°C in the range of a K/Si ratio of 0.25 - 0.5, which is the range in which the bed was in these tests. This kind of a mixture forms immediately a large amount of molten phase (30 - 100 % by weight depending on the K/Si ratio). A small addition of aluminum decreases at first the T_0 to 695°C, but a further increase in the amount of added Al will increase the T_0 to 810°C. An excess amount of aluminum will further shift the T_0 even higher, up to roughly 1030°C. Based on phase diagram analyses of the three dimensional Al-K-Si diagram and using the elemental analyses of the bed particles (EDAX point analyses 1 and 5 presented in Figure 3) we evaluated the T_0 to be 810°C. A small increase in the amount of potassium would decrease the T_0 down to 695°C while a small increase in aluminum would increase the T_0 to 1030°C. The presently available thermodynamic data are not accurate enough for any further melting behavior calculations.

Also the viscosity of the formed melt would change. Potassium is known to lower the viscosity of a silicate melt while aluminium in some cases increases it.

For the Salix combustion these evaluations imply that the bed needs attention. Even if no bed agglomeration was experienced in the tests performed here, already a small shift in bed composition may cause problems. An aging of the bed i.e., no bed material changes during a long period may for example lead to problems. A small change in the ingoing fuel may also lead to bed agglomeration.

In Figure 6 a similar evaluation of two deposits is presented, one with the composition of the type 3 deposit and another with the composition of the type 4 deposit. In this case the available thermodynamic data are accurate enough for melting behavior calculations. In both cases a first melting point, T_0, is reached at approximately 620°C. In the case with the type 3 deposit, the amount increases to a couple of percent immediately but stays below some 15 % melt in the mixture up to roughly 900°C. The type 4 deposit has a very low amount of melt throughout the temperature range. Two implications from these results can be drawn for deposit formation during Salix combustion. One is that neither of the deposits collected from the convective part of the fluegas channel seemed to contain any significant amount melt, conclusively ruling out this as a deposit forming mechanism here. The second implication is, hence, that some other deposit formation mechanism seems to be dominating for the type 4 deposit.

The cause for the type 4 deposits found in the sampling location 3 may be explained with a mechanism involving the reaction of free lime in the fly ash particles with $CO_2(g)$. It has earlier been shown [11] that if CaO is found in a fly ash particle as it hits a surface, it may react at the surface with $CO_2(g)$ and form a deposit. CaO particles recarbonizing to $CaCO_3$ have further in lab scale been shown to cause significant neckgrowth between the particles [12]. In the case with the type 4 deposit some rest CaO in the fly ash may have caused the deposit.

4 Conclusions

Most of the test results imply that Salix can be fired in a circulating fluidized bed with reasonable success. However, some questions need to be addressed as they may cause problems.

The fuel moisture content in these tests (>50 %) was too high, causing high CO emissions from the combustion. This needs to be solved before Salix can be used more generally as a fuel.

Potassium seemed to be somewhat enriched in the quartz bed. With the bed temperatures used in these tests the potassium reacted with the quartz, and formed silicates. Potassium is known to lower both the first melting temperature of silicate mixtures and the viscosity of the formed melt. A significant share of melt in a fluidized bed with a low enough viscosity will cause bed agglomeration. This may be a potential problem especially if the bed is not changed often enough.

Sulfur and chlorine was found in the deposits, in all sampled locations on the back side of the tube surfaces, in the furnace at the primary cyclone inlet and in the hotter convective sampling location also on the front side. Potassium was further found in all sampled deposits. Sulfur and chlorine together with potassium may at suitable concentrations form a significant amount of melt already at 550 - 600°C. For full scale combustion in a CFB this implies possible corrosion of the heat exchanger back sides. Even if no sever deposit build-up was detected during the test runs reported here, a concern needs also to be raised of the fact that a molten phase may be present in the fly ash and through that form deposits.

A recarbonation of the calcium oxide in that fly ash fraction which was transported out through the primary cyclone may have caused the deposits detected in the sampling location after the heat exchangers in the convective part of the fluegas channel. This reaction has earlier been shown to cause deposits.

5 References

1. Dawson, M., Brown, R., C.: Fuel 71, 585 (1991).
2. Baxter, L.: Biomass and Bioenergy 4 (2), 89 (1993).
3. Nordin, A.: Fuel 74, 615 (1995).
4. Nordin, A., Skrifvars, B.-J., Öhman, M., Hupa, M.: Agglomeration and defluidization in FBC of biomass fuels - Mechanisms and measures for prevention, presented at the Engineering Foundation Conference on "Applications of advanced technology to ash-related problems in boilers", Waterville Valley, NH, USA, July 1995.
5. Lyngfelt, A.: Abatement of SO₂ emissions from fluidised bed boilers, Dr. Thesis, Chalmers University of Technology, 1988.
6. Åmand, L-E.: Nitrous oxide emission from circulationg fluidized bed combustion, Dr Thesis, Chalmers University of Technology, 1994.
7. Eriksson, K., Hack, K.: Metallurgical Trans. 21B, 1013 (1990).
8. Backman, R., Hupa, M., Uppstu, E.: Tappi J. 70 (6), (1987).

9. Backman, R.: Sodium and sulfur chemistry in combustion gases, Dr. Thesis, Åbo Akademi University, Åbo/Turku, Finland, 1989.
10. Skrifvars, B.-J., Hupa, M., Backman, R., Hiltunen, M.: Fuel 73 (2), 171 (1994).
11. Skrifvars, B.-J., Hupa, M., Hyöty, P.: J. Inst. Energy 64, 196 (1991).
12. Skrifvars, B.-J.: Sintering tendency of fuel ashes in combustion and gasification conditions, Dr. Thesis, Åbo Akademi University, Åbo/Turku, Finland, 1994.

Acknowledgments

The Salix combustion test runs were performed at the circulating fluidized bed boiler located at the Technical University of Chalmers, Gothenburg, Sweden. The technical support from the operators at the boiler as well as from the scientists at the Department of Energy Conversion, is gratefully acknowledged.

This work was supported by the Swedish National Board for Industrial and Technical Development, Swedish Farmers Foundation for Agricultural Research, Vattenfall AB, the Nordic Energy Research Program and the Finnish National Combustion and Gasification Research Program, LIEKKI-2.

Table 1. Running conditions of the boiler during the Salix combustion tests.

	Start	End	Fuel		Temperatures										Fluegases					
			Salix	Wood	Bed		Cyc. inl.		Cyc. outl.		Conv.hot		Conv.cold		CO		O₂		SO₂	
					min	max	min	max	min	max	min	max	min	max	min	max	min	max	min	max
			kg/s	kg/s	°C	°C	°C	°C	°C	°C	°C	°C	°C	°C	ppm	ppm	%	%	%	%
A	13.12.1994 9:00	14.12.1994 18:55	1.11	0.00	770	784	812	820	803	830	595	630	283	325	1400	3800	6.0	6.1	0.0	0.0
B	15.12.1994 10:50	18.12.1994 8:45	0.81	0.14	831	843	854	858	830	837	605	618	307	332	2200	3900	5.3	6.2	0.0	0.0
C	19.12.1994 12:00	21.12.1994 7.35	0.82	0.17	840	844	860	866	848	857	636	642	358	363	3600	6100	5.1	5.5	0.0	

Table 2. Analyses of the fuels used during test Salix tests.

Fuel analyses		Salix	Wood	Ash analyses		Salix	Wood
Moist	wt-%	55.0	7.0	Ash	wt-% (db)	2.1	0.5
C	wt-% (db)	49.7	50.9	Si	wt-% (ash)	2.8	2.9
H	wt-% (db)	6.2	6.3	Al	wt-% (ash)	0.3	0.7
N	wt-% (db)	0.6	0.1	Ca	wt-% (ash)	27.3	29.6
S	wt-% (db)	0.0	0.1	Fe	wt-% (ash)	0.9	0.7
Cl	wt-% (db)	0.0	0.0	K	wt-% (ash)	14.7	6.8
				Mg	wt-% (ash)	2.3	2.8
HHV	MJ/kg (db)	19.8	20.4	Mn	wt-% (ash)	0.3	2.0
LHV	MJ/kg (db)	18.4	19.0	Na	wt-% (ash)	0.4	0.6
				P	wt-% (ash)	4.3	0.7

Table 3. SEM/EDAX point analyses of bed samples taken during condition C of the Salix tests. Analyses points indicated with corresponding numbers in the image in Figure 3.

Point #	Si	Al	Ca	Fe	K	Mg	Mn	Na	P	S	Cl
						weight-%					
1	32.8	7.2	0.0	0.0	13.4	0.0	0.0	0.0	0.0	0.0	0.0
2	0.8	0.0	22.3	0.7	12.0	9.9	1.1	0.6	8.5	5.4	0.0
3	1.5	0.1	33.5	0.2	1.5	9.0	0.6	0.1	12.9	0.8	0.0
4	14.5	0.0	40.9	0.0	3.9	0.0	0.0	0.0	1.8	1.2	0.0
5	26.6	6.3	5.4	0.0	19.7	0.0	0.0	0.0	0.0	0.0	0.0
6	0.3	0.1	32.8	0.2	15.1	3.8	0.9	0.5	4.2	6.8	0.0

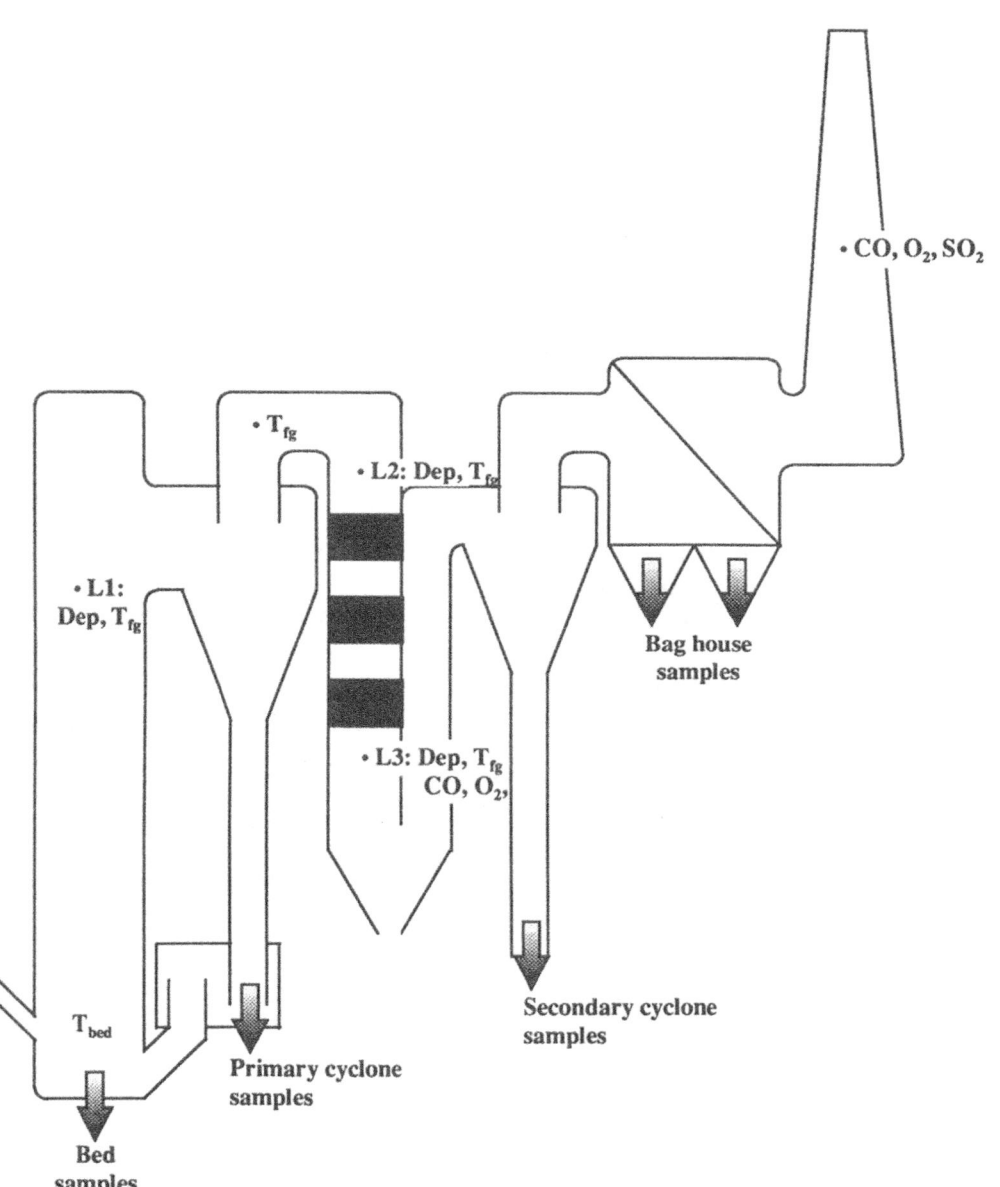

・ CO, O_2, SO_2

・ T_{fg}

・ L2: Dep, T_{fg}

・ L1: Dep, T_{fg}

Bag house samples

・ L3: Dep, T_{fg} CO, O_2,

T_{bed}

Secondary cyclone samples

Primary cyclone samples

Bed samples

Figure 1. A schematic view of the CFB boiler in which the Salix combustion tests were performed. Sampling locations indicated in the figure.

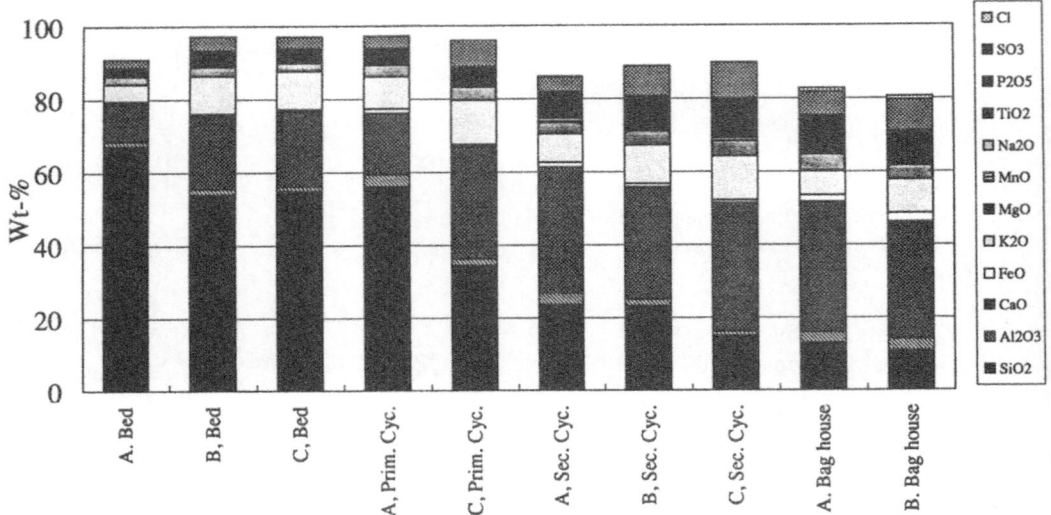

Figure 2. Quantitative wet chemical analyses of bed, primary cyclone, secondary cyclone and bag house samples, taken during the three different running modes A, B and C of the Salix combustion tests. The elements expressed as weight-% of their corresponding oxides, except for Cl. An unamed column has the same sample identification as the first identified column on its left side.

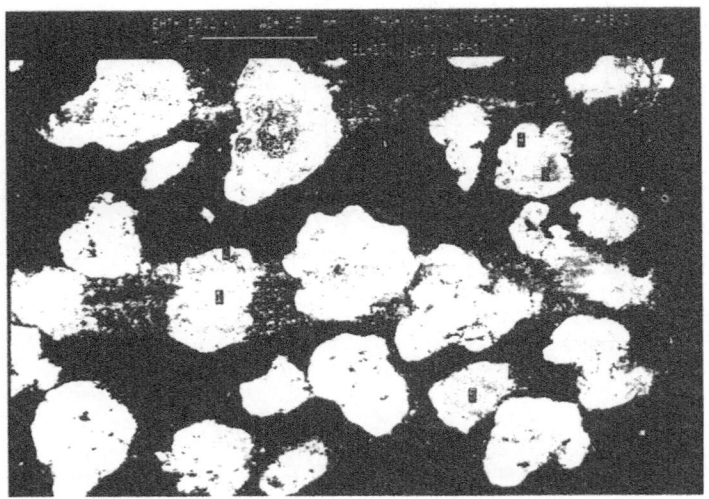

Figure 3. SEM image of a bed sample, collected from the CFB boiler during running condition C of the Salix combustion tests. Magnification: x100. Numbers in the image indicate EDAX point analyses, presented in Table 3.

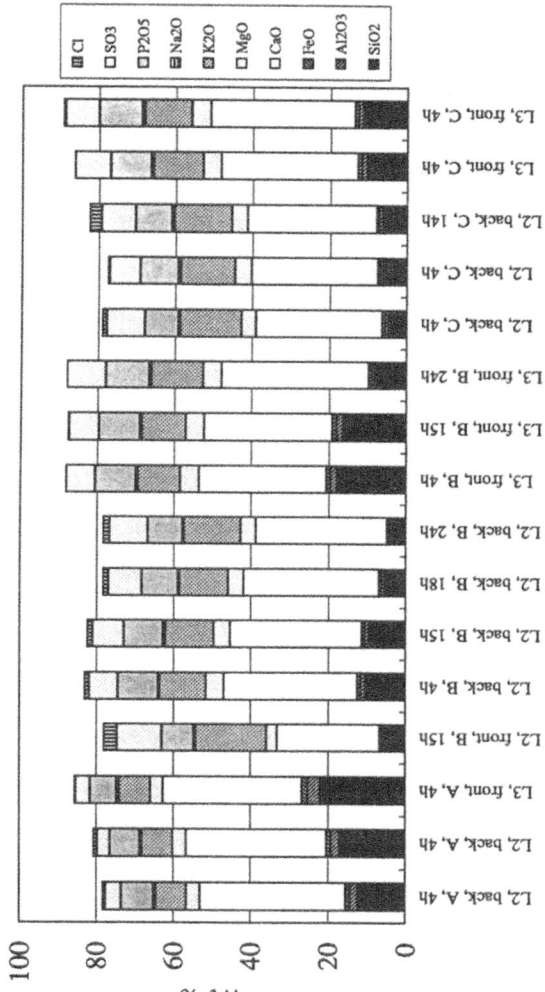

Figure 4. Quantitative wet chemical analyses of front and back side deposits, collected on the deposit probes from the sampling locations 2 and 3 in CFB boiler during the Salix combustion tests. The elements expressed as weight-% of their corresponding oxides, except for Cl. The deposit sampling time indicated on the x-axis.

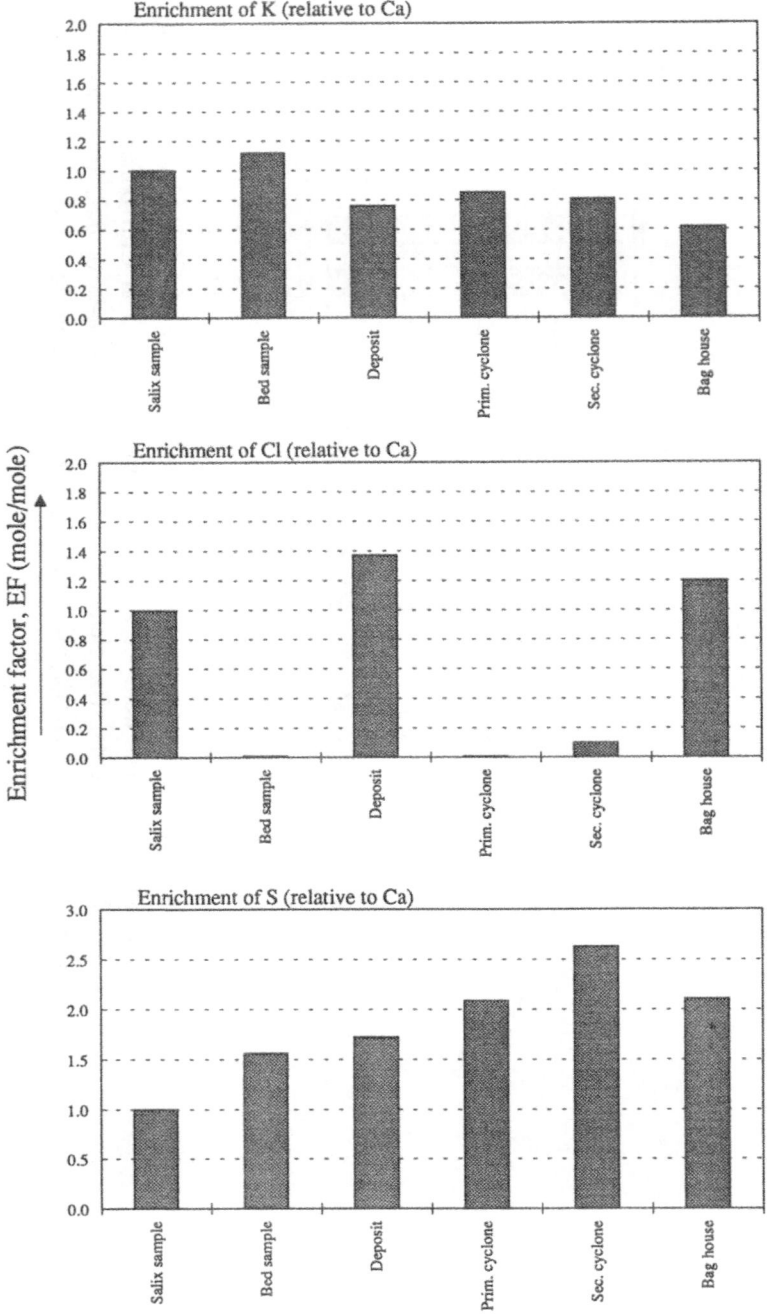

Figure 5. Enrichment factors for potassium, chlorine and sulfur in different locations of the boiler.

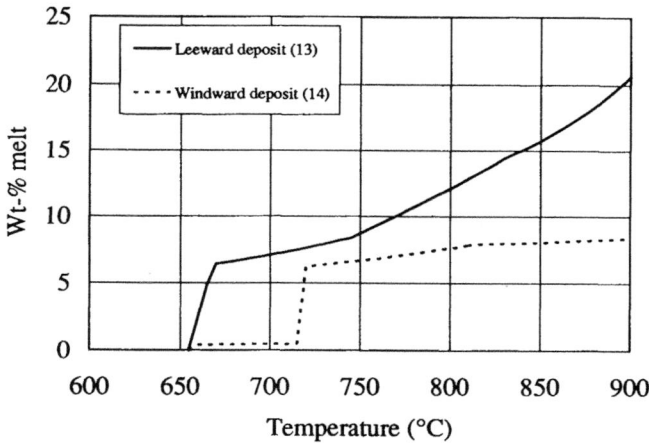

Figure 6. Melting behavior of back side (leeward) deposit, collected from the the sampling locations 2 and front side (windward) deposit collected from location 3 in CFB boiler during the Salix combustion tests. The calculations assume the salts and silicates to be subject for melting. Rest is assumed solid inerts. The amount expressed as weight-%.

THE BEHAVIOR OF INORGANIC MATERIAL IN BIOMASS-FIRED POWER BOILERS: AN OVERVIEW OF THE ALKALI DEPOSITS PROJECT

Inorganic Material Behavior in Boilers

L. L. BAXTER
Sandia National Laboratories, Combustion Research Facility, Livermore, CA, USA
T. R. MILES and T. R. MILES JR.
Thomas R. Miles Consulting Design Engineers, Portland, OR, USA
B. M. JENKINS
University of California, Davis, Biological and Agricultural Eng. Dpt., Davis, CA, USA
T. A. MILNE and D. DAYTON
National Renewable Energy Laboratory, Golden, CO, USA
R. W. BRYERS
Foster Wheeler Development Corp. (Retired), Livingston, NJ, USA
L. L. ODEN
Bureau of Mines, Albany Research Center, Albany, OR, USA

1. Summary

This paper highlights some of the major findings of the Alkali Deposits Investigation, a collaborative effort to understand the causes of unmanageable ash deposits in biomass-fired electric power boilers. A group of interested industrial institutions and the US DOE Energy Efficiency and Renewable Energy Office's Biomass Power Program through the National Renewable Energy Laboratory jointly sponsored the project. The industries contributed both funding and, in most cases, use of facilities to the project and included Mendota Biomass Power Ltd. and Woodland Biomass Power Ltd. (both associated with Thermo Electron Energy Systems), CMS Generation Operating Co. (formerly Hydra-Co Operations Inc.), Wheelabrator, Shasta, and Hudson Energy Cos., Sithe Energy Co., Delano Energy Co. Inc., the Electric Power Research Institute, Foster Wheeler Development Corp., and Elkraft Power Co. Ltd. of Denmark. Research contracts with Thomas R. Miles Consulting Design Engineers, Sandia National Laboratories, and The National Renewable Energy Laboratories provided the government portion of the funding. In addition, the University of California at Davis and the Bureau of Mines performed significant work in close collaboration with the other researchers. This summary highlights the major findings of the project more thoroughly discussed in a recent report (Baxter et al., 1996). We highlight fuel properties, bench-scale combustion tests, a framework for considering ash deposition processes, pilot-scale tests of biomass fuels, and field tests in commercially operating biomass power generation stations.

Detailed chemical analyses of eleven biomass fuels representing a broad cross-section of commercially available fuels reveal their properties that relate to ash deposition tendencies. The fuels fall into three broad categories: (1) straws and grasses (herbaceous materials); (2) pits, shells, hulls and other agricultural byproducts of a generally ligneous nature; and (3) woods and recycle fuels of commercial interest. Woods and wood-derived products represent the most commonly used biomass fuels.

Herbaceous fuels contain silicon and potassium as their principal ash-forming constituents. They are also commonly high in chlorine relative to other biomass fuels. These properties portend potentially severe ash deposition problems at high or moderate combustion temperatures. The primary sources of these problems shown to be: (1) the reaction of alkali with silica to form alkali silicates that melt or soften at low temperatures (can be lower than 700 °C, depending on composition), and (2) the reaction of alkali with sulfur to form alkali sulfates on combustor heat transfer surfaces. Alkali material plays a central role in both processes. The mobility of alkali material, defined as its ability to come in physical contact with other materials, is measured using chemical extractive techniques. Potassium is the dominant source of alkali in most biomass fuels. The analyses below indicate that essentially all of the biologically occurring alkali, in particular potassium, has high mobility. The non-biologically occurring alkali is present as soil contaminants and additives to the fuels, such as clay fillers used in paper production. This non-biologically occurring alkali exhibits far lower mobility than the biological fraction. The relative amounts of biologically vs. non-biologically occurring material depend on fuel type and fuel handling. In the fuels investigated here, the dominant form of alkali was biologically occurring potassium. Some traditional indicators of deposit behavior, most notably ash fusion temperatures, poorly predict ash behavior compared with a more mechanistic interpretation of the data.

Many of the agricultural byproducts also contain high potassium concentrations with equally high potassium mobility. Some woods, on the other hand, contain far less ash overall, differing by as much as a factor of 40 from high-ash straws, for example. In addition, the ash-forming constituents contain greater amounts of calcium with less silicon. The total amount of potassium in wood is much lower than in straws, although this is not necessarily the case when expressed as a fraction of total ash. Calcium reacts with sulfur to form sulfates in ways somewhat analogous to potassium, but the lower mobility and vapor pressure of calcium and the structural properties of the deposits it forms are both more favorable to sustained furnace operation than the ashes formed from straws and grasses.

Chlorine is shown to be a major factor in deposit formation. Chlorine facilitates the mobility of many inorganic compounds, in particular potassium. Potassium chloride is among the most stable high-temperature, gas-phase, alkali-containing species. Chlorine concentration often dictates the amount of alkali vaporized during combustion more strongly than the alkali concentration in the fuel. In most cases, the chlorine appears to play a shuttle role, facilitating the transport of alkali from the fuel to surfaces, where the alkali often forms sulfates. In the absence of sulfur, chlorides often reside on the surface. In the absence of chlorine, alkali hydroxides are the major stable gas-phase species in moist, oxidizing environments (combustion gases).

Bench-scale combustion tests coupled with mass spectrometry techniques reveal the major species evolving from biomass samples during combustion. The fuels subjected to these combustion tests were also examined by chemical analyses. The tests indicate that the major release of alkali material occurs during the char combustion phase and that the primary form of stable alkali-bearing off-gases corresponds with thermodynamic estimates of product stability and vapor pressure. Investigations were performed varying temperature, oxygen concentration, and moisture levels and revealed thermodynamically consistent results; the amount of alkali vaporized increases with temperature and the amount of hydroxide formed increases with increasing moisture content. Details of other species released are also presented. Thermogravimetric and differential thermogravimetric analyses of the samples were also performed, the results of which prove to be marginally useful to predicting ash deposition.

A conceptual framework expresses ash deposition as a combination of four mechanisms: inertial impaction, thermophoretic deposition, condensation, and chemical reaction is presented. This conceptual framework serves to organize the remaining discussion and observations of ash deposition at both pilot and commercial scale. The influences of boiler design, boiler operating conditions, and fuel properties on ash deposit behavior reveal themselves by emphasizing or reducing the role of one or more of these mechanisms and thereby changing deposit composition, phase, and properties.

Pilot-scale investigations on the standard suite of fuels were carried out in an entrained-flow furnace. Isokinetically sampled fly ash samples and deposits collected on instrumented, temperature-regulated probes simulating both waterwalls and convection pass provide the basis for *in situ* and subsequent *ex situ* examination of deposits. Deposit properties reveal spatial dependencies on probe surfaces that are consistent with both the commercial-scale tests and the conceptual framework for deposit growth and property development. In addition, changes in deposit composition with time, temperature, and other operation-relevant variables exhibit the same consistencies with commercial operation and the conceptual framework. Fuels containing high alkali and silica concentrations form alkali silicates that melt or sinter at low temperatures. The rates of deposit growth and sintering/melting increase with increasing temperature and chlorine concentration but are high at all boiler-relevant temperatures and chlorine concentrations for the straws and grasses. Surfaces exposed to impacting particles can accumulate silica and alkali silicates at very rapid rates. Surfaces exposed to combustion gases but less exposed to particle impaction show evidence of thermophoretic accumulation of deposits, vapor condensation, and sulfation of condensed alkali-laden vapors. These processes lead to deposits with markedly different properties (high reflectivity, modest thickness and growth rate) compared to the impacted regions. Fuels containing little alkali or silica indicate far less deposit growth and development of more manageable deposits, by which we mean soot blowers and boiler maintenance techniques are able to sustain operation of a facility for periods of many months without unscheduled shutdowns.

Commercial-scale investigations using nearly every type of commercially significant biomass boiler design provide full scale data for analysis and comparison with pilot-scale results. Bubbling fluid beds, circulating fluid beds, and various grate-based combustors fed by stokers, augers, and a cigar burner provided the data for comparison. Fuel types ranged from wood-derived material with blended agricultural byproducts to straws, sometimes blended with urban wood fuel. Many of the commercial-scale experiments were conducted in the context of commercial operation and employed varying compositions in the fuel. In all cases, comparisons of deposit composition with position in the boiler, type of deposit surface, position on the surface, and fuel properties reveal complete consistency with the fuel analyses, bench-scale combustion results, pilot-scale results, and the conceptual framework. Fuels containing high alkali and silica fractions exhibited the same rapid accumulation of ash deposits with the same sintered/molten character as was observed in the pilot-scale tests. Advanced mineralogical examinations of selected deposits indicate chemical compositions consistent with the conceptual framework and are presented as appendix material. The deposit properties are consistent with the conceptual framework for their formation and the observed bench-scale combustion results. Fuels with less alkali, chlorine, and silica, with less total ash-forming material, and with higher calcium contents exhibit more manageable ash deposits. Wood and non-recyclable paper generated the most manageable deposits.

This report provides highlights of a more detailed analysis of fuel property, operating condition, and boiler design issues that dictate ash deposit formation and property development (Baxter et al., 1996). The span of investigations from bench-top experiments to commercial operation and observations, including both practical illustrations and theoretical background, provides a self-consistent and reasonably robust basis to understand the qualitative nature of ash deposit formation in biomass boilers. While there remain many quantitative details to be pursued to complete our understanding, this project encapsulates essentially all of the conceptual aspects of the issue. It provides a basis for understanding and potentially resolving the technical and environmental issues associated with ash deposition during biomass combustion.

2. Introduction

This document summarizes the principal results and conclusions of the Alkali Deposits Project, a collaborative project designed to identify the causes of ash deposition in biomass-fired combustors used for electric power generation and suggest ways it can be managed (Miles et al., 1996, Baxter et al., 1996). The principal collaborators in this effort include T. R. Miles and T. R. Miles Jr. of T. R. Miles Consulting Design Engineers, B. M. Jenkins of UC Davis, L. Oden of the U.S. Bureau of Mines, L. Baxter of Sandia National Laboratory and a group of ten industrial sponsors. T. R. Miles was principal investigator of a project separately funded by NREL and by industrial sponsors formally entitled Alkali Deposits Found in Biomass Power Plants. Larry Baxter is principal investigator in an ongoing Sandia project complementing that of Miles. In addition, the contributions of D. Dayton and T. Milne of the National Renewable Energy Laboratory are included. This collaborative project combines the resources of industry, design engineers, academia, and government-supported laboratories to address ash deposition, arguably the most critical constraint in burning a wide variety of biomass fuels in existing boilers.

Summarized here are the fuel characteristics, combustion behavior, and ash deposition characteristics for a wide variety of biomass fuels and commercial biomass-fired boilers, only highlights of which can be presented here. Data are derived from laboratory-, pilot-, and commercial-scale operations. Chemical compositions of deposits from the laboratory experiments are reported, and compared with the composition of deposits from commercial boilers and the composition of key feedstocks tested in the laboratory and field.

The scope of the Alkali Deposits Project included fuels characterization, laboratory combustion tests, and field combustion tests at several sites. The focus was on major inorganic species, i.e., silicon, potassium, calcium, aluminum, iron, magnesium, sodium, sulfur, phosphorus, titanium, and chlorine. Information on trace metals in combustion systems is available elsewhere (Tillman, 1994). Fuels characterizations include proximate, ultimate, and ash chemistry analyses of fuels either in use by or of interest to the biomass power generating industry. In addition, modifications to advanced analyses were made that allowed the mode of occurrence of inorganic material to be distinguished. Laboratory combustion tests of a suite of test fuels were conducted in the Multifuel Combustor (MFC) at Sandia. In addition, the MFC was used to perform combustion tests on both non-blended fuels and fuel blends from the industrial collaborators. Fuel samples were collected during the tests and pre- and post-test boiler inspections were performed. During post-test inspections, ash deposits were removed from several locations in the boiler and submitted to a variety of chemical and physical analyses.

3. Fuel Properties

Comprehensive analyses were performed for the series of eleven fuels summarized in Table 1. These are among the 25 fuels that were tested in Sandia's Multifuel Combustor (MFC). Those designated as commercial fuels were also tested at commercial facilities, as described later. The wood/almond shell blend was tested at Mendota and the wood/wheatstraw blend at Hydra-Co, results from which are discussed later. The non-recyclable paper was tested at Wheelabrator primarily for emissions purposes. No ash-related, commercial-scale results were collected from Wheelabrator during these tests as part of this project, although the laboratory results were available to Wheelabrator during their successful bid to obtain a permit to burn such fuels. Samples of many of the fuels were also sent to NREL for mass spectrometry measurements (discussed below), to Foster Wheeler Development Corp. for thermal decomposition and differential thermal decomposition analyses (see appendix), to the US Bureau of Mines for ash chemistry analyses, and to other participating organizations. In this report, non-qualified names of fuels represent different aliquots of the eleven fuels from the MFC laboratory. The scientific names indicated in the table are, in some cases, presumptions. It is not known precisely which woods are included in the commercial fuels, for example, but the most likely candidates include ponderosa pine and douglas fir (two typical sources of lumber).

Fup 1		

Table 1   Summary of fuels used during MFC combustion tests.

Fuel common/ scientific name	Moisture (% fuel)	Ash (% dry fuel)	Higher Heating Value (MJ/kg, daf)
Straws and Grasses (Herbaceous Fuels)/ Family *Gramineae*			
Rice Straw/ *Oryza sativa*	11.22	19.17	18.74
Wheat Straw/ *Triticum aestivum*	8.39	8.08	19.31
Switchgrass/ *Panicum virgatum*	12.98	5.86	19.91
Shells, Pits, and Hulls (Ligneous Fuels)/ Families *Anacardiaceae* (Pistachio), *Oleaceae* (Olive), *Rosaceae* (almond)			
Almond Shells/ *Prunus dulcis*	7.52	2.87	19.83
Pistachio Shells/ *Pistacia vera*	8.08	1.28	19.90
Olive Pits/ *Olea europaea*	6.97	1.83	21.97
Almond Hulls/ *Prunus dulcis*	8.02	5.75	20.00
Wood and Commercial Fuels Families *Pinaceae*			
Wood/Almond Shell Blend (Mendota Fuel)/ *Pinus ponderosa*/ Pseudotsuga menziesii/*Prunus dulcis*	8.62	7.54	19.97
Nonrecyclable Paper i (Wheelabrator Fuel)/ *Pinus ponderosa/ Pseudotsuga menziesii*	5.95	8.21	23.44
Nonrecyclable Paper ii (Wheelabrator Fuel)/ *Pinus ponderosa/ Pseudotsuga menziesii*	5.95	8.21	23.44
Wood/Wheat Straw Blend (Hydra-Co Fuel) *Pinus ponderosa/ Pseudotsuga menziesii/Triticum aestivum*	9.25	7.33	20.56

4.   Bench-Scale Combustion Tests

The evolution of various species during the combustion of biomass was traced using a molecular beam mass spectrometer (MBMS) coupled to a bench scale combustion experiment. The equipment and results are discussed in detail in the larger report (Baxter et al., 1996). An illustration of the interpretation of the results is indicated here, using the gas- and condensed-phase behavior of chlorine as an example. As is illustrated, chlorine and alkali behavior can be coupled and this coupling explains some aspects of how ash deposit structure develops.

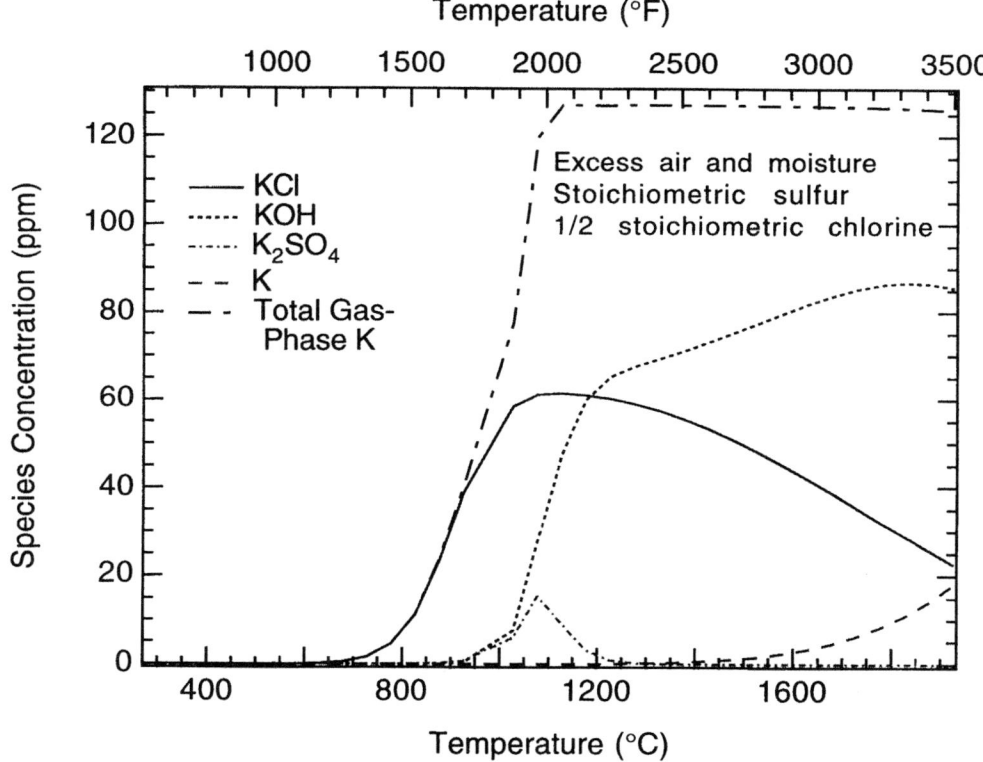

Figure 1   Equilibrium species concentrations for the major potassium-containing, gas-phase species present under typical biomass combustion conditions. Compare with condensed-phase behavior illustrated in Figure 2.

The herbaceous materials present an opportunity to examine the role of chlorine in formation of ash deposits. Chlorides represent among the most stable alkali-bearing species in the gas phase. Essentially all of the fuel chlorine is released early in the combustion process and much of it combines with available alkali to form alkali chloride vapors. In many cases, the amount of alkali vaporized during biomass combustion is determined more by the amount of chlorine available to form stable vapors than by the amount of alkali in the fuel. Equilibrium vapor pressures of representative materials exhibit this tendency. Figure 1 illustrates predicted equilibrium concentrations of gas-phase, potassium-containing species under conditions typical of many biomass combustors. Condensed-phase results are illustrated in Figure 2. The overall stoichiometry of this calculation assumes excess air and moisture and sufficient sulfur to convert all potassium to sulfate. In terms of gas-phase chemistry, few biomass fuels contain this much potassium. (The rationale for choosing these conditions will be explained shortly). Even at these conditions of high sulfur concentration, gas-phase sulfate is seen to play a relatively minor role in potassium chemistry. Peak sulfate concentrations represent about 10 % of the total gas-phase potassium and occur at about 1100 °C. At lower temperatures, potassium sulfate vapor condenses to form liquid or solid sulfate. At higher temperatures, it decomposes.

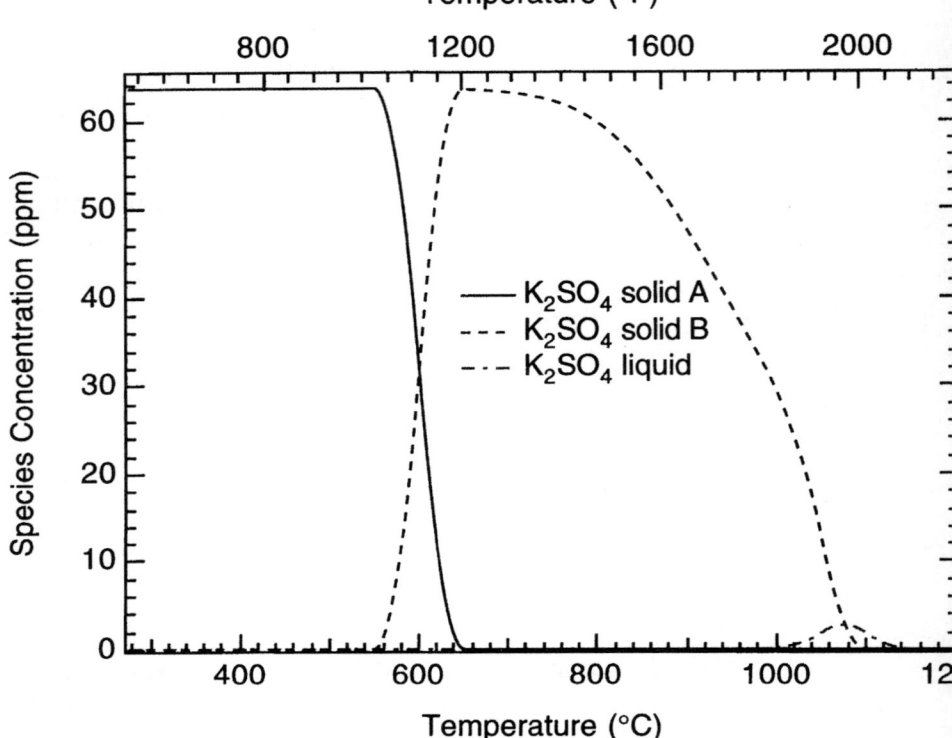

Figure 2   Condensed-phase equilibrium behavior of potassium-containing species as a function of temperature under conditions similar to those in many biomass boilers. Compare with gas-phase behavior illustrated in Figure 1.

The dominant gas-phase, potassium bearing species at flame temperatures (>1400 °C) is potassium hydroxide, followed by the chloride. In the absence of significant chlorine for reaction, only the hydroxide is present. As temperatures cool to convection-pass values (<1000 °C), hydroxides convert to chlorides, the only potassium-bearing species in significant quantities at lower temperatures. In these calculations, there is sufficient total chlorine to convert half of the potassium to chloride. The fact that nearly all of the gas-phase potassium exists as a chloride at intermediate and low temperatures indicates that chlorine is a strong controller of the total gas-phase potassium budget at these temperatures. This is verified by MBMS results, where total potassium release is measured as a function of fuel potassium content and fuel chlorine content. Essentially all of the chlorine in a fuel is released early in the combustion process. This chlorine has a strong influence on the amount of potassium that is found in the vapor.

The condensed-phase behavior of potassium-containing compounds is illustrated in Figure 2 as a function of temperature. In performing these calculations, we have assumed that there is sufficient sulfur to convert all of the potassium to sulfate. Most biomass fuels have less sulfur than the stoichiometric amount required for such a conversion. However, an ash deposit on a surface interacts with a continuous gas stream, providing a continuous source of sulfur. Often, the rate of accumulation of alkali in the deposit is slow compared to the rate of diffusion of sulfur from the bulk gas stream to the deposit surface. Under these conditions, the deposit has opportunity to react with a much larger amount of sulfur than the elemental composition of the fuel may suggest.

As illustrated in the figure, sulfur is completely converted to sulfate at the lower temperatures, even though there are ample reactants to form carbonates and chlorides. Two solid phases of sulfate are indicated, distinguished by A and B. If there is less sulfur in the gas phase than is required to convert all of the alkali to sulfate, chlorides and carbonates form, in that order. The MFC data and the field data reflect these trends in that chlorides and sulfates are commonly found on heat transfer surfaces, and their concentration increases with decreasing temperature.

5. Pilot-Scale Combustion Tests

5.1. Experimental Facilities

Deposit samples were obtained by burning various biomass fuels in the Multifuel Combustor (MFC) at the Combustion Research Facility of Sandia National Laboratories (Figure 3). Gases flow through a series of modular sections that include guard heaters, fuel insertion ports, and thermocouple insertion ports. At the end of these modular heaters is the test section of the combustor. In this section, several air-cooled deposition surfaces that simulate waterwalls and convection pass tubes are inserted in the particle-laden, vitiated flow. In the experiments described here, surface temperatures of these tubes are measured using thermocouples. Surface temperature was held constant in each individual experiment but varied from 350 to 650 °C among the several experiments performed. Advanced diagnostics (e.g., FTIR emission spectroscopy) are used to measure deposit chemical composition and emissivity as a function of time during ash deposition. Deposits are also sampled from the furnace and submitted for SEM-based analyses as well as elemental analyses using atomic emission/absorption spectroscopy with an inductively coupled plasma following digestion. Details regarding the application of the combustor and its advanced diagnostics for coal-related work are available elsewhere (Baxter, 1993). During these investigations, milled biomass fuel passing a 10 mesh sieve or finer was injected pneumatically via a water-cooled lance inserted through the side of the furnace just below the gas burner at the top. The fuel was fired downward from a position about 4 m above the test section, producing a particle residence time of 1 to 2 s, depending of fuel type and other operating conditions. This provided for essentially complete carbon conversion of the fuels within the combustor. For most tests reported here, the furnace wall temperature was set at 900 °C to simulate a typical biomass combustor furnace exit gas temperature ahead of the superheaters. Typically, the gas burner was not used to preheat the air, with some exceptions when simulating very high temperature systems.

The fuels used for the MFC experiments reported here (Table 1) included almond hull, almond shell, olive pits, rice straw, switchgrass, wheat straw, and a blend of urban and agricultural wood fuel with almond shell. The fuels selected provided a wide range of inorganic element concentrations.

Figure 4 is a scanning electron micrograph of a deposit produced during combustion of rice straw in the MFC and illustrates how reactions between deposit components effect deposit structure. The deposit was generated on the wall of the MFC combustor and accumulated over a 3 hour test period. The wall temperature was 900 °C, the gas temperature was 1000 °C, and the gas composition is estimated to contain 6 % oxygen. Most of the deposit has a glassy appearance, with occasional nodules under the otherwise smooth surface.

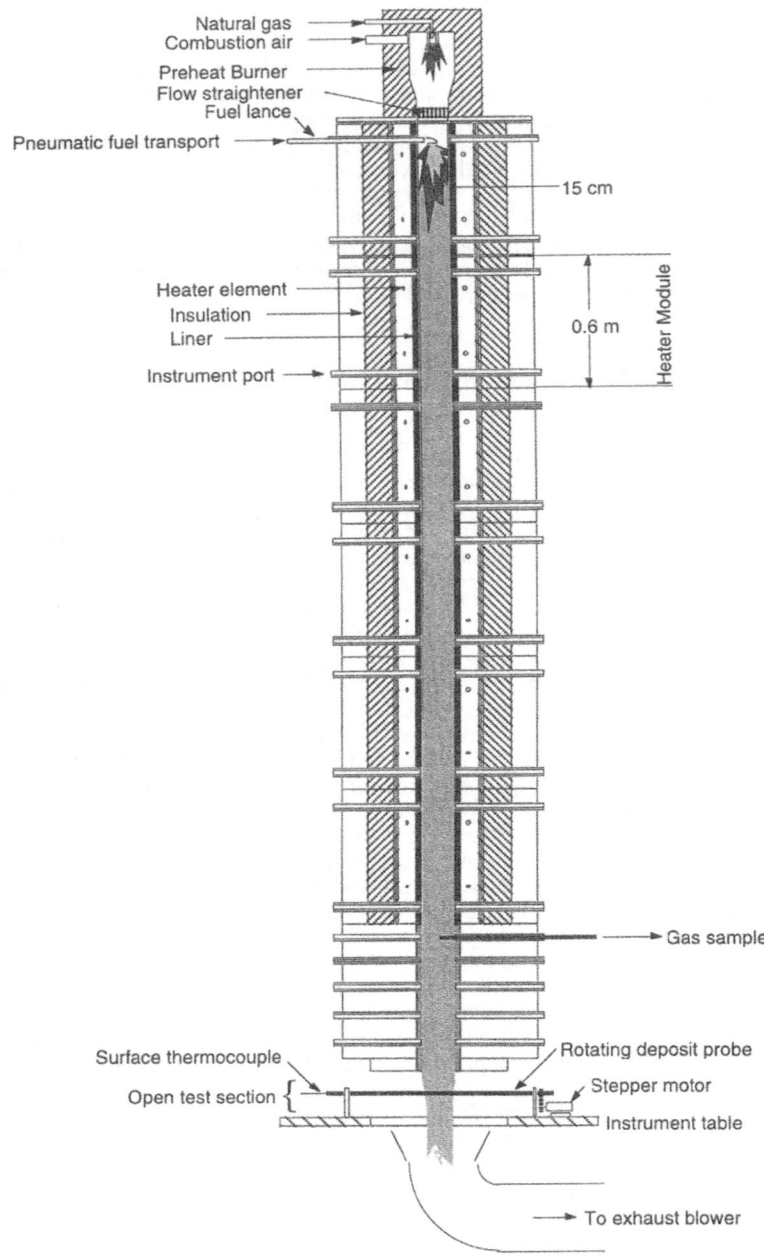

Natural gas
Combustion air

Preheat Burner
Flow straightener
Fuel lance

Pneumatic fuel transport

15 cm

Heater element
Insulation
Liner

0.6 m

Heater Module

Instrument port

Gas sample

Surface thermocouple

Open test section {

Rotating deposit probe

Stepper motor

Instrument table

To exhaust blower

Figure 3   Schematic diagram of the Sandia Multifuel Combustor used in these combustion tests.

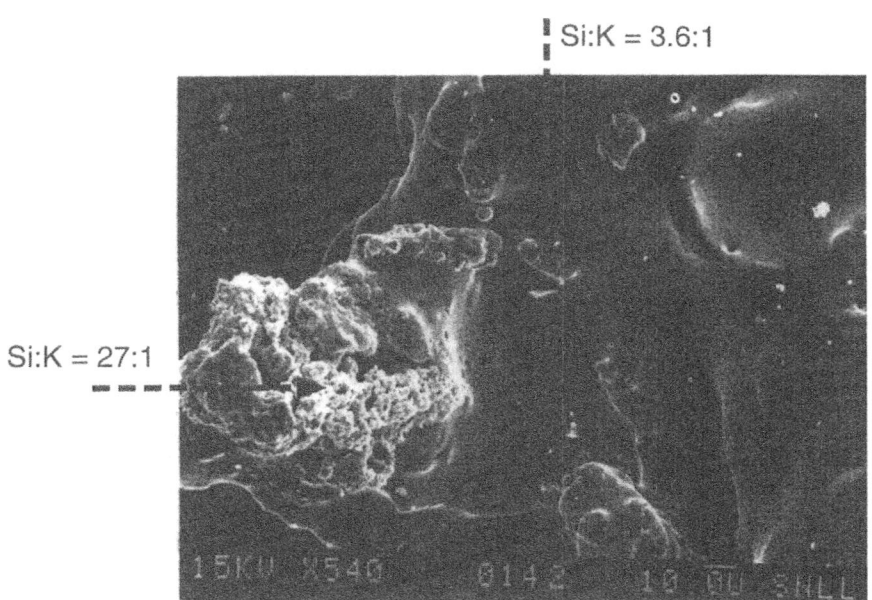

Si:K = 3.6:1

Si:K = 27:1

Figure 4   Scanning electron micrograph of a portion of a rice straw deposit collected from a ceramic surface in the MFC. The porous, silica based material was exposed by fracturing the deposit though one of the nodules evident in many locations on the smooth, glassy surface.

The deposit composition is determined as a function of location on the surface using electron dispersive spectroscopy in a scanning electron microscope. Both the glassy and nodular phases are composed principally of silicon. By comparison with the nodules, the glassy phase contains more nonsilaceous material. More than half of the nonsilaceous fraction is potassium.

Figure 4 illustrates how incorporation of alkalis and other materials changes the physical properties of silica-based deposits. For example, the melting point of silica-containing material decreases from about 1700 °C to less than 750 °C as potassium is introduced to form potassium silicates. Incorporation of additional materials, in particular other alkalis and alkaline earth materials, usually lowers the melting point further still. The silicon to potassium ratio observed in the glassy portions of the deposit illustrated in Figure 4 is about 3.4 on a mass basis, or about 81 % SiO_2 to 19 % K_2O. An equilibrium mixture of such material becomes completely molten at approximately 1300 °C. This is slightly above the temperature of deposit, but the addition of calcium and other heteroatoms to this mixture reduces the melting point significantly (Levin, Robbins, & McMurdie, 1964). The nodular material, on the other hand, has a much higher melting point.

6.   Commercial-Scale Combustion Tests

Field tests were conducted in a range of commercial boilers (Table 2), a few highlights of which are presented here. As with the other results, this paper focuses on one exemplarary data set, with data for the remaining examples used sparingly, but reported in the more comprehensive documents (Miles et al., 1996; Baxter et al., 1996). The Delano combustor, which uses a bubbling fluid bed combustor, will be discussed in detail. During our investigations, the fuels fed to many of the boilers included a large fraction of wood-derived fuel complemented by a large variety of other fuels, mainly agricultural byproducts. The tests conducted at a few of the facilities were more controlled with respect to feed variations, with those conducted at Hydra-Co, Slagelse, and Haslev being the most consistent. Delano operations were somewhere between these extremes. The primary operating characteristics are included in Table 2, with the locations at which ash deposits were sampled indicated in Figure 5.

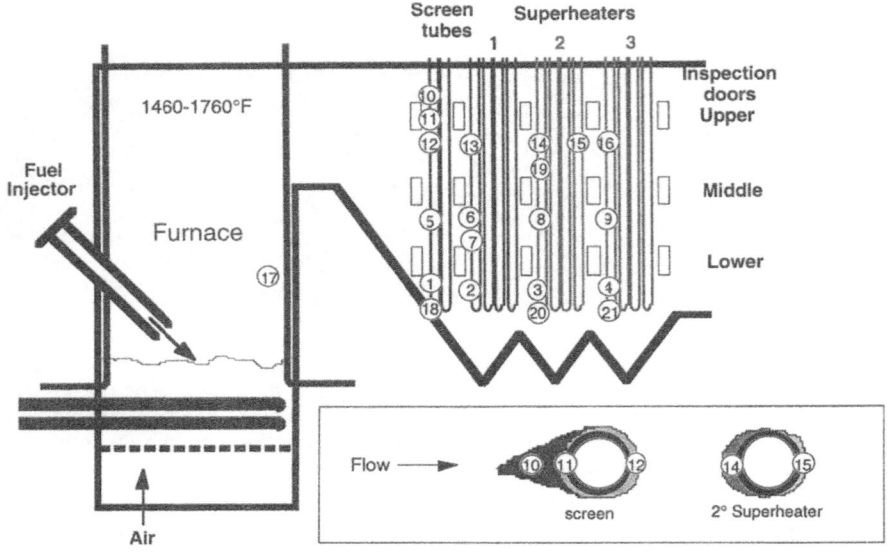

Figure 5   Schematic diagram of the Delano bubbling bed biomass boiler showing major components. This facility was fired primarily by wood during the test burn. Samples were obtained from the regions labeled with numbers, with samples 17-21 being collected after the first shutdown and prior to boiler cleaning (collected on July 10, 1993) and samples 1-16 being collected after the final shutdown (collected on September 4, 1993).

As shown in Figure 6, in regions where mass transfer rates are high, sulfation and alkali enrichment are high. In regions of low mass transfer (back of tube), alkali and sulfate enrichment are less pronounced. A similar trend is seen in the superheater deposits that are spatially resolved. The front deposits exhibit consistent compositions, independent of their vertical height along the tube, whereas the backside deposit is less enriched in potassium, sulfur, and chlorine, the three elements most strongly affected by condensation and chemical reaction mechanisms.

A comparison of the furnace deposit with the convective pass deposit illustrates the substantially different composition of the deposits in different regions of the boiler (Figure 8). These samples show a similar pattern in composition to those illustrated in Figure 7, with the exception that carbonates in the deposits are universally lower in the data illustrated in Figure 8, presumably due to differences in operating conditions. Convective pass deposits with compositions indicated in Figure 8 were collected prior to boiler washing.

Table 2   General information on commercial boilers involved in this investigation.

Name	Delano	Haslev	Hydra-Co	Mendota	Sithe	Slagelse	Wheelabrator	Woodland
Boiler Type	Bubbling Fluid Bed	Auger-fed Grate	Stoker-fired Grate	Circulating Fluid Bed	Circulating Fluid Bed	Cigar-burner-fed Grate	Stoker-fired Grate	Circulating Fluid Bed
Gross Generating Capacity MWe (MWt for cogen plants)	27	5 (13)	20	28	18	12 (28)	55	28
Steam Flow metric t/h	116	26	84	118	77	40	82x3	118
1000 lb/h	256	57	185	260	170	88	180x3	260
Steam Pressure kPa	9308	6701	6378	6240	6206	6701	6206	6240
psi	1350	972	925	905	900	972	900	905
Steam Temperature °C	513	450	421	454	482	450	485	482
°F	955	842	790	849	900	842	905	900
Furnace Exit Gas Temperature °C	960	760	850	882	900	640	650-816	882
°F	1760	1400	1562	1620	1652	1184	1200-1500	1620
Fuel Consumption metric t/h	30	6	22	30	22	8	80	30
1000 lb/h	66	13	48	66	48	18	176	66
Fuel Type	uwf, ag prunings	straw	uwf, straw	wood, ag nuts, shells	uwf, ag	straw	wood, nonrecylabl paper	uwf, ag fuel
Location	Delano	Denmark	Imperial	Mendota	Marysville	Denmark	Shasta, CA	Woodland
Operator	Delano Energy Co., Inc. (Thermo Ecotek)	Elkraft-Midkraft	CMS Generation Operating Co. (Imperial Resource Recovery)	Mendota Biomass Power, Ltd. (Thermo Ecotek)	Sithe Energies, Inc. (shut down)	Elkraft-Midkraft	Wheelabrator Shasta & Hudson Energy Co	Woodland Biomass Power, Ltd. (Thermo Ecotek)

uwf = urban wood fuel (demolition wood, yard clippings, tree trimmings etc.)
ag = agricultural

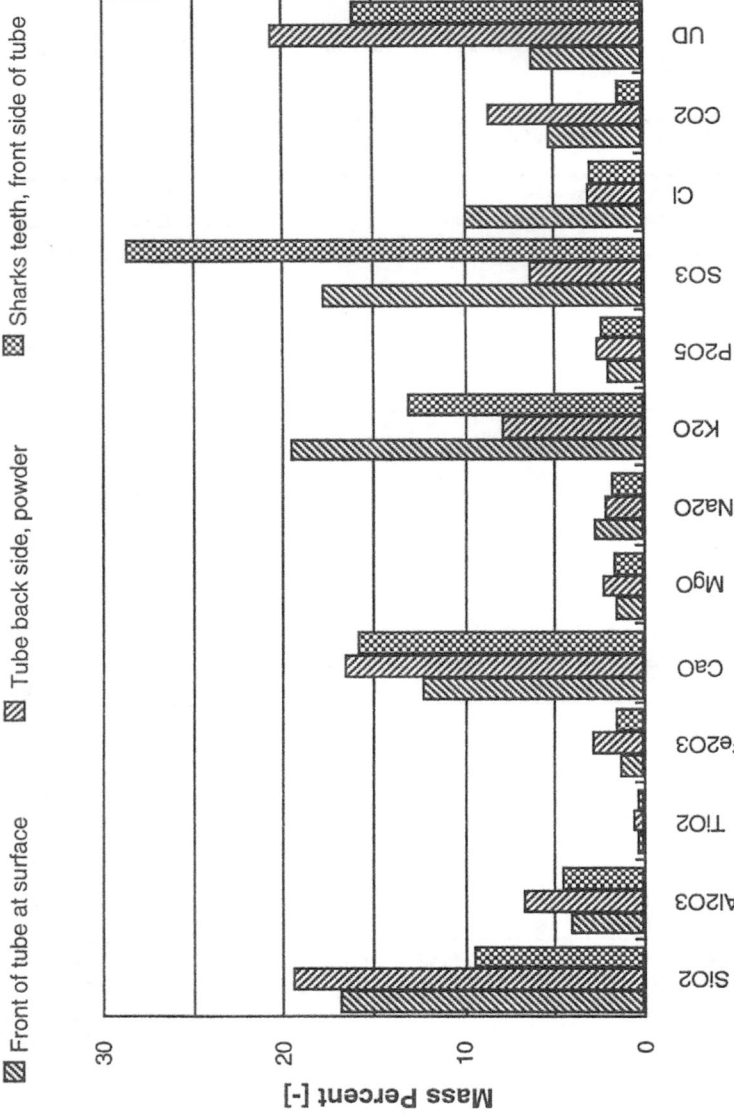

Figure 6 Spatially resolved deposit composition on screen tubes in the Delano boiler. Deposit composition reflects the impact of mass transfer rates on deposit properties. Samples correspond to locations 11, 12, and 10 (left to right) in Figure 5. All values represent percent of total deposit, with UD representing the undetermined fraction as measured by difference between the sum of the oxides and total ash.

The progression of deposit chemistry with temperature is also observable in these samples. At the highest temperatures, chlorides are the most stable form of alkali. As the temperature drops, thermodynamics begin to favor sulfates and finally carbonates. This transition is evident in the samples collected after the boiler shutdown and cleaning (Figure 7). There is also less consistency along the length of a given superheater tube than in the previous results, as is seen by comparing the upper and lower superheater #2 samples.

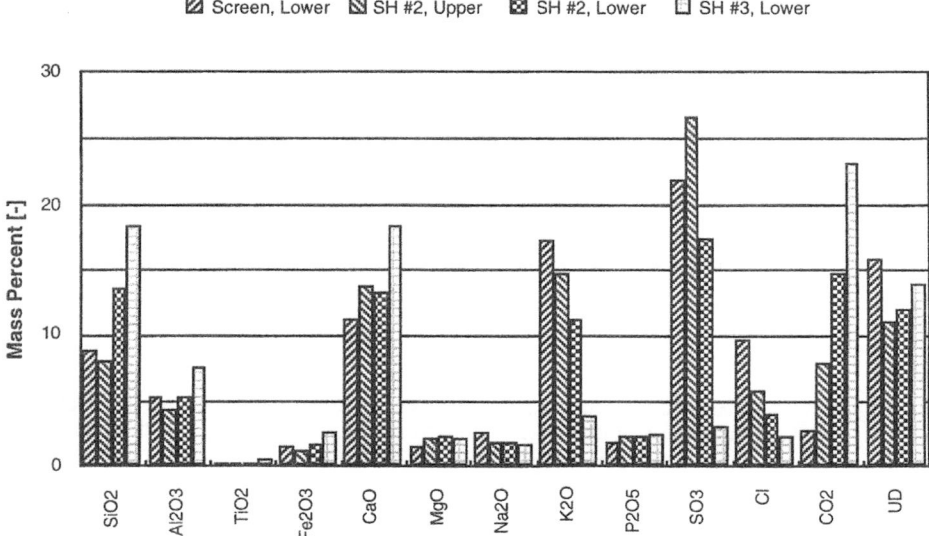

Figure 7   Elemental composition of deposits collected from the Delano boiler at the second shutdown (after boiler washing). Note the progression in deposit chemistry from chlorides to sulfates to carbonates with decreasing gas temperature. Results as displayed left to right correspond to locations 18, 19, 20, and 21 in Figure 5. All values represent percent of total deposit, with UD representing the undetermined fraction as measured by difference between the sum of the oxides and total ash.

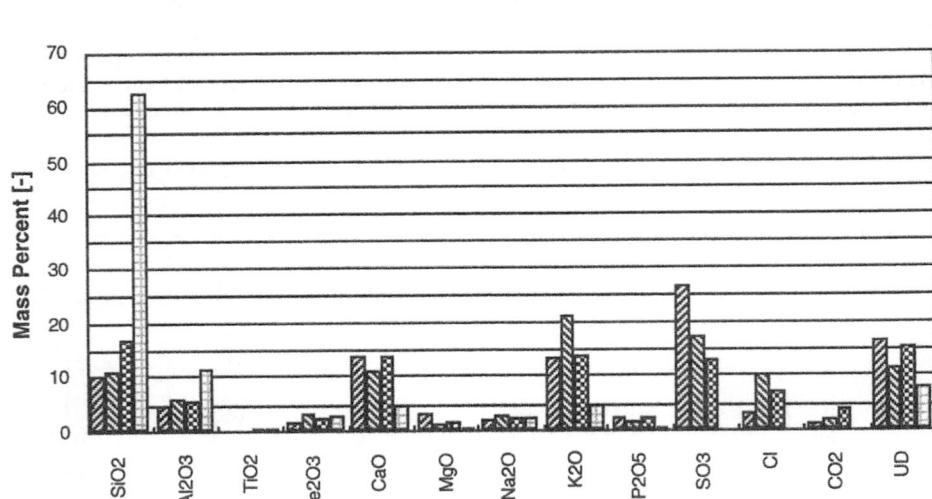

Figure 8   Elemental compositions of superheater deposits collected at midlevel compared with boiler wall deposits. Differences between boiler wall and convective pass deposits are evident and indicate different mechanisms responsible for ash deposition in different regions of the boiler. More subtle differences in convective pass deposits with position and temperature are also evident. Results displayed left to right correspond to samples 5, 6, 8, and 17 in Figure 5. All values represent percent of total deposit, with UD representing the undetermined fraction as measured by difference between the sum of the oxides and total ash.

7. Conclusions

The following represent some of the major conclusions from the overall project, most of which are supported by the highlights discussed above.

Many agricultural fuels have proved unsuitable for use in existing biomass boilers. Straw and other annual herbaceous plant materials cause rapid fouling of heat transfer surfaces, furnace slagging, and agglomeration of fluidized beds. Excessive deposition results from the incompatibility of fuel composition with boiler design and operation. Superheater fouling is perhaps the most critical problem for existing units. Clearly, deposit formation occurs as a result of complex interactions among many compounds, and cannot be described on the basis of the presence and behavior of potassium, silicon, chlorine, or any other element alone.

Full-scale, pilot-scale, and laboratory experiments show the composition of deposits to be consistent with postulated mechanisms of deposit formation and growth. The major alkali species of concern in problematic biomass fuels is potassium. This constrasts with coal, where calcium and sodium are more typically sources of problematic deposits. In herbaceous species, including straws and grasses, canes and stovers, inherent potassium concentrations are about 1% of the fuel dry weight, almost all of which potentially could vaporize during combustion. These same fuels are frequently rich in chlorine, silicon, and sulfur, which in combination with potassium represent the primary fouling agents when these fuels are burned. Chlorine is an important facilitator, leading to the condensation of potassium chloride salt on surfaces, which often reacts with sulfur oxides to form potassium sulfate, and leads to the creation of sticky coatings for enhanced particle attachment following inertial impaction. Potassium, sometimes in combination with alkaline earth metals like calcium, reacts with silicates deposited as fly ash to form molten glassy phases leading to tightly sintered structures. In the furnace, glass reactions lead to the formation of heavy slag deposits resembling fuel ash in composition but substantially depleted in sulfur. On superheater surfaces in boilers firing wheat straw, chlorides represented a significant portion of the deposit mass.

When firing wood fuels, either in combination with agricultural fuels such as hulls, shells, and pits, or woods derived from urban sources with potentially large amounts of adventitious material contamination, the role of potassium may be reduced and that of the alkaline earth metals such as calcium more pronounced. Fuels comprised principally of mature stem wood have relatively low inherent potassium concentrations, around 0.1%, and low inherent silicon and chlorine concentrations. Initial deposition, probably as condensation of hydroxides, is followed by sulfation of alkali and alkaline earth elements, increasing the tenacity of the deposits. If large amounts of adventitious materials in the form of clays or other soil contaminants are present, the role of silicon may still be quite pronounced in secondary deposit growth by particle impaction following the initial formation of condensed layers on surfaces. Complex alkali-alkaline earth-aluminosilicates form or are incorporated into superheater deposits in this manner. Although mature wood is low in silicon, urban and agricultural wood fuels used commercially all have substantial silicon concentrations in the ash. Injection of limestone into fluidized bed combustors leads to the formation of calcium sulfate deposits on superheaters where sulfur is present in the fuel. Wood fuels and blends fired commercially, although generally low in sulfur, generally contain sufficient amounts to cause substantial sulfation of deposits. Calcium sulfate was especially pronounced in the superheater deposits in fluidized beds, compared with the grate units.

Pilot-scale simulations in the Sandia Multifuel Combustor confirm the enrichment of alkali species in the early phases of deposit formation. For straws and grasses, initial deposits containing alkali chlorides can be expected. A wood blend with small amounts of almond shell produce sulfated deposits similar to those observed in commercial units burning similar fuels. Potassium-rich almond hulls and shells generate deposits containing more potassium than could be accounted for as sulfates and chlorides. Sodium rich olive pits produce a similar result. With such fuels, initial deposition as alkali hydroxides or carbonates may be important.

All of the results are consistent with a mechanistic interpretation of ash deposit formation and composition. Changes in deposit composition with position in the boiler and position around individual tubes reflect the changing temperatures, mass transfer rates, heat transfer rates, and gas compoisitions with boiler environment. Changes in deposit composition with fuel type reflect the composition and mobility of the inorganic species in the deposits. Changes in deposit structure and composition with position relative to the surface reflect changing mechanisms of ash deposition.

8.   Acknowledgements

This work was supported through the US DOE Office of Energy Efficiency and Renewable Energy under the financial sponsorship of the Biomass Power Program. In addition, a consortium of industries with interest in biomass power financially contributed to this project. These include Mendota Biomass Power Ltd. and Woodland Biomass Power Ltd. (both associated with Thermo Electron Energy Systems), CMS Generation Operating Co. (formerly Hydra-Co Operations Inc.), Wheelabrator Shasta and Hudson Energy Cos., Sithe Energy Co., Delano Energy Co. Inc., the Electric Power Research Institute, Foster Wheeler Development Corp., and Elkraft Power Co. Ltd. of Denmark. Most of these companies also contributed fuels, use of facilities, and technical expertise in reviewing results of the project. The authors are grateful for the support and interaction with all of the individuals representing these organizations.

9.   Bibliography

1. *Federal Register: Part III; Air Contaminants* (Final Rule No. 29 CFR Part 1910). (1989).
2. Atkins, R. S., & Donovan, C. T. *Wood Products in the Waste Stream: Characterization and Combustion Emissions* (Final Report No. Energy Authority Report 92-8). New York State Energy Research and Development Authority. (1993).
3. Baxter, L. L. Ash Composition Prediction as a Function of Coal Type, Operating Conditions, and Boiler Location. In A. K. Mehta & N. S. Harding (Ed.), *EPRI conference on the Effects of Coal Quality on Power Plants*, (pp. 5/59-5/74). St. Louis, Missouri: Electric Power Research Institute. (1990a).
4. Baxter, L. L., *Progress in Energy and Combustion Sciences, 16*, 261-266. (1990b).
5. Baxter, L. L. Boiler Performance with Blends of Eastern and Western Coals. In *EPRI Conference on The Effects of Coal Quality on Power Plants*, . San Diego, CA: (1992a).
6. Baxter, L. L., *Combustion and Flame, 90*, 174-184. (1992b).
7. Baxter, L. L., *Biomass and Bioenergy, 4*(2), 85-102. (1993).
8. Baxter, L. L. Experimental and Theoretical Comparisons of the Combustion and Ash Deposition Behavior of Blended Coals and that of the Blend Components. In R. W. Bryers & N. S. Harding (Eds.), Coal-Blending and Switching of Low-Sulfur Western Coals (pp. 255-264). New York: ASME. (1994).
9. Baxter, L. L. Ash. In Encyclopecia of Energy Technology and the Environment (pp. 306-321). New York: John Wiley & Sons, Inc. (1995).
10. Baxter, L. L., Abbott, M. F., & Douglas, R. E. Dependence of Elemental Ash Deposit Composition on Coal Ash Chemistry and Combustor Environment. In S. A. Benson (Ed.), *Engineering Foundation Conference on Inorganic Transformations and Ash Deposition During Combustion*, (pp. 679-698). Palm Coast, Florida: The American Society of Mechanical Engineers. (1991).
11. Baxter, L. L., & DeSollar, R. W. Ash Deposition as a Function of Coal Type, Location in a Boiler, and Boiler Operating Conditions: Predictions Compared to Observations. In *International Conference on Environmental Control of Combustion Processes*, . Honolulu, HA: (1991).
12. Baxter, L. L., & DeSollar, R. W., *Fuel, 72*(10), 1411-1418. (1993).
13. Baxter, L. L., & Dora, L., *ASME Paper No. 92-JPGC-FACT-14*. (1992).
14. Baxter, L. L., & Hardesty, D. R. *The Fate of Mineral Matter During Pulverized Coal Combustion* (Quarterly Report No. SAND90-8223). Sandia National Laboratories. (1990).

15. Baxter, L. L., & Hardesty, D. R. *The Fate of Mineral Matter During Pulverized Coal Combustion: April - June* (Quarterly Report No. SAND92-8227). Sandia National Laboratories. (1992a).

16. Baxter, L. L., & Hardesty, D. R. *The Fate of Mineral Matter During Pulverized Coal Combustion: January-March* (Quarterly Report No. SAND92-8210). Sandia National Laboratories. (1992b).

17. Baxter, L. L., Hencken, K. R., & Harding, N. S. The Dynamic Variation of Particle Capture Efficiency During Ash Deposition in Coal-Coal Fired Combustors. In *Twenty-Third Symposium (International) on Combustion*, (pp. 993-999). Orleans, France: (1990).

18. Baxter, L. L., Miles, T. R., Miles, T. R., Jr., Jenkins, B. M., Richards, G. R., & Oden, L. L. Transformations and Deposition of Inorganic Material in Biomass Boilers. In M. G. Carvalho (Ed.), *Second International Conference on Combustion Technologies for a Clean Environment*, 1 (pp. Biomass II: 9-15). Lisbon, Portugal: Commission of European Communities. (1993).

19. Baxter, L. L., Mitchell, R. E., & Fletcher, T. H., *Combustion and Flame*. (1996 to appear).

20. Baxter, L. L., Miles, T. R., Miles, T. R. Jr., Jenkins, B. M., Milne, T., Dayton, D., Bryers, R. W., and Oden, L. L. "The Behavior of Inorganic Material in Biomass-Fired Power Boilers — Field and Laboratory Experiences: Volume II of Alkali Deposits Found in Biomass Power Plants," SAND96-8225 Volume 2 and NREL/TP-433-8142, (1996).

21. Beér, J. M., Monroe, L. S., Barta, L. E., & Sarofim, A. F. From Coal Mineral Matter Properties to Fly Ash Deposition Tendencies; a Modeling Approach. In *The Seventh International Pittsburgh Coal Conference*, . Pittsburgh, PA: (1990).

22. Byers, R. L., & Calvert, S., *Industrial and Engineering Chemistry Fundamentals*, 8(4), 646-655. (1969).

23. Castillo, J. L., & Rosner, D. E., *International Journal of Multiphase Flow*, 14, 99-120. (1988).

24. Chow, O. K., & Lexa, F. F. *Combustion Characterization of the Kentucky No. 9 Cleaned Coals* (Final Report No. CS-4994 Research Project 2425-1). Electric Power Research Institute. (1987).

25. Dayton, D. C., French, R. J., & Milne, T. A., *Energy & Fuels, (to appear)*. (1995).

26. Dayton, D. C., & Wang, D. CID Studies of Inorganic Species Released During BiomassCombustion. In *43rd ASMS Conference on Mass Spectrometry and Allied Topics*, (pp. Poster TPA 034). Atlanta, GA: (1995).

27. Durant, J. F., Kwasnik, A. F., & Lexa, G. F. *Impacts of Cleaning Texas Lignite on "Boiler Performance and Economics* (Final Report No. GS-6517). Electric Power Research Institute. (1989).

28. Evans, R. J., & Thomas A. Milne, T. A., *Energy & Fuels, 1*, 123-127. (1987a).

29. Evans, R. J., & Thomas A. Milne, T. A., *Energy & Fuels, 1*, 311-319. (1987b).

30. FEC Consultants Ltd., O. *Straw Firing of Industrial Boilers* (Contractor Report No. ETSU B 1158). Energy Technology Support Unit: Department of Energy: United Kingdom. (1988).

31. French, R. J., Dayton, D. C., & Milne, T. A. *The Direct Observation of Alkali Vapor Species in Biomass Combustion and Gasification* No. NREL Technical Report (NREL/TP-430-5597)). NREL. (1994).

32. Friedlander, S. K., & Johnstone, H. F., *Industrial and Engineering Chemistry, 49*, 1151-1156. (1957).

33. Fuchs, N. A. The Mechanics of Aerosols. (R. E. Daisley and M. Fuchs, Trans.). New York: Dover. (1964).

34. Gökoglu, S. A., & Rosner, D. E., *International Journal of Heat and Mass Transfer, 27,* 639-646. (1984).
35. Gökoglu, S. A., & Rosner, D. E., *Industrial and Engineering Chemistry Fundamentals, 24,* 208-214. (1985).
36. Gökoglu, S. A., & Rosner, D. E., *AIAA Journal, 24,* 172-179. (1986).
37. Griffith, B. F., Lexa, F. G., & Teigen, B. C. *Pilot-Scale Combustion Characterization of Two Illinois Coals* (Final Report No. CS-6009, Research Project 2425-1). Electric Power Research Institute. (1988).
38. Harb, J. N., Munson, C. L., & Richards, G. H., *Energy & Fuels, 7,* 208-214. (1993).
39. Harding, N. S., & Mai, M. C. Elemental Partitioning during Pilot-Scale Combustion Tests — Effects on Ash Deposit Composition. In R. W. Bryers & K. S. Vorres (Eds.), Mineral Matter and Ash Deposition from Coal (pp. 375-399). Engineering Trustees, Inc. (1990).
40. Hastie, J. W., Zmbov, K. F., & Bonnell, D. W., *High Temperature Science, 17,* 333-364. (1984).
41. Helble, J. J., Neville, M., & Sarofim, A. F. Aggregate Formation from Vaporized Ash During Pulverized Coal Combustion. In *Twenty-First Symposium (International) on Combustion,* (pp. 411-417). Technical University of Munich, Germany: The Combustion Institute. (1986).
42. Helble, J. J., & Sarofim, A. F., *Journal of Colloid and Interface Science, 128*(2), 348-362. (1989a).
43. Helble, J. J., & Sarofim, A. F., *Combustion and Flame, 76,* 183-196. (1989b).
44. Helble, J. J., Srinivasachar, S., & Boni, A. A., *Progress in Energy and Combustion Science, 16,* 267-279. (1990).
45. Hustad, J. E., & Sønju, O. K., *Biomass and Bioenergy, 2*(1-6), 239-261. (1992).
46. Im, K. H., & Chung, P. M., *AIChE Journal, 29*(3), 498-505. (1983).
47. Israel, R., & Rosner, D. E., *Aerosol Science and Technology, 2,* 45-51. (1983).
48. Jacobsen, S., & Brock, J. R., *Journal of Colloid Science, 20,* 544-554. (1965).
49. Jenkins, B. M. Physical Properties of Biomass. In O. Kitani & C. Hall (Eds.), Biomass Handbook New York: Gordon and Breach. (1989).
50. Laitone, J. A., *Jounal of Applied Mechanics, 48,* 465-471. (1981).
51. Levin, E. M., Robbins, C. R., & McMurdie, H. F. Phase Diagrams for Ceramists. Columbus, OH: American Ceramic Society. (1964).
52. Livingston, W. R. *Straw Ash Characteristics* (Contractor Report No. ETSU B 1242). Energy Technology Support Unit: Department of Energy: United Kingdom. (1991).
53. Loehden, D., Walsh, P. M., Sayre, A. N., Béer, J. M., & Sarofim, A. F., *Journal of the Institute of Energy, 62,* 119-127. (1989).
54. Marschner, H. Mineral Nutrition of Higher Plants. (. London: Harcourt Brace Jovanovich. (1986).
55. Martindale, L. P. *The Potential for Large Scale Projects Featuring Straw as a Fuel in the UK* No. ETSU. (1982).
56. Martindale, L. P. *The Potential for Straw as a Fuel in the UK* No. ETSU. (1984).
57. Michelsen, H. P., "Results of Deposit Measurements at Rudkøbing KVV," CHEC Report No. 9603, January 1996.
58. Michelsen, H. P., "Results of Deposit Measurements at Rudkøbing KVV," CHEC Report No. 9604, January 1996.
59. Miles, T. R. Operating Experience with Ash Deposition in Biomass Combustion Systems. In *Biomass Combustion Conference,* . Reno, NV: (1992).
60. Miles, T. R., & Miles, T. R., Jr. Alkali Deposits in Biomass Power Plant Boilers. In *Biomass Power Program,* . Washington, DC: (1993).

61. Miles, T. R., Miles, T. R., Jr., Baxter, L. L., Jenkins, B. M., & Oden, L. L. Alkali Slagging Problems with Biomass Fuels. In *First Biomass Conference of the Americas: Energy, Environment, Agriculture, and Industry*, 1 (pp. 406-421). Burlington, VT: National Renewable Energy Laboratory. (1993).

62. Miles, T. R., Miles, T. R. Jr., Baxter, L. L., Bryers, R. W., Jenkins, B. M., and Oden, L. L. "Alkali Deposits Found in Biomass Power Plants: A Preliminary Investigation of Their Extent and Nature: Volume I," SAND96-8225 Volume 2 and NREL/TP-433-8142, (1996).

63. Milne, T. A., & Klein, H. M., *Journal of Chemical Physics, 33*, 1628-1637. (1960).

64. Quann, R. J., Neville, M., & Sarofim, A. F., *Combustion Science and Technology, 74*, 245-265. (1990).

65. Raask, E. Mineral Impurities in Coal Combustion. (. Washington: Hemisphere Publishing Corporation. (1985).

66. Richards, G. H., Harb, J. N., Baxter, L. L., Bhattacharya, S., Bupta, R. P., & Wall, T. F. Radiative Heat Transfer in PC-Fired Boilers — Development of the Absorptive/Reflective Character of Initial Ash Deposits on Walls. In *Twenty-Fifth Symposium (International) on Combustion*, . Irvine, CA: The Combustion Institute. (1994).

67. Rosner, D. E., *AIChE Journal, 35*(1), 164-167. (1989).

68. Rosner, D. E., & Nagarajan, R., *Chemical Engineering Science, 40*(2), 177-186. (1985).

69. Rosner, D. E., & Tassopoulos, M., *AIChE Journal, 35*(9), 1497-1508. (1989).

70. Salour, D., Jenkins, B. M., Vafaei, M., & Kayhanian, M., *Biomass and Bioenergy, 4*(2), 117-133. (1993).

71. Smouse, S. M., & Wagoner, C. L. Deposit Initiation: Part 2 — Experimental Verification of Hypothesis Using a Simulated Superheater Tube. In S. A. Benson (Ed.), *Conference on Inorganic Transformations and Ash Deposition During Combustion*, (pp. 625-637). Palm Coast, Florida: ASME. (1991).

72. Srinivasachar, S., & Boni, A. A., *Fuel, 68*(7), 829-836. (1989).

73. Srinivasachar, S., Helble, J. J., & Boni, A. A. An Experimental Study of the Inertial Deposition of Ash Under Coal Combustion Conditions. In *Twenty-Third Symposium (International) on Combustion*, (pp. 1305-1312). The University of Orléans, Orléans, France: The Combustion Institute. (1990a).

74. Srinivasachar, S., Helble, J. J., & Boni, A. A., *Progress in Energy and Combustion Science, 16*, 281-292. (1990b).

75. Srinivasachar, S., Helble, J. J., Boni, A. A., Shah, N., Huffman, G. P., & Huggins, F. E., *Progress in Energy and Combustion Science, 16*, 293-302. (1990c).

76. Srinivasachar, S., Senior, C. L., Helble, J. J., & Moore, J. W. A Fundamental Approach to the Prediction of Coal Ash Deposit Formation in Combustion Systems. In *The Twenty-Fourth Symposium (International) on Combustion*, (pp. 1179-1187). The University of Sydney, Australia: The Combustion Institute. (1992).

77. Tillman, D. A. Trace Metals in Combustion Systems. (. San Diego: Academic Press. (1994).

78. Turnbull, J. H., *Biomass and Bioenergy, 4*(2), 75-84. (1993).

79. Wagoner, C. L., & Yan, X.-X. Deposit Initiation via Thermophoresis: Part 1 — Insignt on Deceleration and Retention of Inertially-transported particles. In S. A. Benson (Ed.), *Conference on Inorganic Transformations and Ash Deposition During Combustion*, (pp. 607-623). Palm Coast, Florida: ASME. (1991).

80. Wall, T. F., Baxter, L. L., Richards, G., & Harb, J. N. Ash Deposits, Coal Blends and the Thermal Performance of Furnaces. In R. W. Bryers & N. S. Harding (Eds.), <u>Coal-Blending and Switching of Low-Sulfur Western Coals</u> (pp. 453-463). New York: ASME. (1994a).

81. Wall, T. F., Bhattacharya, S. P., Zhang, D. K., Gupta, R. P., & He, X. The Properties and Thermal Effects of Ash Deposits in Coal-Fired Furnaces. In J. Williamson & F. Wigely (Eds.), The Impact of Ash Deposition on Coal-Fired Plants (pp. 463-477). Washington, D.C.: Taylor & Francis. (1994b).
82. Wibberley, L. J., & Wall, T. F., *Combustion Science and Technology*, *48*, 177-190. (1986).
83. Zygarlicke, C. J., McCollor, D. P., Toman, D. L., Erickson, T. A., Ramanathan, M., & Folkedahl, B. S. *Combustion Inorganic Transformations* (Semiannual Technical Progress Report No. Cooperative Agreement No. DE-FC21-86MC10637). University of North Dakota Energy and Environmental Research Center. (1992a).
84. Zygarlicke, C. J., Ramanathan, M., & Erickson, T. A. Fly Ash Particle-Size Distribution and Composition: Experimental and Phenomenological Approach. In S. A. Benson (Eds.), Inorganic Transformations and Ash Deposition During Combustion (pp. 525-544). New York: American Society of Mechanical Engineers. (1992b).

COMBUSTION

Environment

PRIMARY AND SECONDARY MEASURES FOR THE REDUCTION OF NITRIC OXIDE EMISSIONS FROM BIOMASS COMBUSTION
NO_X reduction in biomass combustion

T. NUSSBAUMER
Verenum Research, Zurich and Swiss Federal Institute of Technology, Zurich, Switzerland

Abstract
Nitric oxide emissions from biomass combustion originate mainly from the fuel bound nitrogen, thermal NO_X are only of minor importance. Since biomass combustion leads to higher NO_X emissions than gas or light fuel oil combustion primary or secondary measures for NO_X reduction are necessary for future combustion plants.

To minimize NO_X emissions by primary measures, the fuel nitrogen must be reduced to molecular nitrogen in zones with an excess air ratio < 1. The following techniques for the reduction of fuel NO_X have been investigated: Air staging with and without separate reduction chamber, fuel staging and flue gas recirculation. It is shown that an NO_X reduction of 40% to 75% can be reached by air staging with separate reduction chamber if the following conditions are met: Primary excess air ratio $\approx 0.7 -$ 0.8, temperature in the reduction chamber $\approx 1'100° - 1'200°C$, residence time $\approx 0.3 -$ 0.5 s. Fuel staging shows a similar potential of NO_X reduction. As for air staging the NOx reduction is mainly influenced by the excess air ratio in the reduction chamber. Fuel staging can only be used for large combustion plants while air staging can be used for automatic wood furnaces from app. 200 kW up to large scale combustion.

If primary measures are not sufficient, secondary measures as the selective catalytic and non-catalytic reduction (SCR, SNCR) through the injection of sal ammoniac, ammonia or urea can be taken. The NO_X reduction in the SNCR process is limited by the ammonia slippage. Further an accurate process control is neceassary to ensure the temperature window of app. 840°C – 920°C. To operate the combustion at low excess air ratio without exceeding the temperature limit, a partial extraction of heat before the SNCR process is necessary. By the SNCR process 60% – 80% reduction can be reached, depending on the fuel composition and the operation conditions. With SCR more than 80% (up to 95%) reduction can be achieved. However there is only little experience with the long term behaviour of catalysts in wood combustion systems.

For large combustion plants combinations of low NO_X and denox-techniques are considered.
Keywords: NO_X, low NO_X, air staging, fuel staging, reburning, staged combustion, ammonia injection, SCR, SNCR, denox.

1 Introduction

Nitric oxides NO and NO_2, summarized as NO_X, from combustion processes are formed in three different reactions:

- Thermal NO_X are formed at high temperature (relevant concetrations can be found > 1'300°C) by the oxidation of nitrogen in the air
- Prompt NO_X can be formed during the combustion of hydrocarbons in reactions of molecular nitrogen with free radicals in the flame
- Fuel NO_X are formed from the nitrogen contained in the fuel.

Fuel NO_X as an oxidation product are usually maximally at high combustion quality. Since typical combustion temperatures in todays wood firings are between 800° and 1200°C only fuel NO_X are of great importance. Therefore nitric oxide emissions typically increase with increasing nitrogen content in the wood (table 1). Furthermore the analysis of flue gas from wood combustion shows that the emission of NO_2 is usually much smaller than the emission of NO.

The NO_X concentration which can be calculated by the Zeldovic mechanism of thermal NO_X formation is neclictible in relation to the measured NO_X emissions from wood firings. This fact was confirmed in different experimental investigations [1] [2] [3]: Wood was burnt in a conventional furnace a) with air and b) with a nitrogen free mixture of oxygen and argon. Since no difference in the emission of NO_X was found the nitrogen in the air is not responsible for the formation of NO_X. Further it was shown in a laboratory furnace that the emission of NO_X from different wood samples are independent of the combustion temperature in the range from 800° – 1300°C.

Table 1. Nitrogen content of pine wood, beech and chip board, concentration of NO_X and conversion rate (NO_X/N) of fuel nitrogen to NO_X resulting from the isothermal combustion in a laboratory furnace (NO_X at 11 Vol.-% O_2) [1].

	Fuel-N [Wt. %]	NO_X [mg/Nm^3]	NO_X/N [%]
Pine wood	0.07	173	67.3
Beech	0.2	231	36.0
Chip board (UF)	2.85	921	8.4

During the pyrolisis of wood approximately 20% of the nitrogen compounds remain in the char and 80% form volatile substances. As a first approximation, the volatile nitrogen conversion to NO and the char conversion to NO can be assumed as independent [4].

The volatile fraction consists of HCN, NH-compounds (NH, NH_2, NH_3) NO and N_2. HCN is converted homogenously to NH_2/NH_3 which can be oxidized to NO or react with NO to N_2 [5] [6] [7]. The overall reaction from fuel nitrogen to NO or N_2 can be influenced by the reaction conditions, especially by the temperature, the concentration of oxidizing agents as O_2 and H_2O, the residence time in the different zones of the combustion and the turbulence.

2 Primary measures for NO_x reduction

2.1 Fundamentals

To ensure combustion with low NO_x emissions the fuel nitrogen must be converted to N_2. Different concepts of staged combustion for NO_x reduction have been investigated for the combustion of coal, oil and gas [8] [9] [10] [11] [12]. For the combustion of wood the following reactions are considered to be important for the NO_x emissions [2] [5]: During the gasification HCN and NH_i radicals as NH_2 and NH_3 are formed. These components can react in different reactions to form molecular oxygen (simplified):

$$NO + NH_2 \rightarrow N_2 + H_2O \tag{1}$$

Where there is a shortage of oxygen NO_x acts as an oxidizing agent for carbon monoxide, methane, hydrocarbons, hydrogen and carbon:

$$NO + CO \quad \rightarrow \quad CO_2 + 0.5\,N_2 \tag{2}$$
$$NO + CH_4 \quad \rightarrow \quad CO + 2\,H_2 + 0.5\,N_2 \tag{3}$$
$$NO + H_2 \quad \rightarrow \quad H_2O + 0.5\,N_2 \tag{4}$$
$$NO + C \quad \rightarrow \quad CO + 0.5\,N_2 \tag{5}$$

Fig. 1 shows the main reactions of the fuel nitrogen during gasification and oxidation. The gas phase reactions are dominant for the NO formation. From HCN which is released from the fuel, NH_i radicals are formed. Under oxidizing atmosphere the NH_i-components are mainly converted to NO, while under reducing conditions, the NH_i-radicals can be reduced to N_2. The already formed NO can react to N_2 in the presence of NH_i or other reducing components such as CH_i, CO etc.. Furthermore NO can react with CH_i to form HCN in the so called NO recycle.

Since in oxidizing and reducing atmosphere different reactions are dominant for the fuel nitrogen, the primary excess air ratio plays an important role for the NO_x formation (Fig. 2, Fig. 3, Fig. 4).

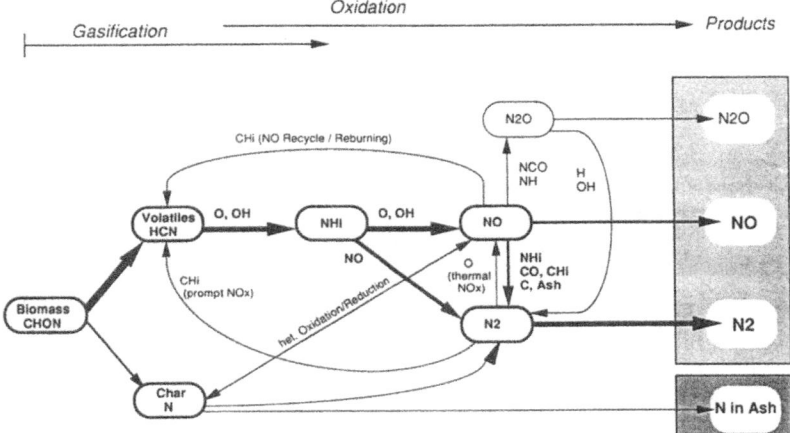

Fig. 1   Reactions of fuel nitrogen during combustion (main paths thick).

2.2 Staged combustion with reduction chamber and air staging

To investigate under which conditions the above mentioned reactions (1) to (5) can significantly reduce NO_x during combustion, a 25 kW test reactor with fixed bed updaft gasification followed by a gas phase combustion was built (Fig. 2). The test reactor can be operated as an under stoker firing, which allows to receive reference NO_x values for conventional combustion. The following concepts of staged combustion have been investigated:

- air staging with separate reduction chamber between the gasification and the combustion chamber
- air staging in the gasification chamber
- flue gas recirculation
- combinations of the listed measures.

Due to the results of different experiments carried out in the fixed bed reactor reaction (1) is supposed to be most important for fixed bed combustion of wood. Reaction (2) might be of importance together with catalytic effects on ash while for reaction (4) and (5) no effects were found during wood combustion in the test reactor. Reaction (5) is supposed to be of major influence during fluidized bed combustion of coal [8]. However for NO_x reduction by air staging in biomass combustion, the heterogenous reactions are of minor importance.

In the test reactor an NO_x reduction > 50% for native wood and > 75% for UF chipboards (N-content app. 3 wt.-%) compared to conventional fixed bed combustion is optained by air staging with a separate reduction chamber between the gasification and the combustion. Table 2 shows the results of different primary measures for native wood and UF chipboards at different operation conditions. However to reach minimal NO_x emissions the reaction conditions must be as follows:

- residence time in the reduction chamber ≈ 0.5 s (> 0.3 s) and sufficient mixing quality (turbulence) to ensure mixing
- reduction temperature $\approx 1'100C° - 1'200C°$
- primary excess air ratio $\approx 0.7 - 0.8$

The primary excess air ratio in the gasification chamber is of great influence on the NO_x emissions (Fig. 4):

- Primary air ratio << 0.7 (gasification): Mainly NH-components (and) HCN are formed during the gasification, while the NO concentration is low. The NH-components are oxidized in the combustion chamber to NO.
- Primary air ratio > 1: The pyrolisis gases are burnt immediately in the gasification chamber (= combustion). The concentration of NH-components to reduce NO in the reduction chamber is close to 0.

An additional NO_x reduction can be achieved by air staging in the gasification chamber (3 stage combustion compared to 2 stage combustion in table 2). Flue gas recirculation was found to have a minor influence on NO_x emissions. The temperature window for the reduction of NO_x by staged combustion with reduction chamber is about 300°C higher than for the SNCR process. It is assumed that methane in the gas consumes OH radicals and therefore inhibits the formation of NH_2 radicals which are necessary for the NO reduction [7]. The experimental results correspond to datas from a kinetic model acc. to Kilpinen et al. [13].

To ensure a constant primary air ratio and a reduction temperature of app. 1'150°C a suitable process design as well as an accurate combustion and gasification control are necessary. As a consequence, a 250 kW grate firing with reduction chamber has been designed (Fig. 5). To reduce heat losses from the reduction chamber, the combustion chamber is surrounding the reduction chamber. In the grate firing similar results as shown in Fig. 4 have been achieved. However measures must be taken to limit the temperature if dry biomass is burnt. Flue gas recirculation, secondary air preheating with convective heat from the primary chamber and wall cooling in the primary zone are considered as measures to limit the temperature < 1'200°C if dry wood is burnt and are investigated in ongoing research.

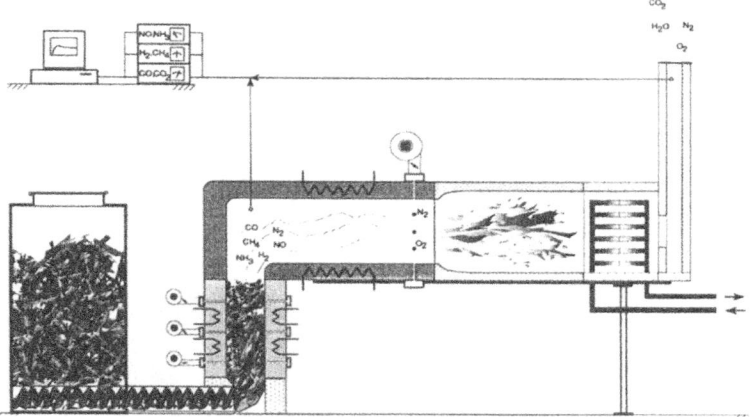

Fig. 2   Laboratory furnace for staged combustion (25 kW).

From left to right: Automatic feeding with under stoker. Updraft-gasification chamber with primary air injection on one level (2 stage combustion) or several levels (staged gasification; 3 stage combustion). Reduction chamber which can be heated for isothermal conditions and with a residence time which can be varied between 0.05 s to 2.0 s. Secondary air injection, combustion chamber and boiler. Option (not shown): Flue gas recirculation. Raw gas analysis after the gasification chamber with FTIR, flue gas analysis by NDIR, CLD and paramagnetic analyzers.

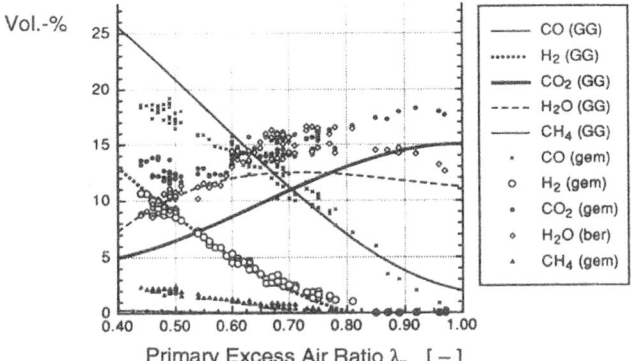

Fig. 3   Composition of the raw gas from the under stoker gasification in the laboratory furnace in function of the primary excess air *(gem)* in comparison to calculated datas *(GG)* [7], [14].

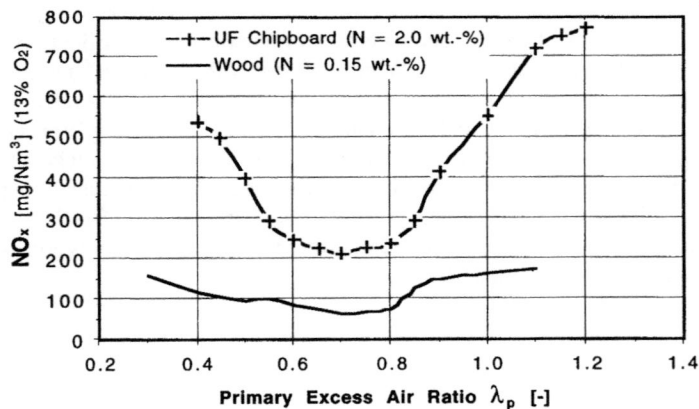

Fig. 4   NO$_x$ emissions in function of the primary excess air ratio during the combustion of wood and UF chipboards in the laboratory furnace with reduction
chamber (2 stage combustion, temperature in the reduction chamber =
1'150°C). Similar results are found in the low-NOx grate firing (Fig. 5).

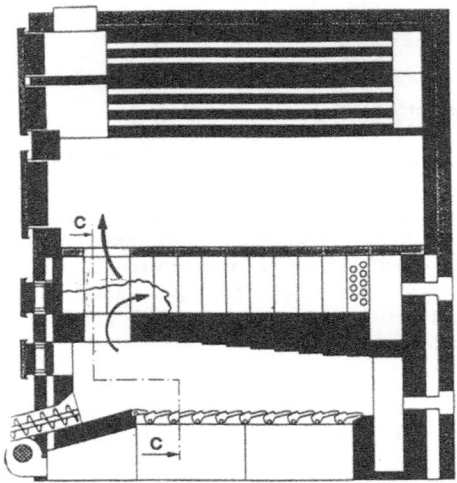

Fig. 5   Prototype grate firing (250 kW) with reduction chamber, which is surrounded
by the combustion chamber to reduce heat losses.

Table 2. NOx emissions and conversion rate of fuel-N to NOx for the combustion of native wood and UF chipboars for different primary air excess ratios [7]. Temperature in the reduction chamber = 1'160°C, residence time = 0.3 – 0.6 s.

Fuel Type	Combustion technology	Primary excess air [-]	NO_x [mg/Nm³] as NO_2 at 13 Vol.-% O_2	Conversion of fuel N to NO_x [%]	NO_x reduction [%]
	2 stage combustion without reduction chamber (ref.)	1 – 1.3	200	39.8	0
Wood N = 0.15 wt.-%	2 stage combustion with understoichiometric operation	0.6 – 0.8	140	28.0	30
	2 stage combustion with reduction chamber	0.7	76	15.1	62
	3 stage combustion with reduction chamber	0.8	60	12.0	70
	2 stage combustion without reduction chamber (ref.)	1 – 1.3	848	15.3	0
UF Chipboard N = 2.0 wt.-%	2 stage combustion with Low-NOx operation	0.6 – 0.8	440	8.1	48
	2 stage combustion with reduction chamber	0.7	216	4.0	75
	3 stage combustion with reduction chamber	0.8	176	3.2	79

2.3 Fuel staging

To reduce NO_x emissions by fuel staging, the primary fuel is burnt with excess air ($\lambda >$ 1). A secondary fuel or reburn fuel is injected in the hot flue gas of the primary fuel. In the mixing zone a reducing atmosphere can be achieved if the excess air ratio is smaller than 1. As in the reactions shown in Fig. 1, NO_x in the flue gas of the primary fuel can be reduced with CH_i, NH_i and other components of the gas from the secondary fuel by the above mentioned reactions. Fuel nitrogen from primary and secondary fuel and also thermal NO_x from the primary fuel can be converted to N_2.

The potential of fuel staging in small scale wood combustion has been investigated in an under stoker furnace [15]. The secondary fuel was introduced on a second grate above the main fuel bed with an energy input ratio of app. 70% primary and 30% secondary fuel. In comparison to unstaged combustion, an NO_x reduction between 52% – 73% has been achieved at a reduction temperature of app. 700°C. In comparison to air staging a NO_x reduction can be achieved at lower temperatures. The influence of the reduction temperature, the residence time and the fuel composition will be investigated in future work.

The application of fuel staging in practice requires an automatic feeding of the primary and the secondary fuel. For this purpose the secondary fuel must be easily dosable. Therefore natural gas, light fuel oil, biomass powder, saw dust or similar fuels can be used. The nitrogen content in the reburn fuel has a positive effect on reducing NO_X emissions. Furthermore the high volatile matter of biomass has a positive effect on the reburn efficiency. However if solid fuels are used as reburn fuel, fine grinding is necessary to achieve high reburn efficiencies and a good burnout [16].

Pyrolisis gas is also a suitable reburn fuel. Beside NH_i, CH_i and other gaseous components, tars in the pyrolisis gas are supposed to improve the reburn efficiency. As air staging (Fig. 4), also fuel staging leads to minimal NO_X emissions for a primary excess air ratio between app. 0.7 – 0.9 [16] [17] [18] (Fig. 6). The heterogenous reactions were found to be of minor importance for the NO_X reduction [19]. Since two feeding systems are necessary, fuel staging is mainly an option for larger biomass combustion systems.

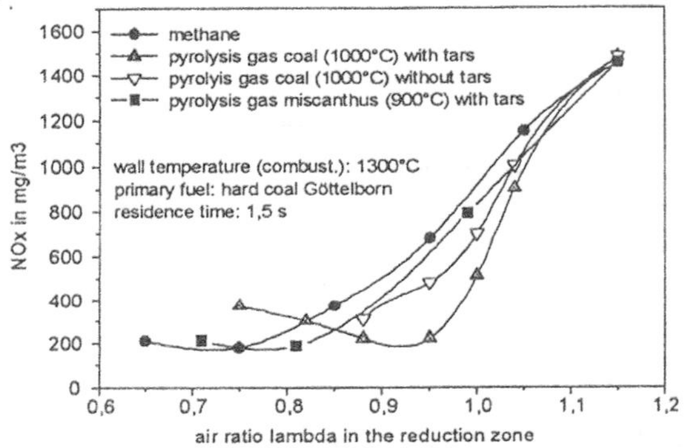

Fig. 6 NO_X emissions in function of the primary excess air ratio for different reburning fuels [16].

3.   Secondary measures for NO_X reduction

3.1 Fundamentals

For native wood the NO_X emissions are usually between 150 and 300 mg/m^3 at 11 Vol.-% O_2 (calculated as NO_2). If chipboards or similar wood residues are burnt, the nitrogen contained in glue and bonding agents (urea) lead to typical NO_X emissions between 500 and 1000 mg/m^3. For biofuels such as straw, miscanthus and grass NO_X emissions are close to those of wood residues since short rotation plants generally show a significantly higher nitrogen content than wood.

In Switzerland secondary measures can be necessary for larger boilers (> 1 – 2 MWth) to fit the emission standard of 250 mg/m^3 of the ordinance on air pollution control (OAPC).

For the secondary measures a reduction agent is injected for a selective NO_X reduction. The most important reduction agents are:

- Ammonia (NH_3) which is gasous at atmospheric pressure and is usually stored in a pressurized tank. NH_3 is toxic and has a low smell limit.

- Sal ammonia (ammonia aq., max. 24 wt.%) is easier to store than ammonia. Smell problems by leackage are similar to ammonia.
- Urea $(NH_2)_2CO$ is a white solid powder which is usually stored as a granulate. To use it as a reduction agent it is dissolved in water. Urea is easy to handle and gives no smell problems.

The reactions in the SCR and SNCR process are similar. The desired reaction is the reduction of NO with NH-reactants, mainly NH_2 (as reaction (1) for primary measures). If urea is used, it has to be decomposed before the reaction with NO_X. Therefore a temperature of at least 250°C – 300°C is necessary.

Desired reaction:

$$4\,NO + 4\,NH_3 + O_2 \longrightarrow \quad 4\,N_2 \quad + \quad 6\,H_2O \tag{6}$$

Side reactions and slippage of reducing agent (4):

$$4\,NH_3 + \quad 5\,O_2 \quad \longrightarrow \quad 4\,NO + \quad 6\,H_2O \tag{7}$$
$$4\,NH_3 + \quad 3\,O_2 \quad \longrightarrow \quad 2\,N_2 \quad + \quad 6\,H_2O \tag{8}$$
$$NH_3 \qquad\qquad \longrightarrow \quad NH_3 \tag{9}$$

In reality reaction (6) is a radical reaction mainly with NH_2 as described for reaction (1) for primary measures: $NO + NH_2 \rightarrow N_2 + H_2O$. Therefore a temperature of about 850°C or the use of a catalyst is necessary for reaction (6).

If the temperature exceeds 950°C the side reactions (7) and (8) become more and more important. Therefore the reaction conditions for the NO_X reduction show an optimum in the temperature window between 850°C to 950°C. Because of the side reaction (8) an over stochiometric amount of reducing agent is necessary. The molar ratio of reducing agent is defined as follows:

$$n = NH_3/NO_X \text{ (mol/mol)} \qquad \text{where } NO_X = NO + NO_2 \text{ (mol)} \tag{10}$$

If a high molar ratio is needed ammonia slippage can lead to high emissions of NH_3 (9). Furthermore N_2O can be formed in undesired side reactions especially when urea is used as reduction agent.

3.2 Selective non-catalytic reduction SNCR

The reaction temperature is most important for the non-catalytic reduction. The highest reduction rate is usually achieved in the temperature window from 850° to 950°C depending on the reducing agent used and other parameters [20]. Own experiments with SNCR with urea in a grate firing with reduction chamber showed an optimum temperature window between 840°C to 920°C [21]. The temperature window can be shifted to lower temperatures by adding H_2 or alcohols which generate OH-radicals. CH_4 can shift the reactions to higher temperatures because it consumes OH-radicals.

Experiments with non-catalytic NO_X reduction in an existing 500 kW grate firing showed unsufficient results. To achieve an NO_X reduction of 50% a highly over stoichiometric amount of reduction agent is needed (n = 3 – 5) which leads to a high NH_3-slippage (> 30 mg/m^3). It was found that the non-ideal mixing of the gases is responsible for the slippage of ammonia [22]. Hence a special reduction chamber with a residence time > 0.3 – 0.5 s is needed for the application of SNCR. To investigate the potential of SNCR processes for biomass combustion, an experimental setup with a pre oven firing and a separate reduction chamber for the SNCR process has been realized (Fig. 7). For wood residues an NO_X-reduction of 60% to 90% was obtained at a molar ratio n = 2 with NH_3-slippage < 10 mg/m^3 at a reaction temperature of app. 900°C. For gaseous ammonia a residence time of app. 0.3 s is sufficent, while for urea and sal ammonia app. 1 s is necessary (Fig. 8).

Laboratory experiments have shown that N_2O can be formed under certain condition when urea is used (Fig. 8) [23]. In several experiments at wood firings no increase of N_2O emissions, which typically were < 5 mg/m^3, has been found [21] [22]. In one case N_2O concentrations > 30 mg/m^3 have been found. The influence on N_2O formation will be studied in future investigations.

From these results several a commercial grate firing was built (Fig. 10). It was shown that an accurate process control is necessary to ensure high NO_X reduction and low NH_3-slippage [22]. Three different control systems to optimize the molar ratio with the aim of minimal NO_X emissions without significant ammonia slippage have been tested:

a   Addition of the reducing agent in function of the load signal
b   Control of the reducing agent in function of the NO_X concentration in the raw gas (in front of the ammonia injection).
c   Control of the reducing agent in function of the NO_X concentration in the flue gas.

Similar results can be achieved with control type a, b and c if the setpoints for type a can be precisely optimized for the actual fuel type. However control type a can only be used in practice if the fuel composition, especially the nitrogen content, is constant. The continous NO_X measurement in the hot raw gas, as needed for control type b, is more complicated than the measurement in the flue gas needed for type c. Since no significant advantages have been achieved with control type b, the measurement of NO_X in the flue gas c is proposed for practical applications.

The test results show that the hot reduction chamber also has a positive effect on the reduction of unburnt pollutants (CO, HC and also PCDD/F) [21]. However the temperature control in the reduction chamber is a major problem in practice. If the grate firing as shown in Fig. 10 is operated at low excess air (< 2), the temperature in the reduction chamber exceeds 950°C and the injected reducing agent immediately reacts to form N_2 and H_2O as described in reaction (8). To ensure high efficiency (low excess air level) and high NO_X reduction efficency, a partial heat extraction from the flue gas is necessary before entering the reduction chamber. Respective design modifications are under investigation.

Fig. 11 shows the design of a large wood combustion plant with combined low-NOx and SNCR process. If saw dust is injected as reburn fuel, a NO_X reduction is achieved by fuel staging. Without saw dust injection, the grate has to be operated with an understoichiometric excess air ratio and the chamber above the grate is used as reduction chamber for air staging. In addition, ammonia or urea can be injected for the SNCR process at different levels in the boiler. At full load operation, the injection is after the first heat exchanger section to avoid temperatures > 950°C, at part load operation, the reducing agent is injected in front of the first heat exchanger section.

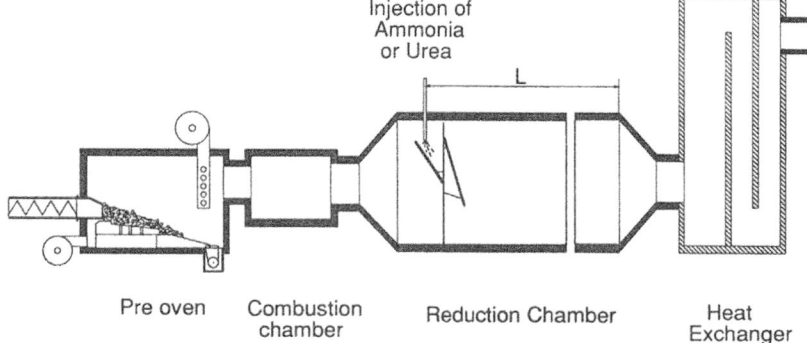

Fig. 7   Pre oven firing with SNCR process in a separate reduction chamber [22]. The residence time can be varied from 0.3 s to 3 s.

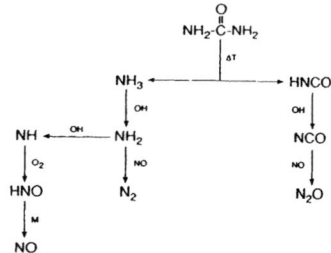

Fig. 8   Reactions during decomposition of urea. Under certain conditions HNCO and N_2O can be formed [23].

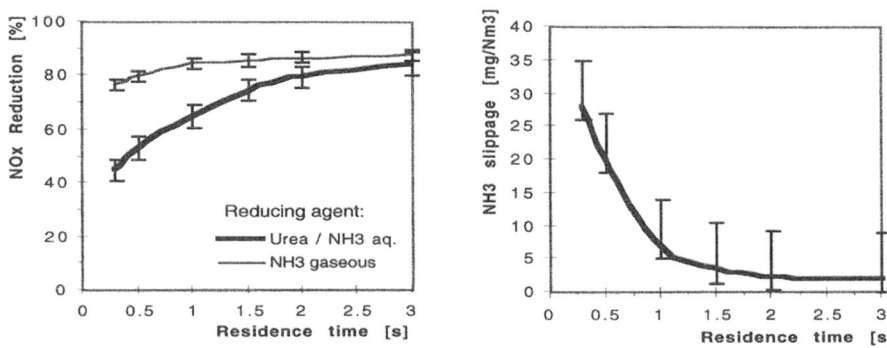

Fig. 9 NO_x reduction by SNCR process with urea and ammonia. Fuel: wood resi- dues, urban waste wood, N content 1 – 2 wt.-%, NO_x level without denox = 700 – 800 mg/Nm3, temperature = 850°C – 930°C°C, n = 1.5 .... 2.2 [22].

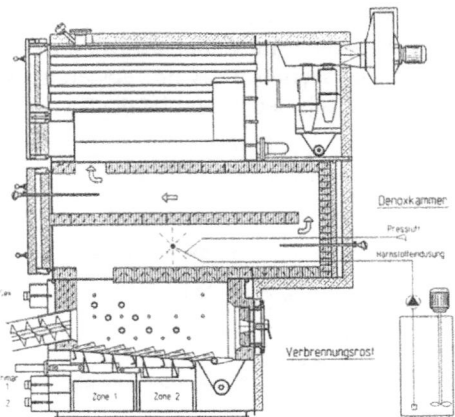

Fig. 10   Grate firing with SNCR process in a separate reduction chamber [20]. Residence time = 0.8 s. At a molar ratio of n = 2.2 up to 90% NOx reduction can be achieved, average reduction in practical operation = 60% – 70%.

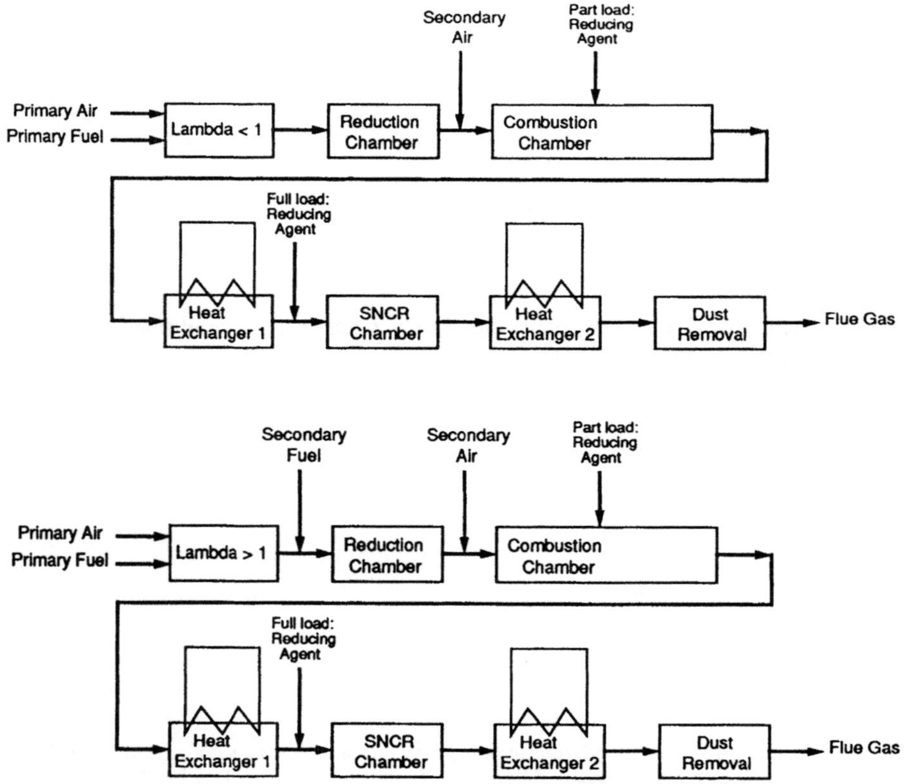

Secondary Air

Part load: Reducing Agent

Primary Air →
Primary Fuel →   Lambda < 1 → Reduction Chamber → Combustion Chamber →

Full load: Reducing Agent

Heat Exchanger 1 → SNCR Chamber → Heat Exchanger 2 → Dust Removal → Flue Gas

Secondary Fuel

Secondary Air

Part load: Reducing Agent

Primary Air →
Primary Fuel →   Lambda > 1 → Reduction Chamber → Combustion Chamber →

Full load: Reducing Agent

Heat Exchanger 1 → SNCR Chamber → Heat Exchanger 2 → Dust Removal → Flue Gas

Fig. 11   Combustion plant with combined low-NOx and denox technology. Above: Low NO_X by air staging, below: low NO_X by fuel staging.

3.3 Selective catalytic reduction SCR

In previous investigations monolithic catalysts ($TiO_2/V_2O_5/WO_3$) have been used with sal ammonia at temperatures between $220° - 270°C$ and with urea at temperatures between $400° - 450°C$. The catalyst was used in low dust and in high dust configuration. For wood residues an NO_x-reduction > 90% was obtained with a molar ratio n = 0.9 – 1 at a catalyst temperature > 250°C. The NH_3-slippage was well below 10 mg/m^3 for 0.7 < n < 2. N_2O emissions were usually low (< 10 mg/m^3), with the exception of one measurement with a catalyst temperature > 350°C [22]. It is assumed that at high catalyst temperature N_2O formation can be significant. Detailed results from the SCR process have been presented in earlier papers [22, 24].

The SCR process shows a high potential for NOx reduction. Similar process control systems as for the SNCR process are necessary to optimize the injection of reducing agents, although ammonia slippage is a minor problem for SCR. The formation of salts in the presence of sulfur and chlorine ((NH_4)HSO_4, (NH_4)SO_4, NH_4Cl etc.) can lead to severe fouling problems in the catalyst, especially at catalyst temperatures below 250°C and in high dust configuration. Due to high costs only few biomass plants have been equipped with SCR and there is only few long term experience. Further development is needed to bring SCR techniques into practice for biomass combustion systems.

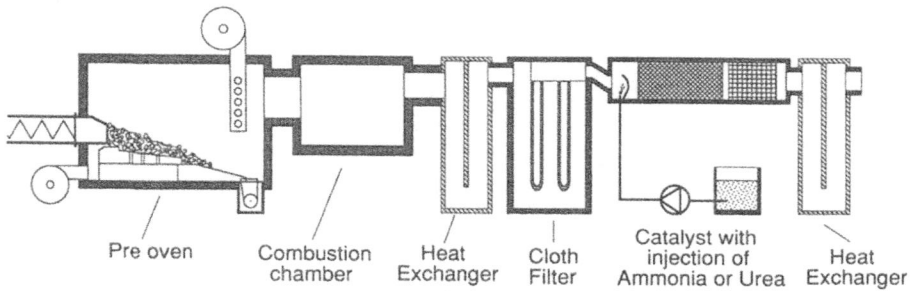

Fig. 12   Wood firing system with NO$_X$ reduction by SCR low dust process.

4.   Conclusions

4.1 Primary measures

Air staging can be considered as a primary measure for NO_x reduction in biomass combustion. By optimizing the excess air ratio towards understoichiometric conditions in the first section of the combustion, a certain NO_x reduction – depending on the furnace design and the fuel – can be achieved in existing combustion systems.

In comparison to such optimized operation of conventional furnaces, air staging with a separate reduction chamber and shows a potential for further NO_x reduction in the order of 40% – 60% for biofuels with low nitrogen content (< 0.5 wt.-%) and 50% – 75% for biofuels with high nitrogen content i.e. UF-chipboards (> 1 wt.-%). For the combustion of natural wood NOx emissions < 150 mg/Nm^3 (calculated as NO$_2$ at 13 Vol.-% O$_2$) were reached in a 250 kW grate firing equipped with air staging and reduction chamber, while for wood residues with a nitrogen content of 2.75 wt.-% NO_x emissions < 250 mg/Nm³ were reached

The NO_x reduction by air staging is mainly influenced by the excess air ratio in the reducing zone, the residence time, the temperature and the mixing quality. Optimal conditions for air staging are: Excess air = 0.7 − 0.8, residence time > 0.3 − 0.5 s, temperature = 1'100° − 1'200°C. To ensure a residence time of 0.3 − 0.5 s in the reducing zone, a separate reduction chamber is needed to achieve high NO_x reduction. Furthermore the realisation of the proposed low-NO_x concept needs an accurate process control to enable a primary excess air ratio below 1 during practical operation. Air staging in the gasification chamber can slightly improve the NO_x reduction rate. However mainly the gas-phase reactions are responsible for the NO_x reduction, while the heterogenous reactions are of minor importance.

Fuel staging shows a similar potential for NO_x reduction as air staging with separate reduction chamber. However fuel staging seems to be effective in a wider temperature range and especially at lower temperatures, while the conditions for the excess air ratio and the residence time are similar to those for air staging. Furthermore a fine grinding of the reburn fuel is needed to achieve significant NO_x reduction. Since two independent automatic fuel feeding systems are necessary for fuel staging, its application is limited to larger biomass plants.

4.2 Secondary measures

To achieve a relevant NO_x reduction without ammonia slippage by the selective non-catalytic reduction (SNCR), a temperature between 850°C and 950°C is needed. To achieve 60% − 70% reduction with ammonia slippage < 10 mg/Nm³ a residence time in the reduction chamber of app. 1 s is needed for urea (aq) or ammonia (aq), while for gaseous ammonia app. 0.3 s is sufficient. For practical application a separate reduction chamber has to be introduced after the combustion zone. A relevant problem for the SNCR process is the very narrow temperature window. Without heat extraction before the reduction chamber, the temperature generally exceeds 950°C at low excess air ratios. Therefore a partial heat extraction before entering the reduction chamber and an accurate process control for the temperature window and the injection of reducing agent are necessary to ensure optimal results in practice. If urea is used N_2O can be formed in the SNCR process under certain conditions. The formation of N_2O will be investigated in future studies.

The selective catalytic reduction (SCR) shows the highest reduction potential of > 80% (up to 95%) at nearly stoichiometric operation with ammonia or urea. Due to high costs only few plants have been equipped in practice with catalysts and only few long term experiences are available. Technical problems have to be considered concerning the possible fouling in the catalyst at low catalyst temperatures (< 300°C), while at high catalyst temperatures the heat management of the combustion plant becomes more complicate.

As an option for large combustion plants low NO_x and denox techniques can be combined in an appropriate way i.e. air staging, fuel staging and SNCR or air staging, fuel staging and SCR.

5. References

1. Nussbaumer, Th.: Stickoxide bei der Holzverbrennung, *Heizung Klima* **12** 1988, AT Aarau, 51-62
2. Nussbaumer, Th.: *Schadstoffbildung bei der Verbrennung von Holz*, PhD Thesis ETH Nr. 8838, Zürich 1989
3. Nussbaumer, Th.: Wood combustion, *Advances in thermochemical biomass conversion*, Vol. I, Blackie Academic 1993, ISBN 0 7514 0171 4, 590 – 604
4. Levy, J.; Longwell, J.; Sarofim, A.; Corley, T.; Heap, M.; Tyson, T.: NO_x abatement in fossil fuel combustion, chemical kinetic considerations, *Proc. 3rd Stationary Source Combustion Symposium*, Vol IV, EPA 600-79-050d, San Francisco, 3-44
5. Nussbaumer, Th.: Grundlagen der Holzverbrennung, *Energetische Nutzung von Holz, Holzreststoffen und Altholz*, Bundesamt für Energiewirtschaft, Bern 1990, 7 – 30
6. Keller, R.; Nussbaumer, Th.; Suter, P.: Untersuchung der Luftstufung mit Reduktionskammer als Primärmassnahmen zur NOx-Minderung bei der Holzverbrennung, *VDI-Bericht Nr. 1090*, 1993, 167-174
7. Keller, R.: *Primärmassnahmen zur NO_x-Minderung bei der Holzverbrennung mit dem Schwerpunkt der Luftstufung*, Diss ETH Nr. 10514, Zürich 1994
8. Beér, J.: Advanced combustion methods for low grade coal utilization, *Low-grade fuels*, Vol. 1, VTT Symposium 108, Espoo (SF) 1990, 83-112
9. Folsom, B.; Bartok, W.: Gas reburning-sorbent injection for simultaneous NO_x and SO_x control, *Energy Technol. XVI*; Conf. 1989, Wash., DC. – Rockville: Gl (1989), 513-527 0-86587-429-8
10. Jansohn, P.; Kolb, Th.; Leuckel, W.: Bildung von Stickstoffoxiden aus Brennstoff-Stickstoff in turbulenten Diffusionsflammen und deren Reduktion durch feuerungstechnische Massnahmen, *Chem.-Ing.-Tech.* **61** 1989, Nr. 11, MS 1801/89
11. Kolb, Th.:: *Experimentelle und theoretische Untersuchungen zur Minderung der NO_x-Emissionen technischer Feuerungen durch gestufte Verbrennungsführung*, Diss TU Karlsruhe 1990
12. Spliethoff, H.: NOX-Minderung durch Brennstoffstufung mit kohlestämmigen Reduktionsgasen, *VDI-Berichte Nr. 765* , 1989, 217-230
13. Kilpinen, P.; Hupa, M.: Homogenous N_2O chemistry at fluidized bed combustion conditions: a kinetic modelling study, *Combustion and Flame*, 85, 94-105
14. Reed, T.: Biomass Gasification, Principles and Technology, Solar Energy Research Institute, Golden, Colorado, 1982
15. Salzmann, R.; Nussbaumer, Th.: *Zweistufige Verbrennung mit Reduktionskammer und Brennstoffstufung als Primärmassnahmen zur Stickoxidminderung bei Holzfeuerungen*, Schlussbericht NEFF 463.1, Laboratorium für Energiesysteme, ETH Zürich 1995
16. Rüdiger, H.; Greul, U.; Spliethoff, H.; Hein, K.: Pyrolisis Gas of Biomass and Coal as a NO_x-Reductive in a Coal fired Test Facility, *Third Int. Conference on Combustion Technologies for a clean Environment*, 3 – 6 July 1995, Lisbon, Portugal
17. Smart, J.; Morgan, D.: The effectiveness of multi-fuel reburning in an internally fuel-staged burner for NO_x reduction, *Fuel* , 1994 vol. 73 no.9, 1437 – 1442
18. Burch, Th., Chen, W.; Lester, Th.; Sterling, A.: Interaction of Fuel Nitrogen with Nitric Oxide During Reburning with Coal, *Combustion and Flame* 98, 1994, 391 – 401
19. Mereb, J., Wendt, J.: Air staging and reburning mechanisms for NO_x abatement in a laboratory coal combustor, *Fuel*, 1994 vol. 73 no.7, 1020 – 1026
20. Schu, G.: *Experimentelle Untersuchungen zur selektiven nichtkatalytischen Reduktion von Stickoxiden in einem Flammrohrkessel*, Diss Universität München 1989
21. Good, J.; Nussbaumer, Th.; Bühler, R.; Jenni, A.: *SNCR-Verfahren zur Stickoxidminderung bei einer Holzfeuerung*, Bundesamt für Energiewirtschaft, Bern 1994
22. Nussbaumer, Th.: Sekundärmassnahmen zur Stickoxidminderung bei Holzfeuerungen, *Brennstoff-Wärme-Kraft*, Vol. 45 (1993) Nr. 11, S. 483-488
23. Köbel, M.: Stickoxidminderung in Abgasen, *Schweizer Ingenieur und Architekt*, Vol. 38, 1992, 693 – 700
24. Nussbaumer, Th.: Selective catalytic reduction and selective non-catalytic reduction of nitric oxides for wood firings, *Advances in thermochemical biomass conversion*, Vol. I, Blackie Academic 1993, ISBN 0 7514 0171 4, 708 – 720

EMPIRICAL NO$_X$-MODELLING AND EXPERIMENTAL RESULTS FROM WOOD STOVE COMBUSTION
Empirical NO$_X$-modelling and experimental results

Ø. SKREIBERG and J. E. HUSTAD
Norwegian University of Science and Technology, Institute of Thermal Energy and Hydropower, Trondheim, Norway
E. KARLSVIK
Foundation for Scientific and Industrial Research at the Norwegian University of Science and Technology, Division of Thermal Energy and Hydropower, Trondheim, Norway

Abstract
Experiments have been performed in three different wood stoves based on batch combustion with natural draught; a traditional stove, a stove equipped with a catalytic afterburner and a staged air unit with downdraught combustion and good insulation of the secondary combustion chamber. The average wood consumption was varied between 0.8-3.8 kg dry wood/h, the average excess air ratio varied between 1.5-5, and the average CO emission level varied between 5-190 g/kg dry fuel (300-12000 ppm). The average NO$_X$ emission level varied between 0.5-1.8 g NO$_2$/kg dry fuel (16-110 ppm), giving a fuel-N to NO$_X$ conversion factor between 0.23-0.55 depending on the wood species (spruce, birch or pine), the operating principle of the wood stove and the operating conditions. The NO$_X$ emission level is traditionally correlated with excess air ratio, temperature, initial fuel-N content and combustion quality. Using the CO emission level as an indicator of the combustion quality, an empirical model has been developed for the total fuel-N to NO$_X$ conversion factor based on average input variables. The excess air ratio and the temperature were found to be the most important input variables, while the CO emission level and the initial fuel-N content only were of minor importance.
Keywords: combustion principles, conversion factors, emissions, empirical modelling, experimental results, NO$_X$, operating conditions, wood stoves.

1 Introduction
In recent years, substantial effort has been put into the study of techniques to reduce NO$_X$ emissions from solid fuel combustion appliances, including biomass combustion appliances [1,2]. Several efficient techniques exist for the reduction of NO$_X$ by the use of secondary measures. Primary measures on the other hand, is usually less efficient and for most practical applications limited to techniques of air-staging and/or fuel-staging, and flue gas recirculation for large scale boilers.

NO$_X$ reduction attempts by the use of primary measures have almost entirely been concerned with continuously fired units of 20 kW and above where the combustion conditions can be assumed to be more or less constant and transient effects can be neglected. For smaller biomass fired batch combustion units with natural draught, such as wood stoves, transient effects are very pronounced. In wood stoves there will be continuously changing combustion conditions with respect to excess air ratio, combustion temperature, fuel composition etc., as illustrated in fig. 1. This makes it more complicated to effectively reduce NO_X emissions from wood stoves.

In wood stoves, typical overall fuel-N to NO_X conversion factors are between 0.2 and 0.55 (this work), dependent mainly on excess air ratio, temperature, initial fuel-N content and combustion quality. The lowest fuel-N to NO_X conversion factors are typically found for wood species with a relatively high fuel-N content, burnt in a traditional stove at low combustion temperatures and consequently resulting in an atmosphere which favours the gas phase reduction of NO_X.

Modern types of wood stoves such as staged air units with downdraught combustion, efficient air preheating and a well insulated combustion chamber are capable of reducing emissions of particles and CO very efficiently, but as a consequence of this, the potential fuel-N to NO_X conversion factor increases. The key to an efficient simultaneous reduction of both CO and NO_X is an air staging which is optimised with respect to a minimum weighted average NO_X emission level in the volatile combustion period. The possibilities of NO_X reduction in the char combustion period are much more limited, due to the predominantly heterogeneous combustion process at high excess air ratios. Most NO_X forms inside the remaining porous char and the possibilities of NO_X reduction in the homogeneous gas phase are low.

This study is aimed at establishing empirical models, based on experimental results, which describes the total, as opposed to the instantaneous, fuel-N to NO_X conversion factor as a function of the most important average variables.

2 Experiments

Experiments have been performed in three different wood stoves based on batch combustion with natural draught; a traditional stove, a stove equipped with a catalytic afterburner and a staged air unit with downdraught combustion and good insulation of the secondary combustion chamber. Schematic drawings of the three different wood stoves are given in fig. 2.

The stoves were placed on an electronic weight for continuos measurement of the weight, as can be seen in fig. 2, and thus the burning rate as a function of time can be calculated. The combustion gases enter the chimney [3] where the emissions of NO_X, CO and unburned hydrocarbons (UHC) in addition to O_2 and CO_2 are measured with continuous analysers. The emissions of NO_X were measured with a chemieluminecent analyser, the emissions of UHC were measured with a flame ionisation analyser, the O_2 level was measured with a paramagnetic analyser and the CO and CO_2 emissions were measured with infrared analysers.

The average wood consumption was varied between 0.8-3.8 kg dry wood/h. The fuel was prepared according to the Norwegian standardised methods, except for the experiments with birch as fuel where standard wood logs, with some bark, were used. Measurements of particle concentrations were done by the use of a dilution tunnel [4].

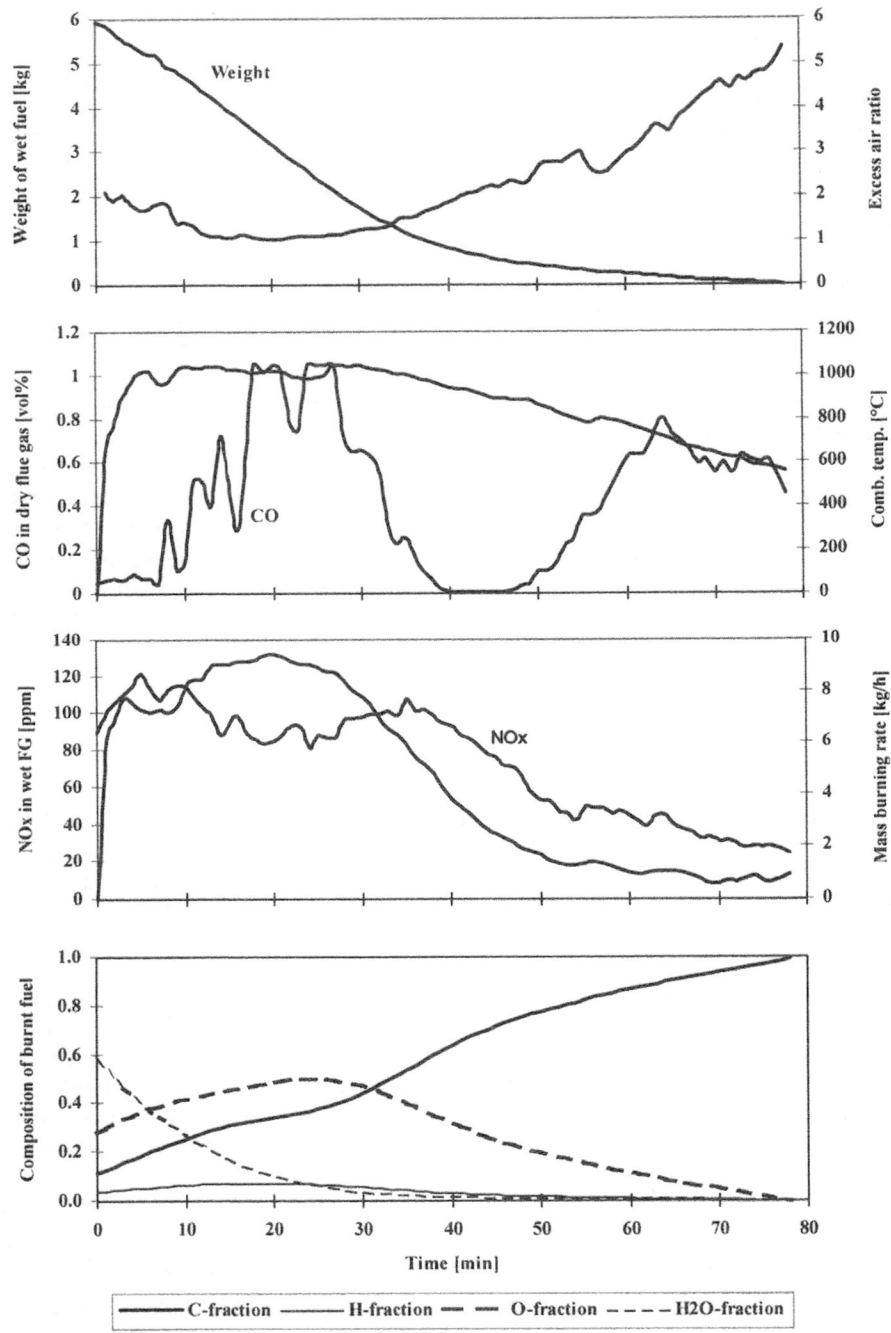

Fig. 1. The complexity of batch combustion.

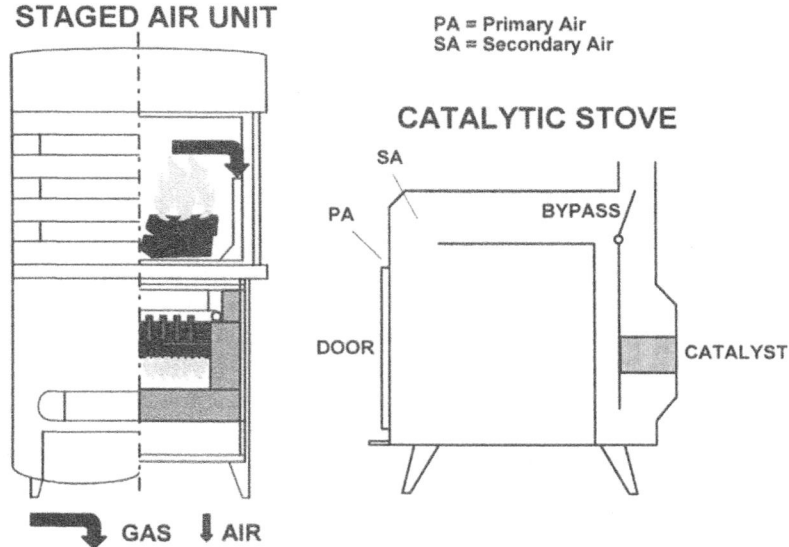

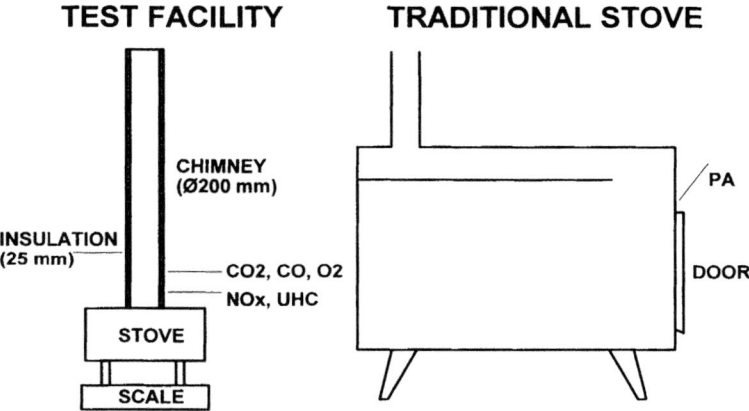

Fig. 2. Schematic drawings of the three types of wood stoves and of the test stand.

Experiments were performed with birch, spruce and pine. The nitrogen content in birch was found to be 0.11% by weight, while the nitrogen content in spruce and pine was found to be 0.07% by weight, using the CHNS-method on an elemental analyser (EA1108 CHNS-O, Carlo Erba Instruments).

3 Empirical modelling

The empirical NO_X modelling approach is based on weighted average values. The NO_X emission level is traditionally correlated with excess air ratio, temperature, initial fuel-N content and combustion quality. Using the CO emission level as an indicator of the combustion quality, and the ratio between the mass burning rate on dry basis and the dry fuel load as an indicator of the combustion temperature, an empirical model has been developed for the total fuel-N to NO_X conversion factor based on these average variables.

The weighted average values which form the basis for the empirical modelling approach are given in table 1, and total weighted average values are given in table 2. Most of the experiments have been done with spruce, but some supplemental experiments have been done with pine and birch, to investigate the effect of fuel type and fuel-N content.

Four variables were used in the empirical models: 1) the average CO emission level (CO) represents the effect of combustion quality and the effect of NO_X reduction by flue gas components formed by incomplete combustion, 2) the average "reactivity" described as the ratio between the average mass burning rate on dry basis and the initial dry fuel load (mB/m) represents the effect of combustion intensity and the influence of temperature, 3) the average excess air ratio (λ) and 4) the initial nitrogen content (N) of dry fuel.

The empirical model describes the conversion of fuel-N to NO_X, N/NO_X, in the form of an exponential expression dependent of the four chosen variables. N/NO_X is the only general and flow independent way to describe the NO_X emission level. N/NO_X can easily be converted to emission values and, from an empirical modelling point of view, it is convenient that it varies between zero and one. Fig. 3 shows the exponential sub-expressions that were chosen to describe the influence of the four chosen variables with respect to N/NO_X. The curve-shape of these sub-expressions represents the general trend in the literature concerning the influence of the chosen average variables on N/NO_X.

During the optimisation process, which minimised the standard deviation between the calculated and modelled N/NO_X, each of the exponential sub-expressions was subjected to boundary conditions, as well as the total expression. The boundary conditions are given in table 3 and the equation for the total expression is given in equation 1.

$$N / NO_X = \exp(-a \cdot CO) \cdot \exp(-b \cdot N) \cdot (1 - \exp(-c \cdot mB / m)) \cdot (1 - \exp(-d \cdot \lambda)) \quad (1)$$

Table 1. Weighted average values
Staged Air Unit (SAU)

Test	Fuel	Fuel-N [w%]	mB/m [1/h]	CO [g/kg]	Excess air ratio	NO_X as NO_2 [g/kg]	N/NO_X
SAU 1	Birch	0.11	0.76	41.4	1.78	1.35	0.373
SAU 2	Birch	0.11	0.67	34.2	2.30	1.77	0.489
SAU 3	Birch	0.11	0.50	24.6	2.03	1.50	0.415
SAU 4	Birch	0.11	0.55	22.6	1.75	1.25	0.347
SAU 5	Pine	0.07	0.56	23.1	1.85	0.77	0.336
SAU 6	Pine	0.07	0.33	18.9	2.16	0.62	0.270
SAU 7	Spruce	0.07	0.24	48.1	2.59	0.54	0.236
SAU 8	Spruce	0.07	0.21	190.9	4.93	0.73	0.319

Catalytic Stove (CS)

Test	Fuel	Fuel-N [w%]	mB/m [1/h]	CO [g/kg]	Excess air ratio	NO_X as NO_2 [g/kg]	N/NO_X
CS 1	Birch	0.11	0.46	4.5	1.61	1.54	0.426
CS 2	Pine	0.07	1.09	10.5	1.76	0.93	0.406
CS 3	Pine	0.07	0.68	36.1	1.80	0.91	0.413
CS 4	Spruce	0.07	0.81	5.8	2.80	1.10	0.479
CS 5	Spruce	0.07	0.74	5.9	2.73	1.21	0.525
CS 6	Spruce	0.07	0.70	10.5	3.53	0.75	0.328
CS 7	Spruce	0.07	0.80	25.4	1.67	0.92	0.402
CS 8	Spruce	0.07	0.36	34.9	2.17	0.88	0.383
CS 9	Spruce	0.07	0.49	23.2	2.23	0.96	0.418
CS 10	Spruce	0.07	0.87	11.9	1.81	0.94	0.408
CS 11	Spruce	0.07	0.72	15.8	2.22	1.02	0.442
CS 12	Spruce	0.07	0.83	23.2	2.19	0.72	0.313
CS 13	Spruce	0.07	0.48	12.2	1.68	0.98	0.426
CS 14	Spruce	0.07	0.36	56.8	1.51	0.71	0.308
CS 15	Spruce	0.07	0.61	16.2	1.75	0.99	0.43
CS 16	Spruce	0.07	0.83	28.3	1.75	0.90	0.392
CS 17	Spruce	0.07	0.48	23.2	1.67	0.81	0.353

Traditional Stove (TS)

Test	Fuel	Fuel-N [w%]	mB/m [1/h]	CO [g/kg]	Excess air ratio	NO_X as NO_2 [g/kg]	N/NO_X
TS 1	Spruce	0.07	0.63	138.4	2.70	0.82	0.358
TS 2	Spruce	0.07	0.79	32.6	2.49	0.63	0.273
TS 3	Spruce	0.07	1.50	9.8	1.58	0.78	0.339
TS 4	Spruce	0.07	0.62	33.0	2.74	0.63	0.275
TS 5	Spruce	0.07	0.57	114.7	4.79	0.68	0.295
TS 6	Spruce	0.07	0.40	92.7	4.20	0.68	0.296
TS 7	Spruce	0.07	0.32	117.2	4.94	0.64	0.28
TS 8	Spruce	0.07	0.39	90.8	2.81	0.56	0.242
TS 9	Spruce	0.07	0.61	37.4	2.36	0.59	0.258
TS 10	Spruce	0.07	0.61	173.4	2.69	0.68	0.297
TS 11	Spruce	0.07	1.62	43.9	2.86	1.26	0.546

mB/m: Ratio between mass burning rate and fuel load [kg/h] / [kg]
N/NO_X: Conversion factor for fuel-N to NO_X

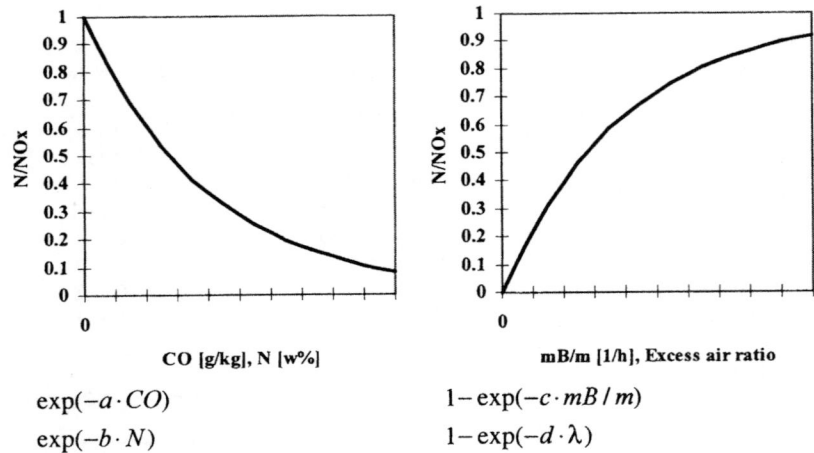

$$\exp(-a \cdot CO)$$
$$\exp(-b \cdot N)$$

$$1 - \exp(-c \cdot mB / m)$$
$$1 - \exp(-d \cdot \lambda)$$

Fig. 3. Exponential sub-expressions.

Table 2. Weighted average fuel-N to NO_X conversion factors

	Birch	Pine	Spruce	Total
Staged Air Unit	0.41	0.30	0.28	0.35
Catalytic Stove	0.43	0.41	0.40	0.40
Traditional Stove	nm	nm	0.31	0.31
Total	0.41	0.36	0.36	0.36

nm = not measured

Table 3. Boundary conditions for sub-expressions and total expression
Sub-expressions:

$CO \rightarrow 0$	$\Rightarrow N / NO_X \rightarrow 1$	$CO \rightarrow \infty$	$\Rightarrow N / NO_X \rightarrow 0$
$N \rightarrow 0$	$\Rightarrow N / NO_X \rightarrow 1$	$N \rightarrow \infty$	$\Rightarrow N / NO_X \rightarrow 0$
$mB / m \rightarrow 0$	$\Rightarrow N / NO_X \rightarrow 0$	$mB / m \rightarrow \infty$	$\Rightarrow N / NO_X \rightarrow 1$
$\lambda \rightarrow 0$	$\Rightarrow N / NO_X \rightarrow 0$	$\lambda \rightarrow \infty$	$\Rightarrow N / NO_X \rightarrow 1$

Total expression:

$$
\left.
\begin{aligned}
CO &\rightarrow 0 \\
N &\rightarrow 0 \\
mB / m &\rightarrow \infty \\
\lambda &\rightarrow \infty
\end{aligned}
\right\} \Rightarrow N / NO_X \rightarrow 1
$$

$$N / NO_X (Average\_values)$$
$$= Average\_ N / NO_X$$

Since mB/m was chosen to represent the influence of temperature, a comparison of the relationship between mB/m and the weighted average chimney inlet temperature was done, as can be seen in fig. 4. The comparison clearly shows that mB/m is a wood stove specific variable and therefore no general empirical model can be made, but it can also be seen that there is a quite good agreement between temperature and mB/m for a specific type of wood stove. Hence, the optimisation process must be performed separately for each type of wood stove.

The optimisation process gave no dependence of the initial fuel-N content on N/NO_X. Nor was there found any significant correlation between the measured CO emission level in the chimney and N/NO_X. For the catalytic stove, there was a clear correlation between the excess air ratio and N/NO_X, but no correlation could be found between N/NO_X and mB/m. The variable mB/m was more important than the excess air ratio for the traditional stove, but for the staged air unit the importance of these two variables was approximately the same. These results are summarised in table 4, where the standard deviation between the calculated and modelled N/NO_X can be found for the optimisation matrix.

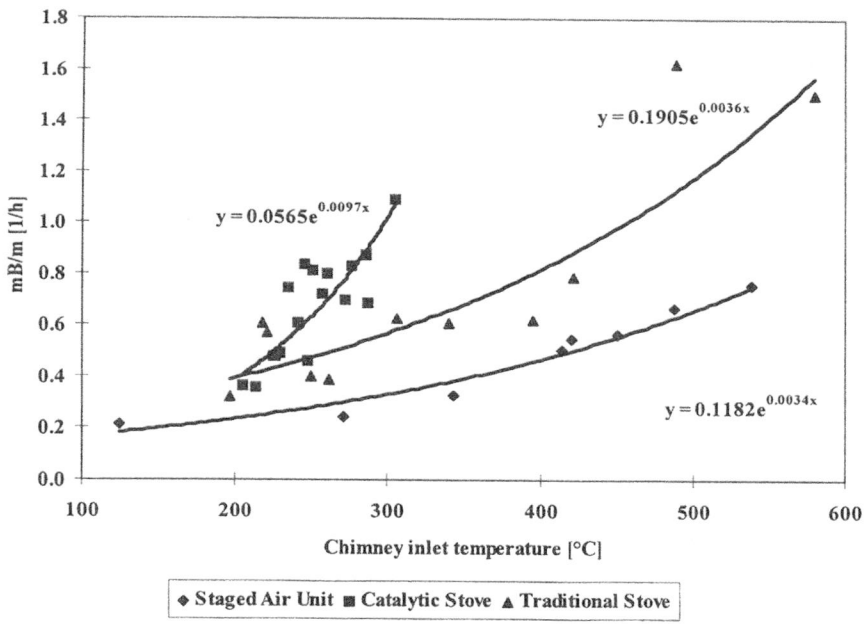

Fig. 4. The relationship between mB/m and the weighted average chimney inlet temperature.

Table 4. Standard deviations in % for the optimisation matrix

Excluding variable(s):	None	N	CO	mB/m	λ	CO&λ	CO&mB/m	λ&mB/m
Staged air unit (SAU)	10.75	10.75	11.22	21.27	18.33	18.33	34.16	58.26
Catalytic Stove (CS)	10.54	10.54	10.54	10.54	40.75	61.99	10.54	40.75
Traditional Stove (TS)	14.05	nm	14.05	28.16	21.56	21.56	29.22	65.47

nm = not measured

N/NO$_X$ can then be described by the following expressions for the three types of wood stoves:

SAU: $\quad N / NO_X = (1 - \exp(-1.6982 \cdot mB / m)) \cdot (1 - \exp(-0.4064 \cdot \lambda))$ (2)

CS: $\quad N / NO_X = (1 - \exp(-0.2515 \cdot \lambda))$ (3)

TS: $\quad N / NO_X = (1 - \exp(-0.7477 \cdot mB / m)) \cdot (1 - \exp(-0.4417 \cdot \lambda))$ (4)

These expressions are shown graphically in fig. 5 for the catalytic stove and in fig. 6 for the staged air unit and the traditional stove.

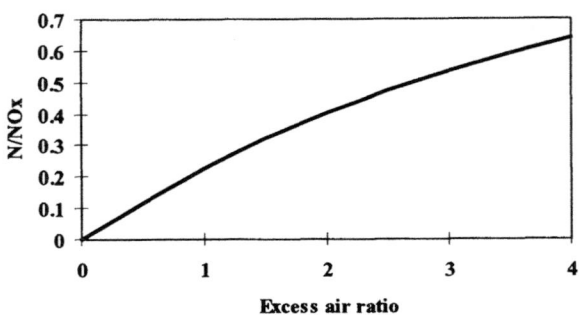

Fig. 5. N/NO$_X$ model for the catalytic stove.

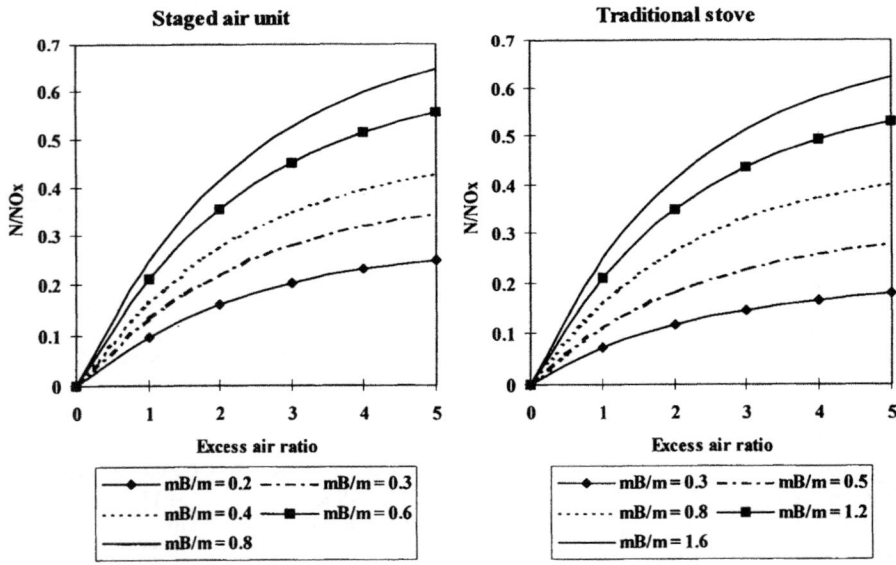

Fig. 6. N/NO$_X$ model for the staged air unit and the traditional stove.

4 Discussions

4.1 Influence of combustion equipment
From table 1, we find the ranges and the average values of the different variables as shown in table 5.

Table 5. Ranges and average values of the different variables

Stove		Fuel-N [w%]	mB/m [1/h]	CO [g/kg]	Excess air ratio	NO_X as NO_2 [g/kg]	N/NO_X
SAU	Min.	0.07	0.21	18.9	1.75	0.54	0.24
	Max.	0.11	0.76	190.9	4.93	1.77	0.49
	Ave.	0.090	0.48	50.5	2.42	1.07	0.35
CS	Min.	0.07	0.36	4.5	1.51	0.71	0.31
	Max.	0.11	1.09	56.8	3.53	1.54	0.53
	Ave.	0.072	0.67	20.3	2.05	0.96	0.40
TS	Min.	0.07	0.32	9.8	1.58	0.56	0.24
	Max.	0.07	1.62	173.4	4.94	1.26	0.55
	Ave.	0.07	0.73	80.4	3.11	0.72	0.31
Total	Min.	0.07	0.21	4.5	1.51	0.54	0.24
	Max.	0.11	1.62	190.9	4.94	1.77	0.55
	Ave.	0.076	0.64	45.3	2.46	0.91	0.36

The ratio between the maximum and minimum N/NO_X varies roughly with a factor of two for all three types of wood stoves. The total N/NO_X is 0.36. The highest average N/NO_X is found for the catalytic stove, about 30% higher than for the traditional stove and 16% higher than for the staged air unit. Recalculating N/NO_X for the staged air unit and the traditional stove by using equation 2 and 4, using the same average excess air ratio as for the catalytic stove, decreases N/NO_X from 0.35 to 0.31 for the staged air unit and from 0.31 to 0.25 for the traditional stove. The average mB/m for each type of wood stove was used in the recalculation. This gives a N/NO_X about 61% higher than for the traditional stove and 28% higher than for the staged air unit at an equal excess air ratio. The large difference in N/NO_X between the catalytic stove and the traditional stove is believed to be caused by the catalytic combustor, since the catalytic stove is comparable with the traditional stove when the catalytic combustor is removed. This was confirmed by Skreiberg [5], by measuring the NO_X emission level both before and behind the catalytic combustor during a single experiment.

The difference between the staged air unit and the traditional stove can be explained partly by a considerably higher combustion chamber temperature (200-300°C) for the staged air unit, and partly by the higher N/NO_X found for birch (see subheading 4.3). Half of the experiments performed in the staged air unit was done with birch.

4.2 Influence of CO emission level
The CO emission level was, in the optimisation process, chosen to represent the effect of combustion quality and the effect of NO_X reduction by flue gas components formed by incomplete combustion. During the course of the optimisation process, no significant correlation between the CO emission level and N/NO_X was found, which

indicates that the CO emission level measured in the chimney is not a suitable indicator for the NO$_X$ reduction in the combustion chamber by flue gas components formed by incomplete combustion, but only a suitable indicator of "combustion quality". This theory is supported by Keller [6] which found no effect of CO on the NO reduction in the reduction chamber of a 25 kW test reactor, comprising a fixed bed gasification chamber, an electrically heated reduction chamber and a combustion chamber. Gas phase kinetical calculations done at sub-stoichiometric conditions confirmed that the effect of CO on the NO$_X$ reduction is insignificant, while the reaction between NH$_2$ and NO is the most important reaction.

4.3 Influence of fuel type

From table 2 we find the average N/NO$_X$ for the three wood species considered. Birch was found to have a N/NO$_X$ about 15% higher than pine and spruce. This difference can be partly be accounted for by the bark layer on the birch wood logs used, which has about five times higher fuel-N content than the trunk [7]. Estimating the relative contribution from the volatile- and char fuel-N content, with the basis in a proximate analysis [8], it was found that the volatile contribution for birch was about 84% and for pine and spruce about 77%. Since the bark layer will be combusted during the volatile period, these results support the theory of a higher N/NO$_X$ for birch due to the presence of a bark layer on the wood logs used.

4.4 Influence of fuel-N content

Due to the higher N/NO$_X$ for birch, no trend of decreasing N/NO$_X$ with increasing fuel-N content could be found. This was reflected in the optimisation process, where no influence of fuel-N content was found. A scan of the literature for N/NO$_X$ reported from wood combustion [9], [10], [11], [12], [13], and including this work, resulted in table 6. As can be seen, depending on fuel-N content, type of combustion equipment and combustion conditions, a wide range of N/NO$_X$ can be found. Plotting the average N/NO$_X$ given in table 6 as a function of initial fuel-N content, as shown in fig. 7, a clear trend can be found, with decreasing N/NO$_X$ with increasing fuel-N content.

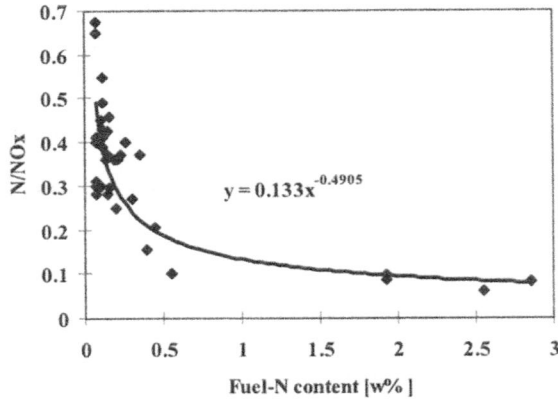

Fig. 7. N/NO$_X$ as a function of fuel-N content.

Table 6. Fuel-N to NO_X conversion factors from wood combustion

Literature Reference	Combustion Equipment	Fuel type	Number of experiments	Particle Size	Fuel-N content [w%, dry basis], Range / Average	Fuel-N to NO_X conversion factor, Range / Average	Comments
[9]	Fixed Bed Reactor	Wood - Acacia	55	200-700 mg	0.21	0.19-0.61 / 0.36	Single particle experiments
		Wood - Beech	48	200-700 mg	0.15	0.11-0.48 / 0.28	Single particle experiments
		Wood - Birch	55	200-700 mg	0.11	0.24-0.86 / 0.55	Single particle experiments
		Wood - Spruce	57	200-700 mg	0.07	0.13-0.70 / 0.40	Single particle experiments
		Wood char - Acacia	9	123-318 mg	0.35	0.28-0.44 / 0.37	Single particle experiments
		Wood char - Birch	6	111-128 mg	0.16	0.38-0.54 / 0.46	Single particle experiments
		Wood char - Spruce	6	93-110 mg	0.11	0.41-0.57 / 0.49	Single particle experiments
This work	Traditional Wood Stove	Wood - Spruce	11	Wood logs	0.07	0.24-0.55 / 0.31	
	Advanced Staged Air Wood Stove	Wood - Birch	4	Wood logs	0.11	0.35-0.49 / 0.41	
		Wood - Pine	2	Wood logs	0.07	0.27-0.34 / 0.30	
		Wood - Spruce	2	Wood logs	0.07	0.24-0.32 / 0.28	
	Wood Stove with Catalytic Afterburner	Wood - Birch	1	Wood logs	0.11	0.43	
		Wood - Pine	2	Wood logs	0.07	0.41-0.41 / 0.41	
		Wood - Spruce	14	Wood logs	0.07	0.31-0.53 / 0.40	
[10]	Laboratory Furnace	Wood - Pine	1	Wood logs	0.07	0.673	
		Wood - Beech	1	Wood logs	0.2	0.36	
		Wood - Chip board	1		2.85	0.084	
[11]	Pyros Stoker Firing	Wood	1	Wood logs	0.16	0.297	
		Wood - Fibre board	1		1.93	0.098	
	Under Stoker Firing	Wood	1	Wood logs	0.23	0.371	
		Wood - Fibre board	1		1.93	0.086	
	Grate Firing	Wood - Fibre board	1		2.55	0.06	
[12]	Laboratory Furnace	Wood - Fir	3	Wood logs	0.065-0.075 / 0.072	0.64-0.67 / 0.65	Data from graph
		Wood - Briquette	2		0.113	0.38-0.40 / 0.39	Data from graph
		Wood - Red Beech	2		0.138	0.36	Data from graph
		Wood - Spruce	2	Wood logs	0.147	0.36-0.38 / 0.37	Data from graph
		Wood - White Beech	3	Wood logs	0.19	0.34-0.37 / 0.36	Data from graph
		Wood - Acacia	3	Wood logs	0.240-0.277 / 0.258	0.38-0.41 / 0.40	Data from graph
[13]	Stationary Fluidized Bed	Wood chips	2		0.4	0.11-0.20 / 0.155	
		Wood chips	2		0.55	0.08-0.12 / 0.10	
	Inclined Moving Grate	Wood chips	NR		0.35-0.55 / 0.45	0.14-0.27 / 0.205	Data from five plants
		Wood chips	NR		0.10-0.20 / 0.15	0.30-0.55 / 0.425	
	Horizontal Fixed Grates	Wood - Briquettes	NR		0.1	0.45	Data from three plants
		Wood - Briquettes	NR		0.1	0.3	
		Wood - Briquettes	NR		0.2	0.25	
		Wood - Briquettes	NR		0.3	0.27	

NR = Not Reported

4.5 Influence of temperature

N/NO_X was found not to be consistently increasing with increasing temperature by Skreiberg [9] in an extensive test matrix for wood particles performed in an electrically heated fixed bed reactor. The N/NO_X consistently increased with temperature until a certain point, where the combustion rate becomes so fast that the limitation of O_2 becomes the controlling parameter, and then decreases. The question is therefore whether the N/NO_X is increasing due solely to an increasing combustion temperature or if this is only an effect of a reduced potential for NO_X reduction by flue gas components formed by incomplete combustion. Nussbaumer [10] found no significant influence of the flame temperature on the NO_X emission level in the range of 800-1300°C in a laboratory furnace, while Keller and Nussbaumer [11] found that with increasing temperature over the range 700-1100°C the fuel NO increased about 20-25% over this range. This indicates that the temperature in itself is not an important controlling variable, but rather a combustion equipment specific indicator of the potential for NO_X reduction by flue gas components formed by incomplete combustion, and which can be described by mB/m as shown in fig. 3.

4.6 Influence of excess air ratio

The excess air ratio is without doubt a very important variable. Increasing N/NO_X with increasing excess air ratio has been found by several investigators including Keller and Nussbaumer [11], Karlsson [13] and Skreiberg [9]. Fig. 8 illustrates the importance of the excess air ratio for the three types of wood stoves using equation 2, 3 and 4. The average mB/m for each type of wood stove was used in the comparison. As can be seen, increasing the excess air ratio from 1.5 to 3.5 increases N/NO_X with about 86% for the catalytic stove, 66% for the staged air unit and 62% for the traditional stove.

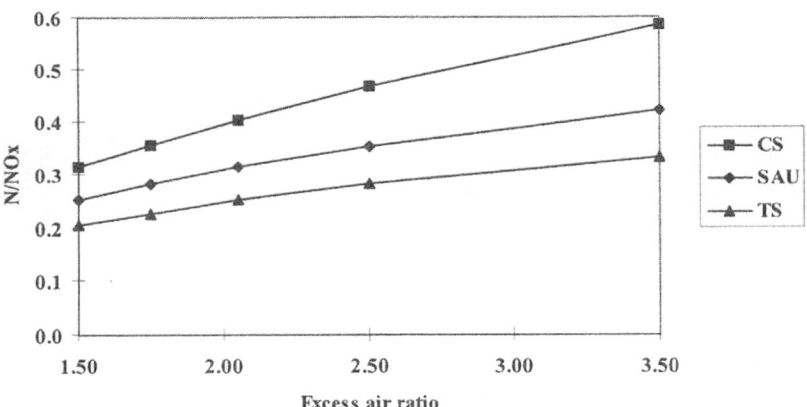

Fig 8. N/NO_X as a function of excess air ratio

5 Conclusions

This work has shown that the excess air ratio and the temperature are the two most important variables controlling the fuel-N to NO_X conversion factor when combusting a specific wood species in a specific combustion appliance. By improved control of the excess air ratio a significant NO_X reduction can be achieved. Also by decreasing the combustion temperature a significant reduction can be achieved. Though, by optimising these to variables a considerable reduction of the fuel-N to NO_X conversion factor can be achieved, even for traditional wood stoves. But care must be taken not to significantly increase the emissions of unburnt components, which are very dependent on the excess air ratio and temperature.

The average fuel-N to NO_X conversion factor for the staged air unit tested in this work was found to be somewhat higher than for the traditional stove. This can partly be explained by a higher combustion temperature. Ideally, staged air combustion should be the most effective means of NO_X reduction by primary measures. But this demands a very effective control of the ratio between the primary air and the secondary air, and of the excess air ratio. Therefore, care must be taken to further optimise the air staging in future development of low-NO_X wood stoves if any significant reduction of the NO_X emission level shall be achieved. For wood stoves equipped with catalytic afterburners, care must be taken to ensure that the catalytic afterburner do not contribute to an increased NO_X emission level, as shown in this work.

6 References

1. Nussbaumer, T. (1993). Sekundärmassnahmen zür Stickoxidminderung bei Holzfeuerungen, *Brennstoff-Wärme-Kraft*, Vol. 45, No. 11, pp. 483-488.
2. Keller, R. (1994). Primärmassnahmen zur NO_X-Minderung bei der Holzverbrennung mit dem Schwerpunkt der Luftstufung, Diss *ETH*, No. 10514, Zürich.
3. Karlsvik, E., Hustad, J.E. and Sønju, O.K. (1993). Emissions from wood stoves and fireplaces, *Advances in Thermochemical Biomass Conversion*, Vol. 1, Blackie Academic & Professional.
4. Karlsvik, E. (1991). Norwegian standards for emissions from wood stoves, *6th European Conference on Biomass for Energy*, Industry and Environment, Athens, Greece, 22-23 April 1991.
5. Skreiberg, Ø. (1994). Advanced techniques for wood log combustion, *COMETT Expert Workshop on Biomass Combustion*, Graz, Austria, 16-17 May 1994.
6. Keller, R. (1994). Stickoxidminderung durch gestufte Verbrennung in einer Unterschufeuerung, Neue Erkenntnisse zur thermischen Nutzung von Holz, *Tagungsband zum 3. Holzenergie-Symposium*, ETH Zürich, 21. Oktober 1994.
7. Hofbauer, H. (1995). Characterisation of biomass fuels and ashes, *IEA Bioenergy Agreement - Task X - Biomass Utilisation - Biomass Combustion*, Project Reports 1992-1994.

8. Grønli, M.G., Sørensen, L.H. and Hustad, J.E. (1994). Thermogravimetric analysis of the components of biomass under nonisothermal conditions - Part I. Experimental, *Nordic Seminar on Biomass Gasification and Combustion*, Trondheim, Norway, 21-22 June 1994.

9. Skreiberg, Ø. (1994). Experimental report on wood combustion experiments in a fixed bed reactor at the Technical University of Denmark, Institute of Thermal Energy and Hydropower, *The Norwegian Institute of Technology*, NTH-ITEV 1994:32.

10. Nussbaumer, T. (1993). Wood combustion, *Advances in Thermochemical Biomass Conversion*, Vol. 1, Blackie Academic & Professional.

11. Keller, R. and Nussbaumer, T. (1993). Fuel nitrogen in wood, *Advances in Thermochemical Biomass Conversion*, Vol. 1, Blackie Academic & Professional.

12. Rath, M. and Hofbauer, H. (1994). Introductory statement to working group 2: Fire Wood Furnaces, *COMETT Expert Workshop on Biomass Combustion*, Graz, Austria, 16-17 May 1994.

13. Karlsson, M.-L. (1994). Emissions from biomass fired plants in the range of 0.5-10 MW, *Seminar on Small-Scale Biomass Combustion Technology*, Jyväskylä, 25 May 1994.

SIMULTANEOUS SO$_x$/NO$_x$ EMISSION CONTROL WITH BIOLIME™ DERIVED FROM BIOMASS PYROLYSIS OIL

Mr. Klaus H. Oehr and Dr. Joe Zhou
DynaMotive Technologies Corporation,
3650 Wesbrook Mall,
Vancouver, B.C., V6S 2L2

Dr. Girard A. Simons
Simons Research Associates
Lynnfield, MA 01940

Dr. Marek Wójtowicz
Advanced Fuel Research, Inc.
87 Church Street
East Hartford, CT 06108

phone: 604-222-5590   FAX: 604-222-5545   E-mail: dmcc@mindlink.bc.ca

ABSTRACT

Recently enacted United States Clean Air Act Amendments have mandated increased control of sulphur dioxide and nitrogen oxide gas emissions from coal fired power stations. Unfortunately, control with conventional lime or limestone sorbents results in high capital and operating costs for SO$_2$ control and little or no NO$_x$ emission control. Destruction of SO$_2$, by lime derived in-situ from thermally degraded organic calcium salts, has exceeded 90% at 1:1 calcium to sulphur dioxide mole ratios. Organic calcium salts, during their thermal decomposition, have also shown an ability to destroy NO$_x$ compounds by 40% or more at 1:1 calcium to sulphur dioxide mole ratios and 3:1 calcium to NO mole ratios. Organic calcium salts are the first combustible species which have shown the ability to simultaneously reduce SO$_x$/NO$_x$ emissions at temperatures typical in pulverized or fluidized bed coal combustors. Previously, the high cost of producing thermally degradable calcium organic salts has precluded their commercialization. However, commercialization of waste biomass pyrolysis reactors has now made a low cost feedstock for organic calcium salts possible. DynaMotive Corporation has developed a proprietary process which combines biomass pyrolysis liquor with lime to produce an organic calcium product called BioLime™. BioLime™ can be produced from a wide variety of waste biomass sources including sawdust.

BACKGROUND

SO_x and NO_x emissions from fossil-fuel combustors are a serious concern because of their adverse impacts on the environment including acid rain production [1]. United States Clean Air Act Amendments mandate the reduction of SO_x emissions by 10 million tons and NO_x emissions by 2 million tons by the year 2000 (vs. 1980 levels). About 80% of SO_x and 56% of NO_x emissions are generated by fuel combustion. Canada and the United States signed an Air Quality Agreement on March 13, 1993 to limit SO_x and NO_x emissions.

For fluidized coal combustors, lime or limestone have been used to reduce SO_x. In theory, 1 mole of calcium can remove 1 mole of sulphur from the combustion gas as calcium sulphite or calcium sulphate. In practice, calcium to sulphur mole ratios of 1.8:1 to 3:1 are required for 90% SO_x removal [2]. Poor calcium utilization is attributed to calcium sulphate or calcium sulphite fouling of unreacted lime [3]. The highly alkaline calcium sulphate/sulphite/lime wastes are difficult to recycle and are considered toxic in many areas [4]. For pulverized bed combustors, representing the majority of current electric utility boilers, the situation is worse. High temperature and low residence time of pulverized bed combustors make in-situ sulphur capture unfeasible, while high capital and operating costs of external SO_x removal systems such as flue-gas desulphurization (FGD) scrubbing systems limit their application. NO_x control is achieved by a combination of burner controls and expensive chemical additives, such as urea and ammonia, which react with NO_x to produce non-toxic nitrogen gas and possible odorous ammonia byproduct emissions.

The use of organic calcium salts such as calcium acetate, calcium magnesium acetate and calcium benzoate, as more suitable calcium oxide precursors, has been suggested [5-7]. These compounds simultaneously reduce SO_x and NO_x emissions at elevated temperatures, typical of coal combustors, at low calcium to sulphur mole ratios. However the high cost of manufacturing these compounds from conventional acetic acid sources ($660/ton U.S.) or benzoic acid ($1200/ton U.S.) renders their commercial use unlikely [8]. The market requires inexpensive sources of organic calcium compounds. Commercialization of waste biomass pyrolysis reactors now makes a low cost feedstock for organic calcium salts possible.

RESULTS

DynaMotive Technologies Corporation has developed a patented process which combines biomass pyrolysis liquor with lime, water and air to produce a mixture of calcium carboxylates and calcium phenoxides called BioLime™ [7]. Additional patents are pending. In preliminary tests partially financed by 3 British Columbia Government, Ministry of Employment and Investment, Technology Assistance Program (TAP) awards, BioLime™ derived from mixed hardwood and softwood waste sawdust achieved a 1.1:1 calcium to sulphur mole ratio during sulphur dioxide adsorption from a simulated fluidized bed combustor exhaust versus 2.5:1 for pulverized limestone [9]. Calcium utilization by sulphur dioxide was 91% for 3 different BioLime™ batches versus only 41% for pulverized limestone. In work partially financed by Energy Mines and Resources Canada, Bio-energy Development Program, a subset of the organic calcium salts in BioLime™ has been shown to simultaneously remove SOx and NOx at temperatures typical of coal combustors [5,10].

Advanced Fuel Research has tested BioLime™ formulations derived from softwood/hardwood sawdust pyrolysis liquors obtained from Ensyn Technologies Inc. which had been diluted with water, filtered through 8 micron polycarbonate filters and injected through a 25 micron aspirator between 900 to 1100°C [11]. The filtered BioLime™ liquors were injected into a 2400 ppm sulphur dioxide and 800 ppm nitric oxide gas stream containing 5% excess oxygen at 1:1 calcium to sulphur mole ratio and 3.5 seconds residence time. BioLime™ achieved 91-92% sulphur dioxide destruction and 38-40% nitric oxide destruction. BioLime™ at 7.3% calcium content by weight has been produced batchwise in a stirred fluidized bed reactor in 5 kg batches in 2 hours at 60-65°C maximum temperature and less than 2 atmospheres pressure using Eucalyptus and Pine derived pyrolysis liquors obtained from Union Fenosa, Spain via RTI Resource Transforms International, Canada blended in a 50:50 weight ratio after pyrolysis. The BioLime™ was diluted with water to form a "cream". This cream was filtered through a 25 micron filter followed by a 12 micron filter. The filtrate, containing 2.9% calcium by weight, was injected through a 25 micron aspirator into a 2400 ppm sulphur dioxide and 800 ppm nitric oxide gas stream containing 5% excess oxygen at 1:1 calcium to sulphur mole ratio and 3 seconds residence time. This BioLime™ filtrate achieved 91-92%, 96-97% and 96-99% sulphur dioxide destruction at 900°C, 1000°C and 1100°C respectively. Nitric oxide destruction was 36-43%, 29-30% and 24-28% at 900°C, 1000°C and 1100°C respectively. The filtrate was injected through a 25 micron aspirator into a 2400 ppm sulphur dioxide and 300 ppm nitrous oxide gas stream containing 5% excess oxygen at 1:1 calcium to sulphur mole ratio and 3 seconds residence time. This BioLime™ filtrate achieved 89-90% and 97% SO_2 destruction at 800°C and 900°C respectively. Nitrous oxide destruction was 31-32% and 60-61% at 800° and 900°C respectively. At the time of this writing, unfiltered BioLime™ with lower moisture content and higher calcium content was being tested for simultaneous SO_2/NOx control.

REFERENCES

1. Bell, A.T., Manzer, L.E., Chen, N.Y., V.W. Weekman, Hegedus, L.L. and C.J. Pereira. "Protecting the Environment through Catalysis", Chemical Engineering Progress, Feb., 1995.

2. Fee, D.C., W. I. Wilson, K.M. Myles, I. Johnson, L.S. Fan, G.W. Smith, S. H. Wong, J.A. Shearer and J.F. Lenc, "Sulfur Control in Fluidized Bed Combustors : Methodology for Predicting the Performance of Limestone and Dolomite Sorbents," Argonne National Laboratory Report no ANL/FE-8-10, NTIS USA, 1982.

3. Vastava, R.D., Campbell, I.M. and Blaustein, B.D., "Coal Bioprocessing: A Research-Needs Assessment," Chemical Engineering Progress, 85(12), 1989, pages 45-53.

4. Foster-Miller Corporation, US Department of Energy, SBIR Solicitation DOE/ER-0629, Feb 27, 1995.

5. Simons, G.A., Parker, T.F., Moore, T.E., C.A. Senior and Y.A. Levendis, "Combined NOx/SOx Control Using a Single Liquid Injection System", Physical Science Inc., TR-1169, February 1992.

6. Levendis, Y. and Wise, D.L., "Use of Aromatic Salts for Simultaneously Removing SO_2 and NOx Pollutants from Exhaust of A Combustion System", United States patent 5,352,423, October 4, 1994.

7. Oehr, K. "Acid Emission Reduction". United States patent 5,458,803. October 17, 1995.

8. Chemical Marketing Reporter, August 15, 1994, page 34.

9. Brereton, C. and A. Ergüdenler, 1995, "Control of Sulphur Dioxide Emissions Using BioLime™ (Calcium Salts of Wood Pyrolysis Product)", unpublished report prepared for Dynamotive Corporation, March 1995.

10. Oehr, K.H. "Biomass Derived Alkaline Carboxylate Road Deicers. Report Prepared for the Bioenergy Development Program, Renewable Energy Division, Energy, Mines and Resources Canada, Ottawa, Ontario. DSS Contract Number:23283-8-60721/01-SZ. 1992.

11. Wójtowicz, M.A. and R. Bassilakis. 1995. Simultaneous SO_2/NO Control by Spray-Drying Solutions of Organic Salts into a Combustion Zone. Report prepared for DynaMotive Corporation by Advanced Fuel Research, East Hartford, Connecticut. August. pages 1-40.

EMISSIONS FROM WOOD COMBUSTION IN FIREPLACES

K.H. ORAVAINEN and J.J. SAASTAMOINEN
VTT Energy
P.O. Box 1603, FIN-40101 Jyväskylä, FINLAND

Abstract

It has been estimated that in Finland even more than 90% of hydrocarbon and PAH emissions from stationary sources are due to small scale wood combustion. Wood log combustion was studied both on field and by laboratory tests. Emissions can be reduced by changing firing habits, using dry wood and developing better combustion appliances, but also by using secondary measurers like catalytic converters. CO measurement is a good indicator for combustion quality.

Keywords: Wood combustion, fireplaces, stoves, emissions

1 Introduction

1.1 Wood log combustion

There are over a million manually fed small heating appliances in Finland where about 5.4 million cubic meters of wood fuel is used annually. In small wood burning appliances, such as kitchen ovens, baking ovens, open fireplaces, heating ovens or stoves, combustion is a batch-type process. Fuel is added according to heat demand as large amounts at a time. By this way the working time used for heating is minimized. The amount of combustion air introduced and the draught in the flue determine the heat output. This type of system works well with fuels having low amount of volatiles. Then combustion is mainly heterogeneous combustion of solid carbon that is more easy to control. Batch-type combustion of fuels having large amount of volatiles leads often to incomplete combustion and high emissions because mixing of combustion air and combustible gases is not sufficient and temperature is not high enough [1,2,3].

Combustion process changes a lot when the share of volatile matter content changes for example from the value of wood (75 - 85%) to that of coke (max. 5%). When burning fuels having high volatile matter content, combustion is quite unsteady-state process. It goes basically through following phases which are partly overlapping: drying of fuel, pyrolysis inside the solid releasing volatile matter to gas phase, combustion of volatiles and residual char.

To illustrate batch-type combustion process of wood, CO_2- and CO -concentrations versus time are plotted in the figure 1. In this experiment birch-wood having moisture content of 13% was used. Unsteady-state combustion can clearly be seen from the figure. Fuel was added several times. Higher CO-content appears after fuel adding.

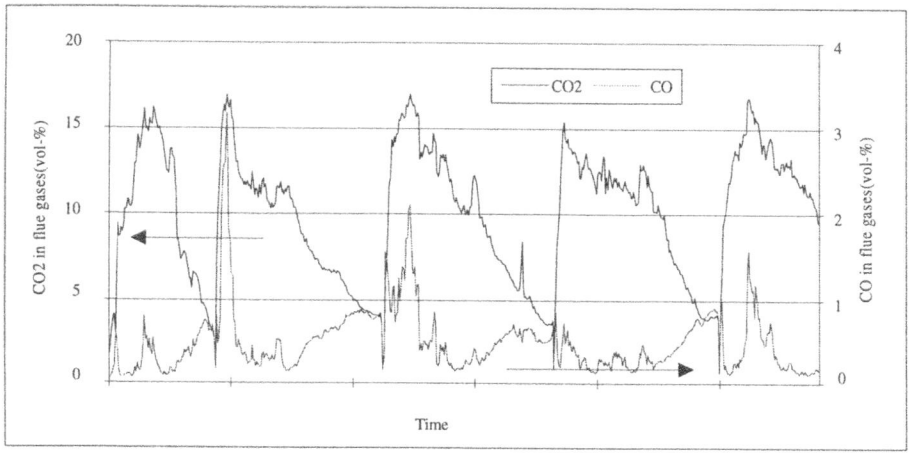

Fig 1. Batch-type wood combustion.

1.2 Emissions from small heating appliances

It has been estimated that in Finland 2/3 of carbon monoxide emissions and even more than 90% of hydrocarbon and PAH emissions from stationary combustion processes (transportation excluded) are due to small scale wood combustion. This is why it has a net effect on greenhouse gas effect [4]. After fuel adding, when volatiles are burned, the releases of carbon monoxide (CO) and other combustible gases are difficult to avoid.

When the CO-content in flue gases is, say over 0.2 - 0.5 %, it is obvious that also other harmful emissions like methane (CH_4) and other VOC's are formed and the amount of polycyclic aromatic hydrocarbons (PAH) compounds can be remarkable. Some PAH-compounds are very carcinogenic. Carbon monoxide and methane emissions from small heating appliances in Finland are compared with emissions from other sources in figure 2.

2 Model for wood log combustion

The combustion rate of wood logs in the primary combustion chamber of a stove depends on different parameters (wood log size, fuel moisture, gas oxygen content, radiative heat exchange between logs and furnace walls, the amount and time interval of succesive fuel inputs) [6]. As an example, the release of volatiles and water vapour (due to drying) from a combusting wood log (0.1 m by diameter, 0.75 m by length, 13% moisture) is shown in Figure 3. It is calculated using a numerical method [7]. The furnace wall temperature is 500°C and the effective gas temperature including the flame effect is 1000°C.

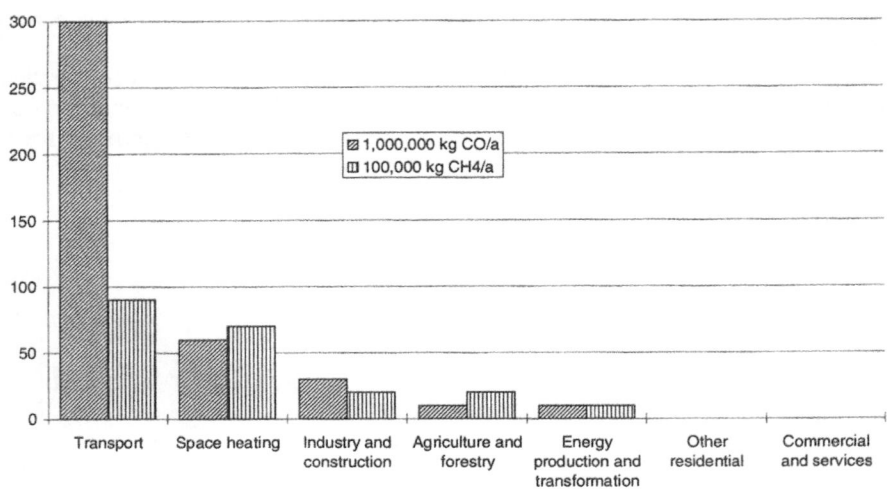

Fig 2. Carbon monoxide and methane emissions in Finland [5].

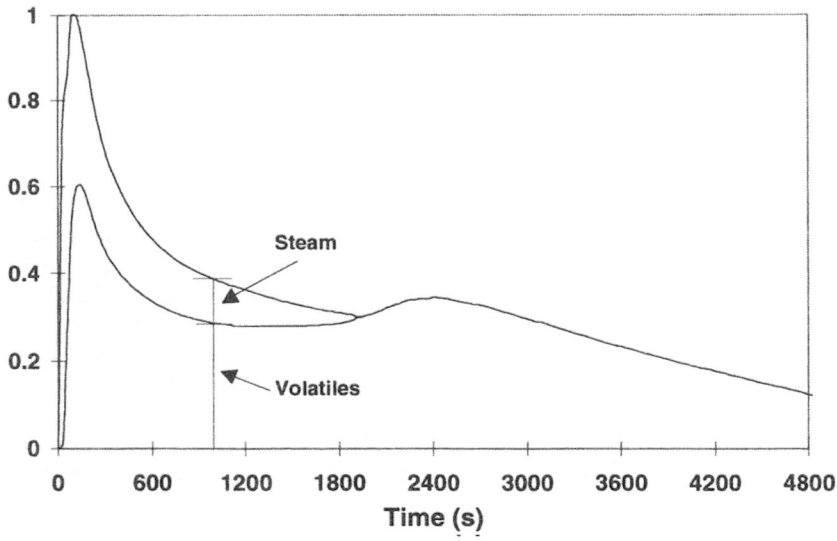

Fig. 3. Release of volatiles and steam from a combusting wood log. The scale is
normalized by dividing the rates (kg/s) with the maximum total rate (kg/s).

Shortly after ignition, the release of volatiles reaches a maximum and then gradually
decreases to lower value. Drying and devolatilization are overlapping about ½ hour in this
case. After that the wood is completely dried and a second maximum for the release of
volatiles is obtained, since the cooling flow of steam from the interior of the log is ceased
due to complete drying. It is clearly seen that the gas release rate from a wood log changes

greatly during the combustion. In order to reduce CO and hydrocarbon emissions, the combustion system should be dimensioned to meet the high release rate soon after fuel input. CO release can also be high during the char oxidation stage.

3 Combustion of carbon monoxide

Westbrook and Dryer [8] give the single step formula for oxidation of CO,

$$\frac{d[CO]}{dt} = k[CO][H_2O]^{0.5}[O_2]^{0.25} , \quad k = 10^{14.6} \exp(-167472 / RT) \tag{1}$$

Time t, temperature T and concentrations in brackets are given in units s, K and mol/cm^3 respectively. R=8.314 kJ/kmolK is the universal gas constant. Equation (1) can be integrated analytically, when oxygen concentration is approximated to remain constant along the flow path. The result for the conversion of CO divided by the initial concentration is

$$X \approx 1 - \exp(-t / \tau) , \quad \tau = t / \left([H_2O]^{0.5}[O_2]^{0.25} \int_0^t k \, dt \right) \tag{2}$$

The time constant $\tau = 1 / \{k[H_2O]^{0.5}[O_2]^{0.25}\}$, when temperature is constant along the flow path. In reality oxygen is consumed along the flow and more exact results presented in figure 4 are obtained by integrating equation (1) numerically.

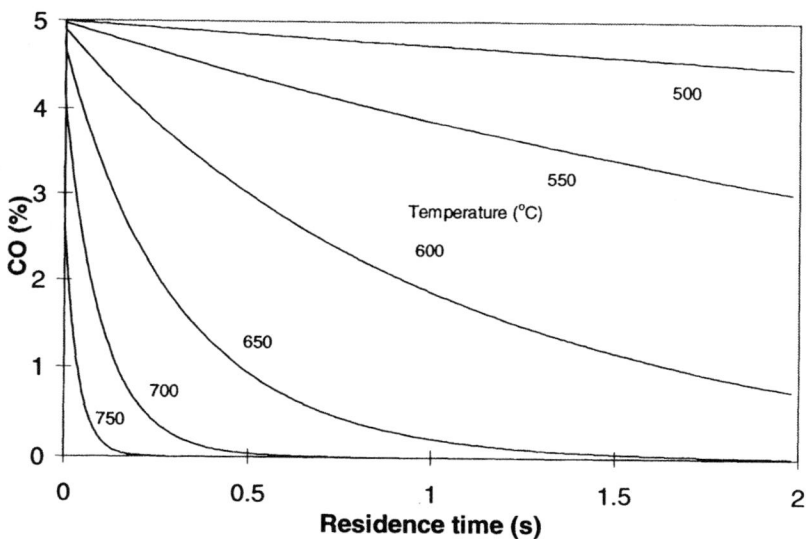

Figure 4. Oxidation of CO in gas containing initially 5% CO , 7 % O_2 and 5% H $_2$O.

However, the use of average oxygen content gives a good accuracy, if the oxygen content is not close to zero, since the exponent 0.25 is small.

The gas volume rate in the combustion chamber can be expressed as

$$\dot{V}_g = [1 + u + \lambda / (fY_{O_2})] \left(\frac{RT}{pM} \right) \dot{m}_f \tag{3}$$

where M is molar weight, p is pressure, $\dot{m}_f$ is dry fuel rate, λ is ratio of actual air rate to stoichiometric one, f is stoichiometric ratio of fuel (daf) to oxygen and $Y_{O_2} = 0.23$ is mass fraction of oxygen in air. In plug flow the gas residence time is related to the volume by $t = V / \dot{V}_g$. The heat effect from combustion is related to the moisture content u (water mass/dry mass) by $P = \eta \dot{m}_f (\Delta h - l_v u)$, where η is the stove efficiency, Δh is heating value of dry fuel and l_v is heat of vaporization. Thus, we obtain the relation betweeen the minimum combustion volume and power P, when a certain conversion X is required

$$V_{min} / P = \frac{1 + u + \lambda / (fY_{O_2})}{\eta (\Delta h - l_v u)} \left(\frac{RT}{pM} \right) \tau \ f(X) \tag{4}$$

where f(X) = - ln(1-X) for a plug flow reactor. In reality, the volume required to reach conversion X is greater due to non-perfect mixing of CO and O_2. Another case is the well-stirred reactor, where the spatial concentration is uniform. Then f(X)=X/(1-X).

The moisture in fuel increases the required volume/power (V/P) in different ways. Firstly, the gas rate is increased due to water vapour decreasing the gas residence time. Secondly the heating value of fuel is decreased with increasing moisture content. Moisture decreases combustion temperature, which reduces the reaction rate of CO. Molar weight and density of the gas mixture decrease with increasing fuel moisture. This also reduces the residence time and increeces V_{min}. A contradictory effect is that water vapour enhances the rate of CO oxidation. The CO oxidation rate may be reduced in the char combustion stage, if the hydrogen has already released during the pyrolysis, since only the water vapour in the combustion air is enhancing the CO oxidation. Increase in the air coefficient $\lambda > 1$ increases the volume V_{min} in two ways. The gas flow rate is increased reducing the residence time and the temperature is reduced due to dilution. A contradictory effect is that the oxygen content is higher, which increases the reaction rate.

The conversion X can be expressed as the function of V/P equation (4). Then it is possible to define a mixing efficiency $\eta = X_m / X_c$ for a combustion chamber, where X_m and X_c are the measured and the calculated conversions respectively.

4 Results from wood log combustion experiments

4.1 Field tests

Field tests were carried out heating two stoves as they are normally heated by house residents. Both stoves were heavy, heat storing stoves (masonry heaters). Stove #1 was a normal roomheater and #2 a baking oven. Flue gas samples were taken from the flue gas coming out of the chimney. CO_2-, O_2- and CO-concentrations were measured with normal continuous analyzers. FTIR was used to measure methane, ethane, ethene and acetylene concentrations, also continuously.

Emissions from stove #1 were much higher than those from the baking oven. Wood log moisture content was about 15% during both tests. Measurement results are given in table 1 and emissions in table 2.

Table 1. Field test results.

Measurement	Stove #1 (roomheater)	Stove #2 (baking oven)
Flue gas temperature in chimney outlet, °C	107	77
CO_2-concentration, %	7.23	4.57
O_2-concentration, %	12.7	15.1
CO-concentration, %	0.91	0.23
Combustion air temperature, °C	24.9	23.7
Chimney draught, Pa	20.2	20.9
Burning rate of wood, g/s	3.272	1.559
Flue gas enthalpy, %	8.2	8.0
Formation enthalpy of CO, %	7.9	3.2
Total flue gas loss, %	16.1	11.2
Fly ash loss, %	1.5	0.1
Total efficiency calculated using flue loss method, %	82.4	88.6

Table 2. Emissions.

Measurement	Unit	Stove #1 (roomheater)	Stove #2 (baking oven)
Particulates	mg/m^3_n	719	43
CO	%	0.91	0.23
Methane	ppm	638	171
Ethane	ppm	78	17
Acetylene	ppm	228	34
Ethene	ppm	354	54
Nitric oxide (NO)	ppm	42	30
Nitric dioxide (NO_2)	ppm	0	0
Nitrous oxide (N_2O)	ppm	0	3

Clearly combustion in stove #2 was much better resulting in lower emissions. When CO concentration is higher, all other emissions are also higher. Especially methane is a strong greenhouse gas. Its GWP-index (Global Warming Potential) is 35 times higher than that of CO_2.

4.2 Laboratory tests to study the effect of primary and secondary air on CO-emission

Tests were carried out with a heat storing roomheater (masonry heater). Secondary combustion chamber was modified so that secondary air could be led through nozzles into the throat between the primary combustion chamber and the secondary combustion chamber. Test parameters were fuel load weight and the amount of primary and secondary air. Birch logs having moisture content of 8% were used as a fuel. Chimney draught was constant (about 20 Pa). During each test several fuel loads were added. Fuel was added when basic firebed was formed. Flue gas temperature was measured along the flue gas flow in the combustion chambers. Parameters of test series are given in table 3.

Table 3. Test parameters.

Test #	Fuel load, kg	Primary air, litr./min	Second. air, litr./min
1	1	600	0
2	4	800	0
3	2	1000	0
4	3	1000	400
5	3	800	200
6	1	800	600
7	1	1000	200
8	4	1000	600
9	2	600	600
10	3	600	400
11	2	800	400
12	4	600	200

Test results were analyzed and statistical models for test results were made using a specific computer code. Combustion was separated to different parts; namely volatile combustion, "normal" combustion and combustion of residual char. An example of the models is given in figure 5. In that CO-emission dependence on the amount of primary and secondary air is modelled for the devolatilization phase.

From this example it can be seen that the effect of secondary air on CO emission is relatively high with 4 kg fuel load. Emissions were lowest using 1Q00 litres/min primary air and 600 litres/min secondary air. With smaller fuel loads similar results were not obtained.

It was clearly seen that this kind of test procedure and statistical test result modelling are not very suitable for an unsteady-state wood log combustion process. Even if the combustion process was divided to separate phases and a statistical model was made separately to them, results were not always satisfactory and were difficult to explain. Furthermore, combustion air in these experiments were introduced with fans, but in practise natural draught is used and the real situation is much different. Batch combustion process of wood logs could perhaps be better studied for example separating different phases from each other. Residual char combustion phase could be studied using wood char coal. Devolatilization phase could be studied using synthetic flue gases and mixing of combustion air and flue gases using visual flow models.

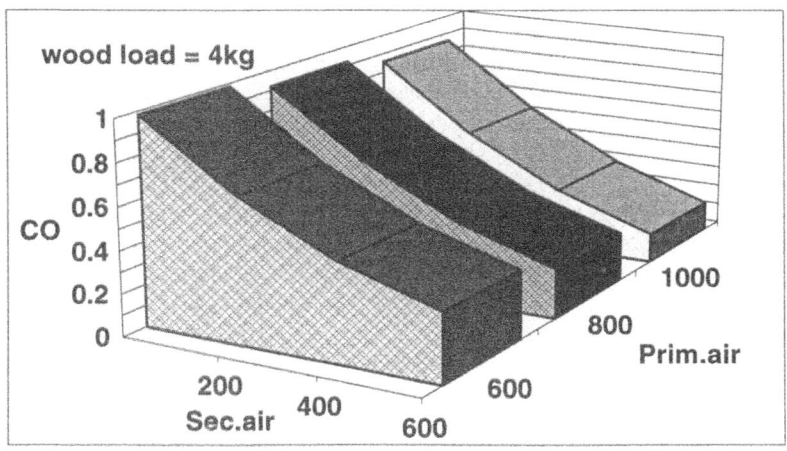

Fig 5. Model for CO-emission during volatile combustion.

4.3 Laboratory tests of a cast iron stove
The effect of fuel load, combustion air control, flue draught, fuel moisture, wood log size and the use of catalysts were studied [9]. A lot of emissions were formed when flue draught was small (4 Pa) and combustion chamber was full of logs. When flue draught was increased

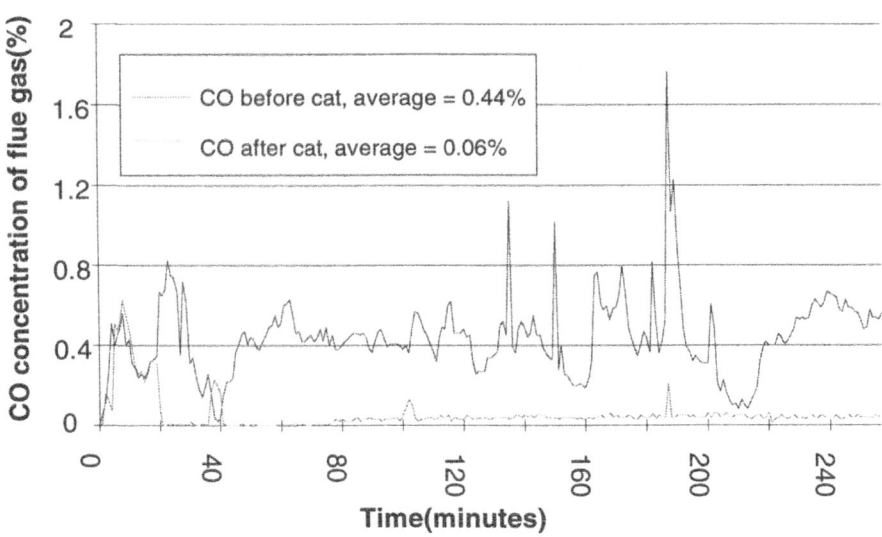

Fig 6. Reduction of CO in wood log combustion using a catalyst.

from 4 Pa to 32 Pa, CO emission decreased to 25% of the original value that was 0.60 % CO at 13% O_2. Higher draught leads to higher heat output. Efficiency of the stove decreases because the flue gas temperature is higher surpassing the effect of better CO oxidation.

When the moisture content of birch wood logs increased from 13% to 40%, CO emission became 5 times higher being 0.52% (13% O_2). At the same time, heat output of the stove decreased about 30%. 1/3 filling of the combustion chamber resulted in lower emission (0.17% CO at 13% O_2) compared to 1/1 filling (0.25% CO at 13% O_2). Heat output of the stove was then 10 - 20% lower. Wood log size (log weight 0.32, 0.70 and 2.0 kg) had very low effect on CO emission and stove efficiency.

Heating procedure and fuel moisture are the main factors affecting the stove efficiency and emissions. Besides those, technical devices like catalysts can also be used to lower emissions. Tests showed that with a suitable catalyst, CO reduction of over 90% can easily be achieved. One example of CO reduction is shown in figure 6 [9].

5 Conclusions

Field and laboratory tests showed that wood log combustion in stoves produce quite high emissions of carbon monoxide and hydrocarbons. Firing habits and selection of wood have great effects on combustion quality. Wood log combustion process is a complicated unsteady-state process. The need of primary and secondary air changes all the time along the burning of fuel load. During volatiles combustion, temperature in the combustion chamber is high which favours complete combustion, but the combustion rate is the highest. Residence time is then much shorter. Combustion control to achieve optimal combustion atmosphere for every phase is difficult with simple heating appliances using natural draught.

Statistical analysis of combustion measurements of a whole wood stove gives an insight to the effect of different parameters on the goodness (emissions and efficiency) of the combustion, but it cannot well be used for dimensioning the stoves and designing optimal control strategy for combustion. Such a work should be carried out with a two stage model in which the burning both in primary and in secondary combustion chambers are simulated separately. The two chambers should be studied using artificial gas mixtures and char coal to test the models in different power ranges and air inlet conditions. The controlling of combustion in the primary chamber determines the power output. The control of combustion in the secondary chamber determines the goodness of combustion. This two stage model could be built by many different ways, for example a simplified model [6] could be used for the primary chamber and a statistical model based on measurements for the secondary chamber. Also a statistical model could be used for the primary chamber. A simplified model of the combustion in the secondary chamber could consist of a simplified gas combustion model with the control of chemical kinetics, which is corrected by using measured mixing efficiencies discussed in this paper. The mixing efficiencies can be measured for a specific configuration and size for different flow conditions and modelled statistically. Also advanced models for turbulent combustion could be used instead of measurements. The stove operation in different control situations in the unsteady state can then be simulated by combining the two models. The output from primary chamber is used as an input for the secondary chamber. Then the optimum control strategy can be studied by calculations and compared to measurements.

The time of the maximum rate of release of volatiles is the most critical for emissions and determines the minimum dimensions of the secondary chamber for a given maximum power. Water vapour in gases enhance the CO oxidation as shown by equation (1). The hydrogen in wood and the moisture in air are the sources of water vapour. The CO release during char combustion stage may increase at cold outdoor temperatures, since moisture content in air is low.

6 References

1. Nussbaumer, T., Wood combustion. In "*Advances in Thermochemical Biomass Conversion*", Vol.1, edited by A.V.Bridgwater, Blackie Academic & Professional, pp.575-589, London 1994.
2. Nussbaumer, T., and Wagner, D., A new method to measure the integrated amount of pollutants from nonstationary wood combustion processes. In "*Advances in Thermochemical Biomass Conversion*", Vol.1, edited by A.V.Bridgwater, Blackie Academic & Professional, pp.563-574, London 1994.
3. Karlsvik, E., Hustad, J.E., Sønju, O.K, Emissions from wood stoves and fireplaces. In "*Advances in Thermochemical Biomass Conversion*", Vol.1, edited by A.V. Bridgwater, Blackie Academic & Professional, pp.690-707, London 1994.
4. Hupa, M. et al. 1988. Energiantuotannon päästöt Suomessa. *KTM Series D:162*. 50 p. + app. 8.
5. Boström, S. et al. 1992. Greenhouse gas emissions in Finland 1988 and 1990. Insinööritoimisto Prosessikemia. 62 p.
6. Saastamoinen, J., Modelling of wood combustion in small stoves. *Nordic Workshop on Combustion of Biomass*. The Norwegian Institute of Technology (NTH), Trondheim, Norway, 7.2.1991. Proceedings, pp.110-132.
7. Saastamoinen, J., and Richard, J.R., Simultaneous drying and pyrolysis of solid fuels. Accepted for publication in *Combustion and Flame*.
8. Westbrook, C.K., and Dryer, F.L., Simplified reaction mechanisms for the oxidation of hydrocarbon fuels in flames. *Combustion Science and Technology*, 1981, Vol.27, pp.31-43.
9. Kouki, J., 1995. Heating procedure and catalyst studies with a cast iron stove (in Finnish). *Työtehoseura ry*, Helsinki. 29 p. + 17 app.

7 Acknowledgments

This work was carried out in research projects funded by the Ministry of Trade and Industry - SIHTI I and BIOENERGY research programmes, Finnish fireplace manufacturers Tulikivi Oy and Upo Valimo Oy, and Kemira Metalkat Oy, for which their support is gratefully acknowledged.

FORMATION AND REDUCTION OF POLYCHLORINATED DIOXINS AND FURANS IN BIOMASS COMBUSTION

Dioxin emissions from biomass combustion

T. NUSSBAUMER and P. HASLER
Verenum Research, Zurich, Switzerland

Abstract

The emission of polychlorinated dibenzo-p-dioxins (PCDD) and polychlorinated dibenzo furans (PCDF) from the combustion of different wood fuels have been determined. Results from different furnaces are discussed and measures for dioxin reduction are proposed.

Emissions from non contaminated wood are usually < 0.1 ng TE/Nm^3 at 11 vol.-% O_2. From the combustion of urban waste wood and demolition wood in grate firings typical dioxin emissions of app. 2 ng TE/Nm^3 were found ranging from $0.03 - 18$ ng TE/Nm^3. Additionaly, two specific fuels were investigated: charcoal and combustible household waste. Charcoal used for grilling meat resulted in the very low PCDD/F emissions of 0.028 ng TE/Nm^3. The emissions from the combustion of household waste in a wood stove were extremely high (114 ng TE/Nm^3).

Dioxin emissions from combustion processes are mainly a consequence of the de novo synthesis. Relevant concentrations of dioxins can be formed in the presence of unburnt carbon, chlorine and catalysts (copper and other heavy metals) in the temperature window between approximately 200°C and 500°C. The dioxin concentrations increase at high levels of oxygen. To avoid dioxin emissions a complete burn out of the fly ash and particles are necessary and the combustion should be operated at low excess air level ($O_2 < 11$ vol.-%) and under stable conditions. For automatic wood furnaces an efficient dust separation is needed and the particle filter must be operated well below the temperature window.

Keywords: PCDD, PCDF, emissions, reduction, de novo synthesis, biomass, combustion.

1   Introduction

The total emissions of polychlorinated dioxins and furans in Switzerland has been estimated in the order of 280 to 370 g TE/a [2] and 100 to 200 g TE/a [3] respectively. The combustion of wood amounts to 1 to 10 TE g/a. However, this estimate is very rough since few data were available at that time. In the following, the Swiss Federal Office of Environment, Forests and Landscape (FOEFL) and the Swiss Federal Office for Energy

(FOE) have supported a number of investigations to improve the knowledge of dioxin emissions originating from the combustion of wood. The results of these studies are presented in this paper.

2 Formation of dioxins and furans in combustion processes

Polychlorinated dioxins and furans belong to a class of polyaromatic ethers with multiple chlorine substitution (figure 1). With varying degrees of chlorine substitution, a total of 75 different dioxin and 135 furan congeners exist. Seventeen of these 210 polychlorinated congeners are considered to exhibit a very high toxicity. All of the highly toxic congeners are substituted at the 2,3,7,8-sites of the dioxin and furan structure. Based on animal experiments and industrial accidents, a toxicity factor is attributed to each of these congeners, e.g. according to NATO-CCMS [1] (I-TEF = International Toxicity Equivalent). The 2,3,7,8-tetra-chlorinated dibenzo-p-dioxin (TCDD; known as Seveso dioxin) has the highest toxicity equivalent factor of 1. Generally, the sum (designated as TE) of the individual toxicities (amount x I-TEF) is used for comparison of different samples.

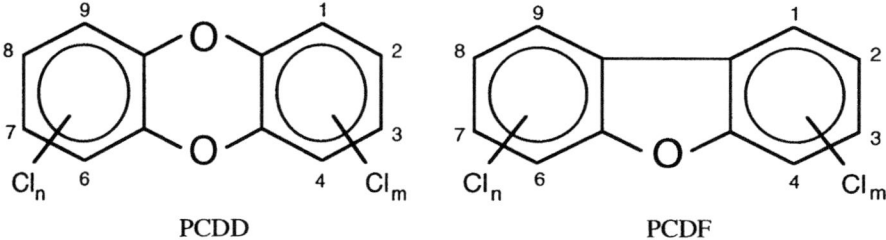

PCDD   PCDF

Figure 1.   Structural formula of polychlorinated dibenzo-p-dioxin (PCDD) and dibenzo-furan (PCDF)

Although the formation mechanism is not yet completely known, important results were found in numerous investigations. Measurements in the hot gas (730°C to 830°C) from municipal solid waste (MSW) incinerators showed that only neglectible quantities of PCDD/F are present in the gas leaving the secondary combustion zone [4], [5]. Therefore, dioxins and furans present in the fuel are completely destroyed. The dominant reaction is the de novo synthesis. The formation takes place in the heat recovery section of the combustion system and requires the presence of chlorine, carbon, oxygen and a solid surface. According to Stieglitz et al. [6], the formation path of chlorinated dioxins/furans and aromatic compounds (ArCl; ArHCl* = adsorbed, chlorinated intermediate product) can be desribed as follows:

a) formation of chlorinated products:

ArH	+	Cu(II)Cl$_2$	$\leftrightarrow$	ArHCl*	+	Cu(I)Cl			(1)

ArH   +   Cu(II)Cl$_2$ $\leftrightarrow$ ArHCl* + Cu(I)Cl   (1)

ArHCl*   +   Cu(II)Cl$_2$ $\leftrightarrow$ ArCl + Cu(I)Cl + HCl   (2)

b) oxidation of the reduced catalyst from reaction (1) and (2):

$$4 \, Cu(I)Cl \quad + \quad 4 \, HCl \quad + \quad O_2 \quad \leftrightarrow \quad 4 \, Cu(II)Cl_2 \; + \; 2 \, H_2O \tag{3}$$

The formation of dioxins and furans by the de novo synthesis occurs in the temperature range from 200°C to 500°C and reaches a maximum rate at approximately 300°C. Inorganic salts of heavy metals, especially of copper salts, exhibit a high catalytic activity for the dioxin formation [7]. The dioxin formation is possible also in the <u>absence</u> of an organic chlorinated compound [8], [9]. The presence of particulate carbon and a chlorine source is sufficient.

The amount of PCDD/F on MSW fly ash exhibits an almost linear correlation with its carbon content [9], [10], [11] (figure 2).

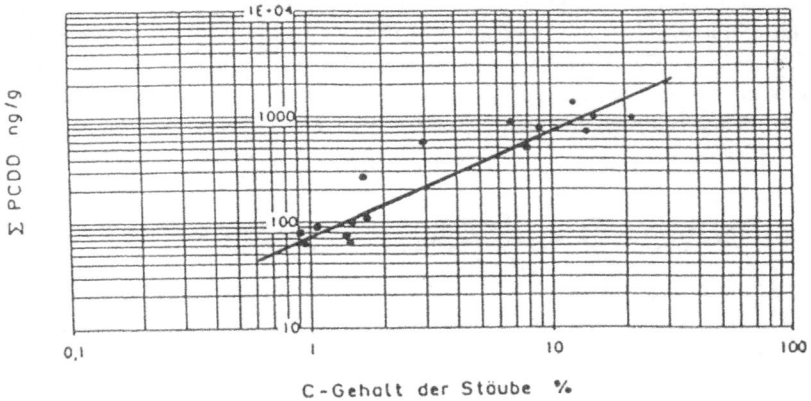

Figure 2.   Amount of PCDD/F on MSW fly ashes as a function of the residual carbon content [10]

A tenfold reduction of the residual carbon content in the fly ash leads to a similar reduction of the PCDD/F content. Therefore, the carbon burn out must be complete and fly ash emission should be minimal. Furthermore, combustion quality must be high to ensure maximal destruction of organic compounds.

The dioxin emissions are also influenced by the residual oxygen concentration in the flue gas [12]. Possible explanations for this effect can be a lower carbon burn out due to lower temperature in the combustion zone and the influence of oxygen on the oxidation of the dioxin formation catalyst (see equation (3) above). In the absence of oxygen, a complete destruction of dioxins and furans is possible at approximately 300°C using copper as a catalyst [7].

The de novo synthesis is a heterogeneous reaction which takes place only in the presence of solid matter which acts as a catalyst. Using synthetic flue gas with HCl or Cl_2 as a chlorine source and dioxin precursors such as 4-chloro phenol, no polychlorinated dioxin and furan congeners could be found in the absence of solid matter [13].

3   PCDD/F emissions from wood combustion systems

Beside the combustion technology and the operation condition, the dioxin formation in wood firings depends also strongly on the fuel type. The combustion of uncontaminated wood (forestry wood chips, log wood or chlorine and heavy metal free chipboards) generally leads to much lower dioxin emissions than the combustion of urban waste wood or demolition wood.

3.1   Dioxin emissions from the combustion of wood, UF chipboards and urban waste wood in a pre oven furnace with denox and ceramic filter

Measurements of dioxin emissions in the raw and clean gas have been made in a 150 kW prototype pre oven furnace [1] during SCR (selective catalytic reduction) and SNCR (selective non catalytic reduction) test runs. The influence of combustion quality, ceramic filter temperature and catalyst temperature has been investigated. The most important results can be summarized as follows [17]:

- Significant dioxin formation can be observed in the ceramic filter. The formation rate is highest at filter temperatures of app. 300°C. When urban waste wood is used as a fuel, an increase in the dioxin emission of 60 ng TE/m^3 can be found (table 1).
- During SCR test runs a high dioxin destruction activity has been observed [17]. The denox catalyst consists of $TiO_2/V_2O_5/WO_3$. After the denox catalyst, a novel metal catalyst (Pt/Pd/Rh) is used to oxidize gases such as ammonia. Most presumably, the dioxin reduction takes place on the novel metal catalyst surface.
- Dioxin emissions from SNCR test runs were found to be much lower than from SCR test runs. Possible explanations are:
  - Air excess ratio is lower and carbon burn out is better.
  - Ammonia acts as a catalyst poison for the dioxin formation on copper chloride.

Table 1.   Formation of PCDD/F in ceramic filters and destruction of PCDD/F in SCR catalyst (Low Dust) during the combustion of uncontaminated wood, UF chipboards and urban waste wood [17]; Values at 11 Vol.-% O_2, dry gas at standard condition; I-TEF acc. to NATO/CCMS

		natural wood	UF chipboard	urban waste wood
Ceramic filter temperature	[°C]	330	300	300
Catalyst temperature	[°C]	240	230	230
PCDD/F in raw gas	[ng TE/m^3]	0.083	0.095	17.76
PCDD/F after filter	[ng TE/m^3]	0.836	2.947	78.25
PCDD/F in clean gas	[ng TE/m^3]	0.052	0.189	8.18
Formation in ceramic filter	[ng TE/ng TE]	10.1	31.0	4.4
Destruction in catalyst	[ng TE/ng TE]	16.1	15.6	9.6
Ratio raw gas/clean gas	[ng TE/ng TE]	0.63	1.99	0.46

---

[1]   The pre oven furnace used for the experiments was designed as a cocurrent grate firing whe the grate section is separated from the combustion/boiler section.

3.2 Uncontaminated wood

Table 2 and figure 3 summarize the results from the combustion of uncontaminated wood. PCDD/F emissions from the combustion of noncontaminated wood range from 0.004 ng TE/Nm^3 to 0.880 ng TE/Nm^3. The highest value was found in a 450 kW moving grate firing (figure 3) which previously was operated for a longer period with urban waste wood. Before the test run with uncontaminated wood, the furnace and the boiler has been thoroughly cleaned and operated for several days with uncontaminated wood. Possibly, the history of the boiler has an influence. Apart from the unexpected high value in the 450 kW grate firing, the dioxin emissions from understoker firings were generally higher than in other firing systems.

Table 2.　　Dioxin emissions in the clean gas from the combustion of uncontaminated wood and from grilling meat on charcoal
CO and PCDD/F values at 11 vol.-% O_2, dry gas at standard condition; I-TEF acc. to NATO-CCMS; λ = excess air ratio ($\lambda = 21/(21-O_2)$)

Combustion system	rating	fuel type	λ [-]	CO [mg/m^3]	PCDD/F in clean gas [ng TE/m^3]	ref.
Wood stove (open)	6 kW	beech wood log	13.1	7400	0.080	[14]
(closed)	6 kW	beech wood log	3.3	4700	0.130	[14]
(open)	n.d.	charcoal	6.0	24000	0.035	[14]
Open chimney	n.d.	beech wood log	50	n.d.	0.090	[15]
Log wood boiler	35 kW	beech wood log	1.7	1000	0.024	[14]
	35 kW	beech wood log	1.4	12800	0.043	[14]
	35 kW	beech wood log	8.2	n.d.	0.080	[15]
Pre oven furnace	35 kW	pine chips	2.0	150	0.015	[16]
	35 kW	pine chips	2.2	190	0.016	[16]
	150 kW	wood chips	3.6	420	0.083*	[17]
	150 kW	wood chips	1.2	4100	0.217*	[17]
	150 kW	Cl-free chipboards	2.9	80	0.095	[17]
	150 kW	Cl-free chipboards	1.2	3800	0.030	[17]
Under stoker firing	110 kW	wood chips	2.3	1700	0.267	[14]
	230 kW	pine chips	6.8	2400	0.303	[18]
	230 kW	pine chips	8.8	3990	0.498	[18]
Moving grate firing	450 kW	wood chips	2.0	70	0.880**	[19]
	2 MW	bark chips	n.d.	n.d.	0.019	[20]
	16 MW	bark chips	n.d.	n.d.	0.006	[20]
Wood dust burner	9.6 MW	pine chips	1.8	49	0.004	[18]
	9.6 MW	pine chips	1.8	37	0.006	[18]

* Sampling at app. 350°C before the ceramic filter
** Furnace was previously operated with urban waste wood. Before the test run with uncontaminated wood, the furnace and the boiler were thoroughly cleaned and operated some days with uncontaminated wood.
n.d. = not determined.

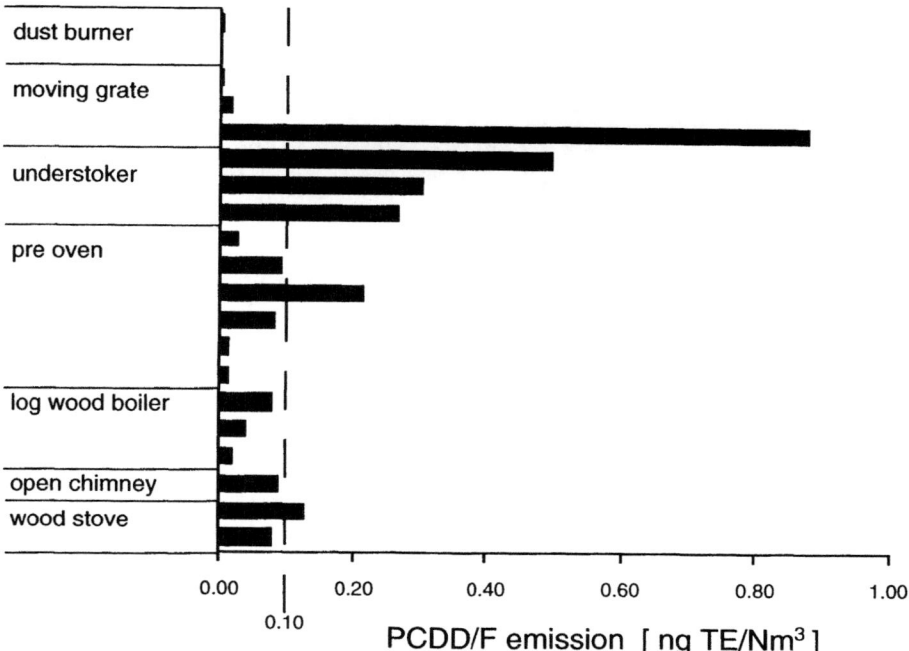

Figure 3. PCDD/F emissions in the clean gas from the combustion of uncontaminated wood (values according to table 2)
Emissions at 11 vol.-% O_2, dry gas at standard condition; I-TEF acc. to NATO-CCMS

3.3 Urban waste wood

Urban waste wood (UWW) originates from the demolition and reconstruction of buildings, from used furnitures and wooden packing materials. In Switzerland, wood wastes such as pressure impregnated wood, wood treated with wood preservatives or wood with halogenorganic coatings are not allowed in the urban waste wood fuel and must be burnt in MSW incinerators. Swiss emission regulations for urban waste wood and waste paper are more stringent than for uncontaminated wood, but are in some cases less severe compared to MSW.

Table 3 shows typical values for the chemical composition and some physical properties of urban waste wood from Swiss origin. In table 4 emission limits according to Ordinance on air pollution control (OAPC) are given for urban waste wood and MSW in Switzerland.

Table 3.   Properties and chemical composition of Swiss urban waste wood [24]
             Except loose density and LHV, all values are based on moisture free wood

		mean	median	min	max
Fine particles (saw dust)	wt.-%	6.5	6.5	4.0	9.0
Loose density	kg/m^3	219	220	180	270
Moisture content	wt.-%	22	18	10	44
LHV (calculated)	MJ/kg	14.8	15.3	12.1	16.6
Ash content	wt.-%	5.3	6.1	1.6	11.1
Nitrogen	N	mg/kg	7 922	7 300	5 600	12 000
Sulfur	S	mg/kg	1 385	450	20	5 800
Chlorine	Cl	mg/kg	849	495	20	4 400
Fluorine	F	mg/kg	42	10	0.01	140
Mercury	Hg	mg/kg	0.3	0.3	0.005	0.7
Arsenic	As	mg/kg	4.1	1.5	0.7	22.0
Cadmium	Cd	mg/kg	3.4	1.0	0.4	24.0
Cobalt	Co	mg/kg	1	1	1	3
Chromium	Cr	mg/kg	32	20	14	93
Lead	Pb	mg/kg	314	320	43	690
Copper	Cu	mg/kg	25	17	11	85
Nickel	Ni	mg/kg	6	5	2	12
Tin	Sn	mg/kg	6	7	0.04	12
Zinc	Zn	mg/kg	535	530	170	960

Table 4.   Swiss emission limits for MSW and urban waste wood (UWW)
             For UWW, some emission limits are valid in cases where the mass flow
             exceeds a fixed value; Emission values at 11 vol.-% O_2, dry gas at standard
             condition.

Compound		MSW	UWW	valid from
Dust	mg/Nm^3	10	50	
Pb+Zn	mg/Nm^3	1	5	
Hg	mg/Nm^3	0.1	0.2	1 g/h
Cd	mg/Nm^3	0.1	0.2	1 g/h
SO$_2$	mg/Nm^3	50	250	2500 g/h
Cl (as HCl)	mg/Nm^3	20	30	300 g/h
F (as HF)	mg/Nm^3	2	5	50 g/h
NO$_x$	mg/Nm^3	80	250	2500 g/h
NH$_3$	mg/Nm^3	5	30	300 g/h
HC	mg/Nm^3	20	50	
CO	mg/Nm^3	50	250	

The combustion of urban waste wood can lead to PCDD/F emissions which are at least one order of magnitude higher than from the combustion of uncontaminated wood (Table 5 and figure 4).

The operation condition and the combustion technology leads to a wide variation of the PCDD/F emissions. The influence of the gas cleaning system can be illustrated with the results from the 150 kW pre oven prototype furnace where PCDD/F levels varied between 0.03 ng TE/Nm^3 and 8.18 ng TE/Nm^3 (at 11 vol.-% O_2). SNCR test runs (denox chamber with 2 seconds residence time at 950°C and static mixer) revealed very low PCDD/F emissions of 0.03 ng TE/Nm^3. Possibly, the low air excess of 1.5 (app. 8 vol.-% O_2) has a major influence. If a ceramic filter is operated at a temperature of 300°C, the PCDD/F emissions are rather high although a catalyst is used after the ceramic filter which destroys more than 90% of the PCDD/F's.

Table 5.   Emission of polychlorinated dioxins and furans in the clean gas during the combustion of urban waste wood in different combustion systems
Emissions at 11 vol.-% O_2, dry gas at standard condition; I-TEF acc. to NATO-CCMS.

Combustion system	Flue gas cleaning system	ref	λ	CO [mg/Nm³]	PCDD/F [ng TE/Nm³]	OAPC
150 kW prototype	cat., cf (SCR High Dust)	[17]	2.6	455	2.83	–
pre oven	cat, cf (SCR High Dust)	[17]	4.1	729	0.74	–
	cf, cat. (SCR Low Dust)	[17]	3.5	105	8.18	–
	cf, cat. (SCR Low Dust)	[17]	1.9	58	2.51	–
	SNCR	[17]	1.5	52	0.03	–
300 kW downdraft gasi-	none	[21]	1.2	< 10	0.12	–
fier with cyclone burner	none	[21]	1.2	< 10	0.15	–
410 kW pre oven	cyclone	[22]	2.1	345	18.0	–
450 kW moving grate	SNCR/cyc. (no DENOX)	[19]	2.4	30	1.87	–
	SNCR/cyc. (with DENOX)	[19]	2.6	50	1.57	–
850 kW moving grate [1]	cyclone/cloth filter (Ca)	[22]	2.0	70	2.70	+
	cyclone/cloth filter (Ca/C)	[23]	2.3	73	2.04	+
1.8 MW moving grate [2]	cyclone/ESP	[22]	2.0	510	9.57	+
12 MW fluidized bed	n.d.	[5]	n.d.	n.d.	0.10	+
17.5 MW circulating	cyclone/ESP	[22]	1.8	2	0.71	+
fluidized bed	cyclone/ESP	[22]	1.8	2	0.73	+

column 'OAPC':   –   system does not meet OAPC regulations (emissions too high; no sufficient dust separation; power rating less than 350 kW)
            +   system meets OAPC regulations

[1]   Carbon contant of fly ash: 4 wt.-%; [2] Combustion system analogeous to 1), on/off operation; Carbon contant of fly ash: 25 wt.-%

Abbreviations: cf = ceramic filter; cat. = catalyst; SCR = Selective Catalytic Reduction; SNCR = Selective Non Catalytic Reduction; Ca = Calciumhydroxide; Ca/C = Calciumhydroxide/activated lignite carbon (7 wt.-%); ESP = electrostatic precipitator; n.d. = not determined.

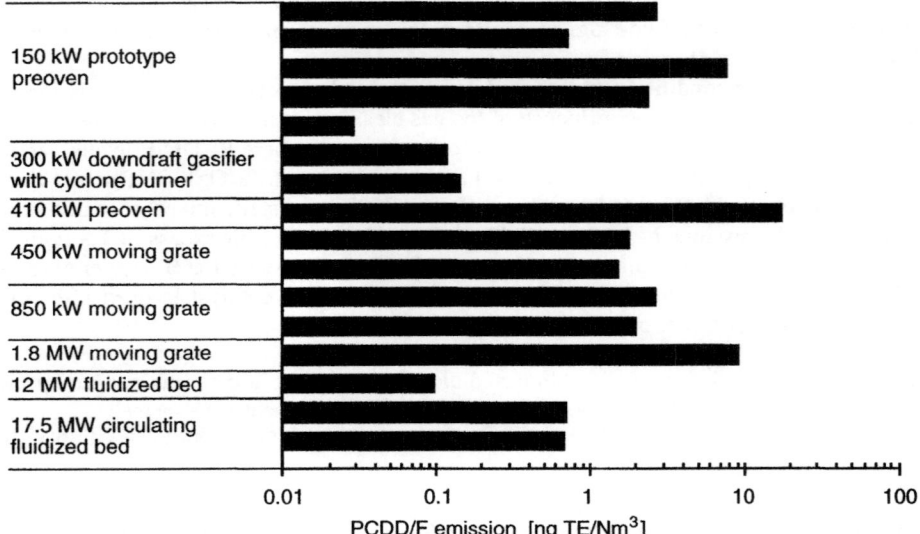

Figure 4.   PCDD/F emissions in the clean gas from the combustion of urban waste wood (values according to table 5)
Emissions at 11 vol.-% O_2, dry gas at standard condition; I-TEF acc. to NATO-CCMS

PCDD/F emissions are reported to be less than 1 ng TE/Nm^3 in fluidized bed combustors and in a downdraft gasifier with cyclone burner. The low CO emissions and the high burn out lead to low carbon contents in the fly as which reduces the de novo synthesis of dioxins.

The lowest PCDD/F emissions of 0.03 ng TE/Nm^3 were found in a pre oven equipped with a SNCR chamber. In moving grate firings (except the pre oven systems) the PCDD/F emissions vary in the range of 1.57 ng TE/Nm^3 to 9.6 ng TE/Nm^3. The highest value was found during intermitting on/off operation. During steady state operation the average PCDD/F emissions are 2 ng TE/Nm^3 approximately. Again, the lowest value of 1.57 ng TE/Nm^3 was found in a system with SNCR reduction chamber. However the value is higher than in the pre oven system reported, possibly due to higher excess air ratio, shorter residence time (0.8 seconds [2]) or due to the absence of a static mixer.

In a 410 kW pre oven furnace in practice, PCDD/F emissions as high as 18 ng TE/Nm^3 were found.

---

[2]   A residence time of 0.8 seconds in the denox chamber is sufficient for a NO_x reduction of 70% Ammonia slippage is less than < 30 mg/Nm^3.

4   Comparison of different fuels in wood combustion systems

Table 6 shows the PCDD/F emissions found in different wood combustion systems using different type of fuels.

Table 6.   Emissions of dioxins and furans from the combustion of different fuels PCDD/F values at 11 vol.-% O_2, dry gas at standard condition; I-TEF acc. to NATO-CCMS; PCP = polychlorinated phenols

Fuel type	ref.	PCDD/F emissions in ng TE/Nm³	
		min	max
Log wood	[14,15]	0.02	0.13
Uncontaminated wood chips	[14,18,19]	0.00	0.88
Urban waste wood	[17,19,22]	0.03 [1),2)]	18.0 [2)]
Chipboard free of chlorine and heavy metals	[17]	0.03	0.10
Chipboards containing PVC or NH4Cl	[16,18,21]	0.05	12.28
Chipboards treated with PCP	[16,21]	0.21	5.14
Combustible household waste in wood stove	[14]	114 [3)]	

[1)] with SNCR, [2)] system does not meet OAPC standards for urban waste wood
[3)] single measurement.

As expected, the combustion of **uncontaminated wood** leads to much lower PCDD/F emissions than of contaminated wood. In most cases with uncontaminated wood, the emissions are less than 0.1 ng TE/Nm³. Higher values are found in understoker firings and in a moving grate firing previously has been operated with urban waste wood.

The PCDD/F emissions are quite variable during the combustion of chipboards. The combustion of chipboards having a comparable composition to uncontaminated wood (except for nitrogen) leads to low PCDD/F emissions.

The combustion of **urban waste wood** and **PVC containing chipboards** leads to comparable PCDD/F emissions. The average level is at least one order of magnitude higher than with uncontaminated wood. The PCDD/F emissions can be as low as with uncontaminated wood if a properly designed SNCR chamber is used. PCDD/F levels below 0.5 ng TE/Nm³ can be expected also for urban waste wood combustion facilities without denox units. This can be achieved by optimizing the combustion process and the heat recovery system as well as with using activated carbon sorbents in cloth filters.

The highest emissions of 114 ng TE/Nm^3 (single measurement) were found during the incineration of **combustible household waste** (food packings, soft drink boxes, paper, magazines) in a hand fed wood stove.

Based on the total consumption of wood and the PCDD/F emissions measured in different type of combustion systems, the annual emissions has been estimated to range from 2.4 g TE to 17.1 g TE [14]. 80% are attributed to the combustion of urban waste wood. The total amount from wood combustion contributes to 2% – 8% of the total dioxin emissions in Switzerland. The illegal incineration of combustible household waste in wood stoves generates annual dioxin emissions between 4.5 and 27 g TE which is more than the amount from all wood combustion systems.

5   Measures for PCDD/F reduction

5.1   Prevention of illegal incineration

The highest PCDD/F emissions can be expected from

    a)   incineration of combustible household waste in manual wood stoves and
    b)   open burning of urban waste wood, e.g. on landfills or
    c)   combustion of urban waste wood in unsuitable combustion systems

Measures to reduce illegal incineration or disposal must be taken with high priority.

5.2   PCDD/F reduction in wood firings especially in urban waste wood firings

The combustion of urban waste wood in Switzerland leads to higher annual emissions than the combustion of uncontaminated wood. The most effective improvements can be realized by reducing the PCDD/F emissions from urban waste wood firings. The following measures will generate lower PCDD/F emissions:

- combustion technology (primary measure; SNCR as a combined measure)
- heat exchanger/boiler (primary measure)
- operation condition (primary measure)
- flue gas cleaning (primary and secondary measure).

The primary measures should be realized first. Secondary measures can be added if the primary measures are not possible or do not lead to a sufficient reduction.

a) Measures in the combustion technology

The primary measures in the combustion technology are aimed to improve the burn out of the gases (HC) and of the fly ash (carbon) as well as in a reduction of the dust content of the raw gas. Possible PCDD/F reduction measures are:

- reduction of the excess air ratio to $\lambda < 2$ (aim: < 1.5)
- high mixing quality of hot gas with secondary air
- increase of residence time in the hot combustion zone before the boiler
- prevention of dust particle removal by minimal disturbance of the glow bed and homogenous distribution of primary air

A further measure is the integration of a SNCR chamber.

b) Measures in the heat exchanger/boiler section

Heat exchangers and boilers should be manufactured with respect to minimal residence time in the temperature range between 200°C and 500°C as well as with minimal dust deposition ability.

c) Measures to improve operation condition

The most important measures to improve the operation conditions are:

- (electronic) combustion control units (temperature-, λ- or CO/λ control system) are able to ensure optimal combustion quality and burn out (HC, carbon)
- steady operation, prevention of intermitting on/off operation
- prevention of rapid load changes.

d) Measures in the gas cleaning system

In the gas cleaning system, the following measures can lead to a PCDD/F reduction:

- prevention of the PCDD/F formation (de novo synthesis)
- destruction of PCDD/F e.g. by catalytic oxidation
- PCDD/F separation in dust separators

Depending on the formation mechanism and gas temperatures the PCDD/F can be adsorbed on particulate matter or can be in a the gas phase. PCDD/F adsorbed on larger particles can be separated by cloth filters or electrostatic precipitators, whereas PCDD/F adsorbed on submicron particles such as aerosols are difficult to separate by conventional filter technologies. Catalytic oxidation can be used primarly for gaseous PCDD/F's. Separation of dioxins in filters leads to highly contaminated fly ashes whereas the catalytic oxidation produces no harmful byproducts. In figure 5 the formation and reduction rates of PCDD/F in several gas cleaning units is illustrated for urban waste wood and MSW. The formation and reduction rate is defined as follows:

$$\text{PCDD/F formation/reduction} = \frac{\text{PCDD/F in clean gas} - \text{PCDD/F in raw gas}}{\text{PCDD/F in raw gas}} 100\%$$

High PCDD/F formation rates are observed in ceramic filters and electrostatic precipitators (ESP) which are operated in the temperature range of the de novo synthesis. No dioxin formation has been detected at temperatures below 160°C using urban waste wood as a fuel. The PCDD/F reduction in a cloth filter using a calcium based sorbent and in a ESP has been determined as 40% and 30% respectively. The PCDD/F reduction in a cloth filter could be increased to 75% by using a sorbent with addition of activated carbon and optimized filter operation [23]. In MSW incinerators or in power stations, PCDD/F reduction of more than 99.9% was observed with activated carbon based sorbents at filter temperatures of app. 100°C [25]. However, filter temperatures below 120°C generally require a previous reduction of acid gases such as sulfur dioxide and hydrogen chloride e.g. in a wet scrubber.

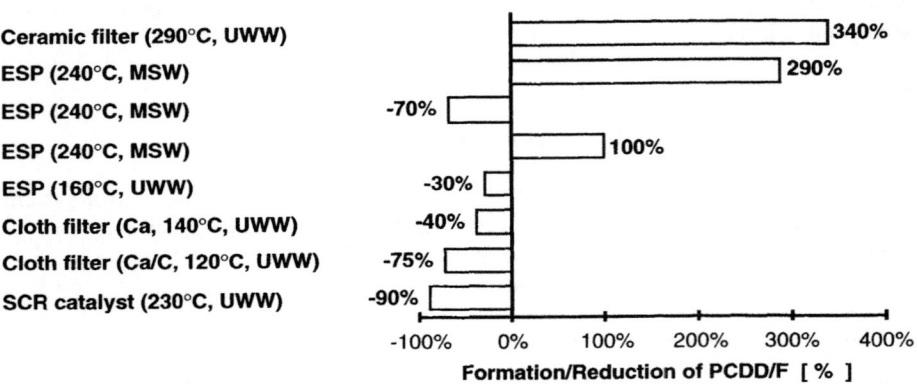

Figure 5.   Formation and reduction of PCDD/F in gas cleaning systems
The amount of PCDD/F in the filter ash is not included in the balance.
ESP = electrostatic precipitator; UWW = urban waste wood; MSW = municipal solid waste; References: MSW [26]; UWW [17], [22], [23].

Catalytic oxidation is a powerful method for PCDD/F destruction in flue gases. Based on actual measurements from urban waste wood combustion, PCDD/F reduction of 90% approximately is possible at catalyst temperatures of 230°C. At higher temperatures and/or longer residence time in the catalyst, higher conversion rates can be expected. Furthermore, catalyst modification may lead to even higher oxidation rates.

The measures for dioxin reduction in the gas cleaning system can be summarized as follows:

- **filter temperatures < 160°C – 180°C**
- cloth filter with activated carbon based sorbents
- catalytic oxidation

6   References

1. NATO-CCMS. 1988: International Toxicity Equivalency Factor Method of Risk Assessment for Complex Mixtures of Dioxins and Related Compounds, Report no. 176 (1988)
2. Schlatter, C: Poiger, H. 1989: Chlorierte Dibenzodioxine und Dibenzofurane - Belastung und gesundheitliche Beurteilung, *UWSF - Z. Umweltchem. Ökotox.* 2 (1989), 11-18
3. BUWAL 1991: Dioxin- und Furanemissionen in die Schweizer Luft, *BUWAL Bulletin* 2/91 (1991)
4. Ballschmiter, K.; Nottrodt, A. 1990: Untersuchungen über Ursachen und Minderung der PCDD/F-Emissionen an einer Hamburger Müllverbrennungsanlage, Teil C: Zusammenfassung der PCDD/F-Bildungswege. *Abschlussbericht, BMFT,* 1990
5. Blessing, R.; Frenkler, W.; Wissmann, G. 1991: Dioxin-Emissionen aus Feuerungsanlagen, *Brennstoff-Wärme-Kraft* 43/3 (1991), E43-48
6. Stieglitz, L.; Vogg, H.; Zwick, G.; Beck, J.; Bautz, H. 1991: On Formation Conditions of Organohalogen Compounds from Particulate Carbon of Fly Ash, *Chemosphere* 23/8 (1991), 1255-1264
7. Hagenmaier, H.P.; Brunner, H.; Haag, R.; Kraft, M. 1987: Copper-Catalyzed Dechlorination/Hydrogenation of Polychlorinated Dibenzo-p-dioxins, Polychlorinated Dibenzofurans and Other Chlorinated Aromatic Compounds, *Environ. Sci. Technol.* 21 (1987), 1085-1088
8. Vogg, H.; Metzger, M.; Stieglitz, L. 1987: Recent Findings on the Formation and Decomposition of PCDD/F in Municipal Solid Waste Incineration, *Waste Management and Research* 5 (1987), 285-294
9. Stieglitz, L.; Vogg, H. 1988: Carbonaceous Particles in the Fly Ash - A Source for the Formation of PCDD/F in Incineration Processes, *ISWA Proc.* Vol. 1 (1988), 331-335
10. Horch, K.; Schetter, G.; Fahlenkamp, H. 1991: Dioxinminderung für Abfallverbrennungsanlagen, *Entsorgungspraxis* 5 (1991), 235-243
11. Kilgroe, J.D. 1990: Combustion Control of PCDD/F Emissions from Municipal Waste Incinerators in North America, *10th Int. Conf. on Organohalogen Compounds*, 10./14. Sept. 1990, Bayreuth
12. Reimann, D. O. 1992: Bilanzierung von Schwermetallen, anorganischen Schadstoffen und Dioxinen/Furanen bei der Restabfallverbrennung sowie deren Verteilung auf Schlacke, Rauchgas und Rückstände. In: *Zukunftsweisende Rauchgasreinigungstechniken*, Handbuch zum VDI-Seminar vom 2./3. April 1992, Düsseldorf
13. Gullet, B.K.; Bruce, K.R.; Beach, L.O. 1990: Formation of Chlorinated Organics During Solid Waste Combustion, *Waste Management and Research* 8 (1990), 203-214
14. Hasler, Ph.; Nussbaumer, Th.; Bühler, R. 1993: Dioxinemissionen von Holzfeuerungen. *Schriftenreihe Umwelt* Nr. 208, Bundesamt für Umwelt, Wald und Landschaft (BUWAL), 1993.

15. Bröker, G.; Geueke, K.J.; Hiester, E.; Niesenhaus, H. 1992: Emission polychlorierter Dibenzo-p-dioxine und -furane aus Hausbrand-Feuerungen, *LIS-Bericht* Nr. 103 (1992), Landesanstalt für Immissionsschutz Nordrhein-Westfalen, Essen
16. Strecker, M.; Marutzky, R. 1993: Untersuchung von Holz und Holzwerkstoffen auf Holzschutzmittel und deren Emissionen bei der Verbrennung, *Forschungsbericht 76-104 03 518* (Umweltbundesamt), Text- und Anlagenband, Wilhelm-Klauditz-Institut, Fraunhofer-Arbeitsgruppe für Holzforschung, Braunschweig, März 1993
17. Nussbaumer, Th.; Bühler, R.; Jenni, A. 1993: Abgasentstickung bei Holzfeuerungen durch selektive katalytische und selektive nicht-katalytische Reduktion SCR und SNCR, Projekt *Altholzfeuerung mit Entstickung - Schlussbericht Phase 2*, Bundesamt für Energiewirtschaft, Bern, 1993
18. Wilken, M.; Kolenda, J.; Gass, H. 1993: Ermittlung und Verminderung der Emissionen von halogenierten Dioxinen und Furanen aus thermischen Prozessen - Holzfeuerungsanlagen.), *UBA-Forschungsbericht 10403365/06*, Umweltbundesamt Berlin (UBA), Juli 1993.
19. Good, J.; Nussbaumer, Th.; Jenni, A.; Bühler, R. 1994: *SNCR-Verfahren zur Stickoxidminderung bei einer Holzfeuerung.* DIANE 8 – Energie aus Altholz und Altpapier, Bundesamt für Energiewirtschaft (BEW), in Vorbereitung.
20. Wurst, F.; Pey, Th.; Boos, R.; Scheidl, K. 1991: Untersuchungen zur PCDD/F-Emission bei der Holzverbrennung, *Organohalogen Compounds* 7 (1991), 197-199
21. Brunner, W.; Pröll, M. 1991: Der Wamsler Thermo Prozessor: Ein zukunftssicherer Weg zur schadstoffarmen Verbrennung, *VDI-Berichte* 922 (1991), 247-257
22. Hasler, Ph.; Nussbaumer, Th. 1994a: *Dioxin- und Furanemissionen bei Altholzfeuerungen.* DIANE 8 – Energie aus Altholz und Altpapier, Bundesamt für Energiewirtschaft (BEW), EDMZ-Nr. 805. 174 d, April 1994.
23. Hasler, Ph.; Nussbaumer, Th. 1995: *Optimierung des Abscheideverhaltens von HCl, SO$_2$ und PCDD/F in einem Gewebefilter nach einer Altholzfeuerungen*, Bundesamt für Energiewirtschaft (BEW), November 1995.
24. Nussbaumer, Th. 1994: Altholzverwertung in 1 – 10 MW-Anlagen. In: Th. Nussbaumer (Hrsg.): *Neue Erkenntnissse zur thermischen Nutzung von Holz*, Bundesamt für Energiewirtschat, Bern 1994, ENET Nr. 30191
25. Mosch, H. 1992: PCDD/F-Minimierung im Abgas und Reststoffe der Müllverbrennung. *Dioxine – Belastung, Quellen, Verbleib*, Symposium 6./7. Oktober 1992, Stuttgart (Veranstalter: Ecoplan, TÜV Stuttgart)
26. Vogg, H.; Merz, A.; Stieglitz, L.; Albert, F.W.; Blattner, G. 1990: Zur Rolle des Elektrofilters bei der Dioxin-Bildung in Abfallverbrennungsanlagen, *Abfallwirtschaftsjournal* 2 (1990), Nr. 9, 529-536

SYSTEM STUDIES

BIOENERGY SYSTEMS

C P MITCHELL
Forestry Department, Aberdeen University, Aberdeen UK
A V BRIDGWATER
Energy Research Group, Aston University, Birmingham, UK
K L MACKIE
NZ Forest Research Institute, Rotorua, New Zealand

Abstract

The objective of this paper is to demonstrate that a bioenergy system has to be considered as an integrated process in which each stage or step interacts with other steps in the overall process. Individual components such as the conversion step must not be considered in isolation and independent of the biomass supply chain or the eventual application.

There are a number of stages in the supply and conversion of woody biomass for energy. Each step in the chain has implications for the next step and for overall system efficiency. The resource can take many forms and will have varying physical and chemical characteristics which will influence the efficiency and cost of conversion. The point in the supply chain at which size and moisture content is reduced and the manner in which it is done is influential in determining feedstock delivered cost and overall system costs. At the conversion plant a pretreatment section is used to modify the incoming feedstock to match the needs of the reactor. To illustrate the interactions within the overall system, the influence of the nature, size and moisture content of delivered feedstocks on costs of generating electricity via thermal conversion processes is examined using a model developed to investigate the inter-relationships between the stages in the supply chain.

There is little information on relationships between steps in the supply chain and considerable potential for system inefficiency. In order for efficient systems to be developed it is essential that both producers and users be aware of the limitations and constraints that each is working under.

Keywords: Bioenergy, supply systems, logistics, short rotation coppice, storage and drying, comminution, bio-electricity, pretreatment, conversion, systems analysis

1 Introduction

The objective of this paper is to demonstrate that a bioenergy system has to be considered as an integrated process in which individual stages such as the conversion step must not be considered in isolation and independent of the biomass supply chain or the eventual application. To illustrate the interactions and necessary integration, the example of bio-electricity production through thermal conversion of woody biomass is used.

The use of biomass to produce heat and generate electricity through combustion and conventional steam cycle technology has been a commercial reality for some time, particularly in the USA and Scandinavia. Bioenergy systems using advanced conversion technologies are currently moving from research and development into commercial operation. For existing systems the supply chain linking the resource through harvesting, transport, storage, conversion and generation of electricity has been largely elaborated. Existing combustion facilities primarily use agricultural, forest or mill residues as a feedstock which in combination are available throughout the year or are available at times of high heat demand, i.e. the winter months in Scandinavia.

New conversion technologies operating at a larger scale of operation to obtain economies of scale, will require larger quantities of feedstock than previously considered and this will be required throughout the year for continuous operation. These advanced conversion processes also require tighter specifications of the fuel supply in terms of particle size and moisture content than conventional combustion. In order to meet these supply needs short rotation coppice (SRC) grown on agricultural land is being developed because it gives higher biomass production and therefore a higher density of supply than forest residues. A major disadvantage of using coppice is its seasonality of supply. SRC can only be harvested during the winter so if year round supply is required, and it is the sole feedstock, efficient storage and drying strategies need to be developed to minimise storage losses.

Gasification has now reached demonstration and small scale commercial operation [1]. There are at least six substantial demonstration scale plants in operation using both atmospheric and pressure gasifiers coupled to a gas turbine for electricity production. Several of these are being built as a commercial operation. In some of these, the biomass supply has already been found to present substantial problems. Pyrolysis for liquid fuels, however, is less well developed and although there are some small scale commercial plants in operation for production of food flavourings, the use of fast pyrolysis liquids for fuel applications is still at the pilot and demonstration stage. In support of this interest in liquid fuels, there is a rapidly growing experience in using the pyrolysis liquids in heat and power applications [2]

2 The Bioenergy Chain

The process of producing energy from biomass is a chain of inter-linking stages which are mutually dependent i.e. a decision at one point has consequences for the next and subsequent stages in the chain.

There are five main stages in the chain of producing bio-electricity which are shown in Figure 1:
1. Production
2. Transport
3. Pretreatment
4. Conversion
5. Generation

Within each of these stages there are a series number of sequential steps - the combination of which determines feedstock form and cost at the end of that stage.

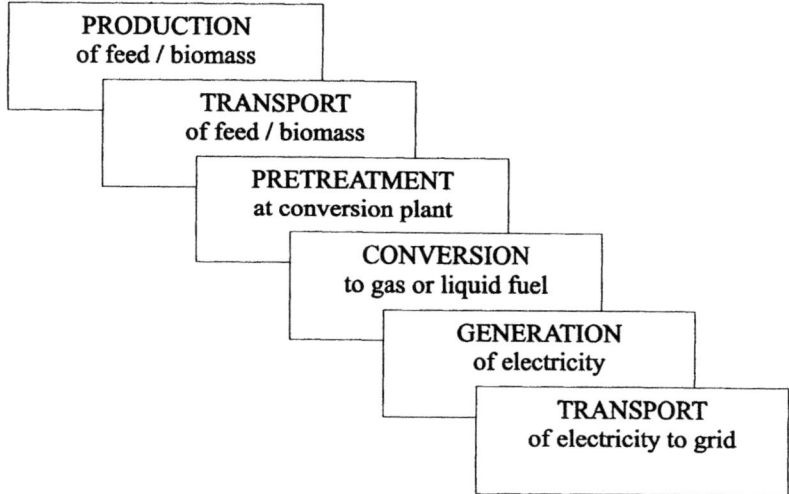

Figure 1   The Bioenergy Chain

The bioenergy chain is analogous to a relay race where the baton carrier passes the baton to the next carrier as smoothly as possible. This entails both players being in the right position for both handing over and receiving. In the bioenergy chain the feedstock is passed from one stage to the next in the form in which the next stage is able to handle it right on through the chain. The aim is to have the whole chain functioning smoothly and cost-effectively with the minimum of losses along the way.

As each conversion technology has specific requirements for the feedstock entering the reactor these requirements must be passed back upstream to the supplier so that the chain can be customised and optimised as much as possible for that end use within the constraints placed on the supplier. A key point to recognise is the interface between production and conversion which is mediated by the pretreatment stage. It is at this stage that the delivered feedstock is modified to optimise its characteristics for feeding to the conversion process.

There are many points along the chain where the properties and suitability of the feed can be controlled, for example sizing or drying. Upstream from the conversion plant the main factors which can be controlled are the form, size, moisture content and

LHV of the feedstock at point of delivery to the plant. These can be controlled at different points in the supply chain [3, 4, 5].

Chemical composition and ash content are other important characteristics, These are not generally controllable in the supply chain other than through selection of appropriate feedstocks.

Size can be controlled through comminution and moisture content can be modified through storage and drying - all of these will occur in both the biomass production area and the conversion facility. Understanding the relationships between particle size, particle shape and behaviour in storage is one key to the development of cost-effective supply systems. The sequence of steps in the supply chain can be changed to suit the overall system and feedstock characteristics required at the conversion plant.

The second stage of drying, comminution and storage occurs in the pretreatment section of the conversion facility. Here, feed material is modified to suit the requirements of the conversion technology in terms of particle size and range and moisture content. Avoidance or minimisation of double processing is essential to optimise the system which is why a complete system has to be considered with proper consideration given to these interfaces between production and conversion.

2.1 Production

The production stage in the bioenergy chain is composed of several steps, the ordering of which can be changed to some degree to meet the needs of the end user. For instance the feedstock can be comminuted before or after storage depending on the supply strategy adopted. The main steps are:

- The resource - growing and managing
- Harvesting
- Comminution (optional according to process specification and optimised requirements)
- Storage
- Drying (optional according to process specification and optimised requirements).

For each step there are numerous options available to choose from whether it be at the level of type of comminution device or individual manufacturers model. Some of the options and issues affecting choice are indicated in Table 1.

2.1.1 Resource

Woody biomass will be available from conventional forestry or short rotation coppice. Conventional forestry offers many opportunities for harvesting woody biomass for energy. The main focus to date has been on recovery of residues left after clearfell and early thinnings for which there is no timber market. The available resource is influenced by harvesting costs and competition from other forest products industries. The density of supply is quite low due to the small proportion of the crop being used for energy, this impacts on the catchment area required to feed a plant of a given size.

Short rotation coppice using fast-growing species such as poplars and willows gives much higher levels of crop productivity than conventional forestry and therefore higher densities of supply can be expected, leading to lower transport distances. The production of SRC on farmland will be influenced by agricultural policy and competing uses for the land.

Production systems for short rotation coppice are still evolving. Mechanisation of operations and development of best management practices to reduce costs of production and minimise environmental impacts are being targeted specifically.

Table 1   Options and Issues for Woody Biomass Production

Step	Options	Issues
Resource	Forestry Short Rotation Coppice	• Species of biomass • Density of supply • Crop productivity • Catchment area • Land area available • Crop area • Harvestable yield • Competition with other industries • Land-use policies
Harvest	Forestry: • Forest residues • Whole tree thinnings • Integrated harvesting Short rotation coppice: • Cut & chip harvesters • Cut & bundle harvesters	• Link with forest products • Harvestable yield • Particle form & size • Moisture content • Storage, drying and transport requirements
Comminution	• Chippers • Chunkers • Hoggers • Tub grinders	• Point of operation • Particle size delivered • Storage, drying and transport requirements
Storage and Drying	• Covered vs. uncovered • Ventilated vs. non ventilated • Active vs. passive drying	• What to store? • When into store? • Where to store? • How long to store? • MC change • DM Change • Storage losses • Fungal spores

2.1.2   Harvesting

Strategies have been developed for harvesting biomass cost effectively from conventional forestry [5]. Forest residues are recovered following clear felling of the stand. There are a range of options available, all of which result in the production of

1513

chips of a size and form similar to those used as feedstocks in the pulp and particleboard industries. Forest residues can be collected and stored in the forest prior to comminution or can be comminuted directly at harvest. Where thinnings are used for energy whole tree harvesting and comminution systems are normally employed. With integrated harvesting systems the energy component is harvested at the same time as the timber. Trees are felled and handled as whole trees with product separation at the edge of the forest. The energy component is separated from the stemwood and comminuted, usually in the same or with a closely coupled operation. In this manner the supply costs are reduced significantly.

The means to harvest SRC crops cost-effectively is still a matter of much concern. Equipment is being developed and some is available commercially. However, they do not necessarily work efficiently, they have the potential to give long term damage to the site, and are costly to operate. SRC harvesting machines are available which cut and produce bundles of cut coppice shoots or which cut and comminute the crop and produce chips, chunks or billets. Modification of fodder, maize or sugar cane harvesters to harvest short rotation coppice has made significant inroads into the reduction of harvesting cost. Crops harvested as shoots can be stored in bundles or after comminution.

2.1.4   Comminution

There is a wide range of equipment which can be used to comminute woody biomass in the field or at the conversion site. Disc and drum chippers are available and can be adjusted to produce different sizes of chip, giving a relatively high degree of homogeneity in the product. Field hoggers (hammermills) and agricultural tub-grinders can be used to produce a more heterogeneous hogfuel which is less useful for most advanced conversion processes. Chunkers are available, although not as commercially well developed, to produce a larger size of particle. Chunks require less power to produce than chips and it is purported that they have better storage characteristics than chips although this has not been demonstrated conclusively [6]. A uniform small particle is generally preferred for advanced conversion processes and if this can achieved in one operation "in the forest" rather than additionally at the conversion plant, there will be a significant cost saving. Chunking is, therefore, unlikely to be optimal.

Generally it is easier to comminute woody biomass at harvest when the moisture content is around 50%. Less power is required and the resultant chip is relatively consistent in size. Once the material starts to dry, comminution becomes harder to perform and there is frequent shattering of the wood and chips. This results in a much wider range of size fractions with often a high fines content. However, if a fine particle size is required by grinding, for example in fast pyrolysis processes, complete drying to the required moisture content is preferred to aid the grinding process.

2.1.5   Storage and drying

With all woody biomass the method of harvesting and the form of the harvested product have significant implications in terms of system efficiency, particularly with respect to storage and drying. If the resource can be harvested all the year round then there will be the minimum of storage. Here the form of the harvested product (chips, chunks or bundles) is a consideration only in terms of transport (solid volume factor) and in-plant handling systems [7].

If short rotation coppice is to be used all the year round then storage for up to 8 months will be necessary. The form and manner in which the biomass is stored will be highly influential in determining moisture content change and dry matter loss. Short rotation coppice chips are more susceptible to biological decomposition through microbial growth than forest residues as they contain greater quantities of readily metabolisable soluble sugars and lower quantities of inhibitory substances such as extractives and phenols. If unchecked the biological activity during storage results in up to 20% losses in dry matter content over six months storage.

2.2   Transport
Transport of feedstock from the forest or field to the conversion facility is one of the simplest handling stages in the chain. The main mode of transport is by road using chip vans, the same as with the pulp and board industry (Table 2).

Table 2 Options and Issues for Biomass Transport

Step	Options	Issues
Transport	• Timber trucks: articulated or rigid • Curtain-siders • Chip vans • Chip bins	• Transport regulations (Gross Vehicle Weight - GVW, dimensions) • Loading capacity (solid volume factor) • Particle size • Bulk density • Moisture content • Transport distance (network inefficiencies) • Environment

Road traffic legislation limits gross vehicle weight and the volume of the truck hence bulk density and solid volume factor of the feedstock are important parameters determining transport efficiency. In some cases, with low bulk density material it may not be possible to achieve maximum payload.

2.3   Pretreatment - "Supply-Conversion Interface"
The pretreatment stage at the conversion plant can be considered the interface between the supplier and user of the biomass. Here, the feedstock is received and treated to meet the requirements of the conversion technology. The number and nature of steps in the pretreatment stage will be determined by the condition of the biomass as delivered relative to the needs of the specific conversion process being used. Table 3 summarises the options and issues for this stage.

A key issue in the bioenergy chain is where the size reduction, storage and drying steps are carried out. The decision hinges on whether these steps are carried out in the forest or at the conversion site or both. The actual situation will vary from case to case. Many of the pretreatment options have been explored by van den Heuvel [8]. The actual equipment which would be used for operations such as drying will be determined by feed throughput rate, degree of drying required and source and quality of heat provided.

Table 3. Steps and Issues for Biomass Pretreatment at the Conversion Plant

Stage	Steps	Issues
Pretreatment	• Unloading • Reception • Storage and reclamation • Screening • Drying • Comminution and grinding • Handling • Buffer storage	• Scale of operation • Dry feed quantity per year • Delivered feed cost • Moisture determination • Feed characterisation • LHV • Size of feed delivered • Shape of feed delivered

An important component of any bioenergy industry will be the means used to monitor the quality and quantity of the delivered feedstock (moisture content, size distribution, LHV, etc.). Whatever system is used it is essential that it is reliable and accurate, particularly if it forms the basis of a payment system.

2.4   Conversion

There are a number of options for the thermal conversion of biomass to an energy carrier as summarised in Table 4 and discussed below. Also included in Table 4 are the issues that require consideration of the interface between production and conversion, which as described above, can be dealt with in the forest or in the conversion plant, or sometimes both.

2.4.1   Combustion

Combustion is widely practised for heat, power and CHP and is therefore reasonably well developed and widely commercially available. An excess of air is used to convert all the chemical energy in the biomass into thermal energy which is then used for steam raising and electricity generation, or heat supply. The efficiency of these systems is limited by the thermodynamics of the Rankine cycle, the limitations of materials of construction, the need to meet performance criteria in operation and to meet environmental regulations. Acceptably high efficiencies can usually only be met in larger scale plants of typically above 20 MWe.

Current concerns relate to ash management in boilers where considerable build-up can occur, and emissions of particulates which are more difficult to control than in the alternative thermal technologies of gasification and fast pyrolysis.

2.4.2   Pyrolysis

Pyrolysis is thermal degradation either on the complete absence of oxidising agent, or with such a limited supply that gasification does not occur to an appreciable extent. In fast pyrolysis for liquids production, relatively low temperatures are employed of around 500 C compared to 800-1200 C in gasification. Although three products are always produced, gas, liquid and char, fast pyrolysis maximises the liquid at around 70 to 75 % yield on dry feed by careful control over temperature, heating rate and vapour residence time.

Table 4 Options and Issues for Thermal Biomass Conversion

Stage	Steps	Issues
Conversion	• Combustion	INTERFACE - FEEDSTOCK
	• Fast Pyrolysis	• Particle size and size range of feed
	• Gasification	• Particle shape
	• atmospheric	• Moisture content of feed
	• pressure	• LHV of feed
		• Chemical composition of feed
		• Ash in feed
		• Delivered feed cost
		OPERATION
		• Scale of operation
		• Performance
		• Reliability
		• Costs
		• Local regulations
		ENVIRONMENT
		• Emissions
		• Noise
		• Health & safety
		• Local regulations

The heat required for pyrolysis can be added indirectly in a variety of ways such as indirect heating through the reactor wall, direct contact with hot gas or hot liquid such as metal or molten salt, or directly by partial gasification with limited addition of oxidising agent such as air to give direct heating [9].

The liquids from fast pyrolysis are known as bio-oil, bio-crude oil or bio-fuel oil and have a range of interesting and peculiar characteristics. They can substitute for fuel oil in many applications such as boilers for heat or electricity and turbines and engines for electricity production. The major problems currently are mostly related to bio-oil stability in applications although considerable experience is being accumulated in their utilisation.

The major operational difference in fast pyrolysis for liquids is that a liquid fuel is produced which can be stored for later use or transported to a distant location. This de-coupling possibility is a significant advantage for decentralised operations.

2.4.3   Gasification

Gasification is the other major thermochemical conversion technology. In this process the primary products from pyrolysis react with oxygen from either air, steam and/or pure oxygen to convert all the solid and liquid products onto gas through partial oxidation and thermal cracking reactions. Gas is the major product although a small amount of tar is also produced from incompletely reacted pyrolysis products which can cause problems in some applications.

The major problems currently are tar cracking or removal to give a sufficiently clean gas for utilisation. Gas turbines are the most demanding in terms of alkali and particulates [1] while close coupling for process heat such as lime kilns in pulp mills is the least demanding. Engine applications are in-between in gas quality requirements, but there is little experience with engines or turbines at large scales of operation in industrialised economies where reliability is important.

2.5 Generation

There are a number of options for the utilisation of products from a biomass conversion process. There are three main products - electricity, which is currently of most interest and is used here to illustrate the importance of considering the complete bio-energy system; heat; liquid fuels such as gasoline and diesel; and chemicals such as alcohols, fertilisers and specialities. The options are summarised in Table 5.

Table 5   Options for Bio-electricity and Issues for Utilisation

Stage	Steps	Issues
Generation	• Gas engine • Diesel engine • Gas turbine • Stirling engine • Boiler • Steam turbine • (Furnace) • (Conversion to chemicals) • (Conversion to fuels)	INTERFACE-UTILISATION • Quality of fuel gas or liquid • Contaminants in fuel product • Reliability of supply • Close-coupling or decoupled requirement ENVIRONMENT • NOx, SOx, particulates emissions • Noise • Health & safety • Local Regulations

3   Planning the supply

Securing a sufficient and constant supply of biomass is essential to any biomass energy system. The supply must be well planned, in particular when SRC is being considered. The main elements which need to be factored into the supply determination are:
- crop productivity,
- catchment area,
- availability of land,
- operational area,
- harvest and handling losses,
- system inefficiencies,
- storage losses,
- delivery of feed to conversion.

In temperate zones production levels of 12 to 15 odt/ha/y are often quoted as being typical. If the average productivities from short rotation coppice plantations grown in Sweden, Denmark or the UK are examined however, it will be seen that productivity levels in the region of 6 to 8 odt/ha/y are more likely to be achieved in practice. These productivity levels could well increase with newer clonal material and better management practice but for now the lower figures are more realistic.

Not all of the SRC crop is harvestable in practice. Cutting height is not always ideal for various reasons and sticks and chips are often dropped onto the ground during harvesting and not recovered later. Not all land in a given hectare will be growing the crop. Up to 20% of a field may be given over to access routes and headlands for machinery to turn and environmental buffers.

It is becoming apparent from studies on SRC harvesting contractors working in Sweden that there are potential problems in the logistics of supply [10]. The difficulty concerns the availability of road transport and the working speed of the harvesters. Often the transport system has difficulty keeping pace with the most productive harvesters resulting in lost production as the harvester is kept waiting with a full load until the lorry or chip bin returns. There are also system inefficiencies with lorries travelling with only a part load or conversely part loads of chips are left in the field.

Planners have tended to be optimistic when determining biomass supply. When each of the above factors are brought into the equation it means that more land and hence longer transport distances than originally calculated are likely to be required.

4   Planning the Conversion and End-use Products

There are four main components in the conversion and end use part of a bioenergy system, continuing to use the example of electricity generation:

1. Pretreatment - in which the characteristics of the feed delivered is adapted to the requirements of the conversion step,
2. Conversion to a bio-fuel - in which prepared solid feed is converted into gas or liquid fuel,
3. Electricity generation - in which the bio-fuel is burned in a high efficiency device and electricity generated,
4. Delivery of electricity to the grid - in which the electricity is metered, adjusted and fed to a suitable local supply.

The first three steps have to be planned together and suitable and compatible alternatives selected. The site will be a significant feature and must be optimised with respect to the biomass supply to not only minimise cost but also to minimise environmental effects including consideration of planning requirements and local regulations. The position of the site with respect to the existing electricity supply is also a significant factor in minimising cost. All these factors mean that site selection is a major consideration and every opportunity will need to be assessed on its merits.

5   Environmental and Health Issues for the Bioenergy Chain

There is a significant interface between the bioenergy chain and the environment which needs to be considered when planning and operating systems. Some of the main environmental issues are listed in Table 6.

Table 6. Environmental Issues and Bioenergy

Stage	Environmental issues
Production	• oil/fuel/hydraulic spillage
	• fossil fuel use
	• atmospheric emissions
	• liquid run-off (leachate) from storage piles
	• soil compaction
	• use of herbicides/pesticides/fertilisers
	• long term site productivity
	• water quality/quantity in catchment
	• noise
	• fungal spores in air
Transport	• oil/fuel/hydraulic spillage
	• fossil fuel use
	• atmospheric emissions
	• noise
	• nuisance
	• road accidents
Pretreatment	• oil/fuel/hydraulic spillage
	• fossil fuel use
	• atmospheric emissions
	• liquid run-off (leachate) from storage piles
	• fungal spores in air
Conversion	• gaseous emissions
	• liquid emissions/spills
	• solids disposal or recycling
	• explosions
	• fires
	• waste water clean-up
Generation	• NOx emissions
	• SOx emissions
	• particulates emissions
	• noise
	• fires
	• electric shock
	• waste water clean-up

Environmental impacts of bioenergy production and use systems have been assessed [11, 12, 13]. The important point to note is that the risks of environmental impact are greater on the supply side than on the conversion side. The conversion technologies have to operate under strict environmental standards for release of hazardous materials and as such any emissions will be within those standards or the plant will not operate.

The importance of the feed reception, storage and handling system to the overall environmental performance has been highlighted [15]. Feed storage and drying both have significant potential problems in terms of emissions and effluents. The risks of accidents such as fires are higher during the pretreatment steps because of the difficulties in handling and containing the feedstock. The low impacts from fuel conversion are balanced by the hazards and waste streams that are introduced by fuel cleaning prior to combustion.

Upstream in the chain Health and Safety legislation will limit exposure levels to hazardous chemicals and materials such as fungal spores. However on the production front legislation is not as well formulated and most production systems are being developed as best practices which seek to minimise environmental impact.

6   Modelling the Bioenergy Chain

The integration of the five main stages in the bioenergy chain for electricity generation has been examined using a spreadsheet-based model as an IEA Bioenergy Activity [14]. The costs of production of electricity using woody biomass from conventional forestry or SRC through four conversion processes (combustion, atmospheric gasification, IGCC and pyrolysis) were examined in various ways and a number of conclusions drawn:

1. Conversion efficiency is important in determining electricity generation cost.
2. Capital cost of the conversion and generating step is very important.
3. Economy of scale of conversion is much more important than transport cost.
4. Electricity generation using conventional forestry feedstocks is generally cheaper than short rotation coppice.
5. For short rotation coppice chips are generally cheaper than whole shoots.

The model can be used to examine the distribution of costs between each of the five stages in the bioenergy chain. As an example the variation in cost distribution with respect to four different SRC supply strategies has been examined for a 30 MWe IGCC facility. There are a number of options available for the harvesting, storage and delivery of short rotation coppice as feedstock for such a process. Material can be harvested as shoots or chips, can be stored in either form and the location of the storage site can vary (this affects transport costs as there are different solid volume factors for shoots and chips). There are different dry matter and moisture content changes for shoots and chips of coppice. The supply strategies differ in the form of the harvested product, the form of the stored feed and where the material is stored (see options in Table 7). The SRC has a productivity of 10 odt/ha/y, the fields are 50ha and 10% of the land in the catchment is growing the crop. The parameters for the conversion technology are as reported in Mitchell *et al.* [16].

The main stages which are affected by the different scenarios are production, transport and pretreatment as would be expected. The production and transport costs vary as a function of dry matter loss during storage (Table 7). The significantly higher production costs for scenarios A and B are due to higher harvesting costs for shoots and the double handling necessary for the separate comminution step. The greater the dry matter loss the more crop that has to be grown and hence greater areas and transport distances required. Pretreatment costs vary primarily as a function of the moisture content of the delivered chips i.e. the wetter the material the more drying is required to bring the feed to the desired level.

Table 7. SRC Feedstock Supply Strategies

	A	B	C	D
Harvested as	Shoots	Shoots	Chips	Chips
Stored as	Shoots	Chips	Chips	Chips
Stored at	Plantation	Conv. Site	Plantation	Conv. Site
Transported as	Chips	Chips	Chips	Chips
Harvest period (months)	4	4	4	4
Delivery period (months)	12	12	12	12
Delivered MC (% wet basis)	32.50	52.50	45.20	53.92
Dry matter loss (%)	6	16	17	17
Costs 000 $/y):				
Production	5137	5672	3968	3587
Transport	770	927	799	931
Pretreatment	2129	3689	2295	3370
Conversion	5863	5863	5863	5863
Generation	4927	4927	4927	4927
Electricity cost (c/kWh)	7.96	8.91	7.55	7.90

These differences are reflected in the costs of electricity produced. The cheapest method is for harvesting as chips and storing at the plantation despite the 17% dry matter loss. Clearly, with different moisture content and dry matter loss relations the picture will change. At the moment there are no strong relationships that can be used as conditions change greatly on a case to case basis.

For the cheapest electricity production cost scenario i.e. Scenario C there are substantial differences in area required and cost of electricity production as a function of SRC productivity. For instance, if the productivity of the crop had been assumed to be 16 odt/ha/y and only 8 odt/ha/y was produced in practice then twice as much land would be required for growing the crop, the annual operating costs for production and transport would rise from 3.5 US$M to 5.7 US$M per year and the cost of electricity production would rise by 13% (Table 8).

Table 8 Influence of Crop Productivity on Costs and Plantation Area

Productivity (odt/ha/y)	8	16
Annual Costs (000s $/y):		
Production	4801	2799
Transport	855	699
Pretreatment	2310	2273
Conversion	5863	5863
Generation	4927	4927
Area required (ha)	20311	10155
Electricity Cost, US c/kWh	7.93	7.00

7   Discussion

There are many problems now facing the developers of bioenergy systems that are poorly appreciated. Those involved with conversion often do not appreciate the limitations and opportunities for production and delivery of feed material and those involved in production have insufficient awareness of the limitations and requirements of conversion processes. A significant barrier to deployment of technologies is the lack of data and experience concerning:

- resource quantification and selection,
- SRC production systems,
- supply strategies i.e. when to harvest and in what form for optimal downstream handling,
- rates of harvesting machine operation for different crops and conditions,
- storage and drying methods with respect to particle size and form and where this should take place: in-forest or on-plant or both,
- biomass handling systems,
- multiple feedstock handling, pretreatment and conversion systems,
- minimum pretreatment requirements for matching biomass feed to conversion technology
- selection of appropriate conversion technology for different sites and opportunities,
- the significant of de-coupled power production,
- management of gas contaminants in gasification,
- management of stability and contaminants in fast pyrolysis liquids,
- management of ash in combustion and gasification,
- reliability of electricity generation systems,
- development of low cost high performance electricity generation devices that are more tolerant of contaminated fuels,
- cost of electricity connection
- assessing and controlling environmental impacts across the bioenergy chain.

From a practical point of view there is a clear need to develop a universally accepted standard classification of woodfuel which can be used to formulate feedstock quality descriptors for use by feedstock suppliers and users.

A key step in the deployment of bioenergy technologies will be the realisation by suppliers and users of biomass as to what each others specifications are and the limitations they are working under.

8   References

1   Bridgwater, A.V., "The technical and economic feasibility of biomass gasification for power generation", Fuel 74 (5) pp 631-653 (1995)
2   Diebold, J.P. and Bridgwater A.V., "Overview of fast pyrolysis of biomass for the production of liquid fuels", These proceedings.
3   Bridgwater, A.V. and Mitchell, C.P. (1987) Interfacing biomass production and biomass conversion, in 4th EC Conference Biomass for Energy and Industry,

(eds. G Grassi, B. Delmon, J-F Molle and H. Zibetta), Elsevier Applied Science. pp 1174 - 1178.

4   Mitchell, C.P. and Bridgwater, A.V. (1990) Supply of biomass and the interface with thermochemical processing, in 5th EC Conference, Biomass for Energy and Industry, (eds. G. Grassi, G. Gosse and G. dos Santos, Elsevier Applied Science. pp. 2.1093 - 2.1097.

5   Mitchell, C.P., Storry, P.G.S., and Hudson, J.B. (1991) Systems analysis of wood fuel supply, in *Economics of wood energy supply systems*, (ed. H. Knutell), Swedish University of Agricultural Sciences, Faculty of Forestry, Department of Operational Efficiency Research Notes No 219, pp. 98 - 109.

6   Mitchell, C.P., Hudson, J.B., Gardner, D.N.A., and Storry, P.G. (1988) A comparative study of the storage and drying of chips and chunks in the UK, in *Production and storage and utilisation of wood fuels*, Swedish University of Agricultural Science, Department of Operational Efficiency Research Notes 133/1988. pp. 226 - 236.

7   Mattsson, J.E. (1990) Basic handling characteristics of wood fuels: angle of repose, friction against surfaces and tendency to bridge for different assortments. *Scandinavian Journal of Forest Research*, vol. 5. pp. 583 - 597.

8   van den Heuvel, E. J.M.T. (1996) Pretreatment technologies for energy crops. Biomass Technology Group. pp 98.

9   Bridgwater, A.V. (1995) Engineering developments in flash pyrolysis technology. In Proceedings of conference on bio-oil production and utilisation, Estes Park, Colorado, USA.

10   Danfors, B. (1996). Personal communication

11   Bridgwater, A.V., Elliott, D.C., Fagernis, L., Gifford, J.S., Mackie, K.L. and Toft, A.J. (1995) The nature and control of solid, liquid and gaseous emissions from the thermochemical processing of biomass. *Biomass and Bioenergy*, Vol. 9, Nos. 1-5. pp. 325-341.

12   Mitchell, C.P. and Bridgwater, A.V. (editors) (1994) Environmental Impacts of Bioenergy. CPL press pp 171.

13   Bridgwater, A.V., Elliott, D.C., Fagernis, L., Gifford, J.S., Mackie, K.L., Rivard, C.R. and Toft, A.J. (1995) The nature and control of solid. liquid and gaseous emissions form the thermochemical processing of biomass. IEA Task X, Activity 7 Environmental Systems Activity. Report no ES 94/2. New Zealand Forest Research Institute.

14   Mitchell, C. P., Bridgwater, A.V., Stevens, D.J., Toft, A. J. and Watters, M. P. (1995) Technoeconomic assessment of biomass to energy. *Biomass and Bioenergy*, Vol. 9, Nos. 1-5. pp. 205-226.

ELECTRICITY GENERATION FROM WOOD-FIRED POWER PLANTS; THE PRINCIPAL TECHNOLOGIES REVIEWED
Wood-fired power plants reviewed

D.R. McIlveen-Wright, B.C. Williams and J.T. McMullan
Centre for Energy Research, University of Ulster,
Coleraine BT52 1SA, UK.

Abstract

The ECLIPSE process simulation package is used to make techno-economic assessments of a range of power generation technologies which are currently available or have been proposed with wood as the fuel. These technologies comprise: (a) combustion with or without wood drying; (b) gasification of the wood followed by power generation using a gas turbine only; (c) Integrated Gasification Combined Cycle or IGCC; (d) Steam Injection Gas Turbine or STIG; (e) Intercooled Steam Injection Gas Turbine or ISTIG; (f) wood gasification followed by a spark ignition gas engine; (g) biomass-fed fuel cells.
With the exception of the STIG, ISTIG and fuel cell systems, each of the technologies is assessed for overall efficiency, break-even electricity selling price and availability of wood supply over a range of plant sizes. The forest size, level of afforestation, transportation distance (and its effect on wood feedstock cost) are also examined and an optimum range of power plant size, in terms of wood feed rate, is suggested for each technology.
Values are assumed for certain parameters in the simulations such as: wood moisture content, harvest yield, afforestation level and discounted cash flow rate. The influence of these parameter values on the system efficiency and economic viability is reported.
Keywords: electricity generation, techno-economic assessments, wood-fired power plant.

1 Introduction

The use of wood, grown in a sustainable fashion, for power generation has undoubted benefits in terms of reduced emissions, security of fuel supply and long-term employment. There are, however, several possible technologies which may be used. The technologies reviewed here are : (a) combustion with or without wood drying; (b) gasification of the wood followed by power generation using a gas turbine only; (c) Integrated Gasification Combined Cycle or IGCC; (d) Steam Injection Gas Turbine or STIG; (e) Intercooled Steam Injection Gas Turbine or ISTIG; (f) wood gasification followed by a spark ignition gas engine; (g) biomass-fed fuel cells.
Several elements in the simulations cannot be precisely determined, for various reasons, and certain values are assumed in all cases. For example, the wood feedstock is taken to be willow from a short rotation forestry plantation with a moisture content of 100% (on a

dry basis), the plantation yield as 10 dry tonnes/ ha/ year, discounted cash flow rate (DCF) as 10% and contingency as 10%. Also, the power plant is assumed to be situated in the centre of a forested area. For the smaller-sized power plants, using less than 1,000 dry tonnes/ day, the level of afforestation around the plant is taken as 100%; for plant sizes between 1,000 and 4,000 dry tonnes/ day the afforestation level is taken as 20%; and for plants which are larger than 4,000 dry tonnes/ day it is taken to be 5%. These values affect the forest size, the transportation distance and, hence, the wood feedstock cost. The effect of taking other reasonable, but less conservative, values for all these elements is also shown.

For all systems it is further assumed that the wood is harvested from a local coppiced plantation, is then chipped and transported to the power plant where it is stored in piles in the open, in sufficient quantities for 5 - 14 days' throughput. Next, the chips are taken from storage, screened (with magnetic separation, size separation and reduction stages) and transferred to buffer storage, ready for use in the combustor or gasifier.

Wood feedstock costs are assumed to be made up of £19.80/ dry tonne for harvesting and chipping plus £0.31/ dry tonne/ km for transportation[1].

The ECLIPSE process simulator[2], which has been successfully used to model a range of power generation systems[3], was used for these assessments. In every case an individual mass and energy balance was optimised for the system chosen. The utilities (electricity, cooling water, etc) used by parts of the plant (e.g. fans, condenser) are then calculated by the following program module, or they can be inputted for non-standard components. Utilities data for some of the biomass-specific machinery (conveyor, chipper, etc) were aquired[4] and entered to complete the technical assessment.

In the economic analysis capital costs are study estimates with an accuracy of ± 30%. Capital costs are difficult to determine for any novel technology, and it is especially difficult to estimate how much they will change after the technology has undergone intensive development and large scale production. The best available capital cost estimates have been used [5],[6], but all the results and conclusions of the economic analyses presented here should be treated with caution.

A more complete description of all the analyses presented here, except for the fuell cell systems, is given elsewhere[7].

2 Combustion

2.1 Process description

A range of plant sizes were examined, using from 10 to 10,000 dry tonnes of wood per day. A simulation was made with and without a drying stage for each plant size. Where a drying stage is used, low pressure steam heats the air used in a rotary dryer where the wood moisture content is reduced to 15% (dry basis). Approximately 15 % excess air, preheated by the exhaust gases, is used in the combustor to ensure complete combustion of the wood. Steam conditions appropriate for each plant size were chosen [8] i.e. 23 bar at 350°C for the 10 dry tonnes/ day size, 60 bar at 480°C for the 100 dry tonnes/ day plants, 80 bar at 520°C for the 500, 1,000 and 2,000 dry tonnes/ day plants and 160 bar at 538°C for the 5,000 and 10,000 dry tonnes/ day plants. Only in the two largest systems was steam

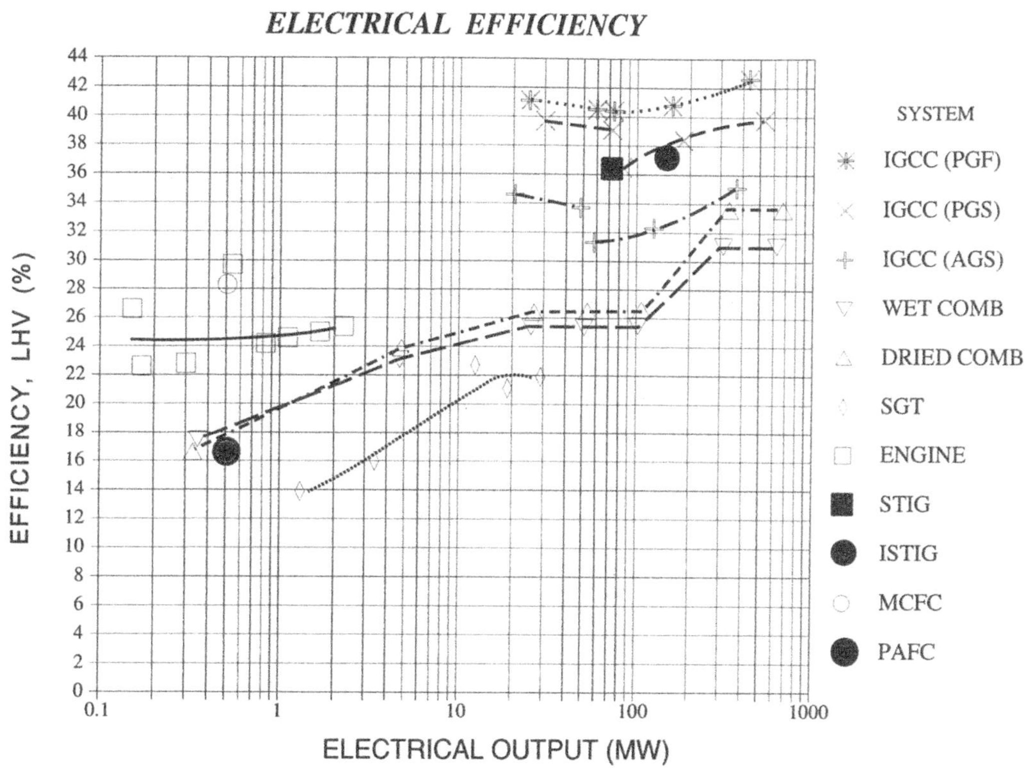

Figure 1 Efficiency versus plant output

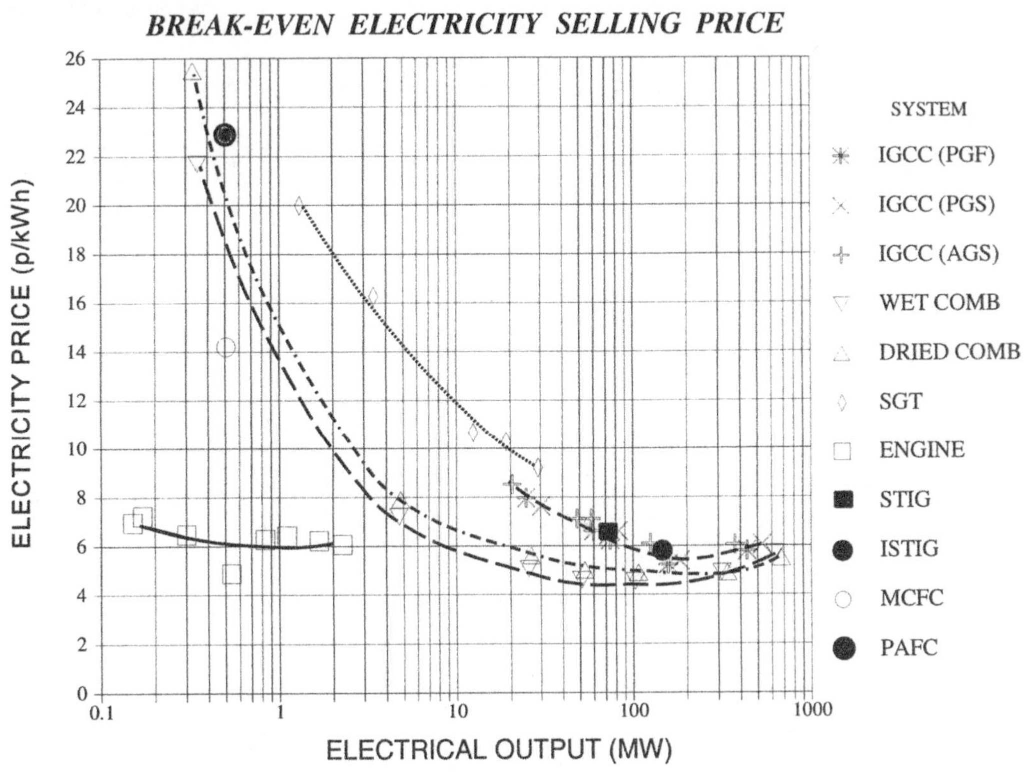

Figure 2 Break-even electricity selling price versus plant output

reheating, together with multiple-stage feedwater preheating, included. For the others a single, low pressure, feedwater heater was used in each case.

2.2 Results and conclusions for combustion plants

For the combustion of wet wood the electrical efficiency (LHV) was found to increase from 17.5% for the 10 dry tonnes/ day plant to 31.1% for the 10,000 dry tonnes/ day plant. For the systems with drying stages the electrical efficiency ranged from 16.6% to 33.6%. The efficiency of the smaller systems is adversely affected by the high base load penalty and the intrinsically lower efficiency of the steam cycle. For example, if the steam cycle used by the 500 - 2,000 dry tonnes/ day plants is also used with the 10 dry tonnes/ day plant, then the efficiency rises from 17.5% to 23.4%, compared with 25.4% for the 1,000 dry tonnes/ day plant. Also, the efficiency of the 5,000 dry tonnes/ day plant falls from 31.1% to 25.5% when it uses the steam conditions of the 1,000 dry tonnes/ day plant. There is no difference in efficiency between the 500, 1,000 and 2,000 dry tonnes/ day plants, which all have the same steam conditions. Efficiencies for all plants are shown in Fig. 1.

Wood Input (dry tonnes/day)	Wet Wood		Dried Wood	
	η_{LHV}	MWe	η_{LHV}	MWe
10	17.5	0.35	16.6	0.33
100	23.1	4.7	23.9	4.8
500	25.4	25.5	26.4	26.5
1,000	25.4	51.2	26.5	53.3
2,000	25.5	102.5	26.6	106.9
5,000	31.1	312.8	33.6	337.2
10,000	31.1	625.5	33.6	675.9

Specific Investment drops sharply from around £7,000/ kWe for the 10 dry tonnes/ day plants to around £2,000/ kWe for the 100 dry tonnes/ day plants. It decreases more slowly to around £1,000/ kWe at 1,000 dry tonnes/ day and down to about £700/ kWe at the 5,000 to 10,000 dry tonnes/ day sizes.

Break-even electricity selling prices, for wet and dried wood combustion respectively, were found to be around 22 p/ kWh and 26 p/ kWh for the 10 dry tonnes/ day plants; around 7 p/ kWh and 8 p/ kWh for 100 dry tonnes/ day plants; around 5 p/ kWh and 5.5 p/ kWh for 500 dry tonnes/ day plants; around 4.5 p/ kWh and 5 p/ kWh for 1,000 dry tonnes/ day plants; around 4.5 p/ kWh and 5 p/ kWh for 2,000 dry tonnes/ day plants; around 5 p/ kWh and 5 p/ kWh for 5,000 dry tonnes/ day plants; and around 5.5 p/ kWh and 5.5 p/ kWh for 10,000 dry tonnes/ day plants. These are shown in Figure 2.

The availability of feedstock may prove to be an unsurmountable problem for the large scale plants. For the 10,000 dry tonnes/ day plants nearly 400,000 ha of forest area would be required. This amount may not be available in a single location in Europe. If the plant is located in a heavily-forested area, say 100% afforestation, then the forest radius would be around 34 km and the average transportation distance 24 km. However, if the plant is located in a lightly-wooded region, say 5% afforestation level, then the average transportation distance rises to nearly 110 km. This has the effect of doubling the feedstock cost from £27.30/ dry tonne to £54.50/ dry tonne and is the reason for the break-even electricity selling price rising again for larger plant sizes.

- Net electrical efficiency, η, was found to be principally determined by the steam conditions and not by the base load linked to plant size. However, larger plants tend to have steam conditions more favourable for high efficiencies. A drying stage generally improves η.
- For large plants, the lack of availability of abundant wood supplies and the high costs of feedstock transportation make them unlikely to be viable.
- For small plants, plentiful wood supplies are available and transportation costs are low, but low η and high capital costs make them uncompetitive.
- Medium- to large-sized combustion plants, using from 500 to 2,000 dry tonnes/ day, (25-110MWe) probably offer the optimum scale for this technology.
- For small systems the effect on BEESP of the efficiency gains in using a drying stage is outweighed by the increase in capital costs. For large systems there is no significant difference in BEESP between systems with or without a drying stage.

3 Integrated Gasification Combined Cycle (IGCC)

3.1 Process description

The harvesting, transportation and handling stages were taken to be the same as for the combustion plants. A drying stage, using heat from the exhaust gases, was used in all cases to reduce the wood moisture content to 15% (dry basis), since wood with a moisture content above 45% (dry basis) produces severely reduced gasification yields.

Data for three industrial gas turbines and two aero-derivatives were used. Each turbine was simulated in three different configurations:- PGF, pressurised gasifier with ceramic filter for hot gas clean-up; PGS, pressurised gasifier with water scrubber for cold gas clean-up; AGS, atmospheric pressure gasifier with scrubber gas clean-up.

A two-stage compressor supplies air to the gasifier at about 10 bar above the pressure in the combustion chamber of the gas turbine in the pressurised gasifier systems. For the systems using atmospheric pressure gasifiers, the gas emerging from the gas scrubber which follows the gasifier is compressed to about 10 bar above gas turbine combustor pressure. The operational temperature of the gasifier is about 900°C. The hot gases leaving the gasifier superheat the steam in the steam cycle and then go to the gas-cleanup system for particulate (and tar, where necessary) removal. The ceramic filter operates at about 420°C, allowing the cleaned gases to reach the gas turbine at a relatively high temperature. The water scrubber operates at about 135°C, which means that the gases are cool when they reach the gas turbine, but they are saturated with water vapour which promotes heat transfer.

Each gas turbine was initially simulated in a stand-alone configuration, with natural gas as the fuel, using the turbine inlet and exhaust temperatures (TIT and TET), compression ratio and mass and air flow given in the Gas Turbine Year Book 92/93 or from the manufacturers' specifications. The turbine blade cooling air bleed pressure and flow rate, as well as the turbine polytropic efficiency, were used as parameters in order to match the simulated TIT and TET to the actual values and to arrive at a power output within 5% of the nominal value. Then the natural gas was replaced by "wood" producer gas, which has a much lower calorific value, and the mass flow of this gas increased until the expected TIT for that turbine was obtained. This increased mass flow through the turbine leads to an increase in the power output of the gas turbine. The increased flow can probably be handled by gas turbines which are based on aircraft engines, but industrial gas turbines may require modifications to their expander sections to accommodate this increased flow. Capital costs for the gas turbines were based on this increased power output i.e. to allow for any costs incurred by the modifications. (It may be preferable to restrict the power output of the gas turbine to the level achieved when using natural gas, but this would lead to a drop in TIT and hence to a drop in efficiency. It would, however, make the capital cost estimate for the gas turbine easier).

Gas Turbine	Configuration	Wood Input (dry tonnes/day)	Efficiency (LHV) (%)	Electrical Output (MWe)
LM1600	PGF	300	41.1	24.7
	PGS	380	39.7	30.2
	AGS	295	34.6	20.5
LM5000	PGF	735	40.5	59.7
	PGS	920	39.1	72.6
	AGS	715	33.7	48.3
MS6001	PGF	915	40.4	74.2
	PGS	1,130	36.5	82.9
	AGS	920	31.3	58.0
MS7001	PGF	1,920	40.8	157
	PGS	2,370	38.4	183
	AGS	1,925	32.3	125
MS9001FA	PGF	5,080	42.6	435
	PGS	6,635	39.8	531
	AGS	5,265	35.1	371

3.2 Results and conclusions for IGCC systems

The two smallest gas turbines modelled, the LM1600 and the LM5000, are aero-derivatives. The three largest, the MS6001, MS7001 and the MS9001FA, are industrial gas turbines. Both types of turbines may have major roles to play in power generation[9]. The aeroderivatives are more efficient, due to their higher compression ratios.

(The LM1600, LM2500, LM5000 and the TYPHOON (all areoderivatives) have been or are being investigated with biomass as fuel.)

The LM1600 is more efficient than the LM5000 because of its higher TIT, even though it is the smaller of the two.Steam conditions were chosen to match the system size i.e. 60 bar at 480°C for the system based on the LM1600 turbine, 80 bar at 520°C for the systems based on the LM5000, the MS6001 or the MS7001 gas turbines, and 160 bar at 538°C for the system using the MS9001FA gas turbine. The relative efficiencies of the IGCC systems based on these gas turbines are in line with the efficiencies of the individual gas turbines. For a given gas turbine the electrical efficiency of the system depends on the gasifier and gas clean-up chosen. The efficiency values are shown on Fig. 1.

Specific Investment decreases from around £2,500/ kWe to around £1,300/ kWe as the system size increases. The AGS versions were the most expensive and the PGS the least expensive, but the difference was only around 10%. The AGS versions also had the highest break-even electricity selling prices (BEESPs), while the PGF and the PGS versions had similar BEESP values. The BEESPs are plotted in Fig. 2.

- Overall system efficiency was found to be improved by using hot gas clean-up and also by using a pressurised gasifier.
- Aero-derivative gas turbines are more efficient than industrial gas turbines; mainly due to their higher compression ratios. For the aero-derivatives, a higher Turbine Inlet Temperature (TIT) leads to higher η.
- Large IGCC systems have the same wood availability and transportation cost problems as the large combustion plants.
- Industrial gas turbines may need modifications to accommodate the low calorific value gas from wood i.e. larger combustor(s) and a larger expander. The costs of such modifications are difficult to assess, so the present economic analysis can only be considered as a guideline estimate. Aero-derivatives may need little or no modifications, since aircraft engines are designed to run under varying load, thrust and atmospheric conditions.
- Medium- to large-sized IGCC plants, using from 500 to 2,000 dry tonnes/ day (electrical output from around 50-160MWe), probably offer the optimum scale for this technology, but smaller IGCC plants based on other aero-derivative gas turbines, and on small gas turbines running in parallel, should be investigated.

4 Gasifier/ gas turbine systems

4.1 Process description

Data from five currently used aero-derivative gas turbines (Hurricane, Typhoon(M), LM1600, LM2500 and LM5000) were used in the Simple Gas Turbine (SGT) configuration (wood drying, atmospheric pressure gasification, filter gas cleaning, gas turbine - no steam raising). A drying stage, using the heat from exhaust gases, was used. A simple hot water generation facility (for CHP) was also investigated. One turbine (LM5000) was also simulated in the STIG and ISTIG configurations.

Similarly to the method used for the IGCC systems, the gas turbine was initially modelled using natural gas as the fuel to ensure that the manufacturers' stated TIT, TET, compression ratio, mass flows and power output were achieved. The high pressure and low pressure steam conditions[10], the power turbine inlet pressure[11] and, for the ISTIG configuration, the intercooler pressure and temperature[12] data were obtained from other sources. These values were also obtained in the simulation, as more fully described elsewhere[13].

4.2 Results and conclusions for gasifier/ gas turbine systems

The efficiency of the smallest SGT system was found to be 13.9%, rising with system size to 21.9% for the largest, based on the LM5000 gas turbine. By contrast, the BEESP values were found to fall from 20 p/ kWh for the smallest system to 9.2 p/ kWh for the largest. For the largest turbine (LM5000) simulations were also made for systems employing the turbine in both the Steam Injection Gas Turbine (STIG) and the Intercooled Steam Injection Gas Turbine (ISTIG) modes. For these two variations steam at two pressures is raised using the gas turbine exhaust heat. The high pressure steam is injected into the turbine combustor and the low pressure steam between the high pressure and low pressure expander stages of the turbine. The ISTIG variant also has an intercooler operating between high and low pressure stages of the turbine compressor.

- The overall electricity generating efficiency was found to increase with plant size, whereas the CHP efficiency remained approximately the same (about 60%) for all the SGT systems. For the LM5000 turbine, electrical η rises from 21.9% (SGT configuration) to 36.4% (STIG) and 37.2% (ISTIG).

Gas Turbine	Wood Input (dry tonnes/day)	Electrical Output (MWe)	Electrical Efficiency (%) (LHV)	CHP Efficiency (%) (LHV)
Hurricane	46	1.3	13.9	61.5
Typhoon(M)	105	3.4	16.0	60.4
LM1600	270	12.4	22.7	61.3
LM2500	445	18.9	21.1	61.4
LM5000	660	29.0	21.9	60.1
LM5000-STIG	970	71.2	36.4	N/A
LM5000-ISTIG	1,950	145.5	37.2	N/A

- For these SGT systems there are no problems with wood availability or transportation costs. This also applies to the STIG system investigated, but the ISTIG system is almost too big (~2,000 dry tonnes/ day).

- Break-Even Electricity Selling Price (BEESP) falls as the system efficiency and system size rises. SGT systems are not competitive for electricity generation (so no optimal size is suggested), but may be viable for CHP. STIG and ISTIG systems have similar BEESPs to IGCC systems of the same size, at present.

5 Gasifier/ gas engine systems

5.1 Process description
In this paper only spark ignition engines were considered. Two engines were simulated in both the Naturally Aspirated (NA) and the Turbocharged Aftercooled (TA) configurations. Four others were considered only in the TA mode. The gasifier was taken to operate at atmospheric pressure. A drying stage, using heat from the exhaust gases, was used. The engine was simulated in a stand-alone configuration initially, using natural gas as the fuel, to replicate the manufacturers' data. An arrangement for providing hot water from the engine cooling was also investigated (for possible CHP applications).

5.2 Results and conclusions for gasifier/ gas engine systems
The electricity generating η was found to be ~23% for systems using NA engines and ~25% for those using TA engines. The CHP η was found to lie between 60 and 70% (these would be higher if the initial wood moisture content were lower).
BEESP values were found to lie between 6 and 7 p/ kWh.

Engine	Wood Input (dry tonnes/day)	Electrical Output (MWe)	Electrical Efficiency (%) (LHV)	CHP Efficiency (%) (LHV)
3408NA	3.8	0.17	22.6	68.7
3516NA	6.6	0.30	22.8	65.6
3408TA	2.8	0.15	26.6	69.7
3516TA	9.0	0.54	29.7	59.2
3606TA	17.1	0.83	24.2	65.2
3608TA	22.3	1.10	24.6	58.8
3612TA	33.1	1.66	25.0	64.3
3616TA	44.5	2.27	25.4	63.8

- Efficiencies are moderate, but plant sizes are small.
- Break-Even Electricity Selling Price (BEESP) was found to be about the same for all but the smallest plant sizes, so almost the whole size range can be considered.
- BEESP values are comparable with those of most IGCC systems, which are much larger.

6 LPO gasifier/ fuel cell systems

6.1 Process description

The system comprises a low pressure oxygen (LPO) wood gasifier, a wood drying stage, cold gas-cleaning and a fuel cell and gives a 500 kWe output. The gasifier is the Koppers-Totzek entrained-flow gasifier, which was originally developed for coal gasification and is considered to be representative of commercially available LPO technology[14]. It has been assessed for biomass[15]. The LPO gasifier is chosen since it gives a gas which is low in methane. This means that no reformer is necessary to "reform" the methane to hydrogen and carbon monoxide. Two fuel cells have been considered, the phosphoric acid fuel cell (PAFC) and the molten carbonate fuel cell (MCFC). The PAFC can only tolerate 1-2% CO at the operating temperature of 200°C, so a "shifter" must be employed to convert the CO to hydrogen. Steam is required for the shift reaction. The MCFC operates at 600°C and uses both hydrogen and CO in electricity production, so it does not require a shifter.

Fuel cell systems have not been in use for a long time, so there is great uncertainty in their operating lifetimes and their capital costs. This makes their economics even more uncertain!

However, for these systems a fuel cell lifetime of 10 years and capital costs of £1,000/kWe for the fuel cell have been assumed. Variations in these values and their effects on the economics of the systems, as well as fuller system descriptions, are discussed elsewhere[16].

6.2 Results and conclusions for LPO gasifier/ fuel cell systems

The 500 kWe PAFC system was found to have an electrical efficiency of 16.6% (CHP η of 68%) with a BEESP of about 16.5 p/ kWh. An electrical efficiency of 28.3% (CHP η of 64%) and a BEESP of about 11p/ kWh were found for the 500 kWe MCFC system.

- The MCFC system has the highest efficiency of the small systems. The PAFC has an efficiency similar to that of wood combustion plants of similar size (i.e. wood input).
- BEESP values are better than for similarly sized wood combustion plants, but are higher than those of the gas engine systems.

- Only one size for each system was investigated, so an optimum size cannot be suggested.

7 Effect of "assumed values" on analysis

The effect of changing the values assumed for certain elements in the analysis to other, less conservative, values has been investigated for the combustion of wet wood [17]. Figure 3

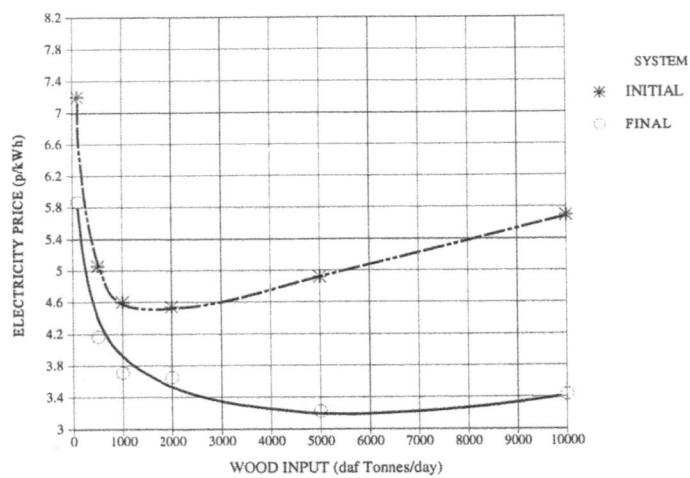

Figure 3 Effect of "assumed values" on break-even selling price

shows the effect of making the following changes:-
- assuming a wood moisture content of 55% (dry basis) instead of 100%;
- assuming a yield of 15 instead of 10 dry tonnes/ha/year;
- an afforestation level of 15% (the approximate level if set-aside land is used) for all systems, instead of a level related to plant size;
- the discounted cash flow rate was taken as 7.5% instead of 10%.
- contingency was taken as 5% instead of 10%.

When these changes in "assumed values"are made, the break-even selling price was found to fall by 18.5% for the 100 dry tonnes/ day plant and by almost 40% for the 10,000 dry tonnes/ day plant.

8 Conclusions

Large-scale plants (> 2,000 dry tonnes/ day) of any technology are unlikely to be built due to the unavailability of sufficient wood feedstock. At the small end of the scale gasifier/ gas engine systems appear to be reasonably efficient and have low BEESP values. The LPO gasifier/ MCFC system is just as efficient, but the economics of such a young technology are not readily quantifiable. Combustion plants in the range 100 - 2,000 dry tonnes/ day have similar efficiencies to the gas engine systems and their BEESP values are also comparable. IGCC, STIG and ISTIG systems are the most efficient, but their BEESP values are no better than those of the larger combustion plants.

Values for certain elements in the simulations were assumed e.g. wood moisture content, yield, DCF, etc. The values used can have a significant effect on the efficiency and economics of the system. Clearly it is important to determine these values as precisely as possible to assess the economic viability of any particular project. As mentioned before, there are considerable uncertainties in the capital costs of major items in the systems, which means that the BEESP values given here can only be considered as guidelines. However, if these BEESP values are considered and compared with the NFFO-3 (3rd tranche of the Non-Fossil Fuel Obligation) average electricity price of 8.65 p/ kWh for biomass systems, most technologies described could be considered viable for some part of their size range.

9 References

1. A study performed by the Department of Agriculture for Northern Ireland (DANI) found the cost of harvesting wood to be £10.80/dry tonne, chipping wood to be £9.00/dry tonne and transportation (up to 20 km) to cost £5.40/dry tonne (equal to £0.27/dry tonne/km if linear). Dawson M. (1992) DANI, private communication. This falls within the range of transportation costs of £0.12 to £0.30/green tonne/km given in the ETSU report *Wood Fuel Supply Strategies*, Vol.1, Report Nr. ETSU B 1176-P1. Another method is to equate the increase in transportation distance with a loss in energy content of the wood fuel i.e 33% drop in energy content after 20 miles, 50% after 40 miles and 70% after 100 miles Method suggested by Foster C., ETSU, private communication, March 1993). The energy content loss/mile can be equated to a drop in weight of the delivered wood/mile and hence converted to an increase in feedstock cost/mile. The value of £0.31/dry tonne/km arose from the latter method, but all three methods gave reasonably similar values.

2. Williams, B.C. and McMullan, J.T. (1996) Techno-economic analysis of fuel conversion and power generation systems - the development of a portable chemical process simulator with capital cost and economic performance analysis capabilities *Int. J. Energy Research*, **(20)**, pp. 125-142,
OR
Williams B.C. (1994) The development of the ECLIPSE simulator and its application to the techno-economic assessment of clean fossil fuel power generation systems. DPhil thesis, University of Ulster, Coleraine, N.Ireland, UK.

3. For example, under the JOULE II programme for R&D in clean coal technology a variety of technologies using coal, lignite, heavy fuel oil, natural gas and wood as fuels, as well as CO_2 sequestration techologies were investigated using ECLIPSE and reported in:
McMullan, J.T., Williams, B.C., Campbell, P., McIlveen-Wright, D. and Bemtgen J.M. (1995) Techno-economic assessment studies of fossil fuel and fuel wood power generation technologies. CEC report, contract nr. JOUF-0017.

4. For example, most of the utilities usages were taken from Bridgwater, A.V. and Double, J.M. (1991) Technical and economic processes for liquid fuel production in Europe. Contract EN3V-0012-UK(RH) for the Non-nuclear programme of the Commission of the Euriopean Communities.

5. Bridgwater, A.V. and Double, J.M. (1991) op. cit.

6. Solantausta, Y., Bridgwater, A.V. and Beckman, D. (1995) An assessment of biomass based power systems. Report for the International Energy Agency Bioenergy Agreement, (VTT).

7. McIlveen-Wright, D. (1995) Electricity Generation from wood. DPhil thesis, University of Ulster, Coleraine, N.Ireland, UK.

8. McIlveen-Wright, D.R., Sloan, P.E., Williams, B.C. and McMullan, J.T., (1995) Wood-fired combustion plants. In the proceedings of the Institute of Energy seminar *Combustion and Emissions Control*, London, pp. 179-188.

9. Williams, R.H. and Larson, E.D. (1989) Expanding roles for gas turbines in power generation. *Electricity, efficient end-use and new generation technologies, and their planning implications*, eds. Johansson, T.B., Bodlund, B. and Williams, R.H., Lund University Press, Sweden.

10. Nielsen, G. (1993) Private communication. Performance tables supplied from GE Marine Diesels Division, USA.

11. Critchley, P. (1994) Private communication from European Gas Turbines, Lincoln, UK.

12. Corman, J.C. (1986) System analysis of simplified IGCC plants. Report prepared for the US Dept. of Energy by the General Electric Reserch and Development, Schenectady, New York, USA.

13. McIlveen-Wright, D. (1995) op. cit.

14. Wyman, C.E., Bain, R.L., Hinman, N.D. and Stevens, D.J. (1993) Ethanol and methanol from cellulosic biomass. Chapter 21 of *Renewable energy, sources for fuels and electricity*, Island Press, Washington DC, pp.865-923.

15. Chem Systems (1990) Assessment of cost of production of methanol from biomass. Report DOE/PE-0097P, Chem Systems, Tarrytown, New York.

16. McMullan, J.T., Williams, B.C., Campbell, P.E., McIlveen-Wright, D.R., Brennan, S. and McCahey, S. (1995) Fuel Cell Optimisation Studies. Final Report of contract JOUL2-CT93-0278 for the European Commission in the framework of the Non-nuclear energy programme.

17. McIlveen-Wright, D.R., Sloan, P.E., Williams, B.C. and McMullan, J.T., (1995) op. cit.

THE PERFORMANCE AND ECONOMICS OF POWER FROM BIOMASS

Y. SOLANTAUSTA, VTT Energy, Espoo, Finland
A.V. BRIDGWATER, Aston University, Birmingham, UK
D. BECKMAN, Zeton Inc., Burlington, Canada

Abstract

The paper presents the results of performance and economic evaluations of advanced biomass power production technologies. The work was carried out as part of the International Energy Agency (IEA) Bioenergy. Models were used to determine performance and to evaluate the cost of new concepts. The modelling tools employed were AspenPlus and GateCycle. The economic models were developed by the IEA working group. Several plant configurations based on gasification and pyrolysis processes with diesel engines, gas turbines and steam turbines were analysed at capacities from 5 to 60 MW$_e$. The study included both technical sensitivities (gasification pressure, dryer type, feed moisture, co-generation of heat, etc.) and economic sensitivities.

Compared to conventional steam cycle power plants, the new systems generally have higher efficiencies (especially combined-cycle based on pressurised gasification, IGCC). The IGCC concepts have potential of becoming an attractive alternative at capacities higher than 30-50 MW$_e$. In smaller scale, gas engines using gasification fuel gas have a relatively high efficiency, but still a high specific investment. The steam injected gas turbine (STIG) cycles analysed did not appear competitive. Pyrolysis systems generally have a lower overall efficiency than gasification systems. However, there is potential for reducing the cost of pyrolysis oil production further. Diesel power plants using pyrolysis oil in peak load and especially in intermittent operation for electricity production appear interesting.

Biomass is often more expensive than fossil fuels. Consequently present commercial applications are generally limited to special cases where biomass residues are available, or when combined-heat and power production is viable. It is difficult to commercialise biomass technologies in larger scale (for example IGCC) in spite of favourable economics, due to the large scale needed for demonstration. It results in a large investment, and a high risk for developers. Development of power production based on pyrolysis systems may be feasible, as it appears viable in small niche projects.

Keywords: biomass, economics, electricity, modelling, system performance

1. Introduction

The paper presents the results of performance and economic evaluations of advanced biomass power production technologies. The work was carried out as part of the International Energy Agency (IEA) Bioenergy during 1992-1994 [1]. The objectives of the activity were:

    - to establish baseline assessments for the performance and economics of power production from biomass,

    - to support national efforts in improving power production systems based on pyrolysis and other conversion methods, and

    - to review research work in the area of liquid fuel production based on pyrolysis.

To satisfy these objectives, models were used to determine the performance and evaluate the cost of new concepts. The modelling tools employed were AspenPlus [2] and GateCycle [3]. The economic models employed were principally developed by the IEA working group and Aston University.

A total of 21 power plant configurations based on gasification and pyrolysis processes with diesel engines, gas turbines and steam turbines were analysed at capacities from 5 to 60 MW$_e$. The study included both technical process sensitivities (gasification pressure, dryer type, gas turbine type, feed moisture, co-generation of heat, etc.) and economic sensitivities. New concepts were compared to existing technology.

2. System description

The following six processes were selected for evaluation. The processes are classified as gasification concepts (four base processes selected) and pyrolysis concepts (two base processes selected). Based on these six base cases, several technical alternatives were studied, bringing the total number of concepts evaluated to 21. The main steps with the six base cases are listed below:

Gasification concepts:

1. Pressurized Gasification Combined-Cycle

    Pressurized fluidized-bed gasifier, hot gas clean-up, combined-cycle.

2. Atmospheric Gasification Combined-Cycle

    Atmospheric fluidized-bed gasifier, dolomite tar cracking, gas scrubbing, combined-cycle.

3. Pressurized Steam Injected Gas Turbine

    Pressurized fluidized-bed gasifier, hot gas clean-up, steam injected gas turbine, STIG.

4. Atmospheric Gasification Diesel Power

    Atmospheric fluidized-bed gasifier, gas scrubbing, combustion engine.

Pyrolysis Concepts:

1. Pyrolysis Diesel Power

    Fast pyrolysis, fuel treatment, diesel engine.

2. Pyrolysis Combined Cycle

Fast Pyrolysis, fuel treatment, gas turbine combined-cycle.

All the 21 process concepts studied are listed in Table 1. The combined-cycles (integration of a gas and a steam turbine), simple cycles (steam injected gas turbines), and internal combustion engines are employed for electricity production. In addition to power production, combined heat and power production is studied. The approximate scale of operation is between 5 to 60 MW$_e$. Note that in the pyrolysis concepts, oil production may be separate from the power plant.

Table 1. Power plant concepts assessed. Cogen = co-generation of heat and power, STIG = steam injected gas turbine, (STIG) = some steam injection, GE = General Electric, EGT = European Gas Turbines.

Fuel conversion	Dryer type	Engine	Product	Nominal capacity MW$_e$
1 Pressurized gasification	Flue gas	GE LM2500	Power	30
2 Pressurized gasification	Flue gas	GE LM2500	Cogen	30
3 Pressurized gasification	Steam	GE LM2500	Cogen	30
4 Pressurized gasification	Flue gas	GE Frame 6	Power	60
5 Atmospheric gasification	Flue gas	EGT Typhoon	Power	5
6 Atmospheric gasification	Flue gas	GE LM2500	Power	30
7 Atmospheric gasification	Flue gas	GE Frame 6	Power	60
8 Atmospheric gasification	Flue gas	GE LM2500	Cogen	30
9 Pressurized gasification	Flue gas	EGT Typhoon STIG	Power	5
10 Pressurized gasification	Flue gas	GE LM2500 STIG	Power	30
11 Pressurized gasification	Steam	GE LM2500 STIG	Cogen	30
12 Pressurized gasification	Flue gas	GE LM2500 STIG	Cogen	30
13 Pressurized gasification	Flue gas	GE Frame 6 (STIG)	Power	40
14 Atmospheric gasification	Flue gas	Combustion engine	Power	5
15 Atmospheric gasification	Flue gas	Combustion engine	Power	25
16 Pyrolysis	Flue gas	Combustion engine	Power	5
17 Pyrolysis	Flue gas	Combustion engine	Power	25
18 Pyrolysis	Flue gas	EGT Typhoon	Power	5
19 Pyrolysis	Flue gas	GE LM2500	Power	30
20 Pyrolysis	Flue gas	GE LM2500	Cogen	30
21 Pyrolysis	Flue gas	GE Frame 6	Power	60

The simplified flowsheet of one of the concepts assessed, an integrated gasification combined-cycle (IGCC) based on pressurised gasification is shown in Figure 1. Detailed descriptions of all the concepts are reported in [1].

In the IGCC, biomass is dried to about 15-25 wt% moisture content in a flue gas dryer, which uses the gas turbine exhaust gases for its energy requirements. Dried biomass is gasified at a pressure of approximately 20-25 bar with air extracted from the gas turbine compressor. Pressure is dependent on gas turbine requirements. Hot fuel gas is cooled in a waste heat boiler, where saturated steam is generated for the steam cycle. Impurities detrimental to the gas turbine (solids, alkali metals) are removed in a ceramic hot gas clean-up unit at around 500 °C. Hot fuel gas is fed to the gas turbine combustor. The flue gases are expanded through the turbine, which generates electricity through a generator. The cooled flue gases are led to the heat recovery steam generator (hrsg). Superheated steam is produced in the hrsg, and the steam is led to a steam turbine, where additional electricity is produced.

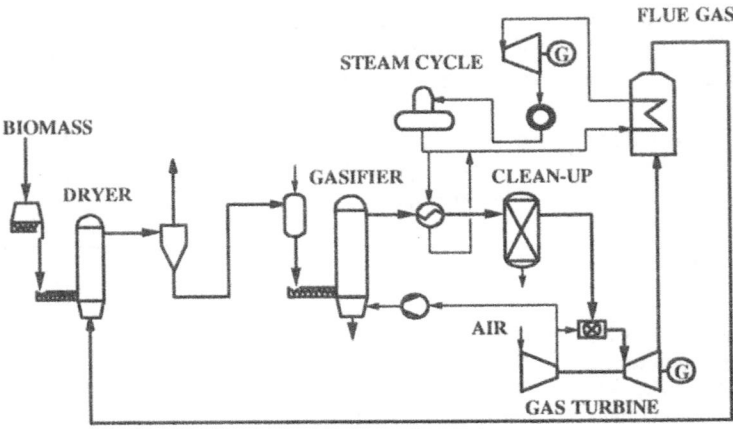

Figure 1. Simplified flowsheet of a pressurised gasification combined-cycle

3. Methods

Simulation models were used to determine performance of new concepts. Modelling tools employed were AspenPlus [2] and GateCycle [3]. The designed models are explained in detail in [1] and [4], [5], respectively. Note that Aspen is fully capable of handling solid materials (including their reactions) in dryers, gasifiers, pyrolysers, and gas cleaning units. Gasification and pyrolysis are unit operations, which are difficult to simulate satisfactorily with less qualified modelling tools. Aspen, with its state-of-the-art physical property estimation procedures, has a major advantage compared to many simulating tools developed for power plant simulation, which typically model rigorously only steam and flue gas.

The gas turbine is the key unit in an IGCC. The entire plant is designed and sized around the gas turbine, and therefore the performance of the gas turbine is critical. The performance of gas turbines employing low heating value (lhv) gases were assessed using the GateCycle simulation programme [4], [5].

The feedstock employed as a fuel was woody biomass with an analysis presented in Table 2.

Table 2. Feedstock analyses

Component	wt-% moisture free wood
C	50.4
H	5.9
N	0.5
O	41.15
S	0.05
Ash	2.0
Higher heating value MJ/kg (m.f.)	20.5
Lower heating value MJ/kg (m.f.)	19.2
Lower heating value MJ/kg (as received)	8.4

Capital costs are conventionally estimated by multiplying equipment costs by a range of factors to account for erection, piping, instruments, electrical, civil, structures, buildings, lagging and incremental materials cost [6]. This gives a direct plant cost (DPC) to which engineering design and management overheads must be added to give an installed plant cost (IPC). Finally there are costs for commissioning, contractors' fee and interest during construction which are added to give a total plant cost (TPC) [6], [7].

This method and typical figures are summarized in Table 3. The factors vary somewhat according to plant size and location, but the values indicated are representative for industrial-scale thermal biomass conversion plants within the accuracy of these cost estimates. The factors are average values derived from detailed analysis of thermal biomass conversion processes [8].

Table 3. Capital cost build-up using typical factors

Basis	Basis	EC basis
Equipment cost (EC)		1.00
Equipment cost (mild steel basis)		
Erection Piping Instruments		
Electrical Civil Structures and buildings		
Lagging		
Incremental materials cost		
Direct plant cost (DPC)		3.03 EC
Engineering design	15 % DPC	
Management overheads	10 % DPC	
Installed plant cost (IPC)		3.79 EC
Commissioning	5 % IPC	
Contractors' fee	10 % IPC	
Interest during construction	10 % IPC	
Total plant cost (TPC)		4.73 EC
Contingency	Vary, typically	10-50 % IPC

The equipment investment costs used have largely been based on the work carried out within the IEA Bioenergy project [1]. Most of the unit costs are correlations derived from a number of sources. For example, the investment cost of the gasification section is based on the data reported in [9]. The correlation employed is based on data collected and analyzed for over 18 complete plants.

The operating costs of the plants were derived based on standard industrial practise. The electricity generation costs were determined with a spreadsheet calculation model developed by Ekono Energy Ltd.

It has been reported by Merrow et al. [10] that early estimates, and even estimates made well into the detailed design of pioneer (first-of-a-kind) plant costs, have been poor predictors of actual costs. Therefore in trying to reduce the uncertainty related to the assessment of new technologies, less emphasis should be placed on the importance of cost estimates, and more emphasis placed on the role of the simulation models. The models are employed in determining mass and energy balances and the process performance. In this study, the models and their accuracy are emphasised along with the performance of the different process concepts. Additional emphasis in the study was made to identify the technical uncertainties existing in the concepts proposed.

4. Results

4.1 Performance

A summary of efficiencies for the studied concepts is shown in Figure 2. Typical efficiencies for conventional steam boiler plants are shown for reference. Note that the values given are design efficiencies. All cases shown are for power production only. Most of the new systems have a higher efficiency than the steam cycles. The highest power production efficiency is achieved with the pressurised IGCC, 45-47 % based on lower heating value (LHV) of wood fuel, depending on the gas turbine. The respective higher heating value (HHV) efficiency is 37-38 %. Gasification combined cycles based on atmospheric pressure gasification range between 37 and 41 %. Concepts where steam injection is applied have lower efficiencies (29-35 %). Combustion engine concepts have an efficiency of 34 %.

The systems based on pyrolysis yield lower efficiencies than the systems with gasification. This is seen in comparing pyrolysis diesel and combined cycle efficiencies to the efficiencies of systems based on gasification. Pyrolysis concepts have LHV-based efficiencies between 30 and 35 %, while systems based on gasification vary between 30 and 45 %.

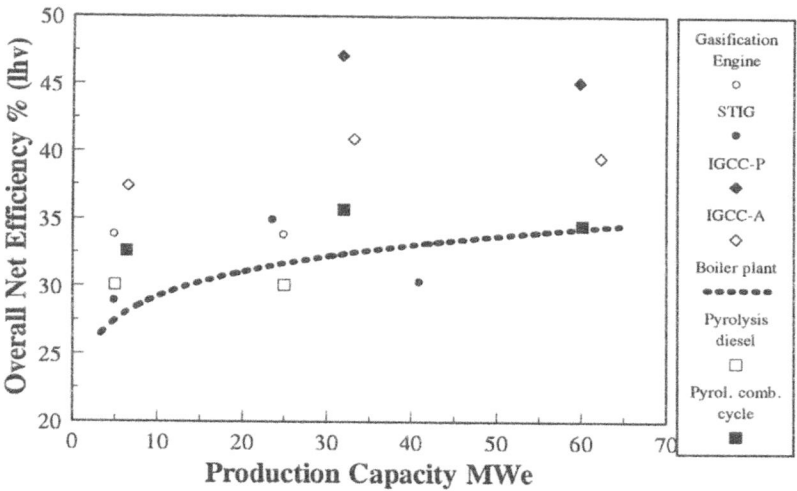

Figure 2. Summary of system efficiencies, power production only

Co-generation of heat and power increases fuel utilization in a power plant considerably. Typically, in a large condensing power plant about 40 % of the fuel energy is converted to electricity. In a co-generation plant, the overall efficiency including electricity and heat is typically 80-90 %. The systems proposed have a higher power-to-heat ratio (PR) than conventional technology. A co-generation IGCC in district heat production has a PR of about 1.0, whereas these plants today have a PR of about 0.5. This may be a considerable advantage, if co-generation is considered favourable in certain locations, as more power may be produced in co-generation service at existing heat loads.

4.2 Capital costs

It is understood that the capital and production costs presented in all figures below should be presented more correctly as wide bands rather than thin lines. However, such a presentation would lead to figures where information would be difficult to perceive.

A summary of specific capital costs for the concepts is shown in Figure 3. An attempt was made to take the "learning effect" into account. The costs are meant to represent mature technology, i.e. specific capital costs of each technology after several commercial plants have been built. New technologies appear to compete with conventional technology in larger scale, approximately above 20 MWe. The higher investments of the new technologies will be off set by their higher efficiencies. It is estimated that the specific capital investment costs of new technologies will be about

US$ 1500 / kW in capacities above 30 MWₑ. This is about the same as presently for boiler power plants.

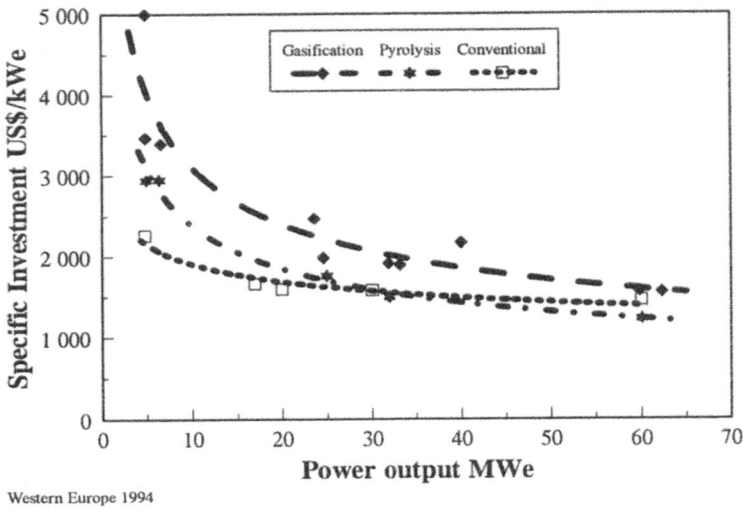

Figure 3. Summary of specific capital costs (nth plant), illustrative correlations shown for systems based on gasification, pyrolysis and boilers.

4.3 Production costs

Some of the parameters employed in the base economic assessment are listed in Table 4. The capital cost share in the production cost was determined with the annuity method. The base cost of woody biomass was set at US$ 25 / wet tonne.

A summary of electricity production costs for all cases is shown in Figure 4. Some tendencies may be pointed out:
- A typical effect of scale may be noticed. At small capacity, the power production cost is about double to that at higher capacities.
- Differences between alternative technologies are not major with the exception of the STIG concepts.
- Co-generation of heat clearly improves the economics. At around 30 MWₑ the improvement is about 20 % with the employed parameters.
- The steam injected gas turbine (STIG) systems do not appear competitive, when lhv fuel gas is employed in the gas turbine. This is mainly due to the smaller steam injection permissible than with natural gas due to surge limits.

Table 4. Bases for economic calculations. TCI = total capital investment

Location	Western-Europe
Time	1994
Rate of interest %	5
Annuity	0.08
Service life years	20
Wood cost US$/MWh	10.6
By-product heat value US$/MWh	20
Service & maintenance % of TCI	1.6
Labour US$/a per person	33 000
Peak operating time h/a	5000

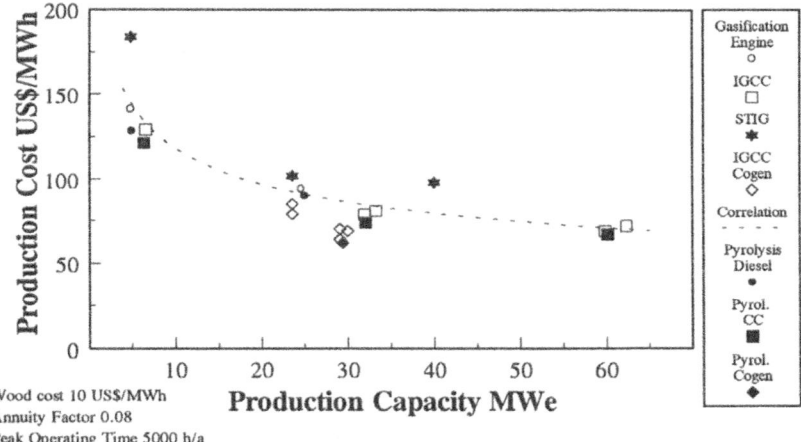

Figure 4. Summary of electricity production costs, an illustrative correlation shown for power production only cases

The cost of biomass was varied in Figure 5 for all the alternatives ± 30 % from the base value of US$ 25 / wet tonne (equals US$ 10.6 / MWh or US$ 2.9 / GJ). Decreasing biomass cost from US$ 14 / MWh to half will reduce power production cost by about 20 %.

Figure 6 shows an attempt to illustrate the "learning effect" on the electricity cost. An assumed reduction of capital cost by 30 %, which is assumed for all concepts, also reduces production cost about 20 %.

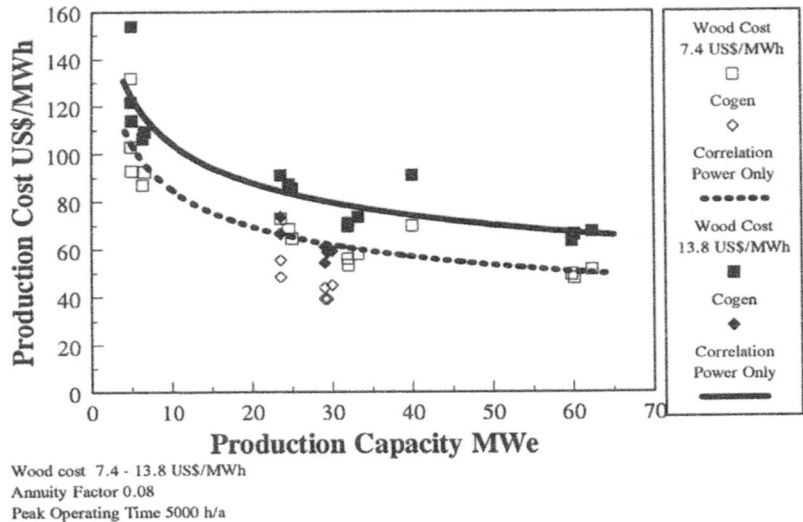

Wood cost 7.4 - 13.8 US$/MWh
Annuity Factor 0.08
Peak Operating Time 5000 h/a

Figure 5. Electricity production costs with different biomass costs, an illustrative correlation shown for power production only cases

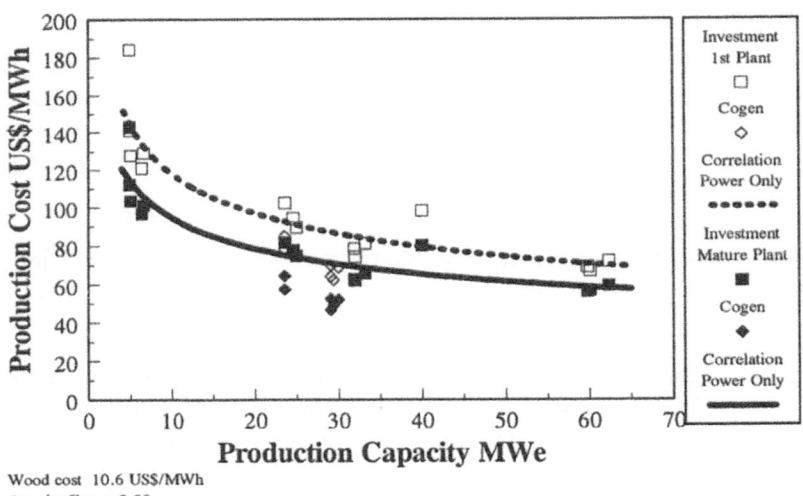

Wood cost 10.6 US$/MWh
Annuity Factor 0.08
Peak Operating Time 5000 h/a

Figure 6. Electricity production costs, effect of reducing capital cost through a learning effect. Illustrative correlations shown for power production only cases.

An attempt was made to compare an atmospheric IGCC to a pressurised IGCC. Although the latter has a clearly higher efficiency, the power production cost of the pressurised system is only marginally lower [1].

The advantage of an IGCC in co-generation is demonstrated in Figure 7, where the IGCC is compared with a conventional boiler plant in district heat production. The annual operating savings compared to the electricity purchase tariff are shown for both alternatives. Both cases have the same heat load, but because of the higher PR of the IGCC, the annual savings of the IGCC at a typical annual operating time of 5000 hours, are about 1.5 million US$. At that point the conventional system will break even.

A fairly small scale co-generation IGCC (30 MWe) is compared to a large-scale condensing coal power plant in Figure 8. Two prices for coal are shown, the present market price of about US$ 10 / MWh and an increased price of US$ 14 / MWh. It may be seen that at present coal price the competitiveness is only reached at fairly high yearly operating terms. However, with a 40 % increase in the coal cost, co-generation biomass IGCC becomes competitive already at above 4000 h/a, which is a realistic peak operating time for a co-generation plant.

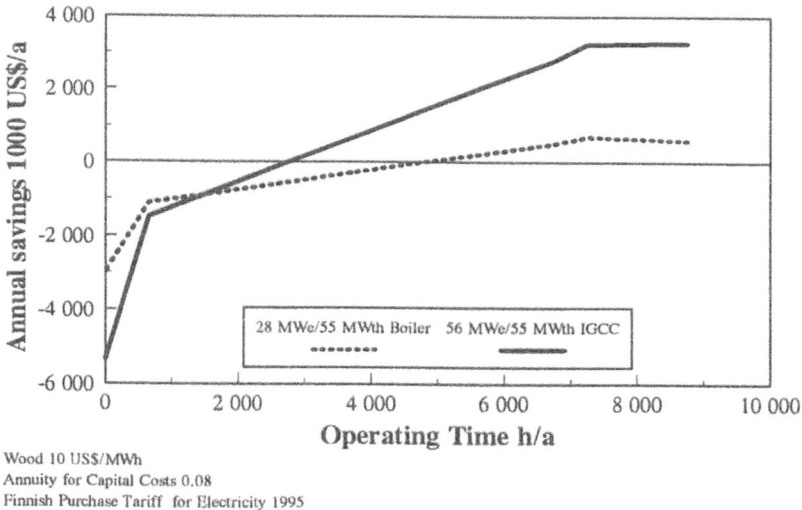

Wood 10 US$/MWh
Annuity for Capital Costs 0.08
Finnish Purchase Tariff for Electricity 1995

Figure 7. Annual savings against the Finnish electricity tariff for a co-generation IGCC and boiler power plants, 55 MJ/s of district heat.

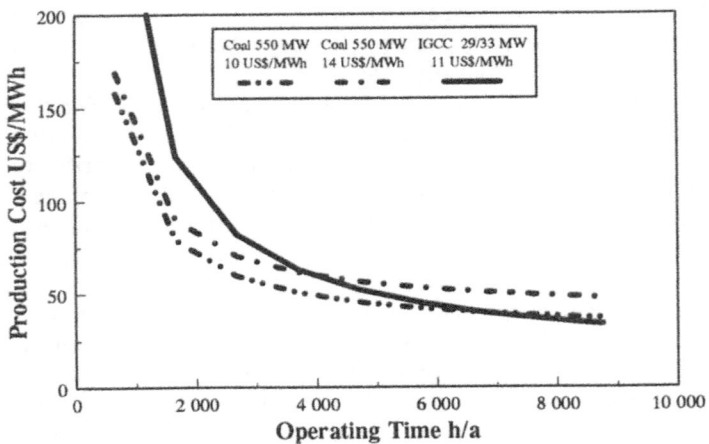

Figure 8. Competitiveness of a wood-fueled co-generation IGCC unit with a condensing coal power plant, base load operation. IGCC 29 MW$_e$ electricity and 33 MJ/s district heat.

In Figure 9, co-generation at a pyrolysis oil diesel and a circulating-fluidized bed (CFB) boiler power plant are compared. The scale is chosen such that the capacity of district heat production is about the same in each case (15 MJ/s for CFB boiler and 14 MJ/s for diesel). Fuel costs have been varied ± 25 % from the base values employed, which are US$ 10.6 / MWh and US$ 26.1 / MWh for wood and pyrolysis oil, respectively. The co-generation diesel is more economic than the CFB boiler in less than about 2000 h/a. Between approximately 2000 and 3000 h/a, depending on the oil cost (less on wood cost), the two concepts compete rather closely. Above 4000 h/a, the CFB boiler yields a lower power production cost. The Finnish electricity purchase tariff is also shown as a reference. The conventional boiler concept begins to compete with the tariff at very high yearly operating hours.

A similar presentation for power production only is shown in Figure 10. In this case the diesel concept is more competitive compared to a conventional technology than in the co-generation case. Diesel has lower electricity production costs below 3000 h/a of operation. Depending on the fuel cost the two systems are fairly close between 3000 and 6000 h/a of operation. The power produced in the CFB boiler power plants is at a lower-cost than that produced in a pyrolysis diesel plant for the very high annual operating hours only.

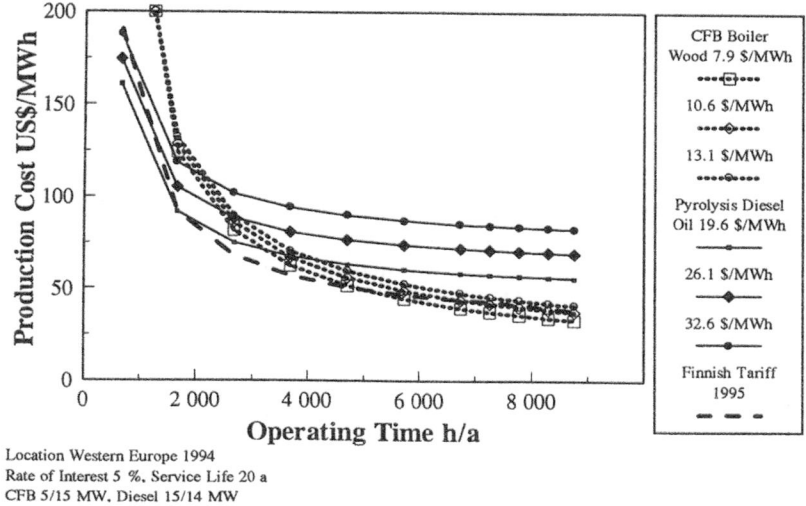

Location Western Europe 1994
Rate of Interest 5 %, Service Life 20 a
CFB 5/15 MW, Diesel 15/14 MW

Figure 9. Comparison of a pyrolysis diesel and a CFB boiler co-generation plants, power production cost as a function of yearly operating costs. The Finnish electricity purchase tariff is shown as a reference.

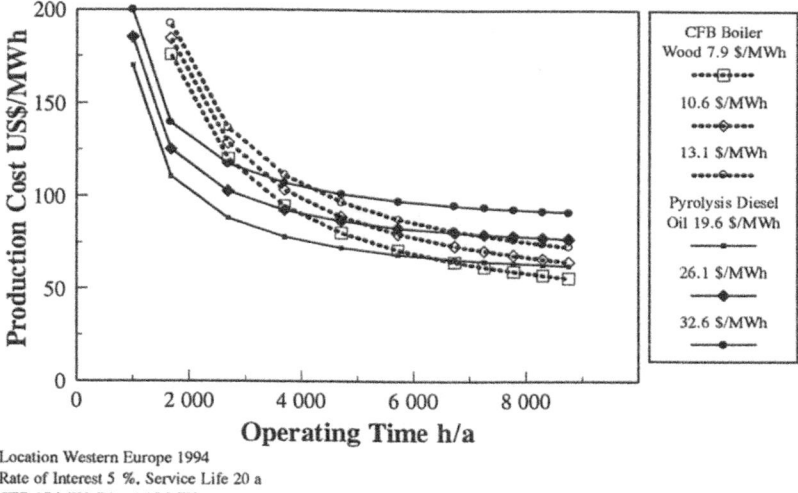

Location Western Europe 1994
Rate of Interest 5 %, Service Life 20 a
CFB 17 MW, Diesel 15 MW

Figure 10. Comparison of a pyrolysis diesel and a CFB boiler plant in power production, the production cost as a function of yearly operating costs.

5. Discussion

The efficiencies of most of the concepts analysed are higher than those of conventional steam power plants. The highest efficiency determined is for a pressurised IGCC based on an aero-derivative (LM2500) engine at 32 MW$_e$, 47 % based on lower heating value of the fuel. The value may be compared with 45 % reported by Salo and Keränen [11] for an IGCC employing an industrial gas turbine in the same size range (39 MW$_e$). Consonni and Larson [12] report an efficiency of 45 % for a system designed around LM2500. An overall efficiency of 45 % was determined in this work for an industrial gas turbine at 60 MW$_e$. The value may be compared with a reported value of 47 % at 69 MW$_e$ [11]. The efficiencies of systems based on atmospheric gasification were determined in this work to be about 5-6 % points lower than those of systems based on pressurised gasification. Consonni and Larson [12] found only a three percentage point difference for these two systems. A conventional state-of-the-art steam power plant in the same capacity range typically has an efficiency of 30 to 35 % based on LHV of the fuel.

Both pressurised and atmospheric pressure IGCCs are being developed [13], [14]. Both approaches include technical uncertainties. However, as the economic analysis showed no major differences between the two, the higher efficiency of the pressurised system may be regarded as a reason to emphasise the pressurised system.

It has been suggested [11] that an IGCC is competitive compared to a conventional steam cycle above 30-40 MW$_e$. The results obtained in this study confirm this [1]. As suggested by among others Salo and Keränen [11], a forest products industry site appears as one of the most feasible sites for an IGCC.

The STIG cycles employing lhv fuel gas do not appear competitive. To be able to fully utilise the increase in power output following from steam injection, the present engines would have to be modified. The increased mass flow resulting from employing a lhv fuel gas in a gas turbine (instead of natural gas) decreases the surge margin. Increasing the mass flow in the turbine with steam injection decreases the surge margin even further. This is especially the case in constant-speed industrial units. No data is available on the use of lhv fuel gas in a gas turbine operated in STIG mode.

Concepts employing internal combustion engines have moderate efficiencies. However, the efficiencies may be improved by integrating a steam cycle to the engine concept [15]. It has been shown that overall efficiencies similar to the IGCC systems may be reached with engine - steam cycle systems employing atmospheric gasification [16]. A similar steam cycle with the pyrolysis - diesel engine concepts would increase their overall efficiency from about 30 to about 36 %. However, it should be borne in mind that to be economic, the capacity has to be higher than the smaller capacity end studied for these concepts (5 MW$_e$).

Technical uncertainties are also related the the use of biomass derived fuels in combustion engines. Lhv fuel gas has been used in engines by several groups. However, at the scale relevant for this work (above 5 MW$_e$), the work carried out by former Studsvik Energiteknik (now TPS) by Waldheim et al. [17] and the work carried out by Wärtsilä Diesel International Ltd. and A. Ahlstrom Corporation [18] are among the most important. The gas clean-up stage has proven to be the critical unit operation [19].

Use of pyrolysis oil in diesel engines is a more recent development. Initially tested in small high speed engines [20], [21], the focus is in developing a medium speed pilot injected engine [22]. Other development efforts and tests are also reported [23], [24]. Durability of the injection equipment is seen as the most critical operation with the concept. Technical uncertainties related to the concept are considerable. A diesel power plant appears quite interesting when it is compared to a conventional CFB boiler, especially in power production only.

Pyrolysis combined-cycles appear fairly interesting both due to their relative high efficiencies and estimated low costs. There is very little data on using pyrolysis oil in gas turbine combustors or in actual engines [25]. At least one active development work aimed at a commercial gas turbine is in progress [26]. However, considerable technical uncertainty exists.

6. Conclusions

Site-specific feasibility studies are needed to demonstrate the competitiveness of new systems. General comparative studies like the present work cannot accurately enough consider distinct features of each site and technology. This underlines the need to carry out the performance analysis accurately in general system studies, as the performance analysis is more generally valid than the economic analysis.

New energy technologies often have initially high investments and technical uncertainties. Therefore demonstration may easily take place only with small-scale systems, thus effectively prohibiting larger scale system commercialization. The development, for example of IGCC systems will be slow for this reason. However, the IGCC system has one of the best development potentials of the systems analysed.

7. References

1. Solantausta, Y., Bridgwater, A., Beckman, D. (1996). Electricity production by advanced biomass power systems. Espoo: Technical Research Centre of Finland. 120 p. + app. 61 p. (VTT Research Notes 1729).
2. ASPEN PLUS™ User Guide Volume 1 (1994). Release 9. Aspen Technology, Massachusetts, USA.
3. Erbes, M., et al. (1989) GATE: A simulation code for analysis of gas turbine power plants. IGTI/ASME Gas Turbine Congress, Toronto, Ontario.
4. Palmer, C. & Erbes, M. 1992. LM2500 Fired on Low Btu Fuels. Final Report for IEA Pyrolysis project. Enter Software, Inc. January 28.
5. Palmer, C. & Erbes, M. 1993. Performance predictions of the Frame 6 fired on low btu fuels. Final Report for IEA Pyrolysis project. Enter Software, Inc. December 14.

6. Anon. 1988. Revised Guide to Capital Cost Estimation. Institution of Chemical Engineers, UK.
7. Garrett, D.E. 1989. Chemical Process Economics. Van Nostrand Reinhold
8. Solantausta, Y., Diebold, J.P., Elliott, D.C., Bridgwater, T. & Beckman, D. (1994). Assessment of liquefaction and pyrolysis systems. Espoo: Technical Research Centre of Finland. 230 p. + app. 79 p. (VTT Research Notes 1573).
9. Bridgwater, A.V. & Evans, G.D. 1993. An assessment of thermochemical conversion systems for processing biomass and refuse. Report to UK DTI. (ETSU B/T1/00207/REP) 254 pp., 1993.
10. Merrow, E., Chapel, S., Worthing, C., A Review of Cost Estimation in New Technologies: implications for Energy Process Plants. Prepared by RAND Corporation for the U.S. Department of Energy. Santa Monica, CA. RAND-R-2481-DOE. July 1979.
11. Salo, K., Keränen, H., Biomass IGCC. Proceedings of the 2nd seminar Power Production from Biomass, March 22-23, 1995, Espoo, Finland. VTT Energy, 1996.
12. Consonni, S., Larson, E., Biomass-gasifier/aeroderivate gas turbine combined cycles, performance calculations and economic assessments. In Proceedings, ASME Cogen Turbo Power 94. American Society of Mech. Eng., NY, 1994, pp. 611-623.
13. Lundqvist, R., The IGCC demonstration plant at Värnamo. Bioresource Technology 46 (1993) 49-53.
14. Maniatis, K., Ferrero, G.L. 1994. The THERMIE Programme: The Energy from Biomass and Waste Sector and Targeted Projects on CHP Production by Biomass Gasification, In: Proc. European Workshop on Market Penetration of Biomass Technologies at Vienna, Austria, October 5, 1994. Linz, Austria, Energiesparverband.
15. Merlo, L., Sadowski, S., Teislev, B., Bottai, G., Giarruso, L., Combined MSW-biomass energy transformation plant for various European conditions. EU DGXII APAS/RENA contractors meeting, November 22-25, 1995, Venice.
16. Solantausta, Y., Ståhlberg, P., Kurkela, E., The competitiveness of the CFB-gasifier gas-engine power plant concept. Intermediate report, VTT Energy. Espoo, February 1996.
17. Waldheim, L., Blackadder, W., Gasification of Wood Fuel and Electricity Production with a Diesel Engine. Studsvik Report EP-90/16. Studsvik 3.12.1990. (in Swedish)
18. Ekono Oy: Competitiveness of gas diesel power plants. Ministry of Trade and Industry, Publications D:52. Helsinki 1984. (in Finnish)
19. Ekström, C., Atmospheric gasification - diesel. Technical evaluation and proposal on future work. Vattenfall June 1991. Vällingby, Sweden. (in Swedish)
20. Solantausta, Y., Nylund, N-O., Westerholm, M., Koljonen, T., Oasmaa, A., Wood pyrolysis oil as fuel in a diesel power plant. Bioresource Technology 46(1993)177-188.
21. Solantausta, Y., Nylund, N-O., Oasmaa, A., Westerholm, M., Sipilä, K., Preliminary tests of wood derived pyrolysis oil as fuel in a stationary diesel engine.

Proceedings Biomass Pyrolysis Oil Properties and Combustion Meeting p. 355, September 26-28, 1994 Estes Park, Colorado. NREL, Golden CO, 1995.)

22. Gros, S., Pyrolysis oil as diesel engine fuel. Proceedings of the 2nd seminar Power Production from Biomass, March 22-23, 1995, Espoo, Finland. VTT Energy, 1996.

23. Casanova, J., Comparative study of various physical and chemical aspects of pyrolysis bio-oils versus conventional fuels, regarding their use in engines. Proceedings Biomass Pyrolysis Oil Properties and Combustion Meeting p. 343, September 26-28, 1994 Estes Park, Colorado. NREL, Golden CO, 1995.

24. Leech, J., Bridgwater, A.V., Development of an internal combustion engine for use with crude bio-oil and evaluation of associated processes. EU DGXII APAS/RENA contractors meeting, November 22-25, 1995, Venice.

25. Grassi, G., Palmarocchi, M., Koeler, J., Trebbi, G., Advanced liquid fuel production from biomass for power generation. Proceedings Second Biomass Conference of the Americas p.1048, August 21-24, 1995 Portland, Oregon. NREL, Golden CO, 1995.

26. Ardy, P., Barbucci, P., Benelli, G., Rossi, C., Zandorlin, S., Development of gas turbine combustor fed with bio-fuel oil. Proceedings Second Biomass Conference of the Americas p. 429, August 21-24, 1995 Portland, Oregon. NREL, Golden CO, 1995.

27. Andrews, R., Patnaik, P., Liu, Q., Thamburaj, R., Firing fast pyrolysis oil in turbines. Proceedings Biomass Pyrolysis Oil Properties and Combustion Meeting p. 383, September 26-28, 1994 Estes Park, Colorado. NREL, Golden CO, 1995.

OPPORTUNITIES FOR FAST PYROLYSIS IN SMALL-SCALE ELECTRICITY GENERATION

Bioenergy systems modelling

A.J. TOFT and A.V. BRIDGWATER
Energy Research Group, Aston University, Birmingham, UK

Abstract

Various systems have been suggested for the generation of electricity from biomass that aim to improve on the low efficiency, capital intensive steam turbine cycles that are currently used. While the integrated gasification combined cycle (IGCC) system using gas turbines has been favoured at large scale (above 30 MW$_e$), diesel engine systems are more suitable at smaller scales because their investment costs and efficiencies are less sensitive to scale. Diesel engines could be driven by liquids produced in fast pyrolysis, a technology that is rapidly gaining acceptance as a means of producing high yields of a liquid fuel. Models have been developed that calculate the costs of producing electricity from biomass via fast pyrolysis and diesel engines. It is expected that such systems would compete with atmospheric gasification and diesel engine systems and these systems have also been evaluated for comparison. Initial analyses have shown that electricity production costs are very similar for the two systems where the conversion (fast pyrolysis or gasification) stage is directly connected to the generating stage. However, fast pyrolysis may be de-coupled from the generator and this could give pyrolysis an advantage over gasification. Three de-coupling options are assessed in this paper to evaluate how de-coupling could be used to improve the economics of fast pyrolysis and hence increase the opportunities for its application in small-scale electricity generation.

Keywords: Fast pyrolysis, diesel engine, electricity generation, economics, system studies

1 Introduction

The sustained use of biomass as an energy resource could have many environmental and social benefits. Biomass use could reduce emissions of carbon dioxide and sulphur dioxide while improving land quality. Socially, biomass production could stimulate rural economies as land no longer used in food production is returned to active service.

None of these benefits will be realised unless biomass can penetrate existing energy markets. The most favourable market for biomass is electricity generation because of the high energy value of electricity; alternative markets such as boiler fuels, transportation fuels and fuel gas are rarely competitive because of the low cost of the fossil fuel alternatives [1].

In the US around 8 GW$_e$ of electricity capacity already uses biomass in conventional steam cycles. Feed availability and feed transport costs tend to constrain plant sizes to typically 25 MW$_e$ and at these capacities generating efficiencies are seldom above 20%* [2]. The high capital cost of small-scale steam cycles and the high feed costs imposed by the low efficiency result in limited opportunities for economic electricity generation.

More efficient generation could be achieved using gas turbines or engines but this requires conversion of the solid biomass to a suitable liquid or gaseous fuel intermediate. Two conversion technologies are widely promoted to achieve this: gasification and fast pyrolysis. The former can produce a low heating value gas by partial oxidation of the feedstock in limited air; the latter produces a liquid fuel by rapidly heating the feedstock in the absence of an oxidant and condensing the organic vapours that arise. Gasification technologies are reviewed by Bridgwater [3], Overend [4] and van Swaaij [5]; fast pyrolysis technologies are reviewed by Bridgwater [6,7].

2 Previous systems analyses

Previous work [8] has evaluated various biomass to electricity systems using the Bioenergy Assessment Model (BEAM) that compared a conventional generating technology with three novel conversion routes:

1. Fluid bed combustion and steam cycle (the current technology);
2. Atmospheric gasification and dual fuel diesel engine;
3. Pressurised gasification and gas turbine combined cycle (IGCC); and
4. Fast pyrolysis and pilot-injected diesel engine.

This work demonstrated the economic advantage of the IGCC system at large capacities due to its better efficiency and decreasing specific capital cost with scale. Engine-based systems were more appropriate at less than 30MW$_e$ because the engines offer relatively high efficiencies at low capacities and the capital costs are less sensitive to scale. Small-scale systems are of particular current interest due to limited biomass

---

* *All energy efficiencies are quoted on the lower heating value of the fuel input.*

1557

availability, high biomass transport costs, and their lower financial risk. It was found that the two engine-based systems produced similar costs and would therefore be competitors in the small-scale electricity generation market.

One of the key features of a fast pyrolysis system is the ability of the conversion technology to produce a liquid fuel that is easy to handle and that could be economically stored or transported. This allows fast pyrolysis to be de-coupled from the engine, so that the conversion and generation stages can operate independently. This option is not available in the atmospheric gasification case: the system must be close-coupled because it is not economic to store or transport the fuel gas produced by the conversion technology. Thus there are potential opportunities to reduce electricity production costs if the system circumstances favour de-coupling. This work examines various small-scale (less than 12 MW_e) electricity generating systems using both fast pyrolysis and gasification conversion technologies and evaluates the opportunities that fast pyrolysis could take advantage of to make it the more attractive technology.

3 Current system models

3.1 Modules
The systems have been modelled by breaking them into stages and modelling each stage as a module. The modules used for each system are listed in Table 1.

3.2 Technology performance parameters
The most essential performance parameters in the system are the efficiencies of the conversion and generation stages. The results will show that a significant portion of the production costs are attributed to purchasing and delivering feed: any change in system efficiency will effect the cost of feedstock.

The fast pyrolysis system operates at a conversion efficiency (the ratio of energy in the feedstock to energy in the liquid product) of 65.2%. This assumes that the total liquids yield is 83% (on dry feed) and that the heating value of the liquids produced is 14.8 GJ/wet tonne. The feed entering the reactor has a lower heating value of 19.2 GJ/odt (oven dry tonne) and a moisture content of 10% (wet basis). No change in conversion efficiency with reactor size has been assumed.

The gasification system operates at a conversion efficiency (the ratio of energy in the feedstock to energy in the cool, clean gas at engine delivery) of 70.5-75.5%. Energy efficiency is assumed to rise with reactor size due to reductions in heat losses.

It is assumed that the engine operates at the same efficiency whether in dual fuel or pilot-injected mode. This efficiency (the ratio of total energy into the engine to electrical energy at the generator terminals) ranges between 35.9% and 38.9%.

3.3 Calculation of costs
All costs are in $US, 1995 basis. Investment costs are total plant costs that have been defined using the same procedures adopted in previous work by Solantausta et al. [9]. Applying total plant costs ensures that all the costs associated with constructing a fully operational plant are considered but it can also produce investment costs that are higher than other published data. All costs are first plant costs and it is expected that these will reduce as experience grows with replication and the technologies are improved.

Table 1 Modules used to model the biomass to electricity systems

Cost Centre	Fast Pyrolysis Modules	Gasification Modules
Feed production	Wood production	Wood production
Feed transport	Wood transport	Wood transport
Pretreatment	Pretreatment	Pretreatment
Conversion	Fast pyrolysis Pyrolysis liquids storage	Gasification
Generation	Pyrolysis liquids transport Fuel storage Pilot-Injected Diesel Engine Grid connection	Dual Fuel Diesel Engine Grid connection

Production costs include capital amortisation; labour, utilities; maintenance and site overheads.

4 Results

Four fast pyrolysis options are evaluated and the basic system concepts are illustrated in Fig. 1. Electricity production costs have been calculated for each system at up to 12 MW_e and the results are presented. Each analysis is compared with an appropriate system that uses atmospheric gasification since previous studies have shown that this technology will compete with fast pyrolysis in the small scale generating market [8].

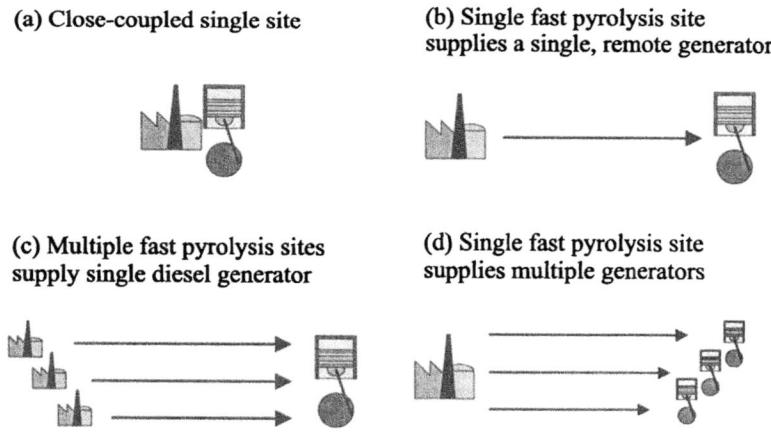

(a) Close-coupled single site

(b) Single fast pyrolysis site supplies a single, remote generator

(c) Multiple fast pyrolysis sites supply single diesel generator

(d) Single fast pyrolysis site supplies multiple generators

Fig. 1 Examples of system configurations using de-coupling

4.1 Close-coupled systems

Fast pyrolysis and gasification systems can be directly compared in systems where the conversion and generation stages operate at the same time and at the same site (Fig. 1(a)). Electricity production costs for a range of capacities are shown in Fig. 2 and results for 4 and 12 MW$_e$ close-coupled systems are summarised in Table 2. The results show that fast pyrolysis is not economic unless low cost feedstocks can be found in sufficient quantities to support a substantial plant (given that electricity generation from conventional plant can be achieved at less than 5 c/kWh). There may however be opportunities for fast pyrolysis where electricity is required in remote communities and the costs of conventional fuels are high.

It can also be seen that the electricity production costs for the fast pyrolysis system and the atmospheric gasification system are very similar, such that the opportunities for fast pyrolysis could be limited by competition from the alternative technology.

Table 2 also illustrates two other issues: the importance of feed costs and hence the need to maintain a high efficiency system and the effect of scale economies. Both factors will be crucial to the de-coupled systems discussed later. At the 4 MW$_e$ scale feed costs (production and transport) account for 25-26% of the total production costs. Fast pyrolysis systems may be able to use de-coupling to reduce feed transport costs and hence the substantial cost of delivered feed. It can also be seen that the specific capital costs of all the systems are reduced by increasing the system capacity, with relatively higher scale economies at the pretreatment and conversion stage. Reduced capital costs in larger plant also increase the proportion of total production costs spent on feed; in the 12 MW$_e$ system feed costs reach 31-36% of the total.

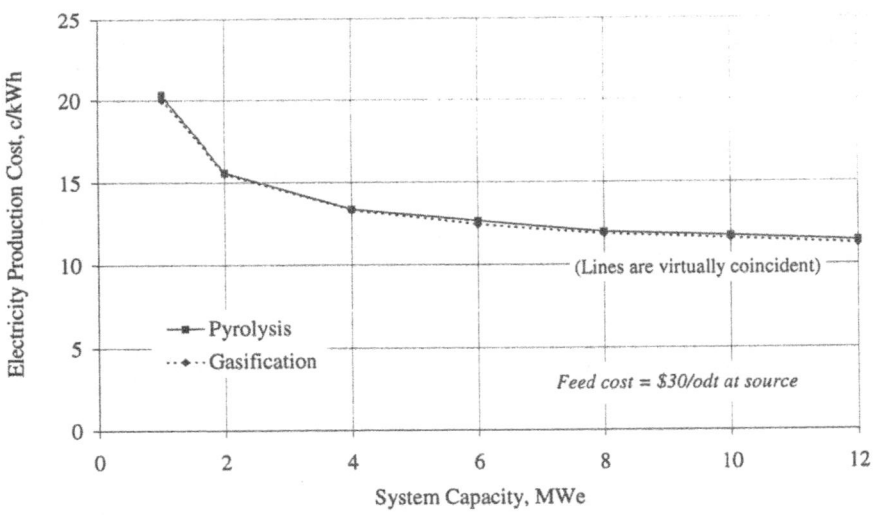

Fig. 2 Electricity Production Costs for Close-Coupled Systems

Table 2 Results for close-coupled systems

System capacity, MWe	4	4	12	12
Conversion route	Pyrolysis and engine	Gasification and engine	Pyrolysis and engine	Gasification and engine
Feed produced, odt/y	29345	26550	88036	77278
Feed production cost, $/odt	30	30	30	30
Moisture content as produced, %	55%	55%	55%	55%
Feed heating value, GJ/odt	19.2	19.2	19.2	19.2
Efficiency, %LHV				
Conversion efficiency	65.2%	73.2%	65.2%	75.5%
Generation efficiency (gross)	38.4%	38.4%	38.4%	38.4%
Overall efficiency*	22.8%	25.1%	22.8%	25.8%
Total plant costs, $/kW$_e$				
Pretreatment	597	436	455	354
Conversion	1868	2289	1217	1609
Generation	1375	1375	1177	1177
Total	3840	4100	2850	3140
Operating costs, $k/y				
Production	880	796	2641	2318
Supply	254	220	1253	1036
Pretreatment	547	333	1234	678
Conversion	1249	1491	2329	2998
Generation	1285	1365	3328	3565
Total	4216	4205	10785	10595
Electricity production cost, c/kWh	13.37	13.34	11.40	11.20

4.2 Remote biomass resource and grid connection

This options examines the case where a biomass resource is available at a location remote from the electricity user or a suitable grid connection. Biomass residues are sometimes cited as examples of low cost material that could be used in biomass to electricity schemes but their high collection and transport costs soon turn a low cost commodity into an expensive one. Fast pyrolysis systems could reduce biomass transport costs by locating the fast pyrolysis plant close to the biomass source and the generator at the grid connection (Fig. 1(b)). The costs of transporting the higher energy density pyrolysis liquid are much less than the costs of transporting biomass.

The alternative gasification system would either incur the cost of biomass transport or extra grid connection costs. In the current analysis it has been assumed that the biomass is transported from source to a close-coupled gasification and generating site located at the grid connection.

---

*   *This is the ratio of net electricity out to the lower heating value of the total feed required*

Fig. 3 shows how electricity production costs vary as the distance between a source of low-cost biomass and a suitable grid connection increase. System capacity is 12 MW$_e$. It is assumed that there is a modest charge for the biomass in all systems of 30 \$/odt. In all pyrolysis cases the biomass residues are transported 5 km to the fast pyrolysis site for processing. The biomass transport costs are 1.26 \$/odt for loading and unloading and 0.59 \$/odt/km for transportation. Pyrolysis liquid loading and unloading costs were 0.1 \$/t and transport costs were 0.3 \$/t. It can be seen that the margin between pyrolysis and gasification system costs increases with distance between biomass source and grid connection. Opportunities for fast pyrolysis are much better at the larger distances. It should be noted that feed transport distances for the 12 MW$_e$ system quoted in Table 2 are 25 km, increasing the delivered feed costs to a level where electricity production costs are almost identical in both systems.

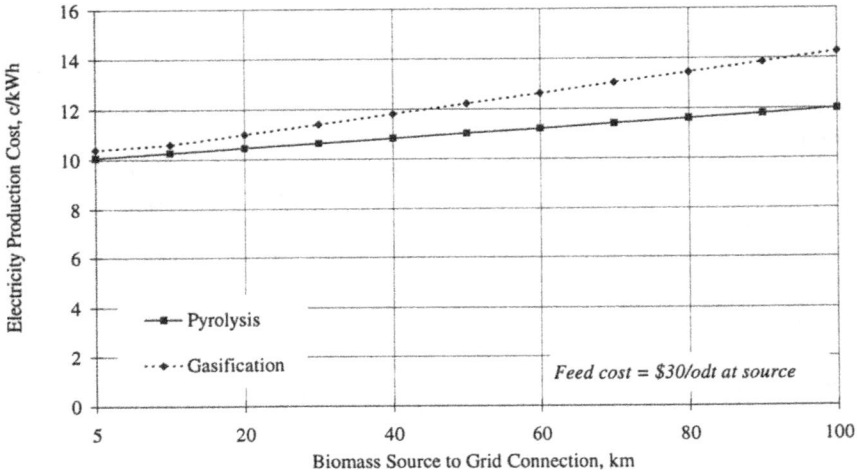

Fig. 3   Electricity production costs for remote biomass resources and grid connections

4.3 Multiple fast pyrolysis sites

It has been suggested that another way of reducing biomass transport costs in a system where the feed is distributed over a large area would be to construct a number of small pyrolysis plants which could then feed a larger, central generator. The disadvantage of such a system is that the total investment costs for several small pyrolysis units would be higher than the cost of one large unit. A brief examination of Table 2 has already shown the significance of scale economies as system capacity changes from 4 MW$_e$ to 12 MW$_e$.

This option has been modelled conceptually as shown in Fig. 4(a). The model assumes that the pyrolysis plants are situated at the point in each sector that minimises the average feed transport distance. A system has been analysed that produces 12 MW$_e$ at the centre of the feed supply area. The electricity demand could be met by using close-coupled systems or by installing three fast pyrolysis plants that each produce 1/3 of the pyrolysis liquid required by the central generator. The results of this analysis are shown in Fig. 5 and system capital costs are presented in Table 3.

It can be seen that the high capital costs incurred at the multiple fast pyrolysis plants (each comprising pretreatment and fast pyrolysis) far outweigh any savings due to lower feed transport costs. Thus this option is not viable at this scale. De-coupling in this way could still be considered for larger scale generators where the transport costs are more significant and the scale effects on investment costs of changing the fast pyrolysis unit capacity are less dramatic.

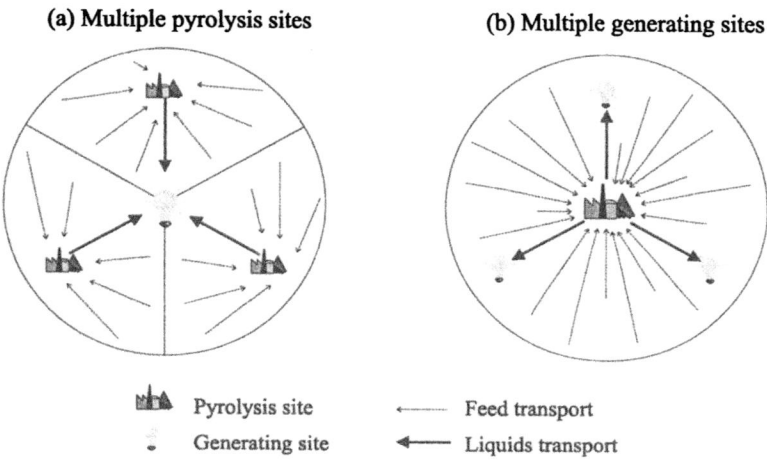

(a) Multiple pyrolysis sites   **(b) Multiple generating sites**

🏭 Pyrolysis site   ←‑‑‑‑‑‑ Feed transport

• Generating site   ◀——— Liquids transport

Fig. 4 Modelling de-coupling using multiple sites

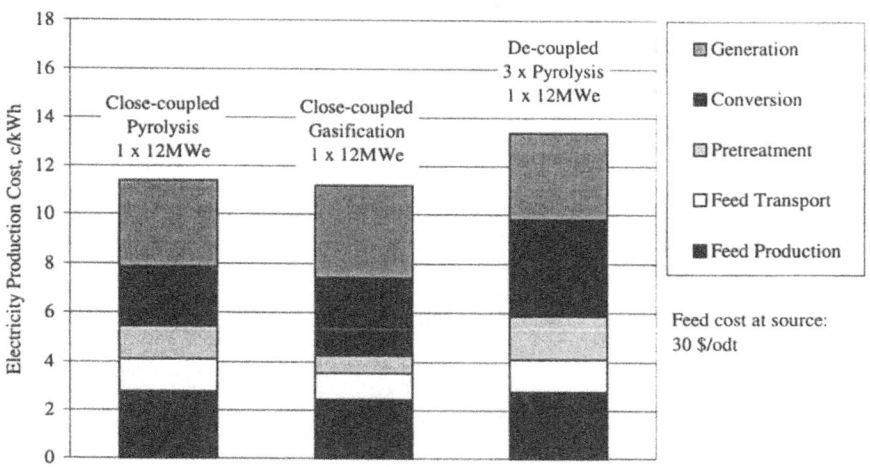

Fig. 5 Electricity production costs at 12 MW$_e$ using single and multiple fast pyrolysis sites compared to a single gasification site.

Table 3 Specific investment costs for 12 MW$_e$ systems using single and multiple fast pyrolysis sites compared to a single gasification site.

Conversion Process	Pyrolysis		Gasification		Pyrolysis	
No. conversion plants	1		1		3	
No. generating sites	1		1		1	
Total plant costs						
Pretreatment, $/kW$_e$ (%)	455	(16.6)	354	(11.3)	597	(16.1)
Conversion, $/kW$_e$ (%)	1217	(42.7)	1609	(51.2)	1868	(50.2)
Generation, $/kW$_e$ (%)	1179	(41.4)	1177	(37.5)	1252	(33.7)
Total, $/kW$_e$	2852		3140		3718	

4.4 Multiple generating sites

This final analysis considers using de-coupling to reduce total system capital costs for systems that require electricity at several points. The concept of the de-coupled system is shown in Fig. 1(d) and the locations of the fast pyrolysis plant and generators used in the model are shown in Fig. 4(b). In this case the positions of the fast pyrolysis plants and generators are reversed from the previous analysis, a simplification that ensures consistency between analyses.

Results for the production of 12 MW$_e$ total electricity at three 4 MW$_e$ generators are shown in Fig. 6. The production of electricity is achieved by installing:

- three complete 4 MW$_e$ fast pyrolysis and diesel engine systems;
- three complete 4 MW$_e$ gasification and diesel engine systems; or
- one large (equivalent to 12 MW$_e$) fast pyrolysis plant and three separate 4 MW$_e$ diesel engines.

As has been seen before, the overall costs of the close-coupled systems are almost identical, although the proportion of costs spent on feed is higher in the pyrolysis case because of its lower efficiency. The de-coupled system however gives an overall electricity production cost 10% lower than the close-coupled systems. This is caused by large capital cost savings at pretreatment and fast pyrolysis. A comparison of specific investment costs is given in Table 4.

Table 4 Specific investment costs for various 12 MW$_e$ systems generating 4 MW$_e$ at 3 sites

Conversion Process	Pyrolysis		Gasification		Pyrolysis	
No. conversion plants	3		3		1	
No. generating sites	3		3		3	
Total plant costs						
Pretreatment, $/kW$_e$ (%)	597	(15.6)	436	(10.6)	455	(14.3)
Conversion, $/kW$_e$ (%)	1868	(48.7)	2289	(55.8)	1217	(38.3)
Generation, $/kW$_e$ (%)	1374	(35.8)	1374	(33.5)	1503	(47.3)
Total, $/kW$_e$	3840		4100		3176	

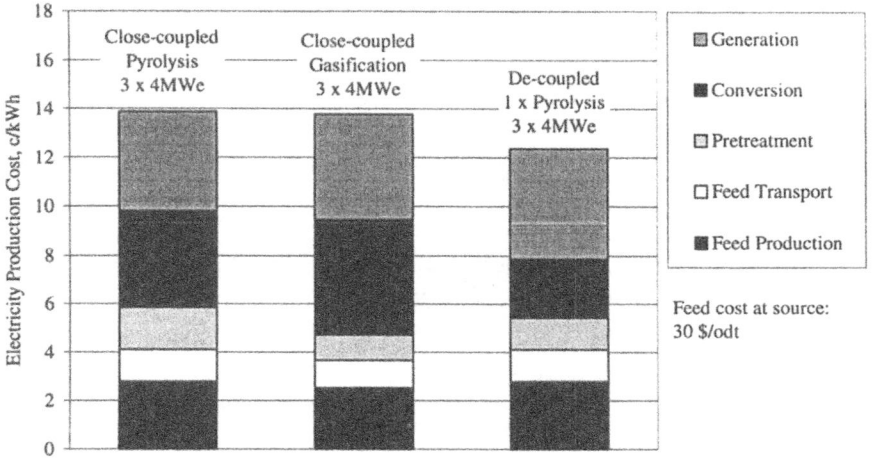

Fig. 6 Electricity production costs for multiple generating sites

5 Conclusions

Techno-economic models have been used to compare the costs of electricity production using fast pyrolysis and diesel engines at capacities of less than 12 MW_e and using various system configurations that illustrate de-coupling options. The results are compared with systems that generate electricity using atmospheric gasification and diesel engines.

It has been shown that the costs of electricity production in close-coupled systems are very similar for both conversion technologies. Gasification systems have higher capital costs but better generating efficiencies so that as the unit feed cost increases or the influence of capital costs decrease they have a slight advantage over fast pyrolysis.

De-coupling gives fast pyrolysis an economic advantage whenever a close-coupled system would require the long-distance haulage of the biomass feedstock. This is because it is less expensive to transport pyrolysis liquids than biomass.

Using multiple pyrolysis sites to supply a large central generator is not cost effective. There is a small saving due to reduced biomass transport costs but these are far outweighed by the increases in system capital costs incurred when constructing several small pyrolysis plants.

If electricity is required at several remote locations then overall electricity production costs can be reduced by using one large pyrolysis plant to produce liquid fuel for all the generators. Investment costs are reduced through economies of scale at the pyrolysis plant.

It has been shown that the costs of electricity production are reduced considerably in specific cases where system de-coupling can be used effectively. The extra flexibility of the fast pyrolysis technology should increase its opportunities to penetrate the small-scale biomass to electricity sector because electricity generating costs can be much lower than competitive technologies that must be used in close-coupled systems.

6 Acknowledgments

This work is part of a current project that is assessing practical applications for fast pyrolysis and diesel engine systems. This project is sponsored by the EC RENA R&D Programme and the UK Department of Trade and Industry.

7 References

1   Elliott, P. (1993) Biomass Energy Overview in the Context of Brazilian Biomass-Power Demonstration. *Bioresource Technology*, Vol.46, pp.13-22

2   Anon (1993) Electricity from Biomass - National Biomass Power Program Five Year Plan (FY1994-1998). US Department of Energy

3   Bridgwater A. V. and Evans, G. D. (1993) An Assessment of Thermochemical Conversion Systems for Processing Biomass and Refuse. ETSU Report No. B/T1/00207/REP, Energy Technology Support Unit, Harwell, UK

4   Overend, R. P. and Rivard, C. J. (1993) Thermal and Biological Gasification. *First Biomass Conference of The Americas*, Vol.1, pp.470-497, National Renewable Energy Laboratory, Golden, Colorado

5   van Swaaij, W. P. M., van den Aarsen, F. G., Bridgwater, A. V., and Heesink, A. B. M., (1994) A Review of Biomass Gasification. Report to the EC DGXII JOULE Program

6   Bridgwater, A .V. and Bridge, S. A. (1991) A Review of Biomass Pyrolysis and Pyrolysis Technologies. *Biomass Pyrolysis Liquids Upgrading and Utilisation* (A. V. Bridgwater and G. Grassi eds.) pp.11-92, Elsevier Applied Science, London

7   Bridgwater, A. V. and Peacocke, G. V. C. (1995) Biomass Fast Pyrolysis. *Proc. Second Biomass Conference of the Americas*, pp.1037-1047, National Renewable Energy Laboratory, Golden, Colorado

8   Mitchell, C. P., Bridgwater, A. V., Stevens, D. J., Toft, A. J. and Watters, M. P. (1995) Technoeconomic Assessment of Biomass to Energy. *Biomass and Bioenergy*, Vol.9, No.1-5, 205-226

9   Solantausta, Y., Diebold, J., Elliot, D. C., Bridgwater, T. and Beckman, D. (1994) Assessment of Liquefaction and Pyrolysis Systems. *VTT Research Notes* No.1573, VTT, Espoo

TECHNOECONOMIC ANALYSIS AND LIFE CYCLE ASSESSMENT OF AN INTEGRATED BIOMASS GASIFICATION COMBINED CYCLE SYSTEM

LIFE CYCLE ASSESSMENT OF BIOMASS-TO-ELECTRICITY

M.K. Mann and P.L. Spath
National Renewable Energy Laboratory
1617 Cole Blvd., Golden, Colorado

ABSTRACT

Life cycle assessment quantifies the environmental impacts of all processes used in transforming a raw material to a final product. Performed in conjunction with a technoeconomic feasibility study, the total economic and environmental benefits and drawbacks of a process can be quantified. This technique can be used in project decision-making, in allocating research and capital dollars, and in comparing the viability of different projects. Additionally, life cycle assessment can distinguish between truly environmentally friendly processes which either mitigate or eliminate upstream emissions and energy consumption, and those that are only environmentally conscious in their final production step.

A biomass gasification combined-cycle power plant, consisting of a low pressure indirectly-heated gasifier integrated with an industrial gas turbine, was simulated using ASPEN Plus.® Economic analyses were then performed to determine the levelized cost of electricity. The economic viability and efficiency of power production from this system appear to be quite attractive, with the cost of electricity near the competitive range of current electricity prices in the United States. To complement this study, a life cycle assessment is being performed to quantify the total benefits and drawbacks of the entire system from biomass crops through power distribution, including emissions and costs from diverse sources such as planting and harvesting, transportation, and power production. A discussion of the economics, efficiency, and environmental benefits of power production from this biomass-based technology, are presented.

Keywords: biomass electricity, dedicated feedstock, impacts, process improvements, life cycle assessment, energy, economic analysis, environment.

1. Introduction

This paper describes the cost and performance potential of a biomass-based integrated gasification combined cycle (IGCC) system, and the method that is currently being used to evaluate the environmental impact of this process. The system chosen for this study was a low pressure indirectly-heated gasifier similar to the Battelle gasifier, coupled to a combined cycle consisting of a utility gas turbine and a two-pressure steam cycle system. A schematic of this system is given in Figure 1. A detailed ASPEN process simulation was developed, and results were used to design, size, and cost the major plant equipment sections. Numerous literature sources and previous biomass and coal studies were used to develop overall plant cost information. Standard economic analyses were performed to determine the levelized cost of electricity. The economic viability and efficiency performance of this biomass-based IGCC generation technology appear, from this study, to be quite attractive.

Life cycle assessment is an analytical tool for quantifying the environmental impacts of all processes used in converting a raw material to a final product. When such an assessment is performed together with a technoeconomic feasibility study, the total economic and environmental benefits and drawbacks of a process can be quantified. A life cycle assessment is being performed on the production of electricity using the biomass gasification combined cycle process described above. The systems considered in the overall analysis consist of the production of biomass as a dedicated feedstock crop,

Figure 1. Low Pressure Indirect BIGCC Schematic

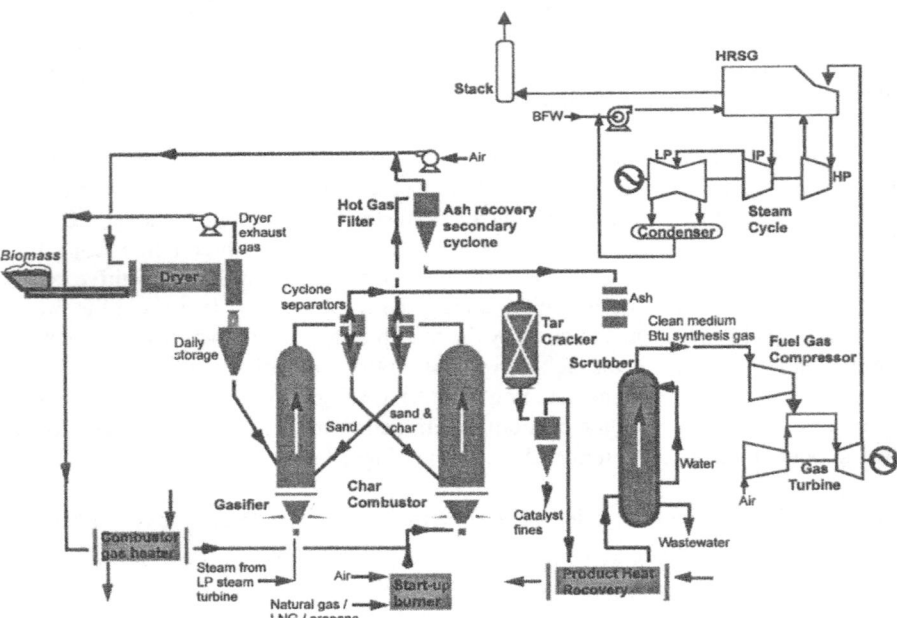

its transportation to the power plant, electricity generation, and all processes associated with intermediate feedstocks. Consistent with a life cycle evaluation, the boundaries of the analysis will include all material and energy streams from these processes. Finally, a sensitivity analysis will be conducted to minimize the risk of imperfect data affecting the results. The life cycle assessment exercise will quantify emissions and material use, and track energy flows within the system. The primary goal of this work is to reduce the environmental impact of this process through design improvements once the impacts of all emissions are assessed. A significant amount of work exists on many parts of this system or similar systems, but very little has been done from a life cycle viewpoint. Unlike previous efforts, this study will serve to assess the life cycle implications of this particular integrated biomass-to-electricity system.

2. Technoeconomic analysis

The integrated gasification combined-cycle plant was simulated in ASPEN Plus© with data from the Battelle Columbus Laboratory (BCL) 9 T/day process development unit [1] [2]. Because the gasifier operates at nearly atmospheric pressure (172 kPa, 25 psi), wood from the rotary dryers is fed to the gasifier using an injection screw feeder. Gasification occurs at 825°C (1517°F), and combustion of the char occurs at 982 °C (1800 °F). Fuel gas from the gasifier is cleaned using a tar cracker to reduce the molecular weight of the larger hydrocarbons, and a cyclone separator to remove particulates. A direct water quench is used to remove alkali species and cool the gas to 97 °C for compression. Although a tar cracker is not necessarily required because the gas is cooled using a direct water quench, one was included in this design to avoid losing the substantial heating value of the tars. Compression of the fuel gas prior to the gas turbine combustor is accomplished in a five-stage centrifugal compressor with interstage cooling. This compressor increases the pressure from 172 kPa to 2,068 kPa (25 psi to 300 psi).

Gas turbine exhaust was ducted to the heat recovery steam generator (HRSG), which incorporates a superheater, high and low pressure boilers, and economizers. The HRSG provides steam superheating and steam reheat, as well as steam for the high and low pressure boilers and economizers. Two percent boiler blowdown was assumed and feedwater heating and deaeration are performed in the HRSG system. All feedwater pumps are motor driven rather than steam turbine driven. In the steam cycle, superheated steam at 538°C and 10 MPa (1000°F, 1465 psia) is expanded in the high pressure turbine. Gasification steam was extracted from the high pressure exhaust, while the remaining steam was combined with steam from the low pressure boiler, reheated, and introduced into the intermediate pressure turbine. Expanded steam quality leaving the low pressure turbine in all cases was 90%.

A mechanical induced-draft cooling tower was assumed in the design. This includes all of the necessary pumps for condenser cooling and makeup water needs. Balance of plant equipment includes plant water supply and demineralization facilities, firewater system, waste water treating, service and instrument air system, and the electric auxiliary systems. General facilities included are roads, administrative, laboratory and maintenance buildings, potable water and sanitary facilities, lighting, heating and air conditioning, flare, fire water system, startup fuel system, and all necessary computer control systems.

A high degree of process integration between the gasifier and combined cycle was incorporated into the ASPEN model to maximize system efficiency. The product gas composition is shown in Table 1. The heat necessary for the endothermic gasification reactions is supplied by sand circulating between a fluidized bed char combustor and the gasification vessel. In addition to supplying heat, the sand acts as the bed material for the gasifier, designed as an entrained fluidized bed reactor. Of the total amount of sand circulating between the gasifier and char combustor, 0.5% is purged to prevent ash build-up in the system. The gasifier operates at nearly atmospheric pressure (172 kPa, 25 psi) and 825°C (1517°F).

Table 1. Indirectly heated gasifier product gas composition, dry basis

Component	H_2	CO	CO_2	CH_4	C_2H_2	C_2H_4	C_2H_6	Tars	H_2S	NH_3
Volume %	21.22	21.22	13.46	15.83	4.62	4.62	0.47	0.40	0.08	0.37

LHV = 13.2 MJ/m^3 (354 Btu/scf)
HHV = 14.2 MJ/m^3 (379 Btu/scf)

The combined cycle investigated was based on the GE MS-6101FA, an advanced gas turbine (high firing temperature, high efficiency). For this analysis, it was shown that the increased gas turbine efficiency offsets the system size increase and keeps the feed requirements within what might be available from a dedicated feedstock supply system (DFSS). Particulate removal prior to the gas turbine was accomplished with cyclones and a fabric filter.

The intent of the technoeconomic study was to evaluate the ultimate potential for application of IGCC technology to biomass-based power systems of large scale (> 30 MW$_e$). Therefore, the plant design was assumed to be for mature, "nth-plant" systems. The aggressive sparing and redundancies typically utilized for "first-plant" designs and the attendant cost associated with such an approach was not applied here. Consistent with the tests at BCL, the biomass used for the simulation is typical of woody biomass such as hybrid poplar. The amount of biomass fed to the plant, as dictated by the fuel requirements of the gas turbine, was 1,486 bone dry Tonnes per day.

The biomass-based IGCC electric generating plant considered in this study consists of the following process sections:

- Fuel receiving, sizing, preparation, and drying: Truck unloading system, wood yard and storage, sizing and conveying system, dryers, live storage area
- Gasification and gas cleaning (Gasification Island): Wood feeding unit, gasifier, char combustion and air heating, primary cyclone, tar cracker, gas quench, particulate removal operation
- Power Island: Gas turbine and generator, Heat Recovery Steam Generator (HRSG), steam turbine and generator, condenser, cooling tower, feed water and blowdown treating unit
- General plant utilities and facilities

Process conditions and system performance for the system examined are summarized in Table 2. Gas turbine output and efficiency based on fuel heating value are greater than those listed in the literature for natural gas fuel. These increases are primarily the result of high fuel gas temperatures and the increased mass flow through the turbine expander (due to lower energy content fuel gas).

Table 2. Process data summary and system performance results

Gasifier Requirements		Fuel Gas Produced	
Wood flowrate, kg/s (lb/hr)	17.2 (136,494)	Fuel gas flowrate, kg/s (lb/hr)	14.5 (114,734)
Air flowrate, kg/s (lb/hr)	0	Fuel gas heating value, LHV,	
Steam flowrate, kg/s (lb/hr)	7.7 (61,346)	wet basis, MJ/m^3 (Btu/SCF)	13.2 (353.9)
Power Island		**Power Production Summary**	
Gas turbine	GE MS-6101FA	Gas turbine output, Mw$_e$	82.1
Turbine PR	14.9	Steam turbine output, Mw$_e$	55.1
Turbine firing temp, °C (°F)	1,288 (2,350)	Internal consumption, Mw$_e$	15.2
Steam cycle conditions,		Net system output, Mw$_e$	122
MPa/°C /°C/	10/538/538	Net plant efficiency, %, HHV	35.4
(psia/°F/°F)	(1,465/1,000/1,000)		

The selling price of electricity in 1990 (the base year for this study) was $0.047/kWh, $0.073/kWh, and $0.078/kWh for industrial, commercial, and residential customers, respectively [3]. By calculating the economics of the processes being studied and comparing the results to the prices within the electricity generating market, the potential profitability can be assessed. The method and assumptions that were used to calculate the cost of electricity are based on those described in the EPRI Technical Assessment Guide (TAG) [4] and reflect typical utility financing parameters. Independent power producers or cogenerators would clearly have different analysis criteria.

Capital costs for the system were estimated using a combination of capacity factored and equipment-based estimates. Capacity factored estimates utilize the ratio of the capacity (flowrate, heat duty, etc.) of an existing piece of equipment to the new equipment multiplied by the cost of the existing equipment to estimate the cost of the new equipment. A scale-up factor particular to the equipment type was applied to the capacity ratio. The equipment-based estimates were determined from more detailed equipment design calculations based on the process conditions and results of the simulations. All costs were estimated in instantaneous 1990 dollars. Where necessary, costs were corrected to 1990 using the Marshall and Swift or Chemical Engineering equipment cost indices. A charge of 20 % of the installed cost of the major plant sections was applied to account for all balance of plant (BOP) equipment and facilities. The major equipment costs were multiplied by standard factors to arrive at the total direct cost of the installed equipment. The results of the economic analysis, including the levelized cost of electricity (COE) are shown in Table 3.

Table 3. Summary of technoeconomic analysis results

Output (Mw$_e$)	122
Efficiency (%, HHV)	35.40
Capital Cost (TCR, $/kW)	1,108
Operating Cost incl. fuel ($1,000/yr)	27,983
COE (¢/kW, Current $)	6.55
COE (Constant $)	5.11

3. Life cycle assessment

3.1 General

Life cycle assessment is an analytical tool for quantifying the environmental impacts of all processes used in converting a raw material to a final product. Performed in conjunction with a technoeconomic feasibility study, the total economic and environmental benefits and drawbacks of a process can be quantified. In performing a life cycle assessment, material and energy balances are used to quantify the emissions, resource depletion, and energy consumption of all processes between transformation of raw materials into useful products and the final disposal of all products and by-products. The results of this inventory are then used to evaluate the environmental impacts of the process so that efforts can be focused on mitigating these effects.

Until recently, life cycle studies were not widely used, and the technique remained in its infancy. In the United States, the Society of Environmental Toxicology and Chemistry (SETAC) has been actively working to advance the methodology through workshops and publications. From their direction, a three-component model for life cycle assessment has been developed, and is considered to be the best overarching guide for conducting such analyses. The three components are inventory, impact analysis, and improvement. The inventory stage involves quantifying all energy and material requirements, air and water emissions, and solid waste from all stages in the life of a product or process. The second element, impact assessment, examines the environmental and human health effects associated with the loadings quantified in the inventory stage. The final component is an improvement assessment in which means to reduce the environmental burden of a process are proposed and implemented. To date, most work in life cycle assessment has been focused on inventory, although the amount of available literature indicates that efforts to advance impact assessment and improvement are significant.

3.2 Life cycle assessment of a biomass power generating system

This following sections serve to establish and disseminate the methodology for conducting a life cycle assessment on the production of electricity using a combined-cycle system based on the BCL indirectly-heated biomass gasifier. The systems considered in the overall analysis consist of the production of biomass as a dedicated feedstock crop, its transportation to the power plant, electricity generation, and all processes associated with intermediate feedstocks. In addition to quantifying emissions, a key aspect of this life cycle project is to evaluate the energy flows within these boundaries to assess the net energy produced. The right-hand boundary of this process will be considered to be at the point of electricity production. Therefore, the nearly infinite ways in which electricity is used will not be taken into account in this study.

It became clear early on in this study that although a significant amount of work has been performed on many parts of this system or similar systems, very little has been done from a life cycle viewpoint. For example, the environmental emissions associated with growing biomass have been assessed, but upstream processes such as fertilizer production were not included in the material and energy consumption inventory.

3.3 System boundaries

The system boundaries for any life cycle assessment should be drawn as broadly as possible. Therefore, in addition to counting the material and energy flows of the primary process of interest, those processes involved in the extraction of raw materials and production of intermediate feedstocks must be included. Furthermore, the means of disposing products, by-products, wastes, and process materials are included within the life cycle boundary. The question of where to stop tracking the energy and material uses of upstream processes is an important one; the analysis is infinite if boundaries are not drawn that encompass the most important impacts to the environment and energy resources. Generally speaking, the impacts of upstream processes become less significant the further you get from the final process, and a situation of diminishing returns becomes apparent past the third level of upstream processes.

Figures 2 and 3 show the processes within the system to be analyzed; the solid lines in these figures represent actual material and energy flows, while the dotted lines indicate the direction of the system, moving between process blocks. In Figure 2, "Other upstream processes" refers to major manufacturing steps needed to produce intermediate feedstocks. The following will be included in the system for the purposes of quantifying material and energy flows: raw material extraction, manufacture of transportation equipment, manufacture of process equipment, manufacture of farm equipment, fertilizer production, pesticide production, herbicide production, processes upstream from production of farm chemicals (e.g., ammonia plant), application of farm chemicals, planting, harvesting biomass, preparation of biomass for use at power plant, all transportation of chemicals, catalysts, biomass, and other materials, and electricity production in a gasification combined-cycle power plant.

Once the system boundary has been defined, the next step in performing a life cycle assessment is to take an inventory of the inputs to and outputs from each process block within the system boundary. The inventory assessment for the biomass-to-electricity process has been divided into the following three process blocks: biomass production, transportation, and power generation. The energy and material balances from all three blocks will be summed to determine the net energy, raw materials, and emissions produced from the biomass-to-electricity system. As the inventory analysis progresses, changes in the process configuration will be made that will reduce emissions, energy consumption, and material use from the original BCL gasification combined-cycle plant design. It is expected that throughout each step in the life cycle assessment, more opportunities will arise that allow further process changes to be made, resulting in a biomass-to-electricity process with minimal impact on the environment.

Figure 2. Biomass Production and Transportation Boundaries for Life Cycle Assessment

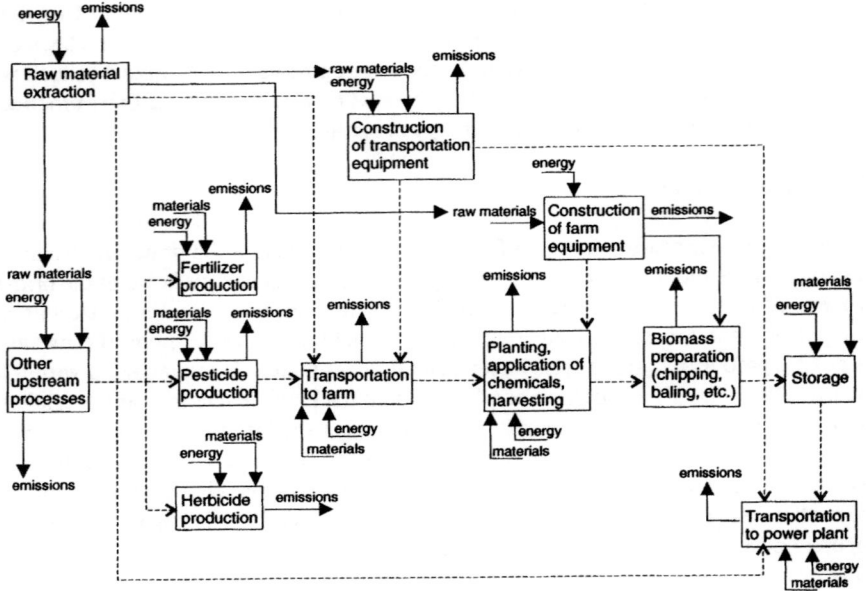

Figure 3. Power Generation and Transportation Boundaries for Life Cycle Assessment

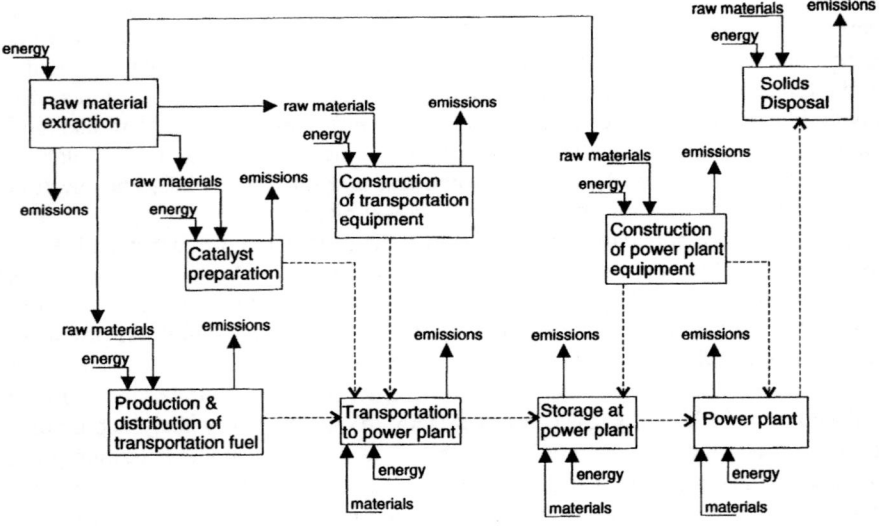

3.4 Biomass production

The environmental emissions and energy use associated with the production of biomass as a dedicated energy crop will be assessed. Processes to be studied include raw material extraction, fertilizer, pesticide, and herbicide production, equipment manufacture, application of chemicals, crop establishment, harvesting, storage, all necessary transportation, and upstream processes to produce intermediate feedstocks. Additionally, soil erosion and net carbon sequestration or emissions will be analyzed. Once validated, much of the data associated with the on-farm practices will be taken from the literature, much of which comes from test plots within the States being studied. Independent work on assessing the impact of this process on biodiversity, wildlife, and habitat will not be performed. However, the available literature and current work within this field will be reviewed.

3.5 Carbon cycle

One of the most talked-about aspects of biomass energy is the potential reduction of atmospheric carbon dioxide (CO_2) per unit of energy produced. In the case of biomass power production, the carbon dioxide absorbed by the growing biomass is equal to or greater than the amount produced from gasification and combustion. On a life cycle basis, however, including all processes required to grow and transport the biomass, produce power, and dispose of wastes, there will be either a net positive or negative flow of carbon within the system. The greater-than portion of this balance is due to carbon that is sequestered in the soil or biomass not used by the power plant or degraded. However, within the life cycle system, carbon dioxide is released in upstream processes used to produce the biomass, thus reducing the total amount sequestered. Additionally, methane can result from decomposition of biomass during storage or biomass that is not used in the process. Because of these factors, it is clear that biomass power is not a zero net CO_2 process, although if there is positive CO_2 production from the system, it will certainly be less than the contribution from a power plant of similar size using fossil fuels. An important distinction between this assessment and other studies that have been performed on biomass crops is that on a life cycle basis, there is no credit for the carbon stored in the biomass even though in order to supply a continuous feedstock to the power plant there is always a mature stand of trees. On a life cycle basis, it must be recognized that eventually the plant will be shut down and the trees will no longer be required, resulting in only short-term carbon sequestration in the biomass standing in the field.

3.6 Transportation

The two forms of transportation that will be considered for the electricity-to-biomass system are trucks and rail cars. However, it is anticipated that the majority of the transportation needs will be met using trucks. The inventory assessment for the transportation block includes the energy required and emissions generated for the transportation of chemicals, biomass, and other items by truck and rail car to and from the boundaries of the biomass production and power generation blocks. All trucks and rail cars are assumed to use diesel as the fuel source. The energy and emissions related to extracting crude oil, distilling it, producing a usable transportation fuel, and distributing the diesel to refueling stations will be assessed.

3.7 Power generation

The inventory assessment for the power generation block begins at the plant gate of the gasification combined-cycle plant and ends with the production of electricity. The material and energy balance for the power generation facility will be examined over the life of the plant then listed for each year of production. The emissions will include those expected at start-up and shut-down, those from normal operation, fugitive emissions and leaks (e.g., from flanges, seals, removal ports from baghouse filters and cyclones, compressors, and pumps), emissions associated with plant improvements or future installations, emissions from accidental releases, and emissions associated with constructing and decommissioning the plant. All of the nonhydrocarbon emissions, except NO_x, will be limited to the amount of sulfur, ash, alkalis, and heavy metals in the feedstock.

Biomass is delivered to the plant and unloaded to a paved storage yard. Before being gasified, the biomass is dried in a rotary kiln dryer. In the current gasification combined-cycle process flow scheme, the gas used to dry the biomass is a mixture of air and flue gas from the HRSG. If this combined stream is not sufficient to meet the drying needs of the plant then a slipstream of combustor flue gas will be mixed in before the dryer. The dryer exhaust gas is used as the combustion air source for the char combustor. This will eliminate any dryer emissions except for fugitive emissions because any wood particulates and VOCs in the exhaust gas will be combusted with the char. If, however, the amount of air necessary to dry the wood exceeds the amount of air required for the char combustor, it will be necessary to purge a slipstream of the dryer exhaust air to the atmosphere. In that case, there will be a small amount of wood particulates and VOCs emitted to the atmosphere. The overall integrated configuration will be examined using the current ASPEN Plus© model.

There will not be any direct air emissions from the gasification step except for fugitive emissions. Most of the emissions from the gasification step will be determined by the elemental composition of the wood.

Once the biomass is gasified, the remaining products will be synthesis gas, char, and ash. In the Battelle gasifier system the synthesis gas will exit the gasifier overhead while greater than 99.5% of the char and ash will be captured with the sand and circulated back to the combustor. The char is burned and any residual sand and ash that is carried overhead with the combustor flue gas captured using a cyclone. Currently, it is anticipated that the sand and ash mixture will be able to be landspread to serve as a nutrient source for the biomass or used in manufacturing asphalt. The emissions and reduction in the need for chemicals associated with the chosen practice will be included in the life cycle assessment.

Energy requirements and emissions associated with catalyst use occur from hauling the fresh catalyst to the plant and disposing it once its useful life is expended. If metals can be reclaimed from the catalyst, then the catalyst will be shipped to a metals-reclaiming facility before being shipped to a disposal site. In this case, the raw materials, energy, and emissions from the metals reclaiming plant will also be included in the analysis.

There are two possible catalysts that will be used in the gasification combined-cycle system. DN-34, a proprietary catalyst, will be used in the tar cracker. A second catalyst that may be required in the plant is for a selective catalytic reduction (SCR) unit, which would be integrated with the HRSG. Synthesis gas from biomass is cleaner than that from coal but is not as clean as natural gas. It has not been determined whether or not the NO_x produced by the gas turbine during the combustion of biomass-derived synthesis gas will be below the allowable NO_x emission level.

Before compressing the synthesis gas to the turbine operating pressure, it must be cooled through heat exchange followed by a water scrubbing step. The water scrubbing step will condense any residual tars that remain after the synthesis gas has passed through the tar cracker. Water from the scrubbing step will be sent to a separation tank where any tars will float to the top and any solid particles will settle to the bottom. Any insoluble tars will be skimmed off of the water and fed back to the char combustor. A portion of the remaining water will be used to rehumidify the synthesis gas prior to combustion in the gas turbine. This will reduce the amount of water that must be treated as well as increase the power output from the plant. The remaining water will then be treated in the wastewater treatment step. The soluble organic compounds will be removed making the wastewater discharge quality and the solids will be collected for disposal.

Because a portion of the flue gas is used to dry the biomass, the gas turbine emissions will be limited to fugitive emissions and those resulting from incomplete combustion, emitted to the atmosphere through the flue gas stack.

Process water will be used throughout the plant as cooling water, potable and utility water, and water for the fire-water system. Once used, this water will be collected and treated in a wastewater treatment step to produce an effluent that can be reused within the plant or discharged without causing serious environmental impacts. There will also be wastewater from the boiler blowdown in the HRSG and rain water which is collected through drainage ducts. Both of these wastewate sources will be sent to the wastewater treatment step to be processed into discharge quality water. There are three types of operations for treating wastewater streams: physical unit operations, chemical unit processes, and biological unit processes. A combination of these processes will be used to treat the various water streams throughout the gasification combined-cycle plant. The treatment steps required will depend on the types of contaminants in each wastewater stream. Gaseous emissions and solid waste products will be produced from the treatment of wastewaters. For the life cycle assessment, all gaseous emissions as well as the disposal of solid waste products produced from the treatment of wastewaters will be included.

The stack is located on the exhaust from the HRSG. The stack gas will be an accumulation of emissions that carry through the process all the way to the HRSG. These emissions will be comprised of gasifier emissions that are not combusted in the gas turbine plus any gas turbine emissions.

3.8 Impact assessment

Under the SETAC directive, the second part of a life cycle assessment is impact assessment. The goals of this section are to describe the principal emissions expected from the system and describe the known results of these chemicals in the environment.

Table 4 summarizes the emissions expected; however not all upstream processes have been characterized. For this life cycle assessment, the majority of the work on impact assessment will be qualitative rather than quantitative.

Table 4. Environmental loadings expected from system
___

Feedstock Production
Gaseous Emissions
 - CO_2, CO, NO_X, SO_2, NH_3, nitric acid, sulfuric acid, terpenes from biogenic sources, fluorine and fluoride compounds from phosphate rock processing
Solids
 - Catalysts requiring disposal from upstream processes
 - Gypsum from the production of phosphate fertilizers
 - Dust and particulates from upstream processes and biomass storage
Liquid Effluents
 - Nitrogen, phosphate, and potassium fertilizer run-off
 - Herbicide and pesticide run-off
 - Brine from the production of potassium fertilizers
Other
 - Microorganisms, spores, and fungi
 - Emissions associated with transportation and farm vehicles (see transportation block)
___

Transportation
 - Carbon monoxide (CO), carbon dioxide (CO_2), methane (CH_4), non-methane hydrocarbons, nitrogen oxides (NO_X), nitrous oxide (N_2O), sulfur oxides (SO_X), particulate matter, ozone (O_3)
 - Spillage & leakage of fuel, oils, and lubricants, evaporative & fugitive emissions, and accidental releases
___

Biomass Power Production
Gaseous Emissions
 - VOC's, PAH's, and terpenes, carbon monoxide (CO), carbon dioxide (CO_2), nitrogen compounds (NO_X and NH_3,), sulfur compounds (SO_X and H_2S), chlorinated compounds, unburned hydrocarbons
 - Fugitive emissions and accidental releases
Solids
 - Sand and ash mixture (may contain alkali and heavy metals), char, catalysts from tar destruction and from selective catalytic reduction of NO_X particulates including aerosols, wood dust
Liquid Effluents
 - Wastewater, organics (dissolved in wastewater and as tars)
 - Leachate of pollutants from feedstock, ash and char piles, and wastewater system
 - Leaks from pumps, water scrubber, and wastewater treatment system
Other
 - Microorganisms, spores, and fungi
 - Emissions associated with transpiration and biomass handling equipment (see transportation block)
___

Classification of inventory data into stressor categories that are potentially linked to ecological health and human health will be the first step in this impact assessment. It must be recognized that discovering and establishing a causal relationship between an emission identified in the inventory and an impact on the environment is not a component of life cycle assessment. That is, the potential impacts on the environment are assessed based on previously identified consequences of environmental emissions. For example, the goal of this study will not be to establish that methane emissions result in global warming. Therefore, our intent is not to prove or disprove that biomass power production via gasification is responsible for any degradation of the environment, but to index expected emissions, energy use, and material consumption with possible impacts. The major impact categories to be looked at are human and ecological health; the stressor groups within each of these categories are shown in Table 5. It is expected that as the life cycle assessment proceeds, additions will be made to this list.

Table 5. Stressor categories associated with biomass power production

Stressor Category	Stressors	Major Impact Category *; Area Impacted **
Toxicants	Pesticides, herbicides, fertilizers	H, E; L
	Tars, diesel fuel, and other hydrocarbons	H, E; L
	SO_2, SO_3, H_2S, NH_3	H, E; L, R, G
	Fluorine and fluorides	H, E; L, R
Photochemical oxidant precursors and photochemical oxidants	Hydrocarbons, non-methane hydrocarbons, VOCs	H, E; L, R
		H, E, L
	Ozone (O_3)	H, E; L, R
Particulates	Wood dust	H, E; L
	Microorganisms, spores, fungi	H, E; L, R
Air pollutants	CO, O_3, NO_x, SO_2, SO_3, H_2S, HC, NH_3	H, E; L
	Chlorinated compounds	H; L
	wood dust, sand dust	H; L
	microorganisms, spores, fungi	H, E; L, R
Solid waste	Catalysts, sand, ash, char, gypsum	H, E; L, R
Physical trauma	Accidents, noise	H; L
	Odor	L
Climate change	CO_2, CH_4, nitrates, sulfates,	E; G
	changes in plant growth	E, L, G
Acidification precursors	SO_2 (H_2SO_4), NO_2 (HNO_3), CO_2 (HCO_4-)	E; R
Nutrients	Nitrates, sulfates	H, E; L, R
Habitat effects	Monoculture, non-native species, flora kill, animal and insect kill	E; L, R
Resource depletion	Fossil fuel use, minerals, ores	E; R, G
	Groundwater	E; L, R
	Topsoil	E; L

* H = Human health, E = Ecological health
** L = Local (county), R = Regional (state), G = Global

In addition to defining stressor categories for each component of biomass power production (feedstock, transportation, power production), characterization of the impact must be performed. Characterization estimates the magnitude of potential impacts for each stressor. Impact characterization descriptors range from simply listing the stressors to conducting a full site-specific exposure and effects assessment. The latter is analogous to risk assessment performed for environmental remediation under CERCLA or an environmental impact assessment under the National Environmental Policy Act. The breadth and depth of the impact assessment should be varied based on the purpose, scope and intended use of the full life cycle assessment. Specific data from the inventory will be listed and grouped according to their potential effect. These data can then be used to identify stages in the life cycle where effects can be reduced. For this level, it is assumed that fewer emissions, less energy use, and less material consumption is better, although the actual effect on the environment of decreased loading cannot be directly assessed. Higher levels of impact assessment will be conducted in future iterations of the analysis.

4.0 Conclusions

Combining technoeconomic and life cycle analyses provides the opportunity to enhance the economic and environmental performance of commercial and concept technologies alike. As impact assessment techniques are advanced, analyses of this type may provide the basis for valuation of sustainable energy systems such as the one analyzed here. The rationale for implementing these technologies, which are now moving toward the demonstration phase, will at last be quantified.

The technoeconomic portion of this study demonstrated that low pressure gasification has the potential to be economically viable. The cost of gas cleanup and compression are more than offset by the substantially lower cost of the gasifier itself (relative to high pressure gasifiers) and the elimination of a hot-gas cleanup unit. The projected cost of electricity falls within the market value of electricity in the United States for the base year of this study. Scale-up and demonstration efforts underway in the DOE Biomass Power Program will help to determine the accuracy of these calculated costs.

The primary goal of the life cycle assessment now being conducted is to reduce the environmental impact of the system through design improvements. The systems considered in the overall analysis consist of the production of biomass as a dedicated feedstock crop, its transportation to the power plant, electricity generation, and all processes associated with intermediate feedstocks. The work is now focusing on an inventory of all environmental emissions and energy flows within the system. This will highlight areas which are responsible for significant emissions and energy consumption. For instance, the integration of the gasifier, char combustor, and dryer has already been changed from the original design to produce fewer air emissions; this was the result of data that was collected in the initial inventory part of the life cycle assessment. Further process improvements will be made based on the assumption that fewer emissions, less energy use, and less material consumption are better. The changes in the process model will be incorporated into the ASPEN model, and the effect on cost of electricity will be determined.

A survey of the expected emissions from this system has been completed, and stressor categories along with possible impacts have been defined. Nearly all stressors from this process will have both human health and ecological effects, although the extent will not be measured directly. With the exception of climate change gases and resource depletion, which will be felt globally, impacts are expected to be local and regional. Because impact assessment is not a fully developed technique, the majority of the initial work on impact assessment will be qualitative rather than quantitative.

5.0 References

1. Feldmann, J.; Paisley, M.A. (May 1988). *Conversion of Forest Residues to a Methane-Rich Gas in a high-throughput Gasifier.* Columbus, Ohio: Battelle Colubus Laboratory.
2. Bain, R. (January, 1992). *Material and Energy Balances for Methanol from Biomass Using Bioamss Gasifiers.* Golden, Colorado: National Renewable Energy Laboratory.
3. Corman, J.C. (September 1986). *System Analysis of Simplified IGCC Plants.* DOE/ET-14928-2233. Morgantown, WV; Morgantown Energy Technology Center. Work performed by General Electric Company Corporate Research and Development, Schenectady, NY.
4. Electric Power Research Institute. (June, 1993). *TAG - Technical Assessment Guide.* EPRI TR-102276-V1R7 Volume 1: Rev. 7. Palo Alto, CA.

EVALUATION OF ELECTRIC POWER GENERATION SYSTEM BY BIOMASS

K. YAMADA, T. AKIYAMA, R. KATO, T. KAWAKAMI, M.SUGIYAMA,

M. TAKASE and M. FUJII

Department of Chemical System Engineering,

The University of Tokyo, Tokyo, Japan

Abstract

Biomass energy systems having 100MW scale of electricity generation were designed and evaluated using thermal efficiency, economics, energy pay-back time and CO_2 emissions as the basis of measure. The systems investigated were of four cases characterized by the combination of liquefaction and gasification processes, and boiler and combined cycles. The evaluation result indicated that each system seemed to be economically competitive and environmentally benign.

1 Introduction

Renewable energy has gained world -wide acceptance as an effective measure in solving global environmental and fossil fuel resource problems. Currently, the main renewable energy resources are hydraulic power generation systems and biomass. The annual biomass consumption is estimated to be 880 ~ 1,080 MTOE (Million Tons Oil Equivalence)[1]~[4]. This amount is not low, however biomass is mainly used as firewood and the thermal efficiency is very low. The establishment of efficient biomass energy systems is very important for global sustainable development. Electric power generation by biomass use is expected to be one of efficient systems.

In this paper, high efficiency processes of electric power generation using biomass are designed and evaluated on the bases of economics, energy pay-back time (EPT) and CO_2 emissions. The processes to produce fuel are by liquefaction[5] and gasification[6] of wood.

2 Design basis

2.1 Capacity of electricity generation

The capacity of a power plant is assumed to be 100MW. Fuel for power plants is produced by the liquefaction and gasification of eucalyptus. Its heat of combustion is assumed to be 16.7MJ/kg-dry basis.

2.2 Biomass production

The annual growth rate of biomass in a plantation area is assumed to be 1.25kg-dry wood/m^2. The plantation - harvesting cycle is 10 years. This means a tree of 500kg (dry basis) can be harvested every 10 years using 40m^2 area. The area of the plantation site required for 100MW plant is calculated to be about 600km^2 which depends on the type of fuel process.

3 Process description

3.1 Liquefaction process

A direct liquefaction process of biomass has been developed by Yokoyama et al.[5].

We design a process as shown in Fig. 1. The liquefaction reaction to produce light oil mainly advanced by de-oxygen of polymers in biomass under hydro thermal conditions.

Characteristics of the process are

(1) reducing gas such as H_2 or CO is not necessary for the liquefaction,

(2) biomass can be fed to the process without being dried,

(3) the contents of S, N and heavy metals in resultant oil are very low.

Raw materials, reaction conditions and products are shown in Tables 1 and 2 respectively. The heat of combustion of the liquid product is 26.4MJ/kg.

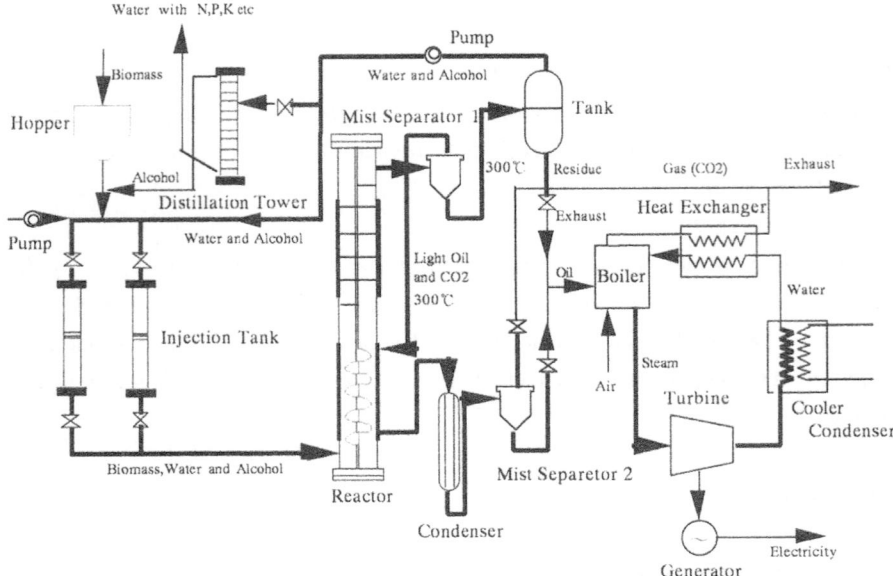

Fig.1.   Process flow diagram of biomass liquefaction - boiler cycle(LST).

Table 1.   Raw material and reaction conditions.

Biomass	Eucalyptus
Solvent	Water, 2 - propanol
Hydrothermal conditions	100 atm, 300°C, 60 min

Table 2.   Products from liquefaction process.

Product	Oil	Char	CO2	other
Weight ratio [%]	46	6	33	15

3.2   Gasification process

Gas is very useful fuel for electricity generation. A gasification process of biomass using a fluidized-bed of pilot scale has been developed by Fujinami et al.[6].

We used their experimental data for our process design. The process flow diagram is shown in Fig.2.

Feedstock wood (25% Water) is fed to the drying step where water content is decreased to 10% using waste gas, and then dried wood is forwarded to the gasification step operated at 850℃. The composition of product gas is shown in Table 3. The heat of combustion of product is 7.24MJ/m^3.

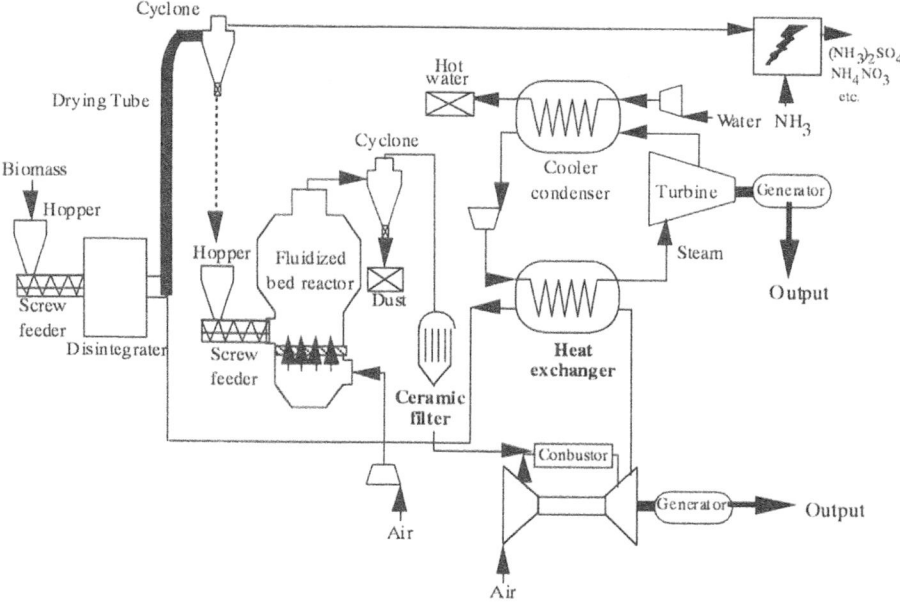

Fig. 2.   Process flow diagram of biomass gasification - combined cycle (GCC).

Table 3.   Composition of product gas.

	N2	CO2	CO	H2	CH4	C2H4
[%]	45	16	22	8	6.5	2.5

3.3   Electricity generation process

Both boiler and combined cycles using the liquid and gas fuels as shown in Table 4 were evaluated.

Table 4. Evaluated processes.

Electricity generation	Steam turbine (boiler)		Combined cycle (gas-steam turbines)	
Fuel	Liquid	Gas	Liquid	Gas
Abbreviation	LST	GST	LCC	GCC

Process flow diagrams of LST and GCC are shown in Figs. 1 and 2 together with the liquefaction and gasification processes.

4 Evaluation results

4.1 Thermal efficiency for electricity generation

Thermal efficiencies for electricity generation by each system were calculated based on the heat of combustion of biomass (HHV) and are shown in Table 5.

The highest efficiency of 30.5% was obtained by the GCC system. This value in very high compared with that (<20%) by a current biomass system because of an efficient heat recovery system.

4.2 Economic evaluation results

Table 5. Thermal efficiency of each process.

Process — Efficiency	LCC	LST	GCC	GST
Biomass conversion [%]	66.9	66.9	83.0	83.0
Generation [%]	43.3	40.0	36.7	32.1
Total [%]	29.0	26.7	30.5	26.7

The cost of biomass production was calculated using a rental cost basis of plantation site in Thailand. The rental cost of hte land is assumed to be $¥24,060 ha^{-1} \cdot y^{-1}$. The labor cost is assumed to be $¥1,500 \cdot day^{-1} \cdot man^{-1}$. The investment cost of the processes and power plants are estimated based on Japanese costs. The annual expenditure was assumed to be 12% of the total investment cost. The calculated electricity costs by each process are shown in Table 6 ($¥10/kWh$ in Japan).

The result indicates that electricity generated by biomass energy systems is competitive to that by existing systems.

The GST system gives the most economical electricity, although the thermal efficiency of GST is lower than that by GCC. The reason of this is a low cost of raw material of biomass and high investment cost the combined cycle.

Table 6.   Investment cost of each process.

[Million Yen]

Process	LCC	LST	GCC	GST
Liquefaction process	2,305	2,305		
Gasification process			480	480
Boiler, turbine, others	5,005	3,977	5,639	4,037
Total	7,310	6,282	6,119	4,517

4.3   EPT and CO2 emissions

Energy input and CO_2 emissions to construct the whole plant from the mining stage of raw materials were calculated for steel and concrete using the weight of equipments. Unit energy consumption and CO_2 emissions are reported to be (1) 25.5GJ/t, 0.59t-C/t for steel, (2) 0.6GJ/t, 0.05t-C/t for concrete. However, herein, we used the values three times higher than the above values.

EPT was calculated by Eq. (1).

$$EPT\,[D] = \frac{EIC}{EG - EOP} \tag{1}$$

where EIC is energy input for the plant construction, EOP is the daily energy consumption for the plant operation (210 GJ/D) and EG is the daily generated electricity (100MW = 8.64×10^3 GJ/D) .

Table 7 shows results on the weights of steel and concrete for power plants, EIC, EOP, CO_2 emissions and EPT. CO_2 emissions from biomass power plants are less than 3g-C/kWh which are very low compared with those from a coal fired power plant (250g-C/kWh) and somewhat lower than those from a photovoltaic (PV) system (10g-C/kWh)[7]. EPT values are low, especially that of GCC is as low as 32 days based on thermal energy. They are lower than that by a PV system (1 year)[7], because unit equipment sizes of biomass systems are much larger than those of the PV system.

Among biomass systems, gasification systems gave lower CO_2 emissions and EPTs than liquefaction systems which required heavy hydro-thermal reactors.

Table 7. Calculation results (material quantities, CO_2 emissions, EPT and cost).

Process	LCC	LST	GCC	GST
Steel (t)	6,100	14,500	3,400	11,800
Concrete (t)	8,700	10,800	6,100	8,200
EIC (GJ)	4.8×10^5	11.3×10^5	2.7×10^5	9.2×10^5
EOP (GJ/d)	200	200	200	200
CO_2 emissions for construction (t-C)	12,000	27,000	7,100	22,000
for operation (t-C/D)	3	3	3	3
Unit CO_2 emissions (g-C/kWh)	2.0	2.9	1.8	2.6
EPT (d)	57	134	32	109
Electricity cost (¥/hWh)	3.9	3.8	3.4	3.3
Electricity cost (¥/kWh)*	6.6	6.9	6.1	6.6

* in the case of labor cost of ¥10,000.d^{-1}·man^{-1}

5 Conclusion

Biomass energy systems having 100MW scale of electricity generation were designed and evaluated using economics, thermal efficiency, EPT and CO_2 emissions as the basis of measure. Systems investigated were of four cases (combination of liquefaction and gasification processes, and boiler and combined cycles). Each system seemed to be economically competitive and environmentally benign. Regarding the gasification - combined cycle system which was the most advantageous, the thermal efficiency, electricity cost, CO_2 emissions and EPT were calculated to be 30.5%, ¥3.4/kWh, 1.8g-C/kWh and 32 days respectively.

6 References

1) K. Yamaji et al., "World Energy Resources : Endowments, Supply / Demand, Economics, and Related Technology Development", Denryoku Tyuo Kenkyusyo Hokoku, Y94001 (1994, Tokyo).

2) World Energy Council, "Renewable Energy Resources : Opportunities and Constraints 1990-2020", World Energy Council (1993).

3) B. Dessus et al., "World Potential of Renewable Energies", Extraits de la Houille Blanchenl (1992).

4) T. B. Johansson et al., "Renewable Energy ; Sources for Fuels and Electricity" Island Press (1992).

5) S. Yokoyama et al., Kagakugogaku Ronbunshu, 17, 326~333 (1991).

6) S. Fujinami et al., "Fluidized-bed gasification of cellulostic wastes", Ebara Jiho No151 (1991, Tokyo).

7) K. Yamada et al., Energy Convers. Mgmt., 36, No.6~9, 819~822 (1995).

The efficiency of heat and power production from combustion and gasification

P. LAMP, A. REICHEL, R. FUNK
Bayerisches Zentrum für Angewandte Energieforschung, Garching, Germany

Abstract

A survey of the state of the art of existing biomass heat and power plants in Bavaria
has been performed and promising technologies utilising combined heat and power
production from solid biomass fuels have been examined. The resulting efficiencies are
compared. Moreover principle considerations comparing heat and power production
using combustion or gasification are presented.

It was found that there are about 200 biomass combustion plants installed in Bavaria,
with a peak in the number of plants in the capacity range of about 1 MW (fuel input).
The vast majority are producing heat only. Combined heat and power production is
done in general by the conventional steam process with a turbine but there is also one
installation using a steam engine. The smallest combined heat and power plant has a
fuel input of 2 MW. The electrical efficiencies range between 10 and 20%.

The prospect of future technologies was examined as well. Gasification of biomass in
combination with a gas engine or gas turbine appears to be most promising but so far
problems with tar content of the gas are not solved satisfactorily. Disregarding
technical problems the combination of the gasification of biomass and an energy
conversion system using the product gas yields higher electrical efficiencies compared
to combustion and a system utilising heat of the flue gas.

There are several options available to realise a combined heat and power plant fired
by biomass. The final choice is determined by the size and the profile of the heat and
power demand.

Keywords: combined heat and power, combustion, gasification

Introduction

The reduction of the emission of CO_2, a gas which contributes essentially to the global warming effect, is one of the most outstanding goals which has to be reached very soon to save our environment. A main source of CO_2-emission is the production of electricity and heat by burning fossil fuels like coal, fuel oil or natural gas. Today there are several renewable energy sources technically available, for example water power, wind energy, solar radiation, geothermal heat or biomass like wood, straw, crops etc. In contrast to the others, the advantage of biomass is, that in principle it can easily be stored and transported in a similar way to coal. Even the conversion of biomass to electrical power and heat can be done in an combustion plant with a steam turbine process, a technique well known in the utilisation of coal or waste. While the burning of biomass is a process emitting CO_2 the net CO_2-balance is zero. During their growth the plants resorb the same amount of CO_2 which is released afterwards.

For this reason the substitution of fossil fuels by biomass seems to be a very promising way to reduce the global CO_2-emissions. However there are features in the fuel which necessitate changes in the technique used as well as in the strategy of power production. The energy density of wood is much smaller than for coal or oil which discourages large distance transport. The kind of available biomass differs from region to region and requires in each case an adapted conversion technique. On a local scale a limited amount of biomass is available freely or at a very low cost. Examples are waste wood from the wood industry or straw from crop production. As a result the use of biomass favours decentralised heat and power plants with smaller capacities, in the range of a few to a few tens of MW. This is a major difference from existing power plants based e.g. on coal which are in the range of hundreds of MW. Therefore the technique of burning the biomass fuel and especially the technique of power production has to be adjusted to this case.

The work presented here is based on a study which investigates the present status of the use of biomass for heat and power production in Bavaria (Germany) and especially new techniques which appear to give high electricity production using biomass in small to medium scale heat and power plants.

This paper reports on the Bavarian situation and focus mainly on some thermodynamic basics and their consequences for the conversion of solid biomass fuels into electricity and heat. Moreover the difference between plants based on combustion or gasification is identified. Throughout this paper capacity denotes the net heating value of the fuel entering the combustion chamber per unit time, otherwise it is mentioned explicitly.

Status in Bavaria

In Germany the amount of biomass (wood and straw) used as primary energy is 110 PJ/a which equals 0,8 % of the total primary energy demand. About 25 PJ/a is attributed to boilers with a capacity lower than 15 kW (individual housings), 50 PJ/a is installed in the capacity range 15 to 1000 kW (multifloor buildings or small enterprises) and 35 PJ/a in heating plants or heat and power plants with a capacity over 1 MW (district heating networks and industry) [1]. A similar distribution can be assumed for Bavaria with 11,2 PJ/a primary energy supplied by biomass which is 0,6 % of the primary energy demand of Bavaria (1886 PJ/a). In the last few years there were installed several biomass combustion plants below 1 MW capacity supplying smaller district heating networks. These installations are supported by the Bavarian government by subsidies which can amount to 50% of the investment costs. In the following we restrict ourselves to the capacity range above 1 MW which is reasonable if combined heat and power production is to be considered. Fig. 1 shows the number of plants installed in Bavaria up to November 1994. They are ordered according to their capacity. Also shown is a sample of 41 plants which were covered by a survey and give an idea of the distribution into heating plants and heat and power plants. One can see that there are only few heat and power plants working at the moment. They are using the steam turbine process for power production except one which works with a steam engine. The electrical efficiencies are rather low (Fig. 2) although one has to state that the plants are in general not designed to yield maximum electricity output but to cover a certain demand of heat and power. The efficiency shows a clear dependence on the size of the plant. The steam engine (400 kW$_{el}$) shows a higher electrical efficiency than steam turbines of similar electricity output.

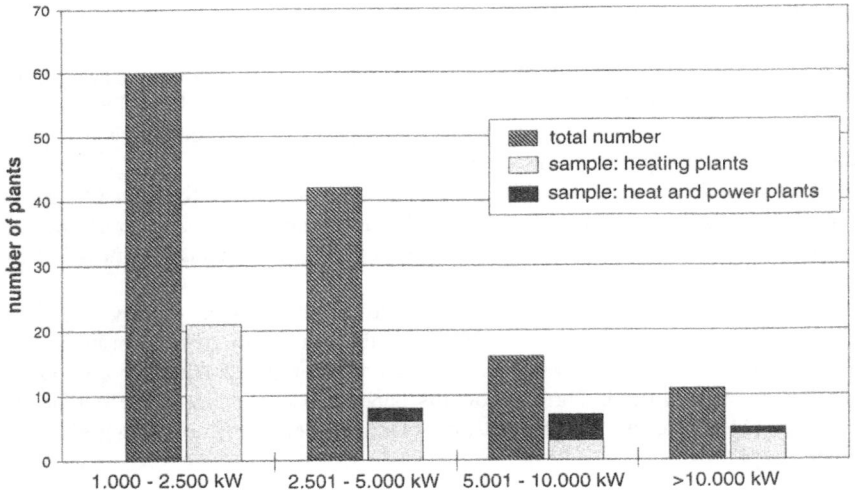

Fig. 1: Number of plants for heat and heat and power production installed in Bavaria categorised according to their capacity.

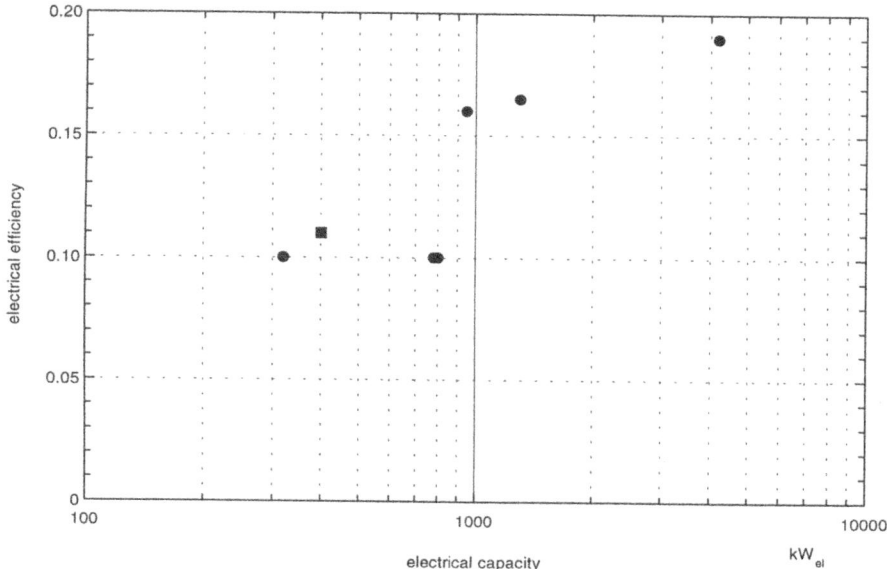

Fig. 2: Electrical efficiencies of existing biomass fired heat and power plants in Bavaria. All are using a steam turbine except the plant with the electrical capacity of 400 kW$_{el}$ which uses a steam engine

Heat and power production using combustion

From a thermodynamic point of view there are different ways to utilise the heat in hot flue gas after a combustion chamber. It can be used to yield steam or hot water for heating as its only product by using a suitable heat exchanger. The production of heat and power can be done using a certain energy conversion process based on a special thermodynamic cycle.

If the electrical efficiency of different concepts of existing or designed heat and power plants have to be compared one has to keep in mind that there are different ways of power production (figure 3). An idealised cycle has been depicted which fits the gliding temperature of the heat released by the flue gas and to the constant temperature at heat rejection.

Mode 1: All of the heat available is put into the energy conversion process and the output heat of the energy conversion process is released at ambient temperature. This yields maximum electricity output.

Mode 2: Part of the heat is put into the energy conversion process. The remaining heat of the flue gas is used to produce process heat or hot water for heating. The output of the energy conversion process is released at ambient temperature.

Mode 3: The output of the energy conversion process is released at a higher temperature to use it for process heat or hot water production (combined heat and power process).

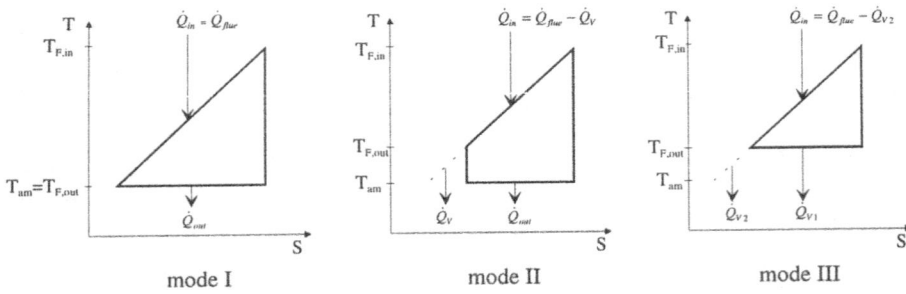

Fig. 3: Different ways of power production shown in a temperature-entropy-diagram. $T_{F,in}$, $T_{F,out}$: temperature of the flue gas before and after the energy conversion process; T_{am}: ambient temperature; $\dot{Q}_{flue}$: total heat flow of the flue gas, $\dot{Q}_{in}$: heat flow going into the energy conversion process; $\dot{Q}_{out}$: waste heat flow; $\dot{Q}_V$: valuable heat flow.

Going from 1 to 3 the electrical efficiency $\eta_{el} = W / \dot{Q}_{flue}$ decreases but the thermal efficiency $\eta_{th} = \dot{Q}_V / \dot{Q}_{flue}$ and also the total efficiency $\eta_{tot} = \eta_{el} + \eta_{th}$ increases. The decision for one of the three possibilities can only be done for a certain application.

As the heat of the flue gas is released not at a fixed temperature but with a large temperature glide, not all possible energy conversion processes are equally suitable. A Stirling process for example (figure 4a) or a steam process without pre- and superheating (for example also an ORC) are not well adapted because the temperature input to these systems is at a fixed temperature. Therefore the working potential of the heat stored in the flue gas at higher temperatures is lost.

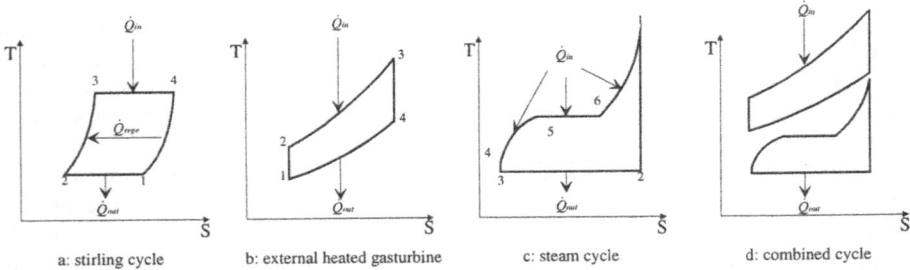

a: stirling cycle   b: external heated gasturbine   c: steam cycle   d: combined cycle

Fig. 4: Different thermodynamic cycles of power production

The external heated gas turbine process (figure 4b) fits much better because its gliding temperature of the heat input. But on the other hand the mean temperature of the reject heat is also raised, thus lowering the electrical efficiency. Figure 4c shows an advanced steam cycle with proper preheating and superheating sections and figure 4d a gas and steam combined cycle. Here the advantage of the gas turbine with the temperature glide of the heat input being very close to the temperatures of flue gas is combined with the possibility to reject the heat at constant low temperature. But in the case of a biomass combustion plant the available flue gas temperature, and therefore the possible gas and steam turbine parameters, are limited. The principle advantage of an increase in efficiency might be counterbalanced by the disadvantage of additional components and therefore higher costs. The steam turbine with proper preheating and superheating possibilities still seems to be a good choice. Please note the following: the above discussion assumes reversible processes, but in practice one has also to take into account irreversibilities appearing in the different components; for a particular application the technical availability and especially the specific investment cost have to be considered which in general favour plants with several tens of MW.

Heat and power production using gasification

Gasification of biomass is at present the most promising alternative conversion technique to combustion. A gasifier is specified by the following parameter: the gasifier efficiency η_{gas}, which is the ratio of the net heating value of the product gas to the net heating value of the biomass input and can range from 50 to 80% depending on the process; the net heating value of the product gas, which is typically about 4 to 6 MJ/m³; the composition of the product gas, which depends on the gasification conditions like temperature, pressure, oxidation medium, external heating.

In general the high content of hydrocarbons and especially tar in the product gas is a problem for the further utilisation of the gas. Disregarding the technical problem of the purity the starting point of the following discussion is the fact that the result of the biomass gasification is a combustible product gas.

One way to use this gas is to burn it in a gas burner resulting in a hot flue gas stream as for combustion. Although this seems to be identical to combustion there is the difference that the gasification process takes place with $\lambda<1$ and the necessary combustion air can be well adjusted in a gas burner ($\lambda \approx 1,2$). The total air supplied and therefore also the mass flow rate of the exhaust gas is reduced resulting in less heat losses by the exhaust gas compared to the combustion process with a typical value of $\lambda>1,5$. Furtheron the reduced exhaust gas mass flow allows for a smaller and less expensive gas cleaning section.

Another advantage is that the gas can be fed into an internal combustion device like a gas engine with a typical electrical efficiency of 40% or into a direct fired gas turbine. In this way all the energy chemically stored in the gas can be transported to the energy conversion process and no heat is lost with the flue gas leaving the burner. The gas turbine, which can be driven by external or internal combustion (see figure 5) can be used for comparison. In all cases the combustion temperature is limited to 1100°C as is usually done in biomass combustion due to the softening of the ashes.

The curve labelled η_{GT} in figure 5 shows the theoretical electrical efficiency of a gas turbine as function of the pressure ratio π of the turbine [2]. Commercial gas turbines using natural gas are operated at pressure ratios between 10 and 30 depending on their size. The real efficiency is mainly determined by the quality of the turbine and the compressor. Typical values for gas turbines with a capacity below 6 MW_{el} are indicated in figure 5 by the filled dots [3].

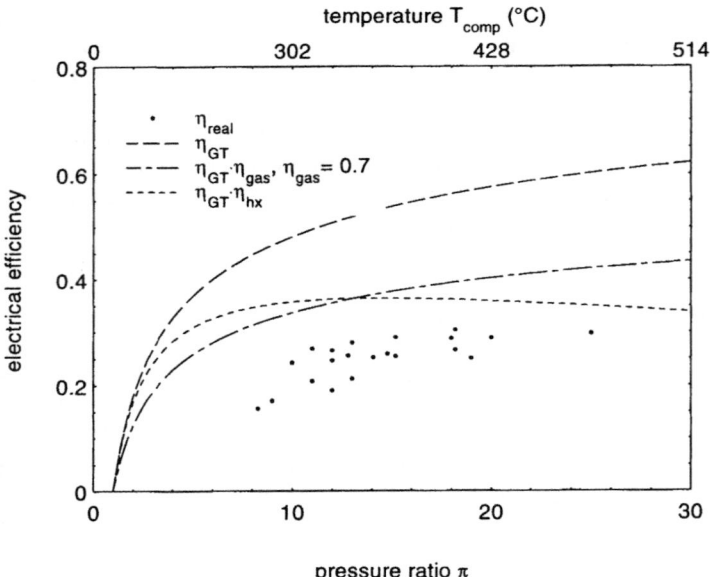

Fig. 5: Electrical efficiency of a gas turbine as function of the pressure ratio. The upper x-axis shows the temperature of the compressed air T_{comp}. η_{GT}: only gas turbine, $\eta_{GT}\cdot\eta_{gas}$: including gasifier efficiency, $\eta_{GT}\cdot\eta_{hx}$: gas turbine heated by flue gas, η_{real}: real efficiencies of gas turbine units using natural gas (0,5 to 6 MW_{el}).

Concerning the use of biomass the comparison must be referred to the net heating value of the solid fuel. Thus the gas turbine efficiency has to be multiplied with the gasifier efficiency η_{gas} which corresponds to the case where the gas is cooled down to room temperature e.g. for gas cleaning purposes. The result is shown in figure 5 by the curve $\eta_{GT}\cdot\eta_{gas}$ for a value of η_{gas}=0.7 . If the product gas enters the burner of the gas turbine at 1100 °C, i.e. without previous cooling, the sensible heat of the product gas is recovered. Neglecting heat losses this corresponds to the theoretical efficiency given in the curve η_{GT}. The difference demonstrates the importance of supplying the product gas as hot as possible to the gas turbine. Suitable ways of hot flue gas cleaning is therefore a major concern of present research.

In the case of combustion the heat input to the gas turbine is achieved by heat exchange between the hot flue gas and the compressed air. The temperature of the air entering the heat exchanger T_{comp}, which is determined by the pressure ratio, gives the minimum temperature to which the flue gas can be cooled down. For a fixed combustion temperature of 1100 °C thus the maximum amount of heat entering the gas turbine is limited. With increasing pressure ratio the amount of heat decreases. To compare with gasification the efficiency of the gas turbine has to be multiplied by the ratio η_{hx} of the heat conveyed to the gas turbine to the total available heat in the flue gas. The result $\eta_{GT}\cdot\eta_{hx}$ is also shown in figure 5. A flat efficiency maximum can be found at a pressure ratio of $\pi=13$. This has to be compared to the case of gasification. If the gas is cooled down completely gasification achieves a higher efficiency only if the pressure ratio exceeds a value of about 15 depending on the gasifier efficiency. But if the product gas can be supplied at a high temperature there is a clear advantage of the gasification process coupled with the internal combustion gas turbine.

A further advantage is obtained if the gas turbine (external and internal) can be coupled with a steam turbine to realise a combined cycle plant.

Existing and designed plants

In the following section some available data on existing or designed plants using the different techniques are presented. In Figure 6 the electrical vs. thermal efficiency is shown. The total efficiency is also given. Although one has to consider that these plants operate at different capacities and are in general not designed for maximum electricity production it can be seen that these values confirm the considerations in the preceeding paragraphs.

The highest electrical efficiency and also total efficiency is reached in an integrated gasification combined cycle plant. In a pilot plant in Värnamo, Sweden, it is tried to prove the feasibility of the technique. The electrical and also total efficiencies are higher than for a usual brown coal cogeneration plant (only steam turbine) which is shown for comparison. A drawback is that economic operation is predicted only for capacities above 50 MW. Figure 6 also shows, that if one is not interested in maximum electrical power output, different techniques can be suitable for different

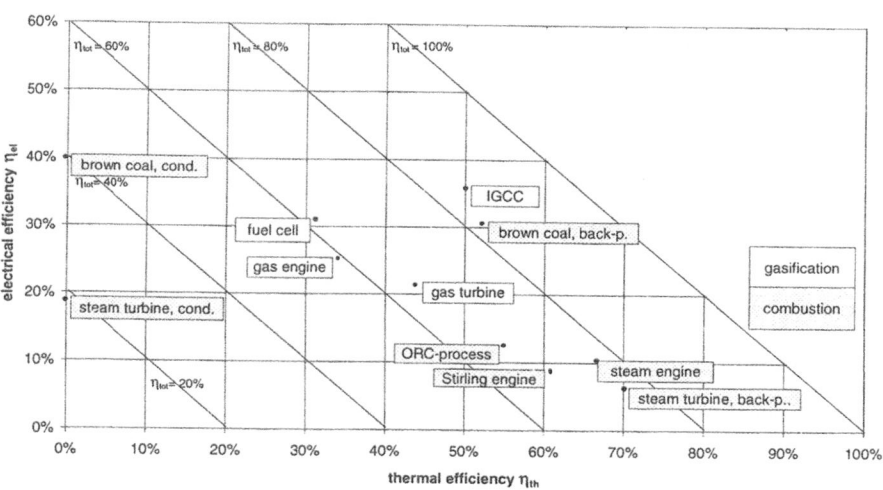

Fig. 6: Electrical and thermal efficiency of existing and designed plants. For comparison a brown coal cogeneration plant is included (back-p.: back-pressure operation, cond.: condensation operation). ORC: organic rankine cycle, IGCC: integrated gasification combined cycle.

applications according to the ratio of heat to power output required. The techniques using combustion in general yield only a smaller amount of electrical power and are thus suitable for applications where heat is the main product. The electrical power can than primarily be used to cover the own consumption of the plant. Regarding efficiency the differences between Stirling engines, steam engine and steam turbine seem to be not significant. It has to be mentioned that the values for the steam engine and steam turbine are data from existing plants while for the Stirling engine and the ORC-process it is only design data. Especially for those techniques using heat only at a high temperature level like the Stirling engine (typical 700°C) preheating of the combustion air by the exhaust gas gives an essential improvement of the electrical efficiency, e.g. from about 10% to about 20% in the case of the Stirling engine.

Summary

Power production with biomass fuels is still rarely found in Bavaria. Nevertheless legal regulations like the high fixed price for electricity produced by biomass (at the moment 153 DM/MWh$_{el}$) which have to be paid by the electric utility company favours plants with a high electrical power output. Starting with this premise a comparison of biomass combustion and gasification technique gives the result that gasification shows a higher potential for electricity production. Using the combustion technique not all possible energy conversion processes are equally suitable. A process with a heat input at gliding temperatures is preferred. For large plants the combination of gasification and combined cycle yields the highest electrical output.

In many applications a fixed ratio of heat to power rather than a maximum power output is required. Depending on this ratio different techniques are suitable. The decision for an installation is determined also by the technical availability and the investment costs rather than by the efficiency.

Acknowledgement

This work was supported by the Bavarian State Ministry of Regional Planning and Environmental Concern under contract no. 8000-722-31233.

References

1   M. Kaltschmitt, S. Becher (1994) Biomassenutzung in Deutschland - Stand und Perspektiven -; im Tagungsband *Thermische Nutzung von Biomasse - Technik, Probleme und Lösungsansätze*; Schriftenreihe „Nachwachsende Rohstoffe", Band 2; Fachagentur Nachwachsende Rohstoffe e.V. (Hrsg.); Stuttgart 1994

2   K. Strauß, *'Kraftwerkstechnik'*, Springer-Verlag, Berlin, 1994

3   Arbeitsgemeinschaft für sparsamen und umweltfreundlichen Energieverbrauch e.V. (ASUE), *'Gasturbinen-Kenndaten'*, Hamburg 1992

AN EVALUATION OF BIOMASS ENERGY POTENTIAL WITH A GLOBAL ENERGY AND LAND USE MODEL

H. YAMAMOTO
Central Research Institute of Electric Power Industry, Tokyo, Japan
K. YAMAJI
The University of Tokyo, Tokyo, Japan

Abstract
The authors evaluate global land use competition and bioenergy potential through developing a global energy and land use model using a SD (System Dynamics) technique. The model describes competition among various uses of biomass such as paper, timber, food, feed, and bioenergy that are produced not only on land but in the water. The model includes food chains from feed (including both crop feed and pasture) to meat and paper recycling process.
Through a simulation study, the authors find that the potential of energy crops is strongly limited due to a large land use requirement for food supply in the developing region of the world, and that effective use of biomass residues, both from food and timber production, may provide a large potential of bioenergy supply.
Keywords: bioenergy, biomass residues, food chain, system dynamics, land use model, recycling.

1 Introduction

Bioenergy is expected to become one of the key energy resources for global sustainable development. Biomass maintained adequately is renewable and free from net CO_2 emission. However, bioenergy cannot be infinite, since land area available for biomass production is limited and a certain amount of biomass must be reserved for food and material.

The purpose of the present study is to evaluate global bioenergy potential in consideration of land use competition. For this purpose, the authors expanded a global energy and land use model developed earlier by the same authors using a SD (System Dynamics) technique [1] [2], which can describe the relation between flow and stock of resources explicitly. Bioenergy potential can be estimated dynamically and comprehensively using this model, which covers a wide range of land uses and food and material chains.

2. The global energy and land use model

2.1 Outline of the model

The global energy and land use model consists of an energy sub-model and a land use sub-model (Fig. 1). The energy sub-model, the basic structure of which is formulated following that of the Edmonds-Reilly model [3], is extended with the addition of a module of chemical material flows including plastics, synthetic rubbers, and synthetic fibers, scrap of which has a certain energy potential in municipal wastes.

The land use sub-model, which consists of two sectors, a food sector and a forest sector, describes land use competition among various uses for biomass applications such as paper, timber, food, feed, and bioenergy. The sub-model includes food chains from feed (including crop feed and pasture) to meat, paper recycling, and various forms of biomass residues.

The time scope of the model is 125 years from 1975 to 2100 with one year time step. In the model, the world is divided into two regions: a developed region and a developing region. The developed region comprises OECD countries (excluding Turkey), former USSR, Israel, and South Africa; the developing region includes all the other countries.

The following sections provide a detailed description of the land use sub-model.

2.2 Biomass balance table of the land use sub-model

The authors developed a "biomass balance table" that shows 33 kinds of biomass in the row and 23 steps of biomass utilization processes in the column (Table 1).

The signs of "+", "–", and "Σ" mean production, consumption, and subtotal of biomass, respectively. The land uses are grouped into four kinds: forest, arable land, pasture, and the other land. Biomass from the water is also dealt with. The kinds of biomass classed with four groups: primary, intermediate, secondary, and scrap *1. Then, the secondary biomass is divided into three kinds: material (paper and timber), food (vegetable and animal), and bioenergy (traditional and modern).

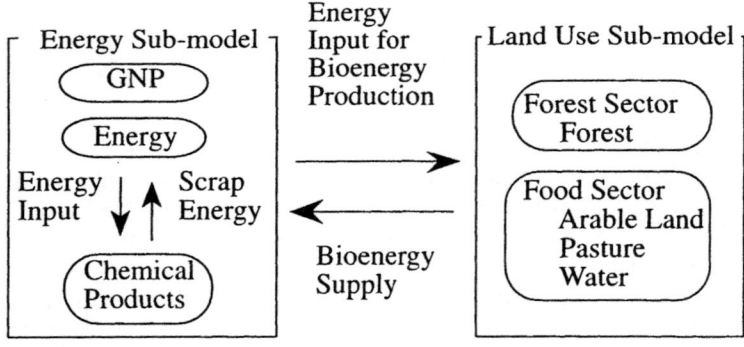

Fig. 1. Outline of the global energy and land use model.

*1 "Primary biomass" is defined as the form of biomass just harvested from land or the sea; "secondary biomass" is the form of biomass supplied to final consumers.

Biomass balances of forest biomass, food biomass, and bioenergy are described in more detail in the following sub-sections.

2.2.1 Forest biomass in the biomass balance table

Uses of forest biomass are shown in Table 1 and Figure 2.
(1) Primary biomass supply
Roundwood is used for wood pulp production and timber production. Fuelwood is used for modern and traditional bioenergy production. In the felling process roundwood residues such as twigs and leaves are left over in the forest. A part of the residues can be utilized for modern bioenergy.
(2) Biomass conversion
Roundwood is converted into timber and woodpulp with mill residues and black liquor, respectively. Most mill residues and black liquor are already used for modern bioenergy . Paper is produced from woodpulp, other fiber pulp (pulp made from straw, bamboo, etc.) and a part of paper scrap.
(3) Biomass consumption
Paper consumed by final consumers is stocked for several years and then scrapped. Timber consumed is stocked for several decades and then scraped. A part of the scrap is recycled; the rest can potentially be used for modern bioenergy.

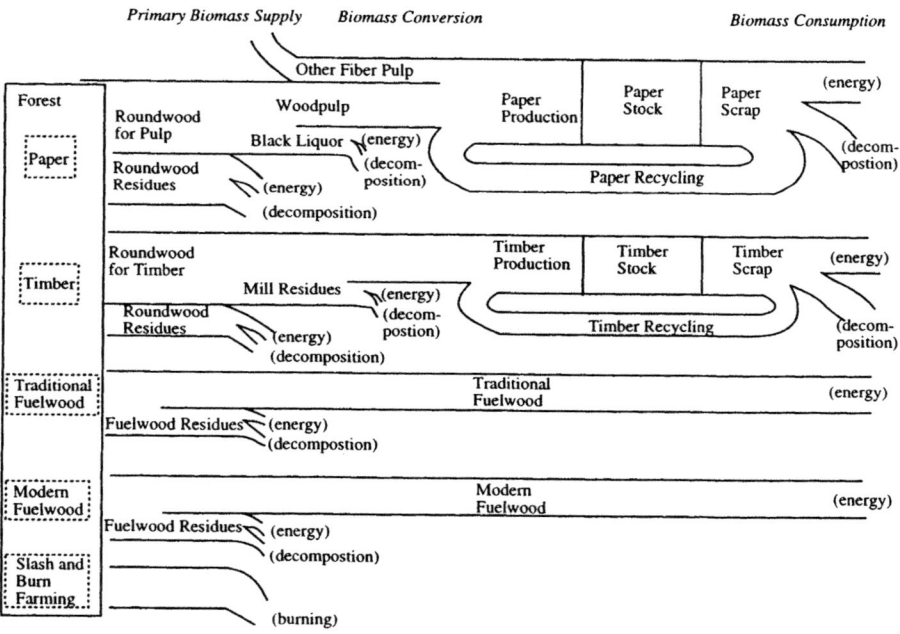

Fig. 2. Flow of forest biomass.

2.2.2 Food biomass in the biomass balance table

Uses of food are shown in Table 1 and Figure 3.

(1) Primary biomass supply

Arable land is used for production of the following five crops: "energy crops", "cereals", "roots", "sugarcane", and "the other crops". Energy crops are used for modern bioenergy exclusively. Sugar is produced from sugarcane, with a byproduct of bagasse that can be used for modern bioenergy. Cereals, roots and the other crops are used for vegetable food production, feed production, and loss (including seed and fertilizer use). A part of residues of cereals and sugarcane can be utilized as modern bioenergy.

On the other hand, pasture is used for feed production exclusively. Fish (including shellfish) caught from the water is used for animal food production except for loss. There is no useful biomass from the other land including dessert, tundra, built-on areas, roads, and barren land.

(2) Biomass conversion

Feed is made from pasture and a part of cereals, roots, sugarcane, the other crops, and meat. In the meat production process livestock produces meat with dung that can be used for modern bioenergy partly.

Vegetable food is made from cereals, roots, the other crops, and sugar. Animal food is the total of edible meat and fish.

(3) Biomass consumption

Kitchen refuse and human faeces are emitted as biomass scrap in the process of food consumption. They can be used for modern bioenergy partly.

The model does not include natural rubbers, natural fibers (e.g., cotton and wool), and aquatic energy crops (e.g., giant kelp) yet; that is the subject of the future study.

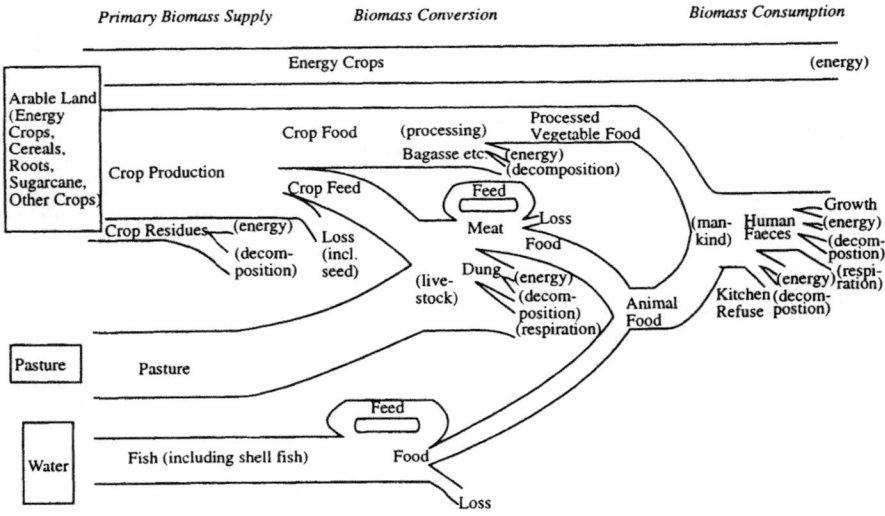

Fig. 3. Flow of food biomass.

Table 1. Biomass balance table (1 of 2).

		Primary (Forest)					Intermediate			Secondary		Scrap		Primary (Arable land)				
		(1) Round-wood	(2) Fuel-wood	(3) Other fiber 4)	(4) Round-wood residues	(5) Fuel-wood residues	(6) Wood-pulp	(7) Black liquor	(8) Mill residues	(9) Paper	(10) Timber	(11) Paper scrap	(12) Timber scrap	(13) Energy crops	(14) Cereals	(15) Roots	(16) Sugar-cane	(17) The other crops
Primary biomass supply	(1) Harvest	+1)	+	+	+5)	+5)								+	+	+	+	+
	(2) Import	+	+	+											+	+	+	+
	(3) Export	-2)	-												-	-		-
	(4) Supply in the region	Σ3)	Σ	Σ	Σ	Σ								Σ	Σ	Σ	Σ	Σ
Biomass conversion	(5) Wood pulp production	-					+	+										
	(6) Paper production			-			-			+		-						
	(7) Timber production	-							+		+							
	(8) Sugar production																-	
	(9) Vegetable food production														-	-	-	-
	(10) Feed production														-	-		-
	(11) Meat production																	
	(12) Animal food production																	
	(13) Trad. bioenergy production		-											-				
	(14) Modern bioenergy production		-									-						
	(15) Loss (incl. seed and fertilizer)				-	-			-			-	-					
Secondary biomass supply	(16) Import									+	+		-					
	(17) Export									-	-							
	(18) Supply in the region									Σ	Σ							
Biomass consumption	(19) Paper consumptn									-		+6)						
	(20) Timber consumption										-		+6)					
	(21) Food consumption														-	-		-
	(22) Trad. bioenergy consumption																	
	(23) Mod. bioenergy consumption																	

1) "+" means that the kind of biomass is produced here.
2) "-" means that the kind of biomass is consumed here.
3) "Σ" means that the subtotal is computed here.

4) Other fibers mean fibrous vegetable material other than wood. They include straw, bamboo, etc.
5) These residues are produced in the processes of harvesting wood and crops.
6) Paper and timber are scrapped after 10 - 100 year stock period.

Table 1. Biomass balance table (2 of 2).

	Primary (the other areas)					Intermediate					Secondary		Scrap		Bioenergy	
	(18) Pasture	(19) The other land	(20) Fish from the water	(21) Cereal residues	(22) Sugarcane residues	(23) Sugar	(24) Bagasse	(25) Feed total	(26) Meat	(27) Dung	(28) Veg. food	(29) Animal food	(30) Kitchen refuses	(31) Human faeces	(32) Trad. bio-energy	(33) Modern bio-energy
Primary biomass supply																
(1) Harvest	+		+	+ 5)	+ 5)											
(2) Import			+													
(3) Export			-													
(4) Supply in the region	Σ		Σ	Σ	Σ											
Biomass conversion																
(5) Wood pulp production																
(6) Paper production																
(7) Timber production																
(8) Sugar production						+	+									
(9) Vegetable food production						-					+					
(10) Feed production	-							+								
(11) Meat production								-	+	+						
(12) Animal food production			-						-			+				
(13) Trad. bioenergy production										-					+	
(14) Modern bioenergy production				-	-		-			-			-	-		+
(15) Loss (incl. seed and fertilizer)			-	-	-		-		-	-			-	-		
Secondary biomass supply																
(16) Import											+	+				
(17) Export											-	-				
(18) Supply in the region											Σ	Σ			Σ	Σ
Biomass consumption																
(19) Paper consumption																
(20) Timber consumption																
(21) Food consumption											-	-	+	+		
(22) Trad. bioenergy consumption															-	
(23) Mod. bioenergy consumption																-

5) These residues are produced in the processes of harvesting wood and crops.

2.2.3 Bioenergy in the biomass balance table

Bioenergy can be classified into two kinds; traditional bioenergy and modern bioenergy [4] [5]. Traditional bioenergy means non commercial bioenergy mainly used in small scale with low energy efficiency (mostly less than 15%). Modern bioenergy means commercial bioenergy mainly used in large scale with high energy efficiency (mostly more than 60%).

A part of fuelwood and a part of dung are used for traditional bioenergy. Modern bioenergy comprises fuelwood, roundwood residues, black liquor, mill residues, paper scrap, timber scrap, energy crops, crop residues, bagasse, dung, kitchen refuse, and human faeces.

2.3 The algorithm of the land use sub-model

2.3.1 The algorithm of the forest sector
Supply and demand of forest biomass are calculated as follows:
(1) Demand of felling area is calculated using population, wood demands per capita (such as paper, timber, traditional fuelwood, modern fuelwood, un-sustainable slash-and-burn farming), and un-sustainable slash-and-farming area *2.
(2) It is assumed that mature forest area *3 can be felled but the growing forest cannot be. If the demand of the felling areas exceeds the supply of the mature forest area, the model allocates felling areas in proportion to the relative demand sizes of the felling areas.

2.3.2 The algorithm of the food sector
A supply potential of energy crops are calculated in the food sector as follows (Fig. 4) :
(1) Food demands (vegetable and animal) are calculated from the population multiplied by the food demand per capita.
(2) The model computes vegetable and animal food supply produced from arable land (except for energy crops and cereals), pasture, and the water.
(3) Cereal demand (for vegetable and animal food production) is equal to the subtraction of the food supply in (2) from the food demand in (1). The demand of animal food is converted into the demand of vegetable feed.
(4) It is assumed that the priorities of the land uses for energy crops and cereals are given as follows:
    1. Cereal production for the own region.
    2. Cereal production for the other region (if food shortage occurs in the other region).
    3. Energy crop production.

---

*2 Slash-and-burn farming is divided into two kinds: sustainable and un-sustainable. Sustainable slash-and-burn farming, defined as cyclic farming repeating slash-and-burn, farming, and natural recovery of the forest, scarecely emits net CO2 through the cycle. Un-sustainable slash-and-burn farming is defined as non-cyclic farming, and emits net CO2.

*3 It is assumed that growing forest, which is forest after felling and reforesting, grows for several decades (see Chapter 3) and is turned into mature forest. In mature forest, the NPP (Net primary product) balances with the rotten biomass.

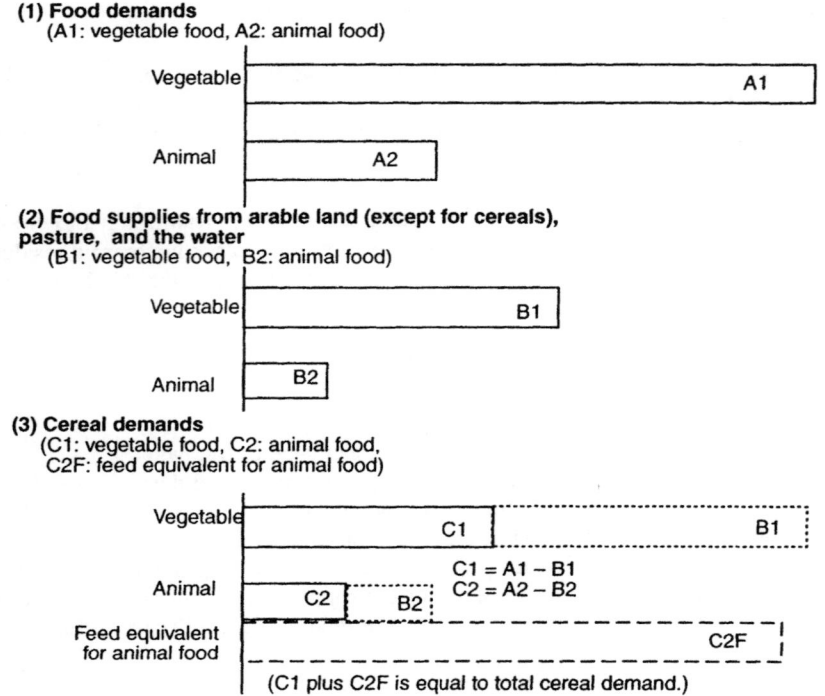

(1) Food demands
(A1: vegetable food, A2: animal food)

Vegetable A1

Animal A2

(2) Food supplies from arable land (except for cereals), pasture, and the water
(B1: vegetable food, B2: animal food)

Vegetable B1

Animal B2

(3) Cereal demands
(C1: vegetable food, C2: animal food, C2F: feed equivalent for animal food)

Vegetable .... C1 .... B1

Animal .... C2 .... B2

C1 = A1 − B1
C2 = A2 − B2

Feed equivalent for animal food C2F

(C1 plus C2F is equal to total cereal demand.)

Fig. 4. The calculation of food supply and demand.

It is assumed that cereals are distributed in proportion to the relative sizes of food and feed demands in a region where food shortage will arise in spite of cereal import. The model does not adjust either demand or supply of food but merely calculates the amount of the food shortage.

3. Input data for the base case

The authors prepared input data for the base case of the model simulation as follows (Table 2, Table 3).

3.1 Population, GDP, and energy

The authors adopted the middle case of World Bank projections for the population scenario [6], and the IS92a of IPCC for the GDP and the energy demand scenario [7]

The model deals with chemical products, scrap of which can be used for energy as well as biomass scrap. The production of the chemical products is calculated on the assumption that the ratio of the feedstock demand to the secondary energy demand in each region is fixed at the level in 1990.

3.2 Land uses

It is assumed that in the developed region the forest area and the pasture area

Table 2. Main input data for the base case.

	Developed Region	Developing Region
Population (million)	1480 (in 2050) 1500 (in 2100)	8580 (in 2050) 10160 (in 2100)
GDP per cap. (thousand US$(1990)/cap.)	13.4 (in 2050) 85.8 (in 2100)	0.8 (in 2050) 14.2 (in 2100)
Biomass demand per cap. (Veg. food, Anim. food, paper, timber, and trad. fuelwood) (Modern fuelwood)	Constant at the 1990 level Not installed	To reach the level of the developed region by 2100 Not installed
Forest protection (Perfect reforestation and extermination of un-sustainable slash-and-burn farming)	Achieved	To be achieved by 2050
Additional arable land area	To change current fallow land (68 Mha) into arable land by 2025	To divert 30% of deforestation area to arable land
Crop productivities (1.0 in 1990)	1.74 (in 2050) 1.77 (in 2100)	2.19 (in 2050) 2.49 (in 2100)
Meat productivities (Meat production per feed in energy, %)	11 (in 1990) 13 (in 2050) 13 (in 2100)	9 (in 1990) 13 (in 2050) 13 (in 2100)
Fish production	Constant at the 1990 level	Constant at the 1990 level
Main forest data • Organism stock (t-C/ha) • Net annual increment of organisum stock (t-C/ha/yr) • Growing period (yr)	100 2.5 40	150 5.0 30
Import and export • Cereals •The others	To be determined by food supply and demand Constant at the 1990 export rates	To be determined by food supply and demand Constant at the 1990 export rates
Biomass uses • Cereals • The others	To be determined by food supplyand demand Constant at the 1990 export rates	To be determined by food supply and demand Constant at the 1990 export rates
Recycling rates • Paper • Timber and chemical products	65 % after 2050 To be assumed zero	65% after 2050 To be assumed zero

will be constant at the 1990 level; fallow land (68 million hectares in 1990 [8]) in the other land will be added to the arable land area by 2025. It is assumed that in the developing region imperfect reforestation and un-sustainable slash-and-burning farming will fade away by 2050. It is assumed that in the developing region seventy percent of the deforestation area will be shifted to pasture and the other will be to permanent arable land. That depends on historical land use changes for the past 20 years [9].

It is assumed that areas for each kind of crop, except cereals and energy crops, will be constant at the level in 1990; areas for cereals and energy crops will vary depending on food supply and demand.

3.3 Biomass demand

It is assumed that each biomass demand (paper, timber, vegetable food, animal food, traditional fuelwood) per capita will be constant in the developed region. Considering that GDP per capita in the developing region reaches the 1990 level of the developed region by 2100 [7], the authors assume that biomass demand per capita in the developing region will reach the 1990 level of the developed region by 2100. It is assumed in the base case that modern fuelwood will not be installed, while biomass residues that does not call for lands for biomass production will be used for bioenergy partly.

3.4 Biomass supply

According to the base scenario of IMAGE 2.0 [10], the productivity of food crops in the developed region will increase by 77% by 2100 compared with that in 1990; similarly, that in the developing region will increase by 149% compared with

Table 3. Discharge rates of biomass residues.

	Process	Discharge rates
(Forest)		
Roundwood residues	Harvesting	39% of forest organisms
Fuelwood residues	Harvesting	20% of forest organisms
Black liquor	Processing	45% of roundwood
Mill residues	Processing	45% of roundwood
Paper scrap	Consuming	26%/yr. of paper storage
Timber scrap	Consuming	3.0%/yr. of timber storage
(Food)		
Cereal residues	Harvesting	1.3 t/t-cereals
Sugarcane residues	Harvesting	0.279t/t-sugarcane
Bagasse	Processing	0.150t/t-sugarcane
Dung	Processing	0.3 J/J-feed
Kitchen refuse	Consuming	0.1 J/J-food supply
Human faeces	Consuming	0.2 J/J-food consumption
(Chemical products)		
Chemical products	Consuming	8.4%/yr. of prod. storage

that in 1990. The productivity of meat (defined as meat per feed in a unit of energy) will increase to 13% by 2100. The catch of fish is assumed to be constant at the level in 1990.

The net annual increments of organisms under growing forest are assumed to be 2.5 t-C/ha/yr. in the developed region and 5.0 t-C/ha/yr. in the developing region [11].

The discharge rates of biomass residues are assumed as shown in Table 3 [8] [12]

It is assumed that export rates (defined as exports per production in an exporting region) and ratios of biomass uses (for pulp, timber, food, feed, loss, etc.) except for cereals are constant at the level in 1990. The export rates and the ratios of cereal uses are calculated in the model according to food supply and demand.

3.5 Recycling rate

It is assumed that the recycling rate of paper scrap will reach 65% –the practical maximum value [11]. The recycling rates of scrap of timber and chemical products are assumed to be zero; and all the scrap is added to the energy potential.

4. Results

4.1 Changes in land use

There will be little changes in land uses in the developed region, since the demand of biomass is stabilized and the reforestation has been done perfectly.

On the other hand, the mature forest area in developing region will decrease from 2.1 billion hectares in 1990 to 1.1 billion hectares in 2050, because of the rapid increase of the biomass demand and the incomplete forest management. The mature forest area in the developing region will continue to decrease to 0.4 billion hectares in 2100, although perfect reforestation is assumed to be achieved by 2050 (Fig. 5).

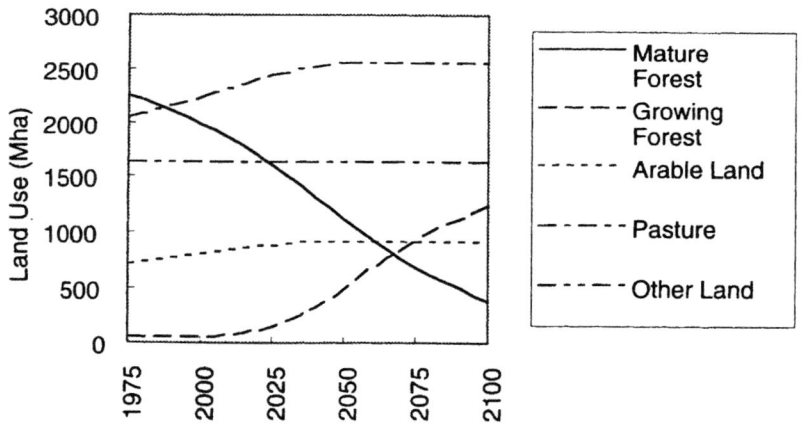

Fig. 5. Land use changes in the developing region.

4.2 Food supply and demand

The food demand will be satisfied completely in the developed region. However, the food demand cannot be met in the developing region, even with food import from the developed region (Fig. 6). This is caused by enormous feed demand originating in the animal food demand in the developing region. Thus, there will be no available land to produce energy crops owing to the food shortage in the world.

4.3 Maximum energy potential

Maximum energy potential is defined as the sum of bioenergy production and all the discharged biomass residues, which include other uses like seed and fertilizer. Figure 7 and Figure 8 show the maximum energy potential of various kinds of biomass and chemical product scrap in the developed region and the developing region, respectively.

As shown in Figure 7 and Figure 8 there will be a large maximum energy potential for roundwood residues, mill residues, timber scrap, cereal residues, and dung; but there will be only a small potential for kitchen refuse, human faeces, paper scrap, and chemical product scrap.

It is important to evaluate practical energy potential in consideration of practical utilization rates of biomass residues, because biomass residues should be substantially consumed for fertilizer and have a large recovery loss. Biomass residues produced at factories such as bagasse, black liquor, and mill residues are exceptions and can be used mainly for energy.

5. Conclusions

The authors developed the global energy and land use model including most kinds of biomass resources and processes (such as harvesting, processing, and consuming) with a SD technique. Through a simulation, the authors obtained the following results.

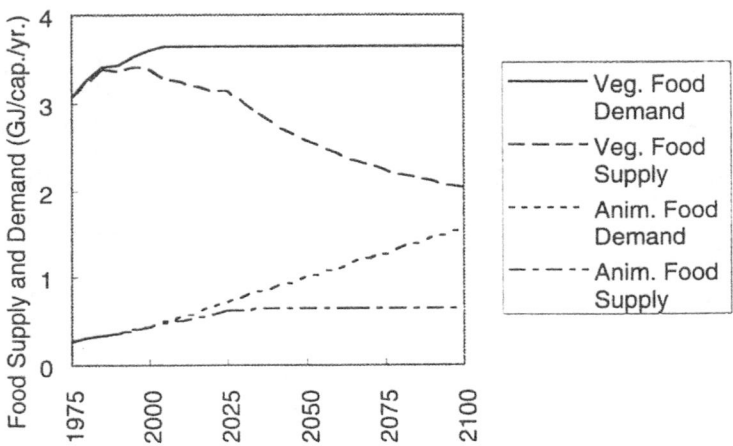

Fig. 6. Food supply and demand (vegetable and animal) in the developing region.

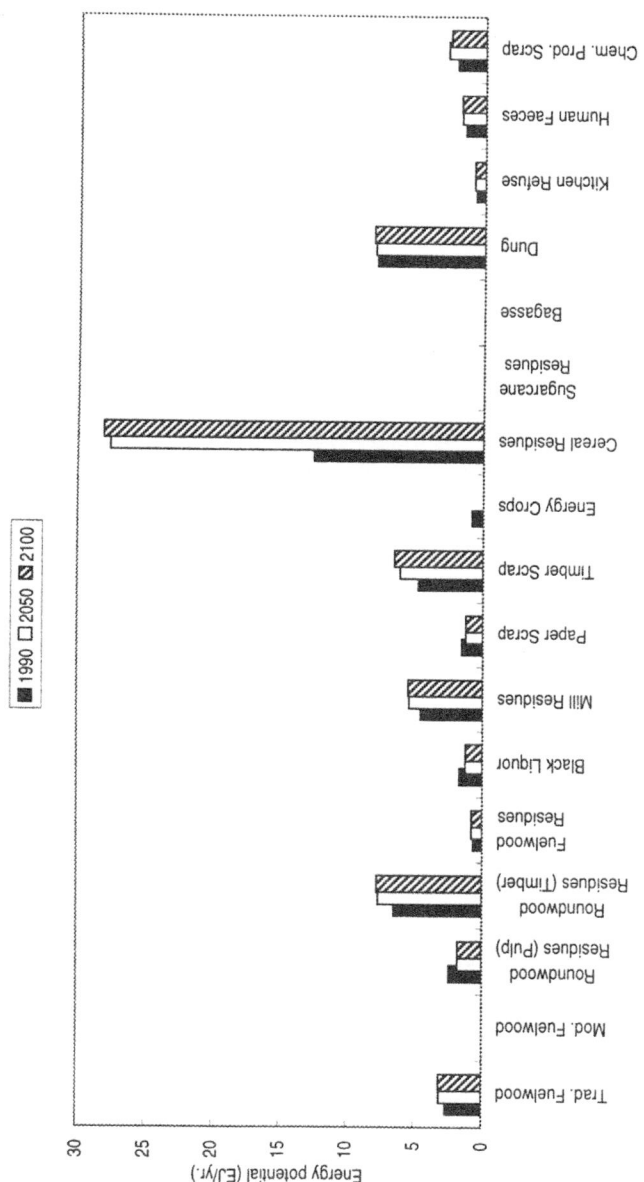

Fig. 7. Maximum energy potential of biomass and chemical products in the developed region.

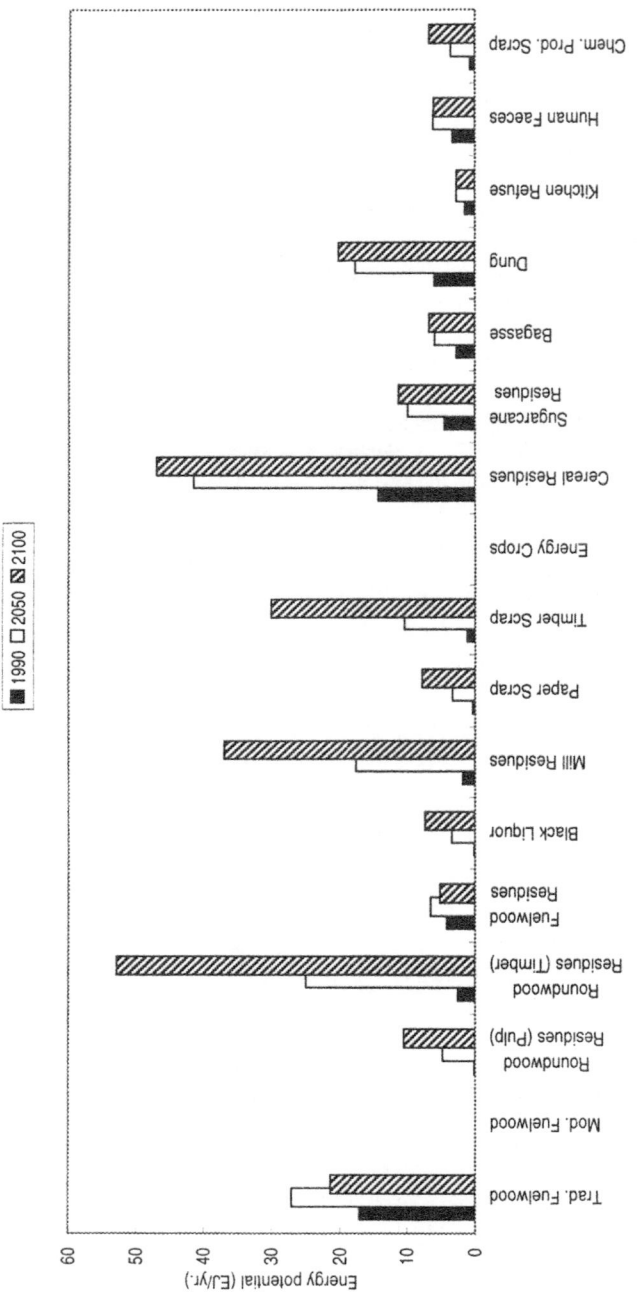

Fig. 8. Maximum energy potential of biomass and chemical products in the developing

1) The mature forest area in the developing region will decrease from 2.1 billion hectares in 1990 to 0.4 billion hectares in 2100, although perfect reforestation is assumed to be achieved by 2050. Thus, it is difficult to fell more fuelwood from the forest in the developing region.

2) Energy potential of energy crops will be strongly limited due to a large land use requirement for food supply in the developing region. Food demand cannot be met in the developing region, even with import of food from the developed region.

3) There will be a large energy potential for cereal residues, dung, roundwood residues, mill residues, and timber scrap. It is important to evaluate practical energy potential in consideration of practical utilization rates of biomass residues, because biomass residues should be substantially consumed for fertilizer and have a large recovery loss.

The authors plan to do sensitivity analysis using the model by varying major parameters such as population, biomass demand, and biomass productivities.

Acknowledgment

This study is supported by a grant from Sumitomo Foundation.

References

1. Yamamoto, H., Yamaji, K., and Sugiyama, T. (1994) *Analysis of Land Use Competition among Productions of Food, Timber, and Biomass Energy with a System Dynamics Model*, Proceedings of International Conference on EcoBalance, Oct. 25-27, Tsukuba, Japan.

2. Yamamoto, H. and Yamaji, K. (1994) *An analysis of Biomass Production Potential with a Global Energy and Land Use Model* (Y94004), Central Research Institute of Electric Power Industry, Tokyo, Japan.

3. Edmonds, J. and Reilly, J. (1983) *A Long-term Global Energy-economic Model of Carbon Dioxide Release from Fossil Fuel Use*, Energy Economics, pp. 283-297, April.

4. Dessus, B., Devin, B., and Pharabod, F. (1992) *World potential of Renewable Energies*, Extraits de la Houille Blanche n 1-1992, Paris.

5. World Energy Council (1993) *Renewable Energy Resources: Opportunities and Constraints 1990 - 2020*, WEC Report.

6. Bos, E. et al. (1993) *World Population Projections 1992-93 Edition*, The John Hopkins University Press.

7. Pepper, W. et al. (1992) *Emission Scenarios for IPCC An Update*, IPCC Working Group 1.

8. Johansson, T. B. et al. eds. (1993) *Renewable Energy*, Island Press.

9. FAO (1993) *FAO Yearbook Production 1992*, vol. 46, FAO, Rome.

10. Alcamo, J. ed. (1994) *IMAGE 2.0 Integrated Modeling of Global Climate Change*, Kluwer Academic Publishers.

11. Yamamoto, H. and Yamaji, K. (1996) *A Comprehensive Evaluation of Biomass Resources with a Global Energy and Land Use Model*, The 10th Conference of Energy Systems and Economics, Tokyo, Japan.

12 FAO (1993) *FAO Yearbook Forest Products 1980-1991*, FAO Forestry Series No. 26, FAO, Rome.

WORKSHOPS

WORKSHOP ON THE FUNDAMENTALS OF THERMOCHEMICAL BIOMASS CONVERSION

Workshop on the fundamentals

M.A. CONNOR
Department of Chemical Engineering, University of Melbourne,
Parkville, 3052, Australia
J.P. DIEBOLD
Consultant, 57 N Yank Way, Lakewood, Co.
K. SJÖSTRÖM
Department of Chemical Engineering and Technology/ Chemical Technology,
Kungl Tekniska Högskolan (KTH), Stockholm

Keywords: workshop, fundamentals, thermochemical, biomass, conversion.

1 Introduction

For the purposes of this workshop the term "fundamentals" was taken to mean the basic science (of a chemical and physical nature) underlying the engineering side of thermochemical biomass conversion. The variety of fundamental aspects of thermochemical biomass conversion still needing further study was made clear by the range of interests of participants in the workshop. These interests encompassed topics as diverse as:

- the influence of biomass properties on patterns of decomposition during pyrolysis.
- gasification processes, including pressure-related effects, catalytic aspects, gas-solid reactions and char reactivity.
- ash-related problems
- nitrogen- and chlorine- containing compounds in biomass and their decomposition pathways.

Such fundamental aspects were seen as having particular relevance to biomass conversion applications in the areas of scale-up and unsteady-state or non-uniform operating conditions. The latter area was seen as encompassing variations in biomass composition or moisture content, changes in demand, e.g. for gas from a gasifier; and desired or undesired changes to process operating conditions.

The Workshop was divided into two sessions. The first of these was a ground-breaking session devoted largely to unstructured discussion of a variety of gasification-related topics. This report covers primarily the more structured discussions of the second session and the areas identified then as needing further study.

To facilitate discussion, topics were assigned to one of three categories:

- intra-particle aspects
- processes at the solid/fluid interface
- aspects linked to applications.

2 Intra-particle aspects

Factors felt to be of relevance to decomposition processes and pathways within heated biomass particles included:

- chemical composition, including moisture content, the percentages present of both the major biomass components (cellulose, lignin and hemicellulose) and the minor components (extractives, ash-forming elements and chlorine and nitrogen compounds).
- biomass structure, reflecting the porous, anisotropic, and chemically non-uniform (at a microstructural level) nature of biomass feedstocks.
- primary decomposition reactions, including kinetic aspects and reaction pathways
- secondary reactions undergone within particles by volatile products of primary decomposition reactions, homogeneous gas phase reactions, heterogeneous gas-solid reactions and catalytic aspects.
- heat transfer
- mass transfer

Discussion of intra-particle aspects largely involved identification of areas where our understanding is as yet incomplete. The first topic addressed was that of ash-forming elements. Of especial interest was how these elements are distributed through the biomass from a microstructural viewpoint, and in what chemical forms they are present.

If non-uniformly distributed, as there appeared to be some evidence to suggest is the case, precisely where they are concentrated could affect the degree to which they might catalyse secondary reactions, for example, as well as the extent to which volatilisation of these elements would be likely in the thermal decomposition process. It could also be expected to affect how well experiments involving impregnation of biomass with specific ash-forming elements prior to thermal treatment could be expected to represent what actually occurs in decomposing biomass.

A second aspect discussed was lignin decomposition and the extent of its influence on char yield. Questions were raised about the relevance of studies using extracted (and therefore chemically and physically modified) lignins to what occurs when naturally occurring lignins undergo decomposition within the biomass particle.

The third topic considered in the intra-particle category was the nature of the secondary reactions that occur within heated biomass particles and that contribute substantially to char-formation. An aspect of particular interest was the role played by inorganic compounds within the biomass.

Finally, there was some discussion as to whether aerosol formation was a process occurring solely through condensation of volatiles in the gas phase or whether processes occurring within particles or at their surfaces during rapid devolatilisation exercised an influence on aerosol production.

3 Processes at the solid/fluid interface

Areas felt to fall into this category included:

- heat transfer between particles and their surroundings.
- mass transfer between particle surfaces and the surrounding gas.
- the nature, and susceptibility to removal by abrasion, of the outer layer of a decomposing particle (ash readily erodes, for example, but the silica skeleton of a rice husk does not).
- surface reactions, and the influence of oxidising and reducing conditions.
- fluid dynamics in the immediate vicinity of the particle surface.
- particle/solid interactions (e.g. collisions).

Discussion of this area focused primarily on char reactivity, which has been observed to change rapidly over comparatively short periods of time. Factors suggested as having the potential to affect char reactivity included the nature and amount of ash-forming elements present, and the oxygen content of the original biomass.

A second topic addressed was that of surface temperature, its importance in any model of thermal biomass decomposition, and the difficulties involved in obtaining accurate measurements of surface temperatures in practice.

4 Aspects linked to applications

Discussion on this class of fundamental aspects centred on problems related to ash and nitrogen. In the case of ash, areas requiring further study were felt to be:

- alkali metals, the stage when and the form in which they leave decomposing particles.
- what are the consequences when chars with a disproportionately high content of potassium are burned.
- the assimilation of ash, formed in the combustion or thermal conversion of biomass, when returned to forested lands.

In the case of nitrogen, it was felt that too little was still known about the molecular forms in which nitrogen is released from decomposing particles and how NO_x is generated during thermal biomass

5 Concluding remarks and recommendation

The Workshop concluded that despite the emphasis at the conference on <u>applications</u> of thermal biomass conversion, further study of a diversity of fundamental aspects of this subject remained extremely important. Part of the reason for the perceived imbalance between papers on fundamentals and papers on applications was felt to lie in the much greater ease with which funding appeared to be obtainable for applications with the potential to generate some future financial return. Concern over this situation led to the formulation of the following recommendation, endorsed at the final Conference plenary session:

"That this meeting strongly supports the importance of continuing research into the fundamental aspects of thermochemical biomass conversion and recommends that funding agencies ensure that adequate funds continue to be made available for fundamental research in the above area".

6 Acknowledgement

The contribution of Dr J. Lédé to the structuring of the second Workshop session is much appreciated.

WORKSHOP ON CHEMICALS FROM BIOMASS

H PAKDEL
Pyrovac Inc.
1560 avenue due Parc Beauvoir, Sillery, QC, G1T 2MT, Canada
J PISKORZ
RTI Ltd
110 Baffin Pl. #5, Waterloo, ON., N2V 1Z7, Canada
A HIMMELBLAU
Biocarbons Corporation
71 Cummings Office Park, Woburn, MA 01801, USA
D CLEMENTS
University of Nebraska
22 L.W. Chase Hall, Lincoln, NE 68583-0726, USA

Abstract

This is a summary of the workshop that discussed and reviewed the production of chemicals from biomass.
Key words: Thermolysis, Pyrolysis biomass derived chemicals, Biomass refinery, Wood smoke.

Introduction - Definition of Chemicals

The main volatile products of biomass thermolysis exit the high temperature zone of all pyrolysers (and gasifiers) in the form of gases, vapours and aerosols, the last ones visible to naked eye as smokes, fumes or fog. Such smokes are formed to a significant degree even in the presence of air. Meat smoking was probably the first chemical use of biomass thermolysis and we are all familiar with the visual effects of tobacco biomass pyrolysis.

The estimated number of biomass pyrolysis "smoke" compounds range from one thousand to several thousands. 10,000 different compounds are possible (1). Hundreds of compounds exist in the condensed, crude biomass pyrolysis oil.

The general public can easily differentiate between so called natural chemicals and synthetic ones.. On the other hand, there is a clear need to put a better label on chemicals derived from biomass. Such a "green" label should emphasise the renewability of resources used in the processing, all aspects related to sustainable development, biodegradability of products, recyclability of those and beneficial effects of biomass conversions on climatic changes, carbon dioxide mitigation and so on.

The discussion concentrated on the following points:
• Pyrolysis liquids supply, availability and eventual commercial price,

- General classification/grouping of biomass derived chemicals,
- Targets of future research: potential of combining selective feedstocks and optimal processing with integrated energy recovery,
- Necessity of long term strategic research perspective, needs for international collaboration and financial support.

Summary

Commercially potential chemicals produced from biomass thermolysis can be classified into two groups. The first one contains "bulk" or "commodity" chemicals for which prices do not exceed, in the first approximation, 1$/kg. Examples of such products are: charcoal, carbon black, adhesives, synthesis gas, hydrogen, "bio-lime", some food additives, silica, fertilizers, etc. The second group of chemicals can be loosely named "speciality" chemicals. The expected market values of those usually exceed 6 $/kg. The list of such compounds can be very long. Only some of them can be extracted from biomass by means other than pyrolysis (solvent extraction of essential oils, for example).

Table 1 (courtesy of C. Roy, Universite Laval) gives some indications of eventual market opportunities. Some recently obtained prices for bulk quantities of specialities were provided such as catechol - 6.17 $/kg, resorcinol - 17.60 $ kg and isoeugenol 13.54 $kg.

Table 1. Typical Wood Derived Phenolic Compounds and their Application

Compound	Price US$/kg	Principal applications
Vanillin	14	Food aroma, Pharmaceutical, Cosmetic, Chemical synthesis
Guaiacol	52	Pharmaceutical, Chemical synthesis
Catechol	55	Food aroma, Pharmaceutical
Resorcinol	55	Food aroma, Pharmaceutical, Chemical synthesis
Isoeugenol	145	Perfume, Chemical synthesis
Syringol	400	Food aroma
Syringaldehyde	400	Food aroma
Allylsyringol	1000	Food aroma

The main hindrance to commercial development of the bulk or commodity chemicals lies in the present low prices of natural gas and hydrocarbons resources in general. Prices of methanol (a natural gas derivative) in the U.S. have stayed in the vicinity of 0.40 $/gallon (1996). Also, a great deal of time and effort must be spent understanding the specific markets and users involved: just providing a commodity chemical at current prices will not produce sales.

Impediments blocking commercialisation of the second group of speciality chemicals are different one and are mostly technical.

There is a strong need to identify the most promising "candidates" from virtually hundreds of speciality chemicals and to find methods to separate and purify them. The resistance of already established food and pharmaceutical industries to "not so natural" ways of production of those chemicals is another problem.

One can speculate that an eventual, successful future development should target simultaneously both classes of chemicals (e.g. extraction of a few high value chemicals, before bulk processing of the oil) and attempt to couple such activity with energy production and/or waste disposal. One of many possible scenarios of integrated biomass conversion is presented in Figure 1, (courtesy of D. Redline, RTI and R Wroth, Dynamotive).

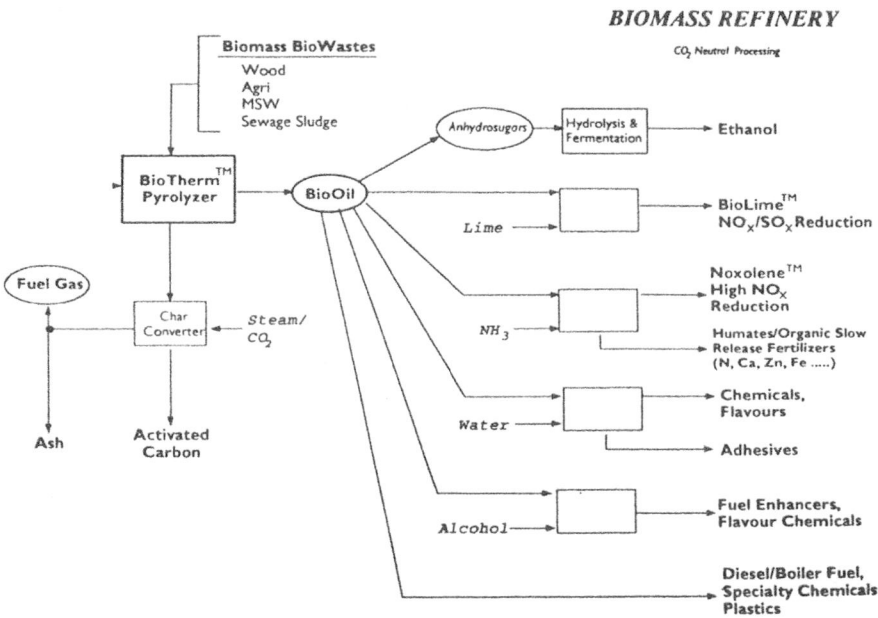

Figure 1. Biomass Refinery

Chemicals; present status; achievements and problems

Recent years have brought significant progress in understanding science of pyrolysis. Specific reaction conditions can be tailored, using selected feeds to obtain improved yields of desired products. Examples of those found in the technical literature vary from steroids to levoglucosan and matol, to electronic grade silica.

But as of now there are no commercial pyrolysis oil suppliers. Apart from good wishes we do not have a pyrolysis industry yet. It is not possible to buy the bio-crude for 75$/ton as some have predicted or announced in the past. Without a 100-300 $/ton range of price for pyrocrude there is almost no opportunity to obtain the finance necessary for specific product and process development. In general, biomass-derived chemicals commercialisation suffers from the present non-availability of bio-oil on a commercial scale. Potential users need to see a source of supply before committing themselves.

Another lesser hindrance is the growing confidentiality of research in those areas of biomass conversion which are the most promising or close to commercial implementation. Such a situation is especially pertinent in North America where government sponsored research conducted in national laboratories and at universities is being tied to private industry involvement and commercial success is only announced after the fact.

The present European Union model of co-operation, partnership and openness augmented by significant funds and targeted specific goals in biomass conversion seems to have a better chance to succeed, at least in accomplishing long term strategy goals.

Conclusions

There are many potential possibilities in solving the world future energy needs. Apart from biomass utilisation we can count on progress in solar energy, wind, waves geothermal, fuel cells, hydrogen and similar options or combination of those. This rather "comfortable" situation does not exist in the chemicals production where the clear choice is a fossil based or non-fossil based industry. There is no doubt that the second option will come. The question is only when.

Meantime there are important educational needs that stress all advantages and benefits of thermochemical biomass conversion. Such educational efforts should target national program administrators and managers, large corporations, media and the general public. This Conference is also a positive step in this direction.

Reference

1   L. Toth, K Potthast, Advances in Food Research, Vol. 29 pp 87-158, (1984)

Appendix

A selection of recent patents in the area of chemicals from biomass by pyrolysis is given below:

E S Grimmet, (Idaho Falls, Idaho), "Production of Charcoal from Sawdust in a Fluidised bed", U S Patent 3,929,585 issued Dec 30, 1975, filed Aug 16, 1972 Assignee: The United States Energy Research and Development Administration

E Chornet, C Roy, "Organic Products and Liquid Fuels from Lignocellulosic Materials by Vacuum Pyrolysis", Canadian patent # 1 163 595, issued March 13, 1984, filed Dec 18, 1980. Granted to University of Sherbrooke, Canada

N K Sowards, "Fluidised Bed Charcoal Particle Production System", U S Patent 4,510,021 issued Apr. 9, 1985, filed Mar 23 1981
Assignee: Energy Products of Idaho, Coeur d'Alene, Id

G Underwood, R G Graham, "Method of using fast pyrolysis liquids as liquid smoke", U S Patent # 4,994,297, issued Fed 19, 1991, filed May 26, 1989
Assignee, Ensyn Engineering Associates, Inc., Ontario, Canada

M Gander, K M Rapp, H Schiweck, "Process for Preparing, 1,6-beta-D-Anhydroglucopyranose (Levoglucosan) in High Purity", U S Patent # 5,023,330, issued Jun 11, 1991, filed Jan 30, 1989
Assignee: Suedzucker-Aktiengesellschaft, Mannheim, Germany
Priority Date: Feb 4, 1988

D A Himmelblau, "Method for Controlling Oil Reservoir Permeability using Biomass Oil", U S Patent 5,115,084 issued May 19, 1992, filed Jan 29, 1991
Assignee: Biocarbons Corporation, Woburn, Mass
Related U S Application, Continuation-in-part of Ser. No 382,232, Jul 19, 1989, Pat no. 5,034,498

K Oehr, D S Scott, S Czernik "Method of Producing Calcium Salts from Biomass", U S Patent # 5,264,623 issued Nov 23, 1993, filed Jan 4, 1993
Patentee, (Assignee): Energy, Mines and Resources Canada

G L Underwood, J A Stradal, "Browning Materials Derived from the Pyrolysis of Sugars and Starches", U S 5,92,541 issued 1994, 03, 08, assigned to Red Arrow Products Company Inc., Manitowoc, WI

J.P. Blanchard, "Smoke generator for food smoking", US Patent # 5,355,782 issued 1994, 10, 18 filed 1993, 02, 18.

L Moens, "Isolation of Levoglucosan from Pyrolysis Oil Derived from Cellulose", U S Patent #5,371,212

P Fransham, J Rasmussen, S Ainslie, "Thermolysis of Pentachlorophenol Treated Poles", U S Patent 5,378,323 assignee: Worthing Industries, Inc. Alberta

J.A. Stradal, G Underwood, "Process from producing hydroxyacetaldehyde", U S Patent 5,393,542, file 1993, 07, 16 issued 1995/02/28
Assigned to Red Arrow Products Company Inc. Manitowoc, WI

D S Scott, J Piskorz, D Radlein, P Redline, "process for the Production of Anhydrosugars from Lignin-and Cellulose-Containing Biomass by Pyrolysis", U S Patent 5,395,455 issued March 7, 1995, filed Mar 9, 1993,
Assigned to Energy, Mines and Resources Canada, Ottawa

M J Antal, "Process for Charcoal Production from Woody and Herbaceous Plant Material", U S 5,435,983 issued 1995/07/25
Assigned to University of Hawaii

K Oehr, "Acid Emission Reduction", U S Patent #5,458,803, issued Oct. 17, 1995, filed Sept 30 1993,
Assigned to Dynamotive, Corporation, Vancouver, Canada.

Workshop Report:
Catalytic Processes

D. C. ELLIOTT
Pacific Northwest National Laboratory, Richland, Washington, USA
R. R. MAGGI
Université Catholique de Louvain, Louvain-la-Neuve, Belgium

1 Introduction

Catalytic processes is a broad subject even within the confines of thermochemical conversion of biomass. The workshop participants expressed interest in a broad range of process technologies involving catalysts. Issues of concern in catalytic processsing of biomass were noted to range across the technology types and impacted on all catalyst applications.

2 Catalytic Processes

A number of catalytic process technologies were identified by the workshop participants. Interest was distributed throughout all technology types.

2.1 Hydrotreating

Catalytic hydrotreating is used to improve the product quality of the oil products produced by pyrolysis or high-pressure liquefaction of biomass. The process is primarily hydrodeoxygenation with hydrocracking and limited hydrogenation. More recently the emphasis has shifted to attempt more economical hydrotreating by limiting the severity of the process. By process modifications of operating temperature, pressure, space velocity or catalyst type, the extent of reaction can be limited to produce more oxygenated products. In this manner hydrotreating can produce a range of upgraded products, and current research focuses on attempts to control the hydrotreating process. In the end, the comparison of processing costs with the upgraded product properties will be the key issue. Much earlier work used sulfided NiMo and CoMo catalysts in a manner similar to petroleum processing. More of the recent research uses metal catalysts which are more active and can be used at lower severity conditions. While results from University of Sassari confirm earlier results showing the high viscosity of the partially deoxygenated products produced at low temperature with sulfided catalysts, new results from the Federal Research Center in Hamburg suggest that nickel on carbon catalyst can be used at temperatures as low as 80°C and 2 bar pressure to cause controlled hydrogenation of biocrude and produce stabilized bio-oils with good handling properties.

2.2 Tar Cracking
Catalytic beds are used to process the product gas from biomass gasification to produce a tar-free fuel. Tar specifics vary with gasification process and collection procedure but most tar cracking work has focused on either nickel metal or dolomites as the catalysts of choice for the tar destruction. The dolomites require higher temperatures than nickel for useful activity but are less subject to coking and deactivation. At the conference, Battelle disclosed that their new catalyst for tar cracking is a metal-free, alumina-based material. The chemistry of tar cracking is primarily steam reforming of the heavier hydrocarbons to light gases (carbon oxides and methane).

2.3 Catalytic Pyrolysis
An interesting new biomass conversion subject is the introduction of catalysts directly into the pyrolysis reactor. By this process the pyrolysis products could be modified as they are produced, such that the collected product has improved properties. Little work has been done in this area. Difficulties in the single reactor catalytic steam-gasification work undertaken at Pacific Northwest Laboratory in the U.S. were identified as indicative of potential problems with catalyst fouling.

2.4 Vapor Cracking / Reforming
Catalytic cracking has been applied at several different institutions to bioliquid products including pyrolysis vapors and condensates, high-pressure liquefaction products, supercritical fluid extractants, vegetable oils and tall oils. In the case of pyrolysis vapors, the main focus of this process application, the current conclusion is that the vapor components are too large to be effectively processed in the pores of the catalyst. Therefore, the activity is limited to the surface of the catalyst and the catalyst is subject to extensive coking. High coke yields limit the economical application of the process.

2.5 Chemicals Production
Chemicals production by catalytic processing of bioliquids is being researched. While work at Aston University attempts catalytic modification of pyrolyzates, the work in the U.S. focuses on catalytic upgrading of bioconversion products from biomass.

2.6 Catalytic Hydroliquefaction
While catalytic hydroliquefaction of biomass (usually at high pressures) has been little researched in recent times, back in the 1970s it was the primary focus of research into the thermochemical means of liquid fuels production from biomass. In the Workshop, interest was raised anew in this type of technology in light of the difficulties experienced in the production of useful fuels by flash pyrolysis. Studies have shown that higher yields of liquid fuels of higher quality are attainable with catalytic hydroliquefaction; however, those same studies concluded that the catalytic hydroliquefaction processes were more expensive.

3 Catalyst Issues

A number of catalyst issues were identified in the workshop as requiring further research. Most of these issues cross-cut all of the technology applications.

3.1 Trace components in biomass - sulfur
While biomass is considered a clean fuel and the small amount of sulfur contained in it is often ignored, sulfur can be a strong catalyst poison even in trace quantities. The sulfur content of pyrolysis oils is usually at the level of several hundred parts per million. At this level most metal catalysts will be poisoned based on conventional petroleum processing results. The use of metal catalysts to date with biomass has been limited in time-on-stream, but sulfur poisoniong has not yet been reported. Much of the processing of biocrudes has been done with pre-sulfided catalysts to avoid the possibility of sulfur poisoning. New research in low-severity processing involves metal catalysts, and sulfur poisoning needs to be evaluated. Sulfur content in gasification tars is seldom reported, but the same type of catalyst poisoning is possible when using nickel metal catalysts and should be investigated.

3.2 Trace components in biomass - alkali
Alkali is more common in biomass than sulfur but is a less serious catalyst poison. The amount of alkali in biomass can vary widely depending on the source, but must be considered in the development of catalytic processing of biomass. Alkali typically can form deposits on the catalyst which block the catalytic sites. However, alkali promoted catalysts are also known and are being studied in hydrotreating of biocrude. Hydrodemetallization of petroleum crudes is well known and removal of metals at levels up to several hundred parts per million (ppm) can be done economically (in which case macropore aluminas are used as the catalyst supports). Similar processing with biocrudes is proposed.

3.3 Activation of catalysts
Activation procedures vary with catalyst type. Sulfiding for CoMo and NiMo catalysts is well-demonstrated; maintenance of the sulfided state is discussed below. Pre-steaming of catalyst support materials was reported as a required step to be able to measure true activity of the catalyst without a significant break-in/deactivation period. This treatment is currently used for cracking catalysts, such as zeolites and clays. Evaluation of this treatment for γ-alumina supported catalysts should also be considered because of its reported reaction with steam at hydrotreating conditions.

3.4 Pretreatment of feedstocks
Hot-gas or hot-vapor filtration is being developed as a means of producing clean feedstocks for further processing. The filtration is aimed at char separation; but, in the process, ash components, including alkali and sulfur, are also removed. Hot vapor filtration of biocrude has produced product with alkali levels at around 10 ppm total alkali while sulfur has remained at 300 ppm.

Water wash and separation of biocrude has also been tested. In this option, the primarily lignin derived heavy tar components settle from the biocrude upon water addition. The heavy tar is envisioned as more appropriate hydrotreating feedstock. Coincidentally, addition of the biocrude to water is reported to produce a solid precipitate as opposed to the heavy tar produced by addition of water to the biocrude. Further treatment of the solid product has not been reported, although its analysis has been reported with results very similar to lignin. The yield of the solid product varies dramatically with the biocrude source.

3.5 Maintenance of the active catalyst
Maintenance in the long term of the active catalyst state is a key consideration for commercialization of any catalytic technology. Research has been ongoing for many years into the sulfided states of the catalysts used in biocrude hydrotreating. Concern has been expressed about the sufficiency of the amount of sulfur in biocrudes in maintaining a sulfided catalyst. There are still disagreements in conclusions drawn from the results, but reduced activity from over-sulfiding has been clearly shown, and no reports yet show a loss of sulfided state and resulting loss of activity.

In the case of gasification tar cracking catalysis, the information is even less definitive. The ongoing research using nickel, dolomites and alumina is mostly short-term operation with limited analytical support to investigate catalytic stability and changes in composition or properties.

3.6 Regeneration of catalysts.
Regeneration of a catalyst is the in-situ chemical processing to reactivate the catalyst as opposed to the removal, disassembly, recovery, and reassembly of the catalyst material at a remote location. Very little work has been done in this area directly with biomass processing catalysts. Because of the limited amount of long-term operation of the catalytic processes, studies of regeneration of degraded catalysts used in biomass conversion processes has not been reported. An important consideration in his issue is that catalysts undergoing a phase change are unlikely to be candidates for regeneration.

3.7 New catalyst supports
A high-surface area support commonly used in petroleum processing is γ-alumina. Research results over the past several years have demonstrated that γ-alumina is not stable in the biomass processing environments containing high levels of water, as discussed further below. In addition, the high acidity of the γ-alumina leads to high coke production and poor selectivity towards useful liquid products because of the unstable nature of biomass pyrolysis products. Alternatives are now being investigated including carbons, zirconia, titania, and spinels. Also, the addition of alkali to limit the acidity of the support is being researched. Research on carbon supports in particular are underway at Université Catholique de Louvain with specific interest in the effects of alkali promoters and variation in pore size distribution.

3.8 Support Stability
As stated above, γ-alumina is not stable in the biomass processing environments containing high levels of water. In fact, most of the conventional support materials (aluminas and silicas) developed for petroleum and gas phase processing are not stable in the presence of high levels of liquid water or steam. The new supports mentioned above are being studied

to address the issue of stability support. However, this issue of liquid water may be only relavant at intermediate temperatures (200-375°C) at elevated pressures where liquid water can exist. Some pillared clays are reportedly stable at up to 700°C temperature in 100% steam atmosphere, while some ultra-stable zeolites are stabilized in steam.

3.9 Mechanical stability of catalysts
The γ-alumina stability issue addressed above is one of chemical stability, primarily, but physical mechanical stability is also an important issue for long term catalyst operation. The chemical change of the γ-alumina to the hydroxide form results in siginifcant loss of surface area and some loss of physical strength in the case of nickel/alumina coprecipitated catalysts. The use of appropriate binders to form catalyst pellets is an art form of the catalyst manufacturing industry. Again, there is limited experience with systems containing high levels of water. Solubilization of various alumina or silica cements renders most conventional binder systems unusable.

3.10 Homogeneous catalysts
New initiatives in homogeneous catalysis of biomass conversion processes have been reported recently. Soluble molybdates have been used with good effect as has iodide. Both systems are operated in aqueous phase. Attempts at recovery of the catalytic material show improvements over earlier reports. Naphthenates were also suggested in the workshop as a potential catalyst form based on coal conversion research. No other organometallic catalytic research was reported, although metallocenes were suggested as a general area of interest.

3.11 Dispersed / Disposable catalysts
Recent results from VEBA OEL suggest a new option in catalytic processing of biocrude fractions. The, as yet, unreported results suggest that a coal-derived additive can, in the presence of very high pressure hydrogen, cause a significant hydrodeoxygenation of the heavy lignin-derived fraction of biocrude (following separation of the water soluble components). The additive, viewed by VEBA as non-catalytic, facilitates the hydrogenation; and a clear light product is recovered by distillation.

4 Conclusions

A wide range of catalytic processes are under development within the field of the Thermochemical Biomass Conversion. A number of issues requiring further research have been identified, most of which cross-cut the range of process applications.

CONCLUSIONS OF THE WORKSHOP "APPLICATIONS FOR THERMOCHEMICAL PROCESSES"

K. MANIATIS
European Commission, DGXVII, Rue de la Loi 200, Brussels, B-1049, Belgium
E. RENSFELT
TPS, Nykoping, S-611 82, Sweden
Y. SOLANTAUSTA
VTT Energy, PO Box 1601, Espoo, FIN-02044 VTT, Finland

1. Introduction

This workshop was held in order to identify the achievements of energy from biomass and to provide future requirements and targets for this renewable energy source to penetrate the various markets such as power, transport fuels, chemicals et al.

The workshop was held over two days and was attended by about 70 participants in each meeting. At the first meeting, the emphasis was placed on identifying the most important areas for applications based on the three thermochemical technologies, Combustion, Pyrolysis and Gasification. The areas of applications were grouped on five sectors being Heating, Power, Chemicals, Transportation and Others. The participants were asked to mark the priorities on the scale 1-3 (3 being the highest score) and the final marks are given in Table 1. From this table it can be seen that the participants identified power generation, industrial heating, waste treatment and co-utilization of biomass with fossil fuels as the most important applications/sectors in which biomass can make a significant contribution and penetrate these markets.

2. Discussions on the five sectors

2.1 Heating
The discussion in this sector was relative short with the main conclusion that there exists readily available technology and the market is developing strongly where the cost of biomass fuels is relative low or there is high concentration of biomass wastes/residues.

2.2 Power
The main long term goal for power generation is hydrogen and this is pursued in the US, Canada and European Union. Already demonstration projects exist with applications in buses and car manufactures are seriously involved in utilizing hydrogen as transport fuel. However, significant amount of work is still required before reliable and economic processes can be developed for widespread utilization. The main process of interest to this conference if the production of syngas by biomass gasification.

In terms of biomass gasification for power generation, the main challenge still remains the gas purification despite all the work and achievements which have been made the last four years. In Canada there is quite some interest for isolated communities in the range of 5-20 MWe based on engines, IGCC or bio-oil. However, these technologies still have to be demonstrated reliably.

Bio-oil has the logistical advantage that it can be produced in large scale production facilities and then used in several de-centralized smaller scale applications.

Finally Stirling engines receive attention in Austria and steam engines in Germany and The Netherlands.

APPLICATIONS FOR THERMOCHEMICAL PROCESSES
Table 1

		Combustion (581)		Pyrolysis (464)		Gasification (699)	
		Untreated 281	Processed 300	Bio-oil 313	Charcoal 151	Low BTU 392	Medium BTU 357
1. Heating	461						
• House Heating	109	28	31	17	12	10	11
• District Heating	169	33	31	32	12	33	28
• Industrial Heating	183	24	31	41	10	32	45
-ovens							
-furnaces							
-kilns							
-dryers							
2. Power	462						
• CHP	249	49	36	50	13	58	43
engines, Stirling engines, steam engines, steam turbines (small scale)							
• IGCC	130	7	17	12	8	55	31
• Fuel cells (hydrogen)	83	12	11	7	8	18	27
3. Chemicals	225						
• Ammonia	51	7	7	7	6	11	13
• Synthesis gas	59	7	7	7	6	16	16
• Ethylene	45	7	7	8	6	8	9
• Special chemicals	70	7	7	26	10	9	11
4. Transportation	303						
• Hydrogen	31	3	3	4	2	6	13
• Methanol	66	7	12	8	8	16	15
• Gasoline	50	7	10	10	7	8	8
• Diesel	48	6	10	16	6	5	5
• DME	45	6	8	9	6	8	8
• Refined bio-oil	63	6	9	25	9	7	7
5. Others	343						
• SNG	52	6	6	5	6	14	15
• Waste treatment	111	12	15	10	5	41	28
• Co-utilization (coal/waste/biomass /natural gas)	180	47	42	19	11	37	24

2.3 Chemicals
There is a general fear that in case syn-gas will be competitively produced by biomass gasification processes, the petroleum industry might intervene to undercut prices. At the present economic situation ethylene, propylene and other chemical can not compete

directly with the chemical industry. An area of interest is the production of phenolic compounds from pyrolysis oils and this could compete directly in the market. Finally, the specialty chemicals market is characterized by low tonnage in which case one or two plants can satisfy the World market.

2.4 Transportation
The economics of transportation fuels from biomass are the main barrier for a wider market penetration. The environmental issues related to carbon dioxide and other emissions can play an important role in new legislation beneficial to biomass derived fuels. It has been proven that methanol - gasoline mixtures reduce NOx emissions and it is generally believed that in the long term methanol from biomass will become economic. Studies indicate that in the case of ethanol, the process is economic in case the residues of the biological process are utilized by thermochemical processes to generate power. In such scenario, about 1/3 of the revenues will originate from the sale of power and will therefore improve the economic viability of an ethanol industry.

2.5 Others
The main interest is in the field of waste treatment and co-utilization of biomass with fossil fuels. In the case of waste, the economics are improved due to the negative value of waste, while in the case of co-utilization, existing infrastructure and plant components can be used, significantly reducing investment costs.

3. Discussions on the applications

3.1 Heating
In Europe the main interest is on district heating either on small or large scale while in the US and Canada the market opportunities are on institutional heating. Wood waste and residues are of primary importance due to their low cost. Nevertheless, the palletized wood industry is growing in Canada and the US while in Europe the interest in this fuel is found in few countries such as The Netherlands, Austria and Sweden. Wood chips are extensively used in Denmark, Germany and Austria. Especially in Austria a 15% support is granted by the government.

There is significant interest in retrofitting applications in the Baltic countries while in China the possibility to distribute producer gas to houses is seriously considered (after special training to avoid CO poisoning).

In addition to heating, district and/or institutional cooling should be considered.

The participants felt that there is reliable technology available in the market for biomass heating applications.

3.2 Power

3.2.1 Emissions

New emission standards are under discussion in EPA while there are prevailing EU regulations for biomass power plants. Public relations are extremely important for the local authorities and or communities in accepting biomass based power plants. A sensitive subject is the definition of "waste" which can have a strong effect on the emission standards. In some countries untreated biomass which is produced as a waste or residue by the wood industry is characterized as waste and not as wood.

An area where little attention has been paid so far concerns the standards for biomass ashes while proper applications for their utilization must be identified.

For the power industry it is important to identify cost effective and reliable technologies for biomass based power generation based on sound environmental issues such as emissions and CO2.

3.2.2 Steam cycles

For the steam cycles the cost of the biomass fuel is crucial due to the relative low efficiencies. In this relation, fuel flexibility is required by the conversion technology in order to overcome problems of seasonal availability and cost of various biomass fuels.

Problems related to fouling of boilers can be minimized by proper selection of biomass fuels. Energy crops can make an important contribution in developing fuels which have low fouling characteristics.

3.2.3 IGCC

The identification of the markets is very important in order to assist the penetration of this technology. In the range of 5-10 MWe, there is a high volume market, however, in this range there is competition from engines. In the US and Scandinavian countries, there are markets for large scale 25 - 100 MWe plants based on the forest and pulp industries.

The development of gas turbines for IGCC cycles still has to progress to reliable and trouble free operation but experience can be gained from blast furnace gases - turbine applications.

3.2.4 Bio-oil

At present there is limited supply of bio-oil for testing work.
An area which needs investigation is the production of bio-oil from MSW, however, the quality of such oils will need special attention to ensure environmentally acceptable utilization.

Work on adapting engines and gas turbines for different fuels instead of adapting the biomass fuels to the engines and gas turbines can make technological breakthroughs. However, this option has not been investigated yet.

3.2.5 Fuel cells

The main problem for biomass based fuel cells remain the gas conditioning for the production of pure hydrogen. However, the participants were encouraged to hear that in 1949 a biomass to hydrogen plant for ammonia synthesis was built and operated successfully in India.

3.2.6 Co-utilization

This was recognized as an important sector for penetration of biomass fuels in the power market. Although the interest is focused on co-utilization of biomass fuels with solid fossil fuels, co-utilization with natural gas (combination of steam lines to increase efficiencies of both cycles) provides certain opportunities. Another area of interest is co-utilization of biomass with sludges and/or fossil fuels.

3.2.7 Financing

In order to create the correct financial environment it is important to collaborate closely with venture capitalists and investment bankers from the very early stages of every project.

4. Conclusions

The participants felt that significant progress has been achieved since the previous conference 'Advances in Thermochemical Biomass Conversion' in Interlaken in 1992 and that at present biomass energy does make a contribution in most sectors discussed in this Workshop.

Work is still required to improve and optimize existing processes and technologies while at the same time identifying new ones.

For the near term, it is important to look for market niches where biomass based energy, fuels and chemicals can be established with cost effective and reliable technologies.

AUTHOR INDEX

SUBJECT INDEX

The index is based on keywords for each paper. The page numbers shown after each index entry refer to the first page of the paper that considers that topic.

Printed by Printforce, the Netherlands